Inorganic Chemistry

PERIODIC TABLE OF THE ELEMENTS

Period \ Group	1 I IA	2 II IIA	3 IIIB	4 IVB	5 VB	6 VIB	7 VIIB	8 VIIIB	9 VIIIB	10 VIIIB	11 IB	12 IIB	13 III IIIA	14 IV IVA	15 V VA	16 VI VIA	17 VII VIIA	18 VIII VIIA
1	1 H hydrogen 1.0080 $1s^1$																	2 He helium 4.00 $1s^2$
2	3 Li lithium 6.94 $2s^1$	4 Be beryllium 9.01 $2s^2$											5 B boron 10.81 $2s^22p^1$	6 C carbon 12.01 $2s^22p^2$	7 N nitrogen 14.01 $2s^22p^3$	8 O oxygen 16.00 $2s^22p^4$	9 F fluorine 19.00 $2s^22p^5$	10 Ne neon 20.18 $2s^22p^6$
3	11 Na sodium 22.99 $3s^1$	12 Mg magnesium 24.31 $3s^2$											13 Al aluminium 26.98 $3s^23p^1$	14 Si silicon 28.09 $3s^23p^2$	15 P phosphorus 30.97 $3s^23p^3$	16 S sulfur 32.06 $3s^23p^4$	17 Cl chlorine 35.45 $3s^23p^5$	18 Ar argon 39.95 $3s^23p^6$
4	19 K potassium 39.10 $4s^1$	20 Ca calcium 40.08 $4s^2$	21 Sc scandium 44.96 $3d^14s^2$	22 Ti titanium 47.87 $3d^24s^2$	23 V vanadium 50.94 $3d^34s^2$	24 Cr chromium 52.00 $3d^54s^1$	25 Mn manganese 54.95 $3d^54s^2$	26 Fe iron 55.84 $3d^64s^2$	27 Co cobalt 58.93 $3d^74s^2$	28 Ni nickel 58.69 $3d^84s^2$	29 Cu copper 63.55 $3d^{10}4s^1$	30 Zn zinc 65.38 $3d^{10}4s^2$	31 Ga gallium 69.72 $4s^24p^1$	32 Ge germanium 72.63 $4s^24p^2$	33 As arsenic 74.92 $4s^24p^3$	34 Se selenium 78.97 $4s^24p^4$	35 Br bromine 79.90 $4s^24p^5$	36 Kr krypton 83.80 $4s^24p^6$
5	37 Rb rubidium 85.47 $5s^1$	38 Sr strontium 87.62 $5s^2$	39 Y yttrium 88.91 $4d^15s^2$	40 Zr zirconium 91.22 $4d^25s^2$	41 Nb niobium 92.91 $4d^45s^1$	42 Mo molybdenum 95.95 $4d^55s^1$	43 Tc technetium (97) $4d^55s^2$	44 Ru ruthenium 101.07 $4d^75s^1$	45 Rh rhodium 102.91 $4d^85s^1$	46 Pd palladium 106.42 $4d^{10}$	47 Ag silver 107.87 $4d^{10}5s^1$	48 Cd cadmium 112.41 $4d^{10}5s^2$	49 In indium 114.82 $5s^25p^1$	50 Sn tin 118.71 $5s^25p^2$	51 Sb antimony 121.76 $5s^25p^3$	52 Te tellurium 127.60 $5s^25p^4$	53 I iodine 126.90 $5s^25p^5$	54 Xe xenon 131.29 $5s^25p^6$
6	55 Cs caesium 132.91 $6s^1$	56 Ba barium 137.33 $6s^2$	57 La lanthanum 138.91 $5d^16s^2$	72 Hf hafnium 178.49 $5d^26s^2$	73 Ta tantalum 180.95 $5d^36s^2$	74 W tungsten 183.84 $5d^46s^2$	75 Re rhenium 186.21 $5d^56s^2$	76 Os osmium 190.23 $5d^66s^2$	77 Ir iridium 192.22 $5d^76s^2$	78 Pt platinum 195.08 $5d^96s^1$	79 Au gold 196.97 $5d^{10}6s^1$	80 Hg mercury 200.59 $5d^{10}6s^2$	81 Tl thallium 204.38 $6s^26p^1$	82 Pb lead 207.2 $6s^26p^2$	83 Bi bismuth 208.98 $6s^26p^3$	84 Po polonium (209) $6s^26p^4$	85 At astatine (210) $6s^26p^5$	86 Rn radon (222) $6s^26p^6$
7	87 Fr francium (223) $7s^1$	88 Ra radium (226) $7s^2$	89 Ac actinium (227) $6d^17s^2$	104 Rf rutherfordium (267) $6d^27s^2$	105 Db dubnium (268) $6d^37s^2$	106 Sg seaborgium (269) $6d^47s^2$	107 Bh bohrium (270) $6d^57s^2$	108 Hs hassium (269) $6d^67s^2$	109 Mt meitnerium (277) $6d^77s^2$	110 Ds darmstadtium (281) $6d^87s^2$	111 Rg roentgenium (282) $6d^{10}7s^1$	112 Cn copernicium (285) $6d^{10}7s^2$	113 Nh nihonium (286) $7s^27p^1$	114 Fl flerovium (290) $7s^27p^2$	115 Mc moscovium (290) $7s^27p^3$	116 Lv livermorium (293) $7s^27p^4$	117 Ts tennessine (294) $7s^27p^5$	118 Og oganesson (294) $7s^27p^6$

Period															
6	58 Ce cerium 140.12 $4f^15d^16s^2$	59 Pr praseodymium 140.91 $4f^36s^2$	60 Nd neodymium 144.24 $4f^46s^2$	61 Pm promethium (145) $4f^56s^2$	62 Sm samarium 150.36 $4f^66s^2$	63 Eu europium 151.96 $4f^76s^2$	64 Gd gadolinium 157.25 $4f^75d^16s^2$	65 Tb terbium 158.93 $4f^96s^2$	66 Dy dysprosium 162.50 $4f^{10}6s^2$	67 Ho holmium 164.93 $4f^{11}6s^2$	68 Er erbium 167.26 $4f^{12}6s^2$	69 Tm thulium 168.93 $4f^{13}6s^2$	70 Yb ytterbium 173.05 $4f^{14}6s^2$	71 Lu lutetium 174.97 $5d^16s^2$	Lanthanoids (lanthanides)
7	90 Th thorium 232.04 $6d^27s^2$	91 Pa protactinium 231.04 $5f^26d^17s^2$	92 U uranium 238.03 $5f^36d^17s^2$	93 Np neptunium (237) $5f^46d^17s^2$	94 Pu plutonium (244) $5f^67s^2$	95 Am americium (243) $5f^77s^2$	96 Cm curium (247) $5f^76d^17s^2$	97 Bk berkelium (247) $5f^97s^2$	98 Cf californium (251) $5f^{10}7s^2$	99 Es einsteinium (252) $5f^{11}7s^2$	100 Fm fermium (257) $5f^{12}7s^2$	101 Md mendelevium (258) $5f^{13}7s^2$	102 No nobelium (259) $5f^{14}7s^2$	103 Lr lawrencium (262) $6d^17s^2$	Actinoids (actinides)

Molar masses (atomic weights) quoted to the number of significant figures given here can be regarded as typical of most naturally occuring sample.

Inorganic Chemistry

An Integrated Approach

FIRST EDITION

AARON L. ODOM
REMI BEAULAC
ADAM R. JOHNSON
MITCH R. SMITH
JAMES K. MCCUSKER
CHIP NATARO

OXFORD
UNIVERSITY PRESS

OXFORD
UNIVERSITY PRESS

Great Clarendon Street, Oxford, OX2 6DP,
United Kingdom

Oxford University Press is a department of the University of Oxford.
It furthers the University's objective of excellence in research, scholarship,
and education by publishing worldwide. Oxford is a registered trade mark of
Oxford University Press in the UK and in certain other countries.

Published in the United States of America by Oxford University Press
198 Madison Avenue, New York, NY 10016, United States of America

British Library Cataloguing in Publication Data
Data available

Library of Congress Control Number: 2024946545

ISBN 9780198905998

Printed in the UK by
Bell & Bain Ltd., Glasgow

The manufacturer's authorized representative in the EU for product safety is
Oxford University Press España S.A. of Parque Empresarial San Fernando de Henares,
Avenida de Castilla, 2 – 28830 Madrid (www.oup.es/en or product.safety@oup.com).
OUP España S.A. also acts as importer into Spain of products made by the manufacturer.

FSC
www.fsc.org
MIX
Paper | Supporting
responsible forestry
FSC® C007785

Preface

This book started off with the desire to modernize inorganic texts while still including all the topics that many instructors might still regard as important. As a result, there are some more up-to-date takes on classic topics in the book. Overall, the result is a text that we think is appropriate for many different levels, if used judiciously for lower-level courses and copiously for higher-level programs.

While some topics could be regarded as advanced, and some are even labeled as such, a title of "Advanced Inorganic Chemistry" would been a little misleading. First, there are many topics that are taught in some high-school and general chemistry classes that are reviewed, such as VSEPR. Second, in reality we do not really believe that such a book in a single, reasonably-sized volume can possibly exist. Inorganic chemistry is too vast a topic to cover all its pendant areas until one is truly in the "advanced" territory. Certainly, entire books, and in many cases series of books, are written on chapter topics in the text, such as spectroscopy, valence-bond theory, organometallic chemistry, kinetics, and others. These key concepts were covered as best as possible then be applied to some examples from the literature.

Overall, there is probably more material than most would cover in two semesters, and the expectation is that instructors will pick and choose topics they prefer to cover in their courses. For exactly these purposes, we include short chapters on topics such as *f*-element chemistry and bioinorganic chemistry.

We hoped to have in one place the materials we wanted our undergraduate and early graduate students to know as part of their research preparation. Therefore, we tried to write a general text that will prepare them well to later specialize in any area of inorganic chemistry. As a result, discussion of topics will tend to simply stop at some point, because to go further was thought to become more specialized than is necessary for most students. Some students will then need to consult more focused texts and the primary or review literature if very in-depth knowledge of the topic is required. That arbitrary line had to be drawn somewhere for every chapter and every topic covered. The instructor or reader may choose to draw that line elsewhere and truncate the discussion, and we tried to make that as easy as possible. Naturally, some instructors may feel we did not go far enough.

Because of the breadth of inorganic chemistry (see Chapter 1), it is impossible to cover a significant percentage of the field in a couple of semesters. In fact, like a lot of fields of science, it is difficult enough to simply keep up with everything being published, much less to catch up from a standing start as a beginning student. As a result, it is necessary to use a large variety of models and theories that may be applicable to thousands or even millions of molecules under certain circumstances, instead of trying to describe excessive numbers of compounds, reactions, and so on. Teaching theories, along with their limitations, provides a course that will be more useful than a totally descriptive approach. However, some descriptive chemistry is a necessity for providing examples of where the theories apply and how to use them. Descriptive chemistry is necessary to show applications of the principles and test their limitations.

There is certainly a place for simply learning as many different chemical transformations and specific compounds as possible, as this helps to build intuition. However, one can argue that the place to build that intuition is by studying the primary literature and reviews, not a general textbook on inorganic chemistry. Even so, we hope there is enough descriptive chemistry in the text to keep things interesting.

In the end, the instructor and student both are encouraged to treat the book like an all-you-can-eat salad bar: feel free to take what you like, in any order you want, leave the rest, and come back as often as you want.

Here are a few notes on how the material could be taught, based on our experience. Obviously, these are merely recommendations, and we do not necessarily follow them ourselves every semester.

Chapter 1 can be taught on the first day of the first semester of undergraduate classes. The purpose of the material is to humanize the science and build interest. This chapter could be skipped for graduate classes, though I usually cover it.

Chapter 2 could be taught for undergraduates as background or review. Advanced topics may not be covered, including mathematical details of LCAO and computational descriptions in that chapter. For graduate-level classes, the advanced topics would be covered, with the other materials assigned for reading as a review.

Chapter 3 would be taught in both undergraduate and graduate classes. For graduate students, one can move quickly through bond dissociation enthalpies, polar covalence theory, definitions for heterocovalent and dative bonds, and bond order.

Chapter 4 can usually just be assigned for reading in undergraduate and graduate classes. Our undergraduate class has a recitation, and VSEPR is reviewed there.

Chapter 5 should be taught in both undergraduate and graduate classes. The matrix algebra may be skipped at the undergraduate level, if desired, but we think it provides a better understanding of the subject when included.

Chapter 6 is covered for undergraduate classes, with the expectation being that they will do projections to derive the correct number of orbitals. Graduate students are required to learn how to orthogonalize the orbitals and show that they are orthogonal. The graduate students may spend more time on molecular orbitals of more complex molecules and produce more examples.

We cover Chapters 7 and 8 in both undergraduate and graduate classes. Advanced topics are reserved for the graduate class. Main-group chemistry is sometimes overlooked in inorganic classes, but there are many lessons from the main group that translate to the transition metals to one degree or other. Furthermore, the students acquire a better understanding of ligand electronic structure, such as phosphines, from a deeper understanding of main-group chemistry.

In many cases, Chapter 9 can be covered in a single lecture with material chosen by the instructor, after which, the chapter is assigned for reading. The general effects, like the inert pair effect, are discussed in class.

Chapter 10 on introductory material for coordination chemistry should be covered for undergraduate and graduate classes. Advanced topics are reserved for the graduate class. Personally, I like to cover some of the historical aspects of coordination chemistry to provide context. Some may prefer to not spend class time on such topics.

Chapters 11 and 12 can be covered in about two lectures, with material chosen by the instructor. The advanced topics are taught in the graduate classes, as desired. There is a lot of reference material included that the instructor can refer to later and that students will hopefully find useful.

Chapter 13 can be taught much earlier if one wishes, such as immediately following group theory. It is here because there are some complexes commonly discussed in this chapter, such as ferrocene, that are not covered in detail until Chapter 12. However, it is unlikely that students will be confused by this. This chapter would fit with our second semester of inorganic chemistry, which in most years includes redox chemistry, kinetics, and solid-state chemistry.

Chapter 14 is treated as advanced material and is taught only to graduate students. However, one could certainly cover most of the material in undergraduate classes without trouble, if desired. The second-order Jahn–Teller discussion is perhaps best reserved for graduate students.

Chapters 15 and 16 may be introduced in the first semester of inorganic chemistry. Usually, the 1-electron model for understanding spectroscopy, with an emphasis on drawing *d*-orbital splittings with different geometries and ligands, is taught in the first semester. Multi-electron term symbols and spectroscopy would be taught in more detail in our second semester course.

Chapter 17 on kinetics is taught in our second semester class on inorganic chemistry. Substitution kinetics is important for the vast majority of processes and can be quite complex, so the kinetics discussion is focused here.

Chapters 18–21 on organometallic, solid-state, bioinorganic, and *f*-element chemistry may be used as 1–2-lecture (each) overviews of the topic in undergraduate and graduate classes, as time permits, or as launching points for more in-depth studies.

Most of all, we hope the text is useful for those looking for a place to begin understanding the field of inorganic chemistry and as a place to find connections to advanced topics.

ALO

Acknowledgments

As with anything in my life, I thank my family for their indulgence of this project, which took a great deal more time than I initially imagined. I want to thank our editors at Oxford University Press for their patience and steadfast belief in the project.

Many people read chapters and offered comments along the way. I would like to thank Clark Landis for his insights on some of the chapters that include natural bond orbital theory, and Peter T. Wolczanski for advice on writing about kinetics. I would also like to thank the many members of the academic community whose generous and insightful feedback helped to shape this text, including:

Dr Arunabhiram Chutia, University of Lincoln

Professor Brian Dolinar, West Virginia University

Professor Mari-Ann Einarsrud, NTNU Norwegian University of Science and Technology

Assistant Professor Srimal Garusinghe, Southeastern Oklahoma State University

Dr Negash Getachew, Addis Ababa University

Professor Katrina Miranda, University of Arizona

Dr Christopher Pask, University of Leeds

Dr John Dominic Smith, Lipscomb University

Professor Jesse Tye, Ball State University

I would also like to thank the many in the community who read various chapters as anonymous reviewers. Your time is much appreciated.

Further, I would like to thank the people who took the time to teach me inorganic (and in some cases organometallic) chemistry over the years in the classroom and in the laboratory. If this book is filled with any insights, they are more probably theirs than mine. Notably, in this regard and in no particular order, I would like to thank Kit Cummins, Dan Nocera, Dick Schrock, Jerry Mills, Bruce Whittlesey, Hanno zur Loye, Dietmar Seyferth, Steve Buchwald, Greg Fu, and Steve Lippard.

I also want to thank my coauthors, who I goaded into this shared endeavor, for their insights, wisdom, and knowledge. I could not have finished it without them. Their excellent help has been invaluable, and any mistakes, typographical errors, and omissions in this book are probably my own.

ALO

Brief Contents

Contents

List of Authors

Aaron L. Odom
Aaron was born in Texas and has a B.S. in chemistry from Texas Tech University, where he carried out undergraduate research with Bruce Whittlesey, Jerry Mills, and Richard Bartsch. He received his Ph.D. from the Massachusetts Institute of Technology with Christopher "Kit" Cummins, on novel chromium, molybdenum, and uranium complexes. Following this, he carried out postdoctoral work with Daniel Nocera at MIT studying light to chemical energy conversion by rhodium complexes. He is currently a Professor in the Chemistry Department at Michigan State University. His research group studies inorganic and organic synthesis, applications of catalysis to pharmacologically active compounds, and modeling of catalytic reactions.

Remi Beaulac
Remi Beaulac has a Ph.D. from the University of Montreal and is currently a Professor of Chemistry at Swarthmore College. His research is in energy processes in inorganic nanomaterials.

Adam R. Johnson
Adam R. Johnson has a B.A. in chemistry from Oberlin College, where he carried out undergraduate research in both organic and inorganic chemistry. He received his Ph.D. from the Massachusetts Institute of Technology, with Christopher "Kit" Cummins on low-coordinate and low-valent complexes of titanium and molybdenum. He undertook postdoctoral work in bioinorganic chemistry with Ken Raymond at U.C. Berkeley. He is currently Professor and Chair of Chemistry at Harvey Mudd College, where his research group studies chiral early metal complexes. He is a founding member of the Interactive Online Network of Inorganic Chemists.

Mitch R. Smith
Smith took what he learned about synthesis, reactivity, and mechanistic studies from his mentors Dick Andersen and Greg Hillhouse, and applied them in the field of catalysis. Researchers from Smith's group and those from groups of his collaborators have made important advances in C–H functionalization, electrocatalysis, and polymer science.

James K. McCusker
James McCusker received a Ph.D. from the University of Illinois at Urbana Champaign and is a Foundation Professor in the Department of Chemistry at Michigan State University. In his research, he studies fundamental aspects and applications of ultrafast processes and magnetic behavior in inorganic complexes.

Chip Nataro
Chip Nataro is the Marshall R. Metzgar Professor of Chemistry at Lafayette College in Easton, PA, USA. He received his B.S. in chemistry from Messiah College. His Ph.D. was earned with Bob Angelici at Iowa State University, where he examined the protonation of metal–metal bonds. He carried out his postdoctoral work at the University of Vermont, with Chris Allen and Bill Geiger. His current research with undergraduates focuses on the synthesis, characterization, and catalytic applications of transition-metal compounds with bis(phosphino)metallocene ligands. He currently serves as a member of the leadership council of the Interactive Online Network of Inorganic Chemists.

CHAPTER 1

What Is Inorganic Chemistry, and Why Study It?

"If you wish to understand the fragrance of the rose, or the tenacity of the oak; if you are not satisfied until you know the secret paths by which the sunshine and the air achieve these wonders; if you wish to see the pattern which underlies one large field of human experience and human measurement, then take up chemistry."

— C. A. Coulson

CONCEPT COVERAGE

1. The definition of inorganic chemistry
2. Application of inorganic chemistry to food production
3. Inorganic chemistry, energy production, and energy storage
4. Inorganic chemistry and human health
5. Connections of inorganic chemistry to other fields

1.1 What Is Inorganic Chemistry?

If you designate something using a negative it often leaves a great deal of breadth for interpretation. Inorganic chemistry is unique among the traditional disciplines of chemistry in that it is named as a negative, literally "not organic". If a field were named "not forks", it could equally well include spoons, poached eggs on toast, and hovercraft (Fig. 1.1). Similarly, *inorganic chemistry is a very broad discipline encompassing topics from structure, bonding, reactivity, and physical properties of molecules bearing atoms from every corner of the periodic table.* The typical definitions of inorganic chemistry can be summarized as the chemistry of molecules not entirely comprised of carbon, hydrogen, oxygen, nitrogen, and monovalent halogens; in some cases, specific categories of sulfur and phosphorus compounds may be considered "organic" as well, depending on the author (though not this one). Some examples of inorganic molecules are shown at the bottom of Fig. 1.1; and, as is most often the case, many of these contain carbon-based substituents on metal centers.

To a beginning student getting their first exposure to the field, inorganic chemistry can look a bit like a large collection of miscellanea, when in reality it is more akin to a few many-sided objects. Learning a handful of different models, such as VSEPR, valence bond theory, Molecular orbital theory, and ligand field theory, and how those models can be applied in different ways to inorganic chemistry as a whole can unify the field from the student's perspective and provide them with the tools for attacking a host of different problems.

While we are attempting to keep the coverage and ordering of topics similar to other texts, the goal is to treat inorganic chemistry using a series of different models. The book will build up a foundation in these different models slowly over the first third of the book or so; in some cases, as many models as possible will be applied to the same problem to show how each arrives at the same answer, though perhaps from different directions. In many cases where multiple models can be applied, one is often much easier to use or more informative than others, making it the more "natural" model to employ. That being said, no single model can capture the whole breadth of inorganic chemistry or universally be applied to all circumstances. A central skill that training inorganic chemists must develop is the ability to recognize the strengths and weaknesses of a variety of approaches, to learn how to look at the same problem from a wide number of perspectives and to use multiple theoretical and experimental tools to attack a given problem, each of which may bring its own nuances and insights.

This breadth makes inorganic chemistry a uniquely complex field. The inclusion of elements from every part of the periodic table, of course, leads to usage of additional orbital types. This, coupled with factors like variations in electronegativity, leads to wide variations in familiar 2-center 2-electron bonds where bond polarity varies widely, and dative bonds are more

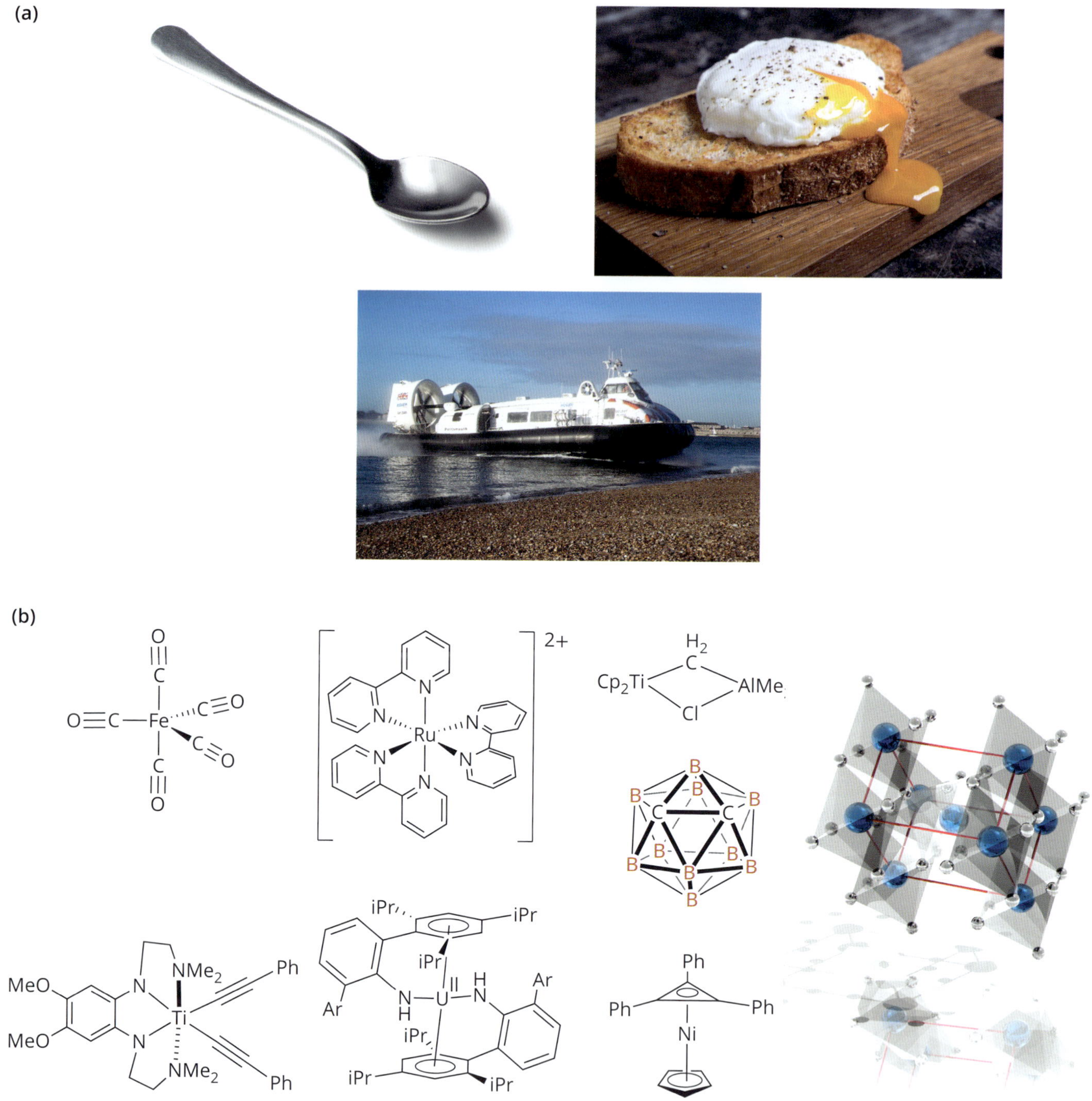

Fig. 1.1 (a) Some things that fit into the category of "not forks". Any group defined as a negative is generally very large. So "not forks" includes a spoon, poached egg on toast, and hovercraft. (b) Some things that are "not organic" and represent some of the large array of inorganic compounds available from organometallic to main group. The extended structure shown is a representation of rutile (TiO_2).

common. The bonding types are richer with δ- and ϕ-bonds, 3-center 4-electron, 3-center 2-electron, and many other bonding types commonly employed to understand the electronic structure of inorganic molecules. These varied bonding situations lead to a wide number of applications related to catalysis, photophysics, and elsewhere. There is a focus in this book on the utility of inorganic chemistry in solving the world's problems through applications, such as catalysis, and integration of inorganic chemistry to fields like biology.

1.2 Application of Inorganic Chemistry to the World's Problems

In the remainder of this chapter, we will briefly discuss the types of problems that inorganic chemists use their craft to address. These will be discussed in three broad categories based on three important topics for society today: sustainability, clean energy, and health. In many cases, some individual researchers will be highlighted to put a human face to the science.

1.2.1 Inorganic Chemistry and Sustainability

One of the many societal challenges that chemistry is uniquely positioned to address is the development of a sustainable modern world. For the sake of this chapter, energy will be dealt with separately, but sustainability touches many different parts of our lives, including food, water, and climate. Let us look at just one of these and discuss how humans have discovered how to improve food growth by taking constituents from the air and making fertilizer. One could argue that the most important reaction that mankind has discovered since the advent of fire is an inorganic reaction used for the generation of ammonia from hydrogen and dinitrogen, the Haber–Bosch reaction.

The only stable allotrope of nitrogen, N_2, is extremely inert towards most chemical species, and dinitrogen makes up about 78% of the air. In fact, dinitrogen is inert enough that it is commonly employed as an inert atmosphere in chemical laboratories to flush reaction vessels in order to avoid unwanted side reactions with oxygen and water. However, nitrogen is a common element found in the biological molecules of every organism on Earth, where it is found in DNA, RNA, amino acids, and so on. Life, as we know it, depends intimately on the existence of nitrogen-based compounds. Obviously, this readily raises a fundamental question: if N_2 is so inert that it does not react, how do living organisms manage to "capture" the nitrogen they need to exist? This problem is referred to as the "activation of nitrogen", which refers to the way unreactive atoms (here, nitrogen) can be successfully made to react. Prior to the work of Haber and Bosch, virtually all of the chemically active nitrogen on Earth came from one source: organisms that contain the enzyme nitrogenase.

Nitrogenases (there are a few different varieties) are metalloenzymes (Fig. 1.2) containing iron, along with vanadium or molybdenum in some cases. These enzymes are found in some bacterial species that often live in the nodules of legumes but also in blue-green algae, some lichens, and some bacteria in the soil. It is estimated that these naturally occurring systems "fix" 99–150 million tons of nitrogen per year—"fixation" is the chemical name for the conversion of dinitrogen to a chemically active form, such as ammonia (NH_3). The most active nitrogenase has molybdenum incorporated into a 7 iron, 9 sulfur cluster with a single carbon atom at its core.[1]

From Haber we have the concept of "bread from air" that avoided famine for billions of people through the production of artificial fertilizers; on the other hand, he also led a central effort to develop chemical warfare in the battlefield during World War I by weaponizing poisonous gases. Haber developed osmium- and ruthenium-based processes for the generation of ammonia from N_2 and H_2 in order to help the Germans produce munitions during the war. On April 25, 1918, Winston Churchill remarked:

> It is a very strange thing to reflect that, but for the invention of Prof. Haber, the Germans could not have continued the War after their original stack of nitrates was exhausted. The invention of this single man has enabled them, utilizing the interval in which their accumulations were used up, not only to maintain an almost unlimited supply of explosives for all purposes, but to provide amply for the needs of agriculture in chemical manures. It is a remarkable fact and shows on what obscure and accidental incidents the fortunes of possibly the whole world may turn in these days of scientific discovery.

Haber (Fig. 1.3) was named as a war criminal by the allies at the end of World War I. Despite being a loyal German who made many contributions during that war, because of his Jewish heritage he had to flee Germany once the Nazis came to power.

Carl Bosch (1874–1940), a German engineer, developed a less expensive iron-based process for the nitrogen fixation reaction that Haber had discovered, and made other changes necessary for industrialization (Fig. 1.4).[2]

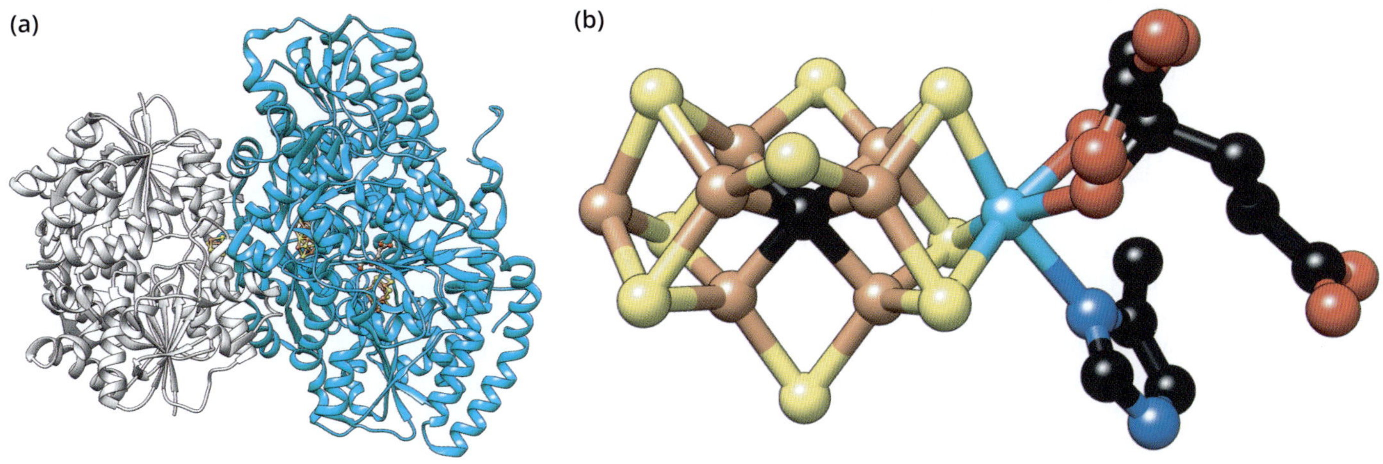

Fig. 1.2 A drawing of the nitrogenase active site discussed in more detail in the chapter on bioinorganic chemistry. Fritz Haber (1868–1934, Fig. 1.3) was one of the more intriguing—as well as controversial—characters of science.[2]

Fig. 1.3 Fritz Haber, inventor of the reaction to generate ammonia from N_2 and H_2.

Fig. 1.4 Carl Bosch was able to use an iron catalyst to generate ammonia, and industrialized what we now call the Haber–Bosch process.

He helped to found and lead IG Farben, a large chemical conglomerate that later was split into now familiar names, such as Bayer and BASF. Bosch was a critic of Nazi policies and the firing of Jewish employees, which led to his dismissal in the 1930s. He died in Heidelberg in 1940.

Each man received a Nobel Prize in Chemistry (Haber in 1918, Bosch in 1931, shared with Friedrich Bergius) for their work on the development of an artificial approach to the fixation of dinitrogen to produce ammonia, Eq. 1.1, which is now known as the *Haber–Bosch reaction*, one of the most famous named reactions in chemistry—two Nobel Prizes for the same chemical reaction, one an optimization of the first! This certainly underscores the importance of this process to society!

$$N_2 + 3\,H_2 \xrightarrow[\substack{300-550^\circ C\\150-250\,atm}]{\substack{\text{iron-based}\\\text{catalyst}}} 2\,NH_3 \qquad 1.1$$

According to the United States Geological Survey Mineral Resources Program, the Haber–Bosch reaction produced 130 million tons of NH_3 worldwide in 2009. This is the mass of ~1500 Washington Monuments or >21 times the weight of the Great Pyramid of Cheops in a single production year (Fig. 1.5).

Nitrogen available for biological purposes comes almost equally from natural and human sources, and our current ability to grow food crops would be greatly curtailed without the Haber–Bosch process. It is estimated that ~50% of the nitrogen in our bodies comes from the Haber–Bosch reaction, or, put another way, the population of the Earth could only be about half of what it is without this reaction: we simply would not be able to grow enough food without the fertilizers produced by the Haber–Bosch process.[2]

1.2.2 Inorganic Chemistry and Clean Energy

Cleaner, more efficient energy production is one of the great challenges for humanity in the twenty-first century. Inorganic chemistry now contributes significantly to cleaner fossil fuel-based energy use by cleaning up emissions. One example is in "three-way catalytic converters" which catalyze the reduction of nitrogen oxides (also called "nox", for NO_x) to N_2 and O_2, the oxidation of CO to CO_2, and the oxidation of any residual hydrocarbons in the emissions to CO_2 and H_2O. These processes are catalyzed by Pt, Pd, and Rh in a solid matrix, which often contains CeO_2, placed in line with the exhaust system of your car, and inclusion of these systems has greatly improved air quality despite increases in vehicle use since they became commonplace in the 1970s.

Fig. 1.5 (*left*) The Washington Monument in Washington, DC, U.S.A. is 554 feet (169 meters) tall and weighs approximately 91,000 tons. (*right*) The Great Pyramid of Cheops is one of the three largest pyramids of Giza and one of the largest buildings ever constructed at 450 feet (137 meters) high and weighing about 5.75 million tons. By comparison, in 2009, 130 million tons of ammonia were produced using the Haber–Bosch reaction.

A topic of great current importance is renewable energy, which is in fact two separate problems: energy conversion/production and energy storage. The only renewable source of energy with the scale to meet all our power needs in the future is solar. It is estimated that the Sun provides energy to the Earth's surface at an average annual rate of about 100,000 TW (terawatts). There is more energy from the Sun that arrives at the Earth in an hour than we need for a year! The problem, of course, is in harnessing that energy.[3–4]

Energy science is a very fast-moving field, and almost anything written will likely be outdated on printing. Here, just a couple of examples are given to demonstrate how inorganic chemistry contributes to energy collection and storage, but the impact of inorganic chemistry is much wider than is presented in what follows here.

The traditional method for collection of solar radiation and its conversion to electricity is by using silicon-based solar panels like those in Fig. 1.6. This technology involves the doping of pure silicon, typically with boron and phosphorus, to produce solid-state systems that convert sunlight to electrical potential.

A quite different method for conversion of sunlight to electricity is the dye-sensitized solar cell (DSSC), sometimes called the Grätzel cell, after Michael Grätzel.[5] For this type of cell (Fig. 1.7), a transparent electrode, like indium tin oxide, is coated with a thin layer of TiO_2. On the TiO_2 is attached a dye to absorb light. One of the most studied dyes was discovered in Grätzel's laboratory and is called "N_3", the ruthenium complex shown. The dye attaches to the TiO_2 surface using the carboxylic acid groups attached to the 2,2′-bipyridine ligands. On excitation, the dye injects an electron into TiO_2, which is collected at the electrode. At this point, the dye is oxidized and must be reduced before it can be excited by another photon. In a popular iteration of the DSSC, the dye regeneration is carried out by I^-, which produces I_2 that in the presence of I^- becomes I_3^-. (The I_3^- molecule is linear and involves a kind of 3-center 4-electron bonding common in the main group that is discussed in detail later in this book.) The I_3^- diffuses back to the counterelectrode, generally platinum, which reduces it back to I^- to complete the circuit.

These decidedly inorganic, electronic devices have reached efficiencies of >15% at this time. Single p-n junction solar panels using highly purified silicon can be expensive but reach efficiencies of 25%, near the Shockley–Queisser theoretical limit of 33.16% for such a cell. Perhaps the hottest materials in energy science at the time of writing are perovskites for solar energy conversion, and these materials quickly reached 15% efficiency.[6] Shown in Fig. 1.8 is the summary of the best solar cell efficiencies from the National Renewable Energy Laboratory (NREL) in Golden, Colorado. Virtually all of these cells involve inorganic materials of some type, which are essential to their operation (Fig. 1.8).

Collecting energy with no method for its storage is often wasteful, and solar energy produced nowadays is often immediately used by placing it onto the electrical grid or stored in batteries, supercapacitors, or other energy-storing technologies. Batteries are important technologies for fast access to electrical potential, but traditional batteries store relatively small amounts of energy relative to their weight compared to chemical storage methods. Shown in Fig. 1.9 is a

Fig. 1.6 Silicon-based solar panels. These cover one of the parking lots at Michigan State University.

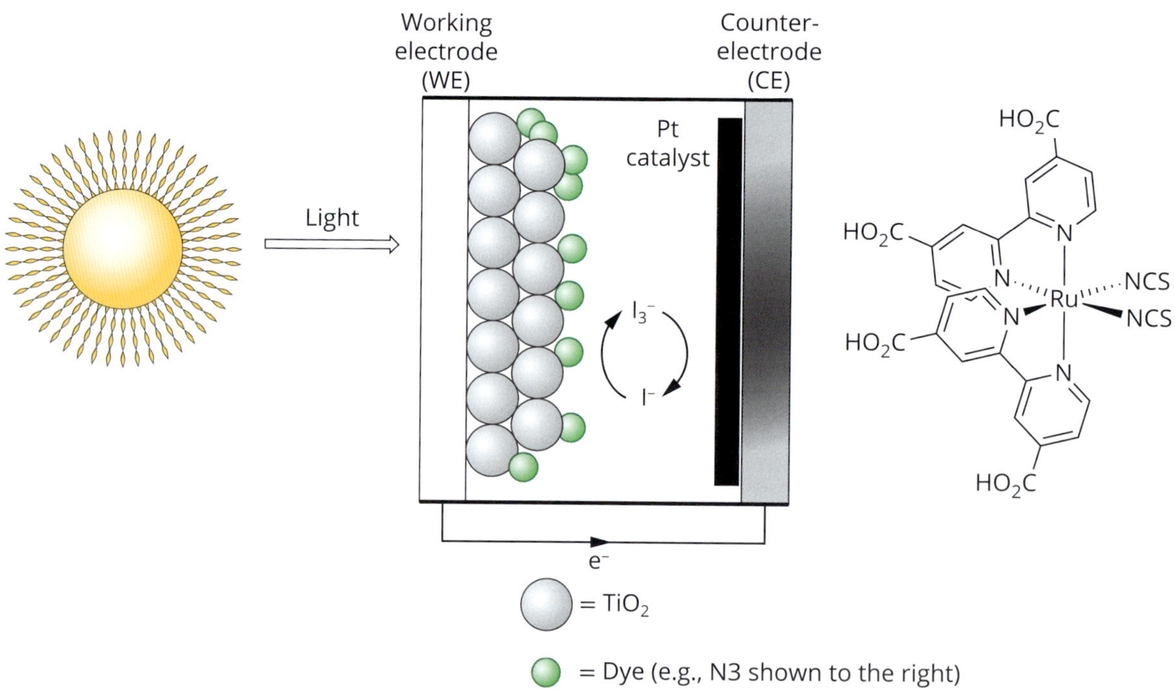

Fig. 1.7 A schematic of the dye-sensitized solar cell. The transparent working electrode (WE) collects electrons from TiO_2 on its surface. Because TiO_2 does not absorb in a very large portion of the solar spectrum, a light-absorbing dye is placed on the TiO_2. Gratzel and coworkers discovered the ruthenium dye shown to the right, the "N_3" dye. Light is shown through the transparent electrode to the dye (green, smaller dots), which absorbs light and injects an electron to TiO_2. The electrons are collected by the WE and travel to the counter electrode (CE). For the dye to be reused, it must be reduced so it can inject another electron. This is done with I^-, which makes elemental I_2. The I_2 reacts with the I^- in solution to make I_3^-, which travels to the CE to be re-reduced to I^-. The maximum energy gained from the light, and process as a whole, is the difference in the energy between the electron injected by the dye and the energy required to re-reduce I_3^-.

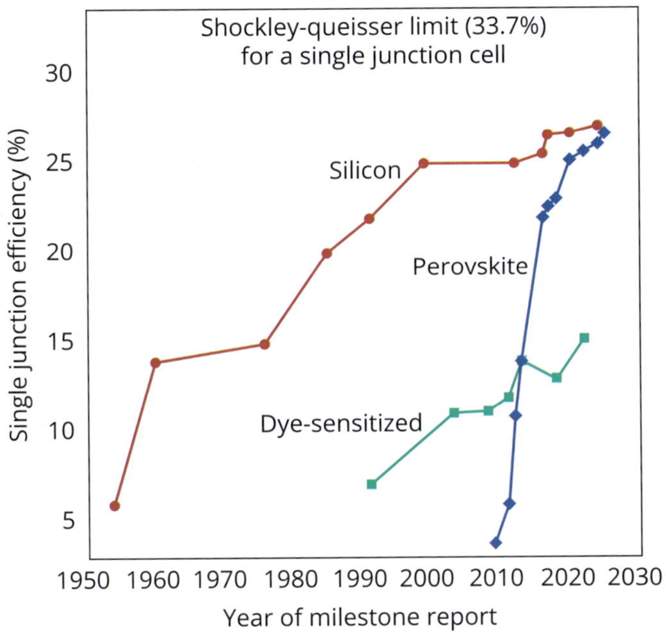

Fig. 1.8 The efficiency of a few, single-junction solar cell types discussed in the text over time as recorded by the National Renewable energy Laboratory (NREL) of the United States. NREL has a far more extensive time line featuring both single and multijunction cells that they maintain on their website.

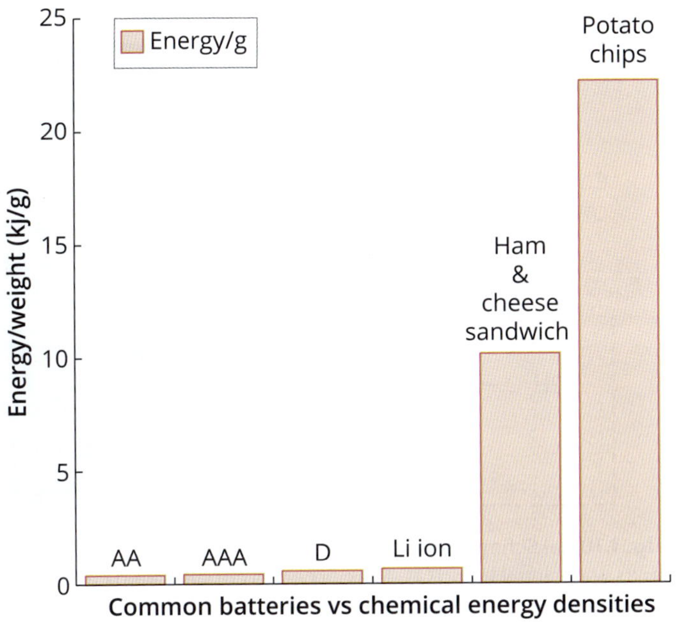

Fig. 1.9 Energy densities of a few common methods of energy storage. Storing energy in chemical bonds (like in food and gasoline) tends to give high energy densities, especially compared to batteries.

chart of the amount of energy per gram for many commonly used systems. In the plot is the amount of energy/weight that common batteries that power our everyday devices (AA, AAA, lithium ion, and so on) store versus chemical sources of energy; namely, the potato chips and ham sandwich you might have for lunch.

Solid-state batteries rely on the reversible storage of electrical charges into electrode materials, and the energy density that can be stored is limited by the maximum number of electrons that each redox-active center of the electrode can accommodate (which is typically not very high), and by the potential required to store each charge (which is generally on the order of 1 volt). Overall, these constraints severely limit the quantity of energy per unit gram of the battery. Comparatively, the energy that can be stored in a chemical bond is typically much higher (higher-energy density), and the energy can be used by feeding the chemical into a fuel cell or simply burned to make heat. Consequently, for large-scale energy storage, like grid-based systems, chemical energy would be far more efficient. Further, energy density is quite important for transportation where heavy energy sources simply require more energy to be used. Of course, a central challenge is to design chemical systems that can be rapidly converted back and forth, without having to waste too much energy to go over the activation barrier, so that energy can be stored and/or extracted quickly, on demand.

From this perspective, H_2 is a very attractive source in many respects as an energy storage medium as it could be produced from water and returned to water when burned. The challenge is that H_2 is a relatively difficult gas to store large quantities of, as it is difficult to liquefy. Nevertheless, hydrogen has been examined extensively as a storage medium for renewable energy sources like solar. Ammonia is easier to liquify and is energy dense, so running the Haber–Bosch reaction backwards could potentially be a way of freeing the H_2 in ammonia.[7–9] Either way, one still has to produce hydrogen gas, which is usually prepared from hydrocarbons but preferably from water.

In 2008, Matt Kanan and Dan Nocera reported the discovery of a cobalt catalyst that could efficiently oxidize water to O_2, which then led the same group to report in 2011 the development of an "artificial leaf", a wireless triple-junction system based on silicon for water splitting. On one side of the artificial leaf, the cobalt phosphate catalyst converts water to O_2 and H^+. On the other side of the thin sheet, H^+ and electrons are converted to H_2 gas by a NiMoZn catalyst. In practice, all one does is place the device in water and shine light on it; the surfaces begin to bubble with H_2 and O_2, storing the energy of the sunlight as the bonds in hydrogen gas (Fig. 1.10).[10–13]

1.2.3 Inorganic Chemistry and Health

Perennially, the two biggest killers in the United States are heart disease and cancer, and inorganic chemistry has made large contributions to the treatment of both. Here we will mention two important cases: cisplatin as a cancer therapeutic, and cardiolite as a diagnostic for heart disease.

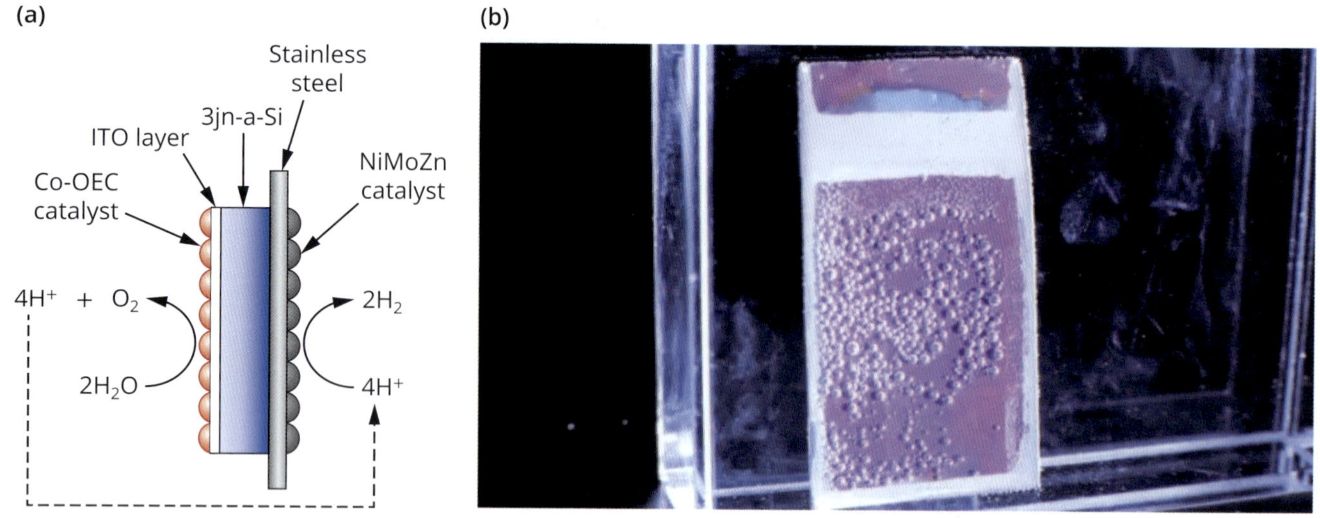

Fig. 1.10 (a) Schematic of the triple-junction cell used in the artificial leaf in 2011. (b) The artificial leaf, when exposed to air and water, splits water into H_2 and O_2. The purple is a cobalt-based catalyst, and the bubbles are O_2. The opposite site has the NiMoZn catalyst for H_2 production.

In the 1970s, Barnett (Barney) Rosenberg (Fig. 1.11), working with Loretta van Camp and Thomas Krigas at Michigan State University, were studying the effects of electromagnetic energy on cell growth. They discovered that effects they were observing were actually due to platinum from the electrodes that combined with chloride and ammonium from the solution. The simple Pt complex formed, "cisplatin" (formally, *cis*-diamminedichloroplatinum(II)), had a dramatic effect on cell growth (Fig. 1.12).[14–16]

Rosenberg spent a great deal of time lobbying for chemotherapy, which was not widely used at that stage in cancer treatment. In 1970, ~90% of patients with testicular cancer died from the disease. Now, cisplatin is very nearly the cure, with a success rate of ~95%, usually in combination with other chemotherapeutics. The more active replacement for cisplatin, carboplatin (Fig. 1.12), which was also developed by Rosenberg and coworkers, is used for a variety of different cancers, including neuroblastoma, brain tumors, lung cancer, and ovarian cancer.

Alan Davison (Fig. 1.13) was a Welshman and an inorganic chemist at the Massachusetts Institute of Technology. He was interested in working with the "new element" technetium (his description to Chris Orvig, one of his graduate students), but the element was, in fact, discovered around the time of his birth in 1936. His laboratory, in collaboration with Prof Alun Jones at Harvard Medical School, developed the imaging drug cardiolite, a homoleptic isonitrile complex of technetium(I).[17–20] The compound is injected into the patient to image the heart. Areas of the heart that remain dark to the technique, Fig. 1.14, are not receiving adequate blood flow (ischemic).[21–22]

It is estimated that about 70% of people diagnosed with heart disease have had a Cardiolite stress test. Other Tc compounds are used for bone, liver, and brain imaging. Some of those were also developed in Alan Davison's laboratories. Prof Davison won many awards, perhaps none more appropriate than the Georg

Fig. 1.11 Barnett Rosenberg (1926–2009), the discoverer of cisplatin.

Cis-platin

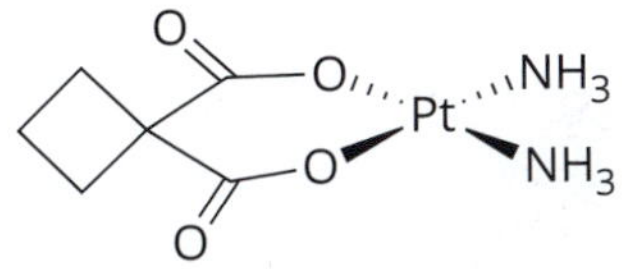

Carboplatin

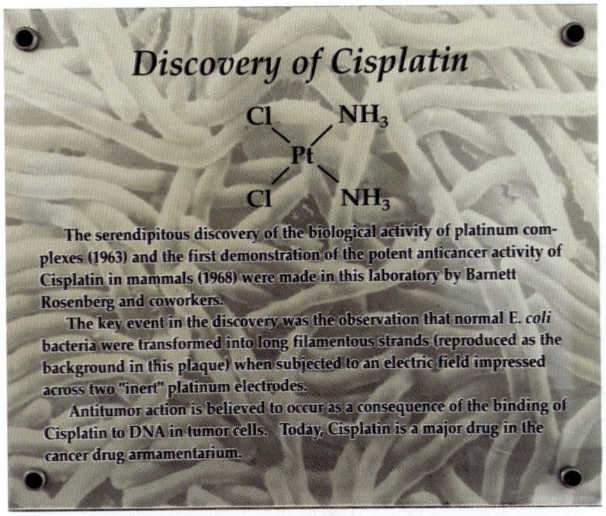

Fig. 1.12 (*top*) Structures of *cis*-platin and carboplatin. (*bottom*) Plaque on the second floor of the chemistry building at Michigan State University. It reads: "The serendipitous discovery of the biological activity of platinum complexes (1963) and the first demonstration of the potent anticancer activity of Cisplatin in mammals (1968) were made in this laboratory by Barnett Rosenberg and coworkers. The key event in the discovery was the observation that normal E. *coli* bacteria were transformed into long filamentous strands (reproduced as the background in this plaque) when subjected to an electric field impressed across two "inert" platinum electrodes. Antitumor action is believed to occur as a consequence of the binding of Cisplatin to DNA in tumor cells. Today, Cisplatin is a major drug in the cancer drug armentarium."

Fig. 1.13 Alan Davison (1936–2015) developed cardiolite for heart imaging.

de Hevesy Nuclear Pioneer Award in 2009 and the American Chemical Society Award for Creative Invention in 2006.

Carolyn Bertozzi (Fig. 1.15) received the Nobel Prize with Morten Meldal and K. Barry Sharpless in 2022 "for the development of click chemistry and bioorthogonal chemistry."[23] Click chemistry is a copper-catalyzed reaction that generates a heterocycle from an azide and alkyne, sometimes called a Husgen 1,3-dipolar cycloaddition.[24–26] Bertozzi's bioorthogonal chemistry involves the use of main-group systems (Chapters 8 and 9) to attach groups, for example, for imaging purposes, to glycans on cells. As the name implies, the strategy uses chemistry foreign to biological systems, which offers advantages in selectivity; that is, biological systems do not typically use groups like azides, so their reactivity can be directed without interference. One example is the use of a Staudinger reaction, reaction of a phosphine with an azide, to attach an imaging group to the cell (Fig. 1.16). The phosphine imide (aza ylide) product of the Staudinger reaction is water sensitive, leading to formation of amine and phosphine oxide. One advance by the Bertozzi group was the use of a nearby ester to accept the amine and make a more water-stable linkage, an organic amide.

1.2.4 Connections and Inorganic Catalysis

Although just a few examples of inorganic chemistry and its applications to societal problems were described above, basic research technologies of this magnitude can cross into applied technologies. Catalysis, especially, is dominated by inorganic compounds and materials and is the center of many of the applications described previously, such as the Haber–Bosch reaction, solar-cell technology, and catalytic converters. Catalysis is used to "crack" petroleum to make gasoline, produce polymers, and make our food.

A single iron-catalyzed reaction, the conversion of H_2 and N_2 to NH_3, which we use to fertilize our fields, expends 1–2% of all the energy we humans use every year. Because of the nature of how the chemistry is done, high pressures and temperatures, the reactions are typically done in large central plants and sometimes shipped long distances. Distributed production schemes that can use renewable energy sources, perhaps on local farms, would be far more efficient and are an active field of study.

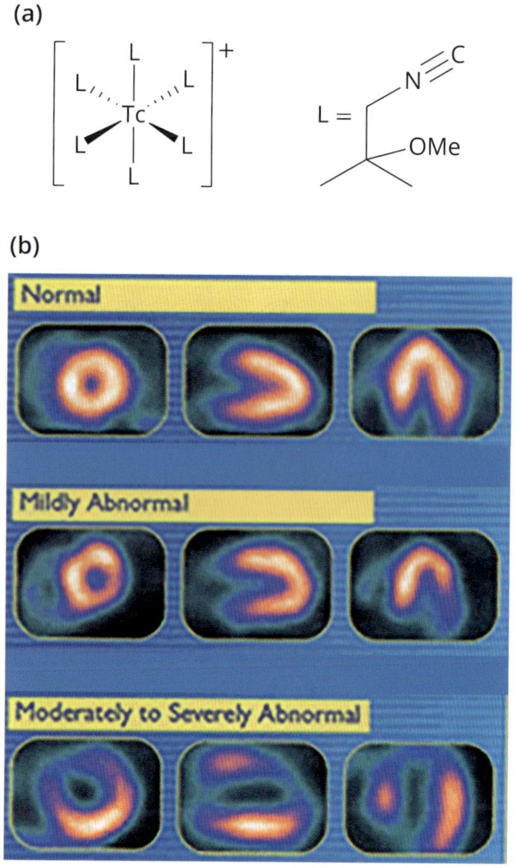

Fig. 1.14 (a) The structure of cardiolite, which uses isonitrile ligands on technetium. (b) Heart images made using cardiolite. The imaging agent is carried in the blood, and dark portions of the heart are not getting sufficient blood flow.

There are many other contributions of inorganic chemistry and places where we can improve. Here are a few:

- Even after the advent of carboplatin and many other chemotherapeutics, cancer remains one of the top killers worldwide. Synthesis and discovery of new drugs are targeted for this and many other maladies, and inorganic compounds are some of the most valuable tools in this arena.
- Inorganic nucleophiles like organolithium and organomagnesium compounds have been employed for well over a century in organic synthesis.
- Palladium-catalyzed reactions of all types (Heck, Suzuki, and Negishi couplings as examples) have become some of the most commonly used methods for new carbon–carbon bond formation in the pharmaceutical industry.
- Asymmetric hydrogenation reactions based on iridium, rhodium, and ruthenium have remained an important method for synthesis since their first applications in the late 1960s. These reactions are some of the most reliable for setting new stereochemical centers.

Fig. 1.15 Carolyn Bertozzi, recipient of the Nobel Prize in Chemistry in 2022.

- Olefin metathesis is an important tool for specialty chemical synthesis, which uses discrete ruthenium or molybdenum catalysts in many cases.

It is estimated that over 20% of the global GDP is contributed by catalysis. While some of the compounds we use every day are inorganic, the field of inorganic chemistry touches our lives in ways that may not be apparent as well, especially through catalysis.

1.3 Specialization and Collaboration in Inorganic Chemistry

Inorganic chemistry, like all of science, is a story about people. By the very nature of the field, no one is an expert in all of inorganic chemistry or the ways that it can be applied. We are all limited, biased creatures. As a result, study in inorganic chemistry, especially with direct applications to areas like energy and health, is often collaborative. This can be seen in many of the previous stories. Alan Davison's work

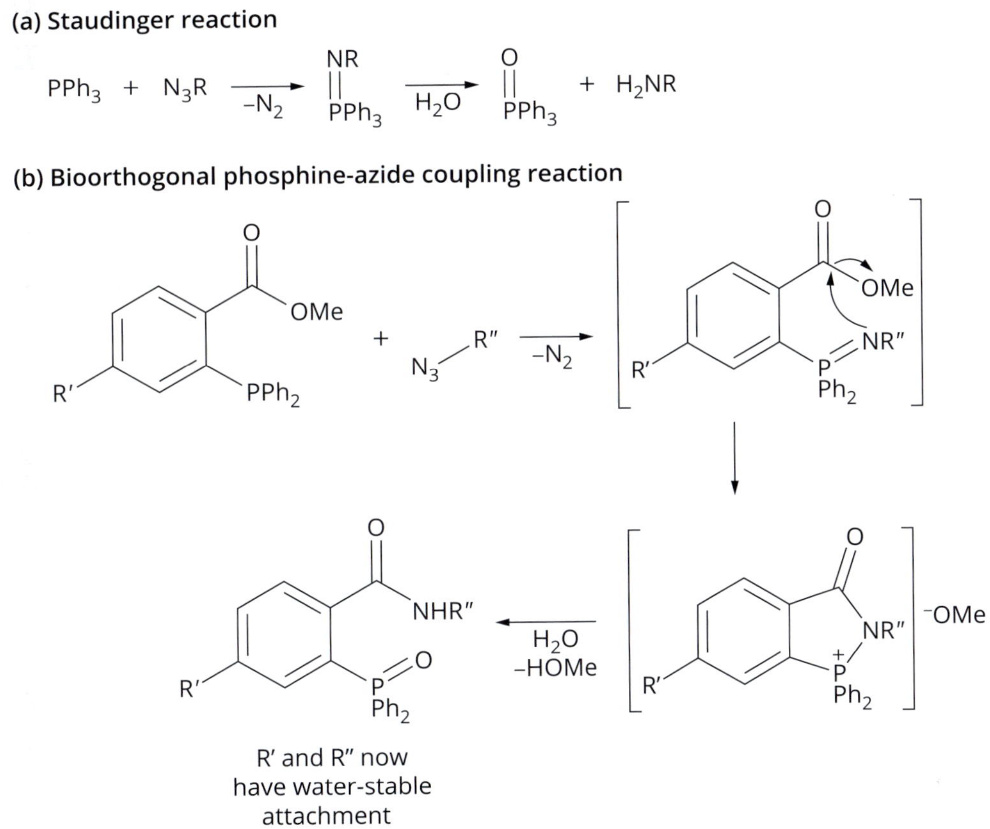

Fig. 1.16 (a) The classic Staudinger reaction. (b) An example of using the Staudinger reaction for bioorthogonal chemistry. The nearby ester traps the reactive phosphine imide, making a stable organic amide.

on imaging with technetium complexes was done with Alun Jones, a professor of radiology at Harvard Medical School. Inorganic chemists collaborate with physicians, engineers, biologists, and others to apply their craft to a variety of different areas.

As a student, it is not unusual to enter a classroom for the first lecture to hear about how the topic you are going to study is one of the most important fields of human endeavor. This seems to be true whether the professor is about to teach law, physics, mathematics, or the writings of Shakespeare. However, one can say with no sense of satire, there are few fields of human endeavor with the breadth, depth, and significance of inorganic chemistry.

One final note: the previous biographical sketches are provided to help humanize the science; it is important to understand that science is done by people, just like you. They have strengths, weaknesses, insecurities, and limitations (just like textbook authors!). Furthermore, while inorganic chemistry has been dominated by men of Jewish and European descent, this is changing rapidly. Today, inorganic chemistry has a young cohort of people who have diversity in viewpoints, ethnicity, and gender, and the field is stronger for it!

Chapter Summary

1. Inorganic chemistry is a very broad field concerning the synthesis, properties, and applications of compounds that are "not organic". This includes most compounds not only containing carbon, hydrogen, nitrogen, and oxygen.
2. Inorganic chemistry has made, and is making, important contributions to key fields like sustainability, energy, and health.
3. Inorganic chemistry, just like all science, is done by people with a variety of goals, strengths, and shortcomings, and they have made important contributions that affect our daily lives.
4. Inorganic chemists often collaborate with scientists in other fields to enable applications of materials and complexes to energy science, biology, and other areas.

Cited References

1. Lancaster, K. M., Roemelt, M., Ettenhuber, P., Hu, Y. L., Ribbe, M. W., Neese, F., Bergmann, U., and DeBeer, S. X-ray emission spectroscopy evidences a central carbon in the nitrogenase iron–molybdenum cofactor, *Science* 2011, *334*, 974–977.
2. Smil, V. *Enriching the Earth: Fritz Haber, Carl Bosch, and the Transformation of World Food Production*. Cambridge, MA: MIT Press, 2001, p. xvii, 338 p.
3. Lewis, N. S., and Nocera, D. G. Powering the planet: Chemical challenges in solar energy utilization, *Proc. Nat. Acad. Sci.* 2006, *103*, 15729–15735.
4. Lewis, N. S., and Nocera, D. G. Powering the planet: Chemical challenges in solar energy utilization, *Proc. Nat. Acad. Sci.* 2007, *104*, 20142.
5. Gratzel, M. Dye-sensitized solar cells, *Journal of Photochemistry and Photobiology C-Photochemistry Reviews* 2003, *4*, 145–153.
6. Kazim, S., Nazeeruddin, M. K., Gratzel, M., and Ahmad, S. Perovskite as light harvester: A game changer in photovoltaics, *Angew. Chem. Int. Ed.* 2014, *53*, 2812–2824.
7. Habibzadeh, F., Miller, S. L., Hamann, T. W., and Smith, M. R. Homogeneous electrocatalytic oxidation of ammonia to N-2 under mild conditions, *Proc. Nat. Acad. Sci.* 2019, *116*, 2849–2853.
8. Little, D. J., Smith, M. R., and Hamann, T. W. Electrolysis of liquid ammonia for hydrogen generation, *Energy & Env. Sci.* 2015, *8*, 2775–2781.
9. Zott, M. D., Garrido-Barros, P., and Peters, J. C. Electrocatalytic ammonia oxidation mediated by a polypyridyl iron catalyst, *ACS Catalysis* 2019, *9*, 10101–10108.
10. Kanan, M. W., and Nocera, D. G. In situ formation of an oxygen-evolving catalyst in neutral water containing phosphate and Co^{2+}, *Science* 2008, *321*, 1072–1075.
11. Kanan, M. W., Surendranath, Y., and Nocera, D. G. Cobalt-phosphate oxygen-evolving compound, *Chem. Soc. Rev.* 2009, *38*, 109–114.
12. Reece, S. Y., Hamel, J. A., Sung, K., Jarvi, T. D., Esswein, A. J., Pijpers, J. J. H., and Nocera, D. G. Wireless solar water splitting using silicon-based semiconductors and Earth-abundant catalysts, *Science* 2011, *334*, 645–648.
13. Surendranath, Y., Kanan, M. W., and Nocera, D. G. Mechanistic studies of the oxygen evolution reaction by a cobalt-phosphate catalyst at neutral pH, *J. Am. Chem. Soc.* 2010, *132*, 16501–16509.
14. Rosenberg, B., Renshaw, E., Vancamp, L., Hartwick, J., and Drobnik, J. Platinum-induced filamentous growth in *Escherichia coli*, *J. Bacteriol.* 1967, *93*, 716–721.
15. Rosenberg, B., Vancamp, L., and Krigas, T. Inhibition of cell division in *Escherichia coli* by electrolysis products from a platinum electrode, *Nature* 1965, *205*, 698–699.
16. Rosenberg, B., Vancamp, L., Trosko, J. E., and Mansour, V. H. Platinum compounds: A new class of potent antitumour agents, *Nature* 1969, *222*, 385–386.
17. Holman, B. L., Jones, A. G., Lister-James, J., Davison, A., Abrams, M. J., Kirshenbaum, J. M., Tumeh, S. S., and English, R. J. A new Tc-99m-labeled myocardial imaging agent, hexakis(*tert*-butylisonitrile)-technetium(I) Tc-99m TBI: Initial experience in the human, *Journal of Nuclear Medicine* 1984, *25*, 1350–1355.
18. Jones, A. G., Abrams, M. J., Davison, A., Brodack, J. W., Toothaker, A. K., Adelstein, S. J., and Kassis, A. I. Biological studies of a new class of technetium complexes: The hexakis(alkylisonitrile)technetium(I) cations, *International Journal of Nuclear Medicine & Biology* 1984, *11*, 225–234.
19. Kronauge, J. F., Noska, M. A., Davison, A., Holman, B. L., and Jones, A. G. Interspecies variation in biodistribution of technetium (2-carbomethoxy-2-isocyanopropane)$_6{}^+$, *Journal of Nuclear Medicine* 1992, *33*, 1357–1365.
20. Piwnicaworms, D., Kronauge, J. F., Holman, B. L. L., Davison, A., and Jones, A. G. Comparative myocardial uptake characteristics of hexakis(alkylisonitrile) technetium(I) complexes effect of lipophilicity, *Investigative Radiology* 1989, *24*, 25–29.
21. Barbarics, E., Kronauge, J. F., Costello, C. E., Janoki, G. A., Holman, B. L., Davison, A., and Jones, A. G. In-vivo metabolism of the technetium isonitrile complex Tc(2-ethoxy-2-methyl-1-isocyanopropane)6^+, *Nuclear Medicine and Biology* 1994, *21*, 583–591.
22. Schibli, R., and Schubiger, P. A. Current use and future potential of organometallic radiopharmaceuticals, *European Journal of Nuclear Medicine and Molecular Imaging* 2002, *29*, 1529–1542.
23. https://www.nobelprize.org/prizes/chemistry/2022/summary/.
24. Kolb, H. C., Finn, M. G., and Sharpless, K. B. Click chemistry: Diverse chemical function from a few good reactions, *Angew. Chem. Int. Ed.* 2001, *40*, 2004–2021.
25. McKay, C. S., and Finn, M. G. Click chemistry in complex mixtures: Bioorthogonal bioconjugation, *Chemistry & Biology* 2014, *21*, 1075–1101.
26. Meldal, M., and Tornoe, C. W. Cu-catalyzed azide-alkyne cycloaddition, *Chem. Rev.* 2008, *108*, 2952–3015.

CHAPTER 2

A Brief Review of Physical Chemistry

"God does not play dice." (displeased with the use of probabilities in quantum mechanics)

— Albert Einstein

"Einstein, stop telling God what to do!" — Niels Bohr

CONCEPT COVERAGE

1. **An overview of the history and theory of quantum mechanics**
2. **The structure of the periodic table**
3. **The shapes and nodal structures of orbitals**
4. **The sizes of orbitals, atoms, and ions**
5. **A brief introduction to *ab initio* methods**

Quantum mechanics is the foundation for much of what will be discussed in this book, and is one of the most important scientific frameworks known. It is estimated that about 30%, and possibly quite a bit higher, of the Gross National Product of the United States is based on inventions made possible to one extent or another by quantum mechanics (transistors, lasers, magnetic resonance imaging, modern chemicals, and so on).

In this chapter we will briefly discuss some of the landmark results in our understanding of atomic structure, much or all of which may be reviewed by students who have completed a physical chemistry course. A course on quantum mechanics would be useful for some of what will be discussed, but is not a requirement. All of the aspects of quantum mechanics used in the book are reviewed here or elsewhere. Those instructing the course may choose to cover some or none of this chapter, depending on how their curriculum is organized, and perhaps assigning it for reading.

Naturally, the material here will be catered to this book's purposes by stressing important points encountered later. The history of quantum mechanics intersects with inorganic chemistry in several places, and the beginnings of the theory are told in a storyline format for ease of reading. Finally, there is a brief discussion of modern computational methods at the novice level that will provide some minimal background for understanding references to computational methods used throughout the book. This last section on computational methods is definitely a "User's Guide" and is not intended as anything approaching complete descriptions of all the techniques discussed. We simply hope to give the student a passing knowledge of a few common methods and some idea of the quality and make-up of basis sets.

2.1 A Brief History of Our Understanding of the Atom and the Periodic Table

As with most science, quantum mechanics began with simple observations without much context. Slowly, the model developed to allow explanation of many decades of results concerning the atom and its structure. To start, we can describe some landmark results, which are summarized in Fig. 2.1.

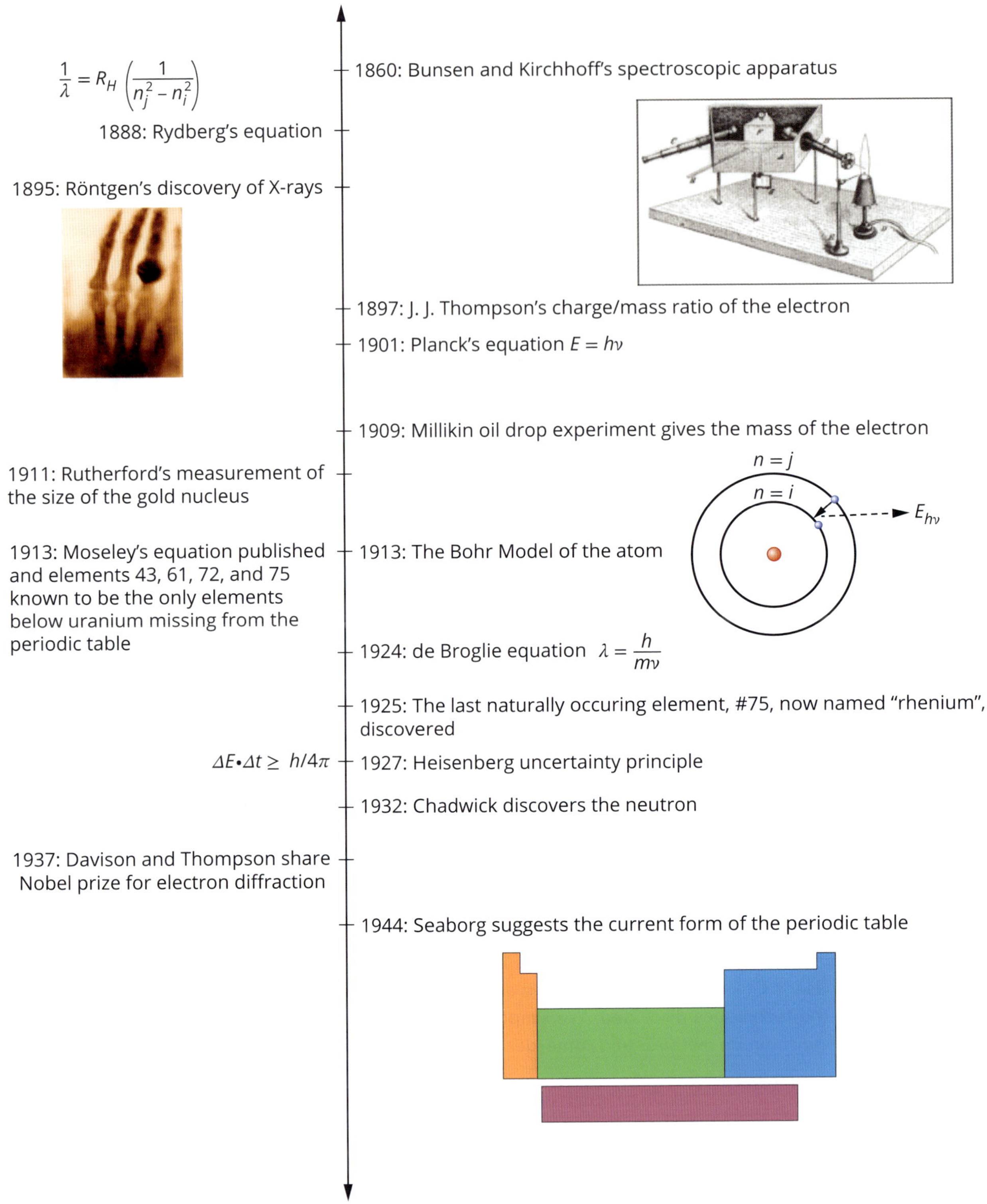

Fig. 2.1 A timeline for the results to be discussed regarding quantum mechanics.

2.1.1 Setting the Stage

Robert Bunsen met Gustav Kirchhoff while they were visiting the University of Breslau in 1851. When Bunsen took a position at the University of Heidelberg in 1852, he arranged for Kirchhoff to teach there as well. Bunsen was an extraordinarily skilled synthetic and analytical chemist who had become interested in the colored flames of elements as a method for the characterization of the many inorganic compounds being prepared in his laboratory.

By 1860, Bunsen had invented his legendary burner, designed to produce a near colorless and extremely hot flame. Kirchhoff suggested that colors resulting from placing compounds containing different elements in the burner's flame could be analyzed by examining their light through a prism (Fig. 2.2).[1] This provided sharp lines characteristic of the emission spectra of the elements in the sample, which were projected onto a wall of the room.

The apparatus was very sensitive. In one reported experiment, 3 mg of sodium chlorate was exploded by exposure to "milk sugar" in the corner of the room farthest from the apparatus.[1] (The room was reported as 60 cubic meters in size.) After a few minutes, a strong, yellow line appeared from the apparatus from the trace sodium in the room's air, which took 10 minutes to dissipate. Many elements are easily identified by their spectral lines in the visible region like Li, K, Sr, Ca, Ba, and others. (For more information, see Table 9.1.)

In the case of sodium two very strong, closely spaced lines at 589.0 and 589.6 nm occur (called the D-lines), along with weaker bands, for example, at 568.8 with 0.7% the intensity of the strongest line. A picture of the two strongest bands in the sodium spectrum is shown in Fig. 2.2, and the color is probably familiar as that of some streetlamps, which in some cases still use sodium bulbs but are rapidly being replaced by LEDs.

Similar excitation spectra for the simplest atom, hydrogen, reveals numerous similar lines. Understanding the line spectra of hydrogen resulted in the real jumping-off point of quantum mechanics: the Bohr model of the atom, which is discussed shortly. The first key development that needs to be discussed related to this occurred in 1888, when Janne Rydberg (1854–1919) developed an equation (based on Balmer's results) for the spectrum of hydrogen.[2] The equation (Eq. 2.1) related the inverse of the wavelength of light (1/λ, which is also called "wavenumbers" and is often denoted in chemistry as cm^{-1}) to a constant, now called the Rydberg constant (R_H = Rydberg constant for hydrogen) of $1.097 \times 10^7\ m^{-1}$. This was a purely empirical observation, and the values n were integers, with $n \geq 1$ and $n_i < n_j$. Spectral analysis was a technique shown to be of great utility for chemical problems by Bunsen and Kirchhoff, and this empirical formula was a tremendous advance for understanding spectroscopy in general and for hydrogen in particular.

$$\frac{1}{\lambda} = R_H\left(\frac{1}{n_j^2 - n_i^2}\right) \qquad 2.1$$

In 1895, Wilhelm Röntgen's laboratory was studying electron beams (cathode rays) emanating from a Crookes tube (Fig. 2.3). William Crookes was an inorganic chemist well known for his studies on main-group compounds such as thallium and selenium, but who became enamored with these electronic devices. Röntgen was following up on the results of Phillip Lenard, who had shown that the rays could be accessed from the Crookes tube if the end were fitted with a thin metallic piece—a 'Lenard window' made of an aluminum sheet thin enough to allow the rays to pass through but still sturdy enough for evacuation of the air from the tube.

Röntgen was attempting to study cathode rays from the tube, which dissipate very quickly in air, with the aid of a phosphor-covered screen and photographic plates. Lenard had reported that the intensity of phosphorescence caused by cathode rays falls off by about half if they travel through about half a centimeter of air at 1 atmosphere of pressure; cathodes rays are quickly dissipated outside of a vacuum.[3]

Röntgen's chosen phosphor was the inorganic complex $Ba\{Pt(CN)_4\}$, which was painted onto a piece of cardboard. The tube, including the Lenard window, had been covered in thick black paper in preparation for the experiments. However, as the instrument was turned on and despite the black covering, the phosphorescent screen would light up *from across the darkened room*! Because of the distance, there was no possible way that cathode rays themselves were causing the screen to phosphoresce.

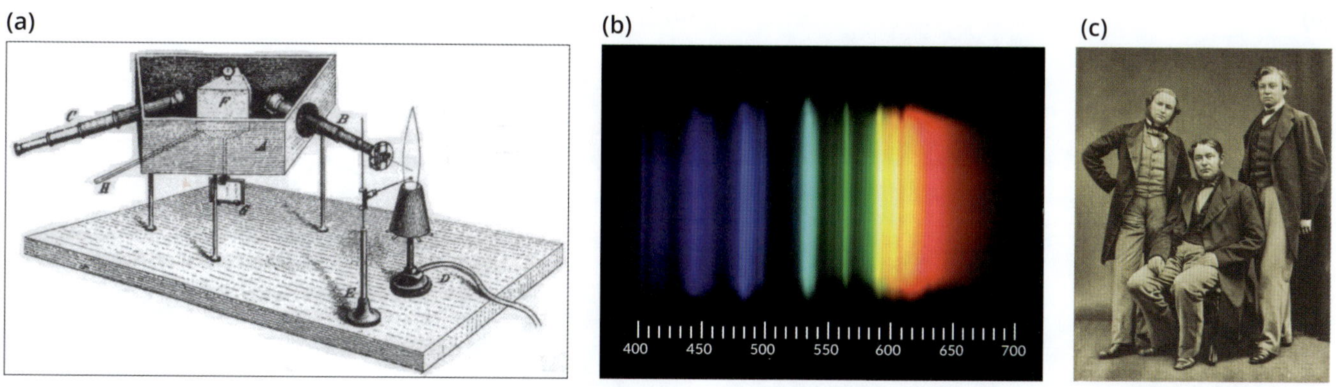

Fig. 2.2 (a) Drawing of Bunsen and Kirchhoff's spectroscopic apparatus from their 1860 publication. The compound to be analyzed was exposed to the burner, and the light from the flame was directed through a prism and focused on a nearby wall.[1] (b) Sodium emission lines due to reaction with water. The very bright lines at 589.0 and 589.6 nm are the familiar yellow D-lines. (c) Photograph taken in 1862 of Bunsen (left), Kirchoff (center), and English chemist Henry Roscoe (right), who was a former student of Bunsen's.

Röntgen worked for eight weeks in isolation to collect his data, discussing nothing with his colleagues, many of whom found out about the discovery after publication of the communication. He discovered that a new type of ray must be the cause and that the new ray had penetrating power through materials roughly proportional with the material's density. Consequently, metals like platinum and lead would stop the rays completely. Imagine his surprise on holding a piece of metal up to the screen and seeing the outline of the bones in his hand, or his wife's surprise on seeing the photographic image he took of her hand, shown in Fig. 2.3.

This discovery is not only important for its practical applications in medicine. Röntgen had discovered the method for the production of X-rays, impinging an electron beam on a metal target, which was a key development in our understanding of atomic structure that informed the research of Henry Moseley (*vide infra*), who made important contributions to the structure of the atom and the periodic table.

In 1897, J. J. Thompson discovered that cathode rays from a Lenard window were composed of particles. He favored "the hypothesis that the carriers of the charges are smaller than the atoms of hydrogen."[3] Thompson had proposed the first subatomic particle, the electron, and went on to determine the charge/mass (e/m_e) ratio of an electron by studying deflection of cathode rays by electric and magnetic fields.

In 1901, Max Planck published his study on "blackbody" radiation. Perhaps the most familiar aspect of radiation due to heating is for hot metals, which can glow red, yellow, and other colors at high heat. Thermal radiation is not isolated to metal objects, as anyone who has seen lava from a volcano can attest. Kirchhoff (yes, the same Kirchhoff mentioned earlier who worked with Bunsen) coined the term "blackbody" for a hypothetical perfect radiator that absorbs all incoming light and, as a consequence, only emits light characteristic of its temperature. As the temperature of the blackbody is increased, the frequency of light emitted increases. Kirchhoff had studied blackbody radiation extensively, and formulated "Kirchhoff's Law of Thermodynamics", which states that at thermal equilibrium the emissivity of an object or surface is equivalent to its absorbance at any given wavelength and temperature.

Planck's studies lead him to propose a new universal constant of nature, h, and the Planck Equation for light, Eq. 2.2, where E is the energy of the light and ν is the frequency. The energy of light was then "quantized" by the constant h. These discrete packets of light were dubbed 'photons' after the Greek word for light, $\phi\omega\varsigma$ (*phos*).

$$E = h\nu \tag{2.2}$$

In 1909, Robert A. Millikin and Harvey Fletcher reported the charge of the electron, measured by examining charged oil drops and the effect of gravity versus an electric field. By combining Thompson's and Millikin's results, one can calculate the mass of the electron, about 9.1×10^{-31} kg, which is astonishingly small at about 1/1800 the mass of the hydrogen atom. As a result, it was concluded that the vast majority of the mass of an atom resides in its nucleus.

In 1911, Earnest Rutherford reported that the size of the nucleus was far smaller than the size of the atom. Using a thin gold foil around 400 atoms thick as a target, α-particles (helium nuclei) were found to largely move through the foil unperturbed but would occasionally be scattered. The assumption is that the high-energy particles easily pass through electron clouds but will be reflected away from a straight path if they strike the heavy nucleus. The experiment will then give a sense of the effective cross-section of the nucleus, σ, which is related to the radius (assuming the nucleus is spherical) by $\sigma = \pi r^2$. If the material is about 400 atoms thick and the nucleus is very small, as they soon found it to be, then the likelihood of there being two nuclei behind one another is very small. Since the particles used (α-particles, which are helium nuclei) only respond to nuclei, the gold foil can be regarded as a sheet of N particles with cross-section σ, and the fraction of the area (A) of the gold occupied by the nuclei is

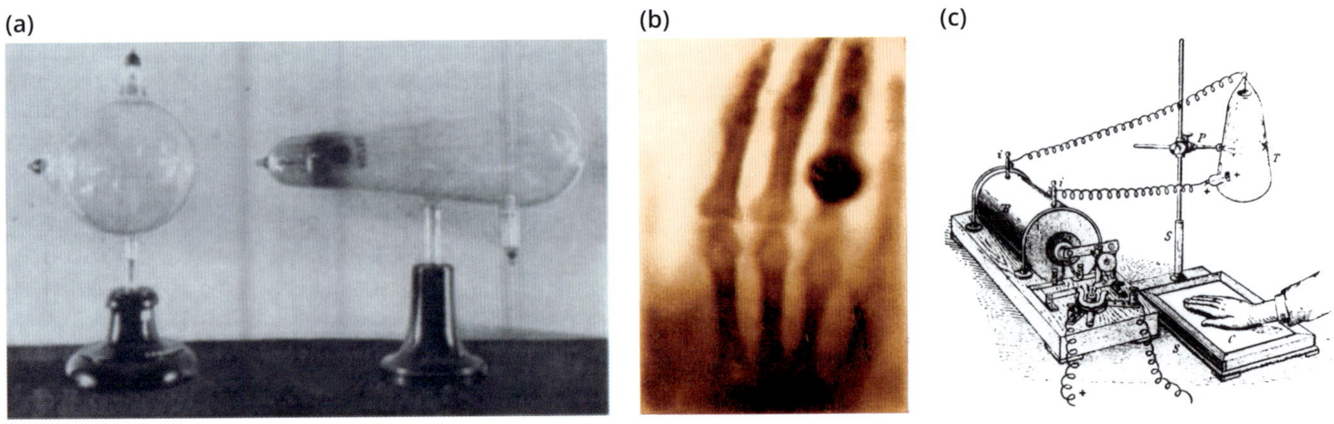

Fig. 2.3 (a) Photograph of Röntgen's Hittorf–Crookes tubes in the Deutsche Museum in Munich. (From *Wilhelm Conrad Röntgen and the Early History of the Roentgen Rays* by Otto Glasser.) (b) The first X-ray image taken by Röntgen of his wife's hand in 1895.[4] (c) Röntgen's drawing of his apparatus for imaging the hand.

then $N\sigma/A$. Because of all of this, a large number of the particles shot at the foil will simply pass through, and let us call the small fraction of particles that happen to strike nuclei n. The radius of the nuclei can then be calculated from this equation.[5]

$$r^2 = n\frac{A}{N\pi}$$

The gold atom was known to be about 10^{-10} m (1 Ångstrom) thick. From the scattering, the size of the nucleus was found to be about 10^{-15} m; that is, 100,000 times smaller than the atom. Consequently, we have a situation where the nucleus contains the vast majority of the mass but occupies very little space relative to the whole atom, a system analogous to our solar system. Indeed, this picture of the atom is sometimes called the "planetary model", which has small electrons orbiting a heavy nucleus, and the physics of the solar system is even valuable as a starting point in describing such a system.

Rutherford is famous for saying, "Physics is the only real science. The rest are just stamp collecting." One wonders if those words rang in his ears as he accepted the Nobel Prize *in Chemistry* in 1908. Rutherford's laboratory spawned many great scientists, including notable chemists such as George de Hevesy, who would win his own Nobel Prize in Chemistry in 1943 for the development of radioactive tracers for studying biological processes.

In 1924, Louis de Broglie, based on Special Relativity and the Planck Equation, postulated wave–particle duality, which in essence suggests that all matter has both wave and particle natures. The derivation is quite simple. Einstein's Special Relativity, which relates energy to mass, is $E = mc^2$, where m = mass and c = speed of light in a vacuum. The Planck Equation is shown in Eq. 2.2. One also needs the relationship $\nu = c/\lambda$, where ν (Greek nu) = the frequency of light and λ = the wavelength.

$$E = mc^2$$

$$E = h\upsilon = \frac{hc}{\lambda}$$

so

$$mc^2 = \frac{hc}{\lambda}$$

$$\lambda = \frac{h}{mc}$$

Simply generalizing the equation to objects that do not move at the speed of light, c is replaced with v (velocity), to produce the de Broglie Wave Equation (Eq. 2.3).

$$\lambda = \frac{h}{mv} \qquad \textbf{2.3}$$

If all matter is a wave, we can take any particle with mass and velocity and find its wavelength using Eq. 2.3. Let us try a macroscopic particle moving very quickly, a Nolan Ryan fastball moving at 100 mph (44.7 m/s). The weight of a regulation baseball should be between 142 and 149 g, so let us round that to two significant figures and use 0.15 kg.

$$\lambda = \frac{h}{mv} = \frac{6.63 \times 10^{-34} m^2 \cdot kg \cdot s^{-1}}{(44.7 \times 0.15) m \cdot kg \cdot s^{-1}} = 9.9 \times 10^{-35} m$$

Using the de Broglie Equation we find that the wavelength is too miniscule, 9.9×10^{-35} m, to be able to observe a baseball's wave properties. In order to diffract a baseball we would need a grating of similar size to the wavelength, and there is no material we can use to make an appropriate grating, because all atoms are far too large (~10^{-10} m).

As a second example we can take an electron moving at 5% of the speed of light, a reasonable speed for a tiny electron. Using Eq. 2.3, we find that the wavelength of the electron is 4.8×10^{-11} m. The d-spacing, the distance between planes of electron density, in a nickel lattice, is 9.1×10^{-11} m, the same order of magnitude. In other words, by altering the velocity of the electrons in a cathode ray and using a nickel crystal as a grating, one can observe diffraction of electrons. This experiment, done by Davison and Germer in 1927, confirmed that electrons are waves. Furthermore, they were able to show that the de Broglie relationship was correctly followed using their experiment by changing the wavelength through changing the velocity of the electrons.

In a related experiment, George P. Thompson used a thin metal sheet that showed rings on a photographic plate when a beam of electrons was passed through; the rings are due to an interference pattern (a distinctly wave-like phenomenon).

Clinton Davison and George Thompson shared the Nobel Prize in Physics in 1937 for the discovery of electron diffraction.

In 1927, Werner Heisenberg, a German physicist, postulated the Uncertainty Principle based on the de Broglie Equation. The Uncertainty Principle places limits on the accuracy of values for atomic and subatomic particles, and is a property that arises due to the wave–particle nature of matter; it has nothing to do with uncertainty due to instrumentation (*vide infra*).

There are two commonly employed expressions of the Uncertainty Principle. The first is that the uncertainty in momentum (Δp) times the error in position (Δx) is greater than $h/4\pi$ (Eq. 2.4). The second relation, Eq. 2.5, says that the product of the uncertainty in energy (ΔE) and the uncertainty in time (Δt) is greater than a constant.

$$\Delta p \cdot \Delta x \geq h/4\pi \qquad \textbf{2.4}$$

$$\Delta E \cdot \Delta t \geq h/4\pi \qquad \textbf{2.5}$$

As an illustration, consider a common problem in many physical chemistry laboratories, the energetic breadth of a short laser pulse. Short laser pulses can be used to measure very fast chemical phenomena, and pulses down to ~100 attoseconds (100 as = 100×10^{-18} s) have been employed. Many different timescales are commonly used in chemistry, such as nano-, pico-, and femtosecond, depending on the problem being studied.

If one is using a 100 fs pulse with an uncertainty of 15 fs with an average wavelength of 450 nm, what is the minimum range of wavelengths (energy) for your pulse? First, you can calculate ΔE using Eq. 2.5.

$$\Delta E \geq \frac{h/4\pi}{\Delta t} = \frac{5.2728 \times 10^{-35}\, J \cdot s}{15 \times 10^{-15}\, s} = 3.515 \times 10^{-21}\, J$$

Now we know the range of error in the energy of the photons. When dealing with energies, using nanometers can make things difficult because this unit is not linearly related to energy. To convert from 450 nm to J, we can use the following relationship, which is the Planck Equation (Eq. 2.2) with the frequency (ν) substituted; this uses the important relation worth memorizing, $\nu = c/\lambda$, where c = speed of light and λ = wavelength.

$$E = \frac{hc}{\lambda} = \frac{(6.626 \times 10^{-34}\, J \cdot s) \times (300 \times 10^{6}\, m/s)}{450 \times 10^{-9}\, m}$$
$$= 4.4173 \times 10^{-19}\, J$$

So, the center of the peak has photons of energy 4.4173×10^{-19} J, and the peak width is at least 3.5153×10^{-21} J. If we say half of the breadth is before the peak and the other half after—that is, the peak is symmetrical in energy—we can calculate the breadth of the peak in Joules. Converting back to nanometers gives a peak from 453 nm to 448 nm. Again, the Uncertainty Principle says the peak will be at least this wide, and this is not a function of spectral width of the source, optics, and so on. If the time frame of the peak is known to within 15 fs, then this is the best spectral width obtainable according to wave mechanics.

What happens if we increase the accuracy of the pulse width? If we have a 100 fs pulse, and we know the temporal error by 5 fs, we can calculate what the minimum pulse width must be around the 450 nm average energy. Now the peak would be from, at least, 445 nm to 461 nm. As we know the time frame of the peak to greater accuracy, the energy broadens.

This behavior is seen in the mathematics of Eqs. 2.4 and 2.5. The differences on the left are equal to a constant ($h/4\pi$). As a result, if one knows momentum more accurately (small Δp), then the position must spread out, giving a large Δx, and vice versa. This is by no means intuitive and is a consequence of the wave properties of matter that are important at the atomic scale.

Illustrations of the momentum-and-position uncertainty relationship are common in quantum mechanics courses. For example, one can show using the mathematics for a "particle in a box" that is the uncertainty in the position of the particle is decreased (the box gets smaller), the uncertainty in the momentum increases.

The Heisenberg Uncertainty Principle is not to be confused with what is known as the Observer Effect. In quantum mechanics, the measurements are on particles of such size that when we attempt to measure them we are often forced to use particles of comparable size or mass; for example, detecting an electron with a photon. This is a little like trying to measure the position of an SUV by crashing into it with a car: one may obtain a measurement of the SUV's position, but it is unlikely to be in the exact same state. These dented fenders (changes in state due to collision) are common in measurements on small particles. It is often the case that making measurements alters the state of the molecule or atom being measured; in fact, many useful measurements commonly employed in chemistry occur at resonant frequencies for the molecule, causing specific changes in the state of the molecule. Again, *this effect is not the uncertainty principle*, but is instead simply a consequence of the nature of observations at the particle level.

Because of these scientific results, a new model for the atom was created, one where very light electrons circled a heavy, small nucleus. The energy of light was quantized. In addition, electrons (and other matter) were found to have wave properties. More often than not, chemists find thinking of electrons as waves more useful than considering them as particles, but this is not always the case, depending on the property or system of interest.

2.1.2 The Bohr Atom[6]

Based on the results just discussed, the atom may be described as being similar to the solar system using classical mechanics and using Kepler's Laws of Planetary Motion. Similar to the Earth orbiting the Sun, the very light electron circles the much heavier nucleus in this "planetary model" of the atom.

2.1.2.1 The Bohr Atom (Advanced Topic)

Niels Bohr was able to derive the first effective model (Fig. 2.4a) of the atom starting from classical arguments. Here, a simple derivation will be given, starting with the simple figure. The radius of the circular path of the electron (r) and the velocity of the electron (v) are shown.

The outward force on the electron (F_o) is as shown in this equation.

$$F_0 = \frac{m_e \upsilon^2}{r}$$

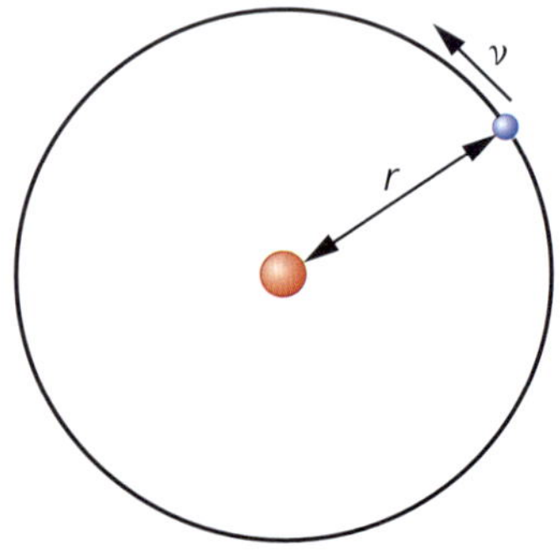

Fig. 2.4a The Bohr model of the hydrogen atom with an electron circling a nucleus at radius r with velocity v.

The electrostatic force holding the electron to the hydrogen nucleus (F_e) is calculated using Coulomb's law, where Z = charge of the nucleus. For the derivation, it is convenient to express the electric charge in a cgs unit "statcouloumbs", sometimes called "esu", for which we will use the symbol e_{sc}. To carry out calculations with the equations, it is easier to convert to SI units. When the formula uses the electric charge in SI, it will be listed as simple e. The conversion will be discussed subsequently, but it involves converting e_{sc} to $e/\sqrt{4\pi\varepsilon_0}$, where ε_0 = permittivity of a vacuum.

$$F_e = -\frac{Ze_{sc}^2}{r^2}$$

For a stable orbit, the outward and inward forces must be equal: $F_o = F_e$. Setting them equal and simplifying gives the equations

$$\frac{m_e v^2}{r} = \frac{Ze_{sc}^2}{r^2}$$

$$m_e v^2 = \frac{Ze_{sc}^2}{r} \qquad \textbf{2.6}$$

Leaving this for a moment, the total energy (E) of any system is the sum of the kinetic (KE) and potential energies (PE).

$$E = KE + PE$$

The kinetic energy of the particle is

$$KE = \tfrac{1}{2} m_e v^2$$

The potential energy, again from Coulombic attraction, is

$$PE = -Ze_{sc}^2/r$$

So, the total energy is

$$E = \tfrac{1}{2} m_e v^2 - Ze_{sc}^2/r \qquad \textbf{2.7}$$

Previously, we derived a different expression that included the term $m_e v^2$ (Eq. 2.6). Substituting into Eq. 2.7 gives Eq. 2.8.

$$E = \tfrac{1}{2} Ze_{sc}^2/r - Ze_{sc}^2/r = -Ze_{sc}^2/2r \qquad \textbf{2.8}$$

What this expression implies is that the electron could continuously give off (negative sign) energy in the form of light to collapse into the nucleus. So why does this not occur? Bohr, thinking back to Planck's results, reasoned that *quantization* is what prevents this collapse. He then assumed that only certain orbits, certain r, are stable.

Since we know that electrons have wave properties (*vide supra*), an assumption can be made that the circumference of the orbit must equal a whole number of wavelengths. The circumference of a circle = $2\pi r$, so the wavelength must be $n\lambda$ (Eq. 2.9), where λ = the wavelength. The value n is a nonzero, positive integer called the "principal quantum number".

$$2\pi r = n\lambda \qquad \textbf{2.9}$$

The angular momentum, l, can then be calculated as well as $l = mvr$. According to the de Broglie relationship (Eq. 2.3), $mv = h/\lambda$. Combining these equations and using $\frac{h}{2\pi} = \hbar$ gives Eq. 2.10. In this, $l = \hbar n$, implying that angular momentum is also quantized (Eq. 2.10).

$$l = mvr = \frac{hr}{\lambda} = \frac{hr}{2\pi r/n} = \frac{nh}{2\pi} = \hbar n \qquad \textbf{2.10}$$

The centripetal force between the electron and nucleus was balanced with the Coulombic attraction earlier as a criterion for a stable orbit (Eq. 2.6). Solving Eq. 2.6 for the radius (r) gives Eq. 2.11.

$$r = \frac{Ze_{sc}^2}{m_e v^2} \qquad \textbf{2.11}$$

While discussing the angular momentum, we developed Eq. 2.10, which includes $mvr=nh/2\pi$. This can be rearranged to Eq. 2.12 for the velocity (v).

$$v = \frac{nh}{2\pi m_e r} \qquad \textbf{2.12}$$

Substitution of Eq. 2.12 into Eq. 2.11 gives

$$r = \frac{Ze_{sc}^2}{m\left[\frac{nh}{2\pi mr}\right]^2} = \frac{Ze_{sc}^2 4\pi^2 m_e^2 r^2}{m_e n^2 h^2}$$

$$r = \frac{Ze_{sc}^2 4\pi^2 m_e^2 r^2}{m_e n^2 h^2}$$

Simplifying and solving for r gives Eq. 2.13.

$$r = \frac{n^2 h^2}{4\pi^2 m_e Ze_{sc}^2} \qquad \textbf{2.13}$$

For the hydrogen atom, $Z = 1$ and $n = 1$, and the radius is generally given the symbol a_0 and is called the Bohr radius.

$$a_0 = \frac{h^2}{4\pi^2 m_e e_{sc}^2}$$

This can be considered a calculation in the non-SI unit of esu (also called statcoulomb). To convert to SI units we replace e_{sc}

in the previous expression with $e/\sqrt{4\pi\varepsilon_0}$, where ε_0 = permittivity of a vacuum (Eq. 2.14).

$$a_0 = \frac{\varepsilon_0 h^2}{m_e e^2 \pi} \qquad \textbf{2.14}$$

Using $\varepsilon_0 = 8.854 \times 10^{-12}\ C^2/N \cdot m^2$,[a] $h = 6.62608 \times 10^{-34}$ Js, $m_e = 9.10939 \times 10^{-31}$ kg, and $e = 1.602177 \times 10^{-19}$ C gives a value for the Bohr radius of 0.531 Å. (1 Å = 1×10^{-10} m).

$$a_0 = \frac{\varepsilon_0 h^2}{m_e e^2 \pi} = \frac{(8.85 \times 10^{-12}\ C^2/N \cdot m^2)(6.63 \times 10^{-34}\ Js)^2}{(9.11 \times 10^{-31} kg)(1.60 \times 10^{-19}\ C)^2(3.14)}$$
$$= 0.531 Å$$

Bohr was able to solve for the energy of the electron as well. Previously, we derived Eq. 2.8 for the energy of the atom. Substituting Eq. 2.13 for the radius into Eq. 2.8 produces Eq. 2.15, where a subscript "i" is added to the energy (E_i) to indicate that this is the energy of the i-th level (n_i).

$$E = -\frac{Ze_{sc}^2}{2r} = -\frac{Ze_{sc}^2}{2\left[\dfrac{n^2h^2}{4\pi^2 m_e Ze_{sc}^2}\right]} = -\frac{4\pi^2 m_e Z^2 e_{sc}^4}{2n^2h^2} \qquad \textbf{2.15}$$

$$E_i = -\frac{2\pi^2 m_e Z^2 e_{sc}^4}{n_i^2 h^2}$$

Niels Bohr's suggestion (Fig. 2.4b) was that the Rydberg description for the emission of the hydrogen atom was describing transitions between energy levels. These energy levels are quantized in line with Planck's findings and are described by the integer values n in the Rydberg expression. In other words, as an electron falls from orbit n_j to an orbit closer to the nucleus and lower in energy, n_i, a photon is released. Using our classical circular orbits, we have the picture in the figure.

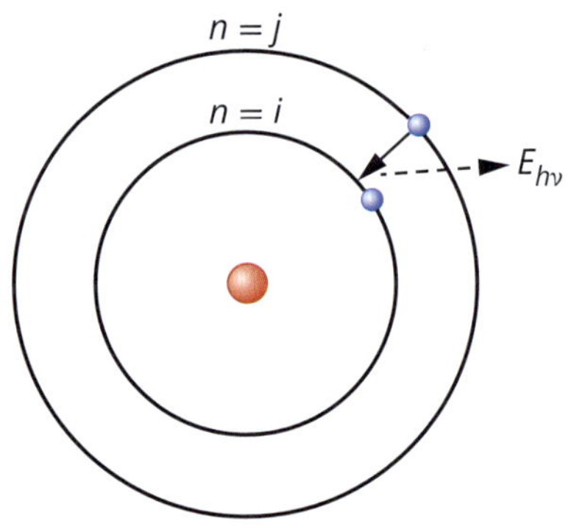

Fig. 2.4b Bohr's rationale for emission from an atom involving a jump from different levels ($n = i$ and j) with release of a photon of energy $E_{h\nu}$.

[a] N is the unit of newtons = kg · m/s². A joule is J = kg · m²/s².

The energy of the photon given off ($h\nu$) in the emission is then the difference in energy between the starting and ending electron energy, the energy required to jump from one quantized level to another. If one is looking at emission, the photon given off is the difference in energy (Eq. 2.16) between the higher-energy j-th level (E_j) and the lower-energy i-th level (E_i).

$$h\nu = E_j - E_i \qquad \textbf{2.16}$$

Substituting the equation for the energies (Eq. 2.15) into Eq. 2.16 gives

$$h\nu = E_j - E_i = -\frac{2\pi^2 m_e Z^2 e_{sc}^4}{n_j^2 h^2} + \frac{2\pi^2 m_e Z^2 e_{sc}^4}{n_i^2 h^2}$$
$$= \frac{2\pi^2 m_e Z^2 e_{sc}^4}{h^2}\left(\frac{1}{n_j^2 - n_i^2}\right)$$

Solving for the frequency gives Eq. 2.17.

$$\nu = \frac{2\pi^2 m_e Z^2 e_{sc}^4}{h^3}\left(\frac{1}{n_j^2 - n_i^2}\right)$$

Again, to convert to SI units, we replace e_{sc} in the previous expression with $e / \sqrt{4\pi\varepsilon_0}$ to obtain Eq. 2.17.

$$\nu = \frac{m_e Z^2 e^4}{8\varepsilon_0^2 h^3}\left(\frac{1}{n_j^2 - n_i^2}\right) \qquad \textbf{2.17}$$

The Rydberg formula was given in Eq. 2.1 and is reproduced here.

$$\frac{1}{\lambda} = R_H\left(\frac{1}{n_j^2 - n_i^2}\right)$$

The Rydberg formula was expressed as "wavenumbers", $1/\lambda$. Using $\nu = c/\lambda$ in Eq. 2.17,

$$\frac{c}{\lambda} = \frac{m_e Z^2 e^4}{8\varepsilon_0^2 h^3}\left(\frac{1}{n_j^2 - n_i^2}\right)$$
$$\frac{1}{\lambda} = \frac{m_e Z^2 e^4}{8\varepsilon_0^2 h^3 c}\left(\frac{1}{n_j^2 - n_i^2}\right)$$

This equation defines the Rydberg constant for hydrogen (R_H) (Eq. 2.8) as a collection of fundamental constants (Eq. 2.18).

$$R_H = \frac{Z^2 e^4 m_e}{8h^3 c\varepsilon_0^2} \qquad \textbf{2.18}$$

2.1.2.2 Overview of the Bohr Model

The Bohr model of the hydrogen atom was a huge success for early quantum mechanics. The model has a few simple features which were described in detail in the previous section and are summarized here.

- The electrons circling the nucleus have specific momenta to ensure the orbit is stable without the electron crashing toward the nucleus or flying away from the atom. Classical physics can take you part of the way to a description of the atom.
- A better description of the atom includes that the electrons cannot have intermediate energy values between stable orbits, because of the wave nature of electrons. In other words, the energies are quantized.
- Bohr was able to rederive the Rydberg equation and describe the fundamental constants found in the Rydberg constant. Electrons "hop" from one stable orbit to the next by absorbing or emitting light.

$$\frac{1}{\lambda} = R_H\left(\frac{1}{n_j^2 - n_i^2}\right) = \frac{m_e Z^2 e^4}{8\varepsilon_0^2 h^3 c}\left(\frac{1}{n_j^2 - n_i^2}\right)$$

- Bohr could calculate what the radius of the hydrogen electron should be, 0.53 Å, which is now called the Bohr radius (a_0) and is a collection of fundamental constants like the mass of the electron and Planck's constant. The Bohr radius is sometimes used to define distances in computational studies.

$$a_0 = \frac{\varepsilon_0 h^2}{m_e e^2 \pi}$$

Although an enormous success, the Bohr model could not be applied to systems with multiple electrons, a result of electron–electron interactions not being in the theory. Addition of these and other effects has led to more quantitative equations for atoms. Some attention will be given to understanding these electron–electron interactions later as they are of great importance in fields like spectroscopy and magnetism.

2.1.3 Henry Moseley and Building the Periodic Table

By the time that the Bohr model was published, chemists had arranged much of the periodic table. The arrangement had taken shape based on the properties of the elements, both chemical and physical. The vast majority of elements can be placed in correct order by simply looking at their atomic masses—values that were fairly readily measured. There are exceptions, however. For example, nickel (element 28) has an atomic weight of 58.693, and cobalt (element 27) has a heavier atomic weight of 58.933. The two elements were correctly grouped so that Ni was with Pd and Pt and so that Co was with Rh and Ir based on their properties by inorganic chemists.

The problem was that no one knew what the atomic numbers actually meant or if they meant anything at all. The elements were simply numbered from approximately lightest to heaviest (pairs like Co and Ni being exceptions to simply using the weight). In fact, Mendeleev, who was magnificent in his success in arranging the periodic table and even anticipating the properties of some undiscovered elements, had postulated two elements lighter than hydrogen, one he called Coronium because he thought it existed in the corona of the Sun. Fortunately, due to lack of evidence they never graced many periodic tables.

Henry Moseley (1887–1915) was an English physicist working at a laboratory in Oxford. Moseley was examining the X-ray spectra of elements using a technique similar to Röntgen's experiment in which the elements were exposed to an electron beam, which caused them to generate X-rays. Moseley found, similar to the Bunsen–Kirchhoff experiments, that the spectrum generated was characteristic of the element. More importantly, *Moseley discovered that the frequency (ν) of what is called the K_α band was related to the atomic number by the empirical formula in Eq. 2.19.* The atomic number was then a measurable quantity, which correlated to the amount of positive charge in the nucleus.

$$\nu(K_\alpha) = 2.47 \times 10^{15}(Z-1)^2\ Hz \qquad \textbf{2.19}$$

X-rays are classified by the element used to create them by impinging an electron beam on a target of the element, such as Cu or Mo, and the nature of the band (Fig. 2.5a). If we classify the principal quantum shells as $n = 1$ (K), 2, (L), 3 (M), and so on, we can begin to classify the X-ray radiation from different target elements. If the electron is in the L shell and moves down to the K shell, a K photon (X-ray) is generated (Fig. 2.5b). If the electron is in the M shell and moves to the K shell, an M photon (X-ray) is generated. If the two quantum levels differ in primary quantum number n by 1, then a subscript α is added, and if they differ by 2 in n then a subscript β is added, and so on. If multiple transitions of the same type are present, then a subscript 1, 2, 3, and so on is added, starting with the highest-energy transition; for example, $K_{\alpha 1}$. A K_α X-ray is formed by relaxation of an electron from the L shell to the K shell.

Moseley's equation was soon examined in light of the Bohr model, which was also published in 1913. The constant that Moseley measured was, like Rydberg's constant, an explicable collection of fundamental constants in the light of the Bohr model.

Using this method, Moseley showed that the gaps in the periodic table of the time were for atomic numbers 43, 61, 72, and 75, which are now known as technetium, promethium, hafnium, and rhenium, respectively. Moseley's method provided a clear method for the identification of elements. This was useful, since some elements, especially the lanthanides, were notoriously difficult to purify and identify.[8]

The lanthanides all have chemical properties that are remarkably (or "unfortunately" is perhaps a better word if you are trying to separate them) similar, which makes their properties

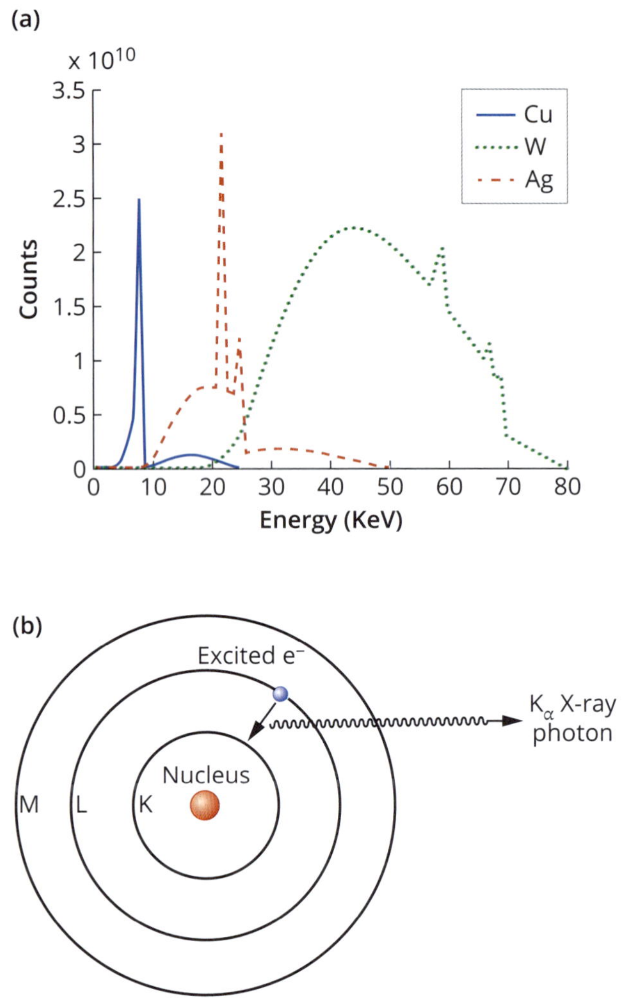

Fig. 2.5 (a) Spectrum of intensity (*I*) vs wavelength (λ) for X-rays made from a beam of electrons impinging on Cu, W, and Ag. (b) An electron from the L shell falls to an unoccupied position in the K shell, producing a K_α X-ray photon.[7]

comparable despite different electron counts. Chemists had identified the elements lanthanum to lutetium as having properties consistent with rare earths (a general name for the lanthanides and Group 3). When Moseley had shown there was a gap in the periodic table at number 72, Georges Urbain claimed that he had found the element and reported it in 1911. Urbain had said the element was found among other rare earth elements, implying that element 72 (for the time named "celtium") had properties similar to the rare earths.

Contrary to Urbain's claim, chemist Charles Bury had suggested in 1921 that element 72 should have similar properties to titanium and zirconium, based on ideas by chemist Irving Langmuir. Niels Bohr, using quantum-mechanical arguments, suggested that the element belonged in Group 4 as well. Furthermore, chemist Fritz Paneth suggested that the element might be found in zirconium ores.

Finally, in 1923, George de Hevesy—a chemist and student of Rutherford who would later earn a Nobel Prize for his work in radiotracers—and physicist Dirk Coster worked together on the discovery of element 72. The compound was discovered, as predicted by Paneth, in zirconium-based minerals and was named "hafnium", after the Latin name for Copenhagen, *Hafnia*.

Georges Urbain was not alone in the assignment of a new element proven later to be incorrect; this occurrence was fairly commonplace at the time. Often, the researcher simply had a mixture of known elements which they mistook for a new single element. In 1909, Masataka Ogawa claimed the discovery of two new elements extracted from samples of thorianite (ThO_2) and molybdenite (MoS_2). The elements based on chemical properties were claimed to be the elusive elements of atomic numbers 43 and 72. Ogawa was working in London with Sir William Ramsay, the discoverer or first to isolate all of the "inert gases"—helium through radon. It was Ramsay who suggested the name "nipponium" for the lighter element that Ogawa had discovered.[9] Unfortunately, no one could reproduce the result, and the claim was discarded. In 1997, re-evaluation of the X-ray spectrum of the sample used by Ogawa confirmed that he had indeed discovered a new element, only it was not element 43; it was the element nestled right below it in the modern periodic chart, now called rhenium.[10]

Rhenium was the last naturally occurring element to be unambiguously identified, and this was accomplished in 1925 by Otto Berg along with the husband and wife team of Walter and Ida Noddack.

In 1932, James Chadwick, another Rutherford student, discovered the neutron, which explains the nature of isotopes—atoms with the same number of protons but different numbers of neutrons. This also provided the modern explanation for why nickel is on average lighter than cobalt even though it has another proton; there are simply more neutrons on the average cobalt atom.

The periodic chart took its modern shape with the lanthanides and actinides below the *s*-, *p*-, and *d*-blocks at the suggestion of Glenn Seaborg in 1944. Seaborg was a graduate student of G. N. Lewis, who shifted his interest toward nuclear chemistry. He was lead discoverer or codiscoverer of ten elements: plutonium, americium, curium, berkelium, californium, einsteinium, fermium, mendelevium, nobelium, and element 106. Seaborg became the first person to have an element named after him while still living, when element 106 was named "seaborgium".

2.2 The Orbitals, Electronic Configurations of the Elements, and the Aufbau Principle

As the discussion of the Bohr model illustrates, understanding the energies and nature of the electrons around an atom or in a molecule allows us to understand the properties of the system. In general, this can be described using the Schrödinger

Equation (Eq. 2.20). The function Ψ is the wavefunction (eigenfunction), a description of electron waves as they encompass the nucleus. The symbol H is the Hamiltonian, an operator to describe the energy, and E is the energy (eigenvalue).

$$H\Psi = E\Psi \qquad \textbf{2.20}$$

The Hamiltonian can be used to describe all of the electron–electron, nuclear–nuclear, and electron–nuclear interactions and motions in a system. In cases where the exact solution of the Schrödinger Equation can be obtained, it has been in excellent agreement with experiment. However, chemical systems typically of interest today are far too complicated for direct solution of Eq. 2.20 without use of approximations. These approximations can generally be divided into three main categories:[11]

1. The nuclear dynamics and electronic motions are separated due to their different time regimes. It is assumed that the nuclei respond to electronic changes relatively slowly, whereas electronic changes occur very rapidly on nuclear displacement. This is the Born–Oppenheimer approximation.
2. Distinct electron–electron interactions are often replaced with an averaged interaction leading to loss of correlation effects in the approximation. This is the 1-electron approximation.
3. Molecular orbitals may be represented by linear combinations of the 1-electron hydrogen-like functions. This is the molecular orbitals (MO) by linear combination of atomic orbitals (LCAO) approximation, which will be used throughout the text and is discussed in more detail in Appendix M.

The shapes and energies of the orbitals in an atom or molecule are of frequent utility to the chemist, and in atoms these are defined by the quantum numbers for the orbitals. Bohr's value n is called the principal quantum number and defines the shell in which the electrons reside. Bohr also proposed angular momentum, l, as a quantum number. (l is sometimes referred to as the azimuthal quantum number.) The other two quantum numbers to be discussed here are m_l, the magnetic quantum number, and m_s, the electron spin. The values for the four quantum numbers are as follows:

$n = 1, 2, 3, 4, \ldots$: main shell.

$l = 0, 1, \ldots, (n - 1)$: subshells contained in the main shell, defines the orbital's shape (s, p, d, f).

$m_l = -l, \ldots, -1, 0, +1, \ldots, +l$: the "magnetic quantum number" defines the angular momentum of the orbital of type l along a particular axis. Linear combinations can be taken to produce different orientations, such as p_x, p_y, or p_z.

$m_s = -1/2$ and $+1/2$: defines the spin of the electron, up or down.

The different l subshells are called: $l = 0 \rightarrow s$-orbitals, $l = 1 \rightarrow p$-orbitals, $l = 2 \rightarrow d$-orbitals, and $l = 3 \rightarrow f$-orbitals. The letters s, p, d, and f denote sharp, principal, diffuse, and fundamental based on descriptions of peaks from spectroscopy. After f, the values become alphabetical: g, h, and so on.

No two electrons in an atom can have the same four quantum numbers (n, l, m_l, m_s). For example, if two electrons occupy the same atomic orbital (n, l, and m_l are identical), they must have different values for spin, m_s. This is called the Pauli Exclusion Principle, after Wolfgang Pauli, who proposed it in 1925. It can be said that these four quantum numbers define an electron's "address" around the nucleus, and no two electrons can have the same address.

For example, in the fourth main shell ($n = 4$), the possible l values are 0, 1, 2, and 3. For each of these l subshells there are $2l + 1$ different possible orientations, m_l values. For the $l = 0$ (called an s-orbital) there is only one possible orientation, $m_l = 0$. The reason for this is obvious; a sphere is the same regardless of how it is oriented. For the $l = 1$ (called p-orbitals) there are three possible orientations, $m_l = 1, 0, -1$; these are generally used as linear combinations directed along the x, y, and z axes (p_x, p_y, and p_z). For the $l = 2$ (called d-orbitals) there are five possible shapes/orientations (d_{xz}, d_{yz}, d_{xy}, d_{z^2}, and $d_{x^2-y^2}$). For the $l = 3$ (called f-orbitals) there are seven possible shapes/orientations; there are different sets of f-orbitals in common use, which are linear combinations of one another, but the cubic set employed here is f_{x^3}, f_{y^3}, f_{z^3}, $f_{x(z^2-y^2)}$, $f_{y(z^2-x^2)}$, $f_{z(x^2-y^2)}$, and f_{xyz}. In general, orbitals are designated by the shorthand nl_{m_l}, so for a d_{xy}-orbital in the fourth shell this would be written $4d_{xy}$.

If one looks at the s-orbitals in different shells, a node is added for each added shell; that is, the 1s-orbital has 0 nodes, 2s-orbital has 1 node, 3s-orbital has 2 nodes, and so on. There are two important points to make about this fact:

(1) *As the number of nodes in an orbital increase, the energy tends to increase.* In other words, the 3s-orbital will be higher in energy than the 1s-orbital.

(2) *The atomic orbitals are arranged so that they are orthogonal (having no net overlap with one another). Orbitals of the same shape must "fit together" so that the orbitals of higher shells use the nodal structure to ensure orthogonality.*

Orbital orthogonality involves another property of waves: *constructive* and *destructive interference*. These interference properties are probably familiar for 1- and 2-dimensional waves (Fig. 2.6a). If two waves moving towards one another have the same phase, they constructively interfere, and the height of the new wave will be the sum of the two waves (*top*). If two waves moving towards one another have different phases, they destructively interfere. Mathematically, one of the waves is written as the negative of the other leading to no wave at all when they mix (*bottom*).

Similar behavior occurs with 3-dimensional waves of electrons. In the case of electrons, the phases are negative or positive. *This is not to be confused with negative or positive charges! All electrons have the same negative charge; however, most*

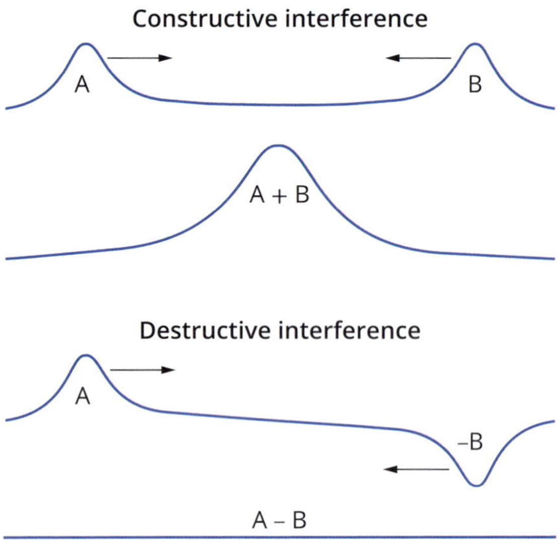

Fig. 2.6a Constructive (*top*) and destructive (*bottom*) interference in 1-dimensional waves.

orbitals, whether they contain electrons or not, have phases that are often designated as positive phase and negative phase. The two different phases are usually designated as shaded or unshaded regions of the orbital or regions of different colors.

To show an example discussed in Chapter 3 in more detail, one can consider the combination of two *s*-orbitals with the same phase (Fig. 2.6b). The two unshaded orbitals combine constructively to give an orbital with more electron density between the nuclei. The different phased orbitals, shown as one unshaded and one shaded, combine destructively to give an orbital with less electron density between the nuclei (bottom of the figure). Phase changes for orbitals are signified mathematically with a change in sign; the bottom orbital combination is *s*–*s* because the second orbital has a different phase from the first.

Another property of waves is that when there is a phase change there is a *node*, a region of zero occupancy of the electron (Fig. 2.6c). This occurs in the bottom picture in Fig. 2.6b, where there is a green (positive phase) to red color (negative phase) change between the two nuclei. We say there is a *nodal*

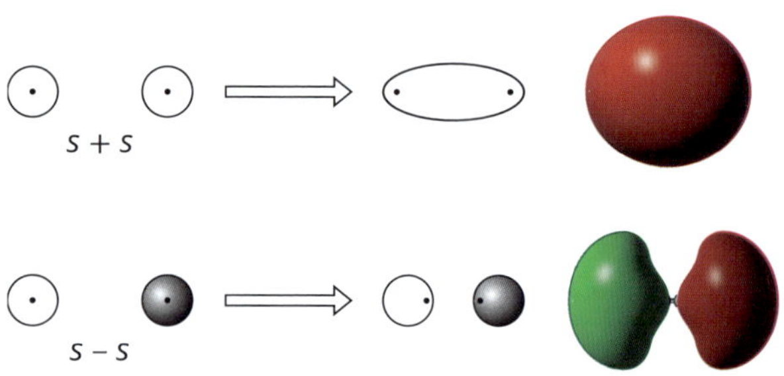

Fig. 2.6b Constructive and destructive interference in 3-dimensional waves of electrons. The dots are intended to represent the positions of the nuclei. The green and red (unshaded and shaded) represent different phases, like above and below horizontal in Fig. 2.6a.

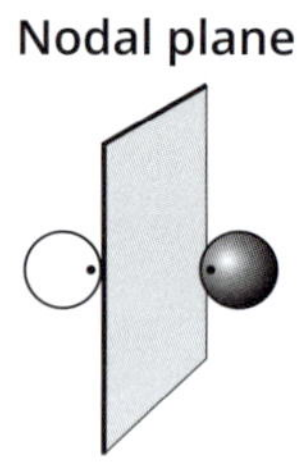

Fig. 2.6c An illustration of the "nodal plane" between the antibonding orbital in Fig. 2.6b where two *s*-orbitals of different phases are interacting. The nodal plane is a plane where no electron density exists and the phase is different on each side.

plane between the two atoms, a position of zero phase, illustrated showing the nodal plane.

Nodal planes can have a variety of different shapes like the planar one above or the spherical one found in a 2*s*-orbital. Fig. 2.6d shows hemispherical cut-outs of the 1*s*- and 2*s*-orbitals. The 1*s*-orbital has the same phase (same orange color) throughout and therefore has no nodes. The 2*s*-orbital has a change from orange to blue signifying a phase change with a spherical region in the center with a different phase.

Orbitals on the same atom must be orthogonal, meaning they sum so that there is no net constructive or destructive interference. In other words, they must have either no shared space (rare when the orbitals are on the same atom) or they must have equal parts constructive and destructive interference. There are two cases to consider: where the two orbitals have the same quantum number l but different principal quantum numbers n, and where the two orbitals have different l.

The overlap between two orbitals on the same atom of the same l (same general shape) but different shell (different n) such as a 1*s*- and 2*s*-orbital, for example, must be zero. Since both orbitals are spherical, this is accomplished by the orbitals having no *net* overlap, and equal amounts of bonding and antibonding interactions. The maximum of the orbital with the lower principal quantum number will generally be at or close to the node of the orbital with the higher principal quantum number. In other words, for the 1*s*- and 2*s*-orbitals to be orthogonal, the amount of overlap between the orange regions on each orbital (bonding) in the figure and the overlap of the blue region of 2*s* and the orange of 1*s* (antibonding) must be equal.

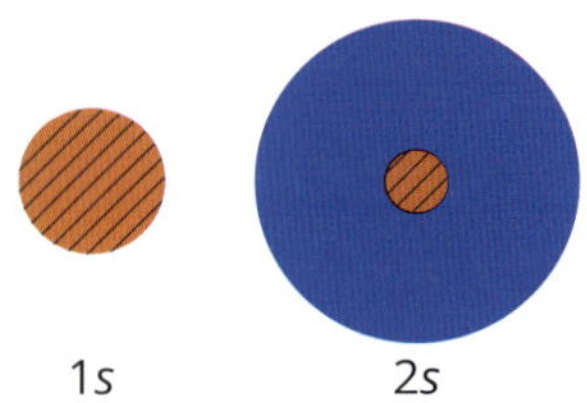

Fig. 2.6d Hemispherical cuts of the 1*s*- and 2*s*-orbitals. The 2*s*-orbital has a spherical node where the phase changes. The two phases are represented by different colors. The lighter, positive phase is also hatched. The size of the orbitals is important as the constructive interference (between the same color) must cancel the destructive interference (different colors) for atomic orbitals.

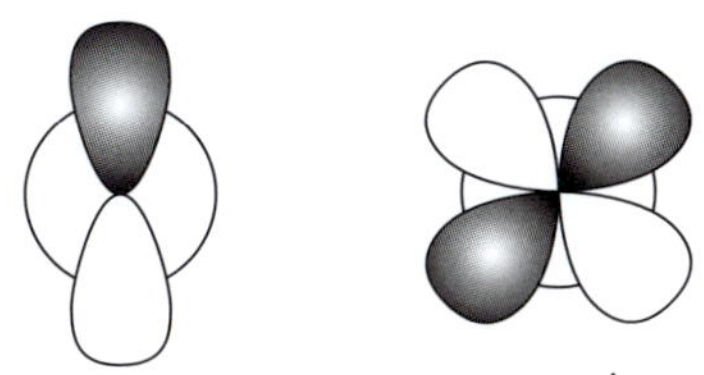

Fig. 2.6e Atomic orbitals with differing angular momentums (l) have no overlap as a consequence of their nodal structure, as well, in this case, due to differences in nodal planes.

If two orbitals have different angular momentum quantum numbers l, it is quite easily seen that they will be orthogonal on the same atom (Fig. 2.6e). Again, orthogonality often involves exact cancelation of constructive and destructive interference between the waves. In this case, rather than orbitals having to be of specific sizes to interact with the spherical nodes as when the orbitals have the same shape, the orbital shape and phase take care of the orthogonality. For example, an *s*-orbital can have no net overlap with the *p*- and *d*-orbitals on the same atom. The unshaded–unshaded areas are exactly the same size as the shaded–unshaded areas regardless of the size of the orbitals due to the shapes and phases of the orbitals themselves.

2.2.1 The Shapes and Nodal Structures of Orbitals in More Detail

The shapes, directions, and nodal positions of the orbitals are of crucial importance throughout chemistry in general. The student should be able to draw, at least, the *s*-, *p*-, and *d*-orbitals with coordinate axes and phases by heart. The *f*-orbitals also have a beauty all of their own, and learning to draw these is worthwhile for chemists working in the fields of lanthanide and actinide chemistry.

In this section we will describe the orbitals in detail, which will obviously be review for many students. However, some undergraduates may benefit from a refresher, especially on the exact shapes, phases, and orientations of the *d*-orbitals. Furthermore, the information in this section is perhaps elementary but also absolutely crucial, which led us to believe that including it here is worthwhile.

An *s*-orbital has no planar nodes, regardless of the primary quantum number (n), due to the spherical nature of the orbital. As the primary quantum number increases, the number of spherical nodes increases so that there are zero nodes in a 1*s*-orbital, 1 node in a 2*s*-orbital, 2 nodes in a 3*s*-orbital, and so on.

Note that as the principal quantum number (n) increases, the size of the orbital also generally increases (Fig. 2.7a). Since each orbital can at most hold two electrons (of opposite spin) regardless of its size, the electron density in these orbitals becomes more diffuse with increasing n.

It is perhaps worth noting at this stage that orbital sizes must be truncated at some arbitrary value, often around where 99% of the probability is contained. Electrons in any orbital have some small probability to be elsewhere, like the Moon or Saturn, but large probabilities to be close to the nucleus.

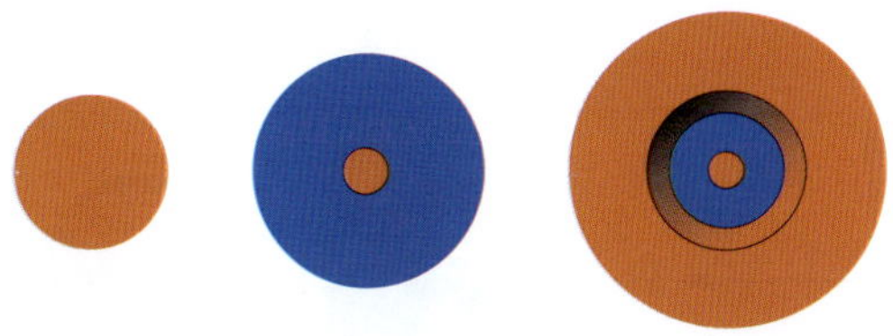

Fig. 2.7a As the principal quantum number increases (n), the size of the orbital generally increases as well.

Unlike an *s*-orbital—which has no planar nodes—a *p*-orbital ($l = 1$) has a single planar node (Fig. 2.7b). The 2*p*-orbital is shown, and the black plane between the two lobes shows the node in the orbital.

As the principal quantum number (n) increases, the *p*-orbital still has a single planar node but adds spherical nodes. For example, the 3*p*-orbital has one spherical node and one planar node. The 4*p*-orbital has two spherical nodes and one planar node. Fig. 2.7c shows the 3*p*- and 4*p*-orbitals with just the orbital and its phases on top (Fig. 2.7c). Immediately below it are the nodes in the orbitals shown as grey-black spheres and planes.

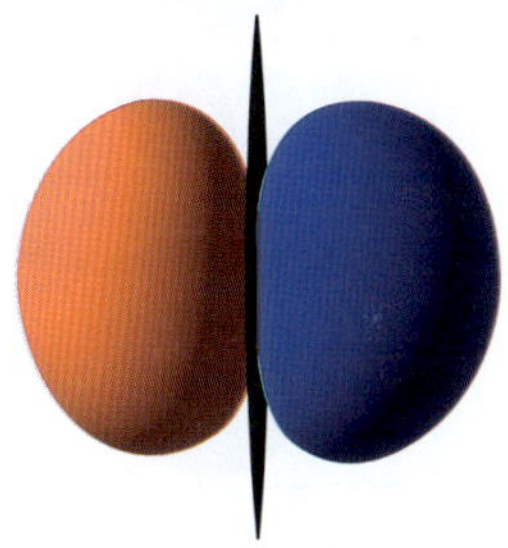

Fig. 2.7b The nodal plane in a *p*-orbital.

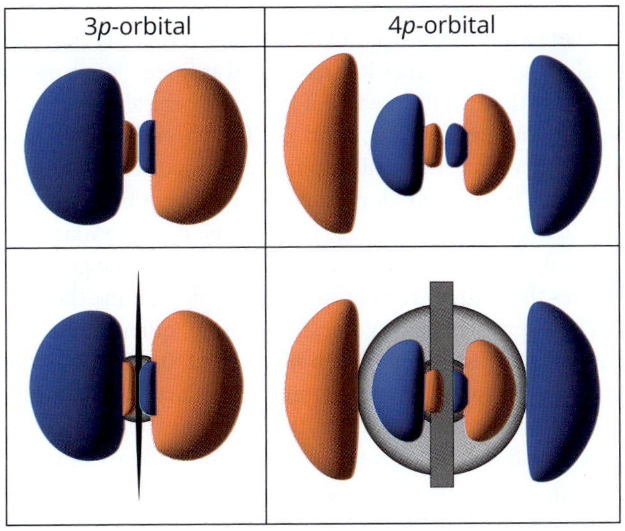

Fig. 2.7c The shapes, with nodal planes and nodal spheres, of the 3*p*- and 4*p*-orbitals.

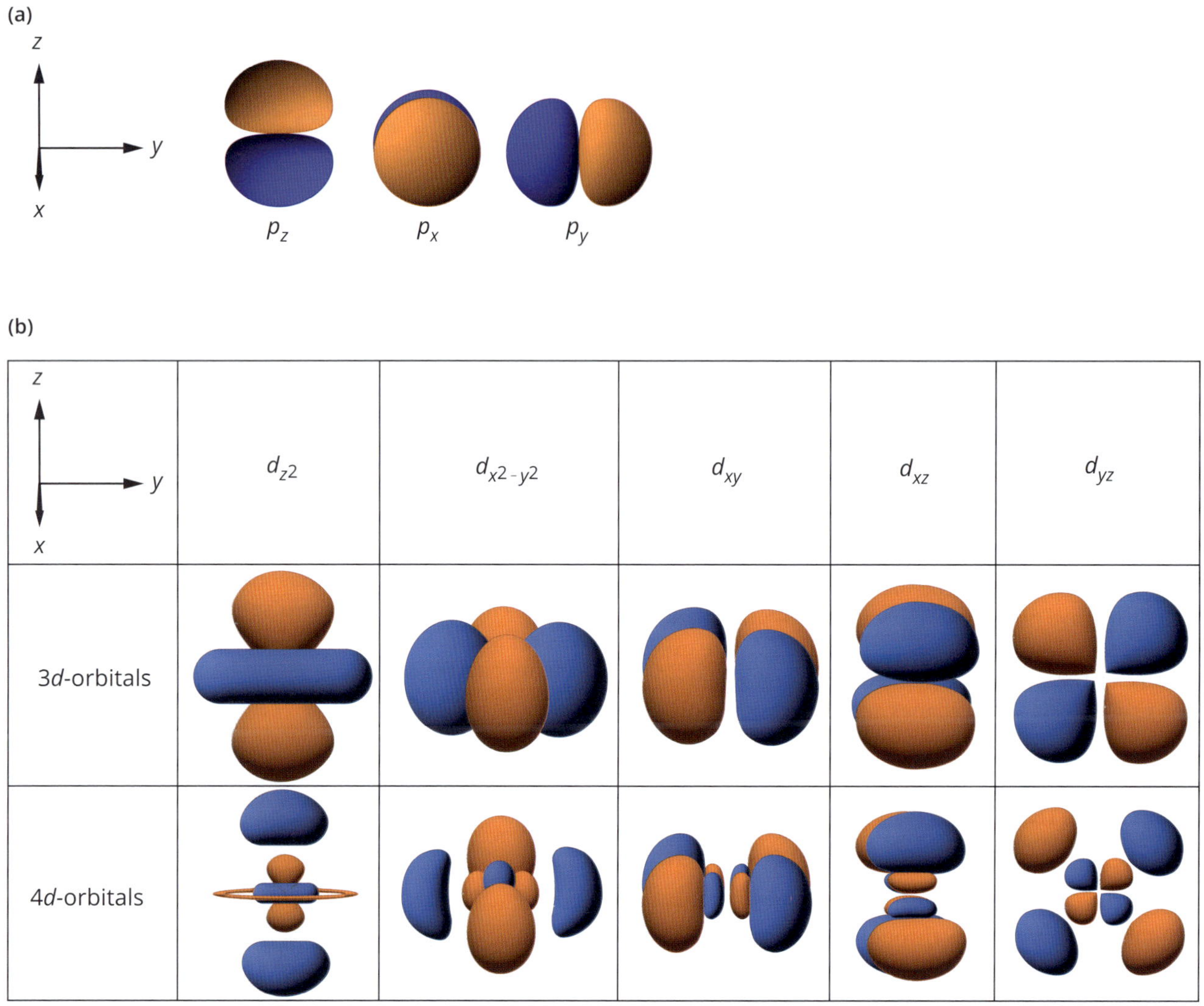

Fig. 2.7d (a) The phases and directions of the p_x-, p_y-, and p_z-orbitals. (b) The shapes of the 3*d*- and 4*d*-orbitals.

There are three *p*-orbitals ($l = 1$; $m_l = +1, 0, -1$) in a given shell. These orbitals are often expressed as linear combinations (Fig. 2.7d) that result in each of the three orbitals being along the Cartesian axes: p_z, p_x, and p_y. The three 2*p*-orbitals are shown. Note that the positive phase (signified with the orange color here) is directed along the positive direction of the axis, and the negative phase (blue) is along the negative axis direction. In other words, the *p*-orbitals have vector-like geometric qualities that mirror the coordinate axes that define them.

The *d*-orbitals of angular momentum quantum number $l = 2$ will have five orbitals with $m_l = +2, +1, 0, -1$, and -2. The usual representation of these orbitals in a Cartesian coordinate system is as linear combinations that give the orbitals d_{z^2}, $d_{x^2-y^2}$, d_{xy}, d_{xz}, and d_{yz}. The five 3*d*- and 4*d*-orbitals are shown in Fig. 2.7d. Note that all of these orbitals have two planar nodes and that the 4*d*-orbitals have an additional spherical node.

Regard the previous table of the 3d-orbitals as absolutely critical information for inorganic chemistry! One must know the shapes, directions, and phases of the d-orbitals in order to learn the chemistry of the transition metals in any depth. These should be set to memory, and the student should be able to reproduce by hand the 3d-orbitals in the previous table, labeling them correctly. Generally speaking, the *d*-orbital planar nodal structure (as represented in the 3*d*-orbitals) is what is usually of importance when we get to topics like crystal/ligand field theory and for most applications. As a result, it is good to understand the radial (spherical) nodes in the 4*d*- and 5*d*-orbitals, but the shapes of the 3*d*-orbitals are more often used for all the transition metals as they contain the more useful planar node information.

Note that the d_{z^2} and $d_{x^2-y^2}$ orbitals have lobes that are along the coordinate axes, and the others (d_{xy}, d_{xz}, and d_{yz}) all have lobes that extend between the axes. All the orbitals have the "four-leaf clover" shape, with lobes of the same phase 180° apart, except for d_{z^2}. The name d_{z^2} is actually a shortening of $d_{2z^2-x^2-y^2}$, meaning that the lobes along the *z*-axis are larger and have a different phase from the smaller torus in the *xy*-plane.

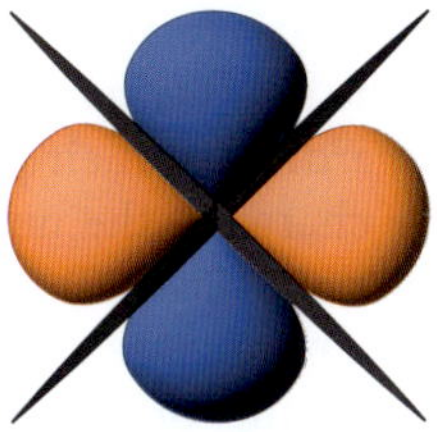

Fig. 2.7e Nodal structure of most *d*-orbitals.

The two planar (also called angular) nodes in the *d*-orbitals with the four-leaf clover shape are at 90°, as in Fig. 2.7e.

The uniquely shaped d_{z^2}-orbital has conical angular nodes, as in Fig. 2.7f. These conical nodes are ~54.73° from the *z*-axis—where the main portion of the orbital resides. This angle is of geometric and chemical significance, as it is half of the tetrahedral angle of ~109.471°.

The number of angular nodes, ν_θ, in an orbital is always equal to the quantum number *l* for the orbital; that is, $\nu_\theta = 0$ for *s*-orbitals, 1 for *p*-orbitals, 2 for *d*-orbitals, and 3 for *f*-orbitals.

$$\nu_\theta = l$$

The number of radial nodes, ν_r in an orbital is equal to the principal quantum number (*n*) minus the angular momentum quantum number (*l*) minus 1 (Eq. 2.21).

$$\nu_r = n - l - 1 \qquad \textbf{2.21}$$

As a result, the total number of nodes (angular + radial) for any orbital is simply the principal quantum number minus one (Eq. 2.22).

$$\nu_\theta + \nu_r = l + n - l - 1$$

$$\nu_\theta + \nu_r = \textit{total nodes} = n - 1 \qquad \textbf{2.22}$$

For example, the total number of nodes in a 4*p*-orbital, where $n = 4$, is 4–1 = 3, as discussed previously. A 5*f*-orbital, as in the valence orbitals of uranium, will have four total nodes (using $n = 5$ in Eq. 2.22, 5–1 = 4).

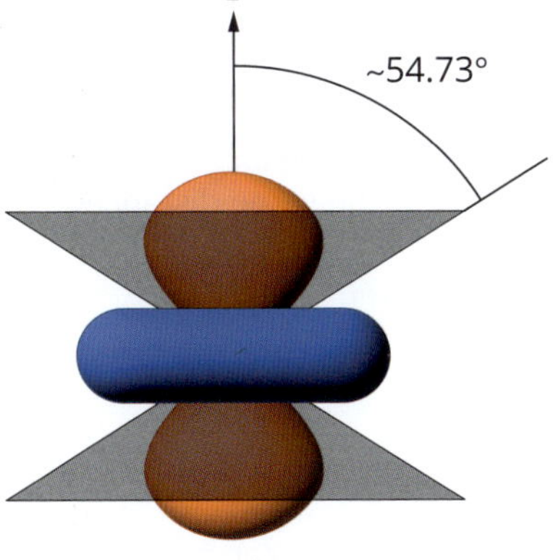

Fig. 2.7f The conical nodes of the d_{z^2}-orbital.

2.2.2 The Mathematical Form of the Hydrogen Wavefunctions and the Effective Nuclear Charge

The orbital shapes as described previously are generally of greater utility to the chemist than the exact formulas for the orbitals, and the mathematical functions for all of the 1-electron wavefunctions will not be discussed here. One can find them in textbooks on quantum mechanics. Nevertheless, we can illustrate an important point by showing only mathematical forms of the wavefunctions for the 1*s*-, 2*s*-, and 3*s*-orbitals, shown in Table 2.1. These are generally divided into a radial, $R_{nl}(r)$, and angular, $Y_{l,m}(\theta, \phi)$ part of the wavefunction.

These equations are only accurate for 1-electron systems like the hydrogen atom and related systems, like He^+. This is because electron–electron repulsion terms are not taken into account. Most *ab initio* calculations (*ab initio* is Latin for "from the beginning", meaning the calculations are done from first principles) do not use these forms of the orbitals, as will be discussed later. These are, however, the starting point for the creation of the commonly employed 3-dimensional orbitals shown in the previous subsection.

In 2-dimensional plots, the radial functions, $R_{n,l}(r)$, are often plotted versus the distance from the nucleus in order to obtain a sense of the size of the orbitals. This can be done using the radial function directly. These plots are shown for the 1*s*- and 2*s*-orbitals in Fig. 2.8. The 1*s*-orbital has no nodes and is positive throughout. The 2*s*-orbital has a single node at ~1 Å, where the plot goes negative. The sum of the 1*s* and 2*s* wavefunctions should cancel; that is, the constructive interference prior to 1 Å should be equal to the destructive interference after 1 Å. As a result, the size of the 1*s* and 2*s* orbitals are intimately related, as will be important in later chapters.

An important point concerning the plots in Fig. 2.8 is the amount of electron density near the nucleus for each orbital. Obviously, the 1*s*-orbital is quite near the nucleus and those electrons "feel" the positive charges there quite well. Another way to think about this is how well the negatively charged electrons in a 1*s*-orbital will "shield" the positive charges of the

Table 2.1 The radial, *R(r)*, and angular, *Y(θ,ϕ)*, forms of the 1-electron wavefunctions for the 1*s*-, 2*s*-, and 3*s*-orbitals, where *Z* = the atomic number and a_0 = 0.529 Å, the Bohr radius.

Orbital	R(r)	Y(θ,ϕ)
1s	$2\left(\frac{Z}{a_0}\right)^{3/2} e^{-Zr/a_0}$	$\left(\frac{1}{4\pi}\right)^{1/2}$
2s	$\left(\frac{Z}{2a_0}\right)^{3/2}\left(2-\frac{Zr}{a_0}\right)e^{-Zr/2a_0}$	$\left(\frac{1}{4\pi}\right)^{1/2}$
3s	$\frac{2}{27}\left(\frac{Z}{3a_0}\right)^{3/2}\left(27-\frac{18Zr}{a_0}+\frac{2Z^2r^2}{a_0^2}\right)e^{-Zr/3a_0}$	$\left(\frac{1}{4\pi}\right)^{1/2}$

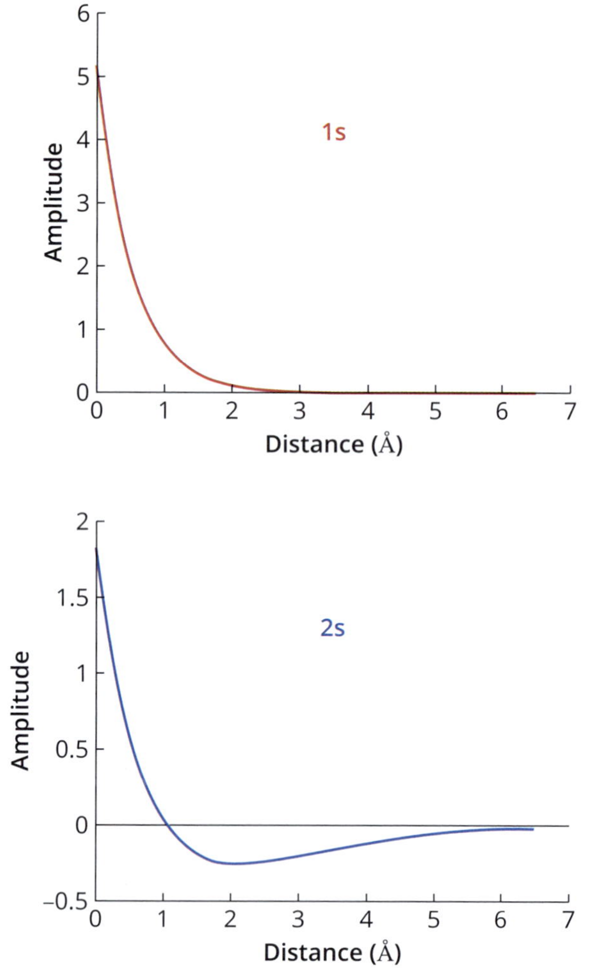

Fig. 2.8 Plots of the radial function, $R_{n,l}(r)$ versus the distance from the nucleus, r, from Table 2.1 with $Z = 1$. (Note the change in scale between orbitals on the y-axis.)

nucleus from some reference electron farther out. If the electrons are very close to the nucleus, as when the orbital is quite small, the electrons regardless of their position in the orbital will shield the positive charge of the nucleus quite well from electrons that are farther out (Fig. 2.9).

The *s*-orbitals always have more electron density near the nucleus as compared with other orbitals in the same shell. This is due to the nodal structure of the *s*-orbitals, which allows relatively close approach of the electrons in these orbitals to the nucleus, thus increasing Coulombic attraction and lowering the energy. This is sometimes called *orbital penetration*, and it causes the shielding of the nuclear charge by electrons in the *s*-orbitals to be higher than electrons in other orbitals. In fact, orbital penetration is inversely related to l; that is, $s > p > d > f$ for orbitals of the same principal quantum number.

If the nuclear charge is Z for a particular atom, we can define the *effective nuclear charge* as being the attraction of a particular electron to the nucleus after the shielding from other electrons is accounted for (Eq. 2.23), which is often given the symbol Z_{eff}.

$$Z_{eff} = Z - S \qquad \textbf{2.23}$$

An approximate but popular method for estimating the shielding term S was developed by Slater, and Slater's Rules are as follows:

(1) The atom's electronic structure is written in order of increasing n and l with the following grouping. Orbitals in the same parentheses are in the same group. The *s*- and *p*-orbitals of the same n are grouped together, while *d*- and *f*-orbitals are in their own groups.

$$(1s)(2s,2p)(3s,3p)(3d)(4s,4p)(4d)(4f)(5s,5p)(5d)(5f)\ldots$$

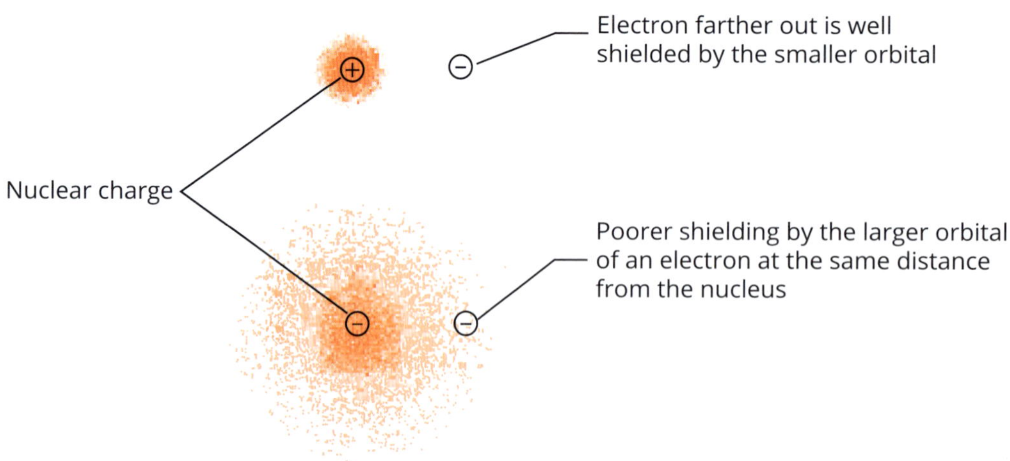

Fig. 2.9 Shielding ability of electrons in smaller versus larger orbitals. The effective nuclear charge, Z_{eff}, can be defined as the nuclear charge of the atom minus a shielding term, S (Eq. 2.23).

(2) Electrons in groups to the right in the previous list do not shield those to their left.

(3) To determine the value of S for a particular electron on the atom the following values are summed:

(a) Each electron in the same group as the electron of interest contributes 0.35, except for a $1s$-electron shielding another $1s$-electron where the contribution is 0.30.

(b) Each electron in $n - 1$ groups contributes 0.85 to S.

(c) Each electron in $n - 2$ groups contributes 1.00 to the value of S.

(d) For nd and nf valence electrons, each electron in the same group contributes 0.35, and each electron in a group to the left contributes 1.00 to S.

For example, let us calculate the Z_{eff} "felt" by, the Coulombic attraction to the nucleus for, any $3p$-electron in the chlorine atom. We start by writing the Slater configurations using the previous list and simply placing two electrons in each orbital starting with $1s$. Chlorine has $Z = 17$, and a neutral chlorine atom therefore has 17 electrons. Keep in mind that there are three p-orbitals that can hold 6 electrons, so each (ns,np) grouping can hold 8 electrons total. Likewise, a d-manifold can hold up to 10 electrons and an f up to 14. The electrons in the group that we are calculating Z_{eff} for are shown in bold.

$$(1s)^2(2s,2p)^8(\mathbf{3s,3p})^7$$

Now we apply Rules 3a–d to this configuration. The valence electrons are in the $(3s,3p)$-group. So, the electrons in $(1s)$, a shell that is two primary quantum numbers lower, contribute 1.00 each. *A common mistake is to not remove the electron you are considering, and electrons should not be considered to shield themselves. In other words, if I am considering the Z_{eff} on one of the electrons in the (3s,3p)-manifold, I need to remove one electron from the total when calculating S. I calculate the shielding based on the other electrons in the manifold and obtain their effect on the reference electron.* The calculation of S for one of the outer electrons in Cl is as follows.

$$\begin{aligned} S &= (2 \times 1.00)_{(1s)} + (8 \times 0.85)_{(2s+2p)} + (6 \times 0.35)_{(3s+3p)} \\ &= 2 + 6.8 + 2.1 \\ &= 10.9 \end{aligned}$$

Using this value for the shielding with $Z = 17$ for chlorine in Eq. 2.23 gives the Z_{eff} for any electron in the valence shell of chlorine as 6.1. In other words, any electron in the valence shell has a Coulombic attraction to the nucleus as if there were no other electrons around the nucleus and the nucleus contained 6.1 protons.

$$\begin{aligned} Z_{eff} &= Z - S \\ &= 17 - 10.9 \\ &= 6.1 \end{aligned}$$

2.2.3 The Filling of Atomic Orbitals and Electronic Configurations

As may be gleaned from the discussion in the previous section, the energies of the various orbitals are not necessarily as simple as the lower quantum numbers being lower in energy. There are several factors that contribute. First, in general, orbitals with fewer nodes are lower in energy. Second, orbital penetration leads to a lowering of orbitals with lower angular momentum quantum numbers, l. Third, an effect that has not been discussed as yet: there are spin multiplicity effects that can contribute to which orbitals are filled with electrons first.

Hund's Rule of Multiplicity can be summarized as (1) electrons do not occupy the same space unless they are required to, and (2), generally, higher multiplicities are lower in energy. The multiplicity (M) of an atom (ion, molecule, and so on) can be defined as the sum of the unpaired electron spins (S = sum of m_s) times 2 plus 1, Eq. 2.24.[b]

$$M = 2S + 1 \qquad \mathbf{2.24}$$

Some common multiplicities are shown in Fig. 2.10. If all the electrons are paired and $S = 0$, then $M = 1$, and the compound is referred to as a singlet. If the compound has one unpaired spin so that the sum of the spins is $S = ½$, then $M = ½(2) + 1 = 2$, and the compound is referred to as a doublet. The names for the multiplicities through eight unpaired spins are singlet, doublet, triplet, quartet, quintet, sextet, septet, octet, and nonet.

To understand why the spin multiplicities are given these names, consider a system of spins of some type in an external magnetic field (B), Fig. 2.10, which is perhaps a situation familiar to many students from NMR. (In NMR you are typically looking at nuclear spin interactions, but the principles are similar for spin ½ electrons). The lower energy configurations for the spin will be where the spin of the electron is oriented with the magnetic field, and a higher energy will result if the spin is oriented against the field. For a singlet, there are equal numbers of spins up and down with no net spin, which obviously gives a single energy level. For a doublet state, the spin may align with the field or against it, two different states with different energies.

For a triplet state, there are two spins in two different orbitals, which may align with the field or against the field in four different ways: ↑/↑, ↓/↓, ↑/↓, and ↓/↑. The two states ↑/↓ and ↓/↑ are degenerate. Both ↑/↑ and ↓/↓ are one-of-a-kind states that have different energies due to the external field. Since up/down and down/up are degenerate, this energy has twice the probability (peak intensity) of ↑/↑ and ↓/↓, which are not degenerate if there is an external field. The quartet and higher states can be rationalized similarly, while predicting the peak intensity based on probability of each energy level.

In Fig. 2.11, all the orbitals of interest are of the same energy, and the lowest energy configuration will have the highest spin

[b] S here is not to be confused with the shielding S term in the previous section. The common use of the same symbol to represent both is an unfortunate coincidence.

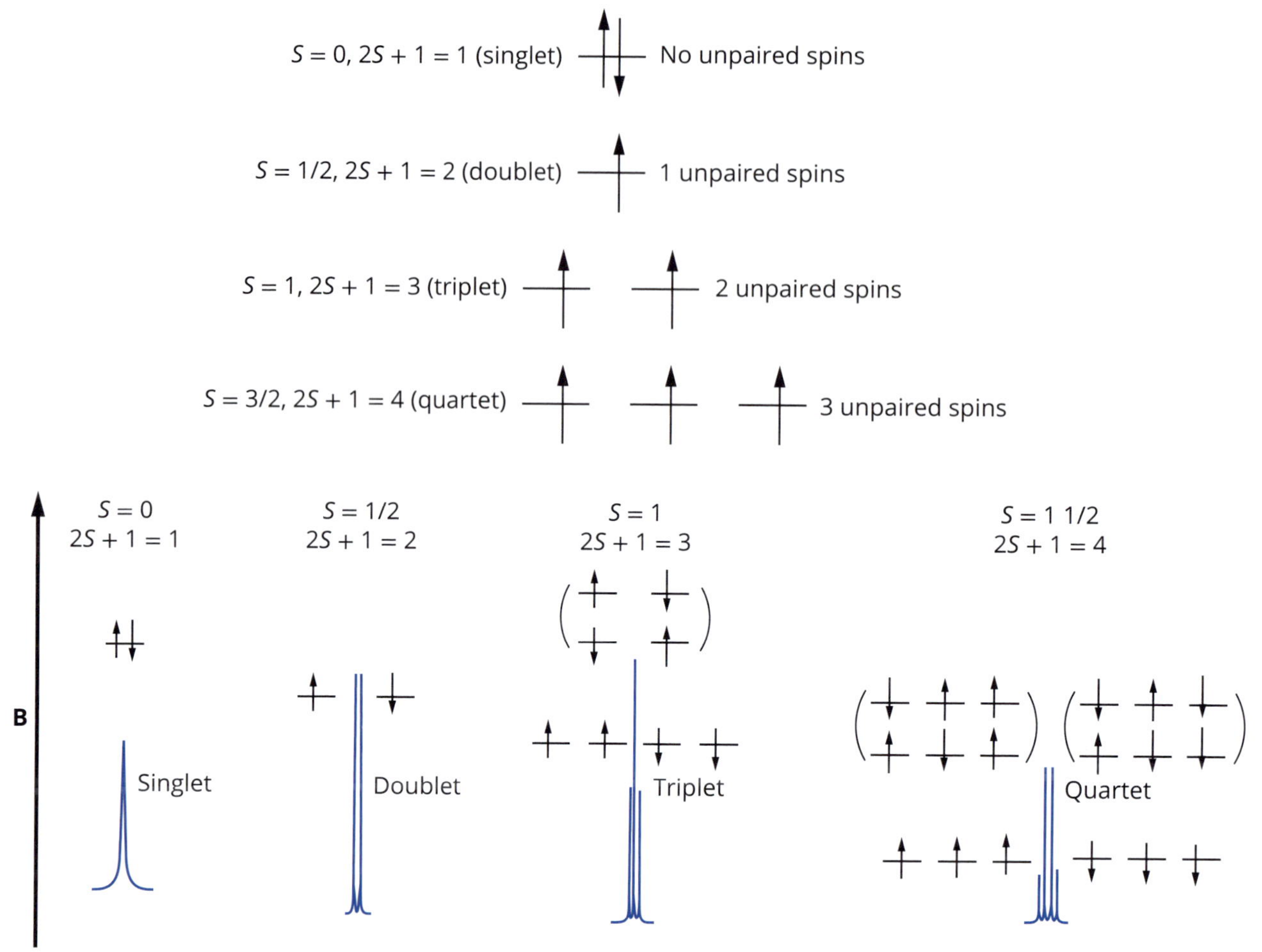

Fig. 2.10 Examples of spin multiplicities (*top*). These are the lowest energy—or in some cases only—configurations for 2 spins/1 orbital, 1 spin/1 orbital, 2 spins/2 orbitals, and 3 spins/3 orbitals situations. (*bottom*) Spins aligned with or against a magnetic field and the resulting energy levels similar to what is found in NMR with spin = ½ nuclei, such as protons. In each case, the lower energy configuration is shown on the left, though this may vary depending on the type of spectroscopy, presentation, molecular symmetry, and other factors.

multiplicity by Hund's Rule. For example, if there are two orbitals of equal energy and two electrons, the lowest energy configuration will have the two electrons in different orbitals with the same spin as shown in Fig. 2.11. *In the triplet state, the two electrons are in different orbitals lowering potential Coulombic repulsions between the two negatively charged particles; in addition, electrons of the same spin tend to stay well separated from one another, which further reduces Coulombic repulsions even when the electrons are in different orbitals.* A common way of expressing this is to say the motions of the electrons in the triplet state will be highly "correlated" so that they do not approach each other often.

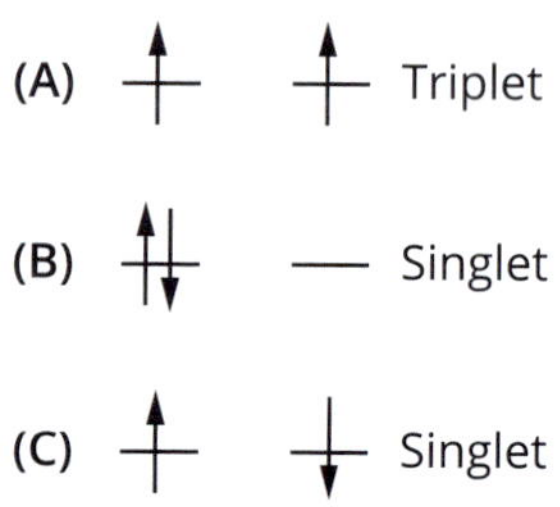

Fig. 2.11 Configurations for the filling of two degenerate orbitals with two electrons. The lowest energy configuration will be A, which has the highest multiplicity, with the two singlet states higher in energy.

Both of the singlet states, configurations B and C in Fig. 2.11, will be higher in energy than the triplet. If the two electrons are paired in the same orbital, they obviously will share all the same space, which increases Coulombic repulsion between the two electrons. Also, if the two electrons reside in different orbitals but have different spin, they will have closer approaches than if they had the same spin, resulting in a higher energy configuration. In other words, a singlet state with electrons in different orbitals is generally higher in energy than a triplet state.

Orbital degeneracy is not a requirement for this effect to be seen. If the orbitals are close in energy, the higher spin state may be preferred if the energy required to pair the electrons in the same orbital (often just called the *pairing energy*) is greater than the difference in orbital energies (configuration A in Fig. 2.12). The singlet state will be preferred in cases where the energy gap between the orbitals is larger than the pairing energy (configuration B).

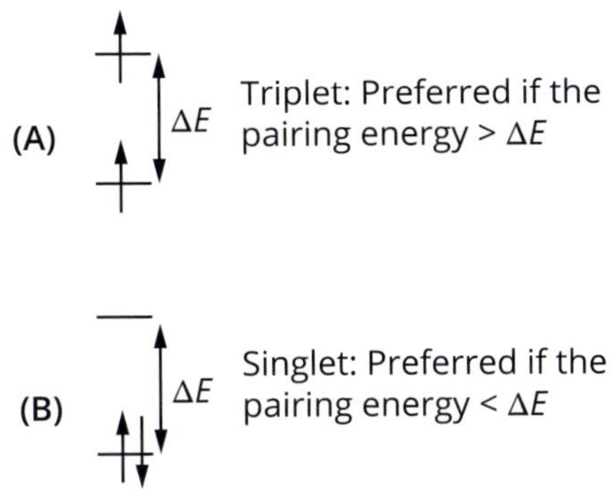

Fig. 2.12 Orbitals that are not degenerate may favor the higher spin state if the energy required to pair the electrons in the same orbital (pairing energy) is less than the energy difference between the orbitals.

If one compiles all of these effects, one can begin to understand the configurations of the elements. First the orbital energies, as determined by orbital penetration and nodal structure, can be remembered based on the following mnemonic suggested by Niels Bohr and Wolfgang Pauli. This has been referred to as the Aufbau principle ("Aufbau" is German for "building up"), because the electronic configurations are built up from hydrogen to the heavier elements. One begins by writing the *s*-orbitals of the shells in one column (Fig. 2.13), followed by the *p*-orbitals in the next, *d*-orbitals in the next, and finally *f*-orbitals in the last column. The orbitals are then filled diagonally. The order of filling is then *1s, 2s, 2p, 3s, 3p, 4s, 3d, 4p, 5s, 4d, 5p, 6s, 4f, 5d, 6p, 7s, 5f, 6d,* and *7p.*

The configurations for the neutral elements are shown in Table 2.2 for the elements through atomic number 112 (copernicium).

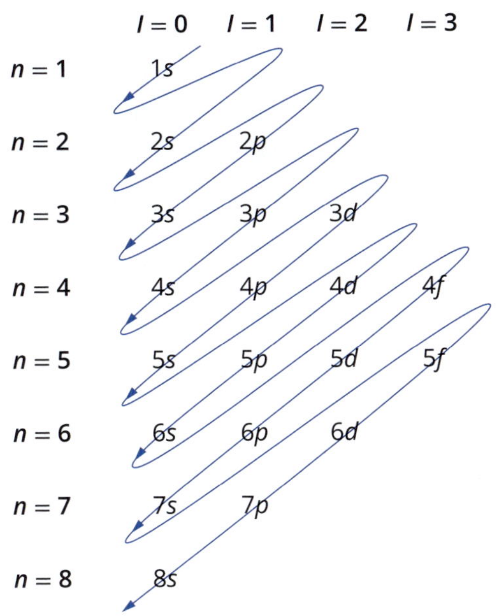

Fig. 2.13 An Aufbau diagram illustrating how orbitals are generally filled, though there are exceptions.

Table 2.2 Configurations of the elements.

Atomic Number	Element	Configuration
1	H	$1s^1$
2	He	$1s^2$
3	Li	$[He]2s^1$
4	Be	$[He]2s^2$
5	B	$[He]2s^22p^1$
6	C	$[He]2s^22p^2$
7	N	$[He]2s^22p^3$
8	O	$[He]2s^22p^4$
9	F	$[He]2s^22p^5$
10	Ne	$[He]2s^22p^6$
11	Na	$[Ne]3s^1$
12	Mg	$[Ne]3s^2$
13	Al	$[Ne]3s^23p^1$
14	Si	$[Ne]3s^23p^2$
15	P	$[Ne]3s^23p^3$
16	S	$[Ne]3s^23p^4$
17	Cl	$[Ne]3s^23p^5$
18	Ar	$[Ne]3s^23p^6$
19	K	$[Ar]4s^1$
20	Ca	$[Ar]4s^2$
21	Sc	$[Ar]4s^23d^1$
22	Ti	$[Ar]4s^23d^2$
23	V	$[Ar]4s^23d^3$
24	Cr	$[Ar]4s^13d^5$
25	Mn	$[Ar]4s^23d^5$
26	Fe	$[Ar]4s^23d^6$
27	Co	$[Ar]4s^23d^7$
28	Ni	$[Ar]4s^23d^8$
29	Cu	$[Ar]4s^13d^{10}$
30	Zn	$[Ar]3d^{10}4s^2$
31	Ga	$[Ar]3d^{10}4s^24p^1$
32	Ge	$[Ar]3d^{10}4s^24p^2$
33	As	$[Ar]3d^{10}4s^24p^3$
34	Se	$[Ar]3d^{10}4s^24p^4$
35	Br	$[Ar]3d^{10}4s^24p^5$
36	Kr	$[Ar]3d^{10}4s^24p^6$
37	Rb	$[Kr]5s^1$
38	Sr	$[Kr]5s^2$
39	Y	$[Kr]4d^15s^2$
40	Zr	$[Kr]4d^25s^2$
41	Nb	$[Kr]4d^45s^1$
42	Mo	$[Kr]4d^55s^1$

(continued)

Table 2.2 Continued

Atomic Number	Element	Configuration
43	Tc	[Kr]$4d^5 5s^2$
44	Ru	[Kr]$4d^7 5s^1$
45	Rh	[Kr]$4d^8 5s^1$
46	Pd	[Kr]$4d^{10}$
47	Ag	[Kr]$4d^{10} 5s^1$
48	Cd	[Kr]$4d^{10} 5s^2$
49	In	[Kr]$4d^{10} 5s^2 5p^1$
50	Sn	[Kr]$4d^{10} 5s^2 5p^2$
51	Sb	[Kr]$4d^{10} 5s^2 5p^3$
52	Te	[Kr]$4d^{10} 5s^2 5p^4$
53	I	[Kr]$4d^{10} 5s^2 5p^5$
54	Xe	[Kr]$4d^{10} 5s^2 5p^6$
55	Cs	[Xe]$6s^1$
56	Ba	[Xe]$6s^2$
57	La	[Xe]$5d^1 6s^2$
58	Ce	[Xe]$4f^1 5d^1 6s^2$
59	Pr	[Xe]$4f^3 6s^2$
60	Nd	[Xe]$4f^4 6s^2$
61	Pm	[Xe]$4f^5 6s^2$
62	Sm	[Xe]$4f^6 6s^2$
63	Eu	[Xe]$4f^7 6s^2$
64	Gd	[Xe]$4f^7 5d^1 6s^2$
65	Tb	[Xe]$4f^9 6s^2$
66	Dy	[Xe]$4f^{10} 6s^2$
67	Ho	[Xe]$4f^{11} 6s^2$
68	Er	[Xe]$4f^{12} 6s^2$
69	Tm	[Xe]$4f^{13} 6s^2$
70	Yb	[Xe]$4f^{14} 6s^2$
71	Lu	[Xe]$4f^{14} 5d^1 6s^2$
72	Hf	[Xe]$4f^{14} 5d^2 6s^2$
73	Ta	[Xe]$4f^{14} 5d^3 6s^2$
74	W	[Xe]$4f^{14} 5d^4 6s^2$
75	Re	[Xe]$4f^{14} 5d^5 6s^2$
76	Os	[Xe]$4f^{14} 5d^6 6s^2$
77	Ir	[Xe]$4f^{14} 5d^7 6s^2$
78	Pt	[Xe]$4f^{14} 5d^9 6s^1$
79	Au	[Xe]$4f^{14} 5d^{10} 6s^1$
80	Hg	[Xe]$4f^{14} 5d^{10} 6s^2$
81	Tl	[Xe]$4f^{14} 5d^2 6s^2 6p^1$
82	Pb	[Xe]$4f^{14} 5d^2 6s^2 6p^2$
83	Bi	[Xe]$4f^{14} 5d^2 6s^2 6p^3$
84	Po	[Xe]$4f^{14} 5d^2 6s^2 6p^4$
85	At	[Xe]$4f^{14} 5d^2 6s^2 6p^5$
86	Rn	[Xe]$4f^{14} 5d^2 6s^2 6p^6$
87	Fr	[Rn]$7s^1$
88	Ra	[Rn]$7s^2$
89	Ac	[Rn]$6d^1 7s^2$
90	Th	[Rn]$6d^2 7s^2$
91	Pa	[Rn]$5f^2 6d^1 7s^2$
92	U	[Rn]$5f^3 6d^1 7s^2$
93	Np	[Rn]$5f^4 6d^1 7s^2$
94	Pu	[Rn]$5f^6 7s^2$
95	Am	[Rn]$5f^7 7s^2$
96	Cm	[Rn]$5f^7 6d^1 7s^2$
97	Bk	[Rn]$5f^9 7s^2$
98	Cf	[Rn]$5f^{10} 7s^2$
99	Es	[Rn]$5f^{11} 7s^2$
100	Fm	[Rn]$5f^{12} 7s^2$
101	Md	[Rn]$5f^{13} 7s^2$
102	No	[Rn]$5f^{14} 7s^2$
103	Lr	[Rn]$5f^{14} 6d^1 7s^2$
104	Rf	[Rn]$5f^{14} 6d^2 7s^2$
105	Db	[Rn]$5f^{14} 6d^3 7s^2$
106	Sg	[Rn]$5f^{14} 6d^4 7s^2$ [a]
107	Bh	[Rn]$5f^{14} 6d^5 7s^2$ [a]
108	Hs	[Rn]$5f^{14} 6d^6 7s^2$ [a]
109	Mt	[Rn]$5f^{14} 6d^7 7s^2$ [a]
110	Ds	[Rn]$5f^{14} 6d^9 7s^1$ [a]
111	Rg	[Rn]$5f^{14} 6d^{10} 7s^1$ [a]
112	Cn	[Rn]$5f^{14} 6d^{10} 7s^2$ [a]
113	Nh	[Rn]$5f^{14} 6d^{10} 7s^2 p^1$ [a]
114	Fl	[Rn]$5f^{14} 6d^{10} 7s^2 p^2$ [a]
115	Mc	[Rn]$5f^{14} 6d^{10} 7s^2 p^3$ [a]
116	Lv	[Rn]$5f^{14} 6d^{10} 7s^2 p^4$ [a]
117	Ts	[Rn]$5f^{14} 6d^{10} 7s^2 p^5$ [a]
118	Og	[Rn]$5f^{14} 6d^{10} 7s^2 p^6$ [a]

[a] Expected electronic configuration

Keep in mind that this (Fig. 2.13) should be regarded as simply a mnemonic for remembering how the orbitals are filled. There are many exceptions for the neutral atoms, some of which will be discussed momentarily. In addition, there are many exceptions for the ions. For example, the neutral scandium atom is correctly predicted by the mnemonic to be $\{1s^2,2s^2,2p^6,3s^2,3p^6,4s^2,3d^1\}$. However, Sc(I) is not $\{1s^2,2s^2,2p^6,3s^2,3p^6,4s^2,3d^0\}$, but is in fact $\{1s^2,2s^2,2p^6,3s^2,3p^6,4s^1,3d^1\}$. In addition, Sc(II) is $\{1s^2,2s^2,2p^6,3s^2,3p^6,4s^0,3d^1\}$, which implies that we have the orbital energy ordering all wrong! Based on this, one may argue that scandium is better written as $\{1s^2,2s^2,2p^6,3s^2,3p^6,3d^1,4s^2\}$ instead. The complexity occurs because the filling is not determined by orbital energy exclusively. Electron spin and also the exact nature of which orbitals are filled can affect the overall energy, and it is the overall energy that nature optimizes.[12]

There are in fact twenty-two exceptions to the expected Aufbau filling for the first 112 neutral atoms. Many of these can be explained satisfactorily by considering the spin state as well. For example, chromium ($Z = 24$) has the configuration $\{Ar\}4s^13d^5$ instead of the expected $\{Ar\}4s^23d^4$ from the Aufbau principle, which is simply due to stabilization of the higher spin state—in this case a septet versus a quintet.

For the heavier elements, relativity begins to become a factor, which can greatly stabilize the *s*-orbital and, to a lesser extent, *p*-orbitals relative to the others. In addition, the effect of *spin–orbit coupling* begins to become significant for the heavier atoms. An electron will reside in an orbital with angular momentum l and has spin m_s; these two different spins sum to give a new quantum number j. This effect too, is more pronounced for heavier atoms and spin–orbit coupling can greatly change the energies (sizes, and so on) of two different orbitals. Both of these effects, discussed in more detail in later chapters, have significant consequences for the properties of atoms.

2.3 Quantifying the Atom: Sizes of Orbitals, Atoms, and Ions

While the details of the atom and the mathematics of the orbitals were discussed previously, there are some empirical properties of atoms that are of utility in chemistry. Some of these properties, such as ionization energies, electron affinities, and so on, are often discussed in general chemistry courses and will not be belabored here. However, it is worthwhile to review some orbital sizes of the elements and factors affecting atomic size due to their importance to the rest of the text.

In many cases the symmetry and shape of the orbitals will be of great utility. In other cases, the exact radial extent of the orbitals will help us greatly in trying to understand the general reactivity patterns of the elements. Table 2.3 shows the principal maxima of the outer orbitals for the vast majority of the elements as calculated by Waber and Cromer.[13] The orbital radii are corrected for relativistic effects.

Note that in all cases there are slight differences between orbitals of the same principle and angular momenta but different j values—different coupling values for l and s derived from $j = l + s$. For example, the 2*p* orbital is listed as $2p_{3/2}$ and $2p_{1/2}$, which occurs because $l = 1$ for a *p*-orbital and $m_s = \pm ½$. The effect is quite small for the lighter elements and very pronounced for the heavier atoms. For example, the radius of the $2p_{1/2}$- and $2p_{3/2}$-orbitals of nitrogen are given as 0.487 and 0.488 Å, a miniscule difference. In contrast, the radii of the $6p_{1/2}$- and $6p_{3/2}$-orbitals of nitrogen's heavier congener bismuth are given as 1.130 and 1.295 Å, a quite significant difference.

The sizes of entire atoms are generally given in the form of radii associated with two different types of interaction: radii when bonding, and radii when in close contact but not bonding. The actual radius of an atom depends on a large number of factors, such as the strength of the interaction, the number of groups attached to the atom, the spin state, and the exact charge.

If there is an ionic component to the bonding, as in most bonds, the anion tends to enlarge and the cation tends to contract somewhat from the neutral atom (Fig. 2.14). In general, radii determined from bond distances are extrapolated towards a purely covalent interaction and are thus termed "covalent radii", removing the size differences due to charges on the atoms. Alternatively, one can develop "ionic radii" for the very electropositive and electronegative elements by extrapolating to a fully ionic structure. For example, the chloride ion, Cl^-, has an estimated radius of 1.67 Å, while neutral chlorine has a much smaller covalent radius of about 0.99 Å. In contrast, making an atom cationic significantly reduces the radius, such as the sodium ion's (Na^+) radius of 1.16 Å relative to its neutral atom of 1.54 Å.

There is no set method for the determination of covalent or ionic radii from experimental data or theoretical calculations; consequently, there are numerous tables of radii in the literature, and they do not always agree.

At several places in this book, how bond distances and atomic radii change with various effects will be discussed. It is important to remember that increasing the charge decreases the radius; that is, more positive atoms are smaller and more negative atoms are larger. In addition, increasing the spin state, which generally means more orbitals are occupied increasing repulsions, increases atomic radii (Fig. 2.14).

Table 2.3 The radii for the principal maxima of the outer occupied orbitals for the elements in Ångstroms as calculated by Waber and Cromer.[13]

He core	He	Li	Be	B	C	N	O	F	Ne
1*s*-1/2	0.291	0.186	0.138	0.110	0.091	0.078	0.068	0.060	0.054
2*s*-1/2		1.586	1.040	0.769	0.620	0.521	0.450	0.396	0.354
2*p*-1/2				0.776	0.596	0.487	0.413	0.359	0.317
2*p*-3/2						0.488	0.414	0.360	0.318

Ne core	Na	Mg	Al	Si	P	S	Cl	Ar	
2*s*-1/2	0.278	0.247	0.221	0.201	0.184	0.169	0.157	0.146	
3*s*-1/2	1.713	1.279	1.044	0.904	0.803	0.723	0.660	0.607	
3*p*-1/2			1.312	1.068	0.916	0.806	0.722	0.655	
3*p*-3/2					0.919	0.810	0.725	0.659	

Ar core	K	Ca	Sc	Ti	V	Cr	Mn	Fe	Co
3*p*-3/2	0.593	0.539	0.500	0.468	0.439	0.416	0.392	0.373	0.355
4*s*-1/2	2.162	1.690	1.570	1.477	1.401	1.453	1.278	1.227	1.181
3*d*-3/2			0.539	0.489	0.449	0.426	0.388	0.363	0.342
3*d*-5/2						0.427	0.389	0.365	0.344

Ar core	Ni	Cu	Zn	Ga	Ge	As	Se	Br	Kr
3*p*-3/2	0.339	0.325	0.311	0.298	0.285	0.274	0.263	0.253	0.244
4*s*-1/2	1.139	1.191	1.065	0.960	0.886	0.826	0.775	0.730	0.691
4*p*-1/2				1.254	1.090	0.982	0.900	0.834	0.779
4*p*-3/2						1.001	0.918	0.851	0.795
3*d*-3/2	0.323	0.310	0.292	0.275	0.260	0.246	0.234	0.223	0.213
3*d*-5/2	0.325	0.312	0.293	0.276	0.261	0.247	0.235	0.224	0.214

Kr core	Rb	Sr	Y	Zr	Nb	Mo	Tc	Ru	Rh
4*p*-3/2	0.735	0.685	0.646	0.613	0.586	0.560	0.535	0.515	0.496
5*s*-1/2	2.287	1.836	1.693	1.593	1.589	1.520	1.391	1.410	1.364
4*d*-3/2			0.865	0.790	0.747	0.696	0.642	0.616	0.584
4*d*-5/2						0.702	0.648	0.622	0.589

Kr core	Pd	Ag	Cd	In	Sn	Sb	Te	I	Xe
4*p*-3/2	0.479	0.462	0.445	0.430	0.416	0.402	0.390	0.378	0.367
5*s*-1/2		1.286	1.184	1.093	1.027	0.969	0.920	0.876	0.837
5*p*-1/2				1.382	1.240	1.140	1.063	0.999	0.944
5*p*-3/2						1.193	1.111	1.044	0.986
4*d*-3/2	0.561	0.530	0.502	0.478	0.456	0.436	0.418	0.401	0.386
4*d*-5/2	0.567	0.536	0.508	0.483	0.460	0.440	0.422	0.406	0.390

Table 2.3 Continued

Xe core	Cs	Ba	Hf	Ta	W	Re	Os	Ir	Pt
5*p*-3/2	0.922	0.869	0.617	0.599	0.582	0.566	0.550	0.536	0.523
6*s*-1/2	2.518	2.060	1.476	1.413	1.360	1.310	1.266	1.227	1.221
5*d*-3/2			1.476	1.413	1.360	1.310	1.266	1.227	1.221
5*d*-5/2						0.734	0.703	0.676	0.657

Kr core	Au	Hg	Tl	Pb	Bi	Po	At	Rn
5*p*-3/2	0.510	0.494	0.485	0.473	0.461	0.450	0.440	0.430
6*s*-1/2	1.187	1.126	1.060	1.010	0.963	0.923	0.885	0.851
6*p*-1/2					1.130	1.065	1.009	0.962
6*p*-3/2					1.295	1.212	1.146	1.090
5*d*-3/2	0.614	0.589	0.566	0.546	0.526	0.509	0.493	0.478
5*d*-5/2	0.634	0.607	0.584	0.562	0.542	0.524	0.507	0.492

Rn core	Fr	Ra	Ac	Th	Pa	U	Np	Pu
6*p*-3/2	1.024	0.972	0.930	0.894	0.884	0.865	0.848	0.842
7*s*-1/2	2.447	2.042	1.895	1.788	1.804	1.775	1.741	1.784
6*d*-3/2			1.274	1.186	1.198	1.171	1.147	
5*f*-5/2					0.567	0.536	0.516	0.504

Rn core	Am	Cm	Bk	Cf	Es	Fm	Md	No
6*p*-3/2	0.826	0.801	0.788	0.775	0.761	0.750	0.739	0.738
7*s*-1/2	1.757	1.657	1.626	1.598	1.576	1.557	1.527	1.581
6*d*-3/2		1.086	1.071	1.054	1.037	1.026	1.015	
5*f*-5/2	0.488	0.469	0.455	0.443	0.431	0.421	0.411	0.403
5*f*-1/2	0.498	0.478	0.464	0.452	0.440	0.429	0.419	0.412

Lanthanides

Xe core	La	Ce	Pr	Nd	Pm	Sm	Eu	Gd
5*p*-3/2	0.827	0.825	0.806	0.788	0.772	0.757	0.742	0.715
4*f*-5/2		0.366	0.351	0.337	0.325	0.314	0.304	0.291
4*f*-7/2							0.307	0.294
5*d*-3/2	1.087							0.960
6*s*-1/2	1.915	1.978	1.942	1.912	1.882	1.854	1.826	1.713

Xe core	Tb	Dy	Ho	Er	Tm	Yb	Lu
5*p*-3/2	0.715	0.702	0.690	0.678	0.667	0.657	0.637
4*f*-5/2	0.286	0.278	0.270	0.263	0.256	0.250	0.242
4*f*-7/2	0.289	0.280	0.273	0.265	0.259	0.253	0.245
5*d*-3/2							0.885
6*s*-1/2	1.775	1.750	1.727	1.703	1.681	1.658	1.553

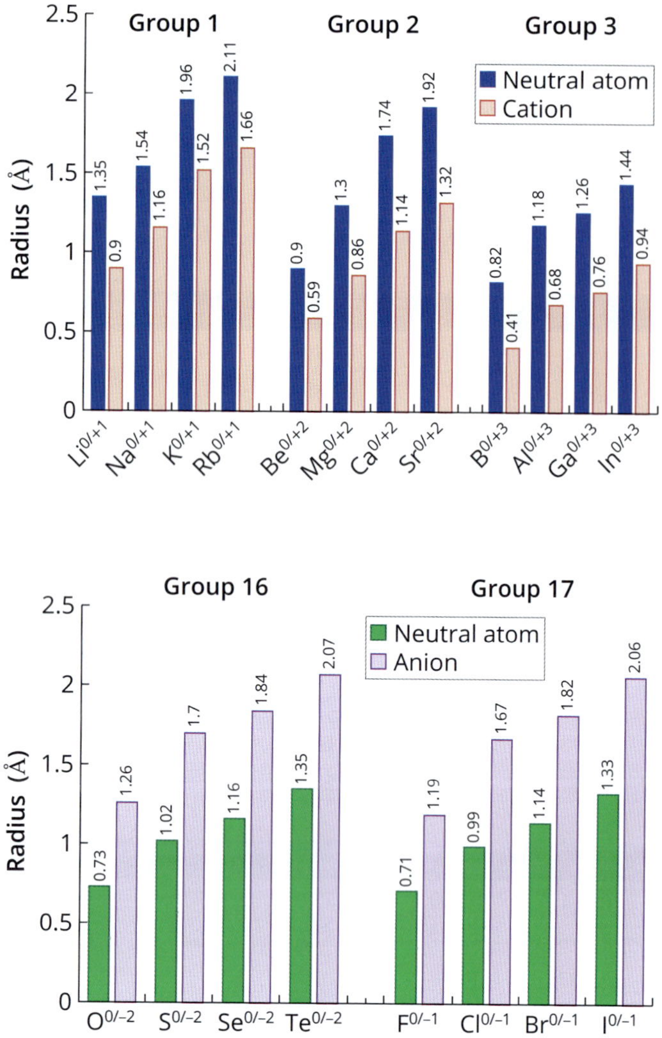

Fig. 2.14 Ionic radii versus radii of the neutral atoms for some of the Group 1, 2, 13, 16, and 17 elements.[14]

2.4 Some Brief Comments on Computational Methods

Here we supply a very brief description of computations methods. Several methods are used throughout the book, and the interested student can turn here for a primer on what the computational notes mean. However, coverage of this section is by no means required to understand the remainder of the text.

The last several decades have seen slow, but steady, incorporation of computational techniques into inorganic chemistry. From the beginning of quantum mechanics there was much bluster about obsolescence of experimental chemistry, first by physicists and then by theoreticians within chemistry due to the new theory. As the decades wore on from the first successes of quantum physics, it was quickly realized that experimental chemistry—which can include solvents, heat, gradients in concentration, paramagnetic species, and a variety of other complicating factors—is exceedingly difficult to model precisely using *ab initio* methods in virtually all cases.

Advances in computational speed and methods have provided some success, and now useful models for some reactions are possible with careful application of theory. Furthermore, understanding the properties of some molecules using models that incorporate the entire molecular systems of moderate to large sizes are possible and are becoming commonplace. Using computational chemistry, one can examine the nature of species that would be impossible to observe experimentally, such as transition states. However, the wise chemist will always treat theoretical results with caution until they can be confirmed by experiment, and even species such as transition states can have their relative energies confirmed experimentally in some cases to lend credence to computational results. Chemistry that is a figment of a computer's imagination is of little utility in the laboratory where compounds can actually be made; good theory should be relatable to experiment.

What can you derive from a quantum-mechanical calculation? The output can be almost anything you are interested in knowing about the molecule. Commonly, one uses *ab initio* calculations to obtain the ground-state geometry, energy, and orbital make-up of a system. For example, a restricted B3PW91 DFT calculation with a small (6–31G**) basis on $Fe(CO)_5$ gives the geometry shown in Fig. 2.15a. (If you have no idea what any of that means, all of it is subsequently described in brief.) This calculation was done on my desktop using the Gaussian16 software package, taking only a few minutes, though I knew the correct geometry for this molecule beforehand and started very close to where the geometry search ended.

The program tells me the energy of the assembly of atoms, −1830.056804 hartrees, which is a massive amount of energy (1.1 million kcal/mol!) required to completely separate all the nuclei and electrons from one another. (Units in calculations are also discussed shortly.) In addition, this calculation gives the entire MO diagram of the molecule, a piece of

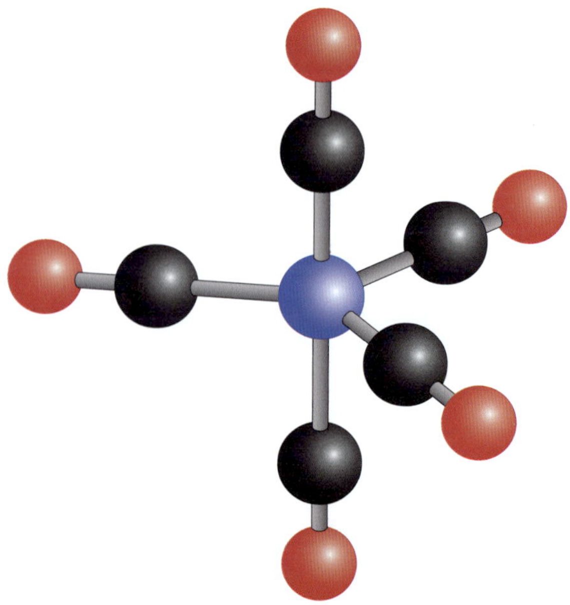

Fig. 2.15a Gas phase geometry of $Fe(CO)_5$ found using density functional theory.

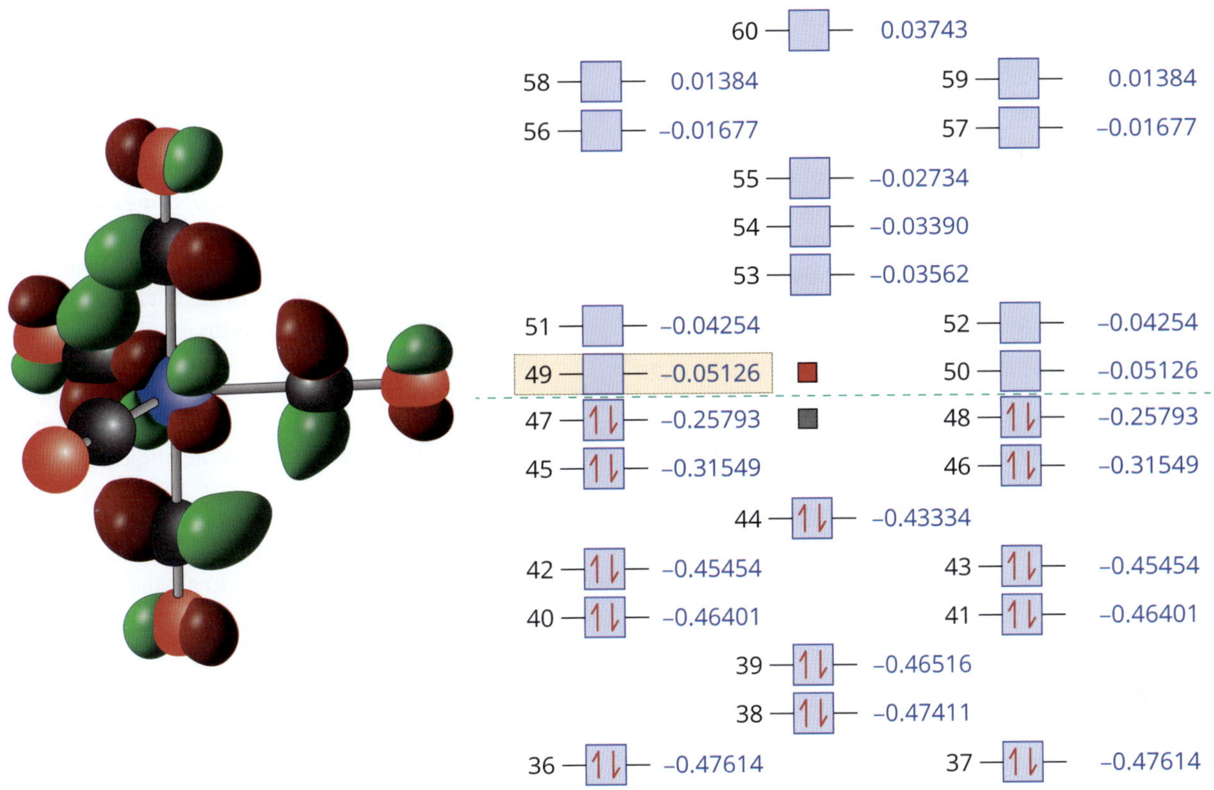

Fig. 2.15b On the left is a picture of the LUMO of $Fe(CO)_5$, and on the right is part of the MO diagram, both from a DFT calculation.

which is shown in Fig. 2.15b, and will calculate what the orbitals look like. The highlighted orbital in the MO diagram is shown. Quantum-chemical calculations can give you vibrational spectra and the nature of the modes, optical spectra, geometries of transition states, and much more. This wealth of information from the calculation, which is relatively easily obtained, is of limited utility without the background to interpret it, which is in part the purpose of this book.

Anything approaching a full accounting of theoretical methods here would be both gratuitous and lengthy, so there will be little in the way of derivation or proof. There are several computational packages in common use with good graphical interfaces that allow easy building of models. In this section we hope to give the student a brief introduction to the terminology of computational methods and a working knowledge of what those terms imply. For more details, we recommend the very accessible text by Frank Jensen, *Introduction to Computational Chemistry*,[18] or the student should consider taking practical computational courses as a part of their studies, which are offered at most universities. It is easily arguable at this stage that some education in computational chemistry should be part of even a synthetic chemist's education.

2.4.1 Basis Sets

Even though we have discussed the mathematical expression for some of the hydrogen orbitals (Table 2.1), these are rarely used explicitly in computational chemistry. First, they are unwieldy mathematically, and doing the (conservatively speaking) hundreds of integrations required for molecules moderate in size with these functions would be very computationally intensive. Second, these are the hydrogen wavefunctions, and the exact forms of the multielectron wavefunctions cannot be derived. In order to combat these problems, relatively simple functions that have similar behavior are optimized as replacements for the exact wavefunction—a process that must be repeated for each atom type. These alternative functions are called *basis sets.* As one might expect, as the basis set becomes larger and more computationally intensive to use, it generally becomes a better approximation for reality.

It is in approximating the multielectron atoms that the division of the wavefunction discussed previously into its radial, $R(r)$, and angular, $Y(\theta,\phi)$, parts really shows its advantage. The angular portion of the wavefunction is much simpler and easily modeled. The radial part in multielectron atoms, unlike in the hydrogen functions, is susceptible to electron–electron repulsion and other multibody effects that are quite difficult to model with great accuracy. The angular parts of the hydrogen wavefunctions are used directly to estimate the orbital shape, but a more easily used version of the radial function is required. In the next subsection, a brief description of how these basis sets are created is given for those interested, but in reality one simply needs to realize that basis sets that are larger generally provide better descriptions of reality but require longer calculations. Small basis sets are typically quicker to use but produce a less complete description of the molecule.

2.4.2 A Brief Description of How Basis Sets Are Generated (Advanced Topic)

One type of function used to approximate the radial part, $R(r)$, of the wavefunctions of multielectron atoms, $\chi(r,\theta,\phi)$, is called a Slater Type Orbital (STO), the form of which is shown in Eq. 2.25. The value of ζ (Greek zeta) is obtained by optimization of the function for the atoms in the basis set. In addition, the parameter ζ is optimized for each type of orbital in the atom as well.

$$\chi(r,\theta,\phi) \propto r^{n-1} \exp^{-\zeta r} Y(\theta,\phi) \tag{2.25}$$

The r^{n-1} factor ensures that the orbitals will become more diffuse, with a larger maximal probability (r_{max}) farther from the nucleus, as one goes down a column of the periodic table. The maximum in the probability is actually related to the formula shown in Eq. 2.26, and STO values should correctly emulate this.

$$r_{\max} = \frac{n^2 a_0}{Z_{eff}} \tag{2.26}$$

These STOs correctly emulate the discontinuity at the nucleus found for the 1*s*-orbital, for example.[17] However, these STOs do not contain radial nodes like the hydrogen wavefunctions with higher principal quantum numbers, *n*, which causes no particular difficulty in their application. While the radial nodes will be discussed often later in the book to understand chemistry of many different systems, apparently, they are not crucial to the calculations.

A more accurate representation of the wavefunction can be obtained if we use multiple parameters in the fit. For example, we can create a "double zeta" wavefunction that can be used to approximate χ (the molecular wavefunction) as follows, where the orbital will be optimized for both ζ_1 and ζ_2.

$$\chi \propto r^{n-1}[c_1 \exp^{-\zeta_1 r} + c_2 \exp^{-\zeta_2 r}]Y(\theta,\phi)$$

A double zeta basis gives the basis the flexibility to expand or contract more precisely to account for more or less electron density at the atom. In order to emulate the radial distributions of the *d*-orbitals well, a double zeta or better wavefunction is required. Triple zeta and larger basis sets are also in common usage.

In order to obtain more accurate results in calculations, it is commonly found that "polarization functions" should be added to the basis set. In isolation, an *s*-orbital containing electron density can reasonably be considered to be spherical. However, placing that *s*-orbital into a nonspherical electric field—such as that found in almost any molecular structure—would be expected to deform the spherical wavefunction. In order to account for this in molecular compounds, wavefunctions of higher *l*-value are added to the basis set. For example, the hydrogen atoms in a molecule will be modeled with not only *s*-orbitals, but commonly *p*-functions are added as well to allow the shape of the orbital to distort. Carbon atoms, for example, may be modeled with 3*d*-functions to account for polarization. Heavy transition metals may be modeled with *f*- or even *g*-functions to account for polarization. *Naturally, none of this implies that those orbitals used to polarize the wavefunctions are actually participating electronically (such as the d-functions in the main-group elements); this is simply a method for better fitting the electron density around the atom.* The polarization functions are of no chemical significance as far as the hybridization of the atoms.

Incidentally, the need for polarization functions in accurate calculations points towards one of the largest approximations in the LCAO MO method in general: the assumption that the orbitals on the atoms in the molecule have the hydrogen-like orbital shapes.

The other commonly employed functions used to estimate the atomic orbitals are Gaussian Type Orbitals (GTOs). In place of using STOs, GTOs can be used to get the same behavior and are more easily integrated in the calculations. In practice, STOs are often estimated using GTOs of the type shown in Eq. 2.27.

$$\exp^{-\zeta r} \approx c_1 \exp^{-\alpha_1 r^2} + c_2 \exp^{-\alpha_2 r^2} + \ldots \tag{2.27}$$

Unlike STOs, GTOs are continuous functions at $r = 0$. In order to get the correct behavior, several different GTO functions are required. Namely, it usually takes at least three GTO functions to correctly emulate an STO. However, the time saved in the calculation is still substantial with GTOs. A common (quite small) basis set still used occasionally is STO-3G, where each Slater Type Orbital (STO) is fit with three Gaussians (3G).

Another type of basis commonly encountered is "3–21G". Here all the orbitals containing the core electrons are approximated using three primitive Gaussian functions (contracted as in Eq. 2.27). The orbitals containing the valence electrons are described by two primitive, contracted Gaussians and another single, more diffuse Gaussian function. The 3–21G basis is of "double zeta quality". Note that STO-3G and 3–21G have the same number of primitive GTOs, but the 3–21G is more flexible in the valence set.

The larger "6–311G(d,p)" basis set, which is sometimes called 6–311G**, includes six functions contracted to one for the core orbitals and a "triple zeta quality" set for the valence orbitals. Furthermore, a *d*-polarization function is added to all the nonhydrogen atoms, and *p*-polarization functions are added to hydrogen atoms.

In some cases, very diffuse functions are added to better ensure the flexibility of the orbital optimization. In the Pople-style sets under current discussion, these are signified by additions of "+" symbols in front of the G. For example, in 6–31++G, where the first + indicates diffuse *s*- and *p*-functions are added to the nonhydrogen atoms and the second + indicates that diffuse *s*-functions are added to hydrogen.

For heavy atoms, what is termed an "effective core potential" (ECP, which is sometimes called a pseudopotential) may be used. In this method for basis development, the exact nature of orbitals below the valence set may be estimated without explicit modeling of the electrons themselves. This is particularly advantageous for large atoms with relativistic effects, as the ECP may model at least the scalar portion of the effect. In addition, there are significant time savings in the use of ECP for heavy atoms, as many, or often most, of the electrons do not need to be dealt with explicitly.

2.4.3 Short Notations for Basis Sets

A "basis set" then approximates the orbitals around an atom and is less complex and easier to calculate than the actual radial functions of the atoms. Here, a few common basis sets will be identified by their shorthand notation. In general, more complex-looking abbreviations signify a larger, more complex basis set; for example, 3–21G is a smaller basis (worse approximation) than 6–311G(d,p). One measure of a basis set quality is its "zeta" value, a description of which can be found in the previous Advanced Topic section; however, "triple zeta" basis sets are better approximations than "double zeta" and so on.

Most basis sets also include polarization functions, which correct for the presence of other atoms in the structure. For example, a hydrogen by itself in a vacuum will have a rigorously spherical *s*-orbital. However, a hydrogen next to another atom will likely have an *s*-orbital that deforms from being rigorously spherical because of the electrons in the other atom. This deformation can be included in the calculation by adding the character of high angular momentum orbitals to the *s*-orbital. For example, *p*- or even *d*-character may be added to the hydrogen *s*-orbital to account for this polarization. Obviously, this does not mean the *s*-orbital of hydrogen has *p*- or *d*-orbital character; polarization is simply a tool for deriving a better description of the electron density.

There are numerous different basis sets available, and the "best" basis for a particular calculation is often a matter for experience, and accuracy of the basis set employed may be limited by the time required for the calculation. Table 2.4 lists common basis sets so that the student may recognize them when seen along with some notes about their quality. However, this is by no means a complete listing. The 3–21G, 6–31G, 6–311G, and other basis sets were developed by Pople and coworkers. On these Pople basis sets, a "G**" implies "G(d,p)" with *d*-functions on nonhydrogen atoms and *p*-functions on hydrogen atoms; for example, 6–31G** is the same as 6–31G(d,p). The basis sets cc-pVDZ, as an example, are the "correlation consistent-polarized" (cc-p) valence only (V) basis sets developed by Dunning and coworkers. If diffuse functions are added to these basis sets they are considered to be "augmented" and become aug-cc-pVDZ, for example. LANL2DZ and SDD are basis sets of double zeta quality that include ECP for many of the heavy atoms. The LANL2DZ ECP was developed by Hay and coworkers at Los Alamos National Laboratory and uses the D95V basis set (Dunning 95 valence). The SDD basis set uses the Stuttgart/Dresden pseudopotential for heavy atoms with the Dunning, Huzinaga, and Hay basis, D95 (the full D95 basis). Table 2.4 lists a few basis sets, but there are many many more, perhaps hundreds in use.

2.4.4 Brief Descriptions of Some Common Computational Methods

Again, even a reasonable description of how each of the computational methods mentioned in what follows would require more space and would carry us far afield. We simply mention a few common methods with a brief description of the technique.

Table 2.4 Some common basis sets with brief descriptions.

Basis Set	Brief Description
3–21G	A double zeta quality set. A common variant is 3–21G**.
6–31G	Perhaps the most commonly employed double zeta basis set. 6–31G**, 6–31++G** and many other variants are in common usage.
6–311G	A very common triple zeta quality basis to which polarization functions are invariably added starting with 6–311G** but adding multiple *d*- and *f*-polarization functions is common.
cc-pVDZ	A fairly large double zeta basis with polarization functions included. If diffuse functions are included, the "augmented" basis is used; that is, aug-cc-pVDZ. ECP-containing basis sets are available for heavier atoms.
cc-pVTZ	A correlation consistent triple zeta basis. ECP-containing basis sets are available for heavier ECP-containing basis sets are available for heavier atoms.
cc-pVQZ	A correlation consistent quadruple zeta basis. ECP-containing basis sets are available for heavier atoms.
cc-pV#Z (where # = 5 or 6)	A correlation consistent quintuple (# = 5) or septuple (# = 6) zeta basis. ECP-containing basis sets are available for heavier atoms.
LANL2DZ	A double zeta quality basis set with ECP that includes most of the elements in the periodic table. Polarization functions are not included.
SDD	A double zeta quality basis set with ECP that includes most of the elements in the periodic table.

An important idea in quantum computing is the *variational principle*, which is defined and derived in all texts on quantum mechanics and is a method of approximating the state of the system (eigenstate). According to the variational principle, an approximation of the eigenvalue (for our purposes usually the energy) will always be higher than the actual value; as a result, a variationally bound method that gives a lower energy is always a better estimate than one with a higher energy. Many common quantum-chemical methods—Hartree–Fock, configuration interaction (CI), coupled cluster (CC) theory—are variationally bound, and one can increase the basis set size (for example) to derive lower and lower energy values for the system.

Here we will briefly mention four types of calculations used to determine the electronic structure of inorganic complexes. Obviously, there are many other methods, but in common practice there are only a few methods that are at all practical for most molecules.

Hartree–Fock

In any *ab initio* calculation there are terms related to nuclear–nuclear repulsion, electron–nuclear attraction, and electron–electron repulsion. On the time scale of electron movement, the nuclei generally can be considered stationary (Born–Oppenheimer approximation), and the input for any quantum-mechanical calculation is the nuclear coordinates. As a result, the nuclear–nuclear repulsion term calculation is simply a geometric problem in which the energy is provided from the charges and Coulomb's law. The electron–nuclear attraction, since the nuclei are considered stationary, is also a relatively easy calculation. The greatest difficulty occurs in the explicit calculation of the instantaneous electron–electron repulsion term, often referred to as "electron correlation".

Hartree–Fock (HF) theory, which is sometimes called Self-Consistent Field (SCF) theory, does not attempt to explicitly calculate electron correlation. Instead, the energy of the electron is calculated as it moves in the average electron field created by the other electrons in the system. Because HF theory is variationally bound, increasing the basis set size lowers the energy as the calculation approaches the "Hartree–Fock Limit", the minimum energy of the calculation using this method obtained with an infinitely large basis set.

HF theory can account for ~99% of the energy of the system; however, in order to describe most chemical phenomena well, the final ~1% must be approximated as closely as possible. The energy difference between the Hartree–Fock Limit and the lowest possible energy is called the "electron correlation energy". In order to lower the energy of the system past the Hartree–Fock Limit, a post-HF method must be used to deal with correlation.

The HF method produces molecular orbitals that are often quite consistent with LCAO MO expectations.

Møller–Plessant (MP) Perturbation Theory

Perhaps the oldest post-HF method is Møller–Plessant perturbation theory, which is used to add electron correlation as a perturbation on the HF result. Perturbation theory involves an infinite Taylor expansion that is truncated after some number of specified terms. The more terms added, generally, the more accurate the approximation. However, adding terms increases the time required for the calculation in a nonlinear fashion because the terms become more and more complex as the order is increased. The commonly encountered methods are MP2 and MP3. The MP method is not variationally bound and approaches (in some cases oscillates above and below)[18] the limiting "exact" value for the calculation as more perturbation terms are added. Fig. 2.16 shows the electronic energies for H_2 as one approaches the Full CI energy (set as zero in the plot).

Configuration Interaction (CI) and Coupled Cluster (CC) Theory

If one starts with a Hartree–Fock (HF) calculation (Fig. 2.17), one method of accounting for electron correlation is called configuration interaction (CI). In this method, the HF calculation is corrected by adding contributions from excited states. The energy corrections to HF can take the form of what are termed singles (S), doubles (D), triples (T), quadruples (Q), and so on. A single (S) is where an excited state is

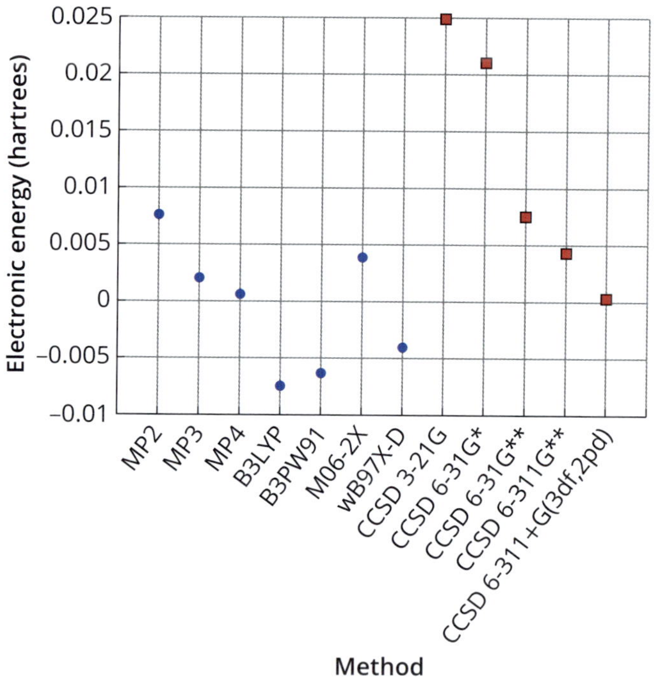

Fig. 2.16 Electronic energy of H_2 with different Møller–Plessant perturbation theory and density functional theory methods with the aug-cc-pVTZ basis set. Energies are given relative to the full CI energy. A variationally bound process will always be higher, which is true for MP here as well. In some cases, MP may oscillate towards the full CI value because it is not variationally bound. Also shown are several values from DFT (*vide infra*), which is also not variationally bound. The red squares are for Coupled Cluster Theory (CCSD) with increasingly larger basis sets, which is variationally bound and approaches the full CI value from higher energy. (All data are from the Computational Chemistry Comparison and Benchmark Database maintained by NIST.)

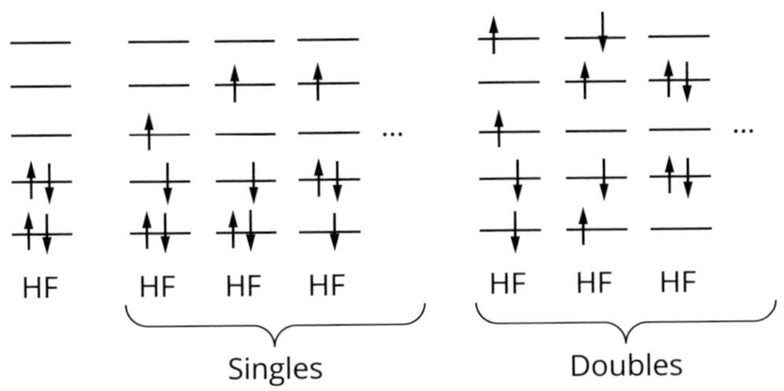

Fig. 2.17 Illustration of the relationship between the Hartree–Fock (HF) calculation and corrections due to singles and doubles.

formed by promoting one electron in the HF configuration to a higher energy orbital as shown in Fig. 2.17. A double (D) involves promotion of two electrons, and so on. The inclusion of these other states is done to lower the overall energy, better accounting for electron correlation.[18]

As can be seen, even when including S only, there are many possible configurations of the system, and the calculation time can be extensive. Full-CI is where every possible excited state for the system is calculated—a technique that is only possible for extremely small systems, such as diatomics. The importance of the correction is generally as S > D > T > Q > and so on; so, the method is often truncated to include the more important terms and reduce computational time to CISD (configuration interaction with singles and doubles) or CISDT. Even then, CI can be onerous in required computer time for the vast majority of chemical systems.

A full-CI calculation extrapolated to infinite basis is sometimes referred to as being the "exact" solution. It should be noted that the term "exact" refers to a solution that still contains approximations and refers to a solution of the Schrodinger Equation under the Born–Oppenheimer approximation and generally without inclusion of effects like relativistic contributions.

Coupled-Cluster (CC) Theory is a method related to CI, which due to the mathematical details is generally much faster than configuration interaction with similar accuracy. Like CI, the terminology of singles, doubles, triples, and so on, is used in coupled cluster theory as well, and the method may be truncated to methods like CCSD. Both CC and CI are variationally bound methods.

Density Functional Theory (DFT)

In DFT, the energy is calculated in a somewhat different fashion from the other methods described previously. The total energy in DFT is determined from the 1-electron density rather than from the wavefunction. An approximate Hamiltonian is used to derive an approximation of the total electron density. The method uses a "functional"—a term that is used for functions of another function—which is used to determine the electron density. The name density functional theory comes from the use of functionals of the electron density.

If the exact nature of "the" functional were known, the functional that describes the actual electron density, DFT could be used to get the exact solutions of the Schrodinger Equation (with the usual caveats of the Born–Oppenheimer approximation, requiring infinite basis, and so on); however, the exact functional that would perfectly describe the molecular electron density is unknown. As a result, there are numerous functionals of various types that have been developed, and it is the accuracy of these functionals, which increased significantly starting in the 1990s, and the speed of the method that has allowed DFT to rise to its current prominence in computational chemistry. Some of these functionals use HF theory (specifically the exchange term) and are termed "hybrid functionals". Some of the functionals may be optimized by using experimental data or more complex calculations as well. Some commonly encountered functionals are B3LYP, B3PW91, LSDA, PBEPBE, WB97XD, M06L, and so on. Unfortunately, there is no precise method for knowing the best functional for any particular calculation. Those who work extensively with DFT are able to make educated guesses as to the best method to use for a specific problem in some cases. In many cases, functionals are optimized for certain types of systems; for example, M06L for transition metals. However, DFT is not variationally bound, and values may be above or below the "real" value (Fig. 2.16); that is, the accuracy of a functional for a new system of interest is essentially unknown.

The advantage of DFT is its relative speed of calculation. If the functional uses HF terms, DFT will scale similarly to the relatively speedy HF but will include electron correlation effects. Nonhybrid methods are, in principle, even faster than HF. The time required generally scales as some power of the number of basis functions and electrons being manipulated; in other words, if the number of basis functions is K, Hartree–Fock may require a time that scales as K^4, but some other method (like CI) may scale as K^{10}, making it much slower.

If a compound of interest is diamagnetic, the calculation may be done in a "restricted" manner, so that all of the orbitals are doubly occupied. If the compound is paramagnetic, an "unrestricted" calculation is required, which will provide "spin orbitals" that are singly occupied. In an unrestricted calculation, the manifold of orbitals with more electrons—by convention those of α spin—will be given separately from those with β spin. These spin orbitals are required to discuss some relatively simple molecules, such as O_2 (Chapter 6).

Calculations on paramagnetic systems are generally regarded as being more difficult to accomplish accurately than restricted calculations on diamagnetic systems. Methods for accurate calculations of paramagnetic transition metal systems are an active area of research in the chemical theory community.

Other additions can be done to improve the accuracy of computational methods, all of which generally require additional computational resources. For example, "dispersion" may be accounted for in calculations. Dispersion is a pure quantum-mechanical phenomenon related to electron cloud fluctuations that may occur. While dispersion effects are energetically much smaller than chemical bonds, in some molecular systems the small energies may add up to be substantial. A brief discussion of the mechanism of dispersion can be found in Section 19.2.1.

2.4.5 Determination of Structure and Thermodynamics

The input for any theoretical calculation of the electronic structure of a molecule is the nuclear coordinates. On the time scale of the calculation, these are considered to be stationary. If one wants to find the structure of a compound using *ab initio* methods, the "best" structure is the lowest energy arrangement of the atoms, and, from there, low-energy molecular motions like vibrations can be considered.

Typically, calculations are done to model the gas phase structure of the molecule, the molecule in isolation, but there are a variety of methods for solution and solid-state models as well. Because the structures are often modeled for the gas phase, comparison with structures from techniques like X-ray diffraction (a solid phase technique), could vary somewhat. However, there is generally very good agreement between accurately calculated structures and structures from techniques such as X-ray diffraction.

The method for optimization of a structure involves calculation of the energy, perturbation of the structure in some way (bond length changes, rotations around bonds, and so on), recalculation of the energy to see if the perturbation lowered the energy, and repeat. The process is stopped when the structure is deemed "converged" based on some set of criteria. The first problem with this process is that each iteration requires calculation of the energy using one of the previous methods, and for large structures this can be quite time-consuming. The second issue is that the program can fall into a local minimum, rather than the desired global minimum. For example, the global minimum of cyclohexane is a chair conformation learned in organic chemistry; however, there may be a local minimum for the boat conformer. By the usual criteria for structure optimization, the program may give the boat cyclohexane structure as "converged" because it is a minimum in the potential, but for most applications one requires the global minimum. In other words, any perturbation of the structure raises the energy from the local minimum until the structure is distorted past some threshold toward the global minimum.

The only way to ensure that the program is in fact finding the global minimum is to use your chemical intuition and to alter the starting structure for the calculation to be near other reasonable conformations, and then to reoptimize to see if the new minimum is of lower energy than the other minima found. It is this intuition that we hope this course will help build for inorganic systems.

Complex structures can be quite time-consuming to optimize (Fig. 2.18). Branching in organics can lead to very long optimizations. For example, where one methyl group is rotated in trying to minimize the energy, but the other nearby structures may no longer be optimized for the new methyl configuration and then need to be reoptimized. For example, consider HCl, MeCl, and *tert*-butylCl. The calculation needs to optimize all the internal degrees of freedom. The internal degrees of freedom are 3n – 5 parameters for a linear system, and 3n – 6 parameters for nonlinear systems, where n is the number of atoms. The only internal parameter for optimization in HCl is the H–Cl internuclear distance, 3(2) – 5 = 1 or a single parameter to optimize. For methyl chloride, $ClCH_3$, there are now 5 atoms and 3(5) – 6 = 9 parameters to optimize including the H–C–H and Cl–C–H angles along with all the internuclear distances. In *tert*-butyl chloride the rotation of one of the methyl groups in the *tert*-butyl group affects its orientation with respect to all the other methyl groups, which will need to have their orientations reoptimized. In other words, *tert*-butyl chloride will have 3(14) – 6 = 46 parameters to optimize simultaneously, many of which are interdependent. Furthermore, as the number of parameters increases, the odds of observing local minima also increase.

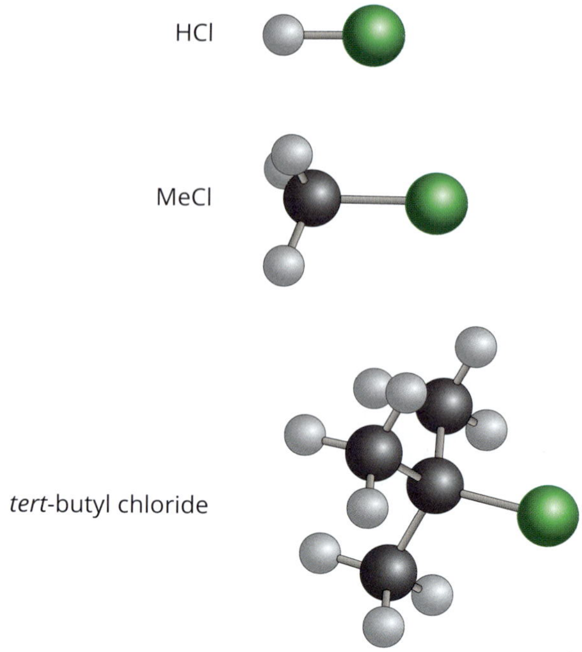

Fig. 2.18 Increasing the complexity of a structure increases the number of parameters that needs to be optimized. For example, HCl only has a single parameter for optimization (a distance). Replacing H with methyl gives MeCl, which has 9 parameters for optimization. Replacing H with *tert*-butyl gives *t*-BuCl, which has 46 parameters to be optimized. Furthermore, changing one parameter results in a need for optimization of all the others; for example, changing the dihedral angle around one methyl group in *t*-BuCl results in needing to reoptimize the other angles on other methyl groups.

As a result of the difficulties in structure optimization and the nonlinear behavior of computational time required with increasing structure size, it is common to see large organic groups replaced with smaller ones or even hydrogen in calculations. The replacement of *tert*-butyl with methyl is likely to be a minor electronic perturbation on the chlorine in *tert*-butyl chloride, for example. As long as the replacement of the larger group with the smaller group does not cause neglect of important steric energies or change the rest of the structure (bond distances and angles) significantly, this is generally deemed acceptable and will save a large amount of computer time. This type of change is less appropriate for comparison with experimental results for very electronically different groups, especially those with π-effects, such as replacement of phenyl with hydrogen.

For very large structures, atoms quite far from, and more or less electronically decoupled from, the center of interest may be modeled with a low level of theory, while the atoms of electronic interest are modeled at a higher level of theory. For example, one may model the metal center and the atoms attached to it at a high level of theory, while using smaller basis sets and lower levels of theory for organic groups distant from the metal center.

All of the material thus far has been concerned with calculation of the electronic energy of the system. In actual chemical systems, the molecules have vibrational, rotational, and other energies that have been ignored. These vibrations are a fundamental property of the molecule and, of course, are also quantized into discrete energy levels.

If the blue curved line in Fig. 2.19 is the electronic energy of a system, which is being approximated as a parabola, then the lowest energy vibrational state is slightly higher in energy than the minimum of the electronic energy. The difference in energy between the ground vibrational state and the minimum of the electronic energy is called the *zero-point energy* (ZPE).

Most computational packages can estimate the vibrational and zero-point energies of molecules as well as electronic energies. As a result, one can calculate the standard free energy (ΔG°), standard enthalpy (ΔH°), and standard entropy (ΔS°) at some levels of theory such as HF, MP2, and DFT. Consequently, the thermodynamics of new molecular systems can be modeled with some degree of accuracy.

Complex hybrid techniques, such as the "G3" and "G4" methods, have been developed which involve a series of optimizations of the structures using HF and MP theory, followed by energy calculations using CI to get an estimate of the thermodynamics for the system, which is then empirically corrected. For many molecule types, methods such as G3 can give very good estimates of thermodynamic parameters for structures but can be somewhat time-consuming for larger molecules.

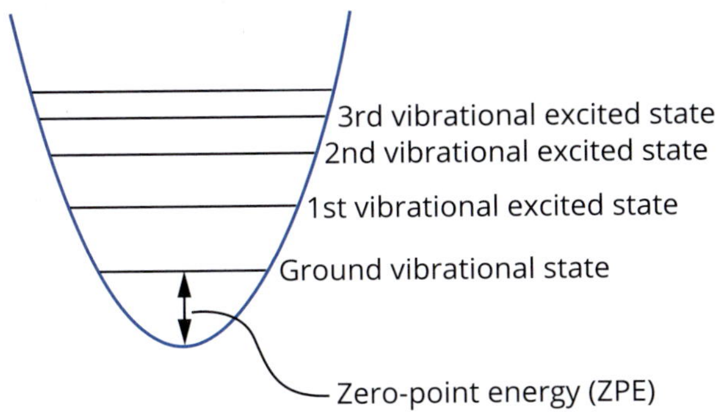

Fig. 2.19 Relationship of the zero-point energy to the vibrational levels. The blue line is the electronic energy with bond distance change, with the curved minimum at the equilibrium distance. The zero-point energy is the energy difference between the electronic energy at the equilibrium distance and the energy of the ground vibrational level.

These methods for the optimization of structures and estimation of energies can be extended to transition states and even entire catalytic cycles to help understand mechanisms of reactions. Furthermore, they can provide the thermodynamics for new reactions that the researcher is considering studying, suggesting if the process is thermodynamically feasible.

2.4.6 Units in Computational Chemistry

The standard unit of energy used in computational chemistry is generally the hartree, E_h, named after physicist Douglas Hartree. It is defined as shown in Eq. 2.28 relative to the Rydberg constant (R_∞), Planck's constant (h), and the speed of light (c).

$$E_h = 2R_\infty hc \qquad 2.28$$

In turn, the Rydberg constant is related to fundamental constants, like the electric charge and mass of the electron, by Eq. 2.18.[c] By including the fundamental constants in place of the Rydberg constant, the hartree becomes Eq. 2.29 and is related to $h/2\pi$, which is abbreviated $\hbar$, the mass of the electron, and the Bohr radius (a_0).

$$E_h = \frac{\hbar^2}{m_e a_0^2} \qquad 2.29$$

The advantage of using the hartree as the unit for computational chemistry is that the computational results become invariant with the current values for the fundamental constants. If a new experimental value for Planck's constant is measured, the computed energy simply shifts with the new experimental result.

The hartree is a relatively large unit for most chemical applications. The conversion factor for the hartree to some common experimental units is as follows.

$$\begin{aligned} 1\,E_h &= 627.509 \text{ kcal/mol} \\ &= 2{,}625.50 \text{ kJ/mol} \\ &= 27.2114 \text{ eV} \\ &= 219{,}475 \text{ cm}^{-1} \end{aligned}$$

2.4.7 Final Comments on Computational Methods

While exact calculation of chemical systems is extremely difficult, the best methods paired with the correct analysis of the problem can derive absolute energies to within a few kcal/mol. For some applications, this type of accuracy is more than adequate, lending great utility to the models. For other applications, 3 kcal/mol can be the difference between a potential

[c] For R_∞ = the heavy atom version of the Rydberg constant where the nucleus is assumed to be infinitely heavier than the electron. $R_\infty = m_e e^4/(8\varepsilon_0^2 h^3 c)$.

compound being the ideal enantioselective catalyst for a transformation, for example, and the compound being of no utility at all.

A fact that often saves computational methods is that the chemist is rarely interested in the absolute energy of the system. Instead, the chemist is usually interested in the energy relative to some starting state. Consequently, the error in the calculation of the initial state is inconsequential to the energy difference if the same magnitude of error occurs in the calculation of the final state. This cancelation of error, when applicable, can lead to calculations with extraordinary accuracy in some cases and may be what saves some methods from being uselessly inaccurate for chemical problems.

In the end, computational methods are of growing utility in their applications to chemical problems. While results from theory should always be applied with some caution to complex chemical systems, some knowledge of theoretical models is highly advisable due to their growing utility.

Chapter Summary

1. Flame color is one of the oldest, but still used, methods for characterizing inorganic compounds, and Bunsen made important contributions to the method.
2. Planck's discovery that photon energy is quantized allowed Bohr to develop the first effective model to understand the spectroscopy of hydrogen, but the Bohr model was difficult to expand to multielectron systems, due to electron correlation.
3. The "planetary" model of the atom, where light electrons circle a heavy nucleus, was very useful but was improved by the realization of wave–particle duality; that is, that the electrons have wave character.
4. The wave character of the electrons causes them to occupy regions of space around the nucleus called orbitals. The shapes and phases of the orbitals are critical to understand for chemical applications. The shapes and phases of the *d*-orbitals should be put to memory for this class (Section 2.2.1).
5. In addition to energy being quantized, the angular momentum of the orbitals and electron spin, for example, are quantized. This gives a specific set of quantum numbers for each electron. Every electron in the atom has a unique set of values.
   a. n: principal quantum number. Gives the shell the electron occupies.
   b. l. angular momentum quantum number. Gives the type of orbital, such as *s*, *p*, or *d*.
   c. m_l: magnetic quantum number. Gives the specific orbital occupied, such as d_{xy}.
   d. m_s: electron spin. Either +1/2 or −1/2.
6. Because electrons can be described as waves, they can constructively and destructively interfere, which are important interactions in bonding.
7. The atomic orbitals of an atom are arranged so as to be orthogonal to one another, meaning there is no net overlap and the orbitals have specific sizes related to placement of nodes in orbitals of the same shape.
8. The multiplicity of a spin system is 2S + 1, where S = sum of the spins. If there is no net spin, S = 0, and 2S + 1 = 1, a singlet state. The names for the multiplicities are singlet, doublet, triplet, quartet, quintet, sextet, septet, octet, and nonet.
9. Quantum calculations can be used to determine a molecule's energy, molecular orbitals, vibrations, ground-state geometry, and other information.
10. The hydrogen-like orbitals have very complex radial functions that would be difficult to calculate for many multielectron atoms. As a result, simpler "basis sets" are used in their place with different levels of approximation.
11. There are several common computation methods in use, such as density functional theory (DFT). There can be a lot of acronyms associated with the methods, but for this course we are more concerned with interpretation of the output than the details of the calculation.

Practice Problems

1. Match the letter corresponding to the correct scientist with their appropriate contribution.
   a. Experimentally determined the charge/mass ratio of electrons and showed that they have a negative charge.
   b. Determined that blackbody radiation is quantized, not continuous.
   c. Electrons move in stable circular orbits around the nucleus with no absorption or emission of energy.
   d. Determined that the mass of the electron is 9.11×10^{-31} kg.
   e. Wrote an equation to fully describe the wave particles of an electron in terms of position, mass, kinetic energy, and potential energy.
   f. Discovered X-rays.
   g. The more accurately one knows the value of the momentum of a particle, the less accurately one knows the particle's position.

h. Discovered that the nucleus is 10^{-15} m in size, much smaller than the atom at 10^{-10} m.

i. Used X-ray spectroscopy to discover that the atomic number is the number of positive charges in the nucleus.

j. Fit the spectroscopy of H atoms to:

$$\frac{1}{\lambda} = R_H\left(\frac{1}{n_l^2} - \frac{1}{n_h^2}\right)$$

(I) Rydberg
(II) J. J. Thompson
(III) G. Seaborg
(IV) Rutherford
(V) N. Bohr
(VI) M. Planck
(VII) E. Schrodinger
(VIII) J. Dalton
(IX) Balmer
(X) de Broglie
(XI) R.A. Millikan
(XII) Heisenberg
(XIII) Moseley
(XIV) E. Fermi
(XV) G. de Hevesy
(XVI) W. Röntgen

2. For each orbital listed here, draw a picture with its orientation. Be sure to include the coordinate axes.

a. d_{z^2}
b. $d_{x^2-y^2}$
c. p_y
d. d_{xy}
e. d_{xz}

3. For each orbital here, give the number of planar and spherical nodes:

a. $3d_{xy}$
b. $5f_{xyz}$
c. $4f_{xyz}$
d. $6p_x$
e. $5s$
f. $3p_z$

4. Explain the difference between a variationally bound calculation and one that is not. Give an example of each.

5. One of the distances, 1.095 or 1.203 Å is the ground state of dinitrogen. The electronic energy of N_2 with a bond distance of 1.095 Å was found to be −109.55593015 hartrees. With a bond distance of 1.203 Å, the energy of the molecule is −109.53153465 hartrees. Which is more likely the distance found in dinitrogen? What is its stability relative to the other geometry in kcal/mol?

6. What are the spin states of atomic:

a. Sc ($4s^2 3d^1$)
b. Ti ($4s^2 3d^2$)
c. V ($4s^2 3d^3$)
d. Cr ($4s^1 3d^5$)
e. Mn ($4s^2 3d^5$)

7. Why does chromium only have one electron in the 4s orbital?

8. If Usain Bolt weighs 94 kg and runs the 100 meter dash in 9.58 seconds, what is his average de Broglie wavelength during that run?

9. If the wavelength of a photon is 320 nm, what is its energy in Joules? What would the same photon be in wavenumbers (cm^{-1})?

10. Answer the following questions about ionization energy (IE) and electron affinity (EA):

a. Define IE.

b. Define EA.

c. Most atoms have a positive EA, but nitrogen does not. Why?

d. All atoms have a positive ionization energy. Why?

e. The first three IE of Be are 9.3, 18.2, and 153.9 eV. Why is the second IE larger than the first? Why is the third almost an order of magnitude larger than the second, and what new effect causes the large change from the second to the third IE?

Cited References

1. Kirchhoff, G., and Bunsen, R. Chemische analyse durch spectralbeobachtungen, *Ann. Phy. Chem.* 1860, *100* (6), 161.
2. Martinson, I., and Curtis, L. J. Janne Rydberg: His life and work, *Nuclear Instruments & Methods in Physics Research Section B-Beam Interactions with Materials and Atoms* 2005, *235*, 17–22.
3. Thompson, J. J. Cathode rays, *The Electrician* 1897, *39*, 104.
4. Sappol, M. *Dream Anatomy*. Bethesda, MD: National Institutes of Health, 2006; p. xii, 180 pp.
5. Feynman, R. P., Leighton, R. B., and Sands, M. L. *The Feynman Lectures on Physics*. Redwood City, CA: Addison-Wesley, 1989.
6. Born, M. *Atomic Physics*, 8th ed. London: Blackie, 1969, p. xiv, 445 pp.
7. Loveland, W. D., Morrissey, D. J., and Seaborg, G. T. *Modern Nuclear Chemistry*, 2nd ed. Hoboken, NJ: Wiley, 2017.
8. Moseley's career was tragically cut short when he was killed at the age of 27 by a sniper during the Gallipoli campaign in the First World War.
9. Howe, J. L. Recent work in inorganic chemistry, *J. Am. Chem. Soc.* 1909, *31*, 1284–1305.
10. Yoshihara, H. K. Nipponium. The element ascribable to rhenium from the modern chemical viewpoint, *Radiochimica Acta* 1997, *77* (1–2), 9–13.
11. Bersuker, I. B., *Electronic Structure and Properties of Transition Metal Compounds: Introduction to the Theory*, 2nd ed. Hoboken, NJ: Wiley, 2010.
12. Exactly what orbitals are filled in degenerate sets can also lead to different electronic configurations, called microstates. These will be discussed in detail when we turn to spectroscopy.
13. Waber, J. T., and Cromer, D. T. Orbital radii of atoms and ions, *J. Chem. Phys.* 1965, *42* (12), 4116–4123.
14. Shannon, R. D. Revised effective ionic-radii and systematic studies of interatomic distances in halides and chalcogenides, *Acta Cryst. A* 1976, *32* (SEP1), 751–767.
15. Cordero, B., Gomez, V., Platero-Prats, A. E., Reves, M., Echeverria, J., Cremades, E., Barragan, F., and Alvarez, S. Covalent radii revisited, *Dalton Trans.* 2008 (21), 2832–2838.
16. Levine, I. N. *Quantum Chemistry*, 5th ed. Upper Saddle River, NJ: Prentice Hall, 2000.
17. Albright, T. A., Burdett, J. K., and Whangbo, M.-H. *Orbital Interactions in Chemistry*, 2nd ed. Hoboken, NJ: Wiley, 2013.
18. Jensen, F. *Introduction to Computational Chemistry*, 2nd ed. Chichester: John Wiley & Sons, 2007.

CHAPTER 3

Basics of Bonding

The 2-Center 2-Electron (2c2e) Bond

CONCEPT COVERAGE

1. Review of Lewis dot structures, formal charge, and formal oxidation state
2. Review of elementary valence bond and molecular orbital concepts
3. Electronegativity
4. Bond dissociation enthalpy: its nature, relationship to bond polarity, and relationship to formal oxidation state
5. The dative vs covalent 2-electron bond
6. Bond order

"I believe the chemical bond is not so simple as some people seem to think."

— Robert S. Mulliken

This Mulliken quote "ought to go down to history as one of the most classic understatements of all time."

— C. A. Coulson[1]

". . . one can still say that quantum mechanics is the key to understanding magnetism. When one enters the first room with this key there are unexpected rooms beyond, but it is always the master key that unlocks each door."

— John Hasbrouck van Vleck[2]

When a chemist hears the word "bond", typically the first thing thought of is the 2-center 2-electron (2c2e) bond which is commonly found in the majority of molecules where covalent bonding is present, such as in H_2, or in heterocovalent compounds like HCl. In this chapter we discuss some basic concepts for understanding 2c2e bonds that most students will have been familiar with from high school, or general chemistry such as Lewis dot structures, formal charges, oxidation states, and electronegativity. Unsurprisingly, considering how early in one's education these concepts are taught, an understanding of many of these fundamental concepts is critical to grasping chemistry in general and inorganic chemistry in particular.

Here, these fundamental concepts are perhaps pushed farther than in previous courses, and procedures for calculating bond energies based on some knowledge of the Lewis structure and electronegativities of the elements will be covered. These empirical bond energy calculations require nothing but a calculator, only take a few minutes to execute, and are applicable broadly to many molecules. Even so, the calculated energies can be stunningly accurate for the majority of cases (typically within 5% of the experimental bond energies), and the approach allows a broad understanding of the nature of the simple 2c2e chemical bond and of the factors influencing energies in such heterocovalent bonds.

The heterocovalent bond will then be contrasted with the dative (Lewis acid–base) bond, the other important type of 2c2e bonding that is commonly encountered in inorganic chemistry. In spite of the absolutely central role that these interactions play in the stability and reactivity of inorganic and organometallic molecules, students often do not have a clear understanding of the nature of these two types of chemical bonding interactions. Very simple MO theory and resonance theory will be introduced, and these theoretical frameworks will be greatly developed in later portions of the book as the tools for exploring molecular electronic structure are expanded.

This chapter is admittedly rather long, but removal of the material to shorten the chapter would have created inconsistences. To counteract the length, the sections have been labeled to help the student and instructor make decisions on coverage.

Key Concepts These are ideas that are critical to understanding inorganic chemistry as a whole or to understand important concepts emphasized in this book.

Expanded Coverage These concepts are not essential but are fairly common in the literature or are referenced by Key Concepts. These sections might have been missed by those looking for more information, and the text would have been incomplete without them. Nevertheless, to save time they can be skipped in formal instruction.

Advanced Topics These are included throughout the book and are generally topics which the authors teach to advanced undergraduates and graduate students.

3.1 Lewis Dot Structures (Key Concept)

The *s*-block (Groups 1 and 2, orange) and *p*-block (Groups 13–18, blue) are often called collectively the "main group" (Fig. 3.1). The main-group elements are those that have their highest valence electron in an *s*- or *p*-orbital. The *d*-block (green) elements, also referred to as the transition metals, have widely varying chemistry that is quite different from the other blocks. In addition, the *f*-block (red) elements have unique chemistry of their own. Group 12 (white) presents a bit of a quandary and is not shaded in the figure. These elements most often only use their *s*-electrons in their chemistry, making them similar in some respects to Group 2; this leaves a d^{10}-shell similar to the *p*-block elements to their right. However, there are isolated instances where mercury may use its *d*-electrons and behave as a transition metal.[3] From a reactivity standpoint, Group 12 compounds typically show more similarities to the main group than the transition metals and, as with every group, some unique features of their own.

Our understanding of the structure of main-group compounds always begins with the Lewis dot structure of the compound. The invention of the Lewis model by G. N. Lewis[4] in 1916 was an insightful and essential first step in the establishment of our modern understanding of the chemical bond, and was predicated on the premise that stable molecules generally require electrons to be grouped in pairs of two. Although the reason for why such pairs of electrons would necessarily lead to stable molecules was not clear to Lewis at that time, a well-known fact was that very few molecules were known to possess odd numbers of electrons (in which case it is mathematically impossible to avoid unpaired electrons), and that, in the few instances where such "free radicals" had been observed, these were invariably reactive. Lewis's insight was to formulate electron pairing as a requirement to stability: whenever two unpaired electrons exist on neighboring atoms, the two electrons on the two centers (2e2c) should lead to a stable chemical bond. In addition to the idea that electrons need to be paired, Lewis conjectured that atoms had to be surrounded by a total of eight electrons, either from their own shell or from sharing with a neighboring atom. This became known as the "Octet Rule", which is so central in the chemistry of the main group that it is still, more than a century later, one of the first concepts taught in introductory chemistry courses.

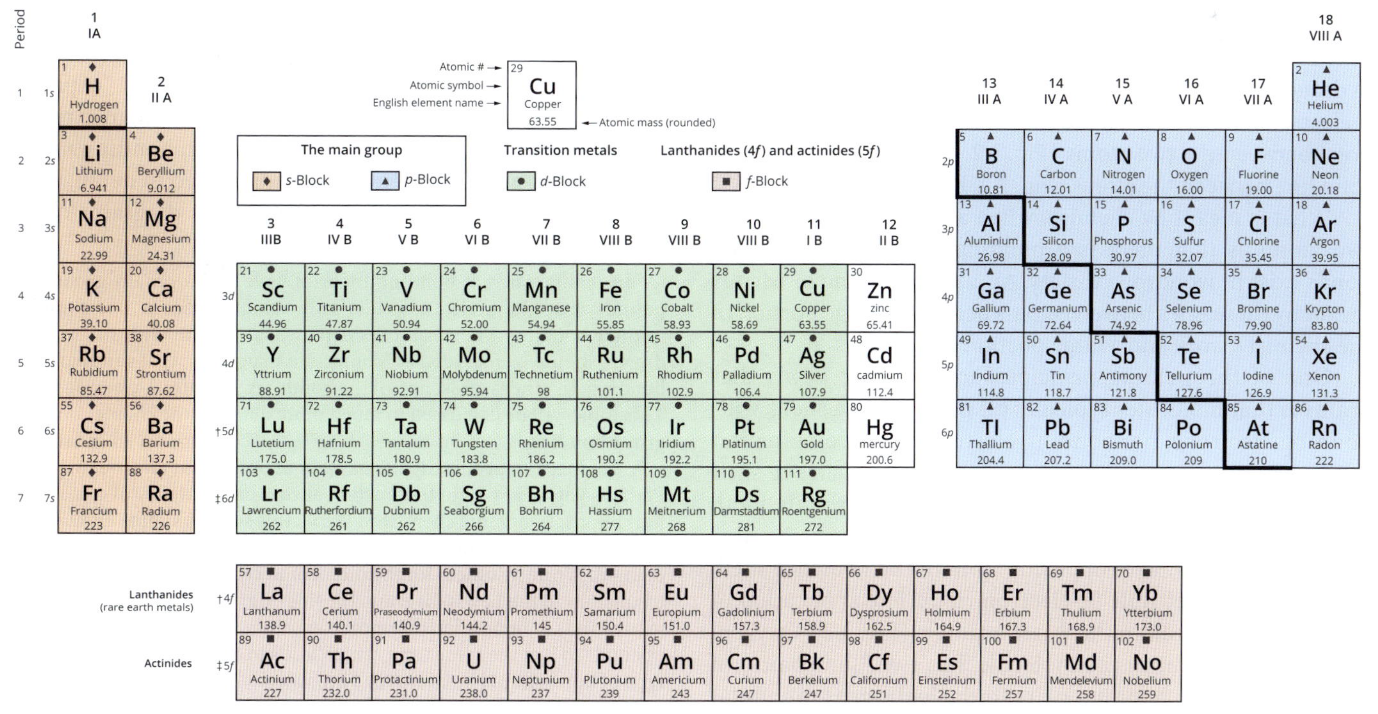

Fig. 3.1 Periodic table colored by block.

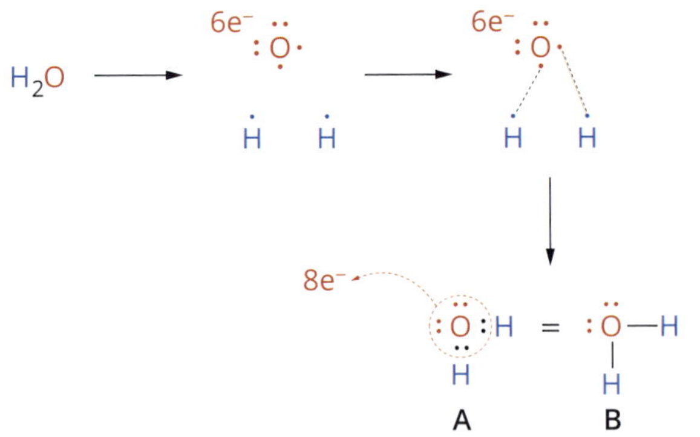

Fig. 3.2 Building the Lewis dot structure of water.

The first step in drawing Lewis diagrams is the representation of each atom in the molecule using simple dot diagrams to represent the electrons that are associated with the atoms, which are typically assumed to be in the neutral charge state. Then, individual atoms are linked by pairing electrons into 2-electron bonds. These diagrams can be regarded, for the moment, as 2-dimensional representations; their relationship to the 3-dimensional structures of molecules will be discussed in Chapter 4 in the context of the Valence Shell Electron Repulsion (VSEPR) theory.

For example, to draw the diagram for water (Fig. 3.2) we begin by drawing the individual atoms with the appropriate number of valence electrons. Oxygen is in Group 16 (Group VIA, Fig. 3.1) and has a $2s^2p^4$-electron configuration in its valence shell. The six valence electrons are shown by using six dots around the oxygen atom. Each hydrogen has a single electron in its 1*s*-orbital and thus is drawn with a single dot adjacent to each hydrogen atom. The single electrons of the two hydrogens can pair with the electrons on the oxygen to form 2-electron bonds. This can be shown as a pair of electrons (two dots) between the two atoms (Fig. 3.2, A) or, equivalently, as a solid line between the atoms (Fig. 3.2, B).

What is meant by chemists when a line is drawn between two atoms is that there is a bonding interaction; however, bonds exist in wondrous variety. An understanding of what is meant by these lines between atoms occupies a large portion of the chemist's work and a large portion of this text.

One of the guiding principles for drawing Lewis diagrams is that molecules typically fill their valence shells with bonding electrons, nonbonding electrons, or a combination of the two. For most main-group elements, the valence shell contains one *ns*- and three *np*-orbitals, four orbitals total. Consequently, elements in the *p*-block often have eight electrons, two per orbital of opposing spin, in their valence shell, fulfilling the Octet Rule (8-electron rule). The application of this rule can be seen in the water example (Fig. 3.2). Oxygen, the central element in the molecule, has two "lone pairs" of electrons and two bonding pairs to the hydrogens for a total of eight electrons. We say that the oxygen of water "obeys" the Octet Rule.

Numerous atoms in molecules, according to their Lewis diagrams at least,[a] do not obey the Octet Rule. Hydrogen atoms are a trivial exception to the Octet Rule simply because their valence shell is comprised of only a single *s*-orbital that can hold two electrons at most. However, there are many examples comprised of molecules containing main-group elements that have Lewis diagrams with atoms (other than hydrogen) below or above an octet of electrons. A more detailed discussion of the bonding in these types of molecules is to follow in later chapters. For now, it suffices to simply draw the requisite number of bonds to the central atom and pair up as many electrons as possible.

Placing more than eight electrons around a first-row *p*-block element (B, C, N, O, F) is not often done in Lewis structures. Also, it is only the central atom of a Lewis diagram that is drawn as having more than eight electrons in its valence shell. Some examples of such compounds are shown in Fig. 3.3.

In the first example, XeF_2, the Lewis structure has three lone pairs and two bonds to fluorine around the central xenon atom. As a result, the xenon atom, according to the Lewis diagram, has ten electrons. In the main group, atoms that exceed an octet in their Lewis structure are often referred to as *hypervalent*.[b] Similarly, in the $OPCl_3$ example, the central phosphorus atom has ten electrons with four electrons coming from the phosphorus–oxygen double bond. In contrast, the chlorine atoms in $OPCl_3$ each have eight electrons around them and are called *normal valent* because they, like the oxygen in water, precisely obey the Octet Rule.

In this book we distinguish between correct Lewis diagrams, bonding schemes with the correct number of electrons and formal charges, and a reasonable bonding model for a molecule where the lines between atoms are well described in terms of their interactions. In other words, it is useful on some levels to be able to draw Lewis diagrams that may exceed the maximum occupancy of the valence orbitals on the central element. (These are sometimes, unfortunately, called "expanded octets"; however, "hypercoordinate" or "hypervalent" are preferable terms to describe such species.) Later, bonding models for the main-group elements will be discussed in detail, which will show that bonds in hypervalent Lewis diagrams can be

[a] In reality, current bonding theories have that all main-group compounds obey the 8-electron rule, only using valence *s*- and *p*-orbitals in their bonding. The structures here, which are the ones still typically drawn, should be regarded as "composite structures" that summarize the bonding. In other words, not all the bonds are 2-center 2-electron bonds. This topic will be discussed in much more detail in the chapter on main-group bonding.

[b] "Valence" is specifically the number of bonds that an atom makes. Perhaps you were taught that carbon always has a valency of 4, makes four bonds, in organic chemistry. While this sounds simple enough, valence for some elements can get quite messy and is somewhat ill-defined. While it suffices for the moment to consider the number of lines to an atom in the Lewis diagram, the lines signify bonds, and there are many kinds of bonds. Furthermore, one atom may have different types of bonds to it at the same time, which leads to decisions about how each type of bond should be counted. The "valency" of an atom is in the common chemistry vernacular, but it is not really a concept we will find extremely useful in this book. Nevertheless, one should be aware of the term's meaning.

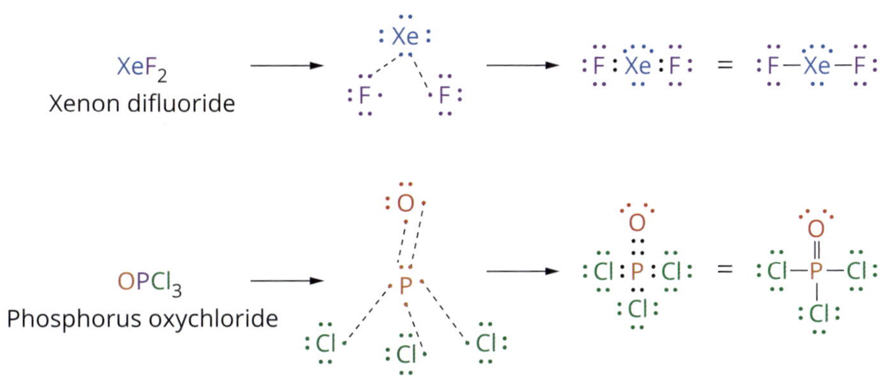

Fig. 3.3 Building the Lewis dot structures of hypervalent XeF_2 and $OPCl_3$.

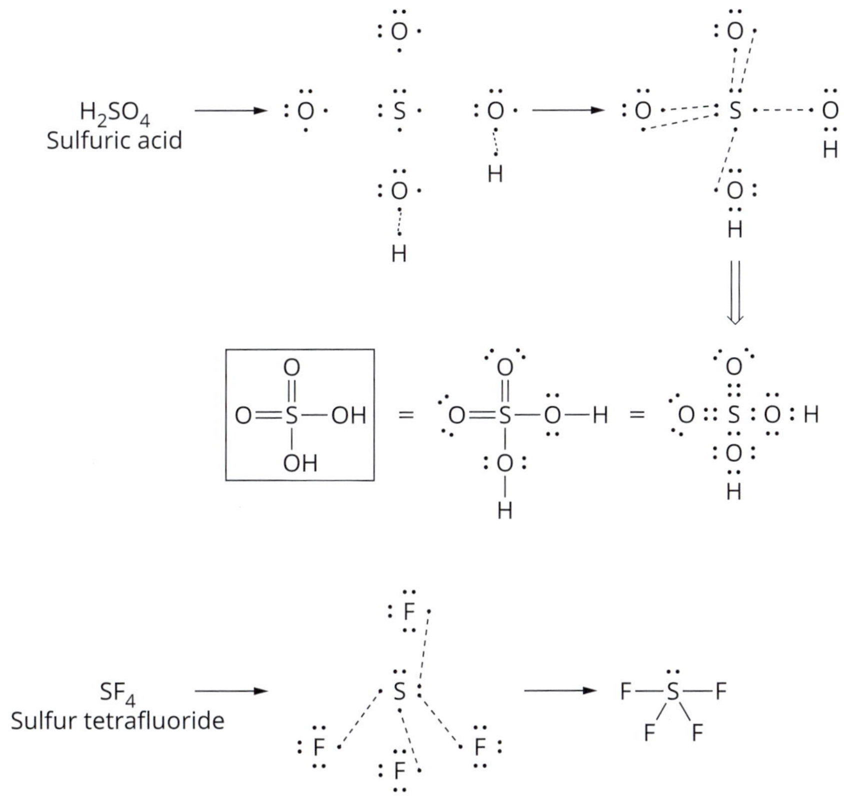

Fig. 3.4 Dropping lone pairs on the periphery of Lewis structures and retaining lone pairs on the central element with H_2SO_4 and SF_4 as examples.

described using only valence *s*- and *p*-orbitals for these compounds. This process will either eliminate some Lewis diagrams as being unreasonable bonding models or will further describe the nature of the bonds in the Lewis diagram.

Some compounds such as the BH_3 monomer have only six electrons around the central atom. The boron has three bonds to hydrogens each with two electrons, a total of six electrons in its valence shell. Main-group atoms in molecules, other than hydrogen, that have fewer than eight electrons in their Lewis structure are called *hypovalent*, and the boron in the BH_3 monomer is an example of a hypovalent atom.

Fig. 3.4 shows a slightly more complex example of a Lewis structure. In sulfuric acid, H_2SO_4, the hydrogens are attached to oxygen, which is in turn attached to the central atom, sulfur. Sulfur and oxygen are both in Group 16 and have six valence electrons as shown. In addition to showing bonding pairs between atoms as a line instead of as pairs of electrons, it is far more common to not draw the lone pairs on the periphery of the molecule. The greatly simplified but still quite clear line-drawing for one bonding scheme of sulfuric acid is boxed in Fig. 3.4.

It is advisable to only drop lone pairs on the periphery—lone pairs on central atoms should be kept because they have an effect on the geometry and bonding of the central atom of the molecule (see Chapter 4 on VSEPR). For example, the Lewis diagram of SF_4 is shown at the bottom portion of Fig. 3.4. The lone pairs on the peripheral fluorines can be dropped without

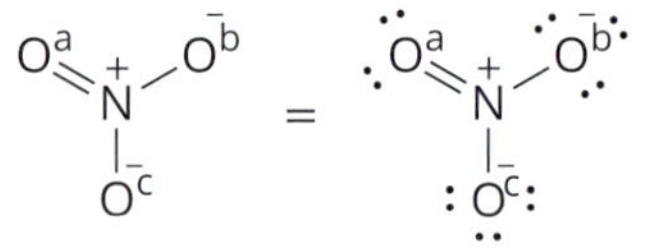

Fig. 3.5 Lewis structure of nitrate, NO_3^-, with formal charges.

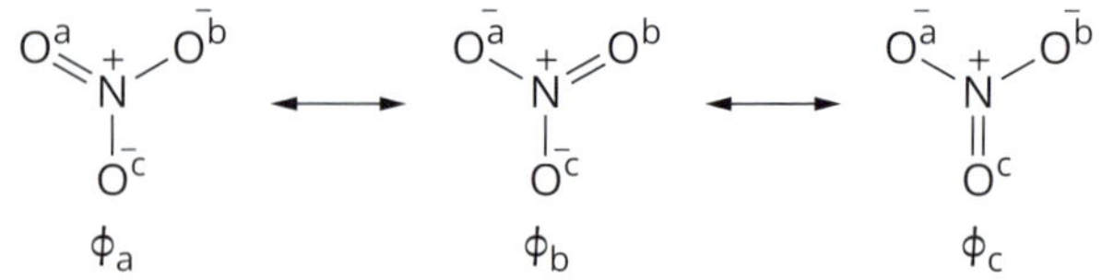

Fig. 3.6 Major contributing resonance forms to the ground state of nitrate.

confusion. However, the lone pair on the central sulfur atom is retained.

Dropping the lone pairs on the periphery is not ambiguous because, if an atom is charged, this should be specified in the structure. The number of electrons on the atom is still explicitly specified if charges are included. There are many stable compounds where there are atoms bearing formal charges in the Lewis structure. These charges should be indicated on the atom in the Lewis structure to indicate that the atom in question is to be considered as charged.

An example of a compound with formal charges on some of its atoms is the nitrate anion, NO_3^-. In the structure (Fig. 3.5), formal charges are found on two of the oxygens (–1) and the central nitrogen (+1).

Once a Lewis structure has been drawn, the formal charge on each atom in the structure can be obtained by using Eq. 3.1:

$$\text{Formal Charge} = \begin{pmatrix} \text{valence} \\ \text{electron} \\ \text{count} \end{pmatrix} - \begin{pmatrix} \text{number of} \\ \text{unshared} \\ \text{electrons} \end{pmatrix} - \begin{pmatrix} \text{number of} \\ \text{2-electron} \\ \text{bonds} \end{pmatrix} \tag{3.1}$$

For instance, the formal charges in the nitrate structure shown in Fig. 3.5 are:

$$\begin{aligned} &\text{For } O^a \quad \text{Formal Charge} = 6-4-2=0\,(\text{neutral}) \\ &\text{For } O^b \quad \text{Formal Charge} = 6-6-1=-1\,(\text{monoanion}) \\ &\text{For } O^c \quad \text{Formal Charge} = 6-6-1=-1\,(\text{monoanion}) \\ &\text{For N} \quad \text{Formal Charge} = 5-0-4=+1\,(\text{monocation}) \end{aligned}$$

As shown in Fig. 3.5, the structure on the left gives all the information of that on the right due to the inclusion of formal charges when the atoms are something other than neutral.

Note that the sum of all the formal charges on the atoms; that is, 0 + (–1) + (–1) + 1 = –1, is necessarily equal to the net charge of the whole nitrate anion (Section 3.4).

3.2 Resonance and an Introduction to Valence Bond Theory (Key Concept)

In the structure of nitrate, it is probably obvious that other Lewis structures can be also be drawn. Some of those structures are shown in Fig. 3.6, where ϕ_a designates the specific bonding pattern (electronic structure) in structure **a**. On the left, the double bond is to O^a, where the entire structure is signified by ϕ_a. In the other structures, the oxygen double bonded to the nitrogen atom is O^b (structure ϕ_b) or O^c (structure ϕ_c), with the requisite changes in formal charges on the oxygens calculated using Eq. 3.1.

In all of these structures, the overall bonding is identical. In other words, if we drop the labels (a, b, c on the oxygens) there is simply one N=O double bond, two N–O single bonds, two anionic oxygens, and a cationic nitrogen in all three structures; without the labels, the structures would be the same if one rotates around the nitrogen atom. As a consequence, we would expect that all of these Lewis diagrams are equally good descriptions of the molecule and, in the case of the nitrate anion, all of these structures contribute equally to the electronic structure of the molecule. The term *resonance* is used to describe the situation where multiple Lewis structures contribute (equally or not) to the stability of a single molecule, and each of the individual Lewis structures is then referred to as a *resonance structure* (or *resonance form*).

A single Lewis structure, more often than not, is not a complete description of a molecule's electronic structure. There are often many different resonance (Lewis) structures that can be drawn for a molecule. Many Lewis diagrams may contribute to one extent or another to the actual ground-state electronic structure of the molecule, and double-headed arrows, like those used in Fig. 3.6, are reserved (exclusively!) to represent resonance between various structures. Although we will come back to this point later, it is important here to not think of resonance as a fast exchange between several individual "real" electronic structures; the actual structure of a species like the nitrate anion does not quickly oscillate between the three forms shown in Fig. 3.6, but rather should be thought of as being the simultaneous superposition of all three forms at the same time. The electronic structure of the actual molecule then is sometimes called the "composite structure", as it is comprised of all the individual resonance forms at once.

In the nitrate resonance forms in Fig. 3.6, labeled ϕ_a through ϕ_c, the three structures are all equivalent in energy because the number and type of bonds between atoms and charges are the same. As a result, all three resonance forms contribute equally to the ground-state structure. In fact, these three resonance forms contribute far more than any other resonance form. As a result, we can approximate the actual ground-state electronic structure of the nitrate anion ($\Phi_{nitrate}$) as shown in Eq. 3.2.

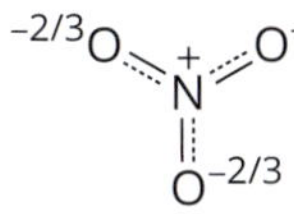

Fig. 3.7 Composite structure of the nitrate anion from Eq. 3.2.

$$\Phi_{nitrate} \approx w_a\phi_a + w_b\phi_b + w_c\phi_c$$
$$\approx w_a\phi_a + w_a\phi_b + w_a\phi_c$$
where $$\sum_i^n w_i = 1 \qquad 3.2$$

where w_a is the weighting factor for the contribution of resonance form ϕ_a. The weighting factor can simply be thought of as the percentage that a particular Lewis form contributes to the overall electronic structure. Because all the energies of the resonance forms are the same, all the weighting factors have the same value, and $w_a = w_b = w_c$.

From these resonance forms, and their related weighting factors, we can generate a resonance *composite structure* (Fig. 3.7) to represent the ground-state electronic structure of nitrate, where the N–O distances are equivalent. Overall, instead of a full anionic charge on two oxygens, the composite structure has a −2/3 charge on three equivalent oxygens. In addition, instead of one double bond and two single bonds, we expect the N–O bonds will be 1 1/3 bonds each with equal N–O bond distances (bond order is discussed in more detail in Section 3.10). The fact that the nitrate anion possesses three identical N–O bonds is directly confirmed in experimental structures obtained from X-ray diffraction of many nitrate compounds, such as potassium nitrate,[5] where the N–O distances are the same within experimental error.

There is no general requirement that all resonance forms contribute equally to the composite structure. Fig. 3.8 shows some additional contributing resonance forms (ϕ_α, ϕ_β, and ϕ_γ) for nitrate anion. In these added forms, one of the oxygen atoms is dianionic and does not share electrons with the central nitrogen. The other oxygens are double bonded to the central nitrogen and are neutral.

$$\Phi_{nitrate} \approx w_a\phi_a + w_b\phi_b + w_c\phi_c + w_\alpha\phi_\alpha + w_\alpha\phi_\beta + w_\alpha\phi_\gamma \qquad 3.3$$

As before, the three resonance forms in the left column contribute equally to the overall structure: $w_a = w_b = w_c$. Similarly, the other three resonance forms in the right column all contribute equally to each other, and $w_\alpha = w_\beta = w_\gamma$. However, the bonding and charges in the resonance forms in the left and right columns are not equivalent. As a result, $w_a \neq w_\alpha$, which means the left and right columns make different contributions to the overall electronic structure. Calculations (natural resonance theory)[6] suggest that $w_a = 0.31$ and $w_\alpha = 0.015$. The ground-state electronic structure for nitrate may now be approximated by using Eq. 3.3, an improved description over Eq. 3.2.

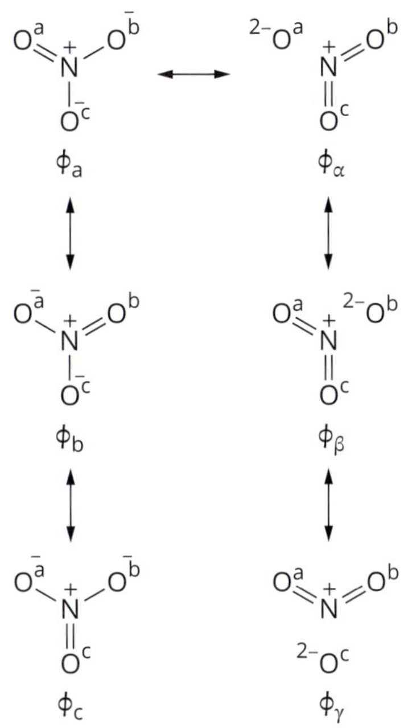

Fig. 3.8 Resonance forms for an improved description of the nitrate anion.

All the weighting factors contributing to the ground-state structure should sum to unity representing 100% of the ground-state electronic configuration. Using the calculated weighting factors from natural resonance theory, we find that 97.5% of the ground-state structure is represented by the six resonance forms in Fig. 3.8. Consequently, these structures, together with their weighting factors, constitute a very good (but not quite complete) picture of the nitrate anion; some other resonance forms not identified must make some very small contributions to the overall electronic structure.

That ground-state structures can be represented as composites of various Lewis structures in resonance is a tenet of valence bond theory, which is one of the most important models used in chemistry to understand the electronic structure of molecules.[7] Valence bond theory (along with a loosely related construct called natural bond orbital theory) will be used, often side-by-side with molecular orbital theory, throughout the remainder of the book as a method for understanding inorganic chemistry in general.

3.3 Heuristic Valence Bond Principles Governing Resonance Forms (Expanded Coverage)

Oftentimes, a variety of different resonance forms can be evaluated based on inspection in order to decide which forms will be the greatest contributors to the overall ground-state electron distribution. There are several principles, most put forward by Pauling, that can be considered when trying to determine which forms will contribute the most to the ground-state wavefunction.[8] Criteria 1 and 2 are strict requirements that *must* be met in order for the structure to be a valid resonance form.

(1) *Contributing structures should have identical atomic positions corresponding to the ground-state geometry of the molecule. Only the electrons should move from one resonance form to another.*

(2) *Principle of Overall Charge Neutrality: The sum of all the ionic charges in a system is zero.* Here, the principle simply points out that in all cases there are counterions associated with any set of ions in solution. A trivial expansion of this would be to state that every resonance form must have the same sum of charges as the system under consideration. For example, if one is considering fully dissociated potassium phosphate, $3\ K^+ + PO_4^{3-}$, all reasonable resonance structures for the phosphate anion will have charges that sum to −3 in order to balance the charges of the three potassium cations.

(3) *A system adopts the highest symmetry consistent with the constraints acting on it.* In essence, this simply suggests that a molecule will be as symmetric as possible. The principle does not suggest that the molecule will attempt to create symmetry where there is none, like making carbon–carbon and carbon–hydrogen bonds similar in something like ethane. Instead, if there are, for example, three C–H bonds around the same carbon, all three C–H bonds will be equal unless there is some electronic force; for example, hyperconjugation involving only one of the hydrogens, making them different. In other words, electronically identical groups are symmetrically distributed.

(4) *Lower-energy structures have charges arranged so that they are surrounded by opposite charges.* Larger contributing structures to the ground state will typically not have charges of the same sign adjacent to one another, but rather will have opposite charges on adjacent atoms. As an example, let us examine hydrazoic acid, HN_3, using natural resonance theory.[6] The results for three common resonance forms drawn for this molecule and their contributions to the ground state are shown in Fig. 3.9. The resonance forms with adjacent opposite charges, ϕ_A and ϕ_B, are the largest contributors. Resonance form ϕ_C, with destabilizing adjacent positive charges, contributes very little.

(5) *Preferred resonance forms have negative charges on more electronegative atoms, and positive charges on more electropositive atoms.*

(6) *Principle of Electroneutrality: Structures with partial charges as close to neutral as possible are usually preferred.* Structures that contribute strongly to the ground-state wavefunction will often have distributed charges rather than localizing charge on one or a few atoms.

Consider as an example for many of the previous criteria, the cyanate ion, NCO^-. The three resonance forms shown in Fig. 3.10 are all reasonable by Criterion 2 because the charges sum to −1 in all cases. The geometry of the molecule is not different in one resonance form relative to another (Criterion 1). The resonance form on the right, ϕ_C, will likely contribute very little because there are two anionic charges on the same atom (Criterion 6) and one positive charge on the electronegative oxygen (Criterion 5). The middle form, ϕ_B, may be a fairly large contributor because it has only a single charge, an anion, on the second most electronegative atom, nitrogen. The resonance form on the left, ϕ_A, has the anionic charge on oxygen, the most favorable situation (Criterion 5).

We would therefore expect the leftmost structure, ϕ_A, to contribute the most to the ground state. Obviously, this does not mean the other resonance structures will not contribute as well; it simply suggests that ϕ_A will contribute the most of these three. Consequently, if we write the ground-state wavefunction, the composite structure for NCO^- (Φ_{NCO^-}), as a linear combination of the electron configurations represented by these resonance forms as shown in Eq. 3.4, we expect that the weighting factors will be $w_A > w_B >> w_C$.

$$\Phi_{NCO^-} = w_A\phi_A + w_B\phi_B + w_C\phi_C \quad (3.4)$$

If one examines the cyanate anion computationally,[6] the following resonance forms are found to contribute with the percentages (weighting factors) shown in Fig. 3.11. As expected, ϕ_A is the largest contributor, with ϕ_B also making a significant contribution. Resonance form ϕ_C makes a minor contribution to the ground state, as does another form with a fully ionized C–O bond, ϕ_D.

We can thus write the ground-state wavefunction for NCO^-, Φ_{NCO^-}, as shown in Eq. 3.5. The weighting factors,

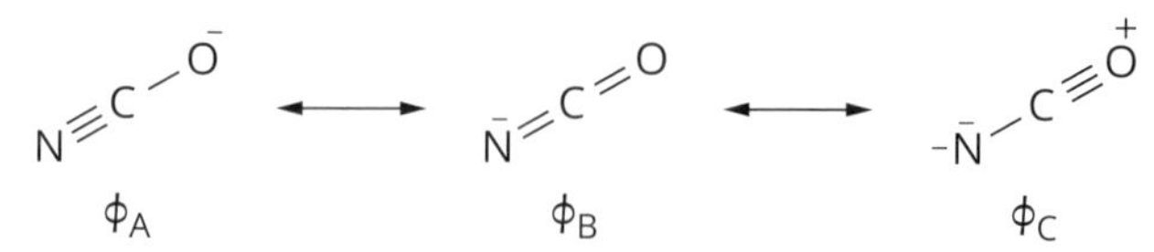

Fig. 3.10 Some resonance forms for cyanate anion.

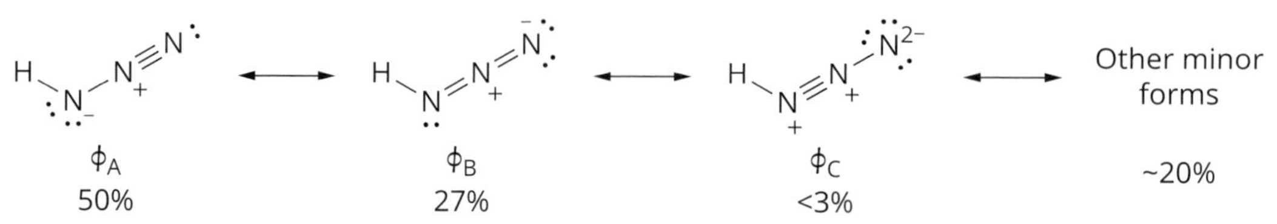

Fig. 3.9 Some resonance forms for hydrazoic acid and their contributions to the ground-state electronic structure, calculated using natural resonance theory (6-311G/B3LYP/NBO5).[6]

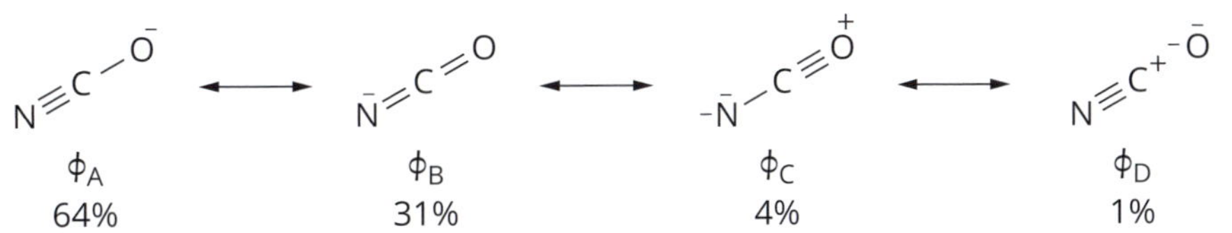

Fig. 3.11 Computed contributions to the ground state for resonance forms of cyanate using natural resonance theory (NBO5).[6]

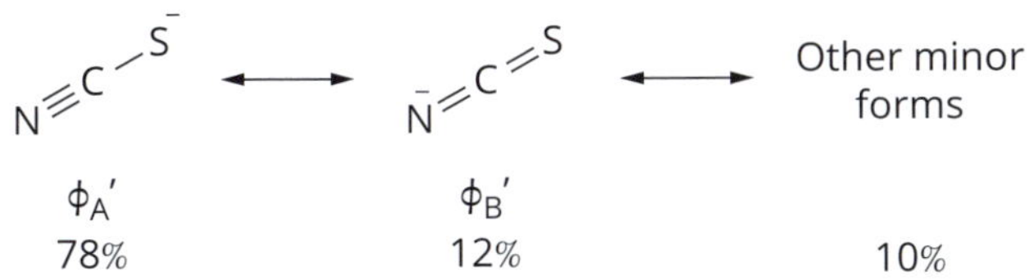

Fig. 3.12 Computed contributions to the ground state for resonance forms of thiocyanate using natural resonance theory (NBO6).[6]

listed previously, sum up to unity, suggesting that the four structures yield a full description of the electronic structure of the cyanate anion, according to NBO theory.

$$\Phi_{NCO^-} = 0.643\phi_A + 0.313\phi_B + 0.035\phi_C + 0.009\phi_D \qquad \textbf{3.5}$$

To these six criteria, an additional (more advanced) criterion can be added regarding overlap.

(7) *Orbital overlap can affect the extent to which a resonance form participates.*

As an example, compare the bonding in thiocyanate, NCS^-, with that of the cyanate anion. What effect will the replacement of O with S have on the electronic structure? We can assume that π-bonding between sulfur and carbon will be weaker than for oxygen, due to worse overlap between the larger 3*p* orbitals of S and the smaller 2*p* orbital of C (relative to O and C). Consequently, we expect the analog of resonance form ϕ_B, with a C=S double bond, will participate even less than in the NCO^-. In fact, form ϕ_B does participate less, which leads to increased participation by resonance form ϕ_A. The O for S replacement, NCS^- instead of NCO^-, causes the ϕ_A bonding pattern (Fig. 3.11) with a triple bond between N–C and a single bond between C–S to represent ~78% of the ground-state structure (Fig. 3.12). This occurs due to overlap considerations and despite the lower electronegativity of sulfur versus oxygen, contrary to Criterion 5.

3.4 Electronegativity Scales (Key Concept)

Once the Lewis structure for a molecule is established, one can begin to think about how the electrons are distributed in the different bonds described by the Lewis structures. When two identical atoms or groups bond to one another, we can expect the electron densities across the bond itself to be symmetrically distributed along the bond axis and around the midpoint between the two atoms. However, if two different elements (or groups of elements) are bonded, the electron density will not necessarily be evenly distributed between the two atoms in the bond.

Fig. 3.13 is a diagram of the H_2 molecule. Around the midpoint of the H–H bond is an even distribution of electron density because each hydrogen atom attracts electrons to the same extent. However, in HF the fluorine atom attracts more of the electron density comprising the two electrons in the bonding orbital than the hydrogen atom, which leads to an uneven, teardrop-shaped distribution of electron density in the σ-bond.

Iodine has a much larger covalent radius (1.333 Å) than fluorine (0.683 Å) because iodine has many additional electrons. However, the electronic distribution in the F–I bond is strongly skewed towards fluorine rather iodine within the σ-bond, which leads to the electron density in that bond being localized mostly on fluorine (Fig. 3.13).

This differential ability of atoms (or groups of atoms) to attract electrons to themselves is called *electronegativity*, for which a reasonable definition is the *ability of an atom to attract electrons in a bond*. In general, larger differences in electronegativity between two atoms in a chemical bond lead to larger partial charges on each atom due to the shifting of electron density in the bond. This effect is called "polarization", and we refer to the bond as being "polarized". The atom attracting more electron density bears a negative partial charge, δ^- ("delta minus"), while the atom attracting less electron density carries a partial positive charge, δ^+. The size of the charge difference and of the bond polarization is proportional to the electronegativity difference in a manner dependent on the electronegativity scale being used.

Although easy to grasp intuitively, there is no unique way to quantitatively define the concept of electronegativity, and the electronegativities of atoms and groups have been examined by a number of different researchers over the years, using

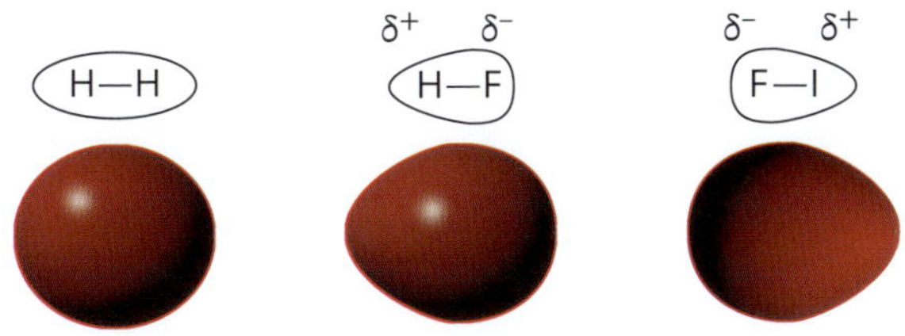

Fig. 3.13 Orbital plot of H_2, HF, and FI, showing the distribution of the two σ-electrons (HOMO for each molecule).

a variety of different approaches, each having their strengths and weaknesses. While a complete analysis of the numerous methods for electronegativity evaluation would be gratuitous, some brief discussion of a few of the most important methods can be used to illustrate some important chemical principles. Electronegativity of the elements on several different scales are provided in Table 3.1. Pearson,[9] Allen,[10–11] Huheey,[12] Gordy,[13] and others[14] have also made important contributions to our understanding of electronegativity.

It should be noted that these scales are not necessarily interchangeable. While it is customary, in some cases, to place the different electronegativities on a scale approximately correlating with the Pauling scale, different applications may require specific definitions of electronegativity: Pauling or Sanderson electronegativity values cannot necessarily be used in place of Allred–Rochow, depending on the problem at hand.

In various courses and textbooks, one is likely to find many of the scales listed in Table 3.1. On occasion, electronegativity will be mentioned in the primary literature, often without mentioning a specific scale, in which case the Pauling scale is usually implied. In this chapter and elsewhere in the book, we will make extensive use of the Sanderson scale to discuss charge distributions and bond energies.

Table 3.1 Electronegativities of the elements.

					Mulliken–Jaffe[e]	
Element	Oxidation[a] State (if Designated)	Pauling[b]	Sanderson[c]	Allred-Rochow[d]	Orbital Hybridization	Electronegativity on Pauling Scale
H		2.20	2.592	2.20	*s*	2.09
Li		0.98	0.670	0.97	*s*	0.94
					sp	0.68
Be		1.57	1.810	1.47	*sp*	1.53
B		2.04	2.275	2.01	*p*	1.47
					sp^3	1.88
					sp^2	2.04
C		2.55	2.746	2.50	*p*	1.85
					sp^3	2.46
					sp^2	2.61
					sp	2.92
N		3.04	3.194	3.07	*p*	2.38
					sp^3	3.01
					sp^2	3.20
					sp	3.56
O		3.44	3.654	3.50	*p*	2.81
					sp^3	3.55
					sp^2	3.77
F		3.98	4.000	4.10	*p*	3.20
					sp^3	4.06
Na		0.93	0.560	1.01	*s*	0.93
					sp	0.66
Mg		1.31	1.318	1.23	*sp*	1.40
Al		1.61	1.714	1.47	sp^2	1.75
Si		1.90	2.138	1.74	sp^3	2.08
P		2.19	2.515	2.06	*p*	1.96
					sp^3	2.51
					sp^2	2.67
					sp	2.99
S		2.58	2.957	2.44	*p*	2.32
					sp^3	2.92

Table 3.1 Continued

Element	Oxidation[a] State (if Designated)	Pauling[b]	Sanderson[c]	Allred–Rochow[d]	Mulliken–Jaffe[e]	
					Orbital Hybridization	Electronegativity on Pauling Scale
Cl		3.16	3.475	2.83	*p*	2.63
K		0.82	0.445	0.91	*s*	0.96
					sp	0.76
Ca		1.00	0.946	1.04	*sp*	
Sc		1.36		1.20		
	Sc(II)		0.64			
	Sc(III)		1.02			
Ti		1.54		1.32		
	Ti(II)		0.73			
	Ti(III)		1.09			
	Ti(IV)		1.50			
V		1.63		1.45		
	V(II)		0.69			
	V(III)		1.39			
	V(IV)		1.89			
	V(V)		2.51			
Cr				1.56		
	Cr(II)	1.66	1.24			
	Cr(III)		1.66			
	Cr(IV)		2.29			
	Cr(V)		2.83			
	Cr(VI)		3.37			
Mn				1.60		
	Mn(II)	1.55	1.66			
	Mn(III)		2.20			
	Mn(IV)		2.74			
	Mn(V)		3.28			
	Mn(VI)		3.82			
Fe				1.64		
	Fe(II)	1.83	1.64			
	Fe(III)	1.96	2.20			
Co				1.70		
	Co(II)	1.88	1.96			
	Co(III)		2.56			
	Co(IV)		3.10			
Ni				1.75		
	Ni(II)	1.91	1.94			
	Ni(III)		2.73			
	Ni(IV)		3.27			
	Ni(V)		3.81			

(continued)

Table 3.1 Continued

Element	Oxidation[a] State (if Designated)	Pauling[b]	Sanderson[c]	Allred–Rochow[d]	Mulliken–Jaffe[e] Orbital Hybridization	Mulliken–Jaffe[e] Electronegativity on Pauling Scale
Cu				1.75		
	Cu(II)		1.98			
Zn	Zn(II)	1.65	2.223	1.66		
Ga	Ga(III)	1.81	2.419	1.82	sp^2	1.94
Ge	Ge(IV)	2.01	2.618	2.02	sp^3	2.16
As				2.20	p	1.92
	As(III)	2.18	2.816		sp^3	2.52
					sp^2	2.70
					sp	3.03
Se				2.48	p	2.23
	Se(II)	2.55	3.014		sp^3	2.70
Br	Br(I)	2.96	3.219	2.74	p	2.48
Rb				0.89	s	0.73
	Rb(I)	0.82	0.312		sp	0.51
Sr	Sr(II)	0.95	0.721	0.99	sp	1.07
Y				1.11		
	Y(II)		0.40			
	Y(III)	1.22	0.65			
Zr				1.22		
	Zr(II)	1.33	0.52			
	Zr(III)		0.79			
	Zr(IV)		0.90			
Nb		1.6		1.23		
	Nb(II)		0.77			
	Nb(III)		1.02			
	Nb(IV)		1.25			
	Nb(V)		1.42			
Mo				1.30		
	Mo(II)	2.16	0.90			
	Mo(III)	2.19	1.15			
	Mo(IV)	2.24	1.40			
	Mo(V)	2.27	1.73			
	Mo(VI)	3.35	2.20			
Tc		1.9		1.36		
Ru		2.2		1.42		
Rh		2.28		1.45		
Pd	Pd(II)	2.20		1.35		
Ag	Ag(I)	1.93	1.826	1.42		
Cd	Cd(II)	1.69	1.978	1.46		
In	In(III)	1.78	2.138	1.49	sp^2	1.83
Sn	Sn(IV)	1.96	2.298	1.72	sp^3	2.16
	Sn(II)	1.80	1.477		p	1.43

Table 3.1 Continued

Element	Oxidation[a] State (if Designated)	Pauling[b]	Sanderson[c]	Allred–Rochow[d]	Mulliken–Jaffe[e]	
					Orbital Hybridization	Electronegativity on Pauling Scale
Sb	Sb(III)	2.05	2.458	1.82	*p*	1.82
Te	Te(II)	2.1	2.618	2.01	*p*	2.10
I	I(I)	2.66	2.778	2.21	*p*	2.25
Cs	Cs(I)	0.79	0.220	0.86	*s*	0.87
					p	0.69
Ba	Ba(II)	0.89	0.651	0.97	*sp*	0.97
La	La(III)	1.10		1.08		
Ce	Ce(III)	1.12		1.08		
Pr		1.13		1.07		
Nd		1.14		1.07		
Pm				1.07		
Sm		1.17		1.07		
Eu				1.01		
Gd		1.20		1.11		
Tb				1.10		
Dy		1.22		1.10		
Ho		1.23		1.10		
Er		1.24		1.11		
Tm		1.25		1.11		
Yb				1.06		
Lu		1.27		1.14		
Hf		1.3		1.23		
	Hf(II)		0.31			
	Hf(III)		0.56			
	Hf(IV)		0.81			
Ta		1.5		1.33		
	Ta(II)		0.44			
	Ta(III)		0.69			
	Ta(IV)		0.94			
	Ta(V)		1.17			
W				1.40		
	W(II)	2.36	0.73			
	W(III)		0.98			
	W(IV)		1.23			
	W(V)		1.48			
	W(VI)		1.67			
Re		1.9		1.46		
Os		2.2		1.52		
Ir		2.20		1.55		
Pt	Pt(II)	2.28		1.44		
Au	Au(I)	1.93		1.42		

(continued)

Table 3.1 Continued

Element	Oxidation[a] State (if Designated)	Pauling[b]	Sanderson[c]	Allred–Rochow[d]	Mulliken–Jaffe[e] Orbital Hybridization	Mulliken–Jaffe[e] Electronegativity on Pauling Scale
Hg	Hg(II)	2.00	2.195	1.44		
Tl	Tl(III)	2.04	2.246	1.44	*sp*^{2}	1.98
	Tl(I)	1.62	0.987		*p*	1.11
Pb	Pb(IV)	2.33	2.291	1.55	*sp*^{3}	2.13
	Pb(II)	1.87	1.900		*p*	1.31
Bi	Bi(III)	2.02	2.342	1.67	*p*	1.43
Po		2.0		1.76	*p*	1.68
At		2.2		1.90	*p*	1.90
Fr				0.86		
Ra				0.97		
Ac				1.00		
Th				1.11		
Pa				1.14		
U		1.38		1.22		
Np		1.36		1.22		
Pu		1.28		1.22		
Am				(1.2)		
Cm				(1.2)		
Bk				(1.2)		
Cf				(1.2)		
Es				(1.2)		
Fm				(1.2)		
Md				(1.2)		
No				(1.2)		

[a] When no value is given for the formal oxidation state, this usually indicates the most common oxidation state or, on some scales, that the neutral atom was used. In any event, no oxidation state was specified in the reference.
[b] Pauling electronegativities are from Allred's corrected values[15] or Pauling.[7]
[c] Main-group values are from Sanderson's *Polar Covalence*,[16] and transition-metal values from Sanderson's study.[17]
[d] Values from Allred and Rochow[18] or from Little and Jones.[19]
[e] Values from the empirical equations of Sen and coworkers under the assumption that the bond in question was covalent,[20] which were then scaled using Equation 9 in the article.[21]

3.4.1 Pauling Electronegativity (Key Concept)

Pauling's electronegativity scale is based on bond energies of compounds containing the elements. *Bond energies* and *bond dissociation enthalpies* (BDEs) are two terms sometimes used interchangeably; however, they each have somewhat different meanings, in agreement with IUPAC recommendations.

BDE, which is also discussed in more detail in Sections 3.6 and 3.9, is the energy required to take a bond of interest from the lowest vibrational state in the electronic ground state and to cleave it homolytically (that is, leaving one electron of the bonding pair on each atom) to a distance where the atoms (or groups) are no longer interacting, as illustrated in this equation.

$$\text{A–B} \longrightarrow \text{A}\bullet + \text{B}\bullet \quad \Delta H = \text{BDE}$$

This can be compared with the bond energy or what is often called more descriptively the *mean bond energy*, which is the average enthalpy of all the bonds in a compound. For example, the bond energy of CH_4 is determined by ¼ the enthalpy of the chemical reaction $CH_4 \rightarrow C + 4\ H$; (397.6 kcal/mol)/4 = 99.4 kcal/mol. For comparison, the BDE of $H{-}CH_3$, the energy required to break the first bond is 104 kcal/mol.

Note that the IUPAC unit often used for bond energies is kilojoules per mole, kJ/mol. However, it is still customary in chemistry circles to use kcal/mol for bond energies and BDE. This is partly due to the traditional use of the kcal/mol unit, but also because of its convenient scale. The H–H bond and C–H bond, to one significant figure, conveniently have bond energies of ~100 kcal/mol, and the strongest hydrogen bonds are around 10 kcal/mol. Typical C–C bonds, again to one significant figure, are about 80 kcal/mol, and the very strong N≡N triple bond BDE is 226 kcal/mol. The IUPAC unit, kJ/mol, which is commonly used as well, can be calculated from the unit used here by using the conversion factor 1 kcal/mol = 4.184 kJ/mol.

Consequently, bond energy and BDE values are the same for diatomic molecules, but can, and usually will, be somewhat different for polyatomic molecules.

Pauling reasoned[7] that a bond between two different atoms, a *heterocovalent bond*, will have an energy related to the geometric mean of the homonuclear bond energies, if the bond between the different atoms were somehow completely covalent. In other words, if A is bound to X with no polarity, then the energy of the purely covalent bond can be calculated using Eq. 3.6, where E_{AA} and E_{XX} are the energies required to break the purely covalent bonds A–A and X–X.

$$E_{AX}^{cov} = \sqrt{E_{AA} \times E_{XX}} \quad 3.6$$

However, the bond will not be purely covalent when two different atoms are bonded together. As illustrated in Fig. 3.13 for HF or FI, any time that two different atoms or groups bond there is an uneven distribution of electron density between the two atoms in the bond. Pauling treated this polarity of the bond by adding an additional term to the previous equation that uses the square of the difference in electronegativity for the two atoms, $(\chi_X^P - \chi_A^P)^2$, times a scaling factor for kcal/mol units, 23.06 kcal/mol,[c] to include the effect of the Coulombic attraction between the charged atoms (Eq. 3.7).

$$E_{AX} = \sqrt{E_{AA} \times E_{XX}} + 23.06(\chi_X^P - \chi_A^P)^2\, kcal/mol \quad 3.7$$

For example, to find the bond energy under this Pauling scheme for HCl, the bond energy for H_2 (E_{H-H}), the bond energy for Cl_2 (E_{Cl-Cl}), and the electronegativities (χ_{Cl}^P and χ_H^P) are employed as shown in Eq. 3.8. The experimental bond energy for HCl is 103 kcal/mol.

$$\begin{aligned} E_{HCl} &= \sqrt{E_{H-H} \times E_{Cl-Cl}} + 23.06(\chi_{Cl}^P - \chi_H^P)^2 \\ E_{HCl} &= \sqrt{104.2 \times 58.0} + 23.06(3.16 - 2.2)^2 \\ E_{HCl} &= 99.0\ \mathrm{kcal/mol} \end{aligned} \quad 3.8$$

The good agreement between calculated and experimental values emphasizes the usefulness of the Pauling scale, and reflects the fact that this scale was derived using experimental bond energies of a large number of common compounds to define the mean values for the electronegativity factors of each elements tabulated in Table 3.1. *Pauling viewed (Eq. 3.7) the ionic component of the bond as something that supplemented and added onto the always present covalent bond energy, a remarkably successful assumption by Pauling that led to the most often used electronegativity scale.* However, this can be contrasted with the approach used later in this chapter (polar covalence theory) where the ionic bond energy *replaces* part of the covalent bond energy.

In order to calculate the electronegativities for the elements in this scheme, it is necessary to set the electronegativity value for an element; Pauling chose to set the electronegativity of fluorine to 4.00. Due to small corrections in the experimental values for the BDEs of fluorides, the value for fluorine's electronegativity has changed slightly from 4 to 3.980 in current tables.[15]

3.4.2 Mulliken (Absolute) and Mulliken–Jaffe Electronegativities (Expanded Coverage)

A quite different reasoning led to the Mulliken electronegativity scale, sometimes called absolute electronegativity.[22] Mulliken proposed a scale based upon the gas-phase abilities of atoms, molecules, ions, and so on, to be ionized by attracting or releasing electrons. More precisely, he proposed that an atom's or group of atom's ability to attract electrons was related to the average of the ionization energy and electron affinity.[d] So, the absolute electronegativity of A, χ_A^{Abs}, is defined as shown in Eq. 3.9, where EA_A and IE_A are the electron affinity and ionization energy of species A, respectively. In principle, any unit could be used, but most often EA and IE are reported in electronvolts, 1 electronvolt (eV) = 96.485 kJ/mol = 23.06 kcal/mol.

$$\chi_A^{Abs} = \frac{EA_A + IE_A}{2} \quad 3.9$$

An advantage of the Mulliken scale is that it can be applied to virtually any species and not just individual atoms. For example, one can measure the ionization energy and electron affinity of triphenylphosphine, PPh_3, and assign an absolute electronegativity value to this collection of atoms. The single number which is obtained for the entire molecule gives a measure of the molecule's ability to attract electrons to itself.

In addition, the charge of an atom or chemical species can be related to the Mulliken electronegativity. An important point about this scale, and electronegativity in general, can be seen by calculating the Mulliken electronegativity of species with different amounts of charge. As an example, we can

[c] The constant 23.06 is derived from Coulomb's law and converts from electron volts to kcal/mol.

[d] The reader will recall that the ionization energy of an atom (or ion) A is the energy of the following reaction: $A \rightarrow A^+ + e^-$, whereas the electron affinity of that same species is the energy of the reaction: $A + e^- \rightarrow A^-$.

calculate the electronegativity of the Sn atom (Eq. 3.10). For this element, $EA_{Sn} = 1.20$ eV and $IE_{Sn} = 7.35$ eV.

$$\chi_{Sn}^{Abs} = \frac{EA_{Sn} + IE_{Sn}}{2} = \frac{1.20 + 7.35}{2} = 4.28\text{eV} \qquad \textbf{3.10}$$

Alternatively, for Sn^{2+} $EA_{Sn(II)} = 15.87$ and $IE_{Sn(II)} = 30.50$. Using these values, one can calculate the electronegativity of the Sn^{2+} ion (Eq. 3.11).

$$\chi_{Sn(II)}^{Abs} = \frac{EA_{Sn(II)} + IE_{Sn(II)}}{2} = \frac{15.87 + 30.50}{2} = 23.19\text{eV} \qquad \textbf{3.11}$$

As might be expected, removing two electrons from tin gives the metal a much higher electronegativity; simply, more electron-deficient species attract electrons much more strongly.

The Mulliken (absolute) electronegativity for an atom A (χ_A^{Abs}) is approximately related to the Pauling scale (χ_A^P) by the relation in Eq. 3.12.[21]

$$\chi_A^P = \left(1.35\sqrt{\chi_A^{Abs}}\right) - 1.37 \qquad \textbf{3.12}$$

As might be imagined, the Mulliken electronegativity of a molecule is dependent to a certain extent on the nature of the orbital being ionized to obtain the experimental or theoretical IE and EA. Jaffé and coworkers examined the idea that electronegativity could be related to the composition of the orbital holding the electrons of interest.[14]

Empirical equations have been reported where the electronegativity of a particular orbital can be calculated based on the partial charge and *s*-orbital character of the orbital for the main-group elements (except the noble gases).[20] Some Mulliken–Jaffé electronegativities for some common hybridizations are shown in Table 3.1, in which the values are normalized to be on the Pauling 4-point scale.

In general, the lower the energy of an orbital, the more electronegative it is, which can be simply understood by the fact that lower-energy orbitals tend to better stabilize electrons. One can see this in the Mulliken–Jaffé orbital electronegativity scale. For example, a pure *p*-orbital on carbon has an electronegativity of 1.85. Mixing this orbital with the lower-energy *s*-orbital should increase the electronegativity; consequently, the more *s*-character that is mixed in, the more electronegative the hybrid orbital must become. As a result, the common hybrids sp^3-, sp^2-, and sp^1-orbitals on carbon have electronegativities of 2.46, 2.61, and 2.92, respectively. The increase in *s*-character from 25% to 33.3% to 50% for sp^3-, sp^2-, and sp^1-hybrids increases the electronegativity—an important effect for molecular orbital and valence bond theory.

3.4.3 Sanderson Electronegativity (Key Concept)

In devising yet another method for the determination of electronegativity, the inorganic chemist R. T. Sanderson (1918–1989) used the observation that, within each period, electronegativity is a function of the electron density of the atom.[16] Sanderson used the expression shown in Eq. 3.13 to calculate the electron density of atom A (ζ_A).[23] In Eq. 3.13, Z is the atomic number of the element, and r is the covalent radius in Ångstroms (1 Å = 10^{-10} m).

$$\zeta_A = \frac{Z}{\left(\frac{4}{3}\right)\pi r^3} e/\text{Å}^3 \qquad \textbf{3.13}$$

Sanderson assumed that the chemical stability of the inert or noble gases was not only associated with their atomic number but also with their uniformly spherical electron density. However, each noble gas has a different electron density, and so does every atom between each pair of noble gases on the periodic table. Sanderson surmised that an ideal electronic density value for any element on the periodic table could be defined by linearly interpolating the data between the adjacent noble gas values. Using Eq. 3.13 and the covalent radii, the average electron densities of the noble gases can be calculated.[23] In order to find the "inert" electron density, ζ^i_A, associated with Li, for example, one finds the equation of the line defined by the points using the atomic number, Z, and the electron density, ζ^i_A (Z_{He} = 2 electrons, $\zeta_{He} = 0.59$ e$^-$/Å^3) and (Z_{Ne} = 10 electrons, $\zeta_{Ne} = 1.06$ e$^-$/Å^3), which is $\zeta^i_{He\text{–}Ne} = 0.4725 + 0.05875\,Z$ (see Fig. 3.14).

This equation can then be used to find the "inert" electron density for any of the elements with atomic numbers from Z = 3–7. For Li, Z = 3 electrons, the equation becomes $\zeta^i_{Li} = (0.4725 + 0.05875 \times 3)$ e$^-$/Å^3, which yields $\zeta^i_{Li} = 0.65$ e$^-$/Å^3, the inert electron density for lithium.

The Sanderson electronegativity value, χ^S, which he originally called the Stability Ratio, is found by dividing the actual (observed) electron density of the atom, ζ_A, by what the density would be in the isoelectronic inert atom, $\chi^S = \zeta_A / \zeta^i_A$. As a result, this unitless term describes how far the atom deviates from its "ideal" electron density.

For example, the experimental value for the covalent radius for fluorine is 0.72 Å. Consequently, for fluorine, Z = 9, the electron density value calculated is 5.75 e$^-$/Å^3 (Eq. 3.14).

$$\zeta_F = \frac{9}{4.19 \times 0.72^3} = 5.75\ e^-/\text{Å}^3 \qquad \textbf{3.14}$$

The ideal electron density for fluorine using the equation relating electron count to the electron densities of He and Ne is $\zeta^i_{He\text{–}Ne} = 0.4725 + 0.05875Z = 0.4725 + 0.05875 \times 9 = 1.00$ e$^-$/Å^3. Consequently, the electronegativity (Stability Ratio) of fluorine is $\zeta_F/\zeta^i_F = 5.75/1.00 = 5.75$.

Again, Sanderson's electronegativity values are derived only from the atomic number and the covalent radius of the atoms. The utility of Sanderson's ideas concerning atomic properties and electron density will be discussed in more detail in Section 3.6 with regards to bond energies. However, there are implications for other fields, such as optical refractometry of oxides,[24] and for charge distributions in proteins (Section 3.6.4).

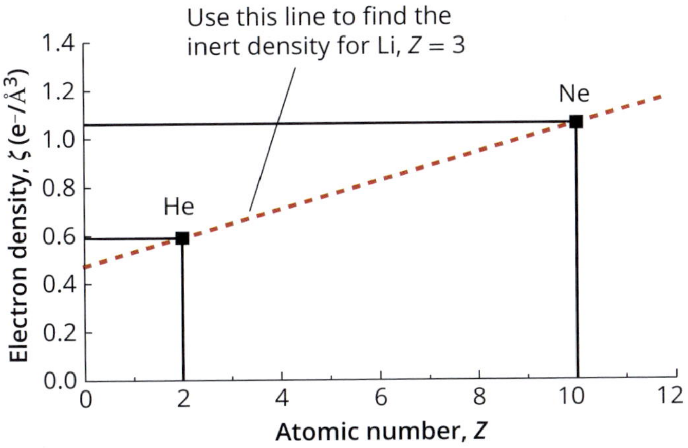

Nobel Gas (Z)	Covalent Radius (Å)	ζ ($e^-/Å^3$)
He (2)	0.93	0.59
Ne (10)	1.31	1.06
Ar (18)	1.74	0.82
Kr (36)	1.89	1.27
Xe (54)	1.41	2.09
Rn (86)	2.29	1.72

Fig. 3.14 (*top*) Plot of electron density (Eq. 3.13) vs atomic number. The line is the "inert density" for the elements between the two noble gases. (*bottom*) Tabulated data of covalent data and calculated (Eq. 3.13) electron densities (ζ).

Later, in order to make the scale more palatable to people accustomed to the Pauling scale where χ_F was originally set to 4.00, Sanderson rescaled his electronegativities to a 4-point scale as well.[16] The nonlinear relationship between Pauling electronegativity and Sanderson electronegativity can be approximated using Eq. 3.15, where χ^P and χ^S are the Pauling and Sanderson values.[25]

$$\chi^P = (0.33\chi^S + 0.66)^2 \qquad \textbf{3.15}$$

Interestingly, Sanderson stated that there was little evidence that his electronegativities were dependent upon orbital hybridization,[16] contrary to the Mulliken–Jaffé scale.

3.4.4 Allred–Rochow Electronegativity (Expanded Coverage)

The Allred–Rochow scale is based on the attractive force between the nucleus and an electron from a bonded atom.[18] The force can be calculated (Eq. 3.16) by using Coulomb's law, if r is the average distance between the electron and nucleus (assumed to reside near the covalent radius of the atom), e is the electron charge, and Z_{eff} is the effective charge from the nuclei, after shielding of the other electrons is considered.

$$Force = \frac{e^2 Z_{eff}}{r^2} \qquad \textbf{3.16}$$

The Allred–Rochow electronegativity scale has been, from its first introduction, scaled to Pauling's Electronegativities.[18] Allred and Rochow did this from the fit to a straight-line plot of their values to Pauling's. The Allred-Rochow values, χ^{AR}, are found using Eq. 3.17, where r^2 is in $Å^2$ and Z_{eff} is in terms of residual electric charge of the atom at the surface of the atom.

$$\chi^{AR} = 0.359\frac{e^2 Z_{eff}}{r^2} + 0.744 \qquad \textbf{3.17}$$

Slater's rules were used to find Z_{eff}. Many times using Slater's rules (Chapter 2) we are concerned with the nuclear attraction to a valence electron; however, it is necessary to remember that Allred and Rochow were considering "the force of attraction between the nucleus and an electron from a bonded atom."[18] In other words, all of the electrons in the atom are counted, and one is concerned with attraction to an additional electron at the covalent radius when calculating the shielding.

Sanderson himself said the Allred-Rochow values would be the best electronegativity scale, if the shielding parameters were precisely known.

3.5 Formal Oxidation State (Key Concept)

One useful convention used throughout inorganic chemistry is the idea of a *formal oxidation state*. Although historically the concept of oxidation state referred to the exact charge that specific atoms were thought to possess in compounds (for instance, the oxidation state of the sodium and chlorine atoms in NaCl would have been assigned as +1 and −1, respectively), the modern concept of the oxidation state does not necessarily have any correlation with the actual charge that is on the atom in question. Rather, it is simply an "electron-bookkeeping" method that is quite useful for inorganic compounds across the periodic table, and the term "formal oxidation state" is often used to remind ourselves of this distinction.

To find the oxidation state of elements in a molecule, a *gedanken* (thought) experiment is carried out by which the bonds are cleaved with the more electronegative atom receiving all of the electrons in the bonds, the extreme of what is referred to as a *heterolytic* bond-breaking process. Whatever formal charge is then found on each atom is then what we refer to as the formal oxidation state of the atom in question. Formal oxidation states are alternatively described using Arabic numerals (+3, +5, −1, −2, . . .) or by affixing the Roman numeral value in parentheses next to the element name or symbol whose oxidation state is described (for example, platinum(IV) ≡ Pt(IV) ≡ Pt^{4+}).

Fig. 3.15 lists several examples for simple main-group compounds. One advantage of this formalism is that it allows us to easily classify molecules that often exhibit related (or divergent) reactivities due to similarities (or differences) in formal oxidation states. For example, tin commonly forms compounds in the +4 and +2 oxidation states, which have different geometries, reactivities, and so on.

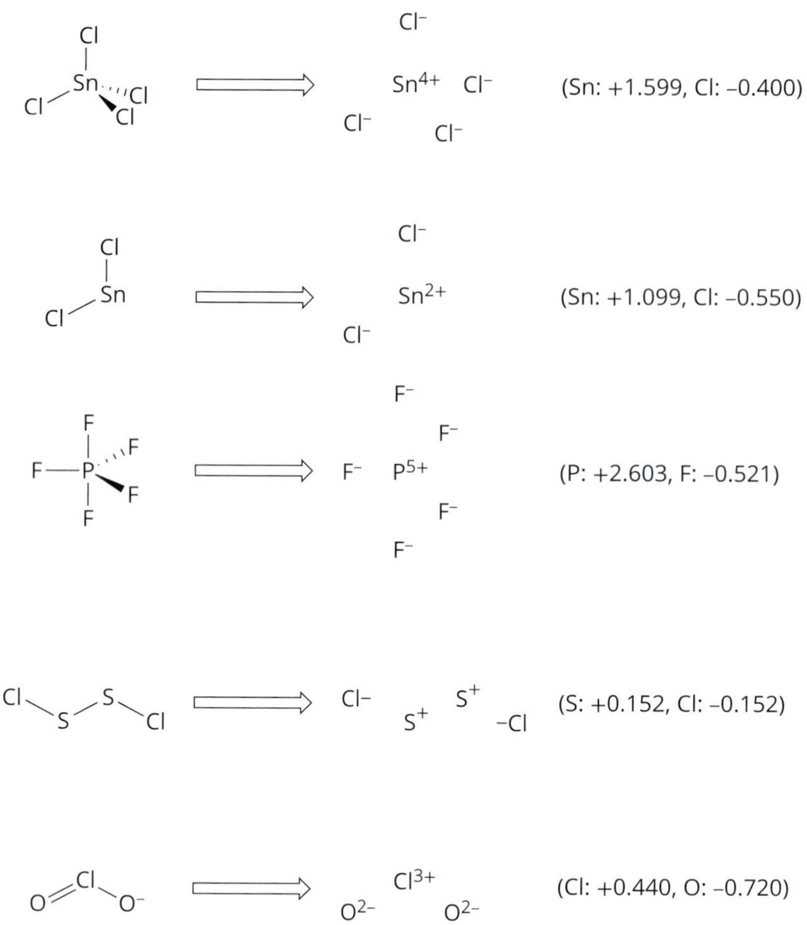

Fig. 3.15 Defining the formal oxidation states of main-group elements by heterolytic bond cleavage. The superscript charges on the atoms on the right are the formal oxidation states of the atoms in those structures. Numbers in parentheses are average charges on the atoms calculated using Natural Population Analysis (B3LYP/LANL2DZ).

Obviously, following the charge-conservation principle, the sum of all the charges in the cleaved-bond state (right side in Fig. 3.15) should be equal to the overall charge on the whole compound (left side in Fig. 3.15). In phosphorus pentafluoride, PF_5, the compound is neutral with a phosphorus(V) center and five peripheral fluorides (F^-); for nitrite, NO_2^-, the sum of the partial charges (+3−2−2 = −1) corresponds indeed to a monoanion.

In breaking homonuclear bonds, like the S–S bond in S_2Cl_2 (Fig. 3.15), where each atom on both ends of the bond are by definition equally electronegative, we simply split each of the two electrons one on each atom; bond-breaking of this sort is called *homolysis*, or *homolytic* bond cleavage. As a result, neither atom accumulates more or less charge than prior to the bond-breaking process, which does not affect the oxidation state. Consequently, the S–S bond does not contribute to the two sulfurs' oxidation state, which is assigned as S(I). The S–Cl bonds are heterolytically cleaved, as usual, with the more electronegative chlorine atom acquiring the two electrons from the S–Cl bond.

The assignment of formal oxidation state to atoms in most compounds is quite simple and yet quite useful as a descriptor with relevance for reactivity and properties. However, that assignment can be somewhat more complex (and in some cases even controversial!) for some ligand sets bound to transition metals. Those issues will be discussed when we review transition metal ligands in detail in Chapters 11 and 12. Again, this is only a useful formalism, and the formal oxidation state has no direct bearing on the magnitude of charge on the elements in a compound. This is illustrated in Fig. 3.15, where the formal oxidation states are given in the structures and the calculated charges are in parentheses to the right; the calculated charges are always smaller in absolute value than the formal oxidation state.

3.6 A Simple Resonance-Based Theory of the 2c2e Bond: Polar Covalence Theory (Key Concept)

From a resonance theory perspective, the 2c2e bond between two atoms A and X with different electronegativities may be described with the two resonance forms shown in Fig. 3.16. In principle, there is an additional polar resonance form with

A^- and X^+; however, we will assume that only one of these polar forms makes a large contribution to the ground state. In other words, we will assume that the more electronegative atom, in this case atom X, gets the anionic charge in the polar form of relevance (Criteria 5, Section 3.3). However, this additional polar form will make a small but nonzero contribution to the ground state in valence bond theory.[26]

In his theory of polar covalence, R. T. Sanderson (whose theory of electronegativity was discussed in Section 3.4.3) assumed that the A–X bond energy could be calculated by considering only the two resonance forms in Fig. 3.16. The structure on the left, represented with a covalent bond between A and X, is assigned a covalent bond energy, E_c. The structure on the right, where the bond is polarized, is assigned an ionic bond energy, E_i. (Compare this with classical Pauling–Wheland VB resonance theory for a 2-electron bond, Chapter 7.) Polar covalence theory is a simple "back-of-the-envelope" empirical method for calculating mean bond energies that is remarkably accurate for many simple (especially binary) compounds and illustrates many important principles of bonding. Some limitations of the theory will be discussed later in this chapter, but in general, *polar covalence theory works best for binary (AX_n) compounds that only contain 2-center 2-electron bonds*, such that the theory works for compounds that do not contain dative bonds (Section 3.8).

Bond energies are an extremely important tool for reaction chemistry. If, as per van Vleck's suggestion quoted at the beginning of this chapter, quantum mechanics is the master key that opens every door, then thermodynamics is the doorman you pay a cover charge to when you want to look inside. Bond energies are often a convenient method for looking at thermodynamics of chemical species as they relate to chemical reactions.

There is no reason to assume that the two resonance forms in Fig. 3.16 will make equal contributions to the ground state, and most often they will not. Consequently, the BDE of A–X will be related to the covalent bond energy (E_c) and ionic bond energy (E_i) by two "blending coefficients" (Sanderson's terminology), t_c and t_i, which are essentially resonance weightings. So, the energy in the bond A–X can be described by Eq. 3.18, where the total bond energy, E_{A-X}, is a weighted sum of the ionic and covalent bond energies.

$$E_{A-X} = t_c E_c + t_i E_i \qquad \textbf{3.18}$$

Since we are assuming that only these two resonance forms are required for a reasonable description of the ground state, according to resonance theory, the blending coefficients should sum to unity (Eq. 3.19):

$$1 = t_c + t_i \qquad \textbf{3.19}$$

Fig. 3.16 Limiting resonance forms for the heterocovalent bond.

This resonance theory-based method for estimating the bond energy of compounds differs from the method used by Pauling when he defined his electronegativities. Pauling (Section 3.4.1) assumed that the bond energy was comprised of an ionic contribution that *added on* to the covalent bond energy. This resonance theory description, polar covalence theory, assumes that the ionic contribution *replaces* part of the covalent bond energy, which occurs through the blending coefficients.

Also keep in mind that Sanderson's electronegativities (Section 3.4.3), unlike Pauling's (Section 3.4.1), are not derived from bond energies for compounds, but rather from ideal electron densities of individual atoms. The fact that Pauling's scale allows calculation of bond energies to a good degree of accuracy is not very surprising; when we use Pauling's scale to calculate bond energies we are simply reversing the process that led to the electronegativity values in the first place. By contrast, Sanderson electronegativities are not derived from bond energies. The fact that Sanderson's polar covalence theory can be used, often with great accuracy, for a wide variety of compounds is a strong validation of the core principles underlying his approach. The theory describes how bond energies are fundamentally related to atomic properties such as electron density, bond distance, and radius. Furthermore, Sanderson discovered some fundamental effects related to atoms containing lone pairs as well as bonds that will be instructive concerning molecular properties and reactivity.

The covalent energy, as discussed for Pauling electronegativity, can be estimated as the geometric mean of the homonuclear covalent bond energies. A valuable correction to this is to compensate for the actual bond distance relative to the sum of the covalent bond radii (Eq. 3.20).

$$E_c = \beta \frac{R_c}{R_0}\sqrt{E_{AA}E_{XX}} \qquad \textbf{3.20}$$

where $R_c = R_A + R_X$, and β = bond order coefficient.

Bond Order	β
single	1.000
double	1.488
triple	1.787

If the A–X bond were purely covalent, the bond distance should be the sum of the covalent radii of A and X, R_A and R_X respectively. However, the real bond distance for a single bond is often slightly different than what it would be for the purely covalent species. This can be accounted for by taking the ratio of the sum of the covalent radii with the actual bond distance, R_0. With this, the covalent bond energy, E_c, can be calculated using Eq. 3.19, where R_c is the covalent bond distance, equal to the sum of R_A and R_X.

In the case of multiple bonds, the A–X bond distance will usually substantially deviate from the covalent radii, and the

R_c/R_0 correction becomes larger. For multiple bonds there is an additional correction, β, which is unity for a single bond. The β-factor becomes larger for double bonds, but is not 2; in other words, a double bond is not twice as strong as a single bond. The factor found empirically for these equations by Sanderson for a double bond was 1.488.[e] For a triple bond, the β-factor is 1.787.

The purely ionic bond energy, E_i, is simply related to the Coulombic attraction between the two charged species making up the ionic bond. In other words, it is inversely related to the experimental bond distance, R_0, as shown in Eq. 3.21. In this equation the constant, 332 kcal • Å/mol (Appendix A), converts to energy in kcal/mol if the value of R_0 is in Ångstroms (1 Å = 1 × 10^{-10} m). This energy is then multiplied by the multiplicity factor β, which equal to 1 for a single bond.

$$E_i = \beta \cdot \frac{332\ kcal \cdot \text{Å}/mol}{R_0 \text{Å}} \quad \textbf{3.21}$$

All that is left is to find the coefficients, t_i and t_c, and the bond energy of A–X can be calculated. The coefficient for the contribution of the ionic resonance form, t_i, can be estimated if one knows the partial charges on A and X in the ground state, δ_A and δ_X respectively. The coefficient can be approximated as the arithmetic mean of the absolute value of the charges across the bond (Eq. 3.22).

$$t_i = \frac{|\delta_A| + |\delta_X|}{2} \quad \textbf{3.22}$$

How can one find the charges in the ground state? The charges on the atoms in the molecule are related to electronegativity differences, as discussed previously (Fig. 3.16). In polar covalence theory, the charges are related to the electronegativities of the individual atoms versus the electronegativity of the whole molecule. The mathematical relationship is shown in Eq. 3.23, where χ_m = molecular electronegativity, χ_A = electronegativity of free atom A, and $\Delta\chi_A = 1.57\sqrt{\chi_A}$ = change of electronegativity in atom A when a unit of charge is added. (The value 1.57 was found empirically.)

$$\delta_A = \frac{(\chi_M - \chi_A)}{1.57\sqrt{\chi_A}} = \frac{(\chi_M - \chi_A)}{\Delta\chi_A} \quad \textbf{3.23}$$

A central aspect of Sanderson's theory is the Principle of Electronegativity Equalization,[27] which states that all atoms *within a molecule* have the same electronegativity (a similar idea was discussed in the context of the Mulliken electronegativity scale, Section 3.4.2). That is, when atoms with their individual electronegativities come together to form a molecule, the entire molecule under examination has the same electronegativity value, χ_m. The basis behind electronegativity equalization can be illustrated with a thought experiment. If atom A approaches atom X, with X being the more electronegative atom, X will slowly start to pull some of the electron density from A and consequently pick up a partial negative charge (and vice versa for A, which will pick up an equal but opposite positive charge). The transfer of charge onto X will gradually decrease its electronegativity (an atom with some anionic charge is less likely to attract additional electron density). Applying the same analysis in reverse, we must conclude that the electronegativity of A will rise as a result of the transfer of charge onto X. This charge transfer and electronegativity change continues until the electronegativity of A and X in the molecule A–X are equivalent. At that point, any further transfer of charge from A to X would make A a more electronegative species than X, and it would pull the charge back. Using similar ideas, we understand that all atoms in a covalent system must either attract or release electron density, picking up partial negative and positive charges, until all atoms reach the same electronegativity.[28–29]

The electronegativity of a molecule is the geometric mean of the electronegativities of the individual atoms. In the case where there are only two atoms (A and X) in the molecule, Eq. 3.24 defines the molecular electronegativity.

$$\chi_m = \sqrt{\chi_A \chi_X} \quad \textbf{3.24}$$

More generally, the molecular electronegativity, χ_m, for a compound is found by using Eq. 3.25, where N is the total number of atoms, and N^A is the number of type A atoms in the molecule. Consequently, $\chi_A^{N^A}$ is the electronegativity of A, χ_A, to the power of N^A, the number of atoms of that type in the molecule.

$$\chi_m = \sqrt[N]{\chi_A^{N^A} \times \chi_B^{N^B} \times \ldots \times \chi_Z^{N^Z}} \quad \textbf{3.25}$$

The relationship between charge and electronegativity used in the polar covalence formalism depends upon the product of a constant, 1.57, and the square root of the atomic electronegativity. The value of $\Delta\chi_A$ in Eq. 3.26 is the electronegativity change on atom A with addition of unit charge. The constant was found through comparison with known bond energies and apparently relates the fundamental charge to Sanderson electronegativity, originally found through electron densities of the atoms (Section 3.4.3). Values for $\Delta\chi_A$ are shown in Table 3.2. Whether the electronegativity increases or decreases with unit charge of course depends on whether the charge being gained is positive or negative.

$$\Delta\chi_A = 1.57\sqrt{\chi_A} \quad \textbf{3.26}$$

[e] It is tempting to think that a double bond is then 1.488 times stronger than a single bond, but that is not quite right either. The shorter double bond also leads to a correction in the R_c/R_0 factor, for example. A double bond is then at least 1.488 times stronger than the corresponding single bond, but generally less than twice as strong.

Table 3.2 Sanderson electronegativities (χ^S), change in electronegativity with unit charge ($\Delta\chi^S$), covalent radii (r_c) in Å (×10^{-10} m), homonuclear fully unweakened (E'''), partially weakened (E''), and fully weakened (E') bond energies in kcal/mol.[a,b,c,d]

Element	χS	$\Delta\chi^s(=1.57\sqrt{\chi^s})$	Rc	E'''	E''	E'
H	2.592	2.528	0.320	104.2		
Li	0.670	1.285	1.336	24.6		
Be	1.810	2.112	0.887	67.6		
B	2.275	2.368	0.822	76.7		
C	2.746	2.602	0.772	85.4		
N	3.194	2.806	0.734	94.9	66.9	38.8
O	3.654	3.001	0.702	104.0	68.8	33.6
F	4.000	3.140	0.681	113.1	76.8	38.2[e]
Na	0.560	1.275	1.539	16.4		
Mg	1.318	1.802	1.373	42.3		
Al	1.714	2.055	1.258	48.2		
Si	2.138	2.296	1.169	54.1		
P	2.515	2.490	1.107	60.0	56.9	53.7
S	2.957	2.700	1.049	65.9	60.4	54.9
Cl	3.475	2.927	0.994	71.8	64.9	58.0
K	0.445	1.047	1.962	13.1		
Ca	0.946	1.527	1.74	30.8		
Sc(II)	0.64	1.256	1.695	31.3		
Sc(III)	1.02	1.586	1.695	31.3		
Ti(II)	0.73	1.341	1.650	31.8		
Ti(III)	1.09	1.639	1.650	31.8		
Ti(IV)	1.50	1.923	1.650	31.8		
V(II)	0.69	1.304	1.605	32.3		
V(III)	1.39	1.851	1.605	32.3		
V(IV)	1.89	2.158	1.605	32.3		
V(V)	2.51	2.487	1.605	32.3		
Cr(II)	1.24	1.748	1.560	32.8		
Cr(III)	1.66	2.023	1.560	32.8		
Cr(IV)	2.29	2.376	1.560	32.8		
Cr(V)	2.83	2.641	1.560	32.8		
Cr(VI)	3.37	2.882	1.560	32.8		
Mn(II)	1.66	2.023	1.516	33.3		
Mn(III)	2.20	2.329	1.516	33.3		
Mn(IV)	2.74	2.599	1.516	33.3		
Mn(V)	3.28	2.843	1.516	33.3		
Mn(VI)	3.82	3.069	1.516	33.3		
Fe(II)	1.64	2.011	1.471	33.8		
Fe(III)	2.20	2.329	1.471	33.8		
Co(II)	1.96	2.198	1.426	34.3		
Co(III)	2.56	2.512	1.426	34.3		
Co(IV)	3.10	2.764	1.426	34.3		
Ni(II)	1.94	2.187	1.381	34.8		

(continued)

Table 3.2 Continued

Element	χS	$\Delta\chi^s(=1.57\sqrt{\chi^s})$	Rc	E'''	E''	E'
Ni(III)	2.73	2.594	1.381	34.8		
Ni(IV)	3.27	2.839	1.381	34.8		
Ni(V)	3.81	3.065	1.381	34.8		
Cu(II)	1.98	2.209	1.336	35.3		
Zn	2.223	2.341	1.292	35.8		
Ga	2.419	2.442	1.256	40.3		
Ge	2.618	2.540	1.223	44.8		
As	2.816	2.635	1.294	49.3	45.1	40.9
Se	3.014	2.726	1.167	53.8	46.0	38.2
Br	3.219	2.817	1.142	58.3	52.2	46.1
Rb	0.312	0.877	2.16	12.4		
Sr	0.721	1.333	1.91	24.6		
Y(II)	0.40	0.993	1.868	25.1		
Y(III)	0.65	1.266	1.868	25.1		
Zr(II)	0.52	1.132	1.826	25.7		
Zr(III)	0.79	1.395	1.826	25.7		
Zr(IV)	0.90	1.489	1.826	25.7		
Nb(II)	0.77	1.378	1.784	26.3		
Nb(III)	1.02	1.586	1.784	26.3		
Nb(IV)	1.25	1.755	1.784	26.3		
Nb(V)	1.42	1.871	1.784	26.3		
Mo(II)	0.90	1.489	1.742	26.9		
Mo(III)	1.15	1.684	1.742	26.9		
Mo(IV)	1.40	1.858	1.742	26.9		
Mo(V)	1.73	2.065	1.742	26.9		
Mo(VI)	2.20	2.329	1.742	26.9		
Cd	1.978	2.208	1.493	30.4		
In	2.138	2.296	1.455	33.1		
Sn(IV)	2.298	2.380	1.420	35.8		
Sn(II)	1.477	1.908	1.420	35.8		
Sb	2.458	2.461	1.389	38.5	35.8	33.0
Te	2.618	2.540	1.360	41.2	39.2	37.2
I	2.778	2.617	1.333	43.9	40.0	36.1
Cs	0.220	0.736	2.35	10.8		
Ba	0.651	1.267	1.98	22.2		
Hf(II)	0.31	0.874	1.80	15.6		
Hf(III)	0.56	1.175	1.80	15.6		
Hf(IV)	0.81	1.413	1.80	15.6		
Ta(II)	0.44	1.041	1.76	14.7		
Ta(III)	0.69	1.304	1.76	14.7		
Ta(IV)	0.94	1.522	1.76	14.7		
Ta(V)	1.17	1.698	1.76	14.7		

Table 3.2 Continued

Element	χS	$\Delta\chi^s(=1.57\sqrt{\chi^s})$	Rc	E'''	E''	E'
W(II)	0.73	1.341	1.73	13.8		
W(III)	0.98	1.554	1.73	13.8		
W(IV)	1.23	1.741	1.73	13.8		
W(V)	1.48	1.910	1.73	13.8		
W(VI)	1.67	2.029	1.73	13.8		
Tl(III)	2.246	2.353	1.490	16.6		
Tl(I)	0.98	1.560	1.49	16.6		
Pb(IV)	2.291	2.376	1.48	24.2		
Pb(II)	1.900	2.164	1.48	24.2		
Bi	2.342	2.403	1.47	32.2		

[a] For a detailed discussion of bond weakening, including the cause of the phenomenon, see Chapter 8.
[b] Main Group values are from Sanderson's 1983 reference.[32]
[c] If no oxidation state is given, the most common oxidation state is implied; that is, F is F(−I), or the oxidation state is not accounted for.
[d] Values for the transition metals are from Sanderson's paper.[17]
[e] Value from Huber and Herzberg.[33]

Admittedly, this is a fairly large series of equations that is used, but in reality these can be used in a few minutes to carry out these simple calculations to give mean bond energies for compounds. Several detailed examples follow, and the method will be used to illustrate some fundamental principles of bonding that become very obvious in the context of these calculations. The following table is a summary of all the polar covalence equations in one place to help with application of the method.

Summary of Equations Used in Polar Covalence Theory

Determination of the A–X bond energy	$E_{A-X} = t_c E_c + t_i E_i$	Eq. 3.17
Sum of the ionic and covalent coefficients is unity	$1 = t_c + t_i$	Eq. 3.18
Covalent bond energy	$E_c = \beta \frac{R_c}{R_0} \sqrt{E_{AA} E_{XX}}$ $R_c = R_A + R_X$ **Bond Order** / β: single 1.000; double 1.488; triple 1.787	Eq. 3.19
Ionic bond energy	$E_i = \beta \cdot \frac{332\ kcal \cdot \mathring{A}/mol}{R_0 \mathring{A}}$	Eq. 3.20
Estimation of the ionic coefficient	$t_i = \frac{\lvert\delta_A\rvert + \lvert\delta_X\rvert}{2}$	Eq. 3.21
Charge on an atom in a molecule	$\delta_A = \frac{(\chi_M - \chi_A)}{1.57\sqrt{\chi_A}} = \frac{(\chi_M - \chi_A)}{\Delta\chi_A}$	Eq. 3.22
Molecular electronegativity	$\chi_m = \sqrt[N]{\chi_A^{NA} \times \chi_B^{NB} \times \ldots \times \chi_Z^{NZ}}$	Eq. 3.24

3.6.1 Calculation of the BDE of the B–F Bonds in BF_3 as an Illustration

Table 3.2 shows most of the numbers necessary to estimate the bond energies for thousands of compounds. The value for R_0 for various species can be found in the experimental literature. For BF_3, R_0 = 1.311 Å. In addition, bond lengths can often be approximated fairly quickly with reasonable accuracy from gas-phase calculations.[30] The bond distances of many representative compounds are shown in Appendix B, or, if all of the previous are inconvenient and accuracy is not paramount, the R_c/R_0 term in Eq. 3.19 could be assumed to be unity for single bonds, in other words one could use R_c in place of R_0 in Eq. 3.19. In addition, bond lengths can be estimated for systems by using charts like that by Pyykkö and coworkers in Section 3.10.1.

We can start by calculating the covalent and ionic bond energies using Eqs. 3.19 and 3.20. We can then plug them into Eq. 3.17 for the bond energy and use the fact that the coefficients must add to unity (Eq. 3.18) to derive the following calculations. In the final line, the coefficient t_i remains as the only unknown required to find the B–F bond energy, E_{BF}.

(Note: For this section we shall focus on single bonds. Therefore, the β-factor will be 1.000 and will be ignored. Some examples of using polar covalence theory with multiple bonds are presented later.)

$$E_c = \frac{R_c}{R_0}\sqrt{E_{BB}E_{FF}} = \frac{0.822 + 0.681}{1.311}\sqrt{76.7 \times 113.1} = 106.8 \text{ kcal/mol}$$

$$E_i = 332/R_0 = 332/1.311 = 253.2 \text{ kcal/mol}$$

$$E_{BF} = t_c E_c + t_i E_i$$

$$E_{BF} = 106.8t_c + 253.2t_i \text{ kcal/mol}$$

but

$$1 = t_c + t_i$$

$$t_c = 1 - t_i$$

So

$$E_{BF} = 106.8 - 106.8t_i + 253.2t_i \text{ kcal/mol}$$

The coefficient t_i is estimated from the partial charges on each atom. We first calculate the molecular electronegativity (Eq. 3.24), χ_m, and, from that quantity, the charges on boron (Eq. 3.22), δ_B, and fluorine, δ_F.

$$\chi_m = \sqrt[4]{\chi_F^3 \chi_B} = \sqrt[4]{4^3 \times 2.275} = 3.474$$

$$\delta_B = \frac{(\chi_m - \chi_B)}{1.57\sqrt{\chi_B}} = \frac{3.474 - 2.275}{2.368} = +0.506$$

$$\delta_F = \frac{(\chi_m - \chi_F)}{1.57\sqrt{\chi_F}} = \frac{3.474 - 4.000}{3.140} = -0.168$$

The charge on boron is +0.506. Since the charge on the molecule overall is zero and, by symmetry, the negative charge should be evenly divided among the three equivalent fluorines, we find that each fluorine atom should have a negative charge that is one third the magnitude of the charge on boron. The value obtained is −0.169, in good agreement with the value calculated using Eq. 3.22, −0.168.

$$\delta_F = -\frac{\delta_B}{3} = -\frac{0.506}{3} = -0.169$$

The blending coefficient t_i is estimated as the average charge across one of the B–F bonds in the molecule (Eq. 3.21).

$$t_i = \frac{|\delta_F| + |\delta_B|}{2} = \frac{0.506 + 0.169}{2} = 0.338$$

Now we can calculate the mean bond energy in the compound.

$$E_{BF} = 106.8 - 106.8t_i + 253.2t_i \text{ kcal/mol}$$

$$E_{BF} = 106.8 - 106.8(0.338) + 253.2(0.338) \text{ kcal/mol}$$

$$E_{BF} = 156.3 \text{ kcal/mol}$$

Often, these calculations on binary compounds are within about ~5% of the experimental values or elaborate quantum calculations. In this case, the experimental BDE of the first bond in BF_3 is 153.5 kcal/mol.[31] Consequently, the polar covalence calculation has an error of about 2% in this case. This method can be used to estimate the energy of one of the bonds; that is, the BDE from F—BF_2 → F + BF_2 with variable results. The value actually calculated by polar covalence theory is the mean bond energy in the compound. Atomization of BF_3—production of B + 3 F from BF_3—should require three times the average bond energy calculated, 468.9 kcal/mol. The experimental atomization energy for BF_3 is 463.0 kcal/mol;[16] consequently, the calculation has an error of 1% in the atomization energy or mean bond energy.

This method of estimating the ionic blending coefficient, t_i, from the charges due to the differences in electronegativity is valuable if attempting to calculate bond energies of compounds that may not be known, or for comparison with literature or computational values. However, another assessment of t_i can be done using Eq. 3.26 if the bond energy (BE) is known (and still under the assumption that there are only two resonance forms of consequence—one ionic and one covalent, Fig. 3.16).

$$t_i = \frac{BE - E_c}{E_i - E_c} \qquad \textbf{3.27}$$

Applying this equation to the previous BF_3 example and using the experimental value for the B–F bond energy (BE = 154.3 kcal/mol), we calculate t_i = 0.325, reasonably close to our previously estimated value of 0.338. In other words, the model suggests that the B–F bond in BF_3 is ~33% ionic under this model.[34]

$$t_i = \frac{BE - E_c}{E_i - E_c} = \frac{154.3 - 106.8}{253.2 - 106.8} = 0.325$$

An important point concerning bond energies is quite apparent from the previous example: *the ionic bond energy is often much larger than the covalent bond energy*. For BF_3, E_i = 253.2 kcal/mol, and E_c = 106.8 kcal/mol. Increasing the ionic character of a bond, which amounts to increasing the ionic blending coefficient t_i, increases the bond energy. This is an important principle for bonding in general and bonding in inorganic complexes in particular: *more ionic bonds are generally stronger*.

As another illustration of the method and demonstrating its validity not just for inorganic compounds but also for organic molecules, let us consider the case of methane CH_4. The equilibrium C–H bond distance in the molecule, R_0, is 1.106 Å by electron diffraction in the gas phase. The mean bond energy is calculated as follows.[35]

$$\chi_M = \sqrt[5]{2.746 \times 2.592^4} = 2.622$$

$$\delta_C = \frac{2.622 - 2.746}{2.602} = -0.048$$

$$\delta_C = \frac{2.622 - 2.592}{2.528} = +0.012$$

$$t_i = \frac{0.048 + 0.012}{2} = 0.030$$

$$E_c = \frac{0.772 + 0.32}{1.106}\sqrt{104.2 \times 85.4} = 93.1\ \text{kcal/mol}$$

$$E_i = \frac{332}{1.206}\ kcal/mol$$

$$E = 93.1 - 93.1(0.03) + 300.2(0.03) = 99.3\ \text{kcal/mol}$$

The experimental BDE of the first H–C bond in CH_4 is 104 kcal/mol, an error of only 4.5% in our calculation. As mentioned, however, the calculated energy using polar covalence is the mean bond energy of all the C–H bonds. The calculated atomization energy is then four times the average, 397.2 kcal/mol. The experimental atomization energy of CH_4 is 397.6 kcal/mol. As a result, the atomization energy from our calculation has an error of only 0.1%!

Note the ionic bond energy for the C–H bond calculated previously for CH_4. The bond energy in the ionic resonance form, E_i, is more than three times stronger than the purely covalent bond energy, E_c. However, the C–H bonds in methane are quite covalent (t_i = 0.03, or about 3% ionicity), and the ionic form contributes very little.

3.6.2 Oxidation States and BDEs in Main-Group Compounds

Some of the main-group elements are commonly found in a single oxidation state, such as magnesium being in the +2 oxidation state most often. However, many main-group elements can access additional oxidation states and form stable compounds, such as Sn(IV) and Sn(II) chlorides. This change in formal oxidation state hints at corresponding changes in partial charge and electronegativity of the metal, and one would expect Sn(IV) to have a higher electronegativity than Sn(II).

The lower oxidation state halides of main-group metals are referred to as the *subhalides*. Typically, stable oxidation state changes in the main group involve changes by a pair of electrons. For example, Ge(IV), Sn(IV), Pb(IV), In(III), Ga(III), and Tl(III) halides are sometimes stable as subhalides Ge(II), Sn(II), Pb(II), In(I), Ga(I), and Tl(I). In general, lower oxidation states become more stable as one proceeds down a particular group—a phenomenon discussed in detail in Chapter 9 under the heading of the inert pair effect.

The larger the difference in electronegativity between two atoms, the more polar the bond, as it is easier for the more electronegative atom to pull electrons to its side of the bond. Consequently, low-oxidation-state main-group metals (which have smaller electronegativities) tend to form more polar bonds with halogen atoms than their higher-oxidation-state (more electronegative) analogs. The higher bond polarity in the lower-valent compounds leads to them having stronger bonds, typically on the order of about 8% but possibly more, relative to the higher oxidation states.

Typically, shorter bonds are thought of as being stronger due to having a higher bond order (Section 3.10). The reasons for longer bonds being generally weaker are easily seen in the equations for E_c (Eq. 3.19) and E_i (Eq. 3.20), and in both cases the actual bond distance, R_0, appears in the denominator. Counterintuitively though, those *stronger* bonds are usually *longer* in these subhalides due to the larger radius of the lower-valent metals relative to their higher-oxidation-state analogs. Nevertheless, the higher polarity of these bonds more than makes up for the bond length increase, leading to overall higher bond energies. This highlights the importance of a holistic approach to understanding chemical bonding, where all aspects of the problem (orbital overlap, bond lengths, partial charges, oxidation states, and so on) are simultaneously considered.

As an example, we can calculate the bond energies of $SnCl_2$ and $SnCl_4$ using Sanderson's polar covalence theory. For $SnCl_4$, R_0 = 2.31 Å, and the other values can be drawn from Table 3.2. We start by calculating the covalent, E_c, and ionic bond energies, E_i. Then, we can use those values in the equations for the bond energy using $1 = t_c + t_i$.

$$E_c = \frac{R_{Sn} + R_{Cl}}{R_0}\sqrt{E_{SnSn}E_{CCCl}} = 53.0\ \text{kcal/mol}$$

$$E_i = 332 / R_0 = 144\ \text{kcal/mol}$$

$$E_{Sn(IV)Cl} = t_cE_c + t_iE_i$$

$$E_{Sn(IV)Cl} = 53.0t_c + 144t_i$$

$$E_{Sn(IV)Cl} = 53.0 - 53.0t_i + 144t_i$$

Now, all that is left is to find the ionic blending coefficient, t_i, which is done using the molecular electronegativity, χ_m, and the charges on tin, δ_{Sn}, and chlorine, δ_{Cl}, as before. Finally, t_i

is calculated as the average of the absolute values of the charge across one of the Sn–Cl bonds.

$$\chi_m = \sqrt[5]{\chi_{cl}^4 \chi_{Sn}} = \sqrt[5]{3.475^4 \times 2.298} = 3.200$$

$$\delta_{Sn} = \frac{\chi_m - \chi_{Sn}}{1.57\sqrt{\chi_{Sn}}} = \frac{3.200 - 2.298}{1.57\sqrt{2.298}} = +0.379$$

$$\delta_{Cl} = \frac{\chi_m - \chi_{Cl}}{1.57\sqrt{\chi_{Cl}}} = \frac{3.200 - 3.475}{1.57\sqrt{3.475}} = -0.094$$

$$t_i = \frac{|\delta_{Cl}| + |\delta_{Sn}|}{2} = \frac{0.094 + 0.379}{2} = 0.237$$

Plugging the blending coefficient into the equation to obtain the bond energy produces the following equations.

$$E_{Sn(IV)Cl} = 53.0 - 53.0t_i + 144t_i$$
$$E_{Sn(IV)Cl} = 53.0 - 53.0(0.237) + 144(0.237)$$
$$E_{Sn(IV)Cl} = 74.5 \text{ kcal/mol}$$

The value obtained for the Sn(IV)–Cl bond energy in $SnCl_4$ is 74.5 kcal/mol, which is within 1% of the experimental value of 75.2 kcal/mol.

We can estimate the Sn(II)–Cl bond energy in $SnCl_2$ similarly using $R_0 = 2.42$ Å and Table 3.2.

$$E_c = \frac{R_{Sn} + R_{Cl}}{R_0}\sqrt{E_{SnSn}E_{ClCl}} = 50.6 \text{ kcal/mol}$$
$$E_i = 332 / R_0 = 137 \text{ kcal/mol}$$
$$E_{Sn(II)Cl} = t_c E_c + t_i E_i$$
$$E_{Sn(II)Cl} = 50.6t_c + 137t_i$$
$$E_{Sn(II)Cl} = 50.6 - 50.6t_i + 137t_i$$

Note that both the ionic and covalent bond energies decreased from the Sn(IV) example because of bond lengthening. However, the molecular electronegativity, the charge on tin, and the ionic blending coefficient show significant differences relative to the higher-oxidation-state species. The tin(II) compound, because of the lower electronegativity of the central metal, is much more ionic and has a much higher ionic blending coefficient: 0.447 versus 0.237 in the Sn(IV) example. In other words, the Sn(II) compound is roughly twice as polar in its bonding to chlorine according to this model.

$$\chi_m = \sqrt[3]{\chi_{Cl}^2 \chi_{Sn}} = \sqrt[3]{3.475^2 \times 1.477} = 2.613$$

$$\delta_{Sn} = \frac{\chi_m - \chi_{Sn}}{1.57\sqrt{\chi_{Sn}}} = \frac{2.613 - 1.477}{1.57\sqrt{1.477}} = +0.595$$

$$\delta_{Cl} = \frac{\chi_m - \chi_{Cl}}{1.57\sqrt{\chi_{Cl}}} = \frac{2.613 - 3.475}{1.57\sqrt{3.475}} = -0.295$$

$$t_i = \frac{|\delta_{Cl}| + |\delta_{Sn}|}{2} = \frac{0.295 + 0.595}{2} = 0.445$$

$$E_{Sn(II)Cl} = 50.6 - 50.6(0.445) + 137(0.445) = 89.0 \text{ kcal/mol}$$

Because more ionic bonds are typically stronger, the bonds in Sn(II) chloride are calculated to be 89.0 kcal/mol, whereas the bonds in Sn(IV) chloride were calculated as 74.5 kcal/mol. The calculated value for the bond strength in $SnCl_2$ is also quite close to the experimental value of 90.2 kcal/mol. Although bond energies can be obtained from experiment directly, it is only upon applying the polar covalence theory analysis that we can conclude that it is the contribution from the ionic resonance form that is responsible for the observed change in the bond energy in $SnCl_2$ vs. $SnCl_4$.

In general, bond energies increase as bond lengths decrease. As a simple example, one can examine the C–C bond lengths and BDEs in the series $H_3C–CH_3$, $F_3C–CH_3$, and $F_3C–CF_3$ (Table 3.3). Here again, polarity is the most important issue. The nonpolar C–C bond lengths in $H_3C–CH_3$ and $F_3C–CF_3$ are essentially identical, while the polar compound, $F_3C–CH_3$, has a shorter and stronger C–C bond. Like most generalities in chemistry, the correlation that shorter bonds are necessarily stronger has exceptions (such as for $SnCl_2$ and $SnCl_4$ discussed previously) and should be used cautiously, especially when comparing compounds that have varying formal oxidation states.

Incidentally, $F_3C–CH_3$—a ternary compound with different environments for the same element (carbon)—is a good example of an instance where the charge scheme in polar covalence fails to give realistic values (see Appendix A).

3.6.3 Group Electronegativity and Bond Energies in More Complex Molecules

While estimating electronegativities in binary molecules can be useful for calculation of charges, bond energies, and other molecular properties, most inorganic molecules contain multiatom ligands bound to a central element. A ligand as simple as hydroxide, OH^-, requires us to develop a model that allows us to have some idea of how the electronegativity of OH^- as a unit is different from either O or H separately.

The Principle of Electronegativity Equalization discussed at the beginning of Section 3.6 serves as a guide in understanding how charge is distributed through chemical groups. By assuming that the electronegativity of all the atoms in a

Table 3.3 Bond distances and bond energies[f] for ethane, hexafluoroethane, and 1,1,1-trifluoroethane.

	$H_3C–CH_3$	$H_3C–CF_3$	$F_3C–CF_3$
C–C Bond distance (Å)	1.533(2)	1.497(3)	1.543(3)
C–C BDE G3 method (kcal/mol)	88.4	103.1	99.6

[f] Bond dissociation enthalpies were calculated using the G3 method in Gaussian09.

BOX 3A

Bond energy and the energy of reaction

The bond energy is the mean energy required to homolytically cleave all the bonds in a binary compound. This quantity is expressed as a positive value with larger energies required for stronger bonds. From a thermodynamic standpoint when thinking of chemical reactions, however, it can be thought of as the energy released when the bond is broken. When energy is released it is given a negative value. Consequently, when using enthalpies of formation, for example, to calculate the enthalpy of reaction the equation takes the following form.

$$\Delta H_{rxn} = \Sigma H_{products} - \Sigma H_{reactants}$$

However, if one wants to use bond energy (BE) to find the enthalpy of reaction, the equation becomes

$$\Delta H_{rxn} = \Sigma BE_{reactants} - \Sigma BE_{products}$$

Example: Is the reaction SnI_2 and I_2 forming SnI_4 exothermic, and what is the enthalpy of reaction?

We can calculate the bond energy in SnI_2 using polar covalence (an R_0 = 2.84 Å value was used from gas-phase calculation using DFT B3LYP/LANL2DZ) as 62 kcal/mol. The average Sn–I bond energy in SnI_4 calculates to (R_0 = 2.76 Å), 46 kcal/mol. The experimental bond energy in I_2 is 36.1 kcal/mol. The ΔH_{rxn} can be calculated with these values as follows.

$$\Delta H_{rxn} = \Sigma BE_{reactants} - \Sigma BE_{products}$$
$$\Delta H_{rxn} = [36.1] - [2 \times 46]$$
$$\Delta H_{rxn} = -56 \text{kcal/mol}$$

This assumes that the only bond that changes during the reaction is the I_2 bond, which is broken, and the Sn–I bonds, which are made. This assumption is often used as a quick method of examining the exothermicity of a bond. In reality, a better method for calculating the energy of the reaction here, where the energies of the starting Sn–I bonds are quite different from the ending Sn–I bonds, would be to atomize the SnI_2 then form four new SnI_4 bonds in the new oxidation state.

$$SnI_2 \rightarrow Sn + 2I\bullet$$

This reacts with I_2 to form SnI_4.

$$Sn + 2I\bullet + I_2 \rightarrow SnI_4$$

The calculation may be done as follows.

$$\Delta H_{rxn} = \Sigma BE_{reactants} - \Sigma BE_{products}$$
$$\Delta H_{rxn} = [2 \times 62 + 36] - [4 \times 46]$$
$$\Delta H_{rxn} = -24 \text{ kcal/mol}$$

The experimental heat of reaction for SnI_2 + I_2 to form SnI_4 from the experimental heats of formation for the compounds is −29 kcal/mol in the gas phase.

covalent molecule have the same electronegativity once electrons have been redistributed upon formation of all bonds, it is possible to design a scheme to calculate the charges associated with every atom or with particular groups in molecules which then directly allows one to quantify the ionic character of each bond. However, the use of polar covalence theory to calculate bond energies in more complex molecules requires compensating for additional effects. For example, in cases where atoms of the same element are in very different environments like the two carbon atoms in CF_3CH_3. In general, use of polar covalence theory is only recommended for binary compounds, should be used with moderate caution for more complex molecules with normal covalent bonds, and should never be used for compounds with dative bonds. More details on polar covalence theory are presented in Appendix A.

3.6.4 Theoretical Models for Calculating Partial Charges (Advanced Topic)

While the Sanderson method outlined previously can be used to estimate partial charges in molecules quite simply, more elaborate methods for the calculation of partial charge have been developed, such as the electronegativity equalization method (EEM)[36–37] and others,[38–41] which rely on various charge schemes in combination with the Principle of Electronegativity Equalization to estimate charges in molecules of great complexity, including proteins.

One drawback to these methods is that such schemes rely upon fixed parameters to describe the charge-equilibration process that occurs in parallel to bond formation, which greatly affect the calculated charges. Those parameters are being developed using various theoretical methods.[42]

More common computational methods for the calculation of charge often rely on "population analysis", by which the electron density in a molecule is calculated, and the partial charge (δ) on each atom is found by taking the nuclear charge (Z) and subtracting the number of electrons (ρ), $\delta = Z - \rho$ assigned to that atom. The main issue at hand, and the fundamental difference between each method, arises from the fact that there are different ways for deciding how electrons should be partitioned between different atoms in a molecule. For details on different charge schemes, the interested reader is referred to relevant texts on computational chemistry.[43] A few common methods are briefly described here.

One method is to partition the electrons based on basis functions; for instance, the valence atomic orbitals used in a

LCAO-MO scheme to partition the molecular orbital densities. One of the most common methods, the Mulliken analysis (also due to Robert S. Mulliken, of the Mulliken electronegativity scale discussed previously) simply partitions the contribution equally among the atoms. For example, for the simple case of a linear combination of atomic orbitals centered on two different atoms, 1 and 2, where c_1 and c_2 are the coefficients for the contribution of the orbitals χ_1 and χ_2:

$$\psi = c_1\chi_1 + c_2\chi_2$$

The electron density is then the square of the function, where S_{12} is the overlap integral between χ_1 and χ_2.

$$\int |\psi|^2\, d\tau = \int |c_1\chi_1 + c_2\chi_2|^2\, d\tau = c_1^2 + c_2^2 + 2c_1c_2S_{12}$$

The term c_1^2 describes an electron density that can unambiguously be assigned to atom 1, and similarly c_2^2 to atom 2. Most charge schemes differ in how the overlap term $2c_1c_2S_{12}$ is treated. The Mulliken scheme takes no sides, and assigns half of the overlap term to each of the two atoms involved in the bond. Consequently, in a Mulliken analysis, atom 1 would be assigned an electron density equal to $c_1^2 + c_1c_2S_{12}$, whereas atom 2 would be assigned a charge density equal to $c_2^2 + c_1c_2S_{12}$. No matter what scheme we use, the sum of all partial densities must necessarily add up to the total density; that is, we cannot "lose" or "create" electrons in the process of dividing the total density on separate atoms, which is obviously the case here.

This approach yields relatively good results in some cases, but the Mulliken analysis has been shown to produce unrealistic charge values in highly polar systems, having been called "wholly unreliable" for use with lithium compounds, for example.[44] Nevertheless, Mulliken analysis is one of the most common schemes used to divvy up electron density among atoms and to approximate partial charges within a molecule.

Another approach which is often used is the atoms in molecules (AIM) method, due to Richard F. W. Bader (1931–2012), a Canadian theoretical chemist.[45] The AIM approach relies on analyzing the total wavefunction directly, rather than relying on the basis functions used to build up the total wavefunction as in a Mulliken analysis. As before, the total electron density in a molecule is determined as the square of the wavefunction, ψ^2, but its decomposition among all atoms is achieved using the topology of that total wavefunction; that is, how it varies across the molecule and where saddle points, minima, or maxima are located, and the electron density is divided into "basins" based on points where the electronic flux is zero. Using this method, the electrons are partitioned around the atoms, and charges can be calculated. Of all the approaches, AIM is arguably the most sound theoretically, but it also depends heavily on an accurate knowledge of the total wavefunction, which we usually do not know in the absence of accurate, high-level quantum-computational studies.

Yet another method for calculation of charge is due to Frank A. Weinhold and coworkers: natural population analysis

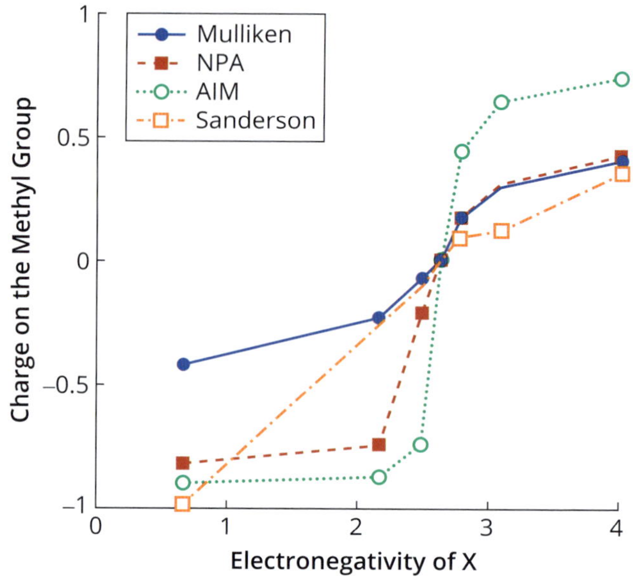

Fig. 3.17 Charge on the methyl group[48] in MeX compounds versus the Sanderson electronegativity of X, where X = Li, HBe, H_2B, Me, H_2N, OH, and F.

(NPA).[46] In NPA, the charge is related to the sum of the electron densities of the "natural atomic orbitals" contributions on the atoms in the molecule. This analysis has the advantage of being largely invariant with basis set and other computational details. This is in stark contrast to many other methods such as Mulliken analysis, which can be quite sensitive to the choice of basis set.[47]

The difference between various charge schemes can be seen if one examines the charge assigned to the methyl group[48] in $Li(CH_3)$, $HBe(CH_3)$, $H_2B(CH_3)$, ethane, $H_2N(CH_3)$, $HO(CH_3)$, and $F(CH_3)$, which is shown in Fig. 3.17 using Mulliken, natural population analysis (NPA), atoms in molecules (AIM), and the Sanderson approach, plotted versus Sanderson electronegativity of the X group in $X(CH_3)$. The charges on the methyl group change significantly depending on the method used, emphasizing the degree of approximation involved in deciding how to apportion electron density in a molecule. The Sanderson method has an approximately linear charge to electronegativity relationship, whereas the other methods have sigmoidal relationships of varying amplitudes. Note that all methods agree on the *sign* but differ in the *amount* of the partial charge assigned to the methyl group.

The more elaborate charge calculations occasionally suggest that the charge scheme used in polar covalence theory is inadequate for certain molecules. This tends to occur, for example, in some molecules that contain the same atom type in different environments. For example, in 1,1,1-trifluoroethane (Table 3.3), F_3CCH_3, the molecular electronegativity is calculated as 3.094 using the usual method (Eq. 3.24), but using Eq. 3.22 to calculate the charge assigned to each of the two carbon atoms yields identical values, which is intuitively unreasonable for a carbon bound to three fluorines compared to one bearing three hydrogens. In this simple implementation, polar covalence assumes that the C–C bond is perfectly nonpolar, which is of course invalid; this inability to account properly for the polarity of the bond leads to a calculated C–C bond energy of 85.5 kcal/mol using polar covalence, which is far lower than the value calculated using reliable theromodynamic calculations[g] of 103.1 kcal/mol. If one uses the bond energy of 103.1 kcal/mol from the quantum-mechanical calculation with E_c and E_i from polar covalence theory with Eq. 3.26, a value for t_i of 0.129 is obtained; in other words, the carbon–carbon bond in $F_3C–CH_3$ is ~13% polar between the two carbons because of the substitution pattern.

Few (if any) theories in chemistry are always accurate, without limitations. Of course, this does not make them completely worthless, but simply means that they need to be used with an understanding of their limitations. For most binary compounds, polar covalence theory is quite accurate, and the theory can be used to illustrate many important bonding principles. However, the theory can fail for compounds containing dative bonds, and some more complex molecules require consideration of reorganization energies of the free radicals or other details. Many of the effects that influence the bond energies illustrate important facets of bonding in molecules; for example, formal oxidation state. Transition metals, like their main-group counterparts discussed previously, have different electronegativities in different oxidation states, which can be accounted for in calculating their bond energies.

Finally, a subtle point on the relationship between formal charges and calculated charges should be noted. Like formal oxidation states, formal charges do not necessarily have any direct relationship to the actual charges on the atoms in the molecule. Formal oxidation states are a method of description and electron-counting. For example, the formal charges in ClSSCl were discussed earlier (Fig. 3.15). There is a formal positive charge on sulfur and a formal negative charge on chlorine. Fig. 3.18 shows the calculated formal charges with the Mulliken charge on the outside and the NBO charge in parentheses.

As alluded to previously, different charge schemes produce quite different values. The absolute magnitude of the charges is uniformly higher in the more sophisticated natural bond orbital method over the Mulliken charges. The Mulliken charge method at least obtains the spirit of the proper charge distributions for the atoms, but the charges are probably more accurate in the natural population analysis.

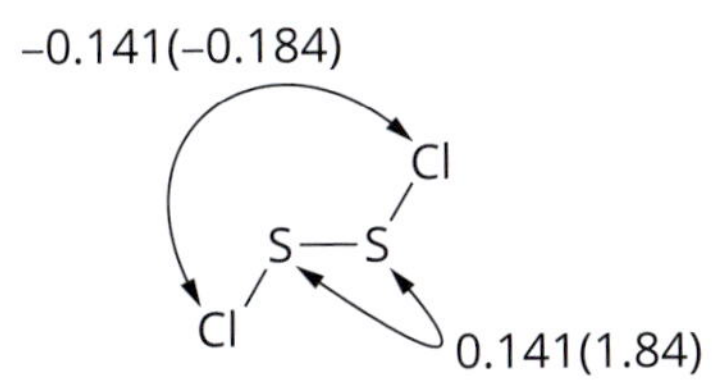

Fig. 3.18 NBO and Mulliken charges on the atoms of ClSSCl.

[g] The G3 method as implemented in Gaussian09 was used to calculated the bond dissociation enthalpy.

3.6.5 Lone-Pair Bond Weakening Effect (Key Concept)

Another complicating factor in polar covalence theory is what is known as the lone-pair bond weakening effect (LPBWE),[16] which can play an extremely important role for molecules that contain atoms with lone pairs. This effect is responsible for the unexpectedly weak bond strength and high reactivity of F_2 and HF, for example.

Here, it is sufficient to state that the LPBWE is more pronounced in bonds where (1) at least one of the bonding atoms contains at least one lone pair and (2) that lone pair does not interact with a π-acceptor atom (an atom with an empty orbital that can form a π-bond with one of the lone pairs). As a result, HF contains a "fully weakened" H–F bond, and the weakened covalent energy (E', Table 3.2) is used to calculate its bond energy. Conversely, in BF_3, the "fully unweakened" covalent bond energy is used because the fluorine lone pairs interact with boron. The weakening is more pronounced for lighter congeners than for heavier ones: the difference between the "unweakened" and "fully weakened" covalent bond energy for iodine is only 8 kcal/mol, compared to 73 kcal/mol for fluorine compounds. The LPBWE has only become understood in any detail relatively recently,[49] and is an important factor in understanding the reactivity of many compounds, such as the halogens. The specific causes from valence bond and molecular orbital models for the LPBWE will be discussed later in the context of main-group bonding and reactivity (Chapters 8 and 9).

It is also known (Section 8.5) that the LPBWE becomes weaker the larger the *s*-character of the lone-pair orbitals. In other words, a large LPBWE is expected for the nitrogen lone pair in ammonia, where the lone pair is nominally sp^3-hybridized (25% *s*-character), compared to the weakening effect in nitriles (R–C≡N), where the lone pair is in nominally an *sp*-hybridized orbital (50% *s*-character). In general, the fully weakened value (E') is used for single bonds, partially weakened (E'') for double bonds, and fully unweakened (E''') for triple bonds in molecules containing elements such as nitrogen and oxygen.

3.6.6 Multiple Bonds and Bond Energy (Key Concept)

Increasing the bond order, the multiplicity of the bond (for example, a single bond has a bond order of 1, a double bond has a bond order of 2, and so on), between two atoms increases the bond strength. However, the bond energy does not increase in direct proportion to the number of electrons involved in the bond; that is, a double bond is not twice as strong as a single bond.

In the context of polar covalence theory,[16] the bond energy is calculated independently of the actual bond order by using the actual (shortened) bond distance; the smaller bond distance, in itself, accounts for some of the additional bond strength of double bonds compared to single bonds. Then, the calculated bond energy is multiplied by a factor of 1.488 to determine the estimated bond energy for the double bond. If the bond in question is a triple bond, the multiplication factor is 1.787. The coefficients (β) were found by Sanderson empirically by considering a large number of compounds, but seem to hold regardless of the atoms involved in the bond; that is, the same multiplication factor (1.787) is used for the triple bond energies of both N_2 and acetylene (H–C≡C–H).

As a simple example, consider the calculation of the N≡N bond energy in dinitrogen. The bond distance is 1.098 Å, and the compound is purely covalent; that is, $t_c = 1$. Since the bond under consideration is a triple bond, the fully unweakened value of the nitrogen covalent bond strength is used, $E''' = 94.9$ kcal/mol (Section 3.6.5). In this case, the β-factor for the triple bond = 1.787. One only needs to calculate the covalent bond energy, E_c.

$$E_c = \beta \frac{R_c}{R_0} \sqrt{E_{NN} E_{NN}}$$

$$E_c = \beta \frac{R_c}{R_0} E_{NN} = \beta \frac{1.468\ \text{Å}}{1.098\ \text{Å}} 94.9 = \beta \times 126.9\ \text{kcal/mol}$$

The value of 126.9 kcal/mol is the "single bond" energy, calculated for equilibrium bond length of the N_2 molecule. This value is then multiplied by the empirical factor 1.787 (β) to derive the triple bond energy:

$$126.9 \frac{kcal}{mol} \times 1.787 = 226.7\ kcal/mol$$

This polar covalence theory calculated value for the dinitrogen bond energy is in excellent agreement (<0.5% error) with the experimental bond energy of 225.9 kcal/mol.

As a second example, let us calculate the bond energy in carbon monoxide, CO. Like N_2, this molecule contains a triple bond, but unlike N_2, CO is polar. The experimental bond distance (R_0) in CO is 1.128 Å, whereas the sum of the covalent radii (R_c, single bond) is 1.474 from the values in Table 3.2. Again, because it is a triple bond we use the fully unweakened value (E''') for E_{OO}.

$$E_c = \beta \frac{R_c}{R_0} \sqrt{E_{CC} E_{OO}}$$

$$E_c = \beta \frac{1.474}{1.128} \sqrt{85.4 \times 104} = 1.787 \times 1.307 \times 94.2$$
$$= 220.1\ \text{kcal/mol}$$

The covalent bond energy alone is an astonishing 220.1 kcal/mol in carbon monoxide! This will mix with an even stronger ionic bond energy.

$$E_i = \beta \cdot 332/R_0 = 1.787 \cdot 332/1.128 = 525.9\ \text{kcal/mol}$$

Since the system is polar, we need to calculate the coefficients t_i and t_c. This can be done by calculating the charge on the carbon and oxygen. The molecular electronegativity in CO (χ_M) is found first.

$$\chi_M = \sqrt{\chi_C \chi_O} = \sqrt{2.746 \times 3.654} = 3.168$$

Then, the charges (δ) on the atoms are found. Again, all the values are shown in Table 3.2. (We could have assumed, since there are only two atoms and the molecule is neutral overall, that the oxygen would have the same charge as carbon with opposite sign, but we did the δ_C calculation as a check.)

$$\delta_C = \frac{\chi_M - \chi_C}{\Delta\chi_C} = \frac{3.168 - 2.746}{2.602} = +0.162$$

$$\delta_O = \frac{\chi_M - \chi_O}{\Delta\chi_O} = \frac{3.168 - 3.654}{3.001} = -0.162$$

The value of t_i is then:

$$t_i = \frac{|\delta_C| + |\delta_O|}{2} = \frac{|+0.162| + |-0.162|}{2} = 0.162$$

The value of $t_c = 1 - t_i = 1 - 0.162 = 0.838$. The polar covalence equation (Eq. 3.17) is then:

$$E_{CO} = t_c E_c + t_i E_i = 0.838 \times 220.1 + 0.162 \times 525.9$$
$$= 269.6\ \text{kcal/mol}$$

The experimental bond energy of CO is 256 kcal/mol. The value calculated by this method of 269.6 kcal/mol is a respectable but unusually large error for polar covalence theory at ~5%.[h] Nevertheless, it illustrates again that, in general, more polar bonds are stronger. Carbon monoxide and dinitrogen are isoelectronic species; however, carbon monoxide has the stronger bond by ~30 kcal/mol. In some cases, students find this surprising, as CO is much more reactive than N_2; however, reactivity is not always related to thermodynamics, as we will discuss later. Kinetic factors play an important role.

3.7 Molecular Orbital Description of the 2c2e Bond: A Brief Review (Key Concept)

We have discussed a couple of different methods for looking at the 2c2e bond already in this chapter. The way the bond in a molecule like HCl is described according to Lewis involves taking two radicals, one of hydrogen and one of chlorine,

[h] Much of the error probably resides in the charge scheme, which is likely where the majority of the error is derived from in most cases. The charge between the C and O in carbon monoxide is probably a bit larger than previously estimated. In addition, CO can be reasonably thought to contain a dative bond, which is known to flummox this simple charge scheme.

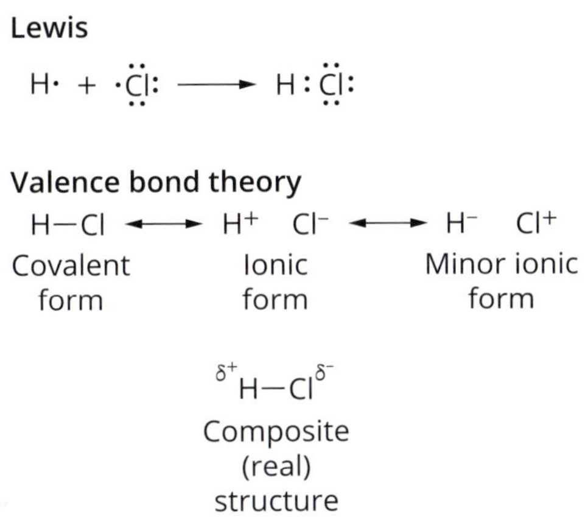

Fig. 3.19 Lewis and simplified valence bond descriptions of the bonding in HCl.

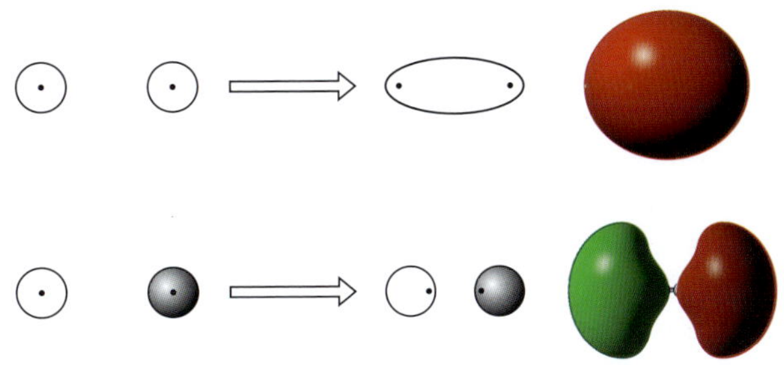

Fig. 3.20 Electron distribution in the bonding and antibonding orbitals of H_2. Dots are used to show the nuclear positions.

which share electrons to make a bond between the two nuclei (Fig. 3.19).

In valence bond theory, the bond in the same molecule may be described using resonance between covalent and ionic contributions. These two forms combine to give a composite structure based on the relative contributions from the different resonance forms. The composite structure, which is a representation of the real electronic structure of the molecule, includes some covalent character to the bonding, as well as an ionic contribution due to the partial charges on the atoms.

Another popular approach for describing chemical bonds is molecular orbital (MO) theory. A common practice is to construct MOs from the atomic orbitals of the individual atoms in the molecule, which leads to the so-called linear combination of atomic orbitals (LCAO) approximation to arrive at the molecular orbitals.

For example, methane, CH_4, is described in the LCAO-MO method by considering the interaction between the atomic 1*s*-orbitals of the four hydrogens and the atomic orbitals of carbon ($1s$, $2s$, $2p_x$, $2p_y$, $2p_z$). Obviously, there are many simultaneous interactions to consider even in as simple a molecule as methane, but the process is greatly simplified by taking advantage of the underlying symmetry of the molecule, which is most easily done using group theory. An extremely powerful mathematical system particularly useful in chemistry, group theory is used in analyzing and understanding symmetry and will be introduced in Chapter 5.

A simple way to represent the essence of the LCAO MO method is to consider the hydrogen molecule, which has only two nuclei and two electrons. The atomic orbitals that are relevant to the bonding are the two 1*s*-orbitals of each hydrogen atom. Atomic orbitals are wave-like objects (they are literally *wavefunctions*, Chapter 2) and, just like any other object with wave-like properties, they can superimpose with each other. A fundamental feature of oscillations is that the combination of two waves can always occur in one of two ways: a superposition is said to be *constructive* if the two waves are "in-phase" (their amplitudes match), or *destructive* if they are "out-of-phase" (their amplitudes cancel). This leads to two possible combinations for the sum of two 1*s* orbitals, depending on whether they have identical or opposite phases, as shown in Fig. 3.20.

As seen in Fig. 3.20, the in-phase combination of the two 1*s* orbitals leads to an increase in electron density between the two nuclei, which in turn leads to shielding of the Coulombic repulsion between the two positive nuclei. Furthermore, the formation of a larger molecular orbital helps to reduce the kinetic energy of the electrons. This effect plays a dominant role in lowering the energy of the in-phase combination of atomic orbitals and can be understood simply by invoking the Heisenberg Principle (Chapter 2). By the Heisenberg Principle, the more delocalized the wavefunction of an electron—the less accurately its position is defined—the lower its momentum needs to be. For these reasons, such in-phase combinations of atomic orbitals are often energetically favorable; that is, more stable than the electrons in individual atomic orbitals on the separate atoms. This more stable combination is called a "bonding" molecular orbital, as it is responsible for the stability of the chemical bond between atoms.

By contrast, the out-of-phase combination of the two atomic orbitals pushes the electron's density to the outside of each nuclei away from the bonding hydrogen, which clearly leads to unfavorable nuclear–nuclear Coulombic repulsions. Furthermore, the presence of a nodal plane in the molecular orbital (a plane across which the phase of the molecular orbitals changes sign and where the probability of finding the electron is strictly zero) implies that the kinetic energy of the electrons is greatly enhanced in such a combination.[i] Electrons in this orbital are less energetically stable than the two separate atomic orbitals before the interaction. Because of these factors, out-of-phase MOs are referred to as "antibonding" MOs, to indicate that they do not favor the stability of the chemical bond.

[i] Recall from Chapter 2 and basic quantum mechanics that kinetic energy is related to the first-derivative of a wavefunction; that is it is the "slope". When a wavefunction changes sign, it necessarily implies that its "slope" must be large, as it has to go from a positive phase value to a negative phase value (a large change). The more nodes a wavefunction has, the higher the kinetic energy. This is a very general feature of quantum mechanical wavefunctions, which is true for the particle-in-a-box, the harmonic oscillator (vibrations), or for orbitals (hence a *p*-orbital is always higher in energy than an *s*-orbital of the same principal value, because of its nodal plane).

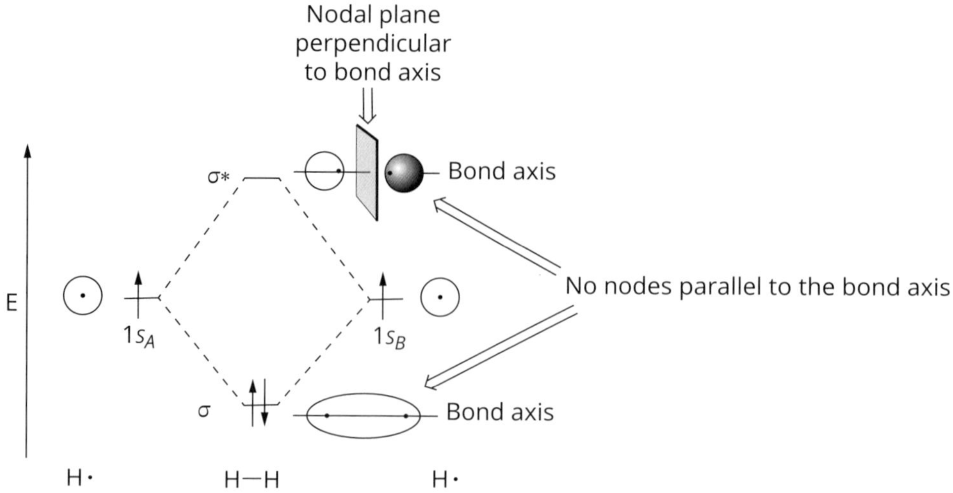

Fig. 3.21 MO diagram for H_2. Dots indicate approximate nuclear positions.

In an LCAO MO diagram, we often draw the atoms (or groups of atoms) being combined on the left and right sides of the diagram, and in the middle the MO combinations of those atomic (or fragment) orbitals. Although it is often not designated as such explicitly, the vertical direction represents the orbital energy.

The MO diagram for H_2 is shown in Fig. 3.21. The two 1*s*-orbitals on the two hydrogens, labeled A and B, start off singly occupied and with identical energy. They can combine in-phase to give what is called the σ-bonding molecular orbital, just labeled σ, or out-of-phase to yield the so-called σ-antibonding MO, labeled σ*. (Antibonding orbitals are always represented with an asterisk.) Electrons that occupy the bonding MO are lower in energy than the electrons in the atomic orbitals, and any electrons in the antibonding MO would be higher in energy than the atomic orbitals. Just like any other orbital (atomic or molecular), these molecular orbitals can accommodate two electrons at most. Since there are two electrons (one from each hydrogen atom), both are assigned to the σ-bonding MO, and the σ is empty. It is this stabilization of the electron energies relative to the isolated hydrogen atoms that leads to the formation of H_2.

In MO theory, bonds are often described based on their symmetry. Specifically, the bonds are designated by the number of nodal planes that run parallel to the axis between the two bonding atoms. A node is a region of zero electron density in the wavefunction where the phase changes on either side of the plane. In the example shown previously, there are no nodal planes for the σ-orbital, but the σ*-orbital has one nodal plane perpendicular to the bond (Fig. 3.21).

σ-bonds are always "cylindrically symmetric" from the chemical bond point of view; this simply means that the orbital looks the same all around the bond. The same type of symmetry would occur if we were to superpose two *p* orbitals on two atoms, each pointing at each other along the chemical bond; this combination would also lead to σ and σ* molecular orbitals (Table 3.4). The situation would be different if we were to superpose two *p* orbitals, but oriented perpendicular to the chemical bond; this combination is no longer cylindrically symmetric, as the phase of the orbital goes from positive to negative around the bond, crossing a nodal plane that includes the chemical bond. Such nodal planes that include the chemical bond are very different from the nodal planes we saw previously for the σ-antibonding orbitals, which are always perpendicular to the bond (they "cut" the bond in half). Molecular orbitals that contain a single nodal plane that includes the chemical bond are called π-bonds (Table 3.4). Both σ- and π-bonding can also arise from the interaction of two *d* orbitals, provided that they are properly oriented (see Table 3.4). A third type of bond can be formed by *d*-orbitals whenever they are aligned face-to-face (Table 3.4), yielding an MO that has two nodal planes including the chemical bond; this is referred to as a δ-bond.

Note that the Greek letters used to designate MOs have a relationship with the Latin letters used for atomic orbitals. For example, *s*-orbitals have no planar nodes, like a σ-bond has no nodal planes containing the bond axis; a *d*-orbital has two planar nodes, like a δ-bond has two nodal planes parallel to the bond axis.

Note that the antibonding orbitals associated with σ-, π-, and δ-bonds have the same number of nodal planes parallel to the bond axis as their bonding analogs. However, they do have one more nodal plane, which is perpendicular to the bond axis. In general, orbital energy rises with the total number (perpendicular and parallel) of nodal planes in a molecular orbital due to the kinetic and Coulombic energy factors discussed previously.

In addition to bonding and antibonding character, molecular orbitals can be nonbonding; typically, nonbonding MOs are mostly centered on one atom, or on atoms that are far apart and that are not interacting with each other. As a result, these nonbonding MOs often have energies similar to the atomic

Table 3.4 Some of the possible atomic orbital combinations to make σ-, π-, and δ-bonds and the number of nodes parallel to the bond axis.

Bonding orbital	Antibonding orbital	Number of nodal planes that include the bond axis	Type of bond
$s + s$ 	$s - s$ 		
$s + p$ 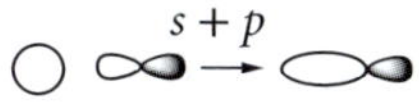	$s - p$ 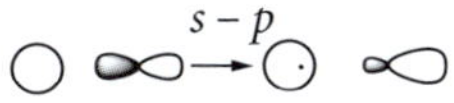		
$p + p$ 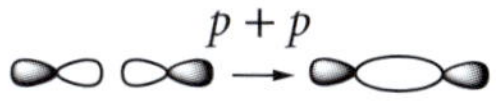	$p - p$ 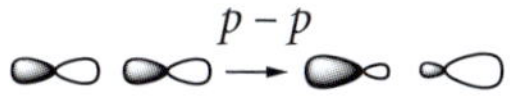		
$s + d$ 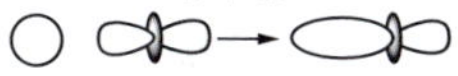	$s - d$ 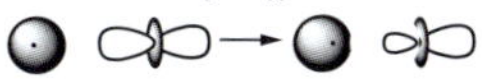	0	σ (σ*) "sigma"
$p + d$ 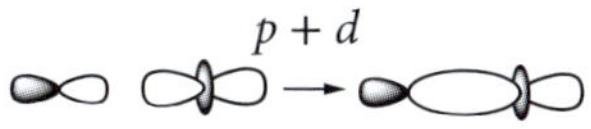	$p - d$ 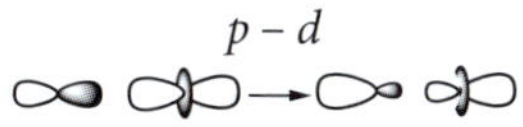		
$d + d$ 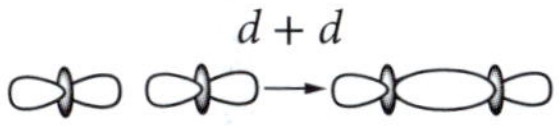	$d - d$ 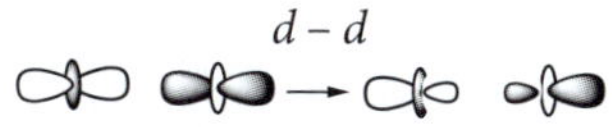		
$p + p$ 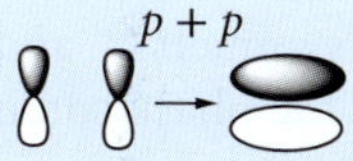	$p - p$ 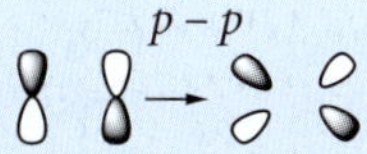		
$p + d$ 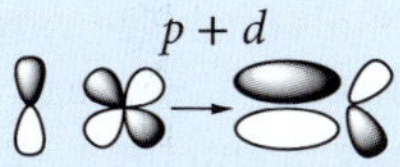	$p - d$ 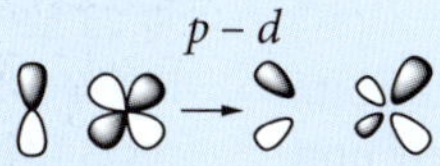	1	π (π*) "pi"
$d + d$ 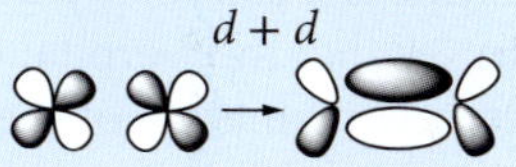	$d - d$ 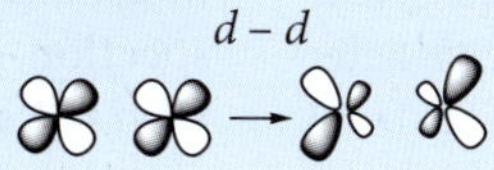		
$d + d$ 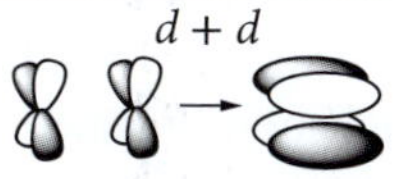	$d - d$ 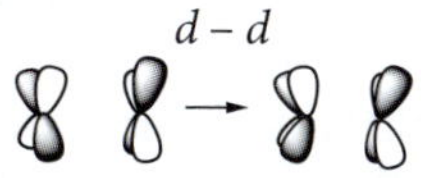	2	δ (δ*) "delta"

orbitals that contribute to the LCAOs. One common type of nonbonding MO is a lone pair of electrons. If one examines the MO diagram of HCl using the LCAO method (Fig. 3.22), the σ-bond between hydrogen and chlorine is formed by the *1s*-orbital on hydrogen and by a mixture of an *s*- and *p*-orbital on chlorine. This mixture of orbitals is called, after valence bond theory, a hybrid orbital. For our purposes here, we simplify things by assuming the bond to hydrogen to be made using a pure *p*-orbital on chlorine. (We will see later when discussing VB theory, that this is a relatively good assumption.) If we designate the H–Cl bond axis as the *z*-axis, then the p_z orbital is the one involved in the σ-bond.

The energies of the orbitals can be assigned, to a first approximation, based on the electronegativity of the atoms or using valence orbital energies. The electronegativities on several different scales are shown in Table 3.1, and Appendix K lists valence orbital energies from ionization energies and from calculations on the atoms. In our MO diagram, the hydrogen *1s*-orbital is drawn higher in energy than the chlorine orbitals. The higher electronegativity of chlorine stabilizes its electrons. Note that we typically only show the valence electrons in an MO diagram; core electrons typically have very poor overlap with other atoms and are very energetically stabilized.

In addition to the bonding orbital formed, a high energy antibonding orbital is also generated, labeled σ*, where the phases of the chlorine $3p_z$-orbital and the hydrogen *1s*-orbital have the opposite phase in the region of overlap. As mentioned previously, the 3*s*-orbital on chlorine has the proper symmetry to participate in the σ-bond; however, for the moment, we will neglect this interaction and postpone to Chapter 6 the

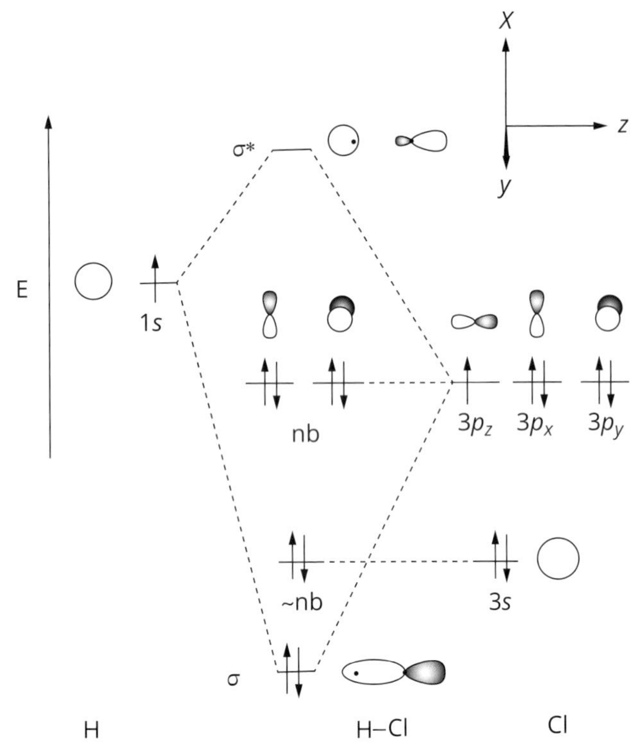

Fig. 3.22 MO diagram for H–Cl. Orbitals below the valence set on Cl are not shown. The dots in some of the orbitals for HCl are meant to give an approximate position for the nuclei.

discussion of 3-orbital interactions. As a result, the 3*s*-orbital remains centered on the chlorine atom as approximately nonbonding, labeled "~nb." (In Chapter 8, we will discuss why this assumption that the lone pair is largely *s* in character and that the H–Cl bond is formed largely by p_z is not far from the truth!)

The remaining two valence *p*-orbitals on chlorine, p_x and p_y, have no way of making a productive interaction with the *s*-orbital on hydrogen. As shown in Fig. 3.23, attempts to mix the hydrogen *s*-orbital with either p_x or p_y leads to an interaction that is as much in-phase as it is out-of-phase. Consequently, the interaction between these orbitals is rigorously nonbonding, labeled "nb." Nonbonding orbitals are typically drawn so that there is no change in their energy versus the starting atomic orbitals. However, this is an approximation. Differences in shielding on the atoms in molecules versus shielding in the isolated atoms or electron correlation effects may lead to slight stabilization or destabilization of the nonbonding orbitals relative to the atomic orbitals.

Under the linear combination of atomic orbitals (LCAO) approach to molecular orbital (MO) theory, bonds between

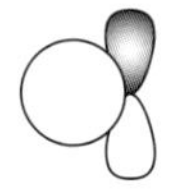

Fig. 3.23 1s-orbital of hydrogen with p_x-orbital on chlorine in HCl produces a net nonbonding interaction.

two atoms can be viewed as a combination of atomic orbitals to form the MOs. It should be kept in mind that the LCAO method is an approximation. It assumes that the MO shapes can be derived from the shapes of the atomic orbitals.

As one final example here, let us examine water. The two hydrogens in water can be regarded as substituents on the oxygen. In MO theory we often decide on the MOs for the ligands and then add them to the central element in cases like this, which often will be seen in the examples throughout the text. So, do we know what the MOs for two hydrogens look like?

The only difference between the molecular orbitals of H_2 and the orbitals for the two hydrogens in water is the distance between the two hydrogens. In H_2, the hydrogens are a distance apart that maximizes their interaction with each other; in H_2O, the H–H distance is determined by the energetics of arranging the orbitals around oxygen, which makes the hydrogens much farther apart.

Just like in H_2, we will have two MOs from the two valence orbitals of the hydrogens in water. Unlike in H_2, because they do not overlap to any significant extent, the MO with the same phase and the MO with different phases on the two hydrogens will have essentially the same energy; because neither of these combination is really "bonding" or "antibonding" in a chemical sense (there is no H–H bond in water!), we will call these two combinations "symmetric" and "antisymmetric" respectively, in relation to the phase combination. These are shown on the left in Fig. 3.24.

The valence orbitals on oxygen are then shown on the right in Fig. 3.24. Note that the axes are drawn so that the *z*-axis bisects the H–O–H angle, and the *x*-axis is perpendicular to the molecular plane (this is a commonly used convention). By inspection, one can see that the *s*-orbital on oxygen can overlap with the symmetric combination of the orbitals on hydrogen, and that p_y can overlap with the antisymmetric combination. These give two delocalized σ-bonding orbitals, the MO versions of the two single bonds drawn as lines between O and H in the Lewis structure.

The next orbital higher in energy from the σ-bonds is labeled as "~nb", approximately nonbonding, and is derived from the 2p_z-orbital of oxygen. It is approximately nonbonding because it is fairly close in energy to its atomic origins on oxygen but does have proper symmetry to overlap with the symmetric combination of hydrogen orbitals. Suggestively, the orbital is drawn slightly lower in energy than the atomic *p*-orbitals of oxygen, giving it some bonding character. These types of (3-orbital) interactions will be discussed in more detail in Chapter 6.

The next orbital higher in energy labeled "nb" is strictly nonbonding and is derived from the 2p_x-orbital on oxygen. This orbital can have no net bonding to the hydrogens.

Finally, the two highest-energy orbitals in Fig. 3.24 are the σ-antibonding, σ*, orbitals in water and are unoccupied.

Both valence bond and molecular orbital theory are developed in more detail in later chapters, and will be applied to a variety of different systems. The VB and MO theories arrive at the same overall answer concerning the electron distributions, but can bring different insights to particular problems.

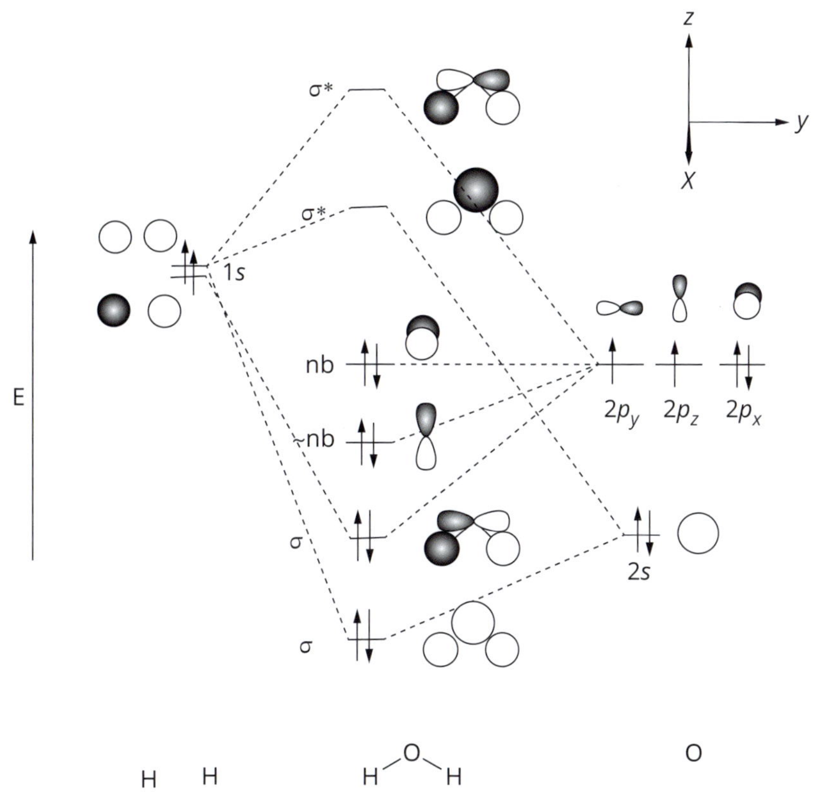

Fig. 3.24 MO diagram for H_2O. The orbital below the valence set on O are not shown.

3.8 Defining 2c2e Bonds: Heterocovalent versus Dative Bonds (Key Concept)

There is a plethora of bond types found in inorganic chemistry. Two of the most common are heterocovalent and dative bonds. The heterocovalent bond has been discussed in some detail in the previous sections of this chapter. It is simply a covalent bond, a bond with some covalent character, between two atoms or groups that are not identical in composition. The definition for the heterocovalent bond particularly relevant to this section is a "bond in which there is a sharing of electrons between the bonding atoms, so that in the resonance form where the electrons are equally shared there are no formal charges across the bond."

For example, if we bring together a hydrogen atom and a chlorine atom we can make a heterocovalent bond. In the resonance form where those electrons are equally shared, there are no formal charges on the hydrogen and chlorine (Fig. 3.25, top left). There is some polarity in the bond, however, and we can consider an unequal sharing of electrons in an ionized resonance form (top right).

Generally speaking, any bond between two different atoms or groups will have some polarity. This can be represented by two different resonance forms (Fig. 3.25)—one with a purely covalent bond (**A**) and another where the bond of interest is ionized (**B**). In a heterocovalent bond, there is no charge on X and Y in the resonance form where the electrons are equally shared, form **A**.

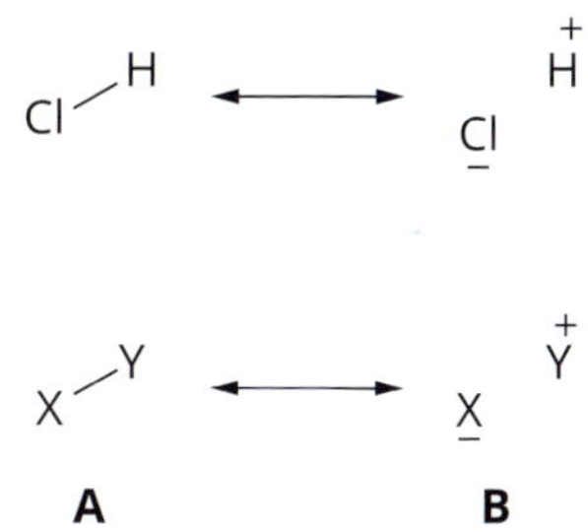

Fig. 3.25 Resonance forms for polar covalent bonds.

Another type of 2c2e bond common in inorganic chemistry is the dative bond (also sometimes called "coordinate" or donor–acceptor bond). The word "dative" (from the latin *dativus*, "given") is used in a grammatical sense to mark the indirect object of a verb. In a sentence like "He gave the gift to me", the word "me" is the dative pronoun because it receives the action of the verb.

In chemistry, a dative bond is one where we can think of both electrons for the 2-electron bond originating from only one of the atoms. Those electrons are then "given" to the receiving atom or group to make the bond. This contrasts with a purely covalent bond where each side donates one electron to the bond, like two hydrogen atoms donating one electron each to make the 2-electron bond in H_2 (Fig. 3.21).

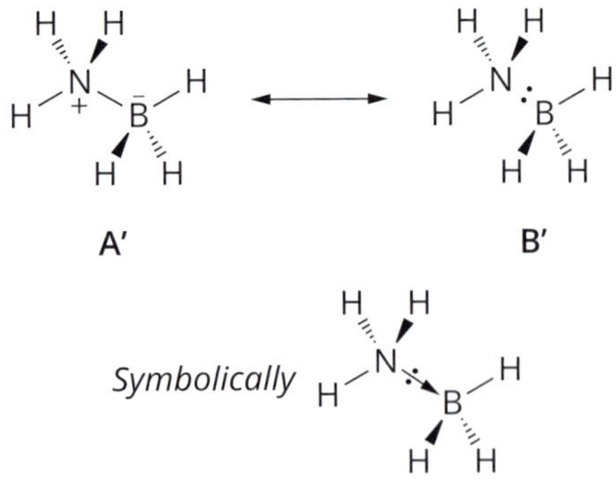

Fig. 3.26 Schematic representation of bonding in amine borane.

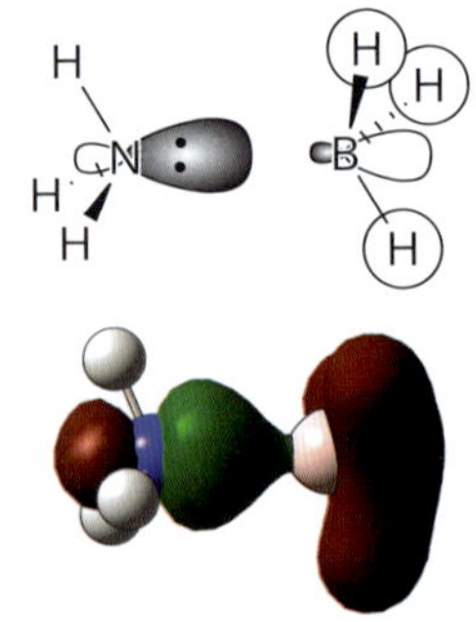

Fig. 3.27 Bonding orbital for the dative interaction between H_3NBH_3 showing the lone pair is still largely on nitrogen (blue).

A fairly simple example of a compound with a dative bond is the ammonia–borane complex (H_3NBH_3, Fig. 3.26). Ammonia (NH_3) has a lone pair of electrons, and borane (BH_3) has an empty *p*-orbital. The new bond between the two can be illustrated with the following two resonance forms. On the left, in form **A′**, there is an equal sharing of electrons between the two atoms, and, across the new N–B bond, new formal charges are generated (positive on nitrogen and negative on boron). In form **B′**, we consider the electrons to reside on the nitrogen with no new charges generated in this unequal sharing. Symbolically, we distinguish a dative bond from a covalent or heterocovalent bond as shown at the bottom of Fig. 3.26, as a lone pair on one atom or group with an arrow going to another atom or group.

Consequently, one definition that could be used for a dative bond is a "bond in which new formal and opposite charges are induced on the participants in the resonance form where the electrons are shared." In essence, there is resonance between a polar bonded form and a nonbonding form in dative bonds. Significant participation by the nonbonding form, which occurs in many compounds, can lead to dative bonds being weaker and longer than common covalent or heterocovalent bonds.

Viewed in this way, the dative bond and the heterocovalent bond are in some respects quite different or, perhaps better, two sides of the same coin. In the heterocovalent bond (Fig. 3.25), there are no charges across the bond when electrons are equally shared (**A**) and there are charges across the bond when the electrons are unequally shared (**B**). However, in the dative bond the opposite is true. In the dative bond (Fig. 3.26) there are formal charges across the bond when electrons are equally shared (**A′**), and no charges across the interatomic axis when the electrons are unequally shared (**B′**).

If one examines the nature of the dative interaction using molecular orbital theory, the stabilization of the electron density and even the shape of the lone pair orbital does not change a great deal for many dative interactions. In the H_3NBH_3 example, it can be clearly seen (Fig. 3.27) from the nature of the orbital for the σ-bonding interaction between the nitrogen (blue) and boron (pink) that relatively little electron density, compared to many covalent bonds (see Fig. 3.13), is being shared by the two nuclei. The "lone pair" still resides largely on the more electronegative nitrogen. In a very polar heterocovalent bond, the electron density resides largely on one of the atoms in the bond, then the bond is very polar, which may give the bond a similar picture to a dative bond like that shown in Fig. 3.27, but the polar bond will have a relatively high bond energy whereas the dative bond will have a low bond energy when there is little sharing.

To illustrate the BDE difference between dative and covalent bonding, we can compare isoelectronic systems. In an isoelectronic system, as the name implies, there are the same number of electrons, but the identity of some of the elements have changed. For example, the ammine borane complex, H_3NBH_3, has one electron from each hydrogen, five electrons from boron, and seven electrons from nitrogen. Consequently, H_3NBH_3 has $(1 \times 6)_H + (5)_B + (7)_N = 6 + 5 + 7 = 18$ electrons. An isoelectronic compound of H_3NBH_3 is ethane, H_3CCH_3, which has $(1 \times 6) + (2 \times 6) = 18$ electrons.[j]

Compare the amine borane complex, H_3NBH_3, with the isoelectronic, covalent compound ethane, H_3CCH_3. The C–C bond in ethane is 1.53 Å, about 0.13 Å shorter than the bond in the amine borane complex. In addition, the covalent C–C bond in ethane is almost three times stronger at 89 kcal/mol, versus 31 kcal/mol in the isoelectronic dative bond. From a BDE perspective, unequal sharing in heterocovalent σ-bonds increases ionic character, which can greatly increase the BDE (Section 3.6.1). Unequal sharing in dative bonding leads to no bond at all between the atoms (Fig. 3.26).

The dative bond as described previously is unlikely to be a new concept for most students; these are Lewis acid–base adducts. However, there are significant differences between seemingly similar Lewis acids like Ca^{2+} and Ba^{2+}, along with significant differences between seemingly similar Lewis bases like NH_3 and PH_3. Dative interactions can be, from an electronic structure standpoint, quite complex and are often mixed with a variety of other bond types.

[j] Compounds also may be "valence isoelectronic", meaning they have the same number of valence electrons. For example, H_3PBH_3 is valence isoelectronic with H_3CCH_3 since the replacement of phosphorus with nitrogen does not change the number of valence electrons, even though it changes the total number of electrons. Compounds that are isoelectronic are also valence isoelectronic, like H_3NBH_3 and H_3CCH_3. However, not all compounds that are valence isoelectronic are isoelectronic; for example, H_3PBH_3 and ethane are valence isoelectronic but not isoelectronic.

3.9 Bond Dissociation Enthalpy in Detail (Advanced Topic)

In the context of electronegativity and valence bond theory, the bond dissociation enthalpy (BDE) was introduced, and here it will be developed in more detail. The BDE can be defined as "the energy required to stretch a bond of interest from the lowest vibrational and electronic state to a distance where the atoms (or groups) no longer interact, resulting in homolytic cleavage of the bond."

The BDE as described here pertains to covalent and heterocovalent bonds as described in Section 3.8. Dative bonds might be considered to heterolytically cleave at long distance; as a result, their bond energies are associated with the Lewis acid–base interaction between the, often closed shell, atoms or groups involved.

The nature of the BDE is illustrated in Fig. 3.28, where the structure on the left is the molecule in the ground electronic and vibrational state. Moving to the right, the distance between A and X is increasing, which causes a concomitant increase in the energy of the molecule until the atoms are no longer interacting and one electron from the single bond resides on A and the other on X. The BDE is the difference in energy between the lowest-energy structure and the bond-stretched derivative where A and X no longer interact. Recombination of the radicals to reform molecule AX often has a small barrier. There is sometimes a small energetic difference between the energy required to break the bond at the transition state, $\Delta H^{\ddagger}$, and the BDE, the energy difference between the ground state and the fully separated radicals.

The energy of the transition state can be understood most easily starting from the dissociated atoms and bringing them together. There are two repulsive and attractive forces at work. As the two dissociated atoms are brought together there is an initial electron–electron repulsion that can occur, raising the energy. As the atoms get closer still, the electrons of one atom begin to "see" the nucleus of the approaching atom and the energy begins to drop.

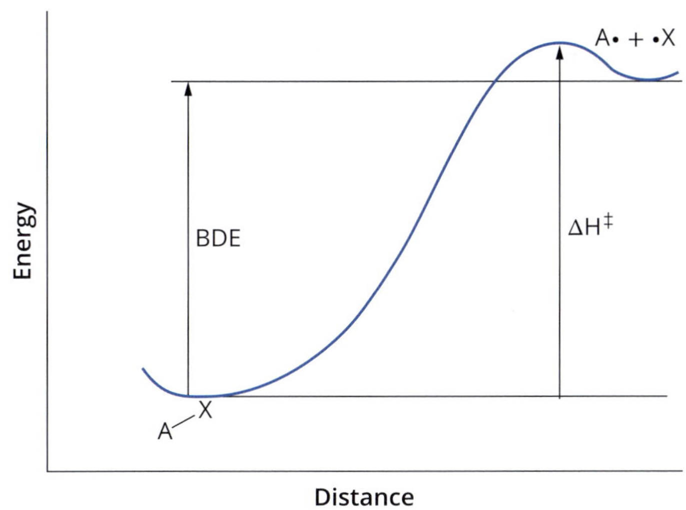

Fig. 3.28 Plot of bond energy versus A–X distance.

In order to understand bond dissociation in more detail, it is necessary to examine how electronic and vibrational energies change with internuclear distance (R). The most common function used for this is the Morse Potential, which has the advantages of being exactly solvable and correctly describing the number of bound vibrational frequencies as finite.[50]

The Morse Function is shown in Eq. 3.27, where $V(R)$ is the energy as a function of the bond distance, and R_e is the equilibrium internuclear distance. In the equation, μ is the reduced mass, $\mu = m_A m_X/(m_A + m_X)$, where m_A is the mass of atom A. The factor D_e is the dissociation energy from the electronic ground state. Finally, ν_e is the fundamental frequency. The vibrational eigenvalues are directly solvable, and the maximum bound vibrational frequency can be estimated with $\nu_{max} \sim 2D_e/\nu_e$.

$$V(R) = D_e(1 - e^{-\alpha(R-R_e)})^2 \quad \alpha = \nu_e\sqrt{\mu / 2D_e} \qquad 3.28$$

This can be applied to simple diatomics easily, and the result for H_2 is shown in Fig. 3.29. The outside contour of the potential well is the electronic energy of the molecule. The electronic energy is the energy of the system in the absence of vibrations. The bottom line internal to the potential well (collinear with the dotted line) labeled $\nu = 0$ is the ground-state vibrational frequency, ν_e. The excited-state vibrational frequencies are then labeled as $\nu = 1, 2, 3, 4, \ldots$ up to where the hydrogen atoms would no longer interact, which occurs at 0 cm^{-1} on the scale of Fig. 3.29.

The difference in energy between the electronic energy, the energy with no atomic motion, and the fully dissociated atoms is labeled as D_e in Fig. 3.29. Naturally, the hydrogen molecule does have vibrational energy. The vibrations increase the energy of the molecule over the electronic energy. Consequently, in order to accurately describe the energy required to break the bond in the hydrogen molecule it is necessary to include the vibrational states. The energy required to be added to hydrogen in order to break the bond from the vibrational ground state is labeled D_0, the bond dissociation enthalpy. For H_2, this is 36,500 cm^{-1} (104.2 kcal/mol, Table 3.2).

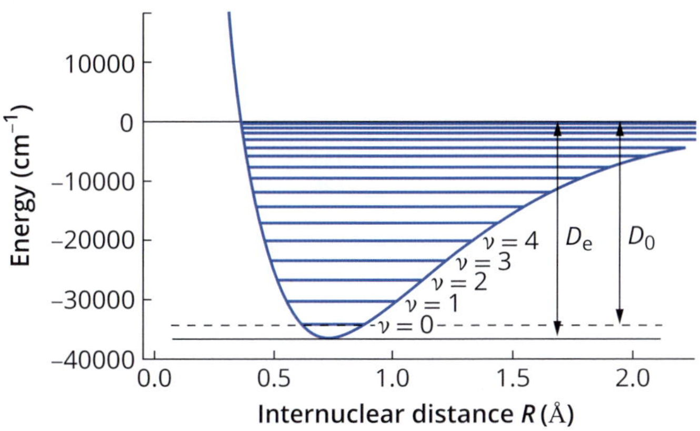

Fig. 3.29 The Morse Function for H_2, showing the vibrational levels.

The relatively small difference in energy between the electronic energy, and the energy of the ground-state vibrational frequency is called the "zero-point energy", which is equal to $|D_e—D_0|$. For reference, some experimental bond dissociation enthalpies are presented in Appendix B.

3.10 Bond Order (Expanded Coverage)

Bond order is a basic concept in chemistry that is necessary for understanding reactivity and properties of molecules. Bond strengths, for example, are intimately tied to bond order in many compounds. However, because of the very large number of possible atom combinations that may be involved in single, double, triple, and so on, bonds in inorganic compounds, evaluating bond orders can quickly become complex.

Molecular orbital theory can be used in one definition of bond order, where a single bond would involve two electrons in a bonding orbital and no electrons in an antibonding orbital, for example. However, not all bonds by this definition are created equal, and an orbital may be technically bonding but involve a very weak interaction between the connecting atoms.

Some of the methods developed to quantify the bond orders in compounds, which vary from simple calculations that can be done by hand to quantum-mechanical techniques, are discussed in what follows. The by-hand methods invariably involve analysis of bond distances, and one should be careful to compare only similar compounds in similar oxidation states using such methods.

3.10.1 Simple Estimation of Bond Orders (Expanded Coverage)

Pyykkö and coworkers have developed estimates of covalent bond radii (Fig. 3.30) for element numbers 1–118 (hydrogen to oganesson). The covalent radii are classified as single, double,

1	2	3	4	5	6	7	8	9	10	11	12	13	14	15	16	17	18
1 H 32 - -																	2 He 46 - -
3 Li 133 124 -	4 Be 102 90 85				Z	Radius, r_n: r_1 r_2 r_3		Symbol				5 B 85 78 73	6 C 75 67 60	7 N 71 60 54	8 O 63 57 53	9 F 64 59 53	10 Ne 67 96 -
11 Na 155 160 -	12 Mg 139 132 127											13 Al 126 113 111	14 Si 116 107 102	15 P 111 102 94	16 S 103 94 95	17 Cl 99 95 93	18 Ar 96 107 96
19 K 196 193 -	20 Ca 171 147 133	21 Sc 148 116 114	22 Ti 136 117 108	23 V 134 112 106	24 Cr 122 111 103	25 Mn 119 105 103	26 Fe 116 109 102	27 Co 111 103 96	28 Ni 110 101 101	29 Cu 112 115 120	30 Zn 118 120 -	31 Ga 124 117 121	32 Ge 121 111 114	33 As 121 114 106	34 Se 116 107 107	35 Br 114 109 110	36 Kr 117 121 108
37 Rb 210 202 -	38 Sr 185 157 139	39 Y 163 130 124	40 Zr 154 127 121	41 Nb 147 125 116	42 Mo 138 121 113	43 Tc 128 120 110	44 Ru 125 114 103	45 Rh 125 110 106	46 Pd 120 117 112	47 Ag 128 139 137	48 Cd 136 144 -	49 In 142 136 146	50 Sn 140 130 132	51 Sb 140 133 127	52 Tc 136 128 121	53 I 133 129 125	54 Xc 131 135 122
55 Cs 232 209 -	56 Ba 196 161 149	La – Lu	72 Hf 152 128 122	73 Ta 146 126 119	74 W 137 120 115	75 Re 131 119 110	76 Os 129 116 109	77 Ir 122 115 107	78 Pt 123 112 110	79 Au 124 121 123	80 Hg 133 142 -	81 Tl 144 142 150	82 Pb 144 135 137	83 Bi 151 141 135	84 Po 145 135 129	85 At 147 138 138	86 Rn 142 145 133
87 Fr 223 218 -	88 Ra 201 173 159	Ac – Lr	104 Rf 157 140 131	105 Db 149 136 126	106 Sg 143 128 121	107 Bh 141 128 119	108 Hs 134 125 118	109 Mt 129 125 113	110 DS 128 116 112	111 Rg 121 116 118	112 122 137 130	113 136	114 143	115 162	116 175	117 165	118 157
		57 La 180 139 139	58 Ce 163 137 131	59 Pr 176 138 128	60 Nd 174 137	61 Pm 173 135	62 Sm 172 134	63 Eu 168 134	64 Gd 169 135 132	65 Tb 168 135	66 Dy 167 133	67 Ho 166 133	68 Er 165 133	69 Tm 164 131	70 Yb 170 129	71 Lu 162 131 131	
		89 Ac 186 153 140	90 Th 175 143 136	91 Pa 169 138 129	92 U 170 134 118	93 Np 171 136 116	94 Pu 172 135	95 Am 166 135	96 Cm 166 136	97 Bk 168 139	98 Cf 168 140	99 Es 165 140	100 Fm 167	101 Md 173 139	102 No 176 159	103 Lr 161 141	

Fig. 3.30 Pyykkö's 2009 covalent radii in pm (1 pm = 10^{-12} m) for the elements.

and triple bonds for the elements, allowing estimation of the vast majority of covalent bond distances by simple addition.

For example, the bond distance in dimethylmagnesium, $MgMe_2$, is 2.13 Å. From Pyykkö's radii, the estimated single bond distance is $r_1(Mg) + r_1(C) = 1.39$ Å + 0.75 Å = 2.14 Å. The Ti=C bond distance in $H_2C{=}TiH_2$ is calculated to be 1.80 Å, which is quite close to the estimated distance of $r_2(Ti) + r_2(C) = 1.17 + 0.67 = 1.84$ Å. It should be emphasized that these are *estimates* of the bond distances, and that covalent radii are known to not be strictly additive as they are influenced by polarity, atom deformation ability, and possibly other factors. More accurate bond distances can be calculated using equations that include electronegativity such as those by Schomaker–Stevenson,[51] Haaland's modified Schomaker–Stevenson equation,[52] and Lang–Smith.[53]

The method from Pyykkö and coworkers will give the researcher an idea of what a Ti–N double bond length, for example, would be; however, more often what one wants to know is the estimated bond order when given an experimental bond length from a technique like X-ray diffraction.

$$B_{AX} = \exp\left(\frac{-(r - r_1)}{b}\right) \qquad 3.29$$

Pauling[54] suggested a simple function to relate bond order to bond length (Eq. 3.28). The bond order between atoms A and X, B_{AX}, is a function of the equilibrium bond distance (r), single bond distance (r_1), and an empirical fitting parameter (b). The bond order then is expected to change exponentially with changes in the A–X bond distance.

Using Pyykkö's values, the single bond distance r_1 can be estimated for any combination of elements that might comprise A and X. Consequently, the only remaining unknown factor to relate bond order to bond distance is the constant b. Pauling suggested that the constant had a value of around 0.3; however, he cautioned that there "is some evidence that the constant varies with the kind of atom and with the type of bond."[54]

In Appendix C, experimental values for b have been collected. For the more than 1,500 values in the chart, the range is 0.26–0.94; however, the average value is 0.377 with a standard deviation of only 0.04. If no further information is available, 0.377 is a good estimate for b.

One method for the evaluation of the constant b is to use Pyykkö's values for the single, double, and triple bonds of interest. To illustrate, let us examine a single complex that has W–C single, double, and triple bonds studied by X-ray diffraction (Fig. 3.31).[55] The compound has the formula $W(C^tBu)(CH^tBu)(CH_2{}^tBu)$(dmpe), where dmpe is the bidentate ligand 1,2-bis(dimethylphosphino)ethane. We can find the b-constant for the W–C bonds by rearranging the Pauling bond order equation to that shown in Eq. 3.29. Then we plot $\ln B_{WC}$ (ln 1, ln 2, and ln 3) versus $-(r-r_1)$. The slope of the straight line fit forced through the origin is equal to $1/b$. For the case of the W–C bond, b is estimated as 0.344 Å using this method.

$$\ln B_{AX} = -(r - r_1) \times \frac{1}{b} \qquad 3.30$$

The experimental bond distance for the W–C single bond is 2.258(9) Å,[55] and using the Pauling equation this is estimated as having a bond order of 0.67. The experimental double and triple bond distances are 1.942(9) and 1.785(8) Å, respectively. These are estimated as having bond orders of 1.65 and 2.65 in this tungsten complex.

One might wonder why the bond orders estimated using this technique are calculated as lower than 1.0 (single), 2.0 (double), and 3.0 (triple). This is often the case and could be due to a variety of factors including steric influences that lengthen the experimental bonds, nonbonding resonance form participation, and so on. In fact, strong participation from nonbonded resonance forms, as in dative bonds, can lead to great disparity between the previous Pauling forms and more complex computational methods. For example, the W–P bonds in the ene-yne-yl complex in Fig. 3.31 using Pauling's method gives W–P bond orders of ~1; conversely, methods that look at electron density between the atoms that might be involved in those bonds (Mayer or natural bond order, Section 3.10.2) typically give much lower bond orders.

The Pauling method works well for many normal valent compounds; however, use of the previous method on many

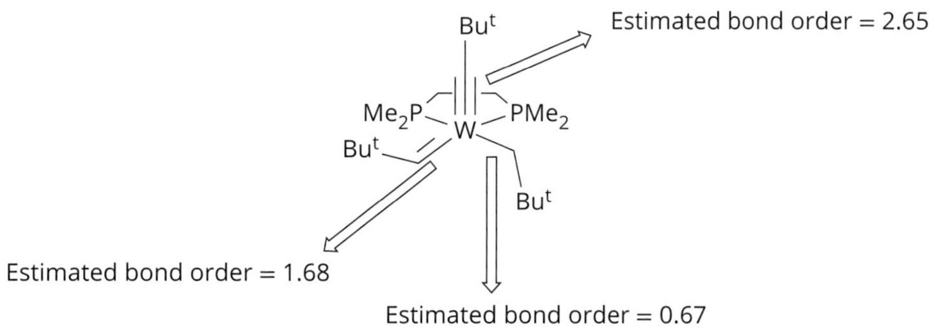

Fig. 3.31 Structure of $W(C^tBu)(CH^tBu)(CH_2{}^tBu)$(dmpe) with the estimated W–C bond orders from Eq. 3.28.

hypervalent main-group compounds produces unrealistic values for the bond order. For example, the bond order in SF_6 can be examined using this technique. The S–F distance in this purely octahedral molecule is 1.561(2) Å from gas-phase electron diffraction.[56] The *b* value for the S–F bond is found to be 0.182 Å using the Pyykkö values, which is quite small and may be erroneous. The S–F bond order using this simple formula is then found to be a quite large, 1.82, which would seem to be (and is) quite unreasonable. However, main-group compounds with hypervalent central atoms often display quite short bonds of this type. The deviations may be due to the high cationic charges on the central atoms and their effect on bond lengths. As noted previously (Section 3.6.2), changes in oxidation states can greatly affect bond distances.

Bond order should correlate with bond strength in most cases, with higher bond orders producing stronger bonds. While bond length affects both the ionic and covalent bond energies, bond length is also affected by factors other than the bond order; for example, bond length can be strongly affected by oxidation state. Bond length, bond order, and bond strength often correlate, but this is not universally true, as shown in a previous example where we calculated the BDEs of $SnCl_2$ and $SnCl_4$ (Section 3.6.2). Tin(II) chloride has longer Sn–Cl bonds of 2.42 Å relative to tin(IV) chloride, which has a bond length of 2.31 Å. However, the bonds in $SnCl_2$ are stronger (90.2 kcal/mol) than in $SnCl_4$ (75.2 kcal/mol) due to the higher polarity of the bonds in the lower oxidation state compound.

Consequently, it is important when deriving bond orders to employ *b* and r_l parameters adjusted for oxidation state whenever possible. Parameters for the Pauling bond order equation have been compiled by several researchers.[8] A tabulated list of available values from the literature is shown in Appendix C. The tabulation is not complete, and Pyykkö's values (as shown previously) can be used to estimate *b* when needed. If the value for *b* derived from Pyykkö's table seems unreasonable, due to oxidation state for example, one can simply use the typical magnitude for *b* of 0.377 as an estimate.

3.10.2 Computational Methods for Bond-Order Determination (Advanced Topic)

The exact details for how various theoretical methods calculate bond order is beyond the scope of this book, and the interested reader is referred to texts on computational chemistry.[43] However, a few general comments will be made here concerning two common methods applied to inorganic compounds and how they compare with the methods for calculating bond orders by hand, discussed previously.

Numerous methods have been applied to the computation of bond orders including an early method by Coulson for examining π-orbitals at the Hückel level.[57] Modern methods often use the density matrix of the molecule of interest as the starting point and often vary in how the density matrix is handled to arrive at the bond orders.[58] One specific method that has received significant attention with regards to inorganic systems is the Mayer Bond Order.[59–60] There are stand-alone programs, such as Mayer's own BORDER,[61] that can take the output from some computational packages—for example, Gaussian—and use that output to calculate bond orders using Mayer's procedure. In addition, Gamess-UK, for example, has the Mayer procedure as an option internal to the code.

Some quantum-mechanical methods, including Mayer's, have a basis set dependence for the computed bond orders. For example, the S–N bond order in S_4N_4 has been calculated by using the density matrix from density functional theory with B3LYP and six different basis functions.[58] The bond order varied from 0.91 to 1.35 with basis set for the molecule. Similarly, the C–H bond order in CH_4 computed from density matrices at the Hartree–Fock level varied from 0.86 to 1.00 with the ten different basis sets examined.[43]

An alternative method for the calculation of bond order uses natural resonance theory (NRT) by Weinhold and coworkers, which is part of the natural bond orbital (NBO) software package.[6] NBO can be added to the source code or comes standard with the code of several common software packages; for example, ADF (Scientific Computing & Modeling N.V.), GAMESS-US (Ames Laboratories, Iowa State University), Gaussian (Gaussian, Inc.), Jaguar (Schrodinger, Inc.), Spartan (Wavefunction, Inc.), and Turbomole (Turbomole GmbH).

The NBO–NRT approach is in many ways unique in its methodology in that it uses the density matrix to identify particular Lewis-like forms that contribute to the overall electronic structure and then quantifies the contributions of each form. From the resonance contributors, delocalization occurs as a weighted average of the resonance forms. Consequently, the NBO–NRT analysis is a resonance theory approach that is friendly to current computational methods. The bond orders produced are a factor of the resonance weights of the Lewis forms involved in a particular structure. A discussion of the differences and similarities between NBO–NRT and traditional valence bond theory can be found in *Valency and Bonding* by Weinhold and Landis, and the two theories are discussed in brief in Chapter 7.[6]

In one of the examples in Section 3.10.1 we found the estimated covalent bond orders for the molecule $W(C^tBu)(CH^tBu)(CH_2{}^tBu)(dmpe)$, which has three different types of W–C linkages in the same molecule, using Pauling's equation and Pyykkö's covalent bond distances. For the nominally W–C triple, double, and single bond distances we found bond orders of 2.65, 1.68, and 0.67, respectively, using the Pauling method (Fig. 3.32).

In order to find the bond order using natural bond orbital analysis, a somewhat simpler model was used to reduce the number of possible resonance forms that have to be calculated. The model used is shown in Fig. 3.31 and has the formula $W(CH)(CH_2)(CH_3)(PH_3)_2$. The bond orders from the NBO analysis are remarkably similar to the those previously calculated by hand.

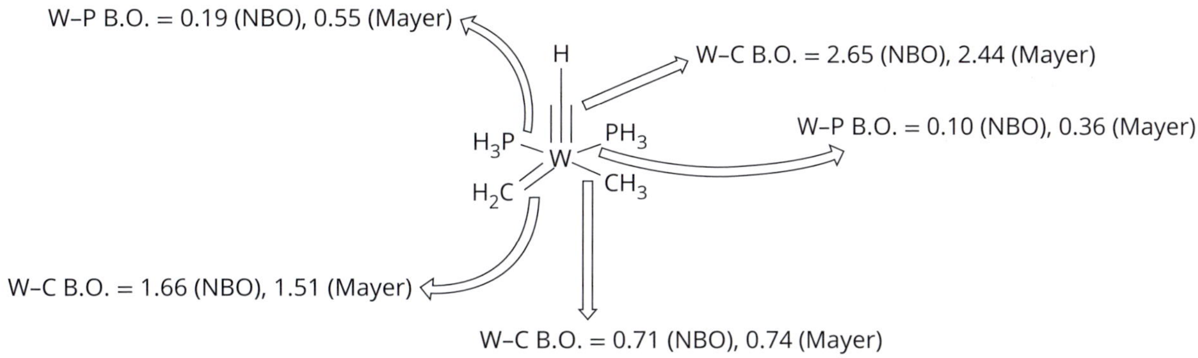

Fig. 3.32 Model for W(C^{t}Bu)(CH^{t}Bu)(CH$_2{}^t$Bu)(dmpe), W(CH)(CH$_2$)(CH$_3$)(PH$_3$)$_2$, and the Mayer and natural bond orders (NRT).

Using Mayer Bond Order analysis on the same model produced values in good agreement. This correlation between calculated bond orders from quantum-mechanical methods and those estimated using the Pauling equation has been noted in the literature, specifically with regard to Mayer Bond Order.[62]

The NBO and Mayer bond analysis, in addition to estimating the heterocovalent bond orders, will provide an estimate for the dative tungsten–phosphorus bond orders, which are comparatively small relative to the W–C bonds. Because the W–P bond is dative, the bond can be considered to have contributions from resonance forms with no bond between these atoms, and there are apparently large contributions from nonbonded resonance forms (Section 3.8). The calculations, again, are looking at the actual electron density in the bonds and recognize the low W–P bond order. The Pauling calculation, which just uses the W–P distance, will find much higher bond orders.

As a final illustration, let us examine the bond order in SF_6. The bond distance in the compound is reported as 1.561 Å. However, the sum of the covalent radii of sulfur and fluorine is 1.73 Å (Table 3.2). Very short bond distances in hypercoordinate main-group compounds are common due to high charges on the central atoms. The unusually short bond distance due to the high oxidation state produced unreasonably high values using the Pauling method for SF_6—a method based exclusively on bond distance (Section 3.10.1).

Using NBO analysis, the bond order in SF_6 is 0.667; in other words, the four valence orbitals ($3s$, $3p_x$, $3p_y$, $3p_z$) on sulfur form bonds with the six fluorines to give 2/3 of a bond to each substituent. (The bonding in such hypercoordinate main-group molecules will be discussed in detail in Chapter 8.)

Application of the Mayer bond order calculation to SF_6 was done by using a variety of basis sets in order to illustrate the dependence of the technique on the computational details. As shown in Table 3.5, the bond order can vary from a very high 1.09 to values very much in line with the NBO analysis of 0.67, depending on the basis set employed. At least in this system, the bond order seems largely insensitive to method and basis set, except when larger basis sets are used. It is advisable to examine basis and method dependence in these (and in many) calculations whenever possible.

Table 3.5 Mayer bond order in SF_6 as a function of method and basis set.

Method	Mayer Bond Order in SF_6
B3LYP/6-311++G	1.09
B3LYP/6-31++G	0.65
B3LYP/6-31G	0.67
B3LYP/3-21G	0.71
WB97XD/6-31G	0.67
PBEPBE/6-31G	0.67
B3PW91/6-31G	0.67
Hartree-Fock/6-31G	0.63

3.10.3 Formal Shortness Ratio (FSR) (Expanded Topic)

While in the previous section we were attempting to correlate bond length with bond order, a simpler approach in some ways is to just correlate bond lengths to other systems. In order to compare different atoms with different atomic radii to one another, one must create a ratio involving the typical covalent radii of the elements involved in the bond. The actual distance between the atoms in the compound, D_{AB}, is then divided by the sum of the covalent radii for the atoms ($R_A + R_B$). Eq. 3.30 for the formal shortness ratio (FSR) was introduced by Cotton and coauthors as a method for examining the "shortness" of metal–metal bonds versus other species.[63] Consequently, an FSR of 1.0 corresponds to a single bond, and values below this to multiple bonds of some type.

$$FSR_{AB} = \frac{D_{AB}}{R_A + R_B} \tag{3.31}$$

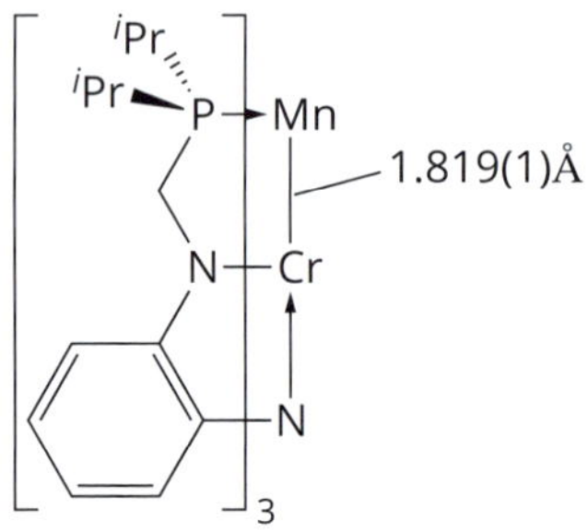

CrMnL

Fig. 3.33 Line drawing of the MnCrL complex.

As an example, consider the following two complexes published by Connie Lu and coworkers.[64] The complex uses a tripodal ligand with a nitrogen-bound chromium center held in close proximity to a phosphorus-bound manganese atom, **MnCrL**. The metal–metal distance is 1.819(1) Å. The covalent bond radii of Cr and Mn are 1.172 and 1.168 Å, respectively. Note that the authors use Pauling's metal radii for the calculation and not the covalent radii found previously in this chapter.[54] This produces an FSR of 0.777, which is in fact quite short. The authors propose that the Mn–Cr bond in the complex is a pentuple (1σ + 2π + 2δ) bond (Fig. 3.33).

Table 3.6 Formal shortness ratios for some representative compounds.

Compound	Bond	FSR
$Cr_2\{2,6\text{-}(MeO)_2C_6H_3\}_4$	Cr≡Cr	0.767
CrMnL	Cr≣Mn	0.777
acetylene	C≡C	0.783
N_2	N≡N	0.786
$Re_2Cl_8^{2-}$	Re≣Re	0.869
ethylene	C=C	0.870

$$FSR_{AB} = \frac{D_{AB}}{R_A + R_B} = \frac{1.819}{1.172 + 1.168} = 0.777$$

These FSR can be compared between any number of compounds to get an idea of bond shortness (Table 3.6). For example, the FSR of N_2 is 0.786, and the original quadruple metal–metal bond in $Re_2Cl_8^{2-}$ has an RSR = 0.869.

Chapter Summary

1. This chapter has focused on the most common type of bond, the 2-center 2-electron (2c2e) bond.
2. There are two main types of 2c2e bond: the (hetero)covalent bond and the dative bond.
3. Covalent bonds between two different atoms or groups, heterocovalent bonds, are greatly strengthened by ionic character. Similar systems with increased ionic character are generally stronger.
4. Dative bonds occur when a molecule with a lone pair of electrons acts as a Lewis base towards a Lewis acid. This gives a system where the bonding can be described as arising from two resonance forms: one where there is a covalent bond between the atoms with induced charges and another resonance form where there is no bond at all. The participation of a resonance form where there is no bond between the atoms makes dative bonds generally weaker than covalent bonds.
5. One method of describing the 2c2e bond is with a Morse Potential, which approximates how the electronic energy changes with bond distance. The vibrational states are then contained within the electronic potential.
6. Simple molecular orbital theory can be used to describe bonds as mixtures of atomic orbitals. Constructive interference gives the bonding interaction with destructive interference occurring in the antibonding orbital.
7. Bond types are defined by the number of nodes parallel to the bond. σ = 0 nodes, π = 1 node, δ = 2 nodes, ϕ = 3 nodes. The antibonding orbital is generally noted with an asterisk (star); for example, π*.
8. Charges on individual atoms in molecular systems are often highly dependent on the method of calculation. Mulliken charges are probably the most common but are poor estimates for polar systems.
9. Bond order as a function of bond distance can be defined using simple equations but should be used cautiously because the details of the bonding are important. One should only compare the distances of the same types of bonds; that is, compare 2c2e heterocovalent bonds in systems with elements in the same oxidation state, when using the Pauling equation based on bond distance.
10. Methods of calculating bond orders are available, such as Mayer Bond Order. These are generally at least self-consistent, but dependency on basis sets should be checked.

Practice Problems

1. For the molecules with the following formulas, draw reasonable Lewis dot Structures. Assign the formal charges and formal oxidation states for all the atoms in your structures. In all cases, the central element is listed first (when there is one in the structure).

   a. SF_4
   b. SO_2
   c. O_3
   d. NO_2^-
   e. NO_3^-
   f. $S(O)Me_2$
   g. S_2Cl_2

2. Draw at least four different resonance forms (Lewis structures) for AlF_3. Of your resonance forms, which is most likely to contribute the largest amount to the ground state structure and why?

3. Calculate the bond dissociation enthalpy of TlF_3 and TlI_3 using the Theory of Polar Covalence. Which halide gives the highest bond energy and, judging from your calculations, why? (Note: You should use the fully unweakened values for the halides from the Sanderson Table. For R_O, use 2.53 Å for TlF and 2.77 Å for TlI.)

4. NF_3 is quite a stable compound and is the only nitrogen halide with a negative enthalpy of formation. NI_3 is a contact explosive. Calculate the bond energies (BDEs) in NF_3 and NI_3. In this case the fully weakened values for both nitrogen and the halides should be used. For R_O, use 1.440 Å for NF_3 and 2.255 Å for NI_3.

5. In Section 3.4, the resonance forms for the cyanate ion, OCN^-, were discussed. An isomer of cyanate, fulminate CNO^- is often used in detonators for explosives as its mercury(II) salt, $Hg(CNO)_2$, which is highly friction and shock sensitive. Draw resonance forms for free CNO^- and for the mercury salt.

6. Dimethyldithiocarbamate, $\{S_2CN(CH_3)_2\}^-$, has the following skeletal structure:

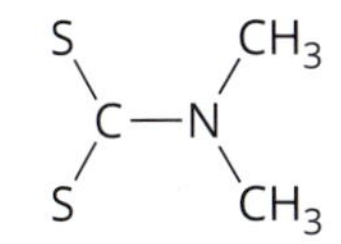

   Provide at least 3 important resonance structures for this ion, including any formal charges where necessary. Indicate which of your resonance structures is most likely to provide the best description of this ion.

7. Xenon forms several compounds of the type XeO_x. Write Lewis dot structures for XeO_2, XeO_3, and XeO_4.

8. Calculate the fully weakened and fully unweakened value for the bond in HF. Which value is closer to the experimental BDE of 135.9 kcal/mol? Is there Lone Pair Bond Weakening (Section 3.6.4) in this molecule and how much? For R_O, use 0.91 Å.

9. The bond distances in Ångstroms (1 Å = 1×10^{-10} m) for the dimethyldithiocarbamate anion, $\{S_2CN(CH_3)_2\}^-$, are shown below. Estimate the S–C, (dithiocarbamate)C–N, and N–C(methyl) bond orders in the compound using the Pauling method. The b can be estimated at 0.377 and r_o values can calculated from Figure 3.30. Compare your bond orders in this problem with the resonance forms you drew in Problem 6. Are the bond orders from the distances consistent with your guess as to which resonance form(s) will participate the most, and why?

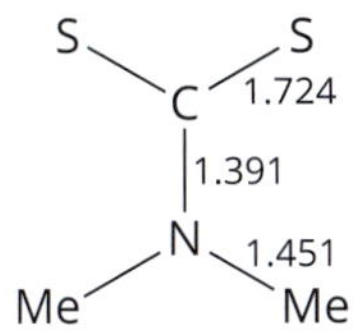

10. What criteria in Section 3.3 suggest that ϕ_D for the cyanate ion (Figure 3.11) will make only a minor contribution to the ground state electron distribution?

Cited References

1. Coulson, C. A., The Present State of Molecular Structure Calculations *Revs. Mod. Phys.* 1960, *32*, 171–177.
2. Van Vleck, J. H., Nobel Prize Lecture in Physics entitled "Quantum Mechanics: the Key to Understanding Magnetism" 1977.
3. Wang, X. F.; Andrews, L.; Riedel, S.; Kaupp, M., Mercury is a transition metal: The first experimental evidence for HgF_4 *Angew Chem. Int. Ed.* 2007, *46*, 8371–8375.
4. Lewis, G. N., The atom and the molecule *J. Am. Chem. Soc.* 1916, *38*, 762–785.
5. Adiwidjaja, G.; Pohl, D., Superstructure of alpha-phase potassium nitrate *Acta Cryst C.* 2003, *59*, I139–I140.
6. Weinhold, F.; Landis, C. R., *Valency and Bonding: A Natural Bond Orbital Donor-Acceptor Perspective* Cambridge University Press: Cambridge, 2005; p ix, 749 p.
7. Pauling, L., *The Nature of the Chemical Bond and the Structure of Molecules and Crystals: an Introduction to Modern Structural Chemistry* 3rd ed.; Cornell University Press: Ithaca, N.Y., 1966; p xx, 644 p.
8. Brown, I. D., Recent Developments in the Methods and Applications of the Bond Valence Model *Chem. Rev.* 2009, *109*, 6858–6919.
9. Pearson, R. G., Electronegativity scales *Acc. Chem. Res.* 1990, *23*, 1–2.
10. Mann, J. B.; Meek, T. L.; Allen, L. C., Configuration energies of the main group elements *J. Am. Chem. Soc.* 2000, *122*, 2780–2783.
11. Mann, J. B.; Meek, T. L.; Knight, E. T.; Capitani, J. F.; Allen, L. C., Configuration energies of the d-block elements *J. Am. Chem. Soc.* 2000, *122*, 5132–5137.
12. Huheey, J. E., Electronegativity of groups *J. Phys. Chem.* 1965, *69*, 3284–&.
13. Gordy, W., A new method of determining electronegativity from other atomic properties *Physical Review* 1946, *69*, 604–607.
14. Mullay, J., Estimation of atomic and group electronegativities *Struct. Bond.* 1987, *66*, 1–25.
15. Allred, A. L., Electronegativity Values from Thermochemical Data *J. Inorg. Nuc. Chem.* 1961, *17*, 215–221.
16. Sanderson, R. T., *Polar Covalence* Academic Press: New York, 1983.
17. Sanderson, R. T., Electronegativity and bonding of transitional elements *Inorg. Chem.* 1986, *25*, 3518–3522.
18. Allred, A. L.; Rochow, E. G., A Scale of Electronegativity Based on Electrostatic Force *J. Inorg. Nuc. Chem.* 1958, *5*, 264–268.
19. Little, E. J.; Jones, M. M., A Complete Table of Electronegativities *J. Chem. Educ.* 1960, *37*, 231–233.

20. Sen, K. D.; Bohm, M. C.; Schmidt, P. C., Electronegativity of atoms and molecular fragments *Struct. Bond.* 1987, *66*, 99–123.
21. Bratsch, S. G., Revised Mulliken Electronegativities .1. Calculation and Conversion to Pauling Units *J. Chem. Educ.* 1988, *65*, 34–41.
22. Mulliken, R. S., A New Electroaffinity Scale; Together with Data on Valence States and on Valence Ionization Potentials and Electron Affinities *J. Chem. Phys.* 1934, *2*, 782–793.
23. Sanderson, R. T., Electronegativities in Inorganic Chemistry *J. Chem. Educ.* 1952, *29*, 539–543.
24. Guo, Y. Y.; Kuo, C. K.; Nicholson, P. S., Structural refractometry of β- and β″-Al_2O_3-type aluminates and gallates *Solid State Ionics* 1998, *110*, 327–333.
25. Bratsch, S. G., A Group Electronegativity Method With Pauling Units *J. Chem. Educ.* 1985, *62*, 101–103.
26. Polar forms are an important contributor even in homopolar bonds in resonance theory, a fact indirectly accounted for in Polar Covalence Theory by using the homopolar bond energies to calculate the covalent bond energy.
27. Sanderson, R. T., An interpretation of bond lengths and a classification of bonds *Science* 1951, *114*, 670–672.
28. Parr, R. G.; Weitao, Y., *Density-Functional Theory of Atoms and Molecules* Oxford University Press: New York, 1994; Vol. 16.
29. Geerlings, P.; De Proft, F.; Langenaeker, W., Conceptual density functional theory *Chem. Rev.* 2003, *103*, 1793–1873.
30. Note that the covalent bond energy of fluorine used here is much higher than the experimental value of about 40 kcal/mol. Here we are using the "fully unweakened" value for the bond energy of fluorine. The experimental value is much smaller due to what is called the "lone pair bond weakening effect", which will be discussed in detail in the chapter on main group chemistry.
31. Haaland, A.; Helgaker, T. U.; Ruud, K.; Shorokhov, D. J., Should gaseous BF_3 and SiF_4 be described as ionic compounds? *J. Chem. Educ.* 2000, *77*, 1076–1080.
32. Sanderson, R. T., Electronegativity and bond-energy *J. Am. Chem. Soc.* 1983, *105*, 2259–2261.
33. Herzberg, G.; Huber, K.-P., *Molecular Spectra and Molecular Structure. IV. Constants of Diatomic Molecules* Van Nostrand: New York, 1979.
34. Charges in Polar Covalence Theory may or may not be the same as other methods for a variety of reasons. See Appendix A and Section 3.6.3 and 3.6.4.
35. There is an additional correction to the covalent bond energy of hydrogen for partial charges that is recommended for hydrogen that is not used for any other element. The correction here is quite small and was neglected. See Appendix B for more details.
36. Mortier, W. J.; Ghosh, S. K.; Shankar, S., Electronegativity equalization method for the calculation of atomic charges in molecules *J. Am. Chem. Soc.* 1986, *108*, 4315–4320.
37. Jirouskova, Z.; Varekova, R. S.; Vanek, J.; Koca, J., Electronegativity Equalization Method: Parameterization and Validation for Organic Molecules using the Merz-Kollman-Singh Charge Distribution Scheme *Journal of Computational Chemistry* 2009, *30*, 1174–1178.
38. Mortier, W. J., Electronegativity equalization and its applications *Struct. Bond.* 1987, *66*, 125–143.
39. Oliferenko, A. A.; Pisarev, S. A.; Palyulin, V. A.; Zefirov, N. S., Atomic charges via electronegativity equalization: Generalizations and perspectives *Adv. Quantum Chem.* 2006, *51*, 139–156.
40. Wu, Y.; Yang, Z. Z., Atom-bond electronegativity equalization method fused into molecular mechanics. II. A seven-site fluctuating charge and flexible body water potential function for liquid water *J. Phys. Chem. A* 2004, *108*, 7563–7576.
41. Lipparini, F.; Barone, V., Polarizable Force Fields and Polarizable Continuum Model: A Fluctuating Charges/PCM Approach. 1. Theory and Implementation *J. Chem. Theory and Comp.* 2011, *7*, 3711–3724.
42. Verstraelen, T.; Bultinck, P.; Van Speybroeck, V.; Ayers, P. W.; Van Neck, D.; Waroquier, M., The Significance of Parameters in Charge Equilibration Models *J. Chem. Theory and Comp.* 2011, *7*, 1750–1764.
43. Jensen, F., *Introduction to Computational Chemistry* 2nd ed.; John Wiley & Sons: Chichester, England, 2007; p xx, 599 p.
44. Collins, J. B.; Streitwieser, A., Integrated spatial electron populations in molecules: Application to simple molecules *Journal of Computational Chemistry* 1980, *1*, 81–87.
45. Bader, R. F. W., *Atoms in Molecules: A Quantum Theory* Oxford University Press: Oxford, 1994.
46. Reed, A. E.; Weinstock, R. B.; Weinhold, F., Natural-Population Analysis *J. Chem. Phys.* 1985, *83*, 735–746.
47. Glendening, E. D.; Landis, C. R.; Weinhold, F., Natural Bond Orbital Methods *Wiley Interdis. Rev. Comp Molec. Sci.* 2012, *2*, 1–42.
48. Wiberg, K. B.; Rablen, P. R., Comparison of atomic charges derived via different procedures *Journal of Computational Chemistry* 1993, *14*, 1504–1518.
49. Lauvergnat, D.; Maitre, P.; Hiberty, P. C.; Volatron, F., Valence bond analysis of the lone pair bond weakening effect for the X-H bonds in the series $XH_n = CH_4$, NH_3, OH_2, FH *J. Phys. Chem.* 1996, *100*, 6463–6468.
50. A simpler function often used for discussions of vibrations is the harmonic oscillator, which will be of great use when discussing vibrational spectroscopy. However, the harmonic oscillator, based simply on a parabolic function, is really only an adequate approximation near the ground state internuclear distance (R_e) and for vibrational frequencies near the fundamental frequency, ω_e. Consequently, the harmonic oscillator model will not serve our purposes here.
51. Schomaker, V.; Stevenson, D. P., Revisions of the covalent radii and the additivity rule for the lengths of partially ionic single covalent bonds *J. Am. Chem. Soc.* 1941, *63*, 37–40.
52. Haaland, A., *Molecules and Models: The Molecular Structures of Main Group Element Compounds* Oxford University Press: Oxford, 2008.
53. Lang, P. F.; Smith, B. C., Electronegativity Effects and Single Covalent Bond Lengths of Molecules in the Gas Phase *Dalton Trans.* 2014, *43*, 8016–8025.
54. Pauling, L., Atomic radii and interatomic distances in metals *J. Am. Chem. Soc.* 1947, *69*, 542–553.
55. Churchill, M. R.; Youngs, W. J., Crystal-structure and molecular-geometry of $W(=CCMe_3)(=CHCMe_3)(CH_2CMe_3)$(DMPE), A mononuclear tungsten(VI) complex with metal-alkylidyne, metal-alkylidene, and metal-alkyl linkages *Inorg. Chem.* 1979, *18*, 2454–2458.
56. Bartell, L. S.; Doun, S. K., Structures of Hexacoordinate Compounds of Main-Group Elements. 3. Electron-Diffraction Study of SF_6 *J. Mol. Struct.* 1978, *43*, 245–249.
57. Coulson, C. A., The electronic structure of some polyenes and aromatic molecules VII. Bonds of fractional order by the molecular orbital method *Proc. R. Soc. Lond. A-Math. Phys. Sci.* 1939, *169*, 0413–0428.
58. Bridgeman, A. J.; Cavigliasso, G.; Ireland, L. R.; Rothery, J., The Mayer bond order as a tool in inorganic chemistry *Dalton Trans.* 2001, 2095-2108.
59. Mayer, I., Charge, bond order and valence in the ab initio SCF theory *Chem. Phys. Lett.* 1983, *97*, 270–274.
60. Mayer, I., Bond order and valence indices: A personal account *Journal of Computational Chemistry* 2007, *28*, 204–221.
61. Mayer, I. *BORDER 1.0*, Chemical Research Center, Hungarian Academy of Sciences, Budapest, 2005.
62. Lendvay, G., On the correlation of bond order and bond length *Theochem-J. Mol. Struct.* 2000, *501*, 389–393.
63. Cotton, F. A.; Murillo, C. A.; Walton, R. A., *Multiple Bonds Between Metal Atoms* 3rd ed.; Springer Science and Business Media, Inc.: New York, 2005.
64. Eisenhart, R. J.; Rudd, P. A.; Planas, N.; Boyce, D. W.; Carlson, R. K.; Tolman, W. B.; Bill, E.; Gagliardi, L.; Lu, C. C., Pushing the Limits of Delta Bonding in Metal-Chromium Complexes with Redox Changes and Metal Swapping *Inorg. Chem.* 2015, *54*, 7579–7592.
65. Bart, S. C.; Lobkovsky, E.; Bill, E.; Chirik, P. J., Synthesis and hydrogenation of bis(imino)pyridine iron Imides *J. Am. Chem. Soc.* 2006, *128*, 5302–5303.
66. Brown, S. D.; Peters, J. C., Ground-state singlet L_3Fe-(μ-N)-FeL_3 and $L_3Fe(NR)$ complexes featuring pseudotetrahedral Fe(II) centers *J. Am. Chem. Soc.* 2005, *127*, 1913–1923.
67. Mahy, J. P.; Battioni, P.; Mansuy, D.; Fisher, J.; Weiss, R.; Mispelter, J.; Morgensternbadarau, I.; Gans, P., Iron porphyrin-nitrene complexes: Preparation from 1,1-dialkylhydrazines, electronic-structure from NMR, mossbauer, and magnetic-susceptibility studies and crystal-structure of the tetrakis(para-chlorophenyl)porphyrinato(2,2,6,6-tetramethyl-1-piperidyl)nitrene iron complexmo *J. Am. Chem. Soc.* 1984, *106*, 1699–1706.
68. Petit, L.; Maldivi, P.; Adamo, C., Predictions of optical excitations in transition-metal complexes with time dependent-density functional theory: Influence of basis sets *J. Chem. Theory and Comp.* 2005, *1*, 953–962.
69. Quinones, G. S.; Hagele, G.; Seppelt, K., MoF_6 and WF_6: Nonrigid molecules? *Chem Eur. J.* 2004, *10*, 4755–4762.

CHAPTER 4

The Shapes of Main-Group Molecules

The VSEPR Model

"The arrangements in space of the covalencies of a polyvalent atom, while they are subject to small variations seldom exceeding 5 or 10°, tend to conform to quite a limited number of types. It is desirable to relate this grouping to some familiar property of the atom. The property here used is the size (in G. N. Lewis's sense) of the valency group of the central atom, and the number of shared electrons which it contains, together with that of the preceding (unshared) electronic group in the atom."

— Nevil V. Sidgwick[1]

CONCEPT COVERAGE

1. Basics of the VSEPR model
2. Estimation of domain sizes with different substituents, central atoms, and electron counts
3. Estimation of 3-dimensional structures using VSEPR
4. The limitations of VSEPR

The Lewis diagrams of main-group molecules were discussed in Chapter 3, which allowed us to discuss the nature of the chemical forces that hold atoms together in molecules. The shapes of molecules within the *s*- and *p*-block can be determined with surprising accuracy using a few simple concepts that collectively are referred as the Valence Shell Electron Pair Repulsion (VSEPR) model. VSEPR is so ingrained into the way chemists view molecular shapes that it is often used without regard to its underlying assumptions and limitations.

In this chapter, the VSEPR model will be developed, followed by a discussion of a few instances where the model fails to predict the correct 3-dimensional shape or at properly estimating changes between related molecules.

Although simple and generally covered in most introductory general chemistry courses, the VSEPR model has a few limitations that are not always well appreciated and that should not be ignored for a proper understanding of the factors influencing the geometry of inorganic compounds. Most important among these limitations, we advise against using VSEPR with transition-metal systems to determine the geometry; nevertheless, VSEPR is often predictive in the main group and is often used as a starting point for discussions of even transition-metal systems, as will be seen throughout this book.

4.1 Basics of the Valence Shell Electron Pair Repulsion Model (VSEPR)[2–3]

The overall geometry of a molecule is an important factor in determining properties and reactivity. The assumption that a molecule has one molecular structure versus another, say trigonal bipyramidal versus square pyramidal, changes expectations for several different properties, such as the number and type of vibrations that the molecule exhibits.

The VSEPR model, which provides an intuitive approach for estimating the ground-state geometry of many main-group molecules from their Lewis diagram, was developed in the 1950s by Ronald Gillespie and Ronald S. Nyholm, based on ideas that Nevil Sidgwick and Herbert Powell had formulated a decade earlier. VSEPR contrasts sharply with other molecular structure models, such as valence bond and molecular orbital theories, in that it is not a quantitative approach: the sole, but important, purpose of VSEPR is to allow quick estimation of molecular geometry.

The foundation of the VSEPR model is the Pauli Exclusion Principle, the most general definition of which is "the total wavefunction for a system must be antisymmetric to the exchange of any two electrons" (see Chapter 16 for further discussion on this important principle). One direct consequence of this principle is that no more than two electrons can occupy the same orbital, and if, *and only if*, they have different spins (another way Pauli's Principle is sometimes formulated for atomic orbitals is that no two electrons can have the same set of quantum numbers). Lone pairs of electrons can exist as long as the two electrons occupying the orbital have different spin quantum numbers, m_s (Section 2.2).

When two electrons occupy the same region of space—as they are bound to if they occupy the same spatial orbital—the Coulombic repulsion between the two is invariably larger than if they were each occupying a different region of space. (Two negative charges always repel one another, the more so the closer they are.) Fermions, such as electrons, cannot be in the same state in the same location; consequently, electrons with the same spin cannot occupy the same space. This Pauli repulsion is not a "force" really, but more an exclusion of the types of possible states that the system can be in. Operationally, there is a strong tendency for electrons of the same spin in a molecule to be as far apart as geometrically and electronically possible, and their overlap must be zero.

Electron density may be regarded as residing in "domains" around the central atom. *These domains will organize themselves around the central atom in the configuration that leads to the least repulsion between the various pairs of electrons (bonding or lone pairs) in the molecule.*

Under the VSEPR model, for example, a silicon with four hydrogens, SiH_4 silane, has four bonding domains around the Si atom (Fig. 4.1). The four bonds of silane will always be organized 3-dimensionally to be as far apart as possible from one another in order to reduce the electronic repulsion between the four bonding pairs of electrons. As it happens, the furthest apart these four pairs can be situated is defined by a tetrahedral geometry.

For the purposes of VSEPR, it is useful to classify the different types of domains. Here, a central atom will be represented by the letter A, a lone pair on the central atom will be assigned an E, and any bonded substituent will be called X (this is sometimes called the AXE classification). For different bonded

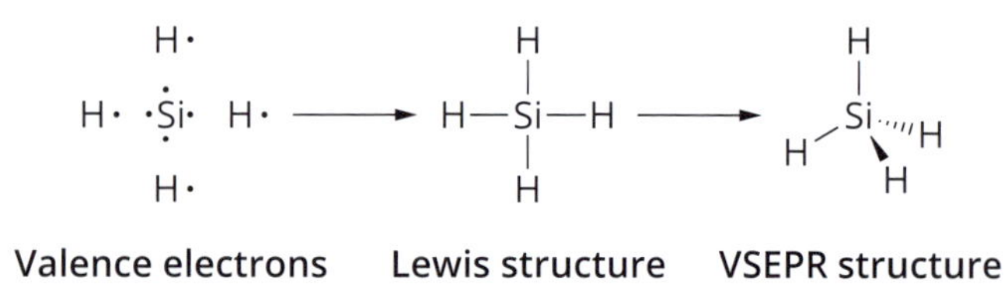

Fig. 4.1 Bonding domains in SiH_4, a tetrahedral compound. The progression from valence electronic structure to Lewis structure to VSEPR 3-dimensional structure.

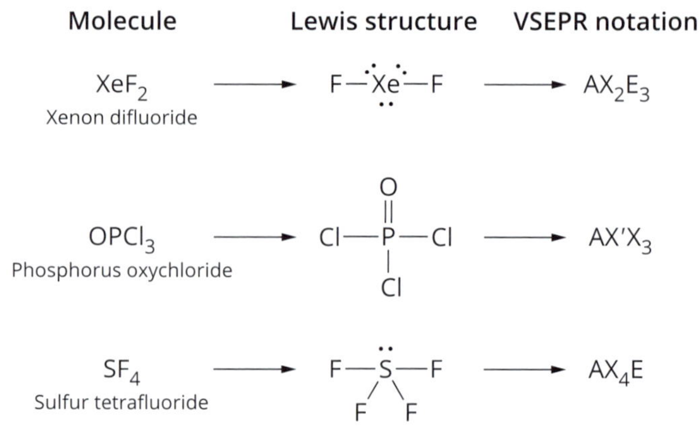

Fig. 4.2 Converting Lewis structures into VSEPR notations.

substituents, primes can be added to the X. Fig. 4.2 shows several examples of this classification, in which lone pairs are omitted in the Lewis structures, except for those on the central atom A. It is perhaps worth noting here, even though such cases will not be covered until Section 4.4, that the unpaired electron of a radical species will be designated with a lower-case "e" and will be treated as an independent VSEPR domain.

The next step in the VSEPR analysis is to determine the number of electron domains in the structure. This is done by adding the number of lone pairs (E), bonds (X), and single (e) electron domains in the VSEPR notation. For example, a molecule represented by AX_2E_3 in VSEPR notation, has $2 + 3 = 5$ domains.

The base geometries (sometimes called "electron domain geometry" or "electronic geometry") corresponding to the number of electron domains in VSEPR are derived from assuming that all substituents on the central atom will be as far apart as possible to minimize like spin–spin repulsions. The base geometries most often encountered are shown in Table 4.1.

To determine the molecular geometry, the central atom A is placed at the center of the base structure with the correct number of domains. Then, the substituents/lone pairs/single-electron domains are placed around the central atom in the shape of the geometry corresponding to that number of domains. Finally, the angles are adjusted to compensate for domain sizes that might be larger or smaller than others.

As examples, CH_4 and SF_6 (Fig. 4.3) have four and six equivalent substituents, respectively. For methane, VSEPR correctly predicts the compound be tetrahedral, and for SF_6, VSEPR correctly predicts an octahedral geometry.

For iodotrifluoromethane (Fig. 4.3), iodine has a much lower electronegativity than fluorine. Consequently, its domain size near the carbon is expected to be larger (*vide infra*), and I–C–F angles are expected to be larger than in the base structure, >109.45°. The larger angles to the less electronegative iodine substituent will require the F–C–F angles to compress to less than the tetrahedral angle.

Although VSEPR does not produce the precise values for the angles in the compound, it nevertheless allows direct

Table 4.1 Base geometries commonly encountered in VSEPR.

Number of domains	Base geometry	
1	Linear	
2	Trigonal planar	
3	Tetrahedral	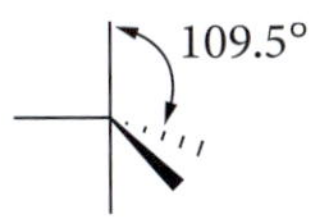
4	Trigonal bipyramidal	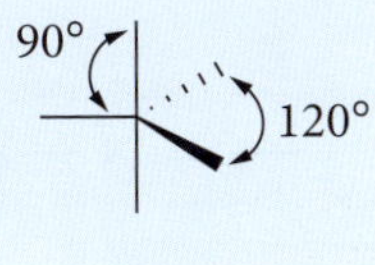
5	Octahedral	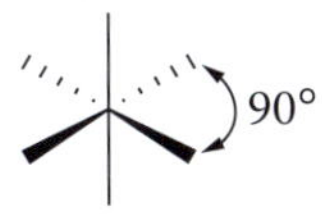
6	Pentagonal bipyramidal	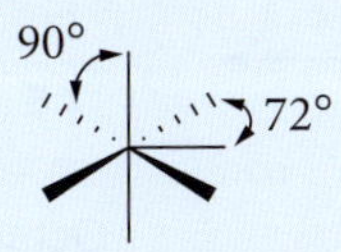

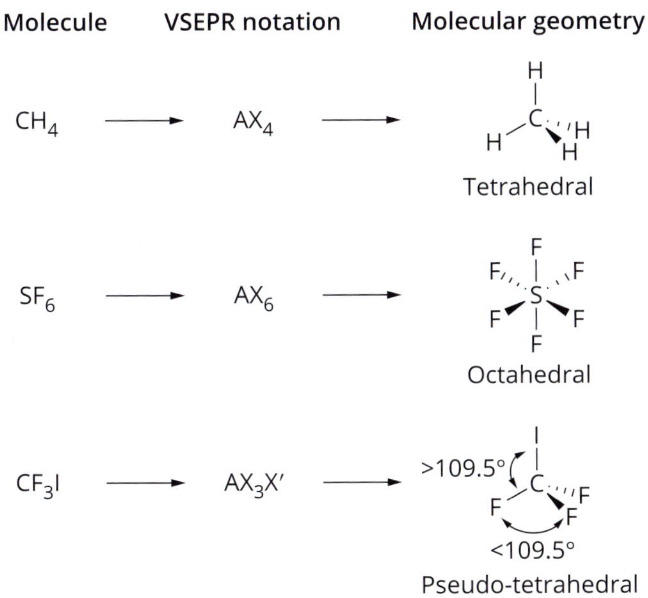

Fig. 4.3 Application of VSEPR to CH_4, SF_6, and CF_3I.

predictions of expected deviations from a particular base structure based on the nature of each substituent. The exact angles in ICF_3 have been found experimentally by using X-ray diffraction to be I–C–F = 111.4° and F–C–F = 107.5°, in agreement with the VSEPR prediction.

Because it is the specific domains occupied by electrons, and not substituents themselves, that dictate the geometry of molecules (at least, according to VSEPR), lone pairs will exhibit an influence on geometry as well. Consider the molecules in Fig. 4.4: SO_2, PH_3, XeF_2, XeF_4.[a] As before, we begin by writing the Lewis structure and then the VSEPR notation. Next, the base geometry according to number of domains (X + E) is found. The molecule is written out with the central atom in the center of the base geometry and with the lone pairs and substituents occupying the vertices. The actual molecular geometry is generally regarded as the shape of the atomic structure only; for example, XeF_4 is generally referred to as "square planar" for the pattern of its atoms and not "octahedral" as implied by the base geometry.

Obviously, an electron pair and a bond to a substituent, such as fluoride, are not equivalent domains, and VSEPR can account for these differences to improve the estimation of the molecular geometry. The lone pair in PH_3 (Fig. 4.4) should occupy much more space than the hydrogens; consequently, the H–P–H angles will be far less than the 109.5° in the base structure (experimentally, the H–P–H angles are found to be 93.5°). It is the electronegativity of the substituents that is the major factor determining domain size, which is the topic of the next section.

4.2 **VSEPR Domains and Electronegativity of the Substituents**

VSEPR can not only be used to predict the geometry of a given molecule, but often it can also successfully account for variations in shapes across series of molecules. For example, one can predict how angles will change (larger or smaller) within similar VSEPR formulas but where substituents vary. *In general, the domain size near a central atom, which is what most greatly affects structure, changes with the electronegativities of the atoms involved.*

Consider the two-electron bonding domains of a hypothetical atom A of electronegativity higher than iodine (χ^P_I = 2.66) as it bonds with either iodine or fluorine (Fig. 4.5). The electrons in the A–I bond will be distributed closer to A, and thus occupy a much larger angle near A, relative to the electrons in the A–F bond, as the higher electronegativity of F (the most electronegative element in the periodic table) will move the electron density shared by the two atoms away from A. As a result, more electronegative atoms can be thought of as forming smaller domains. A lone pair located on a central atom of a molecule can be thought of as an extreme example of the electronegativity effect on the size of the domain. Since there

[a] It is worth noting that structures such as O=PCl_3 in Fig. 4.2 and O=S=O in Fig. 4.4 do not denote 2-center 2-electron bonds in all cases. These are "composite structures" that are commonly drawn. This will become clearer after the discussion of main-group bonding. For VSEPR, one can assume that the lines mean 2-center 2-electron bonds.

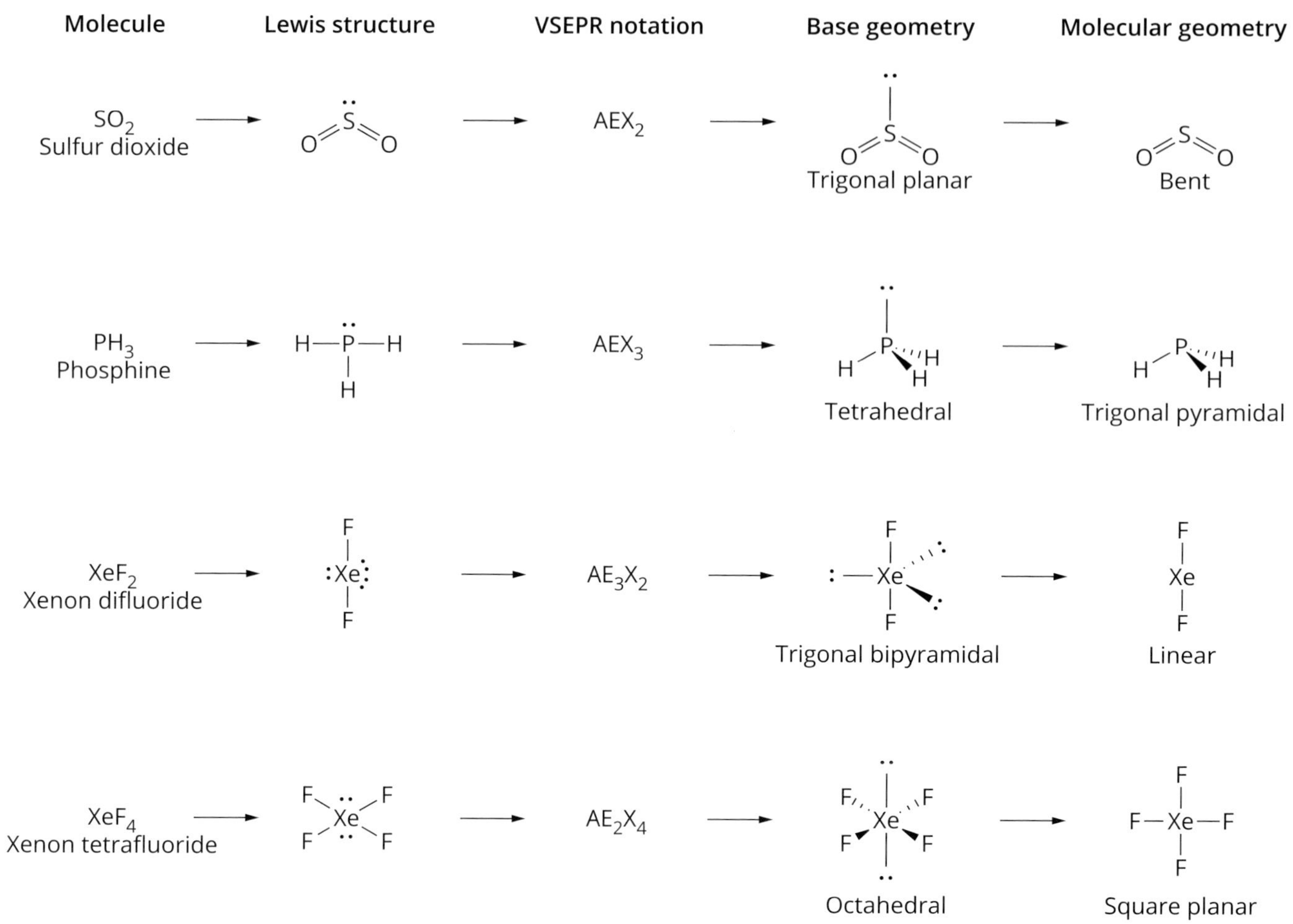

Fig. 4.4 Examples of VSEPR structures where the central atom bears a lone pair of electrons.

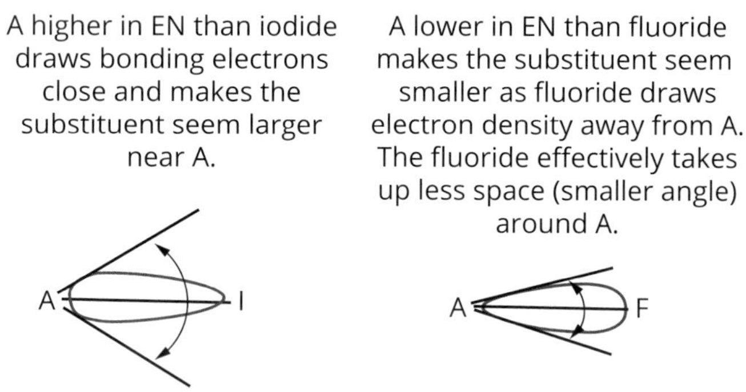

Fig. 4.5 Angle occupied by bonding domains for A–X bonds. Substituents, A, with high electronegativities take up less space around A.

is no atom on the other side of a lone-pair electron domain, the electronegativity of the substituent can be assigned as zero. In other words, *lone pair domains should be larger than the domain of any bonding substituent around a given central atom.*

In the context of the VSEPR model, the primary factors that define the structure of a molecule are the size and orientation of valence-shell electron pairs surrounding the central atom, and the overall size of specific atoms or groups of atoms only play a secondary effect. Any steric effect that the larger atomic radius of iodine, due to its large number of core electrons, would play in defining the position of atoms would be considered as a deviation from VSEPR.

This can be illustrated directly by the angular variations in the series of phosphines PX_3, where X = F, CF_3, Cl, Br, and I, shown in Fig. 4.6. As expected from the previous discussion, these compounds all adopt a trigonal pyramidal geometry. The X–P–X angles generally increase with decreasing electronegativity of the halide, but also increase with increasing covalent radius of the substituent. Are the angular increases due to electronegativity or radius (size) of the group on phosphorus?

The bond angles in $P(CF_3)_3$ are similar to PF_3, and the two substituents have electronegativities much higher than the heavier halides. Using Sanderson's method in Chapter 3, the electronegativity of fluorine is $\chi_H^S = 4.00$, and the electronegativity of the CF_3 group can be calculated as the geometric mean of the atoms, $\chi_{CF_3}^S = \sqrt[4]{4^3 \times 2.746} = 3.64$. In the phosphine series shown here, then, the electronegativies go in the order F > CF_3 > Cl > Br > I, which matches the ordering for the angles in the phosphines.

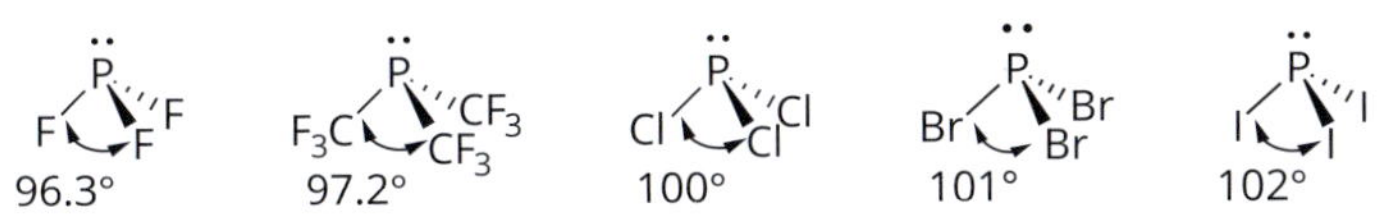

Fig. 4.6 Bond angles in PF_3, $P(CF_3)_3$, PCl_3, PBr_3, and PI_3.

The radius of the CF_3 group in $P(CF_3)_3$ (r_{CF3} ~1.25 Å calculated from the gas phase structure[4]) is between bromine and iodine (r_{Br} = 1.142, r_I = 1.333 Å). The physical size of the groups then goes in the order F < Cl < Br < CF_3 < I, which has no relation to the order of angles observed in the phosphine complexes. Consequently, the electronegativity in these compounds is the more important factor in determining X–P–X bond angles due to the size of the bonding electron domains.

In Fig. 4.6, all of the bond angles, regardless of electronegativity of the substituents, are smaller than the ideal tetrahedral angle (109.5°); in the context of the VSEPR model, this is interpreted as being due to the domain size of the lone pair on phosphorus. Because there is only one nucleus to support the two electrons of the lone pair (unlike bonding pairs, which by definition are shared between two nuclei), the lone-pair domain remains close to the phosphorus nucleus and occupies a large angle around the central atom (Fig. 4.7). Consequently, the angles between the lone pair and the other substituents are larger than 109.5°, forcing the X–P–X angles to close.

Note that in Fig. 4.6 there is good correlation along the series; however, PH_3, which is not included and has lower electronegativity substituents than PF_3, has slightly smaller angles, H–P–H = 94°. Consequently, PH_3 would be an exception to the previous arguments, probably because hydrogen is quite different from the other groups as it has no core electrons.

In comparing congeners containing lone pairs such as OH_2 and SH_2 (Fig. 4.8), VSEPR predicts that lone pair(s) on the central atom with higher electronegativity will occupy less space,

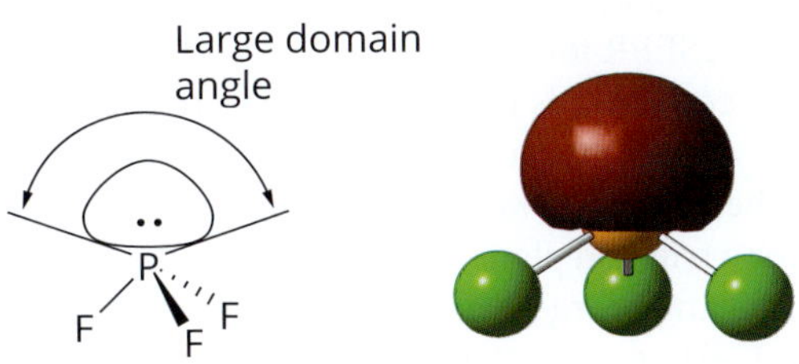

Fig. 4.7 Shape of the lone-pair domain in PF_3. Calculated shape of the lone pair in PF_3, Natural Bond theory, B3LYP/6-311G.

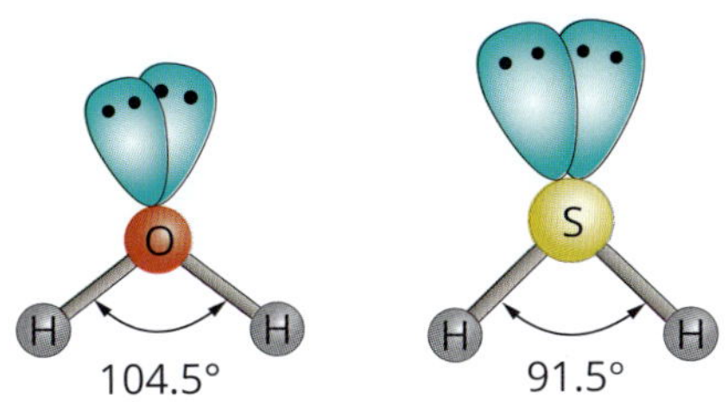

Fig. 4.8 Structures of H_2O and H_2S. The lower electronegativity of sulfur causes an expansion in the lone pairs, which forces the S–H bonds toward lower angles.

allowing substituents to occupy more of the space around the central atom, leading to larger bond angles. For example, the H–E–H angle for E = O (water) is 104.5°, whereas the angle for E = S (hydrogen sulfide) is 91.5°. According to the VSEPR model, the lone pairs in the less electronegative sulfur force the hydrogens to come closer to each other.

4.3 Inequivalent Positions in a Base Structure

In cases where there are inequivalent positions in the base structure (trigonal bipyramidal or pentagonal bipyramidal) or where there are multiple groups with different domain sizes, there will be multiple possible geometrical isomers (molecular species that differ in the way atoms are organized in space) that could be produced. VSEPR predicts that the most stable isomer will be the one where the largest domain sizes are as far apart in space as possible.

In the case of trigonal bipyramidal, the axial sites are 90° from three substituents, while the equatorial sites are 90° from only two substituents. As a result, the axial positions favor smaller domain sizes. For example, in XeF_2 the base geometry is trigonal bipyramidal (Fig. 4.4), and the lone pairs all occupy the less crowded equatorial positions, while the smaller fluorines reside in the axial sites.

The effect of electronegativity on domain size can again be seen in the 5-coordinate phosphorus compounds PF_5, $P(CF_3)F_4$, and $P(CF_3)Cl_4$ (Fig. 4.9). The homoleptic compound, PF_5, has the formula AX_5 by VSEPR notation, suggesting a trigonal bipyramidal structure. The PF_5 trigonal bipyramidal (tbp) structure has been studied in great detail experimentally.[5]

The electronegativity of CF_3 is lower than F and higher than Cl (vide supra). Consequently, the CF_3 group should have a larger domain size than F and will reside in the equatorial plane of $P(CF_3)F_4$. Conversely, CF_3 is more electronegative than Cl and will reside in the axial position of $P(CF_3)Cl_4$. Both of the $P(CF_3)F_4$ and $P(CF_3)Cl_4$ geometries have been shown experimentally to be consistent with VSEPR predictions.[6–7] If a chlorine in $P(CF_3)Cl_4$ is replaced with another CF_3 group, it should also go into the more crowded axial position, leaving the three chlorines in the equatorial plane; this has also been confirmed experimentally.[8]

Stable 7-coordinate compounds are well known in main-group chemistry but are atypical. Central atoms that can accommodate seven different domains to form a stable chemical structure have quite large radii, like iodine.

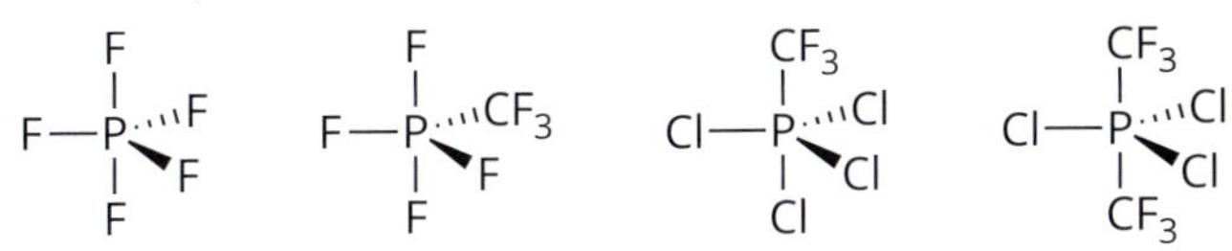

Fig. 4.9 Structures of PF_5, $P(CF_3)F_4$, $P(CF_3)Cl_4$ and $P(CF_3)_2Cl_3$ according to VSEPR.

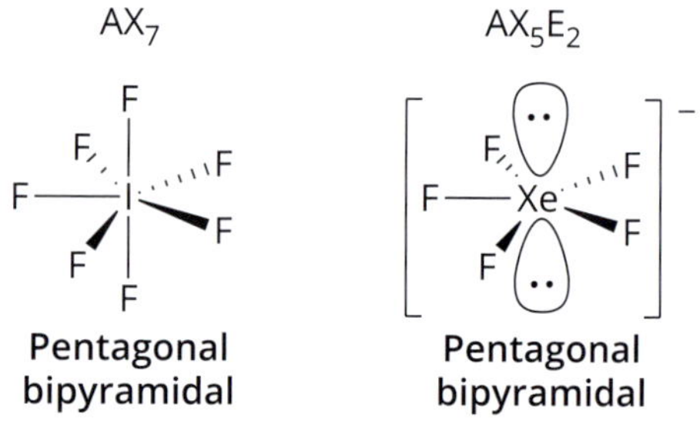

Fig. 4.10 Structures of IF_7 and XeF_5^- according to VSEPR.

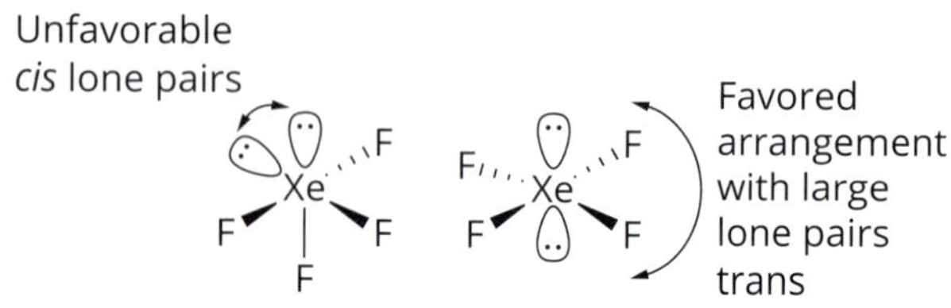

Fig. 4.11 When placing substituents in equivalent positions, such as in the base geometry of octahedral, the larger domains are preferred to be as far apart as possible. The lone pairs are *trans* in XeF_4.

Perhaps the simplest example of a seven-domain compound is binary IF_7 (Fig. 4.10), which has the VSEPR notational formula of AX_7. From Table 4.1, the expectation is that the compound will have a pentagonal bipyramidal structure, which has been confirmed experimentally.[9]

Another example of a seven-domain compound is XeF_5^- (Fig. 4.10), which has an AX_5E_2 VSEPR formula. The less crowded sites, 90° angles to other substituents, are along the axes, whereas in the equatorial plane atoms will be 72° apart. Consequently, the lone pairs will occupy the axial positions and smaller fluorines the equatorial plane, leading to a pentagonal planar molecular geometry.

In some cases a base structure where the substituents are the same may be made inequivalent by addition of different substituents. For example, placing two large domains *cis* to one another in an octahedral compound is less favorable than placing them *trans* to one another (Fig. 4.11). An example of this is XeF_4, which has the VSEPR formula of AE_2X_4. Once one of the large lone pairs is placed into the octahedron, the five remaining sites are no longer equivalent, and VSEPR suggests that the next lone pair will go into the less congested site away from the first lone pair. The result is that XeF_4 has all the fluorines in a single plane, a square planar geometry.

4.4 Single-Electron Domains

As mentioned previously, single-electron domains are generally given the symbol "e" in VSEPR formulas and are generally smaller than lone-pair electron domains, "E". That the single-electron domain is smaller than a lone-pair electron

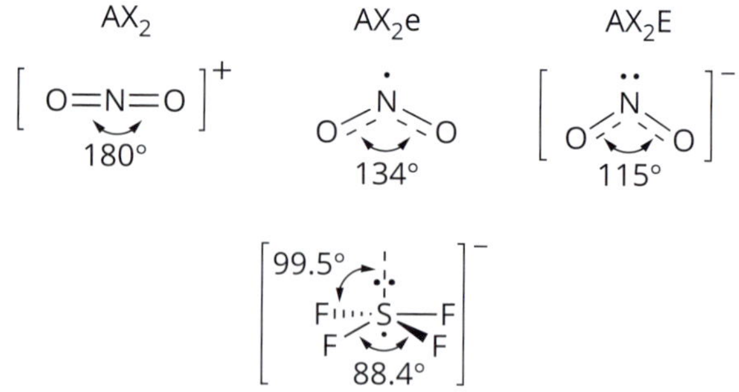

Fig. 4.12 (*top*) Structures of NO_2^+ (nitronium cation), NO_2 (nitrogen dioxide), and NO_2^- (nitrite anion). (*bottom*) Structure of SF_4^-, which has a lone pair and a radical.

domain can be demonstrated with three common compounds: nitronium cation, nitrogen dioxide, and nitrite anion. The VSEPR formulas for the three compounds are (Fig. 4.12), respectively, AX_2, AX_2e, and AX_2E. AX_2 has two domains and a base geometry of linear, as confirmed experimentally for the molecule. AX_2e and AX_2E both have three domains and a trigonal planar base geometry; however, the lone pair will occupy a larger space than the radical (single-electron) domain. As a result, the O–N–O angle is smaller in NO_2^-.

Whether a single-electron domain is larger or smaller than a two-electron domain of a bond seems to be dependent upon the substituents of the bond (*vide infra*). Consider the case of SF_4^-, a radical anion (Fig. 4.12).[10] The lone-pair domain is obviously larger as the fluorines bend away from the lone pair and towards the single-electron domain. However, the single-electron domain is located *trans* to the lone pair, the position expected for the larger substituent. From the VSEPR perspective, we suggest that the single-electron domain is larger than the fluorine-bonding domain and smaller than the lone-pair domain in SF_4^-.

In contrast, consider singlet carbene ($:CH_2$) with the VSEPR formula AX_2E, versus methylene radical cation ($\cdot CH_2^+$) with the VSEPR formula AX_2e (Fig. 4.13). The base geometry for both compounds is trigonal planar. In the case of the compound with a lone pair, singlet carbene in Fig. 4.13 (left), the angles between the two hydrogens are lower than the expected 120° for a trigonal plane, suggesting that the lone-pair domain is significantly larger than the C–H bond domain. However, the H–C–H angle in the CH_2^+ radical (Fig. 4.13 right) is 140°, suggesting that the single-electron domain is significantly smaller than the C–H bond domain.

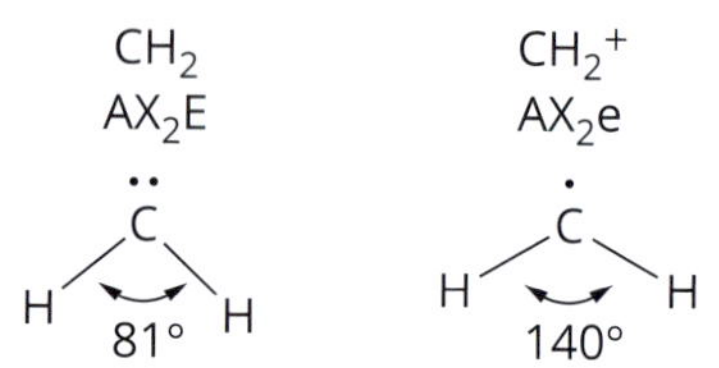

Fig. 4.13 Structures of singlet CH_2 and CH_2^+. (The CH_2^+ geometry was found computationally at the CCSD 6-311++G level using Gaussian09.)

4.5 Multiple Bond Domains

As mentioned previously, the reason that lone pair domains are larger than bonding domains is that there is only one nucleus to support the two-electrons in a lone pair, whereas in a bond both nuclei help to support the electron density. Naturally, as the bond order increases there are more electrons to be supported by the two nuclei involved in the bond, and the increase in electrons swells the bonding domain. As a consequence, double bonds (four electrons for two nuclei) and triple bonds (six electrons for two nuclei) are larger than their single-bond counterparts.

As an example, consider the structure of acetone in the gas phase (Fig. 4.14). Oxygen is more electronegative than the carbons of the methyl groups, and we expect the oxygen to have smaller domains because of this. However, the oxygen has a double bond to the central carbon, increasing the size of its domain beyond that of less electronegative carbon. As a result, the C–C–C angle is only 108° while the O–C–C angle is 126°.[11]

Lowering the electronegativity of the multiple bonded substituent will increase its size even further, as can be seen on comparing $O{=}SF_4$ and $H_2C{=}SF_4$ (Fig. 4.15). The methylene domain appears significantly larger than its oxo counterpart in the equatorial plane where the F(equatorial)–S–F(equatorial) angle significantly shrinks.

BOX 4A

A valence bond description of the bonding in $SeCl_6^{2-}$

In the description of stereochemically inactive lone pairs it was suggested that the lone pair of electrons in $SeCl_6^{2-}$ was going into a spherical *s*-orbital, which would leave all the bonding to be done using pure *p*-orbitals on selenium.

Calculations on the molecule using NBO5.0 show that the lone pair of electrons on selenium is indeed in a pure *s*-orbital. The bonds between Se and Cl involve pure *p*-orbitals on Se and are three-center four-electron interactions (Chapter 8). As a consequence, well over 90% of the ground-state structure is formed by resonance forms such as those shown in the figure, and the bond order in each Se–Cl interaction is ½.

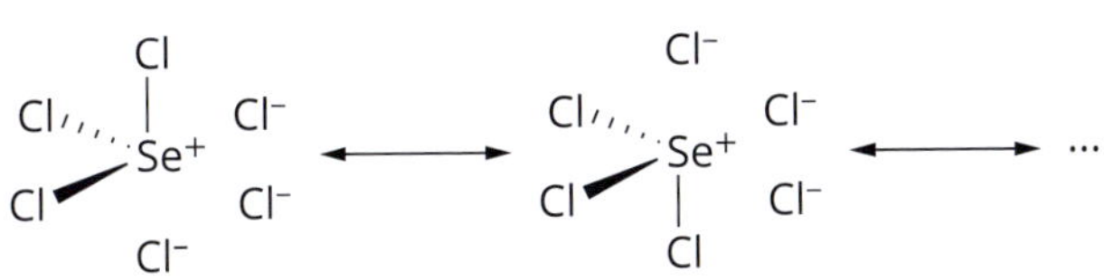

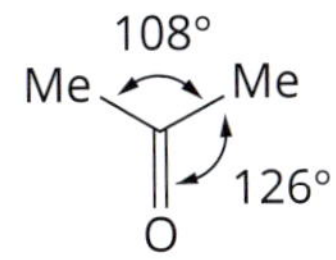

Fig. 4.14 Structure of acetone.

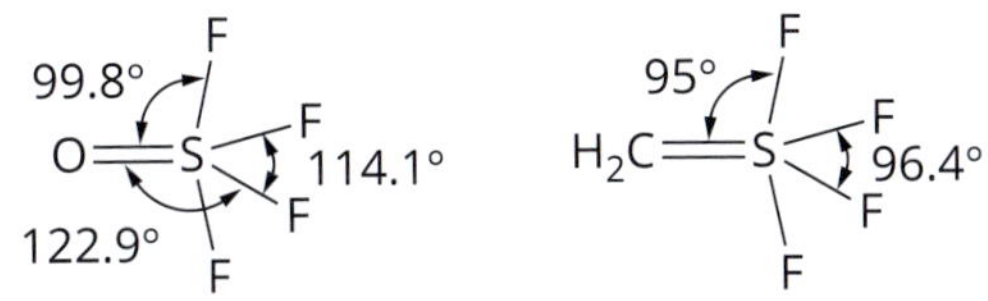

Fig. 4.15 Structures of $O{=}SF_4$ and $H_2C{=}SF_4$, showing the increased size of the less electronegative methylene over oxo.

4.6 Some Causes for Non-VSEPR Geometries[12]

Like all theories, VSEPR has limitations and is not applicable to the entire extraordinary range of inorganic molecules. While attempts have been made to make VSEPR applicable to transition metal systems, we do not recommend relying on the model for these compounds. Other models have been developed for transition-metal-based systems (see Chapter 14).

For transition metals, the assumption made in VSEPR is that the d-electrons reside in orbitals that do not affect the geometry of the molecule and, consequently, do not influence the structure. Only the orbitals bonding to ligands influence structure. In other words, the nonbonding *d*-electrons are thought to reside in a spherical shell, but as we will see later, this is quite wrong. Several examples are shown in Fig. 4.15a. Some molecules

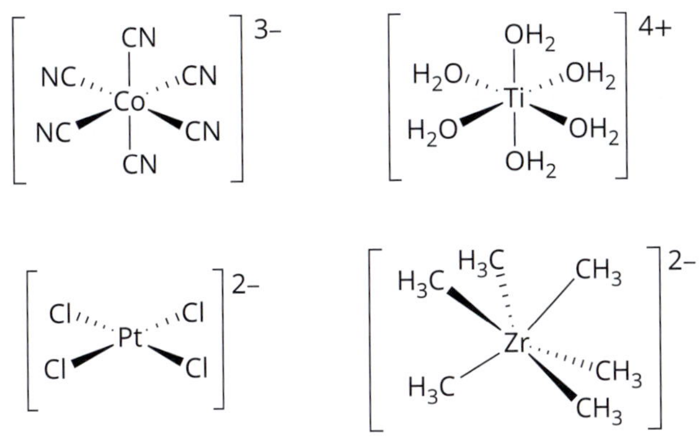

Fig. 4.15a Examples of transition-metal complexes and their geometries. (*top left*) $\{Co(CN)_6\}^{-3}$: According to VSEPR, the *d*-electrons reside in spherical orbitals and do not affect geometry. VSEPR correctly predicts an octahedral geometry. (*top right*) $\{Ti(OH_2)_6\}^{+4}$: VSEPR correctly predicts an octahedral geometry. (*bottom left*) $\{PtCl_4\}^{-2}$: VSEPR predicts a tetrahedral geometry, but the compound is square planar. (*bottom right*) $\{Zr(CH_3)_6\}^{-2}$: VSEPR predicts an octahedral geometry, but the complex is strongly distorted with C-Zr-C angles of 77.5, 78.2, 121.0, 143.5, and 128.2°, for example.[13]

are correctly predicted by VSEPR, such as $\{Co(CN)_6\}^{-3}$ and $\{Ti(OH_2)_6\}^{+4}$, while others are not, such as the square planar $\{PtCl_4\}^{-2}$ and distorted octahedral $\{Zr(CH_3)_6\}^{-2}$.

The *d*-orbitals and the distribution of electrons in the *d*-orbitals quite often lead to non-VSEPR geometries for transition-metal complexes (such as Ligand–Field effects, see Chapter 15). The use of VSEPR for any transition-metal complexes is generally ill-advised. Even for d^0-complexes, where there are no nonbonding *d*-electrons to influence structure, the geometries of compounds can deviate significantly from VSEPR expectations. (See Chapter 14 for examples.)

For transition-metal systems, researchers quite commonly discuss "distortions" of certain observed structures from some expected geometry, and very often the "expected geometry" to which they refer is a highly symmetrical one that VSEPR would predict. For example, one may discuss Jahn–Teller "distortions" of a 6-coordinate metal complex that VSEPR would anticipate being octahedral, but the "distortion" makes the complex elongated along one axis.

VSEPR is quite remarkable in its ability to predict the structures of main-group compounds, being correct far more often than not. Even in the main group there are limitations to the model, which will be briefly discussed subsequently.

4.6.1 Ligand–Ligand Repulsion (Ligand Closest Packing Model)[3]

In VSEPR it is assumed that it is the electron domain sizes around the central atom, not the ligand size, that determine the geometry. In some cases the ligands are in close enough proximity and/or they are large enough that there are steric interactions between the ligands that determine the angles. In some of these cases the compound can be described with the ligand closest packing (LCP) model.

The LCP model centers on ligand repulsions, rather than electron pair repulsions in VSEPR, to predict stable geometries. The lone pairs on a compound are then assumed to occupy as much space as the ligands will allow. The model seems particularly useful for compounds bearing sterically large ligands and when ligands have low electronegativity relative to the central atom.

One simple system where ligand–ligand interactions can be used to explain the structure is HOF, which we can use to illustrate the method. The H–O–F angle in the compound should be determined by the size of the hydrogen and fluorine atoms on oxygen. Because atoms with various amounts of charge have different sizes (Chapter 2), it is best to find the size of the ligand on the central atom of interest, keeping electronegativities involved the same.

For water, the geometry is shown as follows. The O–H bond distances are 0.96 Å (96 pm), and the distance between the two hydrogens is 152 pm (1.52 Å). The ligands are considered to be as close as possible, allowing room for the lone pairs to expand; consequently, the radius of hydrogen on oxygen is ½ of the H–H distance, 152/2 = 76 pm (0.76 Å). The H–O–H angle is 104.7°. This can be found using law of cosines from the bond length data, $C^2 = A^2 + B^2 - 2AB\cos(\theta)$ where C is the H–H distance, A and B are the O–H distances, and cos(θ) is the angle.

96 pm O 96 pm
H- - -H
152 pm

H radius on O = 76 pm

$$C^2 = A^2 + B^2 - 2AB\cos\theta$$
$$152^2 = 96^2 + 96^2 - 2(96)(96)\cos\theta$$
$$\theta = 104.7$$

For OF_2, a similar procedure can be carried out to find the radius of fluorine on oxygen, which is 1.10 Å (110 pm). The F–O–F angle is then 103.6°.

140 pm O 140 pm
F- - -F
220 pm

F radius on O = 110 pm

If one uses VSEPR to predict the geometries of H_2O and F_2O, the expectation is that the more electronegative fluorine will have a smaller domain size than hydrogen. The angle in F_2O is consequently expected to be smaller than in H_2O, which is correct. VSEPR also anticipates that the angle in HOF will be intermediate between H_2O and F_2O.

We can assume that the O–H and O–F bond distances in HOF are the same as in H_2O and F_2O used before, 96 pm and 140 pm respectively. In other words, the radii of the ligands on the oxygen atom are the same as previously calculated. The distance between H and F in the compound will be the sum of the radii, 110 + 76 = 186 pm (1.86 Å). Using these distances and the law of cosines we can calculate the H–O–F angle as 102°, smaller than in either H_2O or F_2O!

96 pm O 140 pm
H- - -F
186 pm

$$C^2 = A^2 + B^2 - 2AB\cos\theta$$
$$186^2 = 96^2 + 140^2 - 2(96)(140)\cos\theta$$
$$\theta = 102^\circ$$

The experimental angle in HOF is actually a bit smaller at 97° because of some small changes in bond distances. The H–O distance in HOF is a little smaller than in H_2O due to the electron withdrawing fluorine in the former. Conversely, the F–O bond in HOF is slightly longer than in F_2O. In essence, however, VSEPR fails in the compound where LCP was successful in predicting the smaller angle in HOF versus F_2O.

Again, this is an instance where VSEPR fails, in essence, because a steric model is more effective. The LCP model is successful in finding the bond distances in BF_3, $O(SiMe_3)_2$, and other examples as well.[3] The LCP model will not be used extensively in this book but is an alternative for molecules where the geometry is sterically determined.

4.6.2 Nonstereochemically Active Lone Pairs

In VSEPR one assumes that all domains, regardless of their number, contribute to the geometry around the central atom. However, for elements in periods 3 and 4, six-electron domains seem to be the maximum that can be accommodated, and angles below 90° are often too small for these elements. For example, SF_6, SeF_6, and BrF_6^+ are all known species, but ClF_7 and BrF_7 have never been observed (Fig. 4.15b).

In periods 3 and 4, if the molecule has six bonding domains and a lone pair, AX_6E compounds such as $SeCl_6^{2-}$, there are seven electron domains total in the period-4 element. Instead of adopting a pentagonal bipyramidal or similar base geometry to accommodate the domains, the compound is octahedral and said to have a "stereochemically inactive lone pair". An analysis is shown in Fig. 4.15c. For the formula $SeCl_6^{2-}$, one can assume momentarily that the charge is localized on selenium, which gives a central atom with eight electrons and four lone pairs. Bonds to the chlorines require six of these electrons with one lone pair remaining; in VSEPR notation the formula is then AX_6E, seven domains. The predicted structure is then based on a pentagonal bipyramid, and the lone pair should be in the least sterically encumbered position—axial. (If the lone pair is in the equatorial position in this case, it would only be 72° from other substituents; in the axial position the lone pair is in the less sterically crowded position where it is 90° from other groups.) However, the experimental structure for the compound is octahedral. In other words, the lone pair has no effect on the stereochemistry, and it is as if the lone pair is not on the atom.

In $SeCl_6^{2-}$, one may consider the lone pair to be forced inside the six bond domains into an *s*-like spherical orbital. In some respects, the lone pair becomes part of the core for the selenium atom and is nonbonding. This results in longer Se–Cl bonds

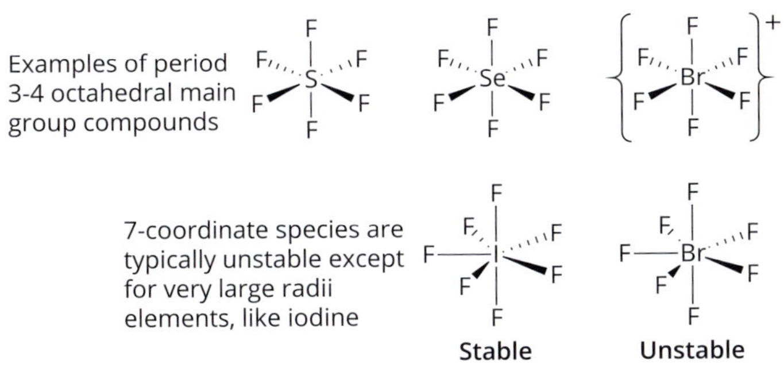

Fig. 4.15b Only main-group elements with very large radii, such as iodine, produce stable species with more than six substituents.

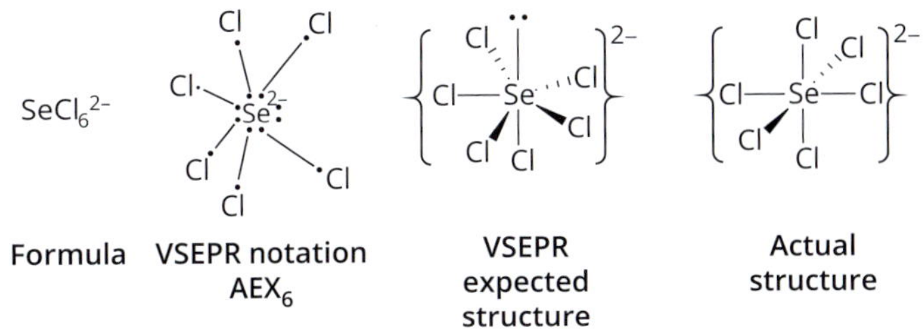

Fig. 4.15c VSEPR analysis of $SeCl_6^{2-}$.

than might be expected, due to increased shielding. The Se–Cl distances are correctly predicted by the LCP model, suggesting that the chlorines are closest packed in this elongated form.

Similarly, BrF_6^- is an octahedral system. However, the period-6 compounds IF_6^- and XeF_6 are not octahedral and can accommodate a lone-pair domain. Nine electron domains cannot be accommodated by even period-6 elements, and XeF_8^{2-}, a rare AX_8E compound, is square antiprismatic like the AX_8 compound IF_8^-.

4.6.3 Incorrect Base Structure Assumption

The base structures in VSEPR (Table 4.1) are assumed to be the most stable geometries accommodating the domains. While electronic or steric effects may cause this assumption to be incorrect, in some cases there are other geometries simply close in energy that become preferred for a variety of reasons. Depending on the exact substituents and other factors, the base structure assumption may be incorrect. Most often, this occurs for 5- and 7-coordinate compounds.

The assumption for 5-coordinate compounds is that trigonal bipyramidal (tbp) is more stable; however, square pyramidal (sp) is often quite close in energy. The classic example of this behavior is for Ni(II) in the complex $\{Cr(en)_3\}\{Ni(CN)_5\} \cdot 1.5\ H_2O$, where the nickel anion crystallizes with two different structures within the same crystal—tbp and sp (Fig. 4.16).[14] (en = $H_2NCH_2CH_2NH_2$, a bidentate ligand[b] that usually binds with both nitrogen lone pairs). Lending further credence to the close energy of the two structures in the complex, another structure done on a slightly more hydrated sample with the formula $\{Cr(en)_3\}\{Ni(CN)_5\} \cdot 2\ H_2O$ in the crystal only showed the square pyramidal structure.[15]

An example of a seven-domain compound is hexafluoroxenon, XeF_6. The VSEPR notation is AX_6E, and the base geometry corresponding to this formula is again a pentagonal bipyramid. In the equatorial plane of the bipyramid the angle between ligands is only 72°; consequently, the substituents with the small domain sizes will occupy the equatorial plane. VSEPR

[b] Transition metal systems will obviously be covered in more detail later. A "ligand" is anything attached to anything else, as originally defined. It is often used in inorganic chemistry to mean a substituent on a metal center. A "bidentate" ligand is one that has two different positions in the same ligand that attach to the metal; in this case, there are two nitrogens that form dative bonds to the metal.

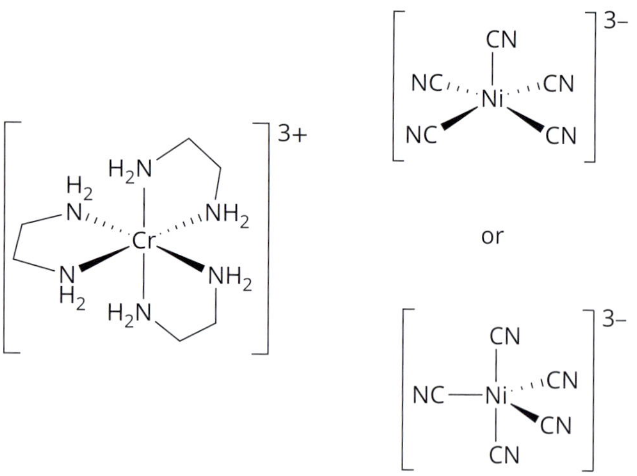

Fig. 4.16 Line drawings of {Cr(en)$_3$}{Ni(CN)$_5$} with both of the observed structures of the Ni anion.

would predict that the structure will have the lone pair in an axial position.

In reality, the structure of XeF_6 is decidedly more complex. Calculations suggest that the lowest energy structure is a distorted "squashed" octahedron (technically, C_{3v} symmetry, see Chapter 5 on group theory).[16] However, the calculations also find that other distorted octahedral structures are within 2 kcal/mol of the proposed ground-state structure. In the proposed octahedral structure, the lone pair becomes stereochemically inactive.

Experimentally, the structure of XeF_6 has been studied by X-ray diffraction, neutron diffraction, solid state NMR, and other techniques.[17] As a consequence of these experiments, at least six different modifications of the structure have been identified.

Chapter Summary

1. Valence Shell Electron Pair Repulsion (VSEPR) is the most common model for finding the 3-dimensional geometry of a molecule from the Lewis structure.
2. In VSEPR, electrons reside in "domains", and the size of the domain depends on whether the electrons are part of a lone pair, bond, or radical.
3. Molecules are given a VSEPR formula where the central atom (A) is surrounded by lone pairs (E), bonds to other groups (X), or single-electron domains (e). For example, H_2O would have a VSEPR formula of AE_2X_2.
4. Depending on the number of domains around the central atom in the VSEPR formula, a base structure is assigned. two domains = linear, three domains = trigonal planar, four domains = tetrahedral, five domains = trigonal bipyramidal, six domains = octahedral, and seven domains = pentagonal bipyramidal.
5. Lone-pair domains are larger than bonds or radical domains. Domains with less electronegative groups are larger than those with more electronegative groups.
6. A large domain tends to reside in positions around the central atom that are farther from the other substituents; for example, a larger domain tends to reside in the equatorial positions in a trigonal bipyramidal base structure.
7. Larger domain sizes may expand the angles around them, often causing angles between smaller domains to contract.
8. Transition-metal structures are often exceptions to VSEPR, but the model is effective for most main-group compounds.

Practice Problems

1. A salt made of an ion having the formula SCl_3 and another ion of formula ICl_4 is prepared and structurally characterized.[18] Which is the cation, and which is the anion in the salt? Using VSEPR, determine the structures of both.
2. In some cases, exceptions to VSEPR can be identified as being differences due to favored resonance forms. In other words, to some extent, identification of the correct VSEPR structure is dependent upon identification of the correct Lewis form. A simple compound where this can be illustrated is nitrosamine, H_2NNO. Two resonance forms for nitrosamine are shown in the figure. Resonance form A shows the perfect pairing form with no formal charges. Resonance form B shows a zwitterionic form. Draw the VSEPR structures clearly showing the 3-dimensional structures for each.

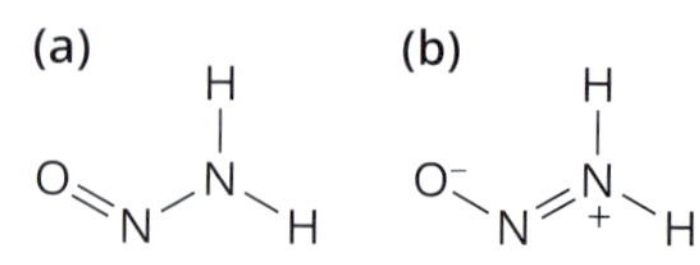

Consistent with NBO calculations resonance form A is the largest participant with a weighting factor of 63%. However, resonance form B is a strong contributor with a weighting factor of 31%. This large contribution from B, which is necessarily flat, cause the molecule to adopt a non-VSEPR geometry. The pyramidalization of the nitrogen bearing the hydrogens would remove the π-interaction with the ON-nitrogen leading to a higher-energy structure. As a result, nitrosamines are planar compounds.

3. Draw a reasonable Lewis structure for the following compounds and find their structures according to VSEPR.
   a. SO_3^{2-}
   b. $SbMe_4^-$
   c. ClF_4^+
   d. BrF_3
   e. BrF_5
   f. CH_3^-
   g. CH_3^+
   h. IO_4^-
   i. XeF_2
   j. CO_3^{2-}
   k. I_3^-
4. Draw the Lewis structures of PF_3 and PF_5. Predict the molecular geometries using VSEPR.
5. For the following compounds, (i) give the VSEPR formula (ii) give the base geometry (iii) draw the 3-dimensional structure (iv) give the deviation in angles from the base geometry, if any, and (v) name the structure of the molecule. Show the angle deviation as >, =, < the standard angle in the base geometry.

Example: SF_2

(i) AE_2X_2

(ii) tetrahedral

(iii and iv)

<109.5°

(v) bent

(a) SbF_6^-
(b) CH_3
(c) SeF_4
(d) XeF_2
(e) XeF_4
(f) $BeCl_2$
(g) SO_2
(h) $O{=}S(CH_3)_2$
(i) ClF_3
(j) BrF_5
(k) OH_3^+
(l) ClO_3^+
(m) $HgBr_2$
(n) BeF_4^{2-}
(o) CO_3^{2-}

Cited References

1. Sidgwick, N. V., Stereochemical types and valency groups *Proc. Chem. Soc. London A* 1940, *176*, 153–180.
2. Gillespie, R. J., The VSEPR model revisited *Chem. Soc. Rev.* 1992, *21*, 59–69.
3. Gillespie, R. J.; Robinson, E. A., Models of molecular geometry *Chem. Soc. Rev.* 2005, *34*, 396–407.
4. Marsden, C. J.; Bartell, L. S., Molecular-structure and vibrational-spectrum of $P(CF_3)_3$: possible example of an altruistic covalent interaction *Inorg. Chem.* 1976, *15*, 2713–2717.
5. Kurimura, H.; Yamamoto, S.; Egawa, T.; Kuchitsu, K., Geometrical Structure of PF_5 as Studied by Gas Electron-Diffraction and Spectroscopic Data *J. Mol. Struct.* 1986, *140*, 79–86.
6. Griffiths, J. E., Molecular Structure of CF_3PF_4 from Infrared Spectroscopy *J. Chem. Phys.* 1968, *49*, 1307.
7. Griffiths, J. E., Molecular Structure of CF_3PCl_4: Infrared and Raman Spectra *J. Chem. Phys.* 1964, *41*, 3510.
8. Griffiths, J. E.; Beach, A. L., Molecular Structure of $(CF_3)_2PCl_3$: Infrared and Raman Spectra *J. Chem. Phys.* 1966, *44*, 2686.
9. Lavilla, R. E.; Bauer, S. H., Molecular structure of iodine heptafluoride *J. Chem. Phys.* 1960, *33*, 182–186.
10. Gillespie, R. J.; Robinson, E. A., Electron domains and the VSEPR model of molecular geometry *Angew. Chem. Int. Ed.* 1996, *35*, 495–514.
11. Allen, P. W.; Bowen, H. J. M.; Sutton, L. E.; Bastiansen, O., The molecular structure of acetone *Trans. Faraday Soc.* 1952, *48*, 991–995.
12. Kaupp, M., "Non-VSEPR" structures and bonding in d^0 systems *Angew. Chem. Int. Ed.* 2001, *40*, 3535–3565.
13. Morse, P. M.; Girolami, G. S., Are d^0 ML_6 complexes always octahedral: The X-ray structure of trigonal-prismatic $[Li(TMED)]_2[ZrMe_6]$ *J. Am. Chem. Soc.* 1989, *111*, 4114–4116.
14. Raymond, K. N.; Corfield, P. W.; Ibers, J. A., Structure of tris(ethylenediamine) chromium(III) pentacyanonickelate(II) sesquahydrate, $[Cr(NH_2CH_2CH_2NH_2)_3][Ni(CN)_5]\cdot1.5\,H_2O$ *Inorg. Chem.* 1968, *7*, 1362.
15. Terzis, A.; Raymond, K. N.; Spiro, T. G., On the Structure of $Ni(CN)_5^{3-}$: Raman, Infrared, and X-ray Crystallographic Evidence *Inorg. Chem.* 1970, *9*, 2415.
16. Peterson, K. A.; Dixon, D. A.; Stoll, H., The Use of Explicitly Correlated Methods on XeF_6 Predicts a C_{3v} Minimum with a Sterically Active, Free Valence Electron Pair on Xe *J. Phys. Chem. A* 2012, *116*, 9777–9782.
17. Hoyer, S.; Emmler, T.; Seppelt, K., The structure of xenon hexafluoride in the solid state *J. Fluorine Chem.* 2006, *127*, 1415–1422.
18. Mishima, M.; Okuda, T., Zeeman effect of Cl-35 and (127)INQR of Form-I of $SCl_3\ ICl_4$ *J. Mol. Struct.* 1997, *412*, 205–211.

CHAPTER 5

Symmetry and Group Theory

CONCEPT COVERAGE

1. Mathematical constructs relevant to symmetry in chemistry
2. Symmetry elements and symmetry operations
3. Definition and properties of a mathematical group
4. Assignment of point groups to molecules
5. Schoenflies symbols for point groups and Mulliken symbols for irreducible representations
6. Reducible and irreducible representations
7. The importance of basis set choice in the reducible representation
8. Vibrational analysis of molecules: introduction to decomposition formula and projection operator
9. Products of irreducible representations

"Symmetry is what we see at a glance" — **Blaise Pascal**

"The chief forms of beauty are order, symmetry, and definiteness" — **Aristotle**

5.1 Symmetry and Nature

Symmetry is a part of our daily lives. Whether we realize it or not, we do numerous symmetry calculations a day to evaluate objects. The reason is that *symmetry affects function*. Almost all cars, for example, have a plane of symmetry where the right and left sides are very similar, a fact our brains recognize immediately (Fig. 5.1). What would you think of an automobile that was quite different on each side, perhaps listing to the left due to larger wheels and tires on the right side? One would definitely think its function would be affected!

We can easily tell there is symmetry in an object, even if we cannot clearly articulate or define that symmetry. For instance, take Fig. 5.2(a), which represents a light-harvesting complex found in some bacteria: most people can easily see that there are nine symmetry-related pieces to the protein. These are, as we will discuss, related by a 9-fold proper rotational axis. Similarly, Fig. 5.2(b) shows a picture prepared from the X-ray diffraction data of the rhinovirus, which has 5-fold symmetry.

Symmetry affects the properties and function of objects in everyday life, and it is therefore unsurprising that symmetry should equally determine the properties and functions of molecules as well. Often, the symmetry of a molecule provides one of the most powerful ways of classifying it, as other molecules that have the same symmetry (or close to it) share some similar properties. In this chapter we will introduce a mathematical formulism for discussing and using symmetry, which will allow us to describe unambiguously the symmetry properties of molecules. We will show that the ensemble of all symmetry operations that describe a molecule constitute a mathematical group, which will allow us to use the powerful machinery of group theory to guide us in our study of the symmetry of molecules.

In chemistry we use symmetry a great deal and in all areas. Below are a few simple examples:

(1) *Symmetry of Orbitals in Atoms and Molecules* All orbitals have symmetry properties determined by their shapes and phases. An *s*-orbital, for example (Fig. 5.3), displays spherical symmetry. It can be rotated, inverted, reflected, etc., without change. If an orbital, like the *s*-orbital, does not change phase on inversion it is labeled "gerade" (/gəˈʁaːdə/, German for "even"). The *p*-orbitals do change phase with inversion through their center point and are thus labeled as "ungerade" (German for "odd"). The *d*-orbitals are gerade. The label goes as the parity, such as being even or odd, of the quantum number l: s ($l = 0$) is gerade/even, p ($l = 1$) is ungerade/odd, d ($l = 2$) is gerade, and f ($l = 3$) is ungerade.

Fig. 5.1 Pictures of (a) a 1967 Pontiac Firebird, and (b) 1969 Dodge Charger. Note that with some superficial exceptions, like the company logos, there is left and right symmetry.

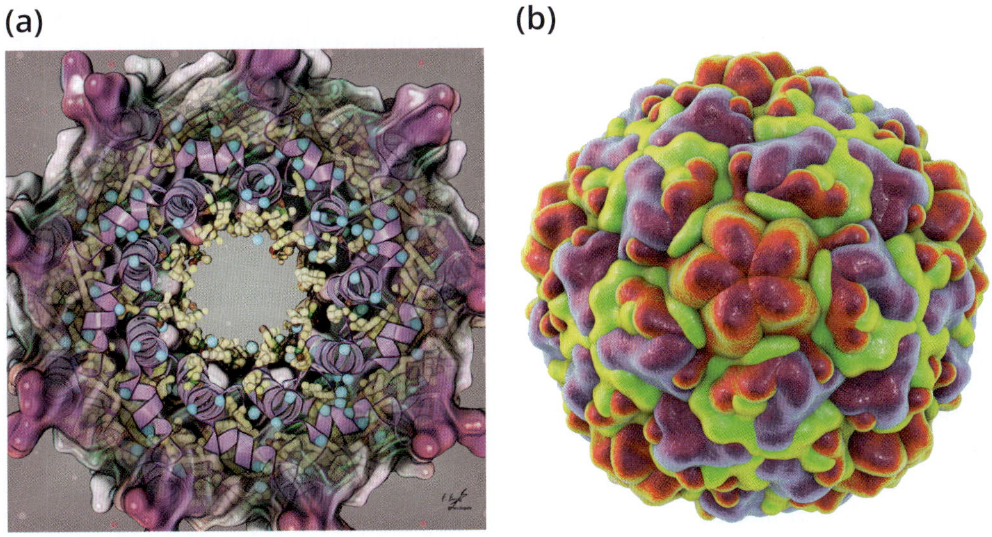

Fig. 5.2 (a) Light-harvesting complex illustrating a 9-fold rotational axis. (b) View of a rhinovirus down its 5-fold rotational axis.[1,2]

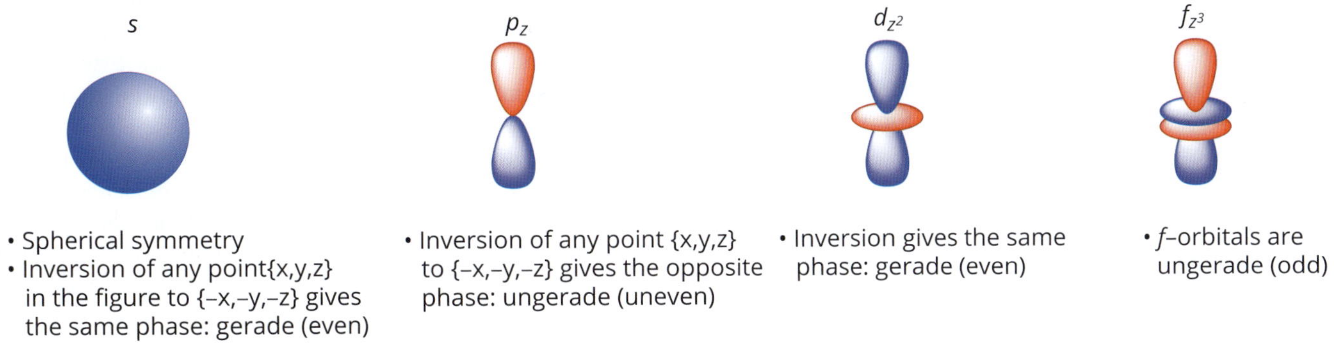

Fig. 5.3 The *s*-, *p*-, d_{z^2}-, and f_{z^3}-orbitals as examples of orbital symmetry. The *s*-orbital (quantum number $l = 0$) is symmetric with respect to inversion (no phase change), which is called "gerade". The *p*-orbitals (quantum number $l = 1$) are antisymmetric with respect to inversion (phase changes on inversion), ungerade. The *d*-orbitals (quantum number $l = 2$) are symmetric with respect to inversion (no phase change), gerade. The *f*-orbitals ($l = 3$) are ungerade.

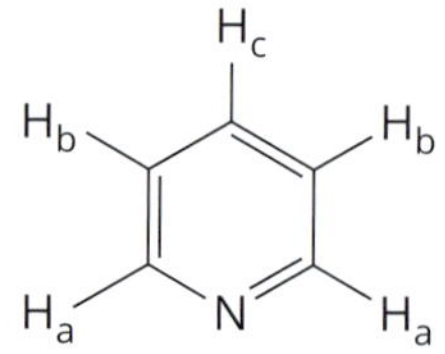

Fig. 5.4. Pyridine with its three different types of protons (H_a, H_b, and H_c), some of which are related by obvious symmetry.

(2) *Computational Chemistry* The consideration of symmetry can be used to great time-saving advantage in calculations of molecular properties. Symmetry is often used to tell very quickly if integrals will be zero or nonzero; in addition, it can greatly reduce the dimensionality of matrices that must be handled and stored in memory by requiring some terms be equivalent due to symmetry.

(3) *Spectroscopy* Symmetry is used both qualitatively and quantitatively in spectroscopy. For example, in the ^1H NMR spectrum of pyridine (shown in Fig. 5.4), the three different types of protons in the molecule each lead to three unique resonances, each with their own spin–spin coupling. In spectroscopy, there are also symmetry-related selection rules for the types of transitions that can be observed. In electronic absorption spectroscopy, for example, excitations of electrons from gerade to other gerade orbitals are "forbidden", which leads to weakly absorbing transitions. In infrared spectroscopy the N_2 molecule does not absorb IR radiation due to its symmetry, and N_2 consequently also does not have any greenhouse effect, in spite of being the dominant species in Earth's atmosphere.

(4) *Chemical Reactivity* Atomic orbitals must be of the same symmetry to combine and form molecular orbitals (see discussion of orthogonal orbitals in Section 3.7). In addition, orbitals must be able to adopt like-symmetries for reactions to occur. Examples of this are the Woodward–Hoffman rules for pericyclic reactions, which are based on orbital symmetry.

In order to use symmetry in chemistry we need to build a vocabulary for its discussion, and we need to understand something of the mathematical basis for the field called group theory.

5.2 Elements and Operations: Building a Vocabulary for the Discussion of Symmetry

In this section we will define the terms *symmetry element* and *symmetry operation*—two closely related ideas that nevertheless are distinct from each other. The central idea that determines what we imply by "symmetry" is the concept of *invariance*, which means exactly what it suggests: "*the property of not changing*". Imagine an object—any object. If you were to close your eyes and have someone else manipulate the object in some definite ways before you were allowed to look at it again, would you be able to know whether the object had been manipulated or not? If the object had absolutely no symmetry at all, then there would be no way to ever manipulate it without your noticing that it had been moved unless they put it back exactly as before. An object that would have some degree of symmetry (say a uniform sphere, or a plain cube) could have been rotated, but for some manipulations (rotating the cube by 180°, for example) there would be no way of knowing for sure whether the object had been moved unless you saw the movement. If you looked away, the action took place, and then if you looked back you would never know the action had occurred. This idea of invariance suggests that symmetry must be related to a specific type of "action"—something that moves the object in a determined manner by taking it to an equivalent orientation in space. Such an action is what we call a *symmetry operation*. As it happens though, symmetry operations cannot be defined directly without making a direct connection to the object itself; we always need a reference frame in order to define what the operation is. For instance, the operation "turn the object by 180°" means nothing unless we specify around which axis that rotation occurred. In this instance, the axis around which the symmetry operation is carried is called a *symmetry element*, which simply refers to the mathematical object (a point, a line, or a plane) around which a symmetry operation occurs. Symmetry operations cannot exist in the absence of symmetry elements, and once a symmetry element is determined to exist, it automatically implies the existence of a specific set of symmetry operations that would leave the object unchanged ("invariant"). Note that although a given symmetry operation is always associated to one, and only one, symmetry element, a given symmetry element can often (but not always) define more than one symmetry operation. For instance, the axis of rotation that goes through the center of a square leads to the 90°, 180°, 270°, and 360° (≡ 0°) rotation operations that leave a square unchanged.

There is a total of four types of symmetry elements that can be found for a 3-dimensional object (there are two types of "lines" of symmetry, as we will see later, in addition to the point and the plane of symmetry). Whenever these elements exist in a given object they will necessarily always cross through the same unique specific point inside the object, which directly implies that there can be only one point of symmetry no matter how high or low the symmetry of an object. (There can be many lines and planes of symmetry, depending on the level of symmetry of the object.) In fact, all the symmetry elements will pass through a single point, giving this the term *point symmetry*. When we discuss the symmetry of objects (including molecules) we will classify them into "point groups", which completely determines the symmetry properties of that object. Although mathematically there is an infinite number of point groups, in practice most molecules fall into maybe a few dozen chemically important groups.

Extending these ideas to include translational symmetry through 3-dimensional space as another possible symmetry operation (which requires that the object repeats over and over in space, as happens in crystal lattices) leads to "space groups", which are used in describing the structures of solids. These will be discussed very briefly in Chapter 19 on solid-state chemistry, and are discussed at length in classes on X-ray diffraction.

5.2.1 The Four Symmetry Elements (C_n, i, σ, and S_n) and Five Symmetry Operations (C_n, i, σ, S_n, and E)

As discussed previously, *a symmetry element is a geometric construct (point, line, plane) about which a symmetry operation is performed.* A *symmetry operation is the manipulation of an object about a symmetry element so that the object is possibly different but is always indistinguishable from the start position.* There are four distinct symmetry elements, and five general types of symmetry operations that can be defined from these four elements.[a]

5.2.1.1 The Identity Operation (E)

The first operation we discuss is the simplest of all operations: the act of doing nothing. Obviously, this "action" necessarily leaves the object unchanged, so technically fits our definition of a symmetry operation. This operation of doing nothing is called the *identity operation* and is assigned the symbol E. The identity operation can be applied to any object, no matter how little symmetry it has otherwise. Although this operation might appear trivial (and perhaps a bit useless at first sight), we will see that it is essential in the mathematical treatment of symmetry operations. For now, you can think of the identity operation as being the symmetry equivalent of the arithmetic operations "adding zero" or "multiplying by one"; they each are meaningful mathematical statements that have the property of leaving the input unchanged.

5.2.1.2 The Proper Rotation (C_n)

A "proper rotation" refers to the usual meaning of a rotation around an axis. As always, to be a symmetry operation a proper rotation must not lead to any noticeable change in the way that the object is perceived for an observer. Generally, the axis around which such a rotation can be performed is labeled using the symbol C_n, where n is called the order of the rotation, which is simply the number of times a given rotation must be carried out to return to the original configuration. A C_n axis of rotation then implies the existence of n rotation operations, each increasing in $360°/n$ increments. For example, a C_2 operation refers to a rotation of $360°/n = 360°/2 = 180°$ around a C_2 axis.[b] Another common way to refer to axes of rotation is to use their order directly, such as 2-fold, 3-fold, and 4-fold axes of proper rotation are C_2, C_3, C_4 axes. In this book we will refer to clockwise rotations with positive angle values, and counterclockwise ones with negative angles. This choice is purely conventional. Other books, such as Cotton's[3] excellent treatise on group theory, use the opposite convention, but whichever convention one chooses must be adhered to consistently and uniformly.

As a simple example, consider the rotational symmetry of an equilateral triangle (Fig. 5.5). Because all the vertices on the triangle have equal lengths and are otherwise indistinguishable, we will label them as 1, 2, and 3 to keep track of their positions. Rotation through 1/3 of 360° in the plane of the triangle does not change the orientation in any perceptible manner (without the artificially added labels); consequently, we say "an equilateral triangle has a 3-fold proper rotational axis." In the top portion of Fig. 5.5, the 3-fold rotation axis is coming out of the plane of the paper (or screen) and is in the precise middle of the triangle. This symmetry element is given the symbol, C_3.

If we rotate clockwise the triangle on the top-middle by 120° clockwise, the labels will move so that we get the triangle on the top-left. This specific operation is called a C_3^1 rotation, because we did a single clockwise rotation of 360°/3: the subscript always refers to the order of the rotation (which also tells you the angle of rotation), whereas the superscript is used to refer to the number of times that operation should be carried out. Most often, the "1" is not written, so this rotation will often be listed simply as "C_3". If we rotate the triangle through a C_3^2 (two clockwise C_3 rotations), the labels in Fig. 5.5 will change

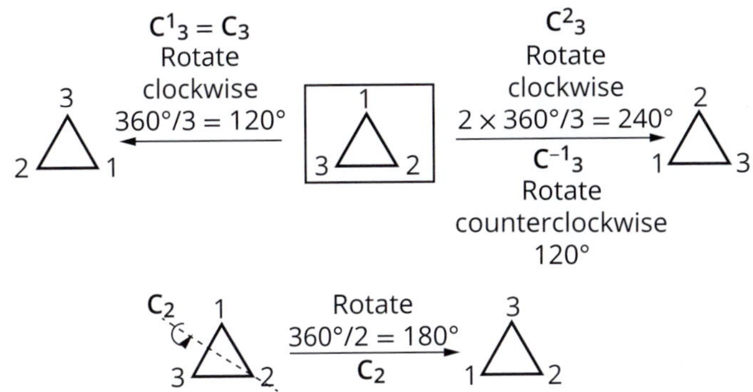

Fig. 5.5 Rotational symmetry in an equilateral triangle.

[a] Strictly speaking, every symmetry operation can be described as one of two types of rotation (proper or improper), so all symmetry elements can be also reduced to rotation axes, either proper or improper. For instance, and as described below, a center of inversion is strictly identical to an improper rotation of order 2 (S_2 axis of rotation), and planes of reflection are, by definition, identical to improper rotations of order one (S_1 rotations) against an axis perpendicular to that plane. Although this is a mathematically more rigorous way of formulating point group theory, we will keep using the more intuitively appealing notions of points and planes of symmetry.

[b] Some texts use distinct symbols to distinguish an operation from the element around which it is defined—say C_2 for a proper axis of rotation of order 2, and $\hat{C}_2$ for the 180° rotation operation around that axis, where the caret symbol is used to explicitly refer to operations. We will not make this distinction here, as the context generally makes it perfectly clear whether the symbol refers to a symmetry element or a symmetry operation.

to yield the configuration shown on the top-right. Alternatively, we could have accessed the same triangle at the top-right by rotating from our starting position 1/3 of a turn counterclockwise, which would be designated as a C_3^{-1} operation.

Finally, if we rotate the triangle three times 120°, C_3^3, the result is a triangle with no changes in label positions, and a return to the original configuration, before any rotation had been performed. A C_3^3 rotation is equivalent to the act of doing nothing in the first place, which you will recall is the *identity operation*. We can thus write $C_3^3 \equiv E$, or, more generally for any proper rotation $C_n^n \equiv E$, operating n times around a C_n axis is formally identical to the identical operation.

These symmetry transformations, like C_3^1, $C_3^2 (\equiv C_3^{-1})$, and $C_3^3 (\equiv E)$, are *symmetry operations* created by the *symmetry element* C_3. In general, a symmetry element may generate many symmetry operations. In the case of a proper rotation axis C_n, the symmetry element will create, at least, n operations, $C_n^1, C_n^2, \ldots, C_n^n$.

An equilateral triangle contains one other type of proper rotation, which involves rotation of the triangle around axes that are in the plane of the triangle and go through one vertex and cut the opposite side in its middle point (bottom portion of Fig. 5.5). There are three such axes, and they are C_2 axes. The existence of three equivalent C_2 axes in the plane of the equilateral triangle is fundamentally linked to the existence of the C_3 axis, and once you define one such C_2 axis then the other two are automatically generated by the C_3 symmetry element.

As an additional example, the C_4 rotational symmetry in a square is described in Fig. 5.6. There are a couple of additional equivalences shown in Fig. 5.6. For example, a C_4^2 rotation is mathematically identical to a C_2 rotation around the same axis ($C_4^2 \equiv C_2$).

5.2.1.3 The Reflection Operation and Mirror Planes (σ)

The reflection operation is defined relative to a plane of symmetry, which is also called a *mirror plane*. Mathematically, a mirror plane exists if every point on one side of the plane corresponds with an identical point on the other side of the mirror located at the exact same distance from the plane, along a line that connects both points and which is perpendicular to the plane. People, for example, contain approximate

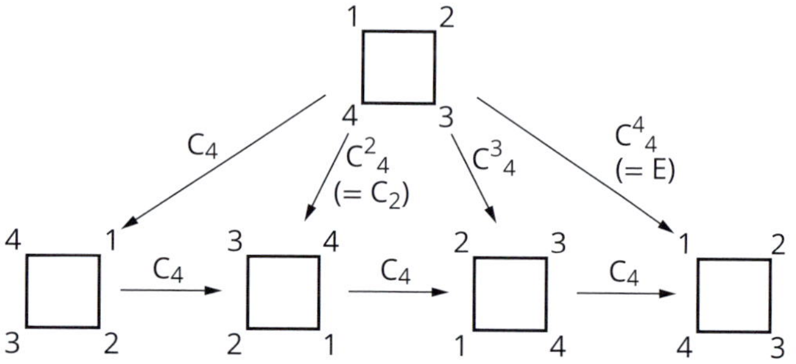

Fig. 5.6 Rotational symmetry in a square.

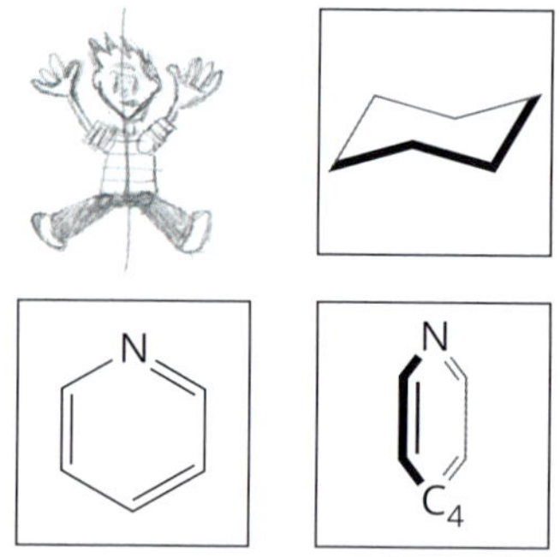

Fig. 5.7 Objects containing a plane of symmetry. (*top*) Approximate plane of symmetry in a cartoon person (drawing by Isaac Odom) and in chair cyclohexane. (*bottom*) Planes of symmetry in pyridine, the plane of the molecule (*left*) and a mirror plane that runs through the nitrogen and C_4 perpendicular to the aromatic ring.

planes of symmetry (bilateral symmetry) on their left and right sides (Fig. 5.7), as do most cars (Fig. 5.1); all polygons and otherwise flat figures (circles, ovals, crescents) have at least one mirror, which is the plane of the figure itself.

Planes of symmetry and the reflection operation are given the symbol σ (Greek letter sigma). Often, it is useful to distinguish between different types of mirror planes that exist in the same object; for example, "horizontal mirror planes" (symbol σ_h) refer to planes that are *perpendicular* to the highest-order rotation axis of the object. The highest-order rotation axis for an equilateral triangle is a C_3 axis, and the mirror plane which is perpendicular to that axis (the plane of the triangle itself) is then called a σ_h plane. A plane of reflection which instead *includes* the highest-order rotation axis is called a "vertical mirror plane" (symbol σ_v); an equilateral triangle has three σ_v planes that each bisect the triangle in two equal halves. Some molecules have more than one type of vertical mirror planes, and the term "dihedral mirror plane" (symbol σ_d) is used to distinguish one type from the other (more about this later). For instance, a square has a total of five distinct planes of reflection: one σ_h plane containing the square itself, and perpendicular to the 4-fold axis of rotation; two σ_v planes, which we can define as being those planes perpendicular to the square and that bisect it along the two long diagonal connecting opposite vertices; and two σ_d planes, which we can define as those two planes that are perpendicular to the square and that bisect it in two equal rectangles along the middle of opposite sides.

A molecule containing all three types of mirror planes is square-planar $\{PtCl_4\}^{2-}$, shown in Fig. 5.8 with a coordinate axis. (The choice of coordinate axes will be discussed in detail shortly.) As discussed previously, the σ_h plane is perpendicular to the highest-order rotation axis; in this case, it contains the plane of the molecule; the σ_v planes run along the Cl–Pt–Cl bond vectors, and the σ_d planes run between bond vectors. (It is customary to refer to the planes that contain the most atoms as vertical planes, and those with the fewest number of atoms at dihedral planes; the distinction is purely conventional.) Importantly, the σ_v (vertical mirror planes) and σ_d (dihedral

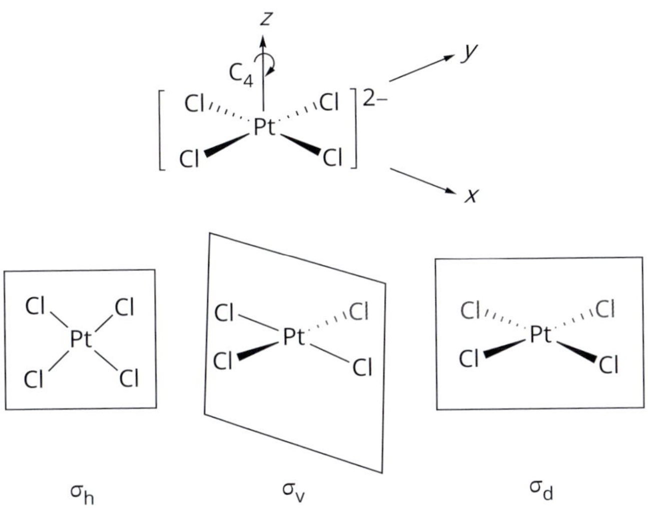

Fig. 5.8 Mirror planes in $PtCl_4^{2-}$.

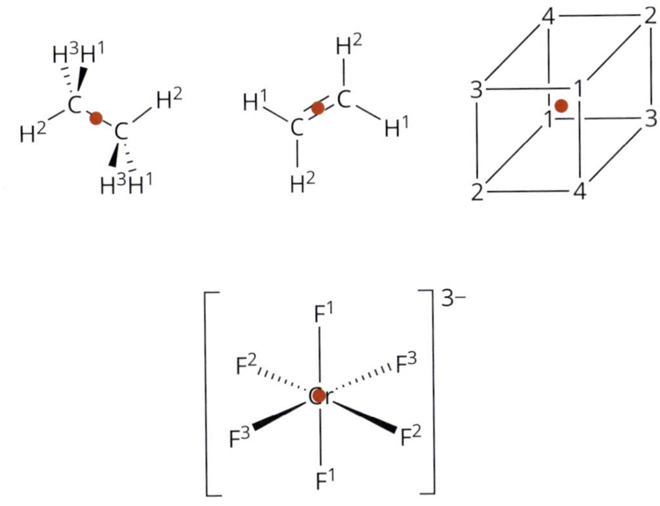

Fig. 5.9 Molecules and objects possessing an inversion center symmetry element. The inversion point is shown in red. Atoms and vertices that are equivalent due to inversion are labeled the same.

mirror planes) are not symmetry-related; whereas one σ_v plane can directly be converted to the other one by a single rotation around the C_4 axis (and similarly for the two σ_d planes), there is no symmetry operation that maps a σ_v into a σ_d.

When considering chemical species, individual atoms in a molecule are always considered to be spherical. As a result, atoms that sit on mirror places are unmoved and unchanged by the reflection. Consequently, the plane of the molecule in pyridine (Fig. 5.7, bottom left) contains a mirror plane symmetry element that reflects the top half of every spherical atom in the molecule into the bottom half.

5.2.1.4 The Inversion Center (i)

All the symmetry elements always meet at a single unique point in a given object under consideration. For example, all planes and axes of rotation of the equilateral triangle and of the square meet at a single point located precisely at the center of the triangle. This observation is a general feature of point symmetry, and it is impossible for symmetry elements not to pass through this point. For example, you will never find two mirror planes that are parallel to each other. Sometimes the point of symmetry is coincident with a symmetry element, a *point of inversion*, which is assigned the symbol *i*. The inversion operation involves taking each point along a line that connects it to the center of inversion, to the same distance on the other side of the center of inversion. Although the point of symmetry of an equilateral triangle is not a center of inversion (if you draw a line from any vertex straight down to the point of symmetry and keep going for the same distance on the other side, you do not end up on another vertex), a square does have a center of inversion, and the inversion operation leaves the square invariant.

Mathematically, if we define the point of symmetry of a molecule (where all symmetry elements meet) as the {0,0,0} coordinate of a Cartesian frame of reference, every point {x,y,z} will be the equivalent to the {−x,−y,−z} point if the molecule has an inversion center. A few molecules and objects having an inversion center symmetry element are shown in Fig. 5.9. The inversion center may be on an atom or between atoms. The inversion-related points and atoms are labeled the same in the figure; for example, F^1/F^1 in CrF_6^{3-}.

An inversion center, by its nature, will often reside on whatever is unique in the molecule, whether it is an element that only appears once or a type of bond not found anywhere else in the molecule. For this reason, $\{CrF_6\}^{3-}$ has an inversion center on the chromium, H_3CCH_3 (staggered conformation) has an inversion center in the middle of the C–C bond, and ethylene has an inversion center in the middle of the double bond.

5.2.1.5 The Improper Rotation (S_n)

The final symmetry element to be discussed is the *improper rotation* axis, given the symbol S_n. An improper rotation can be thought of as a composite operation consisting in a C_n rotation followed by a reflection through a plane perpendicular to the C_n axis of rotation. The existence of an improper axis of rotation only requires that such a combined operation leaves the object unchanged; it is not necessary for either the C_n axis or the plane of reflection used in the combined operation to exist independently on their own.

A compound with an S_6 rotational axis is staggered ethane. As shown in Fig. 5.10, the S_6 axis lies along the C–C bond and converts one of the top hydrogens into one of the bottom, H^1 into H^5, for example. Note here that staggered ethane neither possesses a C_6 axis nor a plane of reflection perpendicular to the C–C bond. Nevertheless, the S_6 axis clearly exists even if none of its individual composite operations do.

Improper rotations of order one (S_1) and two (S_2) are given special names, which we have seen already. An improper rotation of order one (S_1), is a 360° rotation (no rotation) followed by a reflection, which is identical to only doing a reflection operation. Mathematically, all reflections can be thought of

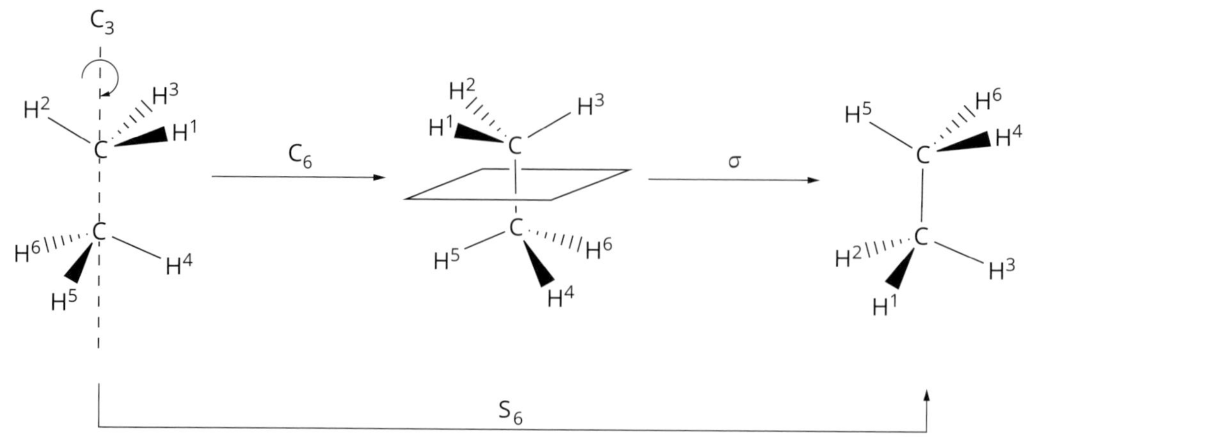

Fig. 5.10 The S_6 improper rotation in staggered ethane.

as being improper rotations of order one; that is, $S_1 = \sigma$. An improper rotation of order two (S_2 rotation) corresponds to a 180° rotation followed by a reflection perpendicular to the axis of rotation, and such a combination of operations is formally equivalent to the inversion operation; that is, $S_2 = i$.

In some cases, the improper rotational axis can be of higher order than the proper rotational axis in the molecule. We saw this in the staggered ethane example, which has an S_6 axis (Fig. 5.10). This also occurs in a tetrahedron, which is the symmetry of methane, CH_4 (Fig. 5.11).

For methane, the highest-order proper rotational axis is a C_3, which resides along one of the C–H bonds. In Fig. 5.11, one of the C_3 axes is shown, the one along the C–H^1 axis. Rotation around this axis clockwise (looking down the H^1–C bond)

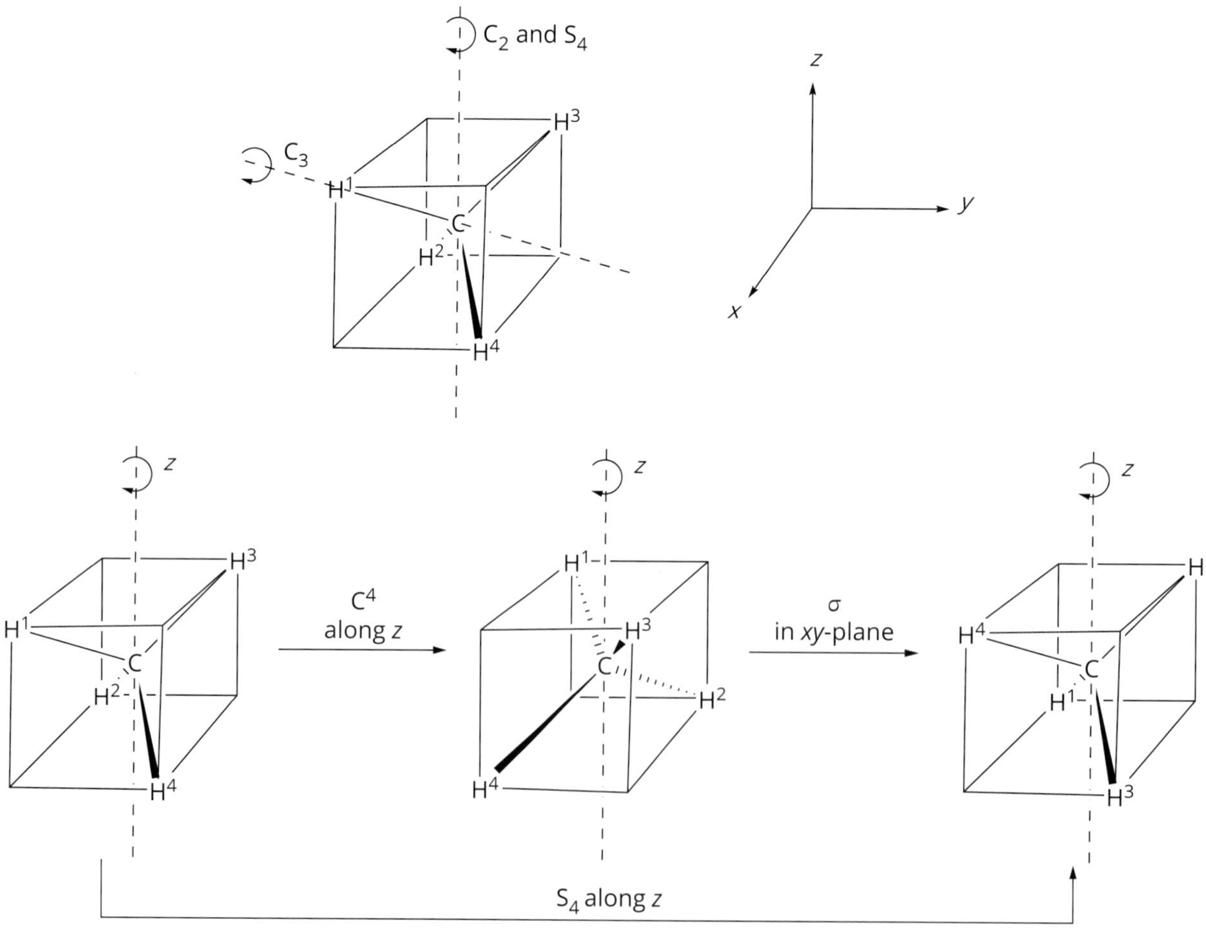

Fig. 5.11 Illustration of proper and improper axes in CH_4.

converts $H^2 \rightarrow H^3$, $H^3 \rightarrow H^4$, and $H^4 \rightarrow H^2$. Counterclockwise rotation around the same C_3 axis converts $H^2 \rightarrow H^4$, $H^4 \rightarrow H^3$, and $H^3 \rightarrow H^2$, which is the same as a $C^2{}_3$ rotation in the clockwise direction. Consequently, there are two different C_3 operations along the H^1–C bond axis, C_3 and $C^2{}_3$; these are different because they convert say H^2 into H^3 in the case of the C_3 and into H^4 in the case of $C^2{}_3$. As a result, there are eight total C_3 operations in a tetrahedral molecule, a C_3 and a $C^2{}_3$ along each bond axis.[c]

There are also C_2 axes bisecting each H–C–H bond. One of these axes is shown in Fig. 5.11 (top). The one shown bisects the H^1–C–H^3 angle, but also bisects the H^2–C–H^4 angle, running through two of the faces of the cube containing the tetrahedron. The C_2 axis shown will convert $H^1 \rightarrow H^3$, $H^3 \rightarrow H^1$, $H^2 \rightarrow H^4$, and $H^4 \rightarrow H^2$. If one places an x, y, and z coordinate axis at the carbon (to the right of the cube in Fig. 5.11), then the C_2 axis drawn would be along the z-axis. One can see that there will also be C_2 axes in the tetrahedral molecule along the y- and x-axes.

Colinear with these C_2 axes are the highest-order rotation axes in the tetrahedron, S_4 axes. In the bottom of Fig. 5.11, one of the S_4 rotations along the z-axis is shown explicitly for methane. The S_4 involves a C_4 around the z-axis (middle bottom) followed by reflection perpendicular to that axis (right bottom). This S_4 operation therefore converts $H^1 \rightarrow H^2$, $H^2 \rightarrow H^3$, $H^3 \rightarrow H^4$, and $H^4 \rightarrow H^1$. An S_4^2 operation is equivalent to doing the S_4 operation twice, which really is the same thing as a single C_2 rotation. An S_4^2 operation is, therefore, not a new symmetry operation in itself.[d] The $S_4^3 (\equiv S_4^{-1})$ operation is distinct. As a result, there are six unique S_4 rotation operations in a methane molecule, S_4 and S_4^3 operations along each of the x-, y-, and z-axes.

5.2.2 The Principal Axis of Rotation

An important issue in symmetry analyses, for the sake of consistency and ease of communication, is the choice of frame of reference for the x-, y-, and z-axes to a molecular system. Although this could be done in a number of equally valid ways, in practice it is desirable to have an agreed upon convention. Most often the highest-order axis of rotation is used to define the coordinate system. This is useful because most molecules have at least one axis of rotation, and many have only one axis of highest order.[e] We will call the axis of rotation (proper or improper) of highest order the *principal rotation axis* and *assign it as coincident with the z-axis*. Once the z-axis has been assigned, the x- and y-axes are generally chosen to lie along other axes of rotation (if they exist) and to go through the most atoms as possible. For example, the highest-order axis of rotation for the square-planar molecule $PtCl_4^{2-}$ is a C_4 proper rotation axis perpendicular to the plane of the molecule and passing through Pt, which we assign as the principal axis of rotation and the z-axis (Fig. 5.12). In $PtCl_4^{2-}$ there are also four C_2 axes perpendicular to the principal axis of rotation, but only two of these axes go through the Cl atoms; that is, they are colinear with Cl–Pt–Cl bonds, so we choose these as the x- and y-axes.[f]

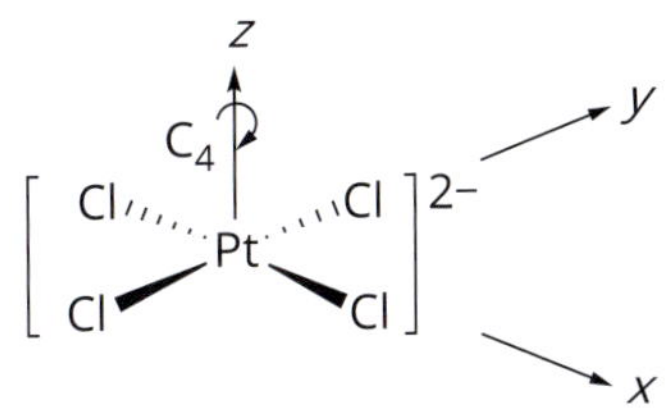

Fig. 5.12 Structure of $PtCl_4^{2-}$ showing the principal rotation axis (C_4) and the assignment of all the coordinate axes.

In methane, the highest-order proper rotational axes are C_3, which are found along any of the four C–H bonds; however, there is a higher-order S_4 improper rotational axis bisecting any H–C–H angle, which defines the principal rotation axis (Fig. 5.11). The three S_4 axes are all equivalent by symmetry; that is, each can each be converted into one of the other two using some of the other symmetry operations of an ideal tetrahedron, so it is immaterial which one we choose as the z-axis. Once we choose one of the three, the other two are fixed using the requirement that the coordinate system be right-handed.

One notable exception to the highest-order rotational axis being the z-axis is for MX_6 octahedral molecules (see Fig. 5.28), where the highest-order axes are S_6 improper axes, which generally do not lie on any of the atoms of the molecule (except the central one). Instead, these molecules have three C_4 axes that all go through the X–M–X bonds, which often makes them much more natural choices for the x-, y-, and z-axes.

5.2.3 Locating Symmetry Elements and Identifying Operations: BF_3 as an Example

As an example, using the previous symmetry elements and symmetry operations, let us examine in some detail the properties of BF_3, which is a planar molecule with equal F–B–F angles as shown in Fig. 5.13a.

[c] All $C^3{}_3$ operations are just the E operation, and not listed as threefold rotations.

[d] Note that this argument only works properly because we can always interchange the sequence order of a reflection and a rotation. Mathematically, the two types of operations are said to "commute" with each other. Not all symmetry operations commute with each other; for instance, two rotation operations that are done around two different axes do not commute. The order in which the operations are carried out does matter in that case.

[e] Special cases for molecules without any axes of rotation or with more than one axis of highest order can be treated separately.

[f] In this book we use a right-handed convention for our Cartesian reference frames, which automatically fixes the choice of the x- and y-axes as shown in Fig. 5.12. As a reminder, a right-handed Cartesian system follows the "right-hand rule": if you point your right-hand thumb in the direction of the positive z-axis and your other four right-hand fingers in the direction of the positive x-axis, then the positive y-axis should be oriented in the direction your four fingers point to when you curl them toward your palm.

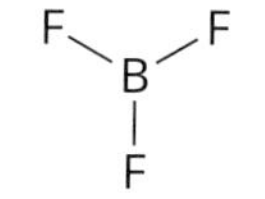

Fig. 5.13a Planar structure of BF_3.

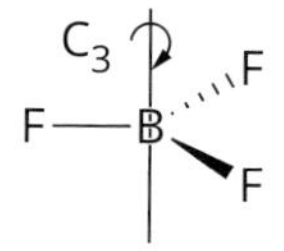

Fig. 5.13b The C_3 in BF_3.

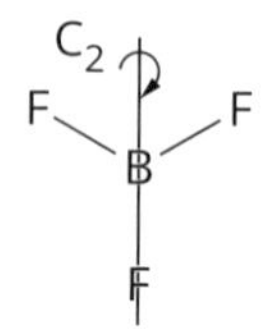

Fig. 5.13c One of the C_2 axes in BF_3.

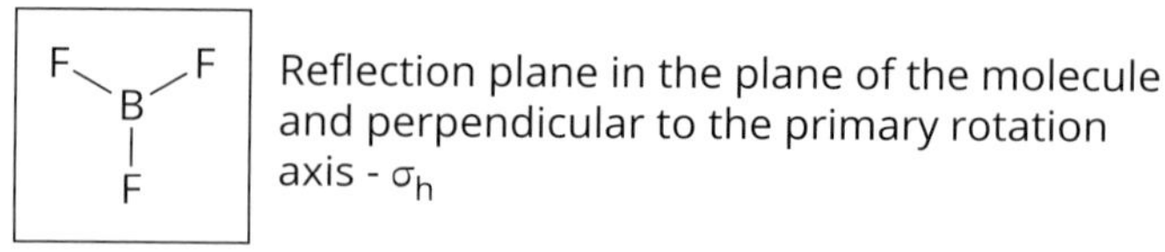

Fig. 5.13e BF_3 contains a plane of symmetry perpendicular to the principal rotation axis, a horizontal reflection plane.

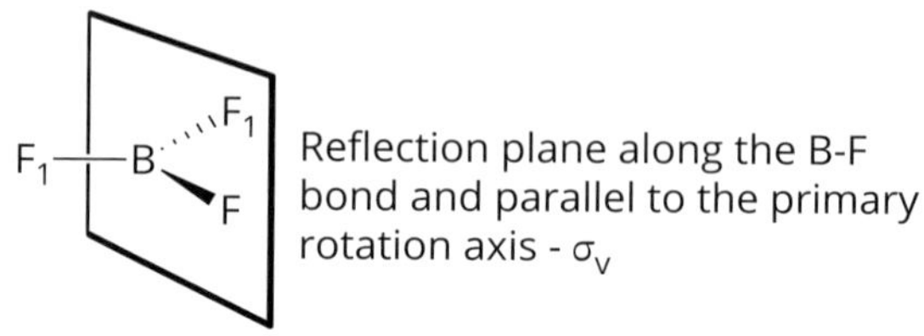

Fig. 5.13f BF_3 contains planes of symmetry that include the principal rotation axis along each bond, vertical reflection planes.

The best way to begin is by identifying some of the most obvious symmetry elements in the molecule.

(1) One obvious symmetry element is the highest-order rotation axis, a C_3 that goes through boron, and which is perpendicular to the plane of the molecule (Fig. 5.13b). This is the highest-order axis of rotation, so it can be automatically designated as the principal axis of rotation. A C_3 axis generates two distinct symmetry operations, which we can label as C_3 and C_3^2.

(2) In addition to the C_3 axis, there are also three C_2 axes that run through each of the three B–F bonds (Fig. 5.13c). Note that this tells us that the boron atom sits on the symmetry point of the molecule, the point where all other symmetry elements meet. Each C_2 axis generates only one operation, so there is a total of three distinct C_2 operations that leave BF_3 invariant.

(3) There is no inversion center, which would have to be located on the boron atom; the inversion operation would *not* leave the molecule invariant (Fig. 5.13d):

(4) The plane of the molecule is a mirror plane. Because it is perpendicular to the principal axis of rotation, we label it as being a horizontal plane of reflection (σ_h). There is only one such plane (Fig. 5.13e).

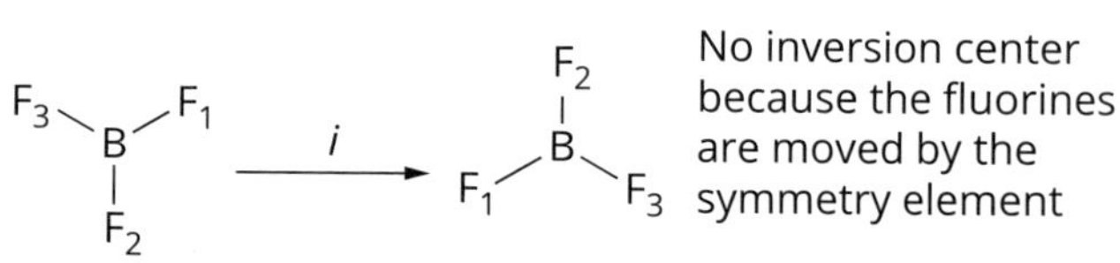

Fig. 5.13d BF_3 moves during inversion, so there is no inversion center.

(5) There are also mirror planes perpendicular to the molecular plane that each contain a B–F bond and which bisect the opposite F–B–F bond angle (Fig. 5.13f). These planes are vertical reflection planes, σ_v. One of these planes is shown that is coincident with one of the B–F bonds and relates the two fluorines labeled "F_1". As always, each plane generates only one distinct symmetry operation, so there is a total of three distinct σ_v operations that leave BF_3 invariant.

(6) Finally, the coexistence of the C_3 axis and σ_h plane imply that there must be an S_3 axis of improper rotation of order 3 coincident with the C_3 axis. This symmetry element generates two distinct operations, S_3 and S_3^2.

Further inspection will reveal that these are all the symmetry elements that define the symmetry of BF_3, and that there is a total of twelve distinct symmetry operations generated from these symmetry elements: two C_3 rotations, one σ_h reflection, three σ_v reflections, three C_2 rotations, two S_3 rotations, and one identity (E) operation. This ensemble of all the symmetry operations that define BF_3 forms what we call a "group" in the mathematical sense of the word, and we will define a group more rigorously in the next section. Because the elements forming that group are all linked by the fact that they share the same point of symmetry, it is customary to call that type of group a "point group". We will see shortly that the specific name of the group of twelve symmetry operations that define BF_3 is assigned the symbol "D_{3h}". Generally, the first thing we will need to do in any problem involving the symmetry of a molecule will be the identification of its unique point group, which requires that we have in hand a list of all the symmetry operations that define the molecule in which we are interested; fortunately, as we will see shortly, we often do not need to identify all the symmetry elements immediately, but only a handful of critical ones, in order to uniquely identify the point group of most molecules.

5.3 Point Groups

As discussed previously, the ensemble of all symmetry operations can be used to classify the symmetry of a molecule; all molecules that contain the exact same set of point symmetry operations are said to belong to the same **point group**.

5.3.1 What Is a Group?

From a mathematical point of view, the concept of a "group" means something very precise. A group is an ensemble of objects (called "elements") that can be combined with each other (through what we call a "product"). In the context of symmetry, the elements of our groups will be symmetry operations, and the "product" of two symmetry operations will simply be the successive application of each of these operations in sequence. Importantly, not every ensemble of elements necessarily forms a mathematical group; a collection of elements must necessarily obey four distinct conditions in order to form a mathematical group:

(1) *The ensemble must be CLOSED.* What this means is that the product of two elements (as well as the product of an element with itself) must always be another element of the group. If the product of two elements in an ensemble generates an element that was not part of the original ensemble to begin with, then that ensemble is not a mathematical group. Mathematically, for every possible pair of elements A and B, $A \times B = C$, where C must be one of the elements of the ensemble. (C could also equal A or B, which is a pretty common occurrence.)

(2) *The ensemble must have an IDENTITY element.* An identity element is an element which always leaves all other elements unchanged. Groups not only have one identity element, but they also cannot have more than one. Mathematically, if we label the identity element as E, then for all elements A of the group: $A \times E = E \times A = A$.

(3) *ASSOCIATIVITY must be preserved.* That is, if we have the product of three elements A, B, C (they do not need to all be different elements), then $A \times B \times C = (A \times B) \times C = A \times (B \times C)$. That is, it does not matter in which order we group the products. In general, communicativity is not preserved; that is, $A \times B$ is not always necessarily the same result as $B \times A$. For example, two noncommutative operations would be "walk five steps north" and "dig a hole"; the order in which two operations are carried matters! We will see that often symmetry operations are noncommutative, but they must always be associative, in the way defined previously. Groups where every element commutes with every other element; that is, groups for which $A \times B = B \times A$ for all A/B pairs of elements, are called *Abelian groups.*

(4) *Every element in the group must have an INVERSE.* The inverse of an element A can be designed symbolically as A^{-1} and has the property $A \times A^{-1} = A^{-1} \times A = E$. In other words, the product of an element with its inverse yields the identity element. An element can be its own inverse. The identity element is an obvious example ($E \times E = E$), but there can be other elements in a group other than the identity that are their own inverse. For example, a mirror plane is typically its own inverse.

It is important to realize that mathematical groups can take a variety of forms and do not have to be groups of symmetry operations. For instance, the group $\{-1, 1\}$ forms a mathematical group if we define the product rule as being the arithmetic multiplication operation, because this ensemble, along with the product rule we chose, obeys the previous four rules:

(1) It is closed; the easiest way to show this is by using a product table[g] to list all possible pairs of products:

	−1	1
−1	1	−1
1	−1	1

This shows that all products generate an element of the group, that this ensemble is closed.

(2) The identity element is "1" as this element never changes the element it multiplies.

(3) Arithmetic multiplication is always associative, so this rule automatically applies here.

(4) The inverse of 1 is 1 ($1 \times 1 = 1$, the identity element); the inverse of −1 is −1 ($-1 \times -1 = 1$), so every element indeed has an inverse.

It is important to not confuse the term "product" for the common meaning it has in arithmetic. The "product" rule can be any operation we choose to define, as long as it leads to a situation where the previous four rules are obeyed. For example, for symmetry operations let us we take one operation as being a C_2 around the z axis and the second operation as inversion; then, the product $C_2 \times i$ would mean: "first rotate the object by 180° around the z axis then do an inversion operation on the object." Conversely, $i \times C_2$ would mean we apply the operations in the reverse order (invert, then rotate).[h] Because symmetry operations do not always commute, being very clear about the order is essential!

As an example of these principles, let us look at the octahedral molecule WF_6 shown in Fig. 5.14. Among other symmetry elements, there is a C_4 axis along the z-axis and a C_2 axis along

[g] A product table such as the one here is also known as a "Cayley table", after the nineteenth-century British mathematician who first described its use.

[h] Mathematically, we say that operators act to their right; if we were to write down the object on which the operator acts (let us call it **O**), then writing $A \times B$ **O** would mean we first need to evaluate the action of B on **O** (which might transform our object into another object, say **O**′), then evaluate the result of A operating on **O**′.

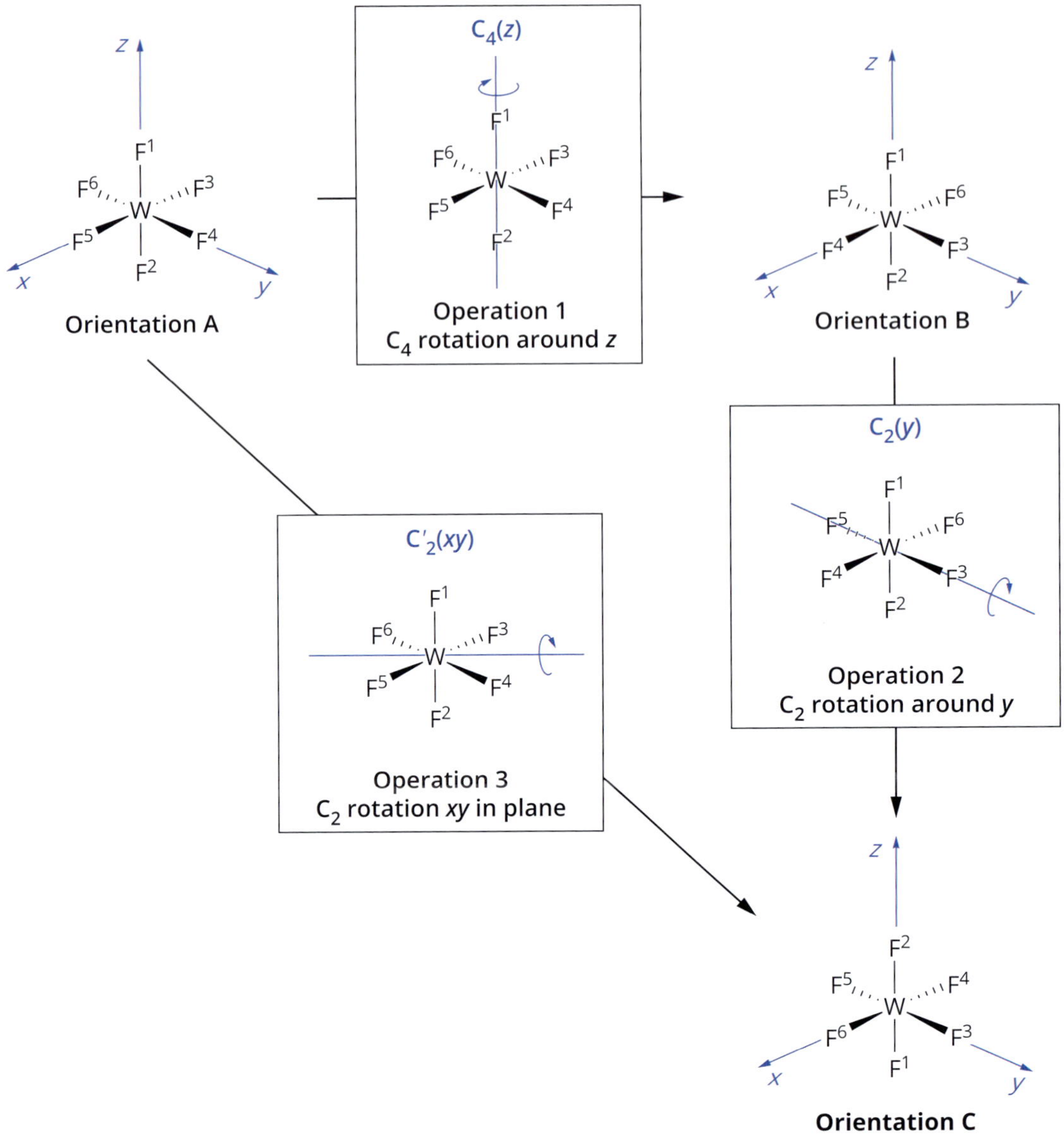

Fig. 5.14 Octahedral molecule WF_6 with the different fluorines labeled 1–6. The $C_4(z)$ operation is conducted first, giving Orientation B with the label pattern shown on the top right. This is then followed with a $C_2(x)$ operation, leading to Orientation C with the label pattern on the bottom right. The same overall transformation can be obtained by doing a single $C'_2(xy)$ operation from the original Orientation A.

the *x*-axis. (Keep in mind that in our approach we rotate the molecule but that axes remain fixed as we carry out each operation.) If we carry out these two symmetry operations in the original molecule, we obtain the result shown at the bottom right, an indistinguishable representation of the original molecule if we ignore the numbering labels for each fluorine atom. In order to have a complete group of symmetry operations for this octahedral species, there must be some symmetry operation that would produce the same outcome as doing a C_4 rotation around the *z* axis followed by a C_2 rotation around the *x* axis; this symmetry operation is a C_2 along an axis that bisects the *x*- and *y*-axes we label $C'_2(xy)$, as shown in Fig. 5.14.

Mathematically, we say that

$$C_2'(xy) = C_2(x) \cdot C_4(z)$$

so $C'_2(xy)$ must also be a member of the group. Importantly, note the way that we represent the order of operations mathematically, by reading the sequence of operations from right to left. A great way to understand the logic behind this approach (and remember it) is to think of each symmetry operation as acting on an object on the right side of the equation. For our sequence of the previous two rotations this would be written as

$$C_2'(xy) \cdot \mathit{Orientation\ A} = C_2(y) \cdot C_4(z) \cdot \mathit{Orientation\ A}$$
$$= C_2(y) \cdot \mathit{Orientation\ B} = \mathit{Orientation\ C}$$

This simply combines rotations, but closure is maintained in a group for any combination of operations.

Any given list of all symmetry operations for a molecule is what we call a "point group". Any two molecules that share the same set of symmetry operations is said to belong to the same point group. The next section will discuss the different types of point groups that exist and outline an efficient procedure to quickly identify the point group of any molecule.

5.3.2 An Overview of Point Groups (Expanded Coverage)

All the point groups are identified using Schoenflies notation, after mathematician Arthur Schoenflies. Table 5.1 lists the major point groups with their Schoenflies symbols. Let us go through the most important ones, beginning with the simplest ones first.

- *Point Group C_1* This group contains only one element: the identity operation. Of course, this is the simplest point group of all. All molecules that have no symmetry elements whatsoever, other than identity (such as C_3^3) belong to this group. An example of a molecule belonging to the C_1 point group is the mixed phosphoryl halide in Fig. 5.15. This molecule contains no axes of rotation (proper or improper), no planes of reflection, nor a center of inversion. It belongs to the C_1 point group. Now is perhaps a good time to point out that the assignment of a point group is much more rigorous than the usual terms we sometimes use to describe the geometry of molecules. Although the structure in Fig. 5.15 has a tetrahedral (or pyramidal) "shape", its formal symmetry is much lower than that of an ideal tetrahedron.
- *Point Group C_s* This group contains only one element symmetry operation (other than the identity operation); namely, a single plane of reflection. The planar molecule formyl fluoride (Fig. 5.16) is an example of a molecule belonging to this point group, and the only symmetry element it has is a reflection in the plane of the molecule.
- *Point Group C_i* This group also has a single element (other than the identity operation), in this case a center of inversion. Few molecules actually happen to have only a center of inversion, and no other symmetry elements. An example is the specific conformer of 1,2-dibromo-1,2-dichloroethane shown in Fig. 5.17 (*anti*-conformer).
- *The Cyclic Point Groups C_n* ($n = 2, 3, 4, \ldots$) This family of point groups is characterized by groups that only have the operations generated by a single C_n axis, and no other symmetry elements. For instance, phosphoric acid, $PO(OH)_3$, in the conformation shown in Fig. 5.18, belongs to the point group C_3, as it has no other symmetry element other than a C_3 axis. (Be careful not to confuse the *point group*'s name with the *symmetry element/operation* here; they are different things that just happen to share the same name.)

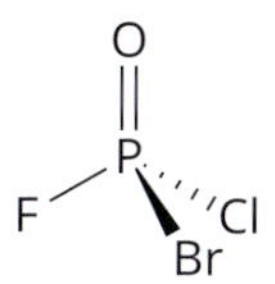

Fig. 5.15 Phosphoryl bromide chloride fluoride, a molecule without any symmetry belonging to the C_1 point group.

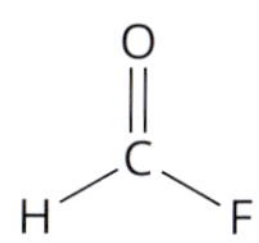

Fig. 5.16 Formyl fluoride, a molecule with a single plane of reflection as the only symmetry element, and thus belonging to the C_s point group.

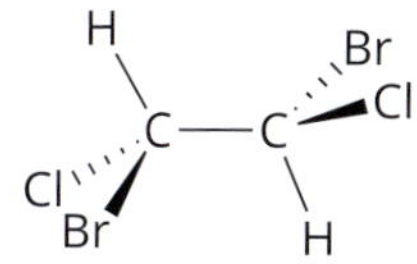

Fig. 5.17 The *anti*-conformer of 1,2-dibromo-1,2-dichloroethane only has an inversion center as a symmetry element; it belongs to the C_i point group.

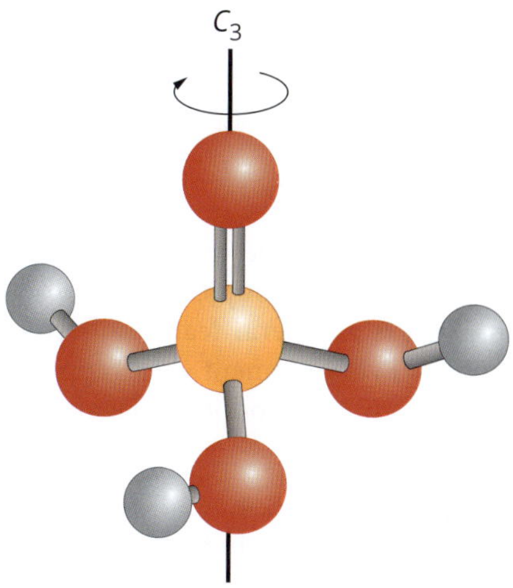

Fig. 5.18 Phosphoric acid, $PO(OH)_3$, contains only one C_3 axis and no other symmetry element: it belongs to the C_3 point group.

The point group C_3 is the first group we encounter that contains more than two operations (including the identity operation). Let us look at it a bit more carefully to ensure that we can identify all the different operations that belong to that point group. One systematic way to do this is to take all the products of the elements we know. In this way we can sometimes generate new symmetry operations that we forgot to list. We know we have the identity operation (always there by default) and the C_3 rotations.

$$E \times E = E$$
$$E \times C_3 = C_3$$
$$C_3 \times C_3 = C_3^2$$
$$E \times C_3^2 = C_3^2$$
$$C_3^2 \times C_3 = C_3 \times C_3^2 = E$$

The list $\{E, C_3, C_3^2\}$ indeed forms a closed group, as no product of two elements in that group yields something which is not already in the group. We also see that every element in there has an inverse (E is its own inverse, and C_3/C_3^2 are each other's inverse), as required for a mathematical group. These properties are all better summarized by showing the product table of the C_3 point group, which simply rearranges the products shown previously in a convenient 3 × 3 table.

C_3	E	C_3	C_3^2
E	E	C_3	C_3^2
C_3	C_3	C_3^2	E
C_3^2	C_3^2	E	C_3

Note that the product table of the C_3 point group always lists each element of the group once, and only once, per row or column. This is a general feature obeyed by all mathematical groups. Every element *must* appear in every row and every column of the product table, and it cannot appear more than once.

- *Point Groups C_{nv} ($n = 2, 3, 4, \ldots$)* This family of point groups is similar to the C_n family in that it also is characterized by having a unique axis of rotation (of order n), but in addition also contains n vertical planes of reflection. (Remember, vertical planes *include* the principal axis of rotation.) The simplest one of these point groups is C_{2v}, which contains four operations; that is, its order is four, $h = 4$: E, C_2, σ_v, and σ_v'. One member of this point group is water (Fig. 5.19).

Perhaps you're wondering whether it is necessary that we have two planes of reflection here; could we not have a group containing a C_2 operation and only *one* plane of reflection? The answer is "No"; a group that has an axis of rotation of order n plus one plane of reflection automatically must have $(n - 1)$ other planes of reflection as well! We say that these two elements "generate" the other symmetry operations, and thus call them the generators of the C_{nv} group.

Let us show that this is indeed the case for C_{2v} by assuming, for now, that there is only one plane of reflection colinear with the C_2 axis (Fig. 5.20). What happens when we use the symmetry operations associated with these two elements to transform a general point in space, say a point located in the positive (x,y,z) sector in Fig. 5.20a? Although the order does not matter here (as it happens, these symmetry operations commute), let us say that we first use the reflection plane (chosen here as the xz plane) to transform our general point, which brings it into the (+, −, +)-octant of the coordinate system, as shown in Fig. 5.20b. We then apply a half-turn operation on that new point, which then takes it all the way to the (−, +, +)-sector (Fig. 5.20b). Remember, the product of two symmetry operations must be a single unique symmetry operation (or else our group is not closed), and we see that the single operation that would take a point in the (+, +, +)-sector into the (−, +, +) -sector is a reflection across the yz plane, which shows that we have to include this plane of reflection as part of our list of symmetry elements. The same type of analysis can be applied to any other of the C_{nv} groups to show that these groups always have n planes of symmetry, no less (otherwise the group would not be closed), no more (otherwise the group would be something other than C_{nv}, see below).

- *Point Groups C_{nh}* These groups are similar to the C_{nv} point groups, except that they are generated by a single proper rotation axis, C_n, combined with a *horizontal* plane of reflection. Obviously, here we do not have to worry about the existence of more than one horizontal plane of reflection; there can only be only one such plane perpendicular to the axis of rotation. (Symmetry elements in the group must all intersect at one unique point, so this automatically rules out the possibility of multiple parallel horizontal planes of reflection.) Nevertheless, the two operations C_n and σ_h always generate other operations. The combination of any proper rotation with a reflection across a plane perpendicular to that axis is the definition of an improper rotation, $C_n \times \sigma_h = S_n$! A closer analysis would reveal that every C_{nh} group has $(n - 1)$ proper rotations and $(n - 1)$ improper rotations; including the identity operation and the plane of reflection. The order (the number of elements it contains) of a C_{nh} group is $h = 2(n - 1) + 2 = 2n$. An example of a molecule belonging to this class of point groups is

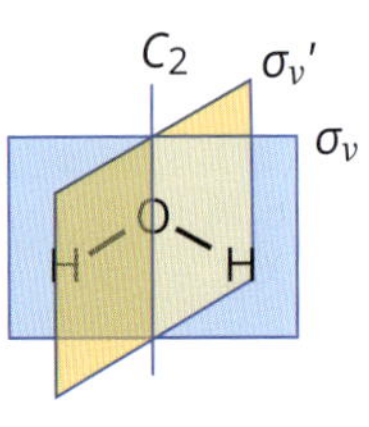

Fig. 5.19 Water belongs to the C_{2v} point group, which contains a C_2 axis and two vertical planes of reflection.

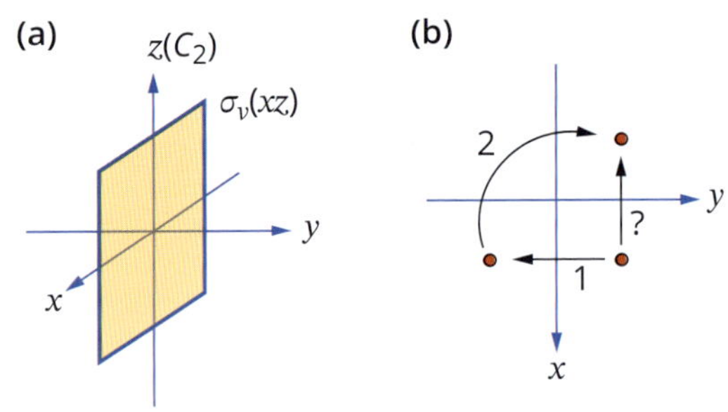

Fig. 5.20 (a) A hypothetical system with a single C_2 axis (colinear with the z axis) and a single plane of reflection (the xz plane). (b) A 2-dimensional view down the z-axis of the coordinate system in (a), and results of a sequence of symmetry operations on a general point initially located in the (+, +, +)-octant of the 3D system ($x > 0$, $y > 0$, $z > 0$). The first operation is a reflection across the xz-plane, the second operation is a C_2 rotation (half-turn) around the z-axis. The product of these two operations is the same as a reflection across the yz-plane, which demonstrates the necessity of a second plane of reflection once a single vertical plane and C_2 axis exist.

trans-dinitrogen difluoride (*trans*-N_2F_2), a planar molecule which belongs to the C_{2h} point group (Fig. 5.21):

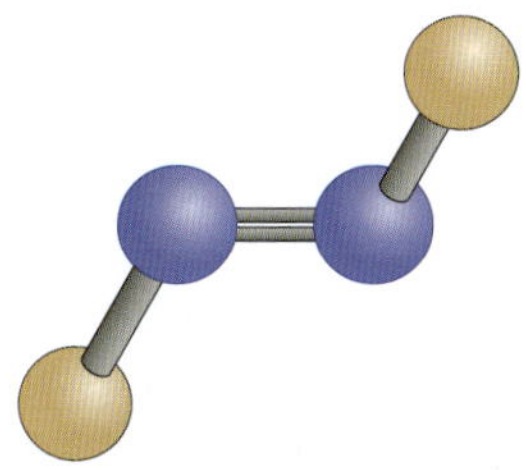

Fig. 5.21 *trans*-dinitrogen difluoride (N_2F_2) belongs to the C_{2h} point group, which is generated by a single C_2 axis and a horizontal plane of reflection (perpencular to the C_2 axis). These two elements also automatically imply the existence of a center of inversion.

Remember that $S_2 = C_2 \times \sigma_h = i$, and *trans*-$N_2F_2$ clearly has a center of inversion. All C_{nh} systems, where n is an even number, will have an S_2 improper rotation axis; that is, a center of inversion. All C_{nh}, where n is an odd number, will automatically contain an improper rotation axis of order n, an S_n axis.

- *The Dihedral Point Groups* D_n ($n = 2, 3, 4, \ldots$). The dihedral family of point groups is generated by the existence of a proper rotation of order n, along with at least one C_2 axis perpendicular to that principal axis. These two elements alone combine to create $(n - 1)$ other C_2 axes of rotation that are in the same plane perpendicular to the principal axis of rotation. A point group belonging to the D_n family always has nC_2 axes perpendicular to its principal C_n axis; its order is then $h = 2n$; that is, the n operations from the C_n principal axis plus nC_2 operations. There are not many molecules that belong to the D_n family, but there are a few important inorganic classes of molecules that are often well approximated ("idealized") by the D_n symmetry (chiral tris-bidentate octahedral complexes notoriously (see Section 5.8).

The biphenyl molecule, in its twisted ground state (Fig. 5.22)—neither flat nor 90° configuration—is an example of a species with a strict D_2 point group symmetry. Note that this molecule indeed has two C_2 axes perpendicular to a C_2 axis. The principal axis of most D_n systems is unambiguously determined, and it will always be the one with $n > 2$, choosing which axis in the D_2 point group should be the principal requires some attention. In the case of the biphenyl molecule, it would make sense to designate the axis of the bond linking the two cycles as the principal axis; as the only axis that contains any atoms, it indeed stands apart from the other two C_2 axes. In the end, the choice in this case is somewhat arbitrary.

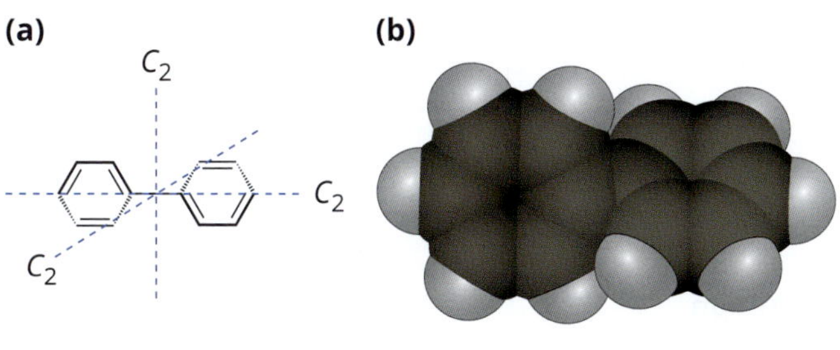

Fig. 5.22 (a) 1,1′-biphenyl $(C_6H_5)_2$, in its twisted conformation. (b) Space-filling model for 1,1′-biphenyl. This molecule contains three C_2 axes all perpendicular to each other, and no other symmetry elements. It belongs to the D_2 point group.

We can check that we listed all the symmetry operations that are generated by the C_n and nC_2 axes. This is readily done by confirming that all possible products of these operations do not lead to new operations; that is, by building a product table. Using the previous biphenyl example, we can build the product table for the D_2 point group the same way we did for the previous C_{2v} group. Take a general point (a general point is a point which does not overlap with any symmetry element) and see what happens to it when pairs of symmetry operations act on that general point; then find the single symmetry operation that brings the original general point to the final position it assumes after the sequence of operations. This yields the Table below for D_2 and confirms that this ensemble of symmetry operations is indeed closed, that no other operations are necessarily generated from this list. The same conclusion would be reached if we were to conduct the same analysis on every other point group belonging to the D_n family.

D_2	E	$C_2(z)$	$C_2(x)$	$C_2(y)$
E	E	$C_2(z)$	$C_2(x)$	$C_2(y)$
$C_2(z)$	$C_2(z)$	E	$C_2(y)$	$C_2(y)$
$C_2(x)$	$C_2(x)$	$C_2(y)$	E	$C_2(z)$
$C_2(y)$	$C_2(y)$	$C_2(y)$	$C_2(z)$	E

- *Point Groups* D_{nd} ($n = 2, 3, 4 \ldots$) D_{nd} point groups are generated by the same elements as the related D_n groups, one C_n axis and n C_2 axes perpendicular to it, but additionally have n planes of symmetry that all contain the principal axis of rotation and bisect the angles separating each pair of C_2 axes. These are called *dihedral* planes, which explains why the Schoenflies symbol is D_{nd} rather than D_{nv}.

An emblematic example is given by allene, $H_2C{=}C{=}CH_2$, which has the structure shown in Fig. 5.23. Each CH_2 group sits in a plane perpendicular to the other group. The hydrogens are labeled "1–4" to help follow where they each go when we apply symmetry operations.

This molecule is often incorrectly assigned to the C_{2v} group, as most people localize the C_2 axis which is colinear with the C=C=C bonds, and the two planes of reflection perpendicular to each other, each one containing one of the two CH_2 groups. The other pair of hydrogen atoms are then reflected, one onto the other. However, there are two more C_2 axes that each convert hydrogens that are on one side of allene to the other side: $H^1 \rightarrow H^3$, $H^1 \rightarrow H^4$, $H^2 \rightarrow H^3$, $H^2 \rightarrow H^3$, and so on. These two axes are located in between each of the two planes of symmetry.

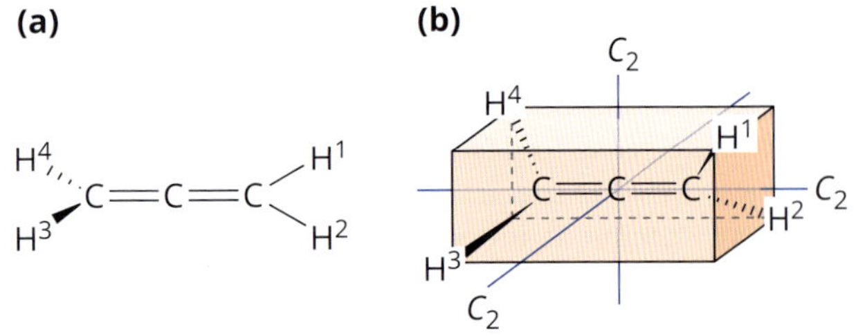

Fig. 5.23 (a) Allene is a molecule that belongs to the D_{2d} point group. (b) Allene, embedded in a square prism, showing the location of the pair of C_2 axes that are perpendicular to the C=C=C bonds. There are also two planes of symmetry that slice the prism along the diagonals of the two square facets.

The D_{2d} nature of the molecule become readily seen if we embed allene in a square prism, as shown in Fig. 5.23b.

One feature of D_{nd} groups is that when the order of the principal axis of rotation is odd, an inversion center is necessarily generated from the combination of the perpendicular C_2 axes and the dihedral planes of reflection elements. When n is odd, there will be a dihedral plane that happens to be perpendicular to one of the C_2 axes, which automatically generates an S_2 axis of rotation—an inversion center. A look at the stable staggered conformer of ethane, which belongs to the D_{3d} point group, reveals the existence of the center of inversion (Fig. 5.24).

Because the center of inversion is automatically implied from the existence of the other elements, it is itself not considered a *generator* element of that family. Besides, n = even members of that family do not have a center of inversion, so the presence/absence of an inversion center cannot be a defining trait for the D_{nd} family.

- *Point Groups D_{nh}* (n = 2, 3, 4 . . .) This family or group is generated by the same elements responsible for the D_n family, with the addition of a horizontal plane of symmetry perpendicular to the principal axis of rotation. This list of elements (C_n axis of rotation, nC_2 axes perpendicular to it, and a horizontal plane) is enough to define the whole family. These elements imply the existence of n planes of reflections that each contain the principal axis of rotation and one of the C_2 axes; that is, vertical planes of reflection. The D_{nh} point groups where n = even always have a center of inversion, necessitated by coexistence of a C_2 operation and the horizontal plane of reflection perpendicular to it. There are many molecules that belong to the D_{nh} family, as most planar species with rotational symmetry in the plane of the molecule fall in that category. A few examples for n = 2, 3, 4, and 6 are shown in Fig. 5.25.

There are also many nonplanar molecules that also belong to the D_{nh} family. A prototypical example of a nonplanar D_{nh} inorganic molecule is the sandwich compound eclipsed bis(benzene)chromium (Fig. 5.26), which belongs to the D_{6h} point group.

- *Point Groups S_n* (n = 4, 6, . . .) This family of point groups is generated by a single axis of improper rotation of order n, and nothing else. We have already covered the case n = 2, which is the C_i point group, the point with no other symmetry elements than an inversion center (remember that $S_2 = i$). When n is odd, then a horizontal plane is always automatically present as well, which then corresponds exactly

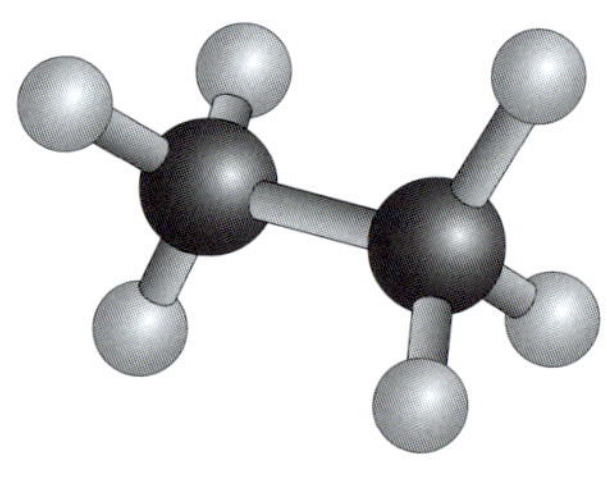

Fig. 5.24 The staggered conformation of ethane, C_2H_6, belongs to the D_{3d} point group, which contains a center of inversion.

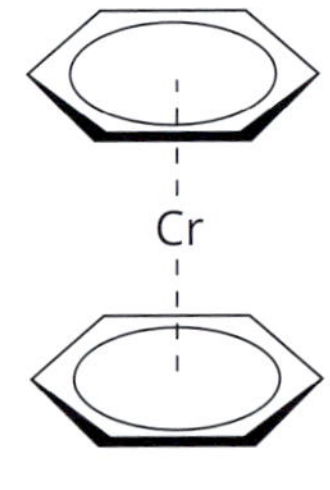

Fig. 5.26 Bis(benzene)chromium, a sandwich compound, belongs to the D_{6h} point group.

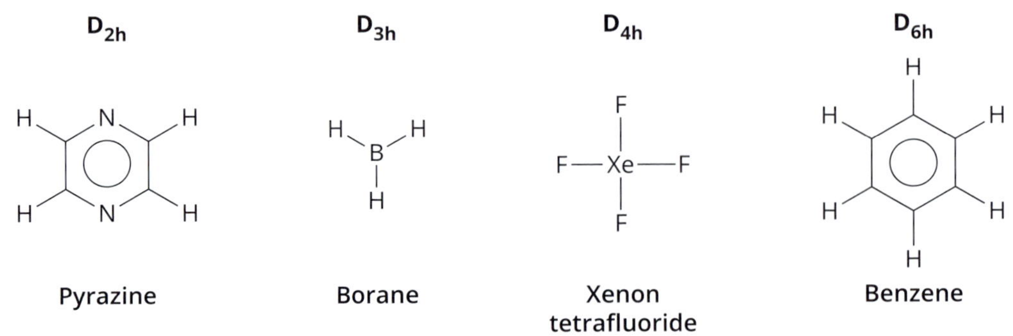

Fig. 5.25 A few examples of molecules belonging to the D_{nh} family of point groups. This family is defined by a principal axis of order n, the existence of nC_2 axes perpendicular to the principal axis, and a horizontal plane of reflection. Among other elements, the D_{nh} groups also contain n vertical planes and, when n = even, a center of inversion.

to the C_{nh} family. We then only need to strictly account for the cases where n is even and greater than 2. This family of point groups is rare in chemistry.

This list of point groups covers the most commonly encountered point groups, of which all were characterized by having, at most, a single axis of rotation of high order (higher than C_2). Of course, this is not a universal requirement; there are cases of very high symmetry where there can be more than one axis of rotation of high order. Point groups of that type are special, and there are a few that are particularly important in chemistry because they describe idealized high symmetries that are commonly found in nature. These special point groups are often identified immediately just because of their very high and very distinctive symmetry.

- *Linear ($D_{\infty h}$ and $C_{\infty v}$)* These are the two point groups for linear molecules, characterized by having a single axis of rotation of *infinite order* (C_∞), which simply refers to the fact that you can always rotate a linear molecule, such as CO_2, by any angle around its molecular axis without changing it. Because our definition of the fraction of a turn of an axis of rotation is $1/n$—that is, a C_2 axis is characterized by having ½-turn invariance around its axis—a C_∞ simply means that you can turn by an infinitesimally small fraction around the axis of rotation without changing the molecule. There are two types of linear molecules—those that have an inversion center, which includes all homodiatomic molecules and CO_2, and those that do not have an inversion center, which includes all heterodiatomic molecules such as HCl. The point group of linear molecules that have a center of inversion is named $D_{\infty h}$, whereas linear molecules that do not have a center of inversion belong to the $C_{\infty v}$ point group.
- *Tetrahedral (T_d)* Tetrahedral symmetry plays a central role in chemistry. A perfect tetrahedron always has the following list of symmetry elements: four C_3 axes (each at 109.47° of each other), three C_2 axes (each at 90° of each other), three S_4 axes that are colinear with the C_2 axes, and six planes of reflection that each contain a pair of C_3 and C_2 axes (Fig. 5.27). The symbol for tetrahedral symmetry is T_d. Methane, the ammonium (NH_4^+) cation, the tetrafluoroborate (BF_4^-) anion, the P_4 allotrope of phosphorus (white phosphorus), and the tetrachloronickelate(II) complex ion ($NiCl_4^{2-}$), are all standard representatives of that important point group.
- *Octahedral (O_h)* Octahedral symmetry plays an important role in inorganic chemistry, as does tetrahedral symmetry. As it turns out, the two symmetries are closely related to each other, each being derived from the symmetric properties of the cube. (They are both sometimes called "cubic symmetries".) An octahedron is a regular polyhedron that has eight equilateral triangular facets—two square pyramids stacked on their square facet. An octahedron shares the same set of symmetry elements that characterizes a cube, which is easily seen by embedding the octahedron inside a cube with each of the six vertices of the octahedron located in the center of a facet of the cube (Fig. 5.28a).

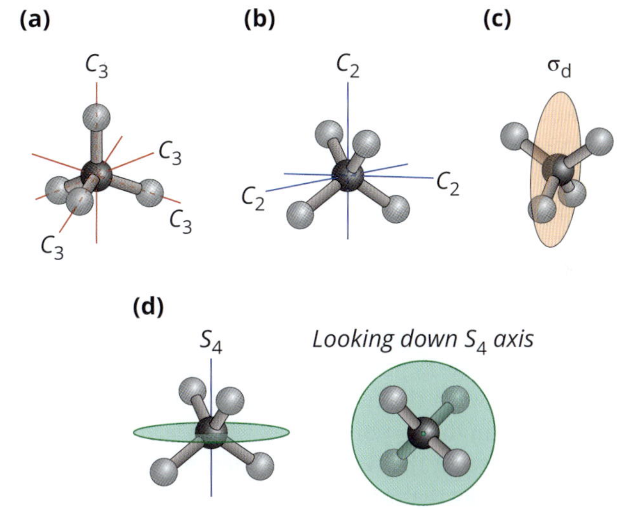

Fig. 5.27 The symmetry elements of an ideal tetrahedron, point group T_d, as illustrated from methane (CH_4). (a) The four C_3 axes in methane. These four axes go through each of the C–H bonds and rotate the other three hydrogens into each other. (b) The three C_2 axes; these axes bisect each pair of H–C–H angles. (c) One of eight equivalent dihedral planes of reflection; each of these planes contain two of the four hydrogens and reflect the other two hydrogen atoms into each other. (d) One of three equivalent S_4 axes; each of these axes is colinear with one of the three C_2 axis. The plane perpendicular to the C_2 axis is shown. Note that this plane is not one of the symmetry elements of the point group T_d.

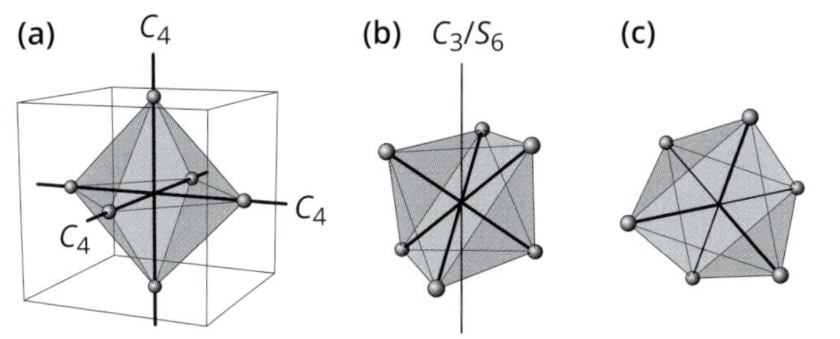

Fig. 5.28 The cubic symmetry of the octahedron. (a) An octahedron can be embedded in a cube, with each vertex of the octahedron positioned in the center of each facet of the cube. The C_4 axes of the cube/octahedron correspond to the *trans*-MX_2 bonds of a MX_6 octahedral complex. (b) The trigonal symmetry of the octahedron can be seen by flipping it on one of its triangular facets; in this orientation, the bottom three vertices can be rotated into each other by a C_3 rotation, and the same for the top three vertices. (c) Another view of the trigonal symmetry of the octahedron, looking straight down the C_3 axis shown in (b); the bold lines represent the M–X bonds that are on top of the octahedron in (b), whereas the thin lines correspond to the bottom three bonds. This perspective helps to highlight the fact that the C_3 axes of the octahedron are in fact S_6 axes; a rotation by a sixth of a turn followed by reflection across a (imaginary) perpendicular plane leaves the octahedron invariant.

The symbol for the octahedral point group is O_h, and contains the following symmetry elements: three C_4 axes of rotation (colinear with the lines that connect the centers of each opposite sides of the cube; these are often chosen as the x,y,z axes of the octahedron), four C_3 axes (colinear with the long diagonals of the cube, the lines that connect each pairs of corners of the cube that do not share the same facet), a center of inversion, three "horizontal" planes of reflections (each one perpendicular to one of the C_4 axes, and containing four of the six vertices of the octahedron), and six "dihedral" planes of reflection (three pairs of planes each containing one of the three C_4 axes, and bisecting the angles between the other two C_4 axes). The four C_3 axes mentioned previously are also S_6 axes, as shown in Fig. 5.28b. All these elements give rise to a grand total of forty-eight symmetry operations; that is, the order of the O_h point group is $h = 48$, which is twice as large as the order of the T_d point group ($h = 24$). A very important distinction between the two groups is the presence of the inversion center in the O_h point group, which leads to important differences between the two symmetries, most notably in the very different way that octahedral and tetrahedral complexes interact with electromagnetic radiation (light).

- *Icosahedral* (I_h) The symbol for the icosahedral point group is I_h. An icosahedron (the polyhedral figure with twenty identical equilateral triangular facets) and dodecahedron (the polyhedral figure with twelve identical equilateral pentagonal facets) are shown in Fig. 5.29a. Such high symmetry is not commonly encountered in chemistry, with a few notable exceptions that include buckminsterfullerene (C_{60}) and the dodecaborate anion ($[B_{12}H_{12}]^{2-}$). The symmetry operations that characterize the I_h point group are six C_5 axes (also S_{10} axes), ten C_3 axes (also S_6 axes), fifteen C_2 axes, fifteen planes of reflection, and a center of inversion. The order of the I_h group is a whopping $h = 120$!
- *Other Cubic Subgroups (O, T, T_h)* There are three other point groups that we have not yet introduced, which are all subgroups of the cubic groups (O_h, T_d) we have already discussed.[i]

The point group O is the group that contains all the proper rotation axes of the octahedron, but nothing else—no plane of reflection, no center of inversion, and no axis of improper rotation. Although the point group O is seldom encountered in chemistry, it is mathematically quite useful as it can directly lead to the full O_h symmetry simply by adding back the center of inversion. In other words, if you take all the elements of the O group (three C_4 axes, four C_3 axes, and six C_2 axes, the same list as for the octahedron) and simply add back the center of inversion, then all the other elements and operations of the O_h point group are recovered. Mathematically, we say that the O_h point group is the *direct product* of the O and C_i point groups. (Direct products are covered in detail later in this chapter.)

$$O_h = O \times C_i$$

A similar situation is found for the point group T_d. We can form a group by considering only the proper rotations (eight C_3 rotations and three C_2 rotations) of tetrahedral, and this group is known simply as T. The full tetrahedral symmetry, point group T_d, can be obtained by taking the direct product of the T subgroup with C_s.

$$T_d = T \times C_s$$

Finally, for the sake of completeness, one more point group can be derived by considering the direct product of the group T with C_i. The resulting point group is called T_h and contains all the same proper rotations as the ideal tetrahedron, plus a center of inversion: four S_6 axes (colinear with the C_3 axes), and three planes of reflection (each perpendicular to one of the C_2 axes).

$$T_h = T \times C_i$$

The T_h group is rare for molecules, but an interesting example is hexanitrocuprate(II) ion, $[Cu(NO_2)_6]^{2-}$, which can be crystallized in the specific arrangement shown in Fig. 5.29b. Although it has a local octahedral geometry near the metal center, this complex clearly does not belong to the O_h point group, as there are no C_4 axes due to the specific arrangement of the nitro ligands. Close inspection would reveal that this complex has all the elements listed previously for the T_h point group.

[i] A subgroup is a subset of a group which itself obeys all the rules of a group. For instance, the point groups C_n are all subgroups of the C_{nv} and C_{nh} point groups, and the point group C_i is a subgroup of any group that has a center of inversion.

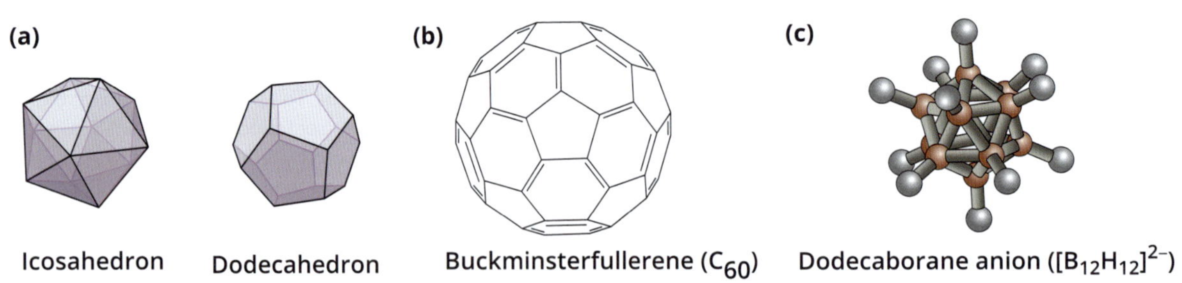

Fig. 5.29 a–c (a) Regular icosahedron and dodecahedron, two polyhedra with a I_h point group symmetry. (b) Buckminsterfullerene (C_{60}) (a truncated icosahedron) and (c) the dodecaborate anion ($[B_{12}H_{12}]^{2-}$) (icosahedron structure) are two molecules that belong to the I_h point group.

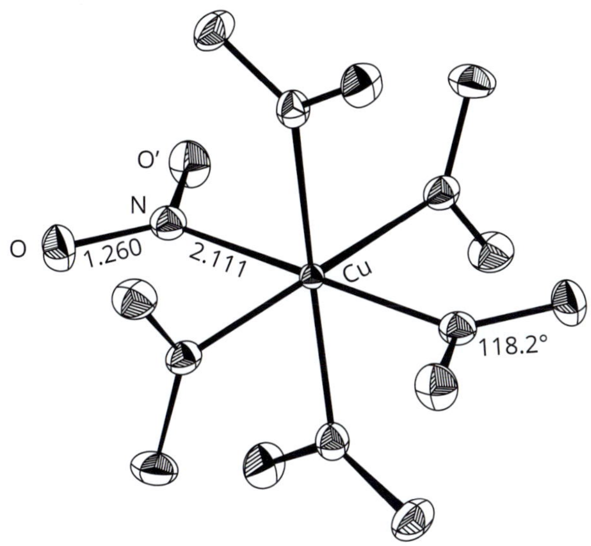

Fig. 5.29d X-ray structure of the hexanitrocuprate(II) ion, as found in crystals of $K_2PbCu(NO_2)_6$. This special arrangement of the planar nitro (NO_2) ligands corresponds to the T_h point group symmetry. Note the presence of the center of inversion and absence of C_4 axes, which characterizes the T_h point group.

A summary of this quick review of point groups is shown in Table 5.1. Most point groups can be defined by the existence of only a few specific symmetry elements, which taken together necessarily imply the existence of all other elements of the point group. These elements are called "generators" and are useful when we need to identify the point group of a molecule.

A systematic procedure to properly identify the point group of any molecule can be summarized in a flow chart, as shown in Fig. 5.30. Flow charts like this are all based on the identification of specific group generators that define each point group. It is therefore rarely necessary to identify every single symmetry operation to classify the symmetry of a molecule.

5.3.3 Point Group Assignment Examples

The assignment of a point group to a molecule is the necessary first step to any symmetry-related analysis, and the best method for becoming proficient is to simply practice. Some symmetry operations are easier to visualize than others; however, in our experience, after a little practice the flow chart is no longer needed, for example, by the final exam in a semester.

Table 5.1 Point groups, generators, and their order (number of symmetry operations, h). Adapted from Douglas and Hollingsworth.[4]

Family	Schoenflies Symbol	Generators	Order (h)
Linear	$C_{\infty v}$	C_∞, σ_v	∞
	$D_{\infty h}$	C_∞, C_2, σ_h	∞
Nonaxial	C_1	no symmetry	1
	C_s	σ	2
	C_i	i	2
Cubic	T	C_3, C_2 (angle between is 54.74°)	12
	T_h	C_3, C_2, i	24
	T_d	C_3, S_4	24
	O	C_3, C_4	24
	O_h	C_3, C_4, i	48
Icosahedral	I	C_3, C_5	60
	I_h	C_3, C_5, i	120
Axial	C_n	C_n	n
	S_{2n}	S_{2n}	$2n$
	C_{nh}	C_n, σ_h	$2n$
	C_{nv}	C_n, σ_v	
Dihedral	D_n	C_n, C_2 (angle between is 90°)	$2n$
	D_{nh}	C_n, C_2, σ_h	$4n$
	D_{nd}	C_n, C_2, σ_d	$4n$

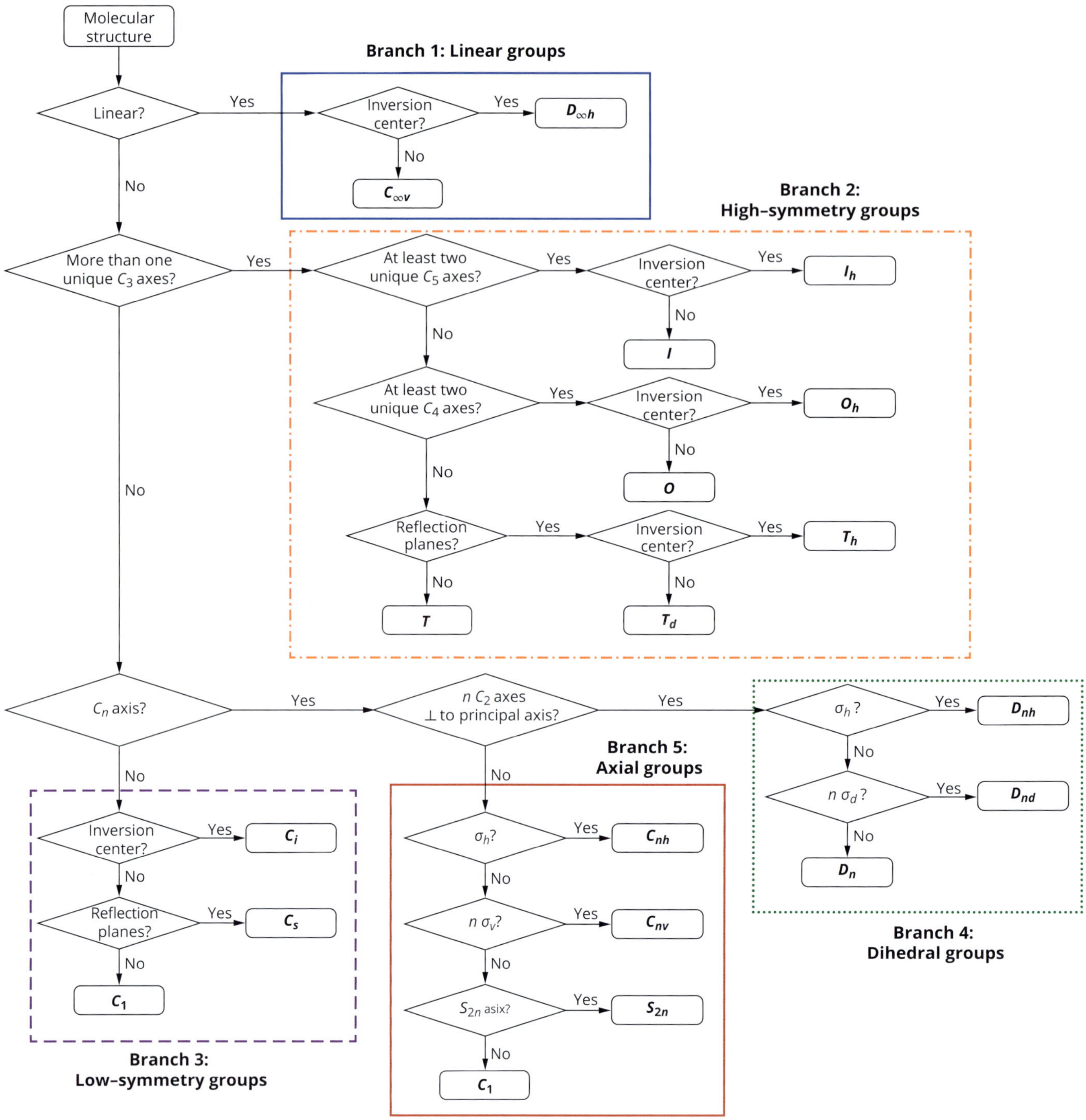

Fig. 5.30 Flow chart for point group assignment.

Following are some details on the point group assignment of some representative molecules. The pathway through the flow chart (Fig. 5.30) is listed.

H_2 (Fig. 5.31a)

- Linear: Yes
- Inversion center: Yes

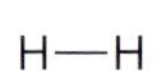

Fig. 5.31a Molecular H_2 ($D_{\infty h}$).

Point group: $D_{\infty h}$

BF_3 (Fig. 5.31b)

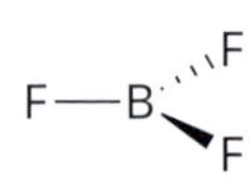

Fig. 5.31b BF_3 (D_{3h}).

- Linear?: No
- Several high-order axes?: No
- Highest-order rotation axis: C_3
- Noncoincident C_n ($n > 2$)?: No
- C_2 perpendicular to C_n?: Yes
- σ_h?: Yes

Point group: D_{3h}

BF_2Cl (Fig. 5.31c)

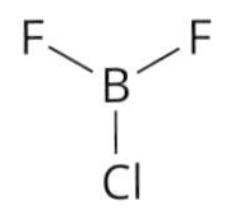

Fig. 5.31c BF_2Cl (C_{2v}).

- Linear?: No
- Several high-order axes? No
- Highest-order rotation axis: C_2
- Noncoincident C_n ($n > 2$)?: No
- C_2 perpendicular to C_n?: No
- σ_h?: No
- σ_v?: Yes

Point group: C_{2v}

Ethylene (Fig. 5.31d)

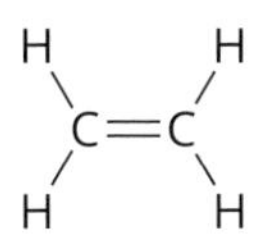

Fig. 5.31d Ethylene (D_{2h}).

- Linear?: No
- Several high-order axes?: No
- Highest-order rotation axis: C_2
- Noncoincident C_n ($n > 2$)?: No
- C_2 perpendicular to C_n?: Yes
- σ_h?: Yes

Point group: D_{2h}

Benzene (Fig. 5.31e)

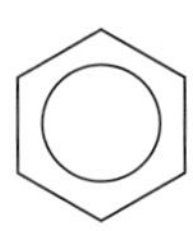

Fig. 5.31e Benzene (D_{6h}).

- Linear?: No
- Several high-order axes?: No
- Highest-order rotation axis: C_6
- Noncoincident C_n ($n > 2$)?: No
- C_2 perpendicular to C_n?: Yes
- σ_h?: Yes

Point group: D_{6h}

$\{PtCl_4\}^{2-}$ (Fig. 5.31f)

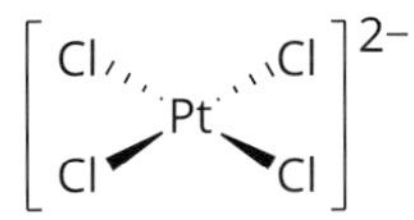

Fig. 5.31f $PtCl_4^{2-}$ (D_{4h}).

- Linear?: No
- Several high-order axes?: No
- Highest order rotation axis: C_4
- Noncoincident C_n ($n > 2$)?: No
- C_2 perpendicular to C_n?: Yes
- σ_h?: Yes

Point group: D_{4h}

Eclipsed ferrocene (Fig. 5.31g)

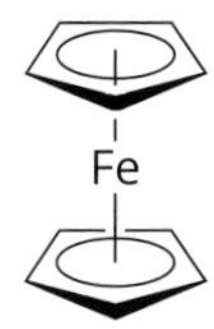

Fig. 5.31g Eclipsed ferrocene (D_{5h}).

- Linear?: No
- Several high-order axes?: No
- Highest-order rotation axis: C_5
- Noncoincident C_n ($n > 2$)?: No
- C_2 perpendicular to C_n?: Yes
- σ_h?: Yes

Point group: D_{5h}

Staggered ferrocene (Fig. 5.31h and i)

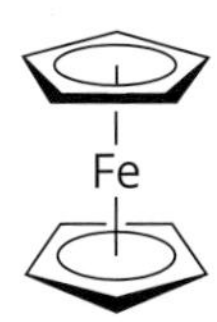

Fig. 5.31h Staggered ferrocene (D_{5d}).

- Linear?: No
- Several high-order axes?: No
- Highest-order rotation axis: C_5
- Noncoincident C_n ($n > 2$)?: No

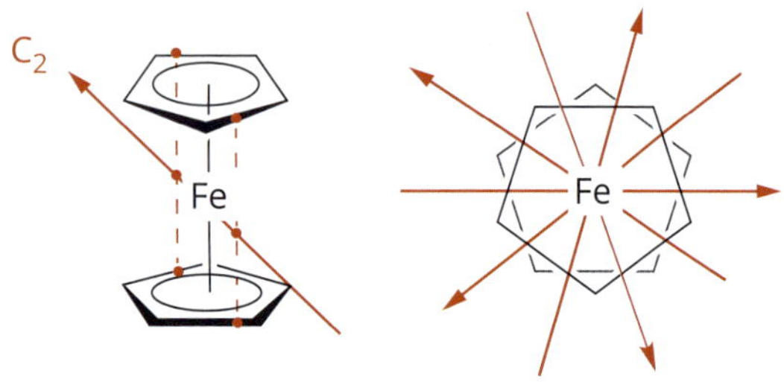

Fig. 5.31i Illustration of the five C_2 axes perpendicular to the C_5 principal axis in staggered ferrocene. Each of these C_2 axes exchanges the two C_5H_5 groups with each other.

- C_2 perpendicular to C_n?: Yes
- σ_h?: No
- σ_d?: Yes

Point group: D_{5d}

5.4 Character Tables and Classes

Appendix D lists the most chemically relevant point groups and their *character tables*. The character table will be the subject of much of the remainder of this chapter. However, at this stage it is useful to simply notice the top row of the character tables. At the top left is listed the Schoenflies symbol for the point group, such as C_{2v} or T_d, as found in the flow chart (Fig. 5.30) and Table 5.1. The rest of the top row lists all of the symmetry operations for that point group. The symmetry operations are grouped into their *conjugate classes*, usually just called *classes*.

Using the flow chart in Fig. 5.30, let us find the point group of ammonia. In detail, we have from the flow chart:

Linear?: No.

Several high-order axes?: No.

Highest-rotation axis: C_3.

Noncoincident C_n ($n > 2$)?: No.

C_2 perpendicular to C_3?: No.

σ_h?: No.

σ_v?: Yes.

→ Point group: C_{3v}

The first row of the character table for C_{3v} is shown as follows. All the character tables include the identity operation, E, as the first operation. For C_{3v}, the other operations are two C_3 rotations (namely C_3^1 and C_3^2) and three σ_v vertical planes of reflection, planes that contain the main rotation axis. These are all the symmetry operations in a C_{3v} molecule such as ammonia.

C_{3v}	E	$2C_3$	$3\sigma_v$

A useful exercise for students learning group theory is to take all the symmetry operations in the character table and identify them for a molecule. For the C_{3v} molecule ammonia, all the symmetry operations are shown in Fig. 5.32. To allow the hydrogens to be distinguished, they are labeled H^1 to H^3 in the figures. For completeness, the identity operation, E, is shown, which does not change the labels. There are two different C_3 rotations sharing one axis, different because they cause the production of different labeling schemes. The two C_3 rotations are in fact a clockwise rotation by 120° and a counterclockwise rotation by 120°, which is equivalent to two clockwise 120° rotations: $C^{-1}{}_3 = C^2{}_3$. There are three mirror planes labeled α, β, and γ in the figure, which are along the N–H^1, N–H^2, and N–H^3 bond axes in the starting configuration.

The three σ_v operations are related in perhaps obvious ways and are said to be in the same "class", which we will now define more rigorously.

Two symmetry operations, A and B, are in the same class if there is another symmetry operation X in the point group that relates the two so that the following relationship is true. The transformation in Eq. 5.1a is called a *similarity transformation*:

$$XAX^{-1} = B \qquad \textbf{5.1a}$$

We say that X transforms A, and that B is the transform of A by X. Because each element in a group must have an inverse, the following similarity transformation also always exists:

$$Y^{-1}AY = B \qquad \textbf{5.1b}$$

where Y is simply the inverse of X: $Y = X^{-1}$. We say that A and B are "conjugate", simply meaning that they are related by other symmetry operations. The complete set of conjugate members in a group is called a "class". Let us look at a couple of these relationships pictorially. From looking at Fig. 5.32 for the symmetry operations in the ammonia molecule, one can see that there is a C_3 and mirror planes that are colinear. Taking any one of the mirror planes and doing a C_3 rotation, one generates a different mirror plane in the molecule. In other words, because of the C_3 operation, if there is one parallel mirror plane there must be two more planes. The three mirror planes are obviously related to one another and behave similarly. This relationship, which can be seen qualitatively, is that the three mirror planes are "conjugate" to one another. We say that the mirror planes belong to the same "class".

While symmetry operations that are in the same class are sometimes obviously directly related to each other, it is still useful to have a more rigorous mathematical definition to show that two symmetry operations are conjugate, especially when their relationship is not as obvious.

Let us look at the similarity transformation pictorially. As transformation operations (X), let us use mirror planes and see their effect on the two different C_3 operations. In the case of mirror planes, they are the inverse of themselves, so $\sigma^{-1} = \sigma$. We will use the same symmetry elements shown in Fig. 5.32, where σ^α is the mirror plane coincident with the N–H^1 bond in the starting structure, and σ^γ is the mirror plane coincident with the N–H^3 bond in the starting structure. For us to be able to keep track of where each element is, keep in mind that as we do symmetry operations on the molecule, the atoms may move but the symmetry elements do not; that is, the plane labeled σ^{γ} in Fig. 5.32 is always the one located in the plane of the page.

If we look at the similarity transform $\sigma^{\alpha}C^{2}_{3}\,\sigma^{\alpha}$, the mirror plane converts the starting geometry into that labeled A1 in Fig. 5.33, exchanging H^2 and H^3. Doing two clockwise rotations around the C_3 yields structure A2. Another reflection through symmetry element σ^{α}, which is now along the N–H^3 bond but still coming out of the plane toward us, gives structure A3, exchanging H^1 and H^2 in structure A2. The structure after doing the operations $\sigma^{\alpha}C^{2}_{3}\,\sigma^{\alpha}$ is identical to what one would obtain by simply doing a 120° clockwise rotation, C_3. As a result, C_3 and C^2_3 are conjugate, making them members of the same class. If one does the same exercise with C_3, using say σ^{γ}, then C^2_3 is obtained (bottom of Fig. 5.33).

Table 5.2 contains all similarity transformations $XAX^{-1} = B$ for the C_{3v} point group, where the transformation operations X are shown along the top row, and the transformed operations A are shown down the first column.

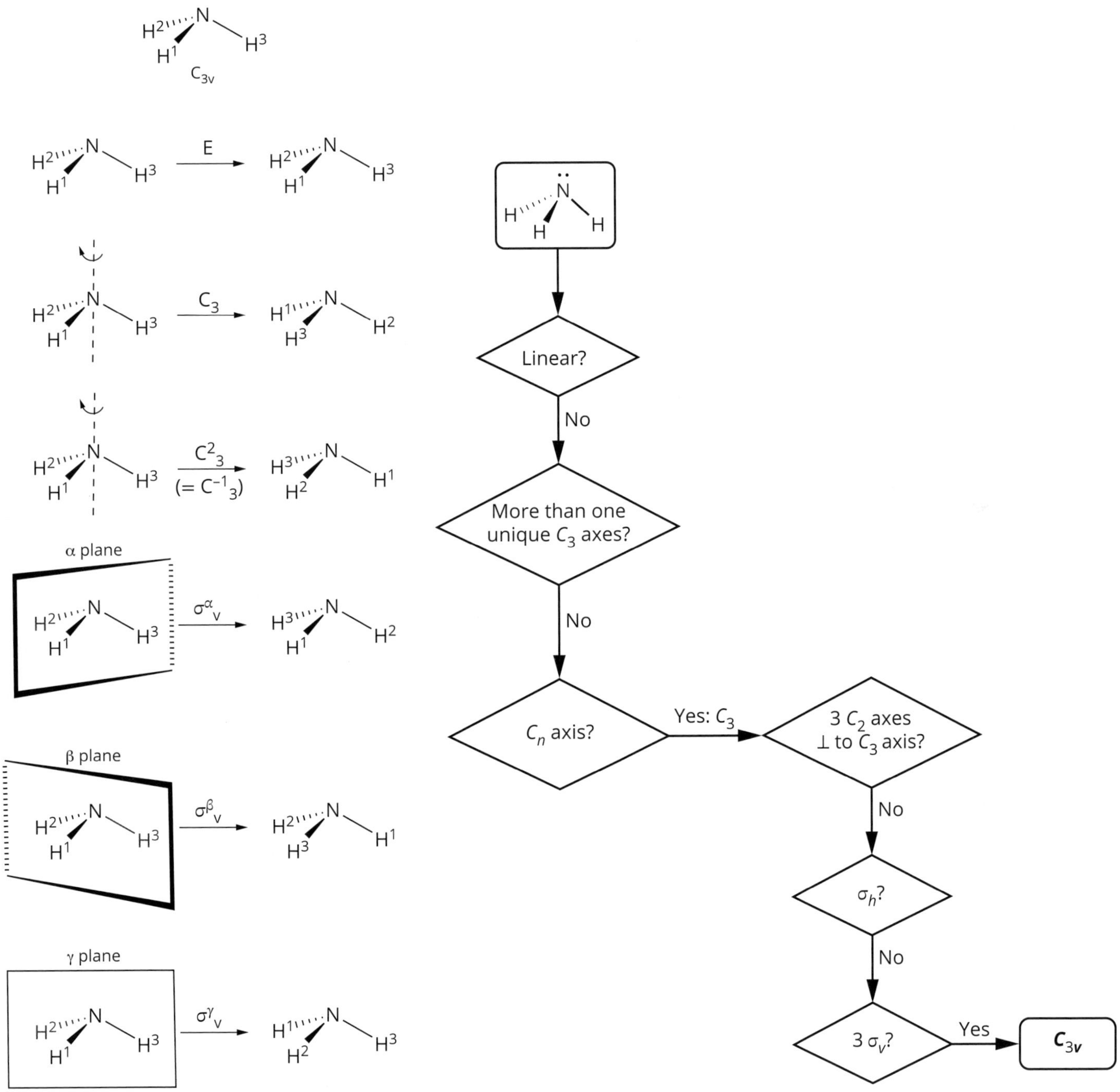

Fig. 5.32 The structure of C_{3v} ammonia and the symmetry operations present. On the right is the path taken through the flow chart in Fig. 5.30.

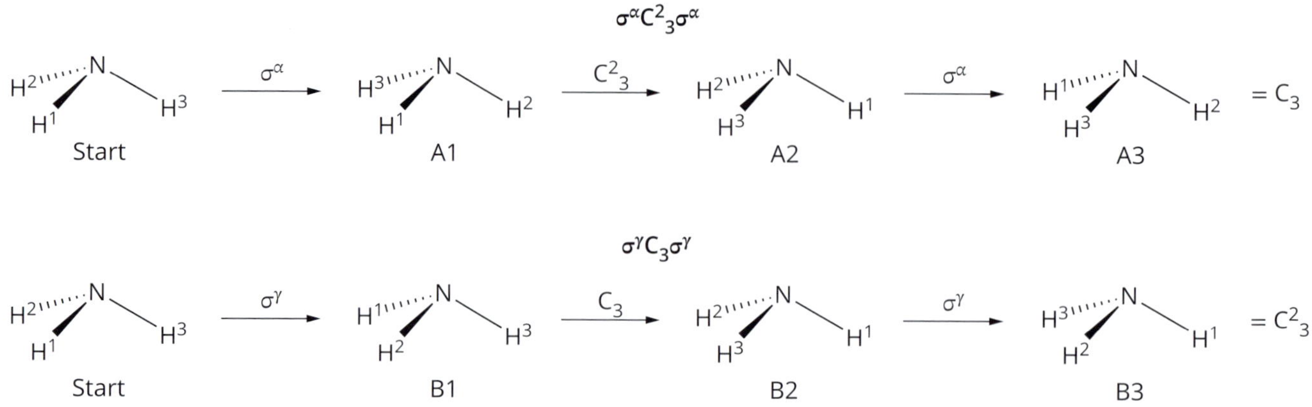

Fig. 5.33 Two examples of pictorially examining complex conjugates for the ammonia symmetry operations. In the first, the C^2_3 is related to C_3 by the mirror plane σ^α. In the second, the C_3 is related to C^2_3 by the mirror plane σ^γ. Since there are other symmetry operations in C_{3v} that related C_3 and C^2_3, these two symmetry operations are in the same class.

A few properties of similarity transformations are worthy of note:

(1) Symmetry operations that appear in the same row will be members of the same class in the character table.

(2) No operation occurs in more than one class.

(3) If one calls the number of members of a class m and h is the order (total number of symmetry operations) of the point group, the total number of symmetry operations, then h/m is the total number of times an operation is transformed into itself. For example, in the row for C_3, C_3 appears three times and C^2_3 appears three times ($h = 6$, $m = 2$, $h/m = 6/2 = 3$). Also, E appears six times because it is in a class by itself ($m = 1$), and $h/m = 6/1 = 6$.

(4) No operation X transforms two different operations A into the same operation B. In other words, each of the operations appears only once in each column.

(5) If the point group is Abelian, then every symmetry operation is in a class by itself, and each row contains only one operation, A. In all cases, $XAX^{-1} = A$.

Because conjugate operations are related to each other, we will generally not have to bother investigating how each member of a class behaves individually. For many applications of group theory, it will be sufficiently often to use one representative operation for each class.

Table 5.2 Similarity transformations $B = XAX^{-1}$ for the symmetry operations in the C_{3v} point group. Operations that appear in the same row together are conjugate and belong to the same class. Remember that in the sequence XAX^{-1}, the operation X^{-1} is applied first, followed by A, and then finally by X.

A \ X	E	C_3	C^2_3	σ^α_v	σ^β_v	σ^γ_v
E	E	E	E	E	E	E
C_3	C_3	C_3	C_3	C^2_3	C^2_3	C^2_3
C^2_3	C^2_3	C^2_3	C^2_3	C_3	C_3	C_3
σ^α_v	σ^α_v	σ^γ_v	σ^β_v	σ^α_v	σ^γ_v	σ^β_v
σ^β_v	σ^β_v	σ^α_v	σ^γ_v	σ^γ_v	σ^β_v	σ^α_v
σ^γ_v	σ^γ_v	σ^β_v	σ^α_v	σ^β_v	σ^α_v	σ^γ_v

To further illustrate these related (conjugate) symmetry operations that form classes, we can briefly examine two additional examples in different point groups, BF_3 (Fig. 5.34) and square planar $\{AuCl_4\}^-$ (Fig. 5.35).

The point group of BF_3 is D_{3h}, and consequently there are three C_2 axes that run along the B–F bonds perpendicular to the C_3 principal rotation axis; the three C_2 operations are all in the same class. Each C_2 axis is associated with a vertical σ_v plane of reflection. All three σ_v operations are also in the same class, as they are related by the C_3. In addition, the plane of the

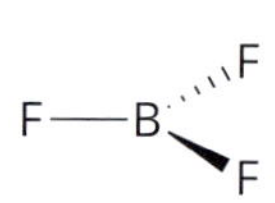

- D_{3h}
- Presence of the C_3 relates the 3 C_2 axes
- 3 σ_v planes related by C_3
- σ_v plane is unique (in a class by itself)

Operations: E, C_3, C^2_3, C^2, C'_2, C''_2, σ_h, S_3, S^5_3, σ_v, σ'_v, σ''_v

Classes: E, $\{C_3, C^2_3\}$, $\{C_2, C'_3, C''_2\}$, σ_h, $\{S_3, S^5_3\}$, $\{\sigma_v, \sigma'_v, \sigma''_v\}$

Fig. 5.34 Some of the symmetry operations of BF_3 and their classes.

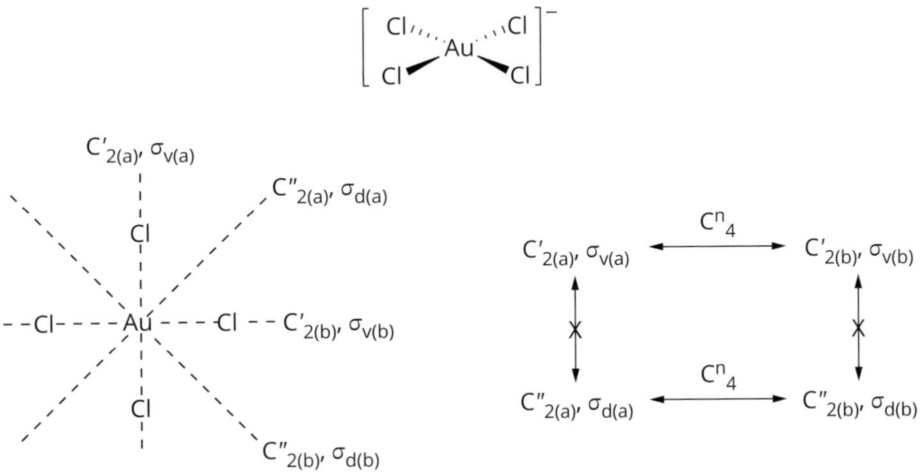

Operations: E, C_4, C^3_4, $C_2(=C^2_4)$, $C'_{2(a)}$, $C'_{2(b)}$, $C''_{2(a)}$, $C''_{2(b)}$, σ_h, S_4, S^3_4, $\sigma_{v(a)}$, $\sigma_{v(b)}$, $\sigma_{d(a)}$, $\sigma_{d(b)}$

Classes: E, $\{C_4, C^3_4\}$, $C_2(=C^2_4)$, $\{C'_{2(a)}, C'_{2(b)}\}$, $\{C''_{2(a)}, C''_{2(b)}\}$, σ_h, $\{S_4, S^3_4\}$, $\{\sigma_{v(a)}, \sigma_{v(b)}\}$, $\{\sigma_{d(a)}, \sigma_{d(b)}\}$

Fig. 5.35 Some of the symmetry operations of $\{AuCl_4\}^-$ and their classes.

molecule resides in a mirror plane, a σ_h. There is no symmetry operation that relates the σ_v to the σ_h, and the horizontal mirror plane is in a class by itself. As always, the identity operation is also in a class of its own.

In all, the BF_3 molecule has twelve symmetry operations: E, C_3, C^2_3, C_2, C'_2, C''_2, σ_h, S_3, S^5_3, σ_v, σ'_v, and σ''_v, where the primes are added to distinguish different symmetry operations of the same type but related to different symmetry elements. These symmetry operations can be grouped into six classes signified by using brackets to enclose multiple operations in the same class: E, $\{C_3, C^2_3\}$, $\{C_2, C'_2, C''_2\}$, σ_h, $\{S_3, S^5_3\}$, and $\{\sigma_v, \sigma'_v, \sigma''_v\}$.

The tetrachloroplatinate anion, $\{AuCl_4\}^-$, is a square planar complex that is in the D_{4h} point group. The principal rotation axis is a C_4. Shown on the left of Fig. 5.35 are four dotted lines that are coincident with C_2 axes and mirror planes in the molecule. The C_2 axes are labeled with primes (C'_2 if along the bond or C''_2 if in between bonds) and "a" and "b" to distinguish them from one another. The mirror planes are given σ_v if along the bond axes and σ_d if in between the bonds. The C_4 axis relates the two C'_2 axes together. Likewise, the C_4 relates the σ_v reflection planes. However, there is no symmetry operation that relates the C'_2 axes to the C''_2 axes, so these belong to different classes. Similarly, the σ_v and σ_d planes belong to different classes of symmetry operations.

In all, there are sixteen symmetry operations in D_{4h}: E, C_4, C^3_4, $C_2(=C^2_4)$, $C'_{2(a)}$, $C'_{2(b)}$, $C''_{2(a)}$, $C''_{2(b)}$, i, S_4, S^3_4, σ_h, $\sigma_{v(a)}$, $\sigma_{v(b)}$, $\sigma_{d(a)}$, and $\sigma_{d(b)}$. These are grouped into ten classes: E, $\{C_4, C^3_4\}$, $C_2(=C^2_4)$, $\{C'_{2(a)}, C'_{2(b)}\}$, $\{C''_{2(a)}, C''_{2(b)}\}$, i, $\{S_4, S^3_4\}$, σ_h, $\{\sigma_{v(a)}, \sigma_{v(b)}\}$, and $\{\sigma_{d(a)}, \sigma_{d(b)}\}$. The C_2 coincident with the C_4 axis is in a class by itself and is labeled $C_2(=C^2_4)$. The identity operation, E, and inversion, i, are always in classes by themselves.

5.5 Symmetry Operations and Matrices

A common method for examining symmetry operations is to use matrix algebra, and there are several important consequences to how the math is done in detail that are worth mentioning. While we will usually employ methods that bypass the full use of matrices in our applications, it is worth demonstrating, with a few examples, how the simplified methods shown later are related to the full mathematics.

There are only a few definitions and operations of matrices that we will need. The main operations on matrices that we will need to carry out are calculation of what are called the character (sometimes called the trace) and matrix multiplication.

A brief tutorial on the multiplication of matrices and matrix algebra is presented in Appendix E. Those unfamiliar with linear algebra or needing a refresher are encouraged to examine this Appendix. There are also a number of examples on multiplication in this section.

The "character" (trace) of a matrix is the sum of the diagonal elements from top left to bottom right. In formula, the character, given the symbol of the Greek letter χ,"chi", is:

$$\chi = \sum_{i=j} a_{ij}$$

A "diagonal" matrix is one where all of the elements not along the diagonal are zero; in other words, where all of the elements of the matrix, a_{ij}, are equal to zero where $i \neq j$; $a_{ij} = 0$ for all $i \neq j$. A special example of such a matrix is the following, where all of the numbers not on the diagonal are zero.

$$\begin{bmatrix} 1 & 0 & 0 & 0 \\ 0 & 1 & 0 & 0 \\ 0 & 0 & 1 & 0 \\ 0 & 0 & 0 & 1 \end{bmatrix}$$

In this case, where all of the elements are "1" in a diagonal matrix, this is called an "identity matrix", specifically the 4 × 4 identity matrix. This is also an example of a "square matrix", with the same number of rows and columns.

We can use matrices to describe symmetry operations in molecules, but to start with a simple example, let's look at the effects of symmetry operations on a point *x*, *y*, and *z*.

To start, consider the effect of inversion on the point (Fig. 5.36a). Pictorially, one can easily see that inversion will take the point from (x, y, z) through the origin to $(-x, -y, -z)$.

We need to find a matrix that will carry out this mathematical operation.

$$\begin{bmatrix} x \\ y \\ z \end{bmatrix} \xrightarrow{i} \begin{bmatrix} -x \\ -y \\ -z \end{bmatrix}$$

The matrix that can be used for the "*i*" symmetry operation is the following.

$$i = \begin{bmatrix} -1 & 0 & 0 \\ 0 & -1 & 0 \\ 0 & 0 & -1 \end{bmatrix}$$

As represented by the following matrix multiplication:

$$i \cdot (x,y,z) = \begin{bmatrix} -1 & 0 & 0 \\ 0 & -1 & 0 \\ 0 & 0 & -1 \end{bmatrix} \cdot \begin{bmatrix} x \\ y \\ z \end{bmatrix} = \begin{bmatrix} -x \\ -y \\ -z \end{bmatrix} = (-x,-y,-z)$$

The character (χ) of the square matrix for *i*—the sum along the diagonal—is $\chi = -3$.

For a reflection in the *xz*-plane coordinates in this plane will not change and *y* will become $-y$ (Fig. 5.36b).

$$\begin{bmatrix} x \\ y \\ z \end{bmatrix} \xrightarrow{\sigma_{xz}} \begin{bmatrix} x \\ -y \\ z \end{bmatrix}$$

The matrix that can carry out this transformation is as follows, with a character, χ, of 1.

$$\sigma_{xz} = \begin{bmatrix} 1 & 0 & 0 \\ 0 & -1 & 0 \\ 0 & 0 & 1 \end{bmatrix}$$

For a proper rotation, the general matrix that can convert any point (x_1, y_1, z_1) to (x_2, y_2, z_2) by clockwise rotation around the *z* axis is shown in Eq. 5.2, where $\theta = 360°/n$:

$$C_n \begin{bmatrix} x_1 \\ y_1 \\ z_1 \end{bmatrix} = \begin{bmatrix} \cos\theta & \sin\theta & 0 \\ -\sin\theta & \cos\theta & 0 \\ 0 & 0 & 1 \end{bmatrix} \begin{bmatrix} x_1 \\ y_1 \\ z_1 \end{bmatrix} = \begin{bmatrix} x_2 \\ y_2 \\ z_2 \end{bmatrix} \quad (5.2)$$

A counterclockwise rotation around *z* is easily defined, remembering that $\cos(-\theta) = \cos(\theta)$ and $\sin(-\theta) = -\sin(\theta)$ (Eq. 5.3):

$$C_n^{-1} \begin{bmatrix} x_1 \\ y_1 \\ z_1 \end{bmatrix} = \begin{bmatrix} \cos(-\theta) & \sin(-\theta) & 0 \\ -\sin(-\theta) & \cos(-\theta) & 0 \\ 0 & 0 & 1 \end{bmatrix} \begin{bmatrix} x_1 \\ y_1 \\ z_1 \end{bmatrix} = \begin{bmatrix} \cos(\theta) & -\sin(\theta) & 0 \\ \sin(\theta) & \cos(\theta) & 0 \\ 0 & 0 & 1 \end{bmatrix} \begin{bmatrix} x_1 \\ y_1 \\ z_1 \end{bmatrix} = \begin{bmatrix} x_2 \\ y_2 \\ z_2 \end{bmatrix} \quad (5.3)$$

Because the rotation matrix is not diagonal, we can see that the *x* and *y* coordinates are mixed into each other by a general rotation around *z*. Applying matrix multiplication gives the

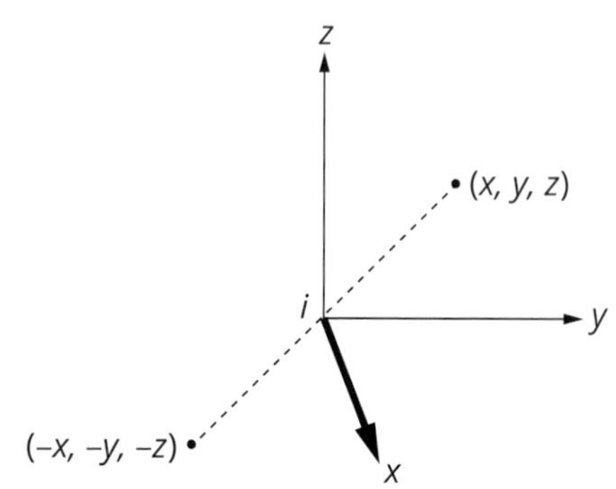

Fig. 5.36a Inversion converts a point (*x*, *y*, *z*) to (−*x*, −*y*, −*z*).

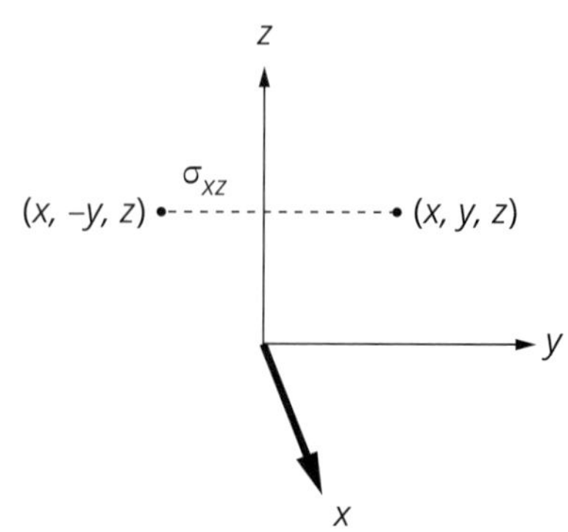

Fig. 5.36b A mirror operation in the *xz*-plane converts a point (*x*, *y*, *z*) to (*x*, −*y*, *z*).

coordinates of the new point (x_2, y_2, z_2) in terms of the original coordinates (x_1, y_1, z_1); for instance, the new coordinates after a C_n rotation are:

$$x_2 = \cos\theta \cdot x_1 + \sin\theta \cdot y_1$$

$$y_2 = -\sin\theta \cdot x_1 + \cos\theta \cdot y_1$$

$$z_2 = z_1$$

Eqs. 5.2 and 5.3 show an important mathematical feature; namely, that a clockwise rotation C_n^1 and a counterclockwise rotation C_n^{-1} are each represented by matrices that are *transpose* of each other. (The transpose of a matrix switches each element a_{ij} with a_{ji}. The element on the second row and third column in matrix A moves to the third row and second column in the transpose matrix of A).

As another example, let us find the matrix that rotates a set of coordinates through a C_4 around the z axis. The general matrix is calculated using $\theta = 360°/4$, which gives the clockwise rotational matrix with a character, χ, of 1. Note that the x and y terms exchange in the matrix with $x \rightarrow y$ and $y \rightarrow -x$, which moves them off the diagonal and gives them no contribution to the character of the matrix.

$$C_4 = \begin{bmatrix} \cos\theta & \sin\theta & 0 \\ -\sin\theta & \cos\theta & 0 \\ 0 & 0 & 1 \end{bmatrix} \xrightarrow{\theta=90°} \begin{bmatrix} 0 & 1 & 0 \\ -1 & 0 & 0 \\ 0 & 0 & 1 \end{bmatrix}$$

90° clockwise rotation

$$\begin{bmatrix} 0 & 1 & 0 \\ -1 & 0 & 0 \\ 0 & 0 & 1 \end{bmatrix}\begin{bmatrix} x \\ y \\ z \end{bmatrix} = \begin{bmatrix} y \\ -x \\ z \end{bmatrix}$$

A C_4 rotation in the counterclockwise rotation, C^{-1}_4 or equivalently C^3_4 if one uses the clockwise rotation matrix, is as follows.

$$C_4^{-1} = \begin{bmatrix} \cos\theta & -\sin\theta & 0 \\ \sin\theta & \cos\theta & 0 \\ 0 & 0 & 1 \end{bmatrix} \xrightarrow{\theta=90°} \begin{bmatrix} 0 & -1 & 0 \\ 1 & 0 & 0 \\ 0 & 0 & 1 \end{bmatrix}$$

90° counterclockwise rotation

As mentioned, the C_4 matrix is the transpose of the C^{-1}_4 matrix, where the columns and rows are exchanged. From a qualitative perspective, if one does a C_4 then a C^{-1}_4 to any set of coordinates one should get back the starting coordinates. In other words, rotating clockwise 90° then counterclockwise 90° should give the symmetry operation E, or more precisely, the matrix corresponding to the operation E, the identity matrix.

$$C_4 C_4^{-1} = E$$

$$\begin{bmatrix} 0 & 1 & 0 \\ -1 & 0 & 0 \\ 0 & 0 & 1 \end{bmatrix}\begin{bmatrix} 0 & -1 & 0 \\ 1 & 0 & 0 \\ 0 & 0 & 1 \end{bmatrix} = \begin{bmatrix} 0+1+0 & 0+0+0 & 0+0+0 \\ 0+0+0 & 1+0+0 & 0+0+0 \\ 0+0+0 & 0+0+0 & 0+0+1 \end{bmatrix}$$

$$= \begin{bmatrix} 1 & 0 & 0 \\ 0 & 1 & 0 \\ 0 & 0 & 1 \end{bmatrix}$$

For an improper rotation around the z-axis, the general matrix involves a rotation of $360°/n = \theta$ followed by reflection perpendicular to z. The reflection changes the z-coordinate from z to $-z$ as shown in the following equation.

$$S_n = \begin{bmatrix} \cos\theta & \sin\theta & 0 \\ -\sin\theta & \cos\theta & 0 \\ 0 & 0 & -1 \end{bmatrix} \quad \mathbf{5.4}$$

5.5.1 Reducible Representation for the Bond Stretches in Water

As an example of applying the previous matrices to a chemical system, let us do a bond stretch analysis (BSA) on water. If we want to know the symmetry properties of the stretches, it makes sense to use vectors between the oxygen and hydrogens, $\vec{r}_1$ and $\vec{r}_2$, as our "basis".

The *basis*, in group theory problems, is "the set of things (orbitals, stretches, coordinates, etc.) whose symmetry characteristics are of interest." The choice of basis is extremely important, as it will define the information obtained from the group theory analysis. In the current example we are interested in the bond stretches, so we are using the bond vectors as our basis. Left out will be any bending modes in the molecule; our basis simply does not include that information.

Fig. 5.37 shows the water molecule with our assigned coordinate system. The highest-order rotation axis is the C_2 through the oxygen, which was assigned as z. We then chose to make the plane of the molecule the yz-plane; this choice is somewhat arbitrary, as making it the xz-plane will produce essentially the same answer with some small changes in the symmetry labels.

The water molecule is in the C_{2v} point group, which has four symmetry operations: E, C_2, $\sigma_v(xz)$, and $\sigma'_v(yz)$. Our basis will be the set of vectors along the bonds, which will be used in a 1 × 2 matrix. The previous four symmetry operations will act on this matrix to provide the result of the symmetry operation. We will determine the form of the matrices for the symmetry operations and their character. *At the moment, this is merely an exercise in matrix algebra, but the characters of these matrices will provide us with important information later.*

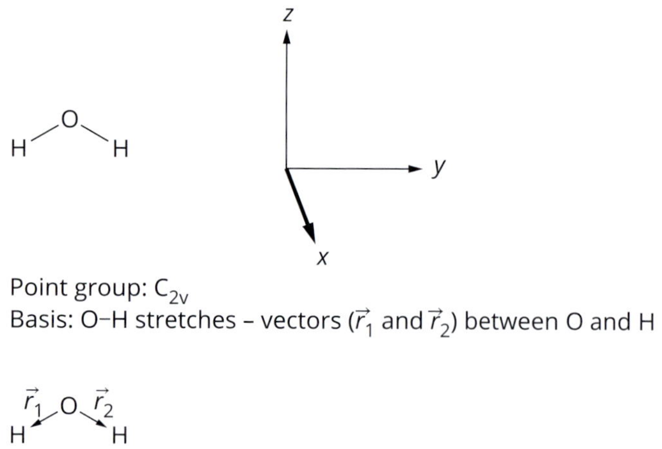

Fig. 5.37 Using the vectors between the bonds in water as a basis.

For the E operation, the starting O–H vectors are left unchanged; as a consequence, the matrix that represents this symmetry operation is the 2×2 identity matrix. The character of the identity matrix is 2. In general, the matrix for the E operation will always be the square identity matrix with a character equal to the number of things (vectors, orbitals, etc.) in our basis. In this case, there are two vectors in our basis, so the character of the E matrix = 2.

$$\begin{bmatrix} \vec{r}_1 \\ \vec{r}_2 \end{bmatrix} \xrightarrow{E} \begin{bmatrix} \vec{r}_1 \\ \vec{r}_2 \end{bmatrix}$$

$$E = \begin{bmatrix} 1 & 0 \\ 0 & 1 \end{bmatrix}$$

(As a side note, this notation implies that the E symmetry operation is being done on the set of vectors. If the basis is written as a column vector as previously, we need to left-multiply the E matrix, see below. We cannot right-multiply just because the number of members of the row does not equal the number of members in a column.)

$$\begin{bmatrix} 1 & 0 \\ 0 & 1 \end{bmatrix} \begin{bmatrix} \vec{r}_1 \\ \vec{r}_2 \end{bmatrix} = \begin{bmatrix} \vec{r}_1 \\ \vec{r}_2 \end{bmatrix}$$

Just by inspection, one can see that rotation around the C_2 axis turns $\vec{r}_1$ into $\vec{r}_2$ and vice versa. The character of this matrix is zero, 0, because all the matrix elements moved off the diagonal.

$$\begin{bmatrix} \vec{r}_1 \\ \vec{r}_2 \end{bmatrix} \xrightarrow{C_2} \begin{bmatrix} \vec{r}_2 \\ \vec{r}_1 \end{bmatrix}$$

$$C_2 = \begin{bmatrix} 0 & 1 \\ 1 & 0 \end{bmatrix}$$

For $\sigma_v(xz)$, the two vectors are converted into one another, and the character is again zero.

$$\begin{bmatrix} \vec{r}_1 \\ \vec{r}_2 \end{bmatrix} \xrightarrow{\sigma_v(xz)} \begin{bmatrix} \vec{r}_2 \\ \vec{r}_1 \end{bmatrix}$$

$$\sigma_v(xz) = \begin{bmatrix} 0 & 1 \\ 1 & 0 \end{bmatrix}$$

Finally, for $\sigma_v(yz)$ in the plane of the molecule, the two vectors are unchanged by the reflection plane.

$$\begin{bmatrix} \vec{r}_1 \\ \vec{r}_2 \end{bmatrix} \xrightarrow{\sigma_v(yz)} \begin{bmatrix} \vec{r}_1 \\ \vec{r}_2 \end{bmatrix}$$

$$\sigma_v(yz) = \begin{bmatrix} 1 & 0 \\ 0 & 1 \end{bmatrix}$$

So, the characters of the matrices with this basis on water are: $E = 2$, $C_2 = 0$, $\sigma_v(xz) = 0$, and $\sigma_v(yz) = 2$. We will return to this example later in this chapter.

5.5.2 Reducible Representation for the Bond Stretches in BF_3

As an example, in a different point group, we can find the symmetry properties of the bond stretches in BF_3. In this case, the molecule resides in point group D_{3h} and our basis will be three vectors along the bonds, $\vec{r}_1$, $\vec{r}_2$ and $\vec{r}_3$.

The character of the matrix representing the E operation will, as always, reflect the size of the basis, in this case 3.

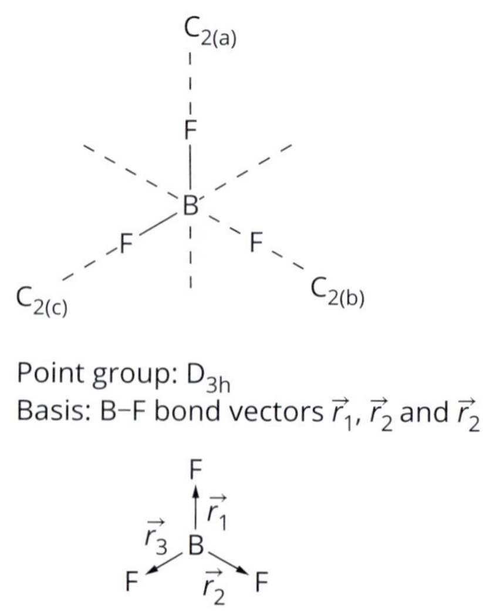

Fig. 5.38 The basis set for BF_3 bond stretches.

$$\begin{bmatrix} \vec{r}_1 \\ \vec{r}_2 \\ \vec{r}_3 \end{bmatrix} \xrightarrow{E} \begin{bmatrix} \vec{r}_1 \\ \vec{r}_2 \\ \vec{r}_3 \end{bmatrix}$$

$$E = \begin{bmatrix} 1 & 0 & 0 \\ 0 & 1 & 0 \\ 0 & 0 & 1 \end{bmatrix} \rightarrow \chi_E = 3$$

Rotation around the C_3 will move all the vectors off diagonal with $\vec{r}_1$ becoming $\vec{r}_2$, etc. Consequently, the character will be zero.

$$\begin{bmatrix} \vec{r}_1 \\ \vec{r}_2 \\ \vec{r}_3 \end{bmatrix} \xrightarrow{C_3} \begin{bmatrix} \vec{r}_2 \\ \vec{r}_3 \\ \vec{r}_1 \end{bmatrix}$$

$$C_3 = \begin{bmatrix} 0 & 1 & 0 \\ 0 & 0 & 1 \\ 1 & 0 & 0 \end{bmatrix} \rightarrow \chi_{C_3} = 0$$

There are three perpendicular C_2 axes, labeled C_{2a-c}. As shown previously, these are members of the same class. *Symmetry operations that belong to the same class will always exhibit the same character (trace) for their matrix.* As shown as follows for C_{2a} and C_{2b}, the nature of the full matrix is slightly different, but the trace is identical for members of the same class.

$$\begin{bmatrix} \vec{r}_1 \\ \vec{r}_2 \\ \vec{r}_3 \end{bmatrix} \xrightarrow{C_{2a}} \begin{bmatrix} \vec{r}_1 \\ \vec{r}_3 \\ \vec{r}_2 \end{bmatrix}$$

$$C_{2a} = \begin{bmatrix} 1 & 0 & 0 \\ 0 & 0 & 1 \\ 0 & 1 & 0 \end{bmatrix} \rightarrow \chi(C_{2a}) = 1$$

$$\begin{bmatrix} \vec{r}_1 \\ \vec{r}_2 \\ \vec{r}_3 \end{bmatrix} \xrightarrow{C_{2b}} \begin{bmatrix} \vec{r}_3 \\ \vec{r}_2 \\ \vec{r}_1 \end{bmatrix}$$

$$C_{2b} = \begin{bmatrix} 0 & 0 & 1 \\ 0 & 1 & 0 \\ 1 & 0 & 0 \end{bmatrix} \rightarrow \chi(C_{2a}) = 1$$

5.6 Choosing a Basis Set: What Do You Want to Learn?

The basis set chosen for a problem is determined by the information you want to learn from the symmetry of the molecule. Choosing a very large basis may give you a lot of information, but it may also be more difficult to interpret and require a lot of mathematical manipulation. A 20-member basis, which is not difficult to reach with even modest-sized molecules, would involve 20 × 20 matrices for each symmetry class in the point group. The best strategy is one that tells you as precisely as possible the information you are really interested in without added manipulations or deconvolution of the information you want from the extraneous details. How we get only the wheat without the chaff requires thoughtful choice of the basis.

In vibrational spectroscopy, perhaps the two most common basis sets used will be for what we will call bond stretch analysis (BSA) and normal mode analysis (NMA). These will be described using $Fe(CO)_5$ as an example.

5.6.1 Bond Stretch Analysis

This is a method of determining the bond stretches in a molecule only. The stretches in a molecule are often higher in energy and easier to identify than bends, and in some molecules, such as metal carbonyls, the vibrations are very easily identified in the infrared spectrum. Group theory does not give us the energy of the observed bands; it only gives the symmetry. One can find all the movements of a molecule using normal mode analysis (discussed next) and then have trouble matching the band in the spectrum to the symmetries obtained. Does this irreducible representation belong to a stretch or a bend? Using BSA, one can identify the symmetry of the stretches exclusively to help with assignment.

To find the number and type of C–O stretching vibrations in $Fe(CO)_4$, for example, we place a vector along the C–O bond (Fig. 5.39b). We will choose to have this vector go from the carbon to the oxygen, although the alternative is just as valid.

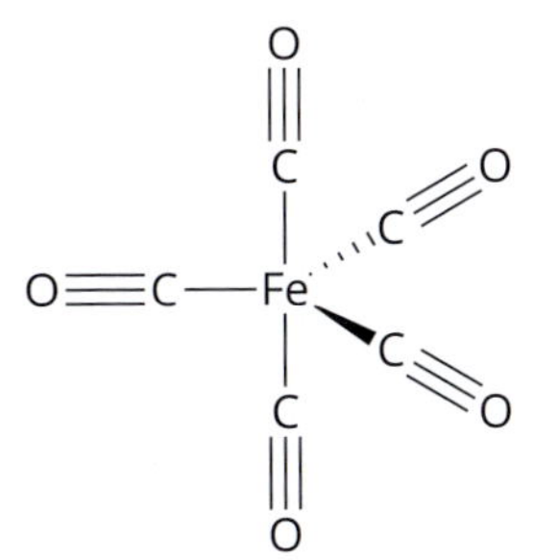

Fig. 5.39a Structure of $Fe(CO)_5$.

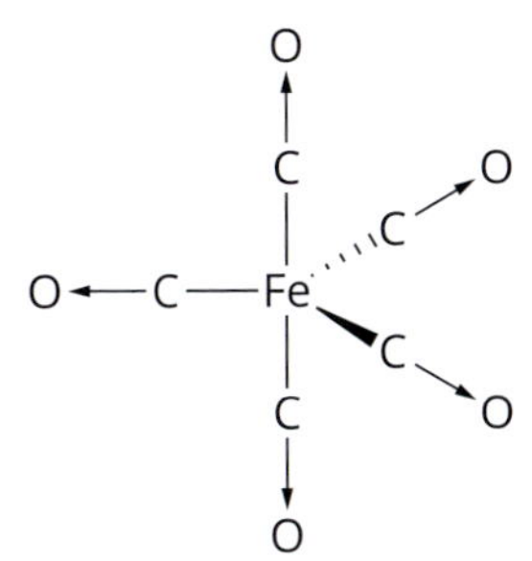

Fig. 5.39b Basis set for examining the C–O bond stretches in $Fe(CO)_5$ is a vector between each C–O bond.

In almost all cases it is advantageous to make the basis symmetric with respect to the principal rotation axis at the very least; this seems to be particularly true for systems with rotation axes higher than C_2. In the previous basis for the D_{3h} example, the vectors of the basis are symmetric with respect to the C_3 and the σ_h, two of the main generators of the group. Keep in mind that this basis is specifically designed to produce information on the stretches. No information will be obtained about any other movement of the molecule.

5.6.2 Normal Mode Analysis

For a normal mode analysis, the basis is an *x*-, *y*-, and *z*-axis on every atom (Fig. 5.39c). In other words, the basis size for a molecule with *n* atoms will be 3*n* in size. In most cases, the problem is greatly simplified by making the basis symmetric with respect to the main rotation axis. Shown below is one possible basis for the $Fe(CO)_5$ molecule for normal mode analysis. In the blue largest box are atoms with axes oriented like the blue topmost *x*-, *y*-, and *z*-vectors shown, a set of vectors on each atom for a total of fifteen in the blue largest box alone. The green, horizontal, box contains atoms with a vector centered on those two atoms with that orientation, and so on. The respective axes are symmetric around the equatorial CO ligands so that the *x*-vector on C^1 will be converted to the *x*-vector on C^2 by the $C^1{}_3$ operation.

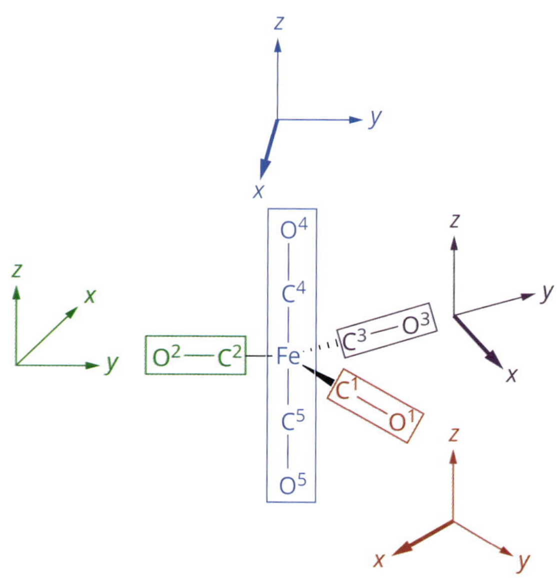

Fig. 5.39c Basis set for examining all the normal modes in $Fe(CO)_5$. Normal mode analysis (NMA) uses a coordinate axis on every atom where each *x*-, *y*-, and *z*-vector is a part of the basis. The different *orientations* of the *x*-, *y*-, and *z*-vectors are shown, but not all the vectors, which would have made for a very crowded figure with a total of 33 vectors. In every case, the *z*-vector should be parallel to the principal rotation axis (PRA). The other vectors are symmetric with respect to the PRA; that is, the C_3 converts the *x*-axis on one carbon in the equatorial plane to another.

The entire basis for this example will be 33-dimensional. Contained within these 33 normal modes are every possible movement of the molecule; however, not every movement is given explicitly. For example, there will be a translation of the entire molecule along *z*, and another translation along *y*. All movements in the *xz*-plane are then linear combinations of these two irreducible representations, even though they are not explicitly listed. In other words, every motion is either one of these normal modes or is a linear combination of them.

5.6.3 Molecular Orbitals and Basis Sets

The use of symmetry in creating molecular orbital diagrams is covered in detail in Chapter 6, and in these cases one may choose to use a full or limited basis, depending on interest. For example, if you want to do a full molecular orbital analysis of PF_5, you need to consider every orbital on phosphorus and all the fluorines. More commonly, we may do a valence orbital analysis where we only consider these as they are of the most chemical relevance.

In some cases it may be advantageous to choose only a specific type of orbital in the molecule for symmetry analysis. If we want to know about the symmetry properties of the π-system of benzene (Fig 5.40), using every valence orbital on every atom will be a 30-dimensional basis set, with the information in which we are interested buried in there somewhere. Alternatively, we could just obtain the information we want by using the p_z-orbitals on each carbon.

One can imagine that the matrices can begin to become unwieldy for larger molecules and larger basis sets, a problem that will now be addressed with a simplified method for finding the characters of a total representation.

5.7 Simplified Method for Generating a Trace for a Matrix and the Total Representation

As the size of a basis becomes quite large, the matrix algebra for group-theoretical analysis can become quite tedious. In applications of group theory to chemistry we will be interested primarily in the trace of the matrix, and fortunately the trace is accessible using simplified methods. First, a very simple common method will be presented where the trace of the matrix is determined by a +1, 0, −1 simplification based on the parts of the basis.

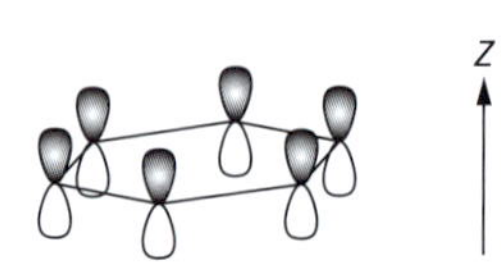

Fig. 5.40 Basis set for examining the π-system of benzene and the orientation of the principal rotation axis.

This method works well for many problems where the parts of the basis either stay the same, or move onto another part of the basis, or into their inverse. However, when the basis does none of these three, such as rotation of a p-orbital by 120°, then one has to find an alternative method to get the contribution to the reducible representation. In general, this (+1, 0, −1)-method will work well for bond stretch analysis. Second, a slightly more mathematical method for finding the contribution of any NMA basis to the reducible representation will be presented. This method may require a calculator, but works well for normal mode analysis of any molecule in any symmetry.

5.7.1 Simplified Method for Generating a Total Representation: +1, 0, −1 Method

The simplified method for determining the trace of the matrix for a symmetry operation involves the following three rules. First we will simply list these rules, and then apply them to a couple of examples.

(1) Any part of a basis that gets mapped onto itself contributes "+1" to the trace. In other words, parts of the basis that do not move under the symmetry operation will contribute +1 each.

(2) Any part of a basis that gets mapped onto its negative contributes "−1" to the trace (Fig. 5.41a). An example of this would be a p_z-orbital undergoing a C_2 around x. The orbital starts at the origin and stays at the origin, but its phases are inverted.

If one is examining bond vectors, these too can invert and contribute negatively to the character. For example, if one is examining a vector between the atoms of H_2, one of the C_2 symmetry operations will invert the vector, causing it to contribute −1.

(3) Any part of the basis that gets mapped onto another part of the basis contributes zero. In essence, this is due to the movement of the basis off diagonal in the matrix, which causes it to not contribute to the trace.

These rules allow the generation of the *total representation* of the basis quite simply. The total representation is the list of characters corresponding to all the symmetry operations acting on a chosen basis. This method is only applicable if all the parts of the basis move into another part of the basis (or its negative) for every symmetry operation.

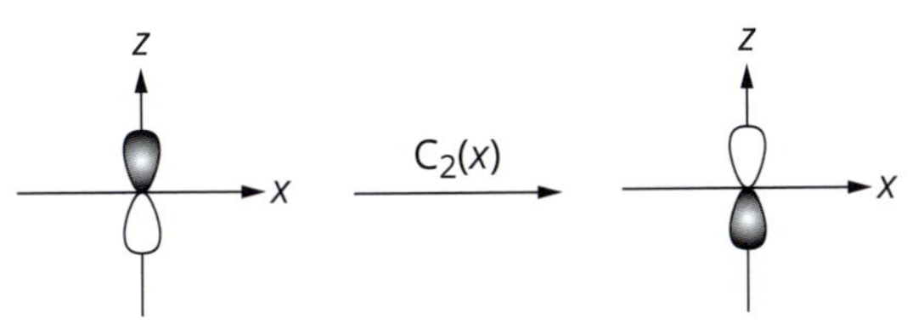

Fig. 5.41a A C_2 perpendicular to a p-orbital leaves the orbital in place but inverts the phase, which causes it to contribute "−1" to the trace.

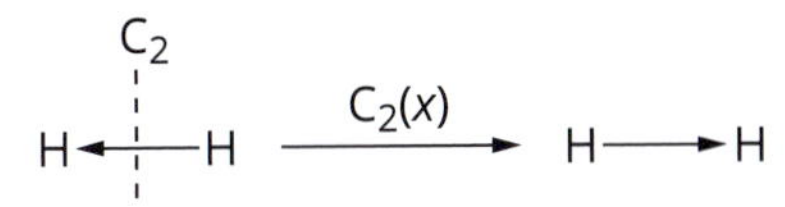

Fig. 5.41b Inverting the direction of a vector in the basis like that in H_2 causes it to contribute "−1" to the trace.

Using this simplified method we can generate the total representation of the bond vectors in BF_3, a continuation of the example in Section 5.3.3. The point group, as before, is D_{3h}, and the basis is the set of vectors (Fig. 5.38).

Since we know that all symmetry operations of the same class have the same trace, we can simply write the classes. If a vector does not move under that symmetry operation it will contribute +1. If the vector moves it will contribute 0. If the vector changes direction by 180° but does not move it will contribute −1. For the E operation, none of the vectors move and each one contributes +1 for a total of +3. A C_3 will move all the vectors, which will give them all a contribution of zero. Each C_2 along the bond vector will leave one unchanged and move the other two for a total of +1. The σ_h is in the plane of the molecule, and none of the vectors move with this operation, +3. Like the C_3, the S_3 will move all the vectors, 0. Like the C_2, one of the bond vectors is along the σ_v and is unmoved, while the other two vectors exchange, +1.

In Table 5.3, the series of characters labeled $\Gamma_{(B-F)}$ is the total representation of the basis, which reflects the symmetry properties of the basis in the point group.

As another example we can examine PCl_5, which has the same point group as BF_3, D_{3h}. In addition, just like we did for BF_3, we will use the bond stretches (bond vectors) as our basis, but because the molecule has changed, the basis has changed (Fig. 5.42). Whereas the bond stretch basis in BF_3 had three members, the basis in PCl_5 will have five members.

Applying the symmetry operations to this basis gives the total representation, $\Gamma_{(P-Cl)}$, shown below. Note that for the S_3 we have to consider the contribution of the vectors after both the C_3 and σ. The axial vectors, $\vec{r}_1$ and $\vec{r}_2$, do not move during the C_3 (note the character in the C_3 operation), but these vectors

Table 5.3 Reducible representation for the bond stretches in BF_3.

D_{3h}	E	$2C_3$	$3C_2$	σ_h	$2S_3$	$3\sigma_v$
$\Gamma_{(B-F)}$	3	0	1	3	0	1

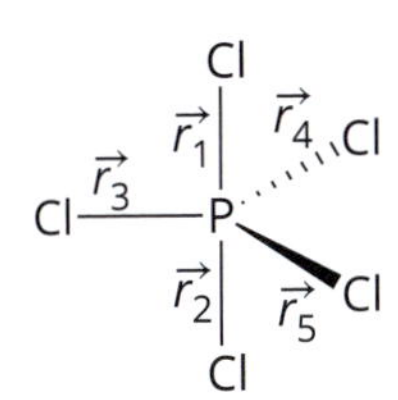

Fig. 5.42 A basis for looking at the bond stretches in PCl_5.

Table 5.4 Reducible representation for the bond stretches in PCl_5.

D_{3h}	E	$2C_3$	$3C_2$	σ_h	$2S_3$	$3\sigma_v$
$\Gamma_{(P-Cl)}$	5	2	1	3	0	3

Table 5.5 Reducible representation for the bond stretches in PCl_5 for just the axial chlorines.

D_{3h}	E	$2C_3$	$3C_2$	σ_h	$2S_3$	$3\sigma_v$
$\Gamma_{(P-Cl\ axial)}$	2	2	0	0	0	2

are moved by the mirror plane. As a result, the character under S_3 is zero (Table 5.4).

If we leave the molecule the same (PCl_5) and change the basis, we will obtain a different total representation as well. For example, if we just want to know the symmetry properties of the axial P–Cl bonds, $\vec{r}_1$ and $\vec{r}_2$, we can use these as our basis. The result will be the total representation (Table 5.5), $\Gamma_{(P-Cl\ axial)}$.

The information we obtain from the group theory analysis is strongly dependent upon the basis chosen. In later chapters there will be a variety of applications where it will be shown how to choose the proper basis. For example, in molecular orbital theory it is common to use the valence atomic orbitals as the basis set for group-theoretical analysis.

5.7.2 Simplified Method for Generating a Total Representation: Proper and Improper Rotation Method for Normal Mode Analysis[5]

To find the reducible representation for a normal mode analysis, we place an x-, y-, and z-axis on every atom, which is then our basis. This simplified method will not require us to even draw out the basis and will only require some simple mathematics. The method will rely on the fact that every symmetry operation can be reduced to either a proper rotation or an improper rotation. For this, we will make a few simplifications to our symmetry operations.

(1) The identity operation, E, will be a proper rotation of 0° (or 360°), C_1.

(2) All proper rotations, C_n, will be treated as normal proper rotations of 360°/n.

(3) An inversion center, i, will be treated as an improper rotation of 180°, S_2.

(4) A mirror plane, σ, will be treated as an improper rotation around the appropriate axis of 0°, S_1.

(5) All improper rotations, S_n, will be treated as normal improper rotations of 360°/n.

What this classification allows us to do is develop two simple equations for getting the contribution of a basis to the reducible representation that will work for any basis in any point group. For a proper rotation (E or C_n) the contribution will be:

$$\chi^C(R) = n_R(2\cos\theta + 1) \quad \textbf{5.5}$$

where $\chi(R)$ is the contribution of that part of the basis to the reducible representation, n_R is the number of atoms retaining their position during the operation, and θ = 360°/n for the rotation.

If the operation is being associated with its equivalent improper rotation (i, σ, and S_n), then we use Eq. 5.6, which is identical to Eq. 5.5 except for a sign change on the term in the parentheses.

$$\chi^S(R) = n_R(2\cos\theta - 1) \quad \textbf{5.6}$$

The trace of the proper rotation matrix (Eq. 5.2) is in parentheses in Eq. 5.5. The trace for the improper rotation matrix (Eq. 5.4) is in parentheses in Eq. 5.6. This character of the matrix for an x-, y-, and z-axis on one atom is then multiplied by the number of atoms that do not move, the number of atoms that contribute to the character. Consequently, these equations get to the character without the matrix algebra.

Previously, we found the bond stretches for BF_3. Here we will find the reducible representations for the normal mode analysis on BF_3. To do this we will place an x-, y-, and z-axis on every atom, whether it be boron or fluorine. Importantly, this basis is *not* well suited for use with the "+1, 0, −1" method learned in the previous subsection.

Consider what happens with just the x-axes on the four atoms when carrying out the C_3 operation. As shown below, and will be mentioned many times through the course of the book, it is generally advisable to use a basis that is symmetric with respect to the main rotation axis whenever possible. Using this principle, the full basis on BH_3, and, the part we will discuss here, just the x-axis, will look as shown in Fig. 5.43.

As the C_3 is carried out, the x-axes on the hydrogens are converted into one another, which, as discussed in the previous subsection, would indicate that they make no contribution to the character of the reducible representation. However, x_1 on boron does not go into another part of the basis on rotation by 120°. It does not go to y_1, which is 90° away. In fact, both x_1 and y_1 will not cleanly go to other parts of the full basis. To find out how much of them is moving we would need to use the general rotation matrix, which essentially tells us the projection of the new position on the old, to determine the contribution.

The details of the calculations are shown in Table 5.6. In the first row, the angle of the proper or improper rotation is given. In the second row, the cosine of this angle is given. The cosine of the angle is then used in Eqs. 5.5 and 5.6, where the plus (+) sign on the 1 is used for a proper rotation, and the minus (–) sign on the 1 is used for an improper rotation. This row is given the

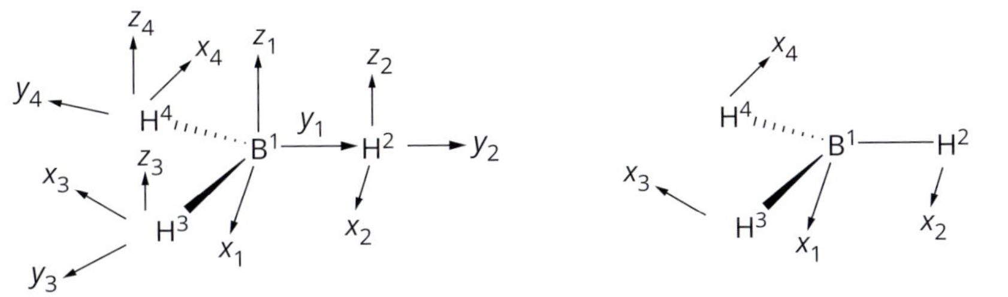

Fig. 5.43 A basis for examining all the possible normal movements of BH_3 on the left, with just the *x*-axes on the right.

Table 5.6 Method for finding the reducible representation for the normal mode analysis of BH_3. (See Fig. 5.43 for a description of the basis of x-vectors.)

	Proper Rotations			Improper Rotations		
D_{3h}	E	$2C_3$	$3C_2$	σ_h	$2S_3$	$3\sigma_v$
θ	0°	120°	180°	0°	120°	0°
cos θ	1	−1/2	−1	1	−1/2	1
±1+ 2cosθ	3	0	−1	1	−2	1
n_R	4	1	2	4	1	2
Γ_{BH3}	12	0	−2	4	−2	2

number of atoms that retain their position during the operation, n_R. Finally, the ±1 + 2cosθ row is multiplied by the n_R row to obtain the reducible representation, Γ_{BH3}.

As a second example of using this method, consider the NMA of trigonal bipyramidal $Fe(CO)_5$, the basis of which was mentioned in Section 5.6.2. The molecule has the same point group as BH_3, D_{3h}. As a result, much of this table can be reused here. The only thing that changes is the number of atoms that do not move with the symmetry operation (Table 5.7), n_R. The results are shown below.

How information can be extracted from these reducible representations will be discussed shortly, but first we need to discuss a few more details about how character tables are constructed and what are called *irreducible representations.*

Table 5.7 Method for finding the reducible representation for the normal mode analysis on $Fe(CO)_5$. See Fig. 5.39c for a description of the basis.

	Proper Rotations			Improper Rotations		
D_{3h}	E	$2C_3$	$3C_2$	σ_h	$2S_3$	$3\sigma_v$
θ	0°	120°	180°	0°	120°	0°
cos θ	1	−1/2	−1	1	−1/2	1
±1+ 2cosθ	3	0	−1	1	−2	1
n_R	11	5	3	7	1	7
$\Gamma_{Fe(CO)5}$	33	0	−3	7	−2	7

5.8 The Great Orthogonality Theorem and Character Table Construction

A **representation** is simply a list of characters that represents the effect of the symmetry operations on a basis. In the previous section we discussed a method for finding the total representation of a basis. In almost all cases, the total representation will be what we will refer to as a *reducible representation*, which can be broken down via similarity transformations into a linear combination of representations having lower dimensionality. As mentioned, the dimensionality of the representation is simply the character under the *E* symmetry operation, which is equivalent to the number of objects (vectors, orbitals, axes, and so on) in the basis.

We can also define an *irreducible representation*, which is a representation where no similarity transformation exists that will lower the dimensionality of the system. The irreducible representations form a closed set found in the character tables for the specific point group.

Because all the symmetry operations of a molecule constitute a mathematical group, we can find the closed set of irreducible representations by using one of the theorems of group theory: the Great Orthogonality Theorem. The consequences of the theory that will be applicable here are divided into five lemmas, each of which will play a role in the construction of the full character tables. As an example, the D_6 character table is shown in Table 5.8, which will be used to illustrate each lemma.

Table 5.8 The D_6 character table used to illustrate the lemmas of the Great Orthogonality Theorem.

D_6	E	$2C_6(z)$	$2C_3(z)$	$C_2(z)$	$3C_2'$	$3C_2''$
Γ_1	1	1	1	1	1	1
Γ_2	1	1	1	1	−1	−1
Γ_3	1	−1	1	−1	1	−1
Γ_4	1	−1	1	−1	−1	1
Γ_5	2	1	−1	−2	0	0
Γ_6	2	−1	−1	2	0	0

D_6	E	$2C_6(z)$	$2C_3(z)$	$C_2(z)$	$3C'_2$	$3C''_2$
Γ_1	1	1	1	1	1	1
Γ_2	1	1	1	1	−1	−1
Γ_3	1	−1	1	−1	1	−1
Γ_4	1	−1	1	−1	−1	1
Γ_5	2	1	−1	−2	0	0
Γ_6	2	−1	−1	2	0	0

$1 + 2 + 2 + 1 + 3 + 3 = 12 = h$

$h = 12 = 1^2 + 1^2 + 1^2 + 1^2 + 2^2 + 2^2$

Fig. 5.44a Illustration of lemma 1 of the Great Orthogonality Theorem: *The sum of the squares of the dimensions (l) of the irreducible representations of a group is equal to the order (h) of the group (Eq. 5.7).*

(1) *The sum of the squares of the dimensions (l) of the irreducible representations of a group is equal to the order (h) of the group (Eq. 5.7).* The *order* (h) of a group is simply the total number of symmetry operations. The sum is over all the irreducible representations, which will be the rows of our character tables.

$$\sum_i l_i^2 = h \qquad 5.7$$

Example: The order (h) of the point group D_6 is 12, $h = 12$ (Fig. 5.44a). What this lemma describes is that the dimensions, which are simply the numbers under the E operation, squared are equal to the order of the group.

(2) *The sum of the squares of the character (χ) of any irreducible representation times the number of operations in the class is equal to the order (h).* As shown in Eq. 5.8, summing all the squares of the characters times the number of operations for a specific irreducible representation will also equal the order. In other words, taking the entries in a row and summing their squares will give h.

$$\sum_R g \times [\chi_i(R)]^2 = h \qquad 5.8$$

Example: The order of the group is equal to the coefficient in front of the symmetry operation at the top of the character table times the square of the character in the irreducible representation, summed over all the classes (Fig. 5.44b).

$h = 12 = (1 \times 1^2) + (2 \times 1^2) + (2 \times 1^2) + (1 \times 1^2) + (3 \times 1^2) + (3 \times 1^2)$

D_6	1E	$2C_6(z)$	$2C_3(z)$	$C_2(z)$	$3C'_2$	$3C''_2$
Γ_1	1	1	1	1	1	1
Γ_2	1	1	1	1	−1	−1
Γ_3	1	−1	1	−1	1	−1
Γ_4	1	−1	1	−1	−1	1
Γ_5	2	1	−1	−2	0	0
Γ_6	2	−1	−1	2	0	0

Fig. 5.44b Illustration of lemma 2 of the Great Orthogonality Theorem: *The sum of the squares of the character (x) of any irreducible representation times the number of operations in the class is equal to the order (h)* (Eq. 5.8).

(3) *All irreducible representations of a group are orthogonal to each other.* In Chapter 6, for example, this will be used explicitly to show two orbitals are orthogonal to one another, an important point for molecular orbital theory, but also for spectroscopy and other areas. For any two irreducible representations to be orthogonal, Eq. 5.9 must be true.

$$\sum_{R} g \times \chi_i(R) \times \chi_j(R) = 0 \qquad \textbf{5.9}$$

This is done by multiplying the character, $\chi_i(R)$ and $\chi_j(R)$, in each irreducible representation (representations i and j) and the number of operations in the class (g), which is summed over all classes, R.

Example: To test if two representations are orthogonal, as they are for the irreducible representations, take the number of operations in a class and multiply by the character of one of the representations, and multiply by the character of the other representation (Fig. 5.44c). The sum of all these terms over all the operations should be zero if the representations are orthogonal.

(4) *The number of irreducible representations in a group is equal to the number of classes in a group.* Since D_6 has six classes, it must have six irreducible representations. In other words, character tables are square matrices.

(5) *The characters of all matrices belonging to operations in the same class are identical.* This has already been discussed in Section 5.5, and it allows us to place the columns of the character tables into classes rather than individual symmetry operations.

As another illustration, let us derive the D_3 character table, which is a common point group for numerous chiral inorganic compounds where three bidentate ligands surround a central metal; for example, $\{Ru(2,2'\text{-bipyridine})_3\}^{2+}$ shown in Fig. 5.44d.

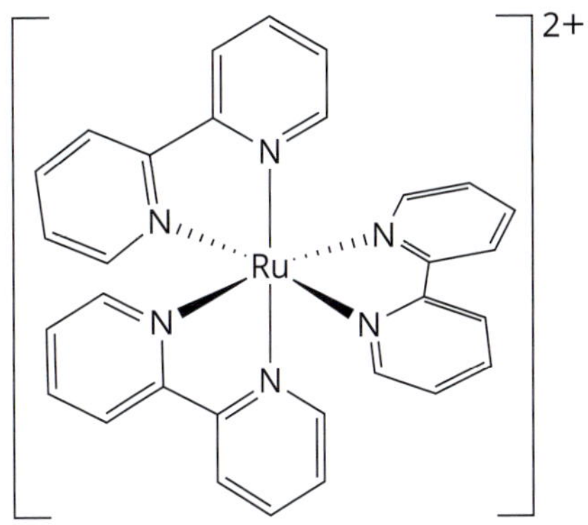

$[Ru(bpy)_3]^{2+}$

Fig. 5.44d Structure of the D_3 molecule $Ru(bpy)_3^{2+}$ (bpy = 2,2′-bipyridine).

The symmetry classes in D_3 are E, $2C_3$, and $3C_2$. These will comprise the heading of the character table. According to Lemma 4, because there are three classes (columns) there will be three irreducible representations (rows). In the character table (Table 5.9a) the irreducible representations are currently labeled Γ_{1-3}, but more descriptive labels will be given later in this section.

Table 5.9a Partial D_3 character table. By Lemma 4, there must be three irreducible representations because there are three classes.

D_3 ($h = 6$)	E	$2C_3$	$3C_2$
Γ_1			
Γ_2			
Γ_3			

(1 × 1 × 1) + (2 × 1 × 1) + (2 × 1 × 1) + (1 × 1 × 1) + (3 × 1 × −1) + (3 × 1 × −1) = 0

Testing if these two are orthogonal representations

D_6	E	$2C_6(z)$	$2C_3(z)$	$C_2(z)$	$3C'_2$	$3C''_2$
Γ_1	1	1	1	1	1	1
Γ_2	1	1	1	1	−1	−1
Γ_3	1	−1	1	−1	1	−1
Γ_4	1	−1	1	−1	−1	1
Γ_5	2	1	−1	−2	0	0
Γ_6	2	−1	−1	2	0	0

Fig. 5.44c Illustration of lemma 3 of the Great Orthogonality Theorem: *All irreducible representations of a group are orthogonal to each other* (Eq. 5.9).

Table 5.9b Partial D_3 character table. The top irreducible representations is always totally symmetric (all "1").

D_3 ($h = 6$)	E	$2C_3$	$3C_2$
Γ_1	1	1	1
Γ_2			
Γ_3			

One row of all the character tables is quite easy to write out (Table 5.9b). In every character table there is what is called a *totally symmetric irreducible representation*, which is invariant with every symmetry operation and is 1-dimensional. This irreducible representation is always written as the first representation listed. In other words, the top row has "1" under every class across the row.

Under the E symmetry operation, all the dimensionalities (characters in each row) must have the sum of their square equal to the order of the group (h); this is first lemma of the great orthogonality theorem. The only set of three numbers whose squares sum to $h = 6$ are {1, 1, 2}. As a result, the numbers down the E column of the character table must be 1, 1, and 2 (Table 5.9c).

$$\sum_i l_i^2 = h$$

$$1^2 + 1^2 + 2^2 = 6$$

Why is there a 2-dimensional representation in D_3? Consider the x- and y-axes in relation to a C_3 around the z-axis. The general matrix to carry out this transformation is not diagonalized. For a C_3, where $\theta = 120°$, the matrix clearly shows that the x and y points are not independent. Instead, they are always intertwined by the symmetry operations. As a result, they can only be reduced to a 2-dimensional representation.

$$C_3\begin{bmatrix} x_1 \\ y_1 \\ z_1 \end{bmatrix} = \begin{bmatrix} \cos 120° & \sin 120° & 0 \\ -\sin 120° & \cos 120° & 0 \\ 0 & 0 & 1 \end{bmatrix}\begin{bmatrix} x_1 \\ y_1 \\ z_1 \end{bmatrix}$$

$$= \begin{bmatrix} -0.5 & 0.866 & 0 \\ -0.866 & -0.5 & 0 \\ 0 & 0 & 1 \end{bmatrix}\begin{bmatrix} x_1 \\ y_1 \\ z_1 \end{bmatrix}$$

$$= \begin{bmatrix} -0.5x_1 + 0.866y_1 \\ -0.866x_1 - 0.5y_1 \\ z_1 \end{bmatrix}$$

$$\theta = 360° / 3 = 120°$$

Table 5.9c Partial D_3 character table. By Lemma 1, the sums of the squares under the E operation must equal the order of the group.

D_3 ($h = 6$)	E	$2C_3$	$3C_2$
Γ_1	1	1	1
Γ_2	1		
Γ_3	2		

Table 5.9d The D_3 character table. The other four characters can be found by using the other lemmas by trial and error.

D_3 ($h = 6$)	E	$2C_3$	$3C_2$
Γ_1	1	1	1
Γ_2	1	1	–1
Γ_3	2	–1	0

So, we have determined all the characters in the D_3 character table, except for the four shown in Table 5.9c. Using the other lemmas, we can determine that the only options for the remaining characters are as shown in Table 5.9d.

The student should confirm that the sum of the squares of the character times the number of operations in the class is equal to the order in any of these representations (Lemma 2). For Γ_3, this is shown as follows.

$$\sum_R [\chi_i(R)]^2 = h$$

$$\sum_R [\chi_{\Gamma_3}(R)]^2 = 6$$

$$(2^2 \times 1)_{\Gamma_3,E} + (-1^2 \times 2)_{\Gamma_3,C_3} + (0^2 \times 3)_{\Gamma_3,C_2} = 6$$

$$4 + 2 + 0 = 6$$

In addition, it can be shown that all the different irreducible representations are orthogonal (Lemma 3). First for Γ_1 and Γ_2, we sum three terms. In the first term we have the number of E operations (always 1) times the character under E in Γ_1 (which is +1) times the number under E in Γ_2 (which is +1). In the second term we have the number of operations under C_3 (which is 2) times the character under C_3 for Γ_1 (which is +1) times the character under C_3 for Γ_2 (which is +1). Finally, in the third term we have the number of operations under C_2 (which is 3) times the character under C_2 for Γ_1 (which is +1) times the character under C_2 for Γ_2 (which is –1). These terms sum to zero, indicating that the two irreducible representations are orthogonal.

$$\sum_R g \times \chi_i(R) \times \chi_j(R) = 0$$

$$(1 \times 1 \times 1)_E + (2 \times 1 \times 1)_{C_3} + (3 \times 1 \times -1)_{C_2} = 0$$

$$1 + 2 - 3 = 0$$

Similarly, the Γ_2 and Γ_3 irreducible representations, for example, can be shown to be orthogonal.

$$\sum_R g \times \chi_i(R) \times \chi_j(R) = 0$$

$$(1 \times 1 \times 2)_E + (2 \times 1 \times -1)_{C_3} + (3 \times -1 \times 0)_{C_2} = 0$$

$$2 - 2 + 0 = 0$$

Typically, instead of simply listing the irreducible representations as $\Gamma_1, \Gamma_2, \Gamma_3, \ldots$, as previously, we use *Mulliken symbols* to provide a more descriptive term. Within the Mulliken symbol is the dimensionality and many of the other symmetry properties of the irreducible representation. Decoding the symbol is achieved as follows:[6]

(1) The dimensionality of the irreducible representation is given by the main symbols A, B, E, or T.

A = single-dimensional representation and symmetric with respect to the principal rotation axis.

B = single-dimensional representation and antisymmetric with respect to the principal rotation axis.

E = 2-dimensional representation.

T = 3-dimensional representation.

(2) A subscript "1" or "2" means that the representation is symmetric or antisymmetric with respect to perpendicular C_2 axes, or, if lacking this, with respect to σ_v.

(3) Primes or double primes are used to indicate that the representation is symmetric or antisymmetric with respect to σ_h.

(4) If the molecule has an inversion center, g (gerade) and u (ungerade) are used to designate that the representation is symmetric or antisymmetric with respect to inversion.

We can apply these to our D_3 character table to replace Γ_{1-3} with the Mulliken symbols. To find the main symbol we simply look first at the dimensionality (Table 5.9e), the character under the symmetry operation E. (E is used for both a 2-dimensional irreducible representation and for the identity symmetry operation. This is unfortunate, but in ordinary usage it does not cause too much confusion.) For Γ_1 and Γ_2 this is "1", so the main symbol is either A or B. For these we then look at the principal rotation axis; if it is even then A, if odd then B. In this case, both are even (a +1 appears under C_3); consequently, both of the first two symbols have A as the main symbol.

In the case of the last representation, Γ_3, it is 2-dimensional, and its main symbol is E.

Going to (2) in our assignment list, we look at the properties under C_2. If it is symmetric with respect to the C_2, as is the first representation, then it is given a subscript "1". If antisymmetric, the subscript is "2". Once all the irreducible representations have different symbols there is no real reason to go further, and the completed character table for D_3 is as shown in Table 5.9f.

Most character tables include a set of commonly used functions to the right of the classes, indicating the irreducible representations to which they belong.

Table 5.9e The D_3 character table with partial Mulliken symbols.

D_3 (h = 6)	E	$2C_3$	$3C_2$
A	1	1	1
A	1	1	−1
E	2	−1	0

Table 5.9f The D_3 character table with full Mulliken symbols.

D_3 (h = 6)	E	$2C_3$	$3C_2$
A_1	1	1	1
A_2	1	1	−1
E	2	−1	0

In the first column to the right of the classes in the character table (Table 5.9g) is often found the rotations around the x-, y-, and z-axes as R_x, R_y and R_z, respectively. In addition, the symmetries of the axes themselves are shown in the same column. Furthermore, because of their phase and symmetry properties, the p_x-orbital has the same properties as the x-axis, for example.

In the next column to the right in the character table (Table 5.9g) are found common important nonlinear function, like those for the d-orbitals.

It should be noted that the s-orbital, because of its spherical symmetry, is always a member of the totally symmetric representation, which is always the top one listed and has all +1 terms under the classes. The totally symmetric representation will have Mulliken symbol A, A′, A_1, A_1', A_g, or A_{1g}.

Adding these to the D_3 character table gives Table 5.9g. All this information is presented in Appendix D for all the commonly encountered point groups in chemistry. For multidimensional representations, as in the E irreducible representation below, the related functions with the same symmetry will be given in parentheses; for example, (x,y). If discussing vibrational modes, orbital energies, and so on, functions in the same parentheses will have the same energy in addition to having the same symmetry properties, and are often regarded as a unit. Functions in different parentheses but in the same representation are not necessarily related by anything other than coincidentally having the same symmetry. For example, d_{xz} and d_{yz} are part of the same E representation and their energies will be required to be identical by symmetry. However, while d_{xy} and $d_{x^2-y^2}$ are symmetry-related and will have the same energy as each other, they do not necessarily have the same energy as the different E group (d_{xz}, d_{yz}), which is in a different set of parentheses.

5.9 Selection Rules for Infrared and Raman Spectroscopy

The total number of vibrations as normal modes can be determined using an internal *degrees of freedom* calculation. If every atom moves independently, then the entire ensemble

Table 5.9g The D_3 character table with full Mulliken symbols and common functions.

D_3 (h = 6)	E	$2C_3$	$3C_2$		
A_1	1	1	1		x^2+y^2, z^2
A_2	1	1	−1	z, R_z	xy
E	2	−1	0	$(x,y),(R_x,R_y)$	$(xz,yz),(x^2-y^2,xy)$

has $3n$ possible ways it can move, in the x, y, and z direction for every atom for n atoms. Three of those movements involve simply translating the entire molecule along the x, y, and z direction. For a nonlinear molecule there are also three rotations of the entire molecule, around the x-, y-, and z-axes. These translations and rotations of the entire molecule are not internal movements in that the atomic positions relative to one another do not change, and as a result they are typically not of interest to chemists. The internal degrees of freedom for a nonlinear molecule, DoF(nonlinear), is then:

$$\text{DoF(nonlinear)} = 3n - 6.$$

If the molecule is linear, then there are only two inequivalent rotations, along the $C_{\infty v}$ and perpendicular to it. Consequently, the internal degrees of freedom for a linear molecule, DoF(linear), are then:

$$\text{DoF(linear)} = 3n - 5.$$

These calculations produce the total number of possible vibrations (stretches and bends) for a molecule; however, depending on the analytical technique, some vibrations will be observable while others are not, depending on the symmetry of the mode. In other words, spectroscopic methods have selection rules that determine the type of vibration that can be observed by using the technique. While a more detailed discussion of selection rules and transition probabilities in spectroscopy will be presented later, a brief discussion of when vibrations are observable and by what technique is useful here to help the student understand some of the utility of group theory.

Infrared (IR) spectroscopy is one of the core analytical methods in chemistry that produces tremendous details about molecular structure. Most IR spectra are run from around 4000–400 cm^{-1}, which is the range for most molecular vibrations of interest. In IR spectroscopy, molecules are excited from the ground vibrational state (v^0) to higher vibrational states, typically from v^0 to v^1.

Badger's rule, named after R. M. Badger, who published it in 1934, asserts that the strength of chemical bond varies directly with its bond length; that is, *longer bonds are weaker bonds*, and vice versa.[7] The bond-force constant, k_O in dynes/cm, is related to the equilibrium distance, r_e in Å, of the bond by the following equation, where d_{ij} = a constant that is dependent upon the atomic elements in the bond which are in the ith row and jth row of the periodic table. For example, in H_2 d_{ij} = 0.025 Å, and in ICl d_{ij} = 1.280 Å.

$$k_0 = \frac{1.86 \times 10^5}{(r_e - d_{ij})}$$

Because the bond distance, r_e, is in the denominator, as the bond becomes shorter the bond force constant increases. Since, generally speaking, as bonds become shorter the bond energy increases (though there are exceptions such as those discussed in Chapter 3),[8] and as the vibrational frequency increases the bond energy increases for similar types of bonds. Vibrational frequencies are often associated with certain functional groups and with certain atom–atom connections.

In all forms of spectroscopy there are *selection rules*, which describe when and what types of transitions will be observed. Symmetry can act as a rough on/off switch for these transitions, describing when a transition may be observed and when it probably will not.

In IR spectroscopy, the absorption depends on changes in dipole moment of the molecule. As a result, the probability of a vibration absorbing light in the IR region requires that the dipole moment changes. The dipoles have the same symmetry as the x-, y-, and z-axes. As a result, a vibration must have the same symmetry as either x, y, or z to be observed, otherwise the transition has an intensity of zero by symmetry.

Raman spectroscopy is an alternative method of observing vibrations of a molecule. Fig. 5.45 illustrates several different possible transitions involving vibrations. In IR (red), the ground vibrational state absorbs a photon of light to produce the first excited vibrational state.

In Rayleigh scattering (blue), a molecule absorbs light to go to a virtual energy state, and then releases a photon of the same energy; this is called elastic scattering, as the energy of the photon is not changed. Rayleigh scattering is the effect that causes the daytime sky to appear blue, the intensity of scattering being inversely proportional to the fourth power of the wavelength. In other words, short-wavelength light (in the blue) is scattered much more efficiently by molecules in the atmosphere than light with longer wavelengths (in the red).

Raman–Stokes scattering and Raman spectroscopy rely on inelastic scattering (green), which occurs when the energy of the photon is not conserved. In the experiment, laser light is used to excite the molecule to a virtual excited state. Many of the excited molecules release light of the same wavelength (elastic scattering), which is filtered out. However, some of the molecules will undergo inelastic scattering to a vibrational excited state, and the difference in the excitation and emission photons produces the energy of the vibration.

In some cases, anti-Stokes scattering can be observed (maroon), which involves excitation of a vibrational excited state, which then emits a photon to return to the ground state.

Raman spectroscopy relies on polarizability of the molecule. As a result, the method is selective for vibrations with the

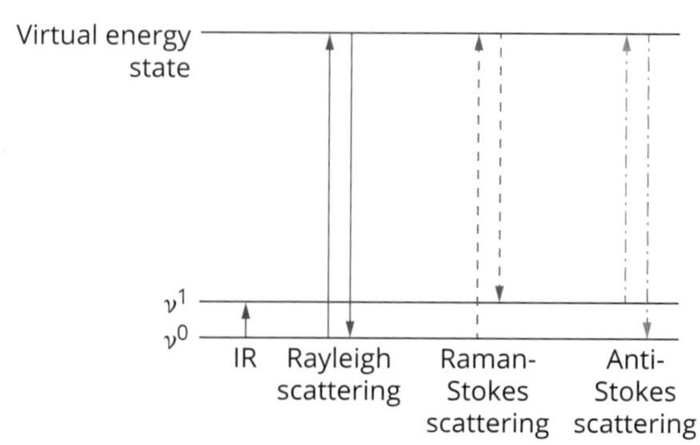

Fig. 5.45 IR, Rayleigh scattering, and Raman–Stokes scattering.

same symmetry as x^2, y^2, z^2, xy, xz, and yz. Observable transitions may also be sums and differences of these functions.

A more detailed discussion of the transition moment integral, which determines the probability of a transition, is in Chapter 16. For our discussion here, *all one needs to remember is that active IR vibrations have the same symmetry as x, y, or z, and that Raman active transitions have the same symmetry as x^2, y^2, z^2, xy, xz, and yz (along with sums and differences of these).*

5.10 Applying Group Theory to Chemical Problems and the Decomposition Formula

In the remaining sections some practical applications of group theory will be discussed by looking at the symmetries of bond stretches and bending modes in simple systems. This is provided as a review of the previous concepts and as an introduction to an important tool, the *decomposition formula*, which will allow the reduction of a reducible representation to its component irreducible representations found in a character table.

All these principles will be applied to additional chemical problems later, such as molecular orbital theory in Chapter 6. As some simple examples here, we will look at how the number and symmetry of vibrations in a molecule may be determined using group theory. One can also find the exact nature of the vibrations and determine whether the resonances are infrared or Raman active.

5.10.1 Finding the Irreducible Representations for Molecular Movements: H_2O as an Example

Our basis will be the bond vectors for this C_{2v} molecule (Fig. 5.46). The plane of the molecule will be placed in the yz-plane.

For these bond vectors we can find the total representation ($\Gamma_{(\vec{r}_1,\vec{r}_2)}$) by applying the symmetry operations, the results of which are as follows.

C_{2v}	E	C_2	$\sigma_v(xz)$	$\sigma_v(yz)$
$\Gamma_{(\vec{r}_1,\vec{r}_2)}$	2	0	0	2

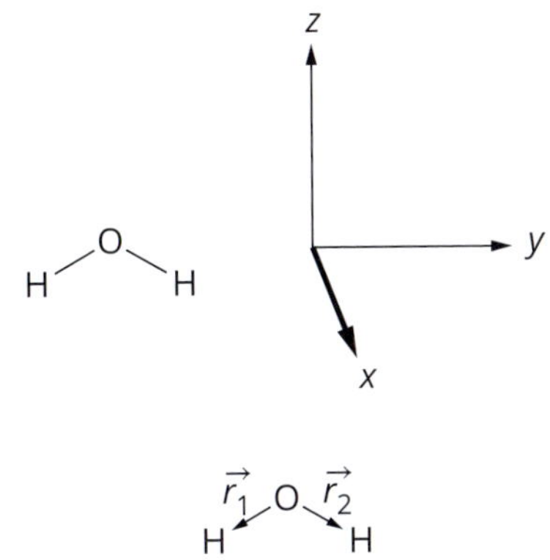

Fig. 5.46 Basis for finding the bond stretches in water.

Qualitatively, we can easily see that there are really only two possibilities for the bond vectors, which may go away and toward the oxygen in unison, or may go with one vector in and one vector out where they alternate in their approach to oxygen (Fig. 5.47). In other words, the vectors may be symmetric, $\vec{r}_1 + \vec{r}_2$, or asymmetric, $\vec{r}_1 - \vec{r}_2$. In addition, the total representation is comprised of these two 1-dimensional representations.

$$\Gamma_{(\vec{r}_1,\vec{r}_2)} = \Gamma_{(\vec{r}_1+\vec{r}_2)} + \Gamma_{(\vec{r}_1-\vec{r}_2)}$$

We can find the symmetry of $\vec{r}_1 + \vec{r}_2$, for example, by using this as our basis. *We are no longer using two separate vectors as our basis. Now our basis is the set of vectors having the same sign, a 1-dimensional basis.* We will give the 1-dimensional representation of this basis the symbol $\Gamma_{(\vec{r}_1+\vec{r}_2)}$, distinct from the previous 2-dimensional basis $\Gamma_{(\vec{r}_1,\vec{r}_2)}$, where the vectors were treated separately. In this case, rotation around the C_2 does not change the basis perceptibly and is given +1 in the representation (Fig. 5.48a). In fact, the symmetric basis $\Gamma_{(\vec{r}_1+\vec{r}_2)}$ does not change with any of the symmetry operations. As a result, we say that it belongs to the totally symmetric representation in C_{2v}, A_1.

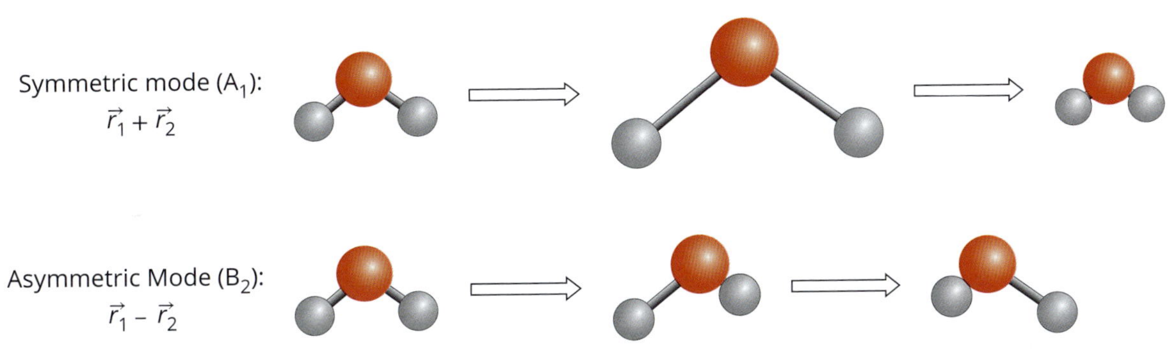

Fig. 5.47 Snapshot pictures of the symmetric and asymmetric modes for the O–H stretches in water.

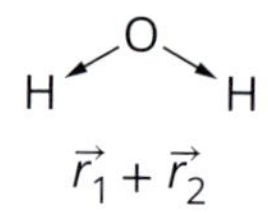

C_{2v}	E	C_2	$\sigma_v(xz)$	$\sigma_v(yz)$
$\Gamma_{(\vec{r}_1+\vec{r}_2)}$	1	1	1	1

Fig. 5.48a The symmetric stretch in water and its irreducible representation.

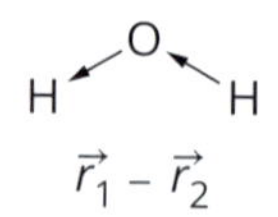

C_{2v}	E	C_2	$\sigma_v(xz)$	$\sigma_v(yz)$
$\Gamma_{(\vec{r}_1-\vec{r}_2)}$	1	−1	−1	1

Fig. 5.48b The asymmetric stretch in water and its irreducible representation.

If we examine the asymmetric combination and its representation $\Gamma_{(\vec{r}_1-\vec{r}_2)}$, the results in Fig. 5.48b are found. Comparing the representation for $\vec{r}_1 - \vec{r}_2$ with the irreducible representations in the C_{2v} character table, we see that this corresponds to B_2.

While the simple representations comprising our original reducible representation $\Gamma_{(\vec{r}_1,\vec{r}_2)}$ are fairly obvious in this case, we need a general method for finding the irreducible representations that make up a reducible representation and their symmetries. The general method applicable to any size basis is called the *decomposition formula*, which as the name implies takes a reducible representation and "decomposes" it to a sum of irreducible representations. The formula is shown in Eq. 5.10, where n_{a_i} is the number of times that irreducible representation (a_i) appears in the total representation, h is the order (number of operations) of the point group, R are the classes, g is the number of members in the class, $\chi_i(R)$ is the character in the irreducible representation a_i under the class R, and $\chi_T(R)$ is the character in the total representation (T) under the class R.

$$n_{a_i} = \frac{1}{h}\sum_R g \times \chi_{a_i}(R) \times \chi_T(R) \qquad \textbf{5.10}$$

To illustrate, we can apply Eq. 5.10 to the reducible representation we found for the bond vectors in water, $\Gamma_{(\vec{r}_1,\vec{r}_2)}$. In Table 5.10 the C_{2v} character table and the reducible representation are reproduced.

Application of the decomposition formula (Eq. 5.10) to each irreducible representation, A_1 to B_2, will tell us the number of times each irreducible representation appears, n_{a_i}, in the reducible representation $\Gamma_{(\vec{r}_1,\vec{r}_2)}$. Beginning with A_1, we have the following equations. First, we have $1/h$ which is the inverse of the order, the number of operations in the point group; in this case $1/h = 1/4$. In the first term in the E operation, we multiply the number of E operations, which is always one, $g = 1$, by the character under E in the A_1 irreducible representation, $\chi_{A_1}(E) = 1$, times the character under E in the reducible representation, $\chi_\Gamma(E) = 2$. The same procedure is carried out for the other operations, and one finds that there is one A_1 in the reducible representation.

Table 5.10 The C_{2v} character table and the reducible representation for the bond stretches in water.

C_{2v}	E	C_2	$\sigma_v(xz)$	$\sigma_v(yz)$
A_1	1	1	1	1
A_2	1	1	−1	−1
B_1	1	−1	1	−1
B_2	1	−1	−1	1
$\Gamma_{(\vec{r}_1,\vec{r}_2)}$	2	0	0	2

$$n_{a_i} = \frac{1}{h}\sum_R g \times \chi_{a_i}(R) \times \chi_T(R)$$

$$n_{A_1} = \frac{1}{4}\begin{bmatrix}(1\times1\times2)_E + (1\times1\times0)_{C_2} \\ +(1\times1\times0)_{\sigma_v(xz)} + (1\times1\times2)_{\sigma_v(yz)}\end{bmatrix}$$

$$n_{A_1} = \frac{1}{4}\left[2+0+0+2\right] = 1$$

The same procedure is followed for the other irreducible representations to see if they contribute, and the result is that there is also a B_2 in the reducible representation, just as we found previously.

$$n_{a_i} = \frac{1}{h}\sum_R g \times \chi_{a_i}(R) \times \chi_T(R)$$

$$n_{A_2} = \frac{1}{4}\Big[(1\times1\times2)_E + (1\times1\times0)_{C_2} + (1\times-1\times0)_{\sigma_v(xz)} + (1\times-1\times2)_{\sigma_v(yz)}\Big]$$

$$n_{A_2} = \frac{1}{4}[2+0-0-2] = 0$$

$$n_{B_1} = \frac{1}{4}\Big[(1\times1\times2)_E + (1\times-1\times0)_{C_2} + (1\times1\times0)_{\sigma_v(xz)} + (1\times-1\times2)_{\sigma_v(yz)}\Big]$$

$$n_{B_1} = \frac{1}{4}[2-0+0-2] = 0$$

$$n_{B_2} = \frac{1}{4}\Big[(1\times1\times2)_E + (1\times-1\times0)_{C_2} + (1\times-1\times0)_{\sigma_v(xz)} + (1\times1\times2)_{\sigma_v(yz)}\Big]$$

$$n_{B_2} = \frac{1}{4}[2-0-0+2] = 1$$

Our total representation can now be presented as a sum of its irreducible representations.

$$\Gamma_{(\vec{r}_1,\vec{r}_2)} = A_1 + B_2$$

C_{2v}	E	C_2	$\sigma_v(xz)$	$\sigma_v(yz)$
A_1	1	1	1	1
B_2	1	−1	−1	1
$\Gamma_{(\vec{r}_1, \vec{r}_2)}$	2	0	0	2

Fig. 5.49 Checking the solutions of the decomposition formula. Summing the characters of the irreducible representations from decomposition should provide the reducible representation.

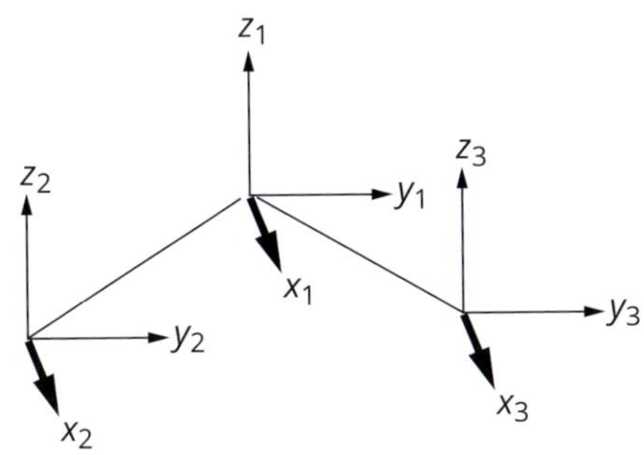

C_{2v}	E	C_2	$\sigma_v(xz)$	$\sigma_v(yz)$
θ(°)	0°	180°	0°	0°
cos θ	1	−1	1	1
±1 + 2cos θ	3	−1	1	1
n_R	3	1	1	3
Γ_{3n}	9	−1	1	3

Fig. 5.50 A basis for normal mode analysis of water and a method for finding the reducible representation.

A useful check on the decomposition formula is to sum the characters in each class of the irreducible representations to reproduce the starting reducible representation (Fig. 5.49). This check is shown below, and summing the characters in A_1 and B_2 does give back $\Gamma_{(\vec{r}_1,\vec{r}_2)}$.

As another example, we can find all the possible movements of the water molecule. To do this, we will use an *x*-, *y*-, and *z*-axis on every atom (Fig. 5.50). This will represent every possible movement of the atoms in the molecule in 3-dimensions, and our basis will be of $3n$ size where n is the number of atoms, nine members total. Included in the total representation will be three translations where all the atoms move in the *x*-, *y*-, or *z*-direction together and three rotations where the molecule rotates around the *x*-, *y*-, or *z*-axis, which are typically of little chemical utility. The internal modes corresponding to stretches and vibrations will then be $3n - 6 = 3$. These are of utility for characterizing molecules by using methods such as infrared and Raman spectroscopy.

The reducible representation is found in the usual way. If a hydrogen atom moves during the operation, the contribution from all of its axes will be zero. During some operations, some vectors will go to their negatives. In this case, either the (+1, 0, −1)-method or the $n_R(\pm1+2\cos\theta)$-method can be used because all the parts of the basis move into another part of the basis (or its negative) for every symmetry operation.

Using the decomposition formula, we can find the irreducible representations that contribute to this 9-dimensional total representation.

$$n_{a_i} = \frac{1}{h}\sum_R g \times \chi_{a_i}(R) \times \chi_T(R)$$

$$n_{A_1} = \frac{1}{4}\Big[(1\times1\times9)_E + (1\times1\times-1)_{C_2} + (1\times1\times1)_{\sigma_v(xz)} + (1\times1\times3)_{\sigma_v(yz)}\Big]$$

$$n_{A_1} = \frac{1}{4}[9 - 1 + 1 + 3] = 3$$

$$n_{A_2} = \frac{1}{4}\Big[(1\times1\times9)_E + (1\times1\times-1)_{C_2} + (1\times-1\times1)_{\sigma_v(xz)} + (1\times-1\times3)_{\sigma_v(yz)}\Big]$$

$$n_{A_2} = \frac{1}{4}[9 - 1 - 1 - 3] = 1$$

$$n_{B_1} = \frac{1}{4}\Big[(1\times1\times9)_E + (1\times-1\times-1)_{C_2} + (1\times1\times1)_{\sigma_v(xz)} + (1\times-1\times3)_{\sigma_v(yz)}\Big]$$

$$n_{B_1} = \frac{1}{4}[9 + 1 + 1 - 3] = 2$$

$$n_{B_2} = \frac{1}{4}\Big[(1\times1\times9)_E + (1\times-1\times-1)_{C_2} + (1\times-1\times1)_{\sigma_v(xz)} + (1\times1\times3)_{\sigma_v(yz)}\Big]$$

$$n_{B_2} = \frac{1}{4}[9 + 1 - 1 + 3] = 3$$

The dimensionality of the sum of the irreducible representations must be equal to the dimensionality of the total representation. In other words (Eq. 5.11), the character under the *E* operation in the total representation, $\chi_T(E)$, must be equal to the sum over all the irreducible representation that comprise the reducible representation, a_i, of the characters under the *E* operation in the irreducible representations, $\chi_{a_i}(E)$, times the number of times that irreducible representation appears, n_{a_i}.

$$\chi_T(E) = \sum_{a_i} n_{a_i} \times \chi_{a_i}(E) \qquad \textbf{5.11}$$

Applied to the problem involving *x*-, *y*-, and *z*-axes on each atom in water, the following equation is obtained. The character under the *E* operation in the total representation, $\chi_T(E)$, is 9 because we are using a 9-dimensional basis. The irreducible

A_1 Bending mode:

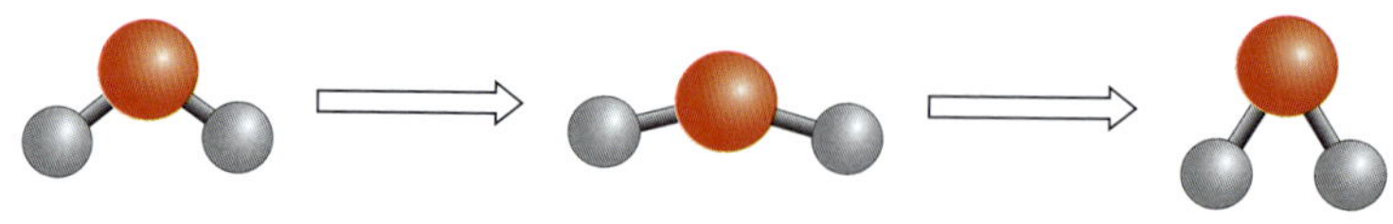

Fig. 5.51 Snapshot pictures of the symmetric bending mode in water.

representations all have a "1" under the character for the E operation because they are all 1-dimensional representations; this is seen from the Mulliken symbol which is either A or B for every irreducible representation in the point group. Finally, the decomposition formula gave us that the total representation is comprised of $3A_1 + A_2 + 2B_1 + 3B_2$. Using Eq. 5.11, we can see that this is a reasonable answer, as far as its dimensionality is concerned.

$$\chi_T(E) = \sum_{a_i} n_{a_i} \times \chi_{a_i}(E)$$
$$9 = (3 \times 1)_{A_1} + (1 \times 1)_{A_2} + (2 \times 1)_{B_1} + (3 \times 1)_{B_2}$$
$$9 = 3 + 1 + 2 + 3$$

In all cases, the decomposition formula, properly employed, should give only positive integers. If one obtains fractions or negative numbers, mathematical errors were made somewhere in the process, if not in the use of the decomposition formula, then previously.

If we look at the full C_{2v} character table (Table 5.11) we can see the linear functions (x,y,z)/rotations (R_x,R_y,R_z) in the column to the right of the last symmetry operation and the nonlinear functions in the final column.

Translations in the x-direction will belong to the same irreducible representation as the x-axis. In this case, translation along x belongs to B_1. Similarly, translation along y belongs to B_2, and translation along z belongs to A_1. Since we are not interested in moving the entire molecule along some axis, in general we can remove one of each of these irreducible representations from our total representation, Γ_{3n}.

$$\Gamma_{3n} = 3A_1 + A_2 + 2B_1 + 3B_2$$
$$\Gamma_{3n-translations} = 3A_1 + A_2 + 2B_1 + 3B_2 - (A_1 + B_1 + B_2)$$
$$\Gamma_{3n-translations} = 2A_1 + A_2 + B_1 + 2B_2$$

We are not often interested in simple rotations of the entire molecule either. Rotation around the x-axis, for example, is shown in the character table as R_x, which in this case is in the irreducible representation B_2. The three rotations have the symmetries of A_2 (for R_z), B_1 (for R_y), and B_2 (for R_z), which can also be removed from our total representation.

$$\Gamma_{3n-translations-rotations} = 2A_1 + A_2 + B_1 + 2B_2 - (A_2 + B_1 + B_2)$$
$$\Gamma_{3n-translations-rotations} = 2A_1 + B_2$$

There are $3n - 6 = 3(3) - 6 = 3$ internal modes in water of symmetry $2\,A_1 + B_2$. There are two stretches that were identified previously (Fig. 5.47), and one bending mode where the H–O–H angle scissors larger and smaller (Fig. 5.51). Note that the stretching and bending modes, labeled A_1, retain the symmetry about the C_2 as required by the "A" designation in the Mulliken symbol. The B_2 stretching mode (Fig. 5.47) is not symmetric about the C_2 during the entire vibrational mode.

The x-, y-, and z-axes have symmetries of B_1, B_2, and A_1, respectively, in C_{2v}. Consequently, all the vibrational modes in water are IR active. Furthermore, all the irreducible representations in C_{2v} have polarization functions of the same symmetry, and, as a result, all vibrations of C_{2v} molecules are Raman active.

5.10.2 Finding the Irreducible Representations for Molecular Movements: *trans*-N_2F_2 as an Example

The molecule *trans*-N_2F_2 has C_{2h} symmetry and the Lewis structure shown in Fig. 5.52a.

The z-axis for the molecule is the highest-order rotational axis, the C_2 that runs through the nitrogen–nitrogen double bond (Fig. 5.52b). We will place the x-axis along the N–N bond. As a result, the plane of the molecule will reside in the xy-plane. The atom labels have been added to help keep track of movements during symmetry operations.

To assign a basis we want to ask a question: what do we want to know? If we want to learn about every possible movement of this simple molecule, we will place an x-, y-, and z-axis on every atom in the molecule. In this case, each atomic axis will be coincident with the molecular axis; in other words,

Table 5.11 The C_{2v} character table.

C_{2v}	E	C_2	$\sigma_v(xz)$	$\sigma_v(yz)$		
A_1	1	1	1	1	z	x^2,y^2,z^2
A_2	1	1	−1	−1	R_z	xy
B_1	1	−1	1	−1	x,R_y	xz
B_2	1	−1	−1	1	y,R_x	yz

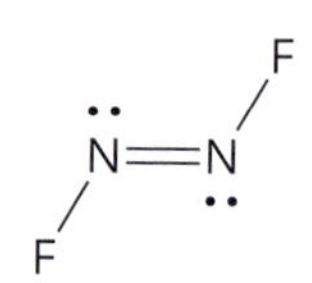

Fig. 5.52a Structure of *trans*-N_2F_2.

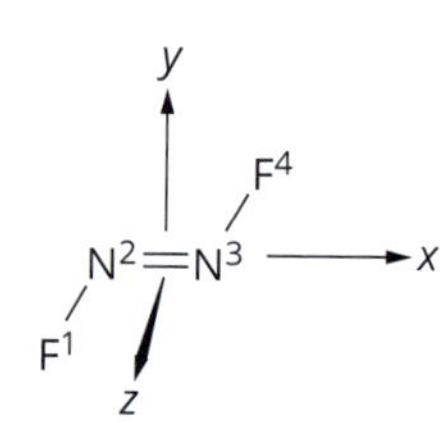

Fig. 5.52b Structure of *trans*-N_2F_2 with a coordinate axis.

there will be a z-axis on nitrogen-3 that will be z_3 that comes out of the plane of the molecule. As a result, our basis size will be twelve vectors: x-, y-, and z-axes on four different atoms.

We examine the effect of the symmetry of the molecule on the basis. To find the effect of the identity operation (E) on the basis, we find the matrix E that fulfills Eq. 5.12a.

$$E\begin{bmatrix} x_1 \\ y_1 \\ z_1 \\ x_2 \\ y_2 \\ z_2 \\ x_3 \\ y_3 \\ z_3 \\ x_4 \\ y_4 \\ z_4 \end{bmatrix} = \begin{bmatrix} x_1 \\ y_1 \\ z_1 \\ x_2 \\ y_2 \\ z_2 \\ x_3 \\ y_3 \\ z_3 \\ x_4 \\ y_4 \\ z_4 \end{bmatrix} \qquad \mathbf{5.12a}$$

The matrix that takes the basis, the set of x-, y-, and z-axes on the four atoms, and returns them unchanged is a square matrix with the same number of rows and columns as the basis with "1" along the diagonal, the 12 × 12 identity matrix (Eq. 5.12b).

$$\begin{bmatrix} 1&0&0&0&0&0&0&0&0&0&0&0 \\ 0&1&0&0&0&0&0&0&0&0&0&0 \\ 0&0&1&0&0&0&0&0&0&0&0&0 \\ 0&0&0&1&0&0&0&0&0&0&0&0 \\ 0&0&0&0&1&0&0&0&0&0&0&0 \\ 0&0&0&0&0&1&0&0&0&0&0&0 \\ 0&0&0&0&0&0&1&0&0&0&0&0 \\ 0&0&0&0&0&0&0&1&0&0&0&0 \\ 0&0&0&0&0&0&0&0&1&0&0&0 \\ 0&0&0&0&0&0&0&0&0&1&0&0 \\ 0&0&0&0&0&0&0&0&0&0&1&0 \\ 0&0&0&0&0&0&0&0&0&0&0&1 \end{bmatrix} \begin{bmatrix} x_1 \\ y_1 \\ z_1 \\ x_2 \\ y_2 \\ z_2 \\ x_3 \\ y_3 \\ z_3 \\ x_4 \\ y_4 \\ z_4 \end{bmatrix} = \begin{bmatrix} x_1 \\ y_1 \\ z_1 \\ x_2 \\ y_2 \\ z_2 \\ x_3 \\ y_3 \\ z_3 \\ x_4 \\ y_4 \\ z_4 \end{bmatrix} \qquad \mathbf{5.12b}$$

The character of this matrix, the sum along the diagonal, is 12 (Table 5.12a). The character under the identity operation is always the size of the basis.

For the C_2 operation we need a matrix that carries out the transformation shown in Eq. 5.12c, where x_1 becomes $-x_4$, etc.

$$C_2\begin{bmatrix} x_1 \\ y_1 \\ z_1 \\ x_2 \\ y_2 \\ z_2 \\ x_3 \\ y_3 \\ z_3 \\ x_4 \\ y_4 \\ z_4 \end{bmatrix} = \begin{bmatrix} -x_4 \\ -y_4 \\ z_4 \\ -x_3 \\ -y_3 \\ z_3 \\ -x_2 \\ -y_2 \\ z_2 \\ -x_1 \\ -y_1 \\ z_1 \end{bmatrix} \qquad \mathbf{5.12c}$$

Writing a 12 × 12 matrix for this transformation produces the result shown in Eq. 5.12d. Notice that the sum of the diagonal is now zero (Table 5.12b), because every part of the basis is moved to a different part during the rotation. Every atom in the molecule moves during the rotation, so no vector on that atom remains the same.

$$\begin{bmatrix} 0&0&0&0&0&0&0&0&0&-1&0&0 \\ 0&0&0&0&0&0&0&0&0&0&-1&0 \\ 0&0&0&0&0&0&0&0&0&0&0&1 \\ 0&0&0&0&0&0&-1&0&0&0&0&0 \\ 0&0&0&0&0&0&0&-1&0&0&0&0 \\ 0&0&0&0&0&0&0&0&1&0&0&0 \\ 0&0&0&-1&0&0&0&0&0&0&0&0 \\ 0&0&0&0&-1&0&0&0&0&0&0&0 \\ 0&0&0&0&0&1&0&0&0&0&0&0 \\ -1&0&0&0&0&0&0&0&0&0&0&0 \\ 0&-1&0&0&0&0&0&0&0&0&0&0 \\ 0&0&1&0&0&0&0&0&0&0&0&0 \end{bmatrix} \begin{bmatrix} x_1 \\ y_1 \\ z_1 \\ x_2 \\ y_2 \\ z_2 \\ x_3 \\ y_3 \\ z_3 \\ x_4 \\ y_4 \\ z_4 \end{bmatrix} = \begin{bmatrix} -x_4 \\ -y_4 \\ z_4 \\ -x_3 \\ -y_3 \\ z_3 \\ -x_2 \\ -y_2 \\ z_2 \\ -x_1 \\ -y_1 \\ z_1 \end{bmatrix} \qquad \mathbf{5.12d}$$

In our short method for finding the reducible representation, if the atom moves, all the parts of the basis on that atom will get zeros. Only atoms that do not move will contribute. If an axis goes to its own negative, it contributes −1. The identity operation leaves every axis on every atom unchanged and

Table 5.12a Partial reducible representation for the normal mode analysis of *trans*-N_2F_2. The trace of the E matrix is 12.

C_{2h}	E	C_2	i	σ_h
$\Gamma_{3n(N_2F_2)}$	12			

Table 5.12b Partial reducible representation for the normal mode analysis of *trans*-N_2F_2. The trace of the C_2 matrix is 0.

C_{2h}	E	C_2	i	σ_h
$\Gamma_{3n(N_2F_2)}$	12	0		

Table 5.12c Full reducible representation for the normal mode analysis of *trans*-N_2F_2.

C_{2h}	E	C_2	i	σ_h
$\Gamma_{3n(N_2F_2)}$	12	0	0	4

Table 5.12d Shorter method for finding the reducible representation applied to *trans*-N_2F_2.

C_{2h}	E	C_2	i	σ_h
θ (°)	0	180	180	0
cos θ	1	−1	−1	1
±1 + 2cosθ	3	−1	−3	1
n_R	4	0	0	4
$\Gamma_{3n(N_2F_2)}$	12	0	0	4

sums to 12. The C_2 moves every atom and sums to zero. The inversion operation moves every atom and also sums to zero. The σ_h inverts all the *z*-axes (−4) and leaves all of the *x*- and *y*-axes the same (+8) and thus sums to +4 (Table 5.12c).

Before we move on, let us apply our shorthand method to the same problem (Table 5.12d). Here, every operation is defined as either a proper rotation (in this case E and C_2) or an improper rotation (in this case i and σ_h). The angles of rotation and cos θ are then shown. If the operation is defined as a proper rotation, then +1 + 2cos θ is calculated; if the operation is defined as an improper rotation, then −1 + 2cos θ is calculated. These are multiplied by the number of atoms that do not move during the operation, n_R. Naturally, this gives us the same reducible representation that we obtained using the previous matrix method (Table 5.12c).

Application of the decomposition formula proceeds as follows.

$$n_{A_g} = \frac{1}{4}[(12)(1) + (0)(1) + (0)(1) + (4)(1)] = 4$$

$$n_{B_g} = \frac{1}{4}[(12)(1) + (0)(-1) + (0)(1) + (4)(-1)] = 2$$

$$n_{B_g} = \frac{1}{4}[(12)(1) + (0)(1) + (0)(-1) + (4)(-1)] = 2$$

$$n_{B_u} = \frac{1}{4}[(12)(1) + (0)(-1) + (0)(-1) + (4)(1)] = 4$$

$$\Gamma = 4A_g + 2B_g + 2A_u + 4B_u$$

Consequently, every possible movement of the molecule can be represented as twelve orthogonal movements represented by the irreducible representations

$$\Gamma_{3n(N_2F_2)} = 4A_g + 2B_g + 2A_u + 4B_u$$

and any other movement that might be imagined is a linear combination of these.

Translations belong to the same irreducible representation as the *x*-, *y*-, and *z*-axes. Rotations around the molecular axes have their symmetry labeled as R_x, R_y, and R_z.

$$\Gamma_{3n(N_2F_2)} = 4A_g + 2B_g + 2A_u + 4B_u$$

$$\Gamma_{translations(N_2F_2)} = A_u + 2B_u$$

$$\Gamma_{rotations(N_2F_2)} = A_g + 2B_g$$

Subtraction of the translations and rotations from the list of irreducible representations comprising the movements give the vibrations alone. The number of vibrations is equal to the internal degrees of freedom = 3n − 6 = 3(4) − 6 = 6, three from the A_g irreducible representation, one from A_u and two from B_u (Eq. 5.11).

$$\Gamma_{vibrations(N_2F_2)} = 3A_g + A_u + 2B_u$$

The vibrational spectroscopy of this simple compound has been studied in great detail by Christe and coworkers.[9] The vibrations of A_g symmetry can be observed using Raman spectroscopy, and the vibrations of A_u and B_u symmetry can be observed in the IR spectrum.

5.10.3 Finding the Irreducible Representations for Molecular Movements: *trans*-$PF_2(CH_3)_3$ as an Example

As an example in yet a different point group, let us examine the P–X stretches in $PF_2(CH_3)_3$. For this example we will ignore the hydrogen atoms on the carbons; consequently, a more appropriate way to view the molecule is as PF_2Me_3, where the methyl groups are treated as a single group or "super-atom".

By VSEPR, the structure of the compound should be that shown in Fig. 5.53, which has D_{3h} symmetry. As our basis, we are using the entire set of P–C and P–F bond vibrations. One could use as the basis the vibrations of the P–C or P–F bonds separately, the sum of which would be the results we are about to obtain using the bigger basis.

An example of the decomposition formula being applied in this case is shown as follows, where the number of A_1'

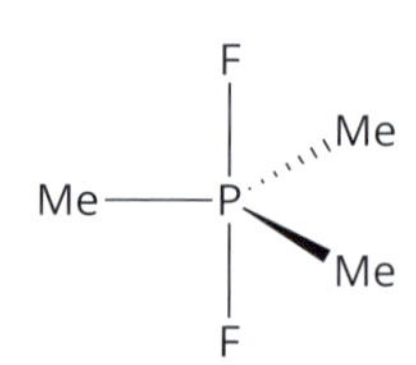

D_{3h}	E	$2C_3$	$3C_2$	σ_h	$2S_3$	$3\sigma_v$
$\Gamma_{(P\text{–}X)}$	5	2	1	3	0	3

Fig. 5.53 Structure of *trans*-PF_2Cl_3 and the reducible representation for the P–X bond stretches.

irreducible representations in the total representation was found to be 2.

$$n_{a_i} = \frac{1}{h}\sum_R g \times \chi_{a_i}(R) \times \chi_T(R)$$

$$n_{A_1'} = \frac{1}{12}\Big[(1\times1\times5)_E + (2\times1\times2)_{C_3} + (3\times1\times1)_{C_2} + (1\times1\times3)_{\sigma_h} + (2\times1\times0)_{S_3} + (3\times1\times3)_{\sigma_v}\Big]$$

$$n_{A_1'} = \frac{1}{12}[5+4+3+3+0+9] = 2$$

Once the calculation is done for the other irreducible representations, the total representation is expressed as a sum of irreducible representations.

$$\Gamma_{P-X} = 2A_1' + E' + A_2''$$

This can be confirmed to have the correct dimensionality by observing that "A"-representations are 1-dimensional and an "E"-representation is 2-dimensional. Alternatively, one can use Eq. 5.8.

$$\chi_T(E) = \sum_{a_i} n_{a_i} \times \chi_{a_i}(E)$$

$$5 = (2\times1)_{A_1'} + (1\times2)_{E'} + (1\times1)_{A_2''}$$

$$5 = 2+2+1$$

5.11 Using Group Theory to Discover the Nature of Vibrations, Rotations, and Translations: The Projection Operator

In the previous section the decomposition formula was introduced, which allowed us to find the irreducible representations that comprise a reducible representation. The reducible representation was, in turn, a symmetry-related construction of the basis chosen. *We will also use this section to further stress a very important point regarding group theoretical analysis: you can only gain information about the molecule that is related to your chosen basis.* One wants to pick a basis as small as possible that gives the information desired; this makes the interpretation as reliable as possible without missing important information.

Once the irreducible representations are known, it is possible to find out the nature of how the basis is related to the irreducible representation. In other words, the decomposition formula will tell you that there is, for example, an A_2' mode in your molecule; the projection operator, discussed here, will tell you what movements of the molecule are involved in that A_2' mode, a "picture" of what the vibration looks like.

5.11.1 Normal Mode Analysis of $\{PtCl_4\}^{2-}$

The anion $\{PtCl_4\}^{2-}$ has D_{4h} symmetry, and normal mode analysis will begin with a decision on the molecular axes and nature of the basis set to be used. The standard molecular axes will place the z-axis along the highest-order rotation axis, C_4. The x- and y-axes will then be along Pt–Cl bonds, as shown in Fig. 5.54a.

Because we want to find all the possible movements of the molecule, we will place an x-, y-, and z-axis on each atom in the molecule (Fig. 5.54a), and we will distinguish each vector in the basis with an atom number as well—for example, x_1, y_1, and z_1 for atom 1—which is the central platinum. The basis on Pt is coincident with the molecular axes in this case; this will be true if there is a central atom in the compound of interest. It is often advantageous to place the vectors on the outer atoms so that they are symmetric with respect to the principal rotation axis. In this case, the x-vectors on atoms 2–5 were set to be along the Pt–Cl bonds, and the y-vectors are 90° from the x-vector to make right-handed axes on each atom. The z-vectors of the basis are all parallel to the molecular z-axis. Using vectors with different directions should give similar results; however, choosing the basis in this way will simplify the analysis.

Applying the symmetry operations in each class to the basis gives the reducible representation (Table 5.13a), $\Gamma_{3n(PtCl_4)}$, that represents all possible movements. In this particular case, all parts of the basis move onto other parts or the negative of other parts of the basis during all of the operations, which means that either shorthand method in Section 5.7 can be used.

The decomposition formula is then applied to determine how each irreducible representation contributes to the reducible representation. The result is shown.

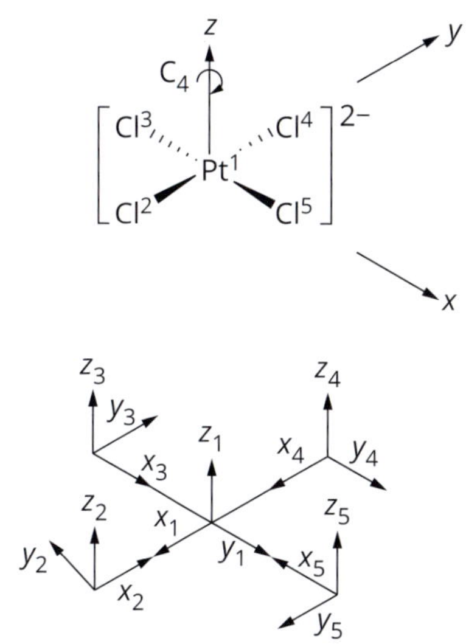

Fig. 5.54a Molecular axes for $\{PtCl_4\}^{2-}$ with atom numbers and basis for finding the normal modes.

Table 5.13a Finding the reducible representation for $PtCl_4^{2-}$ by using the basis in Fig. 5.54a.

D_{4h}	E	$2C_4$	C_2	$2C_2'$	$2C_2''$	i	$2S_4$	σ_h	$2\sigma_v$	$2\sigma_d$
θ (°)	0	90	180	180	180	180	90	0	0	0
cos θ	1	0	−1	−1	−1	−1	0	1	1	1
±1+2cos θ	3	1	−1	−1	−1	−3	−1	1	1	1
n_R	5	1	1	3	1	1	1	5	3	1
$\Gamma_{3n(PtCl_4)}$	15	1	−1	−3	−1	−3	−1	5	3	1

As a check, one should always determine that the dimensionality of the irreducible representation after decomposition is the same as before decomposition (Eq. 5.8). The dimensionality of the basis, number of objects in the basis, in this case vectors, is 15. Each "A" and "B" irreducible representation will contribute one, and each "E" will contribute two because it is doubly degenerate. The size of the decomposed representation is then 1 + 1 + 1 + 1 + (1 × 2) + (2 × 1) + 1 + (3 × 2) = 15, just as it should be.

$$\Gamma_{3n(PtCl_4)} = A_{1g} + A_{2g} + B_{1g} + B_{2g} + E_g + 2A_{2u} + B_{2u} + 3E_u$$

$$n_{a_i} = \frac{1}{h}\sum_R g \times \chi_{a_i}(R) \times \chi_T(R)$$

$$n_{A_{1g}} = \frac{1}{16}\begin{bmatrix} (1\times1\times15)_E + (2\times1\times1)_{C_4} + (1\times1\times-1)_{C_2} \\ +(2\times1\times-3)_{C_2'} + (2\times1\times-1)_{C_2''} \\ +(1\times1\times-3)_i + (2\times1\times-1)_{S_4} + (1\times1\times5)_{\sigma_h} \\ +(2\times1\times3)_{\sigma_v} + (2\times1\times1)_{\sigma_2} \end{bmatrix}$$

$$n_{A_{1g}} = \frac{1}{16}\left[15+2-1-6-2-3-2+5+6+2\right] = 1$$

$$n_{A_{2g}} = \frac{1}{16}\left[15+2-1+6+2-3-2+5-6-2\right] = 1$$

$$n_{B_{1g}} = \frac{1}{16}\left[15-2-1-6+2-3+2+5+6-2\right] = 1$$

$$n_{B_{2g}} = \frac{1}{16}\left[15-2-1+6-2-3+2+5-6+2\right] = 1$$

$$n_{E_g} = \frac{1}{16}\left[30+0+2+0+0-6+0-10+0+0\right] = 1$$

$$n_{A_{1u}} = \frac{1}{16}\left[15+2-1-6-2+3+2-5-6-2\right] = 0$$

$$n_{A_{2u}} = \frac{1}{16}\left[15+2-1+6+2+3+2-5+6+2\right] = 2$$

$$n_{B_{1u}} = \frac{1}{16}\left[15-2-1-6+2+3-2-5-6+2\right] = 0$$

$$n_{B_{2u}} = \frac{1}{16}\left[15-2-1+6-2+3-2-5+6-2\right] = 1$$

$$n_{E_u} = \frac{1}{16}\left[30+0+2+0+0+6+0+10+0+0\right] = 3$$

Again, the total representation is for all possible movements in a normal mode analysis. The molecule can be translated along the *x*-, *y*-, or *z*-axis with all the atoms moving together, which has the same symmetry as the *x*-, *y*-, or *z*-axes in the character table ($\Gamma_{translation(PtCl_4)} = A_{2u} + E_u$). In addition, the molecule can simply rotate around the *x*-, *y*-, or *z*-axis, which is designated in the character table with R_x, R_y, and R_z ($\Gamma_{rotation(PtCl_4)} = A_{2g} + E_g$). If we remove these trivial modes from the possible movements of the molecule, $\Gamma_{3n(PtCl_4)}$, we obtain the irreducible representations that correspond to internal vibrations of the molecule, $\Gamma_{vibrations(PtCl_4)}$.

$$\Gamma_{vibrations(PtCl_4)} = A_{1g} + B_{1g} + B_{2g} + A_{2u} + B_{2u} + 2E_u$$

The degrees of freedom are equal to the number of vibrations in a molecule, 3n − 6 = 3(5) − 6 = 9. The dimensionality of $\Gamma_{vibrations(PtCl_4)}$ should be 9.

The higher-energy stretching modes can be found by investigating the symmetry of the vectors along the $PtCl_4$ bonds, x_2-x_5 in our full basis (Fig. 5.54b). If we find the reducible representation for this piece of the full basis, we will isolate the stretching modes.

The application of the symmetry operations to this subbasis produces the reducible representation in Table 5.13b.

Decomposition of this reducible representation shows that it is comprised of the following irreducible representations.

$$\Gamma_{stretches(PtCl_4)} = A_{1g} + B_{1g} + E_u$$

The bending modes in the complex are thus represented by the following irreducible representations, found by subtracting the stretches from all the vibrations.

$$\Gamma_{bends(PtCl_4)} = B_{2g} + A_{2u} + B_{2u} + E_u$$

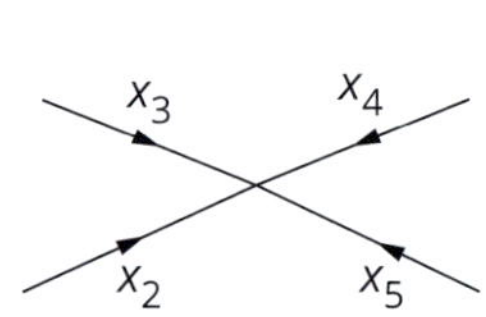

Fig. 5.54b Just the bond stretches in $PtCl_4^{2-}$ as a subset of the full NMA basis in Fig. 5.54a.

Table 5.13b Table for Pt–Cl bond stretches in $PtCl_4^{2-}$ using the basis in Fig. 5.54b.

D_{4h}	E	$2C_4$	C_2	$2C_2'$	$2C_2''$	i	$2S_4$	σ_h	$2\sigma_v$	$2\sigma_d$
$\Gamma_{stretches(PtCl_4)}$	4	0	0	2	0	0	0	4	2	0

In order to discover what the vibrational modes that correspond to these irreducible representations look like, we can use the *projection operator.*

The projection operator takes the function under consideration—for example, movement of the atom along the x-axis—and projects the symmetry of the irreducible representation in the point group onto that part of the basis. We can write this mathematically as shown in Eq. 5.13, where the projection P^{a_i} using a specific part of the basis, β, is $P^{a_i}(\beta)$. This is multiplied by the character of the irreducible representations under that operation, $\chi^{a_i}(r)$, and summed over all the operations r. *In many of the previous equations, such as the decomposition formula, we used the classes of the symmetry operations; here we must use each symmetry operation, r, which may require multiple entries for a class.*

$$P^{a_i}(\beta) = \sum_r \chi^{a_i}(r) \cdot r\beta \qquad \textbf{5.13}$$

As an illustration, let us take a part of the basis used in the previous $PtCl_4^{2-}$ example and apply the projection operator. Keep in mind that we have already removed the possible translations and rotations. The only modes that should be left are vibrations (stretching and bending modes).

To find the stretches we can use the smaller x_2–x_5 basis (Fig. 5.54b). We choose one of the vectors and apply every symmetry operation (Table 5.14a), *not just one symmetry operation from each class, but every operation.* Using x_2 as our starting point, we record the effect of the symmetry operations. Every operation will either leave x_2 as it is (record x_2) or will move it to another vector in the basis (record x_2, x_3, or x_4). Under the E operation column we record the result of applying the identity operation to the x_2 vector and multiplying the character under E in the A_{1g} irreducible representation (1). Under the first C_4 operation (clockwise rotation) we record the result of carrying out this operation on x_2 and multiplying by the character under C_4 in A_{1g} (1), which is x_3. The second C_4 operation is a counterclockwise rotation, giving x_5, and so on.

The results are then summed to produce the result of the projection.

$$P^{A_{1g}}(x_2) = 4x_2 + 4x_3 + 4x_4 + 4x_5$$

What we are interested in is the relative amplitude of the movement, and since "4" appears in front of each term we can drop it. The A_{1g} vibration is simply all Pt–Cl bonds shrinking and expanding together (Fig. 5.55a).

$$P^{A_{1g}}(x_2) = x_2 + x_3 + x_4 + x_5$$

Naturally, the movement with the A_{1g} representation must fit that symmetry label. The "A" in A_{1g} means that the movement is symmetric with respect to the main rotation axis, the C_4; consequently, the movement always has the atoms in positions so that the C_4 is retained. The "1" in A_{1g} implies that the movement is symmetric with respect to the perpendicular C_2's. The irreducible representation is also gerade, symmetric with respect to inversion.

We can carry out the same procedure to discover the nature of the B_{1g} movement (Table 5.14b). In this case, as Cl^2 and Cl^4 are going outward from Pt, Cl^3 and Cl^5 will be moving inward and vice versa. The "B" in B_{1g} implies that the movement will be antisymmetric with respect to the C_4, which is the case with this movement; for example, Cl^2 is moving outward as the chlorine 90° from it (such as Cl^3) is moving inward (Fig. 5.55b).

$$P^{B_{1g}}(x_2) = 4x_2 - 4x_3 + 4x_4 - 4x_5$$

or

$$P^{B_{1g}}(x_2) = x_2 - x_3 + x_4 - x_5$$

Table 5.14a Projections (Eq. 5.13) of A_{1g} on the x_2 part of the basis in Fig. 5.54b. For this, every operation (not just classes) must be used.

D_{4h}	E	C_4	C_4	C_2	C_2'	C_2'	C_2''	C_2''	i	S_4	S_4	σ_h	σ_v	σ_v	σ_d	σ_d
$r\beta = r(x_2)$	x_2	x_3	x_5	x_4	x_2	x_4	x_5	x_3	x_4	x_3	x_5	x_2	x_2	x_4	x_5	x_3
$\chi^{A_{1g}}(r)$	1	1	1	1	1	1	1	1	1	1	1	1	1	1	1	1
$P^{A_{1g}}(x_2)$	x_2	x_3	x_5	x_4	x_2	x_4	x_5	x_3	x_4	x_3	x_5	x_2	x_2	x_4	x_5	x_3

$$P^{A_{1g}}(x_2) = x_2 + x_3 + x_4 + x_5$$

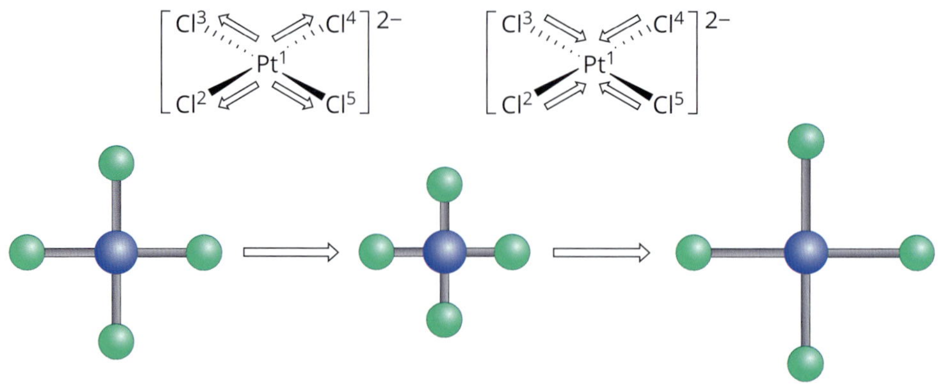

A_{1g}: $x_2 + x_3 + x_4 + x_5$

Fig. 5.55a The $P^{A_{1g}}(x_2)$ projection in mathematical form shows that all the vectors have the same magnitude (1) and the same sign (positive), so the bonds stretch away (or move towards) the metal center at the same time with the same rate, a kind of breathing motion of the atoms around the metal.

Table 5.14b Projections (Eq. 5.13) of B_{1g} on the x_2 part of the basis in Fig. 5.54b.

D_{4h}	E	C_4	C_4	C_2	C_2'	C_2'	C_2''	C_2''	i	S_4	S_4	σ_h	σ_v	σ_v	σ_d	σ_d
$r\beta = r(x_2)$	x_2	x_3	x_5	x_4	x_2	x_4	x_5	x_3	x_4	x_3	x_5	x_2	x_2	x_4	x_5	x_3
$\chi^{B_{1g}}(r)$	1	−1	−1	1	1	1	−1	−1	1	−1	−1	1	1	1	−1	−1
$P^{B_{1g}}(x_2)$	x_2	$-x_3$	$-x_5$	x_4	x_2	x_4	$-x_5$	$-x_3$	x_4	$-x_3$	$-x_5$	x_2	x_2	x_4	$-x_5$	$-x_3$

$$P^{B_{1g}}(x_2) = 4x_2 - 4x_3 + 4x_4 - 4x_5$$

or

$$P^{B_{1g}}(x_2) = x_2 - x_3 + x_4 - x_5$$

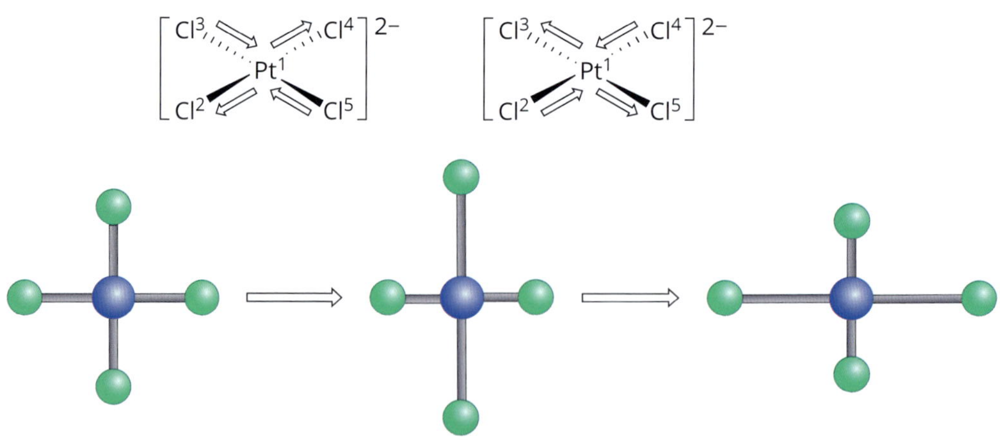

B_{1g}: $x_2 - x_3 + x_4 - x_5$

Fig. 5.55b The $P^{B_{1g}}(x_2)$ projection in mathematical form shows that all the vectors have the same magnitude (1) and have alternating signs, so as the positive parts of the basis are moving away from the metal, the negative parts of the basis are moving toward the metal and vice versa.

The procedure for the degenerate operation is, at least at first, the same. We examine the effect of the E_u operation on x_2 (Table 5.14c, Fig. 5.55c).

$$P^{E_u}(x_2) = 4x_2 - 4x_4$$
or
$$P^{E_u}(x_2) = x_2 - x_4$$

Notice that, in the calculated mode the Pt atom is slightly out of line with the Cl^3–Cl^5 axis (bottom of Fig. 5.55c). The metal moves slightly to ensure that the center of mass does not change during the vibration.

Of course, E_u consists of two symmetry-related modes of equal energy. To find the other mode partnered with $P^{E_u}(x_2)$ to make E_u, we project using another part of the basis. The other mode should be, necessarily, orthogonal to $P^{E_u}(x_2)$. Since $P^{E_u}(x_2)$ contains vectors x_2 and x_4, examining another part of the basis—x_3 or x_5—is advisable. Choosing x_3 for the projection, we obtain the results in Table 5.14d and Fig. 5.55d.

$$P^{E_u}(x_3) = 4x_3 - 4x_5$$
or
$$P^{E_u}(x_3) = x_3 - x_5$$

These two modes described by $P^{E_u}(x_2)$ and $P^{E_u}(x_3)$ in Figs. 5.55c and 5.55d are obviously orthogonal, since they involve different parts of the basis. Procedures for ensuring that

Table 5.14c Table showing the projections (Eq. 5.13) of E_u on the x_2 part of the basis in Fig. 5.39b.

D_{4h}	E	C_4	C_4	C_2	C_2'	C_2'	C_2''	C_2''	i	S_4	S_4	σ_h	σ_v	σ_v	σ_d	σ_d
$r\beta = r(x_2)$	x_2	x_3	x_5	x_4	x_2	x_4	x_5	x_3	x_4	x_3	x_5	x_2	x_2	x_4	x_5	x_3
$\chi^{E_u}(r)$	+2	0	0	−2	0	0	0	0	−2	0	0	2	0	0	0	0
$P^{E_u}(x_2)$	$2x_2$	0	0	$-2x_4$	0	0	0	0	$-2x_4$	0	0	$2x_2$	0	0	0	0

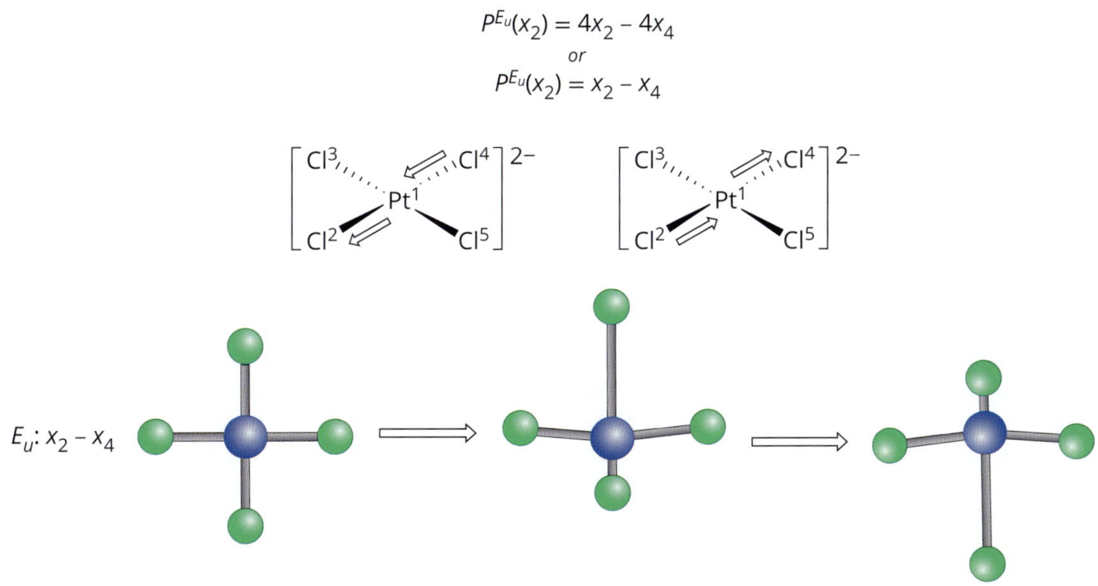

Fig. 5.55c The $P^{E_u}(x_2)$ projection in mathematical form shows that only two vectors, x_2 and x_4, are involved and have different signs. As the positive vector moves away from the metal, the negative vector moves towards the metal, and vice versa. Note the x_3 and x_5 have coefficients of zero and do not move during the vibration. The metal moves slightly to maintain the center of mass.

Table 5.14d Projections (Eq. 5.13) of E_u on the x_3 part of the basis in Fig. 5.55d.

D_{4h}	E	C_4	C_4	C_2	C_2'	C_2'	C_2''	C_2''	i	S_4	S_4	σ_h	σ_v	σ_v	σ_d	σ_d
$r\beta = r(x_3)$	x_3	x_4	x_2	x_5	x_5	x_3	x_2	x_4	x_5	x_4	x_2	x_3	x_5	x_3	x_2	x_4
$\chi^{E_u}(r)$	+2	0	0	−2	0	0	0	0	−2	0	0	2	0	0	0	0
$P^{E_u}(x_3)$	$2x_3$	0	0	$-2x_5$	0	0	0	0	$-2x_5$	0	0	$2x_3$	0	0	0	0

$$P^{E_U}(x_3) = 4x_3 - 4x_5$$

or

$$P^{E_U}(x_3) = x_3 - x_5$$

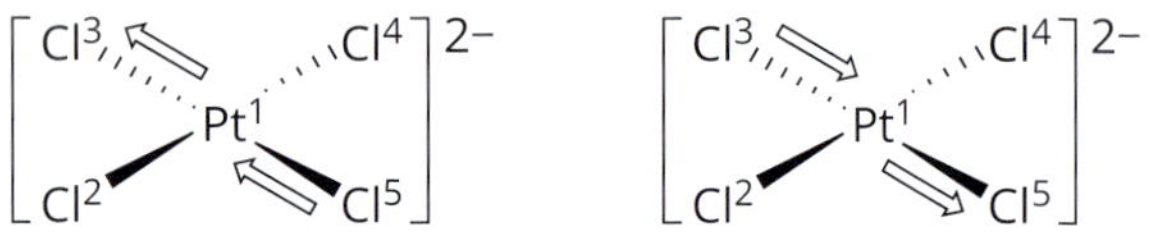

Fig. 5.55d The $P^{E_u}(x_3)$ projection in mathematical form shows that only two vectors, x_3 and x_5, are involved and have different signs. As the positive vector moves away from the metal, the negative vector moves towards the metal and vice versa. Note that x_2 and x_4 have coefficients of zero and do not move during the vibration.

contributors to degenerate irreducible representations are orthogonal are described in Chapter 6.

The nature of the bending modes can be found similarly using the y and z vectors on the chlorines as basis.[j]

5.11.2 Carbonyl Stretches in $Ni(CO)_4$

The nature of the basis determines the nature of the information gained from the analysis. If we are interested in every vibrational mode of $Ni(CO)_4$, then we expect there to be 3n − 6 = 3(9) − 6 = 21 total vibrations. However, for metal carbonyls, the C–O stretching frequency is often in a region of the infrared spectrum uncluttered by other modes and is often diagnostic of the geometry of the compound. As a result, experimentalists are often interested in the possible CO stretching modes alone for carbonyl complexes.

Instead of analyzing the 21 possible vibrations and then trying to decide which are due to the carbonyls, we can use the C–O bonding vectors as our basis to analyze only the CO stretches—an analysis that we have been calling a bond stretch analysis (BSA). Our basis will look like Fig. 5.56a, and application of the symmetry of the T_d point group to this basis provides the reducible representation shown in Table 5.15.

Table 5.15 Reducible representation for the C–O stretches in $Ni(CO)_4$.

T_d	E	$8C_3$	$3C_2$	$6S_4$	$6\sigma_d$
Γ_{CO}	4	1	0	0	2

[j] It can be noted here that this procedure will produce a set of vibrations for the molecule. The "real" vibrations found from spectroscopy may involve mixing of vibrations of the same symmetry—a detail that we shall not be discussing further in this book.

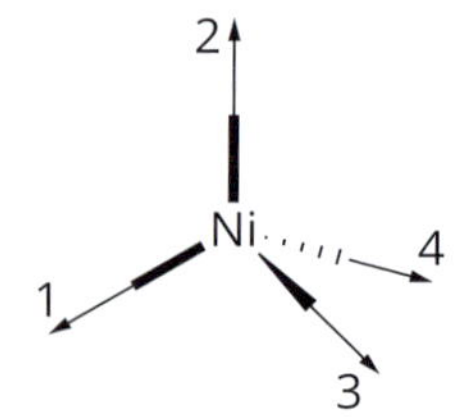

Fig. 5.56a Basis for bond stretch analysis of the C-O stretches in $Ni(CO)_4$.

The decomposition proceeds as follows.

$$n_{A_1} = \frac{1}{24}[(4)(1)(1) + (1)(1)(8) + (0)(1)(3) + (0)(1)(6) + (2)(1)(6)] = 1$$

$$n_{A_2} = \frac{1}{24}[(4)(1)(1) + (1)(1)(8) + (0)(1)(3) + (0)(-1)(6) + (2)(-1)(6)] = 0$$

$$n_E = \frac{1}{24}[(4)(2)(1) + (1)(-1)(8) + (0)(2)(3) + (0)(0)(6) + (2)(0)(6)] = 0$$

$$n_{T_1} = \frac{1}{24}[(4)(3)(1) + (1)(0)(8) + (0)(-1)(3) + (0)(1)(6) + (2)(-1)(6)] = 0$$

$$n_{T_2} = \frac{1}{24}[(4)(3)(1) + (1)(0)(8) + (0)(-1)(3) + (0)(-1)(6) + (2)(1)(6)] = 1$$

$$\Gamma = A_1 + T_2$$

There will be one stretching mode of A_1 symmetry where all of the CO ligands are extending at once and then contracting together (Fig. 5.56b). No other possibility exists for an A_1 symmetry mode.

Similarly, the T_2 mode may be analyzed using projection operators, but there is only one possibility of three orthogonal modes with the four vectors. In each case there will be two vectors that are extending/contracting together and another pair doing the opposite to the keep the center of mass the same. (Again, no net translation or rotation should occur in the mode. Those movements were not in the basis.) In the three degenerate modes we will have (1 & 2)

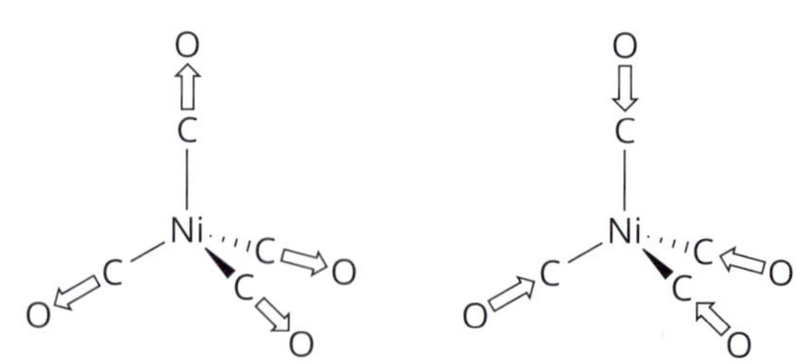

Fig. 5.56b The A_1 symmetry mode for the CO bond stretches in $Ni(CO)_4$. All the CO bonds are all going out or in at the same time.

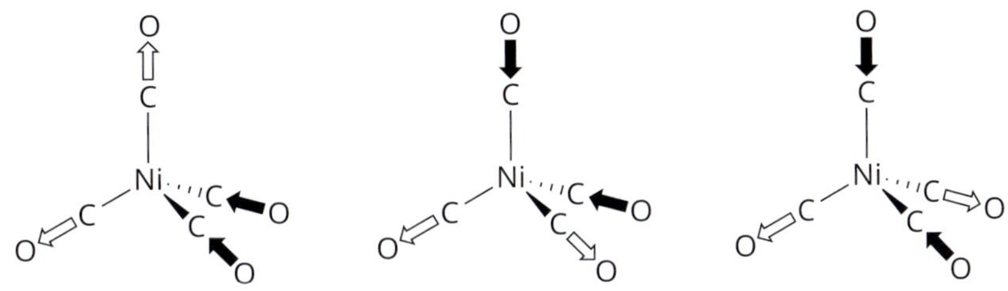

Fig. 5.56c The T_2 symmetry mode for the CO bond stretches in $Ni(CO)_4$.

antisymmetric with (3 & 4), (1 & 3) antisymmetric with (2 & 4), and (1 & 4) antisymmetric with (2 & 3), as shown in Fig. 5.56c.

Again, because of the chosen basis we only gained information about the carbonyl stretches, but if that is our interest then the results are easier to interpret than if we had used a larger basis.

How many CO stretching bands in the IR spectrum are expected for $Ni(CO)_4$? The irreducible representations for the vibrations were found to be A_1 and T_2. In T_d, the *x, y,* and *z* vectors are all part of triply degenerate T_2. Consequently, there will only be one observable band in the IR spectrum for the complex for the CO stretches. Both the T_2 and A_1 vibrations should be observable in the Raman.

5.12 Symmetry in Integration: Multiplying Irreducible Representations (Direct Products)

In some applications, the symmetries of, for example, wavefunctions or operators may be of great importance in determining the strength of a transition, distortion, or other effect. In spectroscopy, for example, the product of the symmetry of a vibration and an operator may determine if the transition in question is "allowed"; that is, if the transition will have high or low intensity. Here we will briefly discuss how irreducible representations may be multiplied and their products discovered.

Table 5.16 Reducible representation of the direct product of $E \times E$ in T_d.

T_d	E	$8C_3$	$3C_2$	$6\sigma_d$	$6S_4$
E	2	−1	2	0	0
E	2	−1	2	0	0
$\Gamma_{prod} = E \times E$	4	1	4	0	0

The product of multiplying a series of irreducible representations ($\Gamma_a, \Gamma_b, \Gamma_c, \ldots$) within a point group can be found by simply multiplying the characters of the irreducible representations under each symmetry class, *R*. The process results in a reducible representation, Γ_{prod}.

As an example, let us multiply the irreducible representation *E* in T_d by itself (Table 5.16); in other words, $E \times E$, which is sometimes referred to as the "direct product" of *E* and *E*.

Subsequently, the decomposition formula (Eq. 5.10) is applied to the reducible representation, Γ_{prod}.

$$n_{A_1} = \frac{1}{24}\sum_R g \times \chi_i(R) \times \chi_T(R) = \frac{1}{24}[4 + 8 + 12 + 0 + 0] = 1$$

$$n_{A_2} = \frac{1}{24}[4 + 8 + 12 - 0 - 0] = 1$$

$$n_E = \frac{1}{24}[8 - 8 + 24 + 0 + 0] = 1$$

Consequently, the product of $E \times E = A_1 + A_2 + E$ in the T_d point group. In general, it can be shown that multiplication of any irreducible representation in any point group by itself includes the totally symmetric representation as a product, and we will be using this fact several times throughout the book. In other words, $E \times E$ in T_d must include A_1 as one of the products, as was shown.

The totally symmetric representation times another representation Γ_x will result in a product of Γ_x; the representation is unchanged by multiplication by the totally symmetric representation.

If the representations are $\Gamma_g \times \Gamma_g$ or $\Gamma_u \times \Gamma_u$ (where again the subscripts g and u represent gerade and ungerade), then the product is Γ_g; if the representations are $\Gamma_u \times \Gamma_g$ or $\Gamma_g \times \Gamma_u$ then the product is Γ_u. Similarly, $\Gamma' \times \Gamma'$ or $\Gamma'' \times \Gamma''$ results in a product of Γ'; if the representations is $\Gamma' \times \Gamma''$, then the product will be Γ''. (Recall that prime is used to indicate that the representation is symmetric with respect to σ_h, and double prime is antisymmetric.)

Chapter Summary

1. There are four symmetry elements: inversion center (*i*), mirror plane (σ), rotation axis (C_n), and improper rotation axis (S_n). A symmetry element is a geometric figure (point, line, or plane) around which a symmetry operation occurs.
2. There are five types of symmetry operation: inversion center, mirror plane, rotation axis, improper rotation axis, and identity (*E*).
3. Symmetry operations happen around a symmetry element, and a symmetry element (such as a rotation axis) may generate several symmetry operations.
4. A few symmetry "generators" may allow determination of all the symmetry in a molecule. Consequently, just finding a few operations can allow molecules to be assigned to point groups.
5. All the symmetry operations of a point group can be found in the top row of a character table, where the operations are divided into classes. The classes are symmetry operations that are related to one another by another symmetry operation.
6. The character tables are square matrices with the classes in the columns and the rows below the classes being formed by the irreducible representations.
7. The character tables obey the Great Orthogonality Theorem, giving them specific properties that can be taken advantage of in chemistry.
8. Group theory is important for several different aspects of inorganic chemistry and is used, for example, to determine the number of observable bands in vibrational spectroscopy.
9. A basis determines the type of information that is gained from a group theoretical analysis. For vibrational spectroscopy we discussed two common bases: normal mode analysis (NMA), which uses an *x, y,* and *z* axis on each atom as the basis; and bond stretch analysis (BSA), which uses a vector between the bonds of interest as the basis.
10. After determining the appropriate basis, the symmetry operations are carried out on the basis to generate the reducible representation, a mathematical description of how the symmetry affects the chemistry problem of interest.
11. The decomposition formula is then used to "decompose" the reducible representation to a linear combination of irreducible representations.
12. Using the projection operator, the symmetry of the irreducible representation can be applied to the basis to determine what the vibration actually looks like, to a first approximation.

Practice Problems

1. Based on the information in Section 5.3.1, show that the set {1, −1, i, −i} constitutes a mathematical group.
2. Using the similarity transformation, Eq. 5.1, show pictorially that the three σ_v planes in ammonia are conjugate; show that one σ_v plane (pick any of them in Fig. 5.33) and show that a different σ_v plane is the transform of the original plane by the C_3. (Hint: this is in some ways the opposite of what was done in Fig. 5.33.)
3. Show mathematically that all the irreducible representations in the C_{3v} point group are orthogonal to one another: A_1 to A_2, A_2 to E, and A_1 to E.
4. What are the conditions that any set must satisfy in order to constitute a mathematical group?
5. What constitutes the "set" with which we can use group theory to solve symmetry-related problems in chemistry?
6. For PF_2Me_3:
   a. Find the irreducible representations for the vibrational modes of the P–F bonds, $\Gamma_{P\text{-}F}$.
   b. Find the irreducible representations for the vibrational modes of the P–C bonds, $\Gamma_{P\text{-}C}$.
   c. Show that $\Gamma_{P\text{-}F} + \Gamma_{P\text{-}C} = \Gamma_{P\text{-}X}$.
7. Draw a real chemical example of a molecule having the following point-group symmetries:
   (i) O_h
   (ii) D_{5d}
   (iii) C_{2h}
   (iv) D_3
   (v) $C_{\infty v}$
8. True or False? Justify your answers.
   a. The order of a group is equal to the number of irreducible representations in that group.
   b. The product of two proper rotations is *always* a proper rotation.
   c. Two representations that are orthogonal to each other are necessarily irreducible.
   d. Groups always have one, and only one, totally symmetric irreducible representation.
9. A *p*-orbital, since it is an odd function, can never have the same symmetry as a *d*-orbital, which is an even function, in a point group with what symmetry operations present? Why?
10. Consider the trigonal bipyramidal $CuCl_5^{3-}$ anion.
    a. Find the total representation of Cu–Cl bond stretches.
    b. Decompose the total representation to a sum of irreducible representations.

Cited References

1. Helliwell, J. R., Synchrotron radiation and crystallography: the first 50 years *Acta Cryst. A* 1998, *54*, 738–749.
2. McDermott, G.; Prince, S. M.; Freer, A. A.; Hawthornthwaite-Lawless, A. M.; Papiz, M. Z.; Cogdell, R. J.; Isaacs, N. W., Crystal-structure of an integral membrane light-harvesting complex from photosynthetic bacteria *Nature* 1995, *374*, 517–521.
3. Cotton, F. A., *Chemical Applications of Group Theory* 3rd ed.; Wiley: New York, 1990.
4. Douglas, B. E.; Hollingsworth, C. A., *Symmetry in bonding and spectra: an introduction* Academic Press: Orlando, 1985.
5. Kunju, A. S.; Krishnan, G., *Group Theory and its Applications in Chemistry* 2nd ed.; PHI Learning: Delhi, 2015.
6. These rules are generally followed; however, there are a few point groups that are exceptions. For example, D_2 and D_{2h} have three perpendicular C_2 axes. The convention is that the "A" irreducible representations in these groups has all of these C_2 axes with +1 characaters. This gives three "B" irreducible representations of the same symmetry with respect to the inversion center, if present. These three B irreducible representations are labeled as 1, 2, and 3 if symmetric with respect to the z, y, and x axes, respectively. .
7. Badger, R. M., A Relation Between Internuclear Distances and Bond Force Constants *J. Chem. Phys.* 1934, *2*.
8. Kaupp, M.; Danovich, D.; Shaik, S., Chemistry is about energy and its changes: A critique of bond-length/bond-strength correlations *Coord. Chem. Rev.* 2017, *344*, 355–362.
9. Christe, K. O.; Dixon, D. A.; Grant, D. J.; Haiges, R.; Tham, F. S.; Vij, A.; Vij, V.; Wang, T.-H.; Wilson, W. W., Dinitrogen Difluoride Chemistry. Improved Syntheses of cis- and *trans*-N_2F_2, Synthesis and Characterization of $[N_2F^+][Sn_2F_9^-]$, Ordered Crystal Structure of $[N_2F^+][Sb_2F_{11}^-]$, High-Level Electronic Structure Calculations of *cis*-N_2F_2, *trans*-N_2F_2, $F_2N{=}N$, and N_2F^+, and Mechanism of the trans-cis Isomerization of N_2F_2 *Inorg. Chem.* 2010, *49*, 6823-6833.
10. Dorta, R.; Stevens, E. D.; Scott, N. M.; Costabile, C.; Cavallo, L.; Hoff, C. D.; Nolan, S. P., Steric and electronic properties of N-heterocyclic carbenes (NHC): A detailed study on their interaction with $Ni(CO)_4$ *J. Am. Chem. Soc.* 2005, *127*, 2485–2495.
11. Popov, I. A.; Jian, T.; Lopez, G. V.; Boldyrev, A. I.; Wang, L. S., Cobalt-centred boron molecular drums with the highest coordination number in the CoB_{16}^- cluster *Nature Comm.* 2015, *6*.
12. Bergendahl, T. J.; Wood, J. S., Crystal and molecular-structure of hexakis(pyridine N-oxide) cobalt(II) perchlorate *Inorg. Chem.* 1975, *14*, 338–342.
13. Boudakian, M. M.; Hyde, G. A., Substitutive aromatic fluorination with chlorine pentafluoride *J. Fluorine Chem.* 1984, *25*, 435–446.
14. Darensbourg, M. Y.; Lyon, E. J.; Zhao, X.; Georgakaki, I. P., The organometallic active site of Fe hydrogenase: Models and entatic states *Proc. Nat. Acad. Sci.* 2003, *100*, 3683–3688.

CHAPTER 6

Group Theory and Molecular Orbital Theory

CONCEPT COVERAGE

1. The symmetry description of orbitals at a central atom
2. Combining the atomic orbitals of ligands to form MOs that can interact with the orbitals of the central atom
3. Using the projection operator to draw pictures of molecular orbitals
4. How to combine orbitals in an MO diagram
5. Examples of MO diagrams for some example molecules
6. X-ray photoelectron spectroscopy (XPS) and MO theory
7. MO theory and diatomics
8. Restricted and unrestricted views of molecules with unpaired electrons
9. Fragment approach for molecular orbitals of complex molecules

"We should regard a single bonding electron as the natural unit of bonding, an anti-bonding electron as a negative unit."

— **Robert S. Mulliken**[1]

"In the LCAO MO method, the MOs are approximated by setting up linear combinations of known AOs (atomic orbitals) of the atoms concerned. The coefficients in these linear combinations are partly, or sometimes wholly, determined by the symmetry of the molecule."

— **Robert S. Mulliken**[2]

In Chapter 5 the framework for group theory was discussed. This mathematical analysis of symmetry and its consequences has applications in numerous areas to chemistry in general and inorganic chemistry in particular. In this chapter, group theory will be applied to the construction of molecular orbital diagrams.

One may question the utility of being able to generate MO diagrams by hand in an age where computers can relatively quickly calculate the MOs for many molecules. However, the practicality comes when one goes to interpret the results from a molecular orbital calculation. Without an understanding of symmetry, orbital mixing, and the origin of the electron density, it is difficult to fully interpret computational results, let alone redesign molecules for new purposes.

The results of a calculation are merely the molecular orbitals as the computer finds them. Identifying the nature of the orbitals requires a detailed understanding of how MOs are derived. An understanding of group theory and symmetry is fundamental to understanding the electronic structure of molecules. In this chapter the procedure for the generation of molecular orbital diagrams by hand, with the aid of group theory, will be discussed.

The treatment here of group theory and molecular orbital theory is by no means complete, and the interested reader is referred to other more specialized texts for more information.[3–5] Some simple background on MO theory was covered in Section 3.7 in the context of basic bonding theory, which one may wish to review before proceeding. In that section, MOs for simple compounds such as water and HCl were covered. Here we will be focusing on how symmetry can be used to generate more complex MOs.

6.1 Constructing Molecular Orbital Diagrams: The Role of Symmetry

The symmetry of the orbitals of an atom sitting on the point origin of a molecule can be found by simply looking at a character table for the point group. For example, examine the character table for O_h. The irreducible representations in the left-most column are correlated with linear and quadratic functions in the two right-most columns. The functions *x*, *y*, and *z* are triply degenerate and of symmetry T_{1u}. The irreducible representation T_{1u} does not correspond to any orbitals

Table 6.1 Character table for O_h.

O_h	E	$8C_3$	$6C_2$	$6C_4$	$3C_2(=C_4^2)$	i	$6S_4$	$8S_6$	$3\sigma_h$	$6\sigma_v$	Linear functions	Quadratic functions
A_{1g}	1	1	1	1	1	1	1	1	1	1		$x^2+y^2+z^2$
A_{2g}	1	1	−1	−1	1	1	−1	1	1	−1		
E_g	2	−1	0	0	2	2	0	−1	2	0		$(2z^2-x^2-y^2, x^2-y^2)$
T_{1g}	3	0	−1	1	−1	3	1	0	−1	−1	(R_x, R_y, R_z)	
T_{2g}	3	0	1	−1	−1	3	−1	0	−1	1		(xy, yz, zx)
A_{1u}	1	1	1	1	1	−1	−1	−1	−1	−1		
A_{2u}	1	1	−1	−1	1	−1	1	−1	−1	1		
E_u	2	−1	0	0	2	−2	0	1	−2	0		
T_{1u}	3	0	−1	1	−1	−3	−1	0	1	1	(x, y, z)	
T_{2u}	3	0	1	−1	−1	−3	1	0	1	−1		

on the central main group or transition metal atom other than p_x, p_y, and p_z; consequently, the label "t_{1u} orbitals" will uniquely identify this set.

As you may have noticed, when we switched from discussing the irreducible representation, T_{1u}, to discussing the orbitals, t_{1u}, the case of the first letter changed. It is customary to distinguish an irreducible representation from the orbitals bearing that symmetry by using lower-case letters for the orbitals.

The valence *s*-orbital on the central atom always has the symmetry of the first irreducible representation listed, the totally symmetric representation: a, a_1, a', a_1', a_g, or a_{1g}, depending on the particular point group.

In the O_h character table (Table 6.1) under E_g, the quadratic (right-most) column has "$2z^2 - x^2 - y^2$". In other symmetries, such as A_1 in C_{3v}, there appears "z^2". Either of these functions can be considered to refer to the d_{z^2}-orbital centered on the point of symmetry.

All orbitals of the same symmetry can, and in many cases will, mix. For example, in the D_{2d} point group (Table 6.2), the (p_x,p_y)- and (d_{xz},d_{yz})-orbitals are both members of the degenerate irreducible representation E. These two different sets of *e*-orbitals may, depending on the compound and orbital energies, mix to some degree when used in bonding by the central element. In the case of the transition metals, the *p*-orbitals are often relatively high in energy with respect to the *d*-orbitals, making mixing less prevalent.

Conversely, atomic orbitals of different symmetry can have no net overlap and are orthogonal to one another. These orbitals will often have bonding and antibonding interactions in the same orbital mixing diagram that exactly cancel due to symmetry.

For example, we can examine a totally symmetric representation of ligand-based orbitals comprised of *s*-orbitals on a D_{2d} molecule all with the same phase (Fig. 6.1). The central element, M, can use an orbital of the same symmetry to interact with this ligand-based orbital. The symmetry of the ligand-based orbital is a_1, and the atomic *s*-orbital on M has the proper symmetry to have net overlap. These ligand- and metal-based orbitals will combine to form bonding and antibonding combinations. If one attempts to overlay the ligand-based a_1-orbital with an orbital on M from a different irreducible representation, such as p_z in irreducible representation B_2, then a nonbonding interaction occurs and there is no net overlap, equal portions of same-phase bonding and different-phase antibonding interactions occur.

It can be noted here that having orbitals of the same symmetry is a necessary but not sufficient criteria for overlap. In other words, having the same symmetry means there may be overlap; however, the amount of overlap could still be zero, depending on the exact sizes and shapes of the orbitals in question. If the orbitals are in different irreducible representations, then the overlap is exactly zero.

In short, the irreducible representations within the point group, from a molecular orbital point of view, describe particular orbitals or combinations of orbitals and their symmetries with respect to phase changes and contributions. On the other hand, if the ligand and metal orbitals are of different

Table 6.2 Character table for D_{2d}.

D_{2d}	E	$2S_4$	C_2	$2C'_2$	$2\sigma_d$	Linear functions	Quadratic functions
A_1	1	1	1	1	1		x^2+y^2, z^2
A_2	1	1	1	−1	−1	R_z	
B_1	1	−1	1	1	−1		x^2-y^2
B_2	1	−1	1	−1	1	z	xy
E	2	0	−2	0	0	$(x, y)(R_x, R_y)$	(xz, yz)

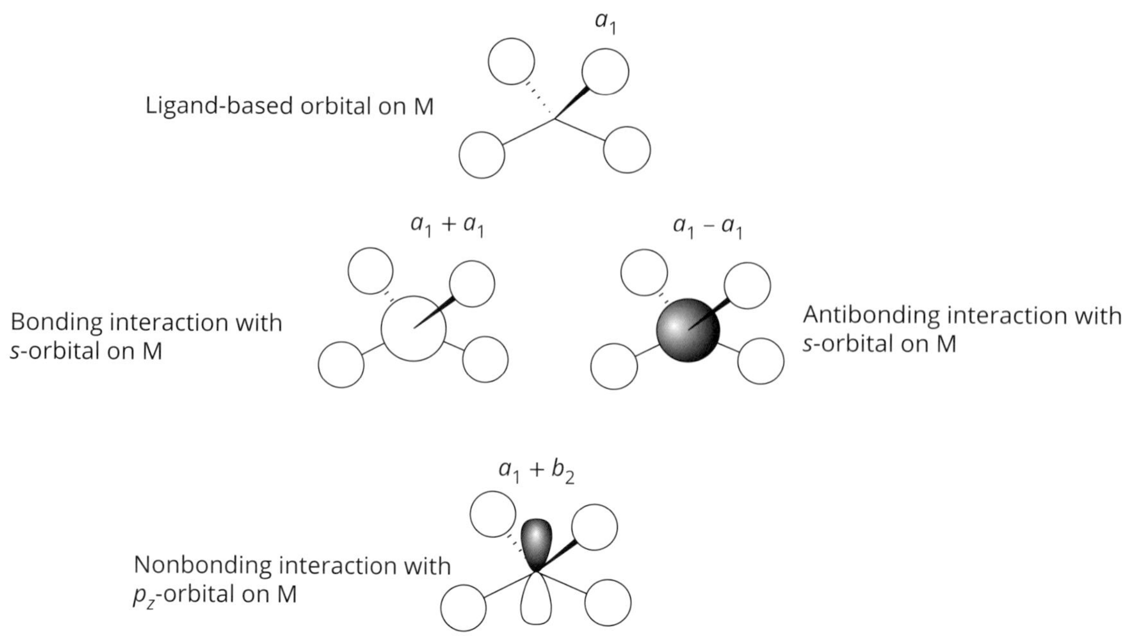

Fig. 6.1 *(top)* Ligand-based orbital a_1 in D_{2d}. *(center)* Combination of ligand-based orbital a_1 with orbital on the central atom of the same symmetry *s* to form bonding and antibonding combinations $a_1 + a_1$ and $a_1 - a_1$, respectively. *(bottom)* Overlay of the ligand-based orbital a_1 with orbital of different symmetry p_z in irreducible representation B_2, which shows that any bonding interaction is exactly canceled by the antibonding interaction.

symmetry (are in a different irreducible representation), the overlap is precisely zero.

As an example, let us look at the valence atomic orbitals of Pt in $PtCl_4^{2-}$ in some detail. The molecule is in the D_{4h} point group, and the Pt atom is sitting on the point origin of the system. What happens to the phase and direction of the orbitals on Pt as we carry out the symmetry operations of the group? If the orbital does not change at all, we will write a "+1". If the phases invert (positive to negative, negative to positive) but the orbital is otherwise unchanged, we will write a "–1". If the orbital moves from its starting position entirely, we will give it a "0" (zero). The results for the nine valence atomic orbitals of Pt are shown in Fig. 6.2.

The 6*s* orbital sitting at the origin will not move or otherwise change during any of the symmetry operations. The spherical *s*-orbital is said to be in the *totally symmetric representation.* Under each class of symmetry operations, we write a "+1". What the +1 under the *E* operation represents is one member of the group being investigated (the set of valence orbitals on Pt); in other words, this is a nondegenerate orbital with a 1-dimensional representation. If we look in the D_{4h} character table, the row with a +1 under each of the classes of symmetry operations is A_{1g}. Consequently, we say that the 6*s* orbital is represented by a_{1g}, the totally symmetric representation.

The $6p_z$-orbital also transforms by itself and is not moved from one member of our set of valence orbitals to another member by any symmetry operation in D_{4h}. As a result, its number under the symmetry operation *E* is also +1. The C_4, and C_2 coincident with it, do not change the p_z-orbital either, which gives them each a +1 assignment. The C_2' and C_2'' operations, perpendicular to the C_4, invert the phase of the orbital, as do the symmetry operations i, S_4, and σ_h, which results in a –1 in our table. The four mirror planes parallel to the C_4 axis do not change the orbital and are assigned a +1. Looking to see which irreducible representation includes these phase changes we see that the p_z-orbital is represented by a_{2u}.

The p_x and p_y-orbitals are different from p_z in their symmetry properties. They do move under some of the symmetry operations, and under some symmetry operations one of these orbitals is converted into the other. *When one orbital is mapped onto another by a symmetry operation it is a sign that they are part of the same irreducible representation*, and these two orbitals comprise a doubly degenerate set. If this is the case, the number under the identity symmetry operation (*E*) is the number of members of our basis set that are in the group, i.e., 2. During the C_4, the orbitals move ($p_x \rightarrow p_y$ and $p_y \rightarrow p_x$), giving them a 0 (zero). For the C_2 coincident with the *z*-axis, the phase of each orbital changes, which is assigned a –2 in our table. For C_2' along the bond axis, one of the orbitals does not move (+1), while the other changes phase (–1), which gives a sum of zero (0). Under C_2'', both orbitals move, producing a 0. Inversion changes the phase of each orbital (–2). The improper rotation moves the orbitals (0). Finally, the σ_h moves no orbitals (+2), and the mirror planes parallel to the main rotation axis produces 0. (Although these parallel mirror planes act on p_x and p_y differently, the result sums to zero.) The 2-dimensional representation for the p_x- and p_y-orbitals is e_u.

The *d*-orbitals can be evaluated similarly. The d_{z^2}-orbital does not exchange with any other orbital with any of the

D_{4h}	E	$2C_4$	C_2	$2C_2'$	$2C_2''$	i	$2S_4$	σ_h	$2\sigma_v$	$2\sigma_d$		
A_{1g}	1	1	1	1	1	1	1	1	1	1		x^2+y^2, z^2
A_{2g}	1	1	1	−1	−1	1	1	1	−1	−1	R_z	
B_{1g}	1	−1	1	1	−1	1	−1	1	1	−1		x^2-y^2
B_{2g}	1	−1	1	−1	1	1	−1	1	−1	1		xy
E_g	2	0	−2	0	0	2	0	−2	0	0	(R_x, R_y)	(xz, yz)
A_{1u}	1	1	1	1	1	−1	−1	−1	−1	−1		
A_{2u}	1	1	1	−1	−1	−1	−1	−1	1	1	z	
B_{1u}	1	−1	1	1	−1	−1	1	−1	−1	1		
B_{2u}	1	−1	1	−1	1	−1	1	−1	1	−1		
E_u	2	0	−2	0	0	−2	0	2	0	0	(x, y)	

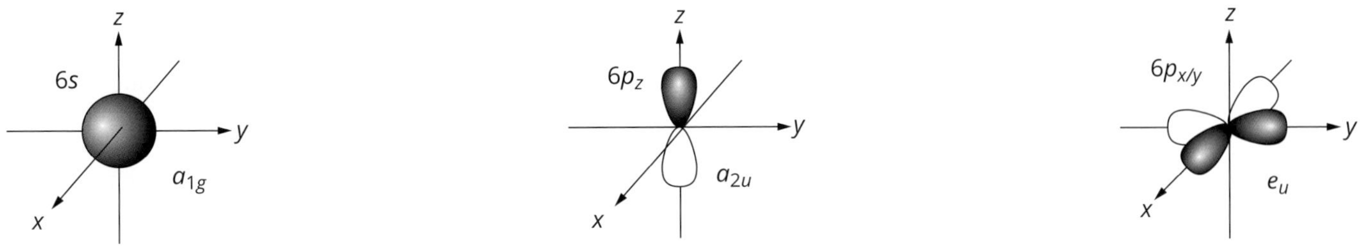

D_{4h}	E	$2C_4$	C_2	$2C_2'$	$2C_2''$	i	$2S_4$	σ_h	$2\sigma_v$	$2\sigma_d$
6s	1	1	1	1	1	1	1	1	1	1

D_{4h}	E	$2C_4$	C_2	$2C_2'$	$2C_2''$	i	$2S_4$	σ_h	$2\sigma_v$	$2\sigma_d$
$6p_z$	1	1	1	−1	−1	−1	−1	−1	1	1

D_{4h}	E	$2C_4$	C_2	$2C_2'$	$2C_2''$	i	$2S_4$	σ_h	$2\sigma_v$	$2\sigma_d$
$6p_{x/y}$	2	0	−2	0	0	−2	0	2	0	0

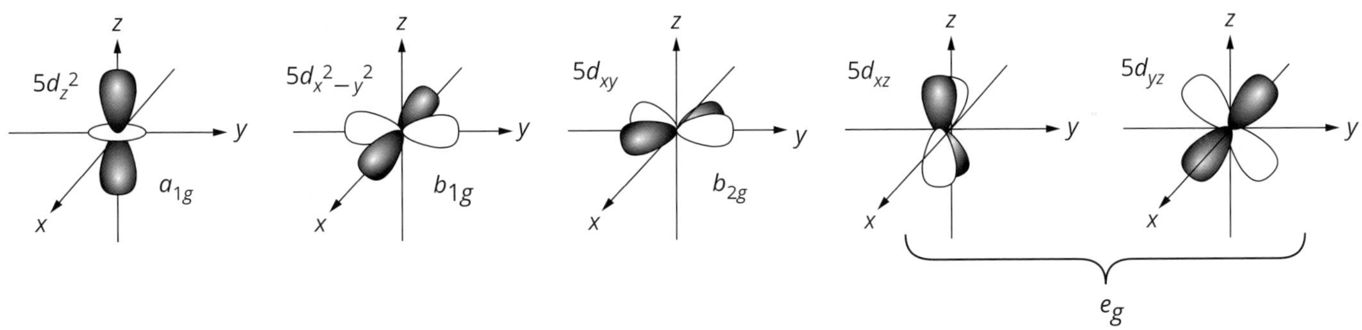

D_{4h}	E	$2C_4$	C_2	$2C_2'$	$2C_2''$	i	$2S_4$	σ_h	$2\sigma_v$	$2\sigma_d$
$6d_{z^2}$	1	1	1	1	1	1	1	1	1	1

D_{4h}	E	$2C_4$	C_2	$2C_2'$	$2C_2''$	i	$2S_4$	σ_h	$2\sigma_v$	$2\sigma_d$
$6d_{xy}$	1	−1	1	−1	1	1	−1	1	−1	1

D_{4h}	E	$2C_4$	C_2	$2C_2'$	$2C_2''$	i	$2S_4$	σ_h	$2\sigma_v$	$2\sigma_d$
$6d_{x^2-y^2}$	1	−1	1	1	−1	1	−1	1	1	−1

D_{4h}	E	$2C_4$	C_2	$2C_2'$	$2C_2''$	i	$2S_4$	σ_h	$2\sigma_v$	$2\sigma_d$
$6d_{xz/yz}$	2	0	−2	0	0	2	0	−2	0	0

Fig. 6.2 The effect of the symmetry operations in the D_{4h} point group on the valence orbitals of Pt in $PtCl_4^{2-}$. The irreducible representations in the character table can be used to describe the transformation properties of the orbitals under the symmetry operations of the molecule.

symmetry operations and, like the *s*-orbital, belongs to the a_{1g} representation. The $d_{x^2-y^2}$- and d_{xy}-orbitals are found in the b_{1g} and b_{2g} representations, respectively. Like the p_x- and p_y-orbitals, the d_{xz}- and d_{yz}-orbitals are sometimes transformed into one another and are parts of an *e*-set; therefore, they are considered together. Unlike the odd-functioned *p*-orbitals, the d_{xz}- and d_{yz}-orbitals are symmetric with respect to inversion and are represented by e_g.

As will be noticed in the character table for D_{4h}, these functions representing the orbitals are in the character table itself when the basis involves an atom on the origin; for example, the valence orbitals on Pt in $PtCl_4^{2-}$. In all cases, if the central atom is at the origin, then the character table will tell us which orbitals fall into which irreducible representations.

What if the atoms, or collection of atoms, are not on the origin, like the Cl atoms in $PtCl_4^{2-}$? How do we find the representations and nature of the ligand orbitals that can overlap with the valence orbitals of the central metal? That is the topic of the next sections. Then, in Section 6.5 we will discuss how to put all these together to generate the MO diagram.

In the case of simple molecules we can try to estimate the molecular orbital diagram simply by looking at which orbitals can overlap well. For example, the MO diagram of water can be drawn with little regard to formal symmetry arguments (Chapter 3).

In H_2O there is a central oxygen with valence *s* and three *p*-orbitals. The two *s*-orbitals on the hydrogens can form only two orbitals: one where they have the same phase, and one where they have different phases (Fig. 6.3). We place the hydrogen orbitals a little higher in energy than the oxygen orbitals because oxygen is more electronegative. Then we simply overlap the oxygen orbitals with the hydrogen orbitals, and this provides a reasonable approximation to the MO diagram.

However, drawing the MO diagrams of more complex systems will be greatly enabled by the use of symmetry. In addition, symmetry provides information about orbital mixing, which can provide important tweaks to MO diagrams, making them much more realistic. This is the case in the MO diagram of water in Fig. 6.3, where the diagram would be more accurate with mixing of *s*- and *p*-orbitals of the same irreducible representation.[a]

[a] A more accurate MO diagram for water is shown later in Chapter 7. An understanding of valence bond theory will be helpful in drawing better MO diagrams for molecules such as ammonia and water.

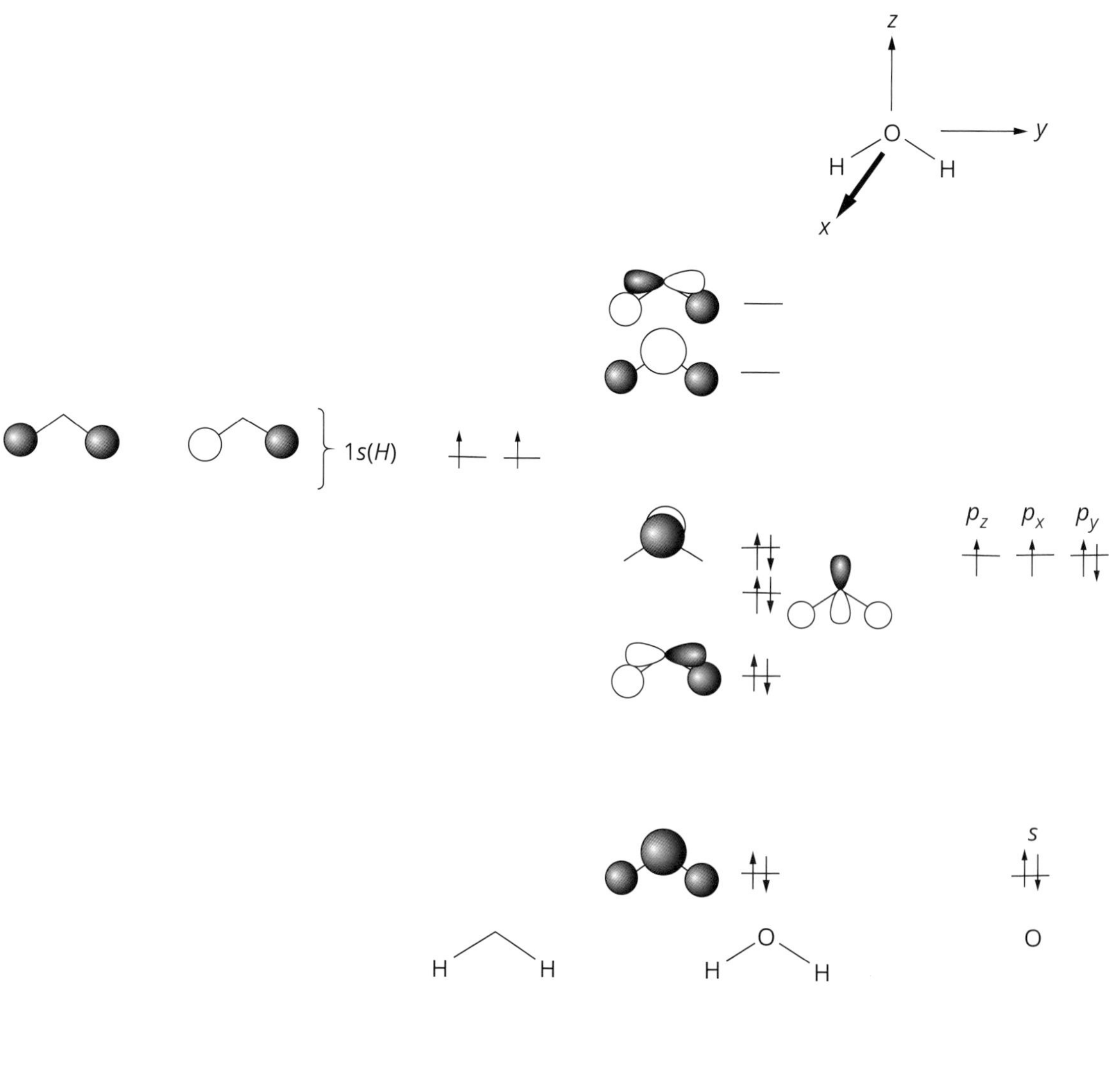

Fig. 6.3 MO diagram of H_2O by inspection of the orbitals.

6.2 Determining the Nature of Ligand-Based Orbitals

As shown, finding the irreducible representations corresponding to the orbitals of the central atom is trivial when a character table is handy. One simply determines the point group and looks in the columns to the right.

How does one find the symmetries of the ligand-based orbitals around the central atom, such as a_1? This is a three-step process.

- *Step 1: Generate the reducible representation* for the orbitals of interest.
- *Step 2: Decompose the reducible representation to its component irreducible representations.* The reducible representation is decomposed to its irreducible representations, providing how the orbitals involved act under the symmetry operation of the point group. These orbitals will most often be combinations of atomic orbitals "adapted" to fit the symmetry of the irreducible representation, and as a consequence are referred to as Symmetry Adapted Linear Combinations of Atomic Orbitals (SALC-AOs, or SALCs).
- *Step 3: Use the projection operator to generate pictures of the SALCs* corresponding to how the atomic orbitals combine in those irreducible representations. In other words, we use the projection operator to determine the nature of the SALCs, enabling us to draw a picture of what the SALC looks like.

Each of these steps will be discussed in turn by using several examples.

6.2.1 Illustrating the Steps for Ligand-Based Orbital Generation: OF_2

As a simple example of how to generate the ligand-based orbitals around a central element, SALCs, let us examine the C_{2v} molecule OF_2, which is a bent molecule similar in structure to water. Both the *s*- and *p*-atomic orbitals on fluorine will contribute to the molecular orbitals and must be accounted for.

The structure, along with its coordinate system, is shown in Fig. 6.4a. The bond angle has been determined to be 103.8 ± 1.5° in the compound.[6] The input for any MO calculation is the geometry, which must be provided from experiment, calculation, or at least approximated by VSEPR or similar methods.

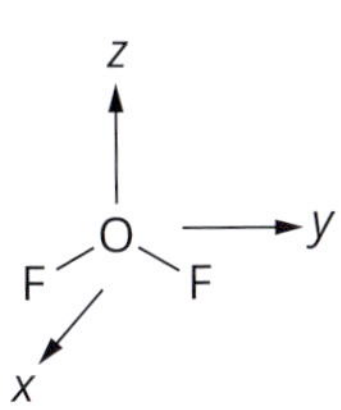

Fig. 6.4a Structure and coordinate system for OF_2.

6.2.1.1 Step 1: Generating the Reducible Representation

The *s*-orbitals on the fluorines can be treated separately from the *p*-orbitals, and we will consider the simpler *s*-orbitals first. Application of the symmetry operations to the basis shown as follows produces the following reducible representation. The *E* operation gives "+2" simply because our basis is comprised of two elements, the two *s*-orbitals on the fluorines. The C_2 operation moves the orbitals and would give zeros along the diagonal of the matrix, which gives a zero in our reducible representation. The $\sigma_v(xz)$ operation also moves both orbitals. Finally, the $\sigma_v'(yz)$ mirror plane is in the molecule and neither *s*-orbitals move, resulting in a +2 in the reducible representation.

To find the symmetry combinations for the *p*-orbitals, we will place an *x*, *y*, and *z*-orbital pointing the direction of each axis on the two fluorines. If the orbital does not change, then it will be given a "1". If the orbital moves entirely it will be given a "0", and if it inverts phase it will be given a "–1". Inversion of phase was, of course, not possible using the *s*-orbitals that we were dealing with previously, and one needs to keep track of phase changes for all other orbital types.

For simplicity, we will treat the p_x-, p_y-, and p_z-orbitals separately. *Our convention used here is to have the phases of the set being used (such as p_z-orbitals on the ligands) such that they are symmetric with respect to the principal rotation axis.* For example, the basis is as shown in Fig. 6.4c for the p_x-orbitals on the fluorines.

For the two p_x-orbitals on the two fluorines, the *E* operation leaves them in the same position with the same phase; consequently, each of the orbitals contributes 1 for a total of 2.

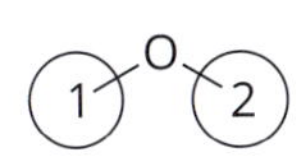

C_{2v}	E	C_2	$\sigma_v(xz)$	$\sigma'_v(yz)$
$\Gamma_s(OF_2)$	2	0	0	2

Fig. 6.4b Basis for the *s*-orbitals on the fluorines of OF_2 and the reducible representation.

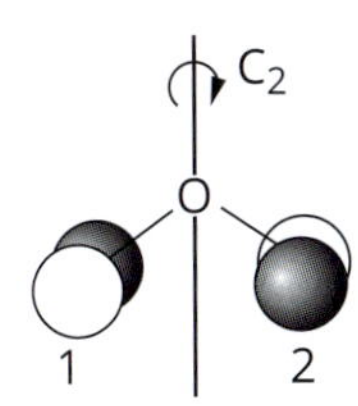

C_{2v}	E	C_2	$\sigma_v(xz)$	$\sigma'_v(yz)$
$\Gamma_{p_x}(OF_2)$	2	0	0	-2

Fig. 6.4c Basis for the p_x-orbitals on the fluorines of OF_2 and the reducible representation.

Basis set for the p_y-orbitals on F: 1 2

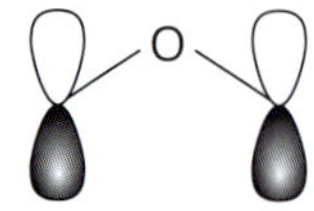

Basis set for the p_z-orbitals on F: 1 2

C_{2v}	E	C_2	$\sigma_v(xz)$	$\sigma'_v(yz)$
$\Gamma_{p_y}(OF_2)$	2	0	0	2
$\Gamma_{p_z}(OF_2)$	2	0	0	2

Fig. 6.4d Bases for the p_y- and p_z-orbitals on the fluorines of OF_2 and the reducible representations.

Carrying out the C_2 operation moves the fluorine atoms. Naturally, all of the *p*-orbitals move as well, and the character under C_2 is zero. As for the $\sigma_v(xz)$ mirror plane, similar to the C_2 operation, this moves the fluorines, and the character is zero. Finally, for $\sigma'_v(yz)$ the mirror plane is coincident with the plane of the molecule (Fig. 6.7), which inverts the phase of the two p_x-orbitals on the fluorines, which gives them a contribution of –2. The reducible representation for the p_x-orbitals on the two fluorines is shown in the following.

A similar process is followed for the p_y and p_z-orbitals. The basis for the two bases are shown in Fig. 6.4d. Application of the symmetry operations produces the following reducible representations for these *p*-orbitals on the fluorines.

The irreducible representations only differ in one of the mirror planes. Fig. 6.4e shows the effect of $\sigma_v'(yz)$ on each of the *p*-orbital basis sets.

6.2.1.2 Step 2: The Decomposition Formula

Next, the decomposition formula is applied to the reducible representations to find the symmetries of the fluorine *s*- and *p*-orbitals in OF_2 under the symmetry of the molecule, C_{2v}.

Decomposing the *s*-orbital-based reducible representation (Fig. 6.4b) shows that it is composed of a_1 and b_2 SALCs.

$$n_{A_1} = \frac{1}{4}[(2\times1\times1)_E + (0\times1\times1)_{C_2} + (0\times1\times1)_{\sigma_v(xz)} + (2\times1\times1)_{\sigma_v(yz)}]$$

$$n_{A_1} = \frac{1}{4}[2+0+0+2] = \frac{1}{4}\times4 = 1$$

$$n_{B_2} = \frac{1}{4}[(2\times1\times1)_E + (0\times-1\times1)_{C_2} + (0\times-1\times1)_{\sigma_v(xz)} + (2\times1\times1)_{\sigma_v(yz)}]$$

$$n_{B_2} = \frac{1}{4}[2-0-0+2] = \frac{1}{4}\times4 = 1$$

$$\Gamma_{s(OF_2)} = a_1 + b_2$$

The p_x-orbital-based reducible representation (Fig. 6.4c) is decomposed similarly.

$$n_{A_1} = \frac{1}{4}[(2\times1\times1)_E + (0\times1\times1)_{C_2} + (0\times1\times1)_{\sigma_v(xz)} + (-2\times1\times1)_{\sigma_v(yz)}]$$

$$n_{A_1} = \frac{1}{4}[2+0+0-2] = \frac{1}{4}\times0 = 0$$

$$n_{A_2} = \frac{1}{4}[(2\times1\times1)_E + (0\times1\times1)_{C_2} + (0\times-1\times1)_{\sigma_v(xz)} + (-2\times-1\times1)_{\sigma_v(yz)}]$$

$$n_{A_2} = \frac{1}{4}[2+0-0+2] = \frac{1}{4}\times4 = 1$$

$$n_{B_1} = \frac{1}{4}[(2\times1\times1)_E + (0\times-1\times1)_{C_2} + (0\times1\times1)_{\sigma_v(xz)} + (-2\times-1\times1)_{\sigma_v(yz)}]$$

$$n_{B_1} = \frac{1}{4}[2-0+0+2] = \frac{1}{4}\times4 = 1$$

$$n_{B_2} = \frac{1}{4}[(2\times1\times1)_E + (0\times-1\times1)_{C_2} + (0\times-1\times1)_{\sigma_v(xz)} + (-2\times1\times1)_{\sigma_v(yz)}]$$

$$n_{B_2} = \frac{1}{4}[2-0-0-2] = \frac{1}{4}\times0 = 0$$

$$\Gamma_{p_x(OF_2)} = a_2 + b_1$$

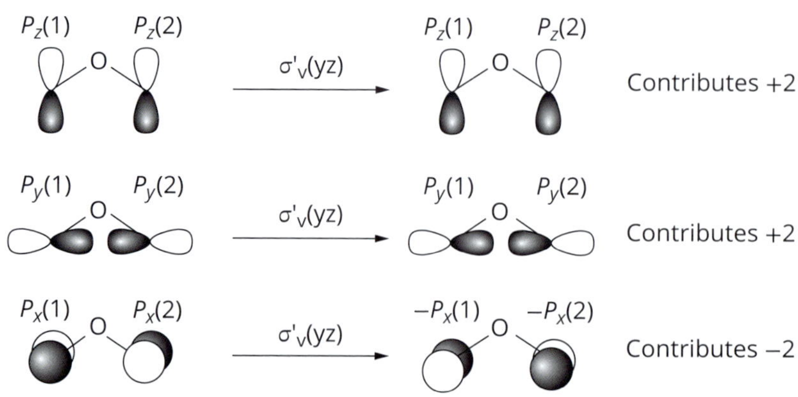

Fig. 6.4e Effect of $\sigma'_v(yz)$ on the pairs of p_x-, p_y-, and p_z-orbitals on the fluorines in OF_2.

Applying the decomposition formula to p_y- and p_z-reducible representations (Fig. 6.4d) produces the following results, which are different orbitals of the same symmetry.

$$\Gamma_{p_y(OF_2)} = a_1 + b_2$$
$$\Gamma_{p_z(OF_2)} = a_1 + b_2$$

6.2.1.3 Step 3: The Projection Operator and Symmetry Adapted Linear Combinations (SALCs) of Atomic Orbitals

The symmetries of the ligand-based orbitals from the previous decomposition formula are useful in the sense that they tell us what irreducible representations on the ligands are available to overlap with the orbitals of the same symmetry on the central atom. However, what would be of greater utility is a picture of how the ligand-based orbitals combine under the symmetry operations. Group theory can provide this as well using what is called the *projection operator*, introduced in Chapter 5.

The projection operator takes the function under consideration—for example, an *s*-orbital on one of the fluorines in OF_2—and projects the symmetry of the irreducible representation in the point group of interest onto it. In short, the projection of the A_1 representation on the *s*-orbital of F^1 in OF_2, $P^{a_1}_{s_1}$, is found by taking the characters of the irreducible representation (χ_{irred}) times the function in question (*s*-orbital on F^1) and summing over all the symmetry operations in the point group. *This is done not just for the classes of symmetry operations but must be done for every symmetry operation.*

Mathematically, this may be written as shown in Eq. 6.1 (reproduced from Eq. 5.13). The projection, $P^{a_i}_{\phi}$, is for a specific orbital that is part of the basis, ϕ, under the irreducible representation, a_i. The projection is the result of carrying out an operation, *r*, on the part of the basis being used in the projection, $r\phi$, times the character of the irreducible representation under that operation, $\chi^{a_i}(r)$, summed over all the operations *r*.

$$P^{a_i}_{\phi} = \sum_r \chi^{a_i}(r) \cdot r\phi \qquad \textbf{6.1}$$

As an illustration, let us return to the *s*-orbitals on the fluorines of OF_2. In the previous section it was shown that the reducible representation is comprised of two orbitals—one belonging to the a_1 irreducible representation, and another belonging to b_2.

$$\Gamma_s = a_1 + b_2$$

It is easy to see that the *s*-orbital on one of the fluorines is being mixed by symmetry with the *s*-orbital on the other fluorine. For example, the C_2 operation moves one orbital into the other. Consequently, if we carry out the projection on one of the *s*-orbitals it will show us how the orbitals are mixing under the irreducible representation to give a_1 and b_2 molecular orbitals. So, let us take the *s*-orbital on F^1, s_1, do the symmetry operations to develop the projection, and multiply by the character in the irreducible representation. If the symmetry operation converts s_1 into s_2, for example, then we record that in the equation. *It is important to note, and bears repeating, that projections are done over all symmetry operations, not the classes.* In this case, however, all the symmetry operations are in a class of their own.

$$\begin{aligned}
P^{a_1}_{s_1} &= \chi^{a_1}(E) \cdot Es_1 + \chi^{a_1}(C_2) \cdot C_2 s_1 + \chi^{a_1}(\sigma_{xz}) \cdot \sigma_{xz} s_1 \\
&\quad + \chi^{a_1}(\sigma'_{yz}) \cdot \sigma'_{yz} s_1 \\
&= (1 \times s_1)_E + (1 \times s_2)_{C_2} + (1 \times s_2)_{\sigma_v(xz)} + (1 \times s_1)_{\sigma'_v(yz)} \\
&= s_1 + s_2 + s_2 + s_1 \\
&= 2s_1 + 2s_2 \\
&\text{or simply} \\
&= s_1 + s_2
\end{aligned}$$

The s_1 function has the operation *E* carried out on it and is multipled by the character under *E* in the A_1 irreducible representation in the character table. Of course, carrying out the *E* operation leaves s_1 as it was, and this is simply multiplied by 1. In the second term, the C_2 operation converts s_1 into s_2, and the result is multiplied by the number under C_2 in the A_1 irreducible representation, +1. This is repeated for all of the operations of the point group. The result is the projection: $2s_1 + 2s_2$. At this point, the a_1 molecular orbital combination of the *s*-orbitals on the fluorines is shown to contain equal portions of s_1 and s_2 with the same phase (the signs are the same). One can at this stage divide to produce the lowest possible coefficient to provide the estimated value of the orbital contributions, or you can go on to simplify the function by *normalization.*

In general, we are most interested in the shape of the wavefunction, and multiplication by a scalar, such as in normalization, is of little significance. However, in some instances it is important to be able to get the normalization correct to show that orbitals are orthogonal, for example, and other operations of potential utility.

For our purposes, *N*, the normalization constant, will be evaluated as shown in the next equation. The wavefunction, ψ, under the Linear Combination of Atomic Orbitals (LCAO) approximation is of the following form with coefficients labeled as $c_1, c_2, \ldots$, in front of atomic orbital contributions. The normalization value, *N*, is chosen so that the probability is unity. The normalization constant is then the inverse of the square root of the sum of the coefficients.

$$\psi = N(c_1\phi_1 + c_2\phi_2 + c_3\phi_3 + \ldots + c_i\phi_i) = N\sum_i c_i\phi_i$$

$$N = \frac{1}{\sqrt{c_1^2 + c_2^2 + c_3^2 + \ldots + c_i^2}} = \frac{1}{\sqrt{\sum_i c_i^2}}$$

If the coefficient in front of the first contributing atomic orbital, c_1, is 1 and the coefficient in front of the second atomic orbital, c_2, is also 1, as in the previous example for $P_{s_1}^{a_1}$ then N is calculated as follows.

$$N = \frac{1}{\sqrt{1^2 + 1^2}} = \frac{1}{\sqrt{2}}$$

If we did not simplify the projection by dividing by the lowest coefficient, we would carry out the normalization as shown in the next equation to obtain the same result. Then once the function is normalized, and we are concerned with the wavefunction instead of simply the projection, we will use $\psi_{s_1}^{a_1}$ in place of $P_{s_1}^{a_1}$. For example, the a_1-molecular orbital for the s-orbitals on the fluorines of OF_2, now normalized, is shown here. In other words, $\psi_{s_1}^{a_1}$ is the normalized SALC-AO found through the projection of s_1 in our basis under the irreducible representation a_1.

$$\begin{aligned} P_{s_1}^{a_1} &= 2s_1 + 2s_2 \\ N &= \frac{1}{\sqrt{2^2 + 2^2}} = \frac{1}{\sqrt{4+4}} = \frac{1}{\sqrt{8}} = \frac{1}{2\sqrt{2}} \\ \psi_{s_1}^{a_1} &= \frac{1}{2\sqrt{2}}(2s_1 + 2s_2) \\ &= \frac{1}{\sqrt{2}}(s_1 + s_2) \end{aligned}$$

If we again use the s_1-orbital on the fluorines in OF_2 to project and find the nature of the b_2 molecular orbital, we find that the result is $s_1 - s_2$, which means that the two s-functions (s_1 and s_2) have opposite phase. Fig. 6.5 shows the nature of the two SALCs and their bonding combinations with orbitals of the same symmetry on oxygen. The a_1 SALC, for example, will interact with the s- and p_z-orbitals on the oxygen.

It is important to realize here that the normalized SALCs, such as a_1 and b_2 in the top portion of Fig. 6.5, are single orbitals. They may have contributions from multiple atoms in the molecule, but they are delocalized single orbitals of a particular symmetry. Similarly, 1 SALC on the ligands overlaps with an orbital on the central atom to give two orbitals—a bonding and antibonding combination formed by changing the phases of the central atom, for example. This will be illustrated extensively once we get to the diagramming of molecular orbitals.

$$\begin{aligned} P_{s_1}^{b_2} &= (1 \times s_1)_E + (-1 \times s_2)_{C_2} + (-1 \times s_2)_{\sigma_v(xz)} + (1 \times s_1)_{\sigma_v(yz)} \\ P_{s_1}^{b_2} &= s_1 - s_2 - s_2 + s_1 \\ &= 2s_1 - 2s_2 \\ \psi_{s_1}^{b_2} &= \frac{1}{\sqrt{2}}(s_1 - s_2) \end{aligned}$$

Applying the projection operator to the p-orbitals is done similarly. Any phase changes on the p-orbital must be taken into account. In addition, the p_x-, p_y-, and p_z-orbitals may transform differently.

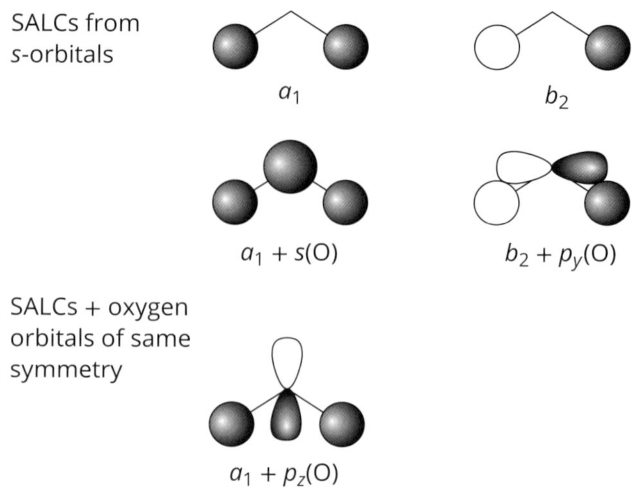

Fig. 6.5 The molecular orbitals due to combinations of the s-orbitals on fluorine in OF_2. *(top)* SALCs on the fluorines. *(bottom)* Bonding combination of the SALCs with oxygen atomic orbitals of the same symmetry.

The irreducible representations corresponding to the combinations of p_x-orbitals of the ligands were found to be a_2 and b_1. Applying the projection operator to the a_2 irreducible representation, then to b_1, produces the following results.

$$\Gamma_{p_x(OF_2)} = a_2 + b_1$$

$$\begin{aligned} P_{p_{x1}}^{a_2} &= (1 \times p_1)_E + (1 \times p_2)_{C_2} + (-1 \times -p_2)_{\sigma_v(xz)} \\ &\quad + (-1 \times -p_1)_{\sigma_v(yz)} \\ &= p_1 + p_2 + p_2 + p_1 \\ &= 2p_1 + 2p_2 \\ \psi_{p_{x1}}^{a_2} &= \frac{1}{\sqrt{2}}(p_1 + p_2) \end{aligned}$$

$$\begin{aligned} P_{p_{x1}}^{b_1} &= (1 \times p_1)_E + (-1 \times p_2)_{C_2} + (1 \times -p_2)_{\sigma_v(xz)} \\ &\quad + (-1 \times -p_1)_{\sigma_v(yz)} \\ &= p_1 - p_2 - p_2 + p_1 \\ &= 2p_1 - 2p_2 \\ \psi_{p_{x1}}^{b_1} &= \frac{1}{\sqrt{2}}(p_1 - p_2) \end{aligned}$$

As shown, the a_2 SALC is a combination where the two p_x-orbitals on the fluorines have the same phase with respect to the principal rotation axis, and the b_1-orbital is a combination where the two p_x-orbitals on the fluorines have the opposite phase. If the reducible representation is comprised of only two orbitals there are really no other possibilities. The same was true for the fluorine s-orbital SALCs, where a_1 was the in-phase combination and b_2 was the out-of-phase combination.

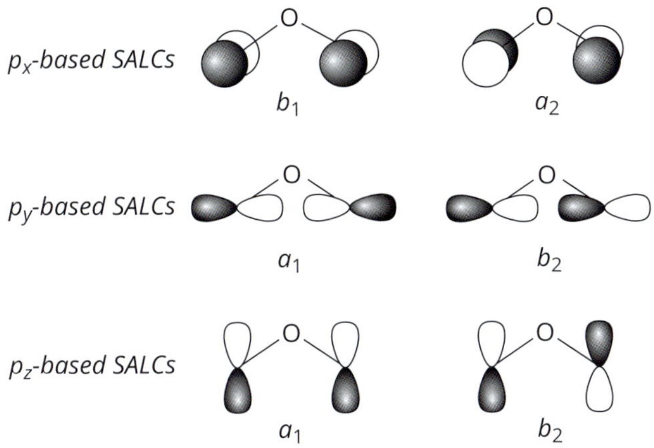

Fig. 6.6 SALCs derived from the *p*-orbitals on the fluorines in OF_2.

The remaining SALCs for the p_y- and p_z-orbitals are again in-phase and out-of-phase combinations with, of course, a_1 being the in-phase molecular orbital. For the p_y-SALC under the a_1 representation, it should be symmetric with respect to the main rotation axis, C_2. The character under the C_2 operation in the a_1 irreducible representation is "1", so there is no phase change. The character under the C_2 operation for b_2 is "–1", so there is a phase change in the orbitals when the C_2 is done. All of the *p*-orbital based SALCs are shown Fig. 6.6.

The results for OF_2 will be used shortly to create an MO diagram for the molecule. Before we move on to the diagram, let us do another example of SALC generation and cover some details about orbital mixing that are important to consider in MO diagram formation.

6.3 Finding the SALCs of SH_4

Step 1: Generating the reducible representation

As an example slightly more complex than OF_2 with some new features, we can discuss SH_4. To find the symmetries of the ligand-based orbitals in the point group of the molecule, we need to generate the reducible representation for the *combinations* of atomic ligand orbitals. As another illustration we can generate the ligand-based orbitals for the hypothetical compound SH_4, which has been examined computationally in some detail.[7] Just as expected from VSEPR, the compound has C_{2v} symmetry. The molecule, with its coordinate axes, is shown in Fig. 6.7a.

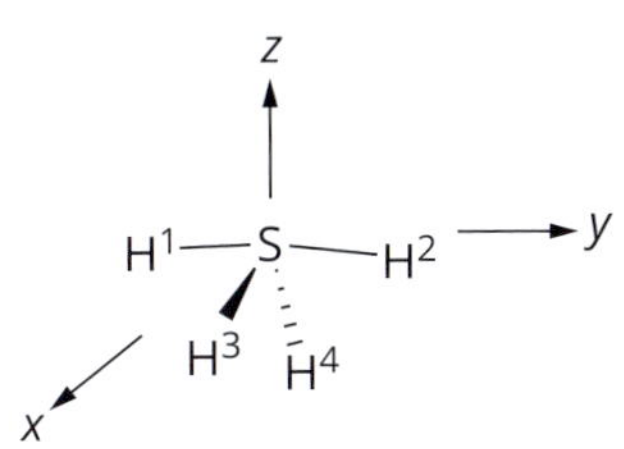

Fig. 6.7a Geometry of C_{2v} SH_4 with its coordinate axes.

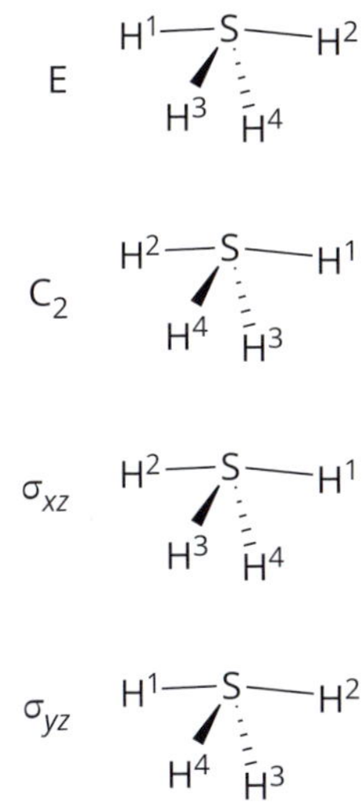

Fig. 6.7b The C_{2v} character table and the results of applying the symmetry operations to the hydrogen *1s*-orbitals on SH_4.

Each *s*-orbital on the hydrogens will combine to form a molecular orbital of specific symmetry, Symmetry Adapted Linear Combinations (SALCs). For SH_4 we will use the hydrogen *1s*-orbitals as our set and apply the symmetry operations to them. In Fig. 6.7b the symmetry operations are applied to the hydrogens labeled 1 to 4 in the figure.

With the operation *E*, none of the hydrogen orbitals move. We count "+1" for the orbitals on the hydrogens that do not change phase and are not transposed to a different hydrogen. In total for the operation *E*, all four hydrogens do not move. For the reducible representation for the hydrogen *s*-orbitals, $\Gamma_s(SH_4)$, we place under the *E* operation a "4", one for each hydrogen. As always, the number under the *E* symmetry operation will correspond to the total number of set members (orbitals in MO theory) being acted upon; in this case, four *s*-orbitals from each of the four hydrogens (Table 6.3a).

Table 6.3a Partial reducible representation for the SALCs in SH_4 with the *E* contribution of 4.

C_{2v}	E	C_2	$\sigma_v(xz)$	$\sigma'_v(yz)$
$\Gamma_s(SH_4)$	4	?	?	?

Next, we can look at the consequences of the C_2 operation on the molecule (Fig. 6.7b). The C_2 along the *z*-axis transposes H^1 to H^2 and H^3 to H^4 (A matrix example will be done momentarily where it will be shown that these move off diagonally.) If the orbital is transposed into a different orbital it is assigned a zero; consequently, the value for $\Gamma_s(SH_4)$ under C_2 is 0 (Table 6.3b).

Table 6.3b Partial reducible representation for the SALCs in SH_4 with the C_2 contribution of 0.

C_{2v}	E	C_2	$\sigma_v(xz)$	$\sigma'_v(yz)$
$\Gamma_s(SH_4)$	4	0	?	?

Table 6.3c Partial reducible representation for the SALCs in SH_4 with the $\sigma_v(xz)$ contribution of 2.

C_{2v}	E	C_2	$\sigma_v(xz)$	$\sigma'_v(yz)$
$\Gamma_s(SH_4)$	4	0	2	?

The xz-plane contains a mirror that converts H^1 into H^2. However, the orbitals sitting in the plane, H^3 and H^4, do not change. Two of the s-orbitals on the hydrogens move, and two do not. Consequently, the character under $\sigma_v(xz)$ is 2 (Table 6.3c).

Similarly, for $\sigma'_v(yz)$ the character is 2 because H^3 is transposed into H^4, but H^1 and H^2 do not move.

The completed reducible representation for the hydrogen s-orbitals is shown in Table 6.3d. We started with four atomic orbitals and must have four molecular orbitals that are generated by using the symmetry operations. The character under the E symmetry operation is the dimensionality and, in the context of MO theory, is equal to the number of orbitals contained in the representation. In the next section we will show that this representation "decomposes" to produce four molecular orbitals.

Table 6.3d Reducible representation for the SALCs in SH_4.

C_{2v}	E	C_2	$\sigma_v(xz)$	$\sigma'_v(yz)$
$\Gamma_s(SH_4)$	4	0	2	2

Before examining the decomposition formula, let's find the reducible representation by using matrix algebra to see mathematically the nature of the characters in the reducible representation. Each s-orbital will have a position associated with it, and for H^1 this location will be h_1, i.e., $h_1 = x, y,$ and z coordinates for hydrogen 1 as a set. As a result, for the four hydrogens each in its position, we start with a column vector containing four members (a 1×4 matrix). If the E symmetry operation is applied to this vector, it should return the same vector. What matrix corresponds to E? Naturally, it is the matrix with 1's on the diagonal shown as follows: a 4×4 identity matrix. (For a review of matrix algebra, see Appendix E.)

$$E\begin{bmatrix} h_1 \\ h_2 \\ h_3 \\ h_4 \end{bmatrix} = \begin{bmatrix} h_1 \\ h_2 \\ h_3 \\ h_4 \end{bmatrix}$$

$$\begin{bmatrix} 1 & 0 & 0 & 0 \\ 0 & 1 & 0 & 0 \\ 0 & 0 & 1 & 0 \\ 0 & 0 & 0 & 1 \end{bmatrix}\begin{bmatrix} h_1 \\ h_2 \\ h_3 \\ h_4 \end{bmatrix} = \begin{bmatrix} h_1 \\ h_2 \\ h_3 \\ h_4 \end{bmatrix}$$

The character (also called the "trace") of a matrix is the sum of the numbers along the diagonal from top left to bottom right, which in this case is $1 + 1 + 1 + 1 = 4$. The character of the E operation for the reducible representation is then 4.

The C_2 operation on the hydrogen atoms is shown as follows in matrix form. During the transformation, h_1 is converted to h_2 and vice versa. In addition, h_3 is converted to h_4. The matrix to carry out that vector transformation has all zeros along the diagonal and a character (trace) of zero.

$$C_2\begin{bmatrix} h_1 \\ h_2 \\ h_3 \\ h_4 \end{bmatrix} = \begin{bmatrix} h_2 \\ h_1 \\ h_4 \\ h_3 \end{bmatrix}$$

$$\begin{bmatrix} 0 & 1 & 0 & 0 \\ 1 & 0 & 0 & 0 \\ 0 & 0 & 0 & 1 \\ 0 & 0 & 1 & 0 \end{bmatrix}\begin{bmatrix} h_1 \\ h_2 \\ h_3 \\ h_4 \end{bmatrix} = \begin{bmatrix} h_2 \\ h_1 \\ h_4 \\ h_3 \end{bmatrix}$$

The matrices for the mirror planes are shown as follows, each with a trace of 2. Putting together the characters of the four matrices representing the symmetry operations of C_{2v} gives the same reducible representation found previously.

$$\sigma_v(xz)\begin{bmatrix} h_1 \\ h_2 \\ h_3 \\ h_4 \end{bmatrix} = \begin{bmatrix} h_2 \\ h_1 \\ h_3 \\ h_4 \end{bmatrix} \qquad \sigma_v(yz)\begin{bmatrix} h_1 \\ h_2 \\ h_3 \\ h_4 \end{bmatrix} = \begin{bmatrix} h_1 \\ h_2 \\ h_4 \\ h_3 \end{bmatrix}$$

$$\begin{bmatrix} 0 & 1 & 0 & 0 \\ 1 & 0 & 0 & 0 \\ 0 & 0 & 1 & 0 \\ 0 & 0 & 0 & 1 \end{bmatrix}\begin{bmatrix} h_1 \\ h_2 \\ h_3 \\ h_4 \end{bmatrix} = \begin{bmatrix} h_2 \\ h_1 \\ h_3 \\ h_4 \end{bmatrix} \qquad \begin{bmatrix} 1 & 0 & 0 & 0 \\ 0 & 1 & 0 & 0 \\ 0 & 0 & 0 & 1 \\ 0 & 0 & 1 & 0 \end{bmatrix}\begin{bmatrix} h_1 \\ h_2 \\ h_3 \\ h_4 \end{bmatrix} = \begin{bmatrix} h_1 \\ h_2 \\ h_4 \\ h_3 \end{bmatrix}$$

Step 2: The decomposition formula

For SH_4 we found the reducible representation (Table 6.3d) for the hydrogen $1s$-orbitals under the symmetry of the point group, C_{2v}.

By using the decomposition formula we can determine how many times the totally symmetric representation, A_1, appears in this reducible representation. For C_{2v}, the order $h = 4$ as the C_{2v} point group has a total of four symmetry operations.

$$n_{irred.rep.} = \frac{1}{h}\sum_{classes} \chi_{red} \times \chi_{irred} \times N$$

$$n_{A_1} = \frac{1}{4}\sum_{classes} \chi_{red} \times \chi_{irred} \times N$$

$$= \frac{1}{4}[(4 \times 1 \times 1)_E + (0 \times 1 \times 1)_{C_2} + (2 \times 1 \times 1)_{\sigma_v(xz)} + (2 \times 1 \times 1)_{\sigma'_v(yz)}]$$

$$n_{A_1} = \frac{1}{4}[4 + 0 + 2 + 2] = \frac{1}{4} \times 8 = 2$$

The decomposition formula suggests that there are two different a_1 orbitals in the reducible representation. Carrying out the same operations with the other irreducible representations in the C_{2v} character table produces the following equations.

$$n_{A_2} = \frac{1}{4}[(4 \times 1 \times 1)_E + (0 \times 1 \times 1)_{C_2} + (2 \times -1 \times 1)_{\sigma_v(xz)} + (2 \times -1 \times 1)_{\sigma_v'(yz)}]$$

$$n_{A_2} = \frac{1}{4}[4 + 0 - 2 - 2] = \frac{1}{4} \times 0 = 0$$

$$n_{B_1} = \frac{1}{4}[(4 \times 1 \times 1)_E + (0 \times -1 \times 1)_{C_2} + (2 \times 1 \times 1)_{\sigma_v(xz)} + (2 \times -1 \times 1)_{\sigma_v'(yz)}]$$

$$n_{B_1} = \frac{1}{4}[4 - 0 + 2 - 2] = \frac{1}{4} \times 4 = 1$$

$$n_{B_2} = \frac{1}{4}[(4 \times 1 \times 1)_E + (0 \times -1 \times 1)_{C_2} + (2 \times -1 \times 1)_{\sigma_v(xz)} + (2 \times 1 \times 1)_{\sigma_v'(yz)}]$$

$$n_{B_2} = \frac{1}{4}[4 - 0 - 2 + 2] = \frac{1}{4} \times 4 = 1$$

Consequently, the decomposition formula allows the reducible representation to be described as SALCs of symmetry b_2, b_1, and two of a_1. None of the orbitals are degenerate, so there are four orbitals total for the four starting atomic orbitals on the hydrogens, as required.

$$\Gamma_s = 2a_1 + b_1 + b_2$$

Note that if we multiply 2 times the row for the A_1 irreducible representation and sum with B_1 and B_2, we recover our original reducible representation. This is always a useful check on the results of the decomposition formula.

	E	C_2	$C\sigma_v(xz)$	$\sigma_v'(yz)$
$2 \times A_1 =$	2	2	2	2
$B_1 =$	1	−1	1	−1
$B_2 =$	1	−1	−1	1
$2A_1 + B_1 + B_2 =$	4	0	2	2

Step 3: Use the projection operator to generate pictures of the SALCs

We can apply the projection operator on the hydrogen *s*-orbital SALCs in SH_4. In the previous section we showed that the orbitals on the hydrogens correspond to the irreducible representations $2a_1 + b_1 + b_2$.

If we pick the *s*-orbital on H^1 (Fig. 6.8), s_1, to apply the projection operator, we find that this a_1 SALC has no information on the s_3 and s_4 orbitals. In other words, there is no symmetry operation that converts s_1 to either s_3 or s_4.

$$P_{s_1}^{a_1} = (1 \times s_1)_E + (1 \times s_2)_{C_2} + (1 \times s_2)_{\sigma_v(xz)} + (1 \times s_1)_{\sigma_v(yz)}$$
$$= s_1 + s_2 + s_2 + s_1$$
$$= 2s_1 + 2s_2$$
$$\psi_{s_1}^{a_1} = \frac{1}{\sqrt{2}}(s_1 + s_2)$$

Obviously, the orbitals on the other hydrogens must be involved in bonding as well, and we need to find their contributions to the molecular orbitals. To acquire some information on the other hydrogens we project on s_3 using the a_1 irreducible representation as well. We find that this too has a totally symmetric representation and could be another a_1 SALC found in our reducible representation.

$$P_{s_3}^{a_1} = (1 \times s_3)_E + (1 \times s_4)_{C_2} + (1 \times s_3)_{\sigma_v(xz)} + (1 \times s_4)_{\sigma_v(yz)}$$
$$= s_3 + s_4 + s_3 + s_4$$
$$= 2s_3 + 2s_4$$
$$\psi_{s_3}^{a_1} = \frac{1}{\sqrt{2}}(s_3 + s_4)$$

Consequently, we have two SALCs of a_1 symmetry in our set that do not include any members of the set in common. In cases like this, the symmetry operations are not giving maximal delocalization of the electron density simply because there are no symmetry operations that convert H^1 to H^3, for example. To increase delocalization and get a better set of SALCs, we should mix these two a_1 SALCs from the projection forming in-phase and out-of-phase combinations (Fig. 6.8). *This mixing can occur only if the SALCs are in the same irreducible representation.* The mixing does not have to be 50/50 in all cases, but we have no reason to assume that there will be significant differences in mixing for these noninteracting SALCs for equivalent *1s* orbitals on hydrogens.[b]

At this stage we do not necessarily need to mix SALCs. The mixing can be done at the diagramming stage, but we introduce it here as a first illustration of mixing within SALCs. All MO theory is a process of deciding how to mix atomic orbitals based on overlap and energy, guided by symmetry.

The creation of the two a_1 SALCs in SH_4 illustrates an important principle. *In all molecules, one can create a SALC on the ligands where all of the orbitals are of the same phase denoted by the totally symmetric representation.* The mixing in Fig. 6.8 gives us that delocalized orbital and another delocalized orbital within the same irreducible representation with a phase change. An additional example of this process will be covered in Section 6.6.4 on PH_5 where this type of mixing is used in SALC generation. Again, this mixing of SALCs of the

[b] The way these SALCs interact with the sulfur will not necessarily be equivalent as the angles between the ligands are different.

same symmetry is not a necessity at this stage, but it does simplify MO diagram formation later.

There are two other SALCs for the s-orbitals on the hydrogens in SH_4, which belong to the irreducible representations b_1 and b_2. Based on the previous account, we know that these will correspond to combinations of s_1 and s_2 or s_3 and s_4 but not all four orbitals, because there are no symmetry operations that relate all four hydrogens and these do not have the same symmetry. Simply looking at the character table shows that the p_x-orbital on sulfur belongs to the b_1 irreducible representation, and H^3 and H^4 both sit on the x-axis. This suggests that we should use s_3 in the projection with b_1. Similarly, we should use s_1 for the projection with b_2.

$$\begin{aligned} P_{s_3}^{b_1} &= (1 \times s_3)_E + (-1 \times s_4)_{C_2} + (1 \times s_3)_{\sigma_v(xz)} + (-1 \times s_4)_{\sigma_v(yz)} \\ &= s_3 - s_4 + s_3 - s_4 \\ &= 2s_3 - 2s_4 \\ \psi_{s_3}^{b_1} &= \frac{1}{\sqrt{2}}(s_3 - s_4) \end{aligned}$$

$$\begin{aligned} P_{s_1}^{b_2} &= (1 \times s_1)_E + (-1 \times s_2)_{C_2} + (-1 \times s_2)_{\sigma_v(xz)} + (1 \times s_1)_{\sigma_v(yz)} \\ &= s_1 - s_2 - s_2 + s_1 \\ &= 2s_3 - 2s_4 \\ \psi_{s_1}^{b_2} &= \frac{1}{\sqrt{2}}(s_1 - s_2) \end{aligned}$$

All of the hydrogen-based SALCs are shown in Fig. 6.8.

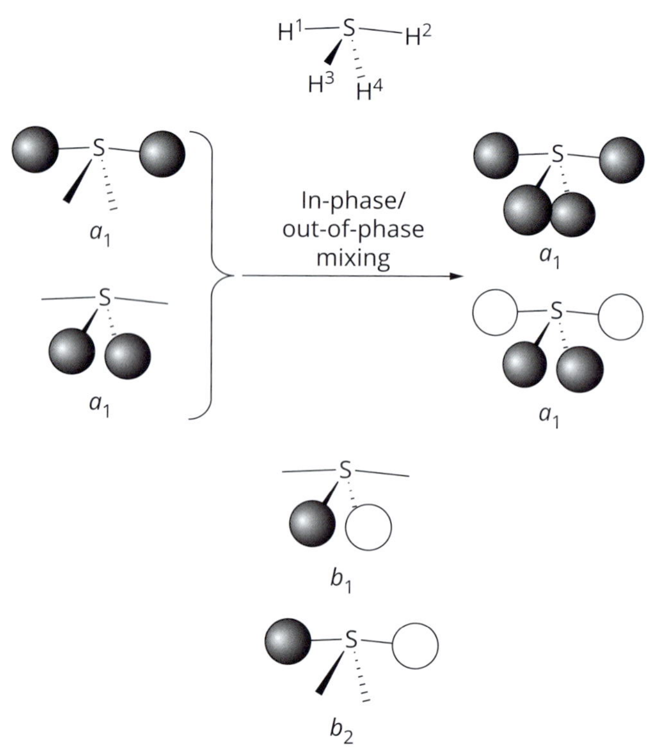

Fig. 6.8 SALCs derived from the hydrogen s-orbitals in SH_4.

6.4 Finding the MOs of Degenerate Representations

In the examples thus far we have restricted ourselves to molecules with no more than a 2-fold rotation axis. Compounds with a 3-fold or higher rotation axis (proper or improper) have degenerate orbitals, sets of orbitals represented by e or t irreducible representations.

Using projection operators while working with a degenerate irreducible representation generates one of the possible molecular orbitals that is part of the set. However, it is required that the other orbital in a degenerate group—for example, an e-set or a t-set—be orthogonal to the first.

Before discussing methods for finding orthogonal degenerate orbitals, it is important to emphasize a few features already covered:[c]

(1) Project normally using one part of the basis to get one projection.

(2) If the orbital is part of an e-set, there should be a second SALC. Use a different part of the basis and project again to get another SALC. If part of a t-set, a third projection is necessary. You should have one SALC for a- and b-symmetry orbitals, two SALCs for e-symmetry orbitals, and three SALCs for t-symmetry orbitals.

(3) When picking a part of the basis to project, try to pick parts that are 90° from the previous projection(s), if possible.

As an example, let us look at the hypothetical molecule PtH_4^{2-} in D_{4h}. The coordinate system for this d^8 square-planar compound is shown in Fig. 6.9. Naturally, the z-axis is the C_4, the highest rotational axis, which places all the hydrogens in the xy-plane.

From the earlier examples and the symmetry of the irreducible representation, the nature of the a_{1g} orbital is probably obvious at this stage, all the s-orbitals on the hydrogens have the same phase and equal contributions. The normalized orbital is as follows.

$$\psi^{a_{1g}} = \frac{1}{2}(s_1 + s_2 + s_3 + s_4)$$

The projection to find the nature of the b_{1g} orbital can be done as shown several times previously.

$$\psi_{s_1}^{b_{1g}} = \frac{1}{2}(s_1 - s_2 + s_3 - s_4)$$

[c] Some instructors may choose to skip the orthogonalization procedure, which does not adversely affect later sections of this book. If so, there are a few things to emphasize with degenerate orbitals and using the projection operator.

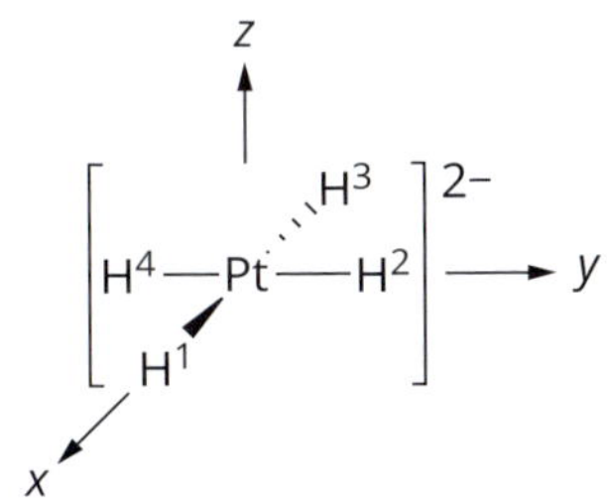

D_{4h}	E	$2C_4$	C_2	$2C_2'$	$2C_2''$	i	$2S_4$	σ_h	$2\sigma_v$	$2\sigma_d$
$\Gamma_\sigma(PtH_4)$	4	0	0	2	0	0	0	4	2	0

$$\Gamma_\sigma = a_{1g} + b_{1g} + e_u$$

Fig. 6.9 Structure of PtH_4^{2-} with the coordinate system used. The basis is essentially the hydrogen positions, an *s*-orbital on each hydrogen. The reducible representation is shown with the decomposition.

Table 6.4a The projection of the E_u irreducible representation on s_1 of the basis for PtH_4^{2-} (Fig. 6.9).

D_{4h}	E	C_4	C_4	C_2	C_2'	C_2'	C_2''	C_2''	i	S_4	S_4	σ_h	σ_v	σ_v	σ_d	σ_d
$\Gamma_{s_1}^{e_u}$	s_1	s_2	s_4	s_3	s_1	s_3	s_2	s_4	s_3	s_2	s_4	s_1	s_1	s_3	s_2	s_4

Next we carry out the projection on the *s*-orbital of H^1 under the E_u irreducible representation in D_{4h}, shown in Table 6.4a.

Tabulating the results and normalizing produces the following orbital.

$$\psi_{s_1}^{e_u} = \frac{1}{\sqrt{2}}(s_1 - s_3)$$

Here we recognize that this is one of the orbitals we need for an e-symmetry orbital. Because the molecular orbitals are of e_u symmetry, we need a second orbital to go with this one. The previous orbital came from projection on s_1 and gave a SALC with only contributions from s_1 and s_3. Let us project on a part of the basis 90° (orthogonal) to this to obtain a new orbital. In other words, we could project with either s_2 or s_4. Using s_2 produces the following result after normalization.

$$\psi_{s_2}^{e_u} = \frac{1}{\sqrt{2}}(s_2 - s_4)$$

If desired, we could show that the these orbitals, parts of the e_u set, are orthogonal in this case without any further manipulation. In this case it is fairly obvious that they are orthogonal, as they do not contain any of the same parts of our basis, and, in the current context, orthogonal is synonymous with having no net overlap of the orbitals.[d]

[d] In our lower-level undergraduate classes, we ask that one recognize that, for example, an *e*-type orbital will produce two orbitals. Furthermore, the projection operator needs to be used successfully on two different parts of the basis. In more advanced classes, we ask that the orbitals be orthogonalized and that they be shown to be orthogonal. The more advanced orthogonalization parts are labeled as "advanced topics" in the sections that follow.

If one wants to learn how to orthogonalize the SALCs constructed and show they are orthogonal, the next section covers how this may be done. There are several procedures one can use to find the orthogonal orbital(s), and here we will discuss perhaps the most general, Gram–Schmidt orthogonalization. Learning to deal with degenerate orbitals should come in two stages: first, recognition that the set is degenerate with generation of the required number of orbitals using the projection operator; and second, orthogonalizing the projected orbitals from the previous step.

6.4.1 Degenerate Orbitals and Gram–Schmidt Orthogonalization (Advanced Topic)

When a molecule contains a rotation axis (proper or improper) that is 3-fold or higher there will be multidimensional irreducible representations in the character table. These multidimensional representations may correspond to degenerate orbitals that transform together under the symmetry of the point group. These orbitals can be thought of as existing as any number of possible linear combinations of one another. However, for most purposes we would prefer to work with orbitals that are orthogonal.

Gram–Schmidt orthogonalization, in the context of molecular orbital theory, is a procedure for extracting an orthogonal SALC, $\psi_\perp^{a_i}$, from a nonorthogonal linear combination, $\psi_1^{a_i}$ and $\psi_2^{a_i}$ found using the projection operator, where a_i is the irreducible representation. If using projection operators, a SALC is generated using ligand orbital L_1 and a SALC $\psi_1^{a_i}$ is obtained, we can obtain a different SALC, $\psi_2^{a_i}$, by using the projection procedure on a different ligand orbital, L_2. *However,* $\psi_1^{a_i}$ *and* $\psi_2^{a_i}$ *are not necessarily orthogonal to one another.* We want to find a different orbital that is orthogonal to $\psi_1^{a_i}$, which will be labeled $\psi_\perp^{a_i}$.

Since $\psi_1^{a_i}$ and $\psi_2^{a_i}$ are not orthogonal, the orbital orthogonal to $\psi_1^{a_i}$, $\psi_\perp^{a_i}$, must be some linear combination of $\psi_1^{a_i}$ and $\psi_2^{a_i}$. Because the orthogonal SALC will be a linear combination of $\psi_1^{a_i}$ and $\psi_2^{a_i}$, Eq. 6.2 should apply, where, again, $\psi_1^{a_i}$ and $\psi_2^{a_i}$ are the nonorthogonal orbitals found using the projection operator and c is a coefficient. In short, as will be shown, Eq. 6.2 simply subtracts the portion of $\psi_1^{a_i}$ that is in $\psi_2^{a_i}$ to make an orthogonal orbital $\psi_\perp^{a_i}$, where c is the overlap between the two projections.

$$\psi_\perp^{a_i} = \psi_2^{a_i} - c\psi_1^{a_i} \qquad \textbf{6.2}$$

One quick note of review: our discussion here will be simplified by using bra-ket notation (see Appendix I on imaginary numbers). This is just a shorthand way of saying an integral with the complex conjugate (denoted with an asterisk "*") in the left-hand (bra) side, with the function on the right-hand (ket) side. In the following integral, a function f is multiplied by its complex conjugate, f^*. For a function of form $f = a + bi$, where i is the imaginary number, $\sqrt{-1}$, the complex conjugate is simply the same function with the imaginary part of the equation having the opposite sign, $f^* = a - bi$. Again, the left side, $\langle f \,|$, is always the complex conjugate and the asterisk is dropped.

$$\int f^* f \, d\tau = \langle f \mid f \rangle$$

Bra-ket notation will shorten the following discussion, where in many cases our functions on the left and right will be assumed to be real ($bi = 0$ in $f = a + bi$), making them even simpler.

The coefficient c in Gram–Schmidt orthogonalization can be found by left-multiplying both sides by $\psi_1^{a_i}$. Since $\psi_1^{a_i}$ and $\psi_\perp^{a_i}$ are orthogonal (by definition, *vide supra*), $\langle \psi_1^{a_i} \mid \psi_\perp^{a_i} \rangle = 0$. This provides the following equations:

$$\begin{aligned}
\left\langle \psi_1^{a_i} \mid \psi_\perp^{a_i} \right\rangle &= \left\langle \psi_1^{a_i} \mid \psi_2^{a_i} - c\psi_1^{a_i} \right\rangle \\
0 &= \left\langle \psi_1^{a_i} \mid \psi_2^{a_i} - c\psi_1^{a_i} \right\rangle \\
0 &= \left\langle \psi_1^{a_i} \mid \psi_2^{a_i} - c\psi_1^{a_i} \mid \psi_1^{a_i} \right\rangle \\
c &= \frac{\left\langle \psi_1^{a_i} \mid \psi_2^{a_i} \right\rangle}{\left\langle \psi_1^{a_i} \mid \psi_1^{a_i} \right\rangle}
\end{aligned}$$

The overlap of a normalized function with itself (as in the previous denominator) is equal to unity, $\langle \psi_1^{a_i} \mid \psi_1^{a_i} \rangle = 1$, and the constant c is simply the overlap between the two nonorthogonal orbitals (Eq. 6.3).

$$c = \langle \psi_1^{a_i} \mid \psi_2^{a_i} \rangle \qquad \textbf{6.3}$$

Once the coefficient c is found, it can be used in Eq. 6.2 to find the SALC orthogonal to orbital $\psi_1^{a_i}$.

6.4.2 Example Using the BH_3 Monomer

The BH_3 monomer is a D_{3h} system with the z-axis coincident with the C_3 rotational axis. The labeling scheme and coordinate axis used are shown in Fig. 6.10a.

6.4.2.1 SALCs of the BH_3 Monomer

The reducible representation for the hydrogen *1s*-orbitals can be found by applying the symmetry operations of the point group to s_1 and assigning it "+1" if it maps onto itself and "0" if it maps onto a different orbital, i.e., s_2 or s_3. Application of the decomposition formula gives that the SALCs have the symmetries $a_1' + e'$.

Finding the nature of the a_1' orbital is trivial using projection on s_1. Following normalization, the SALC is found to be as shown in Fig. 6.10b.

$$\psi_{s_1}^{a_1'} = \frac{1}{\sqrt{3}}(s_1 + s_2 + s_3)$$

Application of the projection operator on s_1 to find the nature of the e SALC produces the following wavefunction after normalization.

$$P_{s_1}^{e'} = 2s_1 - s_2 - s_3 + 2s_1 - s_2 - s_3$$

$$\psi_{s_1}^{e'} = \frac{1}{\sqrt{6}}(2s_1 - s_2 - s_3)$$

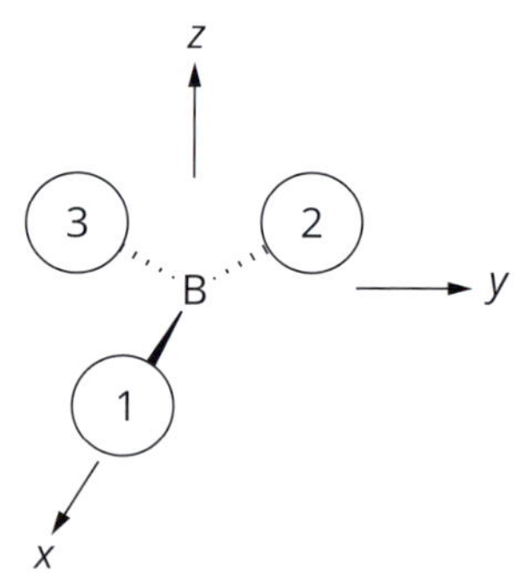

D_{3h}	E	$2C_3$	$3C_2$	σ_h	$2S_3$	$3\sigma_v$
$\Gamma_s(BH_3)$	3	0	1	3	0	1

$\Gamma_{L\sigma} = a_1' + e'$

Fig. 6.10a Basis for BH_3 with the coordinate axes used, reducible representation, and the result of the decomposition.

$$\Psi_{s_1}^{a_1'} = \frac{1}{\sqrt{3}}(s_1 + s_2 + s_3)$$

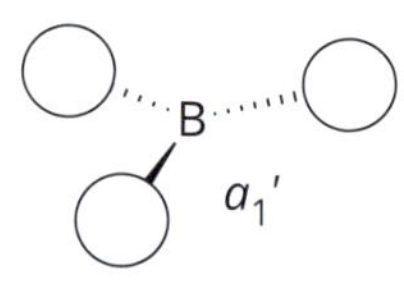

Fig. 6.10b The mathematical form of the a_1' SALC for the hydrogens in BH_3 and a sketch of the delocalized orbital.

Application of the projection operator to a different orbital, s_2, gives:

$$\psi^{e'}_{s_2} = \frac{1}{\sqrt{6}}(2s_2 - s_1 - s_3)$$

Note that the two $\psi^{e'}$ functions are a respectable answer for the question, "What is the nature of the hydrogen SALCs in BH_3?", if one does not want to orthogonalize orbitals. If one wants to orthogonalize these SALCs, the next section shows how this may be done.

6.4.2.2 Orthogonalization of the *e*-Set in the BH_3 Monomer (Advanced Topic)

We can see readily that $\psi^{e'}_{s_1}$ and $\psi^{e'}_{s_2}$ are not orthogonal. If they were, then $\langle\psi^{e'}_{s_1} \mid \psi^{e'}_{s_2}\rangle = 0$, which it does not, as shown in the following. Note that in the equations the cross terms are zero, such as $\langle s_1 \mid s_2\rangle = 0$, and the diagonal terms are unity, such as $\langle s_1 \mid s_1\rangle = 1$.

$$\left\langle\psi^{e'}_{s_1}\middle|\psi^{e'}_{s_2}\right\rangle = \left\langle\frac{1}{\sqrt{6}}(2s_1 - s_2 - s_3)\middle|\frac{1}{\sqrt{6}}(2s_2 - s_1 - s_3)\right\rangle$$
$$= \frac{1}{6}\Big[4\langle s_1|s_2\rangle - 2\langle s_1|s_1\rangle - 2\langle s_1|s_3\rangle - 2\langle s_2|s_2\rangle + \langle s_2|s_1\rangle + \langle s_2|s_3\rangle - 2\langle s_3|s_2\rangle + \langle s_3|s_1\rangle + \langle s_3|s_3\rangle\Big]$$
$$= \frac{1}{6}[4\cdot 0 - 2\cdot 1 - 2\cdot 0 - 2\cdot 1 + 0 + 0 + 0 - 2\cdot 0 + 0 + 1]$$
$$= \frac{1}{6}[-2 - 2 + 1]$$
$$= -\frac{3}{6}$$
$$= -\frac{1}{2}$$

The orbital $\psi^{e'}_{s_2}$ is a linear combination of some orthogonal orbital, $\psi^{e'}_{\perp}$, and $\psi^{e'}_{s_1}$. In other words:

$$\psi^{e'}_{\perp} = \psi^{e'}_{s_2} - c\psi^{e'}_{s_1}$$

The coefficient c is then found from the overlap, $\langle\psi^{e'}_{s_1} \mid \psi^{e'}_{s_2}\rangle$, which we have just solved, and is equal to −½. Placing the values for the variables into the previous equation and solving can be achieved as follows. For this, we will ignore the identical normalization constants on $\psi^{e'}_{s_1}$ and $\psi^{e'}_{s_2}$ then renormalize the orthogonal wavefunction, $\psi^{e'}_{\perp}$, at the end.[e]

[e] For solution of the Gram–Schmidt equation, the form of the projections is typically the same because you are getting them from projection on different parts of the same basis. As a result, the normalization constant added to the projections to make the SALCs are equal. If the coefficients are equal, just as in the previous cases where we had a common factor with which we could divide all the terms (Section 6.2.1.3), it does not affect the final normalized result. If the coefficients are different for some reason, the normalization constants would need to be included.

$$\psi^{e'}_{s_1} = \frac{1}{\sqrt{6}}(2s_1 - s_2 - s_3)$$
$$\psi^{e'}_{\perp} = \frac{1}{\sqrt{2}}(s_2 - s_3)$$

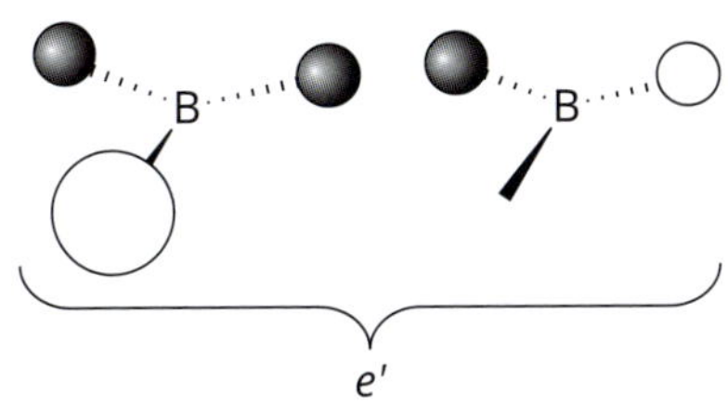

Fig. 6.10c The mathematical form of orthogonal e' SALCs for the hydrogens in BH_3, and a sketch of the delocalized orbitals.

$$\psi^{e'}_{\perp} = \psi^{e'}_{s_2} - c\psi^{e'}_{s_1}$$
$$\text{and } c = -\frac{1}{2} \text{ so}$$
$$= \psi^{e'}_{s_2} + \frac{1}{2}\psi^{e'}_{s_1}$$
$$= 2s_2 - s_1 - s_3 + \frac{1}{2}(2s_1 - s_2 - s_3)$$
$$= 2s_2 - s_1 - s_3 + s_1 - \frac{1}{2}s_2 - \frac{1}{2}s_3$$
$$= 0s_1 + \frac{3}{2}s_2 - \frac{3}{2}s_3$$
...after normalization
$$\psi^{e'}_{\perp} = \frac{1}{\sqrt{2}}(s_2 - s_3)$$

The normalized SALCs that comprise the e' degenerate set are determined and are just labeled $\psi^{e'}_{s_1}$ and $\psi^{e'}_{s_2}$ in Fig. 6.10c. They can be expressed pictorially with white spheres for positive phase and shaded spheres for negative phase.

$$\psi^{e'}_{s_1} = \frac{1}{\sqrt{6}}(2s_1 - s_2 - s_3)$$
$$\psi^{e'}_{\perp} = \frac{1}{\sqrt{2}}(s_2 - s_3)$$

The two orbitals of the e' set should be orthogonal and $\langle\psi^1_{e'} \mid \psi^2_{e'}\rangle = 0$

$$\langle\psi^{e'}_{s_1} \mid \psi^{e'}_{\perp}\rangle = \left\langle\frac{1}{\sqrt{6}}(2s_1 - s_2 - s_3)\middle|\frac{1}{\sqrt{2}}(s_2 - s_3)\right\rangle$$
$$= 0 - \frac{1}{\sqrt{12}} + \frac{1}{\sqrt{12}}$$
$$= 0$$

6.4.3 The SALCs for the Ligands in $\{Co(NH_3)_6\}^{3+}$

For an octahedral complex with all amine donors, we can assume that the ligands are exclusively σ-donors. The compound may be modeled as shown in Fig. 6.11a, with the z-axis positions occupied by ligands 1 and 2.

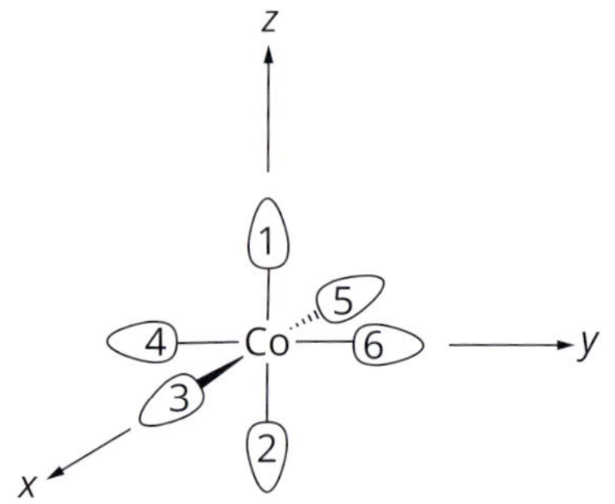

Fig. 6.11a Basis for determining the SALCs of the ammonia lone pairs in $Co(NH_3)_6^{3+}$.

6.4.3.1 The SALCs for the Ligands in $\{Co(NH_3)_6\}^{3+}$

The complex is in the O_h point group if we exclude hydrogens on the amines from consideration, which are electronically isolated from the central Co atom. Note that for the σ-bonds there is no phase change with any symmetry operation; consequently, the orbital being used in the projection such as orbital 1, can only go into itself (and be given a +1) or into another orbital (and be given a 0). The reducible representation is found by using the usual procedure (Table 6.4b), and application of the decomposition formula gives the irreducible representations for the SALCs, $a_{1g} + t_{1u} + e_g$.

$$\Gamma_{L\sigma}(CoL_6) = a_{1g} + t_{1u} + e_g$$

For the application of the projection operator in this case, the pure rotational point group, *O*, can be used. This will save twenty-four operations, speed up the procedure, and still provide all the information on the nature of the SALCs. This is not generally true, but it works here and will expedite the discussion. The gerade labels will be dropped when talking about the *O* representations but will be added back when referring to the SALCs in O_h.

To find the nature of the a_{1g} SALC, one simply needs to apply the projection operator to one of the previous ligand orbitals, say orbital 1 on the *z*-axis (Table 6.4c).

$$\psi^{a_{1g}} = \frac{1}{\sqrt{6}}(\sigma_1 + \sigma_2 + \sigma_3 + \sigma_4 + \sigma_5 + \sigma_6)$$

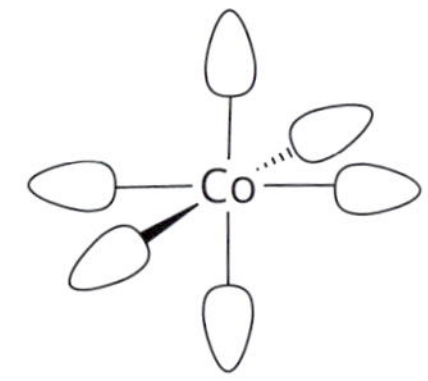

Fig. 6.11b The mathematical form of the a_{1g} SALC for the ammonia lone pairs in $Co(NH_3)_6^{3+}$, and a sketch of the delocalized orbital.

The σ_1-orbital maps onto the other orbitals 1–6 four times each in the sequence. Therefore, the projection of $L_{\sigma 1}$ in the a_1 irreducible representation is as follows.

$$P_{L\sigma 1}^{a_1} = 4(\sigma_1 + \sigma_2 + \sigma_3 + \sigma_4 + \sigma_5 + \sigma_6)$$

Normalization produces the a_{1g} SALC (Fig. 6.11b).

$$\psi^{a_{1g}} = \frac{1}{\sqrt{6}}(\sigma_1 + \sigma_2 + \sigma_3 + \sigma_4 + \sigma_5 + \sigma_6)$$

Applying the projection operator to the *E* irreducible representation on σ_1 produces the following SALC. Note that once the a_{1g} projection is known, the other orbitals can be found by multiplying the characters in the next irreducible representation to be examined, for example e_g, by the orbital labels found during the a_{1g} projection. This gives the following expression for $\psi_{\sigma_1}^{e_g}$.

$$\psi_{\sigma_1}^{e_g} = \frac{1}{2\sqrt{3}}(2\sigma_1 + 2\sigma_2 - \sigma_3 - \sigma_4 - \sigma_5 - \sigma_6)$$

This e_g SALC should have an orthogonal orbital that completes the set. If we pick a ligand orbital orthogonal to ligand 1, like ligand 3, and project, we obtain the wavefunction.

$$\psi_{\sigma_3}^{e_g} = \frac{1}{2\sqrt{3}}(2\sigma_3 + 2\sigma_5 - \sigma_1 - \sigma_2 - \sigma_4 - \sigma_6)$$

Table 6.4b Reducible representation for the ammonia lone pair SALCs in O_h.

O_h	E	$8C_3$	$6C_2$	$6C_4$	$3C_2(=C_4^2)$	i	$6S_4$	$8S_6$	$3\sigma_h$	$6\sigma_d$
$\Gamma_{L\sigma}(CoL_6)$	6	0	0	2	2	0	0	0	4	2

Table 6.4c Projection of σ_1 from Fig. 6.11a onto the a_1 representation in *O*.

O	E	C_3	C_3	C_3	C_3	C_3	C_3	C_3	C_3	C_2	C_2	C_2	C_4	C_4	C_4	C_4	C_4	C_4	C_2	C_2	C_2	C_2	C_2	C_2
$L_{\sigma 1}^{a_1}$	σ_1	σ_3	σ_4	σ_4	σ_5	σ_6	σ_5	σ_6	σ_3	σ_1	σ_2	σ_2	σ_1	σ_1	σ_5	σ_3	σ_4	σ_6	σ_2	σ_2	σ_6	σ_5	σ_4	σ_3

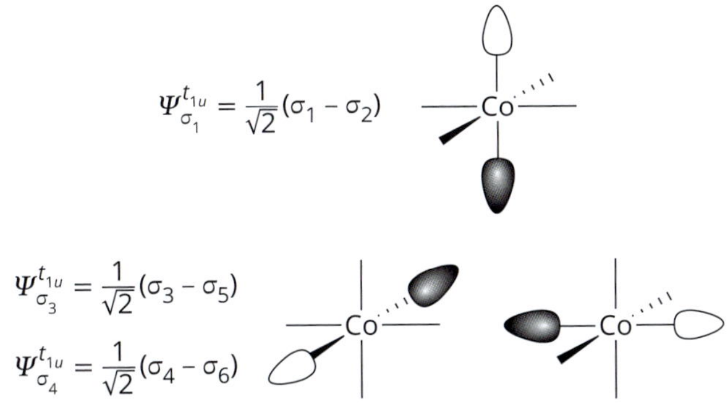

Fig. 6.11c The mathematical form of the t_{1u} SALCs for the ammonia lone pairs in $Co(NH_3)_6^{3+}$ and sketches of the delocalized orbitals.

It turns out that $\psi^{e_g}_{\sigma_1}$ and $\psi^{e_g}_{\sigma_2}$ are not orthogonal. In the next section we will prove they are not orthogonal, and then orthogonalize them. For the moment, let us look at the SALCs for the ligands of t_{1u} symmetry.

The remaining SALCs in the σ-framework are of t_{1u} symmetry. Applying the projection operator on ligand orbital 1 in the t_1 representation of *O*, produces the normalized SALC shown in Fig. 6.11c.

$$\psi^{t_{1u}}_{\sigma_1} = \frac{1}{\sqrt{2}}(\sigma_1 - \sigma_2)$$

$$\psi^{t_{1u}}_{\sigma_3} = \frac{1}{\sqrt{2}}(\sigma_3 - \sigma_5)$$

$$\psi^{t_{1u}}_{\sigma_4} = \frac{1}{\sqrt{2}}(\sigma_4 - \sigma_6)$$

This first function is along the *z*-axis. Considering we are looking for orthogonal orbitals, it seems wise to do projections next along the *x*- and *y*-axes, orthogonal to the first projection. These provide the functions in Fig. 6.11c. These SALCs can be readily reasoned to be orthogonal as they do not contain any overlapping parts of our basis, and, as can be seen by examination of the O_h character table, are of proper symmetry to overlap with the *p*-orbitals on the central metal.

6.4.3.2 Orthogonalization of the e_g SALCs in $\{Co(NH_3)_6\}^{3+}$ (advanced topic)

If $\psi^{e_g}_{\sigma_1}$ and $\psi^{e_g}_{\sigma_2}$ were orthogonal, then the normalized orbitals should fulfill $\langle \psi^{e_g}_{\sigma 1} \mid \psi^{e_g}_{\sigma 3} \rangle = 0$. Consequently, the check for orthogonalization can be achieved as follows. cross terms, such as $\langle 2\sigma_1 \mid 2\sigma_3 \rangle$, which are equal to zero, are not shown.

$$\begin{aligned}
\left\langle \psi^{e_g}_{\sigma_1} \mid \psi^{e_g}_{\sigma_3} \right\rangle &= \frac{1}{12}\big\langle (2\sigma_1 + 2\sigma_2 - \sigma_3 - \sigma_4 - \sigma_5 - \sigma_6) \mid \\
&\qquad (2\sigma_3 + 2\sigma_5 - \sigma_1 - \sigma_2 - \sigma_4 - \sigma_6) \big\rangle \\
&= \frac{1}{12}\left[\begin{array}{l} -2\langle \sigma_1 \mid \sigma_1 \rangle - 2\langle \sigma_2 \mid \sigma_2 \rangle - 2\langle \sigma_3 \mid \sigma_3 \rangle \\ +\langle \sigma_4 \mid \sigma_4 \rangle - 2\langle \sigma_5 \mid \sigma_5 \rangle + \langle \sigma_6 \mid \sigma_6 \rangle \end{array} \right] \\
&= \frac{1}{12}(-2 - 2 - 2 + 1 - 2 + 1) \\
&= -\frac{6}{12} \\
&= -\frac{1}{2}
\end{aligned}$$

Consequently, $\langle \psi^{e_g}_{\sigma_1} \mid \psi^{e_g}_{\sigma_3} \rangle \neq 0$, and the two orbitals are not orthogonal.[f]

For the orthogonalization, we can use the previous two SALCs from the projection operator applications to σ_1 and σ_3 to find the orthogonal orbital.

$$\psi^{e_g}_{\perp} = \psi^{e_g}_{\sigma_3} - c\psi^{e_g}_{\sigma_1}$$

The coefficient *c* is then the overlap between $\psi^{e_g}_{\sigma_1}$ and $\psi^{e_g}_{\sigma_3}$ calculated previously as −½. The orthogonal orbital is found from the previous equation.

$$\begin{aligned}
\psi^{e_g}_{\perp} &= \psi^{e_g}_{\sigma_3} - c\psi^{e_g}_{\sigma_1} \\
&= \frac{1}{2\sqrt{3}}(2\sigma_3 + 2\sigma_5 - \sigma_1 - \sigma_2 - \sigma_4 - \sigma_6) \\
&\quad + \frac{1}{2}\frac{1}{2\sqrt{3}}(2\sigma_1 + 2\sigma_2 - \sigma_3 - \sigma_4 - \sigma_5 - \sigma_6) \\
&= \frac{1}{\sqrt{3}}\sigma_3 + \frac{1}{\sqrt{3}}\sigma_5 - \frac{1}{2\sqrt{3}}\sigma_1 - \frac{1}{2\sqrt{3}}\sigma_2 - \frac{1}{2\sqrt{3}}\sigma_4 \\
&\quad - \frac{1}{2\sqrt{3}}\sigma_6 + \frac{1}{2\sqrt{3}}\sigma_1 + \frac{1}{2\sqrt{3}}\sigma_2 - \frac{1}{4\sqrt{3}}\sigma_3 \\
&\quad - \frac{1}{4\sqrt{3}}\sigma_4 - \frac{1}{4\sqrt{3}}\sigma_5 - \frac{1}{4\sqrt{3}}\sigma_6 \\
&= 0\sigma_1 + 0\sigma_2 + \frac{3}{4\sqrt{3}}\sigma_3 - \frac{3}{4\sqrt{3}}\sigma_4 + \frac{3}{4\sqrt{3}}\sigma_5 - \frac{3}{4\sqrt{3}}\sigma_6
\end{aligned}$$

... after renormalization

$$\psi^{e_g}_{\perp} = \frac{1}{2}(\sigma_3 - \sigma_4 + \sigma_5 - \sigma_6)$$

These normalized orbitals should be orthogonal, but we can check, as they should fulfill $\langle \psi^{e_g}_{\sigma 1} \mid \psi^{e_g}_{\perp} \rangle = 0$.

[f] Note that the overlap integral in the D_{3h} example for BH_3 and for the O_h example here are both −½. This is a coincidence. The overlap integral from projections varies depending on the system.

$$\Psi_{\sigma_1}^{e_g} = \frac{1}{2\sqrt{3}}(2\sigma_1 + 2\sigma_2 - \sigma_3 - \sigma_4 - \sigma_5 - \sigma_6)$$

$$\Psi_{\perp}^{e_g} = \frac{1}{2}(\sigma_3 - \sigma_4 + \sigma_5 - \sigma_6)$$

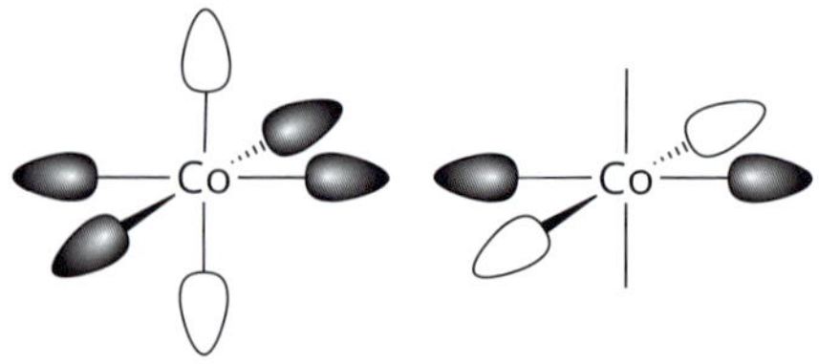

Fig. 6.12 The mathematical form of the orthogonalized e_g SALCs for the ammonia lone pairs in $Co(NH_3)_6^{3+}$, and sketches of the delocalized orbitals.

$$\langle \psi_{\sigma_1}^{e_g} \mid \psi_{\perp}^{e_g} \rangle = \left\langle \frac{1}{2\sqrt{3}}(2\sigma_1 + 2\sigma_2 - \sigma_3 - \sigma_4 - \sigma_5 - \sigma_6) \right| \frac{1}{2}(\sigma_3 - \sigma_4 + \sigma_5 - \sigma_6) \Big\rangle$$

$$= 0 + 0 - \frac{1}{4\sqrt{3}} + \frac{1}{4\sqrt{3}} - \frac{1}{4\sqrt{3}} + \frac{1}{4\sqrt{3}}$$

$$= 0$$

Consequently, the SALCs in the octahedral complex corresponding to the e_g irreducible representation are the normalized functions, $\psi_{\sigma 1}^{e_g}$ and $\psi_{\perp}^{e_g}$, shown in Fig. 6.12. These obviously have the correct symmetry to overlap with d_{z^2} and $d_{x^2-y^2}$ wavefunctions, respectively, on the Co central metal.

$$\psi_{\sigma_1}^{e_g} = \frac{1}{2\sqrt{3}}(2\sigma_1 + 2\sigma_2 - \sigma_3 - \sigma_4 - \sigma_5 - \sigma_6)$$

$$\psi_{\perp}^{e_g} = \frac{1}{2}(\sigma_3 - \sigma_4 + \sigma_5 - \sigma_6)$$

6.5 Constructing the Diagram

By using the reducible representation, decomposition formula, and projection operator one can determine the symmetry labels and even generate pictures of the fragment orbitals for a molecule. All that remains is to use the information to put together the molecular orbital diagram. We will place the SALCs generated from the valence orbitals of the ligands on the left side of the diagram and the valence orbitals of the central atom on the right.

The y-axis, often unlabeled, is energy. *The placement of the SALCs and atomic orbitals of the central atom on the y-axis can be done based on electronegativity of the atoms in question, with valence orbitals on more electronegative atoms lower in energy.* One can also use the ionization potential of the atomic orbitals as an estimate for the energies for placement in an MO diagram. A collection of atomic orbital ionization energies in eV is found in Appendix K.

To illustrate some important principles for the construction of MO diagrams with various numbers of orbitals with specific symmetries, let's discuss three scenarios: one orbital on each side of the diagram (a 2-orbital interaction), 3-orbital interactions, and 5-orbital interactions with two on one side and three on the other. These are certainly not all the possible combinations but were chosen to illustrate some important principles for orbital mixing. Other orbital mixing patterns will be seen in the examples that follow.

6.5.1 One Orbital on Each Side of the Diagram

If there is one orbital on each side of the MO diagram with the same symmetry, then these will combine to form bonding and antibonding combinations. In Fig. 6.13, an orbital of a_1 symmetry on more electronegative ligands, L_n, combines with an orbital of the same symmetry on a less electronegative central atom, M. The MOs of a_1 symmetry are customarily labeled sequentially ($1a_1$, $2a_1$, . . .) from lowest in energy to highest. If the orbital is clearly antibonding, it is often marked with an asterisk, such as, $2a_1$*. The combination closest in energy to the fragment orbital being used typically has more character of that fragment. For example, in Fig. 6.13 the bonding orbital $1a_1$ would have more ligand (L) character, and antibonding orbital $2a_1$* would have more metal (M) character, a common arrangement in metal chemistry.

The strength of the orbital interaction between the two orbitals on L and M determines the energy gap between the antibonding and bonding orbitals, $\Delta_{ab\text{-}b}$ (Fig. 6.13). The strength of the interaction is determined by two independent factors: the overlap and the energy difference between the constituent orbitals.

The overlap is, in essence, determined by how well the orbitals match in size and how well they are directed

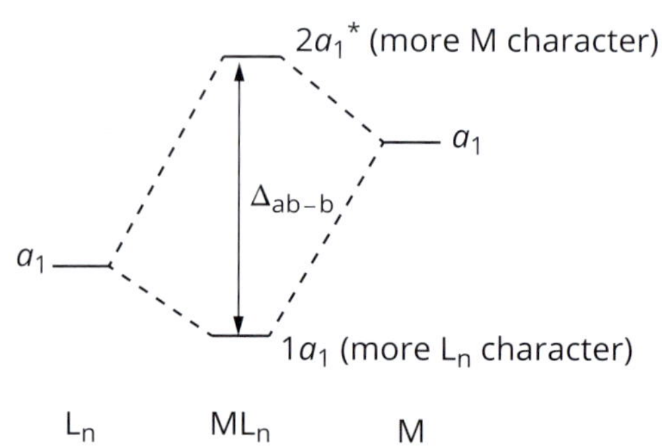

Fig. 6.13 A bonding and antibonding combination where there is one orbital on each side of the diagram with a specific symmetry; for example, a_1. $\Delta_{ab\text{-}b}$ = difference in energy between the antibonding (ab) and bonding (b) orbitals.

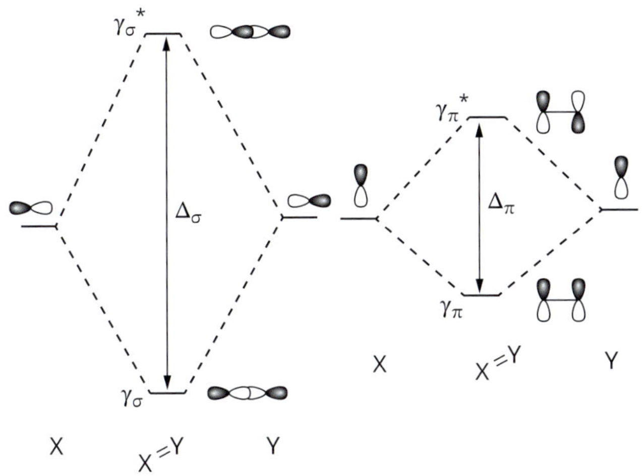

Fig. 6.14 Overlap and the energy gap between bonding orbitals of symmetry γ and antibonding orbitals (γ*). (γ = irreducible representation for the orbitals)

at one another. For example, take a system with a σ- and π-interaction between two fragments X and Y, where the orbitals involved in both types of bonds are of the same energy (Fig. 6.14). This might happen, for example, in a diatomic molecule where p_z along the internuclear axis would be used to make the σ-bond and p_x might be used to make a π-bond. Because the σ-bond orbitals are better directed at one another than the side-on interaction between the *p*-orbitals in the π-bond, Δ_σ is larger than Δ_π.

The effects of energy on the orbital splitting are typically described using perturbation theory (Fig. 6.15). In short, if the overlap is kept constant, the stabilization of the bonding orbital (Δ_b) and destabilization of the antibonding orbital (Δ_{ab}) are inversely proportional to the energy gap Δ_{XY}. The closer the interacting orbitals are in energy, the stronger the interaction at constant overlap. Also note that the stabilization of the bonding and destabilization of the antibonding orbitals relative to their closest contributing orbitals are similar, i.e.,

Fig. 6.15 Relationship between the energy of the interacting orbitals and the stabilization of the bonding orbitals of γ symmetry/destabilization of the antibonding orbital (γ*). (γ = irreducible representation for the orbitals)

$\Delta_b \cong \Delta_{ab}$ (Fig. 6.15). In reality, $\Delta_b = \Delta_{ab}$ only at the limit of zero overlap, and, in general, $\Delta_b < \Delta_{ab}$. However, that $\Delta_b \cong \Delta_{ab}$ is a common approximation; see Appendix M on LCAO for a more detailed discussion of this issue.

6.5.2 3-Orbital Interactions

If there are more than two orbitals involved, then the number of bonding and antibonding orbitals will depend on the number of matching orbitals of the same symmetry on each side and on the nature of the mixing. A 3-orbital interaction will be used to illustrate (Fig. 6.16). If we examine a case where X has two orbitals with the same energy of γ symmetry and Y has a single orbital of the same symmetry, it is obvious that there can only be one traditional bonding interaction between the two species because one side only has one orbital of correct symmetry, but what happens with the third orbital? A reasonable assumption when creating MO diagrams by hand is that the third orbital will be approximately nonbonding (Fig. 6.16A), *2 γ*, and isolated on the fragment on which it originated, X. However, this is an assumption, and the orbital may pick up bonding or antibonding character from either the orbital of the same symmetry above (3 *γ* *) or below (*1 γ*) it. In other words, the orbital of intermediate energy can potentially be anywhere between the bonding and antibonding orbital, mixing nonbonding character with either. If the presumed nonbonding orbital (*2 γ*) mixes with the bonding (*1γ*), then *2γ* will drop in energy as it picks up bonding character (Fig. 6.16B) and *1 γ* will rise in energy as it picks up nonbonding character. Similarly, *3 γ ** will lower in energy and *2 γ* will rise in energy if mixed (Fig. 6.16C).

In addition to possibilities where the two orbitals on X are the same energy, there are many more possible 3-orbital interactions where the orbitals are of the same symmetry but different energies. In Fig. 6.16D there is a stronger interaction between the lower-energy orbital on X (γ^{Xa}) and the orbital on Y (γ^Y). The result is that the bonding orbital (*1γ*) has the character of $\gamma^{Xa} + \gamma^Y$. As a result, the higher-energy orbital on X (γ^{Xb}) is essentially nonbonding, and the antibonding orbital is $\gamma^{Xa} - \gamma^Y$. Alternatively (Fig. 6.16E), there could be a stronger interaction with the higher-energy orbital on X (γ^{Xb}) that results in a bonding MO below the lower-energy orbital where this lower-energy orbital (γ^{Xa}) makes little contribution to the bonding in the system and is essentially nonbonding.

Again, Fig. 6.15 does not contain all the potential 3-orbital interactions but hopefully serves to illustrate some of the many possibilities. In the end, the type of interaction observed depends on the overlap and energy differences between the contributing orbitals. In general, it will be assumed that one of the orbitals on X is approximately nonbonding, generally the one with worse apparent overlap based on the orbital pictures or very different orbital energy, at least until there is additional experimental or computational information gathered to suggest otherwise.

One final principle is illustrated in Fig. 6.16F. In this case, both γ^{Xa} and γ^{Xb} interact with γ^Y to form two bonding

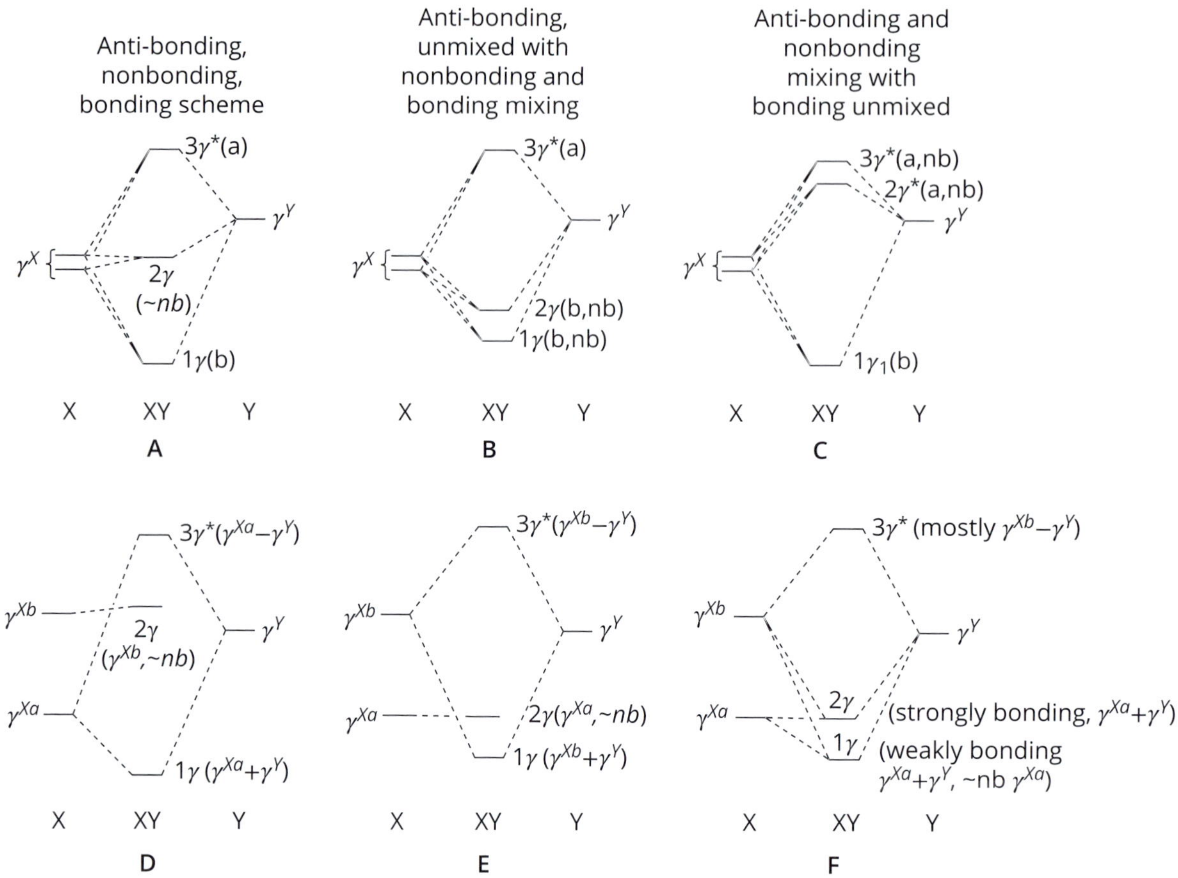

Fig. 6.16 (*top row*) A 3-orbital interaction with two orbitals of the same energy on X and one orbital on Y illustrating the nonbonding assumption (A), mixing with the bonding orbital (B), and mixing with antibonding orbital (C). a = antibonding, b = bonding, nb = nonbonding. (*bottom row*) A 3-orbital interaction with two orbitals of different energy on X and an orbital on Y illustrating a stronger interaction with the lower-energy orbital (D) and a stronger interaction with the higher energy orbital (E). In (F), the lower energy orbital on X is less bonding and mostly nonbonding while the higher-energy orbital on X is more involved in bonding to Y.

interactions (both orbitals are stabilized relative to their starting energies on the left but mixed with nonbonding to produce one bond) and a strongly antibonding orbital. The lower-energy orbital on X, γ^{Xa}, is the major contributor to the lower-energy molecular orbital 1γ; however, this orbital is only weakly stabilized relative to its starting position and is largely nonbonding. On the other hand, γ^{Xb} forms a bonding interaction that is much stronger and is largely responsible for the bonding between X and Y. The starting position and the amount of stabilization are also important features when deciding the amount of bonding being done by an orbital interaction.

6.5.3 The 5-Orbital Interaction in OF_2

If there are five orbitals of the same symmetry—three on fragment X and two on fragment Y—there can be two full bonds with corresponding antibonding orbitals generated when X and Y interact. The additional orbital on X could then be assumed, barring additional information, to be approximately nonbonding. We can return to the OF_2 molecule for an illustration.

The typical Lewis dot structure for OF_2 is in Fig. 6.17a, which we will relate to the MO diagram for OF_2 involving both σ- and π-bonding orbitals shown in Fig. 6.17b. All orbitals of the same symmetry will most likely mix to some degree. The exact energies of the approximately nonbonding orbitals in these multiorbital interactions can be difficult to judge without calculations, and they often have some bonding or antibonding character associated with them due to the mixing.

In the MO diagram for OF_2 (Fig. 6.17b) there are three orbitals of a_1 symmetry (red) in the fluorine SALCs, and two of the same symmetry listed among the atomic orbitals of oxygen, s and p_z. All five of these orbitals will possibly mix. We will estimate that two of the orbitals will be bonding ($1a_1$ and $2a_1$), two will be antibonding ($4a_1$* and $5a_1$*), and one will be approximately nonbonding ($3a_1$). The energy of the

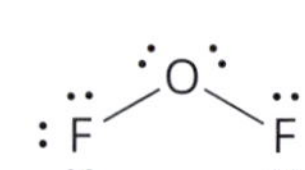

Fig. 6.17a The Lewis dot structure of OF_2 showing all the lone pairs.

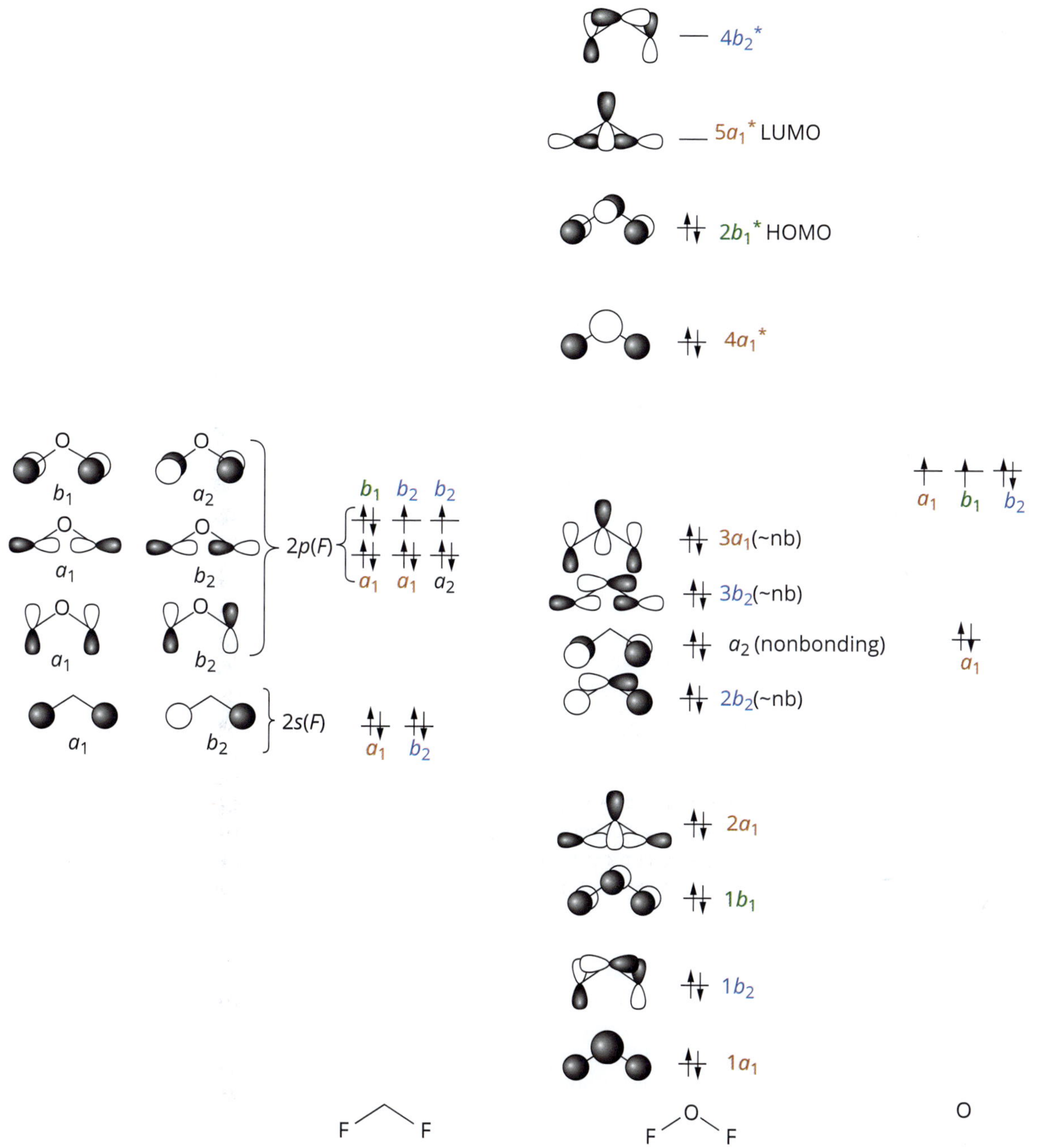

Fig. 6.17b Approximate MO diagram for OF_2. Orbitals in brackets have approximately the same energy, even though they may be drawn at different levels on the *y*-axis.

nonbonding orbital we can estimate, when working by hand, to be similar to the fragment orbitals generating it, and in this case something close in energy to the *p*-orbitals on fluorine is reasonable. Antibonding orbitals should be clearly higher in energy than the fragment orbitals of which they are comprised. Bonding orbitals should be clearly lower in energy than their constituent SALCs and atomic orbitals.

Electrons from the left and right fragments are summed and filled in bottom to top in the center MO diagram. There are two fluorines with a total of fourteen valence electrons and an oxygen with six valence electrons. Consequently, our MO diagram will have twenty electrons in total. This fills the MOs up to and including $2b_1{}^*$, which we assign as the HOMO of the molecule.

From the Lewis diagram for OF_2 (Fig. 6.17a), we expect that there would be two covalent bonds between oxygen and the fluorines. In addition, there should be a total of eight lone pairs—three on each fluorine and two on the oxygen. Where are these "lone pairs" and covalent bonds in the MO diagram?

There are four pairs of approximately nonbonding or precisely nonbonding orbitals in the MO diagram: $3a_1$, $3b_2$, a_2, and $2b_2$. None of these are located on a specific atom. Even

a_2, which is rigorously nonbonding with respect to oxygen, is spread out over the two fluorines. However, these may in some respects be considered "lone pairs", as they have, to a first approximation, no net role in bonding. Similarly, orbitals that have both their bonding and antibonding counterparts filled have no net role in bonding. Consequently, $1a_1/4a_1^*$ and $1b_1/2b_1^*$ may all be considered "lone pairs" in this sense. In the MO description, the lone pairs, to the extent that one may consider them such, are delocalized over the molecule involved in bonding, antibonding, and nonbonding interactions.

The type of interaction is usually classified by the number of parallel nodes between the central element and the individual bonding atoms, i.e., σ-, π-, δ-bonding (Section 3.7). If we examine the four bonding interactions in OF_2 in Fig. 6.17b ($1a_1$, $1b_2$, $1b_1$, and $2a_1$), $1a_1$, $1b_2$, and $2a_1$ have no nodes parallel to the O–F bond axes and are therefore σ-bonding. On the other hand, $1b_1$ does have a node parallel to the bond axis and is a π-bond. As required, the same number of parallel nodes along the O–F bond axes exist in the antibonding interactions as well, perpendicular nodes are added due to the phase changes, however. The filled antibonding interactions correspond to one σ-bond and one π-bond, while there are three σ-bonding and one π-bonding interaction. As a result, there are two net σ-bonds in the compound, corresponding to the two single lines between the oxygen and fluorines in the Lewis diagram (Fig. 6.17a).

This illustrates an important feature of MO theory in general: MO theory stresses delocalization of electrons. In contrast, valence bond theory deals with localized structures that are delocalized through resonance. As a result, MO and VB theory often provide different insights into electronic structure, which will be emphasized in the next chapter.

6.6 Some Additional Applications to Inorganic Systems

As illustrations of the previous procedures, a few additional examples will be worked out in some detail. The first is the bonding in BF_3 to illustrate σ- and π-bonding in compounds of this type. Second, the molecular orbital diagram for the hypothetical O_h compound CrH_6 will be developed. The MO diagrams for octahedral compounds will be discussed under several different circumstances in this book, and this example will provide some background to those discussions. Third, another example involving a d-block complex, PtH_4^{2-}, will be illustrated. Finally, the MO diagram for the hypervalent D_{3h} compound PH_5 will be constructed to demonstrate some features of the bonding in this class. Main-group bonding will be developed in detail in Chapter 8, using both MO and VB models.

6.6.1 σ- and π-Bonding in BF_3

The SALCs involving the s-orbitals in D_{3h} were shown previously in the example on BH_3 (Section 6.4.2). The axes are defined so that the C_3 is along z, and one of the fluorines, F^1, is on the x-axis. The SALCs left to generate are the ones related to the fluorine p-orbitals. For our MO discussions here, the orthogonal orbitals will be used. These can be simply used without derivation, or, as an advanced topic, their generation is shown.

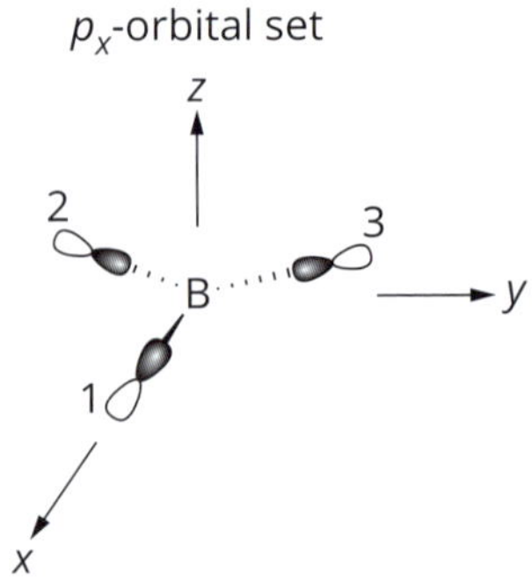

Fig. 6.18a The p_x-orbital basis for the fluorines in BF_3. The basis should be symmetric with respect to the highest-order rotation axis.

6.6.1.1 Generation of the SALCs for the Fluorines in BF_3

It is easiest in this case if we define the p-orbitals relative to the bonding axis (Fig. 6.18a). One of the p-orbitals on each fluorine atom will point towards the boron; let's call this one p_x.

As far as the boron is concerned (Table 6.5a), the p_x-orbital set looks very similar to the s-orbital set in BH_3. No symmetry operation changes the phase of the p-orbitals in question, and the same phase is always directed toward boron. As a result, the reducible representation and its constituents, a_1' and e', are the same as found for s.

$$\Gamma_{p_x} = a_1' + e'$$

Application of the projection operator to discover the nature of the orbitals results in the forms shown as follows for a_1' and e' due to the p_x-orbitals. For the purposes of diagramming, we have orthogonalized the orbitals already and will be used as shown in Fig 6.18b.

$$\psi^{a_1'}_{p_{x1}} = \frac{1}{\sqrt{3}}(p_{x1} + p_{x2} + p_{x3})$$

$$\psi^{e'}_{p_{x1}} = \frac{1}{\sqrt{6}}(2p_{x1} - p_{x2} - p_{x3})$$

$$\psi^{e'}_{p_{x2}} = \frac{1}{\sqrt{2}}(p_{x2} - p_{x3})$$

Table 6.5a The reducible representation for the p_x-basis (Fig. 6.18a) in D_{3h}.

D_{3h}	E	$2C_3$	$3C_2$	σ_h	$2S_3$	$3\sigma_v$
$\Gamma_{p_x}(BF_3)$	3	0	1	3	0	1

$$\Psi^{a_1'}_{p_{x1}} = \frac{1}{\sqrt{3}}(p_{x1} + p_{x2} + p_{x3})$$

$$\Psi^{e'}_{p_{x1}} = \frac{1}{\sqrt{6}}(2p_{x1} - p_{x2} - p_{x3})$$

$$\Psi^{e'}_{p_{x2}} = \frac{1}{\sqrt{2}}(p_{x2} - p_{x3})$$

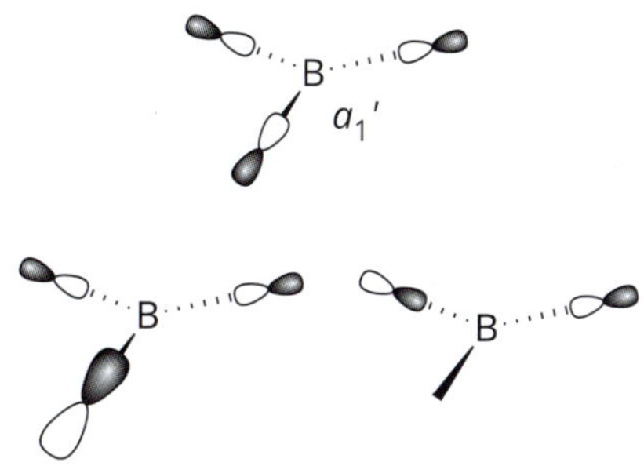

Fig. 6.18b The mathematical form of the normalized fluorine SALCs for the p_x-orbitals on the fluorines in BF_3, and pictorial depictions of the a_1' and e' orbitals.

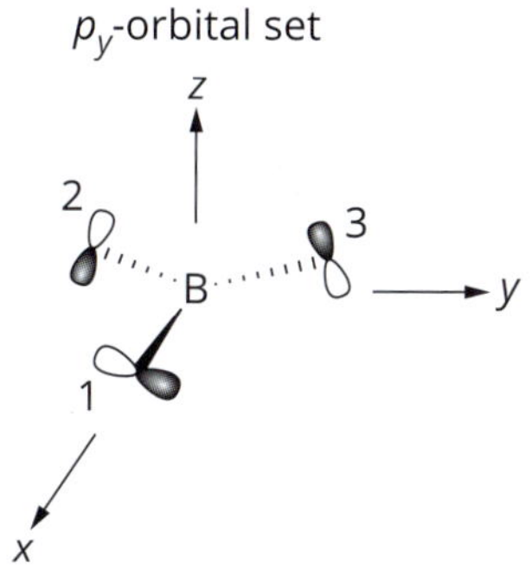

Fig. 6.18c The p_y-orbital basis for the fluorines in BF_3. The basis should be symmetric with respect to the highest-order rotation axis.

We can define the p_y-orbital set as being 90° from the B–F axes in the xy-plane as shown in Fig. 6.18c. The phases are designated so that there is no phase change with respect to the principal rotation axis. The reducible representation (Table 6.5b) and its decomposition result in orbitals of a_2' and e'.

$$\Gamma_{p_y} = a_2' + e'$$

What does the projection of the a_2' orbital look like? To find out, let's do the projection on orbital 1 in Figure 6.5b for the p_y-basis. The result is an orbital with equal and positive contributions from all three fluorines. In other words, the orbital looks much as it does in the figure for our set of p-orbitals. The wavefunctions for the e' orbital will look much as they have in the functions for the s- and p_x-orbitals but applied to the p_y-orbital set. Keep in mind that a positive phase means that the phase does not change with respect to our basis. The following e'-set has already been orthogonalized. The SALCs are collected into Fig. 6.18d.

Table 6.5b The reducible representation for the p_y-basis (Fig. 6.18c) in D_{3h}.

D_{3h}	E	$2C_3$	$3C_2$	σ_h	$2S_3$	$3\sigma_v$
$\Gamma_{py}(BF_3)$	3	0	–1	3	0	–1

$$\psi^{a_2'}_{p_{y1}} = \frac{1}{\sqrt{3}}(p_{y1} + p_{y2} + p_{y3})$$

$$\psi^{e'}_{p_{y1}} = \frac{1}{\sqrt{6}}(2p_{y1} - p_{y2} - p_{y3})$$

$$\psi^{e'}_{p_{y2}} = \frac{1}{\sqrt{2}}(p_{y2} - p_{y3})$$

For the p_z-orbital set on the fluorines, all the orbitals are aligned parallel with the z-axis (Table 6.5c, Fi.g 6.18f).

$$\Psi^{a_2'}_{p_{y1}} = \frac{1}{\sqrt{3}}(p_{y1} + p_{y2} + p_{y3})$$

$$\Psi^{e'}_{p_{y1}} = \frac{1}{\sqrt{6}}(2p_{y1} - p_{y2} - p_{y3})$$

$$\Psi^{e'}_{p_{y2}} = \frac{1}{\sqrt{2}}(p_{y2} - p_{y3})$$

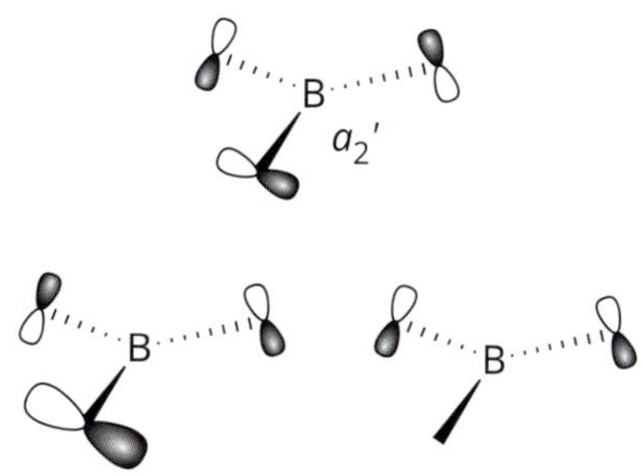

Fig. 6.18d The mathematical form of the normalized fluorine SALCs for the p_y-orbitals on the fluorines in BF_3, and pictorial depictions of the a_2' and e' orbitals.

Table 6.5c The reducible representation for the p_z-basis (Fig. 6.18e) in D_{3h}.

D_{3h}	E	$2C_3$	$3C_2$	σ_h	$2S_3$	$3\sigma_v$
$\Gamma_{pz}(BF_3)$	3	0	−1	−3	0	1

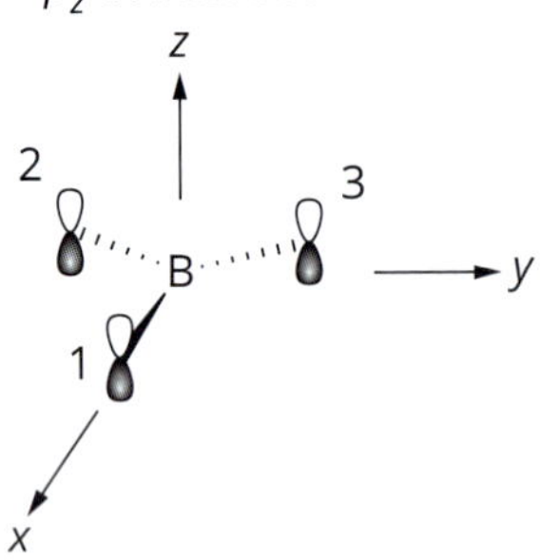

Fig. 6.18e The p_z-orbital basis for the fluorines in BF_3. The basis should be symmetric with respect to the highest-order rotation axis.

$$\Psi^{a_2''}_{p_{z1}} = \frac{1}{\sqrt{3}}(p_{z1} + p_{z2} + p_{z3})$$

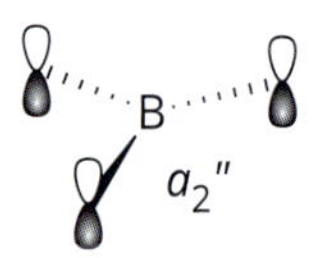

Fig. 6.18f The mathematical form of one of the normalized fluorine SALCs for the p_z-orbitals on the fluorines in BF_3, and a pictorial depiction of the a_2'' orbital.

$$\Gamma_{p_z} = a_2'' + e''$$

$$\psi^{a_2''}_{p_{z1}} = \frac{1}{\sqrt{3}}(p_{z1} + p_{z2} + p_{z3})$$

The projection on the p_z-orbitals of the fluorines under the E'' irreducible representation gives for p_{z1} and p_{z2} the following two functions, which are not orthogonal.

$$\psi^{e''}_{p_{z1}} = \frac{1}{\sqrt{6}}(2p_{z1} - p_{z2} - p_{z3})$$
$$\psi^{e''}_{p_{z2}} = \frac{1}{\sqrt{6}}(2p_{z2} - p_{z1} - p_{z3})$$

6.6.1.2 Orthogonalization of the e'' Orbitals from the p_z Orbitals on Fluorine (Advanced Topic)

As another example of the orthogonalization process, we will apply the Gram–Schmidt equation to the e'' set of the previous p_z-orbitals. We use Eq. 6.3 to find an orbital orthogonal to $\psi^{e''}_{pz1}$.

$$\psi^{e''}_{\perp} = \psi^{e''}_{pz2} - c\psi^{e''}_{pz1}$$

The coefficient c is found as follows from the $\psi^{e''}_{pz1}$ and $\psi^{e''}_{pz2}$ overlap.

$$\begin{aligned}\langle \psi^{e''}_{pz1} \mid \psi^{e''}_{pz2} \rangle &= \frac{1}{6}\langle 2p_{z1} - p_{z2} - p_{z3} \mid 2p_{z2} - p_{z1} - p_{z3} \rangle \\ &= \frac{1}{6}\left[-2\langle p_{z1} \mid p_{z1} \rangle - 2\langle p_{z2} \mid p_{z2} \rangle + 1\langle p_{z3} \mid p_{z3} \rangle\right] \\ &= \frac{1}{6}\left[-2 - 2 + 1\right] \\ &= -\frac{1}{2}\end{aligned}$$

Inserting the coefficient into the equation and solving is accomplished as follows. Not unexpectedly, the function looks very similar to the orthogonal functions for other e-sets that we have generated previously.

$$\begin{aligned}\psi^{e''}_{\perp} &= \psi^{e''}_{pz2} - c\psi^{e''}_{pz1} \\ &= \psi^{e''}_{pz2} + \frac{1}{2}\psi^{e''}_{pz1} \\ &= 2p_{z2} - p_{z1} - p_{z3} + \frac{1}{2}(2p_{z1} - p_{z2} - p_{z3}) \\ &= 2p_{z2} - p_{z1} - p_{z3} + p_{z1} - \frac{1}{2}p_{z2} - \frac{1}{2}p_{z3} \\ &= 0p_{z1} + \frac{3}{2}p_{z2} - \frac{3}{2}p_{z3}\end{aligned}$$

...after normalization

$$= \frac{1}{\sqrt{2}}(p_{z2} - p_{z3})$$

6.6.1.3 Diagramming the MOs of BF_3

Based on the previous results, the fluorine-based SALCs due to the s-atomic orbitals will be comprised of three wavefunctions of a_1' and e' symmetry. The nine p-orbitals on the fluorines will combine to produce nine molecular orbitals represented by the irreducible representations a_1', a_2', a_2'', 2 e', and e''. The orbitals on boron have symmetries of a_1' (s), e' (p_x,p_y), and a_2'' (p_z). Pictures of the fluorine SALCs, all orthogonalized, are shown on the left side of the MO diagram in Fig. 6.18g.

According to our usual approximation when there is a 3-orbital interaction involving a_1' with two orbitals on the SALCs and one orbital on B, this should result in bonding ($1a_1'$), nonbonding ($2a_1'$), and antibonding ($3a_1'$) orbitals. Calculations suggest that the nonbonding orbital is mixing heavily with the bonding so that $2a_1'$ is deep into the bonding manifold. Similarly, $2e'$ was shown computationally to be well into the bonding manifold due to its 4-orbital interaction where there are four orbitals of e' symmetry on the SALCs and one on the boron.

As a result, $2e'$ and $2a_1'$ are drawn with the bonding orbitals in Fig. 6.18g; however, we might have estimated that they were approximately nonbonding without computational data. Again, this shows the utility of group theory analysis in conjunction with calculations as a method for understanding electronic structures, as we can now readily

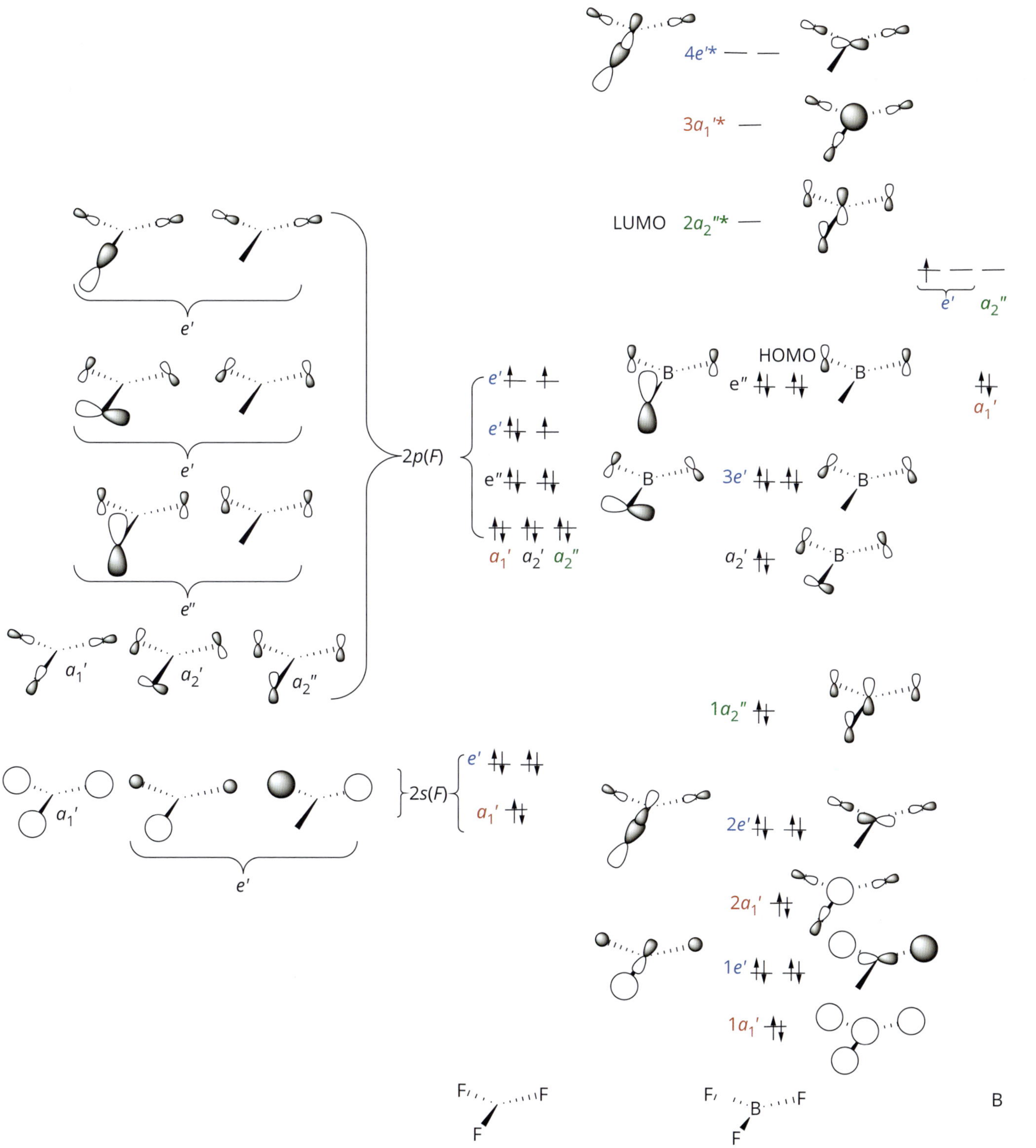

Fig. 6.18g Approximate MO diagram for BF_3 in D_{3h}. Orbitals in brackets have approximately the same energy, even though they may be drawn at different levels on the *y*-axis.

see that these are comprised of nonbonding and bonding mixed interactions.

Two representations (three orbitals) in the diagram are strictly nonbonding, a_2' and e'', as these have no symmetry match on boron. The $3e'$ set seems to have little boron contribution and is approximately nonbonding in the calculation. These are drawn essentially equal in energy with the atomic *p*-orbitals on fluorine.

If for a moment we consider $2e'$ and $2a_1'$ to be approximately nonbonding as these are merely mixing nonbonding character with the bonding orbitals below them, there are four bonding orbitals represented by $1a_1'$, $1e'$, and $1a_2''$. Two of these representations (three orbitals) are σ-bonding, $1a_1'$ and $1e'$; these are the three B–F σ-bonds. The third interaction is π-bonding, $1a_2''$. Consequently, the B–F bond order in BF_3, according to the MO analysis, is 1.33.

The HOMO of the molecule is one of the nonbonding, F-centered orbitals that are essentially delocalized lone pairs. The LUMO is $2a_2''^*$. The LUMO is much closer to boron in energy and will be largely boron-based. Naturally, it is largely this orbital that gives the compound its Lewis-acidic character.

6.6.2 Octahedral Complexes: CrH_6 as an Example

In Section 6.4.3 the ligand-based SALCs for σ-donation to a metal complex of octahedral symmetry were developed. For reference, the functions are reproduced as follows. There is a totally symmetric function represented by the irreducible representation a_{1g}.

$$\psi^{a_{1g}} = \frac{1}{\sqrt{6}}(\sigma_1 + \sigma_2 + \sigma_3 + \sigma_4 + \sigma_5 + \sigma_6)$$

The SALCs contain a pair of degenerate orbitals of e_g symmetry. The orthogonalized orbitals were found to be $\psi_{\sigma_1}^{e_g}$ and $\psi_{\perp}^{e_g}$.

$$\psi_{\sigma_1}^{e_g} = \frac{1}{\sqrt{12}}(2\sigma_1 + 2\sigma_2 - \sigma_3 - \sigma_4 - \sigma_5 - \sigma_6)$$

$$\psi_{\perp}^{e_g} = \frac{1}{2}(\sigma_3 - \sigma_4 + \sigma_5 - \sigma_6)$$

Finally, there is a triply degenerate set of t_{1u} symmetry.

$$\psi_{\sigma_1}^{t_{1u}} = \frac{1}{\sqrt{2}}(\sigma_1 - \sigma_2)$$

$$\psi_{\sigma_3}^{t_{1u}} = \frac{1}{\sqrt{2}}(\sigma_3 - \sigma_5)$$

$$\psi_{\sigma_4}^{t_{1u}} = \frac{1}{\sqrt{2}}(\sigma_4 - \sigma_6)$$

In all, there are six SALCs representing, in this case, symmetry combinations of the six *s*-orbitals on the hydrogens in CrH_6. The valence orbitals on chromium have the symmetries of a_{1g} (*s*), t_{1u} (p_x, p_y, p_z), t_{2g} (d_{xz}, d_{yz}, d_{xy}), and e_g (d_{z^2}, $d_{x^2-y^2}$).

The MO diagram, approximated from the symmetries, is shown in Fig. 6.19. There are, as expected, six bonding interactions, $1a_{1g}$, $1t_{1u}$, and $1e_g$, along with their antibonding orbitals, $2a_{1g}^*$, $2t_{1u}^*$, and $2e_g^*$. The t_{2g} orbital has no symmetry match on the SALCs and is strictly nonbonding, localized on the chromium center. As we will discuss later, the t_{2g} orbitals have the proper symmetry to overlap with π-orbitals on the ligands, but naturally these are absent when hydrogen is the only ligand.

The *d*-orbitals in the CrH_6 complex are largely represented by the nonbonding t_{2g} orbital along with the e_g^* orbital, which will be of mostly chromium character. Later, when discussing ligand field theory, these orbitals will be referred to as the ligand field orbitals. The electronic structure of the *d*-orbitals make large contributions to the properties and reactivity of many transition metal complexes, and from this diagram it is perhaps obvious why: they are typically at or near the frontier orbitals (HOMO and LUMO) of the complex.

6.6.3 Square-Planar Complexes: PtH_4^{2-} as an Example

The SALCs for this d^8 square planar compound were discussed in Section 6.4. Compounds that are 4-coordinate and d^8 commonly adopt a square planar geometry. This exercise will be used to illustrate how MO theory treats compounds in this class to some degree; however, transition-metal electronic structure is covered in much more detail in later chapters. As mentioned, the SALCs are of symmetry:

$$\Gamma_\sigma = a_{1g} + b_{1g} + e_u$$

The a_{1g} orbital has equal contributions, with the same phase, from all of the *s*-orbitals on the hydrogens.

$$\psi^{a_{1g}} = \frac{1}{2}(s_1 + s_2 + s_3 + s_4)$$

The b_{1g} orbital has some phase changes in the contributions. Here we point out that a common method for finding the nature of a SALC is to look at the orbital on the central atom with the same symmetry with which it can overlap. In this case, b_{1g} is also the irreducible representation for $d_{x^2-y^2}$, which has alternating phases on the *xy*-axes. The SALC must have the same series of phase changes giving the formal wavefunction.

$$\psi_{s_1}^{b_{1g}} = \frac{1}{2}(s_1 - s_2 + s_3 - s_4)$$

The normalized wavefunctions for the e_u SALCs are as follows.

$$\psi_{s_1}^{e_u} = \frac{1}{\sqrt{2}}(s_1 - s_3)$$

$$\psi_{s_2}^{e_u} = \frac{1}{\sqrt{2}}(s_2 - s_4)$$

The orbitals on Pt have symmetries of a_{1g} (*s* and d_{z^2}), b_{1g} ($d_{x^2-y^2}$), b_{2g} (d_{xy}), e_g (d_{xz}, d_{yz}), a_{2u} (p_z), and e_u (p_x, p_y); see Fig. 6.2. The estimated MO diagram is shown in Fig. 6.20.

Between the central platinum and the hydrogens there should be a total of four bonds. These bonds are $1a_{1g}$, $1b_{1g}$, and $1e_u$; however, it is important to point out that the e_u bonding interaction is weak due to the high energy of the Pt p_x- and

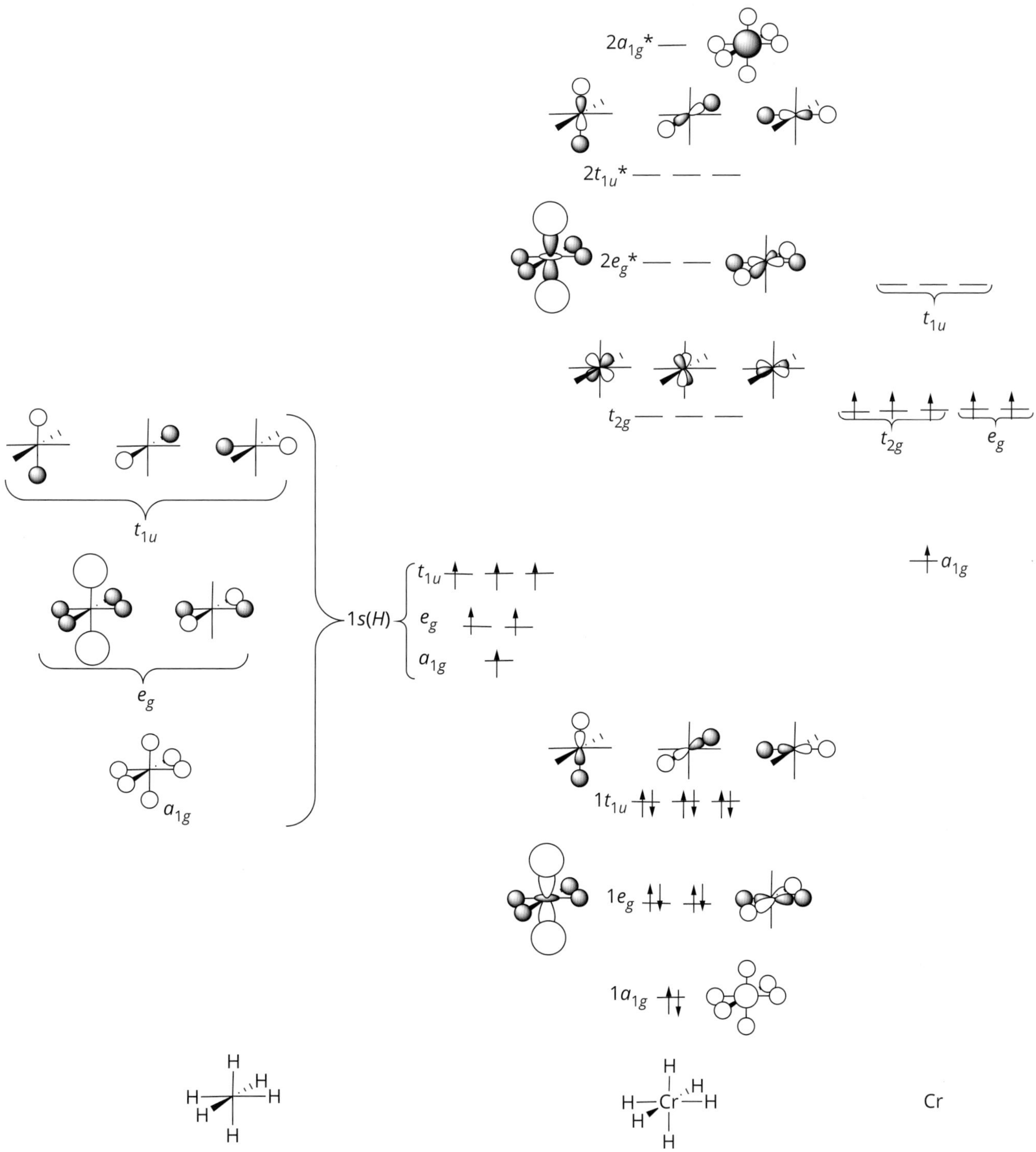

Fig. 6.19 Approximate MO diagram for CrH_6 in O_h. Orbitals in brackets have approximately the same energy, even though they may be drawn at different levels on the *y*-axis.

p_y-orbitals. As a result, the e_u orbitals interact poorly. Because of the delocalized nature of the bonding, these weak bonding interactions are split over all four Pt-H interactions. (See Chapter 14 for an analogous valence bond description.) Also note that the antibonding orbitals $2e_u$ are only slightly destabilized from the atomic orbitals on Pt, which is also consistent with a weak bonding interaction.

There is one 3-orbital interaction involving orbitals of symmetry a_{1g}, and the approximately nonbonding orbital in that interaction is expected to be largely d_{z^2} in character, $2a_{1g}$. The other nonbonding orbitals, which have no symmetry match on the SALCs, b_{2g} (d_{xy}) and e_g (d_{xz}, d_{yz}), will be rigorously nonbonding. One may estimate that these will have approximately the same energy as the Pt atomic d-orbitals.

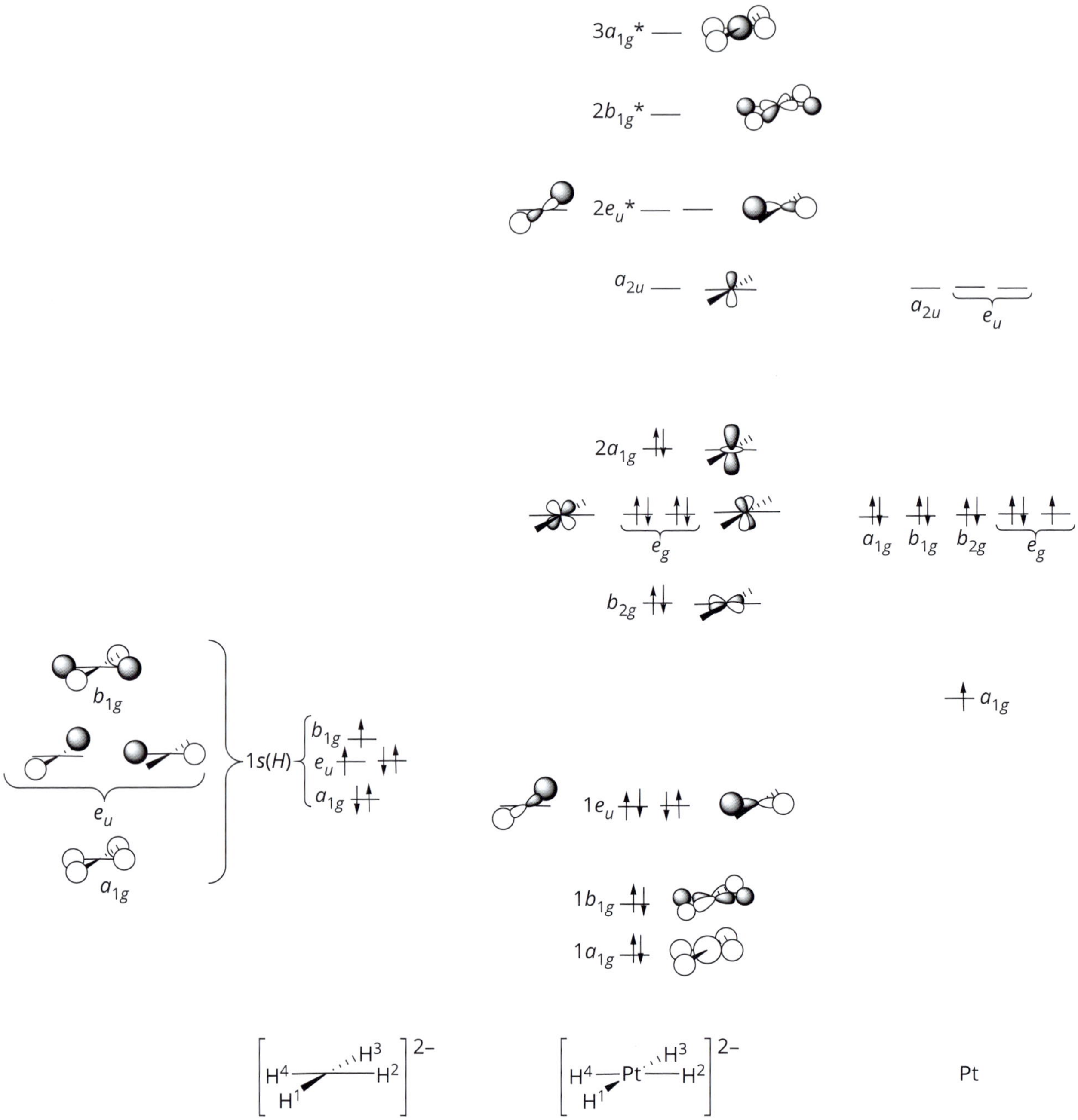

Fig. 6.20 Approximate MO diagram for PtH_4^{2-} in D_{4h}. Orbitals in brackets have the same energy, even though they may be drawn at different levels on the *y*-axis.

However, differing electron correlation effects on the orbitals in different orientations will produce slightly different energies for these orbitals as well.

There will be one valence orbital on Pt, which is empty and rigorously nonbonding, a_{2u}. If the metal orbitals were all filled with electrons or engaged in bonding interactions, the MO diagram using valence orbitals with a central transition metal would contain at least eighteen electrons, or more than eighteen if the ligands have lone pairs not engaged in bonding or if additional antibonding orbitals are occupied. In this case, there are only sixteen electrons around the metal center, and p_z on Pt is the LUMO of the complex.

6.6.4 Main-Group Hypervalency: PH_5 as an Example

We generate the molecular orbital picture of PH_5 "by hand" using group theory, and then compare the molecular orbital shapes with the results from a calculation. The hydrogen labeling scheme and coordinate axes are shown in Fig. 6.21a.

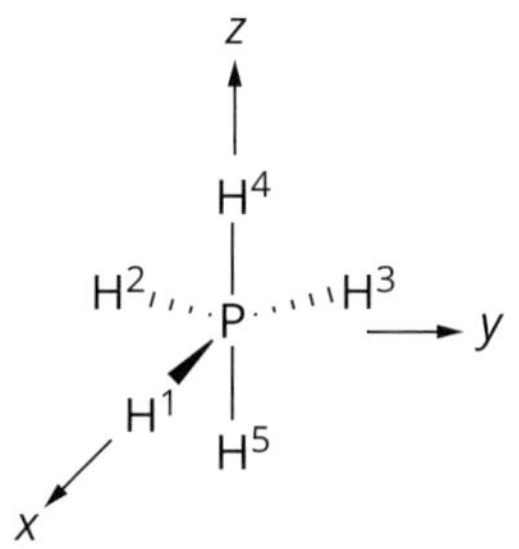

Fig. 6.21a Drawing of PH_5 with coordinate axes.

The molecule has D_{3h} symmetry, and we have generated the SALCs for hydrogen 1*s*-orbitals in this symmetry previously for BH_3. These were orbitals of the form shown as follows in irreducible representations a_1' and e'.

$$\psi^{a_1'} = \frac{1}{\sqrt{3}}(s_1 + s_2 + s_3)$$

$$\psi^{e'}_{s_1} = \frac{1}{\sqrt{6}}(2s_1 - s_2 - s_3)$$

$$\psi^{e'}_{\perp} = \frac{1}{\sqrt{2}}(s_2 - s_3)$$

The difference between the BH_3 system discussed previously and PH_5 from a group-theory standpoint are the two hydrogens along the *z*-axis, and there are no symmetry operations that convert axial and equatorial positions. The axial hydrogens, considering that these are simply combinations of two orbitals, must produce symmetric and asymmetric combinations. These have the symmetries of, by inspection of the character table, a_1' (same symmetry as the *s*-orbital on phosphorus) and a_2'' (same symmetry as the p_z-orbital on phosphorus).

Since there are no symmetry operations that relate the axial hydrogens to the hydrogens in the equatorial *xy*-plane, we generate their SALCs separately. SALCs generated for the axial hydrogens and SALCs generated for the equatorial hydrogens will mix if they belong to the same irreducible representation; Fig. 6.21b. (See Fig 6.8 and accompanying text for a related example.) In this case, the a_1' SALC of the axial hydrogens will mix with the a_1' SALC of the equatorial hydrogens, producing in-phase and out-of-phase combinations as shown in Fig. 6.21b. Note that both of the combinations (symmetric and asymmetric) are still in the a_1' irreducible representation, which can be confirmed by applying the symmetry operations to the orbitals. This operation simply increases delocalization of the SALCs.

The MO diagram for PH_5 is shown in Fig. 6.22. The most interesting aspect of the electronic structure for the compound is the 3-orbital interaction between the two a_1' orbitals on the hydrogen SALCs, and the a_1' (*s*-orbital) on phosphorus. We would expect that we would obtain three orbitals from the 3-orbital interaction: bonding ($1a_1'$), antibonding ($3a_1'$), and nonbonding ($2a_1'$). This expectation is met, and there is an approximately nonbonding orbital with largely hydrogen character in the MO. This orbital is from the antisymmetric combination of the axial and equatorial a_1' SALCs shown in Fig. 6.22 (see top right of Fig. 6.21b), and while it has a_1' symmetry there are no valence orbitals on phosphorus with this shape to their phases. Even though *s* belongs to the correct irreducible representation, its net overlap with this orbital will be quite small.

One interesting aspect of the bonding is that there are only four orbitals with an occupied nonbonding interaction in the MO diagram. However, this nonbonding character is spread out over the five different hydrogens, albeit not necessarily equally. The calculation has the P–H$_{equatorial}$ distance as 1.433 Å, whereas the P–H$_{axial}$ distance is a somewhat longer at 1.489 Å. If we assume that $2a_1'$ is rigorously nonbonding and count bonding interactions, we find that the MO diagram suggests that the P–H$_{equatorial}$ bond order is $(1/5 + 2/3) = 13/15$ (~ 0.87), whereas

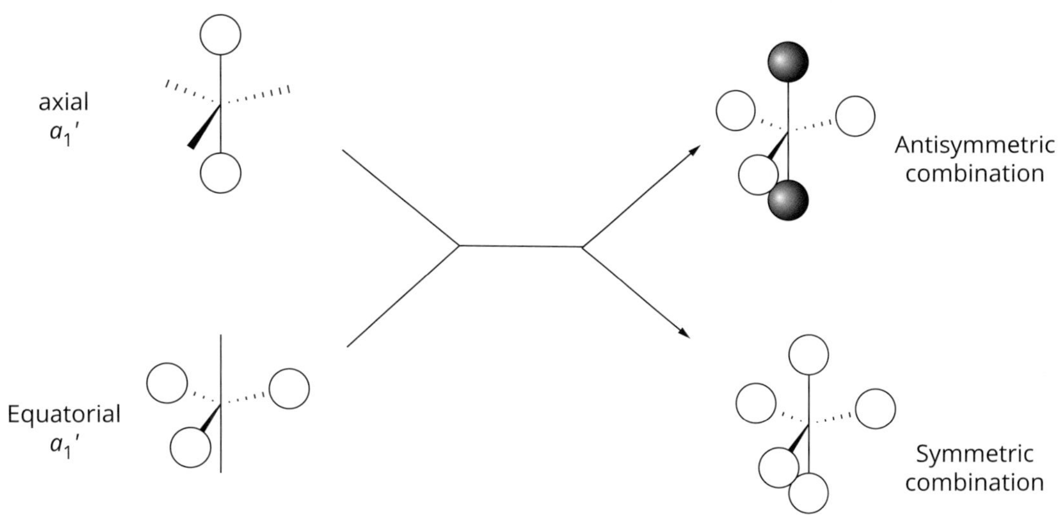

Fig. 6.21b Mixing of the a_1' SALC of the equatorial hydrogens with the a_1' SALC of the axial hydrogens, producing in-phase and out-of-phase combinations.

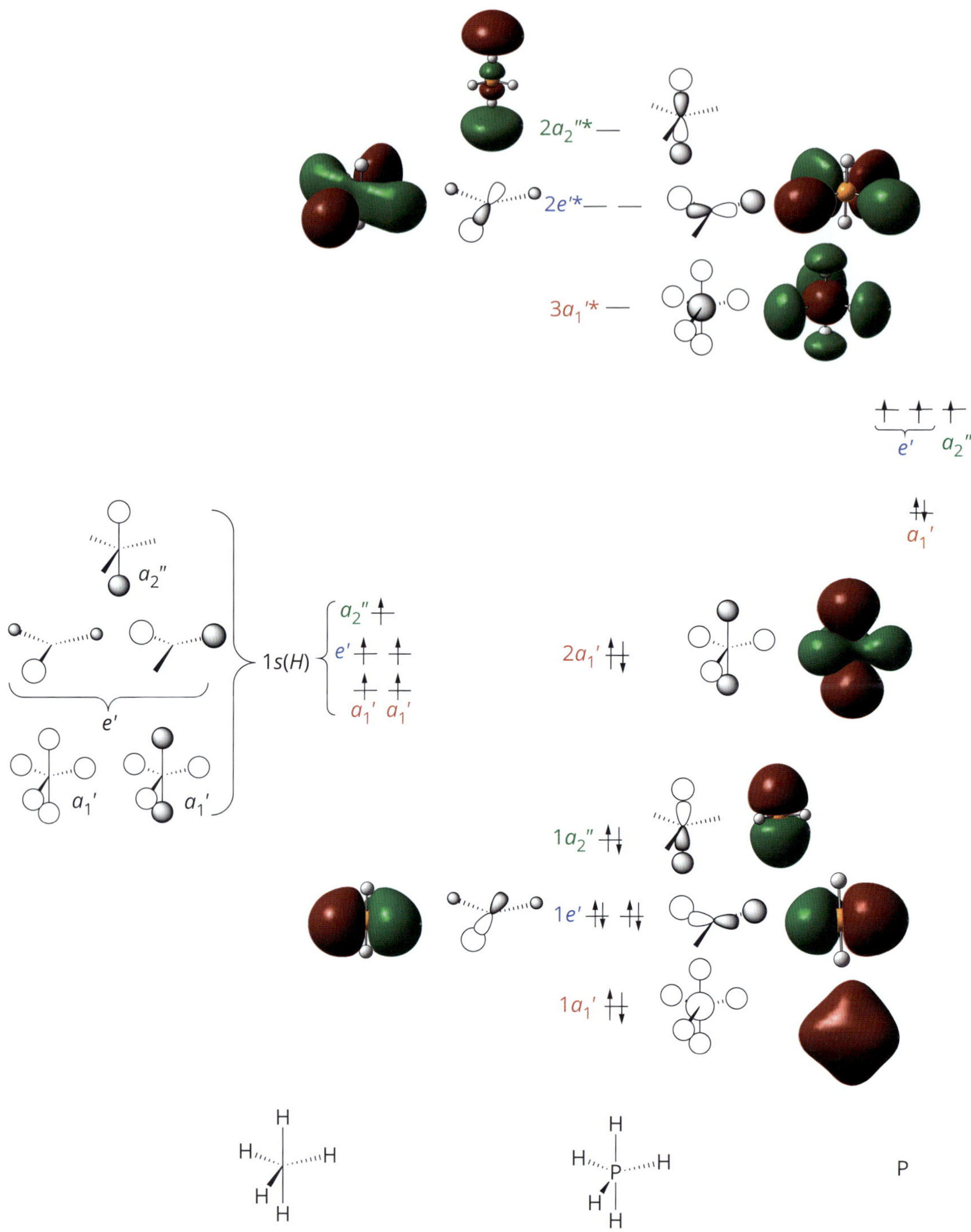

Fig 6.22 Approximate MO diagram for PH_5 in D_{3h}. Orbitals in brackets have approximately the same energy, even though they may be drawn at different levels on the *y*-axis. Included in the diagram are the surface contours from the DFT calculation (B3LYP, 6-311++G), with different phases shown in red and green. The phosphorus atom is orange and the hydrogens are gray.

the P–H_{axial} bond order is (1/5 + 1/2) = 7/10 (or 0.7). This is a common phenomenon in 5-coordinate main-group complexes, and the axial bond distances are often longer than the equatorial due to additional nonbonding character localizing on these axial substituents. These phenomena will be discussed in more detail later, especially in Chapters 8 and 14.

6.7 MO Diagrams of the Diatomics

The diatomics are quite important compounds on their own, but in inorganic chemistry they are exceptionally important as ligands. A detailed understanding of the electronic structure of the first-row p-block diatomics (such as CO, N_2, O_2, NO, F_2) will be invaluable for several areas of inorganic chemistry, such as in Chapter 11, where we will discuss, in detail, diatomics as ligands for metal centers.

6.7.1 MO Diagram of Dinitrogen and Photoelectron Spectroscopy

As an example, let us examine the MO diagram of dinitrogen, N_2. In addition, we will use photoelectron spectroscopy (PES) as an experimental method to examine the electronic structure of the compounds (see Box 6A).

The electronic configuration of nitrogen in the valence shell is $2s^2 2p^3$. If one were to simply draw the expected MO diagram with no s-p mixing, it would look something like Fig. 6.23. The s-orbitals on each nitrogen would combine to produce σ-bonding and σ-antibonding interactions, both filled with electrons and thus resulting in no net bond. The p_z-orbitals would similarly combine to form σ-bonding and σ-antibonding interactions, but only the bonding interaction is occupied with electrons. The p_x- and p_y-orbitals each combine to form two π-bonds. (Only one of the π-orbitals is shown in Fig. 6.23 for $1\pi_u$; the other has the same shape and is rotated 90° along the bond axis.) As would be anticipated, the σ-interaction is stronger, due to larger overlap, than the π-interaction for the p-orbitals, which leads to a larger orbital splitting for the σ-bond (Fig. 6.14).

The two orbitals on each nitrogen capable of forming σ-bonds, $2s$ and $2p_z$, can form sp-hybrids because they are of the same symmetry. The advantage energetically of this atomic orbital mixing is that it allows use of the directionality of the p-orbital to improve overlap, while mixing the lower-energy s-orbital into the bonding to stabilize the electron density better than higher-energy p-orbitals.

The results of sp-mixing are shown in Fig. 6.24. Both $1\sigma_g$ and $1\sigma_u$ are stabilized somewhat. Conversely, $2\sigma_g$ and $2\sigma_u{}^*$ are destabilized. The σ_g orbitals due to s- and p-orbital mixing

BOX 6A

Photoelectron spectroscopy (PES), X-ray photoelectron spectroscopy (XPS), and electron spectroscopy for chemical analysis (ESCA)

PES is the general name typically used to describe the specific techniques of X-ray photoelectron spectroscopy (XPS), which is used to examine core levels, and ultraviolet photoelectron spectroscopy (UPS), which is typically used to examine valence levels.

In PES, a photon is used to eject an electron from the system, leaving a cation. The photon is signified as hν in the following equation because the energy of the photon, E, is Planck's constant times the frequency of light, E = hν.

$$A + h\nu \rightarrow A^+ + e^-$$

The kinetic energy, $E(e^-)$, of the electron ejected is measured and by conservation of energy:

$$E(A) + h\nu = E(A^+) + E(e^-)$$

Rearrangement of this equation produces the following equation, in which the kinetic energy of the electron is equal to the energy of the incident photons minus the difference in the energy between the starting and ending electronic states of the system. The quantity $\{E(A^+) - E(A)\}$ is sometimes referred to as the "binding energy".

$$E(e^-) = h\nu - \{E(A^+) - E(A)\}$$

An assumption sometimes employed is that the binding energy is equal to the negative of the energy of the molecular orbital from which the electron was ejected. This was proposed by Tjalling Koopmans in 1934, specifically for the ionization energy of the highest occupied MO, and is referred to as Koopmans's theorem. It is sometimes helpful in the context of relating PES to MO theory, though it should be used with care as it can lead to erroneous conclusions, even with simple molecules such as N_2. This theorem is generally not helpful in explaining valence bond descriptions of molecules, which are readily explained by using the full binding energy expression. (See Chapter 7 for an example.)

XPS, which examines core orbitals, can be used to determine the concentration of elements on the surface of a sample. Because XPS looks at core electrons, which do not shift much in energy from one compound to the next, the energies are very nearly constant from sample to sample. A specific energy of light is used, typically Mg K_α with hν = 1253.6 eV or Al K_α with hν = 1486.6 eV. Peaks occur at specific kinetic energies, which are then converted to binding energies by using the known energy of the light. The intensities of the peaks are proportional to the concentration of the species. While the peak heights are not directly related to their electron occupation, they can be calibrated versus known concentrations then used for unknown samples. This is called electron spectroscopy for chemical analysis (ESCA), which is a widely used technique for surface characterization.

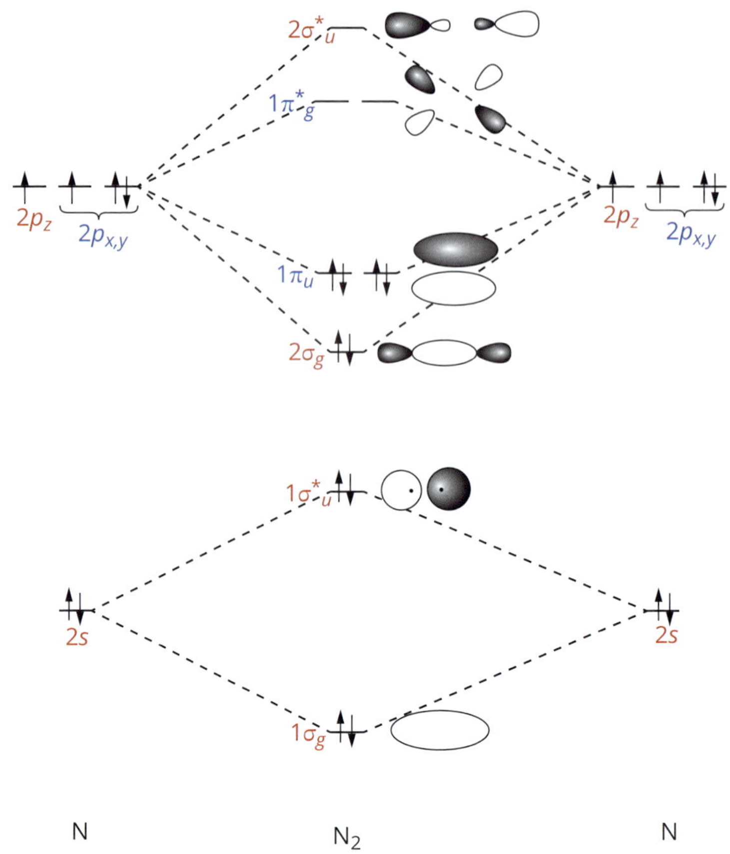

Fig. 6.23 MO diagram for N_2 without *s-p* mixing. This can be used as a starting point for many diatomics but is not the typical orbital ordering without mixing.

may mix constructively (+) and destructively (–), producing orbitals with bonding and antibonding character. In addition, the σ_u orbitals may mix. Due to this mixing, the orbital $2\sigma_g(p)$ picks up some antibonding character and rises in energy.

The result is an electronic structure that is, overall, more stable than without *sp*-mixing. In addition, the HOMO moves from the π-system without *sp*-mixing to an orbital of σ-character.

These orbital ordering arguments can be examined versus experimental results from X-ray photoelectron spectroscopy (XPS) under the Koopmans's theorem assumption (Box 6A). (More generally, photoelectron spectroscopy (PES) is sometimes discussed with excitation photons in the UV or X-ray region.) The bands in the XPS for N_2 can be described as being due to the excitation of the ground-state electronic configuration of N_2 and resulting in the different product configurations to the right of the spectrum in Fig. 6.24. The narrower bands in the XPS with little vibronic structure, such as $^2\Sigma_u^+$ (at ~16 eV), are due to excitation of orbitals with little bonding character that have little effect on the vibrations. Consequently, the experiment suggests that the very broad $^2\Sigma_g^+$ band at the bottom of the spectrum is due to excitation of an electron from the orbital largely responsible for the σ-bonding, and the heavily vibronically coupled $^2\Pi_u$ is from excitation of an electron from the orbital responsible for the π-bonding. The other two orbitals are largely nonbonding in nature and show little vibronic coupling.

Very high-level calculations on the N_2 molecule suggest that the $2\sigma_g$ orbital is actually very slightly lower in energy than $1\pi_u(p)$. This is not necessarily because the calculation or the PES experiment are incorrect. The dinitrogen molecule has a well-known breakdown in Koopmans's theorem (Box 6A). If one uses the actual ionized species (N_2^+) in the calculation of the PES energy, then the calculations can accurately reproduce the PES spectrum.

6.7.2 MO Diagram of Dioxygen: Effects of Unpaired Spin Density

The simplified MO diagram for O_2 is shown in Fig. 6.25. Unlike the previous example with N_2, the *sp*-mixing does not lead to $2\sigma_g$ being close in energy to π_u. This is possibly due to the larger *sp*-orbital energy gap in oxygen relative to nitrogen. The difference in valence orbital ionization energy for 2*s* and 2*p* in oxygen is 16.5 eV; the 2*s* and 2*p* orbitals in nitrogen

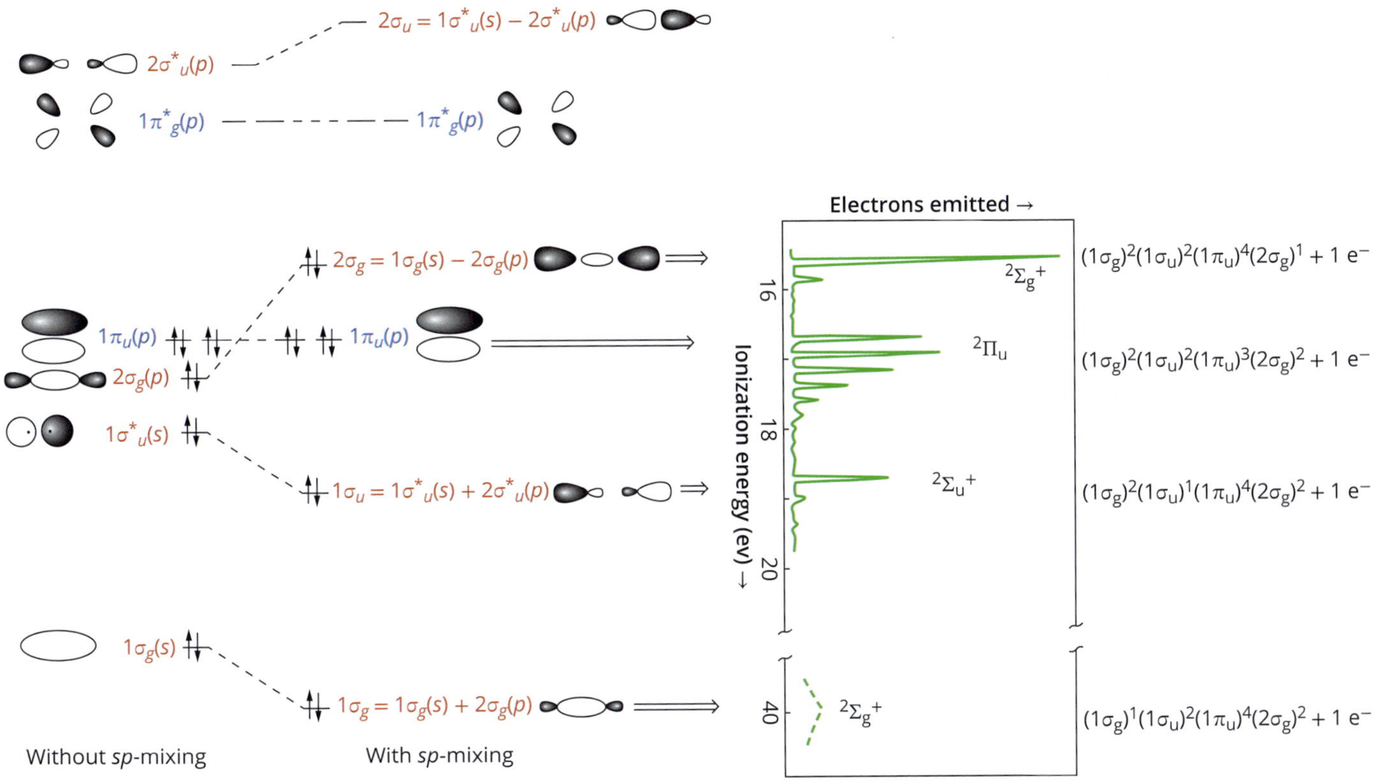

Fig. 6.24 Approximate MO diagram for N_2 with and without *s-p* mixing. On the right is the X-ray photoelectron spectrum of N_2. (Adopted from *Chemical Structure and Bonding* by R. L. DeKock and H. B. Gray.)[8] In reality, the $2\sigma_g$ and $1\pi_u(p)$ orbitals are very close in energy with the $1\pi_u(p)$ orbitals slightly higher in energy than the $2\sigma_g$, according to some calculations, such as CCSDT with cc-pVQZ.

are closer in energy at 12.4 eV. Since the orbital sizes are very nearly the same in oxygen and nitrogen, the energy gap is the main difference between the two atoms. From perturbation theory, the larger energy difference will lead to less mixing for oxygen and explains the MO differences between N_2 and O_2. The $2\sigma_g$-orbital lies well below π_u in energy.

The diagram for O_2 in Fig. 6.25 is "simplified" because the oxygen molecule contains unpaired electron density. The additional α-spin relative to β-spin makes the two spin manifolds different. The diagram in Fig. 6.25 is simplified by averaging spin-orbital energies to obtain the orbital energies for the diagram.

To understand the diagram for O_2 in more detail, we can look at the "unrestricted" calculation on the molecule. A "restricted" calculation is one in which α- and β-spin orbitals are refined together so that they have the same energy, which produces the usual MO picture in which each orbital can contain two electrons. If the molecule contains unpaired spins, by convention more α-spins than β-spins, then the calculation is run "unrestricted" where the usual 2-electron orbitals are divided into two "spin-orbitals", each able to contain one electron. The unrestricted calculation then refines orbital energies for the two spins independent of one another, providing different energies for the same atomic or molecular orbital in different spin manifolds.

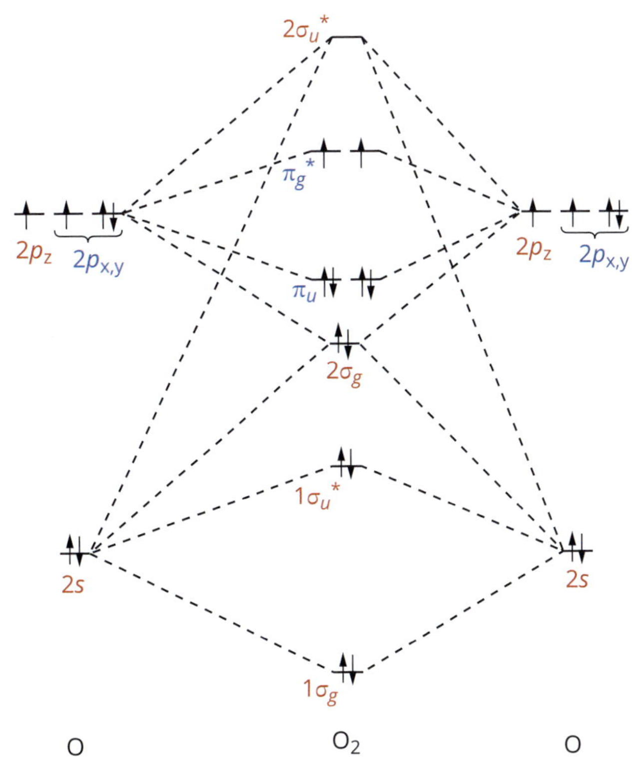

Fig. 6.25 Simplified MO diagram for O_2.

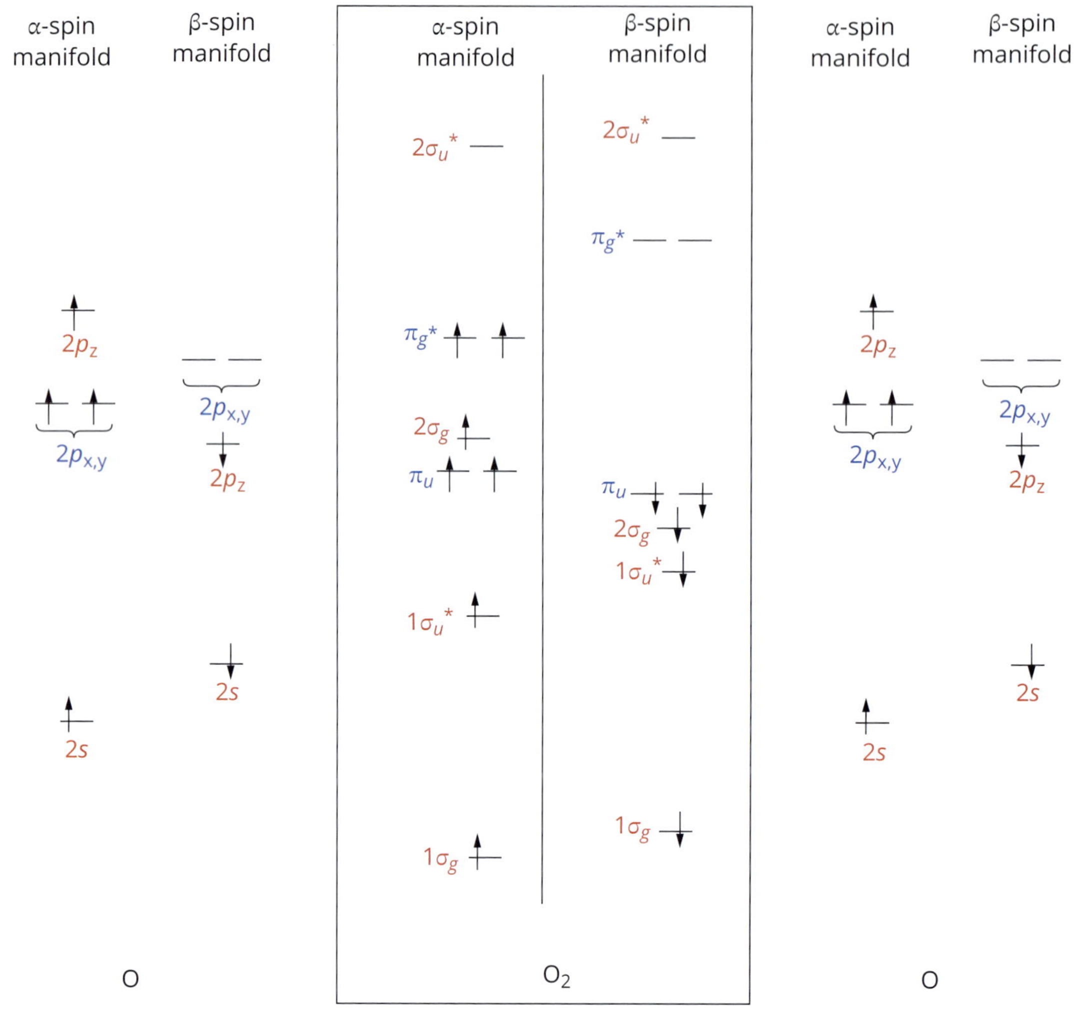

Fig. 6.26 More detailed MO diagram for O_2 showing the α- and β-spin manifolds. On the left and right are the two identical oxygen atoms with the differences in the two spin manifolds shown. The boxed portion is the MO diagram for dioxygen where the orbital energies are taken from a DFT calculation.

In Fig. 6.26 the atomic orbital diagram shows the atomic energies of the oxygen atom, which is also paramagnetic, along with the MO diagram of O_2 divided into α- and β-spin orbitals at their relative energies from the calculation. As can be seen, the energies, and even orbital ordering, changes due to electron correlation differences within the individual manifolds.

This effect is present in any compound with unpaired spins. However, simplified pictures, like those for O_2 in Fig. 6.25, are still of significant utility in discussing and understanding reactivity and properties in some cases, and this type of simplified MO will often be used in this book for molecules that are radicals.

6.8 Fragment Approach to Lower Symmetry Systems (Advanced Topic)

In order to determine the MOs for some systems, such as those with low symmetry or multiple metal centers, a fragment approach is often used. For example, in low-symmetry systems the method may include generation of the MOs for the metal and one hemisphere of the ligands where the symmetry of the whole molecule is imposed on this fragment, with additional orbital mixing as the remainder of the coordination sphere is added.

An example of such a process will be discussed later for the relatively simple molecule CpNiNO. In addition, the molecular orbitals of bent metallocenes will also be discussed, as these are a commonly encountered framework in some fields of inorganic chemistry.

As the symmetry becomes lower, and a full MO approach becomes more unwieldy, it is common for researchers to use crystal field theory (CFT) instead of MO theory for approximating electronic structure and to discuss orbital makeup, especially in laboratory discussions. In the literature, these CFT pictures may then be used and perhaps supplemented with a more rigorous theoretical treatment such as DFT, coupled cluster theory, and so on.

6.8.1 CpNiNO

For CpNiNO, which is of C_{5v} symmetry, one can think of using $CpNi^-/NO^+$ or $Cp^-/NiNO^+$ fragments, and in this case it is really a personal preference. We will use the former fragmentation here (Fig. 6.27a).

To begin, we need to find the MOs for the $CpNi^-$ fragment. In this species, cyclopentadienyl ($C_5H_5^-$) is η^5 ("eta 5") with all five carbons equivalently bound to the nickel. The bonding in ligands of this type will be discussed in detail in Chapter 12; however, the MOs can be looked at here without detailed discussion. Most of what matters at the moment are the symmetries of the orbitals and their mixing. In Chapter 12, the MOs of the π-system of cyclopentadienyl are discussed in the symmetry of the free ligand, D_{5h}. Here we need the equivalent terms in C_{5v} to overlap with the metal, and the labels are adjusted in Fig. 6.27b. The MO of the fragment is built up using the same principles discussed previously for the generation of MOs for full molecules.

Cyclopentadienyl interacts with a metal to form a σ, 2 π, and 2 δ interactions of A_1, E_1, and E_2 symmetry, respectively. The only interaction with significant mixing will be the σ-bond between the Cp and Ni, where there are three a_1 orbitals on the nickel side and one a_1 orbital on the Cp-side. The s and d_{z^2} interaction with the a_1 orbital on the Cp gives a bonding/antibonding/~nonbonding set; in this case there is an orbital that looks very much like a nonbonding d_{z^2}. The bonding interaction in Fig. 6.27b is simply shown as the s-orbital of Ni interacting with the a_1 orbital on Cp. The e_1 orbitals on nickel involving both p- and d-atomic orbitals heavily mix and interact with ligand.

In total, there are five bonding interactions between the metal and the cyclopentadienyl, with an occupied antibonding e_1 set of orbitals.

The molecular orbitals for the $CpNi^-$ fragment are then combined with the frontier orbitals of the NO^+ ligand in the same point group. The N–O single bond is of little consequence, as we are mostly concerned with the metal's electronic structure; similarly, the C–H and C–C single bonds in the cyclopentadienyl were ignored in generating the MOs of the $CpNi^-$ fragment. The nitrosyl cation will interact with the metal by using a lone pair on nitrogen of A_1 symmetry and π*-antibonding orbitals of E_1 symmetry. The approximate MO diagram from the interaction of the $CpNi^-$ fragment with the NO^+ fragment to make neutral CpNiNO is shown in Fig. 6.28. The most important interactions between the two fragments are the e_1 set's stabilization by the interaction with NO π*-orbitals. The a_1 σ-bonding orbital is also shown.

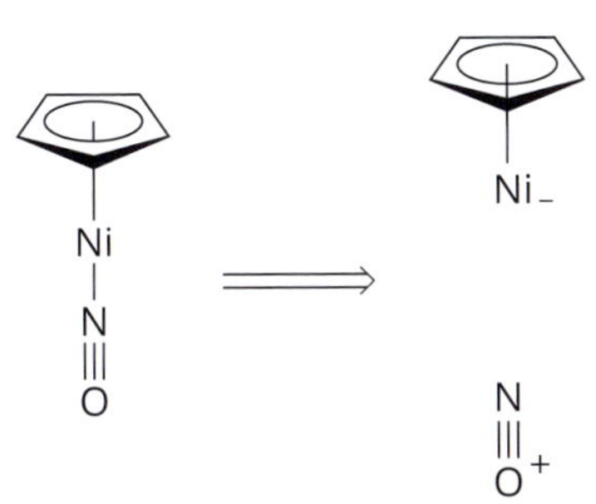

Fig. 6.27a Fragment approach to molecular orbitals for unsymmetrical complexes. In this case we will create fragment MOs for $CpNi^-$ and NO^+.

6.8.2 Bent Metallocenes

Bent metallocenes are commonly encountered in inorganic and organometallic chemistry and are structures of the type Cp_2MX_n, where $n = 1–3$. The most common approach for considering the MOs of such molecules is to start with the MCp_2 fragment. The MOs near the frontier for a structure of this type are shown in Fig. 6.29, and the labels assume C_{2v} symmetry for the molecule overall. The two η^5-cyclopentandienyl ligands will leave three orbitals on the metal for interaction with the X ligands: two orbitals of a_1 symmetry (hybrids of s, d_{z^2}, $d_{x^2-y^2}$, and p_z) and one orbital of b_2 symmetry (mostly d_{yz}). These appear as shown in Fig. 6.29.

As a prototypical example of how these fragments may be used, consider the complex Cp_2TiCl_2. The two cyclopentadienyl ligands are bent to form a cleft in which the chlorides bind, and the molecule is approximately C_{2v} in symmetry. depending on how the Cp rings are rotated. If we just consider σ-bonds between the Ti and Cl atoms and use one of the p-orbitals on the chlorides for simplicity, the two p-orbitals on the chlorides will have an in-phase combination of a_1 symmetry and an out-of-phase combination of b_2 symmetry, which can interact with the fragment orbitals on Cp_2Ti as shown in Fig. 6.29, and these are the two bonds between titanium and chlorine.

In this chapter we have covered methods for the generation of MO diagrams using symmetry, and have discussed common orbital mixing patterns. In later chapters these methods and their results will be used extensively to understand how electronic structure relates to physical properties and reactivity of inorganic compounds. In addition, MO theory will be compared and contrasted with resonance-based models for electronic structure.

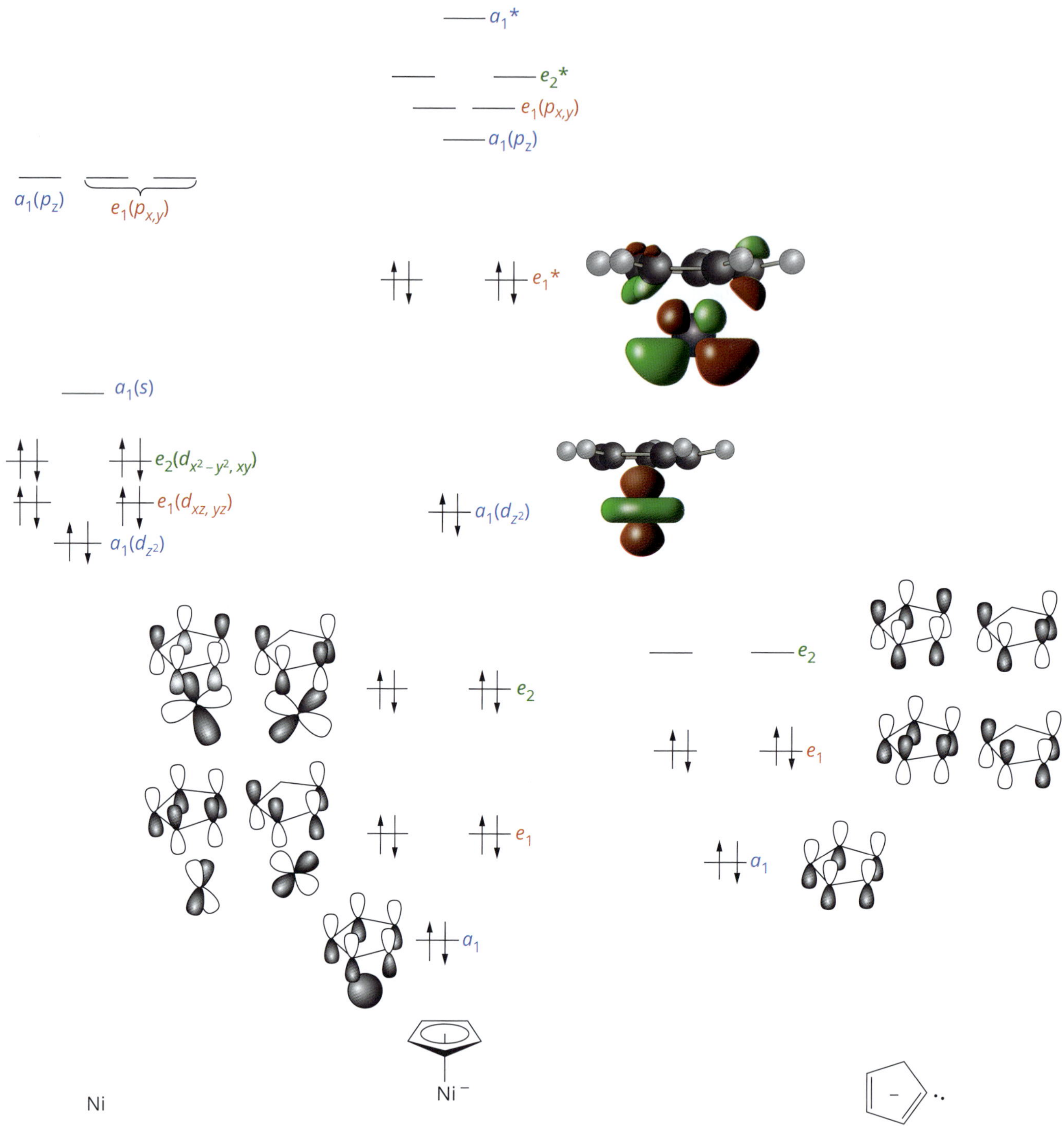

Fig. 6.27b Assembly of the CpNi⁻ fragment.

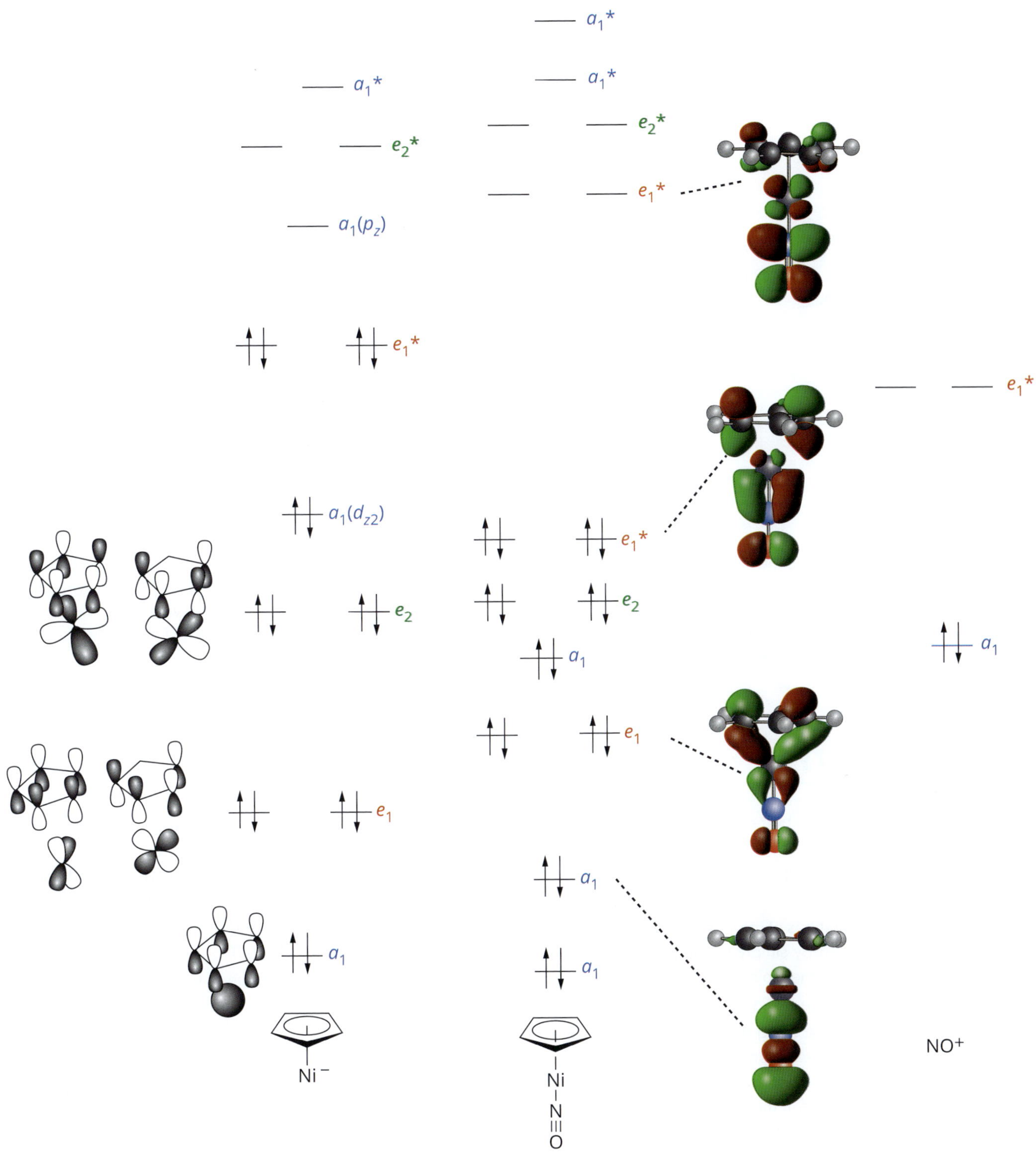

Fig. 6.28 Interaction of the $CpNi^{-}$ fragment with NO^{+} to produce the approximate MOs for CpNiNO.

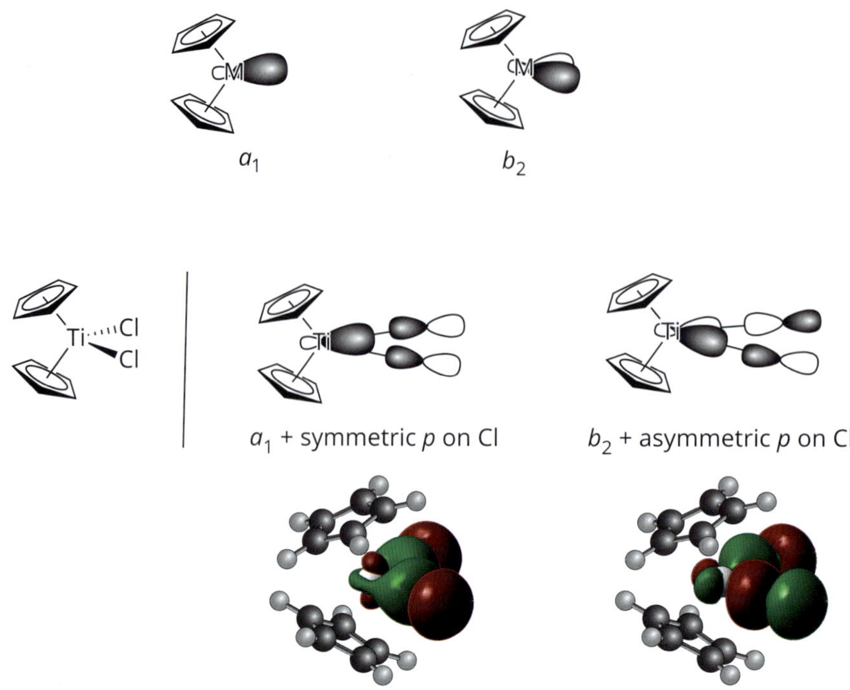

Fig. 6.29 (*top*) The bent metallocene fragment typically leaves three low-lying orbitals with the symmetries shown. There are two totally symmetric orbitals of a_1 symmetry and one orbital of b_2 symmetry. (*bottom*) Structure of Cp_2TiCl_2 with the interaction of the a_1 orbital on the Cp_2Ti fragment and the symmetric combination of *s*-orbitals on the chlorides. The b_2 orbital on the fragment will interact with the asymmetric combination of *s*-orbitals on the chlorides to produce two Ti–Cl bonds. The relevant orbitals from a DFT calculation (M06L/6-31G**) on Cp_2TiCl_2 are shown below.

Chapter Summary

1. Group theory can be used to better understand molecular orbital theory for complex systems. In these cases the orbitals are described by the irreducible representations in the point group of the complex.
2. The irreducible representations of the valence orbitals on a central atom can be obtained by simply looking at the character table for the point group of the molecule.
3. To find the ligand-based orbitals the following procedure is followed: (1) form a basis set using the orbitals on the ligands; (2) find the reducible representation in the point group for your basis; (3) use the decomposition formula to obtain the set of irreducible representations for the symmetry adapted basis; (4) use the projection operator to get a picture of the SALC; and (5) normalize the SALC to get the ligand-based orbital.
4. In cases where there is a rotation axis (C_n or S_n) where $n \geq 3$, there will be representation with dimensions higher than 1, and there could be degenerate orbitals. In cases where the orbital is doubly degenerate, the projection operator should be used twice on different parts of the basis. In cases where the orbital is triply degenerate, the projection operator should be used three times.
5. In cases with degenerate orbitals the projections obtained may not be orthogonal, and procedures such as Gram–Schmidt orthogonalization can be used to identify an orthogonal set of orbitals.
6. In the MO diagram the ligand orbitals are placed at an appropriate energy on one side, and the central atom's valence orbitals are placed on the other side. The energies can be estimated by the electronegativity of the atoms or by examining tables of ionization energies for valence orbitals. The MOs are then generated at the center of the diagram.
7. For orbitals to mix, they must be of the same irreducible representation. Otherwise, they are orthogonal, and no net bonding can occur.
8. When orbital mixing occurs, bonding and antibonding orbitals are generated. Nonbonding orbitals may be brought across at their starting energies. The nonbonding orbitals may mix with the bonding or antibonding orbital(s) of the same symmetry. We recommend leaving the nonbonding orbitals as strictly nonbonding unless additional information is available, such as from a calculation, regarding orbital mixing.
9. The molecular orbitals of diatomics are useful to understand as these will be important later for their abilities to bind metal centers. The MOs of N_2 and O_2 were covered in some detail and feature *sp*-mixing in the σ-framework.
10. In cases with complex molecules, a fragment approach can be used to approximate the molecular orbitals.

Practice Problems

1. The first step in any group theory analysis it to assign the point group. Assign the point groups for the molecules in the figure.

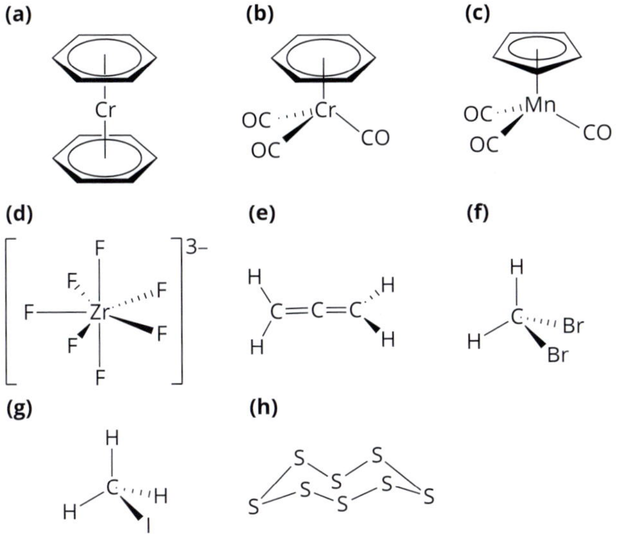

2. In Section 6.3 the SALCs for the hydrogens in SH_4 were generated. Draw the MO diagram for this molecule. Place the hydrogen SALCs on the left side and the sulfur atomic orbitals on the right, with their symmetry labels. In the center, draw all of the molecular orbitals, making reasonable guesses for their relative energies.
3. In 2012 the first organometallic transition metal complex with a pentagonal pyramidal structure was prepared and characterized.[9] The complex was prepared by addition of LiC_6F_5 (a potentially explosive compound) to $NbOCl_3$ in the presence of NBu_4^+ Br^-. Interestingly, the related compound $\{NbOCl_5\}^{2-}$ has the structure expected from VSEPR, pseudo-octahedral, instead of the pentagonal pyramidal structure of the organometallic compound (counterions are not shown).

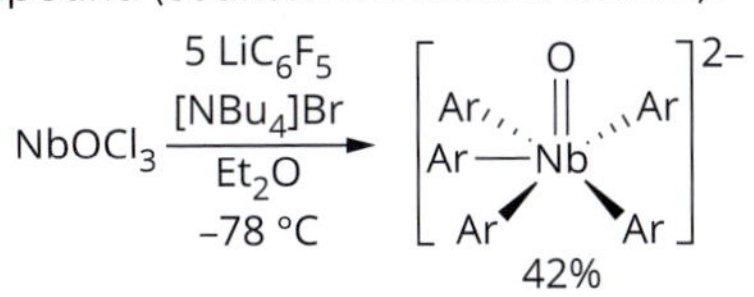

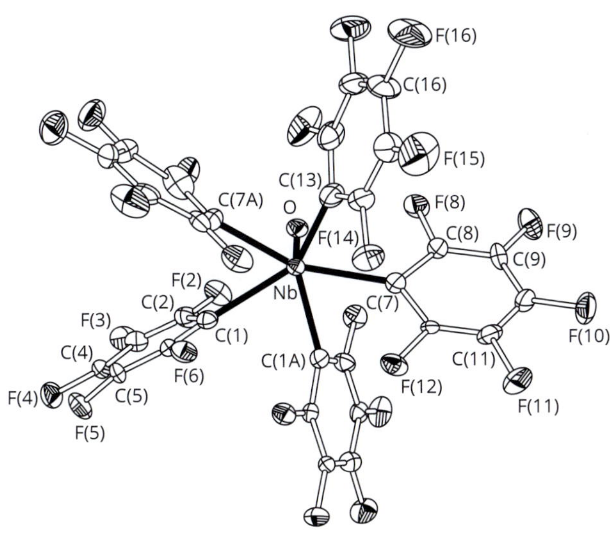

a. What is the point group of the compound ignoring aryl–Nb bond rotational angles?
b. Calculate the SALCs for the carbons giving their symmetries (irreducible representations) in the point group of the complex. You may assume that only σ-interactions are of importance in your analysis of the aryl carbon bonding. In this section, give only one of the orbitals comprising any degenerate sets involved.
c. What Nb-based orbitals are involved in bonding to oxygen? Draw the Nb–O bonding interactions, labeling the atomic orbitals in your diagrams.
d. (Advanced) Calculate the full set of SALCs for the carbons interacting with the metal center, assuming that only the σ-interactions are important, and orthogonalize the orbitals. Draw how these SALCs will interact with metal-based orbitals of the same symmetry.

4. Draw the orbital-splitting diagrams where the orbitals mentioned are in the same irreducible representations. Use the assumption where unmatched interactions will be nonbonding. If we assume that the SALCs (i.e., the ligands) are of the same energy and more electronegative than the central atom (for example, M for metal), identify the major contributor to each orbital. In other words, for each orbital (bond, nonbonding, antibonding) is it mostly metal- or ligand-based?

Example: SALC: 2 orbitals; M: 1 orbital

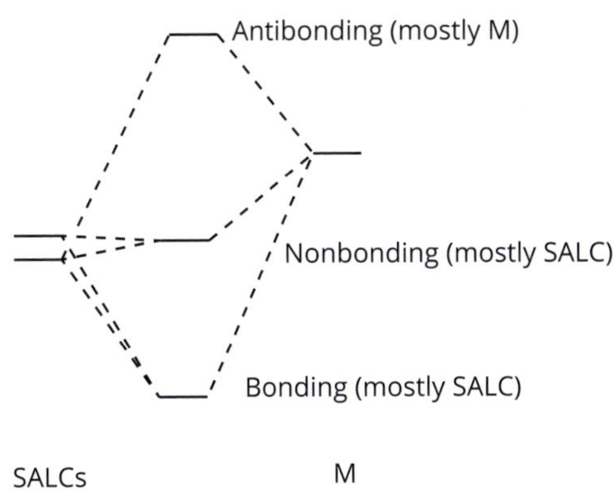

a. SALC: 4 orbitals; M: 2 orbitals
b. SALC: 5 orbitals; M: 3 orbitals
c. SALC: 3 orbitals; M: 3 orbitals
d. SALC: 1 orbital; M: 3 orbitals
e. SALC: 2 orbitals; M: 2 orbitals

5. While many of the molecules of interest in inorganic systems have a central atom, some do not. For example, staggered H_3Si–SiH_3 does not have an atom sitting at the symmetry origin, the inversion center. In this case, two sets of SALCs are generated: one for the Si orbitals, and one for the hydrogen orbitals.

a. Generate the SALCs for the silicon-based orbitals in staggered Si_2H_6.

b. Generate the SALCs for the hydrogen-based orbitals.

c. Draw the MO diagram for the molecule.

6. Answer the following questions regarding SO_2, a product of natural (e.g., volcanos) and man-made (burning of coal) activities.

a. Draw the Lewis structure of SO_2.

b. Based on your Lewis structure, what is the point group of the compound?

c. Generate the SALCs for the oxygen atomic orbitals under the symmetry of the point group.

d. Draw the MO diagram for SO_2.

e. According to your MO diagram, what is the S–O bond order?

7. Draw the molecular orbitals for the π-system in "inorganic benzene", borazine, $B_3N_3H_6$. The *z*-axis will be the 3-fold axis, and the molecule will sit in the *xy*-plane. Use the projection operator to discover their shape. (There is a hydrogen on each of the atoms in the ring, which isn't shown.)

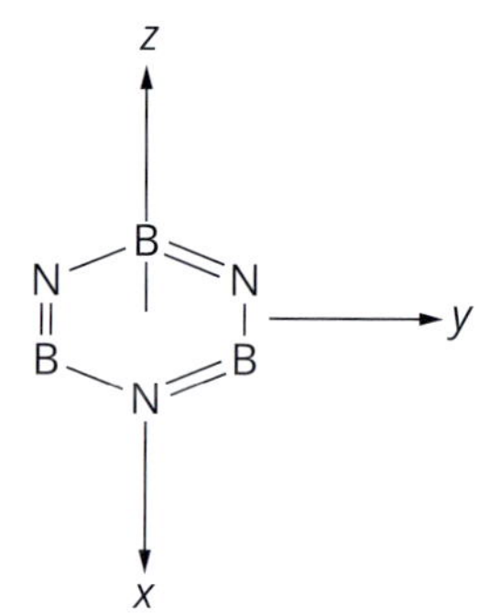

8. Draw the molecular orbitals for the π-systems below. Use Gram-Schmidt orthogonalization to discover the nature of the constituents of degenerate orbitals. Use the projection operator to discover their shape.

a. Benzene

b. Cyclopentadienide, $C_5H_5^-$ (This molecule is aromatic, flat, and a common ligand)

c. Square $C_4H_4^{2-}$ (another aromatic, flat ligand)

9. Tetramethylsilane, $Si(CH_3)_4$, is the standard for ^1H and ^{13}C NMR spectroscopy, where its resonance is set to 0 ppm. The compound is tetrahedral, point group T_d. For the procedure below, assume that the hydrogens on the methyl groups do not influence the symmetry, i.e, the methyl groups act as a spherical group with symbol Me.

a. What are the irreducible representations that symbolize the *s*- and *p*-orbitals on silicon?

b. Find the reducible representation for a basis set comprised of an *s*-orbital on each Me group and use the decomposition formula to find the irreducible representations in the reducible representation.

c. It is obvious from the previous parts that the p-orbitals on silicon, which constitute a t-set, overlap with a set of t-orbitals on the methyl groups. Starting from the p-orbitals on silicon, what do you think the SALCs on the methyl groups must look like? As a hint, you can do the projection for the SALC, which is the sum of one possible set of t-orbitals that will overlap with the p-orbitals on silicon.

d. Draw the equations for the normalized SALCs on the methyl groups.

e. What is the Si–C bond order?

f. Draw each of the σ-bonding C–Si MOs.

10. Find the SALCs for the two cyclopentadienyl ligands in ferrocene and their symmetries. Assume an eclipsed conformation, D_{5h}. In order to do this, take the orbitals of the π-system of one $C_5H_5^-$ ligand (problem 7b) and make in-phase and out-of-phase combinations with another identical π-system orthogonal to the first. In this way, ten orbitals will be generated. Find the irreducible representations to which each orbital belongs by inspection of the character table. The metal will reside between these two π-systems. What metal-based orbitals will overlap with each of your SALCs?

Cited References

1. Mulliken, R. S., Bonding power of electrons and theory of valence *Chem. Rev.* 1931, *9*, 347–388.
2. Mulliken, R. S.; Roothaan, C. C. J., The twisting frequency and the barrier height for free rotation in etylene *Chem. Rev.* 1947, *41*, 219–231.
3. Cotton, F. A., *Chemical Applications of Group Theory* 3rd ed.; Wiley: New York, 1990.
4. Vincent, A., *Molecular Symmetry and Group Theory: A Programmed Introduction to Chemical Applications* 2nd ed.; Wiley: Chichester, 2001.
5. Albright, T. A.; Burdett, J. K.; Whangbo, M.-H., *Orbital Interactions in Chemistry* Wiley: New York, 1985.
6. Ibers, J. A.; Schomaker, V., The structure of oxygen fluoride *J. Phys. Chem.* 1953, *57*, 699–701.
7. Schwenzer, G. M.; Schaefer, H. F., Hypervalent molecules sulfurane (SH_4) and persulfurane (SH_6) *J. Am. Chem. Soc.* 1975, *97*, 1393–1397.
8. DeKock, R. L.; Gray, H. B., *Chemical Structure and Bonding* 2nd ed.; University Science Books: Mill Valley, Calif., 1989.
9. Garcia-Monforte, M. A.; Baya, M.; Falvello, L. R.; Martin, A.; Menjon, B., An Organotransition-Metal Complex with Pentagonal-Pyramidal Structure *Angew. Chem. Int. Ed.* 2012, *51*, 8046–8049.
10. Chisholm, M. H.; Cotton, F. A.; Extine, M. W., Molecular and electronic-structure of tetrakis(dimethylamido)molybdenum(IV) *Inorg. Chem.* 1978, *17*, 1329–1332.
11. Smith, A. E.; Schrauzer, G. N.; Mayweg, V. P.; Heinrich, W., Crystal and molecular structure of $MoS_6C_6H_6$ *J. Am. Chem. Soc.* 1965, *87*, 5798.
12. Kammoun, O.; Rekik, W.; Naili, H.; Bataille, T., Inorganic layered structures in hybrid double sulfates: related phases and thermal reactivity *New J. Chem.* 2015, *39*, 2682–2689.

CHAPTER 7

Hybridization-Based Theories of Bonding

Applications Within the Main Group

"Taken together, MO and VB theories constitute not an arsenal, but a tool kit, simple gifts from the mind to the hands of chemists. Insistence on a journey through the perfervid bounty of modern chemistry equipped with one set of tools and not the other puts one at a disadvantage. Discarding any one of the two theories undermines the intellectual heritage of chemistry."

— Roald Hoffmann[1]

CONCEPT COVERAGE

1. **Brief overview of valence bond (VB) and natural bond orbital (NBO) theories**
2. **The hybridization parameter for main-group compounds (sp^λ)**
3. **The use of general resonance theory equations to evaluate bonding**
4. **Brief description of bond bending**
5. **Brief overview of perceived VB "failures" and misconceptions that led to them**

In hybridization-based theories the electronic configuration for most polyatomic systems is described through combinations of hybrid atomic orbitals, such as sp^λ-hybrids of some type. This hybrid concept, often used in a cursory fashion in organic chemistry as integer values only (sp, sp^2, and sp^3), is one of the hallmarks of hybridization theory. In fact, in organic chemistry, carbons are often described as being "sp^2" and the like, as if all the hybrids around the carbon are identical in molecules like ethylene. However, it is generally understood that this is not the case unless all the substituents on that carbon are identical; for example, as in the middle of a graphite sheet. Most importantly for our purposes, understanding orbital hybridization can lead to insights that are consistent with, and complementary to, a molecular orbital analysis, providing access to additional information on molecular systems. In some cases this hybridization theory analysis is far easier than an MO analysis, and in some cases it is not.

In the following discussions, little will be said about describing the exact energies of orbitals, repulsive states, configuration interaction, various weighting schemes for resonance contributions, and a host of other important issues for quantitative calculation of energies in valence bond theory. This is not unlike the treatment of MO theory, where we spent no time, for example, determining the exact energies of resonance integrals. Like MO theory, the conceptual framework for hybridization theory, and a few simple equations, are enough to provide powerful tools for understanding main-group chemistry.

7.1 Modern Adaptations of Hybridization Theory

Hybridization-based theories use weighted contributions from individual canonical electronic distributions to describe the ground-state electronic configuration for molecules. Each contribution is in resonance, leading to a composite structure lower in energy than any of the contributing forms (Section 3.2).

The main hybridization-based theories to be discussed are valence bond (VB) theory[2–3] and natural bond orbital (NBO)[4–6] theory.

VB theory is the classic hybridization theory for understanding electronic structure, and it was the dominant model for understanding molecular electronics until about the 1950s when molecular orbital theory surged into prominence. While MO theory is still the more used framework today, VB theory has undergone revitalization since the 1980s and is very valuable for understanding the electronic structure of many compounds. Today, the term "VB theory" is used to encompass a large number of related frameworks differing, for example, in how ionic forms are dealt with in the calculations or other aspects.

NBO theory was developed by Weinhold, Landis, and coworkers, and they published a major monograph in 2005 on the subject entitled *Valency and Bonding*.[4] The method starts with a calculation of the total electron density, perhaps by DFT or another method. Electrons are then partitioned as core electrons, lone pairs, bonding electrons, and so on. The theory has evolved into a powerful and often-used tool for understanding chemistry in general and inorganic chemistry in particular.[5] Another feature of NBO is natural resonance theory, which allows the ground-state electronic structure to be expressed as a series of weighted Lewis-structure contributors.

We will discuss the theoretical bases for VB and NBO theory in a very cursory fashion, highlighting some of their similarities and differences. The details for both theories are well beyond the scope of this book, and the interested reader is referred to very accessible texts for both.[2–4] After a brief introduction to the two commonly employed theoretical frameworks and some comments on computational abilities, we will turn to more qualitative uses of hybridization theory and what can be learned from a descriptive hybridization theory analysis of bonding in molecules.

As will be shown in this chapter and the next, hybridization theory is an invaluable complement to MO theory and can provide additional information through simple by-hand calculations. In addition, it can provide very clear understanding of molecule geometries in many cases, and can explain VSEPR geometries through electronic structure arguments. (See problems 7.3–7.5.)

After the brief introduction in the next section, many of the qualitative arguments will apply more or less equally to both NBO and VB theories, which will be used somewhat interchangeably even though they vary in the details of the calculations. In all cases, VB hybridizations are derived from the geometry of the molecule in ways that will be discussed shortly, and NBO hybridizations may be calculated from electron distributions found from methods such as DFT.

In this chapter the discussion will be limited to main-group molecules and *sp*-hybridization. In Chapter 14, theoretical treatments of transition metals using NBO will be discussed and contrasted with MO-based theories for understanding the geometry of transition-metal complexes.

7.1.1 Valence Bond (VB) Theory

In 1927, Heitler and London proposed to treat the hydrogen molecule 2-center 2-electron (2c2e) bond by using an antisymmetrized wavefunction where the orbitals on the hydrogens $1s_1$ and $1s_2$ can be occupied by electrons and shared. With the necessary normalization constant, the wavefunction is as drawn in Eq. 7.1, where the bar over one of the orbitals indicates that the electron spin is down. The second term in the equation is included because the hydrogens and electrons are indistinguishable. This covalent bond is shown pictorially in Fig. 7.1a.

$$\Phi_{HL} = \frac{1}{\sqrt{2}}\left|1s_1\overline{1s_2}\right| - \frac{1}{\sqrt{2}}\left|\overline{1s_1}1s_2\right| \qquad 7.1$$

While this was an extraordinary advance in electronic structure for multielectron species when proposed by Heitler and London, this function is not a reasonable approximation even for the hydrogen molecule's ground state. In order to obtain a more accurate function, it is necessary to include polar terms in the calculation where both electrons may reside on either hydrogen (Fig. 7.1b). The result in Eq. 7.1 is mixed with the polar function, where both electrons in H_2 are in a single *s*-orbital, Eq. 7.2, on H_1 for example.

$$\Phi_{ion(1)} = \left|1s_1\overline{1s_1}\right| \qquad 7.2$$

Naturally, we need a second polar function because the two hydrogens are equivalent. In fact, because they are equivalent we can put them together in a single term with a single coefficient in front of the term to define their contribution, Eq. 7.3. As a result, the full VB wavefunction, $\Phi_{VB\text{-}full}$, is given in Eq. 7.3, where λ and μ are weighting factors such that $\lambda + \mu = 1$.

$$\Phi_{VB-full} = \lambda\Phi_{HL} + \mu(\Phi_{ion(1)} + \Phi_{ion(2)}) \qquad 7.3$$

After Shaik and Hiberty,[2] we can use $H_a\bullet\!-\!\bullet H_b$ to signify the Heitler and London covalent part of the function and $H_a^-\ H_b^+$ and $H_a^+\ H_b^-$ to signify the two polar forms where both electrons reside on one of the hydrogens. Then, the actual bond between the two hydrogens in H_2, H_a—H_b is described as shown in Eq. 7.4, with resonance contributions between covalent and ionic forms. The covalent form contributes about 75%, $\lambda = 0.75$, to the ground-state wavefunction, and the polar forms contribute the remainder of 25% (12.5% for each of the equivalent ionic forms), $\mu = 0.25$. This is shown pictorially in Fig. 7.1c.

$$H_a - H_b \sim 0.75(H_a\bullet\!-\!\bullet H_b) + 0.25(H_a^- H_b^+ \leftrightarrow H_a^+ H_b^-) \qquad 7.4$$

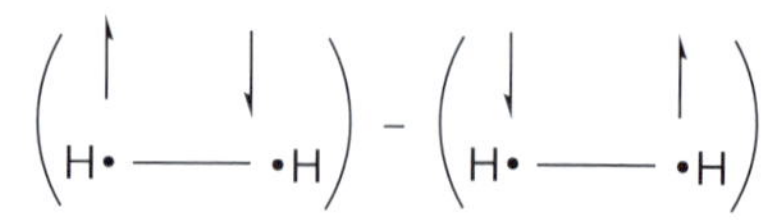

Fig. 7.1a Pictorial representation of Heitler and London's wavefunction for the covalent bond in the hydrogen molecule.

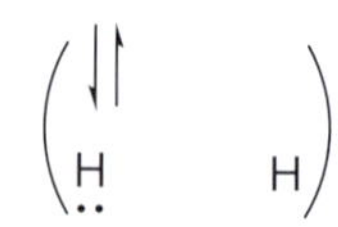

Fig. 7.1b Pictorial representation of one of the ionic bonds in the hydrogen molecule.

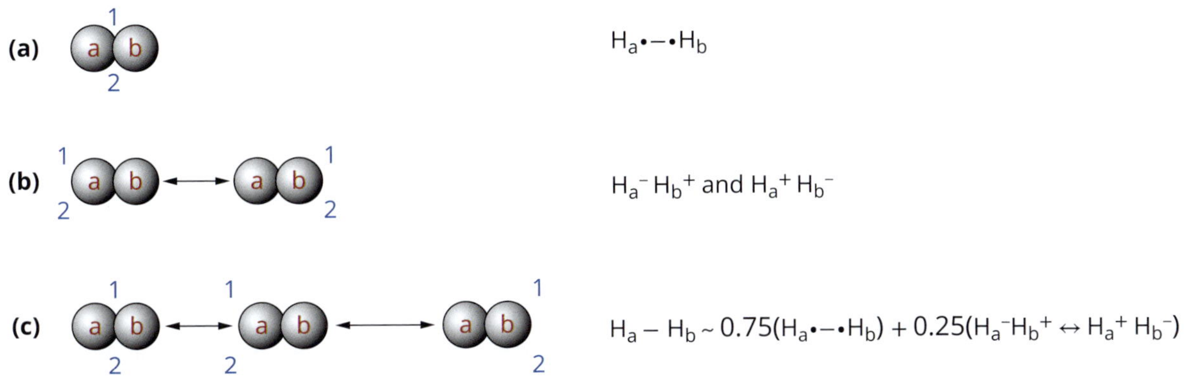

Fig. 7.1c Simplified VB description of the hydrogen molecule: (a) the covalent part of the bond where the electrons (1 and 2) are shared by the two nuclei (a and b); (b) the ionic part of the bond; (c) the complete description involving resonance between covalent and ionic forms.

Inclusion of only the covalent description for the bonding in H_2 produced a computed bond dissociation enthalpy (BDE) of 87 kcal/mol, quite far from the experimentally determined value of 104 kcal/mol. Inclusion of the ionic terms in the description improves the BDE value to 92.7 kcal/mol. Similar to what is done in MO theory calculations, the description can then be improved by inclusion of additional excited state terms and a bigger basis than the minimal 1*s* on each H, to allow for polarization of the 1*s*-function due to the presence of the second nucleus in the molecule. For example, the same Pople-type basis sets (Chapter 2) can be used in VB theory, such as 3-21G.

The advantage of the VB process is that the electron pairs are localized in the space near the atoms, which leads to smaller errors in electron correlation relative to MO theory. The disadvantage of the VB process is that for polyatomic molecules the number of possible resonance forms becomes extremely large rapidly, making implementation of MO theory for precise calculations on large molecules simpler.

An alternative, but closely related, method for describing the 2c2e bond in VB terms was described by Coulson and Fischer.[2,7] In this modification of the VB method, the ionic terms can be removed if we allow the wavefunctions for the individual *s*-orbitals, *a* and *b*, to delocalize slightly onto the adjacent atom; this is done using a mixing coefficient, ε, as shown in Eq. 7.5. The slightly delocalized wave-orbitals, φ_a and φ_b, are then used in a way analogous to the Heitler and London covalent description (Eq. 7.1); however, in this case the Coulson and Fischer slightly delocalized wavefunction Ψ_{CF} is equivalent to the full VB description (Eqs. 7.3 or 7.4) without explicit inclusion of the polar forms.

$$\begin{aligned} \Psi_{CF} &= |\varphi_a\bar{\varphi}_b| - |\bar{\varphi}_a\varphi_b| \\ \varphi_a &= a + \varepsilon b \\ \varphi_b &= b + \varepsilon a \end{aligned} \qquad 7.5$$

From a computational perspective it is fairly safe to say that VB methods are at the moment less accessible than their MO counterparts. However, there are continuous improvements to computational methods being introduced, and there are several quite powerful computational packages available for application of valence-bond techniques. For those interested in computational VB theory, we recommend several quite accessible books written by the theory's practitioners. Some related theoretical constructs, such as generalized valence bond theory,[8] are found in common computational packages such as Gaussian and Gamess US. Others, such as XMVB, can be obtained, often free of charge, from developers.[2]

Perhaps more importantly for our purposes here, VB theory provides a construct for the qualitative understanding of bonding and reactivity that is in many cases informative and powerful, as we will see later in this chapter and in other chapters.

7.1.2 Natural Bond Orbital (NBO) Theory[4–5]

NBO theory begins with the overall electron density of the molecule under study and partitions the electrons into natural atomic orbitals (NAOs) with shapes quite reminiscent of the familiar hydrogen-based orbitals. The NAOs are then orthogonalized and divided into a natural minimal basis for the atoms involved, which are related to the occupied atomic orbitals for isolated atoms. This set of orbitals usually has occupation numbers significantly higher than zero.

During this process, weakly occupied "Rydberg" orbitals are also generated. The number of NAOs generated is related to the number of atomic orbitals, and the number of Rydberg orbitals increases as the basis set used in the calculation increases in size.

The density matrix organized in this way is then searched for atomic blocks with occupation numbers of essentially two electrons, which are then assigned as core orbitals. Then a search is done to find atomic orbitals with an occupancy less than exactly two but very high (~1.90 electrons), which are assigned as lone pairs. Pairs of atoms are then considered with high occupancies, which are then assigned as 2-electron bonds. Finally, if the process has not found enough NBOs to correspond to the NAOs, the program begins to look for other types of bonding patterns, like 3-center bonds discussed in detail in the next chapter. The NBOs can be expressed as linear combinations of the NAOs, producing explicit hybridizations for each NBO.

This description of structures as a Lewis-based system, for the most part, does give NBO a flavor very similar to VB theory. However, while Pauling–Wheland VB theory[2] describes bonds, even homopolar bonds, as resonance hybrids between covalent and ionic contributors (Eqs. 7.3 and 7.4), NBO describes bonds in terms of orbital functions delocalized over the centers with usually near-double occupancy. In some respects this makes it more analogous to Coulson–Fischer VB theory, but even here there are significant differences in the mathematical details.

While NBO theory does not, unlike VB theory, use resonance between ionic and covalent forms within 2-electron bonds, the program does specify polarization coefficients within the bonds and charges on individual atoms, via natural population analysis, during this procedure.

One powerful feature of NBO theory is its expansion to natural resonance theory (NRT). The NBO/NRT takes the electron density and partitions it using canonical Lewis-structure contributors. The output of the technique is a set of contributing resonance forms that describe the bonding in the molecule under study as a set of weighted Lewis structures.

From a computational perspective, NBO/NRT is remarkably accessible and is even relatively insensitive to the method used to obtain the electron density; for example, DFT is adequate as long as a reasonably good basis is employed. (It is preferable to use a triple-ζ basis. NBO seems fairly insensitive to diffuse functions.) NRT is largely applicable to fairly simple compounds or models of more complex compounds as the number of contributing resonance forms can quickly become very large, but the NBO/NRT programs suite is a very powerful hybridization-based computational method that is quite popular in the chemical community. As mentioned in Chapter 3, NBO is available in, or can be added onto, several popular computational programs, including Gaussian and Gamess. An advantage of NBO theory is that it allows the calculation of the exact hybrids being used by a central atom in bonding to its substituents.

Before examining the application of hybrids to molecular systems, let us examine a few important features of hybridization. In this chapter the basics of hybridization theory will be discussed, and applications to hypovalent and normal valent compounds. The bonding schemes here will only involve 2-center 2-electron bonds. In the next chapter, both MO and VB theory will be applied to other bond types. One of the modest goals of this short chapter is to thoroughly explain the hybridization parameter, which is introduced in the next section, and persuade the reader into thinking in terms of hybridizations from sp^0 *to* sp^∞ in incremental values, rather than integer sp^1, sp^2, and sp^3 as taught in organic chemistry.

7.2 The *sp*-Hybridization Parameter

One of the canonical features of VB theory is hybridization of atomic orbitals. In many contexts, these hybrids are used as a simple mathematical construct to generate a specific geometry. In reality, hybrid orbitals are quite sensitive to both geometry *and* the particular substituent to which they are bound in an interdependent fashion, as will be shown.

The formation of *sp*-hybrids can be thought of as taking the form shown in Eq. 7.6. Here we assume that the atom in question is a first-row *p*-block element, such as nitrogen, with 2*s* and 2*p* valence orbitals. The constant *a* is a coefficient for the amount of *p*-orbital character involved in the *sp*-hybrid orbital ϕ_{sp}. The function in the equation is normalized by the fraction in front of the brackets. (Normalization is a method of ensuring that a linear combination of atomic orbitals is just a single orbital. See Chapter 6 for more details.)

$$\phi_{sp} = \frac{1}{\sqrt{1+a^2}}[2s + a(2p)] \qquad 7.6$$

If we want to find the electron density in the hybrid orbital ϕ_{sp}, we square Eq. 7.6, which gives the function in Eq. 7.7. The mixed term, "2a(2*s*)(2*p*)", is zero due to the orthogonality of atomic orbitals. Then, we replace a^2 with a new symbol, λ, where λ is an important coefficient called the *hybridization parameter*. The amount of *s*-character in the hybrid is then $1/(1 + \lambda)$, while the amount of *p*-character is $\lambda/(1 + \lambda)$, the coefficients in front of $(2s)^2$ and $(2p)^2$ in Eq. 7.7.

$$\begin{aligned}
\phi_{sp}^2 &= \frac{1}{1+a^2}[(2s)^2 + 2a(2s)(2p) + a^2(2p)^2] \\
\phi_{sp}^2 &= \frac{1}{1+a^2}(2s)^2 + \frac{a^2}{1+a^2}(2p)^2 \\
\phi_{sp}^2 &= \frac{1}{1+\lambda}(2s)^2 + \frac{\lambda}{1+\lambda}(2p)^2
\end{aligned} \qquad 7.7$$

For any *i*-th hybrid with hybridization parameter λ_i, we can define the hybridization parameter as shown in Eq. 7.8. The coefficients *a* for p_x-, p_y-, and p_z-orbitals labeled as a_{ix}, a_{iy}, and a_{iz}, and a_{i0} is the coefficient for the *s*-orbital. The hybridization parameter is, more simply, the percentage of *p*-character in the hybrid divided by the percentage of *s*-character.

$$\lambda_i = \frac{a_{ix}^2 + a_{iy}^2 + a_{iz}^2}{a_{i0}^2} = \frac{\%p}{\%s} \qquad 7.8$$

Whether most chemistry students are aware of it or not, they are probably quite familiar with the parameter λ. This is the value given in *sp*-hybridizations as a superscript. In other words, *sp*, sp^2, and sp^3 values are *sp*-hybridization parameters where $\lambda = 1$, 2, and 3, respectively; that is, the superscript in sp^λ. However, this is as far as most are taught hybridization. In actuality, λ can run from any value between 0 and infinity,

sp^0 to sp^∞. On this more substantial scale, the "hybrid" sp^0 is a pure *s*-orbital, and sp^∞ is a pure *p*-orbital.

As mentioned previously, the fraction of *s*-character in any *i*-th hybrid (f_s^i) must be as shown in Eq. 7.9, while the fraction of *p*-character (f_p^i) is as shown in Eq. 7.10 (see Eq. 7.7).

$$f_s^i = \frac{1}{1+\lambda_i} \quad \textbf{7.9}$$

$$f_p^i = \frac{\lambda_i}{1+\lambda_i} \quad \textbf{7.10}$$

Furthermore, there is no requirement for a hybrid to be an integer, and a hybrid with parameter $\lambda = 1.5$, $sp^{1.5}$, simply has intermediate amounts of *p*-character relative to *sp* and sp^2, as would be expected. Indeed, by using Eq. 7.9 one can calculate the *s*-character of *sp*, $sp^{1.5}$, and sp^2 hybrids as 50%, 40%, and 33%, respectively.

Naturally, any main-group atom only has one *s* and three *p* valence orbitals available when forming bonds through specific hybrids. Consequently, we can define the *Sum Rules*, which simply state that the sum of all the hybrids around a specific atom will lead to one *s* and three *p*-orbitals. Mathematically (see Eq. 7.7), these can be expressed as shown in Eq. 7.11 and Eq. 7.12, which will be of significant utility in this chapter and the next.

$$\sum_i^n \frac{1}{1+\lambda_i} = 1\,(s\text{-orbital}) \quad \textbf{7.11}$$

$$\sum_i^n \frac{\lambda_i}{1+\lambda_i} = 3(p\text{-orbitals}) \quad \textbf{7.12}$$

7.2.1 Calculating Hybrids in Detail (Advanced Topic)

As an example, we can look at the common problem of generating three equivalent hybrids that point towards three atoms that all reside in the same plane 120° apart, this can be done using combinations of the *s*- and *p*-orbitals. By convention, the 3-fold rotational axis will be along the *z*-axis (Section 5.2.2). Consequently, the three hybrids reside in the *xy*-plane. The equations for the p_x- and p_y-orbitals are as follows. The function in front, $\frac{R(r)}{r}\sqrt{3/4\pi}$, of the direction, for example *x* or *y*, defines the radial extension of the orbital along the *x*- or *y*-axis and does not need to be considered further here. We are mostly concerned with the direction of the hybrid at the moment.

$$p_x = \frac{R(r)}{r}\sqrt{3/4\pi}\cdot x$$

$$p_y = \frac{R(r)}{r}\sqrt{3/4\pi}\cdot y$$

We can consider forming new orbitals where we take linear combinations of the equations for the p_x- and p_y-orbitals and to give *p*-orbitals pointing at the three substituents at $\alpha = 0°$, 120°, and 240°. These ligands (L_1, L_2, L_3) will interact with the new, rotated *p*-orbitals p_1, p_2, and p_3, which will be defined by the following equation.

$$p = \cos\alpha\cdot(p_x) + \sin\alpha\cdot(p_y)$$

Consequently, the three hybrid orbitals are the three linear combinations of p_x- and p_y-orbitals shown as follows. These orbitals are not yet orthogonal. The reader will also readily recognize that they do not obey the Sum Rules, as these are three equivalent hybrids formed from two atomic orbitals . . . we will fix that in a moment.

$$p_1 = \cos(0)(p_x) + \sin(0)(p_y)$$
$$p_1 = p_x$$

$$p_2 = \cos(120)(p_x) + \sin(120)(p_y)$$
$$p_2 = -\frac{1}{2}(p_x) + \frac{\sqrt{3}}{2}(p_y)$$

$$p_3 = \cos(240)(p_x) + \sin(240)(p_y)$$
$$p_3 = -\frac{1}{2}(p_x) - \frac{\sqrt{3}}{2}(p_y)$$

To form the three hybrid orbitals we will need an orbital other than just the two *ps*; that is, we need to mix in the *s*-orbital. The mixing of the spherical *s*-orbital into the rotated linear combinations of p_x- and p_y-orbitals does not change their directions. The hybrid orbitals will have the following forms (see Eq. 7.6).

$$h_1 = \frac{1}{\sqrt{1+\lambda}}[(s) + a(p_1)]$$
$$h_2 = \frac{1}{\sqrt{1+\lambda}}[(s) + a(p_2)]$$
$$h_3 = \frac{1}{\sqrt{1+\lambda}}[(s) + a(p_3)]$$

The only question remaining is the value of the hybridization parameter λ in these equations. (Keep in mind that $a = \sqrt{\lambda}$, so the two constants are related.) A requirement placed on the hybrids on an atom, as well as atomic orbitals, is that they be orthogonal, having no net overlap.

If we orthogonalize the hybrid orbitals (h_1, h_2, and h_3), we can find the value of the hybridization parameter. Operationally, we can take the overlap integral of h_1 and h_2 ($\langle h_1 | h_2\rangle$) and set it equal to zero to orthogonalize; in essence, by setting the equation equal to zero we are finding the value of λ where the orbital fits the definition of "orthogonal". (See Chapter 6 for additional examples of such procedures. The equal normalization factors can be ignored.)

$$0 = \langle h_1 | h_2 \rangle$$
$$= \langle (s) + a(p_1) \,|\, (s) + a(p_2) \rangle$$

Substituting the values of p_1 and p_2 in terms of p_x and p_y, and then multiplying, produces

$$0 = \left\langle (s) + a(p_x) | (s) - \frac{a}{2}(p_x) + a\frac{\sqrt{3}}{2}(p_y) \right\rangle$$
$$= \langle s \mid s \rangle - \frac{a}{2}\langle s | p_x \rangle + a\frac{\sqrt{3}}{2}\langle s | p_y \rangle + a\langle p_x | s \rangle$$
$$-\frac{a^2}{2}\langle p_x | p_x \rangle + a^2\frac{\sqrt{3}}{2}\langle p_x | p_y \rangle$$

The integrals $\langle \phi_1 | \phi_1 \rangle = 1$, such as $\langle s | s \rangle = 1$; in other words, the overlap of an orbital with itself is unity. The overlap of two different atomic functions, $\langle \phi_1 | \phi_2 \rangle = 0$, such as $\langle s | p_x \rangle = 0$, is necessarily zero due to the orthogonality of atomic orbitals. Consequently, it can be greatly simplified.

$$0 = \langle s | s \rangle - \frac{a}{2}\langle s | p_x \rangle + a\frac{\sqrt{3}}{2}\langle s | p_y \rangle + a\langle p_x | s \rangle$$
$$-\frac{a^2}{2}\langle p_x | p_x \rangle + a^2\frac{\sqrt{3}}{2}\langle p_x | p_y \rangle$$
$$= 1 - \left(\frac{a}{2} \times 0\right) + \left(a\frac{\sqrt{3}}{2} \times 0\right) + \left(a \times 0\right)$$
$$-\left(\frac{a^2}{2} \times 1\right) + \left(a^2\frac{\sqrt{3}}{2} \times 0\right)$$
$$= 1 - 0 + 0 + 0 - \frac{a^2}{2} + 0$$

Solving for a^2 and keeping in mind that $a^2 = \lambda$,

$$\frac{1}{2}a^2 = 1$$
$$a^2 = 2$$
$$\lambda = 2$$

In order to make three equivalent hybrids 120° apart in the xy-plane, the hybridization parameter must be 2 in order for the hybrids to be orthogonal to one another. While this is probably not a surprise. as this hybridization is often invoked for planar carbons in, for example, aromatic rings where substituents are ~120° apart, it does illustrate an important principle: *hybrid orbitals, in order to be orthogonal to one another, can only exist at certain angles.*

The three hybrids are then

$$h_1 = \frac{1}{\sqrt{3}}s + \frac{2}{\sqrt{6}}p_x$$

$$h_2 = \frac{1}{\sqrt{3}}s - \frac{1}{\sqrt{6}}p_x + \frac{1}{\sqrt{2}}p_y$$

$$h_3 = \frac{1}{\sqrt{3}}s - \frac{1}{\sqrt{6}}p_x - \frac{1}{\sqrt{2}}p_y$$

For an alternative method based on group theory to derive the same hybrids, see Cotton (3rd ed., p. 225).[9]

7.2.2 The Coulson Directionality Theorem and Relative Energies of Hybrids

If one derives the general expression for the angle between any two hybrids (ω_{ij}) with hybridization parameters λ_i and λ_j, it takes the form shown in Eq. 7.13, often called the *Coulson directionality theorem.*[3]

$$\cos\omega_{ij} = -\frac{1}{\sqrt{\lambda_i\lambda_j}} \qquad \textbf{7.13}$$

For equivalent hybrids, where $\lambda_i = \lambda_j = \lambda$, the equation reduces to

$$\cos\omega = -\frac{1}{\lambda} \qquad \textbf{7.14}$$

and is applicable to any two equivalent *sp*-hybrids.

For example, for the familiar integer values of *sp*, sp^2, and sp^3 along with sp^∞ (two pure *p*-orbitals) the following values are obtained. (Technically, the last is the limit as the equation approaches infinity. Keep in mind that infinity is not a number; it is an idea. Make sure the equation your using gives a value consistent with a pure *p*-orbital, which is what sp^∞ is being used to signify.)

$$\lambda = 1 : \cos\omega = -1, \omega = 180^\circ$$
$$\lambda = 2 : \cos\omega = -\frac{1}{2}, \omega = 120^\circ$$
$$\lambda = 3 : \cos\omega = -\frac{1}{3}, \omega = 109.47^\circ$$
$$\lambda = \infty : \cos\omega = 0, \omega = 90^\circ$$

Note that as the hybridization parameter is increasing, the angle between the equivalent hybrids is decreasing. As a result, the closest approach that the two equivalent *sp*-hybrids can make is 90°, when the hybridization value is itself infinite and the atom is using pure *p*-orbitals (Fig. 7.2).

Naturally, these hybrids do not have the same energies as the atomic orbitals that comprise them, and it can be shown that the energy of the hybrid, E_{hyb}, is as shown in Eq. 7.15, where $\Delta E_{s\to p}$ is the energy gap between the *s*- and *p*-orbitals

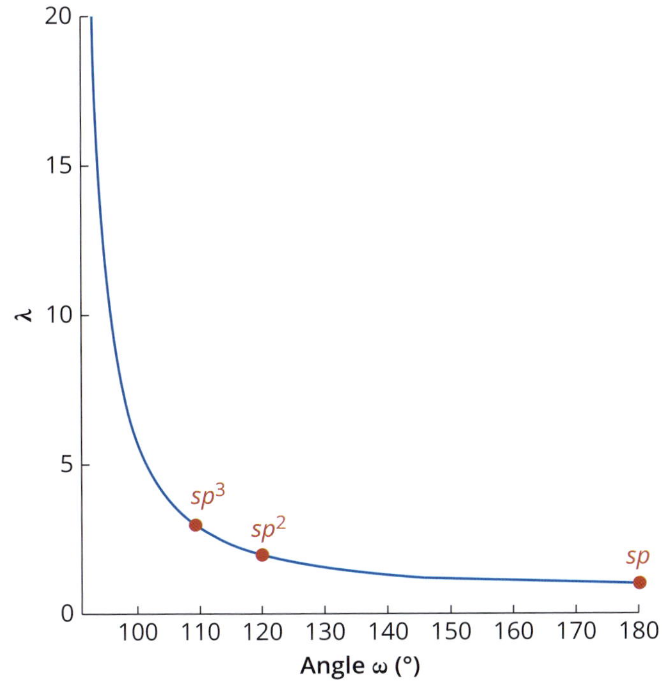

Fig. 7.1d Plot of hybridization parameter (λ) versus angle between hybrids.

in the valence shell (Fig. 7.2). Consequently, the energy of the hybrid is expressed as a fraction of the *s-p* energy gap.

$$E_{hyb} = \frac{\lambda}{1+\lambda}\Delta E_{s\to p} \qquad \mathbf{7.15}$$

The sp^1-hybrid is then halfway between the *s*- and *p*-orbitals in energy, due to equivalent contributions, while sp^2-hybrids are 2/3 of the way up the energy scale between the valence atomic orbitals. *Obviously, but importantly, more s-rich hybrids will be lower in energy.*

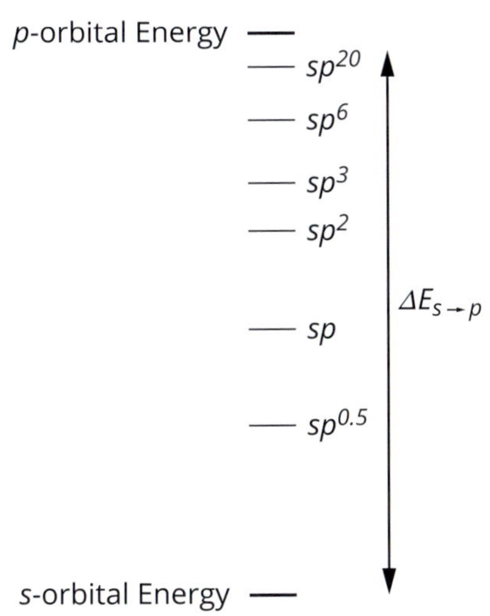

Fig. 7.2 Energies of some hybrids relative to the *s*- to *p*-orbital energy gap.

7.3 Application of Hybridization Theory for Some Simple Compounds and Bent's Rule

The qualitative application of hybridization theory to simple molecules is often achieved by using simple geometries. If the molecule is close to trigonal planar, then the hybridization is assumed to be sp^2, for example. However, using the hybridization parameter as developed previously, we can obtain a bit more detail about the bonding in various compounds, and the only input required is the geometry, the same input as required for MO theory.

Different ligands, lone pairs, single electrons, and so on, bound to a central atom will not reside in identical hybrids as is often assumed. There is no reason to believe that an atom will use the same hybrid to bond both to a fluorine atom and to house a lone pair, and indeed it will not. Even in compounds like ethylene—where we often consider the carbon to be using "sp^2-hybrids"—the carbon will not use exactly the same hybrids to bond to the adjacent carbon that it uses to bond to its hydrogens.

Looking in more detail at the simple example of an olefin, we can use tetrafluoroethylene, in order to make the electronegativity difference between the substituents and carbon more obvious than the ethylene case. Will the carbon use more *s*-rich hybrids to bond to the fluorine or to the adjacent carbon?

As a guide for situations of this type we can use *Bent's rule*, named after Henry Bent: "Atomic *s*-character concentrates in orbitals directed toward electropositive substituents".[10] Expressed in a different way: a central atom tends to direct hybrids of higher *p*-character toward more electronegative groups.

As an illustration we can look at the extreme, and very common, example of a compound with both a lone pair and a substituent of some type, X. The substituent X is bonded to the central atom with a pair of electrons.

Similarly, one might consider the lone pair to be a substituent with an electronegativity of zero. In this sense, a lone pair is a bonding pair with a substituent unable to help support the electron density in the hybrid. Consequently, the central atom will use a hybrid orbital that is best able to support this electron density on its own. The lower-energy (Eq. 7.15) *s*-rich hybrid will be used for the lone pair. In other words, the central atom E will use its most electronegative orbital (orbital with high *s*-character; see Mulliken–Jaffe Electronegativity, Section 3.4.2) to aid the least electronegative substituent A in supporting the bonding electrons.

The substituent X has an electronegativity greater than zero and can help support the bonding pair of electrons through Coulombic stabilization. The central atom can use a higher energy *p*-rich hybrid in order to bond to X because of this aid in stabilizing the pair of bonding electrons.

Returning to the example of tetrafluoroethylene, the computed hybrid (NBO_5) on carbon used to bond to the fluorines is

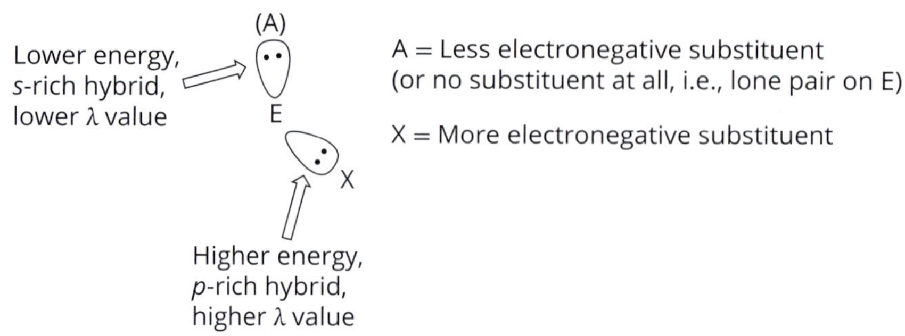

Fig. 7.3 Illustration of hybrid character on central atom E as a function of electronegativity of substituents A and X.

$sp^{2.7}$, while the carbon uses an $sp^{1.2}$-hybrid to bond to the adjacent carbon. Again, this is much as expected with higher-energy p-rich hybrids bonding to more electronegative fluorine.

However, to find the hybridization, all one really needs for many simple molecules is the geometry. The experimental geometry (Fig. 7.4) from electron diffraction[11] for tetrafluoroethylene has F–C–F angles of 113°.

Each carbon will use two equivalent hybrids to form the two equivalent C–F bonds. With the angle of 113° one can calculate the required hybridization of these two C–F bond-forming orbitals on the carbon by using the Coulson directionality theorem (Eqs. 7.13 and 7.14). The hybridization of the orbitals on carbon used to make these bonds is $sp^{2.6}$, nearly identical to that calculated using NBO (*vide supra*).

$$\cos\omega = -\frac{1}{\lambda}$$
$$\cos 113^{\circ} = -\frac{1}{\lambda}$$
$$\lambda = 2.56$$

The carbon will use a pure p-orbital to form the C–C π-bond. Consequently, three of the four hybrids being used by the carbon to form bonds are already described: 2 C–F bonds of hybridization $sp^{2.56}$ and an sp^{∞} hybrid being used for the π-bond. The only unknown orbital composition is the hybrid in use to form the C–C σ-bond. By using the s-orbital sum rule (Eq. 7.11) we can determine this hybridization, shown as follows. (The calculation uses that $\lim_{\lambda\to\infty}\frac{1}{1+\lambda} = 0$.) Again, the result is very similar to the NBO calculation (*vide supra*).

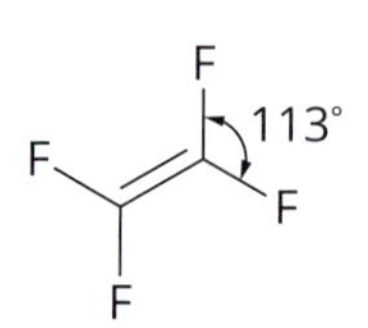

Fig. 7.4 Structure of tetrafluoroethylene with the angle between F–C–F atoms from electron diffraction.

$$\sum_i^n \frac{1}{1+\lambda_i} = 1$$
$$\frac{1}{1+2.56}+\frac{1}{1+2.56}+\frac{1}{1+\infty}+\frac{1}{1+\lambda_{C-C}} = 1$$
$$0.281+0.281+0+\frac{1}{1+\lambda_{C-C}} = 1$$
$$\lambda_{C-C} = 1.29$$

Using the p-orbital sum rule, Eq. 7.12 proceeds as follows, for which $\lim_{\lambda\to\infty}\frac{\lambda}{1+\lambda} = 1$ was used.

$$\sum_i^n \frac{\lambda_i}{1+\lambda_i} = 3$$
$$\frac{2.56}{1+2.56}+\frac{2.56}{1+2.56}+\frac{\infty}{1+\infty}+\frac{\lambda_{C-C}}{1+\lambda_{C-C}} = 3$$
$$0.719+0.719+1+\frac{\lambda_{C-C}}{1+\lambda_{C-C}} = 3$$
$$\frac{\lambda_{C-C}}{1+\lambda_{C-C}} = 0.562$$
$$\lambda_{C-C} = 1.28$$

7.3.1 NH_3 Bonding Analysis: VB and MO Theory Comparison

Ammonia has an experimental H–N–H bond angle of 107° between each of the three equivalent hybrids on nitrogen bonding to hydrogens in the C_{3v} molecule. As would be expected with the angles being slightly smaller than tetrahedral, the hybridization parameter is slightly higher than $\lambda = 3$ (Fig. 7.1d).

$$\cos\omega = -\frac{1}{\lambda}$$
$$\cos 107^{\circ} = -\frac{1}{\lambda}$$
$$\lambda = 3.42$$

The hybridization of the orbital on nitrogen housing the lone pair ($\lambda_{L.P.}$) can be found using the sum rule (Eq. 7.11). The hybrid is significantly higher in *s*-character as predicted by Bent's rule, and is ~sp^2 hybridized. (*Do not be confused by the hybridization of the lone pair!* We are not discussing three equivalent sp^2-hybrids 120° apart as are often invoked for vinyl carbons. We are discussing a single sp^2-hybrid containing a pair of electrons.)

$$\sum_i^n \frac{1}{1+\lambda_i} = 1$$

$$3\left[\frac{1}{1+3.42}\right] + \frac{1}{1+\lambda_{L.P.}} = 1$$

$$0.679 + \frac{1}{1+\lambda_{L.P.}} = 1$$

$$\lambda_{L.P.} = 2.11$$

What is the energy of the $sp^{2.11}$-hybrid holding the pair of electrons? We can determine this as a fraction of the *sp*-energy gap by using Eq. 7.15. In other words, the lone pair is about 2/3 of the way up from the *s*-orbital on nitrogen toward the *p*-orbital in energy.

$$E_{hyb} = \frac{\lambda}{1+\lambda}\Delta E_{s\to p}$$

$$E_{hyb} = \frac{2.11}{1+2.11}\Delta E_{s\to p} = 0.678 \cdot \Delta E_{s\to p}$$

Let us compare this with the results from MO theory in some detail. The SALCs for the hydrogens on NH_3 provide two representations, a_1 and *e*. These orbitals are extremely reminiscent of the totally symmetric and degenerate sets for the hydrogens in BH_3 or the *s*-orbitals on the fluorines in BF_3 (Chapter 6). The atomic orbitals on nitrogen have symmetries a_1 (*s* and p_z) and *e* (p_x, p_y).

The full MO diagram and full VB diagram for NH_3 are shown in Fig. 7.5. In a general sense one can easily see the main difference between the two pictures. MO theory applies symmetry operations to the substituents on the periphery (hydrogens in this case) to generate delocalized orbitals of appropriate symmetry to mix with the pure atomic orbitals on the central atom (nitrogen). Conversely, VB theory modifies the atomic orbitals of the central atom through hybridization to fit the positions of the substituents on the periphery, and the hydrogens simply use their *1s*-orbitals to bind to these hybrids. The MO picture is delocalized due to the symmetry adaptation of the ligand orbitals and orbital mixing with the atomic orbitals on nitrogen. The VB picture is more localized into orbitals containing N–H bonding electrons and lone-pair electrons.

Looking specifically at what the qualitative MO picture reveals that there are several important features. First, one can identify three pairs of bonding electrons residing in $1a_1$ and $1e$, with their corresponding antibonding orbitals $3a_1$* and $2e$*. In the MO diagram there is a 3-orbital interaction of a_1 symmetry, where the SALCs have one orbital and the nitrogen

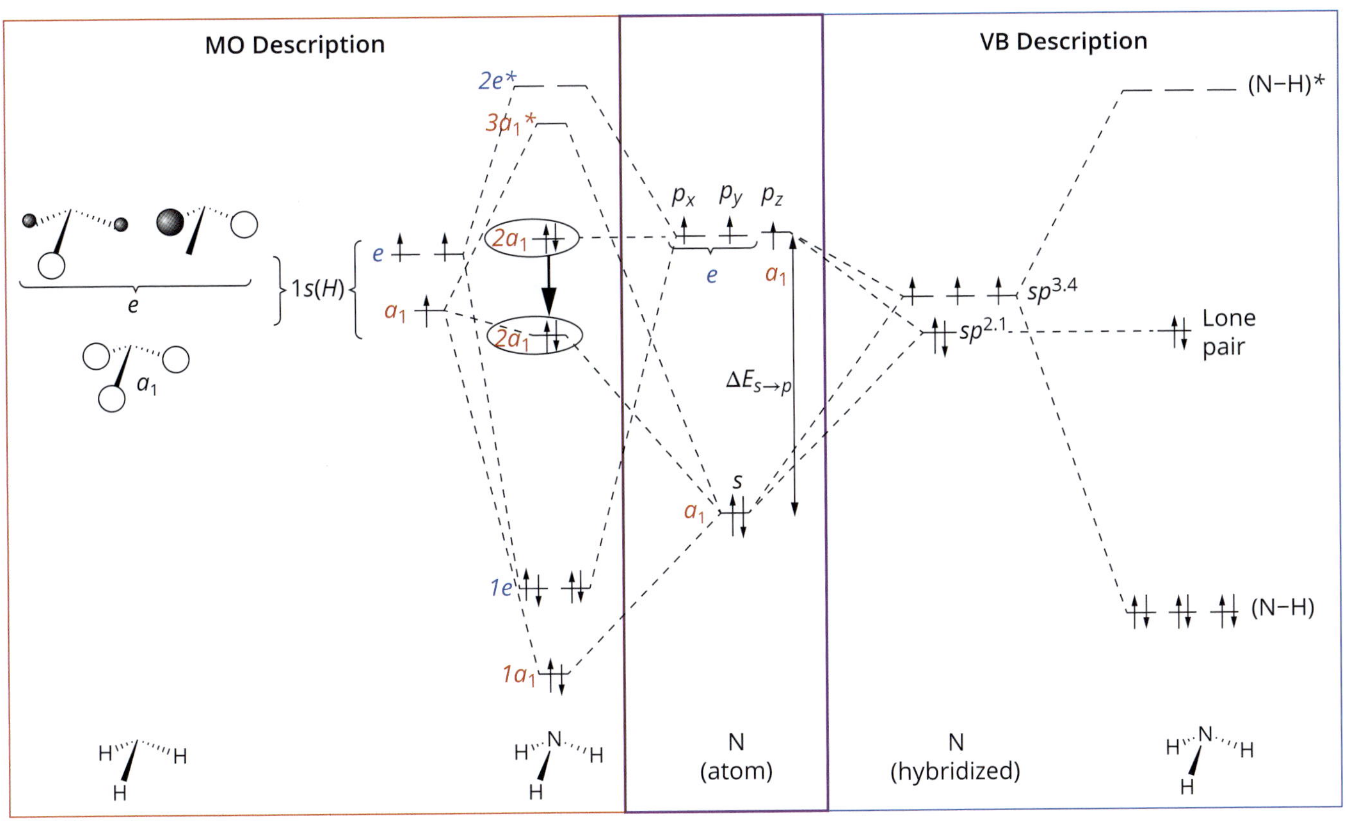

Fig. 7.5 MO (red box) and VB (blue box) diagrams for ammonia. (See the text for discussion.)

has two orbitals of this symmetry. We expect a bonding, antibonding, and perhaps nonbonding (approximately) set of MOs from these orbitals. If we assume that $2a_1$ is nonbonding, we might perhaps glean that the bonding is being done exclusively by the *s*-, p_x-, and p_y-orbitals on nitrogen, and the p_z-orbital contains the lone pair of electrons. Indeed, this is the suggested assumption in Chapter 6, provided you have no additional information.

Qualitative VB theory actually provides explicit information on the nature of this pair of electrons, including its relative energy as a fraction of the *s*-*p* energy gap; however, the lone pair composition in the VB state is an optimized, localized state. In essence, the VB picture is related to the MO picture where the 3-orbital interaction of a_1 representations has been altered so that orbital $2a_1$ is exactly nonbonding with no hydrogen delocalization, one reasonable outcome of the mixing. Will an MO calculation give exactly this composition for the lone pair? Perhaps, but depending on basis optimization and the effect of delocalization on how the program partitions the electrons may provide mixtures of the three a_1 representations other than the one suggested by VB theory. However, at the limit of the theories the VB and MO theories will produce identical pictures of the total electron density of the molecule. The qualitative analysis by MO and VB theory often produces complementary information.

If one adopted the same *sp*-mixing from the VB picture for the qualitative MO case as well, then the composition of the bonding set between the two theories reflects the localized bonding analysis of VB theory versus delocalized for MO theory. The bonding MOs ($1a_1$ and $1e$) may be mixed to get the VB picture with an appropriate energy 2/3 of the way up between the $1a_1$ and $1e$ atomic nitrogen energy levels. The lone-pair energy is then explicitly determined in the qualitative VB model as $0.678\ \Delta E_{s\rightarrow p}$.

This simple example illustrates the power of these methods for understanding orbital hybridization. MO theory will often tell you that two orbitals of similar symmetry will mix, but one is left to guess, based on overlap and energy differences, exactly how much the orbitals will mix. In main-group systems, valence bond theory can provide very accurate values for the hybrids being used by an atom, an interpretation of the amount of orbital mixing. In the MO description the 3N–H bonds are equivalent because of the delocalized nature of the MOs; in the VB description, the 3N–H bonds are electronically identical. *In this case and many others, VB theory and MO theory are more powerful when used in tandem than alone.*

7.3.2 H_2O as a Model for Systems with Two Lone Pairs

For systems such as water, ethers, most halide compounds, and so on, there are multiple lone pairs on a single atom. One can envision the lone pairs equally distributing the *s*-orbital character or placing all the *s*-orbital character into one highly-stabilized lone pair with other lone pairs occupying higher-energy pure *p*-orbitals. These two different scenarios can be analyzed in detail by using hybridization theory for their effects on electron density and energy of the system. As an advanced topic this is presented in the next section. Following this, some recommendations are made on how to analyze such systems.

7.3.2.1 Equal and Unequal Distribution of *s*-Character in Lone Pairs on an Atom: Calculation of Energy and Electron Density (Advanced Topic)

The Lewis dot structure for water suggests two equivalent O–H bonds along with two equivalent lone pairs, and this is a reasonable description of the main contributor to the electronic structure of water from a rough perspective (Fig. 7.6). We can use these tools to calculate by hand if there is an electronic energy difference or a difference in electron density of the two descriptions of the lone pairs in water. This is presented here as an exercise in using the formulas for hybridization theories discussed previously.

The H–O–H angle in water is 104.5° in the gas phase. We can calculate the two equivalent hybrids being used by oxygen to form the OH bonds by using the Coulson directionality theorem (Eq. 7.14) as $sp^{4.0}$.

$$\begin{aligned} \cos\omega &= -\frac{1}{\lambda} \\ \cos 104.5^\circ &= -\frac{1}{\lambda} \\ \lambda &= 3.99 \end{aligned}$$

The hybridization of two equivalent lone pairs in water would then be accessible from the sum rule for the *s*-orbital (Eq. 7.11). The two equivalent lone pairs would then be housed in $sp^{2.3}$-hybrids. As expected from Bent's rule, the hybrids containing

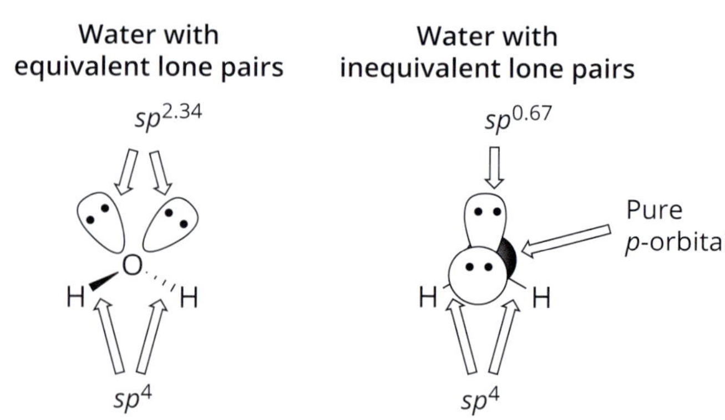

Fig. 7.6 Two different hybridizations for the lone pairs in water: one where the lone pairs reside in equivalent hybrids, and another where one lone pair is a pure *p*-orbital with the other in an *s*-rich hybrid.

the lone pairs have a significantly higher *s*-character than those being used to form the bonds.

$$\sum_{i}^{n} \frac{1}{1+\lambda_i} = 1$$

$$2\left(\frac{1}{1+\lambda_{O-H}}\right) + 2\left(\frac{1}{1+\lambda_{L.P.}}\right) = 1$$

$$2\left(\frac{1}{1+3.99}\right) + 2\left(\frac{1}{1+\lambda_{L.P.}}\right) = 1$$

$$\lambda_{L.P.} = 2.34$$

If we draw the MO diagram for water (Fig. 7.7) we find a major difference with the VB picture described previously. The SALCs for the 1*s*-orbitals on the hydrogens are in representations a_1 (in-phase combination) and b_2 (out-of-phase combination). The atomic orbitals on the oxygen are in a_1 (s and p_z), b_1 (p_x), and b_2 (p_y). There are two bonding orbitals in the MO: $1a_1$ and $1b_2$. The two lone pairs are in the orbitals marked $2a_1$ and b_1. The b_1 orbital is entirely nonbonding with respect to the hydrogens, and contains a pure *p*-orbital lone pair on the oxygen. The $2a_1$ orbital is due to a 3-orbital interaction between s and p_z on oxygen and the in-phase SALC on the hydrogens. This "lone pair" is mixing with $1a_1$ somewhat, and in most MO calculations has some slight bonding character with the hydrogens but is largely nonbonding.

Can we develop a VB picture consistent with the MO diagram? Of course! From Bent's rule we know that the lone pairs will have more *s*-character than the bonding orbitals. Previously, we assumed that the two lone pairs are equivalent. If we change that assumption to be that one of the lone pairs will pick up as much *s*-character as possible and that the other is a pure *p*-orbital, we can use the sum rule to get the hybridization of the *sp*-hybrid holding the lone pair. In this case, the orbitals holding the lone pairs will, on average, have more *s*-character than the bonding orbitals. The molecule would then have one lone pair in a pure *p*-orbital and another in an $sp^{0.67}$-hybrid.

$$\sum_{i}^{n} \frac{1}{1+\lambda_i} = 1$$

$$2\left(\frac{1}{1+\lambda_{O-H}}\right) + \left(\frac{1}{1+\lambda_{L.P.-pure\ p}}\right) + \left(\frac{1}{1+\lambda_{L.P.-sp-hybrid}}\right) = 1$$

$$2\left(\frac{1}{1+3.99}\right) + \left(\frac{1}{1+\infty}\right) + \left(\frac{1}{1+\lambda_{L.P.-sp-hybrid}}\right) = 1$$

$$2(0.200) + 0 + \left(\frac{1}{1+\lambda_{L.P.-sp-hybrid}}\right) = 1$$

$$\lambda_{L.P.-sp-hybrid} = 0.667$$

The average energy of the four electrons in the equivalent lone pairs versus the case with two different lone pairs will be the same. The average energy of the inequivalent lone pairs (pure *p*-orbital and $sp^{0.7}$-hybrid) will be the same as when two equivalent lone pairs are used ($sp^{2.3}$).

We can now define all of the hybrid orbitals on oxygen in water. This is done as in Eq. 7.6, again keeping in mind that $\lambda = a^2$ in this equation. The fraction of *s*- and *p*-orbitals in the

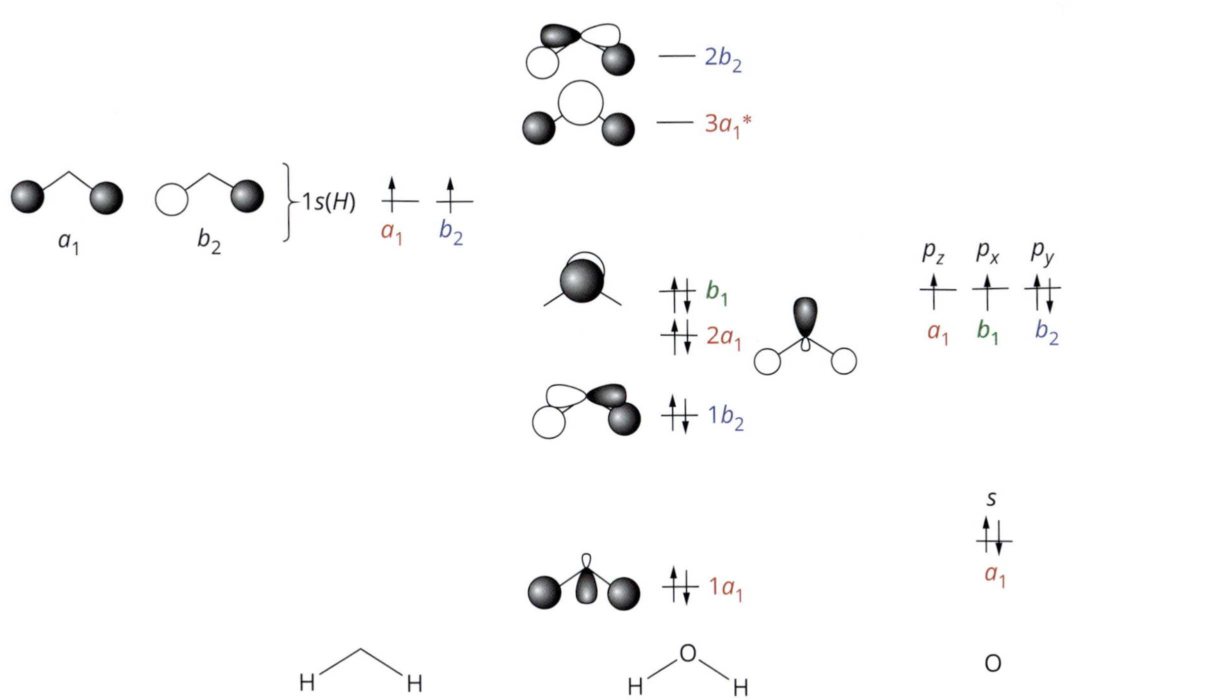

Fig. 7.7 MO diagram of water.

O–H bonding hybrids are found using Eq. 7.9 and Eq. 7.10 to be 0.20 and 0.80, respectively; this produces Eq. 7.16 for the two bonding hybrids. Note that the coefficients in that equation are the square root of the fraction calculated or $\sqrt{(1/(1+\lambda)}$ for the *s*-contribution and $\sqrt{(\lambda/(1+\lambda)}$ for the *p*-contribution, $\lambda_{O-H} \sim 4$.

$$\varphi_{O-H} = \sqrt{0.20}(2s) + \sqrt{0.80}(2p) \qquad 7.16$$

However, which 2*p*-orbitals are being used, and to what extent? If the molecular plane is *yz*, then the hybrids being used to make the bonds are obviously *yz*-hybrids. One can use the geometry to calculate what the contribution should be; however, it is obvious that all the p_y-orbital and some fraction of the p_z-orbital will be used to form the two equivalent O–H bonds. Since they are equivalent, we give them each half (50%) of the p_y-orbital. The remainder of the 80% *p*-character for each hybrid (see Eq. 7.16) is then obviously from p_z: 30% p_z. Furthermore, the two bonds are directed along different directions of the p_y-axis and have different signs. In other words, the hybrids on oxygen being used for the two O–H bonds are as follows.

$$\varphi_{(O-H)}(1) = \sqrt{0.20}(2s) + \sqrt{0.30}(2p_z) + \sqrt{0.50}(2p_y)$$

$$\varphi_{(O-H)}(2) = \sqrt{0.20}(2s) + \sqrt{0.30}(2p_z) - \sqrt{0.50}(2p_y)$$

The lone pairs reside in one pure *p*-orbital ($\varphi_{L.P.(pure-p)}$, Eq. 7.17) and in a hybrid ($\varphi_{L.P.(hybrid)}$, Eq. 7.18) with the following equations. The hybridized lone pair uses the valence *s*-orbital and the remainder of the p_z-orbital, 40%.

$$\varphi_{L.P.(pure-p)} = p_x \qquad 7.17$$

$$\varphi_{L.P.(hybrid)} = (\sqrt{0.60})2s + (\sqrt{0.40})2p_z \qquad 7.18$$

To find the nature of the lone pairs in this scheme in which the two lone pairs were equivalent, we start with the hybridization parameter, 2.34, found earlier and apply Eq. 7.9 and 7.10 to discover how much *s*- and *p*-character the two lone pairs have, which is 0.30 2*s* and 0.70 2*p*. The *p*-character in the orbital is a mixture of p_x and p_z; however, the entirety of the p_x-orbital will be used, and only the remainder of the p_z-orbital not used in O–H bonding is available. Since 80% of the p_z-orbital is already employed, this leaves 20% that may be used for the lone pairs. The nature of the orbitals ($\varphi_{L.P.}$) containing the equivalent lone pairs is then as shown in the following. (The two lone pairs have the same hybridization but are directed differently along the *x*-axis—a fact that does not change any of the discussion. In other words, the sign on the "$2p_x$" term is different. One can check that the two lone pairs are orthogonal as well, which they are.)

$$\varphi_{L.P.} = \left(\sqrt{0.30}\right)2s + \left((\sqrt{0.50}\right)2p_x + \left(\sqrt{0.20}\right)2p_z \qquad 7.19$$

It can be shown that these two schemes, one where the lone pairs are equivalent and one where they are not, describe the electron density exactly the same way. In order to find the electron density in the two equivalent lone pairs, we square the wavefunction $\varphi_{L.P.}$ and multiply by 2. This is done because the electron density (probability density) is the square of the wavefunction, and in this case there are two identical lone pairs (multiplication factor of 2). Squaring the term in brackets will lead to many cross-terms, such as $(0.39)2s2p_x$, only some of which are shown, and all of these cross-terms will be zero due to orthogonality of atomic orbitals.

$$2(\varphi_{L.P})^2 = 2\left[\left(\sqrt{0.30}\right)2s + \left(\sqrt{0.50}\right)2p_x + \left(\sqrt{0.20}\right)2p_z\right]^2$$

$$2(\varphi_{L.P.})^2 = 2\begin{bmatrix}(0.30)2s^2 + (0.50)2p_x^2 + (0.20)2p_z^2 + (0.39)2s2p_x \\ +(0.25)2s2p_z + \ldots\end{bmatrix}$$

$$2(\varphi_{L.P.})^2 = 2\left[(0.30)2s^2 + (0.50)2p_x^2 + (0.20)2p_z^2\right]$$

$$2(\varphi_{L.P.})^2 = (0.60)2s^2 + (1.00)2p_x^2 + (0.40)2p_z^2$$

We find the electron density in the two inequivalent lone pairs by using a similar procedure, by summing the squares of Eq. 7.17 and Eq. 7.18. Again, the cross-terms from squaring the function in brackets will be zero (cross-terms not shown), and the electron density where the lone pairs are inequivalent is exactly the same as when two equivalent lone pairs were employed. *Note that the electron density in the two identical lone pairs is the same as when the lone pairs were inequivalent.* In both cases, the sum of the electron density in the lone pairs is $(0.60)2s^2 + (0.40)2p_z^2 + (1.00)2p_x^2$. As a result, neither description for the electron density in water is preferable to the other in how they describe the electron density nor the energy of the electrons overall.

$$\left(\varphi_{L.P.(hybrid)}\right)^2 + \left(\varphi_{L.P.(pure-p)}\right)^2 = \left[\left(\sqrt{0.60}\right)2s + \left(\sqrt{0.40}\right)2p_z\right]^2 + (1.00)p_x^2$$

$$\left(\varphi_{L.P.(hybrid)}\right)^2 + \left(\varphi_{L.P.(pure-p)}\right)^2 = (0.60)2s^2 + (0.40)2p_z^2 + (1.00)2p_x^2$$

As a result, there is no fundamental reason to prefer a two-inequivalent lone pair versus a two-equivalent lone pair model for water based solely on describing the electron density of the molecule or energies, as the two pictures are indistinguishable.[2] Furthermore, both schemes can be shown to produce orbitals that are orthogonal and are obviously just expressions of different linear combinations of the same orbitals.

7.3.2.2 Recommendations for the Two Lone Pairs in Water and Similar Systems

We argue here that as there is no difference in electron density distribution or energy for placing all the *s*-characters into one lone pair versus distributing it over both lone pairs in water, there is no real reason to prefer one hybridization scheme over the other. However, some do prefer the picture where the lone pairs are different for symmetry reasons. From a symmetry perspective there is no possibility of mixing b_1 (p_x) with $2a_1$ ($s + p_z$ lone pair) in Fig. 7.7; indeed, it is not allowed unless the C_{2v} symmetry is broken in some way.[5] In general, it is recommended that when an atom has multiple lone pairs, one lone pair will reside in a hybrid with as much *s*-character as allowed by the bonding hybrids, and the remaining lone pairs will reside in pure *p*-orbitals, as suggested by Landis and Weinhold.[12]

This recommendation is for systems where there are no other external factors effecting the lone-pair energies. The two lone pairs in water will obviously change if one lone pair is involved in a hydrogen bond or coordination to a metal center and the other lone pair is non-interacting.

7.4 Diatomics

It stands to reason that if hybridization theory and molecular orbital theory are equivalent when taken to their limits but approach molecular electronics in very different ways, then hybridization theory may be simpler than MO theory for some molecules and more complex for others. Indeed, this is the case.

For many of the common diatomics, such as CO, N_2, and H_2, there is little difficulty in understanding their electronic structure by using what is often referred to as "perfect pairing", and the ground-state electronic structure is well described by a simple Lewis form in which the electrons are paired into 2-center 2-electron bonds whenever possible. Using N_2 as an example (Fig. 7.8), the simple Lewis structure is the anticipated σ-bond and two π-bonds.

The two π-bonds will of course be made with pure *p*-orbitals, leaving one *s*-orbital and one *p*-orbital (p_z) on each nitrogen. Consequently, we can try to obtain information on the hybridization of the orbitals used to hold the lone pairs and form the N–N σ-bond.

If we use the sum rule for the *s*-orbital with two pure *p*-orbitals for the π-bonds, and use $\lambda_{L.P.}$ and λ_σ as the hybridization for the lone pair and σ-bonding orbitals respectively, we obtain the set of the following equations that simplify to $\lambda_{L.P.}\lambda_\sigma = 1$.

$$\sum_i^n \frac{1}{1+\lambda_i} = 1$$

$$\left(\frac{1}{1+\lambda_\sigma}\right) + \left(\frac{1}{1+\lambda_{L.P.}}\right) + 2\left(\frac{1}{1+\lambda_{pure-p}}\right) = 1$$

$$\left(\frac{1}{1+\lambda_\sigma}\right) + \left(\frac{1}{1+\lambda_{L.P.}}\right) + 2\left(\frac{1}{1+\infty}\right) = 1$$

$$\left(\frac{1}{1+\lambda_\sigma}\right) + \left(\frac{1}{1+\lambda_{L.P.}}\right) + 2\cdot 0 = 1$$

$$\frac{\lambda_\sigma + \lambda_{L.P.} + 2}{1+\lambda_\sigma + \lambda_{L.P.} + \lambda_\sigma\lambda_{L.P.}} = 1$$

$$\lambda_\sigma\lambda_{L.P.} = 1$$

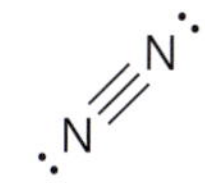

Fig. 7.8 Perfect pairing structure of the dinitrogen molecule.

Our qualitative methods described here do not allow separation of the interrelated hybrids $\lambda_{L.P.}$ and λ_σ and determination of their character; however, we can say from Bent's rule that $\lambda_{L.P.} < \lambda_\sigma$, that is, the lone pair will have more *s*-character. NBO calculation of the hybridization in N_2 gives $\lambda_{L.P.} = 0.6$ and, by the previous equation, $\lambda_\sigma = 1.7$.

While the diatomic N_2 is readily analyzed, one of the perceived "failures" of hybridization theory is in the electronic structure of O_2. Some claim that hybridization theory predicts a singlet, when in actuality correct application of the theory leads correctly to the triplet ground state. The perceived failure rests in attempting to apply "perfect pairing" to O_2, as was done previously for N_2. The perfect pairing structure for O_2 might be drawn as shown in Fig. 7.9a.

Before addressing the hybridization theory picture of O_2, it is worthwhile to draw attention to the simplified MO diagram of the molecule discussed in Chapter 6. The compound has the usual ordering of the MOs and a filling that suggests a ground state configuration of $(1\sigma)^2(2\sigma^*)^2(1\pi_{x,y})^4(3\sigma)^2(2\pi^*_{x,y})^2$. By Hund's rule, in the lower-energy configuration the two electrons in $(2\pi^*_{x,y})^2$ have parallel spins, producing a triplet ground state. In other words, we know to give the two electrons in $(2\pi^*_{x,y})^2$ the same spin in different orbitals because of a rule external to MO theory.

In the Lewis "perfect pairing" form for O_2 in Fig. 7.9a, a pure *p*-orbital will be used for the π-bond, and then an *s*- and two *p*-orbitals remain to form the σ-bond and store the two lone pairs. *Note, however, that all the lone pairs reside in the same plane.* In this singlet perfect-paired state of oxygen there is a tremendous amount of electron density residing in nonbonding orbitals all in the same plane, which leads to very large electron–electron repulsion terms!

Some of the electron density residing in this single plane can be moved to the orthogonal plane by rehybridizing the oxygens (Fig. 7.9b). As we did with N_2, we will use *sp*-hybrids to hold one lone pair on each oxygen and to form the σ-bond. This leaves four pure *p*-orbitals (p_x and p_y on each oxygen) to

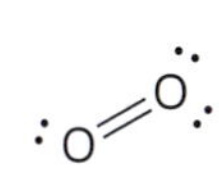

Fig. 7.9a Perfect pairing structure of the dioxygen molecule.

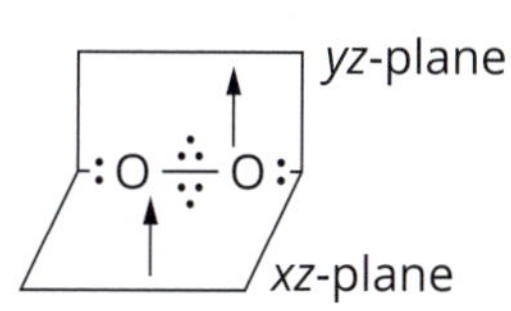

Fig. 7.9b A schematic showing the arrangement of six π-electrons into the four pure *p*-orbitals of O_2 to form a triplet electronic configuration. (Adapted from Shaik and Hiberty.)[2]

be occupied by six electrons. If we place three electrons in each plane, x and y, we obtain two 3-electron bonds. It can be shown quantitatively that the repulsions in this configuration are far less than the perfect-paired singlet configuration.[2] However, it is obvious qualitatively that this arrangement will lower electron–electron repulsions by evening out the electron density in the xz- and yz-planes. This type of 2-center 3-electron bond is discussed in more detail in the next chapter, but there is one full σ-bond and ½ of a π-bond in each of the xz- and yz-planes—a double bond in total between the two oxygens.

Naturally, the 3-electron bonds have unpaired spins, which may be either paired or unpaired to form singlet or triplet states. In order to determine which is preferred, we appeal to Hund's rule, just as in MO theory, to suggest that the triplet state is the ground electronic configuration.[13]

7.5 Bond Bending

For a main-group compound, the closest approach that two hybrids on the same atom can make is at sp^∞ of 90° for pure p-orbitals. However, there are many compounds where the angles between ligands—not angles to the hybrids but angles to the substituents—are less than 90°. Cyclopropane, where the C–C angles are 60°, is one example. Iodine(VII) fluoride, IF_7, is another, more complex, example, with 72° F–I–F angles in the pentagonal plane (Fig. 7.10).

Sometimes the bonding hybrids employed by an atom are not aligned with the interatomic line. Instead of having the substituent at the end of the hybrid, the atom is forced to interact with the hybrid orbital at an angle, θ in Fig. 7.11. This misalignment is referred to as *bond bending*, and there is diminished overlap due to the torsion in the bond. However, the geometry is often required by geometric, steric, or other factors.

Cyclopropane is an apt example, as this is the same effect generally referred to as "ring strain" in organic chemistry where molecular geometry does not allow maximal hybrid overlap. The bond bending here is more than is perhaps expected, in fact. From the gas phase electron diffraction experiment,[14] cyclopropane has H–C–H angles of 114.5(9)°. Assuming there is no bond bending in the C–H bonds, we can calculate the hybrids on carbon used to form these bonds using Eq. 7.14.

$$\begin{aligned} \cos\omega &= -\frac{1}{\lambda_{CH}} \\ \cos 114.5^\circ &= -\frac{1}{\lambda_{CH}} \\ \lambda_{CH} &= 2.41 \end{aligned}$$

The sum rules (Eq. 7.11) can then be used to arrive at the equivalent hybrids on carbon used to form the C–C bonds in cyclopropane.

$$\begin{aligned} 1 &= \sum_i^n \frac{1}{1+\lambda_i} \\ 1 &= 2\left(\frac{1}{1+\lambda_{CH}}\right) + 2\left(\frac{1}{1+\lambda_{CC}}\right) \\ 1 &= 2\left(\frac{1}{1+2.41}\right) + 2\left(\frac{1}{1+\lambda_{CC}}\right) \\ \lambda_{CC} &= 3.84 \end{aligned}$$

The C–C bonds are not formed by sp^∞-hybrids 90° apart. Instead, the carbon uses $sp^{3.8}$-hybrids (NBO calculations suggest similar $sp^{3.5}$ hybrids) to form the C–C bonds.

The amount of bond bending can be calculated geometrically from the hybridizations. Two equivalent hybrids with $\lambda = 3.8$, in order to be orthogonal, must follow the Coulson directionality theorem. The angle between them must be 105.1°, calculated as shown in the following. The bond bending angle is calculated as shown in Fig. 7.12. The internuclear C–C–C angle is 60°, and the hybrids are 105.1° apart. The bond bending angle is then (105.1° – 60°)/2 or 22.6°. NBO calculation produces essentially the same value, 24°, for the bond bending in cyclopropane as this "back-of-the-envelope" number.

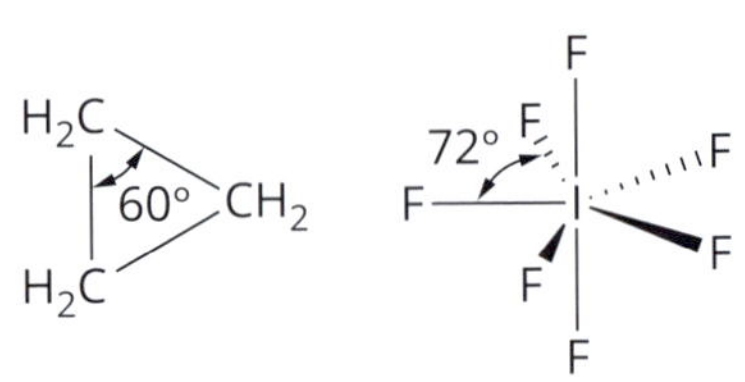

Fig. 7.10 Angles in two compounds containing obvious and severe bond bending.

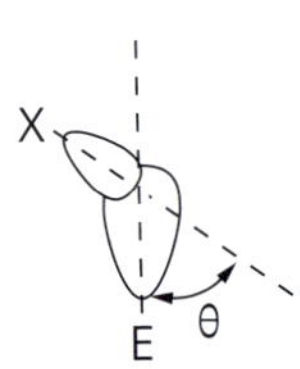

Fig. 7.11 Pictorial explanation of the bond bending angle, θ.

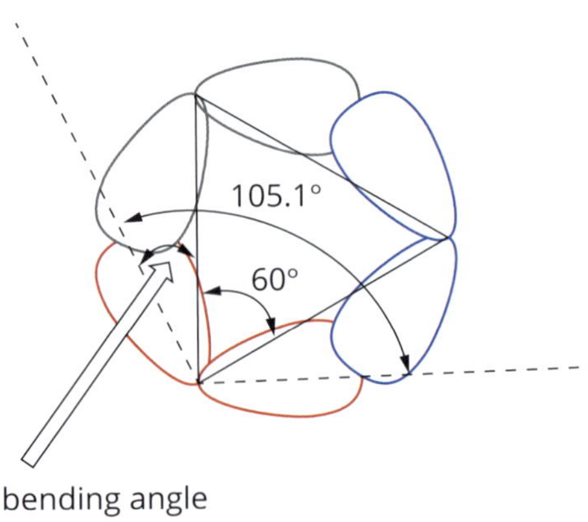

Fig. 7.12 Illustration for the calculation of the bond bending in cyclopropane. The bond bending angle (θ) is ½ the difference in the hybrid-C-hybrid angle (105.1°) and the C–C–C internuclear angle (60°). θ = ½(105.1 – 60) = 22.6°

BOX 7A

OH_3^+ and NH_3 by NBO: apparent anti-bent behavior

While the predictive power of Bent's rule is quite impressive, there are situations that can lead to apparent deviations.[5] Two surprisingly simple compounds that violate the expectation of higher *s*-character in the lone pair than in the bonds, according to NBO theory, are the isoelectronic pair NH_3 and OH_3^+. The usual VB analysis from the geometry of NH_3 implies that the molecule is Bent's rule compliant, with more *s*-character in the lone pair than the bonds. In fact, the molecule was found to be Bent's rule compliant in our simple calculations previously in this chapter. However, more detailed electronic structure calculations under the NBO framework provide a slightly different answer.

According to NBO calculations (NBO_5 implemented in GAMESS US using B3LYP/cc-pVTZ), the N–H bonds in ammonia involve $sp^{2.9}$-hybrids, while the lone pair on nitrogen is in an $sp^{3.4}$-hybrid, slightly more *s*-character in the bonding orbitals than in the lone pair orbital! It should be stressed that *the difference is very small*. A 2° pyramidalization increase or a 0.05 Å bond length increase would make the compound comply with Bent's rule.[5]

The cause of the non-Bent behavior in NH_3 has been ascribed to a lack of cylindrical symmetry in the hybrid orbital bonding to hydrogen.[4] This leads to a small amount of bond bending of about 3.7°. As a result, even though NH_3 appears from its geometry to comply with Bent's rule, NBO calculations suggest that the apparent compliance is deceptive. If the hybrid were cylindrical and there were no bond bending, the H–N–H bond angle would be about 111°.[4]

Interestingly, similar calculations on isoelectronic hydronium, OH_3^+, suggest that there is the same non-Bent behavior. The bonds are formed with $sp^{2.6}$-hybrids, while the lone pair rests in a higher *p*-character $sp^{5.3}$-hybrid. However, the compound exhibits no bond bending and the angles are 112.7°, similar to the angle expected for ammonia with no bond bending.

The cause of this non-Bent behavior in these molecules is perhaps electronic crowding (Pauli repulsion) due the N–H bond electron density being strongly polarized toward the electronegative central nitrogen. Consistent with this, NF_3, in which the more electronegative fluorine will polarize the bond away from the central atom, exhibits behavior consistent with Bent's rule.[5]

Molecular geometry is not the only factor that leads to bond bending in some cases. Hyperconjugation sometimes results in atoms being bent from bonding hybrids, for example. A lack of cylindrical symmetry in the hybrid bonding orbital has also been suggested as a possible cause of bond bending (see Box 7A).

7.6 Photoelectron Spectroscopy and Hybridization Theory: CH_4 as a Case Study (Advanced Topic)

Photoelectron spectroscopy (PES) was employed in Chapter 6 to probe the MO descriptions of molecules. PES is sometimes cited as a reason for preferring MO theory over hybridization theory—which is an oddity, considering that the two theories can be shown to be mathematically equivalent for describing electron densities of molecules. However, there is some utility in discussing the application of PES to molecules and how the experimental technique can be used with hybridization theory. Here we will compare the MO diagram of CH_4 with the hybridization theory description of the same molecule. Then, both will be employed to understand the photoelectron spectrum of the molecular system.

The MO diagram for CH_4 involves the combination of four SALCs comprised of one in the totally symmetric representation, a_1, and another in the triply degenerate representation t_2 in the point group of the molecule, T_d. The two representations for the four orbitals are shown in Fig. 7.13. The atomic orbitals of carbon, not surprisingly, reside in the same two representations with *s* as a_1 and the set of *p*-orbitals as t_2.

The photoelectron spectrum of CH_4 is shown in Fig. 7.14. Two peaks are readily explained by the MO spectrum as arising from ionization of electrons in the filled $1t_2$ and $1a_1$ molecular orbitals. Under the Koopmans's theorem assumption, the energies of the peaks are the negative of the energies of the orbitals. However, more accurately, the peak energies include the energy difference between the cationic, $E(CH_4^+)$, and neutral molecular, $E(CH_4)$, states called the "binding energy", $BE = E(CH_4^+) - E(CH_4)$; see Box 6A. The different peaks are due to electron energy differences on excitation between various ground-state electrons (with various energies) and the energy of the cation formed by instantaneous excitation. In other words, when the high-energy light of the PES experiment hits the molecule, electrons with kinetic energy related to the t_2 level or electrons with kinetic energy related to the a_1 level can be liberated. The peaks in the PES spectrum display these differences as ionization energies (with lower ionization energy for the higher-energy liberated electrons). The t_2 peak is larger due to the larger probability of liberating the more abundant electrons (six electrons) in this level relative to the a_1 level (two electrons)—a simple interpretation of the spectrum.

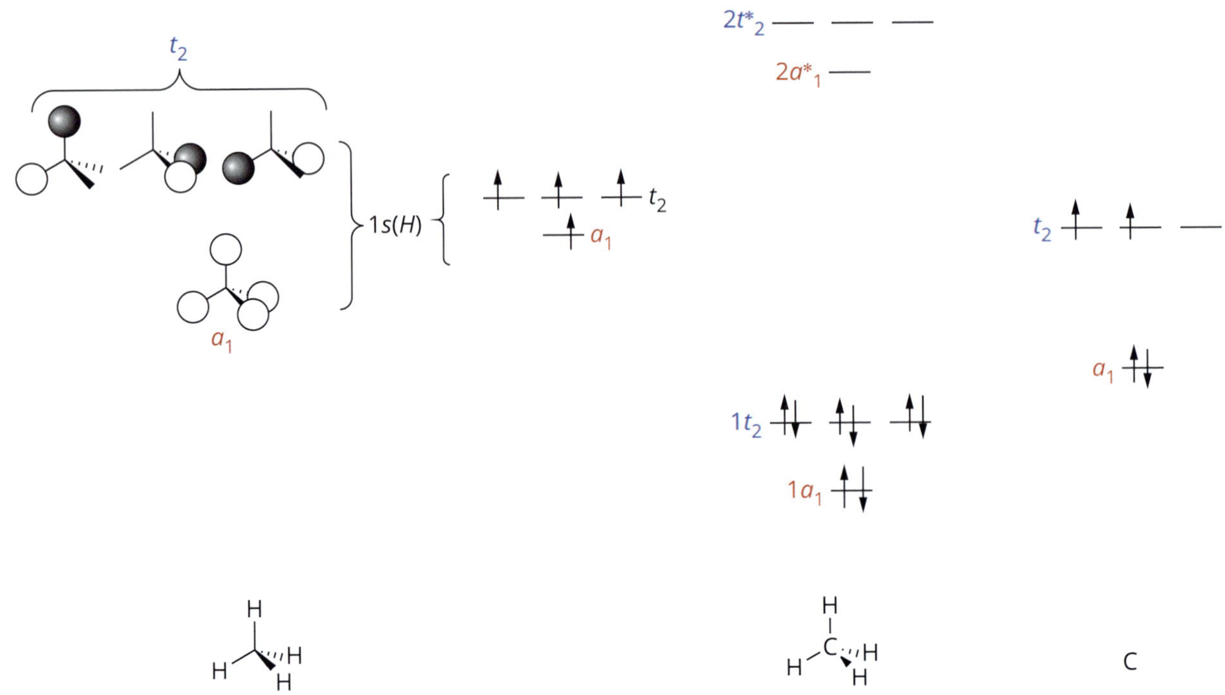

Fig. 7.13 MO diagram for CH_4.

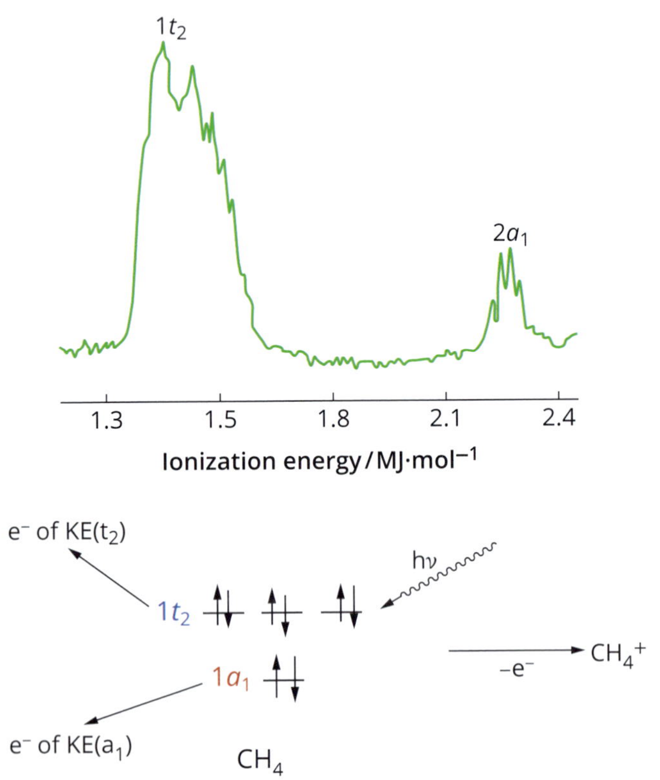

Fig. 7.14 PES spectrum of CH_4, from McQuarrie.[15] There are two main peaks corresponding to two distinct binding energies. The broadening and smaller splittings of the two peaks are due to vibrational structure.

The hybridization theory explanation for the PES spectrum is really no more complex and is due to quite analogous reasoning. The hybridization theory picture of CH_4 is, no doubt, quite familiar. The four equivalent C–H bonds are formed by four exact sp^3-hybrids on carbon interacting with the hydrogen 1*s*-orbitals.

Carbon hybridization can be thought of as changing the ground state electronic configuration of the carbon, $(2s)^2(2p)^2$, through promotion of one of the *s*-electrons into the *p*-manifold to produce a $(2s)^1(2p)^3$ configuration. The orbitals then hybridize to produce the four equivalent sp^3-hybrids (Fig. 7.15), which are then used to form the bonds to the hydrogen atoms.

Photoexcitation of the CH_4 system during the PES experiment, from a hybridization theory perspective, will remove one electron from one of the four equivalent C–H bonds. The kinetic energy of the liberated electrons will all be the same, KE(CH_4), unlike in the MO explanation. However, methane cation can exhibit two different hybridized states.

After Landis and Weinhold,[16] we can focus on the carbon in CH_4, considering that on instantaneous ionization the symmetrically distributed hydrogens will contribute little energetic difference in the two different hybridization theory states. The main differences in energy for cationic CH_4 must be in the rehybridization at carbon.

On instantaneous excitation we are left with a carbocation in an environment of tetrahedrally disposed hydrogens. (It is assumed that the atomic movements in response to the ionization will occur on a much longer timescale). Fig. 7.16 is a state

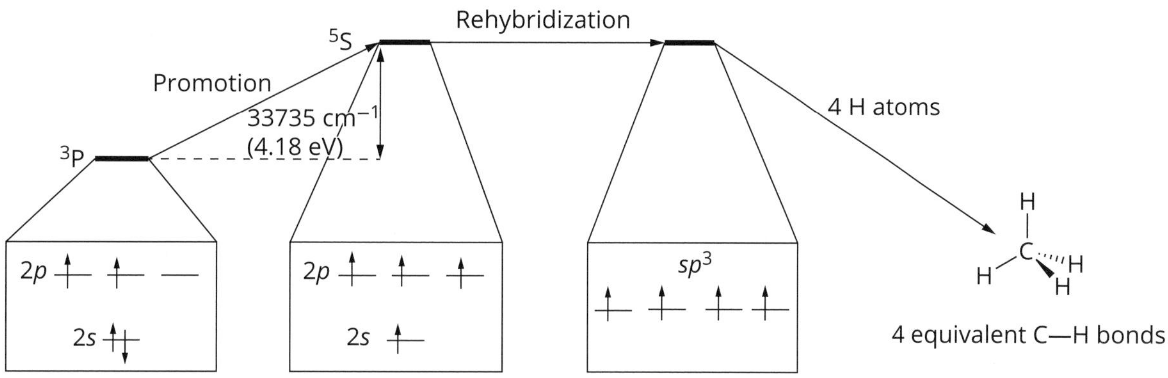

Fig. 7.15 Formation of the sp^3-hybrids from the carbon atomic orbitals. The ground state-configuration of carbon, labeled ^{3}P, has a $(2s)^2(2p)^2$ configuration. An electron is promoted from this ^{3}P ground state to form the state labeled ^{5}S with a $(2s)^1(2p)^3$ configuration. Rehybridization produces the sp^3 hybrids, isoenergetic to the ^{5}S state, which combines with the symmetrically disposed hydrogens to form CH_4.

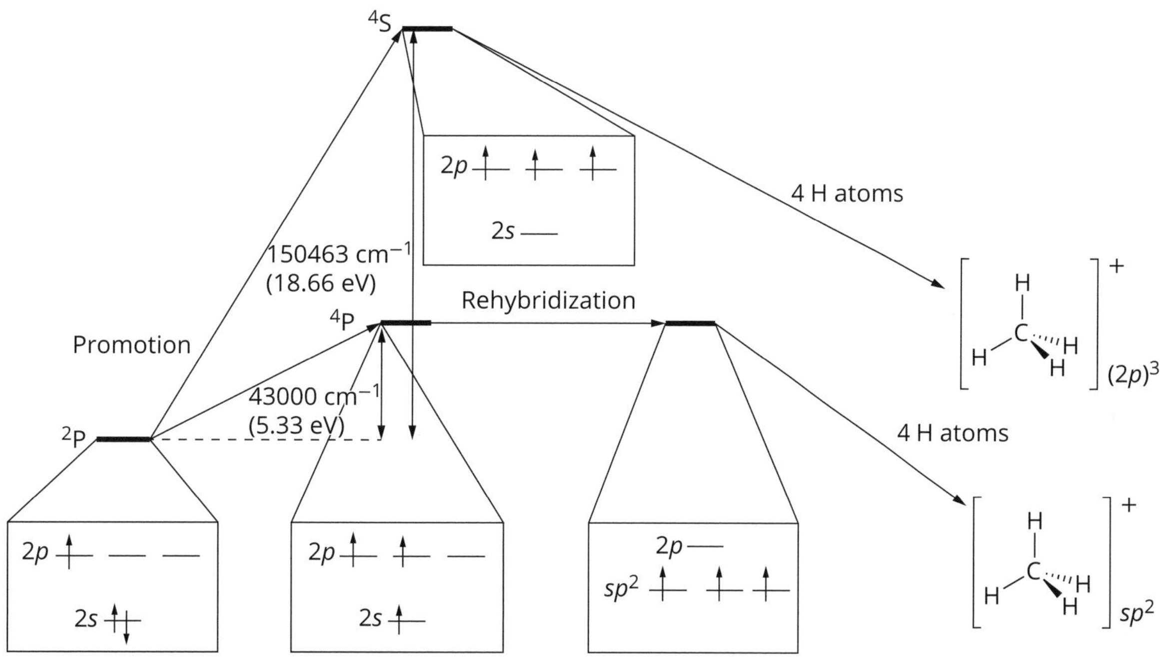

Fig. 7.16 Diagram showing the hybridization of CH_4^+. Since the hybridization of hydrogen cannot change, we focus exclusively on carbon, and the energies shown are excited state energies of a carbon atom; that is, the energies of the CH_4^+ will be different, but the energies shown provide a reasonable starting point to think about relative energies of the different configurations. Starting with a monocationic carbon, the ground state is ^{2}P with a configuration of $(2s)^2(2p)^1$. We can consider two states that could lead to two different hybridizations. In the lower-energy state, ^{4}P, a $(2s)^2(2p)^1$ configuration is found, which can go on to form sp^2-hybrids for bonding. This lower-energy state is, of course, triply degenerate as the vacancy (unoccupied, unhybridized p-orbital in the hybridized system) can equally be p_x, p_y, or p_z. In the higher-energy state, ^{4}S, a $(2p)^3$ configuration is found, which is then placed in the tetrahedral arrangement of hydrogens to produce a methane cation with a different electronic state.

diagram for carbon cation, with its ^{2}P ground state, $(2s)^2(2p)^1$. In order to form 2-electron bonds as well as is possible with the hydrogen atoms in the molecule, the carbon can generate two electronic configurations to combine with its surrounding hydrogen radicals.

In the lower-energy state an electron is promoted from the 2s orbital to the 2p orbital to produce a ^{4}P state. Hybridization of the 2s orbital with the two singly occupied 2p-orbitals will form sp^2-hybrids. This leaves one pure p-orbital on the carbon, which can be either p_x, p_y, or p_z, making this state triply degenerate. Consequently, because there are three isoenergetic ways to draw the state, this low-energy state has a higher probability. Consistent with the PES spectrum there is a lower-energy peak with about three times the intensity of the higher-energy peak.

In the higher-energy state, both 2*s*-electrons are promoted to the 2*p*-orbitals, producing a ^4S state with a $(2p)^3$ configuration. This higher-energy singly degenerate state will then use pure *p*-orbitals in its interactions with the tetrahedrally disposed hydrogens.

On considering both the MO and VB explanations for the PES spectrum of CH_4, the two models differ only in one major respect: the bifurcation leading to two peaks occurs in the neutral CH_4 in the MO model and in the cationic CH_4 molecule in the VB model. Even so, the two models are obviously analogous; for example, the triply degenerate state in both cases leading to the lower-energy and more intense peak in the PES spectrum is derived from the atomic *p*-orbitals on carbon.

The PES measurement deals with, by conservation of energy,

$$E(e^-) = h\nu - \{E(CH_4^+) - E(CH_4)\}$$

where $E(e^-)$ is the kinetic energy of the electrons measured, $h\nu$ is the light source energy (a constant for the experiment), $E(CH_4^+)$ is the energy of the cation, and $E(CH_4)$ is the energy of the methane molecular electrons. Consequently, the hybridization theory and MO models lead to two peaks due to different interpretations of the binding energy, $\{E(CH_4^+) - E(CH_4)\}$.

Chapter Summary

1. Two different types of hybridization theory have been discussed: valence bond (VB) theory, and natural bond orbital (NBO) theory. The two differ in their computational details but are similar in their qualitative aspects. NBO theory is computationally friendly and allows the contributions from different resonance forms to be obtained through natural resonance theory (NRT).
2. Hybridization theory describes the orbitals on a main-group atom in terms of sp^λ hybrids, where λ is the hybridization parameter. The parameter λ varies from 0 to infinity, where sp^0 = pure *s*-orbital and sp^∞ = a pure *p*-orbital. Hybrids may have a value from 0 to infinity, including non-integer values.
3. In the end, there is only one *s*-orbital and three *p*-orbitals on a main-group atom in the valence set. If you know the hybridization parameters, you can set up an equation where all the *s*-characters in the individual hybrids will sum to produce exactly one *s*-orbital on the atom. This sum rule can be used to determine an unknown hybridization parameter for a hybrid if all the other hybrids on the atom are known.
4. The angle between hybrids and the substituents on the hybrids are dependent on each other (in the absence of bond bending). The angle between hybrids is set by the hybridization parameter, and that angle is required for the orbitals to be orthogonal. The relationship is defined by the Coulson directionality theorem.
5. A central element will use a more *p*-rich hybrid to bond to a more electronegative substituent. This is called Bent's rule.
6. The energy of a hybrid relative to the $s \rightarrow p$ energy gap can be calculated if the hybridization parameter is known.
7. In many cases, the Sum Rule, Coulson directionality theorem, and relative energy of the hybrid can be used to supplement the information gained from MO theory, even enabling more precise MO diagrams to be drawn in some cases.
8. If there is more than one lone pair on an atom, such as in H_2O, there is no difference in electron density distribution or energy of the electrons when the lone pairs are equivalent ("rabbit ears" you learned when you first learned to draw the Lewis structure of water) versus when one lone pair is in a pure *p*-orbital and the other lone pair is in an *s*-rich hybrid. We recommend the unequal lone pair configuration on symmetry grounds, however, which gives it an electronic structure similar to the MO diagram for water.
9. The perceived "failures" of hybridization theory, such as in the configuration of O_2 and the PES spectrum of CH_4, are actually due to improper implementation of the theory. Poor hybridization theory will produce the wrong answer, just like improper diagramming of MOs will also produce the wrong answer.
10. In some cases, the substituent attached to a hybrid is not directed to give the best overlap. This is most obvious in cases where the substituents are at angles of <90°, which is the smallest angle that one can achieve with *sp*-hybrids at maximal overlap. This less than optimal overlap occurs in molecules such as cyclopropane, which is described as having "ring strain". A more general term used in hybridization theory is that a molecule can have "bond bending", by which the substituent is not situated for maximal overlap. This occurs in ringed molecules like cyclopropane and cyclobutane, but also in systems such as IF_7, which is pentagonal bipyramidal by VSEPR with 72° F–I–F angles in the equatorial plane. Bond bending can occur in cases where the substituent angles are >90° as well due to a variety of factors. Ammonia, for example, is unexpectedly calculated to have bond bending by NBO.

Practice Problems

1. For the flat carbocyclic systems $(CH_2)_n$, where n = 4-6, find the hybrids of the carbons used to form the carbon–carbon bonds. Assume that the H-C-H angle in all the systems is the same as in the cyclopropane system, 114.5°. How much bond bending would there be in each compound, assuming that all the carbons reside in the same plane?
2. Cyclobutane is not a planar compound. The calculated (B3LYP, aug-pVDZ) structure is shown as follows, with some important metrical parameters. All the H-C-H angles are the same, as are the C-C-C angles. Assuming there is no bond bending in the C-H bonds, calculate the bond bending angle in the C-C bonds.

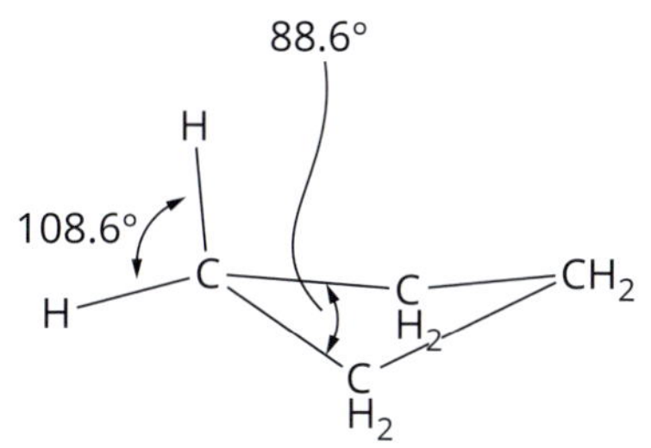

3. Explain the trend in Fig. 4.6 using valence bond theory.

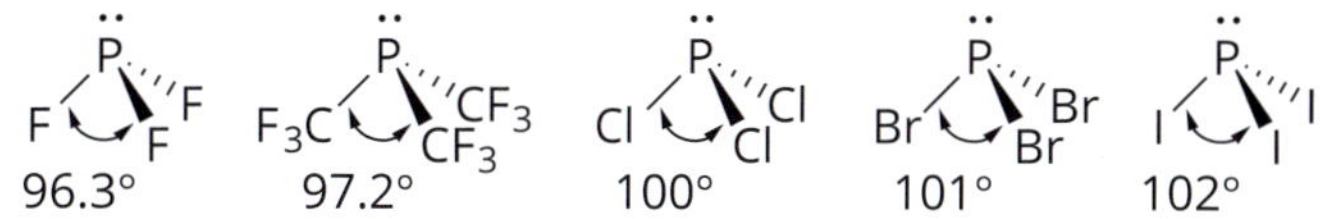

4. Analyze the compounds (see Fig. 4.12) NO_2^+, NO_2, and NO_2^-, using the Coulson directionality theorem and sum rules based on the O-N-O angles of 180°, 134°, and 115° respectively. Without using VSEPR, explain the trend in the angles.
5. The accompanying figure shows the structure of B(H)(F)Cl as found by DFT optimization. Find the hybridization of the boron orbitals used to bind to the three different ligands. Explain your results in terms of Bent's rule.

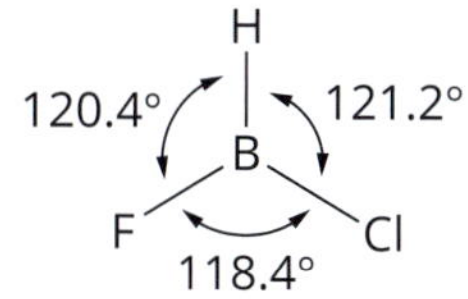

Cited References

1. Hoffmann, R.; Shaik, S.; Hiberty, P. C., A conversation on VB vs MO theory: A never-ending rivalry? *Acc. Chem. Res.* 2003, *36*, 750–756.
2. Shaik, S. S.; Hiberty, P. C., *A Chemist's Guide to Valence Bond Theory* Wiley-Interscience,: Hoboken, N.J., 2008.
3. Coulson, C. A.; McWeeny, R., *Coulson's Valence* 3rd ed.; Oxford University Press: Oxford, 1979.
4. Weinhold, F.; Landis, C. R., *Valency and Bonding: A Natural Bond Orbital Donor-Acceptor Perspective* Cambridge University Press: Cambridge, 2005.
5. Weinhold, F.; Landis, C. R., *Discovering Chemistry with Natural Bond Orbitals* Wiley: Hoboken, N.J., 2012.
6. Weinhold, F.; Landis, C. R.; Glendening, E. D., What is NBO analysis and how is it useful? *Int. Rev. Phys. Chem.* 2016, *35*, 399–440.
7. Coulson, C. A., The electronic structure of some polyenes and aromatic molecules VII. Bonds of fractional order by the molecular orbital method *Proc. R. Soc. Lond. A-Math. Phys. Sci.* 1939, *169*, 0413–0428.
8. Jensen, F., *Introduction to Computational Chemistry* 2nd ed.; John Wiley & Sons: Chichester, England, 2007; p xx, 599 p.
9. Cotton, F. A., *Chemical Applications of Group Theory* 3rd ed.; Wiley: New York, 1990.
10. Bent, H. A., An appraisal of valence-bond structures and hybridization in compounds of the 1st-row elements *Chem. Rev.* 1961, *61*, 275–311.
11. Broun, T. T.; Livingston, R. L., An Electron Diffraction Investigation of the Molecular Structures of Ketene, Carbonyl Flouride and Tetrafluoroethylene *J. Am. Chem. Soc.* 1952, *74*, 6084–6091.
12. Clauss, A. D.; Nelsen, S. F.; Ayoub, M.; Moore, J. W.; Landis, C. R.; Weinhold, F., Rabbit-ears hybrids, VSEPR sterics, and other orbital anachronisms *Chem. Ed. Res. Prac.* 2014, *15*, 417–434.
13. This is something of an oversimplification as some care needs to be taken to ensure the indistiguishability of the electrons and other essential features. For a more rigorous treatment see Shaik and Hiberty, ref. 2.
14. Yamamoto, S.; Nakata, M.; Fukuyama, T.; Kuchitsu, K., Geometrical structure of cyclopropane as studied by gas electron-diffraction and spectroscopic data *J. Phys. Chem.* 1985, *89*, 3298–3302.
15. McQuarrie, D. A.; Simon, J. D., *Physical Chemistry: A Molecular Approach* University Science Books: Sausalito, Calif., 1997.
16. Landis, C. R.; Weinhold, F., Comments on "Is It Time To Retire the Hybrid Atomic Orbital?" *J. Chem. Educ.* 2012, *89*, 570–572.
17. Iijima, T.; Tsuchiya, S.; Kimura, M., Molecular-structure of dimethyl sulfide *Bull. Chem. Soc. Jpn.* 1977, *50*, 2564–2567.

CHAPTER 8

Main-Group Bonding

CONCEPT COVERAGE

1. Explanations for the hybridization differences in first-row main-group compounds versus their heavier congeners
2. The effects of primogenic repulsion on orbital radii throughout the periodic table
3. Relativity and atomic structure
4. π-Bonding down the main group
5. Introduction to bonding other than 2-center 2-electron bonds: the 3-center 4-electron bond (ω-bond), negative hyperconjugation, and the 3-center 2-electron bond (τ-bond)
6. The lone pair bond weakening effect
7. Odd-electron compounds
8. Second-order Jahn–Teller effects in the main group

"One is almost tempted to say, 'at last I can almost see a bond'. But that will never be, for a bond does not really exist at all: it is a most convenient fiction which, as we have seen, is convenient both to experimental and theoretical chemists."

— **C. A. Coulson**

"The general theory of quantum mechanics is now almost complete, the imperfections that still remain being in connection with the exact fitting in of the theory with relativity ideas."

— **P. A. M. Dirac**

In this chapter we will cover many of the important concepts associated with bonding in the main group, including the various types of bonds, other than 2-center 2-electron bonds, that one will commonly encounter. Some important concepts will be introduced, with implications for the entire periodic table.

8.1 First-Row versus Heavier *p*-Block Bonds: NH_3 versus PH_3 as an Illustration

First, we can illustrate some important concepts by examining congeners of normal valent compounds. In particular, it is important to understand the notable differences between the first-row *p*-block elements and the heavier congeners as far as their bonding is concerned. As an example, we will discuss the bonding in NH_3 versus PH_3; however, many pairs of compounds would serve about as well.

8.1.1 VB Analysis of the E–H Bonds in NH_3 and PH_3

Analyzing the molecules in the light of hybridization theory is particularly helpful for our purposes here, but MO theory produces very much the same results.[1] In PH_3, where the H–P–H angles are ~93°, phosphorus uses orbitals with very high *p*-orbital character to bond to hydrogen. In NH_3, where the H–N–H angle is nearer the tetrahedral angle (~107°), the N–H bonds will be made with hybrids having significant *s*-character.

Analysis of NH_3 shows, using Coulson's directionality theorem, that there are $sp^{3.4}$-hybrids interacting with the hydrogens. The lone-pair hybridization is easily determined by using the sum rule as being ~sp^2. Natural resonance theory produces slightly different results, but these really do not change the fact that significantly more *sp*-mixing occurs in NH_3 relative to PH_3. (Due to a small amount of bond-bending, nitrogen uses $sp^{2.9}$-hybrids for bonding to hydrogen and $sp^{3.4}$-hybrids for the lone pair. See Box 7A.)

Analysis of PH_3 using Coulson's directionality theorem (Eq. 7.13) and Sum Rule (Eq. 7.11) shows that the phosphorus is using sp^{19}-orbitals to overlap with the hydrogens, which allows the lone pair to reside in an essentially pure *s*-orbital ($sp^{0.18}$).

From hybridization theory, the phosphorus in PH_3 uses more *p*-orbital character to form the E–H bonds than nitrogen in NH_3; but why?

8.1.2 Explaining the Differences in H–E–H Angles in PH_3 and NH_3

MO theory is in perfect agreement with the hybridization theory analysis, suggesting that there is reduced *s*-orbital character in the P–H bonds relative to the N–H bonds in NH_3. It was found that there is essentially no phosphorus *s*-orbital participation in the bonding of PH_3.[1] However, neither MO nor hybridization theory concepts discussed thus far determine the *reason* for the geometric difference. Here we offer for discussion four possible explanations (and another discussed at the end of this chapter: Jahn–Teller theory):

(1) *VSEPR explains the smaller angles in PH_3 versus NH_3 as being due to the smaller electronegativity of phosphorus.* The lower electronegativity of phosphorus causes the lone pair to have a larger domain size relative to nitrogen, which reduces the H–P–H angles—a plausible explanation under the model.

(2) *The greater hybridization parameter in NH_3 for the E–H bonding orbitals is possibly due to the energetic match between s- and p-orbitals.* This is a perturbation theory argument, which suggests that the closer in energy the orbitals being mixed, the greater their mixing; that is, the difference in orbital energy (ΔE_{ij}) is inversely proportional to the mixing (Section 6.5).

$$mixing \propto \frac{1}{\Delta E_{ij}}$$

This relation is generally true, and the argument is potentially reasonable. *However, it is not the energy gap between the s- and p-orbitals that is responsible in this case.* The energy difference between the *s*- and *p*-orbitals in the first row of the *p*-block is consistently *larger* than for the heavier elements (Fig. 8.1).[2] In other words, if this were the major effect, phosphorus with its smaller *sp*-energy gap would be *more* likely to

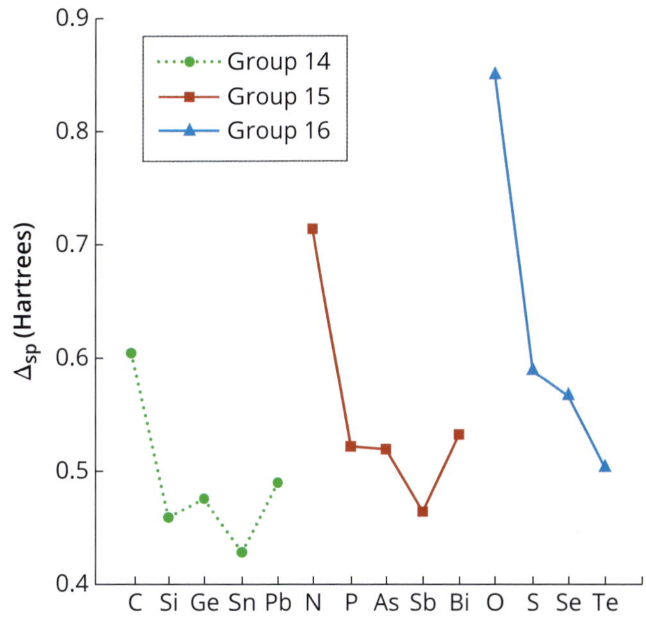

Fig. 8.1 Plot of the calculated (CCSDT using the QZVP basis) energy difference between the valence *s*- and *p*-orbitals (Δ_{sp}) for Groups 14, 15, and 16. In all cases, the energy gap for the first-row element is significantly higher than its heavier congeners.

be sp^3-hybridized than nitrogen, and the opposite is found in the EH_3 compounds.

(3) *Does the smaller radius of nitrogen relative to phosphorus lead to the larger angles in ammonia?* This potentially could be determined using the ligand closest packing model (discussed briefly in Chapter 4), and may explain the result that PH_3 has smaller angles than NH_3.

(4) *Is there an overlap explanation for the angle difference? Do the p-orbitals in PH_3 make a larger contribution than they do to the bonds in NH_3 because there is an orbital size difference?* If one examines the atomic orbital sizes in Group 15 (Fig. 8.2),

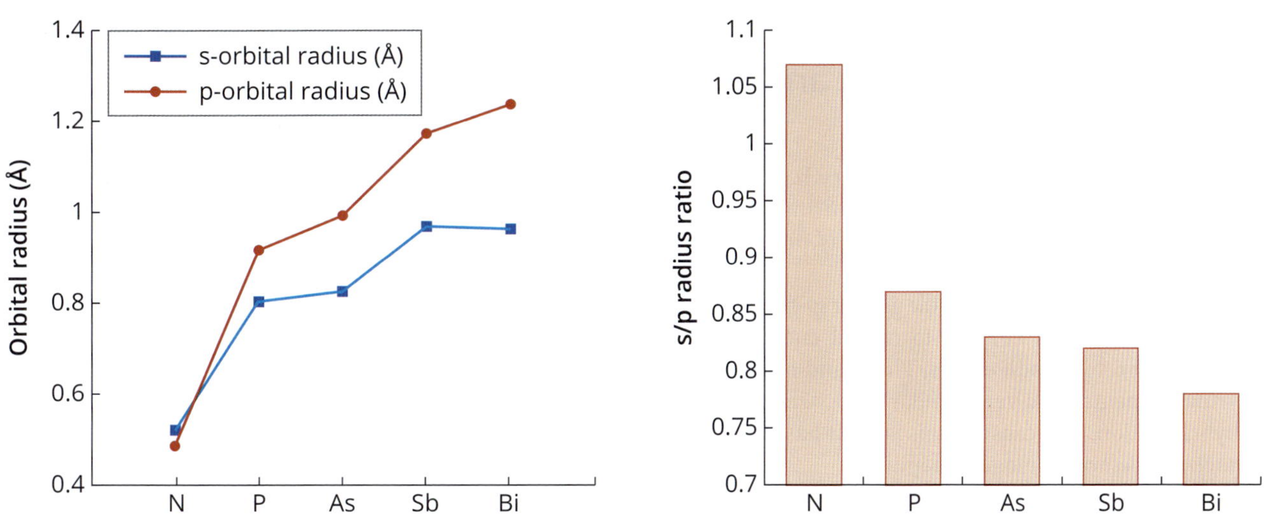

Fig. 8.2 (*Left*) Radii of the *s*- and *p*-orbitals for the Group 15 elements.[4] (*Right*) Bar chart showing the *s*-/*p*-orbital radius ratio versus element. The Sb to Bi difference is due, at least in part, to relativistic effects.

it becomes obvious that there is a notable difference in the ratio of *s*- and *p*-orbital radii for the first row versus the heavier congeners. Nitrogen has *s*- and *p*-orbital radii both near 0.5 Å in size. The other, heavier members of the group have what might be considered the expected relationships between the valence orbital sizes within the same shell where the *s*-orbital is ~80% the size of the *p*; the lower energy *s*-orbital allows the electrons to approach closer to the nucleus and is contracted relative to *p*. The better orbital size match of the *s*/*p*-orbitals in nitrogen leads to better hybridization and explains the higher hybridization in NH_3. The *s*- and *p*-orbitals in phosphorus are of different radial sizes, and the larger *p*-orbitals are extended so that they overlap with the ligands on phosphorus more effectively than the smaller *s*-orbital.

The cause of this disparity in orbital size ratios in the first row versus the heavier elements is an important one in chemistry: the primogenic effect,[3] discussed in the next section.

The last explanation dealing with overlap fits neatly into both hybridization theory and molecular orbital theory and explains the differences using hybridization or overlap with atomic orbitals. Several of these models provide reasonable explanations for the phenomenon in this instance. Of the explanations presented, the focus in this chapter will be on the primogenic effect, which will be a useful concept for understanding geometries and reactivity in inorganic chemistry.

8.2 Some Important Effects in Main-Group Chemistry

At this stage, several important concepts affecting chemistry in the main group will be discussed, though in some cases their influence will not rest only in the *p*-block. Some of the concepts covered will play roles in the reactivity and properties for certain compounds throughout the periodic table.

8.2.1 The Primogenic Effect and Orbital Radii[3]

Any student who has taken organic chemistry has been subjected to a myriad of compounds prepared using nominally *sp*-, *sp*^{2}-, and *sp*^{3}-hybrids on carbon, and has perhaps wondered why such hybridizations are the norm.

In Section 8.1.2 it was shown that the greater *sp*-hybridization in NH_3 relative to PH_3 is *not* due to orbital energies but is well explained by orbital radii effects. The *s*- and *p*-orbitals for the first-row elements have very similar radii, which leads to greater hybridization. The cause of this is what has been called the *primogenic effect* or primogenic repulsion.

Any atomic orbital that lacks a radial node, namely 1*s*, 2*p*, 3*d*, and 4*f*, is more compact than might be expected because there are no inner orbitals of the same angular momentum (Fig. 8.3). A *1s*-orbital, for example, can contract very close to the nucleus. A feature of the 2*s*-orbital is a phase change, a radial node. The 1*s*-orbital will sit in the node of the 2*s*-orbital, giving the two orbitals no net overlap. Because the atomic orbitals are required to be orthogonal (Chapter 2), *this arrangement is required* (Fig. 8.3).

In the first row of the *p*-block, the valence orbitals are 2*s* and 2*p*. The 2*s*-orbital is required to be orthogonal to the inner 1*s*-orbital and is therefore limited in its contraction towards the nucleus. However, the 2*p*-orbital has no such restriction and contracts toward the nucleus to increase the Coulombic attraction between its electrons and the nuclear protons. In other words, the 2*s*-orbital experiences a *primogenic repulsion* and the 2*p*-orbital does not. *As a result, for first-row elements the 2s- and 2p-orbitals have essentially the same radius (Fig. 8.4), which aids in the formation of sp-hybrids. This similarity in valence orbital radii of the first-row elements is not repeated for the heavier congeners, making them less prone to forming sp-hybrids.*

The expected size order for the radii (r) of orbitals around a particular atom would be $r(s) < r(p) < r(d) < r(f)$ for any particular valence shell due to the greater interpenetration and lower energy of the orbitals with lower angular momentum (Chapter 2). From a classical perspective, one can imagine the electrons as balls on a rubber band being swung around the nucleus, where the restoring force of the rubber band is the Coulombic attraction. As the angular momentum increases, from *s* to *p* to *d* to *f*, one would expect the rubber band to stretch and the radius of orbit to increase. All other things

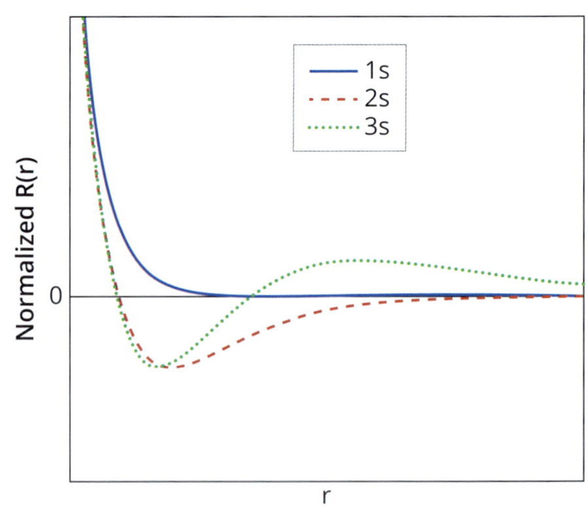

Fig. 8.3 Plot illustrating the importance of the phase difference between orbitals of the same angular momentum, using the radial distributions of the 1s, 2s, and 3s-orbitals, which is necessitated by orthogonality. If the orbitals have the same shape, same angular momentum, the constructive and destructive interference between the different orbitals must exactly cancel, which is done by radial matching. (Note that the plots are normalized to fit on the same axis, so exact cancelation may involve making one curve somewhat larger or smaller, but this makes the nodal structure between the different orbitals easier to see.)

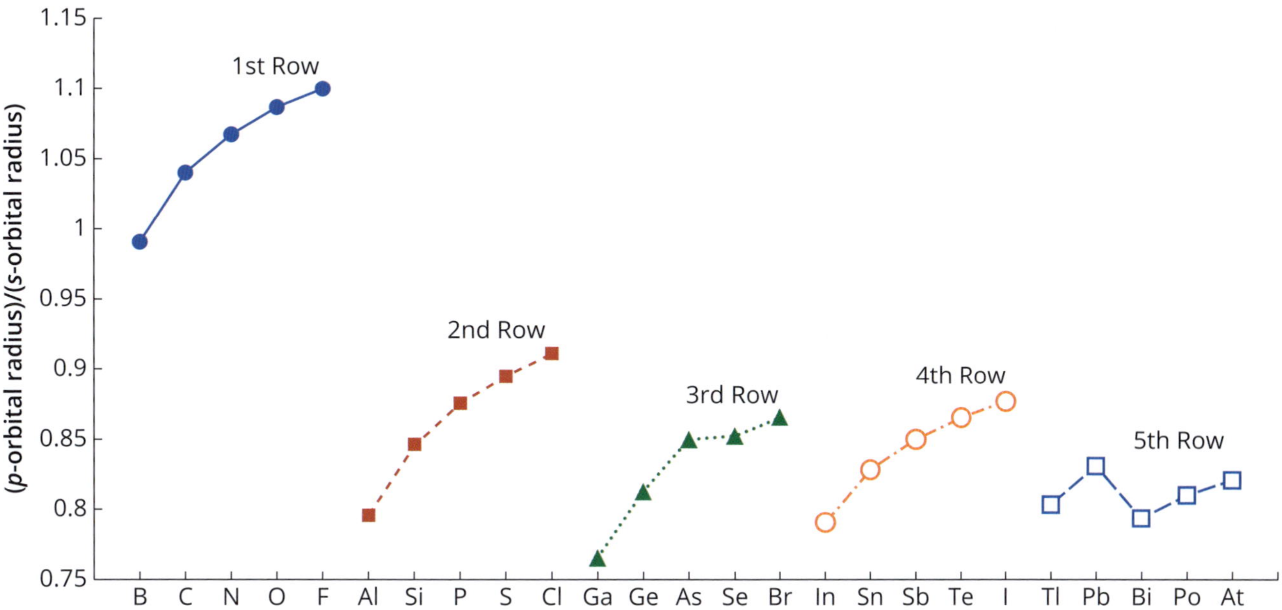

Fig. 8.4 The $r_{p\text{-orbital}}/r_{s\text{-orbital}}$ (p-orbital radius to s-orbital radius) for the different rows of the p-block. (Data taken from Waber and Cromer.[4]) The first-row elements have s- and p-orbital radii that are very nearly identical (ratio ~1), enhancing hybridization, while the other rows have ratios of ~80% with s-orbitals somewhat smaller than p, as expected.

being equal, such as being in the same shell, we would expect s-orbitals to be smaller than p-orbitals, p-orbitals to be smaller than d-orbitals, and so on. If we use the valence s-orbital as a benchmark, we can see how primogenic repulsion leads to small 1s-, 2p-, 3d-, and 4f-orbitals relative to other orbitals of the same angular momentum.

Similar effects for the 3d-orbital sizes can be seen by plotting the valence d-orbital to valence s-orbital ratio versus row in the transition series (Fig. 8.5). This leads to first-row transition metals using their d-orbitals less effectively in bonding than their heavier congeners, and typically smaller ligand fields for first-row transition-metal complexes versus second and third row.

In all cases, the d-orbitals in the valence shell possess a principal quantum number smaller than the valence s-orbitals, which leads to the d-orbitals being smaller than s. For example, the first-row transition-metal valence shell is 3d and 4s. The lack of primogenic repulsion acts to exacerbate the size difference between the d-orbitals and the other valence orbitals of the first-row transition elements. As a result, the average d-orbital/s-orbital radii ratio for the first row is 0.30, while for the second and third rows the ratios are 0.46 and 0.55, respectively.

That there is a poor match in size between 3d and 4s, where the 3d is much smaller, is significant. If one of the orbitals is too large, the diffuse orbital overlaps poorly with ligands at proper distance to interact with the small atomic d-orbitals on the metal. If a valence orbital becomes too small, it may approach the size of filled core orbitals, leading to large Pauli repulsion with electrons of the incoming substituent. Specifically, in the case of first-row transition elements, the 3d orbitals are similar in size to the "inner" filled 3p-orbitals (Fig. 8.6). For example, the 3d orbitals on chromium have a radius of 0.43 Å, whereas the filled 3p-orbitals have a very similar radius of 0.42 Å. These Pauli repulsions contribute further to weaken bonding in the first transition series relative to the second and third series. For example, the 4d-orbitals of molybdenum extend out somewhat (0.70 Å) from the repulsive and filled 4p-orbitals of radii 0.56 Å (Fig. 8.6).

All of this is not to suggest that the d-orbitals are unimportant for bonding in the first row of the transition series; bonding with the 3d-orbitals is crucial in the first transition series. It is simply to say that bonding is typically weaker in the first row than the second and third transition series.

The primogenic effect can also be seen in the f-block elements. In this case, of course, the valence f-orbitals have principal quantum numbers 2 smaller than their valence s. For example, the lanthanides have valence configurations containing the 4f- and 6s-orbitals. The lack of primogenic repulsion, because there is no inner orbital of the same angular momentum with which the 4f must remain orthogonal, leads to a significant additional contraction. The lanthanides have very small f-orbitals relative to their valence s-orbitals; the radius of the s-orbital is around six times larger (Fig. 8.7). The average valence f-orbital/s-orbital radii ratio for the lanthanides is 0.17, while for the actinides the valence f-orbital is 28% as large as the valence s.

Pauli repulsion with filled inner orbitals is a much larger effect for the lanthanides than even for the first-row transition elements. For example, gadolinium has a filled 5p-orbital of radius 0.715 Å, while its valence 4f-orbital only has a radius of 0.29 Å![4] This leads to very large filled–filled repulsions between an occupied orbital of a donor and the core electrons of the metal that prohibits significant f-orbital participation

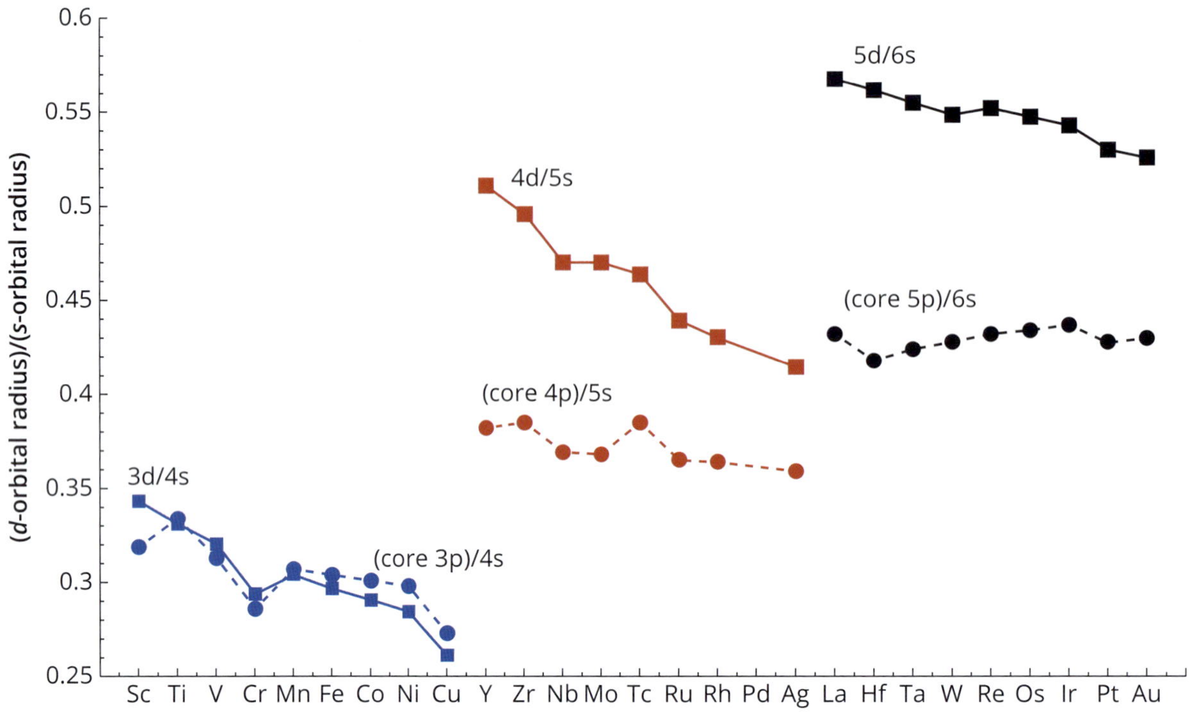

Fig. 8.5 Squares (■) with solid lines are the ratio of average valence *d*-orbital radius to valence *s*-orbital radius for the different rows of the *d*-block. The circles (●) with dotted lines are the ratio of filled core $p_{3/2}$- to valence *s*-orbital ratio. While the second and third rows of the transition series have similar radii ratios, the first row has *d*-orbitals that are significantly smaller relative to their valence *s*-orbitals, which leads to a much smaller *d/s* radii ratio. In addition, large Pauli repulsions may occur when filled core orbitals, such as the $p_{3/2}$-orbital, are of similar size to the valence *d*-orbitals, which occur in the first row. (Data taken from Waber and Cromer.[4]) (Pd data is missing due to lack of occupancy of the valence *s*-orbital in the ground state.)

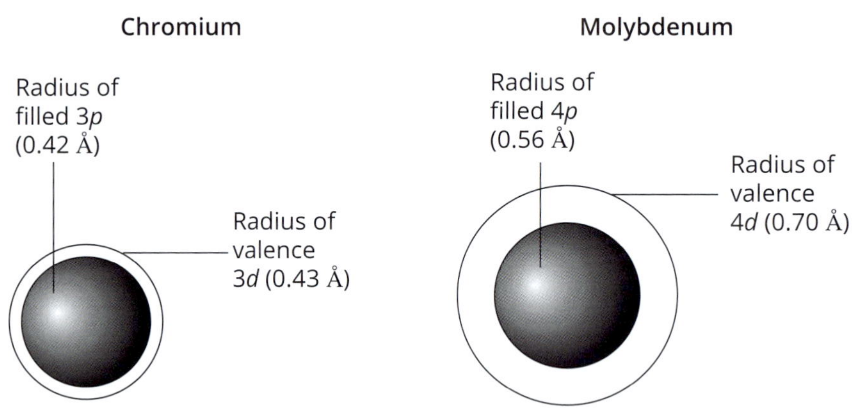

Fig. 8.6 Relative radius of filled, inner *p*-orbitals (shaded circle) to the valence *d*-orbital radii for chromium and molybdenum. Filled–filled electron repulsion can occur when filled ligand orbitals try to interact with first-row elements, like chromium, due to the filled inner *p*-orbitals of similar size. The second- and third-row elements typically have *d*-orbitals that extend beyond these inner orbitals, and these transition metals typically have stronger bonds than their first-row counterparts.

in bonding. As a consequence, 4*f*-orbitals are thought to participate very little in bonding for the lanthanides in most instances.

The consequences of these differences in orbital radii will be discussed in more detail in Chapter 21 on *f*-element chemistry, but it is obvious that the *f*-orbitals are more likely to participate in bonding and reaction chemistry for the actinides than lanthanides.

As can be seen, primogenic repulsion is an important effect throughout the periodic table, but in the context of the main group it leads to far greater *s*- and *p*-orbital hybridization in the first-row *p*-block elements due to the very similar size of the orbitals. The heavier elements of the *p*-block are far more likely to use *p*-rich hybrids in bonding due to the somewhat larger extension of these orbitals, which leads to more effective overlap than if *s* were used significantly.

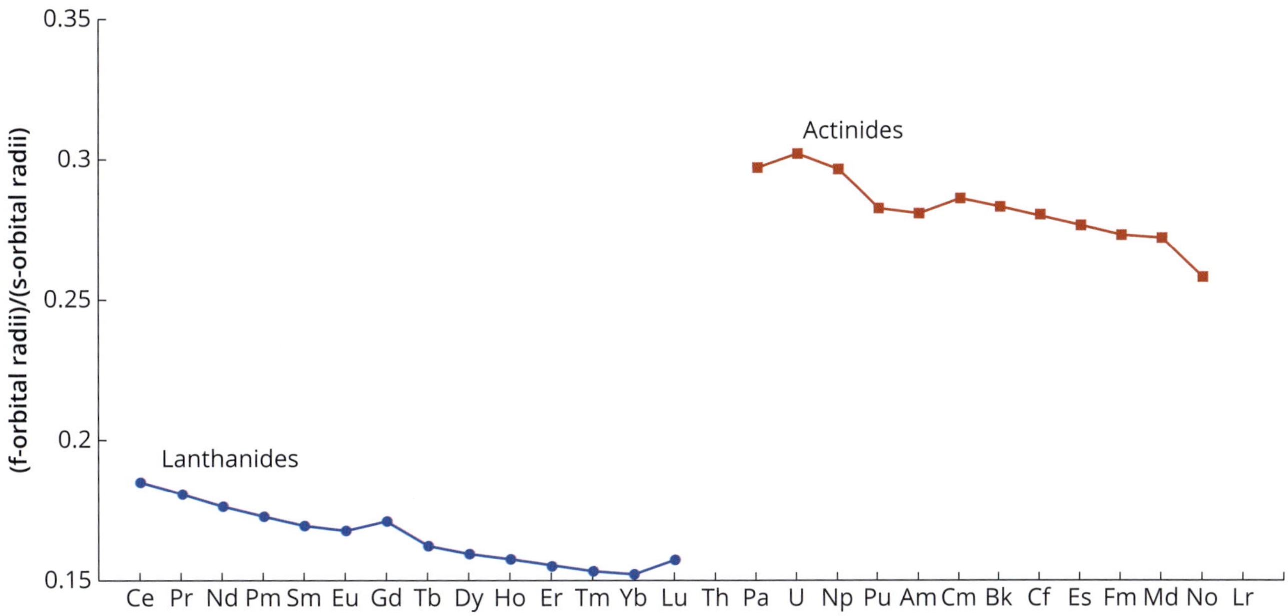

Fig. 8.7 The ratio of valence *s*-orbital radii to the average valence *f*-orbital radii for the lanthanides and actinides. (Data taken from Waber and Cromer.[4]) No data were available for thorium due to lack of *f*-orbital occupancy in the ground state. The actinides have *f*-orbitals closer in radius to the valence *s*-orbitals, while the lanthanides have greatly contracted *f*-orbitals due to the lack of primogenic repulsion. This makes the *f*-orbitals of the lanthanides far less likely to participate in bonding and reaction chemistry than their actinide counterparts.

8.2.2 Relativistic Effects[5,6]

According to the theory of special relativity, the mass of an object is related to its velocity by Eq. 8.1, where m is the mass at velocity v, m_0 is the rest mass, and c is the speed of light.

$$m = m_0[1 - (v/c)^2]^{-1/2} \qquad \textbf{8.1}$$

This "universal speed limit" applies to everything of which we are currently aware, such as light, subatomic particles, macroscopic objects, and so on. However, as shown in Fig. 8.8, relativity has little affect at low-to-medium v/c ratios, where classical mechanics is quite accurate. The majority of calculations still used today ignore relativity; in effect, regarding c as infinite. However, for some elements this can lead to substantial errors.

According to the Bohr classical model of the atom (Chapter 2), the electronic attraction between the electron and the nucleus is exactly balanced in a stable orbit by the momentum of the electron. As the nuclear charge increases, the velocity of the electron also increases; otherwise, the electron would plunge into the nucleus. As a result, according to the simple Bohr model, the velocity of the electron increases linearly with the nuclear charge, as shown in Eq. 8.2, where e is the charge of the electron, n is the principal quantum number, h is Planck's constant, and Z is the nuclear charge.

$$v = Z\frac{2\pi e^2}{nh} \qquad \textbf{8.2}$$

Because the velocity of the electron increases as the nuclear charge (Z) increases, the velocity can become a significant percentage of the speed of light (c) for very heavy nuclei. For example, in bismuth the velocity of the 1s-electron is ~60% of c. As a result, the mass of the 1s-electron in Bi is 25% higher than the electron rest mass, a very significant change.

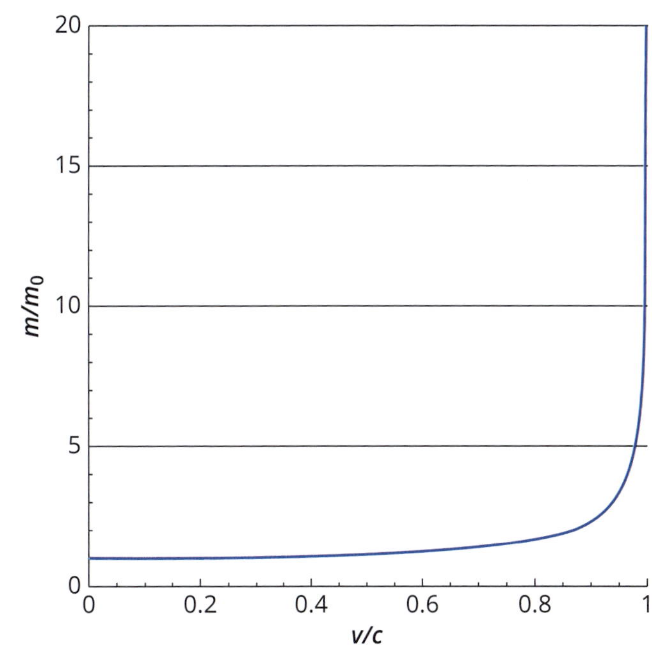

Fig. 8.8 Plot of the ratio of v/c (velocity of the electron over the speed of light) versus $m/m0$ (relativistic mass over the rest mass of the electron). At 99% the speed of light the mass has increased by about seven times.

Table 8.1 Table showing the effect of orbital radius on the estimated velocity and mass of the 1s-electrons down Group 15.

	s-orbital radius (Å)[a]	Average *p*-orbital radius (Å)[a,b]	Velocity of the 1*s*-electron (% of c)	Mass of the electron (m/m_0)
N	0.52	0.48	5	1.001
P	0.80	0.92	11	1.006
As	0.83	1.00	24	1.03
Sb	0.97	1.18	37	1.08
Bi	0.96	1.24	60	1.25

[a] Orbital radii from Waber and Cromer.[4]
[b] Average of $p_{1/2}$- and $p_{3/2}$-radii.

Table 8.1 shows similar calculations for the elements of Group 15. The mass of the 1*s*-electrons in the light element, nitrogen, are 1.0013 times the rest mass of the electron, m_0, whereas for the heaviest element in the group, bismuth, the 1*s*-electrons weighs 1.25m_0. The change due to Eq. 8.1 is not gradual, however. The mass of the 1*s*-electrons in the second-heaviest element of the column, antimony, are increased by only ~8%, and relativistic effects are less important.

The Bohr radius of an orbital is calculated as shown in Eq. 8.3, and changes as the inverse square of the velocity. The decreasing size of the 1*s*-orbital due to the terrific speed of its electrons then affects the other *s*-orbitals. In order to remain orthogonal to 1*s*, the other *s*-orbitals also shrink somewhat. As a result, the heavier bismuth has a slightly smaller valance *s*-orbital than lighter antimony, while average *p*-orbital size increases for the heavier element (Table 8.1). The relativistic effect is very pronounced for bismuth and is far less significant as one ascends the column.

$$r = \frac{Ze^2}{mv^2} \tag{8.3}$$

If the 1*s*-orbital contracts, the outer-shell *s*-orbitals must also contract to maintain orthogonality. In fact, the effect rippling outward to the higher shells—for example, out to the valence 6*s*-orbital—can be quite large. Fig. 8.9 shows a plot of the relativistic contraction of the 6*s*-orbital versus the nonrelativistic radii, from the review by Pyykko and Desclaux. There is additional contraction due to the filling of the 5*d*-subshell, leading to the local minimum at gold, {Xe}$(5d)^{10}(6s)^1$. In fact, the relativistic effect for gold is the largest of any element until fermium (atomic number 100).[7]

The calculated contractions, taking all of the complex effects into account for the 6*s*-orbitals, are, as examples, 4% for Cs, 12% for Re, 18% for Au, and 12% for Rn, versus what the calculated orbital sizes would be without relativity. The (direct) effect is largest for *s*-orbitals, much smaller for *p*, and considered insignificant for *d* and *f*.

One can estimate the relativistic effect by using Eq. 8.4, where again *Z* is the nuclear charge, *c* is the speed of light, and $<v>$ is the mean velocity of the 1*s*-electrons.

$$<v>/c = Z/137 \tag{8.4}$$

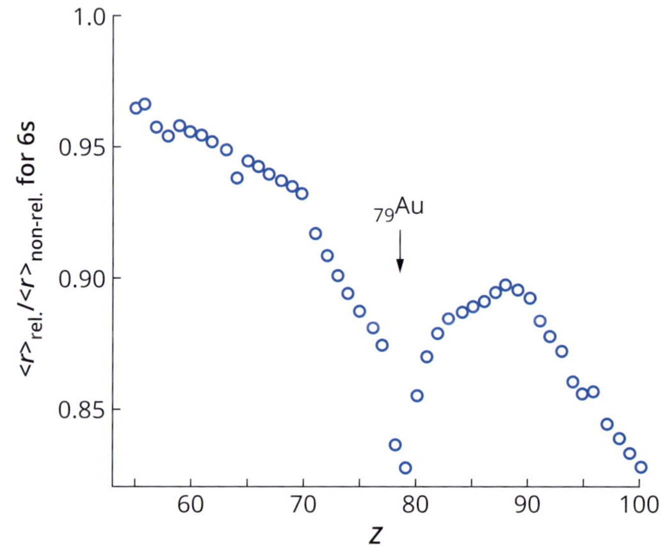

Fig. 8.9 Ratio of relativistic to nonrelativistic radii for the 6*s*-orbitals as a function of atomic number, *Z*.[7]

So, for example, with gold, *Z* is 79, the ratio of the mean velocity over the speed of light is 0.58, 58% of the speed of light. This will lead to an electron mass, by Eq. 8.1, of 1.23 m_0. As a result, the 1*s* orbital contracts, and the other *s*-orbitals contract similarly in order to remain orthogonal to 1*s*.

The relativistic contraction and electronic stabilization of outer *s*-orbitals, such as the 6*s*-orbital in Au, is called the *direct relativistic effect*. The contraction of the *s*-orbitals due to relativity causes a destabilization and expansion of the *d*- and *f*-orbitals due to increased shielding by the smaller *s*-orbital. This destabilization of *d*- and *f*-orbitals is called the *indirect relativistic effect*.

One result of relativity is that the difference in *s*- and *p*-orbital radii for 6*s* and 6*p* tends to be a bit larger than the difference between 5*s*- and 5*p*-orbital radii in the element above it in the periodic table. Fig. 8.2 shows a plot of the *s*/*p*-orbital radii versus element for Group 15; the slightly smaller Bi ratio versus Sb is due, at least in part, to relativistic effects.

Relativity has important consequences on the properties of the heavy elements, including a few notable effects.

(a) *Low melting point of mercury:*[8] Hg and Au are adjacent in the periodic table, but their melting points differ by 1100 °C! (Hg m.p. = −39 °C and Au m.p. = 1064 °C). Relativistic 6*s*-orbital contraction and filled subshells, {Xe}$4f^{14}5d^{10}6s^2$, give Hg weak interatomic interactions. Gold, with its unfilled subshell, {Xe}$4f^{14}5d^{10}6s^1$, has much stronger metallic Au–Au bonds that increase its melting point. The lighter congeners of mercury, with much smaller relativistic effects and identical valence electron configurations ($d^{10}s^2$), melt at much higher temperatures at 420 °C (zinc) and 321 °C (cadmium) because of less stabilized *s*-electrons and strong M–M interactions.

(b) *Color and stability of gold:*[9] Stabilization of the 6*s*-orbital and destabilization of 5*d*-orbitals leads to a lower-energy $\{5d^{10}6s^1\}\ ^2S_{1/2} \rightarrow \{5d^9 6s^2\}\ ^2D_{3/2}$ transition for gold. Absorption in the blue/violet gives the element a yellow color. In contrast, the related transition in silver, where relativistic effects are much smaller, occurs in the ultraviolet. Stabilization of the valence 6*s*-electrons also makes gold more difficult to oxidize by O_2, and Au does not tarnish in air like Ag (Fig. 8.10).

(c) *"{C}ars start due to relativity"*:[10] The lead–acid battery most commonly employed in cars owes 1.7–1.8 V of its 2.1 V to relativistic stabilization of the 6*s*-orbital, which greatly increases the voltage. Most of this additional voltage occurs on the PbO_2 side of the cell.

(d) *Spin–orbit coupling*: In the heavier elements there is coupling between electron spin and orbital angular momentum, which is a result of the relativistic atom.

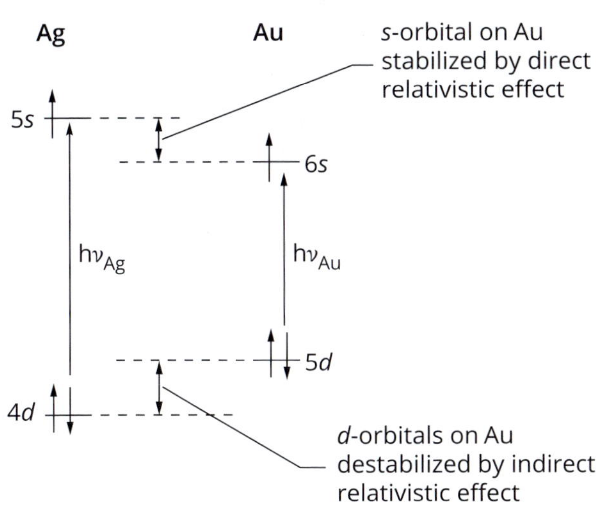

Fig. 8.10 The color of gold is due to a combination of direct and indirect relativistic effects. As related to silver, the *s*-orbital is stabilized in gold (direct effect), and the *d*-orbitals are destabilized in gold (indirect effect). As a result, silver absorbs in the UV, while gold absorbs in the visible.

The relativistic influence on electronic properties of transition-metal compounds is felt even for some first-row complexes of the *d*-block, and profoundly influences the properties of the second- and third-row transition metals.

8.2.3 2-Center 2-Electron π-Bonds Down the *p*-Block: Heavy Alkynes as an Illustration

In Chapter 3, 2-center 2-electron σ-bonds were discussed extensively and for many different elements in the periodic table. However, π-bonds have not been discussed in as much detail, and here we will cover some of the changes that occur in π-bonding for the heavier main-group elements.

As one proceeds down a column in the *p*-block, the frequency of finding π-bonds decreases dramatically, with the first row forming many more double- and triple-bond compounds than even the element immediately below it. Part of the general change in π-bonding stability is due to the lower electronegativities of the elements as one proceeds down the column. In a double or triple bond there are four or six electrons, respectively, that must be supported by only two nuclei. The likelihood is greatly increased that the two nuclei will perform this feat and form a stable compound while supporting so much electron density when they are higher electronegativity elements. Indeed, one finds that there are many compounds with heavier *p*-block elements containing multiple bonds if at least one of the elements involved in the bond has a high electronegativity, such as phosphine oxides, sulfur dioxide, and so on. As we will see later, these compounds often have dominant polar resonance forms with anionic charges on the electronegative element as well.

Another contributing cause to the lower numbers of π-bonds in heavier main-group compounds is the propensity of the heavier congeners to use *s*-orbitals to hold electron pairs. The *s*-orbital extension is relatively low, and its tendency to hybridize is lower than in the first row. As a consequence, placing the large amounts of electron density on the heavier *p*-block elements tends to result in lone-pair formation rather than sp^2- or *sp*-centers that form strong double or triple bonds (*vide infra*). A related issue is the *inert pair effect*, which also leads to the lower oxidation state of the heavy *p*-block elements being more favorable (Chapter 9).

One group of compounds that has been explored in some detail that illustrates many of these points are the *heavy alkynes*—compounds of the general formula Ar–EE–Ar, where E is a Group 14 element and the E–E bond is intentionally vague in bond order.[11] In order to isolate these highly reactive bonds with the heavier elements of the group, it is necessary to use a large substituent as Ar; for example, bis(2,6-(2,6-diisopropylphenyl))phenyl. A typical synthesis is shown in Eq. 8.5, where the divalent germanium compound is reduced using potassium metal to afford the dimeric product.[12]

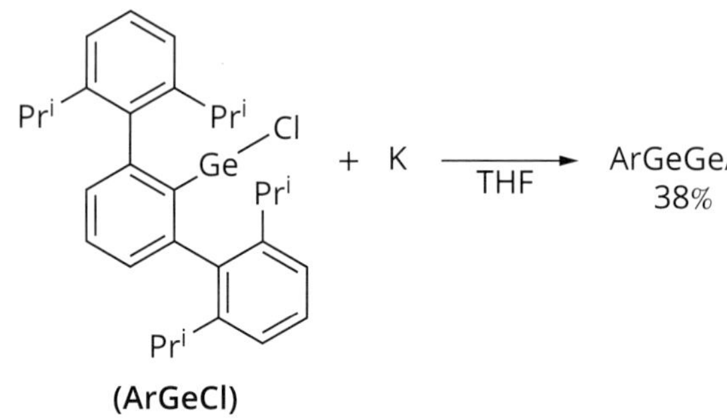

8.5

On examining a series of heavy alkyne compounds, one finds that the approximate bond orders are as shown in Fig. 8.11. When the element is carbon, of course the bond order is 3. By the time the heaviest congener of the group is reached, the bond order is below 2.

One quite noticeable feature of the heavier congeners is their lack of linearity. In order to explain the bent structures and reactivity, five different resonance structures have been proposed as major contributors in this class of compounds. The first resonance form (1 in Fig. 8.12) is the anticipated E–E triple bond. The second is a pair of contributing structures (2 and 2′ in Fig. 8.12) where the electrons of one of the bonds is placed on one atom, ionizing the triple bond to form a zwitterionic structure; naturally, the E–E bond is nonpolar so resonance forms 2 and 2′ contribute equally. The third important resonance form can be thought of as homolysis of one of the π-bonds to form a 1,2-diradical (3). The last is where, instead of a triple bond, a single bond exists between the Group 14 elements and there is a lone pair on each center (**4**).

The diradical resonance form 3 is thought to contribute little. Even in Si and Ge compounds where resonance form 3 makes its largest contribution, the diradical character is thought to be only about 14%.

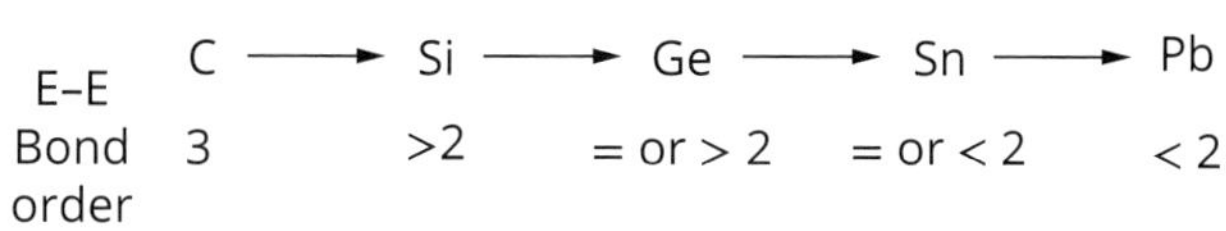

Fig. 8.11 Approximate bond orders in ArEEAr compounds.

As one proceeds from 1 to 4 in the resonance structures in Fig. 8.12, the E–E bond order is decreasing; as a result, it is thought that the decreasing bond orders as one proceeds down the column are due to larger contributions from the later resonance forms in Fig. 8.12. As an example, examination of the model system HPbPbH using natural resonance theory is shown in Fig. 8.12. The largest contributor in this case is resonance form 4 at 35%. The triple-bond form contributes about a quarter of the ground-state electronic distribution, with 2/2′ contributing 7% each for this simple model using a low level of theory (B3LYP/LANL2DZ).

Consistent with this is the reactivity of the series. In the distannyl compound, reactions with oxidants should result in cleavage of the Sn–Sn (approximately) single bond to form Sn(II) compounds (Eq. 8.6). Treatment of ArSnSnAr with the oxygen atom donor nitrous oxide, N_2O, produces the bridging oxo compound. If 4 is the main resonance contributor, one may not expect ArSnSnAr to react with donors, and the compound is unreactive with *tert*-butylisonitrile, $CNBu^t$ (Eq. 8.7)

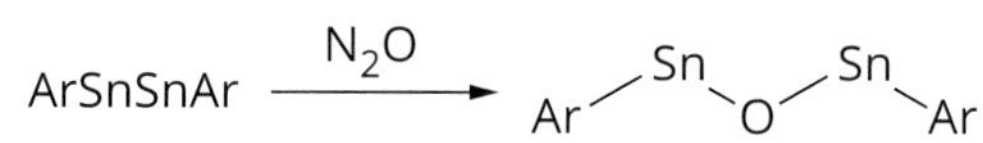

8.6

8.7

The germanium compound has a bond order of about 2, so the main resonance form contributors may be 2/2′ in Fig. 8.12. Treatment of ArGeGeAr with $CNBu^t$ is consistent with this electronic structure being an important contributor, and a stable adduct is formed, Eq. 8.8.

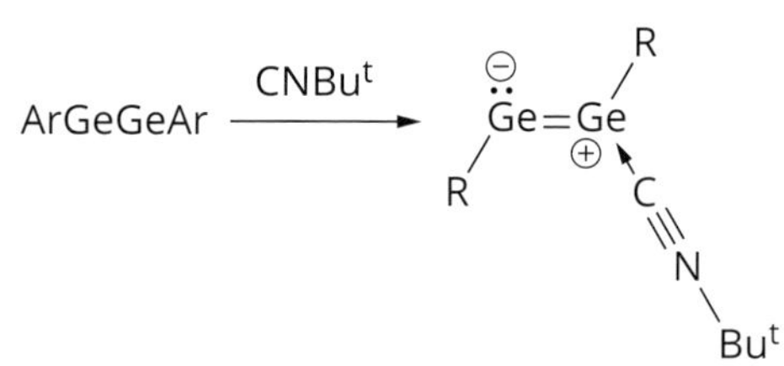

8.8

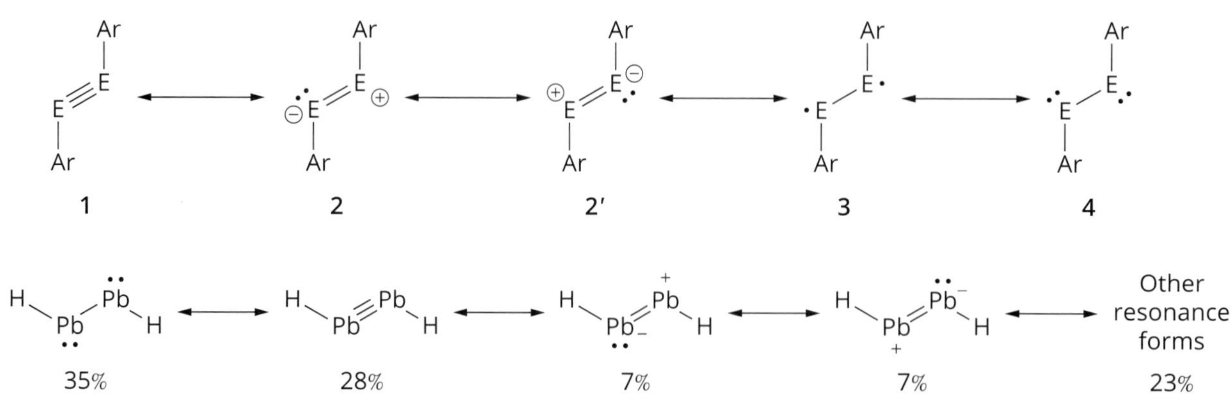

Fig. 8.12 (*top*) Some resonance forms contributing to the bonding in heavy alkynes. (*bottom*) Results of natural resonance theory calculations on HPbPbH as an example.

8.3 Beyond the 2-Center 2-Electron Bond: Common Bonds in Hypovalent and Hypervalent compounds

Under the favored modern models, the preservation of the octet is of paramount importance, and main-group atoms never exceed eight electrons in their valence shell. However, the 2-center 2-electron bond is not always preserved; instead, there are several additional bonding interactions that are of importance in hypovalent and hypervalent compounds.

Using 2-center 2-electron bonds along with three additional interactions—the 3-center 4-electron bond (ω-bond), negative hyperconjugation, and the 3-center 2-electron bond (τ-bond)—the vast majority of main-group compounds can be understood. Each of these interactions is described as follows, using both hybridization and molecular orbital theory.

BOX 8A

The main-group elements and *d*-orbitals

It was at one point thought that bonding in hypervalent main-group compounds involved use of *d*-orbitals on the main-group elements, such as sp^3d-hybrids in trigonal bipyramidal PF_5. The popularity of this theory perhaps peaked in the 1970s, and today most researchers working in main-group chemistry do not believe that the 3*d*-orbitals on phosphorus, for example, are involved significantly in the bonding of hypervalent compounds of the element.

While there are some who still contend that *d*-orbitals play roles in the bonding of hypervalent compounds, the authors do not believe that they play a chemically significant role. We can summarize three reasons for not using this theory in this book.

First, calculations on the *d*-orbitals of phosphorus, even by authors trying to examine the applications of 3*d*-orbitals in hypervalent bonding, found that the radii of these orbitals in the $^6D_0$ term ($3s^13p^33d^1$ configuration) were ~3.6 Å.[51,52] This was the smallest radius found for all the possible terms with this configuration. In contrast, the radii of the valence orbitals on phosphorus are ~0.8 Å (3*s*) and ~1.0 Å (3*p*). In order to hybridize effectively, orbitals need to be close in size (Chapter 3 and Section 8.2.1). At this size for the *d*-orbitals, a typical first-row *p*-block element (C, O, F, and so on) on the phosphorus with a hybrid radius of ~0.5 Å and at a typical P–X bond distance (~1.8 Å) will have very little, if any, overlap. One may reduce the size of the 3*d*-orbitals on phosphorus by increasing the amount of electron density in the orbitals, but this is done at the expense of the promotion energy, which is the next topic.

Second, the promotion energy required for the participation of the *d*-orbitals on phosphorus is extremely high. If one uses a fairly high-level calculation (CCSDT, aug-cc-PVTZ, full electron) to find the energy of the lowest-energy sextet state relative to the quartet ground state ($s2p^3$ configuration), one naturally finds that the lowest-energy sextet has a $3s^13p^34s^1$ configuration as would have been expected, not the $3s^13p^33d^1$ configuration as needed for *d*-orbital hybridization, but this $3s^13p^34s^1$-state is 288 kcal/mol above the ground state. In other words, the promotion energy required to take an electron from 3*s* and put it in the *d*-orbital is >288 kcal/mol. The necessary state can be estimated from experimental atomic spectroscopy as being ~360 kcal/mol over the ground state, but even using the lower 288 kcal/mol value the promotion energy required is >2 times higher than even the highest P–X single bond energies.

The energy to promote from the *s*- to *p*-orbital to give the $3s3p^4$ configuration is 71,000 cm^{-1}. The energy to move an electron from *p* to 3*d* to make $3s^23p^23d^1$ is 72,000 cm^{-1}. To a first approximation, the energy of the $3s^13p^33d^1$ configuration would be the sum of these at ~140,000 cm^{-1} or 400 kcal/mol.[53] Expecting a significant contribution from the sp^3d-configuration is energetically unrealistic.

This was applicable to the neutral phosphorus atom. Going to phosphorus in higher oxidation states seems to exacerbate the promotion energy. Removing one electron, phosphorus(I) gives a ground state of $3s^23p^2$. The first *d*-orbital containing excited state, $3s^23p^13d^1$, is 77,710 cm^{-1}, even higher than in the neutral atom.

Moving to even heavier elements does not decrease the promotion energy substantially enough to make *d*-orbital participation more likely. For example, in iodine the lowest-energy promotion in the neutral atom that includes *d*-orbital participation is the $5s^25p^45d^1$-configuration, which is about 67,000 cm^{-1} (190 kcal/mol) above the ground state.

Third, a final and very compelling reason not to use *d*-orbitals in the bonding of the main group is that they are not required. *The bonding, chemistry and properties of the compounds can be explained by using their valence orbitals, just as we do for the chemistry of all other compounds.*

None of this entirely excludes *d*-orbital participation within the main group; however, if they do participate it is probably at small levels. For example, a molecule where *d*-orbital participation may be near the highest in the main group is SF_6, for which NBO calculations suggest 3*d* orbital occupation as a meager 0.16 electrons, far less than that required for sp^3d^2 hybridization.[54] From a pragmatic standpoint it behooves the student of inorganic chemistry to learn how the bonding in the main group may be explained without *d*-orbitals, as these resonance forms are undoubtedly more important contributors to electron ground state of main-group compounds.

8.3.1 The 3-Center 4-Electron Bond (ω-Bond)

Pimentel and Rundle independently introduced the 3-center 4-electron (ω-bond) bond in 1951.[13,14] Perhaps the simplest example of a compound containing an ω-bond is XeF_2. The geometry according to VSEPR for the AE_3X_2 compound is shown in Fig. 8.13, and, as anticipated, XeF_2 is linear.[15]

If both of the bonds are 2-center 2-electron bonds, the Xe atom has ten electrons (four in Xe–F bonds and six in lone pairs). However, main-group elements have only four valence orbitals and can therefore accommodate only eight electrons.[16] So, how can the bonding in this and many related compounds be understood?

Let's analyze the compound using MO theory first, though we should also consider one important principle from hybridization theory immediately. Bent's rule suggests that lone pairs are best housed in *s*-orbitals, and due to the primogenic and relativistic effects this is especially true for heavier *p*-block elements. Considering there are three lone pairs on Xe to support, the bonding in the molecule is very likely being done by a *p*-orbital on xenon. Since the two fluorines similarly have three lone pairs each, we will consider the bonding by F to be done by a pure *p*-orbital as well.

With the approximation that each atom is using a pure *p*-orbital in place, we realize that there is one orbital on each atom that can participate in bonding because all others are holding electron pairs, and we can draw the three MOs for the bonds by inspection (Fig. 8.14). The lowest-energy combination of three *p*-orbitals will be bonding through the F–Xe–F framework. The next higher-energy molecular orbital will have one node, which will be located on Xe; in this orbital, both positive lobes of the *p*-orbitals on fluorine will point to Xe—an arrangement that cannot bond to a *p*-orbital by symmetry. The highest molecular orbital will have two nodes and will be antibonding between Xe and both fluorines.

Based on how we set up our MO diagram (Fig. 8.14), we can discern how many electrons to include. In setting up the system, we left three lone pairs of electrons on each atom. Consequently, each fluorine has six electrons as lone pairs and one remaining for bonding. Meanwhile, the 8-valence electron Xe has six electrons as lone pairs and two remaining for bonding. In total, there are four electrons in our F–Xe–F bonding MOs.

From the MO diagram there is one filled bonding orbital that extends across the F–Xe–F framework, which is lowest in energy. The other filled orbital is nonbonding with its electron density distributed onto the two fluorines. From MO theory, the Xe–F bond order in XeF_2 is then ½ for each of the bonds.

The VB description for the compound confirms this analysis and is shown in Fig. 8.15. Again, the suggestion is that the compound has a Xe–F bond order of ½.

If we compare the properties of XeF_2 with its closest normal valent compound I–F, we begin to see the consequences of

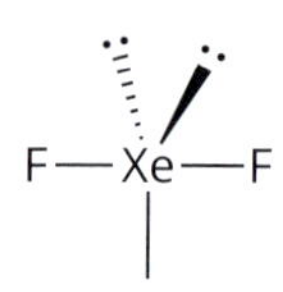

Fig. 8.13 VSEPR structure of xenon difluoride.

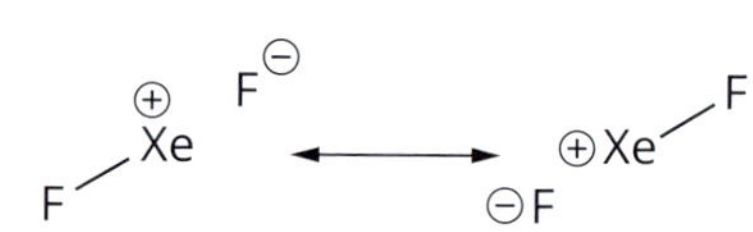

Fig. 8.15 VB description of the bonds in XeF_2.

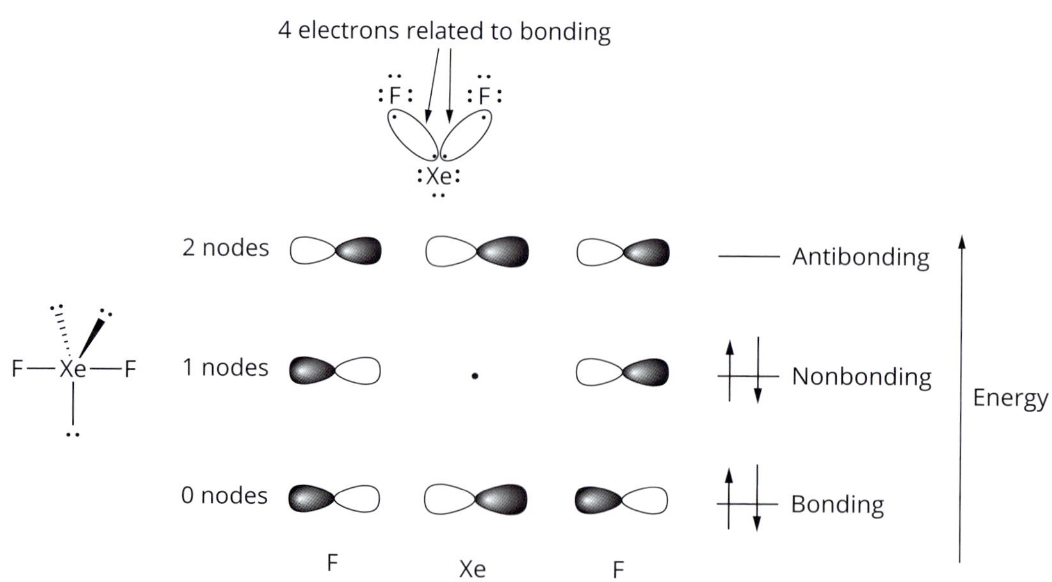

Fig. 8.14 Partial MO diagram for XeF_2, assuming that the bonding is done with *p*-orbitals on each atom. Orbitals holding "lone pairs" are not shown.

ω-bonding. The I–F bond is significantly shorter at 1.93 Å compared to the Xe–F bonds in XeF_2 at 2.14 Å. The bond energy in XeF_2 is 32 kcal/mol, half that in I–F of 60 kcal/mol.[17]

Because 3c4e bonds in the main group will use a pure p-orbital, the two ligands participating in this type of bond will be ~180° apart, like the fluorines in XeF_2. This fact is very useful in identifying bonds of this type and is essential in understanding the bonding in many hypervalent compounds. *In general, if the compound is hypervalent because it contains too many s-bonds, it probably contains one or more 3c4e bonds.*

8.3.1.1 Analyzing Compounds with ω-Bonds: SF_4

With principles discussed so far, we can analyze SF_4 by using hybridization theory to better understand the bonding. The input for a bonding analysis is the geometry of the compound. The experimental geometry is shown in Fig. 8.16a. The compound has two fluorines that are very nearly 180° apart (labeled F^a, axial flourine) that form the axis of the trigonal bipyramidal (tbp) base structure for VSEPR analysis of the AEX_4 compound. In the equatorial plane resides the lone pair and two of the fluorines. The fluorines in the *xy*-plane of the tpb base structure (labeled as F^e in the figure) are only 101.6° apart.

The compound SF_4 as drawn in Fig. 8.16a has ten electrons around sulfur (two from a lone pair and four pairs of bonding electrons to the fluorines) and is hypervalent. It is not possible to form four 2-center 2-electron bonds in this case. Since the compound has two fluorines that are ~180° apart, we can propose that the F^a–S–F^a system involves a 3-center 4-electron (3c4e) bond (Fig. 8.16b), as discussed in Section 8.3.1. As a result, there will be a ½-bond between sulfur and each fluorine along the axis.

For a 3c4e bond to the axial fluorines, a pure *p*-orbital on sulfur is used (Fig. 8.14). As a result, the sulfur has one valence *s*- and two valence *p*-orbitals to form hybrids to house the lone pair and form the S–F^e bonds. If the lone pair and S–F^e bonds were 120° apart, then this could be done with sp^2-hybrids; however, since the lone pair is not equivalent to the fluorines, different hybrids will be used for these different substituents that average to ~sp^2.

The two F^e fluorines are equivalent by symmetry, and orbitals of the same hybridization on sulfur will be used to bond to them. Using the Coulson directionality theorem, we can find the hybridization parameter for any two hybrids if we know the angle between them. The two fluorines are 101.6° apart (Fig. 8.16a), so we assume that there is no bond-bending and use this value in Eq. 7.14, which shows that sp^5 hybrids are used by sulfur to bond to the F^e atoms.

$$\begin{aligned} \cos\omega &= -\frac{1}{\lambda_{SF^e}} \\ \cos 101.6^\circ &= -\frac{1}{\lambda_{SF^e}} \\ -0.201 &= -\frac{1}{\lambda_{SF^e}} \\ \lambda_{SF^e} &= 4.98 \end{aligned}$$

The only hybrid on sulfur that we do not know is the one housing the lone pair. If we use the sum rule, Eq. 7.11, we can find this hybridization as well. In the following equations the hybridization of the sulfur-based orbital bonding to the axial fluorines is a pure *p*-orbital ($\lambda_{SF(a)} = \infty$). The hybrids on sulfur bonding to the equatorial fluorines are sp^5, as discovered previously ($\lambda_{SF^e} = 4.98$). Solving for the hybridization parameter housing the lone pair, $\lambda_{l.p.}$, produces a very *s*-rich orbital with hybridization $sp^{0.5}$. That the lone pair resides in a very *s*-rich orbital could have been guessed beforehand by using Bent's rule (Section 7.3). The hybridization theory analysis is summarized in Fig. 8.16c.

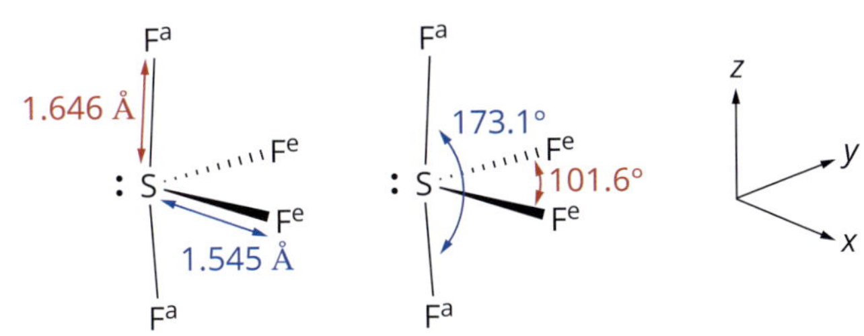

Fig. 8.16a Geometry of SF_4.

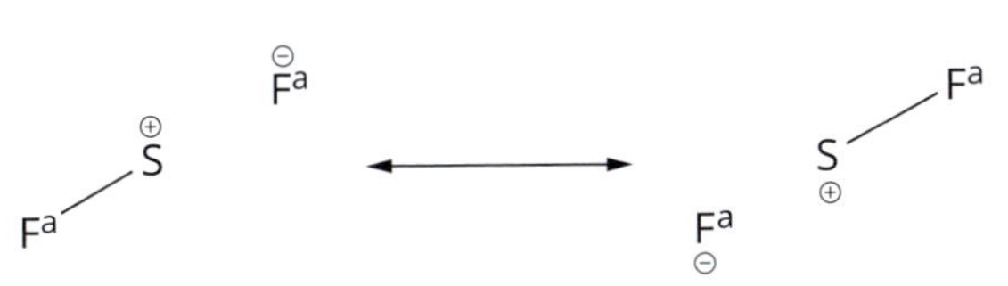

Fig. 8.16b Valence bond diagram of 3-center 4-electron bonds along the SF_4 axis.

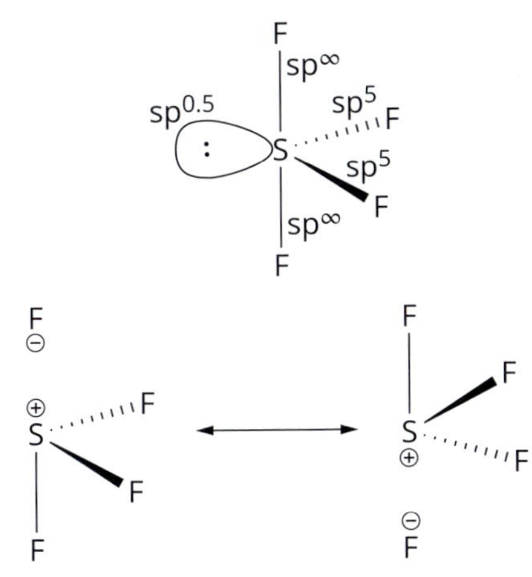

Fig. 8.16c Composite structure from the hybridization analysis showing the hybrids on sulfur, and two of the major contributing resonance forms.

In this calculation, keep in mind that the sulfur is using one orbital (a pure p-orbital) to bind to both of the axial fluorines, so there is only one hybrid in the calculation for both axial fluorines (F^a), an sp^{∞}-orbital.

$$\sum_{i}^{4} \frac{1}{1+\lambda_i} = 1$$

$$\frac{1}{1+\lambda_{SF(a)}} + \frac{1}{1+\lambda_{SF(e)}} + \frac{1}{1+\lambda_{SF(e)}} + \frac{1}{1+\lambda_{l.p.}} = 1$$

$$\frac{1}{1+\infty} + \frac{1}{1+5} + \frac{1}{1+5} + \frac{1}{1+\lambda_{l.p.}} = 1$$

$$0 + 0.167 + 0.167 + \frac{1}{1+\lambda_{l.p.}} = 1$$

$$\frac{1}{1+\lambda_{l.p.}} = 0.666$$

$$\lambda_{l.p.} = 0.50$$

One can analyze SF_4 by MO theory as well (Fig. 8.16d). For this, the SALCs on the fluorines should be more or less identical to those discussed for the hypothetical molecule SH_4 in detail in Chapter 6. In this case we will assume that the fluorines are largely using p-orbitals to form their bonds (Bent's rule); as a result, the s-orbitals used in the SALCs in Chapter 6 for SH_4 are replaced with p-orbitals here, with similar phases pointed at the central sulfur atom. There are two a_1 SALCs: one with all the same phases for the p-orbitals directed at sulfur, and another where the axial p-orbitals have one phase and the equatorial p-orbitals another. The explanation for how these were generated are shown in Fig. 6.8 and surrounding text. There are also SALCs of b_1 and b_2 symmetry.

We can draw a partial MO diagram for SF_4 (Fig. 8.16d). A more detailed diagram would include all of the SALCs for the fluorine lone pairs, which will be nonbonding but may mix with bonding or antibonding orbitals (Chapter 6). For our purposes here, a partial MO diagram will be sufficient to compare with our hybridization theory results. In the MO diagram there are orbitals of a_1 symmetry that comprise an ω-bond set: $1b_2$ (bonding), $2a_1$ (nonbonding), and $2b_2$* (antibonding). The orbital $2a_1$ will be largely nonbonding, but may look different from a calculation due to mixing.

For the MO diagram of SF_4 shown on the left in Fig. 8.16d, a scheme where the energies were largely considered when mixing orbitals was employed. This leads to the s-orbital of sulfur being used in the $1a_1$-orbital, for example. Possible mixing with the $2a_1$ is not shown in the MO diagram in Fig. 8.16d, so that the pure nonbonding MO that is part of the ω-bond could be shown. The sulfur lone pair resides in $3a_1$, which is entirely p-orbital in character, which provides a major clue that something is very wrong with this orbital mixing scheme.

The hybridization theory analysis shows the F(axial)-S bonds being made by sp^5-hybrids and the lone pair as residing in an $sp^{0.5}$-hybrid, which is quite consistent with Bent's rule. The hybridization theory analysis (Fig. 8.16c) suggests that our MO diagram should have substantial mixing of $1a_1$ and $3a_1$. In fact, there is so much mixing of $1a_1$ and $3a_1$ that the lone pair is actually mostly s-character, and the bonding is mostly done with p-orbitals.

If we redraw the MO diagram while keeping the lessons we learned from the hybridization theory analysis in mind, we obtain the diagram for SF_4 shown on the right. Here, sp-mixing occurs to enable overlap of the sulfur orbitals with the fluorines and to stabilize the sulfur-based lone pair ($3a_1$).

We can use natural resonance theory to examine SF_4 and compare with our previous analysis.[18] The calculation finds that there are two major resonance forms that contribute equally and constitute ~70% of the ground-state electronic structure; these are the two resonance forms shown in Fig. 8.16c. Note that other resonance forms may raise the axial bond order slightly above 50%. For example, one can envision using sp^1-hybrids to form two covalent bonds along the axis to the fluorines, then using the two remaining p-orbitals to house the lone pair and bond to one of the equatorial fluorines. This resonance structure will participate relatively little but would exchange bond order in the equatorial fluorines for a higher bond order along the axis. The MO diagram also hints at this, where the nonbonding orbital $2a_1$ may mix with the sulfur lone pair $3a_1$ and $1a_1$ to provide additional bonding along the axis. Again, carrying out a hybridization and MO analysis provides complimentary information that provides more nuanced electronic structures than might be garnered by either method alone.

8.3.1.2 Analyzing Compounds with ω-Bonds: $\{Sb(DMAP)_2(Ph)_3(OTf)\}^+$

Some main-group compounds contain both dative and polar covalent bonds. One such compound is 6-coordinate $\{SbPh_3(DMAP)_2(OTf)\}\{OTf\}$, where DMAP = 4-dimethylaminopyridine and OTf = trifluoromethylsulfonate ($^{-}OSO_2(CF_3)$, triflate).[19] Triflate is a common "pseudohalide"—a monoanionic species that often exhibits reactivity reminiscent of a halide, such as cyanide, cyanate, thiocyanate, and others. DMAP is a common ligand and base that binds through the pyridine nitrogen. The structure of the compound is shown in Fig. 8.17a. If all the bonds are 2c2e bonds, then the compound would have ten electrons around the antimony(V) metal, making the metal hypervalent and meaning that bond types other than 2c2e bonds will be involved.

It is perhaps best to start with the strongest donors in the compound: the relatively low electronegativity phenyl groups. Keeping in mind that the central antimony will make nominally six bonds in the end, one can consider that the best method to interact with the three strongly donating phenyl groups would be to make bonds to the two phenyls 180° apart using ~sp-hybrids and using a pure p-orbital to interact with the third ~90° away. Consequently, we will use 2c2e polar covalent bonds to the phenyl groups employing the s-, p_x, and p_y-orbitals on antimony (Fig. 8.17b).

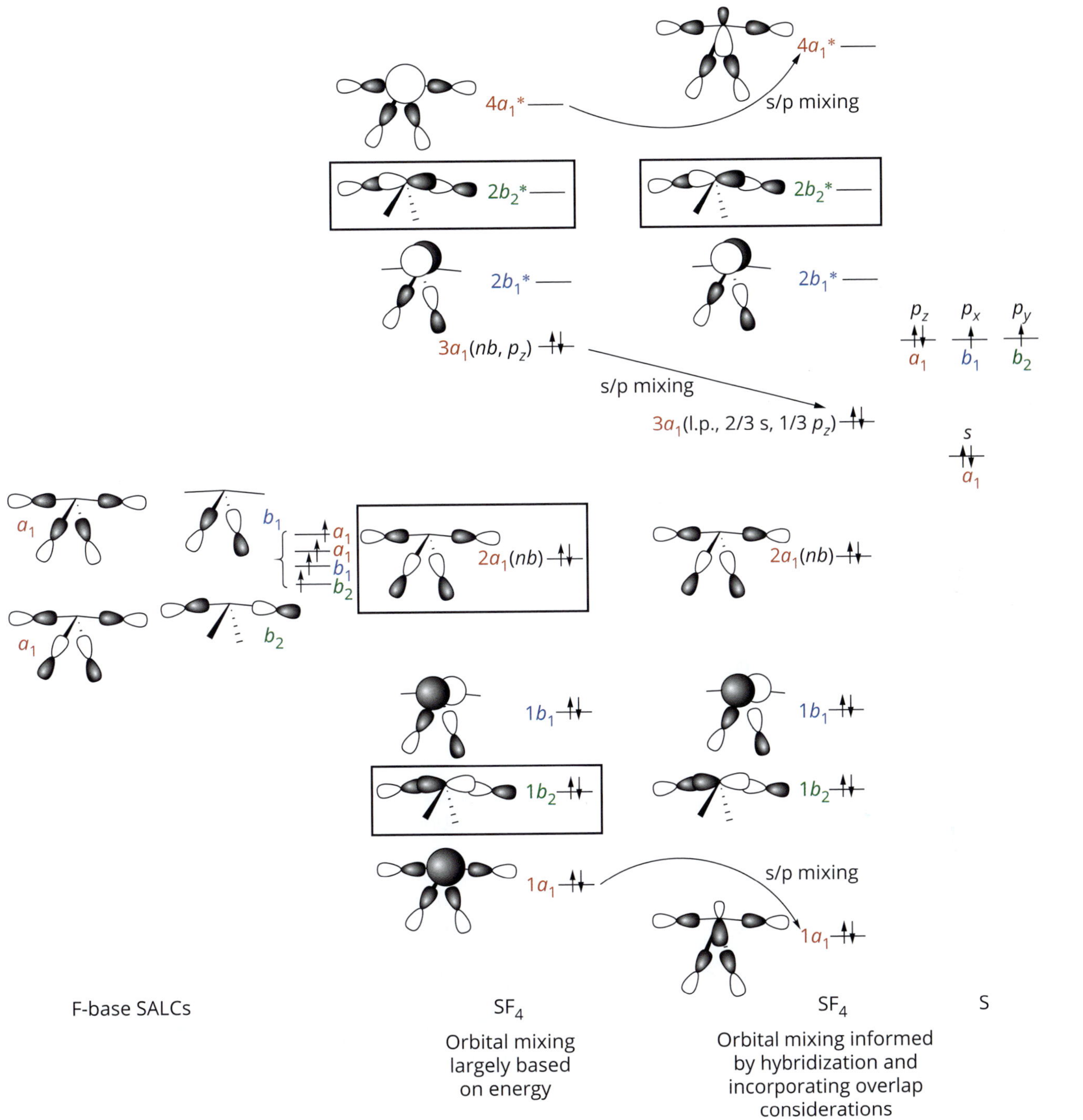

Fig. 8.16d Partial molecular orbital diagram for SF_4. The lone pairs on fluorines are not included. In the boxes are the three orbitals involved in the F–S–F ω-bond: $1b_2$ (bonding), $2a_1$ (nonbonding), and $2b_2$* (antibonding). There are two MO diagrams included for SF_4. The one on the left is where orbital mixing was largely based on orbital energies. The lower-energy *s*-orbital on sulfur was used to bond in $1a_1$. From the hybrization theory analysis, we can tell that this is somewhat incorrect. The bonding from sulfur should be mostly *p*-orbital in character, and the lone pair on sulfur should reside in an orbital that is ~2/3 *s*-character.

Approximately 180° from the phenyl group on the *x*-axis is the triflate. Triflate is a relatively poor donor versus phenyl, but here acts as a ligand to antimony. The bond can only be made in the *xy*-plane using a 3c4e bond with the phenyl group. Because triflate is a poor donor and because its anionic charge is well supported in the resonance form where there is no Sb–OTf bond, the resonance form with an Sb–Ph bond will contribute far more to the ground state (Fig. 8.17c).

The only orbital on antimony that was not used in the previous bonding discussions is p_z. The two dative pyridine ligands along the *z*-axis will bond to the antimony with a 3c4e bond utilizing the p_z-orbital (Fig. 8.17d).

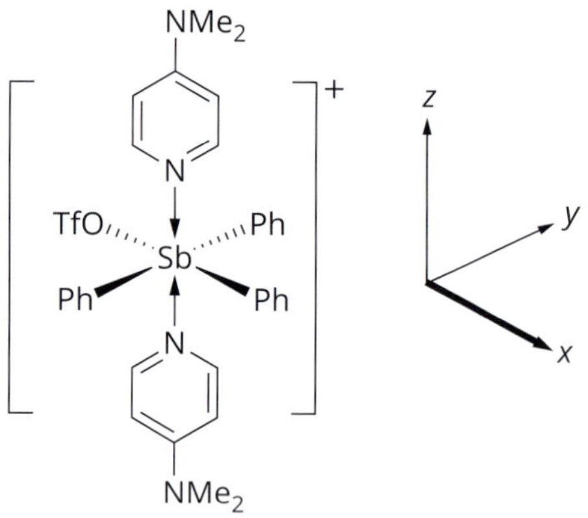

Fig. 8.17a The structure of $\{SbPh_3(DMAP)_2(OTf)\}^+$.

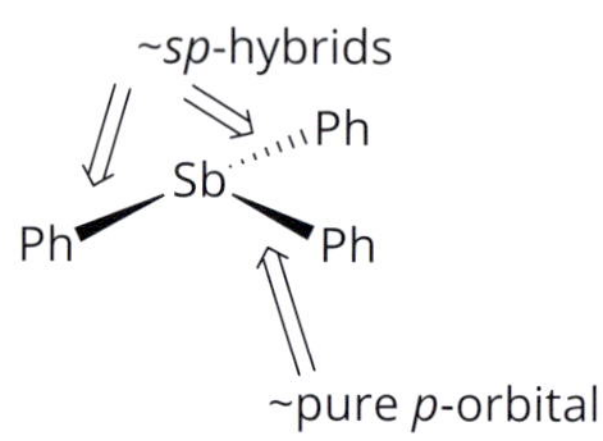

Fig. 8.17b Equatorial phenyl bonds in $\{SbPh_3(DMAP)_2(OTf)\}^+$.

Fig. 8.17c 3-center 2-electron bond in the equatorial plane between phenyl and triflate. The form on the left will be a much larger participant due to the higher electronegativity of triflate.

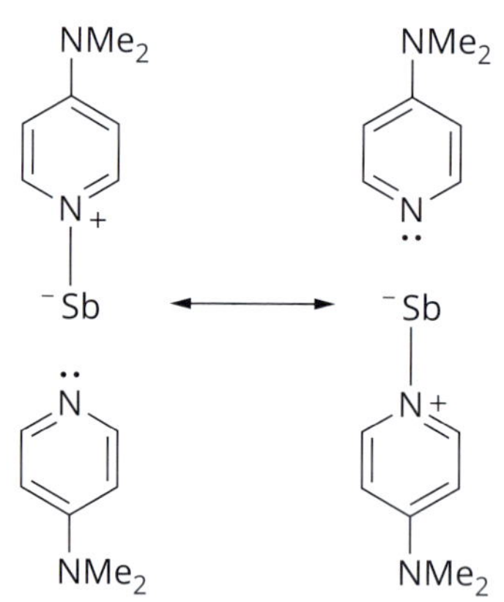

Fig. 8.17d Axial 3-center 2-electron bond in $\{SbPh_3(DMAP)_2(OTf)\}^+$.

The most likely contributing resonance forms are shown in Fig. 8.17e, each of which will use all of the orbitals around antimony and, as is necessary, obey the octet rule. Each of the major contributors on top will have minor contributors where there is an antimony–triflate bond and a C_6H_5 anion (bottom). Consistent with this expectation, an NRT calculation on the model compound $\{SbH_3(NH_3)_2I\}^+$ (Ph → H, DMAP → NH_3, OTf → I) gave the top two structures as the major contributors at >50% of the ground state.

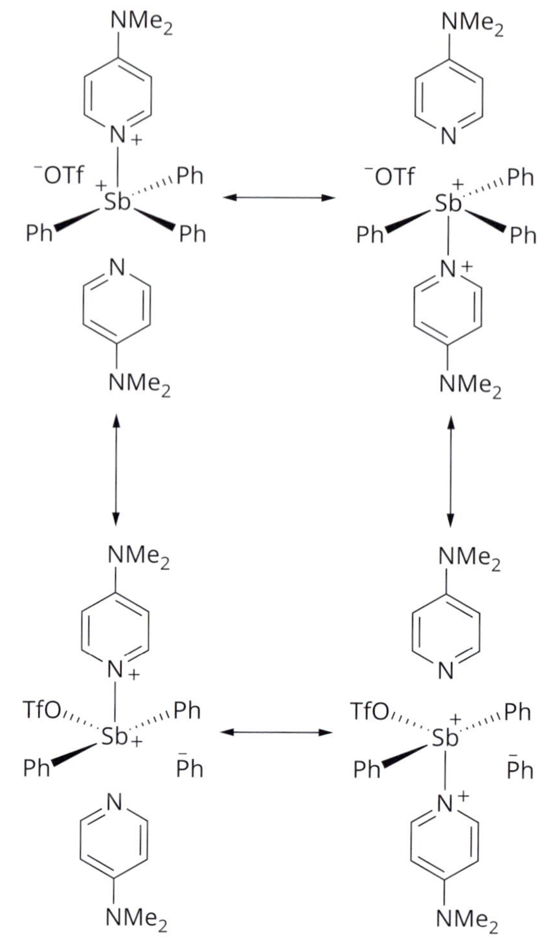

Fig. 8.17e Axial 3-center 2-electron bond in $\{SbPh_3(DMAP)_2(OTf)\}^+$.

8.3.2 Hyperconjugation and Negative Hyperconjugation[20]

Another important bonding interaction in hypervalent main-group compounds is *negative hyperconjugation*, but before discussing this it is useful to briefly review hyperconjugation, which is often taught in sophomore-level organic classes.

Hyperconjugation (positive hyperconjugation) is where electron density is donated from a filled σ-bonding orbital into a nonbonding empty orbital. A commonly discussed example for this type of interaction is found in the ethyl cation. From an MO perspective (Fig. 8.18) this is donation of electron density from a C–H bonding orbital of the CH_3 group into the empty *p*-orbital on the CH_2 carbon. From a hybridization theory perspective, the same interaction involves resonance between the ethyl cation having a CH_2 carbocation and an ethylene with adjacent proton. Hyperconjugation is a major stabilizing effect leading to carbocation stability proceeding as tertiary > secondary > primary in organic chemistry.

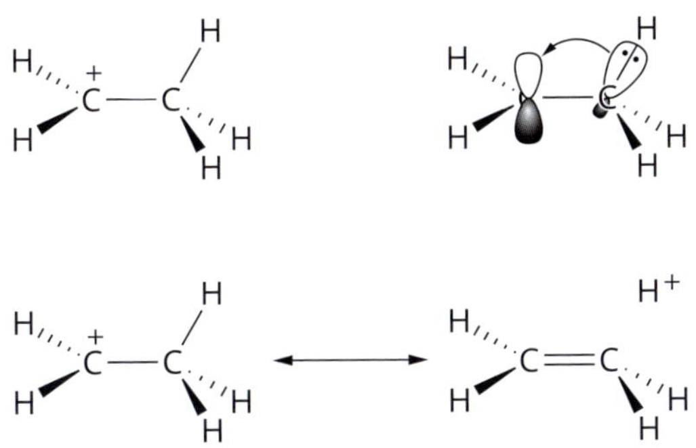

Fig. 8.18 The ethyl cation as an example of positive hyperconjugation with (*top*) an MO description of the interaction and (*bottom*) a hybridization theory analysis.

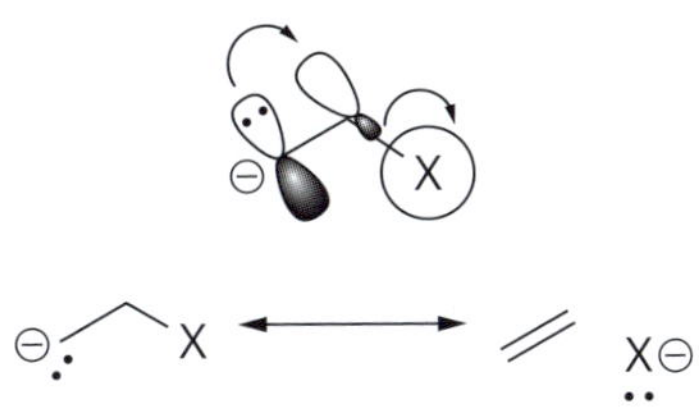

Fig. 8.19 Example of negative hyperconjugation from organic chemistry, with the MO description at the top showing the flow of electron density using the carbanion as a starting point, and with the valence bond description at the bottom. Negative hyperconjugation involves the lone pair on carbon donating into the C–X antibonding orbital, transferring electron density to X.

Negative hyperconjugation involves donation from filled nonbonding orbitals, such as lone pairs, into σ*-orbitals. A simple organic example of such an interaction is shown in Fig. 8.19, in which a carbanion with an electronegative group (X) on the β-carbon may be considered an alkyl carbanion or alkene with adjacent X^-. In this case, the "nonbonding" electrons are from the carbanion and the acceptor is the C–X σ*-orbital.

8.3.2.1 Analysis of Compounds with Negative Hyperconjugation

Negative hyperconjugation interactions are extremely common in hypervalent main-group compounds where, in the hypervalent Lewis structure (Fig. 8.20a), one often draws π-bonds. One very common example is the bonding in phosphine oxides, such as OPH_3. As drawn, the compound has ten electrons around the phosphorus atom, four electrons in the P=O bond, and six in the three P–H bonds. Again, phosphorus has only four valence orbitals and can therefore accommodate interactions involving eight electrons maximum. Consequently, the structure in the figure is not a valid resonance form, even though it is the most common structure drawn for phosphine oxides.

One possible resonance form for OPH_3 that is normal valent is its "ylide" form. An ylide is a compound with adjacent positive and negative charges, a 1,2-dipole. Phosphine oxides in general are quite polar compounds consistent with this resonance form being a major contributor (Fig. 8.20b). This is readily seen in the properties of phosphines versus phosphine oxides. For example, PMe_3 is a nonpolar, colorless liquid that boils at 38–40 °C and melts at −86 °C. Its oxide, $OPMe_3$, is a white solid at room temperature that *melts over 220 °C higher*

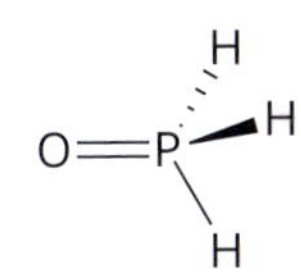

Fig. 8.20a Hypervalent Lewis structure of OPH_3.

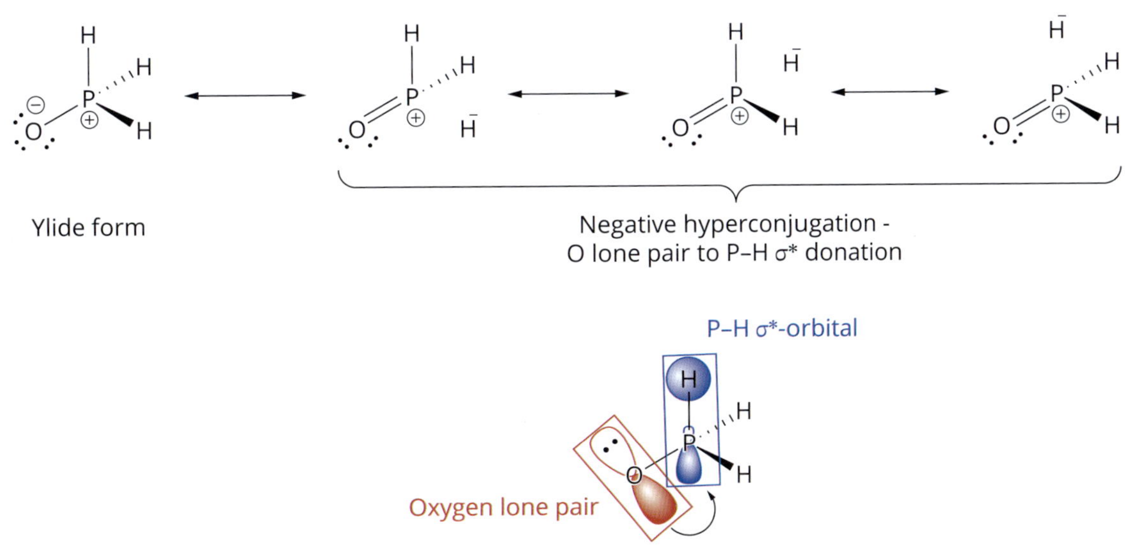

Fig. 8.20b VB (*top*) resonance forms for the bonding in OPH_3, and (*bottom*) the MO description of negative hyperconjugation in the compound.

than PMe₃ at 140–141 °C. It is largely the high polarity of the "O=P" bond that leads to this dramatic change in properties between PMe_3 and $OPMe_3$.

The other major contributing resonance forms can be drawn by considering negative hyperconjugation from the ylide form. An oxygen-based lone pair donates into the P–H antibonding orbital to produce a phosphorus–oxygen double bond and an unbound hydride (Fig. 8.20b). Note that what is not present is the typical structure of a phosphine oxide shown previously with a phosphorus oxygen double bond and three P–H single bonds. However, if one were to draw a shorthand of the composite structure, this would be a reasonable estimate. One should keep in mind, however, that the composite structure usually drawn does not have simple 2-center 2-electron bonds all around.

One natural consequence of the bonding model for phosphine oxides shown in Fig. 8.20b is that, as the substituent X in OPX_3 compounds becomes more electronegative, more negative hyperconjugation should occur. In other words, X groups that stabilize a negative charge should have longer P–X bonds, and conversely, X groups that poorly stabilize a negative charge should have shorter P–X bonds. These effects are clearly borne out in the structures of phosphine oxides, and a couple of examples are given in Fig. 8.20c for comparison. The more electronegative substituent, CF_2CF_3, on phosphorus has a longer P–C bond, while the more electropositive ethyl groups have shorter P–C bonds. Likewise, the P=O bond distance should shorten when the phosphorus contains quite electronegative groups due to increasing negative hyperconjugation, which enhances the P–O bond order. This effect too is readily apparent in Fig. 8.20c.

If the X substituent in OPX_3 has a relatively low electronegativity, such as CH_3 or H, then the ylid resonance form in Fig. 8.20d dominates the ground-state structure, 79%. For example, natural resonance theory can be used to determine the main contributing resonance forms in OPH_3.[18] The three symmetry-related negative hyperconjugation structures contribute 6% each.

Since the three equivalent minor resonance forms in Fig. 8.20d have a negative charge on H, replacing H with a more electronegative substituent stabilizes these resonance forms relative to the ylid structure.

These types of interaction are extremely common in main-group chemistry, including such important and pervasive anions as SO_4^{2-} and PO_4^{-3}. In general, many compounds that are hypervalent with possible π-bonding contain negative hyperconjugation interactions to one extent or other.

The most important resonance forms for some common anions are shown in Fig. 8.21. (Also, see the discussion of NO_3^- in Chapter 3.)

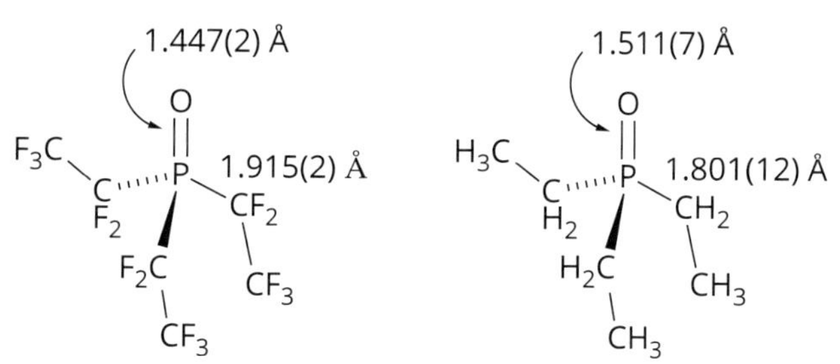

Fig. 8.20c Relevant bond lengths from the X-ray diffraction studies on $OPEt_3$[21] and its fluoroalkyl derivative, $OP(CF_2CF_3)_3$.[22] The fluoroalkyl group is much more electronegative, leading to more negative hyperconjugation in the compound, which results in a longer C–P and shorter P=O bonds for the compound on the left. The P–C bond distances listed are the average of the three chemically equivalent P–C bond distances in the X-ray diffraction studies.

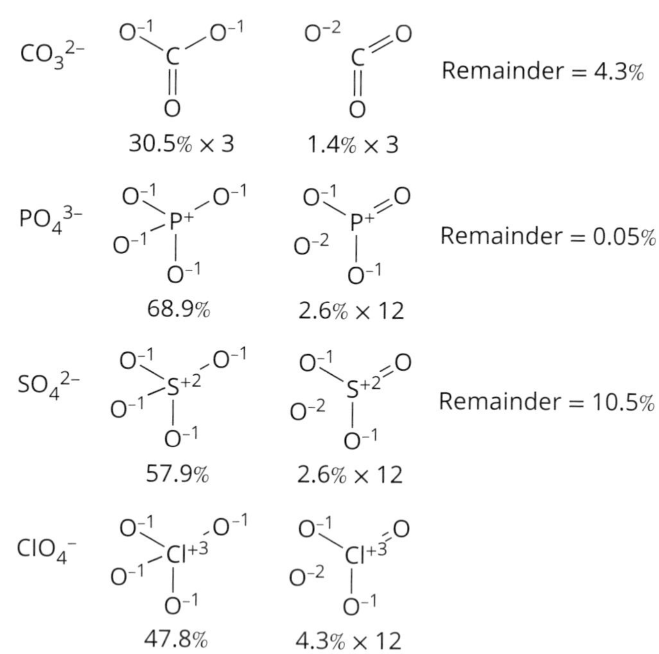

Fig. 8.21 Important resonance forms for some common inorganic anions. (Adopted from Weinhold and Landis.[23])

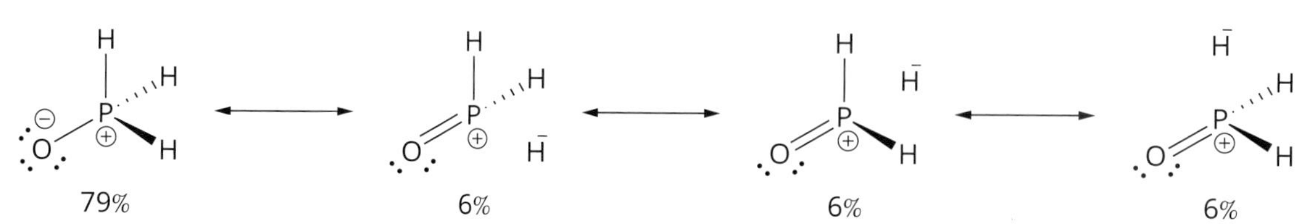

Fig. 8.20d Results of a natural resonance theory calculation on OPH_3, showing the contribution of each Lewis structure to the ground state.

8.3.2.2 Bond Polarity and Negative Hyperconjugation

Since a σ*-orbital is acting as the acceptor, the strength of negative hyperconjugation can be quite sensitive to the nature of this orbital. In order for a strong negative hyperconjugation interaction to be present, the σ*-orbital should largely rest on the central atom to allow good overlap. Consequently, if the central atom, E, has a lower electronegativity than the substituents, X, then the σ-orbital will be largely comprised of X-contributions and the σ*-orbital largely located on E (Fig. 8.22a).

In addition to the orbital argument, if the central atom is of lower electronegativity with higher electronegativity substituents, then the central atom will have a partial positive charge, making it a better acceptor (Fig. 8.22a).

Considering both these orbital and charge effects are enhanced by having a lower electronegativity central atom with higher electronegativity substituents, it should be no surprise that the heavier, less electronegative congeners typically exhibit stronger negative hyperconjugation interactions.

Consider three closely related compounds: ONF_3, OPF_3, and $ONMe_3$. In the first two the central atom is less electronegative than the substituents, and negative hyperconjugation between O–N or O–P should lead to significant multiple-bond character. In the third compound the nitrogen is more electronegative than the methyl groups, and the potentially accepting N–C σ*-orbital will be largely centered on carbon. In addition, since the N–C bonds are polarized to increase the negative charge on nitrogen, the central atom will be a poor acceptor. As a result, there should be little negative hyperconjugation in $ONMe_3$, and the ylid form, $^-O–^+NMe_3$, should dominate.

Examination of the structures and bonding via natural resonance theory calculations on ONF_3 and OPF_3 reveals that they are quite similar.[18] Both compounds exhibit short oxygen–pnictogen bonds with remarkably comparable angles (Fig. 8.22b). The O–N bond length of 1.15 Å is in the double to triple bond range; for example, the N–O double bond distance in O=N(OR) is 1.22 ± 0.02 Å,[24] while the single bond distance in $HO–NH_3^+$ is 1.408(2) Å.[25] The natural resonance theory bond order for the O–N bond is 2.1 in ONF_3. Similarly, NRT gives an O–P bond order of 2.2 for OPF_3. The results of the NRT calculation are summarized in Fig. 8.22b.

By contrast, the N–C antibonding orbital of $ONMe_3$ should be a far worse acceptor than the N–F antibonding orbital. In the case of the N–C bond, the σ*-orbital will largely be located on carbon instead of nitrogen, making overlap poor with the oxygen lone pairs. In addition, nitrogen being more electronegative than carbon will cause the nitrogen to have a more negative partial charge, also decreasing its acceptor ability. As a result, the $O–NMe_3$ bond is quite long at 1.404(5) Å, indicative of a single bond (Fig. 8.22c).[26] NRT suggests that the ylide form contributes 86% of the ground state, while the three negative hypercongugation forms contribute 7% total.

8.3.3 The 3-Center 2-Electron Bond (τ-Bond)

A type of bonding prevalent in many Group 13 and early transition-metal compounds involves two bonding electrons shared between three atoms—a 3-center 2-electron (3c2e) bond. In some cases, this type of bonding is observed

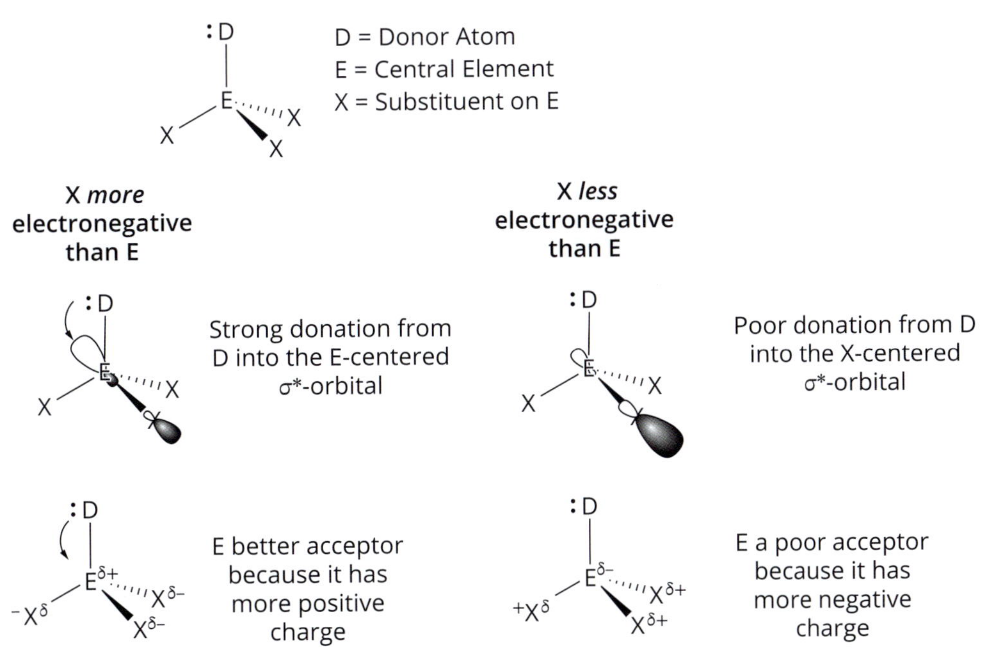

Fig. 8.22a Effects of bond polarity and electronegativity on the strength of negative hyperconjugation. A central element (E) that is more electronegative than the substituents (X) generally leads to poor negative hyperconjugation. When E is less electronegative than X, negative hyperconjugation is more favorable by both charge and overlap considerations.

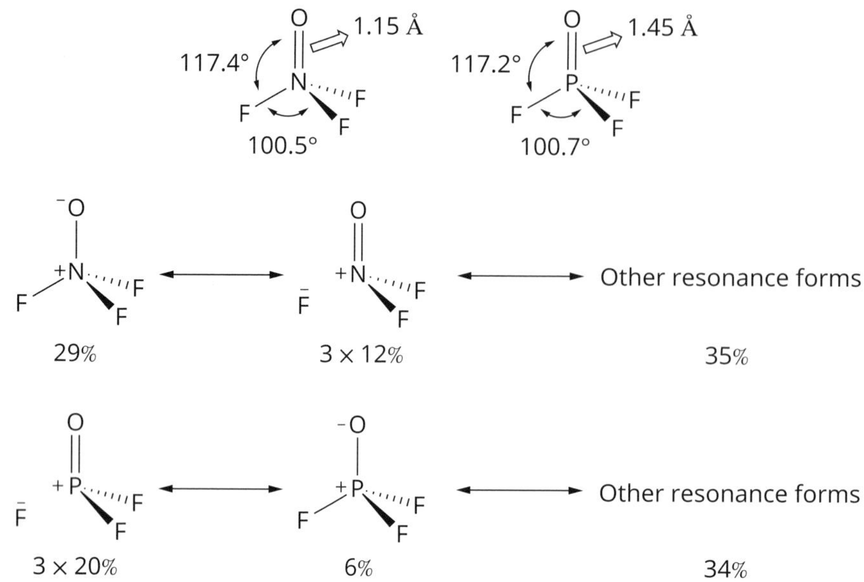

Fig. 8.22b (*top*) Calculated geometries of ONF_3 and OPF_3, using density functional theory. Calculated contributions from the ylide and O=E resonance forms based on natural resonance theory for ONF_3 (*middle*) and OPF_3 (*bottom*).[18]

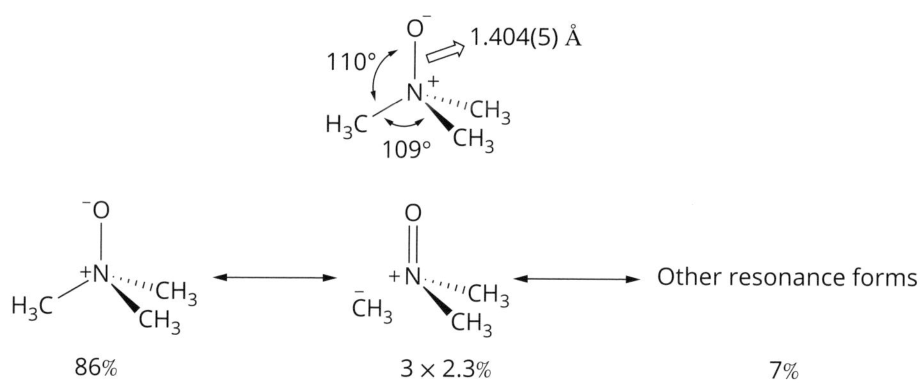

Fig. 8.22c (*top*) Geometries of $ONMe_3$ from X-ray diffraction.[26] (*bottom*) Calculated contributions from the ylide and O=E resonance forms based on natural resonance theory for $ONMe_3$.[18]

in compounds that, if monomeric, would be hypovalent, but instead form dimers with bridging hydride or alkyl groups, such as BH_3 or $AlMe_3$ dimers.

Perhaps the simplest example of this type of bonding is in protonated ethylene, which may be viewed as a proton interaction with the HOMO of ethylene, the π-bond. This is often drawn as shown in Fig. 8.23a.

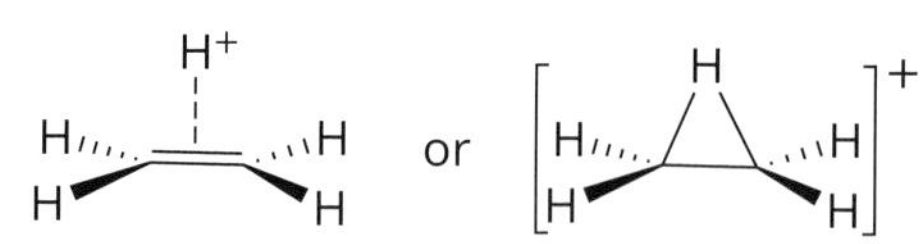

Fig. 8.23a Structure of protonated ethylene demonstrating a 3-center 2-electron bond.

The molecular orbital description of the 3c2e bond (τ-bond) is shown in Fig. 8.23b. On the left is the empty 1*s*-orbital on the bridging hydrogen. Ignoring the σ-bonds in ethylene, which will be little affected by the proton approaching the π-bond, we draw the π- and π*-orbitals on the right side. The 1*s*-orbital on the bridging hydrogen is of proper symmetry to interact with the π-bonding orbital and forms bonding (labeled τ) and antibonding (labeled τ*) orbitals. The π*-orbital is of different symmetry and, being non-interactive with the bridging hydrogen, comes across into the molecular orbitals with little or no energy change. The HOMO of protonated ethylene is the τ-bonding orbital, with significant contributions from both carbons and the bridging hydrogen. The LUMO is in some respects the same as in ethylene, the C–C π*-orbital. Somewhat higher in energy is the τ*-orbital, which is antibonding between the bridging hydrogen and the two carbons.

From a hybridization theory perspective, the bonding in protonated ethylene can be viewed as shown in Fig. 8.23c. There are two symmetry-related resonance forms that will contribute

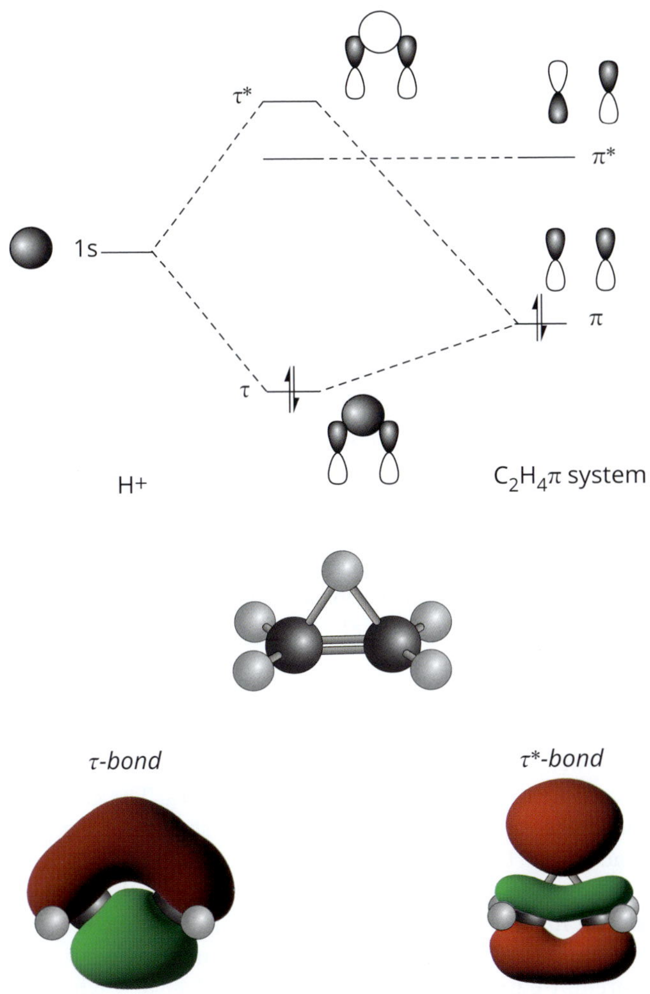

Fig. 8.23b Molecular orbital analysis of the 3c2e bond (τ-bond) in protonated ethylene.

equally. The third form is where there is no bonding between the ethylene and the proton. Natural resonance theory calculations suggest that these three forms contribute about 98% to the ground-state electron density, with the two symmetry equivalent forms (left and middle) contributing about 50% and the nonbonded proton form (right) contributing about 48%.

The quintessential τ-bonded compound is perhaps the borane dimer, B_2H_6 (Fig. 8.24a), the structure of which has bridging and terminal hydrogens. The terminal B–H bonds are 2-center 2-electron (2c2e) bonds, while the bridging hydrogens are held with τ-bonds. Considering that only two electrons are being shared by three atoms or groups, it is perhaps not surprising that τ-bonds are in general weaker than their 2c2e counterparts, and the bridging B–H distance of 1.33 Å is longer than the terminal B–H distances (1.19 Å).

About 96% of the electron density can be described by using the Lewis structure contributors in Fig. 8.24b.

These types of bonds can also be formed with bridging alkyl groups. An example of this is the dimer of trimethylaluminum. The τ-bonds to the bridging methyl groups (Fig. 8.25) are formed with ~sp^3-hybrids on the bridging methyl groups.

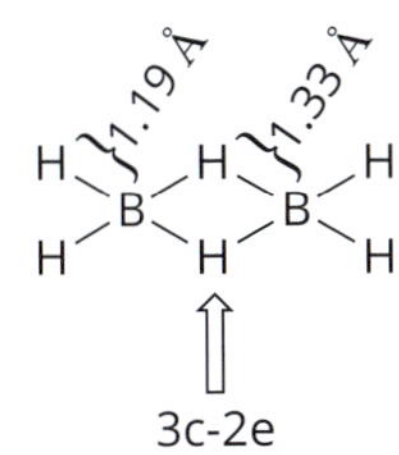

Fig. 8.24a Structure of the borane dimer.

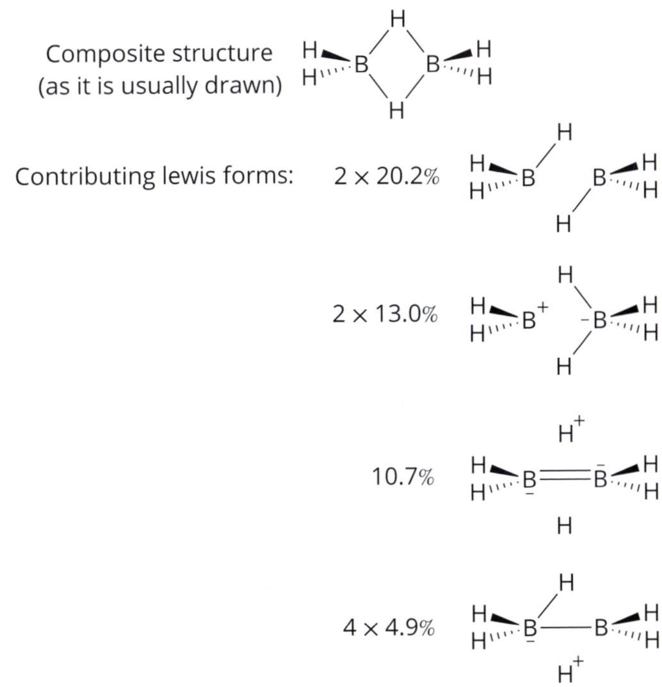

Fig. 8.24b Lewis structures contributing to the ground-state electron density of the diborane dimer according to natural resonance theory. These constitute 96.7% of the ground-state structure. The bond orders (BO) are $BO_{B\text{-}B}$ = 0.432, $BO_{B\text{-}H(terminal)}$ = 0.993, $BO_{B\text{-}H(bridging)}$ = 0.391. (Adopted from *Valency and Bonding* by Weinhold and Landis.[23])

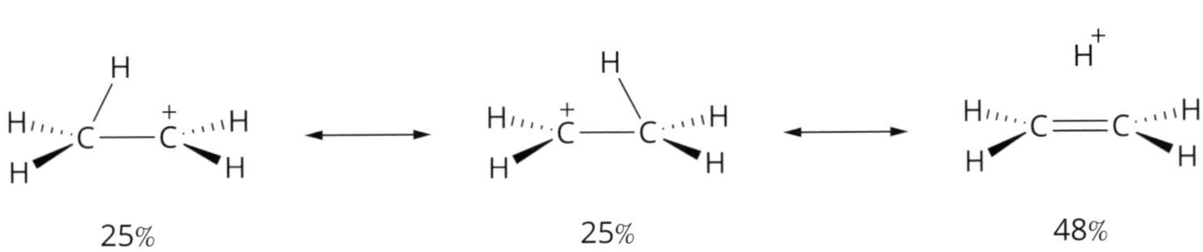

Fig. 8.23c Valence-bond analysis of the 3c2e-bond (τ-bond) in protonated ethylene. The percentages are contributions of that Lewis form by natural resonance theory.[18]

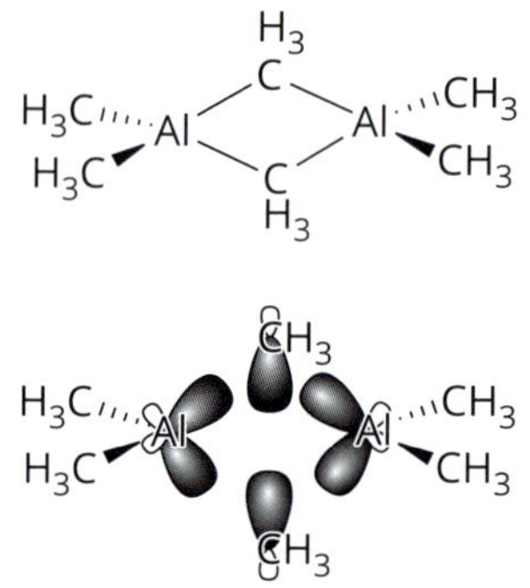

Fig. 8.25 Formation of τ-bonds in Al_2Me_6 using ~sp^3-hybrids on the bridging methyl groups.

The bonding being discussed in this section should not be confused with the types of bonds where the bridging group also has a lone pair, such as bridging halides, alkoxides, or amidos. Bonding with groups containing lone pairs will be better described as being due to covalent and dative resonance. For example, trichloroaluminum (Fig. 8.26) can have a dimeric structure; however, one would not consider the compound to contain τ-bonds. Some of the same resonance forms will contribute to both a τ-bonding and covalent/dative resonance structure, but the covalent/dative structure has many additional stabilizing forms available.

8.4 Odd-Electron Bond Types

Previously, we discussed three bonding interactions commonly found in main-group compounds (3-center 4-electron bonds, negative hyperconjugation, and 3-center 2-electron bonds) that are beyond the 2-center 2-electron bond discussed in detail in Chapter 3. By no means is this a comprehensive list. A chemical bond can be broadly defined as any electronic interaction between two species (ions, atoms, molecules, and so on) with a nonzero orbital overlap that leads to an electronic stabilization of the system with consequences for geometry, atomic distances, electronic structure, or other properties that are measurable. As a result, there are many types of bonds not yet discussed and perhaps not yet discovered.

Furthermore, from our previous discussions one can anticipate 4-center 2-electron and other interactions, which are found in some species. All of these forms that might be encountered can be analyzed with the same models employed in these discussions, such as hybridization and MO theory.

In the previous discussion there was little mention of orbital and bonding effects in radicals, compounds with unpaired electrons. In these common intermediates and occasionally stable species, there is a molecular orbital that is singly occupied. What are the properties of orbitals holding unpaired electrons, and what effects do radicals have on the structure and bonding of the molecule?

8.4.1 Features of Radicals

To further explore the properties of odd electron species, let us examine the two compounds CH_3 and CF_3. The computed geometries for the two compounds are shown in Fig. 8.27. Because the H–C–H angles in CH_3 are 120°, the C–H bonds must be made with sp^2-hybrids on carbon, and the unpaired electron must reside in a pure *p*-orbital. In CF_3, the molecule is pyramidalized somewhat. Using the Coulson directionality theorem, we find that the C–F bonds involve $sp^{2.7}$-hybrids on carbon, and the remaining orbital holding the unpaired electron must reside in an sp^4-hybrid. The angles in CCl_3, with Cl

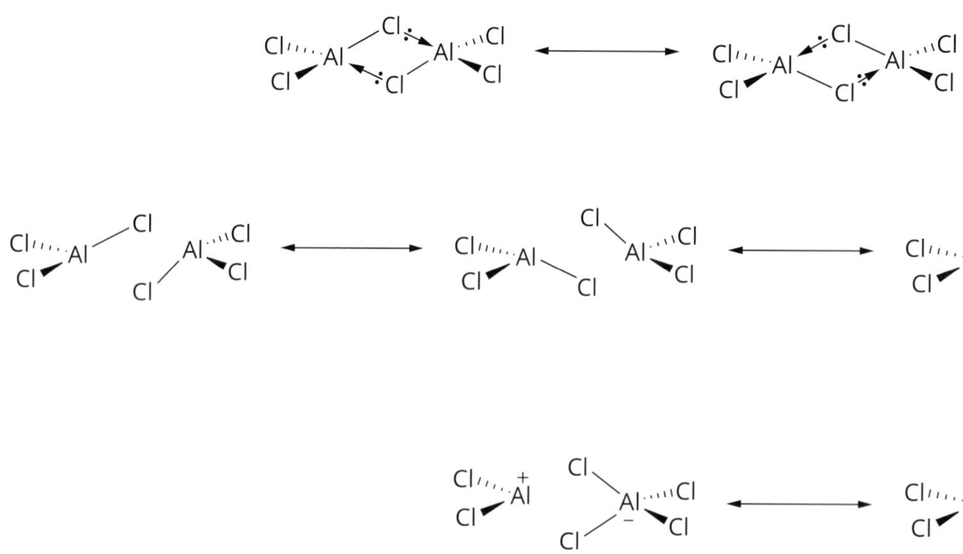

Fig. 8.26 Bonding in the $AlCl_3$ dimer showing covalent/dative bonding and *not* 3-center 2-electron bonds. Roughly speaking, the bonding may be considered resonance between dative and covalent bonds as shown at top. More precisely, the series of resonance forms on bottom should be considered, and other possible contributors are not shown. A covalent/dative resonance system has many stabilizing resonance forms available that are not contributors in τ-bonded systems.

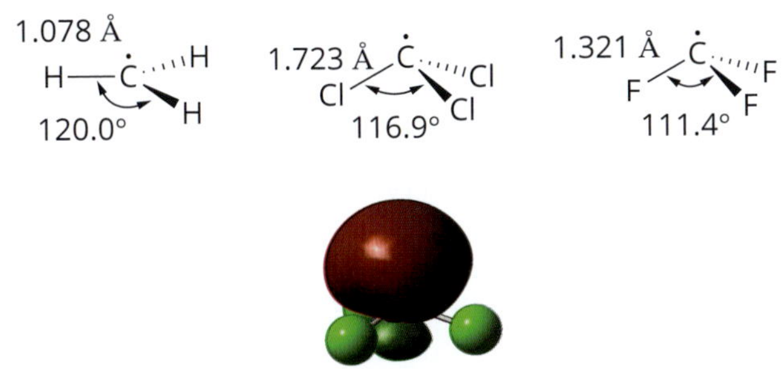

Fig. 8.27 (*top*) Calculated geometries for $\bullet CH_3$ and $\bullet CF_3$.[28] (*bottom*) The singly occupied molecular orbital (SOMO) of $\bullet CF_3$.

of intermediate electronegativity between F and H, are 117°. The bromide derivative has a bond angle of 116.7°.[27]

In the VSEPR description of molecules, single-electron domains were discussed as having an influence on structure. For example, the O–N–O angle in $\bullet NO_2$, nitrogen dioxide, is less than 180°, presumably because the radical resides in a domain of appreciable size.

From VSEPR one would expect that the methyl radical should have angles of less than 120°; however, the molecule does not follow VSEPR's guidelines. The radical, from a VSEPR perspective, in $\bullet CH_3$ is nonstereochemically active and has no influence on the H-C-H angles. By contrast, the $\bullet CF_3$ radical appears VSEPR-compliant with F–C–F angles of less than 120°.

Since the orbital holding the unpaired electron in both $\bullet CH_3$ and $\bullet CF_3$ has higher *p*-character, by Bent's rule, we can suggest that the effective electronegativity of the radical is *higher* than fluorine. Consequently, a half-empty orbital is, or at least acts as if it were, very electronegative. If one looks at the geometries of CX_3 radicals, where X = H, F, Cl, and Br, computationally, one finds that the X–C–X angles are 120°, 111.3°, 116.3°, and 116.7°.[29] This produces hybridization parameters for the C–X bonds of 2, 2.75, 2.26, and 2.23, respectively. Because the F–C–F angle is still greater than 109°, the radical acts as if it were a substituent higher in electronegativity than even fluorine. If one assumes a linear relationship between bond angle and X electronegativity, the carbon-centered radical has an effective electronegativity of ~4.4!

This analysis, however, assumes that the electrophilicity of the radical is a constant regardless of the molecule it is in, which is not usually thought to be the case. Radicals can in practice be made to be electrophilic or nucleophilic, depending on the nature of the molecule they are in. For example, the *tert*-butyl radical is thought to be very nucleophilic and the fluorine radical very electrophilic.[30] Consequently, as is often the true, it is better to view the molecule more holistically, even though analysis of things like "substituent effects" is often useful.

8.4.2 The 2-Center 1-Electron Bond

Perhaps the simplest bond type with an odd electron count is the 2-center 1-electron bond. One instance where such bonds are often discussed in detail is in the $H_2^{+\bullet}$ species in classes on quantum mechanics.

Readily observable species with such bonds can be found in boron chemistry. For example, DuPont and Mills found that electrochemical reduction of BPh_3 in THF results in formation of a dimer with a boron–boron one-electron bond, which was characterized by EPR (Eq. 8.9).[31] More recently, such species have been more fully characterized by building systems that enforce intramolecular B–B interactions.[32,33]

$$[\cdot BPh_3]^- + BPh_3 \xrightarrow[\text{THF}]{} [Ph_3B \cdot BPh_3]^- \qquad \textbf{8.9}$$

8.4.3 The 2-Center 3-Electron Bond

Electrophilic radicals can form what can be regarded as donor–acceptor interactions with lone pairs to produce 2-center 3-electron bonds (2c3e bonds).[34] While this bond type may seem exotic, it is in fact a bond found in many common molecules, as will be seen subsequently, but first let's try to define the interaction more precisely. Diagramatically, a 2c3e bond between E and X may be drawn as E ∴ X.

The simplest example of a 2c3e bond occurs in $He_2^{\bullet+}$. If one ionizes helium, the atom will form a bond with an electron pair from another helium atom. From a molecular orbital perspective (Fig. 8.28), the new bond is of ½ bond order with a half-occupied σ-antibonding orbital. All 2c3e bonds share this fully occupied bonding and half-occupied antibonding orbital layout, though the 2c3e bond may be of any nodal structure; σ, π, δ, and so on.

From a hybridization theory perspective, the 2c3e bond can be thought of in terms of electron transfer between two centers. There are three commonly encountered possibilities (Fig. 8.29): a cation radical interacting with a neutral lone pair,

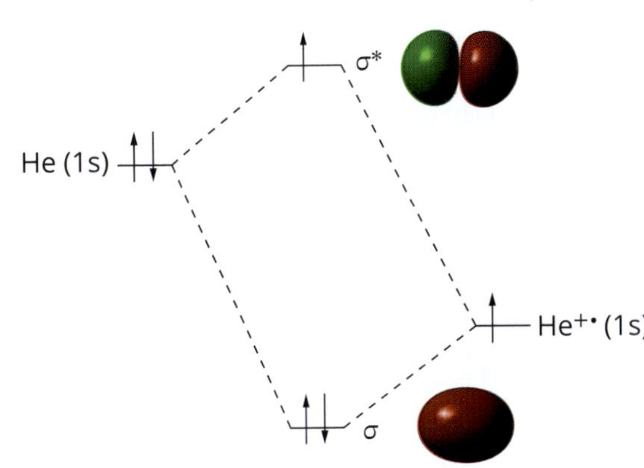

Fig. 8.28 Molecular orbitals of $He_2^{\bullet+}$.

$$^{+}E^{\bullet}\,{:}X \longleftrightarrow E{:}\,{}^{\bullet}X^{+}$$

$$E^{\bullet}\,{:}X^{-} \longleftrightarrow {}^{-}E{:}\,{}^{\bullet}X$$

$$E^{\bullet}\,{:}X \longleftrightarrow {}^{-}E{:}\,{}^{\bullet}X^{+}$$

Fig. 8.29 The 2c3e bonds from a resonance perspective involving electron transfer between E and X atoms.

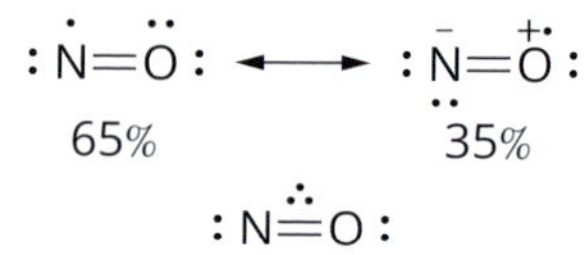

Fig. 8.30 (*top*) The two contributing resonance structures in NO given by Pauling. (*bottom*) A proposed drawing for the composite structure with a σ-bond, a π-bond, and a 2c3e π-bond, for a net bond order of 2 ½.

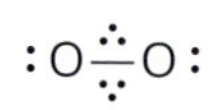

Fig. 8.31 Structure of dioxygen with a bond order of 2 (one full bond and two half bonds).

a neutral radical interacting with an anionic lone pair, and neutral radical interacting with neutral lone pair.

The $He_2^{+\bullet}$ molecule would fall into the first category, a cation radical with a neutral lone pair, with the two resonance forms shown as follows. Thus, the He–He bond order is ½ due to the 2c3e σ-bond.

$$^{+}\mathrm{He}^{\bullet}\mathrm{:He} \leftrightarrow \mathrm{He:}^{\bullet}\mathrm{He}^{+}$$

One commonly encountered molecule with a 2c3e bond is nitric oxide, NO. In this case the 2c3e bond resides in the π-system of the diatomic. The two resonance forms drawn by Pauling for this molecule are shown in Fig. 8.30.[35]

Naturally, the most commonly encountered molecule (every time you take a breath!) with a 2c3e bond is O_2, which has an O–O single bond and two 2c3e π-bonds. The net bond order is then 1 + ½ + ½ = 2. The composite structure may be drawn as shown in Fig. 8.31. Since the two 2c3e bonds have an unpaired electron, singlet and triplet ground states are possible. By Hund's rule, the triplet state will be lower in energy.

See Section 7.4 (hybridization theory analysis) and Section 6.7.2 (molecular orbital analysis) for additional discussion of the bonding in oxygen.

8.5 The Lone-Pair Bond-Weakening Effect

An interesting effect, recognized long ago with regards to the discrepancy in the bond energies of the dihalogen compounds, is the extremely weak F–F bond relative to what might be expected (Fig. 8.32). According to Sanderson, the expected F–F covalent bond energy is ~110 kcal/mol; however, the experimental bond strength of F_2 is a paltry 38 kcal/mol. Sanderson simply ascribed this to a *lone pair bond weakening* effect, but he did not know the origin of, though he did (correctly) ascribe it to a weakening *inherent to the atom* in certain compounds.

Several descriptions of the cause for the weak F–F bond were given. A popular theory, proposed by Pitzer and Mulliken, was that the weak covalent bond in F_2 was due to mutual repulsions between lone pairs (see Fig. 8.33).[36–40] Recent calculations[41,42] suggest that there are large Pauli repulsion terms in F_2; however, valence bond theory suggests that this is not the entire story of why F_2 has a very weak bond.

Using polar covalence theory (Chapter 3), Sanderson suggested that LPBW even occurred in molecules where only one of the atoms had a lone pair, like HF. If one plots the estimated unweakened BDEs versus experiment for HX, where X is halogen, one finds a similar trend to that seen for the X_2 compounds, and there is a larger weakening for F that reduces as one proceeds down the column (Fig. 8.34). Sanderson's weakened values calculated using the E′ numbers in Table 3.2 are reasonable estimates for the experimental values.

Some compounds were "unweakened", and the E‴ values were used. Some compounds had fully weakened bonds, and the E′ values were used. Then the E″ were used for intermediate cases. More simply for elements like nitrogen, E′ was used for single bonds (such as hydrazine, $H_2N{-}NH_2$), E″ for double bonds (such as diazene, HN=NH), and E‴ for triple bonds (such as dinitrogen). What causes this weakening of bonds in some compounds with halogens, and other atoms containing lone pairs, but not others? Why does π-bonding dramatically reduce the effect?

This question of why the bonds in compounds such as F_2 and HF were weakened was addressed from a valence bond

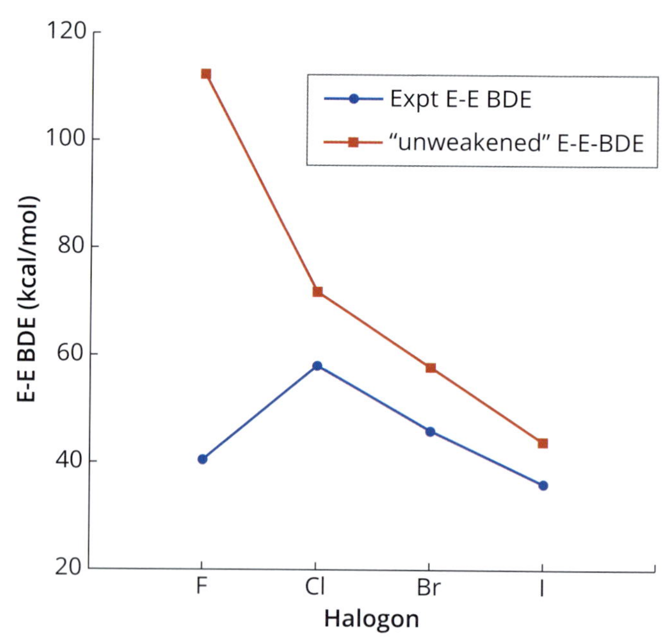

Fig. 8.32 Experimental (blue) and expected "unweakened" halogen–halogen bond energies described by Sanderson.

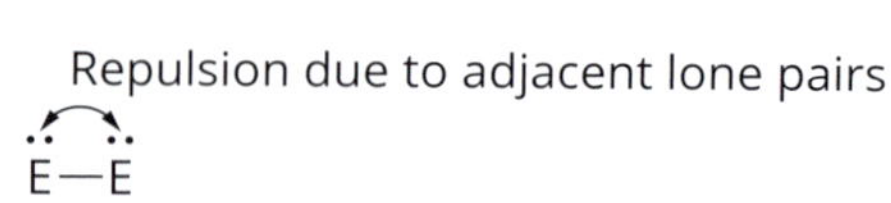

Fig. 8.33 Pitzer and Mulliken lone-pair repulsion as an explanation for weakening.

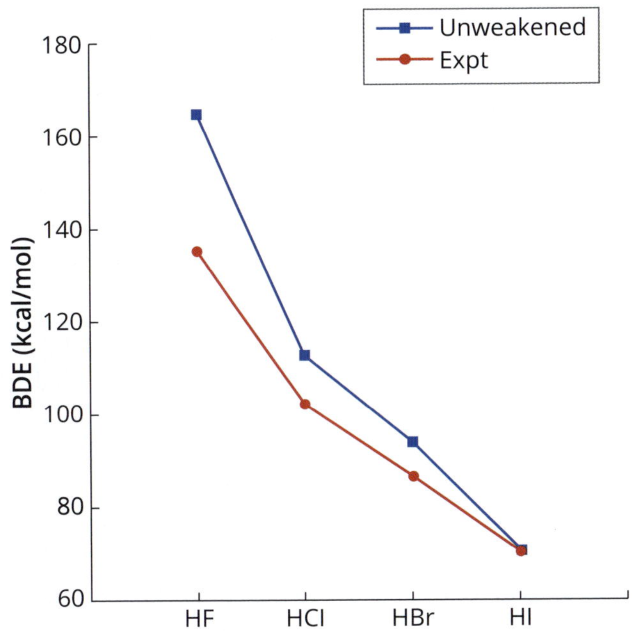

Fig. 8.34 The bond dissociation enthalpies of HX (X is halogen) according to experiment (red) and polar covalence (blue), using unweakened values (E‴ in Table 3.2).

perspective by Volatron, Hiberty, and coworkers[43] in 1996. To illustrate, let us consider the breaking of the N–H bond in ammonia in two steps: one step to break the bond, and another to rehybridize at nitrogen (Fig. 8.35).

On homolyzing the N–H bond, we are left with a singly occupied orbital of hybridization $sp^{3.4}$ and a lone pair in an orbital of hybridization $sp^{2.1}$ on nitrogen. However, the ground state of the NH_2 radical allows the lone pair to pick up an additional *s*-character, and the radical resides in an almost pure *p*-orbital. This rehybridization releases energy of quantity E_R (rehybridization energy) by stabilizing the electron density in the lone pair. Of course, in reality, as the N–H bond is broken the molecule rehybridizes concurrently (solid line in Fig. 8.30). This "stealing" of *s*-character by the lone pair effectively reduces the N–H BDE.

Natural bond orbital calculations on the NH_2 radical confirms this picture, with a hybrid on nitrogen of $sp^{1.2}$ containing the lone pair and a pure *p*-orbital housing the unpaired electron.[18] In other words, the orbital holding the lone pair changes in NH_3 from ~sp^2 to ~sp^1 during the N–H bond-breaking process. Meanwhile, an orbital bonding with hydrogen of ~$sp^{3.4}$ hybridization transforms to an orbital holding a single electron of hybridization ~sp^∞. In essence, LPBW is a type of rehybridization energy loss that occurs during bond-stretching.

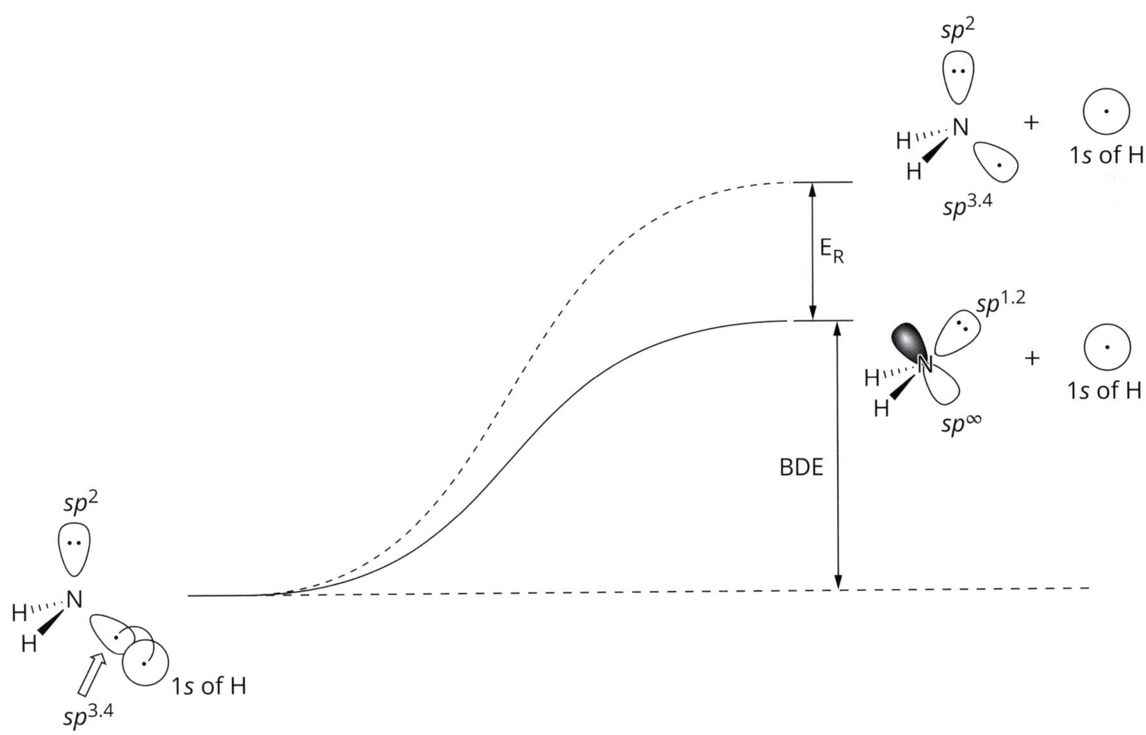

Fig. 8.35 Energy diagram for the homolysis of the N-H bond in ammonia. The top dotted curve represents the homolysis without rehybridization; the N-H bond is broken, but the lone pair and previously bonding electron remain in the same hybrids. The NH_2-radical will rehybridize to place the lone pair in a higher *s*-character orbital and the radical in a pure *p*-orbital. The difference in energy between the NH_2-radical with the NH_3 preferred hybridization and the preferred hybridization of NH_2 is E_R, the rehybridization energy. The solid line is the actual pathway for cleavage of the N-H bond, illustrating that the lone pair will weaken the N-H bond by, in essence, stealing *s*-orbital character as the bond is stretched. (Adopted from Volatron, Hiberty, and coworkers.[43])

Again, this behavior might be expected for any atom having a lone pair that uses hybrids significantly when bonding. In other words, LPBW is going to be more likely to be observed for the first-row *p*-block elements and is likely to be a smaller effect for the heavier congeners of the main group, thus explaining the effect's great significance in fluorine chemistry.

When calculating the bond energy for BF_3 in Chapter 3, the unweakened fluorine bond energy (E‴ in Table 3.2) was used, with great accuracy, to calculate the average bond energy. Generally speaking, most metal-E bonds, where E has one or more lone pairs, do not show bond weakening. Why is there no bond weakening in BF_3 and many other compounds?

The obvious difference between these compounds not exhibiting bond weakening and HX/X_2 compounds is the possibility of π-bonding in the former, either through lone-pair donation into an empty orbital on the central element, or the π-bonding may be in the form of negative hyperconjugation. This is somewhat speculative, but it seems that little π-bonding is required to prevent significant LPBW near the equilibrium bond distance, and the π-bonding may come in the form of dative interactions, negative hyperconjugation, and so on.

Naturally, since LPBW only affects the covalent bond energy (E_c), very polar bonds will see diminished consequences of LPBW as well. Again, this explains why purely covalent F_2 is an extreme case of the effect.

Since mixing of *s*-orbital character into the π-donating lone pair would weaken the π-bonding, perhaps this reduces the *s*-character "stealing" by the lone pairs, and the unweakened bond energy is most appropriate. Consistent with this, it is well known that π-bonding to atoms containing lone pairs reduces the weakening effect. In fact, the intermediate weakening values (E″ in Table 3.2) for elements such as nitrogen were recommended in cases where there is one π-bond to the element, such as the nitrogen in an imine.

Heavier main-group elements, because of the primogenic and related effects, tend to have lone pairs with larger amounts of *s*-character. As a result, rehybridization during bond-breaking in the heavier main-group congeners results in smaller LPBW effects.

8.6 The Jahn–Teller Effect, Bonding, and Geometry in the Main Group[44–46] (Advanced Topic)

A useful tool, especially in conjunction with molecular orbital theory, is the Jahn–Teller effect, which is often discussed as "causing distortions" from some arbitrary, often more symmetrical, geometry. It can be used in a number of ways, such as discovering why the structure does not have an arbitrary geometry, or to discern which vibrational modes may be involved in an intramolecular reaction.

In the first-order Jahn–Teller effect, a molecular distortion occurs due to an electronically degenerate ground state. These distortions with profound effects on the spectroscopy of some transition-metal complexes will be discussed with ligand field theory.

The second-order Jahn–Teller effect, as defined by Pearson and Bader, using a perturbation theory approach (discussed subsequently),[45,47,48] involves coupling of a nondegenerate ground state with a low-lying excited state. From a molecular orbital perspective, this involves the mixing of a relatively high-energy filled orbital with a low-energy unoccupied orbital, most often the HOMO and LUMO of the compound. It should be noted that here we make a rather large approximation in order to employ simple MO theory: we assume that the distortion is due to the interaction of only two orbitals. (For a more detailed discussion, see specialized texts such as Bersuker's excellent treatise.[49])

If we label the arbitrary starting coordinates as Q_0 and small distortion from the starting positions as Q, which are along the reaction- or distortion-coordinate, the new Hamiltonian, *H*, is described relative to the initial Hamiltonian, H_0, by a Taylor–Maclaurin expansion, truncated after the second term (Eq. 8.10). The function *U* describes the nuclear–nuclear and nuclear–electron potential energy.

$$H = H_0 + \left(\frac{\partial U}{\partial Q}\right)Q + \frac{1}{2}\left(\frac{\partial^2 U}{\partial Q^2}\right)Q^2 \qquad \textbf{8.10}$$

Because the Hamiltonian must always be a function that is in the totally symmetric representation, $\left(\frac{\partial U}{\partial Q}\right)$ and Q must belong to the same irreducible representation for the term to be nonzero. Similarly, since Q^2 is totally symmetric, the function $\left(\frac{\partial^2 U}{\partial Q^2}\right)$ must also be totally symmetric to be nonzero. (The square of any irreducible representation, direct product of an irreducible representation with itself, always contains the totally symmetric representation, Section 5.12.)

The energy of the system on distortion is then described using the following equation, where E_0 and E_k are eigenvalues and ψ_0 and ψ_k are electronic eigenfunctions.

$$E = E_0 + Q\left\langle \psi_0 \,|\, \frac{\partial U}{\partial Q} \,|\, \psi_0 \right\rangle + \frac{Q^2}{2}\left\langle \psi_0 \,|\, \frac{\partial^2 U}{\partial Q^2} \,|\, \psi_0 \right\rangle + Q^2 \sum_k \frac{\left[\left\langle \psi_0 \,|\, \frac{\partial U}{\partial Q} \,|\, \psi_k \right\rangle\right]^2}{(E_0 - E_k)} \qquad \textbf{8.11}$$

Notice that the first-order term in Eq. 8.11, after E_0, involves only the initial eigenfunctions (ψ_0), and as a result simply averages the effect of the changing nuclear positions over the original electron distribution. If ψ_0 is nondegenerate, the integral $\left\langle \psi_0 \,|\, \frac{\partial U}{\partial Q} \,|\, \psi_0 \right\rangle \neq 0$ for all points except minima or maxima, and is the slope of a plot of energy vs Q (distortion coordinate). If the wavefunctions in question are degenerate, the coupling to the normal mode (distortion mode) can desymmeterize the molecule in cases where there is an energetic advantage to do so; this is a first-order Jahn–Teller distortion, which is discussed in more detail in Chapter 15 on 1-electron spectroscopy.

If we consider the integral $\left\langle \psi_0 \mid \frac{\partial^2 U}{\partial Q^2} \mid \psi_0 \right\rangle$ in Eq. 8.11, it is found that the term is always nonzero because $\frac{\partial^2 U}{\partial Q^2}$ is totally symmetric and increases the energy. This term is a restoring force, the tendency of the system to return to Q_0 on displacement, because the electron distribution for the original nuclear positions was optimal.

Once the nuclear positions move, the electron distribution will alter to compensate for the change. The change in electron distribution in response to the nuclear movement will always lower the energy. This is illustrated in the last, nonlinear term of Eq. 8.11, which is always negative, because E_0 and E_k are negative with E_0 having the larger magnitude.

The energy decreasing term is the last term in Eq. 8.11, but the integral $\left\langle \psi_0 \mid \frac{\partial U}{\partial Q} \mid \psi_k \right\rangle$ will only be nonzero if the symmetry of the distortion is contained in the direct product of the symmetries of ψ_0 and ψ_k. In other words, Eq. 8.12 must be true for the distortion to be energetically favorable. (The symbol "$\subseteq$" means "subset of or equal to"; in other words, the direct product of $\Gamma_{\psi 0} \times \Gamma_{\psi k}$ is either exactly Γ_Q or contains Γ_Q.)

$$\Gamma_{\psi 0} \times \Gamma_{\psi k} \subseteq \Gamma_Q \qquad \textbf{8.12}$$

The new wavefunction that is obtained is found by using Eq. 8.13. In order for a new wavefunction, ψ, to form as a result of the distortion, the integral $\left\langle \psi_0 \mid \frac{\partial U}{\partial Q} \mid \psi_k \right\rangle$ must be nonzero, and thus Eq. 8.12 must be true. In other words, no electronic mixing can occur unless the product of the symmetries of the wavefunctions contains the symmetry of the distortion.

$$\psi = \psi_0 + \sum_k \frac{Q\left\langle \psi_0 \mid \frac{\partial U}{\partial Q} \mid \psi_k \right\rangle}{(E_0 - E_k)} \psi_k \qquad \textbf{8.13}$$

In applications involving molecular orbital theory, the symmetry rule becomes Eq. 8.14, where ϕ_i is the MO occupied in the ground state and ϕ_f is the changed MO in the excited state. Most commonly, but not always, ϕ_i and ϕ_f are the HOMO and LUMO, which are undergoing mixing. The reaction coordinate, Q, will be in the direction required to increase overlap between ϕ_i and ϕ_f.

$$\Gamma_{\phi_i} \times \Gamma_{\phi f} \subseteq \Gamma_Q \qquad \textbf{8.14}$$

As examples, let us consider three molecules, BH_3, NH_3 and ClH_3. The most symmetric structure that a molecule can adopt is generally taken as the starting geometry (Q_0) for such analyses, which in this case with three substituents on a central atom is trigonal planar (D_{3h}).

In order to ascertain what distortions are possible, and their symmetries, a normal mode analysis on EH_3 in D_{3h} is carried out. The general procedure is shown in Section 5.10. Using x-, y-, and z-axes on all four atoms as the basis (Fig. 8.36a), we find a reducible representation for all of the possible movements of the molecule. The basis includes twelve axes (vectors) total. In this case, we left the y-axes on the hydrogens coincident with their E–H bonds, which simplifies the analysis somewhat. Most of these operations are straightforward, but the C_3 and S_3 character calculations are worthy of special note.

Carrying out the C_3 will move all of the axes on the hydrogens into each other, which makes all of their contributions off diagonal and gives them no contribution to the character; consequently, we can ignore all of the axes except the ones on E. Because E is on the C_3 symmetry element, we need to use the general formula for a C_n rotation, Eq. 5.2, to find the contributions. For a C_3, the angle of rotation is 360°/3 = 120°. Applying the matrix in Eq. 5.2 to the coordinate axes on boron, x_1, y_1, and z_1, the following result is obtained.

$$C_3 \begin{bmatrix} x_1 \\ y_1 \\ z_1 \end{bmatrix} = \begin{bmatrix} \cos 120^\circ & \sin 120^\circ & 0 \\ -\sin 120^\circ & \cos 120^\circ & 0 \\ 0 & 0 & 1 \end{bmatrix} \begin{bmatrix} x_1 \\ y_1 \\ z_1 \end{bmatrix}$$

$$= \begin{bmatrix} -0.5 & \frac{\sqrt{3}}{2} & 0 \\ -\frac{\sqrt{3}}{2} & -0.5 & 0 \\ 0 & 0 & 1 \end{bmatrix} \begin{bmatrix} x_1 \\ y_1 \\ z_1 \end{bmatrix} = \begin{bmatrix} x'_1 \\ y'_1 \\ z'_1 \end{bmatrix}$$

$$\begin{bmatrix} x'_1 \\ y'_1 \\ z'_1 \end{bmatrix} = \begin{bmatrix} -0.5x_1 + \frac{\sqrt{3}}{2} y_1 \\ -\frac{\sqrt{3}}{2} x_1 - 0.5 y_1 \\ z_1 \end{bmatrix}$$

As in all cases, the contribution of x'_1, the new position of part of the basis x_1, is the amount of x_1 that remains in x'_1—the "part that did not move". In this case this is the projection of the x_1 vector onto x'_1 or, more simply, the coefficient in front

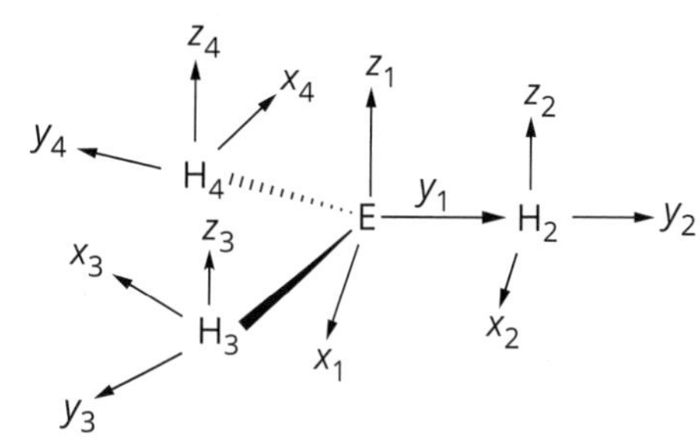

Fig. 8.36a Basis for analysis of EH_3 molecules.

of x_1 in the x'_1 vector. In this case, −0.5 contributes after rotation to x'_1. Likewise, the contribution of y_1 to $y'_1 = -0.5$. The z-axis is unchanged, so the contribution of z_1 to $z'_1 = 1$. The sum of these is the character under the reducible representation, $-0.5 - 0.5 + 1 = 0$ for the C_3 operation.

The S_3 operation starts with a C_3, but then a reflection is done perpendicular to the C_3. The z_1-axis on boron, which does not move during the C_3, is then inverted, giving z_2 a contribution of −1. The x'_1 and y'_1 contributions are the same for C_3 and S_3. As a result, the character under S_3 in the reducible representation is then $-0.5 - 0.5 - 1 = -2$.

So, the reducible representation is then as follows.

D_{3h}	E	$2C_3$	$3C_2$	σ_h	$2S_3$	$3\sigma_v$
$\Gamma_{x,y,z}$	12	0	−2	4	−2	2

Application of the decomposition formula produces the reducible representation for all possible movements as the sum for $\Gamma_{x,y,z}$.

$$\Gamma_{x,y,z} = A_1' + A_2' + 3E' + 2A_2'' + E''$$

As is usually the case, we are only interested in internal modes and not movements that simply result in translation or rotation of all the atoms in the molecule. The translations have symmetries of $E' + A_2''$, same as the x, y, and z in the character table. The rotations have symmetries of $A_2' + E''$ and are marked with R_x, R_y, and R_z in the character table. Removing these provides the possible vibrational movements of a trigonal planar EH_3 molecule, Γ_{vib}.

$$\Gamma_{vib} = A_1' + A_2'' + 2E'$$

We can decide what modes will be able to couple the orbitals simply from the symmetry labels; however, we will get a better understanding of why if we find the motion of the atoms in each mode. This is done using the projection operator.

For the A_1' mode, the only nonzero projections are found for the hydrogen y-axes, which all have the same magnitude and phase; that is, the A_1' vibration, $\upsilon_{A_1'}$, is $\upsilon_{A_1'} = y_2 + y_3 + y_4$. This corresponds to a symmetric mode where all of the hydrogens stretch and contract along the E–H bond vectors (Fig. 8.36b).

For the A_2'' mode, the only nonzero projections are due to the z-axis vectors on the hydrogens, which all have the same phase and magnitude; therefore, $\upsilon_{A_2''} = z_2 + z_3 + z_4$. This corresponds to a bending mode symmetric with respect to the C_3 that pyramidalizes the molecule (Fig. 8.36c).

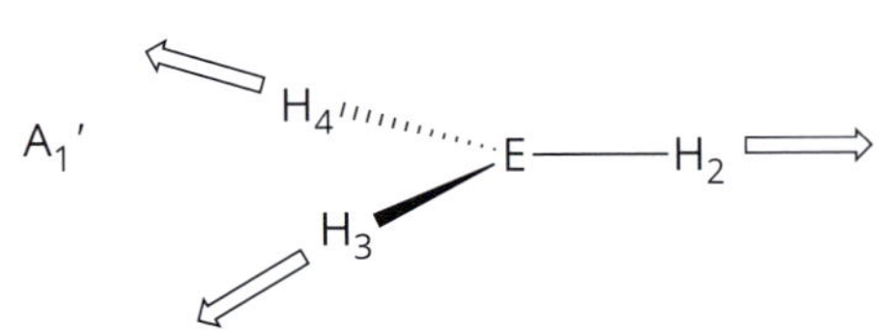

Fig. 8.36b The A_1' stretching mode for EH_3.

Then there are two modes of E′ symmetry. The related modes will involve movement along one particular set of axes. In other words, there is one E′ set of modes that involve movement along the x-axes of the hydrogens, and a separate E′ set of modes that involve movement along the y-axes.

The E′ modes along the x-axes can be found by projection on x_2 and x_3. The two modes then need to be orthogonalized using the Gram–Schmidt procedure (Section 6.4.1). The result is two modes, $\upsilon_{E'(x)} = (2x_2 - x_3 - x_4, x_3 - x_4)$, corresponding to bending motions in the x,y-plane (Fig. 8.36d).

The second E′ mode (Fig. 8.36e) along the y-axes coincident with the E–H bond axes is a stretching mode of form, $\upsilon_{E'(y)} = (2y_2 - y_3 - y_4, y_3 - y_4)$.

Fig. 8.37 shows the MO diagram of EH_3. The electrons, other than the three from the three hydrogen SALCs, are not included, as they change with the nature of the element "E".

The number of electrons and the relative orbital energies will change with element "E" in EH_3. The three molecules to be discussed here have E = B, N, and Cl bearing three, five, and seven electrons respectively. We then fill the MOs with six, eight, and ten electrons from the central atom, plus the three hydrogens. The energy-level diagrams for the three compounds are shown in Fig. 8.38.

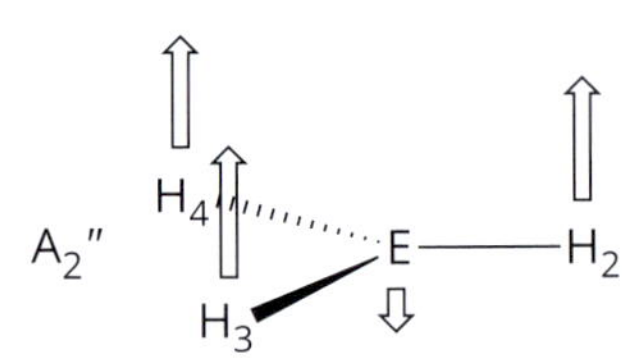

Fig. 8.36c The A_2'' bending mode for EH_3.

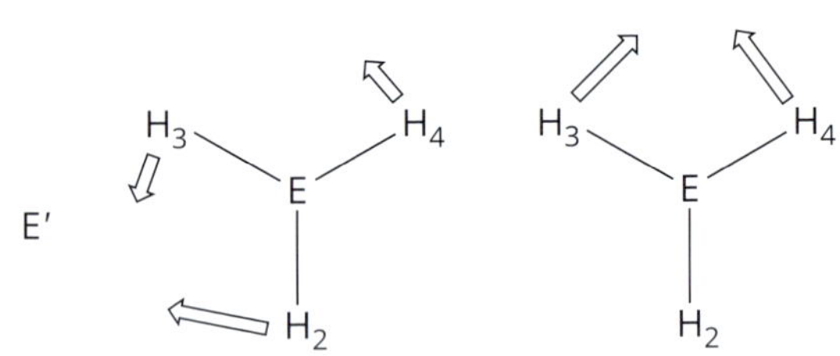

Fig. 8.36d The E′ bending modes for EH_3.

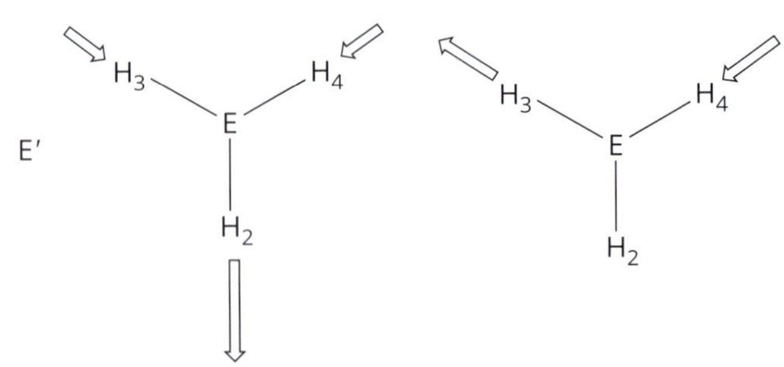

Fig. 8.36e The E′ bending modes for EH_3.

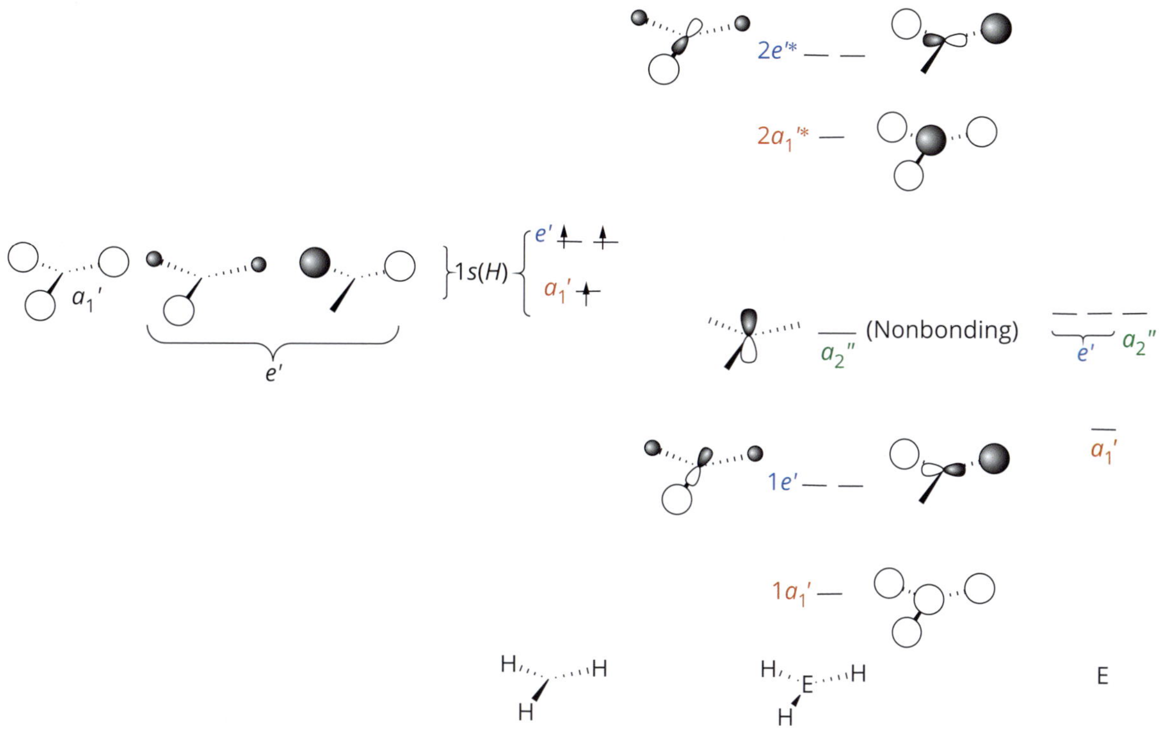

Fig. 8.37 General MO diagram for EH_3 in D_{3h}. Electrons on E are not shown but change with the element in question, with E = B, N, and Cl having three, five, and seven electrons respectively.

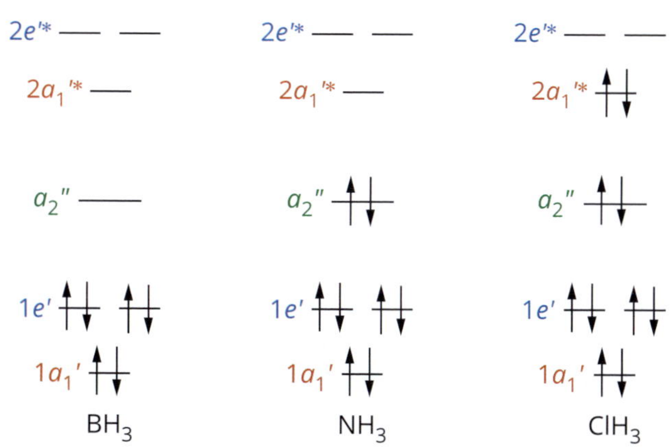

Fig. 8.38 Energy-level diagrams for EH_3 in D_{3h}.

The likelihood that two orbitals will mix is inversely dependent on the energy difference; that is, orbitals close in energy are more likely to mix, which is why we assume that the most probable mixing is between the HOMO and LUMO.

For BH_3, the HOMO is of E′ symmetry while the LUMO is of A_2'' symmetry. The direct product of these is $E' \times A_2'' = E''$. By Eq. 8.14, the direct product of the symmetries of the two orbitals must be equal to or contain the symmetry of the distortion. There are no normal modes in EH_3 of symmetry E″, and as a result, no Jahn–Teller distortion is possible. Consequently, the BH_3 monomer should be, and of course is, trigonal planar.

For NH_3 in D_{3h}, the HOMO is of symmetry A_2'' and the LUMO is of symmetry A_1'. The cross-product of these two irreducible representations is $A_2'' \times A_1' = A_2''$, because the totally symmetric representation (A'_1) does not alter the symmetry of another representation when multiplied. Consequently, the only motion that can couple the HOMO and LUMO is a mode of A_2'' symmetry. As found in our normal mode analysis on EH_3, this mode involves pyramidalization of the molecule, a bending mode. As a result, NH_3 has a pyramidalized, C_{3v} geometry. Note that mixing of the orbital $2a_1'^*$, *s*-character on E, with a_2'', *p*-character on E, results in *sp*-hybridization, consistent with the hybridization theory analysis.

These symmetry-based discussions do not tell us how much of a distortion will occur, which is dependent upon energetics of the new species at specific distortions, but it does tell us that a distortion is likely and the direction in which it will occur. In other words, determining the amount of distortion requires minimizing the energy in Eq. 8.11 and is dependent upon the overlap of ϕ_i and ϕ_f.

For example, valence isoelectronic PH_3 will also "distort" from D_{3h} to C_{3v} and turns out to distort to lower angles. An important consideration in the amount of the distortion that will occur is the relative energy of the HOMO and LUMO orbitals. The smaller the *sp*-energy gap, the larger the mixing that is expected. As mentioned previously and shown in Fig. 8.1, the *sp*-energy gap in nitrogen is much larger than in phosphorus. The Jahn–Teller distortion mixes a_2'' (pure *p*-orbital) and $2a_1'^*$ (σ^*-orbital with *s*-character). Since nitrogen has a primogenically expanded *s*-orbital and contracted *p*-orbitals, the nitrogen *s*-orbital can overlap better than phosphorus with the a_1' SALC (Fig. 8.39). The stronger interaction for nitrogen will

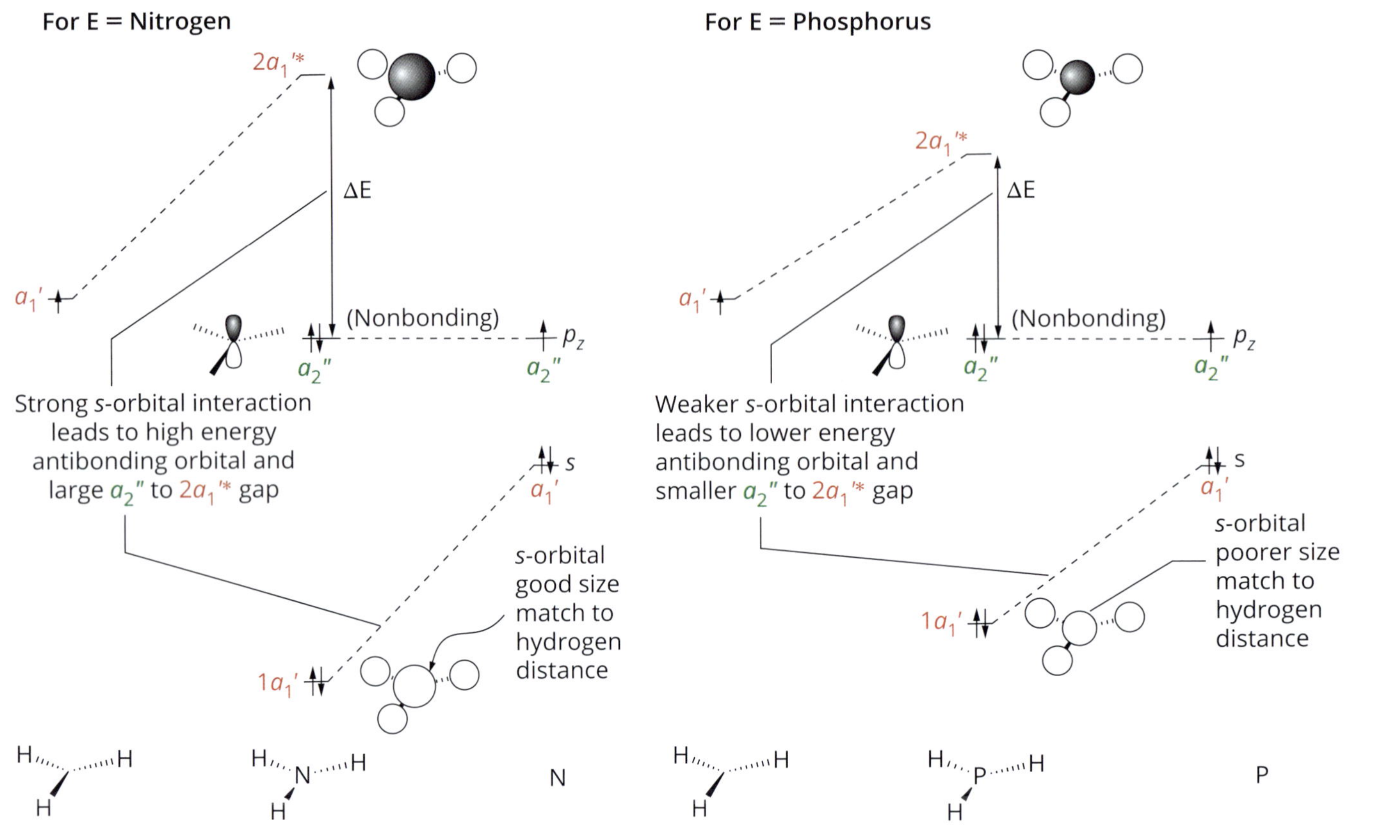

Fig. 8.39 The effect of s-orbital size on the $2a_1'^*$ to a_2'' energy gap in NH_3 and PH_3. The relatively small size of the s-orbital on phosphorus leads to a weaker bonding interaction and a smaller splitting energy. The smaller energy difference will lead to greater HOMO-LUMO mixing and a stronger Jahn–Teller distortion.

lead to a higher-energy antibonding $2a'_1{}^*$ orbital for nitrogen and a larger a_2'' to $2a_1'^*$ energy gap for the lighter congener. Consequently, we expect a smaller a_2'' to $2a_1'^*$ energy gap and a larger distortion for PH_3 than for NH_3.

In other words, like the previous discussion based largely on hybridization theory, molecular orbital-based Jahn–Teller theory suggests that the larger distortion in PH_3 versus NH_3 is due to the smaller relative size of the *s*-orbital on phosphorus with respect to the larger valence *p*-orbitals.

The different mixing gives NH_3 and PH_3 very different properties with respect to inversion, which would occur through the same D_{3h} structure under discussion. The barrier for NH_3 inversion is 5.8 kcal/mol, small enough that inversion occurs rapidly under most conditions. The experimental barrier for PH_3 inversion on the other hand is 31.5 kcal/mol, high enough that inversion is not observed at room temperature. This large barrier to inversion for PH_3 extends to its derivatives, and P(R)(R′)(R″) compounds are often isolable as chiral enantiomers, a useful trait used in asymmetric catalysis.

One-electron oxidation of phosphine derivatives to a cation radical, $^{+\bullet}PR_3$, allows far more facile inversion. As discussed in Section 8.4, radicals often reside in orbitals with far more *p*-character than their lone-pair counterparts, which greatly reduces the energy required to reach the D_{3h} transition state in phosphines. Indeed, calculations suggest the oxidized phosphine has a larger HOMO–LUMO gap than the neutral species due to stabilization of the *p*-orbitals on formation of the cation, which reduces Jahn–Teller mixing and leads to a larger H–P–H angle of 113° (H–P–H angle in neutral PH_3 ~93°).[28]

From a reactivity standpoint it should also be kept in mind that 1-electron oxidation of a phosphine does not necessarily require a strong oxidant; ferrocenium ($FeCp_2^+$) and Cu(II) salts have been shown to oxidize and enable epimerization of chiral phosphines.[50]

Analogously for 7-valence electron $NH_3^{+\bullet}$, the energy gap between the singly occupied a_2''-orbital and the unoccupied $2a_1'$-orbital is larger than in neutral NH_3.[28] The result is lower mixing, and the cation is planar. Likewise, the methyl radical, which is isoelectronic with $NH_3^{+\bullet}$, is planar.

For the hypothetical molecule ClH_3, the HOMO is of A_1' symmetry, and the LUMO is of E′ symmetry. Again, the (spherical) totally symmetric representation does not alter what it is multiplied by, and $E' \times A_1' = E'$. A mode that can couple these two orbitals must have symmetry of E′, of which there are two. From a purely symmetry perspective, either are possible distortion modes; indeed, both may mix in the distortion. The molecule chooses to use the bending modes $\upsilon_{E'(x)} = (2x_2 - x_3 - x_4, x_3 - x_4)$ for the most part. These bending modes could give a Y- or a T-shape to the molecule, and related molecules like ClF_3 have a T-shape.

Note that the distortion in the case of ClH_3 does far more than change the symmetry labels and orbital mixing. The change in symmetry from D_{3h} to C_{2v} results in loss of the C_3 symmetry element. A C_3 rotational axis or higher order is required for degenerate orbitals to be present (Problem 8.7).

One important point one should note is that typically we do not mix orbitals that are not part of the same irreducible representation. This was true in the cases discussed extensively in Chapter 6, where in every case mixing of orbitals required that they be in the same irreducible representation. When we begin to consider the dynamic motions of molecules as we did in this section, specific motions can lead to mixing of orbitals even when in different irreducible representations, but the motion must be specifically suited for the mixing. This coupling of electronic states by vibrational motions is sometimes called "vibronic coupling".

Chapter Summary

1. The lowest shell that an orbital of any angular momentum appears allows the orbital to contract towards the nucleus, because there is no orbital of the same symmetry lower in energy to which it must remain orthogonal. In other words, the 2*s* orbital cannot contract too far towards the nucleus, as it must remain orthogonal to 1*s* below it; however, the 2*p*-orbital can contract towards the nucleus as there is no orbital of the same symmetry below it. We can say that the 2*s*-orbital experiences a primogenic repulsion but the 2*p*-orbital does not, or generally we can call this the primogenic effect, which is important to consider for the 1*s*, 2*p*, 3*d*, and 4*f* orbitals.
2. In the main group, the first row of the *p*-block has *s*- and *p*-orbitals of approximately the same size due to the primogenic effect. It is this similarity in radial size that allows the first-row *p*-block elements, such as carbon, to make *sp*-hybrids effectively. In the heavier congeners of the *p*-block, hybridization does not occur as readily, providing compounds of the heavier elements with different properties from those of the first row.
3. Relativistic effects can occur for any element, but are definitely noticeable once one reaches the elements in the sixth period and beyond. At this point, the electrons in *1s*-orbital of the element have such a high velocity that the orbital contracts due to increased mass of the electron, causing all the *s*-orbitals to contract. The 6*s*-orbital contraction and energetic stabilization is called the direct relativistic effect. The additional shielding by the *s*-orbitals causes some other orbitals to expand, most noticeably the valence *d*- and *f*-orbitals; this is called the "indirect relativistic effect".
4. In the main group, the octet rule is generally maintained, with most atoms having eight electrons, and there is no reason to propose compounds with electron counts higher than eight, even for hypervalent compounds.
5. The bonding in the vast majority of compounds can be explained with 2-center 2-electron bonds plus a few additional bonding effects: 3-center 4-electron bonds (ω-bonds), negative hyperconjugation, and 3-center 2-electron bonds (τ-bonds).
6. MO and hybridization theory are very useful in understanding compounds with various bonding types and were applied to several different hypervalent systems as examples.
7. Radical species are also common in the main group. Some properties of singly occupied radicals were covered, such as that they tend to occupy *p*-rich orbitals. Examples of 2-center 1-electron and 2-center 3-electron bonds were also covered in some detail, and the later type occurs twice in the π-system of the oxygen molecule.
8. The mixing of orbitals in different irreducible representations can occur in an MO diagram, if there is a vibration of the appropriate symmetry to make the mixing allowed. In a first-order Jahn–Teller distortion, a filled and unfilled orbital that are relatively close in energy (typically the HOMO and LUMO) can be mixed. This is a common method for understanding the geometries of compounds where a highly symmetric geometry is considered as a starting point for a distortion to the experimental geometry.

Practice Problems

1. Below is a series of main group compounds. There are arrow(s) in each drawing labeled "a", "b", "c", etc. For each arrow, give the best single description for that bond. List your answer on your paper as "a: {bonding interaction}". Obviously, some compounds have more than one arrow, just follow the labels. If there is a multiple bond, an arrow will specify if it is pointing at the σ- or π-bond. The possible bonding interactions are:

   i. "2c-2e": 2-center 2-electron

   ii. "3c-2e": 3-center 2-electron

   iii. "3c-4e": 3-center 4-electron

   iv. "neg. hyper.": a negative hyperconjugation interaction

   Example

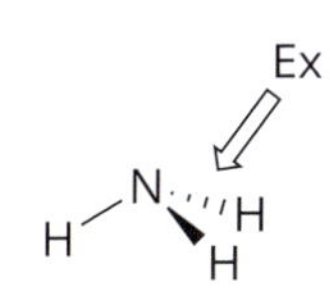

Answer

Ex: 2c2e

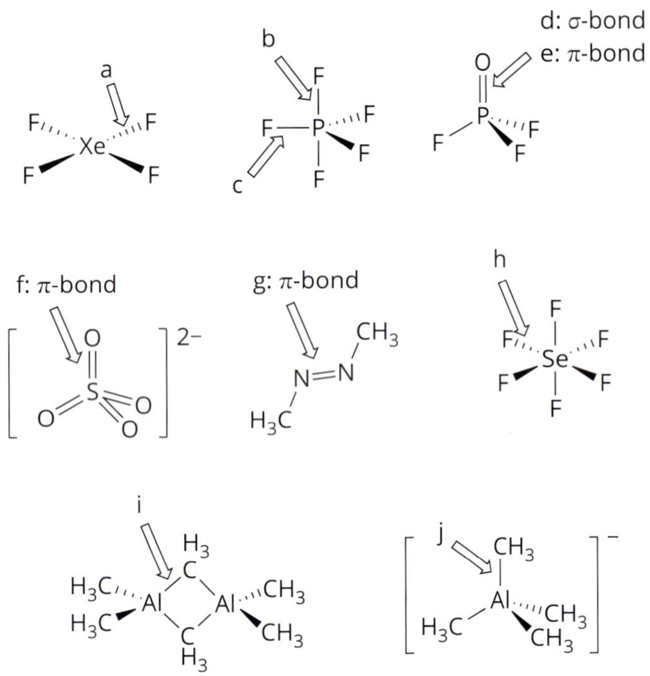

2. Triiodide, I_3^-, is formed by reaction of I_2 with I^-. The compound is linear with two equidistant iodine-iodine bonds. It is used, for example, in certain types of Dye-Sensitized Solar Cells being developed. Describe the bonding in this anion.
   a. What type of bonding is found in I_3^-?
   b. Give a valence bond description of the bonding.
   c. Give a molecular orbital bond description of the σ-bonds. Consider only the p_z-orbital on each atom for your MO diagram.
3. Use VB theory to discuss the bonding in the molecules below. (i) Draw the ground state structure (the composite structure from your resonance forms) with proper stereochemistry and name (what type, i.e., 2c-2e, 3c-4e, dative, etc.) *all* of the bonds. Some bonds may be multiple types, e.g., a mix of dative and heterocovalent 2c-2e bonds. (ii) Draw at least two different resonance forms for each compound.
   a. $\{AlCl_3\}_2$
   b. $\{BH_3\}_2$
   c. XeF_4
   d. SO_2
   e. SO_3
   f. $OSCl_2$
4. Describe the nature of the oxygen π-bonds in OPF_3.
   a. What is the nature of the interaction from an MO picture and draw the donor-acceptor interaction?
   b. What resonance forms are participating from a VB perspective and draw them? (Draw at least two resonance forms.)
   c. Name at least one other compound with a similar type of bond and identify where the bond occurs in the compound either in words or using resonance forms.
5. Analyze the bonding in SF_6 similar to how SF_4 was analyzed in Section 8.3.1. Clearly define the types of bonds to the fluorines in one possible resonance form, then draw all the major resonance forms. Keep in mind that the major resonance form will have as many 2c-2e bonds as possible so as to increase bonding.
6. What is the MO diagram of ClH_3 after Jahn-Teller distortion to a T-shaped molecule? Is this molecule stable to further distortion; in other words, is there a normal mode of proper symmetry to mix the HOMO and LUMO?
7. Below is the geometry of the C_{2v} molecule ClF_3. The molecule is T-shaped as expected form VSEPR.

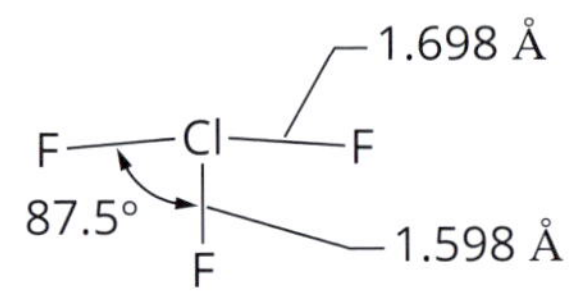

   a. Why is one bond shorter than the other two? Explain this using hybridization theory.
   b. How does the compound manage to make bonds that are <90° between substituents?
8. Consider the hypothetical molecule of "flattened" methane where all the atoms of CH_4 reside in one plane and all the H–C–H angles are 90°.

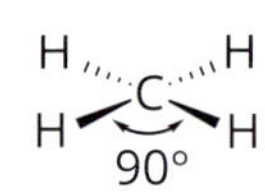

   a. What is the point group of the molecule?
   b. What are the symmetries of the valence orbitals on carbon?
   c. Find symmetries of the SALCs for the hydrogens.
   d. Use the projection operator to determine the shapes of the SALCs from part c.
   e. Draw an MO diagram for flat CH_4.
   f. What is the HOMO of flat CH_4?
   g. Do a VB analysis of flat CH_4. What hybrids on carbon are used to bond to hydrogen?
   h. Using either your VB or MO analysis (or both), what is the C–H bond order in flat CH_4?
9. Douglas Stephan and coworkers at University of Toronto reported the molecule below (**1**) with a pyridyl bis(imine) ligand, which is a dication of phosphorus with an unsymmetrical geometry. The compound is a model for catalysts for allylation of C–F groups by silylallyl compounds. One of the catalysts is the $\{P(Ph)bpy\}^{2+}$ (**2**) compound, and the other is the terpy phenyl phosphine dication **3**. The only one structurally characterized is compound **1**; calculations on **2** give the geometry shown. The authors propose that **3** must uncoordinate one of the pyridines of the terpy to get to the active species, which then is similar to **2**.[55]

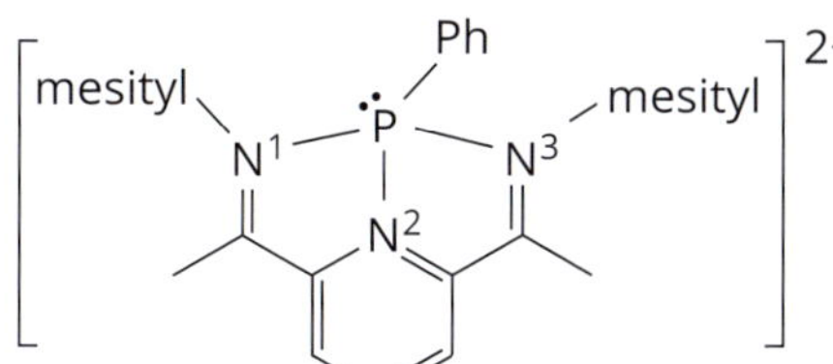

1

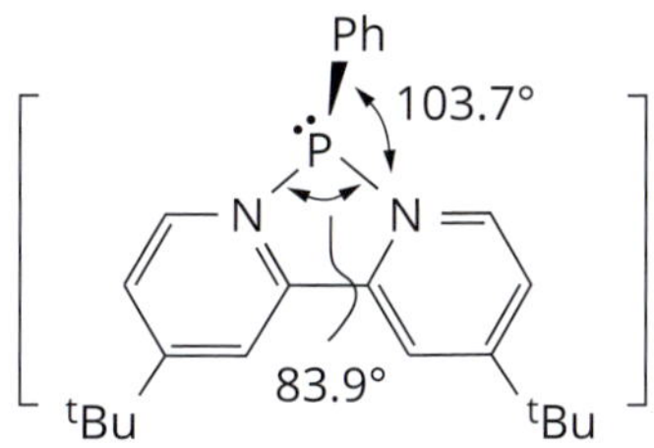

2

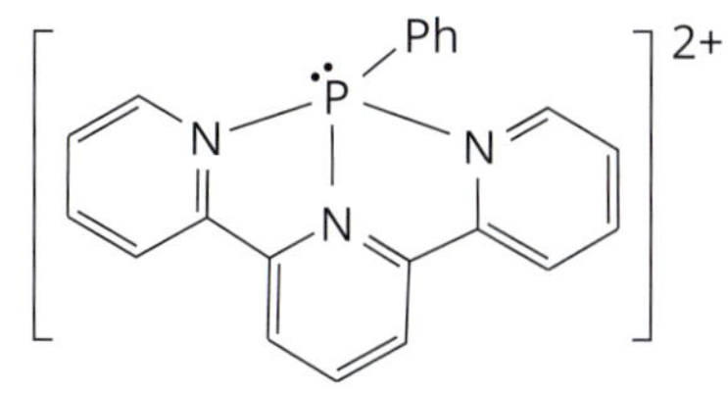

3

a. For **1**, if the N^2–P–C(Ph) angle was 180° and assuming the N^1–P–N^3 bonds are ω-bonds, what would be the hybrid on P holding the lone pair, and why do you think the N^2–P–C(Ph) angle is <180° in the actual compound? Explain in 2 sentences or less.

b. For **2**, NBOs suggests the hybrid on P being used to bond to the nitrogens is $sp^{11.4}$. What should be the angle between two $sp^{11.4}$-hybrids, and why is the angle smaller?

c. For **2**, what hybrid on P is being used to bond to the phenyl?

d. For **2**, what hybrid on P holds the lone pair?

10. Short answer:

 a. What symmetry element is missing in chiral molecules?

 b. What is the indirect relativistic effect?

 c. What is the direct relativistic effect?

 d. In $O{=}PPh_3$, what is the major resonance form, and what type of interaction leads to the O–P π-bond?

 e. What is the VSEPR geometry of XeF_4, and what types of bonds are the Xe–F bonds?

Cited References

1. Gilheany, D. G., No *d*-orbitals but walsh diagrams and maybe banana bonds - chemical bonding in phosphines, phosphine oxides, and phosphonium ylides *Chem. Rev.* 1994, *94*, 1339–1374.
2. Kutzelnigg, W., Chemical Bonding in Higher Main Group Elements *Angew. Chem. Int. Ed.* 1984, *23*, 272–295.
3. Kaupp, M., The role of radial nodes of atomic orbitals for chemical bonding and the periodic table *Journal of Computational Chemistry* 2007, *28*, 320–325.
4. Waber, J. T.; Cromer, D. T., Orbital radii of atoms and ions *J. Chem. Phys.* 1965, *42*, 4116.
5. Pyykko, P., Relativistic effects in structural chemistry *Chem. Rev.* 1988, *88*, 563–594.
6. Thayer, J. S., Relativistic Effects and the Chemistry of the Heavier Main Group Elements. In *Relativistic Methods for Chemists*, Barysz, M.; Ishikawa, Y., Eds. 2010; Vol. 10, pp 63–97.
7. Pyykko, P.; Desclaux, J. P., relativity and the periodic system of elements *Acc. Chem. Res.* 1979, *12*, 276–281.
8. Norrby, L. J., Why is mercury liquid or, why do relativistic effects not get into chemistry textbooks *J. Chem. Educ.* 1991, *68*, 110–113.
9. Li, W.-K.; Blinder, S. M., Introducing Relativity into Quantum Chemistry *J. Chem. Educ.* 2011, *88*, 71–73.
10. Ahuja, R.; Blomqvist, A.; Larsson, P.; Pyykko, P.; Zaleski-Ejgierd, P., Relativity and the Lead-Acid Battery *Phys. Rev. Lett.* 2011, *106*.
11. Power, P. P., Bonding and reactivity of heavier group 14 element alkyne analogues *Organometallics* 2007, *26*, 4362–4372.
12. Stender, M.; Phillips, A. D.; Wright, R. J.; Power, P. P., Synthesis and characterization of a digermanium analogue of an alkyne *Angew. Chem. Int. Ed.* 2002, *41*, 1785.
13. Hach, R. J.; Rundle, R. E., The structure of tetramethylammonium pentaiodide *J. Am. Chem. Soc.* 1951, *73*, 4321–4324.
14. Pimentel, G. C., The bonding of trihalide and bifluoride ions by the molecular orbital method *J. Chem. Phys.* 1951, *19*, 446–448.
15. Kim, M.; Debessai, M.; Yoo, C. S., Two- and three-dimensional extended solids and metallization of compressed XeF_2 *Nature Chem.* 2010, *2*, 784–788.
16. In the past, the dominant model for describing the bonding in compounds of this type involved use of *d*-orbitals on the main group elements. In other words, dsp^3-hybridization would be invoked for this compound. However, in the last 2-3 decades, it has been shown that this *d*-orbital participation is both unlikely and unnecessary. See Box 8A for more details.
17. Tramsek, M.; Zemva, B., Synthesis, properties and chemistry of xenon(II) fluoride *Acta Chim. Slov.* 2006, *53*, 105–116.
18. Calculations were done using cc-pVTZ as the basis set with B3LYP and with NBO5 as implemented in Gamess US.
19. Robertson, A. P. M.; Burford, N.; McDonald, R.; Ferguson, M. J., Coordination Complexes of Ph_3Sb^{2+} and Ph_3Bi^{2+}: Beyond Pnictonium Cations *Angew. Chem. Int. Ed.* 2014, *53*, 3480–3483.
20. Alabugin, I. V.; Gilmore, K. M.; Peterson, P. W., Hyperconjugation *Wiley Interdis. Rev. Comp Molec. Sci.* 2011, *1*, 109–141.
21. Cotton, F. A.; Kibala, P. A.; Miertschin, C. S., 3 New compounds containing Mo_3S_4, W_3S_4 cores prepared from $M_3S_7X_4$ (M = Mo, W, X = Cl, Br) starting materials *Inorg. Chem.* 1991, *30*, 548–553.
22. Lynch, V. M.; Kampa, J. J.; Lagow, R. J.; Davis, B. E., Structure of tris(nonafluoroisobutyl)phosphine oxide *Acta Cryst. C* 1993, *49*, 1549–1551.
23. Weinhold, F.; Landis, C. R., *Valency and Bonding: A Natural Bond Orbital Donor-Acceptor Perspective* Cambridge University Press: Cambridge, 2005.
24. Gordon, A. J.; Ford, R. A., *The Chemist's Companion: A Handbook of Practical Data, Techniques, and References* Wiley: New York, 1972.
25. Andrade, L. C. R.; Costa, M. M. R.; Paixao, J. A.; Moreira, J. A.; Santos, M. L.; Chaves, M. R.; Almeida, A., Crystal structure of diglycine hydroxylammonium chloride, $(C_2H_5NO_2)_2[NH_3(OH)]Cl$Z. *Kristal.* 2002, *217*, 433–434.
26. Caron, A.; Palenik, G. J.; Donohue, J.; Goldish, E., Molecular and Crystal Structure of Trimethylamine Oxide $(CH_3)_3NO$ *Acta Cryst.* 1964, *17*, 102.
27. Calculations were done using SDD as the basis set with B3LYP on G09.
28. Calculations were done using cc-pVTZ as the basis set with B3LYP on G09.
29. Calculations were done using cc-pVDZ as the basis set with B3LYP on G09.
30. De Vleeschouwer, F.; Van Speybroeck, V.; Waroquier, M.; Geerlings, P.; De Proft, F., Electrophilicity and nucleophilicity index for radicals *Org. Lett.* 2007, *9*, 2721–2724.
31. Dupont, T. J.; Mills, J. L., Arylborane Anions: Electrochemical Study *J. Am. Chem. Soc.* 1975, *97*, 6375–6382.
32. Hoefelmeyer, J. D.; Gabbai, F. P., An intramolecular boron-boron one-electron sigma-bond *J. Am. Chem. Soc.* 2000, *122*, 9054–9055.

33. Hübner, A.; Diehl, A. M.; Diefenbach, M.; Endeward, B.; Bolte, M.; Lerner, H.-W.; Holthausen, M. C.; Wagner, M., Confirmed by X-ray Crystallography: The B•B One-Electron Bond *Angew. Chem. Int. Ed.* 2014, *53*, 4832–4835.
34. Fourre, I.; Silvi, B., What can we learn from two-center three-electron bonding with the topological analysis of ELF? *Heteroat. Chem* 2007, *18*, 135–160.
35. Pauling, L., *The Nature of the Chemical Bond and the Structure of Molecules and Crystals: an Introduction to Modern Structural Chemistry* 3rd ed.; Cornell University Press: Ithaca, N. Y., 1966; p xx, 644 p.
36. Mulliken, R. S., Overlap integrals and chemical binding *J. Am. Chem. Soc.* 1950, *72*, 4493–4503.
37. Mulliken, R. S., Structures of the Halogen Molecules and the Strength of Single Bonds *J. Am. Chem. Soc.* 1955, *77*, 884–887.
38. Pitzer, K. S., Repulsive Forces in Relation to Bond Energies, Distances and Other Properties *J. Am. Chem. Soc.* 1948, *70*, 2140–2145.
39. Pitzer, K. S., London Force Contributions to Bond Energies *J. Chem. Phys.* 1955, *23*, 1735–1735.
40. Pitzer, K. S., Intermolecular and Intramolecular Forces and Molecular Polarizability *Adv. Chem. Phys.* 1959, *2*, 59–83.
41. Esterhuysen, C.; Frenking, G., The nature of the chemical bond revisited. An energy partitioning analysis of diatomic molecules E_2 (E=N-Bi, F-I), CO and BF *Theor. Chem. Acc.* 2004, *111*, 381–389.
42. Forslund, L. E.; Kaltsoyannis, N., Why is the F_2 bond so weak? A bond energy decomposition analysis *New J. Chem.* 2003, *27*, 1108–1114.
43. Lauvergnat, D.; Maitre, P.; Hiberty, P. C.; Volatron, F., Valence bond analysis of the lone pair bond weakening effect for the X-H bonds in the series XH_n=CH_4, NH_3, OH_2, FH *J. Phys. Chem.* 1996, *100*, 6463–6468.
44. Opik, U.; Pryce, M. H. L., Studies of the Jahn Teller Effect. 1. A Survey of the Static Problem *Proc. R. Soc. Lond. A-Math. Phys. Sci.* 1957, *238*, 425–447.
45. Pearson, R. G., The 2nd-Order Jahn-Teller Effect *THEOCHEM* 1983, *12*, 25–34.
46. Bersuker, I. B., Pseudo-Jahn-Teller Effect: A Two-State Paradigm in Formation, Deformation, and Transformation of Molecular Systems and Solids *Chem. Rev.* 2013, *113*, 1351–1390.
47. Pearson, R. G., Concerning Jahn-Teller Effects *Proc. Nat. Acad. Sci.* 1975, *72*, 2104–2106.
48. Bader, R. F. W., Vibrationally Induced Perturbations in Molecular Electron Distributions *Can. J. Chem.* 1962, *40*, 1164–1175.
49. Bersuker, I. B., *Electronic Structure and Properties of Transition Metal Compounds: Introduction to the Theory* Wiley: New York, 1996.
50. Reich, K. D.; Ess, D. H.; Radosevich, A. T., Catalyzing Pyramidal Inversion: Configurational Lability of P-Stereogenic Phosphines via Single Electron Oxidation *J. Am. Chem. Soc.* 2013, *135*, 9354–9357.
51. Chandler, G. S.; Thirunam.T, 3d-Radial Functions In sp^3d Configuration of Phosphorus *J. Chem. Phys.* 1967, *47*, 1192.
52. Chandler, G. S.; Thirunam. T; Campbell, J. A.; Chandler, G. S., d Orbitals in sp^3d sp^2d^2 and p^3d^2 Configurations of Phosphorus *J. Chem. Phys.* 1968, *49*, 3640.
53. Martin, W. C.; Zalubas, R.; Musgrove, A., Energy-Levels Of Phosphorus, P(I) Through P(V). *J. Phys. Chem. Ref. Data* 1985, *14*, 751–802.
54. Reed, A. E.; Weinhold, F., On the Role of d Orbitals in SF_6 *J. Am. Chem. Soc.* 1986, *108*, 3586–3593.
55. Chitnis, S. S.; LaFortune, J. H. W.; Cummings, H.; Liu, L. L.; Andrews, R.; Stephan, D. W., Phosphorus Coordination Chemistry in Catalysis: Air Stable P(III)-Dications as Lewis Acid Catalysts for the Allylation of C-F Bonds *Organometallics* 2018, *37*, 4540–4544.
56. Akiba, K. Y., *Chemistry of Hypervalent Compounds* Wiley-VCH: New York, 1999.
57. Baus, J. A.; Laskowski, N.; Tacke, R., A Silacyclopropene with a Six-Coordinate Silicon Atom and an SiN_4C_2 Skeleton: Synthesis and Structural Characterization *Eur. J. Inorg. Chem.* 2016, 5182-5184.
58. Junold, K.; Baus, J. A.; Burschka, C.; Tacke, R., Bis N,N'-diisopropylbenzamidinato(-) silicon(II): A Silicon(II) Compound with Both a Bidentate and a Monodentate Amidinato Ligand *Angew. Chem. Int. Ed.* 2012, *51*, 7020–7023.
59. Lu, J. S.; Su, S. H.; Yang, M. C.; Wen, X. T.; Xie, J. Z.; Su, M. D., Substituent Effects on Boron-Bismuth Triple Bond: A New Target for Synthesis *Organometallics* 2016, *35*, 3924–3931.
60. Sevenard, D. V.; Kolomeitsev, A. A.; Hoge, B.; Lork, E.; Roschenthaler, G. V., Noncyclic [10-S-5] sulfuranide dioxide salts with three S–C bonds: A new class of stable hypervalent compounds *J. Am. Chem. Soc.* 2003, *125*, 12366–12367.

CHAPTER 9

Applying Bonding Principles to Main-Group Reactivity and Properties

"Chemistry is commonly called molecular physics: it is the investigation of the structure and behavior of those little groups of atoms which we call molecules, whose almost infinite repetition constitutes the chemical substances of our ordinary experience."

— Nevil V. Sidgwick[1]

"Such compounds, which result from the union of a metal with one or more organic radicals, are called organometallic compounds. These, with a fairly large number of metals, have been known for a long time, though they do not all lend themselves equally readily to the practical application that I have just mentioned. Some of them, those of the alkaline metals sodium and potassium, are very active but very unstable and difficult to prepare . . . Others, such as those of mercury, err in the other direction – they are too stable . . ."

— Victor Grignard (Nobel Prize Address, 1912)

CONCEPT COVERAGE

1. Allotropy of the elements
2. Common oxidation states of the main-group elements
3. Examples of common reactions and syntheses
4. Inert pair effect
5. Stereochemically active lone pairs
6. Examples of main-group catalysis

In this chapter, a short survey of some main-group chemistry is presented for the purpose of illustrating the utility of the various models discussed and typical reactivity found for the main-group elements. Semester-long classes could easily be taught on the reactivity of all the elements. The examples shown are simply there to illustrate some typical reaction chemistry, to introduce important concepts, and to illustrate some of the bonding principles discussed in previous chapters. Some effort is put in discussion of common or interesting reagents used in the laboratory as well.

The "main group" is generally regarded as being Groups 1, 2, and 12–18 of the periodic table (Fig. 9.1). Common forms of the elements will be discussed first. The organization then is to proceed more or less from left to right across the table with increasing valence electron counts in the elements to review their chemistry. For the elements in a group, only a few features of their reactivity are covered, and there is a brief summary at the beginning of those sections. In contrast, the chapter summary attempts to draw attention to some reactivity trends that cross groups. In much of the book, properties can be described with a few models. This is done here to the extent possible, but reactivity is dependent on many subtle properties and is highly diverse.

9.1 Physical and Chemical States of the Elements

Some elements exist, for all practical purposes, as a single stable form such as nitrogen, which exists almost solely as triply bonded N_2. Some elements, such as carbon and phosphorus, exist in a wondrous array of different structural types, as we will see subsequently. Some of the bonding effects that lead to these different allotropic states are always present within

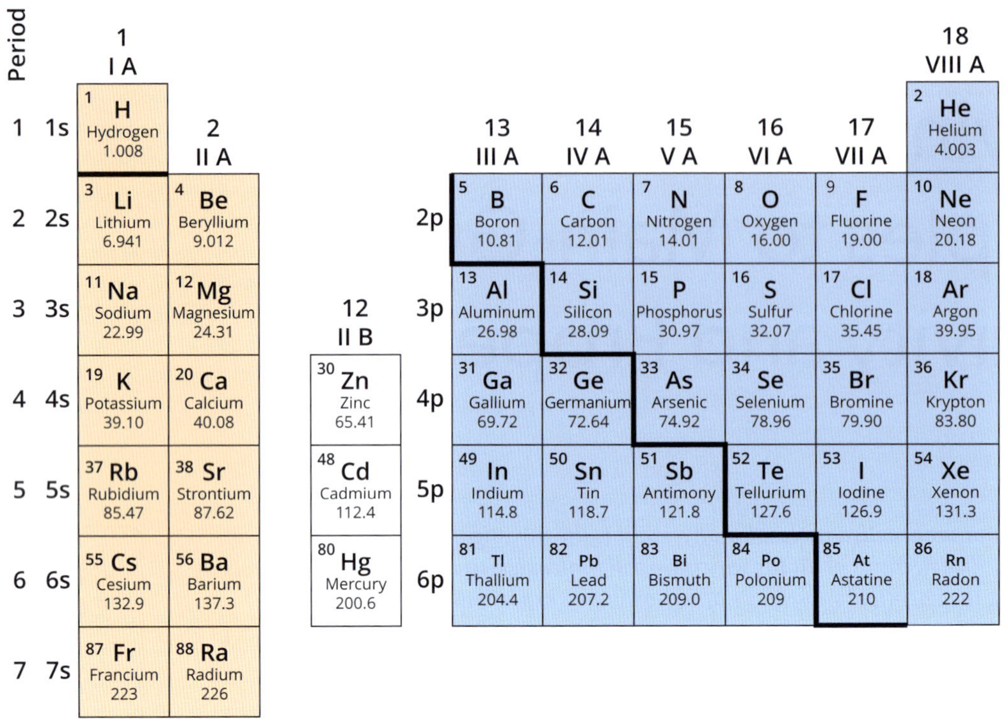

Fig. 9.1 Main-group elements of the periodic table.

the element; consequently, in some cases, bonding in an element's allotropes constitute a baseline on which to build an understanding of the chemistry in binary, tertiary, and other compounds comprised of the main-group elements. Most elements are not found in pure form in nature, however. Nitrogen, oxygen, and a few others are unusual in that their most common form of our experience is as an elemental form. In this subsection we will briefly discuss the chemical states of the main-group elements, their structures, and bonding.

9.1.1 Chemical Forms of the Elements and Allotropy

The word "allotrope" is derived from the Greek words for "other" (*allos*) and "form" (*tropos*). Many elements, such as carbon, can have multiple forms that are kinetically stable, such as diamond and graphite. In all cases, one form will have a higher thermodynamic stability—graphite, for example, is more thermodynamically stable than diamond—but in some cases there is a significant kinetic barrier for conversion between these allotropes. In some cases, specific allotropes are not stable near room temperature and atmospheric pressure, converting to a different form near ambient conditions.

The kinetic barrier that is necessary for a compound to be considered a new allotrope can be subjective. For example, the N_4 compound has been observed in the gas phase and has a structure that is believed to consist of two tightly bound N_2 units with a weaker N–N bond in an open chain structure.[2] The compound has a lifetime of about a microsecond in the gas phase. By some reckoning this could be considered to be a new allotrope of nitrogen; however, we will largely limit our discussions here to compounds that are stable on the order of days or more at ambient temperature and pressure.

The types of stable allotropes for an element are dependent upon its ability and preferred types of linkage, including *catenation*—the linking of the same element into chains. Carbon, boron, phosphorus, and sulfur, along with some of the surrounding elements, are particularly adept at catenation. In some cases there are different allotropes for an element, depending on the type of or degree of catenation.

Table 9.1 shows a selection of common allotropes for the *p*-block elements. Among the main-group elements there exist examples of gases, liquids, and solids under ambient conditions (Fig. 9.2). For solids, the element may cast itself into its chosen form, using covalent bonds or bonds of a highly delocalized metallic nature. If metallic, the atoms pack as spheres in a few standard structures (Fig. 9.3): cubic closest packed (ccp), also called face-centered cubic, hexagonal closest packed (hcp), and body-centered cubic (bcc).

The arrangement of atoms in a crystalline solid is made up of a repeating group called the *unit cell*. There are many different possible types of unit cell, and the atoms within the lattice are described by their symmetry by using *space groups*, which are similar to point groups but include translational symmetry elements as well. The basics of solid-state chemistry are described in a separate chapter, but a brief description of ccp, hcp, and bcc systems are presented here.

For example, the alkali metals (Li, Na, K, Rb, and Cs) all crystallize with bcc structures. Beryllium and magnesium

Table 9.1 Common allotropes, melting and boiling point of the most common allotrope, colors from a Bunsen flame, and other data for the main-group elements.

Element	Stable Forms[a]	Notes	M.P. (°C)[b]	B.P. (°C)[b]
Li	bcc	Has the lowest density of any solid element at room temperature at 0.534 g/cm^{-1}. Flame color: crimson (λ_{max} = 670.8 nm).	181	1347
Na	bcc	Flame color: yellow (λ_{max} = 589.2 nm).	98	881
K	bcc	Flame color: violet (λ_{max} = 766.5 nm).	63	766
Rb	bcc	Flame color: red-violet (λ_{max} = 780.0 nm).	39	688
Cs	bcc	Flame color: blue (λ_{max} = 455.5 nm).	29	705
Be	hcp		1287	~2500
Mg	hcp		649	1105
Ca	fcc	Face-centered cubic at room temperature, with a higher temperature (>450 °C) hcp form. Flame color: orange-red.	839	1494
Sr	fcc	Flame color: red.	768	1381
Ba	bcc	Flame color: green.	727	1850
Ra	bcc	Flame color: red.	700	1700
B	α,β,γ	Some sources list as many as five different forms, but three pure phases are more commonly listed. The most stable form at room temperature and pressure has been contentious, but appears to be the α-rhombohedral structure, which is comprised of B_{12}-icosahedra in a distorted ccp configuration. The common β-rhombohedral form consists of B_{84} polyhedra. The γ-octahedral structure forms at higher temperatures and pressures. All three can be synthesized as pure phases and stored at room temperature.[3] Flame color: green.	2180	~3650
Al	ccp	Metallic.	660	2467
Ga	face-centered tetragonal	The metal adopts a structure very similar to I_2 in the solid state with a pseudo-Ga_2 unit with a Ga–Ga distance of 2.44 Å, with six relatively close neighbors 2.70–2.79 Å away. The interatomic distances are smaller than in Al due to the *d*-block contraction. The unusual structure is thought to be due to a stabilized {Ar}$3d^{10}4s^24p^1$-configuration where the *p*-orbital is used to form the short Ga–Ga bond.	29.8	2403
In	orthorhombic	Metallic with a slightly distorted close-packed structure with four close neighbors at 3.24 Å and 8 at 3.36 Å. Flame color: blue-purple.	156.6	2080
Tl	hcp	Metallic. Flame color: green	303.5	1457
C	graphite, diamond, buckyballs, nanotubes and related structures	The allotropes are quite extensive and useful. Other allotropes will no doubt be discovered for the Group 14 elements, especially C, Si, and Ge.[4]	4100	–
Si	diamond lattice	Semimetallic in nature with a single stable crystalline form. Like many elements, it can be isolated in a quasi-stable amorphous form as well. A β-form is stable only at high pressure.	1420	~3280
Ge	diamond lattice	Semimetallic in nature with a single stable crystalline form, α-germanium. There is a high-pressure form as well; a less dense tetragonal β-germanium with a structure similar to β-tin.	945	2850
Sn	tetragonal β-tin and cubic α-tin	The form stable at room temperature is β-tin, which is white and metallic. Below 13 °C there is a phase change to diamond-like α-tin, which remarkably is less dense by 26% (α: 5.765 g/cm^3; β: 7.265 g/cm^3).	232	2623
Pb	ccp	Metallic element known for its density (11.342 g/cm^3), far denser than Sn, Ge, and Si. However, there are many elements denser still, such as thallium, mercury, and gold. Osmium (22.6 g/cm^3) and iridium (22.4 g/cm^3), for example, are about twice as dense as lead. Flame color: blue-white.	327	1751
N	dinitrogen	A few fleeting allotropes have been reported in the gas phase, but the most stable structure by far is the diatomic, N_2.	−210	−196

(continued)

Table 9.1 Continued

Element	Stable Forms[a]	Notes	M.P. (°C)[b]	B.P. (°C)[b]
P	Molecular P_4, red and black extended structures	P_4 is often the kinetic product of condensation (sublimation) of phosphorus, but is one of the least stable forms thermodynamically. Red phosphorus is made by heating P_4 to form the polymeric, less volatile solid. Black phosphorus is the most thermodynamically stable form and has an orthorhombic structure with interconnected hexagonal sheets of P_6 rings, reminiscent of graphite. Many other allotropes are accessible.	44	281
As	α-As, As_4, and other forms	The most stable α-rhombohedral form is analogous to black phosphorus in structure. In the gas phase, the compound is As_4. There are at least four stable forms. Flame color: blue.	816 (38.6 atm)	615 (subl.)
Sb	α-rhombohedral	The most stable α-form is analogous to black phosphorus in structure. At least five other stable and metastable forms have been prepared. Flame color: pale green.	631	1587
Bi	α-form	The most stable α-rhombohedral form is analogous to black phosphorus in structure. The element does not appear to have other allotropes stable at room temperatures and pressure but some higher=pressure phases are known.	271	1564
O	dioxygen, ozone (O_3)	The only two naturally occurring, stable allotropes are dioxygen, O_2, and ozone, O_3.	−219	−183
S	*cyclic*-S_8, larger and small rings, polymers	Extensive allotropy occurs because of the high catenation ability of sulfur, flexibility in angles, and large range of distances for the S–S bond (1.80–2.60 Å). Even for molecular units of S_8 there are at least three different crystalline forms for the compound. S_8 melts on rapid heating around 107 °C, but is unstable above ~119 °C, where other rings begin to form and depress the melting point. The vapor above *cyclo*-S_8 contains S_2–S_{10} species of various compositions depending on temperatures and pressure. Above 720 °C, the predominant species is triplet S_2, and large amounts (10-20%) of red, ozone-like S_3 can be observed at 440 °C and 10 mm Hg.	~107	See text
Se	*cyclic*-Se_8, grey selenium	There are three forms (α, β, and γ) of Se_8 that differ in intermolecular packing. The metallic grey selenium is the most thermodynamically favored form, which is comprised of helical polymeric chains with a hexagonal structure. Flame color: blue	217	685
Te	grey tellurium	There is one form commonly encountered, which is similar to the hexagonal Se in structure. Flame color: green.	452	990
Po	α- and β-forms	Polonium is the only element known to crystallize as a simple cubic form with six nearest neighbors, the α-form. The β-form has a rhombohedral structure. These two metallic phases can coexist for large temperature ranges near room temperature.	254	962
F	difluorine	The compound has α- and β-forms with layered lattices in the solid state. F_2 has a unique solid-state structure from the other halogens.	−219	−188
Cl	dichlorine	Has a layered lattice with discrete Cl_2 units where each Cl has two weaker interactions at ~180° and ~90° from the Cl–Cl vector. As a result, the layers are made of rectangular units of Cl_2 units, two full Cl_2 molecules, and two halves of Cl_2 units that connect the rectangles.	−101	−34
Br	dibromine	Similar in structure to chlorine in the solid state.	−7	60
I	diiodine	Similar in structure to chlorine in the solid state.	114	185
He	atomic	He is the only liquid that does not freeze to a solid at atmospheric pressure at any temperature.	NA	−269
Ne	atomic		−249	−246
Ar	atomic		−189	−186
Kr	atomic		−157	−153
Xe	atomic		−112	−108
Rn	atomic		−71	−62

[a] ccp = cubic close packed (sometimes called face-centered cubic); hcp = hexagonal close packed; bcc = body-centered cubic.
[b] Data from Greenwood and Earnshaw, which is generally for the first allotrope listed.[5]
[c] Colors for compounds of the element on nichrome wire in a clear to light blue flame. Elements with listed flame colors are generally useful for the flame test.

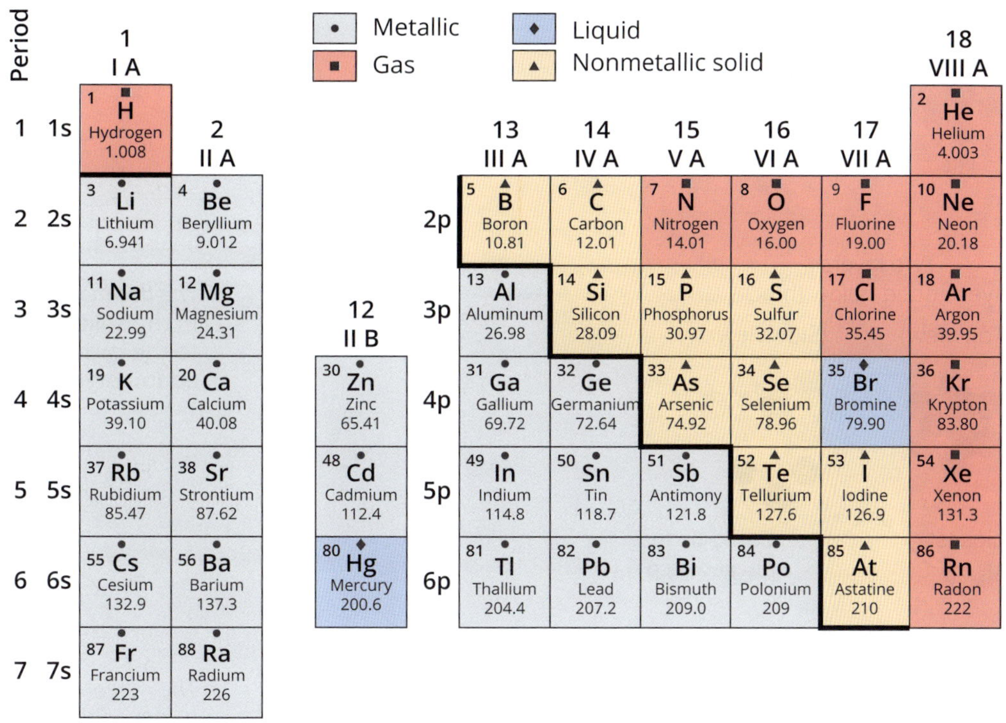

Fig. 9.2 The state of the main-group elements in their most common form: gray (metallic), red (gas), blue (liquid), and orange (nonmetallic solid). The nonmetallic elements touching the dark line in the *p*-block may have metallic properties, and all the elements touching the line may be referred to as "metalloids". Note that francium is shown as a metal, which is a good guess, but the element is exceedingly rare. It is probable that no one has seen enough atoms together in pure form to know its state. It is estimated that there is ~20 g of francium spread throughout the Earth's crust at any one time, as it is an uncommon decay product of uranium and has a half-life of 22 minutes. Astatine has similar issues and is only slightly less rare with a half-life of about 8 hours. The properties of Fr and At are largely determined from either extrapolation of the these elements of from theoretical studies.

have hcp structures. Calcium, at room temperature, and strontium both have fcc crystalline forms. Barium and radium are both bcc, like the alkali metals.

9.1.2 The Many Allotropes of Carbon

It is convenient to consider the allotropes of carbon in two different categories: allotropes containing sp^3-hybridized carbon, and those containing sp^2-hybridized carbon.

There are two closely related allotropes containing sp^3-hybridized carbon: diamond and lonsdaleite. The name diamond comes from the Greek for unbreakable, *adámas*. In 1772 Lavoisier determined that the only product of heating diamonds in air was CO_2, from which he correctly reasoned that diamonds were an allotrope of carbon.

The structure of diamond contains what are referred to as "adamantyl" units. The related hydrocarbon adamantane, $C_{10}H_{16}$, consists of fused-ring chair cyclohexane groups as shown in Fig. 9.4. In the structure of diamond, C–H bonds in adamantane are replaced with C–C bonds to other adamantyl groups.

In the other sp^3-hybridized allotrope, lonsdaleite (hexagonal diamond). Instead of consisting of rings in the chair conformation, the six-membered rings are in the boat conformation (Fig. 9.4).

Both diamond and lonsdaleite can be synthesized from graphite at high temperature and pressure. The two allotropes differ in their solid-state structure with diamond falling into the cubic space group Fd3m and lonsdaleite into the hexagonal space group $P6_3/mmc$.

There are also many allotropes of carbon where the element is sp^2-hybridized. The most common of these is graphite, which is composed of planar sheets of fused, aromatic benzene-like rings (Fig. 9.5a). In graphite, the sheets are translated so that a carbon atom in a sheet below is in the exact center of a ring above it (Fig. 9.5b). Weaker π-stacking interactions hold the sheets together while strong aromatic multiple bonds hold carbons together within the planes.

If graphite is pulled into single or a few layers, small particles can be made with interesting electronic properties; these small particles are often called "graphene" (Fig. 9.5c).

Carbon also forms discrete particulate cages, the most commonly studied of which is C_{60}. To accomplish the curvature, five- and six-membered rings are used in tandem to produce a structure reminiscent of a soccer ball about 7.1 Å in diameter. In Fig. 9.5d, is a line drawing of C_{60}, and in Fig. 9.5e is an STM

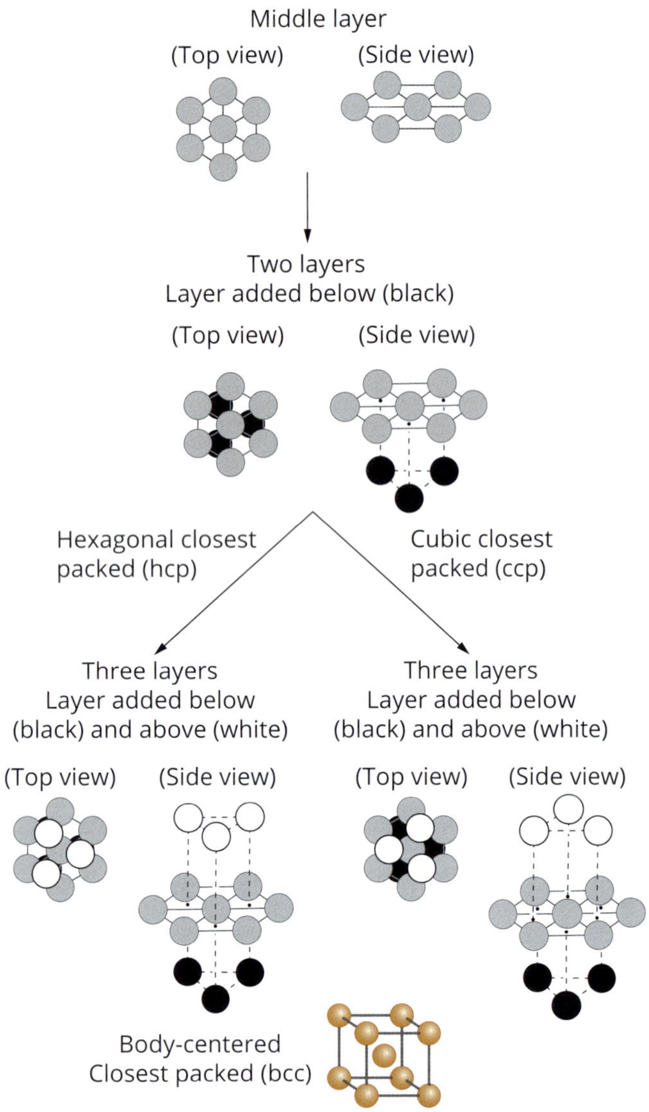

Fig. 9.3 Cubic closest packed (ccp), also called face-centered cubic, hexagonal closest packed (hcp), and body-centered cubic (bcc). In ccp and hcp, each atom has twelve close neighbors. In bcc, each atom has eight close neighbors of equal distance. The diagrams use spheres slightly too small and are expanded from one another to allow clarity of position; naturally, the spheres are packed to touching in the solids like marbles in a jar.

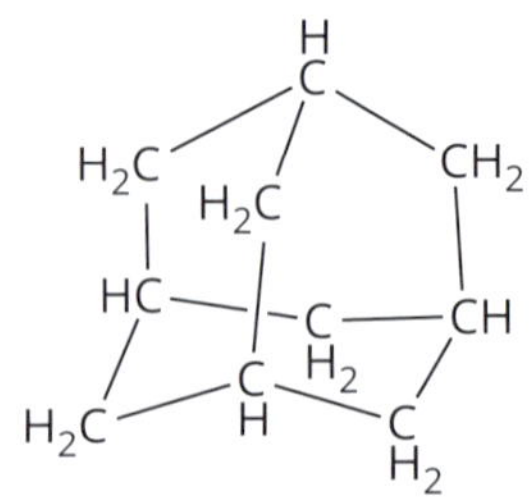

Adamantane

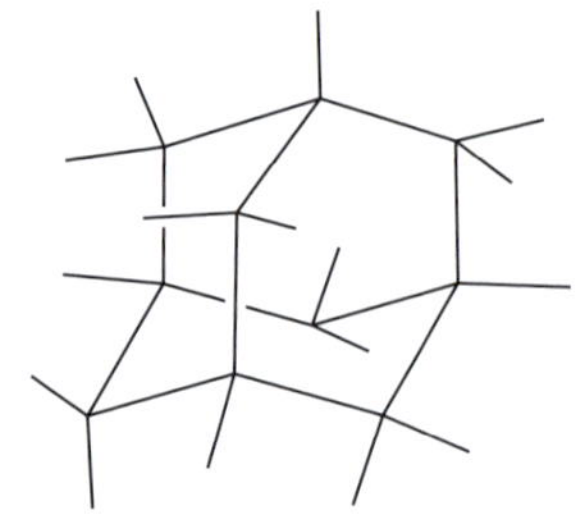

The adamanyl group in diamond. In the solid, each line will lead to the carbon on another adamantyl group.

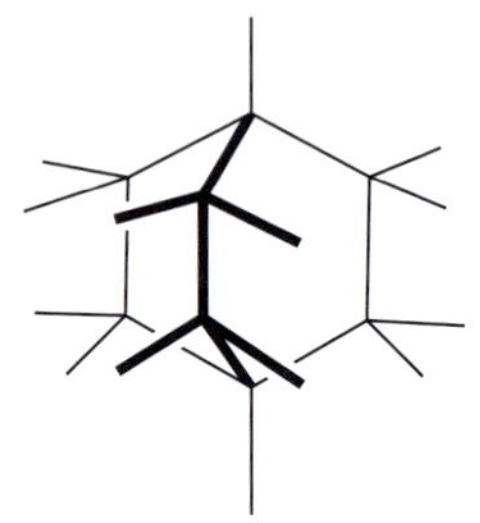

The repeating unit in lonsdaleite. In the solid, each line will lead to the carbon on another repeating unit.

Fig. 9.4 Structure of the hydrocarbon adamantane, diamond, and lonsdaleite.

image of a group of C_{60} molecules on a surface looking very much like soccer balls in a bin. These cages of carbon like C_{60} and C_{70} have been given the moniker "fullerenes" (sometimes "Buckyballs") and Buckminsterfullerene (C_{60} itself), named for the architect of the geodesic dome, Richard Buckminster Fuller. The smallest stable version appears to be C_{20}.

Carbon can also be made into "nanotubes"—graphite-like sheets rolled to forms tubes (Fig. 9.5f) that have typical diameters of 0.5–10 nm. The tubes can be grown to more than 50 cm long, giving them aspect ratios of >100,000,000:1.[8–9] These can be single-walled or multiwalled, capped (ends closed with graphitic hemispheres), or uncapped. There is a rather large variety of nanotubes depending on the angle the graphene sheet is wrapped relative to the large dimension of the tube.

9.1.3 N_2/P_2 and O_2/S_2

Two of the most abundant and important compounds on Earth are dinitrogen and dioxygen. The properties of these two compounds are quite different from their heavier congeners. Dinitrogen and dioxygen are unusual in that these are π-bond-containing molecules, whereas most molecules have their most stable allotropes composed of typically stronger σ-bonds. For example, phosphorus is found as σ-bond-containing P_4, comprised of tetrahedral units and more extended solids (Fig. 9.6). In addition, sulfur is found most commonly as S_8 rings containing only σ-bonds. At atmospheric pressure, P_2 and S_2 only become major portions of the composition of the elements at very high temperatures (~1,800 and 720 °C, respectively).

One explanation that is often proffered for this behavior is that the heavier elements form weaker π-bonds. This is most likely true, but really misses the point! The heavier elements, P and S, are not unusual in preferring σ-bonds to π-bonds; the lighter elements, N and O, are unusual in preferring π-bonds to σ for their most stable form. The question is, "Why are the σ-bonds in N–N and O–O potential allotropes so weak?"

In polar covalence theory, the reason for the weak N–N and O–O bonds is the lone pair bond weakening (LPBW) effect, which is explained in terms of *s*-orbital "stealing" in

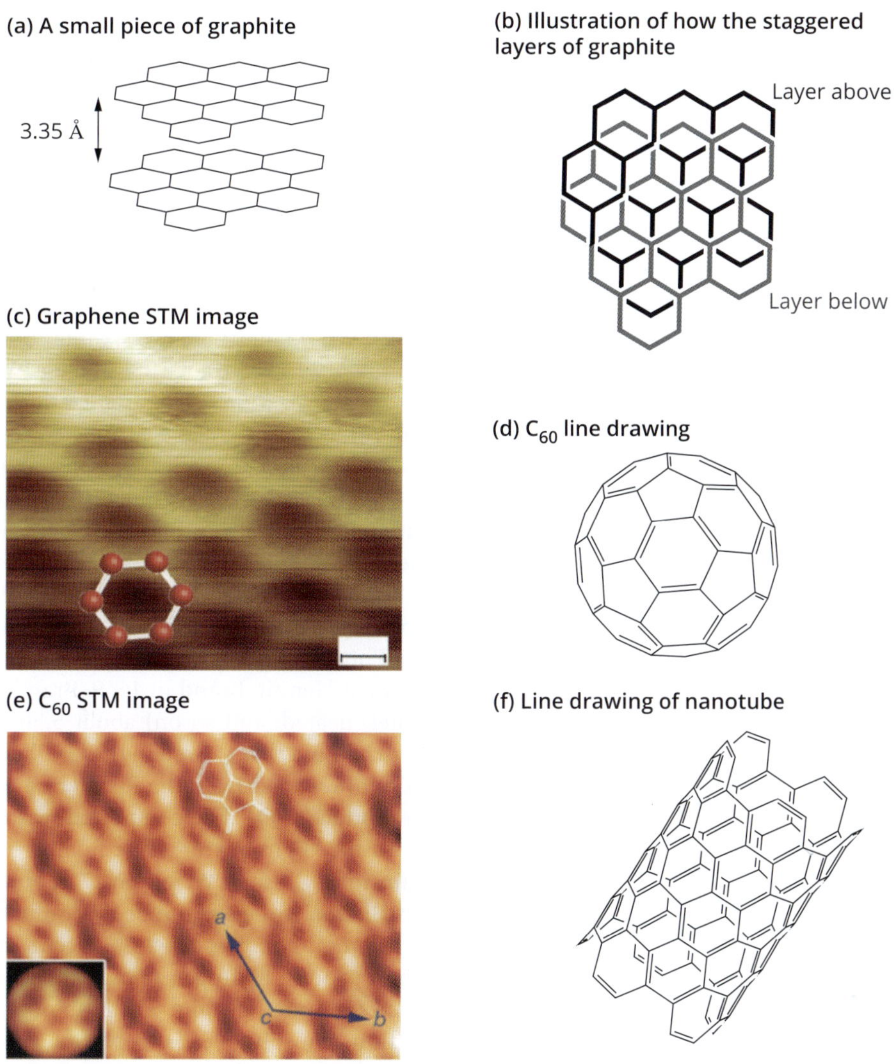

Fig. 9.5 Structures of sp^2-hybridized carbon allotropes: (a) graphite along the planes, which are 3.35 Å apart; (b) graphite perpendicular to the planes showing the carbon of a plane below in the center of the aromatic ring of the plane above it; (c) scanning electron microscope (STM) image of graphene, a single or few layers of graphite. The length bar is 1 Å.[6] (d) Line drawing of C_{60} (e) STM image of a group of C_{60} molecules on a surface looking much like a bunch of soccer balls in a bin;[7] (f) line drawing of a carbon nanotube.

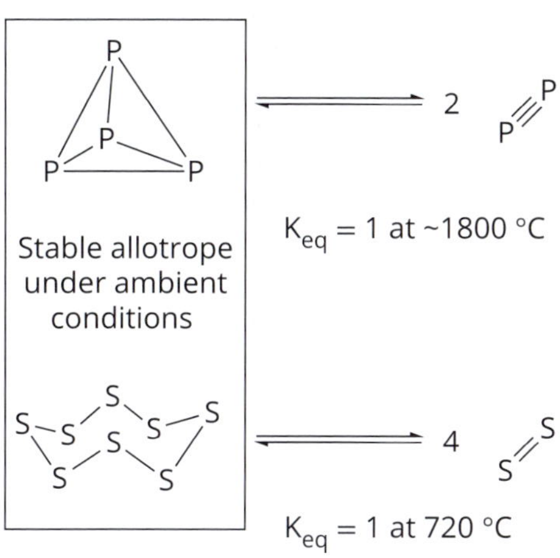

Fig. 9.6 Conversion of room temperature stable P and S allotropes to diatomics.

hybridization theory (Section 8.5). In molecular orbital theory there are repulsion terms that lead to weak σ-bonds and strong π-bonds in these first-row *p*-block compounds.[10–11]

As an illustration, we can calculate the atomization energies (sum of all the bond energies) for two equivalents of N_2 versus tetrahedral N_4, and then do the same for the phosphorus derivatives.

The N–N bond in N_2 has a bond dissociation enthalpy of 225.9 kcal/mol and an experimental interatomic distance of 1.098 Å. Since the bond is, by necessity, purely covalent, only the covalent portion need be calculated and $t_c = 1$. As the number of multiple bonds to an atom with a lone pair increases, the LPBW decreases. Because the π-bonds are generated using entirely *p*-orbitals, the lone pair acquires more *s*-character, lowering the LPBW. By the time the atom has a triple bond, the bond is "fully unweakened", and the E‴ value in Table 3.2 is used. For N_2, the calculation is shown in the following equation for E_c for a N–N single bond with no weakening at the distance of the triple bond (1.098 Å).

$$E_c = \frac{R_c}{R_0} E_{NN}$$
$$= \frac{1.468}{1.098} 94.9$$
$$= 126.9$$

The single-bond energy is multiplied by 1.488 to determine the double bond strength between the same two atoms. The single-bond energy is multiplied by 1.787 to obtain the triple-bond strength between the same two atoms. These bond strengths are also adjusted by the shortening of the bond length in the covalent energy calculation as shown.

Since the unweakened N–N single bond strength is 126.9 kcal/mol, multiplying by 1.787 for a triple bond gives 226.7 kcal/mol, very close to the experimental bond energy for N_2, 225.9 kcal/mol.

For N–N single bonds in hypothetical tetrahedral N_4, the bonds will be fully weakened due to "*s*-orbital stealing" by the lone pairs (Fig. 9.7). If we assume that the bond distances are twice the covalent radius of nitrogen, each N–N bond will be 38.8 kcal/mol. In tetrahedral N_4, there are six total N–N bonds for an atomization energy of $38.8 \times 6 = 232.8$ kcal/mol. Similarly, DFT (B3LPY/aug-PVTZ) calculation of the atomization energy of tetrahedral N_4 gives 270 kcal/mol. Two molecules of dinitrogen have an atomization energy nearly twice as large at $225.9 \times 2 = 453.4$ kcal/mol. (B3LYP/aug-PVTZ gives 453 kcal/mol for two equivalents of N_2 atomization.) Because of LPBW, π-bond containing dinitrogen is far more stable, about twice as stable as all σ-bond containing N_4.

PCT: 6 bonds × 38.8 = 232.8 kcal/mol

(DFT: 270 kcal/mol)

PCT: 2 bonds × 226.7 = 453.4 kcal/mol

DFT: 2 × 226.3 = 452.6

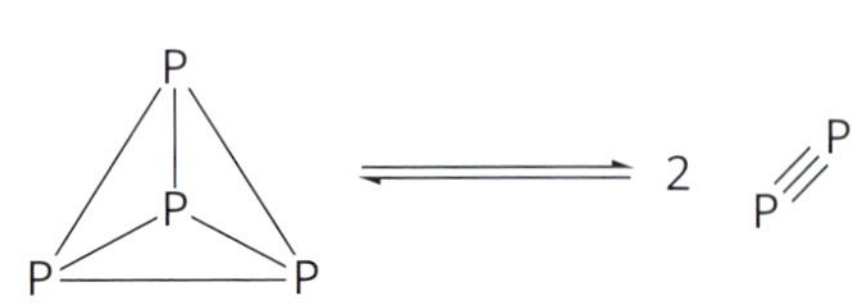

PCT: 6 bonds × 53.7 = 322.2 kcal/mol

(DFT: 273 kcal/mol)

PCT: 2 bonds × 125.1 = 250.2 kcal/mol

DFT: 2 × 114 = 228 kcal/mol

Fig. 9.7 Estimating the differences in energy between N_4/N_2 and P_4/N_2. The high lone pair bond weakening in nitrogen is relieved by the π-bonds in N_2 but leads to relatively weak bonds in N_4. DFT calculations were carried out with (B3LYP/aug-PVTZ).

The same calculation can be carried out for the phosphorus analogues (Fig. 9.7). The interatomic distance in P_2 is 1.89 Å, and the molecule has an experimental bond energy of 125.1 kcal/mol. Polar covalence calculates the P_2 bond energy as 125.6 kcal/mol. (B3LYP/aug-PVTZ gives 114 kcal/mol for P_2 atomization.) The P–P bond energy for the single bonds in P_4 is 53.7, and there are six total bonds for an atomization energy of 322 kcal/mol. (B3LYP/aug-PVTZ gives 273 kcal/mol for P_4 atomization.) The atomization energy of two equivalents of P_2 is only 250.2 kcal/mol; consequently, tetrahedral P_4 is ~50 kcal/mol more enthalpically favorable.

Similar calculations on sulfur and oxygen reveal that the difference in allotropy for the elements is due to lone pair bond weakening, and ultimately, to the lack of primogenic repulsion in the 2*p*-orbitals of the first-row elements, which results in far better *sp*-hybridization in the top row of the *p*-block.

Phosphorus was originally discovered by physician and alchemist Hennig Brand in 1669. Brand processed (distilled off water, heated, and so on) about 5,500 liters of urine to produce a liquid that glowed in the jar. The glow was possibly chemiluminescence from the P_4 reacting with oxygen in the flask to produce P_2O_5. The compound was named from the Greek word for "light-bearer". The final process was depicted in a painting by Joseph Wright in 1717, with Brand overseeing the distillation while surrounded by onlookers (Fig. 9.8a).

Several complex reactions occur in this process. Ammonium sodium hydrogen phosphate liberates ammonia to produce sodium phosphite. Addition of a reductant (graphite) to sodium phosphite leads to phosphorus reduction to P_4, with production of sodium pyrophosphate and CO. In 1680, Robert Boyle improved the procedure by addition of sand (SiO_2) to the second step, which improves the yield of P_4. The distillation is transferred into a receiving vessel with water to prevent the generated P_4 from bursting into flame. (P_4 reacts rapidly with oxygen but not water, and is often stored under water.)

As mentioned previously, white phosphorus, P_4, has a tetrahedral structure and is not the most stable form of the element. Many of the other forms of the element have extended structures. For example, red phosphorus is formed by opening one of the P–P bonds in P_4 to a linear chain.

While white phosphorus is pyrophoric near room temperature, red phosphorus is stable in air to significantly elevated temperatures. As a result, it is used in matches, which are stable until friction due to striking causes a reaction with O_2. Sublimation of red phosphorus (Fig. 9.8b) under vacuum produce white P_4.

Red phosphorus is, in turn, less stable than the black form of the element (Fig. 9.8c), which is composed of chair-conformation rings of P_6 (see below where each vertex is a phosphorus atom). The phosphorus rings are extended into 2-dimensional rippled sheets, and the compound forms black crystalline flakes reminiscent of similarly structured graphite.

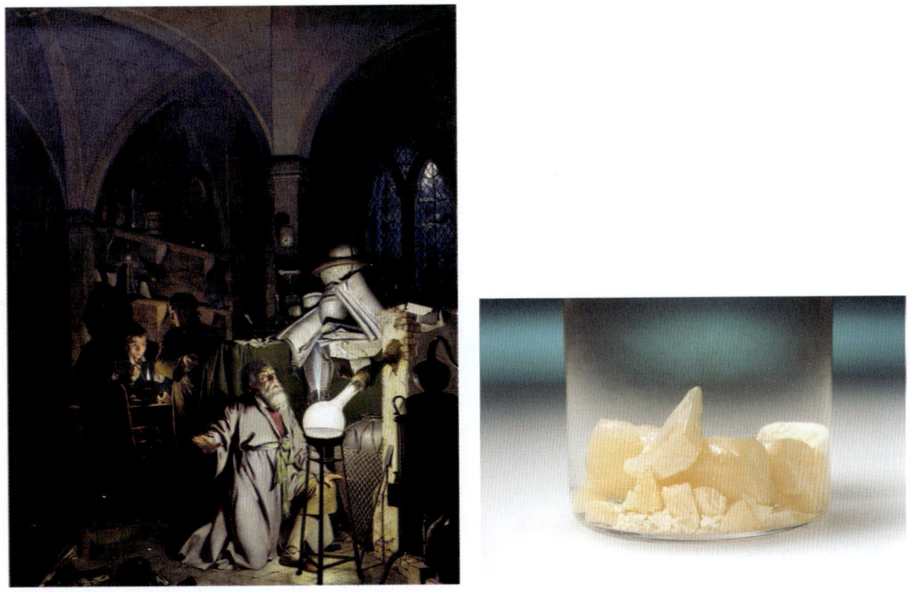

Fig. 9.8a "The Alchemist in Search of the Philosopher's Stone" by Joseph Wright, depicting Brand in his laboratory, isolating phosphorus from processing urine. (*bottom*) The P_4 (called "white" or sometimes "yellow" phosphorus) reportedly produced a chemiluminescent glow.

Red phosphorus

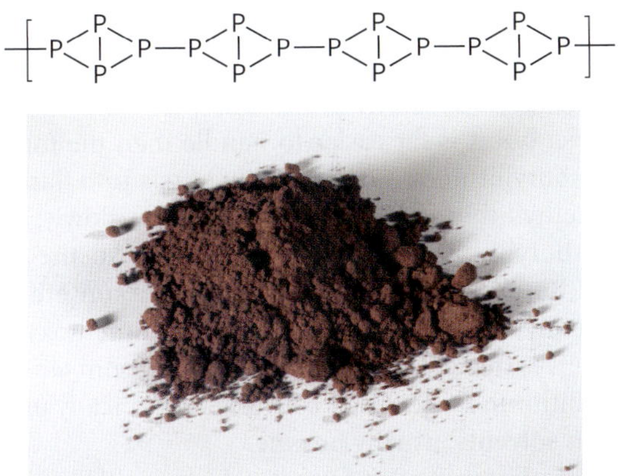

Fig. 9.8b Line drawing and picture of red phosphorus.

Black phosphorus

Fig. 9.8c Line drawing and picture of black phosphorus.

9.2 Group 1: Alkali Metal Chemistry

Features of Group 1 Chemistry

- Usually give up 1-electron to form E^+ and are strong reductants.
- Sodium's reduction potential is often altered by use of an amalgam, a mercury alloy.
- Metal-alkyls, such as methyl lithium, are important reagents.
- Lithium alkyls can be prepared by reaction of lithium metal with alkyl halides, such as $2Li^0 + MeCl \rightarrow LiMe + LiCl$.
- Alkyllithium reagents have cluster structures with bridging alkyls; some clusters use 4-center 2-electron bonds.
- Crown ethers and cryptands can be used to make alkalides, such as "sodide" (Na^-).

All of the alkali metals have an ns^1-configuration, one electron higher than the noble gases with the unpaired electron residing in the valence *s*-orbital. If the one valence electron is lost, the cation has a noble-gas configuration. Consequently, it is not surprising that most of the chemistry of the group is derived from loss of this one electron to form its various compounds.

Although hydrogen has the same ns^1-configuration, H is generally not regarded as a member of the same group (signified by the darkened line between it and the heavier elements of the same column in Fig. 9.1). The chemistry of hydrogen differs greatly from that of alkali metals covered here; although hydrogen does also give up one electron fairly easily as well, its bonding is typically far more covalent than the alkali metals.

The most commonly employed alkali metals in synthetic chemistry are lithium, sodium, and potassium. Sodium and potassium are the seventh and eighth most abundant elements in the Earth's crust, respectively. Lithium is the thirty-fifth most abundant element, but is certainly abundant enough to not be considered "rare."

Rubidium is more abundant than lithium and is ranked as the twenty-third most abundant element, but is used far less than any of the other stable alkali metals. It is not found by itself in any common minerals, but is often found with more abundant potassium. As a result of the lengthy separation of the similar alkali metals and its lack of industrial applications, rubidium and its compounds are often quite expensive, even relative to less abundant cesium.

Cesium, the heaviest stable element in the column, is an order of magnitude less abundant than lithium. Francium is radioactive, with its most stable isotope, ^{223}Fr, having a half-life of only 22 minutes, and is continually being formed by decay of thorium and uranium.

All of the easily observable elements in the group, Li–Cs, are silvery metals. The elements grow softer as one proceeds down the column. Lithium is a silvery-white metal often sold as granules or in a ribbon easily cut with scissors. Sodium and potassium are often purchased as large rods or lumps that are easily cut and scooped out of a larger mass with a dull spatula. Rubidium is an even softer metal with a melting point of 39 °C. Cesium is a silver liquid near room temperature (M.p. 28 °C) and is similar in sheen to mercury, the only metallic element with a lower melting point.

Going down the group, the elements become more reactive with oxygen. Lithium reacts quite slowly with oxygen and water. The reaction generates LiOH and hydrogen gas, often seen as slow bubbling. Sodium and potassium react more violently with water, and the hydrogen gas is ignited with flames shooting from the water or even explosive "popping", especially in the case of the heavier potassium. All three, Li, Na, and K, generate a passivating oxide/hydroxide layer in air, appearing as a white crust and protecting the metal in the inner core of the ribbon, rod, and so on. In contrast, cesium is *pyrophoric*, spontaneously bursting into flame in air.

Both sodium and potassium are essential elements in the human body, making up about 0.1% and 0.2% of our bodies by weight, respectively. Lithium has no clearly understood application in the normally functioning body but is often found in many organisms, including humans in low concentration, and has even been suggested as an essential trace element. Lithium is also used in the treatment of bipolar disorder and related conditions.[12]

The heavier alkali metals, Na-Cs, can be reacted with air to form orange-yellow superoxide salts of formula MO_2. Superoxide, $O_2^{-\bullet}$ is a 1-electron reduced form of oxygen common in nature, and the salts are quite stable in the absence of water. In the presence of water, the superoxide salts are converted to hydroxide salts and oxygen.

Lithium reacts quite slowly with oxygen, but uniquely among the alkali metals it reacts with typically inert dinitrogen. Purified nitrogen gas is often used as the "inert" atmosphere for reactions, but lithium reacts with N_2 to form red Li_3N in an exothermic reaction ($H_f° = 39.4$ kcal/mol).[13] None of the other alkali metals form stable nitrides in this fashion; in fact, lithium can be dissolved in Na metal and reacted with dinitrogen gas to form Li_3N. Sodium is the inert solvent in the reaction!

Addition of lithium to alkyl halides to produce alkyllithium compounds is large-scale industrial process. In the reaction, an alkyl halide (such as methylchloride) is reacted with lithium metal to form LiR and LiCl (Eq. 9.1). One possible side-reaction is Wurtz coupling, where R–R is formed by reaction of LiR with the RX starting material; however, use of less electrophilic alkylchlorides (rather than bromides or iodides) all but eliminates this side-reaction. In the laboratory these reactions are often achieved with lithium granules doped with ~1% sodium in hydrocarbon solvent, such as hexane, under an argon atmosphere.

$$RCl + 2\,Li \rightarrow LiR + LiCl \qquad \mathbf{9.1}$$

These compounds are important reagents in industry and academia. Industrially, they are used, among other purposes, as initiators for polymerizations to form polyisoprene, polybutadiene, and others.

The most commonly available alkyllithium reagents are methyllithium, *n*-butyllithium, *s*-butyllithium, and *t*-butyllithium, with the compounds increasing in reactivity in that order. All of the compounds react violently with water, alcohols, CO_2, and other compounds, and are often added only to cooled solutions for reactions. Methyllithium at typical concentrations, such as 1.6 M in diethylether, can be considered pyrophoric, but it is far easier to handle than the much more reactive *t*-butyllithium, which readily bursts into flame. In addition, methyllithium ring opens THF only slowly, while *n*-, *s*-, and *t*-butyllithium will rapidly polymerize the common solvent. *t*-Butyllithium reacts with the common ethereal solvents—diethyl ether, tetrahydrofuran, and dimethoxyethane—with half-lives that are all within minutes at room temperature. The butyllithium reagents are sold and are typically used in hydrocarbon solvents, usually hexane.

Lithium alkyls are often found in clusters with four or six LiR units per cluster. The alkyl groups reside on the trigonal faces of the polyhedra. For example, in the case of *n*-butyllithium the compound exists in the solid state as a hexamer (Fig. 9.9). The lithiums form a somewhat distorted octahedral core with butyl groups bridging three of the lithium atoms on a trigonal face. The butyl group resides near the center of the trigonal face. The octahedral core is distorted due to uneven occupancy of the faces; an octahedron has eight faces which are occupied by six hexyl groups. The faces not occupied by a hexyl group are wider, while in the hexyl-occupied sides the lithium atoms are more tightly packed near the carbon.

A more symmetrical lithium cluster is found in tetramers, for the simple reason that a tetrahedron has the same number of vertices as faces. Alkyllithium compounds that form tetramers, like methyl- and *tert*-butyllithium, have structures in which there is a tetrahedron of lithium atoms in the center (Fig. 9.10). Each face is capped with an alkyl group symmetrically bound to three lithium atoms, 4-center 2-electron bonds between three lithiums and a carbon.

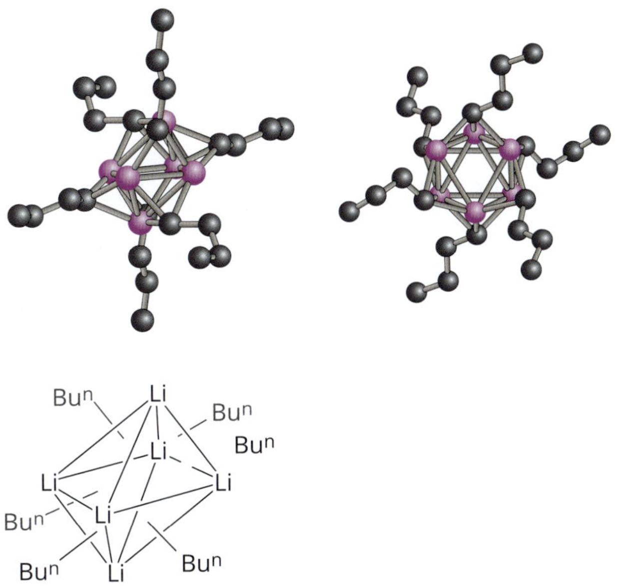

Fig. 9.9 Solid-state structure of the Li-*n*-Bu hexamer with carbons in grey and lithium in lavender. (*top left*) Looking down the edge of the square plane of four lithium atoms. (*top right*) Looking down one of the two unoccupied trigonal faces. (*bottom*) Line drawing of the butyllithium hexamer.

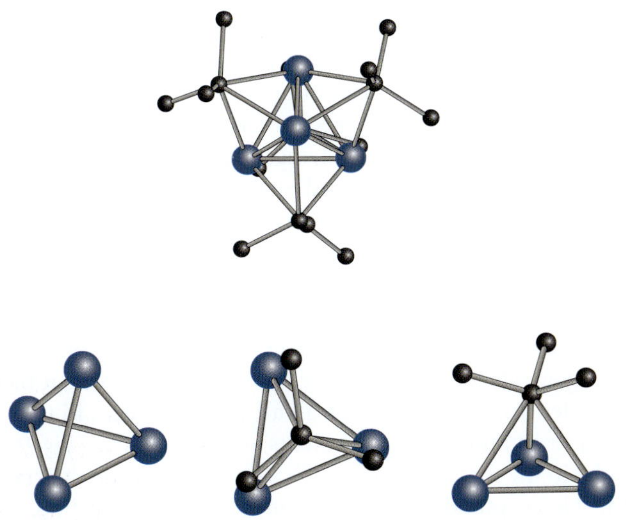

Fig. 9.10 Structure of *t*-butyllithium. Lithium atoms are shown in blue (larger) and carbon in black (smaller). Hydrogens are not shown. (*top*) Full structure. (*bottom left*) The tetrahedron of lithium atoms in the structure. (*bottom middle and right*) Two views of just one of the faces of the tetrahedron with its pendant *tert*-butyl group.

In the case of sp^2-hybridized carbons bearing bromide and iodide, fast halide for lithium exchange occurs (Eq. 9.2). Treatment of arylbromides with *n*-BuLi in hexanes is typically done at –78 °C to limit the amount of Wurtz coupling, reaction of the aryllithium with BrBu to form PhBu. This is a convenient method for the production of many aryllithium reagents, and the aryl lithium can then be used at low temperature as a nucleophile.

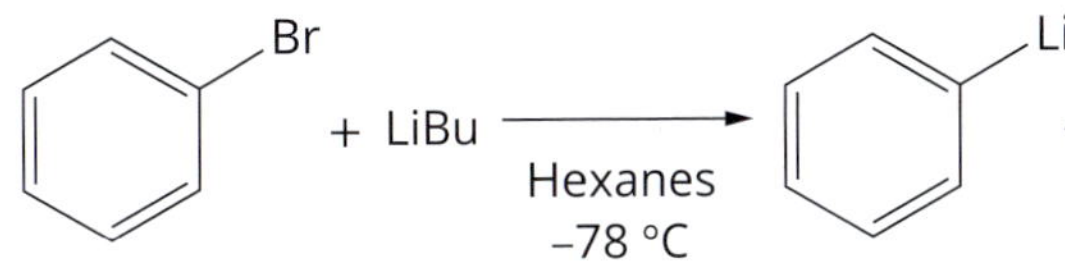

9.2

Alkali metals are also used, because of their tendency towards 1-electron reductions, in the laboratory to create radicals. Perhaps the most commonly seen radical in most synthetic laboratories is prepared from reaction of sodium and benzophenone. The highly delocalized dark purple benzophenone radical is stable for long periods, soluble in ethereal solvents, and reacts with oxygen and water. As a result, benzophenone ketyl is a common drying agent, especially for solvents such as diethyl ether and tetrahydrofuran, Eq. 9.3.

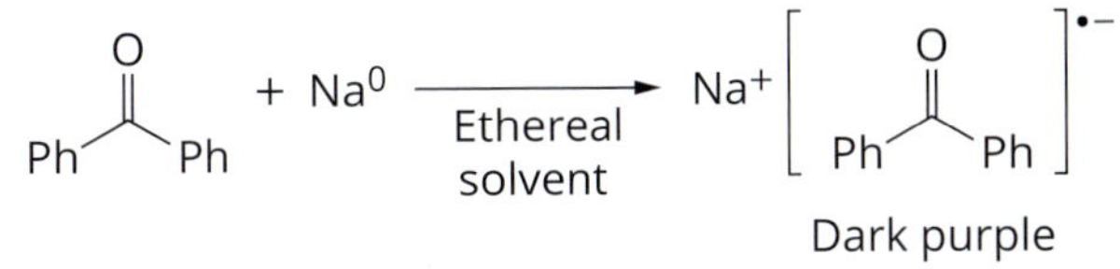

9.3

Benzophenone ketyl solution is also used to ensure that solvents are dry when brought into a glove box. Treatment of the hydrocarbon or ethereal solvent with a few drops of purple sodium ketyl should produce a colored solution or precipitate that retains its color if the solvent is dry. The radical is intensely colored, with an optical absorption at λ_{max} = 630 nm, $\varepsilon = 6 \times 10^3$ in acetonitrile.[14]

Another commonly employed alkali metal radical is naphthalenide, prepared as the lithium, sodium, or potassium salt. Ion pairing in the naphthalenides goes as Li < Na < K, which makes lithium naphthalenide the strongest reductant of the three. In contrast, because of their affinities for oxygen, ion-pairing to benzophenone ketyl (Eq. 9.3) runs in the opposite direction Li > Na > K, and potassium benzophenone ketyl is regarded as the strongest reductant of the three. Naphthalenide anion is one of the strongest reductants commonly employed in the laboratory, at –3.10 V vs ferrocene/ferrocenium; for comparison, sodium amalgam (Section 9.5) in nonaqueous solvent is –2.60 V, and benzophenone ketyl in THF is –2.26 V.[14]

Alkali metals can be used to produce activated metal/alkali halide mixtures, often called Rieke metals (Eq. 9.4). Commonly, the procedure involves metal halide reduction by an alkali metal in xylenes at a temperature exceeding the melting point of the alkali metal. For example, the reaction may be run with potassium at around 100 °C under an argon atmosphere. The product is finely divided metal in a matrix of potassium salt that is far more reactive than bulk metal, which is often lower in surface area and is passivated by surface oxides. Many such Rieke metals have been explored, including magnesium, titanium, manganese, zinc, nickel, palladium, platinum, and others.[15–18]

$$MX_n + xK \xrightarrow{\text{xylenes}} \underset{\text{Rieke Metal}}{\{M, nKX\}} \qquad \textbf{9.4}$$

Alkali metals can also be intercalated into the sheets of graphite. Graphite is composed of atom-thick sheets layered upon one another, and many metals are known to insert into these sheets. A representation of the structure from Bergbreiter and Killough's report is shown in Fig. 9.11.[19] One of the most employed systems has the formula KC_8, which is formed by simply melting potassium over black graphite. The resulting pyrophoric, bronze (or sometimes described as gold) solid is a strong reductant that is a common laboratory reagent. All of the heavier alkali metals (K, Rb, and Cs) produce similar compounds with graphite. The smaller alkali metals provide different stoichiometries, such as LiC_6.

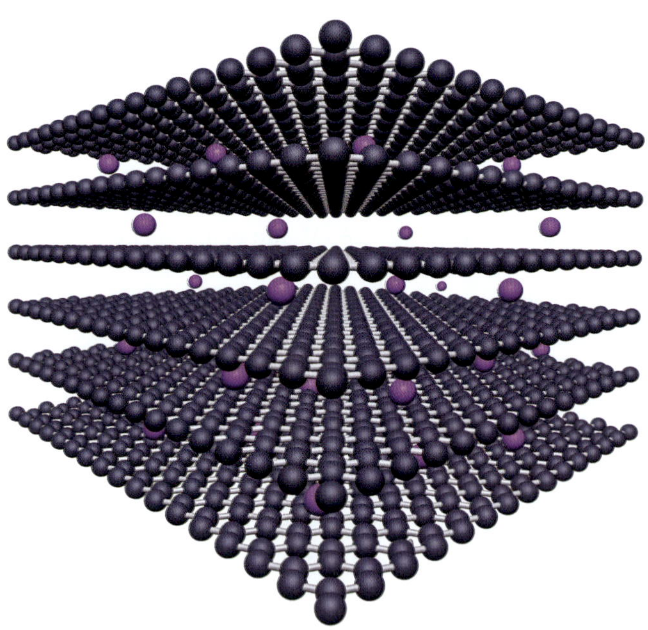

Fig. 9.11 Alkali metals will insert into the layers of graphite. With lithium, a material of composition LiC_6 can be prepared in which the lithium atoms take up residence between the flat carbon planes.

This chemistry is typical of alkali metals where the metal is in the +1 formal oxidation state in stable products. There is also a rich chemistry, discovered by James Dye, where one generates stable compounds through *addition* of one electron to the alkali metals and filling the ns^2 subshell, producing monoanionic *alkalides*. The synthesis of alkalides uses classes of organic compounds called crown ethers and cryptands.

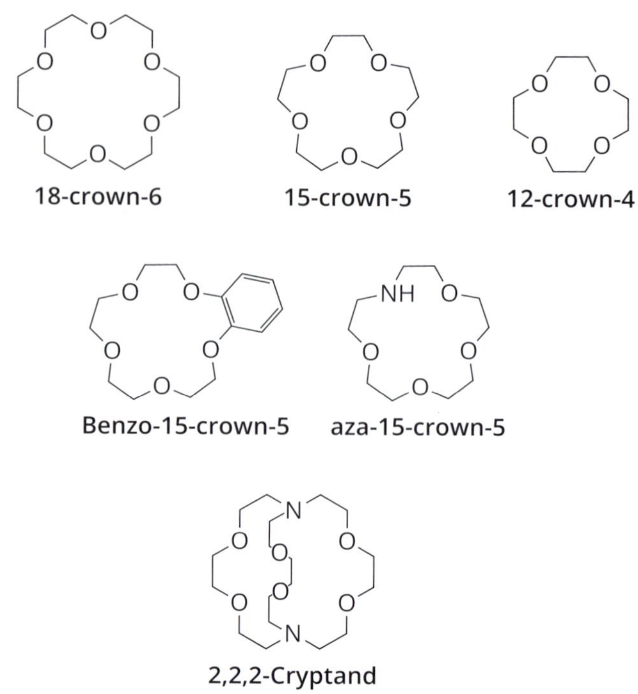

Fig. 9.12 Structure of 18-crown-6, {2,2,2}-cryptand, and some other representative crown ethers.

A *crown ether* is a cyclic polyether. In their simplest form, crown ethers are cyclic oligomers of ethylene oxide with a ($-CH_2CH_2O-$) repeating unit (Fig. 9.12). These compounds can be prepared with a variety of substituents with various ring sizes. The different-sized rings produce the compounds' selectivity for the variable-sized cationic metal atoms. The compounds are named so that the first number is the total atoms in the ring and the second is the number of oxygen (or nitrogen in the case of azacrowns) atoms in the ring. For example, 18-crown-6 is an 18-membered ring compound with 6 oxygen atoms that has a high affinity for the potassium cation. There are many different varieties of crown, including "benzo" crowns with an aromatic ring in place of a ($-CH_2CH_2-$) unit (Fig. 9.12).

Some related complexants are *cryptands*. These contain two nitrogens connected by strings of polyethers that envelop the metal in a pseudo-trigonal environment (Fig. 9.12). Perhaps the most utilized member of this class is commercially available {2,2,2}-cryptand, where there are three strings of 2 ($-CH_2CH_2O-$) repeating units between the nitrogens—an exceedingly strong complexing agent for the potassium cation.

As mentioned, crowns and cryptands are strong complexing agents for alkali metal cations. What happens if these are added to the zero valent alkali metals? The addition of {2,2,1}-cryptand to two equivalents of sodium metal (Fig. 9.13) results in the formation of a salt with sodium inside the cryptand, $\{Na(cryptand)\}^+$, with a counter ion of Na^-, sodide, outside the cage.[20–22] The alkalides of most of the alkali metals (Na^-, K^-, Rb^-, Cs^-) have been prepared by using similar procedures. Lithides are apparently unknown. The driving force for the creation of the alkalides is the stabilization of the alkali metal cation in the crown or cryptand.

Even more remarkably, addition of one equivalent of {2,2,2}-cryptand to sodium metal provides the *electride*—a compound in which the sodium cation is inside the cryptand and a free electron is the "anion".

The cesium-based electride, for example, has been studied in great detail. The cation employed was two equivalents of 15-crown-5 on cesium, $\{Cs(15\text{-crown-}5)_2\}^+$. The cavities between the cations are inhabited by weakly spin-paired electrons (Fig. 9.14).

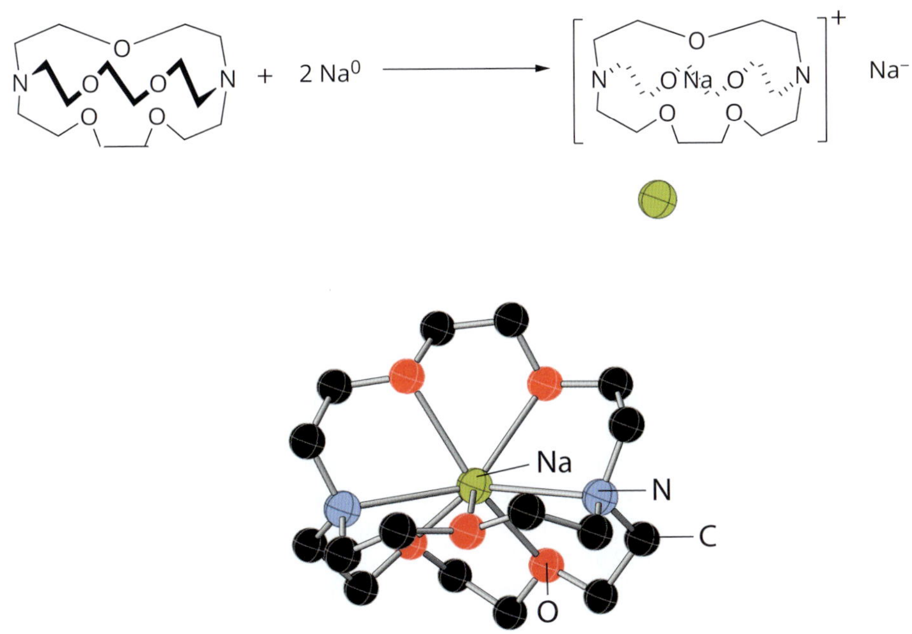

Fig. 9.13 Reaction of {2,2,1}-cryptand with sodium to produce sodide, Na^-. Also shown is the structure of the complex from single-crystal X-ray diffraction with hydrogen atoms excluded. Na = yellow, N = blue, O = red, C = black.[22]

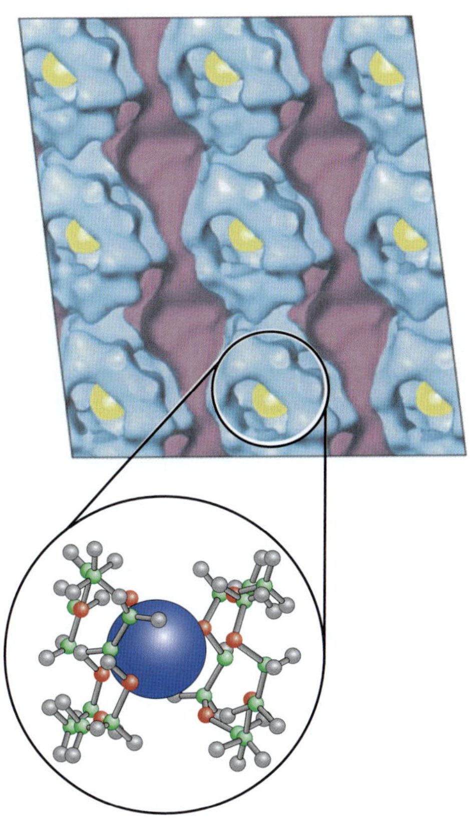

Fig. 9.14 Space-filling representation of the electride, $Cs(15\text{-crown-}5)_2]^+e^-$. In mauve are the cavities containing the weakly spin-paired electrons, the counterions for the complexed cesium cations.[23]

Unsurprisingly, these electride and alkalide materials are extremely reactive and, with typical crown ethers and cryptands, unstable at room temperature. However, Jackson and Dye have published a modified cryptand that results in electrides and sodides that are more thermally stable. The design involves cages formed from bridging piperazine rings and only nitrogen-based donors for the alkali metal cation.[24]

9.3 Chemistry of Group 2 (Alkaline Earth) and Group 12

Features of Group 2 Chemistry

- Both Group 2 and Group 12 commonly act as 2-electron reductants.
- Alkyl magnesium compounds, XMgR, are generally known as "Grignard reagents". Dialkylmagnesium compounds can be prepared through the Schlenk equilibrium; the Grignard reagent (RMgX) reacts with 1,4-dioxane to give $MgR_2 + MgX_2$, where R = organic group and X = halide.
- Edward Frankland prepared one of the first examples of an organometallic compound, pyrophoric $ZnEt_2$, which is a colorless liquid.
- The most common ore of mercury is cinnabar, HgS, which can be converted to mercury metal by heating in the presence of O_2 (or air).

- Mercury catenates quite well, and Hg_2^{2+} is a commonly found unit with a mercury–mercury bond. Mercury(I) compounds invariably contain this unit. In other words, mercury(I) chloride is ClHg–HgCl.
- The other Group 2 and Group 12 elements have metal–metal bond-containing compounds that are known, but they are typically quite reactive.
- A compound in which mercury uses its *d*-electrons is known, so it acts as a transition metal. HgF_4 is a square planar d^8 complex, valence isoelectronic with $PtCl_4^{2-}$.

The electronic configuration of the elements of Group 2 (the alkaline earths) involves two electrons over the noble-gas configuration; that is, ns^2 in the valence shell. Similarly, in Group 12 the metals have a filled *d*-shell and a filled *s*-shell, $(n{-}1)d^{10}ns^2$, and the filled *d*-subshell very rarely participates in the chemistry of these metals (Zn, Cd, and Hg). Not surprisingly, the chemistry of Group 2 and Group 12 are similar in many cases, and we will treat them together here to highlight some of those similarities.

The commonly encountered Group 2 elements are Mg, Ca, Sr, and Ba. Beryllium is of fairly low abundance at 2 ppm in the Earth's crust, making it the fiftieth most abundant element. The other nonradioactive members of the group are some of the most abundant on Earth, with Mg, Ca, Sr, and Ba being sixth, fifth, fifteenth, and fourteenth in abundance, respectively. While small amounts of beryllium are present in the body, the element's similarity to Mg causes it to displace the common metal from enzymes, often with loss of function. As a result, beryllium is quite toxic, especially when inhaled. Fortunately, Be has a relatively low abundance for a light element. Probably as a result of its high toxicity, the chemistry of beryllium is less developed compared to magnesium below it.

Zinc, cadmium, and mercury are the twenty-fourth, sixty-third, and sixty-fourth most abundant elements, respectively. Zinc has a similar abundance to copper. While mercury has about the same abundance as silver, cadmium has concentrations of about twice that of mercury. By far the most common formal oxidation states for the metals are +2, like the Group 2 elements, but mercury is more variable in its oxidation states.

Perhaps the most recognized reaction of any element in the two groups is the reaction of magnesium with an alkyl or aryl halide: the Grignard reaction, developed by Victor Grignard, reported in 1900, and garnering the Nobel Prize in 1912.[25]

In the Grignard reaction, an alkyl (or aryl) halide, RX, reacts with magnesium metal, usually in the form of turnings, to produce RMgX (Fig. 9.15). The alkylmagnesium halide compound is in equilibrium with the dialkylmagnesium and magnesium dihalide, the Schlenk equilibrium. Addition of dioxane, which is thought to bridge magnesium dihalide (MgX_2, such as $MgBr_2$) centers and form insoluble polymers or oligomers, leads to isolation of the dialkylmagnesium compound as the only soluble magnesium species. In other words, addition of dioxane leads to Le Chatelier-driven completion of the equilibrium towards MgR_2 + $MgX_2\downarrow$ by making the magnesium dihalide insoluble.

Invariably, Grignard reagents are prepared in ethereal solvents. In diethyl ether, most RMgX compounds are oligomers at all but the lowest concentrations. In THF, many exist as monomeric $Mg(R)(X)(THF)_2$. Similarly, dialkylmagnesium compounds often retain ether ligands as well.

The reaction of RX with Mg, in which magnesium is oxidized to Mg(II), is exothermic. If the solvent is diethyl ether (b.p. 35 °C), then a solution of RX is often added dropwise to Mg metal slurried in a small amount of solvent. During addition, diethylether often boils due to the exothermic reaction, with no external heating. In many cases, initiation of the reaction, as shown by refluxing of the ether, is rapid, but to ensure initiation, I_2 is added to the magnesium. Alternatively, a small amount, such as 1% or less relative to RX, of 1,2-dibromoethane may be added to the RX solution, which also helps with initiation. Both I_2 and 1,2-dibromoethane are thought to clean the surface of the magnesium, which is coated with surface oxides.

Addition of anthracene to magnesium in THF provides a compound that acts as a soluble surrogate for magnesium metal, the powerful reductant magnesium anthracene, which is often isolated as the orange crystalline tris(tetrahydrofuran) complex (Eq. 9.5).[26]

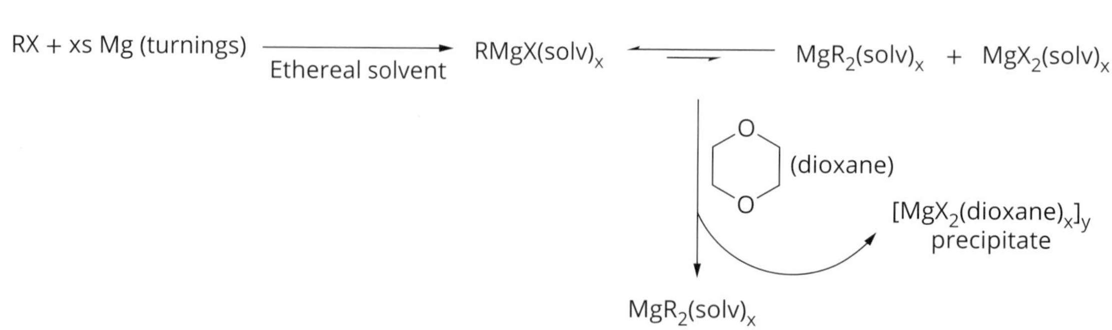

Fig. 9.15 Reaction of magnesium with alkyl halide to form a Grignard reagent, RMgX. The mixed alkyl halide, RMgX, is in equilibrium with MgX_2 and the dialkylmagnesium, MgR_2. Addition of dioxane removes MgX_2 as a solid, forcing the equilibrium to the soluble dialkylmagnesium. (solv = ethereal solvent)

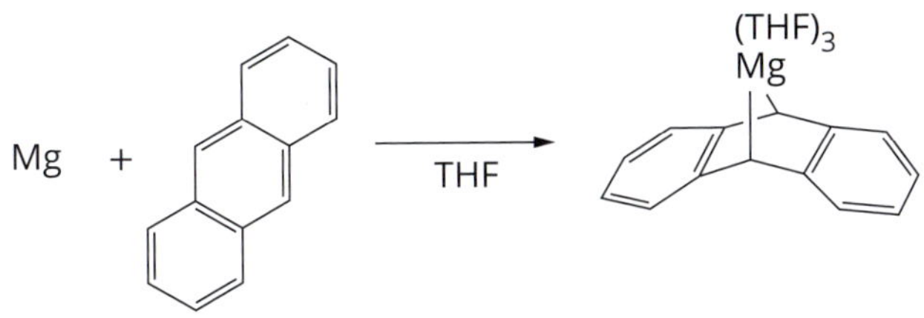

9.5

The dihalides of the alkaline earth metals are, in some cases, interesting structurally and are exceptions to VSEPR, which predicts that the AX_2 compounds should be linear. However, many alkaline earth halides exhibit bent structures to one degree or other, especially for the heavier congeners. Similarly, the pentamethylcyclopentadienyl compounds exhibit very similar angles (Table 9.2). The cause of this bending in some alkaline earth compounds has been explored by using many different models, including covalent, ionic, hybridization, and core polarization explanations.[27–32]

One important zinc reagent commonly encountered is diethylzinc, $ZnEt_2$. The compound was first prepared by Edward Frankland in 1848 at Queenwood College in Hampshire, England, and later more fully characterized while he was working in Robert Bunsen's laboratory at the University of Marburg.[37] The compound, and its methyl derivative, was prepared by reaction of ethyl iodide with zinc metal (Eq. 9.6). The extremely pyrophoric $ZnEt_2$ is a liquid at room temperature, distilled for purification.

$$2\,Zn + 2\,ICH_2CH_3 \rightarrow \underset{\substack{\text{colorless liquid}\\ \text{m.p.: } -28\,^{\circ}C\\ \text{b.p.: } 117\,^{\circ}C}}{Zn(CH_2CH_3)_2} + ZnI_2 \qquad 9.6$$

The heaviest member of Group 12, mercury, is usually found as its sulfide HgS, cinnabar. Cinnabar was commonly used as a red dye in the ancient world. Simply heating cinnabar in oxygen provides mercury metal (Eq. 9.7), named after the messenger of the gods in Roman mythology because of its mobility. Alternatively, reacting HgS with a mild reductant, such as iron(0), will produce the liquid metal. Mercury, also called quicksilver, has a very low melting point of −39 °C, due to a combination of a filled *s*-subshell and relativistic stabilization of these valence electrons (Section 8.2.2).

$$HgS + O_2 \xrightarrow{600\,^{\circ}C} Hg + SO_2 \qquad 9.7$$

Liquid mercury has the ability to form alloys, called amalgams, with many different metals, sometimes with significant liquid ranges. Perhaps the most commonly employed amalgam in the laboratory is with sodium. Addition of sodium metal to mercury is an extremely exothermic reaction that often volatilizes some toxic mercury metal, which should obviously be done with the use of an efficient hood, glove box, or using other precautions. Sodium–mercury amalgam, or Na(Hg) as it is often written, is liquid up to ~1% wt/wt of Na to Hg, with perhaps 0.1% being the most common liquid amalgam employed. Higher-percentage amalgams, say 10%, can be purchased from commercial sources as solid beads.* The advantage of these amalgams is that they are milder reductants than sodium metal itself by ~500 mV[14] and are easier to handle. Obviously, the liquid varieties have the advantage of high surface area as well and no passivating oxide coating that can plague some alkali–metal reactions.

The most common oxidation states of mercury are +1 and +2. The divalent compounds are, like their Zn and Cd counterparts, numerous and of varied types. Mercury(II) halides, HgX_2, and pseudohalides, such as $Hg(NO_3)_2$, are common starting materials, and many are commercially available.

As a safety note, the methyl derivatives of the heavier Group 12 elements can be prepared, as they can for zinc. The compounds $CdMe_2$ and $HgMe_2$ are also volatile liquids like the zinc compound. Dimethylcadmium does fume and sometimes flame in air, but is not as pyrophoric as its lighter congener. $CdMe_2$ is extremely toxic, affecting the liver, kidneys, brain, and nervous system at concentrations of a few milligrams per cubic meter in air. If $CdMe_2$ is "extremely toxic", there are insufficient appropriate adjectives for $HgMe_2$, which is even more dangerous, and very experienced researchers have lost their lives to milligrams of the compound that were absorbed through the skin or lungs. Finally, $HgMe_2$ has been shown by OSHA to permeate many common glove types such as latex, PVC, and neoprene "almost instantly".[38] Butyl gloves, commonly used in glove boxes, were also found to provide inadequate protection.

Mercury(I) compounds, readily available as halides and pseudohalides, are dimeric with a mercury–mercury bond. For example, mercury(I) nitrate is actually $Hg_2(NO_3)_2$ with a stable $\{Hg–Hg\}^{2+}$ unit. Catenation of mercury is not limited to one Hg–Hg bond, and compounds linking several Hg

Table 9.2 Gas phase X–M–X angles (°) in Group 2 compounds.[33–35] Cp* is pentamethylcyclopentadienyl, and the angle is measured from the centroids of the rings. Numbers in parentheses are from the solid phase found by X-ray diffraction.[36]

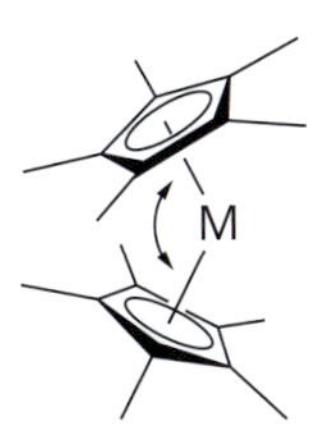

	MF_2 angle	MI_2	MCp^*_2 angle
Mg	180	180	180
Ca	180	180	154 (131)
Sr	144	180	149
Ba	126	157	148 (148)

* Some higher-percentage sodium amalgams become liquid again at room temperature, prepared by melting the lower-percentage solid amalgams and adding more sodium. However, these high-sodium amalgams are pyrophoric and are less commonly employed.

atoms together, such as $Hg_4(AsF_6)_2$, are known, which contain near-linear mercury chains.

Furthermore, M_2^{2+} cations of some of the other Group 2 and Group 12 metals have been prepared, with examples for all of Group 12 and prominent examples for Mg in Group 2 (Fig. 9.16a).[39–44] The solvated dizinc complex $\{Zn_2(THF)_6\}^{2+}$ also has been isolated.[45]

A mixed Group 12 system with unusual Zn–Hg–Zn catenation (Fig. 9.16b) has been reported by Mountford and coworkers.[46]

Although the complex is more appropriately discussed as being of a transition metal rather than a main-group element, mercury(IV) fluoride has been prepared through reaction of mercury with fluorine and trapping in a solid neon matrix.[47] The HgF_4 is a low-spin d^8 square planar complex (Fig. 9.17) similar to valence isoelectronic $\{PtCl_4\}^{2-}$.

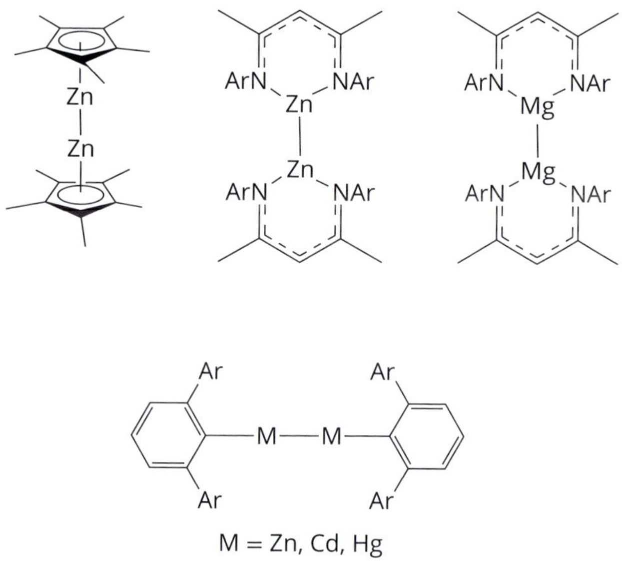

Fig. 9.16a Some examples of $\{M–M\}^{2+}$ compounds for Group 2 and Group 12.

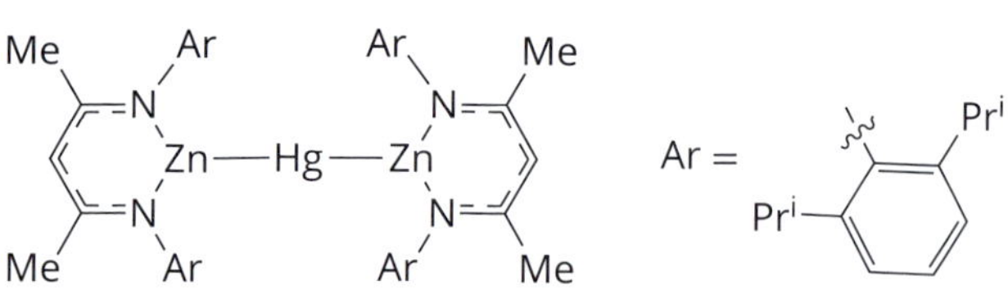

Fig. 9.16b A mixed Group 12 catenated system with zinc and mercury.

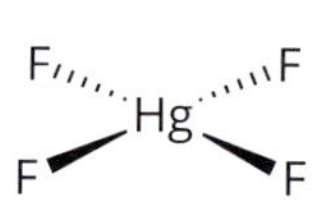

Fig. 9.17 The structure of square planar HgF_4, a low-spin d^8 complex isoelectronic with $\{PtCl_4\}^{2-}$.[47]

9.4 Group 13 Chemistry[48]

Features of Group 13 Chemistry

- Group 13 compounds tend to form hypovalent compounds when monomeric, but often accept Lewis bases or dimerize to occupy all their orbitals.
- Boranes form a vast array of catenated structures. These may be closed or open cages.
- The heavier Group 13 compounds often exhibit two different oxidation states, +1 and +3. The +1 state, such as in thallium(I), may be described as having an "inert pair" of electrons.
- The tendency to have the +1 oxidation state goes as Th >> In > Ga >> Al and B for compounds not having E–E bonds. Al and B compounds in the +1 formal oxidation state are often catenated into clusters of some type and do not have lone pairs on the Group 13 element.
- The tendency for the heavier Group 13 elements to form compounds, such as Group 13 halides, with "inert pairs" of electrons is due to bond strengths of the E–X compounds relative to the X–X (dihalide) bond strengths.

All of the *s*- and *p*-block groups up to and including Group 13 flaunt many electron-deficient compounds with bridging ligands, dative interactions, and other characteristics identified with unfilled valence orbitals, but perhaps no column of elements is identified with electron deficiency as much as Group 13.

The compounds of monomeric formula, MX_3, are often strong Lewis acids. Boron tends to accept one Lewis base to form 4-coordinate compounds, many of which are commercially available, such as $BH_3(THF)$ and $BH_3(SMe_2)$. The heavier congeners, such as $AlCl_3$, can accept more than one base, such as $AlCl_3(THF)_2$, which has a trigonal bipyramidal structure with axial tetrahydrofuran donors.[49] This geometry of axial oxygen donors for $AlCl_3(THF)_2$ is consistent with VSEPR due to oxygen's higher electronegativity; however, VSEPR arguments would be incorrect for $AlCl_3(NHMe_2)_2$, where less electronegative nitrogen is also axial (Fig. 9.18).[50] More accurately, one could easily anticipate that the less donating (relative to Cl^-) dative ligands would be more likely to use the pure *p*-orbital involved in the 3-center 4-electron bonds along the *z*-axis by hybridization theory.

In the absence of an external Lewis base, the electron-deficient monomers often find methods of filling the remaining empty *p*-orbital. In the case of the halogens, they often form dimers or higher oligomers with bridging halogens. The halogen is then σ-bound to one of the metal centers and datively bound to the other (Fig. 9.19), such as $AlCl_3$. Naturally, the dative and covalent bonds are also in resonance across the symmetric Al_2Cl_2 ring (see Section 8.3.3).

If the pendant ligands do not have lone pairs, the electron-deficient monomers often participate in τ-bonded dimers, as in diborane or the $AlMe_3$ dimer (Section 8.3.3).

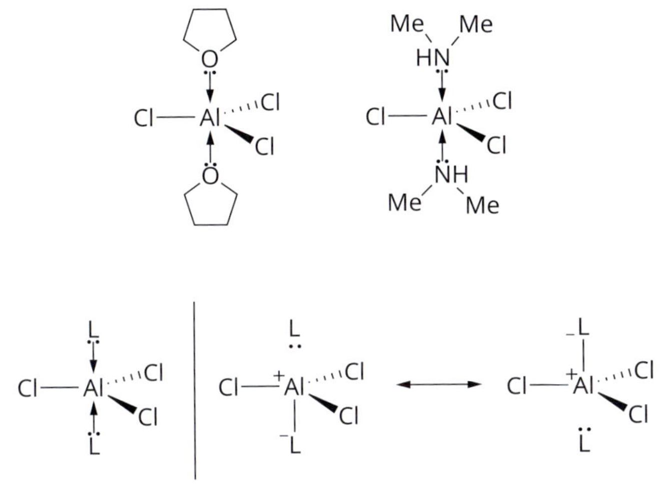

Fig. 9.18 (*top*) Structures of $AlCl_3(THF)_2$ and $AlCl_3(NHMe_2)_2$, one consistent with VSEPR and the other not, but both consistent with expectations from hybridization theory. (*bottom*) General structure of $AlCl_3$ with two dative donors (L) on the left, with some contributing resonance forms on the right. The $AlCl_3$ bonds can be made by sp^2-hybrids 120° apart. The dative ligands would then be attached to aluminum using the p_z-orbital through a 3-center 4-electron bond, the 4-electrons coming from the lone pairs on L.

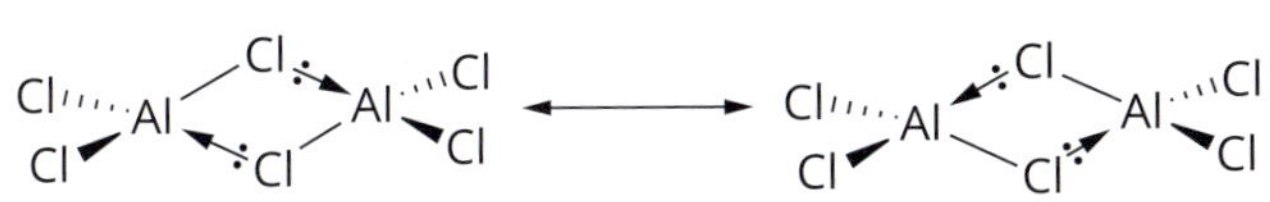

Fig. 9.19 The structure of the $AlCl_3$ dimer. The Al–Cl bridges have resonance between dative and covalent bonds (not τ-bonds).

Boron-containing compounds (Fig. 9.20) with substituents bearing lone pairs of π-symmetry, such as halides, often have monomeric structures such as BF_3, BCl_3, and pinacolborane, HB(Pin). Perhaps these structures are monomeric due to π-donation from the lone pairs of the pendant groups. (See Section 6.6.1 for a complete MO discussion of the bonding in BF_3. The B–F π-bonding orbital is labeled as "$1a_2$".)

There has been extensive disagreement about the Lewis acidity of the boron halides, with different BX_3 compounds being considered stronger acids for various reasons, including orbital overlap, charges on boron, reorganization energy from planar to trigonal pyramidal, and LUMO energy. However, one must be cautious when generalizing about Lewis acidity. The heavier halides on boron are generally thought to provide stronger Lewis acids; for example, BBr_3 is thought to be a stronger Lewis acid than BF_3. However, this is dependent on the Lewis base as well. It is true that BBr_3 is the stronger Lewis acid toward NH_3, but BF_3 is the stronger Lewis acid towards CO. Different effects are no doubt more important for some Lewis acid–base combinations than others, often thwarting efforts at broad generalizations.[28,29]

The halides of boron are all monomeric, and their phase properties at room temperature mimic the properties of their respective molecular halides (X_2): BF_3 (gas, m.p. = −127, b.p. = −100 °C), BCl_3 (gas, m.p. = −107, b.p. = 13 °C), BBr_3 (liquid, m.p. = −46, b.p. = 91 °C), and BI_3 (solid, m.p. = 50, b.p. = 210 °C).[5] The bond in BF_3 is one of, if not, the strongest single bond known (Section 3.6.1) at 154 kcal/mol; no doubt this high bond energy is a combination of high polarity in the bond (~34% by polar covalence) and the high covalent, unweakened bond energies in fluorine (E‴ = 113 kcal/mol) and boron (E‴ = 77 kcal/mol). One can argue, however, that the bond in BF_3 is not truly single with some B–F π-bonding being a necessity to alleviate lone pair bond weakening and reach this bond strength value.

9.4.1 Boranes

The chemistry of boron hydrides is vast, aided by the element's propensity to exhibit multiple bond types and an ability to catenate in unequalled variety. The chemistry of boron hydrides, called "boranes" (compare alkanes), has yielded an exotic menagerie of compounds that requires their own nomenclature, which will only be discussed briefly here. These boron-based polyhedral clusters can also be substituted with carbon, forming "carboranes" in multitudes as well.

Boranes are classified into four general categories based on their formula and structure (Fig. 9.21):

(1) *closo* (from Greek *clovo*, cage). These have the formula $B_nH_n^{2-}$, with n = 6–12 being known. The *closo*-boranes have all trigonal faces around the polyhedral structure. No stable neutral derivatives are known.

(2) *nido* (from Latin *nidus*, nest). *nido*-boranes have a more open structure with one vertex of the parent polyhedron from the related *closo* form left vacant and have the neutral formula B_nH_{n+4}.

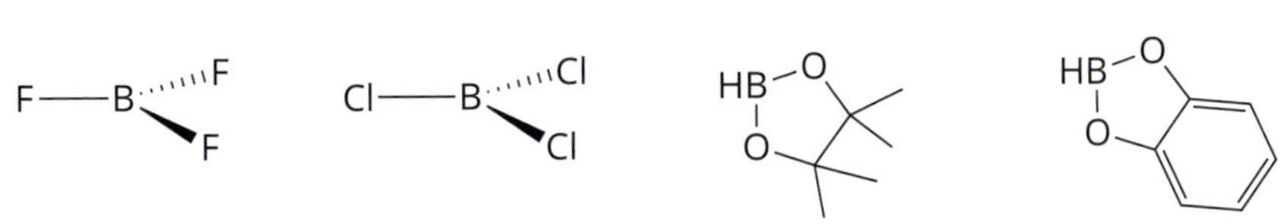

pinacolborane (HBPin) catecholborane (HBCat)

Fig. 9.20 Structures of some common boron monomers: BF_3, BCl_3, pinacolborane "HB(Pin),", and catecholborane "HBCat".

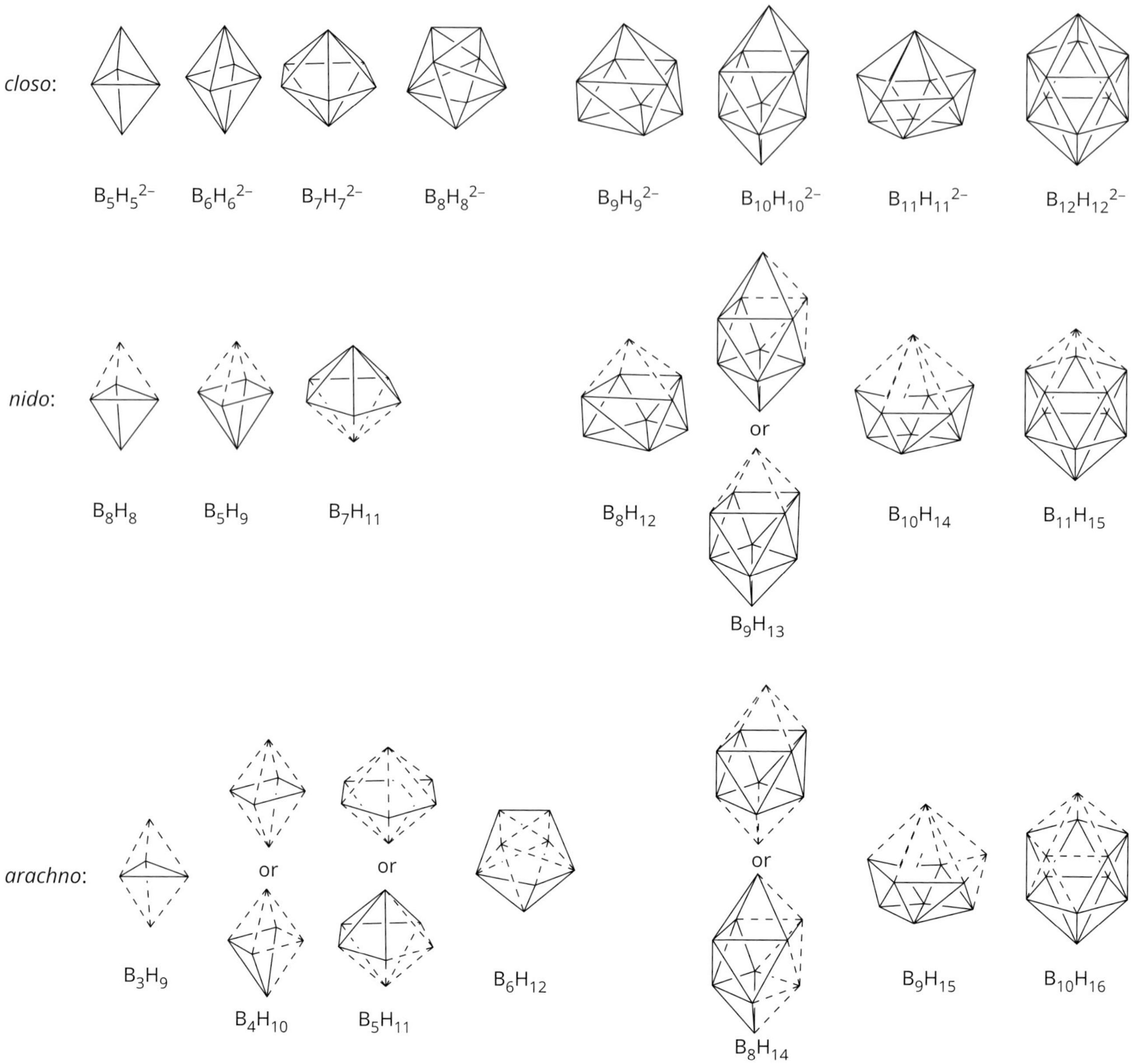

Fig. 9.21 Structures of some *closo-*, *nido-*, and *arachno*-borane structures, illustrating the loss of vertices to generate the more open structures. For the *closo*-boranes, the formulas for the known dianionic compounds are listed. For the *nido-* and *arachno*-boranes, the neutral formulas are given. (Adapted from O'Neill and Wade's chapter on metalloboranes.[51])

(3) *arachno* (from Greek *arachne*, spider's web). These are even more open structures of the neutral formula B_nH_{n+6} with two missing vertices from the *closo* structure.

(4) *hypho* (from Greek *hyphe*, net). A few members of this relatively rare class of boranes are known with the formula B_nH_{n+8}, which have three missing vertices from the *closo* structures. Considering the rarity of the class, *hypho*-boranes will not be discussed further here.

In Fig. 9.21 are shown the *closo*-structures based on the formula $B_nH_n^{2-}$, along with their *nido-* and *arachno*-clusters derived from the same polyhedron.

The bonding in such compounds is analyzed using the topological method of William Lipscomb, whose work on the bonding in boranes garnered the Nobel Prize in 1976. The bonding in the clusters is analyzed based on the number of B–H–B 3-center 2-electron bonds (τ-bonds), B–B–B τ-bonds, B–B 2-center 2-electron (2c2e) bonds, and BH_2 groups. The analysis provides a series of numbers called the *styx*-number for the compound, which is a shorthand for the structure or a small group of structures.

The *styx*-number has each letter defined as:

$$s = \text{B–H–B } \tau\text{-bonds}$$
$$t = \text{B–B–B } \tau\text{-bonds}$$
$$y = \text{B–B 2c2e-bonds}$$
$$x = \text{BH}_2 \text{ units in structure}$$

For example, the *styx* number for diborane = 2 0 0 2, and for B_5H_{11} (which could also be named as *arachno*-pentaborane(11)) the *styx* number = 3 2 0 3. The structures for these compounds are shown in Fig. 9.22.

If we write the formula for a borane as B_nH_{n+m} and assume that each boron atom will have at least one terminal B–H bond, then there are three orbitals and two electrons remaining for every boron atom. Since every skeletal hydrogen not involved exclusively in a B–H unit will contribute one electron, the number of electrons available for the bonding in the polyhedron is $2n + m$ and there are $(2n + m)/2$ skeletal bonds possible; consequently, $(2n + m)/2 = s + t + y + x$.

Since every boron atom has one more orbital than electrons and all the orbitals will participate, every boron will participate in at least one 3-center 2-electron bond. In other words, $t + s = n$. Every hydrogen over and above the n terminal hydrogens (m) must either be part of a BH_2 group (x) or a B–H–B τ-bond (s), $m = s + x$.

As a result of the this analysis, one can define three equations (Eqs. 9.8–9.10) relating the formula subscripts n and m and the s, t, y, and x values:

$$s = n - t \qquad \textbf{9.8}$$
$$x = m - s \qquad \textbf{9.9}$$
$$2y = s - x \qquad \textbf{9.10}$$

Since we have four unknowns (s, t, y, and x) and only three equations (Eqs. 9.8–9.10), unique solutions are not expected; however, one can generally deduce the possible structures for a given formula.

As an example, let us examine the formula B_4H_{10}, where $n = 4$ and $m = 6$. For a formula of $n + 6$, one expects an *arachno*-structure, and there are two possibilities in the polyhedral drawing in Fig. 9.21. If we consider the structure with four borons in approximately one plane, we can begin to calculate the possible *styx* numbers. Since $(2n + m)/2 = s + t + y + x$, our allowable values will have $s + t + y + x = 7$.

One possible solution is *styx* = 4 0 1 2, which would correspond to the structure in Fig. 9.23a.

A different solution (Fig. 9.23b), still with the borons in more or less the same plane, would be *styx* = 3 1 0 3.

The more symmetrical 4 0 1 2 structure in Fig. 9.23a is adopted by *arachno*-B_4H_{10}, which has been studied in great detail experimentally.[52–53]

Isoelectronic substitution into borane clusters can be achieved with a variety of different groups. Most notably, carbon can be substituted into boron clusters to generate

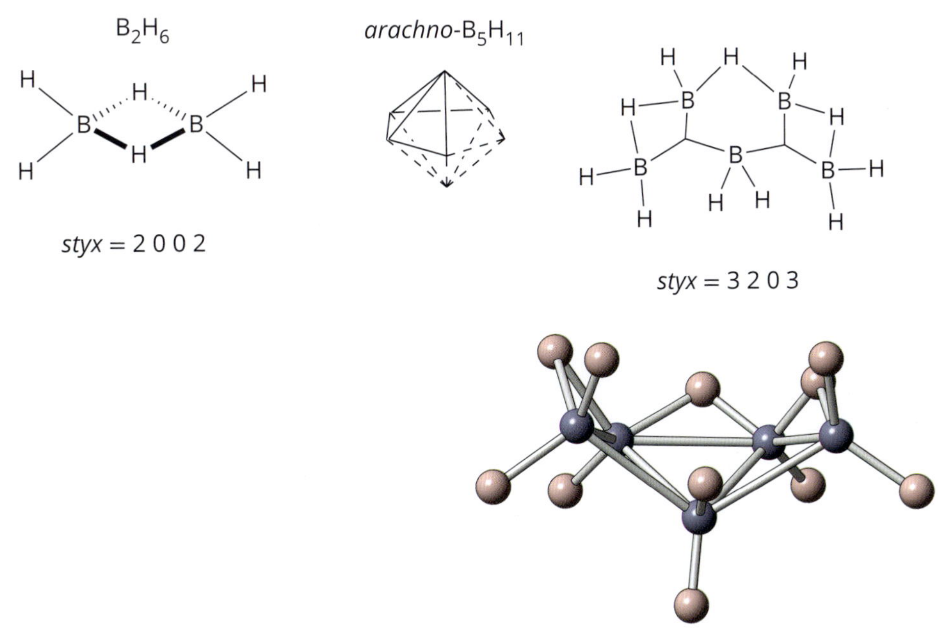

Fig. 9.22 (*left*) Structure and *styx* number for diborane. (*right*) Polyhedral structure, line structure, *styx* number, and rendering of the X-ray diffraction structure[52] of *arachno*-B_5H_{11}.

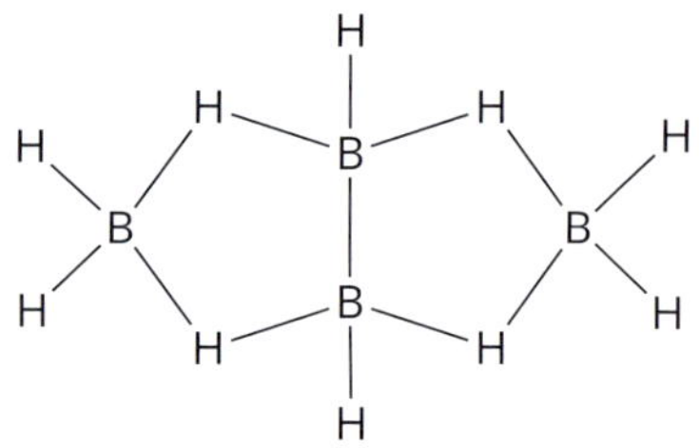

Fig. 9.23a One structural solution to *styx* = 4 0 1 2.

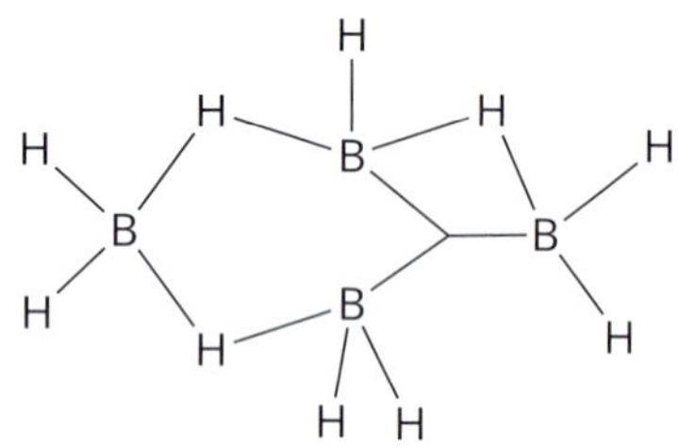

Fig. 9.23b One structural solution to *styx* = 3 1 0 3.

carboranes (Fig. 9.21). Since carbon has one more electron than boron, the isoelectronic structure with one CH substituted for one BH will have one less anionic charge (or one more positive charge). For example, dianionic *closo* structures can be made in a neutral system if two BH units are substituted for CH. The synthetic route to one such compound is shown in Fig. 9.24. Reaction of $B_{10}H_{14}$ with SEt_2 gives $B_{10}H_{12}(SEt_2)_2$, which reacts with acetylene to produce the *ortho-closo*-$C_2B_{10}H_{12}$, a neutral cage isoelectronic with $B_{12}H_{12}^{2-}$. Heating *ortho*-$C_2B_{10}H_{12}$ causes isomerization—first to *meta*- and then *para*-$C_2B_{10}H_{12}$. The driving force for the isomerization is probably charge distribution and moving the more electronegative carbons away from each other.[54]

Substitution with groups other than carbon is quite possible, and appropriate metal centers can be inserted into boron clusters as well. Structures where one vertex is missing from the polyhedron (*nido*-structures) can be used as ligands for transition metals. In some cases, *closo*-structures are quite unreactive anions, which gives them great utility. Monocarba-*closo*-dodecaborate anion, $\{HCB_{11}H_{11}\}^-$, has many known derivatives, such as $\{HCB_{11}Cl_{11}\}^-$ in Fig. 9.24,[38,39,41,53,55–56] which are often employed as weakly coordinating anions.[57–58]

Other isoelectronic replacements have been explored. Computationally, most replacements in the $B_{12}H_{12}^{2-}$-system are destabilizing. Some energies relative to $B_{12}H_{12}^{2-}$ for different replacements are shown as follows. All substitutions,

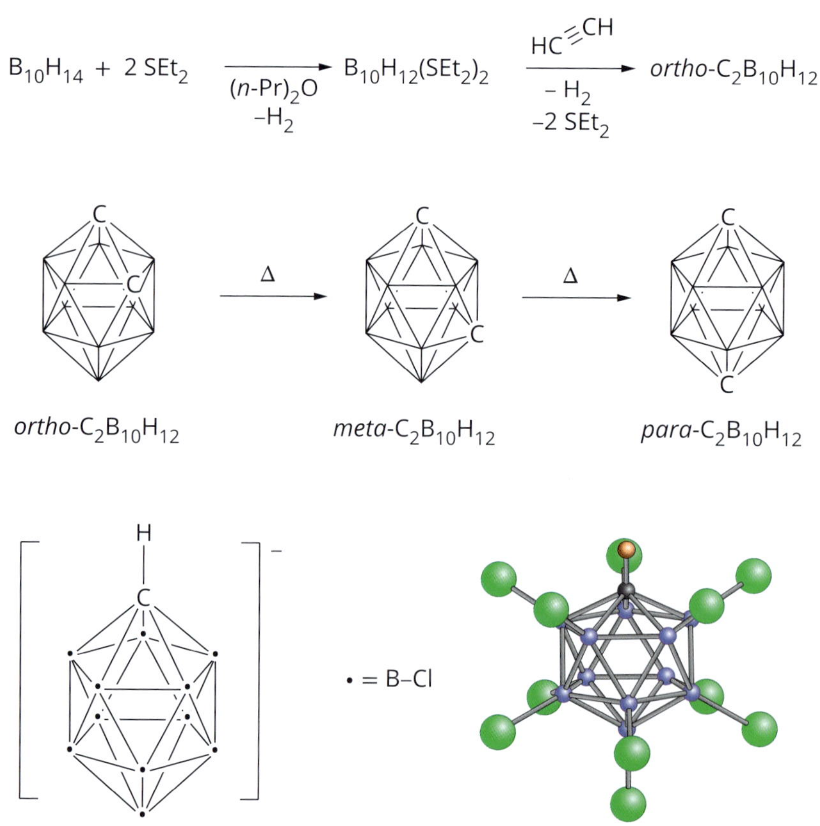

Fig. 9.24 (*top*) Synthesis and isomerization of *ortho*-$C_2B_{10}H_{12}$. Each atom on the polyhedron has a terminal hydrogen atom, and the vertices are BH units. (*bottom left*) Line structure of the $\{HCB_{11}Cl_{11}\}^-$ carborane; the vertices with dots (•) are BCl-units. (*bottom right*) Ball and Stick drawing from the X-ray diffraction study.[55] The cation is not shown.

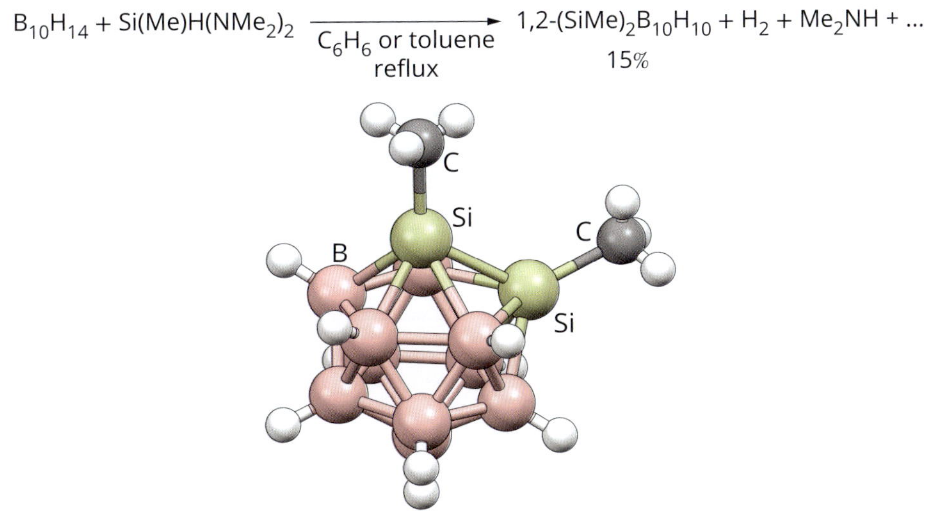

Fig. 9.25 Synthesis and structure of 1,2-$(SiMe)_2B_{10}H_{10}$.

including CH, are positive energies (destabilizing), but X = P and CH have lower relative energies.[59]

$$B_{12}H_{12}^{2-} + 2\,XH_3 \rightarrow para\text{-}X_2B_{10}H_{10} + 2\,BH_4^-$$

X =	CH	SiH	P	N
ΔE(kcal/mol)	56.7	64.6	34.9	94.1

Dietmar Seyferth and coworkers, for example, reported a silicon-based derivative, 1,2-$(SiMe)_2B_{10}H_{10}$ with a pseudo-icosahedral structure (Fig. 9.25). The compound was prepared by reaction of $B_{10}H_{14}$ with $MeSi(H)(NMe_2)_2$.[60–61] The phosphorus compound $P_2B_{10}H_{10}$ is also known.[62]

$$B_{10}H_{14} + Si(Me)H(NMe_2)_2 \xrightarrow[\text{reflux}]{C_6H_6\text{ or toluene}} 1,2\text{-}(SiMe)_2B_{10}H_{10} + H_2 + Me_2NH + \ldots \quad (15\%)$$

For more details on this expansive class of compounds and for a discussion of their 3-dimensional aromaticity, see the many reviews and books on the topic.[63–70] There are numerous applications for carboranes being pursued, including cancer therapies, batteries, heat-resistant polymers, and electronic devices.[71–72]

9.4.2 M(I)/M(III) Monomeric Compounds and the Inert Pair Effect[73–74]

In Group 13 there are two common oxidation states found for monomeric compounds separated by two electrons, E(I) and E(III). The lower oxidation state becomes far more common as one proceeds down the column; consequently, the heavier elements are more likely to have a lone pair. This tendency to adopt the lower oxidation state and have a lone pair of electrons was dubbed by Nevil Sidgwick the "inert pair effect", which is due, at its core, to a variety of effects already discussed related to bond energies (*vide infra*). The term "inert pair effect" should be taken more as a descriptive term rather than as the cause of the lower oxidation state's stability for the heavier members of the group.

As an illustration, Fig. 9.26 shows the calculated energies for the reaction $EX_3 \rightarrow EX + X_2$, where E = B, Al, Ga, In, and Tl with X = H and F. In general, as the electronegativity of X decreases the reaction becomes more endoergic; that is, stability of EX_3 is highest for F > Cl > Br > I > H. All of the Group 13 element trifluorides, EF_3, are thermodynamically stable, while not all of the E(III) iodides and hydrides are isolable.

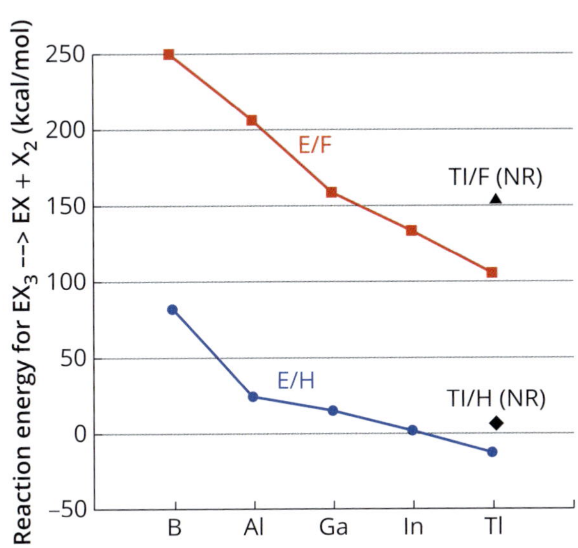

Fig. 9.26 Plot of $EX_3 \rightarrow EX + X_2$ reaction energy versus E, where E is the Group 13 element and X = F and H. NR = Result of nonrelativistic calculation for thallium.

Table 9.3 E–X bond energies in EX_3 for the Group 13 halides.

X_2 E-X bond energies in kcal/mol
Experimental (theoretical: polar covalence)

	B	Al	Ga	In	Tl
F	155.7(158.2)	141.5(144.9)	(101.8)	95.0	(85.5)
Cl	107.4(104.3)	101.8(101.7)	(78.7)	73.8(72.1)	(55.9)
Br	89.3(90.7)	85.7(86.1)	(66.0)	64.2(63.0)	(52.9)
I	66.2(67.0)	68.1(70.0)	(50.9)	50.2(48.8)	(41.7)

Sanderson, R. T. *Polar Covalence.* New York: Academic Press, 1983.

Table 9.4 Thallium halide bond strengths for the +1 and +3 oxidation states.[74]

	Tl(I) BDE (bond length)	Average Tl(III) BDE (bond length)
F	128 (2.061)	106 (2.002)
Cl	95 (2.524)	85 (2.357)
Br	96 (2.267)	73 (2.468)
I	84 (2.871)	64 (2.744)

Schwerdtfeger, P., Heath, G. A., Dolg, M., and Bennett, M. A. *J. Am. Chem. Soc.* 1992, *114*, 7518.

If, for simplicity of discussion, we look only at the halides by using polar covalence theory as a guide, the EX_3 bond energies decrease substantially from B to Tl (Table 9.3). Furthermore, the E–X bond energies decrease substantially, ~50%, as one proceeds from fluoride to iodide.

If one partitions the energy in the bonds of the Group 13 halides by using polar covalence theory into covalent and ionic contributors, the plots in Fig. 9.27 are obtained. Both the ionic and covalent bond energies are dropping for the EF_3 and EI_3 compounds down the series. Even though the ionic character of the bonding is somewhat variable down the row due to changes in electronegativity, the result is increasing stability of the lower oxidation state as one proceeds down the column.

In all cases the lower oxidation state, M(I), has a stronger bond to the halide. As discussed in Chapter 3, this is due to the lower electronegativity of thallium(I), for example, and thus a higher polarity to the bonds. Furthermore, the lower oxidation state has stronger bonds despite having longer M–X distances due to the polarity increase relative to the higher oxidation state (Table 9.4).

As the M(III) bond strengths decrease down the column, the M(I) compounds, which have stronger bonds, become more common; this greater stability of the lower oxidation state down the column has been given the name inert pair effect. Again, inert pair effect is the label given to the phenomenon, but it occurs because of varying bond strengths down the column.

The Group 13 EX_3 halides are all stable for X = F, Cl, and Br. For the iodides, BI_3, AlI_3 and GaI_3 are all apparently quite stable. The indium derivative, InI_3, appears to be stable at low T. There is some evidence that InI_3 is In(I) triiodide (In^+ I_3^-) at high temperatures in the gas phase. Thallium triiodide exists (Fig. 9.28) as the monovalent thallium compound Tl^+ I_3^-.

Considering this where thallium(I) is more stable, it is perhaps not surprising that thallium(III) compounds are commonly employed as oxidants. Aluminum(III) is by no means considered an oxidant and is quite difficult to reduce. While

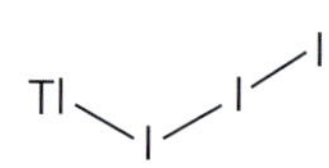

Fig. 9.28 Structure of TlI_3, as Tl^+ I_3^-.

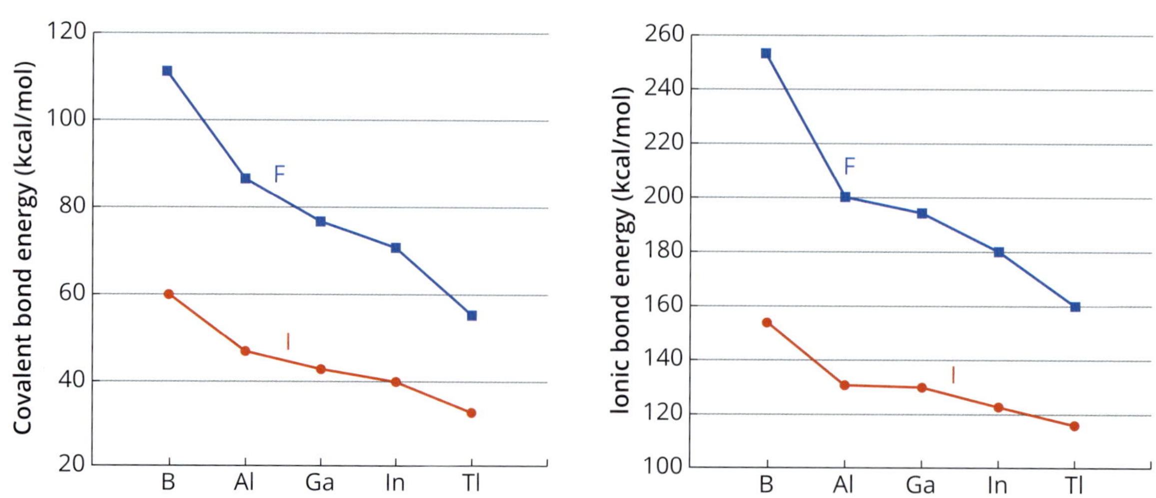

Fig. 9.27 Plots of covalent (*left*) and ionic (*right*) bond energies for EX_3 compounds, where E = Group 13 elements and X = F and I. The values were calculated using polar covalence theory.

thallium(I) compounds are quite common in the laboratory, monomeric aluminum(I) is little known outside of unstable gaseous species. For example, reaction of TlOH with cyclopentadiene produces the C_{5v} compound TlCp. As shown in Eq. 9.11, all five carbons in the Cp ligand bind equally to the metal, an η^5-configuration.

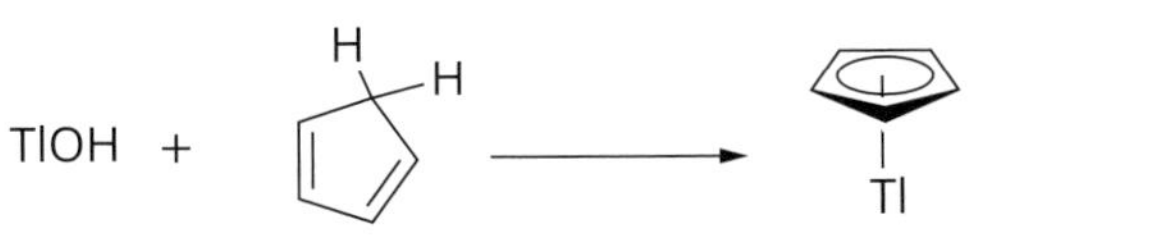

9.11

Many thallium(I) compounds are quite useful as reagents for the metathesis of ligands often to produce thallium(I) halides, which are insoluble in most organic solvents. Thallium(I) is especially useful in cases where other salts tend to reduce or oxidize sensitive metal halide starting materials. The downside of these reagents is their high toxicity and suspected carcinogenicity.[75] Interestingly, Prussian blue—a mixed valent Fe(II)/Fe(III) cyanide complex often used as a dye—has been used as a treatment for thallium poisoning.

Metastable Al(I) chloride solutions can be prepared by reaction of solid aluminum with gaseous HCl at 927 °C to produce AlCl and H_2. The AlCl gas is then cocondensed with toluene and diethyl ether at 77 K to produce a dark red solution of AlCl.[76] In a reaction accelerated by light, heat, and increasing concentration, the aluminum(I) halide valence disproportionates to $AlCl_3$ and aluminum metal, Eq. 9.12. Valence disproportionation is the act of a compound of a specific oxidation state reacting with itself to form species of both higher and lower oxidation states; in this case, Al(I) becomes Al(III) and Al(0). In the converse, valence comproportionation is a compound of one oxidation state reacting with a compound in a different oxidation state to produce two equivalents of a single compound of intermediate oxidation state.

$$3\,AlCl \cdot xOEt_2 \xrightarrow[Et_2O/toluene]{} AlCl_3 \cdot xOEt_2 + 2\,Al \qquad \textbf{9.12}$$

The aluminum(I) chloride solution does in some cases persist long enough at low temperature to undergo reaction to produce species more stable to valence disproportionation. These are generally still quite reactive metal clusters, such as $(AlCp^*)_4$, where Cp* = pentamethylcyclopentadienyl (Fig. 9.29), which is prepared from AlCl·xOEt_2 and $MgCp^*_2$ or, more simply, by

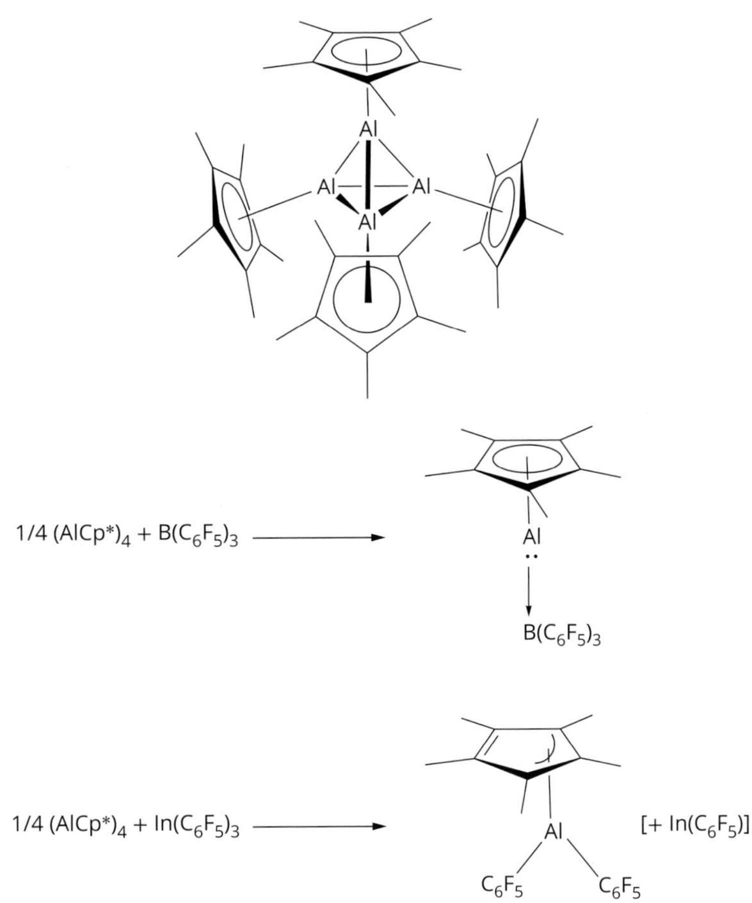

Fig. 9.29 (*top*) Structure of $(AlCp^*)_4$. (*middle*) Synthesis and structure of a borane adduct of AlCp*, $(F_5C_6)_3B{\cdot}AlCp^*$. (*bottom*) Oxidation of $(AlCp^*)_4$ by $In(C_6F_5)_3$ to give $Al(\eta^3\text{-}Cp^*)(C_6F_5)_2$ and (perhaps) $In^I(C_6F_5)$.[80]

reduction of Cp^*AlCl_2 by potassium metal.[77–79] Compare this cluster compound of an aluminum(I) cyclopentadienyl derivative with the monomeric and quite stable TlCp noted previously, where the thallium compound is stabilized by the inert pair effect.

Structurally, isolable aluminum(I) species[78] often exist as clusters with Al–Al bonds; compare boranes. One exception to this is the monomeric adduct of AlCp* with the Lewis acid $B(C_6F_5)_3$ (Fig. 9.29), which forms a stable, monomeric Lewis acid–base adduct.[80] In contrast, reaction of $In(C_6F_5)_3$, where the lower oxidation state In(I) compound is more stable, with $(AlCp^*)_4$ produces $Al(\eta^3\text{-}Cp^*)(C_6F_5)_2$ and presumably $In(C_6F_5)$, which was not isolated (Fig. 9.29).

9.5 Chemistry of the Tetrels (Group 14)

Features of Group 14 Chemistry

- Often found in the +2 and +4 formal oxidation state. Divalent carbon, CH_2 (carbene), can be stabilized with π-donors, as can silenes. The lower oxidation state becomes more stable down the column.
- Stable carbenes with a lone pair on carbon can be prepared with strong amine π-donors on the carbon. Most often these are heterocycles called "*N*-heterocyclic carbenes", or NHCs.
- Pb in its highest oxidation state is oxidizing, such as $Pb(OAc)_4$.
- The compounds in this group in the +2 oxidation state may exhibit "stereochemically active lone pairs" such as $Pb(SPh)_3^-$, which has a pyramidal structure. However, this is a misnomer, as the lone pair is often in a nearly pure *s*-orbital, and the structure is due to almost pure *p*-orbitals used in the bonding.

There are two common formal oxidation states for isolable compounds of Group 14, +2 and +4. Just as they did in Group 13, the monomeric compounds of the lower oxidation state tend to bear a lone pair of electrons. Also, just as for the previous group and for the same reasons, the heavier elements tend to be more stable in the lower oxidation state than the lighter congeners. Most often, CX_2 compounds are unstable towards dimerization and other processes, making them not isolable. However, the chemistry of divalent carbon (carbenes) has seen expansive growth since the late twentieth century, due in large part to interest in these compounds as ligands for transition-metal complexes.

Considering that the chemistry of the tetravalent elements in this group, especially of carbon and to some extent silicon, receive wide-ranging attention in organic courses, little will be said of their chemistry here.

The carbon(II) parent hydrocarbon CH_2, carbene, is a ground state triplet due to the energetic ease of hybridization in the first row of the *p*-block. If one adds π-donors to the divalent carbon, such as OMe, the singlet state becomes energetically preferred (Fig. 9.30) due to π-occupation of a *p*-orbital on carbon from donation of an oxygen lone pair.

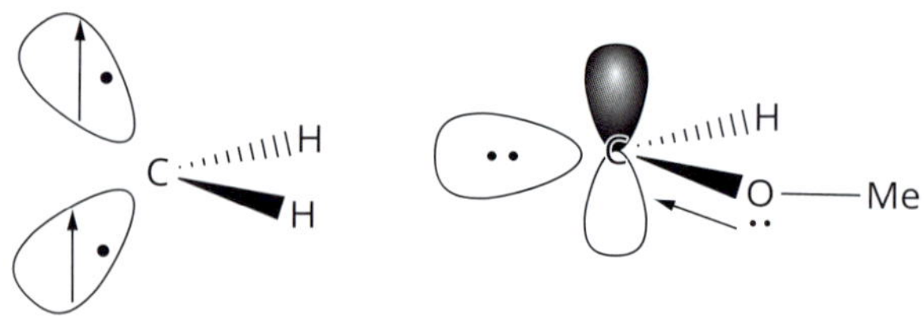

Fig. 9.30 Structure of triplet ground state CH_2 (*left*) and singlet ground state C(H)(OMe) (*right*).

Some common derivatives of singlet carbenes are prepared from imidizolium salts, or similar heterocycles, by addition of base, as shown in Eq. 9.13. There is a multitude of carbenes related to that shown in the equation, with various ring sizes, different R groups (aryl and alkyl), different degrees of saturation in the ring, and so on. As a group, these are often referred to as *N*-heterocyclic carbenes, or NHCs. This general type of carbene and their metal complexes have been studied extensively since initial separate reports by Öfele and Wanzlick in 1968, with key contributions by Lappert, Bertrand, and Arduengo.[81–85]

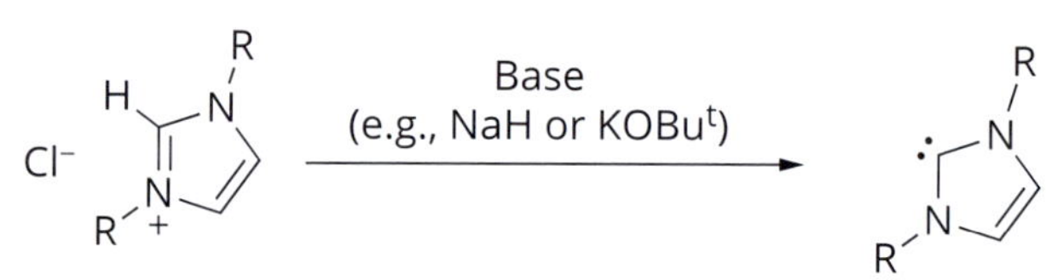

9.13

Very electron-rich olefin tetramines can in some cases undergo rupture of the carbon–carbon double bond to form NHCs. These can be trapped by reactions with a variety of reagents, such as S_8, to form a thiourea, Eq. 9.14.[86]

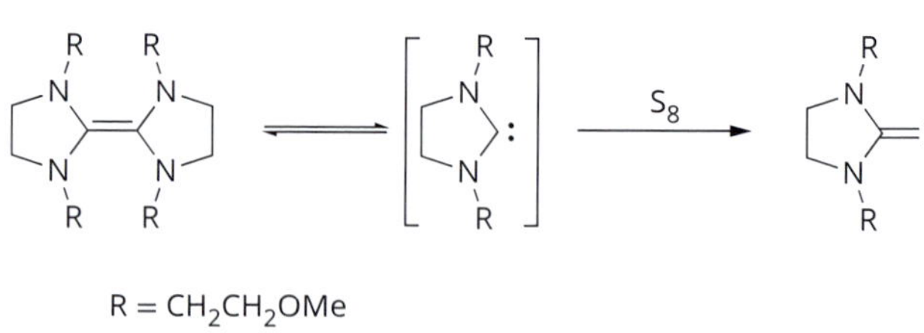

9.14

Natural resonance theory on a model NHC where R = H produced the results shown in Fig. 9.31. The main contributors by far are due to strong π-interactions between the carbene carbon and the adjacent nitrogens. The potential Lewis form shown at the right side of Eq. 9.13, with only two bonds to the carbene carbon, was not found as a resonance contributor in the calculation, even though this is perhaps the most common line-drawing of such compounds. Examination of the highest occupied molecular orbital, not surprisingly shows strong involvement of the carbene carbon lone pair (Fig. 9.31).

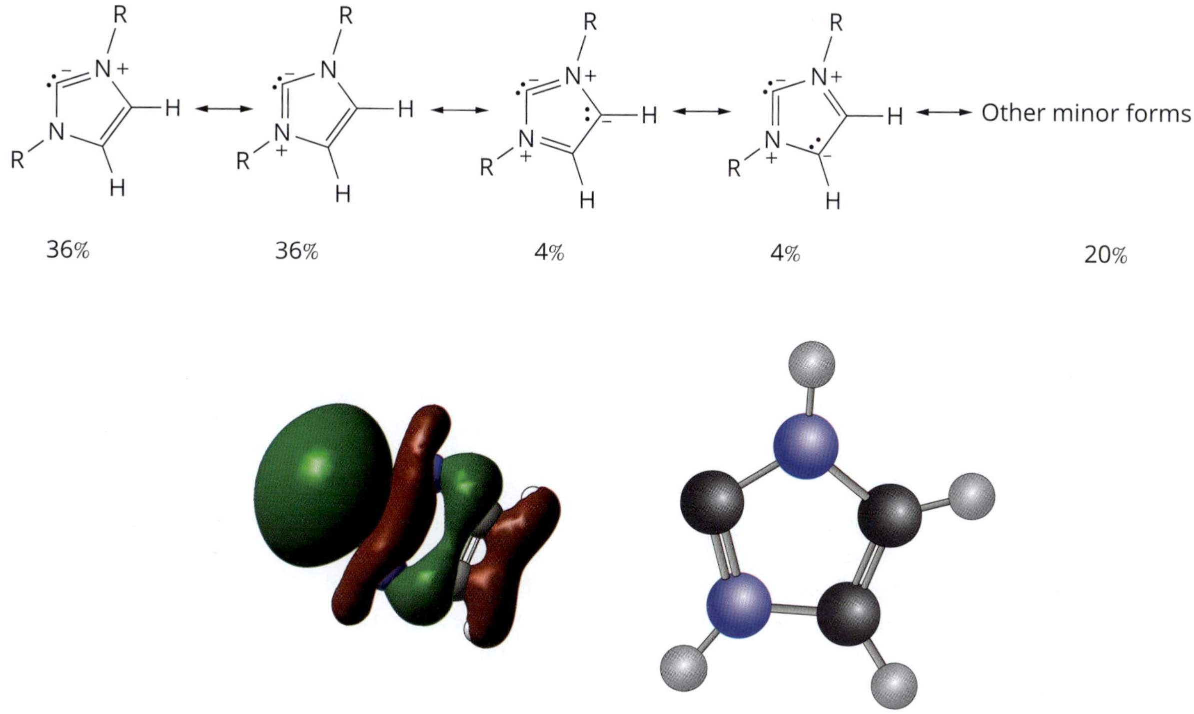

Fig. 9.31 (*top*) Results of the natural resonance theory calculation (NBO5, cc-pVDZ, B3LYP) on a model NHC where R = H. The four highest resonance contributors are shown. (*bottom*) The highest occupied orbital (*left*) and the ball and stick structure in the same orientation (*right*). The carbon-based lone pair is a strong contributor in the HOMO.

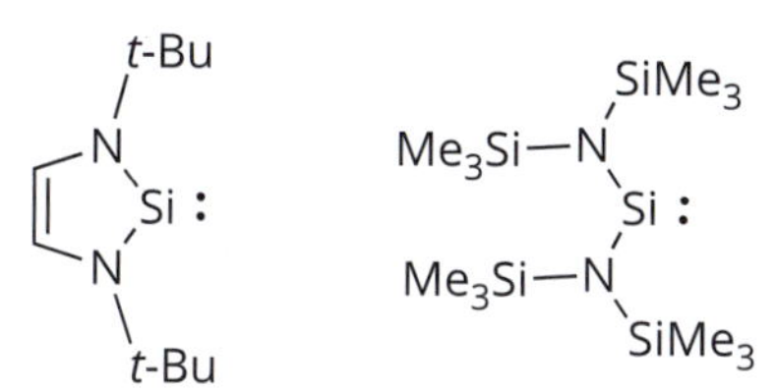

Fig. 9.32 Examples of stable silylenes.

Simple silylenes are generally unstable above 77 K, but through control of ligand structure and careful synthesis, many have been prepared. Many of the stable silylenes bear adjacent donor nitrogens and large steric protection near the Si(II) atom (Fig. 9.32).[79]

In some cases, Group 14 divalent compounds may dimerize reversibly to form alkene-like species. One such case was observed with a silylene bearing very bulky aryl groups.[87] Upon low-temperature reduction of Br_2Si(mesityl)(Tb) (see Fig. 9.33) with lithium naphalinide, two *cis/trans* isomers of disilenes are prepared. On warming, these isomers exchange. The monomeric silylene was also trapped through its reaction with numerous other reagents.

Compounds of germanium(II) are often isolable but still quite reducing. A common starting material is $GeCl_2$(dioxane), which is available from commercial sources. In the solid state, $GeCl_2$(dioxane) exists as linear chains with bridging 1,4-dioxane molecules ((Fig. 9.34, and compare $MgCl_2$•dioxane in Fig. 9.15). The O–Ge–Cl angles are very nearly 90°, and the Cl–Ge–Cl angle is 95°.[88] The O–Ge–O

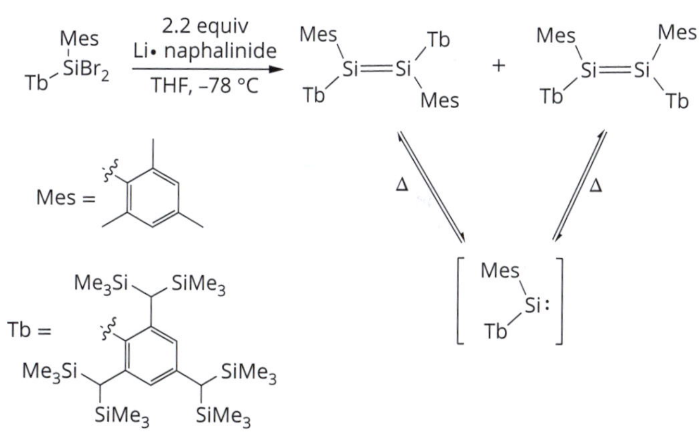

Fig. 9.33 Synthesis of disilenes that thermally convert to the silylene.

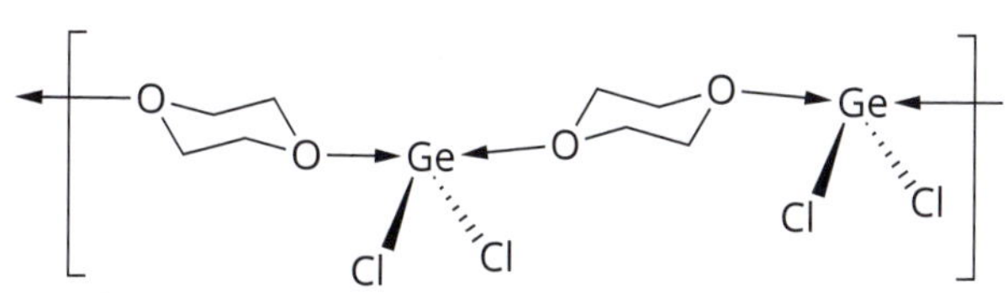

Fig. 9.34 Structure of $\{GeCl_2(dioxane)\}_n$.

angle is about 180°. Consequently, the molecule may be described as having 3c4e bonds using a pure *p*-orbital to interact with the dioxane donors with *p*-rich hybrids on Ge bonding to chloride, leaving the lone pair to reside in a very nearly pure *s*-orbital.

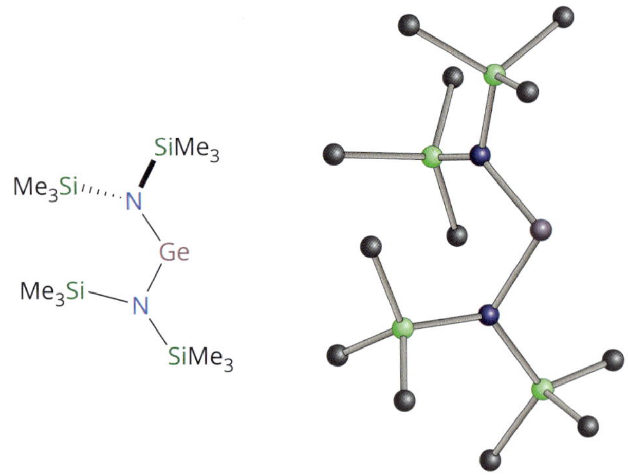

Fig. 9.35 Line drawing of Ge{N(SiMe$_3$)$_2$}$_2$ and its structure from X-ray diffraction.[93] Hydrogens are not shown.

Monomeric Ge(II) 2-coordinate compounds usually have bulky substituents and/or π-donor ligands. For example, Lappert and coworkers have prepared many bisamido germanium(II) compounds,[89–92] such as Ge{N(SiMe$_3$)$_2$}$_2$ shown in Fig. 9.35.[93]

Both tin(IV) and tin(II) compounds are well known and often stable, such as $SnCl_4$ and $SnCl_2$, which are both available commercially. Tin(II) is far less reducing than its lighter congeners of the same oxidation state, but is still used as a mild reductant. For example, $SnCl_2$ is often used for reduction of nitro groups to amines.[94–96]

In general, tin(II) and tin(IV) compounds are not of exceptionally high toxicity. For example, stannous fluoride, SnF_2, is found in some toothpastes and has proven superior to sodium fluoride for dental apatite to fluoroapatite conversion in teeth. However, organotin(IV) compounds can have extremely high toxicity and should be handled with great care. Methyl and ethyltin(IV) compounds are often readily absorbed through the gastrointestinal tract, while the larger compounds less so. Consequently, the bulkier RSn(n-Bu)$_3$ compounds are preferable substrates from a toxicity standpoint for Stille coupling, a common catalysis for forming new carbon–carbon bonds. Once absorbed, the compounds can cause status spongiosus of the central nervous system, including brain lesions.[97–98] Lead is far more toxic than its lighter congeners and has no known safe exposure limit. It causes damage to the nervous system, which if diagnosed can be treated with chelating agents.[99]

Similar to the trend in Group 13, the heaviest element in the group is a good oxidant in the higher oxidation state. Compounds such as Pb(OAc)$_4$ are strongly oxidizing. Lead(IV) oxide is a strong enough oxidant to evolve oxygen on treatment with sulfuric (Eq. 9.15) and nitric acids (Eq. 9.16).[5]

$$PbO_2 + H_2SO_4 \xrightarrow[\text{warm}]{} PbSO_4 + H_2O + 1/2\,O_2 \qquad \textbf{9.15}$$

$$PbO_2 + 2\,HNO_3 \rightarrow Pb(NO_3)_2 + H_2O + 1/2\,O_2 \qquad \textbf{9.16}$$

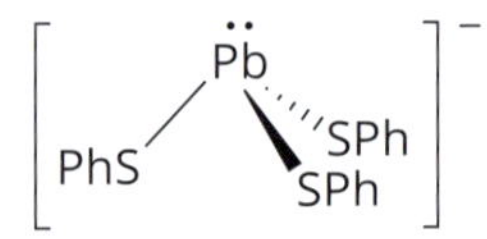

Fig. 9.36 Structure of {Pb(SPh)$_3$}$^-$.

The lower oxidation states of the main group, such as Tl(I), Sn(II), and Pb(II), often exhibit what have been called stereochemically active lone pairs. To pick a molecular example, consider the anion {Pb(SPh)$_3$}$^-$, which has a pyramidal structure and S–Pb–S angles of 95.4° (Fig. 9.36).[100] The lead compound will have a lone pair, which from a VSEPR perspective leads to the pyramidal shape.

However, an MO or hybridization theory analysis on the same compound would suggest that the lone pair is residing in an s-rich hybrid and that Pb is using p-rich hybrids to form the Pb–S bonds. The pyramidalized shape is, at best, indirectly due to the lone pair, which resides in an essentially spherical orbital. The shape is due to the bonding, which is done by very nearly pure p-orbitals on lead.

The halides, like PbI_2, would also exhibit "stereochemically active lone pairs" for the same reasons; that is, the bonding is being done by very nearly pure p-orbitals. As a result, the Pb–I distances in diiodolead(II) are 2.81 Å with an I–Pb–I angle of 102°.[101]

"Nonstereochemically active lone pairs" are said to exist in species such as {PbI$_6$}$^{4-}$, which has an octahedral structure.[102] Typical Pb–I distances in these anions are 3.22 Å, ~0.4 Å longer than in 2-center 2-electron bond containing PbI_2. As a result, the bonding in {PbI$_6$}$^{4-}$ probably involves extensive 3-center 4-electron bonds between I–Pb–I units, resulting in low overall Pb–I bond orders. Like the case with a stereochemically active lone pair, in this molecule the lone pair resides in a pure s-orbital and the p-orbitals on Pb are involved in the bonds.

While the VSEPR-derived nomenclature suggests fundamental differences in the electronic structure at the central element, in many cases, resonance theory suggests that the central element has an s-rich lone pair with bonding being done primarily by p-orbitals through either 2c2e bonds, 3c4e bonds or a combination of the two types. Consequently, the terms "stereochemically active lone pair" and "nonstereochemical active lone pair" should simply be regarded as topological descriptors rather than bonding analyses.

9.6 The Pnictogens (Group 15)

Features of Group 15 Chemistry

- The electronic differences between nitrogen and phosphorus were highlighted in Chapter 8, and here the reactivity differences between the two that result are explored, such as reactions with O_2.

- Compounds with nitrogen–nitrogen bonds are well known and often have interesting reactivity. Hydrazine, H_2N–NH_2, is a high-energy compound with an N–N single bond. Azides, N_3^-, can be explosive. Organic azides are used in the "click" reactions with alkynes and in bio-orthogonal tagging using the Staudinger reaction.
- Nitrogen forms many important oxides as well.
- Phosphorus forms many P–P bonded compounds.
- The syntheses of phosphine compounds can be relatively elaborate, and a very large number of such compounds are known.
- The cacodyls, dimethylarsenic compounds, are historically important and were originally identified by Bunsen.

The nitrogen group is often referred to as the pnictogens, which comes from the Greek *πνιγειν* (*pnigein*) meaning "drown or choke", which may denote the effect of compounds like ammonia on inhalation. The elements of the group have some propensity to catenate, especially phosphorus, and have a very rich chemistry overall.

9.6.1 Contrasting the Chemistry of Nitrogen and Phosphorus

There are notable differences in the chemistry of the first-row main group and the heavier congeners for all of the *p*-block, and we will discuss some of those differences in detail here, using nitrogen and phosphorus to illustrate.

One cause of the differences in NH_3 and PH_3 previously discussed in detail (Section 8.6) is the ease of *sp*-hybridization in nitrogen relative to phosphorus—a result of valence atomic orbitals that are much closer in radii for the nitrogen atom. One result is the much larger relative barrier to PR_3 inversion than NR_3 (Section 8.6)—an effect used to advantage in asymmetric catalysis.

While amines sometimes act as reductants, they are almost invariably stable in oxygen for reasonable periods. Phosphines, on the other hand, have variable reactivity with oxygen; some are stable indefinitely in air, while others are pyrophoric. Triarylphosphines, like PPh_3, often are stable in air. In general, phosphines bearing hydrogens, H_2PR, and only alkyl groups, such as PMe_3, are not stable in air. However, there are exceptions, often for very bulky compounds, like air-stable $H_2P(2,4,5\text{-}Pr^i_3C_6H_2)$. Considering the small size of the reactant, O_2, this is unlikely to be a "steric effect" in the sense of shielding the reagent from being accessed, but is more likely sterics leading to phosphorus rehybridization, which changes the energy of the frontier orbitals. The stability of phosphines in air can be predicted using relatively simple calculations of the singly occupied molecular orbital energy in the radical cation, $^{+\bullet}PR_3$, consistent with Fig. 9.37 for phosphine-to-phosphine oxide conversion.[103]

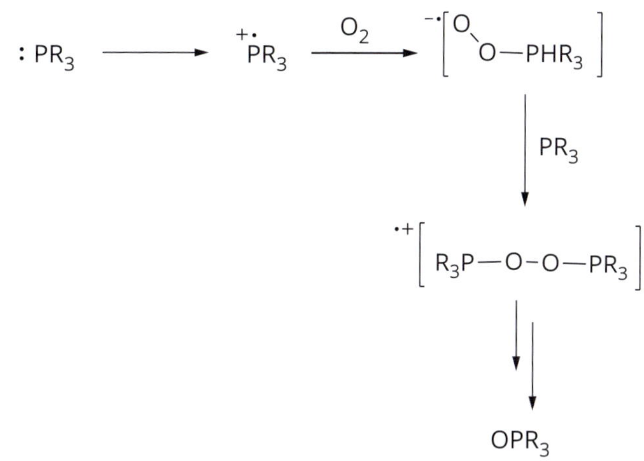

Fig. 9.37 Proposed pathway for reaction of phosphines to form phosphine oxides.[103]

9.6.2 A Sampling of Pnictogen Chemistry

The inorganic and organic chemistry of nitrogen is quite extensive, and the coverage here is limited to some common nitrogen compounds and their oxides. The ability of nitrogen to catenate is limited by the large lone pair bond weakening effect (Section 8.5) in the N–N bond and the high bond dissociation enthalpy of dinitrogen (226 kcal/mol); in fact, N_2 is the common decomposition and combustion product of nitrogen compounds. Even the common reagent hydrazine, H_2N–NH_2, is an endoergic compound ($\Delta G_f° = 35.6$ kcal/mol, $\Delta H_f° = 12.1$ kcal/mol) that is only kinetically stable; hydrazine's stability is improved in the presence of water.[5]

Azide, N_3^-, is accessible from hydrazine by partial oxidation with nitrous acid, nitric acid, hydrogen peroxide, and other oxidants. Azide compounds are often explosive, including pure HN_3, hydrazoic acid. Some salts, such as NaN_3, and compounds, such as Me_3SiN_3, can be handled safely but should always be used with care. The isolated azide anion is linear like isoelectronic carbon dioxide (CO_2), cyanamide (NCN^{2-}), cyanate (NCO^-), fulminate (CNO^-), and nitronium (NO_2^+).

Organoazides undergo 1,3-dipolar cycloaddition reactions with unsaturated substrates like alkynes, Huisgen cycloaddition. However, the thermal reaction often requires heating and can lead to a mixture of isomers. The copper(I)-catalyzed analog of the Huisgen reaction, often called the "click" reaction, in many circumstances can be run at room temperature and usually provides a single isomer. The proposed mechanism for the catalysis and the overall process are shown in Fig. 9.38.[104]

Perhaps more commonly, azides are used as sources of nitrene, N–R, for 2-electron oxidation. As one example, consider the Staudinger reaction, $PR_3 + N_3R$. In the initial step, azide acts as a nucleophile towards the phosphorus atom to provide a phosphazide (Fig. 9.39), which is generally unstable although certain derivatives can be observed at low temperature.[105]

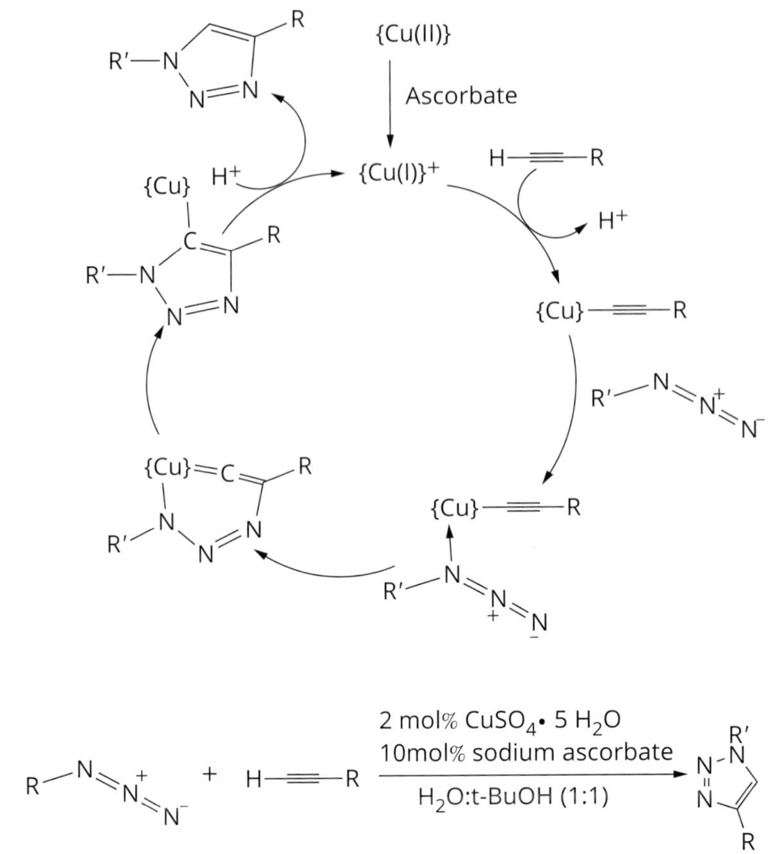

Fig. 9.38 Proposed mechanism (*top*) and overall reaction (*bottom*) of the copper(I) catalyzed reaction of alkynes with organic azides: "click" chemistry. Brackets "{ }" indicate that the other ligands on copper are ill-defined.[104]

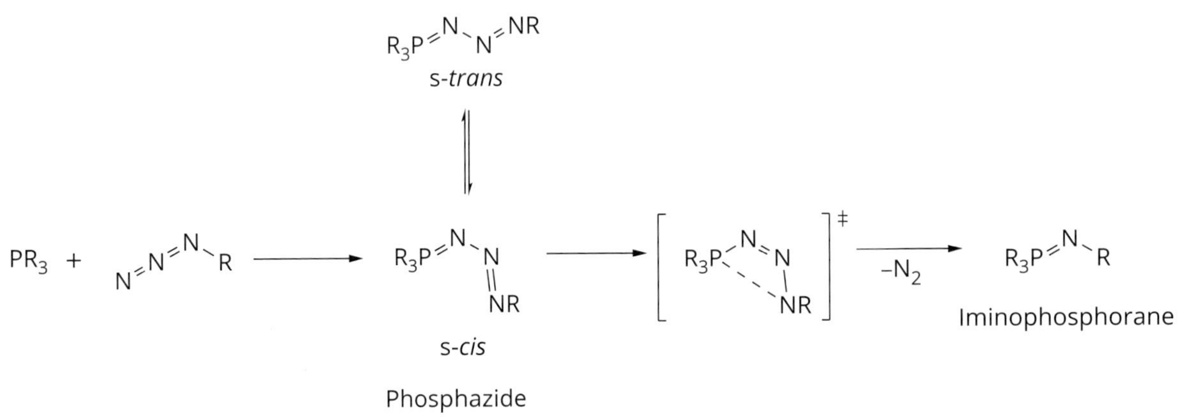

Fig. 9.39 The mechanism of the Staudinger reaction to form iminophosphorane.

In fact, the phosphazide $Ph_2(CN)C–N=N–N=PPh_3$ has been characterized by X-ray diffraction.[106] Phosphazides can exist as either the s-*trans* or s-*cis* isomers. The *s*-cis isomer is thought to utilize a cyclic transition state with loss of N_2 to form the iminophosphorane, formally a 2-electron oxidation of the phosphine from the +3 to the +5 formal oxidation state.[105,107] Phosphazides and iminophosphoranes are electronically similar pseudo-vinylagous species related by –N=N–.

Iminophosphoranes are common reagents in organic chemistry, and they react with a variety of oxygen-containing compounds to produce phosphine oxides (Fig. 9.40).

The Staudinger reaction is sufficiently bio-orthogonal that the "traceless" version has become a valuable tool in biological studies.[108] Since both organoazides and triarylphosphines do not react with most biological substrates but react readily with each other, they can be used to couple biological compounds without risk of side-reaction. In the application shown in Fig. 9.41, the nonfluorescent coumarine-based triarylphosphine reacts with organoazide to produce a fluorescent phosphine oxide. To allow the chemistry to be achieved in water, which also reacts with iminophosphoranes (Fig. 9.41), a carbonyl substituent is placed near the phosphine

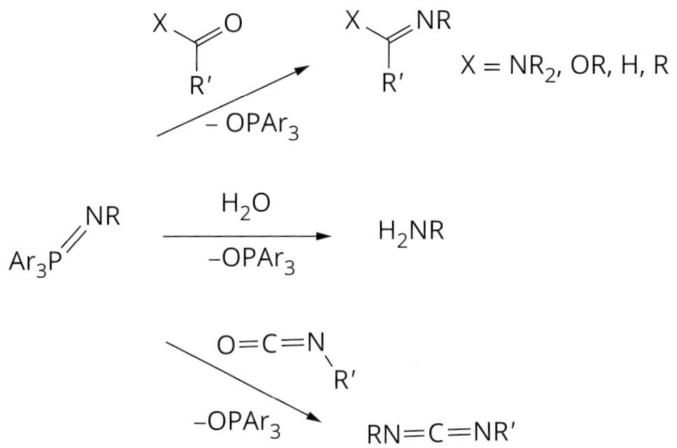

Fig.9.40 Some common reactions of iminophosphoranes.

as a trap. The end product of the phosphine–ester reaction with organoazide is the water-stable phosphine oxide–organoamide. Consequently, treatment of the coumarine-based triarylphosphine results in fluorescent labeling of azide-substituted proteins.[109]

Organoazides react similarly with transition-metal complexes, {M}, to produce metal imido compounds, {M}=NR. Double labeling studies on such systems have yielded results consistent with either unimolecular decomposition of the metal–alkylazide complex, similar to the Staudinger reaction, or bimolecular decomposition.[110–111]

Production of metal azide complexes {M}–N_3, where the metal is at least two below its maximum oxidation state, often leads to loss of N_2 and production of a metal nitrido, {M}≡N. The loss of N_2 is sometimes acid catalyzed.[112]

Because of potential explosion hazards with N_3^-, azide surrogates have been sought. One compound that has shown great efficacy in some cases is 2,3:5,6-dibenzo-7-azabicyclo{2.2.1}hepta-2,5-diene, H(dbabh). Deprotonation of H(dbabh) to form the lithium salt, Li(dbabh), produces a compound with reactivity reminiscent of LiN_3, which instead of N_2 eliminates anthracene to effectively transfer a nitrogen atom (Eq. 9.17).[113]

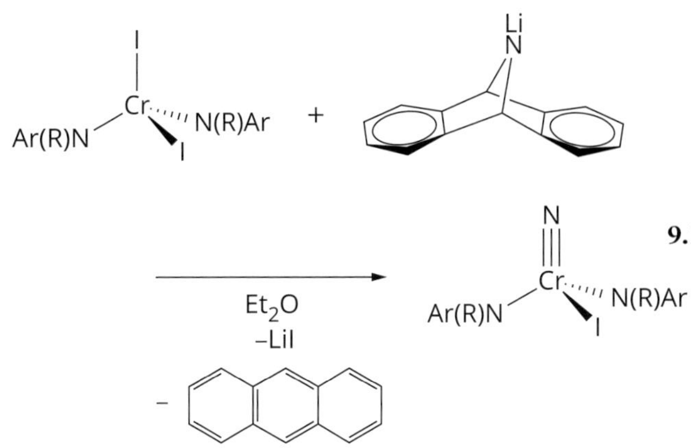

9.17

Nitrogen forms six common oxides, shown in Table 9.5 along with some of their properties.

In common uses, nitrous oxide, N_2O, is employed as a propellant in whipped cream and as an analgesic, "laughing gas". The compound is a kinetically stabilized oxidant that releases

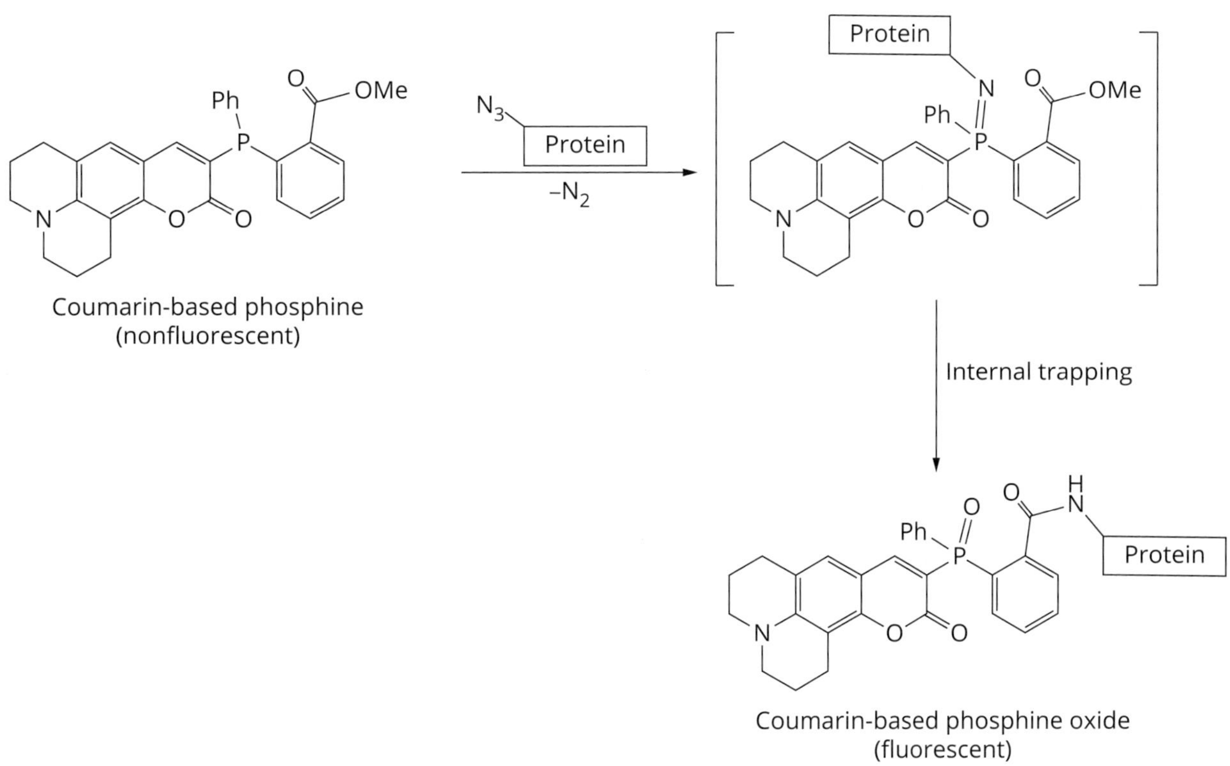

Fig. 9.41 Fluorescent labeling of azido-proteins using the Staudinger reaction.[109]

Table 9.5 Some oxides of nitrogen.[5]

Formula	Name	Some contributing resonance forms	Properties
N_2O	nitrous oxide (dinitrogen monoxide, "laughing gas")		colorless gas (b.p. −88.5 °C)
NO	nitric oxide (nitrogen monoxide)	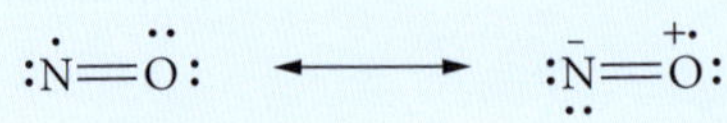	paramagnetic colorless gas (m.p. −163.6 °C b.p. −151.8 °C)
N_2O_3	dinitrogen trioxide	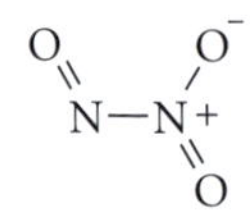	blue solid that dissociates reversibly in the gas phase to NO and NO_2 (m.p. −100.7 °C)
NO_2	nitrogen dioxide	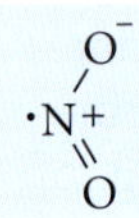	brown paramagnetic gas that dimerizes reversibly to N_2O_4
N_2O_4	dinitrogen tetroxide	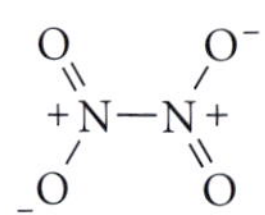	colorless liquid that dissociates reversibly in the gas phase to NO_2. (m.p. −11.2 °C)
N_2O_5	dinitrogen pentoxide (nitric anhydride)	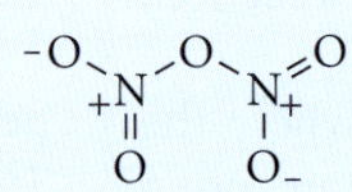	colorless solid that sublimes at 32.4 °C to an unstable gas. The N-O-N angle is near 180°.
NO_3	nitrogen trioxide	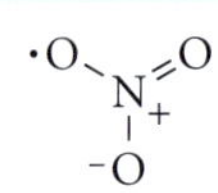	unstable radical

N_2 on O-atom donation. However, it is unreactive even towards some strong reductants, such as the alkali metals, at room temperature. At high temperatures it supports combustion well. Included in its numerous uses as an oxidant, some transition-metal complexes will react with N_2O to form oxides or with oxidation of pendant groups.[114–117] Conversely, there is an unusual example of N_2O acting as a nitrogen atom donor to form a metal nitrido with release of nitric oxide.[118–119]

Nitrous oxide, along with nitric oxide, is on the anaerobic reduction pathway for nitrites, and some organisms possess a copper-containing enzyme, nitrous oxide reductase, for its reduction to N_2. The enzyme contains two different copper centers: a binuclear copper center (CuA) similar to that in cytochrome *c* oxidase, and a tetranuclear copper center (CuZ) with bridging sulfides.[120]

Nitric oxide has a rich inorganic chemistry.[121–122] It is a colorless, paramagnetic gas that reacts readily with O_2, forming brown NO_2. Nitric oxide (NO) is a common ligand for transition metals, and as a radical it can be either oxidized or reduced. The oxidized compounds, such as $\{NO\}^+ BF_4^-$, are stable salts often used as oxidants. Nitric oxide is also an important signaling molecule in biology, which is prepared by nitric oxide synthase—a heme iron protein that oxidizes arginine to form NO.[123–126]

The other oxides of nitrogen in Table 9.5 are unstable in one way or other. Another commonly encountered example is NO_2/N_2O_4, which interconvert reversibly. Nitrogen dioxide is a brown gaseous radical that forms a weak N–N bond in the liquid and solid state to produce N_2O_4. The compound is oxidizing and reduces to nitrite, NO_2^-.

With its lone-pair bond weakening effect lower than nitrogen, catenation in phosphorus chemistry is more common and is less likely to be explosive. Typical P–P bond energies are ~55 kcal/mol, far lower than a typical C–C bond of ~85 kcal/mol; nevertheless, the chemistry of catenated phosphorus is extensive. Substituted *cyclo*-polyphosphanes of general formula P_nR_n are relatively easily prepared with n = 3-6.[127] For example, P_6Ph_6 (m.p. = 190 °C) can be prepared by reaction of $PhPH_2$ with equimolar $PhPCl_2$ (Fig. 9.42). The six-membered phosphane has a boat conformation with equatorial phenyl groups; compare the structure of black phosphorus in Section 9.1.3.[128–129]

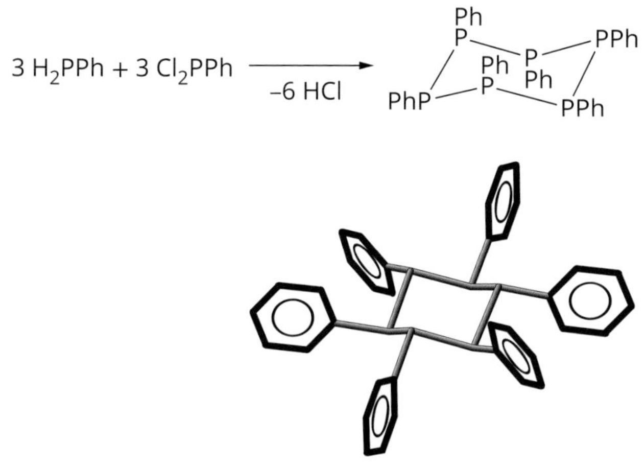

Fig. 9.42 Synthesis and structure of *cyclo*-P_6Ph_6. The structure is from the single crystal X-ray diffraction coordinates with hydrogens on the phenyl groups not shown.[128–129]

In a relatively simple and elegant example, P(3,5-dimethylpyrazolyl)$_3$ reacts with $HPCy_2$, Cy = cyclohexyl, to produce $P(PCy_2)_3$, which contains three P–P bonds (Eq. 9.18).[130]

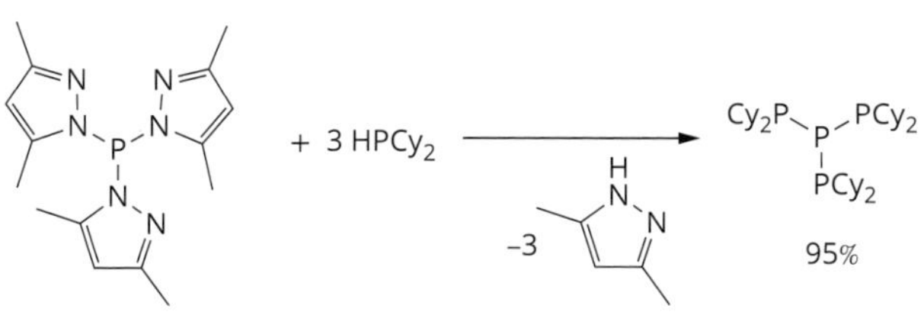

9.18

Phosphorus–phosphorus bonds can in some instances undergo metathesis (exchange of groups) to form new P–P bonds. For example, *cyclo*-$(PCF_3)_{4,5}$ reacts with $HEMe_2$, where E = P or As, to produce $Me_2E–P(CF_3)H$ (Eq. 9.19).[131–132]

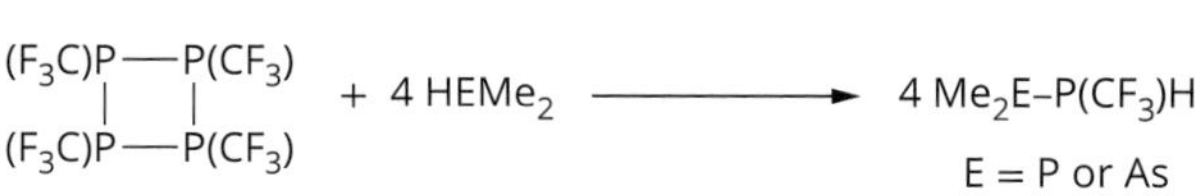

9.19

Diphosphines are also known to undergo rapid and, in some cases, quantitative exchange (Fig. 9.43), such as the reaction of $Me_2P–PMe_2$ with $(F_3C)_2P–P(CF_3)_2$ to produce $Me_2P–P(CF_3)_2$.[132] A possible intermediate would be a four-membered ring where a phosphorus lone pair donates into the P–P antibonding orbital, a form of negative hyperconjugation. Naturally, the largest contributor in the intermediate probably would have the more electron-rich PMe_2 groups donating into the more electron-deficient $P(CF_3)_2$ groups, as shown in Fig. 9.43. The obvious driving force for the reaction is the increased bond strength of the P–P bond in the product relative to the starting materials due to increased polarity in the asymmetric $Me_2P–P(CF_3)_2$ derivative (see Table 3.3 and surrounding discussion for the same effect in a different system).

Complex multistep syntheses are quite common in phosphorus chemistry. For discussion, we can look at the production of the chelating ligand $(t\text{-Bu})_2PCH_2P(t\text{-Bu})_2$ (Fig. 9.44).[133] The synthesis begins with the production of a Grignard reagent by addition of *tert*-butylchloride to magnesium turnings in ether. Grignard reagents are oxygen- and water-sensitive and are generally not isolated, and the presumption is that the production of the reagent is quantitative. If one needs an accurate concentration of a solution of a RMgCl, LiR, ZnR_2, or similar reagent, one can titrate the solution[134–135] or use NMR methods versus an internal standard to determine an accurate concentration.[136]

The *tert*-butylmagnesium chloride solution was then added to PCl_3 in Et_2O to produce $ClP(t\text{-Bu})_2$ with $MgCl_2$ as a by-product. Salts are common by-products in reactions due to their generally quite high heat of formation and ease of separation. *The driving force for many reactions is not found in the reactants versus desired product energies but in the energies of the often ignored by-products.* In this case, the $MgCl_2$ is removed by precipitation using addition of dioxane (Section 9.5). The solution was then filtered and distilled.

The phosphine $ClP(t\text{-Bu})_2$ is a distillable, colorless liquid that is both water- and air-sensitive. The water sensitivity is the result of the hydrolyzable P–Cl bond, which will react with protic sources to release HCl.

Next, $ClP(t\text{-Bu})_2$ is treated with LiMe to replace chloride with methyl, providing the still oxygen-sensitive but water-stable $MeP(t\text{-Bu})_2$. In fact, the alkyllithium reaction is quenched with deoxygenated aqueous NH_4Cl,† and the organic phase is cannulated to a flask containing $MgSO_4$ to remove trace water—a typical work-up for trialkylphosphines.

In the next step, the methyl group of $MeP(t\text{-Bu})_2$ is deprotonated with *tert*-butyllithium to produce $LiCH_2P(t\text{-Bu})_2$. Addition of $ClP(t\text{-Bu})_2$ to $LiCH_2P(t\text{-Bu})_2$ provides the final desired product with elimination of LiCl, which is again isolated after a deoxygenated ammonium chloride quench.

† Aqueous solutions are easily deoxygenated by refluxing for a short while under dinitrogen or argon atmosphere.

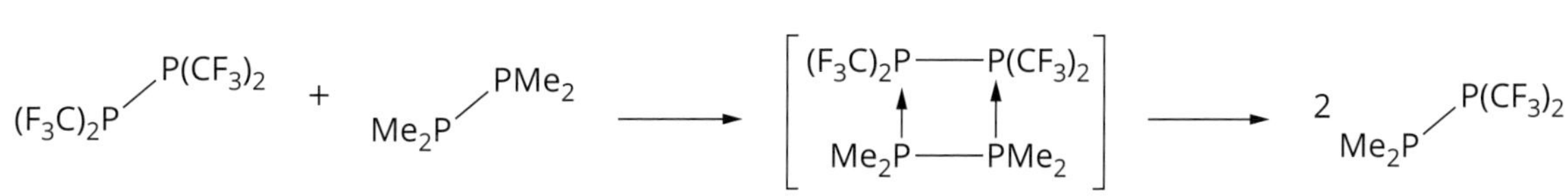

Fig. 9.43 Reaction of $Me_2P–PMe_2$ with $(F_3C)_2P–P(CF_3)_2$ to produce $Me_2P–P(CF_3)_2$ and the proposed intermediate.

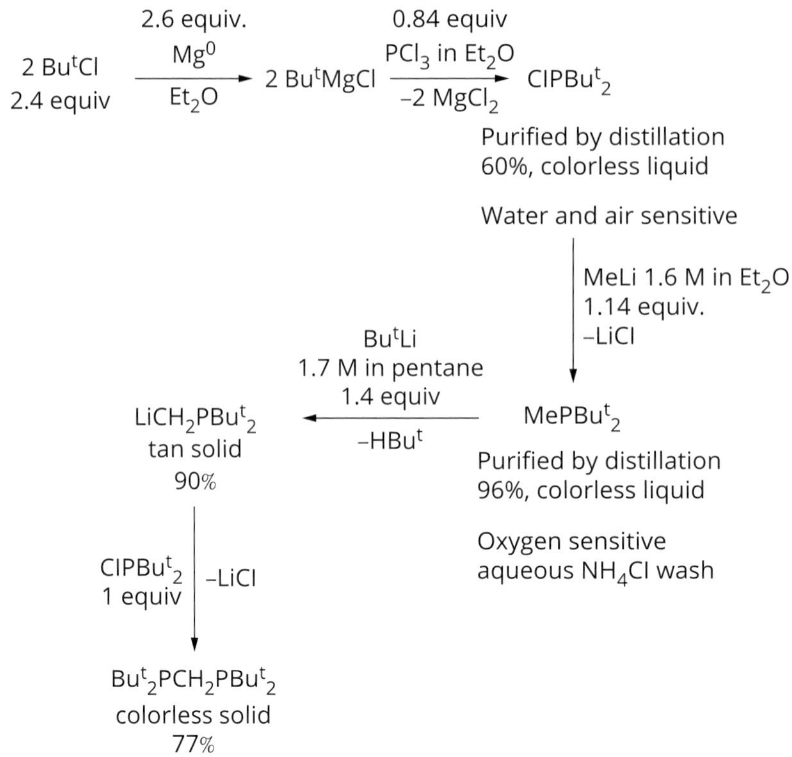

Fig. 9.44 Synthesis of $(t\text{-Bu})_2PCH_2P(t\text{-Bu})_2$.

The deprotonation of $H_3CP(t\text{-Bu})_2$ is enabled by the electronic structure of the phosphine. The deprotonation of the methyl in PMe_2(dodecyl) is an estimated fifteen times faster than NMe_2(dodecyl).[137] The cause for this is probably the difference in negative hyperconjugation strengths of the product anions, which leads to differences in relative stabilities.

If one examines the structures of the anions $\{CH_2–NMe_2\}^-$ and $\{CH_2–PMe_2\}^-$, optimized using DFT (B3LYP, 6-311++G), one finds several differences (Fig. 9.45). Most notably, the H–C–H angle of the deprotonated methyl group of the amine is close to tetrahedral, 108.9°, indicating that there is little interaction between the carbon-centered lone pair and the nitrogen. In contrast, the deprotonated carbon on the phosphine is very nearly planar, 116°, indicating ~sp^2-hybridization and significant π-bonding. Most tellingly, the CH_2–N bond distance is essentially identical to the N–Me bond distances in $\{CH_2–NMe_2\}^-$, whereas the CH_2–P bond distance is ~0.2 Å shorter than the P–Me single bonds.

The P=CH_2 character in $\{CH_2–PMe_2\}^-$ is derived from carbon-centered lone-pair donation into the P–C antibonding orbitals, which can be described in valence bond terms by using the resonance forms in the bottom portion of Fig. 9.45.

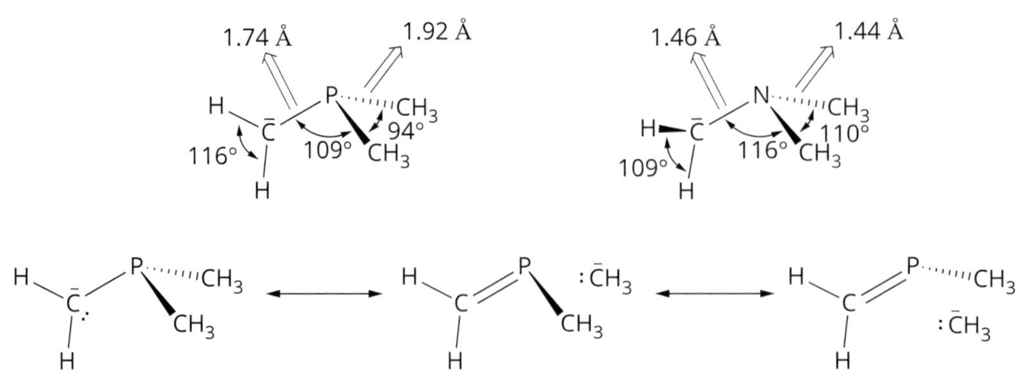

Fig. 9.45 Contrasting the structure and electronics of deprotonated methyl groups on phosphorus and nitrogen. (*top*) The calculated (B3LYP, 6-311++G) geometries of $\{CH_2–NMe_2\}^-$ and $\{CH_2–PMe_2\}^-$. (*bottom*) Some expected resonance forms, including the carbanion structure and P=C structures through negative hyperconjugation.

The stronger π-bonding in the phosphorus analog is due to the greater acceptor ability of the P–Me antibonding orbitals and the polarization of the bonds. In general, P–X bonds will be better acceptors because of the lower electronegativity of phosphorus, making it more likely to bear a partial cationic charge due to bond polarization, $P^{\delta+}$–$C^{\delta-}$. Nitrogen, with its higher electronegativity, will have a more negative charge than the phosphorus analog, $N^{\delta-}$–$C^{\delta+}$, making the central nitrogen a poor electron acceptor (see Fig. 8.20 and accompanying discussion).

In addition, since the electronegativity of phosphorus is lower than carbon, the antibonding orbitals will be centered on phosphorus, increasing overlap with the deprotonated carbon. Since the antibonding orbital in the nitrogen analog will have more carbon character, the overlap between the methylene carbon lone pair and the antibonding orbital of the N–Me bonds will be poor.

While $LiCH_2PMe_2$ can be accessed in straightforward fashion by direct deprotonation of PMe_3 with Li-*t*-Bu, the nitrogen analog is accessible through a more circuitous route involving, in the key step, reaction of $Me_2NCH_2SnBu_3$ with Li-*n*-Bu.[138–139]

The deprotonation of CH groups adjacent to phosphorus in phosphonium salts (Fig. 9.46) would, as you would predict based on their charge, can be accomplished with a variety of bases such as $KOBu^t$, lithium diisopropylamide, and sodium hexamethyldisilazide. The resulting ylides are Wittig reagents, named after Georg Wittig, who shared the Nobel Prize for Chemistry with H. C. Brown in 1979.

Perhaps the most common Wittig reagent employed is from the deprotonation of $MePPh_3^+$, $H_2C{=}PPh_3$. This ylide is a common reagent for the methylenation of carbonyls and is even used industrially for the production of Vitamin A and various carotenoids. On an industrial scale, the $O{=}PPh_3$ by-product is recovered and reused.[140–141] The phosphorus ylide undergoes

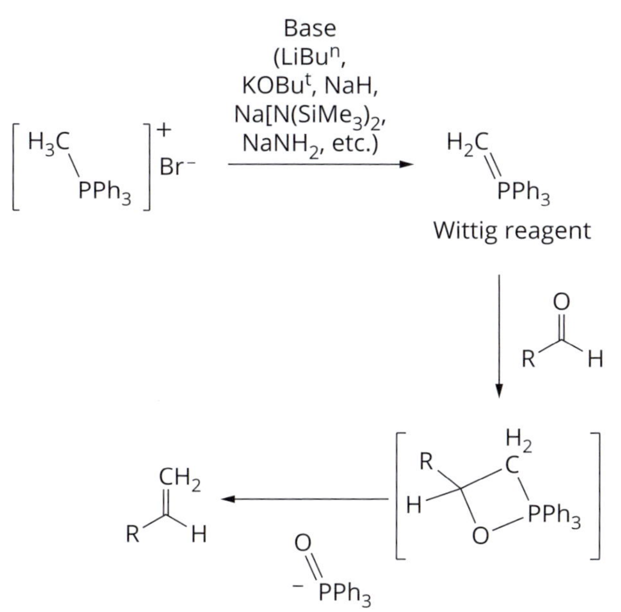

Fig. 9.46 Formation and reactivity of Wittig reagents.

{2 + 2}-cycloaddition with, for example, an aldehyde to produce a four-membered ring, an oxaphosphetane. Retro-{2 + 2}-cycloaddition to form the phosphine oxide and olefin occurs irreversibly (Fig. 9.46).[142]

In general, as the substituents on the phosphine become less electronegative, the Wittig reagent becomes more reactive. Consequently, Wittig reagents based on triphenylphosphine are less reactive than their trialkylphosphine counterparts. Less electronegative substituents on the phosphorus no doubt reduce the negative hyperconjugation and isolate the negative charge on the reactive carbon by encouraging the ylide form.

9.6.3 Bunsen and the Cacodyl-Based Arsenic Compounds

One fascinating account in arsenic chemistry is the discovery of cacodyl, $Me_2AsAsMe_2$. The French pharmacist–chemist Louis-Claude Cadet de Gassicourt carried out the reaction of arsenious oxide, As_2O_3, with potassium acetate, KO_2CMe, in 1757. While heating the mixture in a glass receiver in a furnace, liquids were distilled with a strong odor similar to garlic. When the grease-sealed stopper was opened and the liquid was exposed to air, the liquids "caused immediate inflammation of the grease."[143] At the time, Cadet had little chance of identifying the compound, as chemistry was in its nascency; for example, Lavoisier was aged only 14 at the time, and phlogiston theory was the dominant model for understanding combustion.

Cadet's fuming liquid was investigated by several French scientists in the intervening years, with modest progress, until Robert Bunsen undertook a thorough study of the compound in 1837. Bunsen began with 1 kg of 1:1 As_2O_3 and KO_2CMe and heated the solid mixture. Two liquid layers collected in a receiver flask, the lighter layer containing water, acetone, acetic acid, and some of the arsenic-containing products that were largely in the bottom brown layer. Using his nearly legendary abilities of compound characterization, Bunsen found the main arsenic-containing compound to have the formula $C_4H_{12}As_2$, and Berzelius suggested the name *kakodyl* (derived from Greek for "evil-smelling matter"; in modern usage the "k's" have become "c's").[143]

Arsenic is of extraordinary toxicity in general, and these investigations were quite dangerous. One of the compounds that Bunsen prepared was cacodyl cyanide, NC–$AsMe_2$. In one attempt there was an explosion that nearly took his life from poisoning and permanently lost him partial use of one eye. After recovering, he attempted the synthesis again(!) and isolated the pure compound as colorless crystals. The synthesis was done in an open field while he breathed through a long glass tube to obtain fresh air. The vapor from ~70 mg of the compound in a room was reported to produce sudden numbness of the hands and feet, dizziness, and eventual unconsciousness. Bunsen died decades later at the age of 88.

This work on arsenic chemistry was of great importance because of the properties and structure of the compounds.

The metalloid-containing compounds were distillable oils with properties much like organic compounds, which dramatically changed the way people thought about the chemistry of the metals and metalloids.

9.7 The Chalcogens (Group 16)

Features of Group 16 Chemistry

- The chemistry of sulfur oxides, halides, and ylides.
- Selenium and tellurium compounds in the +4 and +6 formal oxidation states are well known but are somewhat less common than for sulfur and tend to be more oxidizing.
- Se and Te, as in previous groups, favor lower oxidation states.
- Selenium is an essential element for higher life but is toxic at high levels; Se is often found in amino acids as a replacement for sulfur in methionine and cysteine.
- Polonium does not occur naturally and was named by Marie Curie after Poland.

Sulfur is one of the few elements that appears as a pure compound in crustal rocks, which resulted in the element being known throughout recorded history. The compound is mentioned, sometimes translated as "brimstone", in Genesis 19:24, for example. In contrast, sulfur's lighter congner oxygen was finally identified by Priestley, Scheele, and Lavoisier in 1774, although Leonardo da Vinci (1452–1519) reportedly noted that air had several components, one of which supported combustion.

Oxides are known for every stable element of the periodic table except some of the lighter noble gases. For many elements, such as silicon and aluminum, the oxides are the most common mineral source. Metal sulfides are common sources for Mo, W, Mn, Fe, Ru, Co, Ni, Pt, Cu, Ag, Zn, Cd, Hg, Pb, As, Sb, and Bi.

As mentioned previously (Table 9.1), sulfur has extensive catenation abilities, and disulfide bonds are also of biological importance. For example, cysteine-based S–S linkages between alpha-helices can be manipulated to curl or straighten hair.

Like phosphorus, sulfur-based ylides are readily prepared and are also common reagents; however, unlike phosphorus the typical reactivity of sulfur ylides involves redox chemistry at sulfur. For example, the Johnson–Corey–Chaykovsky reaction involves transfer of CHR from a sulfur ylide, $R'_2S{=}CHR$, to a double bond substrate. The sulfur ylide, although isolable in many cases, is often generated *in situ* from the sulfonium or oxysulfonium cation by addition of base. The ylide adds to olefins to produce cyclopropanes, aldehydes to produce epoxides, and imines to produce aziridines. Asymmetric variants are commonly employed, such as the example shown in Fig. 9.47. The chiral sulfur agent can also be used catalytically.[144–147]

The major resonance form for sulfur ylides, just as in phosphorus ylides, as the name implies, is the 1,2-dipole. In fact,

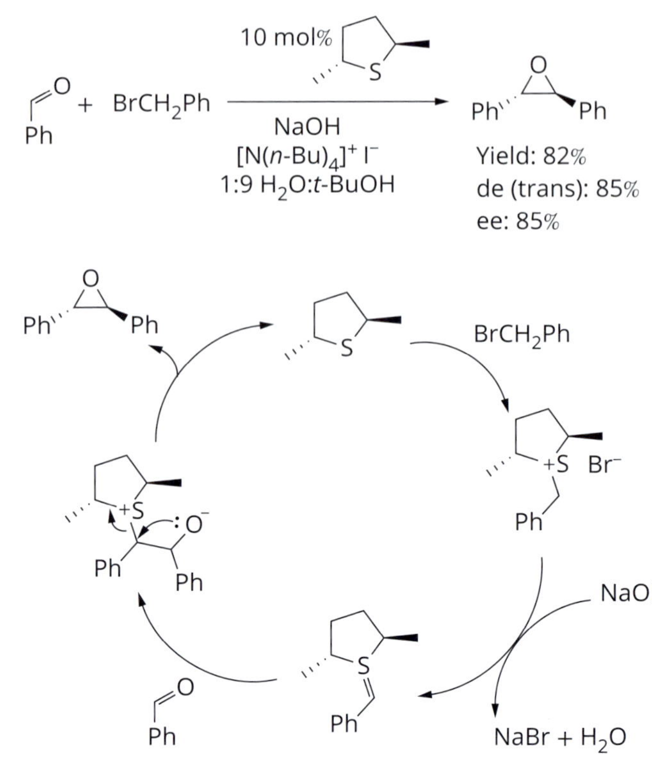

Fig. 9.47 Example of a catalytic asymmetric sulfur ylide epoxidation of an aldehyde.[147]

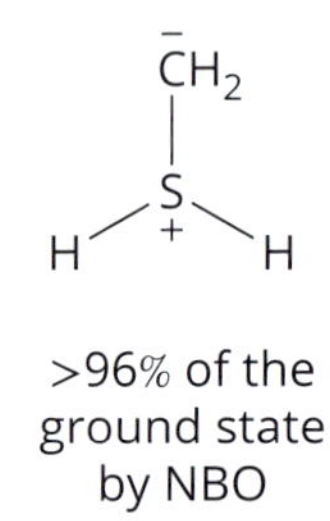

Fig. 9.48 Major resonance form for H_2CSH_2 by NRT.

NBO/NRT calculations[148] suggest that greater than 96% of the ground state for the model compound H_2CSH_2 is comprised of the charge-separated resonance form shown in Fig. 9.48. Oxysulfur ylides, $R'_2(O)S{=}CHR$, and other substituents sometimes greatly alter the reactivity, probably by changing the electronic structure of the S–C ylide bond through hyperconjugation.

Numerous oxides, halides, and oxyhalides of sulfur are used in industry and in the laboratory (Table 9.6). The interesting chemistry of a larger selection of these compounds can be found in Greenwood and Earnshaw's treatise.[5]

One of the largest-scale chemical syntheses in the world is the production of sulfuric acid, which is made by oxidation of SO_2 to SO_3. A catalyst, V_2O_3/K_2O on zeolite support for example, is used to ensure complete oxidation to SO_3. Sulfur trioxide

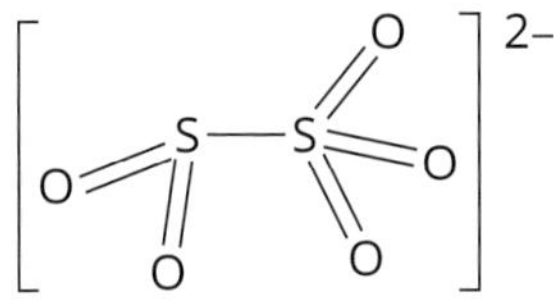

Fig. 9.49 Structure of disulfite.

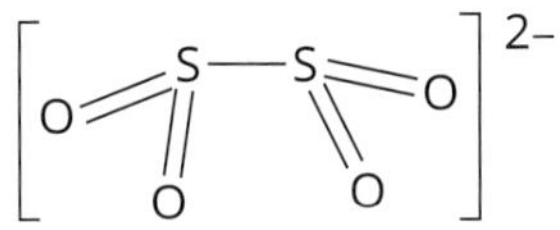

Dithionite

Fig. 9.50 Structure of dithionite.

is highly hygroscopic, and careful reaction with water produces H_2SO_4, which is generally sold as a ~98% pure, viscous oil; the slight impurity of water prevents the compound from solidifying. Sulfuric acid has a melting point of 10.4 °C and decomposes at ~300 °C. Addition of concentrated H_2SO_4 to water is exceedingly exothermic at −210 kcal/mol (−880 kJ/mol) at infinite dilution. Sulfuric acid is a viscous liquid (24 cP), between cream (~17 cP) and vegetable oil (~35 cP).

The starting material for sulfuric acid, SO_2, is obtained by heating sulfur- or sulfide-containing minerals in oxygen. It dissolves in water to form what is generally referred to as "sulfurous acid"; however, there is believed to be little to no H_2SO_3 in solution. Sulfites, salts of SO_3^{2-}, are quite stable and numerous. Hydrogen sulfite salts, HSO_3^-, often eliminate water to produce disulfites, $S_2O_5^{2-}$ (Fig. 9.49). Unlike disulfates and diphosphates, disulfites are not bridging oxides but instead have an S–S bond. Placing disulfites in acidic solution converts them back to HSO_3^-.

Treating hydrogen sulfite, HSO_3^-, with Zn dust or some other reductant produces dithionite, $S_2O_4^{2-}$ (Fig. 9.50). The compound is most often isolated as the sodium salt, $Na_2S_2O_4$ (sodium dithionite). The compound is prepared under an inert atmosphere to avoid reoxidation. Dithionite is a powerful reductant in aqueous solution that converts Ti^{4+} to Ti^{3+} and Cr(VI) oxides to Cr(III). Dithionite is also strong enough to reduce many metal salts (like those of Cu, Ag, Pb, Sb, and Bi) to the metals, which gives the compound applications in removal of heavy metals from polluted water.

Sulfur trioxide exists in a monomeric D_{3h} form, which is in equilibrium with a cyclic trimer S_3O_9 (γ-SO_3). The equilibrium constant at 25 °C is ~1 atm^{-2} between monomer and trimer, and the described properties of SO_3 are often for the equilibrium mixture.

If traces of water are present, the compound can polymerize to form β-SO_3. Trace water can also form the most stable state of the compound α-SO_3, which involves cross-linking of the strands in β-SO_3 into sheets (Fig. 9.51).

Sulfur trioxide is an exceedingly strong Lewis acid and reacts as such with many main-group and transition-metal compounds.

The halides of sulfur are numerous as well, with many formal oxidation states accessible. The simplest, normal valent sulfides

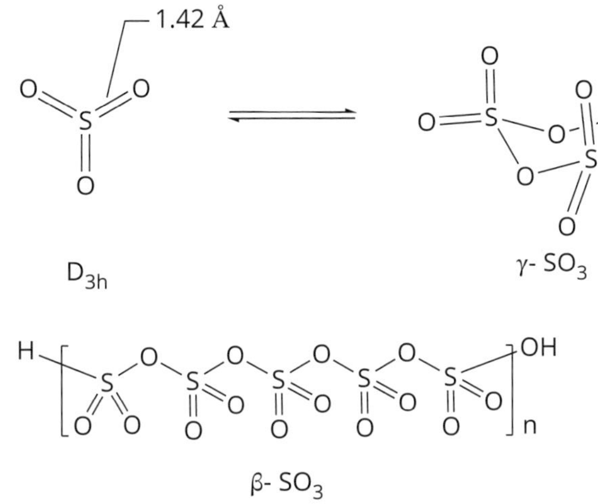

Fig. 9.51 Planar SO_3 is in equilibrium with trimeric γ-SO_3. With trace water, strings or sheets of SO_3 units can be formed.

such as SF_2 and SCl_2 are well studied. While SCl_2 is an easily obtained and stable compound that has been studied by many different techniques, the compound SF_2 is notoriously difficult to prepare in pure form. In one method, SF_2 gas is obtained by reaction of COS with F_2 in the gas phase, which produces largely SF_2; the product over time goes to form FS–SF_3, S=SF_2, and SF_4.[149–150]

Again, catenation in sulfur is far more prevalent than in oxygen due to LPBW effects. So-called subhalides of sulfur can also be prepared as pure compounds. For example, the yellow liquid ClSSCl, sulfur monochloride or dichlorodisulfane, can be prepared by simple addition of Cl_2 to S_8. Other chlorosulfanes ("sulfanes" in analogy to "alkanes") can be prepared in some cases selectively, such as S_7Cl_2. These chlorosulfanes can be used as starting materials to prepare, as one example, new allotropes of sulfur, such as the S_7 synthesis shown in Fig. 9.52. Using related chemistry, S_n rings have been prepared where $n = 6$–20.[151–153]

The fluorides of sulfur are extensive, with the most commonly encountered monosulfur varieties being SF_4 and SF_6. Sulfur tetrafluoride reacts rapidly with water to form HF, and due to this reaction it is exceedingly toxic. Octahedral sulfur hexafluoride, on the other hand, is a colorless and odorless gas that is used as an inert atmosphere in some industrial applications. The heavier halides do not support these higher-sulfur halides, and even SCl_4 decomposes at around −30 °C.

There are numerous oxyhalides of sulfur, including many of industrial importance. In the laboratory as reagents, perhaps the most notable are thionyl chloride, $OSCl_2$, and sulfuryl chloride, O_2SCl_2. Both are chlorinating agents and oxidants.

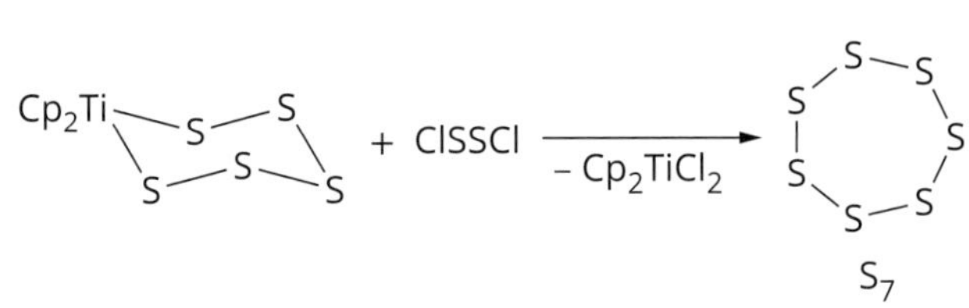

Fig. 9.52 Synthesis of S_7.

Table 9.6 Some oxides, halides, and oxyhalides of sulfur.[5]

Formula	Name	Some contributing resonance forms	Properties
SO	sulfur oxide	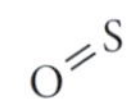	unstable under ambient conditions but known as a ligand. Dimerizes to O=S–S=O
SO_2	sulfur dioxide	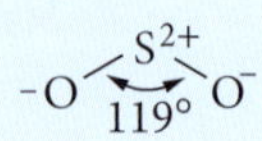	water soluble, toxic gas. Mp. –75.5 °C, B.p. –10 °C.
SO_3	sulfur trioxide	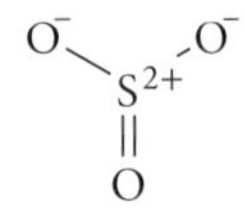	below the M.p. (17 °C) exists as a trimeric solid. B.p. = 44.6 °C. Equilibrium between monomer and trimer in both the liquid and gas phase. Extended structures exist in presence of trace water.
SF_2	sulfur difluoride	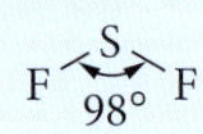	unstable gas that undergoes disproportionation to $FSSF_3$, $S{=}SF_2$, SF_4 and other products
SCl_2	sulfur dichloride	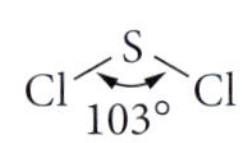	M.p. –78 °C, B.p. 59 °C
ClSSCl	sulfur monochloride	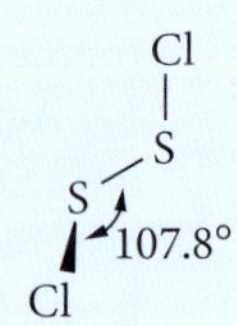	M.p. –80 °C, B.p. 138 °C
SF_4	sulfur tetrafluoride	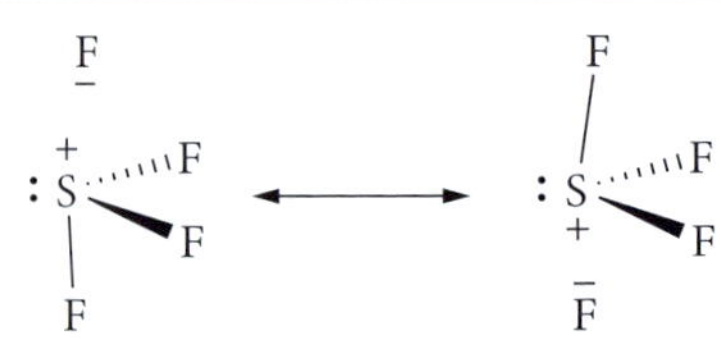	M.p. –121 °C, B.p. –38 °C, toxic colorless gas
SF_6	sulfur hexafluoride	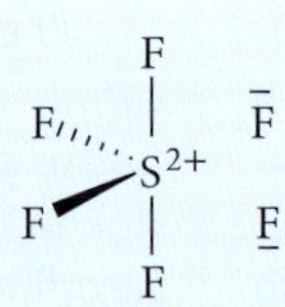	M.p. –51, B.p. –64 °C
$OSCl_2$	thionyl chloride	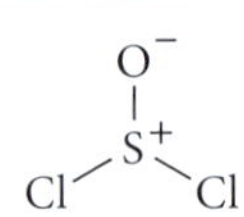	M.p. –101 °C, B.p. 74 °C
O_2SCl_2	sulfuryl chloride	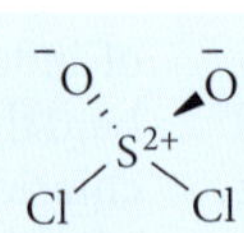	M.p. –54 °C, B.p. 69 °C

Selenium and tellurium have, in some respects, similar chemistry to sulfur. Higher oxidation-state compounds, especially 4- and 6-coordinate compounds, are accessible analogous to sulfur compounds. However, selenium and especially tellurium favor lower oxidation states, and the compounds in higher oxidation states are generally more oxidizing than the sulfur analogs. This is very much in line with Group 13 and Group 14 discussions where lower oxidation states are favored for the heavier elements.

Higher organisms, including humans, require trace amounts of selenium, and a selenium-free diet can cause liver failure, among other ailments. Selenium is found in selenocysteine $HSeCH_2CH(NH_2)CO_2H$ and selenomethionine $MeSeCH_2CH(NH_2)CO_2H$, both of which are replacements of the usual sulfur in the more common amino acids and are found in some enzymes such as glutathione peroxidase, where it is thought that the advantage of selenium over sulfur is that the element "resists permanent oxidation", though the reactions are highly reversible.[154]

None of the twenty-seven known isotopes of polonium are stable. The only isotope known to occur naturally, in very small amounts, is ^{210}Po, which is made through decay of radium. The element was first prepared in 1898 by Marie Sklodowska Curie, who named it after her native Poland.

9.8 The Halogens (Group 17)

Features of Group 17 Chemistry

- Reactivity of F_2.
- Properties of the halogen oxides.
- Some chemistry of hypervalent iodine. Synthesis of Cl_2IPh and OIPh (iodosylbenzene).
- Synthesis and structure of Dess–Martin periodinane (DMP)

Halogen reactivity is so ubiquitous that it was covered in many of the elemental groups discussed previously. Here, a few additional notes on reactivity will be presented. The compounds are all one electron short of a nobel-gas electron configuration, and the −1 oxidation state is by far the most common for all the elements of the group. However, higher-valent compounds are well-known for the heavier elements, especially iodine, where the chemistry is very rich in higher formal oxidation states.

Much of the chemistry of fluorine is determined by its high electronegativity and low F–F bond strength. Fluorine is the only halogen that does not form large numbers of oxides, with OF_2 and some similar derivatives (like HOF) being the only stable ones known.

The lone pair bond weakening (LPBW) effect for F_2 is estimated to weaken the F–F bond by 66% from an unweakened value of 113 kcal/mol to the experimental BDE of 38 kcal/mol (Section 8.5). As a result, fluorine forms stronger bonds to almost any other element than itself, as LPBW only effects the covalent term of the bond and element–F bonds are often quite polar.

Working with F_2 requires special gas-handling equipment, usually made of relatively inert and highly fluorinated plastics such as Teflon. Glass is quickly attacked by the gas and is generally avoided entirely when handling F_2, but carbon steel and Monel piping, made of a corrosion-resistant nickel and copper alloy, sometimes with small amounts of aluminum and titanium, are used to handle the element industrially with little corrosion. The more expensive Monel piping is required at higher temperatures.[155]

9.8.1 Properties of the Halogen Oxides

Chlorine, bromine, and iodine all form extensive oxides and oxyanions in a variety of oxidation states, summarized in Table 9.7.

Table 9.7 Some oxides and oxyanions of chlorine, bromine, and iodine.[5]

Formula E = Cl, Br, or I	E = Cl Notes	E = Br Notes[156]	E = I Notes[157–158]
OE_2	Cl_2O is an explosive brown gas.	Brown solid stable below −60 °C.[159] Available from $HgO + Br_2$.	
^-OE HOE = hypohalous acid ^-OE = hypohalite	Hypochlorous acid and anion hypochlorite (^-OCl) are stable in aqueous solutions and are common chlorinating and oxidizing agents. The sodium salt is found in household bleach.	Hypobromous acid is less stable than the Cl analog but observable in aqueous solutions. Hypobromite, ^-OBr, valance disproportionates to Br_2 and BrO_3^-.	Hypoiodous acid is less stable than the Cl analog but is observable in aqueous solutions. Hypoiodite, ^-OI, valance disproportionates to I_2 and IO_3^-.
E_2O_3	Explosive brown solid.	Yellow solid formed by ozonolysis of Br_2. Formulated as O_2BrOBr.[160]	
EO_2^- HOEO = halous acid EO^- = halite	Chlorous acid, HOClO, is stable in dilute aqueous solution. Many chlorite, ClO_2^-, salts are known. $NaClO_2$ is a stable commercial chemical, while some salts such as Ag, Tl, Cu(II), and ammonium can decompose explosively.	Bromous acid is not a stable compound but is a probable intermediate in the formation of higher oxidation-state acids. Bromites often undergo valence disproportionation.[161–162]	
EO_2	Yellow, paramagnetic gas that can be explosive at pressures above 50 torr but is an important industrial bleaching agent.	Pale yellow solid formed by low-temperature ozonolysis of bromine. Decomposes explosively to the elements around 0 °C. Reacts with alkaline to valence disproportionate: $6\,BrO_2 + 6\,^-OH \rightarrow 5\,BrO_3^- + Br^- + 3\,H_2O$	

(continued)

Table 9.7 Continued

Formula E = Cl, Br, or I	E = Cl Notes	E = Br Notes[156]	E = I Notes[157–158]
EO_3^- $HOEO_2$ = halic acid EO_3^- = halate	Chloric acid, $HOClO_2$, is stable in dilute aqueous solution. Sodium chlorate, $NaClO_3$, is a stable compound used as a powerful oxidant.	$HOBrO_2$ is stable in aqueous solution. Sodium bromate is a commercial compound used as an oxidant. Sodium bromate is used as a hair-perm agent.	Iodic acid is a stable white solid with high water solubility. Many iodate salts are stable.
E_2O_4	$ClOClO_3$: An unstable yellow liquid that decomposes to Cl_2O_6, ClO_2, Cl_2, and O_2 at room temperature.	Unstable above −40 °C and is believed to be a dimer with a Br–Br bond, $O_2Br–BrO_2$. Decomposes to Br_2O_3.	Yellow solid that decomposes to I_2O_5 and I_2 above 85 °C. Formulated as $\{IO\}^+\{IO_3\}^-$.
E_2O_5	The existence of $O_2ClOClO_2$ is apparently unproven, at least in condensed phases.[163]	Ozonolysis of Br_2 at −78 °C forms $BrOBrO_2$ and eventually colorless $O_2BrOBrO_2$, which is unstable above −40 °C.[164]	O_2IOIO_2: White solid. Anhydride of HIO_3, which reforms on addition of water. The related acid with iodine in the +5 oxidation state $H_4I_2O_7$ can also be isolated.
E_2O_6	Dimer of ClO_3 that is in equilibrium with the monomer in the gas phase. Decomposes to ClO_2 and O_2.		I_2O_6 has been reported from decomposition of I_2O_7.
E_2O_7	Distillable colorless liquid	$O_3BrOBrO_3$ is a known but unstable compound.	Has reportedly been formed by dehydration of H_5IO_6 with SO_3, but unlike the oxo-bridged Cl and Br derivatives had a polymeric structure and decomposed to I_2O_6 and O_2.
E_4O_9			Formed by treating HIO_3 with H_3PO_4 or by ozonolysis of I_2. Decomposes at 70 °C in the presence of water it decomposes to I_2O_5. I_4O_9 is formulated as $\{I\}^{3+}\{I^{5+}O_3\}^-{}_3$.
EO_4^- $HOEO_3$ = perhalic acid EO_4^- = perhalate	Perchlorate salts can be made with numerous cations, all of which should be handled as if, and likely are, explosive.	$HOBrO_3$ is stable in aqueous solution. Perbromate, BrO_4^-, is a stronger oxidant than either perchlorate or periodate. Pure $KBrO_4$ is stable even to 275 °C, and NH_4BrO_4 is stable to 170 °C.[165]	Iodine forms stable periodic acid, $HOIO_3$, but related $(HO)_5IO$ with iodine is also in the +7 oxidation state. $(HO)_5IO$ is protonated in strong acids to octahedral $I(OH)_6^+$.[165] Periodanate oxidizes manganous ion to permanganate, and periodic acid and periodanate are common oxidants in organic chemistry.

9.8.2 Hypervalent Iodine Chemistry[157–158]

The chemistry of iodine in oxidation state +3 and higher is extensive and is worthy of note for its utility and diversity. Some of the iodine oxides and oxyanions are shown in Table 9.7. In this section, a few common hypervalent iodine reagents will be discussed.

Addition of a slight excess of peracetic acid to iodobenzene produces iodosobenzene diacetate (Fig. 9.53). As expected from VSEPR, the compound is T-shaped. The probable major resonance forms involve a 3-center 4-electron bond along the O–I–O vector involving the electronegative acetate groups. The O–I–O angle differs somewhat from linear, 164.2(3)°,[166] which may be a result of weak I–O interactions with the carbonyl (C=O) oxygens.[167]

Chlorination of IPh produces the related T-shaped chlorinating agent Cl_2IPh—a yellow solid that is often easier to handle than chlorine gas in the laboratory (Section 13.2.1).[168–169]

Polymeric iodosobenzene, "OIPh", is prepared by basic hydrolysis of iodosobenzene diacetate or dichloride. The structure of iodosobenzene is not known by X-ray diffraction but probably involves bridging oxo groups involved in near-linear 3c4e bonds (Fig. 9.53).

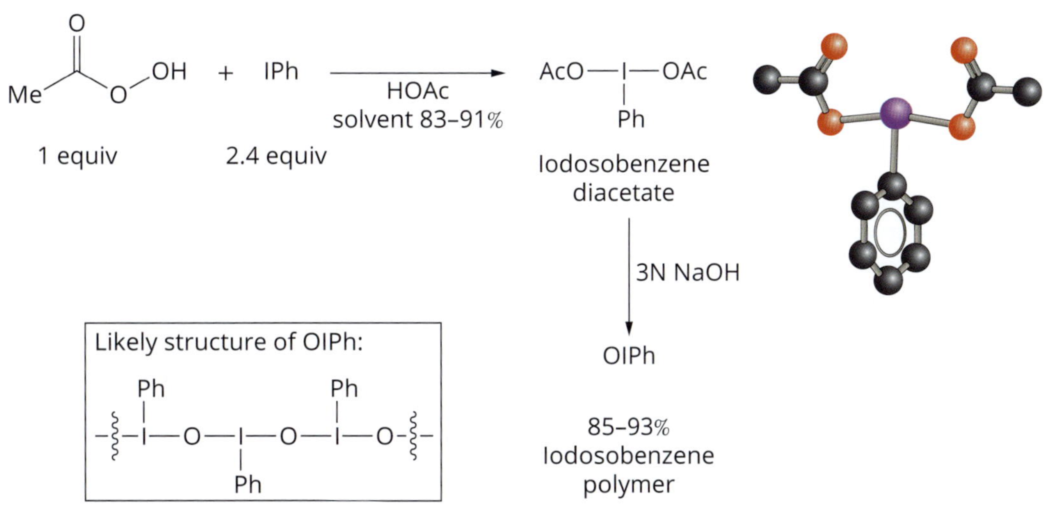

Fig. 9.53 Synthetic pathway to iodosobenzene, the structures of iodosobenzene diacetate from the X-ray diffraction coordinates[166] and the probable structure of iodosobenzene.

A commonly employed oxidant, Dess–Martin periodinane (DMP) is prepared as shown in Fig. 9.54.[170] The synthesis involves the generation of an intermediate hydroxy-oxo iodine(V) compound from the oxidation of 2-carboxylic acid iodobenzene with $KBrO_3$ in acidic solution. The intermediate is *cis*-divacant pseudo-octahedral with the chelating oxygen about 180° from the OH and every other substituent about 90° from each other. Addition of acetic anhydride in acetic acid produces DMP, which is best described as square pyramidal with an axial aryl ring.[171–172]

9.20

DMP is a common reagent for the conversion of alcohols to aldehydes and ketones, due to solubility and clean reactivity (Eq. 9.20). It should be noted that some explosions have occurred with DMP.

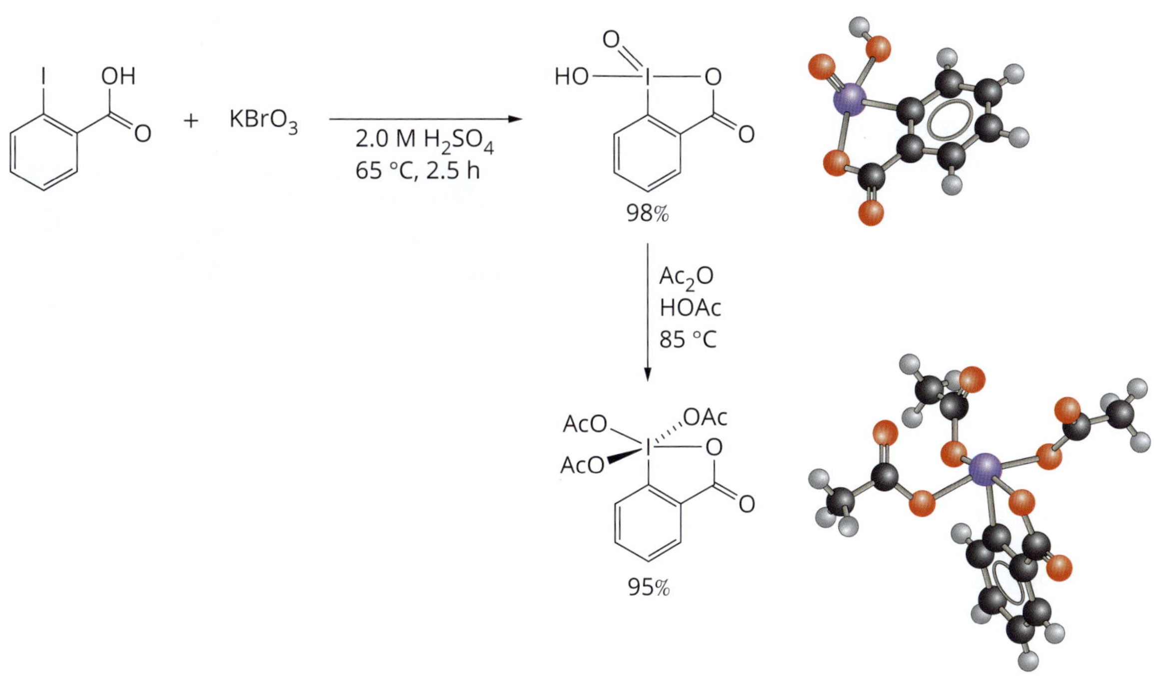

Fig. 9.54 Synthesis of Dess–Martin periodinane and structures of an intermediate and DMP.

9.9 From "Inert" to "Noble" Gases (Group 18)

Features of Group 18 Chemistry

- The Group 18 elements were once called the "inert gases", but Neil Bartlett's discovery of their reactivity led to their renaming as "noble".
- Compounds containing very electronegative elements such as oxygen, chlorine, and fluorine can be used to prepare compounds of the heaviest Group 18 elements, especially xenon.
- XeF_2 can be prepared by photochemical reaction of xenon and F_2 gas.
- Some metal complexes are known where xenon acts as a ligand for the metal center, including homoleptic $AuXe_4^{2+}$.

Group 18 elements came to be known as the "inert gases" and were sometimes referred to as the "rare gases". Indeed, all the elements of the group have relatively weak interatomic interactions due to filled valence shells, and are monoatomic gases under standard conditions. However, argon constitutes about 1% of dry air, making it not really "rare", and Neil Bartlett's seminal discovery of xenon chemistry meant that the group was no longer "inert". The group name has since been changed to "noble" to signify that they only rarely interact with the other elements.

Bartlett's insight should be underscored here. He discovered that PtF_6 was a strong enough oxidant to oxidize dioxygen to form the salt $O_2^+\{PtF_6\}^-$.[173] He surmised that since the ionization potentials of O_2 (12.071 eV) and Xe (12.130 eV) are quite similar, PtF_6 should be a strong enough oxidant to activate the noble gas as well. The reaction of xenon gas with PtF_6 led to a new salt eventually identified as $\{XeF\}^+\{Pt_2F_{11}\}^-$—the first compound of a noble, *née* inert, gas. In fact, within a year of Bartlett's famous experiment a staggering number of noble gas compounds appeared, such as XeF_2, KrF_2, XeF_4, XeF_6, and XeO_3.

One difficulty, especially with the lighter congeners, in activating the noble gases is their extremely high electronegativity. Helium is the most electronegative element, although fluorine is often cited as such. The Allred–Rochow electronegativity of helium is an astonishingly high 5.50, and the electronegativity of neon at 4.84 is still well over that of fluorine (4.10 on this scale). In short, there is simply nothing in the periodic table that will effectively remove electrons from these elements, nor do these closed-shell atoms accept electrons, and, as a consequence, there are no known compounds of He or Ne stable at room temperature and pressure.

Even as the electronegativity of the elements diminishes down the column (Allred–Rochow electronegativity: $\chi_{Ar}^{A-R} = 3.20$, $\chi_{Kr}^{A-R} = 2.94$, and $\chi_{Xe}^{A-R} = 2.40$) one must still generate thermodynamically stable compounds, which is most often accomplished by using elements that tend to have very high bond strengths like oxygen and fluorine. However, compounds containing noble-gas bonds to Cl, Br, I, N, C, B, S, H, all of the first-row transition elements, Rh, Ru, Ag, Au, Re, U, and Hg, among others, are also known, typically with xenon.[174–175]

Argon produces no stable compounds, but some reports of species such as HArF have appeared that suggest that compounds can be prepared at low temperature. KrF_2 can be prepared at low temperature by applying electric discharge to a mixture of Kr and F_2 gas. By far the most common noble-gas element in compounds is xenon. By nature, compounds of xenon are hypervalent, with a rich chemistry often involving 3-center 4-electron bonds and negative hyperconjugation.

Outside of laboratories studying noble-gas chemistry specifically, perhaps the only compound from this class of compounds that might be called a "common laboratory reagent" is xenon difluoride, XeF_2. Xenon difluoride is most easily prepared by irradiating a mixture of xenon and difluorine gas with sunlight, which provides sublimable, colorless crystals with a melting point of 129 °C (Fig. 9.55). However, the compound can be made by heating F_2 and Xe to 400 °C in a nickel vessel and from heating CF_4 and Xe. The compound is a commercially available mild fluorinating agent.[176–177]

Xenon and krypton will also bind to Lewis acidic transition-metal complexes. One fascinating example is square planar (Fig. 9.56) tetra(xenono)gold(II) $\{AuXe_4\}\{Sb_2F_{11}\}_2$, prepared by Seppelt and Seidel in 2000. The structure from X-ray diffraction shows Au–Xe bond lengths of ~2.73 Å and long Au–F contacts to the $Sb_2F_{11}^-$ counterions. The compound is stable to about −40 °C, where it begins to lose gaseous xenon.[178]

Fig. 9.55 Synthesis of XeF_2 from Xe, F_2, and sunlight. (Illustration from Tramšek and Žemva.[177])

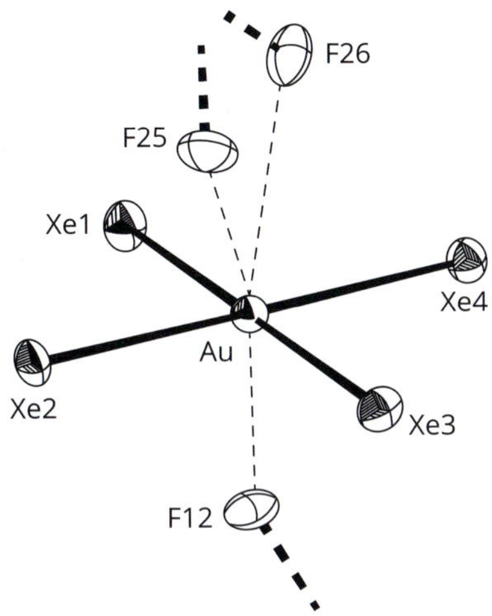

Fig. 9.56 Structure of tetra(xenono)gold(II), $\{AuXe_4\}^{2+}$.[178]

9.10 Concluding Comments on the Frontiers of Main-Group Chemistry

Naturally, since the main group occupies a large portion of the periodic table, the elements discussed in this chapter have almost endless applications in numerous areas. Even so, new utility is constantly being found in the chemistry of the main group. While most chemists may think of the transition metals when considering sophisticated applications, especially in catalysis, the main-group elements are growing in their complexity for a variety of purposes.[179]

9.10.1 "Frustrated" Lewis Pairs[180–181]

The general dogma was that activation of small molecules such as hydrogen required the use of transition metals when it was discovered in 2006 that certain combinations of phosphines, PR_3, and boranes, BR_3, would generate phosphonium borate salts, $\{HPR_3\}\{HBR_3\}$, on addition of H_2. For activity, the phosphine and borane chosen cannot form a strong dative B–P bond, which is prevented by steric encumbrance. The system prevented from formation of a strong acid–base adduct is well positioned for heterolytic activation of hydrogen. For example, addition of $P(t\text{-Bu})_3$ to $B(C_6F_5)_3$ leads to no reaction due to the steric constraints of the system, but addition of H_2 rapidly provides $\{HP(t\text{-Bu})_3\}^+\{HB(C_6F_4)_3\}^-$. These combinations of Lewis acids and Lewis bases that show this type of activity have been called "frustrated Lewis pairs" (FLP), because they are sterically prevented from forming strong acid–base bonds.

The original system where reversible H_2 activation was observed by Stephan and coworkers is shown in Fig. 9.57. The system under study was an arene-bridged phosphinoborane, **A**.[182–183] The phosphinoborane catalyzed the hydrogenation of some substrates like imines.

In principle, any class of Lewis acid and Lewis base could show this activity towards the heterolytic cleavage or activation of small molecules; although boranes are the usual Lewis acid. Such reactivity has been described with phosphine/borane, *N*-heterocyclic carbene/borane, and nitrogen bases/borane systems, for example.

Indeed, it was soon found that some polar substrates such as imines can catalyze their own reduction in the presence of $B(C_6F_5)_3$, where the nitrogen substrate acts as the base in the FLP. The proposed catalytic cycle for such a system is shown in Fig. 9.58. Such systems can be used to reduce imine, nitriles, *N*-heterocycles, and aziridines.[184]

FLPs have been used to activate cyclic ethers, cyclopropanes, olefins, lactones, lactides, dienes, alkynes, CO_2, N_2O, SO_2, and other systems. They constitute a fascinating and active field of catalysis as well. While FLPs are generally not as fast as transition metals systems, do not tolerate impurities as well, and do not have as broad a functional group tolerance as their *d*-block competitors, they have proven to have quite interesting reactivity for a number of purposes since their discovery in 2006.

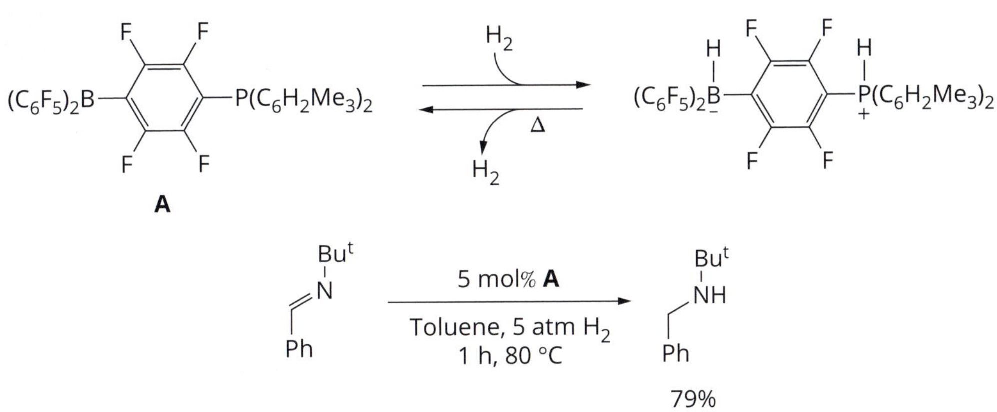

Fig. 9.57 Structure of Stephan and coworkers' initial frustrated Lewis pair system, reversible binding of H_2, and an example of its reactivity, reduction of an imine.

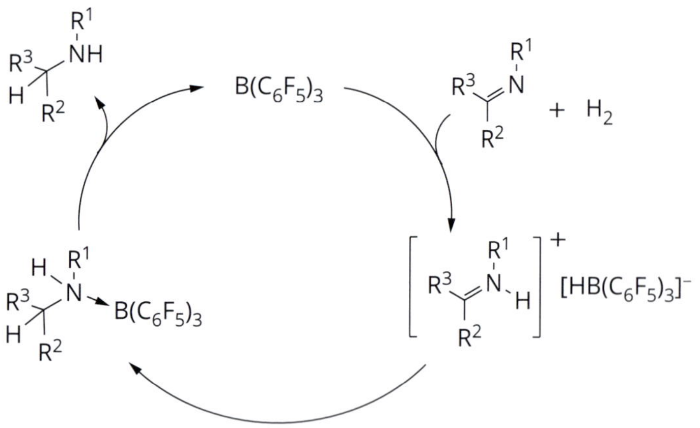

Fig. 9.58 Proposed mechanism for the reduction of an imine by $B(C_6F_5)_3$.

Transition-metal catalysts may be better for most applications at the moment but have undergone decades of development; the future looks quite bright for FLP and related systems.[185]

9.10.2 Oxidative Addition in Main-Group Catalysis

While oxidative addition and reductive elimination catalytic cycles are considered the sole domain of transition-metal systems as well, Radosevich and coworkers have shown that main-group elements will catalyze transfer hydrogenation of azobenzene from amine-borane, H_3NBH_3, using a P(III)/P(V) cycle involving hydrogen oxidative addition to phosphorus (Fig. 9.59).[186–187]

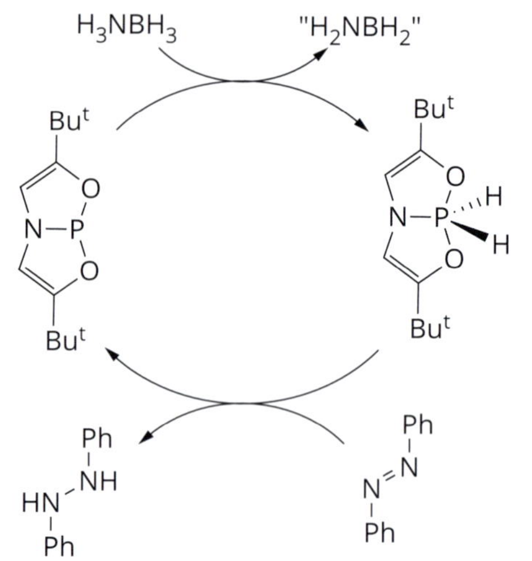

Fig. 9.59 Proposed mechanism for the reduction of azobenzene by a phosphorus center.

As usual in synthetic inorganic chemistry, the ingenuity of the chemist is the only limit to the possible applications within the main group, and almost any reaction that can be conceived may be possible.

Chapter Summary

1. The alkali and alkaline earth elements are common reductants, with Group 1 easily giving up one electron and Group 2 easily giving up two electrons.
2. Reaction of lithium or magnesium with alkyl halides produces metal alkyls that are common reagents in organic synthesis. Common alkyl lithiums are methyl, *n*-butyl, and *t*-butyl lithium, which are clusters with alkyl groups often capping trigonal faces of lithium polyhedra. Many magnesium alkyls are known, and RMgX compounds are generally called Grignard reagents.
3. Boron forms a large class of B_nH_{n+m} compounds called boranes; compare with alkanes. These often form clusters with complex bonding patterns, including τ-bonding in addition to 2-center 2-electron bonds.
4. As one proceeds down Group 13 and Group 14, there is a stabilization of compounds with a lone pair on the heavier elements. While Al(III) is the most common oxidation state for aluminum, thallium(I) is quite common. Similarly, silicon(IV) is the most common oxidation state, but Pb(II) is quite stable. This is sometimes called the "inert pair effect", which is due to changing bond energies as one proceeds down the group.
5. Higher oxidation-state phosphine and sulfur compounds are well known and are often readily synthetically accessible. These compounds often involve π-bonds stabilized by negative hyperconjugation or additional σ-bonds stabilized by 3-center 4-electron bonds. The lighter congeners, nitrogen and oxygen, are rarely hypervalent due to their poor participation in these bonding types.
6. As one proceeds down Group 15 and Group 16, the heavier congeners are less likely to be in high oxidation states than the second-row elements phosphorus or sulfur.
7. Carbon-based ylides such as the Wittig reagent $H_2C{=}PPh_3$ and the sulfur ylide $H_2C{=}SMe_2$ are important reagents for organic synthesis. An ylide is a compound with a 1,2-dipole, and the compounds (although often written as in the sentence previous) have major resonance contributions from $H_2C^{+}{-}^{-}PPh_3$ and $H_2C^{+}{-}^{-}SMe_2$.
8. Wittig reagents often react without redox chemistry. $H_2C{=}PPh_3$ reacts with O=C(H)R to produce phosphine oxide and olefin. In contrast, sulfur ylides often react with oxidation state changes. $H_2C{=}SMe_2$ may react with an aldehyde to produce an epoxide and SMe_2.
9. The vast majority of halogen compounds have the halogen in the −1 oxidation state. However, there is a very rich chemistry of some of the heavier congeners in higher oxidation states, especially iodine.
10. Very electronegative elements, such as F, Cl, and O, can be used to make compounds with some of the heavier noble gases, especially xenon, where many compounds are known.

Practice Problems

1. In the Wittig reaction, a reagent such as $CH_2{=}PPh_3$ is reacted with a carbonyl compound, such as an aldehyde, to form an olefin and $O{=}PPh_3$. However, the double bond in an olefin is weaker than in an aldehyde. For example, the BDE for the double bond in ethylene is 164 kcal/mol, while the double bond in formaldehyde is 175 kcal/mol.[188] Consequently, the change in bond energy in the phosphine compounds must be able to compensate for this unfavorable transformation in the organic products. If one examines the thermodynamics of the model reactions below at the G3 level of theory, one finds that the reaction with the $CH_2{=}PH_3$ reagent is favorable, having $\Delta G_{rxn} = -50$ kcal/mol. In addition, use of the reagent $CH_2{=}PF_3$ would make the reaction even more thermodynamically favorable, $\Delta G_{rxn} = -56$ kcal/mol.
   a. Explain why the OPX_3 (X = H or F) is stronger than the $H_2C{=}PX_3$ bond. In your answer, consider both σ- and π-effects.
   b. Based on your answer in part (a), suggest why the X = F reaction might be slightly more exothermic.
2. There are several important effects in main group chemistry like the:
   a. relativistic effect
   b. primogenic effect
   c. inert pair effect
   d. lone pair bond weakening effect

   For (a)–(d) above:

   (i) *Briefly but clearly* describe what leads to the effect and its consequences on electronic structure, bonding, etc.

   (j) Name at least one important consequence of the effect on the chemistry or properties of a *specific* main group compound, element, etc.
3. a. Using Polar Covalence Theory, calculate the atomization energies of S_8 and four equivalents of S_2. Then, calculate the atomization energies of hypothetical O_8 and four equivalents of O_2. Explain why sulfur is found as single bond containing S_8 while oxygen exists as double bond containing O_2. (O–O distance in O_2 = 1.207 Å, S–S distance in S_2 = 1.887 Å, assume the single bond distances in O_8 and S_8 are twice the covalent radii.)
   b. Should SO gas be more stable as the diatomic or as the S_4O_4 ring with alternating sulfurs and oxygens. (S–O distance in SO = 1.493 Å, assume the O–S distances are the sum of the covalent radii in the ring molecule).
   c. Search for more information on sulfur monoxide. What structure does it adopt and what are its properties?
4. A mild method for the oxidation of a primary or secondary alcohol to the aldehyde or ketone is Swern oxidation. The reagent employed is SCl_2Me_2, which is prepared in situ by reaction of dimethylsulfoxide (DMSO) with oxalyl chloride.

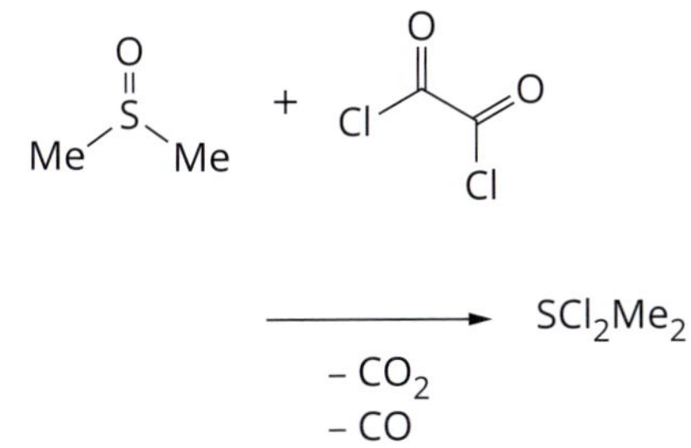

The alcohol is then reacted with SCl_2Me_2 to give the aldehyde/ketone and two equivalents of HCl.

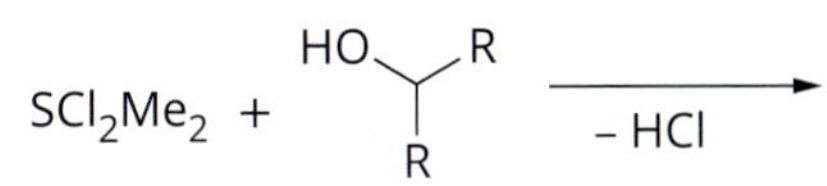

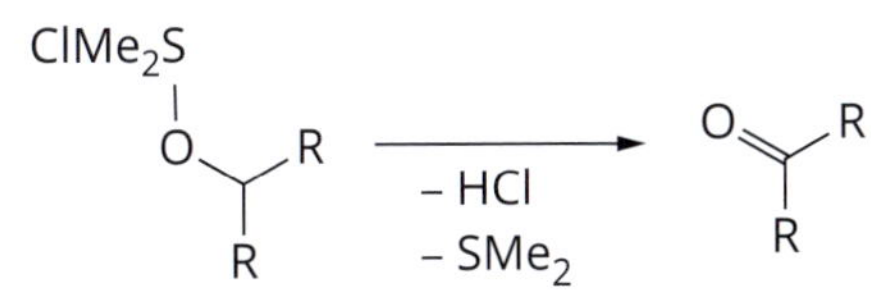

   a. What do you expect for the structure of SCl_2Me_2? Give approximate angles for Me–S–Me, Cl–S–Cl, and Cl–S–Me.
   b. In one short paragraph describe the bonding in SCl_2Me_2.
   c. What is the structure of intermediate $SCl(OR')Me_2$? Why does the compound adopt this structure?
   d. In the intermediate $SCl(OR')Me_2$ why is the S–Cl bond oxidizing?
   e. There is a structurally characterized example in the CCSD database of an SR_2Cl_2 compound.[189] There are no related bromide derivatives. Why might the bromide derivatives be less stable?
5. Use polar covalence theory to show why CO_2 is a molecular gas and SiO_2 is an extended solid under normal conditions. (C=O distance in CO_2 = 1.16 Å, Si=O distance in molecular SiO_2 = 1.52 Å. For single bond distances, use the sum of covalent radii.)
6. You prepare the compound $SeMe_2Cl_2$. What structure do you expect for the compound, and why?
7. For Groups 13 and 14, the heavier congeners tend to have lower oxidation states; however, for the halogens, iodine has the most known hypervalent compounds by far. Explain why.
8. In 2002, Akiba and coworkers published a paper in *Tetrahedron: Asymmetry* on the synthesis of a selenium compound.[190]
   a. The synthesis begins with $O{=}Se(OEt)_2$ as the starting material, which can be prepared from SeO_2 and HOEt. Use VSEPR to determine the geometry of SeO_2 and $O{=}Se(OEt)_2$.
   b. The reaction of $O{=}Se(OEt)_2$ with the Grignard/alkoxide reagent shown below gives the product. The carbon and oxygen bound to magnesium in the starting material are both bound to selenium, and two of these ligands are placed on Se. Both the oxygen and the OEt groups on $O{=}Se(OEt)_2$ are replaced to give $Se(OC)_2$. What geometry do you expect for the product, and why?

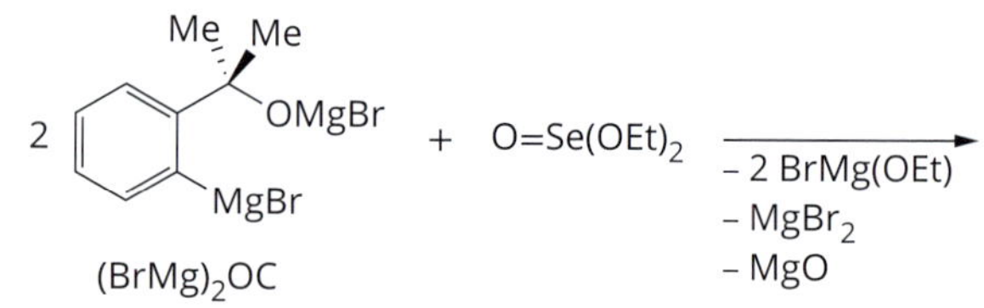

c. As mentioned, this compound (product of the previous part) was published in a journal devoted to asymmetric chemistry. In general, compounds without an improper axis of any kind are chiral; this includes S_1 (mirror planes), S_2 (inversion), and any other order of improper axis. What is the point group of the product, and prove that it is chiral? What are all the symmetry operations of the molecule, and are any of them an improper axis of any kind?

9. BrSePh is a reagent used in organic chemistry to prepare chiral compounds. Many of the reactions involve seleniranium ions, 3-membered rings formed between $PhSe^+$ and an olefin as shown in the example below.[191]

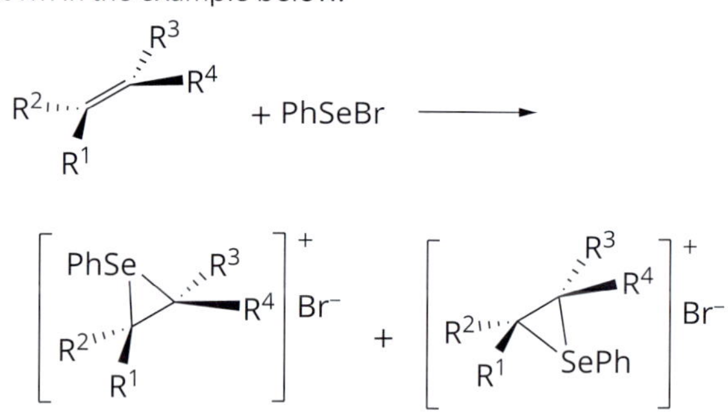

These seleniranium ions are used as intermediates in syntheses like the one below.

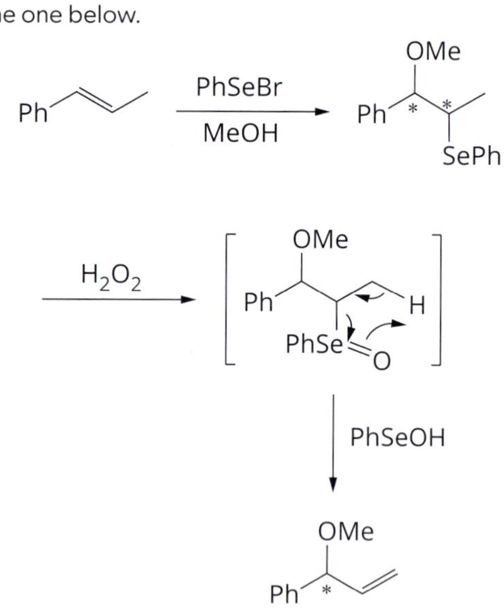

a. Consider a simplified version of the seleniranium ion (shown below) whose geometry was calculated using B3LYP/6-31G**. What is the formal oxidation state of the selenium? What is the geometry around selenium, and how many lone pairs are there? What hybrids on carbon are used to bond to selenium?

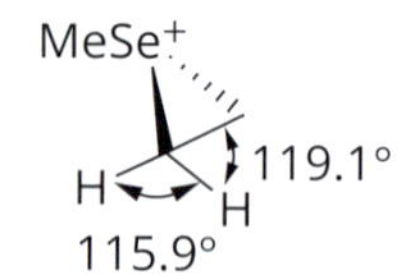

b. The C–C–Se angle is 69°. Calculate the bond bending angle on carbon in the C–Se bond assuming there is no bond bending in the C–H or C–C bonds.

c. BrSePh is made by addition of Br_2 to PhSe–SePh. Calculate the formal oxidation state of Br and Se in the reactants and product. What is being oxidized? What is being reduced?

$$Br_2 + PhSeSePh \rightarrow 2BrSePh$$

d. Thus far, we have been drawing BrSePh as a monomer, and this is commonly how the compound is described in the literature. In actuality, the compound is a tetramer with the interesting solid-state structure shown below in two different views. What is the point group of the compound?[192]

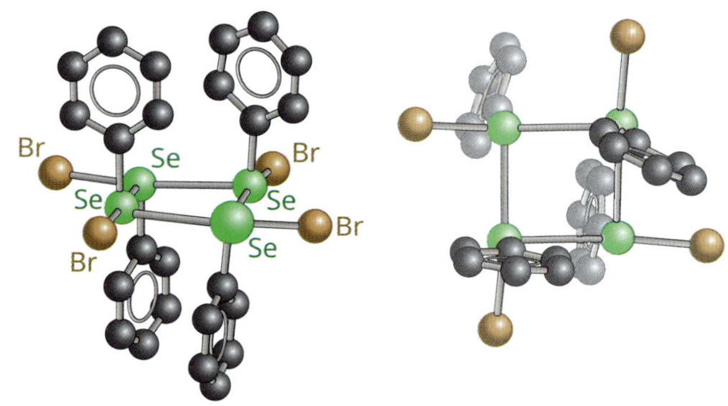

10. The compound SO has a triplet ground state like O_2 but reversibly dimerizes, producing O=S–S=O with an O–S–S angle of 113°. The compound has an O–S–S–O angle of 0°, i.e., the compound is planar. Explain why the compound is planar. (Hint: Are there lone pairs on sulfur? Is there an interaction discussed in the chapter that would lead to S–S multiple bonding?)

Cited References

1. Sidgwick, N. V., Molecules *Science* 1937, 86, 335–340.
2. Cacace, F.; de Petris, G.; Troiani, A., Experimental detection of tetranitrogen *Science* 2002, *295*, 480–481.
3. Parakhonskiy, G.; Dubrovinskaia, N.; Bykova, E.; Wirth, R.; Dubrovinsky, L., Experimental pressure-temperature phase diagram of boron: resolving the long-standing enigma *Scientific Reports* 2011, *1*.
4. Zhao, Z.; Tian, F.; Dong, X.; Li, Q.; Wang, Q.; Wang, H.; Zhong, X.; Xu, B.; Yu, D.; He, J.; Wang, H.-T.; Ma, Y.; Tian, Y., Tetragonal Allotrope of Group 14 Elements *J. Am. Chem. Soc.* 2012, *134*, 12362–12365.
5. Greenwood, N. N.; Earnshaw, A., *Chemistry of the Elements* 1st ed.; Pergamon Press: Oxford 1984.
6. Allen, M. J.; Tung, V. C.; Kaner, R. B., Honeycomb Carbon: A Review of Graphene *Chem. Rev.* 2010, *110*, 132–145.
7. Li, B.; Li, Z.; Yang, J.; Hou, J. G., STM studies of single molecules: Molecular orbital aspects *Chem. Commun.* 2011, *47*, 2747–2762.
8. Wang, X.; Li, Q.; Xie, J.; Jin, Z.; Wang, J.; Li, Y.; Jiang, K.; Fan, S., Fabrication of Ultralong and Electrically Uniform Single-Walled Carbon Nanotubes on Clean Substrates *Nano Lett.* 2009, *9*, 3137–3141.
9. Zhang, R.; Zhang, Y.; Zhang, Q.; Xie, H.; Qian, W.; Wei, F., Growth of Half-Meter Long Carbon Nanotubes Based on Schulz-Flory Distribution *Acs Nano* 2013, *7*, 6156–6161.
10. Kovacs, A.; Esterhuysen, C.; Frenking, G., The nature of the chemical bond revisited: An energy-partitioning analysis of nonpolar bonds *Chem Eur. J.* 2005, *11*, 1813–1825.
11. Forslund, L. E.; Kaltsoyannis, N., Why is the F-2 bond so weak? A bond energy decomposition analysis *New J. Chem.* 2003, *27*, 1108–1114.
12. Schrauzer, G. N., Lithium: Occurrence, dietary intakes, nutritional essentiality *J. Am. Coll. Nut.* 2002, *21*, 14–21.
13. O'Hare, P. A. G.; Johnson, G. K., Lithium nitride (Li_3N) - standard enthalpy of formation by solution calorimetry *J. Chem. Thermodyn.* 1975, *7*, 13–20.
14. Connelly, N. G.; Geiger, W. E., Chemical redox agents for organometallic chemistry *Chem. Rev.* 1996, *96*, 877–910.
15. Furstner, A., Chemistry of and with highly reactive metals *Angew. Chem. Int. Ed.* 1993, *32*, 164–189.
16. Kahn, B. E.; Rieke, R. D., Carbonyl coupling reactions using transition-metals, lanthanides, and actinides *Chem. Rev.* 1988, *88*, 733–745.
17. Rieke, R. D., Preparation of highly reactive metal powders and their use in organic and organometallic synthesis *Acc. Chem. Res.* 1977, *10*, 301–306.
18. Rieke, R. D.; Burns, T. P.; Wehmeyer, R. M.; Kahn, B. E., Preparation of Highly Reactive Metal Powders: Some of Their Uses in Organic and Organometallic Synthesis *ACS Symposium Series* 1987, *333*, 223–245.
19. Bergbreiter, D. E.; Killough, J. M., Reactions of potassium-graphite *J. Am. Chem. Soc.* 1978, *100*, 2126–2134.
20. Dye, J. L.; Ceraso, J. M.; Lok, M. T.; Barnett, B. L.; Tehan, F. J., Crystalline salt of sodium anion (Na^-) *J. Am. Chem. Soc.* 1974, *96*, 608–609.
21. Tehan, F. J.; Barnett, B. L.; Dye, J. L., Alkali anions: preparation and crystal-structure of a compound which contains cryptated sodium cation and sodium anion *J. Am. Chem. Soc.* 1974, *96*, 7203–7208.
22. Huang, R. H.; Huang, S. Z.; Dye, J. L., Syntheses and structures of six compounds that contain the sodium anion *J. Coord. Chem.* 1998, *46*, 13–31.
23. Dye, J. L., Electrons as anions *Science* 2003, *301*, 607–608.
24. Redko, M. Y.; Jackson, J. E.; Huang, R. H.; Dye, J. L., Design and synthesis of a thermally stable organic electride *J. Am. Chem. Soc.* 2005, *127*, 12416–12422.
25. Seyferth, D., The Grignard Reagents *Organometallics* 2009, *28*, 1598–1605.
26. Bogdanovic, B., Magnesium anthracene systems and their application in synthesis and catalysis *Acc. Chem. Res.* 1988, *21*, 261–267.
27. Coulson, C. A., Valence angles in group II dihalide molecules *Isr. J. Chem.* 1973, *11*, 683–690.
28. Guido, M.; Gigli, G., Ion model and equilibrium configuration of gaseous alkaline-earth dihalides *J. Chem. Phys.* 1976, *65*, 1397–1402.
29. Kaupp, M.; Schleyer, P. V., The structural variations of monomeric alkaline-earth MX_2 compounds (M = Ca, Sr, Ba, X = Li, BeH, BH_2, CH_3, NH_2, OH, F): An ab initio pseudopotential study *J. Am. Chem. Soc.* 1992, *114*, 491–497.
30. Weinhold, F.; Landis, C. R., *Valency and Bonding: A Natural Bond Orbital Donor-Acceptor Perspective* Cambridge University Press: Cambridge, 2005; p ix, 749 p.
31. Gillespie, R. J.; Robinson, E. A., Models of molecular geometry *Chem. Soc. Rev.* 2005, *34*, 396–407.
32. Gillespie, R. J., The VSEPR model revisited *Chem. Soc. Rev.* 1992, *21*, 59–69.
33. Andersen, R. A.; Blom, R.; Burns, C. J.; Volden, H. V., Synthesis and thermal average gas-phase molecular-structures of bis(pentamethylcyclopentadienyl)-strontium and bis(pentamethycyclopentadienyl)-barium: The 1st organo-strontium and organo-barium structures *Chem. Comm.* 1987, 768–769.
34. Seijo, L.; Barandiaran, Z.; Huzinaga, S., Abinitio model potential study of the equilibrium geometry of alkaline-earth dihalides: MX_2 (M = Mg, Ca, Sr, Ba, X = F, Cl, Br, I) *J. Chem. Phys.* 1991, *94*, 3762–3773.
35. Kaupp, M.; Schleyer, P. V.; Stoll, H.; Preuss, H., The question of bending of the alkaline-earth dihalides MX_2 (M = Be, Mg, Ca, St, Ba, X = F, Cl, Br, I): An ab initio pseudopotential study *J. Am. Chem. Soc.* 1991, *113*, 6012–6020.
36. Williams, R. A.; Hanusa, T. P.; Huffman, J. C., Structures of ionic decamethyl-metallocenes - crystallographic characterization of $(Me_5C_5)_2Ca$ and $(Me_5C_5)_2Ba$ and a comparison with related organolanthanide species *Organometallics* 1990, *9*, 1128–1134.
37. Seyferth, D., Zinc alkyls, Edward Frankland, and the beginnings of main-group organometallic chemistry *Organometallics* 2001, *20*, 2940–2955.
38. OSHA Hazard Information Bulletins: Dimethylmercury. https://www.osha.gov/dts/hib/hib_data/hib19980309.html.
39. Bonyhady, S. J.; Jones, C.; Nembenna, S.; Stasch, A.; Edwards, A. J.; McIntyre, G. J., beta-Diketiminate-Stabilized Magnesium(I) Dimers and Magnesium(II) Hydride Complexes: Synthesis, Characterization, Adduct Formation, and Reactivity Studies *Chem Eur. J.* 2010, *16*, 938-955.
40. Gondzik, S.; Blaser, D.; Wolper, C.; Schulz, S., Novel Synthetic Pathway for New Zn-Zn-Bonded Compounds from Dizincocene *Chem Eur. J.* 2010, *16*, 13599–13602.
41. Green, S. P.; Jones, C.; Stasch, A., Stable Adducts of a Dimeric Magnesium(I) Compound *Angew. Chem. Int. Ed.* 2008, *47*, 9079–9083.
42. Grirrane, A.; Resa, I.; Rodriguez, A.; Carmona, E.; Alvarez, E.; Gutierrez-Puebla, E.; Monge, A.; Galindo, A.; del Rio, D.; Andersen, R. A., Zinc-zinc bonded zincocene structures. Synthesis and characterization of $Zn_2(\eta^5\text{-}C_5Me_5)_2$ and $Zn_2(\eta^5\text{-}C_5Me_4Et)_2$ *J. Am. Chem. Soc.* 2007, *129*, 693–703.
43. Wang, Y. Z.; Quillian, B.; Wei, P. R.; Wang, H. Y.; Yang, X. J.; Xie, Y. M.; King, R. B.; Schleyer, P. V.; Schaefer, H. F.; Robinson, G. H., On the chemistry of Zn-Zn bonds, RZn-ZnR (R = $[\{(2,6\text{-}(Pr^i{}_2C_6H_3))N(Me)C\}(2)CH]$): Synthesis, structure, and computations *J. Am. Chem. Soc.* 2005, *127*, 11944–11945.
44. Zhu, Z.; Brynda, M.; Wright, R. J.; Fischer, R. C.; Merrill, W. A.; Rivard, E.; Wolf, R. R.; Fettinger, J. C.; Olmstead, M. M.; Power, P. P., Synthesis and characterization of the homologous M-M bonded series Ar'MMAr' (M = Zn, Cd, or Hg; Ar' = $C_6H_3\text{-}2,6\text{-}(C_6H_3\text{-}2,6\text{-}Pr^i{}_2)_2$) and related arylmetal halides and hydride species *J. Am. Chem. Soc.* 2007, *129*, 10847–10857.
45. Banh, H.; Gemel, C.; Seidel, R. W.; Fischer, R. A., A solvated zinc analogue of the calomel-dication *Chem. Commun.* 2015, *51*, 2170–2.
46. Blake, M. P.; Kaltsoyannis, N.; Mountford, P., Synthesis, molecular and electronic structure, and reactions of a Zn–Hg–Zn bonded complex *Chem. Commun.* 2015.
47. Wang, X. F.; Andrews, L.; Riedel, S.; Kaupp, M., Mercury is a transition metal: The first experimental evidence for HgF_4 *Angew. Chem. Int. Ed.* 2007, *46*, 8371–8375.
48. Shapiro, P. J.; Atwood, D. A., *Group 13 Chemistry from Fundamentals to Applications* American Chemical Society: Washington, 2002; Vol. 822.
49. Cowley, A. H.; Cushner, M. C.; Davis, R. E.; Riley, P. E., Crystal and molecular-structure of the 1-2 aluminum trichloride-tetrahydrofuran complex AlCl3•2THF *Inorg. Chem.* 1981, *20*, 1179–1181.
50. Ahmed, A.; Schwarz, W.; Weidlein, J.; Hess, H., Adducts of aluminum-chloride and gallium chloride with secondary-amines: Crystal and molecular-structure of bis(dimethylamine) aluminum trichloride *Zeitschrift Fur Anorganische Und Allgemeine Chemie* 1977, *434*, 207–216.
51. Grimes, R. N., *Metal Interactions with Boron Clusters* Springer: Boston, MA, 1982.
52. Moore, E. B. J.; Dickerson, R. E.; Lipscomb, W. N., Least Squares Refinements of $B_{10}H_{14}$, B_4H_{10}, and B_5H_{11} *J. Chem. Phys.* 1957, *27*, 209.
53. Forster, D.; Hubschle, C. B.; Luger, P.; Hugle, T.; Lentz, D., On the 2-electron 3-center B-H-B bond: Charge density determination of tetraborane(10) *Inorg. Chem.* 2008, *47*, 1874–1876.
54. Bregadze, V. I., Dicarba-*closo*-dodecaboranes $C_2B_{10}H_{12}$ and their derivatives *Chem. Rev.* 1992, *92*, 209–223.
55. Douvris, C.; Stoyanov, E. S.; Tham, F. S.; Reed, C. A., Isolating fluorinated carbocations *Chem. Commun.* 2007, 1145–1147.

56. Gu, W. X.; McCulloch, B. J.; Reibenspies, J. H.; Ozerov, O. V., Improved methods for the halogenation of the $HCB_{11}H_{11}^-$ anion *Chem. Commun.* 2010, *46*, 2820–2822.
57. Reed, C. A., Carboranes: A new class of weakly coordinating anions for strong electrophiles, oxidants, and superacids *Acc. Chem. Res.* 1998, *31*, 133–139.
58. Reed, C. A., H^+, CH_3^+, and R_3Si^+ Carborane Reagents: When Triflates Fail *Acc. Chem. Res.* 2010, *43*, 121–128.
59. Jemmis, E. D.; Kiran, B.; Coffey, D., Ab initio studies on disubstituted closo-icosahedral heteroboranes, $X_2B_{10}H_{10}$ (X = CH, SiH, N, P, and Sb) *Chem. Ber. Recl.* 1997, *130*, 1147–1150.
60. Seyferth, D.; Buchner, K. D.; Rees, W. S.; Wesemann, L.; Davis, W. M.; Bukalov, S. S.; Leites, L. A.; Bock, H.; Solouki, B., 1,2-Dimethyl-1,2-disila-*closo*-dodecaborane(12), a silicon analog of an O-carborane: Synthesis, X-ray crystal structure, NMR, vibrational and photoelectron spectra and bond *J. Am. Chem. Soc.* 1993, *115*, 11656–11656.
61. Seyferth, D.; Buchner, K. D.; Rees, W. S.; Wesemann, L.; Davis, W. M.; Bukalov, S. S.; Leites, L. A.; Bock, H.; Solouki, B., 1,2-dimethyl-1,2-disila-*closo*-dodecaborane(12), a silicon analog of an *o*-carborane - synthesis: X-ray crystal-structure, NMR, vibrational, and photoelectron-spectra and bonding *J. Am. Chem. Soc.* 1993, *115*, 3586–3594.
62. Little, J. L.; Kester, J. G.; Huffman, J. C.; Todd, L. J., Synthesis of polyhedral phosphaboranes: Crystal-structure of 6-$(C_2H_5)_3N$•2-PB_9H_8 *Inorg. Chem.* 1989, *28*, 1087–1091.
63. Chen, Z. F.; King, R. B., Spherical aromaticity: Recent work on fullerenes, polyhedral boranes, and related structures *Chem. Rev.* 2005, *105*, 3613–3642.
64. Grimes, R. N., Boron clusters come of age *J. Chem. Educ.* 2004, *81*, 658–672.
65. King, R. B., Three-dimensional aromaticity in polyhedral boranes and related molecules *Chem. Rev.* 2001, *101*, 1119–1152.
66. Liu, S. A.; Han, Y. F.; Jin, G. X., Formation of direct metal-metal bonds from 16-electron "pseudoaromatic" half-sandwich complexes $Cp''ME_2C_2(B_{10}H_{10})$ *Chem. Soc. Rev.* 2007, *36*, 1543–1560.
67. Martin, D. L., *Borax to Boranes* American Chemical Society: Washington, 1961; Vol. 32.
68. Matteson, D. S., *Stereodirected Synthesis with Organoboranes* Springer: Berlin, 1995.
69. Marder, T. B.; Lin, Z.; Aldridge, S., *Contemporary Metal Boron Chemistry I: Borylenes, Boryls, Borane σ-Complexes and Borohydrides* Springer: Berlin, 2008; Vol. 130.
70. Wade, K., *Electron Deficient Compounds* Springer: Boston, MA, 1971.
71. Grimes, R. N., Synthesis and serendipity in boron chemistry: A 50 year perspective *J. Organomet. Chem.* 2013, *747*, 4–15.
72. Grimes, R. N., Carboranes in the chemist's toolbox *Dalton Trans.* 2015, *44*, 5939–5956.
73. Drago, R. S., Thermodynamic evaluation of the inert pair effect *J. Phys. Chem.* 1958, *62*, 353–357.
74. Schwerdtfeger, P.; Heath, G. A.; Dolg, M.; Bennett, M. A., Low valencies and periodic trends in heavy element chemistry: A theoretical-study of relativistic effects and electron correlation-effects in group-13 and period-6 hydrides and halides *J. Am. Chem. Soc.* 1992, *114*, 7518–7527.
75. Hoffman, R. S., Thallium toxicity and the role of Prussian blue in therapy *Tox. Rev.* 2003, *22*, 29–40.
76. Tacke, M.; Schnockel, H., Metastable AlCl as a solid and in solution *Inorg. Chem.* 1989, *28*, 2895–2896.
77. Dohmeier, C.; Robl, C.; Tacke, M.; Schnockel, H., The tetrameric aluminum(I) compound $(Al(\eta^5\text{-}C_5Me_5))_4$ *Angew. Chem. Int. Ed.* 1991, *30*, 564–565.
78. Roesky, H. W.; Kumar, S. S., Chemistry of aluminium(I) *Chem. Commun.* 2005, 4027–4038.
79. Nagendran, S.; Roesky, H. W., The chemistry of aluminum(I), silicon(II), and germanium(II) *Organometallics* 2008, *27*, 457–492.
80. Gorden, J. D.; Voigt, A.; Macdonald, C. L. B.; Silverman, J. S.; Cowley, A. H., A Lewis acid adduct of an alanediyl: An aluminum(I)-boron donor-acceptor bond *J. Am. Chem. Soc.* 2000, *122*, 950–951.
81. Arduengo, A. J., Looking for stable carbenes: The difficulty in starting anew *Acc. Chem. Res.* 1999, *32*, 913–921.
82. Bohm, V. P. W.; Herrmann, W. A., The "Wanzlick equilibrium" *Angew. Chem. Int. Ed.* 2000, *39*, 4036–4038.
83. Bourissou, D.; Guerret, O.; Gabbai, F. P.; Bertrand, G., Stable carbenes *Chem. Rev.* 2000, *100*, 39–91.
84. Cardin, D. J.; Cetinkay, B; Lappert, M. F., Transition metal-carbene complexes *Chem. Rev.* 1972, *72*, 545.
85. Herrmann, W. A., N-heterocyclic carbenes: A new concept in organometallic catalysis *Angew. Chem. Int. Ed.* 2002, *41*, 1290–1309.
86. Gok, Y.; Cetinkay, E.; Ozdemir, I.; Cetinkay, B.; Lappert, M. F., Synthesis and Characterisation of N-Functionalized Enetetramines, and their properties *Acta Chim. Slov.* 2004, *51*, 437–452.
87. Tokitoh, N.; Suzuki, H.; Okazaki, R.; Ogawa, K., Synthesis, structure, and reactivity of extremely hindered disilenes: The 1st example of thermal-dissociation of a disilene into a silylene *J. Am. Chem. Soc.* 1993, *115*, 10428–10429.
88. Leites, L. A.; Zabula, A. V.; Bukalov, S. S.; Korlyukov, A. A.; Koroteev, P. S.; Maslennikova, O. S.; Egorov, M. P.; Nefedov, O. M., Experimental and theoretical study of vibrational spectra and structure of dihalogermylene and dihalostannylene complexes with 1,4-dioxane and triphenylphosphine *J. Mol. Struct.* 2005, *750*, 116–122.
89. Goldberg, D. E.; Hitchcock, P. B.; Lappert, M. F.; Thomas, K. M.; Thorne, A. J.; Fjeldberg, T.; Haaland, A.; Schilling, B. E. R., Subvalent group-4b metal alkyls and amides .9. Germanium and tin alkene analogs, the dimetallenes M_2R_4 M = Ge or Sn, R = $CH(SiMe_3)_2$: X-ray structures, molecular-orbital calculations for M_2H_4, and trends in the series $M_2R'_4$ (M = C, Si, Ge, or Sn, R' = R, Ph, C_6H_2-2,4,6-Me_3, or C_6H_3-2,6-Et_2 *Dalton Trans.* 1986, 2387–2394.
90. Gynane, M. J. S.; Harris, D. H.; Lappert, M. F.; Power, P. P.; Riviere, P.; Rivierebaudet, M., Subvalent group-4b metal alkyls and amides. 5. Synthesis and physical-properties of thermally stable amides of germanium(II), tin(II), and lead(II) *Dalton Trans.* 1977, 2004–2009.
91. Lappert, M. F.; Slade, M. J.; Atwood, J. L.; Zaworotko, M. J., Monomeric, colored germanium(II) and tin(II) di-t-butylamides, and the crystal and molecular-structure of $Ge(NCMe_2\ [CH_2]_3CMe_2)_2$ *Chem. Comm.* 1980, 621–622.
92. Tokitoh, N.; Okazaki, R., Recent topics in the chemistry of heavier congeners of carbenes *Coord. Chem. Rev.* 2000, *210*, 251–277.
93. Chorley, R. W.; Hitchcock, P. B.; Lappert, M. F.; Leung, W. P.; Power, P. P.; Olmstead, M. M., Subvalent group-14 metal-compounds. 14. The x-ray crystal-structures of 2 monomeric group-14 metal bisamides, $Ge[N(SiMe_3)_2]_2$ and $Sn[NC(Me)_2(CH_2)_3CMe_2]_2$ *Inorg. Chim. Acta* 1992, *198*, 203–209.
94. Bellamy, F. D.; Ou, K., Selective reduction of aromatic nitro-compounds with stannous chloride in non-acidic and non-aqueous medium *Tetrahedron Lett.* 1984, *25*, 839–842.
95. Xing, W. K.; Ogata, Y., Steric acceleration by ortho substituents of the stannous chloride reduction of nitrobenzenes in aqueous ethanol *J. Org. Chem.* 1982, *47*, 3577–3581.
96. Xing, W. K.; Ogata, Y., Effect of meta-substituents and para-substituents on the stannous chloride reduction of nitrobenzenes in aqueous ethanol *J. Org. Chem.* 1983, *48*, 2515–2520.
97. Kimbrough, R. D., Toxicity and health effects of selected organotin compounds: a review *Environ. Health Perspect.* 1976, *14*, 51–6.
98. Sunday, A. O.; Alafara, B. A.; Oladele, O. G., Toxicity and speciation analysis of organotin compounds *Chem. Speciation Bioavailability* 2012, *24*, 216–226.
99. Mason, L. H.; Harp, J. P.; Han, D. Y., Pb Neurotoxicity: Neuropsychological Effects of Lead Toxicity *Biomed. Res. Int.* 2014.
100. Christou, G.; Folting, K.; Huffman, J. C., Mononuclear, 3-coordinate metal thiolates - preparation and crystal-structures of $[NBu^n_4]$ $[Hg(SPh)_3]$ and $[NPr^n_4][Pb(SPh)_3]$ *Polyhedron* 1984, *3*, 1247–1253.
101. Hargittai, M., Molecular structure of metal halides *Chem. Rev.* 2000, *100*, 2233–2301.
102. Vincent, B. R.; Robertson, K. N.; Cameron, T. S.; Knop, O., Alkylammonium lead halides .1. Isolated PBi_6^{4-} ions in $(CH_3NH_3)_4PBI_6$•$2H_2O$ *Can. J. Chem.* 1987, *65*, 1042–1046.
103. Stewart, B.; Harriman, A.; Higham, L. J., Predicting the Air Stability of Phosphines *Organometallics* 2011, *30*, 5338–5343.
104. Himo, F.; Lovell, T.; Hilgraf, R.; Rostovtsev, V. V.; Noodleman, L.; Sharpless, K. B.; Fokin, V. V., Copper(I)-catalyzed synthesis of azoles. DFT study predicts unprecedented reactivity and intermediates *J. Am. Chem. Soc.* 2005, *127*, 210–216.
105. Bebbington, M. W. P.; Bourissou, D., Stabilised phosphazides *Coord. Chem. Rev.* 2009, *253*, 1248–1261.
106. Molina, P.; Lopez-Leonardo, C.; Llamas-Botia, J.; Foces-Foces, C.; Fernandez-Castano, C., Unexpected Staudinger reaction of alpha-azidoacetonitriles alpha-phenyl substituted with triphenylphosphine. Preparation, X-ray crystal and molecular structures of a phosphazine, an aminophosphonium carbanion salt and a phosphazide, with (Z)-configuration *Tetrahedron* 1996, *52*, 9629–9642.
107. Tian, W. Q.; Wang, Y. A., Dynamics of the Staudinger reaction *J. Chem. Theory and Comp.* 2005, *1*, 353–362.

108. Kohn, M.; Breinbauer, R., The Staudinger ligation - A gift to chemical biology' *Angew. Chem. Int. Ed.* 2004, *43*, 3106–3116.
109. Lemieux, G. A.; de Graffenried, C. L.; Bertozzi, C. R., A fluorogenic dye activated by the Staudinger ligation *J. Am. Chem. Soc.* 2003, *125*, 4708–4709.
110. Fickes, M. G.; Davis, W. M.; Cummins, C. C., Isolation and structural characterization of the terminal mesityl azide complex $V(N_3Mes)(I)(NRAr_F)_2$ and its conversion to a vanadium(V) imido complex *J. Am. Chem. Soc.* 1995, *117*, 6384–6385.
111. Proulx, G.; Bergman, R. G., Synthesis and structure of a terminal metal azide complex: An isolated intermediate in the formation of imidometal complexes from organic azides *J. Am. Chem. Soc.* 1995, *117*, 6382–6383.
112. Tran, B. L.; Krzystek, J.; Ozarowski, A.; Chen, C.-H.; Pink, M.; Karty, J. A.; Telser, J.; Meyer, K.; Mindiola, D. J., Formation and Reactivity of the Terminal Vanadium Nitride Functionality *Eur. J. Inorg. Chem.* 2013, *2013*, 3916–3929.
113. Mindiola, D. J.; Cummins, C. C., Deprotonated 2,3:5,6-dibenzo-7-azabicyclo[2.2.1]hepta-2,5-diene as a nitrido nitrogen source by anthracene elimination: Synthesis of an iodide(nitride)chromium(VI) complex *Angew. Chem. Int. Ed.* 1998, *37*, 945–947.
114. Crestani, M. G.; Hickey, A. K.; Gao, X.; Pinter, B.; Cavaliere, V. N.; Ito, J.-I.; Chen, C.-H.; Mindiola, D. J., Room Temperature Dehydrogenation of Ethane, Propane, Linear Alkanes C4-C8, and Some Cyclic Alkanes by Titanium-Carbon Multiple Bonds *J. Am. Chem. Soc.* 2013, *135*, 14754–14767.
115. Harrold, N. D.; Waterman, R.; Hillhouse, G. L.; Cundari, T. R., Group-Transfer Reactions of Nickel-Carbene and -Nitrene Complexes with Organoazides and Nitrous Oxide that Form New C=N, C=O, and N=N Bonds *J. Am. Chem. Soc.* 2009, *131*, 12872-+.
116. Matsunaga, P. T.; Mavropoulos, J. C.; Hillhouse, G. L., Oxygen-atom transfer from nitrous-oxide (N=N=O) to nickel alkyls: Syntheses and reactions to nickel(II) alkoxides *Polyhedron* 1995, *14*, 175–185.
117. Cummins, C. C.; Schrock, R. R.; Davis, W. M., Synthesis of terminal vanadium(V) imido, oxo, sulfido, selenido, and tellurido complexes by imido group or chalcogen atom-transfer to trigonal monopyramidal V{N3N} (N3N = $(Me_3SiNCH_2CH_2)_3{}^{3-}$) *Inorg. Chem.* 1994, *33*, 1448–1457.
118. Cherry, J. P. F.; Johnson, A. R.; Baraldo, L. M.; Tsai, Y. C.; Cummins, C. C.; Kryatov, S. V.; Rybak-Akimova, E. V.; Capps, K. B.; Hoff, C. D.; Haar, C. M.; Nolan, S. P., On the origin of selective nitrous oxide N-N bond cleavage by three-coordinate molybdenum(III) complexes *J. Am. Chem. Soc.* 2001, *123*, 7271–7286.
119. Laplaza, C. E.; Odom, A. L.; Davis, W. M.; Cummins, C. C.; Protasiewicz, J. D., Cleavage of the nitrous-oxide NN bond by a 3-coordinate molybdenum(III) complex *J. Am. Chem. Soc.* 1995, *117*, 4999–5000.
120. Pauleta, S. R.; Dell'Acqua, S.; Moura, I., Nitrous oxide reductase *Coord. Chem. Rev.* 2013, *257*, 332–349.
121. Richter-Addo, G. B.; Legzdins, P., *Metal Nitrosyls* Oxford University Press: New York, 1992.
122. Richter-Addo, G. B.; Legzdins, P.; Burstyn, J., Introduction: Nitric oxide chemistry *Chem. Rev.* 2002, *102*, 857–859.
123. Tota, B.; Trimmer, B., *Nitric Oxide* 1st ed.; Elsevier: Amsterdam, 2007.
124. Ignarro, L. J., *Nitric Oxide: Biology and Pathobiology* 2nd ed.; Elsevier: Amsterdam, 2010.
125. Loscalzo, J.; Vita, J. A., *Nitric Oxide and the Cardiovascular System* Humana Press: Totowa, N.J., 2000; p xi, 601 p.
126. Alderton, W. K.; Cooper, C. E.; Knowles, R. G., Nitric oxide synthases: structure, function and inhibition *Biochem. J* 2001, *357*, 593–615.
127. Henderson, W. A.; Seichter, F. S.; Epstein, M., Some aspects of chemistry of cyclopolyphosphines *J. Am. Chem. Soc.* 1963, *85*, 2462-&.
128. Daly, J. J., Crystal and molecular structure of a triclinic form of phosphobenzene $(PC_6H_5)_6$, referred to as B *J. Chem. Soc. A* 1966, 428.
129. Daly, J. J.; Maier, L., Molecular structure of phosphobenzene *Nature* 1964, *203*, 1167-&.
130. Feldmann, K.-O.; Weigand, J. J., P-N/P-P Bond Metathesis for the Synthesis of Complex Polyphosphanes *J. Am. Chem. Soc.* 2012, *134*, 15443–15456.
131. Avens, L. R.; Cribbs, L. V.; Mills, J. L., Reactions of secondary phosphines with a phosphorus phosphorus bond and related reactions *Inorg. Chem.* 1989, *28*, 205–211.
132. Avens, L. R.; Cribbs, L. V.; Mills, J. L., Exchange-reactions of tetrakis(trifluoromethyl)diphosphine with pnicogen-pnicogen, phosphorus-hydrogen, and phosphorus-chlorine bonds *Inorg. Chem.* 1989, *28*, 211–214.
133. Eisentrager, F.; Gothlich, A.; Gruber, I.; Heiss, H.; Kiener, C. A.; Kruger, C.; Notheis, J. U.; Rominger, F.; Scherhag, G.; Schultz, M.; Straub, B. F.; Volland, M. A. O.; Hofmann, P., Sterically crowded diphosphinomethane ligands: molecular structures, UV-photoelectron spectroscopy and a convenient general synthesis of $(Bu^t{}_2PCH_2PBu^t{}_2)$ and related species *New J. Chem.* 2003, *27*, 540–550.
134. Burchat, A. F.; Chong, J. M.; Nielsen, N., Titration of alkyllithiums with a simple reagent to a blue endpoint *J. Organomet. Chem.* 1997, *542*, 281–283.
135. Gilman, H.; Cartledge, F. K., The analysis of organolithium compounds *J. Organomet. Chem.* 1964, *2*, 447–454.
136. Hoye, T. R.; Eklov, B. M.; Voloshin, M., No-D NMR spectroscopy as a convenient method for titering organolithium (RLi), RMgX, and LDA solutions *Org. Lett.* 2004, *6*, 2567–2570.
137. Peterson, D. J.; Hays, H. R., 3D-orbital resonance in trivalent organophosphines .i. metalation of methylphosphines *J. Org. Chem.* 1965, *30*, 1939-&.
138. Peterson, D. J., N,N-disubstituted aminomethyllithium compounds *J. Am. Chem. Soc.* 1971, *93*, 4027-&.
139. Steinborn, D.; Becke, F.; Boese, R., Synthesis and properties of (aminomethyl) nickel complexes $(Li(OEt_2))_2Ni(CH_2NC_5H_{10})_4$: The first structurally characterized homoleptic aminomethyl complex of a transition-metal *Inorg. Chem.* 1995, *34*, 2625–2628.
140. Valentine, D. H.; Hillhouse, J. H., Alkyl phosphines as reagents and catalysts in organic synthesis *Synth.* 2003, 317-334.
141. Pommer, H.; Thieme, P. C., Industrial applications of the wittig reaction *Top. Curr. Chem.* 1983, *109*, 165–188.
142. Vedejs, E.; Meier, G. P.; Snoble, K. A. J., Low-temperature characterization of the intermediates in the wittig reaction *J. Am. Chem. Soc.* 1981, *103*, 2823–2831.
143. Seyferth, D., Cadet's fuming arsenical liquid and the cacodyl compounds of bunsen *Organometallics* 2001, *20*, 1488–1498.
144. Aggarwal, V. K.; Winn, C. L., Catalytic, asymmetric sulfur ylide-mediated epoxidation of carbonyl compounds: Scope, selectivity, and applications in synthesis *Acc. Chem. Res.* 2004, *37*, 611–620.
145. Li, A. H.; Dai, L. X.; Aggarwal, V. K., Asymmetric ylide reactions: Epoxidation, cyclopropanation, aziridination, olefination, and rearrangement *Chem. Rev.* 1997, *97*, 2341–2372.
146. McGarrigle, E. M.; Myers, E. L.; Illa, O.; Shaw, M. A.; Riches, S. L.; Aggarwal, V. K., Chalcogenides as organocatalysts *Chem. Rev.* 2007, *107*, 5841–5883.
147. Zanardi, J.; Leriverend, C.; Aubert, D.; Julienne, K.; Metzner, P., A catalytic cycle for the asymmetric synthesis of epoxides using sulfur ylides *J. Org. Chem.* 2001, *66*, 5620–5623.
148. Calculations were done using cc-pVTZ as the basis set with B3LYP and with NBO5 as implemented in Gamess US.
149. Gombler, W.; Schaebs, J.; Willner, H., ^{19}F NMR chemical-shift of SF_2 in the gas-phase *Inorg. Chem.* 1990, *29*, 2697–2698.
150. Haas, A.; Willner, H., Lower chalcogen fluorides. 4. Preparation and characterization of pure S_2F_4 *Zeitschrift Fur Anorganische Und Allgemeine Chemie* 1980, *462*, 57–60.
151. Schmidt, M., Elemental sulfur: Challenge to theory and practice *Angew. Chem. Int. Ed.* 1973, *12*, 445–455.
152. Steudel, R.; Hassenberg, K.; Munchow, V.; Schumann, O.; Pickardt, J., Preparation of organic polysulfanes R_2Sn (n = 5, 7, 8, 9) from sulfenyl chlorides, RSCl, and transition metal polysulfido complexes *Eur. J. Inorg. Chem.* 2000, 921–928.
153. Steudel, R.; Munchow, V.; Pickardt, J., Preparation of spiro-cyclic organic polysulfanes of type $C_6H_{10}Sn$ and X-ray structural analysis of $C_6H_{10}S_{11}$, a derivative of cyclo-dodecasulfur *Zeitschrift Fur Anorganische Und Allgemeine Chemie* 1996, *622*, 1594–1600.
154. Maroney, M. J.; Hondal, R. J., Selenium versus sulfur: Reversibility of chemical reactions and resistance to permanent oxidation in proteins and nucleic acids *Free Radical Biol. Med.* 2018, *127*, 228–237.
155. McGuffey, J. R.; Paluzelle, R.; Muldrew, W. E., Handling Gaseous Fluorine in Industry *Ind. Eng. Chem.* 1962, *54*, 46–50.
156. Seppelt, K., Bromine oxides *Acc. Chem. Res.* 1997, *30*, 111–113.
157. Zhdankin, V. V., *Hypervalent iodine chemistry preparation, structure, and synthetic applications of polyvalent iodine compounds* John Wiley & Sons, Inc.,: Chichester, West Sussex, 2014.
158. Wirth, T., *Hypervalent Iodine Chemistry Modern Developments in Organic Synthesis* Springer: Berlin, 2003.
159. Hwang, I. C.; Kuschel, R.; Seppelt, K., Structures of bromine oxygen compounds *Zeitschrift Fur Anorganische Und Allgemeine Chemie* 1997, *623*, 379–383.
160. Kuschel, R.; Seppelt, K., Bromine bromate Br_2O_3 *Angew. Chem. Int. Ed.* 1993, *32*, 1632–1633.
161. Lee, C. L.; Lister, M. W., Decomposition of aqueous sodium bromite *Can. J. Chem.* 1971, *49*, 2822-&.

162. Lee, C. L.; Lister, M. W., Kinetics and mechanisms of some reactions of aqueous sodium bromite *Canadian Journal of Chemistry-Revue Canadienne De Chimie* 1979, *57*, 1524–1530.
163. Quiroga, S. L.; Perissinotti, L. J., Chlorine Oxoacids and Structure od Dichlorine Oxides *Chem. Educator* 2011, *16*, 275–278.
164. Leopold, D.; Seppelt, K., Dibromine pentoxide Br_2O_5 *Angew. Chem. Int. Ed.* 1994, *33*, 975–976.
165. Cotton, F. A.; Wilkinson, G., *Advanced Inorganic Chemistry: A Comprehensive Text* 4th ed.; Wiley: New York, 1980.
166. Lee, C. K.; Mak, T. C. W.; Li, W. K.; Kirner, J. F., Iodobenzene diacetate *Acat Cryst. B:Struc. Sci.* 1977, *33*, 1620–1622.
167. Alcock, N. W.; Countryman, R. M.; Esperas, S.; Sawyer, J. F., Secondary bonding. 5. Crystal and molecular-structures of phenyliodine(III) diacetate and bis(dichloroacetate) *Dalton Trans.* 1979, 854–860.
168. Archer, E. M.; van Schalkwyk, T. G. D., The crystal structure of benzene iododichloride *Acta Cryst.* 1953, *6*, 88–92.
169. Zhao, X.-F.; Zhang, C., Iodobenzene dichloride as a stoichiometric oxidant for the conversion of alcohols into carbonyl compounds; Two facile methods for its preparation *Synth.* 2007, 551–557.
170. Boeckman, R. K.; Shao, P.; Mullins, J. J., The Dess-Martin Periodinane *Organic Syntheses* 2004, *77*, 141.
171. Schroeckeneder, A.; Stichnoth, D.; Mayer, P.; Trauner, D., The crystal structure of the Dess-Martin periodinane *Beilstein J. Org. Chem.* 2012, *8*, 1523–1527.
172. Stevenson, P. J.; Treacy, A. B.; Nieuwenhuyzen, M., Preparation of Dess-Martin periodinane: The role of the morphology of 1-hydroxy-1,2-benziodoxol-3(1H)-one 1-oxide precursor *J. Chem. Soc. Perkin 2* 1997, 589–591.
173. Bartlett, N.; Lohmann, D. H., Dioxygenyl hexafluoroplatinate(V) O^{2+} $[PtF_6]$ *Proc. Chem. Soc. London* 1962, 115.
174. Christe, K. O., Bartlett's discovery of noble gas fluorides, a milestone in chemical history *Chem. Commun.* 2013, *49*, 4588–4590.
175. Hope, E. G., Coordination chemistry of the noble gases and noble gas fluorides *Coord. Chem. Rev.* 2013, *257*, 902–909.
176. Hoppe, R., Valence compounds of inert gases *Angew. Chem. Int. Ed.* 1964, *3*, 538-&.
177. Tramsek, M.; Zemva, B., Synthesis, properties and chemistry of xenon(II) fluoride *Acta Chim. Slov.* 2006, *53*, 105–116.
178. Seidel, S.; Seppelt, K., Xenon as a complex ligand: The tetra xenono gold(II) cation in $[AuXe_4]^{2+}(Sb_2F_{11}^-)_2$ *Science* 2000, *290*, 117–118.
179. Power, P. P., Main-group elements as transition metals *Nature* 2010, *463*, 171–177.
180. Erker, G.; Stephan, D. W., *Frustrated Lewis Pairs I* Springer-Verlag: Berlin, 2013.
181. Erker, G.; Stephan, D. W., *Frustrated Lewis Pairs II* Springer-Verlag: Berlin, 2013.
182. Welch, G. C.; Juan, R. R. S.; Masuda, J. D.; Stephan, D. W., Reversible, metal-free hydrogen activation *Science* 2006, *314*, 1124–1126.
183. Welch, G. C.; Stephan, D. W., Facile heterolytic cleavage of dihydrogen by phosphines and boranes *J. Am. Chem. Soc.* 2007, *129*, 1880.
184. Chase, P. A.; Welch, G. C.; Jurca, T.; Stephan, D. W., Metal-free catalytic hydrogenation *Angew. Chem. Int. Ed.* 2007, *46*, 8050–8053.
185. Hounjet, L. J.; Stephan, D. W., Hydrogenation by Frustrated Lewis Pairs: Main Group Alternatives to Transition Metal Catalysts? *Org. Process Res. Dev.* 2014, *18*, 385–391.
186. Dunn, N. L.; Ha, M.; Radosevich, A. T., Main Group Redox Catalysis: Reversible P(III)/P(V) Redox Cycling at a Phosphorus Platform *J. Am. Chem. Soc.* 2012, *134*, 11330–11333.
187. McCarthy, S. M.; Lin, Y.-C.; Devarajan, D.; Chang, J. W.; Yennawar, H. P.; Rioux, R. M.; Ess, D. H.; Radosevich, A. T., Intermolecular N-H Oxidative Addition of Ammonia, Alkylamines, and Arylamines to a Planar σ^3-Phosphorus Compound via an Entropy-Controlled Electrophilic Mechanism *J. Am. Chem. Soc.* 2014, *136*, 4640–4650.
188. Gordon, A. J.; Ford, R. A., *The Chemist's Companion: A Handbook of Practical Data, Techniques, and References* Wiley: New York, 1972.
189. Baenziger, N. C.; Buckles, R. E.; Maner, R. J.; Simpson, T. D., Crystal structure of chlorine complex of bis(p-chlorophenyl) sulfide *J. Am. Chem. Soc.* 1969, *91*, 5749.
190. Drabowicz, J.; Luczak, J.; Mikolajczyk, M.; Yamamoto, Y.; Matsukawa, S.; Akiba, K. Y., The first enantiomerically pure, C_2-symmetric spiroselenurane: 3,3,3',3'-tetramethyl-1,1'-spirobi 3H,2,1-benzoxaselenole *Tetrahedron-Asymmetry* 2002, *13*, 2079–2082.
191. Wirth, T., Organoselenium chemistry in stereoselective reactions *Angew. Chem. Int. Ed.* 2000, *39*, 3740–3749.
192. Barnes, N. A.; Godfrey, S. M.; Halton, R. T. A.; Mushtaq, I.; Pritchard, R. G.; Sarwar, S., Reactions of $Ph_4Se_4Br_4$ with tertiary phosphines. Structural isomerism within a series of $R_3PSe(Ph)Br$ compounds *Dalton Trans.* 2006, 1517–1523.
193. Hirsekorn, K. F.; Veige, A. S.; Marshak, M. P.; Koldobskaya, Y.; Wolczanski, P. T.; Cundari, T. R.; Lobkovsky, E. B., Thermodynamics, kinetics, and mechanism of $(silox)_3M(olefin)$ to $(silox)_3M(alkylidene)$ rearrangements (silox = (Bu_3SiO); M = Nb, Ta) *J. Am. Chem. Soc.* 2005, *127*, 4809–4830.
194. Bender, J. E.; Holl, M. M. B.; Kampf, J. W., Synthesis and characterization of a novel diarylgermylene containing electron-withdrawing groups *Organometallics* 1997, *16*, 2743–2745.
195. Carr, G. E.; Chambers, R. D.; Holmes, T. F.; Parker, D. G., Polyfluoroaryl organometallic compounds. 17. 2,4,6-tris(trifluoromethyl)phenyllithium, a sterically crowded system *J. Organomet. Chem.* 1987, *325*, 13–23.
196. Kindervater, M. B.; Binder, J. F.; Baird, S. R.; Vogels, C. M.; Geier, S. J.; Macdonald, C. L. B.; Westcott, S. A., The phosphinoboration of 2-diphenylphosphino benzaldehyde and related aldimines *J. Organomet. Chem.* 2019, *880*, 378–385.
197. Zhao, W.; McCarthy, S. M.; Lai, T. Y.; Yennawar, H. P.; Radosevich, A. T., Reversible Intermolecular E-H Oxidative Addition to a Geometrically Deformed and Structurally Dynamic Phosphorous Triamide *J. Am. Chem. Soc.* 2014, *136*, 17634–17644.

CHAPTER 10

Introduction to Transition-Metal Chemistry

"Occasionally, a single man may play such a central role in a particular field of science that his name becomes synonymous with that field . . . [A]lmost a full half century after his death, coordination compounds, particularly metal-ammines, are still known as Werner complexes . . ."

— **George B. Kauffman in the introduction to his biography of Werner**[1]

"I call any geometrical figure, or group of points, chiral, and say that it has chirality, if its image in a plane mirror, ideally realized, cannot be brought to coincide with itself."

— **Lord Kelvin (William Thomson)**

"One of the major goals of inorganic chemistry is to prepare compounds of elements in unusual oxidation states. In 1969, 238 oxidation states were known of the elements with atomic number Z between 21 and 100, that is an average of 3 per element."

— **C. K. Jørgensen**[2]

CONCEPT COVERAGE

1. The beginnings of coordination chemistry and the geometry of metal complexes
2. Common coordination numbers for transition metal complexes
3. The molecular orbital basis for the 18 valence electron rule for *d*-block complexes
4. Charged and neutral methods for electron counting
5. Chelating ligands and the chelate effect
6. Enumerating the various stereoisomers for metal complexes of different coordination numbers
7. The extent and calculation of formal oxidation states for simple transition-metal systems
8. Isomerism and stereochemistry in metal complexes

The transition series has a rich chemistry that will require new models and methods for us to categorize and effectively understand. In this chapter, some basic concepts will be reviewed as an introduction to the more detailed discussion of transition-metal ligands and electronic structure to follow in later chapters. Specifically, the focus here will be on the geometry and isomers of metal complexes, along with introductions to simple ligand types and metal electronic structure.

10.1 Alfred Werner and the Octahedron[1,3]

Werner was born in 1866 in Mulhouse, France, near where Germany, France, and Switzerland are joined. For his graduate work he studied with the well-known organic chemist at University of Zürich, Arthur Hantzsch, and was noted as having a gift for stereochemistry. Hantzsch and Werner published a treatise on the stereochemistry of nitrogen compounds, which explained, for example, the multiple possible isomers in 1,2-diimines. This paper was the first to suggest that unsymmetrical oximes, R(R′)C=NOH, would have two isomers due to the two possible directions in which the nitrogen could point the OH group, and that the two isomers in azobenzene (PhN=NPh) were *cis* and *trans*, besides resolving many other nitrogen-based stereochemical issues. Although Hantzsch's name appeared on the paper, he stated that the work was entirely Werner's.[4] While we may disregard these solutions to these stereochemical issues as obvious, one should keep in mind that this was well before G. N. Lewis's work on chemical bonding, and no one had any notion of what a "lone pair" of electrons was; in fact, the electron would not be discovered for another eight years.

While Alfred Werner was working on lectures for an inorganic chemistry course, he dreamed one night about the stereochemistry of coordination complexes. The next morning he wrote down what would be the foundation for an entire field of chemistry.[5] When he wrote the paper that revolutionized the field, his theory was met with great skepticism, especially from a leader in the area at the time, Sophus Mads Jørgensen at the University of Copenhagen. Werner began experiments on coordination chemistry on a scale that perhaps had not been seen previously. By the end of his career, his more than 200 doctoral students would prepare thousands of compounds, and more than 8,000 of those compounds still rest in carefully labeled glass tubes at the Anorganisch-Chemisches Institut der Universität Zürich.

Of particular intrigue to chemists at the time were cobalt ammine complexes, of which there was a tremendous variety. The bright colors of coordination complexes had always fascinated chemists, artists, and many others. For example, Prussian blue—a deep-blue complex of iron and cyanide—had been used as a pigment in painting for centuries. Complexes of cobalt with amine, chloride, and sometimes water had been prepared in *croceo* (saffron yellow), *luteo* (orange-yellow), *purpureo* (purple), *roseo* (red), *praseo* (green), and *violeo* (violet) varieties. Werner's theory explained the nature of these complexes far differently than the theory of the time, which was due to Swedish chemist Christian Wilhelm Blomstrand and supported by Jørgensen.

The structures of the *luteo*, *purpureo*, *praseo*, and *violeo* complexes proposed by Blomstrand–Jørgensen and Werner are shown in Table 10.1. While the Blomstrand–Jørgensen structures look quite foreign to our eyes today, they were rather innovative at the time, and believe it or not they are a step in the right direction, as will be explained momentarily.

Blomstrand was rebelling against one of the major ideas about bonding theory at the time, Kekule's theory of fixed valency, which was having a stultifying effect on inorganic chemistry. August Kekule (1829–1896), who is most famous for deciphering the correct structure of benzene, suggested that valence (number of bonds to an atom) was "a fundamental property of the atom, which is just as constant and unchangeable as the atomic weight itself."[1] This idea was a boon to organic chemistry, where drawing four bonds to carbon without fail is still taught today, but handcuffed inorganic chemistry.

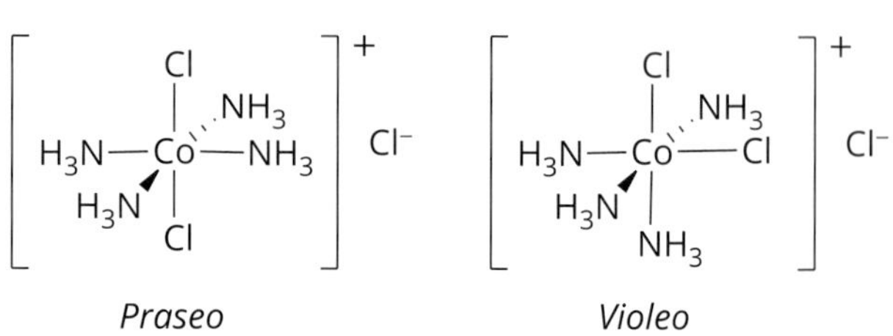

Fig. 10.1 Structures of the *praseo* and *violeo* forms of $\{Co(NH_3)_4Cl_2\}^+ Cl^-$.

In order to account for some commonplace inorganic species, Kekule had to propose two classifications: more stable "atomic compounds" and less stable "molecular compounds", which could not exist in the gas phase. For example, PCl_5 was difficult to explain using fixed valency, as Kekule suggested that phosphorus, like nitrogen, should only be able to reach a valency of 3. Consequently, he suggested that PCl_5 was a "molecular compound" with the formula $PCl_3{\cdot}Cl_2$, because it decomposed to PCl_3 and Cl_2 on heating.[1]

Kekule described cobalt chloride hexamine, for example, as $CoCl_3{\cdot}6NH_3$. Kekule's evasive, and decidedly noncommittal, dot (•) is still used in inorganic chemistry today—a tenacious remnant of his "atomic and molecular compounds" theory. For example, look in any chemical catalog and you are liable to find compounds for sale, such as $CrCl_3{\cdot}6H_2O$, which is probably being sold as a mixture of different isomers, the likes of which will be discussed later in this chapter.

Werner was well aware of Kekule's theory of fixed valency, and at age 17 in his chemistry notes he even identified PCl_5 as a "molecular" compound. Blomstrand's theory involving pentavalent nitrogen (Table 10.1)—compare PCl_5—flew in the face of Kekule's fixed valency, as would Werner's proposals for coordination complexes based on six bonds to central metal centers. In this regard, one could think that Blomstrand's only mistake was in increasing the bonds to the ligand instead of the metal!

Some of the exchanges between Werner and Jørgensen were heated, but Jørgensen conceded defeat when Werner showed that the *praseo* and *violeo* complexes both had the formula $\{Co(NH_3)_4Cl_2\}^+ Cl^-$ and were *cis*- and *trans*-isomers of one another (Fig. 10.1).

Werner expanded his coordination theory by production of enantiomeric complexes, which really took advantage of his talent for stereochemistry. In 1911, he and his student Victor L. King published the resolution of $\{cis\text{-}Co(NH_3)Cl(en)_2\}^+$ as the *d*-α-bromocamphor-sulfonate salts.[6] Two years later, Werner received the Nobel Prize, but his insistence that the metal was the cause of the chirality was still met with skepticism from many because it was believed that carbon was required to imbue this property. It was claimed by some that the chirality was somehow due to the carbons in the chelating ligands. Werner later resolved this issue by isolation of enantiomers of a completely carbon-free complex with inorganic chelating "ligands" comprised of metal complexes, dihydroxyltetraamminecobalt, shown in Fig. 10.2. Over his career, Werner and coworkers would separate enantiomers for more than forty chiral complexes.

In 1874, Jacobus van 't Hoff was the first to suggest that the substituents on carbon pointed towards the center of a tetrahedron. Werner is his inorganic counterpart with his suggestion of the octahedron as the primary geometric component for coordination complexes, but his contributions are more far-reaching than this simple suggestion and in some ways are the basis for the remainder of this book.

Table 10.1 Examples of Blomstrand–Jørgensen and Werner proposals for the structures of cobalt ammine chloride complexes.

Name	Blomstrand-Jørgensen structure	Werner structure
	Cobalt ammine chlorides	
luteo (orange)	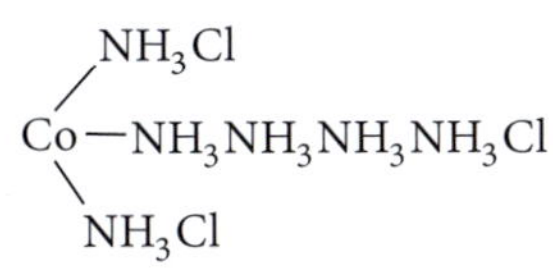	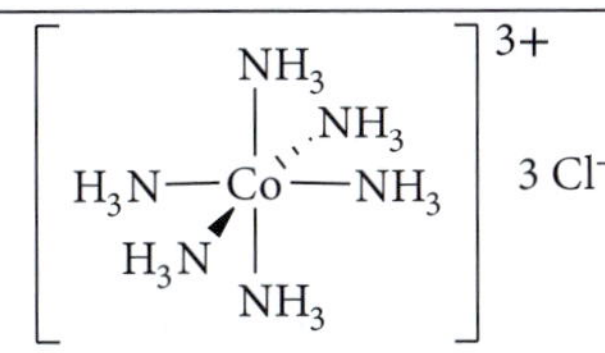
purpureo (purple)	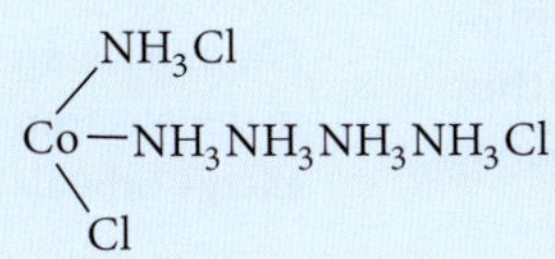	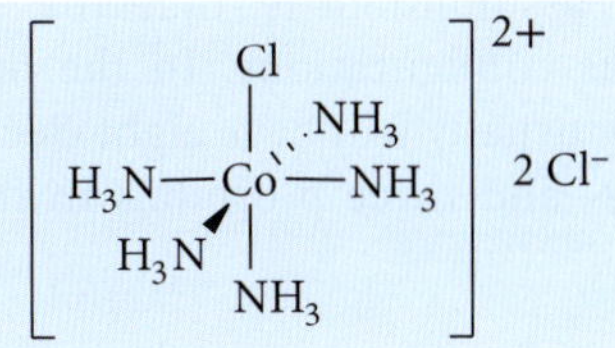
praseo (green)	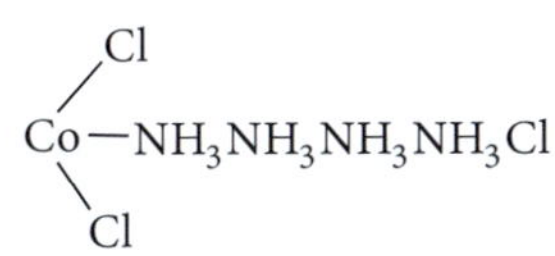	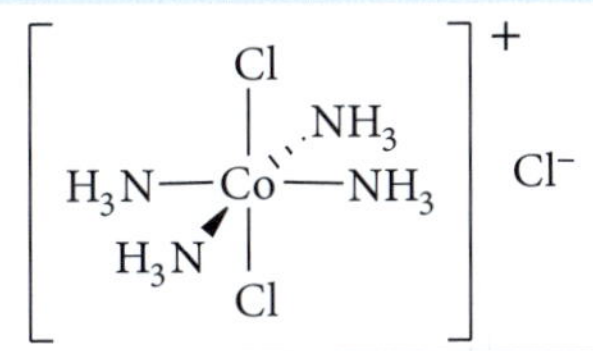
	Cobalt ethylenediamine chlorides	
praseo	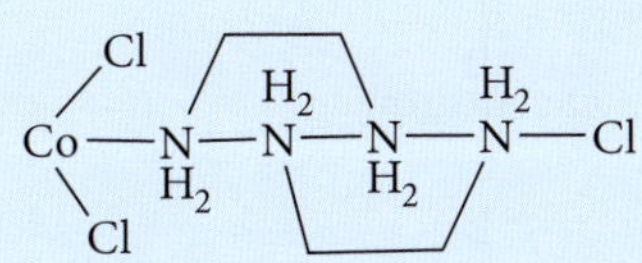	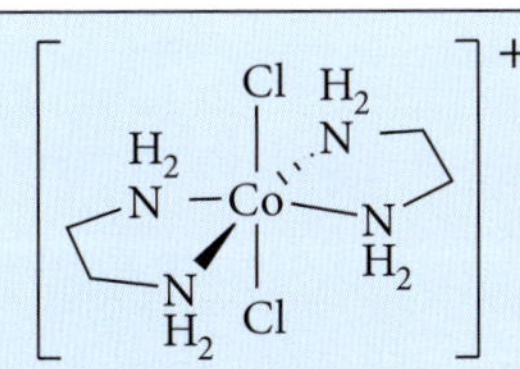 *trans*
violeo	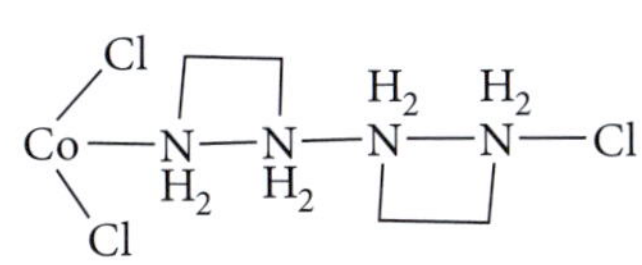	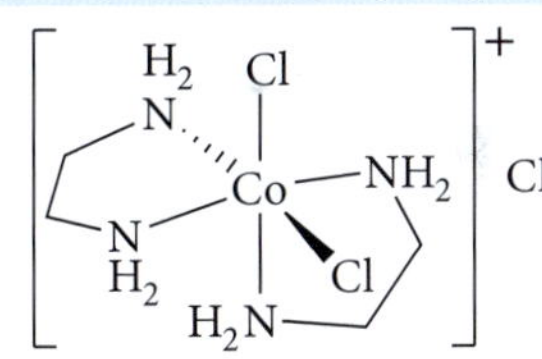 *cis*

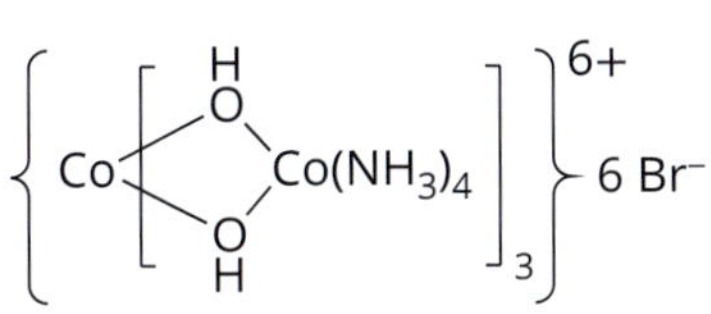

Fig. 10.2 Structure of a carbon-free cobalt complex with the central cobalt as a chiral center.

10.2 Coordination Numbers and Electron Counting in the Transition Series with Simple Ligands

Since Werner, the complexity of ligands surrounding the central metal, still often in a near-octahedral geometry, has increased significantly. One concept of utility in the description of complexes is the **electron count** of the metal center, which is generally a good place to begin to understand electronic structure of a new complex. In this section, electron counting using simple ligands will be discussed. In the following two chapters, more detailed discussions of ligand electronic structure and interactions with the metal will be presented, and ligands of greater complexity will be analyzed. The discussion in this section will serve to introduce some simple ligands and how they may bond to a metal.

One relatively straightforward concept to introduce is the *coordination number*, which is simply the number of ligands around the metal center. The previous discussion concerning Werner centered around octahedral complexes, such as $\{Cr(NH_3)_6\}^{3+}$, which has a coordination number of 6 around chromium. However, Werner was well aware that some metal complexes did not adopt configurations with 6 ligands around the metal. Coordination numbers of 4 and 5 are also quite common, and complexes with these coordination numbers may exhibit several different geometries.

In Table 10.2, some common limiting geometries for transition-metal complexes are organized by coordination number. In cases where the coordination number has more than one limiting form—that is, 4 and 5—actual structures are often somewhere in between, which will be discussed in detail later. The geometries themselves are usually found experimentally; for example, by X-ray diffraction, which directly provides bond distances and angles of the complex. (See Box 10A.)

The assumption is that the elements tend towards noble-gas configurations. In the main group, this implies that they usually will have eight electrons (octet rule) as they fill the *s*- and three *p*-orbitals in their valence shell. Since the transition metals have one *s*-, three *p*-, and five *d*-orbitals in their valence shell, they have nine total orbitals that can hold a total of 18 electrons. Consequently, for the transition metals an "18-electron rule" is discussed; however, while some transition metal complexes fulfill this rule, very many do not. Some early transition metals rarely reach 18 electrons in their complexes, with perhaps $Y\{CH(SiMe_3)_2\}_3$ being one of the largest violators as it only has six electrons around yttrium (Fig. 10.3).[7]

Using electron counting, one can quickly discern some important information about a metal complex that affects its reactivity and properties.

Electron counts ≤16

If the electron count is ≤16 on the metal center, empty low-energy orbitals on the metal center are probably available. This

Table 10.2 Common coordination numbers for transition metals, the most symmetric form, and point group of the most symmetric form.

Coordination number	limiting geometric form	Highest symmetry point group
6	 Octahedron	O_h
5	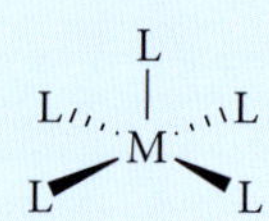 Tetragonal pyramidal (square pyramidal)	C_{4v}
	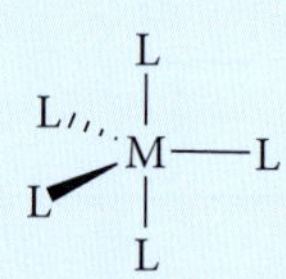 Trigonal bipyramidal (tbp)	D_{3h}
4	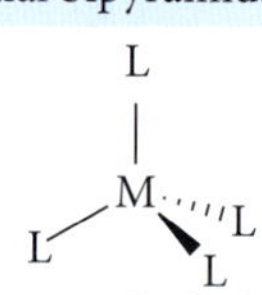 Tetrahedral	T_d
	 Square planar	D_{4h}
3	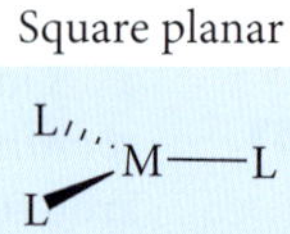 Trigonal planar	D_{3h}
2	L—M—L Linear	$D_{\infty h}$

BOX 10A

Comments on single-crystal X-ray diffraction

X-ray diffraction is one of the most important techniques for characterization of metal complexes. Using this method, bond distances and angles are provided with estimated standard errors. One thing to remember is that not all structures are created equal, and the species that crystallized out is not necessarily the main species in solution.

The accuracy of a structure is measurable using several different statistics of variable utility. From a pragmatic perspective, the most important numbers from the method are bond distances and angles along with their errors, estimated standard deviations (ESDs). ESDs in distances should usually be in the thousandths of Ångstroms or smaller, with ESDs for angles around a tenth of a degree or less for well-behaved structures.

Estimated standard deviations should always be reported with a structural parameter; that is, 1.238(4) Å, and not just 1.238 Å. Ultimately, these determine the utility of the structural parameters. For two metric parameters to be different, three times the ESD of the parameter in one structure and three times the ESD of the parameter in the other structure should not overlap. For three standard deviations, the "confidence limit" on the measurement is 99.73%, and this is by far the most used value. For two standard deviations, the "confidence limit" is 95.45%, but the deviations are estimated, making using this limit more dubious than this figure might suggest. For three estimated standard deviations and above (Table 10A), there is a good chance that the metrics being compared are actually different and are worthy of comment.

Example: A titanium nitrogen distance is determined to be 1.857(6) Å for a Ti–NMe_2 bond and 2.030(5) Å for a Ti–N(pyrrolyl) bond. Is the difference statistically significant?

The three ESD limits for the short distance are 1.839–1.875 Å. The three ESD limits for the longer distance are 2.015–2.045 Å. The limits do not overlap; therefore, the difference in the distances is statistically significant, which means there are probably differences in bonding (such as bond order) between the two nitrogen-based ligands.

Table 10A Confidence limits for standard deviations.

Standard Deviations	Confidence Limit
1 σ	68.27%
2 σ	95.45%
3 σ	99.73%
4 σ	99.994%
5 σ	99.99994%

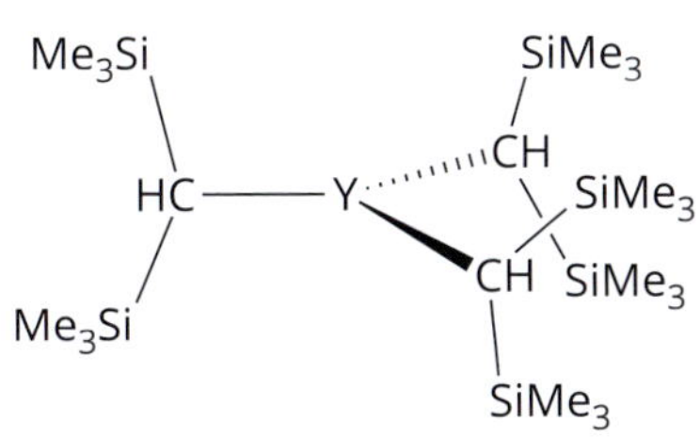

Fig. 10.3 Structure of Y{$CH(SiMe_3)_2$}$_3$, which formally has six valence electrons around Y.

may mean that the complex is Lewis acidic. It may also mean that the complex can undergo rapid substitution reactions.

Electron counts >18

If the electron count exceeds 18, antibonding orbitals are being occupied. Complexes with metal centers having more than 18 valence electrons are not necessarily unstable, but they may give up electrons easily and often undergo rapid ligand substitution by dissociation of ligands.

Odd electron counts

If the metal center has an odd electron count, it will be paramagnetic, which dramatically affects the properties of the compound for any spectroscopy involving a magnetic field, such as EPR, NMR, and magnetometry. In some cases, metal centers with even electron counts can be paramagnetic as well, such as {$Ni(NH_3)_6$}$^{2+}$, shown as follows with a triplet ground state.

From a molecular orbital standpoint, we can return to our MO diagram developed in Chapter 6 for CrH_6, and elaborate (Fig. 10.4). *In all cases, the critical orbitals to watch are the metal-based t_{2g} and e_g (labeled as $2e_g$* in the figure) orbitals. Very often, these two orbital manifolds end up determining the properties and reactivity of transition-metal complexes to a large degree.* As will be discussed next, these orbitals often are, or are near, the frontier molecular orbitals for metal complexes. Many of the chapters that follow will discuss how reactivity and properties of transition metal complexes may be understood by using simple orbital pictures of the t_{2g} and e_g orbitals. In brief, electron counting tells you if the bonding and nonbonding orbitals are filled; this occurs in an octahedral system with a configuration of $(a_{1g})^2(e_g)^4(t_{1u})^6(t_{2g})^6(2e_g^*)^0$, which is 18 valence electrons.

The CrH_6 molecule has a valence electron count on chromium of 12, $(a_{1g})^2(e_g)^4(t_{1u})^6(t_{2g})^0(2e_g^*)^0$. The largely metal-based orbitals t_{2g} and e_g are empty in CrH_6. The empty t_{2g} orbitals on chromium are low-energy nonbonding orbitals, which makes this (hypothetical) complex probably Lewis acidic. An actual compound with a 12-electron count with only σ-donating ligands is WMe_6.[8,a]

In the middle column of Fig. 10.4 is the MO diagram for the Werner complex $\{Co(NH_3)_6\}^{3+}$. The compound has the configuration $(a_{1g})^2(e_g)^4(t_{1u})^6(t_{2g})^6(2e_g{}^*)^0$, which means it has a valence electron count on cobalt of 18.

In the right-hand column of Fig. 10.4 is the MO diagram for $\{Ni(NH_3)_6\}^{2+}$, which has a configuration of $(a_{1g})^2(e_g)^4(t_{1u})^6(t_{2g})^6(2e_g{}^*)^2$. By Hund's rule, the complexes should be a ground-state triplet with two unpaired electrons, and as a result it is paramagnetic. The valence electron count on nickel is twenty electrons. The two electrons in $e_g{}^*$ are in an antibonding orbital, which lowers the bond energy in the complex, making the ligands readily substituted. (See Chapter 17 on kinetics of ligand substitution.)

[a] It should be noted that WMe_6 and CrH_6 are not, in fact, octahedral but are being drawn as such for the sake of argument here. For a detailed discussion, see Chapter 14.

These hopefully illustrate that electron counting gives quick, useful information about a complex. As we proceed further, other models will be introduced to understand the electronic structure of metal complexes, such as crystal field theory. However, molecular orbital arguments will be of great utility throughout most of this book, and electron counting is in some respects a quick method of determining the electron configuration.

There are two common methods for finding the electron count of a transition-metal complex called the *neutral method* and *charged method*. It is advisable to learn how to electron count by using one or both methods. However, the most common error made in finding electron counts is to mix the methods. If you choose to learn one method, it should probably be the charged method, as this one incorporates formal oxidation states, which are often discussed on their own. In this author's opinion, the neutral method is simpler and quicker.

Oxidation states for transition metals in complexes are found much as was shown for main-group compounds in

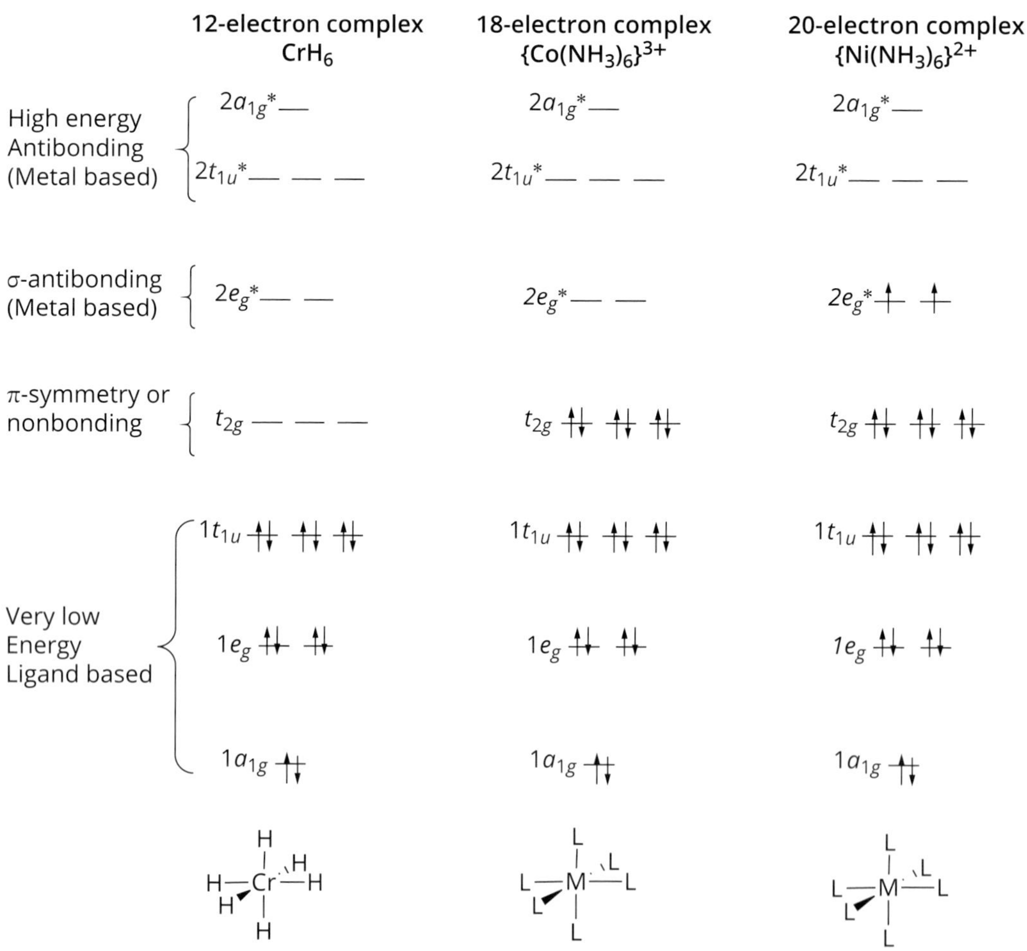

Fig. 10.4 Molecular orbitals and electron counting. The figure provides MO diagrams for σ-only ligands, such as methyl or ammine, in approximately octahedral symmetry, and listed in the caption for each column is a known molecule that will display that electron count. (*left*) CrH_6 is a 12-electron complex, which leaves all of the low-energy metal-based orbitals (t_{2g} and $2e_g{}^*$) empty. (*middle*) The Werner complex $\{Co(NH_3)_6\}^{3+}$ is 18-electron, which exactly fills all the low-energy metal-based orbitals t_{2g} but leaves the antibonding $2e_g{}^*$ orbitals empty. (*right*) $\{Ni(NH_3)_6\}^{2+}$ is a 20-electron system, which fills all the-low energy metal-based orbitals t_{2g} and places two electrons into antibonding $2e_g{}^*$ orbitals.

Chapter 3. The ligands are removed from the metal, and the more electronegative ligands are assumed to take the electrons in covalent bonds. The charge on the metal after heterolytically cleaving the bonds is the formal oxidation state.

The remainder of this section, in addition to teaching these methods for electron counting, will increase your familiarity with common ligands for transition metals, which you will see in the remainder of this chapter and the rest of the book. Table 10.3 lists many simple ligands for metal centers, the name usually given the ligand, and the electron count of the ligand under both methods.

Common ligands to be discussed in this chapter may be covalent or dative in how they bond to the metal center, both bonding types were discussed in Chapter 3. Some ligands may

Table 10.3 Some common simple ligands and their electron counts donated to a metal.

Ligand	Name as ligand	Neutral method electron count	Charge method electron count
Halides (X)	Listed as anions with the "ide" ending, e.g., fluoride and chloride.	•X 1-electron	:X$^-$ 2-electron anion
Alkyls	Named as described in organic chemistry, e.g., methyl, phenyl, butyl, vinyl. An alkyl commonly employed in metal chemistry is neopentyl, $CH_2{}^tBu$.	•R 1-electron	:R$^-$ 2-electron anion
CN	cyanide	•CN 1-electron	:CN$^-$ 2-electron anion
NO_3	nitrate	•ONO_2 1-electron	:ONO_2^- 2-electron anion
SCN	thiocyanate (S-bound) or isothiocyanate (N-bound)	•SCN or •NCS 1-electron	$^-$:SCN or $^-$:NCS 2-electron
NO_2	nitrite, which may be O or N bound	•ONO or •NO_2 1-electron	$^-$:ONO or $^-$:NO_2 2-electron
CO	carbonyl	:CO 2-electron	:CO 2-electron
OH_2	These are referred to as "aquo complexes" if only water ligands are present around the metal.	:OH_2 2-electron	:OH_2 2-electron
NH_3	ammine	:NH_3 2-electron	:NH_3 2-electron
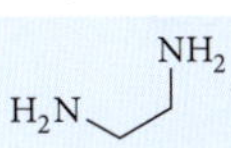	ethylenediamine (en) is "bidentate", attaches in 2 places	Each amine donates 2-electrons, 4-total electrons 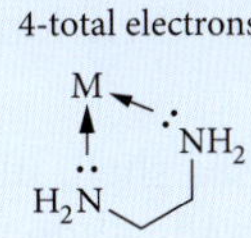	Each amine donates 2-electrons, 4-total electrons 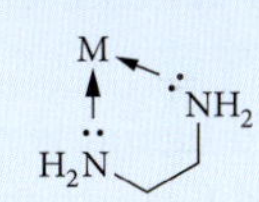
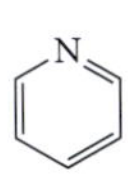	pyridine	2-electrons from nitrogen lone pair 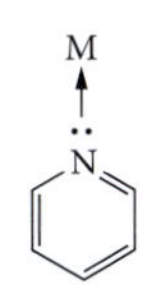	2-electrons from nitrogen lone pair 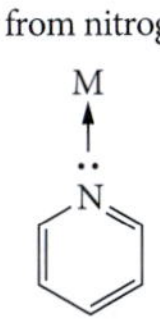
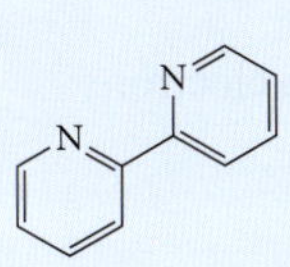	2,2'-bipyridine (bpy or bipy), a bidentate ligand	2-electrons from each nitrogen lone pair 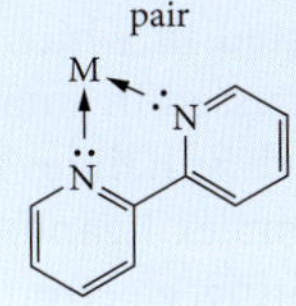	2-electrons from each nitrogen lone pair 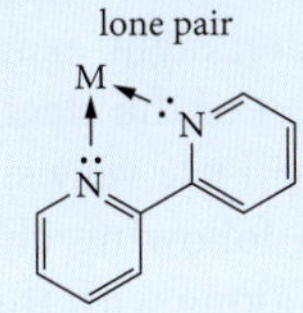
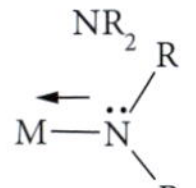	amide The R groups can be most alkyls, e.g., methyl, or even SiMe3. The ligand presents one covalent and one dative bond.	3-electron donor 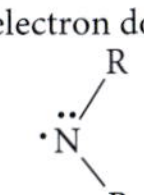	Anionic 4-electron donor 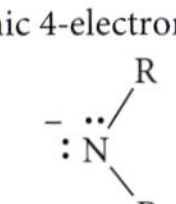
C_2O_4	oxalate (ox) bidentate ligands	Donates 2 × 1 electrons, 2 total electrons 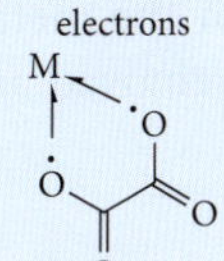	Donates 2 × 2 electrons, 4 total electrons 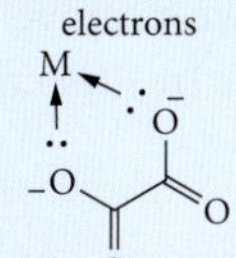

be *multidentate*. Many common ligands, such as halides and cyanide, bond to the metal through a single atom. Some polyatomic ligands bond to metal centers through multiple atoms, and a few examples are shown in Table 10.3. For example, pyridine is a monodentate ligand that datively bonds to metal centers using its nitrogen lone pair; 2,2'-bipyridine is a ligand with two pyridines bound together that acts as a bidentate ligand and datively bonds through both nitrogens.

Multidentate ligands are often favored over multiple monodentate ligands due what is called the "chelate effect", covered in Section 10.3.

10.2.1 Neutral Method for Electron-Counting Simple Complexes

The neutral method stresses covalency in metal complexes. Covalently bound ligands are homolytically cleaved from the metal as radicals. Lewis acid–base interactions occur as well if the ligands are typical Lewis bases, such as ammonia or a phosphine.

For the neutral method:

- All ligands are removed as neutral species.
- The number of electrons contributed by the metal to its own electron count is equal to the column number where the metal resides; for example, iron is in Group 8 and contributes 8 electrons.
- Ligands are generally considered 2-electron donors if neutral and closed shell molecules on their own. For example, NH_3 and PMe_3 are both closed-shell neutral molecules that can act as 2-electron donors using their lone pair of electrons.
- If a listed ligand is not neutral and closed shell, such as halides and CH_3, then it is considered a radical donor to the metal; for example, $\cdot Cl$ or $\cdot CH_3$ are considered 1-electron donors.
- Overall charges are considered to be on the metal and are added to the electron count. As a result, negative charges increase the metal electron count and positive charges reduce it.

To illustrate, eight examples are presented in which the determination of the electron count is discussed in some detail. Here we will use very simple ligands, such as halides and ammonia. In following chapters, more complex ligands will be discussed along with their electron counting, electronic structure, how complexes using the ligands are sometimes prepared, and so on.

(1) VCl_4

To begin, we remove all the ligands from the metal as neutral species. In this case, the ligands are removed as chlorine radicals.

$$VCl_4 \Rightarrow V + 4\,Cl\cdot$$

The number of electrons on the vanadium is simply equal to the group number for the element, 5. Each of the chlorine radicals will donate one electron to the metal. The total number of electrons around vanadium is then nine.

$$VCl_4 \Rightarrow V + 4\,Cl\cdot$$
$$5 + (4 \times 1) = 9\ \text{electrons}$$

(2) TaF_7^{2-}

Again, the halides are removed as radicals. Tantalum is in the same group as vanadium, Group 5. In this case there is an overall charge on the compound, which is considered to reside on tantalum. The –2 charge indicates that the metal has an additional two electrons.

$$TaF_9^{2-} \Rightarrow Ta + 7\,F\cdot + 2\,(\text{from charge})$$
$$5 + (7 \times 1) + 2 = 14\ \text{electrons}$$

(3) $Ni(CN)_4^{2-}$

Removal of the ligands from nickel as neutral species gives CN, which one can tell from its Lewis structure is not a closed shell compound. CN is a radical donor, similar to a halogen for electron-counting purposes. Nickel is in Group 10, and the complex has a negative overall charge.

$$Ni(CN)_4^{2-} \Rightarrow Ni + 4\ \cdot CN + 2\,(\text{from charge})$$
$$10 + (4 \times 1) + 2 = 16\ \text{electrons}$$

(4) $Pt(NH_3)_2Cl_2$

The examples above are *homoleptic*, in that all of the ligands around the metal center are the same. Most complexes under study today are *heteroleptic*, by which the metal center has more than one type of ligand, like platin in this example. In this case, if we remove the ligands as neutral species, NH_3 is removed as a neutral 2-electron donor and $\cdot Cl$ is a radical 1-electron donor. Platinum is in Group 10.

$$Pt(NH_3)_2(Cl)_2 \Rightarrow Pt + 2\ {:}NH_3 + \cdot Cl$$
$$10 + (2 \times 2) + (2 \times 1) = 16\ \text{electrons}$$

(5) $Fe(CO)_5$

Carbon monoxide is a neutral, 2-electron donor that is a common ligand for transition metals. Iron is in Group 8.

$$Fe(CO)_5 \Rightarrow Fe + 5\ {:}CO$$
$$8 + (2 \times 5) = 18\ \text{electrons}$$

(6) $Re(CO)_5(NO_3)$

Nitrate, NO_3^-, is sometimes referred to as a "pseudo-halide", and from its Lewis structure it would be a radical as a neutral compound; compare Cl· in $Pt(NH_3)_2Cl_2$ as an example. Consequently, for the neutral method it is considered $\cdot ONO_2$ and a 1-electron donor. Rhenium is in Group 7.

$$Re(CO)_5(NO_3) \Rightarrow Re + 5\,{:}CO + \cdot ONO_2$$
$$7 + (2 \times 5) + 1 = 18\,\text{electrons}$$

(7) NiF_6^{4-}

This is another example of an anionic halide complex. Nickel is in Group 10, and the electron count around nickel is higher than 18. Since there are twenty electrons and the valence orbitals can only accommodate eighteen, we know that two electrons reside in antibonding orbitals; however, this is a thermally stable complex.

$$NiF_6^{4-} \Rightarrow Ni + 6\,\cdot F + 4\,(\text{from charge})$$
$$10 + (6 \times 1) + 4 = 20\,\text{electrons}$$

(8) $\{Co(NH_3)_6\}^{2+}$

Here is a cationic example, which was reported as the salt $\{Co(NH_3)_6\}^{2+}(BH_4^-)_2$.[9] Since the charge is cationic, we will subtract electrons from the valence set for cobalt at the end.

$$Co(NH_3)_6^{2+} \Rightarrow Co + 6\,{:}NH_3 - 2\,(\text{from charge})$$
$$9 + (6 \times 2) - 2 = 19\,\text{electrons}$$

10.2.2 Charged Method for Electron-Counting Simple Complexes

The charged method stresses the ionic character of metal complexes. Ligands such as halides that are covalently bound to the metal are heterolytically cleaved with the ligand obtaining the bonding electrons. In other words, a metal–halide, M–X, bond is cleaved to M^+ and $X{:}^-$. The charged method also stresses the Lewis acid–base interactions (dative bonding, see Chapter 3) of a metal center with its ligands, and all the ligands could be considered to be acting as Lewis bases towards the metal and donating lone pairs.

In the charged method:

- All ligands are multiples of 2-electron donors; that is, 2-, 4-, or 6-electron donors.
- The oxidation state of the metal is determined by removing ligands similar to what was done with the main group. Neutral closed shell donors do not affect the oxidation state. Other ligands are removed as closed-shell species. If the complex is charged, that charge is considered to reside on the metal. The formal oxidation state is found by removing more electronegative ligands as anions; in other words, the more electronegative atom in the bond, almost invariably considered to be the ligand, obtains the electrons from the heterocovalent bond. For some additional examples, see Section 3.5 for a discussion of formal oxidation state in the main group.

It is common in inorganic chemistry to refer to complexes as having *d*-electron counts equal to their column number minus their oxidation state. For example, Fe(II) is d^6, Ti(IV) is d^0, Os(III) is d^5, and Co(III) is d^6. This convention is the norm in ligand field theory and crystal field theory especially, and will be a useful designation when those models are discussed. One may notice now that these *d*-electron counts correlate with the number of electrons in the t_{2g}/e_g^*-orbitals, which are the metal-based orbitals of low energy in Fig. 10.4. For WMe_6, the metal is in the +6 oxidation state, and tungsten is in Group 6. Consequently, W(VI) is d^0, and there are zero electrons in the t_{2g}/e_g^*-orbitals. For $\{Co(NH_3)_6\}^{3+}$, cobalt is in the +3 oxidations state and d^6; the t_{2g}/e_g^*-orbitals contain six electrons, $(t_{2g})^6(e_g^*)^0$. For the third example in Fig. 10.4, $\{Ni(NH_3)_6\}^{2+}$, nickel is in the +2 oxidation state and d^8; the t_{2g}/e_g^*-orbitals contain eight electrons, $(t_{2g})^6(e_g^*)^2$.

In what follows, the same examples worked previously for the neutral method are examined by using the charged method. Naturally, the results from using the two methods will be the same, although sometimes they arrive there using different routes.

(1) VCl_4

To begin, we remove all the ligands from the metal as closed-shell species; consequently, the Cl ligands are removed as chloride, $:Cl^-$, a 2-electron donor. Since there are four anionic donors and the complex overall is neutral, the vanadium must be in the +4 formal oxidation state and is d^1.

$$VCl_4 \Rightarrow V(IV) + 4\,{:}Cl^-$$

The number of electrons on vanadium is the column number minus the oxidation state; in this case, the metal contributes 5 − 4 = 1 electron. Each of the chlorides donate 2 electrons to the metal. The total number of electrons around vanadium is then 9.

$$VCl_4 \Rightarrow V(IV) + 4\,{:}Cl^-$$
$$1 + (4 \times 2) = 9\,\text{electrons}$$

(2) TaF_7^{2-}

In this case there are seven fluorides. If the complex were neutral, then the Group 5 metal would need to be in the +7 oxidation state—an impossibility, as this is higher than the valence electron count. Placing the overall charge on the complex at the metal center adds two electrons to the metal and produces a formal oxidation state of Ta(V). Consequently, the tantalum center is d^0, has no electrons remaining, and contributes nothing to its electron count. Each fluoride is a 2-electron donor.

$$TaF_9^{2-} \Rightarrow Ta(V) + 7\,{:}F^-$$
$$0 + (7 \times 2) = 14\,\text{electrons}$$

Keep in mind that the overall charge on the complex is accounted for in the oxidation state of the metal.

(3) $Ni(CN)_4^{2-}$

The CN ligands are considered to be cyanide, $^-$:CN. After compensating for the overall charge on the complex, this produces nickel(II), which is d^8.

$$Ni(CN)_4^{2-} \Rightarrow Ni(II) + 4\ ^-{:}CN$$
$$8 + (4 \times 2) = 16 \text{ electrons}$$

(4) $Pt(NH_3)_2Cl_2$

Just like in the neutral method, the ammine ligands are neutral closed-shell 2-electron donors. In the charged method, the Cl-ligands are considered to be chlorides, producing a formal oxidation state from the metal of Pt(II), d^8.

$$Pt(NH_3)_2(Cl)_2 \Rightarrow Pt(II) + 2\ {:}NH_3 + \ ^-{:}Cl$$
$$8 + (2 \times 2) + (2 \times 2) = 16 \text{ electrons}$$

(5) $Fe(CO)_5$

If all of the ligands are neutral closed-shell donors, then the neutral and charged methods calculate the electron count in the exact same way.

$$Fe(CO)_5 \Rightarrow Fe + 5\ {:}CO$$
$$8 + (2 \times 5) = 18 \text{ electrons}$$

(6) $Re(CO)_5(NO_3)$

Nitrate is a pseudo-halide, 2-electron donor, $^-$:ONO_2. The formal oxidation state of the metal is then Re(I), d^6.

$$Re(CO)_5(NO_3) \Rightarrow Re(I) + 5\ {:}CO + \ ^-{:}ONO_2$$
$$6 + (2 \times 5) + 2 = 18 \text{ electrons}$$

(7) NiF_6^{4-}

The central metal is Ni(II) after compensating for the −4 charge on the complex, d^8.

$$NiF_6^{4-} \Rightarrow Ni(II) + 6\ {:}F^-$$
$$8 + (6 \times 2) = 20 \text{ electrons}$$

(8) $\{Co(NH_3)_6\}^{2+}$

The cationic charge is considered to be on the cobalt. Since the other ligands are neutral, the formal oxidation state is Co(II).

$$Co(NH_3)_6^{2+} \Rightarrow Co(II) + 6\ {:}NH_3$$
$$7 + (6 \times 2) = 19 \text{ electrons}$$

10.2.3 Electron-Counting Most Metal–Metal Bond Complexes

Metals can be bound to other metals, and thousands of metal–metal bond-containing complexes have been reported. For electron-counting purposes, the electrons are considered to be shared equally between the metals; in other words, each metal–metal bond will contribute an additional electron. For determination of oxidation state, the metal–metal bond is homolyzed; this is generally practiced whether or not the two metals are the same and have the same electronegativity, but there are exceptions, and metal–metal bonding will be discussed in more detail in Chapter 12.

Metal–metal bond complexes are known from single through pentuple bonds, where the pentuple bond would include σ-, two π-, and two δ-bonds. The following provides several examples of metal–metal bond complexes with different bond orders and types.

(1) $Mo_2(neopentyl)_6$[10] (Fig. 10.5)

Charged Method
The formal oxidation state of the metal is Mo(III), as the metal–metal bond does not affect the oxidation state and there are three neopentyl groups, considered $^-$:CH_2-*t*-Bu for this method. We focus on one of the two equivalent metal centers for electron counting.

$$Mo_2(CH_2\text{-}t\text{-}Bu)_6 \Rightarrow Mo(III) + 3\ ^-{:}CH_2\text{-}t\text{-}Bu + 3\,M{-}M \text{ bonds}$$
$$3 + (3 \times 2) + 3 = 12 \text{ electrons}$$

Neutral Method
The neopentyl groups are considered 1-electron radical donors, ·CH_2-*t*-Bu.

$$Mo_2(CH_2\text{-}t\text{-}Bu)_6 \Rightarrow Mo + 3\ {\cdot}CH_2\text{-}t\text{-}Bu + 3\,M{-}M \text{ bonds}$$
$$6 + (3 \times 1) + 3 = 12 \text{ electrons}$$

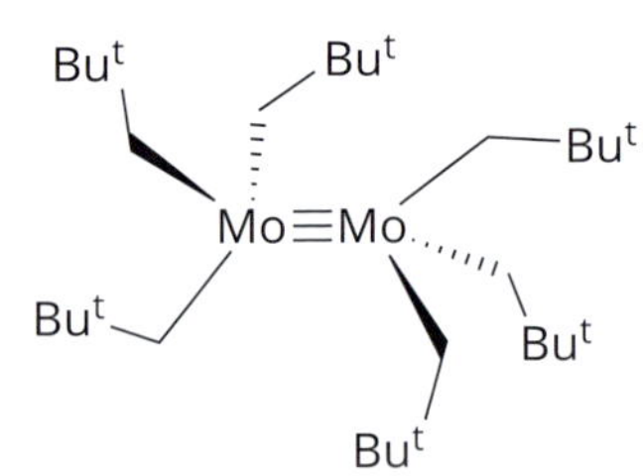

Fig. 10.5 Structure of $Mo_2(neopentyl)_6$ with a metal–metal triple bond.

(2) $Cr_2Ar^*_2$[11] (Fig. 10.6)

Charged Method
The aromatic group is removed as $^-$:Ar*, and the formal oxidation state is Cr(I).

$$Cr_2Ar^*_2 \Rightarrow Cr(I) + ^-{:}Ar^* + 5\,M{-}M \text{ bonds}$$
$$5 + 2 + 5 = 12 \text{ electrons}$$

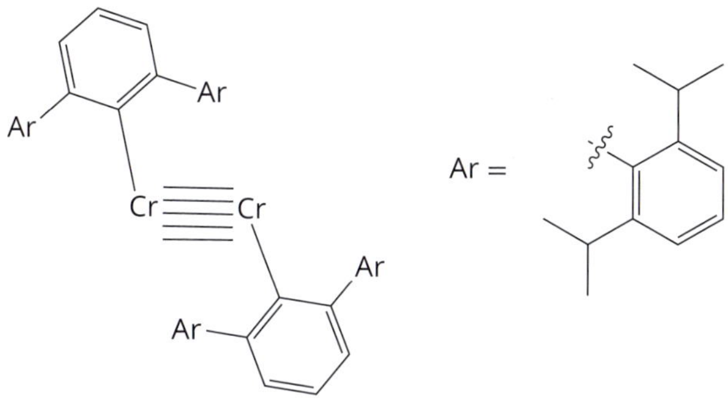

Fig. 10.6 Structure of Cr_2Ar_2 with a metal-metal pentuple bond.

Neutral Method

$$Cr_2Ar^*_2 \Rightarrow Cr + \cdot Ar^* + 5\,M-M \text{ bonds}$$
$$6 + 1 + 5 = 12 \text{ electrons}$$

(3) $(OC)_5Mn{-}Fe(CO)_4{-}Mn(CO)_5$ (Fig. 10.7)

In this cluster there are single bonds between manganese and iron.[12] The two Mn atoms are identical by symmetry.

Since the only ligands on either of the metal types are neutral ligands and metal–metal bonds, the neutral and charged methods obtain identical results. For manganese,

$$Mn(CO)_5(FeCO_4 - Mn(CO)_5) \Rightarrow Mn(0) + 5 :CO + 1M-M \text{ bonds}$$
$$7 + (5 \times 2) + 1 = 18\,e^-$$

The iron is counted similarly, but there are two metal–metal bonds.

$$FeCO_4(-Mn(CO)_5)_2 \Rightarrow Fe(0) + 5 :CO + 2\,M-M \text{ bonds}$$
$$8 + (4 \times 2) + 2 = 18\,e^-$$

This electron-counting is presented as a brief introduction to metals, ligands, and electronic structure. Now we turn to the various ways by which simple ligands may be arranged around a metal center, but first there is an important point regarding ligand structure that should be discussed: the chelate effect.

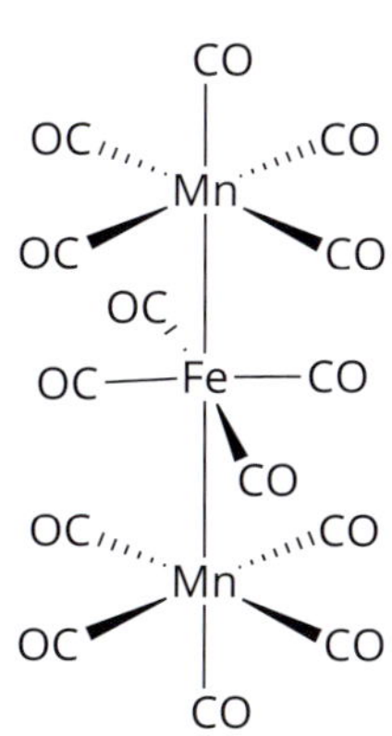

Fig. 10.7 Structure of an Mn_2 and Fe cluster.

10.3 The Chelate Effect[13–15]

The chelate effect is a commonly discussed phenomenon that is used to great advantage in inorganic chemistry. Quite simply, ligands with multiple points of attachment to the metal, "multidentate" or "chelating" ligands, tend to produce more stable complexes than those that are "monodentate", literally "single toothed", having one attachment point to the metal. The word "chelate" is from the same Greek word *chelae*, used to describe the claw of a lobster or other crustacean, and was first used by Morgan and Drew to describe "caliper-like groups which function as two associating units and fasten on to the central metallic atom so as to produce heterocyclic rings."[16]

The chelate effect is, in fact, two different effects. These effects both lead to a complex bearing a single multidentate ligand being favored over a complex with multiple monodentate ligands. There are thermodynamic and kinetic chelate effects, and either or both may be important for any system under study.

10.3.1 The Thermodynamic Chelate Effect

The following often-used relation (the Gibbs equation) concerning the free energy of a system, $\Delta G°$, with enthalpy and entropy, can be interpreted as saying that free energy becomes more negative with increasing exothermicity and increasing number of modes to dissipate that energy (increasing entropy), and a negative $\Delta G°$ is a prerequisite for a spontaneous reaction.

$$\Delta G^\circ = \Delta H^\circ - T\Delta S^\circ$$

To examine how chelation can affect the thermodynamics for a reaction, consider the simple replacement of the ammine ligands in $\{Ni(NH_3)_6\}^{2+}$ with three equivalents of ethylenediamine (Fig. 10.8).[13] In the reaction, NH_3 ligands are replaced with a slightly different donor, NH_2R, and the alkyl group on the ammine does make the amine a slightly stronger Lewis base. This increase in the Ni–N interaction due to the change is the Lewis acid–base interaction would be expected to be reported in the $\Delta H°$ value for the reaction, and the reaction is slightly exothermic.[b]

$$\Delta G^\circ = -13\,\text{kcal/mol}\,(-54\ \text{kJ/mol})$$
$$\Delta H^\circ = -6.9\,\text{kcal/mol}\,(-29\ \text{kJ/mol})$$

$$\Delta S^\circ = 21\,\text{cal/mol}\cdot\text{K}\,(88\,\text{J/mol}\cdot\text{K}) \rightarrow -T\Delta S^\circ_{298K}$$
$$= -6.3\,\text{kcal/mol}\,(-26\ \text{kJ/mol})$$

[b] There may also be differences in solvation and other factors that contribute to $\Delta H°$.

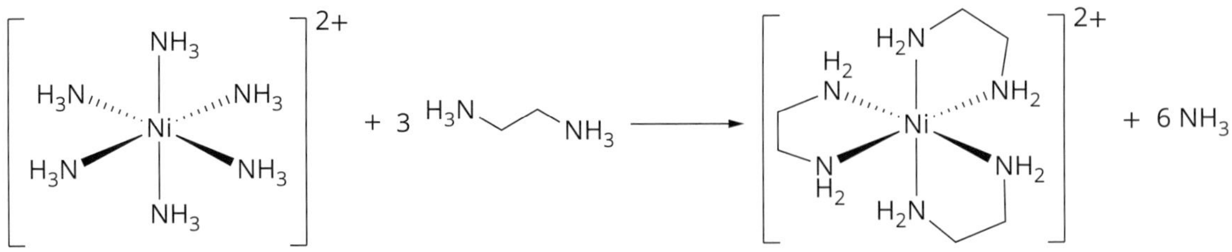

Fig. 10.8 Equation for the reaction of hexamminenickel(II) with ethylenediamine, and the thermodynamic parameters for the reaction.

More importantly for our discussion of the chelate effect, there is a large change in the number of particles during the reaction. The reaction begins with four molecules (one complex and three ethylene diamine ligands) and ends with seven molecules (one complex and six ammonia). The increase in the number of molecules increases the translational and rotational degrees of freedom. There is some smaller loss in rotational degrees of freedom on ring formation in the ethylene diamine ligands between free and bound. In the end, there is a large positive entropy for the reaction of $\Delta S° = 21$ cal/mol·K. The contribution of this term to the reaction is more easily seen in the form of the term "$-T\Delta S°$", and using room temperature gives $-T\Delta S°_{298K} = -6.3$ kcal/mol.

The entropic term in the chelate reaction makes about as large a contribution to $\Delta G°$ as the enthalpy in this case. *Because there is a large increase in the number of molecules during chelation, the reactions of chelating ligands are generally entropically favorable; this is the thermodynamic chelate effect.* The increase in the number of particles leads to a large increase in entropy, and detailed studies on quantifying the chelate effect have been appearing for at least six decades with new studies continuing to appear.[17–22]

In general, metal–ligand rings have preferred sizes of 4-, 5- and 6-members; however, whether this should be attributed to ring strain (bond bending) should be considered carefully on a case-by-case basis. In the main group, the smallest hybrid-to-hybrid angle available is 90°, two pure *p*-orbitals. In the transition series the *ds*-hybrid angles available, depending on the exact nature of the complex, range from as large as 125° to as miniscule as 55°.[23] As a result, one should never assume that, for example, four-membered rings with a transition metal center contain ring-strain (bond-bending). In addition, in three-membered metallacycles there is resonance between the metallacycle and the double bond ligand, such as olefin complex, which must be considered before assigning ring-strain.

An important factor, especially with rings larger than six-members, is loss of rotational degrees of freedom. In quite large rings, the rotational losses may not be compensated by the increases in translational degrees of freedom due to the thermodynamic chelate effect.

For example, consider the replacement of mim, where mim = $MeN(CH_2CO_2^-)_2$, with ethylene diamine tetraacetic acid, Y^{4-} where n = 2 is EDTA, and its derivatives,

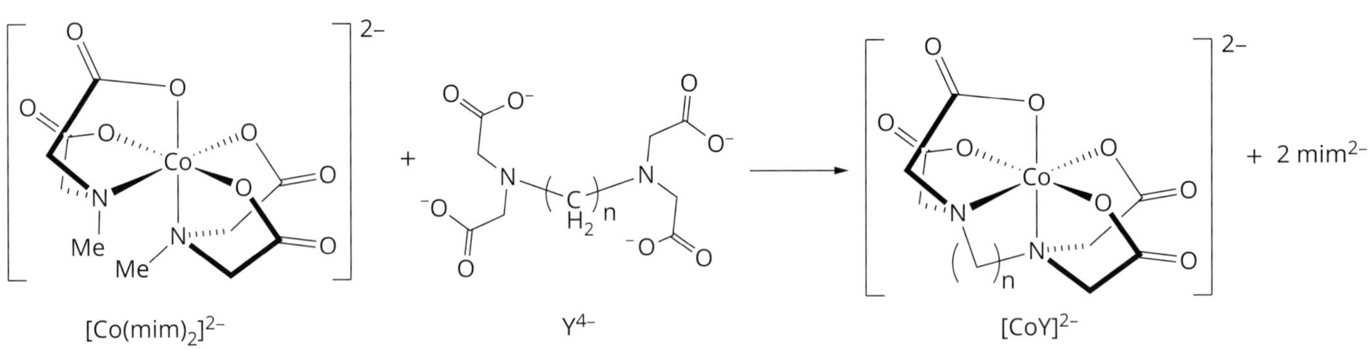

n =	ΔG° kJ/mol (kcal/mol)	ΔH° kJ/mol (kcal/mol)	−TΔS° kJ/mol (kcal/mol)
2	−13.4 (−3.2)	+5.4 (+1.3)	−18.8 (−4.5)
3	−9.2 (−2.2)	+12.1 (+2.9)	−21.3 (−5.1)
4	−9.6 (−2.3)	+16.3 (+3.9)	−25.9 (−6.2)
5	+2.9 (+0.7)	+9.6 (+2.3)	−7.5 (−1.8)

Fig. 10.9 Reaction of $\{Co(mim)_2\}^{2-}$ with Y^{4-} and thermodynamic data as a function of n. $Y^{4-} = (^-O_2CCH_2)_2N(CH_2)_nN(CH_2CO_2^-)_2$. mim = $MeN(CH_2CO_2^-)_2$.

$Y^{4-} = (^-O_2CCH_2)_2N(CH_2)_nN(CH_2CO_2^-)_2$. In this case (Fig. 10.9) there is little difference between the donor on the metal before and after the reaction, and the differences in ΔH° for the reaction may be due to slight deformation from ideal bond angles (bond bending in the attaching ligand atom) due to the heterocyclic ring formation.[21]

First, note that the reactions in Fig. 10.9 would be thermodynamically unfavorable in the absence of the chelate effect; all of the reactions are endothermic (ΔH > 0). Second, as the number of methylenes (n) increases, the loss of rotational degrees of freedom due to the chelate ring increases in importance until at $n = 5$ the reaction is no longer favorable (ΔG° > 0). Eventually, at large ring sizes the increase in entropy due to the increasing number of translational modes from the additional particles in the products is counterbalanced by the loss of rotational entropy in the large ring of the product. This may be an important contributing factor to the stabilization of smaller chelate rings in metal complexes over larger rings.

10.3.2 The Kinetic Chelate Effect[14–15]

For the kinetic chelate effect, a brief explanation will be presented which will be adequate for most purposes. A more detailed explanation will follow in the ensuing subsection by using a more detailed kinetic analysis.

10.3.2.1 Brief Explanation of the Kinetic Chelate Effect

The kinetic chelate effect can be explained using Fig. 10.10. A metal center with two 2 *cis*-ligands has one ligand (L) substituted by one side of the bidentate ligand (**1** → **2**). The reaction to place the first side of the bidentate ligand, A^1, on the metal proceeds with rate constant k_1, and the backreaction of L removing the chelate to go back to starting material (**2** → **1**) occurs with rate constant k_{-1}.

The first thing to consider is the reaction to close the chelate ring (**2** → **3**), which occurs with rate constant k_2. Once A^1 is attached to the metal center, A^2 is structurally required to be close to the metal center by the linking atoms. In other words, the second side of the chelate has a "high local concentration" near the metal. This makes k_2 (**2** → **3**) relatively large, favoring the chelated metal center **3**.

Exacerbating the fast ring closure (k_2) to form **3**, the rate constant for breaking the chelating ring (**3** → **2**) is often slow. One explanation for this is that the chelate requires the atoms to adopt unfavorable momentum vectors for ring breaking, but regardless of the explanation, k_{-2} is often small. As a result, the equilibrium constant $K_2 = k_2/k_{-2}$ will be very large and strongly kinetically favor the chelated system.

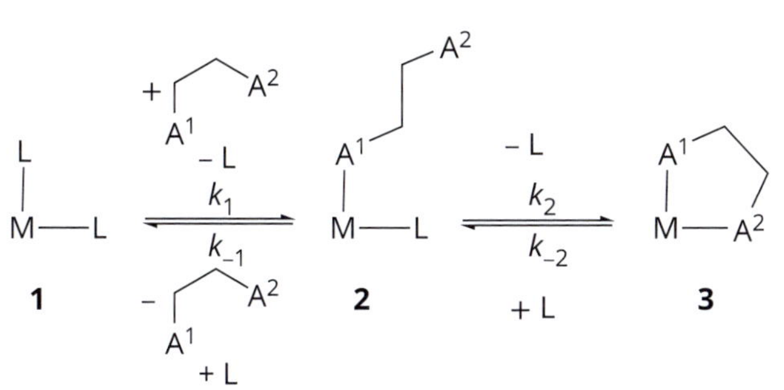

Fig. 10.10 Stepwise replacement of two monodentate ligands with a bidentate ligand.

10.3.2.2 The Kinetic Chelate Effect in More Detail[14–15] (Advanced Topic)

For the kinetics of chelation we can consider the general scheme shown in Fig. 10.11. The starting complex with at least two monodentate ligands (A), presumably a bis(solvate) and shown as two waters, reacts with a chelating ligand (L_2) with a forward rate constant of k_1. Initially, an intermediate (B) will be formed where one of the waters is replaced with one side of the chelate. The ligand L_2 can be removed from B to reform A with rate constant k_{-1}. Intermediate B can also go on to form the ring-closed product C with rate constant k_2. The ring may reopen to reform B with rate constant k_{-2}.

The *rate of chelate formation*, formation of C, is the rate constant k_2 times the concentration of species B, $[B]$. However, C is also removed by ring opening to reform B. As a result, the concentration of C is dependent upon both the forward and back reaction rates, one generating and one consuming C.

Here a common problem arises; we do not know $[B]$, and because it is an intermediate it may or may not be observable. It is often assumed that the concentration of B is independent of time, since it is being generated and consumed by several different processes (steady-state approximation; see Chapters 17).

The *rate of formation of intermediate B* is as shown in Fig. 10.11, and the steady-state approximation says that $\frac{d[B]}{dt} = 0$. Solving for $[B]$ produces the equation shown in the figure, which is substituted into the *rate of chelate formation* equation to obtain an *approximate rate equation* that only contains the concentrations of A, L_2, and C.

The approximated forward (k_f) and backward (k_b) rate constants in terms of the original reaction scheme are then as follows.

$$k_f = \frac{k_1 k_2}{k_{-1} + k_2}$$

$$k_b = \frac{k_{-1} k_{-2}}{k_{-1} + k_2}$$

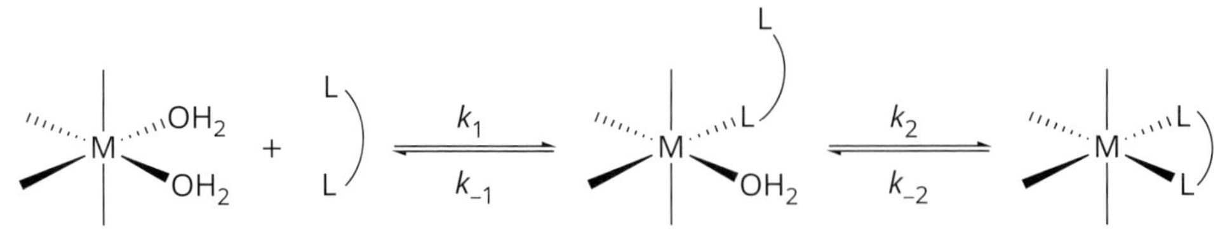

A L_2 B C

Rate of chelate formation	$\frac{d[C]}{dt} = k_2[B] - k_{-2}[C]$
Rate of formation of intermediate B & steady-state assumption to find the concentration of B	$\frac{d[B]}{dt} = k_1[A][L_2] - k_{-1}[B] - k_{-2}[B] + k_2[C] = 0$ $[B] = \frac{k_1}{k_{-1} + k_2}[A][L_2] + \frac{k_{-2}}{k_{-1} + k_2}[C]$
Approximate rate equation	$\frac{d[C]}{dt} = \underbrace{\frac{k_1 k_2}{k_{-1} + k_2}}_{k_f}[A][L_2] - \underbrace{\frac{k_{-1} k_{-2}}{k_{-1} + k_2}}_{k_b}[C]$

Fig. 10.11 Mechanism for chelate formation and kinetic analysis.

One can imagine two limiting cases.

(1) *$k_2 >> k_{-1}$ and small k_{-2}*

In this case $k_f \sim k_1$, indicating that the slow step in the formation of the chelate is the initial replacement. Once one side of the chelate is on the metal, there is a high local concentration of the second ligand and ring closure is relatively fast. This anchimeric assistance (rate enhancement by a neighboring group) is at least one part of the kinetic chelate effect. Formation of the chelate is encouraged kinetically because the second ligand is, by necessity, near the metal center. In addition, some ring sizes have a quite small value for k_2, ring opening, which can also contribute substantially to kinetic stability.

Example 1A The ring-opened complex bearing ethylene diamine (en) can be isolated under acidic conditions where ring-closing by the second nitrogen is prevented by protonation (Fig. 10.12). Then, by controlling pH the protonated nitrogen can be released and the kinetics of ring closure studied.[24]

It was found that the rate constant for the first ring closure, k_2, was 10.4 s^{-1}, and the second ring closure rate, k'_2, was 0.73 s^{-1}. This can be compared with the Cl replacement rate in the absence of the chelate effect (Fig. 10.12) of 5.7×10^{-3} $M^{-1}s^{-1}$, and the second replacement to form $\{Pt(NH_3)_4\}^{2+}$ is about an order of magnitude slower, 5.4×10^{-4} $M^{-1}s^{-1}$. Consequently, the ring closure at the chelate is ~2,000 times faster than the analogous reaction with a monodentate ligand using 1 M NH_3. The greater nucleophilicity of ethylene diamine (en) relative to NH_3 is responsible for some of this as ethylenediamine is about 3.4–5.4 times faster than ammonia for the initial replacement reaction; however, the chelate formation is still hundreds of times faster, even after compensating for this.

It is thought that anchimeric assistance is due, in part, to a high local concentration of the second nitrogen. Cotton and Harris examined this issue computationally using Monte Carlo methods. If the free end of the ligand occupies a hemisphere of radius 3 Å, the effective concentration would then be 0.02 molecule $Å^{-3}$, about 30 M!

The ring opening rate constant, k_{-2}, is often relatively small as well. This has been postulated to be due to steric restrictions to ring opening in one specific direction, and the energy barrier to ring opening results in a decrease in rate constant by ~40 times.[24–26] Consequently, the kinetic chelate effect is due to, one or both, rapid ring closure and slow ring opening.

Example 1B The system in Example 1A was with a square planar complex. Octahedral systems have also been studied in great detail for their kinetics of chelation (Fig. 10.13). In comparing the two related nickel(II) systems $Ni(OH_2)_4(en)^{2+}$ ($K_f = 5 \times 10^7$ M^{-1}) and $Ni(OH_2)_4(NH_3)_2^{2+}$ ($K_f = 1.5 \times 10^5$ M^{-2}), the chelated species shows a formation constant ~300 times larger, ignoring the difference in units.

Detailed kinetics on the system reveals results quite similar to the previous example of platinum, where there is a very large rate constant for the ring closure (k_2) and a very small rate constant for ring opening (k_{-2}).

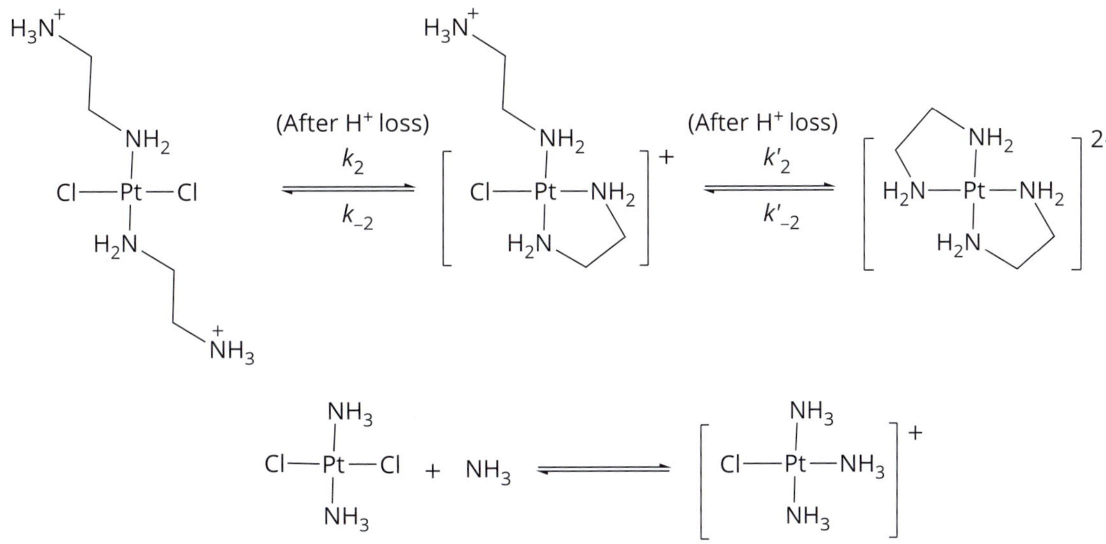

Fig. 10.12 *(top)* Isolable protonated complex used to study ring-closure kinetics by controlling amine protonation using pH. *(bottom)* Similar reaction with monodentate ligands.

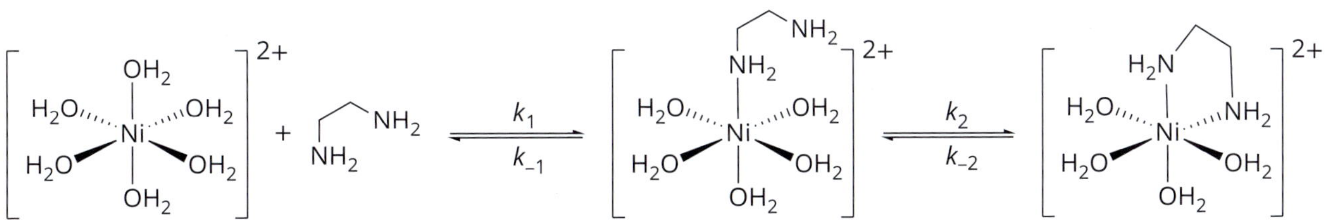

Fig. 10.13 Reaction of $\{Ni(OH_2)_6\}^{2+}$ with ethylene diamine, and the experimental rate constants for the process, which show rapid ring closure (large k_2) and slow ring opening (small k_{-2}).[14]

$$k_1 = 900\ M^{-1}\ s^{-1}$$
$$k_{-1} = 15\ s^{-1}$$
$$k_2 \sim 1.2 \times 10^5\ s^{-1}$$
$$k_{-2} \sim 0.14\ s^{-1}$$

(2) $k_2 << k_{-1}$

While the case where ring closure is rapid is the usual assumption, there are known instances where ring closure is slow relative to the first side of L_2 binding to the metal and $k_f \sim k_1k_2/k_{-1}$. Detailed discussion of such cases is left to more specialized texts, such as R. G. Wilkins' treatise on transition-metal kinetics, but the following is a brief discussion of one example.[15]

As one example (Fig. 10.14) where the ring closure may be slower than initial binding of the chelate, consider the case of cobalt hydroxide complexes with CO_2. The aquo amine complex $\{Co(tren)(OH_2)_2\}^+$ reacts with CO_2 to form the carbonate complex $\{Co(tren)(\kappa^2\text{-}O_2CO)\}^+$. The stoichiometry is a little deceptive, as the reaction involves initial loss of water and deprotonation of coordinated water to form $\{Co(tren)(OH)\}^{+2}$, a 5-coordinate species. This low coordinate species reacts rapidly with CO_2, providing unchelated $\{Co(tren)(OH)(OC(OH)O)\}^+$. Chelation is then comparatively quite slow, as hydroxyl must be protonated and removed. The initial coordination of CO_2 to form carbonate is ~2,500 times faster than chelation because of the required hydroxyl displacement. Note, however, that the fast initial addition is to an empty metal site compared to a ring closure that involves replacement of an excellent nucleophile, OH, which explains the slow chelation event.

10.4 Isomerism in Simple Complexes

As a consequence of the octahedral structure of many coordination complexes, various types of isomerism arise naturally. In this subsection, isomerism is divided into two categories: stereoisomerism and structural isomerism. The relationship between the various isomer types is shown in Fig. 10.15.

The term "isomers" refers to any two or more compounds with the same formula. For metal complexes these will be divided into stereoisomers and structural isomers. Stereoisomers

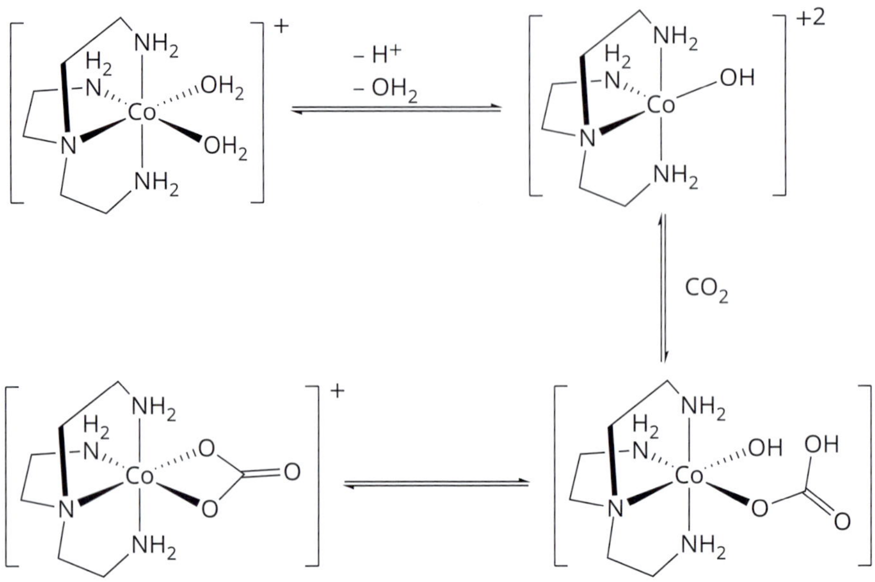

Fig. 10.14 Reaction of $\{Co(tren)(OH_2)_2\}^+$ with CO_2 to form $\{Co(tren)(\kappa^2\text{-}O_2CO)\}^+$. The initial addition of CO_2 does not require ligand displacement, because the reacting species is 5-coordinate, and as a result the reaction is extremely fast. This chelation step requiring ligand displacement is relatively slow.

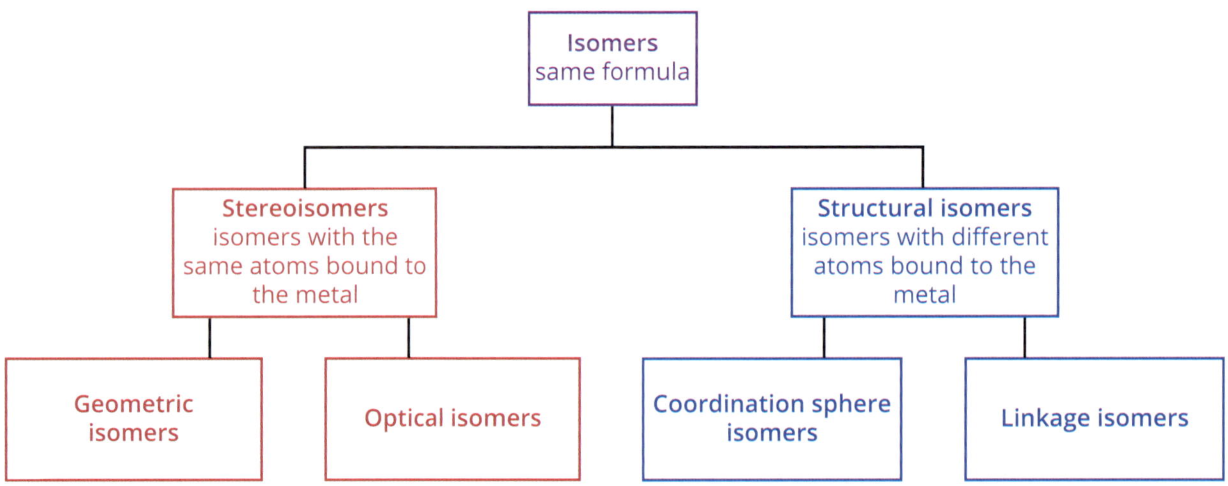

Fig. 10.15 Flow chart for isomers of coordination complexes.

are compounds that have all the same atoms bound to the metal center, but the ligands are arranged differently around the metal; for example, *cis*- and *trans*-isomers.

Stereoisomers can be further divided into geometric and optical isomers. Optical isomers are chiral, enantiomers. Geometric isomers do not necessarily include chiral centers but have different arrangements of ligands around the metal.

Structural isomers have different bonds to the metal center but the same formula, which will be further divided into coordination sphere isomers and linkage isomers. Coordination sphere isomers have ligands that have moved from the primary to secondary coordination sphere and vice versa. Linkage isomers have differences in how one or more ligands are bound to the metal.

10.4.1 Stereoisomers and Geometric Isomers

Stereoisomers are compounds with the same ligands bound to the metal center that have a different arrangement of ligands in 3-dimensional space. For example, *cis* and *trans* compounds of formula MA_4X_2 are stereoisomers (Fig. 10.16a).

Geometric isomers are stereoisomers with a difference in how bonds are arranged around the metal. The *cis* and *trans* compounds are geometric isomers because they differ in the arrangement of ligands; for example, in the *cis* isomer the X ligands are 180° from the A ligands, while they are 180° from X ligands (from each other) in the *trans* isomer.

For MA_4X_2 complexes, where A and X are different monodentate ligands, the number of geometric isomers is two (Fig. 10.16a). The complex is the *trans* isomer if the two X ligands are ~180° apart and is *cis* if the two X ligands are ~90° apart; compare the *praseo* and *violeo* complexes of Werner shown in Fig. 10.1.

If the complex has the general formula of MA_3X_3 then there are again only two isomers (Fig. 10.16b). One of the isomers has ligands of the same identity organized meridionally around the metal center, called the *mer* isomer. The other isomer has ligands of the same identity on the same face of the octahedron, *fac*.

10.4.2 Chirality and Optical Isomers

Chirality is another form of isomerism. The word "chiral" is derived from Greek χειρ (*kheir*), meaning "hand". The IUPAC definition is as follows: "The geometric property of a rigid object (or spatial arrangement of points or atoms) of being nonsuperimposable on its mirror image; such an object has no symmetry elements of the second kind (a mirror plane, $\sigma = S_1$, a center of inversion, $i = S_2$, a rotation reflection axis, S_{2n}). If the object is superimposable on its mirror image the object is described as being achiral." Plainly, an object is chiral if it has no improper axis of symmetry of any kind, S_n where $n = 1, 2$, and so on. Note that this does not mean it has no symmetry, as the object could in principle have many symmetry operations. Keep in mind that inversion is equivalent to an S_2, and a mirror plane is equivalent to an S_1. Chiral molecules do not have inversion centers, mirror planes, or any other S_n improper rotation.

Chiral molecules must belong to certain point groups, and compounds of certain point symmetry can never be chiral. The following are some point groups for chiral and achiral molecules, where n = integer.

Chiral point groups: C_n, D_n, T, O, and I

Achiral point groups: C_s, C_i, S_n, C_{nv}, C_{nh}, D_{nh}, D_{nd}, O_h, T_h, T_d, I_h

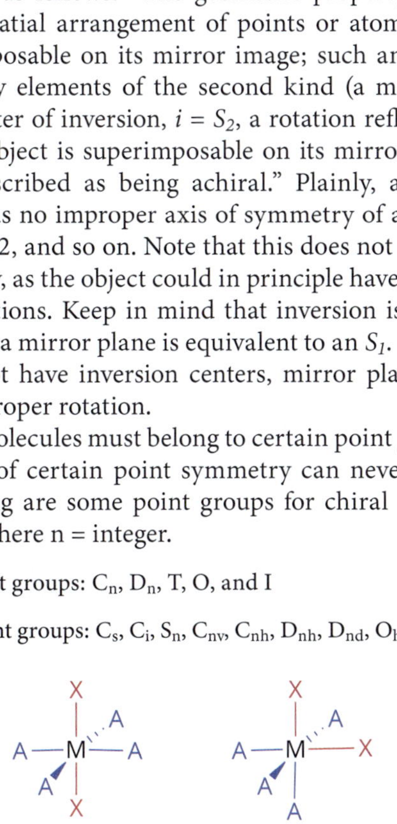

trans *cis*

Fig. 10.16a *cis* vs *trans* isomers.

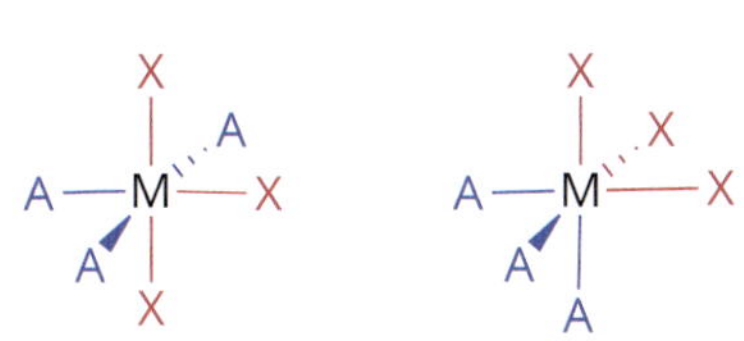

mer *fac*

Fig. 10.16b *mer* vs. *fac* isomers.

Molecules that are stereoisomers have an isomer that is not a geometric isomer. The molecule has an identical bonding pattern (all the ligands are the same angles from one another) but is the mirror image of another isomer. The two isomers related in this way are called enantiomers.

Some inorganic complexes that are chiral contain chelating ligands, such as three bidentate ligands on the metal. In 1911, Werner published a resolution (separation of the two enantiomers) of the chiral complex $\{cis\text{-}Co(NH_3)Cl(en)_2\}^+$. The compound was formed in the reaction as an equal mixture of the two enantiomers (Fig. 10.17).

Enantiomers have identical melting points, boiling points, colors, solubilities, and so on. In fact, enantiomers are identical in pretty much every way, except for the way they interact with plane-polarized light and with how they interact with other chiral molecules. Plane-polarized light is rotated in one direction by one enantiomer and in the other direction by the other enantiomer. If the enantiomer rotates the light clockwise it is called dextrorotatory (*d* or + for short). If the enantiomer rotates the light counter-clockwise the compound is levorotatory (*l* or—for short). For every chiral compound (*d*) there is one compound (*l*) that is its enantiomer, and these always come as pairs.

A chiral molecule will contain one or more chiral centers. A chiral center is defined by IUPAC as "an atom holding a set of ligands in a spatial arrangement which is not superimposable on its mirror image." The simplest example is a carbon atom with four different groups attached, in which the carbon is a chiral center. In the cobalt complexes, the metal is the chiral center.

If a compound contains more than one chiral center, then there can be diastereomers as well as enantiomers. The enantiomer of a compound with multiple chiral centers will have all of the chiral sites with the opposite hand. For example, say a cationic complex is dextrorotary (*d*), and it has a counterion that is also dextrorotary (*d*). For short, we will call this a cation–anion pair (*dd*). This enantiomer (*dd*) has a single enantiomer in which both chiral centers are inverted; this enantiomer is (*ll*), where both centers are levorotary. In contrast, the complexes (*dl*) and (*ld*) are not enantiomers of complex (*dd*) or (*ll*). The compounds (*dd*) and (*dl*), where only one of the chiral centers is inverted, are diastereomers.

The IUPAC definition of a diastereomer is more general: "Diastereoisomers (or diastereomers) are stereoisomers not related as mirror images. Diastereoisomers are characterized by differences in physical properties, and by some differences in chemical behavior towards achiral as well as chiral

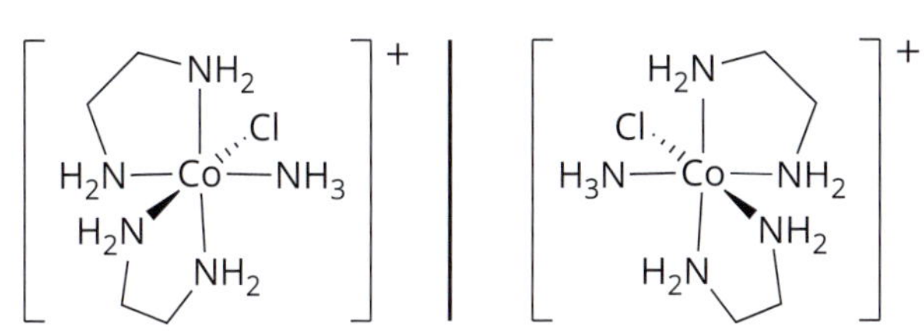

Fig. 10.17 The two stereoisomers of $\{cis\text{-}Co(NH_3)Cl(en)_2\}^+$.

reagents." While enantiomers are two sides of the same coin with many identical physical properties, diastereomers are different compounds altogether, with different melting points, boiling points, solubilities, and so on. Under the general definition, a diastereomer does not necessarily need to contain a chiral center and has different properties. A diastereomer, then, is any stereoisomer of a chiral compound that is not its enantiomer.

Consider the possible isomers of Werner and King's complex $CoCl(NH_3)(en)_2$, where en = ethylenediamine (Fig. 10.18). There are three possible stereoisomers for the complex. There is a *trans*-isomer, which is achiral owing to an internal mirror plane. The other two stereoisomers are *cis*-enantiomers. The *cis*- and *trans*-isomers are diastereomers, since they are not mirror images. The *trans*-derivative is a geometric isomer of the *cis*.

If the counterion for *cis*-$CoCl(NH_3)(en)_2$, shown simply as X^-, is achiral then there is only one chiral center at cobalt. Since the enantiomers have identical properties with regards to solubility, they cannot be separated by crystallization; although, depending on the space group of crystallization, the individual crystals could be enantiopure with a mixture of different enantiomers among the crystals.[c] In some space groups, the two enantiomers will occupy the same crystal lattice and not even be separable by this "Pasteur method" of separating enantiomers, which Louis Pasteur famously applied to separating enantiomers of tartaric acid, one crystal at a time with tweezers while working under a microscope.

However, replacement of X^- with a single enantiomer of a chiral compound, X^{*-}, gives diastereomers. since one chiral site is the same in the two salts and one is different. The two salts at the bottom of Fig. 10.18 are (*dd*) and (*ld*), diastereomers. Diastereomers can have different solubilities and use of a chiral salt allows their separation by crystallization.

Assignment of stereoisomers in inorganic complexes has been systematized using methods such as the *rectus* (R) and *sinister* (S) methods found in organic chemistry. However, these rules are both complex and in some cases inadequate to fully describe chirality in inorganic complexes. A full accounting of those naming rules would be gratuitous here, and interested readers are directed towards more specialized accounts.[27]

If an octahedral complex with three bidentate ligands is generated, then helical chirality results. For example, $\{Co(en)_3\}^{3+}$ is a chiral complex that is kinetically "inert", and has a pair of isolable enantiomers that do not rapidly isomerize. These helical enantiomers are classified as either Δ (delta) or Λ (lambda), depending on the direction the chelating chain takes from one trigonal face to the other (Fig. 10.19). If one looks down a trigonal face and points a thumb down the C_3 axis, the ligands will curl the direction of the right hand if the enantiomer is Δ and the direction of the left hand if the enantiomer is Λ.

The chiral tris(chelate) complexes can have D_3 symmetry. Note that molecules in the D_3 point group contain no

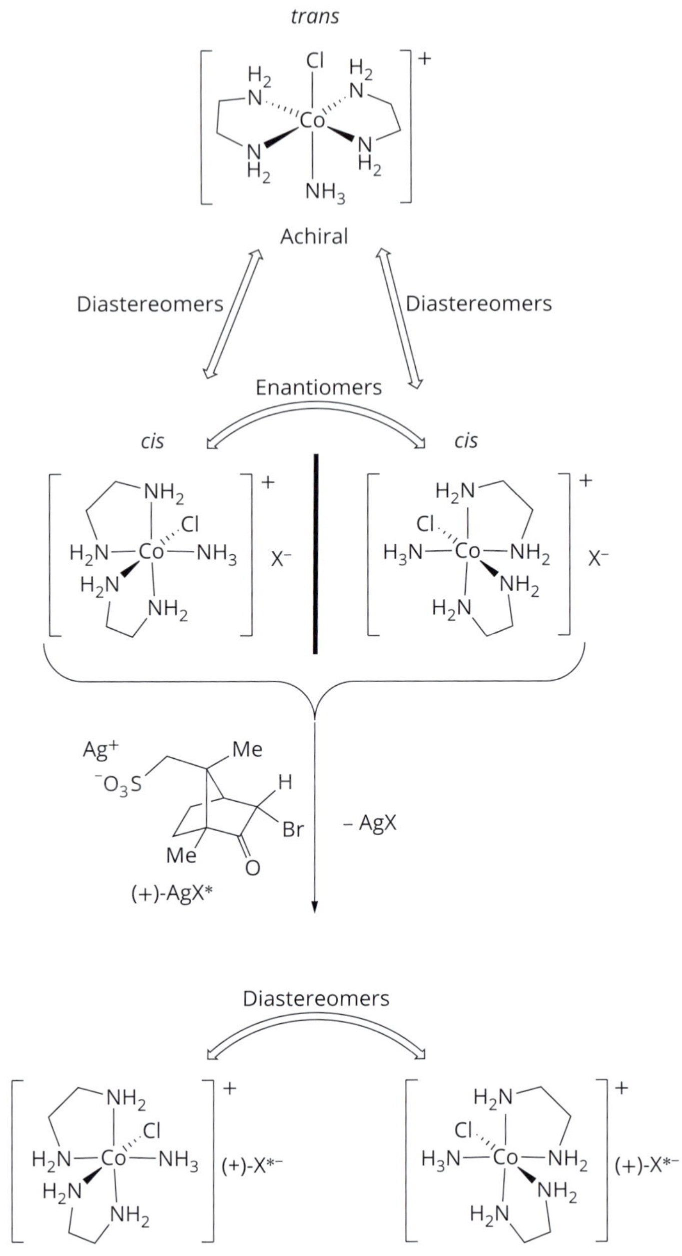

Fig. 10.18 Relationships between different isomeric $\{CoCl(NH_3)(en)_2\}^+$ complexes and production of diastereomers from the enantiomers by salt exchange.[27–28]

improper axes and are therefore chiral. If the two trigonal faces are eclipsed as in a trigonal prismatic arrangement, the complex is D_{3h} and is achiral.

In all cases, two enantiomers will have the same energy; consequently, for the enantiomers to be separated there must be a significant barrier to conversion between the two isomers. If there is a low barrier, the complex will racemize quickly. For example, $\{Co(en)_3\}^{3+}$ has a relatively high barrier to racemization, while the enantiomers of $\{Fe(ox)_3\}^{3-}$, where ox = oxalate (Table 10.3), interconvert quickly and cannot be isolated as separate enantiomers.

[c] For brief discussions of space groups and arrangements of atoms into solids, see Chapters 9 and 19.

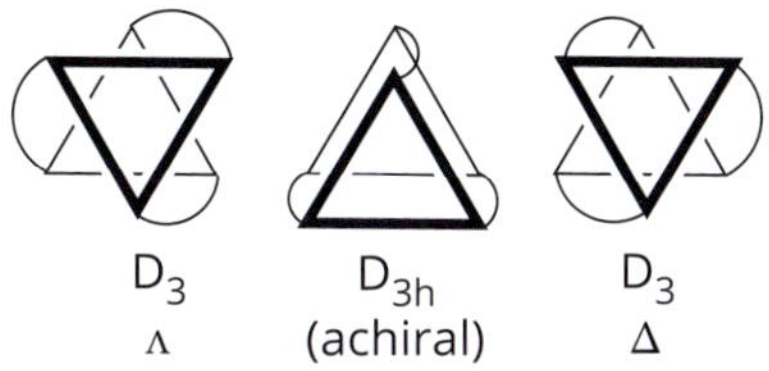

Fig. 10.19 The two D_3 stereoisomers (Δ and Λ) along with the achiral D_{3h} structure that are possible for a complex having three chelating bidentate ligands on the same metal center.

Two stereoisomers that can be separated are those of $\{Ru(bpy)_3\}^{2+}$ (bpy = 2,2′-bipyridyl), shown in Fig. 10.20. Again, one can see that the compound on the left is the Λ-isomer by placing the thumb of the left hand down onto the page/screen and noticing that the fingers curl the direction that the chelates proceed from top to bottom. The compound on the right has a right-handed Δ-isomer configuration.[27]

There are three common mechanisms for interconversion of enantiomers in octahedral complexes.

(1) The first is the *Bailar twist mechanism*, which involves the D_{3h} transition state shown in Fig. 10.19. The trigonal planes of the complex rotate relative to one another to interconvert the enantiomers.

(2) A C_{2v} transition state can also occur as shown in Fig. 10.21, which is called the *Ray–Dutt mechanism*. Here the complex rotates and aligns two different trigonal planes from those used in the Bailar twist mechanism.

(3) The third common mechanism for racemization is for one arm of the chelate to come away from the metal then to reattach to produce the other enantiomer. This is a common occurrence for labile metal complexes. The term "labile" is used for metal complexes that undergo fast exchange of ligands through a variety of different mechanisms, often related to their electronic structure properties, as is discussed in Chapter 17 on kinetics.

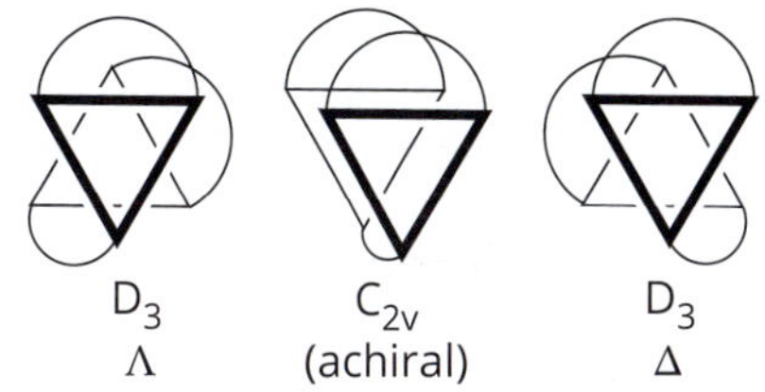

Fig. 10.21 An achiral C_{2v} structure can be used to convert between Δ and Λ isomers, which is the Ray–Dutt mechanism for interconversion.

10.4.2.1 Possible Number of Isomers for Complexes[29–31]

As the number of different ligands around the octahedron increases, it can become far more difficult to draw the possible isomers by hand, and count them. For example, when every ligand of an octahedral system is different, there are thirty possible stereoisomers: fifteen geometric isomers, and fifteen pairs of enantiomorphs (*vide infra*).[30]

The total number of isomers for any system where all of the ligands are different can be calculated readily by using Eq. 10.1. The number of isomers, N, is related to the factorial of the number of ligands, $n!$, and the order of the pure rotational group for the pseudo-geometry, h_{rot}.

$$N = \frac{n!}{h_{rot}} \qquad \textbf{10.1}$$

For example, the pure rotational group for a pseudo-octahedral complex of formula $ML^aL^bL^cL^dL^eL^f$, where L^n are the different ligands, would be the rotational group O, which has an order of 24. Therefore, there are thirty different stereoisomers, as a mixture of enantiomeric pairs and geometric isomers, for a pseudo-octahedral complex with six different ligands.

$$N = \frac{n!}{h_{rot}} = \frac{6!}{24} = \frac{720}{24} = 30$$

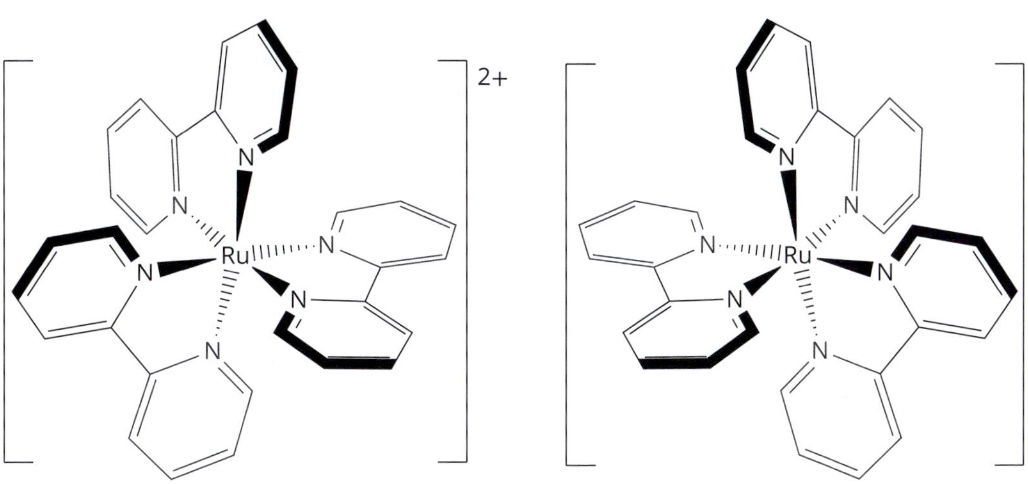

Λ-$\{Ru(bpy)_3\}^{2+}$ Δ-$\{Ru(bpy)_3\}^{2+}$

Fig. 10.20 The two stereoisomers of $\{Ru(bpy)_3\}^{2+}$.

Similarly, one can calculate the number of isomers for a tetrahedron with four different ligands. The pure rotational group is then T, which has an order of 12.

$$N = \frac{n!}{h_{rot}} = \frac{4!}{12} = \frac{24}{12} = 2$$

There are only two isomers possible for a tetrahedron with four different ligands. As learned in organic chemistry, there is one geometric isomer, and in this case it has an enantiomer.

10.4.2.2 Calculating the Number of Regioisomers, Geometric Isomers, and Enantiomeric Pairs (Advanced Topic)

Procedures for calculating the number of isomers with any number of different ligands have been enumerated, and in addition one can even calculate the number of geometric isomers versus enantiomeric pairs for a complex. Here we provide a brief description of how this may be achieved by using the method of Polya.[29–31]

The method relies heavily on group theory. The description here will be as visual as possible, but there is some math involved, which is largely algebra. A symbolic math program, such as Mathematica or Maple, is very helpful in expanding the polynomial expansions, which can become quite tedious for higher coordinate complexes.

Obviously, as the number of different ligands becomes very large, the actual symmetry of a system may become quite low. In this method we will calculate all the possible isomers of various types for every ligand configuration (any number of different ligands) all at once. This is accomplished by using the point group for the highest-possible symmetry of the system (Table 10.2) or the related pure rotational point group. For example, if the compound is 6-coordinate with approximately 90° and 180° angles between ligands, it is "pseudo-octahedral", and for the method here we will use O_h and the pure rotational group O. If the compound is square pyramidal, we would use C_{4v} and C_4.

Table 10.4 lists common point groups for various coordination numbers (compare Table 10.2) and their corresponding pure rotational groups. The pure rotational group is used to find the total number of stereoisomers of any kind, as will be explained subsequently, and the high symmetry point group with mirror planes, and so on, included is used to calculate the number of geometric isomers. From the difference between these two calculations, the number of enantiomeric pairs can be determined. The method requires the calculation of the "vertex interchange" matrix, which is basically a matrix showing how the ligands exchange as the symmetry operations are done. From the vertex interchange matrix, the "permutation groupings" are found. From the permutation groupings, the "cycle index" is found, which is simply a symbol that will be useful momentarily.

Table 10.4 The point groups for some common geometric forms found inorganic chemistry and their corresponding pure rotational groups. For the isomer calculation method described here, the pure rotational group is used to calculate the number of stereoisomers, and the higher symmetry group is used to calculate the number of geometric isomers.[30]

Pure Rotational Group	Higher Symmetry Group
To calculate the number of stereoisomers	To calculate the number of geometric isomers
C_n	C_{nv} C_{nh}
D_n	D_{nd} D_{nh}
T	T_d
O	O_h
I	I_h

The easiest way to explain this is with an example, and a few different examples will be presented in this section. To begin, let's find the number of isomers for a tetrahedral system. The total number of stereoisomers (geometric + enantiomeric pairs) is found by using the pure cyclic group as the point group; for example, for the methane derivative CH(Cl)(Br)(I) use the point group T to find the total number of stereoisomers. The total number of geometric isomers is found by using the group including all the symmetry of the pseudosymmetric point group; for example, for the methane derivative CH(Cl)(Br)(I) use the point group T_d to find the total number of geometric isomers. Consequently, the number of enantiomeric pairs can be found by subtraction of the number found by using the pure cyclic point group (T) and the number found when using the full symmetry (T_d).

If we start in the point group T, this will give us the total number of regioisomers of any type. We will label the ligands in the pseudo-tetrahedral system as 1–4 (Fig. 10.22).

We write the four classes in the T point group (left-hand column, Table 10.5) and make a matrix of how the ligands exchange with the operation. The top line of each vertex interchange matrix is simply a numbering of the ligands—in this case, "1 2 3 4". For E, $1 \rightarrow 1$, $2 \rightarrow 2$, $3 \rightarrow 3$, and $4 \rightarrow 4$, so the second line of the E matrix is "1 2 3 4". There are four C_3 axes, and only one needs to be calculated within each group. For one of the C_3 axes, vertex 1 does not change, while $2 \rightarrow 3$, $3 \rightarrow 4$, and $4 \rightarrow 2$, so the second line of the C_3 matrix is "1 3 4 2". Interchanges are calculated similarly for C^2_3 and C_2.

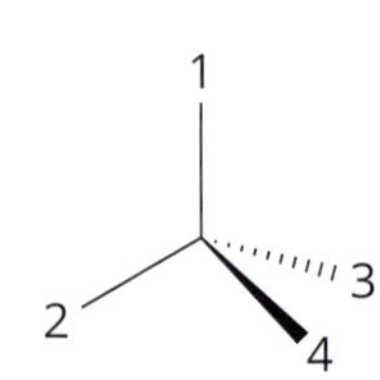

Fig. 10.22 Labels for the T_d system for determination of isomers.

The interchanges are then placed into groups based on the number of vertices exchanging (Table 10.5). For E, nothing undergoes interchange, and everything is in its own group; that is, 1 goes to 1, 2 goes to 2, and so on. For the C_3 there are three vertices in one group (2, 3, and 4) which exchange with one another, and vertex 1 is in a group by itself.

From the permutation groups the "cyclic index", f, is calculated. The number of members in the group is written as a subscript, and the number of groups is the superscript. For E in our example, each group has a single member (column), and there are four groups, f_1^4. For C_3 there is one single-column group and one triple-column group, $f_1^1f_3^1$, or simply f_1f_3. For C_2 there are two groups with two members (columns) each, f_2^2.

To find the number of isomers, Z, we use Eq. 10.2, where h is the order of the point group. All the f-terms are summed over all the different classes R_i in the point group.

$$Z=\frac{1}{h}\sum_i f_x^y\left(R_i\right) \qquad \textbf{10.2}$$

For our example, Eq. 10.2 produces the following expression, where 1/h is one over the order of the point group (T) and the f-terms are from the right-hand column of the Table 10.5.

$$Z^T=\frac{1}{12}(f_1^4+4f_1f_3+4f_1f_3+3f_2^2)$$

This can be simplified to produce Eq. 10.3, for which we simply combine the two f_1f_3 terms. This equation will be general for any 4-coordinate pseudo-tetrahedral system.

$$Z^T=\frac{1}{12}(f_1^4+8f_1f_3+3f_2^2) \qquad \textbf{10.3}$$

We can discover how many stereoisomers there are by assuming that each vertex has a different ligand. In actuality, this process will tell us the total number of isomers for every permutation of different ligands all at once! Each f, in the previous equation will be replaced as shown in Eq. 10.4.

$$f_x^y=(a^x+b^x+c^x+d^x+e^x+\ldots)^y \qquad \textbf{10.4}$$

Table 10.5 Finding the cyclic indices for the classes in T.

Classes in T	Vertex Interchange	Permutation Grouping	Cyclic Index
E	$\begin{matrix}1&2&3&4\\1&2&3&4\end{matrix}$	$\begin{pmatrix}1\\1\end{pmatrix}\begin{pmatrix}2\\2\end{pmatrix}\begin{pmatrix}3\\3\end{pmatrix}\begin{pmatrix}4\\4\end{pmatrix}$	f_1^4
$4C_3$	$\begin{matrix}1&2&3&4\\1&3&4&2\end{matrix}$	$\begin{pmatrix}1\\1\end{pmatrix}\begin{pmatrix}2&3&4\\3&4&2\end{pmatrix}$	$4f_1f_3$
$4C^2_3$	$\begin{matrix}1&2&3&4\\1&4&2&3\end{matrix}$	$\begin{pmatrix}1\\1\end{pmatrix}\begin{pmatrix}2&3&4\\4&2&3\end{pmatrix}$	$4f_1f_3$
$3C_2$	$\begin{matrix}1&2&3&4\\2&1&4&3\end{matrix}$	$\begin{pmatrix}1&2\\2&1\end{pmatrix}\begin{pmatrix}3&4\\4&3\end{pmatrix}$	$3f_2^2$

In our case with four ligands, we truncate Eq. 10.4 after a, b, c, and d. Substitution of Eq. 10.4 into the equation for Z^T (Eq. 10.3) produces the following expression.

$$Z^T=\frac{1}{12}\{(a+b+c+d)^4+8(a+b+c+d)(a^3+b^3+c^3+d^3)+3(a^2+b^2+c^2+d^2)^2\} \qquad \textbf{10.5}$$

At this point, this polynomial is expanded (Eq. 10.6). Again, this can be done by hand, but it is quite tedious. A program like Mathematica or Maple is helpful for the expansion. We can assign the ligands a, b, c, and d to be anything at all, and the following equation tells us directly how many isomers there are by the coefficient! For every term below, the coefficient is "1" except for "2abcd". In other words, there is a single isomer for every substitution pattern except for where all four ligands are different, which has two stereoisomers of some type.

$$\begin{aligned}Z^T&=a^4+a^3b+a^2b^2+ab^3+b^4+a^3c+a^2bc+ab^2c+b^3c\\&+a^2c^2+abc^2+b^2c^2+ac^3+bc^3+c^4+a^3d+a^2bd\\&+ab^2d+b^3d+a^2cd+2abcd+b^2cd+ac^2d+bc^2d\\&+c^3d+a^2d^2+abd^2+c^2d^2+ad^3+bd^3+cd^3+d^4\end{aligned} \qquad \textbf{10.6}$$

Using the pure cyclic group T gave us every stereoisomer. We can now find the number of geometric isomers by using the same procedure but using T_d as the point group (Table 10.6).

This produces a new Z in the following equation.

$$Z^{T_d}=\frac{1}{24}(f_1^4+8f_1f_3+3f_2^2+6f_4^1+6f_1^2f_2^1) \qquad \textbf{10.7}$$

Table 10.6 Finding the cyclic indices for the classes in T_d.

Classes in T_d	Vertex Interchange	Permutation Grouping	Cyclic Index
E	$\begin{matrix}1&2&3&4\\1&2&3&4\end{matrix}$	$\begin{pmatrix}1\\1\end{pmatrix}\begin{pmatrix}2\\2\end{pmatrix}\begin{pmatrix}3\\3\end{pmatrix}\begin{pmatrix}4\\4\end{pmatrix}$	f_1^4
$8C_3$	$\begin{matrix}1&2&3&4\\1&3&4&2\end{matrix}$	$\begin{pmatrix}1\\1\end{pmatrix}\begin{pmatrix}2&3&4\\3&4&2\end{pmatrix}$	$8f_1f_3$
$3C_2$	$\begin{matrix}1&2&3&4\\2&1&4&3\end{matrix}$	$\begin{pmatrix}1&2\\2&1\end{pmatrix}\begin{pmatrix}3&4\\4&3\end{pmatrix}$	$3f_2^2$
$6S_4$	$\begin{matrix}1&2&3&4\\2&4&1&3\end{matrix}$	$\begin{pmatrix}1&2&3&4\\2&4&1&3\end{pmatrix}$	$6f_4^1$
$6\sigma_d$	$\begin{matrix}1&2&3&4\\1&2&4&3\end{matrix}$	$\begin{pmatrix}1\\1\end{pmatrix}\begin{pmatrix}2\\2\end{pmatrix}\begin{pmatrix}3&4\\4&3\end{pmatrix}$	$6f_1^2f_2^1$

Again, we make the substitution $f_x^y = (a^x + b^x + c^x + d^x)^y$, which produces the following equation.

$$Z^{T_d} = \frac{1}{24}\{(a+b+c+d)^4 + 8(a+b+c+d)(a^3+b^3+c^3+d^3) + 3(a^2+b^2+c^2+d^2)^2 + 6(a^4+b^4+c^4+d^4) + 6(a+b+c+d)^2(a^2+b^2+c^2+d^2)\} \quad \textbf{10.8}$$

Expanding this polynomial gives the number of geometric isomers for every possible permutation of ligands in a pseudo-tetrahedral geometry. This time, however, the coefficient for every term is unity, even for *abcd*. Consequently, for every substitution there is a single geometric isomer. The number of enantiomeric pairs (EP) is equal to the number of stereoisomers (SI) minus the number of geometric isomers (GI).

$$SI - GI = EP \quad \textbf{10.9}$$

For an *abcd*-pattern where every ligand is different in a pseudo-tetrahedral system, we obtain:

$$2\text{ stereoisomers} - 1\text{ geometric isomer} = 1\text{ enantiomeric pair}$$

In other words there is one geometric isomer, which has an enantiomer, for a pseudo-tetrahedral system in which all the ligands are different. For any other substitution pattern— for example, ab^2c with three different ligands—there is only one stereoisomer possible.

For a tetrahedral system, the answer was very probably known beforehand. You are taught in organic chemistry that pseudo-tetrahedral systems are chiral if they have four different ligands. Now, let us move to a system that is perhaps less obvious. How many geometric isomers and enantiomeric pairs are there for a 5-coordinate system in a pseudo-trigonal bipyramidal (tbp) coordination geometry with four different ligands; that is, a^2bcd? To find the number of stereoisomers, we use the pure rotational group (D_3, Table 10.7) analogous to D_{3h} (Table 10.8), the point group for tbp. The labels are shown in Fig. 10.23.

$$Z^{D_3} = \frac{1}{6}(f_1^5 + 2f_1^2 f_3^1 + 3f_2^2 f_1^1) \quad \textbf{10.10}$$

Into Eq. 10.10 we substitute $f_x^y = (a^x + b^x + c^x + d^x + e^x)^y$ to obtain Eq. 10.11.[d]

$$Z^{D_3} = \frac{1}{6}\{(a+b+c+d+e)^5 + 2(a+b+c+d+e)^2(a^3+b^3+c^3+d^3+e^3) + 3(a^2+b^2+c^2+d^2+e^2)^2(a+b+c+d+e) \quad \textbf{10.11}$$

[d] Eq. 10.11 and subsequent expansion in Eq. 10.12 could have been made slightly simpler by using only ligands a-d, which is what is in the problem. We decided to show the general result of the expansion with ligands a-e, which includes these.

Table 10.7 Finding the cyclic indices for the classes in D_3.

Classes in D_3	Vertex Interchange	Permutation Grouping	Cyclic Index
E	$\begin{matrix}1&2&3&4&5\\1&2&3&4&5\end{matrix}$	$\begin{pmatrix}1\\1\end{pmatrix}\begin{pmatrix}2\\2\end{pmatrix}\begin{pmatrix}3\\3\end{pmatrix}\begin{pmatrix}4\\4\end{pmatrix}\begin{pmatrix}5\\5\end{pmatrix}$	f_1^5
$2C_3$	$\begin{matrix}1&2&3&4&5\\1&2&4&5&3\end{matrix}$	$\begin{pmatrix}1\\1\end{pmatrix}\begin{pmatrix}2\\2\end{pmatrix}\begin{pmatrix}3&4&5\\4&5&3\end{pmatrix}$	$2f_1^2 f_3^1$
$3C_2$	$\begin{matrix}1&2&3&4&5\\2&1&3&5&4\end{matrix}$	$\begin{pmatrix}1&2\\2&1\end{pmatrix}\begin{pmatrix}3\\3\end{pmatrix}\begin{pmatrix}4&5\\5&4\end{pmatrix}$	$3f_2^2 f_1^1$

Table 10.8 Finding the cyclic indices for the classes in D_{3h}.

Classes in D_{3h}	Vertex Interchange	Permutation Grouping	Cyclic Index
E	$\begin{matrix}1&2&3&4&5\\1&2&3&4&5\end{matrix}$	$\begin{pmatrix}1\\1\end{pmatrix}\begin{pmatrix}2\\2\end{pmatrix}\begin{pmatrix}3\\3\end{pmatrix}\begin{pmatrix}4\\4\end{pmatrix}\begin{pmatrix}5\\5\end{pmatrix}$	f_1^5
$2C_3$	$\begin{matrix}1&2&3&4&5\\1&2&4&5&3\end{matrix}$	$\begin{pmatrix}1\\1\end{pmatrix}\begin{pmatrix}2\\2\end{pmatrix}\begin{pmatrix}3&4&5\\4&5&3\end{pmatrix}$	$2f_1^2 f_3^1$
$3C_2$	$\begin{matrix}1&2&3&4&5\\2&1&3&5&4\end{matrix}$	$\begin{pmatrix}1&2\\2&1\end{pmatrix}\begin{pmatrix}3\\3\end{pmatrix}\begin{pmatrix}4&5\\5&4\end{pmatrix}$	$3f_2^2 f_1^1$
σ_h	$\begin{matrix}1&2&3&4&5\\2&1&3&4&5\end{matrix}$	$\begin{pmatrix}1&2\\2&1\end{pmatrix}\begin{pmatrix}3\\3\end{pmatrix}\begin{pmatrix}4\\4\end{pmatrix}\begin{pmatrix}5\\5\end{pmatrix}$	$f_2^1 f_1^3$
$2S_3$	$\begin{matrix}1&2&3&4&5\\2&1&4&5&3\end{matrix}$	$\begin{pmatrix}1&2\\2&1\end{pmatrix}\begin{pmatrix}3&4&5\\4&5&3\end{pmatrix}$	$2f_2^1 f_3^1$
$3\sigma_v$	$\begin{matrix}1&2&3&4&5\\1&2&3&5&4\end{matrix}$	$\begin{pmatrix}1\\1\end{pmatrix}\begin{pmatrix}2\\2\end{pmatrix}\begin{pmatrix}3\\3\end{pmatrix}\begin{pmatrix}4&5\\5&4\end{pmatrix}$	$3f_1^3 f_2^1$

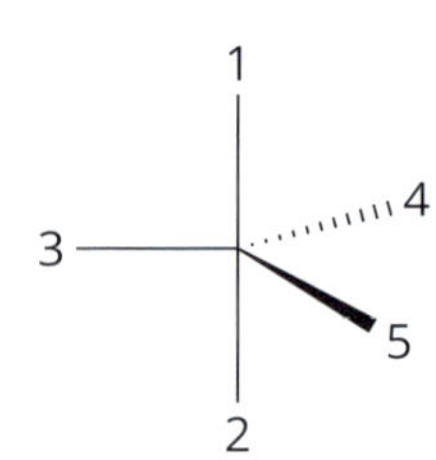

Fig. 10.23 Labels for the trigonal bipyramidal system for determination of isomers.

This gives Z^{D_3} as the long list of terms shown in Eq. 10.12, which tells us the number of isomers for every possible ligand combination. In this case we are interested in an a^2bcd substitution pattern with four different ligands, two ligands the same. This term is found as "$10a^2bcd$", meaning that there are ten different stereoisomers of some type possible in a pseudo-tbp geometry.

$$
\begin{aligned}
Z^{D_3} = {} & a^5 + 2a^4b + 3a^3b^2 + 3a^2b^3 + 2ab^4 + b^5 + 2a^4c + 4a^3bc \\
& +6a^2b^2c + 4ab^3c + 2b^4c + 3a^3c^2 + 6a^2bc^2 + 6ab^2c^2 \\
& +3b^3c^2 + 3a^2c^3 + 4abc^3 + 3b^2c^3 + 2ac^4 + 2bc^4 + c^5 \\
& +2a^4d + 4a^3bd + 6a^2b^2d + 4ab^3d + 2b^4d + 4a^3cd \\
& +10a^2bcd + 10ab^2cd + 4b^3cd + 6a^2c^2d + 10abc^2d \\
& +6b^2c^2d + 4ac^3d + 4bc^3d + 2c^4d + 3a^3d^2 + 6a^2bd^2 \\
& +6ab^2d^2 + 3b^3d^2 + 6a^2cd^2 + 10abcd^2 + 6b^2cd^2 + 6ac^2d^2 \\
& +6bc^2d^2 + 3c^3d^2 + 3a^2d^3 + 4abd^3 + 3b^2d^3 + 4acd^3 \\
& +4bcd^3 + 3c^2d^3 + 2ad^4 + 2bd^4 + 2cd^4 + d^5 + 2a^4e \\
& +4a^3be + 6a^2b^2e + 4ab^3e + 2b^4e + 4a^3ce + 10a^2bce \\
& +10ab^2ce + 4b^3ce + 6a^2c^2e + 10abc^2e + 6b^2c^2e + 4ac^3e \\
& +4bc^3e + 2c^4e + 4a^3de + 10a^2bde + 10ab^2de + 4b^3de \\
& +10a^2cde + 20abcde + 10b^2cde + 10ac^2de + 10bc^2de \\
& +4c^3de + 6a^2d^2e + 10abd^2e + 6b^2d^2e + 10acd^2e + 4bd^3e \\
& +10bcd^2e + 6c^2d^2e + 4ad^3e + 4cd^3e + 2d^4e + 3a^3e^2 \\
& +6a^2be^2 + 6ab^2e^2 + 3b^3e^2 + 6a^2ce^2 + 10abce^2 + 6b^2ce^2 \\
& +6ac^2e^2 + 6bc^2e^2 + 3c^3e^2 + 6a^2de^2 + 10abde^2 + 6b^2de^2 \\
& +10acde^2 + 10bcde^2 + 6c^2de^2 + 6ad^2e^2 + 6bd^2e^2 \\
& +6cd^2e^2 + 3d^3e^2 + 3a^2e^3 + 4abe^3 + 3b^2e^3 + 4ace^3 \\
& +4bce^3 + 3c^2e^3 + 4ade^3 + 4bde^3 + 4cde^3 + 3d^2e^3 \\
& +2ae^4 + 2be^4 + 2ce^4 + 2de^4 + e^5
\end{aligned}
\qquad \mathbf{10.12}
$$

To find the number of geometric isomers, we use the highest point group possible, D_{3h}. This requires a new table (Table 10.8) with the six classes of D_{3h}.

The isomer equation is then as follows.

$$
Z^{D_{3h}} = \frac{1}{12}\left(f_1^5 + 2f_1^2 f_3^1 + 3f_2^2 f_1^1 + f_2^1 f_1^3 + 2f_2^1 f_3^1 + 3f_1^3 f_2^1\right) \qquad \mathbf{10.13}
$$

We again have five vertices. We substitute into this equation $f_x^y = (a^x + b^x + c^x + d^x + e^x)^y$ and expand the polynomial which, once expanded, has many terms. The results of the calculation, and the calculation of the number of stereoisomers, are summarized in Table 10.9.

For a 5-coordinate complex with five different ligands, Mabcde, there are twenty total stereoisomers: ten geometric isomers and ten pairs of enantiomers, for example. The total number of stereoisomers can be checked for the Mabcde-case using Eq. 10.1, so that $N = 5!/6 = 20$, just as found here. The original question regarded an Ma_2bcd in trigonal bipyramidal, and we see there are ten stereoisomers: seven geometric isomers and three pairs of enantiomers.

The cycle indices for some common geometries have been compiled by Brubaker and coworkers,[30] which are shown in Table 10.10. Keep in mind that both the point group and the number of ligands has to be the same before the cycle index is the same. For example, a trigonal planar complex with point group D_{3h} has a different cycle index than a trigonal

Table 10.9 Total number of stereoisomers, geometric isomers, and pairs of enantiomers for a trigonal bipyramidal system with all possible numbers of different ligands.

Formula for pseudo-trigonal bipyramidal system	Stereo-isomers (from D_3)	Geometrical Isomers (from D_{3h})	Pairs of Enantiomers (Difference, Eq. 10.9)
Ma_5	1	1	0
Ma_4b	2	2	0
Ma_3b_2	3	3	0
Ma_3bc	4	4	0
Ma_2b_2c	6	5	1
Ma_2bcd	10	7	3
Mabcde	20	10	10

Table 10.10 Cycle indices for some common structures.[30]

Coordination Number	Geometry	Point Gp	Cycle Index
3	trigonal planar	D_3	$1/8(f_1^3 + 3f_1f_2 + 2f_3)$
		D_{3h}	$1/12(2f_1^3 + 6f_1f_2 + 4f_3)$
4	square planar	D_4	$1/8(f_1^4 + 3f_2^2 + 2f_1^2f_2 + 2f_4)$
		D_{4h}	$1/16(2f_1^4 + 6f_2^2 + 4f_1^2f_2 + 4f^4)$
	tetrahedral	T	$1/12(f_1^4 + 8f_1f_3 + 3f_2^2)$
		T_d	$1/24(f_1^4 + 8f_1f_3 + 3f_2^2 + 6f_1^2f_2 + 6f_4)$
5	trigonal bipyramid	D_3	$1/6(f_1^5 + 2f_1^2f_3 + 3f_1f_2^2)$
		D_{3h}	$1/12(f_1^5 + 2f_1^2f_3 + 3f_1f_2^2 + 4f_1^3f_2 + 2f_3f_2)$
	square pyramid	C_4	$1/4(f_1^5 + 2f_1f^4 + f_1f_2^2)$
		C_{4v}	$1/8(f_1^5 + 2f_1f^4 + 3f_1f_2^2 + 2f_1^3f_2)$
6	octahedral	O	$1/24(f_1^6 + 8f_3^2 + 3f_1^2f_2^2 + 6f_2^3 + 6f_1^2f_4)$
		O_h	$1/48(f_1^6 + 8f_3^2 + 7f_2^3 + 6f_1^2f_4 + 9f_1^2f_2^2 + 8f_6 + 3f_1^4f_2 + 6f_4f_2)$

Table 10.11 The number of stereoisomers, geometric isomers, and pairs of enantiomers for a pseudo-octahedral system with monodentate ligands.[30]

Formula for pseudo-octahedral system	Geometric + Enantiometric pairs(from O)	Geometrical Isomers (from O_h)	Pairs of Enantiomers (Difference)
Ma_6	1	1	0
Ma_5b	1	1	0
Ma_4b_2	2	2	0
Ma_4bc	2	2	0
Ma_3b_3	2	2	0
Ma_3b_2c	3	3	0
Ma_3bcd	5	4	1
$Ma_2b_2c_2$	6	5	1
Ma_2b_2cd	8	6	2
Ma_2bcde	15	9	6
Mabcdef	30	15	15

bipyramidal complex that is also D_{3h}, which is why the listings are by geometry and not point group.

For octahedral systems, the cycle indices are in Table 10.10, and the results are shown in Table 10.11.[30] As shown, as many as three ligands can be the same on the octahedral system with all monodentate ligands when the first pair of enantiomers appears, Ma_3bcd. The $Ma_2b_2c_2$ system can also be chiral (Fig. 10.20 and surrounding discussion).

Here the applications of the method are limited to systems with a central atom, but the method can be applied to virtually any system. For enumerating the structures of the isomers and applications to chelating ligands, Bailar published a systematic procedure.[32]

10.4.2.3 5-Coordinate Complexes: Geometry and Isomerism

For 5-coordinate complexes, there are two limiting geometric forms: square pyramidal and trigonal bipyramidal (tbp). Often, these two structures are very close in energy for 5-coordinate complexes and undergo fast exchange of ligand sites. This is true of main-group compounds, such as the tbp ground state PF_5, and transition metal complexes.

The mechanisms of exchange in 5-coordinate complexes are probably varied, but it seems by far the most important is what is termed *Berry pseudorotation*. The transition state or intermediate in the exchange process between two D_{3h} trigonal bipyramidal structures is a C_{4v} square pyramidal species in the mechanism, as illustrated in Fig. 10.24 for FeL_5, such as 18-electron iron pentacarbonyl, $Fe(CO)_5$. The starting ligands in the tbp structure are labeled as numbers L^1 and L^2, while the equatorial ligands are assigned letters L^{a-c}. The process can be thought of as a distortion of the axial ligands in the tbp structure in the same direction to generate a square pyramidal system where one of the equatorial ligands in the tbp structure, L^a, now becomes the axial ligand in the square pyramidal complex. The originally equatorial ligands in the tbp structure L^b and L^c then become the new axial ligands of the isomerized tbp complex.

In heteroleptic complexes especially, structures between square pyramidal and trigonal bipyramidal are the norm. Actual angles in structures may not be near either of these limiting forms, and a simple method for describing the geometry of 5-coordinate complexes is becoming commonplace.

The method relies on the value of a continuous symmetry parameter, τ, determined by the two largest angles in the complex (Fig. 10.25).[33] The equation for the determination of τ is shown in Eq. 10.14, where α is the largest angle and β is the second-largest angle. For square pyramidal, $\alpha = \beta = 180°$, and $\tau = 0$. For trigonal bipyramidal, $\alpha = 180°$ and $\beta = 120°$, giving $\tau = 1$. From the value of τ one can acquire, at a glance, a sense of whether the structure is closer to trigonal bipyramidal or square pyramidal.

$$\tau = (\alpha - \beta)/60 \qquad \textbf{10.14}$$

In Fig. 10.26 is a structure of a 5-coordinate complex developed by one of the authors as a catalyst for carbon–nitrogen bond formation.[34] The largest and second-largest angles around Ti from the caption are 163.58(12)° and 120.37(12)°.

$$\tau = (\alpha - \beta)/60 = \frac{(163.58 - 120.37)}{60} = 0.72$$

This gives $\tau = 0.72$, and the structure is closer to trigonal bipyramidal.

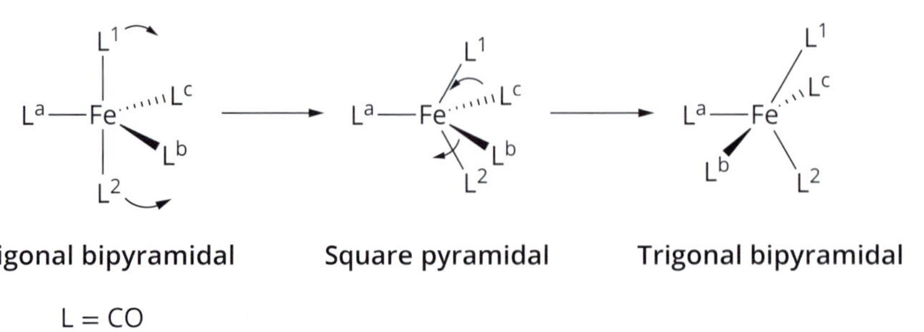

Fig. 10.24 Berry pseudorotation mechanism for exchange of equatorial and axial sites.

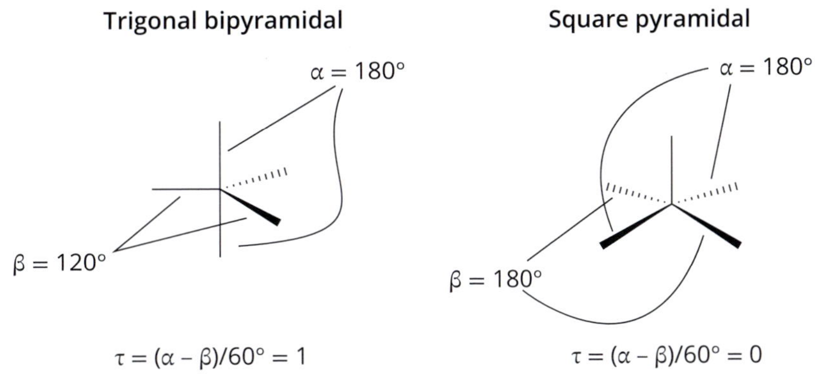

Fig. 10.25 Largest and second-largest angles in trigonal bipyramidal and square pyramidal structures.

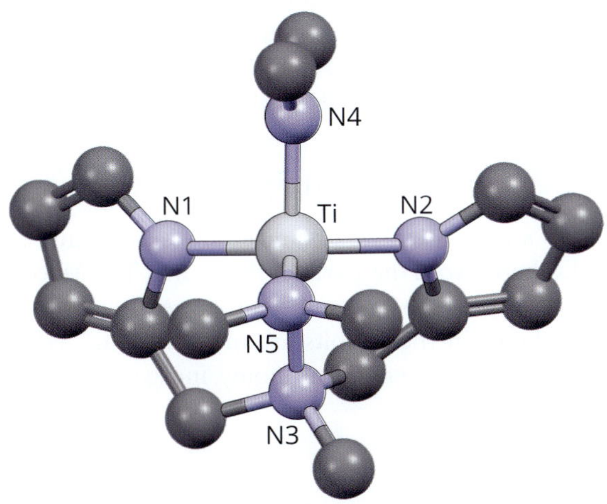

Fig. 10.26 Structure, bond distances, and bond angles of $Ti(NMe_2)_2(dpma)$, where dpma = *N,N*-dipyrrolyl-α-methyl-*N*-methylamine.[34] N(4) and N(5) are dimethylamides, NMe_2. N(1), and N(2) are deprotonated pyrroles, which are similar to amides with weaker π-donating properties. N(3) is a trialkylamine. Selected bond lengths (Å) and angles (°): Ti-N(5) 1.859(3), Ti-N(4) 1.888(3), Ti-N(1) 2.015(3), Ti-N(2) 2.017(3), Ti-N(3) 2.312(3), N(5)-Ti-N(4) 100.74(13), N(5)-Ti-N(1) 115.95(14), N(4)-Ti-N(1) 97.88(14), N(5)-Ti-N(2) 118.21(14), N(4)-Ti-N(2) 94.96(13), N(1)-Ti-N(2) 120.37(12), N(5)-Ti-N(3) 95.59(12), N(4)-Ti-N(3) 163.58(12), N(1)-Ti-N(3) 76.16(12), N(2)-Ti-N(3) 75.67(12). Hydrogens are not shown.

10.4.2.4 4-Coordinate Complexes: Geometry and Isomerism

There are two commonly encountered limiting forms for a 4-coordinate complex, and Alvarez and coworkers developed the t_4 parameter to produce a continuous parameter between square planar and tetrahedral geometries.[35] However, the method was found to be inadequate for some structures containing bidentate ligands. In 2015 the parameter was modified to be more generally applicable.[36] The formula for the Reineke–Kubiak τ_δ parameter is shown in Eq. 10.15, where α and β are the largest and second largest L–M–L angles in the complex, respectively. The new equation is the product of Alvarez's t_4 equation and a new parameter δ.

Table 10.12 The parameter τ_δ as a function of geometric structure in 4-coordinate complexes.[36]

τ_δ parameter	Structure
~0.9–1	tetrahedral
~0.63–0.9	distorted tetrahedral
~0.45–0.63	distorted sawhorse
~0.2–0.45	sawhorse
~0.1–0.2	distorted square planar
~0.1–0.0	square planar

$$\tau_\delta = \frac{360 - (\alpha + \beta)}{141}\delta \text{ where } \delta = \frac{\beta}{\alpha} \quad \textbf{10.15}$$

The new parameter runs from 1.0 (tetrahedral) to 0.0 (square planar) with intermediate structures in between, like a sawhorse (disphenoidal) structure at ~0.5. The structures and the parameters are correlated in Table 10.12.

Kubiak and coworkers were studying a Ni(0) system bearing bidentate *N*-heterocyclic carbene ligands (Fig. 10.27). The α and β angles from X-ray diffraction were found to be 139.86(14) and 119.36(14)°. This gives a τ_δ parameter of 0.61, a sawhorse structure slightly distorted towards tetrahedral.

10.4.3 Structural Isomerism[37]

In addition to stereo and geometric isomers, compounds may be structurally different in their bonds but still have the same chemical formula. There are many possible forms of structural isomerism, and a few common ones will be discussed.

10.4.3.1 Linkage Isomerism

Some ligands have multiple modes of attachment to metal centers, which may be energetically or kinetically favored.

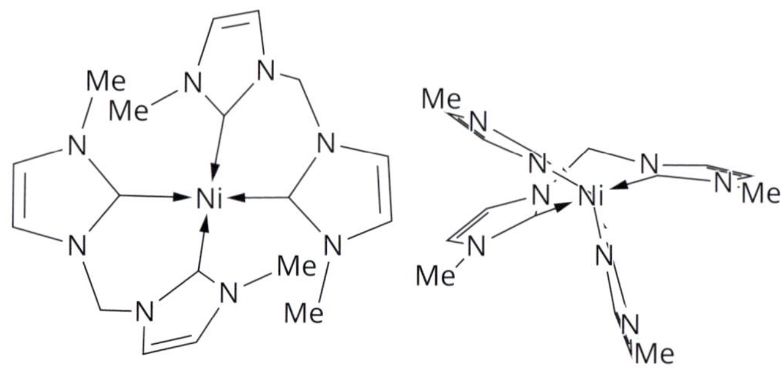

Fig. 10.27 Two views of a nickel(II) complex with two bidentate *N*-heterocyclic carbene ligands prepared by Kubiak and coworkers.[36] The line drawings are from the X-ray diffraction geometry.

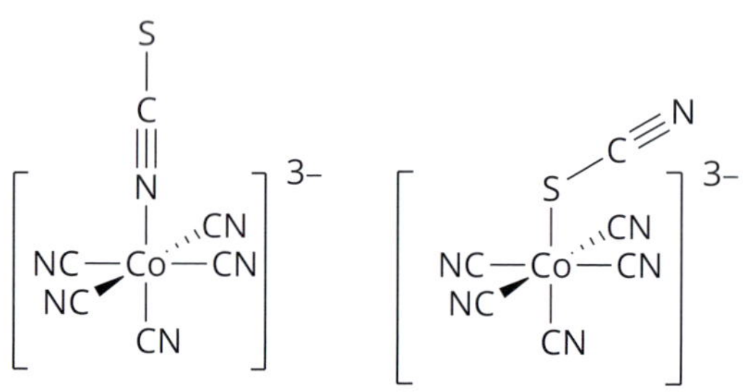

Fig. 10.28 Linkage isomerism in cobalt complexes, illustrated using isothiocyanate (*left*) and thiocyanate (*right*).

Two such ligands, common in coordination chemistry since Werner's time, are nitrite (NO_2^-) and thiocyanate (SCN^-). Nitrite may be N-bound or O-bound to a metal center. Thiocyanate can bind either through the nitrogen or sulfur atom. Examples for thiocyanate are shown in Fig. 10.28 in the form of $\{Co(CN)_5(\kappa^1N\text{-}NCS)\}^{3-}$ and $\{Co(CN)_5((\kappa^1S\text{-}SCN)\}^{3-}$.

$$\left\{Co(CN)_5\left(\kappa^1 N\text{-NCS}\right)\right\}^{3-} \quad \left\{Co(CN)_5\left(\kappa^1 S\text{-SCN}\right)\right\}^{3-}$$

Note that there are several clues as to the structure in the formulae. The atom in the thiocyanate ligand bound to the metal is written closest to the metal in the formula. This is a good habit in general, with carbonyl complexes being written with the bound carbon first, M(CO), and with water being written with the bound oxygen first, $M(OH_2)$. In addition, the κ-notation gives the number of atoms in the ligand bound to the metal center, in this case 1, and the element bound to the metal in italics, *N* or *S*. (See Box 10B.)

10.4.3.2 Coordination Sphere Isomerism

There was a time when we organized the entire universe into celestial spheres, with the Earth at the center. The Moon, Mercury, Venus, the Sun, Mars, and so on, were thought to be carried by spheres in their orbits around the Earth. While this model has been replaced in astronomy, inorganic chemists use

BOX 10B

Symbolic aids for naming inorganic compounds: κ, η, and μ

In inorganic and organometallic chemistry there are a few symbols commonly used in the naming of compounds.

κ (kappa)

The Greek letter κ is used to remove ambiguity in which atoms in a ligand are bound to the metal center, especially when the ligand is multidentate or ambidentate. For example, consider the nitrito ligand. Nitrito is an ambidentate ligand that can bind through either the nitrogen or oxygen, and the symbol κ with the element symbol is used to specify which (Fig. 10B1).

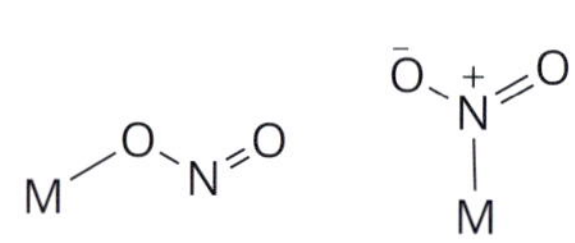

κ*O*-nitrito κ*N*-nitrito

Fig. 10B1 Structures of κ*O*- and κ*N*-nitrito.

Similarly, ligands such as carboxylate may be either κ^1 or κ^2, as shown in (Fig. 10B2), if they bind with one or two oxygens respectively. The superscript 1, κ^1, is assumed when no superscript is present.

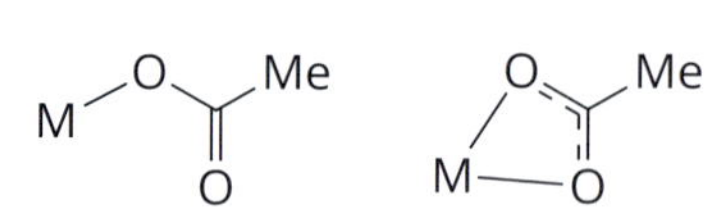

κ*O*-acetate κ^2*O*,*O*′-acetate

Fig. 10B2 Structures of κ*O*- and κ^2*O*,*O*′-acetate.

η (eta)

The Greek letter η is used when there is no ambiguity as far as the type of atom that is binding, but designates the number of atoms in the ligand bound to the metal, especially in organometallic compounds. It is used to signify the number of atoms in a ligand that are bound. For example, the cyclopentadienyl ligand, $C_5H_5^-$, can have 1, 3 or 5 carbons bound to the metal centers designated as η^1, η^3, and η^5 (Fig. 10B3). The symbol η is often used to designate the number of continuous atoms in a chain or ring bound to the metal, differentiating it from κ.

How many atoms in a ligand that are bound to the metal is referred to as the ligand's "hapticity"; for example, the hapticity of the cyclopentadienyl ligands in ferrocene, $Fe(\eta^5\text{-}Cp_2)_2$, is five.

Cyclooctatetraene, C_8H_8, may only bind through some of the carbons in the ring to a metal. If four of the carbons bind as shown in the following, the complex may be designated as $\eta^4\text{-}C_8H_8$ or, to be more specific, 1,2,3,4-$\eta^4\text{-}C_8H_8$ to signify that the carbons bound are contiguous (Fig. 10B4).

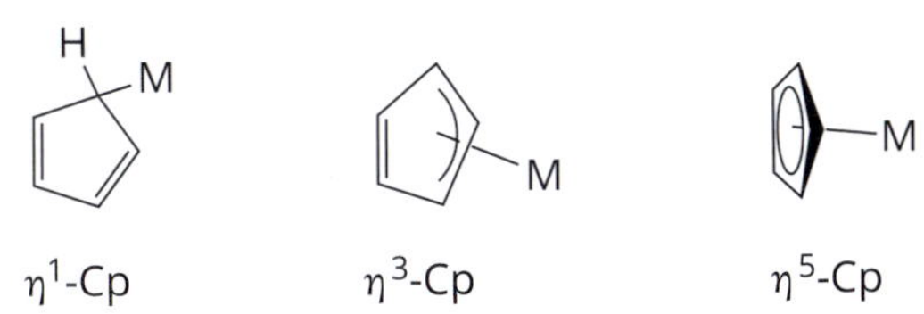

Fig. 10B3 Structures of η^1-, η^3-, and η^5-cyclopentadienyl.

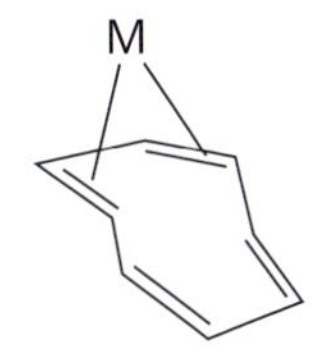

Fig. 10B4 Structure of an η^4-cyclooctatetraene complex.

μ (mu)

The Greek letter μ is used to indicate a bridging ligand and the number of metals that it bridges. A subscript is used to indicate the number of metal bridges. If the ligand bridges only two ligands, the subscript is often omitted (Fig. 10B5).

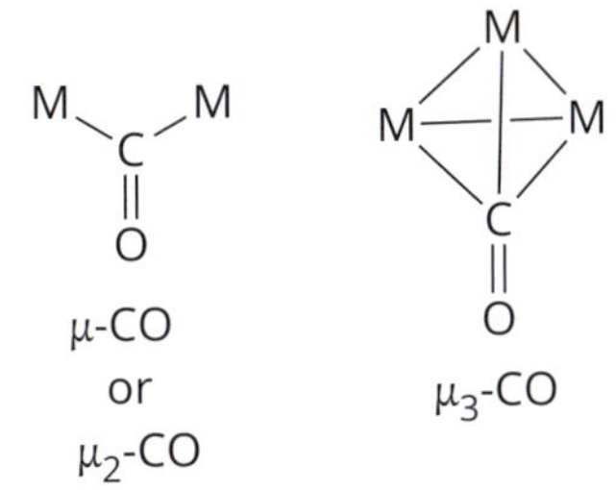

Fig. 10B5 Structures and labels for bridging carbonyl ligands using μ.

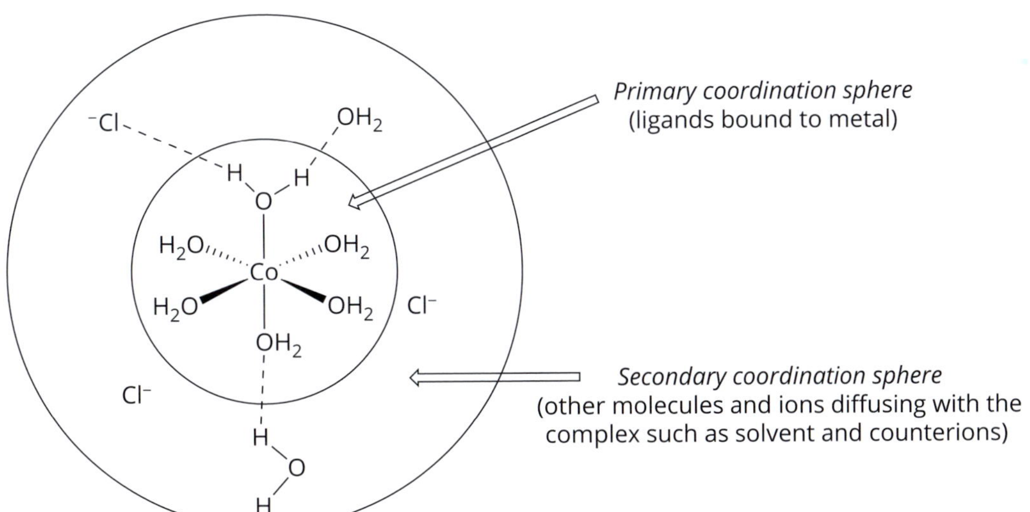

Fig. 10.29 The coordination spheres around the metal. In the primary coordination sphere are the ligand bound to the metal center. In the secondary coordination sphere are solvent, anions, and other species that may diffuse with the complex.

the metal center as the center of their universe, with the rest of the chemical bodies organized around the spherical metal.

The primary coordination sphere is the set of ligands held to the metal through bonding interactions, whether they be dative, heterocovalent, or other bonds (Fig. 10.29). The secondary coordination sphere is the set of ligands that diffuse with the metal but are not directly bound to the metal center. Often, the secondary coordination sphere interacts with the primary coordination sphere through weaker interactions such as van der Waals forces, ionic interactions, or

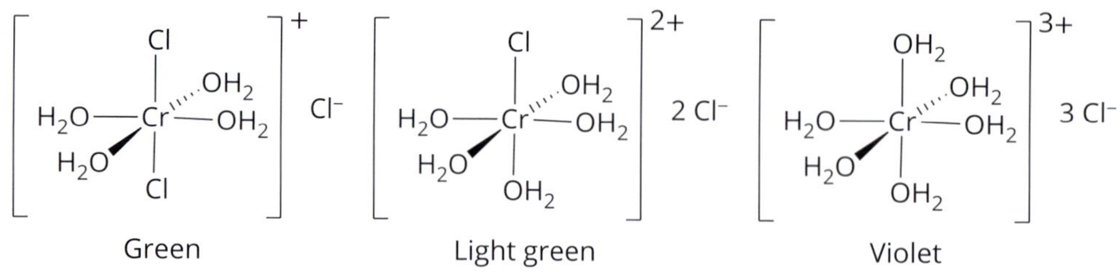

Fig. 10.30 Hydration isomers of $CrCl_3 \cdot 6OH_2$. All the complexes have the same formulae when including the second coordination sphere, but have different ligands bound to the metal.

hydrogen bonds. The secondary coordination sphere often contains things such as solvent (especially water but others as well) and counterions for the complex not directly bound to the metal. Out from the secondary coordination sphere is the bulk solvent.

It is common for salts to crystallize with water in the lattice, and salts involving transition-metal complexes are no exception. The hydrates of chromium(III) chloride of the general formula $CrCl_3 \bullet 6OH_2$ are a good illustration of this type of isomerism (Fig. 10.30).

The most commonly encountered form of chromium(III) chloride hydrate seems to be "Bjerrum green" *trans*-$\{CrCl_2(OH_2)_4\}^+Cl^-$, which often carries two equivalents of water in the lattice, giving it the same formula as violet $\{Cr(OH_2)_6\}Cl_3$. Both of these complexes have been structurally characterized by single-crystal X-ray diffraction (*vide infra*). A third isomer has been reported to be $\{CrCl(OH_2)_5\}Cl_2$, which is often described as light green or blue-green.

Two molecules of water in *trans*-$\{CrCl_2(OH_2)_4\}Cl \cdot 2H_2O$ are in the second coordination sphere, meaning that they are loosely held to the complex by electrostatics and hydrogen bonds but are not directly bound to the metal center. The structure of this complex is shown in Fig. 10.31.[38] The overall formula is $CrCl_3{\cdot}6\ H_2O$, but, as shown in the figure, one of the chlorides and two of the waters are in the lattice joined with each other and the *trans*-$\{CrCl_2(OH_2)_4\}^+$ complex by hydrogen bonds.

In cases where both the anion and cation of a salt are metal complexes, coordination isomers are possible where the metal ligand bonds are different in compounds of the same formula. For example (Fig. 10.32), the salt $CoCr(CN)_6(NH_3)_6$ could be either $\{Co(NH_3)_6\}\{Cr(CN)_6\}$ or $\{Cr(NH_3)_6\}\{Co(CN)_6\}$.

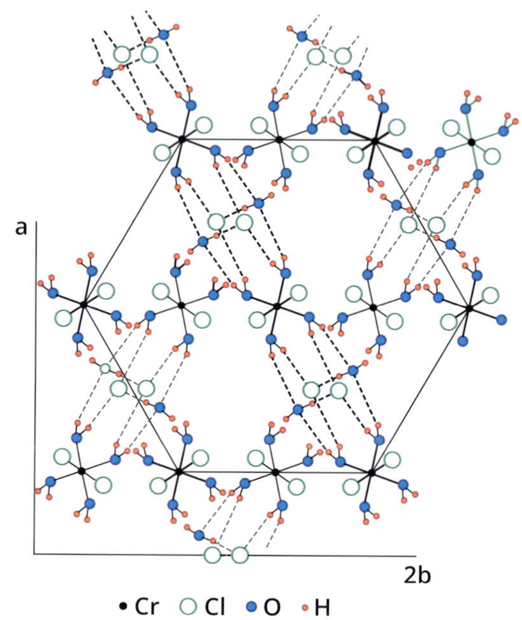

Fig. 10.31 The structure of *trans*-$\{CrCl_2(OH_2)_4\}Cl \cdot 2H_2O$ from X-ray diffraction. In the second coordination sphere of the complex are the chloride and two waters, hydrogen bound to the ligands on chromium.[38]

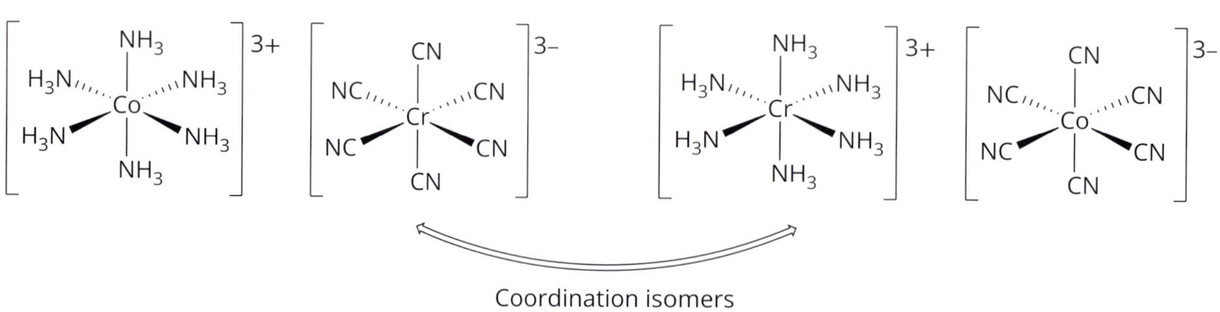

Fig. 10.32 An example of coordination isomers with the formula $CoCr(CN)_6(NH_3)_6$.

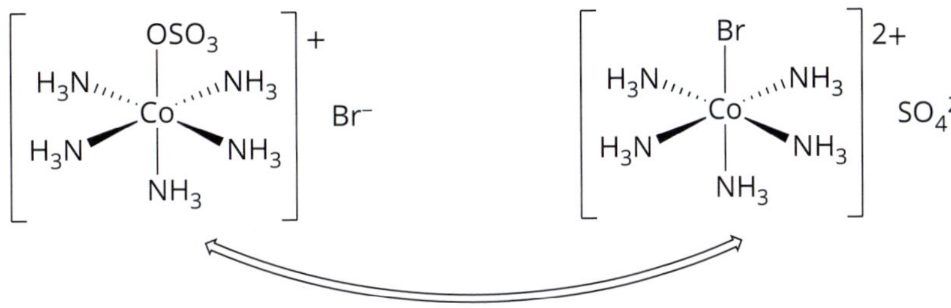

Fig. 10.33 An example of ionization isomers of the formula $Co(NH_3)_5(SO_4)(Br)$, which differ by the ions in the primary and secondary coordination sphere.

An illustration of ionization isomers involves complexes of the formula $Co(NH_3)_5(SO_4)(Br)$, as shown in Fig. 10.33. This cobalt(III) complex can be bound to sulfate with a bromide counterion, $\{Co(NH_3)_5(SO_4)\}Br$, or the complex could exist as the bromide-bound $\{Co(NH_3)_5Br\}SO_4$.[37]

10.5 Generalizations on the Redox Behavior of the Transition Metals

C. K. Jørgensen (not to be confused with S. M. Jørgensen, the rival of Alfred Werner) famously wrote that there were, on average, three oxidation states known for the elements, with $Z = 20$ to 100 (Ca to Fm).[2] That number has expanded significantly since the 1970s when Jørgensen made that observation, and it has been found that the limit of oxidation states for observable systems is, as one may expect, really determined largely by the column number, especially for the heavier elements of the columns. Complexes with formal oxidation states up to the column number are known for elements up to Group 9; for example, IrO_4^+ with an iridium(IX) metal center has been observed.[39] At the time of writing, Pt(X) has been predicted to be stable but has not yet been observed experimentally.[40]

It is useful, to some extent, to think of "common" oxidation states for a particular transition metal. The implication is that metal complexes in this formal oxidation state are neither terribly oxidizing nor reducing towards most common chemicals. The actual utility is made more dubious by the rather large dependence of oxidation state on ligands. Very low oxidation states for very reducing metals can be stabilized by electron-withdrawing ligands like carbonyl, such as titanium(–II) in $\{Ti(CO)_6\}^{2-}$, whereas very high oxidation states can be stabilized by very donating ligands like fluoride, oxide, and nitride, such as chromium(VI) in CrO_3.

All of those caveats aside, it does seem useful to discuss "common" oxidation states for metals, as these may (a) provide clues as to how the metal is interacting with ligands either as a strong donor or acceptor to the ligands, (b) provide some idea of the ease with which the complex will be oxidized or reduced, or (c) provide an indication of the likelihood of certain mechanistic steps, such as oxidative addition and reductive elimination. The question becomes, what is a reasonable method for determining these common oxidation states?

In some cases, standard potentials for the metals are reported as Frost diagrams, potential in volts vs formal oxidation state. However, these are quite sensitive to conditions and are generally reported only in aqueous environments, which limits their applicability somewhat unless those are the conditions of interest. Specific redox potentials, then, are perhaps not the most ideal method of determining the most "common" oxidation state, which depends on the conditions and the system.

One method for examining oxidation state of the transition metals on a more or less even footing is to examine their oxidation states in common minerals. In this case, the stability of the metal towards a common oxidant, namely O_2, is used to determine the most stable oxidation state(s). Furthermore, with these caveats regarding ligands in place, one can assume that complexes involving a metal with an oxidation state at or higher than those in minerals will likely be stable to air, whereas those lower in oxidation state than common minerals will most likely need to be protected from oxygen. (Note that air sensitivity is more than just outer sphere oxidation. O_2 can insert into ligands and other reactions, and water in the air can hydrolyze complexes. Again, stability is ligand dependent.) Table 10.13 shows the Group 4 to Group 11 elements, their commonly encountered minerals, and the formal oxidation states of the metal in those minerals.

A plot of the lowest oxidation state for the transition elements found in common minerals is shown in Fig. 10.34. Again, one may reasonably expect that compounds of the elements of the same oxidation state or higher than the values in the plot will be stable to oxygen. Conversely, complexes containing the metal in a lower oxidation state are likely to react

Table 10.13 Mineral sources of the transition metal elements, their formulae, and the oxidation states of the metals.[41]

Metal(ox. state)	Common Mineral Examples	Formulae	Notes
Group 4			
Ti(IV)	ilmenite	$FeTiO_3$	
	rutile	TiO_2	
Zr(IV)/Hf(IV)	zircon	$ZrSiO_4$	Zircon has been used as a gemstone since ancient times. Baddeleyite has a monoclinic structure. The most often synthetic, cubic zirconia has the same formula and is used as an alternative to diamonds in jewelry. Hafnium is usually found as an impurity in Zr minerals and is extracted.
	baddeleyite	ZrO_2	
Group 5			
V(IV, V)	Variety of vanadates VO_3^- or VO_4^{3-}	For example, $PbCl_2{\cdot}3Pb_3(VO_4)_2$ or $NaVO_3$	The minerals seem to most often be V(V), but OV^{2+} where vanadium is +4 is common in air-stable soluble complexes.
Nb(V)/Ta(V)	columbite or tantalite	$(Fe,Mn)M_2O_6$ where M = Nb and Ta	These metals generally occur together with the most common minerals columbite and tantalite being of the same formula; the name is changed depending on which metal is favored over the other, with columbite for high niobium content and tantalite for high tantalum content. Niobium was also known as columbium until 1950, when IUPAC adopted the current name.
Group 6			
Cr(III)	chromite	$FeCr_2O_4$	Cr_2O_3 is a common pigment with an emerald green color. Some gemstones, most notably ruby, also obtain their color from trace chromium.
	crocoite	$PbCr_2O_4$	
	chrome ochre	Cr_2O_3	
Mo(VI)/W(VI)	wulfenite	$PbMoO_4$	Found together in powellite, but also found as purer sources such as in the deceptively named (one might wrongly assume it is a tungsten mineral) wulfenite, $PbMoO_4$, or scheelite ($CaWO_4$), which have similar structures.
	powellite	$Ca(Mo,W)O_4$	
	scheelite	$CaWO_4$	
Group 7			
Mn(II,III,IV)	pyrolusite	MnO_2	Mn(II) has additional electronic stabilization when high spin (5 unpaired electrons) due to half-filled *d*-orbitals. Higher oxidation states are also found in minerals up to Mn(IV).
	hausmannite	Mn_3O_4	
	rhodochromisite	$MnCO_3$	
Tc	N/A		Not naturally occurring.
Re(IV)	rheniite	ReS_4	Rhenium is most often isolated from Mo and Cu ores as an impurity. Rheniite is quite rare.
Group 8			
Fe(II,III)	hematite	Fe_2O_3	Iron is found in many common minerals in the +2 and +3 oxidation state, even as both in magnetite, which is mixed valent. Pyrite contains iron(II) with S_2^{2-} counterions.
	magnetite	Fe_3O_4	
	limonite	$Fe_2O_3{\cdot}3H_2O$	
	siderite	$FeCO_3$	
	pyrite	FeS_2	
Ru(0)/Os(0)	metallic		Generally found in the metallic state alloyed with other "platinum group" elements (Ru, Os, Ir, Rh, Pt, Pd) and as impurities in nickel and copper sulfide ores.

Table 10.13 Continued

Metal(ox. state)	Common Mineral Examples	Formulae	Notes
Group 9			
Co(II,III)	cobaltite smaltite	CoAsS $CoAs_2$	Cobalt is found in >200 ores, most of which also contain nickel and iron. Cobalt was named after the Germanic *Kobald*, an underground spirit. Weathering the surface of cobaltite produces erythrite, $Co_3(AsO_4)_2{\cdot}8H_2O$, known as red cobalt.
Rh(0)/Ir(0)	metallic		Generally found in the metallic state alloyed with other "platinum group" elements (Ru, Os, Ir, Rh, Pt, Pd).
Group 10			
Ni(II,III)	nickeline or nickelite laterites pentlandite	NiAs $(Ni,Mg)_6Si_4O_{10}(OH)_8$ $(Ni,Fe)_9S_8$	Nickeline was originally given the name *kupfernickel*, German for "Old Nick's copper". The mineral was often mistaken for Cu_2O, from which copper could not be extracted, and which was attributed to a mischievous sprite, Old Nick.
Pd(0)/Pt(0)	metallic		Generally found in the metallic state alloyed with other "platinum group" elements (Ru, Os, Ir, Rh, Pt, Pd).
Group 11			
Cu(0,I,II)	chalcopyrite chalcocite cuprite malachite	$CuFeS_2$ Cu_2S Cu_2O $Cu_2CO_3(OH)_2$	Copper is largely found in a variety of oxide, sulfide, and carbonate ores. "Native" (metallic) copper can also be found, but the metal is more often found in higher oxidation states.
Ag(0,I)/Au (0,I, II)	metallic argentite calaverite	 Ag_2S $AuTe_2$	Silver and gold are found in the metallic state, but also as several sulfide or telluride ores, such as silver(I) argentite and gold(II) calaverite, which contains Te_2^{2-} units.

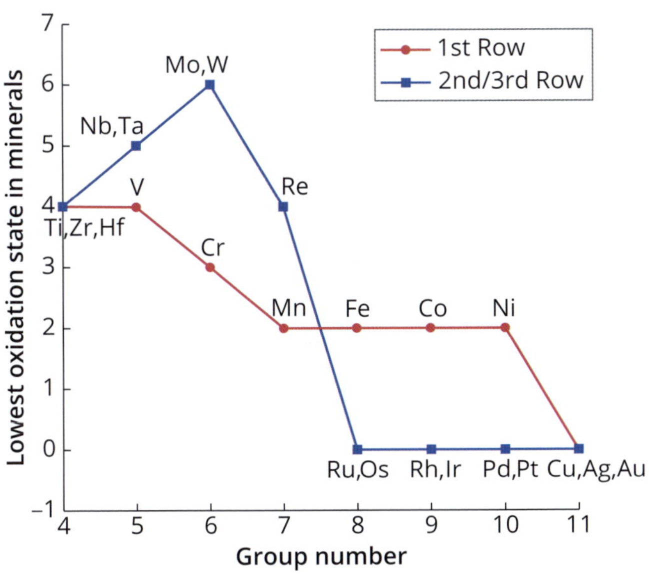

Fig. 10.34 Plot of the lowest oxidation state found in common minerals for the transition elements.

with oxygen. Naturally, there are many exceptions depending on the exact nature of the ligands around the metal, and this is simply offered as a rule of thumb. For example, it is common for Fe(II) complexes to be somewhat air-sensitive, oxidizing to Fe(III), but there also are many air-stable Fe(II) complexes, such as ferrocene.

Another method for examining common oxidation states for the transition metals is through the compounds that have been characterized and added to the Cambridge Crystallographic Structural Database (CCSD). Generally speaking, one would expect that isolated and fully characterized compounds in various oxidation states would favor those states that are more stable. There may be some skewing of the results based on compounds of particular interest due to some property that would tend to be studied more often, but this is a reasonable place to begin.

Fig. 10.35 shows the transition elements from each group of the transition series and the percentage of compounds in the various oxidation states. Many trends can be noted. First, elements from Groups 4, 5, and 6 are in their highest oxidation state more often than those in Groups 7–11. Second, it is somewhat more likely, with some notable exceptions, for second- and third-row elements to be in higher oxidation states. As an exception, note the prevalence of copper(II) relative to Ag(II) and Au(II); many simple Cu(I) salts (such as CuCl) will valence disproportionate to Cu(II) and metallic Cu metal. Third, note that metal centers in negative oxidation

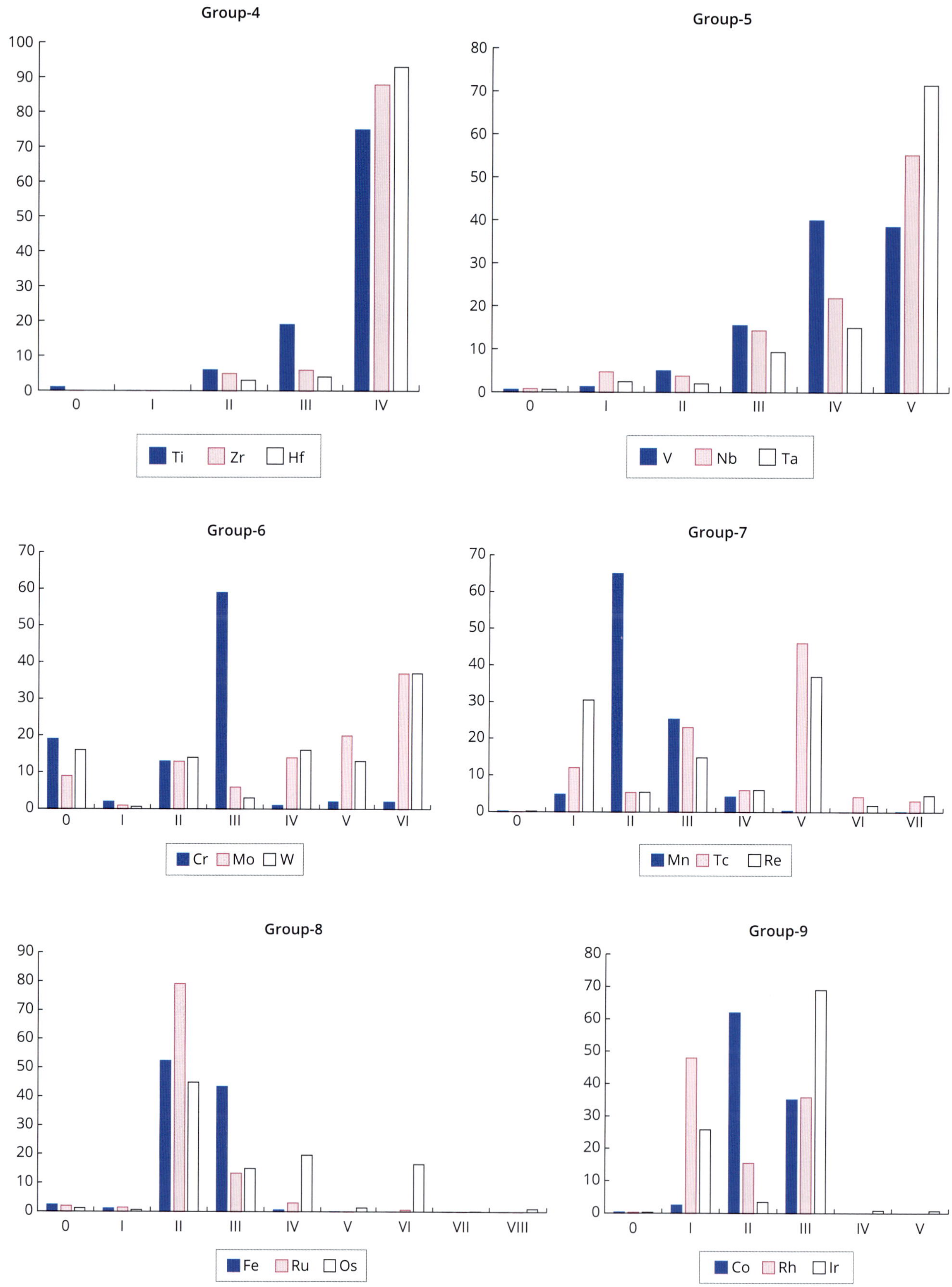
Group-4
100
90
80
70
60
50
40
30
20
10
0
0
I
II
III
IV
Ti
Zr
Hf
Group-5
80
70
60
50
40
30
20
10
0
0
I
II
III
IV
V
V
Nb
Ta
Group-6
70
60
50
40
30
20
10
0
0
I
II
III
IV
V
VI
Cr
Mo
W
Group-7
70
60
50
40
30
20
10
0
0
I
II
III
IV
V
VI
VII
Mn
Tc
Re
Group-8
90
80
70
60
50
40
30
20
10
0
0
I
II
III
IV
V
VI
VII
VIII
Fe
Ru
Os
Group-9
80
70
60
50
40
30
20
10
0
0
I
II
III
IV
V
Co
Rh
Ir

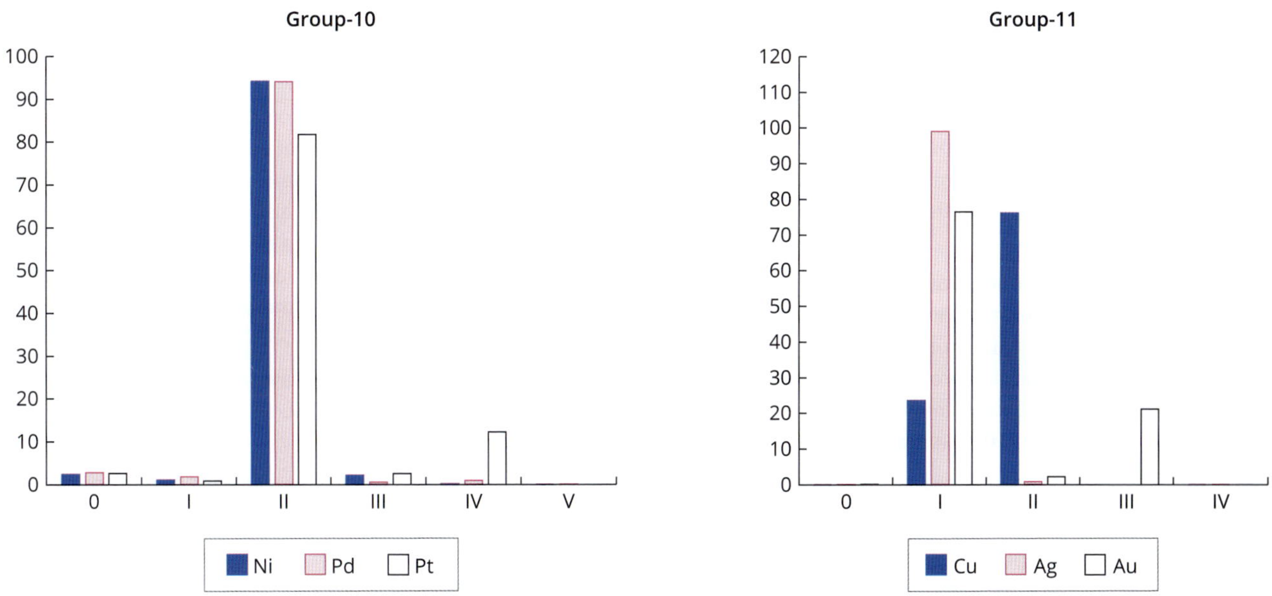

Fig. 10.35 The *y*-axes are the relative percentage of transition metals in certain oxidation states in the CCSD, from December 2016. The *x*-axes are the oxidation states. Oxidation states not shown had very small or no population in the group. No attempt was made to correct or evaluate the formal oxidation states as assigned in the database.

states are not listed. This is not because none are known. Some metal carbonyl anions are well studied, such as Collman's reagent, $\{Fe(CO)_4\}^{2-}$, which is formally iron(−II). The relative number of compounds in negative oxidation states, however, is relatively miniscule. In fact, one should not assume that a particular oxidation state is unknown or unstable simply because there was not a population visible in the plots. The graphs are simply to present an idea of the relative populations of certain oxidation states.

To understand whether a compound is oxidizing or reducing, the ligands need to be considered, and this is best evaluated experimentally by using a technique such as cyclic voltammetry, or chemically by using redox agents (Chapter 13). However, one can start with the assumption that compounds higher in oxidation states than the most common oxidation state for an element are unlikely to be reducing, and compounds lower in oxidation state than that most common for an element are unlikely to be oxidizing.

The common sources for the *d*-block elements are found in the mineral sources shown in Table 10.13, and their abundances are shown in Fig. 10.36. For the most part, the first-row transition metals and the early transition metals have higher abundances. Some of the later, heavier transition metals, such as Re, Os, Ru, Rh, Ir, Pd, Pt, and Au, are some of the rarest elements on Earth. Nevertheless, the chemistry of all of the transition metals is very rich, and even the rare elements listed are commonly employed, especially for applications such as catalysis.

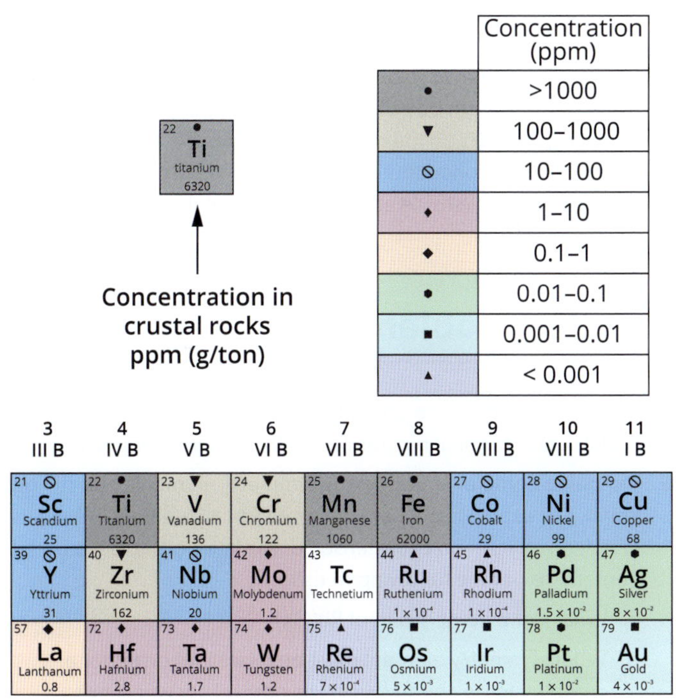

Fig. 10.36 The abundance of the transition metals.[42] The values listed below each element are the number of grams of the element per ton of crustal rocks on average. The data is from Greenwood and Earnshaw.[41] All of the known isotopes of technetium are radioactive, and the element is prepared by nuclear chemistry.

Chapter Summary

1. Alfred Werner convincingly showed that many transition-metal complexes adopt octahedral geometries around the metal. In the process, he made large numbers of well-characterized complexes, including complexes that were chiral at the metal center.
2. Being able to electron-count transition metals can determine whether there are empty orbitals on the metal (electron counts ≤sixteen electrons), if the metal is a radical (odd electron counts), or if antibonding orbitals are occupied (electron counts above 18). There are two common methods for electron counting: the charged method and the neutral method.
3. In the charged method, ligands are closed-shell donors to a metal center. The oxidation state of the metal must be calculated.
4. In the neutral method, ligands are removed as neutral species with dative ligands (such as ammonia) acting as 2-electron donors, and other ligands may be radicals (methyl, for example, is a 1-electron donor). The metal contributes the number of electrons equal to its column number (for example, nickel contributes ten electrons while tungsten contributes six electrons).
5. The chelate effect regards the tendency of chelating, multidentate ligands to replace monodentate ligands on a metal center. There are two components to the effect: thermodynamic and kinetic. The thermodynamic effect is driven by the higher entropy of the system due to a higher number of particles when a chelating ligand is on the metal. The kinetic effect is due, in part, to a high local concentration of the second side of a bidentate ligand after the first side attaches to the metal.
6. Stereoisomers are related compounds with all the same elements bound to the metal but with different bonding patterns. For example, MA_2X_4 complexes have two stereoisomers, *cis* and *trans*. Chiral complexes also fall into this category, and any complex that is a member of a point group without an improper rotation axis of any kind (σ, i, or S_n) is chiral and has an enantiomer that is different only in the direction that it rotates plane-polarized light and interacts with other chiral molecules.
7. Chelating ligands are often favored over monodentate ligands due to the chelate effect. The chelate effect has origins in both thermodynamics and kinetics.
8. Using the Polya method, the number of stereoisomers, geometric isomers, and pairs of enantiomers for compounds can be obtained for any substitution pattern.
9. Structural isomers have the same formula but differ in the atoms bound to the metal center. One example is linkage isomers. Some ligands commonly bind in multiple ways to a metal center, such as thiocyanate (SCN^-), which can bind through sulfur or nitrogen. The two compounds $\{Co(CN)_5(\kappa^1N\text{-}NCS)\}^{3-}$ and $\{Co(CN)_5((\kappa^1S\text{-}SCN)\}^{3-}$ are linkage isomers. The other major type of structural isomerism discussed is coordination sphere isomerism.
10. The transition metals can access wide ranges of oxidations states. In some cases, negative oxidation states up to and including the highest possible oxidation state (column number of the metal) are known. However, there is some utility in discussing common oxidation states. Most metal centers have only a few oxidation states that are commonly found, such as +2 and +3 for iron.

Practice Problems

1. How many isomers are possible when a complex is square pyramidal versus trigonal bipyramidal with 5 different ligands?
2. Inorganic complexes with many symmetry elements can still be chiral. For example, $Rh(Ox)_3^{3-}$, where Ox = oxalate ($C_2O_4^{2-}$), is chiral; an oxygen on each carbon of oxalate bonds to the metal center. Answer the following questions:
   a. What is the point group of $Rh(Ox)_3^{3-}$?
   b. What symmetry element must be missing in order for a complex to be chiral? {Hint: The correct answer is **not** a mirror plane, but mirror planes are a special case of this symmetry element.}
   c. Draw the two enantiomers of $Rh(Ox)_3^{3-}$.
   d. Instead of using the *rectus* (R) and *sinister* (S) notation used in organic chemistry, for tris(chelates) a Δ and Λ notation is typically used. For your two enantiomers in part c, label them as either Δ or Λ.
3. You prepare a homoleptic 5-coordinate transition metal complex, ML_5. From X-ray diffraction the compound is rigorously trigonal bipyramidal; however, when you examine the complex by room temperature NMR, all the ligands look equivalent. Why? Explain in detail how the ligands are being made equivalent on the NMR timescale.
4. For $\{Cr(bpy)_2Cl(NH_3)\}^{2+}$, draw three possible isomers and identify them as geometric or optical isomers.
5. In 2013, the 4-coordinate complex $Ru(NAr)(PMe_3)_3$ was reported, which is neither tetrahedral nor square planar.[43] Examine the paper and determine τ_δ. How does the parameter describe the geometry, e.g., tetrahedral, square planar, see-saw distorted toward square planar, or see-saw distorted toward tetrahedral?
6. For an octahedral molecule with formula $Co(NH_3)_3Cl_3$, what are the two possible stereoisomers? Draw them and for each designate the type of geometrical isomer.

7. What are the two possible geometries for $Fe(bpy)_2Cl_2$? Draw them and for each designate the type of geometrical isomer.

bpy = 2,2′-bipyridine

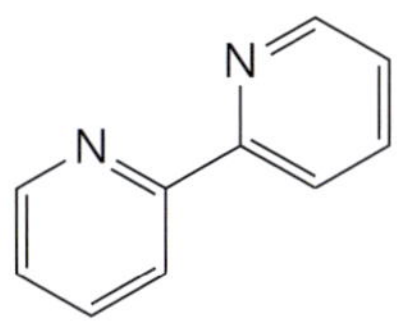

8. For each of the complexes below, clearly draw the 3D structure, give the electron count, oxidation state, and *d*-electron count of the metal.

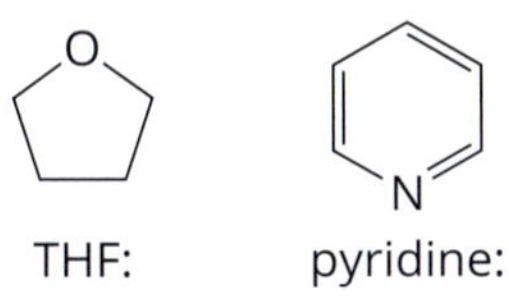

a. *fac*-$\{CoCl_3F_3\}^{3-}$
b. *mer*-$VCl_3(THF)_3$
c. *cis*-$PtCl_2(NH_3)_2$
d. *trans*-$CrCl_2(pyridine)_4$
e. *cis*-$CoCl_2(NH_3)_4$

9. What are the oxidation states and electron counts for the metals in the complexes below?

a. $Ru(P(OMe)_3)_5$
b. ZrF_6^{2-}
c. $(CO)_5Mn–Mn(CO)_5$
d. *trans*-$\{Co(NH_3)_4Cl_2\}^+$

Cited References

1. Kauffman, G. B., *Alfred Werner: Founder of Coordination Chemistry* Springer-Verlag: Berlin, 1966.
2. Jorgensen, C. K., New understanding of unusual oxidation-states in transition groups *Naturwissenschaften* 1976, *63*, 292–292.
3. Kauffman, G. B., Ed., *Coordination Chemistry: A Century of Progress* American Chemical Society: Washington, DC, 1994.
4. Hantzsch, A.; Werner, A., On spatial arrangement of atoms in nitrogen-containing molecules *J. Chem. Educ.* 1966, *43*, 156.
5. Ernst, K.-H.; Wild, F. R. W. P.; Blacque, O.; Berke, H., Alfred Werner's Coordination Chemistry: New Insights from Old Samples *Angew. Chem. Int. Ed.* 2011, *50*, 10780–10787.
6. Kauffman, G. B.; Lindley, E. V., Classic in coordination chemistry: Resolution experiment for inorganic laboratory *J. Chem. Educ.* 1974, *51*, 424–425.
7. Barker, G. K.; Lappert, M. F., Stabilization of transition-metals in a low coordinative environment using bis(trimethylsilyl)methyl ligand: Monomeric Cr(III) alkyl $Cr[CH(SiMe_3)_2]_3$ and related complexes *J. Organomet. Chem.* 1974, *76*, C45–C46.
8. Note, some molecules that appear as if they should be of high symmetry can lower their energy by going to lower symmetry structures. WMe_6 is such a compound. The causes of these distortions are discussed in Chapter 13 for metal centers, but are related to the Jahn-Teller Effects discussed in Chapter 8.
9. Roedern, E.; Jensen, T. R., Ammine-Stabilized Transition-Metal Borohydrides of Iron, Cobalt, and Chromium: Synthesis and Characterization *Inorg. Chem.* 2015, *54*, 10477–10482.
10. Mowat, W.; Wilkinson, G., Elimination stabilized alkyls. 3. Trimethylsilylmethyl and neopentyl alkyls of transition-metals *Dalton Trans.* 1973, 1120-1124.
11. Nguyen, T.; Sutton, A. D.; Brynda, M.; Fettinger, J. C.; Long, G. J.; Power, P. P., Synthesis of a stable compound with fivefold bonding between two chromium(I) centers *Science* 2005, *310*, 844–847.
12. Agron, P. A.; Ellison, R. D.; Levy, H. A., Crystal structure of dimanganese iron carbonyl $Mn_2Fe(CO)_{14}$ *Acta Cryst.* 1967, *23*, 1079-&.
13. Porterfield, W. W., *Inorganic Chemistry: A Unified Approach* 2nd ed.; Academic Press: San Diego, 1993.
14. Jordan, R. B., *Reaction Mechanisms of Inorganic and Organometallic Systems* 3rd ed.; Oxford University Press: Oxford, 2007.
15. Wilkins, R. G., *Kinetics and Mechanisms of Reactions of Transition Metal Complexes* 2nd ed.; VCH Publishers: New York, N.Y., 1991.
16. Morgan, G. T.; Drew, H. D. K., Researches on residual affinity and coordination Part II Acetylacetones of selenium and tellurium *J. Chem. Soc.* 1920, *117*, 1456–1465.
17. Black, D. S. C.; Hartshorn, A. J., Ligand design and synthesis *Coord. Chem. Rev.* 1973, *9*, 219–274.
18. Martell, A. E., Metal chelate compounds *Annu. Rev. Phys. Chem.* 1956, *6*, 239–260.
19. Martell, A. E.; Hancock, R. D.; Motekaitis, R. J., Factors affecting stabilities of chelate, macrocyclic and macrobicyclic complexes in solution *Coord. Chem. Rev.* 1994, *133*, 39–65.
20. Moschetta, E. G.; Gans, K. M.; Rioux, R. M., Elucidating the roles of enthalpy, entropy, and donor atom in the chelate effect for binding different bidentate ligands on the same metal center *J. Catal.* 2014, *309*, 11–20.
21. Schwarzenbach, G., Der chelateffekt *Helv. Chim. Acta* 1952, *35*, 2344–2363.
22. Taylor, R. W.; Begum, R. A.; Day, V. W.; Bowman-James, K., Cooperativity and the Chelate, Macrocyclic and Cryptate Effects. In *Supramolecular Chemistry: From Molecules to Nanomaterials*, John Wiley & Sons: 2012; pp 1–27.
23. Landis, C. R.; Firman, T. K.; Root, D. M.; Cleveland, T., A valence bond perspective on the molecular shapes of simple metal alkyls and hydrides *J. Am. Chem. Soc.* 1998, *120*, 1842–1854.
24. Carter, M. J.; Beattie, J. K., Kinetic chelate effect - chelation of ethylenediamine on platinum(II) *Inorg. Chem.* 1970, *9*, 1233.
25. Ahmed, A. K. S.; Wilkins, R. G., Factors influencing the rates of dissociation of metal complexes. 1. The mechanism of dissociation of mono(ethylenediamine)-nickel(II) ion *J. Chem. Soc.* 1959, 3700–3708.
26. Wilkins, R. G., Mechanisms of ligand replacement in octahedral nickel(II) complexes *Acc. Chem. Res.* 1970, *3*, 408.
27. Amouri, H.; Gruselle, M., *Chirality in Transition Metal Chemistry Molecules, Supramolecular Assemblies and Materials* Wiley: Chichester, U.K., 2008.
28. Kuramoto, M.; Kushi, Y.; Yoneda, H., Structural study of optical resolution. 6. Crystal-structure of (+)589-*cis*-chloroamminebis(ethylenediamine)cobalt(III) d-3-bromocamphor-9-sulfonate monohydrate *Bull. Chem. Soc. Jpn.* 1978, *51*, 3196–3202.
29. Baraldi, I.; Fiori, C.; Vanossi, D., On the cycle index of point groups *J. Math. Chem.* 1999, *25*, 23–30.
30. Kennedy, B. A.; Brubaker, C. H.; McQuarrie, D. A., Group theory and isomerism *Inorg. Chem.* 1964, *3*, 265.
31. McDaniel, D. H., Restatement of Polya's theorem *Inorg. Chem.* 1972, *11*, 2678.
32. Bailar, J. C., The Numbers and Structures of Isomers of Hexacovalent Complexes *J. Chem. Educ.* 1957, *34*, 334–338.
33. Addison, A. W.; Rao, T. N.; Reedijk, J.; Vanrijn, J.; Verschoor, G. C., Synthesis, structure, and spectroscopic properties of copper(II) compounds containing nitrogen sulfur donor ligands: The crystal and molecular-structure of aqua 1,7-bis(N-methylbenzimidazol-2′-yl)-2,6-dithiaheptane copper(II) perchlorate *Dalton Trans.* 1984, 1349-1356.
34. Harris, S. A.; Ciszewski, J. T.; Odom, A. L., Titanium η^1-pyrrolyl complexes: Electronic and structural characteristics imposed by the N,N-di(pyrrolyl-α-methyl)-N-methylamine (dpma) ligand *Inorg. Chem.* 2001, *40*, 1987–1988.

35. Alvarez, S.; Alemany, P.; Casanova, D.; Cirera, J.; Llunell, M.; Avnir, D., Shape maps and polyhedral interconversion paths in transition metal chemistry *Coord. Chem. Rev.* 2005, *249*, 1693–1708.
36. Reineke, M. H.; Sampson, M. D.; Rheingold, A. L.; Kubiak, C. P., Synthesis and Structural Studies of Nickel(0) Tetracarbene Complexes with the Introduction of a New Four-Coordinate Geometric Index, τ_δ *Inorg. Chem.* 2015, *54*, 3211–3217.
37. DeKock, R. L.; Gray, H. B., *Chemical Structure and Bonding* 2nd ed.; University Science Books: Mill Valley, Calif., 1989.
38. Dance, I. G.; Freeman, H. C., Crystal structure of dichlorotetraaquochromium(III) chloride dihydrate: Primary and secondary metal ion hydration *Inorg. Chem.* 1965, *4*, 1555.
39. Wang, G.; Zhou, M.; Goettel, J. T.; Schrobilgen, G. J.; Su, J.; Li, J.; Schloeder, T.; Riedel, S., Identification of an iridium-containing compound with a formal oxidation state of IX *Nature* 2014, *514*, 475.
40. Yu, H. S.; Truhlar, D. G., Oxidation State 10 Exists *Angew. Chem. Int. Ed.* 2016, *55*, 9004–9006.
41. Greenwood, N. N.; Earnshaw, A., *Chemistry of the Elements* 1st ed.; Pergamon Press: Oxford 1984.
42. Odom, A. L.; McDaniel, T. J., Titanium-Catalyzed Multicomponent Couplings: Efficient One-Pot Syntheses of Nitrogen Heterocycles *Acc. Chem. Res.* 2015, *48*, 2822–2833.
43. Singh, A. K.; Levine, B. G.; Staples, R. J.; Odom, A. L., A 4-coordinate Ru(II) imido: unusual geometry, synthesis, and reactivity *Chem. Commun.* 2013, *49*, 10799–10801.
44. Sattelberger, A. P.; Wilson, R. B.; Huffman, J. C., Metal-metal-bonded complexes of the early transition-metals .4. Synthesis and characterization of the tantalum(III) dimer $Ta_2Cl_6(PMe_3)_4$ *Inorg. Chem.* 1982, *21*, 2392–2396.
45. Cruz, S. S.; Amenta, D. S.; Gilje, J. W.; Yap, G. P. A., A re-examination of the reaction of dichloro-tris-(triphenylphosphino)ruthenium(II) with benzonitrile: Crystal structures of triphenylphosphine chloro ruthenium complexes *Polyhedron* 2016, *114*, 179–183.
46. Spingler, B.; Scanavy-Grigorieff, M.; Werner, A.; Berke, H.; Lippard, S. J., Crystal structure determination of a (μ-amido)(μ-hydroxo)(μ-superoxo)dicobalt(III) complex from the Werner collection *Inorg. Chem.* 2001, *40*, 1065.
47. Lummiss, J. A. M.; Perras, F. A.; McDonald, R.; Bryce, D. L.; Fogg, D. E., Sterically Driven Olefin Metathesis: The Impact of Alkylidene Substitution on Catalyst Activity *Organometallics* 2016, *35*, 691–698.

CHAPTER 11

Basics of Transition-Metal Bonding 1

Simple Ligands Including Olefins, Diatomics, Phosphines, and Hydrides

"Formalisms are convenient fictions which contain a piece of the truth—and it is so sad that people spend a lot of time arguing about the deductions they draw, often ingeniously and artfully, from formalisms, without worrying about their underlying assumptions."

— Roald Hoffman (discussing electron counts)[1]

CONCEPT COVERAGE

1. Bonding in olefins and alkynes
2. Complexes involving diatomic ligands
3. Phosphine bonding and parameterization
4. Nitrogen-based ligands
5. *N*-heterocyclic carbenes as ligands
6. Common monoanionic ligands and their donor abilities

The word "ligand", derived from the Latin word *ligandum*, meaning "that which is bound", was perhaps first used by Alfred Stock in 1916, with regards to groups on silicon to mean "atoms and groups bound to another element".[a] The word is used with unchanged meaning today, as anything bound to something else.[2]

There is an almost endless variety of different ligands that are used for transition metals. Some of the more common ones for early and late transition metals will be discussed in this chapter and the next. For ligands not covered, the principles developed will allow students to analyze unfamiliar ligand types.

Determination of electron count and formal oxidation state (Chapter 10) of the metal in transition metal systems is the initial step to developing the ability to discuss properties and reactivity. However, to even determine electron count and oxidation state requires some understanding of electronic structure of the ligands, which is the core of this and the next chapter.

In this book there is no attempt to separate what are sometimes referred to as "organometallic" ligands from "Werner"-type ligands. Especially from a bonding standpoint, the distinction is quite artificial. Carbon-based organometallic ligands are in some ways "special", but all of the elements have their own special chemistry. In many cases, the way we understand Werner complexes has informed our understanding of organometallic complexes, and vice versa.

In many cases, what will be discussed here is the standard, conventional method for finding the electron count and oxidation state. In some cases the electronic structure of the compound will make the convention seem inaccurate; for example, olefins bound to early transition metals in low formal oxidation states. If so, the researcher can and should deviate from the convention but should also consider why the deviation was necessary.

The basics of electron count and formal oxidation state determination were discussed in Chapter 10 using electronically simple ligands. In this chapter and the one following, a more detailed discussion of many common ligand types will be presented. Notice as the

[a] Some claim that the word "ligand" should be used only for substituents on a metal, but the history and IUPAC definition do not support this. "Ligand" is also used for a molecule that binds to proteins for some biological purpose.

discussion proceeds that many of the ligands in this chapter use a similar bonding scheme of σ-donation mixed with π-acceptor ability.

In this chapter, what the authors would refer to as "simple" ligands are discussed, including carbonyl, olefins, dihydrogen, alkyls, and related substituents. In the next chapter, somewhat more complex ligands are covered, including cyclic aromatics, metal–ligand multiple bonds, metal–metal bonds, and redox noninnocent ligands. In undergraduate courses we would recommend the student familiarize themselves with this chapter and then supplementing with the next, using topics which they may find interesting. For example, a very common ligand type covered in the next chapter is cyclopentadienyl, which could be studied as a stand-alone section despite being part of a larger discussion on cyclic aromatics.

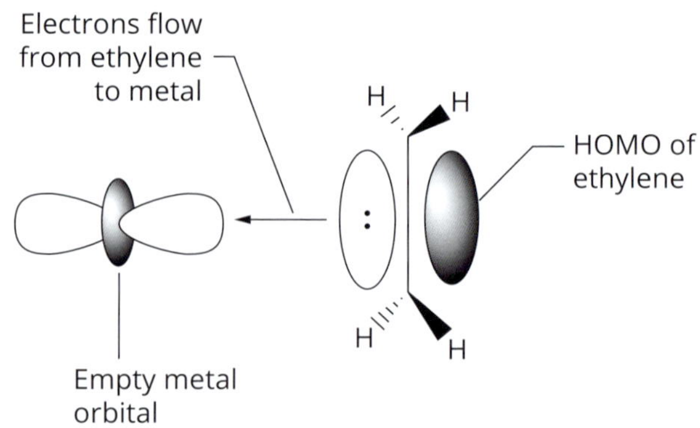

Fig. 11.2 The σ-bond between ethylene and the metal center. The bond involves the donation of the electrons in the HOMO of ethylene, the C–C π-bond, to an empty orbital of appropriate symmetry on the metal.

11.1 Binding of Olefins, Alkynes, and Related Ligands to Metals

Metal centers, especially late transition metals, often bind to double and triple bonds of first-row *p*-block elements. Ethylene is perhaps the simplest example, and a complex of ethylene $\{Pt(ethylene)Cl_3\}^-$, Zeise's salt, has been known since 1831. However, the bonding in the compound was not explained until the Dewar–Chatt–Duncanson model, which was developed in the 1950s.[3–4] This model strongly influenced how the electronic structure of ligands on metals was understood.

Fig. 11.1 shows the drawing from Dewar's original manuscript in which he discussed the interaction of a silver cation with an olefin. The paper derives from a symposium in Montpellier, and the audience responses are appended to the article. The drawing appears as a response to a question by Bartlett.[3]

The Dewar–Chatt–Duncanson model involved two separate interactions between the metal and olefin: a σ- and a π-interaction. The σ-interaction occurs between the π-electrons of ethylene, the HOMO of the isolated olefin, and an orbital of appropriate symmetry on the metal such as *s* or d_{z^2}. Since there is no node parallel to the bond axis, even though the bonding is not between just two atoms and uses π-electrons of the olefin, it is a σ-bond (Fig. 11.2).

The π-bond between ethylene and a metal center involves a filled *d*-orbital on the metal donating into the π*-orbital on ethylene. In this case, electrons can be thought to flow from the metal to the ligand. Some may ordinarily think of ligands as Lewis bases donating to Lewis-acidic metal centers. This flow of electrons from the metal to the ligand runs in the opposite direction and is often termed "backbonding" (Fig. 11.3).

From a valence-bond perspective, one may consider the interactions between the metal and ethylene to vary from a simple dative interaction to a metallacyclopropane (Fig. 11.4), depending on the degree of backbonding.

For a highly reducing metal, the C–C bond length can be similar to a C–C single bond. For example, Sita and coworkers have described the complex ZrCp*(amidinate)(styrene) (Fig. 11.5).[5] Based on our discussion in Chapter 10, we can anticipate that Zr(II) should be very reducing. The backbonding

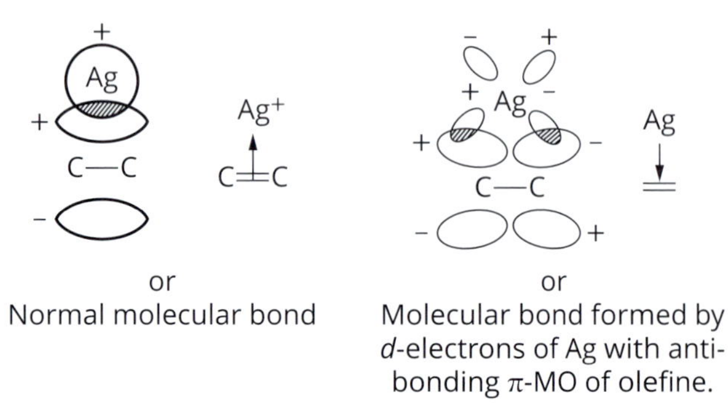

Fig. 11.1 Drawing by Dewar of a silver ion interacting with an olefin. The "+" and "–" in the diagram indicate the phase of the orbital lobes.[3]

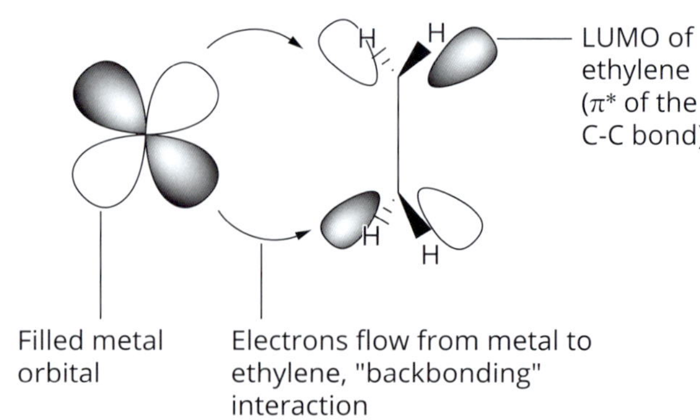

Fig. 11.3 The π-bond between ethylene and the metal center. The bond involves the donation of the electrons from a filled metal-based orbital into the C–C π*-orbital.

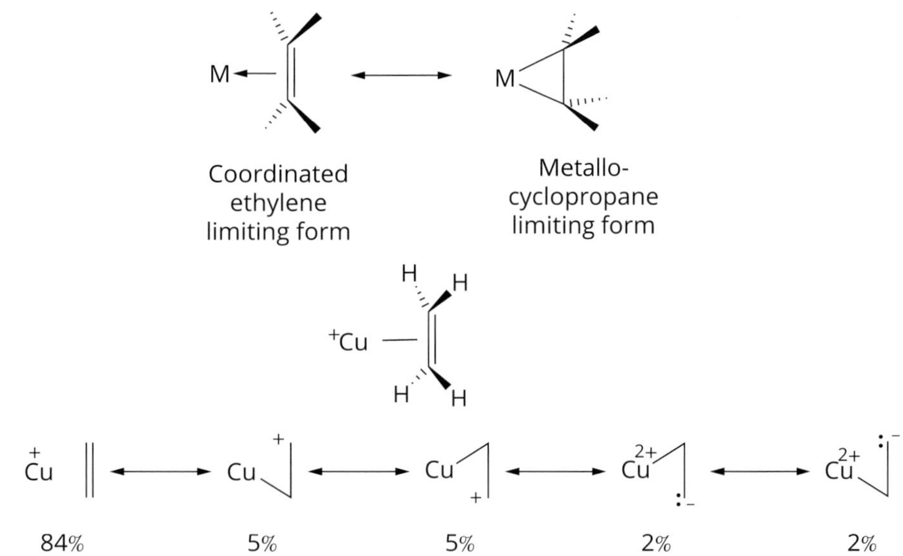

Fig. 11.4 (*top*) Valence-bond descriptions of olefin binding to a metal often consider dative interactions between the olefin π-system and the metal (*top left*) in resonance with a metallacyclopropane structure (*top right*). (*bottom*) The resonance forms in {Cu(ethylene)}$^+$ were examined by using natural resonance theory and using NBO6 from the cc-pvdz/B3LYP optimized structure. The 5-forms below the composite structure constitute 98% of the ground state, and resonance forms contributing <1% are not shown.

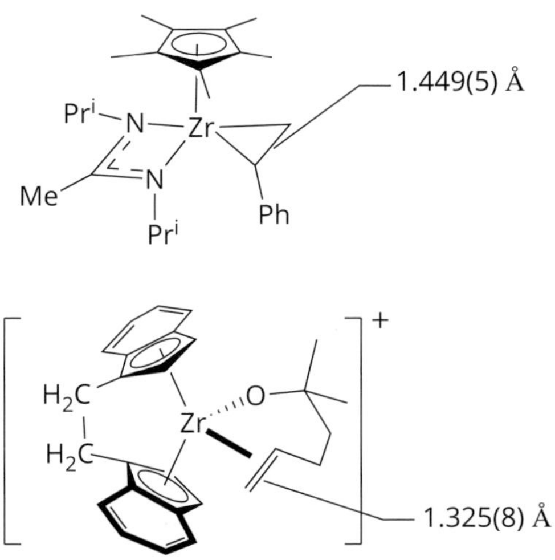

Fig. 11.5 (*top*) Sita and coworkers' zirconium(II) complex ZrCp*(amidinate)(styrene), which is best described as a metallacyclopropane and perhaps more accurately regarded as Zr(IV).[5] (*bottom*) Jordan and coworkers' zirconium(IV) complex {Zr(EBI)(OCMe$_2$CH$_2$CH$_2$CH=CH$_2$)}$^+$, which is best described as a dative olefin complex.[6]

in the complex is so strong, and the metallacyclopropane resonance form is so strongly favored, that the C–C distance is 1.449(5) Å. Pauling's method (Eq. 3.27, b = 0.37, r_1 = 1.54 Å) gives a bond order of 1.2 in this compound. If regarded as a metallacyclopropane, the compound contains Zr(IV).

The π-backbonding interaction is of great importance in the creation of stable metal–olefin complexes. However, very Lewis-acidic d^0-metal centers, which have no filled *d*-orbitals to backbond to the olefin, can bond to olefins, though the interaction is typically much weaker than complexes in which backbonding is possible. Jordan and coworkers prepared the zirconium(IV) complex {Zr(EBI)(OCMe$_2$CH$_2$CH$_2$CH=CH$_2$)}$^+$ with an intramolecular Zr-olefin interaction for structural investigation (Fig. 11.5).[6] The C–C double bond-distance in the complex is 1.325(8) Å—a typical double-bond distance. If the olefin is not appended to the molecule, the zirconium cation {ZrCp'$_2$(OBu^t)(olefin)}$^+$ can be observed, where Cp' = MeC$_5$H$_4$, and the barrier to decomplexation of ethylene has been measured as a very small $\Delta G^\ddagger$ = 11.2 kcal/mol in CD$_2$Cl$_2$.[7]

Considering that metal backdonation decreases the C–C bond order, it is perhaps not surprising that metal centers bind more strongly to ring-strained olefins, in which there is extra impetus for the olefin to act as an acceptor and lower the strain through decreasing the C–C bond order. Metal binding, in addition to the usual M–C bonding generated, relieves some or all of the ring strain. As a result, ring-strained olefins, such as norbornadiene, are frequently encountered ligands. Norbornadiene (NBD), cyclooctene (COE), and cyclooctadiene (COD) are all common examples (Fig. 11.6).

The bonding of carbon–carbon π-bonds to transition metals is not unique, and nitrogen–nitrogen π-bonds, for example, can exhibit similar bonding modes. The nickel(0) complex Ni(PPh$_3$)$_2$(PhN=NPh) has been structurally characterized and contains a π-bound N=N unit with a bond distance of 1.371(3) Å.[8] The formally titanium(II) complex Ti(OAr)$_2$(pyridine)$_2$(PhN=NPh), where Ar = 2,6-Pr$^i_2$C$_6$H$_3$, was prepared by Rothwell and coworkers, and displayed an N–N distance of 1.416(8) Å (Fig. 11.7).[9] Just like in olefin complexes, the bond distances—N–N in this case—increase with increasing metal donation, and the Rothwell complex is better regarded as Ti(IV) with a N–N single bond.

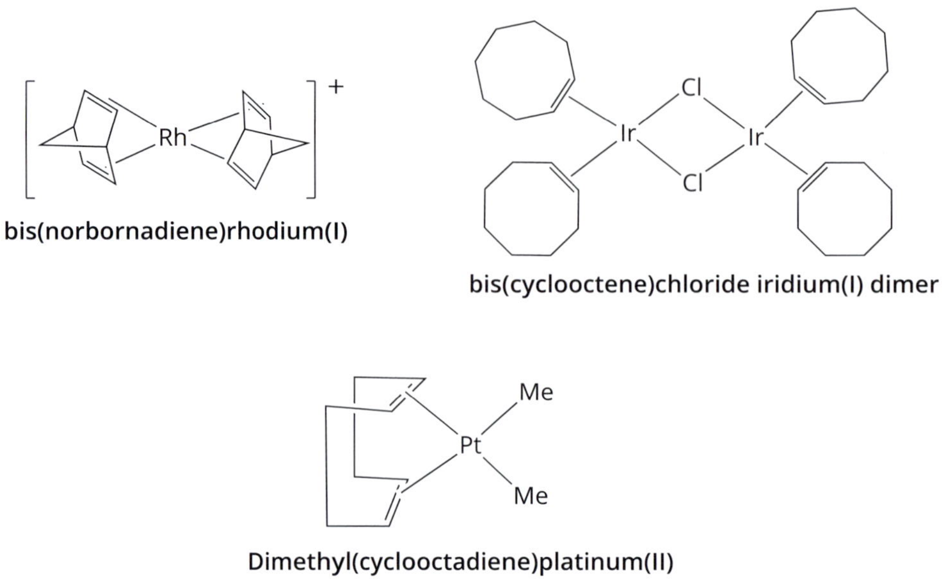

Fig. 11.6 Some commonly encountered olefin ligands: norbornadiene (NBD), cycloctene (COE), and cyclooctadiene (COD). The examples shown are all of commercially available complexes.

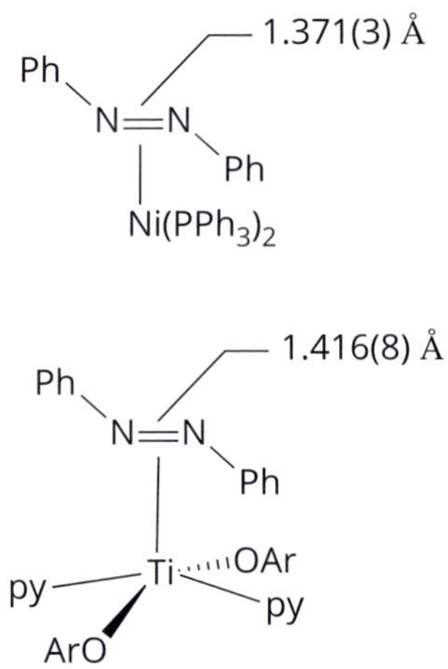

Fig. 11.7 *(top)* A nickel(0) π-bound N=N complex.[8] *(bottom)* A titanium(II) π-bound N=N complex.[9]

The bonding in alkynes has additional possible interactions from those in olefins. In a bound alkyne there are two π-systems (Fig. 11.8): there is the π-bond that is parallel to the M–alkyne vector ($\pi_{\parallel}$), and the π-bond perpendicular to the M–alkyne vector ($\pi_{\perp}$). For each of these π-systems, there is also a π*-orbital that can act as an acceptor orbital for electron density from the metal.[10] As a result, the alkyne may donate up to four electrons to the metal center and have two different acceptor interactions. The symmetries of these four different interactions are σ, two π ($\pi_{\perp}$ and $\pi_{\parallel}$), and δ. In other words, alkynes can be formally quadruply bonded to transition metals.

In general, alkynes interact more strongly than alkenes and will often displace olefins from metal–alkene complexes.

11.2 Diatomic and Related Ligands

Diatomic molecules are some of the most common ligands for transition metals, and this section will provide a brief overview of their properties and discuss some electronically related ligands.

While there will be some brief discussion of ligands that bridge multiple metals, the emphasis will be on monometallic systems and their electronic structure. A diatomic (XY) may bind in one or more of three different limiting geometries to a single metal center: linear, bent, or side-on. In many cases the number of electrons donated depends on the exact nature of diatomic XY and the bonding mode to the metal.

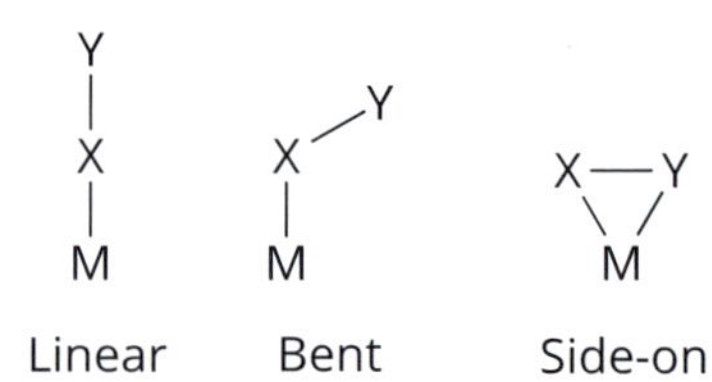

Table 11.1 lists some of the ligands in this class, which will be used to illustrate electron counting and bonding principles. This is not a complete list, and other ligands such as the "heavy nitrosyls" (M–PS and M–PO)[11–13], M–BO,[14] M–BF,[15] and many others have been prepared that are not discussed explicitly. In addition, there are other bonding modes, especially with multiple metal centers, that will not be covered in detail, but the discussion here will probably serve to provide enough information to analyze complexes with bonding modes not examined. Numerous books and reviews have been written on many of these ligands, if more detailed coverage is desired.[14, 16–28]

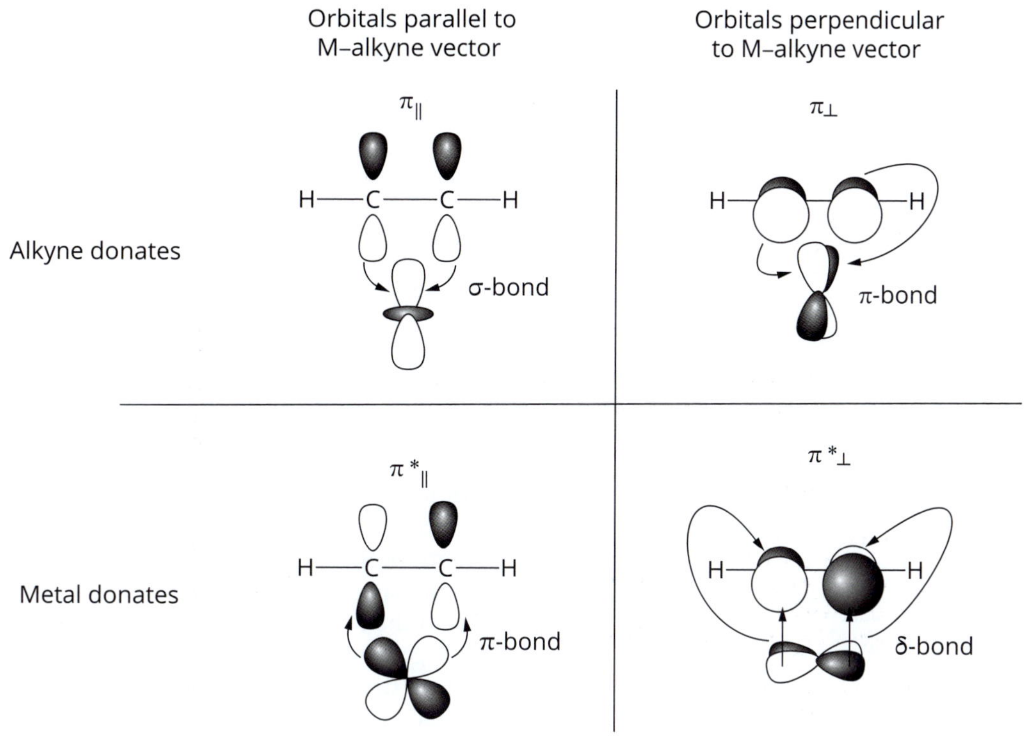

Fig. 11.8 The possible metal–alkyne interactions and their bond types. The alkyne may donate up to four electrons, using the interactions at the top and at bottom left. There could also be two different types of bonds where the alkyne accepts electron density from the metal shown at the top and at bottom right.

Table 11.1 CO and related ligands bonding to a single metal center.

Structure (Neutral/ Charged if different)	Name	Electrons Donated (Neutral Method)	Electrons Donated (Charged Method)	Notes
:CO	carbon monoxide	2	2	Coordinates linearly through carbon.
$\cdot$CN/$^-$:CN	cyanide	1	2	Coordinates linearly through carbon.
:CNR	isonitrile	2	2	R can be a variety of alkyl or aryl groups. M–C–N is linear but C–N–R can bend, depending on donation from metal.
Linear N_2:	dinitrogen	2	2	
Side-on N_2	dinitrogen	2–4	2–4	Bonding model similar to acetylene.
Linear NO: $:\dot{N}O/^+:NO$	nitrosyl	3	2	Ligand is a cation in charged method.
Bent NO: $\cdot NO/^-:NO$	nitrosyl	1	2	Ligand is an anion in the charged method.
Bent O_2: $\cdot O_2/^-:O_2$	superoxide	1	2	Monoanionic, monodentate ligand.
Side-on O_2: $\cdot O–O\cdot/^-:O–O:^-$	peroxo	2	4	Dianionic or neutral ligands, depending on the degree of backbonding, similar to an olefin but generally O_2 is reduced by the metal.

11.2.1 Carbonyl (CO), Cyanide (CN), and Isonitrile (CN–R)[19, 24]

11.2.1.1 Metal Carbonyls

The molecular orbitals of some diatomic molecules were discussed in Chapter 6. In Fig. 11.9 is the molecular orbital diagram of carbon monoxide. The initial expectation is perhaps that the HOMO will be the CO π-bond and that the LUMO will be the π*-orbital, similar to what is found in other simple π-systems such as ethylene. However, the σ_{2p}-orbital has the same symmetry as the mostly carbon-based antibonding σ^*_{2s}. As in dinitrogen (see Fig. 6.24 and accompanying discussion), these two orbitals mix to stabilize σ^*_{2s} and destabilize the σ_{2p}-orbital. This causes the carbon-based lone pair, σ_{2p}, to be the HOMO.

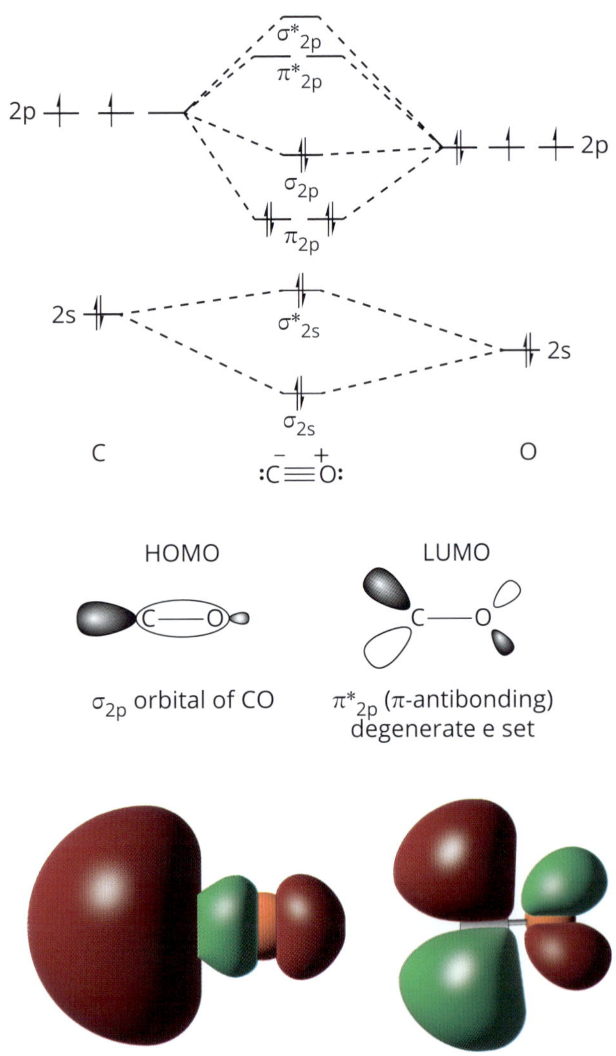

Fig. 11.9 (*top*) Molecular orbital diagram for CO. (*bottom*) Schematic drawing and molecular orbitals calculated for the HOMO (*left*) and one of the two LUMO orbitals (*right*) of CO, both of which are largely based on carbon (gray sphere) rather than oxygen (red sphere).

Due to the mixing, both the σ-based HOMO and the doubly degenerate π-based LUMO are centered on the carbon (Fig. 11.9). When bound to a single metal center, carbon monoxide always bonds through its carbon, which is unsurprising considering that both frontier orbitals are largely based on this atom.

The bonding of CO to a metal center was first discussed, independently in three separate books, by Orgel, Zeiss, and Pauling in 1960.[29–31] The scheme is very reminiscent of the Dewar–Chatt–Duncanson model for olefin bonding to a metal, discussed previously. The HOMO of carbon monoxide acts as a 2-electron donor towards a metal-based acceptor orbital, drawn as d_z2 in Fig. 11.10. If the metal has filled *d*-orbitals of the proper symmetry, the metal can then act as a donor to CO, producing a synergistic flow of electron density both to and from the metal. This donation of electron density from the metal to CO, backbonding, can occur with one or both of the π*-orbitals on CO. The backbonding interaction in many CO complexes is thought to be a larger contributor to the M–CO bond energy than the σ-interaction, and CO is generally regarded as a ligand that greatly reduces the electron density at the metal, stabilizing electron-rich metal centers.

As mentioned, an important factor in the bonding of most CO complexes is the donation of the metal into the C–O π*-orbitals. Naturally, donation in the antibonding orbitals lowers the C–O bond order and increases the M–C bond order. The decrease in C–O bond order can be readily observed in the stretching frequencies of metal-carbonyls, and the more electron-rich the metal center, the stronger the donation and the lower the C–O stretching frequency.

For example, examination of very electron-rich titanium(–II) to relatively electron-poor hexacarbonyl complexes reveals a dramatic change in C–O stretching frequencies. The more electron-rich metals back-donate into the carbonyl π*-orbitals more strongly, lengthening the C–O bonds and reducing the stretching frequency (Table 11.2).

Typical bond dissociation enthalpies for metal carbonyls are in the ~30–50 kcal/mol range, depending on the metal and the other ligands present. A list of some selected first CO BDEs are shown in Table 11.3.

The bonding modes of bridging carbonyls are numerous, and some examples are shown in Fig. 11.11. Carbonyl ligands may bridge two metals, as in the fluxional complex $Fe_3(CO)_{12}$.[38] Also shown is the structure of the heavier congener $Os_3(CO)_{12}$, which has all terminal carbonyls. As these two examples suggest, bridging carbonyls are more commonly observed in complexes with first-row transition metals. Carbonyl ligands bridging two metals typically have smaller C–O bond orders than the linear derivatives, and often appear in the 1720–1850 cm^{-1} range in the IR.

Three metals can also be bridged by CO, as shown in the cluster $\{Fe(\mu_3\text{-}CO)Cp\}_4$ where the irons form a tetrahedral core (shown in red) and a carbonyl bridges each triangular face.[39] Triply bridging CO ligands often have even lower C–O stretching frequencies around 1600–1730 cm^{-1}.

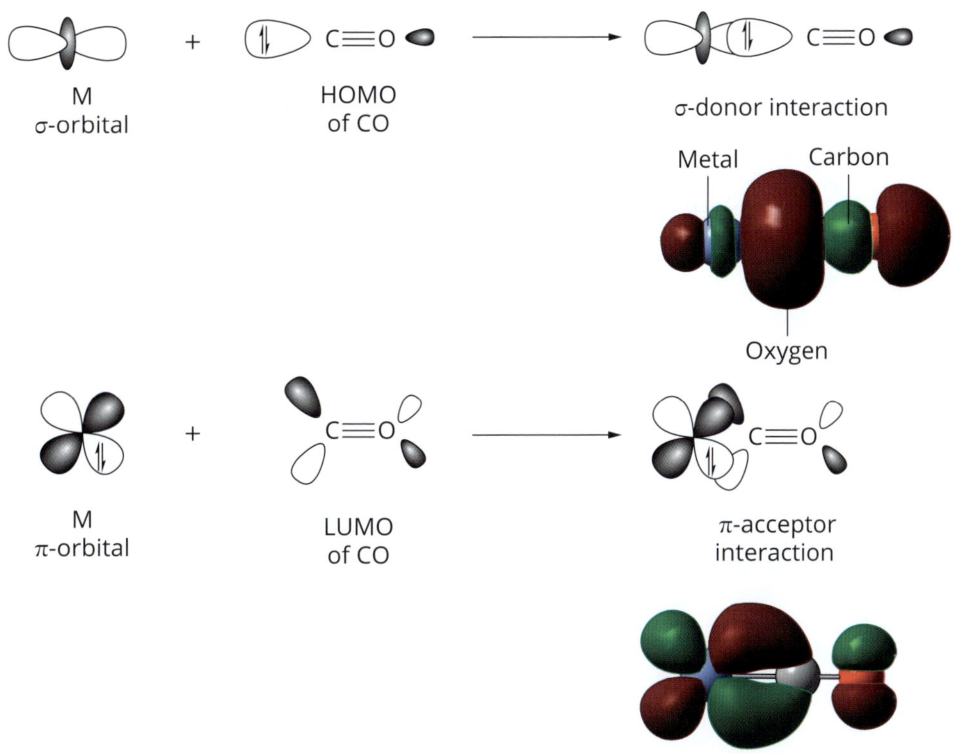

Fig. 11.10 CO bonding to a transition-metal center. Only one of the two equivalent π-acceptor interactions is shown; the other is rotated 90° along the M–CO axis. Metal = blue sphere, Carbon = gray sphere, Oxygen = Red sphere.

Table 11.2 CO-stretching frequencies and IR-active T_{1u} stretch of some isoelectronic (d^6) hexacarbonyl complexes of the first transition series.

Complex	ν_{CO} (cm^{-1})
$\{Ti(CO)_6\}^{2-}$	1748[32]
$\{V(CO)_6\}^{-}$	1860[33]
$Cr(CO)_6$	2003[34]
$\{Mn(CO)_6\}^{1+}$	2106[34]
$\{Fe(CO)_6\}^{2+}$	2204[35]

11.2.1.2 Metal Cyanides[24]

Cyanide, $^{-}$CN, is a pseudohalide counted as a radical in the neutral method and as an anionic 2-electron donor in the charged method (Table 10.3). The anion is isoelectronic with CO and has a very similar bonding scheme to that shown in Fig. 11.10. However, because of the lower electronegativity of nitrogen versus oxygen and the formal anionic charge, $^{-}$CN is a better σ-donor and a weaker π-acceptor than CO. Many complexes with bridging cyanide ligands are known with diverse structures.

There are numerous applications of cyanide complexes. One familiar compound, $Fe_4\{Fe(CN)_6\}_3$, is known as "Prussian blue", which has been used as a pigment for hundreds of years and appears in some well-known paintings, including Vincent van Gogh's *Starry Night*, shown in Fig. 11.12. The

Table 11.3 Bond dissociation enthalpy (BDE) for the loss of the first carbonyl ligand from some selected stable complexes.[36–37]

Complex	BDE (kcal/mol)
$V(CO)_6^-$	30.8 ± 4
$V(Cp)(CO)_4$	35 ± 5
$Cr(CO)_6$	36.8 ± 2
$Mo(CO)_6$	40.5 ± 2
$W(CO)_6$	46.0 ± 2
$Mn(CO)_6^+$	32 ± 5
$Mn(CO)_5^-$	40.6 ± 4
$Mn(Cp)(CO)_3$	55
$Fe(CO)_5$	41 ± 2
$Ru(CO)_5$	27.6 ± 0.4
$Os(CO)_5$	30.6 ± 0.4
$Co(CO)_4^-$	39.7 ± 4
$Rh(Cp)(CO)_2$	46‡
$Ir(Cp)(CO)_2$	57‡
$Ni(CO)_4$	22-25
$Pd(CO)_4$	10‡
$Pt(CO)_4$	13‡
$Ru(dmpe)_2(CO)$	43.0 ± 2

‡ Value from theory. dmpe = 1,2-bis(dimethylphosphino)ethane

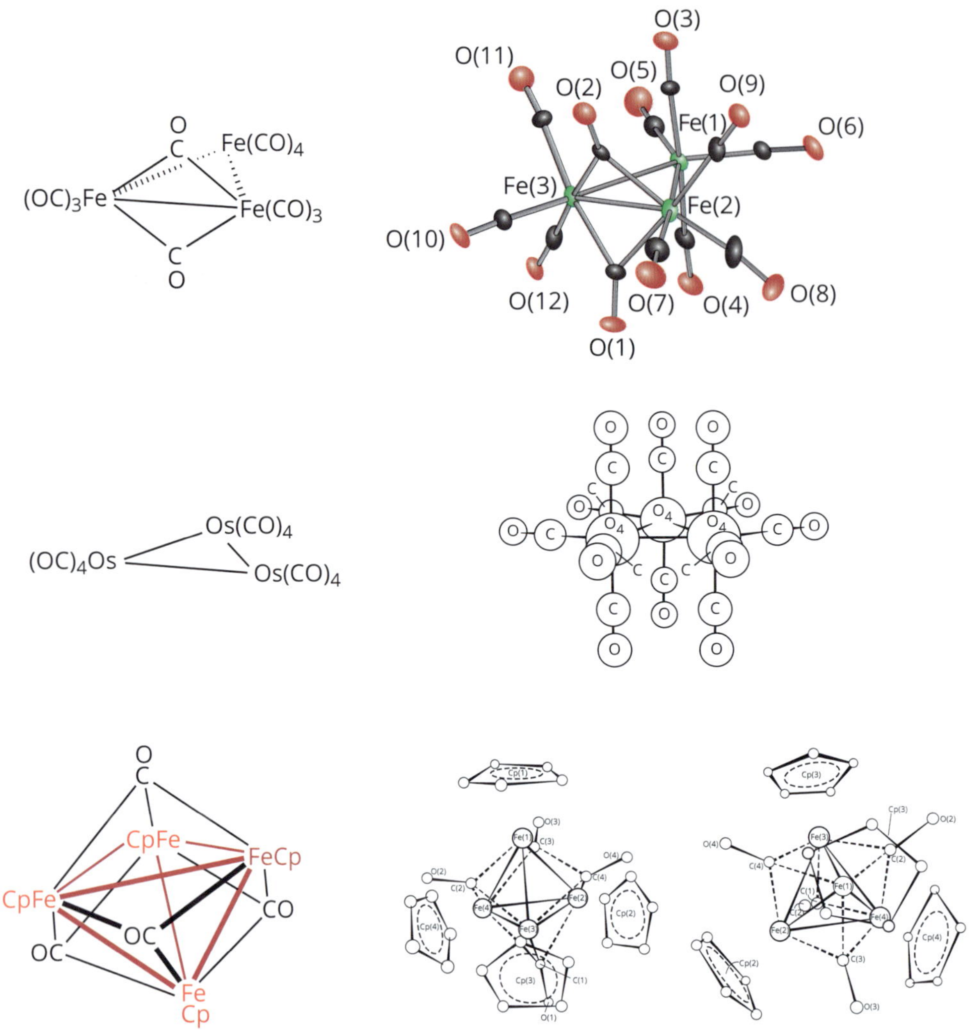

Fig. 11.11 Structures of $Fe_3(CO)_{12}$ and $Os(CO)_{12}$, (*top and middle*). X-ray structures from the review by Dahl and coworkers.[38] (*bottom*) Structure of $\{Fe(\mu_3\text{-}CO)Cp\}_4$, where the irons form a tetrahedral core (shown in red) and a carbonyl bridges each triangular face.[39]

Fig. 11.12 (*left*) Vincent van Gogh's *Starry Night*, which makes extensive use of Prussian blue as a pigment. (*right*) One of the earliest known uses of Prussian blue as a pigment (both the sky and Mary's veil) was in 1709 in the painting *Entombment of Christ*, by Pieter van der Werff.[40]

compound as a pigment may have been first used in 1709 in the painting *Entombment of Christ*, by Pieter van der Werff (Fig. 11.12).[40]

There are a few instances where monometallic isocyanides (M–NC) complexes have been prepared as stable species. From a frontier molecular orbital perspective, it seems that such species should be unlikely to exist. Nevertheless, some isocyanide complexes, particularly of the main group, rare earths, and actinides, have been isolated. It would appear that in such compounds the structure is under charge control, and the metal center is preferentially bonding to the more electronegative nitrogen with the larger amount of anionic character in the CN^- structure. If so, it would be metal centers that poorly backbond with highly polar bonding that might exhibit the structure.[41–43]

11.2.1.3 Metal Isonitriles

Some ligands closely related to carbon monoxide and cyanide, although not themselves diatomics, are isonitriles, CNR, also called isocyanides. The R group on isonitriles can be a large selection of alkyl and aryls. Because the ligand is neutral, as opposed to the anionic ^-CN, isonitriles are generally between cyanide and CO in π-acceptor and σ-donor ability.

The results of backdonation into isonitriles can be seen not only in the bond distances and vibrational frequencies, but also in the C–N–R angle of the ligand (Fig. 11.13). As shown in the resonance forms in the following, as the metal in the n-oxidation state backdonates into the C–N π*-orbitals, the C–N bond order drops and the C–M bond order increases. In the second resonance form, electrons have in essence been transferred from the metal, which is now in the (n + 2)-oxidation state, to the nitrogen to form a lone pair, bending the C–N–R angle.

An extreme example of such bending was reported by Cummins and coworkers in 2004.[44] The very reducing metal complex Mo{N(Bu^t)Ar}$_3$, where Ar = 3,5-$Me_2C_6H_3$, reacts with *tert*-butylisonitrile to give Bu^{t}NC–Mo{N(Bu^t)Ar}. The C–N–Bu^t angle reported was 137.8(7)°, approaching that for an sp^2-nitrogen.[44] Deviation from linearity in isonitrile angles is found in many other examples as well, including Mo(CNBu^t)$_6$, which has a mean C–N–Bu^t angle of ~153°.[45]

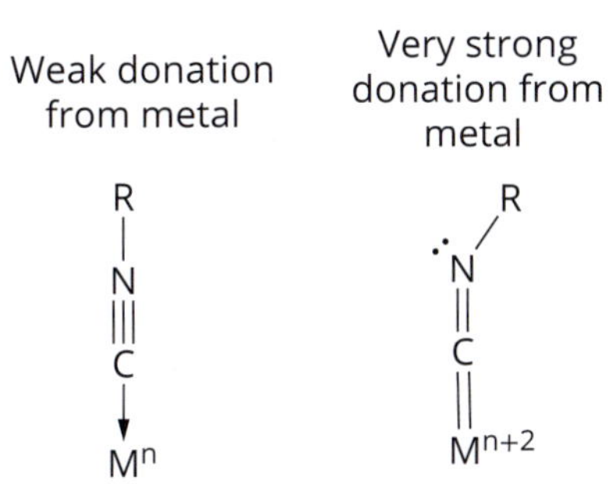

Fig. 11.13 Strong donation to a coordinated isonitrile can lead to bending.

11.2.1.4 Nonclassical Carbonyls and Related Systems (Advanced Topic)

The HOMO of CO has a small amount of antibonding character due to mixing of σ_{2p} with σ^*_{2s} (Fig. 11.9). If the metal center does not π-donate into the π* orbital but interacts in a σ fashion with CO, the interaction with the metal can actually *increase* the CO bond-stretching frequency by stabilization of these electrons with antibonding character. Effectively, adding a σ-only Lewis acid can increase the CO bond strength. Carbonyl complexes of this type have been dubbed "nonclassical carbonyls", defined as a metal complex with $\nu_{CO} > 2143\ cm^{-1}$ and $R_{CO} < 1.128$ Å—a CO stretching frequency greater than and a C–O bond distance less than the values in free carbon monoxide.[46]

A well characterized example of a nonclassical carbonyl complex was described by Strauss and coworkers, $\{Ag(CO)\}^+\{OTeF_5\}^-$. The anion used in the preparation was the weakly coordinating anion "teflate", $\{OTeF_5\}^-$, and some closely related ligands. Attempts to use many other anions, X = Cl^-, ClO^-_4, and so on, lead to displacement of the CO ligand and provided only AgX.

The CO stretching frequency in $\{Ag(CO)\}^+$ versus free CO are shown in Fig. 11.14a. Also observed was the dicarbonyl complex $\{Ag(CO)_2\}^+$.[47] Free carbon monoxide has a $\nu_{CO} = 2143\ cm^{-1}$,

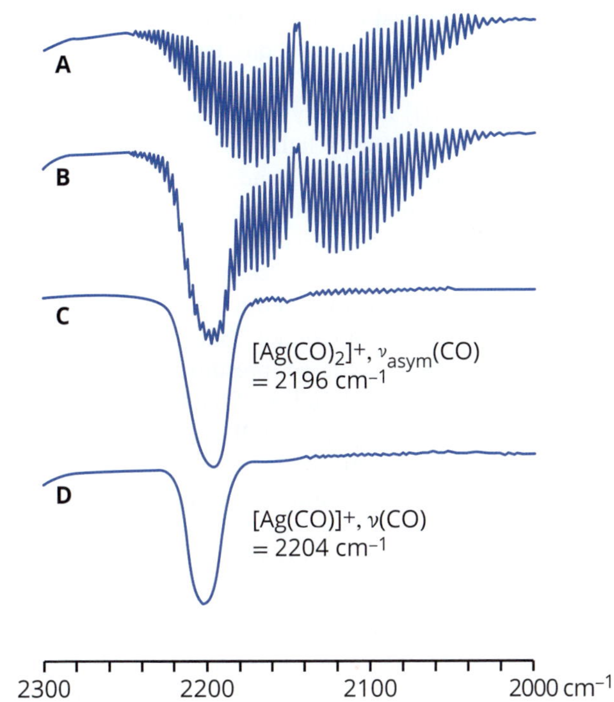

Fig. 11.14a IR spectra taken with a 1.6 cm path length gas-phase cell with AgBr windows: (A) 501 torr of CO, $\nu_{CO} = 2143\ cm^{-1}$; (B) solid $Ag^+\{Nb(OTeF_5)_6\}^-$ deposited on the side of both windows plus 501 torr of CO; (C) spectrum A subtracted from spectrum B; the small peak at 2196 cm^{-1} is assigned to $\nu_{asym}(CO)$ for $\{Ag(CO)_2\}^+$; (D) the sample for spectrum B placed under vacuum for ~10 min—the peak at 2204 cm^{-1} is assigned to ν_{CO} for $\{Ag(CO)\}^+$. Figure from the paper by Strauss and coworkers.[47]

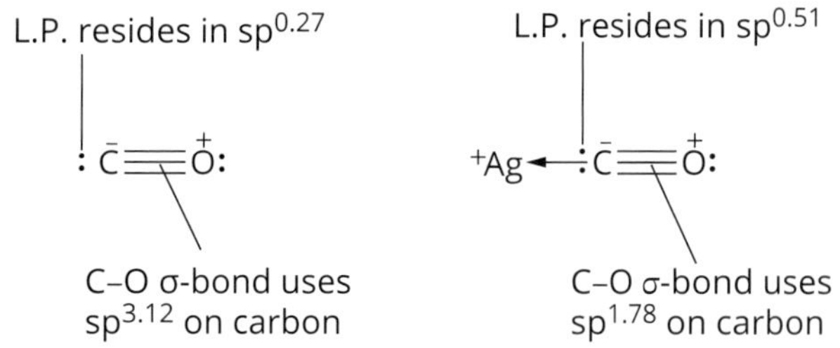

Fig. 11.14b NBO results on CO and $\{Ag(CO)\}^+$. (Both calculations were done using Gaussian 09 with B3LYP/SDD and NBO6.) The CO bond strength increases on coordination to Ag^+ due to increasing *s*-character in the C–O σ-bond.

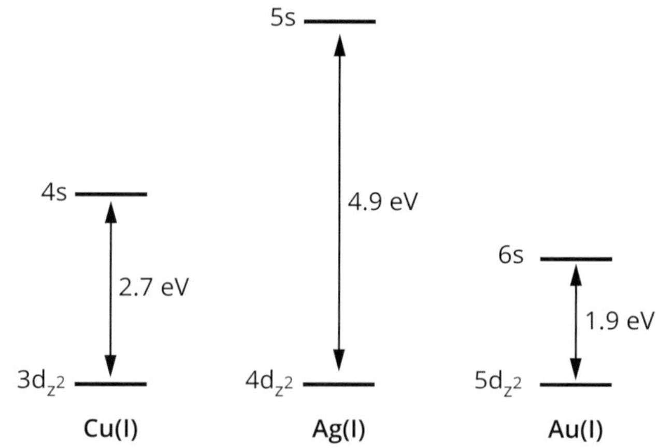

Fig. 11.15 Orbital energy diagram for Cu(I), Ag(I), and Au(I). The *d*-orbital energies are taken from the second ionization potentials of the neutral atoms. The *s*-*d* energy gaps are from the lowest-energy $d^9s^1 \leftarrow d^{10}$ electronic transition energies.

whereas the silver complex $\{Ag(CO)\}^+$ has a significantly higher C–O stretching frequency of 2204 cm^{-1}.

If NBO theory is used to analyze the bonding differences between CO and $\{Ag(CO)\}^+$, the reason for the increased C–O bond strength in the silver salt becomes quite clear (Fig. 11.14b). The lone pair on the carbon of CO is housed in a very *s*-rich orbital, $sp^{0.27}$. Once this lone pair is coordinated to silver(I), the *s*-character can be diminished in the lone pair (Bent's rule), and a larger amount of *s*-character can be used in the C–O σ-bond. As a result, the hybrid on carbon used to bond with oxygen increases in *s*-character from $sp^{3.12}$ in free CO to $sp^{1.78}$ in $\{Ag(CO)\}^+$, making the C–O bond stronger. In other words, the observation of "nonclassical" behavior in carbonyls is a manifestation of the Lone Pair Bond Weakening Effect (Section 8.5).

Why does silver(I) have very low to no π-backbonding to the carbonyl ligand? Silver's lighter and heavier congeners form far stronger bonds with CO, and there were many examples of Cu–CO and Au–CO complexes known when the first silver carbonyl was observed. All of the $\{M(CO)\}^+$, M = Cu, Ag, and Au complexes are nonclassical carbonyls with ν_{CO} = 2180, 2189, and 2179 cm^{-1}, respectively. Both Cu(CO)Cl and Au(CO)Cl are readily isolable compounds, but the corresponding silver salt has not been prepared, presumably due to weaker Ag–CO BDE versus Au–CO. Indeed, Frenking and Veldkamp calculate the M–CO BDEs in $\{M(CO)\}^+$, M = Ag and Au, to be 21 and 45 kcal/mol, respectively.[48]

This difference in CO binding has been ascribed to the ease of *sd*-hybridization. The *d*-orbitals are lower in energy than the *s*-valence orbitals but are filled in the M(I) oxidation state, where M = Cu, Ag, and Au. If the metal can produce an empty *sd*-hybrid, it can use this much lower-energy orbital to bind a donor ligand like CO.[b] The ease with which the metal can form an *sd*-hybrid is, at least partially, determined by the energy gap between the two orbitals. The orbital energies from excitation and ionization data are shown in Fig. 11.15. Silver(I) has the largest *s*-*d* energy gap by far. The smaller *s*-*d* energy gap in gold is due to relativistic stabilization of the valence *s*-orbital (direct relativistic effect) and destabilization of the valence *d*-orbital (indirect relativistic effect), as discussed in Chapter 8.[49]

If the metal can use the lower-energy *d*-orbital to bond with CO through hybridization, a shorter and stronger M–CO bond will result. This will not only strengthen the σ-interaction but also the more distance-sensitive π-backdonation.

This increase in CO bond-stretching frequency due to the σ-interaction should always occur when CO is on a metal center to one extent or another. However, in instances where the metal also π-donates, the π-donation usually reduces ν_{CO} far more than the σ-accepting interaction increases ν_{CO}. Even d^0 complexes may have $\nu_{CO} < 2143$ cm^{-1} if there are filled bonding MOs that can interact with the CO π*-orbitals. Jordan and Shen reported two isomers of the complex $\{ZrCp_2(CO)(COMe)\}^+$—rotational isomers that differ by the acyl positioning.[50] The two d^0 complexes differ substantially in the infrared stretching frequencies for the CO ligands. One can only assume that there is an orbital associated with the acyl that is a much more substantial donor in the right isomer than in the left of Fig. 11.16. In this case, a ligand-based donor is donating electron density to the metal, and the electron density is then being used to backbond to the CO.

Hundreds of nonclassical carbonyls have been reported, and many are isolable. Similar behavior is also observed with the closely related cyanide, CN^-, and isonitrile, CNR, ligands. In the case of cyanide, formation of a strong σ-bond, such as in HCN, is well known to increase the C–N stretching frequency over CN^-. Many "nonclassical" isonitrile complexes are also known, such as *cis*-$TiCl_4(CNBu^t)_2$ (Fig. 11.16), which has a single observed C–N IR stretching frequency at 2226 cm^{-1} (solid, Nujol) or 2210 cm^{-1} (CH_2Cl_2 solution).[51] This is significantly higher in frequency than free $CNBu^t$ at 2067 cm^{-1} (liquid film).[52–53]

[b] See Chapter 14 for an extensive discussion of *sd*-hybridization.

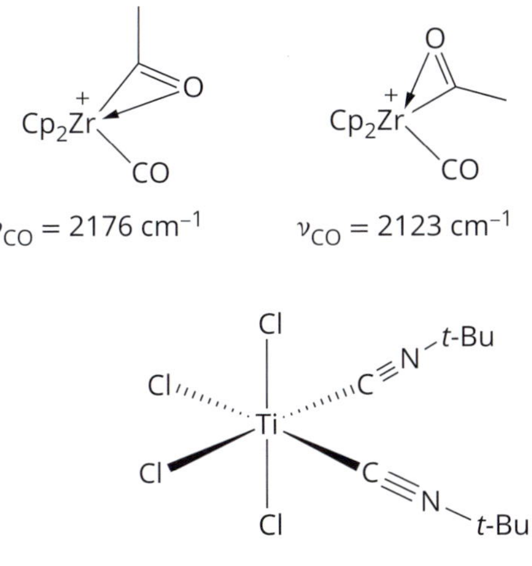

Fig. 11.16 (*top*) Two isomers of the d^0-complex $\{ZrCp_2(CO)(COMe)\}^+$ isolated by Jordan and Shen. (*bottom*) Structure of *cis*-$TiCl_4(CNBu^t)_2$.

11.2.2 Dinitrogen (N_2)

The recommendation, provided no other information (such as structural information of X-ray diffraction), is to regard dinitrogen as a neutral 2-electron donor whether the ligand is bound end-on (linear) or side-on to the metal center for the purposes of formal oxidation state (Fig. 11.17a). In the end-on position, the *sp*-hybridized lone pair acts as a donor to the metal. When side-on, a pair of electrons in a π-bond can donate to the metal, similar to acetylene. However, acetylene is not necessarily a simple 2-electron donor

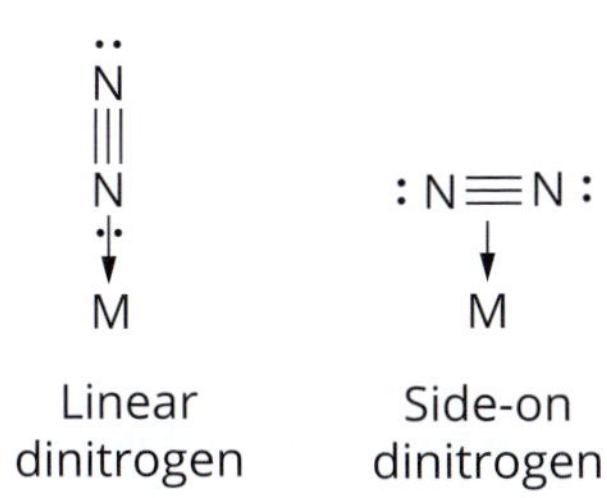

Fig. 11.17a Most common N_2 binding modes to a single metal center.

in all complexes (Section 11.2), and if the electron count permits, side-on nitrogen may be a 4-electron donor as well.

However, dinitrogen binding epitomizes an effect present in all of these CO-related ligands: backbonding changes the N–N bond order, the M–N bond order, and metal oxidation state. This effect is always present, but only becomes significant for very electron-rich metal centers with extremely strong backbonding. This is especially true for dinitrogen bound to multiple metal centers, and is readily observable in bridging carbonyls, which typically have significantly reduced stretching frequencies relative to mononuclear complexes (*vide supra*).

The first metal dinitrogen complex, $\{Ru(N_2)(NH_3)_5\}^{2+}$, was reported by Allen and Senoff in 1965 (Fig. 11.17b), prepared serendipitously through the action of hydrazine on ruthenium(III) chloride. The complex was characterized by conductivity, elemental analysis, evolution of N_2 (detected by mass spectrometry) on decomposition with 95% sulfuric acid, and by infrared spectroscopy, which showed a "strong, sharp band between 2170–2115 cm^{-1}" depending on the counterion (Br^-, I^-, BF_4^-, or PF_6^-).[54–56]

Less than a decade later, another key dinitrogen complex would be reported by Hidai and coworkers, a molybdenum *trans*-dinitrogen complex (Fig. 11.17b).[57–59] Now, dinitrogen complexes are known for most metals in the transition series, like the intriguing Fe(0) example reported in 2013 by Field and coworkers.[60] These complexes are still studied extensively due to their importance in the production of ammonia and hoped-for improvements to the Haber–Bosch reaction.

Dinitrogen that bridges two or more metal centers can either be end-on or side-on. Two examples with uranium are shown in Fig. 11.17c. In the left complex, Roussel and Scott prepared the side-on dinitrogen $(NN'_3)U(\mu_2\text{-}\kappa^2{:}\kappa^2\text{-}N_2)U(N'_3N)$, where $NN'_3 = N(CH_2CH_2NSiBu^tMe_2)_3$.[61] In the example on the right, Cummins and coworkers described the end-on dinitrogen complex $\{Ar(R)N\}_3U(\mu_2\text{-}\kappa^1{:}\kappa^1\text{-}N_2)Mo\{N(Bu^t)Ph\}_3$, where Ar = 3,5-$Me_2C_6H_3$ and R = $C(CD_3)_2CH_3$.[62]

As the metals donate more electron density, the N–N stretching frequency decreases and the N–N bond distance increases. The stretching frequency in free N_2 = 2331 cm^{-1}. Typically, dinitrogen complexes have N–N stretching frequencies that appear from about 1830 to 2150 cm^{-1}.

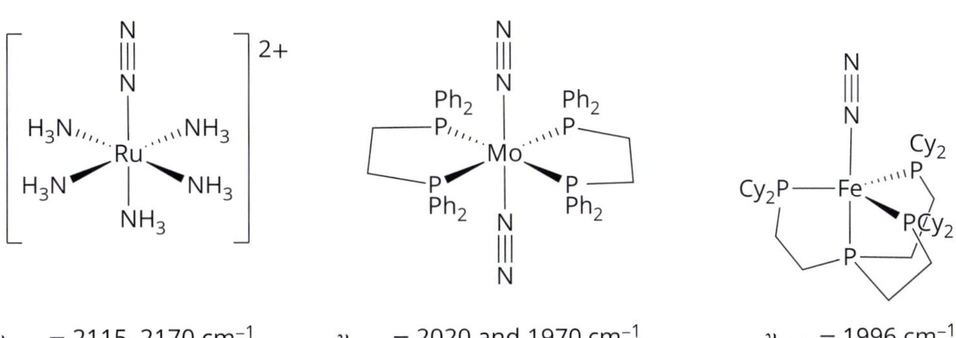

Fig. 11.17b Structures of some selected dinitrogen complexes. (*left*) Allen and Senoff's Ru(II) complex.[54–55] (*middle*) Hidai and coworkers' Mo(0) bis(dinitrogen) compound.[57–59] (*right*) Field and coworkers' Fe(0) dinitrogen complex.[60]

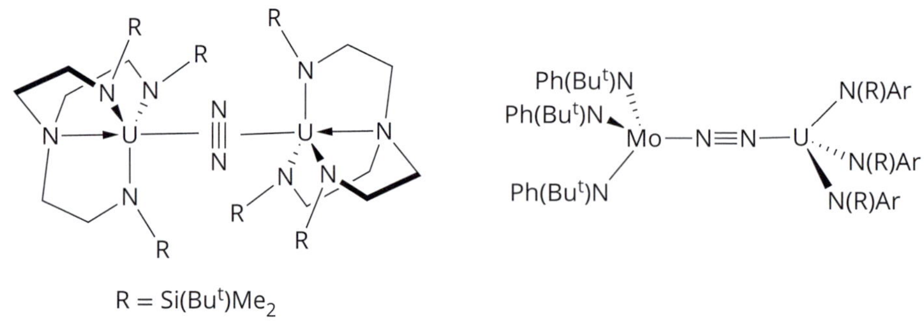

Fig. 11.17c Structures of two uranium-containing dinitrogen complexes. (*left*) Roussel and Scott's side-on[61] diuranium complex, and (*right*) Cummins and coworkers' end-on heterometallic.[62]

The effect of changing ligand donation can be examined with the series *trans*-M(N_2)(Cl)L_2, where M = Ir and Rh. As the phosphines become more donating, the N–N stretching frequency decreases due to increasing reduction of dinitrogen by the metal.[63–66] For example, the N–N stretching frequency in L = PPh_3 is 2105 cm^{-1}, while the more donating P^iPr_3 ligands gives an N–N frequency of 2058 cm^{-1}. In general, the donor ability of transition metals increases down the column, and the Ir metal center in M(Cl)(PPh_3)$_2$(N_2) appears to be much more donating (ν_{NN} = 2105 cm^{-1}) than the Rh derivative (ν_{NN} = 2152 cm^{-1}).

In some cases, metal backbonding may become so strong that treating N_2 as a neutral 2-electron donor becomes obviously incorrect. An example is shown from Fryzuk and coworkers in Fig. 11.18, where the N–N bond distance is 1.548(7) Å, which is longer than the N–N bond in hydrazine of ~1.45 Å. Without N_2, or with N_2 considered a neutral donor, the zirconium centers would be Zr(II). However, the very long N–N bond and relatively short Zr–N(of N_2) bonds belie such significant backbonding that Zr(IV) seems far more likely.[67]

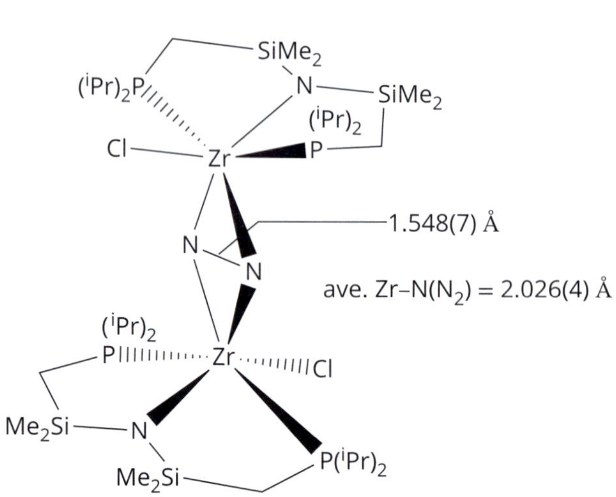

Fig. 11.18 Structures of Fryzuk and coworkers activated N_2 complex.

11.2.3 Nitrosyl (NO)[17–18,68]

The bonding in nitrosyl ligands is somewhat more complex than the previous diatomics due to differences in bonding modes and how they are perceived to affect metal oxidation state. There are two commonly encountered ways that nitric oxide may bind to a metal—linear and bent.

Nitric oxide, NO, is itself a stable radical used in biology as a signaling molecule for vascular relaxation. While NO is a colorless gas, when exposed to oxygen it is rapidly oxidized to higher nitrogen oxides (often producing yellow-brown colors), but NO is quite thermodynamically stable. The compound has an N–O bond order of 2.5 (Section 8.4) due to partial occupancy of the π-bond by one electron. This one unpaired electron can be lost to form a stable cation, and many salts are commercially available, such as NO^+ BF_4^-, and are common oxidants that often release gaseous NO as a by-product.

A linearly bound NO is regarded as a 2-electron donor and a cation for the charged method; in the neutral method, a linear NO is considered a 3-electron donor radical. The reasons for this are shown pictorially in Fig. 11.19. In the charged method, the higher-energy unpaired electron, which resides in an antibonding orbital, is lost to the metal giving NO^+, which is isoelectronic with CO. In the neutral method, the metal pairs one of its *d*-electrons with the unpaired electron of NO to form a single bond and π-bond character results from donation of the lone pair of electrons on nitrogen to an appropriate acceptor orbital on the metal.

As an example of a complex with a linear nitrosyl, Co(NO)$_3$ is shown in Fig. 11.19, which is thought to have three linear nitrosyls.[69] The cobalt complex Co(NO)$_3$ is reported to be a good Lewis base, donating through the cobalt center, that is not displaced from BMe_3 by either NH_3 or NMe_3. A related Lewis acid, {Fe(NO)$_3$}$^+$, has been prepared by Legzdins and coworkers.[70]

A bent nitrosyl is regarded as a 2-electron donor and an anion in the charged method (Fig. 11.20). In the neutral method, nitrosyl is a 1-electron donor. As an example with a bent nitrosyl, the electron counting for Ir(NO)Cl_2(PPh_3)$_2$ is shown.[71–72]

BOX 11A

Harmonic oscillator model for a diatomic and isotopic labeling

The vibrational frequency of a bond is often approximated using a harmonic oscillator model, which might be used to model the movement of two balls connected by a spring. The frequency changes with the reduced mass of the two nuclei (μ) as shown in the following equation, where m_1 and m_2 are the atomic masses of the two bonded atoms.

$$\mu = \frac{m_1 m_2}{m_1 + m_2}$$

The harmonic oscillator model then says that the frequency is related to the reduced mass by *k*, which is related to the bond strength.

$$v = \frac{1}{2\pi}\sqrt{\frac{k}{\mu}} \qquad 11.A1$$

For example, consider the complex $Cr(NO)(NPr^i{}_2)_3$, first prepared by Bradley and coworkers.[156] Most linear nitrosyls, like most carbonyls, are easily identified. However, because the frequency in this complex is extremely low due to very efficient backbonding from the chromium, the resonance appears in the "fingerprint" region with many bending modes of lighter element-containing bonds. Unequivocal identification can be difficult in this region.

Because the NO ligand frequency is substantially different from the Cr–N frequency in energy (the much heavier chromium gives a much lower frequency) the N–O vibration can be treated independently from the rest of the molecule; that is, the N–O and N–Cr vibrations are not strongly coupled. As a result, changing the mass of one or both of the atoms in the nitrosyl ligand will lead to a shift of the NO-related peak with virtually no change to the rest of the spectrum. Subtraction of the natural abundance and isotopically labeled spectra gives the NO frequencies.

The difference in energy for the two peaks will be related by the ratio of inverse square roots of the reduced masses with the frequencies.

$$\frac{v_{vib}^{^{14}NO}}{v_{vib}^{^{15}NO}} = \frac{\sqrt{\mu_{^{15}NO}}}{\sqrt{\mu_{^{14}NO}}} \qquad 11.A2$$

The natural abundance ^{14}NO complex has an N–O stretching frequency of 1643 cm^{-1}. The isotopically labeled ^{15}NO should have a frequency of approximately 1614 cm^{-1}.[157]

$$\frac{v_{vib}^{^{14}NO}}{v_{vib}^{^{15}NO}} = \frac{\sqrt{\mu_{^{15}NO}}}{\sqrt{\mu_{^{14}NO}}}$$

$$v_{vib}^{^{15}NO} = v_{vib}^{^{14}NO}\frac{\sqrt{\mu_{^{14}NO}}}{\sqrt{\mu_{^{15}NO}}} = 1643\frac{\sqrt{7.467}}{\sqrt{7.742}} = 1614$$

The experimentally determined $v_{vib}^{^{15}NO}$ was 1610 cm^{-1}, very close to the harmonic oscillator approximation.

If an adjacent bond has a similar frequency it can couple to the bond of interest, complicating the analysis. This is often the case with M=N–R (imide) complexes where the M=N and N–R bond stretching frequency can be quite similar. Even in these strongly coupled cases, reliable assignments can be made with isotopic changes or with changes to isostructural congeners, though the complexity may increase substantially with such couplings. For example, where this mode coupling problem was addressed in detail, the modes in $M(NAr)X_3L_2$ complexes were assigned using M = Ta and Nb and with X = Cl and Br by Hopkins and coworkers.[158]

The difference between linear and bent in the neutral method is that in the former, the lone pair on nitrogen is donated to the metal giving an approximately *sp*-hybridized nitrogen. In the bent form, the nitrosyl lone pair does not donate to the metal, and as a result one can presume a bent nitrosyl will result when there is no low-energy acceptor orbital on the metal of proper symmetry.

Free nitric oxide has an IR stretching frequency of 1870 cm^{-1}, and on binding to the metal this value may increase or decrease, and there is no steadfast correlation between M–N–O bond angle and the observed stretching frequency. Linear nitrosyls are generally found at higher frequencies; however, there is substantial overlap between the typical linear and bent NO ranges. Linear NO ligands typically appear in the range 1950–1450 cm^{-1}; whereas bent nitrosyls are typically from 1750 to 1400 cm^{-1}. Linear nitrosyls are considered to be more electron-withdrawing than even carbonyls due to the cationic charge in the isoelectronic species (Fig. 11.19).

The best method for linear versus bent determination currently is using single-crystal X-ray diffraction. If the complex will exceed 18-electrons in the linear form, the NO ligand will typically be bent. If the metal has an empty orbital—for example, is 16-electron—when considering the nitrosyl as bent, then a bent nitrosyl is unlikely and the NO ligand will probably be linear.

11.2.4 Dioxygen (O_2)[27–28]

Dioxygen complexes are studied for a host of applications and for understanding numerous enzymatic processes. There are many metalloenzymes that catalyze selective and nonselective oxidation of various substrates, and their study is a vast area

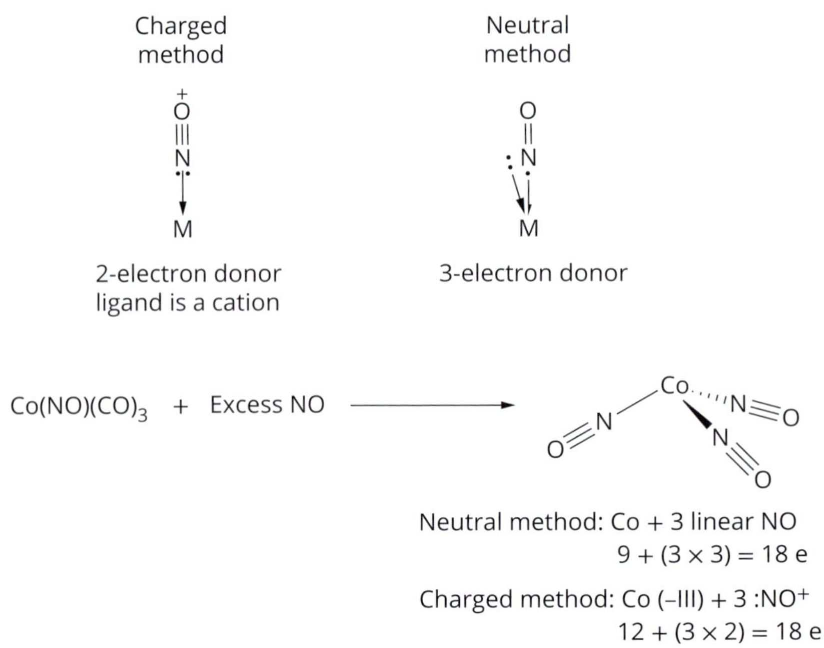

Fig. 11.19 (*top*) Linear nitrosyls are considered 2-electron donor cations in the charged method and 3-electron donors in the neutral method. (*bottom*) $Co(NO)_3$ as an example of a complex with linear nitrosyls.[69]

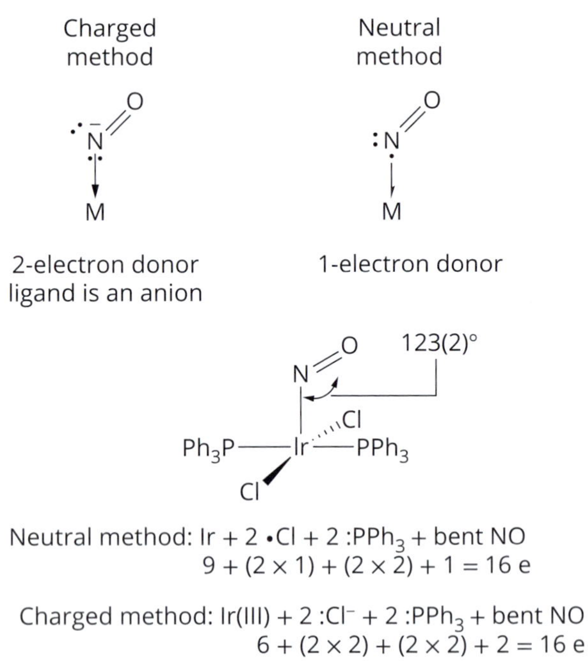

Fig. 11.20 (*top*) Bent nitrosyls are considered 2-electron donor anions in the charged method and 1-electron donors in the neutral method. (*bottom*) The bent nitrosyl complex $Ir(NO)Cl_2(PPh_3)_2$ and its electron counting.[71–72]

in bioinorganic chemistry. Dioxygen on a single metal can act either as a bent ligand, where it is called a superoxo, or as a side-on ligand, where it is called a peroxo. Both are found biologically; as examples, oxygen in oxyhemoglobin is carried as a superoxo, and oxygen in methane monooxygenase goes through a bridging peroxo intermediate, both of which are discussed in more detail in Chapter 20.[73–74]

For electron-counting purposes, a peroxo is regarded as a dianionic, 4-electron donor in the charged method and as a 2-electron donating diradical in the neutral method. An example of a peroxo complex is shown in Fig. 11.21, the homoleptic $\{Ta(\kappa^2\text{-}O_2)_4\}^{3-}$.[75] Using either method, the electron count on tantalum is 16.

Fig. 11.22 shows a relatively simple superoxo complex of platinum, $Pt(\kappa^1\text{-}O_2)_2(NHC)_2$, where NHC is an *N*-heterocyclic carbene (Section 11.3.3), a neutral 2-electron donor similar in some ways to a phosphine.[76] The complex is square planar platinum(II) and 16-electron.

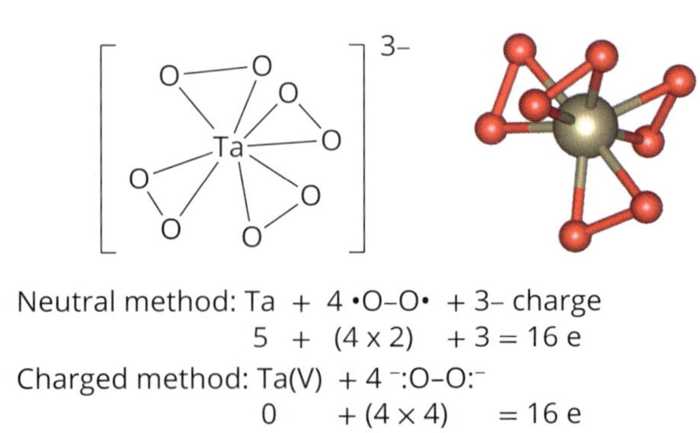

Fig. 11.21 Ball-and-stick structure of the anion from the X-ray diffraction data and electron-counting for the peroxo complex $\{Ta(\kappa^2\text{-}O_2)_4\}^{3-}$. The counterions are guanadinates, $C(NH_2)_3{}^+$.[76]

BOX 11B

Enemark–Feltham notation

Designation of nitrosyls as either bent or linear is in some cases subjective, and a means of categorizing nitrosyl complexes without this assignment has been devised: Enemark–Feltham notation.[159] This notation allows one to compare different nitrosyl–metal complexes and anticipate electronic structure based on known complexes.

The notation takes the nitrosyl(s) and metal as a unit and adds the *d*-electrons on the metal (after accounting for oxidation state while ignoring the nitrosyl) to one electron from each nitrosyl, the high-energy antibonding electron in NO. For the notation, the metal, M, and the number of nitrosyls, x, are placed in brackets with the number of electrons (*d*-electrons on metal plus one electron for every nitrosyl), n, as a superscript outside the bracket, $\{M(NO)_x\}^n$.

For example, $\{V(CN)_5(NO)\}^{3-}$ is a known mononitrosyl with ν_{NO} = 1530 cm^{-1}; the Enemark–Feltham notation for this complex is $\{V(NO)\}^4$. The accompanying complex, $\{Fe(NO)_3(NHC)\}^+$ was reported by Darensbourg and Hsieh, which is an $\{Fe(NO)_3\}^{10}$ complex.[160] Ibers and Mingos described the cationic iridium complex $\{Ir(NO)_2(PPh_3)_2\}^+$, which is an $\{Ir(NO)_2\}^{10}$ compound.[161]

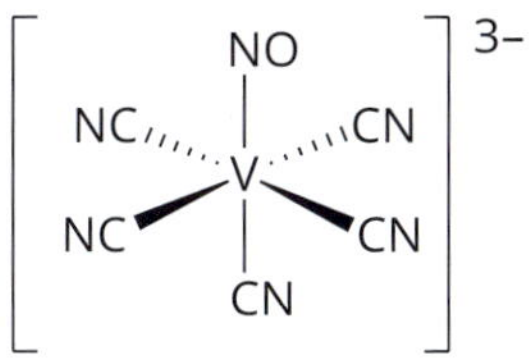

$\{V(NO)\}^4$

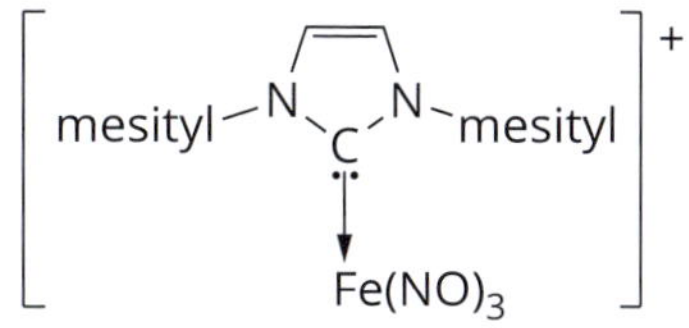

$\{Fe(NO)_3\}^{10}$

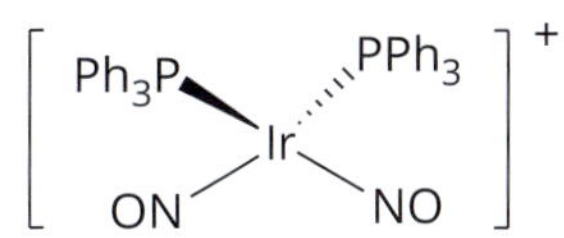

$\{Ir(NO)_2\}^{10}$

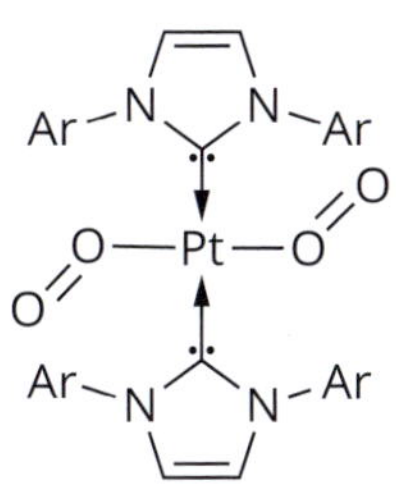

Ar = 2,4,6-Pr^{i}C$_6$H$_2$

Neutral method: Pt + 2 •O=O + 2 :NHC
10 + (2 × 1) + (2 × 2) = 16 e

Charged method: Pt(II) + 2 ⁻:O=O + 2 :NHC
8 + (2 × 2) + (2 × 2) = 16 e

Fig. 11.22 Structure and electron-counting for the superoxo complex $Pt(\kappa^1\text{-}O_2)_2(NHC)_2$, where NHC is an *N*-Heterocyclic Carbene.[76]

11.3 Neutral Donors: Amines, Phosphines, N-Heterocyclic Carbenes (NHCs), and Related Ligands

As with many of the ligand types that will be discussed, there is a dizzying array of amines, phosphines, and NHCs that are employed in coordination chemistry. Rather than providing extensive lists of ligands, although some listing of more common ligands will be of utility, we will discuss models for understanding these ligands generally.

11.3.1 Phosphines as Ligands

Phosphines are perhaps the most common ancillary ligand, a ligand not undergoing reaction, for late transition metals. As such, there is an enormous menagerie of phosphine ligands, which have undergone extensive, almost zoological, classification using a host of different parameters.

First we will discuss the bonding of phosphines to transition metal centers. Second, some common phosphine classes will be described. As an "advanced topic", a brief discussion of classification of ligands by their steric and donor properties is included subsequently.

11.3.1.1 Phosphine Bonding to Metal Centers

For electron-counting purposes, phosphines are simply considered 2-electron donors; however, they can act as σ-donors and π-acceptors. The HOMO of phosphine is the phosphorus-based lone pair, which changes in hybrid composition based on the phosphorus substituents. The LUMO is typically the *e*-set of σ*-orbitals. The MO diagram for PH_3 is shown in Fig. 11.23.

As the X–P–X angles change, the composition of the HOMO will necessarily change as well, and the *sp*-hybridization can be calculated using the Coulson directionality theorem and sum

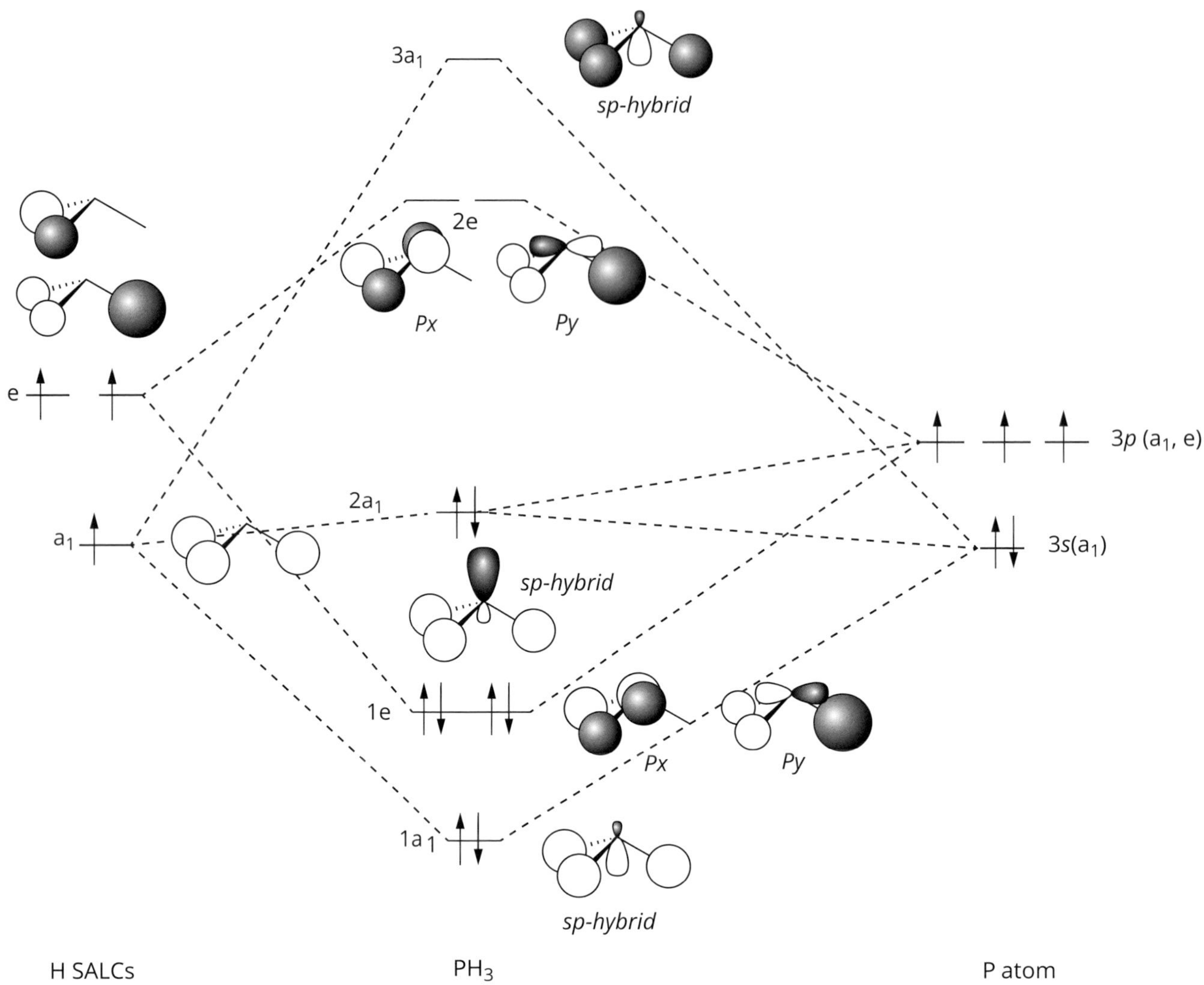

Fig. 11.23 MO diagram form PH_3.

rules (Chapter 7). As the amount of *p*-character in the lone pair increases, the energy of the lone pair will increase making the ligand a better σ-donor. However, one method to increase the X–P–X angle is by increasing the sterics of X, which may also lower donation from the phosphorus if the sterics necessitate a longer P–M bond. Naturally, increasing the electronegativity of the substituents on the phosphorus, which inductively decreases electron density, will also lower the phosphine σ-donor ability.

The backdonation is very similar to π-bonding from oxygen to phosphorus in $^{-}O{-}^{+}PH_3$, negative hyperconjugation (Section 8.3.2). Substituents on the phosphorus that are high in electronegativity increase the phosphorus character of the antibonding orbitals and stabilize anionic charge that builds up on the substituents, both of which increase backbonding. As a result, PF_3 is an extremely strong π-acceptor, thought to be similar to or even better than CO.[77]

Backdonation to the phosphine from the metal should (1) increase the P–X bond distances because donation from the metal occurs into the P–X σ*-orbital and (2) shorten the M–P bond due to increased bond order. These effects can be seen in the structure of *cis*-$PtCl_2(PF_3)(PEt_3)$, for example (Fig. 11.24).[78] The Pt–PF_3 bond is 2.141(3) Å in length, an incredible 0.13 Å shorter than the Pt–PEt_3 bond! This bond shortening runs counter to the expected σ-donor ability of the two ligands, where PF_3 should be a much worse σ-donor, and is directly attributable to backbonding from the metal.

As one additional example (Fig. 11.25), consider the two redox isomers $\{Fe(\eta^3\text{-}C_8H_{13})(P(OMe)_3)_3\}^{0/+1}$. The neutral, iron(I) complex shows shorter Fe–P and longer P–O distances consistent with stronger backbonding than in the cation. Again, keep in mind that the iron(I)–P distance is shorter despite the smaller radius of the higher oxidation state iron(II) metal, which can be seen in the shorter average Fe(II)–C distances.[79–80] This behavior, illustrated by these Fe(I)/Fe(II) phosphine complexes, has been found to be the norm and is quite consistent with metal backdonation into the P–X antibonding orbitals.[81–83]

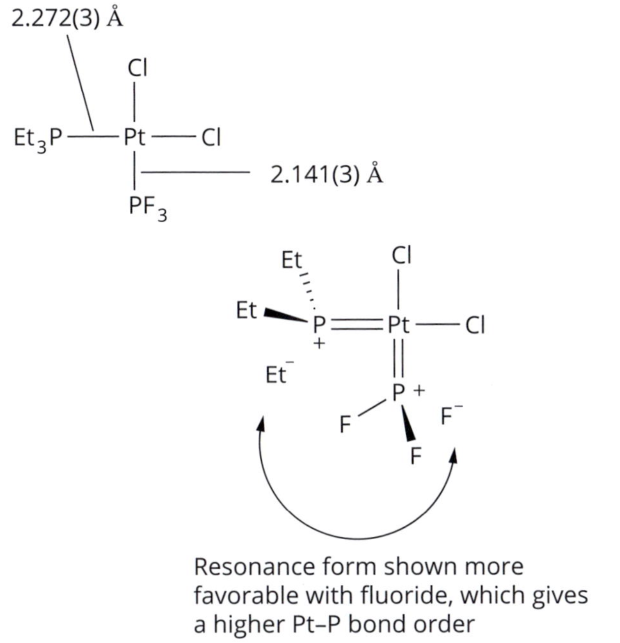

Fig. 11.24 The structure of *cis*-$PtCl_2(PF_3)(PEt_3)$, showing the Pt–P bond lengths.[78] Due to increased negative hyperconjugation, the Pt–P bond order is increased for the higher electronegativity PF_3 due to stabilization of the anion (*bottom*).

11.3.1.2 Some Common Phosphine Varieties

There is a vast and ever-growing array of commercially available phosphines, and a plethora has been reported in the literature. Here we will present a biased, partial list of some phosphines that are commonly encountered. The most common phosphines in a particular field of catalysis, for example, are highly dependent on the specifics of the study, which further emphasizes the remarkable breadth of phosphine ligand properties that are accessible.

Phosphines may be monodentate or multidentate, chiral or achiral, extremely bulky, or quite small. The physical and chemical properties of these phosphines vary widely, some being gases (such as PH_3), others liquids (such as PMe_3), and others relatively high melting solids (such as PPh_3). Some phosphines are quite stable in air, with no special handling requirements, such as PPh_3, while others are pyrophoric, such as PH_3.

Some common monodentate phosphines vary from trimethylphosphine, triphenylphosphine, triethylphosphine, to tri-*tert*-butylphosphine, where there are three identical alkyl groups on the phosphine. Mixed alkyl and aryl phosphines, such as methyldiphenylphosphine, dimethylphenylphosphine, and di(*tert*-butyl)methylphosphine, are also quite common (Fig. 11.26).

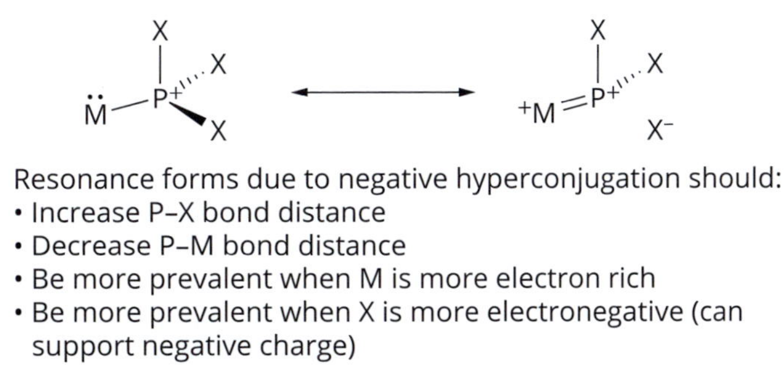

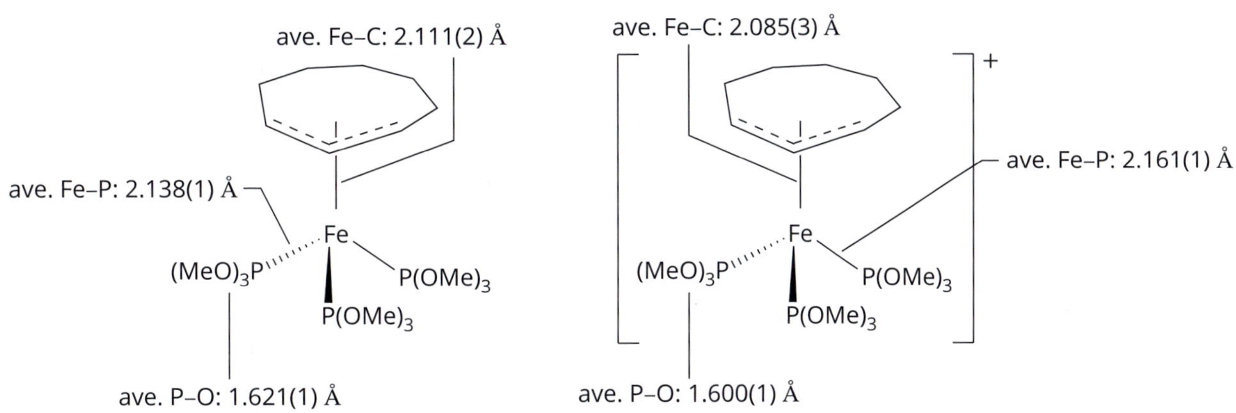

Fig. 11.25 Structures of the redox isomers $\{Fe(\eta^3\text{-}C_8H_{13})(P(OMe)_3)_3\}^{0/+1}$. The higher oxidation state iron(II) complex shows shorter average Fe–C bonds but longer Fe–P bonds, due to decreased backbonding to the phosphines.[c,79–80]

[c] The higher oxidation state complex has an agostic C–H interaction not shown.

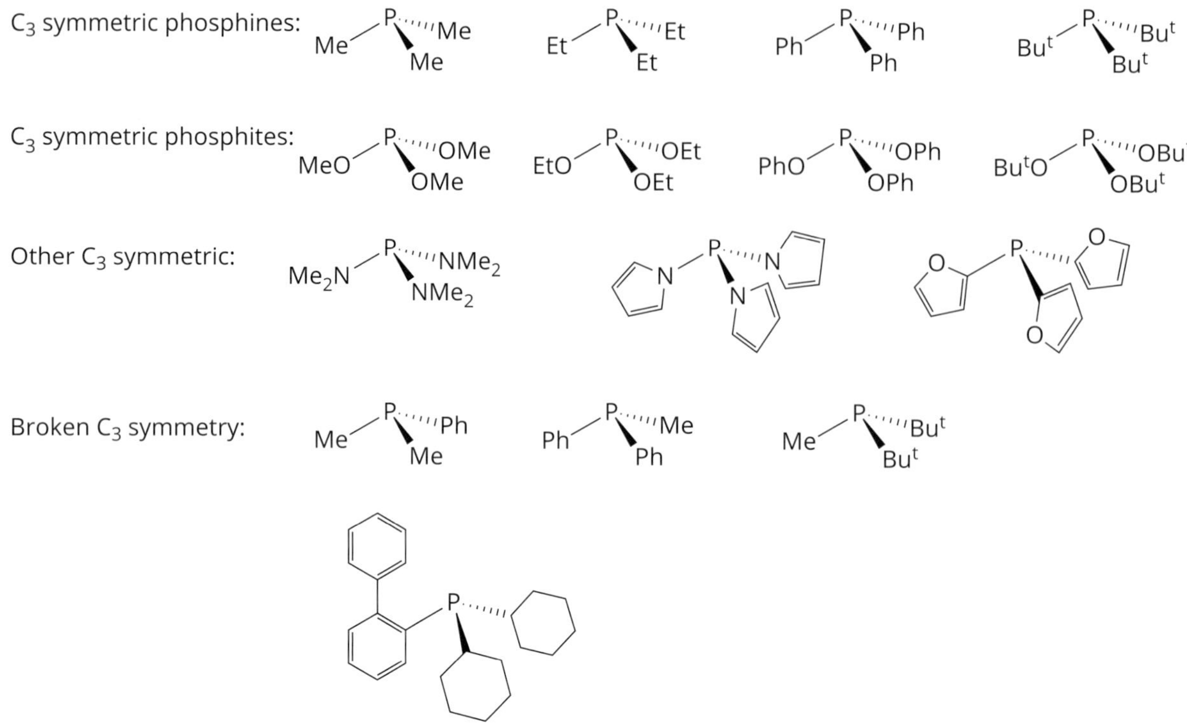

Fig. 11.26 Some examples of commonly encountered monodentate, achiral phosphines. Many multidentate phosphine ligands are also commonly employed to take advantage of the chelate effect and further control the metal's environment. Some of these ligands are shown in Fig. 11.27.

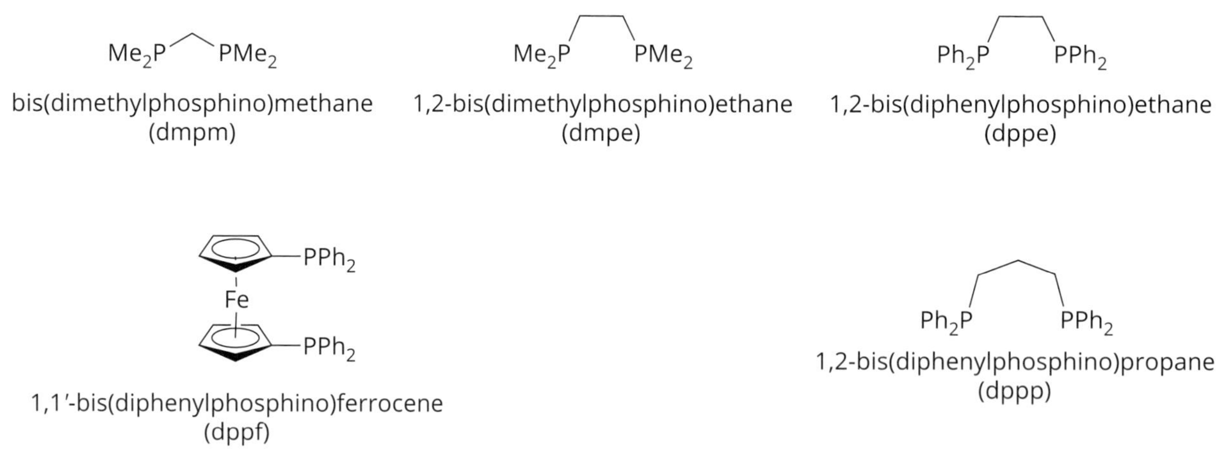

Fig. 11.27 Some examples of commonly encountered multidentate phosphine ligands.

Buchwald and coworkers have popularized a group of monodentate phosphines where the three-fold symmetry around the phosphorus is severely broken with one large, elongated group (biphenyl and derivatives of biphenyl) and two often large but more spherical substituents (cyclohexyl, *tert*-butyl, and so on).[84–86]

Despite the hundreds of commercially available phosphines, the most used derivative, especially on a mass basis, is probably triphenylphosphine, which has remarkable stability, low cost, and utility as a general ligand.

For chiral catalysis, many chiral phosphines are also commercially available. A few common ligands are shown in Fig. 11.28. Because phosphines generally have a fairly high barrier to inversion, chiral phosphines can be divided into two different categories: those with chiral substituents, and those that are chiral at phosphorus.[87–92]

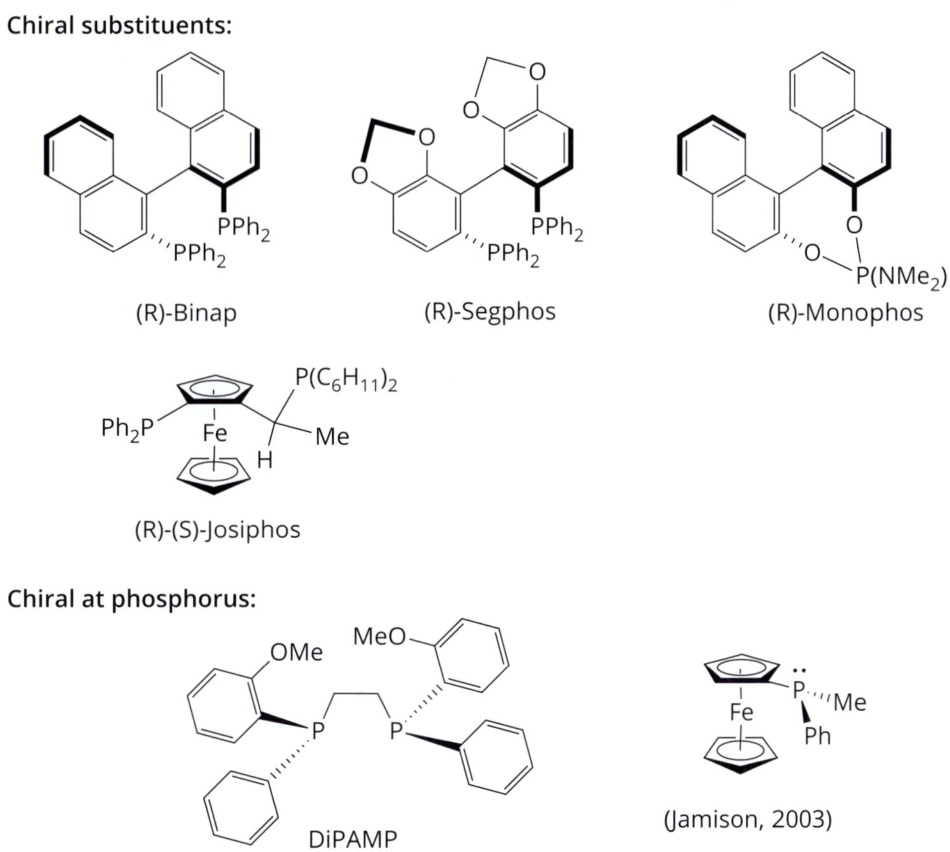

Fig. 11.28 Some selected examples of chiral phosphines.

11.3.1.3 Classifying Ligands: Tolman Electronic Parameter (TEP) and Tolman Cone Angle (TCA)[93–96] (Advanced Topic)

Because of the massive number of phosphines that are available, it is necessary to have some chemical descriptors that allow the classification of the ligands. Some obvious properties that one would want to categorize phosphines are their donor abilities (perhaps separated into σ-donor ability and π-acceptor ability) and sterics. One can imagine many different methods for classification, but addition of every parameter leads to greater and greater complexity in the classification. As a result, generally speaking, the fewer the parameters the better, as long as the properties and reactivity are well described.

There has been almost continual effort in parameterization of phosphines for at least half a century, and new parameterization schemes are still being published. Some of the more basic methods will be discussed here. By far the most common method for electronic description of ligands is to examine CO stretching frequencies in metal carbonyls as phosphines (or related ligands for that matter) are changed in, for example, $Ni(CO)_3(PR_3)$. As the metal center becomes more electron-rich, the C–O vibrational energy measured in wavenumbers (cm^{-1}) drops due to increased backbonding from the ligand. This method has been used since at least the 1950s for evaluation of phosphine ligands by Nyholm and Short, Cotton, Abel and Stone, and others.[97–100]

Perhaps the key paper for the field of phosphine parameterization was published in 1977 by Tolman, in which he summarized both electronic and steric effects for many common phosphorus ligands.[96] In the review are the Tolman electronic parameter (TEP) and the Tolman cone angle (TCA).

The TEP is based on the use of nickel-based complexes by which one carbonyl is replaced with the phosphine, L, of interest, $Ni(CO)_3L$. The symmetric, A_1, stretching frequency of the carbonyls is used as a gauge of the donor ability of L. The infrared spectra are typically taken in dichloromethane or hexane, and can be somewhat solvent dependent. The donor abilities of phosphine ligands are then expressed in wavenumbers based on the carbonyl frequencies, with more donating ligands leading to smaller values. More accurate values were published in 1984 by Bartik and coworkers using FTIR, a selection of which is shown in Table 11.4.[101] In some charts, the TEP is expressed as the difference from the very donating PBu^t_3; in other words, the ligand PBu^t_3 value of 2056.1 cm^{-1} is scaled to zero.

One potential issue with the TEP is that experimentally the measured A_1 vibration in the $Ni(CO)_3$ fragment could couple with vibrations in the ligand L, the Ni–L vibration, or

Table 11.4 Selection of Tolman electronic parameters organized from most to least donating and their pK_a values.

Ligand	TEP (cm^{-1})[a]	χ[b]	pK_a
P(*t*-Bu)$_3$	2056.1	0	11.40
PCy$_3$	2057.5	1.4	9.65
P(i-Pr)$_3$	2059.6	3.45	9.3
PBu$_3$	2061.4	5.25	8.43
PPr$_3$	2061.5	5.4	8.64
P(NMe$_2$)$_3$	2062.1	5.95	3.7
PEt$_3$	2062.4	6.3	8.69
P(mesityl)$_3$	2064.2	8.1	6.0
PMe$_3$	2064.7	8.55	8.65
PMe$_2$Ph	2066.7	10.6	6.50
P(*o*-tolyl)$_3$	2066.8	10.65	3.08
PMePh$_2$	2068.2	12.1	4.77
P(O-t-Bu)$_3$	2069.1	12.95	4.5
PPh$_3$	2069.4	13.25	2.73
PPh(OMe)$_2$	2075.6	19.45	2.64
P(OEt)$_3$	2077.7	21.6	3.31
P(OMe)$_3$	2080.2	24.1	2.60
PPh(OPh)$_2$	2080.2	24.1	−0.4
PH$_3$	2081.0	24.9	−4.3
P(OPh)$_3$	2086.3	30.2	−2.0
P{O-2-(*t*-Bu)C$_6$H$_4$}$_3$	2086.6	30.5	−1.0
P(C$_6$F$_5$)$_3$	2089.7	33.6	
PCl$_3$	2104.1	48	−13
PF$_3$	2110.7	54.6	−9
P(CF$_3$)$_3$	2114.9	58.8	

[a] Values for TEP from Tolman along with Bartik and coworkers' data.[95–96, 101]
[b] Scaled value of the TEP with PBu$^t_3$ used as zero.

modes internal to L. This will alter the nature of the vibration under investigation somewhat, from ligand to ligand. Computationally, one can calculate the vibrational frequency and isolate the Ni(CO)$_3$ fragment in an electronic field equivalent to L, removing this coupling computationally.[102–103]

An alternative experimental method for the determination of donor properties is to use the pK_a values for the phosphines, and some of these are also listed in Table 11.4. Obviously, pK_a values can only tell of the σ-donor ability of the phosphine, and in conjunction with a method such as TEP it can allow separation of σ- and π-donor abilities.

Electronic parameters are now available for many phosphines from both experimental and computational studies. For more details, see one of the many of interesting reviews on this topic.[93–95, 104–106]

The steric effects of phosphines have been studied as much as, if not more than, the electronic effects.[95, 106–107] The first and probably still most used scale is the Tolman cone angle (TCA or θ).[96] The cone angle of a phosphine is the smallest angle of a cylindrical cone required to fully enclose the van der Waals radii of the atoms in the ligand with the Ni–P distance of 2.28 Å with an angle subtended at nickel (Fig. 11.29).

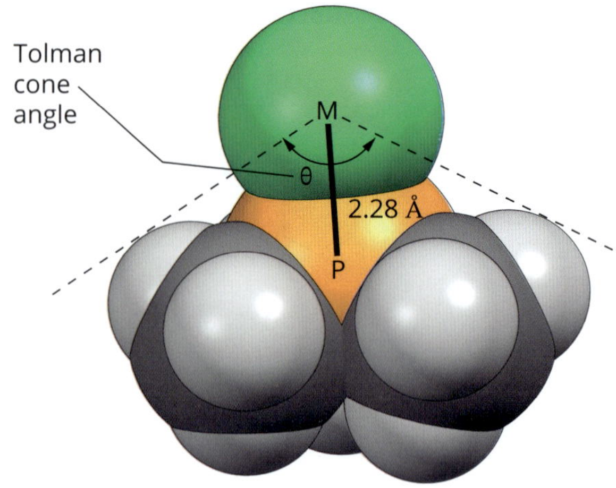

Fig. 11.29 Graphical description of the Tolman cone angle (TCA, θ). The cone angles were originally found for space-filling models of the phosphines where the M–P bond was set to 2.28 Å. The cone angle was then measured with a special ruler.

Tolman's method utilized space-filling models of three-fold symmetric phosphines, PR$_3$. The metal–phosphorus distance was set to 2.28 Å, and the angle between the outermost angular reaches of the van der Waals radii of the atoms in the ligand were measured. For phosphines that were not three-fold symmetric, the cone angle was considered to be the weighted average of the symmetrical phosphines, Eq. 11.1, where θ is the unsymmetrical phosphine cone angle, θ_i = the cone angle of the three-fold symmetric phosphine with the same substituent, and the sum runs from 1 to 3 for the three substituents on the phosphine.

$$\theta = \frac{1}{3}\sum_{i=1}^{3}\theta_i$$
$$\theta = \frac{1}{3}\sum_{i=1}^{3}\theta_i = \frac{1}{3}(145 + 170 + 170) = 162^\circ \qquad \textbf{11.1}$$

To find the cone angle for P(Ph)Cy$_2$, where Cy = cyclohexyl, the cone angles for PPh$_3$ (θ_{PPh_3} = 145°) and PCy$_3$ (θ_{PCy_3} = 170°) are used. The value for PCy$_3$ will appear twice, as two of these groups are in the unsymmetrical ligand to be calculated.

Tolman's cone angles have been modified computationally in numerous ways over the years, and these new methods are worth examination in cases where correlation problems arise.[108–109] However, the original values were used for decades, with great success. A large listing of cone angles can be found in the review by Dias and coworkers.[95]

Cone angle can, in principle, be used as the steric parameter for any ligand. Amine cone angles have been described by Seligson and Trogler, for example.[110]

There are several other notable steric parameters that have been developed, though there are too many to cover in detail here. Brown's ligand repulsive energy (E_R) is a versatile system for steric evaluation with reasonably good correlation with cone angle.[111] Another method, which employs the amount of ligand volume within the first coordination sphere of the metal (often defined as being around 3.5 Å), is percent buried volume (%V_{bur}), developed by Clavier and coworkers and often applied often to *N*-heterocyclic carbenes.[112–113]

It should be noted that steric parameters of this type have been of great utility in understanding reactivity and in the classification of ligands for catalyst development. However, they are quite naturally limited as descriptors. The only 3-dimensional object that one can truly describe with a single parameter is a perfect sphere, and the methods discussed here use a single parameter to describe quite complex structures in some cases. Nevertheless, these parameters will continue to be of tremendous aid to the chemist in the future as simple descriptors of complex shapes and their impact on chemistry. However, one must be cautious, as a lack of correlation between reactivity and a single parameter steric model may say little of value about the importance of sterics in a reaction in some cases.

11.3.1.4 Phosphine Ligands and Modelling Reactivity (Advanced Topic)

The parameters discussed previously for phosphines are often used qualitatively to decide on directions for catalyst system exploration. However, they can in some cases be used quantitatively to explain data such as reaction rate constants. In other words, one can model the reactivity of the system mathematically by using the properties of the ancillary ligands.

The simplest such model perhaps is due to Tolman, who suggested that the property of the complex or reaction could be described by using Eq. 11.2, where *a*, *b*, and *c* are coefficients, χ = TEP or other electronic parameter, and θ = TCA or other steric parameter. The constants are found through a least-squares fit to the data.

$$\text{Property} = a + b(\chi) + c(\theta) \qquad \textbf{11.2}$$

An example of a reaction system in which this equation can be applied is the reaction of $Ir_4(CO)_{12}$ with a phosphine (PR_3) to form $Ir_4(CO)_{11}(PR_3)$. The log of the rate constant for the reaction can be examined versus the steric or electronic parameter independently. In this case, the electronic parameter used was the pK_a of the phosphine, and the steric parameter was the Tolman cone angle (TCA). If one plots TCA (θ) vs (log k + 0.31 × pK_a), the line shown in Fig. 11.30 is obtained, and the data used are shown in the table internal to the figure. The model fits quite well with R^2 for the line shown of 0.96. If π-bonding effects are included, the fit improves even further. The correlation demonstrates a direct relationship between reaction rate and the phosphine size and σ-donor ability.

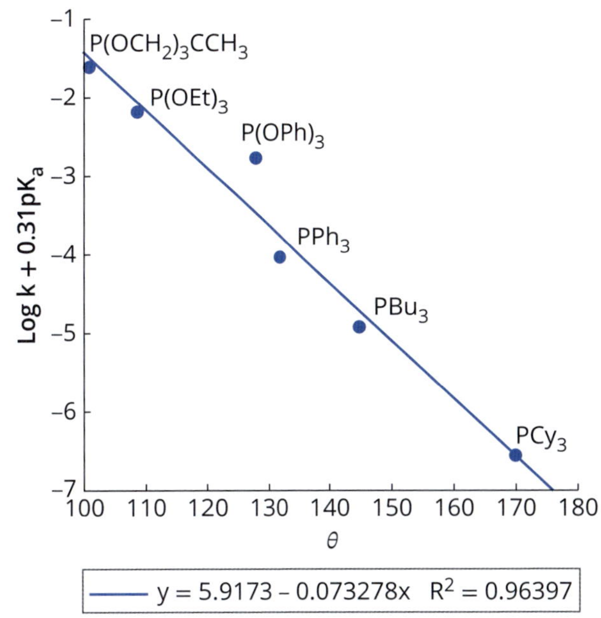

Ligand	log k	pK_a	TCA (°)
PCy_3	−3.567	9.65	170
PBu_3	−1.41	8.43	132
PPh_3	−4.055	2.73	145
$P(OEt)_3$	−1.147	3.31	109
$P(OPh)_3$	−3.382	−2	128
$P(OCH_2)_3CCH_3$	−1.062	1.74	101

Fig. 11.30 Plot of Tolman cone angle (TCA, θ) vs (log k + 0.31 × pK_a), showing good correlation between the log of the rate constant for the reaction $Ir_4(CO)_{12} + PR_3 \rightarrow Ir_4(CO)_{11}(PR_3) + CO$ and a phosphine steric parameter (TCA) and phosphine σ-donor ability as measured by using pK_a.

One highly successful system that has been developed for the quantitative analysis of property and reaction rate variance with different phosphines (and related ligands such as arsines and amines) has been dubbed *quantitative analysis of ligand effects* (QALE). Instead of the simple equation based on one electronic parameter and one steric parameter used in the previous example, it has been found that a somewhat more complex model is required to more generally fit chemical systems involving phosphines.[105,114,115]

11.3.2 Amines and Other Neutral Nitrogen Compounds as Ligands

Amines are common ligands for transition metal systems, though are perhaps not used in as great a variety. The most commonly used monodentate amines contain simple alkyl, aryl, and hydrogen, such as NMe_3 and NH_3. In addition, pyridines and imines are often employed ligands, especially as parts of chelates.

11.3.2.1 Alkyl and Aryl Amines

While many amines can be strong σ-donors, they are generally not regarded as π-acceptors, due to the σ*-orbitals being largely based on, for example, carbon in NMe_3 due to the lower electronegativity of the substituents relative to nitrogen (Section 9.6.2). Similarly, it is often assumed that NH_3 has a π-effect of zero in crystal field theory and the angular overlap model discussed in Chapter 15.

One can confirm the much lower π-acceptor ability of NH_3 relative to the heavier congeners by using NBO calculations (Fig. 11.31). Taking a typical late transition-metal species, in this case $\{PtCl_3(NH_3)\}^-$, NBO can be used to calculate all the Pt lone pair to N–H σ*-interactions. The sum of all these interactions from the second-order perturbation analysis in NBO suggests that these interactions are only about 4 kcal/mol in strength. The same calculations with the heavier congeners PH_3 and AsH_3 produces the Pt lone pair to E–H σ*-interactions as being a much more substantial ~12 kcal/mol.[116]

Similar to phosphines, the cone angles for amines have been calculated (Table 11.5). Seligson and Trogler provided a fairly large listing of cone angles for amines, where a metal–nitrogen distance of 2.2 Å was used.[110]

Because amines are generally not regarded as π-acceptors, σ-only dependent pK_a values are often employed as a measure of their donor ability.

Like phosphines, amines are commonly employed as part of multidentate ligands. Some common prototypical ligands are shown in Fig. 11.32.

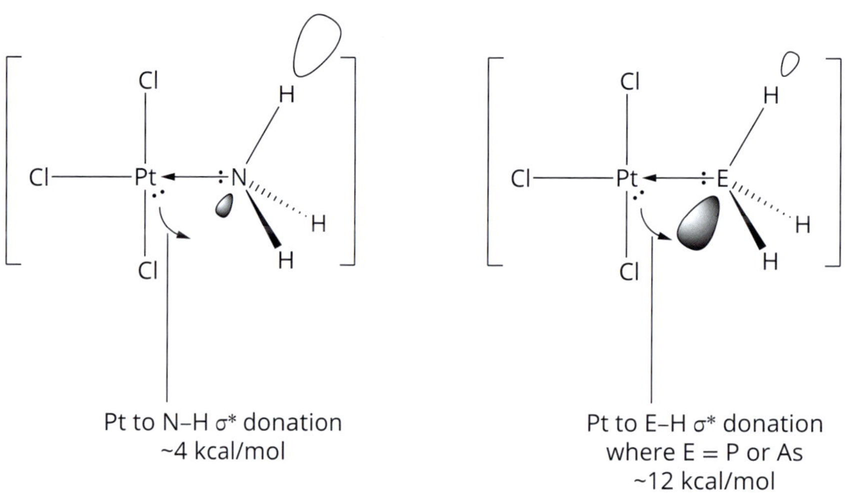

Fig. 11.31 Negative hyperconjugation in $\{PtCl_3(NH_3)\}^-$, where E = N, P, and As, and the calculated Pt to E–H σ* energies.[116]

Table 11.5 Cone angles for some amines with M–N = 2.2 Å.[110]

Amine	Cone Angle (θ)	Amine	Cone Angle (θ)
NH_3	94	$NHPh_2$	136
NHR (R = Me, Et, Pr, *i*-Pr, *i*-Bu, CH_2Ph, NH_2, CH_2-*t*-Bu)	106	NH(*i*-Pr)$_2$	137
NH_2Ph	111	NH(*i*-Bu)$_2$	138
NH_2-*s*-Bu	113	$NH(CH_2Ph)_2$, NMe_2Ph	140
NH_2Cy	115	$NMeEt_2$	145
$NHMe_2$, NHMeEt	119	NEt_3	150
Piperidine	121	NH(*s*-Bu)$_2$	158
NH_2-*t*-Bu	123	NPr_3	160
$NHEt_2$	125	NPh_3	166
NHMePh, NHEtPh	126	NEt_2Ph	170
NH_2(1-adamantyl), $NHPr_2$	127	$N(CH_2Ph)_3$	210
$NHCy_2$	133	N(*i*-Pr)$_3$	220

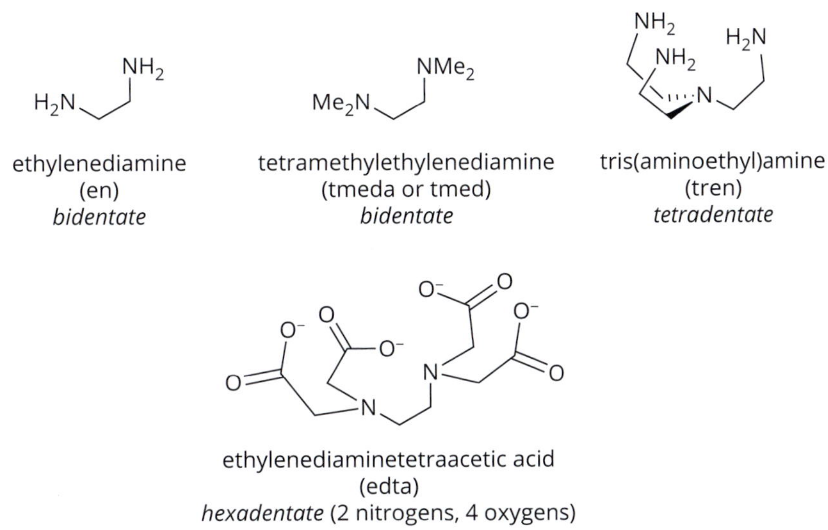

Fig. 11.32 Some common multidentate amines, their abbreviations, and their common denticity.

11.3.2.2 Bipyridine and Related Ligands

Several ligand types based on pyridines are commonly employed, and pyridine itself is a very common ligand across the transition series. There is evidence that it can be either a π-donor or π-acceptor, depending on the metal center, in addition to being a good σ-donor.

Linked pyridines, like 2,2′-bipyridine (bipy or bpy), 2,6-bis(2′-pyridyl)pyridine (terpy or tpy), and their derivatives, are very commonly employed ligands (Fig. 11.33). Phenanthroline is another common, closely-related ligand. All of these are good electron-acceptors, storing the electron in their delocalized π-systems—a property that plays a key role in their photochemical and photophysical properties.

11.3.3 N-Heterocyclic Carbenes (NHCs) as Ligands[117–119]

Often considered an alternative to phosphines, *N*-heterocyclic carbenes (NHCs, see Section 9.5) are often drawn as singlet carbenes. A commonly encountered derivative involving two mesityl groups on the nitrogens is shown in the top portion of Fig. 11.34. Natural resonance theory on a simple NHC was discussed in Chapter 9, and Fig. 11.34 presents calculations on singlet diaminocarbene. Similar to the previous calculations on an imidazole-based NHC, these suggest an ylide structure for the ground state, and the carbene structure was not listed as a contributor to the electronic structure, though this does not necessarily preclude it from participating once the ligand is on the metal center.[120]

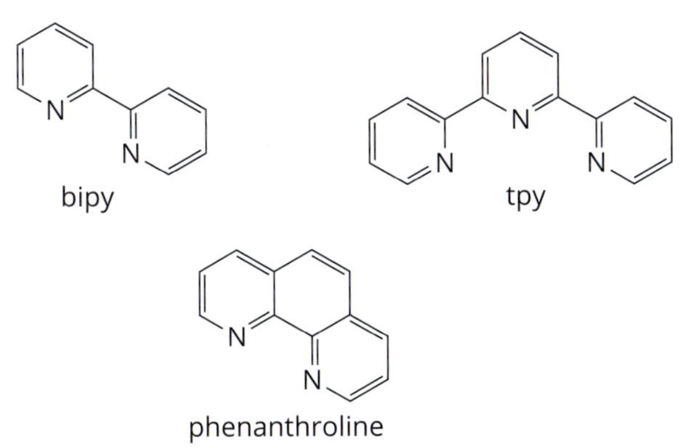

Fig. 11.33 Structures of 2,2′-bipyridine (bipy or bpy), 2,6-bis(2′-pyridyl)pyridine (terpy or tpy), and phenanthroline.

11.3.3.1 NHCs: History and Common Usage

We have discussed extensively the types of bonding found in NHCs, which is similar to phosphines with σ-donor and π-acceptor orbitals. It appears that NHCs are typically stronger donors than phosphines and are typically weak π-acceptors.

Unfortunately, it has become the norm by some researchers using NHCs to simply draw a line between the metal and the attached carbon of the ligand, which, by everything students have been taught about every other chemical system, implies that there is a hydrogen on the carbon that is not actually present in this system. While the authors can state from personal experience that this confuses students learning the field, at this stage there are probably hundreds of papers in print which use this nonstandard presentation, and the reader should be aware of this oddity. A clearer presentation would be to use a carbon lone pair with an arrow, using the carbene or more accurately the ylide structure (Fig. 11.35).

The electronic structure of NHC–metal complexes is an extreme example of that found in Fischer carbenes (Chapter 12), and it would not be entirely inappropriate to draw a metal–carbon double bond in the presentation as well. However, π-bonding to NHCs is probably more limited than in typical Fischer carbenes, due to the very strong π-donors on C_α.

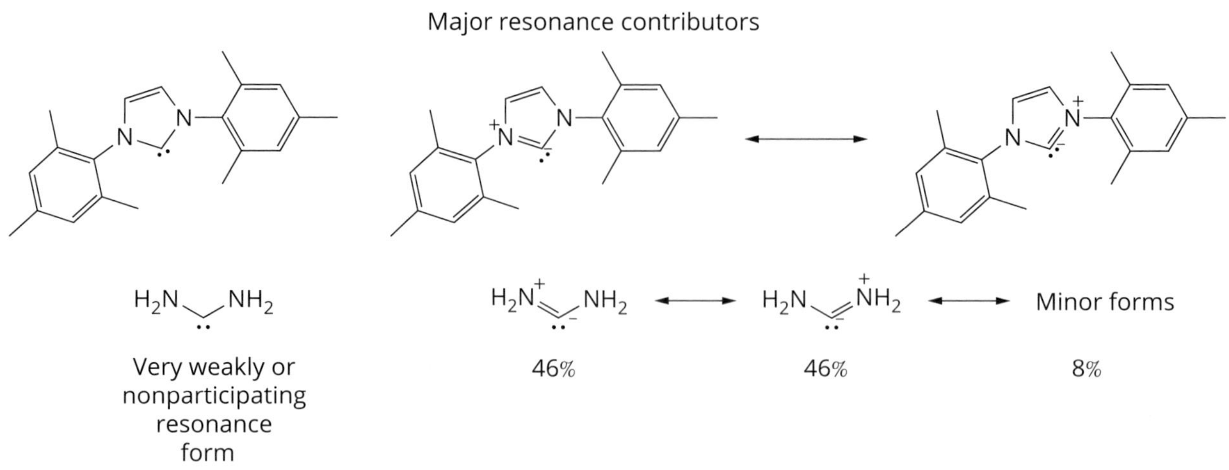

Fig. 11.34 (*top*) A common NHC as it is usually drawn as the free ligand is shown on the left, with its two ylide forms that are its most probably resonance contributors. (*bottom*) Calculations on the model $C(NH_2)_2$ suggest that these ylide forms are by far the largest contributors to the ground-state structure.[121] The carbene was not listed as a contributor, and the minor forms all involved bond-breaking (ionization).

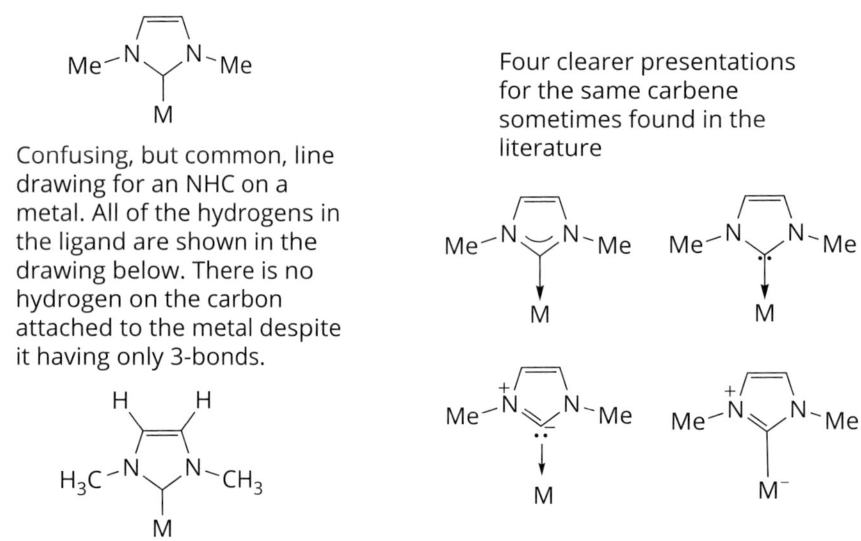

Fig. 11.35 Line drawings for NHCs on a metal center. (*left*) A somewhat confusing, but often encountered line drawing where there is a simple bond between the metal and the carbon. The carbon does not have a hydrogen, despite having only three bonds drawn. At the bottom of the figure is the same line drawing with all of the hydrogens included. (*right*) Four clearer drawings of the NHC bond to the metal.

The first metal–NHC complexes were prepared independently in 1968 by the groups of Öfele[122] and Wanzlick (Fig. 11.36, top and middle),[123] who started from the imidazolium and used a basic ligand on the metal to deprotonate to form the NHC—a common procedure, although sometimes an external base (NaH, $NaOBu^t$, NEt_3, and so on) is added.

Lappert and coworkers then published a very large series of papers in the 1970s and early 1980s in which many metal–NHC complexes are reported. In these papers, electron-rich ene-tetramines were used, dimers of the incipient carbenes. Lappert published reviews on his contributions using electron-rich olefins to generate NHC complexes (Fig. 11.36, bottom).[124–125]

Inexplicably, the field lay dormant until 1991, when Arduengo and coworkers published that increasing the steric bulk around the ylide carbon led to isolable compounds that did not dimerize.[126] After the discovery of these isolable carbenes, a flurry of activity ensued in metal–NHC studies.

Another method for the preparation of metal–NHC complexes is to use a "protected" carbene. For example, addition of NaOMe to saturated NHCs gives the heterocycles with H and OMe added to the would-be carbene carbon. Thermolysis in the presence of a metal complex gives the metal–NHC. In Fig. 11.37 is shown the synthesis of a "second-generation" Grubbs olefin metathesis catalyst using a HOMe-protected carbene.[127–128]

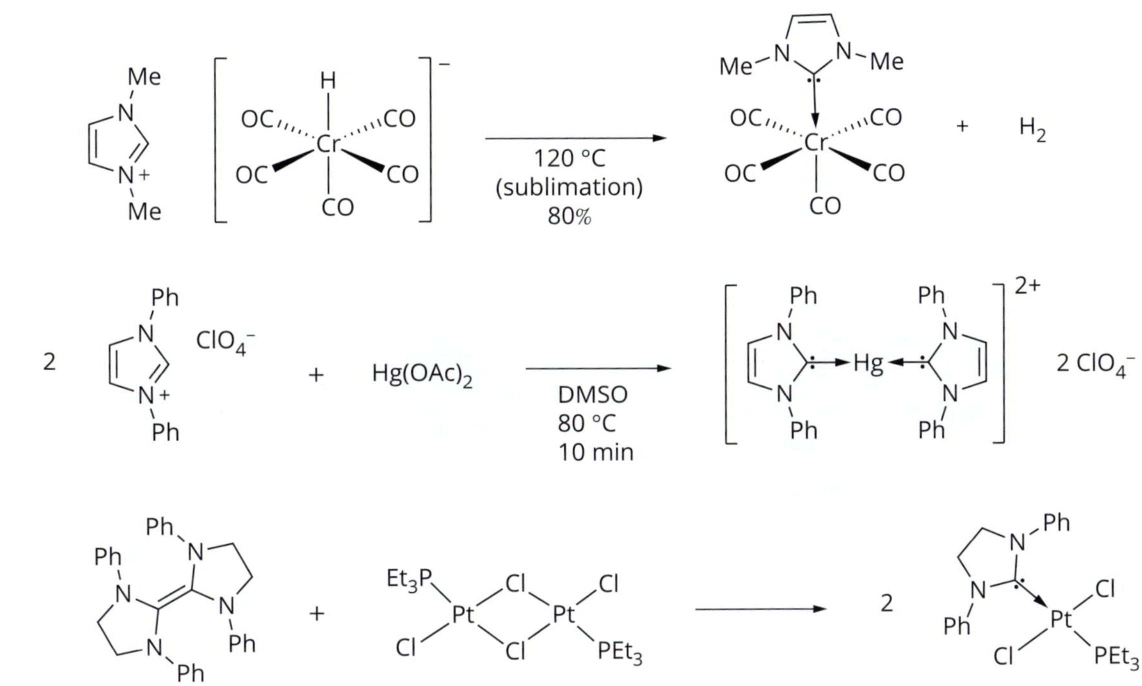

Fig. 11.36 (*top*) Öfele, (*middle*) Wanzlick, and (*bottom*) Lappert's metal–NHC syntheses.

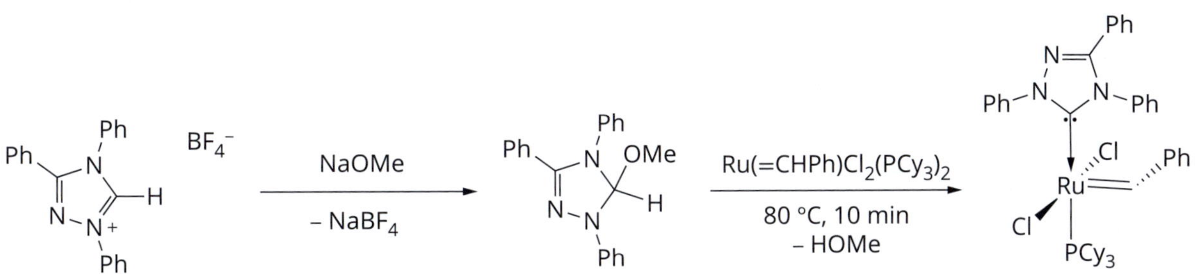

Fig. 11.37 Synthesis of a "second-generation" Grubbs's catalyst from the "first-generation" derivative using a protected carbene.[127–128]

11.3.3.2 Parameterizing NHC Ligands (Advanced Topic)

NHCs have been subjected to similar classification schemes as phosphines. From a steric perspective, the Tolman cone angle (TCA) seems less appropriate for these flatter, but still quite bulky systems. The Tolman electronic parameter is often used for these systems; however, some surrogates of rhodium or iridium carbonyls are quite popular as they linearly correlate with stretching frequencies of $LNi(CO)_3$ but do not require use of quite toxic $Ni(CO)_4$.[118]

For steric characterization, TCA can be used; however, percent buried volume ($\%V_{bur}$), popularized by Clavier, Nolan, and Cavallo, has also been used to great effectiveness with the class.[112–113, 129–130] In steric evaluation using $\%V_{bur}$, it is assumed that there is a specific radius from the metal center, approximating the first coordination sphere, where it is important to evaluate steric effects. Groups outside of this sphere have little or no steric influence. Then, the amount of the first coordination sphere occupied, as a percentage, is calculated for the ligand.[118]

The typical value for the sphere radius used to represent the first coordination sphere of the metal is 3.5 Å—an obviously arbitrary but apparently useful value. Fig. 11.38 shows an example of a molecule, ClAg(NHC), where the 3.5 Å sphere is shown as the transparent red sphere. The $\%V_{bur}$ is the percentage volume of the sphere occupied by the ligand in question.

As far as donor abilities of NHCs, they are generally regarded as stronger σ-donors than phosphines, and in general this appears to be the case. The pK_a values of typical protonated NHCs vary from about 16 to 26 in H_2O,[118] while aqueous pK_a values of phosphoniums are from about 12 to negative

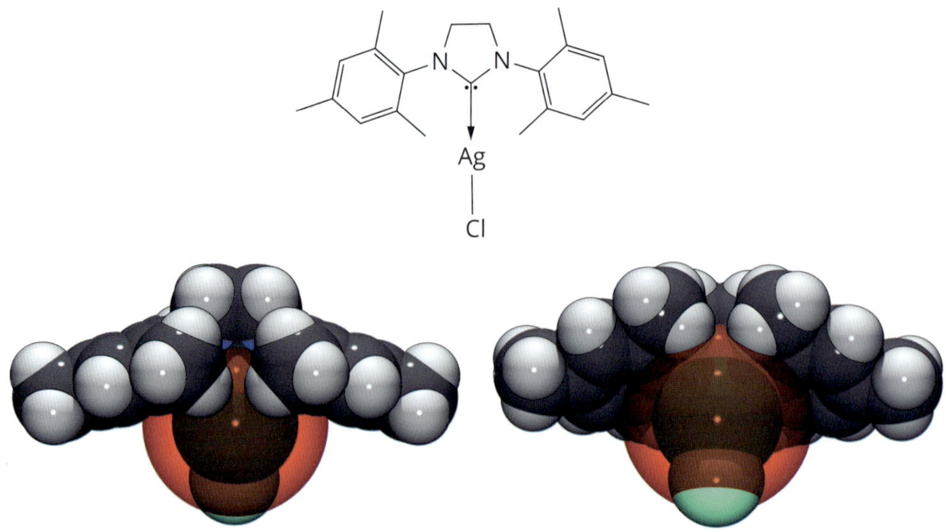

Fig. 11.38 Pictorial explanation of %V_{bur}. (*top*) A line drawing of the molecule being analyzed. (*bottom*) Two views (one orthogonal to the Cl–Ag–C vector and one view 45° from this vector) of the spacefilling model of the molecule with a red transparent sphere centered at the metal with a radius of 3.5 Å. The %V_{bur} is the percentage of the red sphere around the metal center occupied by the ligand of interest.

values.[95] One could argue as to whether the values in aqueous media are the most appropriate, but gas-phase proton affinities (PA) also suggest that a prototypical NHC (the NHC used was derived from *N*-methyl-*N*′-methylimidizolium) has a PA about 10 kcal/mol larger than PCy_3, again consistent with an NHC being a stronger σ-donor.[131]

There is significant evidence building that NHCs can be either π-donors or π-acceptors.[118] One can use the Tolman electronic parameter to evaluate the ligands. A diverse selection of NHC ligands is shown in Table 11.6, along with their estimated TEP value. For reference, the generally accepted TEP value for PBu^t_3, a very donating phosphine, is 2056.1 cm^{-1}, and values lower than this signify a stronger overall donor.[95]

Interestingly, the saturated NHC rings, such as NHC-4, generally have higher TEP values than the unsaturated systems, such as NHC-3, suggesting that the saturated systems are weaker donors. It is thought that the reason for this is that the saturated, less aromatic systems are better π-acceptors, which lowers their overall donor ability.

For some common ligands, like NMe_2 and OBu^t, the number of electrons donated can seem rather ambiguous, depending on how many lone pairs of π-symmetry on the ligand participate in bonding to the metal. In general, if the metal has an empty acceptor orbital of appropriate symmetry for one of these π-lone pairs, it should probably be regarded as being donated to the metal center in a dative sense to some extent. Resonance forms with donation of multiple lone pairs are unlikely to be large contributors to the ground state, as positive charges are generally being added to the donor, which is often electronegative, in those resonance forms as more lone pairs are donated. For example, if an alkoxide donates one lone pair, the electronegative oxygen picks up one formal positive charge; if two lone pairs are donated a 2+ charge on the oxygen is obtained. For alkoxides and (even more so) amidos, π-bonding is important but will be limited by the usual factors that limit dative interactions (Chapter 3).

In later sections, further discussion will be presented for a few classes of ligands discussed in Table 11.7, such as carboxylates and scorpionates. Not discussed are the heavier congeners of many of these common ligands, such as thiolates (^-SR) and phosphides ($^-PR_2$), which are less common but are well known.

11.4 Some Common Monoanionic Ligands

There is a vast array of monoanionic ligands that are commonly employed in inorganic chemistry. Some of the typical monoanionic ligands shown in Table 11.7 are based on the element bound to the metal. In some cases, the ligands will have multiple possible bonding modes; in such cases, the electron counts with increasing number of donor atoms are also given.

11.4.1 Carboxylates and Related Systems

Deprotonated carboxylic acids are quite common ligands for metals in inorganic chemistry in general and bioinorganic chemistry in particular. The ligands can bind either through one oxygen, a κ^1-form, or through both oxygens of the carbonyl carbon, a κ^2-form. At the limit in the κ^2-form, the two C–O bond distances will become equivalent, as may the two M–O distances.

Table 11.6 Some examples of the variety of NHC structures available and their Tolman electronic parameter.[118] In parentheses are the TEP values scaled to P^tBu_3 (2056.1 cm^{-1}) as 0 cm^{-1}; negative values indicate ligands even more donating than this very donating phosphine.

NHC	TEP (cm^{-1})	NHC	TEP (cm^{-1})
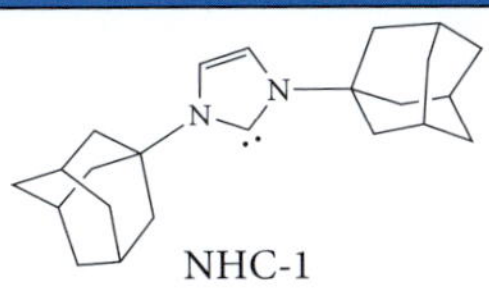 NHC-1	2048.3 (−7.8)	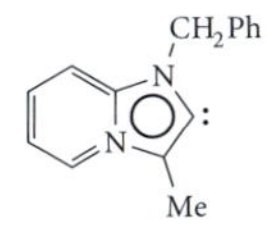 NHC-11	2029.2 (−26.9)
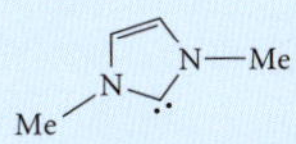 NHC-2	2051.2 (−4.9)	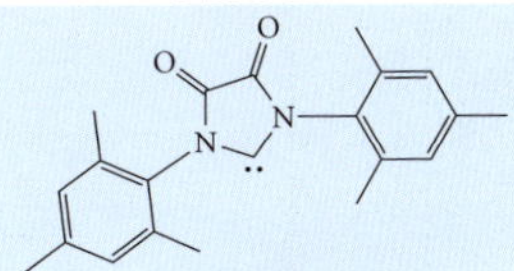 NHC-12	2069.0 (+12.9)
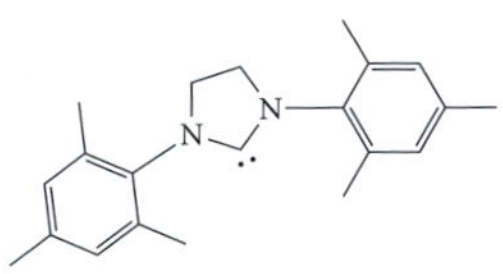 NHC-3	2049.6 (−6.5)	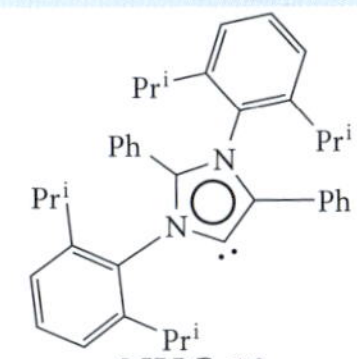 NHC-13	2038.4 (−17.7)
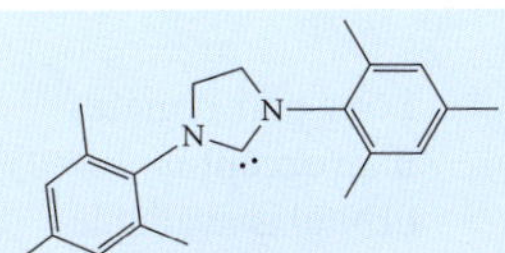 NHC-4	2050.8 (−5.3)	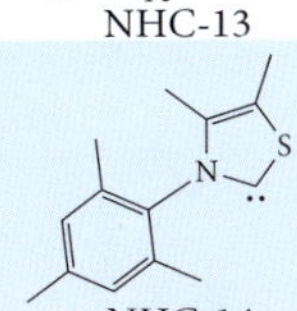 NHC-14	2053.7 (−4.7)
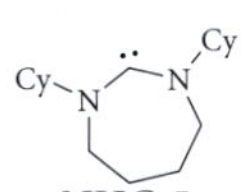 NHC-5	2043.1 (−13)	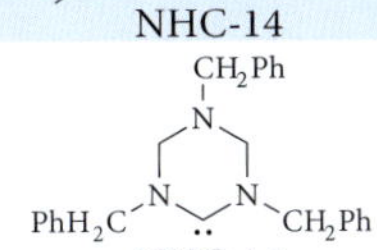 NHC-15	2051.4 (−4.7)
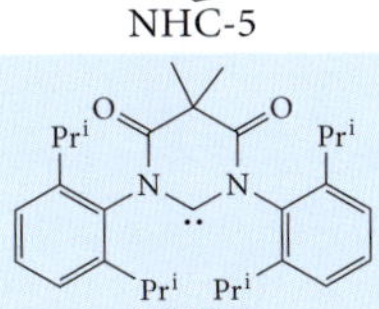 NHC-6	2057.5 (+1.4)	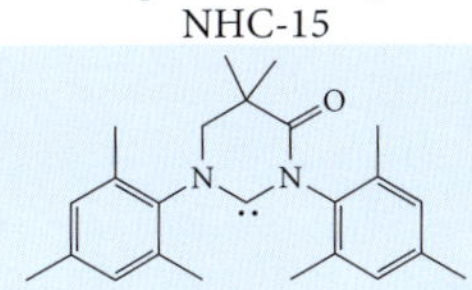 NHC-16	2049.9 (−6.2)
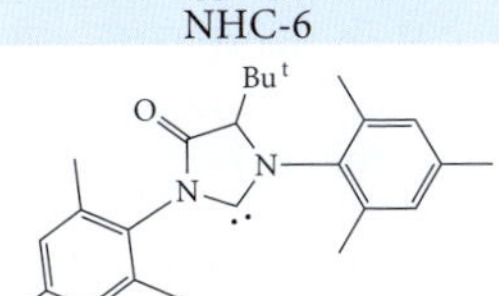 NHC-7	2057.2 (+1.1)	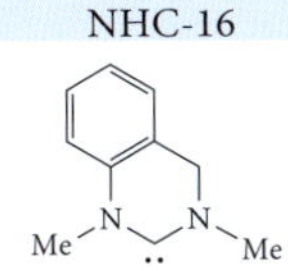 NHC-17	2050.3 (−5.8)
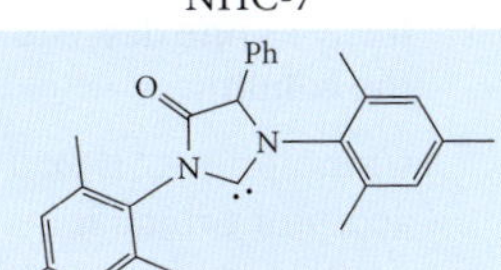 NHC-8	2057.7 (+1.6)	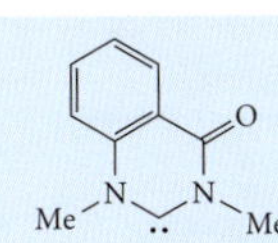 NHC-18	2059.2 (+3.1)
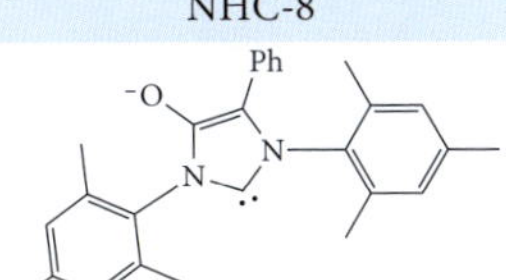 NHC-9	2043.7 (−12.4)	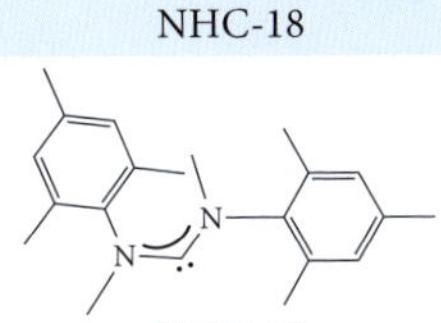 NHC-19	2044.4 (−11.7)
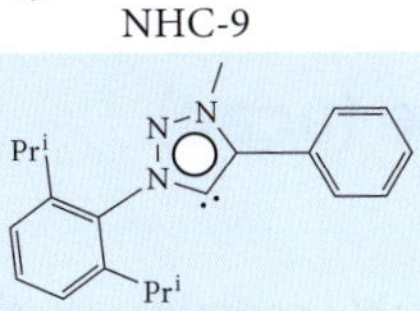 NHC-10	2046.1 (−10.0)	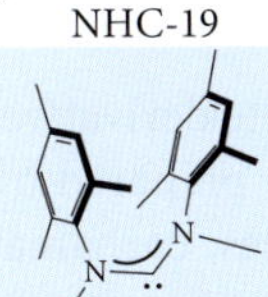 NHC-20	2047.8 (−8.3)

Table 11.7 Common name, structure, and electron counting for some selected monoanionic ligands.

Class Name (abbreviation)	Structure	Electron count Charged method	Electron count Neutral method	Notes
alkyls neopentyl (Np) neophyl (Nph) trimethylsilylmethyl	-R 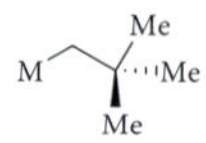 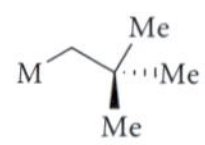 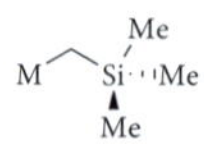	2	1	Other simple alkyls are commonly employed, such as methyl.
aryls (Ar) phenyl (Ph) mesityl	-Ar 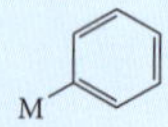 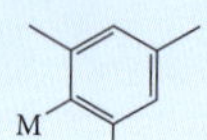	2	1	Two common derivatives shown.
vinyl	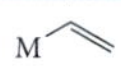	2	1	
alkynyl or acetylide	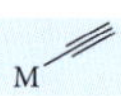	2	1	
amide dimethylamide diisopropylamide hexamethyldisilazide	$-NR_2$ 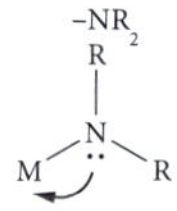 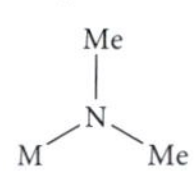 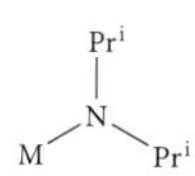 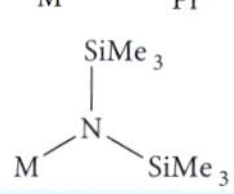	2 or 4	1 or 3	Lone pair donates if an acceptor orbital is available.
alkoxide *tert*-butoxide	-OR 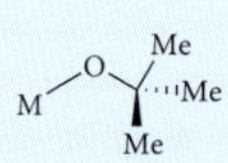	2 or 4 or 6	1 or 3 or 5	Lone pair(s) donate if acceptor orbital(s) are available.
halides	-X (F, Cl, Br, I)	2 or 4 or 6	1 or 3 or 5	Lone pair donation can occur but becomes less likely as one descends the column.
carboxylates	-OC (O)R κ^1: 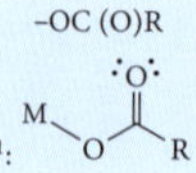 OR κ^2: 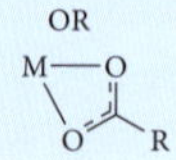	2 for κ^1 or 4 for κ^3	1 for κ^1 or 3 for κ^3	Commonly κ^3 if an acceptor orbital is available on the metal. Many different derivatives (See Section 11.6).
hydride	-H	2	1	

In addition, there is a host of carboxylate derivatives. Carbamates, $^{-}O_2C(NR_2)$, for example, can be prepared by addition of CO_2 to an amide. Like carboxylates, carbamates are most often κ^2, but the distant donor nitrogen can increase the donor ability of the ligand by using resonance forms like those shown in Fig. 11.39. Some other carboxylate-related ligands are also shown in the figure.

11.4.2 Acetylacetonate (acac) and Related Ligands

Acetylacetonate (Fig. 11.40) forms stable complexes with most of the transition metals, including the homoleptic $M(acac)_3$ compounds, many of which are commercially available. In the vast majority of cases, acac acts as a bidentate ligand through

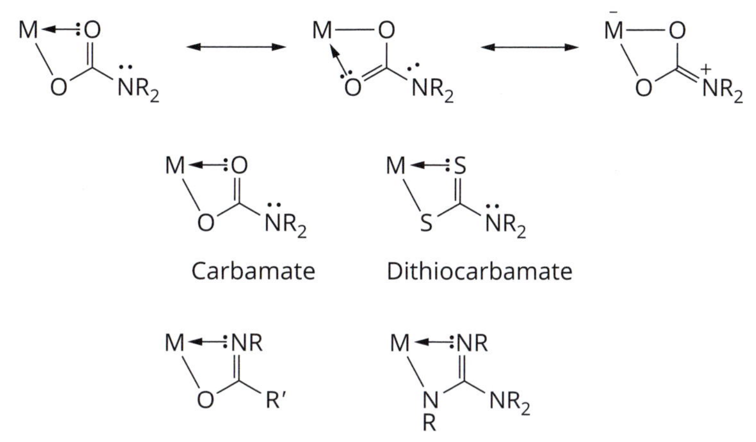

Fig. 11.39 (*top*) Resonance forms in carbamate, including a form where the donor amine causes both oxygens to become covalently bound to the metal. (*bottom*) Some carboxylate-related ligands. In all cases, the ligands can potentially be κ^1 or κ^2 and are monoanionic.

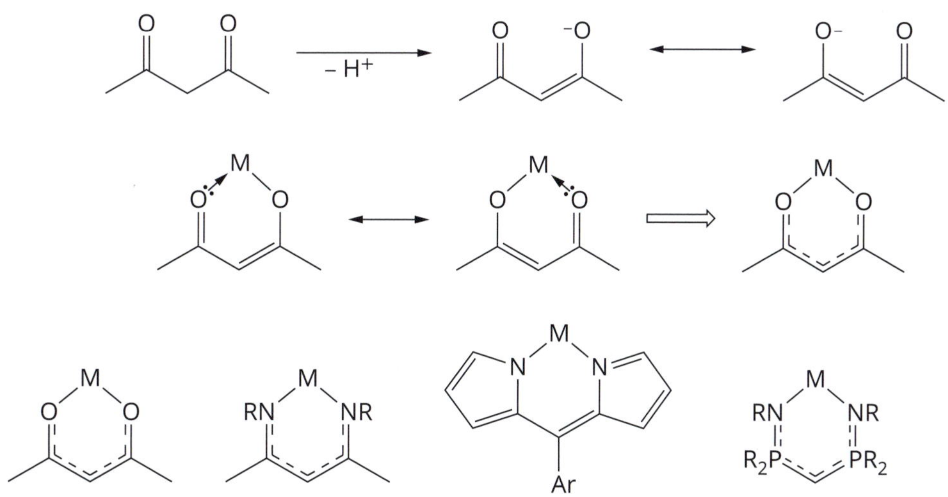

Fig. 11.40 Deprotonation of acetylacetonate at the acidic carbon between the two carbonyl groups produces the acetylacetonate anion, as shown at the top. This acac-anion can act as a ligand to many transition metals and is related to many different monoanionic bidentate ligands like those shown at the bottom.

the two oxygens to form a six-membered metallacycle. One of the oxygens can be thought of as covalently bound, while the other is acting as a dative ligand; naturally, there is resonance between the two structures.

Common derivatives include various alkyl and aryl groups on any of the carbons of the backbone. One or both of the oxygens in acac may be replaced with an NR or NAr group to produce what are sometimes referred to as "nacnac" ligands. There are many related classes of bidentate anionic ligands such as dipyrrolylmethenes, which are somewhat like ½ of a porphyrin ligand, and bis(phosphinimino)methides, which use hypervalent phosphorus.

11.4.3 Macrocyclic Nitrogen-Based Ligands: Porphyrins and Salen

Some of the most common ligands for metal centers in biological systems are derivatives of porphyrin—a large, planar macrocyclic ligand with four nitrogens used to bind the metal. The system is derived from condensation of pyrrole or pyrrole derivatives with aldehydes and is dianionic as a ligand with two neutral donor nitrogens, an 8-electron donor in the charged method and a 6-electron donor in the neutral method.

Porphyrins are such common ligands for iron in biology that iron is regarded as being in one of two categories: heme (bound to a porphyrin derivative), or nonheme (not bound to a porphyrin derivative).

Shown in Fig. 11.41 are a selection of porphyrins and some related ligands. Porphyrin is aromatic, fulfilling the Hückel rule, $4n + 2$, with $n = 4$. Two common, relatively simple porphyrin derivatives, oep and tpp, are also shown. There are many porphyrin-like ligands that are also used, such as the electronically similar pthalocyanine and the trianionic (when deprotonated) ligand corrole.

Porphyrin ligands and their complexes are invariably intensely colored. Most porphyrins are an intense violet or blue color with extraordinarily strong absorptions. In fact, the word "porphyrin" is derived from the Greek *porphura*, meaning purple.

Other ligands similar to porphyrins are "salens" and their many derivatives. The abbreviation "salen" is a contraction of salicylaldehyde and ethylenediamine (Fig. 11.42); however, the term is sometimes used more generally to refer to dianionic, macrocyclic ligands. Also shown in Fig. 11.42 are two other salen-like ligands: one chiral ligand used in asymmetric epoxidation and epoxide ring-opening catalysis[132] and a tetraamido macrocyclic ligand (TAML) developed by Collins and coworkers for peroxide-based water treatment chemistry.[133]

11.4.4 Scorpionates: Tris(pyrazolyl) borates and Related Systems

Another class of common monoanionic ligands is based on the use of a borate-derived chelating ligand. One common variant is tris(pyrazolyl)borate (Fig. 11.43). Bidentate bis(pyrazolyl) borates are also common. These ligands were dubbed "scorpionates" by DuPont scientist Swiatoslaw Trofimenko.[134–135]

For electron-counting purpose, *tris*(pyrazolyl)borate ligands are similar to cyclopentadienyl, covered in the next chapter. In the charged method, the ligands are 6-electron donor anions (top left in Fig. 11.43). In the neutral method they are 5-electron donors with two dative and one covalent bond (top right in Fig. 11.43). The *bis*(pyrazolyl)borate is similar to allyl in its electron counting.

There are many similar ligand types based on borates, such as the *tris*(phosphino)borate shown at the bottom right of Fig. 11.43, popularized by Peters and coworkers.[136] Other variants use different donor ligands, such as NHC ligands.[137]

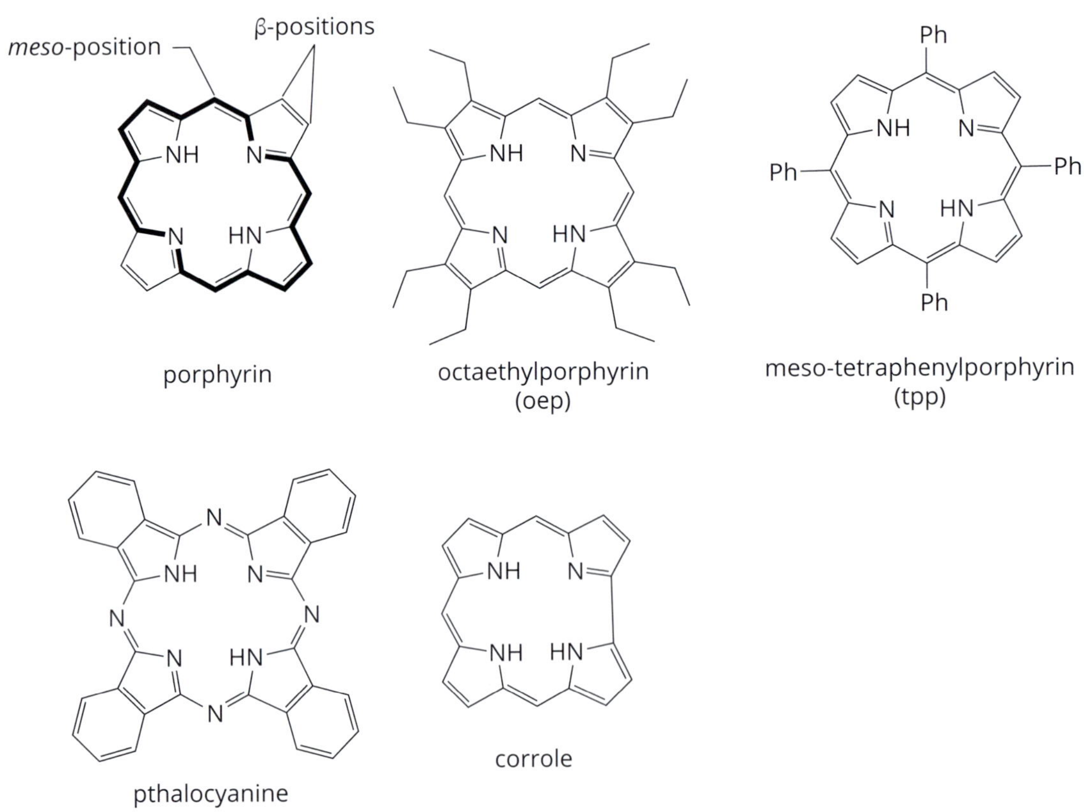

Fig. 11.41 A line drawing of porphyrin, showing the shortest aromatic pathway (dark line in structure at upper left), $4n + 2$ with $n = 4$, and a selection of related macrocyclic ligands. The porphyrin has two positions of common substitution, *meso* and β. Octaethylporphyrin (oep) is a common β-substituted system, and *meso*-tetraphenylporphyrin (tpp) is a common *meso*-substituted system. Pthalocyanine is a related macrocycle that can act as a dianionic ligand, and corrole is a related system that is trianionic when the pyrroles are deprotonated and a metal inserted.

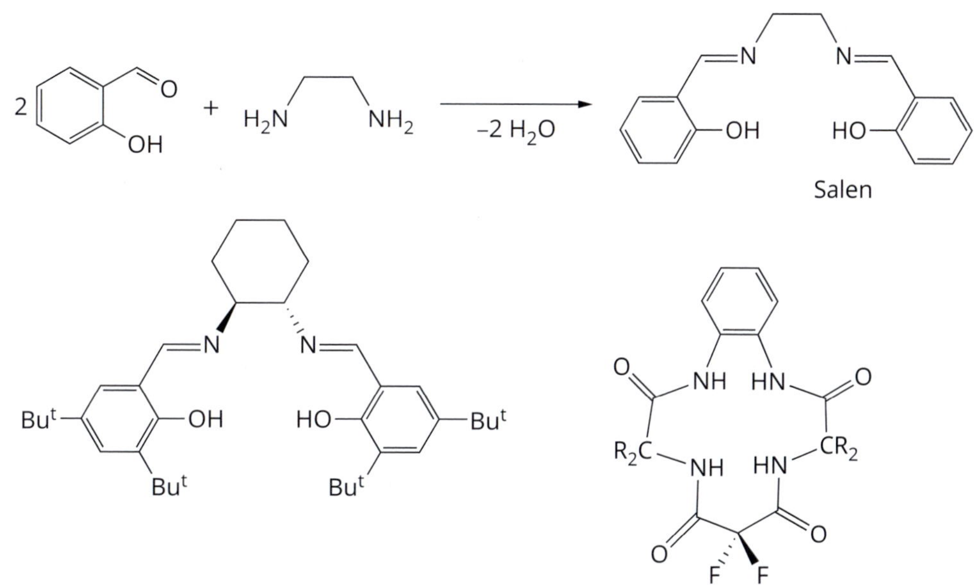

Fig. 11.42 The synthesis of the original salen (salicylaldehyde and ethylenediamine (commonly abbreviated "en")) ligand, and some of the numerous derivatives that have been reported.

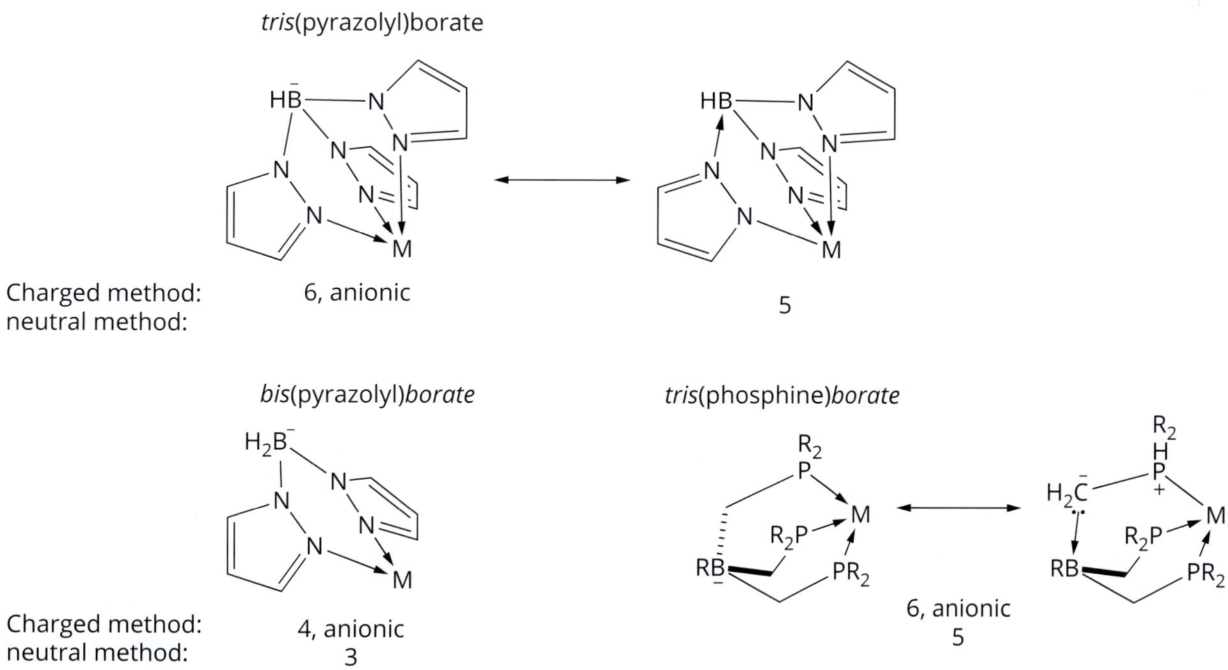

Fig. 11.43 Some common borate-based ligands and their electron counts. The tris(pyrazolyl)borate, often abbreviated Tp, and tris(phosphine)borate are similar to cyclopentadienyl (Cp) in their electronic structure. The bis(pyrazolyl)borate is similar to allyl.

11.5 Hydrogen as a Ligand: Hydride and Dihydrogen

From an electron-counting perspective, hydride ligands are as simple as might be expected: 1-electron donors for the neutral method and 2-electron donors for the charged method. This simplicity in electron-counting hides a great many complexities in hydride's reactivity.

Although all M–H ligands attached to metals are generally referred to as "hydrides", the ligands can vary from being strongly acidic to strongly hydridic (basic), depending on the metal center and coligands. In general, electronegative metal centers with electron-withdrawing ligands that can support the conjugate base's formation are acidic, and complexes with electron-donating ligands, especially of the earlier transition metals, are usually hydridic.[138–143]

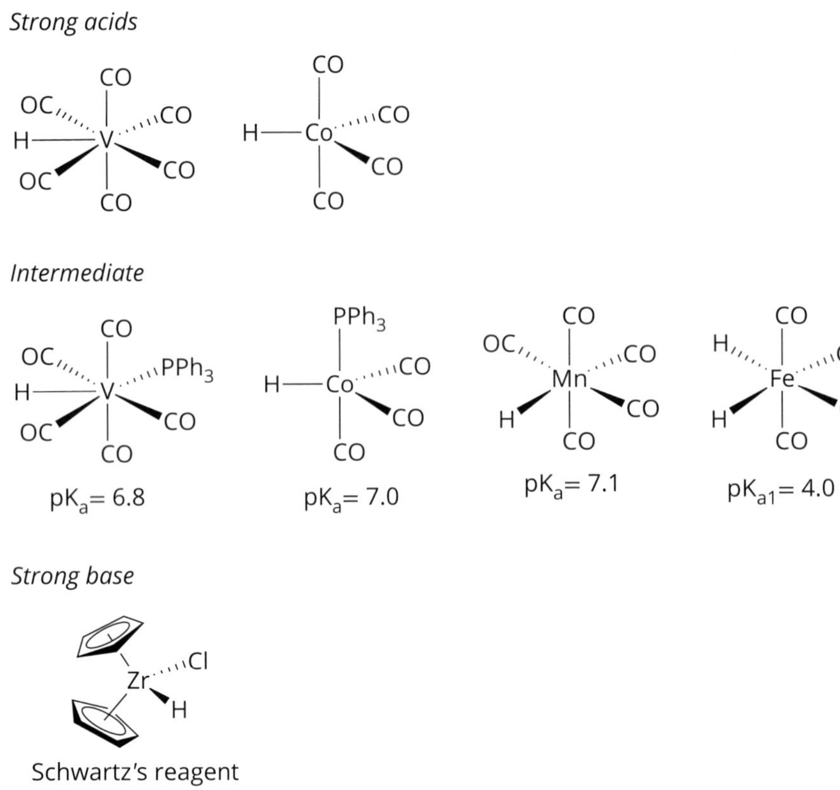

Fig. 11.44 Examples of acidic and basic hydride complexes.

For example, $HV(CO)_6$ and $HCo(CO)_4$ are strong acids in water (Fig. 11.44) due to the electron-withdrawing ability of the carbonyl ligands, which stabilize the anions $V(CO)_6^-$ and $Co(CO)_4^-$. Replacing only one carbonyl in the cobalt complex with a more electron-donating triphenylphosphine, $HCo(CO)_3(PPh_3)$, produces a complex with a $pK_a = 7.0$ in water, greatly mediating the acidity. Similarly, $HV(CO)_5(PPh_3)$ has a $pK_a = 6.8$.

Acidity decreases down a column with $HMn(CO)_5$ having a $pK_a = 15.1$ in methanol, while its heavier congener $HRe(CO)_5$ has a $pK_a = 21$ in the same solvent. This decrease in acidity is presumably due to the decrease in electronegativity, and thus ability to support a negative charge, as one descends a column.

In contrast, Schwartz's reagent $Zr(Cp)_2(Cl)H$ (Fig. 11.44), which has a zirconium center with strongly donating cyclopentadienyl ligands, is strongly hydridic.

There are simple dihydride complexes that are well characterized, like $H_2Fe(CO)_4$ shown in Fig. 11.44. In addition, dihydrogen complexes are known where there is significant H–H bonding as well as interactions with the metal center. The first such complex to be isolated was $W(CO)_3(P^iPr_3)_2(\eta^2\text{-}H_2)$, by Kubas and coworkers at Los Alamos National Laboratory in 1984 (Fig. 11.45).[144–148]

The bonding in H_2 complexes, similar to ethylene, involves σ-donation and π-backbonding (Fig. 11.36). The σ-donation occurs from the H–H σ-bond. The π-backbonding occurs from a filled metal-based orbital into the H–H σ*-orbital. As backbonding increases, the σ*-electron population increases eventually, at the limit, cleaving the H–H bond to form two hydrides. Consequently, there is a continuum of bonding between dihydrogen and dihydrides with varying amounts of H–H and M–H bonding. In the original Kubas complex, the H–H vibrational stretching frequencies were studied in great

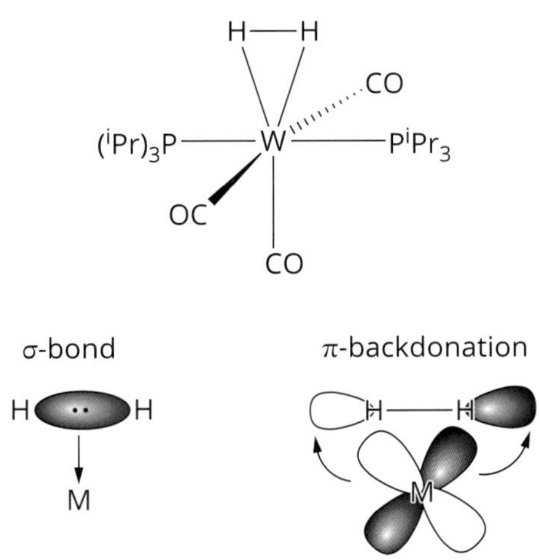

Fig. 11.45 (*top*) The geometry of $W(CO)_3(PPr^i_3)_2(\eta^2\text{-}H_2)$, as drawn in the original manuscript. The positions of the two hydrogen atoms were derived from neutron diffraction data, and the heavy atom position are from X-ray diffraction. (*bottom*) The σ-donation and π-backdonation bonding model for an H_2 complex.[149]

detail (see Problem 11.6), which can provide a measure of where one is on the spectrum from dihydrogen metal complex to dihydride complex.

Hydrides are also exceptionally good bridging ligands, and many hydride clusters are known. One example is "Stryker's reagent" $Cu_6H_6(PPh_3)_6$, which was originally prepared by Osborn and coworkers, but popularized in applications by Stryker and coworkers.[150–153] The related *p*-tolyl cluster has been investigated by neutron diffraction, which allows accurate determination of the hydrides. The hydrides bridge the trigonal faces of the Cu_6 octahedron.

11.6 Donor Abilities of Some Common Anionic Ligands to High-Valent Metals[154]

The donor abilities of ligands towards lower-valent metal complexes are often found by using procedures like that described in Section 11.3.1.3, such as CO-stretching frequencies in $Ni(CO)_3L$ complexes. Late transition metals in low oxidation states often have filled orbitals of π-symmetry on the metal due to higher *d*-electron counts, and the metals rarely act as π-acceptors. In contrast, transition metals in high oxidation states often have empty π-acceptor orbitals, and the ligands often act as both σ- and π-donors. Furthermore, many common anionic ligands for high-valent metals (such as amido, alkoxides, alkyls) all commonly react with CO, making the diatomic a poor tag for looking at the donor properties of many ligands. Couple this with the fact that CO ligands are rare for high-valent metal centers due to weak backbonding, and another method for examining ligand donor abilities is needed for metals in their highest oxidation states.

One system proposed for examining the donor abilities of ligands for high-oxidation-state metals, where both σ- and π-donor abilities of the ligands are important, uses the d^0-nitrido, $NCr(NPr^i_2)_2X$, where X = ligand under interrogation. The complex is shown schematically in Fig. 11.46. The diisopropylamido ligands π-donate to the metal, providing the Cr–N linkage with some double bond character that restricts rotation. The isopropyl group near the nitrido (hydrogen labeled H_{syn}) is in an environment different from that away from the nitrido (hydrogen labeled H_{anti}), and as a result, spin saturation transfer (a straightforward ^1H NMR technique for measuring isomerization rates) can be used to measure the rate of rotation around the $Cr–NPr^i_2$ bond.

The donation from X competes with the NPr^i_2 lone pair for the acceptor orbitals on chromium, and as X becomes a better donor the amido rotates more rapidly. The Eyring equation is used to determine $\Delta G^\ddagger$ for the rotation, which is used to calculate $\Delta H^\ddagger$ with the assumption that the entropy of rotation is constant, regardless of X in $NCr(NPr^i_2)_2X$ complexes. The enthalpy is used as the donor parameter, which has been dubbed the ligand donor parameter (LDP). Because the system is of relatively low symmetry, σ- and π-character orbitals are mixed in the system.

In short, the LDP for X is (with some assumptions) the enthalpic barrier to rotation in $NCr(NPr^i_2)_2X$ and provides a single parameter measure of how σ- and π-donating a ligand is towards a high-valent metal center. Some values for some common ligands are shown in Fig. 11.46.

The LDP for a variety of different X ligands has been published and correlated with spectroscopic and other data for many metals in high oxidation states, such as Ti(IV) and W(VI). Angular overlap model parameters[d] for Cr(III) also correlate linearly with LDP, but, as expected, require σ- and π-donor values from the AOM.[154] These ligand donor values have also been used in conjunction with percent buried volume in modeling catalytic reactions.[130]

Note the halide donor abilities, F >> Cl > Br > I, with fluoride a much stronger donor than its heavier congeners and running counter to electronegativity. Calculations suggest that this is due to better orbital overlap between the 3*d* acceptor orbitals on chromium and the smaller halides.[155]

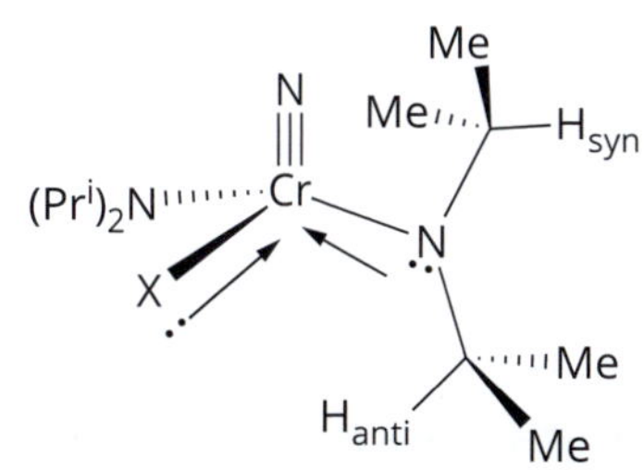

Ligand	LDP(kcal/mol)
NMe_2	9.34
O-*t*-Bu	10.59
OCH_2CF_3	11.63
OPh	11.98
indolyl	13.39
F	13.38
pyrrolyl	13.64
CN	13.98
SPh	13.99
SePh	14.16
TePh	14.20
NCO	14.48
Cl	14.97
Br	15.45
I	15.80

Fig. 11.46 (*top*) Structure of $NCr(NPr^i_2)_2X$. (*bottom*) Representative ligand donor parameter (LDP) values for a variety of monoanionic ligands in kcal/mol. Errors are estimated as ± 0.1 kcal/mol. More donating ligands have smaller LDP values.[154–155]

[d] The angular overlap model is discussed in Chapter 15 on transition-metal spectroscopy.

Chapter Summary

1. The Dewar–Chatt–Duncanson model for understanding how olefins bind to metal centers greatly influenced and informed later models for understanding metal–ligand interactions. In this model the ligand donates in a σ-fashion to the metal, and the metal may donate electrons back to the ligand in a π-fashion. Several other important ligands bond with a similar scheme. For example, carbon monoxide bonds to metals in a similar fashion but can accept electron density through multiple π-acceptor orbitals.
2. There is a host of diatomics that are quite important ligands for metals, and their transformations are of great interest. The diatomics discussed in the chapter are carbon monoxide, cyanide, dinitrogen, dioxygen, and nitric oxide. A related ligand class, isonitriles, which have similar bonding characteristics, was also discussed. Each of these ligands uses a bonding scheme similar, in some respects, to the Dewar–Chatt–Duncanson model, but with important nuances in its interactions with the metal center.
3. The σ-donation and π-acceptor motif continued with the examination of bonding in phosphines. In this case, the σ-donation occurs from the phosphorus lone pair, and the π-acceptor orbitals are through negative hyperconjugation, where nonbonding electrons on the metal donate into the σ*-orbitals of the substituents on phosphorus.
4. The electronics of both the lone pair and acceptor orbitals can be adjusted extensively, as can the steric profile of phosphines, which leads to a large array of different ligands with highly tunable properties. Methods for evaluating and classifying different phosphines have been developed, and then reactivity changes resulting from different ligand sets can be examined, which provides direction to catalyst development efforts.
5. Nitrogen donors are also of great importance. Amines are far poorer π-acceptors than their phosphine counterparts. Pyridines are excellent ligands for transition metals, as are related multidentate ligands such as 2,2′-bipyridine (bpy).
6. There is a large array of anionic ligands commonly employed. Monodentate examples include halides, alkyls, amides, and alkoxides. Carboxylates and some related ligands may be κ^1 or κ^2 to the metal center. Multidentate ligands such as acetylacetonate and pyrazolyl borates can be used to produce a host of stable complexes across the transition series.
7. Hydrogen can bind to many metal centers to produce stable compounds. The complexes may be a strong acid in water or highly hydridic, depending on the metal center and coligands.
8. Dihydrogen complexes can be isolated in some cases. Here again, a σ-donation and π-acceptor motif is observed for the bonding, where the H–H σ-bond donates to the metal and the H–H σ*-orbital accepts electron density from the metal. As a result, there is a continuum of structures for H_2 bound to a metal to where the dihydrogen bond is broken to produce two hydrides, and the position on this continuum is determined by backdonation from the metal center.

Practice Problems

1. In the example used of a strongly bent isonitrile, Bu^{t}NC–Mo{N(Bu^t)Ar} by Cummins and coworkers, the C–N bond distance in the isonitrile was 1.184 Å, and the Mo–C distance = 1.936 Å. Calculate the C–N and Mo–C bond orders using Eq. 3.27 with b = 0.38 Å and using Pyykkö's chart (Fig. 3.30) to calculate r_1. Is the C–N bond order what you would expect from the C–N–C(Bu^t) angle of 138°?
2. Below is a table of tungsten(II) and molybdenum(II) alkyne complexes adopted from the paper by Templeton and Ward.[162] Find the electron count of the metal centers not including the alkyne donors; then, assuming that the complexes reach 18 electrons, determine how many electrons each alkyne donates in each complex. In some cases, the alkyne may not donate a whole number of electrons due to averaging among the triple bond ligands. In the table are the experimentally determined average ^{13}C NMR chemical shifts for the alkyne carbons.

Complex	δ_{ave} ^{13}C NMR alkyne carbons (ppm)
W(CO)(detc)$_2$(HC≡CH)[a]	206.6
W(CO)(detc)$_2$(HC≡CPh)[a]	205.7
W(η^5-Cp)(CO)(CH$_3$)(HC≡CH)	190.0
Mo(CO)(dmtc)$_2$(HC≡CH)[b]	203.7
Mo(CO)(dmtc)$_2$(HC≡CPh)[b]	205.3
W(CO)(EtC≡CEt)$_3$	181.0
{W(η^5-Cp)(CO)(MeC≡CBu)$_2$}$^+$	152.0
Mo(detc)$_2$(HC≡CPh)$_2$[a]	180.2
Mo(η^5-Cp)$_2$(HC≡CH)	117.7
Mo(η^5-Cp)$_2$(MeC≡CMe)	115.3

[a] detc = diethylthiocarbamate.
[b] dmtc = dimethylthiocarbamate.

a. Plot the electrons donated per alkyne (N) in each complex versus the ^{13}C chemical shift of the alkynes. Is there a correlation?

b. If you repeat the plot in part a using separate lines for the Mo and W complexes, is there a stronger correlation? Why or why not?

3. For the known complex $\{Ag(C_2H_4)_3\}^+$,[163-165] which has the structure shown below where all of the carbons reside in the same plane with the silver atom. The C-H bonds reside above and below the plane. Answer the following questions.

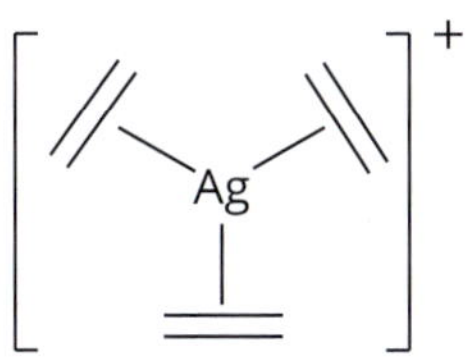

a. What is the point group of the complex as drawn?

b. The average C–C bond length is 1.33 Å in the complex. What is the C–C bond order based on the Pauling equation (Eq. 3.27), using $b = 0.38$ Å, and the single bond distance found using Pyykkö's tables (Fig. 3.30)?

c. Find the symmetries of the SALCs for the ethylene orbitals bonding to silver. Use the projection operator to discover what they look like.

d. Draw an MO diagram for the complex.

4. The trigonal pyramidal complex $Co(NO)_3$ has three linear nitrosyls and overall C_{3v} symmetry. In the infrared spectrum, there are two NO stretches at 1860 and 1795 cm^{-1}. Use group theory to find the symmetries of the stretches and use the projection operator to discover the nature of the stretches. Which stretch can be assigned to which energy, and why?

5. A known tetranitrosyl complex $Cr(NO)_4$ has a single N–O stretch in the IR spectrum at 1716 cm^{-1}. What is the nature of this stretch, including its symmetry and projection?

6. The H–H symmetric stretching frequency in Kubas and coworkers' dihydrogen complex (Fig. 11.45) was found to be 2690 cm^{-1}; however, the resonance was partially obscured by other peaks.

a. The HD derivative was also prepared and had a stretching frequency of 2360 cm^{-1}. Using the HD frequency, show that the 2690 cm^{-1} assignment was correct for the H–H vibration.

b. If the D_2 derivative is prepared, what should the stretching frequency be?

7. Why are CO ligands typically linear for the 2nd and 3rd row transition metals but apparently more likely to bridge for the 1st row transition metals? Use the primogenic effect and experimental CO stretching frequencies from the literature to support your answer.

8. In Chapter 8, the similarity of bonding in O–NF_3 and O–PF_3 was discussed, and the compounds show remarkably similar geometries. PF_3 is a fairly good ligand for transition metals and is an excellent π-acceptor through negative hyperconjugation. Based on the properties of OEF_3 compound, one would expect NF_3 to act as a similar ligand; however, there are dozens of metal–PF_3 structures compared to zero metal–NF_3 structures in the Cambridge Structural Database. Why might this be the case, and if you were going to try to make a metal-NF_3 complex, what properties of the metal would you be concerned with? In other words, what complex would you target for synthesis, and why?

Cited References

1. Saillard, J. Y.; Hoffmann, R., C–H and H–H activation in transition-metal complexes and on surfaces *J. Am. Chem. Soc.* 1984, *106*, 2006–2026.
2. Brock, W. H.; Jensen, K. A.; Jorgensen, C. K.; Kauffman, G. B., The origin and dissemination of the term ligand in chemistry *Polyhedron* 1983, *2*, 1–7.
3. Dewar, M. J. S., A Review of π-Complex Theory *Bull. Soc. Chim. Fr.* 1951, C71.
4. Chatt, J.; Duncanson, L. A., Olefin Co-Ordination Compounds. 3. Infra-red spectra and structure: Attempted preparation of acetylene complexes *J. Chem. Soc.* 1953, 2939–2947.
5. Kissounko, D.; Epshteyn, A.; Fettinger, J. C.; Sita, L. R., Synthesis and formal ring-opening reactions of $(\eta^5\text{-}C_5Me_5)Zr[\eta^2\text{-}CH_2CH(C_6H_5)][N(Pr^i)C(Me)N(Pr^i)]$, a base-free η^2-styrene complex of zirconium, *Organometallics* 2006, *25*, 531–535.
6. Carpentier, J. F.; Wu, Z.; Lee, C. W.; Stromberg, S.; Christopher, J. N.; Jordan, R. F., d^0 metal olefin complexes. Synthesis, structures, and dynamic properties of $(C_5R_5)_2Zr(OCMe_2CH_2CH_2CH{=}CH_2)^+$ complexes: Models for the elusive $(C_5R_5)_2Zr(R)(olefin)^+$ intermediates in metallocene-based olefin polymerization catalysis *J. Am. Chem. Soc.* 2000, *122*, 7750–7767.
7. Stoebenau, E. J.; Jordan, R. F., Nonchelated Zr(IV)-alkoxide-alkyne complexes *Organometallics* 2006, *25*, 3379–3387.
8. Fedotova, Y. V.; Kornev, A. N.; Sushev, V. V.; Kursky, Y. A.; Mushtina, T. G.; Makarenko, N. P.; Fukin, G. K.; Abakumov, G. A.; Zakharov, L. N.; Rheingold, A. L., Phosphinohydrazines and phosphinohydrazides $M(\text{-}N(R)\text{-}N(R)\text{-}PPh_2)_n$ of some transition and main group metals: synthesis and characterization. Rearrangement of Ph_2P-NR-NR- ligands into aminoiminophosphorane, $RN{=}PPh_2$-NR-, and related chemistry *J. Organomet. Chem.* 2004, *689*, 3060–3074.
9. Durfee, L. D.; Hill, J. E.; Fanwick, P. E.; Rothwell, I. P., Formation and characterization of η^2-azobenzene derivatives of titanium containing ancillary aryloxide ligands *Organometallics* 1990, *9*, 75–80.
10. Tatsumi, K.; Hoffmann, R.; Templeton, J. L., Structural and theoretical evidence for participation of the 2nd acetylene π-orbital in transition-metal alkyne complexes *Inorg. Chem.* 1982, *21*, 466–468.
11. Balazs, G.; Green, J. C.; Scheer, M., Terminally coordinated AsS and PS ligands *Chem. Eur. J.* 2006, *12*, 8603–8608.
12. Johnson, M. J. A.; Odom, A. L.; Cummins, C. C., Phosphorus monoxide as a terminal ligand *Chem. Commun.* 1997, 1523–1524.
13. Laplaza, C. E.; Davis, W. M.; Cummins, C. C., A Molybdenum-Phosphorus Triple Bond - Synthesis, Structure, And Reactivity Of The Terminal Phosphido (P^{3-})

Complex Mo(P)(NRAr)$_3$ *Angewandte Chemie, International Edition in English*, 1995, *34*, 2042–2044.

14. Brand, J.; Braunschweig, H.; Sen, S. S., B=B and B E (E = N and O) Multiple Bonds in the Coordination Sphere of Late Transition Metals *Acc. Chem. Res.* 2014, *47*, 180–191.
15. Drance, M. J.; Sears, J. D.; Mrse, A. M.; Moore, C. E.; Rheingold, A. L.; Neidig, M. L.; Figueroa, J. S., Terminal coordination of diatomic boron monofluoride to iron *Science* 2019, *363*, 1203.
16. Tolman, W. B., *Activation of Small Molecules: Organometallic and Bioinorganic Perspectives* Wiley-VCH: Weinheim, 2006; p xvii, 363 p.
17. Richter-Addo, G. B.; Legzdins, P., *Metal Nitrosyls* Oxford University Press: New York, 1992.
18. Richter-Addo, G. B.; Legzdins, P.; Burstyn, J., Introduction: Nitric oxide chemistry *Chem. Rev.* 2002, *102*, 857–859.
19. Dyson, P. J.; McIndoe, J. S., *Transition Metal Carbonyl Cluster Chemistry* Gordon and Breach Science Publishers: Amsterdam, 2000.
20. Sharpe, A. G., *The Chemistry of Cyano Complexes of the Transition Metals* Academic Press: London, 1976.
21. Blackman, A. G.; Tolman, W. B., Copper-dioxygen and copper-oxo species relevant to copper oxygenases and oxidases *Metal-Oxo and Metal-Peroxo Species in Catalytic Oxidations* 2000, *97*, 179–211.
22. Chufan, E. E.; Puiu, S. C.; Karlin, K. D., Heme-copper/dioxygen adduct formation, properties, and reactivity *Acc. Chem. Res.* 2007, *40*, 563–572.
23. Cramer, C. J.; Tolman, W. B., Mononuclear CuO_2 complexes: Geometries, spectroscopic properties, electronic structures, and reactivity *Acc. Chem. Res.* 2007, *40*, 601–608.
24. Dunbar, K. R.; Heintz, R. A., Chemistry of Transition Metal Cyanide Compounds: Modern Perspectives. In *Prog. Inorg. Chem.*, Karlin, K. D., Ed. 1997; Vol. 45, pp 283-391.
25. Fehlhammer, W. P.; Fritz, M., emergence of a CNH and cyano complex based organometallic chemistry *Chem. Rev.* 1993, *93*, 1243–1280.
26. Suzuki, M., Ligand effects on dioxygen activation by copper and nickel complexes: Reactivity and intermediates *Acc. Chem. Res.* 2007, *40*, 609–617.
27. Klotz, I. M.; Kurtz, D. M., metal-dioxygen complexes—a perspective *Chem. Rev.* 1994, *94*, 567–568.
28. Vaska, L., dioxygen-metal complexes: Toward a unified view *Acc. Chem. Res.* 1976, *9*, 175–183.
29. Orgel, L. E., *An Introduction to Transition-Metal Chemistry: Ligand-Field Theory* Wiley: London, 1960; p 180 p.
30. Pauling, L., *The Nature of the Chemical Bond, and the Structure of Molecules and Crystals: an Introduction to Modern Structural Chemistry* 3rd ed.; Cornell University Press: Ithaca, N.Y., 1960.
31. Zeiss, H., *Organometallic Chemistry* Reinhold Pub. Corp.: New York, 1960.
32. Chi, K. M.; Frerichs, S. R.; Philson, S. B.; Ellis, J. E., Highly Reduced Organometallics. 23. Synthesis, Isolation, and Characterization of Hexacarbonyltitanate(2−), $Ti(CO)_6^{2-}$: Titanium NMR-Spectra of Carbonylititanates *J. Am. Chem. Soc.* 1988, *110*, 303–304.
33. Liu, X.; Ellis, J. E., Transition Metal Carbonyl Compounds. 21. Hexacarbonylvanadate(1-) and Hexacarbonylvanadium(0) *Inorg. Synth.* 2004, *34*, 96–103.
34. Reed, Z. D.; Duncan, M. A., Infrared Spectroscopy and Structures of Manganese Carbonyl Cations, $Mn(CO)_n^+$ (n = 1-9) *J. Am. Soc. Mass. Spectrom.* 2010, *21*, 739–749.
35. Bernhardt, E.; Bley, B.; Wartchow, R.; Willner, H.; Bill, E.; Kuhn, P.; Sham, I. H. T.; Bodenbinder, M.; Brochler, R.; Aubke, F., Hexakis(carbonyl)iron(II) undecafluorodiantimonate(V), $[Fe(CO)_6][Sb_2F_{11}]_2$, and -hexafluoroantimonate(V), $[Fe(CO)_6][SbF_6]_2$, their syntheses, and spectroscopic and structural characterization by single crystal X-ray diffraction and normal coordinate analysis *J. Am. Chem. Soc.* 1999, *121*, 7188–7200.
36. Li, J.; Schreckenbach, G.; Ziegler, T., A Reassessment Of The First Metal-Carbonyl Dissociation-Energy in $M(CO)_4$ (M = Ni, Pd, Pt), M(Co)(5) (M = Fe, Ru, Os), And $M(CO)_6$ (M = Cr, Mo, W) by a Quasi-Relativistic Density-Functional Method *J. Am. Chem. Soc.* 1995, *117*, 486–494.
37. Rosini, G. P.; Liu, F. C.; Krogh-Jespersen, K.; Goldman, A. S.; Li, C. B.; Nolan, S. P., Thermodynamics of addition of H_2, CO, N_2, and C–H bonds to $M(P^iPr_3)_2Cl$ (M = Ir, Rh). An unprecedented metal-carbonyl bond strength *J. Am. Chem. Soc.* 1998, *120*, 9256–9266.
38. Campana, C. F.; Guzei, I. A.; Mednikov, E. G.; Dahl, L. F., Sixty-Year Saga (1952–2013) of the Solid-State Structure of Triiron Dodecacarbonyl *J. Cluster Sci.* 2014, *25*, 205–224.
39. Neuman, M. A.; Trinh-Toan; Dahl, L. F., Structural characterization of cyclopentadienyliron carbonyl tetramer, $Fe_4(\eta^5\text{-}C_5H_5)_4(CO)_4$ - cubane-type structure containing a completely bonding tetrametal cluster unit *J. Am. Chem. Soc.* 1972, *94*, 3383–8.
40. Bartoll, J. The Early Use of Prussian Blue in Paintings. 9th International Conference on NDT of Art Jerusalem, Israel, 2008.
41. Ballmann, G.; Elsen, H.; Harder, S., Magnesium Cyanide or Isocyanide? *Angew. Chem. Int. Ed.* 2019, *58*, 15736–15741.
42. Bouzidi, Y.; Belkhiri, L.; Ephritikhine, M.; Halet, J. F.; Boucekkine, A., Cyanide linkage isomerism in cerium(III) and uranium(III) complexes. A relativistic DFT study *J. Organomet. Chem.* 2017, *847*, 82–89.
43. Berthet, J. C.; Thuery, P.; Ephritikhine, M., Advances in f-element cyanide chemistry *Dalton Trans* 2015, *44*, 7727–7742.
44. Stephens, F. H.; Figueroa, J. S.; Cummins, C. C.; Kryatova, O. P.; Kryatov, S. V.; Rybak-Akimova, E. V.; McDonough, J. E.; Hoff, C. D., Small-molecule activation by molybdaziridine hydride complexes: Mechanistic sequence of the small-molecule binding and molybdaziridine ring-opening steps *Organometallics* 2004, *23*, 3126–3138.
45. Carnahan, E. M.; Lippard, S. J., Mechanistic studies of the reductive coupling of isocyanides in 7-coordinate molybdenum complexes: X-ray crystal-structures of $Mo(CNBu^t)_6$ and $[Mo(=CN(Bu^t)(SiMe_2Bu^t))(CNBu^t)_5][O_3SCF_3]$ *Dalton Trans.* 1991, 699–706.
46. Lupinetti, A. J.; Strauss, S. H.; Frenking, G., Nonclassical Metal Carbonyls. In *Prog. Inorg. Chem.*, Karlin, K. D., Ed. 2001; Vol. 49, pp 1–112.
47. Hurlburt, P. K.; Rack, J. J.; Luck, J. S.; Dec, S. F.; Webb, J. D.; Anderson, O. P.; Strauss, S. H., Nonclassical metal-carbonyls: $[Ag(CO)]^+$ and $[Ag(CO)_2]^+$ *J. Am. Chem. Soc.* 1994, *116*, 10003–10014.
48. Veldkamp, A.; Frenking, G., Theoretical-studies of organometallic compounds. 6. Structures and bond-energies of $M(CO)_n^+$, MCN, and $M(CN)_2^-$ (M = Ag, Au; N = 1-3) *Organometallics* 1993, *12*, 4613–4622.
49. Rack, J. J.; Strauss, S. H., Polycarbonyl cations of Cu(I), Ag(I), and Au(I): $M(CO)_n^+$ *Catal. Today* 1997, *36*, 99–106.
50. Shen, H.; Jordan, R. F., Molecular structure and vinyl chloride insertion of a cationic zirconium(IV) acyl carbonyl complex *Organometallics* 2003, *22*, 2080–2086.
51. Carofiglio, T.; Floriani, C.; Chiesivilla, A.; Guastini, C., Isocyanide complexes of titanium(IV) and vanadium(V): Concerning the nonexistent insertion of isocyanides into a metal-chloride bond *Inorg. Chem.* 1989, *28*, 4417–4419.
52. Spectral Database for Organic Compounds, http://sdbs.db.aist.go.jp
53. In principle, *cis*-$TiCl_4(CN\text{-}t\text{-}Bu)_2$ should have two different IR bands corresponding to symmetric and asymmetric vibrations of the isonitriles; however, only one band is observed experimentally. Calculations (B3LYP, 6-31G**) suggest that the two IR bands are very close in energy (<3 wavenumbers), which suggests they may be overlapping in the experimental spectrum. .
54. Allen, A. D.; Bottomley, F.; Harris, R. O.; Reinsalu, V. P.; Senoff, C. V., Ruthenium complexes containing molecular nitrogen *J. Am. Chem. Soc.* 1967, *89*, 5595.
55. Allen, A. D.; Senoff, C. V., Nitrogenopentammineruthenium(II) complexes *Chem. Commun.* 1965, 621.
56. Senoff, C. V., The discovery of $[Ru(NH_3)_5N_2]^{2+}$: A case of serendipity and the scientific method *J. Chem. Educ.* 1990, *67*, 368–370.
57. Hidai, M.; Tominari, K.; Uchida, Y.; Misono, A., A molybdenum complex containing molecular nitrogen *J. Chem. Soc. D-Chem. Comm.* 1969, 814.
58. Hidai, M.; Tominari, K.; Uchida, Y.; Misono, A., A *trans*-dinitrogen complex of molybdenum *J. Chem. Soc. D-Chem. Comm.* 1969, 1392.
59. Uchida, T.; Uchida, Y.; Hidai, M.; Kodama, T., *trans*-bis(dinitrogen)-bis-1,2-bis(diphenylphosphino)ethane molybdenum(0) *Acat Cryst. B:Struc. Sci.* 1975, *31*, 1197–1199.
60. Gilbert-Wilson, R.; Field, L. D.; Colbran, S. B.; Bhadbhade, M. M., Low Oxidation State Iron(0), Iron(I), and Ruthenium(0) Dinitrogen Complexes with a Very Bulky Neutral Phosphine Ligand *Inorg. Chem.* 2013, *52*, 3043–3053.
61. Roussel, P.; Scott, P., Complex of dinitrogen with trivalent uranium *J. Am. Chem. Soc.* 1998, *120*, 1070–1071.
62. Odom, A. L.; Arnold, P. L.; Cummins, C. C., Heterodinuclear uranium/molybdenum dinitrogen complexes *J. Am. Chem. Soc.* 1998, *120*, 5836–5837.
63. Taqui Khan, M. M.; Martell, A. E., *Homogeneous Catalysis by Metal Complexes* Academic Press: New York, 1974.

64. Boettcher, H.-C.; Graf, M.; Mayer, P.; Suenkel, K., Crystal and molecular structure of trans- IrCl(N-2)((PPr3)-Pr-i)(2) *Zeitschrift Fur Anorganische Und Allgemeine Chemie* 2008, *634*, 1241–1243.
65. Collman, J. P.; Kang, J. W., Iridium complexes of molecular nitrogen *J. Am. Chem. Soc.* 1966, *88*, 3459.
66. Collman, J. P.; Kubota, M.; Vastine, F. D.; Sun, J. Y.; Kang, J. W., Iridium complexes of molecular nitrogen *J. Am. Chem. Soc.* 1968, *90*, 5430.
67. Fryzuk, M. D.; Haddad, T. S.; Rettig, S. J., Reduction of dinitrogen by a zirconium phosphine complex to form a side-on-bridging N_2 ligand: Crystal-structure of $(Pr^i_2PCH_2SiMe_2)_2N\ ZrCl)_2(\mu\text{-}\eta^2\text{-}\eta^2\text{-}N_2)$ *J. Am. Chem. Soc.* 1990, *112*, 8185–8186.
68. Wright, A. M.; Hayton, T. W., Recent Developments in Late Metal Nitrosyl Chemistry *Comments Inorg. Chem.* 2012, *33*, 207–248.
69. Sabherwal, I. H.; Burg, A. B., The strong base trinitrosylcobalt *J. Chem. Soc. D-Chem. Comm.* 1970, 1001–2.
70. Hayton, T. W.; McNeil, W. S.; Patrick, B. O.; Legzdins, P., Toward binary nitrosyls: Distinctly bent Fe–N–O linkages in base-stabilized $Fe(NO)_3^+$ complexes *J. Am. Chem. Soc.* 2003, *125*, 12935–12944.
71. Melenkivitz, R.; Hillhouse, G. L., Synthesis, structure, and reactions of a nitroxyl complex of iridium(III), *cis,trans*-$IrHCl_2(NH{=}O)(PPh_3)_2$ *Chem. Commun.* 2002, 660–661.
72. Mingos, D. M. P.; Ibers, J. A., Crystal and molecular structure of dichloronitrosylbis(triphenylphosphine)iridium, $IrCl_2(NO)(P(C_6H_5)_3)_2$ *Inorg. Chem.* 1971, *10*, 1035.
73. Harcourt, R. D., Oxyheme valence bond structures and the oxyheme plus NO reaction *Coord. Chem. Rev.* 2018, *358*, 178–180.
74. Komor, A. J.; Jasniewski, A. J.; Que, L.; Lipscomb, J. D., Diiron monooxygenases in natural product biosynthesis *Natural Product Reports* 2018, *35*, 646–659.
75. Bayot, D.; Tinant, B.; Devillers, M., Spectroscopic and structural characterizations of novel water-soluble tetraperoxo and diperoxo polyaminocarboxylato bis(N-oxido) tantalate(V) complexes *Inorg. Chem.* 2004, *43*, 5999–6005.
76. Cai, X.; Majumdar, S.; Fortman, G. C.; Cazin, C. S. J.; Slawin, A. M. Z.; Lhermitte, C.; Prabhakar, R.; Germain, M. E.; Palluccio, T.; Nolan, S. P., et al., Oxygen Binding to Pd(L)(L') (L = NHC, L' = NHC or PR_3, NHC = N-Heterocyclic Carbene). Synthesis and Structure of a Paramagnetic *trans*-$Pd(NHC)_2(\eta^1\text{-}O_2)_2$ Complex *J. Am. Chem. Soc.* 2011, *133*, 1290-1293.
77. Wolczanski, P. T., Flipping the Oxidation State Formalism: Charge Distribution in Organometallic Complexes As Reported by Carbon Monoxide *Organometallics* 2017, *36*, 622–631.
78. Hitchcock, P. B.; Jacobson, B.; Pidcock, A., *cis*-influences and *trans*-influences in platinum(II) complexes: X-ray crystal-structure analysis of *cis*-dichloro (triethylphosphine)(trifluorophosphine)platinum(II) *Dalton Trans.* 1977, 2043–2048.
79. Brown, R. K.; Williams, J. M.; Schultz, A. J.; Stucky, G. D.; Ittel, S. D.; Harlow, R. L., Delocalized 2-electron 3-center CH-metal interaction: Single-crystal neutron (30 and 110 K) and X-ray (298 K) diffraction study of $Fe(P(OCH_3)_3)_3(\eta^3\text{-}C_8H_{13})^+\ BF_4^-$ *J. Am. Chem. Soc.* 1980, *102*, 981–987.
80. Harlow, R. L.; McKinney, R. J.; Ittel, S. D., Structure of η^3-cyclooctenyl-tris(trimethyl phosphite)iron(I) - bonding of the η^3-alkenyl group to 16-electron, 17-electron, and 18-electron ML_3 systems *J. Am. Chem. Soc.* 1979, *101*, 7496–7504.
81. Orpen, A. G.; Connelly, N. G., Structural systematics: Role of P–A σ^* orbitals in metal-phosphorus π-bonding in redox-related pairs of M-PA_3 COMPLEXES (A = R, Ar, OR, R = Alkyl) *Organometallics* 1990, *9*, 1206–1210.
82. Comas-Vives, A.; Harvey, J. N., How Important Is Backbonding in Metal Complexes Containing N-Heterocyclic Carbenes? Structural and NBO Analysis *Eur. J. Inorg. Chem.* 2011, 5025-5035.
83. Gilheany, D. G., No d-orbitals but walsh diagrams and maybe banana bonds: Chemical bonding in phosphines, phosphine oxides, and phosphonium ylides *Chem. Rev.* 1994, *94*, 1339–1374.
84. Hoshiya, N.; Buchwald, S. L., An Improved Synthesis of BrettPhos- and RockPhos-Type Biarylphosphine Ligands *Adv. Synth. Catal.* 2012, *354*, 2031–2037.
85. Kaye, S.; Fox, J. M.; Hicks, F. A.; Buchwald, S. L., The use of catalytic amounts of CuCl and other improvements in the benzyne route to biphenyl-based phosphine Ligands *Adv. Synth. Catal.* 2001, *343*, 789–794.
86. Tomori, H.; Fox, J. M.; Buchwald, S. L., An improved synthesis of functionalized biphenyl-based phosphine ligands *J. Org. Chem.* 2000, *65*, 5334–5341.
87. Arrayas, R. G.; Adrio, J.; Carretero, J. C., Recent applications of chiral ferrocene ligands in asymmetric catalysis *Angew. Chem. Int. Ed.* 2006, *45*, 7674–7715.
88. Barbaro, P.; Bianchini, C.; Giambastiani, G.; Parisel, S. L., Progress in stereoselective catalysis by metal complexes with chiral ferrocenyl phosphines *Coord. Chem. Rev.* 2004, *248*, 2131–2150.
89. Colby, E. A.; Jamison, T. F., P-chiral, monodentate ferrocenyl phosphines, novel ligands for asymmetric catalysis *J. Org. Chem.* 2003, *68*, 156–166.
90. Drusan, M.; Sebesta, R., Enantioselective C-C and C-heteroatom bond forming reactions using chiral ferrocene catalysts *Tetrahedron* 2014, *70*, 759–786.
91. Knowles, W. S., Asymmetric hydrogenations (Nobel lecture) *Angew. Chem. Int. Ed.* 2002, *41*, 1999–2007.
92. Widhalm, M.; Brecker, L.; Mereiter, K., Rigid P-chiral mono and diphosphines. Configurative stability and P-inversion barrier *Tetrahedron-Asymmetry* 2006, *17*, 1355–1369.
93. Fey, N.; Orpen, A. G.; Harvey, J. N., Building ligand knowledge bases for organometallic chemistry: Computational description of phosphorus(III)-donor ligands and the metal-phosphorus bond *Coord. Chem. Rev.* 2009, *253*, 704–722.
94. Poe, A. J., Pendent group effects, PGEs, in P-donor ligands *New J. Chem.* 2013, *37*, 2957–2964.
95. Dias, P. B.; Depiedade, M. E. M.; Simoes, J. A. M., Bonding and energetics of phosphorus(III) ligands in transition-metal complexes *Coord. Chem. Rev.* 1994, *135*, 737–807.
96. Tolman, C. A., Steric effects of phosphorus ligands in organometallic chemistry and homogeneous catalysis *Chem. Rev.* 1977, *77*, 313–348.
97. Abel, E. W.; Stone, F. G. A., Chemistry of Transition-Metal Carbonyls: Structural Considerations *Quar. Rev.* 1969, *23*, 325–371.
98. Cotton, F. A., Vibrational spectra and bonding in metal carbonyls. 3. Force constants and assignments of CO stretching modes in various molecules: Evaluation of CO bond orders *Inorg. Chem.* 1964, *3*, 702.
99. Nyholm, R. S.; Short, L. N., The structure of nickel tetracarbonyl and some disubstituted derivatives *J. Chem. Soc.* 1953, 2670–2673.
100. Strohmei, W.; Muller, F. J., Klassifizierung phosphorhaltiger liganden in metallcarbonyl-derivaten nach der π-acceptorstarke *Chem. Ber. Recl.* 1967, *100*, 2812.
101. Bartik, T.; Himmler, T.; Schulte, H. G.; Seevogel, K., The effect of substitutions on the basicity of phosphorus ligands in $R_3PNi(CO)_3$ complexes *J. Organomet. Chem.* 1984, *272*, 29–41.
102. Perrin, L.; Clot, E.; Eisenstein, O.; Loch, J.; Crabtree, R. H., Computed ligand electronic parameters, from quantum chemistry and their relation to Tolman parameters, Lever parameters, and Hammett constants *Inorg. Chem.* 2001, *40*, 5806–5811.
103. Kalescky, R.; Kraka, E.; Cremer, D., New Approach to Tolman's Electronic Parameter Based on Local Vibrational Modes *Inorg. Chem.* 2014, *53*, 478–495.
104. Kuhl, O., Predicting the net donating ability of phosphines: Do we need sophisticated theoretical methods? *Coord. Chem. Rev.* 2005, *249*, 693–704.
105. Woska, D.; Prock, A.; Giering, W. P., Determination of the stereoelectronic parameters of PF_3, PCl_3, PH_3, and $P(CH_2CH_2CN)_3$. The quantitative analysis of ligand effects (QALE) *Organometallics* 2000, *19*, 4629–4638.
106. Bunten, K. A.; Chen, L. Z.; Fernandez, A. L.; Poe, A. J., Cone angles: Tolman's and Plato's *Coord. Chem. Rev.* 2002, *233*, 41–51.
107. Brown, T. L.; Lee, K. J., Ligand steric properties *Coord. Chem. Rev.* 1993, *128*, 89–116.
108. Bilbrey, J. A.; Kazez, A. H.; Locklin, J.; Allen, W. D., Exact ligand cone angles *Journal of Computational Chemistry* 2013, *34*, 1189–1197.
109. Smith, J. M.; Taverner, B. C.; Coville, N. J., Cone angle radial profiles *J. Organomet. Chem.* 1997, *530*, 131–140.
110. Seligson, A. L.; Trogler, W. C., Cone angles for amine ligands: X-ray crystal-structures and equilibrium measurements for ammonia, ethylamine, diethylamine, and triethylamine complexes with the (bis(dimethylphosphino) ethane)methylpalladium(II) cation *J. Am. Chem. Soc.* 1991, *113*, 2520–2527.
111. Choi, M. G.; Brown, T. L., A molecular mechanics model of ligand effects. 5. Ligand repulsive energy values for phosphines and phosphites bound to CpRh(CO) and the crystal-structure of $CpRh(CO)(PPH_3)$ *Inorg. Chem.* 1993, *32*, 5603–5610.
112. Clavier, H.; Correa, A.; Cavallo, L.; Escudero-Adan, E. C.; Benet-Buchholz, J.; Slawin, A. M. Z.; Noaln, S. P., Pd(NHC)(allyl)Cl Complexes: Synthesis and Determination of the NHC Percent Buried Volume (%V-bur) Steric Parameter *Eur. J. Inorg. Chem.* 2009, 1767–1773.
113. Clavier, H.; Nolan, S. P., Percent buried volume for phosphine and N-heterocyclic carbene ligands: steric properties in organometallic chemistry *Chem. Commun.* 2010, *46*, 841–861.

114. Bartholomew, J.; Fernandez, A. L.; Lorsbach, B. A.; Wilson, M. R.; Prock, A.; Giering, W. P., Comments on coupling graphical and regression analyses of ligand effect data *Organometallics* 1996, *15*, 295–301.
115. Li, C. B.; Stevens, E. D.; Nolan, S. P., Solution thermochemical study of ligand steric influences on substitution enthalpies in the $L_2Fe(CO)_3$ system *Organometallics* 1995, *14*, 3791–3797.
116. Calculations were done using LANL2DZ with B3LYP and analyzed with NBO6.
117. Hopkinson, M. N.; Richter, C.; Schedler, M.; Glorius, F., An overview of N-heterocyclic carbenes *Nature* 2014, *510*, 485–496.
118. Nelson, D. J.; Nolan, S. P., Quantifying and understanding the electronic properties of N-heterocyclic carbenes *Chem. Soc. Rev.* 2013, *42*, 6723–6753.
119. Peris, E., Routes to N-Heterocyclic Carbene Complexes. In *N-Heterocyclic Carbenes in Transition Metal Catalysis*, Glorius, F., Ed. 2007; Vol. 21, pp 83–116.
120. Boehme, C.; Frenking, G., Electronic structure of stable carbenes, silylenes, and germylenes *J. Am. Chem. Soc.* 1996, *118*, 2039–2046.
121. Calculations were done using cc-pVDZ as the basis set with B3LYP on G09. NBO calculation used NBO6. .
122. Ofele, K., A new transition-metal carbon complexe 1,3-dimethyl-4-imidazolinylid-2-ene pentacarbonylchromium *J. Organomet. Chem.* 1968, *12*, P42.
123. Wanzlick, H. W.; Schonher, H. j., Direct synthesis of a mercury salt-carbene complex *Angew. Chem. Int. Ed.* 1968, *7*, 141.
124. Lappert, M. F., The coordination chemistry of electron-rich alkenes (enetetramines) *J. Organomet. Chem.* 1988, *358*, 185–214.
125. Lappert, M. F., Contributions to the chemistry of carbenemetal chemistry *J. Organomet. Chem.* 2005, *690*, 5467–5473.
126. Arduengo, A. J.; Harlow, R. L.; Kline, M., A stable crystalline carbene *J. Am. Chem. Soc.* 1991, *113*, 361–363.
127. Scholl, M.; Ding, S.; Lee, C. W.; Grubbs, R. H., Synthesis and activity of a new generation of ruthenium-based olefin metathesis catalysts coordinated with 1,3-dimesityl-4,5-dihydroimidazol-2-ylidene ligands *Org. Lett.* 1999, *1*, 953–956.
128. Trnka, T. M.; Morgan, J. P.; Sanford, M. S.; Wilhelm, T. E.; Scholl, M.; Choi, T. L.; Ding, S.; Day, M. W.; Grubbs, R. H., Synthesis and activity of ruthenium alkylidene complexes coordinated with phosphine and N-heterocyclic carbene ligands *J. Am. Chem. Soc.* 2003, *125*, 2546–2558.
129. Falivene, L.; Credendino, R.; Poater, A.; Petta, A.; Serra, L.; Oliva, R.; Scarano, V.; Cavallo, L., SambVca 2. A Web Tool for Analyzing Catalytic Pockets with Topographic Steric Maps *Organometallics* 2016.
130. Billow, B. S.; McDaniel, T. J.; Odom, A. L., Quantifying ligand effects in high-oxidation-state metal catalysis *Nature Chem.* 2017, *9*, 837.
131. Liu, M.; Chen, M.; Zhang, S.; Yang, I.; Buckley, B.; Lee, J. K., Reactivity of carbene center dot phosphine dimers: proton affinity revisited *J. Phys. Org. Chem.* 2011, *24*, 929–936.
132. Jacobsen, E. N., Asymmetric catalysis of epoxide ring opening reactions *Acc. Chem. Res.* 2000, *33*, 421–431.
133. Collins, T. J., TAML oxidant activators: A new approach to the activation of hydrogen peroxide for environmentally significant problems *Acc. Chem. Res.* 2002, *35*, 782–790.
134. Pettinari, C.; Trofimenko, S.; Trofimenko, S., *Scorpionates II : chelating borate ligands* Imperial College Press ; Distributed by World Scientific Pub.: London Singapore ; Hackensack, NJ, 2008; p xxi, 541 p.
135. Trofimenko, S., Scorpionates the coordination chemistry of polypyrazolylborate ligands. Imperial College Press,: River Edge, NJ, 1999.
136. Saouma, C. T.; Peters, J. C., M≡E and M=E complexes of iron and cobalt that emphasize three-fold symmetry (E = O, N, NR) *Coord. Chem. Rev.* 2011, *255*, 920–937.
137. Juarez, R. A.; Lee, W.-T.; Smith, J. M.; Wang, H., Computational evaluation of tris(carbene)borate donor properties in $\{NiNO\}^{10}$ complexes *Dalton Trans.* 2014, *43*, 14689–14695.
138. Belkova, N. V.; Epstein, L. M.; Filippov, O. A.; Shubina, E. S., Hydrogen and Dihydrogen Bonds in the Reactions of Metal Hydrides *Chem. Rev.* 2016, *116*, 8545–8587.
139. Eisenberg, D. C.; Norton, J. R., Hydrogen-atom transfer-reactions of transition-metal hydrides *Isr. J. Chem.* 1991, *31*, 55–66.
140. Morris, R. H., Bronsted-Lowry Acid Strength of Metal Hydride and Dihydrogen Complexes *Chem. Rev.* 2016, *116*, 8588–8654.
141. Perutz, R. N.; Procacci, B., Photochemistry of Transition Metal Hydrides *Chem. Rev.* 2016, *116*, 8506–8544.
142. Wang, D. M.; Angelici, R. J., Metal-hydrogen bond dissociation enthalpies in series of complexes of eight different transition metals *J. Am. Chem. Soc.* 1996, *118*, 935–942.
143. Wiedner, E. S.; Chambers, M. B.; Pitman, C. L.; Bullock, R. M.; Miller, A. J. M.; Appel, A. M., Thermodynamic Hydricity of Transition Metal Hydrides *Chem. Rev.* 2016, *116*, 8655–8692.
144. Kubas, G. J., Fundamentals of H_2 binding and reactivity on transition metals underlying hydrogenase function and H_2 production and storage *Chem. Rev.* 2007, *107*, 4152–4205.
145. Kubas, G. J., Dihydrogen complexes as prototypes for the coordination chemistry of saturated molecules *Proc. Nat. Acad. Sci.* 2007, *104*, 6901–6907.
146. Kubas, G. J., Activation of dihydrogen and coordination of molecular H_2 on transition metals *J. Organomet. Chem.* 2014, *751*, 33–49.
147. Kubas, G. J.; Ryan, R. R.; Wrobleski, D. A., Molecular-hydrogen complexes of the transition-metals. 2. evidence for a new complex, $Mo(CO)(DPPE)_2(H_2)$, and for solution equilibrium between dihydrogen and dihydride forms, M-η^2-H_2 reversible H–M–H, in $W(CO)_3(PR_3)_2(H_2)$ *J. Am. Chem. Soc.* 1986, *108*, 1339–1341.
148. Crabtree, R. H., Dihydrogen Complexation *Chem. Rev.* 2016, *116*, 8750–8769.
149. Kubas, G. J.; Ryan, R. R.; Swanson, B. I.; Vergamini, P. J.; Wasserman, H. J., Characterization of the first examples of isolable molecular-hydrogen complexes, M $(CO)_3(PR_3)_2(H_2)$ (M = Mo, W; R = Cy, *i*-Pr). Evidence for a side-on bonded H_2 ligand *J. Am. Chem. Soc.* 1984, *106*, 451-452.
150. Bezman, S. A.; Churchill, M. R.; Osborn, J. A.; Wormald, J., Preparation and crystallographic characterization of a hexameric triphenylphosphinecopper hydride cluster *J. Am. Chem. Soc.* 1971, *93*, 2063.
151. Churchill, M. R.; Bezman, S. A.; Osborn, J. A.; Wormald, J., Synthesis and molecular geometry of hexameric triphenylphosphinocopper(I) hydride and crystal-structure of $H_6Cu_6(PPh_3)_6$•$HCONMe_2$ *Inorg. Chem.* 1972, *11*, 1818.
152. Mahoney, W. S.; Brestensky, D. M.; Stryker, J. M., Selective hydride-mediated conjugate reduction of alpha,beta-unsaturated carbonyl-compounds using $(Ph_3P)CuH_6$ *J. Am. Chem. Soc.* 1988, *110*, 291–293.
153. Stevens, R. C.; McLean, M. R.; Bau, R.; Koetzle, T. F., Neutron-diffraction structure-analysis of a hexanuclear copper hydrido complex, $H_6Cu_6[P(p\text{-}tolyl)_3]_6$: An unexpected finding *J. Am. Chem. Soc.* 1989, *111*, 3472–3473.
154. DiFranco, S. A.; Maciulis, N. A.; Staples, R. J.; Batrice, R. J.; Odom, A. L., Evaluation of Donor and Steric Properties of Anionic Ligands on High Valent Transition Metals *Inorg. Chem.* 2012, *51*, 1187–1200.
155. Bemowski, R. D.; Singh, A. K.; Bajorek, B. J.; DePorre, Y.; Odom, A. L., Effective donor abilities of E-*t*-Bu and EPh (E = O, S. Se, Te) to a high valent transition metal *Dalton Trans.* 2014, *43*, 12299-12305.
156. Bradley, D. C.; Newing, C. W., Low spin 4-coordinated chromium nitrosyl dialkylamides and alkoxides *J. Chem. Soc. D-Chem. Comm.* 1970, 219.
157. Odom, A. L.; Cummins, C. C.; Protasiewicz, J. D., Nitric-oxide cleavage - synthesis of terminal chromium(VI) nitrido complexes via nitrosyl deoxygenation *J. Am. Chem. Soc.* 1995, *117*, 6613–6614.
158. Heinselman, K. S.; Miskowski, V. M.; Geib, S. J.; Wang, L. C.; Hopkins, M. D., Molecular structures and FT-Raman spectroscopy of luminescent niobium and tantalum arylimido compounds *Inorg. Chem.* 1997, *36*, 5530–5538.
159. Enemark, J. H.; Feltham, R. D., Principles of structure, bonding, and reactivity for metal nitrosyl complexes *Coord. Chem. Rev.* 1974, *13*, 339–406.
160. Hsieh, C.-H.; Darensbourg, M. Y., A $\{Fe(NO)_3\}^{10}$ Trinitrosyliron Complex Stabilized by an N-Heterocyclic Carbene and the Cationic and Neutral $\{Fe(NO)_2\}^{9/10}$ Products of Its NO Release *J. Am. Chem. Soc.* 2010, *132*, 14118–14125.
161. Mingos, D. M. P.; Ibers, J. A., Crystal and molecular structure of dinitrosylbis(triphenylphosphine)iridium perchlorate, $[Ir(NO)_2(P(C_6H_5)_3)_2][ClO_4]$ *Inorg. Chem.* 1970, *9*, 1105–1111.
162. Templeton, J. L.; Ward, B. C., C^{13} Chemical-shifts of alkyne ligands as variable electron-donors in monomeric molybdenum and tungsten complexes *J. Am. Chem. Soc.* 1980, *102*, 3288–3290.
163. Krossing, I.; Reisinger, A., A stable salt of the tris(ethene)silver cation: Structure and characterization of $[Ag(\eta^2\text{-}C_2H_4)_3]^+$ $[Al\{OC(CF_3)_3\}_4]^-$ *Angew. Chem. Int. Ed.* 2003, *42*, 5725–5728.
164. Krossing, I.; Reisinger, A., A stable salt of the tris(ethene)silver cation: Structure and characterization of $[Ag(\eta^2\text{-}C_2H_4)_3]^+$ $[Al\{OC(CF_3)_3\}_4]^-$ *Angew. Chem.* 2003, *115*, 5903–5906.

165. Reisinger, A.; Trapp, N.; Knapp, C.; Himmel, D.; Breher, F.; Rueegger, H.; Krossing, I., Silver-Ethene Complexes $[Ag(\eta^2\text{-}C_2H_4)_n][Al(OR_F)_4]^-$ with n = 1, 2, 3 (R_F = Fluorine-Substituted Group) *Chem Eur. J.* 2009, *15*, 9505–9520.
166. Greenhough, T. J.; Kolthammer, B. W. S.; Legzdins, P.; Trotter, J., Organometallic nitrosyl chemistry. 10. Synthesis, X-ray structural characterization, and properties of dicarbonyl(η^5-cyclopentadienyl)(thionitrosyl) chromium *Inorg. Chem.* 1979, *18*, 3548–3554.
167. Rauchfuss, T. B.; Weatherill, T. D., Roussins red salt revisited - reactivity of $Fe_2(\mu\text{-}E)_2(NO)_4^{2-}$ (E = S, Se, Te) and related-compounds *Inorg. Chem.* 1982, *21*, 827–830.
168. Li, L. J.; Li, L. L., Recent advances in multinuclear metal nitrosyl complexes *Coord. Chem. Rev.* 2016, *306*, 678–700.
169. Cui, X. Y.; Joannou, C. L.; Hughes, M. N.; Cammack, R., The bacteriocidal effects of transition-metal complexes containing the NO^+ group on the food-spoilage bacterium clostridium-sporogenes *FEMS Microbiol. Lett.* 1992, *98*, 67–70.
170. Handy, L. B.; Ruff, J. K.; Dahl, L. F., Structural characterization of dinuclear metal carbonyl anions $[M_2(CO)_{10}]^{2-}$ (M = chromium and molybdenum) and $[Cr_2(CO)_{10}H]^-$: Marked stereochemical effect of a linearly protonated metal-metal bond *J. Am. Chem. Soc.* 1970, *92*, 7312.
171. Roziere, J.; Williams, J. M.; Stewart, R. P.; Petersen, J. L.; Dahl, L. F., A neutron-diffraction study of $[Et_4N]^+[Cr_2(CO)_{10}(\eta_2\text{-}H)]^-$: a $[M_2(CO)_{10}(\mu_2\text{-}H)]^-$ monoanion with a pseudo D_{4h} non-hydrogen geometry together with a disordered, bent symmetric metal-hydrogen-metal bond *J. Am. Chem. Soc.* 1977, *99*, 4497–4499.
172. Vicic, D. A.; Anderson, T. J.; Cowan, J. A.; Schultz, A. J., Synthesis, structure, and reactivity of a dinuclear metal complex with linear M–H–M bonding *J. Am. Chem. Soc.* 2004, *126*, 8132–8133.
173. Beagley, B.; Cruickshank, D. W.; Pinder, P. M.; Robiette, A. G.; Sheldrick, G. M., molecular structure of $Fe(CO)_5$ in gas phase *Acta Cryst. B: Struc. Sci.* 1969, *B 25*, 737.
174. Chin, H. B.; Bau, R., Crystal-structure of $Na_2Fe(CO)_4{\bullet}1.5(C_4H_8O_2)$: Distortion of $Fe(CO)_4^{2-}$ anion in solid-state *J. Am. Chem. Soc.* 1976, *98*, 2434–2439.
175. Teller, R. G.; Finke, R. G.; Collman, J. P.; Chin, H. B.; Bau, R., Dependence of $Fe(CO)_4^{2-}$ geometry on counterion: Crystal-structures of $K_2F3(CO)_4$ AND $[Na(Crypt)]_2[Fe(CO)_4]$ Crypt = $N(CH_2CH_2OCH_2CH_2OCH_2CH_2)_3N$ *J. Am. Chem. Soc.* 1977, *99*, 1104–1111.

CHAPTER 12

Basics of Transition-Metal Bonding 2

Conjugated π-Systems, Metal–Ligand Multiple Bonds, and Other Ligand Types

CONCEPT COVERAGE

1. Open-chain conjugated π-systems
2. Cyclic aromatic π-systems
3. Metal-ligand multiple bonds
4. Redox noninnocent ligands
5. Weakly-coordinating anions
6. Metal–metal bonds

"'Remarkable' seems a pallid word with which to describe this substance."
Marshall Gates (editor of Journal of the American Chemical Society, *regarding Wilkinson and Woodward's article on the structure of ferrocene)*[1]

In the previous chapter, simple ligands commonly encountered in transition-metal chemistry were discussed. Here, a variety of more complex, although many are still common, ligand types will be covered, including conjugated π-systems, carbenes, carbynes, oxos, imidos, and other important ligand classes.

Just as in the previous chapter, there are instances where the standard suggestions for understanding the electronic structure of the system lead to unlikely scenarios. Deviations from the usual suggestions are sensible in these cases, but one should be able to explain why this was done. The electronic structure of Ti(η^7-C_7H_8)Cp is an example that will be discussed, where the typical recommendations would suggest that this is a Ti(0) complex, but a more reasonable rendering would produce Ti(IV). Naturally, the idea is not to provide rules that are never to be broken, but tools to help you understand compounds.

12.1 Open-Chain π-Systems

The simplest of π-systems, olefins and alkynes, were covered in Chapter 11. Here we will cover larger, conjugated systems. While these may, in principle, be of any length, conjugated systems in an open-chain containing more than five members do not often bind to a single metal center all at the same time. As a consequence, the electronic structure of longer chains with dangling olefins may be described as a perturbation on the smaller three- to five-membered systems to be described here.

12.1.1 Common Open-Chain π-Systems

The commonly encountered open-chain π-systems are the simplest: allyl, butadiene, and benzyl. Obviously, these are usually found with a wide variety of substituents in modern examples, but we will cover the most basic forms here.

12.1.1.1 Allyl

The transition-metal chemistry of allyl groups is extensive, especially for late metals such as palladium. The allyl group can have two commonly encountered coordination modes (Table 12.1): η^1 where the ligand is electronically similar to a simple alkyl, and η^3 where all three

Table 12.1 η^1- and η^3-allyl groups and their electron counting.

	Electron count Neutral method	Electron count Charged method
η^1-allyl	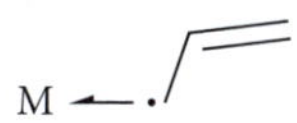 1-electron donor	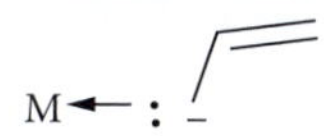 2-electron donor monoanion
η^3-allyl	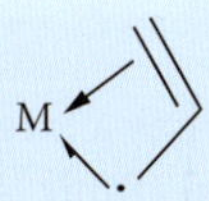 3-electron donor	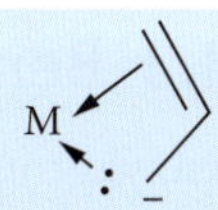 4-electron donor monoanion

carbons bind to the metal. The number of carbons bound is sometimes called the "hapticity" of the ligand. Ligands that undergo changes in hapticity might be said to undergo a "haptotropic shift".

The molecular orbital structure of allyl can be described as shown in Fig. 12.1. The π-system contains three atoms, each contributing one *p*-orbital, and as a result the ligand has three molecular orbitals in the π-system, one each with 1-, 2-, and 3-nodes.

Allyl complexes are prepared through a variety of different processes, including reactions with simple olefins, dienes, allene, and so on, or by simple salt-elimination reactions. Two such examples are shown in Fig. 12.2.[2–3]

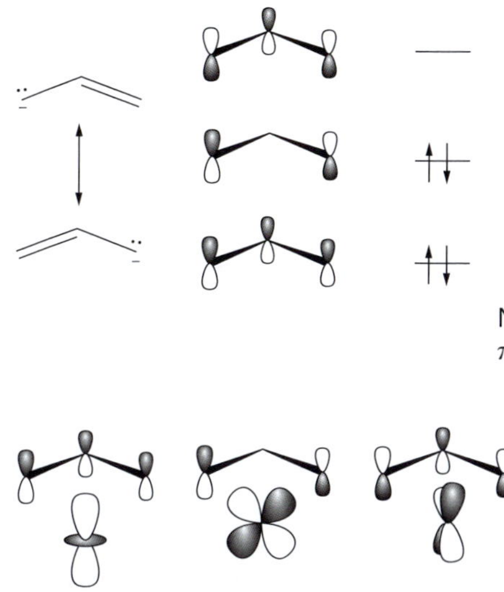

Fig. 12.1 (*top left*) Resonance forms for allyl anion. (*top right*) Molecular orbitals for the allyl anion π-system and (*bottom*) how they might interact with the *d*-orbitals on a transition metal.

12.1.1.2 Butadiene

Butadiene can bind in an s-*cis*- or s-*trans*-conformation to a metal center. In general, s-*cis*-butadiene is a far more common ligand; however, complexes with all four carbons of s-*trans*-butadiene bound to a metal have been characterized. In Fig. 12.3, the two conformations are shown with 17-electron $\{Fe(butadiene)_2\}^-$ as an example of the s-*cis*-butadiene structure, and 18-electron $Zr(1,4\text{-diphenylbutadiene})Cp_2$ as an example of the unusual s-*trans*-butadiene complexation mode.[4–6]

From an electron-counting perspective, butadiene is considered to be much as its line diagram would suggest: two olefins, and therefore a simple 4-electron donor in both the neutral and charged methods. As usual, the ligand is more complex than this simple picture suggests. In order to understand these nuances, molecular orbital (MO) theory can again be applied to the ligand π-system. We will ignore the σ-system and focus exclusively on the interacting π-system. In this case we have four carbons, and each will provide a single *p*-orbital to the π-system. We will then have four MOs of

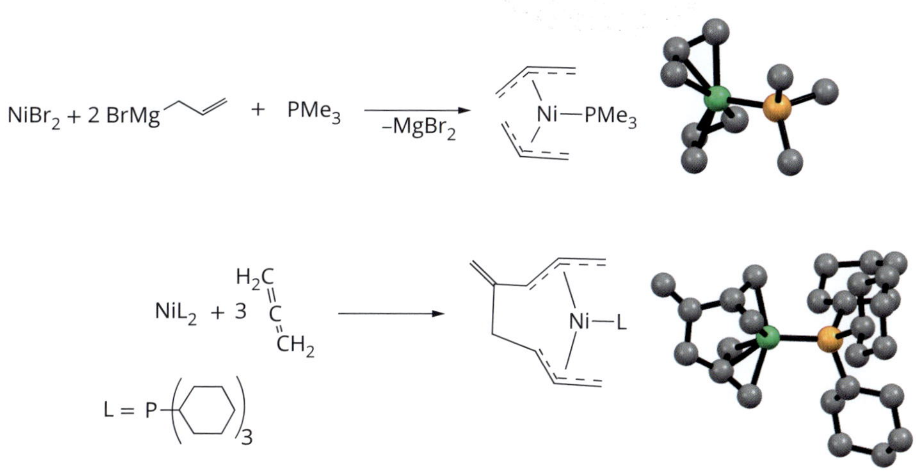

Fig. 12.2 Production of diallylnickel(II) complexes by salt elimination (*top*) and oligomerization of allene (*bottom*).[2–3]

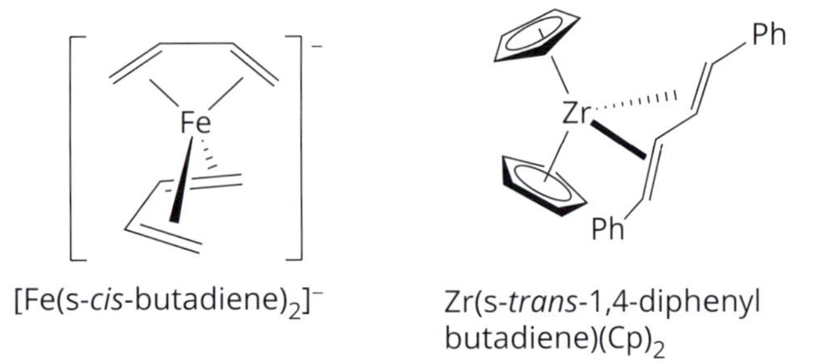

Fig. 12.3 Examples of s-*cis*- and s-*trans*-conformations of butadiene.[4–5]

relevance, which are shown in Fig. 12.4. The occupied orbitals have zero and one node in the π-system of butadiene; the two- and three-node orbitals are unoccupied in the free ligand and may act as acceptor orbitals for metal donation. As a result, one may consider two common bonding scenarios for an η^4-1,3-butadiene ligand, depending on the acceptor and donor properties of the metal to which it is attached (Fig. 12.5).

In any η^4-butadiene complex, both of these resonance forms may participate to one extent or another. For metal centers where the metal is not very reducing, the left-most structure should dominate. If the metal center is very reducing, then butadiene can accept two electrons to raise the oxidation state of the metal. Those two electrons from the metal could be regarded as going into the LUMO of butadiene, which has a C_2-C_3 double bond (Fig. 12.5), making the end carbons of the 1,3-butadiene alkyl groups on the metal. If the right resonance form is favored, butadiene overall acts as a 2-electron acceptor, and the C_2-C_3 double bond will act as a 2-electron donor towards the metal.

The left form is sometimes referred to as the π,η^4-geometry, and the right resonance form is the σ^2,π-geometry. These are obviously limiting forms, and many complexes show characteristics consistent with a mixture of these prototypical bonding patterns.[7–8]

12.1.1.3 Benzyl

The benzyl group, $CH_2C_6H_5$, has a large number of observed binding modes, illustrated in Table 12.2. In 2012, Parkin and coworkers analyzed the structurally characterized benzyl groups and found that most were η^1, 92.6%. The next most common coordination mode was η^2 at 6.1%, followed by η^3 at 0.9%. Furthermore, there is also an example of an η^7-benzyl group.[9]

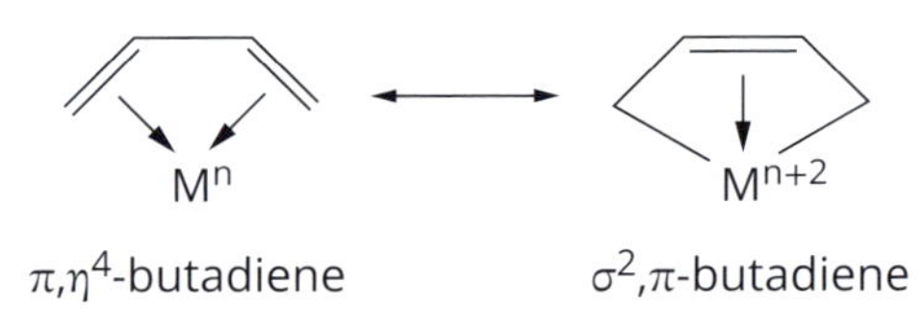

Fig. 12.5 Two different possible resonance forms for η^5-1, 3-butadiene on a metal center, M, which starts in the *n* oxidation state.

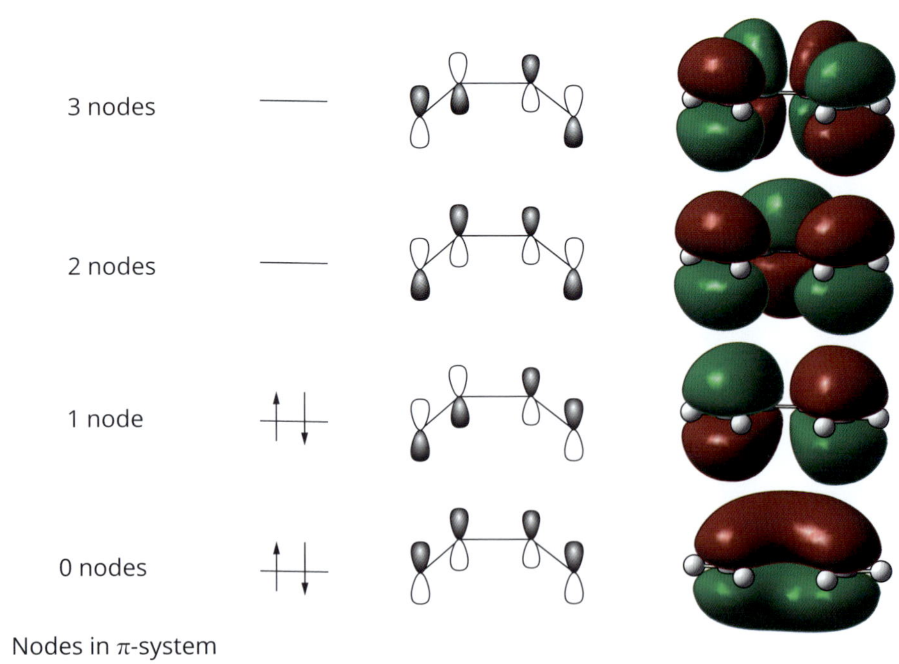

Fig. 12.4 Molecular orbitals for the π-system in 1,3-butadiene.

The η^1-binding mode is basically a simple alkyl and is electron-counted in the same way. The η^2-binding mode, which involves a single bond to the benzyl carbon and a weaker interaction with the *ipso*-carbon, may be regarded as the metal accepting electron density from the aromatic ring, perhaps through the resonance form shown in Table 12.2. An η^3-benzyl is similar electronically to η^3-allyl and is electron-counted in the same way. However, because of rearomatization of the ring, η^3-benzyl generally converts to η^1-benzyl at far lower energies than the η^3- to η^1-allyl conversion. The η^7-benzyl has been observed, which involved η^6-coordination of the phenyl ring to the metal in addition to the σ-bond form the benzyl carbon. An η^4-mode has possibly been observed in some complexes as well.[9]

Table 12.2 Some known binding modes of the benzyl ligand, with examples and suggested donor electron counts in the charged and neutral counting methods. η^1: $Zr(\eta^1\text{-}CH_2Ph)_2(\eta^5\text{-}Cp)_2$.[10] η^2: $Zr(\eta^1\text{-}CH_2Ph)_2(\eta^2\text{-}CH_2Ph)_2$.[9] η^3: $Ni(\eta^1\text{-}CH_2Ph)(\eta^3\text{-}CH_2Ph)PMe_3$.[11] η^7: $\{Zr(\eta^7\text{-}CH_2Ph)(\eta^3\text{-}CH_2Ph)(\eta^5\text{-}Cp)\}^+$.[12]

Binding mode	Electron count	Example	Structure
η^1 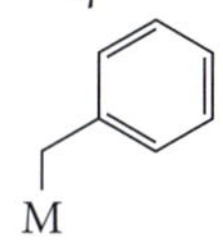	Charged: 2 Neutral: 1	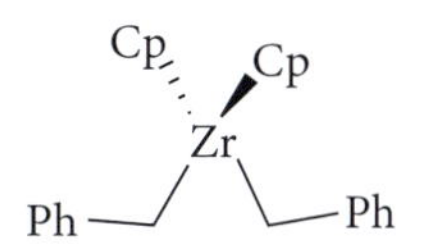 $Zr(\eta^1\text{-}CH_2Ph)_2(\eta^5\text{-}Cp)_2$	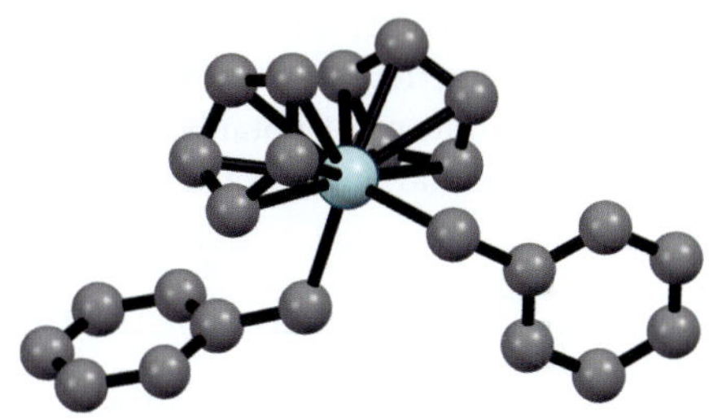
η^2 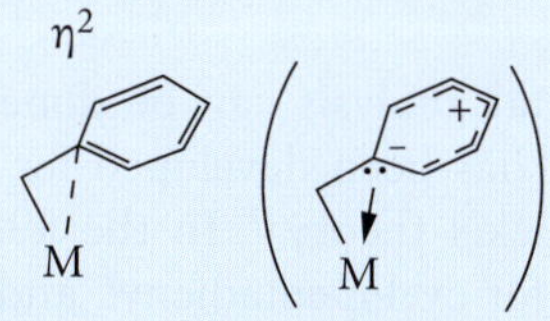	Charged: 4 Neutral: 3	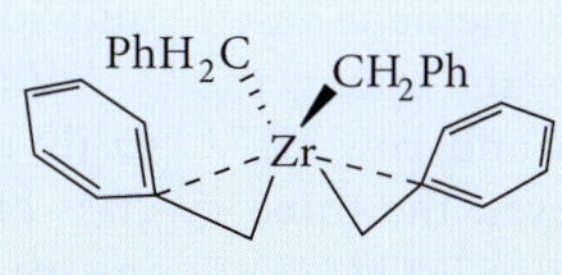 $Zr(\eta^1\text{-}CH_2Ph)_2(\eta^2\text{-}CH_2Ph)_2$	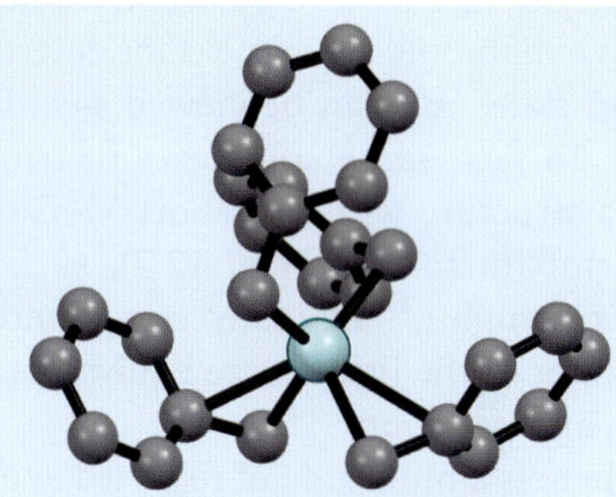
η^3 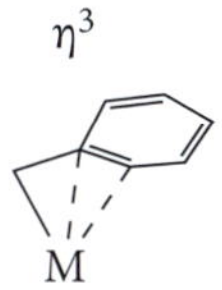	Charged: 4 Neutral: 3	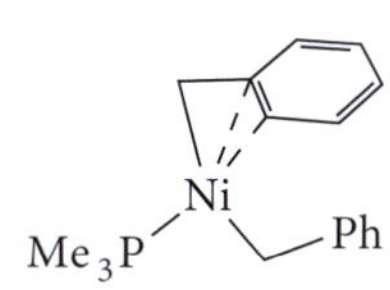 $Ni(\eta^1\text{-}CH_2Ph)(\eta^3\text{-}CH_2Ph)PMe_3$	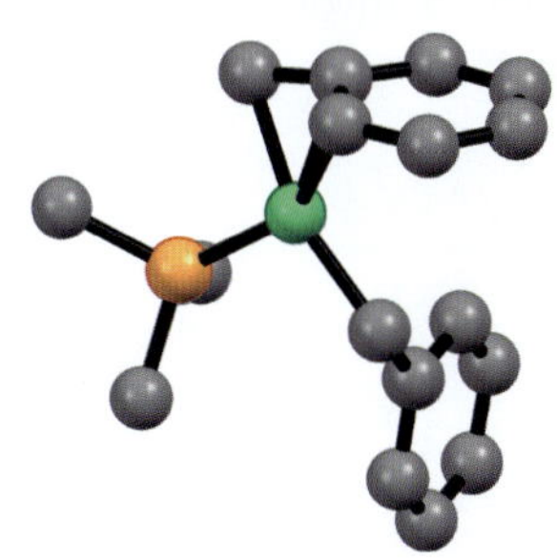
η^7 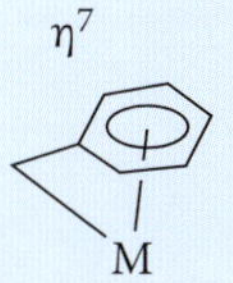	Charged: 8 Neutral: 7	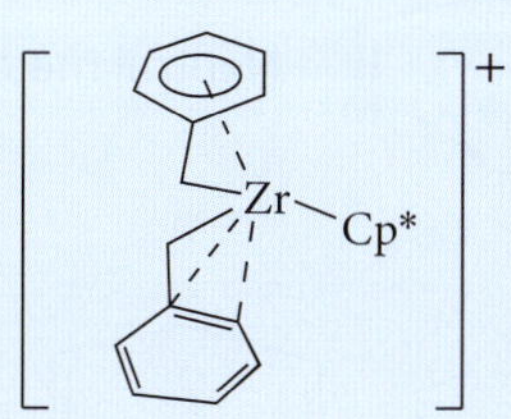 $\{Zr(\eta^7\text{-}CH_2Ph)(\eta^3\text{-}CH_2Ph)(\eta^5\text{-}Cp)\}^+$	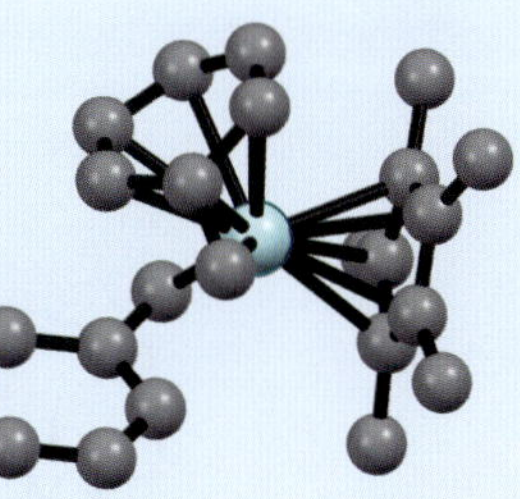

12.1.2 Less Common Open-Chain π-Systems (expanded coverage)

The less common open-chain π-systems to be discussed are pentadienyl and trimethylenemethane.

12.1.2.1 Pentadienyl[13–22]

A surrogate for the more common cyclopentadienyl (Section 12.2.2.1), pentadienyl can also be used as an ancillary ligand and has a fairly large breadth of use across the transition series.

For electron-counting purposes, pentadienyl is identical to cyclopentadienyl: the charged method considers the ligand a 6-electron donor and an anion, while the neutral method considers it a 5-electron (radical + 2 olefins) donor (Fig. 12.6). In some cases, the pentadienyl complex is stable where the cyclopentadienyl is not; most notably, titanocene ($TiCp_2$) is unstable, while Ti(pentadienyl)$_2$ is isolable.[20]

12.1.2.2 Trimethylenemethane (TMM)

A ligand with an intriguing structure related to allyl is trimethylenemethane, a 4-carbon π-bound ligand. The ligand acts as a dianionic substituent for determination of formal oxidation state, and can be viewed as a dialkyl with a pendant olefin. In fact, some complexes of trimethylenemethane exhibit a structure with two short M–C distances and a longer M–olefin interaction (Fig. 12.7), such as Sita and coworkers' Zr complex.[23] In many compounds, however, the three β-carbons of the ligands are equivalent, such as in Fe(TMM)$(CO)_3$ and its cation.[24–25]

There is a multitude of reported syntheses for the ligands, some of which involve predictable reaction sequences and others that involve unusual rearrangements/eliminations. More straightforward routes involve reaction of a 2-chloromethyl-3-trimethylsilylpropene with an appropriate metal precursor; acetoxy is often used in place of chloride as well, such as the example in Eq. 12.1 with Vaska's complex, Ir(Cl)(CO)$(PPh_3)_2$.[26]

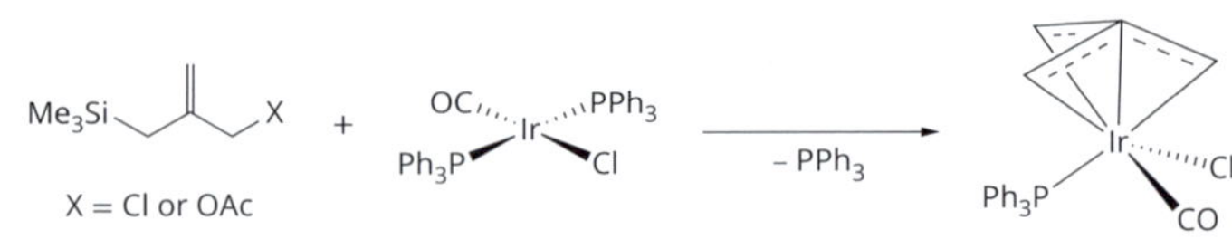

12.1

Anderson, Bergman, and McNeill reported an interesting C–H abstraction, C–C activation, C–H elimination route to a ruthenium TMM shown in Fig. 12.8.[27] Carbon–carbon bond cleavage of this type is very uncommon and may indicate a very high favorability for TMM in this case.

12.2 Cyclic Aromatic π-Systems

A common series of ligands in transition-metal chemistry are cyclic planar π-systems. In order to understand these systems generally, it is convenient to discuss their electronic structure as a class. Following this, the commonly encountered cyclic π-systems will be covered in more detail individually.

While a full molecular orbital analysis can be done, an easily remembered mnemonic[28] has been developed for this class of molecules based on Hückel theory.[29] In the following we illustrate the mnemonic for cyclopentadienyl, and the results for the smallest planar hydrocarbon ring of the class, C_3H_3, through C_8H_8, are shown in Table 12.3.

In the mnemonic (Fig. 12.9a) we consider a circle of radius 2β, where β is a relative energy term called the bonding or

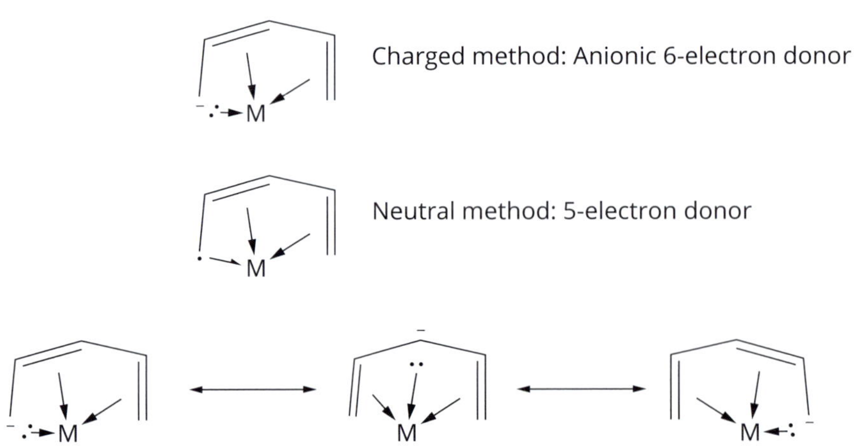

Fig. 12.6 (*top*) Electron-counting using the charged and neutral methods for pentadienyl. (*bottom*) Some important resonance forms for metal pentadienyl complexes.

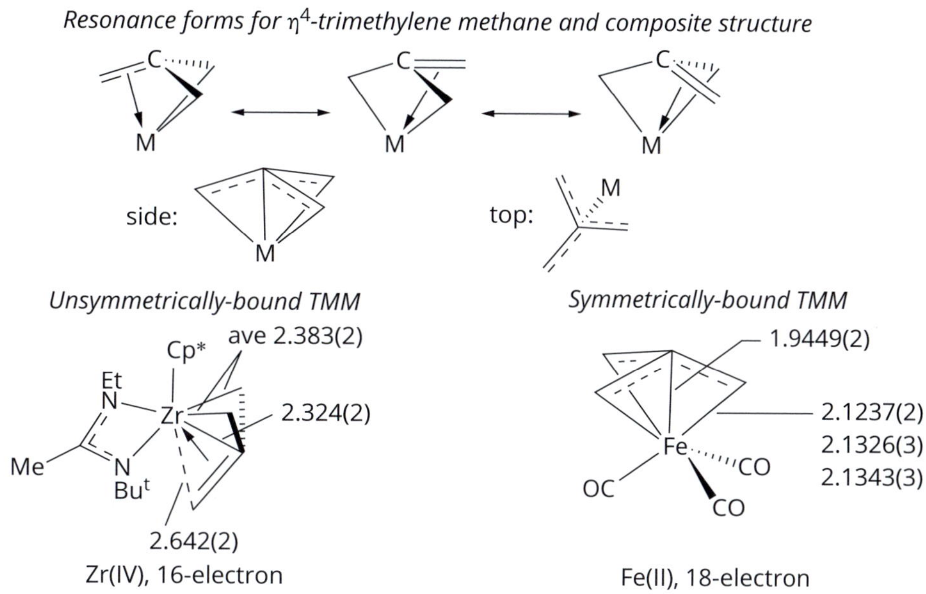

Fig. 12.7 *(top)* Resonance forms for a bound trimethylenemethane (TMM) and often-drawn composite structures. *(bottom)* An example in which all the β-carbons of the TMM ligand are not symmetrically bound *(left)*, and an example where they are *(right)*.[23-25]

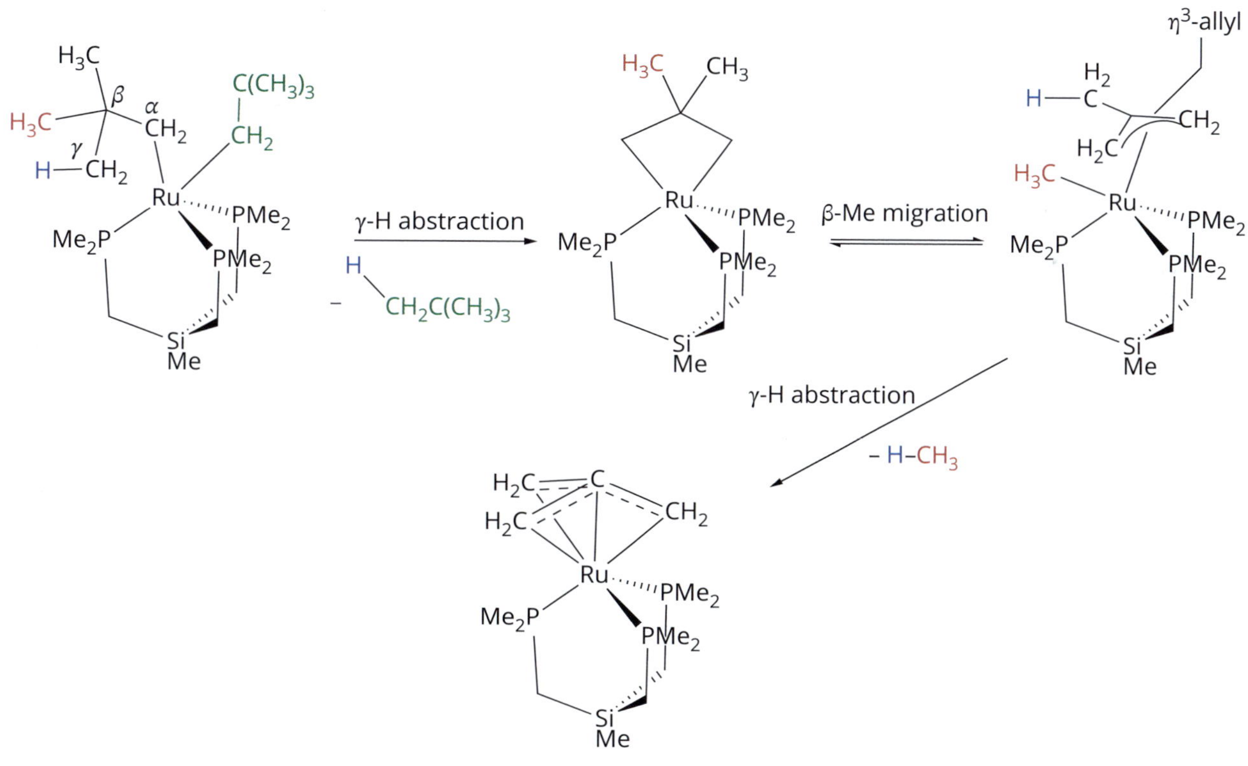

Fig. 12.8 Synthesis of a ruthenium(II) TMM triphosphine complex.[27]

resonance energy (Appendix M). A horizontal line through the center of the circle will be labeled as energy level α, which is the energy of the noninteracting atomic *p*-orbitals in the π-system.

In this case for the cyclopentadienyl ring, we are interested in a fully conjugated ring of $n = 5$, where n = the number of carbons in the ring. The ring of interest is inscribed in the circle with one vertex at the bottom (6 o'clock)—a figure sometimes referred to as a Frost circle.[28] For cyclopentadienyl, we draw a pentagon as shown in Fig. 12.9b.

Anywhere a vertex of the pentagon touches the circle will be an orbital of that relative energy (Fig. 12.9c).

Table 12.3 MO diagrams of the neutral π-systems and electron-counting for conjugated planar hydrocarbon rings from C_3H_3 to C_8H_8.

Number of carbons (n)	Frost circle	MO diagram for π-system of neutral ligand	Neutral method	Charged method
3	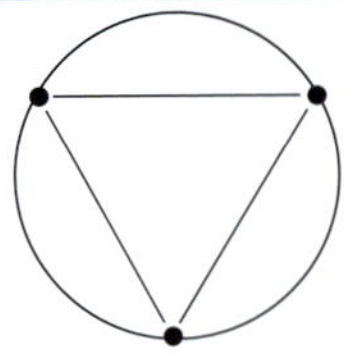	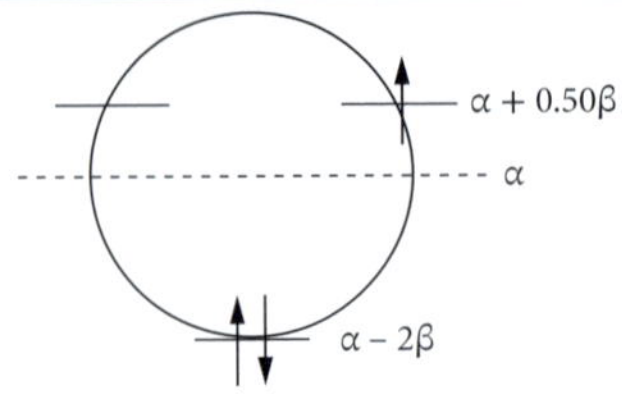	 3-electron donor	 2-electron donor
4	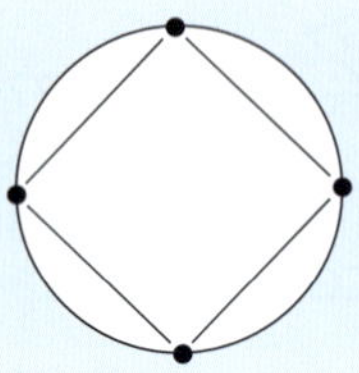	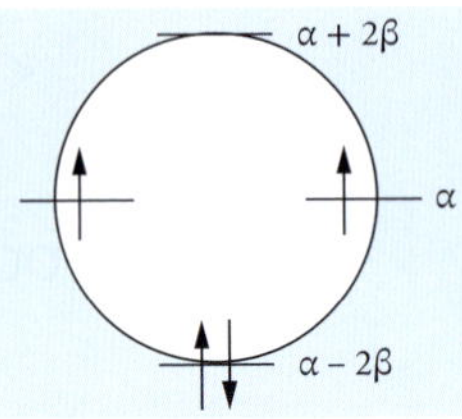	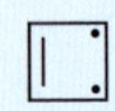 4-electron donor	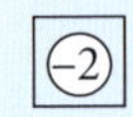 6-electron donor
5	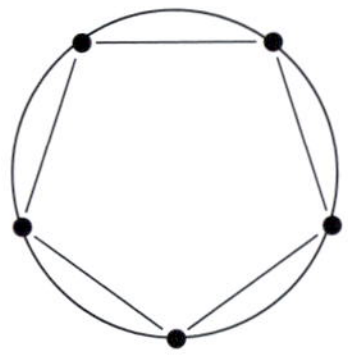	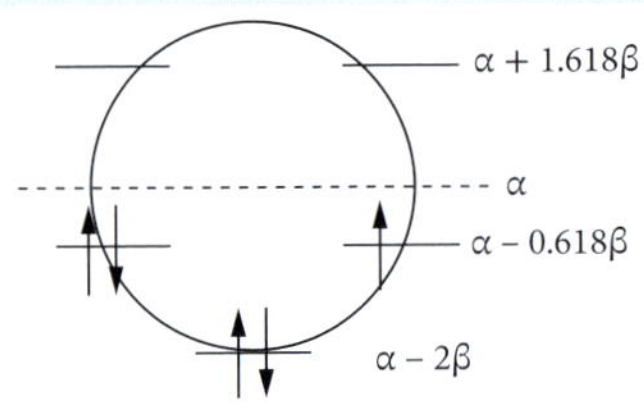	 5-electron donor	 6-electron donor
6	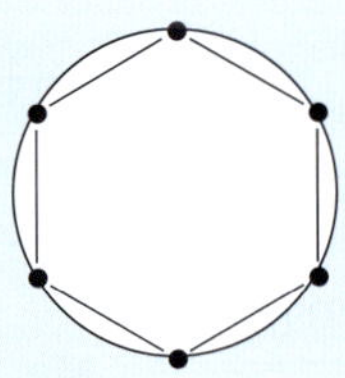	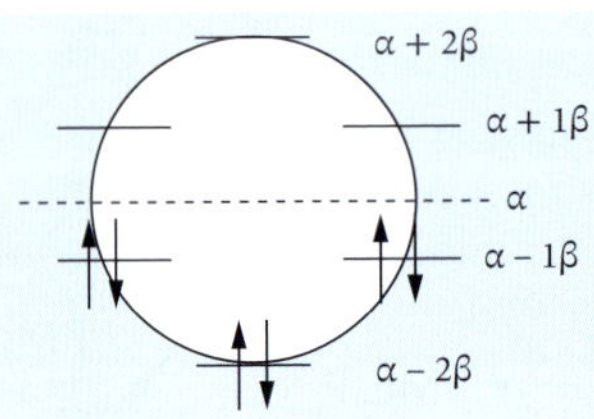	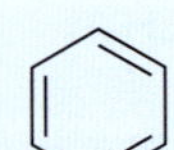 6-electron donor	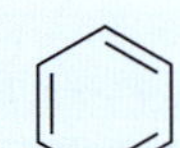 6-electron donor
7		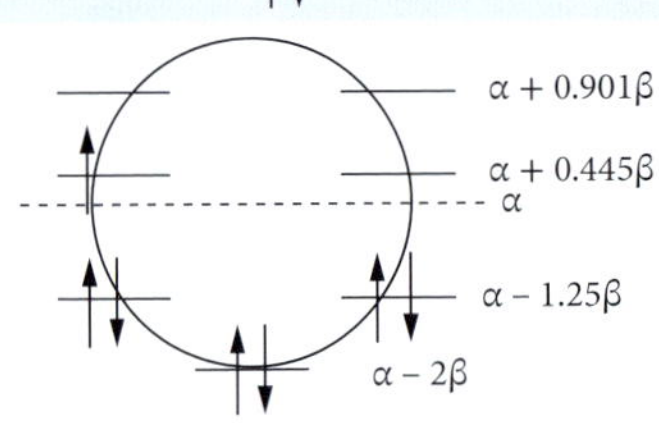	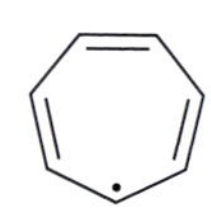 7-electron donor	 6-electron donor
8	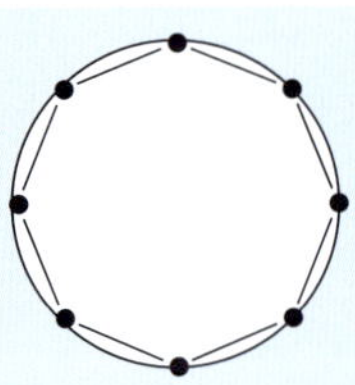	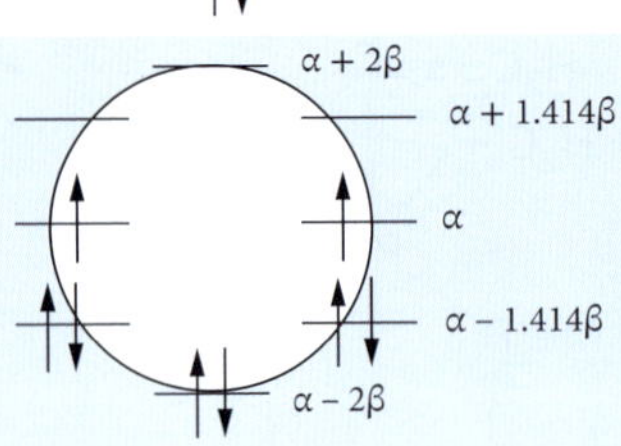	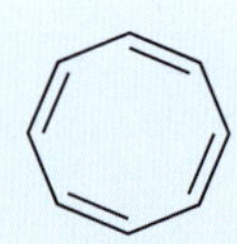 8-electron donor	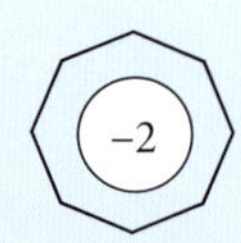 10-electron donor

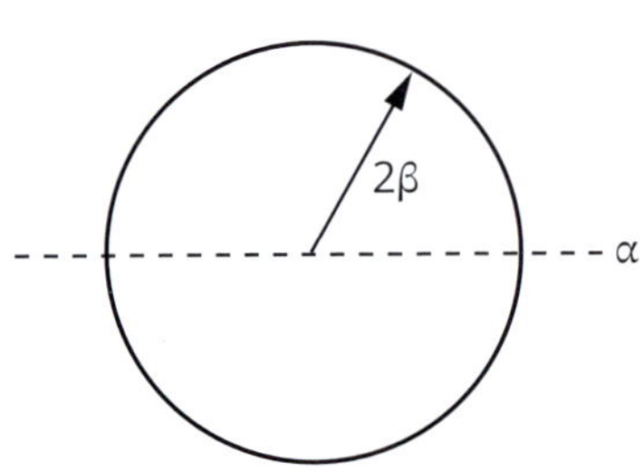

Fig. 12.9a Base Frost circle showing form of α and β.

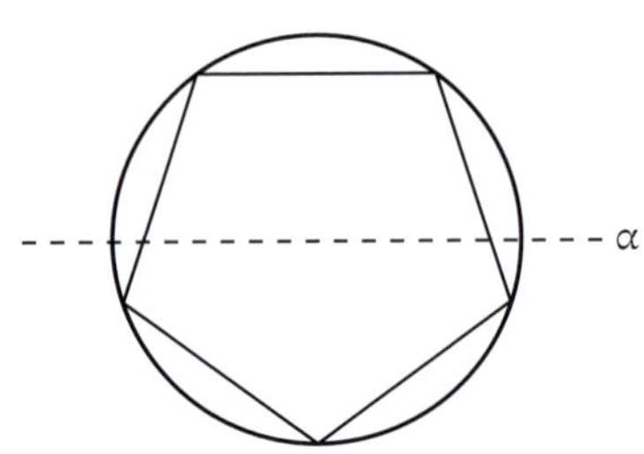

Fig. 12.9b Frost circle for a five-membered ring.

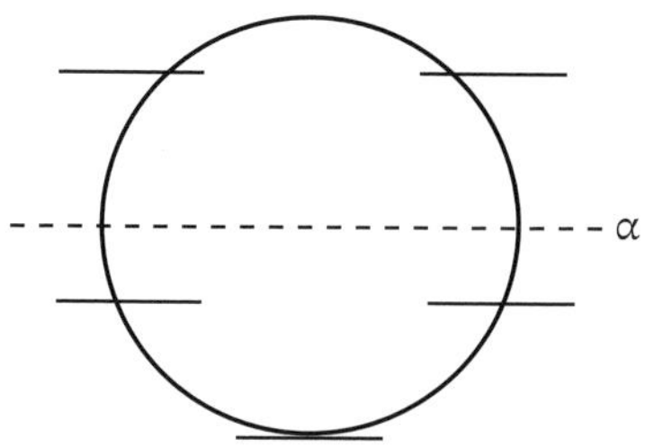

Fig. 12.9c Orbitals for a five-membered ring.

In fact, the relative energies of the orbitals can be calculated at the Hückel level by using Eq. 12.2, where ε^k is the energy of the kth orbital and $k = 0, \pm1, \pm2, \ldots, \tau$.

$$\varepsilon^k = \alpha - 2\beta\cos\frac{2\pi k}{n} \qquad \mathbf{12.2}$$

The factor τ is calculated using Eq. 12.3 if the number of carbons in the ring (n) = odd and using Eq. 12.4 if n = even.

$$\tau = \pm\frac{n-1}{2} \text{ if } n \text{ is odd} \qquad \mathbf{12.3}$$

$$\tau = +\frac{n}{2} \text{ if } n \text{ is even} \qquad \mathbf{12.4}$$

For cyclopentadienyl (n = 5) Eq. 12.3 is used, and $\tau = \pm2$. Consequently, $k = 0, \pm1, \pm2$, which leads to five different orbitals. The relative energies (Eq. 12.2) in terms of α and β are calculated as follows. Because cosine is an even function, the $+k$ and $-k$ energies are the same, producing three different energy levels for the 5-orbitals.

$$k = 0$$
$$\varepsilon^{(0)} = \alpha - 2\beta\cos(0) = \alpha - 2\beta$$

$$k = 1$$
$$\varepsilon^{(\pm1)} = \alpha - 2\beta\cos\left(\frac{360(1)}{5}\right) = \alpha - 0.618\beta$$

$$k = 2$$
$$\varepsilon^{(\pm2)} = \alpha - 2\beta\cos\left(\frac{360(2)}{5}\right) = \alpha + 1.618\beta$$

Each carbon is σ-bound to two other carbons and a hydrogen, using three of its four electrons in the σ-framework. Consequently, each carbon will contribute one electron to the π-system for a total of five electrons in our molecular orbital diagram shown in Fig. 12.9d. This is the MO diagram of the radical shown.

The most stable aromatic electronic configuration has the energy levels up to and including α filled, and those higher in energy than α empty (Fig. 12.9e). In order for cyclopentadienyl

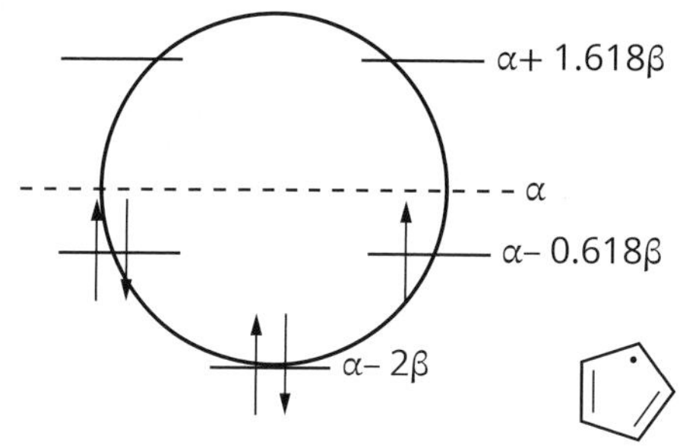

Fig. 12.9d Orbitals with five electrons (one for each carbon) in the π-system. Approximate orbitals for cyclopentadienyl.

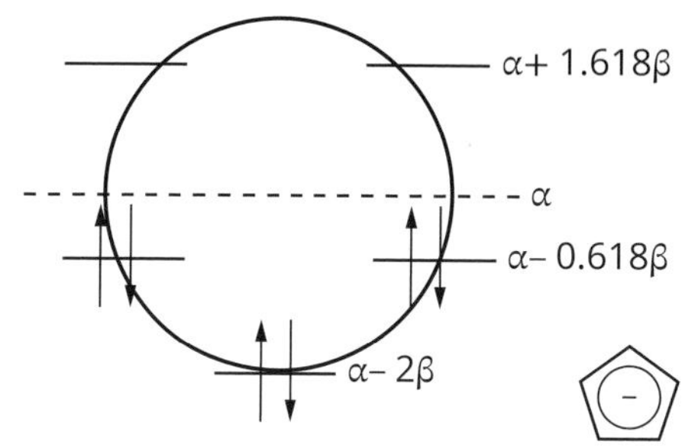

Fig. 12.9e Orbitals for aromatic cyclopentadienide.

to become aromatic, one electron is required to be added to the system into the α – 0.618 level. In other words, cyclopentadienide, $C_5H_5^-$, is aromatic.

As might be expected, the neutral electron-counting system considers these conjugated planar systems as neutral radicals or multiples of 2-electron donors. *In all cases, the number of electrons donated by the ring to the metal in the neutral method is equal to n, the number of carbons in the ring.*

The charged electron-counting system considers these ligands to be aromatic if all the carbons are approximately bound in an equal fashion; as a result, it is sometimes necessary to invoke electron transfer to or from the metal to reach the necessary electron counts on the aromatic rings.

Table 12.3 shows the ring systems with n = 3–8, their molecular orbital diagrams for the π-system with energies in terms of α and β, line drawings demonstrating how the two different electron counting methods view the ligand, and the number of electrons donated in each system.

If one recalls the Hückel rule for aromaticity, $4n + 2$, it is easy to determine the number of electrons required to be added to or removed from the π-system to reach aromaticity.[30–31] The relevant aromatic π-systems should be:

(n = 0) 2 π-electrons: $C_3H_3^+$

(n = 1) 6 π-electrons: $C_4H_4^{2-}$, $C_5H_5^-$, C_6H_6, $C_7H_7^+$

(n = 2) 10 π-electrons: $C_8H_8^{2-}$

The coordination chemistry of these π-systems will be discussed briefly, with examples of electron-counting for each system.

12.2.1 Common Cyclic π-Systems

The two most encountered cyclic π-systems by far are cyclopentadienyl and arenes (or benzene). These will be discussed first, and both are straightforward 6π-electron aromatic systems. These two ligands are discussed in some detail. Following this, some less common ring systems are briefly covered.

12.2.1.1 Cyclopentadienyl (Cp)[32]

A firestorm of activity was created by Pauson and Kealy's report in *Nature* in 1951, detailing the reaction of C_5H_5MgBr with ferric chloride.[33] (It should be noted that Kealy and Pauson were not technically, as is often cited, the first to report ferrocene. Miller, Tebboth, and Tremaine (MTT) published the same complex with an earlier received date, July 11, 1951, versus Kealy and Pauson's August 7, 1951. MTT prepared the compound by passing cyclopentadiene over iron on a support of alumina, potassium oxide, and molybdenum oxide, a Haber–Bosch catalyst, at 300 °C. Indeed, the title of their paper was "Dicyclopentadienyliron". Pauson and Kealy's paper simply appeared in print first, December 1951 versus February 1952.)[34] The anticipated result from Pauson and Kealy's experiment was the production of fulvalene; however, the compound obtained was an orange solid with the formula $FeC_{10}H_{10}$ that was "insoluble in, and apparently unattacked by, water, 10% caustic soda and concentrated hydrochloric acid even at the boiling point." This was in marked contrast to other transition-metal-organics prepared previously, which generally decomposed well below room temperature. The authors proposed a linear dialkyl of iron as the structure, which insightfully was proposed to be stable where other similar derivatives were not, due to "the tendency of the cyclopentadienyl group to become 'aromatic' by acquisition of a negative charge". The authors proposed that strong iconicity in the system, $Cp^- \ Fe^{2+} \ Cp^-$, was perhaps responsible for the stability.

On reading the article, two groups at Harvard set out to examine the compound and further the chemistry of cyclopentadienyl derivatives: the Woodward and Wilkinson groups. Woodward's interest was reportedly in examining the chemistry of the new aromatic system, while Wilkinson set out to expand cyclopentadienyl chemistry to other metals. Wilkinson, Woodward, and coworkers published that the correct structure was the "sandwich" complex shown in Fig. 12.10.[35] On the left is the staggered isomer, and on the right is the eclipsed isomer. The barrier between the two, as might be expected, is quite small: ~3 kcal/mol.[36–37]

The structure of ferrocene has been examined, using a variety of methods including single-crystal X-ray diffraction at various temperatures, neutron diffraction, and even solution-phase neutron diffraction. The compound has phase changes in the solid state with the eclipsed versus staggered, and even structures in between, being favored at different temperatures. In solution, it was found that the eclipsed structure was slightly favored, and in addition the Cp(centroid)–Fe–Cp(centroid) angle of 180° was preferred but varied down to about 150°.[38–42]

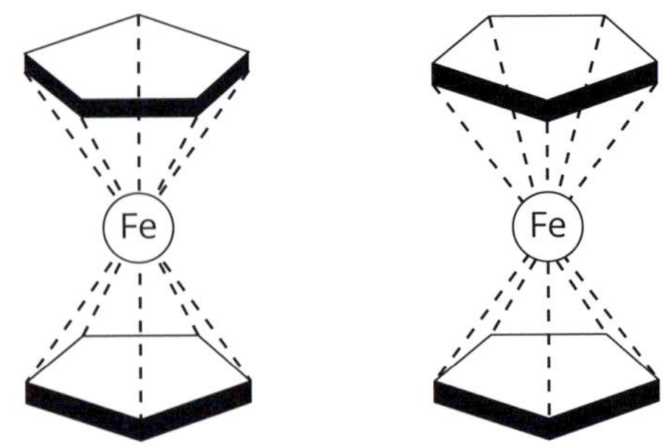

Fig. 12.10 Structure of ferrocene as shown in Wilkinson, Woodward, and coworkers' paper in 1952.[35]

The next two years saw an explosion of reports in metal cyclopentadienyl (and related indenyl, benzocyclopentadienyl) complexes by the groups of Fischer and Wilkinson. In 1973 the two shared the Nobel Prize for "their pioneering work, performed independently, on the chemistry of the organometallic, so called, sandwich compounds."

Cyclopentadienyl (Cp) derivatives are some of the most common ligands in inorganic chemistry, with compounds known across the periodic table. The ligand is commonly found in three different hapticities: alkyl-like η^1, allyl-like η^3, and aromatic η^5. When "Cp" is written without a specified hapticity it is generally assumed to be η^5 (Fig. 12.11).

Fig. 12.11 illustrates an example of a structurally characterized η^1-Cp complex, $Fe(\eta^1\text{-}Cp)(Cp)(CO)_2$, first prepared by Wilkinson and Piper.[43] The compound shows two singlets in the ^1H NMR because of fast migration around the η^1-Cp ring at higher temperature. An η^1-Cp is alkyl-like, making it a 1-electron donor in the neutral method and a 2-electron anionic donor in the charged method.

An η^3-Cp is very much like an η^3-allyl ligand in how it is electron-counted. In the charged method the ligand would be considered an anionic 4-electron donor. In the neutral method it is a 3-electron (1 radical and 1 olefin) donor. Fig. 12.11 shows the seminal example of an η^3-Cp, $W(\eta^3\text{-}Cp)(Cp)(CO)_2$, prepared by Brintzinger and coworkers.[44]

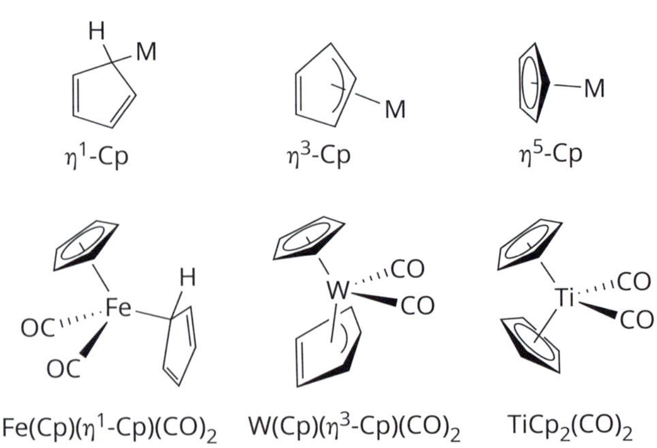

Fig. 12.11 Three compounds of the general formula $MCp_2(CO)_2$ with M = Fe, W, and Ti. All of the compounds are 18-electron at the metal center due to changes in hapticity of the Cp ring.

Distinguishing between hapticities generally relies on M–C bond lengths and the planarity of the five-membered ring. Obviously, the η^1-configuration is characterized by one short and four long M–C bonds with bond alternation in the C_5H_5 ring. Compare the bond distances in the two different Cp rings in WCp(η^1-Cp)(N-mesityl)$_2$, Fig. 12.12; all of the η^5-Cp W–C distances are similar 2.3–2.5 Å) but not necessarily identical due to the asymmetry of the complex, while the η^1-Cp has one distance that is obviously far shorter than the others.[45] Bond alternation should also be observed in the solid state in the cyclopentadienyl ring with typical C–C single and double bond distances.

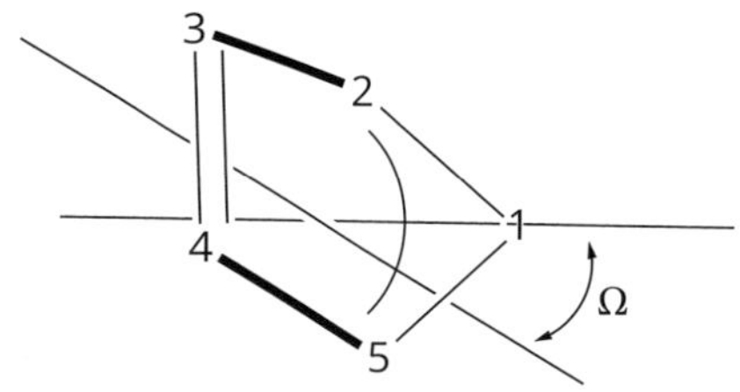

Fig. 12.13 The definition of the Ω angle for an η^3-cyclopentadienyl.

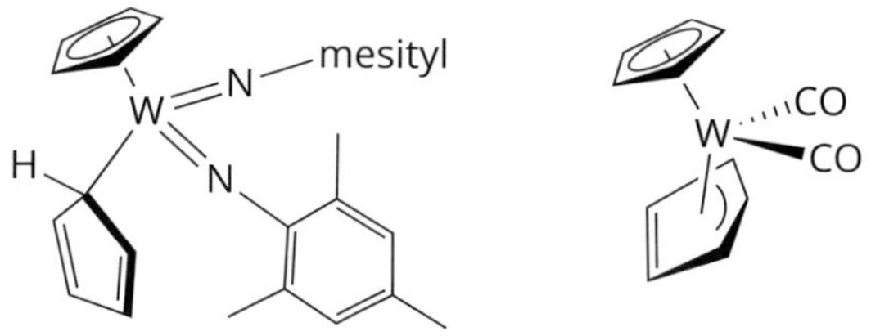

W–C η^1-Cp	W–C η^5-Cp
2.220(6)	2.334(6)
3.144	2.440(6)
3.053	2.502(6)
3.985	2.474(6)
4.034	2.416(6)

W–C η^3-Cp	W–C η^5-Cp
2.28(2)*	2.34(3)*
2.40(2)	2.31(2)
2.98(2)	2.42(2)

Fig. 12.12 Structure of an η^1-Cp example, WCp(η^1-Cp)(Nmesityl)$_2$,[45] and an η^3-Cp example, WCp(η^3-Cp)(CO)$_2$,[44] with W–C bond distances in Ångstroms illustrating the hapticity differences. In WCp(η^1-Cp)(Nmesityl)$_2$, the errors for the nonbonded W–C distances in the η^1-Cp were not reported but are probably similar to the other W–C errors. In WCp(η^3-Cp)(CO)$_2$, only three bond distances are shown because there is a crystallographic mirror plane that runs through the molecule. Consequently, one of the carbons is unique (marked with *) and the other two generate two identical carbons on the other side of the mirror plane with identical W–C distances.

The bond-distance comparison for an η^3-Cp versus an η^5-Cp is also shown in Fig. 12.12. As would be expected, the η^5-Cp has very similar W–C distances from about 2.3 to 2.4 Å. The η^3-Cp has three relatively short W–C distances of 2.28, 2.40, and 2.40 Å with two longer W–C distances of 2.98 Å.[44] In some cases, η^3-Cp rings can fold with the unbounded carbons moving out of plane with the metal-bound carbons. If one plane is defined with the bound carbons (1, 2, and 3) and another with carbons 2, 3, 4, and 5, then the two planes (Fig. 12.13) meet at an angle, Ω. The Ω angle between these two planes in WCp(η^3-Cp)(CO)$_2$ is about 20°.

In some cases the Cp ring may be flat but highly irregularly bound to the metal center with very different M–C distances. These Cp rings are often referred to as "slipped". This is a very common phenomenon in related indenyl complexes, which are far more likely to adopt an η^3-configuration than their Cp analogues. This is due to the fused six-membered ring, which can only be fully aromatic when the Cp ring is not aromatic, and vice versa. A related compound is fluorenyl with two fused benzene rings on a central cyclopentadienyl, which has a strong tendency to slip to η^3- or even η^1-configurations. Examples of these Cp derivatives are shown in Fig. 12.14.

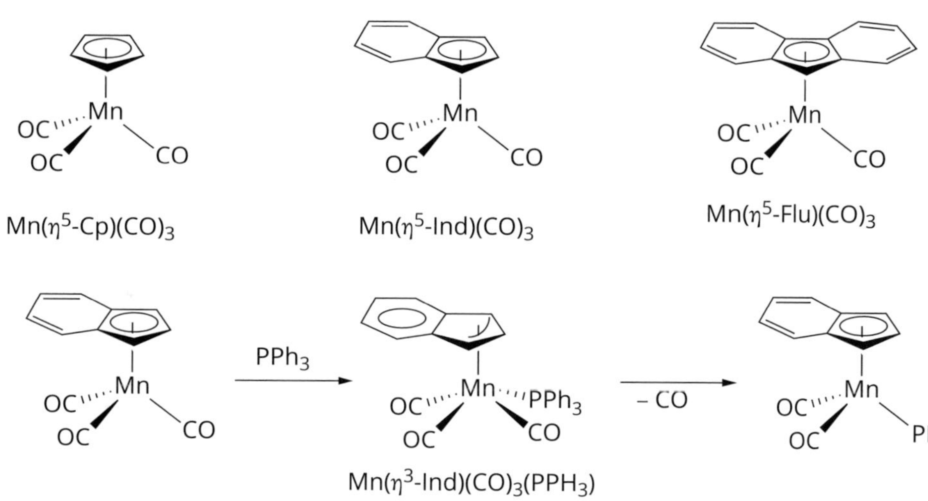

Fig. 12.14 (*top*) Structures of Mn(η^5-Cp)(CO)$_3$, Mn(η^5-Ind)(CO)$_3$, and Mn(η^5-Flu)(CO)$_3$ studied by Basolo.[46–48] (*bottom*) Reaction with PPh$_3$ with the structure of the proposed "slipped" indenyl complex for carbonyl substitution. Note that all of the complexes are 18-electron, and the η^3-Ind shown occurs to avoid a 20-electron intermediate in the substitution reaction.

The ability of indenyl and fluorenyl to slip can lead to dramatic enhancements in rate constants for associative substitution reactions—a phenomenon given the moniker the "indenyl effect" by Basolo. The Basolo group studied the substitution of carbonyls in $Mn(\eta^5\text{-R})(CO)_3$, where R = Cp, indenyl (Ind), or fluorenyl (Flu), with PPh_3. The substitution of CO ligands in this group of complexes with PPh_3 occurs through an associative mechanism, and the ring slips to produce a phosphine-coordinated intermediate. If the Cp or Cp derivative did not slip to an η^3-coordination mode, the complex would reach an unfavorable 20-electron configuration. As a result, systems where ring slippage is favorable have faster substitution rates to form $Mn(\eta^5\text{-R})(CO)_2(PPh_3)$ so that the rate constant for complexes goes as fluorenyl > indenyl >> cyclopentadienyl.[46–48]

A Lewis-like form for an η^5-Cp may be drawn as in Fig. 12.15, with one M–C covalent bond and two metal–olefin interactions. Just as in other olefins, there would be potential backbonding interactions, and resonance forms for metallocyclopropane may participate. In principle, the metal could bond to all five carbons with some covalent character, if orbitals of proper symmetry exist. All of this is in resonance so that all of the M–C bonds are equivalent, or nearly so depending on the specific complex.

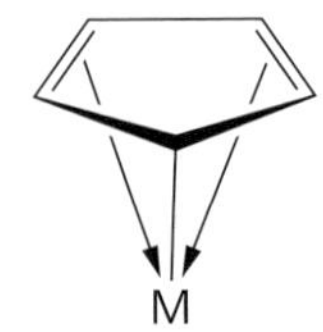

Fig. 12.15 An η^5-Cp may be viewed as an alkyl and two olefins with resonance delocalization.

From an MO perspective, the electronic structure of the π-system of cyclopentadienide is shown in Fig. 12.16. These five molecular orbitals can, in principle, form five bonds to the metal center: one σ, two π, and two δ.

There are numerous related ligands to cyclopentadienyl that contain fewer, or even no, carbon (Fig. 12.17). One common such ligand is pyrrolyl, also called pyrrolide.[49] In addition, there are also many known complexes of P_5^-, including the ferrocene analog, Cp^*FeP_5.[50–53] In Cp^*FeP_5, Cp* (read "Cp star") = pentamethylcyclopentadienyl, $C_5Me_5^-$, a very commonly employed Cp derivative.

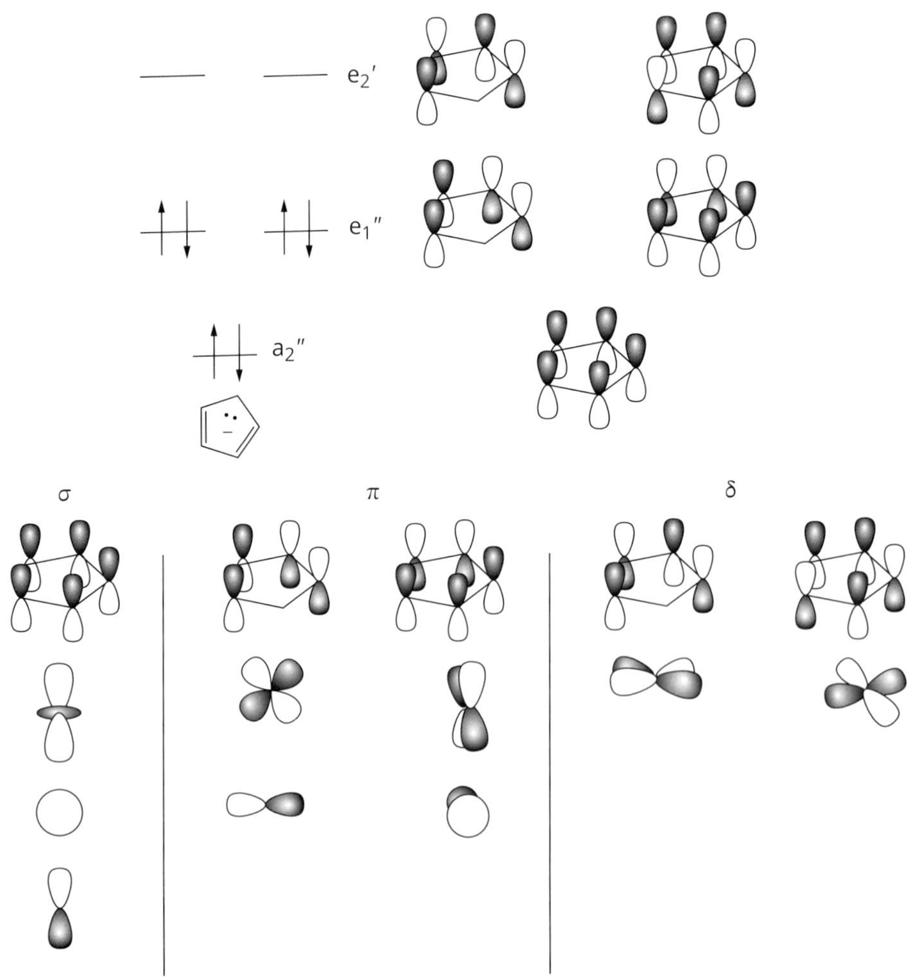

Fig. 12.16 Molecular orbitals of cyclopentadienide (Cp^-) and the metal-based orbitals with which they may interact.

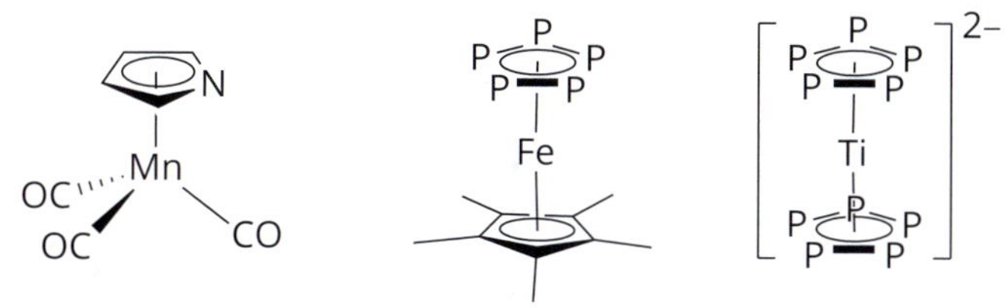

Mn(η^5-pyrrolyl)(CO)$_3$ Fe(η^5-P$_5$)Cp* [Ti(η^5-P$_5$)$_2$]$^{2-}$

Fig. 12.17 A few examples of Cp analogues with fewer or no carbon atoms in the ring.

Another fascinating P_5^- complex was reported by Ellis and coworkers: the entirely carbon-free sandwich complex $\{Ti(P_5)_2\}^{2-}$. Despite being formally titanium(0) and 16-electron, the compound is reported to be "highly stable toward heat and air, both in solution and the solid state." Computationally, the pentaphosphacyclopentadienyl ligand was suggested to be an excellent acceptor, which probably lends the compound its notable stability.[54]

12.2.1.2 Benzene and Arenes as Ligands

In general, benzene coordination to metal centers is weaker than that of anionic cyclopentadienyl; even so, benzene and other arenes can bind with great tenacity to some metal fragments.

The first fully characterized example of an arene complex was reported by E. O. Fischer and Walter Hafner in 1953: Cr(η^6-benzene)$_2$.[55–56] The route to the compound is shown in Fig. 12.18. Reduction of chromium(III) chloride was carried out in the presence of benzene and $AlCl_3$; the aluminum chloride was necessary for solubility. This reaction provides a chromium(I) bis(η^6-benzene) complex. Reduction with dithionite in aqueous base gave the neutral Cr(η^6-benzene)$_2$ as a benzene soluble brown solid that, like ferrocene, was easily sublimed and quite stable, decomposing at >300 °C.

While η^6-arene complexes are by far the most common, examples where only one of the double bonds, η^2, and where two of the double bonds in the ring, η^4, have also been characterized. As might be guessed, the electron-counting for the species is the same for the charged and neutral methods, since no charge is induced by the neutral benzene ligand. The electrons donated depend on the hapticity, with η^2 giving 2-electrons, η^4 giving 4-electrons, and η^6 giving 6-electrons to the metal. Consequently, Cr(η^6-benzene)$_2$ is a chromium(0), 18-electron complex.

Geiger and coworkers reported a very elegant example of several bonding principles of aromatic systems to transition metals (Fig. 12.19).[57] The synthesis begins with the well-known rhodium(I) dimer {RhCp*(μ-Cl)Cl}$_2$. Removal of the chlorides with $AgPF_6$ in the presence of hexamethylbenzene, C_6Me_6, produces {Rh(η^5-Cp*)(η^6-C$_6$Me$_6$)}$^{2+}$, an 18-electron complex.[58–59] A common soluble reductant is cobaltocene, $CoCp_2$, which is 19-electron and easily gives up an electron to produce ferrocene-like cobaltocenium, $CoCp_2^+$. If the dication is reduced with two equivalents of cobaltocene, a neutral rhodium complex is generated; however, if both the Cp* and C_6Me_6 rings stayed aromatic with all of their carbons bound to the metal, a 20-electron complex would be obtained. To avoid making the 20-electron complex, the aromaticity in the benzene ring is lost, and the 18-electron complex Rh(η^5-Cp*)(η^4-C$_6$Me$_6$) is formed. (The structure is shown in Fig. 12.19 with the hydrogens hidden.)[57]

Interestingly, the complex breaks the aromaticity of the benzene ring rather than breaking the aromaticity of the Cp* ring. It has been found that the aromatic stabilization energy of cyclopentadienide is actually higher than benzene, and as a consequence it is less energetically costly to break aromaticity in the C_6Me_6 ring. When aromaticity

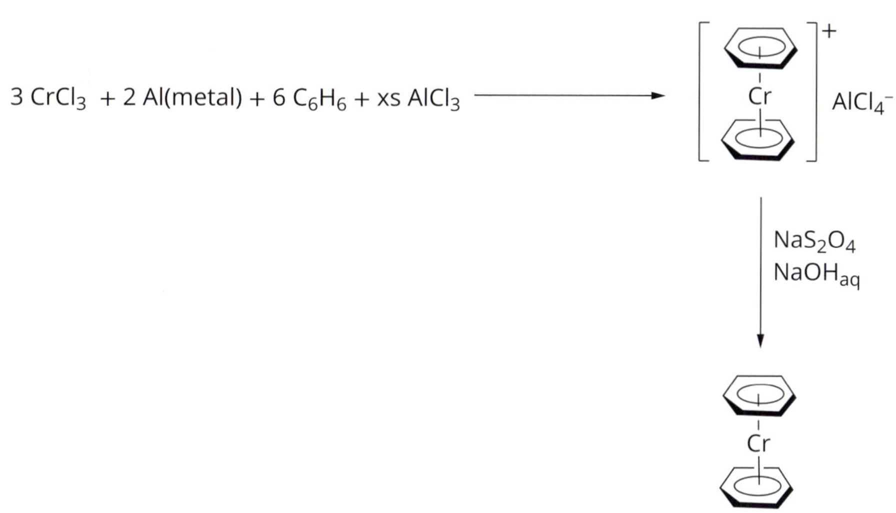

Fig. 12.18 Fischer and Hafner's synthesis of Cr(η^6-benzene)$_2$.

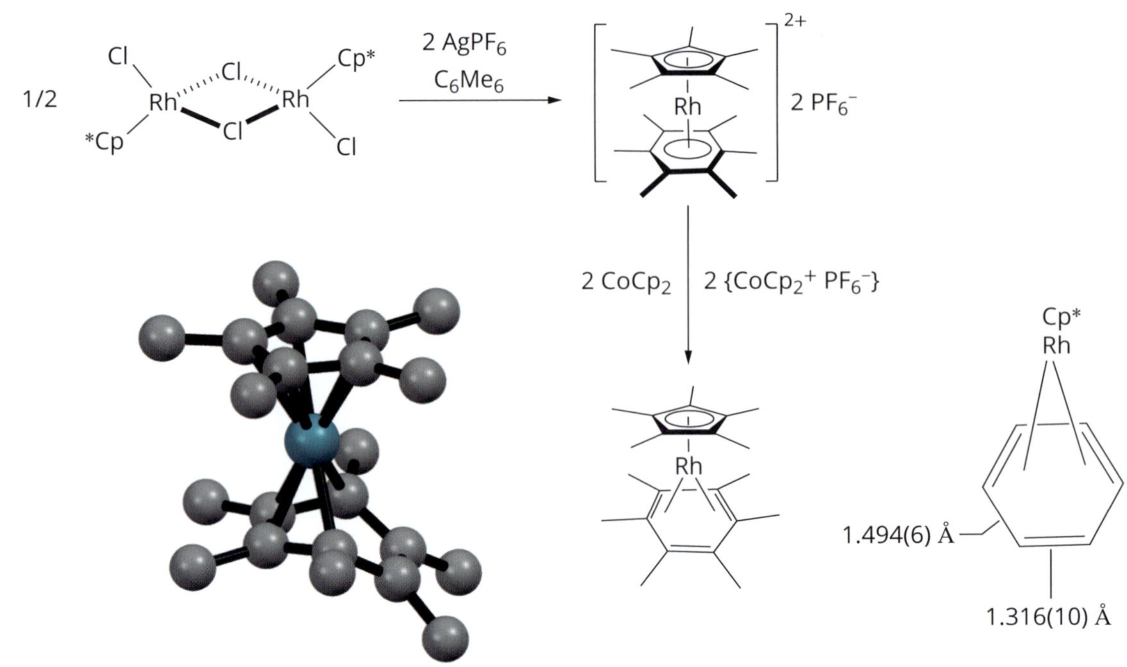

Fig. 12.19 Geiger and coworkers' synthesis of Rh(Cp*)(η^4-C_6Me_6).

in the benzene ring is broken, it is common to observe bond alternation in the dangling part of the ring, which is observed here with a short C–C bond for the side not attached to the metal.[57]

12.2.2 Less Common Cyclic π-Systems (Expanded Coverage)

The less commonly encountered π-systems to be discussed here are cyclopropenyl (three-membered ring), cyclobutadienyl (four-membered ring), cyclohexatrienyl (seven-membered ring), and cyclooctatetrenyl (eight-membered ring). As discussed previously, these build on the same concepts used to understand the five- and six-member ring systems. It is worth noting that cyclooctatetrenyl (COT) ligands are not uncommon for the lanthanides and actinides where, for example, $U(COT)_2$ is known as "uranocene". If the actinides and lanthanides are of interest, these ligands are probably of interest as well.

12.2.2.1 Cyclopropenyl

The cyclopropenyl cation, $C_3Ph_3^+$, was first isolated by Ronald Breslow in 1957,[60] and it is still this triphenyl species that is most often employed as a ligand.[61] However, cyclopropenyl is perhaps one of the least commonly used ligands in this C_nH_n $n = 3$–8 series.

The previously discussed linear nitrosyl and η^3-C_3Ph_3 have strong parallels that have been drawn in the literature, as both are, by the charged method, 2-electron cationic donors to metal centers.[62] Compare, for example, the two complexes CpNiNO and the "mixed sandwich" CpNi(η^3-C_3Ph_3) shown in Fig. 12.20.

Addition of ClC_3Ph_3 to $Ni(CO)_4$ followed by addition of pyridine (py) provides the nickel(0) complex $Ni(C_3Ph_3)(Cl)(py)_2$. The complex is nickel(0) because there is an electronegative chloride, which increases the oxidation state by one, and the C_3Ph_3 moiety, which is aromatic if it gives an electron to the metal to become $C_3Ph_3^+$. The other ligands are neutral pyridines. By the charged method, the system involves 10-electrons (Ni(0)) + 2-electrons ($C_3Ph_3^+$) + 2 electrons (Cl^-) + 2 × 2 electrons (2 pyridines) = 18 electrons. By the neutral method, $Ni(C_3Ph_3)(Cl)(py)_2$ has 10-electrons (Ni(0)) + 3-electrons (C_3Ph_3) + 1 electrons (Cl) + 2 × 2 electrons (2 pyridines) = 18 electrons.[63]

Treatment of $Ni(C_3Ph_3)(Cl)(py)_2$ with thallium(I) cyclopentadienide produces the mixed sandwich complex $Ni(C_3Ph_3)(Cp)$, which is also formally nickel(0) and 18-electron.[64–66] This can be compared with the similar linear nitrosyl complex $Ni(C_3Ph_3)(Cp)$, which is available by treatment of nickelocene ($NiCp_2$) with nitric oxide, and is also formally nickel(0) and 18-electron.[67–68]

In Fig. 12.20 the researchers used thallium cyclopentadienide. This reagent has the advantage of being neither oxidizing nor reducing, as Tl(I) is extremely stable. The alkali metal salts of organometallic reagents are often reducing, and in some cases the thallium reagent can be used instead. Another advantage is that the thallium halide byproduct is extremely insoluble in organic solvents. The main disadvantage, of course, is the toxicity of the thallium reagent and byproduct, which should be handled and disposed of with care.

BOX 12A

Prochiral molecules and planar chirality

Some planar molecules are "prochiral", a kind of 2-dimensional *Flatland* chirality.[209] If an object cannot be superimposed on its mirror image without manipulation in a third dimension to flip the molecule over, it is considered prochiral. For example, consider 2-chlorotoluene and its mirror image in Fig. 12.1A. Through manipulation in the plane, translations and rotations, one cannot superimpose the two molecules shown. However, as is easily seen, these are not true isomers as rotating the molecule out of plane allows superposition of the two structures.

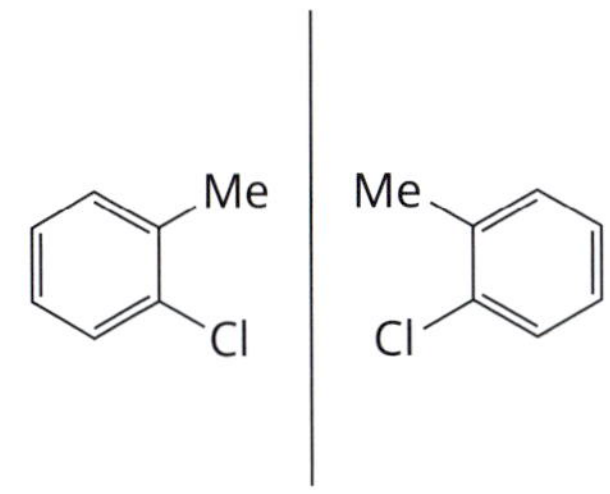

Fig. 12.1A Example of prochiral molecules.

Addition of a substituent to a prochiral molecule out of plane results in a chiral molecule and a pair of enantiomers, thus 2-chlorotoluene is prochiral. For example, if a metal center with its other ligands (ML_n) is added to the π-system of the arene, a molecule that is not superimposable on its mirror image results. This type of chirality in an aromatic system attached to a metal center is sometimes referred to as "planar chirality". As an example, a derivative of Greg Fu's iron DMAP catalysts are shown in Fig. 12.2A, some of which are commercially available.[210]

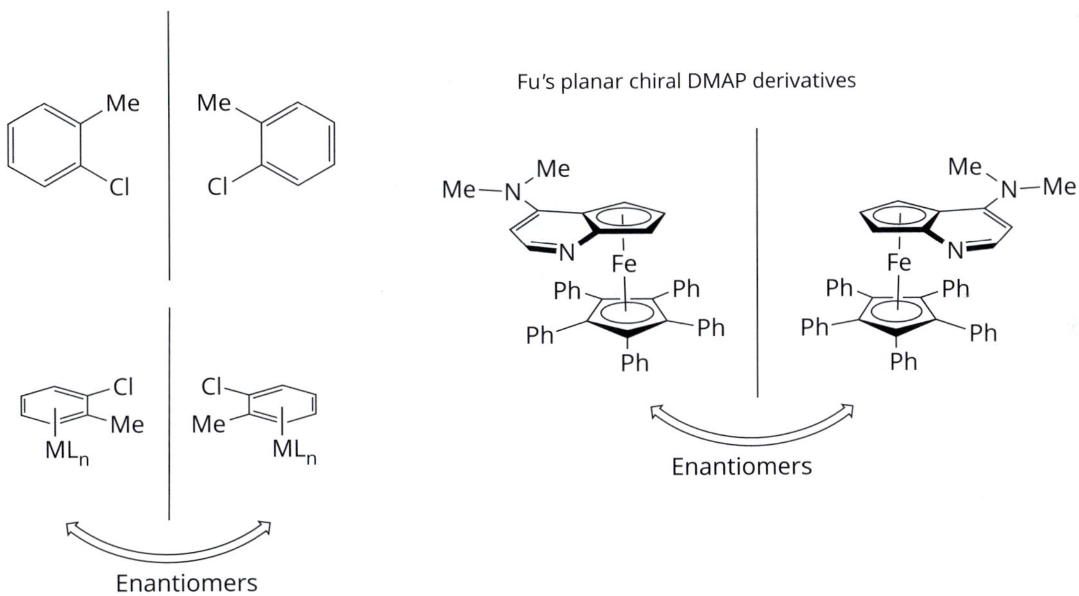

Fig. 12.2A Coordination of prochiral molecules to a metal center creates a chiral center. This was used in Fu and coworkers chiral DMAP derivatives.

The molecular orbital analysis of the free ligand produces the orbital shapes shown in Fig. 12.21 for $C_3H_3^+$. The filled a_2''-orbital can participate as a σ-donor towards the metal center, while the e''-set can act as a π-acceptor from metal-based donor orbitals. Again, strong parallels can be drawn to the bonding in linear NO (NO^+).

Symmetrically bound η^3-C_3R_3 complexes are known for many metals, including smaller, lighter metals such as Co, Fe, and Ni, along with larger, heavier metals such as Rh, Ir, and W.[69–72] In addition to the η^3-mode, η^1-, and η^2-modes have also been observed (Fig. 12.22).

The η^1-C_3Ph_3 ligand can be viewed as a simple alkyl for electron-counting purposes and is observed in Fe(η^1-C_3Ph_3)(η^5-Cp)$(CO)_2$.[73] Based on the Fe–C distances to the C_3Ph_3-ligand, the complex is obviously η^1 with one Fe–C distance of 2.076(1) Å, and the other two nonbonding carbons over

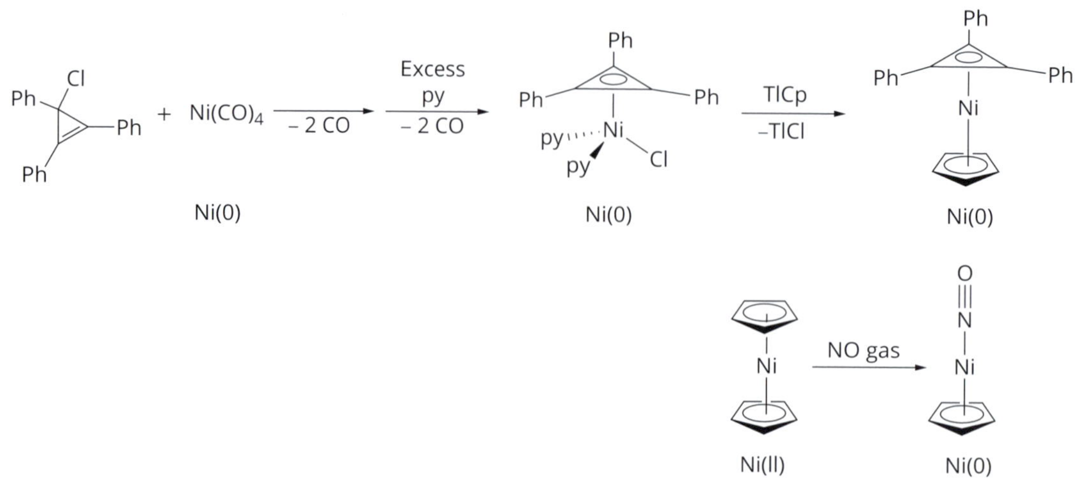

Fig. 12.20 Synthesis of the mixed sandwich $Ni(C_3Ph_3)(Cp)$ and the related CpNiNO complex.

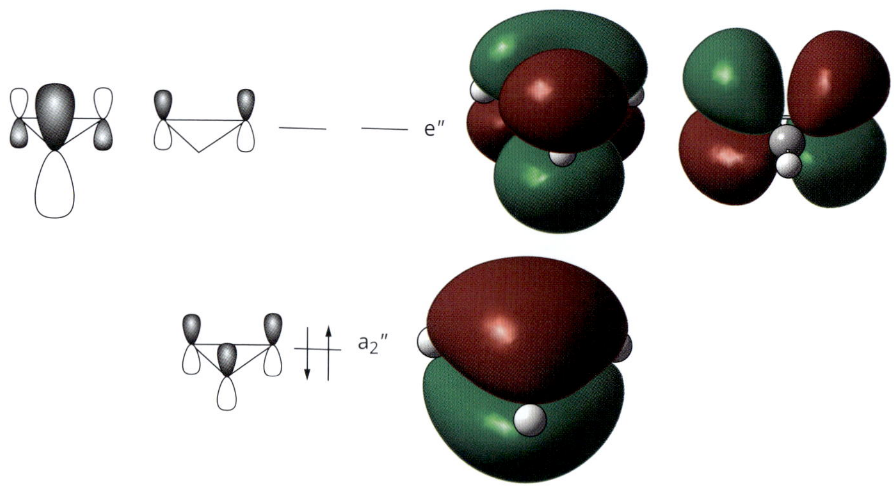

Fig. 12.21 Orbitals in the $C_3H_3^+$ aromatic system. Kohn–Sham orbitals shown are the LUMO and HOMO-1 calculated in Gaussian09 (B3LYP/aug-cc-PVDZ).

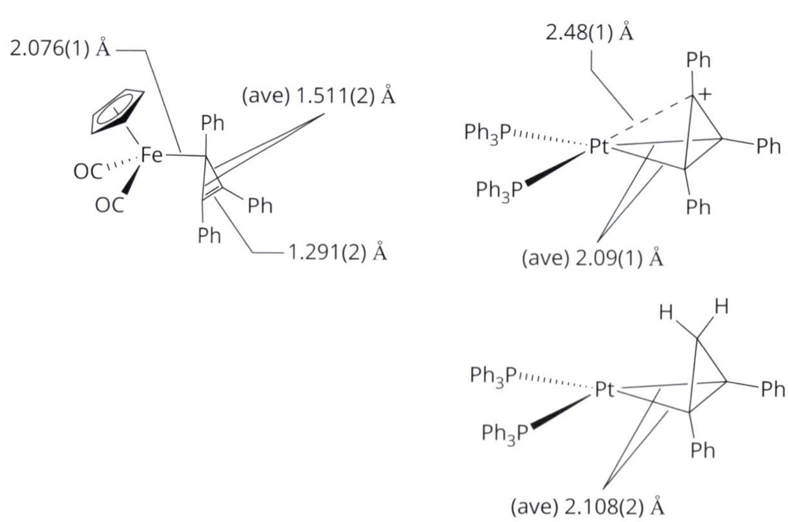

Fig. 12.22 Structures of $Fe(\eta^1\text{-}C_3Ph_3)(\eta^5\text{-}Cp)(CO)_2$ and $\{Pt(\eta^2\text{-}C_3Ph_3)(PPh_3)_2\}^+$.[73–74] In $Fe(\eta^1\text{-}C_3Ph_3)(\eta^5\text{-}Cp)(CO)_2$ the unbound carbons of the C_3Ph_3 ligand are more than 3 Å from the metal.

Table 12.4 Cyclopropenyl versus metallacyclobutadiene isomers.

	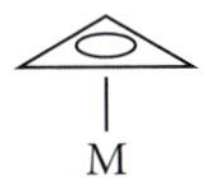	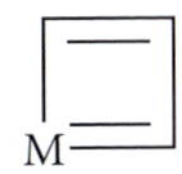
Donor properties	2-electron cationic donor	3-electron donor (neutral method) 6-electron trianionic (charged method)
Metal orbitals required	1 acceptor 2 donors	3 acceptors

3 Å from the metal. The C_3Ph_3 has been observed to act as an η^2-ligand in $\{Pt(\eta^2\text{-}C_3Ph_3)(PPh_3)_2\}^+$[74] The Pt–C distances to the cyclopropenyl ligand are essentially identical to the distances to the neutral cyclopropene ligand complex $Pt(\eta^2\text{-}C_2Ph_2CH_2)(PPh_3)_2$, and the three-membered ring seems to simply act as an olefin with a pendant carbocation. The structures with some pertinent bond distances are shown in Fig. 12.25. Other η^2-C_3R_3 complexes have been prepared, such as Cloke, Scott, and coworkers' $V(C_3Bu^t{}_3)(CO)_4$ system.[75]

Metal cyclopropenyl complexes are isomerically related to metallacyclobutadiene complexes (Table 12.4).[76–77] There is a continuum of structures that have been observed between the two limiting forms with complete C–C bond cleavage, partial C–C bond cleavage, and η^3-C_3R_3 complexes with no carbon–carbon bond cleavage.

The two isomers vary greatly in their donor properties (Table 12.4). The η^3-C_3R_3 ligand is a 2-electron cationic donor (charged system) that requires a metal acceptor orbital and two metal-based donor orbitals. Cyclobutadiene is a 6-electron trianionic donor in the charged system. Unsurprisingly, η^3-C_3R_3 is often encountered on electron-rich, late metal centers, while metallacyclobutadienes are often favored for high-oxidation-state earlier metals. However, as usual, the nature of the ligands and the metal center greatly affect the nature of the bonding.

Schrock and coworkers have reported a series of related complexes of W that show both of the limiting η^3-C_3R_3 and metallacyclobutadiene forms and an intermediate structure (Fig. 12.23).[69–70, 78] Complexes **A** and **B** are formally W(VI) and 12-electron and 16-electron species, respectively. The phosphine complex **C** is then 18-electron and formally W(II). Reversible cyclopropenyl ring opening has been reported by Hughes and coworkers in an iridium system.[79]

As mentioned, cyclopropenyl ligands are somewhat rare as transition-metal ligands. In fact, as of writing there are more structures in the Cambridge Structural Database for the valence isoelectronic cyclotriphosphirane complexes, *cyclo*-P_3—such as the two shown in Fig. 12.24—than cyclopropenyl complexes. The left complex, $Ni(P_3)Cp^*$, is the *cyclo*-P_3 derivative of the nickel(0) complex in Fig. 12.20.[80] The compound at right is formally niobium(I), a very low oxidation state for this reducing metal, and is low spin d^4;[81] the P_n analogs of $(CH)_n$ cyclic systems are stronger electron-withdrawing groups and often support very low formal oxidation states; compare titanium(0) $Ti(cyclo\text{-}P_5)_2^{2-}$ in Fig. 12.17. Considering the P_3-fragment's electron-withdrawing ability and the reducing ability of niobium(I), a better resonance form may be the one at right with covalent bonds to each phosphorus and niobium, complex isolobal with P_4. (See Chapter 14 for a discussion of the isolobal analogy.) The niobium would then be better described as niobium(V); however, the same resonance form would not be used for the nickel molecule on the left as nickel(IV) is highly unlikely. Consequently, one should keep in mind the discussion of common oxidation state in Chapter 10, as the nature of the metal center, not just the ligands, will contribute to the total electronic structure.

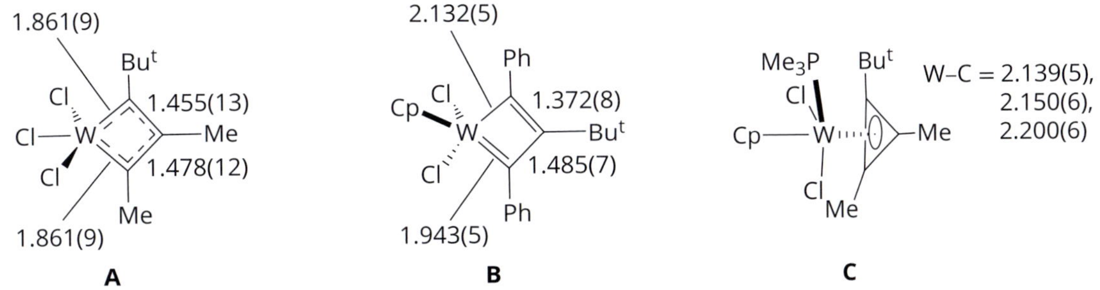

Fig. 12.23 Three related tungsten complexes showing a delocalized metallacyclobutadiene **A**, localized metallacyclobutadiene **B**, and η^3-cyclopropenyl **C**. All distances are in Ångstroms.[134,135,144]

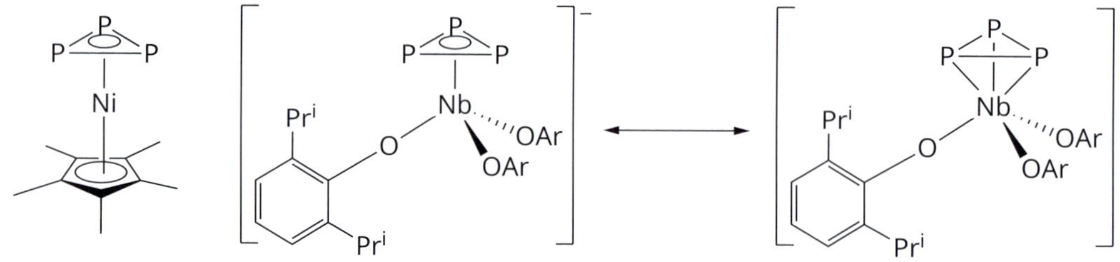

Fig. 12.24 Examples of *cyclo*-P_3 complexes. The nickel complex on the right may be best described as nickel(0), where the P_3-unit is aromatic similar to cyclopropenyl. The niobium complex is perhaps best regarded as Nb(V) with niobium bonding covalently to each phosphorus atom in the ring. The P–P distances are consistent with this, and the nickel complex has a shorter distance within the ring of 2.100(3) Å, while the niobium complex has longer P–P distances of 2.187(5) Å.

12.2.2.2 Cyclobutadiene[82–85]

Free cyclobutadiene is the prototypical antiaromatic compound with $4n$ π-electrons,[86] and as such its electronic structure has been discussed in great detail in the literature. If the compound is restrained to be a square, a triplet ground state is expected (Table 12.4). The singlet, square compound would undergo a Jahn–Teller distortion (Chapter 8) to produce a rectangular structure.[87–89]

Theoretical discussion of aromatic cyclobutadiene metal complexes by Longuet–Higgins and Orgel appeared in 1956. However, the first fully characterized example of a cyclobutadiene transition-metal complex was reported in 1959 by Hübel and coworkers, from the reaction of $Fe_3(CO)_{12}$ with excess diphenylacetylene at high temperatures, and a structure of the complexes by X-ray diffraction appeared soon thereafter.[90] The compound was initially formulated by elemental analysis and by its reaction with $LiAlH_4$ to produce 1,2,3,4-tetraphenylbutadiene. The authors suggested two possible structures for the complex, shown in Fig. 12.25 as **A** and **B**. Dodge and Schomaker's X-ray diffraction experiments showed that the correct structure was **A**;[91] however, **B** is certainly a possible intermediate during the reaction, as ring-contraction of metallacyclopentadienes to cyclobutadienes is now well known.

The first unsubstituted cyclobutadiene complex was prepared by Pettit,[92] and its reactivity was consistent with the C_4H_4 ring being aromatic in that it underwent typical aromatic ring reactions such as Friedel–Crafts acylation—just one of the many reactions that they explored.[93] Further evidence of aromaticity in metallocyclobutadiene is in their structures, which feature flat square rings. One of the early structures of such a compound—in this case the tetramethylcyclobutadiene derivative achieved by Gray and coworkers—is also shown in Fig. 12.26.[94]

From a molecular orbital perspective, η^4-cyclobutadiene may form four bonds to the metal center: one σ, two π, and one δ. The orbitals for the π-system of cyclobutadiene are shown in Fig. 12.27, together with the metal-based orbitals that may in principle bond to those of η^4-C_4H_4.

Metal–cyclobutadiene complexes of hapticity other than η^4 are quite unusual, but some derivatives are known. For example, $Cp_2Zr(\eta^2$-benzocyclobutadiene$)(PMe_3)$ has been characterized (Fig. 12.28),[95] but in this case the η^4-mode would have required a break in aromaticity in the fused six-membered ring, making the η^2-bonding mode far more likely. The reactivity of the complex is what may be expected from a zirconium(II) olefin, zirconacyclopropane.

Every ligand type is unique, but there is nothing particularly special about carbon as a ligand, and there are noncarbon analogues to many metal–aromatic rings that are known. One particularly intriguing example electronically related to cyclobutadiene is P_4^{2-}, tetraphosphacyclobutadiene, which was first observed by Scherer and coworkers by photolysis of a solution of $Nb(C_5Me_4)(CO)_4$ and P_4 (Fig. 12.29). The ligand As_4^{2-} is also known.[96–97]

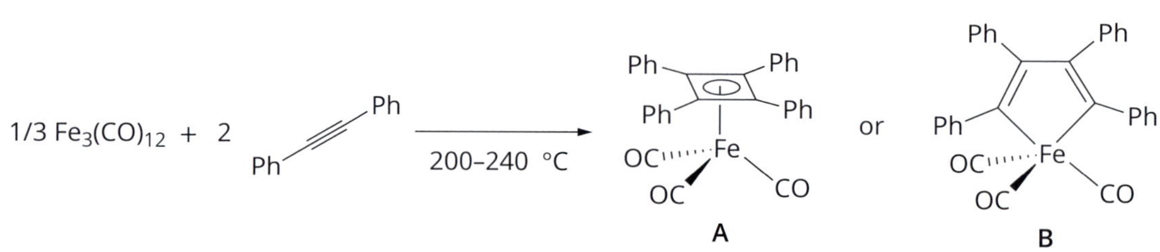

Fig. 12.25 Reaction of $Fe_3(CO)_{12}$ with diphenylacetylene produced $Fe(CO)_3(PhC_2Ph)_2$, which was proposed to have one of the two structures shown. The correct structure **A** was established by X-ray diffraction soon afterwards.

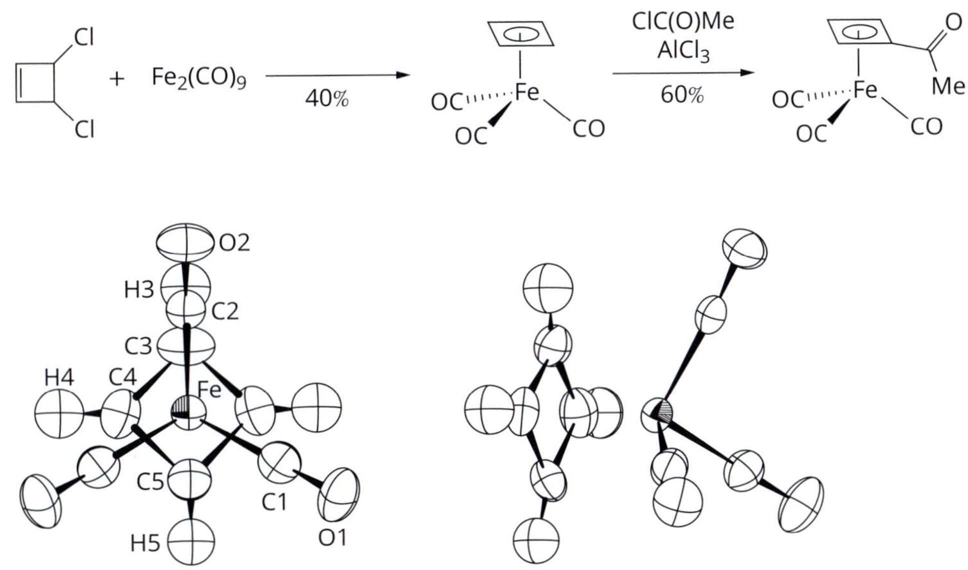

Fig. 12.26 (*top*) Pettit's synthesis of $Fe(C_4H_4)(CO)_3$ and Friedel–Crafts acylation of the aromatic ring. (*bottom*) Two views of the structure of $Fe(C_4Me_4)(CO)_3$ from the paper by Gray and coworkers.[94]

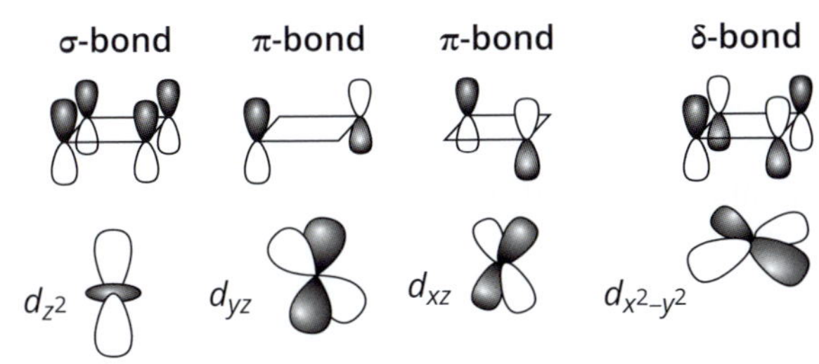

Fig. 12.27 The ligand-based orbitals for η^4-cyclobutadiene and the metal-based orbitals that may interact with them.

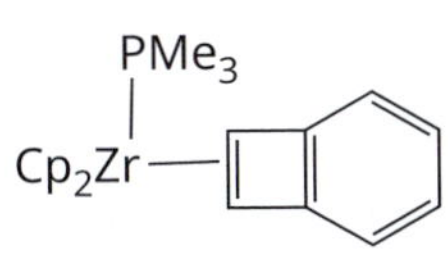

Fig. 12.28 Structure of $Cp_2Zr(\eta^2$-benzocyclobutadiene$)(PMe_3)$.

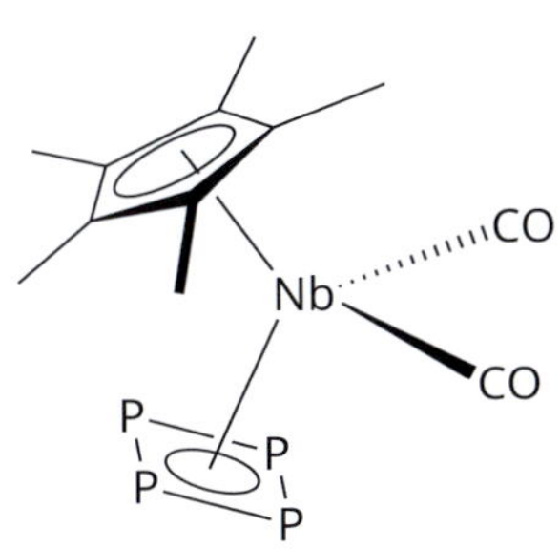

Fig. 12.29 Structure of $Nb(C_5Me_4)(P_4)(CO)_2$.

12.2.2.3 Cycloheptatrienyl

Similar to cyclopropenyl, the cycloheptatrienyl fragment is aromatic as the cation, which is called tropylium ($C_7H_7^+$). Tropylium is a stable aromatic that can be prepared from rearrangement of tolyl cation, $^+CH_2C_6H_5$; in fact, on fragmentation of benzyl groups in mass spectroscopy it is tropylium that is generally observed. The cation is commercially available as the tetrafluoroborate salt.

Tropylium ($C_7H_7^+$) sometimes acts as an outer-sphere oxidant, and perhaps because of this it seems most often placed on a metal center as neutral C_7H_8 rather than as the cation to avoid redox side-reactions. For the deprotonated ligand, the allyl-like η^3-bonding mode, pentadienyl-like η^5-bonding mode and aromatic η^7-bonding mode are all known. All three of these hapticities are demonstrated in Eqs. 12.5–12.7. For electron-counting purposes, aromatic η^7-C_7H_7 is a 7-electron donor in the neutral method, and a cationic 6-electron donor in the charged method, if we follow the suggestion of using the most stable aromatic electron count in the ring.

Eq. 12.5 shows the synthesis of an η^7-C_7H_7 complex from an 18-electron η^7-C_7H_7 molybdenum starting material, $Mo(\eta^6$-$C_7H_8)(CO)_3$. A common reagent for the removal of hydride, trityl cation ($^+CPh_3$) removes H^- to produce a half-sandwich complex $\{Mo(\eta^7$-$C_7H_7)(CO)_3\}^+$ as its tetrafluoroborate salt.[98]

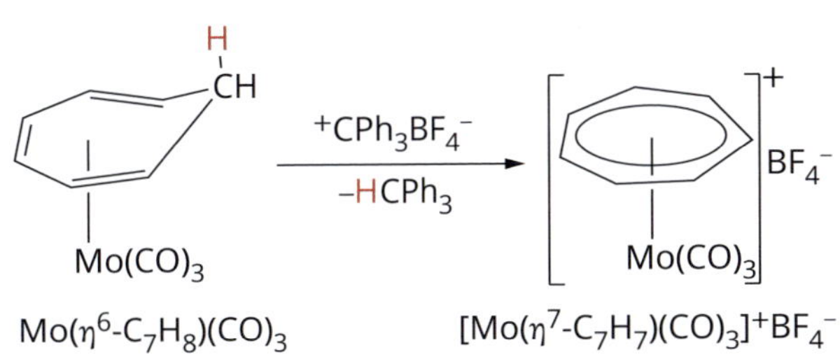

12.5

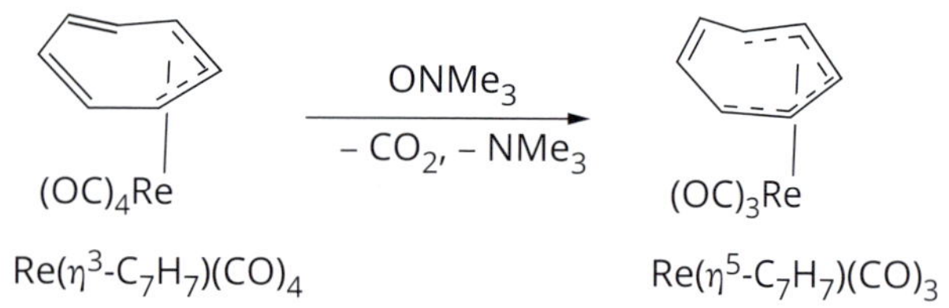

12.6

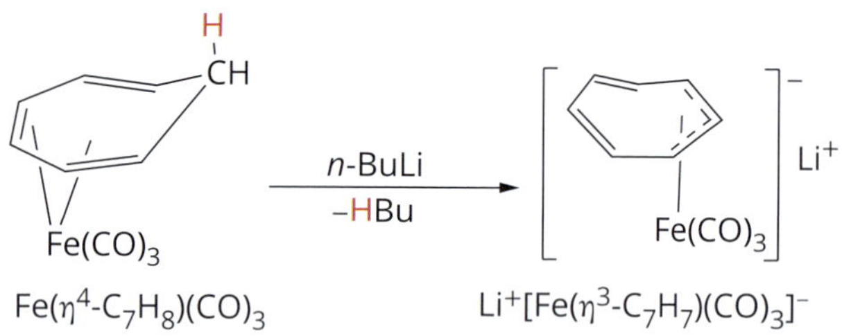

12.7

Eq. 12.6 shows the synthesis of Re(η^5-C_7H_7)$(CO)_3$ by Rest and coworkers. In this case, the seven-membered ring is η^3 in the starting material shown, which is 18-electron. Removal of one carbonyl is accomplished by addition of trimethylamine-*N*-oxide (^-O–$^+NMe_3$). This common strategy for the selective removal of carbonyl, oxidizes one CO to CO_2, which is released. The metal then can accommodate another double bond from the ring to produce the pentadienyl-like η^5-C_7H_7 structure.[99]

The neutral cyclic polyolefin C_7H_8 can act as an η^4-C_7H_8 ligand for iron(0), Eq. 12.7. Deprotonation with a strong base, LiBu, produces the allyl-like η^3-C_7H_7 ligand in a complex that is still formally iron(0).[100]

Again, one should also keep in mind the typical oxidation states of the metal. For example, Ti(η^7-C_7H_8)Cp is a known complex (Fig. 12.30).[101] The cycloheptatrienyl ligand is planar with equivalent C–C bonds around the ring. The aromatic electronic structure for both η^7-C_7H_8 and Cp would produce an unlikely formal oxidation state for titanium of 0, where +4 is by far the most common oxidation state of the metal. An alternative resonance description would be a diallyl/monoalkyl description of the η^4-C_7H_8 ligand, giving the system a −3 charge on the seven-membered ring, and a titanium(IV) metal center, which seems far more likely.

The (2 allyl + 1 alkyl)-resonance form can be recast as a Hückel rule $4n + 2$ case, where $n = 2$; the ten electrons would come from two allyl anions (2 × 4 electrons) and an alkyl anion (2 e). In the context of Table 12.4, this would be electronically equivalent to filling the slightly antibonding ($\sigma + 0.445\beta$)-level of the cycloheptatrienyl ligand to reach the next aromatic state. The 10-electron aromatic π-system should be higher in energy than the 6-electron system, but the metal will contribute to the overall electronic structure and should be considered with the ligand electronics, so that Ti(IV) may be more stable than Ti(0) contributes to the molecular stability.

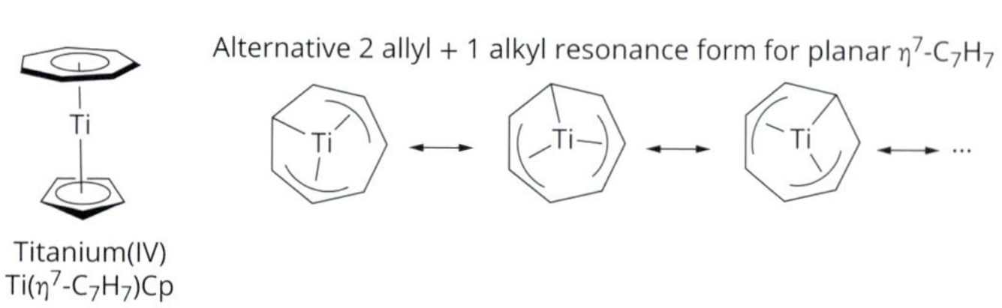

Fig. 12.30 Alternative 2 allyl + 1 alkyl interpretation of the bonding in Ti(η^7-C_7H_8)Cp.

12.2.2.4 Cyclooctatetraene

The largest generally encountered aromatic system for metal centers is cyclooctatetraene, which is aromatic as the dianion when it acts as a 10-electron donor, $4n + 2$, where $n = 2$. In order to reach an aromatic configuration as a transition metal or *f*-element ligand, the metal must bind to all of the carbons of this fairly large ring more or less equally. With some metals, binding modes such as a butadiene-like η^4-mode are more common than an aromatic η^8. However, examples where cyclooctadiene acts as an η^8-aromatic ligand are known, even for the first-row transition metals.

For example, Mountford and coworkers prepared Ti(NBu^t)(η^8-C_8H_8), shown in Fig. 12.31, by replacement of chloride ligands in Ti(NBu^t)$Cl_2(py)_3$[102] with $K_2C_8H_8$.[103] In addition, bis(cyclooctatetraene)titanium is known, Ti(η^8-C_8H_8)(η^4-C_8H_8), with one ring "slipped".[104–106]

Aromatic C_8H_8 is perhaps a more common ligand for the actinides, due to their larger atomic radius, than the transition metals. A quite stable derivative dubbed "uranocene" is the bis(cyclooctadiene) uranium(IV) shown in Fig. 12.31, which was originally reported by Streitwieser and Müller-Westerhoff.[107]

12.3 Metal–Ligand Multiple Bonds

Species with multiple bonds between a transition metal and a first-row *p*-block element are common reactive intermediates in many catalytic cycles. There is a great deal of variety to be found in electronic structure and reactivity for the complexes, depending on the metal center, type of substituents on the multiple-bonded ligand, and ligands attached to the metal.

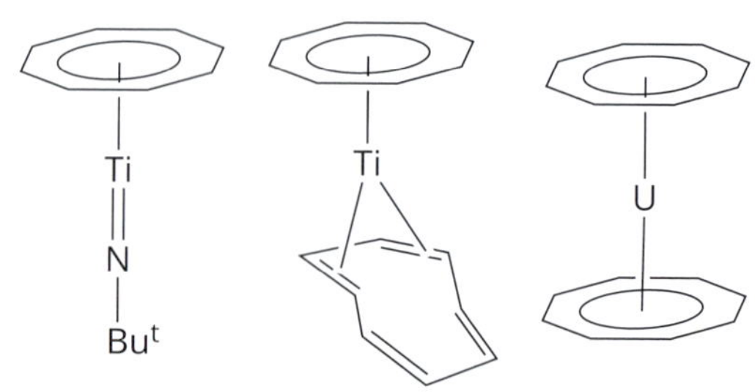

Fig. 12.31 Structure of the complexes Ti(NBu^t)(η^8-C_8H_8),[103] Ti(η^8-C_8H_8)(η^4-C_8H_8), and "uranocene" U$(C_8H_8)_2$.

12.3.1 Carbenes and Alkylidenes

For electron-counting and formal oxidation-state assignment, metal to carbon double and triple bonds are divided into two different categories that vary widely in their properties: Fischer carbenes and Schrock carbenes. The Schrock-type ligands are also called "alkylidenes". To illustrate the electronic difference, it is informative to start with the syntheses of the first examples of each class by Fischer and Schrock.

12.3.1.1 Brief History and Description of the Electronic Structure of Carbenes and Alkylidenes

The synthesis of the first example of a metal carbene was accomplished by Fischer and Maasbol in 1964 (Fig. 12.32).[108] The reaction of LiPh to $W(CO)_6$ "takes places without elimination of CO to yield an orange-yellow, water-soluble, diamagnetic anion, which can be precipitated as its tetramethylammonium salt", Fischer wrote. The salt was identified correctly as $\{NMe_4\}^+\{W(CO)_5(=C(O)Ph)\}^-$. Acidification would give $W(CO)_5(=C(OH)Ph)$, which was methylenated with diazomethane (N_2CH_2) to produce $W(CO)_5(=C(OMe)Ph)$ as a neutral Fischer carbene.

The synthesis begins with $W(CO)_6$, a W(0) complex, and no redox processes occur. In other words, one would expect that the product is also tungsten(0), and this is how the compound is formally assigned, as W(0). The reason for this is found in Fischer's insightful drawing of the molecule in his original paper, which shows a W=C interaction along with a C=O interaction (Fig. 12.32).[108] Application of the Mayer bond-order calculation to Fischer's original carbene shows significant π-bonding between the methoxy group and the carbene carbon, with a bond order of 1.23.

Carbenes that bear heteroatoms on C_α have a singlet ground state with an *sp*-hybridized, filled orbital (HOMO) and an unfilled pure *p*-orbital (LUMO) orthogonal to the C–R–O plane that can bond with the oxygen lone pair (Fig. 12.33). These will act as a σ-donor to the metal and a π-acceptor from the metal, respectively, similar to carbonyl, although with one less π-acceptor orbital available. As one may remember from the discussions in Section 11.3.3 on *N*-heterocyclic carbenes (NHCs), this is functionally equivalent to the metal-bonding discussion for NHCs, which are regarded as neutral 2-electron donors. In other words, the reactions in the top of Fig. 12.32 by Fischer and Maasbol turns one neutral 2-electron donor, CO, into another, a singlet carbene.

The reactivity of typical Fischer carbenes is such that nucleophiles react at C_α (the carbene carbon) and electrophiles may react at the electron-rich metal center. *In other words, in Fischer carbenes C_α is usually electrophilic.* For example, Fischer's original carbene, $W(CO)_5(=C(OMe)Ph)$, was reported by Casey and coworkers (Eq. 12.8) to produce $K\{W(CO)_5(CH(OMe)Ph)\}$, hydride addition to C_α, on reaction with $K^+\{HB(OPr^i)_3\}^-$.[109]

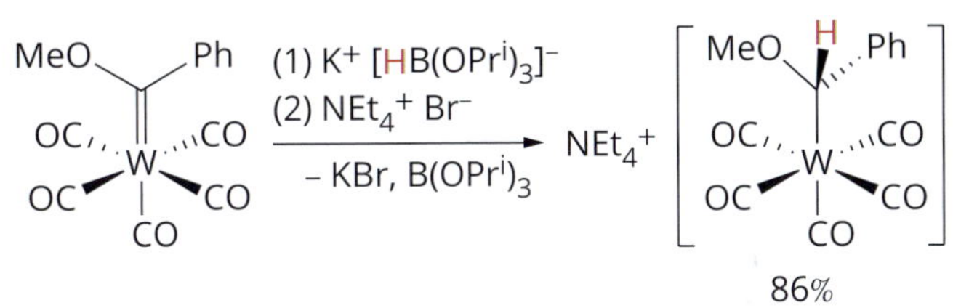

12.8

The first report of a Schrock carbene, otherwise known as an "alkylidene", appeared in 1974.[110–111] The reaction sequence (Fig. 12.34) begins with tantalum(V) chloride, $TaCl_5$. Partial replacement of the chlorides for neopentyl, Np = CH_2Bu^t, is accomplished by using $ZnNp_2$, providing $TaCl_2Np_3$. Addition of LiNp to the dichloride complex produces, as a possible intermediate, $TaNp_5$, which undergoes α-abstraction of a hydrogen by a neighboring neopentyl. This produces the tris(neopentyl)(neopentylidene)tantalum(V) product. The synthesis begins with tantalum(V), and no redox chemistry is proposed; the product is still Ta(V).

Schrock carbenes are often considered to be derived from the interaction of the metal with a triplet carbene (Fig. 12.35). The triplet state can be the ground state for most carbenes

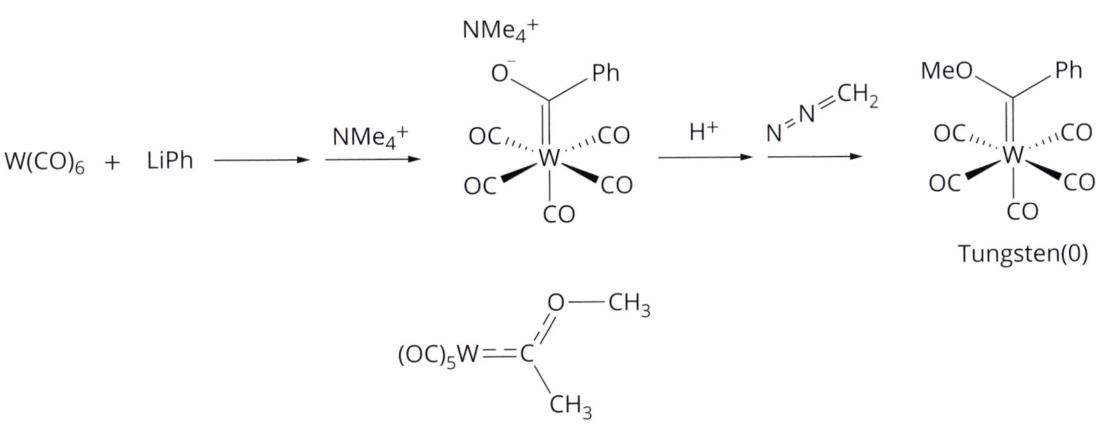

Fig. 12.32 (*top*) Synthesis of the first Fischer carbene, $W(CO)_5(=C(OMe)Ph)$. (*bottom*) The drawing of the structure from their 1965 paper.[108]

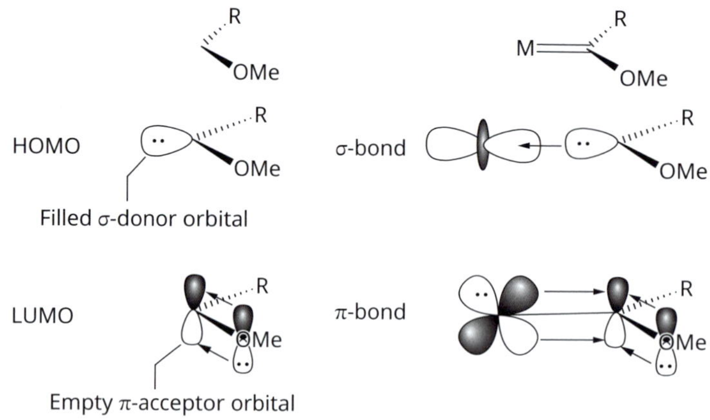

Fig. 12.33 (*left*) The singlet state of an example heteroatom-bearing carbene and (*right*) bonding interactions with a metal center.

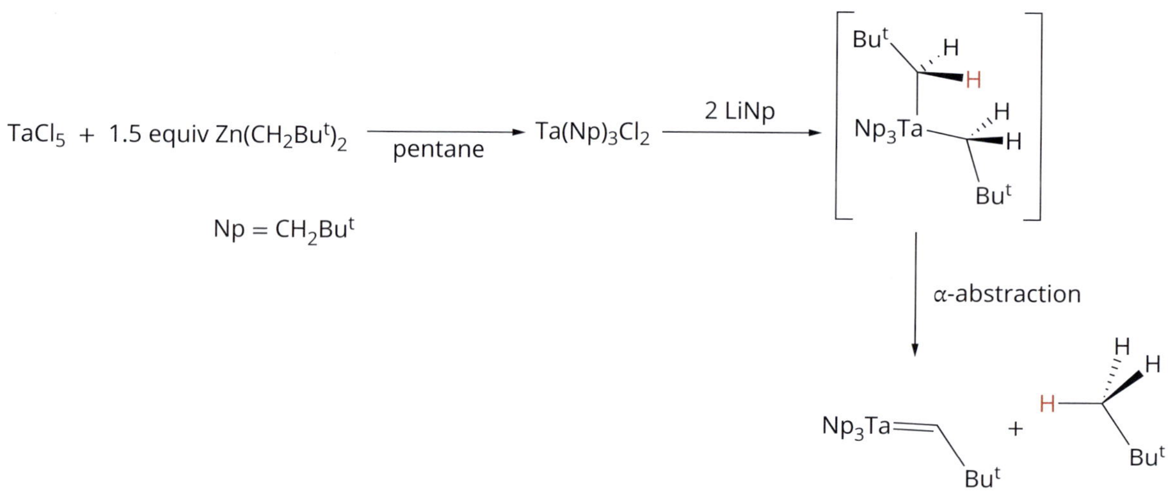

Fig. 12.34 (*top*) Schrock's original synthesis of $TaNp_3(=CHBu^t)$, the first alkylidene complex.[111–112]

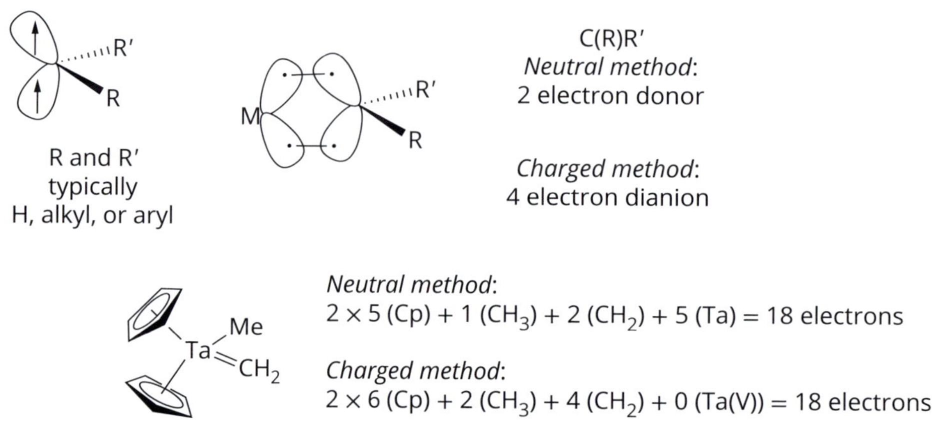

Fig. 12.35 (*left*) The triplet state of an example carbene with no heteroatom and (*right*) bonding interactions with a metal center. (*bottom*) Electron-counting of a typical Schrock carbene.

without a substituent on the carbene carbon that can act as a π-donor; CH_2 has a triplet ground state. However, this is not generally the case; for example, CMe_2 has a singlet ground state due to extensive CH-hyperconjugation.[113] As a result, these hyperconjugative terms are π-repulsive just as a heteroatom's lone pair might be on a Fischer carbene.

While this triplet/singlet carbene ligand argument is pervasive and is of some utility, there must be a more complete definition of a Fischer versus Schrock carbene that includes the transition metal and ancillary ligand properties for there to be any utility at all in these labels. Such a system is discussed shortly that is based on MO theory arguments.

As a general rule of thumb, Schrock carbenes can be identified by having C_α carbons with no heteroatom directly attached, are generally at or near the highest oxidation state of the metal, and typically of the early to middle transition series elements.

For electron-counting purposes, a Schrock carbene is counted quite differently from a Fischer carbene. While the Fischer carbene is regarded as a neutral 2-electron donor in both the neutral and charged methods, the Schrock carbene (alkylidene) is a 2-electron diradical in the neutral system and a 4-electron dianionic donor in the charged method (Fig. 12.35).

Contrary to Fischer carbenes, which are generally electrophilic at C_α, Schrock carbenes are often nucleophilic at C_α. For example, (Eq. 12.9), the tantalum methylidene complex $TaCp_2(Me)(=CH_2)$, reacts with Lewis acids such as $AlMe_3$ at electron-rich C_α.

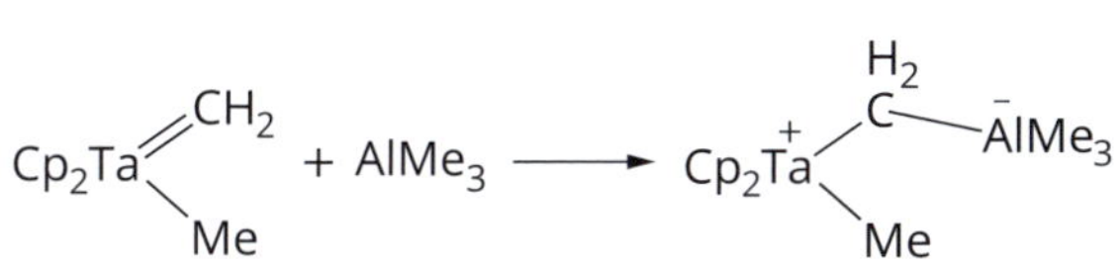

12.9

Along similar lines, Tebbe's reagent is used in organic synthesis and may be regarded as a masked methylidene. The preparation of Tebbe's reagent (Fig. 12.36) involves addition of $AlMe_3$ to Cl_2TiCp_2.

The different electronic structures, from a hybridization theory perspective, may be regarded as in Fig. 12.37. A Fischer carbene has an electrophilic carbon and an electron-rich metal center. A Schrock carbene has a nucleophilic carbon and an electron-deficient metal center.

However, there are now thousands of carbenes known, and their reactivity patterns are quite diverse. *It is best to view Fischer carbenes and Schrock carbenes as two ends of a spectrum of electronic structures, with many compounds having properties in between these limiting forms.*

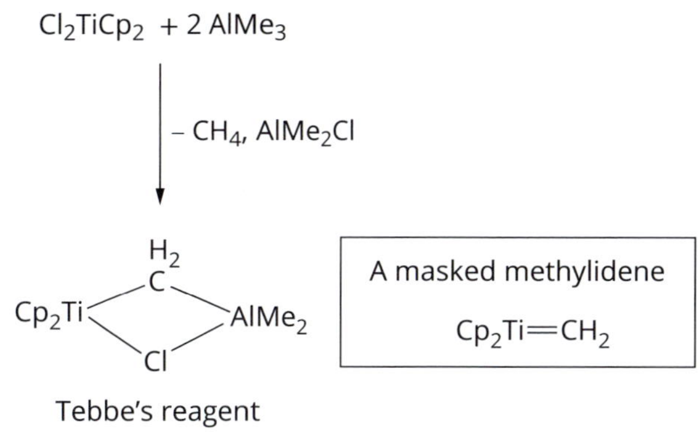

Fig. 12.36 Synthesis of Tebbe's reagent.

12.3.1.2 A Molecular Orbital-Based Unified Picture for Fischer and Schrock Carbenes (Advanced Topic)

As discussed previously, the nucleophilic Schrock carbene and the electrophilic Fischer carbene are two limiting forms. As a result, we should build a unified electronic picture for our description of transition-metal carbenes that allows this continuum to be explored. In order to build this picture we will pick, somewhat arbitrarily, one of the two spin states of the carbene as one of our fragments, the singlet carbene. To bond with the singlet carbene, we will pick typical metal fragments for a Fischer and Schrock carbene. For the Fischer fragment we will use $Cr(CO)_5$, and for the Schrock fragment we will use trimethyltantalum, $TaMe_3$, analogous electronically to the originally reported compounds. For the Fischer fragment, because of their commonality and importance, we will examine the effect of a heteroatom on C_α as well.

We begin with the MO analysis of $R_2C{=}Cr(CO)_5$, and for simplicity we will ignore *p*-orbitals on the metal centers as they are the highest-energy valence orbitals and least likely to participate. We will place the plane of the carbene (R–C–R plane) so that it bisects the OC–Cr–CO vectors in the plane orthogonal to it. The orbitals of the two fragments in this C_{2v} molecule are shown in Fig. 12.38.

The metal-based orbitals are highly stabilized by the presence of carbonyl acceptor orbitals (Fig. 12.38, left). The carbene

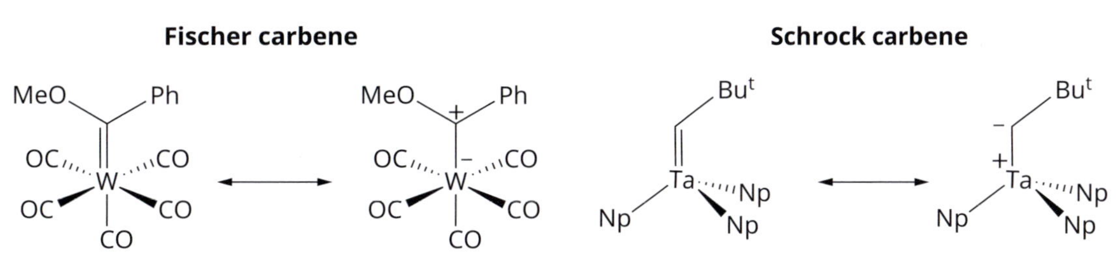

Fig. 12.37 Hybridization theory perspective on typical Fischer and Schrock carbene electronic structure.

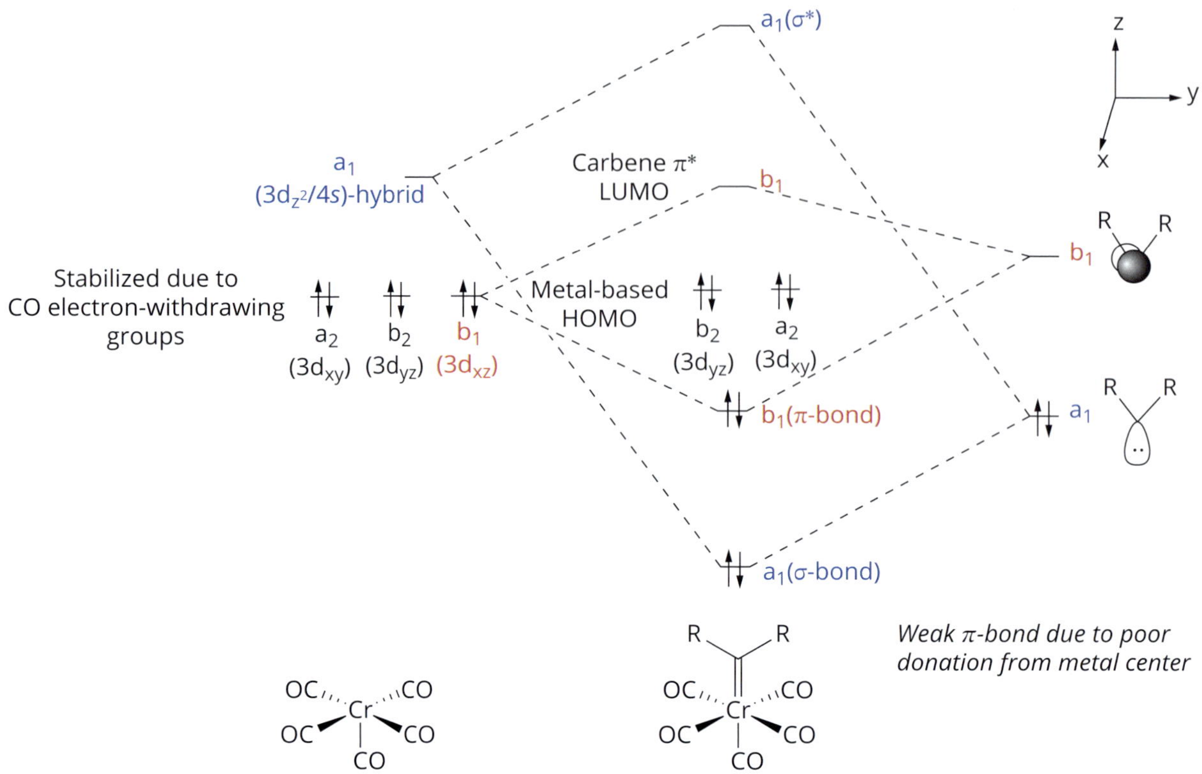

Fig. 12.38 Simplified orbital analysis of $R_2C{=}Cr(CO)_5$.

has a σ-donor orbital, which can interact well with a d_{z^2}/s-hybrid to form bonding and antibonding interactions (blue). There is another orbital on the metal of a_1-symmetry (not shown), $d_{x^2-y^2}$; however, this orbital is rotated so that its interaction with the carbene will be negligible, despite having proper symmetry. The π-acceptor orbital on the carbene (b_1) can interact with $3d_{xz}$, accepting electron density from the metal; however, the donor ability of this metal-based orbital will be tempered by the highly electron-accepting carbonyls on the metal. As a consequence, the π-bond from the metal to the carbene will be quite weak in a typical Fischer carbene. The expectation is for Fischer carbenes to have relatively low M–C bond orders.[114]

How does a donor group on the carbene carbon influence this picture? Addition of a group like OMe in place of R results in donation into the carbene acceptor orbital. The result is to weaken the carbene–metal π-bond even further. This is not unexpected, as an extreme example of metal–carbene bonding is coordination of an NHC, which is viewed as a dative interaction with some small amount of backbonding.

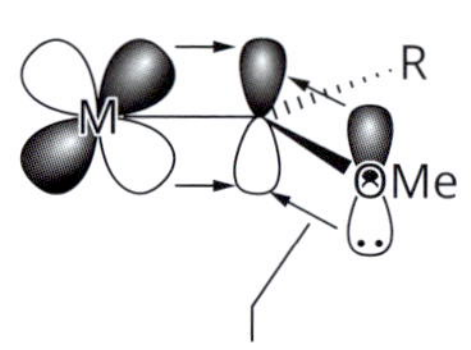

Heteroatom lone pair competes with the metal for the C_α acceptor orbital weakening the M–C π-bond

As shown in Fig. 12.38, the HOMO of a Fischer carbene is typically a metal-based nonbonding orbital(s). The LUMO is typically, the π*-orbital of the M=C bond, which is largely carbon-based. As a result, the C_α of Fischer carbenes is typically electrophilic.

As a model for a Schrock carbene we will use $R_2C{=}TaMe_3$. The overall symmetry of the system is C_s; the low symmetry results in many potential interactions. However, we can reasonably approximate the MO diagram as shown in Fig. 12.39. In this case it is useful to separate the metal-based orbitals into those that will point along the z-axis ($d_z{}^2$ and s, blue), those that reside in the *xy*-plane (approximately nonbonding with respect to the carbene, black), the nonbonding with respect to the carbene d_{yz}-orbital (green), and the carbene π-bonding d_{xz}-orbital (red).

The metal fragment, $TaMe_3$, is a strong donor with a strongly reducing metal in Ta(III) with good σ-donating ligands in methyls. As a result, it will donate very well into the π-acceptor orbital of the singlet carbene fragment, forming a strong Ta–C π-bond. The Lewis-acidic metal center is also has a low electron count and will be a good acceptor for the σ-electron density from the carbene. Consequently, strong Ta–C σ- and π-bonds will be formed. The HOMO of the system is the Ta–C π-bond, which is largely carbon-based, making C_α nucleophilic. The LUMO is an approximately nonbonding metal-based orbital orthogonal to the Ta-carbene vector, though this orbital may be mixed with the carbene π*, also largely on the metal.

Schrock carbenes, alkylidenes, are characterized by strong metal–carbon bonds with relatively high bond orders.

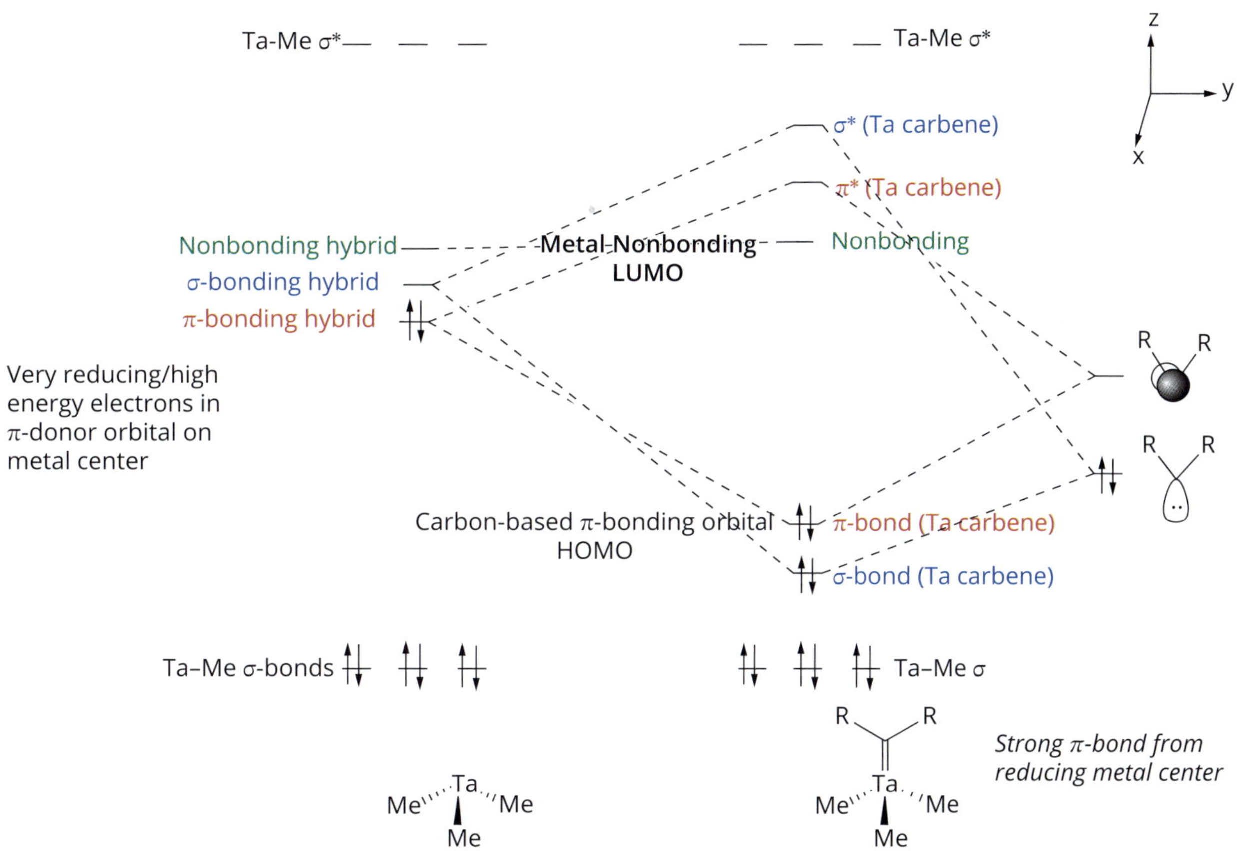

Fig. 12.39 Simplified orbital analysis of $R_2C{=}TaMe_3$.

In contrast, Fischer carbenes are characterized by weaker metal–carbon bonds, especially the π-bond, with relatively low bond orders.

As can readily be imagined, metal carbene systems with donor or acceptor groups on the metal and carbene with metal centers in various oxidation states can be prepared so that the metal–carbon bond strength and bond order changes in an essentially continuous fashion from strong bonds with high bond orders (Schrock-like) to weak bonds with low bond orders (Fischer-like).

Calculations on typical Fischer carbenes have shown them to have BDEs of ~60 kcal/mol. In contrast, typical Schrock carbenes have BDEs of ~90 kcal/mol. The covalent bond order (CBO) calculated from the NBO analysis of $W(=C(H)OMe)(CO)_5$ reportedly had a very low value of 0.70. A similar calculation,[115] albeit with a smaller basis, on $Ta(=CMe_2)Me_3$ gave a CBO = 1.4, a significantly higher covalency as expected.[116–117]

12.3.2 Imido, Hydrazido(2–), and Oxo

Commonly encountered ligands with nominal metal-element double bonds that are similar to carbenes are imido (M=NR),[118] hydrazido(2–) ($M{=}N{-}NR_2$),[119] and oxo (M=O). Certainly, many more are known, such as M=PR (phosphinidines),[120] M=N–OR, M=S (sulfide), and so on, but we will largely limit our discussion to these three groups, with some discussion of bridging sulfides, which are of great importance in biology.

The principles learned in the previous subsection on carbenes versus alkylidenes are quite applicable here, and the discussion of electronic structure will be brief. Generally, for electron-counting purposes the N–X (X = alkyl, aryl, NR_2, OR, and so on) group is counted as a 4- or 6-electron donor in the charged method and as a 2- or 4-electron donor in the neutral method. As an example, consider the complex $Ti(NBu^t)Cl_2(py)_3$, Mountford's reagent,[102] prepared by treating $TiCl_4$ with H_2NBu^t followed by pyridine (Fig. 12.40).[103]

Imide and related ligands can be bent with M=N–X angles down to those expected for approximately sp^2-hybridization, which may occur when the metal center cannot accept the nitrogen lone pair. Severely bent imido ligands are certainly known. The classic example of such a complex is $Mo(NPh)_2(S_2CNEt_2)_2$ by Haymore, Maatta, and Wentworth, which has two different imido ligands in the solid state with Mo=N–C angles of 169.4(4) and 139.4(4)°.[121] The latter, relatively acute angle can be compared with the angle in a typical organic imine (Fig. 12.41), which has a C=N–C angle of 121°. The barriers for organic imines are such that *E*/*Z*-isomers are often observed, and the barriers in inorganic bent imido complexes are often minuscule by comparison. It has been suggested that increased covalency in the M=N bond increases the barrier to changing

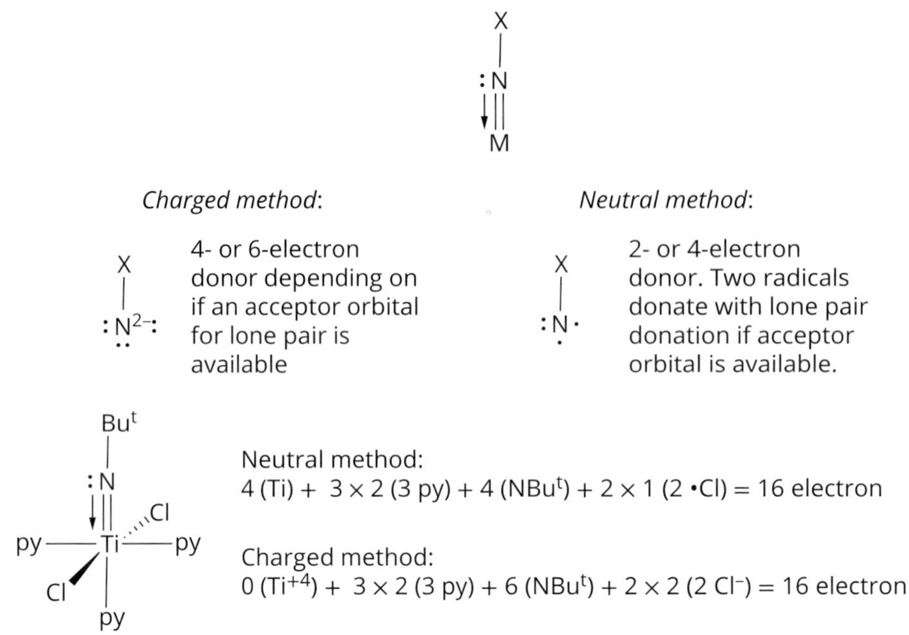

Fig. 12.40 Electron-counting for an N–X metal complex, using Mountford and coworkers' $Ti(NBu^t)(py)_3Cl_2$, where py = pyridine.[102]

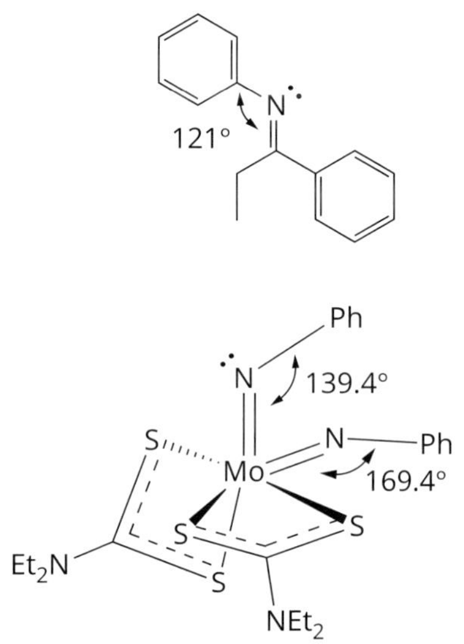

Fig. 12.41 Bond angle in a typical organic imine[123] and in $Mo(NPh)_2(S_2CNEt_2)_2$.[121]

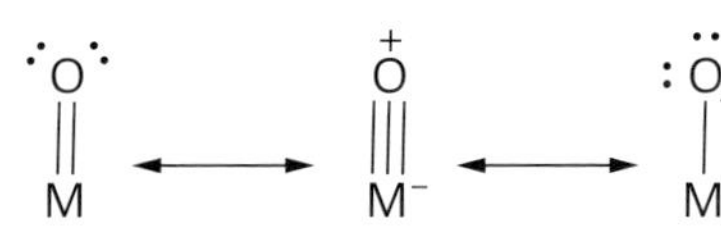

Fig. 12.42 Some of the resonance forms commonly invoked for oxos.

the M=N–C angle; as a result, the largest barriers would be expected when the metal cannot accept the lone pair to produce the bent structure and the metal is quite electronegative.[122]

Imido ligands, NR and NAr, are common across the transition-metal series as both terminal and bridging ligands.[118] Both early[124–130] and late[131–139] transition metals have found applications in catalytic C–N-forming reactions, for example, and have been studied extensively.

Chalcogenide, oxo, and sulfide complexes in particular, are of great importance in industrial, academic, and biological settings.[140–142] From an electronic and electron-counting perspective, they are similar to imido ligands: 4- or 6-electron donor in the charged method and as a 2- or 4-electron donor in the neutral method. However, another resonance form (Fig. 12.42), also present in imido systems but not discussed, may contribute substantially to many oxo complexes. Namely, oxo complexes often can have significant contributions from forms where the π-bond is polarized, leaving an M–O single bond and a full anionic charge on the oxo, and a full cationic charge on the metal.

One interesting and important complex where it is impossible to have triple bonds to all the oxo ligands is OsO_4. This compound is used in dihydroxylation of olefins[143–146] and is quite volatile, with a melting point of 40 °C and a boiling point of 130 °C[a] Because there are four oxo ligands and only nine valence orbitals, there is no possibility of forming three bonds to each oxo. Instead, the single bond resonance form participates strongly, and the Os–O bonds in OsO_4 have a Mayer bond order of about 1½ per bond.[147]

However, many (if not most) oxo and other chalcogenides do form three strong bonds to metal centers when terminal and have calculated bond orders consistent with this. For example (Fig. 12.43), Figueroa and Cummins have prepared the series $ENb\{N(CH_2Bu^t)(3,5\text{-}Me_2C_6H_3)\}_3$, where E = O, S, Se, and Te.[148] Natural resonance theory calculations on the model $ENbH_3$ suggest that the dominant resonance structure in all cases has the chalcogenide triply bonded to the metal center, with this form contributing >90% to the ground-state electronic structure.

[a] OsO_4 is also quite toxic and readily sublimes.

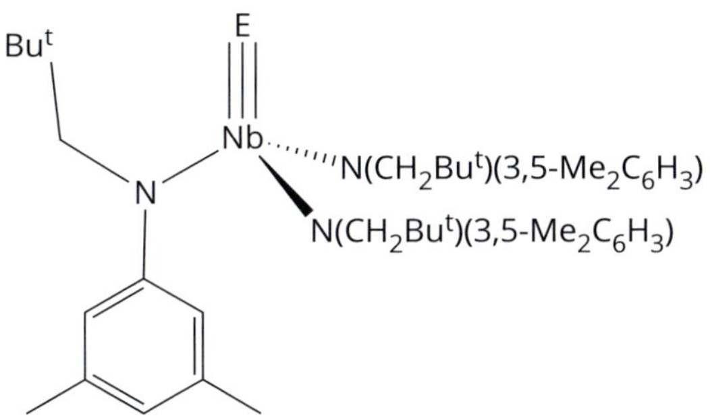

Fig. 12.43 Structures of ENb{N(CH$_2$Bu^t)(3,5-Me$_2$C$_6$H$_3$)}$_3$, where E = O, S, Se, and Te, by Cummins and coworkers.

Especially in the case of first-row late transition metals such as iron, the exact bonding to the metal may vary with the oxygen being best described as anionic, radical, and neutral (Fig. 12.44a). The oxidation state of the metal varies with the nature of the oxygen involved in the resonance form. If the metal acts as a donor to the oxygen atom with a formal −2 charge, then the complex is a "metal oxene". If there is a single covalent bond between metal and oxygen with radical character on the oxygen, the complex is a "metal oxyl"; compare to the 2-center 3-electron bond in O_2. If there is a metal–oxygen double bond with a neutral oxygen, the complex is simply a "metal oxo" as described previously.

A common area where these forms are discussed is iron porphyrin chemistry. If an iron porphyrin chloride, Fe(Por)Cl, is oxidized, the resulting complex may be best described as an oxene, oxyl, or oxo (Fig. 12.44b), depending on the importance of the different resonance forms in calculations or based on the assigned oxidation state from Mössbauer spectroscopy.

Hydrazido(2−) complexes have also been extensively studied for their C–N bond-forming chemistry and for their intermediacy in important dinitrogen reduction chemistry. These ligands are quite varied, with strong parallels to metal carbene/alkylidene electronic structure.[119,149] Two limiting resonance forms can participate when NNR$_2$ binds to a metal, which can be called the "hydrazido(2−)" and "isodiazene"

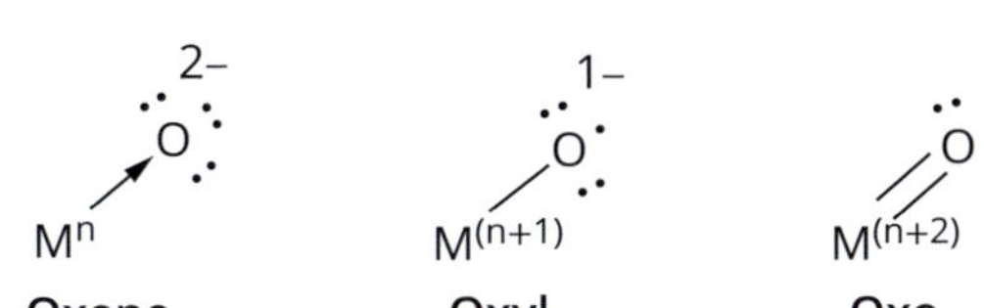

Fig. 12.44a The oxene, oxyl, and oxo resonance forms.

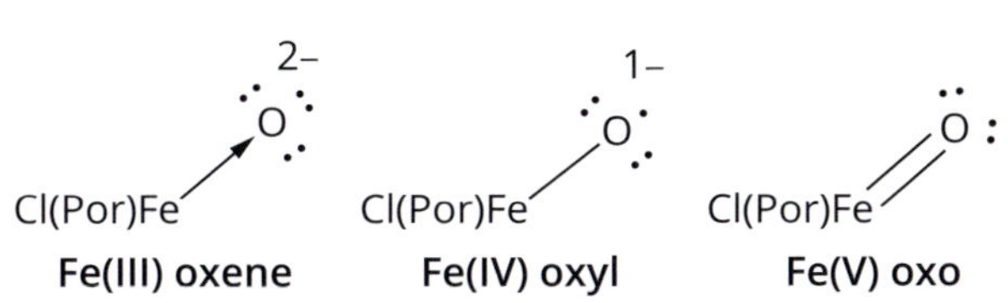

Fig. 12.44b Examples of iron porphyrin structures with the oxene, oxyl, and oxo resonance forms and the effect on formal oxidation state of the metal center.

forms (Fig. 12.45); compare with Schrock and Fischer carbenes. The hydrazido(2−) limiting form is very similar to imido in its electron counting. The isodiazene limiting form is a 2-electron donor that may also act as a π-electron acceptor from the metal center. Notice that if the metal center in M^{n}NNR$_2$ is in the n-oxidation state in the isodiazene it is in the (n + 2)-oxidation state in the hydrazido(2−) resonance form. As a result, the isodiazene resonance structure participates more with less reducing metal centers, such as iron(II). Hydrazido (2−) resonance forms participate strongly in cases where the metal would be very reducing in the M^n oxidation state, such as titanium(II).

Fig. 12.48 shows examples of complexes in which the hydrazido(2−) and isodiazene resonance forms are strongly favored where both utilize the same fragment NNMe$_2$. On the left is Ti(NNMe$_2$)(bpy)(dpma), where dpma = *N,N*-di(pyrrolyl-α-methyl)-*N*-methylamine, which strongly favors a hydrazido(2−) resonance structure.[151] On the right is Chatt, Dilworth, Hursthouse, and coworkers' Mo(NNMe$_2$)(O)(S$_2$CNMe$_2$)$_2$.[150]

12.3.3 Carbyne, Alkylidyne, and Nitride

Transition metal Fischer carbynes and alkylidynes (sometimes called Schrock carbynes) are electronically reminiscent of carbenes and alkylidenes. Fischer carbynes are typically attached to metals in low oxidation states, while alkylidynes are typically associated with metals at or near their highest formal oxidation state.

Electron-counting for Fischer carbynes may appear rather unexpected for beginning students, but is composed to allow understanding of the limiting electronic form associated with Fischer's original species. In accord with our discussion of the electronic structure of Fischer carbenes, these carbynes are expected to have a HOMO that is largely metal-based and a LUMO that is based on C$_\alpha$, making this carbon electrophilic.

For the charged method, the carbyne is similar to linear nitrosyl and is considered to be cationic; in other words, the ligand is called a 2-electron donor that has given up an electron to the metal. A Fischer carbyne is regarded as similar to a linear nitrosyl in the neutral method as well; a 3-electron donor. This can be explained as shown in Fig. 12.46, where the carbyne is either a low-spin doublet if it has not given up an electron, or a singlet if one considers it a cation.

A typical Fischer carbyne synthesis is shown in Fig. 12.46. In this example the tungsten(0) complex W(=C(OMe)Ph)(CO)$_5$ reacts with BBr$_3$, a common reagent for ether cleavage, to replace OMe with Br, However, the bromine ends up attached not to the carbon but to the metal, producing a tungsten bromide and carbyne. Because the bromide is anionic, the carbyne is cationic, and the other ligands are neutral, the formal oxidation state of tungsten remains the same as in the starting carbene, W(0).

Hydrazido(2–)

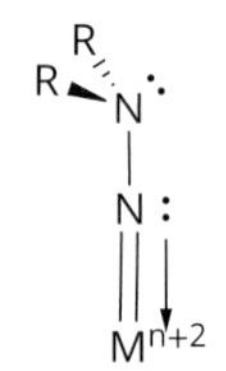

Neutral method:
2- or 4-electron donor

Charged method:
4- or 6-electron dianionic donor

Isodiazene

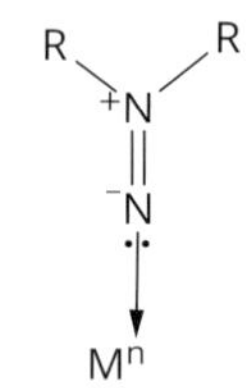

Neutral method:
2-electron donor

Charged method:
2-electron neutral donor

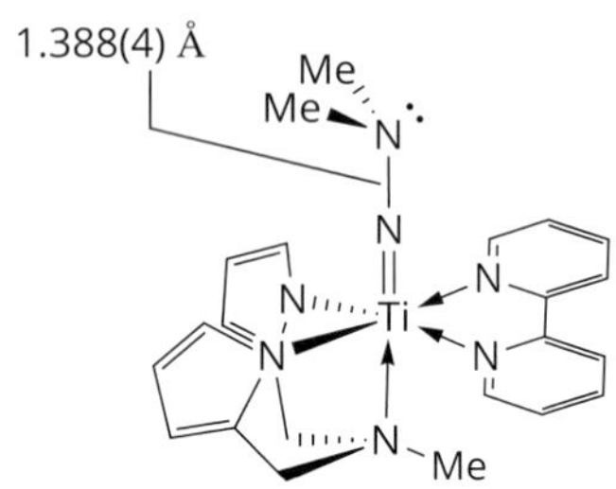

Neutral method:
4 (Ti) + 2 × 2 (bpy) + 4 ($NNMe_2$)
+ 2 × 1 (2 •pyrrolyl) + 2 (NR_2Me) = 16 electron

Charged method:
0 (Ti^{+4}) + 2 × 2 (bpy) + 6 ($NNMe_2$)
+ 2 × 2 (2 ⁻pyrrolide) + 2 (NR_2Me) = 16 electron

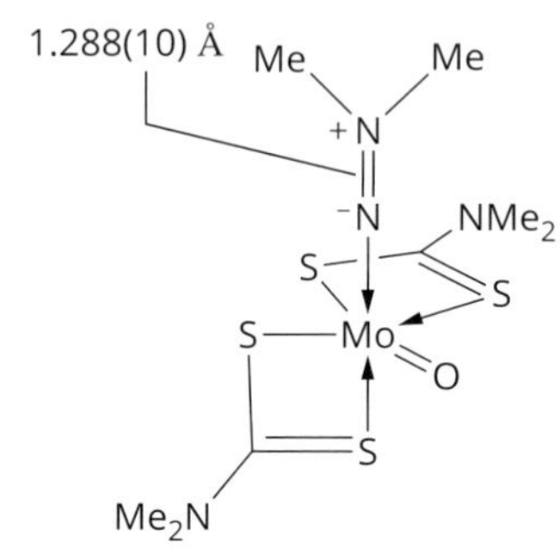

Neutral method:
6 (Mo) + 3 × 2 (S_2CNMe_2) + 4 (oxo)
+2 ($NNMe_2$) = 18 electron

Charged method:
2 (Mo^{+4}) + 4 × 2 (S_2CNMe_2) + 6 (oxo)
+2 ($NNMe_2$) = 18 electron

Fig. 12.45 Electron-counting for hydrazido(2–) and isodiazenes using Ti($NNMe_2$)(bpy)(dpma), where dpma = *N*,*N*-di(pyrrolyl-α-methyl)-*N*-methylamine, and Mo($NNMe_2$)(O)(S_2CNMe_2)$_2$ as examples.[150–151]

Neutral method *Charged method*

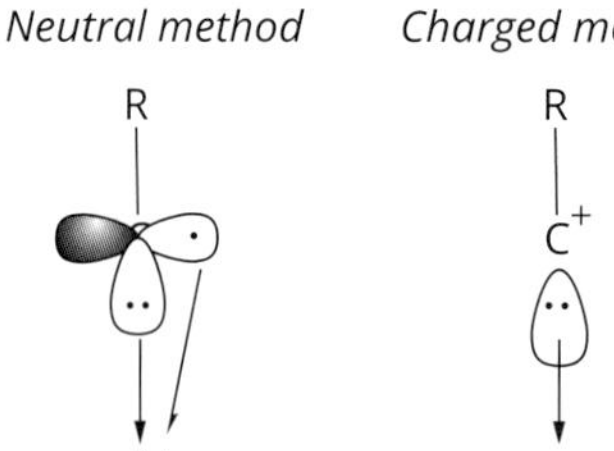

3-electron donor

Cationic 2-electron donor

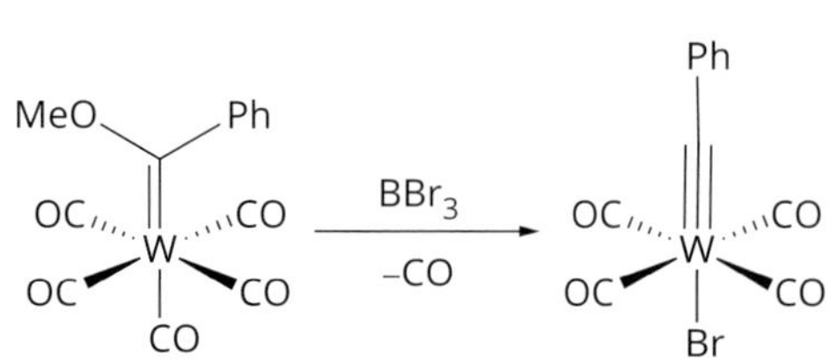

Tungsten(0) Tungsten(0)

Neutral method:
4 × 2 (CO) + 1 (•Br) + 3 (CPh) + 6 (W) = 18 electron

Charged method:
4 × 2 (CO) + 2 (Br^-) + 2 (:CPh^+) + 6 (W(0)) = 18 electron

Fig. 12.46 Fischer carbynes are regarded as low spin and, for determination of formal oxidation state, to have given up an electron to the metal center. In other words, Fischer carbynes are electron-counted the same as linear nitrosyls in both the neutral and charged methods.

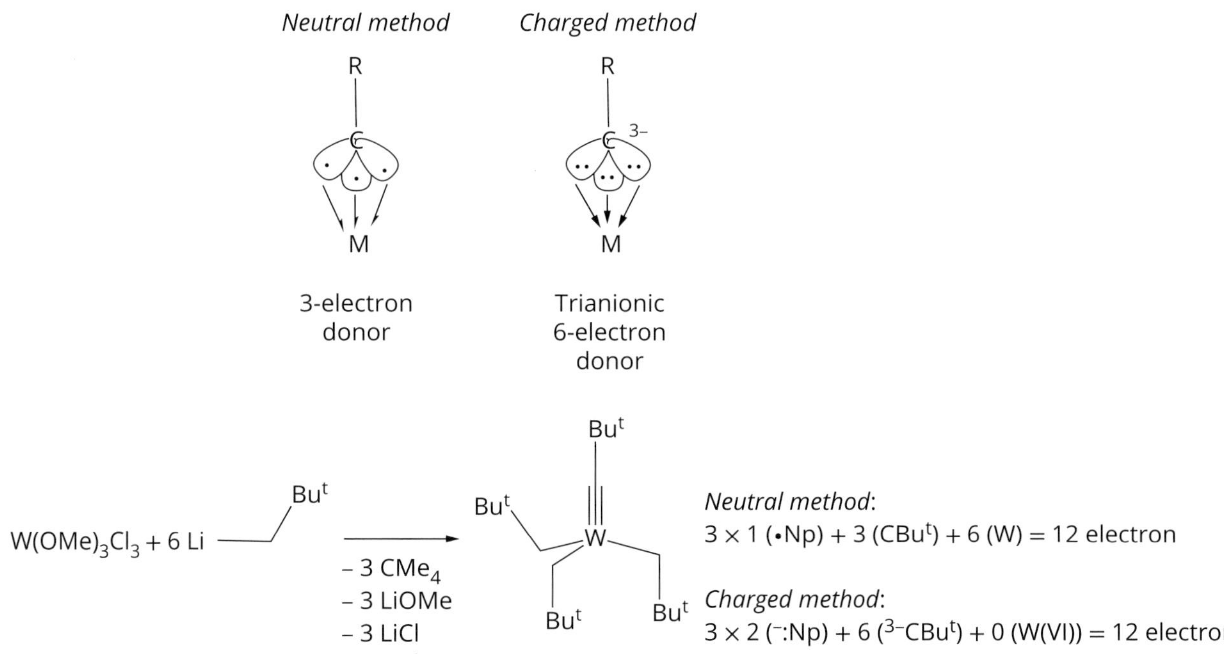

Fig. 12.47 Alkylidynes (Schrock carbynes) are considered triradicals for the neutral method or trianionic 6-electron donors for the charged method. The first alkylidyne was prepared by Schrock, Clark, and coworkers by addition of neopentyllithium (LiNp, $LiCH_2CMe_3$) to $W(OMe)_3Cl_3$.

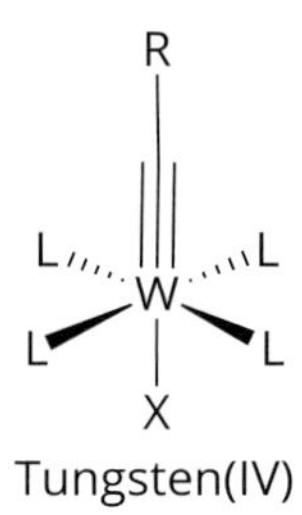

Neutral method:
4 × 2 (L) + 1 (•X) + 3 (CR) + 6 (W) = 18 electron

Charged method:
4 × 2 (L) + 2 (X^-) + 6 ($^{-3}$CR) + 2 (W(IV)) = 18 electron

Fig. 12.48 Electron count on tungsten using the neutral and charged methods regarding the alkylidyne as W(IV), which produces the same value as when using W(0); compare Fig. 12.49.

The electron-counting of Schrock carbynes is more straightforward. As shown in Fig. 12.47, the neutral method regards the ligands as 3-electron donors and the charged method as 6-electron donors. One of the first such complexes to be isolated was prepared by Schrock and Clark from treatment of WCl_6 or $WCl_3(OMe)_3$ with neopentyllithium. The reaction proceeds through repeated α-elimination reactions—a reaction step discussed previously for the synthesis of $Ta(=CHBu^t)Np_3$ (Fig. 12.34).[152–153]

Just as for alkylidenes, the HOMO of an alkylidyne is expected to be the M–C_α π-bond and the LUMO a metal-based nonbonding orbital in most cases. As a consequence, the alkylidyne carbon is often nucleophilic.

Also, as for alkylidenes and carbenes, there is a spectrum of reactivity between the Schrock and Fischer extremes. This high-spin/low-spin character of the free ligand (compare Figs. 12.33 and 12.35) belies a deeper complexity residing in the issue of whether the complex is a Fischer- or Schrock-carbyne, which in actuality is determined by the character of the electronic structure of the entire complexes, as was described for the carbenes using molecular orbital theory.

Furthermore, there are many researchers who choose to regard carbynes like Fischer's species as W(IV) d^2-complexes. In the end, whether the complex is W(0) or W(IV) depends on where along this spectrum from Fischer carbynes to alkylidynes the particular complex resides (Fig. 12.48), and, as can be seen subsequently, has no effect on the formal electron count of the complex. The choice of one formal oxidation state or other depends on how the researcher chooses to regard the nature of the molecular orbitals as mostly ligand-based or mostly metal-based. Either choice, especially in the case of Fischer carbynes it seems, is fine as long as the discussion is clear about the choice being made.

Pragmatically, the real test of whether a complex is a Fischer carbyne or an alkylidyne rests in the compound's reactivity towards the electrophile or nucleophile of interest and the site on

Ar = 2,6-$Pr^i_2C_6H_3$

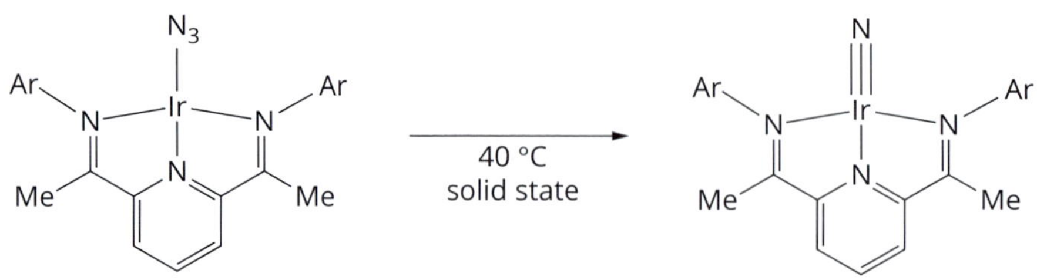

Neutral method:
2 × 2 (NAr) + 2 (pyridine) + 3 (N) + 9 (Ir) = 18 electron

Charged method:
2 × 2 (NAr) + 2 (pyridine) + 6 (N) + 6 (Ir(III)) = 18 electron

Fig. 12.49 An example of electron-counting for a terminal nitrido by Burger and coworkers.[154]

the complex of direct reaction. If C_α is electrophilic, then the complex is best regarded as a Fischer carbyne; if C_α is nucleophilic, then the complex is best regarded as an alkylidyne.

Nitrido complexes are electron-counted similar to alkylidynes and are 3-electron donors in the neutral method and trianionic 6-electron donors in the charged method. Terminal nitrido complexes are known for the majority of transition metals from Groups 5 to 8. Some examples are known for other elements outside of this range. For example, Burger and coworkers have reported a square planar iridium(III) nitrido, prepared by solid-state thermolysis of the iridium(I) azido (Fig. 12.49).[154]

There are many methods for nitrido ligand formation and a few are shown in Fig. 12.50. Thermal or photochemical degradation of an azido ligand is a common method for the production of a nitrido ligand. In addition to the example in Fig. 12.49, *in situ* production of $Mo(N_3)Cl_3$ by addition of Me_3SiN_3 to $MoCl_4$ leads to a solvate of $NMoCl_3$, which may be reacted with a variety of different ligands to produce Mo(VI) nitrido complexes (Fig. 12.50).[155–158]

Also, Cummins and coworkers have pioneered the use of an azido surrogate, Li(dbabh), in nitrido synthesis (Eq. 9.17).[159] The reagent has also been used to prepare an interesting and rare example of an iron terminal nitrido (Fig. 12.50) by Peters, Betley, and coworkers.[160]

Nitrido ligands are also available from oxidation/deprotonation of ammine ligands. For example, Meyer and coworkers have studied the conversion of osmium(II) *trans*-{Os(tpy)$Cl_2(NH_3)$} to osmium(VI) *trans*-$\{Os(tpy)Cl_2(N)\}^+$, where tpy is 2,2′:6′2″-terpyridine, by successive removal of electrons and protons (Fig. 12.50).[161] In addition, some nitrido complexes can be prepared by nitrogen-atom transfer from other nitrido complexes. For example, $NCr(OBu^t)_3$ transfers a nitrogen atom to three coordinate $Cr(NPr^i_2)_3$ to give $NCr(NPr^i_2)_3$.[162]

12.3.4 Bridging Oxides and Sulfides

All of these metal–ligand multiple bonds, instead of being terminal and interacting with a single metal center, can bridge two metal centers. In fact, most binary elemental oxides and sulfides contain mostly bridging ligands. Of particular importance are bridging oxides and sulfides, which are commonly encountered in natural enzymes. As such, a complete structural discussion of bridging oxides and sulfides, even narrowed to iron-containing biological systems, is beyond the scope of this book. We will briefly discuss two biological bridging oxide and sulfide systems: the iron oxide storage system in ferritin and iron-sulfur clusters found in many biological systems.

12.3.4.1 Iron Oxides and Ferritin

Iron is used by all organisms from archaea to eukaryotes to animals for a variety of purposes, such as oxygen transport by hemoglobin and as active sites in enzymes. Storage of iron is done in a diverse class of proteins called ferritin, where the iron is stored as a material with bridging oxides and in the presence of phosphate.

Ferritin is best described as a rhombic dodecahedron—a twelve-sided figure with diamond faces; the figure has O_h point group symmetry. It is composed of twenty-four subunits and is a 450 kDa protein. The interior of the protein is lined with carboxylate groups from glutamic acid and aspartic acid residues, which coordinate to iron as carboxylates. In the cavity about 4,500 irons can be stored. See Fig. 12.51 for a structure of the ferritin protein and for a schematic showing the bonding interior to the core, which involves bridging oxo groups.

The active sites of many proteins include oxo bridges between iron, copper, and other metal centers.

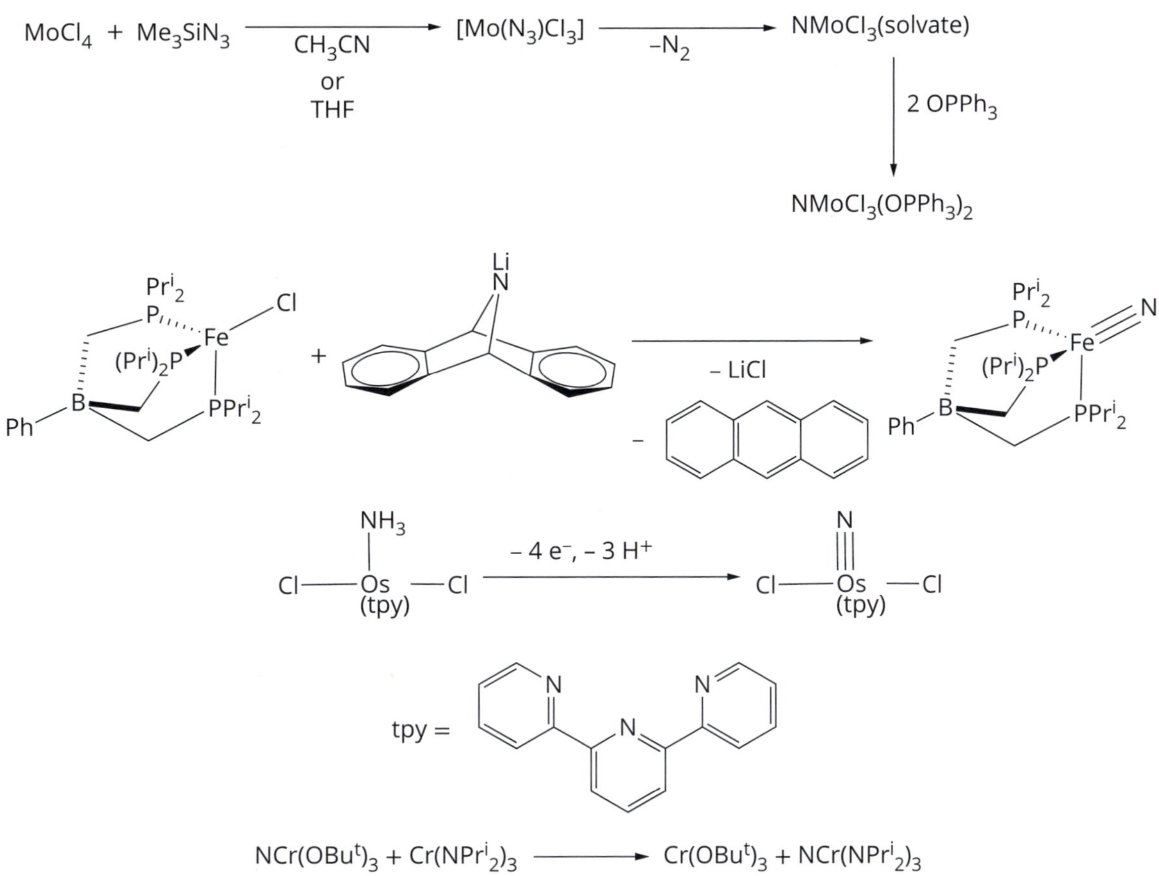

Fig. 12.50 Some examples of nitrido syntheses involving, from top to bottom, azido ligand degradation, Li(dbabh) substitution of halide and anthracene loss, oxidation and deprotonation of a coordinated ammonia, and nitrogen-atom transfer.

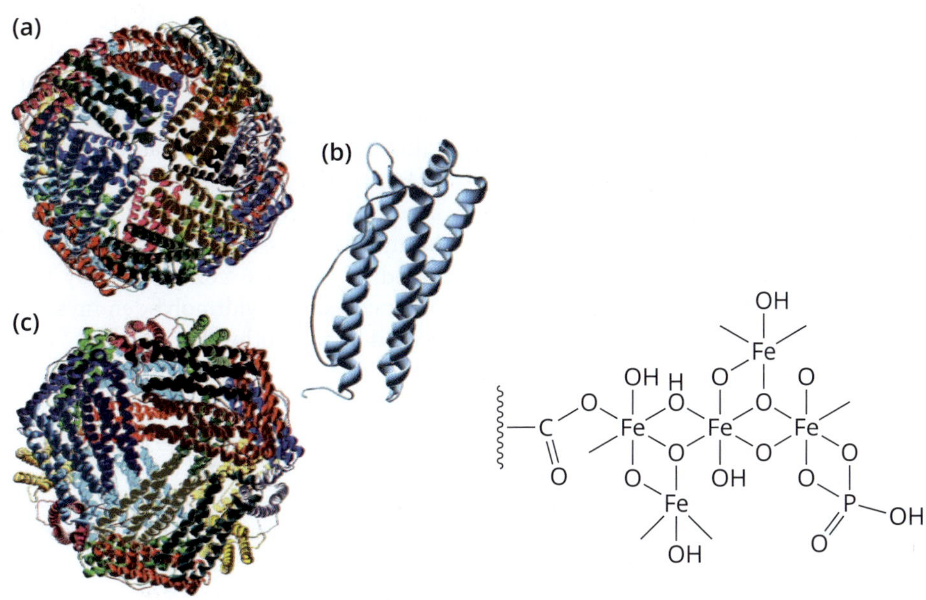

Fig. 12.51 (*left*) A depiction of the structure of the human iron storage protein ferritin: (a) A view down the four-fold symmetric axis (b) The ferritin subunit. The full protein is a 24-mer. (c) A view down the four-fold symmetric axis. (From the review by Le Brun and coworkers.[163]) (*right*) Schematic structure of how the iron is stored in the interior of the protein. The interior of the protein is lined with glutamine and aspartate residues, which contain carboxylates, to bind iron. (From Rehder, *Bioinorganic Chemistry*.[164])

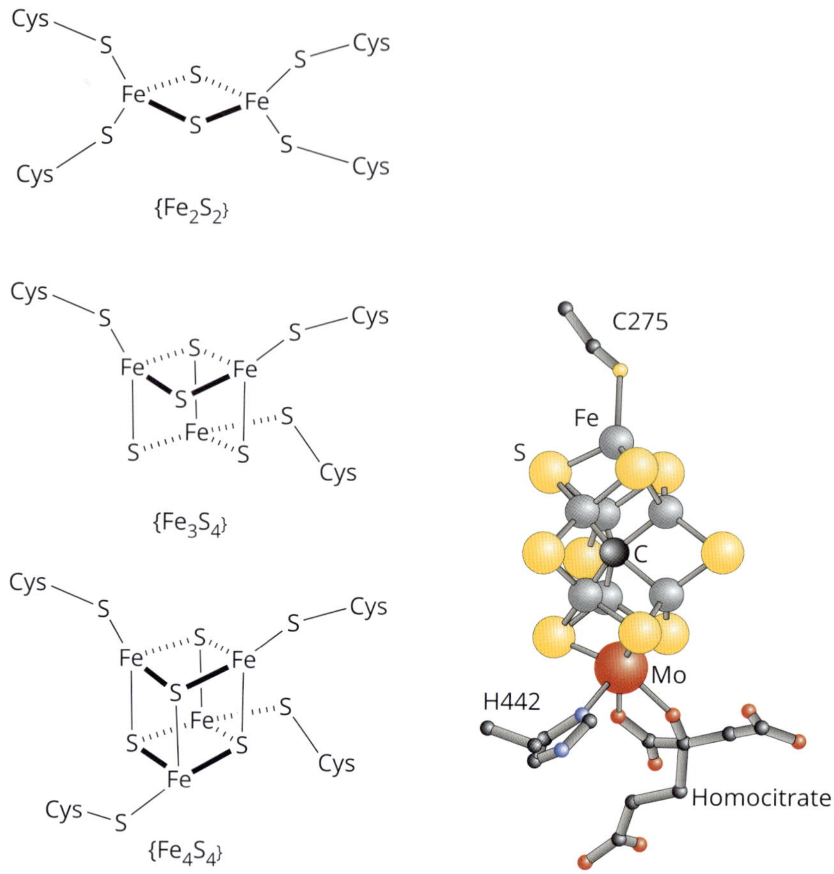

Fig. 12.52 (*left*) The $\{Fe_2S_2\}$-, $\{Fe_3S_4\}$-, and cuboidal $\{Fe_4S_4\}$-structures found in ferredoxin. (*right*) The $MoFe_7S_9C$ bridging sulfide core of nitrogenase supported by a cysteine (*top*) on an iron, and homocitrate with histidine on molybdenum (*bottom*). (Figure on the right from the review by Einsle.[169])

12.3.4.2 Iron–Sulfur Clusters

Sulfide bridges are commonly found in biological systems requiring electron transfer for function, such as respiration, photosynthesis, and nitrogen fixation. For example, the $\{Fe_2S_2\}$-structure is found in ferredoxins where each iron has two alkylthiolates from cysteines as well as sulfide bridges (Fig. 12.52). Also, $\{Fe_3S_4\}$- and cuboidal $\{Fe_4S_4\}$-structures are commonly found in electron-transfer proteins.[165–168] Likewise, the core of nitrogenase, the natural enzyme for the reduction of N_2 to ammonia, has an active site with an $MoFe_7S_9C$-bridging sulfide core.[169] As shown, sulfides commonly bridge more than two metal centers.

12.4 Metal–Metal Bonds[170] (Advanced Topic)

While the better part of two chapters have been spent trying to describe the array of ways that transition metals may bind to ligands, usually comprised of *p*-block elements, metals perhaps bind to each other with even greater diversity. Covalent bond orders between metals are known from single to quintuple, with dative interactions also well established.

Perhaps the simplest transition metal–metal bonded complex is dodecacarbonyldimanganese, $Mn_2(CO)_{10}$, which contains a single covalent bond between the two metals. As an example of a metal–metal double bond, shown in Fig. 12.53 is a Ta=Ta acetylene complex prepared by Cotton and Hall.[171] Many examples of metal–metal triple bonds have been prepared of Group 6 compounds by Chisholm, Rothwell, and others; hexabenzyldimolybdenum shown in the figure was reported by the Rothwell group.[172–174]

F. Albert Cotton and coworkers reported the first quadruple bond (one σ-, two π-, and one δ-bond) complexes in the early 1960s; the first such compounds were salts of $Re_2Cl_8^{2-}$.[175–176] The first pentuple bond (one σ-, two π-, and two δ-bonds) was reported by Power and coworkers in 2005.[177]

There are examples where a metal complex can act as a 2-electron donor (Lewis base) toward another metal complex (Lewis acid). One such example (Fig. 12.54) is $(PMe_3)(CO)_4Os{\rightarrow}Cr(CO)_5$, prepared by reaction of $Os(CO)_4(PMe_3)$ with $Cr(THF)(CO)_5$. The complex is stable in air and even stable to chromatography on Florisil.[178] In this case, the dative bond assignment seems best, as a single bond produces

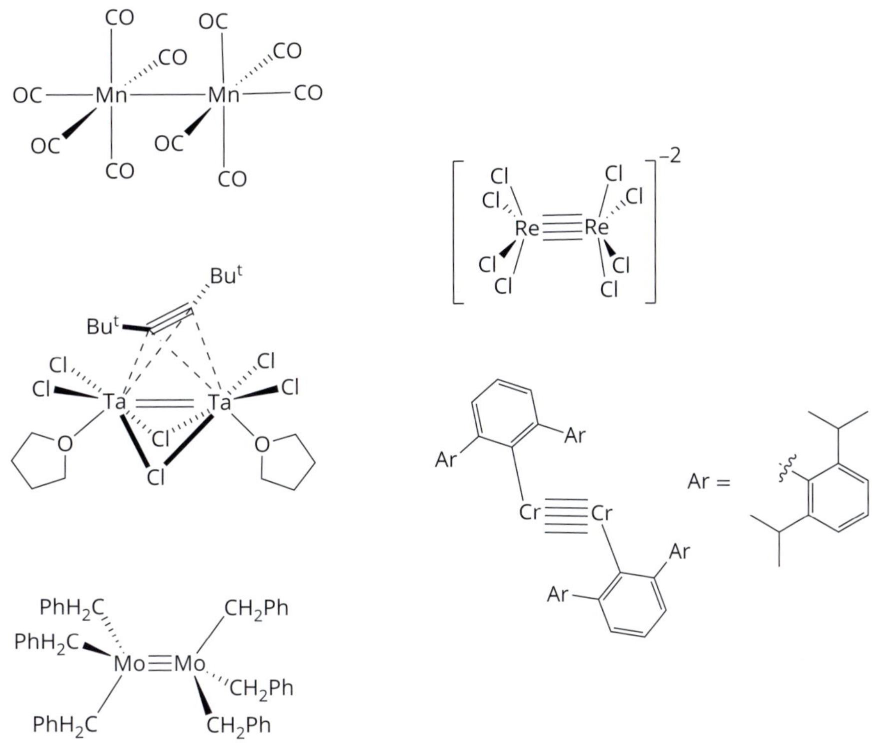

Fig. 12.53 Examples of metal–metal bonds from single to quintuple.

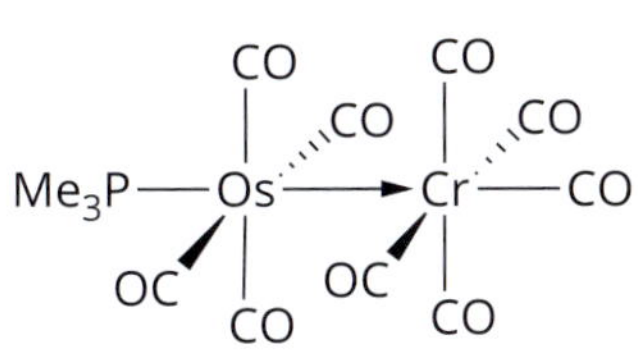

Fig. 12.54 A dative bond between two metal centers.

nineteen electrons on Os and seventeen electrons on Cr, whereas a dative bond suggest two 18-electron metal centers.

Weak metal–metal bonds can also occur due to filled–filled orbital interactions in, for example, d^8-d^8 systems.[179–180] Gray and coworkers extensively investigated such dimers and higher oligomers (Fig. 12.55). In one such system it was found that $\{Rh(CNPh)_4\}^+$ monomer was in equilibrium with dimeric and trimeric structures.[180]

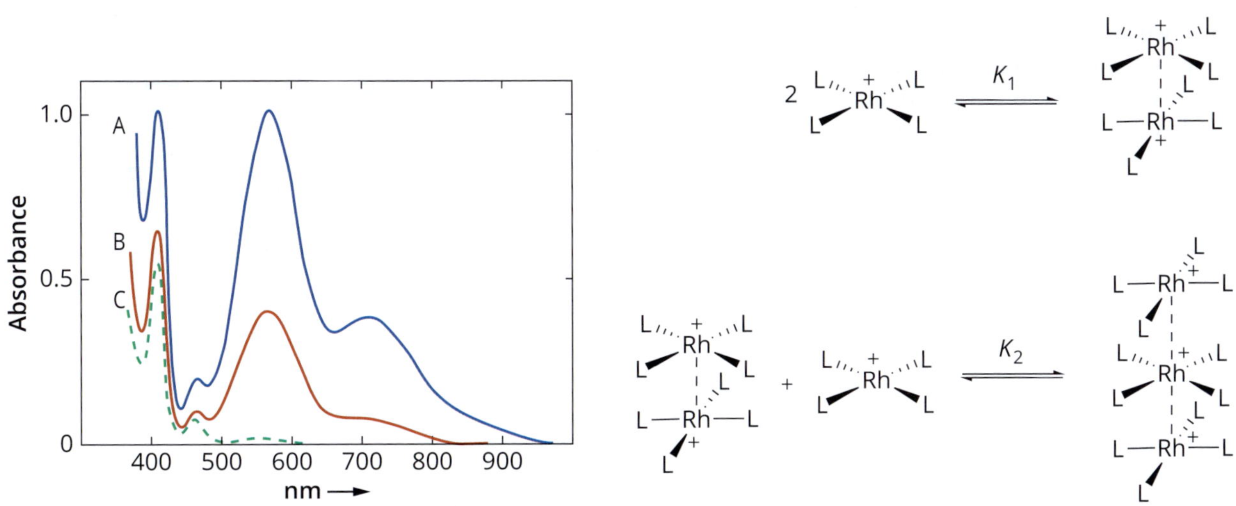

Fig. 12.55 (*left*) Absorption spectrum of $\{Rh(CNPh)_4\}^+$ in acetonitrile at 25 °C at different concentrations (cell path lengths): A, 5.7×10^{-2} M (0.06 mm); B, 2.7×10^{-2} M (0.06 mm); C, 6.3×10^{-4} (0.75 mm). The peaks at 411 nm (ε = 6,000), 568 nm (ε = 14,800), and 750 nm (ε = 16,000) were assigned to the monomer, dimer, and trimer, respectively. (*right*) The equilibria between monomer, dimer, and trimer are shown, and K_1 and K_2 were calculated as 16 ± 1 and 16 ± 3 M^{-1}. L = CNPh. The counterions, not shown, were either BPh_4^- or BF_4^-, which did not impact the spectra. (Absorbance spectra from the reference.[180])

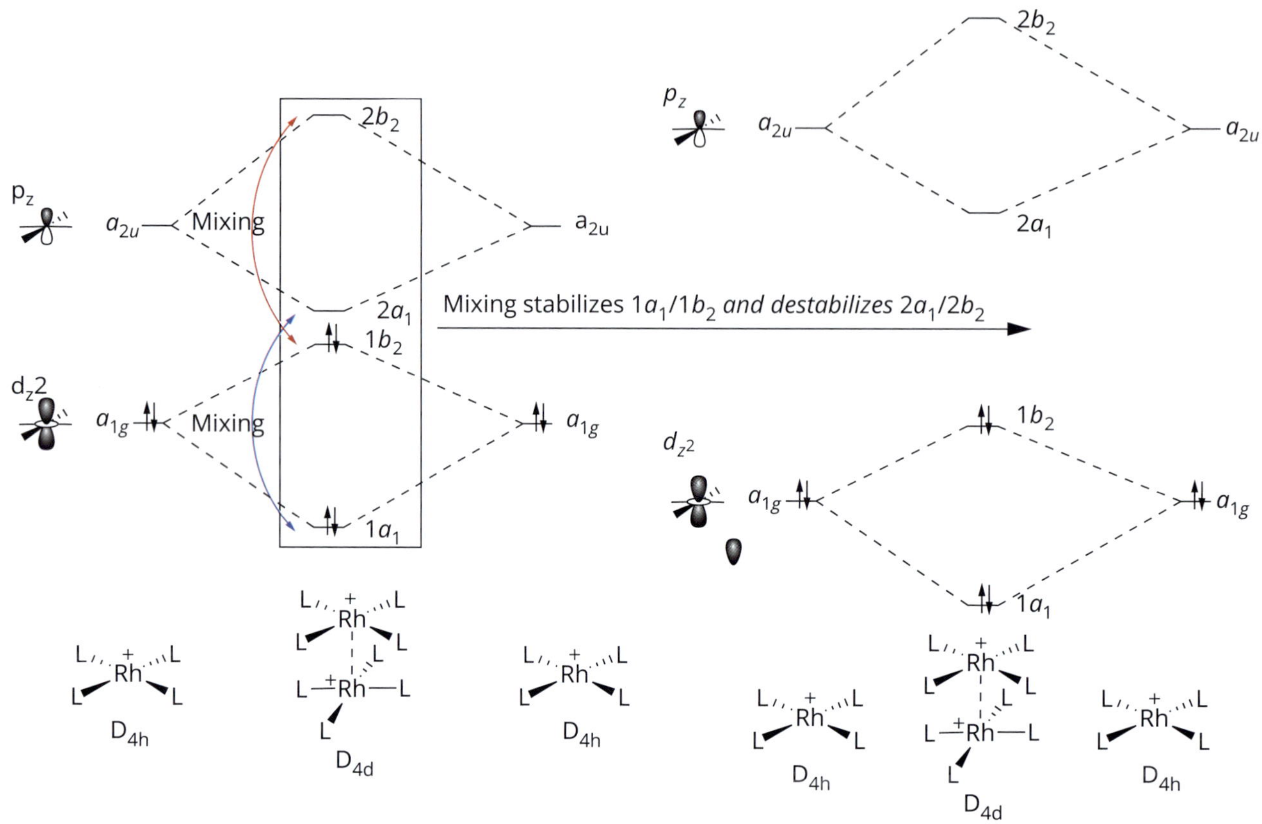

Fig. 12.56 Frontier orbitals and the MO interaction of two $\{Rh(CNPh)_4\}^+$ species. The symmetry change from the fragments to the dimer. The staggered D_{4d} isomer is shown, but the eclipsed D_{4h} isomer is probably present as well. Mixing of the lower-energy filled orbitals with the higher-energy empty orbitals give the dimer some bonding character as antibonding character is shifted to the empty orbitals.[180]

The MO explanation for the formation of a bond between the two filled orbitals involves examination of the frontier orbital interactions (Fig. 12.56). The d_{z^2}-orbitals on the monomers form bonding and antibonding combinations, labeled *1a₁* and *1b₂*. Since both of these orbitals are filled, one would expect no net interaction; however, the LUMO of each monomer (p_z-orbital) will also form a bonding and antibonding combination. Since the orbitals from the HOMO–HOMO interaction and LUMO–LUMO interaction have the same symmetries, these MOs will mix, leading to stabilization of the lower-energy orbitals and destabilization of the empty higher-energy set of orbitals. In all, this orbital mixing leads to net stabilization of metal–metal bond over the monomers. Since the interaction involves mixing of *d*-orbital character with a high energy *p*-orbital, it will be a relatively weak bond.

Weak metal-metal interactions are not limited to square planar d^8-complexes and such bonds are suggested in systems of diverse electronic make-up. One unusual such system involving a ferrocene-derived amido with uranium was published by Diaconescu and coworkers, where they suggested Fe–U bonding interactions (Fig. 12.57). Magnetism

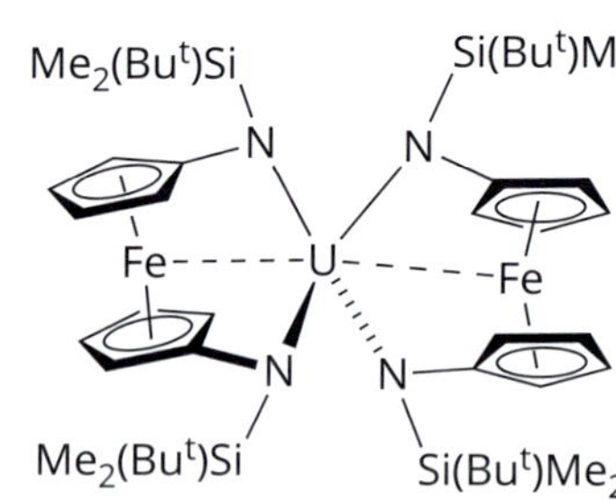

Fig. 12.57 An example of a uranium-iron bond.

studies by Long, Bartlett, and coworkers were consistent with a metal-metal interaction.[181–182]

12.5 Weakly Coordinating Anions[183–187] (Expanded Coverage)

In many cases it is desirable to isolate a cationic complex as a discrete species from its counterion, or as nearly so as possible. Generally, if a complex's reactivity is enhanced by Lewis acidity, the cation will have the highest activity available when it bonds to its anion as weakly as possible. In polar solvents, alleviation of ion-pair effects can be accomplished even with halides in some cases. As one moves to less polar solvents and more reactive cationic metal complexes, specialized anions are required to prevent strong ion-pairing and maximize reactivity. These replacements for halide counterions were often called "noncoordinating anions", but in the solid state and in nonpolar solvents this seems like a misnomer in many cases, and most researchers prefer "weakly-coordinating anions" (WCAs). Some classes of commonly employed WCAs are shown in Table 12.5.

Which ligand is best for a particular application is best decided based on literature precedent for the type of cation to be stabilized; however, the borate-based $B(Ar_F)_4^-$ systems have found extensive use from industry to academia and are perhaps the most generally applicable. There are many known varieties of the $B(Ar_F)_4$-anions, affectionately referred to as "barf" salts, the two most common being $B\{3,5\text{-}(CF_3)_2C_6H_3\}_4^-$ (or BAr_{F24}) and $B(C_6F_5)_4^-$ (or BAr_{F20}).

These $B(Ar_F)_4$-anions are extensively used industrially as counterions to highly Lewis-acidic, cationic catalysts for olefin polymerization. The anions are poor Lewis acids and allow rapid reactivity without quenching by making an adduct to the metal, which would slow substrates' access to the active site. While the anions are weakly coordinating, they do ion pair with cations in nonpolar solutions. Even when tight ion-pairing occurs in nonpolar solution, there is evidence that they

Table 12.5 Some examples of common weakly coordinating anions.

Small anions

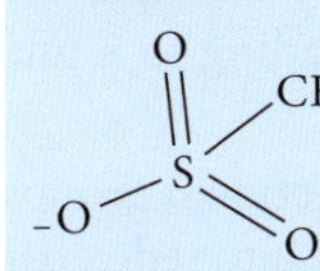

triflate, ^-OTf

BF_4^- ClO_4^- PF_6^- SbF_6^-

Tetraarylborates

$^-BPh_4$

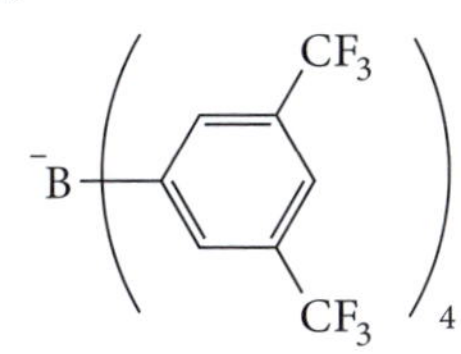

$^-B(Ar_F)_4$ or BAr_{F24}^-

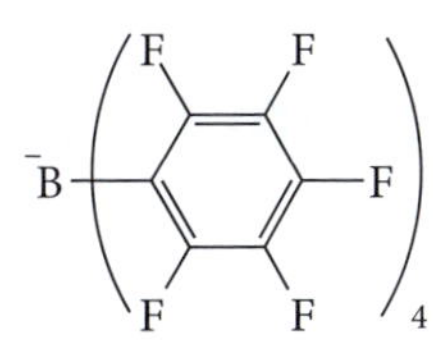

$^-B(Ar_F)_4$ or BAr_{F20}^-

Alkyloxy/aryloxymetallates and teflates

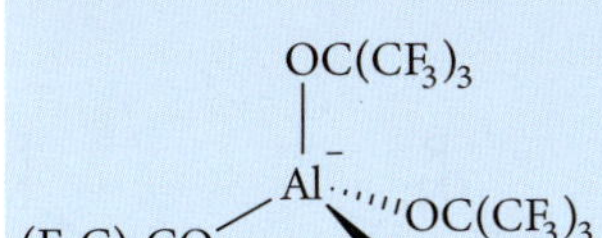

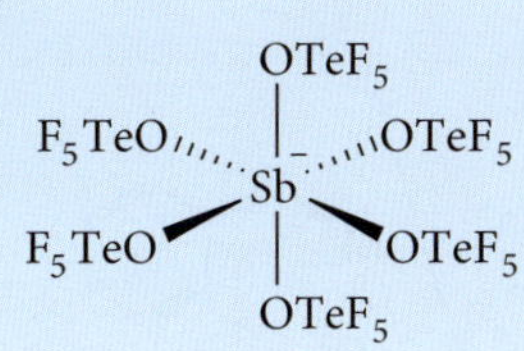

Carboranes

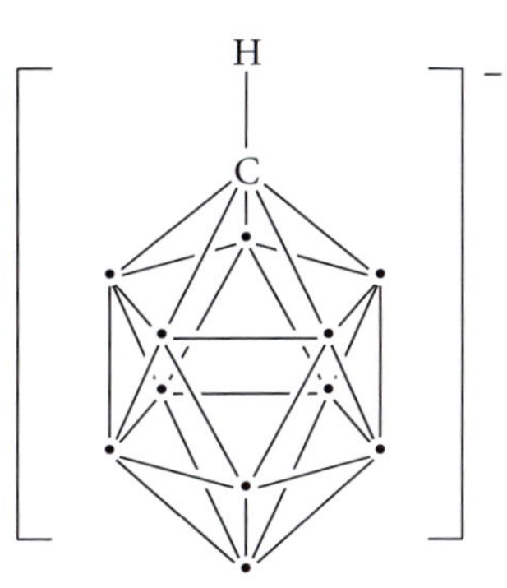

• = B–Cl

are nonspecific in the way they pair to their cation, and, as a result, spend much of their time away from the reactive site.[188]

Some of the anions can coordinate rather well or may decompose, giving up aryl or halide to some metal centers. Triflate is an often-employed WCA, but in high-oxidation-state complexes it appears to donate with a similar ability to iodide.[162] Perchlorate, ClO_4^-, is used relatively rarely, as all of its salts are explosive. Anions like BF_4^- and PF_6^- can release fluoride to metals, and the $B(Ar_F)_4^-$ anions are known to transfer Ar_F in some instances. Of the small anions listed, SbF_6^- may have the best combination of stability, availability, and low coordinating ability. The carboranes $(CB_{11}X_{12})^-$ counterions (where X = H, Me, halides or mixtures of these) may be the most chemically robust anions, but are not the most easily accessed and appear to coordinate strongly in some systems and are the weakest coordinating anions in others.

Commonly employed reagents incorporating WCAs include $^+CPh_3$ (H^- or R^- removal), Ag^+ (oxidation or halide removal), H^+ (strong Brønsted acid, often employed as Et_2O solvate), and ferrocenium (outer sphere oxidant).

Lewis acids, most notably $B(C_6F_5)_3$, can be used with methyl-containing complexes to generate cations with weakly coordinating $\{MeB(C_6F_5)_3\}^-$ as the counterion. Similarly, addition of $AlCl_3$ to chloride complexes may produce cations with $AlCl_4^-$ as the anion.

12.6 Redox Noninnocent Ligands (Advanced Topic)

There is a broad and interesting field where the reactivity and properties of redox interactions between metals and ligands are studied. Some such interactions occur in excited states of the complexes due to metal–ligand or ligand–metal charge transfer, which will be discussed later. Some such redox processes occur thermally, which will be briefly discussed here. The field of redox interactions between ligands and metals is vast and is only briefly mentioned here with a few examples. The interested reader is directed to more detailed reviews.[189–194]

Many ligands (Fig. 12.58), including some of those already discussed, can act as redox noninnocent ligands. There are many examples of such behavior in biological systems as well; for example, porphyrin acts as a redox active ligand in some intermediates of cytochrome P450, and phenoxyl radical is often used as a reducing equivalent carrier when it is, and even when it is not, bound to a metal center.[189,195–197]

There are certain properties found in the majority of redox noninnocent ligands:

(1) Redox noninnocent ligands are, more often than not, multidentate with extended π-systems. The metal center typically has multiple contacts with the donor/acceptor π-system leading to a multitude of stabilizing resonance forms.
(2) Certain functional groups are often employed. These can be classified into redox pairs shown in Fig. 12.58. The vast majority of redox active ligands employ one or more of these redox complexes or closely related systems.

One of the best-known and most studied redox active ligands is derived from *ortho*-catecholate (Fig. 12.59), which on oxidation is known as semiquinone (1-electron oxidation) or quinone (2-electron oxidation).

As one example (Fig. 12.60), McCusker and coworkers have prepared and studied in detail the redox pair of complexes $\{Cr(O_2C_6H_4)(tren)\}^{2+/1+}$, where tren = $N(CH_2CH_2NH_2)_3$ a tetradentate amine ligand. In both cases the chromium center is assigned as Cr(III). Consequently, in the +1 complex the redox

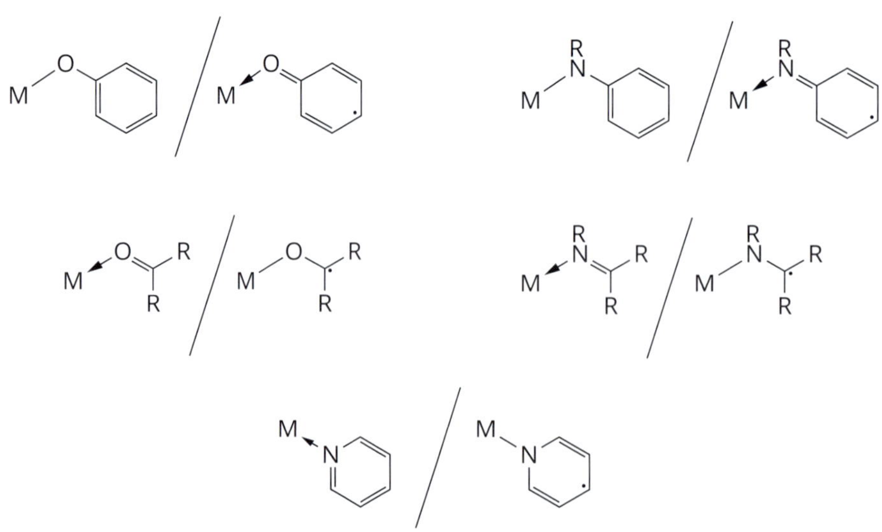

Fig. 12.58 Some common groups found as parts of redox active ligands, which are often chelating. The anionic and dative versions of the ligands are shown with the radical form on the right.

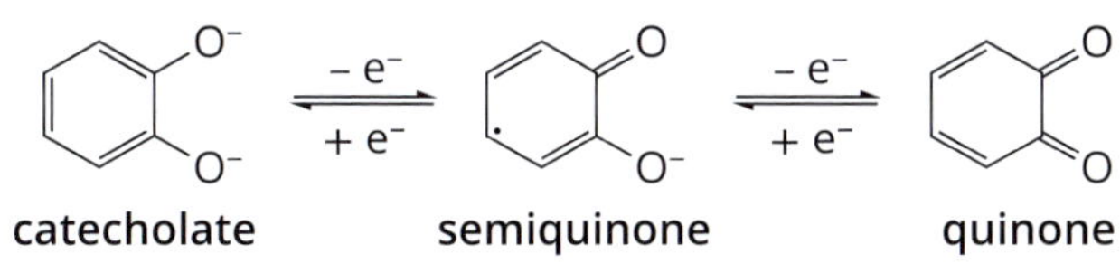

Fig. 12.59 The structures of catecholate, semiquinone, and quinone ligands, a common class of redox-active ligands.

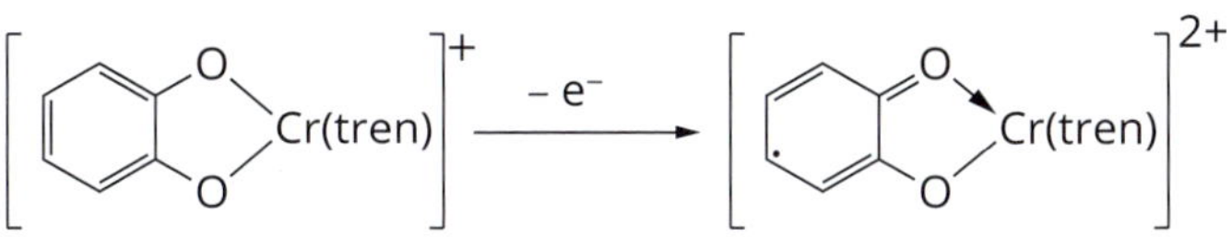

Fig. 12.60 McCusker and coworkers' $\{Cr(O_2C_6H_4)(tren)\}^{2+/1+}$, where tren = $N(CH_2CH_2NH_2)_3$'

active ligand acts as a catecholate, and in the +2 complex the ligand is a semiquinone.[198–199]

The Chirik group has popularized bis(imino)pyridine ligands (Fig. 12.61) for use in cobalt and iron catalysis.[200–202] The catalyses are proposed to involve redox activity at the ligand, metal, or both. The isolated iron species are often ligand-centered radicals as shown in the redox series when in Fig. 12.61. In the dianionic species **A** with an Fe(I) metal center, there is a carbon-centered radical, and the ligand is trianionic towards the metal. Oxidation of **A** gives **B** with a similar ligand electronic structure; the oxidation is iron-centered. Oxidation of **B** gives **C**, which has a dianionic ligand, and the metal oxidation state is unchanged by the ligand-centered redox event.[203]

Redox noninnocent ligands have been explored with early transition metals as well. For example, Heyduk and coworkers have examined imine analogs of the catechol-semiquinone-quinone redox series on Group 4 elements. In one study (Fig. 12.62), a titanium(IV) complex undergoes oxidative addition of Cl_2 (from Cl_2IPh) using reducing equivalents from the catecholate-like ligand. The proposed intermediate is either the quinone-like dichloride (top middle) or dication (bottom middle), which can undergo reductive elimination of two acetylides to form a new carbon–carbon bond.[204]

12.7 Covalent Bond Classification (CBC) Method[205–206] (Advanced Topic)

In organic chemistry, a simple yet powerful test for the validity of any structure is taught: does carbon have four bonds? For the vast majority of organic compounds, carbon will have four bonds involving its four valence electrons and four valence orbitals. Transition metals commonly form six bonds but form a myriad of compounds of lower and higher coordination numbers as well. From all of this arose the system described in this and the previous chapter, in which compounds are analyzed based on electron count and oxidation state.

Classification of compounds using electron count and formal oxidation state does have its shortcomings. For example, the chemistry of the complex $(R_2N)_3MoMo(NR_2)_3$ has a metal–metal triple bond and is formally Mo(III). However, its chemistry is far more like $NMo(NR_2)_3$, which is Mo(VI), than $Mo(NR_2)_3$, which is Mo(III). For example, Mo(III) monometallic complexes must be paramagnetic, whereas "Mo(III)" $(R_2N)_3MoMo(NR_2)_3$ is diamagnetic. A useful method for classification of compounds by their covalent bonds, and other bonding for that matter, has been developed that is generally applicable in order to address such problems, called the covalent bond classification (CBC) method.

The CBC method resolves many of the shortcomings of electron-counting and formal oxidation states, and is a fairly simple approach to understanding the electronic structure of all the elements in the periodic table. Under this method, the molecule is examined as a whole, in contrast to the formal oxidation state in which polarities are assigned based solely on electronegativity while heterolytically cleaving most types of bonds. A CBC analysis is carried out based on the general formula $\{MX_xL_lZ_z\}^{Q\pm}$, where the terms are defined as:

M = element of interest.

X = an atom or group bound to M through a covalent bond.

x = number of X ligands.

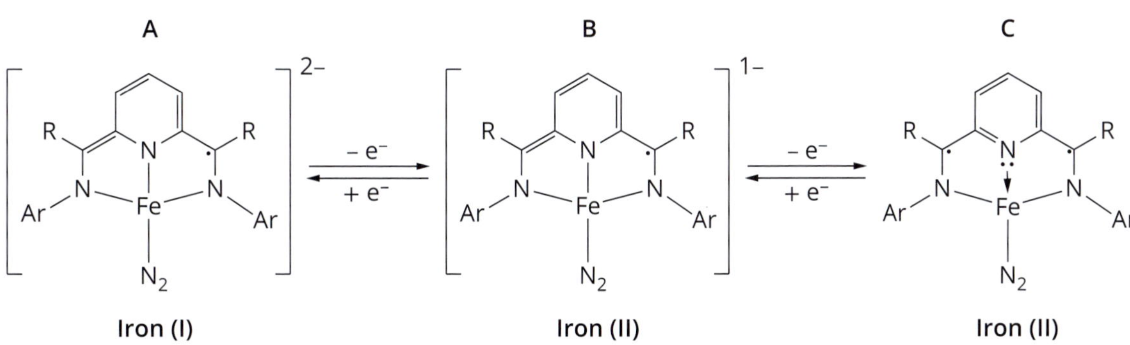

Fig. 12.61 Chirik and coworkers' 2,6-bis(imino)(dinitrogen)pyridine iron complex with some possible redox states of the metal and ligand.

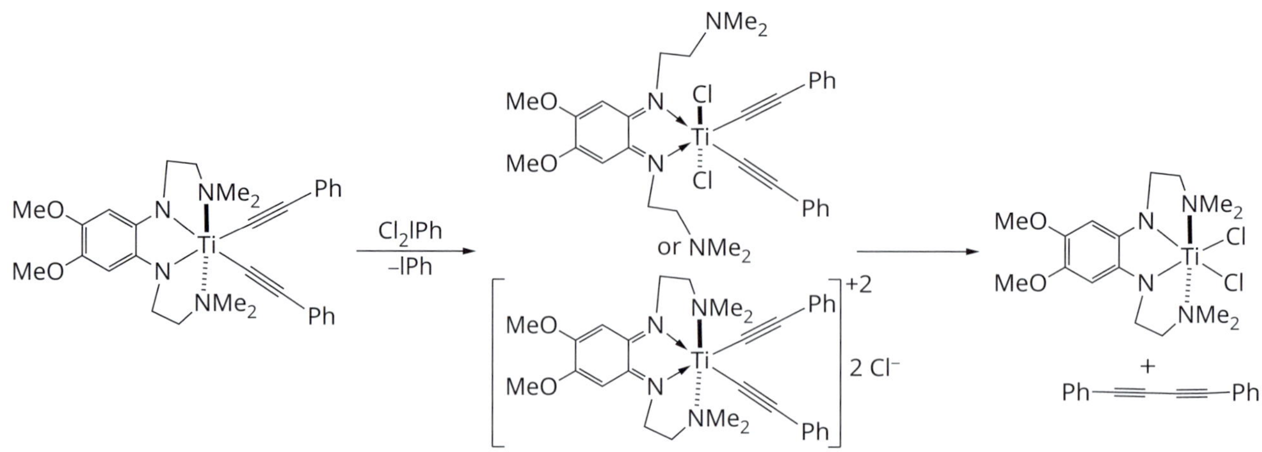

Fig. 12.62 Redox chemistry in a titanium(IV) complex with a redox active ligand.

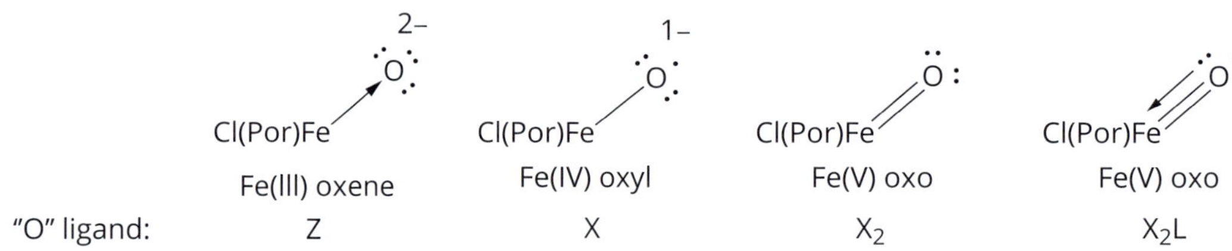

Fig. 12.63 CBC classifications for various types of oxygen-atom interactions with a metal center.

L = a group acting as a donor to M through a dative bond.

l = number of L ligands.

Z = a group acting as an acceptor from M of a dative bond.

z = the number of Z ligands.

Q± = charge on the complex.

A single ligand may contribute more than one X, L, or Z to the metal. For example, a nitride would be described as three X ligands (X_3), and an amide (NR_2) might contribute XL (a covalent bond and a dative bond) to the overall CBC formula. For the descriptions of an oxygen atom on an iron porphyrin chloride described previously, the potential CBC formula describing the oxygen atom's interaction with the metal (Fig. 12.63) can be oxene (Z), oxyl (X), oxo (X_2), or oxo with additional lone-pair donation (X_2L).

Table 12.6 lists many types of ligands with their CBC descriptions and electron donor numbers corresponding to the CBC formulae.

12.7.1 The Equivalent Neutral Class (Advanced Topic)

Once all of the ligands on the metal are classified as described previously, the full CBC formula can be written and the complex classified based on $\{MX_xL_lZ_z\}^{Q\pm}$. The overall charge can then be used to provide the equivalent neutral class for the complex where anionic overall charges result in fewer covalent bonds (fewer X ligands), and cationic overall charges result in additional covalent bonds (more X ligands) in the neutral formula. In other words, L and a positive charge (L^+) is equivalent to X, $L^+ = X$; X and a negative charge (X^-) is equivalent to L, $X^- = L$. This is illustrated in Fig. 12.64 for the three Co(III) complexes $\{Co(NH_3)_6\}^{3+}$, $CoCl_3(NH_3)_3$, and $\{CoCl_6\}^{3-}$, which all belong to the same equivalent neutral class CoL_3X_3.[206]

Furthermore, anionic charges with L ligands are equivalent to LX, and cationic charges with X ligands are equivalent to Z. All of these transformations are shown in Fig. 12.64.[206]

It is important to note that these transformations for anions should be applied sequentially: first use $X^- = L$, then if there are no X apply $L^- = LX$. If a formula contains L and Z, then the following transformation could also be applied: $LZ = X_2$.

12.7.2 Information from the CBC Equivalent Neutral Class (Advanced Topic)

From the formula, several pieces of information can be gleaned, some of which are obviously related to the electron count and formal oxidation state used previously.

Valence number (VN) and d-electron count

$$VN = x + 2z$$

The valence number is the number of electrons used by the atom of interest in bonding. The number of electrons left in

Table 12.6 A short list of common ligands with their CBC descriptions and electron donor numbers. (Taken from Green and Parkin.[206])

Ligand	CBC Description	Electron Donor Number
PR_3	L	2
H	X	1
R	X	1
BR_3	Z	0
η^2-ethylene	L	2
η^2-allyl	LX	3
η^2-butadiene	L_2	4
η^4-C_4H_4	X_2L	4
η^5-C_5H_5	XL_2	5
η^6-C_6H_6	L_3	6
η^7-C_7H_7	L_2X_3[a]	7
η^8-C_8H_8	L_3X_2	8
CO	L	2
NO (bent)	X	1
NO (linear)	XL	3
OR	X, XL, or XL_2	1, 3, or 5
NR_2	X or XL	1 or 3
O	X_2 or X_2L[b]	2 or 4
NR	X_2 or X_2L[b]	2 or 4
CR_2 (Schrock carbene)	X_2	2
CR_2 (Fischer carbene)	L	2
CR	X_3	3
N	X_3	3

[a] This is similar to a diallyl–alkyl view of the C_7H_7 ring discussed.
[b] Most commonly.

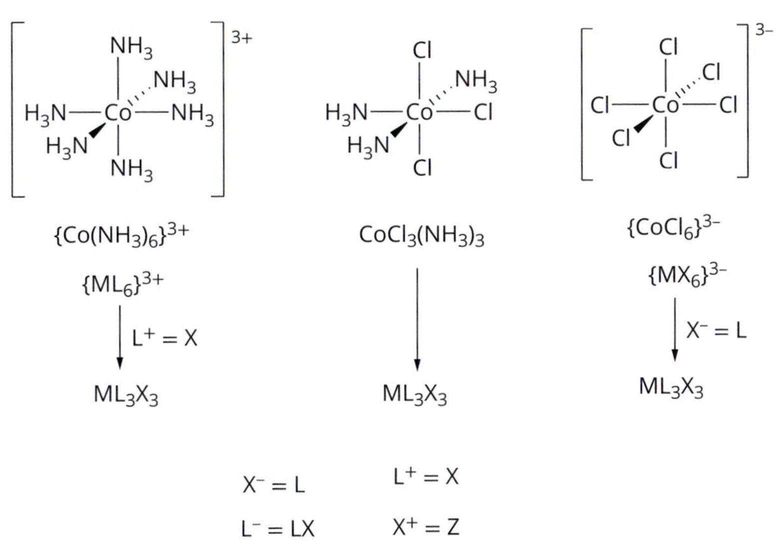

Fig. 12.64 (*top*) The equivalent neutral classes for the Werner complexes shown are all the same, revealing that their electronic structures are similar despite the different charges and (*bottom*) some common exchanges to determine the neutral class.

approximately nonbonding d-orbitals (d^n) is then as shown in the following equation, where m = the column number of the transition metal element (for example, for Fe $m = 8$, for Ti $m = 4$).

$$d^n = m - x - 2z = m - VN$$

Ligand Bond Number (LBN)
The ligand bond number is the effective number of ligands around the metal center in the CBC formula, which is intuitively related to the coordination number. The formula is simply

$$LBN = x + l + z$$

This is related to the number of equivalent monodentate ligands to reach a similar electronic configuration around the metal.

Electron Count (EC)
The electron count under the CBC system is defined based on the starting number of electrons on the metal (m) and the number of X and L ligands in a fashion analogous to the usual method of electron counting.

$$EC = m + 2l + x$$

12.7.3 Some Examples of Molecules Classified by CBC (Advanced Topic)

Consider the complex $\{Mn(CO)_6\}^+$, which is a ML_6^+ molecule. The equivalent neutral class is then ML_5X. The electron count (EC), valence number (VN), d-electron count, and ligand bond number (LBN) are calculated as follows. In all cases, these produce the same results as traditional electron counting.

$$\begin{aligned} EC &= m + 2l + x = 7 + 2(5) + 1 = 18 \\ VN &= x + 2z = 1 + 2(0) = 1 \\ d^n &= m - VN = 7 - 1 = 6 \\ LBN &= x + l + z = 5 + 1 + 0 = 6 \end{aligned}$$

As an additional example, consider the very high oxidation-state complexes $\{ReF_6\}^+ X^-$, where $X^- = SbF_6^-$, $Sb_2F_{11}^-$, AsF_6^-, and IrF_6^- have been prepared.[207–208] The metal is in Group 7, just like the previous example, $m = 7$, and the complex is in equivalent class $\{ReF_6\}^+ \rightarrow MX_6^+ \rightarrow MX_5Z$.

$$\begin{aligned} EC &= m + 2l + x = 7 + 2(0) + 5 = 12 \\ VN &= x + 2z = 5 + 2(1) = 7 \\ d^n &= m - VN = 7 - 7 = 0 \\ LBN &= x + l + z = 5 + 0 + 1 = 6 \end{aligned}$$

Green and Parkin list several advantages to the CBC method:

(1) *All ligands are classified in their neutral form.* Electronegativity is not used to determine which atom acquires the electrons in a bond, which removes some ambiguities.
(2) *Electron-counting is simpler in many cases.* The electron count, EC, which is valuable for determination of empty orbitals or filled antibonding orbitals, is simply the sum of the electron count on the metal (m), twice the donor ligands ($2l$), and the number of covalent bonds (x) for the equivalent neutral class of the complex. Ambiguities with assigning charges to ligands (such as with NO) and determining the metal oxidation state are avoided.
(3) *Assignment of d^n configuration fits with spectroscopy better in some cases.* In a formal oxidation state we typically assign metal–metal bonds as not affecting the oxidation state because the metals are regarded as having the same oxidation state. While this is entrenched in chemistry, it is not particularly accurate as a description of the complex in some cases. For example, the metal–metal triple bond complex (neopentyl)$_3$WW(neopentyl)$_3$ is diamagnetic and formally W(III), which would seem to mean it was a nonsensical d^3. The compound has more in common with NW(neopentyl)$_3$, which is formally W(VI) and diamagnetic. Sensibly, both NW(neopentyl)$_3$ and (neopentyl)$_3$WW(neopentyl)$_3$ are in the same class in the CBC method.

12.7.4 Final Comments on CBC (Advanced Topic)

The method was first presented by Malcolm L. H. Green in the 1990s, and has since been reviewed and expanded several times by Green and Parkin. The $MX_xL_lZ_z$ nomenclature seemed widely adopted almost immediately by the community. However, any time a new method is presented as a way of replacing an old one, the opinions in the community will probably always be mixed, with some seeing the value of the new method and with others unwilling to change. The electron-counting and formal oxidation-state method presented in the previous two chapters have been used for a very long time to good effect, perhaps nearing a century in many aspects. However, "we have always done it this way" is a poor argument in any debate and really falls short in science. The best method should be employed when it comes along; that is how fields move forward.

Even if one adopts the CBC method as their exclusive method for understanding the chemistry of the transition metals, the traditional electron-counting and oxidation-state formalisms have great utility. For example, electron-counting in transition-metal systems can quickly reveal the number of low-energy unoccupied orbitals (electron counts <18) or the occupation of antibonding orbitals (electron counts >18). Formal oxidation state is used extensively in redox reactions, understanding d-electron counts on spectroscopy and magnetism, and so on. The CBC can also provide these things, and in some cases it may even do them better; however, one of the most important jobs of a scientist is to explain their results to others. As CBC is adopted, perhaps it will supplant the usual language of oxidation states. Until then, we recommend those working in inorganic chemistry becoming conversant in both.

Chapter Summary

1. The π-systems of open-chain and cyclic systems can interact in a variety of different ways with transition metals. In some cases, different hapticities to metal centers are common.
2. The most common open π-systems observed are allyl, butadiene, and benzyl, which are used with metals across the transition series.
3. Cyclic π-systems can be understood by using a model where the ring system is regarded as being aromatic. Complexes of 6-π-electron cyclopentadienyl and benzene ligands are commonly encountered and can produce very stable compounds.
4. There are two limiting forms for alkylidenes: metal–carbon double bonds, Schrock carbenes, and Fischer carbenes. Schrock carbenes are generally of early transition metals and have a nucleophilic carbene carbon, while Fischer carbenes generally involve late transition metals and are electrophilic at C_α.
5. The continuum between Schrock and Fischer carbenes can be understood by using an MO-based approach. Schrock carbenes often have a metal-based LUMO and carbon-based HOMO, while Fischer carbenes have a metal-based HOMO and mostly carbon-based LUMO. The M–C bond order in Schrock carbenes is generally higher than in Fischer carbenes.
6. There is a wide variety of metal–metal bonds known from single to pentuple covalent bonds. Dative metal–metal bonds are also known.
7. Weakly coordinating anions are an important part of industrial and academic chemistry that can enable isolation of unique species and observation of unusual reactivity.
8. Redox noninnocent ligands are a field of current and growing importance in catalysis where the ligand may accept or donate electron density to a metal center, producing a redox interplay between different parts of the complex that can greatly enable interesting reactivity.

Practice Problems

1. Many different types of "sandwich" complexes not discussed explicitly in the chapter have been prepared including many triple-decker sandwich complexes like $Ti_2(COT)_3$, where COT = cyclooctatetraene. The structure of the complex from the reference is shown below, where there is a 2-fold crystallographic rotation axis.[105] If we consider the longest of the bonds shown with measurements to be nonbonding (2.57 Å), draw the bonding pattern in the central ring. (In other words, each titanium interacts strongly with three carbons on the central ring.) What is the electron count and oxidation state of the equivalent titanium atoms in your bonding pattern?

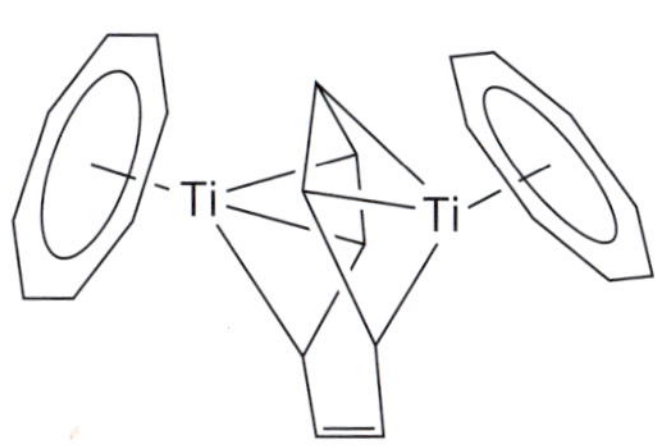

2. Wolczanski and coworkers have shown that, in some systems, the M(olefin) isomer may be less stable than the M(alkylidene) tautomer. For example, an ethylene compound $M(H_2C{=}CH_2)$ may be less stable thermodynamically than an ethylidene M=C(H)Me tautomer. Find this classic paper in the primary literature for the conversion of (olefin)$Ta(OSiBu^t{}_3)_3$ to $R_2C{=}Ta(OSiBu^t{}_3)_3$.[113] What is the proposed mechanism? Name at least one other mechanism that was proposed and how it was eliminated experimentally as a possibility.
3. An interesting complex prepared by Fischer and coworkers is the ruthenium trimethylenemethane complex shown below.[211] Discuss the bonding in the complex between both Ru and C along with Ru and Ga. What is the formal oxidation state and electron count on Ru? What is the formal oxidation state of the Ga atoms? Is the In or Al analog of this complex more or less likely to be stable, and why?

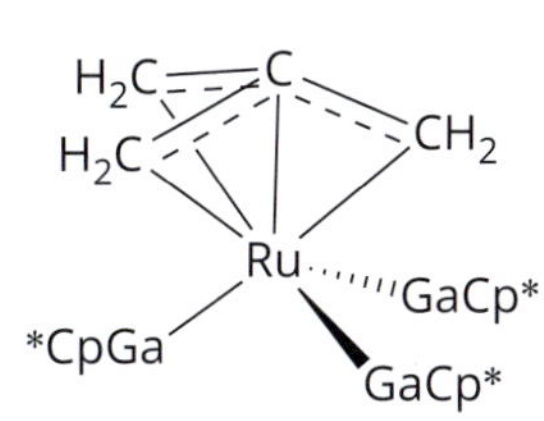

4. Consider the molecule M(pentadienyl). Draw a set of axes for the molecule with the pentadienyl centroid-M as the *z*-axis. Draw the molecular orbitals for the π-system of the free ligand as the anion. What orbitals on the metal, M = transition metal, may overlap with each of the molecular orbitals of the pentadienyl

π-system? Draw the metal-ligand interactions as clearly as possible labeling them with the number of nodes in your ligand MO and the *d*-orbital used on the metal.

5. The fragment Fe(η^5-Cp)(CO)$_2$, often abbreviated Fp (pronounced "Fip"), is a common one that has been bound to many ligands. Fp is observable as the radical Fp•, cation Fp$^+$ and anion Fp$^-$. As discussed in Section 12.2.2.1, cyclopropenyl binds to this fragment in an η^1-fashion. Explain why the complex Fe(η^3-C$_3$Ph$_3$)(η^5-Cp)(CO)$_2$ is not observed.

6. One of the earliest cyclobutadiene metal complexes was Criegee's Complex, prepared by the reaction of Ni(CO)$_4$ with 1,2-dichloro-tetramethylbutadiene. In the reaction, 4 equivalents of CO are lost.[212–213] The formula for the compound is {NiCl(μ-Cl)(η^4-C$_4$Me$_4$)}$_2$.
   a. Give the line drawing for the complex.
   b. What is the formal oxidation state of nickel?
   c. What is the electron count of nickel?

7. Two related complexes Mn(Cp)(CO)$_3$ (**A**) and Mn(η^5-pyrrolyl)(CO)$_3$ (**B**) have been known for many years. The reported CO stretching frequencies in cyclohexane for **A** are 2034, 1947 cm^{-1}. The reported CO stretching frequencies in decalin for **B** are 2037, 1965, 1955 cm^{-1}.[214]
   a. Why does the Cp derivative show two observable CO stretches and the pyrrolyl complex show three? Use group theory in your argument and assign the symmetries for the stretches in your answer.
   b. What do the CO stretching frequencies say about the donor ability of η^5-pyrrolyl versus η^5-Cp?

8. By hand, draw the molecular orbital diagram with symmetry labels and orbital drawings for CpNiNO. In order to accomplish this, first draw the orbitals for the fragments CpNi and NO, labeling the orbitals in the point group of the entire molecule. Next, combine orbitals of proper symmetry to give the orbital picture for the entire molecule.

Cited References

1. Werner, H., At Least 60 Years of Ferrocene: The Discovery and Rediscovery of the Sandwich Complexes *Angew. Chem. Int. Ed.* 2012, *51*, 6052–6058.
2. Barnett, B. L.; Tsay, Y.; Kruger, C., Molecular structure of a bridged cis-bis(π-allyl)nickel system *Angew. Chem. Int. Ed.* 1972, *11*, 137.
3. Henc, B.; Jolly, P. W.; Salz, R.; Stobbe, S.; Wilke, G.; Benn, R.; Mynott, R.; Seevogel, K.; Goddard, R.; Kruger, C., Transition-metal allyls. 4. The (η^3-allyl)$_2$M complexes of nickel, palladium and platinum: reaction with tertiary phosphines *J. Organomet. Chem.* 1980, *191*, 449–475.
4. Brennessel, W. W.; Jilek, R. E.; Ellis, J. E., Bis(1,2,3,4-η^4-anthracene)ferrate(1-): A paramagnetic homoleptic polyarene transition-metal anion *Angew. Chem. Int. Ed.* 2007, *46*, 6132–6136.
5. Kai, Y.; Kanehisa, N.; Miki, K.; Kasai, N.; Mashima, K.; Nagasuna, K.; Yasuda, H.; Nakamura, A., X-ray evidence for a mononuclear s-*trans*-η^4-1,3-diene complex - molecular-structure of Zr(η^5-C$_5$H$_5$)$_2$(s-*trans*-PHCH=CH-CH=CHPH) *Chem. Commun.* 1982, 191–192.
6. Erker, G.; Kehr, G.; Frohlich, R., Some selected chapters from the (butadiene) zirconocene story *J. Organomet. Chem.* 2004, *689*, 4305–4318.
7. Spencer, M. D.; Wilson, S. R.; Girolami, G. S., Butadiene complexes of titanium(II) and titanium(0): Synthesis, butadiene dimerization catalysis, and crystal structures of TiMe2(eta(4)-1,4-C4H4Ph2)(dmpe) and Ti(eta(4)-C4H6)(2)(dmpe) *Organometallics* 1997, *16*, 3055–3067.
8. Green, M. L. H.; Hughes, A. K.; McGowan, P. C.; Mountford, P.; Scott, P.; Simpson, S. J., Cycloheptatriene and cycloheptatrienyl derivatives of niobium *Dalton Trans.* 1992, 1591–1600.
9. Rong, Y.; Al-Harbi, A.; Parkin, G., Highly Variable Zr-CH$_2$-Ph Bond Angles in Tetrabenzylzirconium: Analysis of Benzyl Ligand Coordination Modes *Organometallics* 2012, *31*, 8208–8217.
10. Liu, F. C.; Liu, J. P.; Meyers, E. A.; Shore, S. G., Cyclic hydroborate complexes of metallocenes. IV. Weak unsupported single hydrogen bridges in Cp$_2$Zr{(μ-H)B(X)CH$_2$Ph}{(μ-H)$_2$BX} (X = C$_5$H$_{10}$, C$_4$H$_8$) *Inorg. Chem.* 1999, *38*, 2169–2173.
11. Wasilke, J. C.; Ziller, J. W.; Bazan, G. C., Trimethylphosphine-bis(benzyl) nickel: Synthesis and characterization *Adv. Synth. Catal.* 2005, *347*, 405–408.
12. Pellecchia, C.; Immirzi, A.; Pappalardo, D.; Peluso, A., A novel η^7-coordination mode of a benzyl ligand in a cationic zirconium complex *Organometallics* 1994, *13*, 3773–3775.
13. Reiners, M.; Fecker, A. C.; Freytag, M.; Jones, P. G.; Walter, M. D., Pentadienyl chemistry of the heavy alkaline-earth metals revisited *Dalton Trans.* 2014, *43*, 6614–6617.
14. Ernst, R. D.; Freeman, J. W.; Stahl, L.; Wilson, D. R.; Arif, A. M.; Nuber, B.; Ziegler, M. L., Longer but stronger bonds: Structures of PF$_3$, P(OEt)$_3$, and PMe$_3$ adducts of an open titanocene *J. Am. Chem. Soc.* 1995, *117*, 5075–5081.
15. Gleiter, R.; Hylakryspin, I.; Ziegler, M. L.; Sergeson, G.; Green, J. C.; Stahl, L.; Ernst, R. D., Half-open metallocenes of iron, ruthenium, and osmium: Synthesis, characterization, photoelectron-spectroscopy, and theoretical calculations *Organometallics* 1989, *8*, 298–306.
16. Kralik, M. S.; Stahl, L.; Arif, A. M.; Strouse, C. E.; Ernst, R. D., Open and half-open manganocene chemistry: More associated salts *Organometallics* 1992, *11*, 3617–3621.
17. Newbound, T. D.; Stahl, L.; Ziegler, M. L.; Ernst, R. D., Open and half-open ruthenocenes and osmocenes: protonations, structures, and reactions with carbonyl and phosphine-ligands *Organometallics* 1990, *9*, 2962–2972.
18. Stahl, L.; Ernst, R. D., Synthesis and characterization of bis(pentadienyl) ruthenium compounds *Organometallics* 1983, *2*, 1229–1234.
19. Stahl, L.; Ernst, R. D., Equilibria studies involving ligand coordination to open titanocenes: Phosphine and pentadienyl cone angle influences and the existence of these electron-deficient molecules *J. Am. Chem. Soc.* 1987, *109*, 5673–5680.
20. Stahl, L.; Ernst, R. D., Pentadienyl Complexes of the Group 4 Transition Metals. In *Advances in Organometallic Chemistry* West, R.; Hill, A. F.; Fink, M. J., Eds. 2008; Vol. 55, pp 137–199.
21. Stahl, L.; Ma, H.; Ernst, R. D.; Hylakryspin, I.; Gleiter, R.; Ziegler, M. L., Synthetic, structural and pe spectroscopic studies on bis(pentadienyl) compounds of iron, ruthenium and osmium: The role of the heavy-metal *J. Organomet. Chem.* 1987, *326*, 257–268.
22. Waldman, T. E.; Stahl, L.; Wilson, D. R.; Arif, A. M.; Hutchinson, J. P.; Ernst, R. D., Phosphine adducts of the open metallocenes of zirconium, hafnium, niobium, and molybdenum: Syntheses, structures, and reactions with carbon-monoxide *Organometallics* 1993, *12*, 1543–1552.
23. Keaton, R. J.; Koterwas, L. A.; Fettinger, J. C.; Sita, L. R., Regarding the stability of d^0 monocyclopentadienyl zirconium acetamidinate complexes bearing alkyl substituents with beta-hydrogens *J. Am. Chem. Soc.* 2002, *124*, 5932–5933.
24. Farrugia, L. J.; Evans, C.; Tegel, M., Chemical bonds without "chemical bonding"? A combined experimental and theoretical charge density study on an iron trimethylenemethane complex *J. Phys. Chem. A* 2006, *110*, 7952–7961.
25. Grosselin, J. M.; Lebozec, H.; Moinet, C.; Toupet, L.; Dixneuf, P. H., Electron-rich, hydrocarbon-metal complexes: Synthesis and reversible one-electron oxidation, X-ray structure of a 17-electron iron cation *J. Am. Chem. Soc.* 1985, *107*, 2809–2811.
26. Jones, M. D.; Kemmitt, R. D. W.; Platt, A. W. G.; Russell, D. R.; Sherry, L. J. S., Trimethylenemethane complexes of osmium and iridium: Crystal-structure of η^4-trimethylenemethanechlorocarbonyltriphenylphosphineiridium *Chem. Commun.* 1984, 673–674.
27. McNeill, K.; Andersen, R. A.; Bergman, R. G., C-C and C-H bond activation at ruthenium(II): The stepwise degradation of a neopentyl ligand to a trimethylenemethane ligand *J. Am. Chem. Soc.* 1997, *119*, 11244–11254.

28. Frost, A. A.; Musulin, B., A mnemonic device for molecular orbital energies *J. Chem. Phys.* 1953, *21*, 572–573.
29. Kutzelnigg, W., What I like about Huckel theory *Journal of Computational Chemistry* 2007, *28*, 25–34.
30. Allen, A. D.; Tidwell, T. T., Antiaromaticity in open-shell cyclopropenyl to cycloheptatrienyl cations, anions, free radicals, and radical ions *Chem. Rev.* 2001, *101*, 1333–1348.
31. Wiberg, K. B., Antiaromaticity in monocyclic conjugated carbon rings *Chem. Rev.* 2001, *101*, 1317–1331.
32. Werner, H., A Scientific Revolution: The Discovery of the Sandwich Complexes. In *Landmarks in Organo-Transition Metal Chemistry: A Personal View*, 2009; pp 129–176.
33. Kealy, T. J.; Pauson, P. L., A new type of organo-iron compound *Nature* 1951, *168*, 1039–1040.
34. Miller, S. A.; Tebboth, J. A.; Tremaine, J. F., Dicyclopentadienyliron *J. Chem. Soc.* 1952, 632–635.
35. Wilkinson, G.; Rosenblum, M.; Whiting, M. C.; Woodward, R. B., The structure of iron bis(cyclopentadienyl) *J. Am. Chem. Soc.* 1952, *74*, 2125–2126.
36. Braga, D., Dynamic processes in crystalline organometallic complexes *Chem. Rev.* 1992, *92*, 633–665.
37. Haaland, A.; Nilsson, J. E., Determination of barriers to internal rotation by means of electron diffraction: Ferrocene and ruthenocene *Acta Chem. Scand.* 1968, *22*, 2653.
38. Dunitz, J. D.; Orgel, L. E., Bis-cyclopentadienyl iron: A molecular sandwich *Nature* 1953, *171*, 121–122.
39. Dunitz, J. D.; Orgel, L. E.; Rich, A., The crystal structure of ferrocene *Acta Cryst.* 1956, *9*, 373–375.
40. Maverick, E.; Dunitz, J. D., Rotation barriers in crystals from atomic displacement parameters *Molecular Physics* 1987, *62*, 451–459.
41. Seiler, P.; Dunitz, J. D., Structure of triclinic ferrocene at 101 K, 123 K and 148 K *Acta Cryst. B: Struc. Sci.* 1979, *35*, 2020–2032.
42. Turner, J. F. C.; Soper, A. K., On the determination of the structure of ferrocene in solution *Polyhedron* 2004, *23*, 2975–2983.
43. Piper, T. S.; Wilkinson, G., Alkyl and aryl derivatives of π-cyclopentadienyl compounds of chromium, molybdenum, tungsten, and iron *J. Inorg. Nuc. Chem.* 1956, *3*, 104–124.
44. Huttner, G.; Brintzinger, H. H.; Bell, L. G.; Friedrich, P.; Bejenke, V.; Neugebauer, D., Structure of $(C_5H_5)_2W(CO)_2$, a compound containing a bent trihapto-cyclopentadienyl ligand *J. Organomet. Chem.* 1978, *145*, 329–333.
45. Radius, U.; Sundermeyer, J.; Peters, K.; von Schnering, H. G., Bis(cyclopentadienyl) diimido complexes of molybdenum and tungsten $Cp_2M(NR)_2$ at the limit of π-bond saturation *Eur. J. Inorg. Chem.* 2001, 1617-1623.
46. Ji, L. N.; Rerek, M. E.; Basolo, F., Kinetics and mechanism of substitution-reactions of $Mn(\eta^5\text{-}C_9H_7)(CO)_3$ and $Mn(\eta^5\text{-}C_{13}H_9)(CO)_3$ *Organometallics* 1984, *3*, 740–745.
47. Rerek, M. E.; Basolo, F., Haptotropic rearrangements - kinetics and mechanism of the rearrangement of $Mn(\eta^6\text{-}C_{13}H_9)(CO)_3$ to $MN(\eta^5\text{-}C_{13}H_9)(CO)_3$ *Organometallics* 1984, *3*, 647–649.
48. Rerek, M. E.; Ji, L. N.; Basolo, F., The indenyl ligand effect on the rate of substitution-reactions of $Rh(\eta\text{-}C_9H_7)(CO)_2$ and $Mn(\eta\text{-}C_9H_7)(CO)_3$ *Chem. Comm.* 1983, 1208–1209.
49. Note, the "yl" ending implies a radical bound to the metal in a neutral method-like electronic structure; whereas, the "ide" ending implies an anionic ligand in a charged method-like electronic structure. Both endings are used by different researchers in the literature to refer to the same types of complexes and likely signify subtle differences in how the chemists view the ligand's bonding to the metal.
50. Detzel, M.; Friedrich, G.; Scherer, O. J.; Wolmershauser, G., $\eta^5\eta^2$ coordination of a cyclo-E^5 ligand, E = P, AS *Angew. Chem. Int. Ed.* 1995, *34*, 1321–1323.
51. Mathey, F., The chemistry of phosphacyclopentadienide and polyphosphacyclopentadienide anions *Coord. Chem. Rev.* 1994, *137*, 1–52.
52. Scherer, O. J.; Blath, C.; Wolmershauser, G., Ferrocenes with pentaarsacyclopentadienyl ligands *J. Organomet. Chem.* 1990, *387*, C21–C24.
53. Scherer, O. J.; Bruck, T.; Wolmershauser, G., Pentaphosphametallocenes *Chem. Ber. Recl.* 1988, *121*, 935–938.
54. Urnezius, E.; Brennessel, W. W.; Cramer, C. J.; Ellis, J. E.; Schleyer, P. V., A carbon-free sandwich complex $(P_5)_2Ti^{2-}$ *Science* 2002, *295*, 832–834.
55. Seyferth, D., Bis(benzene)chromium. 2. Its discovery by E. O. Fischer and W. Hafner and subsequent work by the research groups of E. O. Fischer, H. H. Zeiss, F. Hein, C. Elschenbroich, and others *Organometallics* 2002, *21*, 2800–2820.
56. Seyferth, D., Bis(benzene) chromium. 1. Franz Hein at the University of Leipzig and Harold Zeiss and Minoru Tsutsui at Yale *Organometallics* 2002, *21*, 1520–1530.
57. Bowyer, W. J.; Merkert, J. W.; Geiger, W. E.; Rheingold, A. L., Structural consequences of electron-transfer reactions and redox-induced hapticity changes: Effect of substituents on arene bending in a series of rhodium complexes *Organometallics* 1989, *8*, 191–198.
58. Rybinskaya, M. I.; Kudinov, A. R.; Kaganovich, V. S., A new preparative route to cationic arene complexes of ruthenium(II), rhodium(III) and iridium(III) *J. Organomet. Chem.* 1983, *246*, 279–285.
59. White, C.; Thompson, S. J.; Maitlis, P. M., Pentamethylcyclopentadienyl-rhodium and pentamethylcyclopentadienyl-iridium complexes. 12. Tris(solvent) complexes and complexes of η^6-benzene, η^6-napthalene, η^6-phenanthrene, η^6-indene, η^6-indole, η^6-fluorene and η^5-indenyl and η^5-indolyl *Dalton Trans.* 1977, 1654–1661.
60. Breslow, R., Synthesis of the s-triphenylcyclopropenyl cation *J. Am. Chem. Soc.* 1957, *79*, 5318–5318.
61. Komatsu, K.; Kitagawa, T., Cyclopropenylium cations, cyclopropenones, and heteroanalogues - Recent advances *Chem. Rev.* 2003, *103*, 1371–1427.
62. Lichtenberger, D. L.; Hoppe, M. L.; Subramanian, L.; Kober, E. M.; Hughes, R. P.; Hubbard, J. L.; Tucker, D. S., Electron-distribution and bonding in η^3-cyclopropenyl metal-complexes *Organometallics* 1993, *12*, 2025–2031.
63. Gowling, E. W.; Kettle, S. F. A., π-cyclopropenyl derivative of nickel carbonyl *Inorg. Chem.* 1964, *3*, 604.
64. Rausch, M. D.; Tuggle, R. M.; Weaver, D. L., Formation and structure of a new mixed sandwich complex, $(\pi\text{-}C_5H_5)Ni(\pi\text{-}C_3PH_3)$ *J. Am. Chem. Soc.* 1970, *92*, 4981.
65. Tuggle, R. M.; Weaver, D. L., Crystal and molecular structure of mixed sandwich complex π-cyclopentadienyl-π-triphenylcyclopropenylnickel, $(\pi\text{-}C_5H_5)Ni(\pi\text{-}C_3(C_6H_5)_3)$ *Inorg. Chem.* 1971, *10*, 1504.
66. Tuggle, R. M.; Weaver, D. L., Crystal structure and properties of π-triphenylcyclopropenylchlorodipyridinenickel(0)-pyridine, a π-complex of a 3-membered aromatic ring *Inorg. Chem.* 1971, *10*, 2599.
67. Fomitchev, D. V.; Furlani, T. R.; Coppens, P., Combined X-ray diffraction and density functional study of $Ni(NO)(\eta^5\text{-}Cp^*)$ in the ground and light-induced metastable states *Inorg. Chem.* 1998, *37*, 1519–1526.
68. Piper, T. S.; Cotton, F. A.; Wilkinson, G., Cyclopentadienyl-carbon monoxide and related compounds of some transitional metals *J. Inorg. Nuc. Chem.* 1955, *1*, 165–174.
69. Churchill, M. R.; Fettinger, J. C.; McCullough, L. G.; Schrock, R. R., Transformation of a tungstenacyclobutadiene complex into a nonfluxional η^3-cyclopropenyl complex by addition of a donor ligand: The x-ray structure of $W(\eta^5\text{-}C_5H_5)$ $C_3(CMe_3)_2Me(PMe_3)Cl_2$ *J. Am. Chem. Soc.* 1984, *106*, 3356–3357.
70. Churchill, M. R.; Ziller, J. W.; McCullough, L.; Pedersen, S. F.; Schrock, R. R., $W(\eta^5\text{-}C_5H_5)C[(PH)C(CMe_3)C(PH)]Cl_2$: A molecule having a localized, nonplanar, fluxional metallacyclobutadiene ring *Organometallics* 1983, *2*, 1046–1048.
71. Hughes, R. P.; Lambert, J. M. J.; Whitman, D. W.; Hubbard, J. L.; Henry, W. P.; Rheingold, A. L., Reactions of cyclopropenyl cations with tricarbonylnitrosylferrate(1-), tetracarbonylcobaltate(1-), and octacarbonyldicobalt - synthesis and conformational and configurational stabilities of η^3-cyclopropenyl and η^3-oxocyclobutenyl complexes of iron and cobalt: Crystal and molecular-structure of $Fe(\eta\text{-}C_3PH_2\text{-t-Bu})(CO)_2(NO)D$ *Organometallics* 1986, *5*, 789–797.
72. Chiang, T.; Kerber, R. C.; Kimball, S. D.; Lauher, J. W., (η^3-triphenylcyclopropenyl)tricarbonylcobalt *Inorg. Chem.* 1979, *18*, 1687–1691.
73. Gompper, R.; Bartmann, E.; Noth, H., Donor-acceptor substituted aromatic systems. 1. Dicarbonyl(η^5-cyclopentadienyl)(1,2,3-triphenylcyclopropenyl) iron *Chem. Ber. Recl.* 1979, *112*, 218–233.
74. McClure, M. D.; Weaver, D. L., Unusual cyclopropenyl complex - formation and structure of $[Pt(C_3(C_6H_5)_3)(P(C_6H_5)_3)_2][PF6]\cdot C_6H_6$ *J. Organomet. Chem.* 1973, *54*, C59–C61.
75. Blunden, R. B.; Cloke, F. G. N.; Hitchcock, P. B.; Scott, P., $(C_3{}^tBu_3)V(CO)_4$: Partial insertion of vanadium into the cyclopropenyl ring *Organometallics* 1994, *13*, 2917–2919.
76. Jemmis, E. D.; Hoffmann, R., Cleaving CC bonds in cyclopropenium ions *J. Am. Chem. Soc.* 1980, *102*, 2570–2575.
77. Lin, Z. Y.; Hall, M. B., Stabilities of metallacyclobutadiene and metallatetrahedrane complexes *Organometallics* 1994, *13*, 2878–2884.

78. Pedersen, S. F.; Schrock, R. R.; Churchill, M. R.; Wasserman, H. J., Reaction of tungsten(VI) alkylidyne complexes with acetylenes to give tungstenacyclobutadiene and tungsten cyclopentadienyl complexes *J. Am. Chem. Soc.* 1982, *104*, 6808–6809.
79. Hughes, R. P.; Tucker, D. S.; Rheingold, A. L., Reversible insertion of iridium into a cyclopropenyl carbon-carbon bond *Organometallics* 1994, *13*, 4664–4666.
80. Scherer, O. J.; Braun, J.; Wolmershauser, G., Nickelapnicogencubanes *Chem. Ber.* 1990, *123*, 471–475.
81. Cossairt, B. M.; Diawara, M.-C.; Cummins, C. C., Facile Synthesis of AsP3 *Science* 2009, *323*, 602–602.
82. Gleiter, R.; Werz, D. B., Reactions of metal-complexed carbocyclic 4π-systems *Organometallics* 2005, *24*, 4316–4329.
83. Kudinov, A. R.; Mutseneck, E. V.; Loginov, D. A., (Tetramethylcyclobutadiene) cobalt chemistry *Coord. Chem. Rev.* 2004, *248*, 571–585.
84. Seyferth, D., (Cyclobutadiene)iron tricarbonyl: A case of theory before experiment *Organometallics* 2003, *22*, 2–20.
85. Baker, P. K.; Silgram, H., Recent Studies on π-cyclobutadiene transition-metal complexes *Trends Organomet. Chem.* 1999, *3*, 21–33.
86. Breslow, R., Antiaromaticity *Acc. Chem. Res.* 1973, *6*, 393–398.
87. Nakamura, K.; Osamura, Y.; Iwata, S., 2nd-order Jahn-Teller effect of cyclobutadiene in low-lying states: An MCSCF study *Chem. Phys.* 1989, *136*, 67–77.
88. Senn, P., A simple quantum-mechanical model that illustrates the Jahn-Teller effect *J. Chem. Educ.* 1992, *69*, 819–821.
89. Albright, T. A.; Burdett, J. K.; Whangbo, M.-H., *Orbital interactions in chemistry* 2nd ed.; Wiley,: Hoboken, New Jersey, 2013.
90. Hubel, W.; Braye, E. H.; Clauss, A.; Weiss, E.; Kruerke, U.; Brown, D. A.; King, G. S. D.; Hoogzand, C., Organometallic complexes. 1. The reaction of metal carbonyls with acetylenic compounds *J. Inorg. Nuc. Chem.* 1959, *9*, 204–210.
91. Dodge, R. P.; Schomaker, V., Crystal structure of tetraphenylcyclobutadiene iron tricarbonyl *Nature* 1960, *186*, 798–799.
92. Emerson, G. F.; Watts, L.; Pettit, R., Cyclobutadiene- and benzocyclobutadiene-iron tricarbonyl complexes *J. Am. Chem. Soc.* 1965, *87*, 131.
93. Fitzpatrick, J. D.; Watts, L.; Emerson, G. F.; Pettit, R., Cyclobutadieneiron tricarbonyl: A new aromatic system *J. Am. Chem. Soc.* 1965, *87*, 3254.
94. Harvey, P. D.; Schaefer, W. P.; Gray, H. B.; Gilson, D. F. R.; Butler, I. S., Structure of tricarbonyl(η^4-cyclobutadienyl)iron(0) at −45 °C *Inorg. Chem.* 1988, *27*, 57–59.
95. Elschenbroich, C.; Salzer, A., *Organometallics: A Concise Introduction* 2nd ed.; VCH: Weinheim, Federal Republic of Germany ; New York, NY, USA, 1992.
96. Scherer, O. J.; Vondung, J.; Wolmershauser, G., Tetra-arsa-cyclobutadiene as a complex ligand *J. Organomet. Chem.* 1989, *376*, C35–C38.
97. Scherer, O. J.; Vondung, J.; Wolmershauser, G., Tetraphosphacyclobutadiene as complex ligand *Angew. Chem. Int. Ed.* 1989, *28*, 1355–1357.
98. Dauben, H. J.; Honnen, L. R., π-tropenium-molybdenum-tricarbonyl fluoroborate *J. Am. Chem. Soc.* 1958, *80*, 5570–5571.
99. Campen, A. K.; Narayanaswamy, R.; Rest, A. J., Photochemistry of pentacarbonyl(cyclohepta-1,3,5-triene-7-carbonyl)manganese and pentacarbonyl(σ-cycloheptatrienyl)rhenium in frozen gas matrices at ca. 12 K: Infrared spectroscopic evidence for carbon-monoxide ejection with concomitant polyene ring hapticity changes *Dalton Trans.* 1990, 823–832.
100. Maltz, H.; Kelly, B. A., Reaction of tricarbonylcycloheptatrieneiron with base: Tricarbonylcycloheptatrienyliron anion *J. Chem. Soc. D-Chem. Comm.* 1971, 1390.
101. Zeinstra, J. D.; Deboer, J. L., Structure of cyclopentadienylcycloheptatrienyltitanium *J. Organomet. Chem.* 1973, *54*, 207–211.
102. Blake, A. J.; Collier, P. E.; Dunn, S. C.; Li, W. S.; Mountford, P.; Shishkin, O. V., Synthesis and imido-group exchange reactions of tert-butylimidotitanium complexes *Dalton Trans.* 1997, 1549–1558.
103. Dunn, S. C.; Hazari, N.; Jones, N. M.; Moody, A. G.; Blake, A. J.; Cowley, A. R.; Green, J. C.; Mountford, P., Titanium imido complexes of cyclooctatetraenyl ligands *Chem Eur. J.* 2005, *11*, 2111–2124.
104. Breil, H.; Wilke, G., Di(cyclooctatetraene)titanium and tri(cyclooctatetraene) dititanium *Angew. Chem. Int. Ed.* 1966, *5*, 898.
105. Dietrich, H.; Dierks, H., Crystal structure of tri(cyclooctatetraene)dititanium *Angew. Chem. Int. Ed.* 1966, *5*, 899.
106. Dietrich, H.; Soltwisc, M., Crystal structure of bis(cyclooctatetraene)titanium *Angew. Chem. Int. Ed.* 1969, *8*, 765.
107. Streitwieser, A. J.; Muller-Westerhoff, U., Bis(cyclooctatetraenyl)uranium (uranocene): A new class of sandwich complexes that utilize atomic f orbitals *J. Am. Chem. Soc.* 1968, *90*, 7364.
108. Fischer, E. O.; Maasbol, A., On existence of tungsten carbonyl carbene complex *Angew. Chem. Int. Ed.* 1964, *3*, 580.
109. Casey, C. P.; Polichnowski, S. W.; Tuinstra, H. E.; Albin, L. D.; Calabrese, J. C., Synthesis, X-ray crystal-structure, and temperature-dependent nmr-spectrum of tetraethylammonium (alpha-methoxybenzyl)pentacarbonyltungstate *Inorg. Chem.* 1978, *17*, 3045–3049.
110. Schrock, R. R., Transition metal-carbon multiple bonds *Dalton Trans.* 2001, 2541–2550.
111. Schrock, R. R., Recent Advances in High Oxidation State Mo and W Imido Alkylidene Chemistry *Chem. Rev.* 2009, *109*, 3211–3226.
112. Schrock, R. R., Alkylcarbene complex of tantalum by intramolecular α-hydrogen abstraction *J. Am. Chem. Soc.* 1974, *96*, 6796–6797.
113. Hirsekorn, K. F.; Veige, A. S.; Marshak, M. P.; Koldobskaya, Y.; Wolczanski, P. T.; Cundari, T. R.; Lobkovsky, E. B., Thermodynamics, kinetics, and mechanism of $(silox)_3M$(olefin) to $(silox)_3M$(alkylidene) rearrangements (silox = (Bu_3SiO); M = Nb, Ta) *J. Am. Chem. Soc.* 2005, *127*, 4809–4830.
114. Nelson, D. J.; Nolan, S. P., Quantifying and understanding the electronic properties of N-heterocyclic carbenes *Chem. Soc. Rev.* 2013, *42*, 6723–6753.
115. Calculations were done using LANL2DZ with B3LYP and analyzed with NBO6.
116. du Toit, J. I.; van Sittert, C. G. C. E.; Vosloo, H. C. M., Metal carbenes in homogeneous alkene metathesis: Computational investigations *J. Organomet. Chem.* 2013, *738*, 76–91.
117. Occhipinti, G.; Jensen, V. R., Nature of the Transition Metal-Carbene Bond in Grubbs Olefin Metathesis Catalysts *Organometallics* 2011, *30*, 3522–3529.
118. Wigley, D. E., Organoimido Complexes of the Transition-Metals. In *Prog. Inorg. Chem.*, Karlin, K. D., Ed. 1994; Vol. 42, pp 239–482.
119. Kahlal, S.; Saillard, J. Y.; Hamon, J. R.; Manzur, C.; Carrillo, D., Bonding analysis of transition metal NNR end-on complexes and comparison with isoelectronic NNR2 species *New J. Chem.* 2001, *25*, 231–242.
120. Aktas, H.; Slootweg, J. C.; Lammertsma, K., Nucleophilic Phosphinidene Complexes: Access and Applicability *Angew. Chem. Int. Ed.* 2010, *49*, 2102–2113.
121. Haymore, B. L.; Maatta, E. A.; Wentworth, R. A. D., Bis-phenylnitrene complex of molybdenum with a bent nitrene ligand: Preparation and structure of cis-$Mo(NC_6H_5)_2(S_2CN(C_2H_5)_2)_2$ *J. Am. Chem. Soc.* 1979, *101*, 2063–2068.
122. Ciszewski, J. T.; Harrison, J. F.; Odom, A. L., Investigation of transition metal-imido bonding in $M(NBu^t)_2$(dpma) *Inorg. Chem.* 2004, *43*, 3605–3617.
123. Knoepke, L. R.; Spannenberg, A.; Brueckner, A.; Bentrup, U., The influence of substituent effects on spectroscopic properties examined on benzylidene aniline-type imines *Spectroschem. Acta, Part A* 2012, *95*, 18–24.
124. Hannedouche, J.; Schulz, E., Asymmetric Hydroamination: A Survey of the Most Recent Developments *Chem Eur. J.* 2013, *19*, 4972–4985.
125. Lee, A. V.; Schafer, L. L., Modular N,O-chelating ligands: Group-4 amidate complexes for catalytic hydroamination *Eur. J. Inorg. Chem.* 2007, 2243–2255.
126. Mueller, T. E.; Hultzsch, K. C.; Yus, M.; Foubelo, F.; Tada, M., Hydroamination: Direct addition of amines to alkenes and alkynes *Chem. Rev.* 2008, *108*, 3795–3892.
127. Odom, A. L., New C-N and C-C bond forming reactions catalyzed by titanium complexes *Dalton Trans.* 2005, 225–233.
128. Reznichenko, A. L.; Hultzsch, K. C., Early Transition Metal (Group 3-5, Lanthanides and Actinides) and Main Group Metal (Group 1, 2, and 13) Catalyzed Hydroamination. In *Hydrofunctionalization*, Ananikov, V. P.; Tanaka, M., Eds. 2013; Vol. 43, pp 51–114.
129. Hazari, N.; Mountford, P., Reactions and applications of titanium imido complexes *Acc. Chem. Res.* 2005, *38*, 839–849.
130. Odom, A. L.; McDaniel, T. J., Titanium-Catalyzed Multicomponent Couplings: Efficient One-Pot Syntheses of Nitrogen Heterocycles *Acc. Chem. Res.* 2015, *48*, 2822–2833.
131. Berry, J. F., Terminal nitrido and imido complexes of the late transition metals *Comments Inorg. Chem.* 2009, *30*, 28–66.
132. Eikey, R. A.; Abu-Omar, M. M., Nitrido and imido transition metal complexes of Groups 6-8 *Coord. Chem. Rev.* 2003, *243*, 83–124.
133. Holland, P. L., Electronic structure and reactivity of three-coordinate iron complexes *Acc. Chem. Res.* 2008, *41*, 905–914.
134. Mehn, M. P.; Peters, J. C., Mid- to high-valent imido and nitrido complexes of iron *J. Inorg. Biochem.* 2006, *100*, 634–643.
135. Ray, K.; Heims, F.; Pfaff, F. F., Terminal Oxo and Imido Transition-Metal Complexes of Groups 9-11 *Eur. J. Inorg. Chem.* 2013, *2013*, 3784–3807.
136. Saouma, C. T.; Peters, J. C., M E and M=E complexes of iron and cobalt that emphasize three-fold symmetry (E O, N, NR) *Coord. Chem. Rev.* 2011, *255*, 920–937.

137. Sharp, P. R., Oxo and imido ligands in late transition metal chemistry *Dalton Trans.* 2000, 2647–2657.
138. Yi, X.-Y.; Liang, Y.; Li, C., Recent developments in ruthenium nitrido complexes *RSC Adv.* 2013, *3*, 3477–3486.
139. Zhang, L.; Deng, L., C-H bond amination by iron-imido/nitrene species *Chin. Sci. Bull.* 2012, *57*, 2352–2360.
140. Kuchta, M. C.; Parkin, G., Terminal chalcogenido complexes of Group 13 and 14 elements *Coord. Chem. Rev.* 1998, *176*, 323–372.
141. Parkin, G., Terminal Chalcogenido Complexes of the Transition Metals. In *Prog. Inorg. Chem.*, Karlin, K. D., Ed. 1998; Vol. 47, pp 1–165.
142. Trnka, T. M.; Parkin, G., A survey of terminal chalcogenido complexes of the transition metals: Trends in their distribution and the variation of their M=E bond lengths *Polyhedron* 1997, *16*, 1031–1045.
143. Ahrgren, L.; Sutin, L., Sharpless asymmetric dihydroxylation on an industrial scale *Org. Process Res. Dev.* 1997, *1*, 425–427.
144. Andersson, M. A.; Epple, R.; Fokin, V. V.; Sharpless, K. B., A new approach to osmium-catalyzed asymmetric dihydroxylation and aminohydroxylation of olefins *Angew. Chem. Int. Ed.* 2002, *41*, 472–475.
145. Becker, H.; Sharpless, K. B., A new ligand class for the asymmetric dihydroxylation of olefins *Angew. Chem. Int. Ed.* 1996, *35*, 448–451.
146. Zaitsev, A. B.; Adolfsson, H., Recent developments in asymmetric dihydroxylations *Synth.* 2006, 1725–1756.
147. Bridgeman, A. J.; Cavigliasso, G., Density-functional investigation of bonding in tetrahedral MO_4 anions *Polyhedron* 2001, *20*, 2269–2277.
148. Figueroa, J. S.; Cummins, C. C., The niobaziridine-hydride functional group: Synthesis and divergent reactivity *J. Am. Chem. Soc.* 2003, *125*, 4020–4021.
149. Selby, J. D.; Manley, C. D.; Schwarz, A. D.; Clot, E.; Mountford, P., Titanium Hydrazides Supported by Diamide-Amine and Related Ligands: A Combined Experimental and DFT Study *Organometallics* 2008, *27*, 6479–6494.
150. Bishop, M. W.; Chatt, J.; Dilworth, J. R.; Hursthouse, M. B.; Motevalli, M., Reactions of molybdenum oxo-complexes with substituted hydrazines. 3. Preparation and crystal and molecular-structure of an oxo dimethylhydrazido(2-) complex *Dalton Trans.* 1979, 1600–1602.
151. Li, Y. H.; Shi, Y. H.; Odom, A. L., Titanium hydrazido and imido complexes: Synthesis, structure, reactivity, and relevance to alkyne hydroamination *J. Am. Chem. Soc.* 2004, *126*, 1794–1803.
152. Clark, D. N.; Schrock, R. R., Multiple metal-carbon bonds.12. Tungsten and molybdenum neopentylidyne and some tungsten neopentylidene complexes *J. Am. Chem. Soc.* 1978, *100*, 6774–6776.
153. Schrock, R. R.; Clark, D. N.; Sancho, J.; Wengrovius, J. H.; Rocklage, S. M.; Pedersen, S. F., Multiple metal-carbon bonds. 31. Tungsten(VI) neopentylidyne complexes *Organometallics* 1982, *1*, 1645–1651.
154. Schoeffel, J.; Rogachev, A. Y.; George, S. D.; Burger, P., Isolation and Hydrogenation of a Complex with a Terminal Iridium-Nitrido Bond *Angew. Chem. Int. Ed.* 2009, *48*, 4734–4738.
155. Chatt, J.; Dilworth, J. R., Molybdenum(VI) and molybdenum(V) nitrido-complexes and a μ-nitrido-molybdenumrhenium complex *Chem. Comm.* 1974, 517–518.
156. Dietrich, A.; Neumuller, B.; Dehnicke, K., $MoNCl_3(THF)_4$: A nitrido complex with a folded Mo_4N_4 eight-membered ring *Zeitschrift Fur Anorganische Und Allgemeine Chemie* 1999, *625*, 1589–1591.
157. Fehlhammer, W. P.; Beck, W., Azide Chemistry: An Inorganic Perspective, Part I Metal Azides: Overview, General Trends and Recent Developments *Zeitschrift Fur Anorganische Und Allgemeine Chemie* 2013, *639*, 1053–1082.
158. Straehle, J., Reactions of transition metal azido complexes and their reaction products *Zeitschrift Fur Anorganische Und Allgemeine Chemie* 2007, *633*, 1757–1761.
159. Mindiola, D. J.; Cummins, C. C., Deprotonated 2,3:5,6-dibenzo-7-azabicyclo[2.2.1]hepta-2,5-diene as a nitrido nitrogen source by anthracene elimination: Synthesis of an iodide(nitride)chromium(VI) complex *Angew. Chem. Int. Ed.* 1998, *37*, 945–947.
160. Hendrich, M. P.; Gunderson, W.; Behan, R. K.; Green, M. T.; Mehn, M. P.; Betley, T. A.; Lu, C. C.; Peters, J. C., On the feasibility of N_2 fixation via a single-site Fe(I)/Fe(IV) cycle: Spectroscopic studies of Fe(I)(N_2)Fe(I), Fe(IV) N, and related species *Proc. Nat. Acad. Sci.* 2006, *103*, 17107–17112.
161. Meyer, T. J.; Huynh, M. H. V., The remarkable reactivity of high oxidation state ruthenium and osmium polypyridyl complexes *Inorg. Chem.* 2003, *42*, 8140–8160.
162. DiFranco, S. A.; Maciulis, N. A.; Staples, R. J.; Batrice, R. J.; Odom, A. L., Evaluation of Donor and Steric Properties of Anionic Ligands on High Valent Transition Metals *Inorg. Chem.* 2012, *51*, 1187–1200.
163. Lewin, A.; Moore, G. R.; Le Brun, N. E., Formation of protein-coated iron minerals *Dalton Trans.* 2005, 3597–3610.
164. Rehder, D., *Bioinorganic Chemistry* Oxford University Press: Oxford, 2014.
165. Bertini, I.; Luchinat, C.; Rosato, A., NMR Spectra of Iron-Sulfur Proteins. In *Adv. Inorg. Chem.*, Sykes, A. G., Ed. 1999; Vol. 47, pp 251–282.
166. Holm, R. H.; Kennepohl, P.; Solomon, E. I., Structural and Functional Aspects of Metal Sites in Biology *Chem. Rev.* 1996, *96*, 2239–2314.
167. Rao, P. V.; Holm, R. H., Synthetic Analogues of the Active Sites of Iron-Sulfur Proteins *Chem. Rev.* 2004, *104*, 527–559.
168. Zanello, P., The competition between chemistry and biology in assembling iron-sulfur derivatives. Molecular structures and electrochemistry. Part II. {Fe_2S_2(S-Cys(gamma))$_4$} proteins *Coord. Chem. Rev.* 2014, *280*, 54–83.
169. Einsle, O., Nitrogenase FeMo cofactor: an atomic structure in three simple steps *J. Biol. Inorg. Chem.* 2014, *19*, 737–745.
170. Cotton, F. A.; Murillo, C. A.; Walton, R. A., *Multiple Bonds Between Metal Atoms* 3rd ed.; Springer Science and Business Media, Inc.: New York, 2005.
171. Cotton, F. A.; Hall, W. T., Reactions of niobium(III) and tantalum(III) compounds with acetylenes. 2. Preparation and structure of $Ta_2Cl_4(\mu\text{-}Cl)_2(\mu\text{-}Me_3CC{=}CCMe_2)(THF)_2$: The shortest known Ta-Ta bond *Inorg. Chem.* 1980, *19*, 2354–2356.
172. Beshouri, S. M.; Rothwell, I. P.; Folting, K.; Huffman, J. C.; Streib, W. E., Synthesis and structure of hexabenzyl-dimolybdenum (Mo=Mo) *Polyhedron* 1986, *5*, 1191–1195.
173. Chisholm, M. H., Metal-metal bonds and metal-carbon bonds in the chemistry of molybdenum and tungsten alkoxides *Polyhedron* 1983, *2*, 681–722.
174. Chisholm, M. H., Recent advances in the chemistry of metal metal multiple bonds: Introduction *Polyhedron* 1987, *6*, 665–665.
175. Cotton, F. A.; Harris, C. B., Crystal and molecular structure of dipotassium octachlorodirhenate(III) dihydrate $K_2[Re_2Cl_8]\cdot 2H_2O$ *Inorg. Chem.* 1965, *4*, 330.
176. Cotton, F. A.; Harris, C. B., Molecular orbital calculations for complexes of heavier transition elements. 3. Metal-metal bonding and electronic structure of $Re_2Cl_8^{2-}$ *Inorg. Chem.* 1967, *6*, 924.
177. Nguyen, T.; Sutton, A. D.; Brynda, M.; Fettinger, J. C.; Long, G. J.; Power, P. P., Synthesis of a stable compound with fivefold bonding between two chromium(I) centers *Science* 2005, *310*, 844–847.
178. Davis, H. B.; Einstein, F. W. B.; Glavina, P. G.; Jones, T.; Pomeroy, R. K.; Rushman, P., Complexes with unbridged dative bonds between osmium and a group-6 element: Structures of $(Me_3P)(OC)_4OsM(CO)_5$ (M = Cr, W) *Organometallics* 1989, *8*, 1030–1039.
179. Luo, J.; Khusnutdinova, J. R.; Rath, N. P.; Mirica, L. M., Unsupported d^8-d^8 interactions in cationic Pd(II) and Pt(II) complexes: Evidence for a significant metal-metal bonding character *Chem. Commun.* 2012, *48*, 1532–1534.
180. Mann, K. R.; Gordon, J. G.; Gray, H. B., CHARACTERIZATION OF OLIGOMERS OF TETRAKIS(PHENYL ISOCYANIDE)RHODIUM(I) IN ACETONITRILE SOLUTION *J. Am. Chem. Soc.* 1975, *97*, 3553–3555.
181. Monreal, M. J.; Carver, C. T.; Diaconescu, P. L., Redox processes in a uranium bis(1,1′-diamidoferrocene) complex *Inorg. Chem.* 2007, *46*, 7226–7228.
182. Rinehart, J. D.; Harris, T. D.; Kozimor, S. A.; Bartlett, B. M.; Long, J. R., Magnetic Exchange Coupling in Actinide-Containing Molecules *Inorg. Chem.* 2009, *48*, 3382–3395.
183. Beck, W.; Sunkel, K., Metal-complexes of weakly coordinating anions: Precursors of strong cationic organometallic lewis-acids *Chem. Rev.* 1988, *88*, 1405–1421.
184. Krossing, I.; Raabe, I., Noncoordinating anions - Fact or fiction? A survey of likely candidates *Angew. Chem. Int. Ed.* 2004, *43*, 2066–2090.
185. Macchioni, A., Ion pairing in transition-metal organometallic chemistry *Chem. Rev.* 2005, *105*, 2039–2073.
186. Piers, W. E., The chemistry of perfluoroaryl boranes. In *Advances in Organometallic Chemistry, Vol 52*, West, R.; Hill, A. F., Eds. 2005; Vol. 52, pp 1–76.
187. Reed, C. A., H^+, CH_3^+, and R_3Si^+ Carborane Reagents: When Triflates Fail *Acc. Chem. Res.* 2010, *43*, 121–128.
188. Aldrich, K. E.; Billow, B. S.; Holmes, D.; Bemowski, R. D.; Odom, A. L., Weakly Coordinating yet Ion Paired: Anion Effects on an Internal Rearrangement *Organometallics* 2017, *36*, 1227–1237.
189. Chaudhuri, P.; Wieghardt, K., Phenoxyl Radical Complexes. In *Prog. Inorg. Chem.*, 2001; Vol. 50, pp 151–216.
190. Chaudhuri, P.; Wieghardt, K.; Weyhermuller, T.; Paine, T. K.; Mukherjee, S.; Mukherjee, C., Biomimetic metal-radical reactivity: Aerial oxidation of alcohols, amines, aminophenols and catechols catalyzed by transition metal complexes *Biol. Chem.* 2005, *386*, 1023–1033.

191. Ray, K.; Petrenko, T.; Wieghardt, K.; Neese, F., Joint spectroscopic and theoretical investigations of transition metal complexes involving non-innocent ligands *Dalton Trans.* 2007, 1552–1566.
192. Sproules, S.; Wieghardt, K., Dithiolene radicals: Sulfur K-edge X-ray absorption spectroscopy and Harry's intuition *Coord. Chem. Rev.* 2011, *255*, 837–860.
193. Suarez, A. I. O.; Lyaskovskyy, V.; Reek, J. N. H.; van der Vlugt, J. I.; de Bruin, B., Complexes with Nitrogen-Centered Radical Ligands: Classification, Spectroscopic Features, Reactivity, and Catalytic Applications *Angew. Chem. Int. Ed.* 2013, *52*, 12510–12529.
194. Wieghardt, K., The active-sites in manganese-containing metalloproteins and inorganic model complexes *Angew. Chem. Int. Ed.* 1989, *28*, 1153–1172.
195. Limberg, C., The role of radicals in metal-assisted oxygenation reactions *Angew. Chem. Int. Ed.* 2003, *42*, 5932–5954.
196. Fehir, R. J., Jr.; McCusker, J. K., Differential Polarization of Spin and Charge Density in Substituted Phenoxy Radicals *J. Phys. Chem. A* 2009, *113*, 9249–9260.
197. Tommos, C.; Babcock, G. T., Oxygen production in nature: A light-driven metalloradical enzyme process *Acc. Chem. Res.* 1998, *31*, 18–25.
198. Rodriguez, J. H.; Wheeler, D. E.; McCusker, J. K., Density functional studies of a Heisenberg spin coupled chromium-semiquinone complex and its chromium-catechol analog *J. Am. Chem. Soc.* 1998, *120*, 12051–12068.
199. Wheeler, D. E.; McCusker, J. K., Electron exchange and the photophysics of metal-quinone complexes. 1. Synthesis and spectroscopy of chromium-quinone dyads *Inorg. Chem.* 1998, *37*, 2296–2307.
200. Chirik, P. J., Iron- and Cobalt-Catalyzed Alkene Hydrogenation: Catalysis with Both Redox-Active and Strong Field Ligands *Acc. Chem. Res.* 2015, *48*, 1687–1695.
201. Chirik, P. J., Carbon-Carbon Bond Formation in a Weak Ligand Field: Leveraging Open-Shell First-Row Transition-Metal Catalysts *Angew. Chem. Int. Ed.* 2017, *56*, 5170–5181.
202. Obligacion, J. V.; Chirik, P. J., Earth-abundant transition metal catalysts for alkene hydrosilylation and hydroboration *Nature Rev. Chem.* 2018, *2*, 15–34.
203. Tondreau, A. M.; Stieber, S. C. E.; Milsmann, C.; Lobkovsky, E.; Weyhermueller, T.; Semproni, S. P.; Chirik, P. J., Oxidation and Reduction of Bis(imino) pyridine Iron Dinitrogen Complexes: Evidence for Formation of a Chelate Trianion *Inorg. Chem.* 2013, *52*, 635–646.
204. Szigethy, G.; Heyduk, A. F., Steric and Electronic Consequences of Flexibility in a Tetradentate Redox-Active Ligand: Ti(IV) and Zr(IV) Complexes *Inorg. Chem.* 2011, *50*, 125–135.
205. Green, M. L. H., A new approach to the formal classification of covalent compounds of the elements *J. Organomet. Chem.* 1995, *500*, 127–148.
206. Green, M. L. H.; Parkin, G., Application of the Covalent Bond Classification Method for the Teaching of Inorganic Chemistry *J. Chem. Educ.* 2014, *91*, 807–816.
207. Bartlett, N.; Yeh, S.; Kourtakis, K.; Mallouk, T., The fluorobasicities of ReF_7 and IF_7 as measured by the enthalpy change *J. Fluorine Chem.* 1984, *26*, 97–116.
208. Jacob, E.; Fahnle, M., Hexafluororhenium(VII) *Angew. Chem. Int. Ed.* 1976, *15*, 159–160.
209. Abbott, E. A., *Flatland a romance of many dimensions, by a square, with illustration by the author* Little, Brown and company,: Boston, 1899.
210. Fu, G. C., Asymmetric catalysis with "planar-chiral" derivatives of 4-(dimethylamino)pyridine *Acc. Chem. Res.* 2004, *37*, 542–547.
211. Cadenbach, T.; Gemel, C.; Bollermann, T.; Fernandez, I.; Frenking, G.; Fischer, R. A., Organometallic Chemistry of Ga^+: Formation of an Unusual Gallium Dimer in the Coordination Sphere of Ruthenium *Chem Eur. J.* 2008, *14*, 10789–10796.
212. Criegee, R.; Schroder, G., Ein nickel-komplex des tetramethyl-cyclobutadiens *Angew. Chem. Int. Ed.* 1959, *71*, 70–71.
213. Criegee, R.; Schroder, G., Ein derivat des cyclobutadiens *Ann. Chem. Justus Liebig* 1959, *623*, 1–8.
214. Ji, L. N.; Kershner, D. L.; Rerek, M. E.; Basolo, F., Syntheses and carbon-monoxide substitution-reactions of η^5-N-heterocycle manganese tricarbonyls *J. Organomet. Chem.* 1985, *296*, 83–94.
215. Ryan, D. E.; Cardin, D. J.; Hartl, F., eta(3)-Allyl carbonyl complexes of group 6 metals: Structural aspects, isomerism, dynamic behaviour and reactivity *Coord. Chem. Rev.* 2017, *335*, 103–149.
216. Carter, T. J.; Kampf, J. W.; Szymczak, N. K., Reduction of Borazines Mediated by Low-Valent Chromium Species *Angew. Chem. Int. Ed.* 2012, *51*, 13168–13172.
217. Heins, S. P.; Wolczanski, P. T.; Cundari, T. R.; MacMillan, S. N., Redox non-innocence permits catalytic nitrene carbonylation by (dadi)TiNAd (Ad = adamantyl), *Chem. Sci.* 2017, *8*, 3410.

CHAPTER 13

Thermodynamic Considerations in Reactivity

Acid–Base Theory and Redox Reactions

"Ludwig Boltzmann, who spent much of his life studying statistical mechanics, died in 1906, by his own hand. Paul Ehrenfest, carrying on the work, died similarly in 1933. Now it is our turn to study statistical mechanics."

— David Goodstein[1]

CONCEPT COVERAGE

1. Review of thermodynamics
2. Redox reactions in acidic and basic solution
3. Cyclic voltammetry
4. Review of acid–base theory
5. Difficulties with strict Lewis definitions of acidity and basicity for various metals
6. The hard–soft acid–base principle

In this chapter we present some broad generalizations on thermodynamic aspects of inorganic reactivity. After a brief review of thermodynamics that may be reminiscent of general chemistry, two different concepts for discussing reactivity will be covered: redox chemistry and the hard–soft acid–base principle. Some introductory chemistry courses classify all reactivity as either redox or acid–base chemistry. This view does a great disservice to chemistry; to say these are the *only* two reactions feels a bit like coloring the world with only two crayons. There are many reactions, such as some pericyclic reactions like the Diels–Alder dimerization of cyclopentadiene (Eq. 13.1) or radical-chain reactions, which seem to be neither of these two. (However, some may regard even these pericyclic reactions as acid–base reactions with cyclopentadiene acting as both a Lewis acid and base.) In some cases, reactions may be both acid–base and redox.

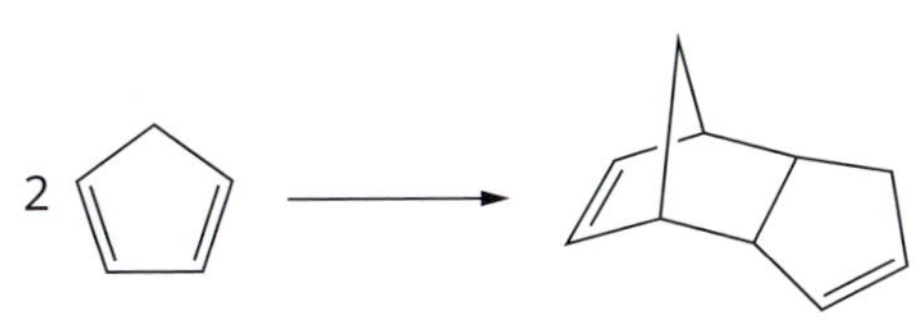

13.1

Nevertheless, acid–base and redox reactions are important components of inorganic chemistry. Redox reactions will be discussed using standard potentials with some emphasis on cyclic voltammetry, as this is currently the most common experimental technique for investigating the redox properties of inorganic complexes. For acid–base chemistry we present the venerable hard–soft acid–base model for understanding some inorganic reactions, especially in aqueous media. The discussion of both types of reactions will largely be restricted to the thermodynamics of the reactions, as kinetics will be discussed in later chapters.

13.1 A Brief Review of Thermodynamic Fundamentals

In this section we provide a very brief review of thermodynamic principles. Many students can skip to Section 13.2 if they have had a thermodynamics course or a general chemistry course covering thermodynamics, but those students may also find this section to be a nice refresher of the material.

Thermodynamics boasts three theories that have risen to the level of being called "laws"—principles found to be true in every measurement thus far and are suspected to remain true under the stated conditions. A law typically does not describe the reason or mechanism for the relationship, simply that the relationship exists. These laws are as follows.

First law of thermodynamics: Energy/matter cannot be created or destroyed.

The first law of thermodynamics refers to the universe as a whole; a particular system may interact with the surroundings (the rest of the universe) so that there is a net change, input or output, of energy. For the purpose of the rest of this discussion we will consider the system to be a chemical reaction. The change in energy (ΔE) is related to any heat change (q) related to the process and work (w) done on or by the reaction in question ($\Delta E = q + w$). The sign conventions for heat going into and out of a system (reaction) are shown in Fig. 13.1. If heat is absorbed by the reaction, then the reaction is said to be endothermic, and q is positive ($+q$). If heat is released by the reaction into the surroundings, then the reaction is exothermic, and q is negative ($-q$).

Most often, chemists are concerned with heat changes resulting from the reaction itself rather than any work being done by the system. As reactions are typically performed at constant pressure, the work term is generally negligible. Under these conditions, the heat is given a different name, "enthalpy," and the symbol used is H. Enthalpy is usually reported in energy/mol—kcal/mol or kJ/mol—and is somewhat arbitrarily defined relative to the most stable forms of the elements set as zero. The change in enthalpy (change in heat at constant pressure) for a reaction under standard conditions is defined in Eq. 13.2. In other words, if the enthalpy of the reactants is higher than the products (if energy of the reactants is higher than the products, just like water flowing downhill), then the reaction is enthalpically favorable.

Reactions for which $\Delta H°$ is negative (exothermic) are often more favorable; however, endothermic reactions ($\Delta H°$ is positive) certainly can occur due to another factor, entropy, which is the next topic.

$$\Delta H^\circ = H^\circ(\text{products}) - H^\circ(\text{reactants}) \qquad \textbf{13.2}$$

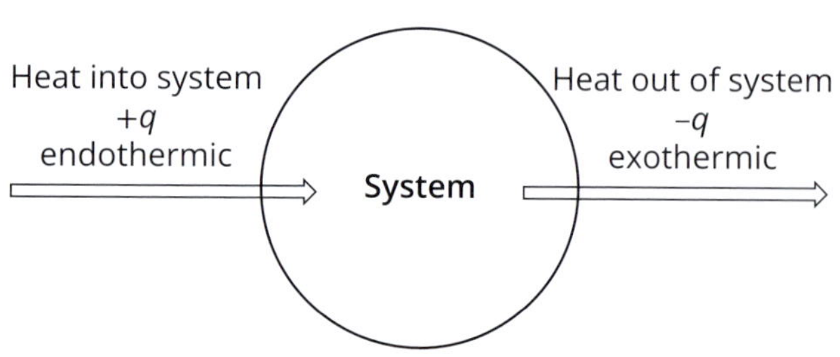

Fig. 13.1 Sign conventions for heat going into and out of a system, such as a reaction. Heat absorbed by the system ("endothermic") is given a positive sign, $+q$, and heat released by the system ("exothermic") is given a negative sign, $-q$.

Second law of thermodynamics: The entropy of the universe is always increasing.

Entropy, given the symbol S, can be viewed as a tendency towards disorder, a definition often useful in determining the most entropically favorable state of a system. Increased translations and rotations, for example, an increase in the number of molecules (especially gases) as products relative to reactants, is entropically favorable. Reactions that increase the orderliness of the system, formation of ring structures or fewer molecules, are often referred to as entropically unfavorable because they reduce the number of possible translations and/or rotations.

From a statistical mechanics viewpoint, the probability of some particular energy state is related to the number of possible arrangements with the same total energy. Placing a hot block of metal next to a cold block of metal results in the objects tending towards the same temperature, heat transfers from hot to cold (Fig. 13.2). Consider blocks of metal that contain two atoms in each block. While these are very small blocks, the results can be extrapolated to larger systems. Now,

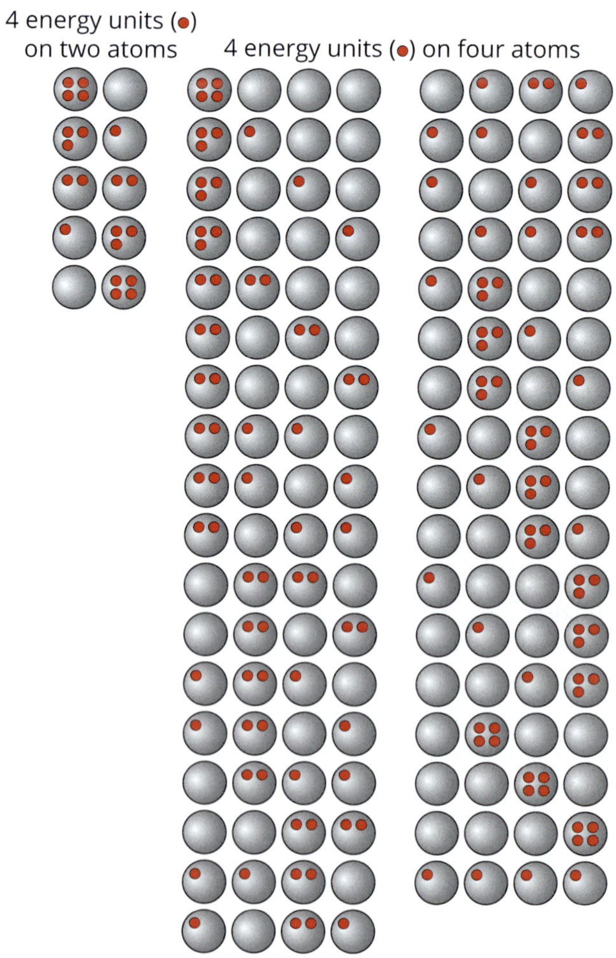

Fig. 13.2 Possible arrangements of four energy units (•) on two (left side) and four (right side) atoms, which illustrates that the spread of energy results in more possible arrangements and is more probable.

let us say there are four indivisible units of energy contained in the hot block and zero in the cold block. The metal atoms, shown as spheres, do not move, but the energy units, shown as red dots, can move between them. For the sake of this illustration, the energy units only move as whole entities and are identical. When the energy is localized on two atoms (left side of Fig. 13.2) there are five ways that the energy can be distributed between the two atoms. If the hot block of one atom is placed in contact with a cold block of three atoms (right side of Fig. 13.2), the four energy units can distribute between all four atoms, and there are thirty-five possible arrangements for the energy of the system. *It is simply more likely that the energy (such as heat) will be distributed between the four atoms rather than stay localized on the original hot atom.*

This relationship was expressed by Boltzmann as shown in Eq. 13.3, where S is the entropy, k_B is the Boltzmann constant, and W is the number of possible arrangements. This expression is chiseled above his rather stern likeness at his grave in Vienna (Fig. 13.3). Consequently, the tendency of energy to be lost as heat and spread out over larger and larger numbers of atoms is simply a tendency towards the most probable state. As such, entropy, sometimes called "time's arrow", always increases on the universal scale.

$$S = k_B \log W \qquad \textbf{13.3}$$

Fig. 13.3 Photograph of Ludwig Boltzmann's grave in Vienna, Austria.

Entropy has units of energy/(mole·Kelvin), which is often cal/mol·K or J/mol·K. The collection of units cal/mol·K is sometimes referred to as "entropy units" or simply "e.u." Using calorimetry, measuring the enthalpy of reaction ($\Delta H°$) as a function of temperature can also be used to obtain the entropy of reaction via the relationship $\Delta S = q/T$.

Third law of thermodynamics: The entropy of a system approaches a constant value as the temperature approaches 0 K.

Zero entropy for any substance is defined as a perfect crystal at 0 K, and as a result, entropy has an absolute scale.

Many of the things discussed in these laws regard the entire universe as the system. We know that $\Delta S_{univ} = \Delta S_{system} + \Delta S_{surr}$ at any time. Likewise, $\Delta S_{system} + \Delta S_{surr} > 0$. However, it is quite hard to measure entropy of a system, and exceedingly difficult to measure the entropy of the rest of the universe! Are there ways of obtaining terms that are more easily measured for the system under study?

One can tell from the units that enthalpy, H (for example, kcal/mol), and entropy, S (for example, kcal/mol·K), are related by the temperature, and the Clausius definition of entropy (under constant pressure) is as follows.

$$\Delta S_{surr} = \Delta H_{surr}/\mathrm{T}$$

Therefore, we can take $\Delta S_{system} + \Delta S_{surr} > 0$ and make the replacement $\Delta S_{surr} = \Delta H_{surr}/T$ to get:

$$\Delta S_{system} + \Delta H_{surr}/T > 0$$

From the first law, when the system loses heat the surroundings gain the same amount of heat, and vice versa. So, $\Delta H_{surr} = -\Delta H_{system}$. Replacing $\Delta H_{surr}/T$ with $-\Delta H_{system}/T$ gives the following equation, which is only in terms of the system (such as a reaction).

$$\Delta S_{system} - \Delta H_{system}/T > 0$$

If we multiply this expression by T, we obtain:

$$T\Delta S_{system} - \Delta H_{system} > 0$$

Here, we will drop the "system" labels, as all the terms are for the system. We then multiply by –1, which switches the direction of the inequality, and rearrange.

$$\Delta H - T\Delta S < 0$$

The term on the left (ΔH–$T\Delta S$) is what is referred to as the "Gibbs free energy" (ΔG), and Eq.13.4 is the "Gibbs equation", perhaps the most important relation in thermodynamics. Because the entropy of the universe must always be increasing ($\Delta S_{univ} > 0$), this expression says that a reaction is only thermodynamically favorable if the free energy is negative, $\Delta H - T\Delta S = \Delta G < 0$. Processes that have $\Delta G < 0$ are said to be "spontaneous". It does not matter how that is accomplished, whether with large positive entropy (ΔS) or with negative enthalpy (ΔH).

$$\Delta G = \Delta H - T\Delta S \quad \mathbf{13.4}$$

Entropy is far more difficult to measure accurately than enthalpy and therefore is less commonly reported. In fact, a common simplifying feature for chemical systems is to look at the enthalpies of specific bonds, often measured versus known enthalpies of bonds being broken, as was done often previously in this book.

A point of emphasis in a variety of chemistry courses (and for good reason) is an understanding of thermodynamic versus kinetic effects. It can be a futile exercise to consider the thermodynamics of a reaction with no thought as to its kinetics, and vice versa. Both exhibit essential controls over the plausibility and utility of chemical processes. If a reaction is thermodynamically favorable ($\Delta G < 0$) and yet kinetically so slow that it will never occur in a hundred lifetimes of the scientist, the thermodynamics are of no practical consequence. If a reaction would be kinetically very fast with a very small barrier to reaction but is not thermodynamically favorable ($\Delta G >> 0$), then the kinetics may be of little consequence.

One can exhibit roughly independent control over either thermodynamics or kinetics in a reaction process because they are determined by distinct energy differences (Fig. 13.4). In the case of thermodynamics, this control often involves changing the reactants or reaction conditions to increase the favorability of the reaction. One common tactic is to increase the thermodynamic favorability of the by-product of the reaction to make a reaction more favorable overall. Increasing kinetic favorability may involve use of a catalyst without necessarily even changing reactants or reaction conditions.

In some cases, a change—for example, raising the temperature—may affect both the thermodynamics and kinetics of a reaction. This may lead to some of the confusion with respect to thermodynamic versus kinetic effects, but the mechanisms of how temperature affects the thermodynamics and kinetics of a process are quite different. The temperature affects thermodynamics largely through the entropic term ($-T\Delta S$). On the other hand, raising the temperature at which a reaction occurs affects the rate by providing more thermal (vibrational, and so on) energy to the starting materials, which makes it more likely that a collision will occur between reactants with sufficient energy to go over the transition state.

All of this must be kept in mind when making chemical arguments. Are the factors influencing a process through their kinetics, thermodynamics, or both? What energies and, usually more importantly, what energy differences are critical to understanding the process? For the remainder of this chapter, two important processes for understanding the thermodynamics of reactions will be discussed: standard potentials and acid–base chemistry.

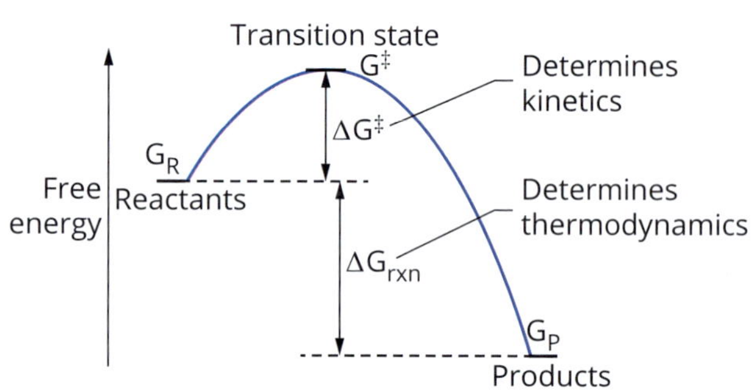

Fig. 13.4 An elementary reaction surface with labeled energies relevant to thermodynamics and kinetics.

13.2 Standard Potentials in Inorganic Chemistry

There are many reactions in inorganic chemistry that involve reduction (addition of electrons thus making the charge more negative) of one species and oxidation (removal of electrons making the charge more positive) of another species. These "redox" reactions can occur with highly variable degrees of complexity, with some reactions involving moving one electron from one compound to another with no bond-breaking or formation occurring, to systems involving multiple electrons being transferred and with many bonds being broken and formed in the process of the redox reaction.

As an example, consider a Clemmensen reduction, which perhaps you learned about in organic chemistry. In this rather complex reaction (Fig. 13.5), a ketone is converted to an alkane by action of zinc amalgam (an alloy of zinc and mercury) in aqueous acid. The mechanism for the reaction is still disputed,

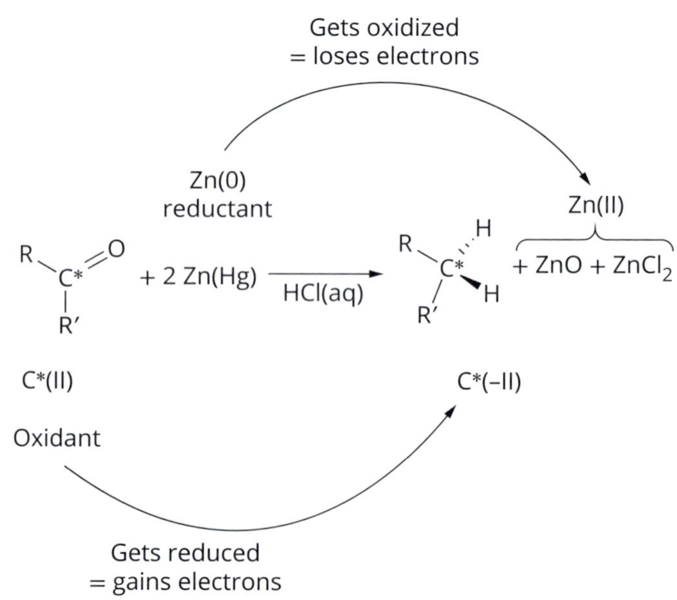

Fig. 13.5 A Clemmensen reduction of a ketone to an alkane.

but from a thermodynamic and redox standpoint this is of little consequence. The oxidant in the reaction is the ketone; the carbonyl carbon, identified with a C*, is in the +2 oxidation state on the reactant side and is reduced to the −2 oxidation state on the product side. The reductant is zinc, which begins the reaction in the 0 oxidation state and is oxidized to the +2 oxidation state. The protons that end up on the carbon come from the aqueous HCl used as the solvent. Overall, carbon is reduced by four electrons, which requires two equivalents of zinc to provide two electrons each.

One can tell several facts about redox reactions from the previous example. First, keeping track of electrons is critical to understanding the reactions. Electrons are usually treated as another species in the reaction and are balanced like other reagents. Second, solvent often participates in the reactions. In the previous example, the solvent is aqueous HCl, which provides protons to carbon and generates $ZnCl_2$. Acidic or basic aqueous solutions are common solvents for redox chemistry, and it is common for species, particularly metals, to lose or gain oxygen atoms and/or protons during the course of redox reactions. Third, it is handy to think about species in terms of what is being oxidized and what is being reduced. As a result, such processes are often divided into two "half-reactions", where one half-reaction has the reductant and the other half-reaction has the oxidant. Summing the two half-reactions gives the overall process. A major advantage of this is that one can combine known half-reactions to readily examine the thermodynamics of new combinations of redox reactions.

For example, consider the reaction of Ca metal with aqueous acid:

$$\mathrm{Ca\,(s) + 2\,H^{+}(aq) \rightarrow Ca^{+2}(aq) + H_2(g)} \qquad \textbf{13.5}$$

The calcium is being oxidized and therefore provides electrons to another species, and so the calcium is serving as the reductant and can be called the reducing agent. The protons are being reduced, making them the oxidant or oxidizing agent. The acid must come with a counter-ion, such as Cl^- or NO_3^-, that is acting as a spectator ion and therefore not participating in the redox reaction. In some cases, spectator ions are not even listed in the reaction, but the charges must, of course, be balanced (Section 3.3).

This reaction can be divided into the two electrochemical half-reactions (Eqns. 13.6 and 13.7).

$$\mathrm{Ca\,(s) \rightarrow Ca^{2+}(aq) + 2\,e^-} \qquad \textbf{13.6}$$

$$\mathrm{2\,H^{+}(aq) + 2\,e^- \rightarrow H_2(g)} \qquad \textbf{13.7}$$

The advantage of writing half-reactions is the emphasis on the redox behavior of the atoms. The disadvantage is that some of the nuances of the chemistry are not explicitly addressed. For example, this redox process takes place regardless of the counter ion of the acid, but if the counter ion were sulfate, the redox reaction would also have resulted in the formation of a precipitate, $CaSO_4$. Many commonly used tables of half-reactions do not actually give the product in the form it is generated in solution; a shorthand notation is given instead of the real chemical species present, especially in water. The chemical nature of the actual products makes a difference in the redox behavior, and it is important to consider exactly what is being reduced or oxidized, as well as the products formed under the reaction conditions.

13.2.1 Thermodynamics and the Standard Potential

The most important thermodynamic quantity to understand in redox processes is the "potential", *E*. *The potential, typically given in volts (V), is directly related to the free energy of the system.* A volt (*V*) is defined as 1 *J/C*, where *C* is Coulombs (a unit of electric charge) and 1 *C* is equivalent to the charge of 6.242×10^{18} electrons. The free energy of the system, $\Delta G°$, under standard conditions (1 bar of pressure, 1 M in reactants, room temperature, and so on) is related to the standard potential for the redox process, $E°$, by Eq. 13.8. In this equation, n = number of electrons in the process, and the Faraday constant, *F*, is the amount of electric charge in a mole of electrons, with 1 *F* = 96.485 kC/mol (9.649×10^4 C/mol). Put another way, a Faraday is the product of Avogadro's constant (N_A) and the elementary charge of the electron (*e*), so that $F = eN_A$. With these units, the free energy is in units of kJ/mol (1 kJ/mol = 0.239 kcal/mol).

$$\Delta G^{\circ} = -nFE^{\circ} \qquad \textbf{13.8}$$

Consider the example of Zn metal in aqueous acid as shown in Eq. 13.9.

$$\mathrm{Zn\,(s) + 2\,H^{+}(aq) \rightarrow Zn^{2+}(aq) + H_2(g)} \qquad \textbf{13.9}$$

The half-reaction in which zinc is oxidized by two electrons is shown in Eq. 13.10. Using Eq. 13.8, we can calculate the standard potential of the half-reaction, $E°(Zn,Zn^{2+})$, which is +0.76 V. Take note of the signs! The free energy for the reaction is favorable, −35 kcal/mol (−147 kJ/mol). Because of the negative sign in Eq. 13.8, standard potential differences, $E°$, that are positive are thermodynamically favorable.

$$\mathrm{Zn\,(s) \rightarrow Zn^{2+}(aq) + 2\,e^-} \quad E^{\circ}(Zn, Zn^{2+}) = +0.76\,\mathrm{V}$$
$$\Delta G^{\circ} = -35\,\mathrm{kcal/mol} \qquad \textbf{13.10}$$

Half-reactions are typically tabulated as reductions. If we are interested in running the reaction in the opposite direction, Zn^{2+} being reduced by two electrons to make Zn metal, the half-reaction is inverted (Eq. 13.11). This results in a sign change in the standard potential. Based on Eq. 13.8, the sign of $\Delta G°$ changes to positive in this direction and thus the reduction is thermodynamically unfavorable; that is, energy must be added for the reaction to proceed.

$$Zn^{2+}(aq) + 2\,e^- \rightarrow Zn(s) \qquad E^\circ(Zn^{2+}, Zn) = -0.76\ V$$
$$\Delta G^\circ = +35\ kcal/mol \qquad \textbf{13.11}$$

The other half of our reaction is the reduction of H^+ to H_2 (Eq. 13.12). The reduction of protons under standard conditions is the reference reaction for electrochemistry with a potential defined as exactly zero V.

$$H^+(aq) + e^- \rightarrow \tfrac{1}{2}H_2(g) \quad E^\circ = 0.000\ V\ (\text{by convention}) \qquad \textbf{13.12}$$

Putting the two half-reactions together requires that we equalize the number of electrons. The overall potential for the reaction is the sum of potentials of the oxidation and reduction half-reactions (Eq. 13.13). As mentioned previously, converting the reduction half-reaction that appears in tables to the oxidation half-reaction requires (1) flipping the sides of the chemical equation and (2) changing the sign of the potential. Consequently, the standard potential for Zn metal reducing protons to H_2 is +0.76 V. It is advisable to use "+" and "–" signs explicitly in electrochemistry to avoid confusion, that is, +0.76 V rather than 0.76 V.

$$E^\circ_{rxn} = E^\circ_{ox} + E^\circ_{red} \qquad \textbf{13.13}$$

oxidation half-reaction:

$$Zn(S) \rightarrow Zn^{2+}(aq) + 2e^- \qquad E^\circ = +0.76\ V$$

reduction half-reaction:

$$2H^+(aq) + 2e^- \rightarrow H_2(g) \qquad E^\circ = +0.00\ V$$

overall redox reaction:

$$Zn(s) + 2H^+(aq) \rightarrow Zn^{2+}(aq) + H_2(g) \qquad E^\circ_{rxn} = +0.76\ V$$

Many elements are capable of accessing a variety of oxidation states and thus are capable of performing different roles in redox reactions. Consider the production of Cl_2 gas (Cl in the zero formal oxidation state) from the reaction of hypochlorous acid, HOCl (Cl in the +1 formal oxidation state), with Cl^- (Cl in the –1 oxidation state) in the presence HCl (Eq. 13.14).

$$HOCl(aq) + H^+(aq) + Cl^-(aq) \rightleftharpoons Cl_2(g) + H_2O(l)$$
$$K = 9.6\text{–}10^2\ M^{-2} \qquad \textbf{13.14}$$

This reaction is of particular interest because it involves chemicals that are obtainable outside of a laboratory setting. Many toilet-bowl cleaners have HCl as the active ingredient. The sodium-containing conjugate base of hypochlorous acid, sodium hypochloride (NaOCl), is the active ingredient in household bleach. Mixing the two household chemicals acidifies the hypochlorite to HOCl (recall that HCl is a strong acid in water, whereas OCl^- is a weak base and therefore some HOCl is produced) producing the previous reaction. The equilibrium constant favors Cl_2 formation. Under nonlaboratory conditions, people have died from mixing bleach and toilet-bowl cleaner in the home. Under safe laboratory conditions, this is a convenient method to prepare small amounts of Cl_2 gas, an important oxidant.

As an aside, a substitute for performing reactions requiring Cl_2 is to use dichloroiodobenzene, which can be prepared without ever isolating Cl_2 by using a variation of the previous reaction. The Cl_2 is generated by addition of household bleach (Clorox® or store-brand bleach) to a solution of iodobenzene (PhI) in aqueous HCl; the chlorine converts the PhI to $PhICl_2$, which precipitates as a yellow solid (Eq. 13.15). Sometimes we think of hypervalent iodine reagents as exotic, but this one is made from bleach, toilet-bowl cleaner, and iodobenzene in near-quantitative yield.[3]

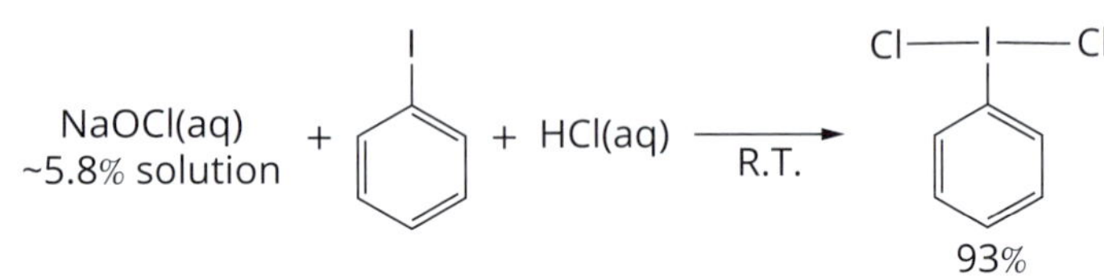

13.15

The two half-reactions for the formation of Cl_2—hypochlorous acid and chloride—are shown with their standard potentials in Eqs. 13.16 and 13.17. *The reaction is performed in the presence of acid, so H^+ appears on the reactant side and water appears as a product to balance the equation*; tabulated redox potentials often specify acidic or basic conditions (Appendix J).

$$2\,HOCl(aq) + 2\,e^- + 2\,H^+(aq) \rightarrow Cl_2(g) + 2\,H_2O(l)$$
$$E^\circ = +1.67\ V \qquad \textbf{13.16}$$

$$\tfrac{1}{2}Cl_2(g) + 1\,e^- \rightarrow Cl^-(aq) \qquad E^\circ = +1.36\ V \qquad \textbf{13.17}$$

The Cl^- is oxidized to produce Cl_2, so we need to invert Eq. 13.17 and change the sign of the potential:

$$Cl^-(aq) \rightarrow \tfrac{1}{2}Cl_2(g) + 1\,e^- \qquad E^\circ = -1.36\ V \qquad \textbf{13.18}$$

Next, we need to balance the number of electrons in the reactions, which can be done by multiplying the oxidation half-reaction by 2 (Eq. 13.19). *Note that the potential does not change. While the number of electrons involved doubles, the electrons are not more energetic, there are just more of them.*

$$2\,Cl^-(aq) \rightarrow Cl_2(g) + 2\,e^- \qquad E^\circ = -1.36\ V \qquad \textbf{13.19}$$

The addition of the chemical reactions and the potentials yields the overall reaction (Eq. 13.20). Note that the potential for the reaction is positive, +0.31 V, which corresponds to a negative free energy (–14.3 kcal/mol) and therefore a spontaneous process.

$$2\,HOCl(aq) + \cancel{2e^-} + 2\,H^+(aq) \rightarrow Cl_2(g) + 2\,H_2O(l)$$
$$E^\circ = +1.67\ V$$
$$2\,Cl^-(aq) \rightarrow Cl_2(g) + \cancel{2e^-} \qquad E^\circ = -1.36\ V$$
$$2\,HOCl(aq) 2\,Cl^-(aq) + 2\,H^+(aq) \rightarrow 2Cl_2(g) + 2\,H_2O(l)$$
$$E^\circ = +0.31\ V \qquad \textbf{13.20}$$

Acidic solution (pH = 0)

$$\underset{+7}{ClO_4^-} \xrightarrow{+1.20\ V} \underset{+5}{ClO_3^-} \xrightarrow{+1.18\ V} \underset{+3}{ClO_2^-} \xrightarrow{+1.65\ V} \underset{+1}{ClO^-} \xrightarrow{+1.67\ V} \underset{0}{Cl_2^-} \xrightarrow{+1.36\ V} \underset{-1}{Cl^-}$$

Fig. 13.6 Latimer diagram for chlorine-containing species in acidic solution.

Reduction potentials for a variety of species containing a specific element are often written in an abbreviated form called a Latimer diagram. An example using the various chlorine oxidation states in acidic solution, generally assumed to be pH = 0, is shown in Fig. 13.6. With the reactions being performed in acidic solution, the equations in the Latimer diagram can be balanced by adding H^+ and H_2O as necessary.

For example, the reduction of ClO_4^- to ClO_3^- can be balanced to produce Eq. 13.21. In acidic solution, anionic species in the list will often be protonated to their neutral form; for example, ClO_3^- in acidic solution will be (at least partially) in the form of $HClO_3$.

$$ClO_4^-(aq) + 2\,H^+(aq) + 2\,e^- \rightarrow ClO_3^-(aq) + H_2O(l) \quad E^\circ = +1.20\ V \qquad \textbf{13.21}$$

The nature of the solution makes a significant difference to the exact species in solution and the reduction potentials. Consequently, in basic solution (pH = 14), the Latimer diagram for the reduction of chlorine-containing species involves different chemical species and therefore different potentials, as shown in Fig. 13.7.

In basic solution, ClO_4^- will actually predominantly exist as this anion, whereas in acidic solution ClO_4^- will exist at least partially as $HClO_4$. Neutral species are easier to reduce than anions, therefore the potential for the reduction of "ClO_4^-" is more positive in acidic solution. The anions in basic solution are more difficult to reduce, and therefore exhibit smaller positive potentials. Note that reduction of Cl_2, a neutral species, in both acidic and basic solutions occurs at the same potential. This is only true of a species if the starting material and the product are in the same state in both solutions, which happens to be the case for this halogen. An example of this is found in the reduction of phosphorus, a neutral species, in acidic and basic solutions. This reduction occurs at different potentials because phosphonium (PH_4^+) is produced in acidic solution whereas the free phosphine (PH_3) is produced in basic solution.

Latimer diagrams provide a picture of the redox landscape of an element under the given conditions, and can be used in

Acidic solution (pH = 0)

$$\underset{0}{P} \xrightarrow{-0.063} \underset{3}{PH_3}$$

Basic solution (pH = 14)

$$\underset{0}{P} \xrightarrow{-0.89} \underset{3}{PH_3}$$

a variety of ways. For example, they can be used to determine whether a species is thermodynamically stable in solution. Some ions will transfer electrons to an ion of the same kind—a process known as "redox disproportionation". The Latimer diagram can be used to determine whether such a redox disproportionation is thermodynamically favorable.

$$M^{n+} + M^{n+} \rightarrow M^{(n+1)+} + M^{(n-1)+}$$

As an example, consider whether Mn^{2+} is stable in acidic solution. As a starting point, a balanced equation for the disproportionation reaction—in this case, the disproportionation of Mn^{2+} to give Mn^{3+} and solid Mn^0—is necessary.

$$3\,Mn^{2+} \rightarrow 2\,Mn^{3+} + Mn^0$$

The Latimer diagram for manganese under acidic conditions is:

$$MnO_4^- \xrightarrow{+0.90\,V} MnO_4^{2-} \xrightarrow{+2.09\,V} MnO_2 \xrightarrow{+0.95\,V} Mn^{3+}$$
$$\xrightarrow{+1.51\,V} Mn^{2+} \xrightarrow{-1.18\,V} Mn^0$$

Next, we can write the half-reactions for the process and sum them.

$$\begin{array}{ll} Mn^{2+} + 2\,e^- \rightarrow Mn(s) & E^\circ = -1.18\,V \\ 2Mn^{2+} \rightarrow 2\,Mn^{3+} + 2\,e^- & E^\circ = -1.51\,V \\ \hline 3\,Mn^{2+} \rightarrow 2\,Mn^{3+} + Mn^0 & E^\circ_{rxn} = -2.69 \end{array}$$

As a result, we can see that manganese(II) should not undergo the redox process described, as it is thermodynamically unfavorable; that is, it has a negative E°_{rxn}.

Basic solution (pH = 14)

$$\underset{+7}{ClO_4^-} \xrightarrow{+0.37\ V} \underset{+5}{ClO_3^-} \xrightarrow{+0.30\ V} \underset{+3}{ClO_2^-} \xrightarrow{+0.68\ V} \underset{+1}{ClO^-} \xrightarrow{+0.42\ V} \underset{0}{Cl_2} \xrightarrow{+1.36\ V} \underset{-1}{Cl^-}$$

Fig. 13.7 Latimer diagram for chlorine in basic solution.

However, if one were to consider the disproportionation of manganese(III) under acidic conditions instead, writing the half-reactions for the process and summing them would show that the redox process is thermodynamically favorable, that it has a positive $E°_{rxn}$.

$$\begin{array}{ll} Mn^{3+} + e^- \rightarrow Mn^{2+} & E° = 1.18\text{ V} \\ Mn^{3+} \rightarrow MnO_2 + e^- & E° = -0.95\text{ V} \\ \hline 2Mn^{3+} \rightarrow Mn^{2+} + MnO_2 & E°_{rxn} = +0.85 \end{array}$$

13.2.2 The Galvanic Cell

In the previous examples, electrons moved directly from one chemical species to another. If the reaction is favorable, the energy of the starting materials is higher than that of the products. That energy is released as the reaction proceeds, perhaps dissipated as heat. One can also use the potential difference between two half-reactions to generate a flow of electrons if the reactions are separated but connected in such a way to allow electrons and ions to flow between them. The overall reaction will still take place, just as it did when the species were together in the same reaction vessel, but the flow of electrons, along with the energy over and above that required for the reaction to occur, can be used for some other process. This, in its simplest form, is referred to as a Galvanic cell.

An example of a simple Galvanic cell is shown in Fig. 13.8. On the left side is a chamber with a zinc electrode and a solution of zinc sulfate. On the right side is a chamber with a solution of copper(II) sulfate ($CuSO_4$) and a copper electrode. The two electrodes are connected by a wire, perhaps with a voltmeter to measure the electrical potential of the cell.

Consider the two reduction half-reactions for $Zn/ZnSO_4$ (Eq. 13.22) and $Cu/CuSO_4$ (Eq. 13.23). The potential for the reduction of Zn^{2+} is negative and far less thermodynamically favorable than the reduction of Cu^{2+}. Consequently, we can surmise that Zn^0 will be the reductant and copper(II) the oxidant.

$$Zn^{2+}(aq) + 2\,e^- \rightarrow Zn(s) \qquad E° = -0.76\text{ V} \qquad \textbf{13.22}$$

$$Cu^{2+}(aq) + 2\,e^- \rightarrow Cu(s) \qquad E° = +0.34\text{ V} \qquad \textbf{13.23}$$

Rewriting the zinc half-reaction as an oxidation and summing the two reactions will produce the overall reaction and its potential, +1.10 V (Eq. 13.24). As the reaction proceeds, Zn^0 (zinc metal) will be dissolved into solution to form aqueous Zn^{2+}, and dissolved Cu^{2+} will precipitate (often called "plating out" when referring to the formation of a solid metal) onto the copper electrode to form additional Cu^0.

$$\begin{array}{ll} Zn(s) \rightarrow Zn^{2+}(aq) + \cancel{2e^-} & E° = +0.76\text{ V} \\ Cu^{2+}(aq) + \cancel{2e^-} \rightarrow Cu(s) & E° = +0.34\text{ V} \\ \hline Zn(s) + Cu^{2+}(aq) \rightarrow Zn^{2+}(aq) + Cu(s) & E° = +1.10\text{ V} \end{array} \qquad \textbf{13.24}$$

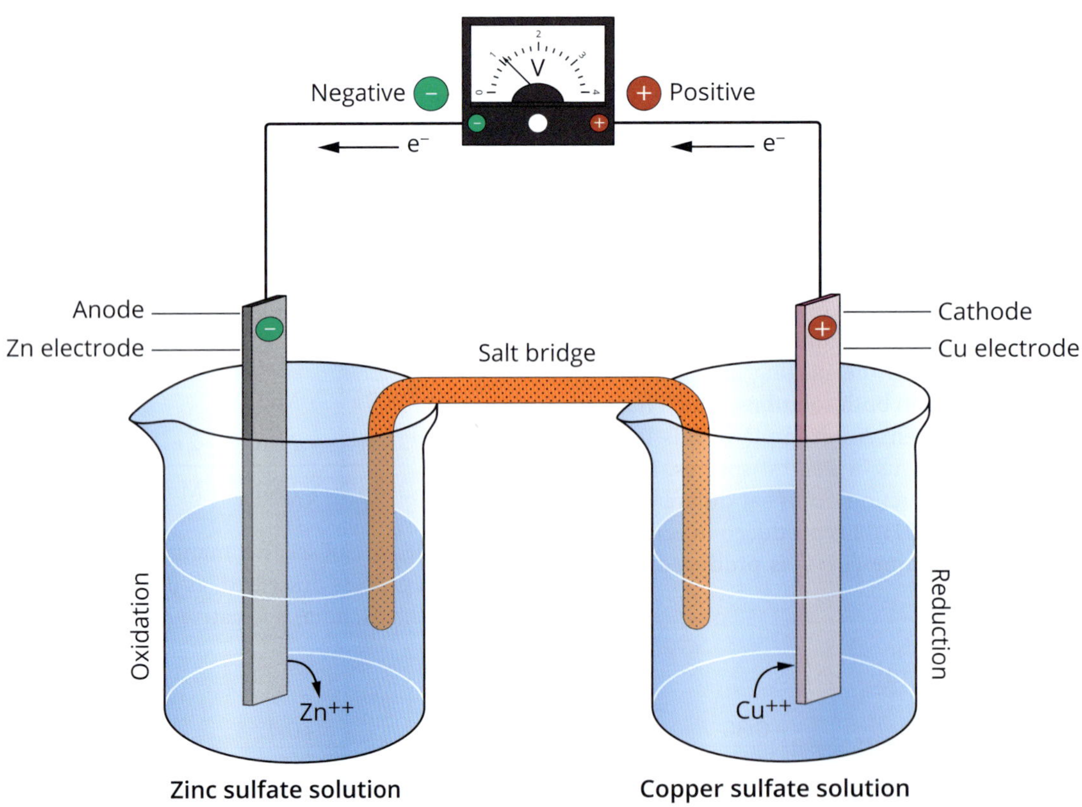

Fig. 13.8 An example of a Galvanic cell. (Wikimedia Commons.)

The zinc side, where the oxidation is occurring, is called the *anode* and is given the symbol "–". One can remember this as being the site where the electrons originate (zinc gives up its electrons) or, as it was originally described, as being the side that anions move toward. Since the zinc is giving up electrons to make Zn^{2+}, sulfate anions must move toward this side with the same rate. The copper side, where the reduction is occurring, is called the *cathode* and is given the symbol "+". Cations, in this case Cu^{2+}, move towards the cathode.

The salt bridge, which allows the transport of these ions, is usually just a glass tube with an aqueous solution and a fritted glass disk (porous glass) or some other material that prevents the solutions from freely mixing but allows ions to pass through. Some may use a U-shaped tube with a salt solution with cotton balls stuffed into the two ends. In simple experiments one can use a wet paper towel with one end in each solution. A salt bridge is used because, if the copper ions make direct contact with the zinc electrode, copper metal will begin to plate out on the zinc; that is, the redox reaction will occur with no flow of electricity. As the reagents are consumed in the cell, the flow of electricity will eventually come to a halt.

Using different half-reactions, many Galvanic cells can be imagined. The Galvanic cell can be seen, at least from a chemical perspective, as the basis for batteries and fuel cells. In order to make modern batteries and fuel cells, much of the technology involved is in controlling the flow of ions precisely (advanced versions of the salt bridge), improving efficiency (which may involve catalysis and lowering kinetic barriers to reactions), and other issues. Battery science is an important specialized field of inorganic chemistry, with an extensive interface with materials science. Batteries are briefly discussed in Chapter 19.

13.2.3 Standard Potentials in Nonaqueous Solution and Cyclic Voltammetry

Until the 1970s the most common method of examining the redox behavior of a species was polarography, but that has since been replaced by cyclic voltammetry (CV). The technique is relatively straightforward experimentally, and yet can provide a wealth of information about the redox properties of the species in solution and the stability of the different redox states formed. Depending on the system, the results from a CV experiment and the interpretation of those results can be rather complex. As such, this chapter will not present an in-depth discussion of all possible results from a CV experiment, for which the reader can turn to one of the many books and review articles devoted to the topic.[2–6] First, there will be a brief description of the species present in solution during a CV experiment. This will be followed by a description of the CV experiment itself, including experimental setup and some background on electrochemistry. For a more detailed discussion of the how to conduct the experiment, troubleshoot, and analyze data, we recommend the articles by Jillian L. Dempsey[7–8] and the short book by Daniel J. Graham.[9]

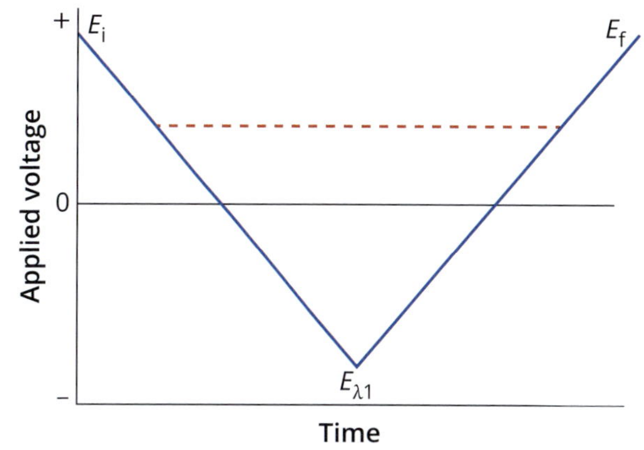

Fig. 13.9 A plot of the variation of applied potential versus time.

In a typical CV experiment the applied potential is varied over time—a plot which might look something like Fig. 13.9. In this example plot, the scan was begun with a positive potential (E_i), but the potential could just as easily be started negative. The scan continues to some set point called the switching potential ($E_{\lambda 1}$), at which the scan will be switched from positive to negative, or vice versa. It is completed upon reaching a final potential (E_f). Ideally, the analyte will undergo at least one redox event (indicated by the red line in Fig. 13.9) within the potential range from E_i to $E_{\lambda 1}$, and will undergo the corresponding reverse process between $E_{\lambda 1}$ and E_f.

Consider the electrochemical process presented in Eq. 13.25. The analyte could start in the oxidized (Ox) form at E_i, and be converted to the reduced (Red) form when the potential reaches the point indicated by the red line in Fig. 13.9. Upon reversing the scan, the reduced form will be converted back to the oxidized form. The cycle can be completed once, or it can be run numerous times. The rate at which the potential is changed is called the "scan rate", and varying this rate can have a significant impact on the redox process. It has a large impact, because the cyclic voltammetry experiment relies on diffusion (as will be discussed later) which is a relatively slow process, and why you stir your reactions (and your coffee after you add cream)—but you do not stir your CV experiments.

$$\text{Ox} \underset{+e^-}{\overset{-e^-}{\rightleftharpoons}} \text{Red} \qquad \textbf{13.25}$$

13.2.3.1 The Solution in Cyclic Voltammetry

CV experiments are typically conducted on solutions of an analyte dissolved in a solvent that ideally does not participate in the redox process. Our discussion of redox properties thus far perhaps gives the impression that most redox chemistry is done in aqueous solution. While aqueous chemistry is still very important and many inorganic chemists do electrochemistry in aqueous solutions, many others do not. One obvious reason for this is that many inorganic species are not soluble in water. In addition, water is rarely an innocent solvent both in terms of its chemical reactivity and redox properties. For example, one could not study the redox chemistry of sodium in aqueous solution as sodium metal would react instantly with the water to form NaOH (aq) and H_2.

To perform CV experiments, one needs a solution that will carry the current. Even pure water does not carry current well as the only ions are due to autoionization, H_2O going to H^+ and OH^-, which has a very small equilibrium constant of 10^{-14}, so acidic or basic solutions are often employed. However, the presence of H^+ and OH^- can often lead to additional chemical complications such as protonating or deprotonating the analyte. To enhance the conductivity of aqueous solutions, the addition of a soluble ionic compound, such as KCl, to the solution can assist in carrying current. These soluble ionic species are known as "supporting electrolytes", or just "electrolytes", in an electrochemical experiment. In nonaqueous solvents that do not undergo autoionization, an electrolyte is also a necessity.

Performing CV experiments in media other than water often simplifies the electrochemistry. In addition, many organic solvents have larger (or at least different) redox windows than water, which allows for the examination of the electrochemically active species that would not be accessible in aqueous solution. Some of the most common conditions for electrochemistry in inorganic laboratories not working with aqueous solutions are to use either acetonitrile or THF (tetrahydrofuran) as the solvent, with supporting electrolytes such as $\{NBu_4\}^+\{PF_6\}^-$ or $\{NBu_4\}^+\{OTf\}^-$ (OTf^- is "triflate," $CF_3SO_3^-$, a very common pseudohalide) at a concentration of ~0.1 M. In the older literature, salts of the perchlorate anion, $\{ClO_4\}^-$, are often seen as supporting electrolytes, but concerns about the explosive nature of perchlorates have greatly reduced their use, and they should be avoided if at all possible.

In CV experiments, a typical analyte concentration is ~1 mM. Oxygen should not be present in the solution, as it is not only redox active but also may react with the analyte or any of the species generated during the CV experiment. This requires that the solution be degassed by some method, and the experiment is typically run under an atmosphere of purified dinitrogen or argon. In addition, the temperature at which the experiments are performed should be recorded. All of the conditions are subject to change, depending on the nature of the analyte, but the conditions just discussed are fairly typical.

13.2.3.2 The CV Experiment

In a typical experiment, the solvent, analyte, and supporting electrolyte are placed within a glass container, usually referred to as an electrochemical cell (Fig. 13.10). The experiment is commonly performed using a "three-wire set-up" (reference electrode, working electrode, and auxiliary electrode), and the auxiliary (sometimes called counter) electrode is typically a platinum wire. The working electrode is where the electrochemistry occurs and is usually made of platinum, gold, or glassy carbon.[a] There are a variety of reference electrodes

[a] Glassy carbon is a form of carbon more similar to graphite than diamond.

Fig. 13.10 An electrochemical cell.

that can be used, with one of the more common ones being made of a silver wire in a silver(I) chloride solution. In some cases a traditional reference electrode is not employed, but a pseudo-reference of platinum or silver wire can be used in solutions with the working electrode in a "two-wire set-up". In either case it is common for a reference compound to be added at the end of the experiment to set the potential; often the ferrocene/ferrocenium couple ($FeCp_2^{0/+}$) is used for this purpose.[10] Values for nonaqueous redox chemistry are often referenced to the $FeCp_2^{0/+}$ couple as 0 V, such as in Connelly and Geiger's redox chemistry review; however, referencing to Ag/Ag^+ or H_2/H^+ are also common.[11] Referencing to H_2/H^+ is most often done indirectly by adjusting the potential from some other electrode, as the H_2/H^+ electrode is physically more difficult to set up and use. Regardless of what reference is used, it is critical that the reference electrode or compound is clearly indicated in any CV experiment.[12]

The redox reaction of interest in a CV experiment is taking place at the surface of the working electrode. This can be thought of in the context of the electrode interacting with a single species referred to here as the oxidized species, "Ox", which can undergo a reduction to the reduced species, "Red" (Eq. 13.25). It is useful to think of the effect changing the potential has on the energy of the electrons at the electrode. As the potential is increased (more positive), the electrons at the electrode are being stabilized. Under these conditions, the electrons at the electrode surface will not have the necessary energy to reduce Ox to Red. As the potential is decreased (more negative), the electrons at the electrode surface are increasing in potential energy, making it more likely that a molecule of Ox will be reduced to Red.

One can imagine that there should be a relationship between the concentration of Ox and Red and the potential of the electrode.[b] Having a large number of Red species near the electrode should make it harder to reduce another Ox to Red, for example. The relationship between the potential of the electrode where the electron transfer occurs and the standard potential, $E°$, discussed in the previous sections, is found by using the Nernst equation (Eq. 13.26). However, this equation only holds under reversible conditions; consequently, we should discuss what reversibility means in this context.

$$E = E^{\circ} - \frac{RT}{nF}\ln\left(\frac{[Red]}{[Ox]}\right) \qquad 13.26$$

The term "reversible" here refers to electrochemical reversibility, which occurs when the electron transfer is very fast. This means that there is a small barrier to electron transfer between the electrode and the analyte. In this context, this means that the electron transfer reaction is fast, setting up an equilibrium between the electrode and Ox/Red and therefore making the system electrochemically reversible. This differs from chemical reversibility in electrochemistry, which refers to the stability of the product of a redox process. For example, if a chemical species undergoes a reduction yielding a species that immediately decomposes or reacts in some way, the process would be said to be a *chemically irreversible* reduction. For the purpose of this discussion, a reversible redox process will constitute both chemical and electrochemical reversibility unless otherwise noted.

One common reason there can be a large barrier to electron transfer is that there is a significant change in the geometry between Ox and Red. Generally, the transfer of an electron from one site to another is fast with respect to nuclear motions, such as vibrations and rotations. If Ox and Red have similar geometries, various molecular motions may allow a significant population of Ox to be in the geometry of Red at any given time. This makes for a low barrier to electron transfer and therefore reversible electron transfer. If Ox and Red have vastly different geometries, there may be very little of Ox that is in a geometry where electron transfer can occur, giving a very high barrier to electron transfer from the electrode to Ox. The high barrier suggests that the electron transfer from Red back to the electrode will likely be slow as well.

Under reversible electron transfer conditions and assuming the diffusion rate of Ox and Red in the solution are similar, one can use the Nernst equation to calculate the potential of the electrode at the point of electron transfer with respect to the Ox/Red equilibrium. The Nernst equation is shown in Eq. 13.26, where R is the gas constant in J/(mol · K), T is the temperature in K, n is the number of electrons in the process, F is Faraday's constant, [Ox] is the concentration of Ox, and [Red] is the concentration of Red. Notice that the ratios of Red and Ox are in a logarithmic function, so if they are in equal concentrations, $\ln([Red]/[Ox]) = 0$, which causes the second term in the Nernst equation to be equal to zero. This gives $E = E°$, where the measured potential is the standard potential. In addition to the relationship between measured and standard potential, the Nernst equation can be used to understand the chemical species present in solution during a CV experiment. The Nernst equation is discussed in more detail later in this chapter.

The results of the application of a potential to a solution are plotted as a current (μA) versus potential (V) plot; a typical cyclic voltammogram is shown in Fig. 13.11. Note that like many other instrumental techniques, the abbreviation "CV" refers to the technique of cyclic voltammetry but is often used for the plot of the results, a cyclic voltammogram. The traditional convention for the x-axis stems from polarography in which potentials become more negative to the right. This differs from the IUPAC convention where potentials become more positive to the right on the x-axis. Consequently, it is important to notice the direction of the potential when examining work from a particular research group to see which convention they follow. However, regardless of the convention used, a more negative potential typically results in a reduction of the analyte and a reductive (cathodic, +) current is observed. When a more positive potential is applied, oxidation of the analyte occurs and an oxidative (anodic, –) current is observed. While the potential is being changed on the x-axis, and current is being observed on the y-axis. You should think of current as a rate, as it is charge per unit time. Current is typically reported in amperes (A, "amps") and 1 A = 1 C/s,

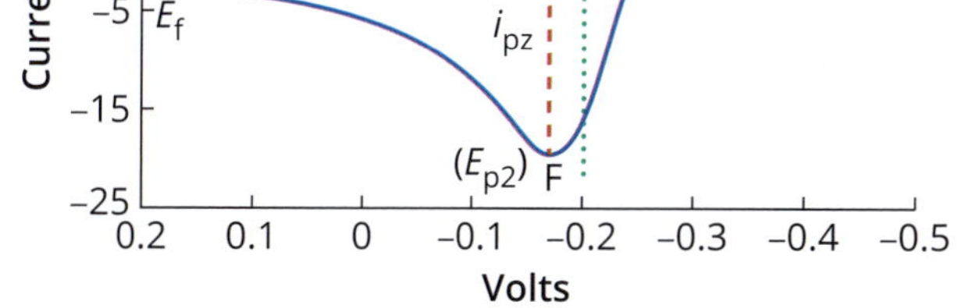

Fig. 13.11 Typical reversible cyclic voltammogram starting at +0.2 V (E_i), switching at −0.5 V ($E_{\lambda1}$) and ending at +0.2 V (E_f). The arrow indicates the direction of the scan. See the text for a description of the regions.

[b] In chemistry it is customary to signify the concentration of a species by putting square brackets around the species. In other words, the concentration of Ox is [Ox] and the concentration of Red is [Red].

where the Coulomb is the charge of 6.242×10^{18} electrons. As the current moves away from zero, a large number of electrons are moving to or from the electrode.

The "duck-shape" of a reversible CV for the reduction of Ox to Red will be described by using the letters in Fig. 13.11 in the context of the Nernst equation to calculate concentrations and diffusion gradients. If there is a large difference in concentration of Ox and Red in solution, the gradient is large and there is a tendency for species to move from regions of high concentration to low concentration. Due to the existence of the electrode in one specific physical location within the solution, concentration gradients will exist and are therefore an essential part of the cyclic voltammetry experiment, and for this reason *the solution is never stirred during a CV experiment.*

Region A

In this region of the voltammogram, the electrons on the electrode surface are relatively stable (high potential) and there is little electron transfer from the electrode to the analyte. In fact, one can calculate the equilibrium [Red]/[Ox] explicitly using the Nernst equation (Eq. 13.26) if $E_{1/2}$ and the potential of the electrode are known. For the 1-electron reduction of Ox to Red, if the potential of the electrode is +1 V relative to $E_{1/2}$ at 298 K, the equilibrium constant will be exceedingly small, $\sim 1 \times 10^{-17}$ (as follows). Because there is such a small amount of Red, there is essentially no concentration gradient, and the solution is essentially 100% Ox.

$$E = E^\circ - \frac{RT}{nF}\ln\left(\frac{[Red]}{[Ox]}\right)$$

$$1V = E - E^\circ = -\frac{RT}{nF}\ln\left(\frac{[Red]}{[Ox]}\right)$$

$$1V = -\frac{\left(8.315\frac{J}{K x mol}\right)(298\,K)}{(1)\left(96{,}485\frac{C}{mol}\right)}\ln\left(\frac{[Red]}{[Ox]}\right)$$

$$-\frac{1V}{0.0257\frac{J}{C}} = \ln\left(\frac{[Red]}{[Ox]}\right)$$

$$e^{-38.9} = \frac{[Red]}{[Ox]}$$

$$\frac{[Red]}{[Ox]} = 1.26 \times 10^{-17}$$

Region B: "foot of the wave"

As the potential approaches $E_{1/2}$, the current begins to increase. The potential where the current begins to increase, sometimes called the "foot of the wave", occurs where small but significant amounts of Ox are reduced to Red. Here, a concentration gradient begins to be created between the solution (~100% Ox) and the solution near the electrode, which has a small amount of Red. If we choose the point **B** at −0.06 V as our foot of the wave, and estimate $E_{1/2}$ for the reduction of Ox to Red as −0.20 V, we can use the Nernst equation to calculate the [Red]/[Ox] at this point, shown as follows. At this point there are about four Red for every thousand Ox, which is still a small amount of reduced species but enough that Red has a small concentration gradient and begins to move away from the electrode.

$$E = E^\circ - \frac{RT}{nF}\ln\left(\frac{[Red]}{[Ox]}\right)$$

$$1V = (-0.06) - (-0.20) = -\frac{RT}{nF}\ln\left(\frac{[Red]}{[Ox]}\right)$$

$$-\frac{0.14V}{0.0257\frac{J}{C}} = \ln\left(\frac{[Red]}{[Ox]}\right)$$

$$e^{-5.45} = \frac{[Red]}{[Ox]}$$

$$\frac{[Red]}{[Ox]} = 4.30 \times 10^{-3}$$

Point C: $E_{1/2}$

The $E_{1/2}$ occurs halfway between the two waves, **D** and **F**, in the voltammogram, and for this reversible redox couple $E_{1/2} = E^\circ$, assuming that Ox and Red diffuse at the same rate. At this point, the Nernst equation shows that [Ox] = [Red]. The concentration of Red near the electrode is much higher than in the bulk of the solution, which is still essentially 100% Ox. This means that the analyte, Ox, can still diffuse more quickly towards the electrode to be reduced to Red, which is why the current has not reached its maximum.

Point D

In methods such as NMR and absorption spectroscopy, the peak maximum is what is reported, but in cyclic voltammetry the wave maxima (E_{p1} and E_{p2}) are not of great significance. Here the concentration gradients are very large between the region close to the electrode and the bulk. The analyte, Ox, is reduced as quickly as allowed by diffusion, which leads to the highest current. The difference in potential for E_{p1} and E_{p2} in a reversible system with a slow scan rate should be 59 mV. As a result, **D** (E_{p1}) should occur ~29.5 mV higher than $E_{1/2}$. In addition, for a reversible process the ratio of the peak currents (i_{pa}/i_{pc} indicated in red in Fig. 13.6.) should be ~1.

Region E

After the wave, the analyte, Ox, near the electrode surface has mostly been converted to Red, and thus the concentration of Ox is extremely low. In addition, the concentration of Red is increasing farther and farther away from the electrode surface. Thus, the observed current is due to Ox diffusing in from the bulk solution through this region of high Red concentration in order to become reduced.

Point F

At $E_{\lambda 1}$ the scan is reversed, and the reverse reaction in which Red is oxidized to Ox occurs. Initially, there is a large amount of Red near the electrode, and the current is high. Close to $E_{\lambda 1}$, the potential is insufficient to oxidize Red, but as the potential is increased, the Red near the electrode surface will begin to be oxidized. A diffusion gradient similar to what was observed in going to point **D** is established. Eventually the system goes through a point at **F** where Red is being oxidized as quickly as allowed by diffusion of Red to the electrode surface. As a result, **F** (E_{p2}) should occur ~29.5 mV lower than $E_{1/2}$.

If a system is electrochemically reversible, $E_{1/2}$ should not change with scan rate. However, the peak separation, $E_{p1} - E_{p2} = \Delta E_p$, will increase with increasing scan rate. A plot of the peak potential of one of the waves, say E_{p1}, versus the log of the scan rate should give a slope of −59 mV/decade, where a decade is one unit on the logarithmic scale. Therefore, only at very slow scan rates is one likely to observe $\Delta E_p = 59$ mV.

Another method for looking at reversibility is to examine the square of the peak current (i_p^2) as a function of scan rate (ν). Shown in Fig. 13.12 is a table with CV data for ferrocene in acetonitrile at room temperature. For a reversible process, a plot of i_p versus $\nu^{1/2}$ (or i_p^2 versus ν) should produce a straight line. Such a plot is shown in the bottom of Fig. 13.12. This plot is based on the Randles–Sevcik relation at room temperature, $i_p = (2.69 \times 10^5)n^{3/2}AD^{1/2}C_i\nu^{1/2}$, where n is the number of electrons transferred, D is the diffusion coefficient in cm²/s, C is the concentration in mol/cm³, and A is the electrode area in cm². Note that $E_{1/2}$ does not change with scan rate for this reversible system. The peak difference (ΔE_p) is about 59 mV at slow scan rates, and increases as the scan rate increases.

Until now, we have considered the electron transfer between the electrode surface and the analyte to be fast. If a redox process between the electrode and analyte is slow, the system is *electrochemically irreversible*. In the case of an electrochemically irreversible wave, the surface concentrations at the electrode are only dependent on the heterogeneous rate constant for the electron transfer from the electrode surface to the analyte. Under these conditions, little thermodynamic information is obtained from the experiment. In electrochemically irreversible systems, the irreversible waves will shift, depending on scan rate, and $E_{1/2} \neq E°$, and the ΔE_p values can be quite large. If the CV for the system appears reversible, or nearly so, to the eye, but the i_p^2 versus scan rate plot is nonlinear, the system may be referred to as *quasi-reversible*. Thermodynamic quantities such as $E_{1/2}$ should be cautiously interpreted in such cases.

A full discussion of mechanistic information that can be gleaned from a cyclic voltammogram is more related to kinetics and thus is beyond the scope of the current discussion. In addition, the variety of mechanisms that can be investigated by CV,

Scan rate (mV/s)	$E_{1/2}$ (V)[a]	ΔE_p	i_p (μA)[b]
5	0.45	59	10.3
50	0.45	71	32.5
100	0.45	85	46.2
200	0.45	104	65.2

[a]Versus a silver pseudo-reference. [b]Peak current for the oxidation.

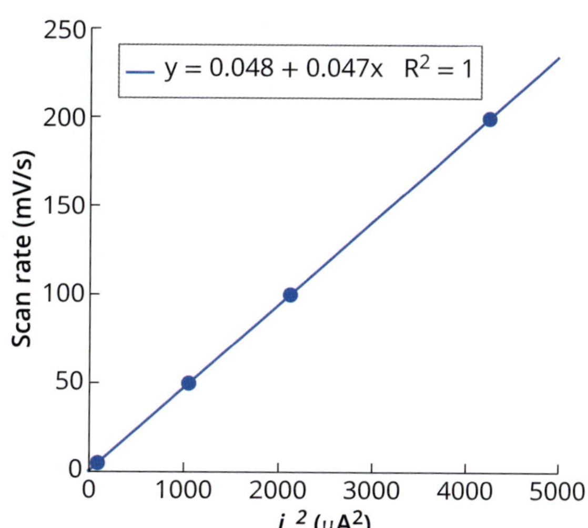

Fig. 13.12 (*top*) Tabulated cyclic voltammetry data for $FeCp_2^{0/+}$, using 3.3 mM ferrocene in acetonitrile with 0.1 M $\{NBu_4\}^+\{PF_6\}^-$ with a 0.4 mm diameter platinum disk working electrode at 25 °C. (*bottom*) Plot of the square of the oxidation peak current (i_p^2) versus the scan rate (*v*), which produces a straight line, showing that $FeCp_2^{0/+}$ is reversible.

such as electrochemical–electrochemical (EE), electrochemical–chemical–electrochemical (ECE), and chemical-electrochemical (CE) are quite numerous and are beyond the scope of this text. A detailed discussion of these mechanisms has been presented by Geiger.[13] Electrocatalytic reactions can also be examined by CV.[8] In these studies, a substrate and a catalyst are added to the solution. Large currents are observed as the catalyst reacts with a substrate, and they interact with the electrode. In such cases, the rate of the catalysis is directly related to the current observed.

A good report for a CV experiment will include the following.

(1) A plot of the current versus potential
(2) Solvent
(3) Supporting electrolyte
(4) Concentrations (supporting electrolyte and analyte)
(5) Temperature
(6) Atmosphere
(7) Potentials (initial, switching, and final)
(8) Reference for the reported potentials
(9) Scan rate (typically in mV/s)
(10) Number of sweeps
(11) Changes with sweep number
(12) Types of electrodes used

In some cases it is also helpful to describe the procedure used for cleaning the electrodes between runs. One of the classic references for redox chemistry in nonaqueous solutions was published in *Chemical Reviews* by Neil Connelly and Bill Geiger[10] in 1996. The review has extensive listings of oxidizing and reducing agents with their standard potentials, usually reported in nonaqueous solvents. All of the potentials of these redox agents are listed versus the $FeCp_2^{0/+}$ couple as 0 V, which is 0.40 V versus the normal hydrogen electron (NHE).

13.3 Acid–Base Reactions

Some of the first reactions presented in many general chemistry courses are acid–base reactions. One of the reasons for this is that they provide simple models for illustrating bonding in relation to reactivity. The reaction of H^+ with OH^- is frequently presented soon after the introduction of Lewis structures and the octet rule, with electron pair donation leading to a heterocovalent 2c2e bond. In this section we will discuss acid–base models very briefly, and then move on to the hard–soft acid–base (HSAB) principle, which was created in an attempt to overcome some limitations of understanding inorganic chemistry as reactions between simple Lewis acids and bases. HSAB is most often used in a qualitative sense, although quantitative methods have been developed. Applications and limitations of HSAB will also be discussed.

In general, one can divide HSAB into aqueous and nonaqueous applications. Originally, HSAB was designed to be used in water, and the solvent effects are large and often dominant. In nonaqueous solutions there still exists a preference of hard for hard and soft for soft, but the origins of the effect are sometimes different, being derived from preferential ionic versus covalent bonding between species, and generally not as large as in aqueous solutions.

13.3.1 Brief Review for Definitions of Acidity

It was Arrhenius who proposed that acids are substances that dissociate in aqueous solutions to produce H^+ and bases dissociate to produce OH^-. However, a free proton (H^+) is not the major species in aqueous solution; acid dissociation is most often considered to generate hydronium, H_3O^+. Even so, H^+ is frequently used as shorthand for acid in water.

In 1923, Johannes Brønsted and Tomas Lowry expanded on Arrhenius's definition. In Brønsted–Lowry theory, an acid (HA) and base (E) react to form a conjugate acid (HE^+) due to loss of proton and conjugate base (A^-), shown as follows.

$$HA + E \rightarrow A^- + HE^+ \quad \textbf{13.27}$$

For example, hydrochloric acid, HCl, fully dissociates in water to generate H^+ (in the form of hydronium) and Cl^- (Eq. 13.28). Chloride is the "conjugate base" of hydrochloric acid. Hydronium is the conjugate acid of water, which is acting as the base.

$$HCl + H_2O \rightarrow Cl^- + H_3O^+ \quad \textbf{13.28}$$

Hydrochloric acid is an example of a "strong acid"—a species that fully dissociates in water. Many acids do not fully dissociate in water and are therefore considered to be "weak acids". Acids are then classified by their acid dissociation constant, K_a. A capital K is used to represent an equilibrium constant, and K_a represents the equilibrium constant for the dissociation of a proton in a specific solvent, typically water. For the chemical equation of an acid dissociating in water, the equilibrium K_a is defined as shown in the following equation.

$$HA + H_2O \rightleftharpoons A^- + H_3O^+$$

$$K_a = \frac{[A^-][H_3O^+]}{[HA][H_2O]} \quad \textbf{13.29}$$

The concentration of water in aqueous solution is approximately constant and may be ignored. As a result, Eq. 13.29 is sometimes written as in Eq. 13.30 without the water being shown explicitly. Values for K_a are most easily presented as their negative base 10 log, $-\log_{10} K_a$, abbreviated as pK_a (Eq. 13.31).

$$K_a = \frac{[A^-][H^+]}{[HA]} \quad \textbf{13.30}$$

$$pK_a = -\log_{10} K_a \qquad \textbf{13.31}$$

The general trend is that the smaller (less positive) the pK_a value, the stronger the acid. A selection of pK_a values in common solvents (water and DMSO—dimethylsulfoxide, $OSMe_2$) as well as isolated in the gas phase (near-vacuum) are presented in Table 13.1. Note that even though there is some correlation between varying conditions, the values shift significantly as solvent effects can greatly change how acidic certain compounds are relative to others. Expectedly, water is a much stronger acid than toluene (H–CH_2Ph) in water and DMSO; however, water is a weaker acid than toluene in the gas phase! This dependence on solvent is a common theme in acid–base chemistry, regardless of the framework employed to understand the reactions. This solvent dependence is greatly complicated by the relative lack of precision in our understanding of solvent processes and their effect on reactivity. Nevertheless, some solvent-related arguments are presented in Table 13.1.

One fact often dwelled upon in general chemistry courses is that HF is a weak acid while the rest of the hydrohalides are strong acids. This is due to a combination of enthalpic and entropic effects. The H–X bond strengths go as HF > HCl > HBr > HI with gas phase acidities (ΔG° of heterolytic cleavage) of 364, 327, 317, and 309 kcal/mol, respectively. In other words, it is more energetically costly to break the HF bond in the gas phase relative to the other HX compounds. In addition, F^- has a larger charge density (larger charge/radius), which leads to a larger solvent response to the anion to counterbalance the charge. On top of this, the orbital size in fluoride leads to stronger hydrogen bonding with water. The hydrogen bonding and solvent polarization lead to an entropically costly structuring of the water around the fluoride anion, which makes HF dissociation less favorable than the other hydrohalides. As a result, acidity goes as HF << HCl < HBr < HI both in the gas phase and in aqueous solution.

In aqueous solution where H^+ is the most common acid, the order of base strength is the order of the pK_a of the conjugate acids, with the stronger bases having the higher pK_a of the conjugate acids. For example, aqueous ammonia will react with HF in an equilibrium reaction to form ammonium and fluoride (Eq. 13.32). The pK_a of HF is 3.45, whereas the pK_a of NH_4^+ is 9.25. Consequently, the right side of the following equation is strongly favored because F^- is a weaker base than NH_3 and NH_4^+ is a weaker acid than HF.

$$_{pK_a}\ NH_3(aq) + \underset{3.45}{HF(aq)} \rightleftharpoons \underset{9.25}{NH_4^+(aq)} + F^-(aq) \qquad \textbf{13.32}$$

The equilibrium constant for the reaction can be found by using Eq. 13.33, which indicates that the equilibrium lies considerably towards formation of products.

$$\log K_{eq} = pK_a(NH_4^+) - pK_a(HF) \qquad \textbf{13.33}$$

$$\log K_{eq} = 9.25 - 3.45 = 5.80$$
$$K_{eq} = 6.3 \times 10^5$$

This general-chemistry view of acid and base theory begins to experience difficulties when confronted with all the elements of the periodic table. In 1923, G. N. Lewis proposed a definition of acids and bases devoid of reliance upon the proton. He suggested that an acid is any species that accepts a pair of electrons, and a base is any species that donates a pair of electrons. The resulting product of the reaction of a Lewis acid with a Lewis base is sometimes called an "adduct". This Lewis definition encompasses the previous definitions; in the previous equation, ammonia serves as the lone-pair donor (base) and the proton as the lone-pair acceptor (acid). However, the Lewis definition is much broader and would also include the reaction of thiocyanate, SCN^-, (base) with Cd^{2+} (acid) to form the complex $[Cd(SCN)_4]^{2-}$ (Eq. 13.34).

$$Cd^{2+}(aq) + 4\ SCN^-(aq) \rightleftharpoons [Cd(SCN)_4]^{2-}(aq) \qquad \textbf{13.34}$$

As was seen in the previous example with ammonia and HF, an acid–base reaction will favor the formation of the weaker acid and base. By applying this principle to the far more general definition of Lewis acids and bases, we can determine that a stronger Lewis base (B_{strong}) will displace a weaker Lewis base (B_{weak}) from an acid (A) (Eq. 13.35).

$$:B_{strong} + A:B_{weak} \rightarrow :B_{weak} + A:B_{strong} \qquad \textbf{13.35}$$

The K_f of $[Cd(SCN)_4]^{2-}$ (aq) is 1.0 × 10^3, while that of $[Cd(CN)_4]^{2-}$ (aq) is 7.7 × 10^{16}. Thus, the addition of CN^- to $[Cd(SCN)_4]^{2-}$ should substantially favor the formation of $[Cd(CN)_4]^{2-}$ (Eq. 13.36)

$$[Cd(SCN)_4]^{2-}_{(aq)} + 4\ CN^-_{(aq)} \rightleftharpoons [Cd(CN)_4]^{2-}_{(aq)} + 4\ SCN^-_{(aq)} \qquad \textbf{13.36}$$

$$\log K_{eq} = pK_f([Cd(CN)_4]^{2-}) - pK_f([Cd(SCN)_4]^{2-})$$
$$\log K_{eq} = 16.9 - 3.0 = 13.9$$
$$K_{eq} = 7.9 \times 10^{13}$$

Table 13.1 Some selected pK_a values for common compounds in water.[10–13]

Compound	pK_a (water)	pK_a (DMSO)	pK_a (near vacuum)
HI	-5.2		314
HBr	-4.7	0.9	324
HCl	-2.2	1.8	333
HF	3.2	15	371
H-OC(O)Me	4.8	12.6	~345
HOMe	16	29	381
H_2O	15.7	31.4	391
H-CH_2Ph	41	43	379
H-alkyl	~50	~56	~420

Using the Lewis definition, we are no longer restricted to just protons as our acid. This provides the opportunity to compare the reactivity of vastly different acids with bases, which is where things can become interesting. For example, a series of stability constants for the reactions of Ag^+, H^+, and Ca^{2+} with a variety of different Lewis bases in water indicates that the Lewis acids have different affinities for a given Lewis base. In addition, the relative affinity of these Lewis acids for a Lewis base does not remain consistent as the Lewis base is varied. The addition of ammonia to CaF^+ does not result in the formation of $Ca(NH_3)^+$, suggesting that fluoride is the stronger base. In contrast, the fluoride in AgF is displaced by ammonia to make $Ag(NH_3)^+$. Examining similar series with I^- produces yet a different ordering of Lewis acidity for H^+, Ag^+, and Ca^{2+} as summarized here. In other words, ammonia is the strongest base towards H^+ and is the weakest base for Ca^{2+}, but I^- is a stronger base toward Ag^+ than it is for H^+.[14]

$$H^+ > Ag^+ > Ca^{2+} \text{ toward } NH_3(aq)$$

but

$$H^+ > Ca^{2+} > Ag^+ \text{ toward } F^-(aq)$$

and

$$Ag^+ > Ca^{2+} > H^+ \text{ toward } I^-(aq)$$

In order to explain these differences, Ralph Pearson proposed classification of acids and bases as being either "hard" or "soft", and developed the hard–soft acid–base (HSAB) principle.

13.3.2 Hard–Soft Acid–Base (HSAB) Principle in Aqueous Solution

To explain similar series in which certain acids bind with higher thermodynamic favorability to certain bases, all chemical species were classified as being "hard" or "soft". A "hard" species was one with low polarizability, whereas "soft" species were easily polarizable. In general, hard acids and bases have an atom with a high electronegativity as the center of reactivity. So, F^- is a very hard base, and Al^{3+} would be an example of a hard acid, since the metal has a very high positive charge and is relatively electronegative. Conversely, systems with low electronegativity are soft, such as the soft base SH^- and the soft acid Tl^+. In some cases, it is a specific orbital that is considered; for example, ethylene is considered a soft base since the electrons in the π-system are used in bonding and are relatively polarizable despite being on an electronegative atom, carbon. Finally, the charge on an acid or base also plays a role in the hard or soft nature of an acid or base. For example, Co^{3+} will be a harder acid than Co^{2+} due to the higher charge.

Here it is important to mention an important fact: the HSAB Principle was designed for and works best in water or other similar polar solvents. The reasons for this will be discussed later, and HSAB in water is still used to understand effects in inorganic complexes. In particular, this principle may be applied to bioinorganic systems in general and in the understanding of toxicology.

What Pearson has described as the hard–soft acid–base principle is that *hard acids display a thermodynamic preference for binding to hard bases* and *soft acids display a thermodynamic preference for binding to soft bases.* In the exchange reaction, Eq. 13.37, if either the acids (A/A′) or bases (B/B′) have the same strength, there is a thermodynamic preference for formation of A:B′ and A′:B, if A and B′ are hard (*h*) while A′ and B are soft (*s*).

$$A : B + A' : B' \rightleftharpoons A : B' + A' : B \qquad \textbf{13.37}$$

$$hs + sh \rightleftharpoons hh + ss \qquad 0 > \Delta H^\circ$$

Through looking at equilibria, Pearson designated acid and bases as either hard, soft, or borderline (Fig. 13.13). Note that increasing the oxidation state increases the hardness of the species. Numerous examples are listed, such as Fe^{3+} (hard) versus Fe^{2+} (borderline), SO_3 (hard) versus SO_2 (borderline), and Cu^{2+} (borderline) versus Cu^{1+} (soft).

13.3.3 Solvation Effects, Complex Ions, and the Hard–Soft Acid–Base (HSAB) Principle

Typically, the equilibrium constants in water are used to classify Lewis acids, and as a result, hydration energies can affect the values. In the gas phase, fluoride generally forms stronger bonds than iodide. However, in solution, iodide can bind more strongly than fluoride to some soft metals, such as Ag^+, due to the high solvation energy of F^- relative to I^-. Water is both a hard acid and a hard base. Consequently, it will moderate some of the acidity of hard acids more so than soft acids, and hard bases more than soft bases.

In solvents less polar than water, the solvation energies are typically smaller and thus have a less substantial impact. Formation constants suggest that binding goes as Cl > Br > I even for soft ions like Rh(I) and Pt(II) in less polar solvents. Pearson gives as another example, the reaction of $Ag(NH_3)^+$ with $Li(PH_3)^+$ (Eq. 13.38). This reaction is unfavorable in the gas phase but is exothermic, as expected from the HSAB principle, when run in water due to heats of hydration of NH_3 and PH_3. A more in-depth discussion of solvation effects and HSAB is included in Pearson's book, *Chemical Hardness*.[14]

$$Ag(NH_3)^+ + Li(PH_3)^+ \rightleftharpoons Ag(PH_3)^+ + Li(NH_3)^+ \qquad \textbf{13.38}$$

$$\Delta H^\circ = -16\,\text{kcal/mol (in water)}$$

$$\Delta H^\circ = +11\,\text{kcal/mol (gas phase)}$$

Hard acids	Borderline	Soft acids
H^+, Li^+	Na^+	K^+
Be^{2+}, Mg^{2+}, Ca^{2+}		
BF_3, BCl_3, $B(OR)_3$	BBr_3, BMe_3	BH_3, Tl^+, Tl^{3+}
Al^{3+}, $Al(CH_3)_3$, $AlCl_3$, AlH_3		
Cr^{3+}, Mn^{2+}, Fe^{3+}, Co^{3+}	Fe^{2+}, Co^{2+}, Ni^{2+}, Cu^{2+} Zn^{2+}, Rh^{3+}	Cu^+, Ag^+, Au^+ Cd^{2+}, $Hg_2{}^{2+}$, Hg^{2+} Pd^{2+}, Pt^{2+}, Pt^{4+}
SO_3	SO_2	

Hard bases	Borderline	Soft bases
F^-, Cl^-	Br^-	H^-, I^-
H_2O, OH^-, O^{2-}		H_2S, SH^-, S^{2-}
ROH, RO^-, R_2O, ^-O_2CR		RSH, RS^-, R_2S
$NO_3{}^-$, $ClO_4{}^-$, $SC\underline{N}^-$	$\underline{N}O_2{}^-$, $N_3{}^-$, N_2	$\underline{S}CN^-$, CN^-, RNC, CO
$CO_3{}^{2-}$, $SO_4{}^{2-}$, $PO_4{}^{3-}$	$SO_3{}^{2-}$	
NH_3, RNH_2	$PhNH_2$, pyridine	PR_3, π-systems (e.g., ethylene, benzene)

Fig. 13.13 Qualitative classification of some common Lewis acids and bases as hard, soft, or borderline.

Classifying new ligands and some metal centers as hard or soft acids or bases is not without complications. For example, one must somehow account for differences in solvation effects. In addition, hardness changes with oxidation state, and some metals will then oxidize or reduce the ligands being used to classify them. For example, I^{7+} and Cr^{6+} are probably hard, but using I^- to test this will result in redox chemistry occurring as opposed to a Lewis acid–base interaction. The choice of the reference acid or base can significantly impact the ordering of hardness for different acids or bases, and as such, a different, quantitative methods for the determination of hardness were desirable, which will be discussed later.

13.3.4 Applications of the HSAB Principle to Aqueous Metal Complexes and Toxicology

The HSAB concept was originally used to describe, and perhaps still works best for, aqueous systems. In water, solvation energies can make a large contribution to the thermodynamics of the acid–base reaction, especially the entropic term. Consider the hard–hard reaction of an Al^{3+} ion with F^- in water. Naturally, the aluminum ion is solvated by water and probably exists largely as $[Al(OH_2)_6]^{3+}$. The fluoride ion perhaps makes the strongest hydrogen bonds of any known species to water. Because of the fluoride ion's charge, small size, and substantial hydrogen bonding, it structures the water around it. In other words, solvated fluoride lowers the entropy of the solution. When fluoride coordinates to aluminum, the charge on fluoride lessens due to electronegativity equalization, and the water becomes less structured. Further, to coordinate fluoride, aluminum releases water. Consequently, F^- coordination to Al^{3+} is very entropically favorable. In fact, *this entropic effect due to solvation is the driving force for F^- to bind to Al^{3+} in water. The reaction itself is slightly endothermic* despite its being difficult to imagine a more canonical example of a "hard–hard" interaction than Al^{3+} interacting with F^-. The thermodynamic values for the reaction are shown in Eq. 13.39, and is the entropic term ($-T\Delta S°$) that results in the favorable free energy.

$$Al^{3+}(aq) + F^-(aq) \rightarrow AlF^{2+} \qquad 25\,°C \tag{13.39}$$

$$\Delta G^\circ = -8.4\,\text{kcal/mol}$$
$$\Delta H^\circ = 1.1\,\text{kcal/mol}$$
$$-T\Delta S^\circ = -9.5\,\text{kcal/mol}$$

Contrast this with the soft–soft reaction of Hg^{2+} with I^- in aqueous solution. The solvation of iodide by water is quite weak due to its large size and poor hydrogen-bonding ability. Consequently, solvation of iodide is probably unfavorable, as it disrupts the strong water–water interactions and forms few interactions with water itself. The entropy for the reaction of soft Hg^{2+} with soft I^- allows more structured water–water interactions, as the two soft ions interact with each other and take up less solvent volume, and thus $-T\Delta S°$ is slightly positive (unfavorable). However, the enthalpy is relatively large and exothermic (Eq. 13.40). Part of this is the additional water–water interactions that can occur as the soft species interact with one another.[c]

$$Hg^{2+}(aq) + I^-(aq) \rightarrow HgI^+ \qquad 25\,°C \qquad \textbf{13.40}$$

$$\begin{aligned}\Delta G^\circ &= -17.5\,\text{kcal/mol}\\ \Delta H^\circ &= -18.0\,\text{kcal/mol}\\ -T\Delta S^\circ &= +0.5\,\text{kcal/mol}\end{aligned}$$

As can be readily gleaned from the previous discussion, understanding acid–base interactions in water *requires* consideration of solvation effects. In aqueous solution, the HSAB principle seems to work quite well and has been applied successfully to fields such as toxicology and bioinorganic chemistry, which are concerned largely with aqueous systems.

One can see these effects when examining the physiological activity of thallium,[15–17] for example. Thallium(I) has an ionic radius very similar to another soft metal, potassium(I), at 1.47 and 1.52 Å, respectively. As a result, thallium can take the place of potassium in certain enzymes, such as pyruvate kinase and succinate dehydrogenase. Both of these enzymes are involved in glucose metabolism, and the inclusion of thallium inhibits enzymatic activity. The metal interferes with ATPase by taking the place of potassium and has a high affinity for sulfhydryl (SH) groups from many proteins and other biologically relevant molecules. Furthermore, thallium can interact with disulfide bonds leading to disruption of cysteine cross-links in keratin, which leads to alopecia (hair loss) and Mees' lines (white lines across the fingernails). The disruption of cysteine cross-links also affects glutathione production and results in accumulation of lipid peroxides in the brain. Thallium also damages the ribosome, interfering with protein synthesis. These are only some of the known effects, and perhaps needless to say, the eventual outcome is death. Note that the vast majority of these effects can be traced to soft–soft interactions between thallium and sulfur or thallium and the typical binding sites of potassium.

The HSAB principle has also been applied to organic molecules and their interactions within the body with great success. In general, all the same arguments apply, and the HSAB principle seems quite useful for these systems as well.

[c] Another important factor is that mercury does not alleviate lone pair bond weakening in water perhaps even as well as will hydrogen bonding to water. LPBW will be discussed shortly in the context of HSAB principle.

13.3.5 Parameterizing Hardness, Electronegativity, and Lewis Acidity in Nonaqueous Media[15] (Advanced Topic)

Originally, HSAB was used as a qualitative principle for understanding reactivity. Initial attempts at quantifying HSAB were based on equilibrium constants, which were in many cases difficult to measure, or were reference-dependent. Other, more easily obtained, methods for measuring Lewis acidity and basicity were sought. In response, Pearson developed scales for "absolute hardness" based on ionization energies and electron affinities. There are several related methods for measuring Lewis acidity and basicity, and here one additional method, the Guttmann–Beckett acceptor number, will be briefly described.

While the HSAB principle was originally designed for aqueous solutions, these methods move the species to other media in most cases. As a result, solvation effects become less important, and the details of the bonding between the acids and bases becomes paramount to the discussion.

13.3.5.1 Absolute Hardness and Electronegativity[15] (Advanced Topic)

A typical plot of the energy of a chemical system (atom or molecule) as a function of the relative number of electrons is depicted in Fig. 13.14. The points shown are experimentally observable, and the line is simply extrapolation between points. The addition of an electron to a chemical system often has a small and favorable effect on the energy. In contrast, removal of an electron often causes the energy to rise more sharply.

It was proposed that the slope of the line in Eq. 13.41 at any point provides what is known as the "chemical potential" (μ)—a measure of the tendency for a molecule (or atom, ion, and so on) to lose or donate electrons. Due to electronegativity equalization, the principle that atoms in a molecule have equal electronegativity values at their equilibrium bond distances,

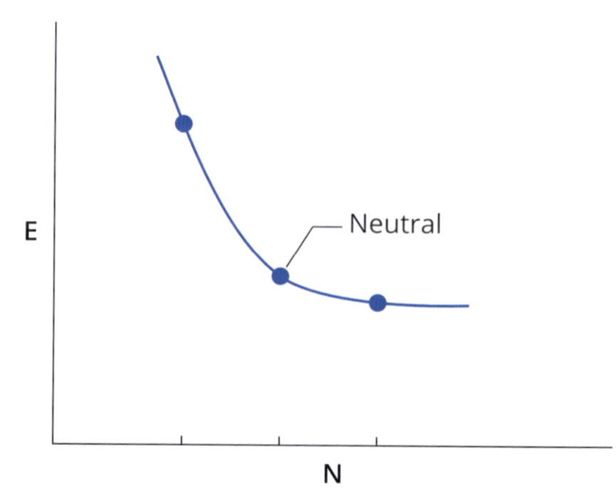

Fig. 13.14 Plot of electronic energy (E) versus number of electrons (N) for a chemical species.[18]

the chemical potential is the same everywhere in a molecule. Thus, the substitution of an electronegative group, such as replacing a hydrogen in a molecule with a fluorine, increases the electronegativity of the entire molecule.

$$\mu = \left(\frac{\partial E}{\partial N}\right) \tag{13.41}$$

Consider the general acid–base reaction in Eq. 13.42, where a Lewis acid, A, reacts with a Lewis base, B, to produce an adduct, A:B. The reaction occurs only if $\mu_B > \mu_A$; in other words, the reaction will occur if B is more willing to donate electrons. For example, consider the Lewis acid BH_3, which will quickly react with a source of H^- such as NaH to yield BH_4^-. However, no reaction will take place between BH_3 and the isoelectronic He atom, as the He atom is less willing to donate electrons than H^-.

$$A + :B \rightarrow A:B \tag{13.42}$$

A serious problem with this approach is that we do not know μ, the slope of the plot, by any experimental method. However, the mean slope for the change (N – 1) to N is the negative of the ionization energy for the species, –IE. In addition, the mean slope from N to (N + 1) is the negative of the electron affinity, –EA. There is a mathematical trick for dealing with such a situation called the *finite differences method*, which is often presented in courses on differential equations. Application of this to the problem of solving for the slope at *N* from two related slopes produces Eq. 13.43, in which the slope at point *N* is the average of IE and EA for the species. This is the definition of Mulliken (sometimes called "absolute") electronegativity (Chapter 3). The absolute electronegativity, χ, is the propensity of a chemical species to add or keep electrons. The larger the value of χ, the more likely a species is to accept electron density, and the more exothermic the reaction when adding a pair of electrons.

$$-\mu = -\left(\frac{\partial E}{\partial N}\right) \approx \frac{(IE + EA)}{2} = \chi \tag{13.43}$$

Consider the two-step reaction by which a pair of electrons is added to a Lewis acid, A^+, one at a time. The absolute electronegativity is a measure of the attraction of A^+ for a pair of electrons, which can also be thought of as the average of the attraction of both A^+ and A for a single electron. In the first step, A^+ is reduced to A, which has an enthalpic barrier of –IE, as it is the reverse of the ionization energy. In the second step, A gains an electron to yield A^-, which has a barrier of –EA. As a result, the overall energy required is –(IE + EA), Eq. 13.44. In other words, absolute electronegativity is related to half the energy released on addition of a pair of electrons.

$$\begin{array}{ll} A^+ + e^- \rightarrow A & \Delta H = -IE \\ A + e^- \rightarrow A^- & \Delta H = -EA \\ \hline A^+ + 2e^- \rightarrow A^- & \Delta H = -(IE + EA) \end{array} \tag{13.44}$$

Returning to the plot in Fig. 13.14 and the hard–soft concept, the rate of change of the slope is defined as the "absolute hardness" (η)—a measure of how difficult it is to change the electron density (Eq. 13.45). Again, using the finite differences method, it is possible to estimate the second partial derivative. In this case, the hardness is one-half the difference between the ionization energy and the electron affinity. The absolute hardness is the propensity to keep electron density as it is; in other words, a resistance to deformation or change in total electron density. A larger η means that there is a higher propensity to keep the electron density the same, and this tendency to stay the same means that the species in question is "hard". A small η means the species readily undergoes electron transfer and polarization and therefore is "soft".

$$\eta = \frac{1}{2}\left(\frac{\partial^2 E}{\partial N^2}\right) = \frac{1}{2}\left(\frac{\partial \mu}{\partial N}\right) = \frac{(IE - EA)}{2} \tag{13.45}$$

A short list of absolute electronegativity and absolute hardness values are shown in Table 13.2. From this discussion, we can reach the following general conclusions:

Small atoms like fluorine often:

- hold electrons tightly
- have unfilled orbitals to accept electrons (small EA)
- have high values of η
- are referred to as "hard"

Large atoms like iodine often:

- hold outer electrons loosely
- have small differences in energy between currently occupied orbitals and incoming electrons (relatively high EA)
- have low values of η
- are referred to as "soft"

Due to their relationship, if both χ and η are known, the amount of charge transferred between an acid and a base can be estimated. The hardness of the acid (η_A) and the base (η_B) can be calculated in the adduct A:B by using Eq. 13.46, in which ΔN is the partial charge transferred, and $\eta_A°$ and $\eta_B°$ are the values for the hardness of the isolated acid and base before the reaction.

$$\begin{aligned} \mu_A &= \mu_A^\circ + 2\eta_A \Delta N \\ \mu_B &= \mu_B^\circ + 2\eta_B \Delta N \end{aligned} \tag{13.46}$$

The concept of electronegativity equalization tells us that in a Lewis acid–base reaction, the formation of the adduct will result in an increase in μ_A and a decrease in μ_B until they reach the point at which they are equal, μ_{AB}. This allows us to derive the equation for ΔN presented in Eq. 13.47. As may be expected, the difference in the absolute electronegativity values for the isolated species ($\chi_B° - \chi_A°$) is what drives the electron transfer during adduct formation, with larger electronegativity differences between species resulting in larger charge transfer. The amount of charge transferred is tempered by the

Table 13.2 Examples of absolute electronegativity and hardness.[15]

Species (ion, atom or molecule)	$\chi°$, eV	η, eV
Na^+	26.2	21.1
Ag^+	14.6	6.9
I^+	14.8	4.3
Mn^{2+}	24.4	9.3
Ca^{2+}	31.6	19.7
Fe^{2+}	23.4	7.3
Zn^{2+}	28.8	10.8
Pd^{2+}	26.2	6.8
Hg^{2+}	26.5	7.7
Al^{3+}	74.2	45.8
Sc^{3+}	49.3	24.6
Tl^{3+}	40.3	10.5
SF_6	7.9	7.4
BF_3	7.8	7.8
SO_3	7.2	5.5
N_2	7.0	8.6
Cl_2	7.0	4.6
SO_2	6.7	5.6
H_2	6.7	8.7
O_2	6.3	5.9
PF_3	5.7	6.7
CO	6.1	7.9
I_2	6.0	3.4
Pt	5.6	3.5
CH_3I	4.9	4.7
HCl	4.7	8.0
C_5H_5N	4.4	5.0
C_2H_4	4.4	6.1
H_2S	4.3	6.4
C_6H_6	4.0	5.2
PH_3	4.0	6.0
H_2O	3.1	9.5
CH_3Cl	3.8	7.5
V	3.6	3.1
NH_3	2.9	7.9
$As(CH_3)_3$	2.8	5.5
$P(CH_3)_3$	2.8	5.9
$S(CH_3)_2$	2.7	6.0
$O(CH_3)_2$	2.0	8.0
$N(CH_3)_3$	1.5	6.3

sum of the hardness values for the acid and base ($\eta_A + \eta_B$), and electron transfer is smaller regardless of whether the acid or base is hard.

$$\Delta N = \frac{(\chi_B^\circ - \chi_A^\circ)}{2(\eta_A + \eta_B)} \qquad \textbf{13.47}$$

As an example of this charge transfer in a Lewis acid–base reaction, consider the interaction of I_2 with benzene. Iodine is a brown color in some solvents, such as in EtOH or Et_2O, but in some solvents such as C_6H_6, I_2 dissolves to give violet solutions (Fig. 13.15) due to interactions between solvent and halide: a "halogen bond".[d] There is substantial evidence that I_2 interacts with the π-system of benzene so that the I–I bond is nearly perpendicular to the plane of the benzene ring (shown in the top right of Fig. 13.15). Calculations (B3LYP/SDD) show an interaction between the I_2 σ-bonding orbital and the π-system. The HOMO−9, which shows this interaction, and the LUMO are at the bottom of Fig. 13.15.

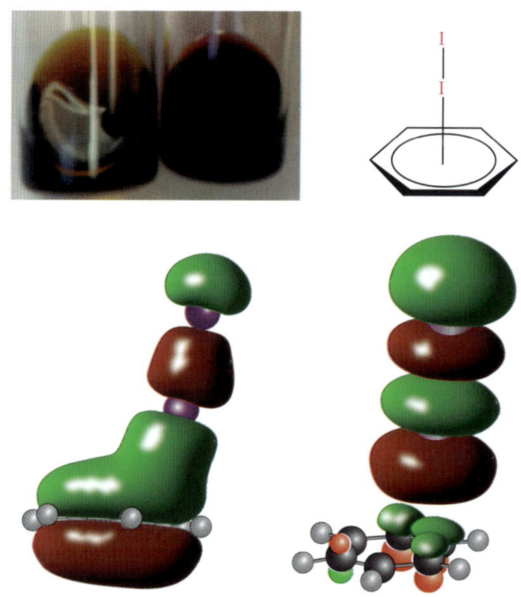

Fig. 13.15 (*top left*) Iodine solutions of EtOH (left) and C_6H_6 (right). (*top right*) Geometry of the I_2/benzene complex. (*bottom left*) The HOMO-9 of the I_2/benzene complex showing an interaction between the benzene π-system and the σ-bonding orbital of I_2, which is largely based on the benzene ring. (*bottom right*) The LUMO of the I_2/benzene complex, which is strongly antibonding and based on I_2.

---

[d] A "halogen bond" is essentially where a donor of some sort donates into the σ* of an X–E bond, where X is a halide and E is another halide or some other atom. For example, an amine, NR_3, may donate into I–Cl to give a stable complex, $R_3N{:}\!\rightarrow$I–Cl. There is a large and expanding literature on the exact nature of such interactions. For example, see Robertson, C. C., Perutz, R. N., Brammer, L., and Hunter, C. A. *Chem. Sci.* 2014, *5*, 4179–4183. Also see Moghadasnia, M. P., Eckstein, B. J., Balaich, G. J., and McGuirk, C. M. *Cryst. Growth Des.* 2023, *23*, 5066–5077.

First, which species is the electron donor, and which is the acceptor? In short, the species with the higher absolute electronegativity will act as the electron acceptor. In order to determine this, we need the ionization energy (IE) and electron affinity (EA) for the two species, which are available on the NIST website. Benzene is one of the atypical systems where it actually costs energy to add an electron, and therefore it has a negative value for EA.

$$I_2 : IE = 9.31\,eV, EA = 2.52\,eV$$
$$C_6H_6 : IE = 9.24\,eV, EA = -0.78\,eV$$

Using these values, the absolute electronegativity for these two species can be calculated (Eqs. 13.48 and 13.49). The I_2 molecule is more electronegative, and electrons will flow towards this molecule, so it is behaving as the Lewis acid while benzene serves as the Lewis base.

$$\chi_{I_2} = \frac{(IE + EA)}{2} = \frac{(9.31 + 2.52)}{2} = 5.92\,eV \quad \textbf{13.48}$$

$$\chi_{C_6H_6} = \frac{(IE + EA)}{2} = \frac{(9.24 - 0.78)}{2} = 4.23\,eV \quad \textbf{13.49}$$

Using absolute hardness and electronegativity, one can calculate the amount of electron density transferred from benzene to I_2. The difference in electronegativity causes electron density to flow from one compound to the other until equalization occurs. However, the amount of electron density that actually moves between the molecules is deterred by the hardness of the two species. The harder the molecules involved, the more energy required to change their electron density. The calculation suggests that about 0.10 electrons will be transferred to iodine (Eq. 13.50). This is somewhat smaller than the value suggested experimentally (^{129}I Mössbauer spectroscopy), which suggests that about 0.26 electrons are transferred from benzene to I_2.[19,e]

$$\Delta N = \frac{(\chi^{\circ}_{benzene} - \chi^{\circ}_{iodine})}{2(\eta_{iodine} - \eta_{benzene})} = \frac{(4.23 - 5.92)}{2(3.39 + 5.01)} = -0.10 \quad \textbf{13.50}$$

Pearson has suggested that a relationship exists between MO theory and the HSAB principle that utilizes the expanded Koopmans's theorem approximation (Box 6A). For this, the ionization energy (IE) is equivalent to the negative of the energy of the HOMO, $-E_{HOMO}$, and the electron affinity (EA) is the energy of the LUMO, E_{LUMO}. This means that the HOMO-LUMO gap, E_{HL}, is equivalent to the difference between the IE and EA (Eq. 13.51). As previously defined, the absolute hardness is this difference divided by 2, therefore the hardness is simply one-half the HOMO-LUMO gap (Eq. 13.52). Likewise, the absolute electronegativity is the average energy of the HOMO and LUMO (Eq. 13.53).

$$E_{HL} = E_{LUMO} - E_{HOMO} = IE - EA \quad \textbf{13.51}$$

$$\eta = \frac{(IE - EA)}{2} = \frac{(E_{LUMO} - E_{HOMO})}{2} \quad \textbf{13.52}$$

$$-\chi = \frac{(IE + EA)}{2} = \frac{(E_{LUMO} + E_{HOMO})}{2} \quad \textbf{13.53}$$

Pictorially, the absolute electronegativity and hardness are shown in Fig. 13.16, with the electronegativity being the midpoint between the HOMO-LUMO gap and the hardness being the range in energy from the midpoint to either the HOMO or LUMO energy. A soft molecule, atom, or ion will have a small HOMO-LUMO gap and may be more prone to mixing of low-lying excited states. Hard molecules, atoms, or ions conversely will have large HOMO-LUMO gaps.

Can this MO-based picture of the HSAB principle be used to explain why "hard should prefer hard" and "soft should prefer soft"? In part, yes. To illustrate, consider how two species of the same electronegativity but different values for hardness might interact (Fig. 13.17). By the previous rationale, the "hard" species has a large HOMO-LUMO gap and the "soft" species has a small HOMO-LUMO gap, but the frontier orbitals will be centered around the same midpoint because they have the same electronegativity. We can define three different interactions that will occur:

(1) *There will be partial transfer of electron density from the HOMO of each species to the LUMO of the other* (shown with "1" in Fig. 13.17). This leads to electron delocalization and an energy stabilization that, by perturbation theory, decreases with the difference in energy between the HOMOs and LUMOs of the two species. In other words, as the hardness decreases and the HOMO-LUMO gap decreases, there will be more delocalization. This means that soft–soft interactions have more delocalization and therefore will be more covalent.

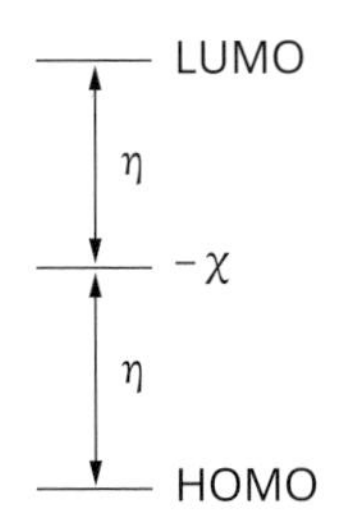

Fig. 13.16 Definition of absolute electronegativity (χ) and absolute hardness (η) relative to the HOMO-LUMO gap.

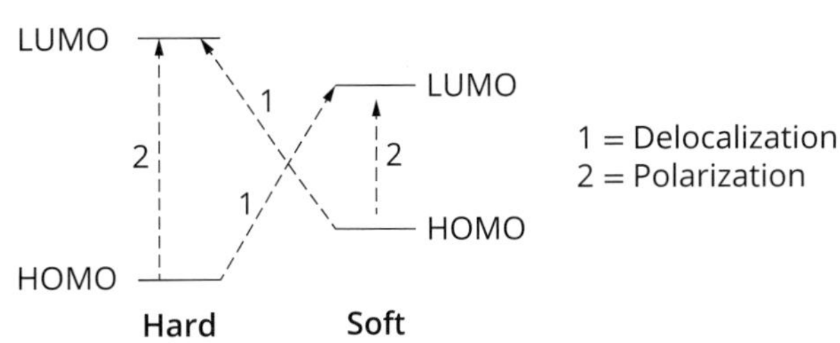

Fig. 13.17 HOMO-LUMO orbitals in hard and soft species.

[e] There is reason to be somewhat dubious of the precision of electronegativity equalization for dative systems, and the calculation here uses that assumption. See Chapter 3 for a detailed discussion.

(2) *There will be mixing of the HOMO with the LUMO on the same species to polarize the system* (shown with a "2" in Fig. 13.17). In some cases, this polarization can be used to lower repulsive forces between the two species coming into contact. Again, this is encouraged by a small HOMO–LUMO gap for both interacting species and is strongest for the soft–soft interaction.

(3) Now leaving Fig. 13.17, what if the electronegativities are very different, and the two species are near equal hardness? Naturally, this leads to large electron transfer (Eq. 13.47) between the two compounds and ionic bonding. It is common for hard cations, such as lithium, to have low electronegativities and for hard anions, such as fluoride, to have high electronegativities, which leads to very polar and very strong bonding between these species. This is the hard–hard interaction, which is often more polar. Note, the hardness tempers electron transfer; overcoming large hardness requires large electronegativity differences.

The use of absolute hardness to classify HSAB has the advantages of being more quantifiable for a larger number of compounds and not depending upon measuring equilibria. In many cases, absolute hardness seems to capture the spirit of the HSAB principle as well aiding in predicting reactivity. However, absolute hardness does use many approximations. Koopmans's theorem was used to develop the MO picture discussed previously, which relates the ionization energy and electron affinity of a species to the energy of the HOMO and LUMO, respectively. Furthermore, the entire system is only energy-based, while bond energies depend heavily on both energy *and* overlap parameters. Nevertheless, absolute hardness and electronegativity are concepts often discussed in the chemical literature that are used as a basis for discussing reactions, and all chemists should be aware of these concepts.

13.3.5.2 Gutmann–Beckett Method and the Acceptor Number (Advanced Topic)

A quite different method for examining the Lewis acidity of chemical species was originally proposed by Viktor Gutmann in 1975.[20] Gutmann had previously published a method for the determination of the *donor number* (DN) of solvents based on the ΔH of the reaction between the solvent and the strong Lewis acid $SbCl_5$. Gutmann's interest was in quantifying the acidity and basicity of solvents, but his method was generalized to other systems, first by Beckett and now by many other research groups. The Gutmann–Beckett method, as it is now generally called, examines the ^{31}P NMR chemical shift of $OPEt_3$ in the presence of a compound that can act as a Lewis acid towards the oxygen lone pairs. Initially, Gutmann found that for a given solvent there were definite effects on the chemical shift based on the concentration of $OPEt_3$. Based on his definition of donor numbers, Gutmann chose to scale his "acceptor numbers" (AN) by setting the chemical shift of the phosphorus in the adduct $Et_3PO–SbCl_5$ as 100.

In 1996, Beckett and coworkers published a simplified method for determination of AN based directly on the ^{31}P NMR chemical shift (δ_{sample}) of $OPEt_3$ in the presence of an excess of the Lewis acid in $CDCl_3$.[21] Beckett and others used this method with many different Lewis acids; the values for boron-based compounds are particularly numerous.[22] A few examples are shown in Table 13.3. Comparison of these numbers depends on whether the measurements were made in the same solvents and whether the sample was corrected for changes in the chemical shift of $OPEt_3$ under the experimental conditions. A reasonable estimate of the errors, depending on the exact nature of the measurements, is perhaps 5%. However, additional error in the measurements can occur, especially if the Lewis acid is very bulky.[22]

The Gutmann–Beckett measurements are based on interactions with a hard base, the oxygen atom of $OPEt_3$. A soft version of the Gutmann–Beckett measurement was introduced by Britovsek in 2005. in which the difference in the ^{31}P chemical shift for free $SPEt_3$ was compared to the same compound in the presence of a Lewis acid.[23] By comparing, for example, $AlCl_3$ and $GaCl_3$ with the hard base, $OPEt_3$, and the soft base, $SPEt_3$, one can determine that the gallium compound is softer than the aluminum (Table 13.4). The larger chemical shift difference suggests a stronger interaction. Specifically, aluminum has a larger chemical shift difference with $OPEt_3$ (Al: 36 versus Ga: 34.9 ppm), while gallium has a larger chemical shift difference with $SPEt_3$ (Ga: 10.5 versus Al: 9.2 ppm).

13.3.5.3 Additional Means of Quantifying the HSAB Principle (Advanced Topic)

The previous two methods (absolute electronegativity/hardness and Gutmann–Beckett) represent two ways that have been developed to quantify HSAB relationships. While both methods have their limitations, their basis on relatively easy-to-obtain experimental data is a mutual strength. Other methods for quantifying the HSAB principle have been developed. Pearson also developed a "class a and b" system for acids and bases that is not synonymous with hard and soft because it includes additional parameters other than just

Table 13.3 Acceptor numbers (AN) for a selection of Lewis acids using $OPEt_3$.

Compound	AN
BF_3	84.0
BCl_3	105.7
BBr_3	109.3
BI_3	114.9
$B(OMe)_3$	23.1
$B(NMe_2)_3$	9.1
$B(C_6F_5)_3$	81.8
$AlCl_3$	87
$TiCl_4$	70

Table 13.4 Comparison of δ_{OPEt3} and δ_{SPEt3} for some Lewis acids.[23]

Lewis Acid	δ_{OPEt3}	$\Delta\ \delta_{OPEt3}{}^{a}$	δ_{SPEt3}	$\Delta\ \delta_{SPEt3}{}^{a}$
None	46.2	0	53.5	0
$AlCl_3$	82.2	36.0	62.7	9.2
$GaCl_3$	81.1	34.9	64.0	10.5
$B(C_6F_5)_3$	76.0	29.8	58.0	4.5
$Al(C_6F_5)_3$	76.0	29.8	61.2	6.7
$Ga(C_6F_5)_3$	73.1	26.9	61.3	7.8
$AlMe_3$	66.7	20.5	58.1	4.6
$AlEt_3$	65.9	19.7	58.3	4.8
BEt_3	53.7	7.5	53.5	0
BPh_3	72.0	25.8	53.5	0

[a] $\Delta\ \delta = \delta(\text{Lewis acid}) - \delta(\text{none})$

hardness.[24] Drago developed an alternative system called the ECW model.[25] This method uses E and C parameters as approximations for the ionic and covalent contributions to bonding and relates these parameters to the ΔH_{rxn} for a Lewis acid–base reaction.

13.3.6 Predictions of the HSAB Principle in Nonaqueous Media (Advanced Topic)

Keep in mind that the HSAB principle is not, in reality, a "theory" by Pearson's own definition.[24] It is merely an observation that often fits the data. In general, we recommend applying the HSAB principle liberally in aqueous systems, for which it was originally designed, and cautiously elsewhere. The HSAB principle does not explain many chemical reactions well, especially outside of aqueous systems, and yet it often seems to be applied to these systems. As mentioned in Chapter 10, early transition metals are often oxophilic (prefer to bind to oxygen), while many later metals can be said to be thiophilic (prefer to bind to sulfur and, more generally, any of the heavier chalcogens such as selenium or tellurium). This can be seen in the nature of minerals of the transition elements. Hard-acid titanium(IV), for example, is always bound to oxygen in naturally occurring minerals, while soft-acid silver(I) is more often associated with sulfur and other heavier chalcogens.

One is tempted to apply the HSAB concept to oxophilicity and, in fact, many have in the scientific literature. However, as has been shown by Kepp, oxophilicity does not correlate with absolute hardness, and in fact there is a slight *negative* correlation between oxophilicity and hardness for the transition metals as a whole![26]

In order to examine oxophilicity in a quantitative fashion, one needs to develop a scale for the oxophilic character of the transition metals.[26] Using M=O bond energies (D_O) is inadequate since the metal–oxygen bonds are almost universally stronger than the bond energies of their M=S (D_S) counterparts. Since we are interested in the relative bond energies of M=O and M=S systems, one can define ΔD_{OS} for each metal (M) (Eq. 13.54). The ΔD_{OS} is most often positive but is negative in a few cases, such as gold.

$$\Delta D_{OS}(M) = D_O(M) - D_S(M) \qquad \textbf{13.54}$$

The oxophilicity (Θ) for a metal is defined relative to gold (Eq. 13.55); the equation contains a normalization constant, *a*, chosen so the scale runs from 0 (low oxophilicity/high thiophilicity) to 1 (high oxophilicity), with 1 being set by hafnium, the metal with the highest oxophilicity.

$$\Theta(M) = (\Delta D_{OS}(M) - \Delta D_{OS}(Au))/a(\Delta D_{OS}(Au)) \qquad \textbf{13.55}$$

These oxophilicities were calculated for most of the elements of the periodic table, although we will largely focus on the transition metals here, based on the bond energies of the diatomics, metal–oxo (MO) and metal–sulfide (MS). Experimental values were used when available, with the remaining values calculated using DFT. Shown in Fig. 13.18 is Kepp's plot of Θ versus absolute hardness (η), which shows essentially no correlation between hardness and oxophilicity.

Instead of hardness one can apply the general principles of bonding discussed throughout this book, especially in Chapter 3, to oxophilicity with excellent results. For example, it is expected that bond energies will increase with increasing polarity. High-polarity metal–oxo species will probably have higher bond strengths relative to lower-polarity systems. Indeed, there seems to be a strong correlation between oxophilicity and ionic character in the metal–oxo bond as shown in Fig. 13.19.

Perhaps of greater interest is the *difference* between metal–sulfide and metal–oxo complexes. Why do some metal centers prefer oxygen strongly over sulfur, while others prefer sulfur over oxygen? To analyze, let's look at one monoanionic soft ligand, SH^-, and one hard ligand, OH^-, with one hard metal, Ti(IV) in a TiH_3 fragment, and one soft metal, Pb(II). We will examine exactly what is gained and lost energetically through all possible combinations in the context of concepts already discussed in detail in this book. Doing this, one discovers that HSAB concepts are a shortcut that often produces the correct result without quantifying the specific interactions (*vide infra*). Four compounds, $HSTiH_3$, $HOTiH_3$, $Pb(OH)_2$, and $Pb(SH)_2$, were examined using DFT, NBO, NRT, and polar covalence theory.

The information for analyzing these systems is shown in Fig. 13.20. The geometry from the DFT calculations is shown at the top, with the M–E–H angle and M–E bond distance. Below this is shown the major resonance form and its percent participation from the natural resonance theory calculation. In every case, a single resonance form dominates the electronic structure. If there are lone pair(s) on the oxygen or sulfur atom in the natural bond orbital calculation, the hybridization of

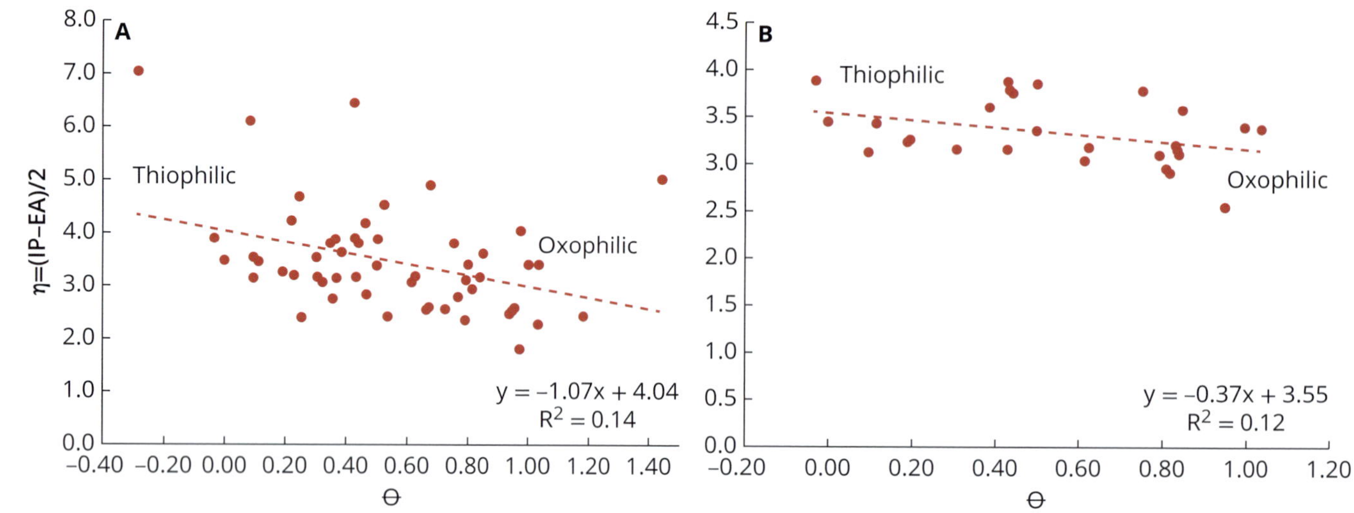

Fig. 13.18 Plot of Θ versus absolute hardness (η), where the hardness is calculated from experimental IE and EA for all elements where experimental data is available (A) and for the *d*-block elements (B). As shown, the correlation between oxophilicity and hardness is negligible (R^2 = 0.14 and 0.12), and, if anything, the oxophilicity increases with decreasing hardness.[26]

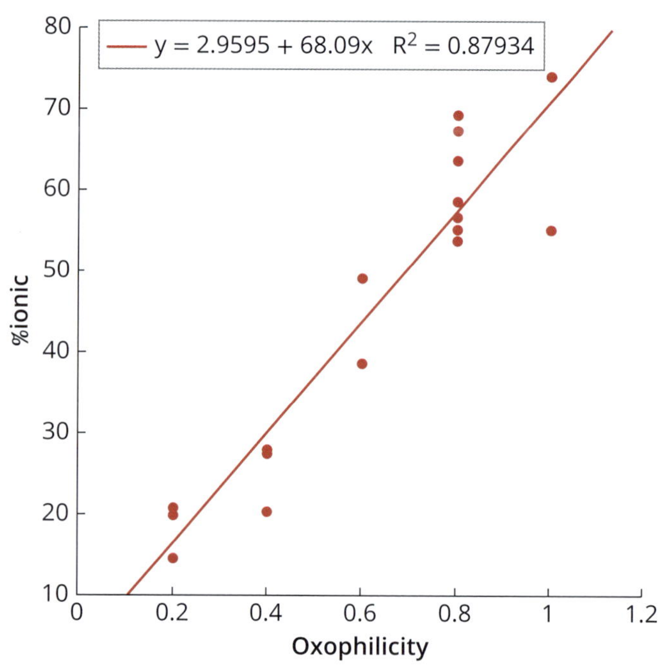

Fig. 13.19 Plot of oxophilicity versus %ionic character in a metal–oxo bond calculated using polar covalence theory for the transition metals.

the lone pair(s) is provided. Finally, the results from the polar covalence (PC) calculations are tabulated: %ionic for the bonds, the covalent bond energy (E_c), ionic bond energy (E_i), R_c/R_0 for the M–E bond, and the bond dissociation enthalpy (BDE). The value of R_c/R_0 provides a rough measure of the deviation from single bonding; systems with a single bond will have a R_c/R_0 value ~1, while deviations from 1 may suggest some π-bonding. Overall, the R_c/R_0 results are consistent with the NRT calculations; for example, $HOTiH_3$ shows significant multiple bonding and the highest R_c/R_0 value, 1.33. Using HSAB principles, there should be additional, energetically favorable interactions for the hard–hard and soft–soft interactions that will be lost for the complexes with hard–soft interactions. Let us analyze the systems to identify these interactions in each ligand–metal combination.

Hard Metal–Hard Ligand (TiH_3–OH)

For any first-row *p*-block element, lone pair bond weakening (LPBW) is an important effect to consider. LPBW is a hybridization effect where lone pairs "steal" *s*-character from bonds; the effect is most pronounced for the first-row *p*-block elements, where the valence orbitals are very nearly the same size (due to the primogenic effect) and hybridize easily. As the number of π-bonds to an element such as oxygen increase, the LPBW effect tends to decrease. This is due to the symmetry of π-bonds, which are made exclusively by *p*-orbitals leaving the *s*-orbital more available to stabilize any lone pair on the *p*-block element. *Consequently, LPBW will decrease for compounds containing metals with empty orbitals of π-symmetry. The earlier the metal center and the higher its oxidation state, the more likely it is to have empty orbitals of π-symmetry to accept additional lone pairs from oxygen* (see the following). With a triple bond, LPBW tends to be negligible, and double bonds are "partially weakened". In short, π-bonds to a first-row *p*-block element such as oxygen greatly increase the covalent bond strength, not just by increasing the number of bonds between the atoms, but also by increasing the overall bond strengths as well through alleviation of LPBW.

Almost invariably, "hard" Lewis acids have empty, low-energy π-orbitals that can act as acceptors for lone pairs. These low-energy orbitals may be *d*-orbitals, like for high valent transition metals, or may take the form of low-energy σ*-orbitals, such as SiH_3^+, which accepts lone pairs via negative hyperconjugation. Note that this is only dependent on the

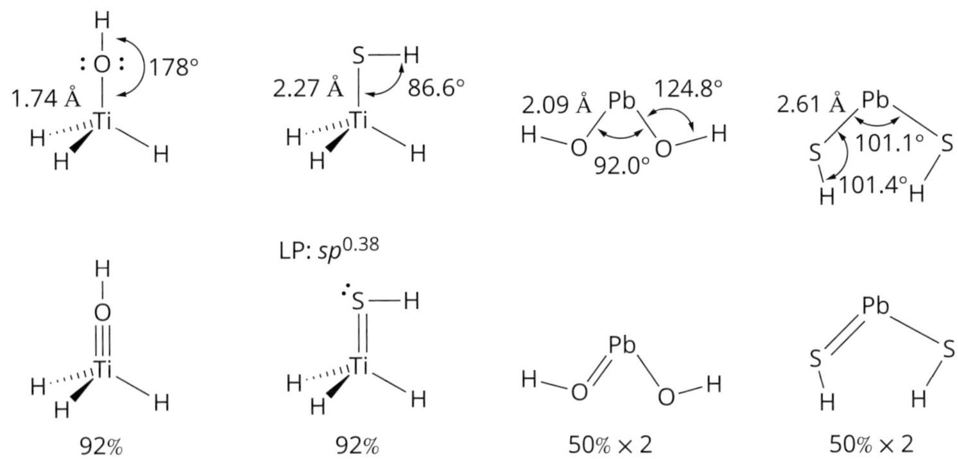

Polar covalence analysis[a]	TiH_3–OHb[b]	TiH_3–SH[c]	Pb–OH[d]	Pb–SH[d]
%ionic (t_i)	45.3	33.9	35	22
E_c (kcal/mol)	76.2	51.2	36.7	36.2
E_i (kcal/mol)	190.8	146.3	158.9	127.3
R_c/R_0[e]	1.33	1.17	1.04	0.97
BDE_{M-E} (kcal/mol)	128.1	83.4	79.5	56.2
$BDE_{M-O} - BDE_{M-S}$	44.7 kcal/mol		23.3 kcal/mol	

Fig. 13.20 Geometries are from DFT (B3LYP/cc-pvdz for Ti complexes, and B3LYP/SDD for Pb complexes). Lone-pair hybridizations are from NBO analysis. The lower structures are the main resonance form found by NRT with the percentage participation below it. Below the structures is a tabular summary of the polar-covalence analysis of the compounds. The values for the polar covalence calculations were taken from Table 3.2. [a]All distances are from gas-phase DFT calculations. [b]The fully unweakened oxygen covalent bond energy was used because of the Ti-O triple bond. [c]The partially unweakened sulfur covalent bond energy was used because of the Ti-S double bond. [d]The average of the fully weakened and partially weakened covalent bond energies were used because the bond order is 1.5 to each ligand. [e]The sum of the covalent radii (R_c) over the bond distance from the DFT optimization (R_0).

number of empty acceptor orbitals on the metal, and explains why a metal like hafnium, which by absolute electronegativity should be relatively soft, is a better oxophile than titanium, which should be relatively hard.

The NRT calculation (Fig. 13.20) suggests that there is a Ti–O triple bond in TiH_3OH, which should completely alleviate LPBW in the covalent part of the bond energy (E_c). Consequently, the fully unweakened O–O bond energy (E‴) of 104.0 kcal/mol is used in the calculation of E_c.

$$E_c = \frac{R_c}{R_0}\sqrt{E_{TiTi}E'''_{OO}} = \frac{2.307}{1.74}\sqrt{31.8 \times 104.0} = 76.2$$

If there were no acceptor orbitals on Ti, and therefore no π-bonding, the fully weakened O–O covalent bond energy (E‴) would only be 33.6 kcal/mol. This would cause the covalent bond energy to drop from 76.3 to 43.4 kcal/mol in the compound. Since the bond is roughly 54% covalent, LPBW would cause the Ti–O bond energy to decrease by 24 kcal/mol!

$$E_c = \frac{R_c}{R_0}\sqrt{E_{TiTi}E'_{OO}} = \frac{2.307}{1.74}\sqrt{31.8 \times 33.6} = 43.4$$

The second striking value in Fig. 13.20 is the %ionic of the hard–hard interaction. The ionic bond energy is always greater than the covalent bond energy ($E_i > E_c$), and therefore more ionic bonds are generally stronger. The %ionic in TiH_3–OH is 45%, as compared to the hard–soft interaction of TiH_3–SH, which is only 34% ionic, which greatly increases the Ti–O bond strength.

Hard Metal-Soft Ligand (TiH_3-SH)

The lower bond energy in this compound relative to the hard–hard interaction in TiH_3OH is due to a number of factors. (1) Since the NRT suggested that the Ti–S bond was a double bond, the LPBW in this system is also diminished relative to if it were a single bond. Thus the "partially weakened" S–S covalent bond energy was used in calculations; however, for sulfur, LPBW is greatly diminished relative to oxygen due to poorer *s-p* hybridization. The lone pair that remains on sulfur is very *s*-rich at $sp^{0.38}$. In the end, there is a lower covalent bond energy in the Ti–S system. (2) There is also a lower ionic bond energy here due to the larger radius of sulfur relative to oxygen, which gives a lower E_i. (3) In addition, the electronegativity difference between sulfur and titanium is lower than between sulfur and titanium, which gives a lower %ionic for the Ti–S bond. Overall, the Ti–S bond energy is then about 45 kcal/mol weaker than its hard–hard Ti–O counterpart.

Soft Metal–Soft Ligand (Pb–SH)

The Pb–S bond is less polar than the Pb–O bond, 22% and 35% respectively. More ionic bonds are typically stronger, and this system is not an exception, with the Pb–O bond being about 23 kcal/mol stronger than the Pb–S. However, the softer lead does seem to have a higher preference for sulfur than titanium, where the Ti–O bond is nearly 45 kcal/mol stronger than the Ti–S bond; but why?

In the absence of LPBW, sulfur has a fully unweakened (E‴) S–S bond energy of 65.9 kcal/mol. In a fully weakened (E′) S–S bond with no π-bonding, the bond is 54.9 kcal/mol. A bond-energy difference of 11 kcal/mol is not insignificant, but it is much smaller than the difference in oxygen where $E''' - E' = 70.4$ kcal/mol. This is of great importance for the interaction with soft atoms such as lead, where empty π-symmetry orbitals are less common and their large orbitals lead to poor π-overlap. Consequently, the fully weakened S–S and O–O bond energies will often be used to calculate the covalent bond energy (E_c). In this case, natural resonance theory does suggest a 1.5 bond to each side, so the average of the fully weakened and partially weakened values was used. Because LPBW is very large for oxygen relative to sulfur, forming a bond with sulfur when no or little π-bonding occurs, gives a lower LPBW penalty.

Soft Metal–Hard Ligand (Pb–OH)

The soft-hard Pb–O bond is actually a bit (23 kcal/mol) stronger than the soft–soft Pb–S bond in the gas phase and nonpolar solvents. The Pb–O bond is 13% more ionic than the Pb–S counterpart, which leads to the somewhat stronger lead–oxygen bond. In this case, the ionic contribution to the bond energy for Pb–O is higher than the covalent ($t_iE_i = 55.6$ versus $t_cE_c = 23.9$ kcal/mol); in contrast, the covalent contribution is slightly higher than the ionic for Pb–S bond strengths ($t_iE_i = 28.0$ versus $t_cE_c = 28.2$ kcal/mol), which contrasts with the titanium system in which the covalent *and* ionic bond energies are both higher for Ti–O as compared to Ti–S. In the gas phase or nonpolar solvents, it is quite rare to find lower polarity soft–soft interactions that are favored over soft–hard interactions. As previously discussed, HSAB principles work best in aqueous solution in which solvation effects can aid the soft–soft preference.

Summary of TiH_3EH versus $Pb(EH)_2$

The point of discussing these four compounds in such detail is two-fold. First, we wanted to show that the concepts discussed in this text can be used effectively to analyze HSAB problems. Second, the HSAB principle of "soft prefers soft and hard prefers hard" is more of a guideline than an explanation. It provides none of the "why" and only the "what." In order to explain the "why" here, we used a combination of theory and experiment, but largely used hybridization theory and polar covalence. No doubt, molecular orbital theory could be used instead, perhaps in conjunction with energy decomposition analysis.

Using this previous analysis, one can explain the preference of the hard–hard interaction by (1) alleviation of LPBW due to the empty π-symmetry orbitals on titanium and (2) the increase in ionic nature of the bonding. For the titanium system, the Ti–O bond has a higher E_c, E_i, and %ionic; everything points to a stronger Ti–O bond than Ti–S bond. The soft–soft interaction in Pb–SH is less unfavorable than might be expected, due to the lower importance of LPBW for sulfur than for oxygen, when the metal has no π-acceptor ability. However, in the Pb–O versus Pb–S case, the soft–soft "preference" is more convoluted. The Pb–O bond is more ionic, and ultimately slightly stronger than the Pb–S bond. However, the Pb–S bond has a higher covalent bond energy than Pb–O, and both bonds are largely covalent. Since the Pb–S bond has a higher E_c and Pb–O has a higher E_i, the difference in BDE for the two is relatively small. If one were to set up an equilibrium with $TiH_3SH + Pb(OH)_2$, it is probable, from the previous analysis, that the $TiH_3(OH) + Pb(SH)_2$ species (hard–hard and soft–soft) would be favored over $TiH_3(SH) + Pb(OH)_2$ (Fig. 13.20), meaning that the HSAB principle is a useful shorthand in this and many cases.

Ending this section with Ralph Pearson's quote on HSAB explanations seems very appropriate: "It must be emphasized again that the HSAB principle is intended to be phenomenological in nature. This means that there must be underlying theoretical reasons which explain the chemical facts which the principle summarizes. It seems certain that there will be no one simple theory. To explain the stability of acid–base complexes, such as A:B, will require a consideration of all the factors which determine the strength of chemical bonds."[30] In this chapter we have applied the simple bonding principles outlined previously in this book, and more detailed analyses using more advanced methods will probably be enlightening. However, this suffices to provide the student with an idea of the utility of HSAB and when it may be applicable, such as in water or when there are large LPBW effects in ligands that will be alleviated effectively on one side of the equilibrium.

13.3.7 Ambidentate Ligands and HSAB

There are some ambidentate ligands that can bind to metals either through a hard atom or a soft atom. Thiocyanate, SCN^-, is the archetype of the ligand class. The HSAB principle suggests that soft metals should preferentially bind to the *S*-side of SCN^- and hard metals should bind to the *N*-side. Often, such discussions start with the author showing two molecules like those in Fig. 13.21, and for the homoleptic complexes HSAB does quite well. Cu(I) is a relatively soft metal, and Cu(II) is harder. One expects that Cu–SCN linkages are more common in copper(I) and Cu–NCS linkages more common in Cu(II).

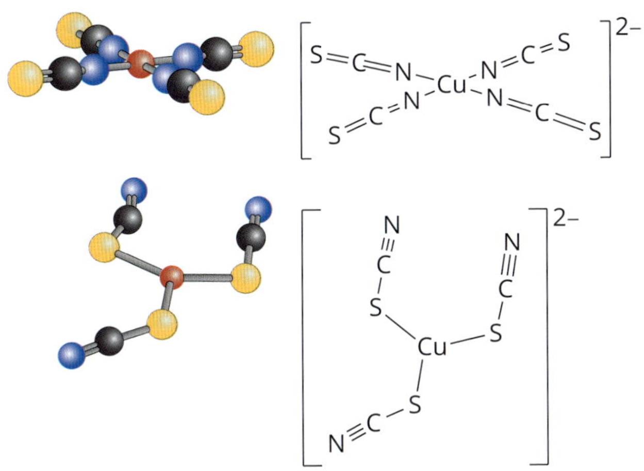

Fig. 13.21 Structures of copper(II) $Cu(NCS)_4^{2-}$ and copper(I) $Cu(SCN)_3^{2-}$ from their X-ray diffraction studies with line drawings.[31,32]

The 3-coordinate Cu(I) tris(thiocyanate) complex is known with the soft metal bound to the soft side of the ligand.[31] In addition, the 4-coordinate Cu(II) tetrakis(isothiocyanate) is known with the harder metal bound to the harder side of the ligand.[32]

When one wants to use the HSAB principle to predict the structure of a complex, in all likelihood you will not be concerned with just the homoleptic compounds. How well does the principle work for the larger selection of structurally characterized complexes? Fig. 13.22 shows the number of structures with Cu–NCS versus Cu–SCN linkages in the Cambridge Crystallographic Data Centre (CCDC) for copper(I) and copper(II).[33] There are far more Cu(II) structures than Cu(I), which is unimportant for our present discussion. What we want to know is: For what percentage of the time does the HSAB principle work for these two metal centers with SCN^-? For copper(I), the metal does prefer the sulfur-side of the ligand around 65% of the time. For a significant portion of the time, 35%, the nitrogen is bound to Cu(I). For harder Cu(II), for most (83%) of the time the HSAB principle correctly predicts that the nitrogen side will be bound. It should also be noted that there are more than 150 reported structures in which a copper center is bound to the sulfur of one SCN^- ligand and the nitrogen of a second SCN^- ligand. The majority of these compounds are dimeric or polymeric in nature, and the SCN^- ligands serve to bridge two copper centers.

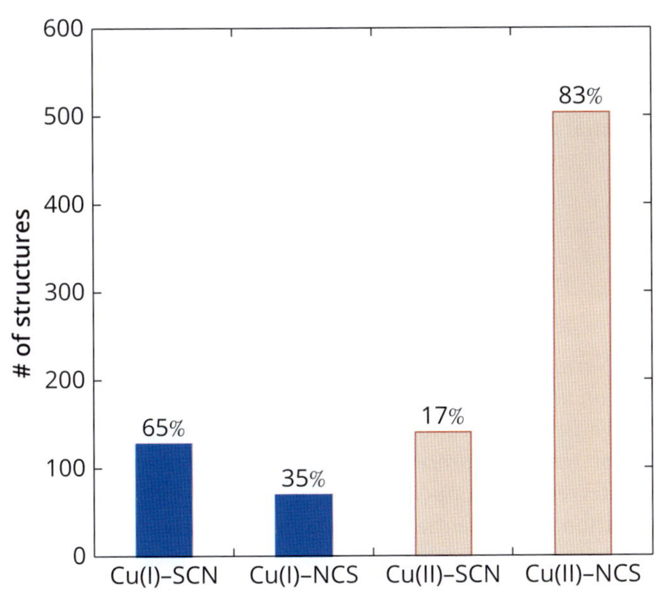

Fig. 13.22 Number of nitrogen- versus sulfur-bound thiocyanates for Cu(I) and Cu(II) in the CCDC.

Why may HSAB work somewhat well to explain the binding preference for NCS^- to copper? The charge distribution in NCS^- has an almost neutral carbon with a slightly higher negative charge on the nitrogen (Fig. 13.23). The NBO charges are shown in the following from a B3LYP/cc-pvTZ optimized structure. Consequently, there is not a large, charge-specific influence to thiocyanate binding. The explanation as to why the HSAB principle works (or not) in a particular case would require analysis similar to that in Section 13.3.6, where the bonding energies are partitioned by using some chosen theoretical framework. We will not present such an analysis here. Suffice it to say that the HSAB principle seems to work for a majority of the time and is a good starting point for investigations, but it is more of a tool for anticipating reactivity than an explanation as to why a reaction provides a specific product.

Unlike copper, there are many cases, especially in the area of the periodic table around cobalt, in which the preferred binding of thiocyanate is very sensitive to the conditions.[27–28] Cases have been identified in which "linkage isomerism" occurs—movement of the metal from one site on the ligand to another, depending on conditions and counterions. For example, in the closely related selenocyanate ligand, $SeCN^-$, Gutterman and Gray reported that $\{Co(CN)_5(NCSe)\}^{3-}$ was the preferred complex when K^+ was the counterion, and $\{Co(CN)_5(SeCN)\}^{3-}$ was the preferred complex when NBu_4^+ was the counterion. In aqueous solutions, $\{Co(CN)_5(SCN)\}^{3-}$ sets up an equilibrium of 70% S-bound and 30% N-bound complexes, which converts entirely to $\{Co(CN)_5(NCS)\}^{3-}$ in CH_2Cl_2.

There are many possible explanations for why the HSAB principle may be violated in individual cases, such as differences in solubility of the isomers, changes in metal hardness with other ligands on the metal, and production of a kinetic product that does not easily isomerize to the thermodynamic complex. However, in general the HSAB principle provides good guidance for the isomer likely to be produced. This is especially true for very hard metals like those in Group 3 and Group 4, where currently every thiocyanate complex in the database is nitrogen-bound (~80 structures). For one of the softest metals, gold, virtually all of the thiocyanate complexes in the database are sulfur-bound (forty-three structures), and the only exception is a complex with two thiocyanates—one sulfur- and the other nitrogen-bound.

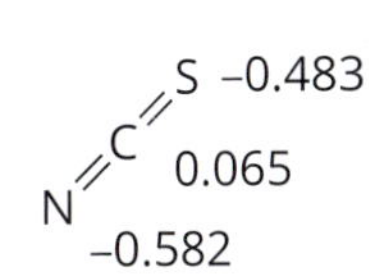

Fig. 13.23 Natural charges for NCS^-.

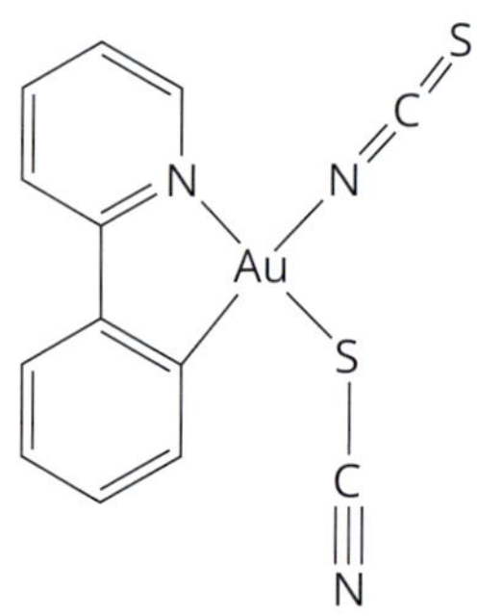

Fig. 13.24 Structure of pseudo-square planar Au(*N*,*C*-2-phenylpyridine)(NCS)(SCN).

The N-bound thiocyanate in this structure has a weak Au–S interaction with the gold of a different molecule, which may be contributing to this isomer's stability (Fig. 13.24).[29]

13.4 Integrated Examples

Thus far, we have examined electrochemistry and the HSAB principle. As they are separate sections of this chapter, one might be tempted to think of them as unrelated topics. However, this chapter began with a review of thermodynamics, which can serve as a link for these concepts, and there are many important links in redox and acid–base chemistry.

13.4.1 The Nernst Equation and Pourbaix Diagrams

The expression for how ΔG° is related to the standard potential has been discussed in some detail and is defined in Eq. 13.8. However, more often than not, the reaction will not be under standard conditions. How do you calculate the potential then, and how is it related to the standard potentials you can look up for many different systems?

First, just consider that Eq. 13.8 will still hold if not under standard conditions. In other words, we can just rewrite the expression as Eq. 13.56.

$$\Delta G = -nFE \qquad \textbf{13.56}$$

Next, recall from general chemistry that there is an expression for the free energy of a system away from equilibrium (Eq. 13.57), where Q is the reaction quotient (equilibrium expression with initial concentrations instead of the equilibrium concentrations).

$$\Delta G = \Delta G^{\circ} + RT\,ln\,Q \qquad \textbf{13.57}$$

Substituting Eq. 13.56 (and 13.8) into Eq. 13.57 produces the following expression.

$$-nFE = -nFE^{\circ} + RT\ln Q$$

Simplifying this expression produces the Nernst equation, which was briefly mentioned previously (Eq. 13.26) and describes how the standard potential and equilibria are related to the actual potential. The equation is written in a slightly different form in Eq. 13.58.

$$E = E^{\circ} - \frac{RT}{nF}\ln(Q) \qquad \textbf{13.58}$$

On occasion, different forms of the Nernst equation are useful, and Eqs. 13.59 and 13.60 are forms using base-10 log rather than natural log, and where the temperature is assumed to be room temperature (298 K).

$$E = E^{\circ} - \left(2.303\frac{RT}{nF}\right)\log(Q) \qquad \textbf{13.59}$$

$$E = E^{\circ} - \left(\frac{0.0591V}{n}\right)\log(Q)(\text{at } 25\,^{\circ}\text{C}) \qquad \textbf{13.60}$$

As an example of using Eq. 13.60, consider a cell in which a zinc electrode is submerged in an acidic 0.80 M zinc(II) solution, which is connected by a salt bridge to a 1.30 M silver(I) solution containing a silver electrode. What is the initial voltage of the cell at 298 K?

$Zn(s) \rightarrow Zn^{2+} + 2e^-$	$E^{\circ} = +0.76$ V
$2Ag^+ + 2e^- \rightarrow 2Ag(s)$	$E^{\circ} = +0.80$ V
$2Ag^+ + Zn(s) \rightarrow Zn^{2+} + 2Ag(s)$	$E^{\circ} = +1.56$ V

As written, this is the standard potential, but the cell is not at standard conditions. The Nernst equation can help us calculate what the voltage will be under the conditions listed. To use the equation we need to find Q, which has the same form as an equilibrium constant, [products]/[reactants]. The solid metals do not contribute to Q.

$$Q = [Zn^{2+}]/[Ag^+]^2 = 0.80/(1.30)^2 = 0.47$$

Then, using Eq. 13.49 and substituting values, we obtain the potential under these conditions.

$$E = E^{\circ} - \left(\frac{0.0591V}{n}\right)\log(Q)$$

$$E = 1.56 - \left(\frac{0.0591V}{2}\right)\log(0.47) = 1.57\,V$$

The Nernst equation (Eq. 13.60) actually means that the half-cell potential will change by 59/n millivolts for every ten-fold change in concentration of a substance involved in an oxidation or reduction process, where *n* is the number of electrons transferred. For copper dissolving in water with concomitant oxidation, the potential becomes more positive as the copper ions in solution decrease. In other words, as one might predict from Le Chatelier's principle, the dissolution of copper is more favorable the less copper that is already present in solution (Fig. 13.25).

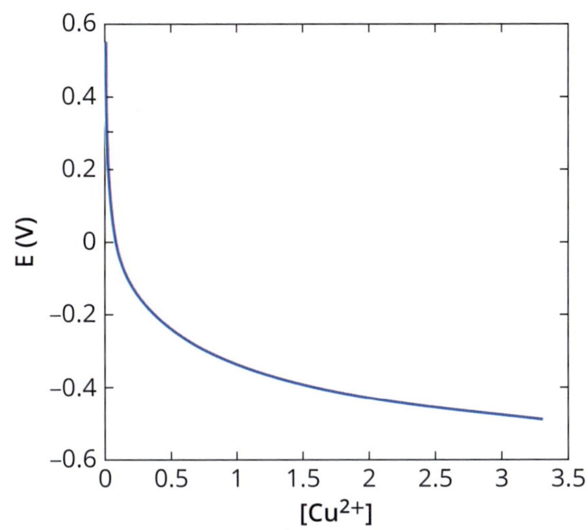

Fig. 13.25 Plot of potential (electromotive force) versus copper concentration, using the Nernst equation. As the copper concentration increases, dissolving more copper becomes thermodynamically less favorable.

$$Cu(s) \rightarrow Cu^{2+} + 2\,e^- \qquad E^\circ = -0.337$$

$$E = -0.337 - \left(\frac{0.0591V}{2}\right)\log[Cu^{2+}]$$

It is worth noting here that the Nernst equation works best at low concentrations. As the concentration increases, ion pairs become more common, and deviations from values predicted by the Nernst equation become larger and larger.

From the previous discussion of redox processes it is obvious that water, protons, and hydroxide often play an important role in aqueous solutions. As you probably also know, water can be reduced or oxidized under some conditions and by some compounds. For example, alkali metals react with water to release H_2 gas and form hydroxide. The reaction in acidic solution can be written as follows.

$$2\,H^+ + 2\,e^- \rightarrow H_2(g)$$

If the solution is basic or neutral, then the following reaction may occur.

$$H_2O + 2\,e^- \rightarrow H_2(g) + 2\,OH^-$$

For either reaction, the Nernst equation that it should follow is simplified (Eq. 13.61a) if one assumes that the H_2 pressure is 1 atm and that the temperature is 25 °C. Furthermore, $E^\circ_{H^+/H_2} = 0$ by definition.

$$E_{H^+/H_2} = E^\circ_{H^+/H_2} - \frac{RT}{nF}ln\left(\frac{[H^+]^2}{P_{H_2}}\right)$$

$$E_{H^+/H_2} = E^\circ_{(H^+/H_2)} - \frac{0.059V}{2}2 \times pH = -0.059\,pH \qquad \textbf{13.61a}$$

If one considers the oxidation of water, a similar equation can be derived. Again, we assume that the pressures are 1 atm and the temperature is 25 °C.

$$2\,H_2O \rightarrow O_2(g) + 4\,H^+ + 4\,e^-$$

$$E_{O_2/H_2O} = E^\circ_{O_2/H_2O} - \frac{RT}{nF}ln(P_{O_2}[H^+]^4)$$

$$E_{O_2/H_2O} = 1.23V - 0.059V \times pH \qquad \textbf{13.61b}$$

These equations describe the stability of water to redox processes as a function of pH. A plot of the potential, *E*, versus pH for a substance like this is called a "Pourbaix diagram", and the plot for water is shown in Fig. 13.26. The red and blue lines were drawn with Eqs. 13.61a and 13.61b, respectively. Water at 1 atm and 25 °C is thermodynamically stable between the two lines. Above the red line it is thermodynamically possible to oxidize water, and below the blue line, water can be reduced. As one might expect, H^+ is easier to reduce, and the thermodynamic favorability of reduction increases with decreasing pH; that is, the more H^+ in solution, the easier it is to reduce. As you probably also suspect, it is easier to oxidize water at higher pH, where more OH^- is present.

Pourbaix diagrams can be drawn for far more complex systems as well. Let us consider how pH affects the potentials of manganese ions from formal oxidation states of +7 to 0 in

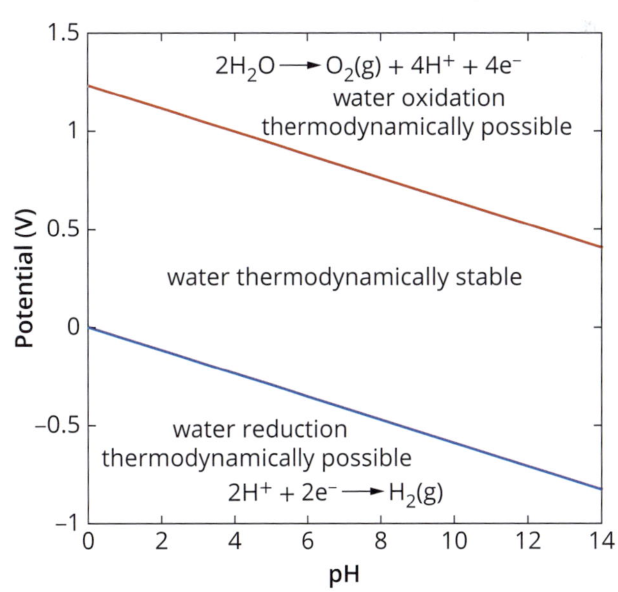

Fig. 13.26 Pourbaix diagram for water. The red line was drawn by using Eq. 13.61a, and the blue line was drawn by using Eq. 13.61b. Between the two lines, water is thermodynamically stable. Water reduction becomes thermodynamically more favorable as the solution becomes more acidic, but water oxidation becomes less thermodynamically favorable by the same amount with increased acidity.

aqueous solution. The Latimer diagram under acidic conditions (pH = 0) was shown previously but is reproduced here.

$$MnO_4^- \xrightarrow{+0.90\,V} MnO_4^{2-} \xrightarrow{+2.09\,V} MnO_2 \xrightarrow{+0.95\,V} Mn^{3+} \xrightarrow{+1.51\,V} Mn^{2+} \xrightarrow{-1.18\,V} Mn^0$$

The first reduction is very simple, as H^+ is not used in the balanced equation. It is simply reduction of Mn(VII) to Mn(VI). As a result, the potential is +0.90 V at all pH values.

$$MnO_4^- + 1e^- \rightarrow MnO_4^{2-} \qquad E^\circ = +0.90\,V$$

Next, let us consider the 2-electron reduction of Mn(VI) to Mn(IV).

$$MnO_4^{2-} + 2e^- \rightarrow MnO_2 \qquad E^\circ = +2.09\,V$$

This equation from the Latimer diagram is obviously not balanced, and if we want to understand its pH dependence, we need to balance it.

$$4H^+ + MnO_4^{2-} + 2e^- \rightarrow MnO_2(s) + 2H_2O$$

At pH = 0, E = +2.09 V. We can use the Nernst equation to calculate the potential at any other pH, so let us use pH = 7. The value of Q can be found by using the previous equation. The compound MnO_2 is a solid under the conditions and thus does not contribute to Q (value of 1); water is the solvent and thus does not contribute either. We will assume that the other manganese species are at standard conditions—that their concentrations are unity.

$$E_{Mn(VI)/Mn(IV)} = E^\circ_{Mn(VI)/Mn(IV)} - \frac{RT}{nF} ln(Q)$$

$$E_{Mn(VI)/Mn(IV)} = E^\circ_{Mn(VI)/Mn(IV)} - \frac{RT}{nF} ln\left(\frac{[MnO_2(s)][H_2O]}{[H^+]^4[MnO_4^{2-}]}\right)$$

$$E_{Mn(VI)/Mn(IV)} = E^\circ_{Mn(VI)/Mn(IV)} - \frac{RT}{nF} ln\left(\frac{1}{[H^+]^4}\right)$$

$$E_{Mn(VI)/Mn(IV)} = +2.09 - \frac{0.059}{2} log\left(\frac{1}{[H^+]^4}\right)$$

At pH = 7, $[H^+]^{-4} = (10^{-7})^{-4}$. Substituting in this equation produces

$$E_{Mn(VI)/Mn(IV)} = +2.09 - \frac{0.059}{2} log(10^{28}) = +1.26\,V$$

At pH = 14, the same equation produces

$$E_{Mn(VI)/Mn(IV)} = +2.09 - \frac{0.059}{2} log(10^{56}) = +0.44\,V$$

For Mn^{4+} to Mn^{3+}, the Latimer diagram suggests the following standard potential, but again this is not a balanced equation, which involves the addition of protons.

$$MnO_2 + e^- \rightarrow Mn^{3+} \qquad E^\circ = +0.95\,V$$

$$4H^+ + MnO_2 + e^- \rightarrow Mn^{3+} + 2H_2O$$

From this equation we can write the relevant Nernst equation. The water concentration and solid MnO_2 can be dropped.

$$E_{Mn(IV)/Mn(III)} = E^\circ_{Mn(IV)/Mn(III)} - \frac{RT}{nF} ln\left(\frac{[Mn^{3+}][H_2O]^2}{[H^+]^4[MnO_2(s)]}\right)$$

$$E_{Mn(IV)/Mn(III)} = E^\circ_{Mn(IV)/Mn(III)} - \frac{RT}{nF} ln\left(\frac{[Mn^{3+}]}{[H^+]^4}\right)$$

The value in the Latimer diagram under acidic conditions is for pH = 0. We can now write the equations for any pH, where we will assume the manganese concentration to be unity. At pH = 7, we have

$$E_{Mn(IV)/Mn(III)} = 0.95 - \frac{0.059}{1} log\left(\frac{1}{[10^{-7}]^4}\right) = -0.70V$$

Similarly, at pH = 14, we have

$$E_{Mn(IV)/Mn(III)} = 0.95 - \frac{0.059}{1} log\left(\frac{1}{[10^{-14}]^4}\right) = -2.35V$$

For the 2-electron reduction of MnO_2 to Mn^{2+}, we can write the following equation. Here our Latimer diagram does not contain this standard potential, but it can be calculated from the steps by using the following equation.

$$E^\circ(multistep) = \frac{Sum\ of\ potentials\ for\ each\ step}{number\ of\ electrons} \qquad \textbf{13.62}$$

For MnO_2 to Mn^{2+}, the Latimer diagram is

$$MnO_2 \xrightarrow{+0.95\,V} Mn^{3+} \xrightarrow{+1.51\,V} Mn^{2+}$$

Consequently, the standard potential for this reaction under acidic conditions is

$$E^\circ_{MnO_2/Mn^{3+}} = \frac{+0.95 + 1.51}{2} = 1.23\,V$$

The Nernst expression in terms of pH is then

$$E_{MnO_2/Mn^{3+}} = E^\circ_{MnO_2/Mn^{3+}} - \frac{0.059}{2} log\left(\frac{1}{[H^+]^4}\right)$$

Finally, we will include a couple of reactions that should not be pH-dependent.

$$Mn^{3+} + e^- \rightarrow Mn^{2+} \qquad E^\circ = +1.51\,V$$

$$Mn^{2+} + 2e^- \rightarrow Mn^0 \qquad E^\circ = -1.18\,V$$

All of these Nernst equations for the manganese redox reactions just discussed are summarized in Fig. 13.27. However,

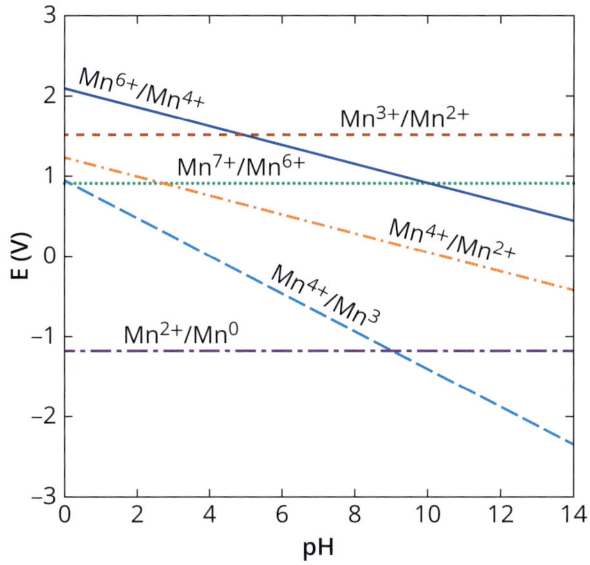

Fig. 13.27 Plot of some of the Nernst equations describing manganese redox chemistry in aqueous solutions at pH = 1–14.

to really obtain the complete picture of the redox behavior of manganese in solution, we would need to do many more calculations. For example, Mn^{7+}/Mn^{6+} redox couple is shown in blue. However, Mn^{7+} can probably oxidize many of the lower oxidation states of the metal, and we have not examined any of these redox disproportionation reactions and their proton dependencies. We would expect that Mn^{7+} would only be stable at higher potentials than the other species listed. Similarly, the metal, Mn^0, is probably only going to be present at very low potentials after all the possible disproportionation reactions are considered. In addition to the species listed in the Lattimer diagram for manganese, there are also multinuclear (containing multiple manganese metal centers) species that can be generated.

The full Pourbaix diagram for manganese is shown in Fig. 13.28, which contains many of the species that we have been discussing, along with some protonated species and multinuclear compounds that we have not discussed. The diagram shows the dominant species in solution at a given potential and pH. At points on lines, the species on either side of the line are in equilibrium.

13.4.2 Metal–Hydride Bond Energies from pK_a and Redox Measurements

A series of M–H bond dissociation energies (BDEs) for cationic metal hydrides has been estimated by Angelici, using cyclic voltammetry and thermochemical measurements of acid–base reactions.[38]

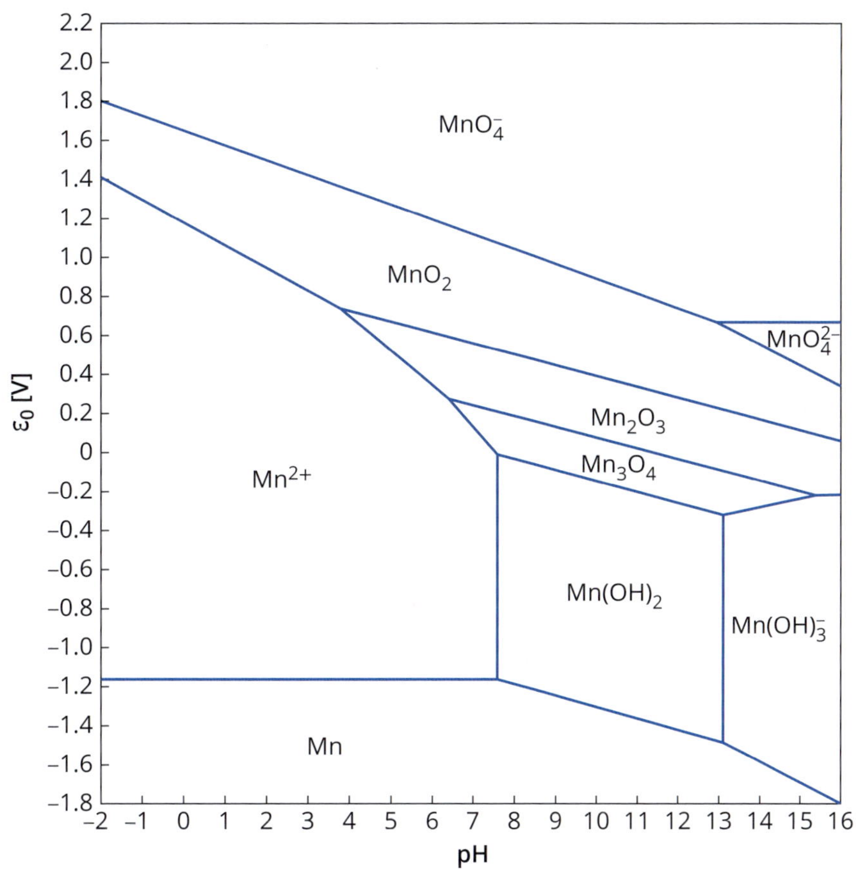

Fig. 13.28 Pourbaix diagram for manganese in aqueous solution.

While a variety of metal and ligand combinations were examined, we will focus on two compounds, $\{HFe(CO)_3(PPh_3)_2\}^+\{CF_3SO_3\}^-$ and $\{HFe(CO)_3(PMe_3)_2\}^+\{CF_3SO_3\}^-$, which vary by the phosphine used. These metal "hydrides" are, in fact, acidic (see Chapter 11); consequently, the hydrogen is lost from the complex as a proton, H^+ or to a different base. In this scheme, the proton is lost to the very weak base triflate, $CF_3SO_3^-$.

Generally speaking, when one refers to the bond dissociation enthalpy (BDE) it is assumed that the homolytic bond energy is being referenced. For a metal hydride $L_nM–H$, the homolytic bond energy is the difference between $L_nM^• + H^•$ and the starting complex.

$$M{-}H \longrightarrow M^• + H^• \quad \text{BDE} \qquad \textbf{13.63}$$

An acidic compound losing a proton is a heterolytic cleavage (Eq. 13.64); that is, the charges on the species are different from BDE (Eq. 13.63). Angelici and coworkers determined the BDE of acidic metal hydrides using a combination of acid–base and redox chemistry. In this scheme, Angelici used titration calorimetry to determine the enthalpies of protonation (ΔH_{HM}) of various metal complexes with the strong acid CF_3SO_3H (triflic acid) in 1,2-dichloroethane (DCE) at 25 °C (Eq. 13.65). As the magnitude of ΔH_{HM} increases, the metal complex is becoming more basic, and the resulting acid $H–ML_n^+$ is a weaker acid. As a result, ΔH_{HM} is related to the pK_a.[30]

$$M{-}H \rightleftharpoons M^- + H^+ \quad pK_a \qquad \textbf{13.64}$$

$$ML_n + CF_3SO_3H \xrightarrow[25^\circ C]{DCE} [H-ML_n^+][CF_3SO_3^-] \quad \Delta H_{HM} \qquad \textbf{13.65}$$

The general scheme for finding the BDE of M–H bonds by using this method is shown in Fig. 13.29. Hess's law states that thermodynamic quantities are independent of pathway—that is, state functions—and consequently we can determine the BDE using the alternative pathway involving proton transfer determined by the negative of the energy $-\Delta H_{HM}$ (compare Eq. 13.63), $E_{1/2}$, and *B*. In other words, we can approximate the BDE if we know the enthalpy of protonation with triflic acid, standard potential for L_nM/L_nM^+, and *B*. The constant

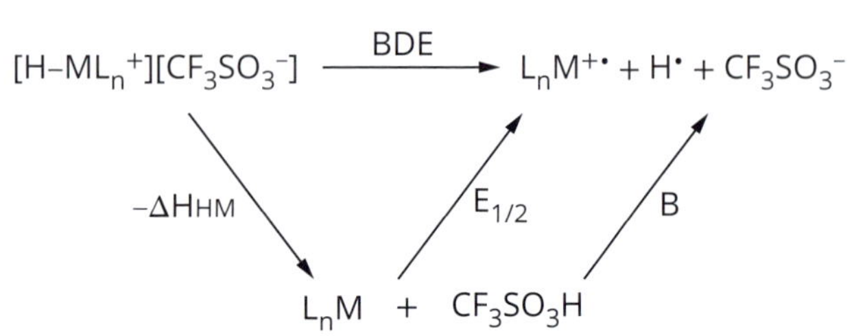

Fig. 13.29 Scheme for approximating the BDE of transition metal hydrides.

Table 13.5 $-\Delta H_{HM}$ and $E_{1/2}$ used to calculate the *BDE* for iron hydride phosphine complexes $\{H–Fe(CO)_3(PR_3)_2\}^+$.[38]

Complex	ΔH_{HM} (kcal/mol)	$E_{1/2}$ (V)	BDE (kcal/mol)[a]
$Fe(CO)_3(PPh_3)_2$	−14.1	0.55	60.1
$Fe(CO)_3(PMePh_2)_2$	−17.6	0.49	62.2
$Fe(CO)_3(PMe_2Ph)_2$	−21.2	0.45	64.9
$Fe(CO)_3(PMe_3)_2$	−23.3	0.41	66.1
$Fe(CO)_3(dppp)_2$[b]	−21.1	0.31	61.5
$Fe(CO)_3(dppm)_2$[c]	−24.0	0.40	66.5

[a] Calculated by using Eq. 13.66.
[b] dppp = $Ph_2P(CH_2)_3PPh_2$.
[c] dppm = $Ph_2PCH_2PPh_2$.

B was found experimentally and is related to proton reduction from triflic acid in the solvent.

Once the thermochemical value for $-\Delta H_{HM}$ is measured by calorimetry and the electrochemical value for $E_{1/2}$ is measured by cyclic voltammetry, one can estimate the BDE by using Eq. 13.66.[38]

$$\text{BDE}(M - H^+) = -\Delta H_{HM} + 23.06(E_{1/2}) + 33.3 \text{ (in kcal/mol)} \qquad \textbf{13.66}$$

Now consider the specific examples $[HFe(CO)_3(PPh_3)_2]^+[CF_3SO_3]^-$ and $[HFe(CO)_3(PMe_3)_2]^+[CF_3SO_3]^-$ (Table 13.5). The value for $-\Delta H_{HM}$ is the enthalpy (heat at constant pressure) found by using titration calorimetry of $[Fe(CO)_3(PPh_3)_2]$ or $[Fe(CO)_3(PMe_3)_2]$ reacting with CF_3SO_3H, the energy released (negative) when the complex is protonated. The value of $E_{1/2}$ is found by using cyclic voltammetry to determine the potential at which the neutral metal complexes are oxidized to yield the 17-electron cations, $[Fe(CO)_3(PPh_3)_2]^+$ or $[Fe(CO)_3(PMe_3)_2]^+$.

For the compounds of interest, $[HFe(CO)_3(PPh_3)_2]^+[CF_3SO_3]^-$ and $[HFe(CO)_3(PMe_3)_2]^+[CF_3SO_3]^-$, the Fe–H BDE of the PMe_3 compound was found to be higher than the PPh_3 compound. The PMe_3 was proposed to be a better electron donor (see Section 11.3.1.3) to the iron, thus making the iron center more basic, which is observed in the ΔH_{HM} values. This same trend was observed in the electrochemistry as the potential at which oxidation of the PMe_3 compound occurred was 0.14 V less positive than the PPh_3 compound; that is, $Fe(CO)_3(PMe_3)_2$ is easier to oxidize than its PPh_3 analog. The combination of these data suggested that the M–H BDE for $\{HFe(CO)_3(PMe_3)_2\}\{CF_3SO_3\}$ was 6 kcal/mol greater than that of $\{HFe(CO)_3(PPh_3)_2\}\{CF_3SO_3\}$.

13.4.3 Proton Reduction Catalysis

Another area where redox and acid–base chemistry overlap is in the important area of water reduction. For renewable energy to be practical there must be methods for storing energy when the Sun goes down and the wind does not blow.

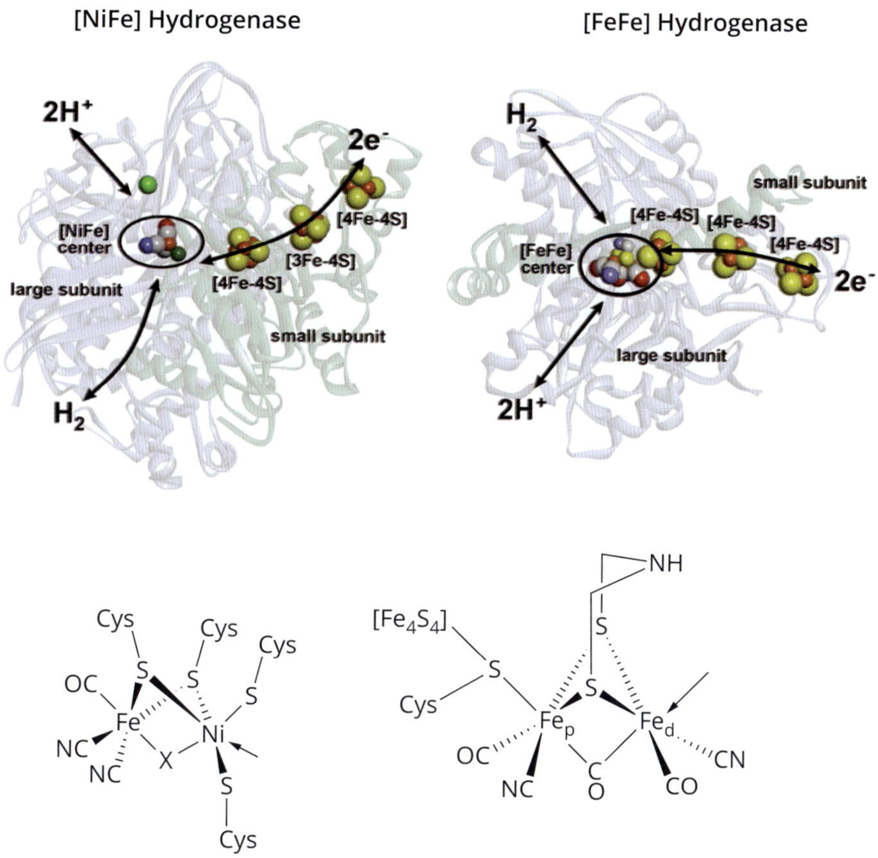

Fig. 13.30 The structures of the Ni–Fe- and Fe–Fe-hydrogenases. The active metal cores are shown below the ribbon structures of the proteins. (Figure from the review by Lubitz et. al.[32])

One of the most energy-dense methods for storing energy is in chemical bonds, and water splitting to H_2 and O_2 very cleanly stores a tremendous amount of energy, while burning H_2 only remakes water.

Water-splitting presents two different problems: proton reduction to H_2, and production of O_2. The synthesis of O_2 turns out to be the much more difficult problem as it is a 4-electron process; however, there are relatively inexpensive and efficient catalysts for the reaction, such as the cobalt phosphate catalyst discovered by Nocera and coworkers.[31] The proton-reduction problem would seem more straightforward, but to be practical it needs to be done with little energy loss and very rapidly.

In nature there are exceedingly good catalysts for H_2 production from protons and for production of protons from H_2. In fact, the same catalysts go in either direction—a trick that nature has perfected in this case, but which we as chemists still find difficult. The remarkable enzymes that carry out these reactions are generally known as "hydrogenases", and there are a few different varieties: nickel–iron, iron–iron, and iron.[32] The structure of the Ni–Fe- and Fe–Fe-hydrogenases are shown in Fig. 13.30. One interesting feature that you may notice first is that these contain cyanide and carbon monoxide as ligands—two famously toxic molecules appear in this natural system!

The next feature of these fantastic chemical systems that interests us here is the pendant amine group in the Fe–Fe hydrogenase system at right in Fig. 13.30. This is proposed to act as a Lewis basic site to move protons in the catalysis. A very active unnatural system for proton reduction has been developed by Bullock and coworkers, the structure of which is shown in Fig. 13.31.[33] The catalyst contains a nickel center and very similar pendant amines bound through chelating phosphine ligands. These catalysts show rates for proton reduction of up to $4.5 \times 10^7\ s^{-1}$!

Water-splitting for energy storage is a very active field in which researchers are constantly balancing acid–base and redox behavior to maximize outcomes.

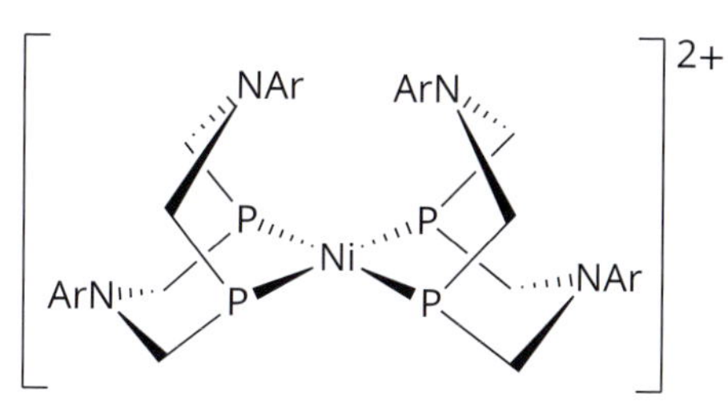

Fig. 13.31 Structure of Bullock and coworkers' proton reduction catalyst.

Chapter Summary

1. Redox reactions involve the addition or subtraction of electrons from a chemical species, and, in this chapter only the thermodynamics of these processes were considered.
2. The standard potential ($E°$) is directly related to the free energy of the process by the relation $\Delta G° = -nFE°$. The standard potential typically has units of volts and is scaled so that the half reaction $H^+ + e^- \rightarrow ½ H_2$ has $E° = 0.000$ V.
3. A Latimer diagram gives the standard potentials for one chemical species being converted into another in aqueous media. The diagrams can be different under acidic (pH = 0) and basic (pH = 14) conditions.
4. The most common electrochemical experiment to examine redox behavior is probably cyclic voltammetry, which can be used to measure oxidation and reduction potentials and the reversibility of redox events under the experimental conditions.
5. Acidity and basicity have several useful definitions varying from the Brønsted, Lewis, and other constructs you have likely encountered in previous coursework. One principle that is useful under some conditions in inorganic chemistry is hard–soft acid–base, which is very effective in aqueous media for predicting reaction outcomes. The principle is used in nonaqueous media in some cases, but is often less predictive.
6. There are many points at which redox behavior and acid–base theory meet. One is in the Nernst equation, which allows the calculation of potentials outside of standard conditions.
7. A Pourbaix diagram plots potential in Volts versus the pH of the solution, producing the chemical species present under those conditions.
8. An important field where redox and acid–base chemistry meet is in water-splitting for energy storage. There are natural enzymes for proton reduction—for example, hydrogenase—that are known, and chemists are making unnatural complexes with very high activities with related structures.

Practice Problems

1. Based on the redox potential calculations in this chapter, what is the free energy of the $Zn + 2 H^+$ to $Zn(II) + H_2$ reaction in kcal/mol under standard conditions?
2. Explain the driving forces for the following reaction:

   $$AsF_3 + PI_3 \rightarrow AsI_3 + PF_3$$

   Using:

   a. The HSAB principle.

   b. Physical properties (look up boiling points for each reactant).

   c. Polar covalence bond energies. For this, break down why the reaction is favorable, e.g., increasing polar bond strengths, increasing covalent bond strengths, and so on.
3. Use the HSAB principle to discuss the following reaction.

   $$MgS + BaO \rightarrow MgO + BaS$$

4. Verkade's bases are a class of compounds with a C_{3v} geometry where a nitrogen donor is situated trans to a phosphine lone pair and greatly increases the basicity of the phosphorus atom. The compounds contain a hard nitrogen site and a soft phosphorus site. It was reported that Verkade's base (R = *iso*-propyl) reacts with different Lewis acids at different sites, depending on their hardness.[34] For example, $GaCl_3$ reacts at P, and $AlCl_3$ reacts at N. The acids that react with Verkade's base without alteration of the base (some acids lead to P–N bond cleavage and other reactions) were H^+, tropylium ($C_7H_7^+$), AlX_3 (X = Cl, C_6F_5, Me, Et, *i*-Bu), GaZ_3 (Z = Cl, C_6F_5), BF_3, and BH_3.

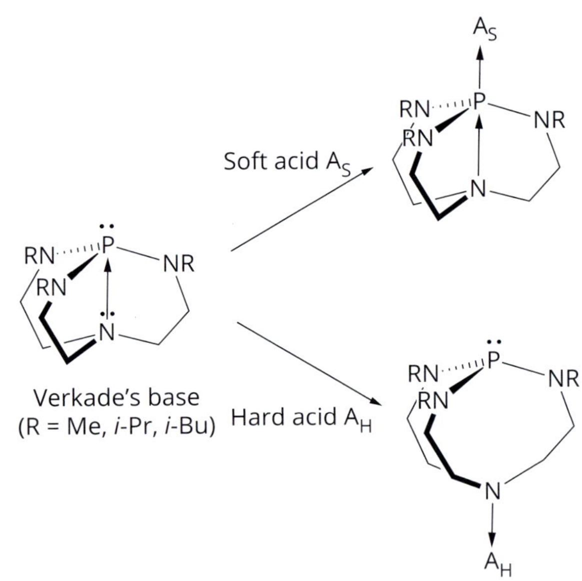

   a. Based on the Gutmann–Beckett-like numbers for $OPEt_3$ and $SPEt_3$, where will $AlCl_3$ and $GaCl_3$ attack Verkade's base? Explain your reasoning based on the data in the table.

   b. For the reaction with H^+, do you expect this to be a hard or soft acid, and why? Based on this expectation, how should H^+ interact with Verkade's base?

   c. What about the π-system of tropylium: is tropylium hard or soft? Why? Based on this prediction, how should tropylium interact with Verkade's base?

d. Now examine the paper by Krempner and coworkers.[34] Can you explain the site of attack for the full list of acids (H^+, tropylium, AlX_3, GaZ_3, BF_3, and BH_3) by using HSAB and Gutmann-Beckett acceptor numbers? Why, or why not?

e. In the cases where the Lewis acid interacts with the phosphorus in Verkade's base, the nitrogen also interacts with phosphorus. However, when the Lewis acid interacts with the nitrogen there is no P–N interaction. Why?

5. Use the Nernst equation to show that at $E_{1/2}$ the concentration of [X] and [X^-] are equal. (Note: Do not forget to discuss the difference between $E_{1/2}$ and $E°$ and under what conditions they are the same.)

Cited References

1. Goodstein, D. L. *States of Matter*. Englewood Cliffs, N.J: Prentice-Hall, 1975.
2. Heinze, J. Cyclic voltammetry: electrochemical spectroscopy. *Angew. Chem. Int. Ed.* 1984, *23*, 831–847.
3. Marken, F., Neudeck, A., and Bond, A. M. Cyclic voltammetry. In Scholz, F. (ed.) *Electroanalytical Methods: Guide to Experiments and Applications*, 2nd edn., 2010; pp 1–359.
4. Rusling, J. F., and Suib, S. L. Characterizing materials with cyclic voltammetry. *Adv. Mater.* 1994, *6*, 922–930.
5. Workentin, M. S., Maran, F., Chiba, K., and Savéant, J. M. *Organic Electrochemistry: Proceedings of the Fifth International Manuel M. Baizer Symposium in Honor of Professor Jean Michel Savéant*. Pennington, NJ: Electrochemical Society, Organic and Biological Electrochemistry Division, 2002, p. ix, 162 pp.
6. Connelly, N. G., and Geiger, W. E. Chemical redox agents for organometallic chemistry. *Chem. Rev.* 1996, *96*, 877–910.
7. Elgrishi, N., Rountree, K. J., McCarthy, B. D., Rountree, E. S., Eisenhart, T. T., and Dempsey, J. L. A practical beginner's guide to cyclic voltammetry. *J. Chem. Educ.* 2018, *95*, 197–206.
8. Rountree, E. S., McCarthy, B. D., Eisenhart, T. T., and Dempsey, J. L. Evaluation of homogeneous electrocatalysts by cyclic voltammetry. *Inorg. Chem.* 2014, *53*, 9983–10002.
9. Graham, D. J. *Standard Operating Procedures for Cyclic Voltammetry*, 2018. www.sop4cv.com
10. Reich, H. J. *Bordwell pKa Table* https://www.chem.wisc.edu/areas/reich/pkatable/.
11. Bordwell, F. G., Equilibrium acidities in dimethyl-sulfoxide solution. *Acc. Chem. Res.* 1988, *21*, 456–463.
12. Bordwell, F. G., and Zhang, X. M. From equilibrium acidities to radical stabilization energies *Acc. Chem. Res.* 1993, *26*, 510–517.
13. Goumont, R., Magnier, E., Kizilian, E., and Terrier, F. Acidity inversions of alpha-NO2 and alpha-SO2CF3 activated carbon acids as a result of contrasting solvent effects on transfer from water to dimethyl sulfoxide solutions. *J. Org. Chem.* 2003, *68*, 6566–6570.
14. Douglas, B. E., McDaniel, D. H., and Alexander, J. J. *Concepts and Models of Inorganic Chemistry*, 3rd edn. New York, NY: Wiley, 1994, p. xviii, 928+65 pp.
15. Galvan-Arzate, S., and Santamaria, A. Thallium toxicity. *Toxicol. Lett.* 1998, *99*, 1–13.
16. Hasan, M., and Ali, S. F. Effects of thallium, nickel, and cobalt administration on the lipid-peroxidation in different regions of the rat-brain. *Toxicology and Applied Pharmacology* 1981, *57*, 8–13.
17. Hoffman, R. S. Thallium toxicity and the role of Prussian blue in therapy. *Tox. Rev.* 2003, *22*, 29–40.
18. Pearson, R. G. Recent advances in the concept of hard and soft acids and bases. *J. Chem. Educ.* 1987, *64*, 561–567.
19. *Mössbauer Effect Methodology*. New York, NY: Plenum, 1970, Vol. 5.
20. Mayer, U., Gutmann, V., and Gerger, W. Acceptor number: Quantitative empirical parameter for electrophilic properties of solvents. *Monatshefte Fur Chemie* 1975, *106*, 1235–1257.
21. Beckett, M. A., Strickland, G. C., Holland, J. R., and Varma, K. S. A convenient NMR method for the measurement of Lewis acidity at boron centres: Correlation of reaction rates of Lewis acid initiated epoxide polymerizations with Lewis acidity. *Polymer Comm.* 1996, *37*, 4629–4631.
22. Sivaev, I. B., and Bregadze, V. I. Lewis acidity of boron compounds. *Coord. Chem. Rev.* 2014, *270*, 75–88.
23. Britovsek, G. J. P., Ugolotti, J., and White, A. J. P. From $B(C_6F_5)_3$ to $B(OC_6F_5)_3$: Synthesis of $(C_6F_5)_2BOC_6F_5$ and $C_6F_5B(OC_6F_5)_2$ and their relative lewis acidity. *Organometallics* 2005, *24*, 1685–1691.
24. Pearson, R. G. *Chemical Hardness*. Weinheim and New York, NY; Wiley-VCH, 1997, p. x, 198 pp.
25. Vogel, G. C., and Drago, R. S. The ECW model. *J. Chem. Educ.* 1996, *73*, 701–707.
26. Kepp, K. P. A quantitative scale of oxophilicity and thiophilicity. *Inorg. Chem.* 2016, *55*, 9461–9470.
27. Gutterman, D. F., and Gray, H. B. Cation-induced linkage isomerism of thiocyanatopentacyanocobaltate(III) complex. *J. Am. Chem. Soc.* 1969, 91, 3105.
28. Gutterman, D. F., and Gray, H. B. Electronic structures of thiocyanatopentacyanocobaltate(III) and related complexes. *J. Am. Chem. Soc.* 1971, *93*, 3364.
29. Fan, D. M., Yang, C. T., Ranford, J. D., Vittal, J. J., and Lee, P. F. Synthesis, characterization, and biological activities of 2-phenylpyridine gold(III) complexes with thiolate ligands. *Dalton Trans.* 2003, 3376–3381.
30. Angelici, R. J. Basicities of transition-metal complexes from studies of their heats of protonation: A guide to complex reactivity. *Acc. Chem. Res.* 1995, *28*, 51–60.
31. Kanan, M. W., Surendranath, Y., and Nocera, D. G. Cobalt-phosphate oxygen-evolving compound. *Chem. Soc. Rev.* 2009, *38*, 109–114.
32. Lubitz, W., Ogata, H., Rudiger, O., and Reijerse, E. Hydrogenases. *Chem. Rev.* 2014, *114*, 4081–4148.
33. Cardenas, A. J. P., Ginovska, B., Kumar, N., Hou, J., Raugei, S., Helm, M. L., Appel, A. M., Bullock, R. M., and O'Hagan, M. Controlling proton delivery through catalyst structural dynamics. *Angew. Chem. Int. Ed.* 2016, *55*, 13509–13513.
34. Mummadi, S., Kenefake, D., Diaz, R., Unruh, D. K., and Krempner, C. Interactions of Verkade's superbase with strong Lewis acids: From labile mono- and binuclear lewis acid–base complexes to phosphenium cations. *Inorg. Chem.* 2017, *56*, 10748–10759.
35. Duff, C. M., and Heath, G. A. Stepwise ligand-additivity effects on electrode-potentials and charge-transfer spectra in hexahalide, mixed halide nitrile, and hexakis(nitrile) complexes of ruthenium(IV), ruthenium(III), and ruthenium(II). *Inorg. Chem.* 1991, *30*, 2528–2535.
36. Grenon, N., and Baumgartner, T. Exploration of hypervalent Lewis acid–base interactions in 2-(2′-thiazolyl)-3-thienylphosphanes *Inorg. Chem.* 2018, *57*, 1630–1644.

CHAPTER 14

Understanding Transition-Metal Complexes Using Hybridization-Based Theory

CONCEPT COVERAGE

1. Foundations of the isolobal analogy
2. How to develop an isolobal series
3. Using the isolobal analogy to understand reactivity
4. *sd*-hybrids and consequences for molecular geometry
5. Applications of *sd*-hybridization to hypo-, hyper-, and normal valent systems
6. Relationship between the hybridization explanation for the geometry of WMe_6 and the MO theory explanation using Jahn–Teller distortions

"The isolobal analogy is a model. It is the duty of our scientific craft to push it to its extremes, and being only a model it is certain to fail somewhere. For any model, as ingenious a construction as it might be, is bound to abstract only a piece of reality."

— Roald Hoffmann

"The development of reliable quantum chemical computations has aided the detailed analyses of the geometries and electronic structures of many simple metallo-hydride and -alkyl molecules that so far have eluded empirical characterization. Surprisingly, we find that such molecular structures are most simply described by using the concepts of directed valency first proposed by Pauling and Slater at the onset of modern bonding theory."

— Clark Landis[1]

Understanding a host of different models and knowing how to apply them gives the chemist tools for explaining new results and designing new experiments or computational studies. In this chapter, some models outside of full MO or hybridization theory explanations for transition-metal systems will be developed. The models under discussion are only related in a cursory fashion to one another, but at the same time make an interesting contrast. In addition, both models have been found useful for understanding geometry, bonding, and reactivity of *d*-block compounds.

This chapter should be regarded as an advanced topic; however, the isolobal analogy and hybridization-based bonding approach to transition-metal bonding are, in reality, not too complex to be taught at the undergraduate level. Nevertheless, they are perhaps best reserved for more advanced classes that move through the material quickly. While the models here are related to hybridization theory, there is a connection drawn in the last section to molecular orbital theory-related Jahn–Teller distortions in WMe_6 to allow one to see how the concepts described cross into MO-based approaches.

14.1 Isolobal Analogy

One very useful model for analyzing the bonding in a transition-metal complex or fragment is the isolobal analogy, which was presented by Hoffman in 1982. Molecular fragments are isolobal "if the number, symmetry properties, approximate energy and shape of the frontier orbitals and the number of electrons in them are similar."[2] In other words, the isolobal analogy allows the chemist to draw conclusions about the frontier orbitals of a transition-metal fragment and relate it to other systems, whether they contain transition metals or only main-group elements. The two molecules or fragments need to have frontier orbitals with

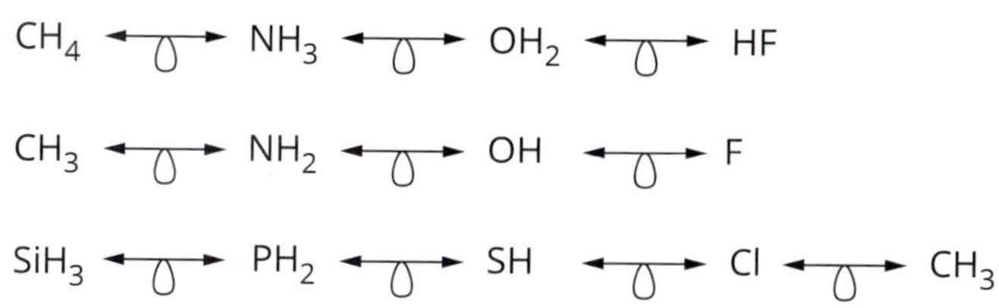

Fig. 14.1 Some isolobal main-group molecules and fragments.

similar: (a) numbers of frontier orbitals, (b) symmetry properties, (c) occupation by electrons, (d) energetic arrangement, and (e) relative radial extent.

In essence, the model allows one to relate seemingly exotic metal complexes quite often to “run-of-the-mill” organic fragments. For example, $H_3C–Mn(CO)_5$ becomes, from an electronic point of view, a fancy form of mundane ethane, $H_3C–CH_3$.

An isolobal group is signified with a double-headed arrow with a lobe (teardrop shape) below the arrow. For example, for some main-group fragments and molecules we can draw the following isolobal groups (Fig. 14.1). Starting with the closed-shell main-group hydride CH_4, we can draw isolobal compounds by removing hydrogens and adding electrons to make similar closed-shell compounds NH_3, OH_2, and HF. If we start with the radical CH_3 we can make analogous main-group radicals NH_2, OH, and F. As you might guess, moving to the next row of the *p*-block and starting with SiH_3 produces the isolobal series PH_2, SH, and Cl, which is also isolobal with the lighter fragments of the *p*-block just discussed, such as CH_3.

14.1.1 Development of the Isolobal Analogy for Transition-Metal Systems

For a transition-metal fragment, an octahedral geometry is the usual basis. We begin by creating a greatly simplified and generalized MO diagram for O_h. Transition metals have one *s*, five *d*, and three *p* in the valence shell. We will hybridize those orbitals to a 3-below-6 pattern, as shown in Fig. 14.2. The three low-energy orbitals in the pattern are the t_{2g} set (d_{xz}, d_{yz}, d_{xy}) in octahedral symmetry, which are not pointed directly at the vertices of an octahedron.

The usual method for the production of octahedral symmetries in transition-metal systems is the use of sp^3d^2-hybrids in a hybridization theory approach. This approach has the

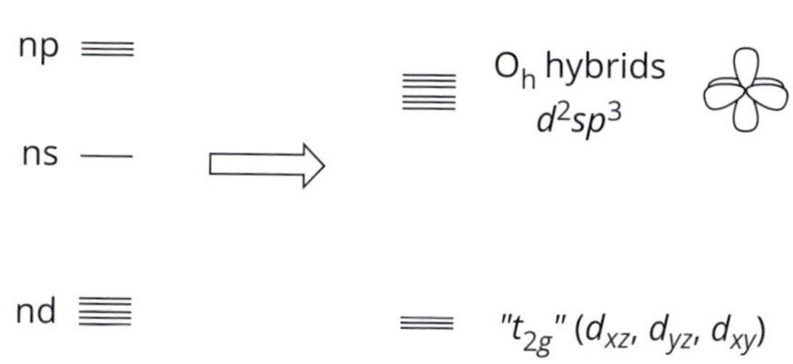

Fig. 14.2 Rehybridization of a transition metal to a 3-below-6 pattern with sp^3d^2-hybrids.

$$\Psi_1 = \frac{1}{\sqrt{6}} s + \frac{1}{\sqrt{2}} p_z + \frac{1}{\sqrt{3}} d_{z^2}$$

$$\Psi_2 = \frac{1}{\sqrt{6}} s - \frac{1}{\sqrt{2}} p_z + \frac{1}{\sqrt{3}} d_{z^2}$$

$$\Psi_3 = \frac{1}{\sqrt{6}} s + \frac{1}{\sqrt{2}} p_x - \frac{1}{\sqrt{12}} d_{z^2} + \frac{1}{2} d_{x^2-y^2}$$

$$\Psi_4 = \frac{1}{\sqrt{6}} s - \frac{1}{\sqrt{2}} p_x - \frac{1}{\sqrt{12}} d_{z^2} + \frac{1}{2} d_{x^2-y^2}$$

$$\Psi_5 = \frac{1}{\sqrt{6}} s + \frac{1}{\sqrt{2}} p_y - \frac{1}{\sqrt{12}} d_{z^2} - \frac{1}{2} d_{x^2-y^2}$$

$$\Psi_6 = \frac{1}{\sqrt{6}} s - \frac{1}{\sqrt{2}} p_y - \frac{1}{\sqrt{12}} d_{z^2} - \frac{1}{2} d_{x^2-y^2}$$

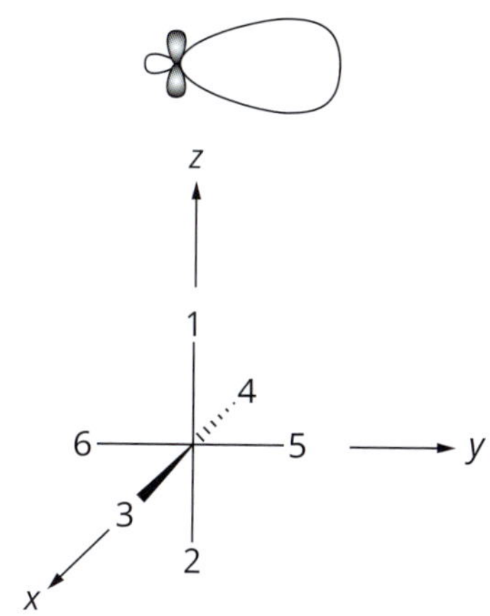

Fig. 14.3 (*top*) The mathematical form of the six sp^3d^2-hybrids. (*middle*) The general shape of the hybrids. The hybrid shape is rotated depending on its atomic orbital composition. (*bottom*) The direction of the large lobe for each of the wavefunctions (ψ) at the top points with ψ_1 towards position "1", and so on.

weight of intuition behind it, as the *s*-orbital and the orbitals pointed at the ligands (p_x, p_y, p_z, $d_{x^2-y^2}$, d_{z^2}) are expected to be involved in the bonding. Indeed, Pauling derived the nature of such hybrids (in work extended by Hultgren), and the composition is shown in Fig. 14.3; their general shape is also shown and there are six such hybrids pointed along the vertices of the octahedron.[4]

In analogy with the main group, we can discuss hybrids including *d*-orbitals as well as *s* and *p* as $sp^\lambda d^\mu$-hybrids, where λ is the ratio of *p* to *s* participation as before, and μ is the *d* to *s* participation. Similar to the treatment in Chapter 8, we can then write the normalized hybrid, *h*, as Eq. 14.1.[5]

$$h = \frac{1}{\sqrt{1 + \lambda + \mu}}(s + \lambda^{1/2} p + \mu^{1/2} d) \qquad \textbf{14.1}$$

If we look at this expanded Coulson directionality equation including *d*-orbitals, the following equation is found for the angle(s) between hybrids.[5]

$$\cos\alpha = \frac{-\lambda \pm \left[\lambda^2 - 3\mu(2-\mu)\right]^{1/2}}{3\mu} \qquad \mathbf{14.2}$$

For sp^3d^2-hybrids there are two possible solutions of $\cos^{-1}(0) = 90°$ and $\cos^{-1}(-1) = 180°$. This simply confirms that the hybrids point at the vertices of an octahedron, and each hybrid is then either 90° or 180° apart.

$$\cos\alpha = \frac{-\lambda \pm [\lambda^2 - 3\mu(2-\mu)]^{1/2}}{3\mu} = \frac{-3 \pm [9 - 6(2-2)]^{1/2}}{6}$$
$$= \frac{-3 \pm 3}{6} = 0 \;\&\; -1$$

If we take the hybrid orbital picture in Fig. 14.2 and remove one ligand from the octahedron, the fragment will have one hybrid that remains at the energy of the d^2sp^3-hybrids, while the other five hybrids will form bonding and antibonding combinations as shown in Fig. 14.4. We are only concerned with the frontier orbitals, which are in the box in Fig. 14.4, a 1-above-3 pattern.

Similarly, for the removal of two and three adjacent ligands from an octahedron, we obtain orbital patterns as shown in Fig. 14.5.

Consider the transition-metal fragment $Mn(CO)_5$. There is only one ligand missing from the octahedral vertices, and there will be one low-energy hybrid. The compound is Mn(0) and therefore has seven *d*-electrons. If we take the frontier orbital picture in Fig. 14.4 (the boxed portion) and fill in 7-electrons, we obtain the picture in Fig. 14.6. Six electrons are used to fill the t_{2g} orbitals, and one electron occupies the hybrid. The complex is a radical with filled lower energy orbitals. This orbital pattern is quite similar to that in CH_3 (Fig. 14.6), and these two fragments are isolobal.

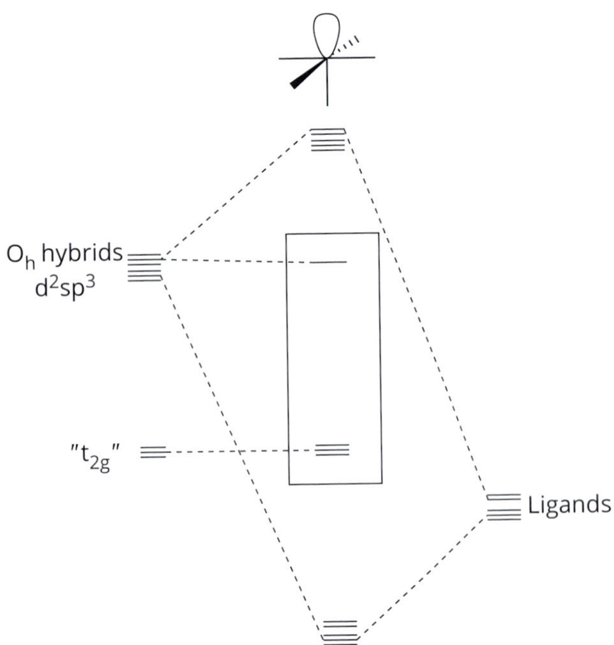

Fig. 14.4 Simplified MO diagram for a 5-c`oordinate transition-metal fragment based on an octahedron. The three t_{2g} orbitals remain unchanged, as does one d^2sp^3-hybrid. The other hybrids form bonding and antibonding orbitals with the ligands approaching in a square pyramidal geometry. The frontier orbitals are shown in the box.

Fig. 14.5 Frontier orbital patterns for the removal of 2- and 3-adjacent ligands from the octahedron with the three low-energy orbitals of the t_{2g} orbitals and two and three hybrids, respectively.

Because $Mn(CO)_5$ is isolobal with CH_3 (Fig. 14.7a), for any compound containing CH_3 we might anticipate a stable derivative with $Mn(CO)_5$. For example, ethane, $H_3C{-}CH_3$, contains two CH_3 fragments, and we would expect that both $H_3C{-}Mn(CO)_5$ and $(CO)_5Mn{-}Mn(CO)_5$ are stable complexes, which they are.

If $Mn(CO)_5$ is isolobal with CH_3, we can also draw the analogies in Fig. 14.7b with the heavier elements of the same group and with the cation of the next group $\{Fe(CO)_5\}^+$, as examples.

What is $Cr(CO)_5$ isolobal with? Chromium has one electron fewer than manganese, so the hybrid orbital in Fig. 14.6 will be empty. Consequently, neutral $Cr(CO)_5$ is isolobal with the methyl cation, CH_3^+ (Fig. 14.8).

By similar reasoning, $Fe(CO)_5$ in a square pyramidal arrangement is isolobal with CH_3^- and NH_3 (Fig. 14.9). Note that $Fe(CO)_5$ is a stable complex on its own, but has a trigonal bipyramidal structure in the ground state, not the square pyramidal arrangement here that leads to a structure isolobal with methide. Furthermore, while CO is being used as the default ligand here, there is nothing particularly special about this ligand in this context, and many others, such as PMe_3, would do just as well.

For our generalization, ligands such as Cp^- are converted to a collection of monodentate ligands with possible charges. The cyclopentadienide ligand will occupy three

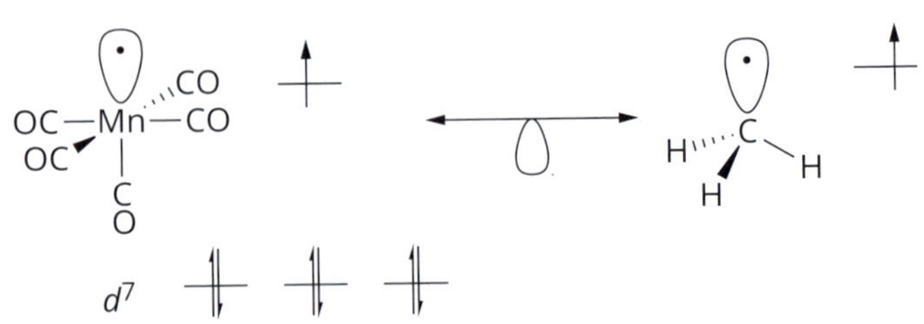

Fig. 14.6 Frontier orbital diagram for the isolobal radicals $Mn(CO)_5$ and CH_3.

Fig. 14.7a Some isolobal relationships with $Mn(CO)_5$.

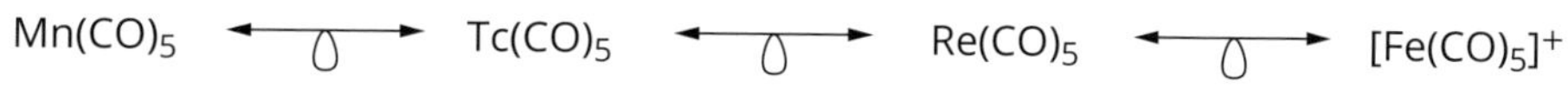

Fig. 14.7b Some isolobal relationships with $Mn(CO)_5$ and other transition-metal complexes.

Fig. 14.8 $Cr(CO)_5$ is isolobal with methyl cation.

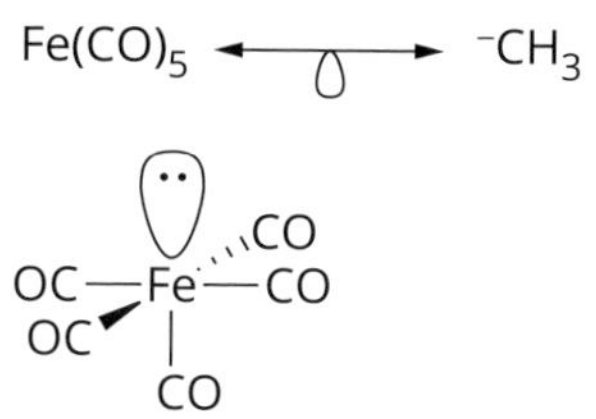

Fig. 14.9 $Fe(CO)_5$ is isolobal with methyl anion.

orbitals and induce a +1 charge on the metal (Fig. 14.10). A cyclopentadienide-containing fragment will be isolobal with a similar system with three neutral ligands in place of Cp and a cationic charge or one metal to the left (Cp = 3 CO and +1 charge, for example). So, the common fragment $CpFe(CO)_2$ (Fp) is isolobal with $Mn(CO)_5$, both are radicals when neutral. Replacing CO with phosphines makes no difference for the isolobal analogy. In addition, anionic ligands can replace neutral ligands as long as the *d*-electron count on the metal remains the same; for example, $Mn(CO)_5$ is isolobal with $\{Mn(PMe_3)_4Cl\}^-$.

The *cis*-divacant fragment $Fe(CO)_4$ is d^8, and placing eight electrons in the frontier orbital diagram from the left side of Fig. 14.5 produces the picture in Fig. 14.11. This metal fragment is isolobal with triplet carbene.

The fragment $Fe(CO)_4$ should be exchangeable in compounds containing CH_2 (Fig. 14.12). Such a series of known compounds is shown. Cyclopropane is isolobal with the $Fe(ethylene)(CO)_4$ (shown as the metallacyclopropane resonance form), which is isolobal with the μ-carbenediiron cluster and $Os_3(CO)_{12}$.

The fragment $Co(CO)_3$ is an example of a *fac*-trivacant octahedron. Placing nine electrons into the frontier orbital diagram on the right side of Fig. 14.5 produces the diagram in Fig. 14.13, which is isolobal with quartet CH.

A known series where CH is repeatedly replaced with $Co(CO)_3$ is shown in Fig. 14.14, beginning with well-known tetrahedrane (R = ^{t}Bu, for example). The next member of the series is the cyclopropenyl cobalt complex. These are isolobal with the acetylene complex of $Co_2(CO)_6$, the Co_2C_2 cluster.

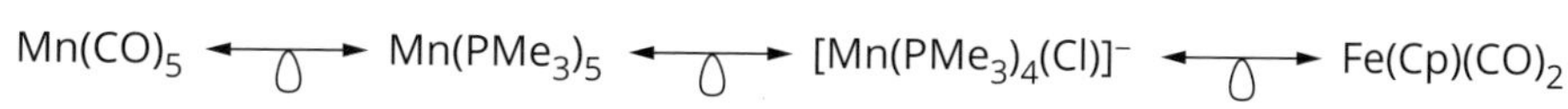

Fig. 14.10 Some isolobal relationships with $Mn(CO)_5$, including exchange of Cp for monodentate ligands.

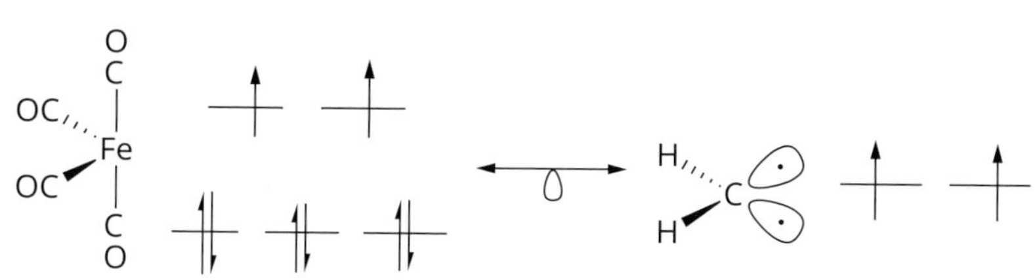

Fig. 14.11 $Fe(CO)_4$ is Isolobal with triplet CH_2.

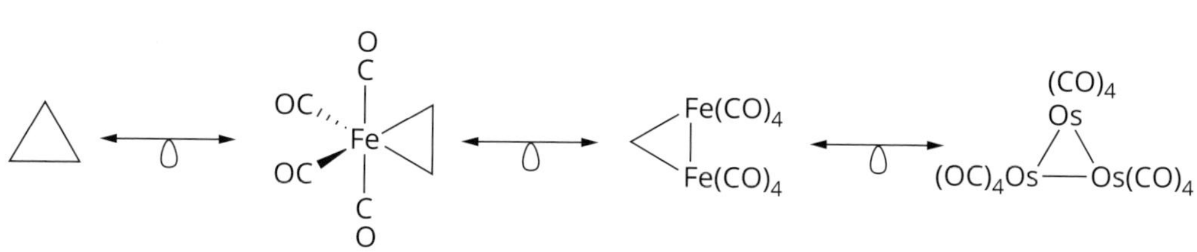

Fig. 14.12 Some isolobal relationships with $Fe(CO)_4/CH_2$ exchanges to generate known compounds.

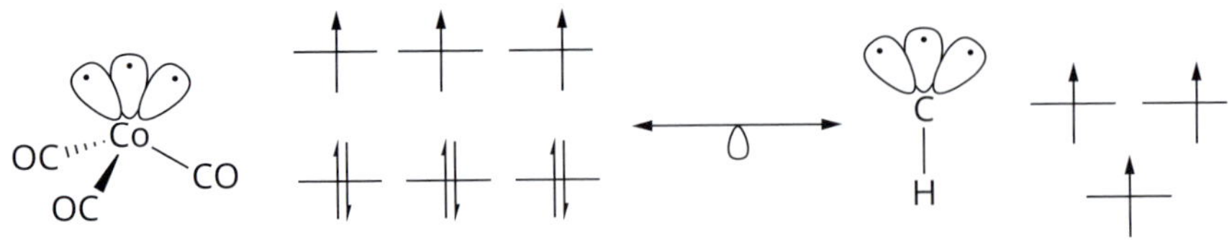

Fig. 14.13 $Co(CO)_3$ is isolobal with quartet CH.

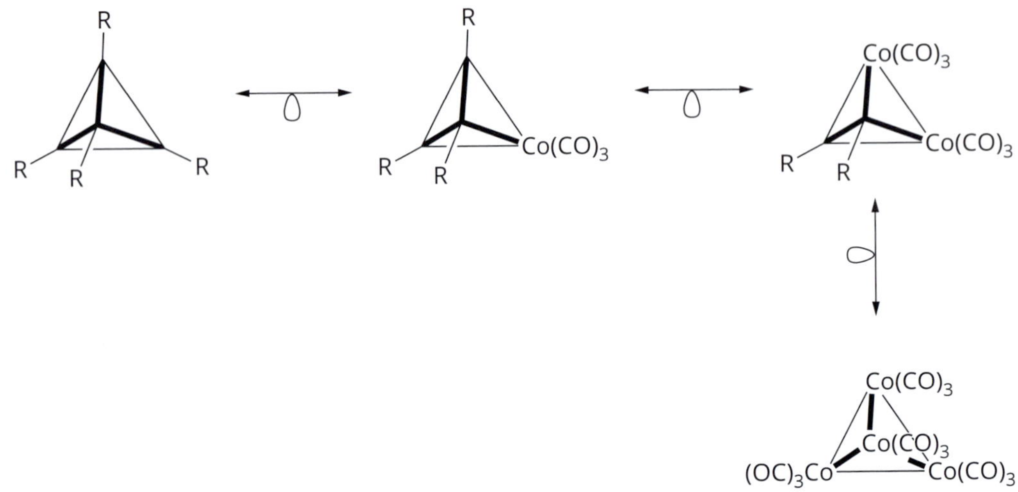

Fig. 14.14 Some isolobal relationships with $Co(CO)_3$/CH exchanges to generate known compounds.

One of the stable cobalt carbonyls is tetrahedral $Co_4(CO)_{12}$. The example with two carbons and two cobalt atoms in the tetrahedron is prepared from an alkyne and $Co_2(CO)_9$, and is a method of protecting alkyne groups, for example. These isolobal exchanges are summarized in Fig. 14.15.

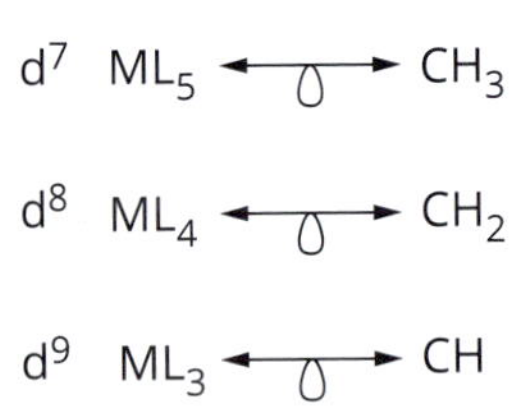

Fig. 14.15 Summary of isolobal exchanges.

A second method for the formation of isolobal fragments is by removing both ligands along the z-axis. This causes the loss of hybridization in ψ_1 and ψ_2 in Fig. 14.3, and as a result it drops in energy to the nonbonding manifold with t_{2g} (Fig. 14.16). A p-orbital is also extracted from the hybrids, p_z, but this orbital is relatively high in energy after removing hybridization, and is ignored. Now, we have four low-energy nonbonding orbitals, d_{xz}, d_{yz}, d_{xy}, and d_{z^2}. To make a fragment isolobal with the octahedral-based hybrids, we need two additional electrons. As shown, a monovacant d^n O_h-fragment will be isolobal with a monovacant d^{n+2} D_{4h} fragment.

For example, d^6 $Cr(CO)_5$ is isolobal with d^8 $PtCl_3^-$, which are both isolobal with the methyl cation (Fig. 14.17).

Similarly, the d^8 fragment *cis*-divacant $Fe(CO)_4$ is isolobal with d^{10} $Ni(PR_3)_2$. Placing ethylene on both fragments

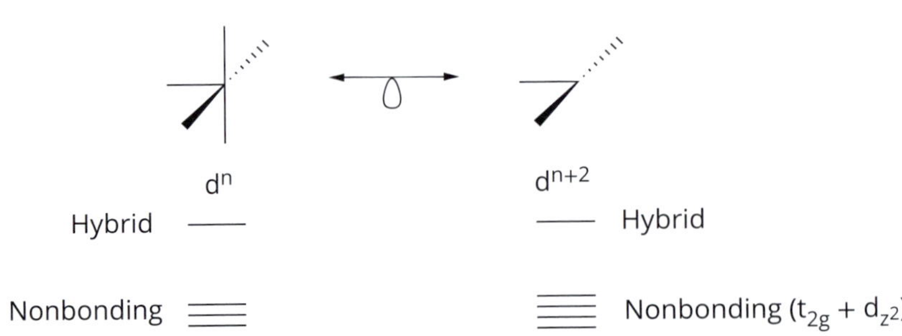

Fig. 14.16 Removal of axial ligands causes d_{z^2} to drop into the nonbonding set. As a result, a monovacant octahedral d^n complex is isolobal with a T-shaped d^{n+2} complex.

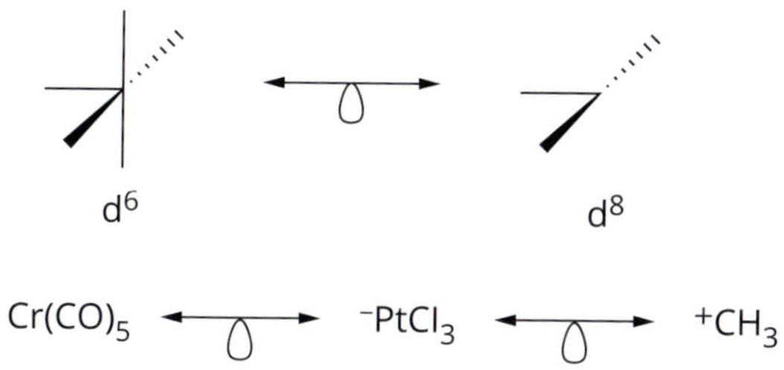

Fig. 14.17 Isolobal relationship between $Cr(CO)_5$, $PtCl_3^-$, and $^+CH_3$.

produces isolobal systems as well, as shown in Fig. 14.18 as metallacyclopropanes.

A third feature of the isolobal analogy is the ability to dip into the t_{2g} electrons to form isolobal fragments. Through this, the analogies in Fig. 14.19 will be explained.

Previously, we showed that $Cr(CO)_5$ is isolobal with $^+CH_3$. Therefore, $^+Cr(CO)_5$ is isolobal with $^{2+}CH_3$, but how is $^+Cr(CO)_5$ isolobal with CH as shown previously? We start with our simplified MO diagram for $^+Cr(CO)_5$ and allow it to go high spin (Fig. 14.20); that is, we promote an electron from the lower-energy t_{2g} set into the empty hybrids. *In this generalization, we are not necessarily placing restrictions on spin state. In the end, the important thing is not to blindly follow a set of rules, but to use the model in a logical fashion to gain insight into the problem at hand!*

From these arguments we create the isolobal analogies shown in Fig. 14.21. On the right, it is seen that (for the purposes of the isolobal analogy) we can indiscriminately remove protons from the organic fragment to obtain an equally good analogy. This is because we are only interested in the gross electronic structure.

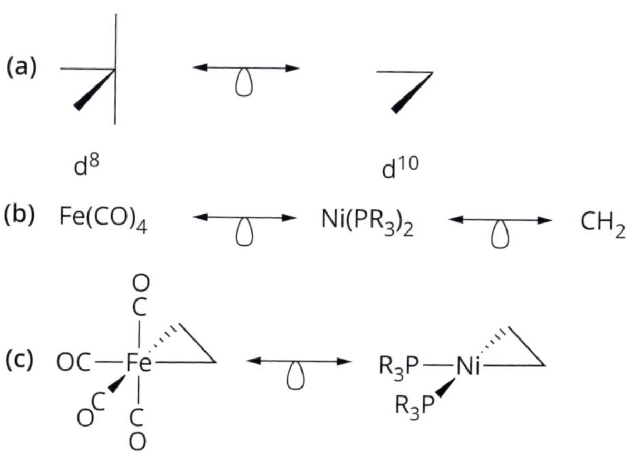

Fig. 14.18 (a) Divacant octahedral d^8 complexes are isolobal with 2-coordinate d^{10} complexes. (b) As examples of fragments, $Fe(CO)_4$ is isolobal with $Ni(PR_3)_2$ and triplet CH_2. (c) Compounds that are isolobal in this class include $Fe(CO)_4$(ethylene) and $Ni(PR_3)_2$(ethylene) complexes.

14.1.2 Applications of the Isolobal Analogy

Since its introduction, the isolobal analogy has been applied to hundreds if not thousands of chemical systems.[6] Many of the applications are in cluster chemistry, where we have already shown several analogies. Next we will discuss two other applications, one of which concerns whether there is an expected barrier to rotation for a metal–olefin bond, while the other will be to the reactivity of MX_3 fragments of Group 6.

14.1.2.1 Rotational Barriers by the Isolobal Analogy

Olefin orientation can, in some circumstances, greatly affect the rate of reactions such as 1,2-insertion of an olefin into a metal–hydride—a key step in catalyses like olefin hydrogenation. We can use the isolobal analogy to provide an indication of whether or not an electronic preference for a specific orientation is expected in a metal–olefin complex.

Take the olefin complex Cr(ethylene)$(CO)_5$ in Fig. 14.22. The isolobal analogy for the metal fragment $Cr(CO)_5$ suggests that there is one low-energy hybrid for the olefin to donate into. The chromium can backbond by using the filled t_{2g} orbitals. If we call the Cr–olefin vector the z-axis, the orbitals in t_{2g} of proper symmetry to interact with the olefin in the position on the left is d_{xz}; when the olefin in the position on the

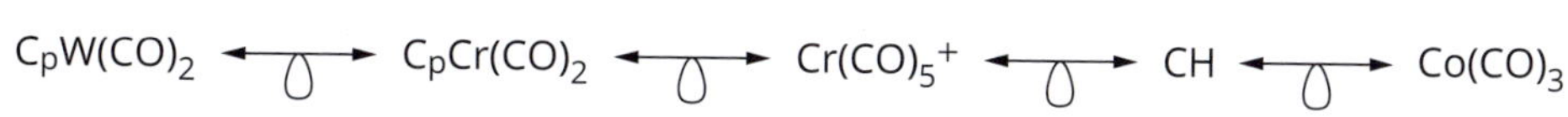

Fig. 14.19 Some isolobal fragments related by using the t_{2g} electrons in some cases.

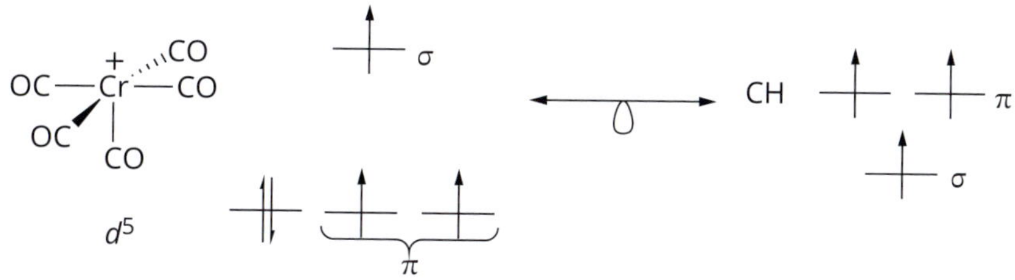

Fig. 14.20 One can promote electrons from the t_{2g} set to a hybrid, which is another way to generate isolobal systems. In this way, $^+Cr(CO)_5$ is isolobal with quartet CH.

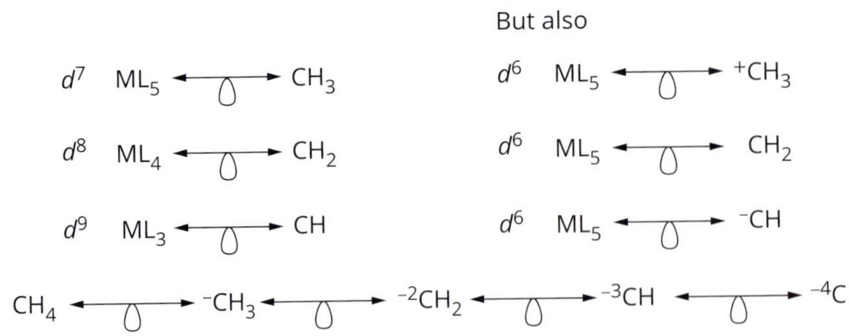

Fig. 14.21 Some isolobal systems that were discussed, including the systems on the right where protons are being removed to make isolobal fragments. This is further expanded at the bottom of the figure, where removing protons from methane is shown to produce isolobal fragments.

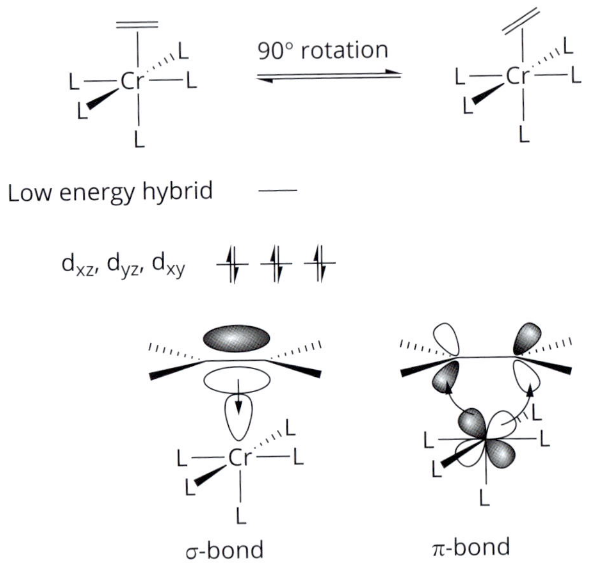

Fig. 14.22 The $Cr(CO)_5$ fragment on an olefin. No barrier to rotation is expected around the Cr–olefin σ-bond because the π-bond on the right has a partner rotated 90° of the same energy.

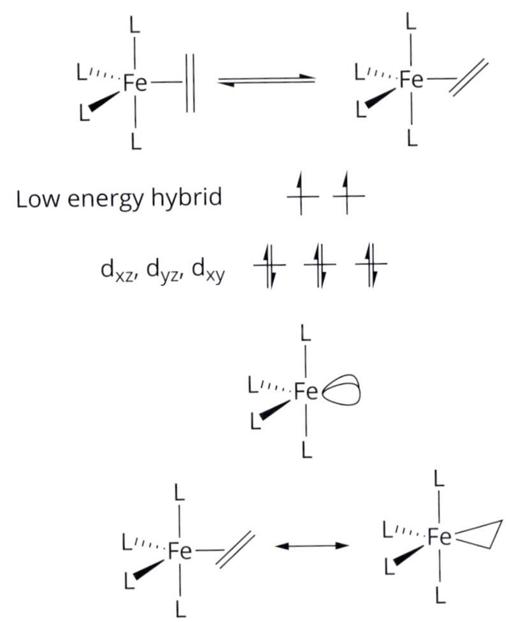

Fig. 14.23 The $Fe(CO)_4$ fragment on an olefin.

right, d_{yz} will backdonate into the olefin. Because both of these orbitals have identical energies, we expect no preference for either position. In fact, the olefin may reside between the ligands as well, or any other angle without an electronic barrier, since the e-symmetry set of orbitals provides a torus of equal electron density with which the olefin can interact.

Now let us examine an olefin bound to the *cis*-divacant $Fe(CO)_4$ fragment. Do we expect an orientational preference for the olefin in this complex? The complex with the olefin in each limiting rotational position is shown in Fig. 14.23. For the $Fe(CO)_4$ fragment we expect two half-filled low-energy hybrids. These hybrids point at the empty vertices of the octahedron.

Because the higher-energy hybrids are better donors than the lower-energy t_{2g} electrons,[a] the olefin will prefer to interact with these orbitals. Indeed, if the olefin complex is drawn as a metallacyclopropane, one sees a clear preference for this orientation over the one rotated 90° (top left top Fig. 14.8) where lower-energy, less basic t_{2g} orbitals will donate into the olefin π^*-orbital. Furthermore, realizing the $Fe(CO)_4$ is isolobal with CH_2 suggests a clear preference for how the olefin will be oriented.

[a] Electron pairs in low-energy orbitals are stabilized, making them poor donors. By the same token, electrons in high-energy orbitals are poorly stabilized, making them good donors.

14.1.2.2 Structure–Reactivity Relationships in MX_3 Complexes

One of the most powerful applications of the isolobal analogy is in structure–reactivity relationships between complex metal groups and simple main-group fragments. The syntheses of many organometallic complexes have been guided by isolobal analogies to common organic molecules. One series of molecules where the isolobal analogy has been employed

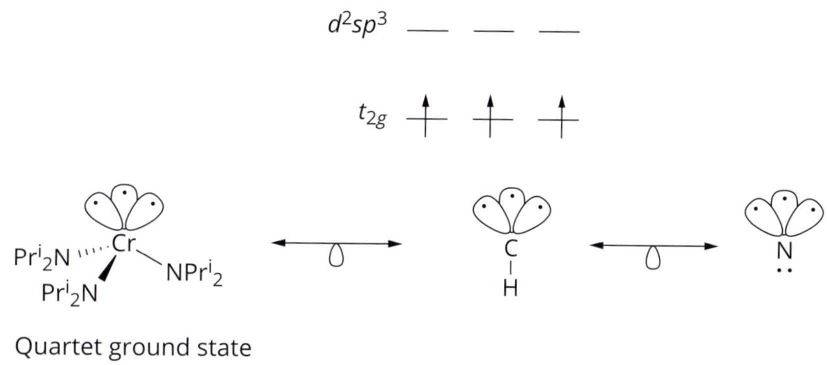

Fig. 14.24 The quartet $Cr(NPr^i_2)_3$ species is isolobal with CH due to the half-filled t_{2g} set. Naturally, quartet CH is also isolobal with the nitrogen atom.

is in chromium and molybdenum tris(amido) complexes in chemistry performed by Cummins and coworkers.[7] The complex $Cr(NPr^i_2)_3$, originally prepared by Bradley,[8] has a quartet ground state and is d^3. The complex is isolobal with CH. This is a case that we will look into the low-energy t_{2g} orbitals (compare Fig. 14.20), which will all be half-occupied. More usefully, we will draw analogies between compounds of $Cr(NPr^i_2)_3$ and the nitrogen atom, which is isolobal with CH.

A reaction of $Cr(NPr^i_2)_3$, reported by Bradley and coworkers, was with NO, where the complex $ONCr(NPr^i_2)_3$ was prepared.[7] The mononitrosyl chromium compound is isolobal with N_2O, nitrous oxide, which is a common oxidant that liberates N_2 after O-atom transfer. Because N_2O is an oxidant, it was reasoned that $ONCr(NPr^i_2)_3$ could also be deoxygenated, which would produce $NCr(NPr^i_2)_3$. Terminal nitrido complexes of chromium(VI) were, at the time, unknown, despite a plethora of fully characterized Mo(VI) and W(VI) terminal nitrido complexes.[9]

A search for the proper oxygen-atom acceptor led to $V(mesityl)_3 \cdot THF$, which worked well (Fig. 14.25) to extract the oxygen, producing $O[V(mesityl)_3]_2$ and the N_2 analogue $NCr(NPr^i_2)_3$. In related chemistry, complexes such as $OCr(OMe)(NPr^i_2)_2$ were also generated—chromium(V) complexes that are isolobal with the radical small molecule nitric oxide, NO.[10–12]

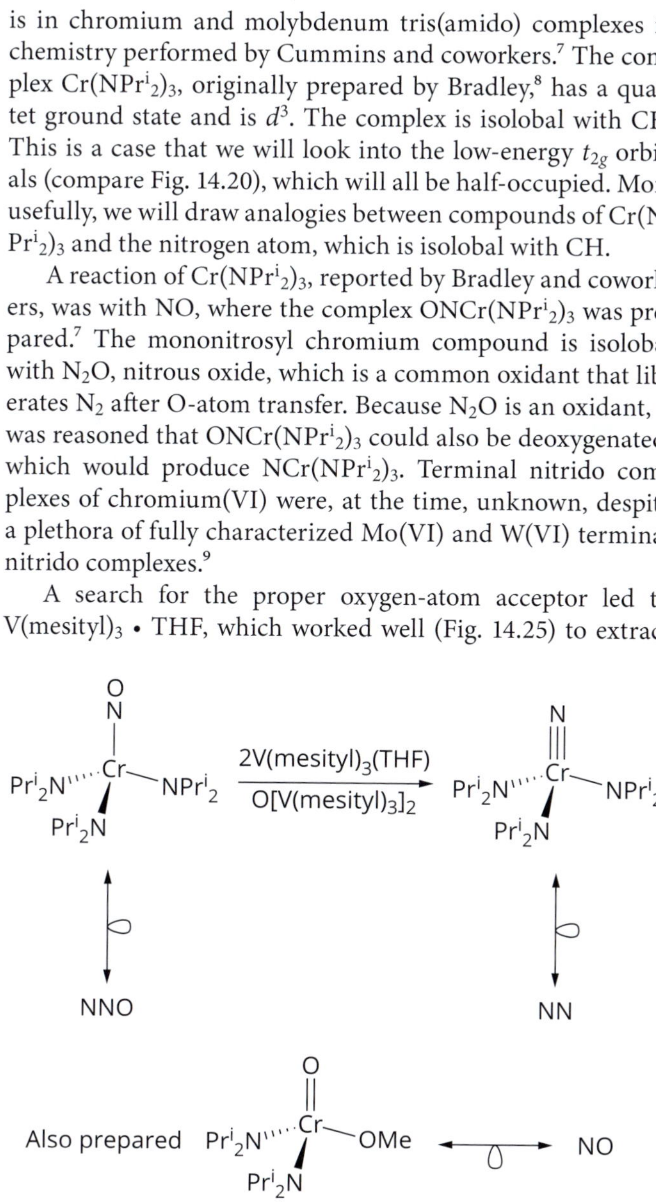

Fig. 14.25 Chromium complexes isolobal with nitrous oxide (N_2O), dinitrogen (N_2), and nitric oxide (NO).

14.2 A Hybridization-Based Bond Approach to Transition-Metal Geometries and Electronic Structure

The strength of a bonding interaction is inversely proportional to the energy difference of the orbitals involved in the participating atoms of the bond. As the energy difference between orbitals of the interacting parts increases, the weaker their bonding interaction, all else being equal. Similarly, a mismatch in atomic orbital energy or size on the atom can lead to poor hybridization of orbitals. The energy of the *p*-orbitals of the transition metals is quite high compared to the valence *s*- and *d*-orbitals. As a result, the expectation that most metal centers will use d^2sp^3-hybrids with half of the hybrids comprised of higher-energy *p*-orbitals seems unreasonable.

A natural bond orbital calculation, NBO6, on low-spin $\{Co(NH_3)_6\}^{3+}$, a prototypical Werner complex, suggests that the hybrids used by cobalt to form the bonds have far less than 50% *p*-character. Indeed, the calculated hybrids used by cobalt are $sp^{0.01}d^{2.07}$; that is, sd^2-hybrids. The metal is using exclusively its lower energy *s*- and *d*-orbitals for bonding, but why sd^2, and how do we now describe the bonding?

Landis and coworkers have developed a hybridization theory method for examining transition-metal bonding. Typically, the method ignores *p*-orbital contributions. In many cases this leads to intuitive understanding of geometries. The method has drawn the ire of some who feel that *p*-orbitals should be included in the valence set,[13–15] but like all theories this NBO-based system can be modified in cases where needed with *p*-orbital participation. This is no different from what we do with MO theory: we add layers of complexity to improve the model. The hybridization method for transition-metal geometries is included here, as in many cases it provides intuitive results useful for understanding transition-metal chemistry. More precise measurements beyond our intuition may need the inclusion of the smaller amounts of valence *p*-orbital participation.

14.2.1 The *sd*-Hybrids and Preferred Geometries

Landis and coworkers have examined transition-metal bonding in some detail, with the assumption that there is minimal or no *p*-orbital participation in transition-metal bonding. In this hybrid-orbital approach, the bonding by the metal is achieved exclusively with *sd*-hybrids. In the model, the bonding in transition metals is explained by using Lewis-like electronic structures based on the shapes of the *sd*-hybrids.

The *d*-block molecules will be "normal valent" if they occupy six orbitals, "hypovalent" if they occupy fewer than six, and "hypervalent" if it appears that the complex has more than six orbitals occupied. Normal valent and hypovalent molecules will be treated together in the ensuing discussion, as their bonding will involve combinations of nonbonding electrons (if any), 2-electron 2-center bonds between the metal and ligand, and perhaps empty orbitals on the metal. Hypervalent molecules will be treated by using some of the same tools developed previously for understanding main-group hypervalent molecules, such as 3-center 4-electron bonds.

The first question to be addressed is: "What do the sd^{μ}-hybrids look like?" Fig. 14.26 shows the shapes of these hybrids. If we start from the bottom right, the d_{z^2}-orbital is shown. The angles from the largest part of the orbital along the *z*-axis to the nodes near the torus are 55° and 125°. In other words, if we had two d_{z^2}-shaped orbitals occupied with electrons and want to arrange them to minimize repulsion, we would place them rotated at either 55° or 125° to minimize electron–electron repulsion between these occupied orbitals.[1]

In the bottom middle of Fig. 14.26 is shown the sd^5-hybrid. The mixing of some *s*-character into the d_{z^2}-orbital reduces the size of the torus, and the smaller nodal angle increases to 63°,

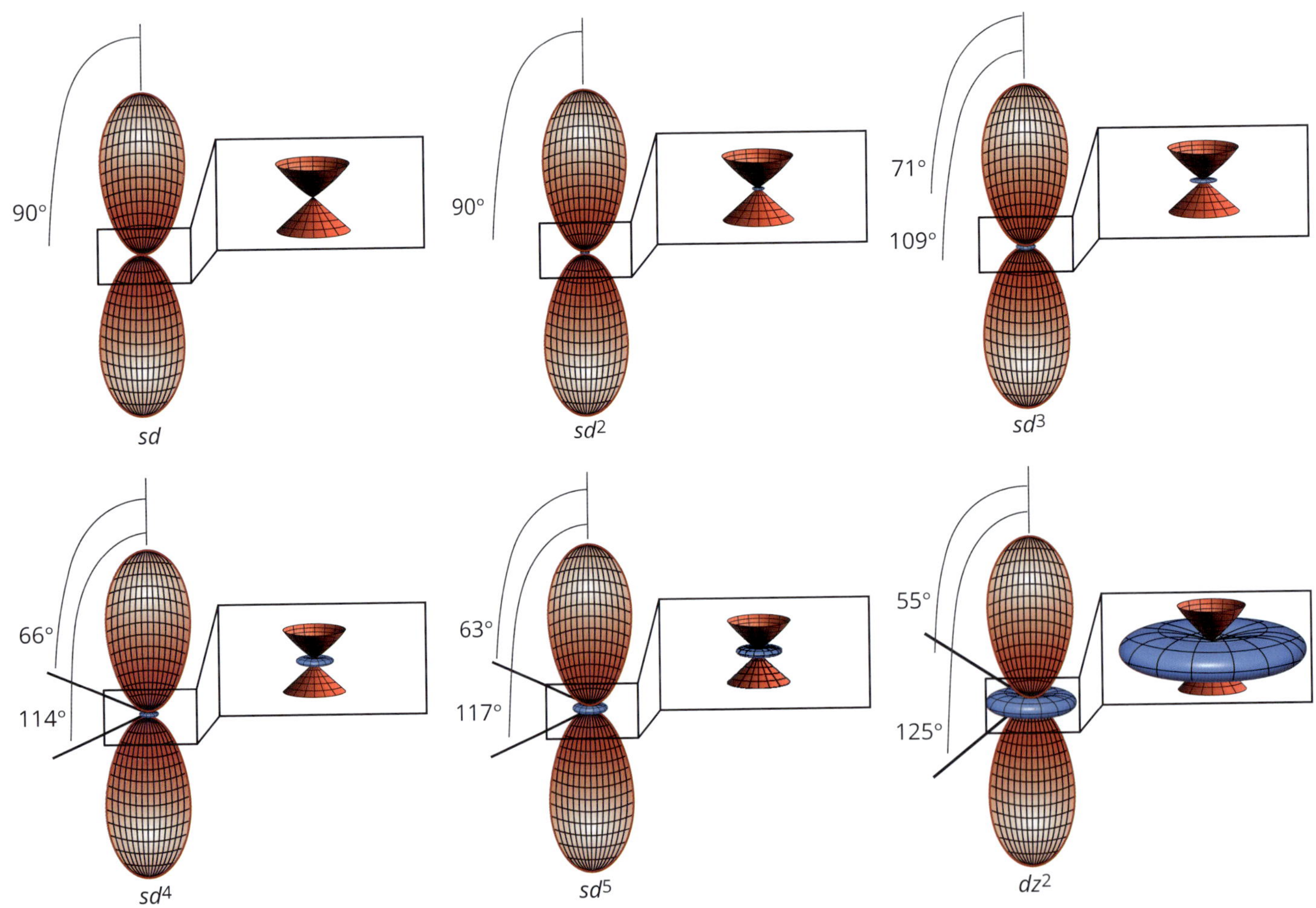

Fig. 14.26 The shapes of the sd^{μ}-hybrids and the angle from the largest part of the orbital to the nodes. (Figure from the article by Landis and coworkers.[1])

while the larger nodal angle decreases to 117°. As more *s*-character is added to the orbital, the size of the torus continues to shrink until at sd^2 and *sd* there is a single node at 90°.

The requirement that hybrids be orthogonal was used for sp^λ-hybrids to produce an equation relating the hybridization factor λ and the angle (ω_{ij}) between hybrids *i* and *j*, which we called Coulson's directionality theorem. A similar exercise with the sd^μ-hybrids produces the following equations between the angles of the hybrids and the hybridization factor μ.[5]

$$\cos\omega_{ij} = \pm\sqrt{\frac{1-\frac{2}{\sqrt{\mu_i\mu_j}}}{3}} \quad for\ \mu > 2$$
$$= 0 \quad for\ \mu \le 2$$

Using these sd^μ-hybrids, one can calculate the resulting molecular shapes due to energetic minima (Table 14.1). For some of the hybridizations, multiple minima are found; however, in all cases there is typically one of these geometries—seemingly the one that will minimize ligand–ligand (steric) repulsions—that is found for molecules with that hybridization.

Before we discuss the shapes of molecules by using this method, a few rules should be kept in mind when deciding the hybridization of the metal center:

(1) Nonbonding electrons on the metal center will occupy pure *d*-orbitals, which will prevent the orbital from hybridizing. For example, if the complex is a singlet d^2 with one pair of electrons in a *d*-orbital, then the highest possible hybridization will use the four remaining *d*-orbitals, sd^4.

(2) Any substituents with π-bonds to the metal center must use *d*-orbitals to bond. One of the assumptions was that the bonding was done with only *s*- and *d*-orbitals. Any π-bonds must, by symmetry, be done with the *d*-orbitals. If there is one π-bond, then one *d*-orbital is not available for forming hybrids.

We will now examine experimental complexes that are examples of these various types of hybridization.

14.2.1.1 sd^5-Hybridization

The complex WMe_6 was reported by Shortland and Wilkinson in 1973 (Fig. 14.27).[16] The synthesis, which looks quite simple but hides significant complexity, involves the addition of LiMe to WCl_6 in diethyl ether at low temperature to provide the pure complex as red crystals. On its surface, no one necessarily expects that the complex would require a flurry of activity on the part of theoreticians to explain its structure. By VSEPR, and indeed in accord with many other W(VI) complexes such as the WCl_6 starting material, one would expect WMe_6 to be octahedral. However, the compound exhibits a C_{3v} geometry that is estimated by calculations to be 142 kcal/mol more stable than O_h![17–18] The compound is not only less

Table 14.1 Molecular shapes of the arrangement of sd^μ-hybrids around a single metal center.[1]

Idealized hybridization and angles	Idealized molecular shape and its point group	Experimental examples with specified hybridization
sd^5 117 and 63°	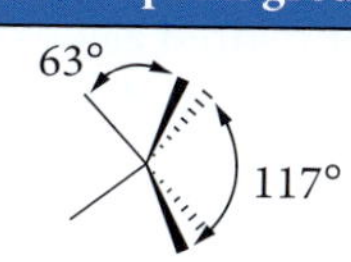 C_{3v}	WMe_6,[16–18] $\{ZrMe_6\}^{2-}$,[19] W(*o*-xylidene)$_3$[20]
sd^4 114 and 66°	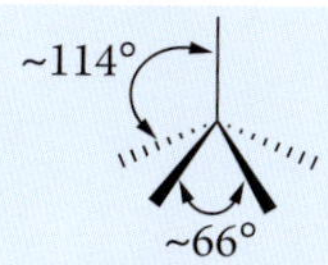 C_{4v}	$TaMe_5$,[21–22] Ta(CH_2-*p*-MeC_6H_4)$_5$[23], $MoMe_5$[24]
sd^3 109 and 71°	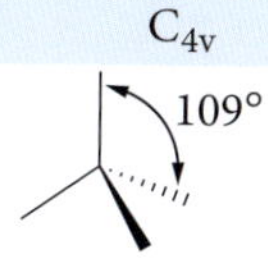 T_d	$OsPh_4$, M(*o*-tolyl)$_4$ (M = Os, Re, Ru, Mo), M(cyclohexyl)$_4$ (M = Ru, Os, Cr), M(mesityl)$_4$ (M = Ru, Ir)
sd^2 90°	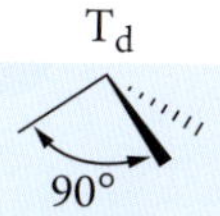 C_{3v}	Rh(mesityl)$_3$, Ir(mesityl)$_3$
sd 90°	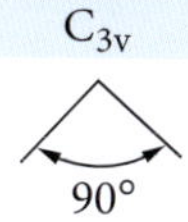 C_{2v}	Cr{N(Ar)BAr_2}$_2$[25] (Ar = mesityl), $Cr_2Ar'_2$ (Ar′= 2,6-(2,6-$Pr^i{}_2C_6H_3$)C_6H_3)

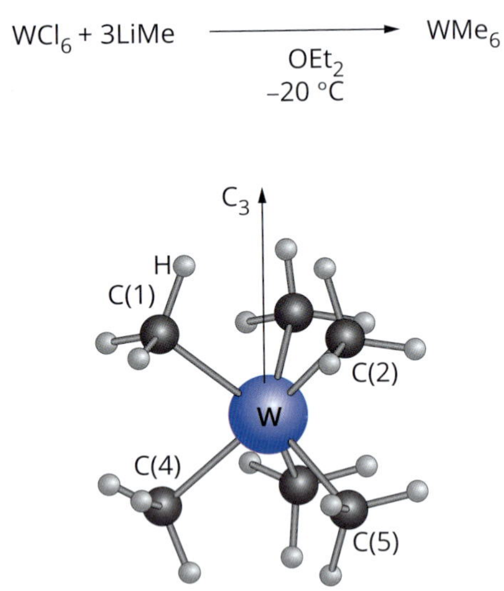

Fig. 14.27 Synthesis and structure of WMe_6. The structure by electron diffraction in the gas phase is shown, which was described as being "deformed to C_{3v} symmetry".[18] (Figure from Haaland and coworkers.[18])

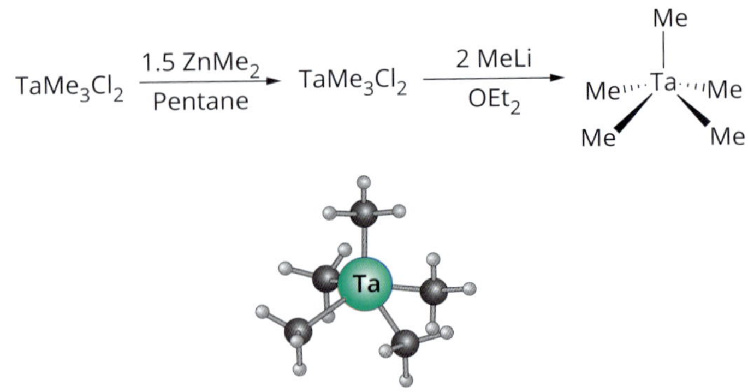

Fig. 14.28 Synthesis and electron-diffraction structure of $TaMe_5$.[17,18]

symmetrical than octahedral but profoundly so, especially from an energetic standpoint. In contrast, WF_6, $W(OMe)_6$, and $W(NMe_2)_6$ all have pseudo-octahedral coordination around the metal center. (Note that each of those ligands has π-symmetry lone pairs that can donate to the metal.) The structure from electron diffraction was described as "deformed to C_{3v} symmetry",[18] but considering the molecule always has C_{3v} symmetry, one can wonder what symmetry it was "deformed" from. The answer, of course, is the expectation of octahedral symmetry from VSEPR theory (or the highest symmetry possible of O_h).

The complex WMe_6 is a 12-electron compound, which will be normal valent under this scheme. Since the compound has no π-bonding and is d^0, all of the d-orbitals on tungsten can be hybridized with s to form sd^5-hybrids. In fact, the idealized geometry for this hybridization shown in Table 14.1 is very close to the experimental geometry for WMe_6.

The compound WF_6, for example, is octahedral because of hypervalence due to π-bonding from the fluorines. The calculated (SDD/B3LYP NBO6) hybrids being used by tungsten to form the W–F bonds are sd^2, because some d-orbitals—namely, the t_{2g} set—are occupied in π-interactions. Any complex of general formula WX_6, where X does not have π-effects and is anionic, is expected to produce the C_{3v} structure, while any system where X includes filled π-donor orbitals should be O_h.

14.2.1.2 sd^4-Hybridization

The 5-coordinate complex $TaMe_5$, originally reported by Schrock and Meakin in 1974, has a square pyramidal geometry rather than the trigonal bipyramidal arrangement expected by VSEPR.[21–22] The complex is 10-electron and therefore hypovalent with an empty unused d-orbital remaining on tantalum. The gas-phase electron-diffraction structure of $TaMe_5$ is quite consistent with the C_{4v} structure anticipated from sd^4-hybridization (Fig. 14.28).[17,18]

Another molecule with obvious sd^4-hybridization is $MoMe_5$ prepared by Seppelt and coworkers, in which the d-orbital not used in bonding is half-filled.[24]

14.2.1.3 sd^3-Hybridization

One of the most common hybridizations for 4-coordinate molecules is sd^3, which has a tetrahedral disposition of ligands around the central metal. When the metal center is sd^3-hybridized and hypo- or normal valent, the compound is almost invariably tetrahedral.

Examples of this class of compound are numerous, such as high-spin d^2 $Cr(CH_2Bu^t)_4$, which was first reported by Wilkinson's research group and structurally characterized by Gambarotta and coworkers. The complex exhibits a tetrahedral structure consistent with sd^3-hybridization and with the two unpaired electrons residing in pure d-orbitals (Fig. 14.29). The compound is prepared from the action of neopentylmagnesium

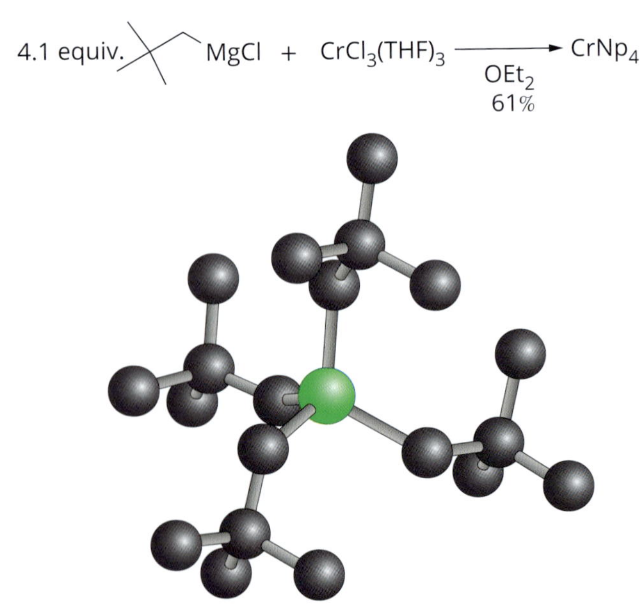

Fig. 14.29 Synthesis and X-ray diffraction structure of $CrNp_4$. Hydrogens are not shown.

chloride on $CrCl_3(THF)_3$ in ether, which provides pyrophoric, dark brown tetrakis(neopentyl)chromium(IV) after valence disproportionation (Fig. 14.29).[26–28]

14.2.1.4 *sd*²-Hybridization

The classic examples of normal valent sd^2-hybridization are two complexes prepared by Wilkinson and coworkers, M(mesityl)$_3$, where M = Rh and Ir. These two complexes have low-spin d^6-metal centers, meaning that three pure *d*-orbitals are filled with electrons. Consequently, the highest hybridization possible is sd^2.[29–30]

From Table 14.1, sd^2-hybrids are 90° apart, and one expects that these complexes will be pyramidalized 3-coordinate species with 90° angles. The complexes are in fact strongly pyramidalized with angles of 105° and 108° for the rhodium and iridium derivatives, respectively, from X-ray diffraction (Fig. 14.30).

This pyramidalized geometry is predicted to be preserved for all normal valent complexes with metals having sd^2-hybridization with only σ-effects to the ligands. DFT calculations (B3LYP, SDD) on d^6 RhH_3 and d^0 ZrH_3^+ predict H–Zr–H angles of 84° and 103°, respectively. Furthermore, the structure of d^0 La{CH(SiMe$_3$)$_2$}$_3$ has C–La–C angles of 109°.[31]

14.2.1.5 *sd*-Hybridization

The preferred angle for *sd*-hybrids is 90°, and there are few normal valent or hypovalent compounds that use this hybridization (though it is common for hypervalent compounds, as we shall see). An example of a complex that may use *sd*-hybrids is Cr{N(Ar)BAr$_2$}$_2$, where Ar = mesityl, prepared by Power and coworkers.[25] The compound is high-spin with a $\mu_{eff} = 4.91\ \mu_B$, close to that expected for four unpaired electrons. These four unpaired electrons will reside in pure *d*-orbitals, leaving the highest possible hybridization as sd^1.

One would expect that the N–Cr–N angle will be ~90°, and the complex has a significantly bent structure of 112.3(3)° (Fig. 14.31). The hypervalent and related complexes

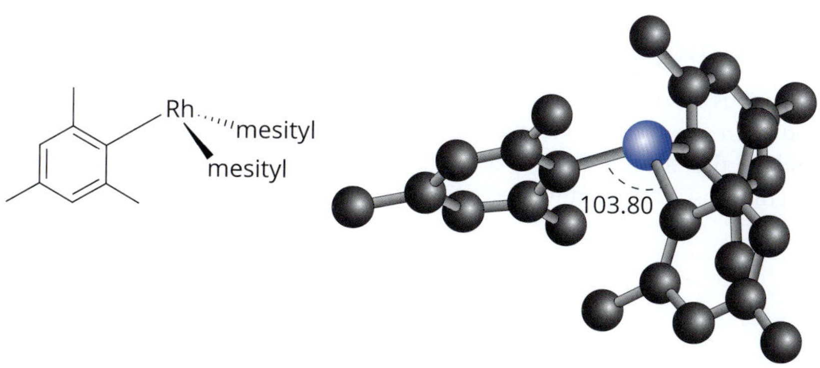

Fig. 14.30 Structure of Rh(mesityl)$_3$ along with a ball-and-stick drawing from the X-ray coordinates. Hydrogens in calculated positions are not shown in the X-ray structure.[29]

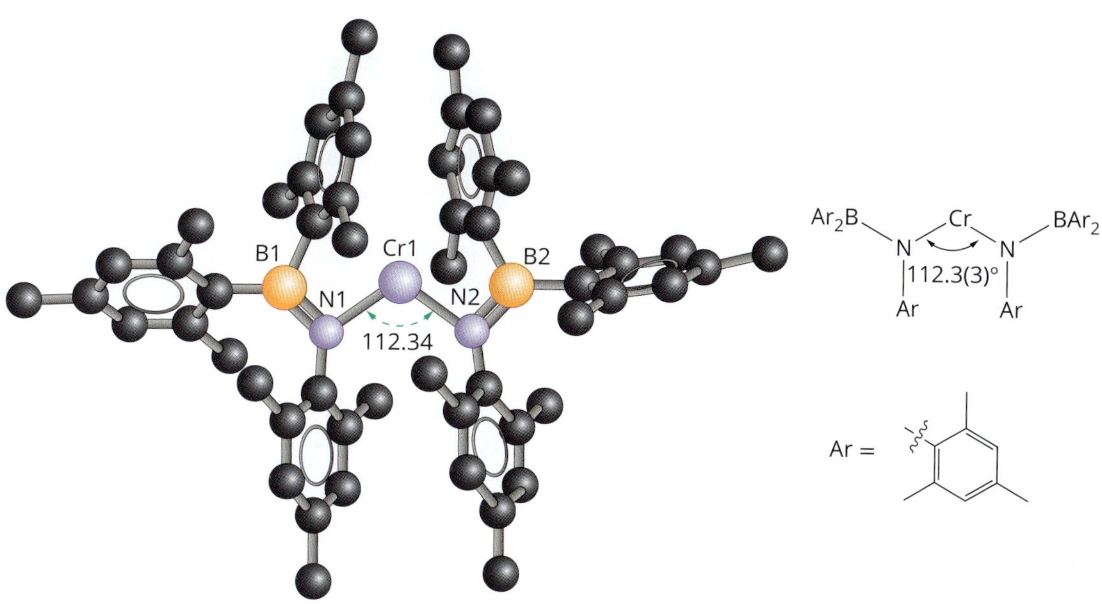

Fig. 14.31 The structure of Cr{N(Ar)BAr$_2$}$_2$, where Ar = mesityl, prepared by Power and coworkers.[25] The ball-and-stick structure is based on the X-ray diffraction coordinates. Hydrogens and a molecule of pentane in the lattice are not shown.

M{N(Ar)BAr$_2$}$_2$, where M = Mn, Fe, Co and Ni, are high-spin with μ_{eff} = 5.98, 4.92, 4.36, and 2.9 μ_B, respectively, and have structures far closer to linear with N–M–N angles of 160.4°, 166.6°, 168.4°, and 167.9°.

The bending in Cr{N(Ar)BAr$_2$}$_2$ was ascribed to possible Cr–C(*ipso*) interactions, which may contribute to the geometry; however, the electronic nature of the bending is consistent with calculations on the geometry of CrH_2 in the quintet state (four unpaired electrons). The calculated angle in CrH_2 is 111°, which is remarkably similar to the angle in Cr{N(Ar)BAr$_2$}$_2$, and NBO6 suggests $sd^{1.22}p^{0.02}$-hybrids are being used to form the bonds on chromium,[32] very nearly the expected sd^1. The slightly higher *d*-character in the bonds is due to the preference to place some electron density from the unpaired electrons in the higher-energy orbital; in this case, *s*.

Another example of a metal complex displaying *sd*-hybridization[33] is the pentuple bond complex of Power and coworkers, $Cr_2Ar'_2$, Ar′ = 2,6-(2,6-$Pr^i_2C_6H_3$)C_6H_3, shown in Fig. 14.32.[34] The chromium center should use sd^1-hybrids in its σ-bonds with angles of 90°, and the angles are very nearly at 103°.

14.2.2 Hypervalent Complexes

All of the complexes in the previous subsection were "normal valent" or "hypovalent", and there were ample orbitals for the metal-based electrons and for the metal to form 2-center 2-electron (2c2e) bonds to all of the substituents. Many transition metals are "hypervalent" and do not have enough orbitals (as *sd*-hybrids) to hold all the *d*-electrons and form 2c2e bonds to all of the attached groups; the bonding in these complexes can often be explained by forming the maximal sd^μ-hybrids allowed after orbitals are used to hold *d*-electrons (single or doubly occupied orbitals) then using 3-center 4-electron (3c4e) bonds.

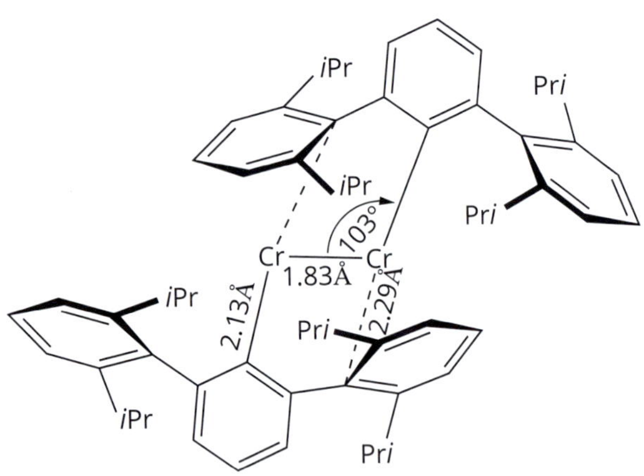

Fig. 14.32 Structure of $Cr_2Ar'_2$, Ar′ = 2,6-(2,6-$Pr^i_2C_6H_3$)C_6H_3, by Power and coworkers. (Drawing from the paper by Landis and Weinhold.[33-34])

A very common example is that of d^8 square planar complexes, such as $PtCl_4^{2-}$ (Fig. 14.33). These are low-spin complexes in which *d*-electrons occupy four orbitals in pairs. Consequently, there is only one *d*-orbital that remains for bonding, which is used to make sd^1-hybrids. The *sd*-hybrids have maximal overlap with the chloride ligands when the Cl–Pt–Cl angles are at 90° (Table 14.1), but there are four Cl–Pt bonds to be made and only two hybrids to form the bonds. In all cases, 3c4e bonds are ~180° apart; consequently, there are two covalent bonds 90° apart with 3c4e bonds to the chlorides 180° from each of these.

From a hybridization theory perspective, the two *sd*-hybrids will bond to two adjacent chlorines, while the remaining two chlorines are unbound to the metal. By rotating through 90-degree increments, a set of resonance forms is obtained with a ½ bond order in each Pt–Cl linkage.

From an MO perspective, the *p*-orbitals on Pt are quite high in energy and participate very little in the bonding, even though they are of appropriate symmetry. Examination of the MO diagram for PtH_4^{2-} in Fig. 6.20 shows that the e_u-orbital is stabilized very little relative to the SALC, indicating very little bonding interaction. Disregarding this weak interaction with the *p*-orbitals, as is done in this approximation, produces the *s*-orbital interaction ($1a_{1g}$) and a *d*-orbital interaction ($1b_{1g}$) as the only bonding to the four substituents, completely in agreement with the previous picture.

A similar effect will occur when the orbitals are occupied by one electron instead of two. The crystal field (*d*-orbital) picture for a square planar d^8 complex is shown in Fig. 14.34. Also shown is the crystal field picture for a high-spin 4-coordinate d^4 complex such as $CrCl_2$(2,6-lutidine)$_2$. Because four *d*-orbitals are occupied with electrons, the highest hybridization is *sd*, the same as the $PtCl_4^{2-}$ case.[35]

Note that an intimate knowledge of MO theory or crystal field theory was not necessary to apply this VB-based model to determining molecular shape. This model is quite independent, and we are simply pointing out that the different models produce similar results. Here, all one needs to realize is that the maximum hybridization in a molecule such as $PtCl_4^{2-}$ was

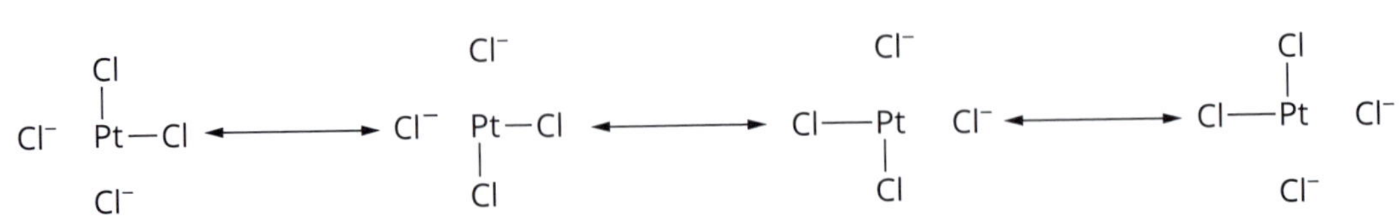

Fig. 14.33 Resonance forms for $PtCl_4^{2-}$. The bonding of each *trans*-pair of chlorides is a 3-center 2-electron bond. The covalent bonds in each resonance form are made with *sd*-hybrids on platinum that are 90° apart.

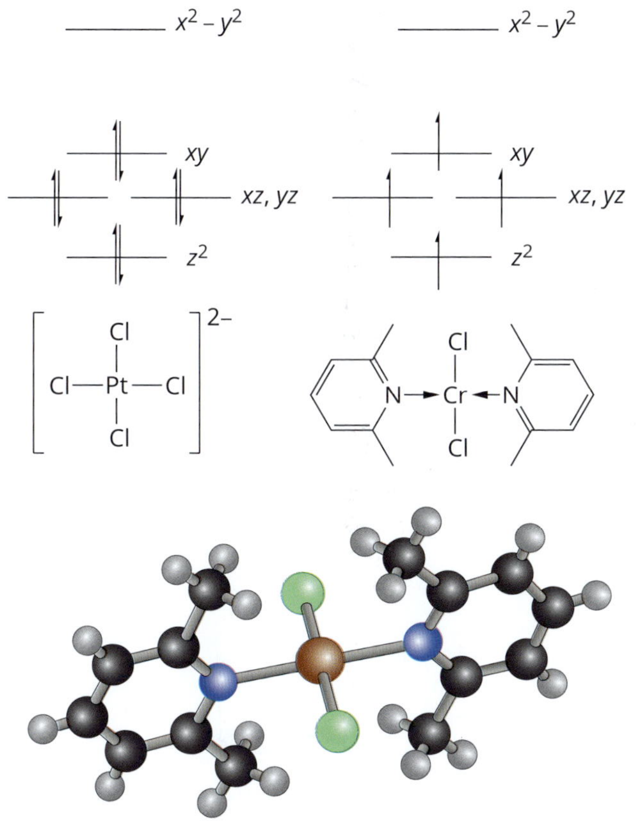

Fig. 14.34 Crystal field (*d*-orbital splitting) diagrams for low-spin d^8 $PtCl_4^{2-}$ and high-spin d^4 $CrCl_2$(2,6-lutidine)$_2$, showing that in each case four *d*-orbitals are used in simply holding electrons. The hybridization at each metal is *sd*, which produces the square planar structures shown. The ball-and-stick diagram from the X-ray diffraction study on the chromium complex is also shown.[35]

sd, that the hybrids were 90° apart, and that the bonds involve 3-center 4-electron bonds that are 180° from one another. From that information the square planar geometry is the only possible result.

14.2.3 A Jahn–Teller Explanation for the Structure of WMe_6: A Complementary Approach[36]

In the hybridization theory-based explanation of the geometry of transition-metal complexes with σ-only ligands, the assumption was that only the *s*- and *d*-orbitals participate in the bonding. One of the great triumphs of this approach is the correct assignment of the structure of WMe_6—a molecule that has a C_{3v} geometry instead of the VSEPR expected O_h. How does molecular orbital theory explain the geometry of WMe_6, and is the result similar to the hybridization theory approach?

Albright and coworkers used a second-order Jahn–Teller approach to explain the geometry of WMe_6, which is what we will briefly recount here.[36] For this method, we will begin with the highest-symmetry geometry, octahedral, and will then allow distortions from this high symmetry to lower symmetry.

As was shown in Section 14.1.1, in order to obtain six hybrids 90° from one another around a transition-metal center for 2-center 2-electron bonds to each methyl group, the required hybridization at the metal is d^2sp^3. Since the ligands in WMe_6 have no π-component and there are no *d*-electrons remaining on the metal center, there is only the σ-bonding to the ligands with which to be concerned.

For simplicity, we will discuss the model complex WH_6, which will simplify the calculations somewhat and increase the actual symmetry. The initial MO diagram in O_h must be consistent with d^2sp^3-hybridization and is shown in Fig. 14.35. This diagram is identical to the diagram for CrH_6 discussed in Chapter 6. Note that the occupied valence orbitals have character attributable to the tungsten *s* ($1a_{1g}$), two *d*-orbitals ($1e_g$), and three *p*-orbitals ($1t_{1u}$), consistent with the d^2sp^3-hybridization anticipated by VB theory for a purely octahedral system.

The MO diagram in Fig. 14.35 is required by symmetry, but does it make the most chemical sense? The lowest-energy orbitals on tungsten are the *s*- and *d*-orbitals. In this diagram, the *p*-orbitals are being used more than either the *s*- or *d*-orbitals in bonding, sp^3d^2-hybridization, which seems like an energetic disadvantage for this high-symmetry structure.

There are several normal modes in an octahedral system that can mix the HOMO and LUMO to cause a second-order Jahn–Teller distortion from this structure. A full understanding of how the normal modes may be used to find the end structure is beyond the scope of this book but is discussed in detail by Bersuker.[37] However, Albright and coworkers' paper on the topic provides a very clear, stepwise diagram for why the distortion occurs, which is reproduced in Fig. 14.36.[36]

As in the detailed MO diagram in Fig. 14.36, the HOMO in the O_h geometry is a filled t_{1u} orbital, and the LUMO is of t_{2g} symmetry. To obtain the orbitals in the middle of Fig. 14.36, we first mix the orbitals in the *xy*-plane, and a structure of D_{3h} symmetry is obtained. During this, *d*-character is mixed into t_{1u}, producing a lower-energy orbital of e′ symmetry (mostly *d*-orbital) and a higher energy, mostly *p*-orbital, of e′ symmetry. Furthermore, to obtain this D_{3h} structure, p_z and d_{z^2} are not mixed and are of different symmetry, a_2'' and a_1' respectively.

However, D_{3h} symmetry does not provide a structure that is immune to Jahn–Teller mixing of the orbitals either. This symmetry too is susceptible to HOMO-LUMO mixing through normal modes. This leads to mixing of p_z and d_{z^2} and a lowering of the overall symmetry to C_{3v}.

Both stages of this mixing take p-orbital(s) in the HOMO and mix them with d-character in the LUMO. In other words, the end result is approximate sd^5-hybridization around tungsten—a result that was readily obtained by using the hybridization theory method of Landis and coworkers.

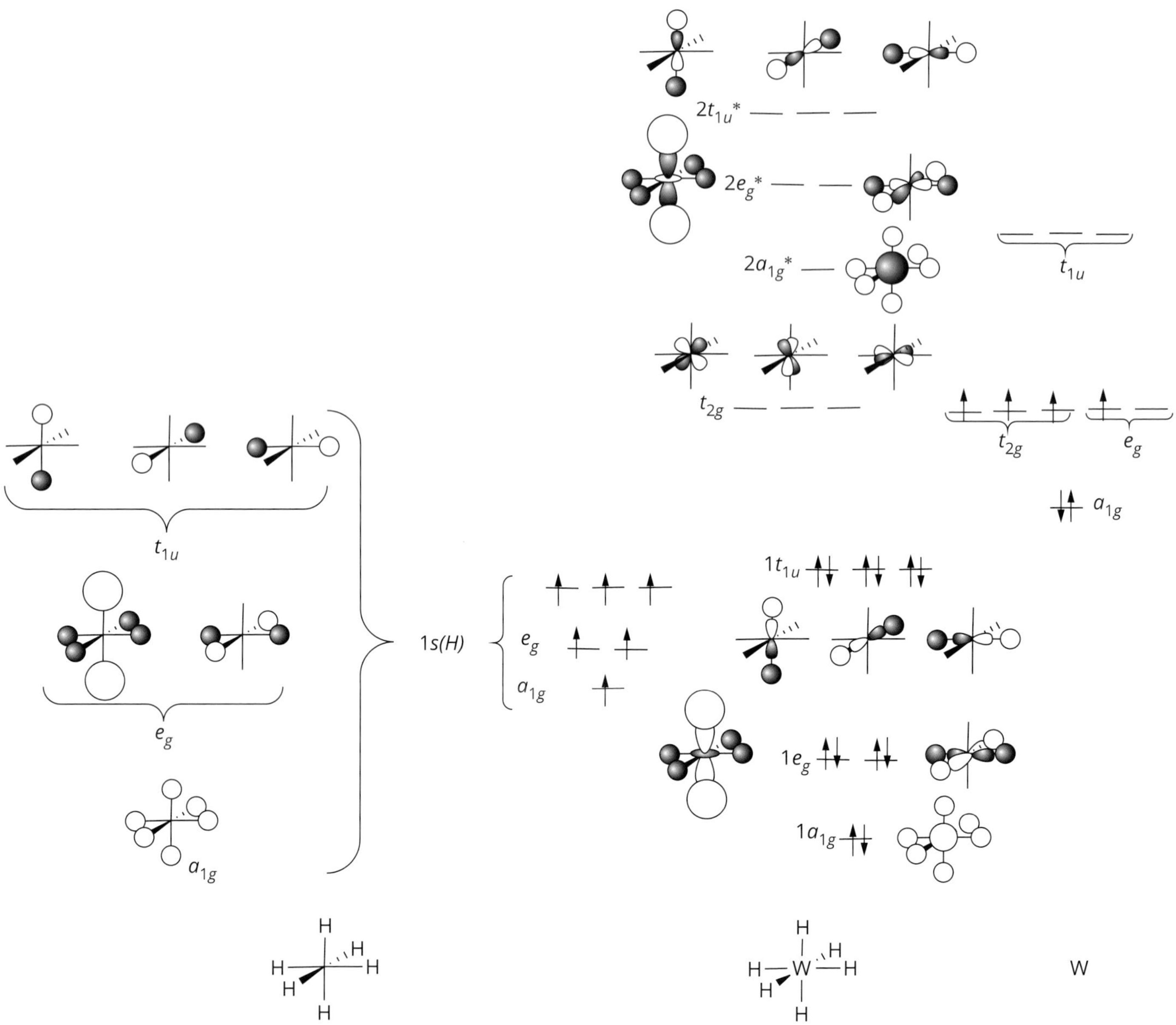

Fig. 14.35 MO diagram of WH_6 in O_h.

How, then, does one explain the octahedral geometry of WF_6? Fluoride is a strong π-donor, and the orbitals of appropriate symmetry to interact with the π-donor orbitals on fluorine are the t_{2g} orbitals on tungsten. As a result, a much larger energy gap exists between the t_{1u} and t_{2g} orbitals in WF_6 than in octahedral WH_6. The amount of mixing is inversely proportional to the energy gap, which leads to a much smaller (in this case unobserved) distortion. The calculated energy gap in octahedral WF_6 from the highest-energy p-character orbital to the lowest-energy empty d-character orbital is 354 kcal/mol (0.564 Hartree), whereas the same gap in octahedral WH_6 is only 212 kcal/mol (0.338 Hartree). (Geometry was optimized by using B3LYP/SDD, and orbital energies were calculated by using CCSD/SDD from that geometry.)

Similar types of Jahn–Teller distortions can be used to explain why the structures of $TaMe_5$ is not trigonal bipyramidal, why Ir(mesityl)$_3$ is not trigonal planar, and other structural effects due to metal rehybridization.

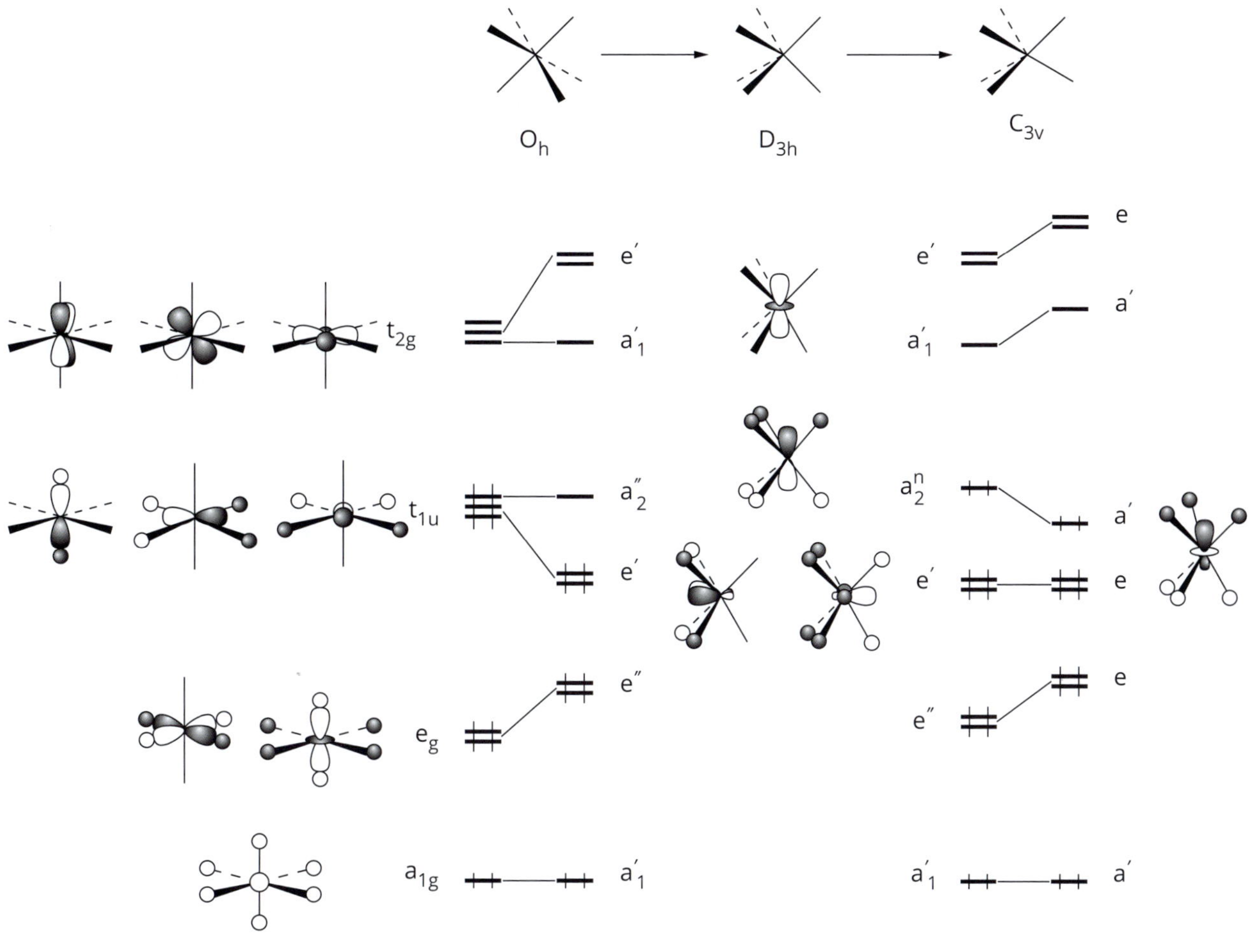

Fig. 14.36 Albright and coworkers' MO diagram for HOMO-LUMO mixing in WH_6. (Figure from the article by Albright and coworkers.[36])

Chapter Summary

1. The chapter presents a few different methods for understanding transition-metal complexes by using hybridization-based methods.
2. The isolobal analogy can be used to relate fragments based on the number and shape of the frontier molecular orbitals and their electron occupancy.
3. The isolobal analogy provides a method for understanding transition-metal reactivity and geometry in the context of perhaps simpler fragments, often atomic species from the main group.
4. A hybridization method for understanding the geometry of transition-metal complexes was proposed by Landis and coworkers, in which the metal center uses *sd*-hybrids to bond to the ligands. This approximation allows a grasp of many unusual transition-metal complexes based on the shapes of these hybrids.
5. From this *sd*-hybrid perspective, compounds then can be hypovalent or normal valent, which requires the number of *sd*-hybrids be equal to the number of ligands. Alternatively, the complex could be hypervalent, and 3-center 4-electron schemes may be used.
6. For *d*-orbitals to be available to hybridize, they must not be occupied by nonbonding electrons or with ϖ-bonds to the ligands.
7. An example of a complex easily understood by using *sd*-hybrids is WMe_6. The metal can be viewed as having sd^5-hybridization, which gives it a C_{3v} geometry. The alternative MO theory method by using Jahn–Teller distortions for understanding this geometry was reported by Albright and coworkers, who came to the same conclusion: the C_{3v} geometry is due to low *p*-orbital participation in the bonding.

Practice Problems

1. From a VSEPR perspective, the complex $TaMe_5$ should be trigonal bipyramidal.
   a. Draw an MO diagram for TaH_5, a simplified model for $TaMe_5$.
   b. What are the symmetries of the HOMO and LUMO of the complex?
   c. Find the symmetries of the normal bending modes for $TaMe_5$ by first calculating all of the vibrations and then removing the stretches.
   d. Should trigonal bipyramidal $TaMe_5$ undergo a Jahn-Teller distortion? If so, use the projection operator to identify the type of distortion that might occur.
2. Show that the following compounds are isolobal: P_4, $Co_4(CO)_{12}$, $As_2Co_2(CO)_6$, $PCo_3(CO)_9$ and $S_2Fe(CO)_4$. Are all of these compounds known? Find references for the syntheses of the compounds.
3. Analyze the $TiCp_2$ fragment in C_{2v}. Use the projection operator, or simply draw by inspection, the SALCs on the two Cp ligands. How will these overlap with the metal-based orbitals?
4. Which compound below is not isolobal to the other two?[2]

(a)

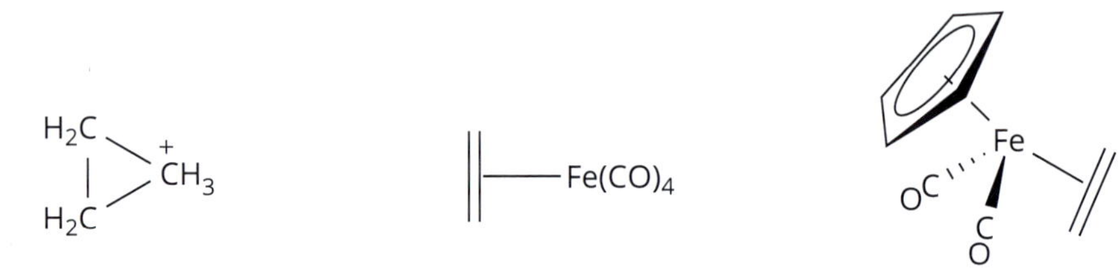

(b)

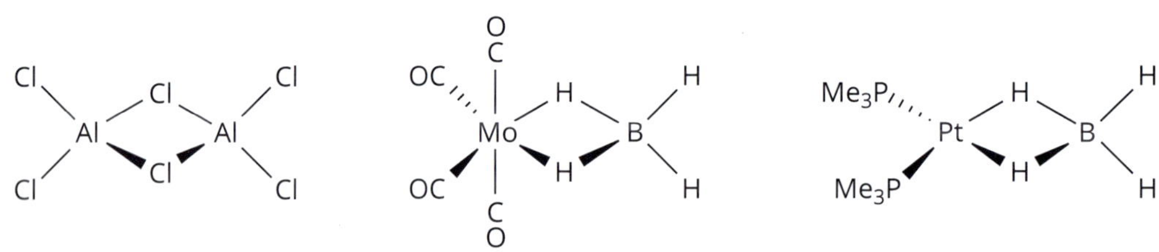

(c)

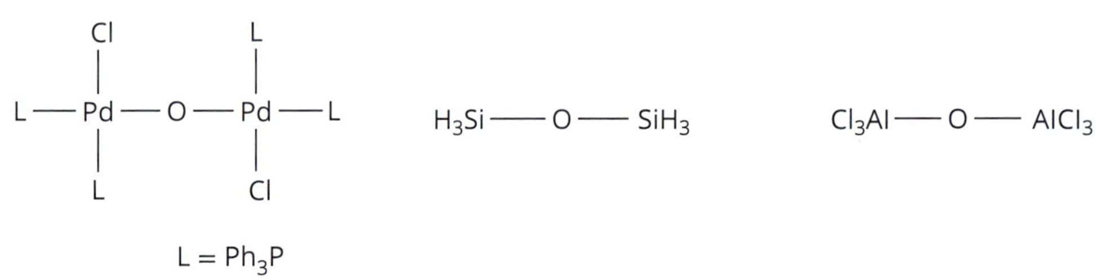

(d)

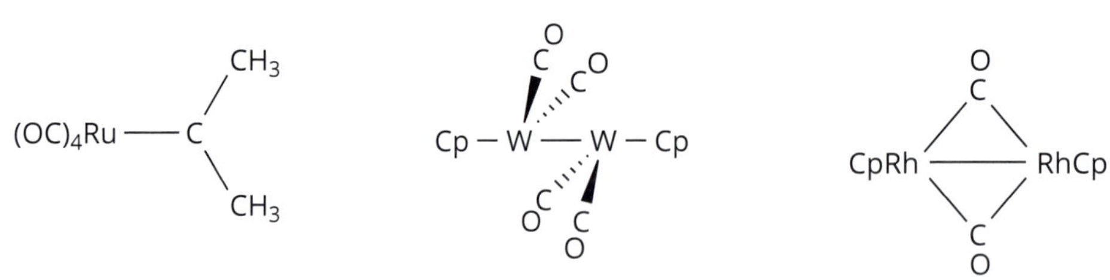

5. Draw an all-organic analog of the following molecules.[2]

(a)

(b)
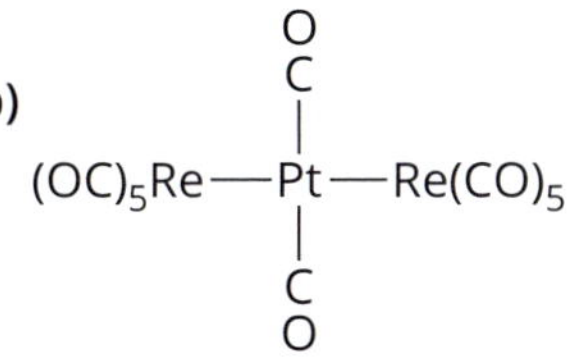

(c)
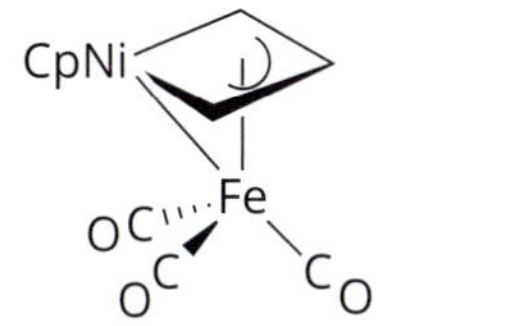

(d)
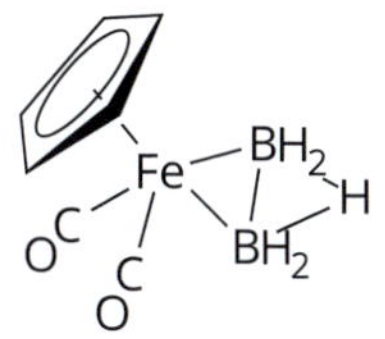

(e)
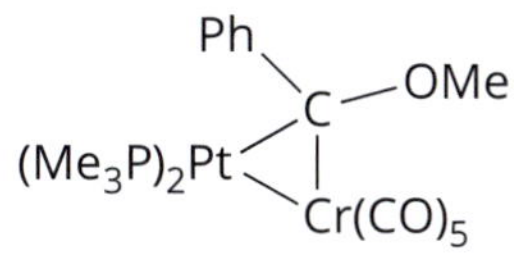

(f)
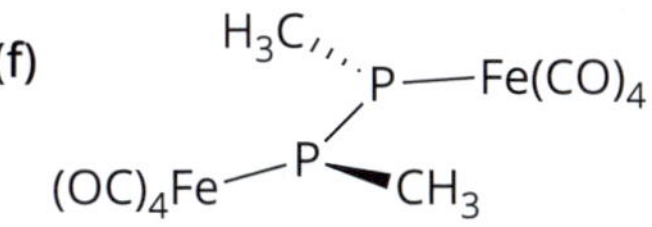

(g)
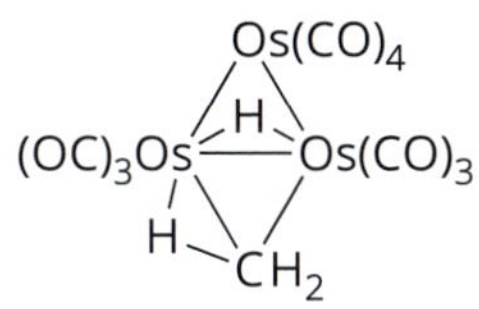

(h)
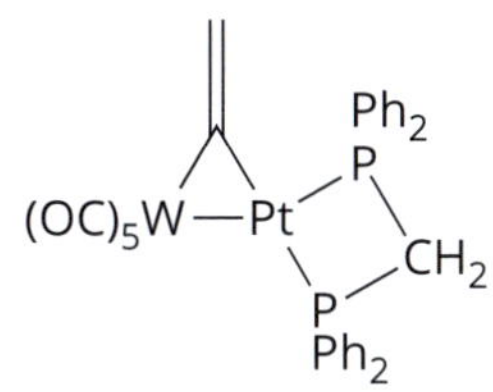

(i)
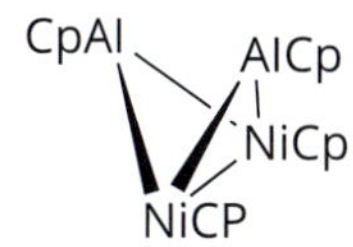

(j)
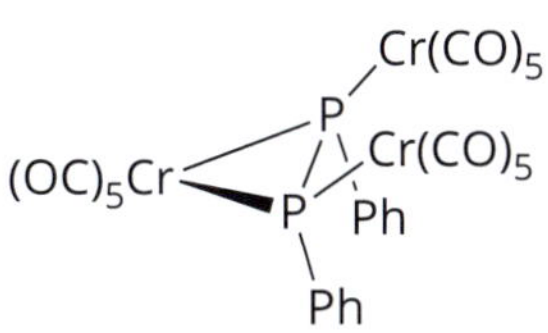

(k)
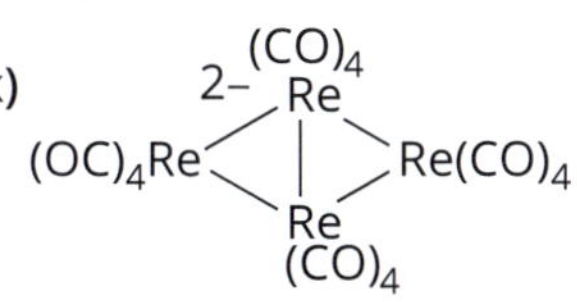

(l)
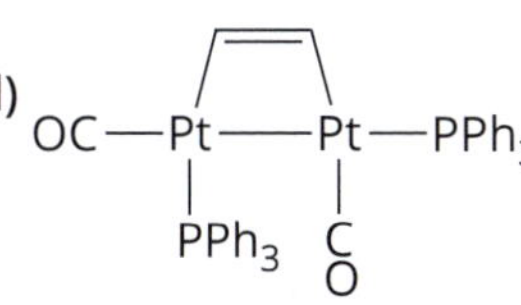

(m)
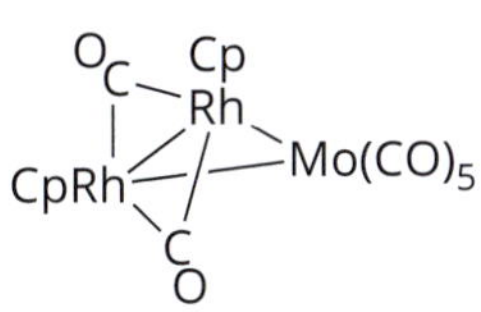

(n)
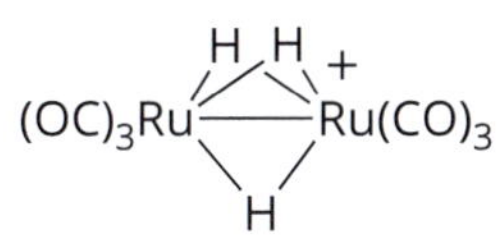

(o)
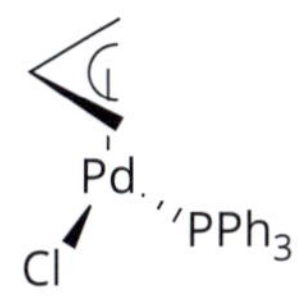

(p)
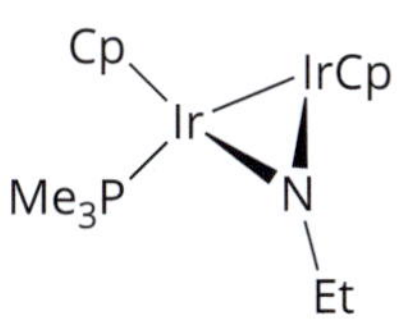

(q)
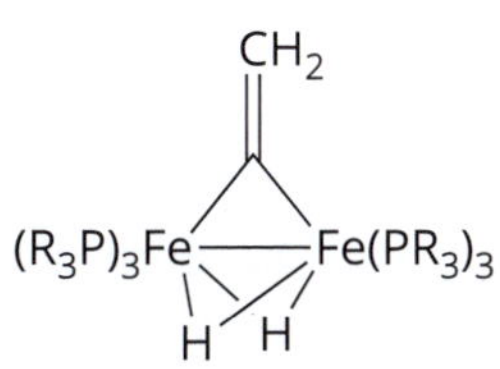

(r)
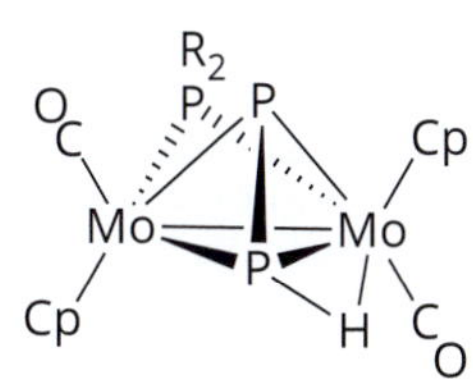

(s)
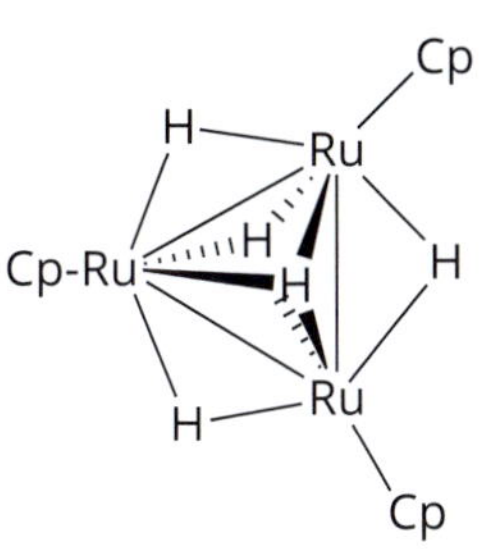

(t)
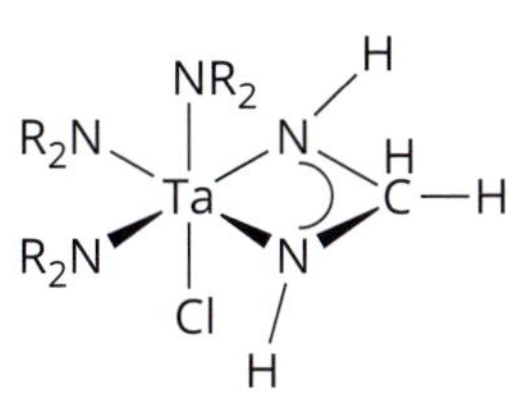

6. Using valence bond theory, predict the shapes of the molecules $FeMe_4^{2-}$ and $FeMe_4^{1-}$. Assume that the former is diamagnetic with no unpaired spins, and assume that the later has 3 unpaired electrons.

Cited References

1. Landis, C. R.; Firman, T. K.; Root, D. M.; Cleveland, T., A valence bond perspective on the molecular shapes of simple metal alkyls and hydrides. *J. Am. Chem. Soc.* 1998, *120*, 1842–1854.
2. Albright, T. A.; Burdett, J. K.; Whangbo, M.-H.; Albright, T. A.; Burdett, J. K.; Whangbo, M. H., The Isolobal Analogy. *Orbital Interactions in Chemistry, 2nd Edition* 2013, 616–652.
3. Hoffmann, R., Building bridges between inorganic and organic-chemistry (nobel lecture). *Angew. Chem. Int. Ed.* 1982, *21*, 711–724.
4. Hultgren, R., Equivalent chemical bonds formed by s, p, and d eigenfunctions. *Physical Review* 1932, *40*, 0891–0907.
5. Weinhold, F.; Landis, C. R., *Valency and Bonding: A Natural Bond Orbital Donor-Acceptor Perspective*. Cambridge University Press: Cambridge, 2005.
6. Stone, F. G. A., Metal-carbon and metal-metal multiple bonds as ligands in transition-metal chemistry: The isolobal connection. *Angew. Chem. Int. Ed.* 1984, *23*, 89–99.
7. Bradley, D. C.; Newing, C. W., Low spin 4-coordinated chromium nitrosyl dialkylamides and alkoxides. *J. Chem. Soc. D-Chem. Comm.* 1970, 219.
8. Bradley, D. C.; Hursthouse, M. B.; Newing, C. W., Crystal structure and electron spin resonance spectrum of *tris*-diisopropylaminatochromium(III). *J. Chem. Soc. D-Chem. Comm.* 1971, 411–2.
9. Dehnicke, K.; Strahle, J., Nitrido complexes of transition-metals. *Angew. Chem. Int. Ed.* 1992, *31*, 955–978.
10. Odom, A. L.; Cummins, C. C., Chromium(VI) nitrido complexes: reactions with Bronsted acids and synthesis of organometallic derivatives. *Polyhedron* 1998, *17*, 675–688.
11. Odom, A. L.; Cummins, C. C.; Protasiewicz, J. D., Nitric-oxide cleavage - synthesis of terminal chromium(vi) nitrido complexes via nitrosyl deoxygenation. *J. Am. Chem. Soc.* 1995, *117*, 6613–6614.
12. Odom, A. L.; Mindiola, D. J.; Cummins, C. C., A nucleophilic chromium(V) dioxo radical anion. *Inorg. Chem.* 1999, *38*, 3290–3295.
13. Bayse, C. A.; Hall, M. B., Prediction of the geometries of simple transition metal polyhydride complexes by symmetry analysis. *J. Am. Chem. Soc.* 1999, *121*, 1348–1358.
14. Diefenbach, A.; Bickelhaupt, F. M.; Frenking, G., The nature of the transition metal-carbonyl bond and the question about the valence orbitals of transition metals. A bond-energy decomposition analysis of $TM(CO)_6{}^q$ ($TM^q = Hf^{2-}, Ta^-$, W, Re^+, Os^{2+}, Ir^{3+}). *J. Am. Chem. Soc.* 2000, *122*, 6449–6458.
15. Maseras, F.; Morokuma, K., Application of the natural-population analysis to transition-metal complexes: should the empty metal p-orbitals be included in the valence space. *Chem. Phys. Lett.* 1992, *195*, 500–504.
16. Shortland, A. J.; Wilkinson, G., Preparation and properties of hexamethyl-tungsten. *Dalton Trans.* 1973, 872–876.
17. Pfennig, V.; Seppelt, K., Crystal and molecular structures of hexamethyltungsten and hexamethylrhenium. *Science* 1996, *271*, 626–628.
18. Haaland, A.; Hammel, A.; Rypdal, K.; Volden, H. V., Coordination geometry of gaseous hexamethyltungsten: Not octahedral. *J. Am. Chem. Soc.* 1990, *112*, 4547–4549.
19. Morse, P. M.; Girolami, G. S., Are d^0 ML_6 complexes always octahedral: The X-ray structure of trigonal-prismatic $[Li(TMED)]_2[ZrMe_6]$. *J. Am. Chem. Soc.* 1989, *111*, 4114–4116.
20. Lappert, M. F.; Raston, C. L.; Skelton, B. W.; White, A. H., A metallotricycle - synthesis, one-electron reduction, and crystal-structure of the thermally robust complex tris-ortho-xylidenetungsten(VI), $W(CH_2C_6H_4CH_2\text{-ortho})_3$ *Chem. Comm.* 1981, 485–486.
21. Schrock, R. R.; Meakin, P., Pentamethyl complexes of niobium and tantalum. *J. Am. Chem. Soc.* 1974, *96*, 5288–5290.
22. Pulham, C.; Haaland, A.; Hammel, A.; Rypdal, K.; Verne, H. P.; Volden, H. V., The structures of pentamethyltantalum and pentamethylantimony: One square pyramid and one trigonal bipyramid. *Angew. Chem. Int. Ed.* 1992, *31*, 1464–1467.
23. Piersol, C. J.; Profilet, R. D.; Fanwick, P. E.; Rothwell, I. P., Crystal and molecular-structure of penta(4-methylbenzyl)tantalum, $Ta(CH_2C_6H_4\text{-}4Me)_5$. *Polyhedron* 1993, *12*, 1779–1783.
24. Roessler, B.; Kleinhenz, S.; Seppelt, K., Pentamethylmolybdenum. *Chem. Commun.* 2000, 1039–1040.
25. Chen, H.; Bartlett, R. A.; Olmstead, M. M.; Power, P. P.; Shoner, S. C., Series of 2-coordinate and quasi-2-coordinate transition-metal complexes: Synthesis, structural, and spectroscopic studies of sterically demanding borylamide ligands-$NRBR'_2$ (R = PH, R′ = Mes, XYL - R = R′ = Mes), their lithium-salts $Li(Et_2O)_2NRBR'_2$, and their transition-metal derivatives $M(NPHBMes_2)_2$ (M = Cr, Co, Ni), $Co(NPhBXYL_2)_2$, and M(NMeSBMes2)2 (M = Cr-Ni). *J. Am. Chem. Soc.* 1990, *112*, 1048–1055.
26. Mowat, W.; Shortlan, A.; Wilkinso, G.; Yagupsky, G.; Yagupsky, M.; Hill, N. J., Elimination stabilized alkyls. 1. Chromium, molybdenum, tungsten, and vanadium. *Dalton Trans.* 1972, 533.
27. Mowat, W.; Shortlan, A.j.; Hill, N. J.; Wilkinso.G, Elimination stabilized alkyls. 2. neopentyl and related alkyls of chromium(IV). *Dalton Trans.* 1973, 770–778.
28. Schulzke, C.; Enright, D.; Sugiyama, H.; LeBlanc, G.; Gambarotta, S.; Yap, G. P. A.; Thompson, L. K.; Wilson, D. R.; Duchateau, R., The unusual stability of homoleptic di- and tetravalent chromium alkyls. *Organometallics* 2002, *21*, 3810–3816.
29. Haymotherwell, R. S.; Hussainbates, B.; Hursthouse, M. B.; Wilkinson, G., Synthesis and X-ray crystal-structure of trimesitylrhodium(III). *Chem. Comm.* 1990, 1242–1243.
30. Haymotherwell, R. S.; Wilkinson, G.; Hussainbates, B.; Hursthouse, M. B., Synthesis and structure of iridium tetramesityl. *Polyhedron* 1991, *10*, 1457–1458.
31. Hitchcock, P. B.; Lappert, M. F.; Smith, R. G.; Bartlett, R. A.; Power, P. P., Synthesis and structural characterization of the 1st neutral homoleptic lanthanide metal(III) alkyls LnR_3 Ln = La or Sm, $R = CH(SiMe_3)_2$. *Chem. Comm.* 1988, 1007–1009.
32. Calculations were done using cc-pVDZ as the basis set with B3LYP on G09.
33. Landis, C. R.; Weinhold, F., Origin of trans-bent geometries in maximally bonded transition metal and main group molecules. *J. Am. Chem. Soc.* 2006, *128*, 7335–7345.
34. Nguyen, T.; Sutton, A. D.; Brynda, M.; Fettinger, J. C.; Long, G. J.; Power, P. P., Synthesis of a stable compound with fivefold bonding between two chromium(I) centers. *Science* 2005, *310*, 844–847.
35. Gibson, V. C.; Newton, C.; Redshaw, C.; Solan, G. A.; White, A. J. P.; Williams, D. J., Low valent chromium complexes bearing N,O-chelating pyridylenolate ligands $[OC(Bu^t)(=2\text{-}CHN_5H_3Me\text{-}x)]^-$ (x = 3-6). *Dalton Trans.* 2003, 4612–4617.
36. Kang, S. K.; Tang, H.; Albright, T. A., Structures for d^0 ML_6 and ML_5 complexes. *J. Am. Chem. Soc.* 1993, *115*, 1971–1981.
37. Bersuker, I. B., *Electronic Structure and Properties of Transition Metal Compounds: Introduction to the Theory*. 2nd ed.; Wiley: Hoboken, N.J., 2010.
38. Carter, T. J.; Kampf, J. W.; Szymczak, N. K., Reduction of Borazines Mediated by Low-Valent Chromium Species. *Angew. Chem. Int. Ed.* 2012, *51*, 13168–13172.
39. Hoffmann, R.; Alvarez, S.; Mealli, C.; Falceto, A.; Cahill, T. J., III; Zeng, T.; Manca, G., From Widely Accepted Concepts in Coordination Chemistry to Inverted Ligand Fields. *Chem. Rev.* 2016, *116*, 8173–8192.

CHAPTER 15

The 1-Electron Picture of Transition-Metal Electronic Structure

"This theory [crystal field theory] gives a survey of the effects of electric fields of definite symmetries on an atom. Accordingly it will be possible to apply it to advantage in the treatment of symmetrical molecules."

— Hans A. Bethe (1929)[1]

"As soon as I got to Copenhagen, I knew crystal field theory was going down . . . I started working on what we now call the ligand field theory, and that was to hybridize Robert Mulliken's molecular orbital theory with the best parts of crystal field theory."

— Harry B. Gray (interview with Richard Eisenberg in 2011)

". . . one can still say that quantum mechanics is the key to understanding magnetism. When one enters the first room with this key there are unexpected rooms beyond, but it is always the master key that unlocks each door."

— John H. Van Vleck[2]

CONCEPT COVERAGE

1. Introduction to crystal field theory at the 1-electron approximation
2. Introduction to ligand field theory at the 1-electron approximation
3. Ligand and metal properties affecting ligand fields
4. High-spin versus low-spin complexes and pairing energies
5. First-order Jahn–Teller effects
6. Angular overlap model
7. Magnetism

Our discussion of electronic structure and spectroscopy for transition-metal complexes will be divided into two parts. In this chapter we will consider a 1-electron picture of spectroscopy with a focus on how orbital energies change with structure. In the next chapter, this framework of ligand effects will be used as a basis for discussing multielectronic terms and the spectroscopy of metal complexes in more detail. Furthermore, the theory and applications of Curie magnetism will be discussed here, which fits well with this 1-electron picture.

This division between 1-electron and multielectron models allows students to understand some critical aspects of how ligand donor properties affect *d*-orbital splittings and relative energies before beginning to deal with term symbols, perhaps for the first time. Then in the next chapter, the goals are to provide the student with an idea of what term symbols are, define microstates, and present a few tools for understanding the optical spectra of d^2–d^8-complexes.

15.1 Background and the Crystal Field Theory Approximation for Octahedral Systems

The deep color and beauty of many compounds has been a source of fascination for people throughout history, and chemists are no exception. When Tyrian purple (Fig. 15.1), a dye extracted from sea snails, was worth its weight in silver and only available to royalty, color was a sign of power and status. For a young William Henry Perkins, for whom the Royal Society of Chemistry journal for organic chemistry once was named, the purple color

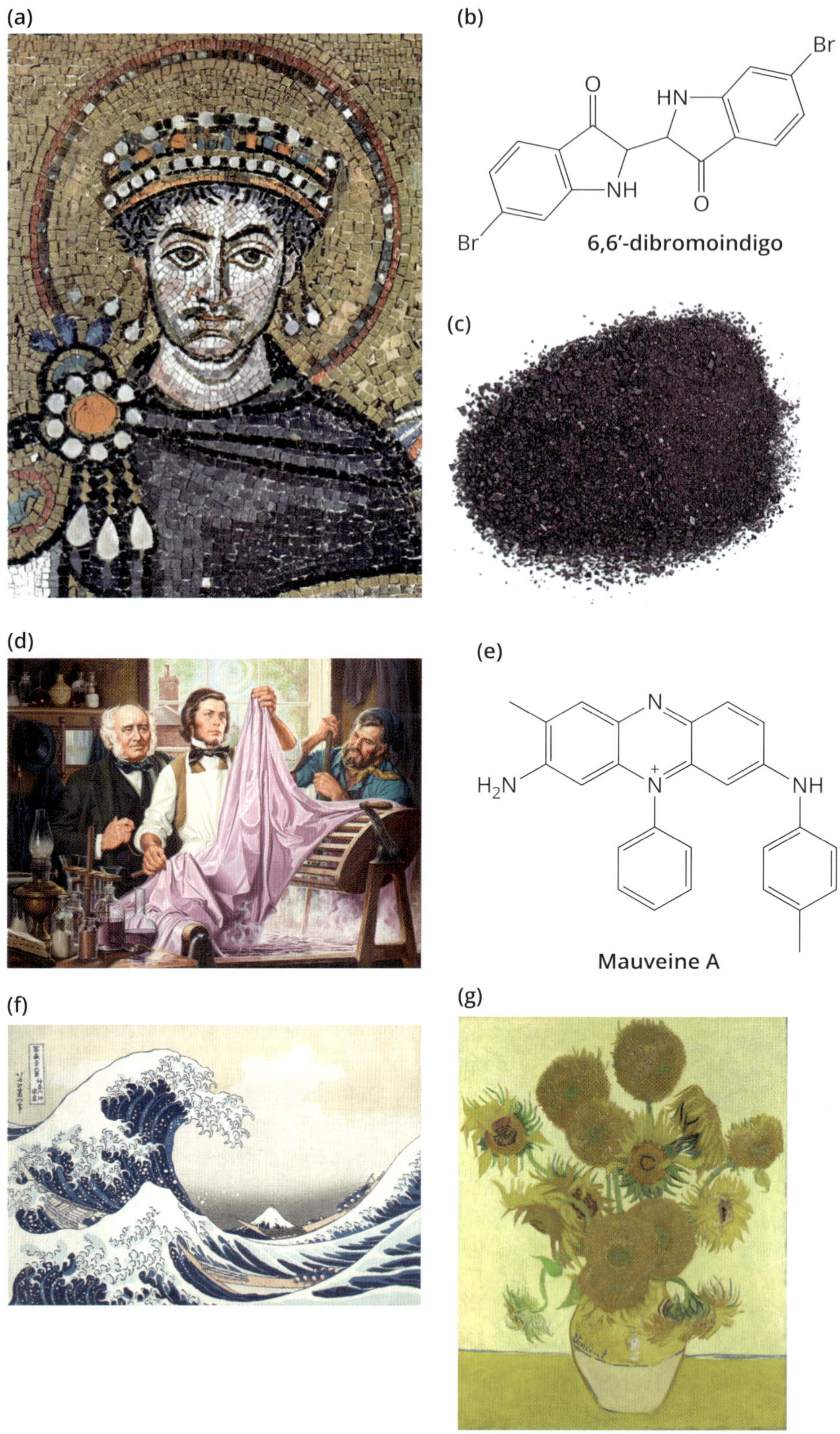

Fig. 15.1 (a) Mosaic portrait of Eastern Roman emperor Justinian I (AD 482–565), clad in Tyrian purple garment. (b) Structure of 6,6′-dibromoindigo, the principal dye of Tyrian purple. (c) Powdered Tyrian purple, a dye obtained from marine snails.[3] (d) Portrait of William Henry Perkins (center), the inventor of an important new dye in the nineteenth century. (e) Structure of Perkins' mauveine A dye.[4–5] (f) *The Great Wave off Kanagawa*, by Hokusai (1831), which uses Prussian blue ($Fe_7(CN)_{18}$). (g) Vincent van Gogh's *Sunflowers* (fourth version), which uses chrome yellow ($PbCrO_4$). Due to its toxicity, chrome yellow has largely been replaced by another inorganic pigment, cadmium orange (CdS), which also is not entirely nontoxic.

in his flask after a failed attempt to make quinine by oxidizing aniline, *p*-toluidine, and *o*-toluidine with potassium chromate was a sign of opportunity in the textile business. Inorganic dyes and pigments are of great importance to art because they are generally more "color fast" than organic dyes and were used in important paintings such as Monet's *Bathers at La Grenouillère* (Prussian blue as well as cobalt blue) and van Gogh's *Vase with Twelve Sunflowers* (chrome yellow). Today, for chemists, color is a sign of opportunity in characterization of electronic structure in a newly prepared complex. For some fields, understanding and controlling color is essential for key features of the device, such as spectral coverage for solar-cell design or the main characteristics of light-emitting devices (LED) and solid-state lasers, for example.

Crystal field and ligand field theories—two related theoretical frameworks—are the most commonly used tools to describe the electronic structure of transition-metal systems, and in particular their optical spectra (responsible for their color in the visible range) and magnetic properties. In this chapter, the basis for the two models will be developed using a "1-electron" description. In other words, we will not concern ourselves yet with interactions between different electrons on the same metal but will instead examine the role of metal–ligand interactions and other factors that affect the energies of the *d*-orbitals. This will allow a rudimentary discussion of absorption spectroscopy and magnetism—two of the most important experimental techniques for understanding electronic structure. In essence, we will discuss some of the features of spectroscopy for d^1 and d^9 systems, complexes with either one electron (or one hole)[a], along with metal–ligand effects in this chapter. In the next chapter, electron–electron repulsion effects will be introduced, leading to multielectronic terms and to a more detailed coverage of absorption spectroscopy for transition-metal systems.

While inorganic spectroscopy in general perhaps started with Bunsen and Kirchhoff, the genesis of modern inorganic spectroscopy occurred in 1929 when the French physicist Jean Becquerel[b] suggested that the electronic structure of transition-metal centers could be understood on the basis of electric fields created by the atoms surrounding them in a crystal lattice. That same year, Hans A. Bethe[c] published a seminal paper that singlehandedly established what is now known as "crystal field theory" by directly applying group theory (which was then newly developed for physical science applications by the mathematicians/physicists Eugene Wigner and John Von Neumann) to take advantage of the high symmetry of crystalline environments in order to simplify what were then impossibly complex calculations.

John H. Van Vleck (1899–1980) was among the first to apply Bethe's theory to actual chemical systems (including molecules, not just crystalline salts), resulting in important advances in the theory of the electronic spectroscopy and magnetism of transition-metal compounds. Furthermore, whereas Bethe's theory was initially strictly set up only for crystalline (extended) structures, Van Vleck recognized early on that the same formalism could also be applied to molecular inorganic complexes—a paradigmatic shift that was not at all obvious to make at the time. For his multiple contributions in solid-state physics, Van Vleck shared the Nobel Prize in Physics with Sir Nevill F. Mott and Philip W. Anderson in 1977. Important pioneering contributions to the electronic theory of transition-metal ions also includes the work of H. Hartmann, L. E. Orgel, J. S. Griffith, W. Moffitt, C. K. Jørgensen, C. J. Ballhausen, D. S. McClure, C. E. Schäffer, M. Gerloch, H. B. Gray, and many others.

15.1.1 Pictorial Descriptions of Orbital Splitting Symmetries

In crystal field theory (CFT), ligands are represented by point charges (or sometimes point dipoles/multipoles) located at precise locations around the central transition metal. The symmetry of the charge arrangement around the metal center leads to the splitting of the *d*-orbitals (different energies for the *d*-orbitals), and therefore plays a primary role in determining the electronic structure of the metal center. Within the limits of CFT, the main factor modulating the properties of an inorganic complex is the *relative* energies of the *d*-orbitals; factors that raise or lower the energies of the *d*-orbitals as a group are generally disregarded. Another postulate of CFT—and this is by far the most curious one from a chemical perspective—is that we should consider the transition-metal center as an isolated ion with the same charge as the one associated with the formal oxidation state of the metal center in the molecule or materials under consideration. That is, CFT makes the chemically bold assumption that metal–ligand bonds are purely *ionic*: within the CFT framework, there are no covalent contributions to the overall stability of the complex. Of course, this assumption is not very good (in fact, it is more *wrong* than right), and as expected, CFT calculations are notoriously quantitatively inaccurate. Nevertheless, because the overall theory is deeply rooted

[a] An "electron hole" is a position that an electron could occupy but is currently not occupied. For example, an orbital with a single electron has one electron and one hole because each orbital can hold two electrons of opposite spin.

[b] Jean Becquerel (1878–1953) was a member of an important scientific dynasty in nineteenth- and early twentieth-century France. His father, Henri Becquerel (1852–1908), shared the Nobel Prize in Physics in 1903, along with Marie and Pierre Curie for their discovery of radioactivity (1897). Henri's father, Edmond Becquerel (1820–1891) discovered the photoelectric effect and made many important contributions in the study of the emission of light by incandescence, phosphorescence, and bioluminescence. Finally, Edmond's father, Antoine César Becquerel (1788–1878), made a number of significant discoveries in electrochemistry, and his pioneering work directly inspired better-known contemporaries such as Michael Faraday and John Frederic Daniell.

[c] Hans A. Bethe (1906–2005), a German-born American physicist, was awarded the Nobel Prize in 1967 for his contributions to the theory of nuclear reactions in stars. He led the theoretical division of the Manhattan Project in Los Alamos during the Second World War, and then spent most of his career at Cornell University.

in symmetry, it actually performs quite well in predicting qualitatively the energetic ordering of metal-centered orbitals and, as will be soon made clear, can be easily extended to accommodate a more reasonable description of the metal–ligand bonding interactions using a molecular orbital theory approach.

Going back to the crude CFT approach, the collection of point charges surrounding a transition-metal ion forms, as a whole, a specific electric field that affects the energies of the *d*-orbitals by generating regions in space where electrons centered on the metal ion (generally, electrons located in the *d*-orbitals of the metal) are more or less likely to be found. Because ligands are generally considered as Lewis bases (electron donors), ligands are represented in CFT as point *negative* charges, whether or not the ligand itself is formally anionic. The collection of these point charges around a metal center therefore creates a specific electric potential[d] that modulates the energy of electrons located on the metal center. The potential energy of an electron on the metal increases (becomes less favorable) in directions that point directly toward the ligands (negative charges repel each other), and consequently those orbitals on the metal center that point directly toward ligands will always be less favorable to occupy than those that are oriented between ligands. This statement is, in a nutshell, the essence of CFT. The practical implementation of the theory then requires that we obtain a method to qualitatively estimate the energy shift of each type of orbital on the metal center.

Importantly, the symmetry of the so-called "crystal-field" potential is directly and uniquely dictated by the symmetry of the ligand arrangement around the metal central. If one starts with a free metal ion—that is, a metal ion isolated in space—the spherical symmetry of the ion leads to a fully degenerate *d*-orbital manifold; there is no reason for electrons to favor one or the other directions around the metal (Fig. 15.2). If we bring in an electric field due to a set of ligands first in a spherical symmetry, then all of the *d*-orbitals will rise in energy due to Coulombic repulsion of the electron density in the orbitals and the electron density in the spherical ligand field, although no

[d] Recall that the electric potential is a spatial function that tells you the potential energy that a unit test charge would have at every position in space: the SI unit of the electric potential, the volt, is formally equal to 1 joule per coulomb, 1 V = 120 J/C.

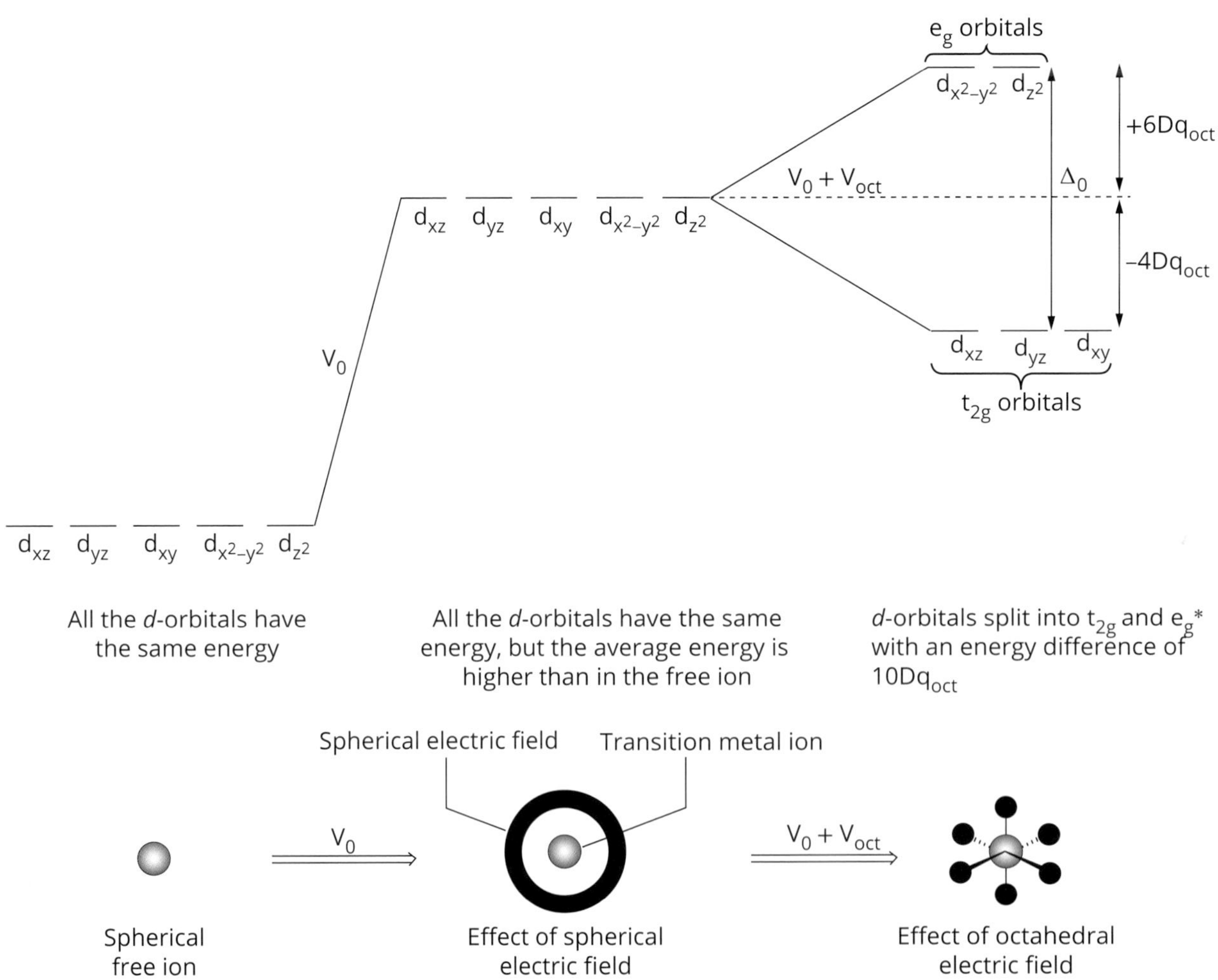

Fig. 15.2 The crystal field theory progression from a spherical ion to a system with a spherical electric field to a metal ion in an octahedral ligand field. Note that the black dots in the diagram on the right are not orbitals but represent electric fields (regions of anionic charge).

d-orbital (crystal-field) splitting is observed. This electric field of spherical symmetry, which leaves all of the orbitals of the same energy, will be called V_0. Then, we allow the electric field to deform and pool at the positions of the ligands, changing the symmetry of the field affecting the d-orbitals from spherical to octahedral, for example. This introduces an additional perturbation to the spherical field V_0, which we will call V_{oct}. The effect of V_{oct} will be to raise the energy of those orbitals that are oriented closer to the ligands and lower the energy of those orbitals that point away from the ligands. Although the total effect of the crystal field is the sum of both V_0 and V_{oct}, in practice we are mostly interested in the energy difference between orbitals rather than their absolute energy, which is poorly described by CFT, and we therefore often focus only on the effects due to V_{oct} (or lower-symmetry crystal fields when needed).

Let us identify the potential energy stabilization or destabilization of the d orbitals in an octahedral ligand arrangement by V_{oct}. This requires that we first define the coordinate system, which boils down to deciding where each of the six point charges associated with each ligand of the octahedral complex will be located. A very useful convention is to position pairs of ligands at equal distances on each of the three Cartesian axes (x, y, and z). Because this positions the C_4 axes of the octahedron colinearly with the Cartesian axes, this is sometimes called the "tetragonal" convention. Keep in mind that the highest-order rotation axis is usually chosen to be the z-axis (Chapter 5), and octahedral does have an S_6 axis that is not coincident with the C_4. However, this tetragonal convention of picking the C_4 to be the z-axis is usually preferable as it simplifies analysis. (Later we will discuss the "trigonal" convention as an advanced topic, where one of the S_6/C_3 axes of the octahedron is chosen to coincide with the z-axis.) In the tetragonal convention, the orbitals that point along the axes, and which are therefore closest to the point charges of the ligands, are the $d_{x^2-y^2}$ and d_{z^2} orbitals, and transform as the e_g irreducible representation in the O_h point group. The orbitals between the point charges are the d_{xz}, d_{yz}, and d_{xy} orbitals, which as a set belong to the t_{2g} representation in O_h.

The potential V_{oct} therefore leads to a uniform lowering in energy of the t_{2g} set of orbitals and a uniform rise of the e_g orbitals. The energy shift of each group of orbitals can be calculated easily by invoking the so-called *barycenter* principle (from the Greek, meaning "heavy center"), which states that the sum of all destabilizing orbital energy shifts must be equal to the sum of all stabilizing shifts. That is, the three t_{2g} orbitals must each drop by a smaller amount ($\Delta E_{t_{2g}}$) than the positive destabilizing rise of the two e_g orbitals (ΔE_{e_g}). Mathematically, the barycenter principle simply requires that $3\Delta E_{t_{2g}} + 2\Delta E_{e_g} = 0$. If we call the total energy gap between the t_{2g} and e_g orbitals Δ_o ($\Delta_0 = \Delta E_{e_g} - \Delta E_{t_{2g}}$), then we have two equations with two unknowns, which can be solved directly to yield the following equations (Eq. 15.1–2).

$$\Delta E_{t_{2g}} = -\frac{2}{5}\Delta_0 = -4Dq \quad \textbf{15.1}$$

$$\Delta E_{e_g} = \frac{3}{5}\Delta_0 = 6Dq \quad \textbf{15.2}$$

This confirms that the destabilization of the e_g orbitals is larger than the stabilization of the t_{2g} orbitals.[e] Historically, the CFT parameter Δ_o was called 10 Dq ("ten-dee-cue"), and this name still is often used in the literature; using this convention, the stabilization of the t_{2g} orbitals would be −4 Dq and the destabilization of the e_g orbitals would be +6 Dq. The energy term Dq can be calculated from the radial part of the d-orbital wavefunctions and the details of those calculations can be found in more specialized texts.[6–8] Here, the results of those calculations will be used without proof. In practice, accurate *ab initio* calculation of crystal-field parameters can be a nontrivial undertaking even with modern computational power and models,[9–14] and as a result, the CFT and LFT are practically always used semiempirically by determining (or fixing) experimentally the value of Δ_o (=10 Dq) of specific inorganic complexes, as discussed in the following.

15.2 The Tetrahedral Ligand Field

The second most common ligand arrangement in inorganic chemistry after octahedral is perhaps tetrahedral (T_d); interestingly, although the tetrahedral arrangement might appear to be very different than the octahedral one, there is a deep connection between the two symmetries. Fundamentally, both the octahedron and tetrahedron are related to the symmetry of a cube, and in fact both O_h and T_d belong to the "cubic groups" family. In parallel to what we observed for octahedral symmetry, the d-orbitals in the T_d point group are split into two groups labelled as t_2 and e irreducible representations in T_d, and the crystal field splitting of tetrahedral complexes can therefore be described by a single parameter which we will call Δ_t. Note that while both T_d and O_h are cubic groups, the center of inversion is lost when going from O_h to T_d; consequently, atomic orbitals do not have the gerade (g) and ungerade (u) designations that are used for the octahedral symmetry.

As we did previously for the octahedral case, we first need to define a coordinate system for the tetrahedral arrangement. It is customary in most systems to orient the molecule with the z-axis along the highest-order rotation axis (octahedral is a notable exception, where the C_4 instead of the S_6 is used as the primary axis in the tetragonal scheme). The highest-order rotation axes of a tetrahedron are three equivalent S_4 axes, which each bisect the angle between any pair of metal–ligand bonds. Perhaps the easiest way to see the ligand positions is by positioning them on the opposing apexes of a cube, with the x-, y-, and z-axes (the three S_4 axes of the

[e] Remember that we are talking about the relative shifts of the d-orbitals, as shown in Fig. 15.2. All five d-orbitals shift up in energy relative to the free ion, but the t_{2g} orbitals are not destabilized by as much as the e_g orbitals.

tetrahedron) extending through the center of the cube's faces (Fig. 15.3). In this coordinate system, the off-axis t_2 orbitals, d_{xy}, d_{xz}, d_{yz}, point towards the ligands—not completely directly, but not completely between the ligands either. As a result, the t_2 set will be destabilized in tetrahedral systems, which is the opposite of the situation for the octahedral arrangement. Analogously, the on-axis e orbitals, $d_{x^2-y^2}$ and d_{z^2}, will be stabilized relative to the t_2 set, as they are oriented between the ligands.

The energy gap in the tetrahedral orbital splitting, Δ_t, can be related directly to the octahedral orbital splitting using the angular properties of the crystal-field and of the *d*-orbitals (see Ballhausen for derivation).[6] Because the tetrahedral arrangement has only four ligands versus the six of the octahedral arrangement, not only is the orbital splitting inverted, but the overall splitting (assuming an otherwise identical crystal-field strength for each ligand) is smaller by a factor of 4/9 (Eq. 15.3) for the tetrahedral crystal-field relative to the octahedral one (Fig. 15.4). Mathematically:

$$\Delta_t = -(4/9)\ \Delta_0 \qquad \textbf{15.3}$$

where the negative sign simply reflects the inversion of the orbital pattern, t_2 over e for tetrahedron versus e_g over t_{2g} for the octahedron.

The factor of 4/9 for the tetrahedral field strength versus the octahedral field strength should be a good approximation for the same metal ion in the same oxidation state with the same ligands. For example, Δ_t in $FeCl_4^-$ is about 5,000 cm^{-1}, whereas Δ_o in $FeCl_6^{3-}$ is ~14,000 cm^{-1}. The crystal-field approximation suggests that tetrahedral splitting should be 14,000 cm^{-1} × 4/9 = 6,222 cm^{-1} (instead of 5,000)—a fairly reasonable estimate considering the crude assumptions of the model.

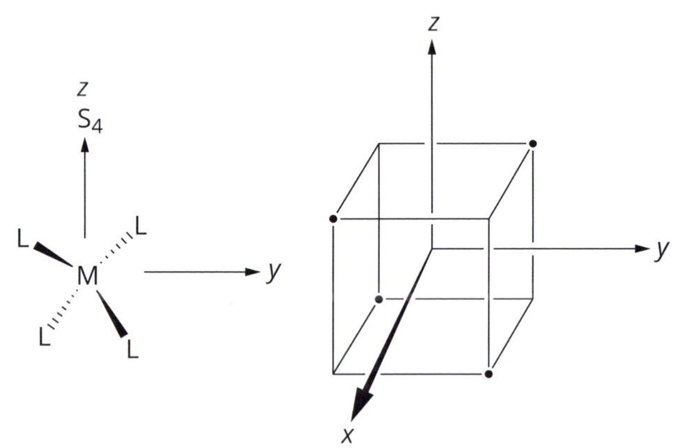

Fig. 15.3 The S_4 axes of a tetrahedron

15.3 CFT and Some Lower-Symmetry Systems

For the cubic systems discussed previously, O_h and T_d, one can model the field with a single parameter such as 10 Dq, Δ_o, or Δ_t. The crystal field splittings for other ligand arrangements can be obtained similarly as for the cubic symmetries, but the number of parameters needed to fully characterize the *d*-orbital splittings generally increases

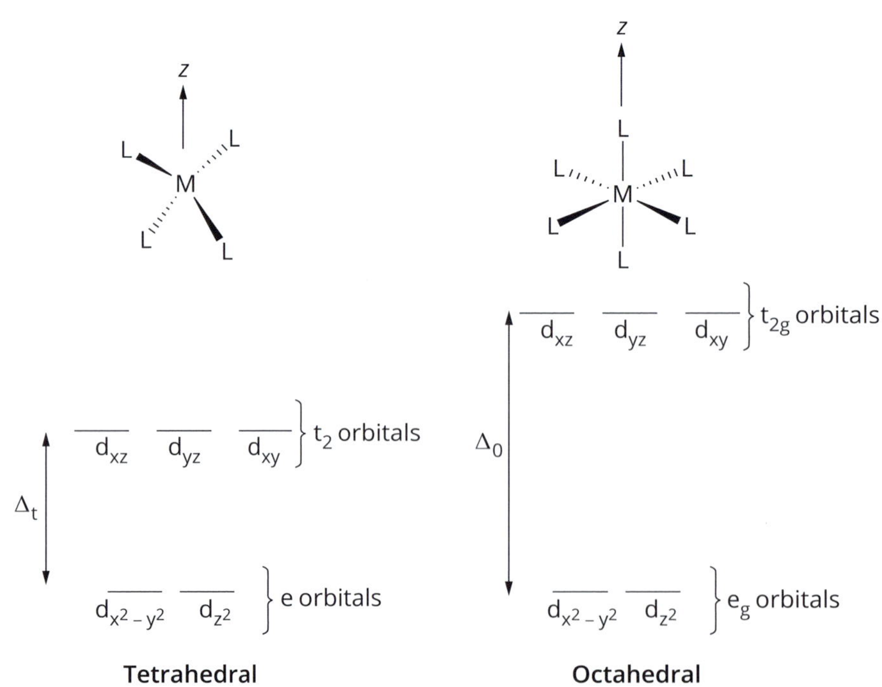

Fig. 15.4 Comparison of the tetrahedral and octahedral ligand fields. With identical ligands and metal, $\Delta_t \sim -4/9\,\Delta_o$.

the lower the symmetry of the ligand arrangement around the metal. For instance, in square-planar symmetry (D_{4h} point group), the five *d*-orbitals split into four groups (one two-fold degenerate set for d_{xz} and d_{yz}, and the three non-degenerate d_{z^2}, $d_{x^2-y^2}$, and d_{xy} orbitals), requiring three independent parameters to fully describe the orbital splitting pattern using the barycenter principle. These extra parameters are not always easy to extract from experimental data, and the gains from using a more accurate symmetry description are often not matched by the extra complications in the analysis and the uncertainty in knowing exactly the value of these extra parameters. Consequently, one often tries to use the highest symmetry that still properly describes a complex, at least to a first approximation, in order to avoid unnecessarily complicated analyses. The lower-symmetry cases are then treated as a perturbation on the higher symmetry, as will be shown later. It is nevertheless useful to first consider simple methods for quickly drawing crystal field splitting diagrams for other, lower-symmetry systems. These diagrams are often useful first steps to understanding transition-metal electronic structure, whether it be to discover the frontier molecular orbitals for applications to reactivity or for discerning ground and excited term states for multielectron systems.

In the remainder of this section we will discuss how orbital energies change as various “distortions” occur; that is, how orbital energies change from some supposed high-symmetry structure as one moves the ligands to the actual, lower-symmetry structure. Why these distortions occur will be discussed in a later section, and that discussion will be greatly enabled by an understanding of orbital structure covered here.

15.3.1 CFT and Tetragonally Distorted Complexes, Including Square-Planar Complexes

We begin our discussion of the crystal field splitting diagram for square-planar complexes by using a specific example, $\{PtCl_4\}^{2-}$. As always, symmetry and group theory are useful tools for sorting the *d*-orbitals. The character table of the D_{4h} point group of a square-planar complex reveals both d_{xz} and d_{yz} transform as the e_g representation, and therefore form a set of degenerate orbitals. The remaining *d*-orbitals all belong to different irreducible representations in D_{4h} (Fig. 15.5).

To qualitatively assign the *d*-orbital splitting, the directionality and amplitude of the orbitals towards each ligand of the square-planar complex need to be evaluated. In general, the closer a *d*-orbital is to a donor center, the higher the energy of that orbital; quantitatively, stronger donors (stronger electric field) will shift the metal orbitals more than weaker donors, but at this point we are only interested in the order of the orbital splitting, not in quantitative energy splittings.

A fruitful approach in thinking about low-symmetry problems, such as the square-planar symmetry versus octahedral symmetry, in the CFT framework is to imagine a distortion that continuously transforms the higher-symmetry complex into the lower-symmetry one. In the case of a square-planar complex, we consider a perfect octahedron (here, a hypothetical $\{PtCl_6\}^{4-}$ complex) that is continuously distorted by pulling the two chloride ligands on the *z*-axis away from the metal center to infinite distance. Because this type of distortion retains the C_4 axis of rotation on the *z*-axis, it is conventional to designate this as a “tetragonal distortion”. By extension, complexes with four

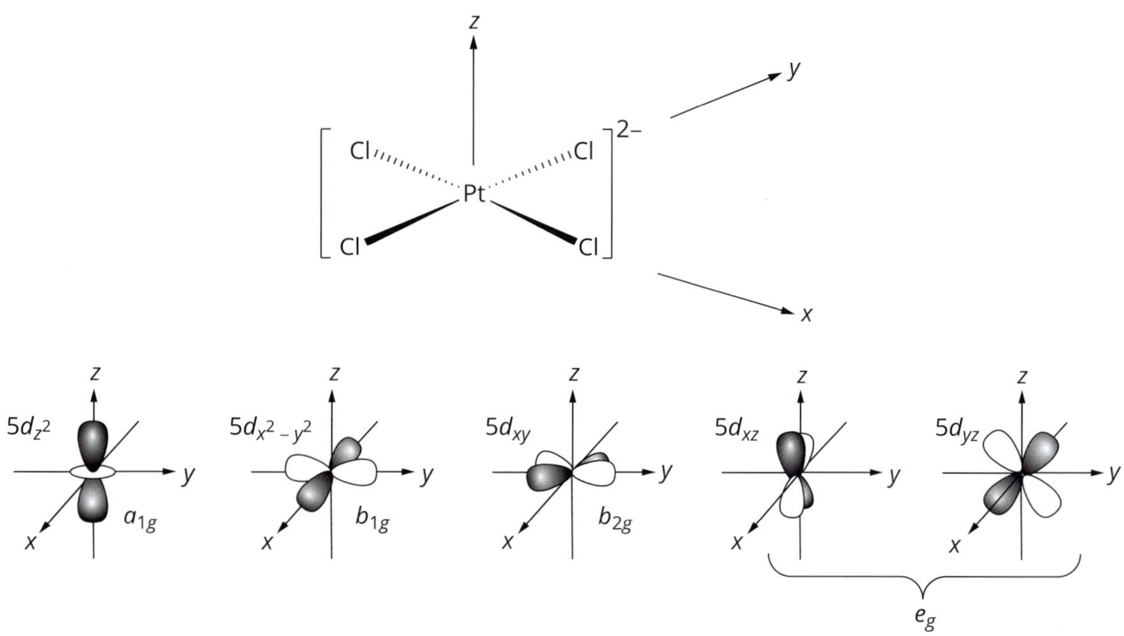

Fig. 15.5 The axes for $\{PtCl_4\}^{2-}$, the orientations of the five *d*-orbitals, and the irreducible representations to which they belong.

identical ligands in the *xy*-plane and two other ligands *trans* to each other on the *z*-axis—for example, *trans*-$\{CrCl_2(H_2O)_4\}^{3+}$—are also called "tetragonal" complexes or "tetragonally distorted" complexes, depending on the context. These all are related to the D_{4h} point group, at least in the idealized limit.

First, let us see what happens to the ($d_{x^2-y^2}$, d_{z^2}) set (the e_g orbitals in the O_h group) as we pull the two ligands on the *z*-axis slightly away from the metal center. Clearly, the $d_{x^2-y^2}$ orbital (b_{1g} symmetry in D_{4h}) will be destabilized relative to the d_{z^2} orbital (a_{1g} symmetry in D_{4h}), which has most of its amplitude on the *z*-axis away from the four ligands that stay close to the metal. The barycenter principle tells us that the amount of destabilization of the $d_{x^2-y^2}$ orbital should be equal to the energetic stabilization of the d_{z^2} orbital.

Next, we investigate the case of the (d_{xy}, d_{xz}, d_{yz}) orbitals (the t_{2g} set in O_h) with ligands along the *z*-axis slightly elongated. There, simple inspection shows us that the d_{xy} orbital (b_{2g} symmetry in D_{4h}) must behave differently than the (d_{xz}, d_{yz}) orbitals (e_g symmetry in D_{4h}), as the former is fully located in the *xy*-plane, whereas the other two are oriented in the *z*-direction where the two axial ligands (negative point charges in CFT) have been pulled away. Consequently, the (d_{xz}, d_{yz}) set must go down in energy, whereas the d_{xy} orbital must go up; we can apply the barycenter principle, which leads us to conclude that the destabilization of the d_{xy} orbital must be twice as large as the stabilization of the two (d_{xz}, d_{yz}) orbitals. This gives us the full diagram of a tetragonally distorted complex, as shown in Fig. 15.6.

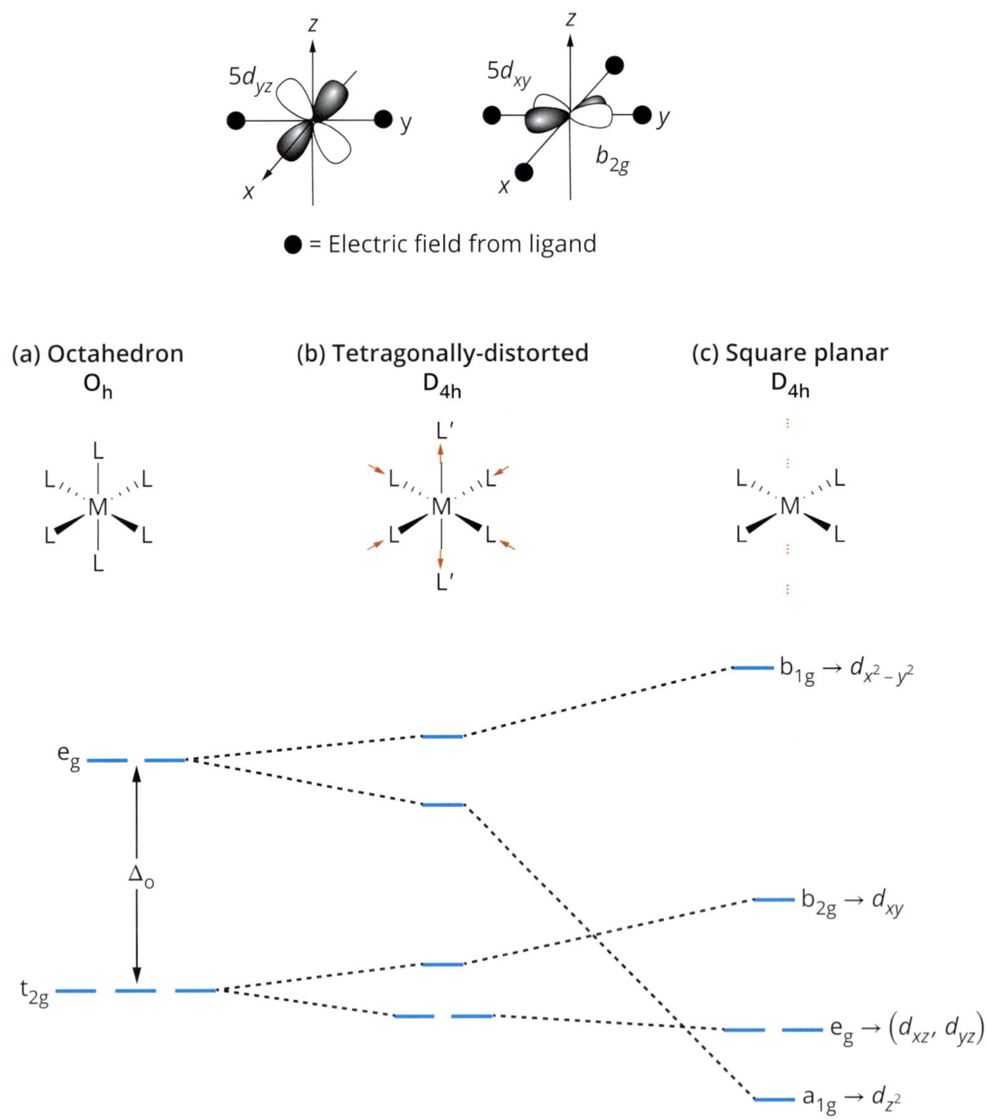

Fig. 15.6 A generalized view of crystal field theory applied to an octahedral system (*a*) which is then distorted by removal of ligands along one axis in two stages; (*b*) the *z*-axis ligand slightly lengthened and the equatorial ligands respond by contracting; and (*c*) the *z*-axis ligands are removed entirely. Depending on the details of the ligand donor properties, the exact ordering of the bottom four orbitals in the square-planar geometry (b_{1g}, e_g, and a_{1g}) may change somewhat.

To obtain the orbital diagram corresponding to an actual square-planar complex, we simply need to consider the case where the tetragonal distortion is infinitely large, corresponding to the situation where the two axial ligands are so far away that they do not interact with the metal orbitals any longer. In this case, all orbital shifts will keep going in the same direction as observed in the limit of a small tetragonal distortion, but now some orbitals can cross energetically. A detailed quantitative CFT analysis (not included here) would show that in fact the only unknown here is the precise position of the d_{z^2} orbital relative to the d_{xy} and (d_{xz}, d_{yz}) orbitals. If the four ligands of the square-planar complex are weak enough donors (small electric field in the *xy*-plane), then a reasonable estimate would be to assume that the d_{z^2} orbital, which has most of its amplitude lying on the *z*-axis, is more stable than the d_{xy} orbital which is concentrated in the *xy*-plane where the four ligands are. A less immediately obvious question is whether the d_{z^2} orbital eventually crosses the (d_{xz}, d_{yz}) orbitals to become the most stable of the five *d*-orbitals, as this really depends on the nature of the ligands and of the metal forming the square-planar complex. If the ligands provide a sufficiently weak crystal field—if Δ_o for the imaginary 6-coordinate complex is small enough to begin with—then it is probable that the d_{z^2} orbital is the most stable orbital of the square-planar configuration, as the destabilization due to the small in-plane amplitude of the orbital is more than adequately compensated by the strong amplitude on the *z*-axis that completely avoids the four ligands. This is not the case for the (d_{xz}, d_{yz}) set, which orients the electronic density towards the ligands, albeit not directly. If the four ligands of the square-planar complex are strong enough, the direct interaction between these four ligands and the torus of the d_{z^2} orbital, as measured by the large Δ_o for the imaginary 6-coordinate complex, will lead to enough electrostatic repulsion to prevent the d_{z^2} from crossing the energy of the (d_{xz}, d_{yz}) set, which then remains the most stable one. Fig. 15.6 illustrates these different scenarios and underlines the richness of square-planar complexes. Where octahedral complexes differ only by the size of the energy gap between the t_{2g} and e_g orbitals, square-planar complexes can organize the orbitals in many different ways, each one leading to unique chemical and electronic properties depending on the specific ligands and the nature and oxidation state of the metal center. No matter what, a common feature of tetragonal complexes, including square-planar ones, is that the $d_{x^2-y^2}$ orbital will always have the highest energy, but the exact energetic ordering of d_{z^2}, d_{xz}/d_{yz}, and d_{xy} will vary from complex to complex. This ambiguity in the orbital ordering in the CFT analysis of $\{PtCl_4\}^{2-}$ is a difficult one that has been discussed extensively in the literature, and it has been established that the d_{z^2}-orbital is the lowest in energy in $\{PtCl_4\}^{2-}$ (Fig. 15.6).[10,16,17]

15.3.2 Quickly Drawing Crystal Field Diagrams

The groundwork as to how and why CFT works for describing the *d*-orbital splittings in transition-metal complexes was laid out in detail in the previous section. Being able to quickly draw qualitative pictures of the orbital splittings critical to the properties and reactivities of transition metals is a very powerful tool. At this point, we felt it would be worthwhile to provide a brief tutorial on drawing crystal field splitting diagrams with a few examples. This will review the previous information and provide the reader with a quick guide to using CFT in qualitative fashion. For the following examples, X will be a ligand of some type that donates more strongly than L.

(a) *trans*-MX_4L_2

(i) First draw the system with a set of coordinate axes and assign the point group (Fig. 15.7a).

(ii) Look up how the orbitals are grouped in the character table (Fig. 15.7b). The orbitals in brackets are degenerate. In parentheses are the irreducible representations for the *d*-orbitals.

(iii) Draw the crystal field splitting diagram with the orbitals pointed directly at the ligands highest in energy (Fig. 15.7c). If there are multiple orbitals pointed at the ligands, the orbitals directed at the most donating ligands are highest in energy. (Keep in mind that we said X is a better donor than L.)

Explanation: $\{d_{z^2}\ (a_{1g})\}$ and $\{d_{x^2-y^2}\ (b_{1g})\}$ are pointed directly at the ligands, and $\{d_{x^2-y^2}\ (b_{1g})\}$ is pointed directly at the better donating X ligands, making it the orbital highest in energy. The other three orbitals are between ligands, but $\{d_{xy}\ (b_{2g})\}$ is in the same plane as the better-donating X ligands, making it higher in energy than the degenerate $\{d_{xz}, d_{yz}\ (e_g)\}$.

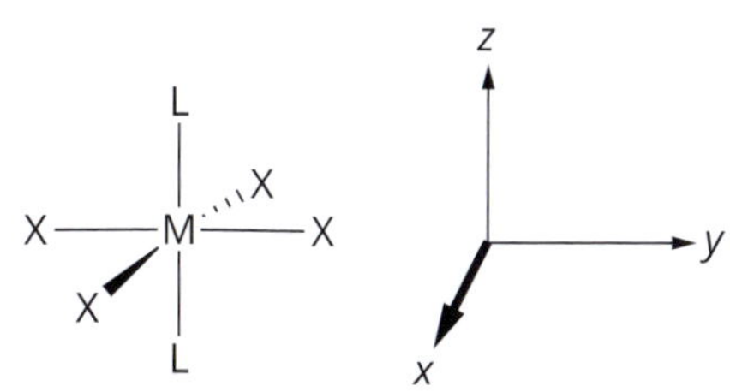

Point group: D_{4h}

Fig. 15.7a Drawing of *trans*-MX_4L_2 and the coordinate axes in use.

$$\{d_{z^2}(a_{1g})\}\ \{d_{x^2-y^2}(b_{1g})\}\ \{d_{xy}(b_{2g})\}\ \{d_{xz}, d_{yz}(e_g)\}$$

Fig. 15.7b Groupings of the *d*-orbitals by symmetry found by inspection of the D_{4h} character table.

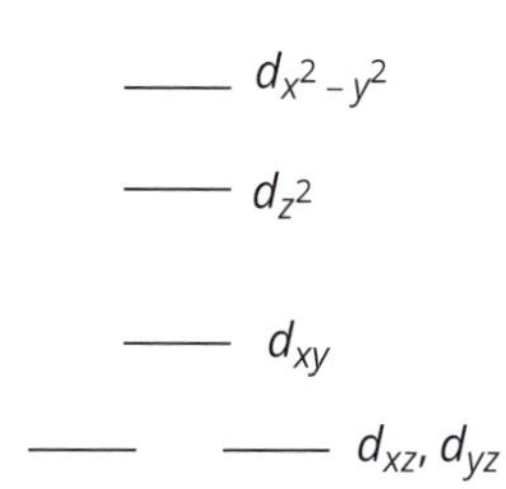

Fig. 15.7c Crystal field splitting diagram of a *trans*-MX_4L_2 complex.

(b) MXL_3

(i) First draw the system with a set of coordinate axes and assign the point group (Fig. 15.8a).

(ii) Look up how the orbitals are grouped in the character table (Fig. 15.8b). The orbitals that are degenerate are in brackets. There are two different sets of degenerate orbitals in the E irreducible representation; while the two orbitals in each set have the same energy, the two degenerate sets as a whole do not necessarily have the same energy.

(iii) Draw the crystal field splitting diagram (Fig. 15.8c).

Explanation: This one is a bit complicated to know the ordering of the two e-sets. $\{d_{z^2}\ (a_1)\}$ will likely be the highest in energy as it is directed at the most donating ligand, X. None of the orbitals in the two e-sets are pointed directly at the ligands. We place the $\{d_{xz}, d_{yz}\ (e)\}$ set higher in energy as it directed more towards the strongly donating X group and, depending on the X–M–L angle, may be strongly involved in bonding to the L ligands.[f]

A couple of brief examples are presented, but the student should be able to reason through any complex given to them to quickly draw out a crystal field splitting diagram. While diagrams can be drawn from just the rough geometry of the molecule, additional data (absorption spectra, magnetism, X-ray diffraction structure, *ab initio* calculations, and so on) can then be used to refine the picture.

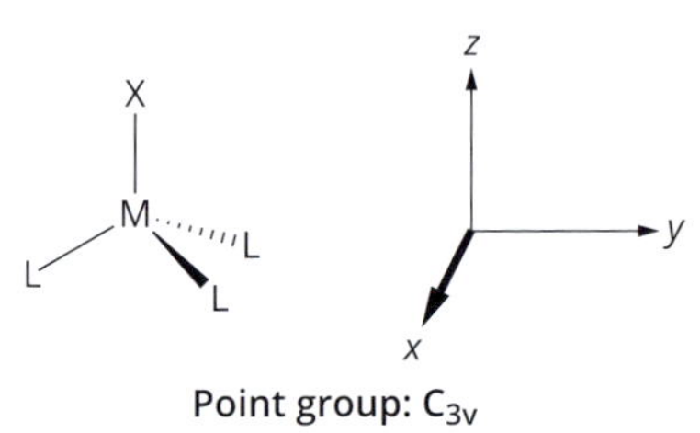

Fig. 15.8a Drawing of MXL_3 and the coordinate axes in use.

$$\{d_{z^2}(a_1)\}\ \{d_{x^2-y^2}, d_{xy}(e)\}\ \{d_{xz}, d_{yz}(e)\}$$

Fig. 15.8b Groupings of the *d*-orbitals by symmetry found by inspection of the C_{3v} character table.

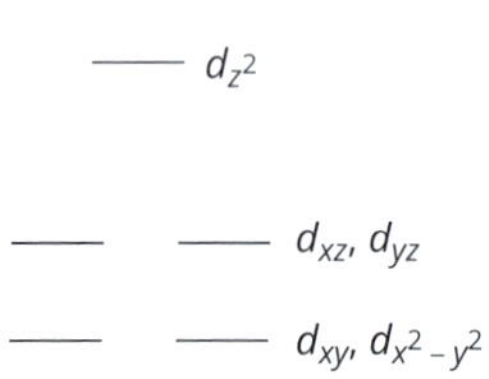

Fig. 15.8c Crystal field splitting diagram of an MXL_3 complex.

[f] In actuality, the two e-sets of the same symmetry can mix and will probably produce a strongly bonding set directed at the L ligands in the base along with a mostly nonbonding set. In any event, a 1 above 2 above 2 pattern would be expected, even if designation of the orbitals comprising the e-sets is somewhat ambiguous.

15.3.3 CFT and Trigonal/Trigonally Distorted Complexes (Advanced Topic)

In general, the term "trigonal" complex is used to refer to any complex in which the highest-order proper rotation axis is a C_3 axis. There are two important distinct classes of trigonal complexes that we will discuss here. The first class of trigonal complexes that are sometimes encountered are trigonal bipyramidal, with three identical (or similar) ligands in the *xy*-plane (the equatorial ligands) and two others located on the *z*-axis (the axial ligands); the idealized point group of this class is D_{3h}. The complex may be C_{3v} if the two axial ligands are not identical; cases where the three equatorial ligands are very different are more complicated. We can use an analysis similar to that used for the previous tetragonal complexes to determine qualitatively the crystal field splitting pattern of trigonal bipyramidal compounds. First, of all five *d* orbitals, the d_{z^2}-orbital (a_1' in D_{3h}) alone interacts strongly with all five ligands, and as such we can expect it to have the highest energy. The d_{xy}- and $d_{x^2-y^2}$-orbitals in D_{3h} are equivalent (e' irreducible representation), and although neither of the two orbitals points directly at all three equatorial ligands simultaneously (Fig. 15.9a), as a set they do strongly interact (and, by symmetry, equally) with these ligands (point charges in the CFT), and are consequently destabilized, albeit not as much as for the d_{z^2}-orbital. Finally, the other two *d*-orbitals (d_{xz}, d_{yz}) are also equivalent in D_{3h} (e" irrep), and because they never point directly at any of the five ligands, they are the most stable set of the trigonal bipyramid configuration.

The probable relative energies of the three orbital levels of the trigonal bipyramidal configuration are shown in Fig. 15.9b.

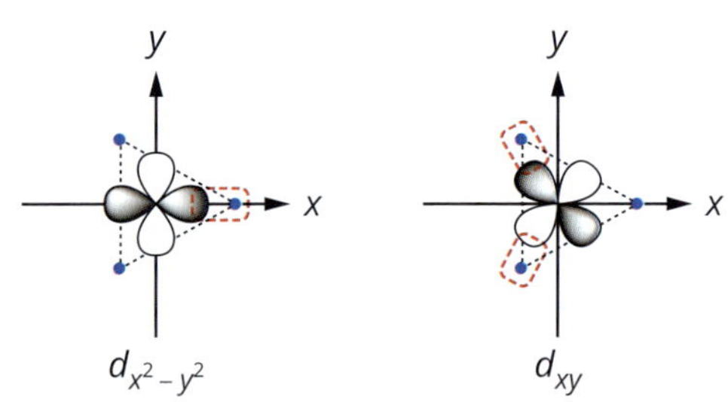

Fig. 15.9a Interaction of the d_{xy}- and $d_{x^2-y^2}$-orbitals in D_{3h} with the ligands in the equatorial plane of a trigonal bipyramidal system. While these orbitals do not point directly at all the ligands at once, they interact strongly, and by symmetry equally, with the ligands.

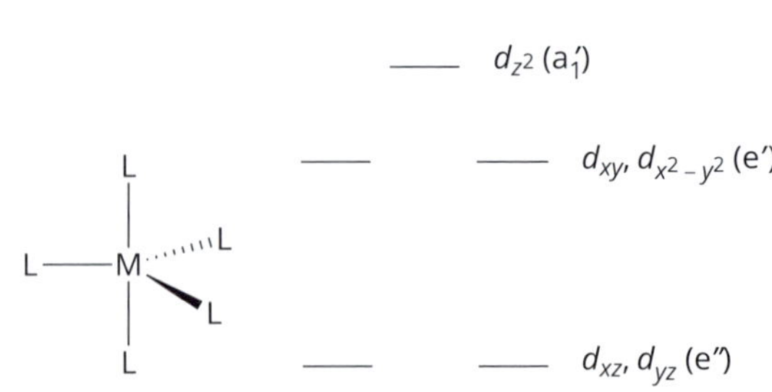

Fig. 15.9b The crystal field splitting diagram for a trigonal bipyramidal complex.

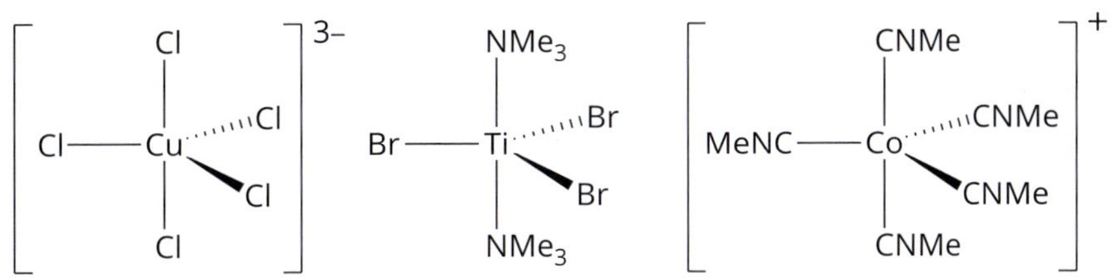

Fig. 15.9c Examples of trigonal bipyramidal transition metal complexes.

There are many trigonal bipyramidal complexes for the transition metals known,[15–16] such as d^9 $\{CuCl_5\}^{3-}$,[17–19] d^1 $TiBr_3(NMe_3)_2$,[20–21] and d^8 $\{Co(CNMe)_5\}^+$, shown in Fig. 15.9c).[22]

Related 3-coordinate complexes are also well known with appropriate, generally large, ligands (Fig. 15.10a).[23–26] Removal of the axial ligands also leads to a trigonal planar system (still in D_{3h}), but with a dramatic lowering of the d_{z^2}-orbital relative to the trigonal bipyramidal geometry. The ordering between the e′, e″, and a_1' orbitals is determined by the nature of the ligands (Section 15.4.2).

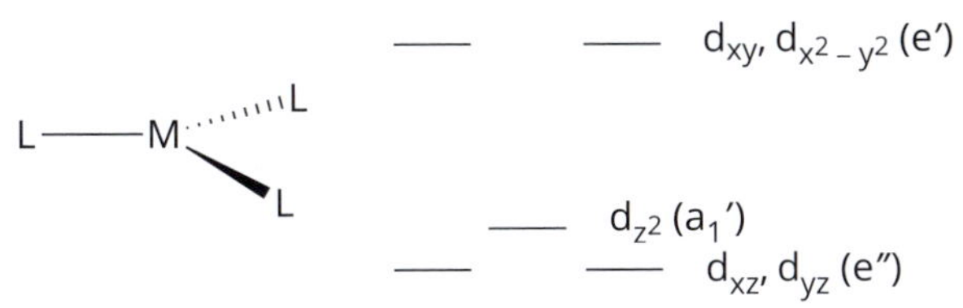

Fig. 15.10a The crystal field splitting diagram for a trigonal planar complex.

A second important class of trigonal complexes, not directly related to the trigonal bipyramid (or trigonal 3-coordinate systems), are 6-coordinate, pseudo-octahedral complexes, that are distorted in a way that retains the C_3 axis. This type of distortion is commonly found in many crystal structures, and this is often achieved by pulling/pushing the two triads of ligands of an octahedral complex that are related by the three-fold axis (trigonal elongation/compression, see Fig. 15.10b), yielding the so-called “trigonal antiprismatic” geometry.

Figuring out the crystal field splitting of a trigonally distorted 6-coordinate complex requires a little more care than we

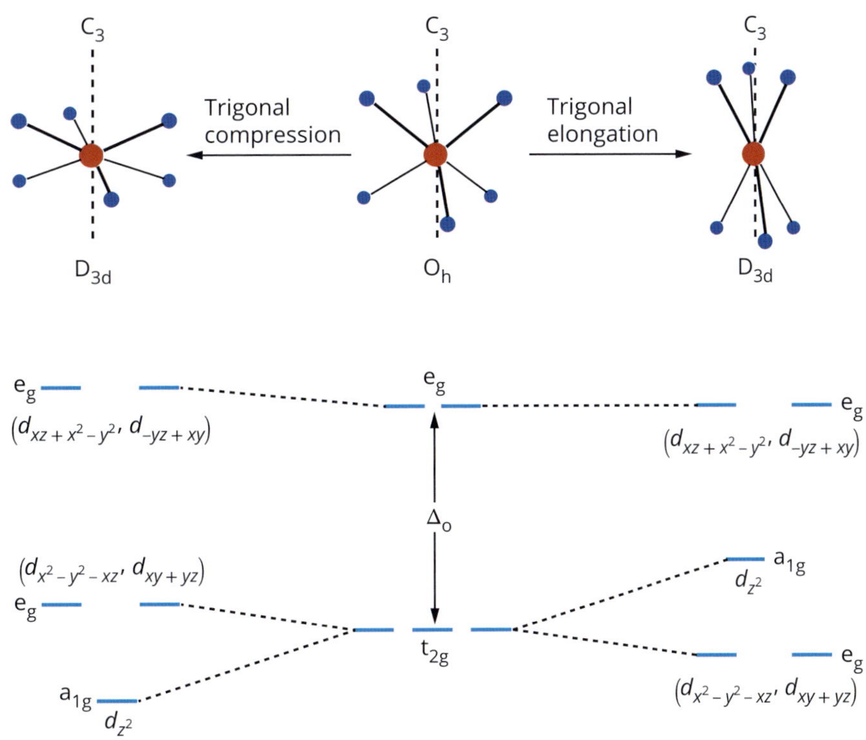

Fig. 15.10b (top) Trigonal distortions that convert an ideal octahedron (center) into trigonal antiprisms. In a perfect octahedron there are four equivalent C_3 symmetry elements; a trigonal distortion conserves only one of these C_3 symmetry elements. Note that the extent of the distortions shown here have been severely exaggerated for emphasis; most trigonal distortions are often so small that the actual deviation from ideal octahedral geometry is hard to detect without analyzing quantitatively the bond lengths and angles. (bottom) Crystal field orbital splitting for trigonal distortion of an octahedral complex (O_h to D_{3d}). (*Left*) trigonal compression; (*right*) trigonal elongation.

had to use for the tetragonally distorted complexes, which has to do with the coordinate system choice. Up to this point, we always used a coordinate system in which the z-axis was chosen to coincide with one of the C_4 axes of the octahedron. This is a *conventional* choice, not a necessary one; none of our conclusions can ever depend on the specific coordinate system that we decide to choose. In practice, though, not all coordinate systems are equally easy to use, even if they are strictly equivalent mathematically. Choosing a proper coordinate system matters a great deal in how easily we can solve a specific problem. Because tetragonal distortions retain the C_4 axis—that is, the one which is used to define the z-axis—the conventional tetragonal system we used up to this point was the most rational one, and it is by far the one you are likely to see employed most often. So much so, people commonly talk about the tetragonal d-orbitals—the d_{xy}, d_{xz}, d_{yz}, $d_{x^2-y^2}$, and d_{z^2} orbitals—as if they were *the* d-orbitals, whereas they are simply one convenient set among an infinite number of possible sets!

For trigonal distortions, using the tetragonal d-orbitals is a very *inconvenient* choice, because none of the C_4 axes of the ideal octahedron remain after the distortion. Instead, trigonal distortions retain one of the C_3 axes of the parent O_h symmetry, which are colinear with S_6 axes in O_h. This leads to very important changes in how each orbital will be labeled and even how they will appear, and we need to pay special attention to this aspect. That this is not a trivial issue can be readily appreciated by looking at the symmetry of the d-orbitals of the trigonally distorted complexes. The ideal point group of a trigonal antiprism, either compressed or elongated, is D_{3d}. In this point group, the five d-orbitals transform as a_{1g} (d_{z^2}) and two e_g sets (($d_{x^2-y^2}$, d_{xy}) and (d_{xz}, d_{yz})). Notice that if we were not careful with the labels, we would quickly get into trouble; in D_{3d}, $d_{x^2-y^2}$ and d_{xy} must have the same energy (they are part of the same set of degenerate orbitals), whereas in perfect octahedral symmetry they seemingly must have very different energies, separated by an energy difference of $\Delta_o = 10$ Dq. This cannot be true, as it would imply that an infinitesimally small trigonal distortion would instantaneously lead to a large discontinuous change in the energy of the orbitals, which is of course physically nonsensical.

The problem is that the conventional definition of the xyz-axes in O_h is different than in D_{3d}, which means that the label we give for each orbital means something different in each point group. For instance, what we call a "d_{z^2}" orbital in O_h is *not* the same orbital as the one we call "d_{z^2}" in D_{3d}. As shown in Fig. 15.10c, in the O_h point group the d_{z^2} orbital along one of the C_4 axes (which we designate as the z-axis), but in D_{3d} the "d_{z^2}" orbital is instead located in a plane perpendicular to a C_3 axis, which acts as the z-axis in that point group. Although the two orbital shapes are identical, the same mathematical function but rotated in space, it is clear that the two are quite inequivalent chemically. The d_{z^2} orbital oriented with the C_4 axis directly points at the ligands on the z-axis and is destabilized by the crystal field, but the d_{z^2} orbital oriented with the C_3 axis points exactly *between* all of the ligands and is stabilized. *You see then that strictly it is incorrect to refer to a "d_{z^2}" orbital (or any other label) without first making it clear what coordinate axis system we used to define that orbital.* In practice, the use of the tetragonal system is so engrained in everyday inorganic chemistry parlance that it is generally very safe to assume that the d-orbitals, if nothing else are specified about them, are described by using axes along the ligands.[g]

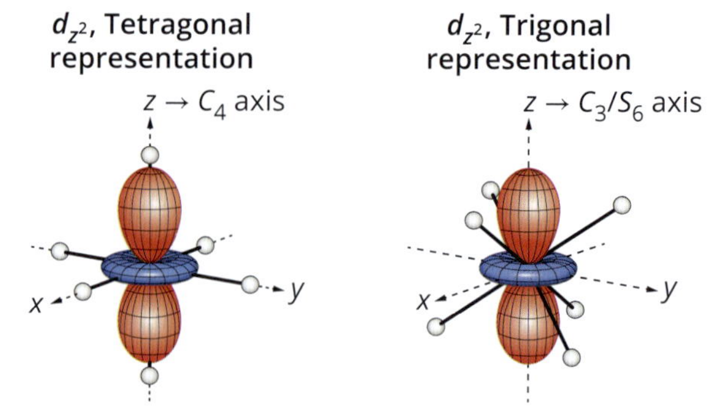

Fig. 15.10c The specific coordinate system we use impacts how we define each orbital. Here, two different "d_{z^2}" orbitals are shown for the same octahedral complex—one defined with the z-axis oriented on one of the C_4 axes of the octahedral complex ("tetragonal representation", *left*) and the other one defined by using one of the C_3 axes of the octahedral complex as the z-axis ("trigonal representation", *right*). In the tetragonal representation, the d_{z^2} orbital is destabilized by the crystal field (oriented toward the point charges), whereas in the trigonal representation the d_{z^2} orbital is stabilized by the crystal field (oriented between the point charges).

What would the other d-orbitals look like, and what would their names be, in the trigonal representation? We could use the exact same five labels we used in the tetragonal representation (d_{xy}, d_{xz}, d_{yz}, $d_{x^2-y^2}$, d_{z^2}), which would have the benefit of keeping things simple in terms of not having to learn new names (not a bad reason in itself), but it turns out that there is a better way to define the d-orbitals in the trigonal representation, which leads to orbitals that are oriented in a more "natural" way relative to the ligands and simplifies how we think about crystal-field effects. This "natural" way, which is only yet another conventional choice, we can think of as coming from mixing (hybridizing) the d_{xy}, d_{xz}, d_{yz}, and $d_{x^2-y^2}$ orbitals together to conserve the symmetry properties of the O_h point group. The procedure to achieve this is outlined in more advanced texts, but here we present the end result:[6]

$$t_{2g}:$$

$$\sqrt{\frac{2}{3}}d_{x^2-y^2} - \sqrt{\frac{1}{3}}d_{xz} \equiv d_{x^2-y^2-xz}$$

$$\sqrt{\frac{2}{3}}d_{xy} + \sqrt{\frac{1}{3}}d_{yz} \equiv d_{xy+yz}$$

$$d_{z^2}$$

[g] In quantum-mechanical terms, we say that the angular momentum basis functions are quantized relative to the C_4 axis.

$$e_g:$$

$$\sqrt{\frac{1}{3}}d_{x^2-y^2} + \sqrt{\frac{2}{3}}d_{xz} \equiv d_{xz+x^2-y^2}$$

$$\sqrt{\frac{1}{3}}d_{xy} - \sqrt{\frac{2}{3}}d_{yz} \equiv d_{-yz+xy}$$

As can be seen directly here, in the trigonal representation, the d_{xy}, d_{xz}, d_{yz}, and $d_{x^2-y^2}$ orbitals are not "pure" orbitals of the system. The d_{xy} and d_{yz} orbitals mix together to form one of the t_{2g} and one of the e_g orbitals, and $d_{x^2-y^2}$ and d_{xz} mix together to also one of the t_{2g} and one of the e_g orbitals. In O_h, the mixing coefficients are precisely dictated by symmetry, with the values shown previously, but as we lower the symmetry the amount of mixing becomes variable and will be dictated by the nature of the ligands used and the exact geometry of the complex, among other things.

Do not be overly bothered by the more complex appearance of the equations that define the trigonal *d*-orbitals in this trigonal setting. The simpler labels are a very good reason why, when we can, we prefer to use the tetragonal basis; however, remember that even in the tetragonal basis the names are shortcuts for more complex functions. For instance, the "z^2" in d_{z^2} is in fact "$2z^2-x^2-y^2$".

In order to be able to keep track of things, we need to assign names to each of the trigonal *d*-orbitals, but there is not a unique convention for how this is done. What we do here is use the names of the individual components, listing first the dominant member. For instance, a $d_{x^2-y^2-xz}$ orbital would refer to the trigonal *d*-orbital that contains more $d_{x^2-y^2}$ character and is combined with a different phase (hence the minus symbol) of the d_{xz} orbital. As shown in Eq. 15.4, under this convention the five trigonal *d*-orbital names would be $d_{x^2-y^2-xz}$, d_{xy+yz}, $d_{xz+x^2-y^2}$, d_{-yz+xy}, and d_{z^2}.

The trigonal *d*-orbitals are shown in Fig. 15.11, which clearly demonstrates that the three t_{2g} orbitals defined previously point between the ligands and are therefore stabilized, whereas the two e_g orbitals point directly at the ligands and are therefore destabilized. This is the same result that we found before for the crystal field splitting of the d-orbitals in an ideal octahedral symmetry. This is as it should be, of course, as the only thing we changed here is the orientation of the *xyz*-axes! That being said, note how the *shapes* of some of the trigonal *d*-orbitals differ from that of the tetragonal representation. In the tetragonal representation, there are four "four-leafed

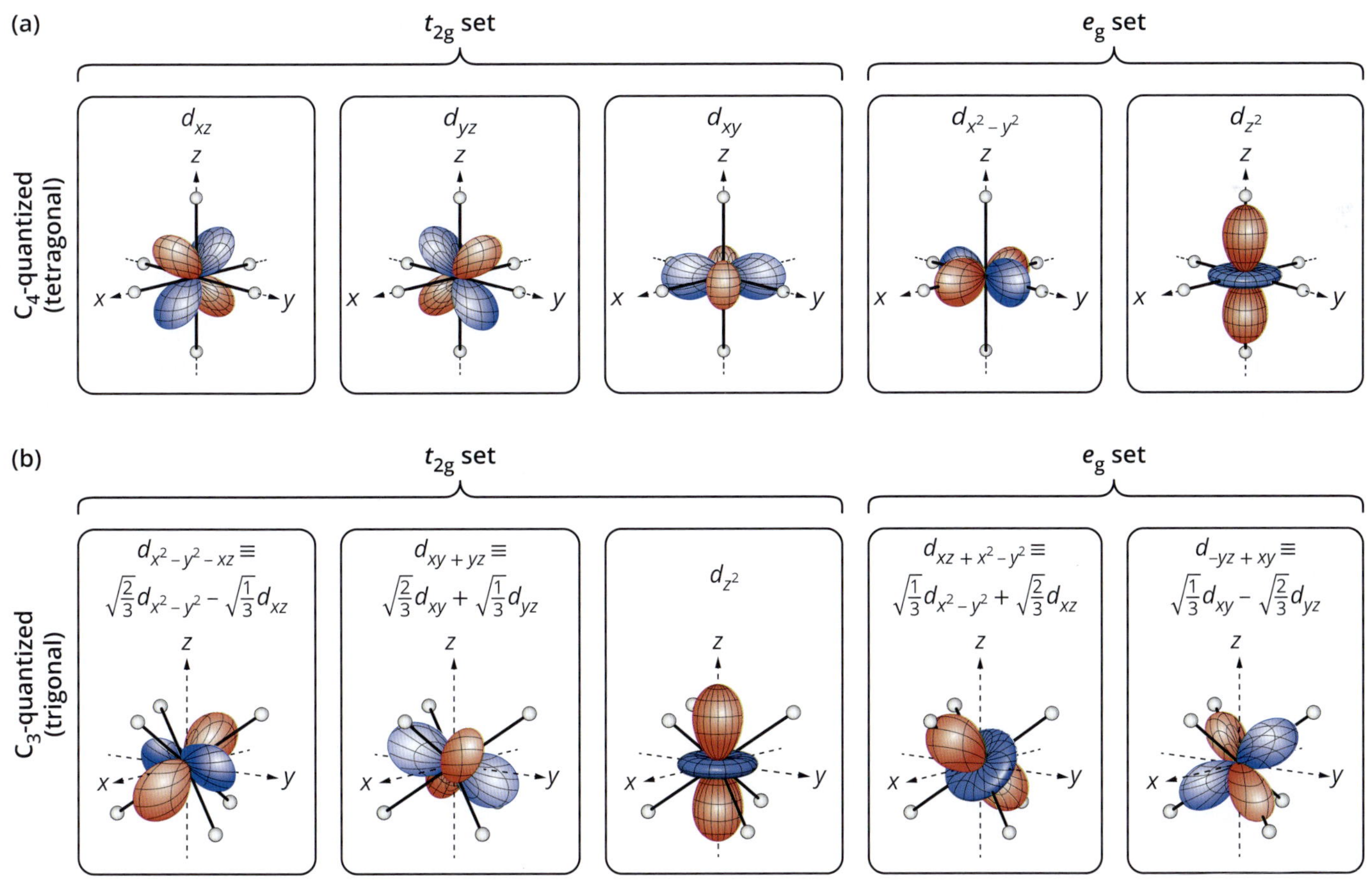

Fig. 15.11 The five *d*-orbitals in the O_h point group, as defined in the (a) tetragonal (*z*-axis = C_4 axis), and (b) trigonal (*z*-axis = C_3 axis) representations. Note that since the coordinate systems are different, the *xyz* labels mean different things in each representation; that is, the d_{z^2} orbital in the tetragonal representation is *not* the same orbital as the d_{z^2} orbital in the trigonal representation.

clover" *d*-orbitals and one "axial-plus-torus" orbital, but the trigonal representation we chose here has only two "four-leafed clover" orbitals, two "axial-plus-torus" orbitals, and a fifth one which is a sort of hybrid between the two shapes (the one we call $d_{x^2-y^2-xz}$). Again, remember that this is simply the result of choosing a coordinate system and of mixing the five orbitals together until they are oriented in a specific manner. These are arbitrary choices that we make based on what we think will help us simplify the problem on which we are working.[h]

With this new information in hand, let us go back to the problem of figuring out the crystal field splitting for a trigonally distorted octahedral complex. For the sake of simplicity, we will assume that the ideal point group of this complex is D_{3d} (Fig. 15.10b), but the same analysis would apply just as well for other lower-symmetry configurations, such as D_3 or C_3, which are two subgroups of D_{3d}. If the complex is trigonally elongated, then the negative point charges are all moved closer to the *z*-axis, and away from the *xy*-plane (see Fig. 15.10c). Clearly, then, the trigonal d_{z^2} orbital, which transforms as a_{1g} in D_{3d}, will be destabilized. The behavior of the other four orbitals is less clearly defined, at least without doing any quantitative calculations, as they are made up of a mixture of orbitals that are located in the *xy*-plane and in the *z*-direction. Nevertheless, if we assume that the trigonal distortion is small compared to 10 Dq, then we can invoke the barycenter rule to propose that the pair ($d_{x^2-y^2-xz}$, d_{xy+yz}), which becomes one of two e_g sets in D_{3d}, will be stabilized by about half the amount of destabilization of the d_{z^2} orbital. To a first approximation, we then assume that the other pair of orbitals, the e_g set in O_h—this also remains a degenerate e_g set in D_{3d}—does not move much. In reality, because the two e_g sets are both of the same symmetry they can mix together, and this amount of mixing ultimately determines the exact position of these orbitals. Unfortunately, this effect varies from case to case; that is, it is not fixed by the symmetry of the complex but by the specific geometrical and chemical parameters that define each complex and cannot be predicted from a simple qualitative analysis.[i] Nevertheless, the overall crystal field splitting will be as shown in Fig. 15.10b.

[h] Note also that this feature is not something you see for lower angular-momentum orbitals, such as *s*-orbitals. Being nondegenerate by default, they have only one orientation and one shape, independent of the coordinate system that you choose. The *p*-orbitals can be oriented in any way we want, depending on the coordinate system that we choose, but their shapes cannot be altered by combining them. Any sum of two or three *p*-orbitals always yields another orbital with the same shape but with a different orientation.

[i] The e_g–e_g mixing not only affects the energy of the orbitals, but also the "shape" of these two sets. If the mixing is very strong, then the proper orbitals that define the two e_g sets in D_{3d} will be closer to the tetragonal orbitals: for trigonal elongations, the low-energy e_g set will look more like the ($d_{x^2-y^2}$, d_{xy})-orbitals—more localized in the *xy*-plane—whereas the high-energy e_g set will look more like the (d_{xz}, d_{yz})-orbitals—more localized along the *z*-axis. The opposite mixture arises for trigonal compressions. Again, there is no direct way to predict exactly what the actual composition of the trigonal *d*-orbitals will be, except for the a_{1g} orbital, which always remains purely d_{z^2} for all trigonal distortions.

As always, the behavior of trigonal compressions will be the inverse of that obtained for trigonal elongations, but in this case, because the bottom e_g set will be moving energetically toward the higher e_g set, we can infer that the mixing between the two sets should be larger than for trigonal elongations, because quantum-mechanical mixing between eigenfunctions is always larger the smaller the energy difference in their energies, everything else being equal, which would lead to stabilization of the bottom-most e_g set and destabilization of the top-most one, as shown in Fig. 15.10b.

15.4 Crystal Fields to Ligand Fields

Considering the number and crudeness of the approximations on which it relies, crystal field theory (CFT) is a remarkably successful approach for describing the electronic structure of transition-metal complexes. The CFT approach is a powerful first step that allows us to quickly draw qualitative orbital splitting patterns that are generally very close to the actual ones that more advanced theoretical and/or experimental methods provide, but on a quantitative basis there is no denying that the crystal field approach is flawed. Historically, it was quickly realized that CFT systematically predicted optical transition energies that were very far from the experimental values, requiring the use of unreasonable values for the physical parameters of the theory.[87] Of course, in hindsight this is not surprising. We know that ligands are not point charges or dipoles, and the interaction between the electrons of a metal center and those of a ligand are not described by simple classical electrostatics but rather by quantum mechanics.

These deficiencies of CFT were recognized not so long after its conception, and various approaches to improve the crude approximations of the CFT were sought. Two particular aspects were thought to be especially important, which are both related to the fact that metal–ligand bonds are covalent interactions. First, the covalency of the bond implies that what we call a "*d*-orbital" is actually an admixture of a pure metal-centered orbital with ligand-centered orbitals, which this has the important impact that "*d*-electrons" are much more delocalized in practice than what the crude CFT would lead us to believe. Second, unlike classical electrostatic repulsions, covalent interactions are governed by the overlap of orbitals on different centers, and these overlaps can take different forms, depending on the type and orientations of the overlapping orbitals (σ-, π-, δ-interactions, and so on). Each of these specific types of interactions will play a different role in the overall orbital splitting and needs to be evaluated separately.

A simple approach that successfully describes these chemical effects is molecular orbital theory, and the relationship between crystal field theory and molecular orbital theory for metal complexes was first established in a classic paper by van Vleck in 1935.[28] (He also pointed out the relationship between MO theory and VB theory for good measure.)[28–29] The incorporation of MO theory into CFT is what we now refer to as ligand field theory (LFT).

Like crystal field theory, ligand field theory focuses only on the *d*-orbitals of the metal center, but it recognizes that the energies of these orbitals are affected by the way they interact with the ligand-based orbitals. For instance, Δ_o, the energy difference between the t_{2g} and e_g orbitals, which CFT calculates as a difference in electrostatic interactions between electrons in each type of orbital and the ligands, will be interpreted in the LFT framework as the combined sum of all interactions between the metal and the ligands orbitals (σ, π, δ, … bonding interactions). As we will see shortly, this new perspective allows LFT to tell us something specific about the nature of the metal–ligand bonding from a chemical point of view, which is why LFT remains to this day such a powerful tool in the modern inorganic chemistry toolbox.

First, let us begin by justifying the focus on the *d*-orbitals in both CFT and LFT, which can be readily seen by examination of the molecular orbital diagram of a simple octahedral complex, say $\{Co(NH_3)_6\}^{3+}$ (Fig. 15.12). In this complex,

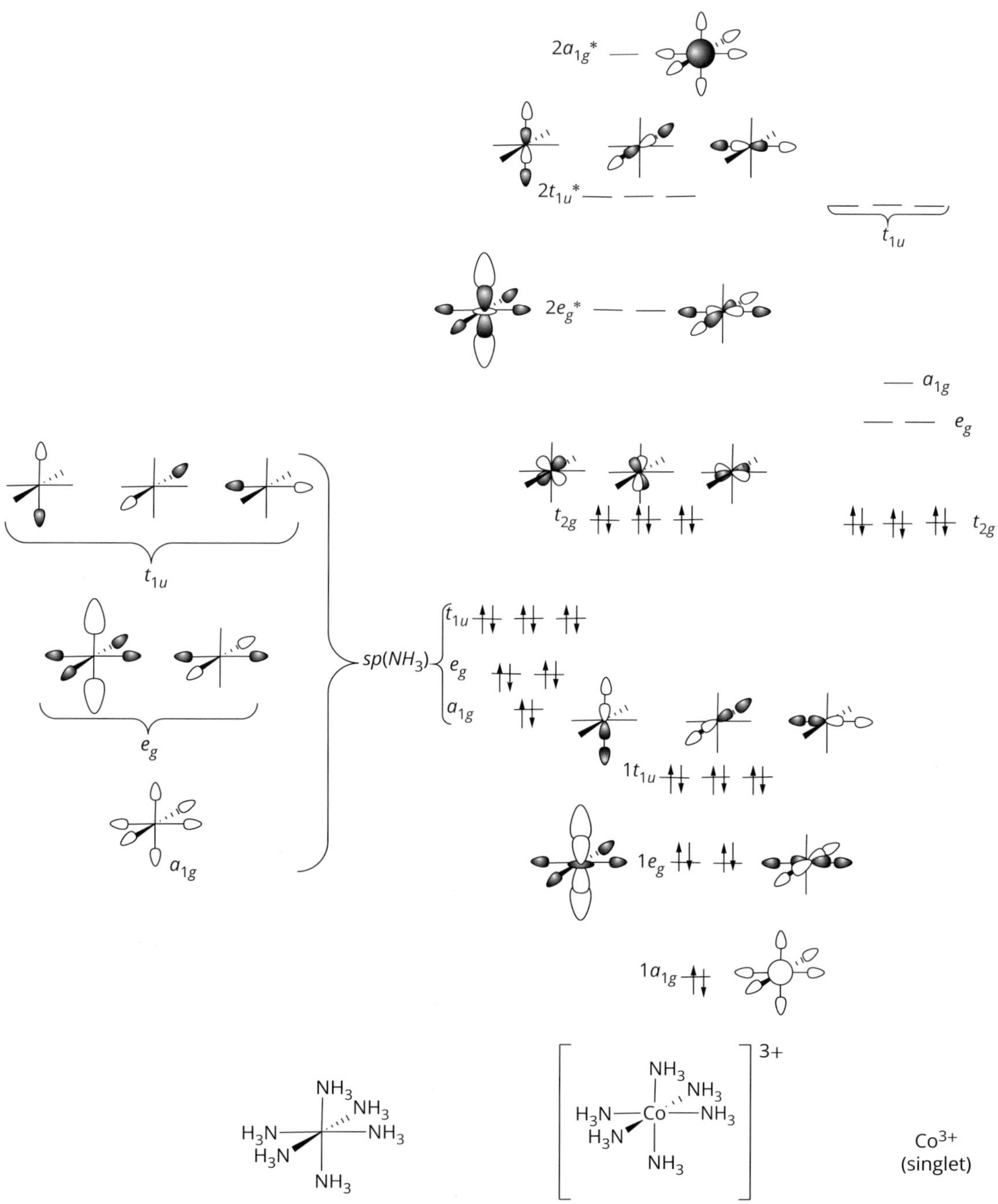

Fig. 15.12 MO diagram for $\{Co(NH_3)_6\}^{3+}$.

(Note: We typically think of all the d-orbitals being equivalent prior to interaction. Here the energies are separated as might be seen in a calculation of the Co(III) ion designated as a singlet, with the occupied orbitals slightly lower in energy, accentuating the π-nonbonding nature of amines on interaction.)

the Co^{3+} ion has a *d*-electron count of six electrons (the nine the element would have, minus three to form the ion). An important feature of transition-metal ions is that the order of the atomic orbitals is altered from that of the neutral species; where the ground-state configuration of Co^0 is {Ar}$4s^2 3d^7$, the higher oxidation states invariably push the energy of the *s* orbital up in energy and the ground-state configurations are always built up by filling *d*-orbitals. As a result, we generally put the valence *s* and *p* orbitals of the metal higher in energy in a MO diagram of an inorganic complex, and fill up the *d*-orbitals first. For the MO diagram in Fig. 15.12, we show the relative energies of the Co(III) free ion in a singlet state, which is most relevant to the problem at hand.

The ligand-based orbitals consist of symmetry-adapted combinations of the nitrogen-based lone-pair donor orbitals with a_{1g}, e_g, and t_{1u} symmetry. These ammine ligands contribute a total of twelve electrons—a pair of electrons each, yielding a total valence electron count of eighteen for the complex.

The eighteen valence electrons fill the MOs as shown in Fig. 15.12, up through the nonbonding metal-centered t_{2g} orbitals. Notice from the MO diagram that the orbitals that will largely be of *d*-orbital character, t_{2g} and e_g^*, are where any "*d*-electrons" in the system will be housed. The student should convince themselves that different *d*-electron counts, d^0 to d^{10}, found by subtracting the oxidation state of the metal from the electron count of the neutral transition element, correspond to the number of electrons that reside in these predominately *d*-character t_{2g} and e_g^* orbitals in the molecular orbital diagram. In the case of $\{Co(NH_3)_6\}^{3+}$, the eighteen electrons produce a $(1a_{1g})^2(1e_g)^4(1t_{1u})^6(t_{2g})^6(e_g^*)^0$ configuration; all six *d*-electrons of Co(III) reside in the metal-based t_{2g}-orbital.

The t_{2g}- and e_g^*-orbitals will always be, or at least be near, the frontier molecular orbitals. For some *d*-electron configurations, such as low-spin d^6 systems like $\{Co(NH_3)_6\}^{3+}$, t_{2g} and e_g^* will be the HOMO and LUMO, respectively, of the complex; in other cases, at least one of the two will be either the HOMO or LUMO. Consequently, the properties and reactivity of transition-metal complexes are often closely tied to the nature and energies of the *d*-orbitals and the number electrons that fill them, which is the justification for the "*d*-orbital-centricity" of LFT and CFT.

In LFT, the ligands, which are typically more electronegative than the metal to which they are attached, are considered to be the major contributors to the bonding orbitals. The *d*-orbitals in the MO picture of a simple octahedral complex are then of t_{2g} and e_g symmetry and are largely metal-based, often making them nonbonding or antibonding in the MO diagram. The e_g^* orbitals, the σ-antibonding set of ($d_{x^2-y^2}$, d_{z^2})-orbitals, corresponds to the same set in crystal field theory. The same orbitals for t_{2g} (d_{xz}, d_{yz}, d_{xy}) are directly below the e_g^* in the octahedral field and have correct symmetry to interact with any π-donating or π-accepting orbitals on the ligands. In the simple case discussed previously, $\{Co(NH_3)_6\}^{3+}$, the ammine ligands do not have any such π-donating or π-accepting orbitals, so the t_{2g} set remains purely nonbonding.

15.4.1 Square-Planar Geometries and Ligand Field Theory

Under the context of crystal field theory we examined the effect of elongating the bonds to the ligands along the *z*-axis of an octahedral complex (Fig. 15.6). At the extreme, where the ligands along the *z*-axis are moved infinitely far from the metal, a square-planar complex is formed. With LFT, one can discuss the orbital effects in rather more detail and how ligand overlap is affecting the *d*-orbital splittings within square-planar complexes, which provides some additional insights and prompts a more in-depth discussion of orbital mixing.

Square-planar complexes are the norm for d^8 4-coordinate complexes, especially for the second- and third-row elements such as Pt^{2+}, Pd^{2+}, Au^{3+}, Rh^{1+}, Ir^{1+}, and so on. These compounds share a splitting pattern similar to that shown in Fig. 15.6, with one high-energy orbital ($d_{x^2-y^2}$) and the other four *d*-orbitals occupied with the eight metal-based electrons. In some cases, further stabilization is gained by mixing of the d_{z^2}-orbital with the valence *s*-orbital, the two orbitals having the same symmetry (a_{1g}) in the D_{4h} point group (see Section 15.9.8). This d_{z^2}-stabilization prevents new ligands from coming in along the z-axis to produce an isolable octahedral system, and encourages lower coordination numbers.

The lighter congeners in the same valence configurations, such as Ni^{2+} and Co^{1+}, can adopt square-planar geometries; however, higher coordination numbers are often stable as well, such as octahedral $\{Ni(NH_3)_6\}^{2+}$. One possible explanation for this is that the *d*-orbitals of the first row are contracted due to the primogenic effect, and consequently do not hybridize as effectively with the valence 4*s*-orbital, which limits the d_{z^2}-stabilization obtained from *s*–*d* mixing (Section 15.9.8).

Furthermore, square-planar configurations are almost always found for low-spin d^8-complexes of the second- and third-row metals. The energy required to pair spins for the second- and third-row complexes is smaller for these larger radii congeners (Section 15.7). The pairing energy for the first row is higher, and it seems that these complexes are more likely to adopt octahedral (or tetrahedral) geometries with high-spin ground states.

For example, NBO calculations on DFT optimized $\{Ni(NH_3)_4\}^{2+}$ produce the expected four lone pairs on the d^8 nickel center, with one (d_{z^2}) being very slightly hybridized with the *s*-orbital to give a sd^{35}-hybrid (~3% *s* character, ~97% *d* character). In contrast, the same metal-centered pair of electrons in $\{Pt(NH_3)_4\}^{2+}$ resides in an sd^{11}-hybrid (that is, ~8% *s* character, 92% *d* character). The combined absence of the primogenic effect and of increased *s*–*d* mixing for the heavier congener leads to a more effective stabilization of the orbital along the *z*-axis that stabilizes the square-planar geometry. In fact, square-planar $\{Ni(NH_3)_4\}^{2+}$ is not isolable, presumably because the higher-energy d_{z^2}-orbital in that complex favors the binding of two additional ammine ligands along the *z*-axis to form octahedral $\{Ni(NH_3)_6\}^{2+}$, that is, the d_{z^2}-orbital and the electrons in it are not stabilized enough to give a stable complex.

Some d^9-complexes also adopt square-planar geometries. For example, copper(II) ions in water give tetragonally distorted octahedral hexa(aqua)copper(II) in water. Addition of ammonia produces blue, square-planar tetra(ammine)copper(II), which has been characterized in the solid state as an independent species in a few cases (Eq. 15.4); however, $Cu(NH_3)_4^{2+}$ usually has at least one, and more often two, ligands along the z-axis, such as in $\{Cu(OH_2)(NH_3)_4\}^{2+}$ or $\{Cu(OH_2)_2(NH_3)_4\}^{2+}$.

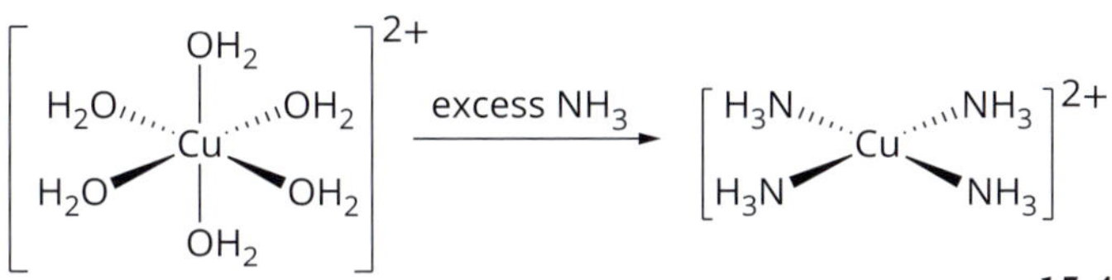

15.4

Here we have discussed elongation of one axis of the octahedron; however, one can consider a different tetragonal distortion where the z-axis has a stronger crystal field than that in the xy-plane. As before, the stronger electric field (stronger donors) will cause the orbitals directed at them to be higher in energy. As a result, a very different crystal field splitting diagram will be obtained (see Problem 15.1).

15.4.2 Trigonal Planar Geometries and Ligand Field Theory

This molecular orbital-based LFT can be applied to systems of any other geometry as well; for example, trigonal planar complexes previously discussed briefly under CFT. The effects of various π-donors and -acceptors on the crystal field diagram of planar 3-coordinate complexes are shown in Fig. 15.13.[j] The figure shows the splitting diagrams for donor ligands that are σ-only (such as Me), ligands that have two π-oriented lone pairs (such as methoxide), and ligands with one lone pair in the xy-plane (such as an NMe_2).

As an example, consider the spectrum of *tris*(hexamethylsilazide)titanium(III), $Ti\{N(SiMe_3)_2\}_3$, a bright blue complex originally prepared by Bradley and coworkers.[27] The complex shows *d–d* transitions (crystal field transitions between *d*-orbitals). The ground state of the compound was shown experimentally (by EPR) to be $(d_{z^2})^1$, which is unsurprising, as d_{z^2} is one of the lower-energy orbitals in all the configurations in Fig. 15.13. Amides, as mentioned, may act as π-donors, because nitrogen has a lone pair of electrons in an orbital perpendicular to the C–C–N–M plane. Fig. 15.13 shows an example in which the amide is rigorously aligned with the z-axis, where the complex would have D_{3h} symmetry. In $Ti\{N(SiMe_3)_2\}_3$, the amide ligands are tilted similar to a boat propeller, giving the complex D_3 symmetry,[28] which causes the amides to π-donate into both e-sets, acting similar to a donor with two lone pairs in that both e' and e" have some π-antibonding character (middle of Fig. 15.13).

The ligand field splitting diagram can be approximated as shown in Fig. 15.14. The lower-energy transition at 4,800 cm^{-1} is from configuration $(d_{z^2})^1 \rightarrow (d_{xz}, d_{yz})^1$. The higher-energy transition at 17,400 cm^{-1} is from configuration $(d_{z^2})^1 \rightarrow (d_{x^2-y^2}, d_{xy})^1$.

[j] In these diagrams we used the tetragonal labels for the *d*-orbitals, which work fine for a trigonal planar system.

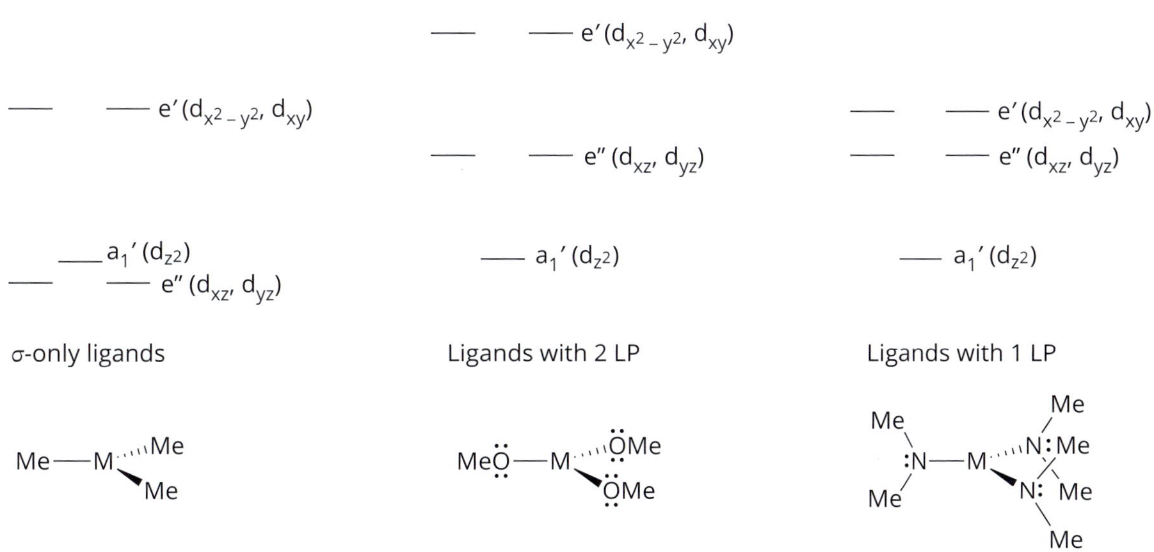

Fig. 15.13 Qualitative crystal field splitting diagrams for 3-coordinate complexes with different donor properties.[23] (*left*) For a complex with σ-only donor ligands, such as Me, we expect a high-energy e' (σ antibonding) and low-energy a₁' and e" orbitals (approximately nonbonding; some antibonding character with d_{z^2} due to interaction with the torus). (*middle*) For a complex with two lone pairs on each donor ligand, such as methoxide, where the lone pairs will be directed both in the xy-plane and along the z-axis, both e-sets will be higher in energy than a_1'. The e' orbitals will be highest as it is both σ- and π-antibonding, while the e" orbitals are π-antibonding. (*right*) If there are lone pairs on the ligands only in the xy-plane, such as NMe_2 ligands oriented with the C–N–C plane parallel to the z-axis, then the e" orbital will not be raised due to π-donation (since there are none), but the e' will be higher in energy than a_1' because it is both σ- and π-antibonding.

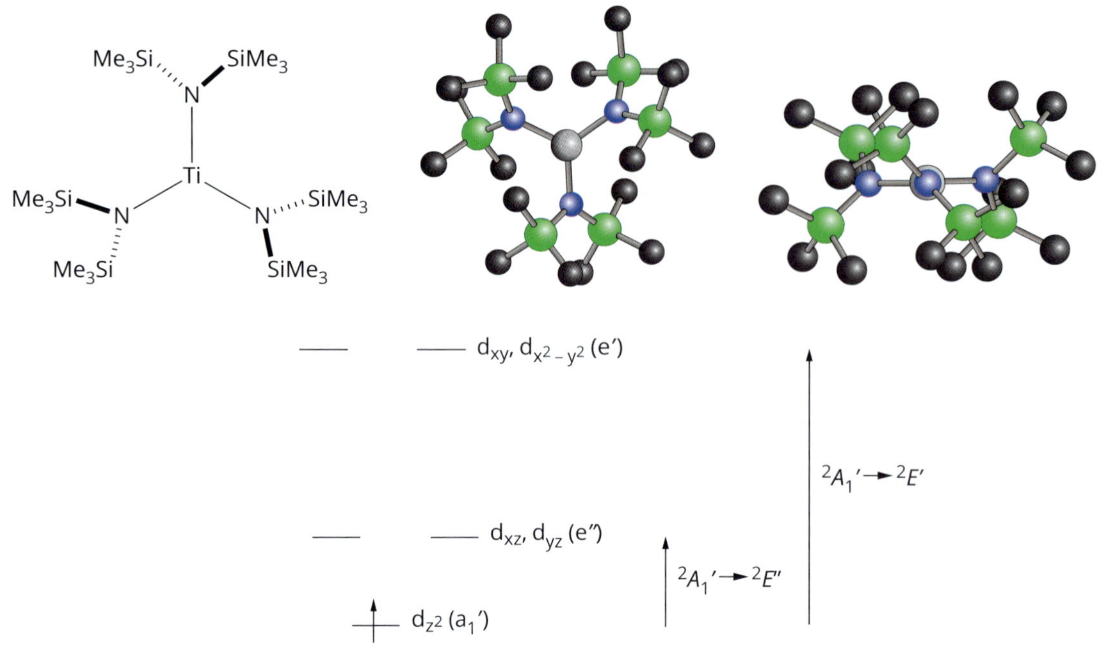

Fig. 15.14 *(top)* The structure of d[1] Ti{N(SiMe$_3$)$_2$}$_3$, with one view down the C_3 axis and another down one of the Ti–N bonds showing the tilt of the Si–N–Si plane relative to the C_3. Overall, the structure has D_3 symmetry.[28] *(bottom)* The crystal field splitting diagram for d[1] Ti{N(SiMe$_3$)$_2$}$_3$ under the 1-electron approximation. The transitions occur experimentally at $^2A_1' \rightarrow {}^2E''$: 4,800 cm^{-1} (10) and $^2A_1' \rightarrow {}^2E'$: 17,400 cm^{-1} (122), where the values in parentheses are the molar extinction coefficients in units of M^{-1}cm^{-1}. The state assignments for the transitions are based on the smaller extinction coefficient for the electric dipole forbidden $^2A_1' \rightarrow {}^2E''$ transition.[27–28]

Further confirmation of this assignment comes from the intensity of the bands (Fig. 15.14).[k]

It will be advantageous, especially for multielectronic systems discussed in Chapter 16, to designate an electronic state by its symmetry. For systems with one unpaired electron that are either d^1 or d^9, the state will be a doublet (Section 2.2.3), and the term symbol will be the symmetry of the orbital containing the unpaired electron. Consequently, Ti{N(SiMe$_3$)$_2$}$_3$ will have a $^2A_1'$ ground state. The electronic transition corresponding to $(d_{z^2})^1 \rightarrow (d_{xz}, d_{yz})^1$ comes from states $^2A_1' \rightarrow {}^2E''$, and the transition from $(d_{z^2})^1 \rightarrow (d_{x^2-y^2}, d_{xy})^1$ will be from state $^2A_1'$ to $^2E'$.

15.5 The Idealized Spectrum of {Ti(OH$_2$)$_6$}$^{3+}$

Consider the d[1]-system, $(t_{2g})^1(e_g{}^*)^0$, hexa(aqua)titanium(III), {Ti(OH$_2$)$_6$}$^{3+}$. Presently, we will discuss the absorption spectrum of a purely octahedral Ti(III) ion in water; the experimental system has an added complication, a Jahn–Teller distortion, that will be discussed later.[l]

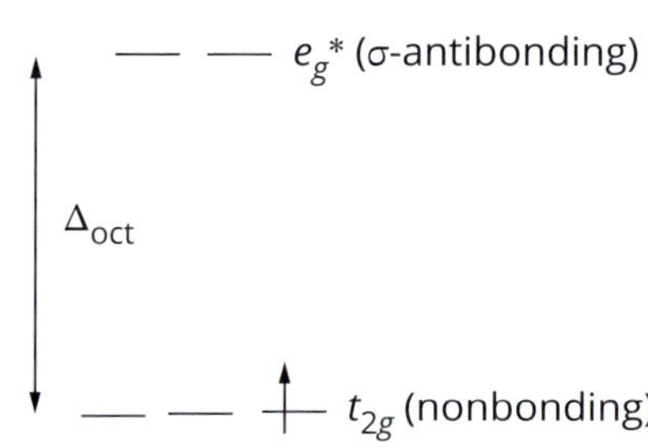

Fig. 15.15a Idealized O_h Ti(OH$_2$)$_6{}^{3+}$ crystal field splitting diagram.

Octahedral {Ti(OH$_2$)$_6$}$^{3+}$ will have a ligand field diagram as shown in Fig. 15.15a, with a single electron occupying the t_{2g} orbital set, a $^2T_{2g}$ state.

On excitation with white light, {Ti(OH$_2$)$_6$}$^{3+}$ will absorb the impinging light with energy approximately equal to Δ_o (Fig. 15.15b).[m]

The idealized spectrum of {Ti(OH$_2$)$_6$}$^{3+}$ is shown in Fig. 15.15c. The absorption band is centered at ~20,000 cm^{-1}. The band in question would be reported as λ_{max} = 20,300 cm^{-1}, or as λ_{max} = 493 nm, and with ε_{max} = 5 M^{-1} cm^{-1} (Box 16A). The absorption of blue-green light leads to a violet-red color to the solution (Box 16C).

Similarly, octahedral Cu(II) can be expected to have a single possible ligand-field transition (Fig. 15.15d). Copper(II) is d^9 with a $(t_{2g})^6(e_g{}^*)^3$-configuration.

---

k Band intensity and the mechanisms that determine it are discussed in Chapter 16.

l It turns out that all d^1 systems will have the complication of a Jahn–Teller distortion, but it is useful to discuss an electronic transition of this type without this added feature first. That is why we are discussing an "idealized" version of the spectrum, which allows us to build the complexity of the discussion more slowly.

m To a first approximation, the light absorbed would be exactly equal to the energy gap; however, there are factors that lead to absorption line breadth, such as vibrations. These will be discussed in Chapter 16.

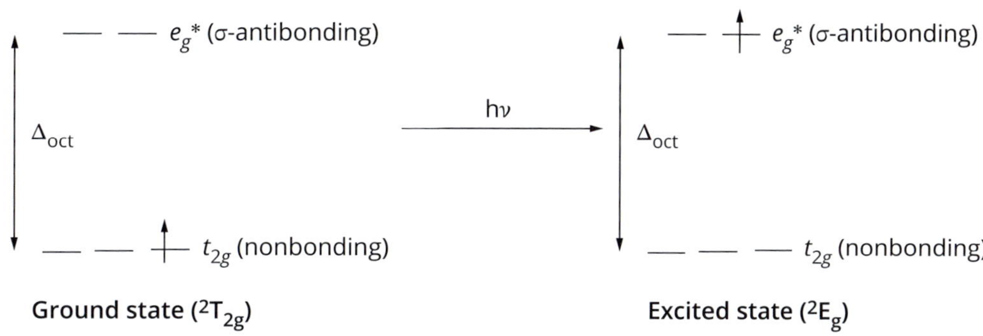

Fig. 15.15b Electron promotion due to light absorption in an idealized O_h $Ti(OH_2)_6^{3+}$ system.

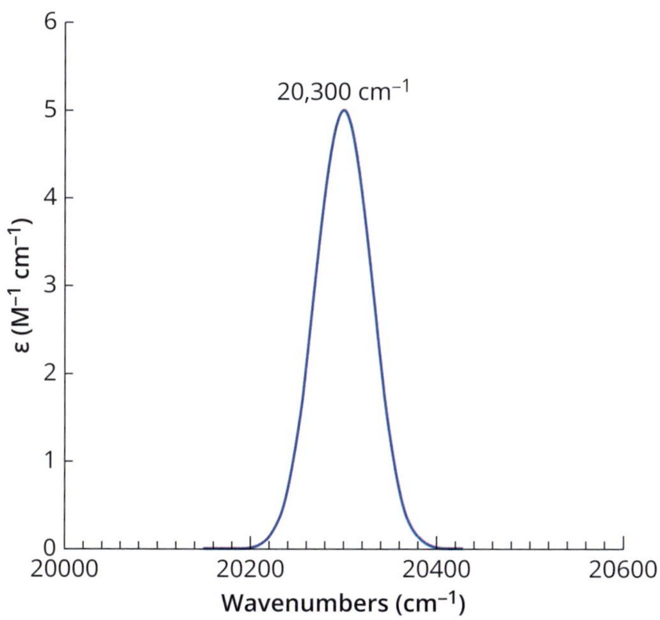

Fig. 15.15c An idealized spectrum of $\{Ti(OH_2)_6\}^{3+}$.[31] (The actual spectrum includes a complication, a Jahn–Teller distortion, that will be added a little later.)

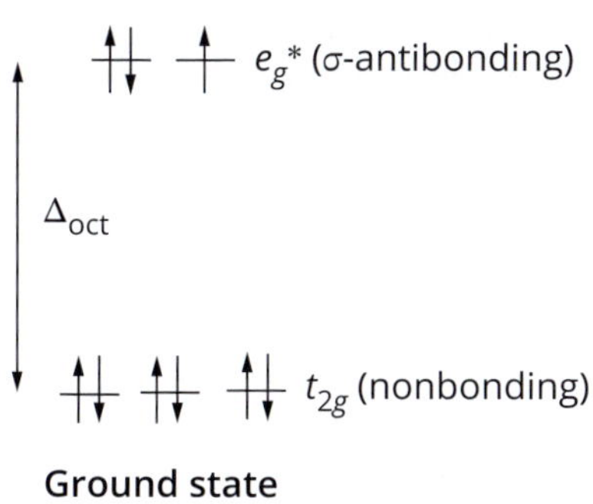

Fig. 15.15d Idealized O_h Cu(II) crystal field splitting diagram.

The one transition possible is where a spin-down electron in the t_{2g} orbital manifold is excited into the e_g* orbital that is unfilled (Fig. 15.15e), which produces the $(t_{2g})^5(e_g{}^*)^4$-configuration.

In the hexaaquacopper(II) complex, $\{Cu(OH_2)_6\}^{2+}$, this transition occurs with λ_{max} ~ 12,000 cm^{-1} (~830 nm) and with ε = 11 M^{-1} cm^{-1}, but the absorption band tails well into the visible to ~17,000 cm^{-1}. The red absorption gives the complex a pale blue color from transmitted white light.

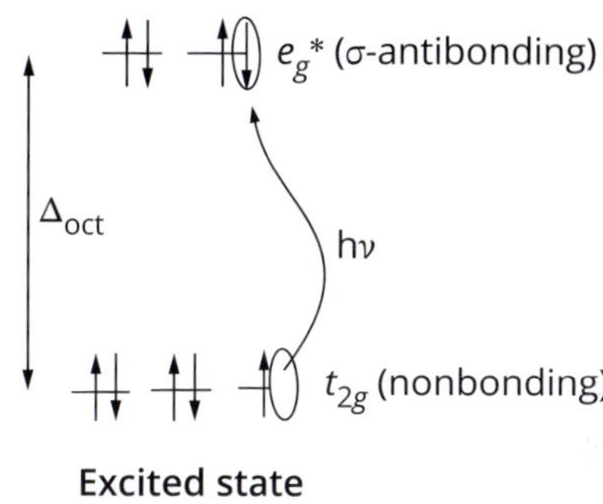

Fig. 15.15e Idealized O_h Cu(II) crystal field splitting diagram for excited state species.

Visible light is often of the appropriate energy to promote electrons across the t_{2g}–e_g* energy gap in octahedral complexes making many octahedral transition-metal complexes colored.

15.6 Ligand Field Strength

The "ligand field" is what leads to the splitting of the *d*-orbitals due to the ligand symmetry and donor properties of the substituents. The size of the splitting is determined by the geometry of the complex, the nature of the metal center, and the properties of the ligands. The previous sections have discussed the geometric factors affecting the splitting. In this section we will discuss the properties of metal centers and ligands that affect the energetic spread of the *d*-orbitals by the ligand field.

15.6.1 Ligand Effects

While ligands have variable electric fields in the CFT model, with stronger fields raising the energy of the orbitals higher in energy, in LFT the stronger donating ligands will raise the energy of the e_g* (through σ-donation) and t_{2g} (through π-donation) orbitals because these orbitals are antibonding in character.[32] If the ligands are π-accepting, the metal-based t_{2g} orbital set will be stabilized (Fig. 15.16). All of these orbital

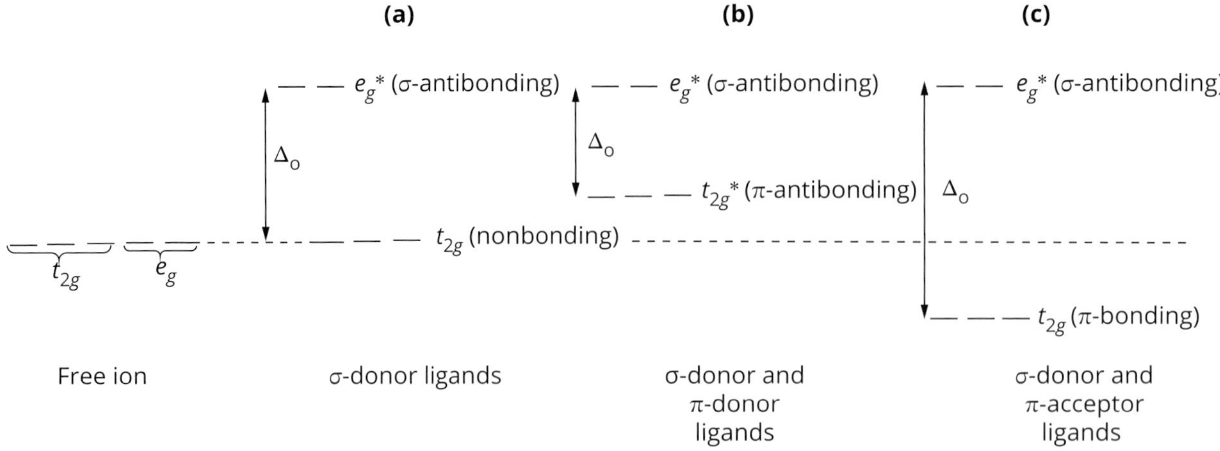

Fig. 15.16 Starting with the free ion where all the *d*-orbitals are the same energy (*far left*), the figure shows the effect of (a) -only donors, (b) - and π-donating ligands, and (c) -donating and π-accepting ligands on the relative *d*-orbital energies in an octahedral field. Note that the t_{2g} orbitals are nonbonding in case (a) but antibonding in case (b), where they are labeled as "t_{2g}*". In the case of (c), we accentuate the fact that the important interaction is the electrons in the metal-based t_{2g} donating into the ligand antibonding acceptor orbitals, and describe t_{2g} as bonding.

energy changes are natural consequences of forming the molecular orbitals.[n]

For ligands that are only σ-donors, the ligand field splitting, Δ_o, increases with increasing ligand σ-donor ability (Fig. 15.16a), which is mainly determined by the energy of the ligand donor orbital and how well it overlaps with d-orbitals. Ligands that are strong π-donors, with equivalent σ-donation as before, will generally decrease Δ_o (Fig. 15.16b) by raising the energy of the t_{2g}-orbitals, whereas ligands that are strong π-acceptors will generally lead to large values for Δ_o (Fig. 15.16c) by lowering the energy of the t_{2g}-orbitals.

These trends with π-donor and π-acceptor abilities can be understood from an MO perspective as shown in Fig. 15.17. The majority of ligands are more electronegative than the metal to which they are attached; as a result, the filled valence orbitals that might act as π-donors to the metal are lower in energy than the metal-based acceptor orbitals (t_{2g} orbitals in O_h) as shown in Fig. 15.17a. When these filled ligand orbitals mix with metal-based π-acceptor orbitals, a lower-energy bonding set is created, which is mostly ligand in character, and a higher-energy antibonding set of orbitals is created. The antibonding orbitals have mostly metal character, and the t_{2g}* set of orbitals is effectively raised in energy, which decreases Δ_o.

[n] One may notice that in CFT, the energy of the *d*-orbitals only goes up (Fig. 15.2); negative point charges can only make the energy of *d*-electrons higher, never more stable. In contrast, in the MO-based LFT approach the orbital energies change from the atomic free-ion energy on ligand interaction, and sometimes these interactions can lead to stabilization effects; that is, π-accepting interactions. If, however, we concern ourselves with only the relative *d*-orbital energies, which is what CFT limits itself to, then this difference is of little overall consequence, but it is important to appreciate the new insights that the MO picture of LFT provides.

In Fig. 15.17b there are ligand-based π*-orbitals that may act as acceptors, and electrons in the t_{2g}-set of orbitals on the metal that can act as donors. This situation is commonly encountered in ligand types such as CO, CN^-, olefins, and so on. Because the acceptor orbitals on the ligands are antibonding in nature, they are relatively high in energy, and when they overlap with the metal, a bonding set of orbitals is created that is stabilized and has principally metal character. Because the metal's t_{2g}-orbitals are stabilized by the interaction, the energy gap Δ_o increases.

The effect of the ligands on the size of the energy gap, Δ_o, has been determined empirically by the measurement of absorption spectra for many different complexes. From this, the ligands can be arranged sequentially into what is called the *spectrochemical series* (Fig. 15.18). Homoleptic complexes of ligands at the beginning of the series result in relatively small Δ_o compared to those bearing ligands later in the series. Ligands at the beginning of the series are commonly referred to as being *weak-field ligands*, whereas those that lead to large Δ_o are referred to as *strong-field ligands*.

Based on Figs. 15.16 or 15.17, the Spectrochemical Series (Fig. 15.18) can be explained as variations in the σ- and π-donor ability of each ligand. Ignoring π-effects and simply focusing on σ-donation for the moment, we can say that ligands that are stronger σ-donors should raise the level of e_g*, and result in a larger Δ_o. For example, I^- is a weaker Lewis base than Br^- and appears earlier in the series. Ligands that are π-donors will appear earlier in the series (they tend to decrease Δ_o), whereas π-acceptors show up later (they lead to larger Δ_o). For example, OH^- is a very good π-donor and appears earlier in the series than CO and CN^-, which are both good π-acceptors.

Because the traditional spectrochemical series arranges all ligands in a single sequence (so-called *one-dimensional* series),

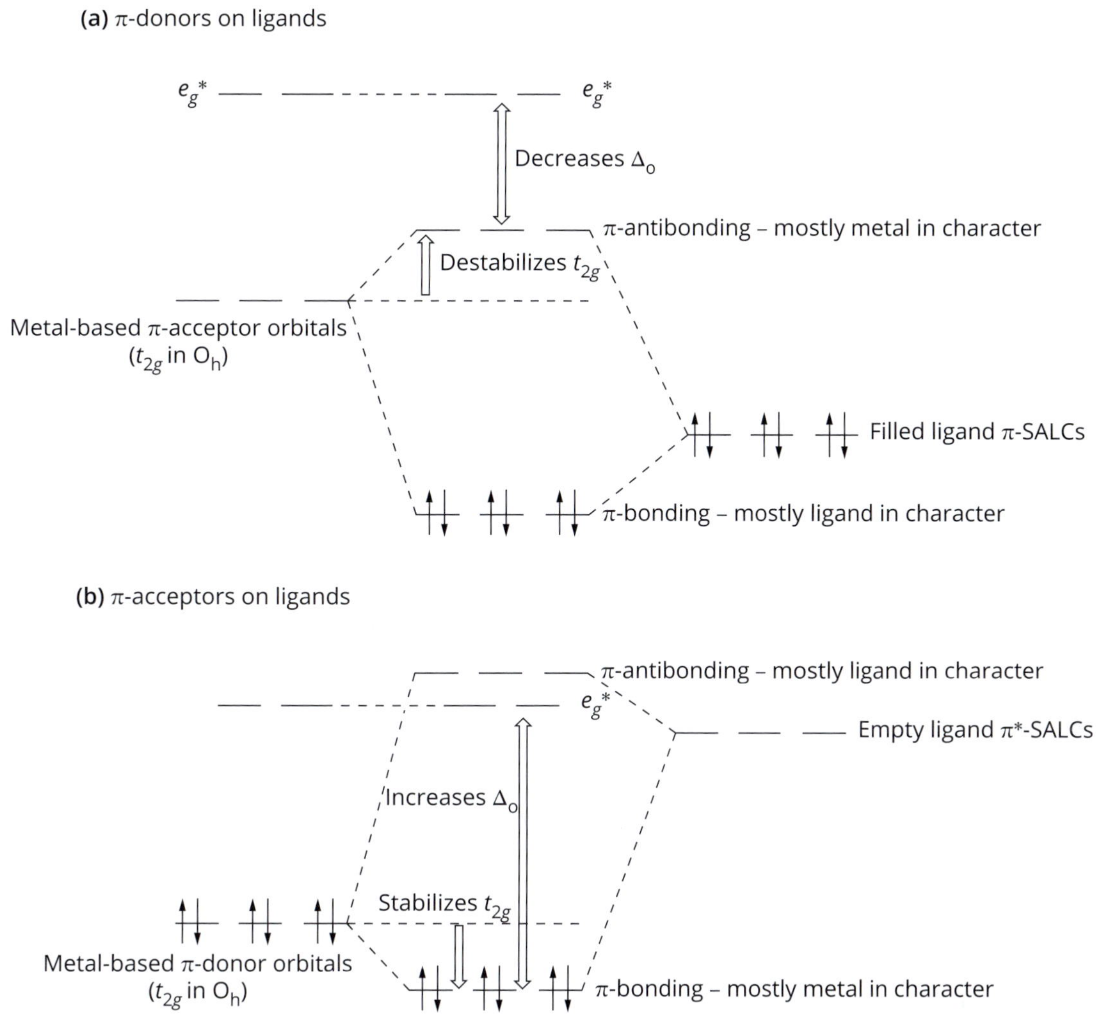

Fig. 15.17 (*a, top*) Ligand-based π-donors typically raise the energy of the metal-based t_{2g} orbitals and decrease Δ_o. (*b, bottom*) Ligand-based π-acceptors typically lower the energy of the metal-based t_{2g} orbitals and increase Δ_o.

it is impossible to completely separate σ- and π-effects. For example, why does the spectrochemical series rank CO at a similar field strength as CN^- given that the former ligand is a better π-acceptor? One reason is that CO is a better π-acceptor but CN^- is a better σ-donor than CO, such that the two ligands end up at a similar position in this one-dimensional series. Based exclusively on the spectrochemical series and CFT, this is pure speculation. Using the angular overlap model (AOM), discussed a little later in this chapter, we will be able to build a two-dimensional spectrochemical series that separates both σ- and π-effects, which will clarify some of the reasons for the ordering in Fig. 15.18.

The effect of changing ligands across the spectrochemical series can be quite dramatic. In the series of d^3 Cr(III) complexes $\{CrCl_6\}^{3-}$, $\{Cr(OH_2)_6\}^{3+}$, $\{Cr(NH_3)_6\}^{3+}$, and $\{Cr(CN)_6\}^{3-}$, the Δ_o values are 13,200, 17,400, 21,600, and 26,600 cm^{-1}.[31] A much larger selection of ligand field splitting parameters and their variance with metal and ligands, adapted from Figgis and Hitchman, are shown in Table 15.1.[7]

15.6.2 The Effect of the Metal on Orbital Splitting

As previously discussed with the ligand σ-effects, as the strength of the metal–ligand (M–L) interaction increases, the ligand (crystal) field splitting, Δ_o, increases. In general, first-row M–L interactions are weaker than second- and third-row. One reason for this is the primogenic effect discussed in Section 8.2.1.

Fig. 8.5 shows the valence *d*/*s*-orbital radii. For the first-row transition metals, the *d*/*s*-orbital radii ratio is quite small at ~0.3. The *d*-orbitals are contracted because there are no orbitals of the same angular momentum for them to remain orthogonal to, and as a consequence they can stabilize by moving closer to the nucleus. In the second and third row, the orbitals have core orbitals of the same angular momentum, and they are larger relative to their valence *s*-orbitals with ratios of around 0.5. The larger *d*-orbitals of the second and third row interact with the ligands more strongly than the contracted

Weak-field ligands
Poor σ-donors and/or good π-donors ←

I^- < Br^- < $^-NCS^*$ < Cl^- < F^- < OH^- < $RC(O)O^-$ < ONO^{*-} < OH_2 < SCN^{*-} < edta <

$NCCH_3$ < py ~ NH_3 < en < bpy < phen < $^*NO_2^-$ < PPh_3 < CN^- ~ CO < NO^+ (linear)

→ Good σ-donors and/or good π-acceptors
Strong-field ligands

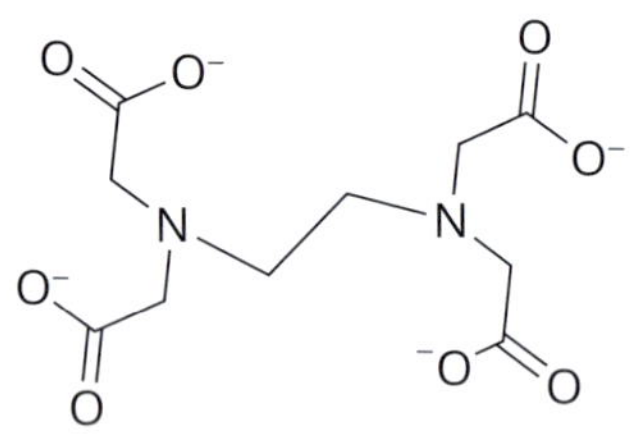

edta = ethylenediamine tetraacetic acid

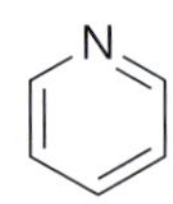

py = pyridine

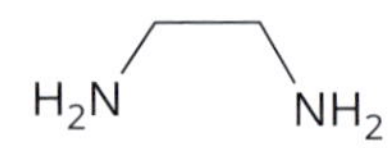

en = ethylene diamine

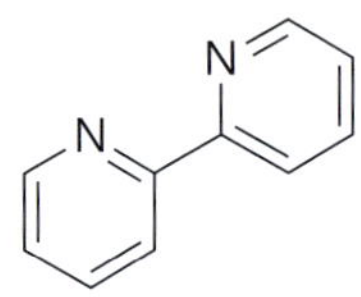

bpy = 2,2′-bipyridine

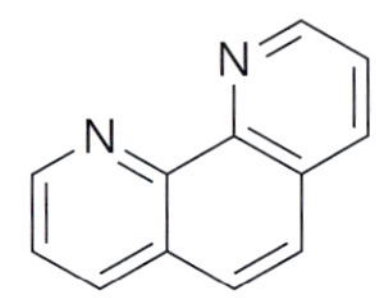

phen = 1,10-phenanthroline

Fig. 15.18 The one-dimensional spectrochemical series. The asterisk (*) designates the site of metal binding in ambidentate ligands. Homoleptic complexes of ligands at the beginning of the series are expected to have smaller Δ_o with respect to those later in the series. The ligands at the beginning of the series are commonly referred to as being *weak-field ligands*, whereas those that lead to large Δ_o are referred to as *strong-field ligands*.

Table 15.1 Ligand field splitting for some complexes with ligands across the spectrochemical series. Values are given in 1,000 cm^{-1} (for example, "13" is 13,000 cm^{-1}). (Table adapted from Figgis and Hitchman.[7])

d-electrons	Ion	6F	6Cl	6Br	$6OH_2$	3Ox	$6NH_3$	3en	6CN	4Cl
1	Ti(III)		13		20		17?			
2	V(III)	16	13		19	18	18?			
3	Cr(III)		13	17	17	22	22	26		
	Re(IV)	32	33							
4	Cr(II)		13[a]		14[a]			18[a]		
	Mn(III)	22[a]	20[a]		21[a]				30[b]	
5	Mn(II)	8	8		8			10		2
	Fe(III)	14			14	14			35[b]	5
6	Fe(II)				10				33[b]	4
	Co(III)	13[c]			19	18	24	24	34	
	Rh(III)		20	19	27	26	34	35		
	Ir(III)		25	23				41		
	Pt(IV)	33	29	25						
7	Co(II)				10	11	11	11		3.2
8	Ni(II)		7	6	9		11	12		3.5
9	Cu(II)				13[a]		15[a]	16[a]		

[a] Value refers to the in-plane ligands of a tetragonally elongated octahedron complex.
[b] Low-spin complex and other values for this metal are for the high-spin complexes.
[c] High-spin complex and the other values for this metal are for the low-spin complexes.

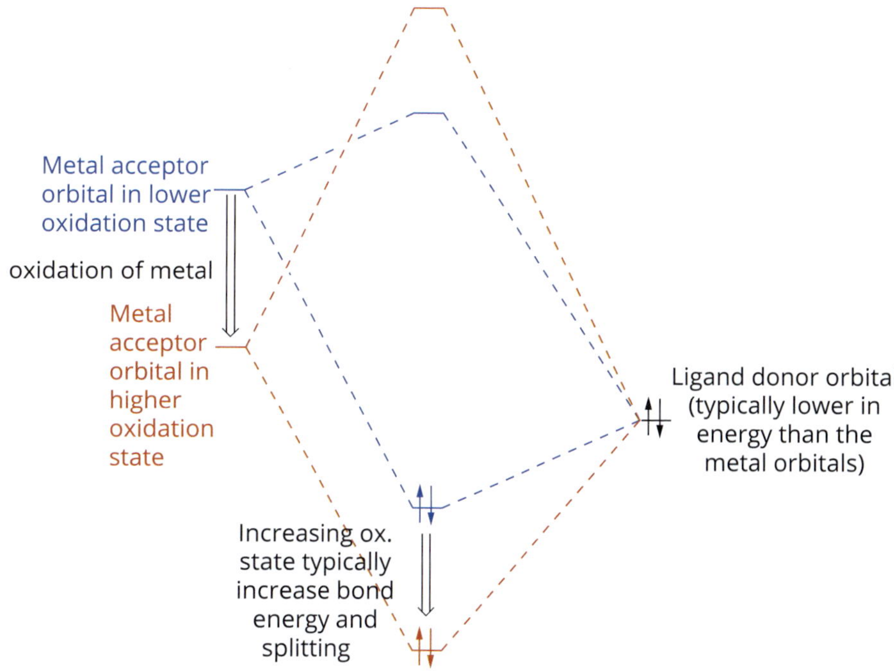

Fig. 15.19 Increasing the metal oxidation state lowers the energy of the *d*-orbitals, increasing the interaction with the ligands.

d-orbitals of the first row. *It is certainly not that the d-orbitals are unimportant for bonding in the first row, but the contraction weakens the M–L interactions relative to the more expanded d-orbitals of the second- and third-row transition elements.*

Another important effect that limits M–L interaction strengths in the first row is that the *d*-orbitals are quite close in size to the core 3*p*-orbitals. The relative sizes of these orbitals can be seen in Fig. 8.5. Because there are filled core orbitals of about the same size, there is some Pauli repulsion associated with the use of the 3*d*-orbitals (Fig. 8.6), which serves to further weaken M–L bonds in the first row. For the second and third row, the core *p*-orbitals are substantially smaller than the valence *d*-orbitals due to primogenic repulsion acting on the *d*-orbitals.

The strengthening of the M–L bonds in the second and third rows versus the first row can be seen in almost any series down a transition-metal column. For example, in the complexes $M(CO)_6$, where M = Cr, Mo, and W, the bond dissociation energies for the loss of the first CO ligand are 36.8 ± 2, 40.5 ± 2, and 46.0 ± 2 kcal/mol, respectively. For an example with heterocovalent bonds, we can examine the M–F bond energies in MF_4 complexes, where M = Ti, Zr, and Hf. The calculated BDEs of the compounds are 127, 144, and 146 kcal/mol, respectively.[29] Again, the heavier elements in the column display stronger metal–ligand bonds; this difference seems to generally be preserved whether the ligands form dative or heterocovalent bonds.

The effect is also observed in the Δ_o values for analogous complexes as well. For example, $\{M(NH_3)_6\}^{3+}$, where M = Co, Rh, and Ir, have Δ_o values of 22,900, 34,100, and 40,000 cm^{-1}, respectively.[30] The stronger bonds in the heavier congeners lead to larger ligand fields.

Another factor that can increase Δ_o is the metal oxidation state. *As the metal oxidation state increases, the ligand field splitting tends to increase as well.* Generally speaking, ligand orbital energies are typically below those of the metal orbital energies. As the metal's oxidation state increases, the orbital energies on the metal decrease. The better energy match between the metal and ligand orbitals leads to stronger M–L interactions (Fig. 15.19) and larger ligand fields.

Increasing the oxidation state of the metal generally shortens the M–L distances as well, leading to stronger interactions. Also, metals in higher oxidation states may be viewed as being more Lewis-acidic, which would strengthen bonds with Lewis basic ligands. For example, in the iron cyano complexes $\{Fe(CN)_6\}^{3-}$ and $\{Fe(CN)_6\}^{4-}$, which are in the +3 and +2 oxidation states respectively, the ligand field splitting is higher for the more oxidized metal (35,000 cm^{-1}) relative to the less oxidized (33,800 cm^{-1}) metal.[31]

15.6.3 Summary of Metal and Ligand Effects on Δ_o

The effects of the metal and the ligand on the crystal (ligand) field splitting, Δ_o, can be summarized as follows:

- Ligands higher on the spectrochemical series increase Δ_o, with complexes containing ligands with strong σ-donation and π-acceptors having a larger Δ_o and complexes containing ligands with weak σ-donation and strong π-donors having smaller Δ_o values.
- In general, metal–ligand interactions are typically larger for second- and third-row metals than for first row. As a result, second- and third-row metals have larger Δ_o energies.
- Increasing the oxidation state of the metal generally leads to an increase in the strength of the metal–ligand interactions, increasing Δ_o.

15.7 High-Spin versus Low-Spin Complexes and the Pairing Energy

In an octahedral field, metal centers with d^4 to d^7 electron counts can have two different ways of filling the orbitals, depending on the ligand field and the energy required to pair electrons in the same orbital, Π (the "pairing energy"). The possible configurations in an octahedral field and their unpaired electronic spins are shown in Fig. 15.20. If the energetic cost of adding an electron to the e_g* orbital is lower than the cost of pairing electrons in the same orbital, then the complex will be high spin. If the ligand field splitting, Δ_O, is larger than Π, then the complex will be low spin.

Since the valence *d*-orbital radii of the second- and third-row transition metals are larger, their pairing energies are significantly smaller; consequently, *complexes of the second- and third-row transition metals tend to be low spin.*[34]

Whether a metal ion is high spin or low spin dramatically affects the ionic radius of the metal. High-spin complexes begin to occupy antibonding e_g* orbitals to a larger degree, leading to longer bonds and larger effective ionic radii (Fig. 15.21). Low-spin complexes have invariably shorter radii than high-spin complexes for d^4 to d^7 ions of the same metal in the same oxidation state.[7,31]

As an example of the effect of oxidation state on bond length, consider the two high-spin ions $\{Fe(OH_2)_6\}^{2+}$ and $\{Fe(OH_2)_6\}^{3+}$. The Fe(III)–O distances are ~1.97 Å due to the smaller size of the highly charged iron(III) ion and stronger Lewis acidity towards the ligands. The bond distances in $\{Fe(OH_2)_6\}^{2+}\{closo\text{-}B_{12}H_{12}\}^{2-}$ average a much longer 2.115 Å.[32]

To see the effect on bond distances from the change in spin state, it is advantageous to examine a *spin-crossover complex*—a compound that exhibits temperature-dependent concentrations of high-spin and low-spin complexes. For *tris*(dimethyldithiocarbamate)iron(III), $Fe(S_2CNMe_2)_3$ shown in the following, the low-spin Fe–S distances average 2.302 Å, while in the high-spin state the Fe–S distances are ~2.45 Å.[33]

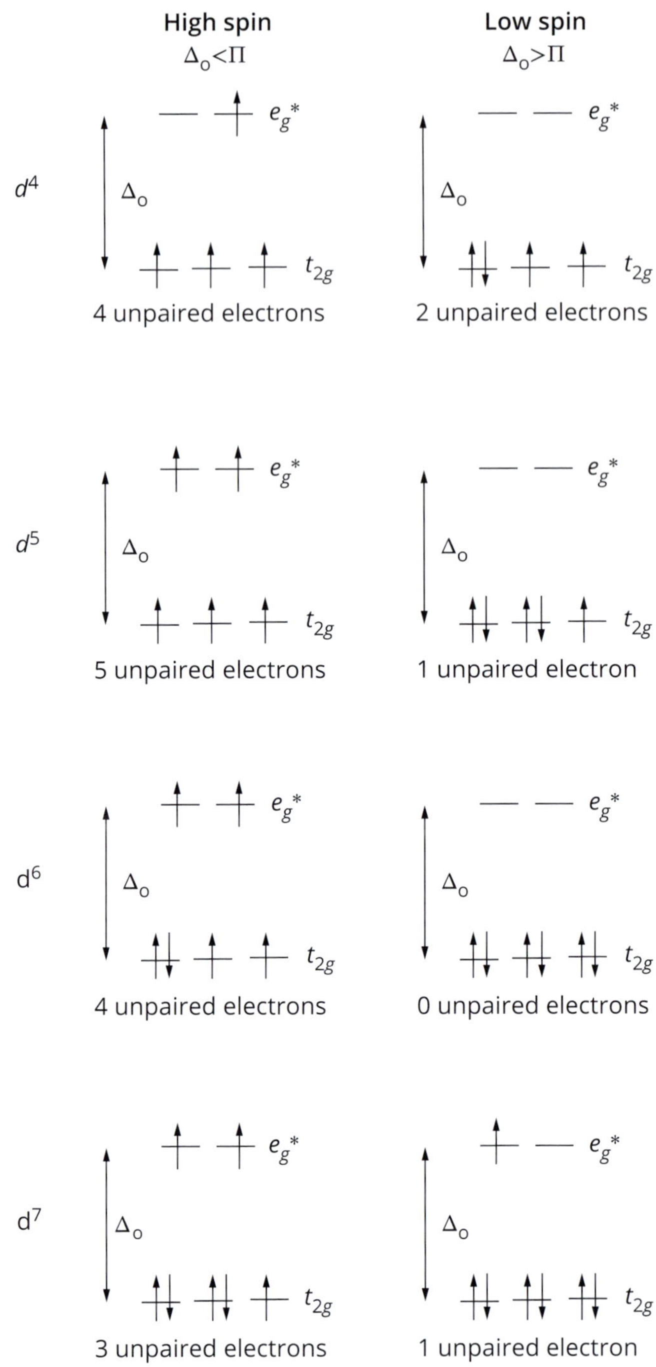

Fig. 15.20 Complexes with *d*-electron counts from d^4 to d^7 can have two different configurations, high spin and low spin. The high-spin configuration is found when $\Delta_o < \Pi$, where Π is the pairing energy.

15.8 Introduction to First-Order Jahn–Teller Distortions

Jahn–Teller (J–T) effects can influence chemical properties and constitute a fairly complex area that we only plan to wade into barely above the ankles. More detailed and generalized descriptions of the J–T effect can be found in more specialized texts and reviews.[36–40] A discussion of related second-order J–T effects on simple main-group molecules can be found in Chapter 8, and a discussion of second-order J–T effects and d^0 transition-metal systems can be found in Chapter 14.[41–42]

In Chapter 8, the energetics and symmetry of a second-order J–T distortion from some high-symmetry geometry was discussed. The energy of the second-order distortion was nonzero if the second-order term, the last term in Eq. 8.11, was nonzero, which involves both the wavefunction of the starting state and the new wavefunction associated with the distortion coordinate *Q*.[43] This equation is fairly intimidating, but in reality we will only need to focus on a small portion here. A more detailed discussion is an Advanced Topic in Chapter 8.

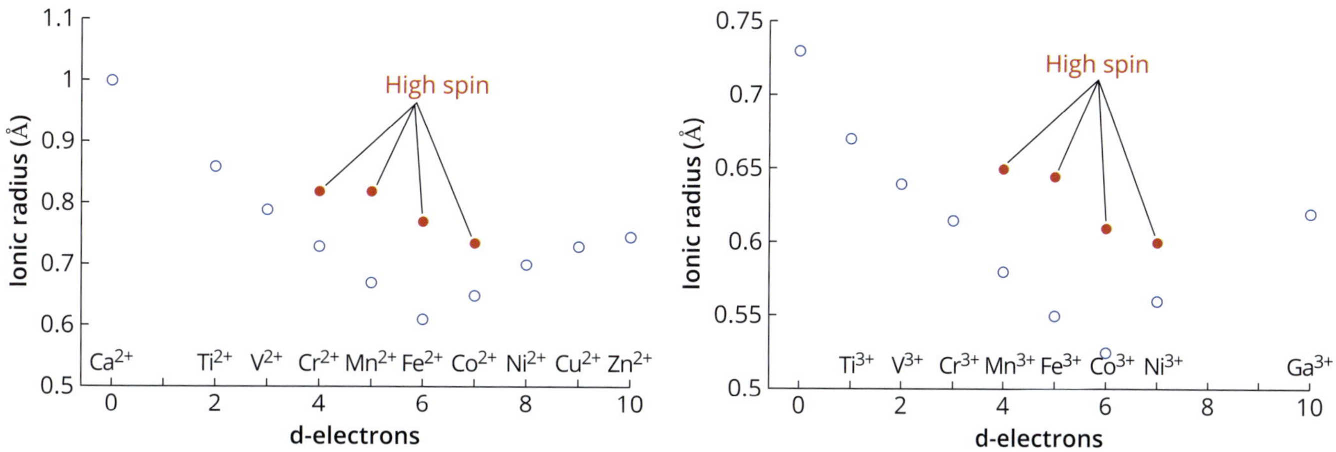

Fig. 15.21 Effective ionic radius versus the number of *d*-electrons for (*left*) divalent and (*right*) trivalent metal ions including the first-row transition metals. High-spin complexes are shown as red filled dots, and low-spin as blue outline dots. (Data from Shannon and Prewitt.[31])

$$E = E_0 + Q\left\langle \psi_0 \left| \frac{\partial U}{\partial Q} \right| \psi_0 \right\rangle + \frac{Q^2}{2}\left\langle \psi_0 \left| \frac{\partial^2 U}{\partial Q^2} \right| \psi_0 \right\rangle + Q^2 \sum_k \frac{\left[\left\langle \psi_0 \left| \frac{\partial U}{\partial Q} \right| \psi_k \right\rangle \right]^2}{(E_0 - E_k)} \qquad \textbf{8.11}$$

The first-order J–T effect, as might be guessed, involves the first-order ψ_0 term in Eq. 8.11. For a small displacement Q there is a mixing of electronic and vibrational terms that can lead to a lowering of the molecular symmetry, although fluxionality involving the distortion coordinate is probable (*vide infra*). In the first-order term, the wavefunction for the ground state, ψ_0, appears twice in the integral, and as a result the symmetry of the distortion (a normal mode of the system) will necessarily be included in the direct product of the ground-state function with itself.

The first-order J–T effect occurs due to unsymmetrically filled degenerate orbitals, such as a single electron in a doubly or triply degenerate set. The distortion will break the degeneracy of the orbitals with some orbitals descending in energy and the others rising. For example, if there is one electron in a degenerate pair of orbitals, the J–T distortion will lower the energy of one orbital and raise the other, leading to an obvious electronic stabilization (as in the following). Like our similar discussions on crystal field theory previously, the energies of the orbitals changed by the distortion are barycentered. The energetic difference between the starting geometry and the distorted system being the electronic stabilization due to distortion (Q), ε_Q.

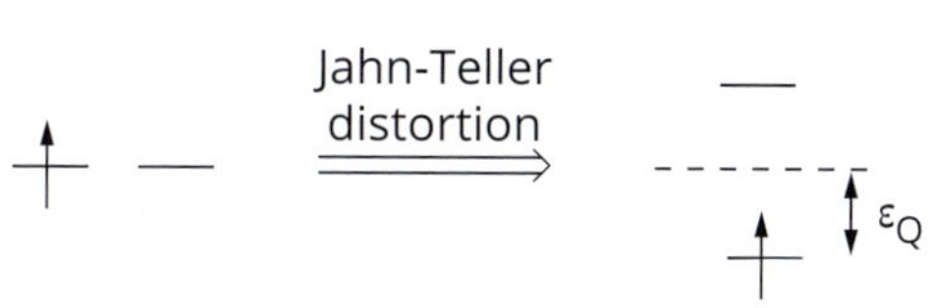

Contrary to the second-order J–T effect, such as HOMO-LUMO mixing, which can result in very large energies in some cases, the first-order J–T effect is typically smaller and on the same energetic order as vibrations. Nevertheless, it is a very important effect for understanding the chemistry and spectroscopy of many inorganic systems.

The J–T effect derives from degenerate states and their coupling to normal modes. Coupling to a vibrational mode of appropriate symmetry breaks the degeneracy, lowering the symmetry of the system. To illustrate, we can draw all the possible ways to fill two degenerate orbitals, such as an $e_g{}^*$ set for an octahedral complex, with various electron counts zero through a maximum of four, labeled $\delta^0 - \delta^4$. Only two of those configurations have alternative methods for the filling of the orbitals: those where there is one electron in the degenerate set (δ^1), and another where there are three electrons (δ^3).

The two configurations in Fig. 15.22 with two different ways of filling the orbitals (δ^1 and δ^3) are equivalent under the 1-electron approximation—that is, they are the same discounting electron–electron interactions—but these two systems are subject to J–T distortions to break the degeneracy and lower the energy of the system. We can illustrate this by saying that the degenerate orbitals in Fig. 15.22 are the $e_g{}^*$ orbitals of a low-spin octahedral system. As a result, the generalized δ^1 configuration becomes, for example, the low-spin d^7 $(t_{2g})^6(e_g{}^*)^1$ configuration for a transition-metal system. The one electron in the $e_g{}^*$ orbital can be either in the $d_{x^2-y^2}$ or d_{z^2} orbital. If the $e_g{}^*$ electron resides in $d_{x^2-y^2}$, from a crystal field perspective, the electron will repel the equatorial donors, weakening their bonds. If the $e_g{}^*$ electron resides in d_{z^2}, the ligands along the z-axis will be significantly weakened relative to those in the equatorial plane.

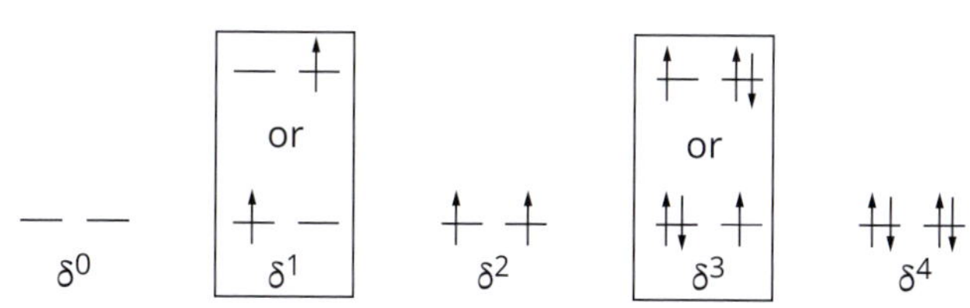

Fig. 15.22 Different possible ways to fill two degenerate orbitals with zero to four electrons. Two of the configurations can be filled in two different ways; that is, have degenerate states.

The distortion coordinate must involve a normal mode of the system that is a direct product of the irreducible representation for the wavefunction with itself (*vide supra*). The ground-state function for a $(t_{2g})^6(e_g{}^*)^1$ configuration has the symmetry of the orbital containing the unpaired electron, E_g. The direct product of this irreducible representation with itself is $E_g \times E_g = A_{1g} + A_{2g} + E_g$. The normal modes of an octahedron belong to representations $A_{1g} + E_g + 2T_{1u} + T_{2g} + T_{2u}$. Consequently, the only vibrational modes that can couple with an E_g symmetry system are A_{1g} and E_g.[o] The totally symmetric representation cannot lead to a molecular distortion as it only involves a "breathing" mode of the ligands in which all the bonds expand and contract together. This mode cannot break the symmetry of the *e*-set and allow the orbitals to have different energies. The only possible mode that can be involved in the J–T distortion is the E_g normal mode. The nature of these vibrations (Q_ε and Q_θ) are shown in Fig. 15.23. The result is a lowering of the symmetry from O_h via a tetragonal distortion to D_{4h}—a result that is often easily measured by absorption spectroscopy and other techniques.

During the distortion, the d_{z^2}- and $d_{x^2-y^2}$-orbitals become inequivalent (Fig. 15.6). In D_{4h}, the d_{z^2}-orbital belongs to irreducible representation A_{1g}, and $d_{x^2-y^2}$ belongs to B_{1g}. From a CFT standpoint, as the ligands along the *z*-axis are elongated by the distortion, the electric field along that axis will weaken, and the d_{z^2} orbital will be stabilized (Fig. 15.23). Alternatively, the ligands along the *z*-axis could be compressed relative to the equatorial ligands, leading to a rise in energy for

[o] Experienced readers will notice that we are attempting to avoid multielectronic terms here before their introduction in the next chapter. The $(t_{2g})^6(e_g{}^*)^1$ configuration has the symmetry of the ^2E_g electronic state, but we instead simply declare that the configuration has the symmetry of the orbital with the unpaired electron.

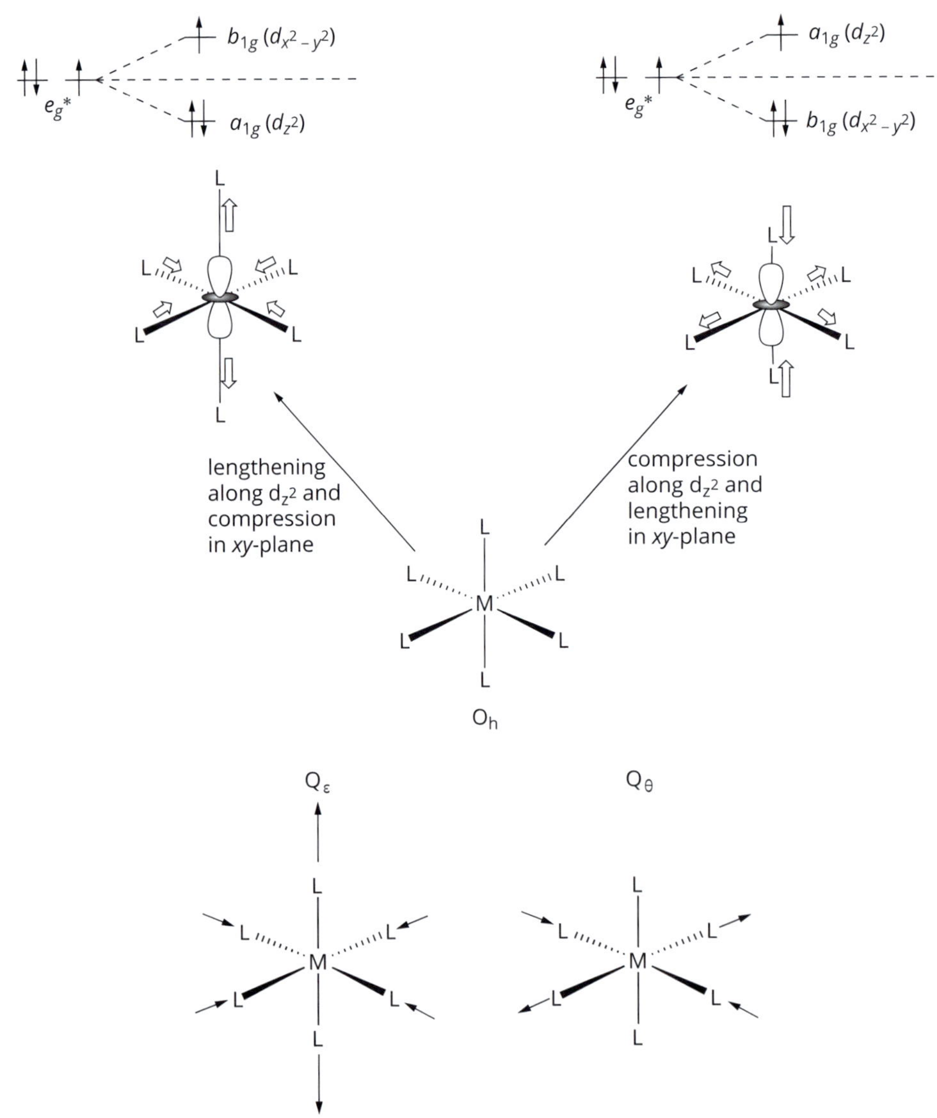

Fig. 15.23 (*top*) Effect of elongating or compressing along the *z*-axis on the energies of the $d_{x^2-y^2}$ and d_{z^2}-orbitals. (*bottom*) The E_g two-fold degenerate vibration in O_h to which the electronics are coupling; the two degenerate modes are labeled as Q_ε and Q_θ. The arrow lengths represent the relative size of displacement for that atom.

the d_{z^2}-orbital. Because the result of the distortion is a lowering of two electrons in energy and a rise in energy of one electron by the same amount, there is a net stabilization of the system as a whole, the driving force for the distortion.

In general, *nonlinear systems with unsymmetrically filled degenerate orbitals are susceptible to J–T distortions.* The energetics and size of the distortion are determined by the nature of the degenerate state. Without a distinct electronic advantage for distortion, higher symmetries with degenerate orbitals are preferred, as degenerate orbitals generally allow greater electron delocalization. If the unsymmetrically filled state is important energetically to the bonding, then a larger distortion occurs. As a result:[43]

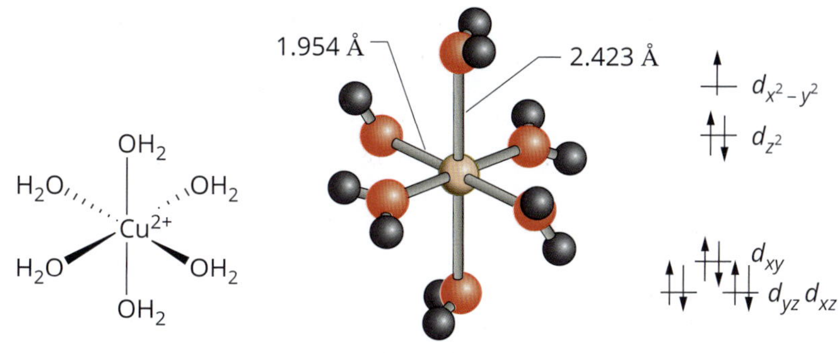

Fig. 15.24 The structure of the Cu^{2+} complex in $\{Cu(OH_2)_6\}\{O_3S\text{-}p\text{-tolyl}\}_2$ from X-ray diffraction with hydrogen atoms in refined positions, and the CFT splitting diagram. (For simplicity, we approximate the complex as D_{4h}.)[34]

- Nonbonding orbitals produce little or no distortion. For example, unsymmetrically filled *f*-orbitals of the lanthanides do not lead to significant J–T distortions.
- Weakly bonding or antibonding orbitals will induce a small to modest distortion. For example, an unsymmetrically filled t_{2g} orbital of an octahedral transition-metal complex usually lead to a small distortion, as these are π-bonding only.
- Unsymmetrically filled, strongly bonding or antibonding orbitals will induce large distortions. For example, one or three electrons in the σ-antibonding, $e_g{}^*$ orbital of an octahedral transition-metal complex.

For octahedral systems, distortions can occur for d^1, d^2, d^4, low-spin d^5, high-spin d^6, d^7, and d^9. As mentioned previously, distortions involving orbitals participating in more energetic bonds will often be larger; that is, J–T distortions due to unsymmetrically filled σ-bond-related $e_g{}^*$ orbitals are typically larger than those involving the π-bond-related t_{2g} orbitals. As a result, complexes with high-spin d^4, low-spin d^7, and d^9 configurations usually exhibit very noticeable tetragonal distortions.

For example, the distortion of the Cu^{2+} ion in $\{Cu(OH_2)_6\}$ $\{O_3S\text{-}p\text{-tolyl}\}_2$, with a $(t_{2g})^6(e_g{}^*)^3$ configuration, has been subjected to an X-ray diffraction study.[44] In the *xy*-plane of the molecule, the unique Cu–O distances are 1.953(2) and 1.954(2) Å.[p] Along the *z*-axis, the Cu–O distances are much longer at 2.423(2) Å—a difference of almost half an Ångstrom, an enormous difference! The structure and CFT splitting of this complex are shown in Fig. 15.24.

In principle, the distortion observed could be an axial elongation or compression (Fig. 15.23, top). In practice, most complexes exhibit tetragonal elongations. An elongation of the ligand distances along one axis will stabilize the d_{z^2} orbital, which is of the same symmetry as the *s*-orbital on the metal with which it can mix, leading to additional stabilization (Section 15.9.8). However, ligand compressions are observed for some systems—notably, some high-spin Mn(III) complexes.[37]

[p] The copper sits on a crystallographic inversion center.

In practice, a J-T-induced *z*-axis elongation can occur along any axis of the molecule, and we can call any axis "*z*". The energetic barriers[38] for conversion between these different minima are typically small, on the order of vibrations. For example, consider again the structure of $\{Cu(OH_2)_6\}$ in Fig. 15.24.[34] Note that this distortion occurs with an unsymmetrical anion in a relatively low-symmetry crystal system, $P2_1/c$. As a result, one particular distortion is enforced in the crystal by the asymmetry. It is estimated that the energy required to change the Cu(II) ion from one geometry to another in solution, such as switching the elongated axis from being $x \to y \to z$, is ~600 cm^{-1} (~1.7 kcal/mol), which is such a small energy that one would expect rapid interconversion.[7] However, the tetragonally distorted form is regularly observed in the solid state, and the lower symmetry is consistent with spectroscopic results in solution and frozen solids as well, such as EPR and EXAFS.[35]

Now, let us return to the d^1-system, $Ti(OH_2)_6{}^{3+}$. If we approximate the system as O_h— that is, if we ignore the impact of the hydrogens on the symmetry—the now familiar crystal field splitting of $(t_{2g})^1(e_g{}^*)^0$ results. The t_{2g} orbital set, composed of d_{xz}, d_{yz}, and d_{xy}, can house the single *d*-electron three different ways. The complex is susceptible to a Jahn–Teller distortion resulting from this degeneracy.

Again, we say the symmetry of the electronic system is the symmetry of the orbital with the unpaired electron, T_{2g}. The normal mode(s) that can couple into this electronic configuration must be included in the symmetry of the direct product of the configuration with itself $T_{2g} \times T_{2g} = A_{1g} + E_g + T_{1g} + T_{2g}$. Consequently, the possible normal modes that can couple by symmetry are $A_{1g} + E_g + T_{2g}$. Again, choosing the E_g mode for discussion, an elongation along the *z*-axis due to the J–T distortion will produce an approximate D_{4h} structure (ignoring the water hydrogens' effect on the symmetry).

As a result, the complex has not one transition in the visible but two (Fig. 15.25) about 2000 cm^{-1} apart. The two peaks result from excitation from the occupied $(d_{xz}, d_{yz})^1$-orbital to either $d_{x^2-y^2}$ or d_{z^2}. The third possible peak, excitation from $(d_{xz}, d_{yz})^1$-orbital to d_{xy} is of quite low energy and is not observed here.

Any other system with a degenerate ground state (unsymmetrically filled degenerate orbitals) can undergo a J–T

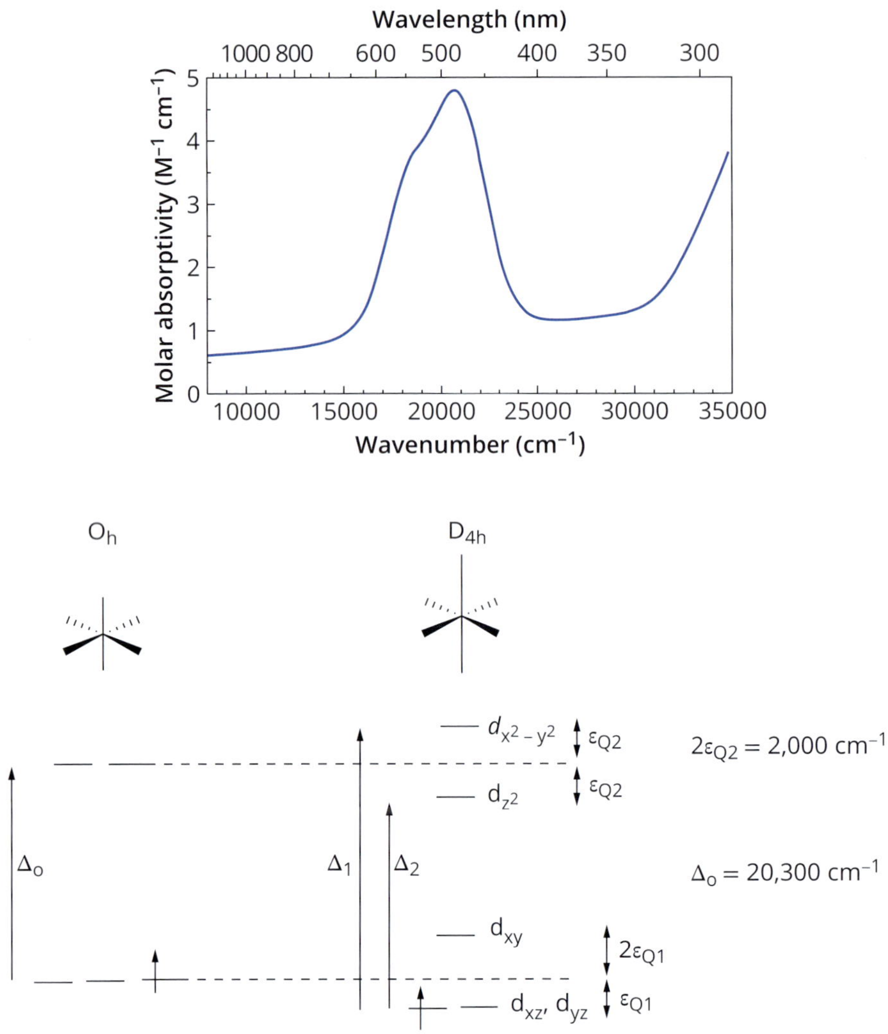

Fig. 15.25 Absorption spectrum of $\{Ti(OH_2)_6\}^{3+}$ and the crystal field splitting diagram. The two transitions (Δ_1 and Δ_2) are due to excitation from the singly occupied degenerate orbital set at the bottom to either $d_{x^2-y^2}$ or d_{z^2}. The energy difference between the two peaks in the spectrum at the top are of energy $2\varepsilon_{Q2}$ due to the splitting in the $e_g{}^*$ set in O_h on J-T distortion by coupling with the E_g vibration.

distortion. Consequently, tetrahedral systems with, for example, nine *d*-electrons would be expected to undergo large distortions from a T_d structure. The configuration of the tetrahedral system would be $(e)^4(t_2)^5$—a $^2T_2$ state (Fig. 15.26). The direct product of the ground state with itself is then $T_2 \times T_2 = A_1 + E + T_1 + T_2$, which must contain the normal mode responsible for the distortion. The normal modes for a tetrahedral system have symmetries of $A_1 + E + 2T_2$. It is again the E normal mode that is generally the one that couples, which involves a flattening of the tetrahedron, resulting in D_{2d} or even D_{4h} structures as shown in the figure. As a result, d^9 4-coordinate transition-metal complexes, such as $CuCl_4{}^{2-}$, have D_{2d} or square-planar (D_{4h}) structures, depending on the extent of the distortion from tetrahedral.

15.9 The Angular Overlap Model

The angular overlap model (AOM) is a method for obtaining more precise energies of the *d*-orbitals if the geometry is known, and the method lends itself to the precise parameterization of the ligand donor properties. Using this method, the *d*-orbital splittings can be identified with more exact relative energies. The donor abilities, separated into σ- and π-donation, can also be measured.

The method is fairly involved when applied precisely to complexes in the laboratory, which is best done on complexes for which the precise geometry is known, such as from X-ray diffraction. In the next subsection there is a fairly detailed development of the model as an Advanced Topic. If desired, these sections may be skipped for the relatively simple description that follows it in Section 15.9.3.

15.9.1 A Brief Development of the Angular Overlap Model (Advanced Topic)

The angular overlap model provides a semiempirical method for the quantification of *d*-orbital energies and ligand donor abilities based on spectroscopic data. Consider the interaction of one *d*-orbital, d_i, with a single σ-orbital on ligand L_j, σ_j. As is typically the case, the ligand donor orbital will be assumed to be lower in energy, E_L, than the metal acceptor orbital, E_M ($E_L < E_M$). A bonding orbital will be generated of

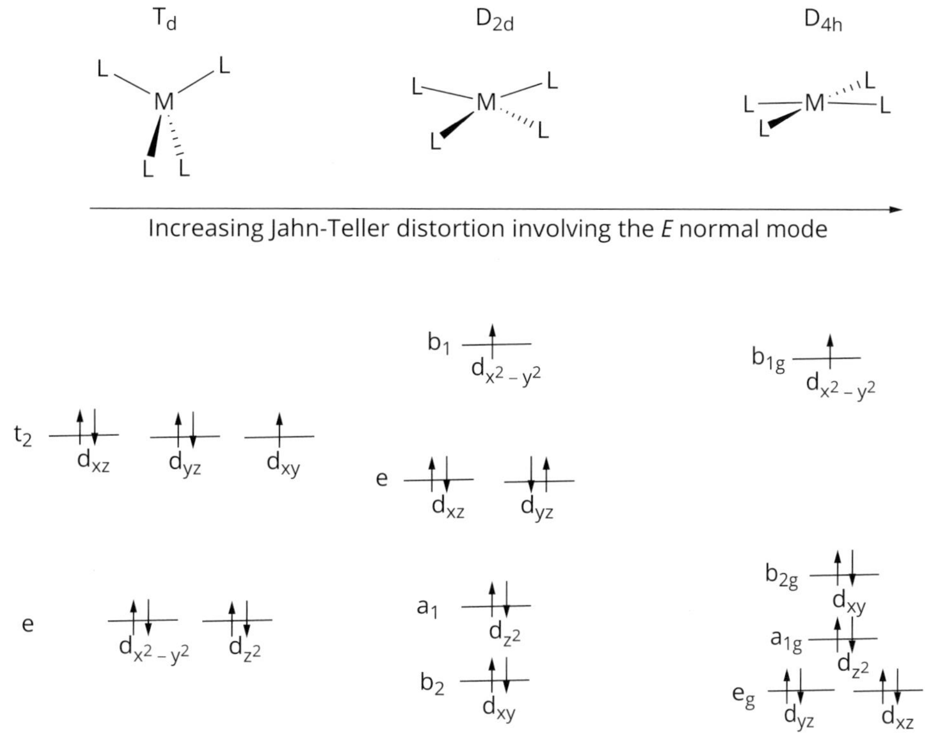

Fig. 15.26 (*top*) Distortions from T$_d$ (tetrahedral) to D$_{2d}$ to D$_{4h}$ (square-planar). (*bottom*) Crystal (ligand) field splitting diagrams for the different geometries.

energy E_{ij} that is largely ligand-based. An antibonding orbital of largely metal character will also be created of energy E_{ij}^* (Fig. 15.27).

The energy difference between the starting ligand orbital energy (E_L) and the energy of the bonding, largely ligand-based bonding orbital (E_{ij}) will be based on the stabilization energy (ε_{ij}). In turn, the difference in energy between the starting *d*-orbital energy (E_M) and the mostly metal-based antibonding orbital will be ε_{ij}^*. Here we will assume that the stabilization of the bonding orbital (ε_{ij}) will be equivalent to the destabilization of the metal-based antibonding orbital; $|\varepsilon_{ij}|=|\varepsilon_{ij}^*|$.[q]

$$\varepsilon_{ij} = E_L - E_{ij}$$
$$\varepsilon^*_{ij} = E^*_{ij} - E_M$$
$$\varepsilon_{ij} = \varepsilon^*_{ij}$$

From the *variational principle*, the expectation value for an orbital energy is given by the following expression. The top integral involves the wavefunction for the orbital (ψ), its complex conjugate (ψ^*), and the Hamiltonian (H) integrated over all space ($d\tau$). The bottom integral involves only the wavefunction and its complex conjugate, the overlap integral.

$$E = \frac{\int \psi^* H \psi \, d\tau}{\int \psi^* \psi \, d\tau}$$

If we assume that the wavefunctions are real—that they have no imaginary components—then $\psi = \psi^*$.

$$E = \frac{\int \psi H \psi \, d\tau}{\int \psi^2 d\tau}$$

Using the *linear combination of atomic orbitals* (LCAO), the nature of the wavefunction, ψ, can be approximated by using the following equation, where c_M and c_L are coefficients, ϕ_M is the metal *d*-orbital wavefunction, and ϕ_L is the ligand orbital (Appendix M).

$$\psi = c_M \phi_M + c_L \phi_L$$

For simplicity, the integrals will be abbreviated. The integral involving the metal orbital only will be labeled as α_M, and the orbital involving the ligand orbital only will be α_L. There are two orbitals involving the metal and ligand, β_{ML} and β_{LM}, which because we assumed $\psi = \psi^*$ are equal and will be labeled as simply β. The integral α is called a "Coulomb integral" and is a measure of the atomic orbital in the environment of the molecule, approximately the starting energy of the orbital. The integral β is called a "resonance integral" and is a measure of the strength of the bonding interaction; that is, how much the orbital is stabilized or destabilized from α.

$$\alpha_M = c_M^2 \int \phi_M H \phi_M \, d\tau$$
$$\alpha_L = c_L^2 \int \phi_L H \phi_L \, d\tau$$
$$\beta_{ML} = c_M c_L \int \phi_M H \phi_L \, d\tau$$
$$\beta_{LM} = c_L c_M \int \phi_L H \phi_M \, d\tau$$

[q] See Appendix M for a discussion, but this is only strictly true at zero overlap.

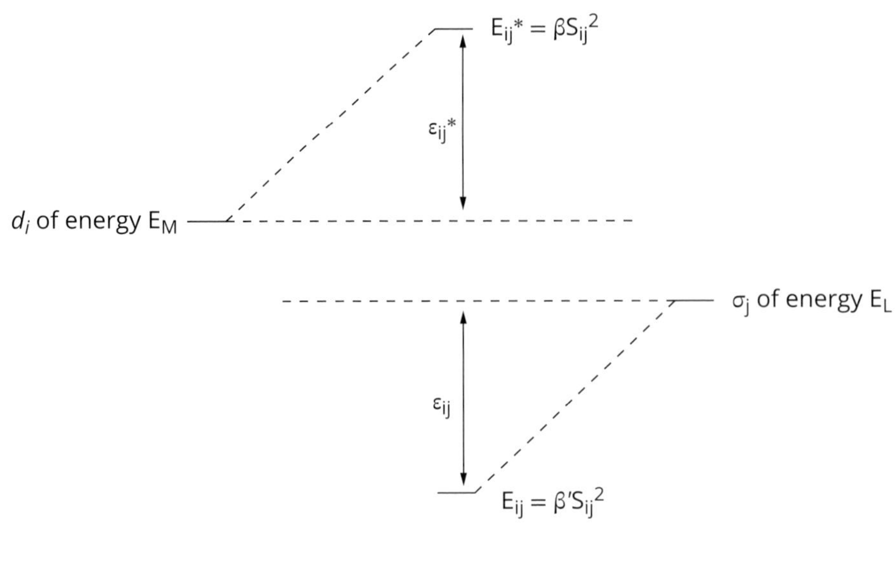

Fig. 15.27 MO diagram for the interaction of one metal *d*-orbital, d_i, of energy E_M with a ligand σ-orbital, σ_j, of energy E_L.

The secular determinant is as follows.

$$\begin{vmatrix} \alpha_M - E & \beta - ES_{ML} \\ \beta - ES_{ML} & \alpha_L - E \end{vmatrix} = 0$$

Solving the determinant requires several approximations: α_M and α_L are well separated in energy, $\alpha_M > \alpha_L$, and S_{ML} is small. We also make use of the Wolfsberg–Helmholtz approximation to eliminate the variable β from the equation—an approximation often employed in extended Hückel theory. After this, the energies for the bonding, E_b, and antibonding, E_a, orbitals can be written as follows.[7]

$$E_a \approx \alpha_M + \frac{\alpha_L^2 S_{ML}^2}{\alpha_M - \alpha_L} \qquad E_b \approx \alpha_L - \frac{\alpha_M^2 S_{ML}^2}{\alpha_M - \alpha_L} \qquad 15.5$$

Here the antibonding orbital is raised in energy from the starting metal orbital energy, α_M, by an amount proportional to the square of the overlap, $S_{ML}{}^2$, and inversely proportional to the metal–ligand energy difference, $\alpha_M - \alpha_L$. Since we are assuming that the rise in energy of the metal orbital is equivalent to the fall in energy of the ligand orbital on bonding, $|\varepsilon_{ij}|=|\varepsilon_{ij}^*|$, we can simply write the energy change due to the interaction of this one orbital on the metal with a σ-orbital on the ligand.

$$\varepsilon_\sigma = \beta K S_{ML}^2 \text{ where } K = \frac{\alpha_L^2}{\alpha_M - \alpha_L} \qquad 15.6$$

The σ-interaction energy, ε_σ, is then proportional to the square of the overlap. The proportionality constant K is dependent upon the ligand and metal energies. The value β is a coefficient associated with the expansion that is always present and, for our purposes here, can be ignored as it will simply be carried along.[r]

In order to determine ε_σ for a σ-bonding interaction, or ε_π for a π-bonding interaction, the overlap integral, S, in Eq. 15.6 needs to be determined. The overlap can be factored into a distance-dependent overlap, $S(r)$, and an angular factor, $F(\theta,\phi)$. We will then collect the angularly independent parameters $S(r)$ and K into one constant e_σ (or e_π for π-effects), Eq. 15.7.

$$\varepsilon_\sigma = e_\sigma \beta F^2(\theta,\phi) \text{ where } e_\sigma = KS^2(r) \qquad 15.7$$

Similarly, a π-interaction energy, e_π, can be derived. With axially asymmetric ligands, such as pyridine, the π-interaction energy will be different if the orbital is aligned with the π-system of the ligand or edge along the N–C bonds. In these cases, an $e_{\pi x}$ and $e_{\pi y}$ parameter will be used. For such planar ligands, the *x*-direction is generally chosen to be the plane of the pyridine. If the ligand is axially symmetric, such as halide, then only e_π is necessary.

The interaction energies e_σ, $e_{\pi x}$, and $e_{\pi y}$ are left as parameters in the AOM calculations and are typically found experimentally. Note that the value K in Eq. 15.7 is largely dependent

[r] For more details on the origin of β, see Appendix F. Unfortunately, it is the same symbol as that used for the resonance integral in the secular determinates, but they are not the same. Many of these symbols are traditional throughout most of the literature on the topics. We retain them here for consistency, but it can be confusing. Be aware that there are several different βs in this chapter that are not the same: resonance integral, coefficient in a Taylor expansion in AOM, and Bohr magneton. We will try to make it as clear as possible in the text what each symbol refers to in every equation.

upon the metal and ligand orbital energies; if the metal stays the same and the same ligand is investigated, the AOM suggests that the interaction should be the same, regardless of the other ligands. Indeed, there is some indication that parameters may be ported from one system to another. Attempts to port e_σ and e_π from one metal system to another are not uncommon and are sometimes successful but should be tried with caution.

If the metal complex contains the same ligand at two different distances, a common occurrence in J–T susceptible systems, the change in distance will affect the e_σ and e_π parameters. Typically, changing the distance *r* affects the parameter exponentially as $\sim r^{-5}$. It should also be mentioned that changing the distance affects π-overlap more dramatically than σ.[7] As a result, the parameters in the axial and equatorial positions would be different in a J–T distorted complex due to changes in S(r) in Eq. 15.7.

For ligands such as carbonyl, CO, the acceptor orbital on the ligand, which is C–O π* in character, is often higher in energy than the metal-based donor orbital. For such ligands, the term *K* in Eq. 15.7 will change sign, and the e_π term will be negative. *As a result, e_σ is used as a positive value, and e_π will be positive for ligands that are π-donors and negative for π-acceptors.*

Since the terms e_σ and e_π in Eq. 15.7 are found experimentally, the only factor left to calculate to determine the change in energy due to the interaction (ε) is the angular factor, *F*, which is dependent upon the position of the ligand orbital in spherical coordinates (*θ*, *ϕ*) and the shape of the *d*-orbital.

A spherical coordinate system is shown in Fig. 15.28. The angle from the axis designated as the *z*-axis is θ, and the angle from the *x*-axis is ϕ. Because the parameter e_σ (or e_π) is angle-independent, the overlap of the ligand orbital with the *d*-orbital being considered will only be due to changes in the shape of the *d*-orbital with angle. In other words, as the ligand is rotated around the metal, the angular factor, *F(θ,ϕ)*, will change from 0 to 1, depending on the position. This is illustrated in Fig. 15.29. Note that because the *F(θ,ϕ)* term is then squared (Eq. 15.7), the phase dependence will not affect the sign of the parameters e_σ or e_π as the angle changes.

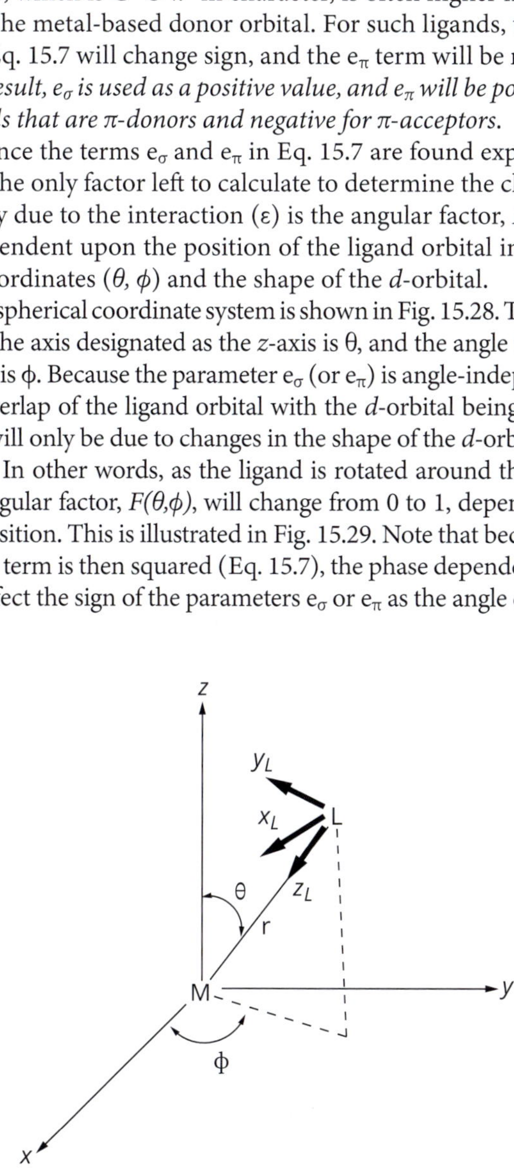

Fig. 15.28 (*top*) Spherical coordinates for a ligand L relative to the metal center M at the origin.

For an *s*-orbital, the angularly independent shape would lead to a value of 1 for *F*, regardless of θ and *ϕ* (Fig. 15.29). The dependence of a p_z-orbital in the AOM would be cos θ. The value for the d_{z^2}-orbital is based on the angular function for the orbital, ($3cos^2\theta - 1$). In all cases, the angular factors, *F*, are scaled so that the maximum is 1.

All of the angular dependence factors, *F(θ,ϕ)*, for *σ*, π_x, and π_y are shown for the *d*-orbitals in Table 15.2.[7]

15.9.2 The General Angular Dependence Factors (Advanced Topic)

In this section, some examples utilizing the *general angular dependence factors* in Table 15.2 will be worked out in detail. While the equations in the table may appear intimidating, they generally reduce to 0, 1, or a simple fractions for symmetrical structures. If one only wants to use the AOM with common geometries for transition-metal complexes as pure forms, there is a simplified chart in Section 15.9.3 (Table 15.4). It is generally recommended by practitioners in the field that the exact angles be used. (See specialized texts and reviews for more details.[7, 36])

For simplicity, we will assume that the ligand(s) being used are cylindrically symmetric (such as halide, CN, CO, and so on) and $e_{\pi x} + e_{\pi y} = e_\pi$, which will be a constant for any angle of *ψ* used. As a result, for cylindrically symmetrical ligands, *ψ* can be set to an arbitrary value, such as *ψ* = 0. The positions for the ligand labels are shown in Fig. 15.30.

The energy of the bonding orbital will be the sum of the ligand–metal orbital interactions (L) scaled by the square of the angular factor $F_i^2(\theta,\phi)$, which is expressed in terms of e_σ(L) and e_π(L). For a *σ*-interaction, the relationship may be expressed as in Eq. 15.8, where *i* is the *d*-orbital in question, such as d_{z^2}, and *j* identifies the specific ligand.

$$\varepsilon_{\sigma,i} = \sum_j F_j^2(\theta,\phi,\psi) \cdot e_\sigma(L_j) \qquad \textbf{15.8}$$

If all the ligands are identical, then Eq. 15.8 simplifies to Eq. 15.9, with identical e_σ values.

$$\varepsilon_{\sigma,i} = e_\sigma \sum_j F_j^2(\theta,\phi,\psi) \qquad \textbf{15.9}$$

For example, for the d_{z^2}-orbital ($i = z^2$) we can take the ligand coordinates for an octahedral system, Table 15.3, and use these for this orbital in the angular expressions in Table 15.2. For the *σ*-interactions between the metal and ligand, we square the expression in the table and apply the *θ*-coordinates of the ligands; the *ϕ*-coordinate does not appear in the expression for $i = z^2$.[s] The energy of the bonding orbital (largely ligand in character) due to interaction with the d_{z^2}-orbital for the *σ*-bonds, ε_{σ,z^2}, is then found by summing the terms F_j^2 (F_1^2 to F_6^2) and multiplying by e_σ(L).

[s] The d_{z^2} orbital is symmetric around the *z*-axis. As a result, there are no changes in energy due to changing the *ϕ*-coordinate.

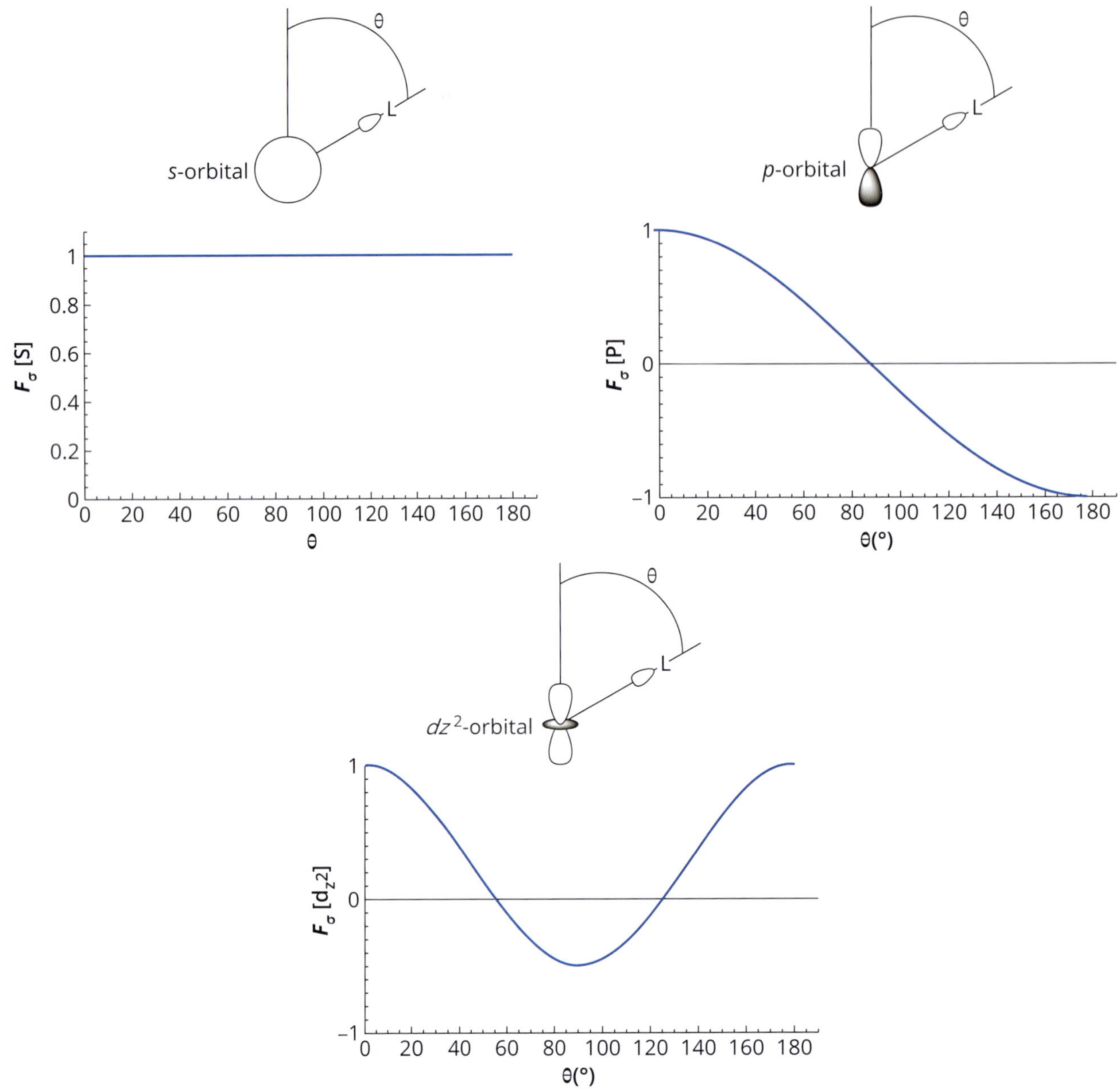

Fig. 15.29 The angular factor, $F(\theta,\phi)$, traces the shapes of the orbitals, changing the overlap with the orbital shape from 1 to 0. The plots shown are in the *xz*-plane with an axially symmetric ligand so that $\phi = \psi = 0$.

	$F^2_{\sigma,z^2,L}(\theta) = \left[\frac{1+3\cos(2\theta)}{4}\right]^2$	
Ligand Position	**θ (°)**	$F^2_{\sigma,z^2,L}(\theta)$
1	90	0.25
2	90	0.25
3	90	0.25
4	90	0.25
5	0	1
6	180	1

$$\varepsilon_{\sigma,z^2} = e_\sigma(L)\sum_L F^2_{\sigma,z^2,L}(\theta)$$

$$\varepsilon_{\sigma,z^2} = (0.25 + 0.25 + 0.25 + 0.25 + 1 + 1)e_\sigma(L)$$

$$\varepsilon_{\sigma,z^2} = 3e_\sigma(L)$$

As shown, the energy of the d_{z^2}-orbital due to σ-interactions is 3 $e_\sigma(L)$, where $e_\sigma(L)$ reflects the effect of the σ-donor ability of L on the energy of the d_{z^2}. This energy is usually tabulated in wavenumbers, cm^{-1} for various ligands with various metal centers (*vide infra*).

Table 15.2 General angular dependence factors for the *d*-orbitals.[7]

	$F_\sigma\{d_i, L(\theta,\phi,\psi)\}$	$F_{\pi x}\{d_i, L(\theta,\phi,\psi)\}$	$F_{\pi y}\{d_i, L(\theta,\phi,\psi)\}$
$d_{x^2-y^2}$	$(\sqrt{3}/4)\cos 2\phi(1-\cos 2\theta)$	$-\sin 2\phi \sin\theta \cos\psi - (\cos 2\phi \sin 2\theta \sin\psi)/2$	$-\sin 2\phi \sin\theta \sin\psi + (\cos 2\phi \sin 2\theta \cos\psi)/2$
d_{z^2}	$(1+3\cos 2\theta)/4$	$(\sqrt{3}/2)\sin 2\theta \sin\psi$	$(\sqrt{3}/2)\sin 2\theta \cos\psi$
d_{xy}	$(\sqrt{3}/4)\sin 2\phi(1-\cos 2\theta)$	$\cos 2\phi \sin\theta \cos\psi - (\sin 2\phi \sin 2\theta \sin\psi)/2$	$\cos 2\phi \sin\theta \sin\psi - (\sin 2\phi \sin 2\theta \cos\psi)/2$
d_{xz}	$(\sqrt{3}/2)\cos\phi \sin 2\theta$	$-\sin\phi \cos\theta \cos\psi - \cos\phi \cos 2\theta \sin\psi$	$-\sin\phi \cos\theta \sin\psi - \cos\phi \cos 2\theta \cos\psi$
d_{yz}	$(\sqrt{3}/2)\sin\phi \sin 2\theta$	$-\cos\phi \cos\theta \cos\psi - \sin\phi \cos 2\theta \sin\psi$	$-\cos\phi \cos\theta \sin\psi - \sin\phi \cos 2\theta \cos\psi$

The energy of the π-bonding interaction with the d_{z^2}-orbital is found similarly. In this case, all of the π-interactions are found to be zero, and the d_{z^2}-orbital is, by symmetry induced necessity, not involved in any π-bonding when ligands are at the coordinates for an octahedron.

If we now turn to a different orbital, d_{xy}, we find that all of the σ-interactions are zero. Calculation of the π-interactions goes as shown as follows. In this case, all of the x-axes on the individual ligands are coincident with the xy-plane, and the ψ-values will be 0 and 90° for π_x and π_y, respectively. As a result, all of the ligands in the xy-plane (L = 1, 2, 3, and 4) can interact with d_{xy} with their π_x-orbital. The π_y-orbitals on the ligands are all directed parallel to the z-axis, orthogonal to the d_{xy}-orbital, and do not contribute to the energy, $0e_{\pi y}$. Because the ligands are cylindrically symmetric and $e_{\pi x} = e_{\pi y} = e_\pi$, the energy of the bonding orbital (mostly ligand in character) is stabilized by $4e_\pi$, while the antibonding orbital (mostly metal in character) is destabilized by $4e_\pi$.

Ligand Position	$F^2_{\pi x,xy,L}(\theta,\phi,\psi) = [\sin\theta \cdot \cos 2\phi \cdot \cos\psi - (\sin\theta \cdot \sin 2\phi \cdot \sin\psi)/2]^2$ θ (°)	ϕ (°)	ψ (°)	
1	90	0	0	1
2	90	90	0	1
3	90	180	0	1
4	90	270	0	1
5	0	0	0	0
6	180	180	0	0

$$\varepsilon_{\pi x,xy} = e_{\pi x}(L)\sum_L F^2_{\pi x,xy,L}(\theta)$$

$$\varepsilon_{\pi x,xy} = (1+1+1+1+0+0)e_\sigma(L)$$

$$\varepsilon_{\pi x,xy} = 4e_{\pi x}(L)$$

Table 15.3 The ligand coordinates for an octahedral system in spherical coordinates. (See Fig. 15.30 for the labeling scheme.)

Ligand	θ	ϕ	ψ
1	90	0	0
2	90	90	0
3	90	180	0
4	90	270	0
5	0	0	0
6	180	0	0

For the effect of the πy-component on the d_{xy} orbital, the contribution is found to be zero. Note that in Fig. 15.30 the $y_{1\text{-}4}$ vectors are out of the xy-plane and along z, so do not affect the energy of the d_{xy}-orbital.

$$\varepsilon_{\pi y,xy} = e_{\pi y}(L)\sum_L F^2_{\pi y,xy,L}(\theta,\phi,\psi)$$

$$\varepsilon_{\pi y,xy} = 0e_{\pi x}(L) = (0+0+0+0+0+0)e_{\pi y}(L) = 0e_{\pi y}(L)$$

One of the starting assumptions was that the ligands were axially symmetric so that the total π-donor ability was a constant, regardless of angle ψ. Consequently, if we choose any angle for ψ (such as ψ = 0°) and sum the $e_{\pi x}$ and $e_{\pi y}$ values, that will give e_π. So, the energy of $d_{xy} = 4e_{\pi x} + 0e_{\pi y} = 4e_\pi$. Calculation of the other orbital energies proceeds similarly to produce the following results.

$$\varepsilon_{xy} = 4e_\pi$$
$$\varepsilon_{xz} = 4e_\pi$$
$$\varepsilon_{yz} = 4e_\pi$$
$$\varepsilon_{x^2-y^2} = 3e_\sigma$$
$$\varepsilon_{z^2} = 3e_\sigma$$

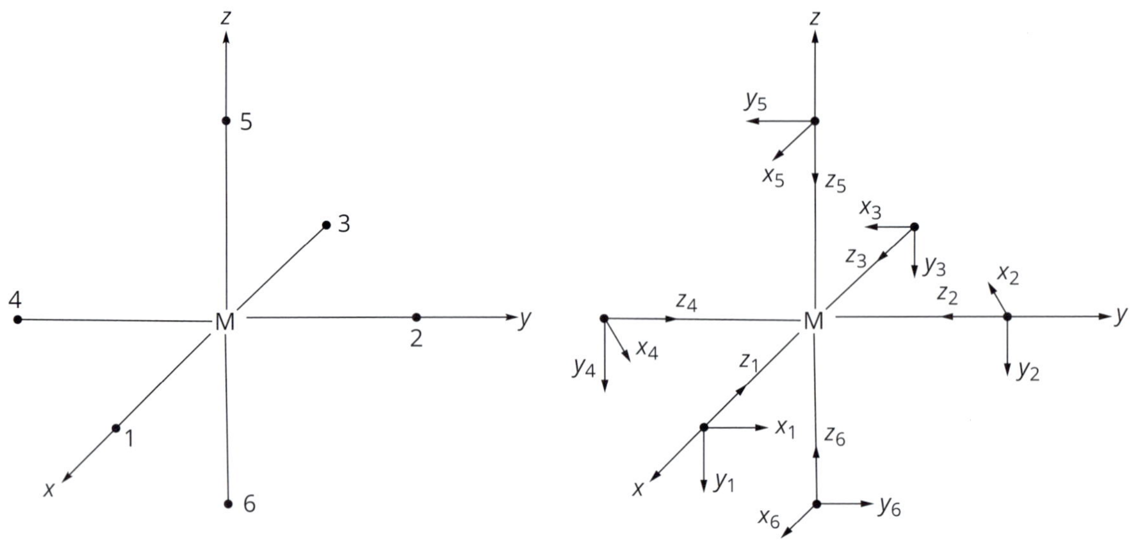

Fig. 15.30 (*left*) Ligand numbering scheme for octahedral. (*right*) Chosen coordinates on the ligands for an octahedral system. The ligand coordinates were chosen, somewhat arbitrarily, so that they are right-handed axes symmetric with respect to the highest-order proper rotation, C_4. If the ligand is planar, the *x*-axis for each ligand should be, by convention, rotated to be along the plane of the ligand.

The MO diagrams with energies listed as their *changes in energy relative to the starting energy of the majority contributor* are shown in Fig. 15.31. Shown on the left side of the figure is the situation when π-donor ligands are present, which has $e_\pi > 0$ and raises the energy of the *d*-orbitals. On the right side of the figure is shown the situation when π-acceptor ligands (such CN^-) are present on the metal, where $e_\pi < 0$ and the *d*-orbitals are stabilized.

One advantage of the AOM is that the energy gaps are parameterized specifically based on ligand–metal properties. The t_{2g} to e_g energy gap, Δ_o, is $3e_\sigma - 4e_\pi$, regardless of the metal or ligand type. The expression for Δ_o is related only to the symmetry. The actual energy is adjusted using the e_σ and e_π parameters determined for the ligand–metal combination.

15.9.3 Solutions of the Angular Dependence Factors for Common Geometries

As can be seen from the solutions to the octahedral systems, high-symmetry geometries often result in relatively simple values for the orbital (de)stabilization energies—ones, zeros, and simple fractions. In this section we will discuss some properties of the AOM model with respect to some high-symmetry systems, which come from solving the general $F(\theta,\phi,\psi)$ equations for the spherical coordinates for the ligand positions. Table 15.5 shows the $F^2(\theta,\phi,\psi)$-values that allow the calculation of the AOM destabilization energies of molecules in many high-symmetry systems, including linear, trigonal planar, square-planar, tetrahedral, trigonal bipyramidal, and octahedral.

We can use the F^2 values in Table 15.4 to calculate the octahedral splitting energy. Since in an octahedral complex all the ligands are equivalent, we have only one ligand type with σ- and π-effects. By the nature of the tabulated values, the calculated π-effects will be for a ligand with two equivalent π-donating orbitals, and the calculated ε_π-value will be from $\varepsilon_{\pi x} + \varepsilon_{\pi y}$. The destabilization energy of the *d*-orbital in question derives simply from adding the appropriate rows (those corresponding to ligand positions 1–6) down the column for that *d*-orbital. For example, the destabilization energy due to σ-interactions for the d_{z^2}-orbital times the e_σ value for the ligand.

$$\begin{aligned}\varepsilon_{\sigma,z^2} &= e_\sigma \sum_L F_L^2(\theta,\phi) \\ &= e_\sigma(1/4 + 1/4 + 1/4 + 1/4 + 1 + 1) \\ &= 3e_\sigma\end{aligned}$$

In the case of the d_{z^2}-orbital, the contribution to the orbital's destabilization due to π-interactions is zero. In contrast, the energy of the d_{xz}-orbital is not influenced by σ-interactions, and the π-interactions are calculated by taking the rows corresponding to the F^2 values for the π-contributions of ligands 1–6 and multiplying by e_π.

$$\begin{aligned}\varepsilon_{\pi,xz} &= e_\pi \sum_L F_L^2(\theta,\phi) \\ &= e_\pi(1 + 0 + 1 + 0 + 1 + 1) \\ &= 4e_\pi\end{aligned}$$

Calculating for the remainder of the *d*-orbitals produces the octahedral splitting.

$$\varepsilon_{z^2} = \varepsilon_{x^2-y^2} = 3e_\sigma$$

$$\varepsilon_{xz} = \varepsilon_{yz} = \varepsilon_{xy} = 4e_\pi$$

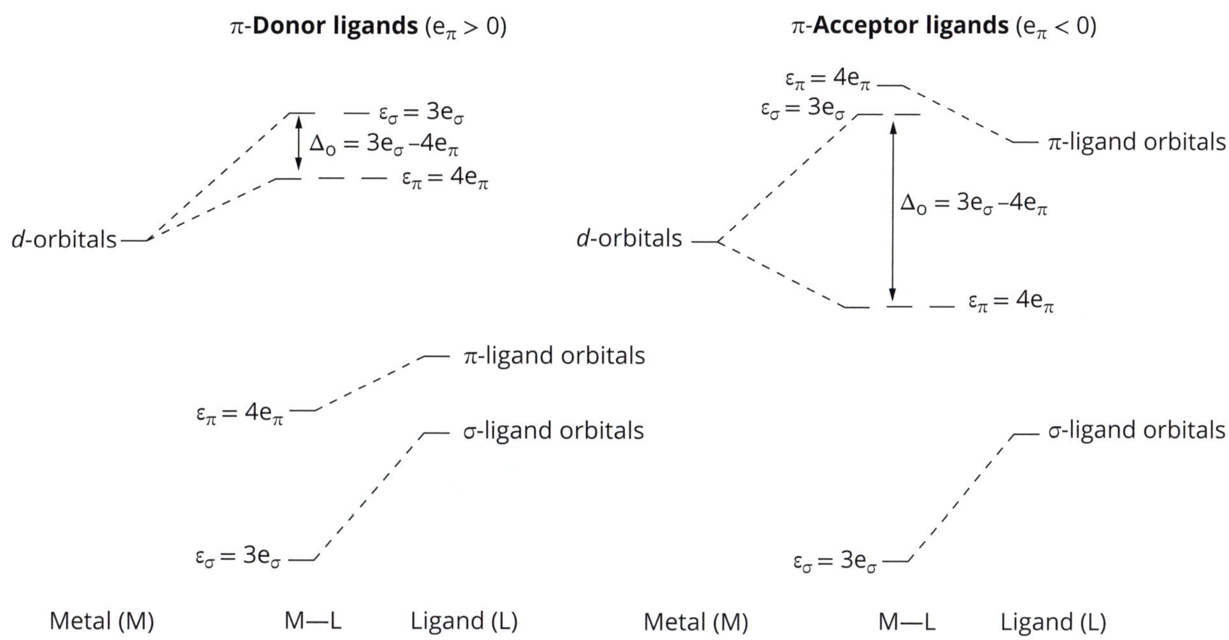

Fig. 15.31 MO diagrams with energies in terms of orbital energy changes from the major contributing orbital. (*left*) Where the ligand is a π-donor ($e_\pi > 0$). (*right*) Where the ligand is a π-acceptor ($e_\pi < 0$).

If we have a system in which the ligand angles are the same as the octahedral system but have a mixture of ligands (or the same ligand at different distances due to J–T distortion), then we can still use ligands 1–6 in Table 15.4. However, we need to use the most general equation for the destabilization energy, ε, Eq. 15.9, keeping track of the individual interaction energies for the different ligands, such as $e_\sigma(L_1)$ versus $e_\sigma(L_2)$. As an example, let's calculate the orbital energies for *trans*-ML_4X_2 and *cis*-ML_4X_2, where L and X are two different ligand types.

For *trans*-ML_4X_2 we will assume that the X ligands are along the *z*-axis, positions 5 and 6. For orbitals in the *xy*-plane they will be unaffected by the X ligands in the AOM model, and their energies will be identical to the octahedral case with contributions exclusively from the L ligand.

$$\varepsilon_{x^2-y^2} = 3e_\sigma(L)$$
$$\varepsilon_{xy} = 4e_\pi(L)$$

The other *d*-orbitals will have energy contributions from both L and X. Calculating the energy of the d_{z^2}-orbital proceeds as follows.

$$\begin{aligned}\varepsilon_{\sigma,z^2} &= \sum_L F_L^2(\theta,\phi,\psi)\cdot e_\sigma(L)\\ &= 1/4\cdot e_\sigma(L) + 1/4\cdot e_\sigma(L) + 1/4\cdot e_\sigma(L) + 1/4\cdot e_\sigma(L)\\ &\quad + 1\cdot e_\sigma(X) + 1\cdot e_\sigma(X)\\ &= e_\sigma(L) + 2e_\sigma(X)\end{aligned}$$

The d_{xz}- and d_{yz}-orbitals are degenerate.

$$\varepsilon_{\pi,xz} = \varepsilon_{\pi,yz} = 2e_\pi(L) + 2e_\pi(X)$$

The *d*-orbital energies for the *trans*-ML_4X_2 system are then as shown below, which are the same splittings expected for a J-T distortion of an octahedral complex.

$$\begin{aligned}\varepsilon_{\sigma,z^2} &= e_\sigma(L) + 2e_\sigma(X)\\ \varepsilon_{x^2-y^2} &= 3e_\sigma(L)\\ \varepsilon_{\pi,xz} &= \varepsilon_{\pi,yz} = 2e_\pi(L) + 2e_\pi(X)\\ \varepsilon_{xy} &= 4e_\pi(L)\end{aligned}$$

For *cis*-ML_4X_2 we will assume that the X ligands are in positions 1 and 5. The resulting orbitals, which just come from adding the products of the F^2 terms in Table 15.5 with $e_{\sigma/\pi}$(L or X), are as follows.

$$\begin{aligned}\varepsilon_{\sigma,z^2} &= 1\tfrac{3}{4}e_\sigma(L) + 1\tfrac{1}{4}e_\sigma(X)\\ \varepsilon_{x^2-y^2} &= 2\tfrac{1}{4}e_\sigma(L) + \tfrac{3}{4}e_\sigma(X)\\ \varepsilon_{\pi,xz} &= 2e_\pi(L) + 2e_\pi(X)\\ \varepsilon_{\pi,yz} &= 3e_\pi(L) + 1e_\pi(X)\\ \varepsilon_{xy} &= 4e_\pi(L)\end{aligned}$$

15.9.4 Calculation of the Stabilization Energy for Complexes

In the previous section it was shown that the crystal field splitting in an octahedral system can be parameterized so that the e_g*-orbitals are at energy $3e_\sigma$ and the t_{2g}-orbitals are

Table 15.4 $F^2(\theta,\phi,\psi)$-values necessary for the calculation of AOM d_i-orbital energies for many high-symmetry systems. The table is for axially symmetric ligands where $e_{\pi x} + e_{\pi y} = e_{\pi}$.

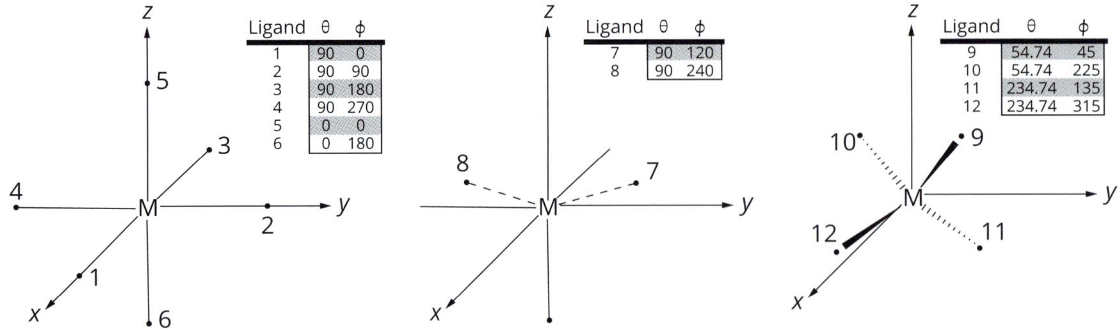

Position	F_i^2	i = z²	x² – y²	xz	yz	xy
1	F^2_σ	1/4	3/4	0	0	0
	F^2_π	0	0	1	0	1
2	F^2_σ	1/4	3/4	0	0	0
	F^2_π	0	0	0	1	1
3	F^2_σ	1/4	3/4	0	0	0
	F^2_π	0	0	1	0	1
4	F^2_σ	1/4	3/4	0	0	0
	F^2_π	0	0	0	1	1
5	F^2_σ	1	0	0	0	0
	F^2_π	0	0	1	1	0
6	F^2_σ	1	0	0	0	0
	F^2_π	0	0	1	1	0
7	F^2_σ	1/4	3/16	0	0	9/16
	F^2_π	0	3/4	1/4	3/4	1/4
8	F^2_σ	1/4	3/16	0	0	9/16
	F^2_π	0	3/4	1/4	3/4	1/4
9	F^2_σ	0	0	1/3	1/3	1/3
	F^2_π	2/3	2/3	2/9	2/9	2/9
10	F^2_σ	0	0	1/3	1/3	1/3
	F^2_π	2/3	2/3	2/9	2/9	2/9
11	F^2_σ	0	0	1/3	1/3	1/3
	F^2_π	2/3	2/3	2/9	2/9	2/9
12	F^2_σ	0	0	1/3	1/3	1/3
	F^2_π	2/3	2/3	2/9	2/9	2/9

Geometry	Ligand Positions
Octahedral	1–6
Square pyramidal	1–5
Square planar	1–4
Linear	5, 6
Trigonal bipyramidal	1, 5–8
Trigonal-planar	1, 7, 8
Tetrahedral	9–12

at energy $4e_\pi$. The stability of a specific geometry is dependent upon the number of electrons and the relative energies of the orbitals which those electrons occupy. Unoccupied orbitals do not affect the energy.

Since the (de)stabilization energy (ε or ε^*) is relative to the starting position of the orbitals, different geometries can be evaluated relative to a common starting position, the beginning energies of the noninteracting ligand and *d*-orbitals (Fig. 15.27, $\varepsilon^* = 0$ and $\varepsilon = 0$). The *AOM stabilization energy (Φ)* is related to the sum (Eq. 15.10) of the product of the bonding orbital stabilizations (ε_i) times the number of electrons in each orbital minus the product of the antibonding orbital destabilization energies (ε_i^*) times the number of electrons in those orbitals, n_i. *If comparing two geometries, the larger stabilization energy, Φ, is more stable.* The total number of electrons will be those required to fill the antibonding orbitals (10-electrons) plus the number of *d*-electrons in the complex (the number of electrons in antibonding orbitals).

$$\Phi = \sum_i n_i\varepsilon_i - \sum_i n_i\varepsilon_i^* \qquad 15.10$$

As an example, let us examine the high-spin d^5 complex, $\{Fe(OH_2)_6\}^{3+}$. The complex is octahedral, so its orbital energies are as shown in Fig. 15.32. The total number of electrons in our diagram will be fifteen—10-electrons to fill the bonding orbitals derived from the water lone pairs, and 5-electrons on the metal center that occupy the largely antibonding orbitals in a high-spin configuration.

From Eq. 15.10, the AOM stabilization energy (Φ) is calculated as follows. We start with the bonding orbitals and find the stabilization due to the bonding electrons. Then we subtract the destabilization due to occupation of the antibonding orbitals.

$$\begin{aligned}\Phi &= \sum_i n_i\varepsilon_i - \sum_i n_i\varepsilon_i^* \\ &= (4\times 3e_\sigma + 6\times 4e_\pi) - (3\times 4e_\pi + 2\times 3e_\sigma) \\ &= (12e_\sigma + 24e_\pi) - (12e_\pi + 6e_\sigma) \\ &= 6e_\sigma + 12e_\pi\end{aligned}$$

More simply, we can realize that for every hole in an antibonding orbital, *h*, there is a filled bonding orbital with a stabilized electron. Consequently, the sum of the number of holes in each orbital times the orbital energy will provide the stabilization energy.

$$\Phi = \sum_i h_i\varepsilon_i^* \qquad 15.11$$

For our example using high-spin $\{Fe(OH_2)_6\}^{2+}$ in Fig. 15.32, there are five holes in the antibonding orbitals, and calculation of the stabilization energy (Φ) using Eq. 15.11 is shown in the following. There are three holes in t_{2g}, and that value is multiplied by the orbital energy, $4e_\pi$. This is summed with the number of holes in e_g (4) times the energy of e_g ($3e_\sigma$). The simplification of the AOM stabilization energy in Eq. 15.11 allows us to focus on the *d*-electron count exclusively as $h = 10 - (d\text{-electrons})$ and on the *d*-orbital splittings, making the method more akin to crystal field theory.

$$\begin{aligned}\Phi &= \sum_i h_i\varepsilon_i^* \\ &= (3\times 4e_\pi) + (2\times 3e_\sigma) \\ &= 6e_\sigma + 12e_\pi\end{aligned}$$

We can use Eq. 15.11 to determine whether or not a complex is susceptible to a J–T distortion. Let us compare two complexes

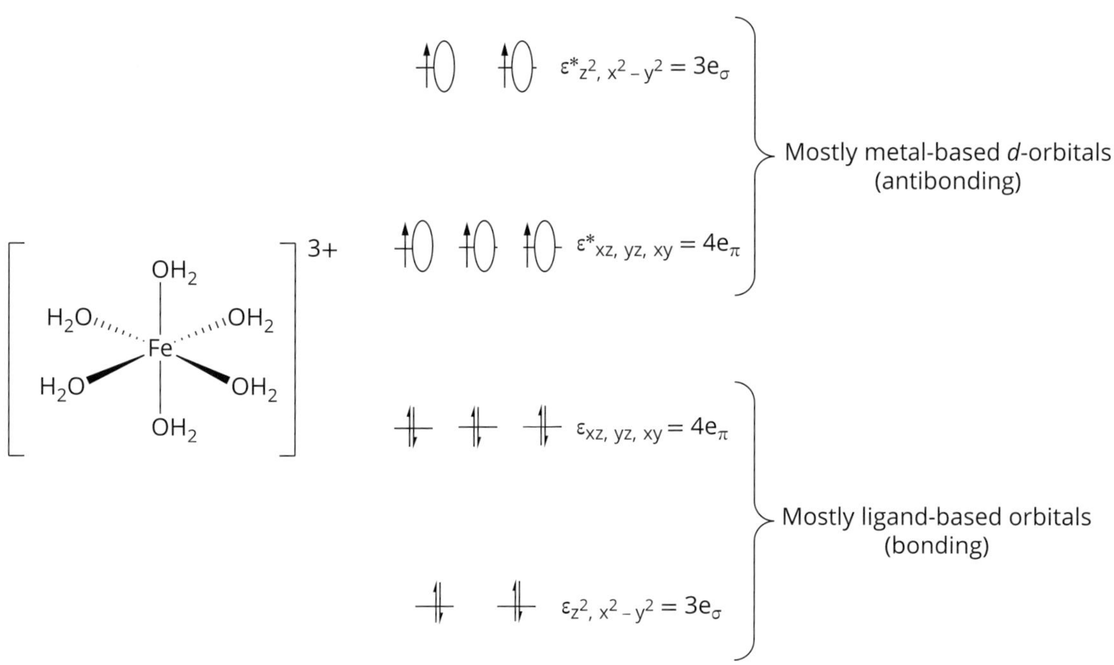

Fig. 15.32 AOM (de)stabilization energies for O_h, high-spin $Fe(OH_2)_6^{2+}$. The ovals represent the five holes, *h*, in the antibonding orbitals.

d^8 $\{Ni(OH_2)_6\}^{2+}$ and d^9 $\{Cu(OH_2)_6\}^{2+}$ in O_h and axially elongated (tetragonally distorted) D_{4h}.

For $\{Ni(OH_2)_6\}^{2+}$, the splitting picture for the two geometries is as follows. The splitting in the e_g orbital due to the axial elongation will not be large enough for the complex to become low-spin, and both geometries will have a triplet ground state (Fig. 15.33).

Calculation of the stabilization energy in the two geometries (O_h and D_{4h}) proceeds as follows.

$$\Phi_{Ni(II),O_h} = \sum_i h_i \varepsilon_i^* = 2 \times 3e_\sigma = 6e_\sigma$$

$$\Phi_{Ni(II),D_{4h}} = \sum_i h_i \varepsilon_i^* = 1 \times 3e_\sigma(L_{eq}) + 1 \times [2e_\sigma(L_{ax}) + e_\sigma(L_{eq})] = 4e_\sigma(L_{eq}) + 2e_\sigma(L_{ax})$$

What is the relationship between e_σ and the Jahn–Teller distorted ligand parameters? As shown in Fig. 15.25, the stabilization of the d_{z^2}-orbital is by ε_Q, making $e_\sigma(L_{ax}) = e_\sigma - \varepsilon_Q$. Conversely, the $d_{x^2-y^2}$-orbital is raised in energy by energy ε_Q, making $e_\sigma(L_{eq}) = e_\sigma + \varepsilon_Q$. If the d_{z^2}-orbital and $d_{x^2-y^2}$-orbital are raised and lowered by the same amount we can set the two ε_Q-values equal and solve the following equations, which produces the J–T distorted parameters in terms of the octahedral parameter, e_σ.

$$3e_\sigma + \varepsilon_Q = 3e_\sigma(L_{eq})$$
$$3e_\sigma - \varepsilon_Q = e_\sigma(L_{eq}) + 2e_\sigma(L_{ax})$$

Solving both equations for ε_Q and setting equal to each other:

$$\varepsilon_Q = 3e_\sigma(L_{eq}) - 3e_\sigma$$
$$\varepsilon_Q = -e_\sigma(L_{eq}) - 2e_\sigma(L_{ax}) + 3e_\sigma$$
$$3e_\sigma(L_{eq}) - 3e_\sigma = -e_\sigma(L_{eq}) - 2e_\sigma(L_{ax}) + 3e_\sigma$$

Solving for e_σ:

$$e_\sigma = (2/3)e_\sigma(L_{eq}) + (1/3)e_\sigma(L_{ax})$$

Multiplying both sides of this equation by 6 (for the six ligands) shows that the two geometries, tetragonally distorted O_h and O_h, are of equal energy for $Ni(OH_2)_6^{2+}$. Consequently, the higher-symmetry geometry is chosen, since the distortion leads to no energy stabilization.[t]

$$6e_\sigma = 4e_\sigma(L_{eq}) + 2e_\sigma(L_{ax})$$

AOM diagram and energies for d^9 $Cu(OH_2)_6^{2+}$ are shown in Fig. 15.34.

[t] This choice of the higher-symmetry structure could be viewed as a manifestation of one of Pauling's heuristic principles for choosing between structures in Chapter 3: "A system adopts the highest symmetry consistent with the constraints acting on it." Higher-symmetry systems are favorable for electron delocalization.

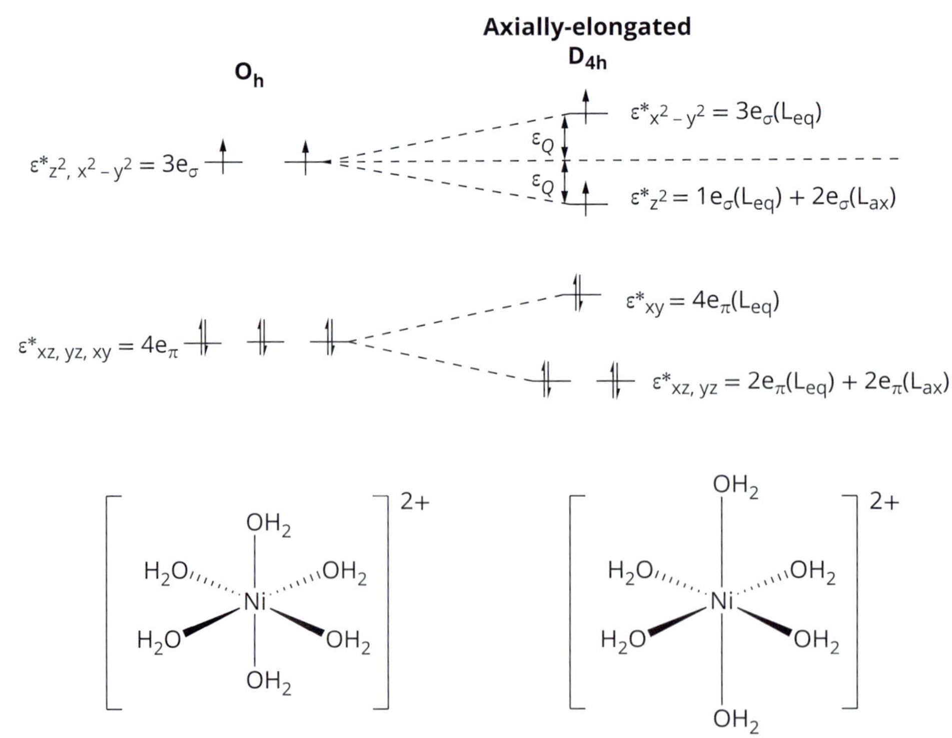

Fig. 15.33 AOM energies and crystal field splitting diagram for octahedral $Ni(OH_2)_6^{2+}$ and the J–T distorted complex.

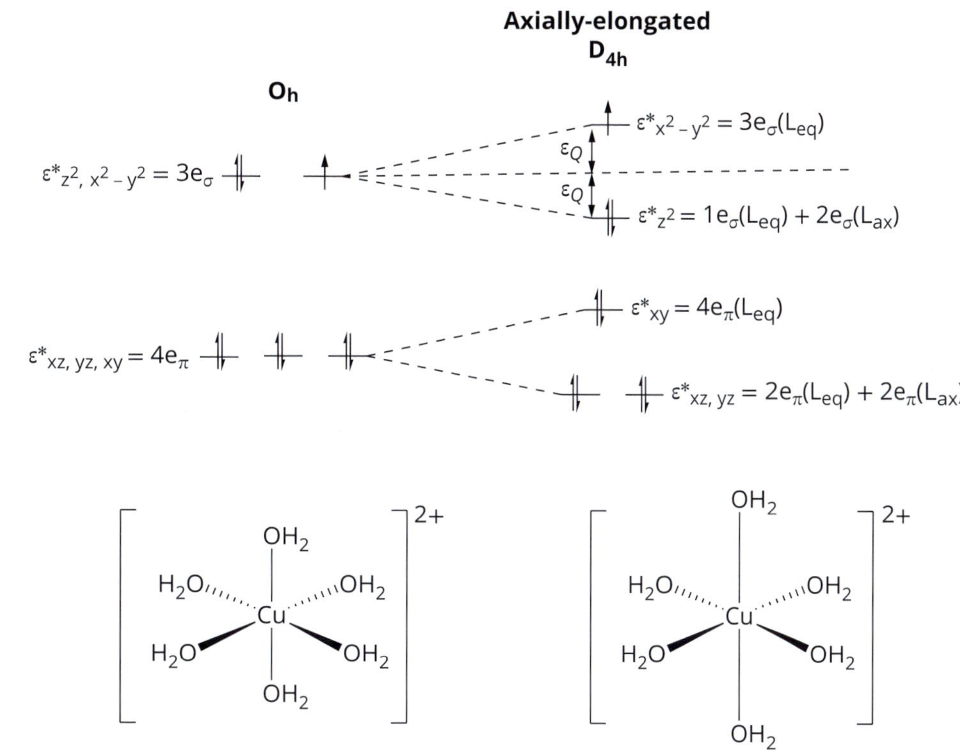

Fig. 15.34 AOM energies and crystal field splitting diagram for octahedral $Cu(OH_2)_6^{2+}$ and the Jahn–Teller distorted complex.

Calculation of the stabilization energy, involving only a single hole in an orbital for each geometry, O_h and D_{4h}, is simply the energy of the highest-energy orbital.

$$\begin{aligned}\Phi_{Cu(II),O_h} &= \sum_i h_i \varepsilon_i^* \\ &= 1 \times 3e_\sigma \\ &= 3e_\sigma\end{aligned}$$

$$\begin{aligned}\Phi_{Cu(II),D_{4h}} &= \sum_i h_i \varepsilon_i^* \\ &= 1 \times 3e_\sigma(L_{eq}) \\ &= 3e_\sigma(L_{eq})\end{aligned}$$

Since $e_\sigma(L_{eq}) = e_\sigma + \varepsilon_Q$, the tetragonally distorted geometry, D_{4h}, will be more stable by ε_Q (where ε_Q = any positive value), corresponding to an axial elongation, and $Cu(OH_2)_6^{2+}$ is susceptible to Jahn–Teller distortion.

15.9.5 Tetrahedral Fields in the AOM

For a tetrahedral system, the ligands 9–12 are used in Table 15.4 to determine the orbital energies according to the AOM. The results are shown as follows. The energy-ordering is reversed, as discussed under CFT, with respect to octahedral complexes with the t_2-orbitals higher in energy than the e-orbitals. In stark contrast to the orthoaxial octahedral system, the σ- and π-effects are mixed in the t_2-orbitals.

$$\begin{aligned}\varepsilon_{xz,yz,xy} &= \frac{4}{3}e_\sigma + \frac{8}{9}e_\pi \\ \varepsilon_{x^2-y^2,z^2} &= \frac{8}{3}e_\pi\end{aligned}$$

From this, it is obvious that the Δ_t energy gap is significantly smaller than the Δ_o energy gap.

$$\Delta_t = \varepsilon_{xz,yz,xy} - \varepsilon_{x^2-y^2,z^2} = \left(\frac{4}{3}e_\sigma + \frac{8}{9}e_\pi\right) - \frac{8}{3}e_\pi$$

$$\Delta_t = \frac{4}{3}e_\sigma - \frac{16}{9}e_\pi$$

The relationship between the octahedral and tetrahedral energy gaps can be calculated by using the previous values. The value obtained is Δ_t=(4/9) Δ_0 as discussed previously, but obviously this assumes that e_σ and e_π are identical for the tetrahedral and octahedral complexes in question. One could add the negative sign here, signifying the inversion of the orbitals, to get Eq. 15.3.

$$\begin{aligned}\frac{\Delta_t}{\Delta_o} &= \frac{\frac{4}{3}e_\sigma - \frac{16}{9}e_\pi}{3e_\sigma - 4e_\pi} \\ \frac{\Delta_t}{\Delta_o} &= \frac{\frac{4}{9}(3e_\sigma - 4e_\pi)}{3e_\sigma - 4e_\pi} \\ \frac{\Delta_t}{\Delta_o} &= \frac{4}{9} \\ \Delta_t &= \frac{4}{9}\Delta_o\end{aligned}$$

15.9.6 Example of Using AOM Stabilization Energy to Find the Preferred Geometry: 4-Coordinate Complexes

Using the limiting ligand positions in Table 15.4, the orbital stabilization energies can be found for different geometries. For a specific *d*-electron count, the stabilization energies (Φ) for the different isomers can be calculated as a function of the donor parameters e_σ and e_π.

As an example, following that of Burdett and also DeKock and Gray,[30,37] let us calculate the *d*-orbital energies for three different 4-coordinate geometries: tetrahedral, square-planar, and *cis*-divacant octahedral (also called disphenoidal or "see-saw"). We can then calculate the preferred geometry based on the stabilization energy.

First we need to calculate the orbital energies as a function of e_σ and e_π. These were done using ligands {9–12} for tetrahedral, {1–4} for square-planar, and {1,4–6} for *cis*-divacant octahedral. The results are shown in Fig. 15.35.

Using these orbital energies, the stabilization energies can be calculated. A few examples will be worked out in detail, but Table 15.5 shows the results for many of the possible configurations.

As the first example, the stabilization energy of a low-spin d^8 complex will be calculated in each of the three geometries under investigation. As an abbreviation, a low-spin d^8 configuration will be written as 22220, signifying that the lowest four orbitals are doubly filled and the highest has two vacancies. The three geometries will be described by the Mulliken symbols for their point-group symmetries: tetrahedral → T_d, square-planar → D_{4h}, and *cis*-divacant octahedral → C_{2v}. For example, the stabilization energy of the *cis*-divacant octahedral system in low-spin d^8 is symbolized by $\Phi^{C_{2v}}_{22220}$.

The results of the calculations are shown as follows. Recall that the only contributors to the stabilization energy are the holes, *h*, and the energy of the orbitals "occupied" by the holes (Eq. 15.11). Since e_σ is typically much larger than e_π, we will ignore e_π.[u] The stabilization energies are then $\Phi^{D_{4h}}_{22220} > \Phi^{C_{2v}}_{22220} > \Phi^{T_d}_{22220}$. The largest stabilization energy indicates the most stable geometry, and low-spin 4-coordinate d^8-complexes are expected to be (and are) square-planar, such as $PtCl_4^{2-}$.

$$\Phi^{T_d}_{22220} = \sum_i h_i \varepsilon_i^\star = 2 \times \left(\frac{4}{3} e_\sigma + \frac{8}{9} e_\pi \right) = \frac{8}{3} e_\sigma + \frac{16}{9} e_\pi$$

$$\Phi^{D_{4h}}_{22220} = \sum_i h_i \varepsilon_i^\star = 2 \times 3e_\sigma = 6e_\sigma$$

$$\Phi^{C_{2v}}_{22220} = \sum_i h_i \varepsilon_i^\star = 2 \times \frac{5}{2} e_\sigma = 5e_\sigma$$

What is the preferred geometry of a d^{10} complexes, such as $Ni(CO)_4$? If the complex is d^{10} then there are no holes in any of the *d*-orbitals, and Eq. 15.11 is necessarily zero, regardless of geometry: $\Phi^{D_{4h}}_{22222} = \Phi^{C_{2v}}_{22222} = \Phi^{T_d}_{22222}$. If two or more geometries have the same stabilization energy due to overlap, the higher-symmetry geometry is preferred. This is an expression of Pauling's statement that molecules adopt the highest symmetry consistent with the constraints acting on the system. Quantum-mechanically, higher-symmetry geometries often have degenerate wavefunctions, which increase electron

[u] This is quite a good assumption considering that e_π would need to be ~$2e_\sigma$ before tetrahedral would be the most stable geometry in this case—a very unlikely scenario.

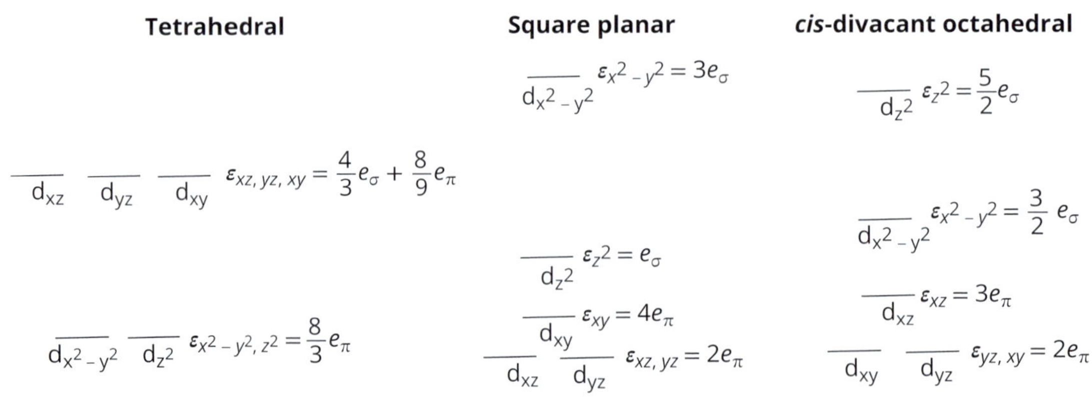

Fig. 15.35 Orbital energies calculated using the AOM for tetrahedral, square-planar, and *cis*-divacant octahedral.

Table 15.5 Stabilization energies (Φ) in units of e_σ for many of the possible electronic configurations for 4-coordinate tetrahedral (Φ^{T_d}), square-planar ($\Phi^{D_{4h}}$), and *cis*-divacant octahedral ($\Phi^{C_{2v}}$) geometries.

d-electrons	Electronic Configuration	Φ^{T_d}	$\Phi^{D_{4h}}$	$\Phi^{C_{2v}}$	Observed Geometry
10	22222	0	0	0	Tetrahedral[a]
9	22221	1 1/3	3	2 1/2	Square-planar
8	22220	2 2/3	6	5	Square-planar
	22211	2 2/3	4	4	*cis*-divacant octahedral[b]
7	22210	4	7	6 1/2	Square-planar
	22111	4	4	4	Tetrahedral[a]
6	22200	5 1/3	8	8	*cis*-divacant octahedral[b]
	22110	5 1/3	7	6 1/2	Square-planar
5	22100	6 2/3	8	8	*cis*-divacant octahedral[b]
4	22000	8	8	8	Tetrahedral
	11110	1 1/3	3	2 ½	Square-planar

[a] Preferred geometry determined by higher symmetry.
[b] Preferred geometry determined by inclusion of higher order overlap terms (Appendix F).

delocalization.[42] In addition, the preferred geometry in these cases is often (but not always, see Chapter 14) consistent with VSEPR expectations. Finally, the most symmetric structure often places ligands as far apart as possible, lowering steric interactions. Any or all of these may come into play for a particular system.

The third configuration to be discussed in detail is low-spin d^6, Φ_{22200}. The calculation of the stabilization energies proceeds as follows. The tetrahedral stabilization energy appears to be lowest by a significant margin, and the complex is unlikely to adopt this geometry. According to our calculations, the square-planar and *cis*-divacant octahedral (disphenoidal) geometries have equal stabilization energies, 8 e_σ.

$$\begin{aligned}\Phi_{22200}^{T_d} &= \sum_i h_i \varepsilon_i^* \\ &= 4 \times \left(\frac{4}{3}e_\sigma + \frac{8}{9}e_\pi\right) \\ &= \frac{16}{3}e_\sigma + \frac{32}{9}e_\pi\end{aligned}$$

$$\begin{aligned}\Phi_{22200}^{D_{4h}} &= \sum_i h_i \varepsilon_i^* \\ &= 2 \times 3e_\sigma + 2 \times 1e_\sigma \\ &= 8e_\sigma\end{aligned}$$

$$\begin{aligned}\Phi_{22200}^{C_{2v}} &= \sum_i h_i \varepsilon_i^* \\ &= 2 \times \frac{5}{2}e_\sigma + 2 \times \frac{3}{2}e_\sigma \\ &= 8e_\sigma\end{aligned}$$

In our development of the AOM we, as is typically done, truncated the overlap terms at second order, S^2. However, if one repeats the calculation while including the fourth-order terms, S^4, one finds that the *cis*-divacant octahedral geometry is preferred for low-spin d^6. A discussion of S^4 terms in the AOM can be found in Appendix F.[41] One such complex is FeR_4^{2-}, which perhaps has as close as is sterically possible to a *cis*-divacant octahedral geometry.[38–39]

Collected in Table 15.5 are the total stabilization energies (Φ) for 4-coordinate complexes of various geometries in units of e_σ (again, π-effects were ignored, $e_\pi \sim 0$) for many of the possible electronic configurations based on 4-coordinate tetrahedral (Φ^{T_d}), square-planar ($\Phi^{D_{4h}}$), and *cis*-divacant octahedral ($\Phi^{C_{2v}}$) geometries. Again, the structure with the highest Φ value is the most stable, and ties are broken either by resorting to F^4 terms or the structure with the higher symmetry when truly energetically equal.

15.9.7 The Two-Dimensional Spectrochemical Series

Using the angular overlap model, ligands on a variety of different metal centers can be parameterized according to their σ-donating and π-donating/π-accepting abilities. While we were able to define a linear spectrochemical series earlier, in which Δ_o changed σ- and π-donor ability (Fig. 15.18), we can now quantify those effects individually by using the AOM to develop a two-dimensional spectrochemical series. One such series is shown in Fig. 15.36 for ligands on chromium(III), which is perhaps the largest such series known for a single metal center.[40]

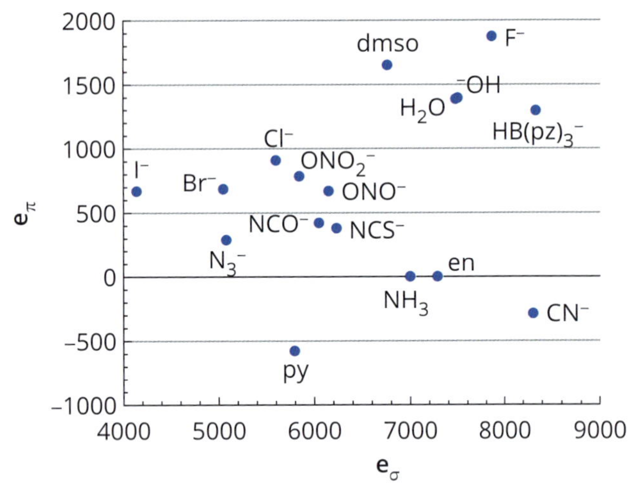

Fig. 15.36 A two-dimensional spectrochemical series for ligands on chromium(III). Values are in wavenumbers (cm^{-1}).[40]

Using the two-dimensional spectrochemical series, it is now possible to explain several relationships in the linear series that may have been puzzling. These are a few examples:

- In the linear series, pyridine and NH_3 have essentially the same position in the series, but ammonia is a stronger σ-base. One would expect, in the absence of π-effects, for the stronger σ-base to be higher on the spectrochemical series. One can see in the 2-D series that ammonia is indeed a stronger σ-donor, but pyridine is π-electron withdrawing, which results in the two ligands being about the same on the linear series.
- Fluorine is actually a stronger σ-donor than the vast majority of ligands. The only reason it is so low on the linear spectrochemical series is because it is also one of the strongest π-donors.
- Cyanide does not seem to be a better π-acceptor than pyridine for Cr(III). Cyanide is near the top of the linear spectrochemical series, mainly due to its extremely large σ-donor ability.

It should be kept in mind that these are the values for Cr(III). Other metals and other oxidation states tend to have somewhat different e_σ and e_π values for the same ligands.

15.9.8 *ds*-Mixing in Metal Complexes

While the focus in ligand field theory, crystal field theory, and the angular overlap model is on the *d*-orbitals generally to the exclusion of all else, there are instances where the next-closest orbital in energy, the valence *s*-orbital, can affect the system in ways that cannot be ignored. The mechanism for the *s*-orbital influencing the LFT or AOM model is through mixing with a *d*-orbital in the same irreducible representation as the *s*-orbital in the point group of the molecule; that is, a *d*-orbital in the totally symmetric representation.

In cubic symmetries, octahedral and tetrahedral, all five *d*-orbitals are involved in two different degenerate sets. Since the *s*-orbital always has the symmetry of the totally symmetric representation, there can be no mixing between the *s*- and *d*-orbitals in these symmetries.

In lower symmetries, most notably trigonal (D_{3h}), tetragonally distorted octahedral (D_{4h}), square-planar (D_{4h}), and Jahn–Teller distorted tetrahedral (D_{2d}) the d_{z^2}-orbital is represented by the totally symmetric representation along with the *s*-orbital. As a result, *sd*-mixing will occur, resulting in a stabilization of the d_{z^2}-orbital.

Consider as an example what may occur in 4-coordinate complexes. If a tetrahedral system contains an unsymmetrically filled degenerate orbital, it will be susceptible to a Jahn–Teller distortion similar to what was discussed previously for octahedral complexes. The nature of the distortion often results in systems intermediate between T_d and square-planar D_{4h}, that is, D_{2d} geometries.

Whether the system is D_{4h} or D_{2d}, the d_{z^2}-orbital has the same symmetry as the *s*-orbital once any distortion occurs, which allows *sd*-mixing. The *sd*-mixing will lead to a net stabilization of the d_{z^2}-orbital—a fact that is represented by a negative term, $-E_{ds}$, added to the AOM expression for the orbital energy. In D_{4h}, the d_{z^2}-orbital destabilization energy, ε_{z^2}, then has energy $\varepsilon_{z^2} = e_\sigma - E_{ds}$, and the AOM splitting diagram would be as shown in Fig. 15.37.[7]

Obviously, the orbital ordering involving d_{z^2} is then variable, and the parameter E_{ds} is generally found experimentally.[v] It has been suggested for square-planar platinum(II) complexes that $e_\sigma \sim E_{ds}$, and the d_{z^2}-orbital will be the lowest-energy orbital.[40]

The incorporation of *ds*-mixing is commonly discussed in square-planar systems like those previously, but may occur in many systems in which the d_{z^2}-orbital belongs to the same irreducible representation as *s*. For example, trigonal systems may have greatly stabilized d_{z^2}-orbitals due to *ds*-mixing, which in many cases can greatly stabilize the complex.

[v] In some schemes for square-planar complexes, the *ds*-mixing with the d_{z^2}-orbital is parameterized as $-4e_{ds} = E_{ds}$. (see Figgis and Hitchman)

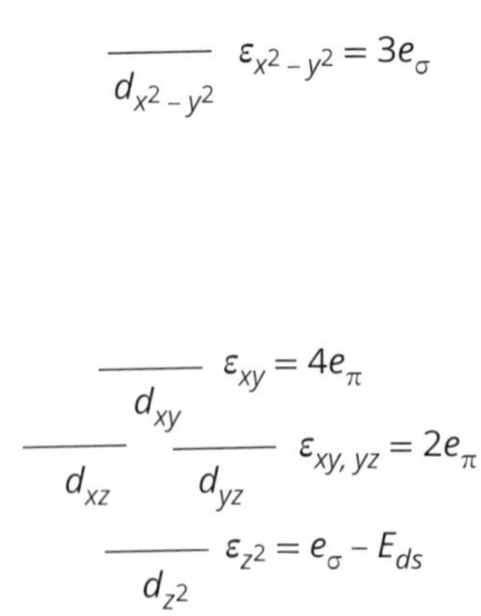

Fig. 15.37 AOM splitting diagram for a D_{4h} complex with *sd*-mixing.

In some cases, *dp*-mixing can occur due to symmetry, and accurate modeling of the system in a few cases requires inclusion of these parameters. However, *dp*-mixing is generally unlikely to be significant.[7]

15.9.9 Metal-Oxo Complexes and the Oxo Wall

An interesting application of the ideas presented in this chapter can be found in metal-oxo complexes. The metal-oxo motif is not only fascinating to inorganic chemists, but is also an important feature of a variety of critical biological systems such as hemoglobin (oxygen transport in animals), cytochrome oxidase (cellular respiration), P450 enzymes (monooxygenases), and photosystem II (photosynthesis). Whereas metal-oxos are commonly formed with early transition metals, they are less common for late transition metals. This interesting observation was first made by at Caltech by Harry Gray, who generalized the phenomenon and named it the "oxo wall". In all appearance, there seems to be a "wall" separating Group 8 elements (the iron family) and Group 9 elements (the cobalt family). Beyond that wall, tetragonal (pseudo-octahedral or square pyramidal) metal-oxos do not exist as stable species (Fig. 15.38).[41–42]

It turns out that it is relatively straightforward to rationalize the existence of the oxo wall, at least once the foundations of ligand-field theory are settled.[w] A prototypical example of an metal-oxo complex is given by the so-called "vanadyl ion", $\{VO(H_2O)_5\}^{2+}$, which has an idealized C_{4v} symmetry (neglecting the hydrogen atoms). The LFT MO arrangement for this tetragonal species can be easily constructed by using the approaches previously outlined to produce the diagram shown in Fig. 15.38b.[41–42]

The key feature of this diagram is the fact that the lowest metal-centered MO is a nonbonding orbital; any configuration with two (or fewer) *d* electrons would not impact the stability of the metal-oxo complex. On the other hand, the next available metal-centered orbitals are the d_{xz}/d_{yz} set, which is predominantly involved in the π-bond between the metal and the oxo ligand. The bond order of the metal-oxo bond is therefore linked directly to the nature of the metal center; any configuration above d^2 will gradually weaken the metal-oxo bond as these electrons populate the antibonding $e_g{}^*$ set. The highest M–O bond order is clearly obtained for the lowest electron counts, and metal-oxo complexes formed out of the d^0, d^1, or d^2 configurations should be ideally represented with a formal triple bond between the metal center and the oxo ligand—a feature that is often neglected in the representation of these species. At higher electron counts, say d^3 or d^4, the M–O bond order decreases down to a "simple" double bond; at higher counts, the metal-oxo bond order is lower than two, and the oxo ligand begins to accumulate too high an electron density, without the benefits of a stable π-bond. Such species are probably much too basic/nucleophilic to exist, as the oxo ligand would probably become protonated or attacked by electrophilic species. This low electron-count constraint is the basis for the oxo wall. The line between Group 8 and Group 9 forms a rather rigid barrier beyond which the d^4 configuration is extremely difficult to generate, at least not without forcing very high oxidation states such as Co(V) or Ni(VI), which would themselves be much too oxidizing to coexist with water ligands (or most other solvents), further reinforcing the structural integrity of the oxo wall. It is important to note that the apparently stringent restrictions imposed by the oxo wall all arise from the tetragonal symmetry used to construct the MO diagram in Fig. 15.38b.

There are some known late-metal oxo complexes, such as Wilkinson's 4-coordinate oxotrimesityliridium(V) complex, which is a d^4 metal–oxo complex with a pseudo-C_{3v} symmetry (Fig. 15.38c), but they all invariably feature low coordination numbers and nontetragonal symmetries.[43–45]

[w] As a matter of a fact, Harry Gray began his illustrious career by collaborating with one of the leading figures of ligand field theory, Carl Ballhausen, in Denmark, which led to a seminal paper published in the very first issue of *Inorganic Chemistry* titled "The electronic structure of the vanadyl ion". This paper helped establish ligand field theory as a clearly superior approach over crystal field theory, and basically contains all the elements needed to rationalize the existence of the oxo wall.

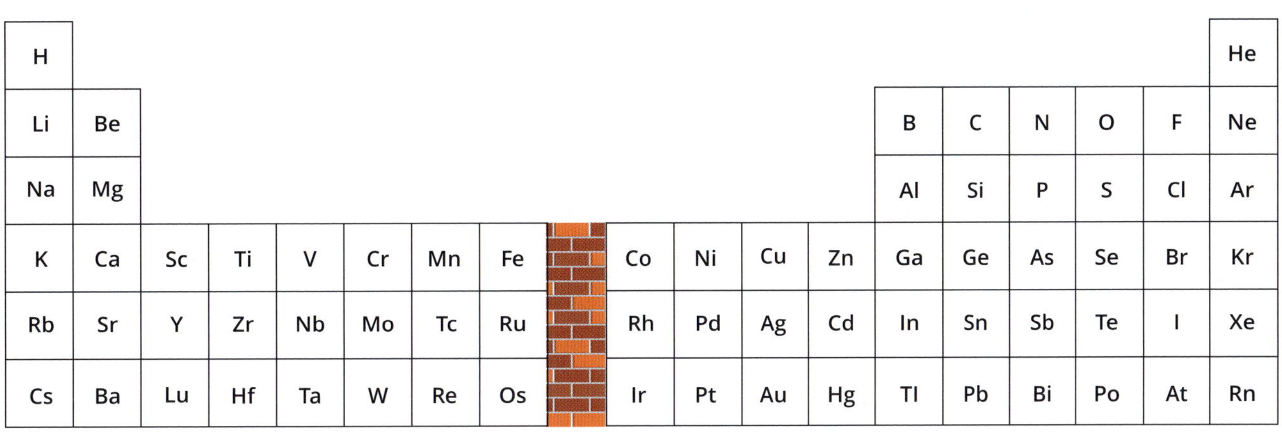

Fig. 15.38a The "oxo wall". Beyond Group 8 (iron family), no stable tetragonal metal-oxo species have been observed.

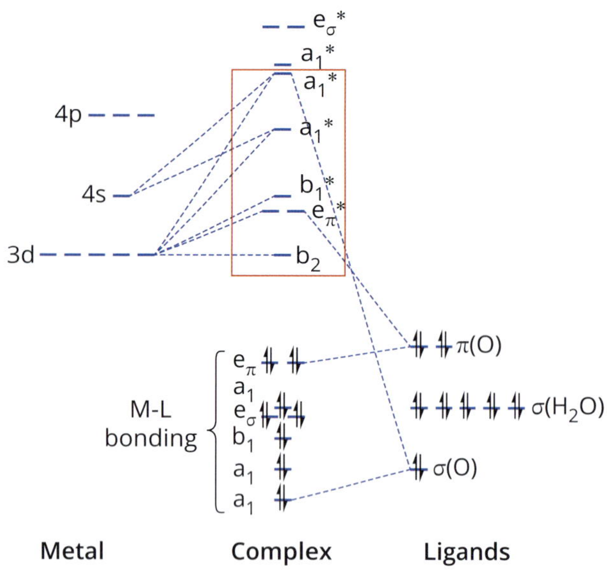

Fig. 15.38b Ligand-field MO diagram for a prototypical $MO(H_2O)_5^{n+}$ metal-oxo ion. The red rectangle encloses the MO with a predominant 3d character. The lowest metal-centered orbital is nonbonding (b_2, d_{xy}), but the next one up, e_g (d_{xz}, d_{yz}), is a π*(M–O) orbital.

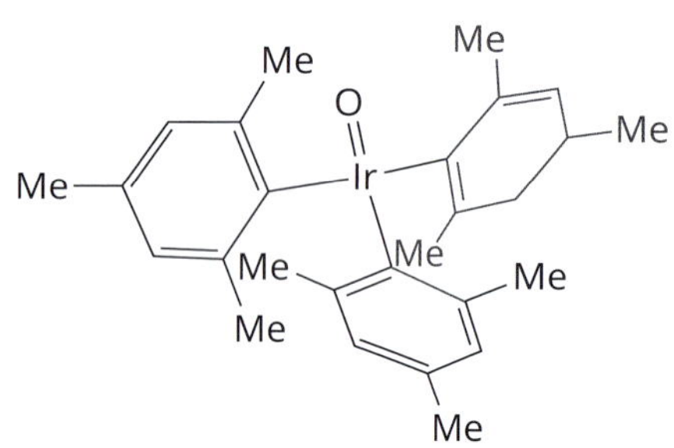

Fig. 15.38c Wilkinson's (trimesityl)oxoiridium(V), a d^4 metal–oxo complex with a pseudo-C_{3v} symmetry.

15.10 Characterizing Electronic Ground States using Magnetism

As we have seen, crystal field theory by Hans Bethe is remarkably successful at helping us to understand the electronic structures of transition-metal complexes, illustrating the various perturbations that can arise in response to changes in geometry, electronic configuration, ligand donor properties, and so on. Insights into electronic structure are also afforded by examining the electronic absorption spectra of metal-containing compounds—a topic that will be discussed in greater detail in Chapter 16.

One of the less heralded successes of crystal field theory was its ability to provide a simple framework to rationalize, if not completely explain, the magnetic properties of chemical systems. The notion that substances can exhibit magnetic behavior is as old as civilization itself. Indeed, in ancient Greece the material lodestone—a relatively rare form of the mineral magnetite (Fe_3O_4)—was known to possess certain properties that allowed it to attract iron. By the eleventh century AD, the Chinese had already developed the first compass by recognizing that a suspended fragment of "high-quality" lodestone oriented in preferred directions—a notion that was later codified in Western civilization in the sixteenth century by William Gilbert, who established the concept of "poles" and the tendency of these materials to be attracted or repelled, depending on their relative orientations.

Long before Bethe developed CFT, the idea that molecules could exhibit magnetic behavior under the right conditions was already established, though the underlying reason for this behavior was not yet fully understood. Take, for example, two simple molecules: $\{Fe(NH_3)_6\}^{2+}$ and $\{Fe(2,2'\text{-bipyridine})_3\}^{2+}$ (Fig. 15.39). These two compounds are Fe(II) and d^6, with six nitrogen atoms ligated to the metal center. Rationalizing why $\{Fe(NH_3)_6\}^{2+}$ is paramagnetic, possessing four unpaired electrons, while $\{Fe(bpy)_3\}^{2+}$ is diamagnetic, with no unpaired electrons, was perplexing to say the least. The formalism put forth by Bethe, and later expanded by van Vleck, provided a framework with which to explain this experimental observation: for Fe(II), NH_3 is a weak-field ligand, thereby resulting in a high-spin configuration, whereas 2,2′-bipyridine is a strong-field ligand, which for d^6 means that all of the electrons in the valence shell are paired. This simple example illustrates the intimate connection that exists between the magnetic properties of molecules and an understanding of their electronic structure, and it is through this lens that we now take a more detailed look at the electronic structure of transition-metal complexes by using their magnetic properties as our guide.

What follows are several sections in which the magnetic behavior of chemical systems is derived in some detail by using, for the most part, classical arguments. This is followed by a summary of basic magnetism that can be used as a stand-alone shorter treatment if desired.

15.10.1 Basic Physics of Magnetism (Advanced Topic)

The magnetic properties of all substances, be it lodestone or a transition-metal complex, ultimately derive from two fundamental physical facts: (1) electrons are (negatively) charged, and (2) electrons have angular momentum due to their movement. As we know from simple empirical observations and can describe quantitatively by using Maxwell's equations, a moving charge creates a magnetic field. Thus, the movement of electrons is the source of all magnetism. Prior to the development of quantum mechanics, magnetism in molecular systems was treated classically using Maxwell's equations within the context of classical electromagnetic theory. A quantum-mechanical analog of this treatment—quantum electrodynamics (QED)—

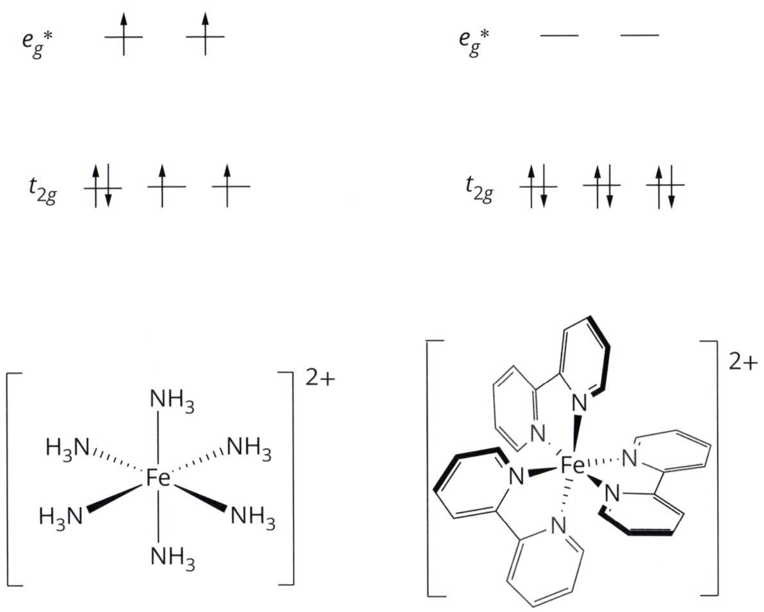

{Fe(NH$_3$)$_6$}$^{2+}$ {Fe(bpy)$_3$}$^{2+}$

Fig. 15.39 Structures of {Fe(NH$_3$)$_6$}$^{2+}$ and {Fe(bpy)$_3$}$^{2+}$ and their O$_h$ splitting diagrams.

has its origins with the British physicist Paul Dirac; important contributions from the likes of Heisenberg, Wigner, Pauli, and in particular Enrico Fermi, followed soon thereafter. The modern formulation of QED is largely attributed to Richard Feynman, Julian Schwinger, and Shin'ichiro Tomonaga, who together shared the 1965 Nobel Prize in Physics for their work. Although electron spin is an inherently quantum-mechanical phenomenon—the m_s quantum number fell out of the Schrodinger equation after Dirac incorporated Einstein's concepts of relativity into Schrodinger's original work—much of what we need to know about magnetism for understanding the electronic structure of transition-metal complexes can be gleaned from classical physics. As a result, this is the approach we will take here.

15.10.1.1 The Orbital Magnetic Dipole (Advanced Topic)

The Bohr model revolutionized how people conceptualized the atom (Section 2.1.2). Consider an electron of mass m, charge e, and velocity v, moving in a circular orbit of radius r around a nucleus, as shown in Fig. 15.40a. The movement of the charge can be thought of as a circular current whose magnitude is defined by the ratio of the magnitude of the charge (e) to that of its orbital period (T) via the mathematical relation shown in Eq. 15.12.

$$i = \frac{e}{T} = \frac{ev}{2\pi r} \qquad \textbf{15.12}$$

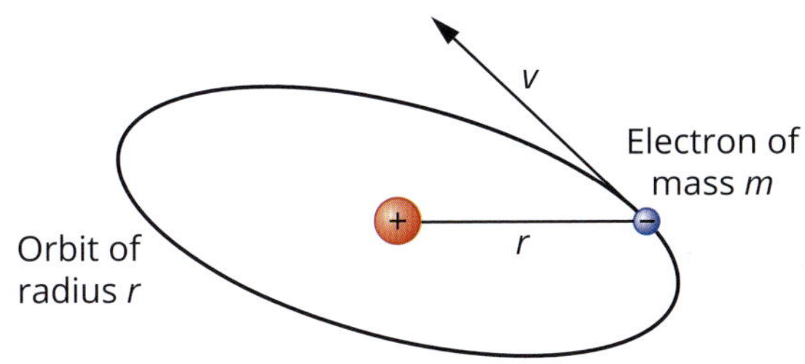

Fig. 15.40a Bohr model of the hydrogen atom.

As with any flowing current, a magnetic field is created as the electron precesses around the nucleus (Fig. 15.40a). In the case of a current loop in which the current is flowing counterclockwise, the direction of this field is determined by the so-called "right-hand rule". This can be visualized by curling four fingers of your right hand into a fist, and the direction of the field is given by your thumb, perpendicular to the plane defined by the loop. This physical situation is identical to what one would have if, instead of a current flowing around the loop, you placed a magnetic dipole at the center of the loop (Fig. 15.40b).

The magnitude of the magnetic dipole, μ_L, can be expressed as the product of the current (i) and the area of the loop (A), as in Eq. 15.13.

$$\mu_L = i \bullet A = \frac{ev}{2\pi r} \bullet \pi r^2 = \frac{evr}{2} \qquad \textbf{15.13}$$

Due to the negative charge of the electron, the direction of this dipole is actually opposite to the direction of the field, as

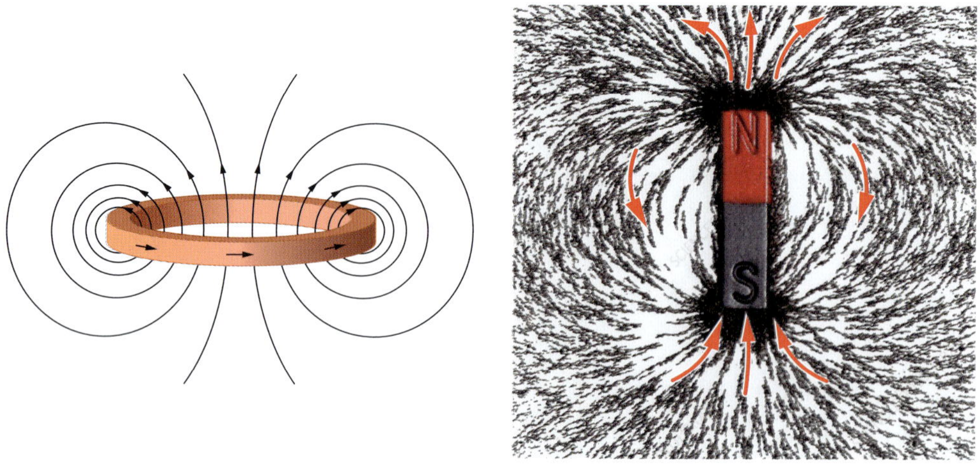

Figure 15.40b *(left)* The magnetic field lines created by a current that is flowing in the direction indicated. *(right)* The magnetic field lines from a simple bar magnet.

depicted in Fig. 15.40b. The angular momentum (***L***) is defined as the product of an object's linear momentum (*mv*) and the radius of its motion (*r*), $L = mvr$, giving rise to the following equation.

$$\frac{\mu_L}{L} = \frac{evr}{2mvr} = \frac{e}{2m} = \frac{2\pi g_L}{h}\beta \qquad \textbf{15.14a}$$

where β is the electron Bohr magneton,[x]

$$\beta = \frac{eh}{4\pi m} = 0.927 \times 10^{-23}\ A \bullet m^2 \qquad \textbf{15.14b}$$

and g_L is called the orbital *g*-factor, which in this case is equal to 1. What is interesting about Eq. 15.14a is that the ratio of the magnitude of the magnetic dipole to its angular momentum is not dependent upon the size of the loop (the "orbit") or the orbital period, but rather on a set of fundamental physical constants. This suggests that the magnetic properties associated with an orbiting electron in an atom may not be dependent upon the model used to describe it, but rather reflects a fundamental property of nature. Indeed, a result analogous to Eq. 15.14a is obtained from quantum electrodynamics when the problem is treated quantum-mechanically, as opposed to the classical approach used here.

[x] Magnetism, perhaps as much or more than any field in science, suffers from confusion due to units, constants, and symbols. Here we are using "electron Bohr magneton (β)" to represent the physical constant developed by Bohr, which has the value and units shown (Equation 15.14b, Box 15B). In some cases, this same quantity is designated as μ_B or B.M. Unfortunately, the symbol μ_B has been used as the "unit" for effective magnetic moment (μ_{eff}), but μ_{eff}, as it is usually defined, is unitless. Consquently, we decided to avoid the use of the somewhat confused symbol μ_B and used β to mean the Bohr magneton instead, a convention also in common usage. The IUPAC convention, however, is to use μ_B for the Bohr magneton. See Box 15B for additional information concerning units in magnetism.

How large is this magnetic dipole? This can easily be gauged by considering how much energy it would take to re-orient this dipole in the presence of an applied field, *H*—an exercise similar to the force one feels on a small magnet when brought under the influence of a larger magnet. From classical physics, we know that a dipole in an applied magnetic field will experience a torque, τ (Eq. 15.15a).

$$\boldsymbol{\tau} = \boldsymbol{\mu}_L \otimes \boldsymbol{H} \qquad \textbf{15.15a}$$

This torque gives rise to a potential energy associated with the orientation of the dipole relative to the applied field given by Eq. 15.15c, where θ is the angle between the two vectors that define the orientations of $\boldsymbol{\mu}_L$ and ***H***.

$$\Delta E = -\mu_L \cdot \boldsymbol{H} \qquad \textbf{15.15b}$$

$$= -\mu_L \cdot H \cdot cos \qquad \textbf{15.15c}$$

Assuming $\theta = 0$ and a magnetic dipole of 1 β (Bohr Magneton) in a field of 1 T (10,000 gauss), the energy difference between parallel ($+\mu_L$) and antiparallel ($-\mu_L$) alignment of the magnetic dipole is about 1 wavenumber (cm^{-1}):

$$2\mu_L H = 1.16 \times 10^{-4}\ \text{eV} = 0.93\ \text{cm}^{-1}$$

Given that thermal energy at room temperature is ~200 cm^{-1}, this is a relatively small amount of energy. Yet, as we will see, this small energetic perturbation is enough to give rise to many of the magnetic properties of molecular systems.

15.10.1.2 Electron Spin: The Stern–Gerlach Experiment (Advanced Topic)

The preceding discussion never mentioned the concept of electron spin; all of the physics we just summarized described the electron as a simple, negatively charged particle moving in an orbit. Indeed, Schrodinger's original formulation of quantum mechanics yielded only three quantum numbers: a principle quantum number (n), and two angular quantum numbers (l and its projection in space, m_l)—not the four that we are accustomed to thinking about. It would take Dirac's reformulation of quantum mechanics and the incorporation of Einstein's theory of relativity in 1929 for the fourth quantum number—m_s, the spin of the electron—to fall out of the math. Prior to this, many researchers were interested in understanding the origins of the physical phenomena that were being documented by many research groups around the world. One of the most consequential experiments was conceived in 1921 by Otto Stern, who at the time was an assistant to the famous physicist Max Born at the Institute for Theoretical Physics at the University of Frankfurt.[y] The experiment was carried out by Stern and Walther Gerlach, who was in the Institute for Experimental Physics at the same university in 1922.

Stern and Gerlach designed an experiment to measure the magnetic dipole of an atom that again relied on the influence of magnetic fields on magnetic dipoles. Eq. 15.15 expresses the torque that a magnetic dipole will feel in the presence of an applied field. The hypothetical situation described previously, in which the dipole aligns perfectly parallel to the direction of the field ($\theta = 0$), is never fully realized, and in practice the dipole will orient itself close to the directionality of the field lines, but not precisely parallel ($\theta \neq 0$). If the applied magnetic field is uniform—in other words, if the density of field lines is constant for any cross-section of the field—the substance possessing the dipole will precess much like a top at a frequency proportional to the magnitude of the field (**B**) and that of the dipole.

$$\omega = \frac{2\pi g_L \mu_B}{h} \bullet \boldsymbol{H} \qquad 15.16$$

This precession is called the *Larmor precession*, with ω being the *Larmor frequency*. It is a consequence of the fact that a uniform field does not produce any net translational force on the object and is thereby the mechanism by which the system as a whole maintains the principle of conservation of energy—the same reason why an ice skater spins faster and faster as he/she pulls their arms closer into their bodies. If, however, the applied field is nonuniform the dipole will be subject to a translational force ($\boldsymbol{F}_z$) proportional to the gradient of the field in the direction of the translation ($\frac{\partial \boldsymbol{H}_z}{\partial z}$) and the magnitude of the dipole projected along that same direction (μ_{L_z}) as shown in Eq. 15.17.

$$\boldsymbol{F}_z = \frac{\partial \boldsymbol{H}_z}{\partial z} \bullet \mu_{L_z} \qquad 15.17$$

Stern and Gerlach's idea was to use this basic concept from physics as a means of experimentally determining the magnitude of μ_L. The experimental design is shown in Fig. 15.41. They proposed to pass a beam of vaporized silver atoms through the nonuniform magnetic field. The gradient of the field was designed to be transverse to the axis along which

[y] In 1954, Max Born, together with Walther Bothe, was awarded the Nobel Prize in Physics for his contributions to quantum mechanics, such as showing that the Schrödinger equation provided statistical predictions rather than exact solutions. He is also famously the grandfather of pop singer Olivia Newton-John.

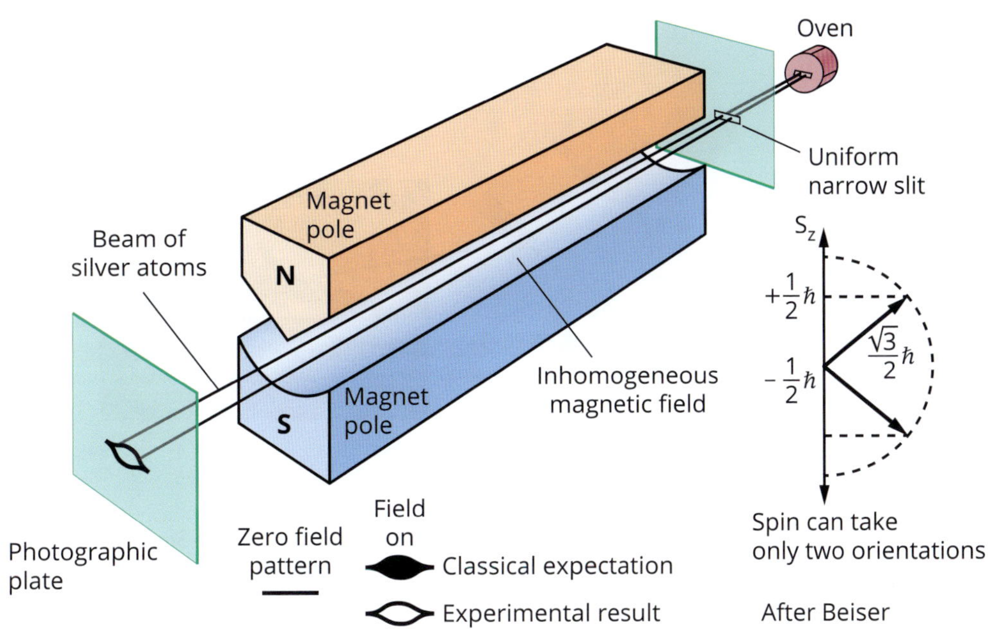

Fig. 15.41 Schematic of the Stern–Gerlach experiment.

the Ag atoms were propagating. After passing through the magnet, the Ag atoms would hit a target, where they would solidify, and, since it was silver, they would be easily observable to the naked eye. Since Stern and Gerlach knew the magnitude of the gradient of the magnetic field ($\partial \boldsymbol{H}_z/\partial z$), the magnitude of μ_L could be determined simply by measuring how far from the propagation axis the atoms had been deflected.

At the time that Stern and Gerlach carried out this experiment, the notion of quantization of angular momentum as we understand it today from quantum mechanics did not exist. Based on classical physics, the magnitude of μ_L could therefore take on a continuum of values anywhere between $+\mu_L$ and $-\mu_L$, which led to the expectation of a circular impact image of silver with a radius proportional to μ_L. Instead, Stern and Gerlach observed the image shown in the right panel of Fig. 15.42.[46]

Instead of a continuous distribution of spots, the experiment produced only two regions of deposited Ag atoms. If that was not surprising enough, the deposits were equidistant apart, above and below, from the center axis of the experimental setup, suggesting that the magnetic dipole that was being measured possessed only two values that were equal in magnitude but opposite in sign. Stern and Gerlach reconfigured the experiment multiple times—redefining the relative orientation of the magnetic field and the beam of atoms—but observed the same result every time, indicating that the observation was independent of the orientation of the magnet. Analogous experiments using other easily vaporizable metals yielded the same observation: two values for the magnetic dipole, equal in magnitude and opposite in sign.

In 1927, Phipps and Taylor repeated the Stern–Gerlach experiment by using a stream of hydrogen atoms, and obtained the same result. This was an extremely important experiment because it narrowed down the possible origins of the magnetic dipole to either the proton or the electron. Theoretical work that was inspired in part by the Stern–Gerlach experiment had already yielded an expression for the expected magnitude of the deflection (Eq. 15.18),

$$\Delta d \propto \pm \frac{eh}{4\pi m} \qquad \textbf{15.18}$$

where Δd is the deflection distance from the center axis, e is the magnitude of the charge on the particle, h is Planck's constant, and m is the mass of the particle. By this time, the mass of both the proton and electron had been experimentally determined; the factor of ~2,000 difference between the two made it easy to determine that the origin of magnetic dipole being measured was that of the electron.

Around the same time, several other developments in the nascent field of quantum mechanics occurred that, when combined, brought the entire picture into focus. Of particular note is the contribution made by two graduate students, Goudsmit and Uhlenbeck, who in 1925 were trying to understand why certain lines in the optical spectrum of hydrogen (as well as the alkali metals) were actually two lines that were closely spaced together. Sommerfeld had previously observed this and attributed it to relativistic effects associated with the mass of the electron, but his model failed when applied to anything other than hydrogen. It was Goudsmit and Uhlenbeck who postulated that the electron itself has angular momentum and therefore a magnetic dipole, so the splitting of the lines in the hydrogen spectrum could be attributed to the energetics associated with the intrinsic dipole of the electron coupling to the dipole created by the electron moving in its orbit. We now refer to this phenomenon as spin–orbit coupling, which is qualitatively similar to the forces felt when bringing two magnets into close proximity. The Stern–Gerlach experiment, coupled with

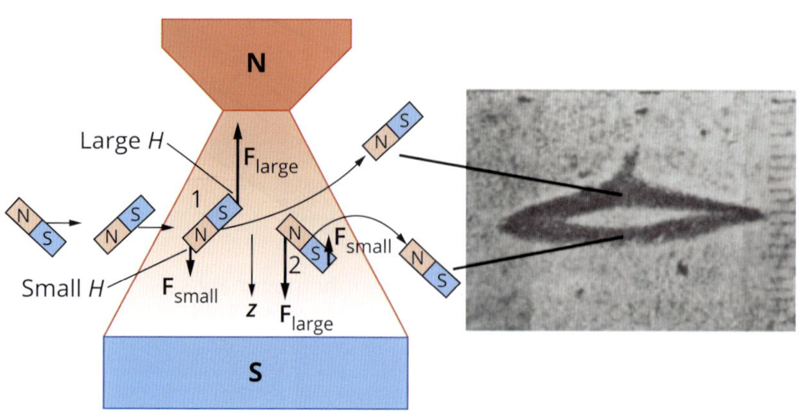

Fig. 15.42 (*left*) A schematic representation of the physical principles giving rise to the results of the Stern–Gerlach experiment. (*right*) A photograph of the actual target that Stern and Gerlach obtained after passing vaporized silver atoms through their inhomogeneous magnetic field.[46]

the more refined measurements of Phipps and Taylor and the postulate of Goudsmit and Uhlenbeck, thus lead us to the now well understood conclusion that the electron itself has spin and is quantized to two values that are equal in magnitude but opposite in sign.

It is interesting to note that in the years between the development of Schrodinger's wave mechanics and Heisenberg's (mathematically equivalent) matrix mechanics, Wolfgang Pauli introduced a fourth quantum number having values of ± 1/2, but he never connected this quantum number with a physical characteristic. It was not until Dirac introduced relativity into quantum mechanics that the m_s quantum number naturally arose from the math—an indication that *electron spin is a wholly relativistic phenomenon.* Of course, all we really needed to do was look back to the musings of the famous British physicist A. H. Compton. In 1921, as the last sentence of a research paper dealing with the scattering of X-rays by atoms, Compton wrote: "May I then conclude that the electron itself, spinning like a tiny gyroscope is probably the ultimate magnetic particle." By all accounts, he was simply speculating on this point and never pursued the idea, but he was correct.

15.10.2 Types of Magnetic Behavior (Advanced Topic)

All materials, regardless of their composition, will respond in some way to the application of a magnetic field. The exact nature of this response, which arises due to specific interactions between the electrons comprising the material and the applied field, serve to phenomenologically differentiate the various types of magnetic behavior that can be encountered in substances ranging from molecules to bulk materials

With the exception of a certain class of superconducting materials (called "Type I" superconductors), a magnetic field will penetrate into any material. The field within the material—its magnitude and orientation to be specific—will generally be different from the field outside the material. This situation is depicted schematically in Fig. 15.43, where $\boldsymbol{H}$ is the field within the substance, $\boldsymbol{H_0}$ is the applied field, and $\Delta \boldsymbol{H}$ is the response of the substance to the applied field. More formally, this response is called *magnetic induction* (B) and is defined as in Eq. 15.19, where $\boldsymbol{M}$ is the magnetization (the degree to which the field "magnetizes" the material).

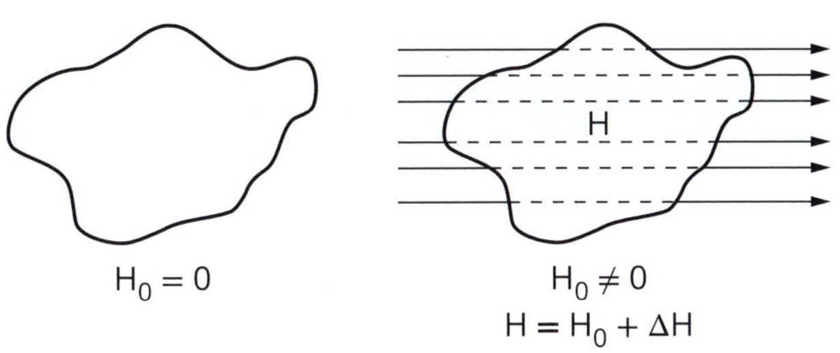

Fig. 15.43 (*left*) A generic material in the absence of an applied field ($\mathbf{H_0}$ = 0). (*right*) The same substance in an applied field ($\mathbf{H_0} \neq 0$). The field within the sample, represented by the dashed lines, is **H**, which will be larger or smaller than the applied field, depending on the nature of the material, which will determine the sign and magnitude of **ΔH**.

$$\boldsymbol{B} = \boldsymbol{H_0} + \boldsymbol{M} \qquad 15.19$$

Empirically, it is found that for most (but not all) substances, the magnitude of the magnetization is directly proportional to the magnitude of the applied field (Eq. 15.20).

$$\boldsymbol{M} \propto \boldsymbol{H_0} \qquad 15.20$$

As with any mathematical proportionality, the relation in this equation can be expressed as an equality by including a proportionality constant, χ, as in Eq. 15.21.

$$\boldsymbol{M} = \chi \boldsymbol{H_0} \qquad 15.21$$

This proportionality constant, χ, is referred to as the magnetic susceptibility. It can be thought of as the nature and degree to which a given material responds—is "susceptible"—to the application of a magnetic field. The magnitude and sign of χ is specific to a given material, and in this sense, all the chemical information about a material that is relevant for defining its response to an applied magnetic field (such as high spin versus low spin) is contained within this parameter. Accordingly, this is where we need to focus our attention if we are to understand how magnetism folds into our broader discussion of the electronic structure of inorganic compounds.

The magnetization response of most substances falls into two regimes. If the magnitude of the field within the sample is smaller than the magnitude of the applied field—if $|\boldsymbol{B}|<|\boldsymbol{H_0}|$—then Eq. 15.19 requires that $|\boldsymbol{M}| < 0$. Phenomenologically, given the fact that the applied field, magnetic induction, and magnetization are all vector quantities—they have direction as well as magnitude—this condition is telling us that the application of the magnetic field is somehow inducing a response within the material such that the magnetic dipoles making up the material are being oriented to oppose the applied field. Conversely, if the field within the material is larger than the applied field ($|\boldsymbol{B}|>|\boldsymbol{H_0}|$), then $|\boldsymbol{M}| > 0$ and the application of the field is creating a response such that the magnetic dipoles within the material are being aligned with the field. These two conditions can be distilled further by using Eq. 15.19 as a guide, as shown in the following equations.

$$|\boldsymbol{B}| < |\boldsymbol{H_0}| \Rightarrow |\boldsymbol{M}| < 0 \Rightarrow \chi < 0 \qquad 15.22a$$

$$|\boldsymbol{B}| > |\boldsymbol{H_0}| \Rightarrow |\boldsymbol{M}| > 0 \Rightarrow \chi > 0 \qquad 15.22b$$

Eq. 15.22a, with a value of the magnetic susceptibility (χ) that is negative, serves to define what is referred to as *diamagnetism*, whereas a positive value for χ (Eq. 15.22b) is a characteristic of *paramagnetism*. As it turns out, the alignment of the dipoles by the magnetic field—whether that alignment is parallel (paramagnetic) or antiparallel (diamagnetic) to the direction of the applied field—creates a force within the material that can be measured, such as the Stern–Gerlach experiment. This force provides us with the experimental handle that we need to actually quantify the magnetic susceptibility and link it to the ideas developed earlier in this chapter with regard to electronic structure. Some general features of χ for diamagnetism and paramagnetism, along with a summary of common parameters used in describing magnetism (and the units employed to quantify these parameters, which turns out to be a particularly problematic aspect of the field of magnetism), are provided in Box 15A and 15B, respectively.

15.10.3 Diamagnetism (Advanced Topic)

As mentioned previously, the results of the Stern–Gerlach experiment suggested that there were two possible values of the magnetic dipoles associated with the atoms of Ag that they passed through their magnetic field; Phipps and Taylor used

BOX 15A

Types of magnetic behavior

Diamagnetism ($\chi < 0$)	Paramagnetism ($\chi > 0$)
very weak (~10^{-6} emu cm^{-3})	stronger than diamagnetism (10^{-4}–10^{-1} emu cm^{-3})
field-induced	intrinsic but detection requires an applied field
field-proportional (linear)	field-proportional (can be linear or nonlinear)
additive	additive (for noninteracting components)
temperature-independent	temperature-dependent

Although not discussed in this chapter, the presence of multiple paramagnetic centers in a given molecule or material can give rise to interactions among these paramagnetic centers. This can lead to patterns of magnetic behavior quite distinct from that observed for so-called "simple" paramagnets:

Ferromagnetism ($\chi > 0$)
arises due to the parallel alignment of the magnetic dipoles associated with two (or more) distinct paramagnetic centers
can be comparable to or much stronger than paramagnetism in molecular-based and bulk materials, respectively
can be intrinsic or field-induced; detection can require the application of a field, especially for molecular systems, but in the case of certain materials, the magnetic response does not require a field, so-called "permanent magnets"
field dependence is usually nonlinear
temperature-dependent

Antiferromagnetism ($\chi > 0$, but can have a very complex mathematical form)
arises due to the antiparallel alignment of the magnetic dipoles associated with two (or more) distinct paramagnetic centers
can be comparable in strength to paramagnetism, but can also be so strong as to produce what appears to be a diamagnetic response
intrinsic (the coupling does not require the application of a field), but detection is field-induced
field dependence can be complex
temperature dependence can be complex

Ferrimagnetism ($\chi > 0$)
essentially the same as antiferromagnetism; that is, antiparallel alignment of the magnetic dipoles on distinct centers, but the dipoles differ in magnitude, leading to net residual paramagnetism

BOX 15B

Common units in magnetism

Property	CGS Units	S.I. Units	Conversion (CGS to SI)
magnetic induction (B)	gauss (G)	tesla (T)	$\times 10^{-4}$
applied field (H)	oersteds (Oe)	$A\ m^{-1}$	$\times 10^3/4\pi$
magnetization (M)	$emu\ cm^{-3}$	$A\ m^{-1}$	$\times 10^3$
magnetization ($4\pi M$)	gauss (G)	—	—
magnetic polarization	—	tesla (T)	—
specific magnetization (σ)	$emu\ g^{-1}$	$J\ T^{-1}\ kg^{-1}$	$\times 1$
permeability (μ)	dimensionless	$H\ m^{-1}$	$\times 4\pi/10^{-7}$
relative permeability	—	dimensionless	—
susceptibility (χ)	$Emu\ cm^{-3}\ Oe^{-1}$	dimensionless	$\times 4\pi$
electron magnetic moment or Bohr magneton (β or μ_B)	0.927×10^{-20} erg/G	$9.27 \times 10^{-24}\ A{\cdot}m^2$	$\times 10^{-3}$

This is by no means a complete list, but it does contain some of the more common properties that are of interest in describing magnetic systems. The physics community generally uses the SI formalism, and while the chemistry community is moving toward this system, a significant fraction of the literature still reports data in cgs units.

a stream of hydrogen atoms to refine this notion and confirm that the magnetic dipole in question was that of the electron. The fact that the atoms in both of these experiments were deflected the exact same distance but opposite in direction indicates that the two possible values of the magnetic dipole of the electron are equal in magnitude but opposite in sign; that is, $m_s = \pm\ ^1/_2$. *Another way to view this is that the magnetic dipoles of two electrons that are paired cancel each other out exactly; this is the physical condition that gives rise to diamagnetism.* Since the ground electronic states of all substances having more than one electron will possess paired electrons, diamagnetism is a property of virtually all materials (except the hydrogen atom, of course . . . and He^+. . . and U^{91+} . . . you get the idea!).

The origin of diamagnetism—which also allows us to understand the phenomenological effect of it giving rise to a reduction in the magnitude of the magnetic field within a given material—is a reflection of a principle known as Lenz's Law, which states:

> The application of a magnetic field on a wire loop results in the creation of a current flowing in a direction such that the resulting magnetic dipole opposes the applied field.

In reality, Lenz's Law is a manifestation of how electromagnetic circuits obey Newton's Third Law:

> If an object A exerts a force F_A on object B, then B exerts simultaneously a force F_B on A such that $F_A = -F_B$.

This arises due to the concept of conservation of energy. For chemists, the most familiar consequence of diamagnetism is not coupled to the field of magnetism *per se*, but rather NMR spectroscopy. It is well known that protons associated with aromatic groups are typically observed downfield from those attached to aliphatic groups. For example, the single proton resonance for benzene occurs at circa 7.3 ppm (relative to the usual standard, the hydrogens of tetramethylsilane at 0 ppm), whereas the corresponding resonances for cyclohexane are found at ~1.5 ppm under the same conditions. Described as being "deshielded", the larger chemical shift for benzene reflects the fact that the local magnetic field surrounding its protons is slightly larger than the corresponding environment for cyclohexane. The physical origin of this phenomenon is basically Lenz's Law. If one envisions the electrons of the π system of benzene as a loop, the applied field from the magnet of the NMR spectrometer can be thought of as inducing a current within that loop. Lenz's Law says that the current created by the applied field will have a magnetic moment that opposes the applied field. Since the physics of magnetic field lines dictates that they exist as closed loops, this induced current will result in a magnetic field that is antiparallel to the applied field at the center of the benzene ring but becomes parallel in the vicinity of the proton (Fig. 15.44). This additional field that the proton feels is the reason for the more downfield (higher ppm) position of the aromatic resonances and is a physical manifestation of diamagnetism. If one places protons near the center of an aromatic system, one sees

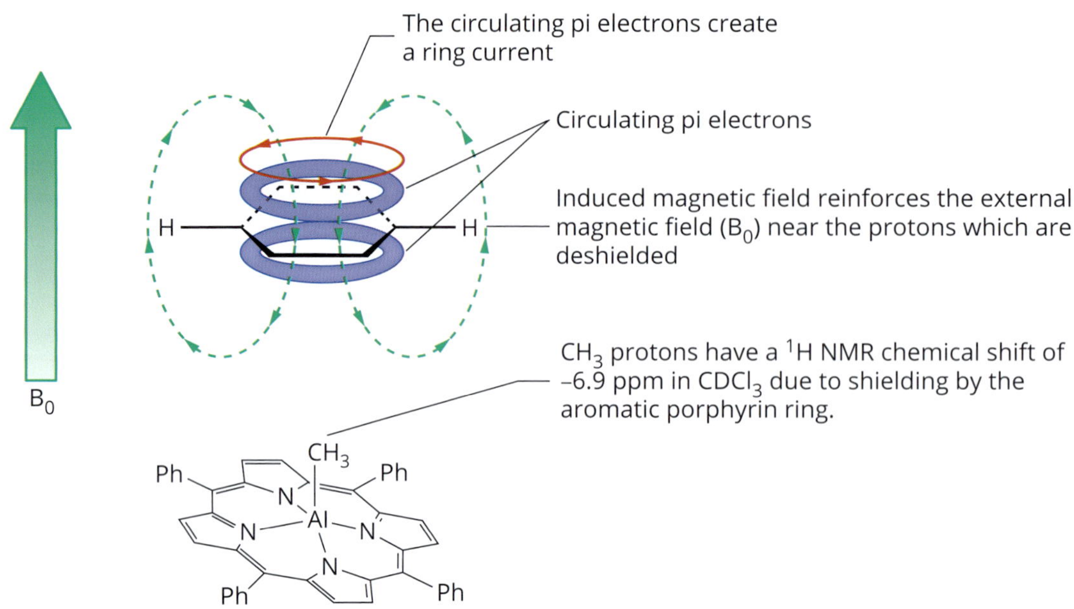

Fig. 15.44 (*top*) The induced magnetic field created by the inductive effect as the result of the application of an external field. The resulting field lines—a consequence of Lenz's Law—create an added field at the proton, which is why its resonance occurs further downfield than what is found for a compound such as cyclohexane. (*bottom*) The same effect is seen in MeAl(TPP), where TPP is tetraphenylporphyrin, except that the methyl group in the center of the large aromatic ring is strongly shielded, occurring at −6.9 ppm. This is due to the ring current induced in the aromatic porphyrin ring.

the protons "shielded*" (move to smaller ppm); this can be seen in a molecule such as MeAl(TPP), where TPP is tetraphenylporphyrin. The TPP ligand is also aromatic, but the methyl group sits in the center of the ring and is strongly shielded with a resonance at −6.9 ppm.[47]

The current loop model, although very much a classical concept, is nevertheless useful enough to allow us to calculate how much of a diamagnetic response we can expect for atoms, functional groups, and so on, that comprise a molecule. For example, one can readily derive an expression for the magnetic susceptibility of a given atom, χ_{atom} (Eq. 15.23), where r_i is the effective radius of the orbit of the electron—the "size" of the current loop.

$$\chi_{atom} = -\frac{Ne^2}{6mc^2}\sum_{i=1}^{n} r_i^2 = -2.83 \times 10^{10}\sum_{i=1}^{n} r_i^2 \qquad \textbf{15.23}$$

The complete diamagnetic response, χ_{dia}, will reflect the additive influences of all of the paired electrons of the system on an atom-by-atom basis, as well as any effects due to field-induced electron mobility associated with various functional groups such as double and triple bonds, aromatic residues, and so on.

$$\chi_{dia} = \sum_i \chi_{atom} + \sum_j \chi_{bond} \qquad \textbf{15.24}$$

The magnetic susceptibility one measures experimentally, χ_{obs}, is going to reflect the combined influence of any unpaired electrons in the system (χ_{para}), as well as these diamagnetic contributions.

$$\chi_{obs} = \chi_{para} + \chi_{dia} \qquad \textbf{15.25}$$

Generally speaking, the information we are looking for from a magnetic measurement—which is what informs us about the electronic structure of the compound in question—is contained in χ_{para}, which means that we must be able to account for the diamagnetic contributions so that the value of χ_{para} can be solved from Eq. 15.25. The complicating factor in all of this is that although the absolute magnitude of χ_{dia} is a factor of 100 or 1,000 smaller than χ_{para}, the effects are additive, and there are usually many more paired electrons in a given chemical system than unpaired electrons. For example, suppose we wanted to measure the paramagnetism of a small iron-containing metalloprotein like ferric myoglobin (such as Fe^{III}-Mb). The iron center in this compound is low-spin d^5, so this system as a whole possesses one unpaired electron that

* There are additional contributions to the observed magnetic susceptibility such as temperature-independent paramagnetism (TIP), as well as other, higher-order effects that can influence the magnetic properties of a given material. These are generally much smaller in magnitude than the two terms listed in Eq 15.24, so for the sake of simplicity we are not including these contributions in this discussion.

will give rise to χ_{para}. However, myoglobin has a molecular weight of nearly 17,000 Daltons. Even if the absolute magnitude of the paramagnetic susceptibility of the low-spin Fe(III) center were 1,000-fold larger than any one of the lighter atoms in the protein backbone, the additive nature of the diamagnetic response will completely overwhelm the paramagnetic signal and give rise to a negative value of χ_{obs}; that is, the system will look experimentally as though it is diamagnetic.

The example of a metalloprotein containing a single unpaired electron is extreme, but it nevertheless highlights the fact that the diamagnetic contributions from all of the atoms in a given molecule must be accounted for in order to obtain reliable experimental data on the paramagnetic response in which we are usually interested. The most common method for doing this is through the use of what are called Pascal's constants. These are tabulated values of the diamagnetic contributions for atoms, unsaturations, functional groups, and so on, using expressions similar to Eq. 15.24. An example of how to carry out one of these calculations is shown in Box 15C, and a rather extensive list of Pascal's constants for all elements of the periodic table as well as a large number of functional groups and commonly used ligands for inorganic compounds can be found in Appendix N.

Pascal's constants provide a reasonable estimate for χ_{dia} in the case of simple coordination compounds, but for very large systems, such as the myoglobin example discussed previously, the approximations inherent in expressions like Eq. 15.24 lead to sufficiently large errors—errors that tend to be additive—such that more precise approaches become necessary. If a diamagnetic analog of a paramagnetic system of interest is available or can be prepared, such as a Zn(II) version of Fe^{III}-Mb or perhaps the apo (metal-free) form, this usually represents the best alternative in that one can experimentally determine χ_{dia} for the entire system, then use Pascal's constants only to account for the paired electrons of the metals. Solution-phase magnetic measurements present a different problem because of the high intrinsic concentration associated with a neat solvent. These more specialized circumstances notwithstanding, for the majority of simple coordination compounds, Pascal's constants provide a sufficiently accurate accounting of the inherent diamagnetism to allow one to focus on χ_{para}—a quantity to which we now turn our attention.

BOX 15C

Diamagnetic corrections

Since diamagnetism is a property of paired electrons—and virtually all molecules, materials, and substances possess paired electrons—there will almost always be a diamagnetic contribution to the observed magnetic susceptibility. For most molecules, the use of Pascal's constants is usually sufficient to account for χ_{dia}, which then allows for the paramagnetic portion of the experimentally measured susceptibility to be extracted.

Diamagnetic contributions are strictly additive, which makes the calculation very simple to perform. Let's take the simple example of potassium ferricyanide, $K_3\{Fe(CN)_6\}$. Cyanide is a strong-field ligand, so we expect the Fe(III) ion to be in its low-spin state (so, S = ½, one unpaired electron for low-spin d^5). If we refer to the table of Pascal's constants in Appendix N, we find the following:

Ion	χ (emu cm^{-3})
K^+	-14.9×10^{-6}
CN^-	-13.0×10^{-6}
Fe(III)	-10.0×10^{-6}

Note that the constant for Fe(III) is only accounting for the paired electrons of the core, not the valence electrons from which the paramagnetic response originates.

Because the diamagnetic contributions are additive, we need only look at the stoichiometry of the compound and carry out the calculation accordingly. So:

$$\begin{aligned}\chi_{dia}(K_3\{Fe(CN)_6\}) &= 3\chi_{dia}(K^+) + 6\chi_{dia}(CN^-) + \chi_{dia}(Fe^{3+})\\ &= 3(-14.9\times10^{-6}) + 6(-13.0\times10^{-6})\\ &\quad + (-10.0\times10^{-6})\\ &= -132.7\times10^{-6} = -1.327\times10^{-4}\ \text{emu cm}^{-3}\end{aligned}$$

Notice that although the magnitude of the diamagnetic contribution from each individual component is on the order of 10^{-5}—an order of magnitude smaller than the low end of the range expected for the paramagnetic response—the additive nature of χ_{dia}, coupled with the fact that there are so many paired electrons in a typical molecule, results in an overall contribution that is a significant correction to the signal produced by the unpaired electron in this system. This is why accounting for diamagnetic contributions is such an important (if not particularly interesting) aspect of studying the magnetic properties of molecules.

15.10.4 Simple Paramagnets: The Curie Law (Advanced Topic)

Correcting the experimentally observed magnetic susceptibility for the diamagnetic contributions to χ_{obs} allows us to extract the underlying paramagnetic response, χ_{para}, via rearrangement of Eq. 15.25. The parameter χ_{para} contains all the information we need to determine how many unpaired electrons that a molecule possesses, which in turn can be combined with other information such as metal oxidation state, optical properties, and so on, for characterizing the electronic structure of a compound. In order to do this, however, we need to consider what a given value of χ_{para} means with regard to the number of unpaired electrons giving rise to its value. For that, we need to develop a model that links the physical response of an unpaired electron in an applied magnetic field to an actual value of χ_{para}. We will now derive such a model based on some very simple physical principles. The specific derivation will be carried out for a system comprised of a single unpaired electron, but the result is easily generalized (subject to certain restrictions that will become apparent) to a system with an arbitrary number of unpaired electrons.

As discussed previously, paramagnetism corresponds to a circumstance in which the magnetic field within a given material is larger than the magnitude of the applied field. For this to occur, the vectoral nature of a magnetic field means that the magnetic dipoles within the material must be positioned such that there is a net alignment of these dipoles parallel to the applied field. We can express this condition mathematically as in Eq. 15.26,

$$M = N\sum_{N}\mu_n P_n \qquad 15.26$$

where M is the magnetization (Eq. 15.21), N is Avogadro's number (included to put the result on a molar basis), μ_n is the magnetic moment (the magnitude of the magnetic dipole) of the nth state, and P_n is the population of that state. The index n runs over all the magnetic states available to the system, which in this case is 2, corresponding to the $m_s = -^1/_2$ and $m_s = +^1/_2$ states of a single electron. Eq. 15.26 states that the magnetization is given by a simple sum: for all the states which the system possesses (n), how big is the magnetic dipole of each of those states (μ_n), and how many particles are in that state (P_n).

As a simple illustration of this, suppose we had 100 particles with dipole magnitudes of either +1 or −1. If all 100 of these were in the +1 state, the magnetization would be 100. If, however, 75 of those particles were in the +1 state and the remining 25 were in the −1 state, then the *net* magnetization would be 50. Now we simply need to define the parameters in Eq. 15.26 as it applies to the magnetic dipole of an electron in a magnetic field.

The magnetic moment is defined in a manner reflective of the condition specified by Eq. 15.15; that is, how much does the energy of a particular orientation of the dipole change with respect to a given change in the magnitude of the applied field. This can be expressed mathematically as follows,

$$\mu_n = -\frac{\partial E_n}{\partial H} \qquad 15.27a$$

where H is the applied field; a negative sign is introduced because the negative charge on the electron causes the orientation of the magnetic dipole to be opposite that of its angular momentum vector (Figure 15.45). The expression for the energy associated with a given orientation of a magnetic dipole interacting with an applied field is based directly on Eq. 15.15 and is shown in Eq. 15.27b,

$$E = +m_s g\beta H \qquad 15.27b$$

where g is electron g-factor ($\approx$ 2) and β is the electron Bohr magneton (Eq. 15.14).

Combining Eqs. 15.27a and 15.27b, we have Eq. 15.27c.

$$\mu_n = -\frac{\partial E_n}{\partial H} = -\frac{\partial(+m_s g\beta H)}{\partial H} = -m_s g\beta \qquad 15.27c$$

For a single electron, this situation gives rise to the creation of two energetically distinct states—one corresponding to $m_s = -^1/_2$, the other to $m_s = +^1/_2$—whose energy will be directly proportional to the magnitude of the applied field. This is depicted schematically in Fig. 15.45.

The energy difference between these two states ($\Delta E = 2m_s g\beta H = g\beta H$ for $m_s = \pm^1/_2$) is what ultimately serves to determine the population of that particular state (P_n); since the magnitude of the splitting with which we are concerned is on the order of thermal energy, we can describe the population distribution across these states in terms of simple Boltzmann statistics. This allows us to specify Eq. 15.26 in more detail, as follows,

$$M = N\sum_{n}\mu_n P_n = N\left[\frac{\sum_{m_s=-1/2}^{m_s=+1/2}(-m_s g\beta)\exp\left(-\frac{m_s g\beta H}{k_b T}\right)}{\sum_{m_s=-1/2}^{m_s=+1/2}\exp\left(-\frac{m_s g\beta H}{k_b T}\right)}\right] \qquad 15.28a$$

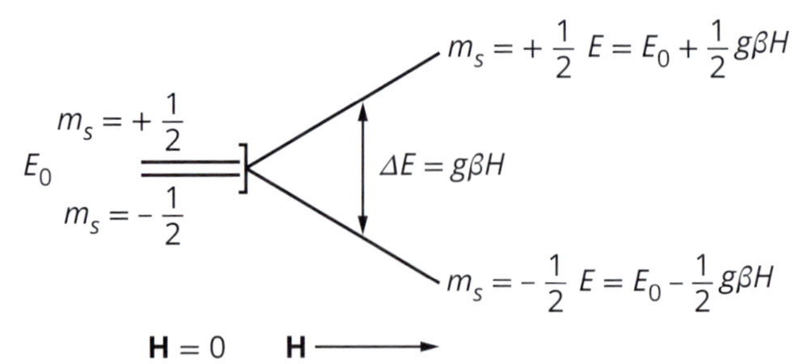

Fig. 15.45 Energy-level diagram for the splitting of the $m_s = \pm^1/_2$ states of a single electron. The two states are degenerate in the absence of a field (**H** = 0), but that degeneracy is lifted in an applied field. Because of the negative charge of the electron, the magnetic dipole of the $m_s = -^1/_2$ state is oriented with the field and is therefore stabilized.

$$= N\left[\frac{\left\{\frac{1}{2}g\beta\, exp\left(\frac{g\beta H}{2k_bT}\right)\right\} + \left\{-\frac{1}{2}g\beta\, exp\left(-\frac{g\beta H}{2k_bT}\right)\right\}}{exp\left(\frac{g\beta H}{2k_bT}\right) + exp\left(-\frac{g\beta H}{2k_bT}\right)}\right] \quad \textbf{15.28b}$$

$$= \frac{Ng\beta}{2}\left[\frac{exp\left(\frac{g\beta H}{2k_bT}\right) - exp\left(-\frac{g\beta H}{2k_bT}\right)}{exp\left(\frac{g\beta H}{2k_bT}\right) + exp\left(-\frac{g\beta H}{2k_bT}\right)}\right] \quad \textbf{15.28c}$$

where k_b is Boltzmann's constant. This expression can be written in an equivalent form by recalling the mathematical equality:

$$\tanh(x) = \frac{\exp(x) - \exp(-x)}{\exp(x) + \exp(-x)}$$

with

$$x = \frac{g\beta H}{2k_bT}$$

which gives

$$M = N\sum_n \mu_n P_n = \frac{Ng\beta}{2}\tanh\left(\frac{g\beta H}{2k_bT}\right) \quad \textbf{15.28d}$$

Admittedly, converting a sum of exponentials into an expression containing a hyperbolic tangent function may look as if we have complicated (as opposed to simplified) the situation. However, we can take advantage of a property of the tanh function that is true under certain conditions. The tanh function can also be expressed as

$$\tanh(x) = \left[\frac{(1 + x + \ldots) - (1 - x + \ldots)}{(1 + x + \ldots) + (1 - x + \ldots)}\right]$$

which, for sufficiently small values of X (X << 1), collapses to

$$\tanh(x) \approx x$$

Applying this concept to the mathematical description for the magnetization represented by Eq. 15.28, we can now simplify our expression. Specifically, if the splitting between the m_s levels depicted in Fig. 15.45 is small with respect to the available thermal energy at a given temperature; that is, if

$$m_s g\beta H \lll k_bT$$

then we can rewrite this as Eq. 15.29.

$$M = N\sum_n \mu_n P_n = \frac{Ng\beta}{2}\tanh\left(\frac{g\beta H}{2k_bT}\right) \approx \frac{Ng\beta}{2}\left(\frac{g\beta H}{2k_bT}\right) = \frac{Ng^2\beta^2 H}{4k_bT} \quad \textbf{15.29}$$

We know phenomenologically that the magnetic susceptibility is the proportionality constant that connects magnetization to the strength of the applied field (Eq. 15.21). Rearranging this expression and combining it with Eq. 15.29, we have Eq. 15.30.

$$\chi = \frac{M}{H} = \frac{Ng^2\beta^2}{4k_bT} = \frac{C}{T} \quad \textbf{15.30}$$

where

$$C = \frac{Ng^2\beta^2}{4k_b}$$

This is known as the Curie Law for one unpaired electron, and C is called the Curie constant. It describes the temperature-dependence of the magnetic susceptibility of a system consisting of a single unpaired electron in an applied magnetic field. Notice that the value of the Curie constant does not depend on the strength of the applied field. Application of the field removes the degeneracy between the two m_s states as shown in Fig. 15.45, and allows us to carry out a measurement to assess the value of χ, but it does not *create* the susceptibility, which is an intrinsic property of the system.

This derivation was carried out specifically for the case of a single unpaired electron. Accordingly, Eq. 15.30 applies only to compounds with one unpaired electron. This result can be generalized to a system with an arbitrary number of unpaired electrons; while the math becomes rather more complicated, no new physical principles are introduced in this approach. The total number of unpaired electrons in a molecule can be described by Eq. 15.31,

$$S = \sum_n |m_s^n| = \frac{1}{2} \times (\#\ of\ unpaired\ electrons) \quad \textbf{15.31}$$

where S is defined as the total spin of the system (the so-called total spin quantum number, Section 2.2.3) and is simply the sum of all of the m_s values for all of the unpaired electrons of the system; for example, a system of two unpaired electrons corresponds to $S = 2/2 = 1$, three unpaired electrons $S = 3/2$, and so on.

Using this definition—and without going through the details of the derivation—Eq. 15.30 can be recast in a more general form, as follows.

$$\chi = \frac{Ng^2\beta^2}{3k_bT}S(S+1) = \frac{C}{T}\text{(Curie Law)} \quad \textbf{15.32}$$

where

$$C = \frac{Ng^2\beta^2}{3k_b}S(S+1)$$

For a single unpaired electron, S = ½, when we plug this value into Eq. 15.32, we see that this expression collapses to Eq. 15.30.

The Curie Law describes an inverse dependence between temperature and magnetic susceptibility: as the temperature increases, the observed magnetic susceptibility decreases. This makes qualitative sense with our expectations concerning the magnetization. We described that in terms of a net polarization of magnetic dipoles aligned with the applied field. The lowest-energy state ($m_s = -^1/_2$) is oriented with the field, while the higher-energy state ($m_s = +^1/_2$) is aligned against the field. The maximum magnetization, and therefore the largest magnetic susceptibility, will therefore be observed at the lowest temperature when all the particles are in the lowest-energy state. If the higher-energy state is thermally accessible, the net magnetization will decrease with increasing temperature as the state characterized by an antiparallel alignment of the magnetic dipoles with respect to the applied field is populated. The $^1/_T$ dependence is a direct consequence of the Boltzmann nature of thermal distributions—in this case as it applies to the various magnetic states available to the system.

We can represent the Curie Law graphically in two ways, both of which are depicted in Fig. 15.46. On the left is a plot of Eq. 15.32; that is, the magnetic susceptibility as a function of temperature. The curvature of this plot is reminiscent of a plot of volume as a function of pressure—indeed, the Curie Law is sometimes referred to as the pressure–volume analog in magnetism. The panel at the right of the figure casts Eq. 15.32 in a slightly different format, linearizing the susceptibility by plotting it as a function of $^1/_T$. The information content of these two plots is exactly the same—it is simply two ways of plotting the exact same equation—but the linear plot helps to more clearly illustrate the information that we can glean out of experimental data acquired on a sample. By plotting χ as a function of $^1/_T$, Eq. 15.32 predicts a straight line whose slope should correspond to the Curie constant. We can see from Eq. 15.32 that the Curie constant is comprised of a collection of physical constants; that is, g, β, and k_b, as well as the total spin of the system, S. Therefore, by plotting the magnetic susceptibility as a function of $^1/_T$, we can simply fit the data to a straight line and directly obtain a value for S, which in turn tells us how many unpaired electrons the system possesses.

Of course, since both plots in Fig. 15.46 derive from the same mathematical expression, one can fit the curve from the

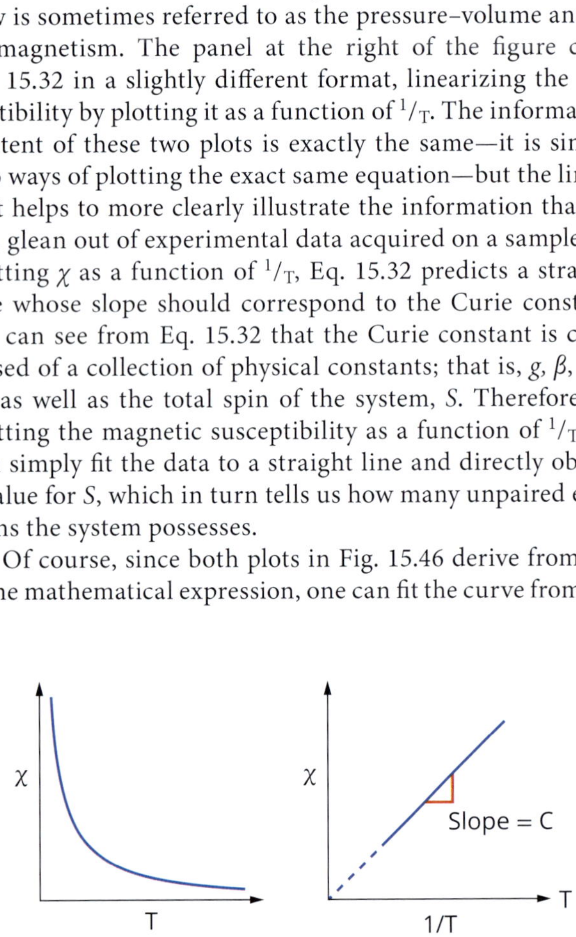

Fig. 15.46 (*left*) Qualitative plot of χ versus T. (*right*) Qualitative plot of χ versus $^1/_T$

χ versus T plot just as easily.[z] That said, there is a pedagogical use for linearization of the data that relates back to the applicability of the Curie Law. In the derivation outlined, we made a critical approximation that allowed us to obtain the relatively simple closed-form expression of Eq. 15.30 (and by extension Eq. 15.32). That approximation is reflected in Eq. 15.29: the assumption that $m_s g\beta H << k_b T$. It was this mathematical condition that allowed us to replace the *tanh(x)* function by the argument of the function itself (tanh(x) ~ x), but it tells us something about the circumstances under which the Curie Law can be applied. As with any model, the Curie Law can be used provided that any assumptions, approximations, and so on, that went into the derivation of that expression are valid. In the present case, *the Curie Law can only be expected to apply if the field-induced splitting of the m_s levels of a given system is much smaller in magnitude than the amount of thermal energy available at a given temperature. If this condition is not met, then we should not expect the Curie Law to accurately describe the magnetic properties of a given system under those conditions.* There is not a hard-and-fast cut-off for this circumstance, as it relates to how valid it is to approximate *tanh(x)* as x itself; the more divergent the discrepancy between the value of x as compared to *tanh(x)*, the worse we can expect the Curie Law to accurately describe the magnetic susceptibility. Going back to the plot of χ versus $^1/_T$, this divergence will manifest as a deviation from linearity, particularly toward the right portion of the plot where data acquired at lower temperatures (and thus less thermal energy) will reside. The human eye is remarkably adept at identifying deviations from linearity (some estimates place it at better than $^1/_{64}$ of an inch), hence the utility of the linearized plot in assessing whether the Curie Law is being adhered to in a given system.

Based on the conditions necessary for applicability of the Curie Law, the threshold for where the model would be expected to break down is going to be different for different systems and will depend on a combination of the magnitude of the magnetic dipole in question (m_s), the temperature ($k_b T$), and the strength of the applied field (H). For example, in an applied field of 1 T (10,000 gauss), the energy-splitting of the two m_s states for a $S = ^1/_2$ system is ~0.9 cm^{-1}. At a temperature of 10 K there is ~7 cm^{-1} of thermal energy available, so in this case the Curie approximation should not present too much of a problem. If, however, we were to measure data for a high-spin d^5 complex ($S = ^5/_2$) in a field of 3 T, the energy splitting would be 14 cm^{-1}. This would clearly invalidate the use of Eq. 15.32 at 10 K, but would be expected to start to recover above 50 K. So, as with any model, it is important to understand the limitations of the Curie Law in order to use it properly.

The temperature-dependence of paramagnetic susceptibility—which, to reiterate, arises due to changes in the population distribution among accessible magnetic states in

[z] Linearization of the data was more common many years ago before the advent of fitting software that is readily available today.

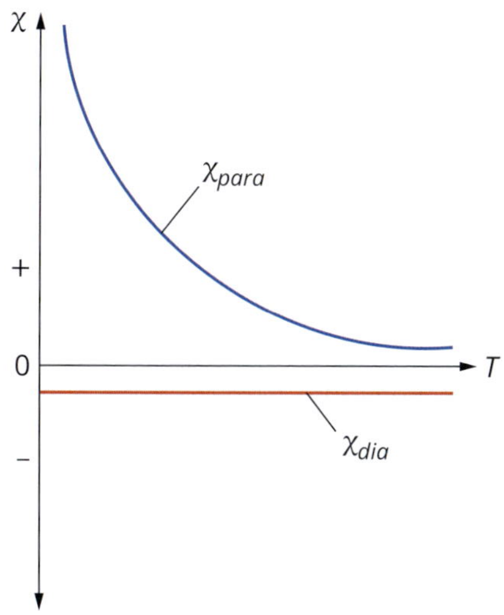

Fig. 15.47 Qualitative plot of magnetic susceptibility versus *T*. The inverse temperature-dependence of χ_{para} results in a substantial difference in its contribution to the overall observed susceptibility (χ_{obs}) at low temperature as compared to high temperature. The corresponding insensitivity of χ_{dia} to changes in temperature means that the diamagnetic response is an increasingly significant contribution to χ_{obs} with increasing temperature. This can have significant implications for the interpretation of experimental data under certain circumstances.

response to changes in the available thermal energy—stands in stark contrast to the temperature-independence of diamagnetic susceptibility. The physical origin of χ_{dia} does not rely on thermal energy and so is therefore essentially constant with changes in temperature. This difference is graphically illustrated in Fig. 15.47.

This plot illustrates an important aspect of magnetic susceptibility from an experimental perspective. It was mentioned previously in connection with Eq. 15.25 that what one measures is the total susceptibility, χ_{obs}, and that this reflects contributions from both the paramagnetic and diamagnetic responses of the material. We also mentioned that although the absolute magnitude of χ_{para} can be orders of magnitude larger than χ_{dia}, the additive nature of χ_{dia} coupled with the fact that there is usually a far greater number of paired electrons in a system than unpaired electrons means that χ_{dia} can be quite significant under certain circumstances. Fig. 15.47 illustrates that this situation is likely to be much more problematic at high temperature, where the thermal distribution across magnetic states that serve to attenuate the net magnetization can lead to a reduction in the value of χ_{para} to a point where it can be comparable in magnitude to χ_{dia}. Since these two contributions are opposite in sign, it is entirely possible for χ_{obs} to be close to or equal to zero in some cases. Accordingly, measuring the temperature-dependence of the magnetic susceptibility can be an important experimental protocol for assessing the magnetic properties of some (but not all) molecular systems. Changing the strength of the applied field, which is relatively easy to do using modern magnetometers, can also be a useful tool due to the fact that the diamagnetic response typically has a weaker field dependence than the linear response characteristic of simple paramagnets.

Prior to the development of instrumentation for measuring magnetic properties available today, carrying out magnetic measurements over a wide temperature range was not a trivial experiment. Even today, although magnetometers that enable the acquisition of magnetization data as a function of both field and temperature are available from several companies, these instruments are quite expensive. There are methods for measuring magnetic susceptibility at room temperature that are relatively simple and inexpensive, which makes recasting the expression for χ_{para} into a form that only requires data at a single temperature a useful exercise. Accordingly, one can rewrite the general expression for the Curie Law in a manner shown as follows.

$$\chi = \frac{Ng^2\beta^2}{3k_bT}S(S+1) = \frac{N}{3k_bT}\mu_{eff}^2 \qquad \textbf{15.33}$$

which can be rearranged to yield Eq. 15.34.[aa]

$$\mu_{eff} = \sqrt{\frac{3k_b}{N\beta^2}\chi T} = 2.828\sqrt{\chi T} \qquad \textbf{15.34}$$

The effective magnetic moment shown in Eq. 15.33 is defined by Eq. 15.35:

$$\mu_{eff} = g\beta\sqrt{S(S+1)} \qquad \textbf{15.35}$$

If we take Eq. 15.35 and simplify it by assuming a value for the *g*-factor equal to that of the free electron (≈2), we get Eq. 15.36,

$$\mu_{eff} = \sqrt{n(n+2)} \qquad \textbf{15.36}$$

where *n* corresponds to the number of unpaired electrons in the system. Eq. 15.36 provides us with what amounts to a theoretical expression for the value of μ_{eff},[ab] which can now be compared to the experimental value afforded by Eq. 15.35. We have already shown through our derivation of the Curie Law that the magnetic susceptibility is inversely proportional to temperature, $\chi = C/T$. The temperature-dependence in Eq. 15.35 cancels out, so the value of μ_{eff} is a temperature-independent parameter and allows us to take a data point at single temperature and determine the number of unpaired electrons in the system (Eq. 15.36).

[aa] The form shown in Eq. 15.35 is the most commonly used, and χ has cgs units of $cm^3\ mol^{-1}$ K. In SI units, χ has units of $m^3\ mol^{-1}$ K, and $\mu_{eff} = 797.727\sqrt{\chi T}$. The value of the effective magnetic moment, μ_{eff}, is unitless.[48–49]

[ab] In some cases, effective magnetic moment is given in units of Bohr magneton. However, effective magnetic moment, as it is usually defined, is unitless.

While convenient, the accuracy of this model also has its limitations. First, the temperature-independence of μ_{eff} is predicated on the validity of the Curie Law, so any deviations from the latter will directly impact the accuracy of μ_{eff} for describing the magnetic properties of the system. In addition, Eq. 15.36 is only arrived at by assuming that the only source of magnetism in the system arises from the spin of the electrons, m_s. While this is a reasonable assumption for many systems, this is certainly not universally true. The nature of this approximation is incorporated into the specific description of Eq. 15.36 as the "spin-only" approximation, and is usually given by the term $\mu_{eff}^{s.o.}$, where the superscript "s.o." refers to "spin-only". Table 15.6 provides a list of values for $\mu_{eff}^{s.o.}$ calculated for n = 1 through 5, which for simple transition-metal-based complexes containing a single metal ion represents the full range of values one can expect to encounter. Over the past decade or so, the magnetism community has begun to move away from the use of effective magnetic moments, and more commonly describes magnetic properties in terms of the product χT, mainly because these are the two experimental parameters one actually measures in addition to the fact that the spin-only bias endemic to $\mu_{eff}^{s.o.}$ is removed. To reflect this trend, corresponding values of χT at 300 K are provided.

The preceding section focused on what we can learn about the electronic structure of compounds from their magnetic properties by using some relatively straightforward concepts from classical mechanics. Much like crystal field theory, this sort of model has its shortcomings but is nevertheless an extremely useful model in a wide range of circumstances. Of course, like CFT, a more complete description requires an acknowledgement that molecules are quantum-mechanical entities, and this perturbation can and does impact our ability to use simpler constructs to describe in greater detail the magnetic properties of molecules. A complete treatment of this topic is beyond the scope of this chapter but is definitely necessary if a more detailed description of a wider range of chemical systems and experimentally observed properties is desired.

Table 15.6 Values for $\mu_{eff}^{s.o.}$ (unitless), the "spin-only" moment, for various numbers of unpaired electrons (*n*) from Eq. 15.36. in addition to corresponding values for XT.[7]

n	$\mu_{eff}^{s.o.}$	XT
1	1.73	0.375
2	2.83	1.00
3	3.87	1.87
4	4.90	3.00
5	5.92	4.38

15.10.5 A Brief Description of the Magnetic Behavior of Simple Paramagnets

In the preceding sections a detailed description of simple magnetic behavior was laid out with derivation of many of the important equations by using classical mechanics. Here, a far briefer description is presented as either a review or a stand-alone, terse section on using magnetism to characterize metal complexes. Equations have the same numbers as in the previous sections, but are reproduced here for readability. For those interested in the origin of the equations, we recommend reading the advanced sections prior to this.

All materials are affected by a magnetic field, and, depending on the nature of the material, can be attracted (+) or repelled (–) by the magnetic field. When a substance is placed in a magnetic field, the response of the material, the magnetization (***M***), is proportional (for most substances) to the magnitude of the applied field (***H₀***).

$$\boldsymbol{M} \propto \boldsymbol{H}_0$$

As with any mathematical proportionality, this relation can be expressed as an equality by including a proportionality constant, χ_{obs}, shown as follows. This proportionality constant, χ, is referred to as the "magnetic susceptibility", and all the chemical information about the material with regard to its response to the applied magnetic field is contained within this term.

$$\boldsymbol{M} = \chi_{obs}\boldsymbol{H}_0$$

The observed susceptibility (χ_{obs}) from the experiment is a sum of the diamagnetic (χ_{dia}) and paramagnetic (χ_{para}) susceptibilities, as shown in Eq. 15.37. The paramagnetic effect is due to unpaired electrons in the system and goes with (positive) the magnetic field, and the diamagnetic effect is due to paired electrons and runs counter (negative) to the applied field. The paramagnetic effect is generally much larger than the diamagnetic for any particular atom with unpaired spins, but the diamagnetic effect is additive. In general, a compound will have many sets of paired electrons, and as the molecule becomes larger, so χ_{dia} scales with the size of the molecular system. For very large molecules with only a few paramagnetic centers, such as metalloproteins with paramagnetic transition metals, the diamagnetic effect can be as large or larger than (χ_{para}).

$$\chi_{obs} = \chi_{para} + \chi_{dia} \tag{15.37}$$

In order to acquire an accurate accounting of χ_{para}, which contains information about the number of unpaired electrons, one needs to obtain a reasonable accounting of χ_{dia}. This can be done a few different ways, but one of the most common is to use Pascal's constants (Appendix M, Box 15C). As mentioned, the diamagnetic susceptibility is additive, and as a result one can add the diamagnetic contributions for all the atoms and bonds in a molecule by using tables like those in Appendix M to obtain an estimate of χ_{dia}.

Once the paramagnetic susceptibility is in hand, one wants to be able to determine what types of complexes lead to specific values of χ_{para}, allowing characterization of new compounds. One way to do this is by examining how χ_{para} changes with temperature. One can show that the sum of the electron spins, S, is related to χ_{para} by Eq. 15.38. (S is simply the number of unpaired electrons times the spin of the electron, ½. In other words, if the complex has five unpaired electrons then $S = {}^5/_2$.) One can fit a χ_{para} versus T curve to determine the number of unpaired electrons in the complex, in the form of S (Eq. 15.38).

$$\chi_{para} = \frac{Ng^2\beta^2}{3k_bT}S(S+1) = \frac{C}{T} \tag{15.38}$$

$$C = \frac{Ng^2\beta^2}{3k_b}S(S+1)$$

If one wants to perform a single temperature measurement, which can involve far less expensive instrumentation, related information on the number of unpaired electrons can be obtained. In order to do this, the equations are most often reformulated to be expressed in "effective magnetic susceptibility", μ_{eff}, which is a unitless number.[48–49] Eq. 15.39 shows the relationship between the paramagnetic susceptibility, χ_{para}, and μ_{eff}.

$$\mu_{eff} = \sqrt{\frac{3k_b}{N\beta^2}\chi T} = 2.828\sqrt{\chi T} \tag{15.39}$$

The approximate value of μ_{eff} as a function of the number of unpaired electrons can be calculated by using Eq. 15.40,

Table 15.7 Values for $\mu_{eff}^{s.o.}$ (unitless), the "spin-only" moment, for various numbers of unpaired electrons (*n*) from Eq. 15.36, and examples of complexes with the given number of unpaired spins.[7]

n	$\mu_{eff}^{s.o.}$	Example/experimental μ_{eff} at 300 K
1	1.73	Cu(II), $K_2Cu(SO_4)_2 \cdot 6H_2O$/1.91
2	2.83	Ni(II), $K_2Ni(SO_4)_2 \cdot 6H_2O$/3.23
3	3.87	Mn(IV), $BaMnF_6$/3.80
4	4.90	Cr(II), $CrSO_4 \cdot 6H_2O$/4.82
5	5.92	Fe(III), $KFe(SO_4)_2 \cdot 12H_2O$/5.89

where *n* is the number of unpaired electrons. Table 15.7 shows the values for unpaired electrons relevant to transition-metal complexes; that is, 1–5 unpaired electrons.

$$\mu_{eff} = \sqrt{n(n+2)} \tag{15.40}$$

As can be seen, simple Curie magnetism, derived largely from classical mechanics, can be used to understand the behavior of paramagnetic systems. With ligand (crystal) field theory, the relative energies of the property-defining *d*-orbitals in transition-metal complexes can be found by simply knowing the general shape of the complex. The *d*-electrons can then be filled into these orbitals if one knows some details about the magnetism of the complex, which provides critical information as to whether the metal center is high spin or low spin.

It should be noted that this is most relevant to first-row transition-metal complexes, where spin-orbit coupling is minimal. While Curie magnetism can act as a guide for the heavier transition elements, large deviations are often observed. Even so, for the creation of a working crystal field splitting diagram, this is less often a problem for second- and third-row transition metals as they typically lead to complexes that are low spin; that is, ambiguity in electron-filling is less of an issue for these heavier congeners.

Chapter Summary

1. In crystal field theory (CFT) the ligands are considered to be point charges that raise the energy of the metal center's *d*-orbitals based on the ligands' location relative to the orbital: the more the orbital points at the ligand, the higher in energy it goes.
2. Using CFT, one can get the relative energies of the *d*-orbitals on a particular metal center in the environment of the ligands, which produces a tremendous amount of information about the complex that ranges from reactivity to spectroscopic properties.
3. While CFT is quite useful, it fails to give quantitative values for the energies, largely because the ligands are not point charges as assumed; the ligands bond to the metal with some covalency. Incorporation of covalency into CFT provides ligand field theory (LFT)—a merger of molecular orbital theory and the symmetry-based arguments of CFT.
4. In LFT, the centricity of the *d*-orbitals in their determination of properties is seen as being due to the fact that the high metal-character *d*-orbitals tend to be (or be near) the frontier orbitals of the complex.

5. CFT and LFT provide *d*-orbital "splittings", pictures of the relative energies of the orbitals and their occupancies. The size of the splitting is determined by the metal center and the donor properties of the ligands. In an octahedral (O_h) complex, the difference between the higher-energy e_g* set and the lower-energy t_{2g} set of orbitals is Δ_o. In a tetrahedral (T_d) complex, the difference between the higher-energy t_2 and lower-energy *e* orbitals is Δ_t.
6. For the same metal in the same oxidation state with the same ligands, $\Delta_t \sim (4/9)\ \Delta_o$. The smaller splitting in the tetrahedral system is due to the smaller number of ligands relative to octahedral and their orientation.
7. If the "pairing energy" of the electrons on the metal is larger than Δ_o, one will get a "high-spin" complex with the fourth electron going into the higher-energy e_g orbital. If the pairing energy is lower than Δ_o, the complex is low spin, and electrons do not occupy the higher-energy e_g orbitals until the t_{2g} set is full.
8. First-row metal complexes produce both high-spin and low-spin complexes depending on the metal center and ligands, while second- and third-row complexes tend to have lower pairing energies and are generally low-spin.
9. Metal centers in higher oxidation states tend to have larger Δ_o.
10. Ligands that induce larger Δ_o values are called "strong field" ligands, which include very strong σ-donors and π-acceptors. Conversely, ligands that produce complexes with small Δ_o values are called "weak field ligands", which are generally weak σ-donors and strong π-donors. The spectrochemical series shows how Δ_o changes with ligand structure.
11. The angular overlap model quantizes the CFT picture and allows calculation of orbital energies in terms of the σ- and π-donor abilities of the ligands towards a particular metal ion.
12. Magnetism, which should be viewed as a way of probing electronic structure, can be used to determine the number of unpaired electrons of a metal complex; for example, to determine whether the complex is high-spin or low-spin.

Practice Problems

1. For a complex ML_6, which is tetragonally distorted from octahedral geometry, where all the ligands are identical but the M–L distances are not, answer the following:
   a. Draw a crystal field diagram where the ligands along the *z*-axis are *elongated*. Label all the orbitals by their tetragonal *d*-orbital designation (e.g., d_{xy}) and by the label for the irreducible representation to which they belong in the point group.
   b. Draw a crystal field diagram where the ligands along the *z*-axis are *compressed*. Label all the orbitals by their tetragonal *d*-orbital designation (e.g., d_{xy}) and by the label for the irreducible representation to which they belong in the point group.
   c. Discuss how the relative energies are found for specific geometries in crystal field theory and the differences between the two diagrams in parts a and b.
2. Answer the following regarding complexes of different geometries with various ligand structures. Your *z*-axis should be defined by the highest order rotation axis (except for O_h, where you should use the C_4).
   a. For a (pseudo-)octahedral complex with six ligands all 90° from one another with the following formulas what is the point group, and to which irreducible representations do the *d*-orbitals belong? Draw a crystal field splitting diagram assuming that X is a greater donor than L.
      i. ML_6
      ii. ML_5X
      iii. *trans*-ML_4X_2
      iv. *cis*-ML_4X_2
      v. *fac*-ML_3X_3
      vi. *mer*-ML_3X_3
   b. For a (pseudo-)trigonal bipyramidal complex with the following formulas what is the point group, and to which irreducible representations do the *d*-orbitals belong? Draw a crystal field splitting diagram assuming that X is a greater donor than L.
      i. ML_5
      ii. ML_4X (X axial)
      iii. ML_4X (X equatorial)
      iv. *trans*-ML_3X_2
      v. *cis*-ML_3X_2
3. What leads to the diamagnetic susceptibility (χ_{dia})?
4. Why are most Zn^{2+} complexes colorless?
5. What does the ligand field splitting diagram for $W(CO)_6$ look like? What is the metal's oxidation state and *d*-electron count?
6. Using the Angular Overlap Model (AOM), find the energies of the *d*-orbitals for a 3-coordinate complex in terms of e_σ, $e_{\pi x}$, and $e_{\pi y}$ in three different geometries: trigonal planar, trigonal pyramidal, and T-shaped. Using the axes below. (Nonstandard axes are being used for the T-shaped molecule so that it will better correlate with the other two geometries.) Using your AOM diagrams for the orbital energies predict the lowest energy geometries for the complexes below.

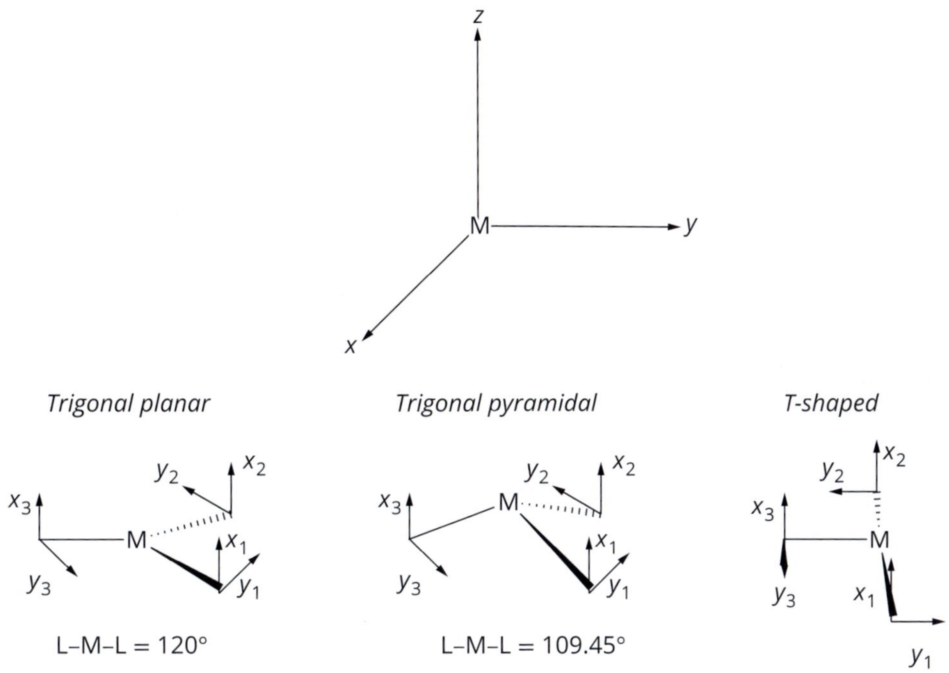

a. $Cr(NPr^i_2)_3$ (Assume all the amido ligands are parallel to the global *z*-axis, with $e_{\pi x} = 0$ and $e_{\pi y} = ¼\ e_\sigma$).

b. $PtCl_3^-$ (Assume $e_\sigma = 4\ e_{\pi x} = 4\ e_{\pi y}$.)

7. For $Cr(NH_3)_6^{3+}$, the value of 10 Dq (Δ_o) = 19599 cm^{-1}. It is often the practice in the AOM to assume that NH_3 has no π-component. If we make that assumption here, what is e_σ for NH_3 in a Cr(III) complex?

Cited References

1. Bethe, H. A. *Selected Works of Hans A. Bethe: With Commentary*. Singapore: World Scientific Publishing Co, 1997.
2. Van Vleck, J. H. Nobel Prize lecture in physics entitled "Quantum mechanics: The key to understanding magnetism", 1977.
3. Papanastasiou, M., Allen, N. S., McMahon, A., Naegel, L. C. A., Edge, M., and Protopappas, S. Analysis of Indigo-type compounds in natural dyes by negative ion atmospheric pressure photoionization mass spectrometry. *Dyes and Pigments* 2012, *92*, 1192–1198.
4. Holme, I. Sir William Henry Perkin: A review of his life, work and legacy. *Coloration Technology* 2006, *122*, 235–251.
5. Sir William Perkin and the 150th anniversary of the discovery of mauveine. http://www.rsc.org/Chemsoc/Activities/Perkin/2006/.
6. Ballhausen, C. J. *Introduction to Ligand Field Theory*. New York, NY: McGraw-Hill, 1962, 298 pp.
7. Figgis, B. N., and Hitchman, M. A. *Ligand Field Theory and Its Applications*. New York, NY: Wiley-VCH, 2000.
8. Gerloch, M., and Slade, R. C. *Ligand-Field Parameters*. Cambridge: Cambridge University Press, 1973.
9. Deeth, R. J. Ligand field and density functional descriptions of the *d*-states and bonding in transition metal complexes. *Faraday Discuss.* 2003, *124*, 379–391.
10. Anthon, C., Bendix, J., and Schaffer, C. E. Mimicking the two-dimensional spectrochemical series using density functional computations. *Inorg. Chem.* 2004, *43*, 7882–7886.
11. Anthon, C., Bendix, J., Schaffer, C. E. Elucidation of ligand-field theory: Reformulation and revival by density functional theory. *Optical Spectra and Chemical Bonding in Transition Metal Complexes* 2004, *107*, 207–301.
12. Reinen, D., Atanasov, M., Kohler, P., and Babel, D. Jahn–Teller coupling and the influence of strain in T_g and E_g ground and excited states: A ligand field and DFT study on halide $M^{III}X_6$ model complexes [M = Ti^{III}-Cu^{III}; X = F^-, Cl^-]. *Coord. Chem. Rev.* 2010, *254*, 2703–2754.
13. Schaffer, C. E., Anthon, C., and Bendix, J. Kohn–Sham DFT results projected on ligand-field models: Using DFT to supplement ligand-field descriptions and to supply ligand-field parameters. *Coord. Chem. Rev.* 2009, *253*, 575–593.
14. Atanasov, A., Daul, C. A., and Rauzy, C. A DFT based ligand field theory. In Schoenherr, T. (ed.), *Optical Spectra and Chemical Bonding in Inorganic Compounds*, 2004, Vol. 106, pp. 97–125.
15. Hoskins, B. F., and Whillans, F. D. Geometry of pentacoordinate complexes. *Coord. Chem. Rev.* 1973, *9*, 365–388.
16. Morassi, R., Bertini, I., and Sacconi, L. 5-Coordination in iron(II), cobalt(II) and nickel(II) complexes. *Coord. Chem. Rev.* 1973, *11*, 343–402.
17. Mori, M., Saito, Y., and Watanabe, T. The crystal structure of $[Cr(NH_3)_6]$ $[CuCl_5]$. *Bull. Chem. Soc. Jpn.* 1961, *34*, 295–296.
18. Smith, D. W. Chlorocuprates(II). *Coord. Chem. Rev.* 1976, *21*, 93–158.
19. Stnikolov, G. Geometry of the pentachlorocuprate(II) ion. *Inorg. Chem.* 1994, *33*, 1144–1152.
20. Greene, P. T., Russ, B. J., and Wood, J. S. Crystal and molecular structures of trimethylamine complexes of titanium tribromide and chromium trichloride. *J. Chem. Soc. A* 1971, 3636.
21. Russ, B. J., and Wood, J. S. Molecular structure of bistrimethylaminetitanium tribromide: An example of Jahn-Teller effect in pentacoordinate molecules. *Chem. Commun.* 1966, 745.
22. Cotton, F. A., Dunne, T. G., and Wood, J. S. Crystal and molecular structure of pentakis(methylisonitrile)cobalt(I) perchlorate. *Inorg. Chem.* 1965, *4*, 318.

23. Alvarez, S. Bonding and stereochemistry of three-coordinated transition metal compounds. *Coord. Chem. Rev.* 1999, *193–5*, 13–41.
24. Bradley, D. C., and Chisholm, M. H. Transition-metal dialkylamides and disilylamides. *Acc. Chem. Res.* 1976, *9*, 273–280.
25. Cummins, C. C. Three-coordinate complexes of "hard" ligands: Advances in synthesis, structure and reactivity. *Progress in Inorganic Chemistry* 1998, *47*, 685–836.
26. Eller, P. G., Bradley, D. C., Hursthouse, M. B., and Meek, D. W. 3 Coordination in metal-complexes. *Coord. Chem. Rev.* 1977, *24*, 1–95.
27. Alyea, E. C., Bradley, D. C., Copperthwaite, R. G., and Sales, K. D. 3-coordinated transition-metal compounds.2. Electronic-spectra and magnetism of tris(bis-trimethylsilylamido)derivatives of scandium, titanium, vanadium, chromium, and iron. *Dalton Trans.* 1973, 185–191.
28. Covert, K. J., Wolczanski, P. T., Hill, S. A., and Krusic, P. J. Ketyl complexes of $(silox)_3Ti$ (silox = $^tBu_3SiO^-$). *Inorg. Chem.* 1992, *31*, 66–78.
29. Calculations were done using SDD as the basis set with B3LYP on G09.
30. DeKock, R. L., and Gray, H. B. *Chemical Structure and Bonding*, 2nd ed. Mill Valley, CA: University Science Books, 1989.
31. Shannon, R. D., and Prewitt, C. T. Effective ionic radii in oxides and fluorides. *Acta Cryst. B: Struc. Sci.* 1969, *B 25*, 925.
32. Zimmermann, L. W., and Schleid, T. Crystal structures and properties of divalent transition metal decahydro-closo-decaborate hydrates $M(H_2O)_6\ B_{10}H_{10} \cdot 2H_2O$ (M = Mn, Fe, Co, Ni, Zn) *Z. Kristal.* 2013, *228*, 558–564.
33. Albertsson, J., Oskarsson, A., Stahl, K., Svensson, C., and Ymen, I. Compounds with intermediate spin.5. X-ray study of tris(N,N-dimethyldithiocarbamato) iron(III) at the 2 extreme temperatures 25 and 400 K. *Acta Cryst. B* 1981, *37*, 50–56.
34. Couldwell, C., Prout, K., Robey, D., Taylor, R., and Rossotti, F. J. C. Crystal and molecular-structures of hexaaquacopper(II) benzenesulfonate, toluene-4-sulphonate and d-camphor-10-sulphonate. *Acat Cryst. B:Struc. Sci.* 1978, *34*, 1491–1499.
35. Persson, I., Persson, P., Sandstrom, M., and Ullstrom, A. S. Structure of Jahn–Teller distorted solvated copper(II) ions in solution, and in solids with apparently regular octahedral coordination geometry. *Dalton Trans.* 2002, 1256–1265.
36. Hoggard, P. E. Angular overlap model parameters. *Optical Spectra and Chemical Bonding in Inorganic Compounds* 2004, *106*, 37–57.
37. Burdett, J. K. New method for determination of geometries of binary transition-metal complexes. *Inorg. Chem.* 1975, *14*, 375–382.
38. Bazhenova, T. A., Lobkovskaya, R. M., Shibaeva, R. P., Shilova, A. K., Gruselle, M., Leny, G., and Deschamps, E. Structure of the naphthyl iron(II) complex formed in the reaction of $FeCl_3$ and $C_{10}H_7Li$. *J. Organomet. Chem.* 1983, *244*, 375–382.
39. Landis, C. R., Firman, T. K., Root, D. M., and Cleveland, T. A valence bond perspective on the molecular shapes of simple metal alkyls and hydrides. *J. Am. Chem. Soc.* 1998, *120*, 1842–1854.
40. Vanquickenborne, L. G., and Ceulemans, A. On the ligand-field spectra of square-planar platinum(II) and palladium(II) complexes. *Inorg. Chem.* 1981, *20*, 796–800.
41. Winkler, J. R., and Gray, H. B. Electronic structures of oxo-metal Ions. In Mingos, D. M. P., Day, P., and Dahl, J. P. (eds.), *Molecular Electronic Structures of Transition Metal Complexes I* 2012, Vol. 142, pp. 17–28.
42. Ballhausen, C. J., and Gray, H. B. The electronic structure of the vanadyl ion. *Inorg. Chem.* 1962, *1*, 111–122.
43. Fortner, K. C., Laitar, D. S., Muldoon, J., Pu, L. H., Braun-Sand, S. B., Wiest, O., and Brown, S. N. Ultrafast and ultraslow oxygen atom transfer reactions between late metal centers. *J. Am. Chem. Soc.* 2007, *129*, 588–600.
44. Jacobi, B. G., Laitar, D. S., Pu, L. H., Wargocki, M. F., DiPasquale, A. G., Fortner, K. C., Schuck, S. M., and Brown, S. N. Stoichiometric and catalytic oxygen activation by trimesityliridium(III). *Inorg. Chem.* 2002, *41*, 4815–4823.
45. Haymotherwell, R. S., Wilkinson, G., Hussainbates, B., and Hursthouse, M. B. Synthesis and x-ray crystal-structure of oxotrimesityliridium(V). *Polyhedron* 1993, *12*, 2009–2012.
46. Gerlach, W., and Stern, O. Der experimentelle Nachweis der Richtungsquantelung im Magnetfeld *Z. Phys.* 1922, *9*, 349–352.
47. Li, D., Gao, B., and Duan, Q. Preparation of star-shaped functionalized polylactides by metal porphyrin complexes as both catalysts and cocatalysts. *J. Porphyrins Phthalocyanines* 2019, *23*, 1020.
48. Hoppee, J. I. Effective magnetic-moment. *J. Chem. Educ.* 1972, *49*, 505.
49. Kettle, S. F. A. *Physical Inorganic Chemistry: A Coordination Chemistry Approach*. Berlin, Heidelberg: Springer, 1996.

CHAPTER 16

Optical Spectroscopy and Excited-State Electronic Structure

> *"Your experience with the things that you have seen before is incomplete. The behavior of things on a very tiny scale is simply different. An atom does not behave like a weight hanging on a spring and oscillating. Nor does it behave like a miniature representation of the solar system with little planets going around in orbits. Nor does it appear to be somewhat like a cloud or fog of some sort surrounding the nucleus. It behaves like nothing you have seen before."*
>
> —Richard Feynman (1967)[1]

CONCEPT COVERAGE

1. Why we need a multielectron picture
2. Multielectronic states
3. Microstates
4. Atomic term symbols
5. Correlation, Orgel, and Tanabe-Sugano diagrams
6. Racah parameters
7. Term symbols in octahedral and tetrahedral fields
8. Selection rules

In the previous chapter we focused mostly on what is often referred to as the "1-electron picture" of inorganic complexes. In spite of what this name implies, we did not limit ourselves to just the d^1 configuration. The term "1-electron" referred to the simplification of treating the *d*-electrons as if they did not interact with each other. Of course, whenever there is only one electron in the system, then this is a valid point of view; there are no other electrons to interact with, so treating this electron as an independent entity is entirely appropriate! For instance, we were able to interpret the optical spectrum of $\{Ti(OH_2)_6\}^{3+}$ directly by looking at the orbital picture; the spectrum exhibits a single electronic transition, which corresponds to the promotion of the lone electron from a t_{2g} to e_g orbital. Furthermore, we also were able to rationalize the asymmetry of the absorption band by invoking the Jahn–Teller effect, which breaks the symmetry of the orbitals of an ideal octahedron. Although the "1-electron" picture still provides important insights in explaining the behavior of configurations with more than one electron, *the repulsive interactions between each d-electron are very important and cannot be ignored.* This leads us to go from the "1-electron" picture that relies primarily on orbital considerations to a "multi-electron" description that fully takes into account the role of electron–electron repulsions and spin–orbit coupling in discussing the electronic structure of transition-metal compounds.

16.1 Breakdown of the 1-Electron Picture

Fig. 16.1 shows the absorption spectrum of the $\{Ni(OH_2)_6\}^{3+}$ ion, which is a d^8 complex. The spectrum shows three broad bands at about 8,000, 14,000, and 23,000 cm^{-1}, which all are due to excitations of the *d* electrons. But why are there three peaks? Based on the 1-electron picture, there should only be, *at most*, two ways to excite the *d* electrons: one transition corresponding to the promotion of a single electron from the t_{2g} set to the e_g set (which we would write $t_{2g}{}^6e_g{}^2 \rightarrow t_{2g}{}^5e_g{}^3$ using 1-electron configurations), and a second one corresponding to the excitation of two electrons from t_{2g} to e_g (written as $t_{2g}{}^6e_g{}^2 \rightarrow t_{2g}{}^4e_g{}^4$). Not only are there too many bands in the spectrum, but there is a further issue with the 1-electron

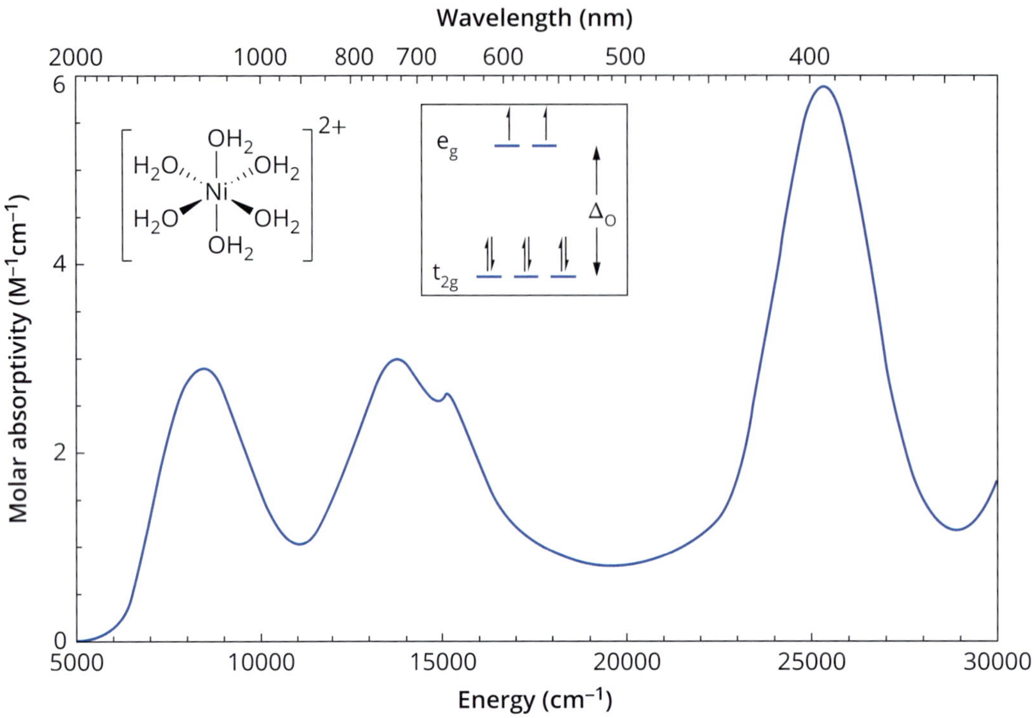

Fig. 16.1 (*top*) Absorption spectrum of $\{Ni(OH_2)_6\}^{3+}$ in water. There are three broad transitions in the spectrum: one around 8,400 cm^{-1}, a second ~14,300 cm^{-1}. (with a shoulder ~15,200 cm^{-1}), and a third around 25,300 cm^{-1}. (*inset*) Crystal field splitting diagram for an octahedral d^8 complex.

picture: quantum mechanics teaches us that single photons[a] cannot excite more than one electron at a time, so the second transition written previously (which is formally a 2-electron excitation) should strictly not even be observable using a standard spectrometer. So, we now have *two* absorption bands in the spectrum of $[Ni(H_2O)_6]^{2+}$ whose origin we cannot explain by using simple orbital configurations! Clearly, there must be something important missing here that is not provided by the 1-electron (orbital) picture. What that simple picture fails to do is to take into account the fact that electrons are negatively charged particles that interact with each other through Coulombic repulsion forces. Every time we reorganize the distribution of electrons in space—when we promote them from one orbital to another—we need to carefully consider how electron repulsion changes in our system in order to predict how many independent ways there are to reorganize the electrons. Let us look at a simpler case to illustrate how this works.

The hexaaquavanadium(III) ion, $\{V(OH_2)_6\}^{3+}$, has a d^2 ground-state configuration (Fig. 16.2), and the complex exhibits two absorption bands at 17,800 and 25,700 cm^{-1}. Let us suppose, for now, that in the ground state—prior to any excitation—the two d-electrons are located in the d_{xy} and d_{xz} orbitals; by symmetry, all three t_{2g} orbitals have the same energy, so it should not matter out of which orbital we excite an electron. Let us then assume that a photon of the right energy and momentum hits the complex, leading to the promotion of the d_{xz} electron to the e_g level, say to the $d_{x^2-y^2}$ orbital.[b] Let us call this excited configuration **A**; since the $d_{x^2-y^2}$ orbital (an e_g orbital) is located at an energy Δ_o above the t_{2g} orbitals, we might expect the photon energy required for this excitation to be roughly equal to Δ_o.

Of course, this is only one possibility, and an equally as good option would be to promote the same electron also from the d_{xz} orbital, but this time into the d_{z^2} orbital, which is also located at an energy Δ_o. Let us label this excited configuration with the letter **B**.

Are these two excited-state configurations identical? Well, in **A** the two electrons both reside in the xy-plane, whereas in **B** the excited electron is concentrated along the

[a] Multi-photon excitations (so-called "nonlinear" optical effects) are also possible, but only under very intense light irradiation of the sort provided by very strong laser beams. In a typical absorption spectrum, the intensity of the light beam is far too weak to lead to significant two-photon absorption processes.

[b] If you know your quantum mechanics well, then you perhaps realized that since electrons are indistinguishable particles, it is not possible to label individual electrons as we are doing here. Strictly speaking, statements such as "*promoting a d_{xy} electron to the $d_{x^2-y^2}$ orbital*" are quantum-mechanically nonsensical, at least once there is more than one electron in the system. In a way, this is also part of the problem we are running into with the orbital representation, and we will ultimately stop using this sort of language once we manage to define multi-electron states.

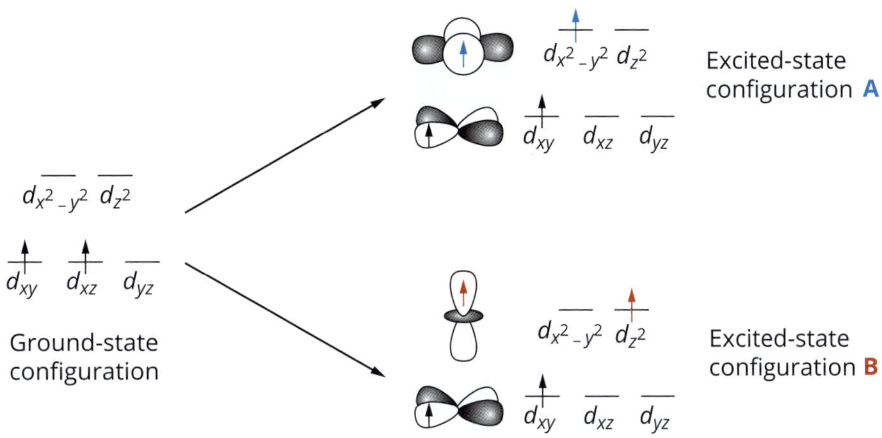

Fig. 16.2 Electronic configurations corresponding to possible excitations of a d^2 configuration. One possible arrangement for the ground-state configuration puts electrons in the d_{xy} and d_{xz} orbitals is shown at left. Two different 1-electron-excited-state electronic configurations can be considered, each leading to different electron–electron repulsion interactions. In excited-state configuration **A** (blue), both electrons reside in the *xy*-plane. In excited-state configuration **B** (red), the two electronic wavefunctions are distributed along different directions in the molecule. Obviously, electron repulsion will not be the same in configurations A and B, which gives the two states different energies.

z-axis, away from the *xy*-plane where the other "nonexcited" electron is left. Clearly, these are not identical configurations once we include the electron–electron interactions; the electrons in **A** must experience a larger repulsive force than in **B**, where they better avoid occupying a similar region of space. We are thus led to acknowledge that when we write an electronic configuration such as $t_{2g}{}^1e_g{}^1$ we do not provide all the information necessary to know how the electrons behave. Clearly, it matters how the electrons are organized inside these degenerate orbitals. Our first take-home message then is that electron–electron repulsion ultimately dictates not only the energy of excited states, but also the number of distinct excited states that exist. In order to accurately predict the number of allowed transitions inside a degenerate set of orbitals (like the *d*-orbital manifold), we must then learn how to account for the effect of electron–electron repulsion. Fortunately, there is a way to do this systematically, as we will discuss next.

BOX 16A

Light properties and Beer's Law

Table 16.B1 lists the pertinent ranges for visible-light absorption in both nanometers (nm) and wavenumbers (cm^{-1}). Light can be identified uniquely by its wavelength (λ), frequency (ν), or energy (E). The relationships between λ, ν, and E are shown in Eqs. 16.B1 and 16.B2, which involve the fundamental constants: the speed of light (c = 299,792,458 m/s ~ 3 × 10^8 m/s), and Planck's constant (h = 6.626 × 10^{-34} J·s).

$$\nu = c/\lambda \qquad 16.B1$$

$$E = h\nu \qquad 16.B2$$

Combining these two produces Eq. 16.B3. The energy of the light, which will be related to the energy of the electronic transition later, is proportional to the inverse of the wavelength. While most instruments report data in nanometers, many inorganic spectroscopists prefer the use of wavenumbers (cm^{-1}), because the unit is linearly proportional to the energy.

$$E = hc/\lambda \qquad 16.B3$$

To convert from nanometers to wavenumbers ($\bar{\nu}$), Eq. 16.B4 can be used.

$$\bar{\nu}\,(\text{in cm}^{-1}) = \frac{1\times 10^7}{\lambda\,(\text{in nm})} \qquad 16.B4$$

The intensity of light absorbed is given from the instrument most often as absorbance (*A*), which depends upon the experimental conditions and the property of the compound under

Table 16.B1 Approximate spectral ranges often used in light absorption.

	cm^{-1}	nm	eV
Ultraviolet (UV)	1,000,000–26,300	10–380	124–3.26
Visible Light			
Violet	26,300–23,000	380–435	3.26–2.85
Blue	23,000–19,200	435–495	2.85–2.38
Green	19,200–17,700	495–565	2.38–2.20
Yellow	17,700–16,900	565–590	2.20–2.10
Orange	16,900–16,000	590–625	2.10–1.98
Red	16,000–14,000	625–750	1.98–1.65
Near Infrared (NIR)	13,300–4,000	750–2,500	1.65–0.50

study through Beer's Law (Eq. 16.B5). The absorbance is the base-10 logarithm of the ratio of the incident light intensity (I_0) to the transmitted light intensity (I). The path length of the cell in centimeters (d) and the molar concentration (c) provide values for the molar extinction coefficient (or molar absorptivity, ε) of M^{-1} cm^{-1} (equivalently L mol^{-1} cm^{-1}).

$$A = \log_{10}\left(\frac{I_0}{I}\right) = \varepsilon dc \qquad 16.B5$$

The molar extinction coefficient (ε) is a property of the absorption band in the compound independent of the experimental conditions (concentration and pathlength) and is what is generally reported. Be aware that the molar extinction coefficient depends strongly on the specific wavelength/energy at which a transition appears; compounds absorb strongly at certain wavelengths (large ε) and poorly (or not at all) at other wavelengths (zero or very small ε). It is commonplace to report the molar extinction coefficient for a specific transition as the value it takes at the maximum of the peak associated with the transition. Absorption spectra often use molar absorptivity ε as the y-axis and either wavelength (nm) or wavenumber (cm^{-1}) for the x-axis.

16.2 The Multi-Electron Picture: From Orbitals to States

First, let us define some of the terminology used to describe electrons. An electronic *configuration* is a specific list of the population of the available orbitals of the system in which we are interested. Typically, the valence orbitals are all listed, but sometimes only a subset is chosen—say, only the *d*-orbitals of a complex (leaving the ligand orbitals out), or by listing only some of the ligand orbitals that interact directly with the metal, either the frontier orbitals or any other subset deemed useful. Generally, only the occupied orbitals are listed as part of the configuration, but sometimes it is also useful to list a few of the lowest-energy unoccupied orbitals as well, especially in comparing two configurations together that differ in the transfer of an electron from an occupied to an unoccupied orbital.

Often, if the symmetry is high enough there will be more than one way to organize electrons inside the orbitals of a given configuration. Each of these separate arrangements of the electrons is called a *microstate* of the electronic configuration. Any configuration where all orbitals are fully occupied is called a *closed-shell* configuration, which can have one, and only one, microstate. Electronic configurations that have at least one singly occupied orbital are called *open-shell* configurations, and generally possess many microstates. Often, many of these microstates will have the same *observable* properties and therefore cannot be distinguished from an outsider's point of view. We will call the group of all those microstates that share the same observable properties an electronic *state*.[c] Technically, if we know the electronic state of a system, then we know everything we can possibly know about the electrons of the system we want to describe: the overall energy stored in the electronic degrees of freedom, the angular momentum (orbital and spin) of the electrons, and the spatial distribution of the electronic density, to name but a few.[d] In practice, the number of individual *states* that a given electronic configuration gives rise to is often not that high (this is all the more true the higher the symmetry of the system is), so it is useful to be able to make a list of all the possible states that exist for a given configuration, and to assign names/labels to each state of the system. For instance,

[c] Atomic spectroscopists sometimes use the name "**term**" to refer to the same concept.

[d] In this way, the use of the term "state" almost parallels the thermodynamics concept of "state". A thermodynamic "state" corresponds to a definite (often implicit) list of all the macroscopically observable properties of a system.

when previously we were trying to make sense of the absorption spectrum of $\{Ni(H_2O)_6\}^{2+}$, what we really needed to know is the number of electronic *states* that this *d*-orbital system has, as this would directly give us the number of possible transitions we should expect to observe in the spectrum. Furthermore, having labels for each of the excited *states* of the complex would allow us to label the electronic transitions and, as we will see shortly, will also allow us to begin answering questions related to the precise energy, intensity, and width/shape of the observed transitions in the spectrum.

Now that we have these definitions in hand, let us figure out the number of microstates that are included in a specific configuration—a number which we are going to call N_{MS} here. (This is also called the *degeneracy* of that configuration.) Counting microstates simply boils down to counting how many independent ways you can organize N_e electrons in N_o degenerate orbitals, which is given by the binomial coefficient:

$$N_{MS} = \frac{(2N_o)!}{(2N_o - N_e)!N_e!} \quad \textbf{16.1}$$

Let us begin with the easiest case: a single electron in a single orbital, say the $1s^1$ configuration of the hydrogen atom; this configuration has $N_e = 1$ (one electron) and $N_o = 1$ (one orbital) and plugging this into the previous equation gives $N_{MS}(1s^1) = 2$. We see that the $1s^1$ (or any s^1 configuration for that matter) has two independent microstates attached to it—one for each of the spin orientations (spin up or spin down) of the electron. Of course, matters become more interesting the more orbitals we have. For instance, for the d^1 configuration there are five equivalent *d*-orbitals ($N_o = 5$) and one electron ($N_e = 1$), and so:

$$N_{MS}\left(d^1\right) = \frac{(2 \cdot 5)!}{(2 \cdot 5 - 1)! \cdot 1!} = \frac{10!}{9!} = 10$$

There are ten microstates in a d^1 configuration. This makes sense, as there two independent ways to put one electron in an orbital (spin up or spin down), and there are five different *d* orbitals. For a t_{2g}^2 configuration, $N_o = 3$ (three t_{2g} orbitals) and $N_e = 2$ (two electrons), so:

$$N_{MS}\left(t_{2g}^2\right) = \frac{(2 \cdot 3)!}{(2 \cdot 3 - 2)! \cdot 2!} = \frac{6!}{4! \cdot 2!} = \frac{6 \cdot 5}{2 \cdot 1} = 15$$

This also is intuitively correct. There are six ways to put two electrons in three orbitals (spin up or down in each), and once you put the first electron in, there are only five other ways left for the second electron, because Pauli's principle dictates that you cannot put two electrons with the same spin in the same orbital. This yields thirty combinations, but because the two electrons are indistinguishable we need to divide that number by 2, which produces the fifteen independent microstates that we deduced from Eq. 16.1. Table 16.1 lists the degeneracies of many of the important configurations that are most often encountered in inorganic chemistry. (A good exercise is to double-check these numbers yourself!)

Table 16.1 Number of microstates (N_{MS}) included in different electronic configurations.

d^1/d^9	10	p^1/p^5 $t_{2g}{}^1/t_{2g}{}^5$	6	f^1/f^{13}	14
d^2/d^8	45	p^2/p^4 $t_{2g}{}^2/t_{2g}{}^4$	15	f^2/f^{12}	91
d^3/d^7	120	p^3, $t_{2g}{}^3$	20	f^3/f^{11}	364
d^4/d^6	210	$e_g{}^1/e_g{}^3$	4	f^4/f^{10}	1001
d^5	252	$e_2{}^2$	6	f^5/f^9	2002
				f^6/f^8	3003
				f^7	3432

As Table 16.1 shows, the d^2 configuration gives rise to 45 independent microstates, which perhaps looks terrifying. Things are not so bad in reality, because very rarely can one distinguish all of them experimentally. Many of them will possess the same physical observables and are therefore, for practical purposes, identical to other listed microstates. What matters is how many independent *states* exist for a given configuration; that is, how many separate subgroups with distinguishable observable properties are in the 45 microstates of the d^2 configuration. An important rule that we will often go back to is *the total degeneracy of a configuration—the total number of microstates arising from the configuration—is equal to the sum of the degeneracy of all the states that belong to it—the number of microstates clustered into each state.*

For instance, the total degeneracy of the p^2 configuration is 15. (Table 16.1; we assume here that the system is perfectly spherically symmetrical, as would apply for a free atom/ion.) After carrying out the type of analysis we will discuss later (that is, once we include the effects of electron–electron repulsion) we will find these fifteen microstates group themselves into three, and only three, distinct states: one state that has a total degeneracy of five (5), a second state with a total degeneracy of nine (9), and a third state with a total degeneracy of one (1). As predicted by our rule, the sum of the degeneracy of all these three states adds up to the total degeneracy of the configuration as a whole: 5 + 9 + 1 = 15. This is absolutely required because microstates are neither created nor destroyed in the process of grouping them into observable states, so you can think of the electron–electron repulsion as simply breaking down the fifteen-fold degeneracy of the p^2 configuration into three groups: one five-fold state, a nine-fold state, and a nondegenerate state.

In order to learn more about how states are defined and what their degeneracy is, we will return to the behavior of free ions, which are the simplest systems to analyze due to their spherical symmetry; once we know how the states of a free ions are defined, we will show how we can go from there to the case of transition-metal complexes with lower-symmetry crystal/ligand fields applied to the free ion.

16.3 Atomic Spectroscopy and Term Symbols

16.3.1 Total Spin Angular Momentum

The states of a multi-electron free ion are characterized by their total orbital angular momentum and total spin angular momentum, which can be obtained simply by coupling the angular momenta components of each individual electron. For instance, if two spin angular momenta are coupled together (s_1 and s_2), then the magnitude of the total spin angular momentum (S_{tot}) can take all the values included in Eq. 16.2.

$$S_{tot} = s_1 + s_2, |s_1 + s_2 - 1|, \ldots, |s_1 - s_2| \qquad \textbf{16.2}$$

The absolute-value brackets are used to emphasize that the magnitude of a vectorial quantity can never take negative values; that is, the length of a vector is necessarily a positive number. In calculating lists such as that one in Eq. 16.2, always start by calculating the first and last terms, then add in all the terms that go between these two limits by systematically subtracting 1 from the maximum value. For instance, if we have two electrons ($s_1 = s_2 = ½$) and want to determine the possible total spin momenta for the pair, then we obtain $S_{tot}{}^{max} = (½ + ½) = 1$ and $S_{tot}{}^{min} = |(½ - ½)| = 0$, and there can be no other values between these two, since $S_{tot}{}^{max} - 1 = S_{tot}{}^{min}$. Consequently, the total spin angular momentum of a pair of electrons can be either $S = 1$ or $S = 0$, and nothing else. In this chapter we will follow the well-established convention where upper-case letters are used for *total/multielectronic* symbols and lower-case letters for *individual/single electron states* ("*orbitals*") symbols.

How do we go about coupling spins of more than two electrons? Let us first consider the case of a state formed by the coupling between three electrons. We begin by coupling two of the three electrons together just as we just did previously, and then we couple the total spin of the first two, which can be $S_{12} = 1$ or 0 as shown previously, to the spin of the third one, which is always ½ for a single electron, in order to obtain the total spin functions. Starting with $S_{12} = 1$, we obtain $S_{tot}{}^{max} = (1 + ½) = 3/2$ and $S_{tot}{}^{min} = |(1 - ½)| = ½$; for $S_{12} = 0$, we obtain $S_{tot}{}^{max} = (0 + ½) = ½$, which is the same value as $S_{tot}{}^{min} = |(0 - ½)| = ½$. So, coupling three electron spins yields three separate total spin functions: one function of magnitude $S = 3/2$, and two functions that each have $S = ½$. Obviously, we can keep applying the same approach to calculate the *possible* total spin values for any arbitrary number of electrons. We say "*possible*" here because these are all the possibilities if we assume that all the individual electrons are free to adopt any spin orientation independent of the orientations which the other electrons adopt, which is rarely the case in practice.

Generally, given a specific orbital configuration, Pauli's principle (see Section 16.3.4) will actually prohibit certain spin alignments, so not all previous possibilities can necessarily occur. This is most easily seen by considering two electrons occupying the same *s* orbital, such as for the ground-state of the helium atom. In this case, the only acceptable spin function is the one with $S = 0$ total spin ("spin-paired"): the $1s^2$ configuration can only have a state with total spin of $S = 0$. On the other hand, two electrons in a $1s^12s^1$ configuration (an excited state of the helium atom) are free to adopt any spin orientation they can, given that the two electrons are not in the same spatial orbital, and therefore not limited by Pauli's exclusion principle. This specific configuration can then lead to all possible spin states for two independent electrons: either $S = 0$ or $S = 1$.

Just as an individual electron spin can take different orientations that we label using the quantum number $m_s = ½$ or $-½$, the total spin momentum of a multielectron system can also take specific orientations, which by analogy with the single-electron symbol we will label using the symbol M_S. (Again, upper-case symbols for multielectron terms.) For a specific total spin value S, the possible orientations are associated with the quantum numbers $M_S = S, S - 1, S - 2, \ldots, -S$; for instance, $S = 0$ has only one orientation ($M_S = 0$), $S = ½$ has two possible orientations ($M_S = ½, -½$), $S = 1$ has three orientations ($M_S = \pm1, 0$), and so on (Table 16.2). It is not too hard to see that the number of orientations allowed for each total spin value is $2S + 1$; this is an important quantity, which is called the spin *multiplicity*. When we refer to the multiplicity of a state, we are simply specifying its total spin degeneracy; how many equivalent ways can the total spin of that state be oriented?[e] The multiplicity is in fact such an important quantity that we assign each type of multiplicity a specific name: a *singlet* is a state with a multiplicity of 1, a *doublet* a state with a multiplicity of 2, and so on.

[e] We really could have called this quantity *spin degeneracy*—which is really all it is—but the term *multiplicity* is the one that was assigned historically and is still commonly used.

Table 16.2 The names associated with each type of spin multiplicity, up to $S = 4$.

S	0	½	1	1 ½	2	2 ½	3	3 ½	4
Multiplicity (= 2S + 1)	1	2	3	4	5	6	7	8	9
Name	singlet	doublet	triplet	quartet	quintet	sextet	septet	octet	nonet

16.3.2 Total Orbital Angular Momentum

The second important quantity that defines the electronic properties of a free ion is the orbital angular momentum of the system. Here again, the total angular momentum of a group of electrons can be defined by coupling the angular momenta of each individual electron. By analogy with the coupling of spin momenta, the magnitude of the total orbital angular momentum is as follows.

$$L_{tot} = l_1 + l_2, |l_1 + l_2 - 1|, \ldots, |l_1 - l_2| \qquad \textbf{16.3}$$

The magnitude of the individual orbital momenta depends on the type of orbital occupied by a given electron: an *s* electron has $l = 0$, a *p* electron has $l = 1$, a *d* electron has $l = 2$, and so on. It is therefore straightforward to determine the magnitudes of the allowed total orbital momentum once the electronic configuration of a free ion is specified. For example, an s^2 configuration has both electrons in the same *s* orbital, $l_1 = l_2 = 0$, and this configuration leads only to one possible total orbital angular momentum state with $L = 0$. A d^2 configuration has more possibilities: $L_{tot}^{max} = 2 + 2 = 4$, and $L_{tot}^{min} = 2 - 2 = 0$, so the possible values are $L = 4, 3, 2, 1, 0$. Similar to the approach used for the total spin momentum described previously, the total orbital angular momentum of a system of more than two electrons would be given by first coupling two of the electrons, then coupling their allowed total orbital momenta with a third electron, and so on until all electrons have been coupled.

By analogy with the system used to label atomic orbitals, we use upper-case letters to label each type of total orbital momentum: an "S" state has $L = 0$ (just as an "*s*" orbital has $l = 0$), a "P" state has $L = 1$, a "D" state has $L = 2$, and so on. Here again, every type of state can take up to $2L + 1$ orientations, which defines the *orbital degeneracy* of that state. Note that although electronic orbitals are generally limited to $l \leq 3$ (*f* orbitals are the highest ones typically encountered in chemistry), the multielectron states of free ions (or other spherically symmetrical species) can often possess much larger orbital angular momenta than that.

16.3.3 Atomic Term Symbols

Once we know the total spin and orbital angular momenta of a system, we generally know almost everything we need to label the state of that system.[f] We do this by combining the symbol for the total orbital angular momentum with the spin multiplicity of the state by writing it this way "$^{2S+1}\Gamma$", where you should recognize the number $2S + 1$ as the spin multiplicity, and the Greek letter Γ (gamma) standing in for the specific total orbital angular momentum symbol (Table 16.3). Such a symbol that combines the spin and orbital information of a state is called a "*term symbol*", and an *atomic term symbol* if the system is an atom or a free ion. For example, a ^{2}S state ("doublet S") is a state that has a spin multiplicity of 2 (hence $S = ½$), and $L = 0$; a ^{4}H state ("quartet H") is a state that has a spin multiplicity of 4 (hence $S = 3/2$), and $L = 5$. Because spin and orbital momenta are free to take any value independent of the other (in the absence of spin–orbital coupling), the total degeneracy of a state is simply the product of the spin and angular degeneracies in Eq. 16.4. Using this equation, we find that a ^{2}S state has a degeneracy of $2 \times 1 = 2$, whereas a ^{4}H state has a degeneracy of $4 \times 11 = 44$, and so on for any other atomic terms.

$$\text{Total state degeneracy} = (2S + 1)(2L + 1) \qquad \textbf{16.4}$$

Table 16.3 Total orbital momentum symbols.

L	0	1	2	3	4	5	6
$2L + 1$	1	3	5	7	9	11	13
Symbol	S	P	D	F	G	H	I

16.3.4 The Electronic States Associated with a Configuration

We are now able to start working out the states that belong to a specific electronic configuration. There are a few simple methods for finding the state that one can follow, depending on the electronic configuration. We will divide these into a few different possibilities, and the discussion here will assume a transition-metal complex with octahedral or pseudo-octahedral symmetry. Here at the beginning we will summarize these somewhat, while the remainder of this section will expand upon these possibilities.

- *Possibility 1: All the valence orbitals on the metal are filled, which is often the case for an 18-electron configuration.* In this case, the electron density will be spherical, and the term symbol will be $^1A_{1g}$ in an octahedral field or ^{1}S in atomic spectroscopy.
- *Possibility 2: There is 1 electron in a nondegenerate orbital.* An example would be a d^9 Jahn–Teller distorted octahedral system with axial elongation, now with symmetry D_{4h}. The highest-energy orbital would be expected to be $d_{x^2-y^2}$, which is in irreducible representation b_{1g} and is singly occupied. As a result, the ground state would be $^2B_{1g}$—doublet because there is one unpaired electron with spin = ½ (Table 16.2) and B_{1g} because that is the symmetry of the occupied orbital.
- *Possibility 3: Two unpaired electrons in two different non-degenerate orbitals.* In this case, the term will be singlet or triplet (depending on how the spins are assigned in the two

[f] Strictly speaking, we also need to eventually characterize the amount of *coupling* between the two types of angular momenta (this is called *spin-orbit* coupling) which would break the degeneracy of the states, forcing us to use one more label to distinguish each spin–orbit component as well. In practice, these effects are much smaller than the electron–electron repulsion effects in some cases, and we will neglect them here for now.

orbitals) and the direct product (Section 5.12) of the orbitals' irreducible representations. For example, let us say the configuration might be $b_{1g}{}^{1}a_{1g}{}^{1}$. The direct product of B_{1g} and $A_{1g} = B_{1g}$. (Anything times the totally symmetric representation is itself.) As a result, the terms from this configuration would be $^{1}B_{1g}$ and $^{3}B_{1g}$, with the higher spin state ($^{3}B_{1g}$, Hund's rule, Section 2.2.3) being the ground state.

- *Possibility 4: There are two or more electrons in degenerate orbitals.* In this case, the symmetries will still be related to the direct products. In other words, in a d^2 octahedral system with configuration $(t_{2g})^2$, the terms will be $T_{2g} \times T_{2g} = A_{1g} + E_g + T_{1g} + T_{2g}$. However, not all of the possible spin states (singlet and doublet) will be possible for all of these, due to the Pauli exclusion principle. How can we find the allowable spin states and, furthermore, the microstates that are related to each term symbol for degenerate systems? This question will occupy much of the remainder of this section.

Let's expand on these possibilities and discuss the details behind them, starting with the degenerate state problem.

Finding all the states associated with degenerate configurations can be somewhat more difficult, generally because there are many microstates, and one must make sure to account for the indistinguishability of electrons and spin restrictions. To keep things simple at first, we will work out the procedure for a simple configuration, p^2, but the process is not very different for other configurations. We found previously that the p^2 configuration contains fifteen microstates (see Table 16.1), but we now want to figure out how to group these microstates, which are not directly observable, to form the different observable states of the system. The simplest way to approach this problem is to begin by writing out a *microstate table*, in which we organize each microstate according to its total spin orientation (M_S) and total orbital angular momentum (M_L), which we can always easily figure out by directly summing the individual spin and orbital angular momenta of each electron in that microstate. To do this efficiently we need to adopt a convention to quickly write out each microstate in a way that makes it easy to see what these individual quantum numbers, m_s and m_l, are for all electrons. There are four things to note:

(1) Because we need the quantum numbers m_l, we want to use the complex orbitals here, not the real ones, so instead of the p_x, p_y, p_z orbitals, we will use the p_{-1}, p_0, p_1 ones.[g]

(2) Instead of writing down the symbol of the orbital every time, we will simply use the m_l value: the three p orbitals will be labeled "1", "0", and "–1", the five d orbitals "2", "1", "0", "–1", and "–2", and so on. One possible drawback from this approach is that we lose the information on the type of orbital (p_0 and d_0 are both labeled "0" in this system), but in practice we seldomly work with configurations containing different types of orbitals at the same time, so this is not really a problem.[h]

(3) We will use a superscript "+" or "–" to label whether each electron is in a "spin up" ($m_s = +½$) or "spin down" ($m_s = -½$) orientation. A spin-up electron in a d_0 orbital will be written "0^+".

(4) Our microstate symbols will be written by listing all electrons; to avoid repeating the same microstates (*vide infra*), we will always list the electrons in decreasing m_l values, and when there are two electrons with the same m_l value, we will always list the spin-up electron first. $(2^+,-1^+)$ is an acceptable microstate in our convention; $(-1^+,2^+)$ is really the same microstate, but it does not follow the convention so we do not list it. (0^+0^-) and (0^-0^+) are the same microstate, but only the former symbol obeys our convention.

As an illustration, one of the microstates of the p^2 configuration might be written symbolically as $(1^+,0^-)$: this would correspond to a microstate where the first electron is put in a p_1 orbital with a spin up, and the second electron in a p_0 orbital with a spin down. A useful feature of our system is that we can always easily define the total angular momentum orientation properties (M_L and M_S values) of any microstate. For instance, our $(1^+,0^-)$ microstate clearly has a total orbital angular momentum orientation $M_L = 1 + 0 = 1$ and a total spin momentum orientation $M_S = +½ - ½ = 0$. Before going on and listing all fifteen microstates of the p^2 configuration, we need to be aware of two fundamental rules that rigidly dictate what can and cannot be written as a microstate:

(1) *Pauli's Exclusion Principle*: Two electrons cannot have the same set of quantum numbers. *A microstate such as $(1^+,1^+)$ is not physically allowed. We cannot put two electrons with the same spin in the same orbital.*[i]

(2) *Electrons are indistinguishable*: Exchanging two electrons does not create a new microstate. *The microstate $(1^+,1^-)$ is really the same one as $(1^-,1^+)$.* Similarly, *$(1^+,0^+)$ and $(0^+,1^+)$ are also not distinguishable microstates.* The thing to look out for is whether you can get the same label simply by rearranging the order of the spin-orbitals listed in the microstate; if you can, then the two labels are just different ways of writing out the same microstate. What about $(1^+,0^-)$ and $(1^-,0^+)$? These are *not* identical microstates; there is no way we can rearrange the two spin-orbitals to get the same label. The approach described in the previous note 4 helps to avoid repeating identical states by systematizing the labeling of the microstates.

[g] Recall from Chapter 2 that complex orbitals have well-defined angular momentum values, which is not the case with the real orbitals; the two sets are equivalent, so we are perfectly free to choose whichever works best for us.

[h] Furthermore, the states that arise from mixed configurations can be directly obtained by taking the direct product of the states from each separate "sub-configurations" (Section 16.4.2.2).

[i] Unless, of course, the two $m_l = 1$ orbitals belong to a different set of orbitals, say $3d_1$ and $3p_1$. As mentioned previously, here we consider only configurations where all orbitals are from the same set.

Table 16.4a Microstate table for the p^2 configuration.

$M_L \backslash M_S$	+1	0	−1
+2		$(1^+,1^-)$	
+1	$(1^+,0^+)$	$(1^+,0^-)$ $(1^-,0^+)$	$(1^-,0^-)$
0	$(1^+,-1^+)$	$(1^+,-1^-)$ $(1^-,-1^+)$ $(0^+,0^-)$	$(1^-,-1^-)$
−1	$(0^+,-1^+)$	$(0^+,-1^-)$ $(0^-,-1^+)$	$(0^-,-1^-)$
−2		$(-1^+,-1^-)$	

All fifteen microstates of the p^2 configuration can now be written and organized in a microstate table, as shown in Table 16.4a. The number of rows/columns for the table is directly linked to the type of orbitals used as well as to the number of electrons in the configuration. With two electrons located in *p*-orbitals, M_L can take integer values between $-(1 + 1)$ and $(1 + 1)$, so $M_L = -2, -1, 0, 1, 2$, and the microstate table has five distinct rows for each M_L value. The total spin orientation can take all values between $-(½ + ½)$ and $+(½ + ½)$, so $M_S = -1, 0, +1$, and our table should have three distinct columns for each of these values. The table is then systematically populated with each possible microstate.

You can convince yourself that we did not leave out any microstate, or that we did not violate the convention established in note 4. A systematic approach to building a microstate table begins by placing both electrons in the orbital with the highest m_l value ($m_l = +1$), which forces us to pair the spins in order to not violate Pauli's principle. This microstate is labeled $(1^+,1^-)$; by adding up the m_l and m_s values, we conclude that this microstate belongs in the $M_L = 2$, $M_S = 0$ cell in our microstate table. Given that we are out of options by keeping both electrons in the $m_l = +1$ orbital, we go on next by moving one of the two electrons into the next-highest m_l valued orbital, which is $m_l = 0$. It turns out that there are four distinct ways to arrange the spins of these two electrons in two distinct orbitals, which we write as $(1^+,0^+)$, $(1^+,0^-)$, $(1^-,0^+)$, and $(1^-,0^-)$; again, each of these microstates is easily assigned to the proper cell of the table by summing the orbital and spin orientation values. By moving both electrons from orbital to orbital and organizing their spins according to Pauli's exclusion principle, we can write out the ten remaining microstates and put them in the proper position in the microstate table.

At this point we have figured out how many microstates were included in the p^2 configuration (15), learned how to assign each one a specific label, and organized these microstates in a table according to their unique total orbital and spin angular momentum properties. We are now one step away from grouping these microstates into states.[j] The trick to do this is to carefully examine our microstate table. Notice how there is a microstate with a total orbital momentum projection of $M_L = 2$, *but nothing larger than that value.* This is a very significant observation, as this implies that $M_L = 2$ *must* be the maximum angular momentum orientation of one of the states included in the p^2 configuration, which tells us directly that there *must* be a state with total orbital momentum of magnitude $L = 2$. Any state with *less* angular momentum would not have an $M_L = 2$ projection, and any state with *more* angular momentum would need to have a $M_L > 2$ component, but there is not one listed in our table. So, simply by looking at our table for a few seconds, we directly conclude that there must be a "D" state ($L = 2$) in the p^2 configuration.

Now, what is the spin multiplicity of that "D" state? What we need to find is the *maximum* M_S value consistent with the *maximum* M_L value of our state. Here, we find out that there is no microstate that has $M_L = +2$ and $M_S = 1$; that is, there is no microstate in the $M_L = +2/M_S = 1$ cell of our microstate table. Given that $M_S = 0$ is the only possible spin orientation for our "D" state, this means that this state *must* be a $S = 0$ state—^{1}D a "*singlet D*" state.

Using Eq. 16.4, we know that this state has a total degeneracy of $(2S + 1) \times (2L + 1) = 1 \times (2 \cdot 4 + 1) = 5$, so we can now account for five of the fifteen microstates in the p^2 configuration, leaving ten that still need to be assigned. To keep track of the fact that we have already assigned some of the microstates, we will strike out five microstates from our table (Table 16.4b) to remind us that we already found a ^{1}D state. Of course, we need to be clever about this; we need to count one microstate for each of the M_L orientations of our ^{1}D state. This is a singlet state, so $M_S = 0$ for all five microstates of the ^{1}D state. When there is more than one microstate in a M_L box, we will simply strike out any one of them (it does not matter which one) as a way to keep track of the numbers.[k]

The next step proceeds by using the exact same approach as was used to find the ^{1}D state. We begin by identifying the most extreme M_L/M_S component in the table. Now that we

Table 16.4b Updated microstate table for the p^2 configuration; microstates that are struck through have been removed to account for the five-fold degeneracy of the ^{1}D state.

$M_L \backslash M_S$	+1	0	−1
+2		~~$(1^+,1^-)$~~	
+1	$(1^+,0^+)$	~~$(1^+,0^-)$~~ $(1^-,0^+)$	$(1^-,0^-)$
0	$(1^+,-1^+)$	~~$(1^+,-1^-)$~~ $(1^-,-1^+)$ $(0^+,0^-)$	$(1^-,-1^-)$
−1	$(0^+,-1^+)$	~~$(0^+,-1^-)$~~ $(0^-,-1^+)$	$(0^-,-1^-)$
−2		~~$(-1^+,-1^-)$~~	

[j] Remember that "states" are groups of microstates that all have the same energy, so these "groups" are the only thing we usually get to observe experimentally, not the individual members, the microstates, of these groups.

[k] At this point, you might be tempted to ask *which* microstate should be assigned to the ^{1}D when there is more than one microstate in the same box; for example, the $M_L = +1$, $M_S = 0$ box has two microstates listed: $(1^+,0^-)$ and $(1^-,0^+)$. Which one of those two should we pick to assign to the ^{1}D state? The short answer is that a combination of both microstates will be used. Finding out that specific combination is a problem that goes beyond the scope of this book, but relies on using elementary angular momentum operations on the microstates. For our purposes, we will not need to find the exact wavefunctions of each microstate, so we will leave this problem aside. See Ballhausen or Griffith for a discussion of that problem.

Table 16.4c Updated microstate table for the p^2 configuration; microstates that are struck through have been removed to account for the five-fold degeneracy of the ^{1}D state and the nine-fold degeneracy of the ^{3}P state.

$M_L \backslash M_S$	+1	0	−1
+2		~~$(1^+,1^-)$~~	
+1	~~$(1^+,0^+)$~~	~~$(1^+,0^-)$~~ ~~$(1^-,0^+)$~~	~~$(1^-,0^-)$~~
0	~~$(1^+,-1^+)$~~	~~$(1^+,-1^-)$~~ ~~$(1^-,-1^+)$~~ $(0^+,0^-)$	~~$(1^-,-1^-)$~~
−1	~~$(0^+,-1^+)$~~	~~$(0^+,-1^-)$~~ ~~$(0^-,-1^+)$~~	~~$(0^-,-1^-)$~~
−2		~~$(-1^+,-1^-)$~~	

have accounted for the ^{1}D state, we find that the next-highest angular momenta values must be $L = 1$ and $S = 1$ (top left corner of the table), which means that we have found a ^{3}P state. That state has a total degeneracy of $3 \times (2 \cdot 1 + 1) = 9$, so there is now only one microstate left to assign (Table 16.4c).

Of course, there is only one way to assign the last remaining microstate. It must belong to a state that has $M_L = 0$ and $M_S = 0$ as the only possible orbital/spin momenta orientations, which can only happen for a state with $L = S = 0$, which corresponds to the nondegenerate ^{1}S state.

We have now found all states that arise from the p^2 configuration: ^{1}D (5 microstates), ^{3}P (9 microstates), ^{1}S (1 microstate). These states can be observed experimentally, for instance, from the gas-phase spectroscopy of atomic carbon (Fig. 16.3), which has a $2s^22p^2$ configuration. As can be seen directly here, the energy ordering of the three states of the p^2 configuration follows well the first two of Hund's rules:

(1) Within a given configuration, the ground state (the one with the lowest energy) is the one with the *highest multiplicity*.

(2) For states that have the same multiplicity, the one with the largest orbital angular momentum has the lowest energy.

The C(0) example emphasizes an important point: filled shells (or subshells) do *not* contribute to the angular momentum (spin or orbital) of the atom, which is why we are able to directly compare the states of the $2s^22p^2$ configuration of atomic carbon with those we got out of the p^2 configuration. The $2s^2$ set plays no role here, at least not in determining the angular momentum properties of the states, which is what we need to determine states and affix labels. You should be able to convince yourself that this is always true by trying to write out the microstates of any filled-shell configuration, where you would quickly realize that there is always one and only one way to arrange electrons, which means that a full shell can only have $L = S = 0$. For filled orbitals, spins are always going to be all paired, and every time you put an electron in an m_l orbital, you will always put one in the corresponding $-m_l$ orbital, so M_L will always add up to zero as well. Concretely, we say that filled sets of atomic orbitals provide a spherically symmetric density. At most they shift uniformly the energies of all electronic states, but they neither change the number nor angular momentum properties of these states.

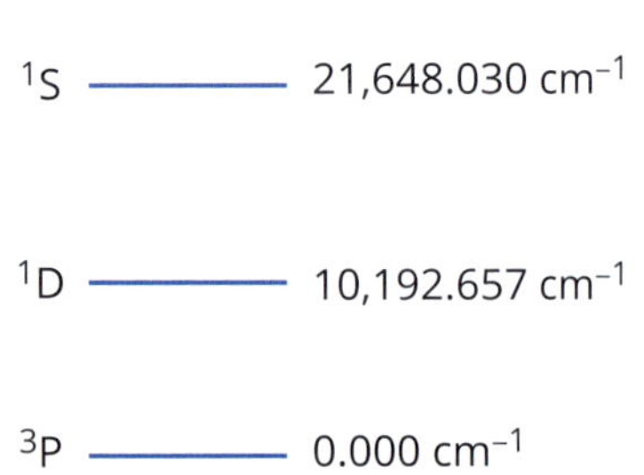

Fig. 16.3 Energy levels of free carbon atom ($2s^22p^2$ configuration). Note the high precision of the data, which is typical for atomic spectroscopy. (Data acquired from the NIST database.)

Configurations that have only one electron ($s^1, p^1, d^1, f^1 \ldots$) are easy to figure out, as mentioned previously; the state arising from this configuration (it will always be just one state) will necessarily have the same orbital and spin momenta as the orbital populated, so we can just write $S = ½$—a doublet state—and use the same upper-case letter as the orbital to designate the state, which would produce ^{2}S, ^{2}P, ^{2}D, ^{2}F . . . respectively for the four 1-electron configurations listed previously.[1]

Interestingly, we can also show that configurations that are almost full *except for missing a single electron* (p^5, d^9, $f^{13} \ldots$) lead to *precisely the same states* as those that have only one electron. Consider the d^1/d^9 case shown in Fig. 16.4, both of which provide a ^{2}D state.

As can be seen, when a shell is nearly full, most of the individual m_l values cancel out. Every time an electron is located in an m_l orbital there is another one with the value $-m_l$. In the d^9 configuration this happens for every electron *except for the last one*. This one electron alone then dictates the orbital momentum behavior of the configuration. At the same time, all spins cancel out except for the last unpaired electron, so the spin state of the configuration is also ultimately determined by the last electron. In the end, both configurations are dictated by the behavior of a single electron in a set of *d*-orbitals, which is why the states that arise are the same ones.

This correspondence comes from a very fundamental principle, called the *electron-hole equivalence*. A *hole* is an abstract construct used to represent the *absence of an electron*. This might appear somewhat disconcerting if this is the first time you encounter this concept (how can "nothing" be "something"?!), but this is not as counterintuitive as it might appear. Holes are simply a convenient way of quickly talking about where electrons are *not* located, which is a useful thing to do

[1] Do not be confused by the fact that we use upper-case symbols, which are reserved for multi-electron states, to a state that comes from a "1-electron". We are trying to distinguish orbitals, which are not observable directly, and the electronic states of a system, which are measurable directly (usually spectroscopically). Furthermore, except for the notable case of the hydrogen atom, remember that when we talk about a "single-electron" configuration, we do not imply that the system has only one electron; we are simply choosing to ignore those other electrons that are in filled shells that do not contribute to the angular momentum of the system.

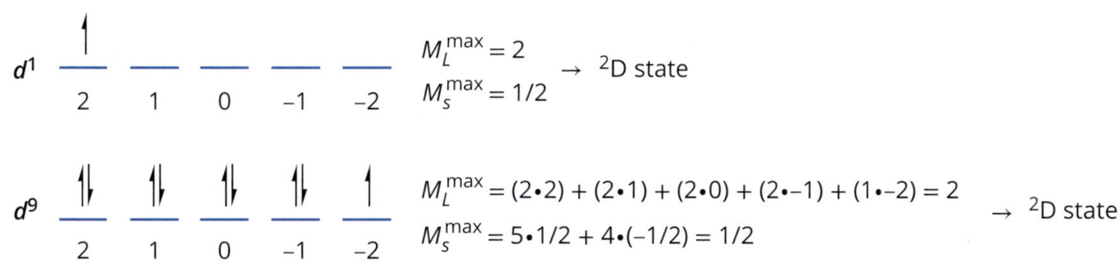

Fig. 16.4 The electron-hole analogy: the states arising from the d^n configuration are the same as those arising from the d^{10-n} configuration, illustrated here for the $n = 1$ scenario. Both d^1 and d^9 configurations have the same electronic ground-state, ^{2}D.

whenever there are more electrons than places to store electrons; that is, when a level is more than half-filled. Consider a large amphitheater as an analogy. If the place is almost sold out, it makes little practical sense to list all occupied seats just to figure out where you can still sit one more spectator; you are much better off by listing the few *unoccupied* seats. Conversely, if everything is open but for a few seats, then it is much easier to list those few seats that are *occupied* as a way to describe the state of occupancy of the amphitheater. Each approach—listing the occupied or the unoccupied seats—contains the same amount of information on where you can sit someone, but the point of reference has been switched. Because holes represent missing electrons, we need to treat them like particles that behave in the exact opposite way; that is, as positively charged particles. An important consequence of this is that although electrons tend to fill the lowest-energy orbitals first, holes act the opposite way; they always populate the highest-energy orbitals first.[m]

The spin of a hole is also the opposite of the spin that an electron would have in the same orbital; if you remove a spin-down electron from a filled shell, then the hole that is left behind will be spin-up, reflecting the fact that the unpaired electron left behind is indeed spin-up. We have already seen a similar effect when we considered the free-ion d^1 and d^9 configurations; the configuration associated with one electron spin-up in the d_{+2} orbital is the same as the d^9 configuration with the hole in the d_{-2} orbital, as shown in Fig. 16.4. That is, the physics of a system of n negative electrons defined by the set of quantum numbers L, M_L, S, M_S is completely identical to that of a system of n positive holes defined by the set $L, -M_L, S, -M_S$. Paul Dirac—one of the great pioneers of quantum mechanics—originally conceived the idea of a particle named the "positron" (the electron's antiparticle in nuclear physics), which he also associated with the concept of a "hole" in a "sea" of electrons. Nevertheless, whereas positrons are now known to be real particles, it is important to realize that what we call holes in our context are not real particles, but convenient constructs to avoid listing large numbers of electrons when we write down microstates, by instead focusing on the orbitals occupied by holes; that is, orbitals where electrons are missing.[n]

The electron-hole equivalence can also be used to show that any configuration that misses n electrons from the maximum number contains the same states as the analog configuration that has only n electrons in it. For instance, we worked out the states for the previous p^2 configuration, but without any more work we can invoke the electron-hole equivalence to show that the configuration with two missing electrons ($p^{6-2} = p^4$) also gives rise to ^{3}P, ^{1}D, and ^{1}S states. By the same token, the d^1/d^9, d^2/d^8, d^3/d^7, and d^4/d^6 configurations are all deeply connected, as once you know the states for one, you obtain the other for free. (The same applies to the configurations of O_h complexes, as we will see shortly.) We already knew previously that these pairs of configurations contained the same number of microstates (Eq. 16.3 and Table 16.1), but now we also know that they also possess the same states. The idea is illustrated in Fig. 16.5 for the d^1/d^9 configurations in O_h symmetry, and we will soon demonstrate the connection between these seemingly different scenarios.

Using these principles and methodology, we can obtain all atomic states (also known as atomic terms) that arise from the different d^n configurations, which play a central role in inorganic chemistry (Table 16.5).

You can see that the number of states quickly increases as we increase the number of d electrons up to the half-filled

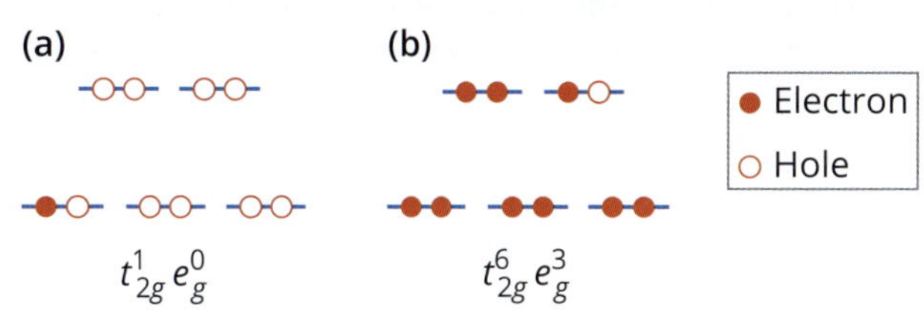

Fig. 16.5 The electron-hole analogy for transition-metal ions in an octahedral (O_h) environment. (a) d^1 ground-state configuration (O_h symmetry). The single electron goes into the lowest-energy orbital. (b) d^9 ground-state configure (O_h symmetry). The single hole goes into the highest-energy orbital.

[m] You can think, if you will, of holes as "bubbles" that rise to the top of a liquid, and electrons as particles that sink to the bottom of the liquid.

[n] Solid-state physicists make the distinction between the two concepts by referring to holes as "virtual" particles, emphasizing the fact that the concept of a hole is simply a convenient way of describing the behavior of a collection of particles as a single object.

Table 16.5 Term symbols from the states that arise from each atomic d^n configuration. The total degeneracy of each state is shown in parentheses.

d^1/d^9	^{2}D (10)
d^2/d^8	^{3}F (21), ^{3}P (9), ^{1}G (9), ^{1}D (5), ^{1}S (1)
d^3/d^7	^{4}F (28), ^{4}P (12), ^{2}H (22), ^{2}G (18), ^{2}F (14), 2 × ^{2}D (10), ^{2}P (6)
d^4/d^6	^{5}D (25), ^{3}H (33), ^{3}G (27), 2 × ^{3}F (21), ^{3}D (15), 2 × ^{3}P (9), ^{1}I (13), 2 × ^{1}G (9), ^{1}F (7), 2 × ^{1}D (5), 2 × ^{1}S (1)
d^5	^{6}S (6), ^{4}G (36), ^{4}F (28), ^{4}D (20), ^{4}P (12), ^{2}I (26), ^{2}H (22), 2 × ^{2}G (18), 2 × ^{2}F (14), 3 × ^{2}D (10), ^{2}P (6), ^{2}S (2)

shell; this tracks the increase in the number of microstates included in the d^n configurations (see Table 16.1). A good exercise is to double-check that the degeneracy of each state corresponds to the term to which it is assigned, and that the sum of all the state degeneracies adds up to the number of microstates included in each configuration, as it should.

16.3.5 The Energies of Atomic States: Electron–Electron Repulsion Parameters

Each individual electronic state that arises from a given configuration has its own set of physical observables, and most importantly a specific energy associated with it. The energy of a multi-electron state is modulated in large part by the amount of electron–electron repulsion associated with the specific spatial organization of the electrons in that state. For an atom (or an ion in free space, a "free ion"), these electron–electron repulsion terms are dictated by the radial and angular distributions of the electronic wavefunction. Interestingly, a closer analysis of the theory of atomic spectra shows that the amount of electron–electron repulsion energy for any state of a d^n configuration can be described by a set of three parameters only. The exact value of these three parameters can (and will) change, depending on the transition-metal ion (its identity, oxidation state, chemical environment, and so on), but these three parameters alone, once they are known, are enough to predict the electron–electron repulsion energies of all d^n states of that transition-metal ion. In fact, because we always set the energy of the ground-state to zero (generally, we are interested in the difference in energy between the ground and excited states, not their absolute energies), we really need only two parameters to define the energy of any d^n states (Fig. 16.6).

There are various schemes to define these parameters, but the most versatile and commonly used ones are the so-called *Racah parameters*, which we owe to the work of the Israeli-Italian physicist Giulo Racah (1909–1965). The three Racah parameters for the d^n configurations are simply called A, B, and C, with the parameter A often omitted because the energy of all d^n states arising from the same configuration share a

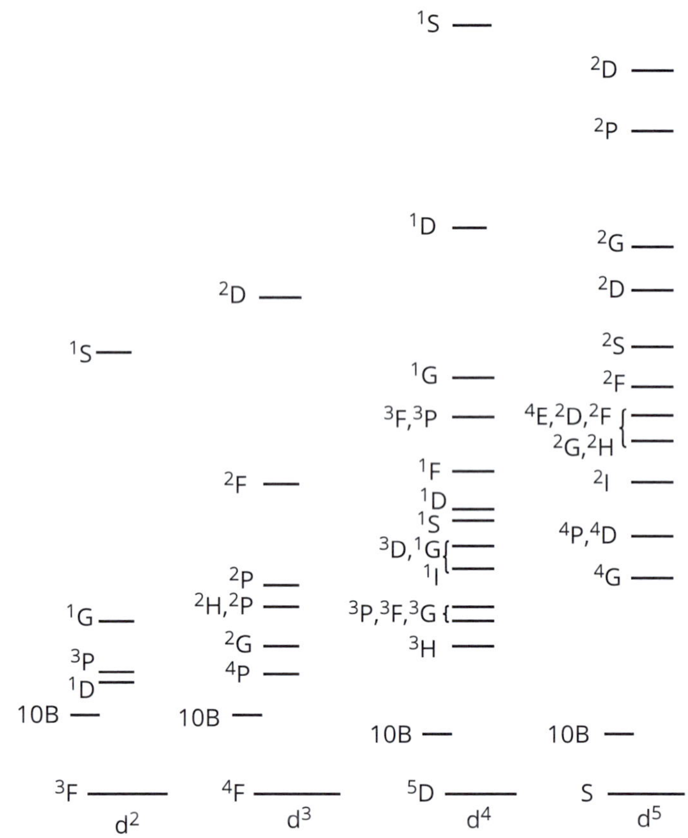

Fig. 16.6 Relative energies of terms in various d^n configurations. The diagram is constructed with C/B = 4.7. Only d^2–d^5 are shown as diagrams while d^{10-n} are the same. (Adapted from Figgis and Hitchman.[2])

common A. Whereas the A term represents a uniform shift in energy for all states of a given d^n configuration, the B term in the Racah formalism has the useful property that the energy differences between states that have the same spin multiplicity as the ground state depend only on B. Table 16.6 lists the energy of all free-ion d^n states in terms of the Racah parameters, and Table 16.7 provides empirical estimations of the free-ion values of the Racah parameters B and C for many common transition-metal ions in various oxidation states.

Several factors contribute to the exact value of the electron–electron repulsion parameters, but the general behavior is closely associated with the radii of the d-orbitals. That is, the values of the Racah parameters increase with increasing nuclear number Z and with increasing oxidation state; that is, decreasing ionic radii. For example, titanium has the largest radius and nickel the smallest for the elements along the first row, and the Racah parameter values in Table 16.7 increase fairly steadily across the series. As the oxidation state for any element increases, electrons are forced to occupy smaller orbital volumes due to radial contraction, leading to an increase in the Racah parameters. We will see later that delocalization of electrons onto ligand orbitals due to metal–ligand bond formation also leads to decreases of the effective values of the Racah parameters in transition-metal ion complexes: the nephelauxetic effect.

Table 16.6 Free-ion energies of d^n states.

d^n Configuration	State Energies
d^2 (for d^8, add $27A - 42B + 21C$ to all terms)	$E(^1S) = A + 14B + 7C$ $E(^1G) = A + 4B + 2C$ $E(^3P) = A + 7B$ $E(^1D) = A - 3B + 2C$ $E(^3F) = A - 8B$
d^3 (for d^7, add $18A - 28B + 14C$ to all terms)	$E(^2P) = 3A - 6B + 3C$ $E(^2_{\pm}D) = 3A + 5B + 5C \pm \sqrt{193B^2 + 8BC + 4C^2}$ $E(^2F) = 3A + 9B + 3C$ $E(^2G) = 3A - 11B + 3C$ $E(^2H) = 3A - 6B + 3C$ $E(^4P) = 3A$ $E(^4F) = 3A - 15B$
d^4 (for d^6, add $9A - 14B + 7C$ to all terms)	$E(^1_{\pm}S) = 6A + 10B + 10C \pm 2\sqrt{193B^2 + 8BC + 4C^2}$ $E(^1_{\pm}D) = 6A + 9B + \frac{15}{2}C \pm \frac{3}{2}\sqrt{144B^2 + 8BC + C^2}$ $E(^1F) = 6A + 6C$ $E(^1_{\pm}G) = 6A - 5B + \frac{15}{2}C \pm \frac{1}{2}\sqrt{708B^2 - 12BC + 9C^2}$ $E(^1I) = 6A - 15B + 6C$ $E(^3_{\pm}P) = 6A - 5B + \frac{11}{2}C \pm \frac{1}{2}\sqrt{912B^2 - 24BC + 9C^2}$ $E(^3D) = 6A - 5B + 4C$ $E(^3_{\pm}F) = 6A - 5B + \frac{11}{2}C \pm \frac{3}{2}\sqrt{68B^2 + 4BC + C^2}$ $E(^3G) = 6A - 12B + 4C$ $E(^3H) = 6A - 17B + 4C$ $E(^5D) = 6A - 21B$
d^5	$E(^2S) = 10A - 3B + 8C$ $E(^2P) = 10A + 20B + 10C$ $E(^2_{\pm}D) = 10A - 3B + 11C \pm 3\sqrt{57B^2 + 2BC + C^2}$ $E(^2D) = 10A - 4B + 10C$ $E(^2_{\pm}F) = 10A - 17B + 9C \pm (8B - C)$ $E(^2_{\pm}G) = 10A - 8B + 9C \pm (5B + C)$ $E(^2H) = 10A - 22B + 10C$ $E(^2I) = 10A - 24B + 8C$ $E(^4P) = 10A - 28B + 7C$ $E(^4D) = 10A - 18B + 5C$ $E(^4F) = 10A - 13B + 7C$ $E(^4G) = 10A - 25B + 5C$ $E(^6S) = 10A - 35B$

Table 16.7 Some Racah parameters for the free ions of the first-row transition elements. B is shown in bold in units of cm^{-1}. The second value is the C/B ratio.[2]

	Charge				
Element	0	+1	+2	+3	+4
Ti	**380**, 2.4	**583**, 3.4	**714**, 3.7		
V	**436**, 2.4	**585**, 4.2	**760**, 3.8	**886**, 4.0	
Cr	**790**, 3.2	**655**, 4.1	**796**, 4.2	**933**, 4.0	**1038**, 4.1
Mn	**720**, 4.3	**680**, 4.6	**859**, 4.1	**950**, 4.3	**1088**, 4.1
Fe	**805**, 4.4	**764**, 4.5	**897**, 4.3	**1029**, 4.1	**1122**, 4.2
Co	**780**, 5.3	**798**, 5.5	**989**, 4.3	**1080**, 4.2	**1185**, 4.2
Ni	**1025**, 4.1	**1040**, 4.2	**1042**, 4.4	**1149**, 4.2	**1238**, 4.2
Cu		**1220**, 4.0	**1240**, 3.8		

BOX 16B

The composition and colors of gemstones

The color of gemstones may arise from three different origins. In some cases, the color of the crystal is a natural property of the components; such minerals are called *idiochromatic*. In many cases, the origin of the color is a transition-metal trace impurity; these are referred to as *allochromatic*. In a few cases, the color is a result of defects in the crystal structure, referred to as a *color center*.

Cinnabar

Gemstones that are idiochromatic are relatively uncommon compared to allochromatic ones. The mineral *cinnabar*—a crystalline form of mercury(II) sulfide, HgS—is a well-known example. This material was used historically to produce the intense red pigment called *vermillon* that was known from antiquity and used by painters from the Renaissance to the twentieth century. Because of the high toxicity associated with the production of vermillon, the pigment was eventually replaced with cadmium red—a solid solution of cadmium sulfide and selenide; a safer material to work with (although by no means completely inoffensive). The blue-green mineral known as *turquoise*, with the formula $CuAl_6(PO_4)_4(OH)_8 \cdot 4H_2O$, is another well-known example of idiochromatic gemstone, by which the blue-green color we now associate with the name itself arises from copper(II).

Turquoise

Ruby

Citrine

Peridot

Emerald

Amethyst

Comparatively, there is a large variety of allochromatic gemstones. In fact, many of the best-known and sought-after precious gemstones belong to the allochromatic family. For example, *citrine* is a yellow or gold-colored gemstone involving trace impurities of iron(III) in quartz (SiO_2). Amethyst is a deep-purple gem, also of quartz, that contains trace iron(IV). The olive-green gem *peridot*, also known as *olivine*, is another example. Olivine is a magnesium iron silicate mineral with the chemical formula $(Mg,Fe)_2SiO_4$; the Fe(II) centers impart the green color associated with peridot. The amount of Fe(II) is variable from sample to sample, and the formula does not imply a 1:1 ratio between the +2 ions. (If only iron cations are present—that is, Fe_2SiO_4—the mineral is intensely colored and is called *fayalite*, technically an idiochromatic gem.) If only magnesium cations are present, the mineral is the colorless *forsterite*. Rubies and emerald both owe their distinctive colors to the presence of chromium(III) impurities, which are both oxide lattices. Red rubies are chromium-doped *corundum* crystals (aluminum oxide, Al_2O_3); whereas green emeralds are chromium-doped *beryl* crystals, $Be_3Al_2(SiO_3)_6$. The striking difference in color arises simply from the change of ligand-field strength in each environment.

Diamond is an allotrope of carbon, which is naturally colorless. However, impurities can replace carbon in the lattice and create what are called "color centers"; an important example is single substitutional nitrogen atoms in diamond, which are also called "C-centers". C-center-containing diamonds are generally more or less deeply yellow-colored, depending on the nitrogen defect concentration. Color centers may also be due to the absence of an atom in the lattice, called *vacancies*; an example of this occurs in the purple *fluorite* (CaF_2), which acquires its color from vacant fluorine sites.

Fluorite

The following table lists a few additional examples of gemstones with colors derived from ligand-field splittings within an inorganic framework. In all cases, the transition metal is in an approximately octahedral geometry.

Gemstone	Formula (Mineral)	Color	Trace Impurity
Ruby	Al_2O_3 (Corundum)	Red	Cr(III)
Emerald	$Be_3Al_2(SiO_3)_6$ (Beryl)	Green	Cr(III) for Al(III)
Alexandrite	Al_2BeO_4 (Chrysoberyl)	Red to Green	Cr(III) for Al(III)
Tourmaline	$Na_3Li_3Al_6(BO_3)_3(SiO_3)_6F_4$	Pink	Mn(II) for Li(I) and Al(III)
Turquoise	$Al_6(PO_4)_4(OH)_8 \cdot 4H_2O$	Blue-Green	^-OH and OH_2 ligated Cu(II)
Sapphire	Al_2O_3 (Corundum)	Blue	Intervalence transition from Fe(II) to Ti(IV)
Aquamarine	$Be_3Al_2(SiO_3)_6$ (Beryl)	Blue	Intervalence transition from Fe(II) to Fe(III)
Amethyst	SiO_2 (Quartz)	Blue-Violet	Fe(III) and Fe(IV)

16.4 Multielectronic States of Octahedral Crystal Fields

As beautiful as the theory of atomic spectroscopy is, inorganic chemists are rarely concerned with individual atoms in spherical fields! Chemically relevant systems invariably possess lower symmetries on account of the ligands surrounding metal centers, and this leads to critical effects on the electronic structure of inorganic complexes. By far, the most common geometry in inorganic chemistry is, as Werner surmised, the 6-coordinate octahedral case (or at least, pseudo-octahedral), and we will discuss that specific case first. There are two classical ways in which we can approach the problem of determining the electronic states of a transition-metal complex in a low-symmetry environment:

(1) We begin with the free-ion states of the metal ion, which we obtained in the previous section, and then consider the impact that an octahedral crystal/ligand field will have on these states. This is known as the *weak-field scheme*, because the crystal/ligand fields are considered as small perturbations on the free-ion wavefunctions.

(2) We start with the orbitals of the metal center located in an octahedral electric field populated with n noninteracting electrons, which was the approach of the previous chapter, and then include the Coulombic electron–electron repulsions between these electrons as an additional perturbation. This is known as the *strong-field scheme*, because the electron–electron interactions are taken as being small perturbations compared to the effects of crystal/ligand field forces on the d-orbitals.

In real life, both contributions are actually close enough in magnitude that the only proper way to solve this problem is to include both terms in the same order of perturbation, referred to as an *intermediate-field scheme*. Historically, the cost of computing resources was sufficiently prohibitive that it was necessary to simplify calculations by taking one of the two extreme schemes described previously, but nowadays this is not so much of an issue, as any modern personal computer can easily carry any ligand-field calculation in the full limit of the intermediate-field scheme. Nevertheless, the weak- and strong-field schemes are still relevant in modern inorganic chemistry in that they provide useful starting points (prior to doing any calculations) to help us understand and predict, qualitatively at least, the behavior of electrons in inorganic complexes, and we will proceed to explore each scheme individually.

16.4.1 The Weak-Field Scheme

The most direct impact that the crystal/ligand-field exerts on the electronic states of a free ion is due to the lowering of the symmetry. The spherical symmetry of free atoms/ions plays a major role in the behavior of these species. For instance, the maximum degeneracy of spherical free atoms/ions can be extremely high. This is seen by the existence of states such as ^{1}I in d^4 (Table 16.5), which has $L = 6$ (Table 16.3). In lower symmetries, the maximum degeneracies that can occur are typically much smaller. For instance, in O_h (a very high symmetry scenario), the highest possible degeneracy is 3, which, as we saw in Chapter 15, is why the five *d*-orbitals cannot have the same energy and therefore need to be broken into two different sets: the threefold t_{2g} and the twofold e_g sets of orbitals. Although we can arrive at this conclusion simply by looking at the O_h character table, it is informative to understand from a more rigorous perspective how this arises, as the same principles will be used to predict how each of the spherical free-ion terms (the atomic S, P, D, F, G, H states, and so forth) transform under the influence of a low-symmetry ligand-field.

Remember that the term symbols that we obtain are really representing the total orbital angular momentum behavior of the electronic part of the wavefunction of the free ion. When we referred to an "S" state, we meant to indicate that the angular momentum of the overall electronic density of the free ion was described by $L = 0$, and so on for all other term symbols as shown in Table 16.3. Our first task, then, in the weak-field scheme, is to deduce how each of the orbital momentum functions associated with the quantum number L transform ("*what their irreducible representation is*") in the O_h point group or any other point group of interest. That is, we need to calculate the *character* of specific angular momentum functions under each symmetry operation of that point group.

At first this might seem like a daunting task. Not only are there many (48) elements in the O_h point group—so that potentially we need to carry out 48 independent operations to find out the total representation we need and then further break it down into a sum of irreducible representations[o]—but how do we even calculate the character of a symmetry operation on an angular momentum function? First, it can be shown that we significantly reduce the scope of the problem by focusing only on the proper rotations of the group, which cuts the number of operations in half for O_h. Effectively, we can use the point group O instead of O_h. We then only need to determine the behavior of the angular-momentum functions under the inversion operation. These two steps are always easy for angular-momentum functions[p] as shown in Eq. 16.5, where L is the angular-momentum operator.

$$\hat{L}Y_l^{m_l}(\theta,\phi) = \hbar\sqrt{l(l+1)}Y_l^{m_l}(\theta,\phi) \qquad \textbf{16.5}$$

Spherical harmonics have a well-known mathematical form (the atomic shapes, Chapter 2) and can always be written as the product of two independent functions: one that depends on the angle θ (the angle between any specific point in space in which we are interested and the *z*-axis), and another that depends only on ϕ (the angle between that point and the *x*-axis). Because we are always free to choose our *z*-axis as being anywhere in space, we can always define the axis of any proper rotation as the *z*-axis. Practically, this means that the only angle of relevance for symmetry purposes is the angle ϕ. We say that rotations in a cubic group are invariant under the angle θ, and only the part of the spherical harmonics that depends on ϕ matters due to symmetry. As it happens, the part of the spherical harmonics that depends on ϕ is actually quite simple (Eq. 16.6),

$$\Phi_{m_l}(\phi) = e^{-m_l\phi} \qquad \textbf{16.6}$$

where the values for m_l, as usual, run between $-l$ and $+l$. For example, for *d*-orbitals ($l = 2$), $m_l = 2, 1, 0, -1$, or -2. This defines a basis for the five *d* orbitals, which can be represented by a column matrix, as shown in Eq. 16.7. The advantage of this way of representing the *d* orbitals (or any other eigenfunction of the angular momentum operator) is that it is easy to represent what happens when the functions are rotated by some angle α around the *z* axis:

$$\hat{R}_\alpha^{l=2} \cdot \begin{bmatrix} e^{-2i\phi} \\ e^{-i\phi} \\ e^{0\phi} \\ e^{i\phi} \\ e^{2i\phi} \end{bmatrix} = \begin{bmatrix} e^{-2i(\phi+\alpha)} \\ e^{-i(\phi+\alpha)} \\ e^{0(\phi+\alpha)} \\ e^{i(\phi+\alpha)} \\ e^{2i(\phi+\alpha)} \end{bmatrix} \qquad \textbf{16.7}$$

The matrix that represents mathematically the "rotation-by-α-around-the-*z*-axis" operator is directly seen as being Eq. 16.8.

[o] You may remember that all operations that belong to the same class have the same character, so we really have only to carry out one calculation for every class in a point group; in O_h, this implies ten independent calculations, not 48.

[p] This is easily done, because the eigenfunctions of the angular momentum operator $\hat{L}$ are the spherical harmonics.

$$\hat{R}_{\alpha}^{l=2} = \begin{bmatrix} e^{-2i\alpha} & 0 & 0 & 0 & 0 \\ 0 & e^{-i\alpha} & 0 & 0 & 0 \\ 0 & 0 & e^{0} & 0 & 0 \\ 0 & 0 & 0 & e^{i\alpha} & 0 \\ 0 & 0 & 0 & 0 & e^{2i\alpha} \end{bmatrix} \quad \textbf{16.8}$$

This is the rotation matrix for the set of $l = 2$ angular momentum functions, but it is easy to generalize this result to any other values of l, as in Eq. 16.9.

$$\hat{R}_{\alpha}^{l} = \begin{bmatrix} e^{-li\alpha} & 0 & \cdots & 0 \\ 0 & e^{-(l-1)i\alpha} & \cdots & 0 \\ \vdots & \vdots & \ddots & \vdots \\ 0 & 0 & \cdots & e^{l\alpha} \end{bmatrix} \quad \textbf{16.9}$$

The general equation for the character of this matrix (the sum of the diagonal elements) can be found to be Eq. 16.10, for $\alpha \neq 0$.

$$\chi_l(\alpha) = \sum_{m_l=-l}^{m_l=+l} e^{im_l\alpha} = \frac{\sin\left[\frac{1}{2}(2l+1)\alpha\right]}{\sin\frac{\alpha}{2}} \quad \textbf{16.10}$$

This equation might seem complicated, but applying it is straightforward. An octahedral molecule has three types of proper rotations: C_2 (rotation by 180°, or $\alpha = \pi$ radians), C_3 (rotation by 120°, or $\alpha = 2\pi/3$), and C_4 (rotation by 90°, or $\alpha = \pi/2$). Using Eq. 16.10, the characters for the rotation of the $l = 2$ set of angular momentum functions are:

C_2: $\alpha = \pi$

$$\chi_{l=2}(C_2) = \frac{\sin\left[\frac{1}{2}(2\cdot 2+1)\pi\right]}{\sin\left(\frac{\pi}{2}\right)} = \frac{\sin\left[\frac{5\pi}{2}\right]}{\sin\left(\frac{\pi}{2}\right)} = \frac{1}{1} = 1$$

C_3: $\alpha = 2\pi/3$

$$\chi_{l=2}(C_3) = \frac{\sin\left[\frac{1}{2}(2\cdot 2+1)\frac{2\pi}{3}\right]}{\sin(\frac{\pi}{3})} = \frac{\sin\left[\frac{5\pi}{3}\right]}{\sin(\frac{\pi}{3})} = \frac{-\sqrt{3}/2}{\sqrt{3}/2} = -1$$

C_4: $\alpha = 2\pi/4 = \pi/2$

$$\chi_{l=2}(C_4) = \frac{\sin\left[\frac{1}{2}(2\cdot 2+1)\frac{\pi}{2}\right]}{\sin\left(\frac{\pi}{4}\right)} = \frac{\sin\left[\frac{5\pi}{4}\right]}{\sin\left(\frac{\pi}{4}\right)} = \frac{-\sqrt{2}/2}{\sqrt{2}/2} = -1$$

This equation is indeterminate when $\alpha = 0$ (division by zero); however, we can apply L'Hôpital's rule directly to Eq. 16.10, to obtain the character of a rotation by $\alpha = 0$; that is, the character of doing nothing, otherwise known as the identity operation, E, is shown as follows.

$$\chi_l(E) = \lim_{\alpha\to 0} \frac{\mathrm{d}\sin\left[\frac{1}{2}(2l+1)\alpha\right]/\mathrm{d}\alpha}{\mathrm{d}\sin(\alpha/2)\,/\,\mathrm{d}\alpha}$$

$$= \lim_{\alpha\to 0} \frac{\frac{1}{2}(2l+1)\cos\left[\frac{1}{2}(2l+1)\alpha\right]}{\frac{1}{2}\cos[\alpha/2]} = 2l+1$$

$$\chi_l(E) = 2l+1$$

That is, the character under the identity operation is simply the total degeneracy of the angular momentum function, for the $l = 2$ orbital angular function, $\chi(E) = 2l + 1 = 5$. (In other words, the character under E is the number of functions involved, as usual.) Using these results, we have a full set of characters for the proper rotations of an octahedron. The reducible representation of the $l = 2$ orbital angular momentum basis set is:

O	E	$8C_3$	$6C'_2$	$6C_4$	$3C_2(=C_4^2)$
$l = 2$	5	−1	1	−1	1

Using the decomposition formula, we find that this spans the $(e + t_2)$ irreducible representations. To relate our analysis to the O_h point group, we simply need to determine the behavior of these angular-momentum functions relative to the inversion operation, i. Here things vary, depending on whether the angular-momentum functions that we characterize are single-electron (orbitals) or multi-electron wavefunctions. The single-electron case is simple, and we already know the result from our knowledge of the symmetry of the atomic orbitals: an $l = 0$ function has the parity of an s orbital (gerade, subscript "g"), an $l = 1$ function has the parity of a p orbital (ungerade, subscript "u"), and so on. Clearly, for even l values the angular momentum functions are gerade (the character under the inversion operation has the same value as that under the identity operation), whereas for odd values the angular functions are ungerade (the character under the inversion operation is the negative of the character value under the identity operation). Combining this with the previous result, we thus conclude that the d orbitals transform like $e_g + t_{2g}$ in the O_h point group, which is, of course, as expected from Chapter 15. Table 16.8 shows how any of the single-electron angular-momentum functions transform in the O_h symmetry, up to $l = 6$.

For multi-electron angular-momentum functions, such as the free-ion terms derived in the previous section, parity is determined by looking at the direct product of the parities of the individual 1-electron components. For instance, any term that arises from a d^n configuration necessarily has an even parity (gerade), because it is made up of a product of

Table 16.8 Symmetry of single-electron angular-momentum functions.

Single-electron angular momentum function	l	Degeneracy $\chi(E) = 2l + 1$	O_h irreducible representations[a]
s	0	1	a_{1g}
p	1	3	t_{1u}
d	2	5	$e_g + t_{2g}$
f	3	7	$a_{2u} + t_{1u} + t_{2u}$
g	4	9	$a_{1g} + e_g + t_{1g} + t_{2g}$
h	5	11	$e_u + 2\,t_{1u} + t_{2u}$
i	6	13	$a_{1g} + a_{2g} + e_g + t_{1g} + 2\,t_{2g}$

[a] For the terms in the O point group, simply drop the g or u subscript.

individually even functions. (The d-orbitals are gerade, and any product of even functions yields an even function.) On the other hand, the parity of the terms that arise from a p^n configuration depends on the value n; if n is even (p^2, p^4, or p^6), then the multielectron terms are all gerade. (The product of an even number of odd functions is itself even.) However, if n is odd (p^1, p^3, p^5), then the parity of the terms is ungerade (odd).

Using this procedure, one can determine how states split under different low-symmetry ligand-fields as well. Obviously, the number of microstates does not change from the free ion to the lower symmetry case: states are neither created nor eliminated when symmetry is lowered. However, the number of individual terms generally increases substantially the lower the symmetry of the system as the lower symmetry splits the states; that is, there are fewer degeneracies in the states.

As an example, let us find the symmetry of the states for d^2 in O_h. The free-ion terms (Table 16.5) are ^3F, ^3P, ^1G, ^1D, and ^1S, some of which will split, and all of which will change labels in O_h. The ^3F term, for example, transforms like an f-orbital angular-momentum function in O_h, which span the $a_{2g} + t_{1g} + t_{2g}$ irreducible representations (Table 16.8); these labels will correspond to the orbital parts of the O_h system. Furthermore, since we are dealing with a d^n configuration, all terms must have the gerade parity (subscript g). Finally, lowering the symmetry never changes the spin part of the wavefunction, so a ^3F must yield three separate states in the O_h point group with the same spin; that is, $^3A_{2g}$, $^3T_{1g}$, and $^3T_{2g}$. Similarly, the ^3P becomes $^3T_{1g}$, ^1G becomes $^1A_{1g} + {}^1E_g + {}^1T_{1g} + {}^1T_{2g}$, and so on.

$$^3F = {}^3A_{1g} + {}^3T_{1g} + {}^3T_{2g}$$
$$^1D = {}^1E_g + {}^1T_{2g}$$
$$^3P = {}^3T_{1g}$$
$$^1G = {}^1A_{1g} + {}^1E_g + {}^1T_{1g} + {}^1T_{2g}$$
$$^1S = {}^1A_{1g}$$

In this way, any free-ion term can always be broken down for any symmetry; one simply has to calculate the character of the angular-momentum functions for any of the symmetry operations of the point group of interest, and then reduce the resulting representation. In practice, once we know the symmetry in one high-order point group, it is often simpler to use a correlation table, which compiles the transformation of all irreducible representations between different point groups. Table 16.9 gives the correlation properties from O_h to many common lower symmetries. As an example of how this works out, the same ^3F free-ion state that gives $^3A_{2g} + {}^3T_{1g} + {}^3T_{2g}$ in O_h symmetry would produce six independent states in D_{4h} symmetry; that is, the symmetry for a MX_4 square-planar complex, with the labels $^3B_{1g} + {}^3A_{2g} + {}^3B_{2g} + 2E_g$. Obviously, the total degeneracy of all these states is conserved.

At this point we have not yet discussed the energy of each of these weak-field terms, which obviously will vary, depending on the crystal/ligand-field strength. This will be covered later, but first we discuss how to obtain the symmetry of the multielectron terms in a strong-field scheme approach. Importantly, because the number and symmetry of the states that a given metal ion has in a certain ligand-field environment do not depend on the approach (weak-field or strong-field) we take to derive them, we already know what the answer will be!

Table 16.9 Correlation table from O_h.

O_h	O	T_d	D_{4h}	C_{4v}	C_{2v}	C_{2h}	D_3	D_{2d}
A_{1g}	A_1	A_1	A_{1g}	A_1	A_1	A_g	A_1	A_1
A_{2g}	A_2	A_2	B_{1g}	B_1	A_2	A_g	A_2	B_1
E_g	E	E	$A_{1g} + B_{1g}$	$A_1 + B_1$	$A_1 + A_2$	$2A_g$	E	$A_1 + B_1$
T_{1g}	T_1	T_1	$A_{2g} + E_g$	$A_2 + E$	$A_2 + B_1 + B_2$	$A_g + 2B_g$	$A_2 + E$	$A_2 + E$
T_{2g}	T_2	T_2	$B_{2g} + E_g$	$B_2 + E$	$A_1 + B_1 + B_2$	$A_g + 2B_g$	$A_1 + E$	$B_2 + E$
A_{1u}	A_1	A_2	A_{1u}	A_2	A_2	A_u	A_1	B_1
A_{2u}	A_2	A_1	B_{1u}	B_2	A_1	A_u	A_2	A_1
E_u	E	E	$A_{1u} + B_{1u}$	$A_2 + B_2$	$A_1 + A_2$	$2A_u$	E	$A_1 + B_1$
T_{1u}	T_1	T_2	$A_{2u} + E_u$	$A_1 + E$	$A_1 + B_1 + B_2$	$A_u + 2B_u$	$A_2 + E$	$B_2 + E$
T_{2u}	T_2	T_1	$B_{2u} + E_u$	$B_1 + E$	$A_2 + B_1 + B_2$	$A_u + 2B_u$	$A_1 + E$	$A_2 + E$

Correlations assume the same coordinate axes between O_h and lower symmetries.

Table 16.10 Strong-field direct products for two electrons occupying the t_{2g} and e_g orbitals in O_h symmetry.

	E	$8C_3$	$6C_2$	$6C_4$	$3C_2$	i	$6S_4$	$8S_6$	$3\sigma_h$	$6\sigma_d$	
t_{2g}	3	0	1	−1	−1	3	−1	0	−1	1	
e_g	2	−1	0	0	2	2	0	−1	2	0	
$(t_{2g})^2 = t_{2g} \otimes t_{2g}$	9	0	1	1	1	9	1	0	1	1	$A_{1g} + E_g + T_{1g} + T_{2g}$
$(t_{2g})^1(e_g)^1 = t_{2g} \otimes e_g$	6	0	0	0	−2	6	0	0	−2	0	$T_{1g} + T_{2g}$
$(e_g)^2 = e_g \otimes e_g$	4	1	0	0	4	4	0	1	4	0	$A_{1g} + A_{2g} + E_g$

16.4.2 The Strong-Field Scheme

The strong-field scheme begins by considering the individual 1-electron levels (the orbitals) of the metal center in the low-symmetry ligand field, then considers how the inclusion of electron–electron interactions will affect the behavior of electrons populating these orbitals. As previously, let us show how this works by using the simple example of a d^2 configuration in O_h symmetry. We know that the possible electron configurations for a d^2 center (O_h point group) are $(t_{2g})^2$, $(t_{2g})^1(e_g)^1$, and $(e_g)^2$: these are called *strong-field configurations* to distinguish them from the free-ion configurations (d^2, d^3, and so forth). To find the terms included in each of these strong-field configurations, one takes the direct products (Section 5.12) of each orbital listed in the configuration (Table 16.10).

This table does not include the spin states for the terms. When two electrons are occupying different orbital sets, as in the $(t_{2g})^1(e_g)^1$ configuration case, then each electron is free to take any spin orientation available; the spins can either be aligned (triplet state) or paired (singlet state), so each strong-field configuration listed in the table gives rise to both types of spin states.

$$(t_{2g})^1(e_g)^1 = {}^1T_{1g} + {}^3T_{1g} + {}^1T_{2g} + {}^3T_{2g}$$

When two (or more) electrons are located in the same strong-field set of orbitals, then not every conceivable combination of spins is necessarily allowed, due to restrictions imposed by Pauli's exclusion principle. One easy way to see that this must be so is to consider the number of microstates that are included in the $(t_{2g})^2$ strong-field configuration, which we calculated previously (Table 16.1 and Eq. 16.1). There are fifteen independent microstates in the $(t_{2g})^2$ configuration. If all terms obtained from the $t_{2g} \otimes t_{2g}$ direct product in Table 16.10 were allowed to form singlet and triplet states, then we would largely exceed the known degeneracy of the $(t_{2g})^2$ configuration:

$$\underbrace{{}^1A_{1g}}_{1} + \underbrace{{}^3A_{1g}}_{3} + \underbrace{{}^1E_g}_{2} + \underbrace{{}^3E_g}_{6} + \underbrace{{}^1T_{1g}}_{3} + \underbrace{{}^3T_{1g}}_{9} + \underbrace{{}^1T_{2g}}_{3} + \underbrace{{}^3T_{2g}}_{9} \rightarrow 36\,\text{microstates}$$

So clearly, most of these possibilities correspond to unphysical states that do not exist in the $(t_{2g})^2$ configuration; that is, some of these possibilities must be unphysical. The origin of this issue is because we did not account for the fact that two electrons cannot occupy the same orbital *and* have their spins aligned; we also did not account for the fact that electrons are indistinguishable particles as well, and the direct product $t_{2g} \otimes t_{2g}$ yields repeated identical states—but which of these states should be eliminated as unphysical? There are two basic approaches to solve this problem: (1) "descent in symmetry", using a correlation table, and (2) use of the symmetric/antisymmetric direct products. The first method (descent in symmetry) is straightforward based on what we already know, but requires a fair amount of work; the second method (symmetry/antisymmetric direct products) hinges on some new physical principles, but requires far less effort to use in general.

16.4.2.1 Finding the Spin States in the Strong-Field Scheme: Descent in Symmetry

The idea behind this approach is to recognize that there are issues only in those cases where more than one electron is located in a degenerate set of orbitals. If we dealt exclusively with nondegenerate orbitals, then there would never be any ambiguity. One electron in a nondegenerate orbital can always adopt either spin orientations (spin-up or spin-down), and two electrons in a nondegenerate orbital necessarily must be spin-paired. It is only when we have degenerate orbitals that we need to be particularly careful not to double-count states as well as keeping track of the spin restrictions on orbital occupancy. The descent-in-symmetry approach solves this issue by artificially lowering the symmetry of the system such that no orbital degeneracy remains. We then determine the allowed spin states in this lower symmetry, then move back up to the higher symmetry to determine the actual states that are allowed in the presence of orbital degeneracy. Let us see how this works by using the $(t_{2g})^2$ strong-field configuration.

First, we begin by identifying a symmetry where the threefold degeneracy of the t_{2g} orbitals would be entirely lifted; a quick glance at a correlation table (Table 16.9) suggests two

Table 16.11 Descent-in-symmetry configurations for the $(t_{2g})^2$ configuration in the C_{2v} point group. The spatial symmetry is obtained from the direct product of the C_{2v} orbitals. Two electrons in the same spatial orbital can only yield singlet spin states, whereas two electrons in two independent orbitals yield both singlet and triplet spin states.

	C_{2v} configuration		C_{2v} direct product		Allowed Spin States
i)	$(a_1)^2(b_1)^0(b_2)^0$	→	$a_1 \otimes a_1 = A_1$	→	$^1A_1$
ii)	$(a_1)^0(b_1)^2(b_2)^0$	→	$b_1 \otimes b_1 = A_1$	→	$^1A_1$
iii)	$(a_1)^0(b_1)^0(b_2)^2$	→	$b_2 \otimes b_2 = A_1$	→	$^1A_1$
iv)	$(a_1)^1(b_1)^1(b_2)^0$	→	$a_1 \otimes b_1 = B_1$	→	$^1B_1 + {}^3B_1$
v)	$(a_1)^1(b_1)^0(b_2)^1$	→	$a_1 \otimes b_2 = B_2$	→	$^1B_2 + {}^3B_2$
vi)	$(a_1)^0(b_1)^1(b_2)^1$	→	$b_1 \otimes b_2 = A_2$	→	$^1A_2 + {}^3A_2$

options: both C_{2h} and C_{2v} points groups transform the t_{2g} set into three nondegenerate orbitals:

O_h	C_{2h}	C_{2v}
t_{2g}	$a_g + 2b_g$	$a_1 + b_1 + b_2$

It does not make a difference which of these two subgroups of O_h we choose to work with; we are only temporarily adopting a lower symmetry to solve our problem. Since we need to choose one, let us decide on C_{2v}, where all three single orbitals have a distinct symmetry label as a more convenient way of keeping track of the change in symmetry. The O_h configuration $(t_{2g})^2$ then gives rise to six low-symmetry (C_{2v}) configurations, depending in which of the three C_{2v} orbitals (a_1, b_1, or b_2) the two electrons are placed (Table 16.11).

The states of each of these six C_{2v} configurations are then easily obtained by taking the proper direct product, and the spin states are determined directly by considering the allowed possibilities that are consistent with Pauli's principle (two electrons in the same orbital can only lead to a singlet state; two electrons in two independent orbitals can form both a singlet and triplet state). At this point, note that the total degeneracy of all our C_{2v} states listed in Table 16.11 for the $(t_{2g})^2$ configuration indeed adds up to 15, as required. Now that we have labeled all our C_{2v} states, we need to go back ("reascend") to our original high symmetry (O_h). We can do this unambiguously because we already know what the spatial symmetry of the O_h states must be; we listed them from the direct product $t_{2g} \otimes t_{2g}$ in Table 16.10. Furthermore, we can use our correlation table (Table 16.9) to link the O_h states to the C_{2v} states. We know, for instance, that a T_{1g} state in O_h is related to $A_2 + B_1 + B_2$ in C_{2v}, whereas a T_{2g} (O_h) state is related to $A_1 + B_1 + B_2$ states in C_{2v}. This allows us to draw the correlation diagram shown in Fig. 16.7.

From here, the only thing we need to do is to assign the spin for each of these states, which is easily done from the

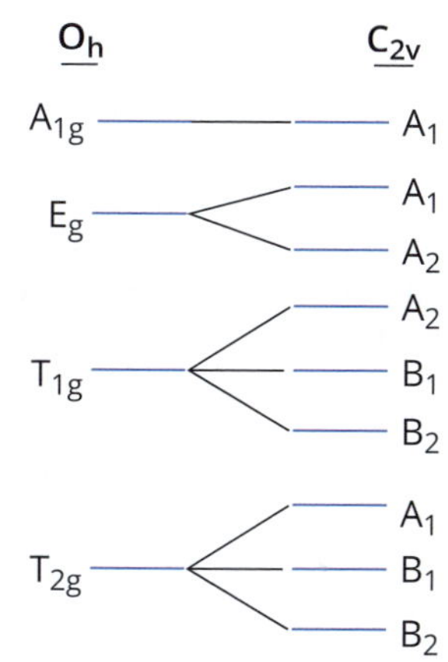

Fig. 16.7 Correlation diagram between the O_h strong-field states arising from the $t_{2g} \otimes t_{2g}$ direct product and the C_{2v} states obtained in a descent-in-symmetry analysis of the $(t_{2g})^2$ configuration.

fact that that lowering the symmetry never changes the spin state; that is, *all C_{2v} components of the same O_h state must have the same spin state.* A quick glance at the list of C_{2v} states in Table 16.11 shows that all "A_1" states are singlet states, but there are no $^3A_1$ states in our list. This solves our problem directly, as it tells us that the A_g, E_g, and T_{2g} (O_h) states must all be singlet states, as they all have an A_1 component in the C_{2v} point group. This leaves the T_{1g} (O_h) state as the only possibility for a triplet state, and we can directly check that, indeed, the $A_2 + B_1 + B_2$ (C_{2v}) are T_{1g} (O_h) in Table 16.11. In summary, the $(t_{2g})^2$ strong-field configuration must give rise to the following multielectron states in O_h symmetry: $^1A_{1g} + {}^1E_g + {}^1T_{2g} + {}^3T_{2g}$. Using the same method, we would find that the $(e_g)^2$ strong-field configuration (degeneracy of six microstates, Table 16.1) gives rise to $^1A_{1g} + {}^1E_g + {}^3A_{2g}$ states.

Although straightforward, the descent in symmetry method requires a fair amount of work, as well as some judicious thinking to properly assign the spin states within the strong-field scheme. Notably, this approach becomes much harder to use systematically for configurations with large numbers of states. In these cases, the symmetric/antisymmetric direct product method provides a general and unambiguous way to solve these problems.

16.4.2.2 Finding the Spin States in the Strong-Field Scheme: Symmetric/Antisymmetric Direct Products

This method relies on a fundamental symmetry property that all electronic wavefunctions must possess, which is related to the *exchange* of electrons. The general form of the Pauli principle states that *the total wavefunction must always be antisymmetric with respect to exchange of any two electrons.*

As indicated here, this statement is in fact another version of Pauli's principle, which in its simplest form states that no two electrons can have the same set of quantum numbers. If we were to have a system of *n* electrons defined by a wavefunction $\Psi(1,2 \ldots n)$ and we were able to magically interchange two electrons with one another, then obviously none of the observable properties would change, and in particular the total electronic density, $|\Psi|^2$, would remain unchanged. (This is, by definition, what symmetry operations do—their actions leave the system invariant. It is in that general sense that the operation *interchanging two electrons* is really a fundamental symmetry operation.)

Now there are two ways that this could work out, and let's use the symbol $\hat{X}_e$ ("exe hat sub e") to designate the operator for the exchange of two electrons. It must be true that $\hat{X}_e\Psi = \pm\Psi$, since $|\Psi|^2$ cannot change when we exchange two electrons; Ψ itself must either go into either $+\Psi$ or $-\Psi$ after the exchange operation so that the square of the magnitude of the total wavefunction remains unchanged. Pauli's principle states that it is only the latter ($\hat{X}_e\Psi = -\Psi$) that applies for electrons, and we say that the wavefunction is *antisymmetric* (changes sign) with respect to the exchange of two electrons.[q]

How does that help us here? Well, we know that the total electronic wavefunction of any system can generally be factorized into spatial and spin components (Eq. 16.11).[r]

$$\Psi_{total} = \Psi_{space} \times \Psi_{spin} \qquad \textbf{16.11}$$

[q] This is true for particles with half-integral spin, such as electrons. Half-integer spin particles are referred to as *fermions*, and wavefunctions are always antisymmetric with respect to exchange of two fermions. Bosons, particles with integer spin have the opposite behavior; that is, Ψ is always symmetric with respect to exchange of two bosons. Photons (spin-1 particles) are important representatives of the boson family, and many of the distinguishing properties of light come from the fact that photons are bosons rather than fermions.

[r] Strictly speaking, this is only true in the absence of spin–orbit coupling, but we are trying here to find out the spin state of strong-field terms, which are defined in the first place by assuming no spin–orbit coupling.

Pauli's principle requires that the *total* wavefunction be antisymmetric with respect to electron exchange, which means that only one of the two parts of the total wavefunction can be antisymmetric while the other one is symmetric relative to electron exchange. (Remember: antisymmetric × antisymmetric = symmetric, so the spatial and spin parts of the total wavefunction cannot both be antisymmetric at the same time.) Let us look at the spin part of the wavefunction first, and let us use the symbols α for a "spin-up" electron and β for a "spin-down" electron. If we have two electrons, then we already know that there are only two types of spin states/wavefunctions, which we named singlet ("spin-paired") and triplet ("spin-aligned"). As a first guess, we might write the spin wavefunction for a singlet state as $\Psi_{singlet} = \alpha(1)\beta(2)$, where we assigned the spin-up part to electron 1 and the spin-down part to electron 2. Another equally as good first guess would be to assign the opposite spins to each electron $\Psi_{singlet} = \alpha(2)\beta(1)$. Which guess is the appropriate one? Well, neither one, as choosing one or the other would require that we can distinguish each electron precisely (by forcing its spin state), which is not allowed quantum-mechanically given that electrons are fundamentally indistinguishable particles. Rather, we should take a linear combination (a superposition) of these two possibilities. Physically, it turns out that the combination in the equation

$$\Psi_{singlet} = \frac{1}{\sqrt{2}}\left[\alpha(1)\beta(2) - \alpha(2)\beta(1)\right] \qquad \textbf{16.12}$$

is the appropriate one for a singlet state (the $1/\sqrt{2}$ constant is a simple normalization factor), showing that singlet-spin wavefunctions must be antisymmetric relative to electron exchange.

$$\hat{X}_e\Psi_{singlet} = -\Psi_{singlet} \qquad \textbf{16.13}$$

Let us explore the case of the triplet wavefunctions; there should be three independent triplet functions, as implied by the name. Two obvious possibilities are:

$$\Psi_{triplet,1} = \alpha(1)\alpha(2)$$
$$\Psi_{triplet,2} = \beta(1)\beta(2)$$

where both electrons take the same spin orientation, both up or both down. These are perfectly fine wavefunctions, as neither violates the indistinguishability requirement, since both electrons have the same spin, that property could not be used to tell which electron we would be looking at. We still need a third triplet wavefunction though, which is provided by the other linear combination of spin-up/spin-down states (Eq. 16.14).[s]

[s] If you are confused as to why a triplet spin function would be represented by a spin-up/spin-down combination, just notice that a triplet spin state ($S = 1$) must have three component functions assigned to $M_S = -1, 0, +1$. It is the $M_S = 0$ component of the triplet state which is defined by the spin-up/spin-down combination of Eq. 16.14.

$$\Psi_{\text{triplet},3} = \frac{1}{\sqrt{2}}\left[\alpha(1)\beta(2) + \alpha(2)\beta(1)\right] \quad \textbf{16.14}$$

As is immediately clear here, triplet spin state wavefunctions are *symmetric* relative to the exchange of two electrons:

$$\hat{X}_e \Psi_{\text{triplet}} = +\Psi_{\text{triplet}}$$

The fact that singlet spin functions are antisymmetric (having opposite spin) imposes that the spatial part associated with them must be symmetric; conversely, given that the spin part of triplet states is symmetric (having the same spin), the spatial part of a triplet state must be antisymmetric for the total wavefunction to be antisymmetric. This strict requirement is what ultimately allows us to assign the components in a strong-field scheme. To better understand what a symmetric/antisymmetric spatial function might look like, let us consider the case of the hydrogen molecule, H_2. In the simplest MO treatment, the ground-state of the H_2 molecule is represented by the $(\sigma_g)^2(\sigma_u^*)^0$ configuration, where σ_g and σ_u^* are the lowest-energy σ-bonding and σ-antibonding MOs, respectively. The spatial part of the ground-state electronic wavefunction would then be represented by assigning both electrons to the σ-bonding MO, $\Psi_{0,spatial} = \sigma_g(1)\sigma_g(2)$. This spatial wavefunction is clearly symmetric against electron exchange, requiring that this spatial wavefunction be coupled with an antisymmetric spin function; that is, a singlet function. This shows directly how Pauli's exclusion principle (*an orbital can accommodate at most two electrons, each with opposite spins*) naturally arises from the requirement that the total wavefunction be antisymmetric; two electrons occupying the same orbital define a symmetric spatial wavefunction, which necessarily implies a spin-paired singlet state, an antisymmetric spin wavefunction. Since there are only two possible spin states that individual electrons can adopt (up or down), then it is not possible to construct a total wavefunction that would be antisymmetric if more than two electrons were to occupy the same spatial wavefunction/orbital. The lowest-energy triplet state in H_2 occurs for the first excited electronic configuration corresponding to the 1-electron promotion to the σ-antibonding orbital: $(\sigma_g)^1(\sigma_u^*)^1$. In order to retain the indistinguishability requirement, the spatial part would need to be written as shown in Eq. 16.15, which is itself an antisymmetric function, a function can be multiplied with any one of the three symmetric triplet spin functions. (Remember that in the absence of a magnetic field or of spin–orbit coupling effects, all three spin functions of a triplet state are degenerate).

$$\Psi_{\text{T1, spatial}} = \frac{1}{\sqrt{2}}\left[\sigma_g(1)\sigma_u^*(2) - \sigma_g(2)\sigma_u^*(1)\right] \quad \textbf{16.15}$$

The case for H_2 is sufficiently simple that we can easily write down the symmetric and antisymmetric spatial wavefunctions, but in general a more robust and systematic method is needed to obtain these combinations. Remember what we did when we tried to find out the states that arise from the $(t_{2g})^2$ strong-field configuration; we first took the direct product $t_{2g} \otimes t_{2g}$, which we found to be equal to $A_{1g} + E_g + T_{1g} + T_{2g}$. These are the symmetry representations of the *spatial* part of the wavefunction, and we used the descent in symmetry method to figure out which spin function (triplet or singlet) went with each of these spatial wavefunctions, which lead us to the result $t_{2g} \otimes t_{2g} = {}^1A_{1g} + {}^1E_g + {}^3T_{1g} + {}^1T_{2g}$.

If you look at this result in light of the discussion just presented, then you might conclude that when we took the direct product $t_{2g} \otimes t_{2g}$, the three components $(A_{1g} + E_g + T_{2g})$ were *symmetric* outputs of the direct product; that is, they ended up being identified as singlet states, so they necessarily represented symmetric spatial parts, whereas the T_{1g} component of $t_{2g} \otimes t_{2g}$ must necessarily have been an antisymmetric output of that same direct product; that is, it ended up being associated with the triplet spin functions, which implies that the T_{1g} component must be a symmetric spatial function. The key here is that if we had a method to directly compute *symmetric/antisymmetric* direct products, then we would always automatically know with which spin state to pair each component of the direct product. Precisely this method exists, and the characters of the symmetric and antisymmetric direct products of a representation with itself are as shown in Eq. 16.16.[t]

$$\chi^+(R) = \frac{1}{2}\left\{\left[\chi(R)\right]^2 + \chi(R^2)\right\}$$
$$\chi^-(R) = \frac{1}{2}\left\{\left[\chi(R)\right]^2 - \chi(R^2)\right\} \quad \textbf{16.16}$$

These equations might be intimidating at first sight, but they are no more difficult to use than the group-theory equations you learned in Chapter 5. These equations each contain two terms, which have very different meanings. The first one, which we wrote $[\chi(R)]^2$, represents "*the square of the character of the irreducible representation under operation R*". For example, if we are trying to compute the direct product of $t_{2g} \otimes t_{2g}$, then $[\chi(R)]^2 = 3^2 = 9$ when $R = E$.

The other term seems similar but represents something very different; $\chi(R^2)$ is to be interpreted as "*the character of the irreducible representation for the square of operation R*". As we saw in Chapter 5, the square of a symmetry operation simply means "*applying that operation twice in a row*," so for $t_{2g} \otimes t_{2g}$, then $\chi(R^2) = 3$ when $R = \sigma_h$, because $(\sigma_h)^2 = E$ (two reflections back-to-back yield the same outcome as the identity operation), and the character of t_{2g} under E is 3.

Let's look at a full example to see how easy it is to compute the symmetric and antisymmetric direct products once we understand the previous equations (Table 16.12). Let's again use the $t_{2g} \otimes t_{2g}$ case, which we know from the previous section should give us ${}^1A_{1g} + {}^1E_g + {}^3T_{1g} + {}^1T_{2g}$. A convenient way to work this out is to work directly from the character

[t] The direct product of two independent irreducible representations is never an issue. You can always form both symmetric and antisymmetric combinations of every component of the direct product. We already covered this when we discussed the previous $(t_{2g})^1(e_g)^1$ case.

Table 16.12 Calculation of the symmetric/antisymmetric direct products for $t_{2g} \otimes t_{2g}$.

Row #	Term	E	$8C_3$	$6C_2$	$6C_4$	$3C_2\ (\equiv C_4^2)$	i	$6S_4$	$8S_6$	$3\sigma_h$	$6\sigma_d$	Interpretation of row
1	$\chi_{t_{2g}}(R)$	3	0	1	−1	−1	3	−1	0	−1	1	*the characters of the t_{2g} irrep in O_h for each operation R*
2	R^2	E	C_3^2	E	C_4^2	E	E	C_4^2	C_3	E	E	*the square of each operation in O_h*
3	$\chi_{t_{2g}}(R^2)$	3	0	3	−1	3	3	−1	0	3	3	*the characters of the t_{2g} irrep in O_h for each operation R^2*
4	$\left[\chi_{t_{2g}}(R)\right]^2$	9	0	1	1	1	9	1	0	1	1	*the square of the characters of the t_{2g} irrep in O_h*
5	$\chi^+_{t_{2g}}(R)$	6	0	2	0	2	6	0	0	2	2	$\chi^+(R) = \frac{1}{2}\left\{\left[\chi(R)\right]^2 + \chi(R^2)\right\}$
6	$\chi^-_{t_{2g}}(R)$	3	0	−1	1	−1	3	1	0	−1	−1	$\chi^-(R) = \frac{1}{2}\left\{[\chi(R)]^2 - \chi(R^2)\right\}$

table for the t_{2g} irreducible representation in the O_h point group and compute the values for three intermediates rows: one for R^2 (the operation that represents the squares of the symmetry operation R), one for $[\chi(R)]^2$, and one for $\chi(R^2)$. Here is what such a table would look like for the t_{2g} irrep, where details of what each row represents are given in the last column.

The representations for both symmetric and antisymmetric direct products will, in general, not be irreducible, but we can apply the decomposition formula (Eq. 5.10) to determine what irreducible representations they are each comprised of, as shown in Table 16.13.

So, $t_{2g} \otimes t_{2g}$ yields three symmetric spatial functions (A_{1g} + $E_g + T_{2g}$), which are to be assigned as singlet states, and one antisymmetric spatial function (T_{1g}) which is necessarily a triplet state. This is, of course, the same result that we obtained using the descent in symmetry method previously. Although the equations might appear intimidating at first, the calculation of symmetric/antisymmetric direct products is much more straightforward, time-efficient, and less prone to error than the comparatively cumbersome descent-in-symmetry approach.

But what if we have more complex cases, such as the ground-state strong-field configuration of a d^3 system, $(t_{2g})^3$? Three electrons can define two types of spin state: doublets (two spin paired, third free to take any orientation, total spin $S = 1/2$) and quartets (all three spins aligned, total spin $S = 3/2$). Although we will not write out these functions explicitly here, doublet states are necessarily antisymmetric spin functions, whereas quartet states are necessarily symmetric. To compute the symmetric/antisymmetric direct products of the spatial parts, we use Eq. 16.17.

$$\chi^{d} = \chi(\text{doublet}) = \frac{1}{3}\left\{\left[\chi(R)\right]^3 - \chi(R^3)\right\}$$
$$\chi^{q} = \chi(\text{quartet}) = \frac{1}{6}\left\{\left[\chi(R)\right]^3 - 3\chi(R)\chi(R^2) + 2\chi(R^3)\right\} \qquad \mathbf{16.17}$$

Where everything is to be interpreted the same way as before, R^3 represents the outcome of operating three times in a row with operation R, and so on. The calculation for the $t_{2g} \otimes t_{2g} \otimes t_{2g}$ is shown in Table 16.14.

This tells us directly that the $(t_{2g})^3$ strong-field configuration yields ^2E_g + $^2T_{1g}$ + $^2T_{2g}$ + $^4A_{2g}$ states in O_h symmetry; from there, we could easily find the states of a lower-symmetry system related to O_h by using a correlation table.[u]

More complicated cases can always be evaluated, but in reality we rarely need to go beyond this level of complexity because the degeneracy of orbitals in the vast majority of molecular systems never exceeds three. (Notable exceptions are icosahedral systems, but these are rare.) What about $(t_{2g})^4$; do we need to compute the quartic direct product $t_{2g} \otimes t_{2g} \otimes t_{2g} \otimes t_{2g}$? Well, no; we can use the electron-hole analogy to show that the states that come out of the $(t_{2g})^4$ configuration

[u] Remember that lowering the spatial symmetry never changes the spin state.

Table 16.13 Analysis of the symmetric/antisymmetric direct products of $t_{2g} \otimes t_{2g}$.

	E	$8C_3$	$6C_2$	$6C_4$	$3C_2\ (\equiv C_4^2)$	i	$6S_4$	$8S_6$	$3\sigma_h$	$6\sigma_d$	Irr. Rep.
$\chi^+_{t_{2g}}$	6	0	2	0	2	6	0	0	2	2	$A_{1g} + E_g + T_{2g}$
$\chi^-_{t_{2g}}$	3	0	−1	1	−1	3	1	0	−1	−1	T_{1g}

Table 16.14 Analysis of the symmetric/antisymmetric direct products of $t_{2g} \otimes t_{2g} \otimes t_{2g}$.

	E	8C$_3$	6C$_2$	6C$_4$	3C$_2$(≡C$_4^2$)	i	6S$_4$	8S$_6$	3σ$_h$	6σ$_d$
$\chi_{t_{2g}}(R)$	3	0	1	−1	−1	3	−1	0	−1	1
R^2	E	C$_3^2$	E	C$_4^2$	E	E	C$_2$	C$_3$	E	E
R^3	E	E	C$_2$	C$_4^3$	C$_2$	i	S$_4^3$	i	σ$_h$	σ$_d$
$\chi_{t_{2g}}(R^3)$	3	3	1	−1	−1	3	−1	3	−1	1
$\left[\chi_{t_{2g}}(R)\right]^3$	27	0	1	−1	−1	27	−1	0	−1	1
$\chi_{t_{2g}}(R^2)$	3	0	3	−1	3	3	−1	0	3	3
$\chi^{d}_{t_{2g}}(R)$	8	−1	0	0	0	8	0	−1	0	0
$\chi^{q}_{t_{2g}}(R)$	1	1	−1	−1	1	1	−1	1	1	−1

Table 16.15 Multielectron states associated with O$_h$ strong-field configurations

Strong-Field Configuration	Multielectron States
$(t_{2g})^1 \equiv (t_{2g})^5$	$^2T_{2g}$
$(t_{2g})^2 \equiv (t_{2g})^4$	$^1A_{1g} + {}^1E_g + {}^1T_{2g} + {}^3T_{1g}$
t_{2g}^3	$^2E_g + {}^2T_{1g} + {}^2T_{2g} + {}^4A_{2g}$
$(e_g)^1 \equiv (e_g)^3$	^2E_g
e_g^2	$^1A_{1g} + {}^1E_g + {}^3A_{2g}$

are going to be the same ones as those out of the $(t_{2g})^2$ configuration. Instead of considering the case of four electrons in t_{2g} orbitals, think of the problem as being two holes in the t_{2g} orbitals, which we solved previously. So really, there are only a few "key" configurations that we need to work out. These key configurations are listed in Table 16.15, along with the multielectron states that they give rise to.

Any other mixed configuration—those that are products of those listed in Table 16.15, such as $(t_{2g})^2(e_g)^1$ or $(t_{2g})^4(e_g)^2$—can be obtained by taking the direct products of the states of each individual configuration, taking into consideration the way that spin states multiply together, which are obtained from the rules of angular momentum coupling, as shown in Table 16.16.

For instance, the $(t_{2g})^3(e_g)^2$ configuration of a d^5 ion contains 120 microstates, the product of twenty states from the $(t_{2g})^3$ configuration and six states from the $(e_g)^2$ configuration, using Eq. 16.1. These 120 microstates are lumped into twenty-one independent states, making this case a particularly onerous one to solve using the descent-in-symmetry approach. By comparison, the list of states is determined in a straightforward manner by employing the direct-product method and using Tables 16.15 and 16.16 as shown as follows.

$$(t_{2g})^3(e_g)^2 = \begin{array}{l}\left({}^2E_g + {}^2T_{1g} + {}^2T_{2g} + {}^4A_{2g}\right)\\ \otimes\left({}^1A_{1g} + {}^1E_g + {}^3A_{2g}\right)\end{array}$$

$$= \left\{\begin{array}{l}
{}^2E_g \otimes {}^1A_{1g} = {}^2E_g\\
{}^2E_g \otimes {}^1E_g = {}^2A_{1g} + {}^2A_{2g} + {}^2E_g\\
{}^2E_g \otimes {}^3A_{2g} = {}^2E_g + {}^4E_g\\
{}^2T_{1g} \otimes {}^1A_{1g} = {}^2T_{1g}\\
{}^2T_{1g} \otimes {}^1E_g = {}^2T_{1g} + {}^2T_{2g}\\
{}^2T_{1g} \otimes {}^3A_{2g} = {}^2T_{2g} + {}^4T_{2g}\\
{}^2T_{2g} \otimes {}^1A_{1g} = {}^2T_{2g}\\
{}^2T_{2g} \otimes {}^1E_g = {}^2T_{1g} + {}^2T_{2g}\\
{}^2T_{2g} \otimes {}^3A_{2g} = {}^2T_{1g} + {}^4T_{1g}\\
{}^4A_{2g} \otimes {}^1A_{1g} = {}^4A_{2g}\\
{}^4A_{2g} \otimes {}^1E_g = {}^4E_g\\
{}^4A_{2g} \otimes {}^3A_{2g} = {}^2A_{1g} + {}^4A_{1g} + {}^6A_{1g}
\end{array}\right.$$

You can check that the list does contain twenty states, and that the total degeneracy indeed adds up to 120 microstates, as required.

16.4.3 The Energy of Multielectronic States in Octahedral Fields

Now that we know how to identify the symmetries of the states that arise from any given electronic configuration, we can tackle the harder issue of assigning the energy associated

Table 16.16 Product rules for spin states.

Separate Terms	Product	$^{x}A_{1g} \times {}^{y}A_{1g}$ Example
Singlet × Singlet	Singlet	$^{1}A_{1g} \times {}^{1}A_{1g} = {}^{1}A_{1g}$
Singlet × Doublet	Doublet	$^{1}A_{1g} \times {}^{2}A_{1g} = {}^{2}A_{1g}$
Singlet × Triplet	Triplet	$^{1}A_{1g} \times {}^{3}A_{1g} = {}^{3}A_{1g}$
Singlet × Quartet	Quartet	$^{1}A_{1g} \times {}^{4}A_{1g} = {}^{4}A_{1g}$
Doublet × Doublet	Singlet + Triplet	$^{2}A_{1g} \times {}^{2}A_{1g} = {}^{1}A_{1g} + {}^{3}A_{1g}$
Doublet × Triplet	Doublet + Quartet	$^{2}A_{1g} \times {}^{3}A_{1g} = {}^{2}A_{1g} + {}^{4}A_{1g}$
Doublet × Quartet	Triplet + Quintet	$^{2}A_{1g} \times {}^{4}A_{1g} = {}^{3}A_{1g} + {}^{5}A_{1g}$
Triplet × Triplet	Singlet + Triplet + Quintet	$^{3}A_{1g} \times {}^{3}A_{1g} = {}^{1}A_{1g} + {}^{3}A_{1g} + {}^{5}A_{1g}$
Triplet × Quartet	Doublet + Quartet + Sextet	$^{3}A_{1g} \times {}^{4}A_{1g} = {}^{2}A_{1g} + {}^{4}A_{1g} + {}^{6}A_{1g}$

with each of these states. This is a harder problem, in part because there is always an element of quantitative approximation associated with this process; that is, the answers we obtain depend invariably on our starting point and on the various approximations. This is where our choice between the strong- versus weak-field scheme will have the most dramatic impact. Whereas both approaches lead to the same number of states and state symmetries, the approximate energies assigned to each of these states depend strongly on the scheme that we use to define these states in the first place. Of course, we can always correct these "first-order" energies to the point where both schemes converge to the same answer, but the amount of work required to get to that point is often dramatically different. To illustrate, let's use the d^2 configuration again.

The first thing we do is to define the *zeroth-order energies*—the energies that the state would have if we were to completely ignore the effect of the perturbation. Each of the two schemes we use have a different picture of what the zeroth-order states are, and what the "perturbation" acting on these is.

For the weak-field scheme, our zeroth-order states are the free-ion wavefunctions, and the perturbation acting on them arises from the crystal field forces due to the ligands around the central ion. We have already tabulated these energies in Table 16.6. All states that arise from the same free-ion state will have the same energy in zeroth-order when we are assuming that the ligand-field is infinitely small. For instance, the energy of the ^{3}F state of a d^2 ion is $E(^3F) = -8B$, which is also going to be the zeroth-order weak-field energy of the three O_h components ($^3T_{1g}$, $^3T_{2g}$, $^3A_{2g}$) that arise out of it.

For the strong-field scheme, our zeroth-order states are the strong-field orbital configurations, and the perturbation is the electron–electron interaction. The zeroth-order energies in the strong-field scheme are easy to obtain, as they are simply the sum of the orbital energies for each electron, and in Chapter 15 we obtained these as $E(t_{2g}) = -4$ Dq and $E(e_g) = 6$ Dq. The zeroth-order energy of any strong-field configuration with m electrons in the t_{2g} set and n electrons in the e_g set is then:

$$E^{(0)}\left(t_{2g}^{m}e_{g}^{n}\right) = (6n - 4m)Dq \qquad \textbf{16.18}$$

For example, the zeroth-order energies of the three d^2 strong-field configurations are $E^{(0)}(t_{2g}^{2}) = -8$ Dq, $E^{(0)}(t_{2g}^{1}e_{g}^{1}) = 2$ Dq, and $E^{(0)}(e_{g}^{2}) = 12$ Dq.

We then need to draw a *correlation diagram*—a figure that connects the states from the weak-field side to the strong-field side. We can do this unambiguously whenever there are only one of a given type of state symmetry in the whole configuration (say, for the $^3A_{2g}$, $^3T_{2g}$, or $^1T_{1g}$ states of the d^2 configuration), but when there is more than one state with the same symmetry, then we need to abide by the *noncrossing rule*: *When a perturbation is applied on a system, the energies of two states that have the same symmetry never cross* (Fig. 16.8). When the energy difference between two states of the same symmetry becomes sufficiently small, they will mix to form low-energy and a high-energy components.

With this principle in hand, a rough correlation diagram for the d^2 configuration can be obtained, where, for instance, the lowest-energy $^3T_{1g}$ component on the weak-field side (the one that comes out of the free-ion ^{3}P state) correlates with the lowest-energy $^3T_{1g}$ component on the strong-field side (the one associated with the t_{2g}^{2} configuration), and the same for the high-energy $^3T_{1g}$ states (the one from the ^{3}P free-ion state correlating with the one from the $t_{2g}^{1}e_{g}^{1}$ configuration), as shown in Fig. 16.9.

An immediate question that might be asked at this point is whether the exact correlation that links both limits together is necessarily a linear one (as implied by our previous rough diagram). As it happens, when there is only one state of a given type and symmetry, then that state's energy does change linearly, as shown. For the three states that occur only once in the d^2 configuration, then we know everything we need to know to define their energies! For instance, the energy of the $^3T_{2g}$ state must be

$$E\left(^{3}T_{2g}\right) = -8B + 2Dq \qquad \textbf{16.19}$$

which works for both limits: in the weak-field limit (where $Dq \sim 0$) $E(^3T_{2g}) = -8B$, but in the strong-field limit (where $B \sim 0$), $E(^3T_{2g}) = 2Dq$. The energies of the other two

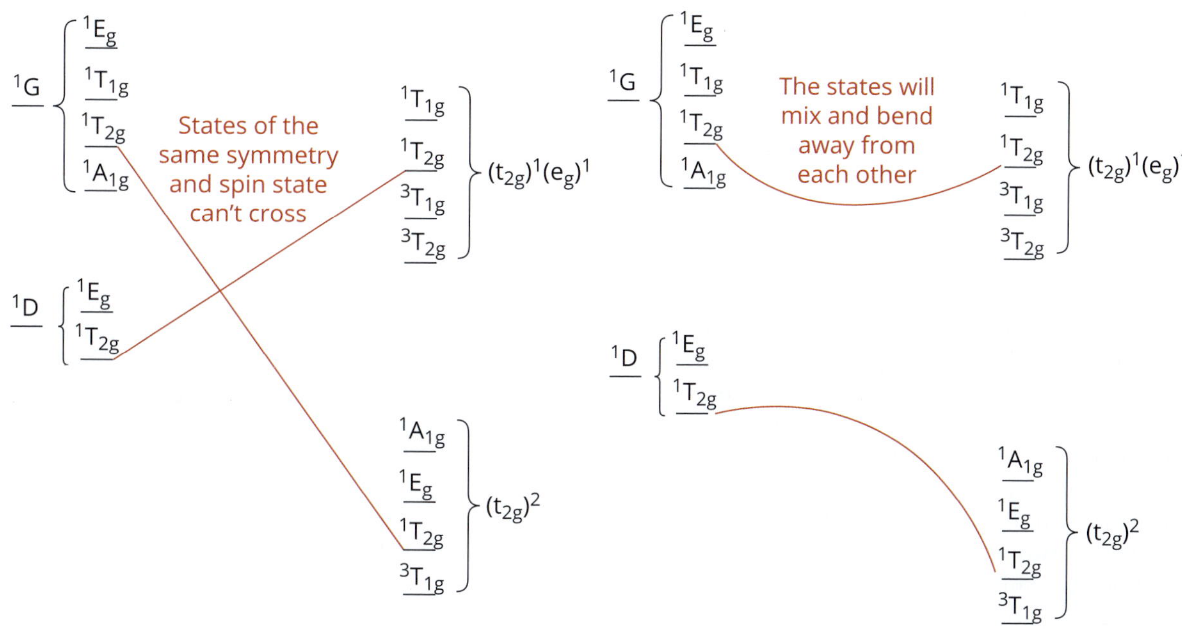

Fig. 16.8 The *anticrossing* rule. The $^1T_{2g}$ state derived from the ^{1}G free-ion term in the weak-field limit is associated with the $(t_{2g})^1(e_g)^1$ configuration in the strong-field limit, in order to avoid crossing the $^1T_{2g}$ state associated with the ^{1}D free-ion state (weak-field limit)/$(t_{2g})^2$ configuration (strong-field limit).

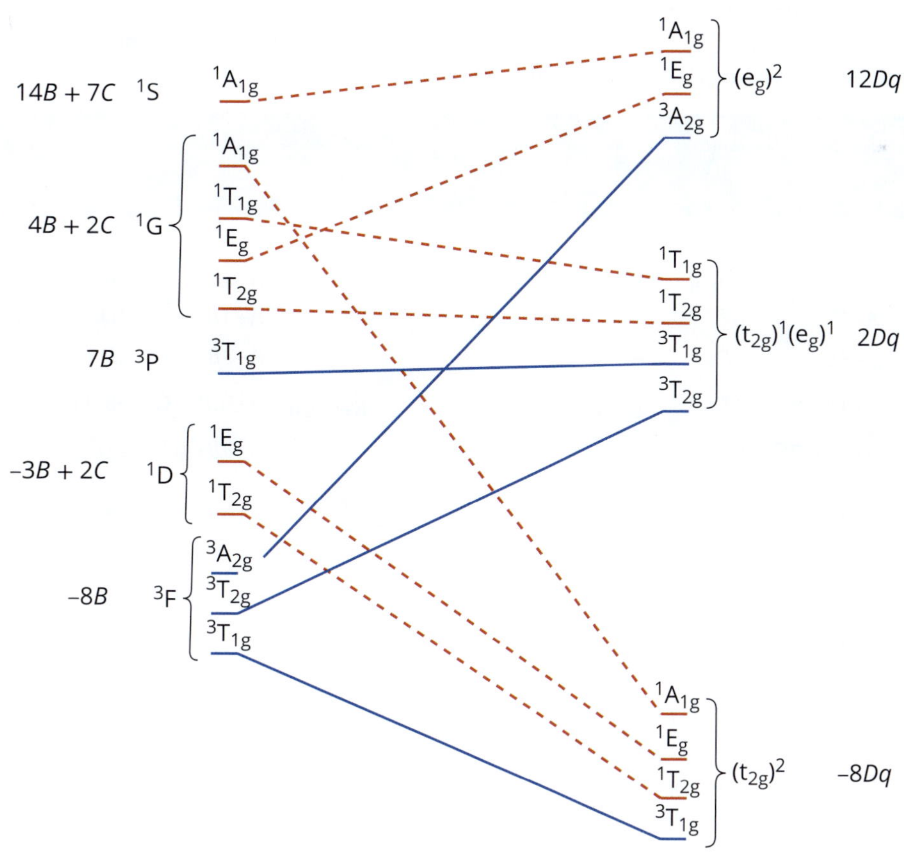

Fig. 16.9 Correlation diagram for the d^2 configuration. The zeroth-order energies are indicated next to each group of states, either in the weak-field limit (left side) or in the strong-field limit (right side).

states that occur only once in the d^2 configuration are similarly obtained (Eq. 16.20–21).

$$E\left({}^3A_{2g}\right) = -8B + 12Dq \tag{16.20}$$

$$E\left({}^1T_{1g}\right) = 4B + 2C + 2Dq \tag{16.21}$$

Things are a little more complex when there is more than one type of state of a given symmetry, because then the correlation is *not* linear, and the energy change will depend on the mixing between the two components, which in turn depends on how close the two components are in energy. A direct way to handle this mixing is by solving the so-called secular determinant, by which the first-order energy of each individual state of a given type and symmetry is given as a diagonal element, and the coupling term responsible for the mixing between two components is inserted in the off-diagonal position that links the two states. These first-order energies and coupling terms depend on our starting point (strong- versus weak-field schemes), and while their derivation goes beyond the level of this text, their values have now been tabulated for the most common symmetries; the values for all states of the d^2 configuration in O_h symmetry are given in Table 16.17.

To obtain the energies of coupled states, we solve the secular determinant obtained by putting the first-order energy of each state—the sum of the zeroth-order energy and the first-order correction—as diagonal elements, and the coupling element in the nondiagonal positions. For instance, the secular determinant for the ${}^3T_{1g}$ states in O_h symmetry, in the weak-field limit, is given by Eq. 16.22, where we subtracted the common Racah parameter A from both diagonal elements, since we are only interested in energy differences between states.

$$\begin{matrix} {}^3T_{1g}({}^3F) \\ {}^3T_{1g}({}^3P) \end{matrix} \begin{vmatrix} -6Dq - 8B - E & 4Dq \\ 4Dq & 7B - E \end{vmatrix} = 0 \tag{16.22}$$

This secular equation yields a quadratic equation that can be solved to yield two energies (Eq. 16.23), in which the

Table 16.17 Weak–field and strong-field zeroth-order and first-order energies of all states of the d^2 configuration in O_h symmetry, along with coupling elements. In the weak-field approach, the zeroth-order energies are given by the energy of the free-ion term associated with the state (compare Table 16.6). In the strong-field approach, the zeroth-order energies are given by the energy of the configuration, with orbital energies $E(t_{2g}) = -4Dq$ and $E(e_g) = +6Dq$.

Weak-Field Perturbation					Strong-Field Perturbation				
State	Free-ion term	$E^{(0)}$	$E^{(1)}$	Coupling element	Coupling element	$E^{(1)}$	$E^{(0)}$	Configuration	State
${}^3A_{2g}$	${}^3F$	$A-8B$	$12Dq$	–	–	$A-8B$	$12Dq$	$(e_g)^2$	${}^3A_{2g}$
${}^3T_{1g}$	${}^3F$	$A-8B$	$-6Dq$	$4Dq$	$-6B$	$A-5B$	$-8Dq$	$(t_{2g})^2$	${}^3T_{1g}$
	${}^3P$	$A+7B$	0			$A+4B$	2Dq	$(t_{2g})^1(e_g)^1$	
${}^3T_{2g}$	${}^3F$	$A-8B$	$2Dq$	–	–	$A-8B$	$2Dq$	$(t_{2g})^1(e_g)^1$	${}^3T_{2g}$
${}^1A_{1g}$	${}^1G$	$A+4B+2C$	$4Dq$	$4\sqrt{6}\,Dq$	$(2B+C)\sqrt{6}$	$A+10B+5C$	$-8Dq$	$(t_{2g})^2$	${}^1A_{1g}$
	${}^1S$	$A+14B+7C$	0			$A+8B+4C$	$12Dq$	$(e_g)^2$	
${}^1E_g$	${}^1D$	$A-3B+2C$	$\frac{24}{7}Dq$	$\frac{40\sqrt{3}}{7}Dq$	$2B\sqrt{3}$	$A+B+2C$	$-8Dq$	$(t_{2g})^2$	${}^1E_g$
	${}^1G$	$A+4B+2C$	$\frac{4}{7}Dq$			$A+2C$	$12Dq$	$(e_g)^2$	
${}^1T_{1g}$	${}^1G$	$A+4B+2C$	$2Dq$	–	–	$A+4B+2C$	$2Dq$	$(t_{2g})^1(e_g)^1$	${}^1T_{1g}$
${}^1T_{2g}$	${}^1D$	$A-3B+2C$	$-\frac{16}{7}Dq$	$\frac{20\sqrt{3}}{7}Dq$	$-2B\sqrt{3}$	$A+B+2C$	$-8Dq$	$(t_{2g})^2$	${}^1T_{2g}$
	${}^1G$	$A+4B+2C$	$-\frac{26}{7}Dq$			$A+2C$	$2Dq$	$(t_{2g})^1(e_g)^1$	

superscripts *lower/upper* have been used to indicate the high-/low-energy root of Eq. 16.22.

$$E\left({}^3\mathrm{T}_{1g}^{lower}\right) = \frac{-6Dq - B - \sqrt{100Dq^2 + 180B \cdot Dq + 225B^2}}{2} \qquad \textbf{16.23a}$$

$$E\left({}^3\mathrm{T}_{1g}^{upper}\right) = \frac{-6Dq - B + \sqrt{100Dq^2 + 180B \cdot Dq + 225B^2}}{2} \qquad \textbf{16.23b}$$

If we put $Dq = 0$ in Eq. 16.23—that is, assume that there is no ligand-field effect—we obtain the zeroth-order energy in the weak-field limit, as expected. Perhaps more interestingly, you can verify for yourself that if we look at the limit $B \to 0$—the strong-field limit—then both equations converge to the strong-field zeroth-order energy values of the ${}^3\mathrm{T}_{1g}$ states. Although the result is the same, we can also build a secular determinant for the same two ${}^3\mathrm{T}_{1g}$ states starting from the strong-field limit as well (Eq. 16.24).

$$\begin{matrix} {}^3\mathrm{T}_{1g}\left(\mathrm{t}_{2g}^2\right): \\ {}^3\mathrm{T}_{1g}\left(\mathrm{t}_{2g}^1\mathrm{e}_g^1\right): \end{matrix} \begin{vmatrix} -8Dq - 5B - E & -6B \\ -6B & 2Dq + 4B - E \end{vmatrix} = 0 \qquad \textbf{16.24}$$

Even though the determinant is different, the two roots of the quadratic in Eq. 16.24 are exactly the same as those given previously for the weak-field determinant, Eq. 16.22. This is necessarily so because we are solving both determinants with exactly a full basis; that is, we did not exclude any ${}^3\mathrm{T}_{1g}$ component from the d^2 configuration here. In the limit where the secular determinants are solved exactly, there is no difference between the strong-field or weak-field starting points.

Using this approach, quantitatively accurate energy diagrams can be drawn for the ligand-field splitting of any d^n configuration, as shown in Fig. 16.10.

We have completed the d^2 correlation diagram in O_h, but we mentioned previously that the d^8 configuration should be closely related to the d^2 configuration (electron-hole analogy), so this similarity must somehow appear in their respective energy diagrams. For starters, we know that their free-ion terms are identical (Table 16.1), and the d^8 energy diagram must therefore be identical to the d^2 diagram when $Dq = 0$. Of course, the strong-field side of the d^8 diagram will look different from d^2 in that the energy ordering of the states has to be "inverted". For d^2, the $(t_{2g})^0(e_g)^2$ configuration; that is, zero electrons in t_{2g} and two electrons in e_g, is highest in energy and produces terms ${}^1\mathrm{A}_{1g}$, ${}^1\mathrm{E}_g$, and ${}^3\mathrm{A}_{2g}$; whereas for d^8 the $(t_{2g})^6(e_g)^2$ configuration—zero holes in t_{2g} and two holes in e_g—is the lowest in energy. Using the electron-hole analogy, we must have the same set of states as the $(t_{2g})^0(e_g)^2$ configuration. Consequently, the strong side of the correlation diagram is flipped for d^8 relative to d^2; states that are lowest in energy in d^2 become the highest in energy in d^8. This is simply another representation of the fact that the energetic stability of holes is reversed from that of electrons which tend to occupy orbitals that are the lowest in energy, whereas holes tend to occupy orbitals that are the highest in energy.

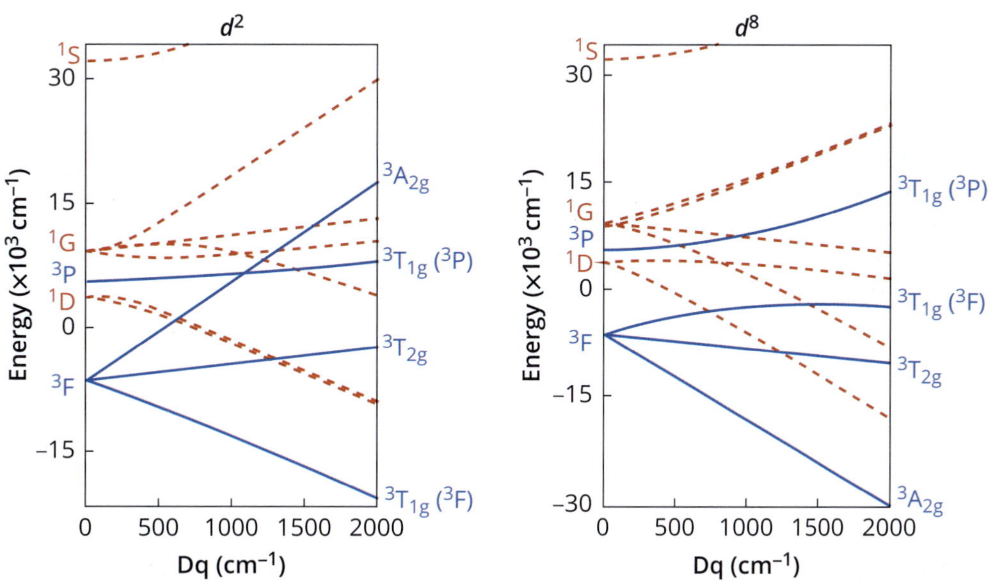

Fig. 16.10 (*Left*) Energy diagram of the multielectron states of the d^2 configuration, calculated by using $B = 800$ cm^{-1} and $C = 3{,}000$ cm^{-1}. Triplet states are shown and labeled in blue, and singlet states are shown by dashed red lines. (*Right*) Energy diagram of the multielectron states of the d^8 configuration, calculated by using $B = 800$ cm^{-1} and $C = 3{,}000$ cm^{-1}. Triplet states are shown and labeled in blue, and singlet states are shown by dashed red lines.

A very convenient way to represent the intrinsic symmetry between d^2 and d^8, or any other pair of configurations related through the electron-hole analogy, is through the use of *Orgel diagrams*.[v] Fundamentally, Orgel diagrams (Fig. 16.11) are nothing more complicated than the energy diagram shown in Fig. 16.10, but with the following distinguishing features:

1) The x-axis of an Orgel diagram represents the ligand-field strength in units of Dq, but this value is now allowed to take both positive *and* negative values. The zero of the x-axis ($Dq = 0$, the free-ion case) is always shown in the center of the diagram.
2) The negative Dq values are used to represent the switch from the electron representation to the equivalent hole representation. (Remember, holes behave opposite to electrons in terms of their energy stabilization.) If the positive Dq values (right side of the Orgel diagram) are used to show the diagram of the d^2 configuration, then the negative Dq values (left side of the Orgel diagram) would represent the hole-equivalent of the d^2 configuration; that is, the d^8 configuration.
3) Traditionally, Orgel diagrams show only those states that have the same multiplicity as the ground state. For the d^2/d^8 case, this means that singlet states are generally omitted. This helps for the clarity of the diagram but is by no means an absolute requirement.

The Orgel diagrams for the d^2/d^8 and the d^3/d^7 configurations are shown in Fig. 16.11. The strength of Orgel diagrams is that they nicely show the intimate connection between the d^n/d^{10-n} configurations, visually underlying the fundamental role played by the electron-hole analogy. In particular, notice how much clearer the configuration interaction that mixes the T_{1g} states now appears as an unmistakable avoided crossing in the Orgel diagrams, compared to what can be perceived from the individual energy diagrams.

The idea of using negative Dq values to join configurations that are partners through the electron-hole analogy can be extended even further. Recall that the energy ordering of d-orbitals of a tetrahedral complex (t_2 above e) is the inverse of that found for octahedral complexes (e_g above t_{2g}). This implies that we could think of the negative Dq value as representing the change from O_h to T_d, as shown in Fig. 16.12.

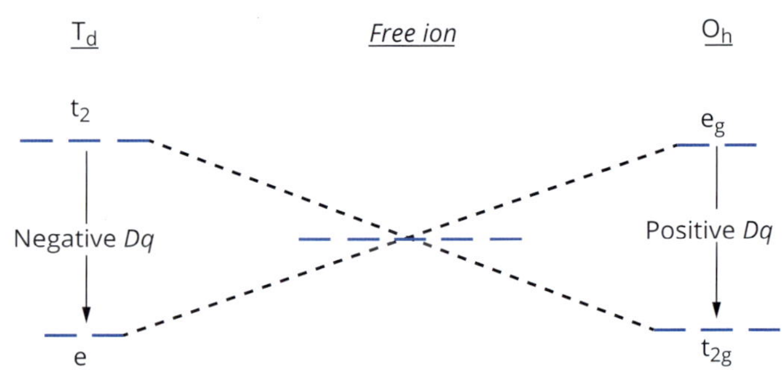

Fig. 16.12 The intimate relationship between the octahedral (O_h) and tetrahedral (T_d) symmetries from the point of view of the ligand-field splitting of the d-orbital manifold. The T_d orbital splitting can be related to the O_h splitting by using a negative-valued ligand-field splitting parameter Dq.

[v] These are named after Leslie E. Orgel (1927–2007), a British chemist and evolutionary biologist who introduced them in *An Introduction to Transition Metal Chemistry* (1959)—one of the first textbooks that incorporated ligand-field theory as part of the standard curriculum of inorganic chemistry.

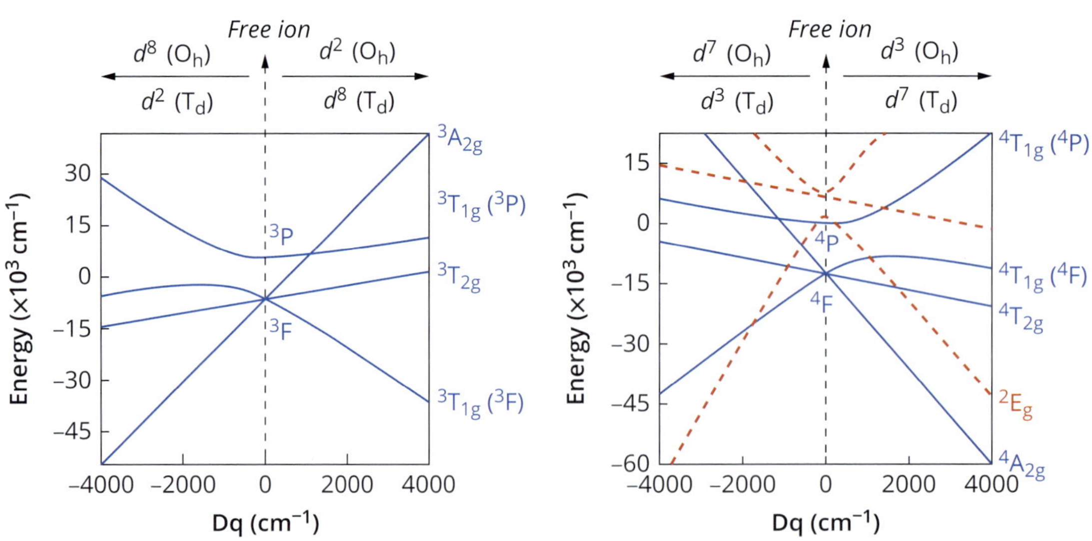

Fig. 16.11 (*Left*) Orgel diagram for the d^2/d^8 configurations, calculated using $B = 800\ cm^{-1}$. (*Right*) Orgel diagram for the d^3/d^7 configurations, calculated using $B = 850\ cm^{-1}$. For tetrahedral configurations, the subscripts "g" should be omitted for the state labels. The lowest-energy doublet states are shown as red dashed lines in the diagram, showing that the d^7 (O_h)/d^3 (T_d) configurations exhibit a high-spin-to-low-spin ground-state transition at high-enough ligand-field strength.

Orgel diagrams show how the energy diagram for a tetrahedral complex of a d^n ion is related to the O_h diagram of the d^{10-n} configuration. Of course, we need to be careful with the interpretation of "10*Dq*" (recall that $\Delta_{oct} = -9/5\ \Delta_{tet}$ for the same ligand-field environment, see Chapter 15), but as long as we use the diagram qualitatively to see how many states there are and what their relative energy ordering is, this is not a problem.[w]

16.5 Tanabe–Sugano Diagrams and Racah Parameters

While Orgel diagrams are extremely useful for illustrating the underlying relationships of d^n/d^{10-n} configurations or O_h/T_d symmetries, there are more useful ways of representing the energies of the *d*-states of transition-metal compounds. For one thing, Orgel diagrams are computed for one given set of Racah parameters, which means that every time we want to study a different ion with different electron–electron repulsion parameters, we should recalculate a new Orgel diagram. Furthermore, although it is sometimes nice to see how the ground-state energy changes in absolute terms when we change the ligand-field strength, in real life we never measure the absolute energy of the ground state; we can only measure the energy difference between states. Practically, we generally like to think of the ground state as defining the zero of our energy scale and define the energy of the other excited states in reference to that of the ground state. Tanabe–Sugano diagrams[x] solve these issues and are now the standard representation used to discuss the electronic structure of transition-metal ions.

A Tanabe–Sugano diagram is constructed in the same way as any other ligand-field energy diagram: the *x*-axis is used to represent the strength of the ligand field, and the *y*-axis is used to represent the energy of the states. In contrast to Orgel diagrams, the energy of the ground state—the state with the lowest energy—is always defined as the zero of the energy scale, so the energy of that state is subtracted from the energy of every other state. Consequently, the line associated with the ground-state energy is always superimposed on the abscissa in a Tanabe–Sugano diagram.

Another significant characteristic of Tanabe–Sugano diagrams is that energies are scaled relative to the Racah parameter *B*, which means that both the abscissa and ordinate axes in a Tanabe–Sugano diagram are dimensionless. The *y*-axis is the energy of each state in units of *B*, and the *x*-axis is the ligand-field strength parameter *Dq* divided by *B*. Of course, in order to fully calculate every state energy, we still need to define the value of the C Racah parameter; in a Tanabe–Sugano diagram, this problem is resolved by calculating the diagram by assuming a given *C*/*B* ratio. This ratio should therefore always be given for each Tanabe–Sugano diagram. The *C*/*B* ratios that are generally used to compute generic Tanabe–Sugano diagrams are typically around four. Fixing this ratio allows one to obtain "semi-universal" diagrams, which are valid for any compound as long as the ratio *C*/*B* is not too different than the one that was used to plot the diagram. In practice, although both *B* and *C* change from compound to compound, their ratio is more or less constant (but by no means invariant), so the approach taken in Tanabe–Sugano diagrams is a good solution to the problem of plotting energy functions that depend on three independent variables (*Dq*, *B*, and *C*). Because Tanabe–Sugano diagrams are often used qualitatively to acquire a sense of the energy ordering between states, or to approximately obtain an idea of the relative energy differences between states, this is often not a significant limitation.

The Tanabe–Sugano diagrams for the d^2–d^8 configurations are shown in Figs. 16.13a–g, and in the next sections

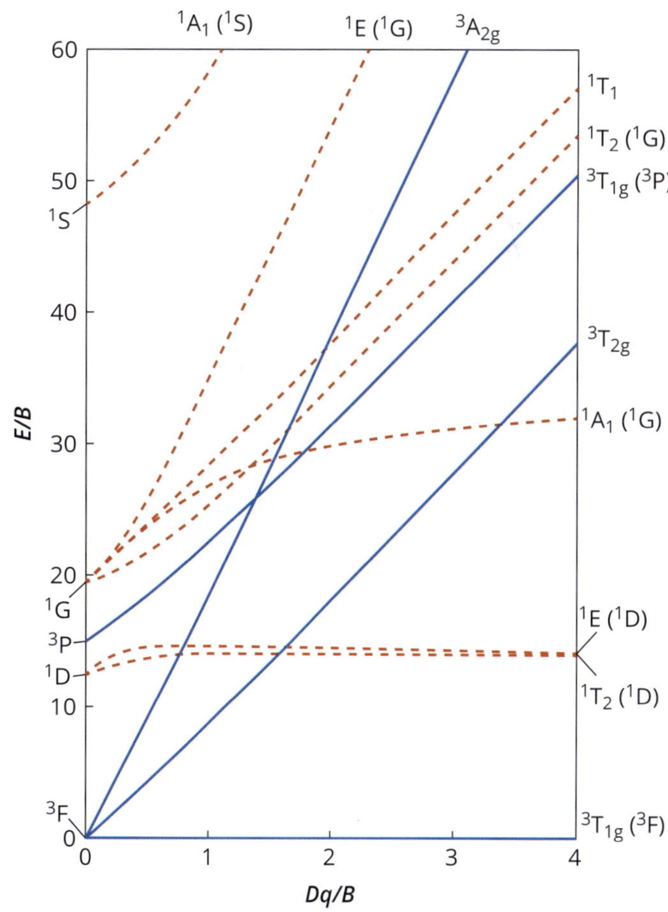

Fig. 16.13a Tanabe–Sugano diagram for the d^2 configuration, calculated with $C/B = 3.75$.

---

[w] Remember also that there is no center of inversion in the T_d point group, so you need to ignore the parity symbols "g" of the symmetry labels that are given in the O_h energy diagrams of the d^{10-n} configuration.

[x] Named after Yukito Tanabe and Satoru Sagano, two Japanese theoretical physicists who developed ligand-field calculations that led to these diagrams in a series of landmark papers in the mid-1950s.

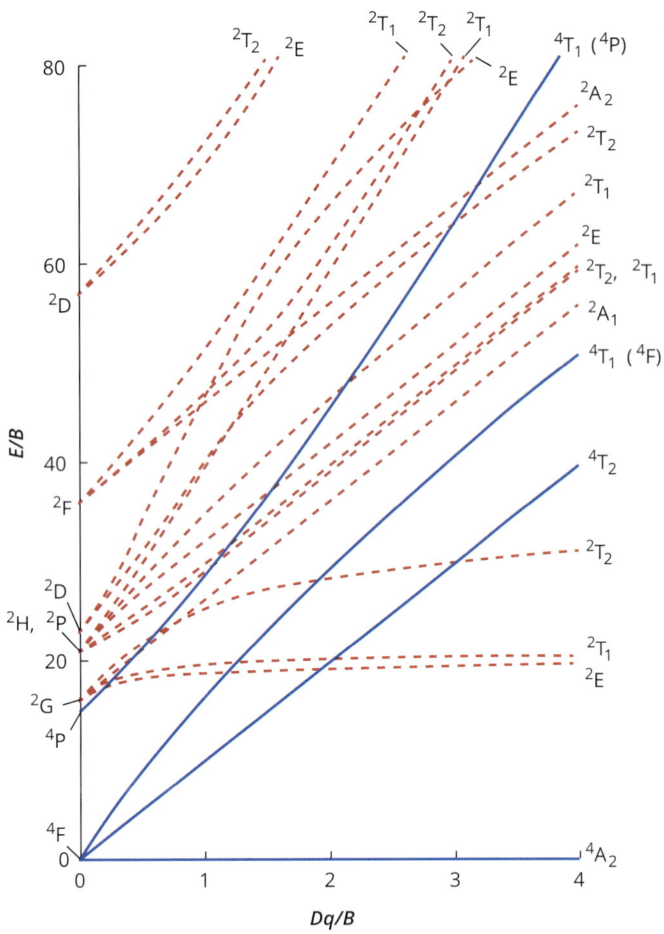

Fig. 16.13b Tanabe–Sugano diagram for the d^3 configuration, calculated with $C/B = 4.0$.

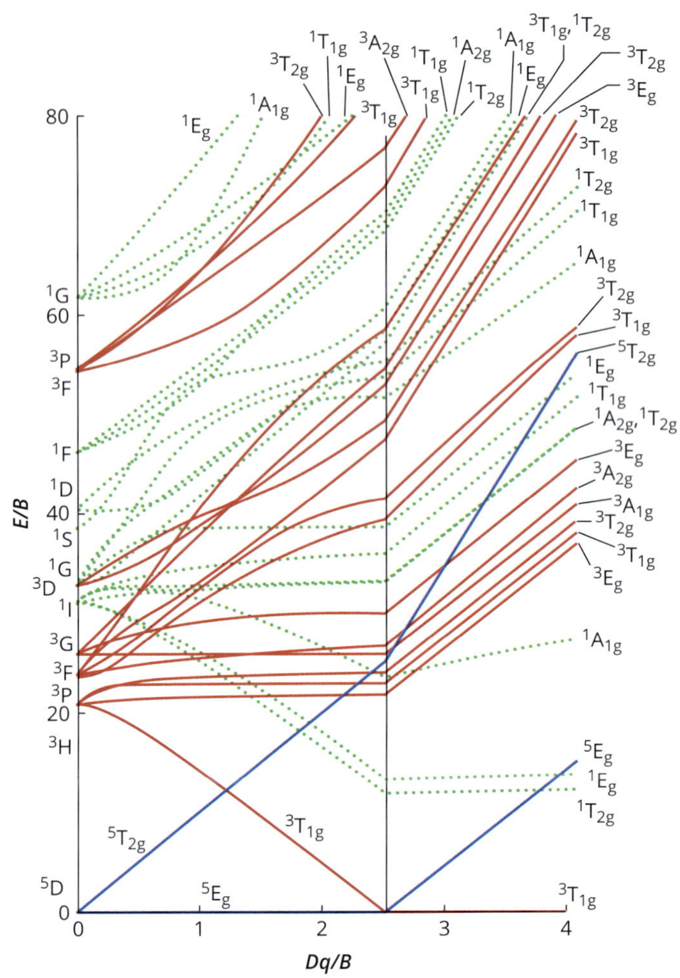

Fig. 16.13c Tanabe–Sugano diagram for the d^4 configuration, calculated with $C/B = 4.2$.

we will show how these diagrams can be used to analyze the electronic spectra of transition-metal compounds. Before doing so, note how the Tanabe–Sugano diagrams for the d^4, d^5, d^6, and d^7 differ from that of the d^2, d^3, and d^8 diagrams; all of the former have a vertical line that clearly separates the diagram into two different regimes. This vertical line represents the point where the ground-state changes from high-spin (to the left of the vertical line) to low-spin (to the right of the vertical line). Because d^2, d^3, and d^8 have only one unique ground state, which is independent of the ligand-field strength; that is, there is no distinction between high and low field for these three configurations, and the Tanabe–Sugano diagrams are simpler. Realize, however, that the suddenly discontinuous nature of the Tanabe–Sugano diagrams of the d^4, d^5, d^6, and d^7 configurations is just an artefact of choosing to put the energy of the ground state equal to zero.

16.6 The Intensity of Electronic Transitions: Transition Moment Integrals and Symmetry-Based Selection Rules

The intensity of a peak in the absorption spectrum can provide valuable information about the nature of the transition between the two states being observed. The intensity of a band can be calculated by using the *transition moment integral*, shown in Eq. 16.25. In the equation, I is the intensity of the band, ψ_i is the initial wavefunction, ψ_f is the final wavefunction, and μ is the transition-moment operator. In other words, the intensity of a transition is dependent upon the initial and final states and an operator that represents the light–matter interaction, μ.

$$I = \int \psi_f^* \cdot \mu \cdot \psi_i \, d\tau \qquad \textbf{16.25}$$

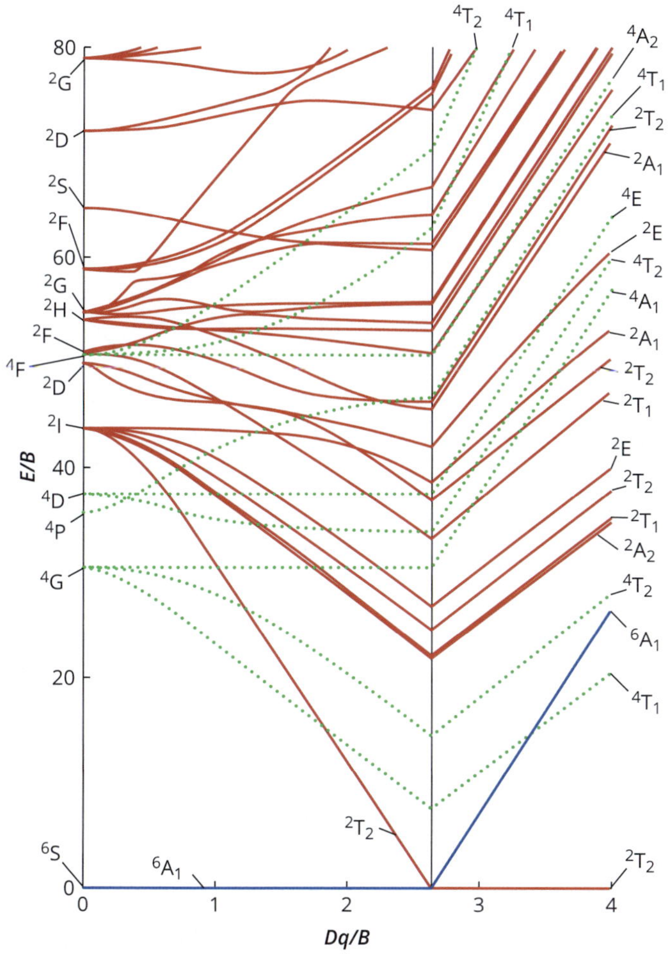

Fig. 16.13d Tanabe–Sugano diagram for the d^5 configuration, calculated with $C/B = 4.1$.

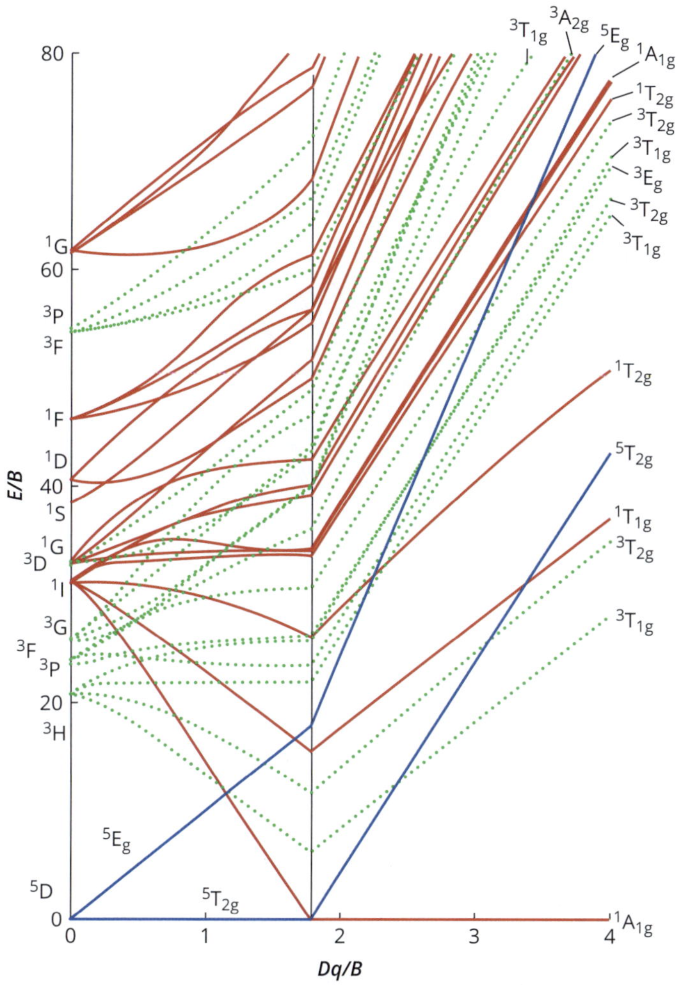

Fig. 16.13e Tanabe–Sugano diagram for the d^6 configuration, calculated with $C/B = 4.2$.

Among a number of different ways that light can interact with matter, a major contribution to μ is the so-called *electric-dipole transition moment*, in which μ represents the magnitude of the oscillating electric dipole associated with the electronic excitation between the two electronic states. A general feature of many standard optical spectroscopic methods—such as infrared, UV-visible, near-infrared spectroscopy, and so on—is that the principal mechanism of the light–matter interaction, which is responsible for the absorption of energy from the light beam by molecules (or vice versa in the case of emission), occurs through the direct action of the oscillating electric field of the light-beam and the oscillation of electric charges (electrons and/or protons) that constitute matter. Although there are many ways by which charges can interact with light, by far the most important is oscillation of electric dipoles. These dipoles can exist either permanently, such as in polar molecules, or they can be generated transiently as the molecule spontaneously distorts through one of its vibrational modes or under the influence of other fields, due to other surrounding molecules, for instance.

The integral in Eq. 16.25 can in principle be calculated exactly to derive the actual strength of the coupling (the "intensity") between an oscillating electromagnetic beam of light and a specific dipole oscillation within a molecule corresponding to the transition between the states ψ_i and ψ_f, but this is generally a difficult process that requires highly accurate wavefunctions that are not easy to obtain for most molecular systems. That being said, one can still extract extremely important information about the transition intensity by simply considering the symmetry constraints on the transition-moment integral. Such considerations are absolutely independent of any quantum-mechanical calculations and provide us with details that must apply absolutely, no matter what the exact wavefunctions happen to be. A short summary of the symmetry requirement on the transition-moment integral is that *the intensity, I, is nonzero if, and only if, the direct product*

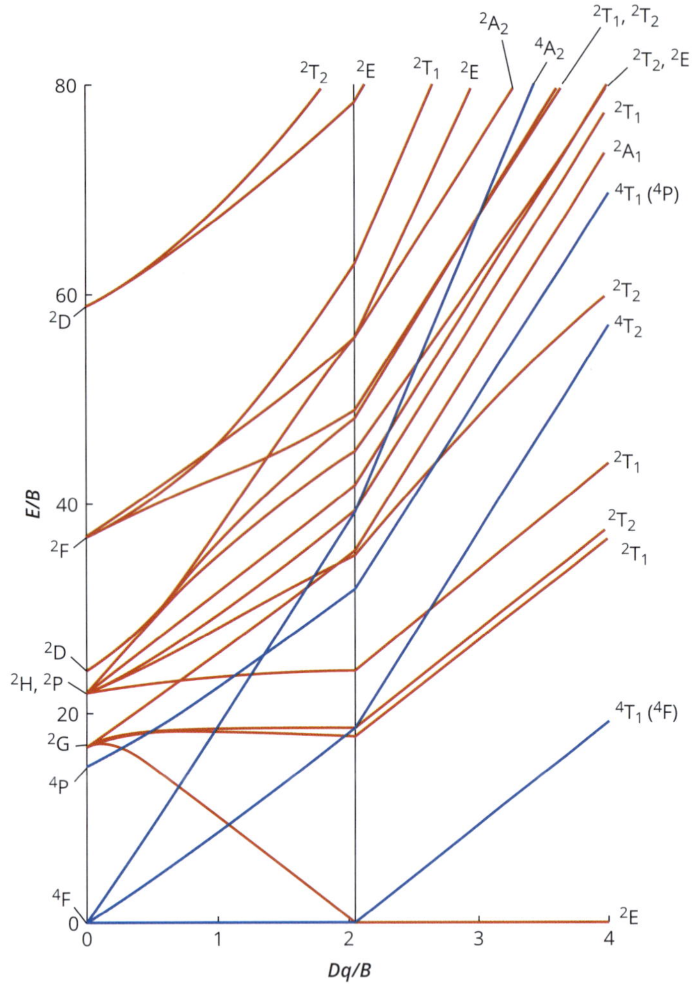

Fig. 16.13f Tanabe–Sugano diagram for the d^7 configuration, calculated with $C/B = 4.3$

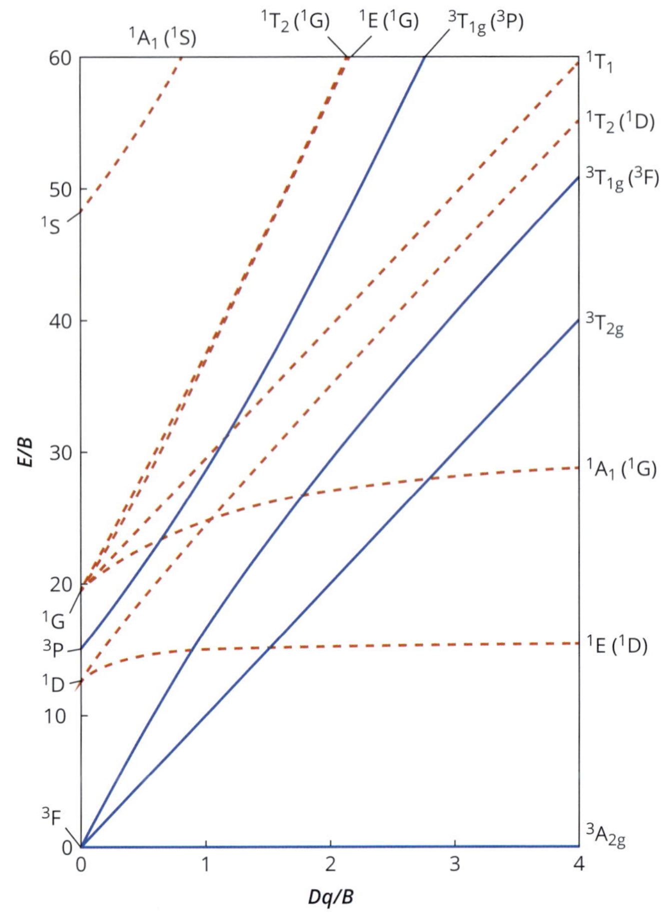

Fig. 16.13g Tanabe–Sugano diagram for the d^8 configuration, calculated with $C/B = 3.75$.

of the representations of the functions in the integral in Eq. 16.25 contains the totally symmetric representation. That is, if the symmetry of the initial wavefunction is Γ_i, the final wavefunction is Γ_f, and the transition-moment operator is Γ_μ, the direct product of the three representations must include the totally symmetric representation:

$$\Gamma_f \otimes \Gamma_\mu \otimes \Gamma_i \supseteq \textit{totally symm. rep.} \qquad \textbf{16.26}$$

The integration over all space of any function that does not transform as the totally symmetric irreducible representation must always be strictly zero, as such a function must always have an equal magnitude of positive and negative amplitude distributed around the point of symmetry. Take for instance an *s*-orbital centered on the point of symmetry as an example of a totally symmetric function. Such an s-orbital always remains unchanged against all symmetry operations of the point group, and thus always belong to the A_{1g} irreducible representation. (We are assuming for simplicity that the point group is O_h, but the argument works just as well for any other point group, as long as you translate "A_{1g}" to the proper totally symmetric irreducible representation of the group.) It should be obvious that because all points in space of an s-orbital share the same phase at any given point in time, the integration over all space has to yield a nonzero value, unless the amplitude at every point in space is exactly zero.[y] Conversely, a function such as a p_z-orbital, which does not belong to the totally symmetric representation (in O_h at least), has precisely counterbalancing phases that integrate to zero over all space, and by symmetry considerations there has to always be an equal amount of negative and positive amplitudes around both sides of the *p*-orbital. These simple examples illustrate how we can use symmetry to determine whether some integration over all space vanishes, or not, by symmetry constraints.

Note that symmetry alone can only determine whether a certain integral is necessarily zero or not. For instance, take the same p_z orbital considered previously, but in the C_{2v} point group. Obviously, the integral value remains the same, no matter what the point group (it is the same mathematical

[y] If this is not immediately obvious to you, recall that integrations are simply additions of infinitesimally small quantities. If all these quantities all have the same sign, then you cannot obtain a total sum equal to zero, unless all these small quantities are mathematically equal to zero.

function!), but p_z now transforms at the totally symmetric representation in C_{2v}. Clearly, the requirement in Eq. 16.26 is a necessary but not sufficient condition to guarantee that the integration does not equal zero.

Any transition that obeys Eq. 16.26 is said to be *symmetry-allowed*; in turn, if the transition moment integral vanishes by symmetry—if the direct product in Eq. 16.26 does not contain a totally symmetry component—then the transition is said to be *symmetry-forbidden*.

Forbidden transitions cannot physically occur under the mechanism of the electric-dipole operator, and generally will therefore be associated with very weak intensities of absorption. (However, there are physical mechanisms other than the electric dipole interaction that can lead to nonzero transition probabilities.) Conversely, symmetry-allowed electronic transitions are generally associated with more intense absorptions than forbidden ones. Symmetry alone does not inform us on the actual strength of the light–matter interaction, and some symmetry-allowed transitions are associated with rather weak light–matter interactions.

For a moment, consider vibrational IR absorption spectroscopy (Chapter 5). In IR absorption, the usually relevant excitations are those that occur between the ground vibrational state (ψ_i) to the first excited vibrational state (ψ_f).[z] The symmetries of vibrational levels (ν) can be determined by using $(\Gamma)^\nu$, where Γ is the the irreducible representation of the vibration, and ν being the vibrational quantum number (ν = 0 being the ground state, ν = 1 the first excited state, and so on).

In water, we determined using group theory (Section 5.10.1) that there should be three vibrations: one symmetric stretching mode of A_1 symmetry, one asymmetric stretching mode of B_2 symmetry, and one bending mode of A_1 symmetry. At that time, we simply stated that only excitations of those vibrational modes that belong to the same irreducible representation as the x, y, and z axes can be observed in IR spectroscopy. Using the transition moment integral, we can now put this on firmer ground. To see why, let us first consider the mathematical form of the vibrational wavefunctions of a harmonic oscillator. For the ground state of any vibrational level Eq. 16.27 should obtain.

$$\psi_0 = N_0\, e^{-\alpha Q^2} \qquad \mathbf{16.27}$$

Here, N_0 is a simple normalization constant (a number), α is a constant that depends on the mass of the atoms and on the force constant of the bonds involved in the vibrational mode, and Q is the normal coordinate of the mode. The term Q is the part that dictates what symmetry the wavefunction will have. For example, for B_1 modes, Q transforms like the B_1 irreducible representation, and so on for all other modes. The important part here is that the ground-state vibrational wavefunction given by Eq. 16.27 involves only Q^2, and no other powers of Q.[aa] The symmetry of the ground-state wavefunction is therefore the same as the direct product of the symmetry of the mode with itself, which always includes or is equal to the totally symmetric representation. So the initial state of any IR vibrational excitation (as long as the temperature is not so high as to populate states other than the ground state) always has the totally symmetric representation.

What about the symmetry of the first vibrational excited-state? In this case, the general form of the vibrational wavefunction in the harmonic oscillator approximation is:

$$\psi_1 = N_1\, Q e^{-\alpha Q^2}$$

which is almost identical to the ground-state vibrational wavefunction except for a difference in the normalization constant (irrelevant for symmetry considerations) and the presence of the function Q in front of the totally symmetric Gaussian part. For obvious reasons, then, the symmetry of the first excited state (the final state in most IR absorption excitations) thus transforms exactly the same as the function Q; if the mode has a B_2 symmetry, then the wavefunction for its ground state (ν = 0) transforms like A_1 (as all ground-state vibrational wavefunctions do) and the wavefunction for its first excited state (ν = 1) transforms like the mode itself, so B_2.

Knowing how to determine the symmetry of the vibrational wavefunctions, we now only need the symmetry of the light–matter interaction, μ in Eq. 16.25. If we assume that the important mode of coupling occurs through the electric dipole mechanism, then the symmetry of μ is simply the same as any dipole vector would have; that is, μ transforms as any of the functions x, y, or z.[bb] Taking the case of water as an example, the function μ in Eq. 16.25 can belong to one of three irreducible representations of the C_{2v} point group: B_1 (μ oriented along the x axis), B_2 (μ oriented along the y axis), or A_1 (μ oriented along the z axis).

We now have in hand all we need to determine the selection rules that govern the intensity of spectroscopic transitions in IR absorption. First, consider the question of whether the B_2 vibrational mode of water can be observed with x-polarized light in IR spectroscopy. As always, the ground-state vibration is totally symmetric: A_1 in C_{2v}. The symmetry of the first excited state has the symmetry of the mode, B_2. The symmetry of the operator corresponding to an x-polarized electric dipole

[z] The energy of vibrational modes is generally large enough that few excited states are thermally populated at room temperature, which prevents excitations from these states. Furthermore, within the approximation that vibrational modes behave like harmonic oscillators (a good approximation most of the time), only excitations between adjacent vibrational quantum states are allowed. Overtones and combination bands, which involve the simultaneous excitation of multiple quanta of vibration, are almost always much weaker than fundamental lines in IR spectroscopy.

[aa] You might recognize function in Eq. 16.27 as being nothing else than the well-known Gaussian, or bell-curve, function.

[bb] We assume here that the beam of light is unpolarized and that molecules are free to take any orientation relative to the beam of light. Otherwise, only specific orientations (*polarizations*) of the electric dipole mechanism should be considered.

mechanism is B_1. Taking all this together, the transition dipole integral (Eq. 16.25) transforms like direct product $A_1 \otimes B_1 \otimes B_2 = A_2$, which is not the totally symmetric representation; consequently, the B_2 vibrations of water cannot be observed by infrared absorption spectroscopy along the x direction of the molecule.

The asymmetric stretching mode can nevertheless be excited by interaction with y-polarized light with the molecule, in which case the transition dipole moment integral transforms as $A_1 \otimes B_2 \otimes B_2 = A_1$. It is not hard to deduce the following general selection rule regarding vibrational spectroscopy.

The IR spectroscopy selection rule can be summarized as *the only vibrational modes that are symmetry-allowed (electric-dipole mechanism) in IR spectroscopy are those that have the same symmetry as one of the three cartesian coordinates: x, y, or z.*

Returning to optical spectroscopy, we can use the transition-moment integral to determine if a specific transition is allowed. First, in an octahedral system or any system with an inversion center, one may realize that *d*–*d*-transitions will never be allowed if the beginning state and ending state have the same symmetry with respect to inversion. This is because the totally symmetric representation is, of course, gerade in symmetry, and the x, y, or z axes are always ungerade. The result is an integral that must have a final state that is ungerade to have a nonzero transition-moment integral. The *d*-orbitals are always gerade, and the transition moment integral always produces zero.

$$I_{d-d} = \int \psi_i \mu \psi_f \, d\tau \propto g \cdot u \cdot g = u$$

Transitions between *d*-orbitals in octahedral symmetry are, for this reason, not allowed. Transitions that are forbidden by symmetry are often referred to as being Laporte-forbidden, after Otto Laporte, who first noted the phenomenon and published it in cooperation with William Meggers in 1925.[3] In tetrahedral symmetry there is no inversion center, and *d*–*d* transitions are Laporte-allowed.

At this point it is useful to note that the designation "forbidden transition" does not imply that such transitions cannot be observed; such statements always simply refer to a specific mechanism and/or set of approximations. In practice, few transitions are ever completely and strictly forbidden. Transitions that are electric-dipole forbidden can sometimes arise through other light-matter interaction mechanisms, the electric quadrupoles and magnetic dipole mechanisms being the usual alternatives. Since the physics behind these mechanisms is operationally completely different, the symmetry associated with each is generally different than the symmetry that governs electric dipole transitions, leading to different selection rules. However, everything else being equal, completely "allowed" electric quadrupole or magnetic dipole transitions are much less intense than electric-dipole allowed transitions, and consequently electric-dipole forbidden transitions are markedly less noticeable than those that are electric-dipole allowed. So, the term "forbidden", even without the "electric-dipole" qualification, is still generally meaningful in designating very weak bands from strong ones.

Another source of intensity for formally electric-dipole forbidden transitions, which generally is responsible for most of the intensity of Laporte-forbidden *d*–*d* transitions is the so-called vibronic-coupling mechanism. The transition-moment integral used previously, Eq. 16.25, is a very simplified form of the actual integral that controls light-matter interactions, where only the electronic part of the wavefunction has been included.

Another form of forbidden-ness mentioned previously occurs in transitions between states with different spin. In a typical absorption experiment, a chemical species will only absorb a single photon before relaxing back to the ground state—a 1-photon experiment. Quantum-mechanically, one can absorb a photon to change the spin state of a molecule, or absorb a photon to complete a transition between two electronic states; and a single photon cannot do both. As a result, absorption transitions between two states with different spins are forbidden. but there are also ways of relaxing this restriction through spin-orbit coupling.

Through these different mechanisms that limit the intensity of transitions, one can often guess as to the nature of the transition. Table 16.18 summarizes the typical intensities, in terms of extinction coefficient ε, for all major classes of electronic transitions.

While the focus in this chapter has been on transitions between different electronic states related to the transition metal in complexes, *d*–*d* transitions, another type of band is often observed in metal complexes. These are charge-transfer (CT) bands where charge is moved from one part of the molecule to another on light absorption. These CT bands are both spin- and Laporte-allowed, making them more intense than most *d*–*d* transitions.

Charge-transfer bands are often assigned by listing the electron-rich donor first and electron-poor acceptor second; namely, as metal-to-ligand charge transfer (MLCT), ligand-to-metal charge transfer (LMCT), and ligand-to-ligand charge transfer (LLCT). The energy of the bands are nominally related to the ease of oxidizing the donor and reducing the acceptor. For example, for LMCT bands involving halides the energy of the band is typically I < Br < Cl < F, and

Table 16.18 Typical order of magnitudes for extinction coefficients for electronic transitions of transition-metal complexes, in $M^{-1}cm^{-1}$ units.[2]

Electronic Transitions	ε
d–*d* band in O_h, spin- and Laporte-forbidden	0.1
d–*d* band in T_d, spin-forbidden and Laporte-allowed	1
d–*d* band in O_h, spin-allowed and Laporte-forbidden	10
d–*d* band in T_d, spin- and Laporte-allowed	100
Charge transfer, spin- and Laporte-allowed	1000

Table 16.19 Solvatochromism in $LFe(CN)_5^{2-}$ complexes. Peak maxima are given in cm^{-1}.[4]

Complex / Solvent	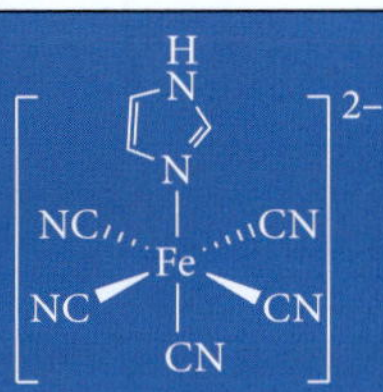	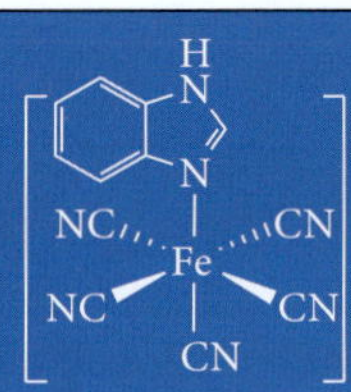	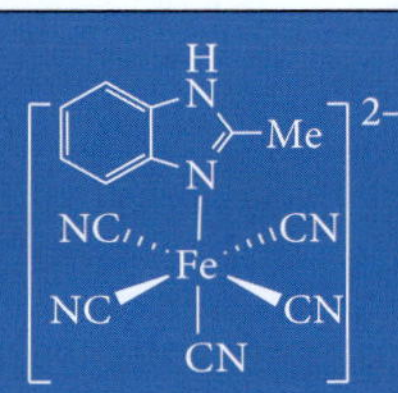	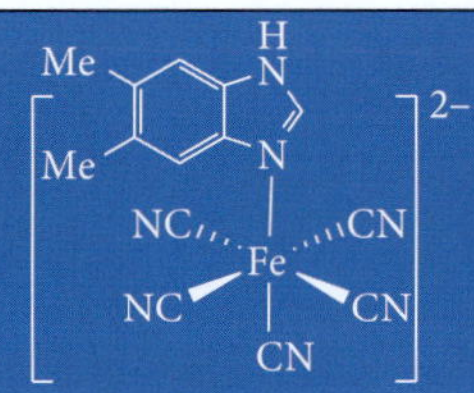	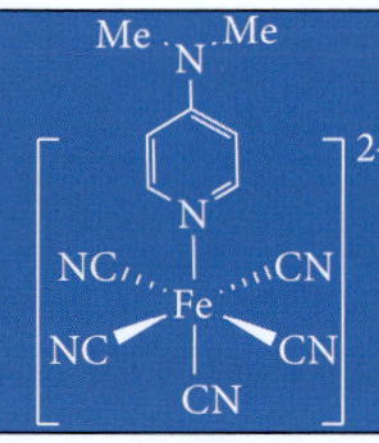
H_2O	20000	19802	19417	18762	15106
CH_3OH	21882	21645	21645	20661	18182
C_2H_5OH	22371	22026	24938	20790	19011
$OS(CH_3)_2$	23095	20661	24213	22831	20367
$NCCH_3$	23256	23041	23753	23041	20080
DMF	23419	23256	24213	23148	21368

the band energy is lower for oxidizing metals in high oxidation states; for example, an Fe(III) complex would probably have a lower energy band than a similar Fe(II) complex for an LMCT.

Because charge is transferred from one place in the complex to another, the polarity often changes dramatically from the ground state to the excited state. For example, Table 16.19 shows the LMCT band maxima for $LFe(CN)_5^{2-}$ complexes, L = heterocycles as shown, in various solvents. Charge-transfer bands often show solvatochromism, whereas *d*–*d* bands (ligand field transitions) will show little to no shift with solvent polarity. In the case of $LFe(CN)_5^{2-}$, all the complexes shift substantially from water to DMF (dimethylformamide), >3000 cm^{-1}.

The complexes in Table 16.19 undergo a bathochromic shift (red, lower-energy shift) in higher-polarity solvents, which is most often what is observed. This is what might be expected as the charge transfer is increasing the polarity from the ground state, and the more polar solvent will stabilize the excited state leading to a lower energy. Compounds with hypsochromic shifts (blue, higher-energy shifts) in higher-polarity solvents are well known, however.[5–10]

16.7 Electronic Spectroscopy I: The Energy of Electronic Transitions

An important use of Tanabe–Sugano diagrams is in the interpretation of electronic spectra of transition metals, to estimate values for the ligand field splitting parameter Δ and the interelectronic repulsion parameter B. Metal complexes result in a remarkable array of different spectra. Some examples for relatively simple transition-metal salts in aqueous solutions and hydrated ions in the solid state are shown in Fig. 16.14. These salts span the entire transition series from titanium to copper, and can all be explained using ligand field theory as a model, with varying degrees of difficulty.

In this section, several examples will be discussed to illustrate how this may be accomplished.

16.7.1 Example 1: $\{V(H_2O)_6\}^{3+}$

Fig. 16.15 shows the absorption spectrum of $\{V(OH_2)_6\}^{3+}$ in water. The ground state of V(III), a d^2 complex, is a triplet spin state, $^3T_{1g}(F)$. Two peaks are readily apparent at 17,635 and 26,290 cm^{-1}, each with molar absorptivities (ε) below 10, as expected for spin-allowed, Laporte-forbidden *d*–*d* transitions. From the d^2 Tanabe–Sugano diagram, these two transitions are readily assigned to the lowest two spin-allowed transitions, $^3T_{1g}(F) \rightarrow {}^3T_{2g}$ and $^3T_{1g}(F) \rightarrow {}^3T_{1g}(P)$.

We can use the equations for the energy of each excited state (Eq. 16.19 for $^3T_{2g}$, and Eq. 16.23b for $^3T_{1g}$), along with the energy of the $^3T_{1g}$ ground state, Eq. 16.23a, to solve for the two unknowns, *Dq* and *B*, which yields

$$Dq = 1900\ \text{cm}^{-1}$$
$$B = 669\ \text{cm}^{-1}$$

These values imply that $Dq/B = 2.84$, which is indicated as a vertical dashed line in the Tanabe–Sugano diagram in Fig. 16.15. A rescaled version of the spectrum of $\{V(H_2O)_6\}^{3+}$ is superimposed next to the vertical axis of the Tanabe–Sugano diagram, dividing the energy axis of the spectrum by $B = 669$ cm^{-1}, showing that the predicted energy of the first two triplet–triplet transitions in the Tanabe–Sugano diagram matches well with the observed positions of the two peaks.

The ligand-field strength value obtained here ($10Dq \sim 19{,}000$ cm^{-1}) is about what we would expect for water ligands,

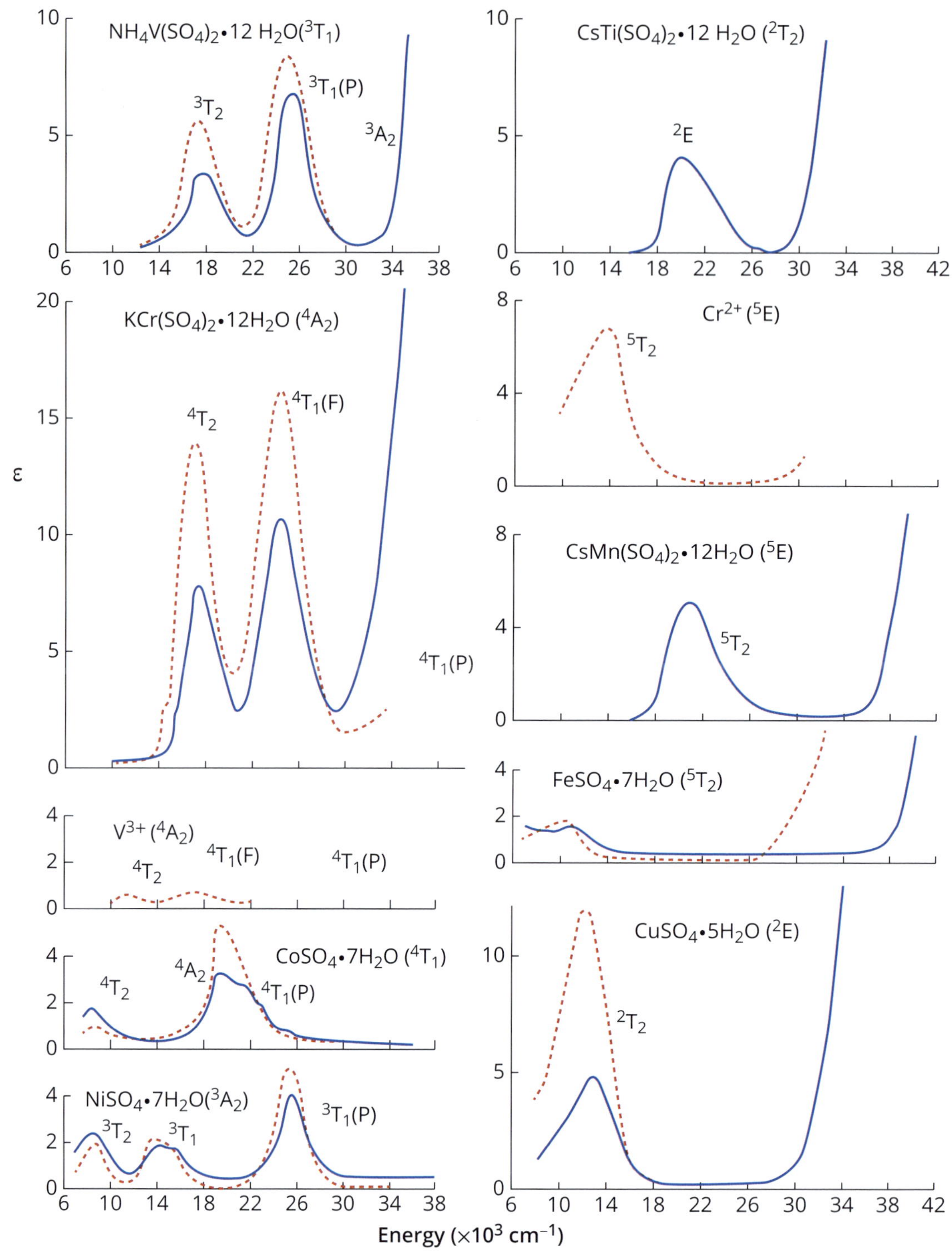

Fig. 16.14 Optical spectra of aqueous 3*d* transition-metal complexes. The spectra taken in aqueous solution are shown as dotted lines, while the solid lines are of the crystalline hydrates. The ground state symmetries (O point group) are given in parentheses next to each sample. (Figure adopted from McClure.[11])

given that water lies about in the middle of the spectrochemical series (see Table 15.1). The value obtained for B (669 cm^{-1}) is significantly smaller than the free-ion Racah parameter for V(III), 886 cm^{-1} (Table 16.7). This is in fact a very general observation; the magnitude of *d*–*d* electronic repulsions in molecular complexes is always reduced from the value observed for the free ion of the same metal centers. This effect is known as the *nephelauxetic effect* and arises from covalency of the metal–ligand bonds, which effectively increases the volume that *d*-electrons occupy in space, and consequently increases the average separation between each electron, thereby decreasing their mutual repulsion. The nephelauxetic parameter (β)

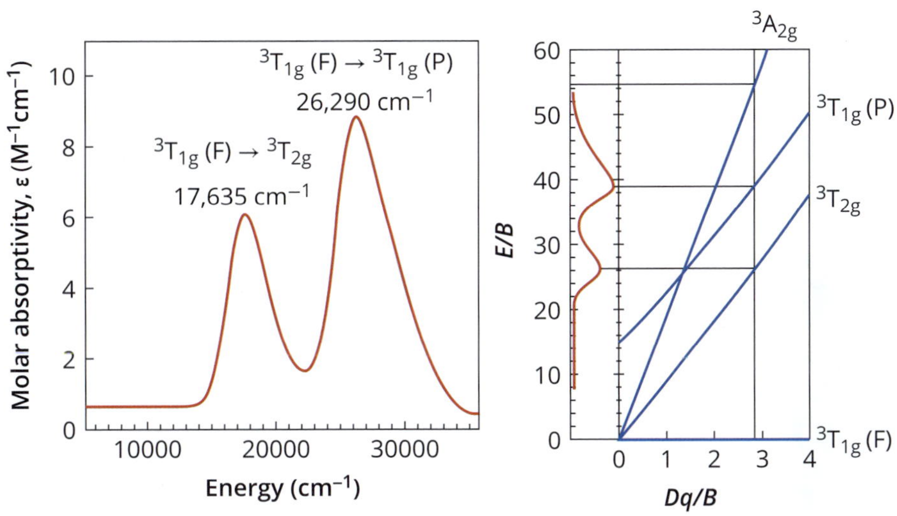

Fig. 16.15 (*Left*) Spectrum of $\{V(OH_2)_6\}^{3+}$ in water. (*Right*) Vertical spectrum of V(III) matched to the d^2 Tanabe–Sugano diagram; the energy axis of the spectrum was scaled using $B = 669$ cm^{-1}.

measures the strength of the nephelauxetic effect and is a measure of the covalency of the metal–ligand bonds that involve the *d*-orbitals of the metal.

$$\beta = \frac{B_{complex}}{B_{free\,ion}} \tag{16.28}$$

It is interesting to note that the third spin-allowed transition, $^3T_{1g}(F) \rightarrow {}^3A_{2g}(F)$, is not observed for $\{V(H_2O)_6\}^{3+}$. Based on the Tanabe–Sugano diagram, this transition should be observed around ~36,000 cm^{-1} for Dq/B ~ 2.8. This can actually be explained easily by considering the d^2 strong-field configurations; the $^3A_{2g}$ excited state is associated with an $(e_g)^2$ strong-field configuration (Fig. 16.9), which differs from the ground-state electronic configuration $(t_{2g})^2$ by the simultaneous excitation of two electrons. Two-electron excitations are strictly forbidden quantum-mechanically (at least for one-photon absorption processes), so the only way the $^3T_{1g}(F) \rightarrow {}^3A_{2g}$ transition can be observed is by mixing in some other configuration in either state to introduce 1-electron excitation character. Such configuration interaction can only be done between states that share the same symmetry. Given that there is only one $^3A_{2g}$ state in the d^2 manifold, it is not possible for it to adopt any other strong-field configurations as long as the O_h symmetry is maintained, and it must remain purely associated with the $(e_g)^2$ configuration at all ligand-field strengths. There are, however, two $^3T_{1g}$ states, and any coupling between the two will introduce some $t_{2g}{}^1e_g{}^1$ strong-field character into the $^3T_{1g}(F)$ ground state, and it is the $t_{2g}{}^1e_g{}^1$ character of that state that ultimately would allow any transition to the $^3A_{2g}$ state to occur. (Note that both $^3A_{2g}$ and $^3T_{1g}(F)$ are derived from the ^3F term on the weak-field side of Fig. 16.9.) As it happens though, the two $^3T_{1g}$ states of V(III) are very far in energy from each other (almost 26,000 cm^{-1} apart), which reduces the coupling to a negligeable value. Consequently, the $^3T_{1g}(F)$ ground state of V(III) remains practically purely of $t_{2g}{}^2$ character, and the transition to the $^3A_{2g}$ excited state remains a very good approximation of a purely 2-electron excitation with almost no intensity.

16.7.2 Example 2: Cr(III) in Rubies and Emeralds

Chromium(III) forms many stable octahedral systems with distinctive photophysical properties. Among many well-known examples, the green color of emeralds is due to Cr(III) replacing some of the aluminum sites in beryl, $Be_3Al_2Si_6O_{18}$, whereas the red color of ruby is due to Cr(III) impurities in corundum, Al_2O_3. The hexaaquachromium(III) complex, $\{Cr(OH_2)_6\}^{3+}$, yields violet crystals. Interestingly, in all these cases the chromium center is directly coordinated to six oxygen atoms, albeit with different chemical environments. Clearly, the perceived color of metal complexes is highly sensitive to the specific nature of the ligand field, especially when there are multiple transitions in the visible range of the spectrum, as is the case for d^3 centers. Fig. 16.16 compares the electronic absorption spectra of both emerald and ruby, superimposed over the visible spectrum. You can see that the first two electronic transitions of these d^3 spectra overlap substantially over most of the visible range, but that the absorption peaks of emeralds are slightly lower in energy (longer wavelength, or "red-shifted") compared to ruby. Emeralds absorb most of the red in the spectrum. Importantly, the absorption gap between the two peaks in the visible falls clearly in the green part of the spectrum for emeralds, which explains the green color associated with these gems.

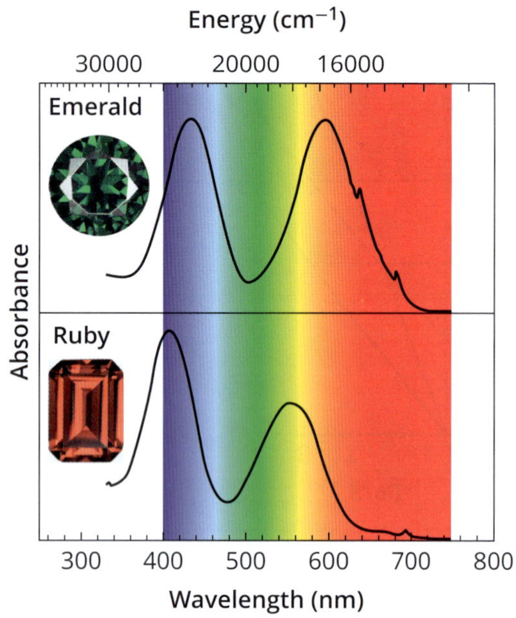

Fig. 16.16 Electronic absorption spectra of green emerald and red ruby. In both cases, the color of the gemstone is due to Cr^{3+} impurities in an otherwise colorless lattice.

We can use our knowledge of ligand field theory to extract some more detailed information about the electronic structure responsible for these optical properties. As it happens, d^3 is quite simple to handle, and will provide us with a convenient way to see how to approach ligand field analyses in general.

A more detailed presentation of the electronic spectrum of ruby (Cr^{3+}:Al_2O_3) is shown in Fig. 16.17. A quick look at the d^3 Tanabe–Sugano diagram (Fig. 16.13b) shows that the ground state of octahedral d^3 centers is $^4A_{2g}$ (*x*-axis of the plot), and that there are three excited quartet states, $^4T_{2g}$, $^4T_{1g}$(F), and $^4T_{1g}$(P), in increasing order of energy. Transitions to each of these states are spin-allowed, and these transitions can be readily assigned, with corresponding energies, as follows:

$$^4A_{2g} \rightarrow {}^4T_{2g} \qquad E = 18{,}050 \text{ cm}^{-1}$$
$$^4A_{2g} \rightarrow {}^4T_{1g}(\text{F}) \qquad E = 24{,}550 \text{ cm}^{-1}$$
$$^4A_{2g} \rightarrow {}^4T_{1g}(\text{P}) \qquad E \approx 40{,}000 \text{ cm}^{-1}$$

The peak molar absorptivity of these three bands is around 30 $M^{-1}cm^{-1}$, consistent with (if somewhat larger than) the value expected for Laporte-forbidden spin-allowed transitions. Another set of peaks, much weaker and narrower than the spin-allowed transitions, is also observed, as shown in the inset in Fig. 16.17. These peaks are associated with a spin-forbidden transition, the lowest-energy doublet, ^2E_g.

$$^4A_{2g} \rightarrow {}^2E_g \qquad E = 14{,}420 \text{ cm}^{-1}$$

The origin of the two peaks, instead of just one, for the ^2E_g transition is the low symmetry of the ligand field (lower than O_h) in ruby.

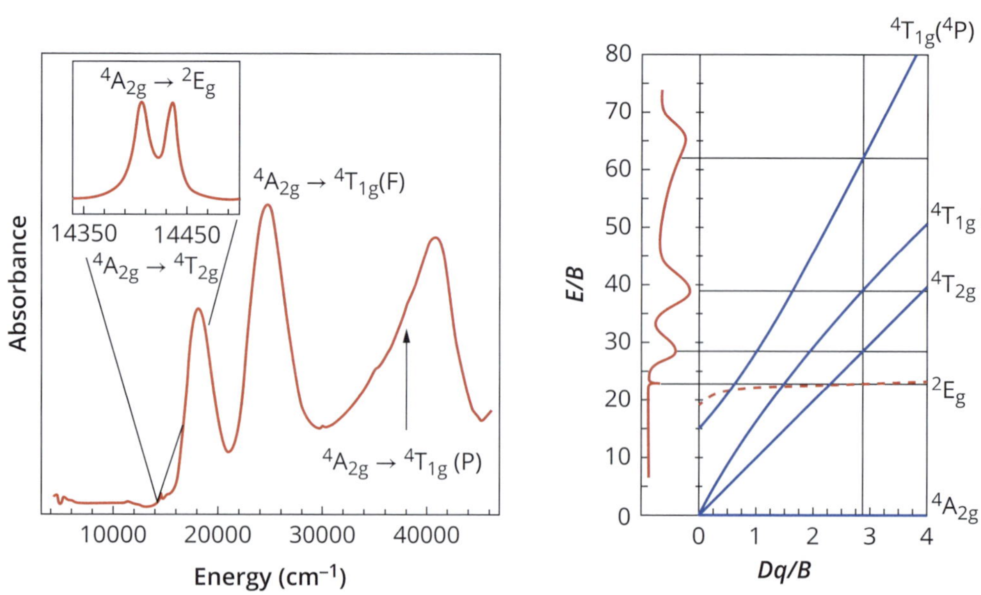

Fig. 16.17 A detailed overview of the electronic spectrum of ruby (*left*), showing the spin-forbidden $^4A_{2g} \rightarrow {}^2E_g$ transition in the inset. A simplified Tanabe–Sugano diagram for this system, calculated with C = 3230 cm^{-1} (see the text) is shown at right, along with the vertically oriented absorption spectrum; the vertical line at Dq/*B* = 2.86 crosses each electronic state at the observed energy.

Using these energies, we can deduce the values of the three ligand field parameters, Dq, B, and C, that define the electronic structure of Cr^{3+} in ruby. From Table G2 (in Appendix G), we see that the energy of the lowest excited quartet state ($^4T_{2g}$) relative to the ground state ($^4A_{2g}$) is simply equal to $10Dq$:

$$E\left({}^4T_{2g}\right) - E\left({}^4A_{2g}\right) = 18050\ \text{cm}^{-1}$$
$$= (-15B - 2Dq) - (-15B - 12Dq) = 10Dq$$

This directly gives:

$$Dq(\text{ruby}) = 1805\ \text{cm}^{-1}$$

The secular determinant for the $^4T_{1g}$ state is (Table G2):

$$\begin{vmatrix} -2Dq - 3B - E & 6B \\ 6B & 8Dq - 12B - E \end{vmatrix} = 0$$

This is solved for the energy, $E(^4T_{1g})$ (taking the solution with the negative sign before the root).[cc]

$$E({}^4T_{1g}) = 1/2\left(-15B + 6Dq - \sqrt{225B^2 - 180B \cdot Dq + 100Dq^2}\right)$$

We then subtract this energy from the energy of the ground state, $E(^4A_{2g}) = -12\ Dq - 15\ B = -\frac{1}{2}\ (30\ B + 24\ Dq)$.

$$E\left({}^4T_{1g}\right) - E\left({}^4A_{2g}\right)$$
$$= 1/2\left(-15B + 6Dq - \sqrt{225B^2 - 180B \cdot Dq + 100Dq^2}\right)$$
$$+ 1/2(30B + 24Dq)$$

$$E\left({}^4T_{1g}\right) - E\left({}^4A_{2g}\right)$$
$$= 1/2\left(15B + 30Dq - \sqrt{225B^2 - 180B \cdot Dq + 100Dq^2}\right)$$

Using the value of Dq obtained previously and the experimental energy of the transition, $E({}^4T_{1g}) - E({}^4A_{2g}) = 24{,}550\ \text{cm}^{-1}$ in the previous equation, we can solve this for B.

$$B(\text{ruby}) = 630\ \text{cm}^{-1}$$

Finally, the value of C can be obtained from the observed energy of the ^2E_g. Because there are four independent ^2E_g states in $d^3(O_h)$, this is a much harder problem to solve by hand, requiring the handling of a fourth-order polynomial equation, corresponding to the lowest root of the following secular determinant:

$$\begin{vmatrix} -12Dq - 6B + 3C - E & -6B\sqrt{2} & 3B\sqrt{2} & 0 \\ -6B\sqrt{2} & -2Dq + 8B + 6C - E & -10B & (2B + C)\sqrt{3} \\ 3B\sqrt{2} & -10B & -2Dq - B + 3C - E & -2B\sqrt{3} \\ 0 & (2B + C)\sqrt{3} & -2B\sqrt{3} & 18Dq - 8B + 4C - E \end{vmatrix} = 0$$

We nevertheless can solve such equations easily by using modern computational tools. Here the solution involves finding the lowest eigenvalue of the secular determinant shown,[dd] yielding:

$$C(\text{ruby}) = 3230\ \text{cm}^{-1}$$

For emerald, the Cr(III) has three reported transitions at 16,300 cm^{-1}, 23,700 cm^{-1}, and 37,700 cm^{-1}. There are also weaker bands, assigned to spin-forbidden transitions, at ~15,000 cm^{-1}. The Tanabe–Sugano diagram for d^3 is shown in Fig. 16.13b. These transitions can be assigned to:

$$^4A_{2g}(F) \rightarrow {}^4T_{2g}(F),\ 16{,}300\ \text{cm}^{-1}$$
$$^4A_{2g}(F) \rightarrow {}^4T_{1g}(F),\ 23{,}700\ \text{cm}^{-1}$$
$$^4A_{2g}(F) \rightarrow {}^4T_{1g}(P),\ 37{,}700\ \text{cm}^{-1}$$
$$^4A_{2g}(F) \rightarrow {}^2E_{g}(F),\ {\sim}15{,}000\ \text{cm}^{-1}$$

There is a crossover point for ^2E_g and $^4T_{2g}$ at about $\Delta/B = 20$ (or $Dq/B = 2$), and these two states are quite close in energy in emerald, with the $^4T_{2g}$ slightly higher in energy. Consistent

[cc] The determinant provides two solutions—one with a negative sign and one with a positive sign. We are looking for the one with the lower energy, the $^4T_{1g}(F)$ state. The other root of the equation is for the higher-energy $^4T_{1g}(P)$ state.

[dd] Concepts such as these are often taught in courses on linear algebra. Unfortunately, the math here is a little more than can be demonstrated in a quick review.

with this, Δ for emerald is assigned as being at 20.4 *B* in the literature.[12] The *E/B* values for the observable spin-allowed states are then:

$$^4A_{2g}(F) \rightarrow {}^4T_{2g}(F),\ E/B = 20.4 = \Delta_o$$
$$^4A_{2g}(F) \rightarrow {}^4T_{1g}(F),\ E/B = 29.7$$
$$^4A_{2g}(F) \rightarrow {}^4T_{1g}(P),\ E/B = 46.1$$

These produce B = 800 cm^{-1} for the emerald system. Note that Δ/B = 20.4 with this value of B produces Δ = 16,300 cm^{-1}, exactly the energy of the $^4A_{2g}(F) \rightarrow {}^4T_{2g}(F)$ transition. In other words, the energy of the $^4A_{2g}(F) \rightarrow {}^4T_{2g}(F)$ transition is the ligand field splitting in this case.

16.7.3 Example 3: $\{Ni(en)_3\}^{2+}$

The kinetics and properties of the species $\{Ni(en)_x(OH_2)_{6-2x}\}^{2+}$ were studied in some detail by Sun, Marceau, and Che.[13] The spectra of all three species, x = 1, 2, and 3, are shown in Fig. 16.18. Focusing on the $Ni(en)_3^{2+}$ complex here, there are three peaks at ν_1 = 888 nm (11, 300 cm^{-1}), ν_2 = 543 nm (18,400 cm^{-1}), and ν_3 = 342 nm (29,200), ascribable to the three spin-allowed *d–d* transitions of a d^8 center. The assignment of these three peaks is trivial in the d^8 Tanabe–Sugano diagram in the figure.

Again, we assign the Δ/B value of the system by looking at the ratios of the peak energies versus the ratios of peak energies in the Tanabe–Sugano diagram. The ratios of the first two and last two peak energies are shown in the following equation. Here it is essential to recall that the wavelength in nm

Box 16C

Light, color, and human perception

Our eyes have two different kinds of light receptors, called rods and cones, shown in the accompanying figure. The rods are responsible for vision in low light and see in black-and-white, while the cones are responsible for color vision. Our eyes, in fact, have three different receptors for colored light, with different absorbances to cover the "visible" region of the spectrum, which are simply called short, medium, and long receptors, depending on the wavelength of light to which they are most sensitive.

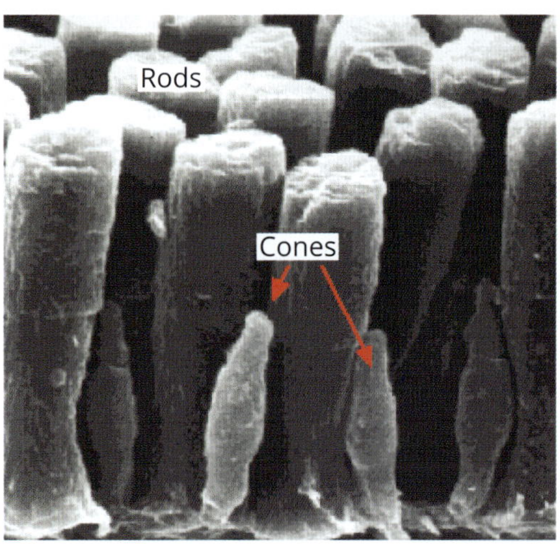

While these receptors may be thought of as organic analogs of the detector in an absorption spectrometer, our brains have sophisticated automatic data-analysis capabilities dealing with optical input to make our daily activities easier, though unfortunately it is not necessarily ideal for scientific analysis of the color of compounds, as in many cases it results in optical illusions.

For example, our eyes try to retain "color constancy", meaning that they try to preserve the color of an item whether it is in bright light or in dull light—an amazing trait of the optical processing in our brains! The accompanying figure is a common optical illusion that takes advantage of this processing to fool the mind. The bottom square looks much lighter than the top square, but the two squares are in fact the same color. To observe this, use your finger to cover the illuminated portion between the two squares! Your eyes receive the color information correctly, and the two squares are the same color. However, the specific lighting suggested by the whiter section between the two blocks suggests to your brain that the bottom figure is not directly in the light, and as a result your brain knows that objects in the shade should be darker than those in the light. Since to your eyes the bottom square appears the same color as the top but is in a shaded area according to your perception, your brain interprets that it must be lighter (or more reflective), which causes you to observe a lighter color. This is not a problem with our eyes; it is a problem with our perception.

As we age, our eyes tend to become less sensitive to blue light as well, which can affect the color of compounds; that is, an older researcher may not see the same color for a compound as a younger student might. This is possibly part of the white/gold versus blue/black dress controversy that occurred on the Internet. The accompanying two photographs are of the same dress, but some see the one on the left as white and gold. Poor lighting in the picture on the left is also probably a contributor to the controversy.

Similar difficulties with perception can occur with color in transmitted versus reflected light. If the light coming to the eye is of a specific frequency, then that is interpreted based on the color sensors in our eyes (the rods) as a specific color. However, there is another way in which the same color may be observed. If white light is transmitted through a substance and all but one color is absorbed, then the color is interpreted by the eye as the complementary color to the one transmitted. For example, if white light is shown through a crystal of a substance that absorbs in the blue and violet regions, the compound will appear yellow-orange. In other words, yellow light reflected off a compound may be perceived as the same color as light transmitted through a substance that absorbs in the blue.

A "color wheel" can be used to estimate what color will be observed when a specific color is absorbed; the color observed is 180° from absorption (see figure). For example, complexes that absorb in the red will appear green or blue when observed with transmitted white light.

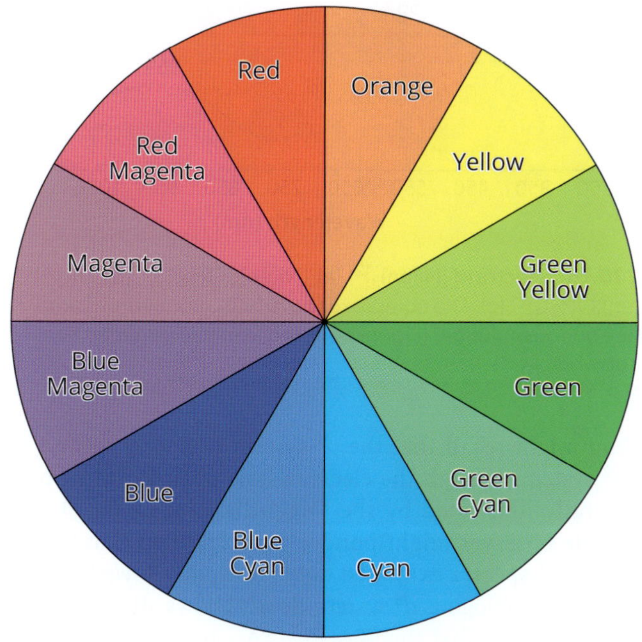

In scientific papers it is commonplace to describe the color of compounds perceived by the eye, and this is quite proper. The color of an inorganic complex can be quite distinctive as a quick piece of information as to the compound's identity. However, the researcher should be aware that "eyeball spectroscopy" is influenced by our perception and can be influenced by multiple transitions, shading, and other factors

is *not* proportional to energy, but the wavenumber value is. In other words, do not take the ratios of the wavelengths; always convert to cm^{-1}.

$$\nu_1/\nu_2 = 0.614$$
$$\nu_1/\nu_2 = 0.63$$

From Table G1 (in Appendix G), the lowest-energy transition at 11,300 cm^{-1} should be equal to

$$^3A_{2g}(F) \rightarrow {}^3T_{2g}(F) = 12Dq - 8B - (2Dq - 8B) = 10\,Dq$$

and so $Dq = 1{,}130$ cm^{-1}.

From the Tanabe–Diagram one can find that these ratios correspond to $10\,Dq/B = 12.2$. Using this value produces $B =$ 930 cm^{-1} for the system.

$$^3A_{2g}(F) \rightarrow {}^3T_{2g}(F)\text{: } E/B = 12.2 = 11{,}300\text{ cm}^{-1}/B = \Delta_0$$
$$^3A_{2g}(F) \rightarrow {}^3T_{1g}(F)\text{: } E/B = 19.8 = 18{,}400\text{ cm}^{-1}/B$$
$$^3A_{2g}(F) \rightarrow {}^3T_{1g}(P)\text{: } E/B = 31.4 = 29{,}200\text{ cm}^{-1}/B$$

Note that the first spin-allowed transition $^3A_{2g}(F) \rightarrow {}^3T_{2g}(F) =$ Δ_o for d^8. This is true for a few transition-metal systems like d^3 and d^8.

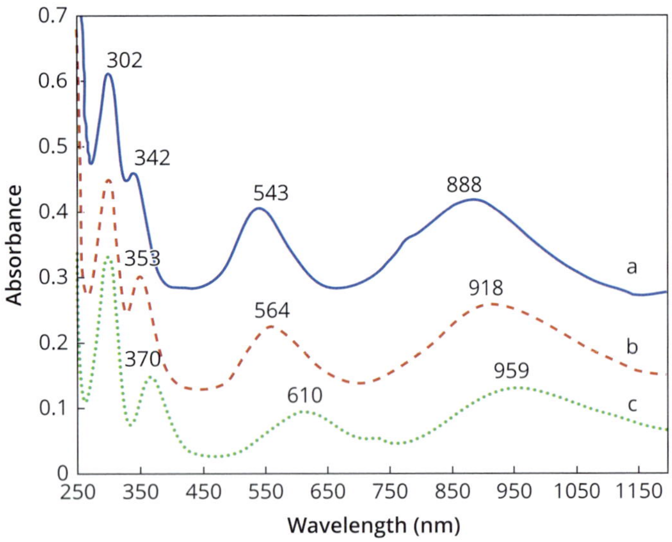

Fig. 16.18 Spectra of $\{Ni(en)_3\}^{2+}$ (a), $\{Ni(en)_2(OH_2)_2\}^{2+}$ (b), and $\{Ni(en)(OH_2)_4\}^{2+}$ (c). The complexes have NO_3^- counterions and are in aqueous solution. (Figure adopted from the paper by Sun and coworkers.[13])

You might recall that the d^8 system is closely related to the d^2 system through the electron-hole analogy; this point is centrally illustrated by the Orgel diagram in Fig. 16.11. An astute observer might ponder why it is then that the d^2 $\{V(OH_2)_6\}^{3+}$ spectra exhibited only two spin-allowed bands, whereas the analogous d^8 system clearly shows all three transitions. Why is it that the $^3T_{1g}(F) \rightarrow {}^3A_{2g}$ transition of V(III) carries practically no intensity, but the reversed $^3A_{2g} \rightarrow {}^3T_{1g}(F)$ transition of Ni(II) is practically as intense as the other two transitions?

There are two important reasons for this striking difference in behavior. First, remember that for V(III) we found that the coupling between the $^3T_{1g}(F)$ ground state and the $^3T_{1g}(P)$ excited state was negligeable because of the large energy difference between them, which led us to associate the $^3T_{1g}(F) \rightarrow {}^3A_{2g}$ with a practically pure 2-electron $t_{2g}{}^2 \rightarrow e_g{}^2$ strong-field excitation, which cannot be achieved directly by one-photon excitation (the normal situation for low-intensity beams of light such as used in absorption spectrometers). The situation is quite different for Ni(II); the two $^3T_{1g}$ states are now both excited states, and are therefore closer to each other than for the d^2 case, as can be seen directly in the d^2/d^8 Tanabe–Sugano diagrams. For the $\{Ni(en)_x(OH_2)_{6-2x}\}^{2+}$ examples, the energy difference is about 10,000 cm^{-1}, which allows for stronger mixing between the two states. Second, the larger molar extinction coefficients of the $^3A_{2g} \rightarrow {}^3T_{1g}(F)$ transition of Ni(II) complexes is the coupling between the formally pure *d–d* transitions and so-called charge-transfer transitions that formally correspond to electronic excitations from the metal to the ligand orbitals (metal-to-ligand-charge-transfer, or MLCT) or vice versa (ligand-to-metal-charge-transfer, LMCT). These charge-transfer transitions typically have extremely large molar extinction values, and any mixing with the Laporte-forbidden *d–d* states can lead to significant enhancement in the intensity of these transitions. The onset of strong charge-transfer transitions is clearly observed in the spectrum of the Ni(II) species, very close in energy to the $^3A_{2g} \rightarrow {}^3T_{1g}(F)$ transition, and probably significantly contributing to the high intensity of the latter.

16.7.4 Example 4: Tetrahedral $\{CoCl_4\}^{2-}$

Tetrahedral $\{CoCl_4\}^{2-}$ is readily formed in aqueous solution and has a bright blue color. Its Laporte-allowed *d–d* transitions are higher in absorption coefficient than the octahedral complexes discussed previously. The energies of the bands are lower in energy due to the smaller ligand field—four ligands instead of six. There are intense bands at 5,800 cm^{-1} ($\varepsilon = 64$) and 15,000 cm^{-1} ($\varepsilon = 600$). In addition, there are weaker transitions ($\varepsilon \sim 5$) at 17,000 and 23,000 cm^{-1}.

There are two differences to keep in mind when using Tanabe–Sugano diagram to interpret tetrahedral complexes versus octahedral. First, the energies of the orbitals (*e* is lower in energy than t_2 in T_d) and the states are inverted relative to octahedral systems. *As a result, one should use the* d^{10-n} *Tanabe–Sugano diagram, where n is the number of d-electrons, for a tetrahedral system.* For example, for d^7 $CoCl_4{}^{2-}$ we will use the diagram for d^3 in an octahedral system. Second, tetrahedral complexes sometimes have ligand fields small enough that their lowest-energy transitions are out of the visible range and into the near-infrared (400–4,000 cm^{-1}), which turns out to be the case for $\{CoCl_4\}^{2-}$. Typically, near-infrared transitions require a different detector. Currently, this is an extra feature on spectrometers that may or may not have been added to the instrument being used.

One cannot fit the experimental peaks to the Tanabe–Sugano diagram at any Δ/B value while assuming the intense bands at 5,800 cm^{-1} and 15,000 cm^{-1} are due to $^4A_2 \rightarrow {}^4T_2(F)$ and $^4A_2 \rightarrow {}^4T_1(F)$ transitions. If we assume that there is a lower-energy band than 5,800 cm^{-1} in the near-infrared that we are not detecting, the two observed spin-allowed transition can be fit to the $^4A_2 \rightarrow {}^4T_1(F)$ and $^4A_2 \rightarrow {}^4T_1(P)$ transitions.

Chloride is very low on the spectrochemical series (Chapter 15), and tetrahedral systems have relatively low ligand fields in general. As a result, the best fit occurs at $\Delta/B = 4.4$—a very low ligand field. One can find the electron repulsion term *B* by using the data and this ligand field, which produces B = 730 cm^{-1}. One can then calculate that the lower-energy-allowed transition, $^4A_2 \rightarrow {}^4T_2(F)$, should occur at 3,200 cm^{-1}, which is also Δ_t.

$$^4A_2 \rightarrow {}^4T_1(F)\text{: } E/B = 7.68$$

$$^4A_2 \rightarrow {}^4T_1(P)\text{: } E/B = 20.5$$

16.7.5 Example 5: Iron(III) and Mn(II) in Aqueous Solution and Peak Shape

There are some very simple complexes for which this process of determining their ligand field spectral parameters is quite difficult. An example is d^5 $\{Fe(OH_2)_6\}^{3+}$. On dissolving iron in aqueous solution at most pHs, a yellow color is observed which is due to charge-transfer bands resulting from hydroxide formation.[2] $\{Fe(OH_2)_6\}^{3+}$ complex is quite acidic and readily releases protons from coordinated water ligands. As a result, low pHs are required to stabilize $\{Fe(OH_2)_6\}^{3+}$ in solution. The speciation of iron(III) in aqueous solution is shown in Fig. 16.19.[14]

Iron(III) in aqueous solution at pH = 0, $\{Fe(OH_2)_6\}^{3+}$, shows a single band in the visible at 41,700 cm^{-1} (ε = 4).[14] A full analysis of the ion in aqueous solution is difficult, but the Δ_o value has been estimated as ~14,000 cm^{-1}.[2]

The aqueous spectrum of d^5 manganese is much easier to study. The manganese(II) ion, because of its larger radius and smaller charge, is far less Lewis-acidic than iron(III), which in turn leads to a reduced acidity of coordinated water. The ion is a very pale pink in solution, and the spectrum shows a series of low-intensity peaks, some of which are quite sharp relative to any spectrum that we have discussed so far (Fig. 16.20).

The breadth of the peaks in the absorption spectrum can provide useful information regarding the complex, but several factors influence band width and shape, some or all of which may be operational at the same time.[2] For example, the width of the band may be influenced by the sensitivity of the band to the ligand field. In real complexes, the ligands are not stagnant but instead vibrate with some frequency. The electronic transition caused by light impinging on the sample occurs far more rapidly than the nuclear motions (the Franck–Condon principle), and as a result, the light strikes complexes in various stages of the vibration, all with slightly different ligand fields due to slightly different bond distances. The ensemble produces a peak with a breadth sensitive to how strongly the state is influenced by the ligand. This sensitivity can readily be seen in the Tanabe–Sugano diagram as the slope of the line for the state near the transition. A large slope indicates that the state is very sensitive to small changes in ligand field, which will lead to a broad peak for that transition. A small slope indicates that the state is not very sensitive to small changes in ligand field, which will produce narrower peaks. One of the very narrow peaks in Fig. 16.20 is associated with the spin-forbidden $^6A_{1g} \rightarrow {}^4A_{1g}$ transition, which is very nearly horizontal on the low-field (high-spin) side of the Tanabe–Sugano diagram for d^5 (Fig. 16.13d). This can be contrasted with the $^6A_{1g} \rightarrow {}^4T_{1g}(G)$ transition, which has a relatively large negative slope and a broader peak.

There is a wealth of peaks in the spectrum of $\{Mn(OH_2)_6\}^{2+}$. The complex is high spin, and as a result there can be no bands that are spin-allowed, which leads to all of the d–d transitions having very low intensities (Table 16.18), ε < 0.05. The first band at 18,600 cm^{-1} can be assigned to $^6A_{1g} \rightarrow {}^4T_{1g}(G)$, and the

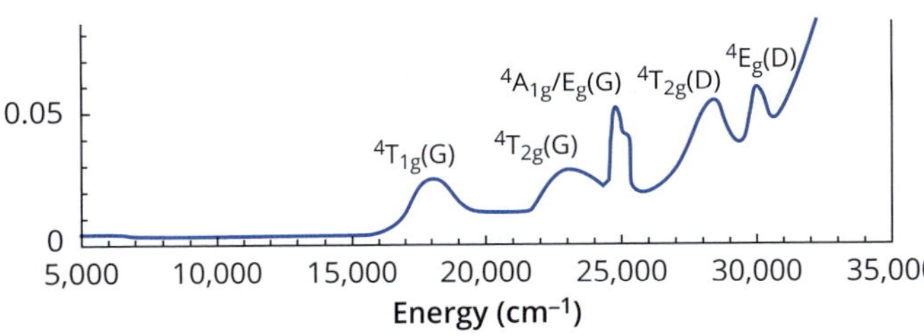

Fig. 16.20 Absorption spectrum for $Mn(OH_2)_6^{2+}$. The y-axis is in ε. Above the absorptions are the states to which the $^6A_{1g}$ ground state excites to on irradiation.[2]

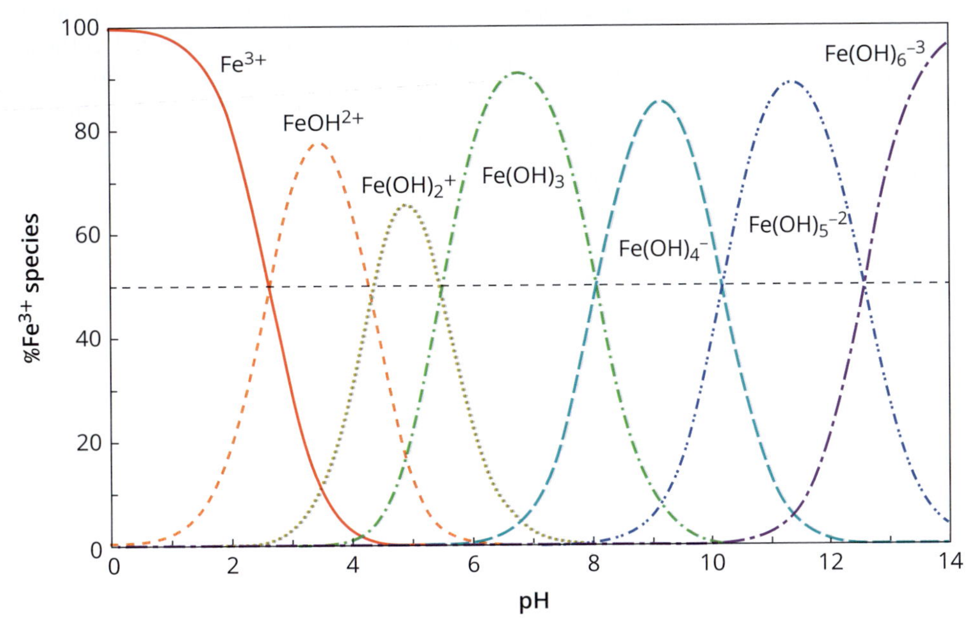

Fig. 16.19 Speciation of iron(III) in aqueous solution as a function of pH. (Adapted from Perera and Hefter.[14])

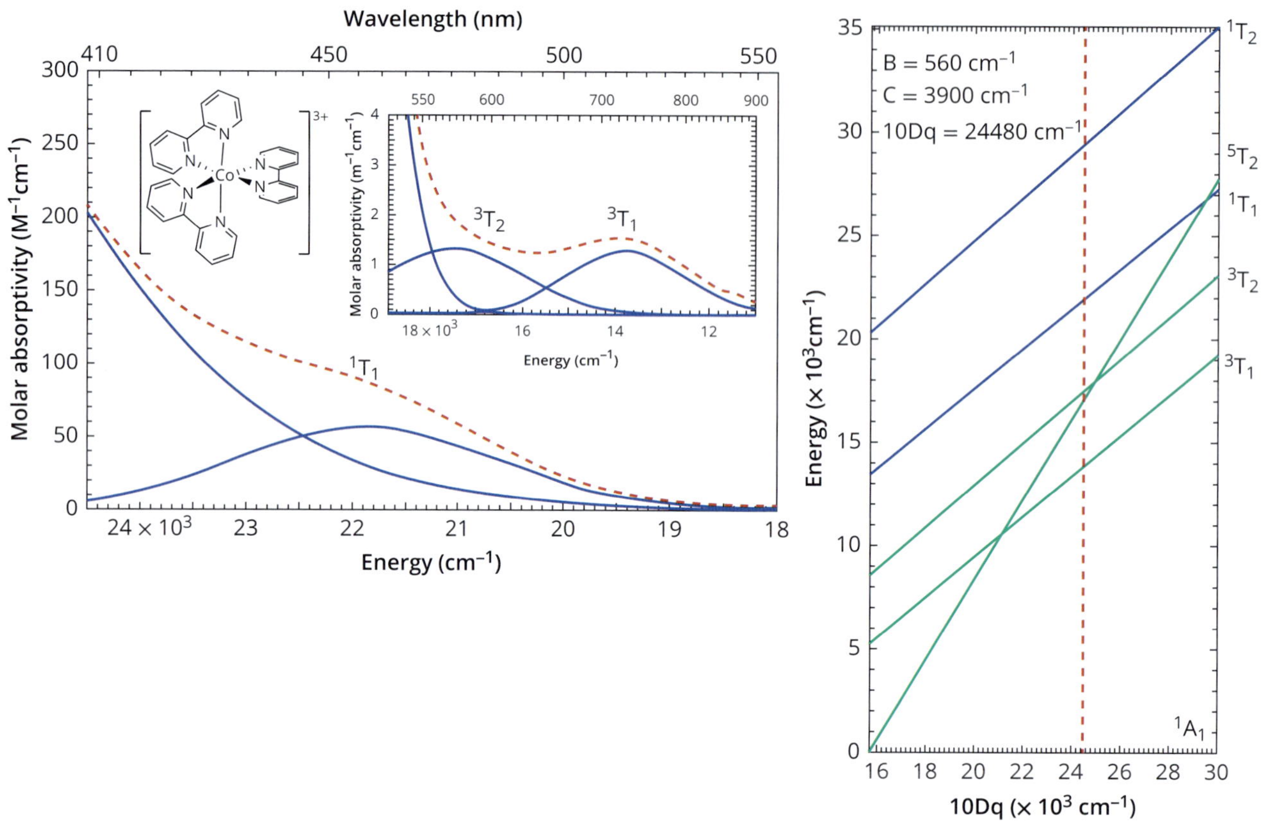

Fig. 16.21 Structure of $\{Co(bpy)_3\}^{3+}$, absorption spectrum with the excited state assigned, Tanabe–Sugano diagram in the point group O for this d^6 complex with the values for the ligand-field parameters shown. (Figure from reference.[15])

spectrum fits well at $\Delta/B = 11$. This peak occurs at $E/B = 24.5$ in the Tanabe–Sugano diagram, which gives $B = 760$ cm^{-1} and $\Delta_o = 8400$ cm^{-1}. The other bands fit exceptionally well to the d^5 Tanabe–Sugano diagram. In all cases, only the weak quartet bands are observed, and the doublet terms in the diagram would require two spin flips and are doubly spin-forbidden, so to speak.

All of the peaks are relatively narrow, but the peaks labeled $^4A_{1g}/E_g(G)$ especially so. In the ligand-field treatment, these states have the same energies, $-25B + 5C$ (Table G4).[ee] Because the peaks are so narrow, one can see that they have slightly different energies, a small deviation from the LFT model.

16.7.6 Example 6: Low-spin d^6 Co(III) Pseudo-Octahedral Complexes[15]

While these discussions have focused on Werner-type ligands and naturally occurring minerals with transition metals, it is far more common in modern applications to have more complex ligand structures, often involving multidentate ligands. Nevertheless, the same types of analysis can be applied to these systems, often while approximating 6-coordinate systems as octahedral. As an example, Yarranton and McCusker recently analyzed a series of 6-coordinate cobalt(III) complexes, most of them using three bidentate ligands in the coordination sphere. The 2,2′-bipyridine (bpy) complex, $\{Co(bpy)_3\}^{3+}$, could be regarded as the archetypal member of the class (Fig. 16.21). In all, a dozen complexes were examined in the work, and for each complex B, C, and Dq were assigned.

The values for 10 Dq are perhaps the simplest to understand based on 1-electron argument put forward in Chapter 15 and the spectrochemical series. The largest value of 10 Dq was found for $\{Co(CN)_6\}^{3-}$ (34,160 cm^{-1}) and the smallest for $Co(acac)_3$ (19,210 cm^{-1}). As part of the study, the authors discussed the effects of the various ligands as well. As discussed in Section 16.7.1 above, β (see eq. (16.28)) measures the strength of the nephelauxetic effect, which is related to the decrease in the electron–electron repulsion term B in the complex relative to the free ion. Covalency in the metal–ligand bonds involving the d-orbitals leads to this lowering of the repulsion term. The B values for the complexes studied varied from 370 to 580 cm^{-1}, which gives β values of 0.34 to 0.53. In essence, increased covalency delocalizes the electrons in the d-orbital reducing repulsion.

[ee] The energy of the $^4A_{1g}$ state can be read directly from Table G4 in Appendix G. The energy of the ^4E_g state is found by determining the eigenvalue of the 2 × 2 matrix for that state, which turns out to be identical in energy to the $^4A_{1g}$ state.

Chapter Summary

1. If a system has more than one *d*-electron (or more than one hole), then electron–electron repulsion necessitates the use of multielectronic states to understand the electronic structure, rather than the simple orbital pictures we used in Chapter 15. The orbital energies still play a critical role in our understanding, but the electron–electron repulsions are of similar energy and need to be concomitantly considered.
2. The specific fillings of the orbitals with electrons are called "microstates", and many microstates of the same energy may comprise an electronic state or term.
3. The term symbols have the spin multiplicity and the angular momentum included. For example, a term symbol of ^3F ("triplet f") term has microstates where the system has two unpaired electrons with the spins parallel ($S = ½ + ½ = 1$). The spin multiplicity is then $2S + 1 = 2(1) + 1 = 3$ (triplet). The angular momentum takes values from $2L + 1$ to $-2L + 1$. For an F term, just like an *f*-orbital where the *l* angular momentum value is 3, values of 3, 2, 1, 0, –1, –2, and –3 are present; a total of seven values. The total degeneracy is then:

$$\text{Total state degeneracy} = (2S + 1)(2L + 1) = (3)(7) = 21 \textit{ microstates}$$

4. Assigning the microstates to terms, in some cases, is fairly easy, but not in all.
   a. *Possibility 1: All the valence orbitals on the metal are filled, which is often the case for an 18-electron configuration.* In this case, the electron density will be spherical, and the term symbol will be $^1A_{1g}$ in an octahedral field or ^1S in atomic spectroscopy.
   b. *Possibility 2: There is one electron in a nondegenerate orbital.* An example would be a d^9 Jahn–Teller distorted octahedral system with axial elongation, now with symmetry D_{4h}. The highest-energy orbital would be expected to be $d_{x^2-y^2}$, which is in irreducible representation b_{1g} and is singly occupied. As a result, the ground state would be $^2B_{1g}$ doublet because there is one unpaired electron with spin = ½ (Table 16.2) and B_{1g} because that is the symmetry of the occupied orbital.
   c. *Possibility 3: Two unpaired electrons in two different nondegenerate orbitals.* In this case, the term will be singlet or triplet (depending on how the spins are assigned in the two orbitals) and the direct product (Section 5.12) of the orbitals' irreducible representations. For example, let us say the configuration might be $b_{1g}{}^1a_{1g}{}^1$. The direct product of $B_{1g} \times A_{1g} = B_{1g}$. (Anything times the totally symmetric representation is itself.) As a result, the terms from this configuration would be $^1B_{1g}$ and $^3B_{1g}$, with the higher spin state ($^3B_{1g}$, Hund's rule, Section 2.2.3) being the ground state.
   d. *Possibility 4: There are two or more electrons in degenerate orbitals.* In this case, the symmetries will still be related to the direct products. In other words, in a d^2 octahedral system with configuration $(t_{2g})^2$, the terms will be $T_{2g} \times T_{2g} = A_{1g} + E_g + T_{1g} + T_{2g}$. However, not all the possible spin states (singlet and doublet) will be possible for all of these, due to the Pauli exclusion principle. This is where some difficulties can occur, systematic methods of finding the terms were examined. On the strong-field side, two different methods were discussed for removing these unphysical states: 1. "descent in symmetry" using a correlation table, and 2. use of the symmetric/antisymmetric direct products.
5. The electronic states in an octahedral field can be assigned at the weak-field (free-ion) limit and at the strong-field limit. In so doing, we can extrapolate the energies and state splittings in between to develop crystal field-splitting diagrams. The most common of these is the Tanabe–Sugano diagram, which has the ground state as the *x*-axis.
6. The energies in ligand (crystal) field theory are often given in Racah parameters. The three Racah parameters for the d^n configurations are simply called *A*, *B*, and *C*, with the parameter *A* often omitted because the energy of all d^n states arising from the same configuration share a common *A*. Whereas the *A* term represents a uniform shift in energy for all states of a given d^n configuration, the *B* term in the Racah formalism has the useful property that the energy differences between states that have the same spin multiplicity as the ground state only depend on *B*. Table 16.6 lists the energy of all free-ion d^n states in terms of the Racah parameters, and Table 16.7 provides empirical estimations of the free-ion values of the Racah parameters *B* and *C* for many common transition-metal ions in various oxidation states.
7. Racah parameters change from complex to complex and are experimentally determined in most cases. The ratio of the **B** value for the complex to the *B* value for the free metal ion is called the nephelauxetic ratio (β). The nephelauxetic effect is that the ***B*** parameter, which is related to electron–electron repulsion in the orbital, changes with the nature of the complex. This is due to covalency in the bonding that effective increases the size of the metal orbital due to mixing with the ligand-based orbital(s). As a result, the higher the covalency the smaller the electron–electron repulsion parameter will be relative to the free ion.

$$\beta = \frac{B_{complex}}{B_{free\,ion}}$$

8. Using the tools in this chapter, the spectra of some common complexes were assigned, finding their ***Dq*** and relevant Racah values, with the help of Tanabe–Sugano diagrams.

Practice Problems

1. Calculate the number of microstates for the following:
   a. d^2
   b. p^4
   c. d^7
2. Explain why there cannot be any F state arising from a np^3 configuration. Is that also the case for a $np^2(n+1)p^1$ configuration?
3. a. How many microstates are included in a p^1d^1 configuration?
   b. Build a microstate table for this configuration and derive all the electronic states that arise from it.
   c. For each state, give the spin multiplicity and orbital degeneracy.
4. Determine the electronic states that arise in the O_h point group from:
   a. The ^3F state of a d^2 ion.
   b. The ^2G state of a d^3 ion.
   c. The ^1D state of a p^2 ion.
5. Repeat the exercise in Question 4 above, but for the D_{4h} point group, without using a correlation table. Confirm your assignment using the correlation table given in Table 16.9.
6. Answer the following:
   a. Indicate the spin, orbital, and total angular momentum quantum numbers and their corresponding degeneracies for the following free-ion term states.
      i. ^1I
      ii. ^6S
      iii. ^2G
      iv. $^4P_{3/2}$
   b. For each of the following total spin and angular momentum quantum numbers, indicate the appropriate term symbol (including all total angular momentum terms).
      i. L = 5, S = 1
      ii. L = 1, S = 3/2
      iii. L = 0, S = 1/2
      iv. L = 2, S = 2

Cited References

1. Feynman, R. P. *The Character of Physical Law.* Cambridge, MA: MIT Press, 1967, 173 pp.
2. Figgis, B. N., and Hitchman, M. A. *Ligand Field Theory and Its Applications.* New York, NY: Wiley-VCH, 2000.
3. Laporte, O., and Meggers, W. F. Some rules of spectral structure. *Journal of the Optical Society of America and Review of Scientific Instruments* 1925, *11*, 459–463.
4. Shepherd, R. E., Hoq, M. F., Hoblack, N., and Johnson, C. R. Solvatochromism of the LMCT transition of pentacyanoferrate(III) complexes. *Inorg. Chem.* 1984, *23*, 3249–3252.
5. Lees, A. J. The luminescence rigidochromic effect exhibited by organometallic complexes: Rationale and applications. *Comments Inorg. Chem.* 1995, *17*, 319–346.
6. Moya, M. L., Rodriguez, A., and Sanchez, F. Specific cation solute interactions as a major contributor to the salt effects on charge-transfer transitions. *Inorg. Chim. Acta* 1992, *197*, 227–232.
7. Patel, S., Li, Y., and Odom, A. L. Synthesis, structure, and LLCT transitions in terminal hydrazido(2-) bipyridine complexes of titanium. *Inorg. Chem.* 2007, *46*, 6373–6381.
8. Pieters, P. J. J., Vanbeek, J. A. M., and Vantol, M. F. H. A method for the prediction of metallocene-type catalyst activity in olefin (co)polymerisation reactions. *Macromol. Rapid Commun.* 1995, *16*, 463–467.
9. Stasyuk, A. J., Stasyuk, O. A., Sola, M., and Voityuk, A. A. Electron transfer in a Li^+-doped Zn-porphyrin-10 CPP superset of fullerene junction and charge-separated bands with opposite response to polar environments. *J. Phys. Chem. B* 2020, *124*, 9095–9102.
10. Zalis, S., Sieger, M., Greulich, S., Stoll, H., and Kaim, W. Replacement of 2,2'-bipyridine by 1,4-diazabutadiene acceptor ligands: Why the bathochromic shift for $(NN)IrCl(C_5Me_5)^+$ complexes but the hypsochromic shift for $(NN)Ir(C_5Me_5)$? *Inorg. Chem.* 2003, *42*, 5185–5191.
11. McClure, D. S. Electronic spectra of molecules and ions in crystals. 2. Spectra of ions in crystals. *Solid State Phys.* 1959, *9*, 399–525.
12. Konig, E. The nephelauxetic effect calculation and accuracy of the interelectronic repulsion parameters. I. Cubic high-spin d^2, d^3, d^7, and d^8 systems. *Struc. Bond.* 1971, *9*, 175–212.
13. Sun, K. Q., Marceau, E., and Che, M. Evolution of nickel speciation during preparation of Ni-SiO_2 catalysts: effect of the number of chelating ligands in $Ni(en)_x(H_2O)_{(6-2x)}^{2+}$ precursor complexes. *Physical Chemistry Chemical Physics* 2006, *8*, 1731–1738.
14. Perera, W. N., and Hefter, G. Mononuclear cyano- and hydroxo-complexes of iron(III). *Inorg. Chem.* 2003, *42*, 5917–5923.
15. Yarranton, J. T., and McCusker, J. K. Ligand-field spectroscopy of Co(III) complexes and the development of a spectrochemical series for low-spin d^6 charge-transfer. *J. Am. Chem. Soc.* 2022, *144*, 12488–12500.

CHAPTER 17

Kinetics and Mechanisms of Ligand Substitutions

"The goal for those of us interested in descriptive chemistry is not kinetics as an end in itself but is rather to understand reactivity . . . it [kinetics] being essential to developing the subject of reactivity as a science."

— Henry Taube[1]

"No matter how thorough, a kinetic study does not really determine *a mechanism in the same sense that single-crystal X-ray diffraction measurements determine a structure. The reaction mechanism is but a scientific postulate that is open to revision when new data, new insights, or new theories of reactivity emerge."*

— James H. Espenson[2]

A mechanism, like a scientific theory, can never be proven correct, but it can be proven wrong. One can test a mechanism in various ways, and if it always holds up to the test, then one believes that the mechanism may be correct."

— Fred Basolo[3]

CONCEPT COVERAGE

1. **Brief descriptive review of kinetics**
2. **Inorganic substitution rate laws**
3. ***Trans*-effect and *trans*-influence**
4. **Linear free energy relationships**
5. **How *d*-electron count, charge, and other factors affect mechanism**

In Chapter 13, some aspects of the thermodynamics of reactions were discussed. Thermodynamics can be considered the "why" of reaction chemistry. Thermodynamics is concerned with the energies of, for example, bonds made and broken in a reaction; that is, why a reaction is energetically feasible and occurs. Kinetics, in many respects, deals with the "how" of reaction chemistry. The rate of a reaction can be used to produce information on mechanism for a chemical transformation to describe how bonds are made or broken, which is often the primary goal of kinetic analysis.

In this chapter, the mechanisms of substitution reactions in inorganic complexes are discussed. This specific class of reactions garners its own chapter because of its general importance in inorganic chemistry. Very few reactions of interest do not involve putting on or taking off a ligand from the metal at some point during the mechanism; consequently, whatever else may be occurring, substitution reactions are usually involved.

17.1 History and Importance

Werner and Jørgensen utilized substitution chemistry to synthesize compounds and lay the foundation of coordination chemistry. Their efforts were the first demonstrations that ligands influenced the reactivity of metal complexes. Later, Taube systematized reaction rates of 6-coordinate inorganic complexes as functions of their electronic structure.[4] In the decades that followed, numerous studies deepened our understanding with particularly noteworthy early contributions coming from the laboratories of Basolo, Chatt, Pearson, Gray, Merbach, and others.

The reaction mechanisms of transition-metal complexes have broad impact on chemistry ranging from biological function to homogeneous catalysis. There are some mechanisms that resemble those for organic reactions, while others are unique to inorganic chemistry. There will also be a focus on introductory kinetic theory of substitution reactions and how one experimentally distinguishes between mechanistic possibilities.

17.2 Introduction to Kinetics

Kinetics is usually covered in several different courses during a student's career, including general chemistry and physical chemistry courses. Consequently, the introduction here will largely be conceptual.

17.2.1 Reaction Coordinate Diagrams

Chemical reactions are often described by plots of energy of reacting species as they proceed along reaction coordinates where reactants are converted to products. For the simplest process that might be called a chemical reaction, forming or breaking a bond between two isolated atoms, the reaction information can be plotted in two dimensions, with the *x*-axis being the interatomic distance and the *y*-axis being energy. However, as soon as one reaches three-atom systems, such as $OH^- + H^+$, there are more parameters than can be plotted explicitly in three dimensions. For example, for $OH_a^- + H_b^+$ there are minimally four important parameters: O–H_a distance, O–H_b distance, H_b–O–H_a angle, and energy. Consequently, it is necessary to describe the many-parameter space of the molecular geometry as one (or sometimes two) parameters for visualization. Generally, the parameters are simply regarded as some arbitrary distance along what is called the "reaction coordinate"—a collection of all the atomic rearrangements required to progress from reactants to products. An example of a 3-dimensional surface is shown in Fig. 17.1 (left), where one may plot some important parameters in the reaction coordinate versus other important parameters versus their energy for visualization. However, reaction mechanisms are typically depicted as 2-dimensional plots of the energy versus a simplified reaction coordinate as shown in Fig. 17.1.

The choice of reaction coordinates in Fig. 17.1 could be interatomic distances between atoms that are forming or breaking bonds, for example. There is a saddle point in the figure on the left that corresponds to the maximum energy on the lowest-energy pathway connecting products and reactants. This point is called the "transition state" or activated complex for the reaction, and energy terms related to the transition state are often designated with a superscript "double dagger", such as $\Delta G^{\ddagger}$. Since the transition state is not a local minimum, its lifetime is fleeting—typically on the order of a bond vibration.[a]

17.2.2 Kinetic versus Thermodynamic Control

For the reaction coordinate in Fig. 17.1 (right), there are two important energy features that govern chemical reactivity. The first is the energy difference between the ground state and the transition state. This corresponds to the activation energy, E_a in the Arrhenius equation, or the free energy,

[a] Typical bond vibrations require on the order of femtoseconds.

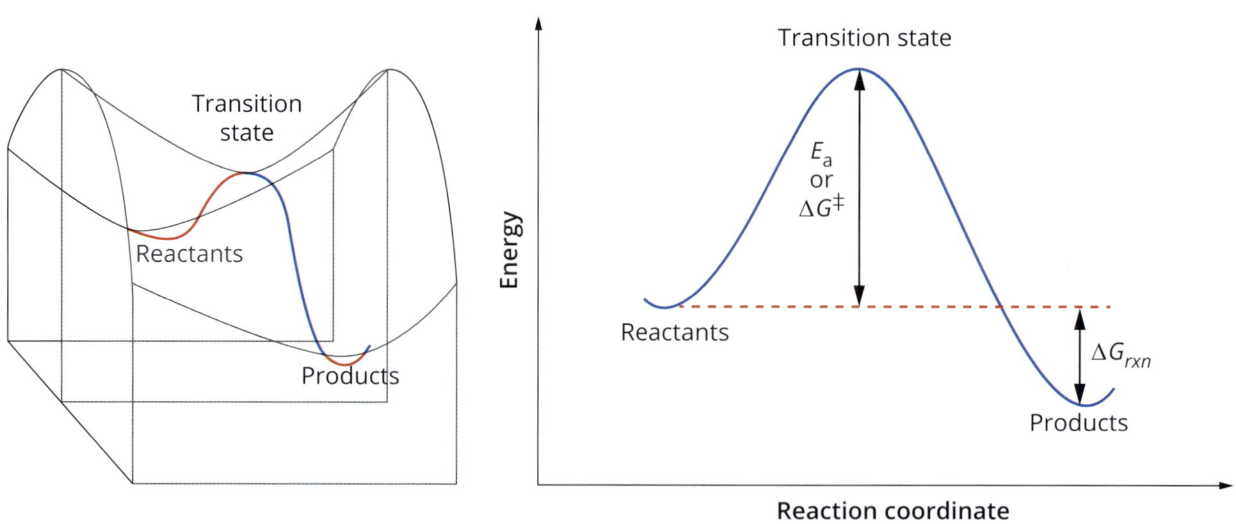

Fig. 17.1 (*left*) A 3-dimensional reaction coordinate diagram relating reactants and products. (*right*) A 2-dimensional representation of the multidimensional energy surface shown on the left depicting the lowest-energy pathway (dark line) connecting reactants and products.

$\Delta G^{\ddagger}$ in the Eyring equation, required to reach the transition state.[5] The barrier height determines the overall rate of the reaction denoted by the rate constant, k. (Note that a small k is always used for a rate constant, and a large K is used for an equilibrium constant.) We use k to describe the kinetic parameters of the reaction. In Arrhenius and Eyring equations, k is proportional to E_a and $\Delta G^{\ddagger}$ as shown in Eq. 17.1. (See Box 17A for the relationship between the Arrhenius and Eyring equations.)

$$k \propto e^{-E_a/RT} \propto e^{-\Delta G^{\ddagger}/RT} \qquad \textbf{17.1}$$

The second feature of importance is the energy difference between the reactants and the products. We denote this as ΔG, the free energy of the reaction, which determines the equilibrium constant, K_{eq}, or just K:

$$K_{eq} = e^{-\Delta G/RT} \qquad \textbf{17.2}$$

Thermodynamics determine whether the reaction is energetically feasible, while kinetics determine how fast a reaction can occur. There are two limiting regimes for a chemical reaction. We say a reaction is under "kinetic control" when a product distribution is determined by the fastest reaction. Conversely,

BOX 17A

The Arrhenius and Eyring equations

Both the Arrhenius and Eyring equations describe the relationship between the rate constant and temperature. The Arrhenius equation, which was found empirically, is usually written as

$$k = A\exp\left(-\frac{E_a}{RT}\right)$$

where k is the rate constant, A is simply called the "pre-exponential factor", E_a is the activation energy, R is the gas constant, and T is the temperature. For plotting purposes, the equation can be reorganized to

$$\ln k = \ln A - \frac{E_a}{R}\left(\frac{1}{T}\right)$$

so that a plot of ln k versus $1/T$ yields a straight line with slope $-E_a/R$ and intercept ln A.

The Eyring equation is derived from transition state theory and can be expressed as

$$k = \frac{\kappa k_B T}{hc^n} e^{\frac{\Delta S^{\ddagger}}{R}} e^{-\frac{\Delta H^{\ddagger}}{RT}}$$

where k_B is the Boltzmann constant, h is Planck's constant, and the factor c^n is the concentration where n = 0 for first-order reactions, and n = 1 for second-order reactions. Since the factor $1/c^n = 1$ for first-order systems, here it will be dropped for the remainder of the discussion. However, one should be cautious comparing activation parameters between reactions of different order.

The factor κ is called the transmission coefficient, which classically varies from 0 to 1, and is related to the number of molecules that come to the transition state and then go on to product. Often, it is assumed that $\kappa = 1$ as species that come to the transition state often have atoms with momentum vectors that lead to product, and it is unlikely that they are reflected backward towards the starting material. In certain cases, κ can be much higher than 1 where quantum-mechanical tunneling occurs; that is, the system does not even reach the transition state energy but goes on to product, which is most likely where motions to reach the transition state only involve light atoms such as hydrogen and at very low temperatures. In practice, for many reactions near room temperature not involving movement of light atoms the assumption that $\kappa \sim 1$ is reasonable.

Again, for data analysis, the equation can be reorganized as

$$\ln\left(\frac{k}{T}\right) = \frac{\Delta S^{\ddagger}}{R} + \ln\left(\frac{k_B}{h}\right) - \frac{\Delta H^{\ddagger}}{RT}$$

so that a plot of ln(k/T) vs $1/T$ has slope m and intercept b, which are related to the thermodynamic parameters as shown.

$$\Delta H^{\ddagger} = -mR$$
$$\Delta S^{\ddagger} = R\left[b - \ln\left(\frac{k_B}{h}\right)\right]$$

Since the Arrhenius and Eyring equations are both expressions relating the rate constant to temperature, the Arrhenius parameters of activation energy, E_a, and pre-exponential, A, can be related to $\Delta S^{\ddagger}$ and $\Delta H^{\ddagger}$.

$$E_a = \Delta H^{\ddagger} + RT$$
$$A = \left(\frac{k_B T}{h}\right)\exp\left(\frac{\Delta S^{\ddagger}}{R}\right)$$

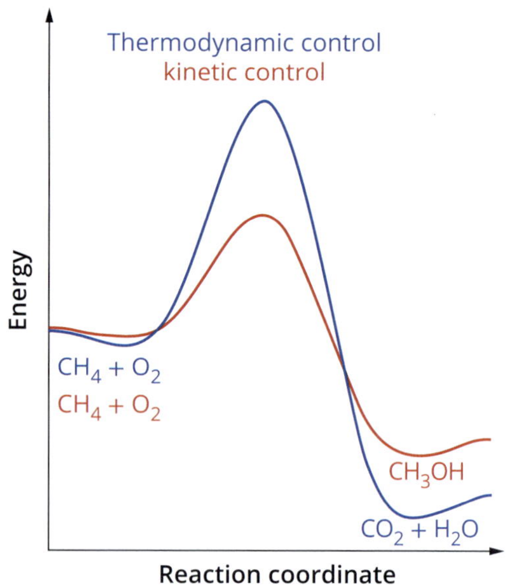

Fig. 17.2 Reaction coordinates for oxidation of methane to methanol (red), the kinetic product, and methane to CO_2 and H_2O (blue), the thermodynamic products.

a reaction is under "thermodynamic control" when the most stable products are formed.

To illustrate this point, let us consider the oxidation of methane by oxygen. The thermodynamic products of the reaction are the products of complete combustion, CO_2 and H_2O. If we can control the rates of oxidation by using a catalyst, it is possible to selectively oxidize methane to methanol. This is a process of considerable interest because the energy density per volume of methanol, a liquid, is much greater than that of methane, a gas. By using a catalyst, we can lower the barrier of the reaction for making methanol such that its rate is faster than subsequent oxidations that lead to CO_2 and H_2O. This is represented in the reaction coordinate diagram in Fig. 17.2, where the multi-step oxidation to CO_2 and H_2O is represented simplistically as a single step. If the reaction is carried out under conditions where methanol is formed selectively, the reaction is under kinetic control. If CO_2 and H_2O are formed, the reaction is under thermodynamic control.

17.2.3 Reactive Intermediates, Concerted Reactions, and Rate-Determining Steps

For the generic reaction

$$A + B \rightarrow D + E$$

consider the two putative reaction coordinates shown in Fig. 17.3. In the blue reaction coordinate, the reaction path connecting reactants and products has a maximum energy. This point of highest energy corresponds to the "transition state", or equivalently "activated complex". The red pathway

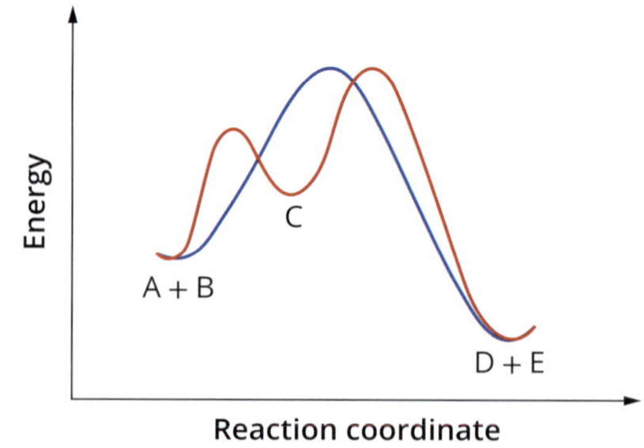

Fig. 17.3 Reaction coordinate diagrams connecting reactants A and B with products D and E. On the blue pathway, A and B combine directly to form products D and E. On the red pathway, A and B react to generate an intermediate C with some finite lifetime, which then proceeds on to D and E.

has two local maxima that are connected by a local minimum, which corresponds to an intermediate, C. Ultimately, the points of highest energy along these two paths are the same. Hence, their reaction rates would be identical.

If we consider the red pathway in Fig. 17.3, where the reaction between A and B proceeds through an intermediate C, there are cases where one can infer information about C. In rare cases, intermediate C may be observable. *Even when this occurs, discretion must be exercised because it can be difficult to distinguish between a true reactive intermediate and a species that is in equilibrium with the reactants but does not lie on the path to products (Fig. 17.4).* In cases where C cannot be detected, its intermediacy can sometimes be inferred from reaction rates. This generally requires that C is formed before the slow step of the reaction.

The slow step of a reaction is commonly referred to as the *rate-determining step* (RDS). In normal experience we often think of each step of a process requiring some amount of time. For example, we think of going to the grocery store (30 minutes), picking up the groceries (15 minutes), making dinner (1 hour), and eating dinner (30 minutes) as taking 2 hours and 15 minutes, which of course is true. For chemical processes, there are two major differences.

First, typical chemical processes involve a staggeringly large number of molecules all of which start the reaction at slightly different times. Consequently, we do not necessarily perceive (using Fig. 17.4 (a)) each molecular step occurring as A + B to C to D + E; instead, we perceived, for example, that step A + B was relatively fast because many of the molecules stayed in form C for enough time that it had a large enough population to be observable.

Second, the variability in reaction rates is far larger than the variability in rates of our usual experience. Even so, we would not consider any one 30-minute trip to the grocery store when we consider how we spent our year; we would think of things like work and sleep as taking up our time for the most part. This is because 30 minutes is only about 0.006% of a year.

If a reaction takes an hour to reach 50%, a process that takes a microsecond contributes little to how long the reaction takes.

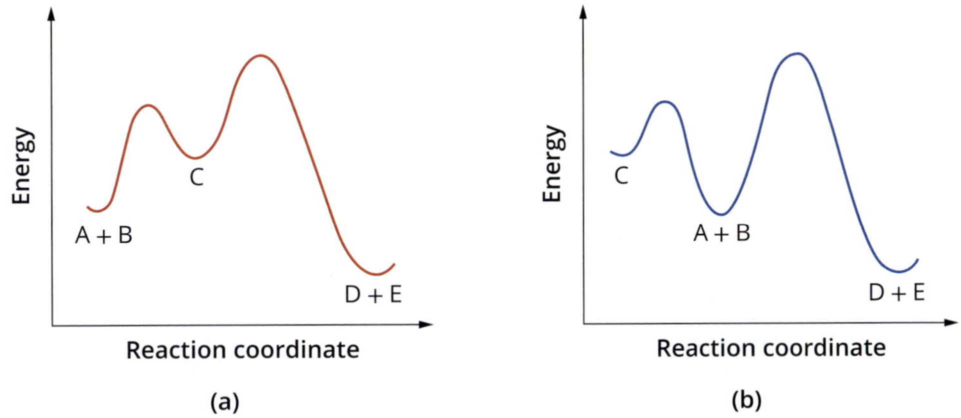

Fig. 17.4 Examples of equilibria between reactants A, B, and C (a) where C is an intermediate and (b) where C is not an intermediate in the reaction between A and B to produce products D and E.

Unimolecular reaction rates may occur on the order of a bond vibration (femtoseconds, 10^{-15} s) to billions of years (10^{13} s). Bimolecular reactions in solution are limited (on the fast side) by diffusion. It is possible and perhaps common for individual steps in a mechanism to occur millions of times faster or slower than one another. Consequently, a reaction rate is determined largely by the slow step, the rate-determining step (RDS), and the steps that lead up to it.

As an example, consider the simple replacement of water surrounding a metal center with water in the solvent, a $\Delta G_{rxn} = 0$ process that will be discussed in great detail in this chapter. On an aqueous Co^{2+} ion, the water exchanges rapidly, while in Ir^{3+} the reaction is exceedingly slow. The rate constants for these exchange reactions range from about 10^{10} to 10^{-10} s^{-1}, changes in magnitude that are often reserved for astronomers (Fig. 17.5). The cause of this large difference in rate constants,

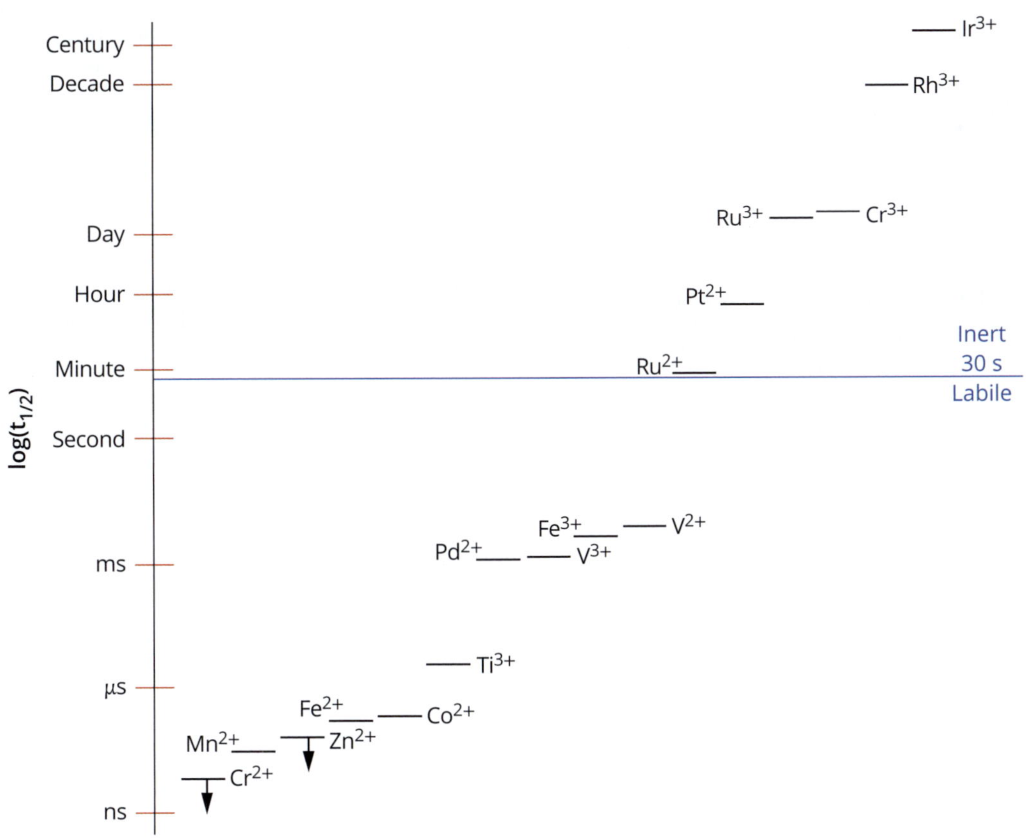

Fig. 17.5 Range of water exchange rate constants (log $t_{1/2}$ in s) for first coordination sphere water exchange with solvent water at 25 °C.[6] The "inert" and "labile" break in the rate constants at 30 s is discussed later in this chapter. The downward arrow (Zn^{2+} and Cr^{2+}) indicates that the value given is an upper limit for the half-life.

for what on the surface looks like the same transformation occurring at different metal cations, will be discussed in due course. However, the degenerate water substitution example illustrates the large range of possible rate constants that can be obtained. This is true not only from changing species in a reaction but also for different steps in the same reaction.

17.2.4 The Importance of the Rate-Determining Step (RDS)

An important consequence of the position of the slow step within the reaction sequence is that *information about the chemical reaction can only be obtained for steps that precede the rate-limiting step*. An example of a reaction coordinate diagram for a generic reaction is shown in Fig. 17.6. By monitoring the course of the reaction, it is possible to glean information about the formation of C and D, but we cannot intuit details about the conversion of D to E + F. This is simply because the population of D is very small relative to the other species in solution.

The reason why intermediate D is difficult to observe is for the same reason that there are fewer cars after you pass a crash on the highway, as opposed to the slow creeping traffic before: when the crash is in front of you, the rate of travel is slow as people cautiously pass the wrecked cars. The slow rate causes a buildup of concentration of vehicles in the "before we reach the crash" state. After cars pass the crash, both safety and curiosity sated, they go back to highway speeds, and the concentration of cars per mile of road decreases as the rate increases.

Likewise, molecules along a reaction pathway tend to spend more time as the species before the rate-determining step, and intermediates that occur after the RDS tend to have little impact on the kinetics.

17.2.5 Microscopic Reversibility

One principle that can be useful for elucidating mechanisms of reactions is the principle of microscopic reversibility, which simply states that the lowest-energy pathway for forward and reverse reactions is the same. It is important that rate laws describing reactions do not violate this principle. One useful corollary of the principle of microscopic reversibility is that we can sometimes gain mechanistic insight by running a reaction in reverse, since we know that any information obtained is relevant to the forward reaction.

Say we have a reaction where A + B goes to make C + D, and there is a transition state $X^{\ddagger}$ through which the reaction proceeds in the forward direction (Fig. 17.7). If the reaction is run backwards, C + D → A + B, the reaction will follow the same pathway through $X^{\ddagger}$.

It should be noted that the principle of microscopic reversibility cannot be used to eliminate mechanisms of a different kinetic order. For example, if the rate law for a reaction in the forward direction is second-order, this does not necessarily mean that the lowest-energy pathway in the backward direction is of the same order. However, if the backward reaction has the same kinetic form, it must occur by the microscopic reverse of the forward reaction.

One place where microscopic reversibility has been extensively used as a method to study the reverse reaction is in C–H activation. Functionalization of C–H bonds catalyzed by metal complexes is an important topic of current research. However, direct observation of a metal breaking a C–H bond through insertion, although known, is more often studied by placing the CH_3 and H groups on the metal center and watching the C–H bond form. By the principle of microscopic reversibility, the same pathway must be followed by the reaction when the metal inserts into the C–H bond as when the C and H eliminate from the metal.

Bercaw and coworkers have studied the microscopic reverse of C–H activation in Pt complexes by protonation of metal–alkyl complexes. One can consider two limiting mechanisms where addition of a proton removes a metal alkyl: Fig. 17.8 (Path 1) protonation of the metal center to form an intermediate metal alkyl hydride complex followed by C–H bond formation, and (Path 2) direct protonation of the alkyl group to form a metal

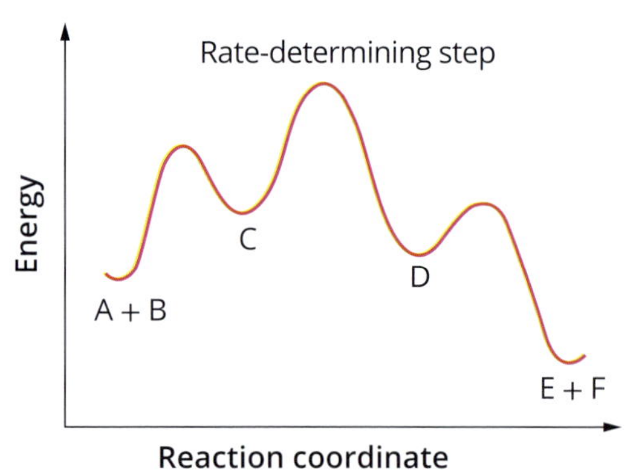

Fig. 17.6 A reaction coordinate diagram for A + B reacting to form E + F via intermediates C and D. Kinetics experiments can provide information about intermediate C because its formation precedes the rate-determining step. Intermediate D will be kinetically "invisible" as it appears after the slow step in the reaction.

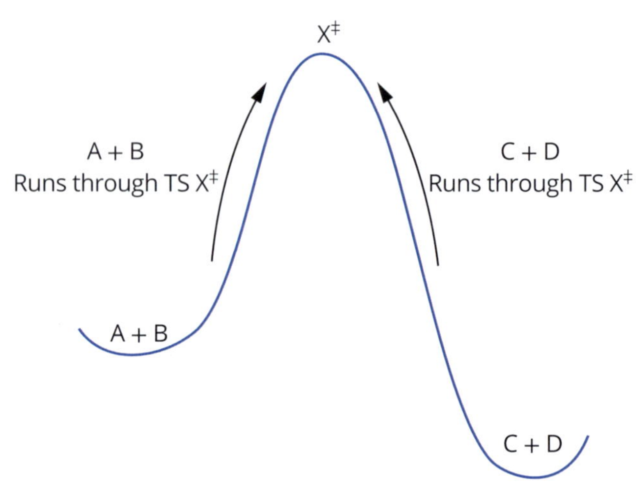

Fig. 17.7 By microscopic reversibility, the reaction A + B to make C + D that goes through a transition state labeled $X^{\ddagger}$, if run backwards, will also follow the same path through $X^{\ddagger}$.

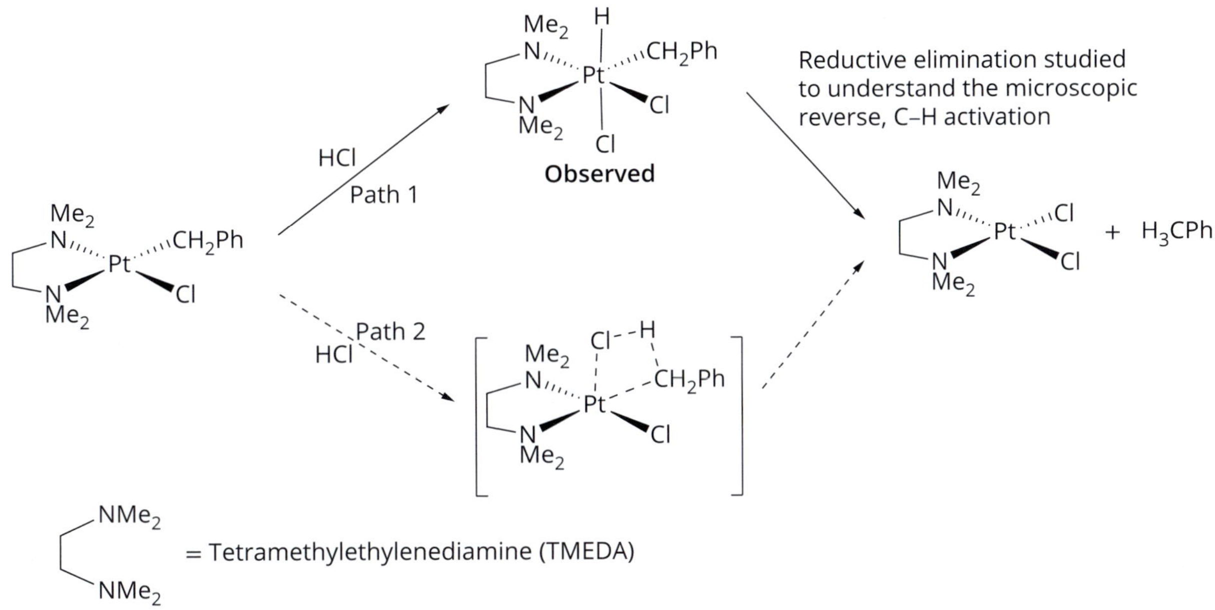

Fig. 17.8 Two limiting paths for removal of a CH_2Ph group from Pt by addition of HCl.

cation that is then trapped by the counterion of the acid. In the first pathway, a 6-coordinate Pt(IV) complex is expected as an intermediate. In the second pathway, the alkyl is directly protonated off the metal and no intermediate hydride is formed, perhaps through a transition state like the one shown where chloride binds to the metal concomitant with C–H bond formation. Stahl, Labinger, and Bercaw were able to directly observe the 6-coordinate Pt(IV) complex and observe kinetically its decomposition to $PtCl_2$(TMEDA) and toluene[7–8] Knowing this, one can study this reaction and learn about the C–H activation of H_3CPh by $PtCl_2$(TMEDA), the microscopic reverse.

17.2.6 Reaction Rates, Reaction Orders, and Pseudo-First-Order Reactions

Determining the pathway for a reaction is important because the detailed information provides deeper insight into reactivity. There are a number of approaches to determining a mechanism. In extremely rare cases, the intermediates, like C in Fig. 17.4, are observed. More typically, mechanisms are determined by studying how reaction rates are influenced by concentrations of reactants and products, temperature, pressure, and so on.

The starting point of any kinetic study is the determination of the rate law. Consider the generic reaction in which two molecules of A and one molecule of B react to produce three molecules of D and one of E.

$$2A + B \rightleftharpoons 3D + E$$

The rate law is measured by determining how the concentrations of reactants and/or products change as a function of time. The overall rate of the reaction—sometimes abbreviated with the Greek letter ν (nu)—is related to the disappearance of starting materials, which then have a negative sign (Eq. 17.3). Since the overall reaction requires two equivalents of A, the disappearance of A is written as $-\frac{1}{2}d[A]/dt$.[b] The same rate for the overall reaction just requires the consumption of one equivalent of B, $-d[B]/dt$. The products D and E increase in concentration over time, and have positive values for the derivatives.

$$rate = \nu = -\frac{1}{2}\frac{d[A]}{dt} = -\frac{d[B]}{dt} = +\frac{1}{3}\frac{d[D]}{dt} = +\frac{d[E]}{dt} \qquad \textbf{17.3}$$

The rate equation for a mechanism relates this change due to formation of a product with a rate constant, k, and concentrations of reactants and products as shown in Eq. 17.4. *Importantly, superscripts in the rate law are rarely equal to the coefficients in the balanced reaction.* In other words, it can never be assumed that $a = 2$, $b = 1$, $d = 3$ or $e = 1$ in Eq. 17.4. The rate equation, or rate law of the reaction as it is sometimes called, is related to the rate-determining step and the steps prior to it, and not necessarily the stoichiometry.

$$\upsilon = k[A]^a[B]^b[D]^d[E]^e \qquad \textbf{17.4}$$

The reason for this difference between the stoichiometry of a reaction and the rate equation is that some of the equivalents of, for example, A, may be used in the reaction after the RDS. Even though two equivalents are necessary for the reaction to occur, only one equivalent may affect the rate. The second equivalent may be used in very fast steps that occur after the RDS.

[b] Note that square brackets denote the concentration of the species within.

The superscripts in the rate law are called the order of the reaction for each species in the rate law. For example, when $a = b = 1$ and $d = e = 0$, the reaction is said to be first-order in the reactants, A and B, and zero-order in the products, D and E (Eq. 17.5). These reaction orders are determined experimentally by varying the concentrations of A, B, D, and/or E and measuring how these changes affect the reaction rate.

$$\upsilon = k[A][B][D]^0[E]^0 \qquad \textbf{17.5}$$

For this equation, doubling [A] will double the initial reaction rate, while doubling [D] (which has an order of zero, $[D]^0$) will have no effect on the rate. In common practice, zero-order terms are not included in the rate law, given $a = b = 1$ and $d = e = 0$, and Eq. 17.5 would instead be written as 17.6.

$$\upsilon = k[A][B] \qquad \textbf{17.6}$$

The formal definition of reaction order with respect to the concentration of species A, [A], is shown in Eq. 17.7. In the equation, the order of species A, lowercase "*a*", is the partial derivative of the log of the rate, ν, with respect to the log of [A], *while the concentration of all other species,* [∞], *are held constant.*

$$a = \left(\frac{\partial \log v}{\partial \log[A]}\right)_{[\infty]} \qquad \textbf{17.7}$$

Let us define the concentration at time t of species A as $[A]_t$, at the start of the reaction ($t = 0$) as $[A]_0$, and the final concentration after the reaction is complete as $[A]_f$. In the previous example, the reaction rate depends on the concentrations of A and B. If [A] = [B], the rate of disappearance of A will appear to be second-order. In other words, if [A] = [B] in k[A][B], then this is indistinguishable from an alternative rate law, $k[A]^2$. For this and other reasons, finding the order of a reactant involves observing the rate as a function of individually changing concentration of each of the species involved in the reaction.

After Eq. 17.7, rates are often measured under conditions where the concentration of one species in the rate equation is limiting and the concentrations of the other species are held in excess, ≥10 times the concentration of limiting reagent, which means their concentration will change an insignificant amount over the course of the reaction. For instance, if $[B]_0 = 10[A]_0$ in the previous example, the concentration of B at the end of the reaction, $[B]_f$, will be 90% of $[B]_0$ and warrant the approximation $[B]_f \approx [B]_i$. In other words, [B] is effectively constant.

If we rearrange Eq. 17.6 as shown in Eq. 17.8 for the situation where $[B]_0 = 10[A]_0$, the approximately constant [B] is factored into the observed rate constant, k_{obs}. Under these pseudo-first-order conditions, $k_{obs} \cong k[B]_0$. The rate law in Eq. 17.8 is a simple first-order rate law. Since the reaction is not really first-order, we refer to this as a *pseudo-first-order rate law*, and the conditions where all but the limiting species in the rate law are held in excess are commonly referred to as pseudo-first-order conditions.

$$\upsilon = k[B][A]$$
$$\upsilon = k_{obs}[A] \qquad \textbf{17.8}$$

Notice that in this example, what we have done is hold all the other species at constant concentration so that we can examine how [A] affects the reaction rate. This allows us to find the order in A by using the formal definition described in Eq. 17.7.

Rearranging Eq. 17.8 and integrating the differential equation (remember that v is as shown in Eq. 17.3) produces the following.

$$\ln([A]_t/[A]_0) = -k_{obs}t \qquad \textbf{17.9}$$

The concentration of A at any time t, $[A]_t$, is given as follows.

$$[A]_t = [A]_0 e^{-k_{obs}t} \qquad \textbf{17.10}$$

Idealized plots of Eqs. 17.9 and 17.10 as a function of time are shown in Fig. 17.9. From Eq. 17.9, the slope gives the value of $-k_{obs}$. Since the slope is always negative, k_{obs} is always a positive number. All rate constants are always positive.

17.2.7 Equilibria and Unobserved Intermediates: Using the Steady-State Approximation

A common situation that occurs in inorganic systems is an equilibrium between generic species A and C that proceeds through a common intermediate, B. Inferred but not detected intermediates are often signified with curly brackets, such as {B} in Eq. 17.11.

$$\mathrm{A} \underset{k_{-1}}{\overset{k_1}{\rightleftharpoons}} \{\mathrm{B}\} \underset{k_{-2}}{\overset{k_2}{\rightleftharpoons}} \mathrm{C} \qquad \textbf{17.11}$$

We will begin with Eq. 17.12, which gives [A] as a function of time. Reactions that affect the concentration of A are the forward and back equilibrium reactions with B. The reaction that reduces [A] over time, $-\frac{d[A]}{dt}$, is the forward reaction with rate constant k_1 dependent on the concentration of A. Because the reaction involving k_1 involves a step that moves the reaction forward, a term $k_1[A]$ with a *positive sign* is included in Eq. 17.12. The back reaction from intermediate B to A runs counter to the decreasing rate being described, it generates A, and is given a *negative sign* with its rate constant k_{-1}.

$$\frac{-d[A]}{dt} = k_1[A] - k_{-1}[B] \qquad \textbf{17.12}$$

In Eq. 17.13, the rate of formation of intermediate B is described. B is formed from A, which is described by adding a

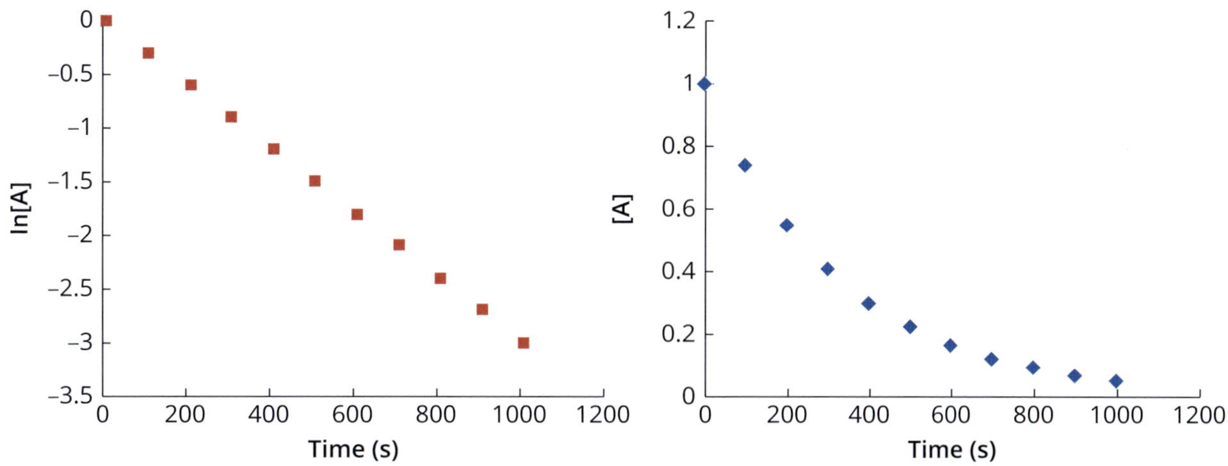

Fig. 17.9 Plot of ln[A] versus time (Eq. 17.9), which produces a straight line for the first-order equation (*left*), and [A] vs time (Eq. 17.10) (*right*).

term $+k_1[A]$. The intermediate B is also formed from C, which is described by adding a term $+k_{-2}[C]$. Compound B is removed from the reaction (negative sign) by forming A and C, $-(k_{-1} + k_2)[B]$. The form of d[C]/dt can be found similarly from the proposed mechanism in Eq. 17.11.

$$\frac{d[B]}{dt} = k_1[A] + k_{-2}[C] - k_{-1}[B] - k_2[B] \quad \textbf{17.13}$$

$$\frac{d[C]}{dt} = k_2[B] - k_{-2}[C] \quad \textbf{17.14}$$

If we are describing the reaction at $t = 0$, the only species present is A, whose concentration is $[A]_0$. Mass balance requires that Eq. 17.15 be true at all times.

$$[A]_0 = [A] + [B] + [C] \quad \textbf{17.15}$$

If the intermediate B is fleeting, we can assume that its rate of formation and disappearance are about the same, which means that its concentration is essentially constant. In other words, Eq. 17.16 will be true.

$$\frac{d[B]}{dt} \approx 0 \quad \textbf{17.16}$$

Thus, [B] will be in a "steady state", and Eq. 17.16 is referred to as the *steady-state approximation*. Combining Eqs. 17.13 and 17.16 we can derive an expression for the concentration of the unobserved intermediate B.

$$\frac{d[B]}{dt} \sim 0 = k_1[A] + k_{-2}[C] - k_{-1}[B] + k_2[B]$$

$$k_1[A] + k_{-2}[C] = (k_{-1} + k_2)[B]$$

$$[B] = \frac{k_1[A] + k_{-2}[C]}{(k_{-1} + k_2)} \quad \textbf{17.17}$$

Furthermore, under the steady-state approximation, when $[B] \sim 0$ we can express [C] in terms of [A] and $[A]_0$. (compare Eq. 17.15)

$$[C] = [A]_0 - [A] \quad \textbf{17.18}$$

Substituting this value for [C] into Eq. 17.17 produces

$$[B] = \frac{k_1[A] + k_{-2}[A]_0 - k_{-2}[A]}{(k_{-1} + k_2)} \quad \textbf{17.19}$$

Substituting this expression for [B] into Eq. 17.12 produces

$$\frac{-d[A]}{dt} = k_1[A] - k_{-1}\frac{k_1[A] + k_{-2}[A]_0 - k_{-2}[A]}{(k_{-1} + k_2)} \quad \textbf{17.20}$$

Collecting terms and rearranging yields Eq. 17.21, which suggests that the rate of disappearance of A to form C is dependent on only [A] and $[A]_0$.

$$\frac{-d[A]}{dt} = \frac{(k_1k_2 + k_{-1}k_{-2})}{(k_{-1} + k_2)}[A] - \frac{k_{-1}k_{-2}}{(k_{-1} + k_2)}[A]_0 \quad \textbf{17.21}$$

When the system reaches equilibrium, the concentrations of A and C can be designated as $[A]_{eq}$ and $[C]_{eq}$. The ratio of concentrations at equilibrium defines the equilibrium constant K_{eq}, which is simply the product of the forward-rate constants k_1 and k_2 divided by the product of the reverse-rate constants, k_{-1} and k_{-2}. This produces Eq. 17.22, which in turn gives $[C]_{eq}$ in terms of the rate constants and $[A]_{eq}$ (Eq. 17.23).

$$\frac{[C]_{eq}}{[A]_{eq}} = K_{eq} = \frac{k_1k_2}{k_{-1}k_{-2}} \quad \textbf{17.22}$$

$$[C]_{eq} = \frac{k_1k_2}{k_{-1}k_{-2}}[A]_{eq} \quad \textbf{17.23}$$

Mass balance requires that

$$[A]_0 = [A]_{eq} + [C]_{eq} \quad \textbf{17.24}$$

so combining Eqs. 17.23 and 17.24 relates $[A]_0$ and $[A]_{eq}$ to rate constants

$$[A]_0 = \frac{k_{-1}k_{-2} + k_1k_2}{k_{-1}k_{-2}}[A]_{eq} \quad \textbf{17.25}$$

which can be rewritten as

$$k_{-1}k_{-2}[A]_0 = (k_{-1}k_{-2} + k_1k_2)[A]_{eq} \quad \textbf{17.26}$$

Substituting for $k_{-1}k_{-2}[A]_0$ in Eq. 17.21 and rearranging, we obtain

$$\frac{-d[A]}{dt} = \frac{(k_1k_2 + k_{-1}k_{-2})}{(k_{-1} + k_2)}([A] - [A]_{eq}) \quad \textbf{17.27}$$

Collecting [A] terms on the left and integrating over time we obtain

$$\int_{[A]_0}^{[A]_t} \frac{1}{([A] - [A]_{eq})} d[A] = -k_{obs}\int_0^t dt \quad \textbf{17.28}$$

Solving the integral produces Eq. 17.29, which describes how the system approaches equilibrium.

$$ln([A]_t - [A]_{eq}) - ln([A]_0 - [A]_{eq}) = -k_{obs}t \quad \textbf{17.29}$$

where k_{obs} is the experimentally determined rate constant.

$$k_{obs} = \frac{k_1k_2 + k_{-1}k_{-2}}{k_{-1} + k_2} \quad \textbf{17.30}$$

This is an example of the types of mathematical manipulations common to kinetics, and the steady-state approximation is a common method for eliminating intermediates from rate laws so that the mathematical expression related only to

BOX 17B

Validity of the steady-state approximation

The steady-state approximation is a common assumption in kinetics that allows the removal of an intermediate concentration from the rate law; this is done because in the vast majority of cases intermediates are not observed. As with all assumptions, there are circumstances where it is valid and others where it is a poor approximation.[56]

If we consider a simple reaction involving a starting material A in equilibrium with an intermediate B that then converts irreversibly to product C, we can apply the steady-state approximation to species B.

$$A \underset{k_{-1}}{\overset{k_1}{\rightleftharpoons}} B \xrightarrow{k_2} C$$

First, we write the expression for the concentration of species B over time as shown in the following equation. There is one process that produces B, which is written as the positive term $k_1[A]$. There are two processes that decrease the concentration of B, which are the negative terms $-k_{-1}[B]$ and $-k_2[B]$. The steady-state approximation says that the concentration of B does not change significantly over the course of the reaction; consequently, the derivative with respect to time is approximately zero.

$$\frac{d[B]}{dt} = k_1[A] - k_{-1}[B] - k_2[B] \approx 0$$

If we solve this expression for [B], we find the following.

$$\frac{d[B]}{dt} = k_1[A] - k_{-1}[B] - k_2[B] \approx 0$$

$$[B] = \frac{k_1}{k_{-1} + k_2}[A]$$

The rate of formation of product C is simply k_2 times the concentration of B, but we have no way of measuring the concentration of intermediate B in most cases, as intermediates are rarely observed. Using the steady-state approximation produced an expression for the concentration of B, however. Substitution goes as follows, which provides an expression for the rate of formation in C in the starting material A.

$$\frac{d[C]}{dt} = k_2[B]$$

$$\frac{d[C]}{dt} = \frac{k_2k_1}{k_{-1} + k_2}[A]$$

It has been found that this rate law is a good approximation if two criteria are met:

(Criterion 1) $k_1 \ll k_{-1} + k_2$

In other words, the rate constant for the formation of intermediate B should be much smaller than the sum of the rate constants for the intermediate's consumption. This will make the overall concentration of the intermediate low.

(Criterion 2) $t \gg 1/(k_{-1} + k_2)$

The variable t is the timescale of the experiment. As a result, the timescale of the reaction should be much longer than the inverse of the rate constants for the consumption of B.

We can look at these criteria in three different kinetics simulations. The starting concentration of A is set to "100" (unitless) so that we can easily talk of the percentage of A, B, and C in the reaction.

In the first simulation (Fig. 17.1B) we show the reaction where Criteria 1 and 2 are met. In this simulation, $k_1 = 1$, $k_{-1} = 15$, and $k_2 = 15$. The concentration of B—obviously zero at the beginning of the reaction—builds up slightly but never exceeds 3% of the reaction mixture. As a consequence, [B] only varies from 0% to 3% and is relatively constant throughout the course of the reaction.

If we change the rate constants so that Criterion 1 is not fulfilled with $k_1 = 2$ and $k_{-1} = k_2 = 1$ the simulation looks quite different (Fig. 17.2B). As can be seen, the concentration of B changes substantially over the course of the reaction, varying from 0% to about 40%.

Finally, Criterion 2 is not fulfilled when the inverse of the sum of the rate constants consuming B are similar to the timeframe of the reaction (Fig. 17.2B). A simulation with $k_1 = 10$ and $k_{-1} = k_2 = 0.5$. With these rate constants, the reaction takes ~10 time units, and [C] has not quite reached 100% at this time, $1/(k_{-1} + k_2) = 1$. As can be seen, even with an order of magnitude difference between k_1 and $k_{-1} + k_2$, the intermediate species B is at some times during the experiment the most abundant species and its concentration varies significantly from 0 to ~70%.

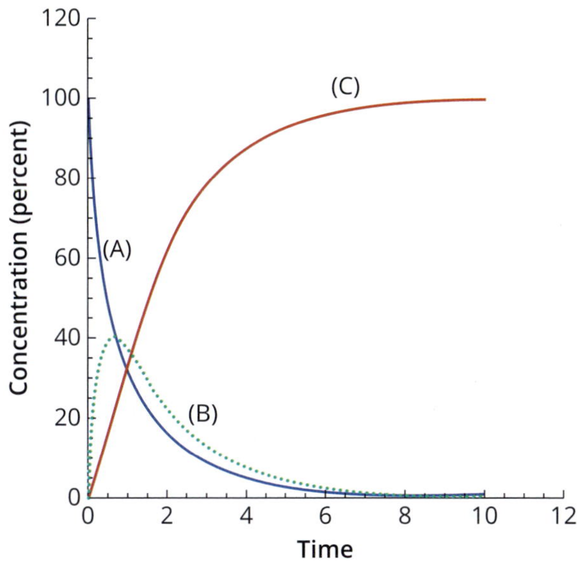

Fig. 17.2B A plot of the concentrations of A, B, and C over time when both Criterion 1 is not met for the steady-state approximation, $k_1 = 2$ and $k_{-1} = k_2 = 1$.

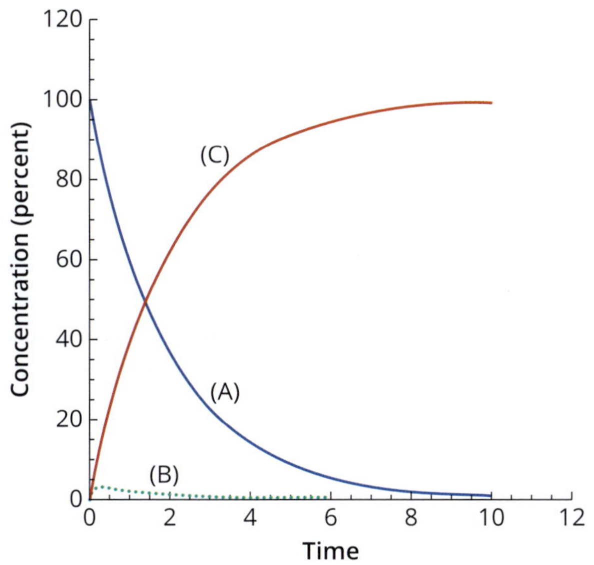

Fig. 17.1B A plot of the concentrations of A, B, and C over time when both Criteria 1 and 2 are met for the steady-state approximation, $k_1 = 1$, $k_{-1} = 15$, and $k_2 = 15$.

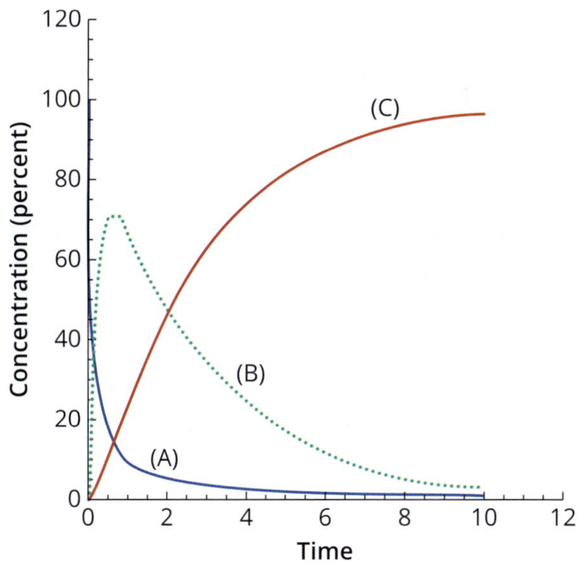

Fig. 17.3B A plot of the concentrations of A, B, and C over time when Criterion 2 is not met for the steady-state approximation.

observables can be obtained. This method (and some of the results presented here) will often be used in this chapter.

17.3 Interrogating the Transition State

By measurement of the observed rate constant as a function of temperature, one can gain information about the thermodynamics of the transition state relative to the ground state. This information can be used to form hypotheses about the nature of the key transition state for a transformation.

17.3.1 Activation Parameters and the Eyring Equation

One piece of information that the rates provide is the energy separating the reactants and the highest-energy transition state. This relationship was originally posited by Arrhenius, who developed an empirical relationship between the reaction rate, k, the energy of activation, E_a, and temperature, T (Box 17A).

$$k = Ae^{-E_a/RT} \quad \textbf{17.31}$$

In Eq. 17.31, R is the gas constant and A is a proportionality constant that is related to the collision frequency and entropy of activation. Later, Eyring, Polanyi, and Evans developed transition-state theory where the rate is related to temperature, fundamental constants, and the free energy of activation, $\Delta G^{\ddagger}$. Eq. 17.32 is generally referred to as the "Eyring equation" (sometimes the Eyring–Polanyi equation), where k_B is Boltzmann's constant, h is Planck's constant, and R is the gas constant.[9] Rearranging Eq. 17.32 by dividing by T and taking the natural log of both sides, produces Eq. 17.33, and Eq. 17.34 can be derived by using the same process in conjunction with the Gibbs equation.

$$k = \frac{k_B T}{h} e^{-\Delta G^{\ddagger}/RT} = \frac{k_B T}{h} e^{-\Delta H^{\ddagger}/RT + \Delta S^{\ddagger}/R} \text{ (Eyring equation)} \quad \textbf{17.32}$$

$$ln\left(\frac{k}{T}\right) = ln\left(\frac{k_B}{h}\right) - \frac{\Delta G^{\ddagger}}{RT} \quad \textbf{17.33}$$

$$\ln\left(\frac{k}{T}\right) = -\frac{\Delta H^{\ddagger}}{RT} + \frac{\Delta S^{\ddagger}}{R} + ln\left(\frac{k_B}{h}\right) \quad \textbf{17.34}$$

By experimentally determining reaction rates over a range of temperatures where the Eyring relationship is valid, a plot of ln(k/T) versus 1/T (Eq. 17.34) will produce a line with slope = $-\Delta H^{\ddagger}$/R and y-intercept = $\Delta S^{\ddagger}/R + \ln(k_B/h)$. The enthalpy of activation, $\Delta H^{\ddagger}$, is the difference in enthalpy between the transition state, $H^{\ddagger}$, and the reactants, H_r, $\Delta H^{\ddagger} = H^{\ddagger} - H_r$. Likewise, the entropy of activation ($\Delta S^{\ddagger}$) is the differences in entropy between the transition state and the reactants, $\Delta S^{\ddagger} = S^{\ddagger} - S_r$. By measuring the rate as a function of temperature, both $\Delta H^{\ddagger}$ and $\Delta S^{\ddagger}$ can be measured experimentally. This provides insight into the roles that bond-making, bond-breaking, and entropy play in a chemical reaction. The interpretation of these results will be discussed later in this chapter and the next.

From statistical analysis of the Eyring equation, the error in the entropy of activation is directly related to the error in the enthalpy through Eq. 17.35, where $\sigma(\Delta S^{\ddagger})$ is the error in the measured entropy of activation, $\sigma(\Delta H^{\ddagger}$ is the error in measured enthalpy of activation, and T_{ave} is the average of the temperature-range used. For most solution phase studies, $\sigma(\Delta S^{\ddagger}) \sim 0.003\ K^{-1} \times \sigma(\Delta H^{\ddagger})$.[10–12]

$$T_{ave} \times \sigma(\Delta S^{\ddagger}) = \sigma(\Delta H^{\ddagger}) \quad \textbf{17.35}$$

An example for the substitution of H_2O by ethylene in $Ru(H_2O)_6^{2+}$ is shown in Fig. 17.10.[13] The rate constants for substitution were measured between 296 and 325 K. Using Eq. 17.34, we can plot the second-order rate data from the paper. For this, ln(k/T) vs 1/T is plotted. The slope, m, contains information about the enthalpy of activation, $\Delta H^{\ddagger}$, which is found by using the following equation.

$$\Delta H^{\ddagger} = -mR \quad \textbf{17.36}$$

From the least-squares fit, m = –9627, with units of Kelvins. Using the gas constant in kcal/mol ($R = 1.987 \times 10^{-3}$ kcal/mol • K) proceeds as follows, providing $\Delta H^{\ddagger}$ = 19.1 ± 0.5 kcal/mol.

$$\Delta H^{\ddagger} = -(-9627\,K)(1.987 \times 10^{-3}\,kcal/mol \cdot K) = 19.1\,kcal/mol$$

The entropy of activation, $\Delta S^{\ddagger}$, is related to the intercept of the plot, b, through Eq. 17.37.

$$\Delta S^{\ddagger} = \left(b - ln\frac{k_b}{h}\right)R \quad \textbf{17.37}$$

From the least-squares fit, b = 19.78 (unitless). If we use the gas constant in calories (R = 1.987 cal/mol K), the equation appears as follows, and $\Delta S^{\ddagger}$ = –7.8 ± 1.7 e.u.[c]

$$\Delta S^{\ddagger} = \left(+19.78 - ln\left(\frac{1.30 \times 10^{-23}\,J/K}{6.626 \times 10^{-34}\,J/Hz}\right)\right) 1.987\,cal/mol \cdot K$$
$$= -7.8\,e.u.$$

Note that the errors in the values from the least-squares fit obey Eq. 17.35 to a high degree. The error in $\Delta S^{\ddagger}$, $\sigma(\Delta S^{\ddagger})$ = 1.74 e.u., and the error in $\Delta H^{\ddagger}$, $\sigma(\Delta H^{\ddagger})$ = 0.542 kcal/mol = 542 cal/mol. The average temperature used is (325+296)/2 = 310.5 K, so $1/T_{ave} = 0.0032\ K^{-1}$.

[c] e.u = entropy units = cal/mol·K.

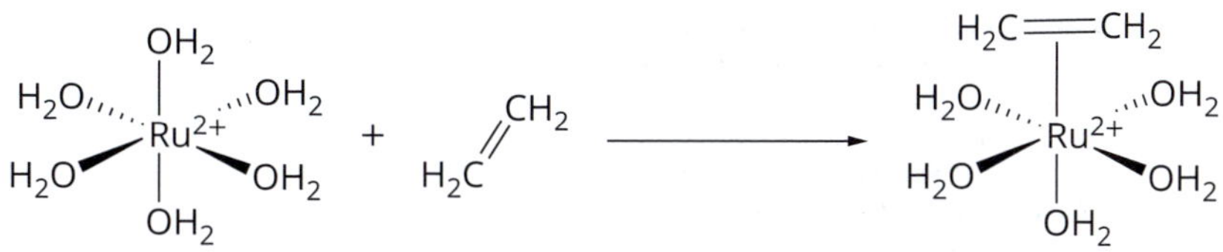

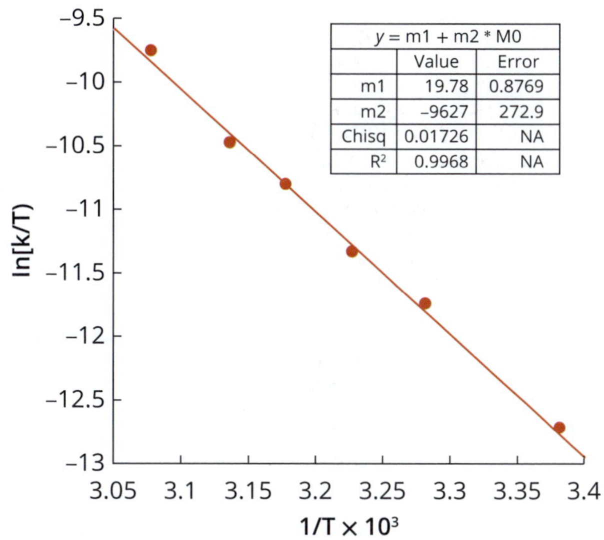

y = m1 + m2 * M0		
	Value	Error
m1	19.78	0.8769
m2	-9627	272.9
Chisq	0.01726	NA
R^2	0.9968	NA

T	k	ln(k/T)	1/T x 10^3
296	8.80E-04	-12.73	3.378
305	2.41E-03	-11.75	3.279
310	3.74E-03	-11.35	3.226
315	6.37E-03	-10.81	3.175
319	9.03E-03	-10.47	3.135
325	1.88E-02	-9.758	3.077

Fig. 17.10 (*top*) Equation for the substitution reaction $[Ru(H_2O)_6]^{2+} + CH_2{=}CH_2 \rightarrow [Ru(CH_2{=}CH_2)(H_2O)_5]^{2+} + H_2O$. (*middle*) An example Eyring plot for the reaction with the tabulated data. The negative entropy is consistent with net bond breaking in the transition state relative to the ground state. (*bottom*) Tabulated data for the plot.

$$\sigma(\Delta S^{\ddagger}) = \frac{1}{T_{ave}}\sigma(\Delta H^{\ddagger})$$

$$1.74 = 0.0032(542)$$

$$1.74 = 1.74$$

17.3.1.1 Interpretation of $\Delta H^{\ddagger}$

The magnitude of $\Delta H^{\ddagger}$ generally describes the size of the enthalpy change between the transition state and the reactant(s). Any interaction that contributes to the overall enthalpy may contribute, including differences in bonding, solvation, dipolar interactions, and so on. In some cases, bond-making and bond-breaking processes make the largest contributions to the term.

As a first approximation, then, an endothermic $\Delta H^{\ddagger}$ indicates bond-breaking in the transition state relative to the reactants. An exothermic $\Delta H^{\ddagger}$ is rare, but would indicate stronger bonds in the transition state than in the starting materials.[d]

It should be kept in mind, however, that $\Delta H^{\ddagger}$ will describe the difference in energy between the entire transition-state structure and entire structure of the reactant(s). As a result, not only the bond being made or broken in the transition state will contribute, but other bonds in the molecule responding to the change in electronic structure by changing bond distances and angles may also contribute to the enthalpy difference.

[d] Heat changes associated with solvation differences, among other factors, could contribute to $\Delta H^{\ddagger}$ as well.

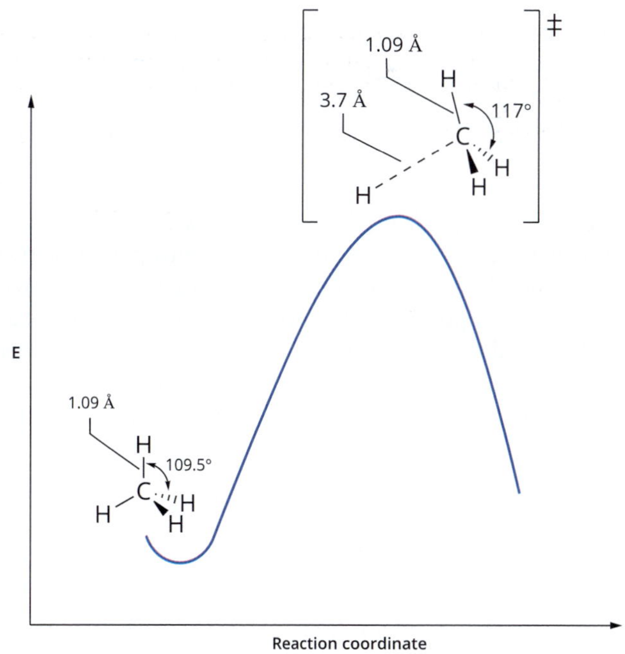

Fig. 17.11 Some metrical parameters for methane and the transition state for C–H bond homolysis calculated using MP2/6-311G(d,p).

As an illustration, consider the important reaction of cleaving a C–H bond in methane. In Fig. 17.11 the calculated (MP2/6-311G(d,p)) ground state and transition state for methane C–H bond cleavage and their energies are shown. The reaction involves stretching one of the C–H bonds until there is no net interaction between C and one H. However, the reaction coordinate plotted on the *x*-axis is in fact a little more complicated. As one C–H bond is stretched, the tetrahedral symmetry is broken and the other C–H bonds respond by changing their angles. In the transition state, an approximately planar CH_3 radical is obtained where the carbon uses ~sp^2 hybrids to form its C–H bonds. The enthalpy and entropy are not just responding to one bond changing but to all the changes in the molecule.

Similarly, in a metal complex, as one bond to the metal is made or broken in the transition state, the other metal–ligand bonds will respond to help stabilize the system. The measured $\Delta H^\ddagger$ and $\Delta S^\ddagger$ respond to all these changes between the ground state and excited state.

Examples of typical values of $\Delta H^\ddagger$, and how the values are typically interpreted, for inorganic substitution reactions will be discussed later in this chapter.

17.3.1.2 Interpretation of $\Delta S^\ddagger$

The entropy of activation is the difference in entropy (disorder) in the transition state relative to the reactant, $\Delta S^\ddagger = S^\ddagger - S_r$. A negative value implies that the reactant is more disordered than the transition state; that is, the transition state is more ordered than the reactant. In many cases this implies steric crowding near the metal as additional ligands enter to restrict modes, formation of rings that restrict bond rotations, and so on. In some cases, a negative entropy of activation is interpreted as indicating net bond formation, which increases crowding near the metal, in the transition state.

Conversely, a positive entropy of activation implies a transition state that is less ordered relative to the starting material. This is often interpreted in substitution reactions as indicating net bond-breaking during the transition state.

Often times it is useful to compare entropies of activation between different reactions, using a reaction of “known” mechanism as a point of reference. A note of caution is warranted when doing this, however. In order to be comparable, the reactions need to be of the same standard state; in addition, the comparisons are most useful if extremely similar and should be of the same molecularity.

17.3.2 Volume of Activation

Another parameter that can change between the ground state and transition state is the volume occupied by the reactants and the activated complex. This difference is the volume of activation, $\Delta V^\ddagger = V_{ac}–V_{gs}$, where V_{ac} is the volume of the activated complex (transition state) and V_{gs} is the volume of the ground state. Because volume and pressure are related, we might expect that the rate of the reaction would be influenced by pressure, depending on the sign of $\Delta V^\ddagger$. For example, when the volume of the activated complex is less than the volume of ground-state reactants, intuition tells us that increasing the pressure should accelerate the rate. Conversely, when the volume of the activated complex is greater than the volume of the reactants, we expect that increasing the pressure would retard the reaction rate. The reaction rate at a given pressure, $(k_T)_P$, the volume of activation, $\Delta V^\ddagger$, and the pressure, *P*, are related as shown in Eq. 17.38.

$$\ln(k_T)_P = \ln(k)_0 - P\Delta V^\ddagger / RT \qquad \textbf{17.38}$$

Thus, by measuring reaction rates at constant temperature and various pressures and plotting $\ln(k_T)_P$ vs. *P*, we obtain a line whose slope is $\Delta V^\ddagger/RT$ with an intercept of $ln(k)_0$, natural log of the rate constant at zero pressure.[14]

17.4 Ligand Substitution Mechanisms[6,15]

Some of the most important reactions in transition-metal chemistry are ligand substitutions, which provide pathways for substrates to interact with metal centers in synthetic and biological systems. The mechanisms of ligand substitution are classified, after Langford and Gray,[16] in terms of the relationships between the entering and leaving ligands. As shown in Fig. 17.12, there are three types of mechanism classifications: associative, dissociative, and interchange, which define how the entering and leaving ligands approach and depart the coordination sphere of the complex.

In the associative mechanism (***A***), the entering ligand coordinates to the metal center and generates an intermediate where the coordination number is increased by one. The leaving ligand then dissociates from the intermediate to generate the product. In a dissociative mechanism (***D***), the leaving ligand dissociates from the starting complex to generate an intermediate where the coordination number is decreased by one, and the entering ligand reacts with this intermediate to form the product. Although the definition of associative and dissociative mechanisms requires intermediates to form, it is important to note that these intermediates are rarely observed.

The interchange mechanism (***I***) is a concerted process in which the metal–ligand bond of the leaving ligand is broken as the metal–ligand bond of the entering ligand is formed. In this case, no intermediate complex is generated. The interchange mechanism is usually broken down into two subclasses: associative interchange (I_a), and dissociative interchange (I_d) mechanisms. In the I_a mechanism there is a greater degree of bond-making from the entering ligand than bond-breaking from the exiting ligand in the transition state. In an I_d mechanism the transition state reflects a greater degree of bond-breaking from the exiting ligand than bond-making from the entering ligand. In principle, the amount of bond-breaking and bond-making in the transition state could be equal, which

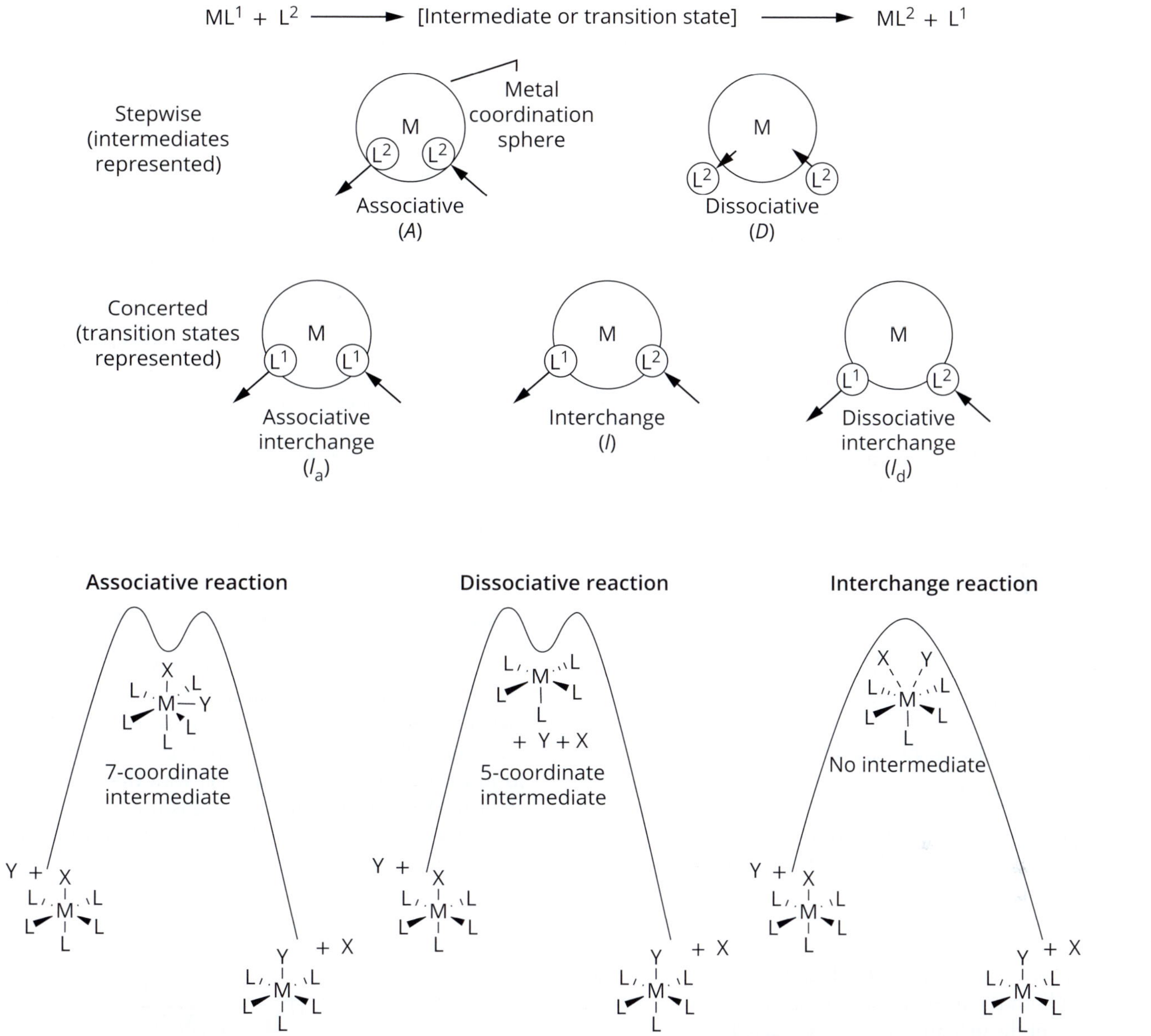

Fig. 17.12 (*top*) A schematic representing commonly designated substitution mechanisms based on the nature of their intermediates (***A*** and ***D***) or transition state (***I*$_\mathbf{a}$**, ***I***, and ***I*$_\mathbf{d}$**). (Adapted from Wilkins's book.[17]) (*bottom*) Representative reaction diagrams for substitutions on 6-coordinate complexes.

is simply an interchange (***I***) mechanism, which is relatively rare but is known to occur, as will be seen shortly.

17.4.1 Rate Law for Associative Mechanisms

Consider substitution of Y for X in a generic metal complex, $L_nM(X)$. In an associative pathway, an intermediate, $\{L_nM(X)(Y)\}$, forms prior to loss of X. This is represented in the following scheme with an equilibrium followed by an irreversible step where X leaves.

$$\mathrm{LnM(X)} + \mathrm{Y} \underset{k_{-1}}{\overset{k_1}{\rightleftharpoons}} \{\mathrm{LnM(X)(Y)}\} \xrightarrow{k_2} \mathrm{LnM(Y)} + \mathrm{X}$$

The rate of formation of the product, YML_n, is then simply the following equation with the rate constant times the concentration of the intermediate.

$$\upsilon = k_2[\{L_nM(X)(Y)\}]$$

Using the steady-state approximation, we can eliminate the intermediate in the rate equation as follows, which is advantageous, as we do not know the concentration of the intermediate $\{L_nM(X)(Y)\}$, and intermediates usually are not observable.

$$\frac{d[\{L_nM(X)(Y)\}]}{dt} \sim 0 = -k_2[\{L_nM(X)(Y)\}] + k_1[L_nM(X)][Y] - k_{-1}[\{L_nM(X)(Y)\}]$$

$$[\{L_nM(X)(Y)\}] = \frac{k_1[L_nM(X)][Y]}{k_{-1} + k_2}$$

Substituting this expression for $[\{L_nM(X)(Y)\}]$ in the rate equation produces the following expression.

$$\upsilon = \frac{k_1k_2[Y][L_nM(X)]}{k_{-1} + k_2} \qquad \textbf{17.39}$$

Note that experimentally, this equation is a simple second-order rate law; first order in [Y] and first order in $[XML_n]$, with $k_{obs} = k_1k_2/(k_1 + k_2)$.

In the limiting case of a purely associative (***A***) reaction with irreversible loss of X, $k_{-2} = 0$, the rate is relatively insensitive to the nature of the ligand that is leaving. The reaction would, however, be very sensitive to the nucleophilicity of the incoming ligand, which affects k_1.

17.4.2 Rate Law for Interchange Mechanisms

Interchange mechanisms (***I***) involve formation of a second coordination sphere complex between the incoming ligand, outgoing ligand, and metal complex. The strength of the interactions between the incoming and outgoing ligands determines whether the mechanism is I_a, I, or I_d.

Associative Interchange (***I_a***): Relatively strong interactions between both the incoming and outgoing ligands. A stronger influence of the incoming group is expected.

Dissociative Interchange (***I_d***): Relatively weak interactions between both the incoming and outgoing ligands. A stronger influence of the leaving group is expected.

Distinguishing between different interchange mechanisms generally requires examination of the effects of incoming and leaving groups, but information can also be obtained from the enthalpy ($\Delta H^{\ddagger}$), entropy ($\Delta S^{\ddagger}$), and volume ($\Delta V^{\ddagger}$) of activation. From a kinetic standpoint, the forms of the rate laws are the same between different interchange mechanisms.

Following our previous discussion on equilibria, we can derive an expression for the rate law under these conditions. The complex $\{L_nMX,Y\}$ is often referred to as the "encounter complex". In $\{L_nMX,Y\}$, X is still bound to the metal, but Y has some interaction with the metal as well.

$$L_nMX + Y \overset{K_1(fast)}{\rightleftharpoons} \{L_nMX,Y\} \quad K_1 = k_1/k_{-1}$$

$$\{L_nMX,Y\} \xrightarrow{k_2} L_nMY + X$$

The rate of formation of L_nMY is related to the interchange of X for Y in the encounter complex.

$$\frac{d[L_nMY]}{dt} = k_2[\{L_nMX,Y\}] \qquad \textbf{17.40}$$

However, we do not know $[\{L_nMX,Y\}]$, and we need an expression for this concentration. This can be done[18] by realizing that the species in equilibrium can be defined according to the total concentration of metal-containing species in equilibrium, [E].

$$[E] = [L_nMX] + [\{L_nMX,Y\}]$$

The total concentration of metal complexes $[M]_0$ is defined as follows, which is equal to the initial concentration of the metal starting material $[L_nMX]_0$.

$$[M]_0 = [L_nMX] + [\{L_nMX,Y\}] + [L_nMY] = [E] + [L_nMY]$$

Alternatively, [E] can be expressed in terms of the product concentration and the concentration of metal complexes.

$$[E] = [L_nMX]_0 - [L_nMY] \qquad \textbf{17.41}$$

Getting the rate expression involves some mathematical manipulation. Starting with the equilibrium expression, simply form its inverse. We will leave [Y] with the equilibrium constant in the following manipulations (multiply both sides by [Y] in the standard equilibrium expression), which will simplify things later.

$$K_1 = \frac{[\{L_nMX,Y\}]}{[L_nMX][Y]}$$

$$K_1[Y] = \frac{[\{L_nMX,Y\}]}{[L_nMX]}$$

$$\frac{1}{K_1[Y]} = \frac{[L_nMX]}{[\{L_nMX,Y\}]}$$

Adding 1 to both sides produces the following equations. On the right, we use $[\{L_nMX,Y\}]/[\{L_nMX,Y\}] = 1$. We then get each side over a common denominator.

$$\frac{1}{K_1[Y]} + 1 = \frac{[L_nMX]}{[\{L_nMX,Y\}]} + \frac{[\{L_nMX,Y\}]}{[\{L_nMX,Y\}]} = \frac{[L_nMX] + [\{L_nMX,Y\}]}{[\{L_nMX,Y\}]}$$

$$\frac{1 + K_1[Y]}{K_1[Y]} = \frac{[L_nMX] + [\{L_nMX,Y\}]}{[\{L_nMX,Y\}]}$$

Notice that the numerator on the right side is now [E], the equilibrium concentration.

$$\frac{1+K_1[Y]}{K_1[Y]} = \frac{[E]}{[\{L_nMX,Y\}]}$$

Now, substituting [E] for Eq. 17.41 to express things in terms of the total concentration:

$$\frac{1+K_1[Y]}{K_1[Y]} = \frac{[L_nMX]_0 - [L_nMY]}{[\{L_nMX,Y\}]}$$

$$[\{L_nMX,Y\}] = \frac{K_1[Y]}{1+K_1[Y]}([L_nMX]_0 - [L_nMY])$$

Substitution of this result into Eq. 17.40, which involves multiplication of the previous by k_2, gives the rate expression for the production of L_nMY.

$$\frac{d[L_nMY]}{dt} = \frac{k_2K_1[Y]}{1+K_1[Y]}([L_nMX]_0 - [L_nMY])$$

One of the general assumptions of the steady-state approximation is that the concentration of the intermediate is low. Consequently, we will generalize this equation by assuming that the concentration of the starting material, $[L_nMX]$, is the initial concentration of the starting material, $[L_nMX]_0$, minus the concentration of product that has been produced, $[L_nMY]$. Again, this mass balance should be true if the concentration of the intermediate is not significant with respect to $[L_nMX]_0$ during the reaction (Box 17B).

$$[L_nMX] \sim [L_nMX]_0 - [L_nMY]$$

Using this approximate expression produces the rate law for an interchange reaction (Eq. 17.42). This is the rate law found in many standard substitution reactions; for example, water substitution in $[Ni(OH_2)_6]^{2+}$ by SO_4^{2-}.[19]

$$\frac{d[L_nMY]}{dt} = \frac{k_2K_1[Y]}{1+K_1[Y]}[L_nMX] \qquad \textbf{17.42}$$

Under these conditions there are two limiting regimes for k_{obs}. When $K_1[Y] << 1$, k_{obs} reduces to $k_2K_1[Y]$, and the rate will depend on [Y]. When $K_1[Y] >> 1$, k_{obs} reduces to k_2 and the rate will be independent of [Y]. At this point, the rate is said to be "saturated", approaching its maximum value of k_2. Fig. 17.13 shows plots of k_{obs} for Eqs. 17.39 (associative mechanism) and 17.42 (interchange mechanism) as a function of [Y]. At low [Y], k_{obs} is a linear function of [Y]. At high [Y], the mechanism involving pre-equilibrium association in the interchange mechanism is distinguishable as k_{obs} approaches a constant value of k_2.

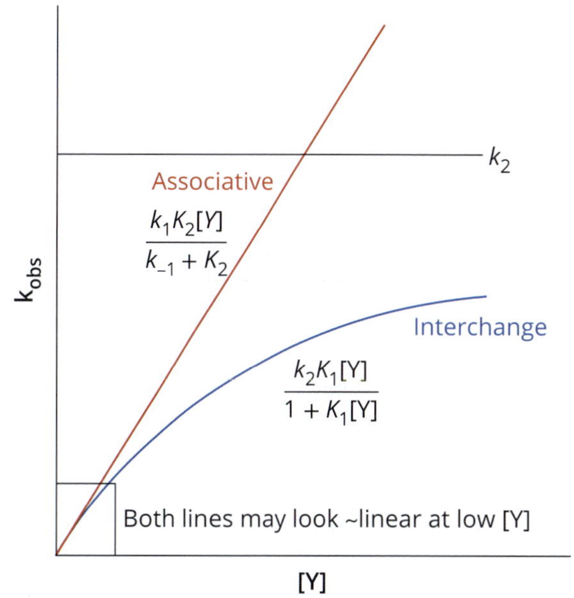

Fig. 17.13 Plots of k_{obs} versus ligand concentration for associative (*A*) and interchange (*I*) mechanisms. Expressions for k_{obs} are from Eqs. 17.39 and 17.42.

17.4.3 Rate Law for Dissociative Mechanisms

Consider the dissociative pathway shown as follows. Note that the path involves production of lower-coordinate intermediate ML_n, which is in solution with both X and Y. The two ligand X and Y then compete for the open coordination site on the metal center with rate constants k_{-1} and k_2, respectively.

$$L_nM(X) + Y \underset{k_{-1}}{\overset{k_1}{\rightleftharpoons}} \{L_nM\} + X + Y \xrightarrow{k_2} L_nM(Y) + X$$

The following differential equations describe $[L_nM(X)]$ and $[L_nM]$ as functions of time.

$$\frac{-d[L_nM(X)]}{dt} = k_1[L_nM(X)] - k_{-1}[X][L_nM] \qquad \textbf{17.43}$$

$$\frac{d[L_nM]}{dt} = k_1[L_nM(X)] - k_{-1}[X][L_nM] - k_2[Y][L_nM] \qquad \textbf{17.44}$$

This produces a rate law analogous to Eq. 17.27, repeated here.

$$\frac{-d[A]}{dt} = \frac{(k_1k_2 + k_{-1}k_{-2})}{(k_{-1}+k_2)}([A] - [A]_{eq})$$

In this case, we adjust the equation by making $[L_nM(X)]_{eq} = 0$, $k_{-2} = 0$, $k_2 = k_2[Y]$, and $k_{-1} = k_{-1}[X]$, which produces the rate law for the dissociative mechanism.

$$\frac{-d[L_nM(X)]}{dt} = \frac{k_1k_2[Y]}{k_{-1}[X] + k_2[Y]}[L_nM(X)] \qquad \textbf{17.45}$$

By varying [X] and [Y] it is often possible to reach limiting regimes where $k_{-1}[X] >> k_2[Y]$ and $k_{-1}[X] << k_2[Y]$.

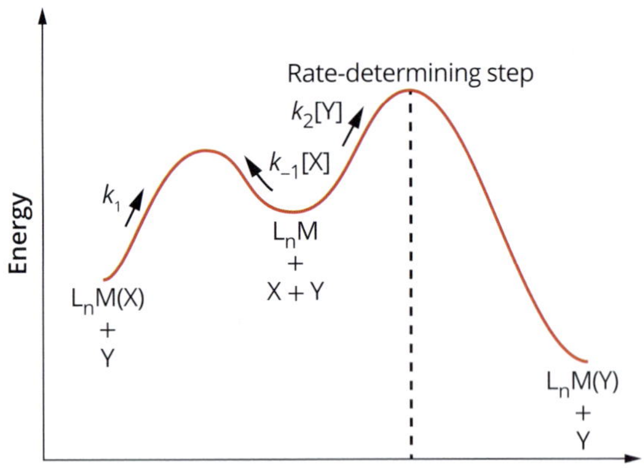

Fig. 17.14 Reaction coordinate diagram for dissociative ligand substitution in complex $L_nM(X)$ where $k_{-1}[X] >> k_2[Y]$. All terms before the rate-determining step contribute to k_{obs} (Eq. 17.42). When $k_{-1}[X] << k_2[Y]$, the rate expression Eq. 17.45 reduces to Eq. 17.47.

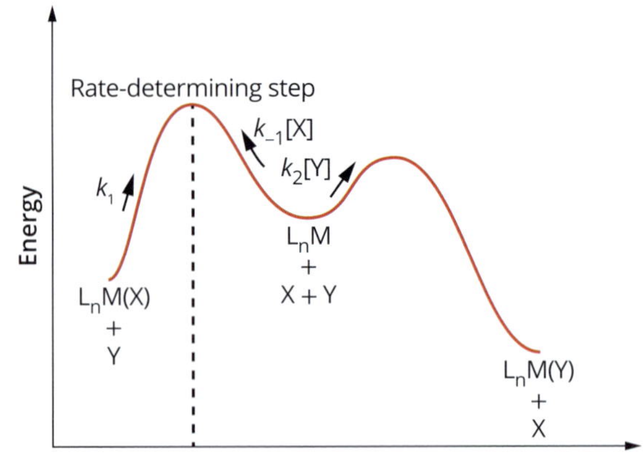

Fig. 17.15 Reaction coordinate diagram for dissociative ligand substitution in complex $L_nM(X)$ where $k_{-1}[X] << k_2[Y]$. In this regime, dissociation of X is the rate-determining step, and k_{obs} is given by Eq. 17.47.

When $k_{-1}[X] >> k_2[Y]$,[e] the rate expression Eq. 17.45 reduces to Eq. 17.46.

$$\frac{-d[L_nM(X)]}{dt} = \frac{k_1k_2[Y]}{k_{-1}[X]}[L_nM(X)]$$
$$(\text{Eq. 17.45 if } k_{-1}[X] \gg k_2[Y]) \qquad \textbf{17.46}$$

Under these conditions the reaction is first-order in concentration of the incoming ligand Y, and inverse first-order in the concentration of the departing ligand X. If a reagent appears in the denominator, like X in Eq. 17.46, this means that it inhibits the progress of the reaction; that is, increasing its concentration will slow the reaction. *This inhibition by X is a hallmark of a dissociative process.* When $k_{-1}[X] >> k_2[Y]$, the second step of the reaction is rate-limiting, so the rate behavior reveals participation of X in the dissociation and Y in trapping of the intermediate, L_nM. Relative barriers for the transition states might be as shown in Fig. 17.14.

[e] For this type of assumption, $k_{-1}[X] >> k_2[Y]$ implies that $k_{-1}[X] + k_2[Y] \cong k_{-1}[X]$.

When $k_{-1}[X] << k_2[Y]$, the rate expression Eq. 17.45 reduces to Eq. 17.47.

$$\frac{-d[L_nM(X)]}{dt} = \frac{k_1k_2[Y]}{k_2[Y]}[L_nM(X)]$$
$$(\text{Eq. 17.45 if } k_{-1}[X] \ll k_2[Y])$$

$$\frac{-d[L_nM(X)]}{dt} = k_1[L_nM(X)] \qquad \textbf{17.47}$$

Under these conditions, the slow step is dissociation of X corresponding to the reaction coordinate in Fig. 17.15.

17.4.4 Summary of Common Rate Expressions for Substitution

Table 17.1 summarizes rate behaviors for ***I***, ***A***, and ***D*** mechanisms for the substitution of Y for X in L_nMX for different

Table 17.1 Limiting rate laws for substitution mechanism for the reaction $L_nMX + Y \rightarrow L_nMY + X$ at different values of [Y] and [X].

$L_nMX + Y \rightarrow L_nMY + X$

Mechanism	Rate (ν) Low [Y]	Rate (ν) Medium [Y]	Rate (ν) High [Y]
A	$\frac{k_1k_2[L_nM(X)]_0[Y]}{k_{-1}+k_2}$ pseudo-first order in [Y]	$\frac{k_1k_2[L_nM(X)][Y]}{k_{-1}+k_2}$	$\frac{k_1k_2[L_nM(X)][Y]_0}{k_{-1}+k_2}$ pseudo-first order in $[L_nM(X)]$
I	$k_2K_1[Y][L_nM(X)]$ $(K_1[Y] \ll 1)$ small build-up of encounter complex	$\frac{k_2K_1[Y][L_nM(X)]}{1+K_1[Y]}$	$k_2[L_nM(X)]$ $(K_1[Y] \gg 1)$ large build-up of encounter complex
D	$\frac{k_1k_2[Y]}{k_{-1}[X]}[L_nM(X)]$ $(k_{-1}[X] >> k_2[Y])$	$\frac{k_1k_2[Y]}{k_{-1}[X]+k_2[Y]}[L_nM(X)]$	$k_1[L_nM(X)]$ $(k_{-1}[X] << k_2[Y])$

[X] and [Y]. Distinction between an ***I*** and ***A*** mechanism can be made only if saturation of the rate is observed at high [Y], which establishes an ***I*** pathway. The ***D*** mechanism can usually be distinguished from ***I*** and ***A*** mechanisms because inhibition by the leaving ligand X is usually observed at high [X].

17.5 Ligand Substitution in 6-Coordinate Complexes

Henry Taube, in a large treatise published in *Chemical Reviews* in 1952, summarized the substitution reactions of metal complexes and their reaction rates.[4] In his descriptions, he called ions that quickly substituted "labile", and those that required longer times "inert" to substitution. The cutoff chosen between the two classifications was $t_{1/2} \sim 30$ s. In this study, he brought attention to the fact that ligand substitution was dependent upon the *d*-electron configuration of the metal ion.

By simple examination of the *d*-orbital configuration, one can make a reasonable guess as to the type of mechanism expected for substitution for an octahedral or pseudo-octahedral complex. Consider three different metal ions, all with a +3 charge (Fig. 17.13): d^2 vanadium(III), d^3 chromium(III), and d^5 high-spin iron(III).

In vanadium(III) complexes there is an empty orbital in the t_{2g} set (Fig. 17.16). As a result, a pair of electrons from an incoming ligand can easily interact with the metal center to form a new bond, and an associative (***A***) or associative interchange ($\boldsymbol{I_a}$) mechanism is expected. In general, associative character is expected for any metal center that has fewer than three *d*-electrons.

In chromium(III), the t_{2g} set is half-filled. Electrons from an incoming ligand will experience electron–electron repulsions if directed towards these orbitals. Consequently, the system is more inert to substitution than the vanadium(III) case. In fact, reactions with chromium(III) complexes are often many orders of magnitude slower than their vanadium(III) counterparts (*vide infra*). Occupation of all the t_{2g} orbitals, without occupation of the $e_g{}^*$ orbitals, produces inert complexes.

In high-spin iron(III), the metal has five unpaired electrons—one in each of the *d*-orbitals. Occupation of the t_{2g} orbitals produces no low-energy orbitals on the metals to act as acceptors, but occupation of the (antibonding) $e_g{}^*$ orbitals leads to destabilization of the metal–ligand interactions. This favors a mechanism with dissociative character (***D*** or $\boldsymbol{I_d}$). Metal complexes with occupied $e_g{}^*$ orbitals are usually labile to ligand exchange.

The discussion here is focused on octahedral complexes, but the same arguments can be extended to other coordination numbers as well. A crystal field theory analysis can be used to discover the nature and occupation of the orbitals, and empty low-lying

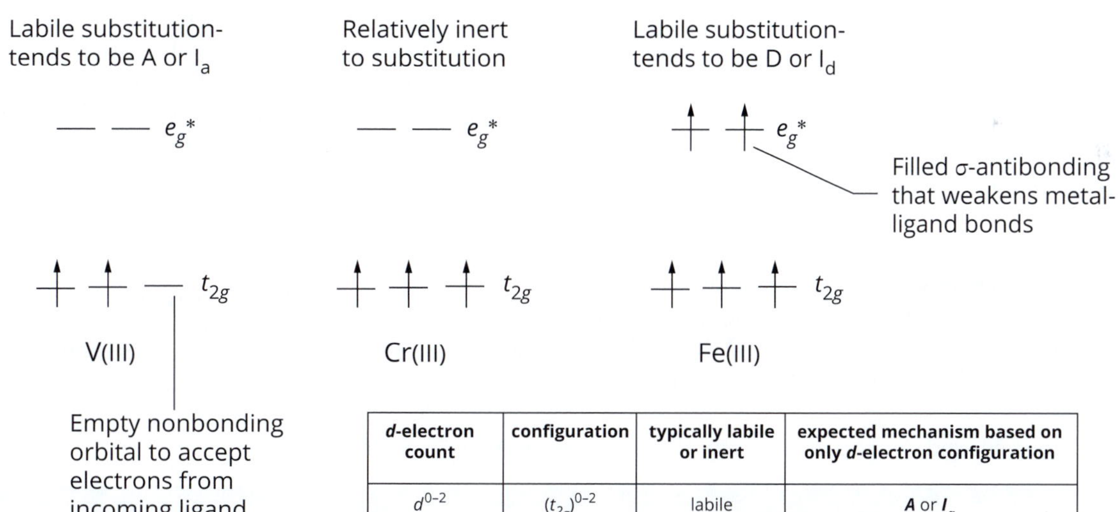

d-electron count	configuration	typically labile or inert	expected mechanism based on only *d*-electron configuration
d^{0-2}	$(t_{2g})^{0-2}$	labile	***A*** or $\boldsymbol{I_a}$
d^3	$(t_{2g})^3$	inert	—
d^{4-5} low spin	$(t_{2g})^{4-5}$	inert	—
d^{4-5} high spin	$(t_{2g})^3(e_g)^{1-2}$	labile	***D*** or $\boldsymbol{I_d}$
d^6 low spin	$(t_{2g})^6$	inert	—
d^6 high spin	$(t_{2g})^4(e_g)^2$	labile	***D*** or $\boldsymbol{I_d}$
d^{7-10}	$(t_{2g})^7(e_g)^{1-4}$	labile	***D*** or $\boldsymbol{I_d}$

Fig. 17.16 (*top*) Some prototypical metal configurations in the +3 oxidation state. The metals are labile with associative or associative interchange in their mechanism (V^{+3}), relatively inert to substitution (Cr^{+3}), or labile with dissociative or dissociative interchange mechanisms (Fe^{+3}) depending on the electronic configuration. Jahn–Teller distortion in the case of V(III) is being ignored. (*bottom*) Table correlating *d*-electron count, electron configuration, lability of ligands, and probable mechanism of substitution.

orbitals or occupied antibonding orbitals will typically produce labile complexes. Obviously, steric factors can also contribute. Coordination numbers greater than 7 will often lead to mechanisms with dissociative character, simply because bringing yet another ligand into a crowded complex for an associative mechanism is energetically costly. Coordination numbers of less than 6 may favor mechanisms with associative character; although here some care is warranted. For example, there are 4-coordinate complexes that are well known to use dissociative mechanisms.[20–21]

Obviously, these rationalizations without data should be used cautiously. However, first considering associative and associative interchange mechanisms for complexes of low coordination number and/or with empty t_{2g}-orbitals is common. If the complex contains occupied antibonding e_g*-orbitals or is >6 coordinate, dissociative and dissociative interchange mechanisms are often reasonably considered first. If the configuration is "inert", arguments of this type cannot really be employed, and simply observing the ground-state electronic configuration does not make a convincing argument for a specific mechanism in these cases. In these inert systems, effects of spectator ligands and charge effects seem to make large contributions, as will be discussed in later sections.

17.5.1 Activation Parameters for Water Exchange in $[M(OH_2)_6]^{3+}$ Complexes

The interchange between bulk solvent and coordinated water in metal aquo complexes is one of the most fundamental examples of ligand exchange. Consequently, it is also one of the most widely studied. Remarkably, the rate for exchange between bulk solvent water and bound water, k_{H2O}, spans 20 orders of magnitude with the mean lifetimes at 298 K ranging from 300 years for $[Ir(H_2O)_6]^{3+}$ to 200 ps for $[Eu(H_2O)_7]^{3+}$![14] For transition-metal aquo complexes, k_{H2O} can be conveniently measured by various NMR techniques, and some values of k_{H2O} are listed in Table 17.2 for trivalent hexaaquo complexes, $[M(H_2O)_6]^{3+}$.[22] The electronic structures of these compounds have been studied thoroughly, and their crystal structures have been determined. Consequently, the ground states are well characterized.[23–25]

The rates for exchange in Table 17.2 span 10 orders of magnitude. Because the charge is equal in each instance, factors such as electronic configuration and ionic radius must have important roles in determining rates.

17.5.2 Connecting Activation Parameters to Substitution Mechanisms

The entropy of activation, $\Delta S^‡$, is often interpreted as reflecting the extent of associative or dissociative character in a reaction. In a purely associative reaction, the transition state will be more ordered due to loss of translational (and possibly other) entropy. Hence, $\Delta S^‡ < 0$ would be expected. Conversely, a less ordered transition state might be expected in a purely dissociative reaction, where it could be anticipated that $\Delta S^‡ > 0$. This simple view has several pitfalls. For example, we are not considering the electronic states that are populated in the ground state versus the transition state.[f] Likewise, our simple model for associative and dissociative transition states ignores contributions from the solvent. This is particularly dangerous for water, because ordering of the solvation sphere in the ground versus transition states can contribute significantly to $\Delta S^‡$ (see Chapter 13 on Hard-Soft Acid Base). These effects can mask the details of the comings and goings of the ligands during the reaction in the entropy of activation; thus a more reliable predictor of mechanism is desirable.

To understand how $\Delta V^‡$ sheds light on mechanism, we will consider an associative reaction between H_2O and $[M(H_2O)_6]^{3+}$ and a dissociative reaction of H_2O from $[M(H_2O)_6]^{3+}$ in Fig. 17.17. In the associative (or I_a) mechanism, the volume of the activated complex is $V\{M(H_2O)_7{}^{3+}\}^‡$ and $\Delta V^‡ = V\{M(H_2O)_7{}^{3+}\}^‡ - V\{M(H_2O)_6{}^{3+}\} - V\{H_2O\}$. Because $V\{M(H_2O)_7{}^{3+}\}^‡ > V\{M(H_2O)_6{}^{3+}\}$, $V\{M(H_2O)_7{}^{3+}\}^‡ - V\{M(H_2O)_6{}^{3+}\} > 0$.

[f] This is the "electronic entropy". Entropy is related to the probabilities of occupying different states (Section 13.1). This also applies to electronic states. If the molecule has low-lying excited states that can be occupied, this can affect the entropy.

Table 17.2 Exchange rates and kinetic parameters for H_2O exchange in $[M(H_2O)_6]^{3+}$ complexes.

ElectronicConfiguration	$(t_{2g})^1$	$(t_{2g})^2$	$(t_{2g})^3$	$(t_{2g})^3\ (e_g)^2$	$(t_{2g})^6\ (e_g)^4$
$[M(H_2O)_6]^{3+}$	Ti^{3+}	V^{3+}	Cr^{3+}	Fe^{3+}	Ga^{3+}
ionic radius, Å	0.67	0.64	0.61	0.64	0.62
$M–O_{exp}$ (average, Å)[b]	2.025	1.991	1.961	1.994	1.962[c]
$M–O_{calcd}$ (average, Å)[b]	2.094	2.049	2.011	2.05	
$k_{H2O}{}^{298}$, s^{-1} [a]	1.8×10^5	5.0×10^2	2.4×10^{-5}	1.6×10^2	4.0×10^2
$\Delta H^‡$, kcal mol^{-1}	10.4	11.8	26.0	15.3	14.8
$\Delta S^‡$, cal K^{-1} mol^{-1}	0.3	-6.6	2.8	2.9	7.2
$\Delta V^‡$, cm^3 mol^{-1}	-12.1	-8.9	-9.6	-5.4	+5

[a] $k_{H_2O}^{298}$ = rate constant for water exchange at 298 K.[17]
[b] Transition-metal distances, from Kallies and Meier.[23]
[c] Ga–O distance, from Raston and coworkers.[26]

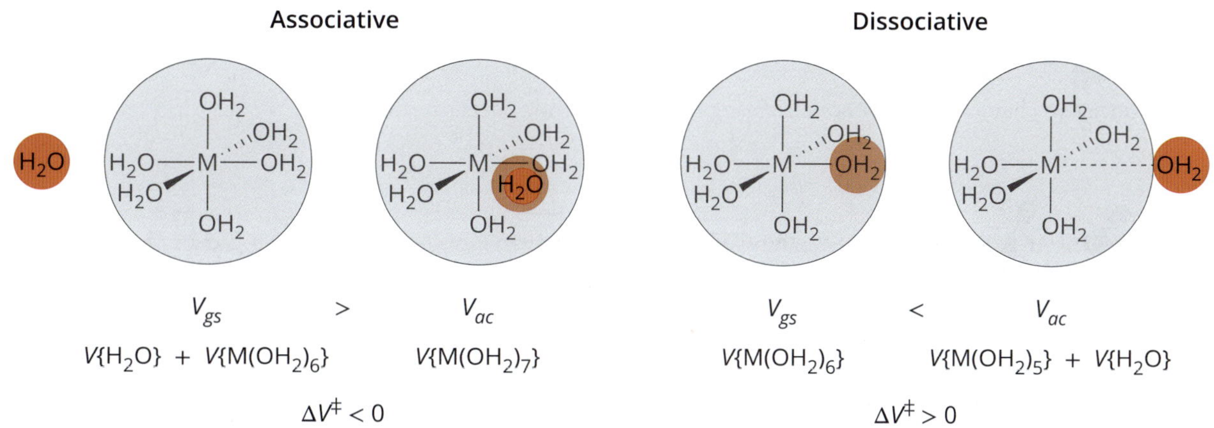

Fig. 17.17 Representations of $\Delta V^‡$ in associative and dissociative mechanisms for H_2O exchange in $M(H_2O)_6^{3+}$ complexes. The side with free water plus the complex has a larger volume because the water becomes partially buried in the complex volume on coordination.

However, because a significant portion of the H_2O ligand in $\{M(H_2O)_7^{3+}\}^‡$ can be considered to be buried within the coordination sphere of $\{M(H_2O)_6^{3+}\}$, the difference in volume between the activated metal complex and the ground-state aquo complex, $V\{M(H_2O)_7^{3+}\}^‡ - V\{M(H_2O)_6^{3+}\}$, will be less than $V\{H_2O\}$. This means that the sign of $\Delta V^‡$ will be negative with the magnitude $|\Delta V^‡| < V\{H_2O\}$. For a pure ***A*** mechanism, $\Delta V^‡$ is estimated to be approximately -13 cm^3 mol^{-1}, which is ~70% of the molar volume for H_2O of 18 cm^3 mol^{-1}.

For a dissociative process, the volume of the activated complex $V_{ac} = V\{M(H_2O)_5^{3+}\} + V\{H_2O\}$ will be larger than the volume of the ground-state complex $V\{M(H_2O)_6^{3+}\}$, V_{gs}. Therefore, in a dissociative mechanism, $\Delta V^‡ > 0$. Based on similar arguments to those for the associative reaction $|V[M(H_2O)_5^{3+}] - V[M(H_2O)_6^{3+}]| < V\{H_2O\}$, and $\Delta V^‡$ will reach an upper limit of approximately +13 cm^3 mol^{-1} for a pure ***D*** mechanism for H_2O dissociation.[14, 23]

Considering the data in Table 17.2 in light of Fig. 17.17, we can use $\Delta V^‡$ to classify the mechanisms for H_2O exchange for a given aquo complex. For example, $\Delta V^‡ = -5.4$ cm^3 mol^{-1} for $[Fe(H_2O)_6]^{3+}$, so its mechanism can be classified as $\boldsymbol{I_a}$. As we move to the left in the periodic table the amount of associative character increases, with $[Ti(H_2O)_6]^{3+}$ reaching what is nearly the value for a pure ***A*** mechanism. The only value of $\Delta V^‡$ in Table 17.2 that indicates dissociative character is for $[Ga(H_2O)_6]^{3+}$.

Exchange rates have been measured, and values for $\Delta H^‡$, $\Delta S^‡$, and $\Delta V^‡$ have been extracted, for the divalent hexaaquo complexes $[M(H_2O)_6]^{2+}$. These data are tabulated in Table 17.3. Again, we see a trend where $\Delta V^‡$ suggests $\boldsymbol{I_a}$ character for the lower d^n configurations, V^{2+} and possibly Mn^{2+}, with a switchover to $\boldsymbol{I_d}$ at Fe^{2+} and increasing dissociative character moving to Co^{2+} and Ni^{2+}.

The most important correlation is between mechanism and electronic structure. In ***A*** and $\boldsymbol{I_a}$ mechanisms, the incoming H_2O ligand will attack the metal by coordination of its lone pair to spatially accessible empty or half-filled metal orbitals. Consider the electronic structures for the sequence Ti-Cr for $[M(H_2O)_6]^{3+}$ (Fig. 17.18).

The lowest unfilled MOs for Ti and V are orbitals of t_{2g} symmetry and approximately nonbonding.[g] Because the coordinated H_2O ligands reside in the nodal planes of the t_{2g} orbitals, the incoming H_2O molecule can interact favorably with these orbitals in the activated complex. Consequently, a mechanism with associative character is favored for Ti^{3+} and V^{3+}. The mechanism for Cr^{3+} is also associative in this case, but it

[g] The t_{2g} orbitals are of correct symmetry to π-overlap with the ligands, which is likely to be fairly weak with water as a ligand. They are labelled as nonbonding (nb).

Table 17.3 Exchange rates and kinetic parameters for H_2O exchange in $M(H_2O)_6^{2+}$ complexes.[17]

Electronic Configuration	$(t_{2g})^3$	$(t_{2g})^3(e_g)^2$	$(t_{2g})^4(e_g)^2$	$(t_{2g})^5(e_g)^2$	$(t_{2g})^6(e_g)^2$
$[M(H_2O)_6]^{2+}$	V^{2+}	Mn^{2+}	Fe^{2+}	Co^{2+}	Ni^{2+}
ionic radius, Å	0.79	0.83	0.78	0.74	0.69
$k^{298}_{H_2O}$, s^{-1}[a]	87	2.1×10^7	4.4×10^6	3.2×10^6	3.2×10^4
$\Delta H^‡$, kcal mol^{-1}	14.8	7.9	9.9	11.2	13.6
$\Delta S^‡$, cal K^{-1} mol^{-1}	−0.1	1.4	5.1	8.9	7.6
$\Delta V^‡$, cm^3 mol^{-1}	−4.1	−5.4	3.8	6.1	7.2

[a] $k^{298}_{H_2O}$ = rate constant for water exchange at 298 K.

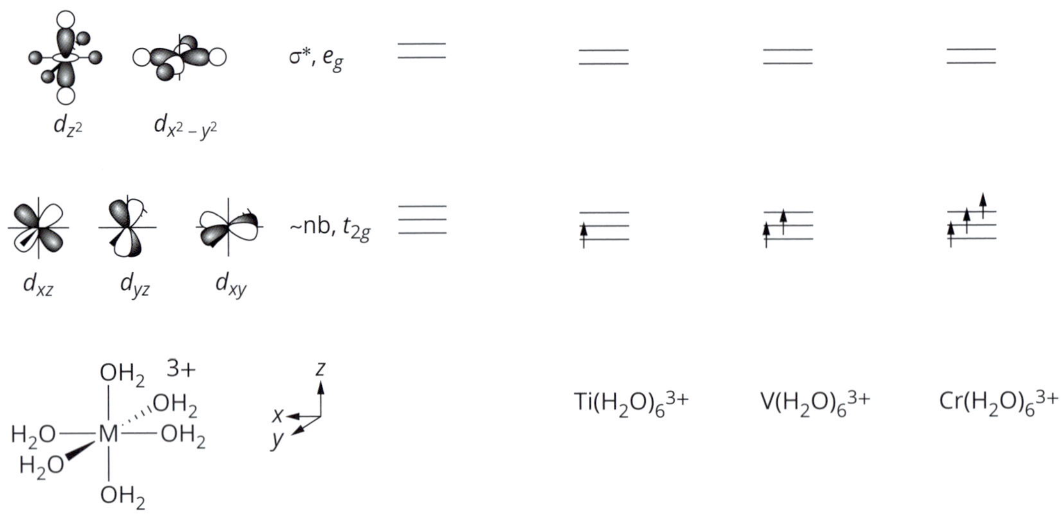

Fig. 17.18 Primitive MO diagrams for hexaaqua complexes of Ti-Cr in the 3+ oxidation state. Jahn–Teller distortions for Ti and V are being ignored.

is not obvious from the electronic structure that this would be true.[h]

From the data in Table 17.2, note that Ti has the more negative volume of activation and a faster overall rate of substitution, $k_{H2O}{}^{298}$. Two factors contribute to this. The first concerns electron–electron repulsions between the incoming H_2O ligand and the metal *d*-electrons. Since the V complex has an additional *d*-electron, electron–electron repulsions will be greater than for Ti, and the relative rate of substitution will be diminished. The second factor is the M–O bond distances. The Ti–O bonds in $[Ti(H_2O)_6]^{3+}$ are 2% longer than the V–O distances in $[V(H_2O)_6]^{3+}$. This means that the Ti coordination sphere is slightly more open. Therefore, both the steric repulsion between the incoming H_2O ligand and the coordinated H_2O ligands, along with the electron–electron repulsion between the H_2O electrons and the *d*-electron(s), will be smaller for Ti, making its rate of substitution faster than for V.

The situation for Cr^{3+} differs substantively from that for Ti^{3+} and V^{3+}. Phenomenologically, the rate of substitution is far slower—about 7 orders of magnitude slower. The fundamental origins of this difference can be understood from the differences in bonding and electronic structure for chromium. In Fig. 17.19 we can see that each MO in the t_{2g} set is singly occupied. The $\Delta V^{\ddagger}$ reflects significant associative character because the interaction between a filled oxygen lone-pair orbital in the incoming H_2O ligand and the singly occupied t_{2g} orbital has net electronic stabilization. However, electron–electron repulsions are unavoidable because the accessible t_{2g} orbitals each have one electron. Another factor that contributes to the diminished rate for Cr is that Cr–O distances in $[Cr(H_2O)_6]^{3+}$ are shorter than those in $[Ti(H_2O)_6]^{3+}$

[h] The associative character may be ascribed to the high positive charge. Mechanism of "inert" complexes and the factors that control their substitution are discussed in Section 17.5.3.3.

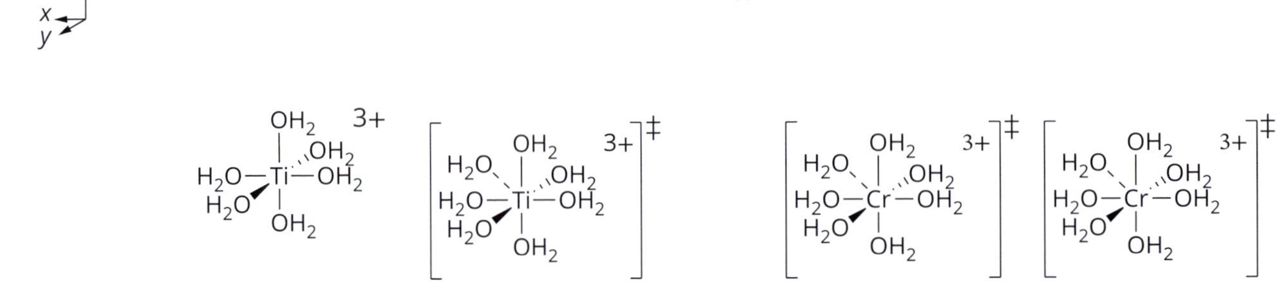

Fig. 17.19 Illustration of electronic stabilization in activated complexes for I_a substitution mechanisms for $[Cr(H_2O)_6]^{3+}$ and $[Ti(H_2O)_6]^{3+}$.

and $[V(H_2O)_6]^{3+}$. Therefore, both steric repulsions between the incoming water molecule and the coordinated H_2O ligands, along with the electronic repulsions between the H_2O electrons and the t_{2g} electrons, are exacerbated for Cr relative to Ti and V. This substantially increases the electronic barrier for Cr relative to Ti and V, which is reflected by the large increase in $\Delta H^‡$.

As we move further to the right for the 3*d* $[M(H_2O)_6]^{3+}$ complexes, we see that the reaction rate for exchange increases relative to that in $[Cr(H_2O)_6]^{3+}$ and $\Delta V^‡$ decreases. The change in $\Delta V^‡$ can be understood by again considering differences in electronic structures between Cr and Fe. Specifically, high-spin $[Fe(H_2O)_6]^{3+}$ has a $t_{2g}{}^3e_g{}^{*2}$ configuration where two electrons reside in M–O σ^* orbitals, which destabilizes the Fe–OH_2 bonds. While the net mechanism is associative, population of σ^* orbitals introduces bond-breaking character. In other words, the reaction is less associative in nature, I_a, which is reflected by a less negative $\Delta V^‡$.

Similar arguments apply for divalent hexaaqua complexes in Table 17.3. For example, the slowest exchange rate is for $[V(H_2O)_6]^{2+}$, which has a $(t_{2g})^3$ configuration like Cr^{3+}. Its $\Delta V^‡$ value is consistent with an $\boldsymbol{I_a}$ mechanism. Despite having identical electronic configurations, the magnitude of $\Delta V^‡$ is less for V^{2+} than Cr^{3+}. The size of $\Delta V^‡$ is diminished for V^{2+} because there is less charge on the complex. Therefore, the attraction between the H_2O lone pair and the metal will be lowered, and the reaction will be less associative. This means that both inter-ligand steric repulsion and electron–electron repulsion between the $(t_{2g})^3$ electrons and the lone pair of the approaching water molecule are greatly diminished for V^{2+} relative to Cr^{3+}. This leads to a reduction of $\Delta H^‡$ by 11.2 kcal/mol for V^{2+}. This lower enthalpy of activation for V^{2+} versus Cr^{3+} accounts for the seven orders of magnitude increase in k_{H2O} and much faster reaction rate for vanadium, despite the same electronic configuration.

Proceeding to the right in the periodic table, we note that the divalent Mn complex has the same electronic configuration as the trivalent Fe complex, and $\Delta V^‡$ for the two complexes are identical, consistent with an $\boldsymbol{I_a}$ mechanism. The ionic radius of Mn^{2+} is 30% larger than that of Fe^{3+}. Therefore, the same rationale that differentiates V^{2+} and Cr^{3+} applies, and k_{H2O} for Mn^{2+} is considerably faster than for Fe^{3+}.

As we move further to the right, the t_{2g} set continues to fill with electrons, and we see a change in sign for $\Delta V^‡$, signifying a switch in mechanism to $\boldsymbol{I_d}$ at Fe^{2+}. The crossover in mechanism occurs because further population of t_{2g} and decreases in the ionic radius make the $\boldsymbol{I_a}$ mechanism unfavorable relative to the $\boldsymbol{I_d}$ pathway. This is consistent with increasing $\boldsymbol{I_d}$ character in the order $Fe^{2+} < Co^{2+} < Ni^{2+}$.

Associative and dissociative pathways are two limiting routes for H_2O exchange, with $\boldsymbol{I_a}$, $\boldsymbol{I}$, and $\boldsymbol{I_d}$ mechanisms falling between these two extremes. Two reaction coordinates can be represented in two dimensions by creating More O'Ferrall–Jencks plots. As shown in Fig. 17.20, the origin of the plot is in the upper left corner. The horizontal axis represents one

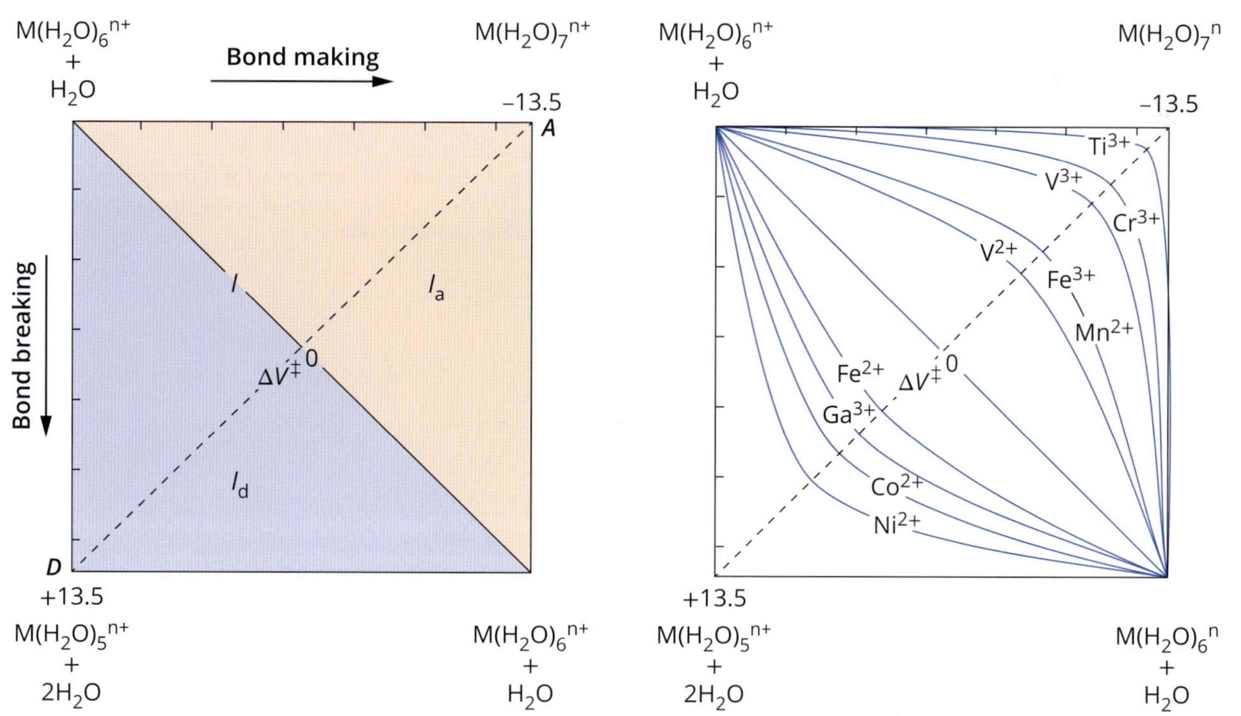

Fig. 17.20 More O'Ferrall–Jencks plots for associative (horizontal) and dissociative (vertical) H_2O exchange mechanisms. (*left*) The dashed diagonal is the axis for $\Delta V^‡$. The solid diagonal is the ***I*** pathway, which intersects the dashed diagonal at $\Delta V^‡ = 0$. The ***D*** and ***A*** mechanism transition states are at the lower left and upper right corners, respectively. All pathways in the blue shaded area are $\boldsymbol{I_d}$ mechanisms, and all pathways in the red shaded areas are $\boldsymbol{I_a}$ mechanisms. (*right*) Lowest-energy pathways for H_2O exchange of various $[M(H_2O)_6]^{n+}$ ions. For a particular pathway, the transition states lie at the intersection of the curved and dashed lines. At this point the dashed line shows the value of $\Delta V^‡$.[14]

reaction coordinate—in this case the ***A*** pathway, which is a bond-making process. The vertical axis represents the bond-breaking, ***D***, pathway. The solid diagonal that connects the upper left and lower right corners has equal ***A*** and ***D*** character. This line represents the concerted, ***I***, pathway. Any pathway that lies below this line corresponds to an I_d mechanism, and pathways that lie above the line are I_a mechanisms. This is represented in Fig. 17.20 (*left*). Reaction pathways for the ions in Tables 17.2 and 17.3 are depicted in Fig. 17.20 (right), with $[Ti(H_2O)_6]^{3+}$ having the strongest ***A*** character and $[Ni(H_2O)_6]^{2+}$ showing the highest ***D*** character.

17.5.3 Entering and Leaving Group Effects

Thus far, we have concerned ourselves with the rates at which reactions occur and considered only degenerate solvent exchange. Most inorganic compounds have more than one ligand type. In complexes like these, multiple pathways are available for ligand substitution. Consider a generic complex that contains *n* nonlabile spectator ligands, L, and two substitutable ligands X and Y. In a reaction with an external, third ligand, Z, two substitutions are possible corresponding to the equilibria.

$$L_nM(X)(Y) + Z \underset{k_{-1}}{\overset{k_1}{\rightleftharpoons}} L_nM(Z)(Y) + X \quad K_1 = \frac{k_1}{k_{-1}}$$

$$L_nM(X)(Y) + Z \underset{k_{-2}}{\overset{k_2}{\rightleftharpoons}} L_nM(X)(Z) + Y \quad K_2 = \frac{k_2}{k_{-2}}$$

The rates of the reactions will be determined by the magnitudes of k_1 and k_2, but the relative stabilities are governed by the equilibrium constants K_1 and K_2. Consequently, it is possible for product distributions to change as a function of time and temperature, depending on whether a reaction is under kinetic or thermodynamic control. In the reaction coordinate diagram in Fig. 17.21, we see that while the barrier for substitution favors initial substitution of X ($k_1 > k_2$), ultimately, thermodynamics will favor substitution of Y ($K_2 > K_1$), because $k_{-2} << k_{-1}$.[i] Initially, $L_nM(Z)(Y)$, the kinetic product, will form. When the system reaches equilibrium, the thermodynamic product, $L_nM(X)(Z)$, will be the final product.

Generally, the more basic ligand will be thermodynamically preferred, but the thermodynamics can vary as a function of ligand and metal. For example, the preference for halide binding for early transition metals follows the trend for Brønsted basicity, F > Cl > Br > I. However, for late transition metals the trend can reverse to I > Br > Cl > F in water. Many of these issues were discussed in Chapter 13 with hard–soft acid–base theory, but in short the difference is often associated with solvation effects and differences in lone pair bond weakening for the ligands.

17.5.3.1 Effects of Entering Groups (Advanced Topic)

The nature of the incoming ligand may influence the rate of reaction for all substitutions. This is even the case for certain I_d mechanisms that proceed via intimate ion pairs. An example that is thought to proceed via an ***A*** or I_a mechanism is the substitution of water in $[Ti(H_2O)_6]^{3+}$ by various ligands. The data in Table 17.4 show that the rate increases with the basicity of the entering ligand, as would be expected for an associative reaction.[27] Associative character in the substitution is also confirmed by a large negative $\Delta V^{\ddagger}$ for H_2O exchange.[28]

[i] Just as a reminder, the rate constant is inversely proportional to the barrier. The arrow with k_{-2} is associated with a large barrier, and therefore the rate constant is small.

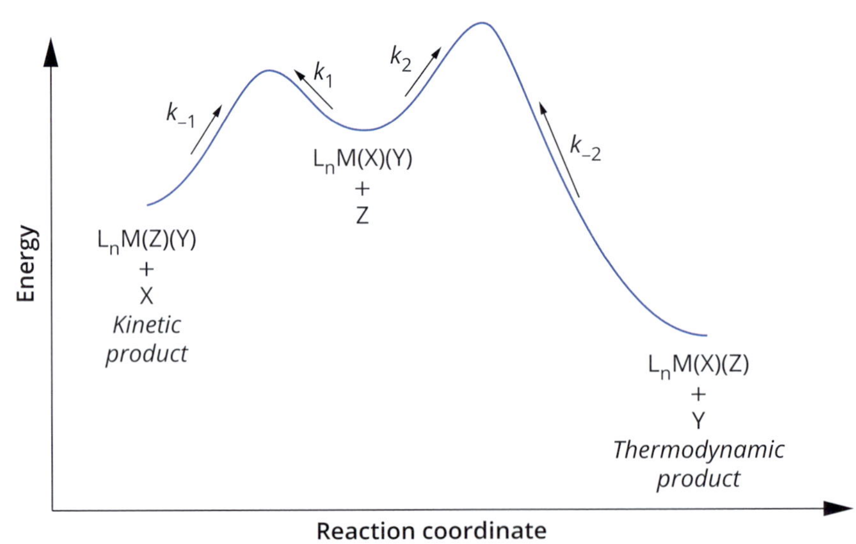

Fig. 17.21 An energy profile depicting kinetic and thermodynamic products.

Table 17.4 Entering group effects of substitution rates $[Ti(H_2O)_6]^{3+} + L \rightarrow [Ti(L)(H_2O)_5]^{3+} + H_2O$ in H_2O at 15 °C.[27]

L	k_{obs}, M^{-1} s^{-1}
$ClCH_2CO_2H$	0.7×10^3
CH_3CO_2H	0.7×10^3
NCS^-	8×10^3 [a]
H_2O	8.6×10^3 [b]
$Cl_2CHCO_2^-$	1.1×10^5
$ClCH_2CO_2^-$	2.1×10^5
$HO_2CC(H)CH_3CO_2^-$	3.2×10^5
$HO_2CCH_2CO_2^-$	4.2×10^5
$CH_3CO_2^-$	1.8×10^6

[a] Rate measured at 8–9 °C.
[b] Rate measured at 12 °C.[15]

The H_2O ligands in $[Ni(H_2O)_6]^{2+}$ are also labile and undergo substitution with various ligands. Some examples with rate constants, k_{obs}, are shown in Table 17.5. Unlike the previous Ti example, there is no correlation between rate and basicity of the entering ligand. Instead, the rates vary as a function of charge on the entering ligand. The data have been interpreted as being consistent with pre-equilibrium ion-pairing with H_2O loss, and ligand coordination proceeding from the ion-paired complex. This mechanism is depicted in Scheme 17.1.

$$Ni(OH_2)_6^{2+} + L^{n-} \underset{}{\overset{K_0}{\rightleftharpoons}} \{Ni(OH_2)_6, L\}^{(2-n)+} \xrightarrow{K_1} Ni(OH_2)^5 L^{(2-n)+} + H_2O$$

Scheme 17.1 The Eigen–Wilkins dissociative ion pairing mechanism for substitutions of $[Ni(H_2O)_6]^{2+}$.

Table 17.5 Values of K_O and k_I for the Ni(II) substitution reactions shown in Scheme 17.1.[29]

Ligand	K_O[a]	$k_I \times 10^{-4}$ (s^{-1})
$PO_3(OMe)^{2-}$	40[b]	1.5
$CH_3CO_2^-$	3	0.7
SCN^-	1[c]	3
F^-	1[c]	0.8
HF	0.15	2
H_2O		3[d]
NH_3	0.15	3
pyridine	0.15	3
imidazole	0.15	4
$NH_2(CH_2)_2NMe_3^+$	0.02	2

[a] Calculated using the Eigen–Fuoss equation.
[b] Experimentally determined.
[c] Different from $CH_3CO_2^-$ because of a different ionic strength.
[d] Included for reference. All of the experimentally determined exchange rates are near that for water, and the equilibrium constant K_O is the main factor differentiating the different ligands.

A crucial factor in the rate of these reactions is the equilibrium constant for the encounter complex formation. Eigen and Fuoss have developed a model to estimate the equilibrium constant based on the solvent, ionic strength of the solution, and the charges of the species involved, shown in Eq. 17.48. In these equations, N_A is Avogadro's number (6.022×10^{23}), a is the distance of center-to-center contact for the ions in cm, k_B is the Boltzmann constant (1.38×10^{-16} erg/K), e is the charge of the electron (4.803×10^{-10} esu), ε is the dielectric constant of the medium, and μ is the ionic strength.[18]

$$K_0 = \frac{4\pi N_A a^3}{3000} exp\left(-\frac{U}{k_B T}\right) \qquad 17.48$$

$$U = \frac{z_1 z_2 e^2}{\varepsilon}\left(\frac{1}{a(1-\kappa a)}\right)$$

$$\kappa = \sqrt{\frac{8\pi N_A e^2 \mu}{1000\varepsilon k_B T}}$$

While these equations may seem intimidating at first glance, the expression is not terribly complex on close inspection. The denominator in the parentheses for the expression defining U becomes the inverse of the distance ($1/a$) as κ goes to zero, which occurs as the ionic strength of the solution (μ) goes to zero. The form of the equation for U suggests that this is a Coulombic term with charges (z_1 and z_2) over the distance (simplifying to just a as the ionic strength goes to zero) and the dielectric (ε) of the medium. The equilibrium constant, K_O, contains terms for the volume of a sphere, ($4\pi r^3/3$), and an exponential containing the Coulombic energy term; compare $K_{eq} = exp(-\Delta G/RT)$.

An examination of how K_O changes as a function of the medium and sum of the charges, $z_1 z_2$, is informative (Table 17.6). Here, we assume $\mu = 0.01$, $a = 5 \times 10^{-8}$ cm (that

Table 17.6 Calculated (Eigen–Fuoss equation) formation constants for the ion pair, K_O, for some common charges in water, methanol, and dichloromethane.[a]

z_1	z_2	H_2O ($\varepsilon = 78.3$)	CH_3OH ($\varepsilon = 32$)	CH_2Cl_2 ($\varepsilon = 9.1$)
+1	−1	1.1	5.1	1300
+2	−1	3.7	83.2	5.3×10^6
+3	−1	12.6	1350	2.1×10^{10}
+1	−2	3.7	83.2	5.3×10^6
+2	−2	4.3	2.2×10^4	8.7×10^{13}
+3	−2	506	5.8×10^6	1.5×10^{21}
+1	−3	12.6	1350	2.1×10^{10}
+2	−3	506	5.8×10^6	1.45×10^{21}
+3	−3	2.0×10^4	2.5×10^{10}	9.9×10^{31}
0	0	0.32	0.32	0.32
+1	+1	0.09	.02	7.7×10^{-5}

[a] Assuming $\mu = 0.01$, $a = 5 \times 10^{-8}$ cm (5 Å), and $T = 298$ K.

is, 5 Å), and $T = 298$ K. If we assume that both species are charged, a +3 metal and a −1 ligand, $z_1 z_2 = -3$. The dielectric constant makes an enormous contribution, with the equilibrium constant in water being about 13 and in dichloromethane 21 billion under the same conditions for the same species in solution!

The rate law for substitution for the reaction shown in Scheme 17.1 is presented in Eq. 17.49. In the case of $PO_3(OMe)^{2-}$, the ion-paired intermediate can be detected, and values of K_0 and k_1 can be determined.[30] For the other ligands, the value of K_0 was calculated by using the Eigen–Fuoss equation and used to determine k_1 from the experimental data. As seen in Table 17.5, the values of K_0 reflect decreasing Coulombic attraction as the amount of negative charge on the entering ligand is decreased. The values for k_1 are all similar to the water self-exchange rate for Ni^{2+} in Table 17.3, 3.2×10^4 s^{-1}, which is consistent with rate limiting H_2O dissociation from an ion-paired complex. Consequently, as one may expect for a ***D*** or $\boldsymbol{I_d}$ mechanism, the actual ligand exchange is not strongly affected by the type of incoming ligand, but the formation of the encounter complex prior to ligand exchange is affected by the incoming ligand, largely through the charge.

$$\frac{-d[\{Ni(H_2O)_6\}^{2+}]}{dt} = \frac{K_0 k_1[L]}{1+K_0[L]}[\{Ni(H_2O)_6\}^{2+}] \qquad \textbf{17.49}$$

17.5.3.2 Effects of Leaving Groups (Advanced Topic)

In the case of complexes $[Co(NH_3)_5X]^{n+}$ (Eq. 17.50), reaction rates and equilibrium constants for acid hydrolysis have been measured for $X^- = F^-, H_2PO_4^{2-}, Cl^-, Br^-$, and NO_3^-.

$$Co(NH_3)_5X^{2+} + H_2O \underset{k_{-1}}{\overset{k_1}{\rightleftharpoons}} Co(NH_3)_5(H_2O)^{3+} + X^-$$
$$K_{eq} = k_1/k_{-1} \qquad \textbf{17.50}$$

If the mechanism is truly dissociative, the transition state, where the negatively charged ligand X is dissociating, should resemble the product where the ligand is completely dissociated. Thus, we would expect $\Delta G^‡$, the energy difference from the transition state to the starting material, and ΔG, the energy difference from the product to the starting material, to be correlated in a linear fashion where the change in $\Delta G^‡$, $\Delta\Delta G^‡$, is equal to the change in ΔG, $\Delta\Delta G$, times a constant β, where β varies from zero to one, $\beta = 0 \rightarrow 1$ (Eq. 17.51). When $\beta = 0$, there is no correlation between the transition state and the product. When $\beta = 1$, there is a complete correlation between the transition state and the product.

$$\Delta\Delta G^‡ = \beta\Delta\Delta G \qquad \textbf{17.51}$$

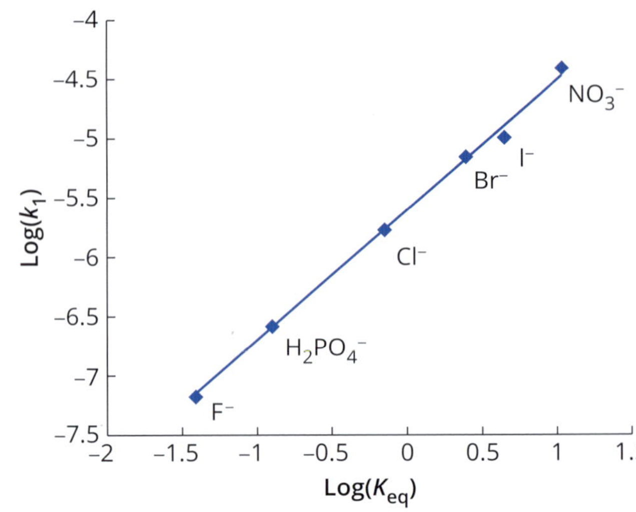

Fig. 17.22 Linear free energy relationship for rates and thermodynamics for acid-hydrolysis of $[CoX(NH_3)_5]^{2+}$.[16]

$\Delta G^‡$ is related to the rate constant k by Eq. 17.1, and ΔG is related to K_{eq} by Eq. 17.2. Thus, a plot of $\log(k_1)$ versus $\log(K_{eq})$ for the equilibria in Eq. 17.50 will have the form of Eq. 17.51. If the linear free energy relationship holds, the data can be fit to a line whose slope is β. Fig. 17.22 shows this relationship for k_1 and K_{eq} measured at 25 °C for replacement of X in $[CoX(NH_3)_5]^{2+}$ by H_2O.

As we can see from Fig. 17.22, the linear relationship between $\log(k_1)$ versus $\log(K_{eq})$ is excellent. The slope of the line gives $\beta = 1$, therefore the transition state resembles the product in having no X ligand, consistent with a ***D*** mechanism.[16] We note that as the Brønsted basicity of X^- decreases, the rate of substitution increases, and the equilibrium favors $[Co(NH_3)_5(H_2O)]^{3+}$. The trend in equilibria is reasonable for a dissociative substitution where the hard Co^{III} center will bind more basic ligands tightly. Likewise, the trend in rates is consistent with species that have weaker Co–X bonds being more reactive.

17.5.3.3 Case Study on Substitution Reactions on Inert Rhodium and Iridium Complexes: Arguments from Charge and Ligand Donation[31–32] (Advanced Topic)

We have discussed how electronic structure is a major driver in controlling rates in transition-metal complexes overall, but what factors affect substitution rates and mechanisms in "inert" complexes? The rate constants for water exchange in these complexes can vary over about 10 orders of magnitude. What affects the rates of substitution and the mechanism for these complexes?

As an example of substitution in octahedral, inert complexes, consider water exchange in $[Ir(OH_2)_6]^{3+}$ by labeled water, O*H$_2$. Obviously, the reaction is approximately thermoneutral, and the iridium center is d^6 low-spin. By Taube's criterion, the complex is inert, and in this case inert is a particularly good descriptor, as this metal ion has one of the slowest rates known for water exchange. The mechanism of the reaction has been somewhat contentious but seems to be resolved for the moment. Merbach's 2004 study was quite exhaustive in this case, and many of the details will be presented here as an example of how substitution mechanisms may be studied.[32]

For the first-row transition metals, the mechanism of substitution in 6-coordinate complexes changes from $\boldsymbol{I_a}$ to $\boldsymbol{I_d}$ as one moves from left to right and the number of d-electrons increases and as the atomic radius decreases. Studies on second- and third-row transition metals have proven less predictable. Investigation of anation (Cl^-, Br^-, and I^-) of low-spin $[Ru(OH_2)_6]^{2+}$ gave $\Delta V^\ddagger = -0.4 \pm 0.7$ cm^3/mol. For low-spin d^6 metals of this type, the filled t_{2g} set is expected to strongly repel any ligand approaching the open faces of the octahedron, while the empty $e_g{}^*$ set implies stable metal–ligand bonds unlikely to dissociate. Apparently, $[Ru(OH_2)_6]^{2+}$ solves this problem with a nearly pure interchange mechanism that is neither associative nor dissociative, as indicated by a $\Delta V^\ddagger \sim 0$.

The $[Rh(OH_2)_6]^{3+}$ and $[Ir(OH_2)_6]^{3+}$ substitution mechanisms have a more colorful history. In the 1960s and 1970s it was decided that these metals used a $\boldsymbol{D}$ mechanism in their substitutions. This was decided based largely on identical rate constants and similar enthalpies of activation for $[Rh(OH_2)_6]^{3+}$ reacting with both Cl^- and Br^-.[j] Later, Merbach and coworkers found that the volumes of activation for the reactions showed a negative value for $[Rh(OH_2)_6]^{3+}$ (−4.2 cm^3/mol), and an even more negative value for $[Ir(OH_2)_6]^{3+}$ (−5.7 cm^3/mol), which led to the conclusion that the complexes use an $\boldsymbol{I_a}$ mechanism.

They also carried out computational studies on these systems at the Hartree–Fock level and found that the $\boldsymbol{I_a}$ mechanism was lower in energy than the $\boldsymbol{D}$ mechanism for Rh^{3+} and Ir^{3+}. The calculated transition state is shown in Fig. 17.23. In the bottom structure, the arrows show the direction of motion of the negative frequency. Because the top of the transition state is a peak, there is one negative frequency showing the direction of motion along the reaction pathway. In this case, the interchange mechanism shows one water entering as the other leaves.

Merbach and coworkers computed the pathways for the rhodium(III), iridium(III), and ruthenium(III) substitutions. In the case of the Rh^{3+} and Ir^{3+} systems, they found transition states for interchange, like those shown in Fig. 17.23. The dissociative mechanistic path was also calculated and had a significantly higher energy, Fig. 17.24.

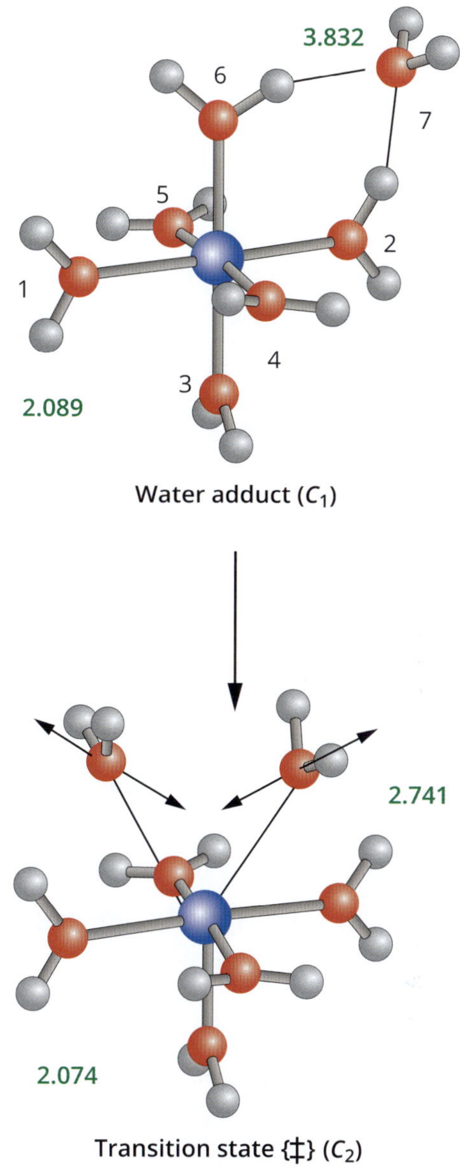

Fig. 17.23 Calculated (Hartree–Fock) geometry of the water adduct (*top*) and the transition state for ligand interchange (*bottom*), with arrows showing the nature of the negative frequency in the transition state for water exchange in $[Ir(OH_2)_6]^{3+}$. The green numbers on the right show the distance between the incoming water and Ir, while the green numbers on the left show the average bond distance of the spectator ligands.

In the case of ruthenium(II), no interchange pathway was found to compare with experiment. Experimentally, the Ru^{2+} system clearly has more dissociative character than the Rh^{3+}/Ir^{3+} systems ($\boldsymbol{I}$ vs $\boldsymbol{I_a}$). This was attributed to the higher charge of the Group 9 metals, leading to stronger metal–ligand bonds than the ruthenium(II) complex.

In general, for "inert" complexes of the second and third row, one finds that complexes bearing only weakly donating

[j] As discussed previously, insensitivity to the incoming ligand is a hallmark of a dissociative mechanism.

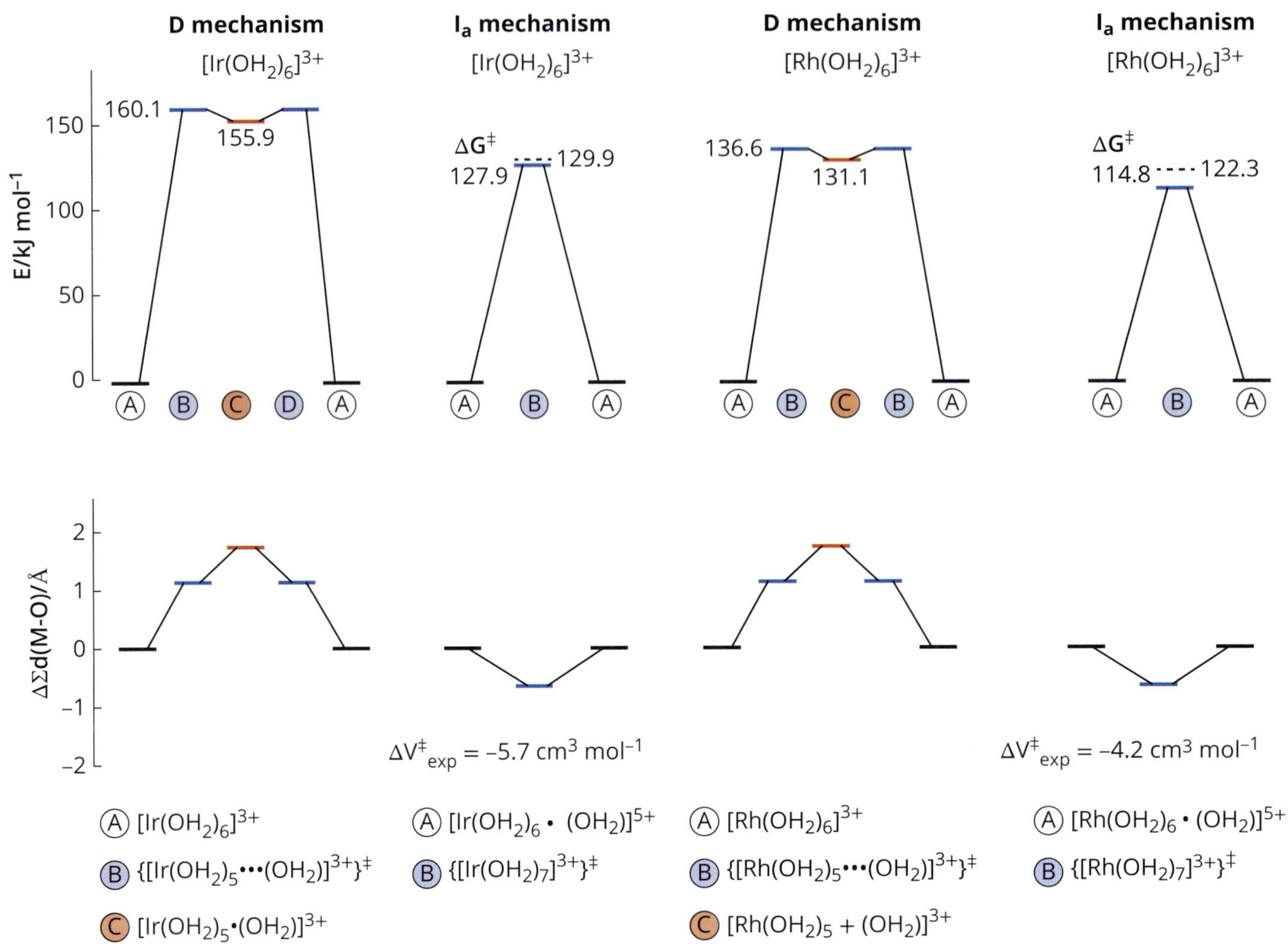

Fig. 17.24 Energies and bond distances for reaction profiles for water substitution in $[Ir(OH_2)_6]^{3+}$ and $[Rh(OH_2)_6]^{3+}$. The I_a mechanisms are clearly lower in energy for iridium(III) and rhodium(III), consistent with experiment.

ligands (neutral ligands such as water) tend to favor ***I*** or I_a mechanisms, and those bearing more strongly donating ligands (anionic ligands such as halides and cyclopentadienyl in addition to water) tend to favor I_d or ***D*** mechanisms for weak ligand (water, DMSO, and so on) exchange (Table 17.7). In addition, one finds that increasing the oxidation state of the metal seems to make the reaction more associative by increasing metal–ligand interaction strengths. For example, the volume of activation indicates a more associative reaction for exchange in $[Ru(OH_2)_6]^{3+}$ than in $[Ru(OH_2)_6]^{2+}$, and the latter is an almost pure interchange reaction. Consequently, for complexes with the same ligands, the higher oxidation-state complex tends to have a more associative reaction mechanism, and at least in this case a smaller rate constant.

For Ru(II), the η^6-benzene dication complex $[Ru(\eta^6\text{-}C_6H_6)(NCMe)_3]^{2+}$ has nearly a pure interchange mechanism for acetonitrile exchange, and the complex containing the more donating cyclopentadienyl anion $[Ru(\eta^5\text{-}Cp)(NCMe)_3]^+$ has a clearly dissociative mechanism. Similar observations can be made in the Ir(III) and Rh(III) complexes; the more electron-rich complexes tend to undergo more dissociative exchange mechanisms, often with reaction-rate constants many orders of magnitude larger.

In these inert complexes, general effects of the ancillary ligands in changing substitution rates can be seen, while more electron-deficient complexes due to either high oxidation state or less donating ligands can be seen to clearly affect reaction rate and mechanism. In the next section, more specific effects related to spectator ligands are discussed.

Table 17.7 Mechanisms of solvent exchange for some "inert" complexes as a function of spectator ligands.[a]

Complex	Metal Oxidation State	Configuration	k_{exch} (298 K, s^{-1})	$\Delta V^{\ddagger}$ (cm^3/mol)	Proposed Mech.
$[Ru(OH_2)_6]^{2+}$	Ru(II)	$(t_{2g})^6$	1.8×10^{-2}	−0.4	I
$[Ru(OH_2)_6]^{3+}$	Ru(III)	$(t_{2g})^5$	3.5×10^{-6}	−8	I_a
$[Ru(\eta^6\text{-}C_6H_6)(NCMe)_3]^{2+}$	Ru(II)	$(t_{2g})^6$	4.07×10^{-5}	+2.4	I_d
$[Ru(\eta^5\text{-}Cp)(NCMe)_3]^{+}$	Ru(II)	$(t_{2g})^6$	5.6	+11.1	D
$[Rh(OH_2)_6]^{3+}$	Rh(III)	$(t_{2g})^6$	2.2×10^{-9}	−4.2	I_a
$[RhCp^*(NCMe)_3]^{2+}$	Rh(III)	$(t_{2g})^6$	37	+0.8	I
$[RhCp^*(DMSO)_3]^{2+}$	Rh(III)	$(t_{2g})^6$	3600	+3.3	I_d
$[Ir(OH_2)_6]^{3+}$	Ir(III)	$(t_{2g})^6$	1.1×10^{-10}	−5.7	I_a
$[Ir(NH_3)_5(OH_2)]^{3+}$	Ir(III)	$(t_{2g})^6$	6.4×10^{-8}	−3.2	I_a
$[Ir(Cp^*)(OH_2)_3]^{2+}$	Ir(III)	$(t_{2g})^6$	25000	+2.4	I_d

[a] Data from Helm and Merbach's review.[14]

17.5.4 The Roles of Spectator Ligands in Substitution Reactions

Spectator ligands (also called ancillary ligands—those inert to substitution under the reaction conditions) can influence the reaction rates of labile ligands in the metal coordination sphere. Consequently, the choice of spectator ligands is important when designing reaction chemistry of inorganic compounds. Let us consider H_2O substitution chemistry for Ru aquo complexes of the formula $[Ru(L)(H_2O)_5]^{2+}$, where L = H_2O, MeCN, Me_2SO, $CH_2{=}CH_2$, CO, and $CH_2{=}CF_2$ (Fig. 17.25).[33] In the hexaaquo complex (L = H_2O), all of the H_2O ligands are identical. For the complexes where L ≠ H_2O there are two chemically distinct H_2O environments as one water is bound *trans* to L, while the remaining four occupy *cis* sites. The rates of H_2O exchange have been monitored by using ^{17}O NMR as functions of L, temperature, and pressure.[33] From these studies, it has been shown that ligand substitutions occur via an ***I*** or ***I*****d** mechanism, and that the relative and absolute rates for *trans*- and *cis*-substitution vary as a function of L. The rates for substitution of the *trans*-ligand are much higher, and we will confine our discussion to this process.

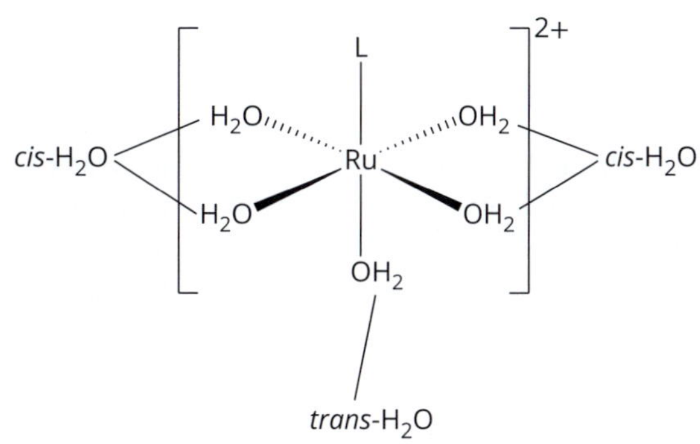

	MeCN	< H_2O	< CO	< $OSMe_2$	< $H_2C{=}CH_2$	< $F_2C{=}CH_2$
k_{trans}/k_{H2O}	0.3	1	2.5	45	173	373

	H_2O	< MeCN	< CO	< $F_2C{=}CH_2$	< $H_2C{=}CH_2$	< $OSMe_2$
$d(Ru\text{–}O_{trans}) - d(Ru\text{–}O_{hexaaquo})$	0	0.015	0.046	0.054	0.061	0.078

Fig. 17.25 The *trans*-effect and *trans*-influence in $[Ru(L)(OH_2)_5]^{2+}$ complexes. The *trans*-influence is measured by the change in Ru-OH_2 distance to the water *trans* to L minus the distance in $[Ru(OH_2)_6]^{2+}$. The distances (*d*), listed in Ångstroms, were calculated by using DFT (M06L/SDD), as only a couple are structurally characterized: L = OH_2 and CO.[34-35] The energy of the *O*-bound DMSO complex was significantly lower (~24 kcal/mol) than the *S*-bound DMSO complex in the calculations, so the former's structure was used. The relative *trans*-effect is the experimental rate constant for substitution of the *trans*-water relative to substitution in $\{Ru(OH_2)_6]^{2+}$.[33]

As shown in the following, the rates span three orders of magnitude. The substitution rates increase as the calculated binding energies of the *trans*-H_2O ligand decrease (L = H_2O > CO > CH_2=CH_2) and as the calculated π-backdonation increases (L = H_2O < CO < CH_2=CH_2) for this system.

Strong σ-donors generally weaken bonds that are *trans*-to them, because they compete for the same σ-bonding metal orbital (Fig. 17.25). This thermodynamic phenomenon of the *trans*-ligand weakening, a bond is termed "*trans*-influence" and is typically estimated by how much metal–ligand bond distances change as a function of the ligand in the *trans*-position. In the previous example, where the Ru–OH_2(*trans*) distance was determined computationally (M06L/SDD) since only a couple of the compounds in the series have been structurally characterized, (O-bound) DMSO has the strongest *trans*-influence and H_2O the weakest, which suggests that DMSO is the strongest σ-donor and H_2O is the weakest.

Weakening of the *trans*-H_2O bond being substituted can be viewed as destabilizing the ground state, which reduces the energy of activation and can increase the rate of substitution. However, σ-and π-bonding can also affect energies of the transition state, and *trans*-influence is not the only factor that affects rate of substitution (*trans*-effect), as can easily be seen by examining the two series in Fig. 17.25, which have quite different orders.

The "*trans*-effect", a kinetic phenomenon related to the influence of a ligand *trans* to the substituting ligand on reaction rate, and *trans*-influence, a thermodynamic effect where the *trans* ligand destabilizes the substituent *trans* to it, have also been studied systematically in detail for square-planar compounds, which are discussed separately. Consequently, we will return to these spectator ligand effects later in this chapter.

17.5.5 Stereochemical Probes of Mechanism

The importance of stereochemistry in elucidating mechanisms in organic chemistry is well documented. Six-coordinate metal complexes have an even richer stereochemistry first recognized by Werner, who prepared carbon-free optical isomers to debunk the notion that optical activity was a special property of carbon (Chapter 10).[36] Consider the Co^{III} complexes of the formula $[Co(en)_2AX]^{n+}$, where A ≠ X and n = 1 or 2, depending on whether A is negatively charged or neutral. (X will be considered to have a negative charge as the free ligand, such as a halide.) Three stereoisomers are possible: *trans*-$[Co(en)_2AX]^{n+}$, and a pair of *cis*-$[Co(en)_2AX]^{n+}$ enantiomers (Fig. 17.26).

Now consider substitution of X by H_2O under acidic conditions. In an associative pathway an external H_2O ligand would attack a triangular face of the octahedron. The outcomes for associative substitution pathways are shown in Fig. 17.27. Both *trans*- and *cis*- isomers of $[Co(en)_2AX]^{n+}$ would be expected to produce a mixture of *cis*- and *trans*-$[Co(en)_2A(H_2O)]^{n+2}$ products. For *trans*-$[Co(en)_2AX]^{n+}$, the *cis*-product that would result from attack at an A-N-N face will be racemic. In the case of *cis*-$[Co(en)_2AX]^{n+}$, two distinct pathways to *cis*-products are possible. Attack at an A–N–N face will invert the stereochemistry at the metal center, whereas attack at an X–N–N face will preserve it. It is difficult to predict *a priori* whether a retention or inversion pathway would be favored.

In a dissociative pathway, 5-coordinate intermediates will be generated. The geometry of the 5-coordinate intermediates can be either square pyramidal or trigonal bipyramidal. As shown in Fig. 17.28, dissociation of X to generate a stereochemically rigid square pyramidal structure will give products in which the stereochemistry is preserved. If a trigonal bipyramidal structure is preferred, both *cis*- and *trans*-$[Co(en)_2AX]^{n+}$ are expected to produce mixtures of *cis*- and *trans*-products. In the event that identical trigonal bipyramidal intermediates are formed, the isomer ratios of products will be the same, regardless of whether *cis*- or *trans*-$[Co(en)_2AX]^{n+}$ starting materials are used.

The fate of the substitution has been examined for a range of A and X ligands. In general, the reaction is found to be stereospecific with *trans*-reactants giving *trans*-products and *cis*-reactants giving *cis*-products. For enantiomerically pure *cis*-isomers, the reaction proceeds with retention of stereochemistry at the metal.[37] These results strongly support a ***D*** or $\boldsymbol{I_d}$ mechanism for substitution proceeding through a stereochemically rigid, square pyramidal $[Co(en)_2A]^{n+2}$ intermediate. It is thought that many other $[Co^{III}(L)X]^{n+}$ complexes undergo substitution of X by similar pathways under acidic conditions.[38]

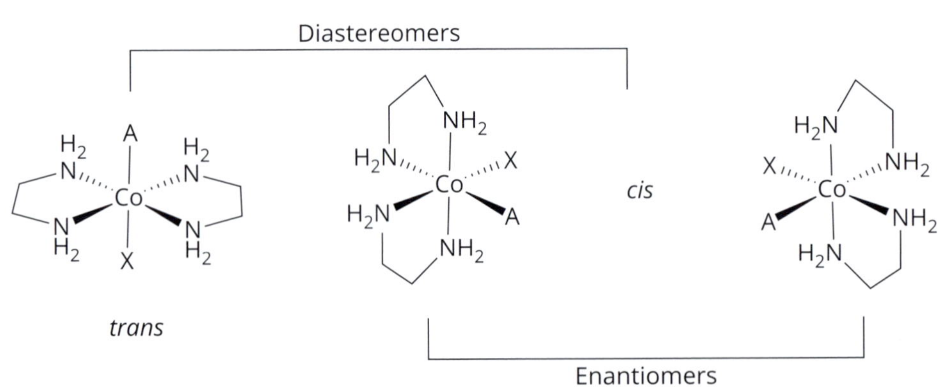

Fig. 17.26 Possible stereoisomers for the formula $[Co(en)_2AX]^{n+}$ (en = $NH_2CH_2CH_2NH_2$).

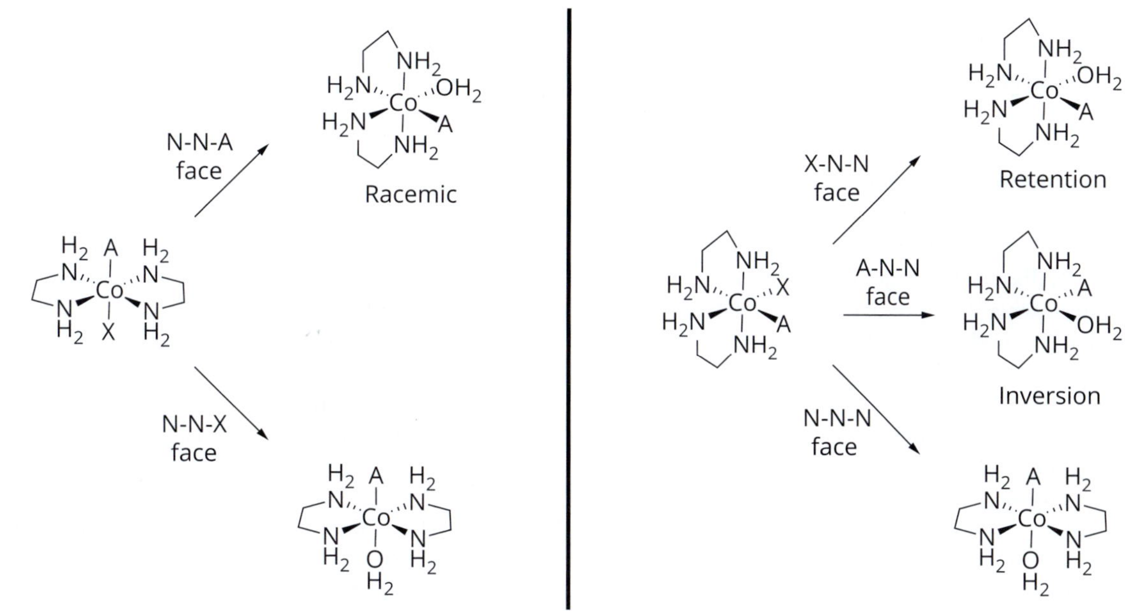

Fig. 17.27 Stereochemical outcomes for associative pathways for water substitution on *trans*- and *cis*-[Co(en)$_2$(A)(X)] complexes.

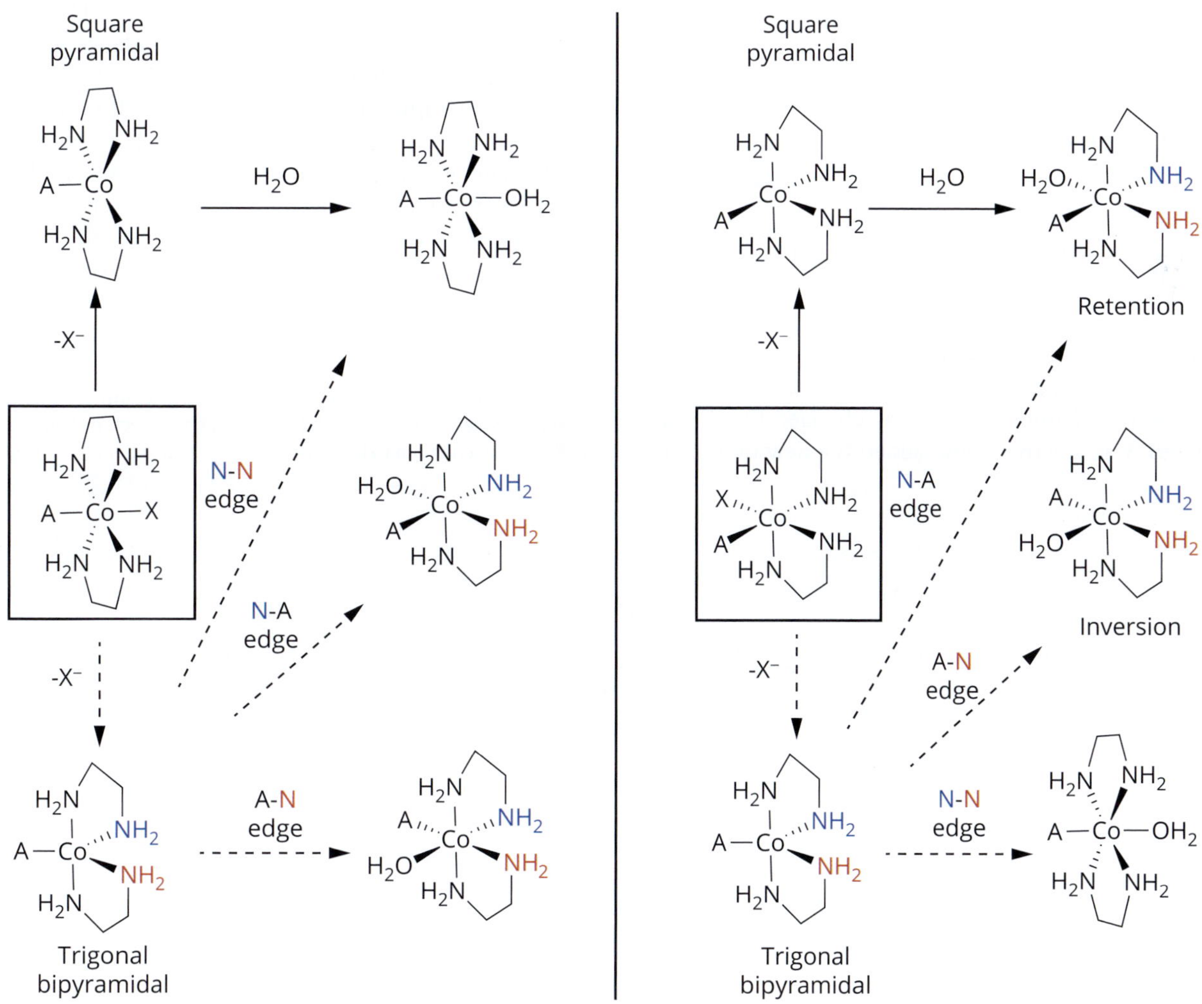

Fig. 17.28 Stereochemical outcomes for associative pathways for hydrolysis of *trans*- and *cis*-[Co(en)$_2$(A)(X)] complexes. The solid arrows indicate pathways that are consistent with experimental results.

17.5.6 Acid- and Base-Catalyzed Substitution Reactions (Advanced Topic)

Rates and mechanisms of substitution reactions are known to be affected by acids and bases. In the case of acids, protonation of a leaving group X can accelerate the rates for substitution of both associative and dissociative reactions. This effect tends to be more pronounced when the product HX is a relatively weak Brønsted acid. For example, the substitution of the acetate ligand in $[Cr(H_2O)_5(OAc)]^{2+}$ follows a two-term rate law.[39] Any time there are two terms in a rate law, such as $k_1 + k_2[H^+]$ in Eq. 17.52, it indicates two competing pathways for the reaction. The first term of the rate law in Eq. 17.52 is first-order in metal. The second term is first-order in metal concentration *and* $[H^+]$. This pathway corresponds to protonation of the unbound oxygen of the acetate ligand to generate an intermediate with a coordinated acetic acid ligand, which is much less basic than coordinated acetate due to loss of anionic charge. Taube found that $k_1 = (0.41 \pm 0.15) \times 10^{-6}\ s^{-1}$ and $k_2 = (7.70 \pm 0.30) \times 10^{-6}\ s^{-1}M^{-1}$ at ionic strength = 1.00 and 25 °C.

$$(AcO)Cr(OH_2)_5^{2+} \xrightarrow[OH_2]{k_1} Cr(OH_2)_6^{3+} + {}^{-}OAc \qquad \textit{1st mechanism}$$

$$H^+ + (AcO)Cr(OH_2)_5^{2+} \xrightarrow[k_2]{OH_2} Cr(OH_2)_6^{3+} + HOAc \qquad \textit{2nd mechanism}$$

$$\frac{-d[\{Cr(H_2O)_5(OAc)\}^{2+}]}{dt} = (k_1 + k_2[H^+])[\{Cr(H_2O)_5(OAc)\}^{2+}] \qquad \mathbf{17.52}$$

Hughes and Ingold created the S_N1 and S_N2 labels for organic chemistry,[40] and originally the same labels were used in inorganic chemistry. Langford and Gray changed the S_N1 and S_N2 to ***D*** and ***A***, respectively, and added the new interchange mechanism (***I***) to make the labels more descriptive and account for the subtleties observed in the inorganic systems.[41]

Cobalt ammine complexes have a history that predates even Werner, and their substitution reactions have been studied in great detail. The reaction of $[Co(NH_3)_5Cl]^{2+}$ with various bases to substitute chloride had been extensively studied. Brown, Nyholm, and Ingold had declared that "the mechanism is well established" and "is certainly a bimolecular nucleophilic substitution S_N2."[42] Basolo had noticed that the reaction was sluggish with some bases when properly buffered, and to account for this he proposed the mechanism shown in Fig. 17.29, which he called an "S_N1CB" (CB = conjugate base). Later, Basolo and coworkers disproved the S_N2 mechanism espoused by Ingold and coworkers through a series of elegant experiments, while all the data were consistent with his "S_N1CB" route.[k,3]

Basolo's rate law for the mechanism is shown in Eq. 17.53. Because $[H^+]$ and $[OH^-]$ are inversely related, rate laws involving water that have an inverse dependence on $[H^+]$ are often catalyzed by OH^-. For the substitution of X^- in $[Co(NH_3)_5X]^{2+}$ at high pH (Fig. 17.29), substitution follows a two-term rate law. The first term in Eq. 17.53 is first-order in $[Co(NH_3)_5X^{2+}]$ and corresponds to a typical dissociative pathway. The second term is first order in $[Co(NH_3)_5X^{2+}]$ and $[OH^-]$ (or equivalently, inverse in $[H^+]$). The $[OH^-]$ dependence is due to pre-equilibrium deprotonation of $[Co(NH_3)_5X]^{2+}$ to generate the amido tetraamine intermediate, $[Co(NH_2)(NH_3)_4X^+]$. Dissociation of X^- from this species forming the 5-coordinate intermediate $[Co(NH_2)(NH_3)_4]^{2+}$ is highly favorable; amides (NH_2^-) are much stronger donors than ammines, strongly favoring ligand dissociation. Furthermore, the rate for this pathway is accelerated relative to the uncatalyzed dissociative process because the NH_2 ligand can form a π-bond to Co, stabilizing the 5-coordinate intermediate.

[k] Note that this is the typical strategy in mechanistic chemistry. One cannot prove a particular mechanism to be correct, but you can prove a different mechanism to be incorrect.

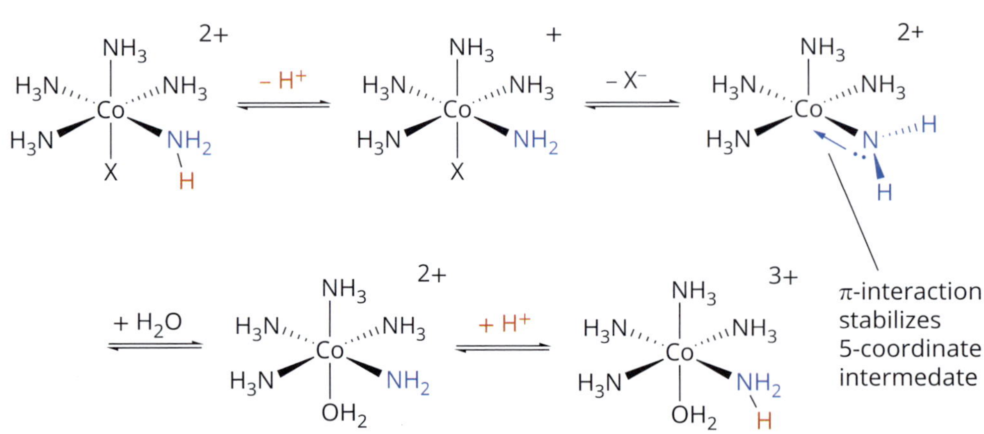

Fig. 17.29 Mechanism for base-catalyzed X substitution in $[Co(NH_3)_5X]^{2+}$ complexes, with the proton undergoing movement in red and the NH_2 group being protonated/deprotonated in blue.

$$\frac{-d[\{Co(NH_3)_5X\}^{2+}]}{dt} = (k_1 + k_2[H^+]^{-1})[\{Co(NH_3)_5X\}^{2+}] \quad \mathbf{17.53}$$

A similar mechanism for ruthenium(III) has been observed for water substitution in $[Ru(OH_2)_6]^{3+}$.[43] The ruthenium(II) case was discussed previously and is assigned as being very nearly a pure ***I*** mechanism. Ruthenium(III) is still a Taube "inert" system with an electron configuration of $(t_{2g})^5$, but the additional charge might suggest that a more associative mechanism would be found, similar to Rh(III) and Ir(III). In fact, this turns out to be the case, and with $\Delta V^{\ddagger} = -8.3 \pm 2.1$ cm^3/mol, the mechanism was assigned as $\boldsymbol{I_a}$. The magnitude suggests a relatively large amount of associative character. As reference, the volume of activation for titanium(III) seems to be one of the largest negative values for a hexa(aquo) ion at –12.1 cm^3/mol, which is near where some predict the limit is for water exchange (–13.9 cm^3/mol) in a purely associative process.[44] Consequently, the $\boldsymbol{I_a}$ label was given to the Ru(III) case.[43]

The rate law for exchange in $[Ru(OH_2)_6]^{3+}$ was found to involve a two-term rate law (Eq. 17.54). There is a direct exchange with the parameters discussed previously and rate constant $k_1 = (3.5 \pm 0.3) \times 10^{-6}$ s^{-1} in water at 298 K. The second term has an inverse in [H$^+$] term; again, this usually indicates a process involving catalytic base. In this case, a "conjugate base" mechanism is proposed where one water ligand is deprotonated to form $[Ru(OH)(OH_2)_5]^{2+}$; that is, the "S_N1CB" mechanism in Fig. 17.29, identified by Basolo. For this second pathway, $\Delta V^{\ddagger} = +0.9$ cm^3/mol was calculated from the experimental data, which suggests a far more dissociative mechanism for this more electron-rich ruthenium hydroxide species.[43]

$$v = (k_1 + k_2 / [H^+])[Ru(OH_2)_6]^{3+} \quad \mathbf{17.54}$$

Table 17.8 shows some data for several main-group and transition-metal complexes in the +3 oxidation state. In all cases, a two-term rate law was identified involving direct water exchange, and a "conjugate base" mechanism involving the metal hydroxide. In all cases, the CB mechanism has a significantly larger rate constant and consequently is faster but dependent on the pH of the experiment. The conjugate base mechanisms are always significantly more dissociative according to the volume of activations.

17.6 Ligand Substitution in Square-Planar Complexes

Another class of compounds whose substitution chemistry has been intensely studied is square-planar ML_4 d^8 complexes. The most common members of this family are compounds of Ni^{II}, Pd^{II}, Pt^{II}, Rh^{I}, and Ir^{I}. As we have seen in previous chapters, ML_4 d^8 complexes have sixteen valence electrons—two electrons short of the 18-electron noble-gas configuration. Since these complexes are electron-deficient, they might be expected to react via ***A*** or $\boldsymbol{I_a}$ mechanisms. Typically, these complexes undergo associative mechanisms, but this cannot be assumed, as there are known exceptions.[21] In this section, some prototypical cases will be examined.

17.6.1 Associative and Dissociative Substitutions

Consider the reaction of $[(dien)PtCl]^+$ {dien = $(NH_2CH_2CH_2)_2NH$} (A) with pyridine to form B in Eq. 17.55.

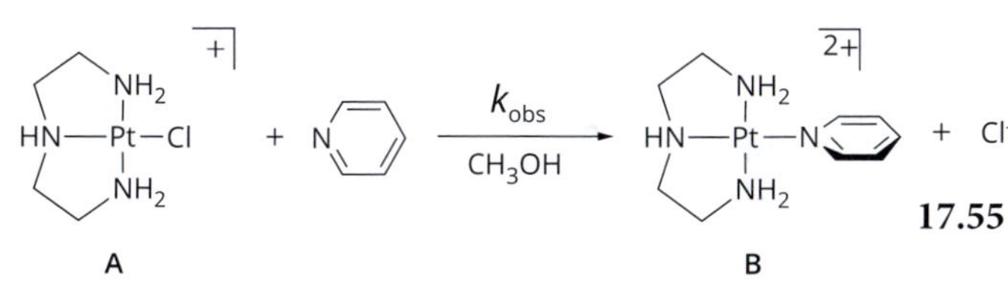

17.55

By varying concentrations of A, pyridine, and chloride ion, the rate law can be formulated as follows.

$$\frac{-d[A]}{dt} = (k_1 + k_2[py]^1)[A]^1[Cl^-]^0$$

Table 17.8 Data for M(III) aqueous cations undergoing water exchange.[a,b]

	k/s^{-1}	k_2/k_1	$\Delta H^{\ddagger}$ (kcal/mol)	$\Delta S^{\ddagger}$ (cal/mol K)	$\Delta V^{\ddagger}$ (cm^3/mol)	pK_a
Ga^{3+}	4.0×10^2	275	16.0	7.20	+5.0	~3.9
$Ga(OH)^{2+}$	1.1×10^5		14.1		+6.2	
Fe^{3+}	1.6×10^2	750	15.3	2.89	–5.4	2.9
$Fe(OH)^{2+}$	1.2×10^5		10.1	1.3	+7.0	
Cr^{3+}	2.4×10^{-6}	75	26.0	2.77	–9.6	4.1
$Cr(OH)^{2+}$	1.8×10^{-4}		26.3	13.3	+2.7	
Ru^{3+}	3.5×10^{-6}	170	21.5	11.5	–8.3	2.7
$Ru(OH)^{2+}$	5.9×10^{-4}		22.9	3.56	+0.9	

[a] All the systems have rate laws of the type $(k_1 + k_2/[H^+])[M(OH_2)_6]^{3+}$ with two pathways: one involving direct exchange related to k_1, and the other in which a coordinated water is deprotonated to form $M(OH)(OH_2)_6{}^{2+}$ related to k_2. The temperature is 298 K.
[b] (Data from Rapaport, Helm, Merbach, Bernhard, and Ludi.[43])

The rate law tells us that the reaction is first-order in A. The k_1 term arises from a solvent (MeOH) assisted pathway, and the k_2 term is first-order in [py]. The rate is unaffected by chloride ion concentrations, and an absence of any effect of leaving group on the rate is suggestive of an associative mechanism. These findings are consistent with an $\boldsymbol{I_a}$ or $\boldsymbol{A}$ mechanism whose reaction coordinate diagrams are shown in Fig. 17.30. As mentioned, the difference between the $\boldsymbol{I_a}$ and $\boldsymbol{A}$ mechanisms is that an intermediate—a species that is a local energy minimum on the reaction surface—forms in the latter case. The only way to definitively distinguish between these two pathways experimentally is to observe the intermediate. Unfortunately for the reaction in Eq. 17.55, the intermediate has not been observed. Several computational studies have failed to locate intermediates on the reaction coordinate for ligand substitutions of this class of ML_4 complexes, so for the sake of discussion[l] we will assume that an $\boldsymbol{I_a}$ mechanism operates.[45–47]

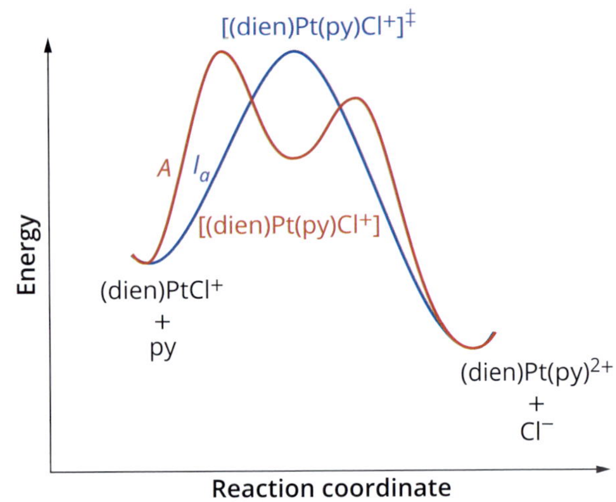

Fig. 17.30 Reaction coordinate diagrams for concerted ($\boldsymbol{I_a}$, blue) and stepwise associative ($\boldsymbol{A}$, red) mechanisms for the reaction in Eq. 17.55.

To understand how a substituting ligand would approach an ML_4 complex, let us consider its electronic structure as depicted in the MO diagram in Fig. 17.31.[m] Typically, an incoming ligand will interact with the LUMO. In the case of ML_4 d_8 complexes, this is the p_z-orbital.[47] The p_z-orbital is rigorously nonbonding in D_{4h}; however, as the incoming ligand comes in axially, the symmetry changes to C_{4v}, which allows mixing of the p_z-orbital with the σ-antibonding d_{z^2}-orbital.[n] As the incoming ligand approaches closer, one ligand in the square plane bends, proceeding to a transition state that has a pseudo-trigonal bipyramidal geometry, which lowers the symmetry further. We will return to this energy diagram later

[l] Failure to find an intermediate in calculations should not be used as an indication that one does not exist, just as not observing one experimentally is not an indication that an intermediate does not exist. Both are negative evidence.

[m] For a detailed discussion of the MO diagram for a square planar complex, see Section 6.6.3.

[n] Both orbitals are A_1 symmetry in C_{4v}.

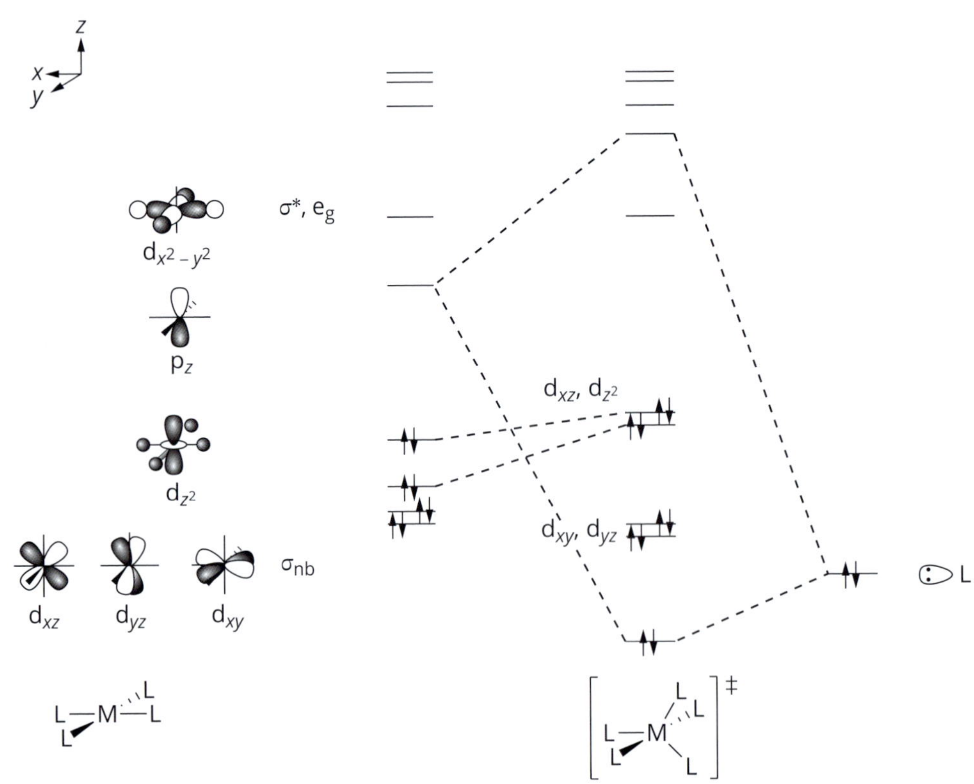

Fig. 17.31 Interaction between an entering ligand and an ML_4, d^8 complex to generate an ML_5 transition state.

when we discuss the ligand effects on substitution reactions. In an $\boldsymbol{I_a}$ mechanism, the Cl^- ligand begins to lengthen in the transition state and is eventually cleaved in the formation of the product.

If we consider a more general reaction,

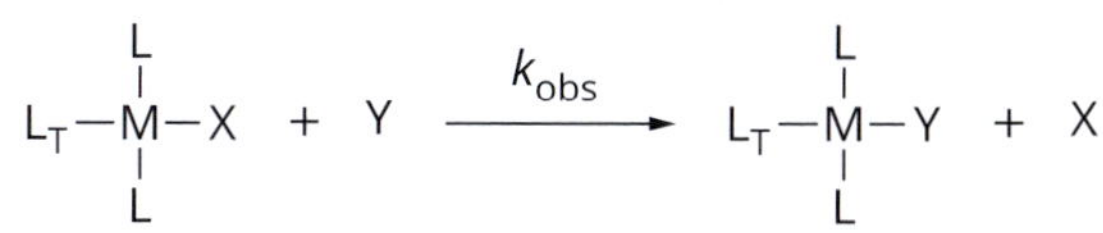

we note that the reaction is stereospecific because the incoming ligand Y has the same *trans*-relationship to L_T that the leaving group X did. We expect that rates of substitution will be influenced by X, Y, L, and L_T. The roles that these variables play on relative rates are outlined in the following.

Fundamentally, a faster rate arises when the energy separation between the reactants and the transition state of the rate-determining step is decreased. As shown in Fig. 17.32, if we start with the black reaction coordinate curve, this can be accomplished by changing the relative energies of the ground state (blue curve), which is a thermodynamic effect: the *trans*-influence (Section 17.5.4). Ground-state destabilization brings the reactants closer to the energy of the transition state, increasing the rate. Alternatively, the rate can be increased by lowering the energy of the transition state (red curve). Transition-state stabilization brings the energy closer to the ground state, thereby increasing the rate. The *trans*-effect is a kinetic effect on the rate of substitution that could be a consequence of either ground-state destabilization (blue curve), transition-state stabilization (red curve), or both.

Historically, ground-state effects have been easier to identify. For example, relative bond distances for a particular chemical bond in a family of complexes often correlate to relative bond dissociation enthalpies. In contrast, transition states have fleeting (on the order of a vibration, in the range of femtoseconds) lifetimes that make their experimental characterization

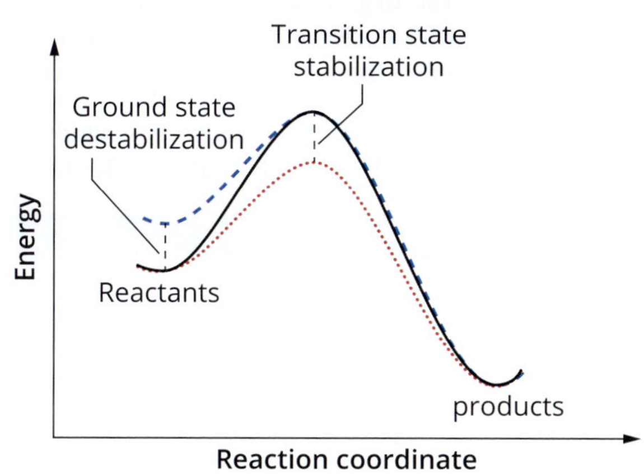

Fig. 17.32 Pathways where a reaction rate (black) is increased by ground-state destabilization, which is *trans*-influence (blue dashed line), and transition-state stabilization (red dotted line). The two effects together— more specifically their effect on reaction rate—is called the *trans*-effect.

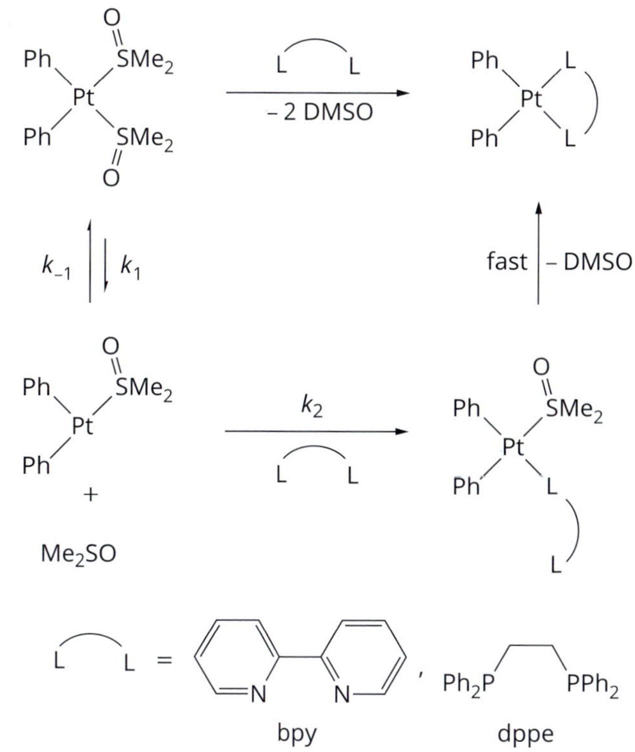

Fig. 17.33 Dissociative substitution of DMSO ligands in the reactions of *cis*-$[PtPh_2(DMSO)_2]$ with bidentate ligands.

virtually impossible. However, computational studies can provide insight to transition-state structures and have more recently contributed to our understanding of their role in determining the rates of ligand substitution reactions.

While many substitution reactions of square-planar complexes are associative in nature, there are examples where dissociative mechanisms operate. Pathways of this type are particularly important in catalytic reactions of Pd phosphine complexes. An early example for dissociative substitution involves displacement of coordinated DMSO ligands in *cis*-$[PtPh_2(DMSO)_2]$ by chelating ligands, L_n (Fig. 17.33).[48–49]

The rate law for this reaction was established through kinetics experiments as

$$\frac{-d[PtPh_2DMSO_2]}{dt} = k_{obs}[PtPh_2DMSO_2] \qquad \textbf{17.56}$$

where

$$k_{obs} = \frac{k_1k_2[L_2]}{k_{-1}[DMSO] + k_2[L_2]} \qquad \textbf{17.57}$$

At relatively high $[DMSO]:[L_2]$ ratios, the term $k_2[L_2]$ in the denominator may be ignored, and k_{obs} reduces to

$$k_{obs} = \frac{k_1k_2[L_2]}{k_{-1}[DMSO]} \qquad \textbf{17.58}$$

Under these conditions, the reaction rate is inhibited by DMSO. At high $[L_2]$, $[DMSO]:[L_2]$ is small; ignoring the $k_{-1}[DMSO]$ term in the denominator of Eq. 17.54 gives simply

$$k_{obs} = k_1 \qquad \textbf{17.59}$$

In other words, the rate is determined exclusively by the rate of DMSO dissociation. Experimentally, the same value for DMSO dissociation, k_1, was found for reactions in which L_2 was bpy (Fig. 11.33) or dppe (Fig. 11.27)—two quite different chelating ligands. This is consistent with a dissociative mechanism, as the nature of incoming L_2 does not factor strongly into the dissociation step.

From the experimentally determined value of k_1, the ratio k_2/k_{-1} can be calculated under conditions where Eq. 17.58 is valid. This gives the ratio of trapping rates of the 3-coordinate intermediate by DMSO and L_2. The data show that the efficiency in trapping follows the order bpy < DMSO < dppe. This indicates a kinetic preference of the Pt^{II} intermediate for larger radii sulfur and phosphorus ligands, as opposed to the smaller radius and sterically hindered nitrogen ligand.

17.6.2 Effects of Leaving Groups

The first variable we will consider is the effect that the leaving group X has on the reaction rate. This has been evaluated for the substitution of X in the following reaction.[50]

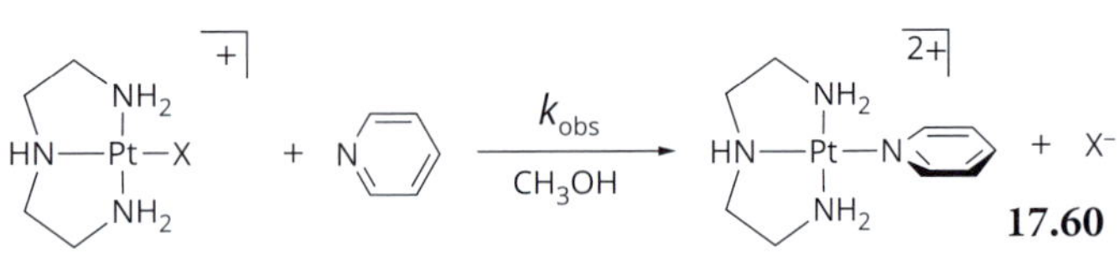

17.60

Table 17.9 lists k_{obs} values for substitution reactions in methanol at 25 °C. The relative ordering is $NO_3^- > H_2O > Cl^- > Br^- > I^- > N_3^- > SCN^- > NO_2^- > CN^-$. Ligands that can π-accept undergo slow substitution, and the halide relative rates follow the order $Cl^- > Br^- > I^-$. This is the opposite trend seen in the organic analog of this reaction, S_N2 displacements, and substitution of halides in $[Co(NH_3)_5X]^{2+}$, but it is consistent with the relative ordering of formation constants $[Pt(dien)I]^+ > [Pt(dien)Br]^+ > [Pt(dien)Cl]^+$, which can be rationalized on the preferential binding of a soft Pt^{II} center to the softest halide. (Alternatively, the rate could be reporting on the π-donor ability of the leaving group. Better π-donors are faster, and π-acceptors are slower.) As discussed previously, stabilization of ground states will decrease relative rates of a chemical reaction.

Associative reactions are *relatively* insensitive (relative to dissociative reactions) to leaving group effects. However, stabilization of the ground state, reported with a higher formation constant, can affect substitution rates.

Table 17.9 Leaving group effects of substitution rates for $[Pt(dien)(X)]^+ + py \rightarrow [Pt(dien)(py)]^{2+} + X^-$ in H_2O at 25 °C.[50]

X	$k_{obs} \times 10^8$, s^{-1}
NO_3^-	Too fast to measure
Cl^-	3500
Br^-	2300
I^-	1000
N_3^-	83
SCN^-	30
NO_2^-	5.0
CN^-	1.7

17.6.3 Effects of Entering Groups

Another important consideration in square-planar substitution is to consider the effect that the entering group has on the substitution rate. This has been evaluated for a number of complexes, with the broadest range studied being for displacement of chloride from *trans*-$Pt(py)_2Cl_2$.[51]
Table 17.10 lists k_{obs} values for substitution reactions in H_2O

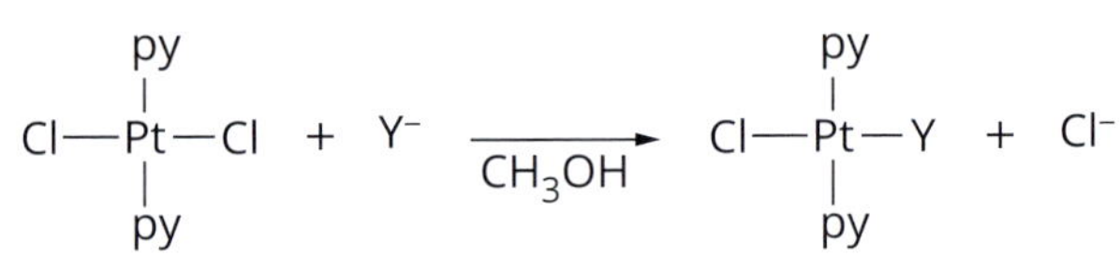

17.61

at 25 °C. The relative ordering is $PhS^- > SeCN^- > SCN^- > I^- > PhSH > Br^- > N_3^- > NO_2^- > py > Cl^- > MeO^-$. For the halides, the relative rates follow the order $I^- > Br^- > Cl^-$, which is the reverse of the leaving group trend (Table 17.9). The comparison between PhS^- and PhSH shows that increased negative charge, not surprisingly, accelerates the rate of substitution.

In general, the softer the incoming nucleophile, the faster the rate of substitution in this case. This can be rationalized as the softer nucleophiles being more polarizable, which means that the electron density can be adjusted to minimize electron–electron repulsion between the approaching ligand's electron pairs and the metal HOMO d_{z^2}-electrons. An alternative explanation concerns the degree of solvation of the incoming nucleophile. The harder nucleophiles are probably better solvated, such as stronger hydrogen bonds, which may act to slow

Table 17.10 Entering group effects of substitution rates for $[Pt(py)_2Cl_2] + Y \rightarrow [Pt(py)_2(Cl)(Y)] + Cl^-$ in $HOCH_3$ at 30 °C.[51]

X	$k_{obs} \times 10^4$, M^{-1} s^{-1}
MeO^-	1
$^{35}Cl^-$	4.5
py	5.5
NO_2^-	6.8
N_3^-	16
Br^-	37
PhSH	57
I^-	1,100
SCN^-	1,800
$SeCN^-$	52,000
PhS^-	60,000

nucleophilic reactions by delocalizing any anionic charge on the incoming nucleophile and by making a sterically encumbered ligand–solvent complex. On the other hand, "softer" ligands such as thiolates tend to hydrogen-bond less strongly and are probably less stabilized by solvation in methanol.

When Basolo, Pearson, and coworkers plotted the reaction rate versus the concentration of the incoming ligand, [Y], for a variety of ligands the series of lines did not converge at the origin (Fig. 17.34, top). In other words, extrapolating to infinite dilution of Y, the reaction still continues with a nonzero rate constant. This was interpreted as there being a background reaction, where instead of the Pt-complex interacting with Y, the complex interacts with methanol solvent first to form $[Pt(py)_2Cl(HOMe)]^+$. The methanol complex was shown to react rapidly with Y displacing methanol to produce $[Pt(py)_2Cl(Y)]$, the same product as direct reaction with Y.

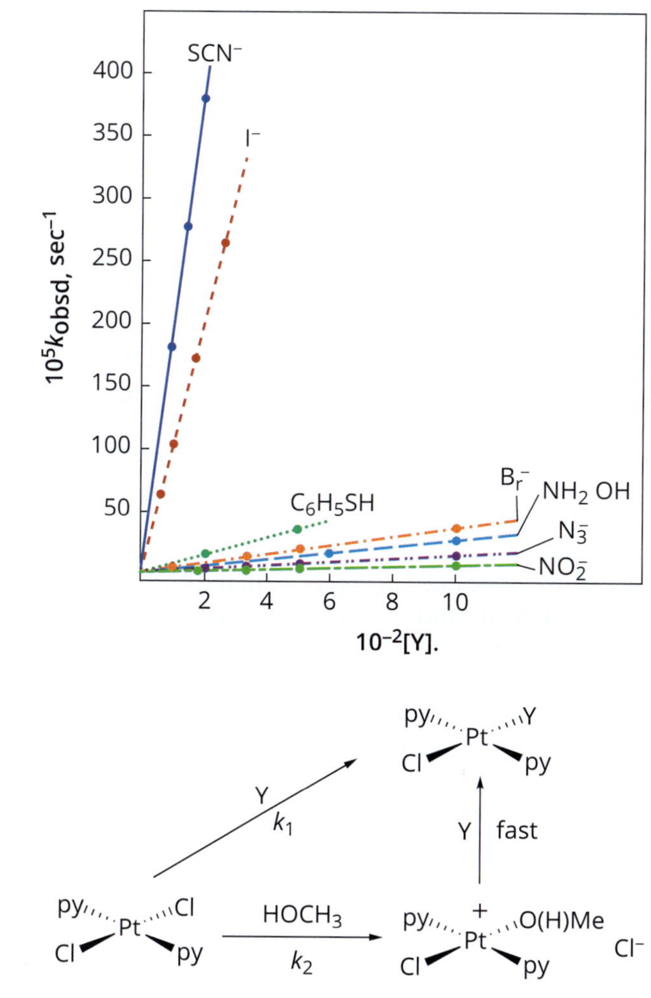

Fig. 17.34 (*top*) Plot of the observed rate constant versus [Y]. (Plot from Basolo and coworkers study.[51]) (*bottom*) The two different pathways to $[Pt(py)_2Cl(Y)]$ identified by Basolo, Pearson, and coworkers. The first is direct reaction with Y with rate constant k_1. The second is reaction with methanol solvent to form an intermediate solvent complex, which is rate-determining in this route, with rate constant k_2. Then, the solvated complex reacts rapidly with Y to produce $[Pt(py)_2Cl(Y)]$.

Consequently, a two-term rate law (Eq. 17.62) was found for these two different routes (Fig. 17.34, bottom). The rate is dependent upon the complex reacting directly with Y with rate constant k_1 and on the complex reacting with MeOH with rate constant k_2 followed by fast reaction with Y. As is often the case, the solvent does not appear in the rate law, as it is always in excess, that is, its concentration does not change significantly and becomes part of the rate constant. The incoming ligand Y does not appear in the second term related to k_2, because its reaction with the intermediate $[Pt(py)_2Cl(HOMe)]^+$ is beyond the rate-determining step, the RDS being replacement of Cl with the solvent on this pathway.[51]

$$v = (k_1[Y] + k_2)[PtCl_2(py)_2] \qquad \textbf{17.62}$$

As mentioned, substitution of $[Pt(py)_2Cl_2]$ has probably been studied more than any other platinum system. Can we use this system as a standard to understand reaction rates in other Pt(II) square-planar complexes? Basolo and coworkers did exactly this through a linear free-energy relationship. In this case, what was desired was how discriminating different complexes were towards incoming nucleophiles. This was analyzed by taking the ratio of rate constants k_1 and k_2. The logarithm was taken to make the large numbers common to kinetics more manageable, and a factor relating the standard system to other systems being interrogated, *s*, was added. The standard system, $[Pt(py)_2Cl_2]$, then has the ratio $\log(k_1/k_2)_0$, which was further abbreviated to n_{Pt}.

$$s \cdot log\left(\frac{k_1}{k_2}\right)_0 = s \cdot n_{Pt} = log\left(\frac{k_1}{k_2}\right)$$

This equation can be linearized as follows. A plot of $log(k_1)$ vs n_{Pt} for various platinum systems gave lines with slopes = *s*. Values for *s* and log k_2 are shown in Table 17.11. There is a loose inverse correlation between k_2 and *s*, which suggests quite reasonably that the complex becomes more reactive towards substitution as the discrimination for different ligand types decreases.[51]

$$\log(k_1) = s \cdot n_{Pt} + log(k_2)$$

Table 17.11 Nucleophile discrimination parameters (*s*) and intercepts for some Pt(II) complexes.[51]

Complex	*s*	log(k_2)	Solvent, Temp. (°C)
trans-$[Pt(PEt_3)_2Cl_2]$	1.43	-6.83	MeOH, 30
trans-$[Pt(AsEt_3)_2Cl_2]$	1.25	-5.75	MeOH, 30
trans-$[Pt\{S(s\text{-}Bu)_2\}_2Cl_2]$	0.57	-4.95	MeOH, 55
trans-$[Pt(SEt_2)_2Cl_2]$	1.05	-4.67	MeOH, 30
trans-$[Pt(pip)_2Cl_2]$	0.91	-4.56	MeOH, 30
$[Pt(en)Cl_2]$	0.64	-4.33	H_2O, 35
$[Pt(dien)Br]^+$	0.75	-4.06	H_2O, 25
$[Pt(dien)Cl]^+$	0.65	-3.61	H_2O, 35
$[Pt(dien)(OH_2)]^{2+}$	0.44	-0.44	H_2O, 25

17.6.4 The *trans*-Effect (Advanced Topic)

In the reaction

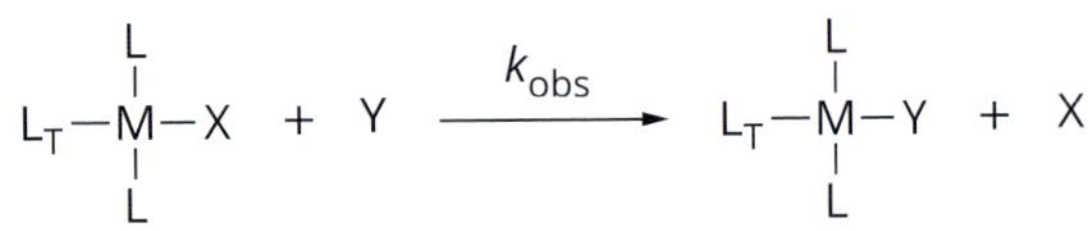

L_T is a spectator ligand—it holds its position throughout the course of the reaction. The nature of L_T has been shown to have a profound influence on reaction rate, and this phenomenon has been termed the *trans*-effect. The relative rates of substitutions as functions of L_T can vary over several orders of magnitude. When the relative rate of substitution is high, L_T is said to have a large *trans*-effect. Rates for the reactions in the top part of Table 17.12 have been measured and are listed in the table.[52] These data establish a *trans*-effect ordering H > Me > C_6H_5 > *o*-tolyl > Cl^- > mesityl. (See Practice Problem 7.) For the ligands where σ-bonding effects dominate, strong σ-donors have high *trans*-effects. Historically, this has been rationalized by strong σ-donors destabilizing the ground state. As we will see, computational studies indicate that strong σ-donors can also accelerate rates by transition-state stabilization. For π-acceptor ligands, such as phosphines, acceleration is attributed to transition-state stabilization.

Table 17.12 Effects of L_T on substitution rates of Cl^- by pyridine in ethanol of *trans*-$[Pt(PEt_3)_2(L_T)(Cl)]$.[52]

$$\frac{-d[Pt-Cl]}{dt} = k_{obs}[Pt-Cl] \qquad k_{obs} = k_1 + k_2[py]$$

L_T	k_{obs},[a] s^{-1}	T (°C)
H	4×10^{-2}	0
Me	5×10^{-4}	25
Ph	1×10^{-4}	25
o-tolyl	2×10^{-5}	25
Cl	3×10^{-6}	25
mesityl	6×10^{-6}	25

[a] k_{obs} is calculated for [py] = 0.005.

A generally accepted[53] trend for a broader range of ligands for the *trans*-effect is

$$CH_2{=}CH_2, CO, CN^- > PR_3, H^- > CH_3^- > C_6H_5^-, NO_2^-, SCN^-, I^- > Br^-, Cl^- > py, NH_3, OH^-, H_2O$$

Factors that give rise to the net effect have been dissected into σ-effects and π-effects. Computational studies have been carried out to provide a better understanding of the *trans*-effect. For example, the exchange of complexes *cis*-$[PtH(Cl)(NH_3)_2]$ and *cis*-$[RhH(CO)(NH_3)_2]$ with free NH_3 (Fig. 17.35) has been evaluated. In the ground-state structures, there are two chemically inequivalent amine ligands. Therefore, we would expect that the amines would have different rates of exchange, and the energy of their transition states would be different. Because these transition states are reached from the same reactants, any energy difference calculated will arise solely from transition-state geometry effects.

In the Pt and Rh complexes, the two pathways that lead to two different transition states are depicted in Fig. 17.36. Calculations show that for the Pt complex, the transition state for exchange of the NH_3 *trans* to H is lower than that for exchange *trans* to Cl. Likewise, the transition states for exchange in the Rh complexes show that the transition state for the NH_3 *trans* to CO is of lower energy than the exchange for the NH_3 *trans* to H. Thus, the Pt and Rh calculations produce a relative *trans*-effect ordering CO > H > Cl, which parallels the experimental trend.

A more detailed examination of bond distances and angles in the transition-state structures in Fig. 17.36 reveals some unusual features illustrated in Fig. 17.37. In the Rh structures, we note the Rh–N distance is considerably shorter for the structure corresponding to substitution of the more labile NH_3 *trans*- to CO (Structure A) than the Rh–N distance in the less favored transition state for substitution of the NH_3 *trans*- to H (Structure B). In other words, the transition state with the shorter Rh–N distance leads to faster replacement.

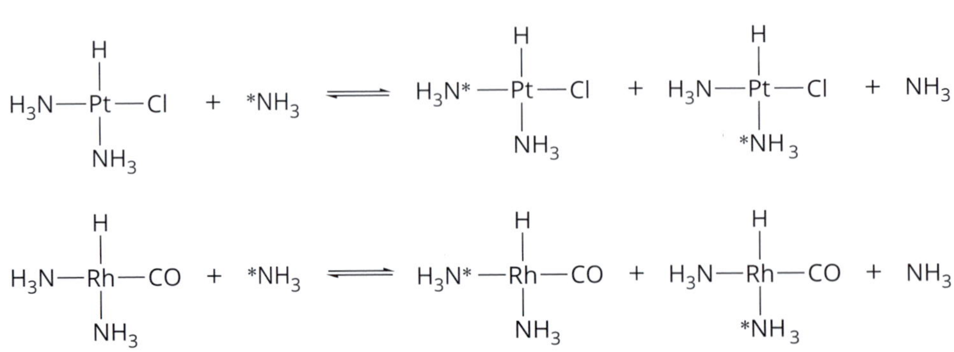

Fig. 17.35 Exchange of free NH_3 with coordinated NH_3 ligands in chemically distinct environments.

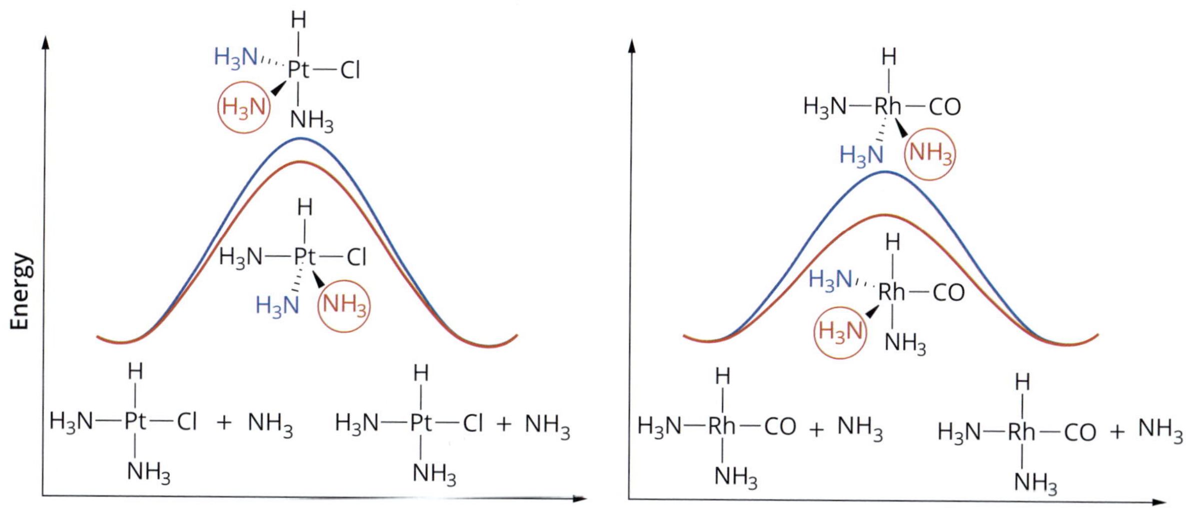

Fig. 17.36 Calculated reaction coordinates for NH_3 substitution in *cis*-PtH(Cl)$(NH_3)_2$ and *cis*-RhH(CO)$(NH_3)_2$. The incoming NH_3 is labeled red, circled and outgoing blue in the transition states. The energy differences arise exclusively for differences in transition-state structures. The relative rates for substitution of NH_3 as a function of *trans*-ligand follows the order CO > H > Cl.

In the Pt complex, the Pt–N distance for substitution *trans* to Cl (Structure C) is shorter than the Pt–N in the transition state for substitution *trans* to H (Structure D). In this case, the transition state with the shorter M–N distance is *less* favored.

To understand this apparent contradiction we will consider the electronic interactions between the incoming NH_3 ligand and the square-planar complex as the reaction proceeds towards the transition state. We will begin the discussion by recalling the electronic structure of ML_4, d^8 complexes. In the MO diagram at the left in Fig. 17.31, we see that four of the five metal-based frontier molecular orbitals are filled with electrons. The $d_{x^2-y^2}$-orbital is strongly σ-antibonding and is unpopulated. The d_{xy}, d_{xz}, and d_{yz} orbitals are σ-nonbonding and the lowest-lying metal orbitals. The d_{z^2}-orbital is the weakly σ-antibonding HOMO, and the LUMO has primarily p_z-character.

The lowest-energy calculated trajectory for an initial NH_3 approach is along the *z*-axis. This is consistent with interaction of the NH_3 lone pair with the metal LUMO. As the M–N distance becomes shorter, the M–N vector cants towards the *xy*-plane, and the NH_3 that will be substituted bends below the *xy*-plane. The principal barrier to the approach is electron–electron repulsions between the NH_3 lone pair of electrons and the electrons in the metal *d*-orbitals. Initially, the strongest repulsion is with the d_{z^2} HOMO because its electron density is aligned on the *z*-axis. As the M–N vector begins canting towards the *xy*-plane, repulsions between the nitrogen lone-pair electrons and d_{xz} increase. In the transition state this repulsion dominates and is reflected by destabilization of the d_{xz} orbital in the interaction diagram in Fig. 17.31. By understanding this interaction we can rationalize the relative transition-state energies in Fig. 17.36.

In the transition state for substitution of the NH_3 *trans* to CO in *cis*-[RhH(CO)$(NH_3)_2$], d_{xz} has the proper symmetry to π-backbond to the CO ligand. This causes electron density from the metal center to be delocalized onto the ligand (Fig. 17.38). Thus, by drawing electron density from the metal, repulsion between the metal electrons and the incoming NH_3 lone pair is minimized. Consequently, the NH_3 can make a closer approach as reflected by the Rh–N distance of 2.50 Å in the transition state. Since H cannot engage in π-bonding, the M–N distance is considerably longer in the higher-energy substitution pathway.

For the Pt complex, neither the Cl nor H can π-accept, and the repulsion with d_{z^2} is unavoidable. Because H is a much stronger donor than Cl, the M–N distances in the transition state for substitution of the NH_3 *trans* to H are considerably longer. These longer distances minimize electron–electron repulsions between the NH_3 and the metal electrons relative to

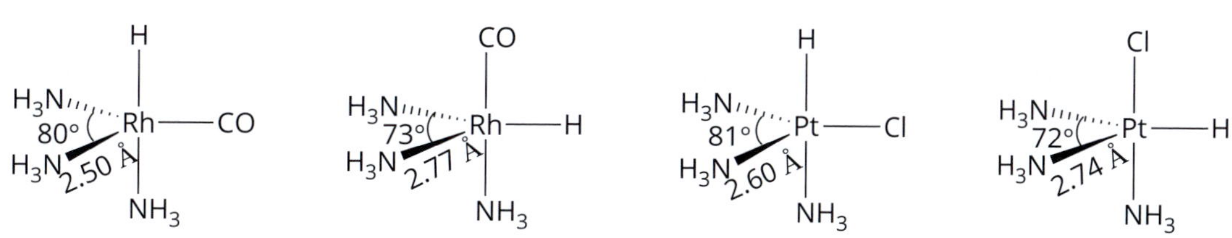

A B C D

Fig. 17.37 Transition states for ammonia exchange in square-planar d^8 complexes.

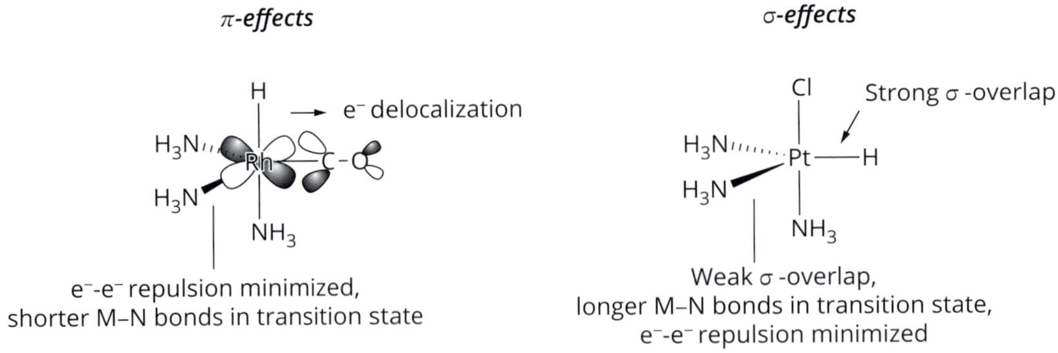

Fig. 17.38 Rationalization of π- and σ-effects in transition-state stabilization for I_a ligand exchange in square-planar d^8 complexes.

BOX 17C

The *trans*-influence and *sd*-hybridization

The *trans*-influence can be quite profound, with large effects on bond distances in, for example, square-planar and octahedral compounds. The reason for this is quite obvious if one analyzes the compounds by using hybridization theory, such as NBO theory (see Section 14.2).

As an illustration, we can examine the resonance forms in Pt square-planar complexes. If we start with homoleptic $[PtCl_4]^{2-}$, the contributing resonance forms are shown in the following. The complex is d^8, and the metal will be *sd* hybridized. Since the compound is 16-electron, it is hypervalent under this system. The two sd^1-hybrids will be 90° apart, and the bonds to the chlorine ligands 180° away will be made with 3-center 4-electron bonds. Since the compound is homoleptic, each of the following resonance forms will contribute equally to the ground state. The calculated, B3PW91/SDD/NBO6, Pt–Cl bond order was 0.5.

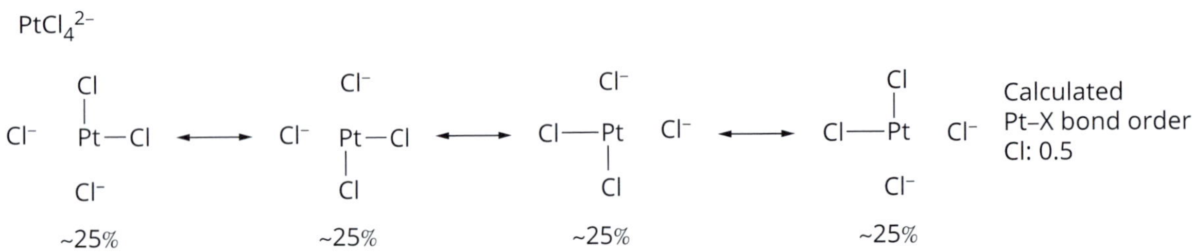

If we replace one of the chlorides with a much stronger donor like hydride, which is one of the strongest known *trans*-influencing ligands, the platinum center will need to use more of the hybrid to stabilize the more strongly basic H^- ligand. This can be seen in the calculations for this complex, summarized in the following.

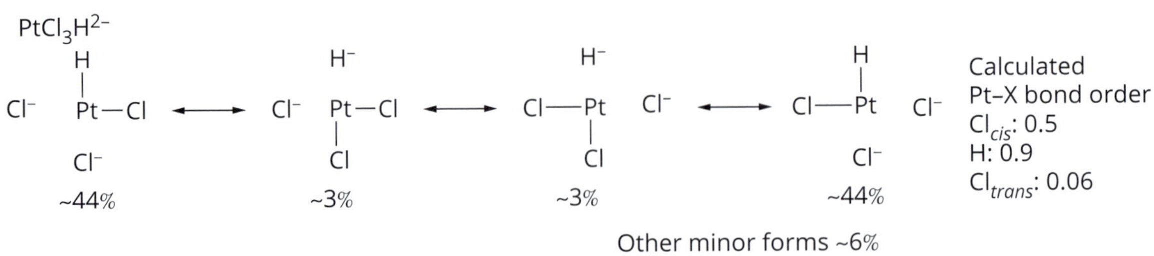

In this complex, the resonance forms where the hydride has a bond to platinum contribute much more than those where there is no Pt–H bond; as a result, the Pt–H bond order is very high at 0.9. Since the two *sd*-hybrids are 90° apart, the Pt–Cl_{cis} bonds still have relatively high bond orders of 0.5. In order to have the very high bond order to hydride, the bond order to the *trans*-chloride will be low and was calculated as 0.06.

The *trans*-chloride as a result has a much longer calculated distance as well at 2.60 Å versus 2.44 for the *cis*-chlorides. This *trans*-influence is a result of the lower bond order to the *trans*-substituent caused by the strongly donating hydride and its effect on the resonance structure of the complex.

those in the substitution for the NH_3 *trans* to Cl. Hence, the barrier for substitution *trans* to H is lower. These studies show that both σ- and π-effects are important for transition-state stabilization.

17.6.5 Metal Effects on Substitution Rates

For the Group 10 metals, substitutions of the chloride ligand in the complexes *trans*-[M(PEt_3)$_2$(*o*-tolyl)(Cl)] (M = Ni, Pd, Pt) with pyridine have been measured, and values of k_1 are listed in Table 17.13. The rates span 7 orders of magnitude with an ordering Ni > Pd >> Pt. The high reactivity of Ni relative to Pd and Pt is consistent with the greater propensity of Ni^{II} to form 5-coordinate complexes. Consequently, the transition state for an $\boldsymbol{I_a}$ mechanism should be stabilized relative to those for Pd and Pt. Due to the lanthanide contraction, bond distances for *4d* and *5d* third-row complexes are quite similar. Thus, the difference in rates between Pd and Pt complexes in Table 17.13 has primarily electronic origins. If there is a significant increase in antibonding character in the transition state, we would expect less destabilization for Pd because it has a weaker ligand field than Pt. Alternatively, a recent experimental and theoretical examination of olefin substitution in Nb and Ta complexes attributes the greater reactivity of the *4d* Nb complex to a higher density of electronic states at the transition state when compared to the *5d* Ta complex.[54] Regardless of the origin, reaction rates for *4d* metals are often faster than those of their *5d* congeners.

Table 17.13 Effects of metal on substitution rates of Cl^- by pyridine in ethanol of *trans*-[M(PEt_3)$_2$(o-tolyl)(Cl)] (M = Ni, Pd, Pt) at 25 °C.

M	k_1, s^{-1}
Ni[a]	33
Pd[a]	0.6
Pt	7×10^{-6}

[a] Estimated from rates measured between −80 to 0 °C.[55]

Chapter Summary

1. Reaction progress is usually thought of in terms of distance along a reaction coordinate. The reaction coordinate is a conglomeration of all the geometric changes that occur during the reaction from starting materials to products.
2. The rate for an elementary reaction—a reaction with a single step—is determined by the barrier height—the difference in energy between the starting materials and the transition state (also referred to as the activated complex). Parameters associated with the transition state are often marked with a superscripted "double dagger", ‡. For example, the enthalpy difference between the starting materials and the transition state would be denoted as $\Delta H^{‡}$.
3. A reaction is under kinetic control if the product is determined by the lowest energy barrier and under thermodynamic control if the most thermodynamically stable products are formed.
4. For multistep reactions, the step with the highest barrier is often referred to as the "rate-determining step", RDS. Only steps that occur before the rate-determining step are observable kinetically.
5. For substitution reactions, Taube determined that complexes with empty t_{2g} or occupied $e_g{}^*$ orbitals are often labile, undergoing rapid ligand exchange. Other configurations, such as low-spin d^6, lead to complexes that are usually inert to substitution. Taube arbitrary called complexes with $t_{1/2} > 30$ s inert, and those with half-lives less than this mark labile.
6. There are three general mechanisms for substitution: associative (**A**), dissociative (**D**), and interchange (**I**). The **I** mechanism is further divided into dissociative interchange ($\mathbf{I_d}$) and associative interchange ($\mathbf{I_a}$). The activation parameters and kinetic behavior can sometimes be used to distinguish these different mechanisms.
7. Substitutions in inert complexes are often controlled by factors such as ligand donor abilities and charges of the interacting species.
8. The *trans*-influence is the effect of a ligand's donor properties on the ground-state geometry of a complex. This thermodynamic effect may lead to long metal–ligand bonds *trans* to a particular substituent, for example.
9. The *trans*-effect is a kinetic effect by which substitution occurs more or less rapidly, depending on a ligand *trans* to the site of substitution. This could be due to ground-state effects, with similar causes as the *trans*-influence, or the observed kinetic effect could be due to transition state (de)stabilization, which would be completely unrelated to the *trans*-influence. A series of ligands organized from those with the largest effect to the least shows that there is generally little correlation between ligands that are strong *trans*-influencers versus those that cause a strong *trans*-effect.

Practice Problems

1. The rates for acetonitrile ligand substitutions in $[(\eta^6\text{-}C_6H_6)Ru(NCCH_3)_3]^{2+}$ and $[(\eta^5\text{-}C_5H_5)Ru(NCCH_3)_3]^{+}$ have been measured as functions of temperature and pressure.
    a. Using the data in the table below, calculate $\Delta H^‡$, $\Delta S^‡$, and $\Delta V^‡$ for acetonitrile exchange in both complexes. The variable pressure experiments are at 304.6 K for $[(\eta^6\text{-}C_6H_6)Ru(NCCH_3)_3]^{2+}$ and 314.7 K for $[(\eta^5\text{-}C_5H_5)Ru(NCCH_3)_3]^{+}$.

$[(\eta^6\text{-}C_6H_6)Ru(NCCH_3)_3]^{2+}$				$[(\eta^5\text{-}C_5H_5)Ru(NCCH_3)_3]^{+}$			
T (K)	10^5k (s^{-1})	P (MPa)	10^5k (s^{-1})	T (K)	k (s^{-1})	P (MPa)	k (s^{-1})
295.2	3.0	0.1	12.2	287.6	1.5	0.1	40.4
298.2	3.9	40.0	12.2	295.1	4.1	35.0	34.1
303.2	7.7	120.0	10.5	304.6	10.8	50.0	32.9
308.2	14.8	158.0	9.8	309.6	22.8	77.0	29.1
313.2	32.5	200	10.3	313.7	33.2	100.0	27.4
318.2	62.2			321.0	75.3	102.0	27.1
						126.0	23.8
						150.0	21.2
						170.0	18.8
						181.0	19.0

    b. Based on the activation parameters, propose mechanisms for substitution.

2. Using $PtCl_4^{2-}$ as a starting material, use the *trans*-effect data and propose syntheses for the complexes below.

a. 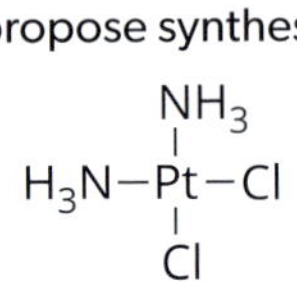

b. 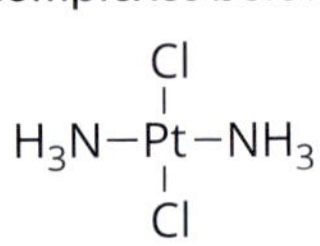

c. 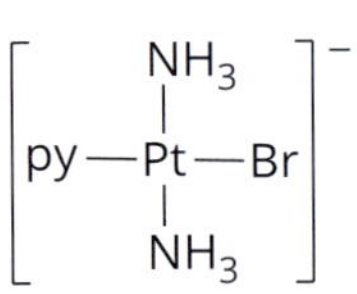

d. 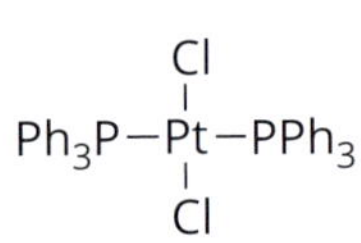

3. The rate law for F^- displacement from $[Co(NH_3)_5F]^{2+}$ by H_2O is shown below

$$\frac{-d[Co(NH_3)_5F^{2+}]}{dt} = (k_1 + k_2[H^+] + k_3[H^+]^{-1})[Co(NH_3)_5F^{2+}]$$

where k_{obs} at constant $[H^+]$ is

$$k_{obs} = k_1 + k_2[H^+] + k_3[H^+]^{-1}$$

Propose mechanisms for the reactions corresponding to k_1, k_2, and k_3.

4. The substitution of Cl^- by pyridine (py) in the complex [Pt(dien)Cl]Cl in H_2O (**1**) to give $[Pt(dien)(py)]Cl_2$ follows the rate law given below:

$$\frac{-d[\mathbf{1}]}{dt} = (k_1 + k_2[py])[\mathbf{1}]$$

where $k_1 = 1.7 \times 10^{-5}\ s^{-1}$ and $k_2 = 5.3 \times 10^{-3}\ M^{-1}{\cdot}s^{-1}$.

    a. The k_1 term is believed to arise from displacement of Cl^- by the solvent, H_2O, to give an intermediate $[Pt(dien)(OH_2)]Cl_2$. If $[H_2O] = 55$ M, calculate the second-order rate constant for substitution of Cl^- by H_2O. How does it compare to k_2 for the py reaction?
    b. The complex $[Pt(dien)(OH_2)]Cl_2$ is not observed under the reaction conditions. Explain this in terms of the rate of substitution of Cl^- by H_2O in [Pt(dien)Cl]Cl and H_2O by py in $[Pt(dien)(py)]Cl_2$.

5. In the displacement reactions of Cl^- by NO_2^-, the neutral complex *trans*-$[PtCl_2(NH_3)_2]$ reacts more than 100 times faster than the cationic complex $[PtCl(NH_3)_3]^+$. Offer an explanation for this behavior.

6. The table below shows H_2O exchange rates and activation parameters for the complexes $[M(OH)(H_2O)_5]^{2+}$ (M = Cr, Fe, Ga). Assign mechanisms for the exchange and account for differences in mechanisms, rates, and activation parameters of the hexaaquo complexes in Table 17.2.

Electronic Configuration	$(t_{2g})^3$	$(t_{2g})^3(e_g)^2$	$(t_{2g})^6(e_g)^4$
$M(OH)(H_2O)_5^{2+}$	Cr^{3+}	Fe^{3+}	Ga^{3+}
k_{H2O}^{298} (s^{-1})	1.8×10^{-4}	1.2×10^5	1×10^5
$\Delta H^‡$ (kcal·mol^{-1})	26.3	10.0	14.0
$\Delta S^‡$ (cal·K^{-1}·mol^{-1})	13.1	1.2	—
$\Delta V^‡$ (cm^3·mol^{-1})	2.7	7.40	6.2

7. Steric effects can influence ligand substitution rates. The table below lists rates for Cl^- displacement by py for a series of Pt complexes.

$$Pt(PEt_3)_2(R)Cl + py \longrightarrow Pt(PEt_3)_2(R)(py)^+ + Cl^-$$

	Rate Constant ($M^{-1}s^{-1}$)	
Ligand	R *trans* to Cl[a]	R *cis* to Cl[b]
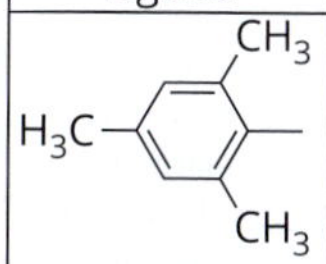 Mesityl	3.4×10^{-6}	1.0×10^{-6}
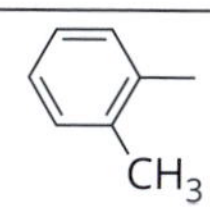 *o*-Tolyl	1.7×10^{-5}	2.0×10^{-4}
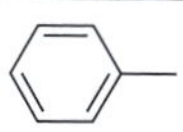 Phenyl	1.2×10^{-4}	$8.0 \times 10^{-2(a)}$

[a] rates measured at 25 °C.
[b] rates measured at 0 °C

a. Propose a mechanism for the substitution reaction.

b. Explain the difference in reaction rates of the *trans*-isomers compared to the *cis*-isomers.

c. Offer an explanation for the rate differences for phenyl, *o*-tolyl, and mesityl ligands.

d. The relative rates for the *cis*-isomers span a broader range than those for the *trans*-isomers. Using your proposed mechanism, account for this difference.

8. Base hydrolysis of isotopically labeled *trans*-$[Co(^{15}NH_3)(NH_3)_4Cl]^{2+}$ gives a 50:50 mixture of *cis*- and *trans*-$[Co(^{15}NH_3)(NH_3)_4(H_2O)]^{3+}$. If the reaction proceeds via a five-coordinate intermediate, $[Co(NH_2)(NH_3)_4]^{2+}$, use the principle of microscopic reversibility to predict the outcomes for deprotonation of a *cis*-NH_3 vs. *trans*-$^{15}NH_3$ to generate the five-coordinate intermediate. Which pathway is more consistent with the observed product ratio?

Cited References

1. Creutz, C., Ford, P. C., and Meyer, T. J. Henry Taube: Inorganic chemist extraordinaire. *Inorganic Chemistry* 2006, *45* (18), 7059–7068.
2. Espenson, J. H. *Chemical Kinetics and Reaction Mechanisms.* New York, NY: McGraw-Hill, 1981.
3. Basolo, F. Retrospective on studies of ligand substitution-reactions of metal-complexes. *Coord. Chem. Rev.* 1990, *100*, 47–66.
4. Taube, H. Rates and mechanisms of substitution in inorganic complexes in solution. *Chem. Rev.* 1952, *50* (1), 69–126.
5. Eyring, H. The activated complex in chemical reactions. *J. Chem. Phys.* 1935, *3* (2), 107–115.
6. Richens, D. T. Ligand substitution reactions at inorganic centers. *Chem. Rev.* 2005, *105* (6), 1961–2002.
7. Stahl, S. S., Labinger, J. A., and Bercaw, J. E. Homogeneous oxidation of alkanes by electrophilic late transition metals. *Angew. Chem. Int. Ed.* 1998, *37* (16), 2181–2192.
8. Stahl, S. S., Labinger, J. A., and Bercaw, J. E. Exploring the mechanism of aqueous C-H activation by Pt(II) through model chemistry: Evidence for the intermediacy of alkylhydridoplatinum(IV) and alkane sigma-adducts. *J. Am. Chem. Soc.* 1996, *118* (25), 5961–5976.
9. Laidler, K. J., and King, M. C. Development of transition-state theory. *J. Phys. Chem.* 1983, *87* (15), 2657–2664.
10. Lente, G., Fabian, I., and Poe, A. J. A common misconception about the Eyring equation. *New J. Chem.* 2005, *29* (6), 759–760.
11. Perrin, C. L. Relation between errors in enthalpy and entropy. *J. Phys. Org. Chem.* 2013, *26* (3), 269–270.
12. Note that it is not uncommon to be told that the error in the entropy of activation is unduly large due to extrapolation to the intercept. However, this assertion is incorrect, and the error between enthalpy and entropy of activation are simply related by the average temperature of the experiments. It is true, however, that as one approaches closer to absolute zero, the error drops as might be expected by the extrapolation claim.
13. Aebischer, N., Churlaud, R., Dolci, L., Frey, U., and Merbach, A. E. Mechanism of complex formation of ruthenium(II) aquacomplexes with $H_2C{=}CH_2$, MeCN, Me_2SO, and CO: Metal–water bond rupture as rate-determining step. *Inorg. Chem.* 1998, *37* (22), 5915–5924.
14. Helm, L., and Merbach, A. E. Inorganic and bioinorganic solvent exchange mechanisms. *Chem. Rev.* 2005, *105* (6), 1923–1959.
15. Langford, C. H., and Gray, H. B. *Ligand Substiution Processes.* Reading, MA: W. A. Benjamin, 1966.
16. Langford, C. H. On acid hydrolysis of $(Co(NH_3)_5X)^{+2}$ ions and mechanism of interchange. *Inorg. Chem.* 1965, *4* (2), 265.
17. Wilkins, R. G. *Kinetics and Mechanisms of Reactions of Transition Metal Complexes*, 2nd edn. New York, NY: VCH Publishers, 1991, p. xv, 465 pp.
18. Jordan, R. B. *Reaction Mechanisms of Inorganic and Organometallic Systems*, 3rd edn. Oxford: Oxford University Press, 2007, p. viii, 521 pp.
19. Ducommun, Y., Newman, K. E., and Merbach, A. E. High-pressure NMR kinetics. 11. High-pressure O-17 NMR evidence for a gradual mechanistic changeover from IA to ID for water exchange on divalent octahedral metal-ions going from manganese(II) to nickel(II). *Inorg. Chem.* 1980, *19* (12), 3696–3703.
20. Plutino, M. R., Scolaro, L. M., Romeo, R., and Grassi, A. To what extent can cyclometalation promote associative or dissociative ligand substitution at platinum(II) complexes? A combined kinetic and theoretical approach. *Inorg. Chem.* 2000, *39* (13), 2712–2720.
21. Romeo, R., Grassi, A., and Scolaro, L. M. Factors affecting reaction pathways in nucleophilic-substitution reactions on platinum(II) complexes: A comparative kinetic and theoretical study. *Inorg. Chem.* 1992, *31* (21), 4383–4390.
22. Hugicleary, D., Helm, L., and Merbach, A. E. High-pressure NMR kinetics. 30. Water exchange on hexaaquagallium(III): High-pressure evidence for a dissociative interchange exchange mechanism. *Journal of the American Chemical Society* 1987, *109* (15), 4444–4450.
23. Kallies, B., and Meier, R. Electronic structure of 3d $M(H_2O)_6^{3+}$ ions from Sc(III) to Fe(III): A quantum mechanical study based on DFT computations and natural bond orbital analyses. *Inorg. Chem.* 2001, *40* (13), 3101–3112.
24. Figgis, B. N., Sobolev, A. N., Young, D. M., Schultz, A. J., and Reynolds, P. A. X-ray-neutron study of the electron density in diammonium hexaaquairon(II) disulfate-d_{20} at 10.5 K and the success of ligand field modeling. *J. Am. Chem. Soc.* 1998, *120* (34), 8715–8723.
25. Beattie, J. K., and Best, S. P. Structures and spectroscopy of hexaaquametal(III) ions. *Coord. Chem. Rev.* 1997, *166*, 391–415.
26. Dalgarno, S. J., Hardie, M. J., and Raston, C. L. pH-dependent formation of molecular capsules and coordination polymers. *Crystal Growth & Design* 2004, *4* (2), 227–234.

27. Chaudhuri, P., and Diebler, H. Kinetics and equilibria of 1:1 complex formation of hexa-aquatitanium(III) with malonic and methylmalonic acid. *J. Chem. Soc., Dalton Trans.* 1986, *0* (8), 1693–1695.
28. Hugi, A. D., Helm, L., and Merbach, A. E. High-pressure NMR kinetics. 29. Variable-temperature, variable-pressure, and variable-frequency O-17 NMR study of water exchange on hexaaquatitanium(III): A limiting associative mechanism. *Inorg. Chem.* 1987, *26* (11), 1763–1768.
29. Wilkins, R. G. Mechanisms of ligand replacement in octahedral nickel(II) complexes. *Acc. Chem. Res.* 1970, *3* (12), 408-&.
30. Brintzinger, H., and Hammes, G. G. Inner- and outer-sphere complex formation in aqueous solutions of nickel(II)-methyl phosphate. *Inorg. Chem.* 1966, *5* (7), 1286.
31. De Vito, D., Sidorenkova, H., Rotzinger, F. P., Weber, J., and Merbach, A. E. Can octahedral t_{2g}^6 complexes substitute associatively? The case of the isoelectronic ruthenium(II) and rhodium(III) hexaaquaions. *Inorg. Chem.* 2000, *39* (24), 5547–5552.
32. De Vito, D., Weber, J., and Merbach, A. E. Calculated volume and energy profiles for water exchange on t_{2g}^6 rhodium(III) and iridium(III) hexaaqualons: Conclusive evidence for an I_a mechanism. *Inorg. Chem.* 2004, *43* (3), 858–864.
33. Aebischer, N., Sidorenkova, E., Ravera, M., Laurenczy, G., Osella, D., Weber, J., and Merbach, A. E. *Trans-* and *cis*-water reactivities in d^6 octahedral ruthenium(II) pentaaqua complexes: Experimental and density functional theory studies. *Inorg. Chem.* 1997, *36* (26), 6009–6020.
34. Bernhard, P., Burgi, H. B., Hauser, J., Lehmann, H., and Ludi, A. Syntheses and crystal and molecular-structures of hexaaquaruthenium(II) p-toluenesulfonate and hexaaquaruthenium(III) p-toluenesulfonate, $Ru(H_2O)_6(C_7H7SO_3)_2$ and $Ru(H_2O)_6(C_7H_7SO_3)\bullet3.3H_2O$. *Inorg. Chem.* 1982, *21* (11), 3936–3941.
35. Meier, U. C., Scopelliti, R., Solari, E., and Merbach, A. E. Structure and reactivity of aquacarbonylruthenium(II) complexes: An X-ray and oxygen-17 NMR study. *Inorg. Chem.* 2000, *39* (17), 3816–3822.
36. Werner, A. Information on asymmetric cobalt atoms XII: Optic activity with carbon compounds. *Ber. Deutsch. Chem. Ges.* 1914, *47*, 3087–3094.
37. Ingold, C., Nyholm, R. S., and Tobe, M. L. Orienting effects in octahedral aquation. *Nature* 1960, *187* (4736), 477–480.
38. Jackson, W. G., McGregor, B. C., and Jurisson, S. S. On the existence and lifetime of the $Co(NH_3)_5^{3+}$, intermediate in the substitution-reactions of $(NH_3)_5CoX^{N+}$. *Inorg. Chem.* 1987, *26* (8), 1286–1291.
39. Deutsch, E., and Taube, H. Studies on acetatopentaaquochromium(III) ion. *Inorg. Chem.* 1968, *7* (8), 1532.
40. Hughes, E. D., and Ingold, C. K. Mechanism of substitution at a saturated carbon atom. Part IV. A discussion of constitutional and solvent effects on the mechanism, kinetics, velocity, and orientation of substitution. *J. Chem. Soc.* 1935, 244–255.
41. Langford, C. H., and Gray, H. B. *Ligand Substitution Processes*. New York, NY: W.A. Benjamin, 1965.
42. Ingold, C. K., Nyholm, R. S., and Tobe, M. L. Mechanism, kinetics, and sterochemistry of octahedral substitutions. 4. Bimolecular basic hydrolysis and aquation of some halogen oisothiocyanato bis(ethylenediamine) cobalt(III) ions. *J. Chem. Soc.* 1956, 1691–1707.
43. Rapaport, I., Helm, L., Merbach, A. E., Bernhard, P., and Ludi, A. High-pressure NMR kinetics. 34. Variable-temperature and variable-pressure NMR kinetic-study of solvent exchange on $Ru(H_2O)_6^{3+}$, $Ru(H_2O)_6^{2+}$, and $Ru(CH_3CN)_6^{2+}$. *Inorg. Chem.* 1988, *27* (5), 873–879.
44. Swaddle, T. W. A model for the prediction of the molar volume effects accompanying changes in the primary coordination-number of aqueous metal-ions. *Inorg. Chem.* 1983, *22* (19), 2663–2665.
45. Banerjee, S., and Mukherjee, A. K. A detailed quantum-chemical study of the interactions of $Pt(dien)Cl^+$ with a series of S-donor ligands: A computational approach. *Computational and Theoretical Chemistry* 2012, *991*, 116–123.
46. Chval, Z., Sip, M., and Burda, J. V. The trans effect in square-planar platinum(II) complexes: A density functional study. *J. Comput. Chem.* 2008, *29* (14), 2370–2381.
47. Lin, Z. Y., and Hall, M. B. Theoretical studies of inorganic and organometallic reaction mechanisms. 2. The *trans* effect in square-planar platinum(II) and rhodium(I) substitution reactions. *Inorg. Chem.* 1991, *30* (4), 646–651.
48. Frey, U., Helm, L., Merbach, A. E., and Romeo, R. High-pressure NMR kinetics. 41. Dissociative substitution in 4-coordinate planar platinum(II) complexes as evidenced by variable-pressure high-resolution proton NMR magnetization transfer experiments. *J. Am. Chem. Soc.* 1989, *111* (21), 8161–8165.
49. Lanza, S., Minniti, D., Romeo, R., Moore, P., Sachinidis, J., and Tobe, M. L. Dissociative substitution in 4-co-ordinate planar platinum(II) complexes: The kinetics of sulfoxide exchange and its displacement by bidentate ligands in the reactions of *cis*-di(aryl) bis(dimethyl sulfoxide) platinum(II) in chloroform and benzene. *J. Chem. Soc.-Chem. Commun.* 1984, (8), 542–543.
50. Basolo, F., Gray, H. B., and Pearson, R. G. Mechanism of substitution reactions of complex ions. 17. Rates of reaction of some platinum(II) and palladium(II) complexes with pyridine. *J. Am. Chem. Soc.* 1960, *82* (16), 4200–4203.
51. Belluco, U., Cattalini, L., Basolo, F., Pearson, R. G., and Turco, A. Nucleophilic constants and substrate discrimination factors for substitution reactions of platinum(II) complexes. *J. Am. Chem. Soc.* 1965, *87* (2), 241.
52. Basolo, F., Pearson, R. G., Chatt, J., Gray, H. B., and Shaw, B. L. Kinetics of reaction of alkyl and aryl compounds of nickel group with pyridine. *J. Chem. Soc.* 1961, 2207.
53. Basolo, F., and Pearson, R. G. The *trans* effect in metal complexes. *Prog. Inorg. Chem.* 1962, *4*, 381–453.
54. Hirsekorn, K. F., Hulley, E. B., Wolczanski, P. T., and Cundari, T. R. Olefin substitution in $(silox)_3M(olefin)$ (silox = $^tBu_3SiO$; M = Nb, Ta): The role of density of states in second vs third row transition metal reactivity. *J. Am. Chem. Soc.* 2008, *130* (4), 1183–1196.
55. Cotton, F. A., and Wilkinson, G. *Advanced Inorganic Chemistry: A Comprehensive Text*. New York, NY: Interscience Publishers, 1962, 959 pp.
56. Gellene, G. I. Application of kinetic approximations to the A-reversible-arrow-B-C reaction system. *J. Chem. Educ.* 1995, *72* (3), 196–199.

CHAPTER 18

Reactivity in Organometallic Chemistry

"In a recent publication, Kealy and Pauson described the preparation of a compound, $C_{10}H_{10}Fe$, which was formulated as dicyclopentadienyl iron. This substance was found to be very stable toward acids and bases, and its isolation stands in striking contrast to the failure of previous investigators to prepare stable organoiron compounds. These circumstances led us to consider whether an alternative to the structure might be more nearly in accord with the unique character of Kealy and Pauson's compound."

— Geoffrey Wilkinson and Robert Woodward in their 1952 paper that described the correct structure of ferrocene and launched the field of organometallic chemistry[1]

CONCEPT COVERAGE

1. History of organometallic chemistry
2. Organometallic ligands
3. Common mechanistic steps
4. Catalysis and catalytic cycles

Organometallic chemistry is the study of the metal–carbon bond, and the field impacts both inorganic and organic synthesis. For organic chemists, organometallic reagents are critical in many modern chemical transformations, and for inorganic chemists organometallic complexes may be prepared and studied for their unusual reactivity, structure, or properties. One, if not the, major thrust of the organometallic field is catalysis, and many different catalytic cycles will be analyzed as a consequence. These figures can take some time to digest at first, but learning a few common steps makes understanding these cycles much easier.

Organometallic chemists are usually generous in their consideration of not only the "organo" portion of the name, but also the "metallic" portion. They study a variety of ligands with similar reaction chemistry to metal alkyls, including such ligands as hydride (H^-), amide ($^-NR_2$), alkoxide (^-OR), and supporting ligands that help to stabilize the complexes such as phosphines (PR_3). Many would also include in the field the study and characterization of metalloid compounds such as boron, aluminum, germanium, or silicon.

This chapter begins with a brief history of the field, which will have some overlap with previous discussions concerning ligands but is included for context, followed by a discussion of the mechanistic steps often found in the reactions of organometallic compounds. The chapter ends with a discussion of how these elementary steps can be combined into catalytic cycles that are used in academic and industrial laboratories.

18.1 A Short Introduction and History

Organometallic chemistry has become central to catalysis and fine-chemical production, but it is a relatively young discipline in comparison with organic chemistry or coordination chemistry. This means that its birth is well-documented. In late 1951 and early 1952, two research groups independently published the synthesis of "a new type of organoiron compound", dicyclopentadienyl iron, that was reported to contain two iron–carbon σ-bonds (Fig. 18.1).[2–3]

Previously, many researchers had attempted to prepare organo–iron complexes by reacting Grignard reagents with ferric chloride with no success; in fact, metal complexes with simple alkyl ligands were essentially unknown.[4] The new compound, however, was stable in boiling hydrochloric acid, sublimed readily at 100 °C, and was stable under pressures of hydrogen and

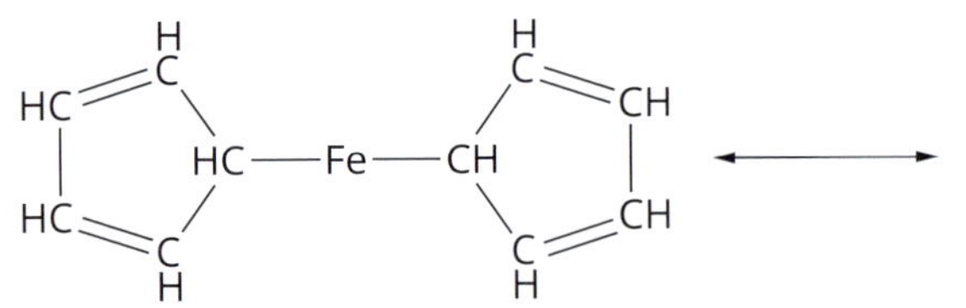

Fig. 18.1 The initially proposed resonance structures of dicyclopentadienyl iron.[2]

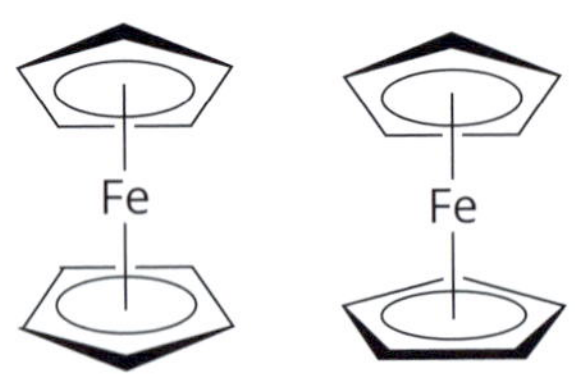

Fig. 18.2 The correct structure of dicyclopentadienyl iron showing the "iron sandwich" with two planar cyclopentadienyl rings bound above and below an iron atom. On the left is the "eclipsed" and on the right the "staggered" form.[5]

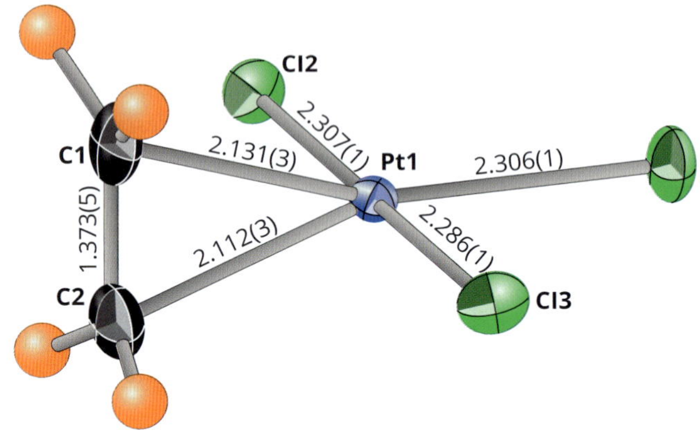

Fig. 18.3 The X-ray crystal structure of Zeise's salt, $K[(C_2H_4)PtCl_3]$, the first well-characterized transition-metal organometallic complex. Figure adapted from data deposited at the Cambridge Structural Database; only the anion is shown.[9]

nitrogen up to 250 °C. Geoffrey Wilkinson—a new assistant professor at Harvard at the time—read the Pauson report, and upon seeing the structure, exclaimed "It can't be that."[5] In a short time he had proposed the new and ultimately verified structure of the "iron sandwich", later called ferrocene, which opened the floodgates of research into these novel compounds (Fig. 18.2). The story of the discovery of ferrocene was documented by those present and later by chemical historians, providing a fascinating insight into the discovery process and the construction of history.[6]

Although responsible for the surge in interest in the field of organometallic chemistry, ferrocene was by no means the first organometallic complex ever to have been prepared. In 1825, William Zeise at Copenhagen began his study of the reactions of platinum chloride with ethanol. While $PtCl_2$ reacted with ethanol to form a black solid (finely divided metallic platinum with organic impurities), $PtCl_4$ reacted with ethanol in the presence of KCl to produce yellow crystals of $K[(C_2H_4)PtCl_3]$, "Zeise's salt". Zeise determined the correct chemical composition, though the nature of the bonding was not understood, and its structure was left unverified for more than 120 years until the X-ray diffraction crystal structure was reported in 1954,[7] followed by a neutron diffraction study in 1975.[8] The first well-characterized transition-metal organometallic compound was not a σ-complex, an interaction between a metal and a carbon with a σ-bond, but instead a π-complex, in which the interaction between the metal and carbon is through a π-interaction from the organic ligand (Fig. 18.3).

The preparation of organometallic compounds even predates that of Zeise's salt. The first compound that many chemists would consider to be "organometallic" was Cadet's "fuming and ill-smelling" arsenic compound $[Me_2As]_2$ first made in 1760. Later, Mond and Berthelot synthesized the first binary metal–carbonyl complexes. In the mid- to late 1800s, Frankland synthesized air-sensitive dialkylzinc reagents, while Grignard and Sabatier synthesized alkyl magnesium compounds. These alkylating reagents are perhaps the most familiar to those beginning their study of organometallic chemistry, due to their extensive use in organic synthesis. Only a few transition-metal organometallics were prepared before the discovery of ferrocene, including: $PtMe_3I$, the first stable transition-metal alkyl; $Fe(butadiene)(CO)_3$, the first diene complex; and $FeH_2(CO)_4$, the first metal–hydride complex.

Grignard reagents were found to be of great importance in the field of organic synthesis, and the reaction of $CrCl_3$ with PhMgBr was carried out with the goal of forming phenyl chromium complexes by Hein at the University of Leipzig in 1918. The reaction led to a large number of products, and further study of the reaction of chromium with alkylating reagents led to the discovery of the bis(benzene)chromium cation by Tsutsui and Zeiss in 1953 and bis(benzene)chromium by Fischer and Hafner in 1956. The structure of bis(benzene)chromium is also a sandwich structure, and description of the bonding in that molecule was clearly informed by the structure of ferrocene.

For additional reading, detailed historical accounts of the molecules $(C_6H_6)_2Cr$,[10–11] $Fe(butadiene)(CO)_3$ and $Fe(cyclobutadiene)(CO)_3$,[12] Zeise's salt,[13] zinc[14] and magnesium[15] alkyls, among others, were published in the journal *Organometallics* by its founding editor, Dietmar Seyferth, from 2001 to 2009.

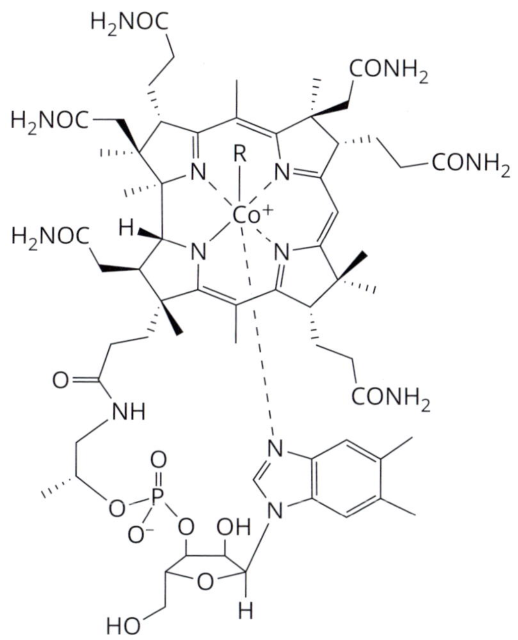

Fig. 18.4 The structure of cyanocobalamin (synthetic vitamin B-12, R = CN) with a Co-CN linkage, methylcobalamin (R = CH_3) with a Co-CH_3 linkage, and 5′-deoxyadenosylcobalamin (R = adenosyl) containing a cobalt-sugar linkage. These are all biologically relevant organometallic complexes.

Although organometallic compounds have been known since the late 1700s, organometallic chemistry and organometallic reagents have been present in biological systems since prehistoric times. As just one example, consider Vitamin B-12. The synthetic version of this vitamin, first prepared by Woodward and Eschenmoser in 1972,[16–17] contains a Co–CN linkage (Fig. 18.4), which can be converted *in vivo* into biologically relevant 5′-deoxyadenosylcobalamin and methylcobalamin, each containing a Co–C bond. These two coenzymes catalyze free-radical mechanisms and are used as methylating agents (post-translational methylation of proteins), reducing agents (reducing ribonucleotides to deoxyribonucleotides), and as a tool to carry out skeletal rearrangements during protein synthesis and metabolism. The coenzyme contains a cobalt atom within a corrin ring with a σ-bond to carbon. This coenzyme is made by only a few specialized bacteria and must be ingested by higher organisms.

There are many types of ligands common to what is generally referred to as "organometallic chemistry", and common types, along with some descriptions of electronic structures, have been discussed in Chapters 11 and 12. All of the previous discussions concerning Werner complexes are applicable here, but oftentimes the focus is more synthetic than electronic in this subfield, which will be the crux of this discussion. Olefins, aromatic π-complexes, carbon monoxide, and alkyls are of particular importance to understanding organometallic chemistry, and the student may wish to review these ligands discussed in previous chapters before continuing.

18.2 Common Organometallic Mechanistic Steps

Perhaps the four most common reactions in organometallic chemistry are oxidative addition, reductive elimination, insertion, and elimination. These four reactions form two pairs of reactions that are the microscopic reverse of each other, meaning that the forward and reverse direction of the reaction follows the same pathway. Oxidative addition is the microscopic reverse of reductive elimination, and insertion is the microscopic reverse of elimination. Understanding the energies of molecular orbitals and steric requirements for one direction of the reaction gives insights towards the understanding of the reverse reaction. General trends about these transformations are relatively straightforward to discern, but the specifics often become quite complicated, challenging, and interesting. In addition to the four main reactions, there are others such as cycloadditions and oxidation/reduction events. Understanding individual mechanistic steps is the first part of understanding catalysis. Once one understands how a chemical reaction proceeds, one can design complexes that can carry out the chemical reaction more efficiently, more rapidly, or by using less expensive reagents.

18.2.1 Oxidative Addition and Reductive Elimination

The first pair of reactions that will be considered are the oxidative addition and reductive elimination reactions. Oxidative addition is a very important mechanistic step, as it is a commonly observed way to coordinate (add) a substrate to a metal center. As such, it is often written as the first, or one of the first, steps in a series showing the conversion of a substrate to a product (Eq. 18.1). Reductive elimination involves the removal (elimination) of a product from the metal center, and therefore it is often proposed as one of the last steps of a chemical reaction (Eq. 18.2). An excellent tutorial on the oxidative addition reaction was published in 2015.[18]

$$X - Y + ML_n \rightarrow M(X)(Y)L_n \quad \text{oxidative addition} \qquad \textbf{18.1}$$

$$M(X)(Y)L_n \rightarrow X - Y + ML_n \quad \text{reductive elimination} \qquad \textbf{18.2}$$

Oxidative addition is a chemical reaction in which a substrate binds, and simultaneously oxidizes, a metal in the x-oxidation state in the starting material that will be in the $(x + 2)$-oxidation state after addition. The metal center reaches a higher oxidation state and increases its coordination number. As a result, oxidative addition is encouraged for metals in a low oxidation state with low coordination number. Furthermore, addition to the metal is encouraged by ligands that are small and electron donating. The substrate for oxidative addition is typically a neutral molecule that is reduced by two electrons into two formally anionic parts that then each bind to the metal center. Since the electrons come from the metal center, the substrate

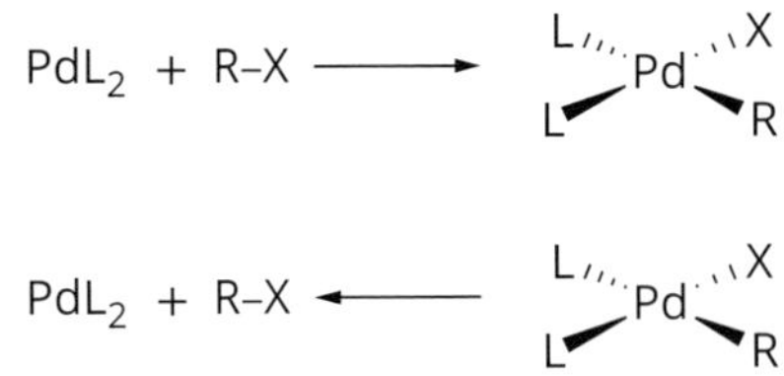

Forward: Oxidative addition
Encouraged by small, electron-rich L

Reverse: Reductive elimination
Encouraged by large, electron-deficient L

Oxidation state:	0	+2
Coordination number:	2	4
Valence electron count:	14	16
d^n count:	d^{10}	d^8

Fig. 18.5 Electron-counting as a method of classification for oxidative addition and reductive elimination reactions.

is reduced, and the metal center is oxidized. Fig. 18.5 shows the oxidative addition (moving left to right) of an alkyl halide to a palladium(0) starting material, and reductive elimination (moving right to left) of a palladium(II) alkyl halide complex.

During the reductive elimination, the oxidation state, coordination number, and valence electron count of the metal all decrease by 2, while the d^n count of the metal increases by 2. As a result, reductive elimination is more favorable for metals in high oxidation states with high coordination numbers. Further, reductive elimination from the metal can be encouraged by ligands that are sterically encumbering and electron withdrawing.

Substrates (X–Y) for the reaction are either electrophilic (polar), or nonelectrophilic (nonpolar). Electrophilic substrates are polar (or polarizable) molecules such as the halides CH_3X, X_2, Ar–X, or H–X, where X is Cl, Br, I or a pseudohalide such as triflate. Nonelectrophilic substrates include H_2; the H-element bond in H–SiR_3, H–BR_2, H–NR_2; the S–S bond in disulfides; and other related molecules. One substrate that currently receives a great deal of attention is the H–C bond in hydrocarbons. While C–H bond activation is an extremely important target for the organometallic community, selective activation of many types of these bonds is difficult, although some variants are becoming practical for synthetic applications (Section 18.3.11).

Typical starting materials for the oxidative addition reaction include 16-electron square planar complexes, and typical products are 18-electron octahedral complexes. The geometric considerations for the reaction depend very highly on substrate, metal center, and ligands. The mechanism of oxidative addition also depends on the substrate type (electrophilic or nonelectrophilic, Section 18.2.1.1).

Before diving deeply into mechanistic aspects of the reaction, several example reactions will be described. The platinum diphosphine dimethyl complex activates the polar substrate methyl iodide (Fig. 18.6, dppe = 1,2-bis(diphenylphosphino) ethane, a bidentate diphosphine). The reaction converts a 4-coordinate, 16-electron platinum(II) complex into a 6-coordinate, 18-electron platinum(IV) complex. Iodomethane is reduced to a methyl group and iodide, and each of these ligands ends up coordinated to the Pt center at the end of the reaction. Labeling of the methyliodide (ICD_3) shows that the reaction occurs in such a way that the I- and CD_3-groups are *trans* in the product. This suggests an S_N2-like mechanism with a $[PtMe_3(dppe)]^+ I^-$ intermediate.[19–20]

Wilkinson's catalyst, $RhCl(PPh_3)_3$, is commonly used for the hydrogenation of olefins. This complex reacts with the nonpolar substrate dihydrogen (Fig. 18.7). The 4-coordinate 16-electron rhodium(I) center is converted into a 6-coordinate 18-electron rhodium(III) center. Dihydrogen is reduced to two

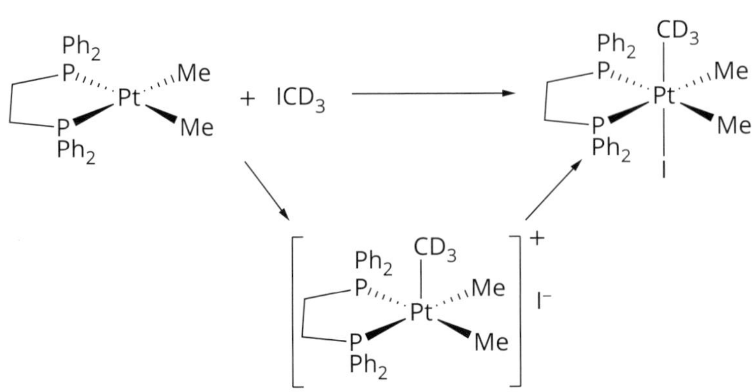

Fig. 18.6 Oxidative addition of CD_3I to $PtMe_2(dppe)$ occurs to give the CD_3- and I-groups mutually *trans*, presumably through an S_N2-like mechanism with an intermediate like the one shown.

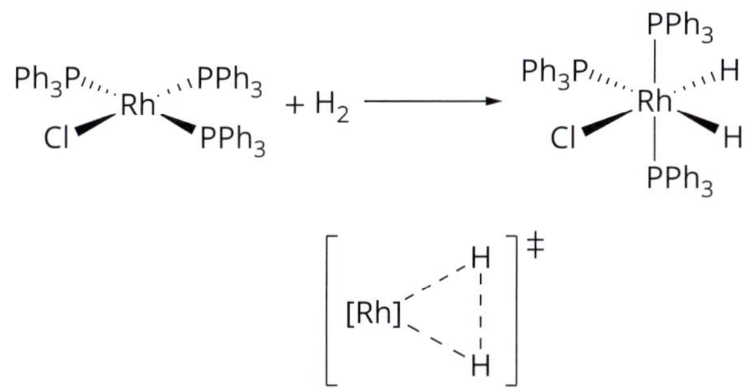

Fig. 18.7 Addition of H_2 to $ClRh(PPh_3)_3$, Wilkinson's catalyst, occurs through a 3-center intermediate where the new Rh–H bonds are made as the H–H bond is broken. The kinetic product of this type of reaction places the two hydrides *cis*.

hydrides, and each is bound to the metal in a *cis*-relationship at the end of the reaction. This reaction is proposed to occur through a 3-center reaction, and the proposed transition state is also shown in Fig. 18.7. Reactions of this type have been studied using *para*-H_2 by NMR spectroscopy, a technique popularized by Eisenberg and coworkers.[21–24]

Another and more complex example is one in which the oxidative addition occurs across two metal centers, but the net changes in oxidation state, coordination number, and electron counts still hold when considering both metal centers together (Eq. 18.3). Two Mn(0) centers in metal–metal bond complex $(OC)_5Mn–Mn(CO)_5$ are converted into two 18-electron Mn(I) centers. The bromine is (formally) reduced to two bromide ions, each of which coordinates to a single manganese.[25–26]

$$(OC)_5Mn - Mn(CO)_5 + Br_2 \rightarrow 2\,MnBr(CO)_5 \qquad \textbf{18.3}$$

Reductive elimination is a chemical reaction in which two ligands that start out coordinated to a transition-metal center bind to one other, allowing them to be removed from the metal center with a simultaneous reduction of the metal by two electrons and reduction of coordination number by 2. The ligands, which are typically formally anionic, are oxidized and joined together as they eliminate from the metal center. The oxidation state, coordination number, and valence electron count of the metal all decrease by 2, while the d^n count increases by 2.

When considered in the reverse direction, the examples shown for oxidative addition are reductive elimination reactions. In the first example (Fig. 18.6) the 6-coordinate 18-electron platinum(IV) complex undergoes a reaction to generate the 4-coordinate 16-electron platinum(II) complex and iodomethane. This is, in fact, the direction that the reaction was studied by Goldberg to elucidate the mechanism (*vide infra*). In the second example (Fig. 18.7) the 6-coordinate 18-electron rhodium(III) complex, $RhH_2(Cl)(PPh_3)_3$, is reduced to the 4-coordinate 16-electron rhodium(I) complex, $RhCl(PPh_3)_3$, and dihydrogen. The reaction of dihydrogen with Wilkinson's catalyst is, in fact, reversible, though the equilibrium lies to the right. The third example (Eq. 18.3), if it were to run in reverse, would perhaps result in the transformation of the 6-coordinate 18-electron manganese(I) bromide into the 5-coordinate 17-electron manganese(0) radical $\bullet Mn(CO)_5$, followed by reformation of the $Mn_2(CO)_{10}$ dimer and bromine.

18.2.1.1 Concerted versus Stepwise Additions

The two broad mechanistic categories for oxidative addition and reductive elimination are stepwise and concerted (Fig. 18.8). Polar, electrophilic substrates tend to react in a stepwise fashion (Fig. 18.8a), while nonpolar, nonelectrophilic substrates usually undergo a concerted addition (Fig. 18.8b). The nature of the addition dictates the geometrical placement of the added ligands.

The stepwise mechanism of oxidative addition and reductive elimination for an electrophilic substrate is very similar to

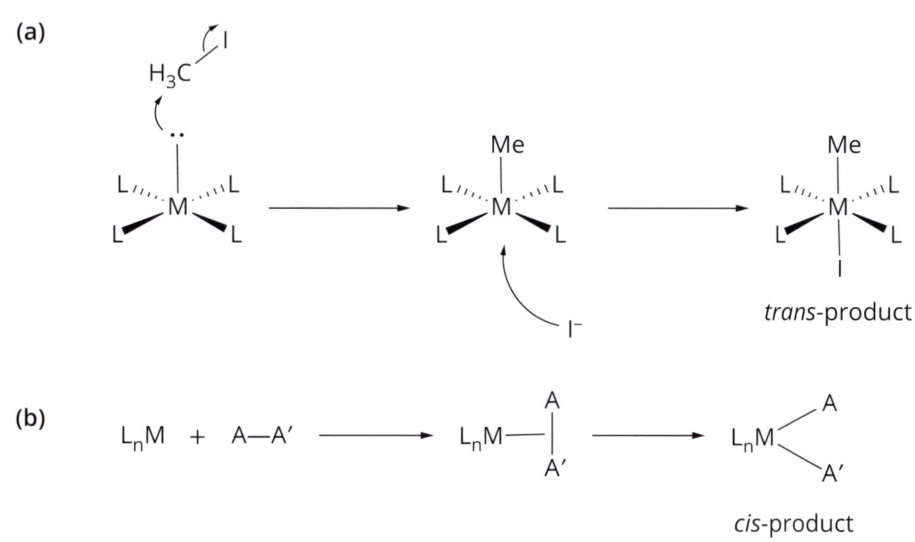

Fig. 18.8 (a) Stepwise versus (b) concerted oxidative addition mechanisms.

that of the S_N2 reaction. An electron pair from the transition-metal center acts as a nucleophile on the substrate, displacing the electronegative substituent as a counterion. This counterion can then coordinate to the metal center, typically in a *trans*-position to the original substrate addition site. In the example shown in Fig. 18.6, square-planar Pt(dppe)Me_2 reacts initially to form a transient 5-coordinate species [Pt(dppe)$Me_3]^+$. This species then reacts with external iodide to form Pt(dppe)Me_3I.[27–28]

The mechanism for the nonelectrophilic substrates is quite different, proceeding in a concerted manner by which the bonds from the metal to both parts of the substrate form simultaneously. As such, the geometry of the addition must be *cis*. This mechanism requires the presence of an empty orbital on the metal for coordination of the substrate. Once the substrate binds to the metal by using the electron pair from the σ-bond, the σ-bond is partially weakened, activating the substrate. Removal of electrons from the metal center and reducing the substrate results in the formation of two new M–E bonds, and the oxidative addition reaction is complete.

The competitive reductive elimination of either CH_3I or C_2H_6 from Pt(dppe)Me_3I was studied by Goldberg and coworkers, and in this one example both possible pathways (concerted or stepwise) can take place. Through examination of the relative rates of formation of the two products, a complete thermodynamic reaction coordinate diagram was prepared (Fig. 18.9). The reductive elimination to form methyl iodide proceeds by initial loss of iodide, with concomitant reduction in the electron count at the metal center, followed by external attack of iodide on the apical methyl group. This reaction has an overall thermodynamic barrier of 104 kJ/mol (25 kcal/mol) and is endothermic by 66 kJ/mol (16 kcal/mol). However, methyl iodide is a reactive substrate and can be captured by the 4-coordinate Pt complex to regenerate the starting material. Reductive elimination to form ethane also proceeds by initial loss of iodide, but this is followed by a concerted reductive elimination by using the nonpolar mechanism of two of the *cis*-methyl groups to generate the 3-coordinate intermediate [Pt(dppe)Me]$^+$, which captures iodide to form Pt(dppe)MeI. This reaction has a larger thermodynamic barrier but is exothermic by 105 kJ/mol (25 kcal/mol). As the barrier for methyl iodide is lower, at early reaction times the reaction favors the formation of that product. As the reaction to form ethane is irreversible, over long periods of time (days) the formation of ethane is favored.

There are other types of oxidative addition mechanisms known. For example, radical pathways for net oxidative addition have been identified, and interested readers are directed to more specialized sources.[29]

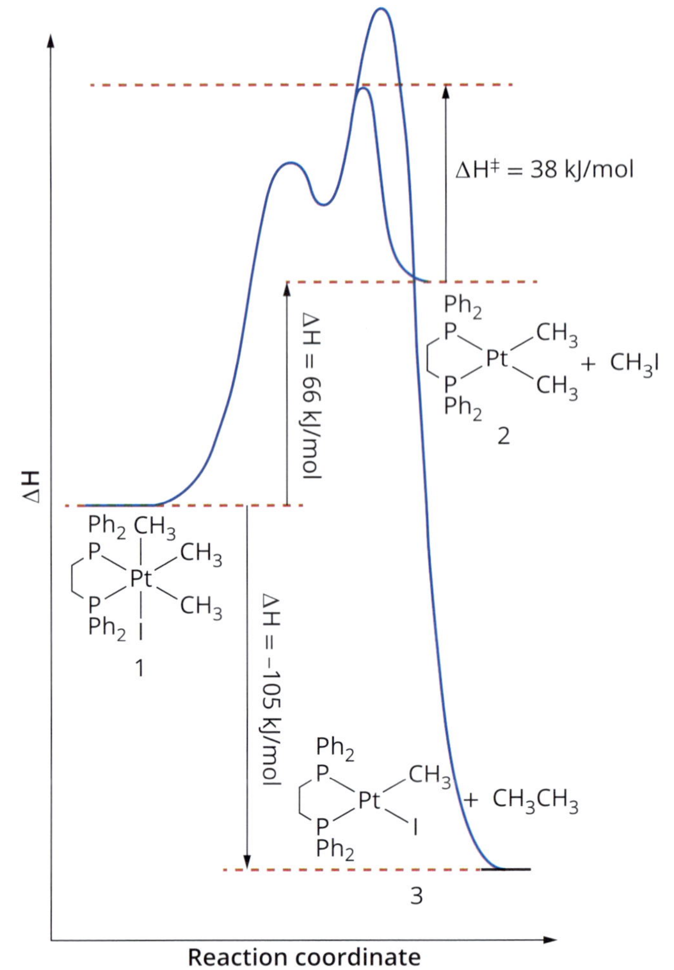

Fig. 18.9 Reaction-coordinate diagram for the two possible reductive elimination reactions starting from Pt(dppe)Me_3I. Both reactions start by loss of iodide to generate the 5-coordinate intermediate [Pt(dppe)$Me_3]^+$. The stepwise process gives Pt(dppe)Me_2 and CH_3I; the concerted process gives Pt(dppe)MeI and ethane.

18.2.1.2 Periodic Trends and Ligand Effects

As the oxidative addition reaction requires the removal of electrons from a metal center, increased electron density at the metal can ease the transformation. Strong σ-donating ancillary ligands and the absence of π-acceptors would favor oxidative addition. Alternatively, to prevent oxidative addition, electron-withdrawing ligands can be used. This trend can be seen in rate data for the reaction of IrX(CO)$(PPh_3)_2$ with H_2 and O_2 (Table 18.1). The interpretation here is that the more electron-withdrawing chloride exhibits a 10–100-fold rate decrease of the oxidative addition relative to the less electron-withdrawing iodide, while bromide has intermediate reactivity. The $\Delta H^\ddagger$ values are unremarkable, but the $\Delta S^\ddagger$ values in the range of −20 e.u. suggest an increase in polarity at the transition state or loss of degrees of freedom at the transition state. The reaction of IrX(CO)$(PPh_3)_2$ with CH_3I shows the opposite trend in terms of rate where the chloride complex reacts two to three times faster than the bromide or iodide. A steric argument was proposed in which the larger halogens interact with the phenyl rings of the PPh_3 ligands, causing them to block the axial interaction with the incoming substrate.[30] This argument was

Table 18.1 Kinetic data for oxidative addition reactions of $IrX(CO)(PPh_3)_2$

X	Reactant	Rate Const (M^{-1} sec^{-1})	$\Delta H^‡$ (kcal/mol)	$\Delta S^‡$ (cal/mol · K = e.u.)
Cl	H_2	0.67	10.8	−23
Br		10.5	12.0	−14
I		>100	—	—
Cl	O_2	3.4×10^{-2}	13.1	−21
Br		7.4×10^{-2}	11.8	−24
I		30×10^{-2}	10.9	−24
Cl	CH_3I	3.5×10^{-3}	5.6	−51
Br		1.6×10^{-3}	7.6	−46
I		0.9×10^{-3}	8.8	−43

Data from Chock and Halpern.[30] Reactions were run in benzene at 25 °C.

based on the significantly larger negative $\Delta S^‡$ for the reaction. From these data, the authors proposed an S_N2-reaction type rather than a concerted addition for CH_3I undergoing oxidative addition with the iridium complexes.

The opposite trends would be expected to hold for reductive elimination. Reduced electron density due to the presence of strong π-acidic ligands would be expected to encourage the reaction. Also, large ligands would be expected to promote reductive elimination. As might be expected, ligands have been developed over the years to enhance either the forward or back reaction. Electron-rich, bulky phosphine ligands encourage oxidative addition by their increased electron density and encourage reductive elimination by their steric bulk. Careful balance between both steric and electronic effects can enhance either the forward or backward rate of the reaction. Other reagents must sometimes be added to encourage the desired chemistry as well. To encourage reductive elimination, sometimes an oxidizing agent is added, such as $CuCl_2$ or F^+.[31] The reduced electron density at the metal then allows the reductive elimination to proceed more smoothly.

Although it is generally accepted that electron-donating ligands favor oxidative addition, this has recently been shown to not be a general principle. A recent report suggests that under certain circumstances, less electron-donating ligands strongly favor oxidative addition. Specifically, the study involved pincer complexes of iridium where the ligand trans to the addition site was altered. As has been mentioned many times, it is quite possible that any generalization regarding chemistry has an exception. Goldman and coworkers offered an orbital-based rationale employing sd^n-hybridization (see Section 14.2 and Box 17C).[32]

18.2.1.3 Electronically Noninnocent Ligands

A good rule of thumb to remember is that d^0 metals cannot do oxidative addition reactions. By definition, a d^0 metal has no valence electrons at the metal center; it is already in its highest oxidation state and cannot be further oxidized without the removal of core electrons. Given this information, examine the following reaction reported by Heyduk in 2005.[33]

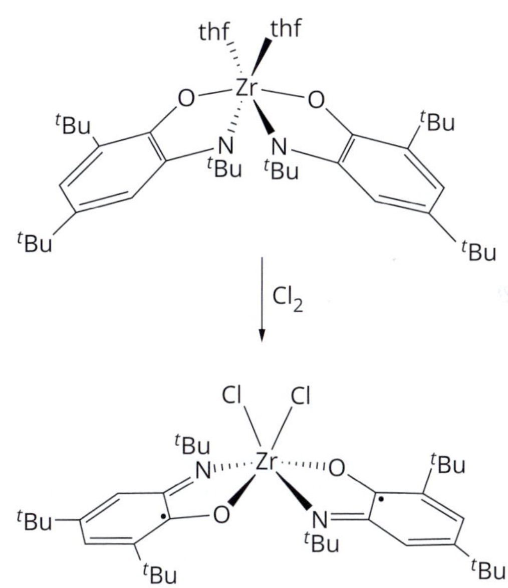

18.4

Addition of chlorine to a metal complex to form two chloride ligands is an addition, and a reduction of the chlorine, but oxidation of Zr(IV) to "Zr(VI)" is not possible. Instead, the ligands are oxidized from amidophenolates, dianionic X_2 ligands, to iminosemiquinones, monoanionic XL (one covalent and one dative) ligands with a radical located on each aromatic ring. This method of allowing an oxidative addition reaction to occur for d^0 metals illustrates the concept of using redox active ligands in organometallic chemistry. Redox-active ligands are finding use across organometallic chemistry and challenge the assumption that redox events only take place at a metal center. This is a bourgeoning research area, and there is much yet to be explored.

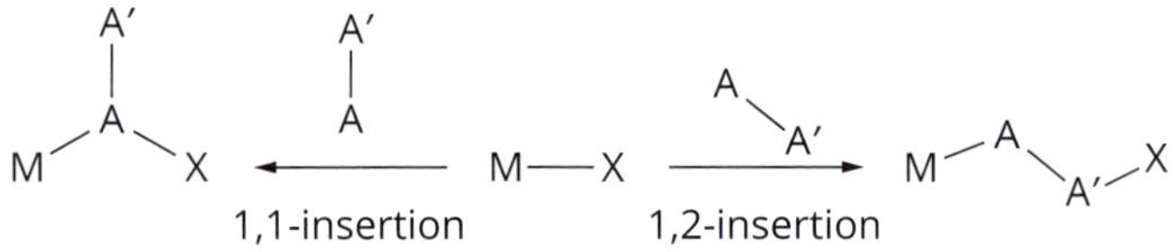

Fig. 18.10 (*left*) A 1,1-insertion where both the metal and ligand end up on the 1-position of the inserting substrate, and (*right*) a 1,2-insertion where the metal and ligand end up on the 1- and 2-positions of the inserting substrate.

18.2.2 Insertions and Eliminations

Once substrates are bound to a metal center, chemistry at the ligands has to happen in order for a transformation to take place. Oxidative addition immediately followed by reductive elimination, described in the previous section, would result in no net chemical reaction. The second common pair of reactions in organometallic chemistry that we will discuss are insertions and eliminations, which are again the microscopic reverse of one another. The terms "insertion" and "elimination" refer to reactions that change the chemical composition of the ligands bound to a metal center. These reactions are useful for transforming one chemical species into another, and are often proposed as steps in the middle region of a longer chemical transformation or catalytic cycle. There are two broad classes of insertion reactions commonly encountered: 1,1- and 1,2-insertion. Formally, these reactions are the insertion of an unsaturated substrate, such as carbon monoxide or ethylene, into a M–X bond. Since the metal and ligand end up on the 1- and 1-positions in carbon monoxide, or on 1- and 2-positions of ethylene, these are called 1,1- or 1,2-insertions, respectively (Fig. 18.10).

18.2.2.1 Carbonyl 1,1-Insertions

The classic example of an insertion reaction is the reaction that transforms $Mn(CH_3)(CO)_5$ into $Mn[C(=O)CH_3](CO)_5$ under an atmosphere of CO. The methyl group on the manganese is transformed into an acyl group, and the original name for the reaction was the carbonyl insertion reaction, because in the product it appears that a CO ligand has inserted itself into the $Mn–CH_3$ bond. Since other substrates, such as NO, can achieve this reaction, it is more correctly referred to as the migratory insertion reaction. Both of these names appear to imply something about the mechanism of the reaction, but it was not until careful experiments were carried out that the details of the chemical transformation were known. It turns out that the mechanism of insertion can be different with different metals and substrates.

In short, an example of the 1,1-insertion of carbon monoxide is shown in Eq. 18.5. In this case it was shown that the reaction occurs by migration of the methyl to one of the pendant CO ligands. The details of that interesting investigation are presented in the next section as an Advanced Topic, but these types of migratory insertions are very common, especially for CO and ligands electronically related to CO, such as isonitriles.

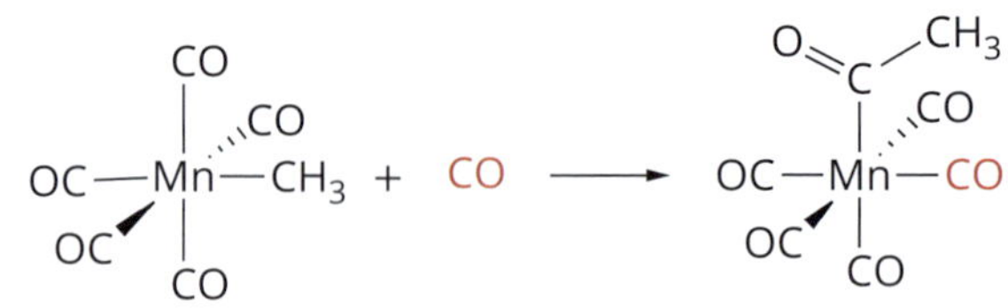

18.5

18.2.2.2 Determination of the Mechanism of 1,1-Insertions (Advanced Topic)

There are three readily conceivable mechanisms for the insertion of CO into a metal–alkyl bond: alkyl migration, direct insertion, and carbonyl migration. To determine the mechanism of the reaction, ^{13}C-labeled CO was used in a set of subtle but powerful experiments by Fausto Calderazzo and coworkers.[34] This labeled CO was easy to follow by IR spectroscopy (Box 11A), and careful experimental design allowed the mechanism to be elucidated.[35]

First consider the reaction of $Mn(CH_3)(CO)_5$ with ^{13}CO. The stereochemical outcomes expected for the three mechanisms are shown in Fig. 18.11. The top pathway results from methyl migration to one of the four equivalent *cis*-CO ligands, followed by trapping with the labeled ^{13}CO. The middle pathway is direct insertion of the ^{13}CO into the $Mn–CH_3$ bond. The bottom pathway results from CO migration from one of the four equivalent *cis*-positions to the methyl group, followed by trapping with the labeled ^{13}CO. The transient intermediate for the top and bottom reactions is postulated to be the solvent adduct $Mn[C(=O)CH_3](CO)_4(sol)$ based on the fact that a solvent adduct can sometimes be detected by IR spectroscopy.

The products of the top and bottom mechanisms are identical (although drawn in different orientations), and only the middle path results in acyl with ^{13}C label. Since no ^{13}C labeled acyl ligand was ever detected, the middle mechanism (direct insertion) can be ruled out. However, it is possible that the direct insertion mechanism does occur, but just at a slower rate than one of the other two. This is an example of "negative evidence", and one should be cautious to base a mechanistic proposal on this type of evidence. It is better (but of course, not always easy) to design an experiment where seeing a product informs the researcher on the mechanism rather than drawing conclusions from *not* seeing a product. The results of this study also confirm that rearrangement, at least for the timescale and temperature of this reaction, does not occur to a large extent, otherwise a statistical distribution of the ^{13}C labelled products would be expected in both *cis*- and *trans*-positions.

Next, consider the reaction of *cis*-$Mn(CH_3)(^{13}CO)(CO)_4$ with unlabeled CO. As before, three different product distributions are possible (Fig. 18.12). For now, the direct insertion mechanism is retained as a possible candidate, since we

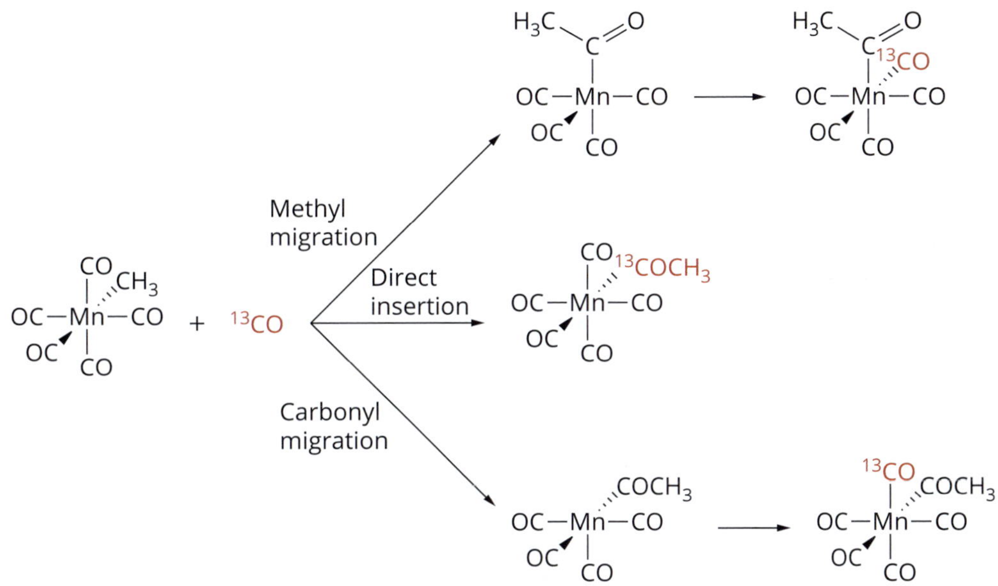

Fig. 18.11 The three possible products for methyl migration, direct insertion, and carbonyl migration (top, middle, and bottom, respectively) in the reaction of $Mn(CH_3)(CO)_5$ with ^{13}CO.

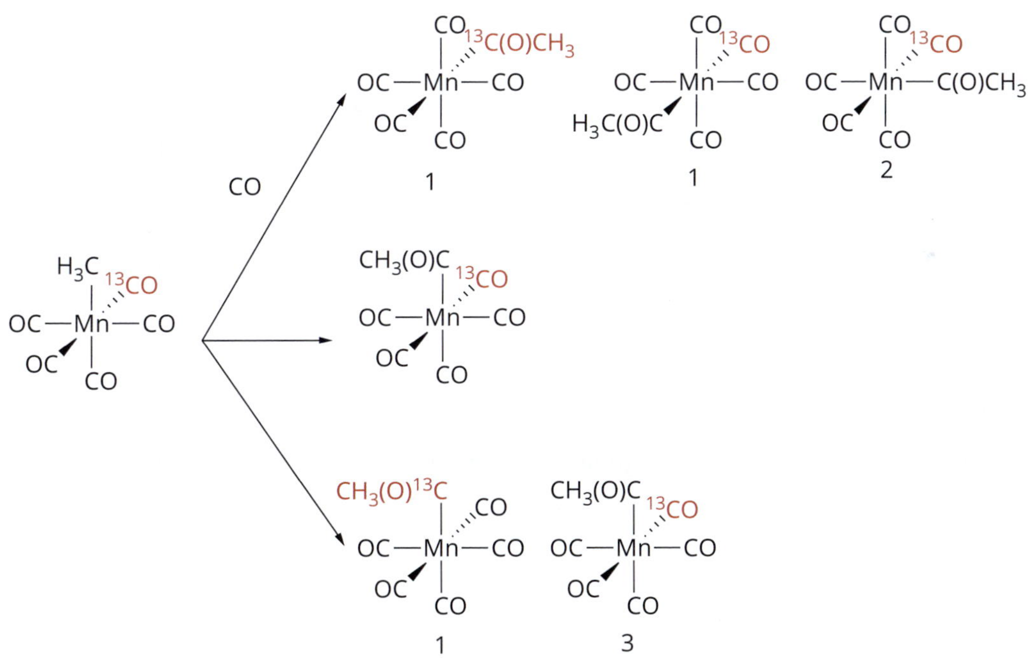

Fig. 18.12 The three possible product distributions for alkyl migration direct insertion, and carbonyl migration (top, middle, and bottom, respectively) in the reaction of $Mn(CH_3)(^{13}CO)(CO)_4$ with CO.

only had indirect evidence against it. If the methyl group migrates according to the top pathway, it would migrate to one of the four *cis*-carbonyls present, leaving an empty coordination site. In a second step, the incoming CO adds to the vacant position. This mechanism would produce a 1:1:2 ratio of ^{13}C labeled acyl, *trans*, and *cis*-products. The middle pathway, direct insertion, would result in only *cis*-product. The carbonyl migration pathway would produce a mixture of one part ^{13}C-labeled acyl to three parts *cis*-product. This experimentally determined product distribution was a 1:1:2 ratio as predicted from the methyl migration pathway.

Finally, consider the product distribution for the *decarbonylation*, the reverse of 1,1-insertion of CO, of *cis*-Mn[C(=O)CH_3](^{13}CO)$(CO)_4$, shown in Fig. 18.13. The top mechanism of CO migration can occur only when one of the three *cis*-CO or *cis*-^{13}CO ligand is lost, followed by migration of the methyl group to the new position. This would result in a 1:1:2 ratio of unlabeled product to *trans*-product to *cis*-product. Direct

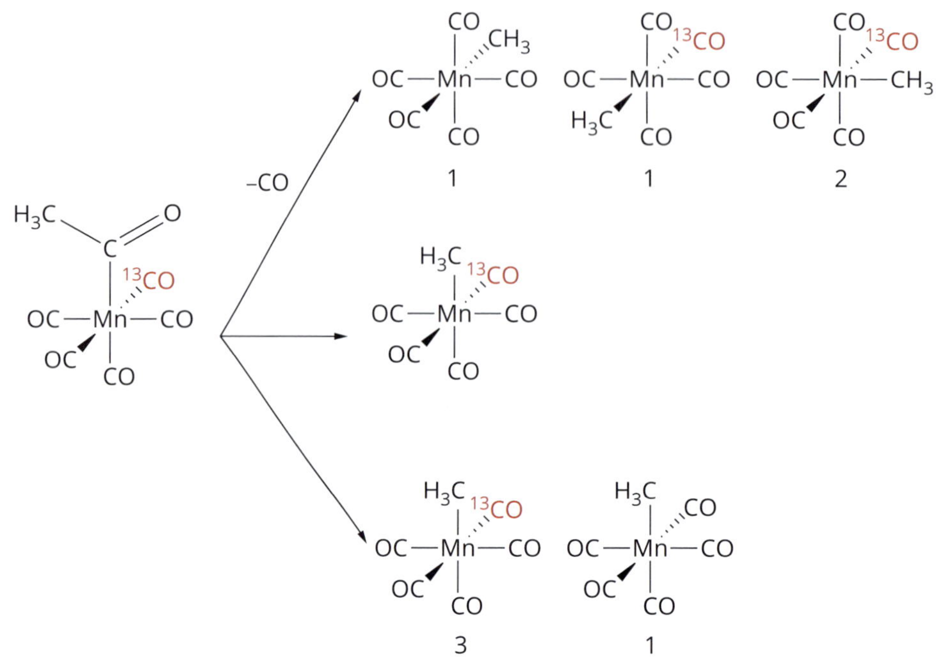

Fig. 18.13 The three possible products for alkyl migration direct insertion, and carbonyl migration (top, middle, and bottom, respectively) in the decarbonylation reaction of $Mn[C(=O)CH_3)](^{13}CO)(CO)_4$.

decarbonylation by the middle pathway of direct deinsertion can only result in the *cis*-product. Decarbonylation by CO migration would produce a 3:1 ratio of *cis*-product to unlabeled product. Experimentally, a 1:1:2 ratio matching that of the alkyl migration mechanism is observed.

Taken together, these experiments show several important features of the carbonyl insertion reaction. First, the forward and backward reactions are occurring by the same mechanism, as they must according to the *principle of microscopic reversibility.* Second, at least in this manganese system, alkyl migration is the observed mechanism. Third, there is a *cis*-relationship between the position of the alkyl group in the starting material and the acyl group in the product. This *cis*-relationship would be true even for the carbonyl-migration mechanism. These experiments also showcase the manner in which carefully designed labeling experiments can be used to elucidate mechanism.

Not all insertion reactions go by this mechanism, however. In fact, in some cases the mechanism can switch, depending on solvent. This was examined by using the chiral complex $CpFe(CO)(PPh_3)R$ by Flood and coworkers, and it was found that in nitromethane, alkyl migration was observed, while in donor solvents, such as DMSO or HMPA, CO migration was observed (Fig. 18.14).[36] The stereochemistry at the chiral iron

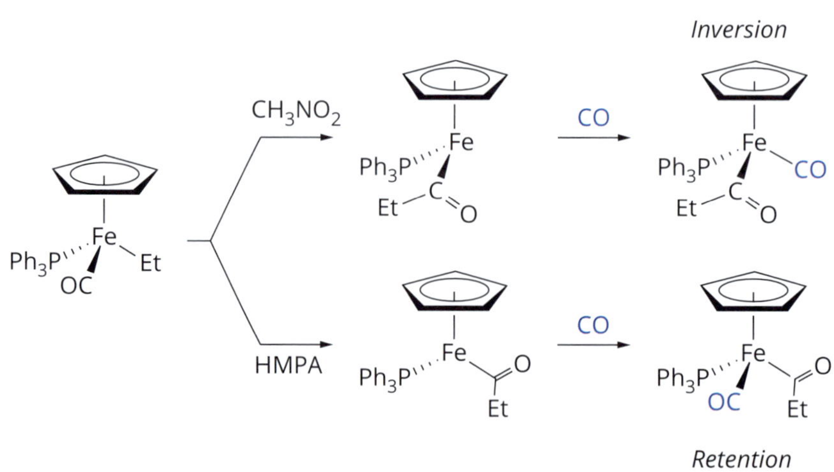

Fig. 18.14 A chiral iron complex that undergoes alkyl migration in nitromethane and carbonyl insertion in donor solvents.[36]

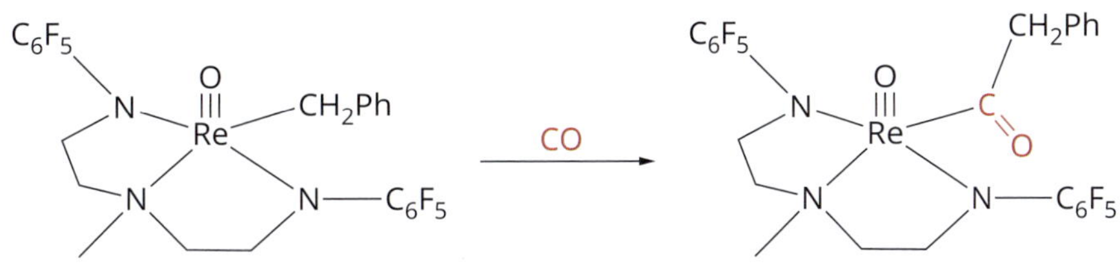

Fig. 18.15 Direct insertion of a carbon monoxide unit into a rhenium alkyl bond.

center showed inversion or retention, respectively, for the two mechanisms.

It is also possible to observe direct carbonyl insertion in some cases, as illustrated recently in a series of papers by Ison.[37–40] An oxo rhenium complex with a tridentate nitrogen ligand was postulated to either first undergo weak CO binding followed by CO insertion, or to undergo direct insertion. The kinetics and labeling studies, including some quantum-mechanical calculations, were strongly supportive of a mechanism that proceeded by direct CO insertion into the Re–C bond (Fig. 18.15).

Carbon monoxide is by far the most studied ligand that undergoes 1,1-insertion, and while the previous focus was on insertion into alkyls, CO will insert into many different ligand types, often reversibly. However, CO is not the only ligand that will 1,1-insert. Many of the ligands with similar electronic structure to CO, such as NO and isonitriles, will do similar reactions.

18.2.2.3 Ethylene 1,2-Insertions

The 1,2-insertion mechanism is carried out on an immense scale in the polymerization of ethylene (around 80 million tons per year globally), and propylene (around 50 million tons per year globally). Currently, we are producing ~34 lbs of polyolefin per person on Earth per year! The hydrogenation of olefins is an example of 1,2-insertion that may be familiar from organic chemistry. The reaction begins by coordinating a reactive, unsaturated ligand, such as ethylene, to a metal complex that contains an X ligand. In the case of hydrogenation, the X ligand is a hydride; in the case of polymerization, the X ligand is the growing polymer chain, an alkyl group.

Coordination of the unsaturated bond to the metal center activates it for nucleophilic attack by the X ligand. The metal and X ligand, both being on the same side of the unsaturated substrate, end up in a *syn*-arrangement in the final product. This *syn*-addition is a hallmark of a 1,2-insertion reaction, differentiating it from other possible mechanism types such as external attack of nucleophile onto a coordinated substrate, which would proceed in an *anti*-fashion. The mechanism of the reaction contains a planar, four-membered transition state with simultaneous M–X bond breaking, M–C bond formation, and C=C π-bond breaking (Fig. 18.16). Triple bonds, such as alkynes and nitriles, can insert as well, and are often favored over olefins for insertion reactions.

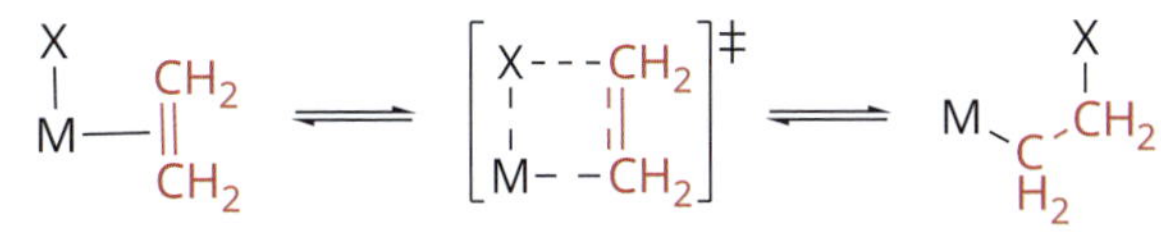

Fig. 18.16 Mechanism of 1,2-insertion reaction showing four membered transition state. (Image source: the authors.)

18.2.2.4 Agostic Interactions

Alkyl complexes can sometimes engage in either an α- or β-agostic interaction.[41] This term derives from the Greek *agostos*, "to draw towards",[42] and is used to indicate that a hydrogen atom is covalently bonded to both a carbon and to a transition-metal atom.

This distortion brings a C–H bond into close proximity with the metal center and results in significant changes in both the NMR and IR spectra. The chemical shift of the proton and carbon involved in the agostic interaction can shift from that of an alkane to more that of a hydride. The C–H coupling constant ($J_{C\text{-}H}$) can shift to lower values: 70–100 Hz rather than 125–130 Hz normally seen for an sp^3-hybridized alkane. The C–H stretching frequency in the IR spectrum can sometimes be observed to decrease: 2,200–2,600 cm^{-1} for the C–H involved in the interaction rather than 2,850–2,950 cm^{-1} for an alkane. X-ray crystallography and neutron diffraction reveals significant distortions at the α-carbon undergoing the interaction, with M–C–C angles increasing from 109° to more than 160°.

If the metal producing the interaction is d^0 and has no electrons to backbond, this is an interaction discussed previously in Chapter 8 on main-group bonding, a 3-center 2-electron bond or τ-bond. If the metal has *d*-electrons, there could possibly be backbonding into the C–H σ*-orbital, in which case the bonding arrangement could be similar to a metal dihydrogen complex.

The tantalum carbene $Ta(CHCMe_3)Cl_3(PMe_3)$ has a C–H stretching frequency of 2,605 cm^{-1}, significantly reduced from the typical value for aliphatic C–H bonds of 2,900–3,000 cm^{-1}. The carbon–hydrogen coupling constant is observed to be 101 Hz, significantly reduced from a typical sp^2-carbon J_{CH} value of 150–200 Hz. These data suggests a significant reduction in C–H bond strength and is attributed to an agostic interaction between the C–H bond and an adjacent empty orbital on tantalum (Fig. 18.17). Tantalum here is in its highest oxidation state, and a τ-bond results where the metal is involved in a 3-center 2-electron bond with H and C.

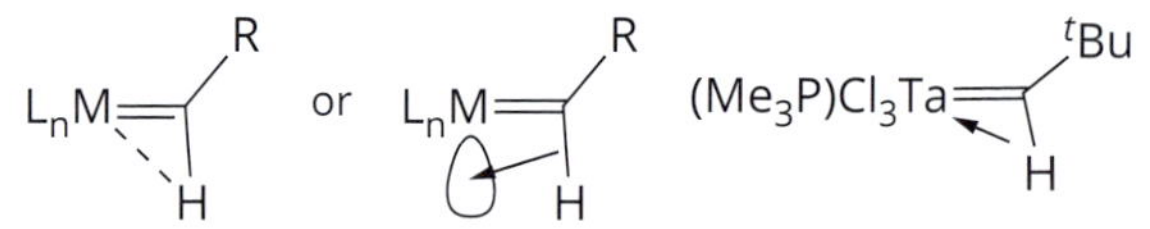

Fig. 18.17 An α-agostic interaction, a type of τ-bond, found in $Ta(CHCMe_3)Cl_3(PMe_3)$: δ = 5.3 ppm, J_{CH} = 101 Hz, ν_{CH} = 2605 cm^{-1}.[43]

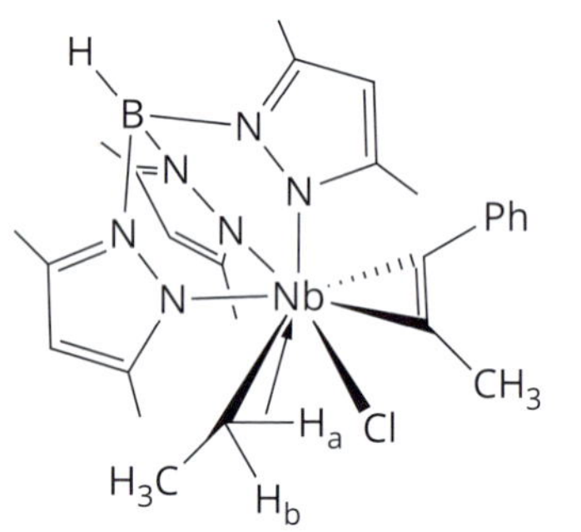

Fig. 18.18 A niobium ethyl complex that undergoes an α-agostic interaction. The two α-hydrogens have very different chemical shifts and C-H coupling constants. $NbC\underline{H_a}H_bCH_3$: δ = 3.84 ppm, J_{CH} = 108 Hz, and $NbCH_a\underline{H_b}CH_3$: δ = 0.37, J_{CH} = 129 Hz; Nb-C-C = 126°.[44]

The niobium ethyl complex shown in Fig. 18.18 also undergoes an α-agostic interaction. The two α-hydrogens on the ethyl group have very different chemical shifts and coupling constants. For the normal C–H bond, the chemical shift is 3.84 ppm and the $^1J_{CH}$ is 129 Hz, while for the C–H bond that is undergoing the agostic interaction, the chemical shift is 0.37 ppm and the $^1J_{CH}$ is 108 Hz. X-ray crystallography shows that the Nb–C–C bond angle is 126°, which is opened substantially from the expected tetrahedral angle of 109.5°.[44] This is an example of a static agostic interaction; in many cases, there is fluxionality between the protons and H_a and H_b would then appear equivalent, with intermediate chemical shifts and coupling constants represented by the two different actual environments. In this case, one of the two diastereotopic hydrogens is perhaps favored for coordination by the chiral metal center.

The β-agostic interaction looks to be a trapped transition state for the 1,2-insertion reaction (or its microscopic reverse, the β-hydrogen elimination; Section 18.2.2.6). The intermediate can sometimes be observed on the pathway between coordinated alkene hydride and metal alkyl, shown as a transition state in Fig. 18.16. As in the α-agostic interaction, molecules undergoing this interaction have significant changes in NMR chemical shifts, C–H coupling constants, and bond angles. In Fig. 18.19, a titanium ethyl complex has a β-agostic interaction.[41] The Ti–C–C angle of 85.9° is significantly reduced from the tetrahedral angle in order to bring the β–H closer to the metal center. The C–C–H angle in the ethyl group undergoing the interaction. 124.4°, is significantly increased from the tetrahedral angle. The ethyl group is part-way transformed into a titanium–ethylene–hydride complex.

18.2.2.5 Kinetics and Thermodynamics of Insertion Reactions

Metal–carbon bonds and metal–hydrogen bond dissociation enthalpies vary widely across the periodic table, but M–C bonds are typically 15–30 kcal/mol weaker than the corresponding M–H bonds (Table 18.2).[45] Insertion of a coordinated alkene into a metal hydride might then be expected to be thermodynamically unfavorable, though the large-scale polymerization of ethylene and propylene suggests otherwise. To understand the full picture, one must consider the thermodynamics of the whole process. The major additional factor is the formation of a new C–H bond at the expense of the C=C π-bond.

The C–C single bond enthalpy in ethane is 90 kcal/mol, while the C=C bond enthalpy in ethylene is 174 kcal/mol. We can thus estimate the C=C π-bond as 174 − 90 = 84 kcal/mol.

Table 18.2 Some selected M–C and M–H bond enthalpies (kcal/mol)[45] and the difference between pairs of complexes that have the same coligands. For example, the Zr–H bond strength of 81 kcal/mol is stronger than the Zr–C bond strength of 62 kcal/mol by 19 kcal/mol.

Fragment	Fragment-Me BDE(X = Me) (kcal/mol)	Fragment-H BDE(X = H) (kcal/mol)	Difference(kcal/mol)
$X–Zr(X)Cp^*_2$	62	81	19
$X–Hf(X)Cp^*_2$	67	83	15
$X–Mo(X)Cp_2$	34	61	27
$X–Mo(CO)_3Cp$	43	67	25
$X–W(X)Cp_2$	47	74	27
$X–Mn(CO)_5$	39	59	20
$X–Ir(Cl)(CO)(PMe_3)_2I$	—[a]	—[a]	30
$X–Pt(PEt_3)_2$(*trans*-Cl)	47	73	27

[a] The Ir–C and Ir–H bond strengths are not known, but the difference is known.

Table 18.3 Thermochemical analysis of the insertion of ethylene into a Zr–H or Zr–C bond.

Reaction	Bonds broken	Enthalpy(kcal/mol)	Bonds formed	Enthalpy(kcal/mol)	Difference(kcal/mol)
Zr–H + C_2H_4	M–H	81	M–C	−62	
	C=C	84	C–H	−101	
Total		165		−163	+2 kcal/mol
Zr–R + C_2H_4	M–C	62	M–C	−62	
	C=C	84	C–C	−90	
Total		146		−152	−6 kcal/mol

The C–H bond strength in ethane is 101 kcal/mol, while the C–H bond strength in ethylene is 110 kcal/mol.[46–47]

Using these data, we can describe the thermodynamics for the insertion of ethylene into a Zr–H or a Zr–C bond (Table 18.3). A similar analysis could be carried out for other M–H/C bonds. The two factors that contribute most to the favorability of the insertion of ethylene into an M–H bond are the increased C–H bond strength relative to a C=C bond (favors insertion) and the decreased M–C bond relative to the M–H bond (disfavors insertion). As the difference in M–H/M–C bond strength increases beyond about 20 kcal/mol, the reaction passes from exothermic to endothermic and would be expected to become less favorable. For the insertion of ethylene into an M–C bond, the major enthalpic consideration is the increased C–C bond strength relative to the C=C bond, which makes the reaction slightly exothermic in all cases. One can use similar analysis to show that the β-hydrogen elimination can be enthalpically favorable in some cases, but the β-carbon elimination is not.

18.2.2.6 Elimination Reactions

An elimination reaction is the microscopic reverse of an insertion reaction. As such, eliminations corresponding to both 1,1- and 1,2-insertions can occur. The reverse of the carbonyl or 1,1-insertion reaction is typically called a deinsertion rather than an elimination, and it could conceivably occur by direct deinsertion, carbonyl migration, or alkyl migration, just as the insertion reaction had three pathways. The elucidation of the mechanism of the carbonyl-insertion reaction used deinsertion chemistry, the microscopic reverse, as evidence (Section 18.2.2.2).

The term “elimination” is typically reserved for the reverse of the 1,2-insertion reaction. This type of reaction is seen in organic chemistry, where an alkyl halide eliminates the hydrogen halide to form a carbon–carbon double bond. In organometallic chemistry, the metal–alkyl complex can react to form a metal–olefin complex with another X-type ligand, such as hydride. Previously, an agostic interaction was described as a trapped 1,2-insertion reaction, but of course, since the reverse reaction is also trapped it can be considered as a trapped β-hydrogen elimination (Fig. 18.19). This reaction requires an empty orbital on the metal and sufficient lack of steric crowding such that the alkyl group can bend towards the metal to engage in an agostic interaction. The reaction could occur directly without a β-hydrogen agostic intermediate, which is illustrated in Eq. 18.6 as direct conversion of an alkyl group to an alkene and a hydride with the transition state shown.

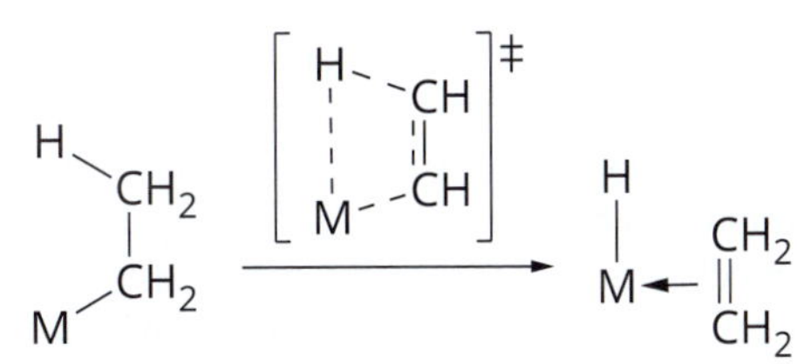

18.6

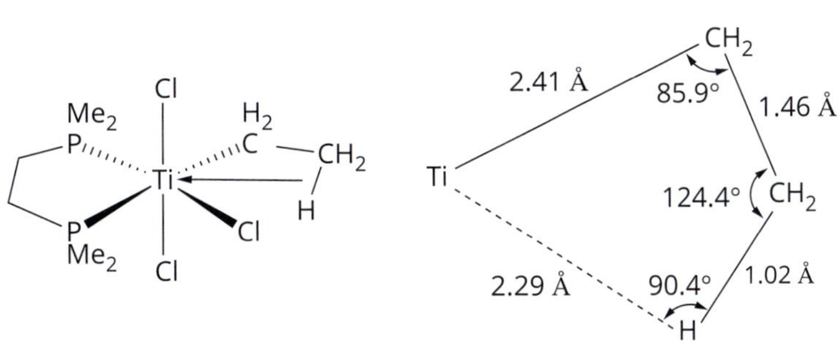

Fig. 18.19 A Ti complex undergoing a β-agostic interaction, showing significant bond distortions from that of a noninteracting ethyl group.

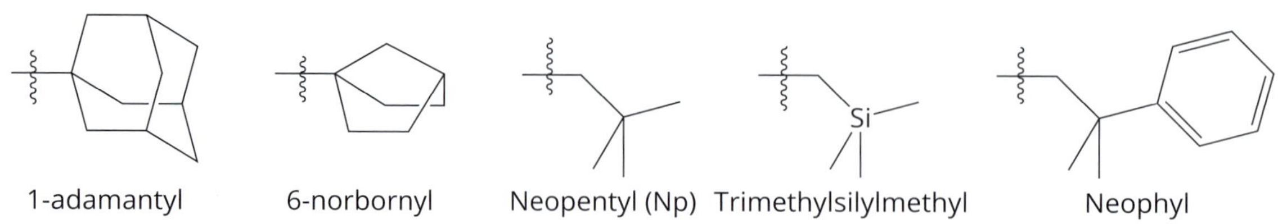

Fig. 18.20 Alkyl groups that often form stable metal complexes.

The favorability of the β-hydrogen elimination reaction was a principal reason for the inability of observing simple transition-metal alkyl complexes during the early years of organometallic chemistry. Simple metal alkyls, such as a metal ethyl, can decompose by β-hydride elimination, typically very rapidly. This decomposition reaction leaves the metal's oxidation state unchanged if the hydride is stable, but the electron count increases by two. However, transition metals with low d^n counts, especially early metals, do not form strong bonds to the resulting alkene, and the olefin dissociates. The resulting metal hydride is stable in some cases, but most often undergoes oligomerization and reduction through elimination of H_2 in a uni- or bimolecular process.

We now know that β-hydride elimination can be inhibited by careful alkyl group selection (Fig. 18.20). Cyclic alkyls such as 1-adamantyl or 6-norbornyl would form unfavorable alkenes, including a bridgehead carbon on elimination (Bredt's rule), while alkyls such as CH_2CMe_3 (neopentyl) or CH_2SiMe_3 (trimethylsilylmethyl) do not contain β-hydrogen atoms at all.

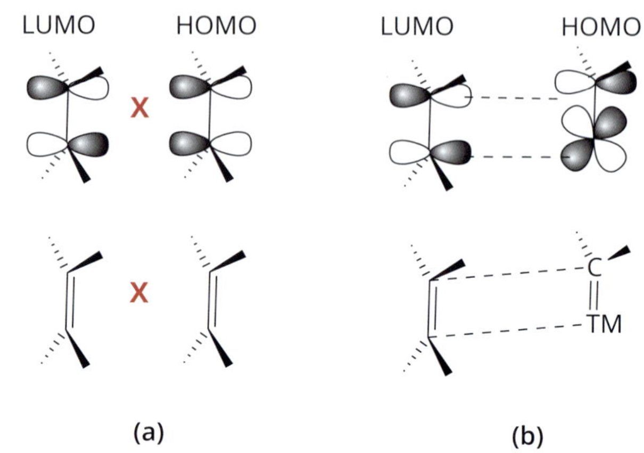

Fig. 18.21 Orbital symmetries of the HOMO and LUMO for (a) an olefin–olefin interaction and (b) an olefin-transition metal (TM) carbene interaction. The olefin–olefin interaction obviously has the wrong symmetry, and the [2 + 2]-cyclization is forbidden. The gerade symmetry of the *d*-orbital in the π-system of the TM carbene allows the reaction.

18.2.3 Cycloadditions

A cycloaddition is a chemical reaction in which two or more unsaturated functional groups come together and react, usually in a concerted fashion, to make and break several chemical bonds to give the final product, which has fewer multiple bonds. A cycloaddition familiar from organic chemistry is the Diels–Alder reaction, where a diene and a dienophile react to form a cyclohexene. There are many examples of cycloadditions in organometallic chemistry. The [2 + 2]-cycloaddition is found in the mechanism of the olefin metathesis reaction (Section 18.3.6) and one possible mechanism in hydroelementation reactions (Section 18.3.8). The Pauson–Khand reaction (Section 18.3.5) can be formally thought of as a [2 + 2 + 1]-cycloaddition. Cycloadditions are important reaction types across the transition-metal series and for many types of transformations.

18.2.3.1 *d*-Orbitals and Conservation of Orbital Symmetry

In organic chemistry some cyclization reactions are symmetry-allowed, which means that there is not an additional barrier to the reaction due to the symmetry of the atomic orbitals involved in the process. When there is an aromatic number ($4n + 2$) of π-electrons participating in the organic cyclization reaction, the reaction is symmetry allowed, and the process can be conducted thermally. The [2 + 2]-cycloaddition is disallowed, however, because there are 4π-electrons participating in the reaction, and the symmetry characteristics of *p*-orbitals make the reaction forbidden (Fig. 18.21 (a)). In order to carry out the chemical transformation, high-energy UV photons are usually required, so the reaction may proceed through an excited state.[48]

Unlike the *p*-orbitals, the *d*-orbitals are symmetric with respect to inversion; in other words, they have gerade symmetry. The presence of *d*-orbitals in a cyclization reaction involving a transition metal thus removes the symmetry-imposed barrier to the reaction (Fig. 18.21 (b)). As a result, transition-metal complexes catalyze a number of reactions that can be formally thought of as [2 + 2] or other seemingly symmetry-forbidden reactions.[49] As is sometimes in the case in organic chemistry as well, cycloaddition reactions are not necessarily concerted processes but may instead occur in a stepwise manner.[50]

18.2.3.2 [2 + 2]- and [2 + 2 + 1]-Cycloadditions

A [2 + 2]-cyclization reaction involves a metal–element double bond (Fig. 18.22) that reacts with an unsaturated substrate. A metal carbene (M=CR$_2$) would react to form a metallacyclobutane, while a metal imide (M=NR) would react to form

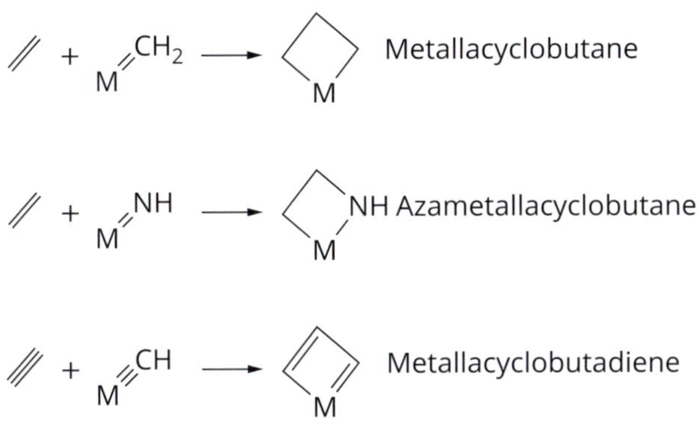

Fig. 18.22 A few examples of [2 + 2]-cyclization reactions with metal–ligand multiple bonds. Olefins can react with carbenes and imides to produce metallacyclobutanes and azametallacyclobutanes, respectively. Triple bonds can also undergo [2 + 2]-cyclization reactions, such as an alkyne reacting with a carbyne to produce a metallacyclobutadiene.

an azametallacyclobutane. Reactions of alkynes are also known, which may react with species such as alkylidynes to produce metallacyclobutadienes.

18.3 Some Common Reactions and Catalyses

Once some of the important chemical reactions that take place in organometallic chemistry had been discovered, it was found that many interesting chemical transformations could be drawn by combining them into catalytic cycles, which is what we will discuss next. A metal complex reacts with a substrate, various other transformations take place, and finally the product is released from the metal catalyst. Some of these reactions are industrially relevant, and some of them are useful for organic transformations to prepare fine chemicals, but all of them can be understood as ensembles of reaction steps, like those just discussed, into cycles that regenerate the initial catalyst.

The catalyses described are in a modular format so that one may pick the reactions of interest for study and class coverage. Many of the examples chosen are industrially important classic cases for discussion. However, there are also several examples taken from the recent literature, such as titanium-catalyzed multicomponent couplings and C–H activations. Furthermore, we tried to provide a good mix of both early and late transition-metal examples, as they often have different mechanistic steps.

It is not unusual for students, coming into courses such as this, to be confused as to what a catalyst actually is. The dictionary definition of catalyst often says something along the lines of "a complex that affects the rate of the reaction without being consumed." Perhaps a better definition that will become clear on studying some mechanisms of catalytic reactions is "a molecule that makes other molecules, often giving the chemist some control of the rate and chemical outcome."

18.3.1 Hydrogenation

Hydrogenation is the chemical reaction that takes a carbon–carbon double bond and reduces it to a carbon–carbon single bond. While conceptually simple and thermodynamically favorable, the reaction requires the use of a catalyst to be practical and selective. Typical reaction conditions in organic chemistry include finely divided palladium on carbon under an atmosphere of hydrogen gas. The metal serves to activate the hydrogen and the substrate to give the desired product. These conditions are heterogeneous, involving a solution-phase substrate, solid-phase catalyst, and gas-phase hydrogen. These conditions make the reaction difficult to study using standard solution-phase instrumentation. Detailed mechanistic experiments required the preparation of a soluble metal complex that could catalyze the reaction.

In 1866,[51] such a complex was realized when Wilkinson, Osborn, and coworkers examined the chemical reactivity of $RhCl(PPh_3)_3$ and discovered that it was highly active for the hydrogenation of organic substrates. The complex $RhCl(PPh_3)_3$ is now referred to as "Wilkinson's catalyst". This homogeneous catalyst was relatively easy to study by standard physical measurements, allowing a detailed investigation of the mechanism of the reaction.

First, before we consider hydrogenation by Wilkinson's catalyst in more detail, it is useful to consider a simple scheme for understanding hydrogenation reactions, which are generally classified in two categories (Fig. 18.23): *hydrogen first* and *olefin first*, both of which are established to occur in various systems. To illustrate, we will use a generalized square planar 16-electron catalyst ML_4, which may have all the same or different ligands, L. In the hydrogen first mechanism (Fig. 18.23a) H_2 oxidatively adds to ML_4 to produce the pseudo-octahedral 18-electron complex H_2ML_4, which then undergoes ligand exchange with the olefin. The olefin then 1,2-inserts into the hydride to produce a *cis*-alkyl hydride, which can undergo reductive elimination to give product and reform catalyst with addition of L back to the metal. The general scheme for the olefin first mechanism (Fig. 18.23b) has olefin exchanging with a ligand, which then oxidatively adds H_2.

These are simplified schemes, and it is commonplace to find such schemes for new reactions being drawn, even without any mechanistic data. These are, in essence, not mechanisms as much as mnemonics for understanding the reaction. The actual mechanisms of reactions are generally far more complex, with additional steps and off-cycle intermediates. To illustrate, we will discuss the mechanism of hydrogenation with Wilkinson's catalyst in more detail, as this reaction has seen extensive investigation.

Wilkinson's catalyst can be prepared by treatment of inert $RhCl_3(OH_2)_3$ with an excess of PPh_3 in ethanol (Eq. 18.7). The initial phosphine product proposed was Cl_2PPh_3, which would react with water in the reaction mixture to produce $OPPh_3$ and HCl.[52] Consequently, in this case phosphine is the proposed reductant for the metal center.

$$RhCl_3(OH_2)_3 + 4\,PPh_3 \rightarrow RhCl(PPh_3)_3 + Cl_2PPh_3 \qquad \mathbf{18.7}$$

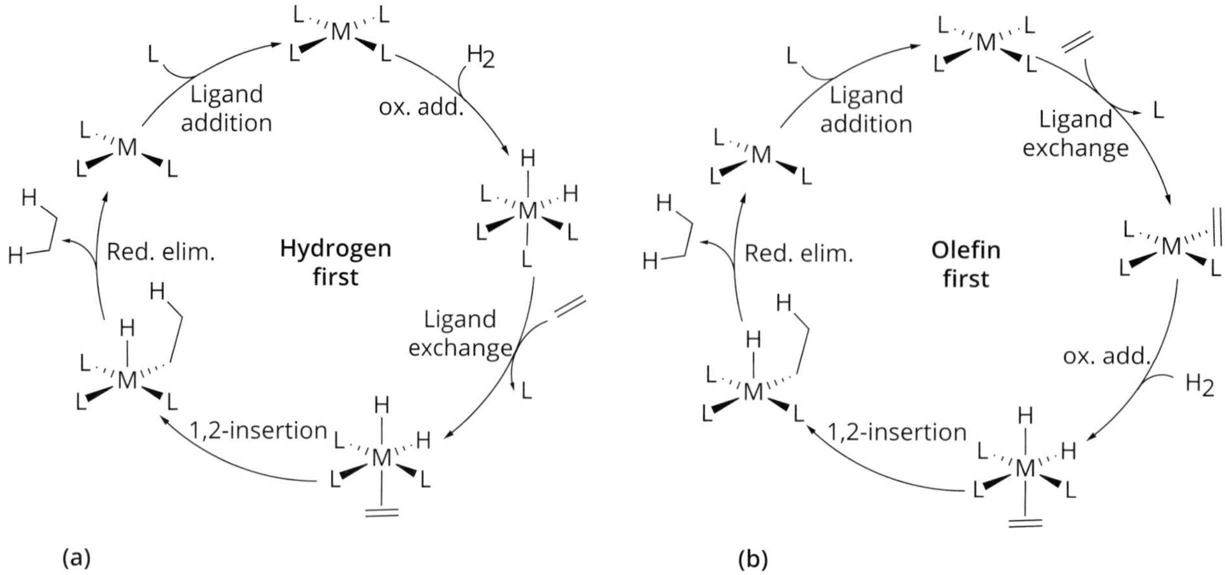

Fig. 18.23 Two generalized mechanisms for hydrogenation of ethylene. The two mechanisms are distinguished by which substrate adds to the metal first: (a) hydrogen, or (b) olefin, and both are believed to occur in known catalytic systems.

It was found that using a mixture of H_2 and D_2 leads to, by and large, all D_2 or all H_2 addition to the olefin with very little scrambling. This implies that dihydrogen is activated, and both hydrogen atoms stay on the metal during the entire catalytic cycle as a dihydride. Furthermore, the isotope effect was small, with $k_H/k_D = 1.15$, which is consistent with a rate-determining step that is something other than cleavage of the H–H/D–D bond, but does involve hydride/deuteride. For example, this is consistent with 1,2-insertion of the olefin into the hydride being rate-determining.[53]

The rate of the reaction was found to decrease in the presence of excess PPh_3, suggesting that ligand loss is an important step of the reaction. The rate increased in the presence of polar solvents, suggesting that a solvent interaction is important for the catalysis. Strong π-acids that do not undergo the hydrogenation reaction (such as ethylene, which binds more strongly than other olefins) "poison" the catalyst, shutting down the reaction. Increased steric bulk on the olefin leads to a decrease of the hydrogenation rate. Successful hydrogenation of these more difficult substrates is therefore likely to require activation and coordination of hydrogen before substrate binding.

A proposed mechanism that is consistent with all the results of these investigations is shown in Fig. 18.24. The main catalytic cycle is shown in the green circle. A solvated 16-electron rhodium complex, $RhCl(PPh_3)_2(solv)$, undergoes an oxidative addition with H_2 to give the *cis*-dihydride product. Loss of solvent and addition of the alkene produces the 18-electron octahedral organometallic complex $RhCl(PPh_3)_2(H)_2$ (η^2-$R_2C{=}CR_2$). The alkene can undergo a 1,2-insertion to produce an alkyl complex, which can be trapped with solvent coordination. The backward reaction, β-hydrogen elimination, is prevented by the coordination of the solvent. Reductive elimination of the alkyl hydride provides the product alkane and regenerates the active solvated catalyst.

The mechanism of the hydrogenation reaction with Wilkinson's catalyst has been widely examined by utilizing a number of other techniques, and all point to this general picture of the reaction. This system exhibits a large number of side reactions and inert resting states of the catalyst that can feed into the catalytic cycle in some cases, making the full analysis of the mechanism more difficult. The starting rhodium complex reacts with dihydrogen to produce $Rh(H)_2(PPh_3)_3Cl$, which can lose phosphine and solvate to enter the cycle. The starting Wilkinson's catalyst can also react with the olefin to lose phosphine and produce $Rh(Cl)(PPh_3)_2$(olefin), which can lose the olefin and dimerize. These species can all be observed or even isolated, suggesting that they are of significantly lower energy and reactivity than the species participating in the catalytic cycle.

This mechanistic picture illustrates an important point: if a structure can be isolated or observed spectroscopically, it is probably not the catalyst. Catalytic reactions tend to be relatively rapid, at least for good catalysts such as Wilkinson's, and resting states are often off of the main cycle of reactivity. Additionally, it can be seen from this complicated mechanism that simple mechanisms that might be predicted are not always in operation. For example, one might have proposed the oxidative addition of dihydrogen to Wilkinson's catalyst itself to generate the *cis*-dihydride as the first step of the reaction. This species certainly does form, but the loss of ligand from $Rh(PPh_3)_3(H)_2(Cl)$ to enter the catalytic cycle is slower. Activation of the metal complex through ligand loss results in a faster oxidative addition reaction.

The discovery of homogeneous hydrogenation led the way to the industrial synthesis of L-Dopa—a Parkinson's disease

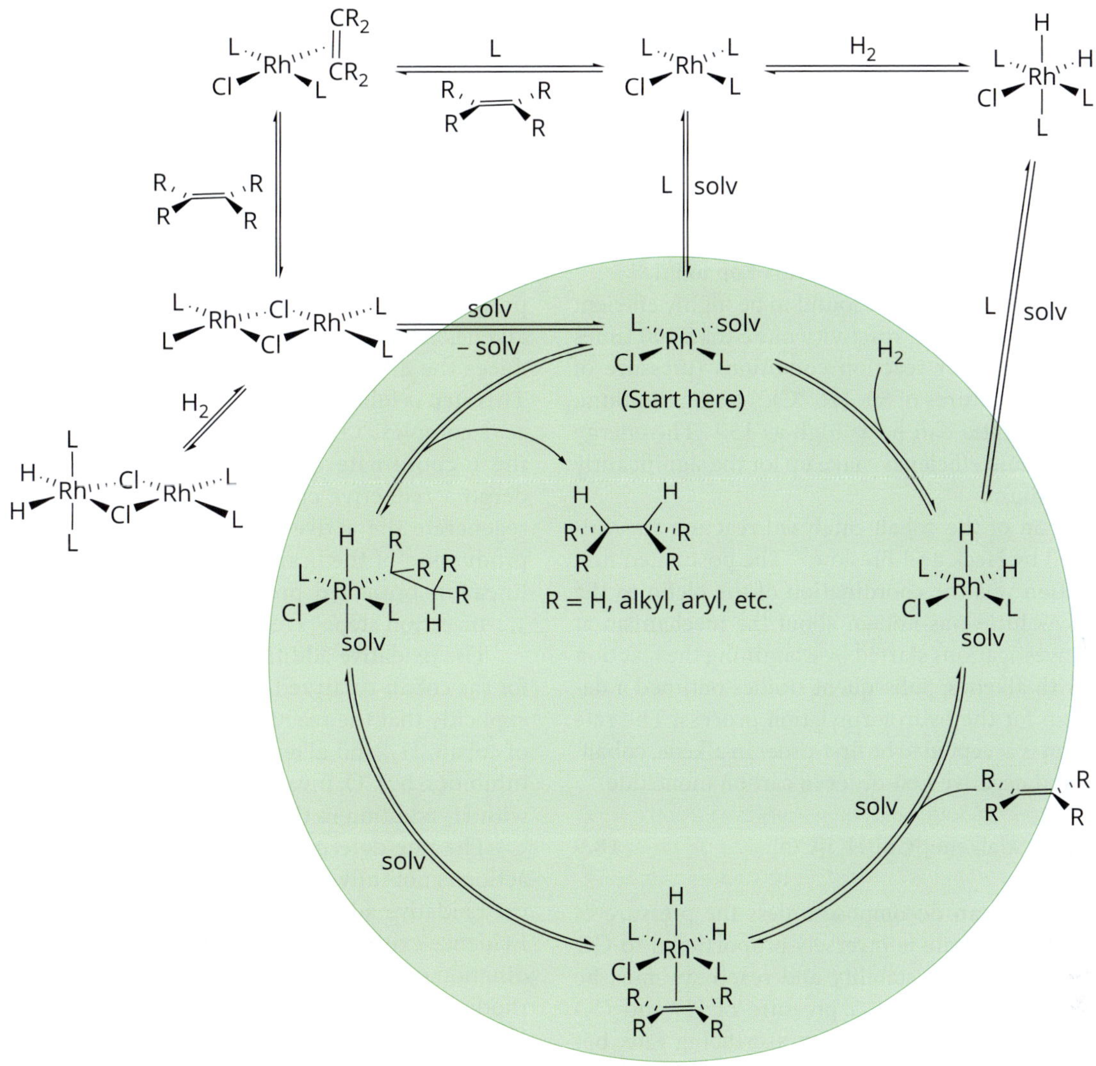

Fig. 18.24 The proposed mechanism of catalytic hydrogenation by $RhCl(PPh_3)_3$. The green circle contains the productive hydrogenation cycle, and other reactions are side reactions that feed into the cycle at various points. L = PPh_3, and solv = solvent.

drug—by the Monsanto corporation. The development of solution-phase catalysts allowed the use of chiral phosphine ligands, which controlled the addition of H_2 to a single face of prochiral substrates. This was one of the first large-scale organometallic reactions carried out industrially.

18.3.2 Hydroformylation

The hydroformylation reaction, also known as the oxo process, is a reaction carried out on industrial scale for the conversion of n-alkenes into $(n + 1)$-aldehydes. The reaction is the net addition of a formyl group, C(O)H, so the resulting aldehyde has one more carbon atom than the initial alkene substrate (Eq. 18.8). Aldehydes are important industrial chemicals, as they can be converted into a number of high-value commodities such as plasticizers, detergents, and specialty chemicals.

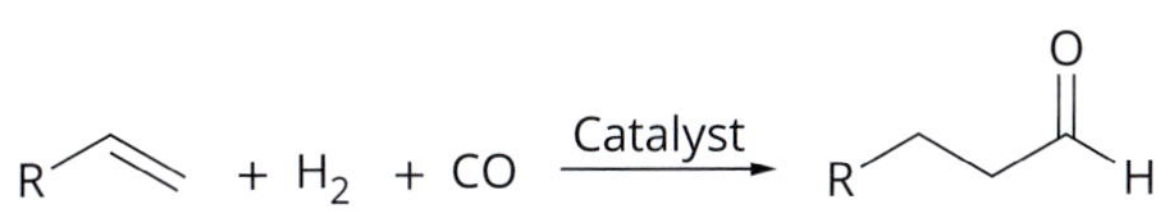

18.8

The reaction was discovered by Otto Roelen in 1938 while he was investigating the Fischer–Tropsch process, which converts carbon monoxide and hydrogen to liquid hydrocarbons. When ethylene was incorporated into the reaction mixture, aldehydes and ketones were observed. It was discovered that $HCo(CO)_4$ was an efficient catalyst for the reaction, though requiring pressures of 200–300 atm and temperatures ranging from 110 °C to 180 °C. The process is still carried out on the order of 10^6 tons per year, making this the oldest homogeneous catalysis reaction that is still in use.

One of the most important considerations for development of hydroformylation is based on improving the selectivity for linear instead of branched aldehydes, such as Eq. 18.8, showing the linear isomer as the product. Catalysts that favor the formation of linear aldehydes are desired because these products are generally preferred for applications. The initial cobalt catalyst, $HCo(CO)_4$, provides a linear:branched ratio of around 4:1. An early improvement to the process was the phosphine-modified catalyst by which bulky trialkylphosphine ligands such as PEt_3, PPh_3, and PBu_3 were used. These catalysts had an increased rate and an improved linear:branched ratio of up to 9:1.[54]

Rhodium complexes were later found to be highly efficient catalysts for the reaction, with reactivity more than 100 times higher under much milder reaction conditions (pressure of 15–25 atm and temperatures of 80–125 °C).[55] Using rhodium, the linear:branched ratio can be as high as 15:1. The energy savings, selectivity, and efficiency make up for the significantly higher cost of rhodium.

The mechanism of the cobalt-catalyzed reaction was initially investigated by Heck and Breslow.[56] The postulated first step of the reaction was the coordination of the olefin to the cobalt hydride. As little was known about the mechanism at the time, their investigations started by examining the reaction of $HCo(CO)_4$ with alkenes. Subsequent studies outlined a detailed mechanism for the hydroformylation process. The rate law of the reaction is accepted to be first-order in alkene, cobalt, and hydrogen, and negative first-order in carbon monoxide.[57]

$$\text{Rate} = k[\text{alkene}][\text{Co}][\text{H}_2][\text{CO}]^{-1} \qquad \textbf{18.9}$$

Since cobalt catalysts can decompose unless the pressure of CO is kept high but the rate is inversely proportional to CO pressure, a balance between stability and reactivity must be maintained. Maintaining an equal pressure of H_2 and CO allows stability to be preserved without sacrificing rate, because the rate becomes independent of pressure with large excess pressures of H_2 and CO.

Wilkinson studied the mechanism of hydroformylation by using rhodium catalysts. After an induction period during which there is no catalytic activity, the rate of the reaction was found to be pseudo-first-order with respect to catalyst concentration, inversely proportional to the partial pressure of CO, and proportional to temperature. A significant ligand effect was observed in which the rate is proportional to the degree of electron donation from the phosphine ligands.[58] Rapid exchange of alkenes was observed during the catalytic process, though the rate of isomerization of the alkenes is much slower. The results imply that an alkyl intermediate is formed reversibly, and isomerization can occur by β-hydrogen elimination to produce the starting alkene or isomerized alkene. This alkyl complex was not observed spectroscopically. The rate-determining step of the catalysis was postulated to be oxidative addition of dihydrogen to a square planar Rh(I) complex, as this step will be much more sensitive to the electronic properties of the ligands, with better donors producing faster rates as observed. The induction period was reported to be dissociation of a phosphine from $Rh(H)(CO)_2(PR_3)_2$ to produce square-planar $Rh(H)(CO)_2(PPh_3)$.

Recent kinetic studies and a rate law were published in 2000,[59] but even then not all of the details of the ligand effects on the reaction were fully understood. However, a general mechanistic picture is agreed upon, and it agrees with the work published by Wilkinson in 1968, (Fig. 18.25). Both the cobalt- and rhodium-catalyzed reactions are believed to follow similar mechanisms. The starting precatalyst, dicobaltoctacarbonyl, reacts with dihydrogen under the reaction conditions to generate $HCo(CO)_4$. Dissociation of CO produces an unsaturated complex that can coordinate the alkene. Alkene 1,2-insertion into the cobalt hydride produces the alkyl, which can then coordinate additional CO. This step is followed by a 1,1-insertion to produce the cobalt–acyl complex. Oxidative addition of dihydrogen generates the 6-coordinate acyl–dihydride complex, which can undergo a reductive elimination to generate the product and regenerate the active catalyst, $HCo(CO)_3$. Fig. 18.25 shows production of the branched aldehyde; the selectivity for the linear or branched product occurs, to some extent, in the 1,2-insertion step.

The oxidative addition of H_2 is the rate-determining step for the cobalt-catalyzed reaction. The rate law, Eq. 18.9, shows explicitly that the reaction is dependent on the concentration of cobalt, H_2, and alkene, as would be expected. There is also inhibition by CO, inverse first-order, which may be competing with H_2 addition in the reaction.

The rate-determining step for the rhodium-catalyzed reaction is not fully understood. It was initially postulated to be the oxidative addition of H_2, but later studies disputed this. Evidence exists that the rate-determining step is alkene coordination or CO dissociation. Several of the major steps in the rhodium catalytic cycle may have very similar rate constants. Mechanistic investigations, especially of asymmetric variants, are an active field of study.[60–61]

18.3.3 Monsanto Acetic Acid Process

Acetic acid is a highly important and relatively inexpensive commodity chemical with current production on the order of 5 million metric tons per year. It is used in the synthesis of vinyl acetate, which is used to prepare cellulose acetate, polyvinyl acetate, and synthetic fibers.

The Monsanto acetic acid process was reported in 1976.[62] Although earlier work suggested that various rhodium complexes reacted with iodide compounds to catalyze the carbonylation of acetic acid, no evidence existed for the catalytically active species before Monsanto published their findings. In this same paper they presented the mechanism for the reaction. The reaction rate is independent of the pressure of CO or the methanol concentration, and is first-order in both the rhodium complex and methyl iodide.

Prior work made it clear that the formation of a Rh(I) species that was capable of undergoing an oxidative addition of methyl iodide was necessary for reaction to occur. A number of Rh(III)-halide starting materials can be reduced under the reaction conditions, which include excess carbon monoxide and

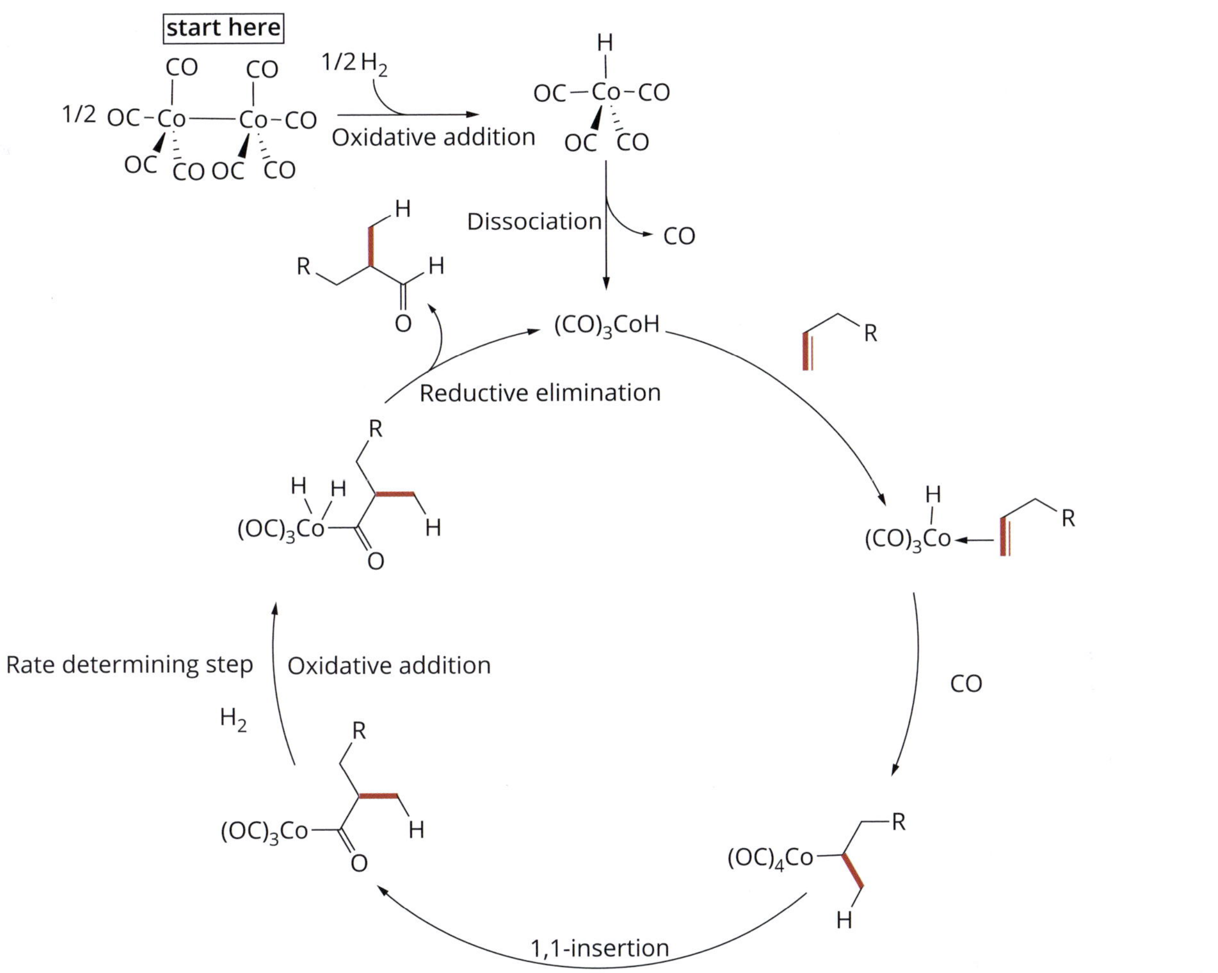

Fig. 18.25 The proposed mechanism for cobalt-catalyzed hydroformylation. To help follow the sites of addition, the two carbons starting off as the olefin are shown in red and bold.

iodide. Halogen exchange from the starting materials is complete on the timescale of the reaction, and the reactive starting material is the anionic–diiodide complex $[Rh(CO)_2I_2]^-$.

When the diiodide is treated with methyl iodide at room temperature, the IR spectrum exhibits a strong peak at 1711 cm^{-1}, which is attributable to an acyl group on the metal. The product of the oxidative addition, $[RhI_3(CH_3)(CO)_2]^-$, is unstable and not observed; the complex rapidly converts to the acyl $[RhI_3(CO)(C(O)CH_3)]^-$ by a migratory insertion reaction (Fig. 18.26).

When the acyl complex is treated with excess carbon monoxide, a new species forms which contains two CO-stretching frequencies and a single acyl frequency (Eq. 18.10). This species decomposes slowly to regenerate the starting diiodide with the formation of acyl iodide.

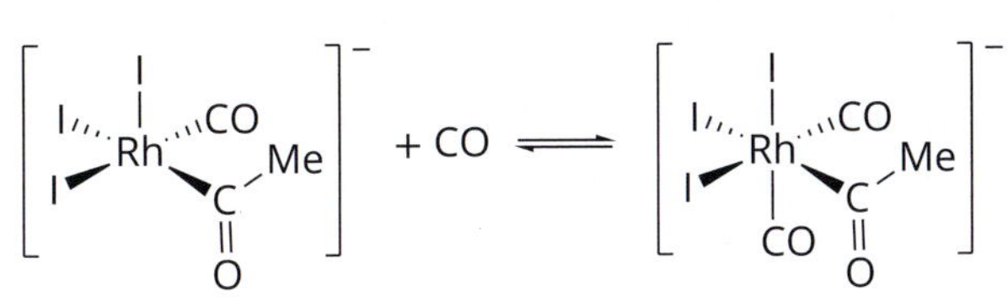

18.10

These steps assembled into a catalytic cycle are shown in Fig. 18.27. The organometallic steps include oxidative addition of CH_3I, 1,1-insertion, and reductive elimination of acyl

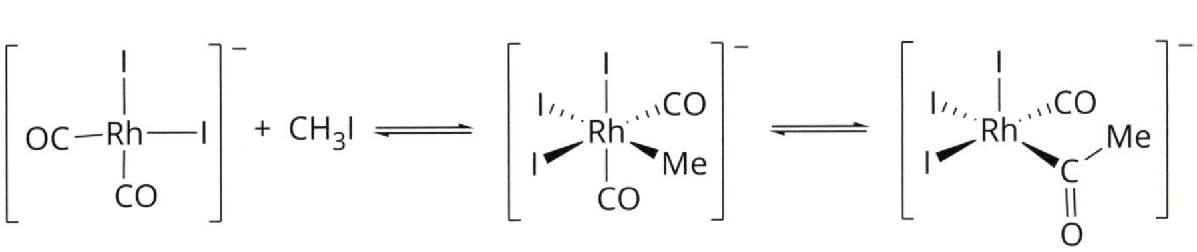

Fig. 18.26 Oxidative addition of MeI by $RhI_2(CO)_2^-$ and the 1,1-insertion mechanism.

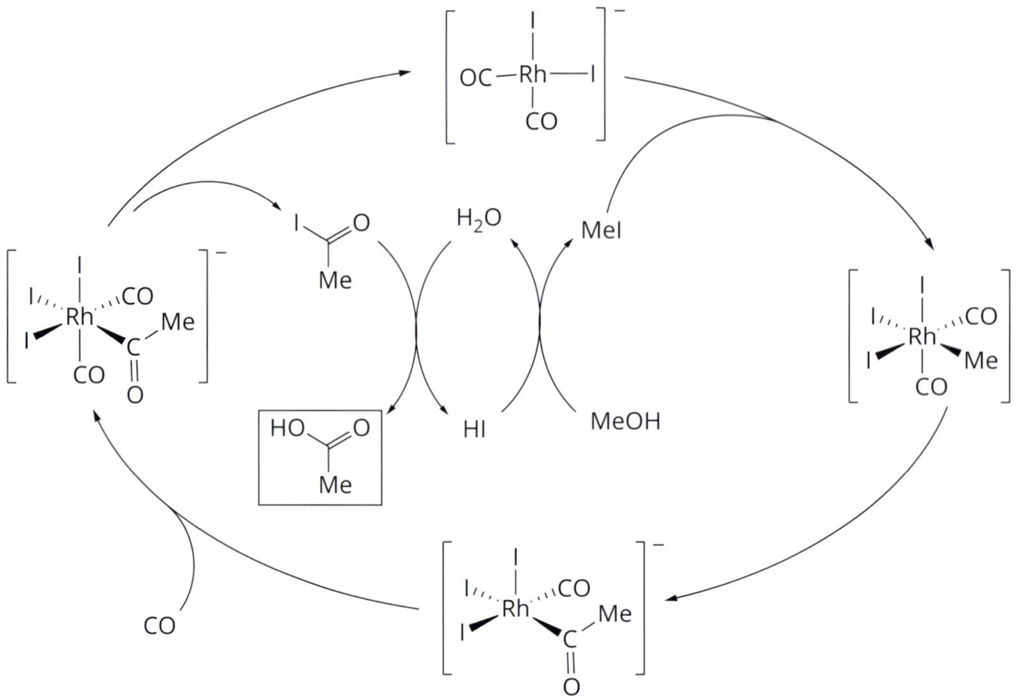

Fig. 18.27 The proposed mechanism for the Monsanto acetic acid process.

iodide. In nonmetal-catalyzed reactions, acyl iodide reacts with water to produce HI and the acetic acid product, which reacts with methanol to regenerate methyl iodide.

Although the Monsanto process was once the primary pathway for the industrial production of acetic acid, it has been replaced by the mechanistically similar Cativa process developed by BP Chemicals in the 1990s. This process, which uses iridium instead of rhodium, is slightly more efficient and has fewer side-reactions. These properties are both due to the use of less water in the catalytic system, which shuts down the water–gas shift reaction, which leads to consumption of CO to form CO_2 and H_2 (Eq. 18.11).

$$H_2O(g) + CO(g) \rightarrow CO_2(g) + H_2(g) \qquad \textbf{18.11}$$

18.3.4 Coupling Reactions[63–66]

The formation of carbon–carbon bonds is of great importance and interest to organic and organometallic chemists. One major area of research is finding milder and more selective carbon–carbon bond-forming reactions. In the early 1970s a variety of palladium-catalyzed coupling reactions were discovered. The great importance of these reactions was recognized by the awarding of the Nobel Prize to Akira Suzuki, Richard Heck, and Ei-ichi Negishi in 2010, for their discovery and development of these reactions in organic synthesis. Chronologically, the Heck coupling was developed in 1972 (based on earlier work by Mizoriko), the Negishi coupling in 1976, Stille coupling in 1976–77 (with a milder and more broadly applicable procedure in 1978), and then Suzuki coupling in 1979. All four reactions have one coupling partner that is an alkyl or aryl halide. The other coupling partner in the Heck reaction is an unsaturated C=C double bond, while the other three reactions couple to alkyl stannanes (Sille coupling: typically $RSnMe_3$ or $RSnBu_3$), alkyl boronates (Suzuki coupling: $RB(OR')_2$), or alkyl zinc reagents (Negishi coupling: R_2Zn or RZnX). The mechanism for the Heck reaction is different from the other three because of the different reactivity of the alkene coupling partner, but the other three coupling mechanisms are similar.

18.3.4.1 Heck Reaction

The Heck reaction couples an unsaturated organohalide or organotriflate with an alkene. The reaction uses a palladium catalyst and requires the addition of base, and the product is a substituted alkene. This was the first example of a carbon–carbon bond-forming reaction that follows the now commonly recognized Pd(0)/Pd(II) catalytic cycle.

Early research on the reaction was used to determine some of the important mechanistic parameters, and these facts were later incorporated during the development of the proposed mechanism. The isolated palladium complex $Pd(PPh_3)_2(Ph)(Cl)$ reacts with styrene to produce *trans*-stilbene, and yields were improved if a base, such as triethylamine, was added to neutralize the HCl byproduct that formed. Excess triphenylphosphine shuts down the oxidative addition of the aryl halide, by preventing the formation of the open coordination site that is required. Electron-donating phosphines improve

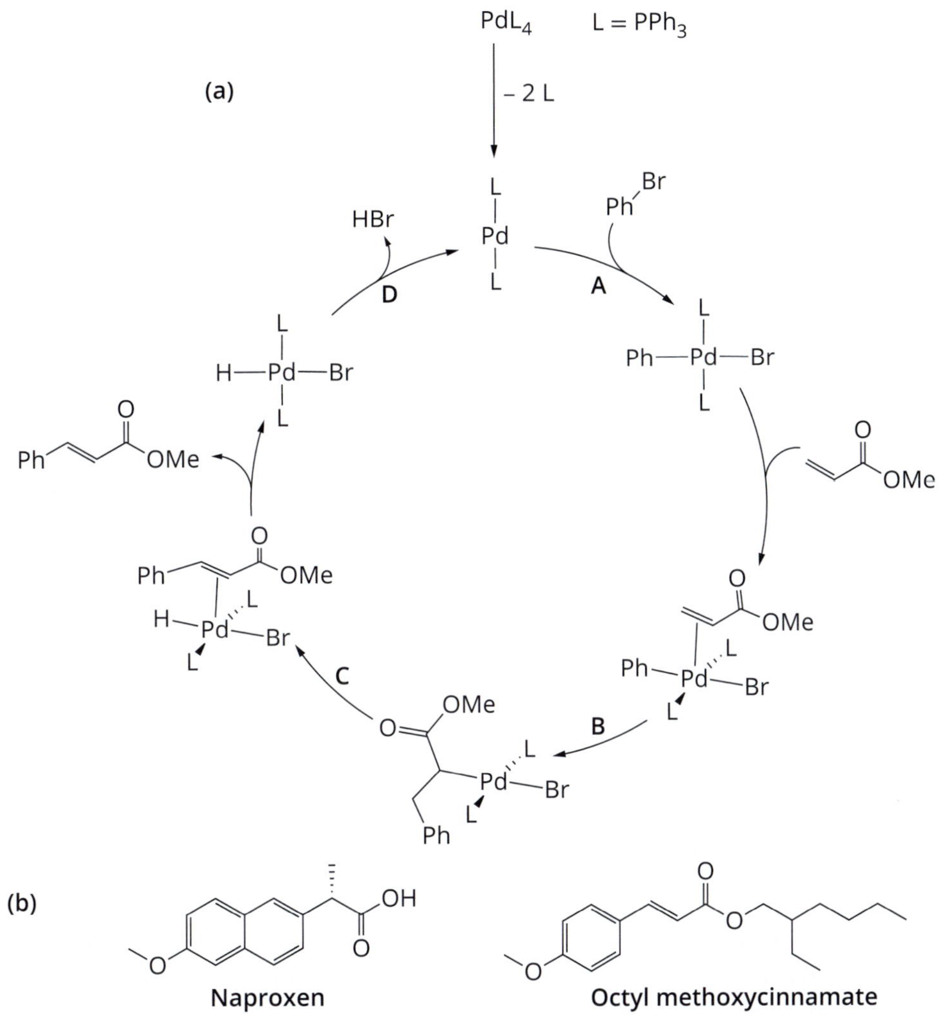

Fig. 18.28 (a) The mechanism of the Heck reaction has four common organometallic steps: oxidative addition of ArBr (Step A), 1,2-insertion (Step B), β-hydrogen elimination (Step C), and reductive elimination (Step D) that work in concert to form an aryl substituted alkene. (b) Two industrial products synthesized by using the Heck reaction.

the rate of the reaction by presumably enhancing the rate of the slow oxidative addition step. The oxidative addition is believed to take place by concerted addition.

A mechanism for the Heck reaction that is consistent with these facts is shown in Fig. 18.28 (a). Starting with an 18-electron palladium(0) species such as $Pd(PPh_3)_4$, the first step is ligand loss under the reaction conditions to produce the unsaturated 14-electron species PdL_2. In Step A, oxidative addition of an aryl bromide gives the 16-electron Pd(II) phenyl bromide compound. Coordination of the incoming alkene to the palladium is followed by a 1,2-insertion reaction to generate the palladium alkyl (Step B). A β-hydrogen elimination step (Step C) forms the product, which remains coordinated to the metal initially. The *trans*-stereochemistry is strongly favored. Reductive elimination of HBr (Step D) regenerates the Pd(0) species. The base is present to react with the HBr that forms to prevent side reactions. The Heck reaction is used industrially in the synthesis of the anti-inflammatory drug Naproxen and the sunscreen compound octyl methoxycinnamate, shown in Fig. 18.28 (b).

18.3.4.2 Negishi Reaction

The Negishi and Stille coupling reactions share a mechanism, differing only in the identity of the second coupling partner (Fig. 18.29). For the Negishi reaction, the coupling partner is an alkyl zinc reagent. Since the zinc reagents are air- and water-sensitive, the reaction must be carried out under rigorously anhydrous conditions. This limits the applicability and functional group tolerance of the reaction, but since zinc reagents are very reactive the reaction rates are faster. Unlike the Heck reaction, the Negishi reaction allows for coupling between an aryl or alkenyl halide and an *sp*-, *sp*^{2}-, or *sp*^{3}-carbon. The increased reactivity of the zinc reagent allows for coupling of complex or difficult-to-activate intermediates. In addition, the zinc reagents are more environmentally friendly than the tin reagents in the Stille coupling.

The first step of the mechanism is oxidative addition of R–X to Pd(0), producing the Pd(II) alkyl complex. The second step is transmetallation of the newly formed Pd–X bond by

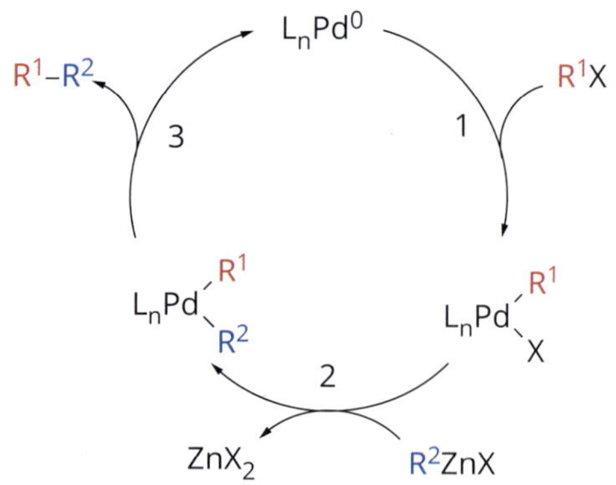

Fig. 18.29 The mechanism of the Negishi cross-coupling reaction. Step 1 is oxidative addition of R_1X to the Pd^0 center. Step 2 is transmetallation with RZnX. Step 3 is reductive elimination of product.

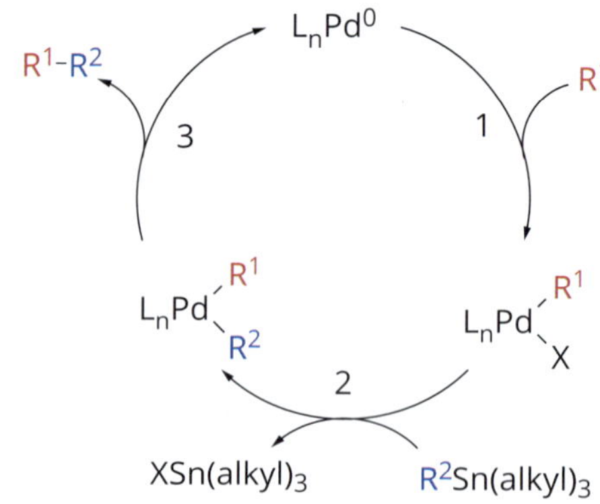

Fig. 18.30 The mechanism of the Stille cross-coupling reaction. Step 1 is oxidative addition of R^1X to Pd^0. Step 2 is transmetallation with $RSn(alkyl)_3$ to produce the byproduct $XSn(alkyl)_3$. Step 3 is reductive elimination of product R^1-R^2 from dialkylpalladium.

using the incoming XZn–R and loss of the side product, ZnX_2. The complex can undergo a rearrangement if necessary, and the last step is reductive elimination of the R–R′ species to give the desired product and regenerate the Pd(0) catalyst.

For the Negishi reaction, the oxidative addition may proceed via concerted addition to produce the *cis*-product or by a stepwise mechanism to produce the *trans*-product. In any case, the *cis*-product can isomerize to the *trans*-product under the reaction conditions. Transmetallation is the rate-determining step of the transformation, though the exact mechanism of this step is not well known. Aggregation of the zinc reagent into a 4-coordinate species (either a zincate or a solvate) is believed to occur with alkyl zinc species, though not for aryls. As in the previous step, there is evidence for both *cis*- and *trans*-dialkyls, but only the *cis*-product can progress through the mechanism via reductive elimination of the alkyl.

18.3.4.3 Stille Reaction

The mechanism of the Stille coupling has been studied extensively, and as such, the mechanism is known to be extremely complex with a number of possible pathways. However, there is general agreement in the overall reaction profile (Fig. 18.30). As in the Negishi coupling, oxidative addition of the alkyl halide to a palladium(0) starting material is the first step. Transmetalation with a trialkylstannane, usually R^2SnBu_3, replaces the halide with the second alkyl group. Reductive elimination of the two alkyls groups gives the final product.

The rate-limiting step of the reaction is usually found to be the oxidative addition step, and the relative reactivity for a wide class of substrates decreases in the order from RI ≥ ROTf > RBr >> RCl. The oxidative addition proceeds with retention of configuration at carbon for alkenyl halides and inversion of stereochemistry for allylic and benzylic halides. Alkyl halides that contain β-hydrogens are not often useful for the coupling reaction, because β-hydrogen elimination from the palladium alkyl complex is competitive. The oxidative addition step is believed to usually follow a concerted process to give the *cis*-product, although in some cases, usually when an sp^3-substrate is employed, the step-wise oxidative addition occurs to give the *trans*-product. The *cis*-product rapidly isomerizes to the more stable *trans*-isomer.

Transmetallation is known to proceed by a number of different mechanistic pathways that depend highly on the substrate and reaction conditions. However, the most common transmetallation mechanism involves the organostannane C–Sn bond reacting with the Pd–X bond in a 4-center cyclic pathway to produce the cyclic C–X–Sn–C species, which can lose the Sn–X species (Fig. 18.31). Alternatively, the palladium complex can attack the carbon directly in an S_N2-type pathway. In order for the reductive elimination step to occur, the *trans*-species must isomerize to the *cis*-species.

The tin reagents and catalyst are air-stable, giving this method the benefit of being much easier procedurally than the Negishi coupling. The reaction is generally quite fast, occurring in minutes. The mechanism of the reaction that was postulated in the original report was essentially correct[67] though highly simplified, in that it does not address the *cis*/*trans*-isomerization steps that are now known to occur, and the multifaceted pathways known for the transmetallation step. The use of Stille coupling in synthesis often follows empirical rather than logical rules, and while this has led to substantial use of the reaction over the years, many specific details of the mechanism remain to be learned. Certainly, the mechanism is highly variable and depends on reaction conditions and substrates.

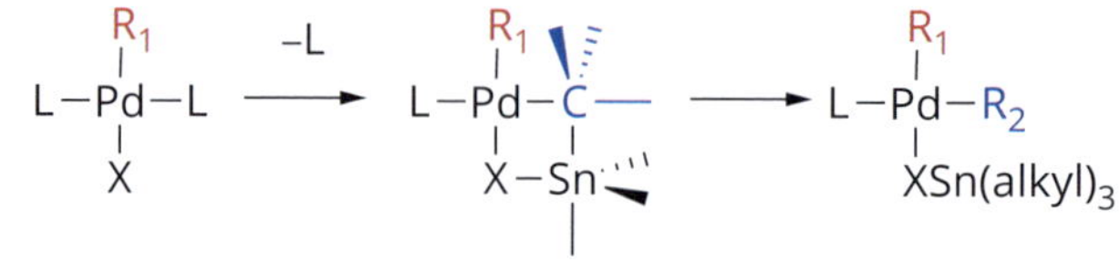

Fig. 18.31 Four-centered mechanism for transmetallation in the Stille coupling.

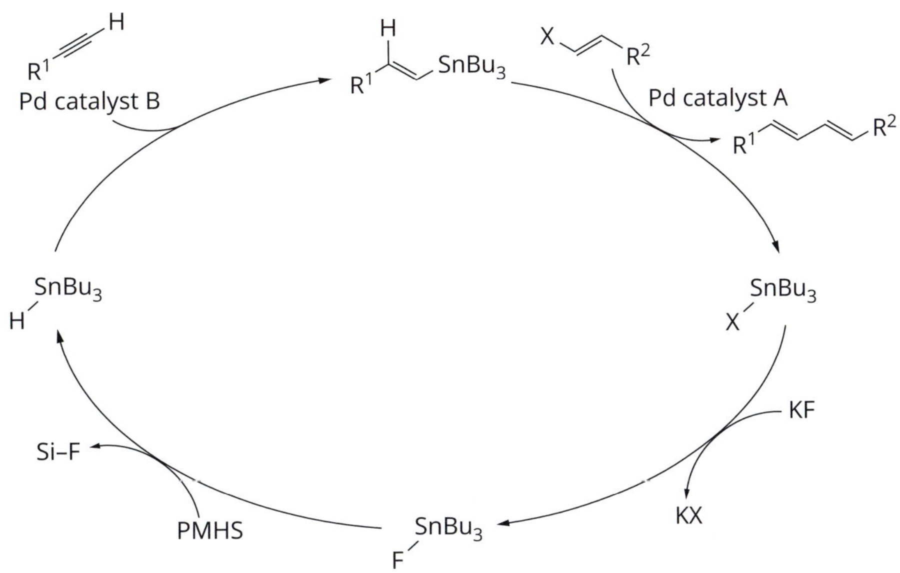

Fig. 18.32 The reaction mechanism for Stille couplings catalytic in tin.

The greatest limitation of Stille coupling is the toxicity of the tin-based starting materials and byproducts. The $RSnMe_3$ reagents are reactive but extremely toxic. The toxicity can be lowered somewhat by using the tributyl compounds, $RSnBu_3$, but even so, the compounds must be handled carefully and can be expensive to dispose of safely. One interesting method of obviating the use of these reagents is to use the tin catalytically, which is a reaction developed by Maleczka and coworkers (Fig. 18.32). Two different palladium-catalyzed processes run concurrently during the reaction: a Stille coupling and a catalysis to regenerate the tin reagent. For these two different catalyses, two different catalysts were required. The scheme involves hydrostannation by $HSnBu_3$ of alkynes to give vinyl coupling partners. After coupling, the halide in $XSnBu_3$ is converted back to hydride by polymethylhydrosiloxane (PMHS) in the reaction mixture.[68–69]

There has been substantial work on using silicon in place of tin in cross-coupling reactions, but silicon is a far less active transfer agent under the same conditions as Stille couplings. However, $RSiMe_3$ reagents can be used if activated with fluoride ion, which is called Hiyama coupling. In a somewhat different strategy, Denmark and coworkers have pioneered the use of silanols in palladium-catalyzed couplings.[70–72]

18.3.4.4 Suzuki Reaction

Suzuki coupling is the reaction in this category that has found greatest use in organic synthesis. Due to the relative ease of preparing the boronic acid coupling partners, and the high reactivity, nontoxic byproducts, mild conditions, and stereo- and regioselectivity of the coupling partner, this reaction is regularly used by researchers to prepare new organic molecules. Mechanistically, Suzuki coupling has very much in common with both Negishi and Stille couplings. The major difference is the use of an external base to promote the transmetallation from the boron compound. The base was originally hydroxide or alkoxide, but a host of different bases can be used.

Fig. 18.33 is a representative mechanism—one of several currently thought possible, depending on substrates and conditions. As seen in the other two coupling reactions, Step 1 of the mechanism is oxidative addition of R^1–X to Pd(0), producing the Pd(II) alkyl complex. In Step 2, the palladium halide is replaced with a *tert*-butoxide by metathesis, which reacts more readily with the incoming boronic ester. In Step 3, $R^2B(OR)_2$ may interact with the Pd–O^tBu species, followed by transmetallation and formation of $^tBuOB(OR)_2$, Step 4 in the complex can undergo a rearrangement if necessary, and then the last step is reductive elimination of the R^1–R^2 to regenerate Pd(0) catalyst.

Organoboron reagents are generally unreactive to a variety of Pd(II) complexes, due to the low nucleophilicity of the carbon on boron. The alkoxide base is used to form the boronate complex $RB(OR)_3^-$ from the neutral alkylborate $RB(OR)_2$. This quaternization of boron increases the nucleophilicity of the organic group.

The reaction allows aryls, alkyls, alkenyls, and alkynyls to be coupled under some conditions, and aromatic substrates are by far the most common, including heterocycles. The advantages of the reaction over other coupling reactions are the ease of use and preparation of boronic acids, mild reaction conditions, and substantially less toxic reagents, especially compared to tin reagents. The Suzuki reaction can be scaled up significantly and has seen much use in the pharmaceutical industry. One large-scale example, by Merck, is to prepare more than 1,000 tons/year of the fungicide boscalid (Fig. 18.34 (a)). Clariant uses the reaction to prepare o-tolylbenzonitrile (OTBN) (Fig. 18.34b) as an intermediate in the preparation of a family of Sartan derivatives—drugs for lowering blood pressure. In both of these syntheses, the crucial aryl–aryl bond is prepared by a Suzuki coupling.

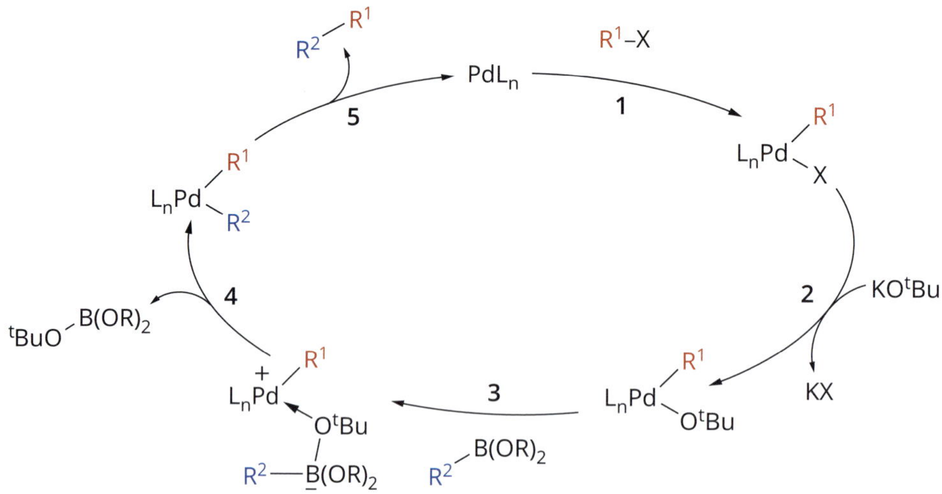

Fig. 18.33 A representative Suzuki coupling mechanism.

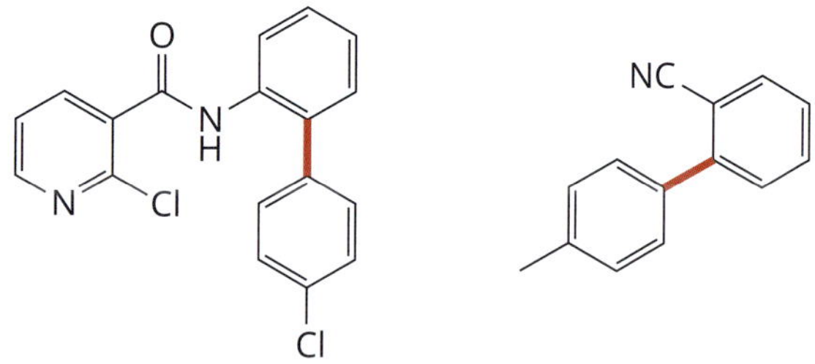

Fig. 18.34 (a) Boscalid, and (b) o-tolylbenzonitrile (OTBN)—two important drugs prepared by using Suzuki coupling. The carbon–carbon bond formed is highlighted in red and bold.

18.3.5 The Pauson–Khand Reaction

Moving away from simple carbon–carbon bond-forming reactions, the Pauson–Khand reaction is a straightforward route into cyclopentenone compounds as a three-component coupling reaction using cobalt carbonyl complexes. The reaction can be thought of as a [2 + 2 + 1]-cycloaddition reaction between CO, an alkene, and an alkyne to produce a α,β-cyclopentenone. The reaction can be made catalytic in $Co_2(CO)_8$, though the original reaction conditions (initially reported in 1973) were stoichiometric in metal. The reaction leads to useful yields (40–60% are typical) and predictable regioselectivity based on sterics. Intramolecular variants of the reaction are much more regioselective than intermolecular ones.

The reaction begins by coordination of alkyne to $Co_2(CO)_8$ to produce the organometallic metallatetrahedrane $[(CO)_3Co]_2(CH)_2$ (Fig. 18.35). This is a reaction that can be used as a protecting group strategy for an alkyne. The resulting tetrahedrane complex can be isolated and then used as prepared in the remaining steps of the cyclization reaction, suggesting its mechanistic importance. $Co(CO)_3$ is isolobal to a CH fragment (Chapter 14). So, the initial product is isolobal with tetrahedrane $(CH)_4$.

Addition of the alkene to the dimetallatetrahedrane occurs by displacement of a CO ligand. Steric repulsion of the R group on the alkene positions the group for a somewhat selective 1,2-insertion into the unsubstituted Co–C bond coming from the alkyne. Carbonyl insertion into the newly formed Co–C bond produces the cobalt acyl, which then undergoes a reductive elimination to produce the coordinated alkene (or metallacyclopropane), which can dissociate to produce the cyclic enone.

The Pauson–Khand reaction works with terminal and internal alkynes, though it suffers from decreased yields for internal alkynes. The reaction does not work well with tetrasubstituted alkenes or alkenes that contain strong electron-withdrawing groups.

The steps of the proposed mechanism after the formation of the tetrahedrane complex are all reasonable, though the exact mechanism is not known with great certainty. A detailed kinetic study was carried out in the coupling of norbornadiene with trimethylsilylacetylene (Figure 18.36), and a complex rate law was obtained.[73]

18.3.6 Olefin Metathesis

Olefin metathesis is another reaction resulting in the award of the Nobel Prize, in 2005, to Yves Chauvin, Robert Grubbs, and Richard Schrock, for their discovery of this reaction, investigating its mechanism, and developing catalysts to improve its selectivity and efficiency, and a detailed historical account of the reaction was published soon afterwards.[74] It interconverts alkenes by cleavage and recombination of carbon–carbon *double* bonds. The chemistry leading up to the discovery of olefin metathesis began in the mid-1950s when a number of industrial laboratories accidentally discovered

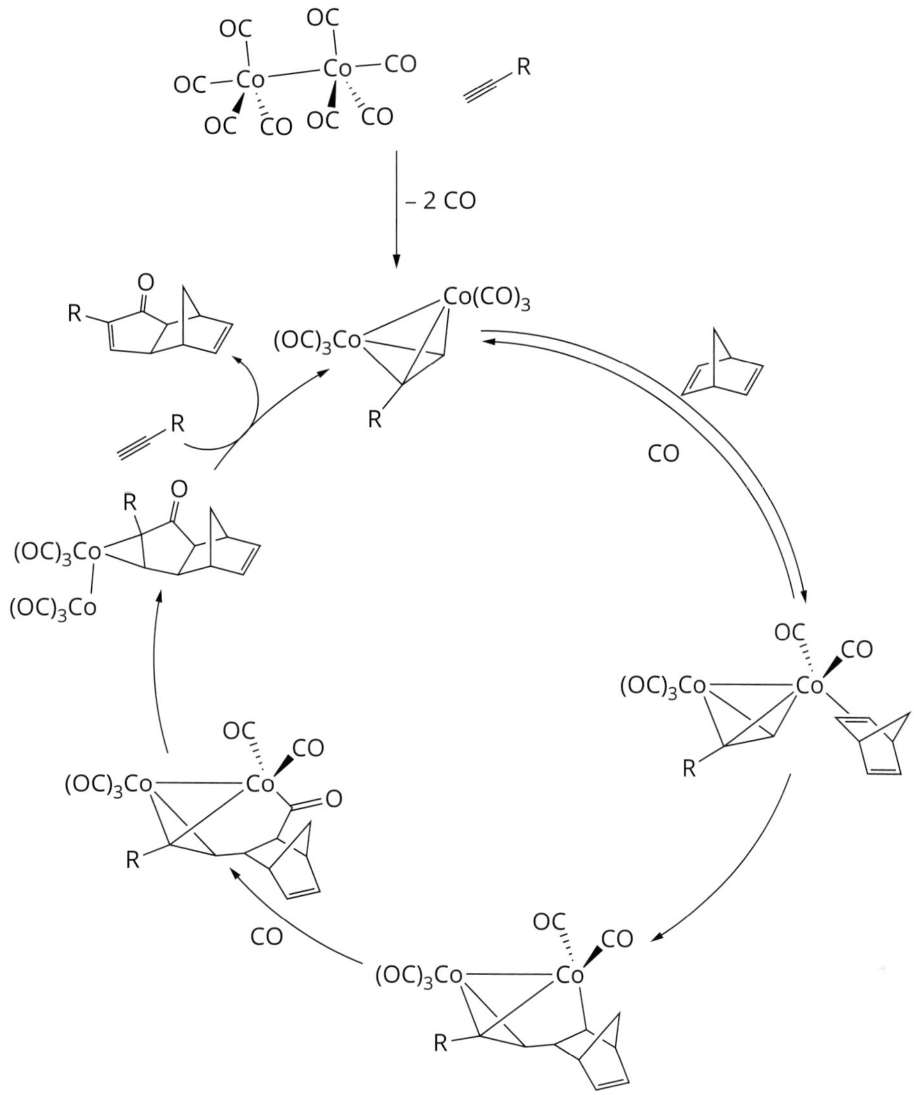

Fig. 18.35 The Pauson–Khand reaction for the synthesis of cyclopentenones from alkenes, alkynes, and carbon monoxide.

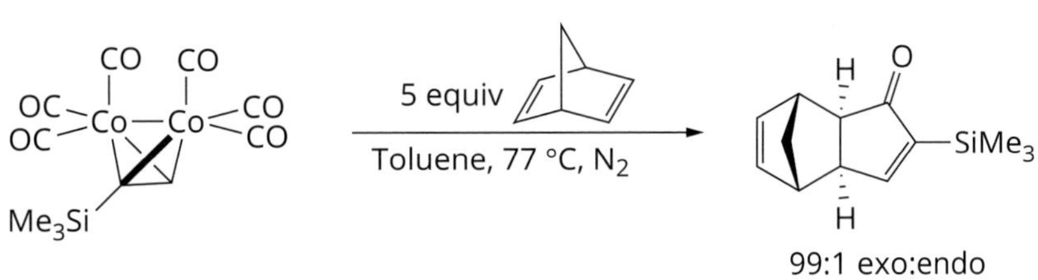

Fig. 18.36 Pauson–Khand reaction between norbornadiene and trimethylsilylacetylene.

chemical reactivity that would later be classified as olefin metathesis. For example, norbornene was found to polymerize into polynorbornene (Eq. 18.12) by a mechanism postulated to occur by concerted C–C and M–C bond-breaking steps. Later, in the mid-1960s, a reaction described as *olefin disproportionation* was observed with a number of Mo and W catalysts. This reaction converted propylene into ethylene and 2-butene.

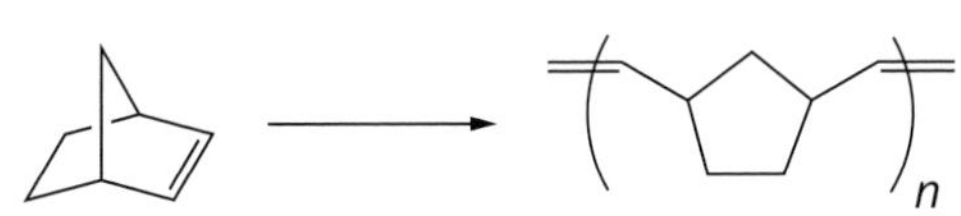 **18.12**

The regiochemistry of two different chemical reactions was influential in providing data to help with determining the mechanism. The first study was carried out at the Goodyear Tire and Rubber Company.[75] This experiment was carried out

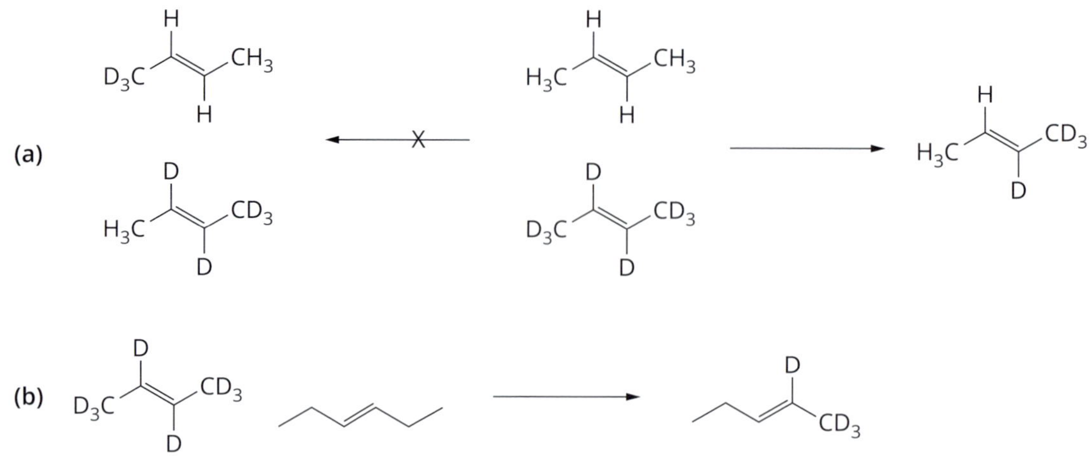

Fig. 18.37 Two chemical reactions that show only cleavage of the carbon–carbon double bond during an olefin metathesis reaction.

by using isotopic labeling and was designed to probe whether it was the C–C single bonds or the C=C double bonds that were reacting. A mixture of 2-butene with 2-butene-d_8 was subjected to the catalytic reaction conditions, and the only products observed were due to exchange through cleavage of the C=C bond (Fig. 18.37a). Similarly, 2-butene-d_8 was treated with 3-hexene, and the only observed product maintained the halves of the molecule on either side of the double bond, demonstrating that the C=C bond is the reactive site under metathesis conditions (Fig. 18.37b).

The second influential study was carried out by Chauvin.[76] Cyclopentene was found to react with 2-pentene in the presence of a W(VI) catalyst activated with tetrabutyltin. The initial product that formed was a mixture of C_9, C_{10}, and C_{11} products in a 1:2:1 ratio, consistent with the calculated statistical distribution expected for reaction at the double bond (Eq. 18.13). At all reaction times and conversions, a mixture of products was found in a ratio consistent with the [2 + 2]-cyclization mechanism that he proposed.

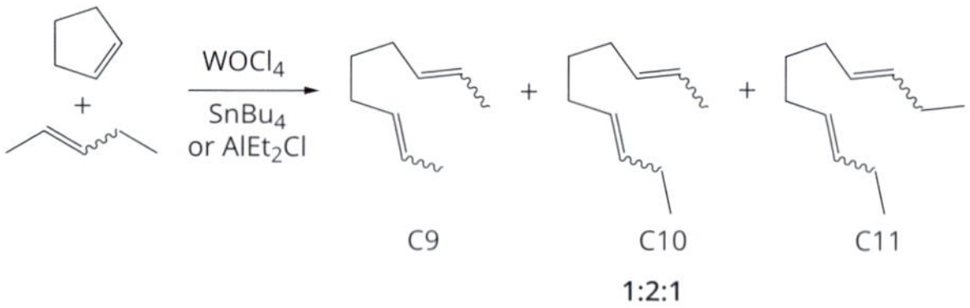

18.13

The mechanism that Chauvin proposed involved a metal-mediated [2 + 2]-cycloaddition to produce a metal–cyclobutane complex that could cyclorevert to give the new products (Fig. 18.38). While organic [2 + 2]-cycloadditions are symmetry forbidden according to the Woodward–Hoffman rules, the metal-mediated reaction is symmetry-allowed (Fig. 18.21) due to the presence of the metal *d*-orbitals, which lack an inversion center and are of the correct symmetry to interact with the olefin π-orbitals.

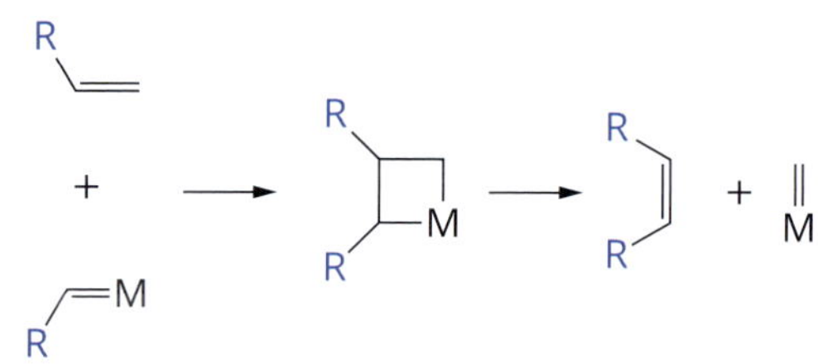

Fig. 18.38 The [2 + 2]-cycloaddition and cycloreversion mechanism proposed by Chauvin for the olefin metathesis reaction.

In the 1970s, Grubbs developed the so-called "pairwise" mechanism that was also consistent with previous data.[77] However, his own later experiments, which involved isotopic labels, were eventually used to disprove his mechanism and provide further evidence for Chauvin's mechanism.[78–79]

A typical catalyst precursor for the olefin metathesis reaction is $RuCl_2(PCy_3)_2(=CHPh)$, which is sometimes referred to as the "Generation 1" Grubbs's (pre)catalyst. A mechanism consistent with the available data begins by phosphine ligand dissociation from the 16-electron precatalyst to generate a coordinatively unsaturated 14-electron complex (Fig. 18.39), which can react with substrate and cyclorevert to generate the active catalyst, $RuCl_2(=CH_2)PCy_3$. A series of [2 + 2]-cycloadditions can occur, and only the productive pathway is shown in Fig. 18.39 (a). All of the steps are, in principle, reversible, but the reaction is often driven to completion by allowing a volatile olefin, such as ethylene gas, to escape from the reaction.

Significant effort has been placed into the improvement of the catalyst, which has been realized by replacement of one of the phosphine ligands in the so-called "first-generation" catalysts with an *N*-heterocyclic carbene ligand in the "second generation", and replacement of the second phosphine with a pyridine ligand in the "third generation" (Fig. 18.39b). These subsequent generations are each more reactive, with shorter induction periods due to the weaker coordination

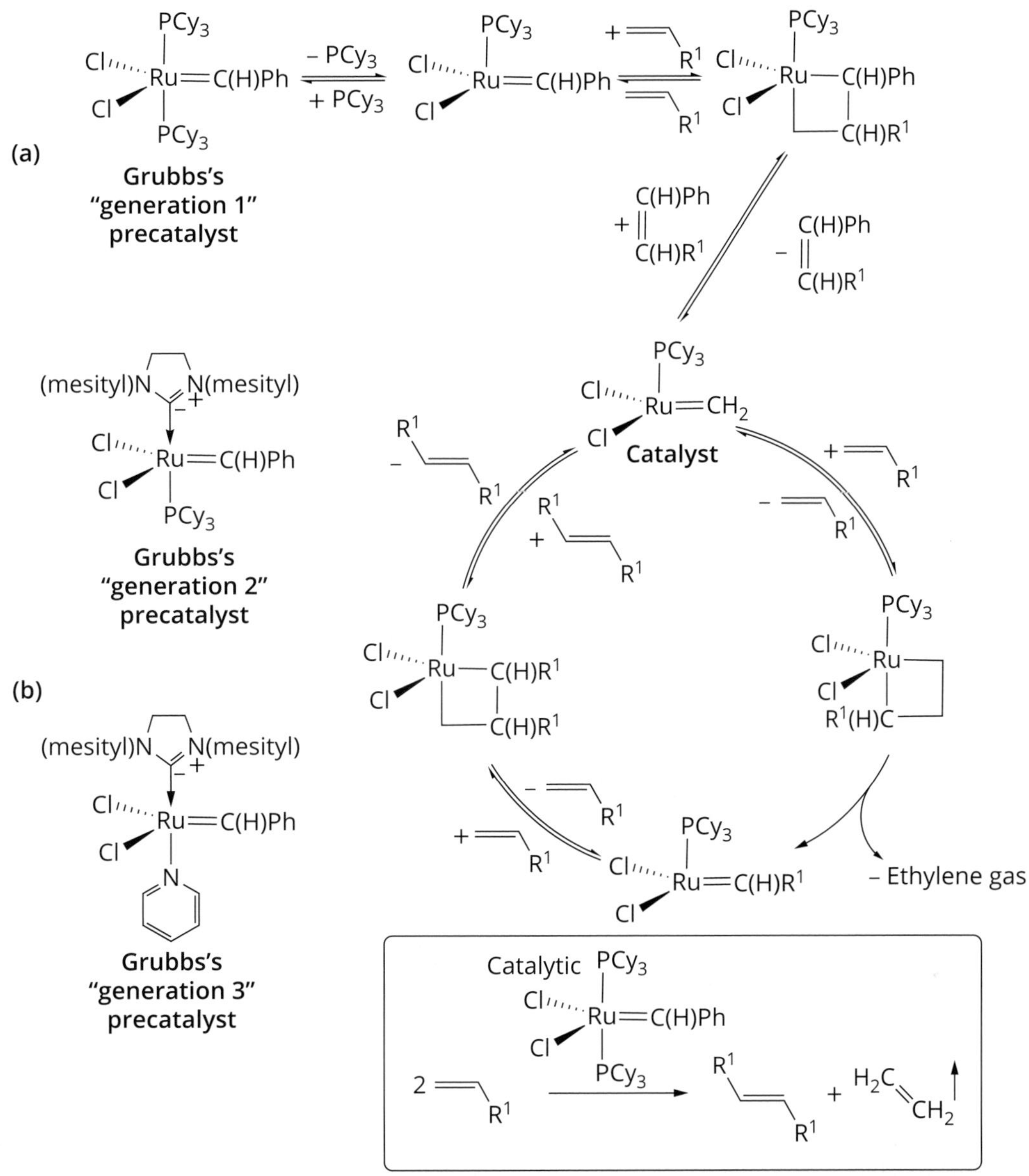

Fig. 18.39 (a) The mechanism of olefin metathesis for Grubbs' first-generation ruthenium catalysts, and (b) the structure of the second- and third-generation catalysts.

or increased steric requirements of the supporting ligand environment.

Differences between the early-metal (Mo or Ta) and later-metal (Ru) catalyzed systems are mostly concerned with ease of synthesis and moisture sensitivity, with the earlier-metal catalysts requiring more rigorous air and water exclusion.

In recent years, a major focus of the metathesis reaction community has been the development of catalysts selective for formation of *Z*-olefins. The olefin metathesis reaction generally favors the more thermodynamically stable *E*-isomer. Current efforts in Mo metathesis begun by Schrock and Hoveyda have led to monoaryloxide pyrrolide (MAP) catalysts that selectively provide *Z*-vinyl halides. This occurs even when using a mixture of *E*:*Z* 1,2-dihaloethene as the coupling partner (Eq. 18.14).[80]

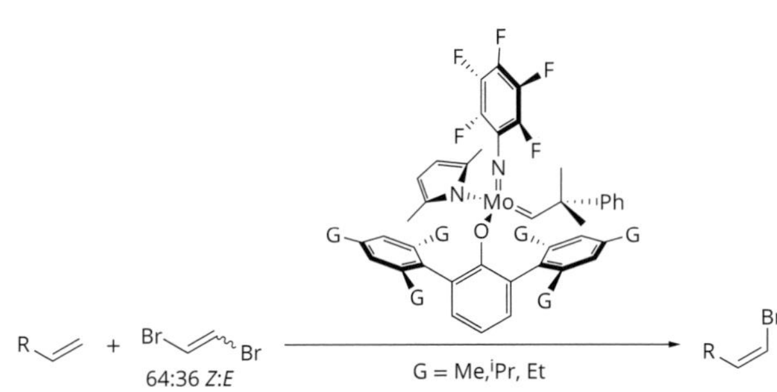

18.14

The addition of bulky substituents to the ruthenium catalyst structure has enabled the formation of the less favorable *Z*-isomer with up to 98% selectivity. Calculations show that the cause of the selectivity for the Ru catalysts is derived from

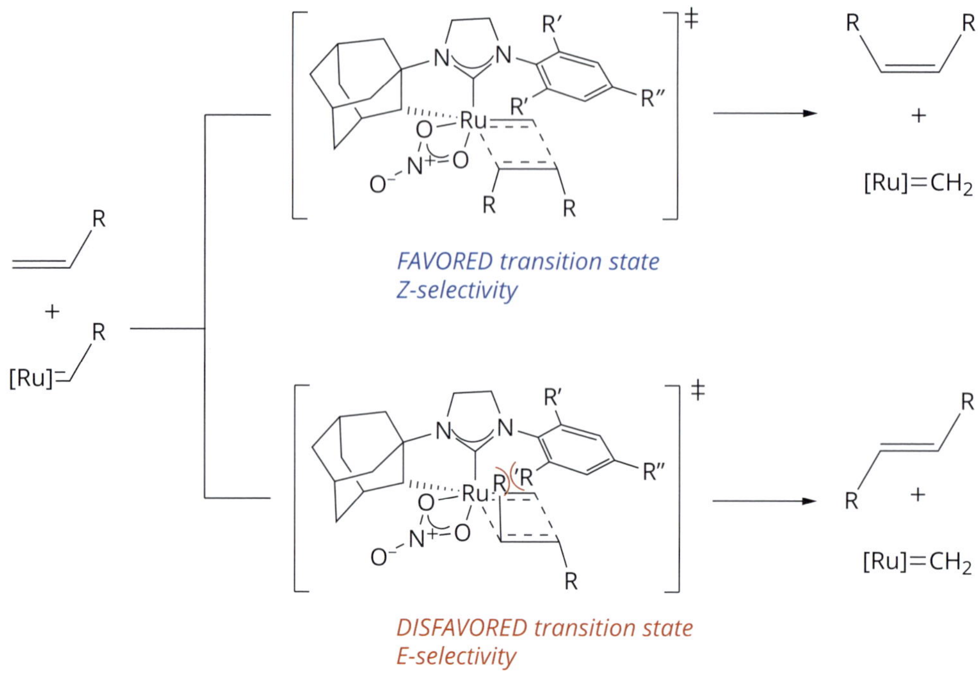

Fig. 18.40 The origin of *Z*-selectivity on ruthenium-catalyzed olefin metathesis stems from steric interactions of the substrate with bulky ligands.

a change in mechanism where the incoming olefin binds "sideways" *cis* to the axial NHC ligand rather than *trans* as is typically observed (Fig. 18.40).[81] Exploitation of this selectivity has been used in the synthesis of a variety of natural products.

18.3.7 Sharpless Epoxidation

The Sharpless epoxidation reaction is an enantioselective chemical reaction to prepare 2,3-epoxyalcohols from primary and secondary allylic alcohols. The 2001 Nobel Prize was partially awarded to Sharpless for its discovery and application. This transition-metal-catalyzed asymmetric reaction has found wide use for a number of reasons, including the cheap and readily available starting materials, ease of conversion of the resulting epoxide to diols, amino alcohols, and other valuable functional groups, and the high enantioselectivity of the reaction. A related reaction that also produces chiral epoxides, but from alkenes, is the manganese-catalyzed Jacobsen–Katsuki epoxidation.[82–83]

Sharpless epoxidation, first reported in 1980,[84] provided extensive procedural details and showed the high yield (>70%) and enantioselectivity (>90% ee) that was possible for a number of di- and trisubstituted allylic alcohols. No mechanistic information was revealed in that publication. The reaction is very easy to set up, adding to its synthetic utility.

Detailed reports of the mechanism through both kinetics and structural studies were published eleven years later.[85–86] Mixing the commercially available titanium tetrakisisopropoxide with one equivalent of dialkyltartrate affords a dimeric titanium complex and releases two equivalents of isopropanol. This complex reacts with *tert*-butyl hydroperoxide, releasing one additional isopropanol, and in a subsequent step it reacts with the substrate allylic alcohol, releasing another equivalent of alcohol and forming the active species that can carry out the oxidation of the substrate (Fig. 18.41). Internal oxygen-atom exchange from the hydroperoxide to the substrate produces the coordinated epoxide. Addition of isopropanol regenerates the initial catalyst species by protonolysis of the product. Alternatively, the product can be released by the addition of the next equivalent of *tert*-butyl hydroperoxide and substrate alcohol.

Under pseudo-first-order conditions, where catalyst and oxidant concentrations were in excess of the substrate concentration, the rate was found to be first-order in the concentration of substrate, titanium complex, and oxidant. The alcohol that is released during catalyst formation inhibits the reaction, with a negative second-order dependence.

The catalytic reaction is inhibited by the addition of excess dialkyltartrate due to the formation of the catalytically inactive [Ti(tart)$_2$] complex, but addition of a substoichiometric amount of the tartrate ligand leads to reduced enantioselectivity. For this reason, a slight (20%) excess of tartrate ligand is recommended in order to fully coordinate to the titanium center.

The rate of reaction of one substrate with both enantiomers of the tartrate catalyst was determined and found to have a difference in rate constant of more than seventy-fold. The rate of epoxidation of the enantiomer of (*E*)-1-cyclohexyl-2-buten-1-ol (Fig. 18.42) is 73 ± 4 times faster for the (–)-tartrate catalyst compared to the (+)-tartrate-derived catalyst. A steric interaction that prevents the reaction of the mismatched pair was used to explain the widely disparate rates.

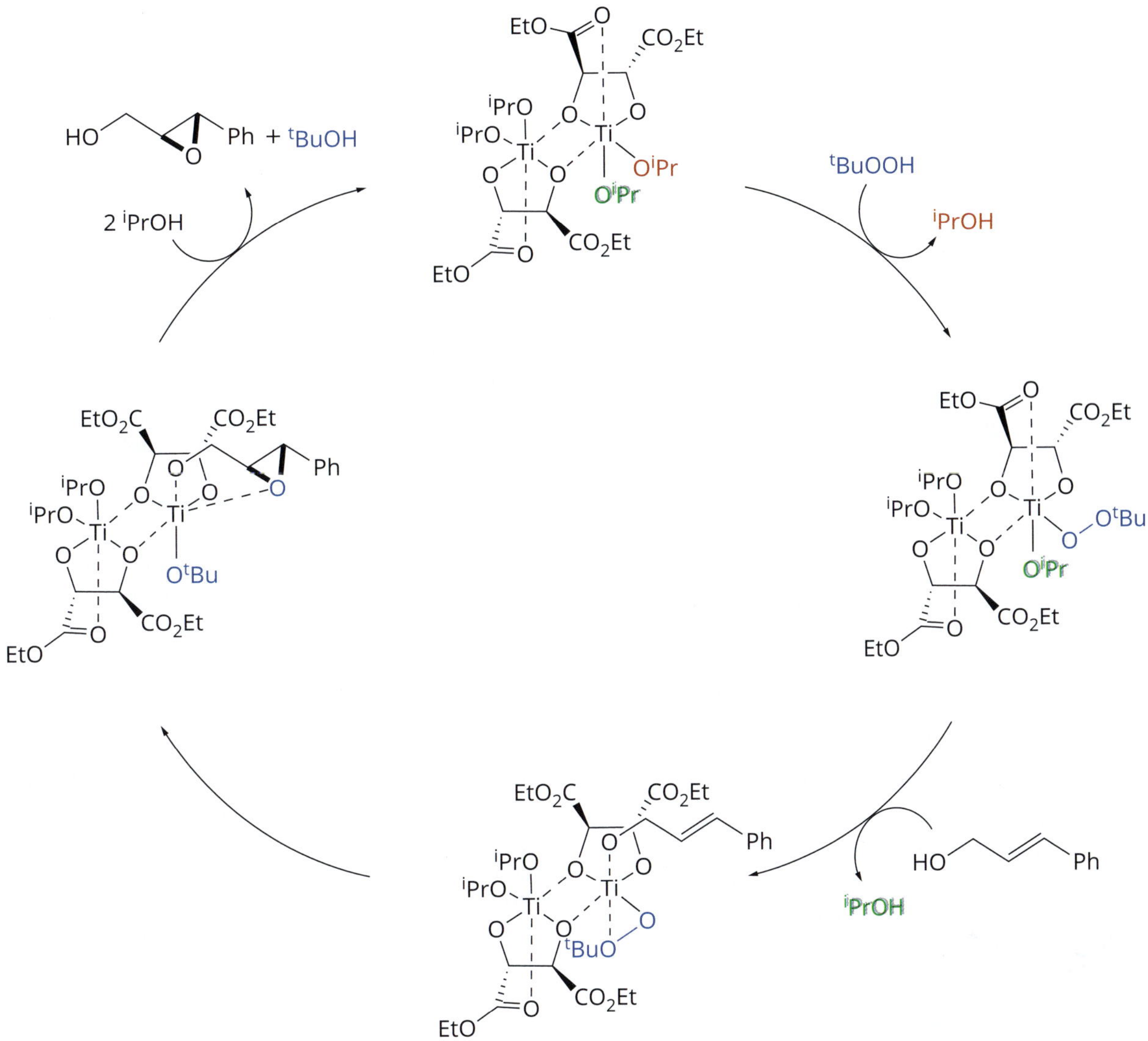

Fig. 18.41 Mechanism of asymmetric oxidation by using titanium tartrate complexes.

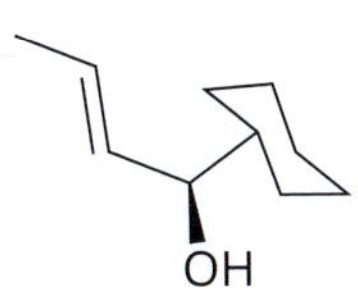

Fig. 18.42 (*E*)-1-cyclohexylbut-2-en-1-ol.

A more synthetically useful way of using this mismatch in rate is to take a racemic mixture of substrate and treat it with a single enantiomer of catalyst. An example of this used in synthesis is the kinetic resolution of racemic allylic alcohols, as seen in Eq. 18.15, which employs diisopropyltartrate (DIPT) and *tert*-butyl hydrogen peroxide (TBHP). The matched pair of reagents reacts more rapidly, leading to the epoxide product, while the mismatched pair remains as the unreacted allylic alcohol. If the reaction is stopped at approximately 50% conversion, an easily separable mixture of two different chiral molecules—one starting material and one product—can both be obtained with high enantiomeric excess.[87]

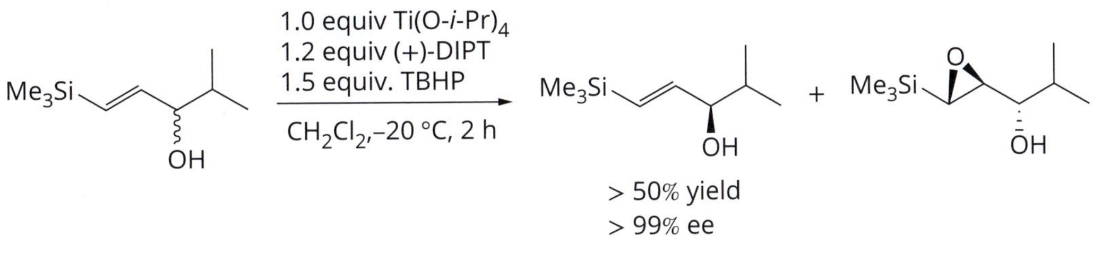

18.15

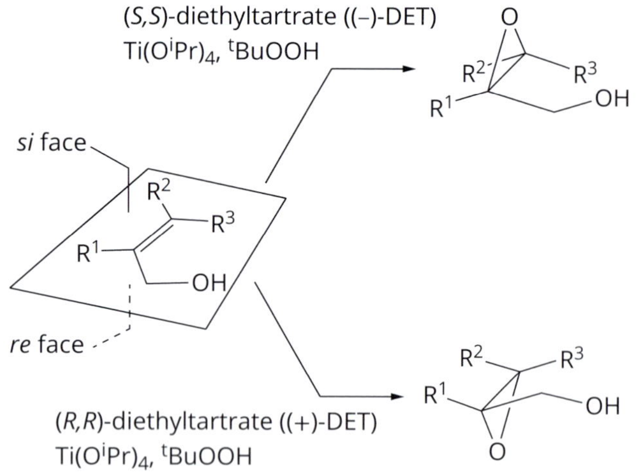

Fig. 18.43 A steric mnemonic for predicting the stereochemistry of the Sharpless epoxidation.

How the titanium complex reacts with the allylic alcohol substrate is understood as including hydrogen-bonding networks that position the substrate and oxidant with subtle steric interactions (both 1,2- and 1,3-diaxial interactions) between the substrate and chiral catalyst to rationalize the observed product distribution.[88] Related work on vanadium oxidation of allylic alcohols postulated a similar model based on sterics.[89–90] A mnemonic allows for very good prediction of the product of the reaction by using either the (+)- or (–)-tartrate-derived catalyst; the (–)-tartrate catalyst reacts more rapidly with the *si* face of the olefin, while the (+)-tartrate reacts more rapidly with the *re* face (Fig. 18.43).

18.3.8 Hydroelementation

Hydroelementation is a class of reactions by which an H–E bond is formally added across a multiple bond. Typically, this involves addition across C–C double or triple bonds, but the addition can also occur with C–N or C–O multiple bonds. There are many examples, including hydrosilylation (addition of H–Si), hydroboration (addition of H–B), and hydroamination (addition of H–N). In general, these reactions have high synthetic utility and atom economy. Hydroboration can be used to prepare boranes that can be converted to boronates suitable for the Suzuki reaction. The hydroboration reaction can be carried out by using simple borane reagents, but more often it is a metal-catalyzed reaction to enhance rate and/or selectivity. Hydrosilylation can be used to prepare industrially useful silicones and organosilanes that are intermediates in organic synthesis. Hydroamination can be used to prepare nitrogen heterocycles of potential pharmaceutical utility. All three of these reactions are highly atom-economical, as all the atoms of the starting material end up in the product. However, the hydroboration reaction is often used in the synthesis of alcohols, as the boron substituent can be transformed into an alcohol after the hydroelementation reaction. In the hydroamination and hydrosilylation reactions, the Si–H and N–H addition products may be the desired final products.

18.3.8.1 Hydroboration

The hydroboration reaction was first reported in early 1956, and in 1979, Herbert C. Brown received the Nobel Prize for developing boron chemistry. The reaction can take place via a nonmetal-catalyzed pathway by the direct addition of B–H bond to the olefin. The organometallic version of the reaction is used to enhance the reactivity and selectivity of the hydroboration reaction, allowing the use of substrates that are slow to react themselves or giving products not usually found from the uncatalyzed process.

The mechanism of the reaction has been studied by using Wilkinsons's catalyst, for example. The 16-electron precatalyst $Rh(PPh_3)_3Cl$ loses a phosphine to form the more reactive 14-electron complex, which undergoes an oxidative addition with R_2BH to produce $R_2BRh(H)(PPh_3)_2Cl$ (Fig. 18.44). Substitution of a triphenylphosphine with an alkene produces the coordinated alkene substrate, which can undergo a 1,2-insertion with the hydride. The carbon–boron bond is formed during a reductive elimination step that releases product and regenerates catalyst. A competing mechanism that operates in some cases does not require loss of phosphine.

An advantage of the metal-catalyzed reaction is more control of selectivity. For the uncatalyzed addition, the anti-Markovnikov product is usually obtained.[91] This addition

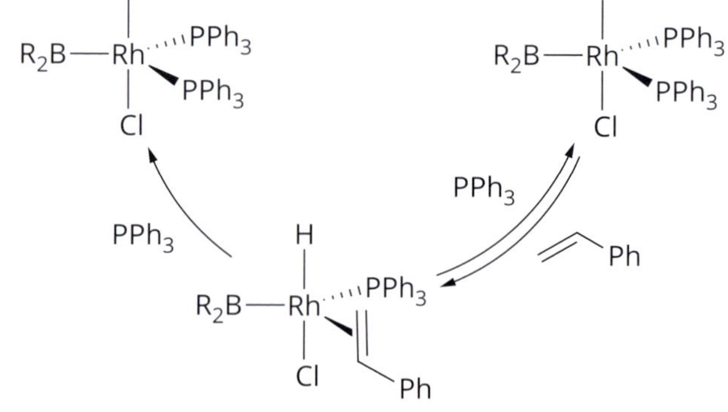

Fig. 18.44 The proposed mechanism for rhodium-catalyzed hydroboration.

proceeds by a concerted four-membered transition state, with the boron and hydrogen atoms adding to the same side of the alkene. The boron develops a slight buildup of positive charge at the transition state, so the slightly negative hydrogen adds to the more substituted and therefore more stable partial carbocation that forms. On the other hand, with the metal-mediated reaction the phosphine ligands can be used to control the addition to form either the Markovnikov or anti-Markovnikov product, and chiral phosphines can be used to prepare chiral products. Treatment of the borane with basic hydrogen peroxide results in an alcohol with retention of configuration at the carbon, making this method very useful for the synthesis of chiral alcohols.

18.3.8.2 Hydrosilylation[92–93]

The hydrosilylation reaction is the formal addition of a trialkylsilane ($HSiR_3$) across a carbon–carbon multiple bond. Although first reported in 1947, this version of the reaction was believed to go through a radical mechanism ($R_3Si\cdot$) via decomposition of a peroxide coreagent. The metal-catalyzed process can be thought of as very similar to hydrogenation, and typical catalysts contain platinum. Oxidative addition of the Si–H bond to the Pt(0) center produces the Pt(II) silyl hydride (Fig. 18.45). The olefin can coordinate and undergo a 1,2-addition with the hydride to produce the Pt(II) alkyl-silyl complex. Reductive elimination affords the desired product and regenerates the Pt(0) center. There are some examples in which the order of insertion and elimination are reversed, resulting in the metal–(β-silyl)alkyl–hydride complex. The asymmetric reaction can be carried out by using suitable chiral phosphine ligands on the catalytic center.

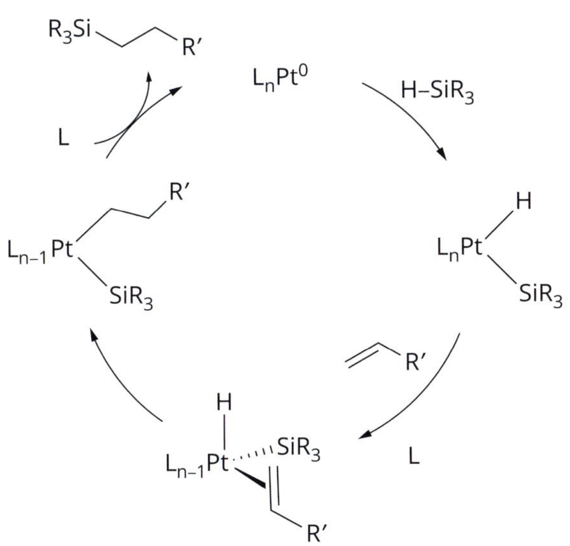

Fig. 18.45 Mechanism of hydrosilation by using a metal catalyst.

18.3.8.3 Hydroamination[94]

Hydroamination is the formal addition of an N–H bond across a C–C multiple bond. The reaction is approximately thermoneutral in some cases, but it has a high-energy barrier in all cases where a strongly electron-withdrawing group is not present on the olefin or alkyne (Michael addition), thus requiring a catalyst or stoichiometric reagent to carry out the reaction. The reaction has been studied extensively, since many pharmaceuticals contain nitrogen heterocycles and there is much interest in the catalytic formation of new carbon–nitrogen bonds. Hydroamination is highly atom-economical, as all the atoms from the starting material end up in the product.

There are both inter- and intramolecular variants of the reaction. In the intermolecular version, much work has been done to favor the anti-Markovnikov product, a regiochemical consideration, while in the intramolecular version, both regio- and enantioselectivity have been examined through careful catalyst design.

Late-metal catalysts (such as gold or platinum) often have a mechanism by which the olefin is activated before the amine (Fig. 18.46). Coordination of the olefin to the metal center makes it more susceptible to external nucleophilic attack by an amine. Proton transfer from the resulting ammonium complex releases the product and regenerates the active metal catalyst.

The reaction has different mechanisms of action, depending on the identity of the metal catalyst. The first is activation of the amine. The most extensive mechanistic studies were carried out by Marks, who studied lanthanide-catalyzed intramolecular hydroamination. Generically, $L_2Ln–N(SiMe_3)_2$ reacts with an incoming aminoalkene to liberate $HN(SiMe_3)_2$ and form the amido complex of the substrate $L_2Ln–NHR'$ (Fig. 18.47). A 1,2-insertion reaction generates the N–C bond, and subsequent protonolysis by incoming substrate releases the product and regenerates the catalyst.[95–97]

Titanium and zirconium complexes have been found to use yet a different mechanism, first discovered by Bergman and coworkers.[98–99] L_2TiR_2 precatalysts react with an amine

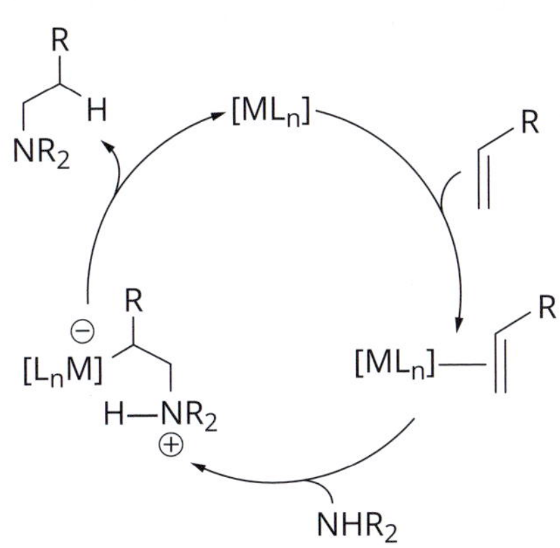

Fig. 18.46 Mechanism of late metal-catalyzed hydroamination.

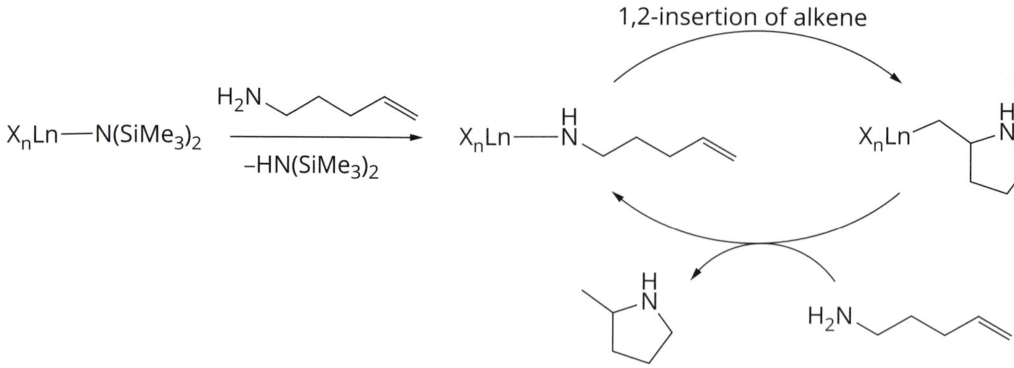

Fig. 18.47 Mechanism of lanthanide-catalyzed hydroamination of an aminoalkene.

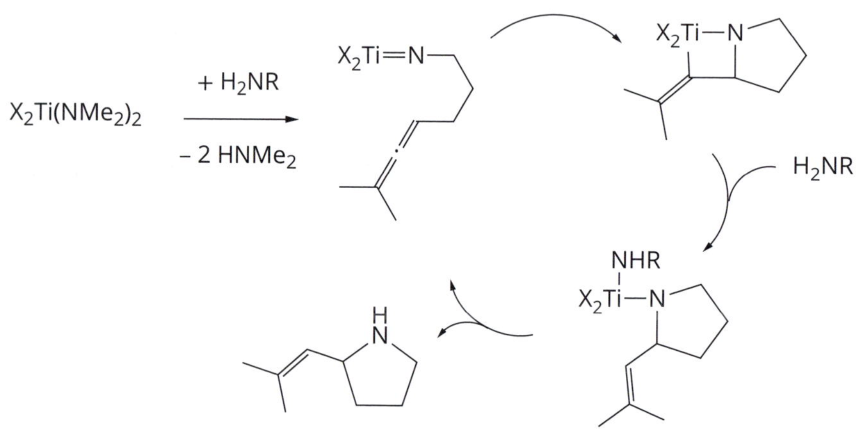

Fig. 18.48 Mechanism of early metal-catalyzed, intramolecular hydroamination of an aminoallene.

(or aminoalkene) to generate the imido complex ($L_2Ti{=}NR$). This undergoes a [2 + 2]-cycloaddition to form the azametallacyclobutane. Protonolysis by incoming substrate regenerates the imido catalyst (Fig. 18.48).[100–103]

18.3.9 Titanium-Catalyzed Multicomponent Coupling Reactions

There is a large and growing group of reactions related to titanium hydroamination chemistry that uses titanium imides and amides as intermediates in the generation of new carbon–nitrogen bonds. Some of these reactions are multicomponent couplings that make carbon–carbon as well as carbon–nitrogen bonds in the same catalytic cycle. The advantage of multicomponent coupling reactions is that they form complex products from relatively simple starting materials in a single step.

One straightforward example is titanium-catalyzed iminoamination, which is a three-component coupling reaction where an amine, alkyne, and isonitrile are used to form tautomers of 1,3-diimines. The proposed catalytic cycle is shown in Fig. 18.49, which starts similarly to hydroamination (Fig. 18.48), but the azatitanacyclobutene is trapped by 1,1-insertion of the isonitrile. The unsymmetrical 1,3-diimines accessed have been applied in one-pot syntheses of many different types of nitrogen heterocycles, such as pyridines, pyrroles, isoxazoles, pyridines, and pyrazoles.[104]

A very interesting reaction type that can be catalyzed by titanium, zirconium, and other metals, but perhaps first became practical with tantalum, is hydroaminoalkylation of olefins.[105–106] In 2007, Herzon and Hartwig published examples of using $Ta(NMe_2)_5$ as catalyst to add substrates such as HN(Me)Ph across olefins, as shown in Fig. 18.50. Since then, there have been great advances by workers such as Schafer, Doye, Rovis, and others on developing the reaction with a variety of catalyst structures.[107–110]

In related chemistry also involving titanium imide intermediates, Tonks and coworkers have developed multicomponent reactions to produce pyrroles, among other compounds.[111–112] An example is shown in Fig. 18.51. The precatalyst is Mountford's complex $Ti({=}NPh)Cl_2py_3$,[113] which can be used to produce pyrroles by coupling two alkynes and half an equivalent of azobenzene. One unique feature of the reaction is that it employs a Ti(II)/Ti(IV) redox couple similar to late-metal coupling mechanisms discussed previously but unusual for early transition metals.[112] Reductive elimination produces the proposed Ti^{II} complex, which reacts with azobenzene to release the pyrrole.

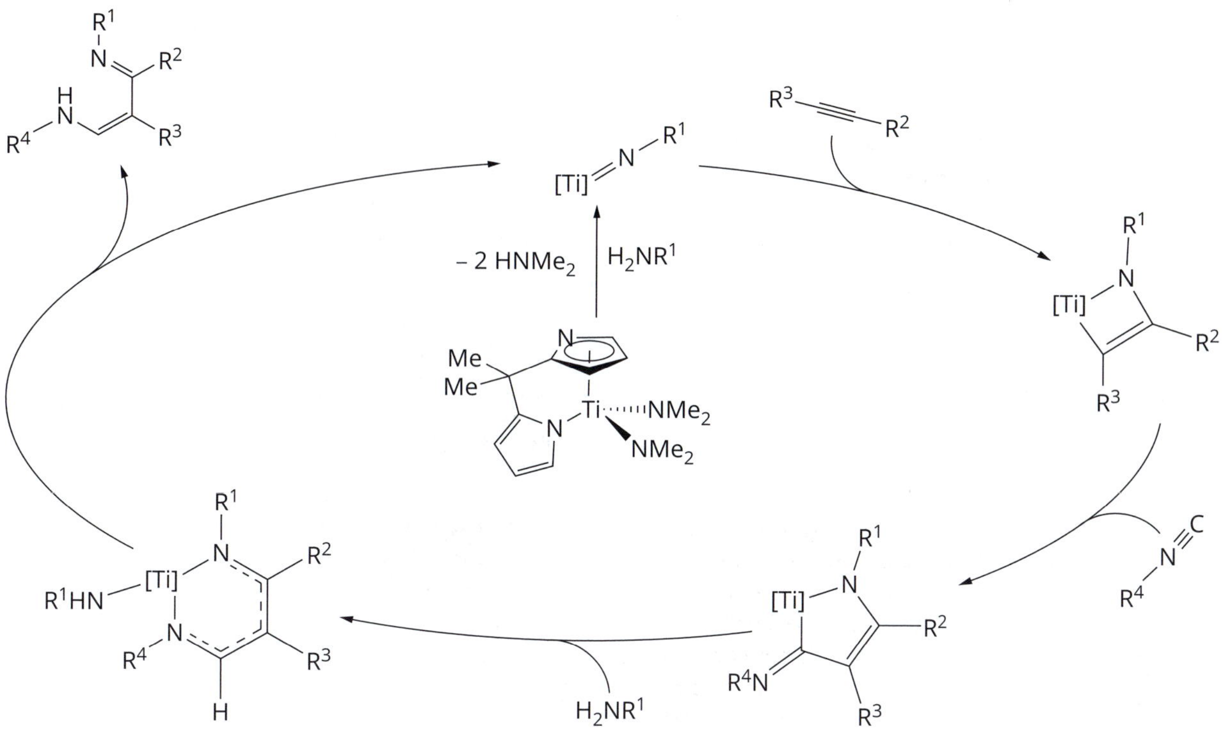

Fig. 18.49 In the iminoamination catalytic cycle, the titanium catalyst (center) reacts with primary amine to form the active titanium imide. The imide can undergo [2 + 2]-cycloaddition with the alkyne to form the azatitanacyclobutene, which is trapped by 1,1-insertion of the isonitrile. Protolytic cleavage of the Ti–C and Ti–N bonds leads to the product and reforms the imide.

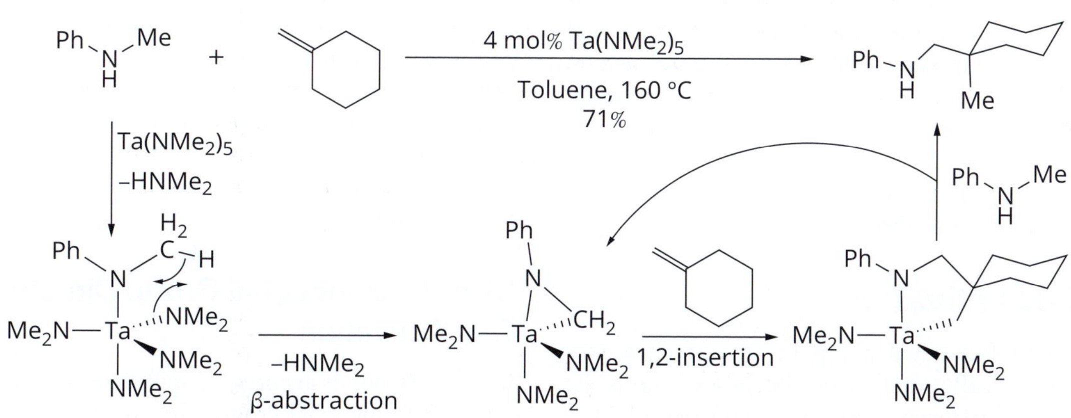

Fig. 18.50 A plausible mechanism for hydroaminoalkylation involving $Ta(NMe_2)_5$ as precatalyst. The cycle involves transamination of NMe_2 with the HN(Me)Ph substrate. The β-hydrogen on the substrate undergoes β-abstraction by another dimethylamide group to produce a metallaziridine that can 1,2-insert the olefin substrate. The azatantallacyclopentane is then protolytically cleaved to give the product and reform the metallaziridine.[105]

18.3.10 Nickel-Catalyzed Multicomponent Coupling Reactions

There is a large and growing list of late-transition-metal multicomponent coupling reactions that are also known.[114–121] The advantage of multicomponent coupling chemistry is that it allows the generation of complex products from simple starting materials in a single step. As an example, consider the nickel-catalyzed multicomponent coupling chemistry developed by the Jamison group (Fig. 18.52).[122]

The proposed mechanism for the reaction uses many of the steps that we have covered in other sections of this chapter. The catalysis begins with coordination of alkyne to a $Ni(PR_3)_2$ precursor of some type in solution. If the alkyne complex is

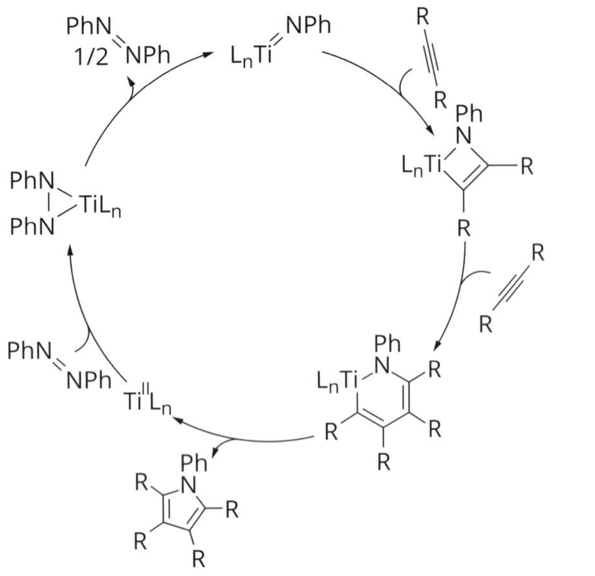

Fig. 18.51 Tonks and coworkers multicomponent coupling reaction to generate pyrroles. The reaction shares some features with titanium hydroamination and iminoamination. However, the system is unusual in that a Ti(II)/Ti(IV) redox reaction is proposed.

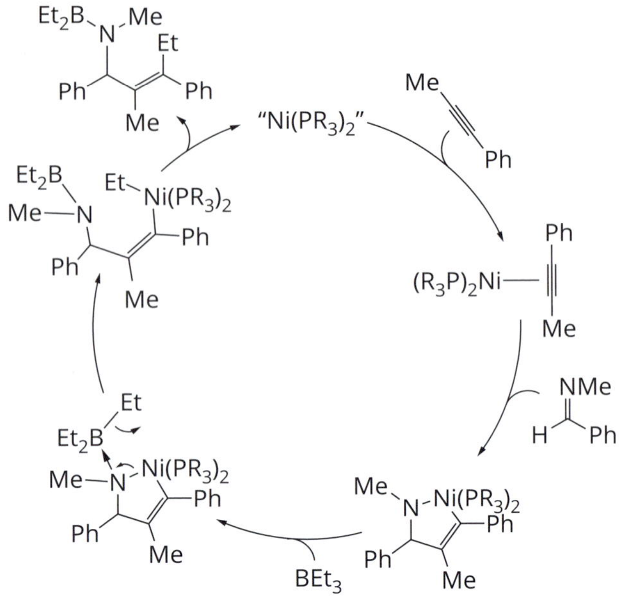

Fig. 18.52 A nickel-catalyzed multicomponent coupling between an alkyne, imine, and trialkylborane.

viewed as a metallacyclopropene, the next step could be called a 1,2-insertion of the imine into the Ni–C bond; alternatively, it might be referred to as an oxidative cyclization of the alkyne, imine, and nickel to produce a five-membered ring. Triethylborane may then coordinate to the nitrogen of the nickel amide. Then, transmetallation from boron to nickel occurs with the amide as the leaving group, which results in a nickel center with vinyl and ethyl groups. Reductive elimination of the two carbon substituents on nickel gives the product and reproduces the "$Ni(PR_3)_2$" catalyst. In this process, two new carbon–carbon bonds are formed.

18.3.11 C–H activation

The most common functional group in organic chemistry (if one can agreeably call it that for our purposes here) is the C–H bond. Carbon–hydrogen bonds are usually described as inert, and the chemistry of alkanes is very limited. A major challenge in the field of organometallic chemistry is the selective functionalization of the C–H group into more valuable functionality. Carbon–carbon and carbon–hydrogen bonds are strong, and simple alkanes do not have low-lying empty orbitals or high-lying filled orbitals that could more readily undergo chemical transformations. One molecule in particular, methane, has attracted significant attention for pragmatic reasons.

Methane is a gas at standard temperatures and pressures, and thus is relatively difficult and expensive to compress, transport, and store. As a result, in 2012 approximately 3.5% of the world's natural-gas supply (143 billion cubic meters) was burned at the source rather than used, resulting in 350 million tons of carbon dioxide being released for want of a better use for the molecule. Methane has a strong C–H bond at 104 kcal/mol. A significant challenge in methane activation is not finding a reagent to break that bond, but to stop at only breaking the first bond. Biological organisms have evolved enzymes to selectively activate and use methane as their only carbon source: methanotrophs. The synthetic chemist is still early in the journey towards selective C–H bond activation.

18.3.11.1 Functional Group Directed C–H Activation

Some C–H bonds in organic molecules are readily activated due to the functional groups already present in the molecule. For example, the carbanion formed by deprotonating the α-hydrogen in a β-diketone is strongly resonance-stabilized into both oxygen atoms. This stabilization makes the α-hydrogen quite acidic, with a pKa around 9, and it is capable of being deprotonated by bases such as hydroxide, carbonate, or even ammonia.

Directed C–H bond functionalization can be carried out by transforming a functional group in an organic molecule into a suitable ligand for a transition-metal complex. The ligand internal to the substrate directs—usually by encouraging a five- or six-membered transition state—the oxidative addition of the C–H bond.

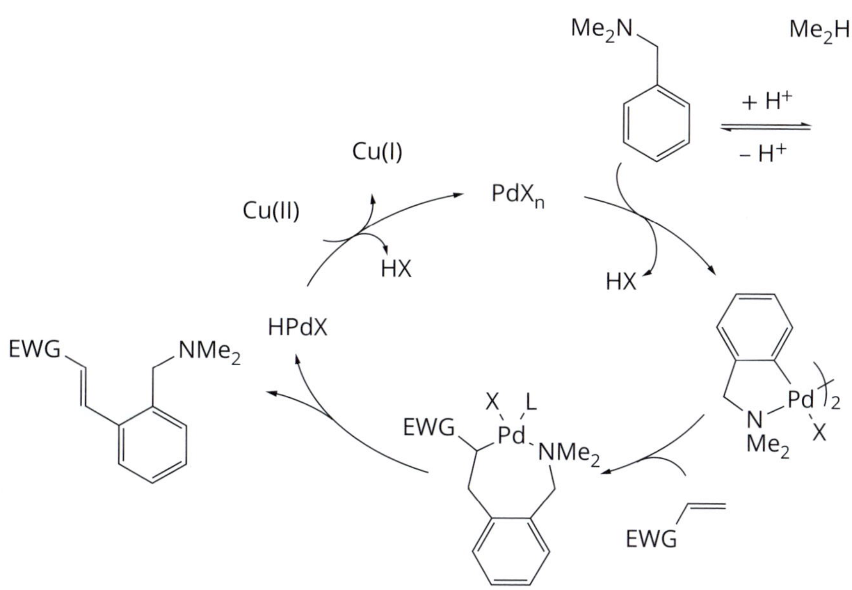

Fig. 18.53 The proposed mechanism for the directed activation of an aromatic C–H bond by Pd(II).[123]

The first example of a palladium-catalyzed amine-directed C–H functionalization was carried out in 2007.[123] A number of substituted *N,N*-dimethylbenzylamines were derivatized at the *ortho*-position of the arene. The dimethylamine group acted as a ligand for a Pd(II) species, which was brought into close proximity of the *ortho*-hydrogen, allowing for its activation by hydrogen-atom abstraction by X to form HX (Fig. 18.53). (One may be tempted to draw a C–H oxidative addition mechanism, but while Pd(IV) complexes are known,[124–126] this is a fairly rare oxidation state for the palladium organometallic complexes.) The Pd–C bond undergoes a 1,2-insertion with an olefin substrate, and subsequent β-hydrogen elimination generated a Pd(II) hydride. The Pd(II) hydride undergoes reductive elimination to form a second equivalent of HX, and oxidation by Cu(II) regenerates the active Pd(II) species. The solvent system was critical for the success of this system as it moderated the relative concentration of the tertiary amine.

The development of catalysts capable of activating a C–H bond at a *meta*- or *para*-position is more difficult, as the five- or six-membered metallacycles required for *ortho*-functionalization would have to be expanded to activate C–H bonds further away from the directing group. The use of coordination chemistry to enforce a *meta*-activation for sulfonation was first realized by using a Ru(cymene) catalyst.[127] The work was then extended to allow for the formation of C–C bonds, and detailed mechanistic experiments were used to postulate a mechanism.[128] Coordination of 2-phenylpyridine to the ruthenium center activates one of the *ortho*-C–H bonds in the phenyl ring (Fig. 18.54). However, instead of functionalization at that carbon, which was the initially expected result, the Ru–C bond exhibits a strong *para*-directing effect, and the sulfonation or alkylation reaction occurs at the carbon that was initially *meta* to the chelating pyridyl group. Cleavage of the product from the metal can be achieved by using bicarbonate as the proton source. This is a young field that is rapidly growing. A recent review also showcases several examples of *para*-direction in the C–H functionalization reaction.[129]

These directed C–H activation reactions can be used to generate a variety of different products. For example, the imine in an α,β-unsaturated imine can be used as a directing group to add an alkyne in a rhodium-catalyzed process (Fig. 18.55). The product of the C–H activation/insertion reaction then undergoes a 6-electron cyclization reaction to produce the dihydropyridine product, which can be isolated or converted to a pyridine derivative by deprotection.[130–131]

Directed C–H activation is now a valuable tool in the synthesis of complex natural products. For example, Potter and Ellman reported the use of rhodium-catalyzed C–H activation and insertion as the key step in the synthesis of (+)-pancratistatin (Fig. 18.56).[132]

18.3.11.2 Undirected C–H Activation

Without a directing group to position the metal center, activation of a C–H bond is difficult, and selective activation is even more challenging. The catalyst must coordinate to the bond of interest with good selectivity for a particular C–H bond. Tertiary C–H bonds are weaker and therefore more

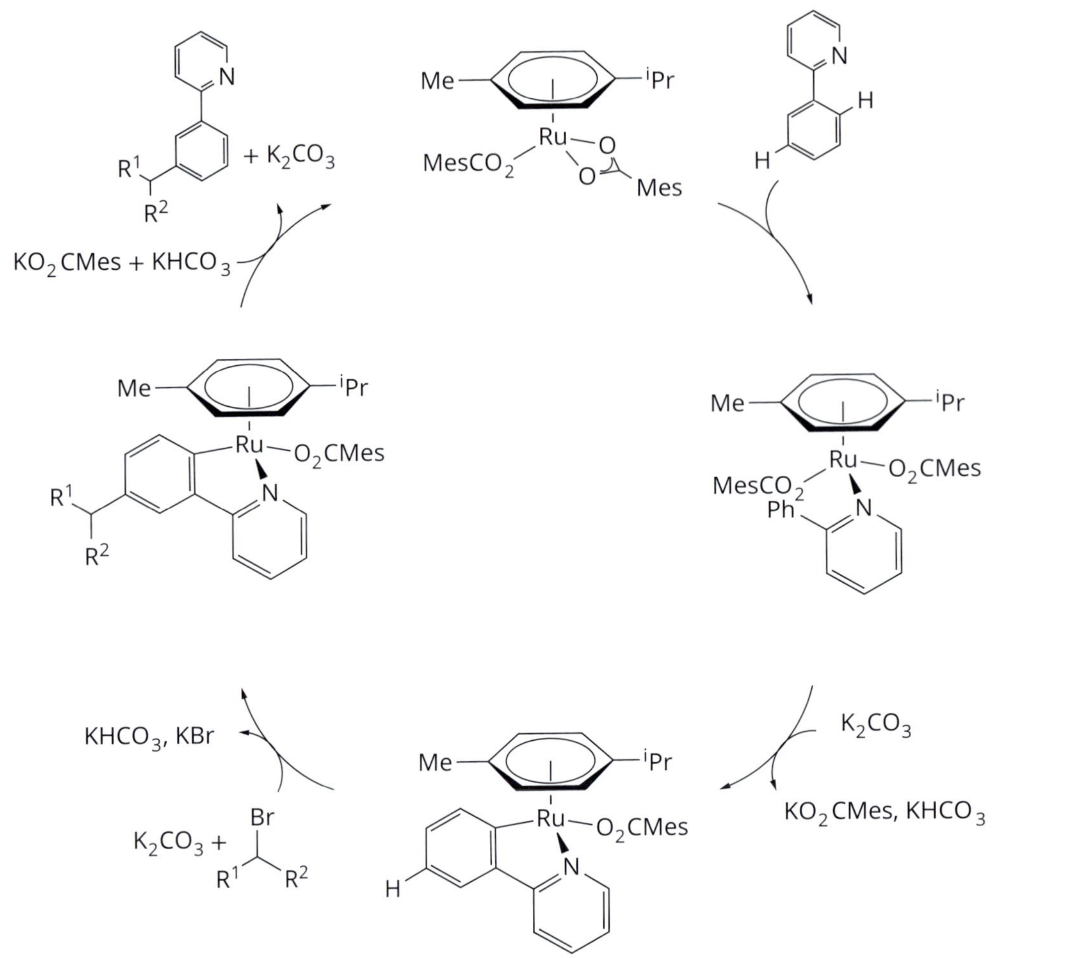

Fig. 18.54 Proposed mechanism for C–H activation with Ru(cymene) complex that produces the *meta*-substituted product.[127]

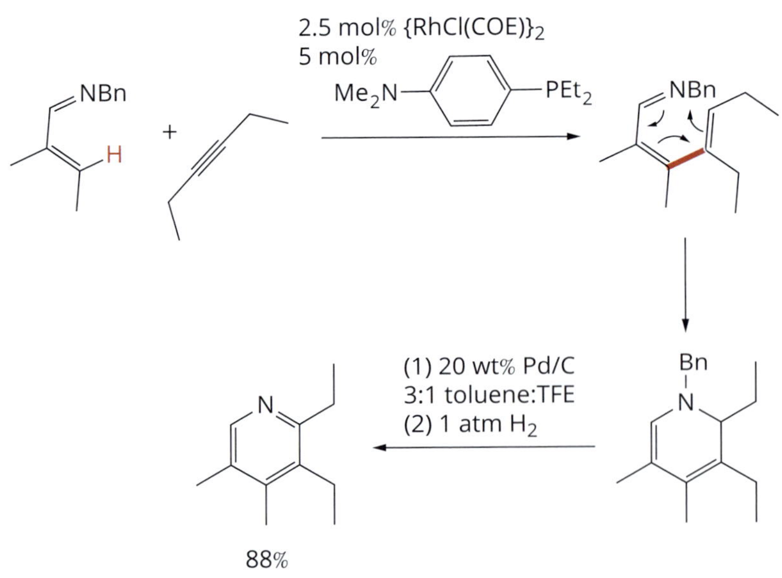

Fig. 18.55 Example of pyridine synthesis by imine-directed C–H activation by rhodium.

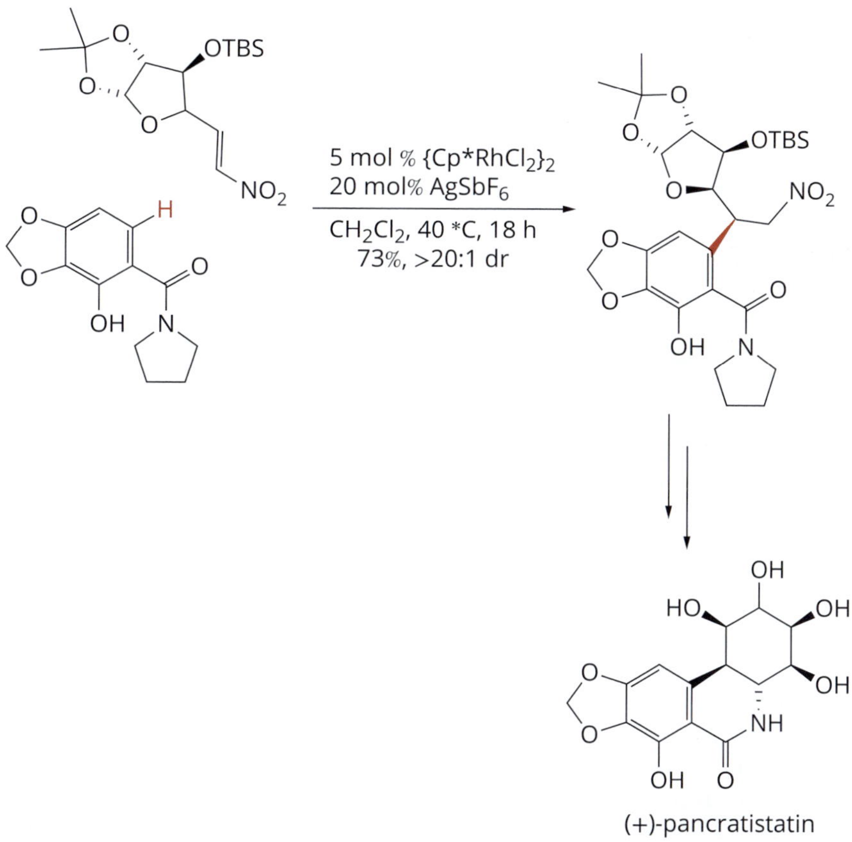

Fig. 18.56 An example of using directed C-H activation in organic synthesis. In this case, the Rh-catalyzed reaction is directed by the adjacent amide group on the aromatic ring, which leads to olefin insertion and reductive elimination.[132]

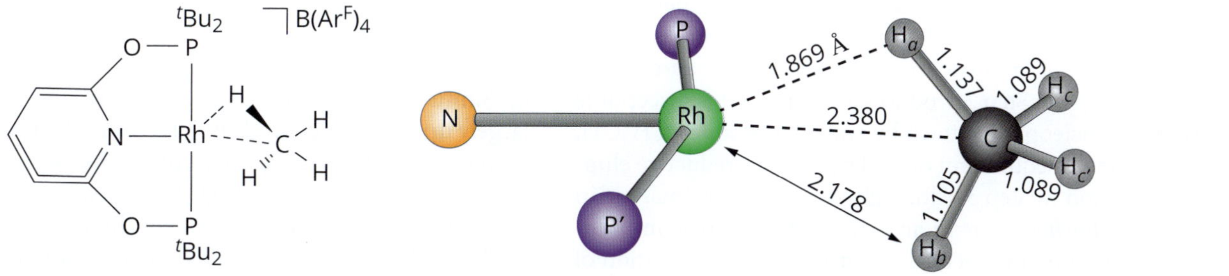

Fig. 18.57 The calculated structure of the first metal–methane complex that was spectroscopically characterized. The metal is interacting unequally with two of the four C–H bonds in methane.[133]

reactive than secondary or primary C–H bonds, but activation of a terminal (primary) bond is often the desired reaction. The C–H bonds in methane are all equivalent, so there is no selectivity issue.

Adding to the challenge, the C–H bond is essentially nonpolar and does not easily bind to a transition metal. However, prior sections of this chapter provide a clue as to how chemists are attempting to activate the C–H bond. It is possible to spectroscopically characterize metal–methane complexes that are engaged in one or even two interactions of C–H bonds with the metal center. The complex shown in Fig. 18.57 was prepared by protonation of a metal–methyl complex—the reverse of a C–H activation process.[133]

Beginning in the 1970s, it became apparent that these C–H bonds could be activated, because deuterium–hydrogen exchange was observed to take place in a number of systems. The resulting organometallic species were believed to be unstable, and careful design of complexes resulted in the first

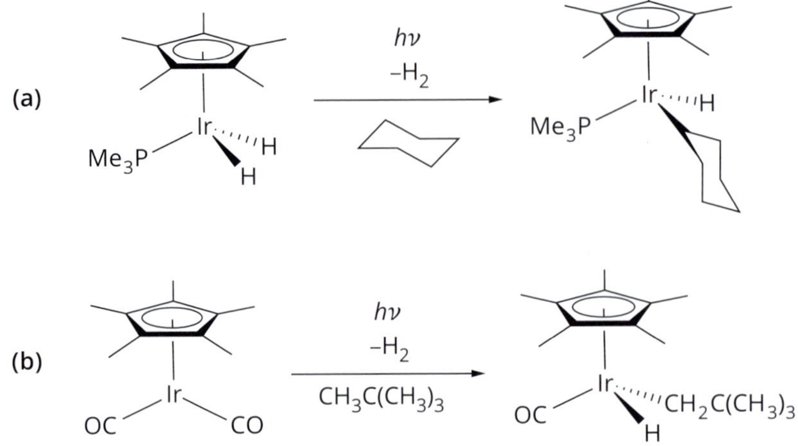

Fig. 18.58 Two of the first examples of C–H bond activation leading to well-characterized products.

demonstrated observations of intramolecular C–H bond activation in 1982 (Fig. 18.58).[134–135] Interestingly, in this system, use of a PPh_3 ligand resulted in directed C–H bond activation of the phenyl group, and this is not uncommon, as aromatic group activation is often more kinetically favorable than alkyl–H activation. Two alkyl activation reactions are shown in Fig. 18.58, in which the active species in both cases is believed to be Cp*IrL (L = PMe_3 or CO), which is capable of coordinating to a C–H bond of an alkane and then activating it by oxidative addition.

Only a few years later, Shilov reported the direct activation of methane by using a Pt(II)/Pt(IV) oxidative-addition reaction sequence (Fig. 18.59a).[136] In Step 1, methane undergoes an oxidative addition to the Pt(II) center, and then HCl is lost by reductive elimination. Oxidation of the Pt(II) by an external oxidant, $PtCl_6^{2-}$, in Step 2 produces a species that is susceptible to external nucleophilic attack by water to produce CH_3OH, and then another equivalent of HCl by another reductive elimination reaction in Step 3. Although the reaction is catalytic in Pt(II), it is *stoichiometric* in an external Pt(IV) oxidant, rendering the reaction impractical but inspiring to a generation of chemists willing to take up the challenge of making the chemistry amenable to large-scale use.

Much effort has been spent to find external oxidants besides Pt(IV), and a success in this area was realized in the system developed by researchers at Catalytica Advanced Technologies.[137] In this reaction, methane is activated in hot sulfuric acid to form the protected product methylbisulphate, H_3COSO_3H (Fig. 18.59b). In this system the oxidant is SO_3, which is reduced to SO_2. Unfortunately, the protected product has little value and would need to first be separated from the reaction mixture, and then converted to methanol in an additional step.

More recently, oxidation of terminal C–H bonds in substituted alkanes has been realized. For example, Sanford and coworkers have developed a method for direct hydroxylation of C–H bonds in amines using platinum catalysis, in some cases with excellent selectivity (Eq. 18.16, image by the authors).[138]

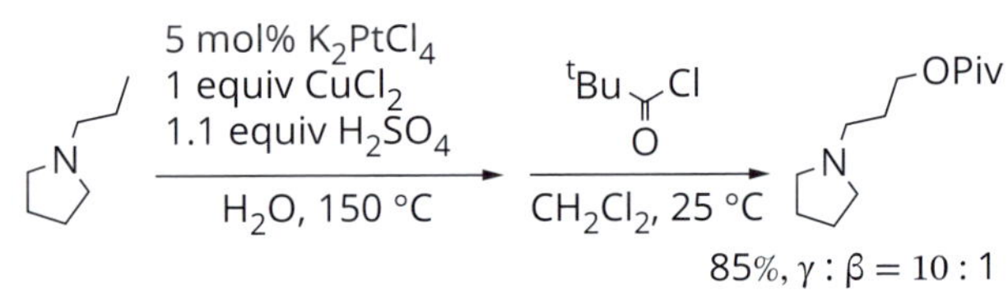

18.16

Perhaps one of the most reliable and well-developed areas of C–H activation is in iridium-catalyzed borylation of aromatic rings—an area pioneered independently by the groups of Smith and Hartwig.[139–142] Originally, the reactions were largely under steric control, and, for example, 1,3-disubstituted arenes could be used as selective substrates in borylation meta to both substituents (Fig. 18.60). Now, however, there is an extensive literature with known directing effects for controlling substitution patterns. These borylation reactions give products that are then ideal for Suzuki coupling and other modifications in far fewer steps than previous methods. The reactions are also applicable to many aromatic heterocycles. Pyrroles, for example, are typically borylated in the α-position unless the nitrogen is protected to sterically favor the β-position,[143] but there are "traceless" β-borylation procedures for pyrroles now that use nitrogen borylation to encourage the selectivity change.[144] There is now significant evidence that the catalytic cycle for these useful iridium-catalyzed borylations involves an iridium(III)/(V) cycle.[145]

18.3.11.3 Radical Pathways

Porphyrins are common ligands in biology and bioinorganic chemistry, and rhodium porphyrin complexes are a bridge

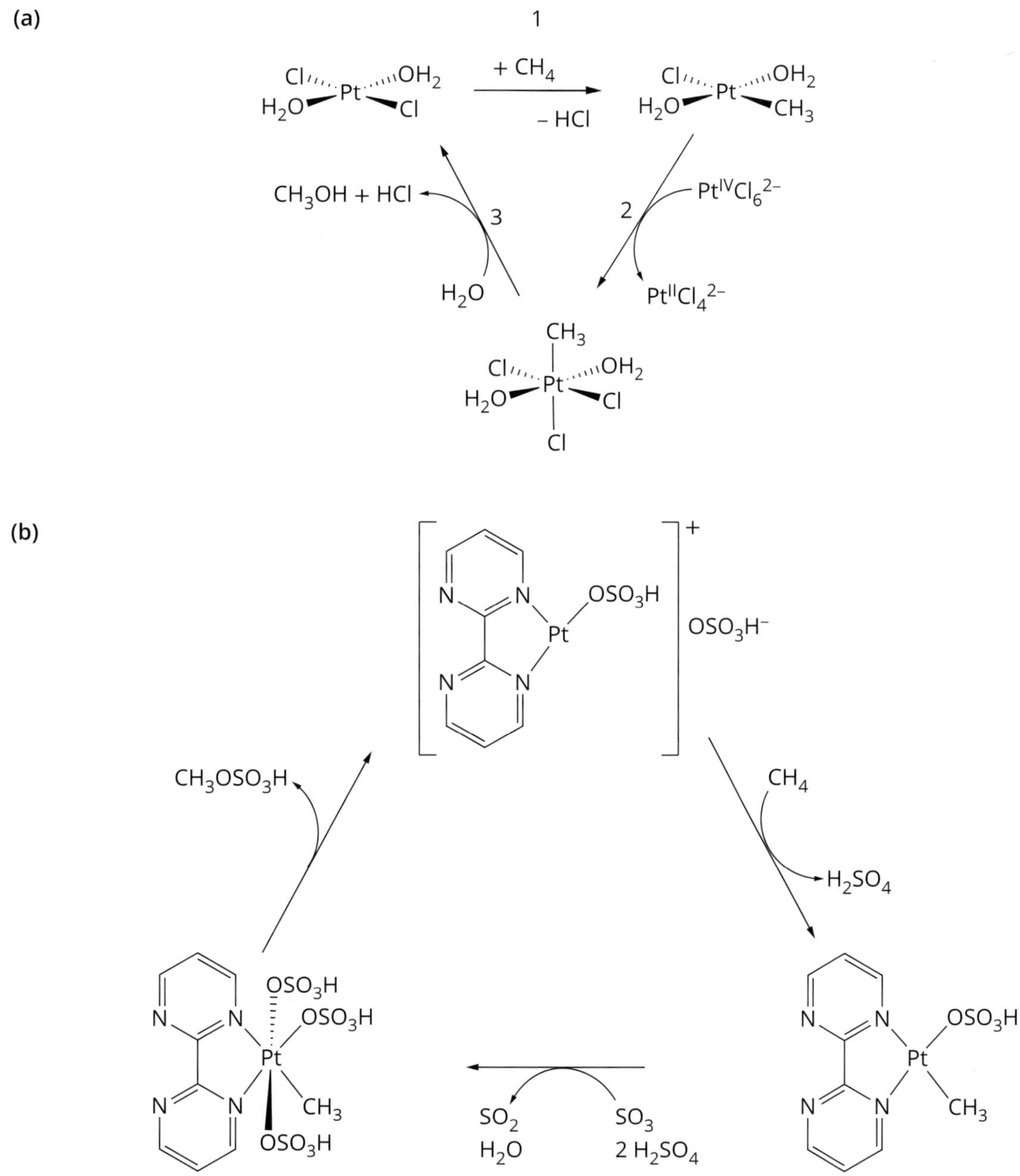

Fig. 18.59 (a) The Shilov system, and (b) the catalytic system for direct conversion of methane to methanol by using catalytic Pt(II) and either Pt(IV) (a) or H_2SO_4 (b) as the stoichiometric oxidant.

between the field of organometallic C–H bond activation and biological systems. Sterically bulky porphyrins (tetramesitylporphyrin, TMP) discourage dimerization of the two rhodium centers, and the resulting monomeric radical complex can react with methane selectively in benzene solution to form one equivalent each of the (TMP)Rh–H and (TMP)Rh–CH_3 complexes.[146]

The rate of the reaction is first-order in methane and second-order in rhodium, has a kinetic isotope effect of 8.6, and has $\Delta S^\ddagger = -37 \pm 5$ cal/mol K. The rate data suggest a termolecular reaction, the KIE suggests a high degree of C–H bond breaking at the transition state, and the entropy of activation implies a high degree of order in the transition state for the reaction. A mechanism that is compatible with the data is the formation of a metal–methane complex that then interacts with a second metal complex to form a 4-centered linear Rh···C···H···Rh interaction (Fig. 18.61).[147]

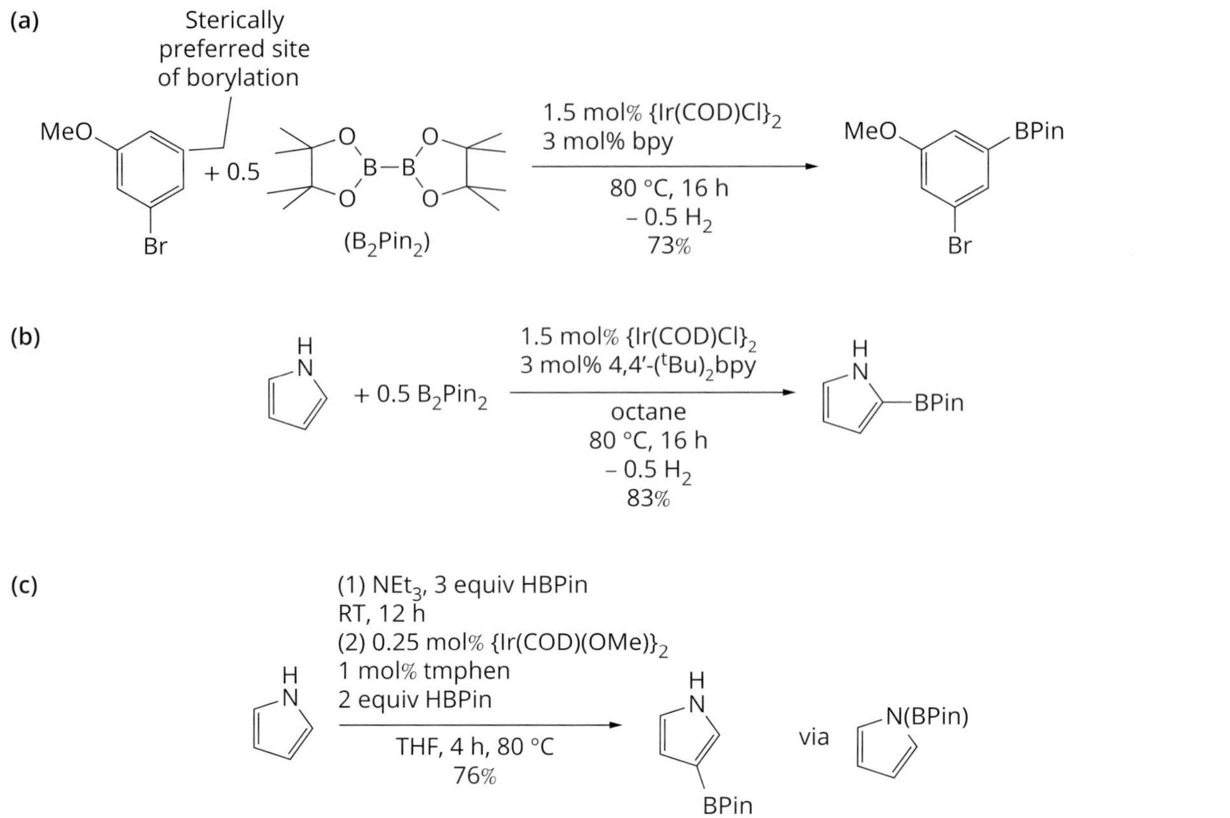

Fig. 18.60 Some examples of iridium-catalyzed borylation of aromatic rings. In many cases, sterics can be used to control the selectivity, although directing effects are becoming better understood. (a) Borylation of a 1,3-disubstituted arene produces selective borylation at the least hindered site meta to both groups. (b) For pyrrole, borylation is favored at the α-position, but (c) placing a group on the nitrogen can lead to β-borylation, which can now be achieved in a traceless nitrogen–borylation reaction. The bottom reaction initial forms an N–BPin pyrrole, and the large BPin-group on nitrogen directs C–H borylation away from the α-positions, allowing selective production of the β-BPin pyrrole. The BPin-group on the nitrogen is removed during workup of the reaction.

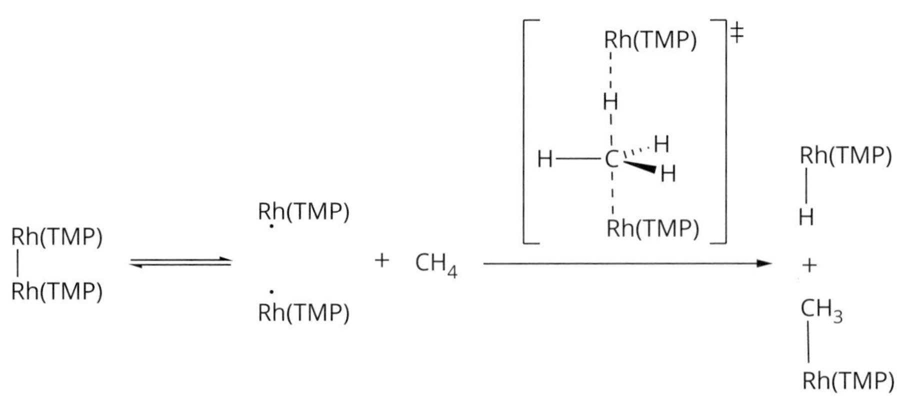

Fig. 18.61 The linear, 4-centered transition state for the diradical activation of the C–H bond in methane by rhodium porphyrin complexes.

Chapter Summary

1. The addition of many small alkyl groups, such as ethyl or butyl, to metal centers leads to unstable products and rapid decomposition. As a result, the field can arguably be said to have started in earnest with the report of very stable ferrocene, $FeCp_2$.
2. Now, organometallic chemistry is a critical field for production of many molecules we rely on, such as polymers and pharmaceuticals. Some of the catalyses described in this chapter, for example, are very common methods for making carbon–carbon bonds in the pharmaceutical industry.
3. The key to unlocking the field of organometallic chemistry was in understanding the common reaction steps taken by transition-metal centers with common organic fragments as ligands. Understanding a few common mechanistic steps allows the understanding of many different catalytic cycles. Catalyses usually work best when the catalytic species stays around the same energy to prevent large barriers along the chemical pathway, which often means that many of the steps will be reversible to some extent.
4. Oxidative addition and reductive elimination were the first pair of mechanistic steps discussed, and reductive elimination is the reverse of oxidative addition. In an oxidative addition, a metal in the *x* oxidation state adds some single-bond species (A–X) to give a product with the metal in the *x* + 2 oxidation state and new M–A and M–X bonds. Reductive elimination is the reaction in the opposite direction.

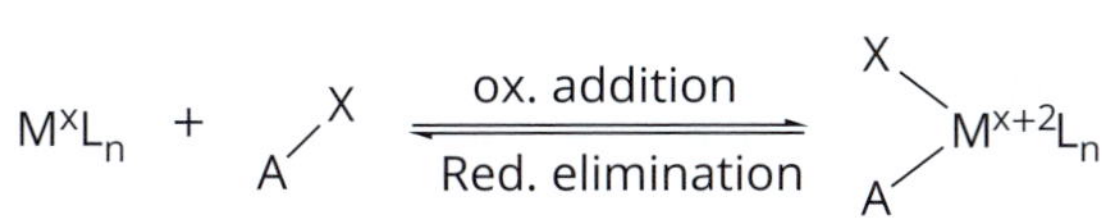

5. Another set of common transformations are insertion/deinsertion reactions. These may occur in a variety of different ways, but two commonly observed reactions are 1,1-insertions and 1,2-insertions, along with their microscopic reverse.
6. Carbon monoxide, and some close derivatives such as isonitriles and NO, often undergo a 1,1-insertion reaction. In a 1,1-insertion, the metal center and one of its covalently bound ligands adds to the carbon in CO. The reverse is 1,1-deinsertion or α-elimination.

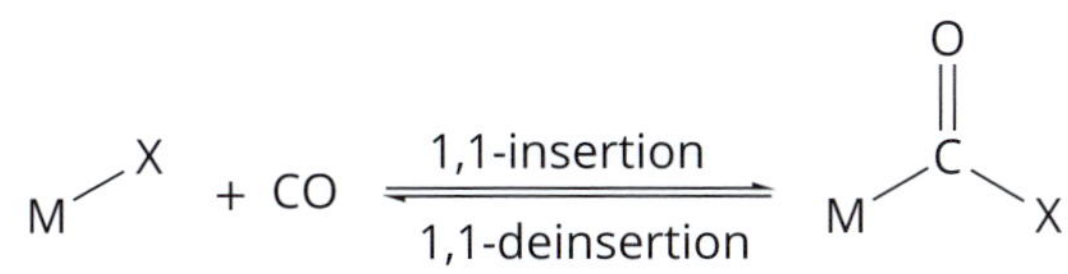

7. Species with multiple bonds, possibly ethylene, may insert into metal–ligand bonds in a 1,2-insertion. The 1,2-deinsertion reaction, usually referred to as β-elimination, is also quite common. A particularly common variant is β-hydrogen elimination, shown in the figure where a hydrogen from, for example, an ethyl group, may add to the metal to produce a coordinated ethylene.

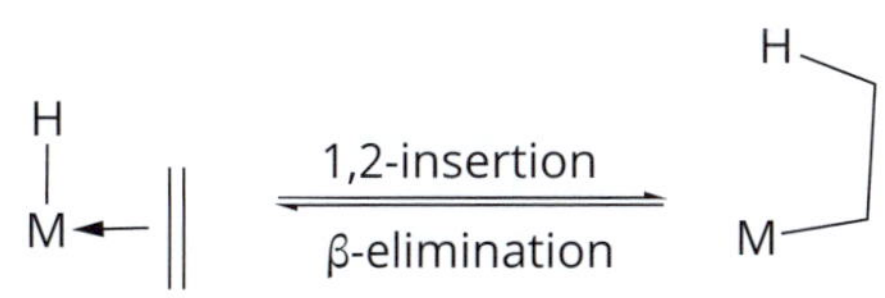

8. Cycloaddition reactions are also quite common with metal centers. In particular, [2 + 2]-cycloaddition is the key mechanistic step in the important olefin metathesis reaction.
9. Most catalytic cycles involve addition of ligands to the metal center (a reaction covered in great detail in Chapter 17), and the mechanistic steps discussed previously. Depending on the catalysis, other steps may also be involved, such as transmetallation from one metal to another. Using these ideas, chemists can modify C–H bonds and hydrogenate olefins, and even carry out multicomponent coupling reactions to generate several bonds in a single step.

Practice Problems

1. Use an analysis similar to that shown in Section 18.2.2.5 to show that the β-hydrogen elimination can be enthalpically favorable in some cases, but the β-carbon elimination is not. Explain the entropic reasons why a β-hydrogen elimination might make the reaction favorable, even if it is almost thermoneutral.
2. Draw a catalytic cycle that produces pentanal from 1-butene under an atmosphere of hydrogen and carbon monoxide catalyzed by $Rh(H)(CO)_2(PR_3)_2$.
3. Propose a catalytic cycle for hydrogenation of $RCH{=}CH_2$ by using $RhH(PPh_3)_3$. Name all mechanistic steps, and assign the oxidation state, valence electron count, and d^n count for each species.
4. Predict the products for the 1,1-insertion reaction, starting with one enantiomer CpFe(Me)(CO)(NO) under an atmosphere of ^{13}C labeled CO for each of the three possible mechanisms. Under which mechanisms is stereochemistry retained or inverted at iron?
5. Propose a reason why M-Ph or M-CMe_3 are stable metal alkyls that do not undergo the β-hydrogen elimination by drawing the expected products of that elimination reaction.

Cited References

1. Wilkinson, G.; Rosenblum, M.; Whiting, M. C.; Woodward, R. B., The structure of Iron bis-Cyclopentadienyl *J. Am. Chem. Soc.* 1952, *74*, 2125–2126.
2. Kealy, T. J.; Pauson, P. L., A new type of organo-iron compound *Nature* 1951, *168*, 1039–1040.
3. Miller, S. A.; Tebboth, J. A.; Tremaine, J. F., Dicyclopentadienyliron *J. Chem. Soc.* 1952, 632–635.
4. Cotton, F. A., Alkyls and aryls of transition metals *Chem. Rev.* 1955, *55*, 551–594.
5. Wilkinson, G., Iron sandwich: Recollection of 1st 4 months *J. Organomet. Chem.* 1975, *100*, 273–278.
6. Laszlo, P.; Hoffmann, R., Ferrocene: Ironclad history or Rashomon tale? *Angew. Chem. Int. Ed.* 2000, *39*, 123.
7. Wunderlich, J. A.; Mellor, D. P., A note on the crystal structure of Zeise's salt *Acta Crystallographica* 1954, *7*, 130.
8. Love, R. A.; Koetzle, T. F.; Williams, G. J. B.; Andrews, L. C.; Bau, R., Neutron-diffraction study of structure of zeises salt, $KPtCl_3(C_2H_4)\bullet H_2O$ *Inorg.Chem.* 1975, *14*, 2653–2657.
9. CCDC 185104: experimental Crystal Structure Determination, 2002, http://doi.org/10.5517/cc66m33
10. Seyferth, D., Bis(benzene)chromium. 2. Its discovery by E. O. Fischer and W. Hafner and subsequent work by the research groups of E. O. Fischer, H. H. Zeiss, F. Hein, C. Elschenbroich, and others *Organometallics* 2002, *21*, 2800–2820.
11. Seyferth, D., Bis(benzene) chromium. 1. Franz Hein at the University of Leipzig and Harold Zeiss and Minoru Tsutsui at Yale *Organometallics* 2002, *21*, 1520–1530.
12. Seyferth, D., (Cyclobutadiene)iron tricarbonyl: A case of theory before experiment *Organometallics* 2003, *22*, 2–20.
13. Seyferth, D., $(C_2H_4)PtCl_3^-$, the anion of Zeise's salt, $K[(C_2H_4)PtCl_3]\bullet H_2O$ *Organometallics* 2001, *20*, 2–6.
14. Seyferth, D., Zinc alkyls, Edward Frankland, and the beginnings of main-group organometallic chemistry *Organometallics* 2001, *20*, 2940–2955.
15. Seyferth, D., The Grignard Reagents *Organometallics* 2009, *28*, 1598–1605.
16. Maag, H.; Obata, N.; Holmes, A.; Schneider, P.; Schilling, W.; Schreiber, J.; Eschenmoser, A., Total synthesis of vitamin-B_{12}: The final steps *Chimia* 1972, *26*, 320.
17. Woodward, R. B., The total synthesis of vitamin B_{12} *Pure Appl.Chem.* 1973, *33*, 145.
18. Labinger, J. A., Tutorial on Oxidative Addition *Organometallics* 2015, *34*, 4784–4795.
19. Seddon, K. R.; Upton, C. E. E., Infrared and raman studies of halogeno-trimethylplatinum(IV) complexes and related compounds *Dalton Trans.* 1974, 2415–2419.
20. Appleton, T. G.; Bennett, M. A.; Tomkins, I. B., Effect of chelate-ring size on spectroscopic and chemical properties of methylplatinum(II) complexes of ditertiary phosphines $Ph_2P(CH_2)_nPPh_2$(n = 1, 2, or 3) *Dalton Trans.* 1976, 439–446.
21. Duckett, S. B.; Newell, C. L.; Eisenberg, R., More than inept: Parahydrogen and inept give unprecedented resonance enhancement to ^{13}C by direct ^1H polarization transfer *J. Am. Chem. Soc.* 1993, *115*, 1156–1157.
22. Duckett, S. B.; Newell, C. L.; Eisenberg, R., Observation of new intermediates in hydrogenation catalyzed by wilkinsons catalyst, $RhCl(PPh_3)_3$, using parahydrogen-induced polarization *J. Am. Chem. Soc.* 1994, *116*, 10548–10556.
23. Eisenberg, R.; Eisenschmid, T. C.; Chinn, M. S.; Kirss, R. U., Parahydrogen-induced polarization and polarization transfer in hydrogenation and oxidative addition-reactions: mechanistic probe *Advances in Chemistry Series D* 1992, 47–74.
24. Duckett, S. B.; Newell, C. L.; Eisenberg, R., Observation of new intermediates in hydrogenation catalyzed by wilkinsons catalyst, $RhCl(PPh_3)_3$, using parahydrogen-induced polarization: Addition *J. Am. Chem. Soc.* 1997, *110*, 2068.
25. Abel, E. W.; Wilkinson, G., Carbonyl halides of manganese and some related compounds *J. Chem. Soc.* 1959, 1501–1505.
26. Kuninobu, Y.; Nishina, Y.; Takeuchi, T.; Takai, K., Manganese-catalyzed insertion of aldehydes into a C-H bond *Angew. Chem. Int. Ed.* 2007, *46*, 6518–6520.
27. Goldberg, K. I.; Yan, J. Y.; Breitung, E. M., Energetics and mechanisms of carbon-carbon and carbon-iodide reductive elimination from a Pt(IV) center *JAm. Chem. Soc.* 1995, *117*, 6889–6896.
28. Procelewska, J.; Zahl, A.; Liehr, G.; van Eldik, R.; Smythe, N. A.; Williams, B. S.; Goldberg, K. I., Mechanistic information on the reductive elimination from cationic Trimethylplatinum(IV) complexes to form carbon-carbon bonds *Inorg. Chem.* 2005, *44*, 7732–7742.
29. Elschenbroich, C., *Organometallics* 3rd ed.; Wiley-VCH: Weinheim, Germany, 2006.
30. Chock, P. B.; Halpern, J., Kinetics of addition of hydrogen oxygen and methyl iodide to some square-planar iridium(I) complexes *J. Am. Chem. Soc.* 1966, *88*, 3511.
31. Engle, K. M.; Mei, T. S.; Wang, X. S.; Yu, J. Q., Bystanding F+ Oxidants Enable Selective Reductive Elimination from High-Valent Metal Centers in Catalysis *Angew. Chem. Int. Ed.* 2011, *50*, 1478–1491.
32. Wang, D. Y.; Choliy, Y.; Haibach, M. C.; Hartwig, J. F.; Krogh-Jespersen, K.; Goldman, A. S., Assessment of the Electronic Factors Determining the Thermodynamics of "Oxidative Addition" of C-H and N-H Bonds to Ir(I) Complexes *J. Am. Chem. Soc.* 2016, *138*, 149–163.
33. Blackmore, K. J.; Ziller, J. W.; Heyduk, A. F., "Oxidative addition" to a Zirconium(IV) redox-active ligand complex *Inorg. Chem.* 2005, *44*, 5559–5561.
34. Maitlis, P. M.; Dell'Amico, D. B., Fausto Calderazzo: Pioneer in Mechanistic Organometallic Chemistry *Organometallics* 2014, *33*, 6989–7006.
35. Noack, K.; Calderazzo, F., Carbon monoxide insertion reactions. V. Carbonylation of methylmanganese pentacarbonyl with ^{13}CO *J. Organomet. Chem.* 1967, *10*, 101.
36. Flood, T. C.; Campbell, K. D., CO insertion into the iron-carbon sigma-bond in optically-active CpFe(CO)(L)R: Solvent dependence and acid catalysis *J. Am. Chem. Soc.* 1984, *106*, 2853–2860.
37. Lilly, C. P.; Boyle, P. D.; Ison, E. A., Synthesis of Oxorhenium Acetyl and Benzoyl Complexes Incorporating Diamidopyridine Ligands: Implications for the Mechanism of CO Insertion *Organometallics* 2012, *31*, 4295–4301.
38. Robbins, L. K.; Lilly, C. P.; Smeltz, J. L.; Boyle, P. D.; Ison, E. A., Synthesis and Reactivity of Oxorhenium(V) Methyl, Benzyl, and Phenyl Complexes with CO: Implications for a Unique Mechanism for Migratory Insertion *Organometallics* 2015, *34*, 3152–3158.
39. Smeltz, J. L.; Boyle, P. D.; Ison, E. A., Mechanism for the Activation of Carbon Monoxide via Oxorhenium Complexes *J. Am. Chem. Soc.* 2011, *133*, 13288–13291.
40. Smeltz, J. L.; Webster, C. E.; Ison, E. A., Computational Investigation of the Mechanism for the Activation of CO by Oxorhenium Complexes *Organometallics* 2012, *31*, 4055–4062.
41. Brookhart, M.; Green, M. L. H.; Parkin, G., Agostic interactions in transition metal compounds *Proc. Nat. Acad. Sci.* 2007, *104*, 6908–6914.
42. Brookhart, M.; Green, M. L. H., Carbon-hydrogen-transition metal bonds *J. Organomet. Chem.* 1983, *250*, 395–408.
43. Schultz, A. J.; Williams, J. M.; Schrock, R. R.; Rupprecht, G. A.; Fellmann, J. D., Interaction of hydrogen and hydrocarbons with transition-metals: Neutron-diffraction evidence for an activated C-H bond in an electron-deficient tantalum-neopentylidene complex *J. Am. Chem. Soc.* 1979, *101*, 1593–1595.
44. Etienne, M., Steric and electronic probes for α-agostic interaction in a niobium ethyl complex *Organometallics* 1994, *13*, 410–412.
45. Simoes, J. A. M.; Beauchamp, J. L., Transition metal-hydrogen and metal-carbon bond strengths: the keys to catalysis *Chem. Rev.* 1990, *90*, 629–688.
46. Ervin, K. M.; Gronert, S.; Barlow, S. E.; Gilles, M. K.; Harrison, A. G.; Bierbaum, V. M.; DePuy, C. H.; Lineberger, W. C.; Ellison, G. B., Bond strengths of ethylene and acetylene *J. Am. Chem. Soc.* 1990, *112*, 5750–5759.
47. Ruscic, B., Active Thermochemical Tables: Sequential Bond Dissociation Enthalpies of Methane, Ethane, and Methanol and the Related Thermochemistry *J. Phys. Chem. A* 2015, *119*, 7810–7837.
48. Turro, N. J.; Ramamurthy, V.; Scaiano, J. C., *Modern Molecular Photochemistry of Organic Molecules* University Science Books: Sausalito, Calif., 2010.
49. Nugent, W. A.; Mayer, J. M., *Metal-Ligand Multiple Bonds: the Chemistry of Transition Metal Complexes Containing Oxo, Nitrido, Imido, Alkylidene, or Alkylidyne Ligands* Wiley: New York, 1988.
50. Symmetry restrictions may also be relaxed if the reaction proceeds in a step-wise rather than concerted fashion.
51. Osborn, J. A.; Jardine, F. H.; Young, J. F.; Wilkinson, G., Preparation and properties of tris(triphenylphosphine)halogenorhodium(I) and some reactions thereof including catalytic homogeneous hydrogenation of olefins and acetylenes and their derivatives *J. Chem. Soc. A* 1966, 1711.

52. Osborn, J. A.; Wilkinson, G., Chlorotris(triphenylphosphine)rhodium(I) (Wilkinson catalyst) *Inorg. Synth.* 1990, *28*, 77–79.

53. Halpern, J.; Okamoto, T.; Zakhariev, A., Mechanism of chlorotris(triphenylphosphine)rhodium(I)-catalyzed hydrogenation of alkenes: Reaction of chlorodihydridotris(triphenylphosphine)rhodium(III) with cyclohexene *J. Mol. Catal.* 1977, *2*, 65–68.

54. Slaugh, L. H.; Mullineaux, R. D., Novel Hydroformylation catalysts *J. Organometal. Chem.* 1968, *13*, 469–477.

55. Osborn, J. A.; Wilkinson, G.; Young, J. F., Mild Hydroformylation of Olefins using Rhodium Catalysts *J. Chem. Soc., Chem. Commun.* 1965, 17.

56. Heck, R. F.; Breslow, D. S., The Reaction of Cobalt Hydrotetracarbonyl with Olefins *J. Am. Chem. Soc.* 1961, *83*, 4023–4027.

57. Jordan, R. B., *Reaction Mechanisms of Inorganic and Organometallic Systems* 3rd ed.; Oxford University Press: Oxford, 2007.

58. Evans, D.; Osborn, J. A.; Wilkinson, G., Hydroformylation of alkenes by use of rhodium complex catalysts *J. Chem. Soc. (A)* 1968, 3133–3142.

59. Kranenburg, M.; Van der burgt, Y. E. M.; Kamer, P. C. J.; Van Leeuwen, P. W. N. M.; Goubitz, K.; Fraanje, J., New diphosphine ligands based on heterocyclic aromatics inducing very high regioselectivity in rhodium-catalyzed hydroformylation: Effect of the bite angle *Organometallics* 1995, *14*, 3081–3089.

60. Brezny, A. C.; Landis, C. R., Recent Developments in the Scope, Practicality, and Mechanistic Understanding of Enantioselective Hydroformylation *Acc.Chem. Res.* 2018, *51*, 2344–2354.

61. Brezny,A.C.;Landis,C.R.,DevelopmentofaComprehensiveMicrokineticModel for Rh(bis(diazaphospholane))-Catalyzed Hydroformylation *ACS Catalysis* 2019, *9*, 2501–2513.

62. Forster, D., Mechanism of a rhodium-complex-catalyzed carbonylation of methanol to acetic-acid *J. Am. Chem. Soc.* 1976, *98*, 846–848.

63. Biffis, A.; Centomo, P.; Del Zotto, A.; Zeccal, M., Pd Metal Catalysts for Cross-Couplings and Related Reactions in the 21st Century: A Critical Review *Chem. Rev.* 2018, *118*, 2249–2295.

64. Hazari, N.; Melvin, P. R.; Beromi, M. M., Well-defined nickel and palladium precatalysts for cross-coupling *Nature Rev. Chem.* 2017, *1*.

65. Maluenda, I.; Navarro, O., Recent Developments in the Suzuki-Miyaura Reaction: 2010–2014 *Molecules* 2015, *20*, 7528–7557.

66. Roy, D.; Uozumi, Y., Recent Advances in Palladium-Catalyzed Cross-Coupling Reactions at ppm to ppb Molar Catalyst Loadings *Adv. Synth. Catal.* 2018, *360*, 602–625.

67. Milstein, D.; Stille, J. K., A General, Selective, and Facile Method for Ketone Synthesis from Acid Chlorides and Organotin Compounds Catalyzed by Palladium *J. Am. Chem. Soc.* 1978, *100*, 3636–3638.

68. Gallagher, W. P.; Maleczka, R. E., Stille reactions catalytic in tin: A "Sn-F" route for intermolecular and intramolecular couplings *J. Org. Chem.* 2005, *70*, 841–846.

69. Maleczka, R. E.; Gallagher, W. P., Stille couplings catalytic in tin: A "Sn-F" approach *OrgLett.* 2001, *3*, 4173–4176.

70. Denmark, S. E.; Ambrosi, A., Why You Really Should Consider Using Palladium-Catalyzed Cross-Coupling of Silanols and Silanolates *Org. Process Res. Dev.* 2015, *19*, 982–994.

71. Hiyama, T., Selective Organic Synthesis Through Generation and Reactivity Control of Hyper-Coordinate Metal Species *Chem. Rec.* 2008, *8*, 337–350.

72. Tymonko, S. A.; Smith, R. C.; Ambrosi, A.; Denmark, S. E., Mechanistic Significance of the Si-O-Pd Bond in the Palladium-Catalyzed Cross-Coupling Reactions of Alkenylsilanolates *J. Am. Chem. Soc.* 2015, *137*, 6192–6199.

73. Cabot, R.; Lledó, A.; Revés, M.; Riera, A.; Verdaguer, X., Kinetic Studies on the Cobalt-Catalyzed Norbornadiene Intermolecular Pauson–Khand Reaction *Organometallics* 2007, *26*, 1134–1142.

74. Casey, C. P., 2005 Nobel Prize in Chemistry. Development of the Olefin Metathesis Method in Organic Synthesis *J. Chem. Educ.* 2006, *83*, 192.

75. Calderon, N.; Ofstead, E. A.; Ward, J. P.; W. Allen Judy; Scott, K. W., Olefin Metathesis. I. Acyclic Vinylenic Hydrocarbons *J. Am. Chem. Soc.* 1968, *90*, 4133–4140.

76. Hérisson, J.-L.; Chauvin, Y., Catalyse de transformation des oléfines par les complexes du tungstène *Makromol. Chem.* 1970, *141*, 161–176.

77. Grubbs, R. H.; Brunck, T. K., A Possible Intermediate in the Tungsten-Catalyzed Olefin Metathesis Reaction *J. Am. Chem. Soc.* 1972, *94*, 2538–2540.

78. Grubbs, R. H.; Burk, P. L.; Carr, D. D., Consideration of the Mechanism of the Olefin Metathesis Reaction *J. Am. Chem. Soc.* 1975, *97*, 3265–3267.

79. Grubbs, R. H.; Carr, D. D.; Hoppin, C.; Burk, P. L., Consideration of the Mechanism of the Metal Catalyzed Olefin Metathesis Reaction *J. Am. Chem. Soc.* 1976, *98*, 3478–3483.

80. Koh, M. J.; Nguyen, T. T.; Zhang, H. M.; Schrock, R. R.; Hoveyda, A. H., Direct synthesis of Z-alkenyl halides through catalytic cross-metathesis *Nature* 2016, *531*, 459–465.

81. Herbert, M. B.; Grubbs, R. H., Z-Selective Cross Metathesis with Ruthenium Catalysts: Synthetic Applications and Mechanistic Implications *Angew. Chem. Int. Ed.* 2015, *54*, 5018–5024.

82. Katsuki, T., Some Recent Advances in Metallosalen Chemistry *Synlett* 2003, *2003*, 0281–0297.

83. Jacobsen, E. N., Asymmetric Catalysis of Epoxide Ring-Opening Reactions *Accounts of Chemical Research* 2000, *33*, 421–431.

84. Katsuki, T.; Sharpless, K. B., The First Practical Method for Asymmetric Epoxidation *J. Am. Chem. Soc.* 1980, *102*, 5976–5978.

85. Woodard, S. S.; Finn, M. G.; Sharpless, K. B., Mechanism of Asymmetric Epoxidation. 1. Kinetics *J. Am. Chem. Soc.* 1991, *113*, 106–113.

86. Finn, M. G.; Sharpless, K. B., Mechanism of Asymmetric Epoxidation. 2. Catalyst Structure *J. Am. Chem. Soc.* 1991, *113*, 113–126.

87. Martin, V. S.; Woodard, S. S.; Katsuki, T.; Yamada, Y.; Ikeda, M.; Sharpless, K. B., Kinetic Resolution of Racemic Allylic Alcohols by Enantioselective Epoxidation. A Route to Substances of Absolute Enantiomeric Purity? *J. Am. Chem. Soc.* 1981, *103*, 6237–6240.

88. Sharpless, K. B.; Woodard, S. S.; Finn, M. G., On the mechanism of titanium-tartrate catalyzed asymmetric epoxidation *Pure & Appl. Chem.* 1983, *55*, 1823–1836.

89. Mihelich, E. D.; Daniels, K.; Eickhoff, D. J., Vanadium-CatalyzedEpoxidations. 2. Highly StereoselectiveEpoxidations of Acyclic Homoallylic Alcohols Predicted by a Detailed Transition-State Model *J. Am. Chem. Soc.* 1981, *103*, 7690–7692.

90. Mihelich, E. D., Vanadium-Catalyzed Epoxidations. I. A New Selectivity Pattern for Acyclic Allylic Alcohols *Tet. Lett.* 1979, *49*, 4729–4732.

91. Markovnikov addition of hydrogen-element bonds is where the hydrogen adds to the side of an olefin that already has the most hydrogens. Markovnikov was studying HBr addition to, for example, terminal olefins. Proton addition to the terminal carbon of a substrate like propene generates the more stable secondary carbocation, where bromide traps to give 2-bromopropane. The term anti-Markovnikov is used to signify that E would add to the less stable carbocation. These terms are still used to signify regiochemistry even though the mechanim may not involve a carbocation.

92. Marciniec, B.; Gulinski, J., Recent advances in catalytic hydrosilylation *J. Organomet. Chem.* 1993, *446*, 15–23.

93. Roy, A. K., A Review of Recent Progress in Catalyzed Homogeneous Hydrosilation (Hydrosilylation). In *Adv. Organomet. Chem.*, West, R.; Hill, A. F.; Fink, M. J., Eds. 2008; Vol. 55, pp 1–59.

94. Mueller, T. E.; Hultzsch, K. C.; Yus, M.; Foubelo, F.; Tada, M., Hydroamination: Direct addition of amines to alkenes and alkynes *Chem. Rev.* 2008, *108*, 3795–3892.

95. Reznichenko, A. L.; Hultzsch, K. C., Early Transition Metal (Group 3–5, Lanthanides and Actinides) and Main Group Metal (Group 1, 2, and 13) Catalyzed Hydroamination. In *Hydrofunctionalization*, Ananikov, V. P.; Tanaka, M., Eds. 2013; Vol. 43, pp 51–114.

96. Li, Y. W.; Marks, T. J., Organolanthanide-catalyzed intramolecular hydroamination/cyclization of aminoalkynes *J. Am. Chem. Soc.* 1996, *118*, 9295–9306.

97. Hong, S.; Marks, T. J., Organolanthanide-catalyzed hydroamination *Acc. Chem. Res.* 2004, *37*, 673–686.

98. Duncan, A. P.; Bergman, R. G., Selective transformations of organic compounds by imidozirconocene complexes *Chem. Rec.* 2002, *2*, 431–445.

99. Walsh, P. J.; Baranger, A. M.; Bergman, R. G., Stoichiometric and catalytic hydroamination of alkynes and allene by zirconium bisamides $Cp_2Zr(NHR)_2$ *J. Am. Chem. Soc.* 1992, *114*, 1708–1719.

100. Hoover, J. M.; Petersen, J. R.; Pikul, J. H.; Johnson, A. R., Catalytic intramolecular hydroamination of substituted aminoallenes by chiral titanium amino-alcohol complexes *Organometallics* 2004, *23*, 4614–4620.

101. Bytschkov, I.; Doye, S., Group-IV metal complexes as hydroamination catalysts *Eur. J. Org. Chem.* 2003, 935–946.

102. Manna, K.; Xu, S. C.; Sadow, A. D., A Highly Enantioselective Zirconium Catalyst for Intramolecular Alkene Hydroamination: Significant Isotope Effects on Rate and Stereoselectivity *Angew. Chem. Int. Ed.* 2011, *50*, 1865–1868.

103. Odom, A. L., New C-N and C-C bond forming reactions catalyzed by titanium complexes *Dalton Trans.* 2005, 225–233.

104. Odom, A. L.; McDaniel, T. J., Titanium-Catalyzed Multicomponent Couplings: Efficient One-Pot Syntheses of Nitrogen Heterocycles *Acc. Chem. Res.* 2015, *48*, 2822–2833.

105. Herzon, S. B.; Hartwig, J. F., Direct, catalytic hydroaminoalkylation of unactivated olefins with N-alkyl arylamines *J. Am. Chem. Soc.* 2007, *129*, 6690.
106. Herzon, S. B.; Hartwig, J. F., Hydroaminoalkylation of Unactivated Olefins with Dialkylamines *J. Am. Chem. Soc.* 2008, *130*, 14940.
107. Ryken, S. A.; Schafer, L. L., N,O-Chelating Four-Membered Metallacyclic Titanium(IV) Complexes for Atom-Economic Catalytic Reactions *Acc. Chem. Res.* 2015, *48*, 2576–2586.
108. Edwards, P. M.; Schafer, L. L., In Situ Generation of a Regio- and Diastereoselective Hydroaminoalkylation Catalyst Using Commercially Available Starting Materials *Org. Lett.* 2017, *19*, 5720–5723.
109. Chong, E.; Garcia, P.; Schafer, L. L., Hydroaminoalkylation: Early-Transition-Metal-Catalyzed α-Alkylation of Amines *Synth.* 2014, *46*, 2884–2896.
110. Bielefeld, J.; Doye, S., Dimethylamine as a Substrate in Hydroaminoalkylation Reactions *Angew. Chem. Int. Ed.* 2017, *56*, 15155–15158.
111. Davis-Gilbert, Z. W.; Tonks, I. A., Titanium redox catalysis: insights and applications of an earth-abundant base metal *Dalton Trans.* 2017, *46*, 11522–11528.
112. Davis-Gilbert, Z. W.; Wen, X. L.; Goodpaster, J. D.; Tonks, I. A., Mechanism of Ti-Catalyzed Oxidative Nitrene Transfer in 2+2+1 Pyrrole Synthesis from Alkynes and Azobenzene *J. Am. Chem. Soc.* 2018, *140*, 7267–7281.
113. Blake, A. J.; Collier, P. E.; Dunn, S. C.; Li, W. S.; Mountford, P.; Shishkin, O. V., Synthesis and imido-group exchange reactions of *tert*-butylimidotitanium complexes *Dalton Trans.* 1997, 1549–1558.
114. Arndtsen, B. A., Metal-Catalyzed One-Step Synthesis: Towards Direct Alternatives to Multistep Heterocycle and Amino Acid Derivative Formation *Chem Eur. J.* 2009, *15*, 302–313.
115. Lessing, T.; Mueller, T. J. J., Sequentially Palladium-Catalyzed Processes in One-Pot Syntheses of Heterocycles *Applied Sciences-Basel* 2015, *5*, 1803–1836.
116. Levi, L.; Mueller, T. J. J., Multicomponent Syntheses of Fluorophores Initiated by Metal Catalysis *Eur. J. Org. Chem.* 2016, 2907–2918.
117. Montgomery, J., Nickel-catalyzed reductive cyclizations and couplings *Angew. Chem. Int. Ed.* 2004, *43*, 3890–3908.
118. Pellissier, H., Recent Developments in Enantioselective Metal-Catalyzed Domino Reactions *Adv. Synth. Catal.* 2016, *358*, 2194–2259.
119. Shen, C.; Wu, X.-F., Palladium-Catalyzed Carbonylative Multicomponent Reactions *Chem Eur. J.* 2017, *23*, 2973–2987.
120. Vlaar, T.; Ruijter, E.; Orru, R. V. A., Recent Advances in Palladium-Catalyzed Cascade Cyclizations *Adv. Synth. Catal.* 2011, *353*, 809–841.
121. Zhang, M.; Neumann, H.; Beller, M., Selective Ruthenium-Catalyzed Three-Component Synthesis of Pyrroles *Angew. Chem. Int. Ed.* 2013, *52*, 597–601.
122. Patel, S. J.; Jamison, T. F., Catalytic three-component coupling of alkynes, imines, and organoboron reagents *Angew. Chem. Int. Ed.* 2003, *42*, 1364–1367.
123. Cai, G.; Fu, Y.; Li, Y.; Wan, X.; Shi, Z., Indirect ortho Functionalization of Substituted Toluenes through ortho Olefination of N,N-Dimethylbenzylamines Tuned by the Acidity of Reaction Conditions *J. Am. Chem. Soc.* 2007, *129*, 7666–7673.
124. Canty, A. J., Development of organopalladium(IV) chemistry: Fundamental-aspects and systems for studies of mechanism in organometallic chemistry and catalysis *Acc. Chem. Res.* 1992, *25*, 83–90.
125. Dick, A. R.; Kampf, J. W.; Sanford, M. S., Unusually stable palladium(IV) complexes: Detailed mechanistic investigation of C-O bond-forming reductive elimination *J. Am. Chem. Soc.* 2005, *127*, 12790–12791.
126. Xu, L.-M.; Li, B.-J.; Yang, Z.; Shi, Z.-J., Organopalladium(IV) chemistry *Chem. Soc. Rev.* 2010, *39*, 712–733.
127. Saidi, O.; Marafie, J.; Ledger, A. E. W.; Liu, P. M.; Mahon, M. F.; Kociok-Köhn, G.; Whittlesey, M. K.; Frost, C. G., Ruthenium-Catalyzed Meta Sulfonation of 2-Phenylpyridines *J. Am. Chem. Soc.* 2011, *133*, 19298–19301.
128. Hofmann, N.; Ackermann, L., meta-Selective C–H Bond Alkylation with Secondary Alkyl Halides *J. Am. Chem. Soc.* 2013, *135*, 5877–5884.
129. Zhang, M.; Zhang, Y.; Jie, X.; Zhao, H.; Li, G.; Su, W., Recent advances in directed C-H functionalizations using monodentate nitrogen-based directing groups *Org. Chem. Front.* 2014, *1*, 843–895.
130. Colby, D. A.; Bergman, R. G.; Ellman, J. A., Synthesis of dihydropyridines and pyridines from Imines and alkynes via C-H activation *J. Am. Chem. Soc.* 2008, *130*, 3645–3651.
131. Colby, D. A.; Tsai, A. S.; Bergman, R. G.; Ellman, J. A., Rhodium Catalyzed Chelation-Assisted C-H Bond Functionalization Reactions *Acc. Chem. Res.* 2012, *45*, 814–825.
132. Potter, T. J.; Ellman, J. A., Total Synthesis of (+)-Pancratistatin by the Rh(III)-Catalyzed Addition of a Densely Functionalized Benzamide to a Sugar-Derived Nitroalkene *Org. Lett.* 2017, *19*, 2985–2988.
133. Bernskoetter, W. H.; Schauer, C. K.; Goldberg, K. I.; Brookhart, M., Characterization of a Rhodium(I) σ-Methane Complex in Solution *Science* 2009, *326*, 553–556.
134. Janowicz, A. H.; Bergman, R. G., Carbon-hydrogen activation in completely saturated hydrocarbons: Direct observation of M + R-H -> M(R)(H) *J. Am. Chem. Soc.* 1982, *104*, 352–354.
135. Hoyano, J. K.; Graham, W. A. G., Oxidative addition of the carbon-hydrogen bonds of neopentane and cyclohexane to a photochemically generated iridium(I) complex *J. Am. Chem. Soc.* 1982, *104*, 3723–3725.
136. Shilov, A. E.; Shul'pin, G. B., Activation and Catalytic Reactions of Alkanes in Solutions of Metal Complexes *Russ. Chem. Rev.* 1987, *56*, 442.
137. Periana, R. A.; Taube, D. J.; Gamble, S.; Taube, H.; Satoh, T.; Fujii, H., Platinum catalysts for the high-yield oxidation of methane to a methanol derivative *Science* 1998, *280*, 560–564.
138. Lee, M.; Sanford, M. S., Platinum-Catalyzed, Terminal-Selective C(sp^3)-H Oxidation of Aliphatic Amines *J. Am. Chem. Soc.* 2015, *137*, 12796–12799.
139. Cho, J. Y.; Iverson, C. N.; Smith, M. R., Steric and chelate directing effects in aromatic borylation *J. Am. Chem. Soc.* 2000, *122*, 12868–12869.
140. Cho, J. Y.; Tse, M. K.; Holmes, D.; Maleczka, R. E.; Smith, M. R., Remarkably selective iridium catalysts for the elaboration of aromatic C-H bonds *Science* 2002, *295*, 305–308.
141. Ishiyama, T.; Takagi, J.; Hartwig, J. F.; Miyaura, N., A stoichiometric aromatic C-H borylation catalyzed by iridium(I)/2,2 -bipyridine complexes at room temperature *Angew. Chem.-Int. Edit.* 2002, *41*, 3056–3058.
142. Mkhalid, I. A. I.; Barnard, J. H.; Marder, T. B.; Murphy, J. M.; Hartwig, J. F., C-H Activation for the Construction of C-B Bonds *Chem. Rev.* 2010, *110*, 890–931.
143. Takagi, J.; Sato, K.; Hartwig, J. F.; Ishiyama, T.; Miyaura, N., Iridium-catalyzed C-H coupling reaction of heteroaromatic compounds with bis(pinacolato) diboron: regioselective synthesis of heteroarylboronates *Tetrahedron Lett.* 2002, *43*, 5649–5651.
144. Preshlock, S. M.; Plattner, D. L.; Maligres, P. E.; Krska, S. W.; Maleczka, R. E., Jr.; Smith, M. R., III, A Traceless Directing Group for C-H Borylation *Angew. Chem. Int. Ed.* 2013, *52*, 12915–12919.
145. Vanchura, B. A., II; Preshlock, S. M.; Roosen, P. C.; Kallepalli, V. A.; Staples, R. J.; Maleczka, R. E., Jr.; Singleton, D. A.; Smith, M. R., III, Electronic effects in iridium C-H borylations: insights from unencumbered substrates and variation of boryl ligand substituents *Chem. Commun.* 2010, *46*, 7724–7726.
146. Sherry, A. E.; Wayland, B. B., Metalloradical activation of methane *J. Am. Chem. Soc.* 1990, *112*, 1259–1261.
147. Wayland, B. B.; Ba, S.; Sherry, A. E., Activation of methane and toluene by rhodium(II) porphyrin complexes *J. Am. Chem. Soc.* 1991, *113*, 5305–5311.
148. Hebrard, F.; Kalck, P.; Saussine, L.; Magna, L.; Olivier-Bourbigou, H., Unexpected reduction pathway of a Co^{2+} salt to $HCo(CO)_4$ via $Co_2(CO)_8$ in an ionic liquid *Dalton Trans.* 2007, 190–191.
149. Haibach, M. C.; Kundu, S.; Brookhart, M.; Goldman, A. S., Alkane Metathesis by Tandem Alkane-Dehydrogenation-Olefin-Metathesis Catalysis and Related Chemistry *Acc. Chem. Res.* 2012, *45*, 947–958.

CHAPTER 19

Solid-State Chemistry

CONCEPT COVERAGE

1. The structure of solids in three dimensions with some cursory discussion of crystal symmetry
2. Forces important in molecular solids, ionic solids, and covalent solids
3. From MOs to band structure in extended systems
4. Phase changes and electronic structure
5. Synthetic strategies in solid-state chemistry using oxides as examples
6. Silicon-based solar cells as examples of crystallization, impurities, and doping
7. Lithium-ion batteries as an example of solid-state chemistry applications in battery technology
8. Metal–organic frameworks (MOF)
9. Intermetallics from the view of metal atoms to clusters to extended materials
10. Quantum dots as an example of a material showing quantum confinement, with a brief discussion of electronic structure and synthesis

"The chemists are a strange class of mortals, impelled by an almost insane impulse to seek their pleasures amid smoke and vapor, soot and flame, poisons and poverty; yet among all these evils I seem to live so sweetly . . ."

— **Johann Joachim Becher**

"Given the incredibly rich heritage of solid-state chemistry, it is probably hard for a newcomer (a young doctoral student, for example) to see the forest for all the trees."

— **Richard Dronskowski, Shinichi Kikkawa, and Andreas Stein, in the Preface to the *Handbook for Solid-State Chemistry*[1]**

Solid-state chemistry is a vast field of its own, often populated with its own community of scientists from a variety of different fields. Material scientists, for example, are often in an engineering college. Condensed-matter physicists often study in glorious and gaudy detail the properties of solid-state compounds. The position of chemistry in this field is in many ways highly advantageous; it is often the chemists that are required to produce the new materials for study, which gives them the opportunity to study the properties in which they are most interested, often through collaborations.

One can argue that materials define us as humans, and in some cases we define ourselves by the materials that we use. The ages of human history are often defined by the materials of the time, such as the Stone Age, Bronze Age, and Iron Age. Today we live in an Information Age made possible by materials chemistry and physics applied to computers. Without concrete to build structures, ceramic to make dishes, steel to make tools, and other materials we use daily and with little thought, we would be scraping an existence together like the wild animals we now hold dominion over, in large part because of materials to which we have access. Without modern materials, like all those listed above and plastics made using inorganic catalysts, life would be quite different.

Just to choose one example, stainless steel is one remarkable material with very sophisticated chemistry that was discovered largely through trial and error. Steel is obtained by adding typically <2% carbon to iron, which greatly enhances the strength over pure iron. Stainless steel is prepared by addition of typically ~10% chromium to steel and greatly increases the resistance of the material to rusting. Typical steel reacts with oxygen to form hematite, Fe_2O_3, which does not bond to the steel, but rather flakes away. The chromium in stainless steel is thermodynamically easier to oxidize than iron and produces a thin oxide coating that adheres to the metal, which protects the surface from further degradation. This oxide coating has the effect of making the metal tasteless because it does not interact with saliva, making it ideal for flatware and cutlery. Before 1913, when stainless steel was invented, metal utensils probably altered the flavor of foods. Stainless steel is now used extensively for many applications where a pristine metallic look and longevity are required. A notable example in art is *Cloud Gate*—a stainless steel sculpture found in Millennium Park in Chicago, that can reflect the impressive city skyline and the faces of pedestrians (Fig. 19.1).

Fig. 19.1 *Cloud Gate* by Sir Anish Kapoor, located in Millennium Park in Chicago, Illinois.

The story of stainless steel is greatly dependent upon serendipity—first in the amount of carbon to make steel, then in the addition of chromium to slow rusting, and this is not unusual for materials (or science in general, for that matter). The properties of materials that make them useful are often dependent upon subtleties that are not easily anticipated, such as chromium oxide adhering to steel while iron oxide does not. These properties are often the product of relatively weak interactions that sum to enormous effects as one goes from the chemical scale to the macroscale. Some properties are more easily anticipated, such as the ionic character of the material, because they are bulk properties relatable to molecular properties which we are more accustomed to considering.

The language of solid-state chemistry is very much its own and is foreign to many inorganic chemists, as it has become populated with terms from materials science and condensed-matter physics. Nevertheless, chemical concepts that were used throughout this book apply just as readily to the solid-state. However, some of the conceptual framework will need to be modified to accommodate the large arrays of atoms to be discussed. In addition, one needs to consider intermolecular interactions on a level greater than in any of the other topics covered previously to understand molecular solids.

The importance of solid-state inorganic chemistry in many applications is undeniable, and as we proceed through this chapter, some of these applications will be mentioned. This chapter will begin with the structure of solids as both arrays of molecules and extended covalent systems, including a brief description of space groups and some common arrangements of atoms in the solid-state. Next, forces important for the solid-state will be discussed, from weak interactions, such as van der Waals interactions, to ionic bonding. Then, the electronic structure of solids will be introduced. Some common methods to produce solid-state compounds will be presented, including some methods for surface modification. Finally, a few applications of solid-state chemistry and important current areas will be discussed, such as batteries, superconductors, metal–organic frameworks, and quantum dots.

19.1 The Structure of Solids

One can effectively divide all solids into three discrete categories: (1) *molecular solids* composed of discrete molecules held together using forces often much weaker in isolation than bonds; (2) *solid-state compounds* that involve traditional bonds (ionic or covalent) between the repeating units that comprise the solid; and (3) *low-dimensional materials* that are, to a first approximation, 0-, 1-, or 2-dimensional in that they have structures similar to solid-state compounds but are relatively small (such as CdSe quantum dots) or are extended systems much bigger in one or two dimensions than another (such as nanotubes and graphene). As far as how the crystal is arranged, there is little difference between the types, and there are definable symmetry properties of all crystalline systems, as will be introduced briefly. Furthermore, the weak forces in molecular solids are worth discussing, as these are present even in the heterocovalent frameworks of the solid-state compounds and low-dimensional materials. In this section, the symmetry and make-up of crystalline frameworks is discussed, and in the next section some of the forces relevant to solid-state compounds—outside the well-detailed ground of heterocovalent bonding found in previous chapters—will be covered.

Most solids are composed of what may be regarded, from a molecular perspective, as an infinite array of molecules, atoms, or ions. Crystalline solids are comprised of *unit cells*, the repeated building-block of the solid that is related to all the other building-blocks by translational symmetry. If there is no regular repeating unit, the solid is *amorphous*, from the Greek for "without form, shapeless." By and large, only crystalline solids will be discussed here, although some amorphous solids are commonly encountered, such as borosilicate glass. The details of crystallography are not required to appreciate much of the chemistry to be discussed, but a brief Advanced Topic section is included to provide those interested in a primer with a more detailed understanding of the arrangement of molecules in solids.

19.1.1 A Brief Description of Crystal Symmetry (Advanced Topic)

The crystal is sometimes called a *lattice*, as it can be considered to be comprised of a set of repeating unit cells with a specific structure. The unit cell can generally be described as shown in Fig. 19.2. The sides of the cell are assigned lower-case letters (a, b, c), the faces of the cell are labeled with upper-case letters (A, B, C), and the angles are named with Greek letters (α, β, γ). By convention, α is the angle between the sides b and c, β is the angle from a and c, and γ is the angle from a to b. Similarly, the faces are defined by the sides not on that face; for example, face A has sides b and c as its edges. (Perhaps better, side B is perpendicular to the b axis.)

Within the unit cell, however, will be a specific number of what is called the "asymmetric unit". It is called "asymmetric" because it is contains no internal crystallographic symmetry, and as a consequence it is the unit that is repeated by the crystal symmetry to build up the unit cell and, by extension, the solid. The asymmetric unit may contain one molecule or part of a molecule, or it may contain many chemically equivalent molecules. There is not necessarily a relationship between the number of chemical species (molecules, ions, and so on) and the asymmetric unit in the crystal.

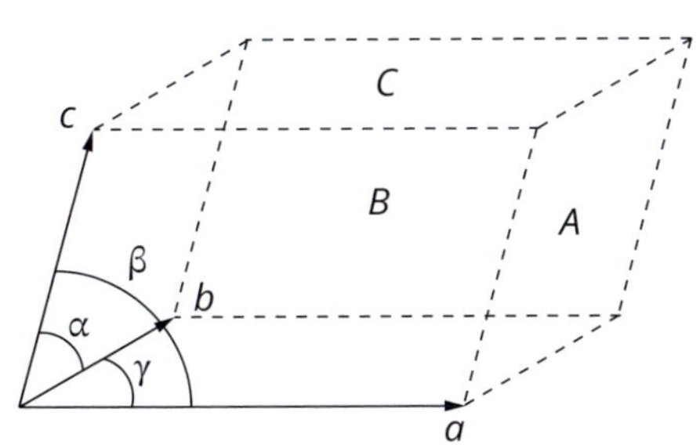

Fig. 19.2 Defining the sides (a, b, c), angles (α, β, γ), and faces (A, B, C) of the unit cell.

Unit cells have different symmetries related to their sides and angles as well. All of the fourteen possibilities shown in Fig. 19.3 are called the Bravais lattices. Here we will not attempt to describe all of crystal symmetry and will simply point out a few features that are commonly encountered. A full course in X-ray diffraction is offered at most major universities, and is advisable for many synthetic chemists.

The points in Fig. 19.3 are called "lattice points" and are mathematical constructs around which everything is the same in the crystal. In the least symmetric crystal system, triclinic, none of the angles are equal to 90° and none of the sides are equivalent. There is a single lattice point in the triclinic cell, $^1/_8$ of the point residing in each corner. In addition to lattices with points at the corners, one may have points on faces (face-centered) or in the center of the unit cell (body-centered).

The unit cell in triclinic is defined by the maximum number of parameters—a total of six from three different sides and three different angles. The most symmetric unit cell possible is cubic, where $a = b= c$ and $\alpha = \beta = \gamma = 90°$, which has a cell fully defined by a single parameter, the length of one side.

In space groups we add translations to the symmetry elements of the point groups, such as rotation, reflection, and inversion. The combination of a rotational axis with a translation parallel to the axis of rotation is called a "screw axis". A 3_1 screw axis, for example, involves a three-fold rotation and a translation of $^1/_3$ of the unit cell; likewise, a 3_2 screw axis involves a three-fold rotation and translation $^2/_3$ of the unit cell. These are shown in Fig. 19.4. Notice that one must translate into the next unit cell as well. Also note that 3_2 and 3_1 have different handedness to their rotations within the unit cell.

Another translational symmetry element found in space groups is called a *glide plane*—a combination of a mirror plane and translation along the plane. The glide plane's translation can be along an edge (a-glide, b-glide, c-glide), along the diagonal of a face (n-glide), or along the diagonal of the cell through the center (d-glide, "diamond" glide). The length of the translation is generally ½ the length of the dimension being translated down; for example, the translation length for an a-glide would be $a/2$.

The combination of the thirty-two point groups with the fourteen Bravais lattices gives produces unique space groups, which are the only arrangements of a repeating unit to fill space. The normal method for describing a specific space group is by using the Hermann–Mauguin notation. The space groups are first identified as to where the lattice points are arranged. If the lattice only has points at the corners it is called "primitive" and given the symbol P. The other lattice types, all identified by upper-case letters I, F, C, and so on, are shown in Table 19.1.

In the least symmetric Bravais lattice, triclinic, there are only two space groups, $P1$ and $P\bar{1}$ (read "p-one-bar"). The point group $P1$ has no symmetry at all except that of

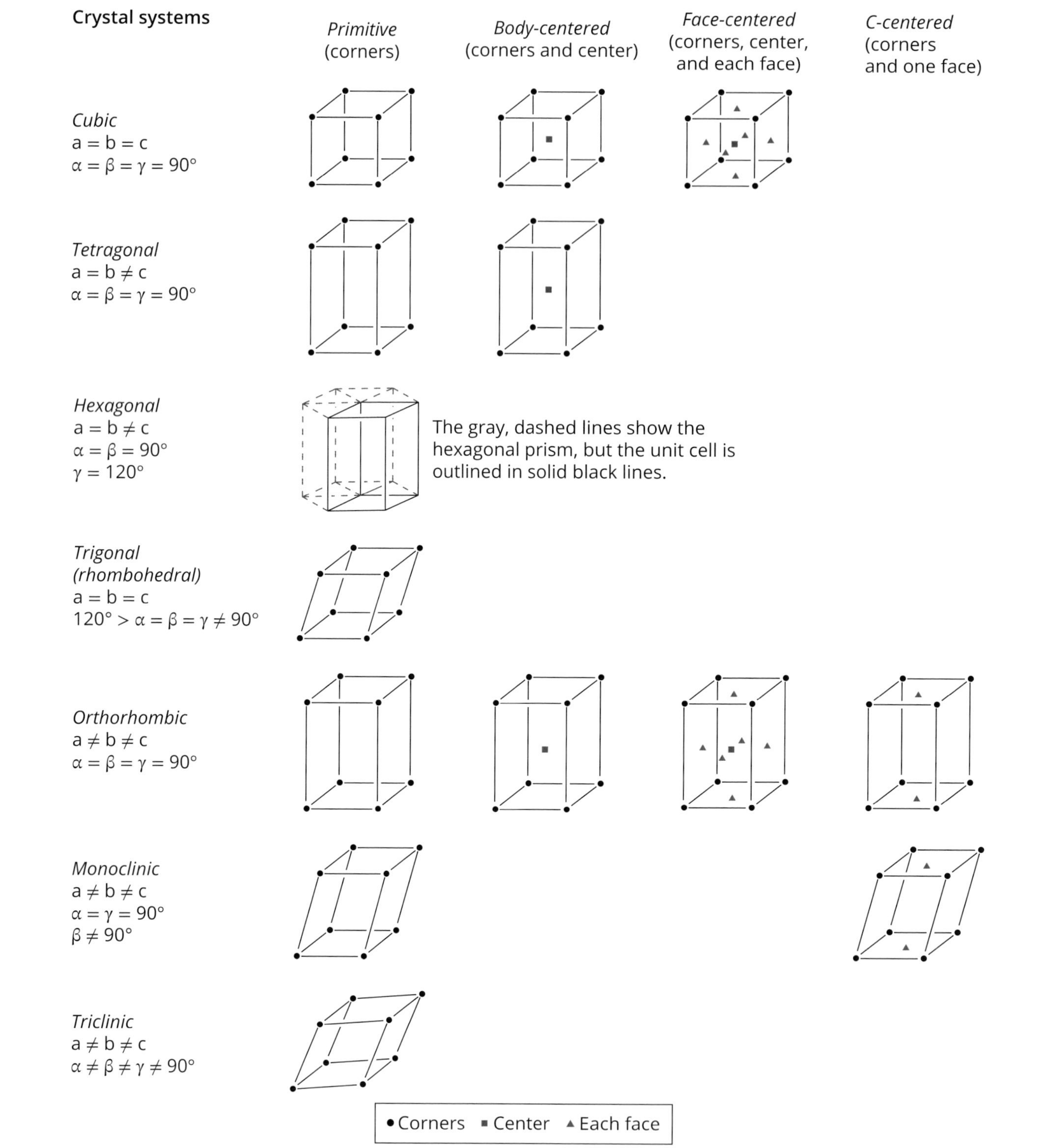

Fig. 19.3 The seven crystal systems (left column) and the fourteen Bravais lattices. The dots represent lattice points and not necessarily atomic positions. All the Bravais lattices have lattice points at the corners (black) and some have point at the center (red) and faces (blue).

translation necessary for building up of the crystal, and as a result is the lowest symmetry possible for a crystalline solid. The *P1* point group is actually very rarely found experimentally, and typically there is some symmetry. The point group $P\bar{1}$ has an inversion center in the middle of the cell as the only additional symmetry. As the symmetry of the crystal increases, less and less data are necessary to fully describe the crystal.

Table 19.2 shows all of 230 space groups organized by their crystal systems. While the list of point groups may seem

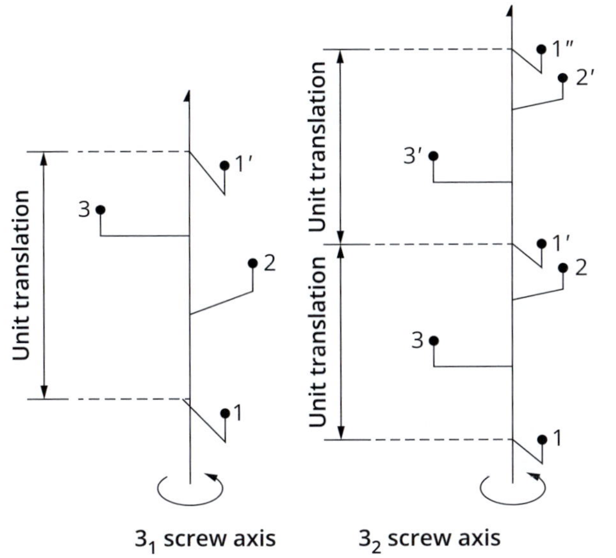

Fig. 19.4 3_1 and 3_2 screw axes. From point 1, point 2 is generated by the first application of the symmetry element, and so on. (Adopted from Stout and Jensen.[2])

daunting, in reality only a few of these are commonly found. The most common space group by far is in the monoclinic setting $P2_1/c$ ("p-2-1 on c"), which is sometimes reported in a nonstandard setting of $P2_1/n$ and has a 2_1-screw axis and a glide plane. About 36% of all compounds studied by X-ray diffraction crystallized in this space group. The next most common is the triclinic group $P\overline{1}$, which occurs about 19% of the time. The next three most common are orthorhombic $P2_12_12_1$ (9%), monoclinic $C2/c$ (7%), and monoclinic $P2_1$ (6%). Around three-quarters of all structures reported are in this handful of space groups.

Again, for our purposes here, it is worth noting that a firm grasp of what these symmetry elements mean is less important than understanding that crystals are comprised of translations plus the types of symmetry operations covered in Chapter 5 on group theory. This leads to specific arrangements of atoms in the solid-state that are related by symmetry operations and translations. Good courses in crystallography will teach these symmetry concepts in more detail, along with the practical aspects of determining the structure of crystalline solids.

Table 19.1 Lattice types and their symbols.

Lattice type	Symbol	Location of lattice points	Diagram
Primitive	P	Lattice points at corners only	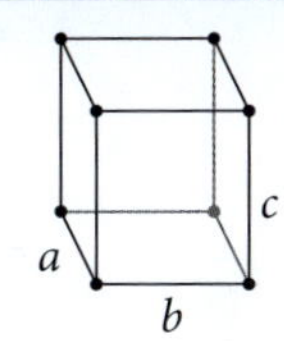
Body centered	I	Lattice points at the corners and in the center of the cell	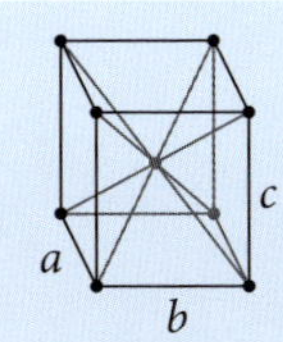
Face centered	F	Lattice points at the corners and in the center of each face	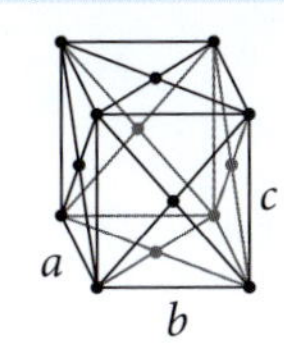
Base centered	A, B, or C	Lattice points at the corner and at the center of one face	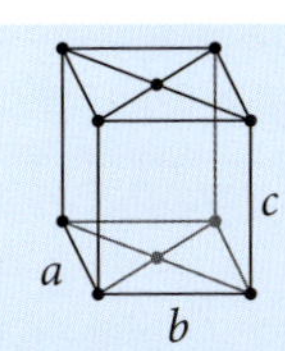 (*C*-centered shown)
Rhombohedral	R	Lattice points at corners and along diagonal of cell	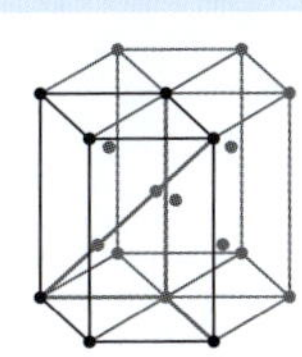

Table 19.2 The 230 space groups organized by crystal system.[a] The last column is the fraction required to collect all the unique data by symmetry. (Adopted from Stout and Jensen.[2])

System	Point Group	Space Group						Fraction
Triclinic	1	$P1$						1/2
	$\bar{1}$	$P\bar{1}$						
Monoclinic	2	$P2$	$P2_1$	$C2$				1/4
	m	Pm	Pc	Cm	Cc			
	$2/m$	$P2/m$	$P2_1/m$	$C2/m$	$P2/c$	$P2_1/c$	$C2/c$	
Orthorhombic	222	$P222$	$P222_1$	$P2_12_12$	$P2_12_12_1$	$C222_1$	$C222$	1/8
		$F222$	$I222$	$I2_12_12_1$				
	$mm2$	$Pmm2$	$Pmc2_1$	$Pcc2$	$Pma2$	$Pca2_1$	$Pnc2$	
		$Pmn2_1$	$Pba2$	$Pna2_1$	$Pnn2$	$Cmm2$	$Cmc2_1$	
		$Ccc2$	$Amm2$	$Abm2$	$Ama2$	$Aba2$	$Fmm2$	
		$Fdd2$	$Imm2$	$Iba2$	$Ima2$			
	mmm	$Pmmm$	$Pnnn$	$Pccm$	$Pban$	$Pmma$	$Pnna$	
		$Pmna$	$Pcca$	$Pbam$	$Pccn$	$Pbcm$	$Pnnm$	
		$Pmmn$	$pbcn$	$pbca$	$pnma$	$Cmcm$	$Cmca$	
		$Cmmm$	$Cccm$	$Cmma$	$Ccca$	$Fmmm$	$Fddd$	
		$Immm$	$Ibam$	$Ibca$	$Imma$			
Tetragonal	4	$P4$	$P4_1$	$P4_2$	$P4_3$	$I4$	$I4_1$	1/8
	$\bar{4}$	$P\bar{4}$	$I\bar{4}$					
	$4/m$	$P4/m$	$P4_2/m$	$P4/n$	$P4_2/n$	$I4/m$	$I4_1/a$	
	422	$P422$	$P42_12$	$P4_122$	$P4_12_12$	$P4_222$	$P4_22_12$	1/16
		$P4_322$	$P4_32_12$	$I422$	$I4_122$			
	$4mm$	$P4mm$	$P4bm$	$P4_2cm$	$P4_2nm$	$P4cc$	$P4nc$	
		$P4_2mc$	$P4_2bc$	$I4mm$	$I4cm$	$I4_1md$	$I4_1cd$	
	$\bar{4}2m$	$P\bar{4}2m$	$P\bar{4}2c$	$P\bar{4}2_1m$	$P\bar{4}2_1c$	$P\bar{4}m2$	$P\bar{4}c2$	
		$P\bar{4}b2$	$P\bar{4}n2$	$I\bar{4}m2$	$I\bar{4}c2$	$I\bar{4}2m$	$I\bar{4}2d$	
	$4/mmm$	$P4/mmm$	$P4/mcc$	$P4/nbm$	$P4/nnc$	$P4/mbm$	$P4/mnc$	
		$P4/nmm$	$P4/ncc$	$P4_2/mmc$	$P4_2/mcm$	$P4_2/nbc$	$P4_2/nnm$	
		$P4_2/mbc$	$P4_2/mnm$	$P4_2/nmc$	$P4_2/ncm$	$I4/mmm$	$I4/mcm$	
		$I4_1/amd$	$I4_1/acd$					
Trigonal/ rhombohedral	3	$P3$	$P3_1$	$P3_2$	$R3$			1/6
	$\bar{3}$	$P\bar{3}$	$R\bar{3}$					
	32	$P312$	$P321$	$P3_112$	$P3_121$	$P3_212$	$P3_221$	1/12
		$R32$						
	$3m$	$P3m1$	$P31m$	$P3c1$	$P31c$	$R3m$	$R3c$	
	$\bar{3}m$	$P\bar{3}1m$	$P\bar{3}1c$	$P\bar{3}m1$	$P\bar{3}c1$	$R\bar{3}m$	$R\bar{3}c$	
Hexagonal	6	$P6$	$P6_1$	$P6_5$	$P6_2$	$P6_4$	$P6_3$	1/12
	$\bar{6}$	$P\bar{6}$						
	$6/m$	$P6/m$	$P6_3/m$					
	622	$P622$	$P6_122$	$P6_522$	$P6_222$	$P6_422$	$P6_322$	1/24
	$6mm$	$P6mm$	$P6cc$	$P6_3cm$	$P6_3mc$			

Table 19.2 Continued

System	Point Group	Space Group						Fraction
	$\bar{6}m2$	$P\bar{6}m2$	$P\bar{6}c2$	$P\bar{6}2m$	$P\bar{6}2c$			
	$6/mmm$	$P6/mmm$	$P6/mcc$	$P6_3/mcm$	$P6_3/mmc$			
Cubic	23	$P23$	$F23$	$I23$	$P2_13$	$I2_13$		1/24
	$m3$	$Pm3$	$Pn3$	$Fm3$	$Fd3$	$Im3$	$Pa3$	
		$Ia3$						
	432	$P432$	$P4_232$	$F432$	$F4_132$	$I432$	$P4_332$	1/48
		$P4_132$	$I4_132$					
	$\bar{4}3m$	$P\bar{4}3m$	$F\bar{4}3m$	$I\bar{4}3m$	$P\bar{4}3n$	$F\bar{4}3c$	$I\bar{4}3d$	
	$m3m$	$Pm3m$	$Pn3n$	$Pm3n$	$Pn3m$	$Fm3m$	$Fm3c$	
		$Fd3m$	$Fd3c$	$Im3m$	$Ia3d$			

19.1.2 Some Common Structures Found in Solid-State Chemistry

Some common structures tend to occur repeatedly in solid-state chemistry. In general, these structure types are often referred to by a common compound that has that structure. For example, one may hear about a compound that has the "rock salt" structure of NaCl, which has octahedral sodium and chloride ions, for any number of other compounds with a similar arrangement of atoms. For example, magnesium oxide (MgO) has the rock-salt structure with a = 4.211 Å, and titanium carbide (TiC) has a = 4.329 Å. The cell is cubic, so the only dimension that needs to be specified to fully describe the cubic cell is the length of one side, a. A few of these prototypical structures are shown in Fig. 19.5. These prototypical structures are often in high-symmetry space groups, but the structures of other compounds may not be in the same space group but merely have a similar arrangement of atoms.

Another feature of some solid-state compounds is partial formulae. In extended structures it is common for a single phase to not have a discrete whole-number formula. In some cases this is accomplished by intentionally doping an ion similar to one in the pure structure. An example of such solids are rubies, which, in addition to occurring naturally, are prepared synthetically. A typical ruby has chromium(III) ions (0.0725 Å or 72.5 pm ionic radius)[3] that replace a small amount of the aluminum(III) ions (0.0675 Å or 67.5 pm ionic radius)[3] in alumina (Al_2O_3) for a molecular formula of $Al_{2-x}Cr_xO_3$, where $x \sim 0.03$. It is this small chromium impurity that gives ruby its characteristic deep red color, and it is common for small impurities to have large effects on properties. In the X-ray diffraction structure, these impurities may be so small as to be difficult to identify, and are often randomly distributed throughout the structure.

One can often replace an ion of similar charge and size in a lattice, as was shown in the ruby example, with Al^{3+} and Cr^{3+}. Consequently, tables on ionic radii for the elements as a function of charge have been created that are of common use in solid-state laboratories. Such a listing is provided in Appendix H.[a]

19.2 Forces Important in the Solid State

In this section, the forces that hold a solid together will be discussed, with special attention to those other than covalent bonds. First, the forces commonly seen in molecular solids will be reviewed. Then, the lattice energies of ionic solids will be discussed. A more detailed look at electronic structure, especially of covalent systems, is then presented.

19.2.1 Van der Waals Forces and Molecular Solids

In molecular solids, species with relatively strong forces—the bonds between the atoms in the molecule—are held together by far weaker forces. While each of the interactions holding the individual molecules together are typically <10 kcal/mol, these forces sum to significant energies that greatly affect properties. Collectively, the weak forces between molecules are known as van der Waals interactions, which are typically taught in general and physical chemistry courses and will be discussed only briefly here.

There are two principle intermolecular forces to be considered for all molecular solids, regardless of their structure. The first is a repulsive term between filled orbitals due to Pauli

[a] Ionic radii in general deserve a note of caution. Depending on how the values are calculated in the source, the radii can vary substantially.

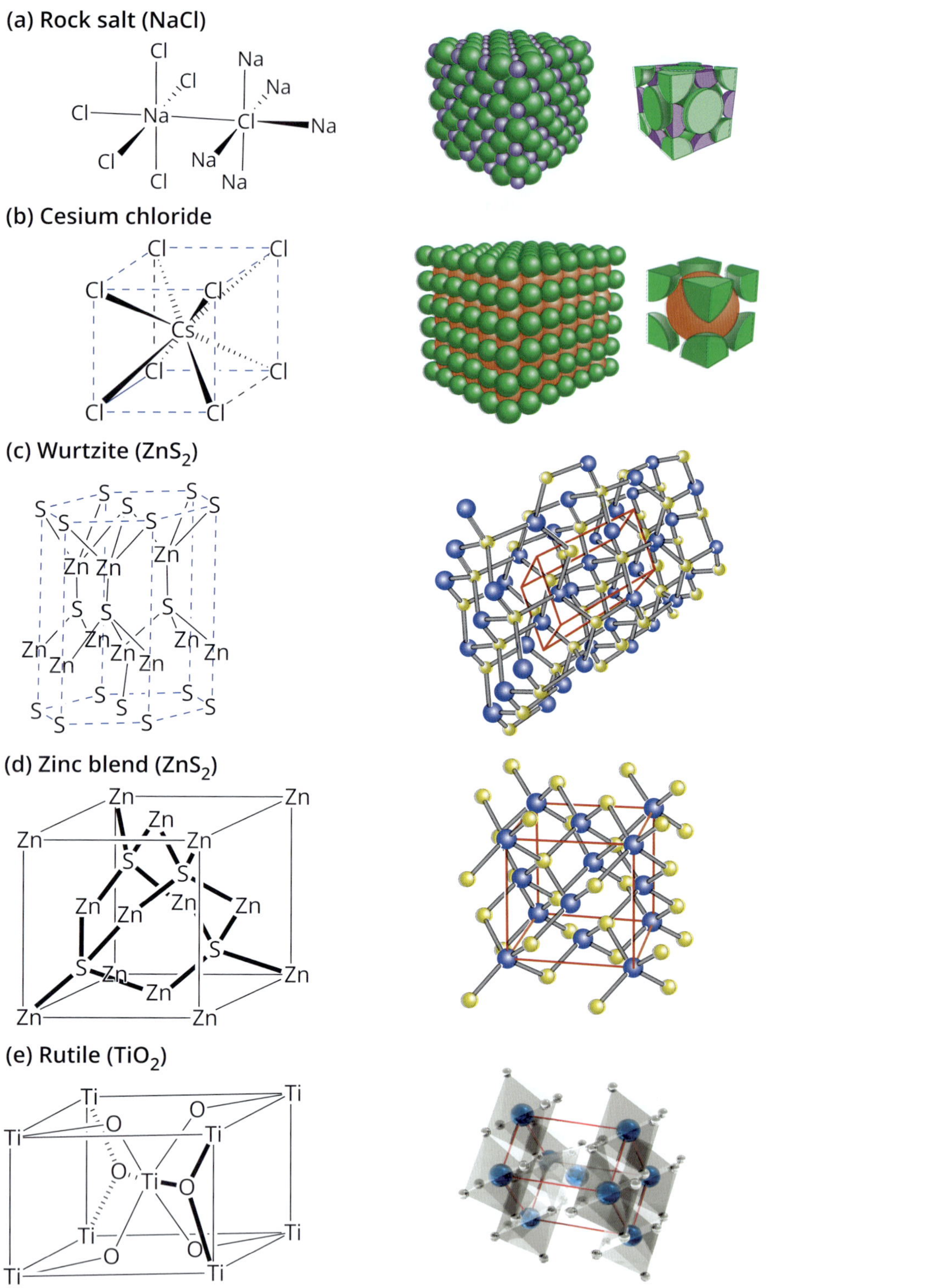

Fig. 19.5 Some common structure types. (a) Rock salt (NaCl, cubic $Fm\overline{3}m$) has alternating cations and anions. Every atom is in an octahedral site, with each sodium surrounded by six chlorides and each chloride surrounded by six sodiums. (b) Cesium chloride (cubic $Pm3m$) has an 8-coordinate metal ion. (c) Wurtzite (ZnS in hexagonal $P6_3mc$) has tetrahedral Zn and S sites. (d) Zinc blend has a structure reminiscent of linked adamantyl groups (ZnS in cubic $F\overline{4}3m$). (e) Rutile (TiO_2 in tetragonal $P4/mmm$) has 3-coordinate oxygens around an octahedral metal center.

repulsion. The second is an attractive force called "London dispersion" or just "dispersion".

The repulsive term (R) is related to close approach of electrons in orbitals, similar to what are generally called "steric interactions" in chemistry. As shown in Eq. 19.1, the repulsive force drops off exponentially with the distance between the interacting bodies, r. The two constants in the equation, a and b, are typically found from experiment, such as deviations from ideal gas behavior.

$$R = be^{-ar} \qquad \textbf{19.1}$$

The attractive force, dispersion (D), is attraction due to synchronization of the electrons in the two different bodies, but is a wholly quantum-mechanical phenomenon. To simplify the discussion somewhat, we will consider two helium atoms interacting, but the arguments may be extrapolated conceptually to molecular ensembles. In short, all atoms and molecules are undergoing constant fluctuations in their electron density. When one helium atom fluctuates, it gains a nonspherical electron density on a very short timeframe (possibly attoseconds). A nearby helium atom can respond to this fluctuation, and both atoms will place more electron density between the two nuclei (Fig. 19.6).[4] This arrangement leads to greater electron–electron repulsion but also greater electron–nuclear interactions (similar to covalent bonding),[b] and the net result is that the electrons experience no net additional forces, due to interaction of the two helium nuclei. However, the helium nuclei experience attraction to the electrons residing between the two nuclear centers—a net attractive force between the atoms. The electron clouds in molecules are in constant fluctuation in this way, which leads to an attractive interaction at long distances relative to covalent bonds. The force drops off as r^{-6}, quickly dissipating as the bodies are moved apart. As shown in Eq. 19.2, the sign of the force is negative, because the interaction is attractive.

$$D = -d/r^6 \qquad \textbf{19.2}$$

The total potential energy (U) associated with these two interactions are then as shown in Eq. 19.3. The constants a, b, and d can be found from nonideal gas behavior, for example. This is sometimes referred to as the Buckingham potential.

$$U = R + D = be^{-ar} - d/r^6 \qquad \textbf{19.3}$$

More often, these van der Waals interactions are described by using a Leonard-Jones potential, which has the general form shown in Eq. 19.4. This form of the potential is a little simpler and adequate for our purposes here. The potential energy, U, is described as an attractive force, $-(\sigma/r)^6$, and a repulsive force, $(\sigma/r)^{12}$, where σ is the inter-particle distance where the potential is zero, r is the inter-particle distance, and ε is the depth of the potential well. The minimum in the potential, r_m, is related to the inter-particle potential by $r_m = 2^{1/6}\sigma$.

$$U = 4\epsilon\left\{\left(\frac{\sigma}{r}\right)^{12} - \left(\frac{\sigma}{r}\right)^{6}\right\} \qquad \textbf{19.4}$$

[b] One can, however, draw clear distinctions between the dispersion force and covalent bonding. Dispersion is an electron correlation effect, and no molecular orbital is generated. For example, two spin-up hydrogen atoms can have a dispersion force attracting them and will place additional electron density between the nuclei, even though to form a covalent bond a spin-flip on one of the hydrogens would have to occur for molecular orbital formation and occupation.

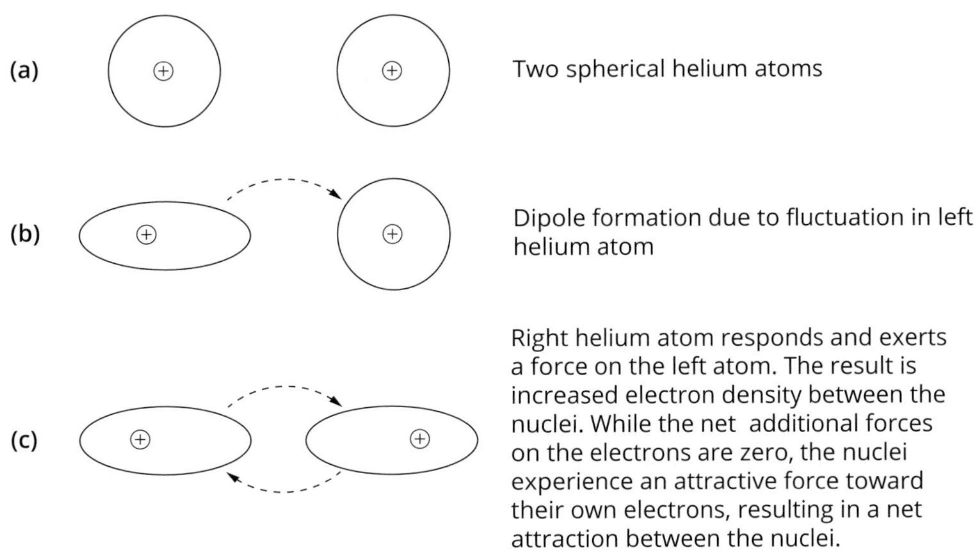

Fig. 19.6 The dispersion or London force. Atoms (and molecules) are constantly undergoing rapid fluctuations in electron density. In a two-atom system (a) the mean electron density is spherical on each atom; however, (b) a fluctuation on the left He atom can lead to a response (c) by the right atom. The result is to put additional electron density between the two nuclei.

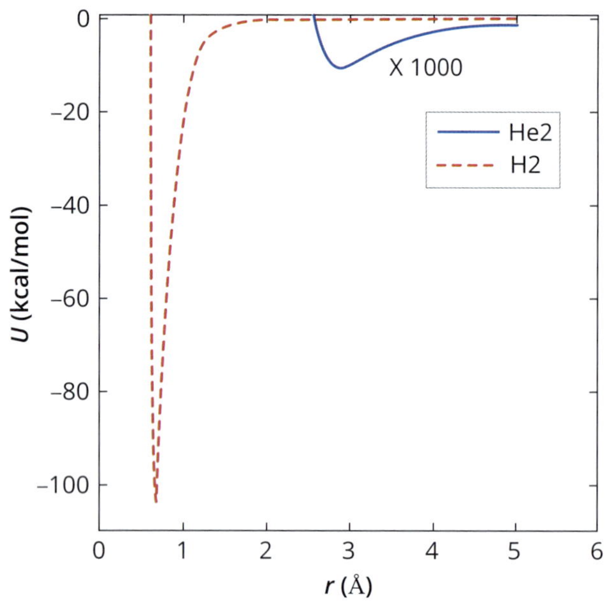

Fig. 19.7 Leonard-Jones potential energy (*U*) plot for the He–He "molecule" (blue) and for H_2 (red) versus distance (*r*) in Å. Note that the energy of the He–He interaction is in cal/mol and the H–H bond is in kcal/mol; the helium interaction had to be multiplied by 1,000 to place it on the same scale. Obviously, the van der Waals interactions between the helium atoms are thousands of times smaller than the bond holding the hydrogen atoms together.

For the He_2 "molecule"—two very weakly interacting helium atoms—a plot of these forces versus interatomic distance is shown in Fig. 19.7. For reference, the plot of energy versus distance for the hydrogen molecule is also shown. The attraction between two helium atoms is strongest, 18.2 cal/mol, at about 3 Å. Bringing the atoms closer than this causes the energy to rise exponentially as the repulsive electron clouds interact. As the atoms are moved apart, the attractive interaction drops off as r^6. For reference, Fig. 19.7 shows the Leonard-Jones potential energy plot for two hydrogen atoms forming a covalent bond, which has an energy of 104 kcal/mol at 0.74 Å. Note the change in units! The He–He interaction is 18 *cal*/mol, while the bond in H_2 is in *kcal*/mol; in the figure, the energies of the helium interactions were multiplied by 1,000 to make them viewable on the same scale!

In the case of helium, a single vibrational frequency, the ground state, can be observed,[5] but any excitation leads to dissociation. The zero-point energy is almost precisely equal to the well depth. Helium is a liquid to absolute zero at 1 atm of pressure, never becoming a solid unless higher pressures are added.

Increasing the number of electrons around the nucleus makes the atom more polarizable and significantly increases the dispersion force. For example, Ar_2 has a minimum at about 4 Å with an interaction energy of 250 cal/mol. This stronger, though still not strong by typical bonding terms, interatomic energy gives argon a melting point of 89 K (−184 °C).

In general, as a molecule becomes larger and has more electrons, the volatility decreases and the melting point increases due to increased intermolecular interactions. This is true, whether or not the atoms are large and polarizable; however, large atoms that are polarizable dramatically increase the effect. For example, ethane is a gas, while longer chain pentane is a liquid and even longer-chain hydrocarbons can be solid waxes. The diatomic of poorly polarizable nitrogen atoms, N_2, boils at −196 °C, while the diatomic of highly polarizable iodine atoms, I_2, is a solid at room temperature (albeit a volatile one).

In addition to these interactions, some molecules may have specific structural features that lead to intermolecular interactions. These include a permanent dipole, hydrogen bonding, aryl–aryl stacking interactions, C–H/aryl interactions, and others. Again, none of these interactions are strong relative to bonding interactions, with hydrogen bonding being the strongest of these with energies up to ~10 kcal/mol. However, it is these interactions that determine the important properties of molecular solids, such as boiling and melting points. These interactions also determine the orientations of the molecules for various applications, such as liquid crystals.

19.2.2 Ionic Solids, Lattice Energies, and the Madelung Constant

Ionic solids are, as the name implies, composed of atoms or ions having quite different electronegativity leading to large ionic components to the bonding. As is usually the case, the high ionic character leads to very large bond energies. In the case of an ionic solid, the bond energy must be considered in all three dimensions with cations and anions in the lattice. This energy that holds the ions together is called the lattice energy (*U*), defined as the energy *released* when one mole of a substance is formed from its constituent gas-phase ions, and this quantity is always positive.

The lattice energy of a solid cannot be measured directly but can be gleaned from the properties of the species involved by using a Born–Haber cycle, an example of which, for KCl, is shown in Fig. 19.8. Starting at the left bottom with KCl(s), we go upward to the standard state of the constituent elements K(s) and $1/2Cl_2(g)$ by the reverse (opposite sign) of the enthalpy of formation, $\Delta H_f°$, which is +104.6 kcal/mol. The energy required to take solid potassium to gas-phase potassium, the sublimation energy, is +21.3 kcal/mol. Next, the ionization energy of potassium is 101.5 kcal/mol, which makes $K^+(g)$ and a free electron. One half the energy of dissociation of Cl_2 is then 29.1 kcal/mol. Chlorine then picks up the electron from potassium, which stabilizes the system by the electron affinity of the chlorine atom, 84.8 kcal/mol. Now, the system has $K^+(g)$ and $Cl^-(g)$, and the energy released when these combine to form the solid is the lattice energy, *U*, which can be calculated as 171.7 kcal/mol (719 kJ/mol).

Table 19.3 lists lattice energies for some common salts found experimentally by using Born–Haber cycles. As can be seen in Fig. 19.8, a tremendous amount of data are required to calculate the lattice energy from thermochemical measurements, and

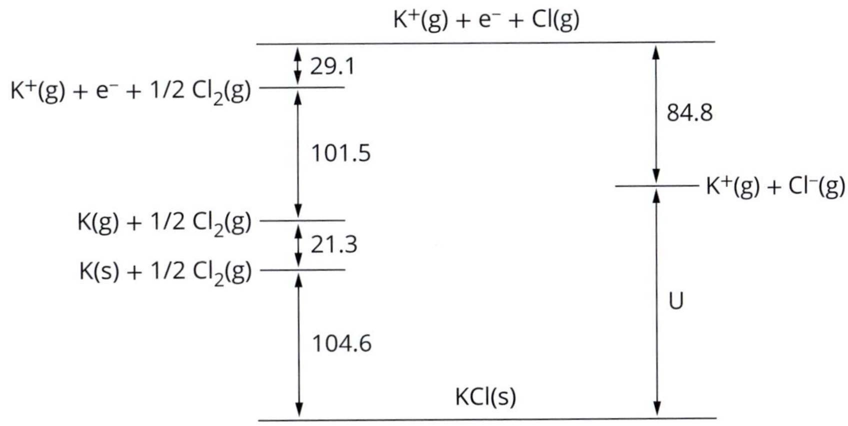

Fig. 19.8 Born–Haber cycles for solid KCl formation with enthalpies in kcal/mol.

Table 19.3 Experimental lattice energies of some common salts.[6]

Compound	Lattice Energy in kcal/mol (kJ/mol)
NaO_2CCH_3 (sodium acetate)	182 (763)
NaN_3 (sodium azide)	175 (731)
$NaHCO_3$ (sodium bicarbonate)	196 (818)
K_2CO_3 (potassium carbonate)	444 (1858)
LiF (lithium fluoride)	248 (1036)
LiCl (lithium chloride)	204 (853)
LiBr (lithium bromide)	193 (807)
LiI (lithium iodide)	181 (757)
NaCl (sodium chloride)	188 (786)
KCl (potassium chloride)	171 (715)
RbCl (rubidium chloride)	165 (689)
CsCl (cesium chloride)	158 (659)
CuI (copper iodide)	231 (966)
InI (indium(I) iodide)	175 (732)
CrF_2 (chromium(II) fluoride)	697 (2917)
$TiCl_3$ (titanium(III) chloride)	1227 (5134)
$FeCl_3$ (iron(III) chloride)	1281 (5359)
$FePO_4$ (iron(III) phosphate)	1745 (7303)
$Ca_3(PO_4)_2$ (calcium phosphate)	2504 (10478)

there has been significant work on the estimation of lattice energies based on the structures and charges, which requires the calculation of individual ion interactions summed over the crystal.

If we do not have all of this data available, we can estimate the lattice energy theoretically. At the limit of a fully ionized system, with actual cations and anions in the lattice and no covalency, the energy of interaction (V_{AB}) is defined by Coulomb's law as shown in Eq. 19.5, where z_A is the charge on species A, e is the electron charge, ε_0 is the permittivity of a vacuum, and r_{AB} is the distance between cation and anion.

$$V_{AB} = \frac{(z_A e) \times (z_B e)}{4\pi\varepsilon_0 r_{AB}} \tag{19.5}$$

This equation is for an isolated cation and anion. What happens if we put the cation and anion into a lattice? What is the energy of the ionic interactions then?

As a simple example, let us look at a 1-dimensional string of alternating cations and anions as shown in Fig. 19.9. The string is, for all intents and purposes in chemistry, infinitely long. Consequently, we need to start by choosing a reference somewhere in the string arbitrarily, shown by an arrow. The distance between each cation and anion will be d; that is, $r_{AB} = d$.

If we look at the reference point, there is a cation with an anion to the right at distance d. Using Eq. 19.5, we calculate the potential energy of that one interaction as follows, where

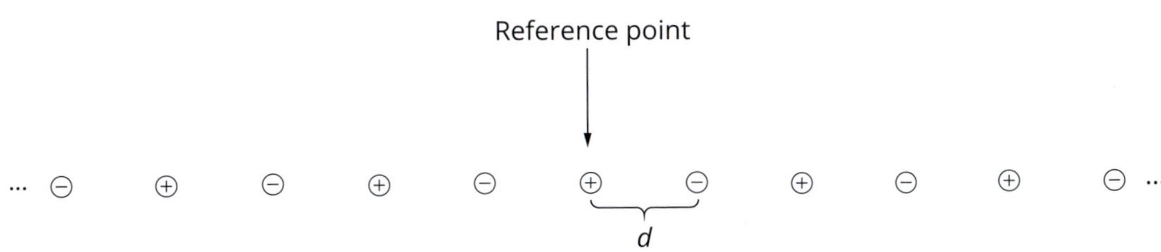

Fig. 19.9 An infinite string of cations and anions equally spaced by distance d.

we have assumed that $z_A = z_B = z$; that is, the charges on the cation and anion are equivalent in absolute value.

$$V_{AB} = \frac{(z_A e) \times (z_B e)}{4\pi\varepsilon_0 r_{AB}} = \frac{z^2 e^2}{4\pi\varepsilon_0 d}$$

At the reference point we first looked to the right, but if we then look to the left we have another anion with the same charge. As a result, we double the previous result and obtain interaction of our reference cation with the two anions on either side. The energy is negative because it is attractive. The constants in the following equation have been collected together outside the parentheses.

$$V_{AB} = \left(-\frac{2z^2}{d}\right)\frac{e^2}{4\pi\varepsilon_0}$$

We can then consider the interaction of the cation with the cations on the other side of the closest anions. This interaction is repulsive because it has the same charge as our reference, and occurs at a distance of $2d$. One can continue calculating the energies between the reference and the cations and anions down the chain. The terms would continue on to infinity, but become smaller due to the larger distances as one moves away from the reference.

$$V_{AB} = \left(-\frac{2z^2}{d} + \frac{2z^2}{2d} - \frac{2z^2}{3d} + \frac{2z^2}{4d} - \frac{2z^2}{5d}\cdots\right)\frac{e^2}{4\pi\varepsilon_0}$$

but

$1 - 1/2 + 1/3 - 1/4 + 1/5\ldots$ converges at $\ln 2$

so

$$= \left(-\frac{2z^2}{d}\right)\cdot \ln 2 \cdot \frac{e^2}{4\pi\varepsilon_0}$$

The series 1 – ½ + 1/3–1/4 . . . conditionally converges at natural log 2, so the energy can be concisely described as shown. The first term, $-2z^2/d$, depends on the charges and distances, while the last term, $e^2/4\pi\varepsilon_0$, is simply a collection of constants. The middle term, ln 2, depends on the symmetry of the system, which in this example is alternating charges in a straight line, equally spaced.

The symmetry-related value, ln 2, is called a "Madelung constant", which is often assigned the symbol A. This constant, just as in the previous example, is independent of the charges and interionic distances; the value changes only with the symmetry of the lattice.

The potential energy of the ionic system, V_{AB}, can be described as shown in Eq. 19.6, where N_A is Avogadro's number and the other terms are as defined previously. Some examples of Madelung constants for some common crystal systems are shown in Table 19.4. The energy V_{AB} is proportional to the Madelung constant, A, and inversely proportional to the distance between ions, d. The distance d often increases with increasing coordination number, and the increased distance may offset the increase in the Madelung constant.

Table 19.4 Some examples of Madelung constants (A).

Structure Type	Coordination #	A
CsCl	8	1.763
Fluorite (CaF_2)	8	2.519
Rock Salt	6	1.748
Rutile (TiO_2)	4	2.408
Wurtzite (ZnS)	4	1.641

$$V_m = N_A \frac{-e^2}{4\pi\varepsilon_0} \times \frac{z_A z_B}{d} \times A \qquad \textbf{19.6}$$

The attractive forces in ionic solids due to polarization and dispersion, discussed previously, are typically much smaller than the Coulombic attraction due to charges. The repulsive force due to Pauli repulsion, which prevent the ions from collapsing into one another, decrease exponentially as one proceeds from the nucleus. The repulsions can be expressed by using the following equation, where it is positive to show repulsion and where C' and d* are constants. The constant C' will disappear in the following treatment. The value of d* is found experimentally to be ~0.345 Å.

$$V = +N_A C' e^{-d/d*}$$

The total energy of the ionic crystal on a molar basis (V_m) is then as shown in Eq. 19.7, where the Coulombic attraction is summed with the repulsive term.

$$V_m = N_A \frac{-e^2}{4\pi\varepsilon_0} \times \frac{z_A z_B}{d} \times A + N_A C' e^{-d/d*} \qquad \textbf{19.7}$$

If we ignore the kinetic motion (vibrations) of the system, which is equivalent to being at absolute zero, then we can take the derivative of the previous expression with respect to the interionic distance, dV_m/dd, and set to zero, which will provide where the potential energy is at a minimum with respect to d.

$$N_A C' e^{-d/d*} = N_A \frac{e^2}{4\pi\varepsilon_0} z_A z_B \times \frac{Ad^*}{d^2} \qquad \textbf{19.8}$$

Substituting the expression in Eq. 19.8 into Eq. 19.7 produces, after some algebraic rearrangement, Eq. 19.9, which is known as the Born–Mayer equation. This equation can be used to estimate the lattice energy of an ionic solid if one knows the value of d and the type of lattice, which determines the Madelung constant (A). The approximations inherent in Eq. 19.9

are that the system is entirely ionic and that there is no vibrational motion (temperature = 0 K).

$$U = \frac{N_A z_A z_B e^2}{4\pi\varepsilon_0 d}(1 - \frac{d^*}{d})A \qquad \textbf{19.9}$$

In 1956, Anatoli Kapustinskii published a modification of the Born–Mayer equation that simplifies things greatly.[7] He noticed that the Madelung constant, A, divided by the number of ions per formula unit and the distance, d, was approximately constant, $A/nd \sim$ constant, where n is the number of ions in each formula. He hypothesized that there is a rock salt structure (Fig. 19.5) energetically equivalent to the true structure with distance d. The result is that we do not need to know the structure and the Madelung constant to estimate the lattice energy. In addition, we can estimate the interionic distance, d, as the sum of the ionic radii for the constituent ions, $d = r_+ + r_-$, the ionic radius of the cation and anion respectively.

The Kapustinskii expression for the lattice energy is shown in Eq. 19.10. The term $d^* = 0.345$ Å and the Kapustinskii constant $K = 290.1$ kcal·Å/mol (1214 Kj·Å/mol). Again, n is the number of species in the formula; for example, $n = 2$ for KCl for one potassium ion and one chloride ion. Values for ionic radii of the elements with various coordination numbers, and ionic radii of some complex anions, are shown in Appendix H.

$$U = \frac{n z_A z_B}{r_- + r_+}\left(1 - \frac{d^*}{r_- + r_+}\right)K \qquad \textbf{19.10}$$

As an example, we can estimate the lattice energy in potassium nitrate, KNO_3. The radius of the potassium cation (6-coordinate, as in the rock salt structure) is 1.52 Å. The complex anion nitrate, NO_3^-, has a radius of 1.89 Å. The use of Eq. 19.10 is illustrated as follows, and the result is a lattice energy of 153 kcal/mol. The thermochemical lattice energy from a Born–Haber cycle is 164 kcal/mol, and our error from the Kapustinskii equation is ~7%.

$$U = \frac{n z_A z_B}{r_- + r_+}\left(1 - \frac{d^*}{r_- + r_+}\right)K$$

$$U = \frac{2(1)(1)}{1.89\,\text{Å} + 1.52\,\text{Å}}\left(1 - \frac{0.345\,\text{Å}}{1.89\,\text{Å} + 1.52\,\text{Å}}\right)290.1\,kcal{\cdot}\text{Å}/mol$$

$$U = \frac{2}{3.41\,\text{Å}}\left(1 - \frac{0.345\,\text{Å}}{3.41\,\text{Å}}\right)290.1\,kcal{\cdot}\text{Å}/mol$$

$$U = 0.587\,\text{Å}^{-1}(0.899)290.1\,kcal{\cdot}\text{Å}/mol$$

$$U = 153\,kcal/mol$$

19.3 A Brief Introduction to the Electronic Structure of Solids[8]

In Section 19.2, molecular solids and ionic solids were quickly reviewed. In both of these types of solids, the electronic structure is typically quite isolated with molecules interacting only weakly and ions only interacting through Coulombic forces and van der Waals interactions. In this section, the electronic structure of solids with significant covalent bonding throughout the extended system will be described, requiring modification and expansion of our ideas on molecular orbital concepts. Specifically, to understand the electronic structure of solids we need to introduce "bands", which are used in place of discrete orbitals.

To illustrate the necessity for bands in the electronic structure, consider a 1-dimensional string of lithium ({He}$2s^1$) atoms (Fig. 19.10). Starting with two atoms, a bonding and antibonding pair of orbitals is generated highly reminiscent of the MOs for H_2. With three atoms, a bonding,

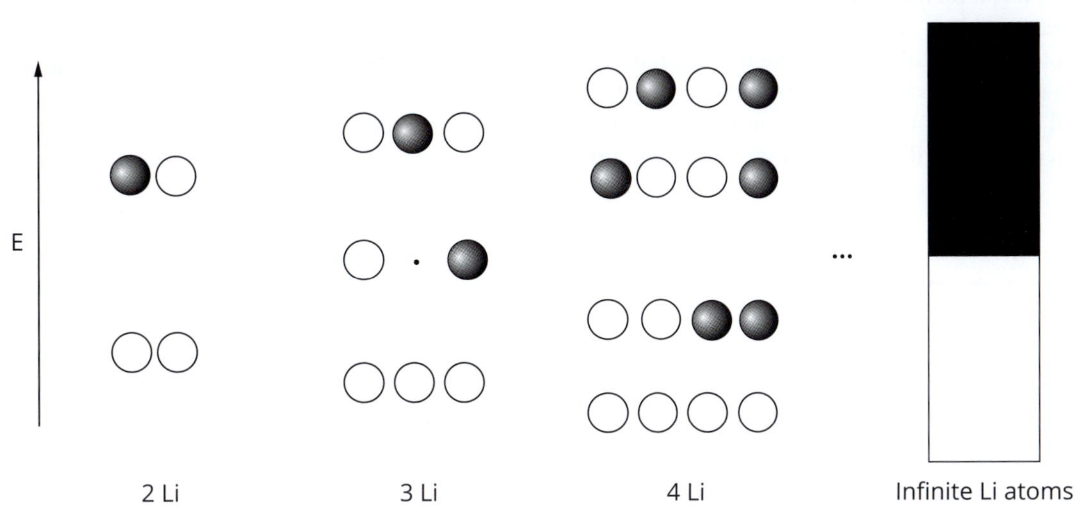

Fig. 19.10 Using lithium atoms as an example, the figure starts with the MOs of two atoms and then builds up a band of orbital energies. Because the band was built from lithium, which has a half-filled 2s-orbital, the band is half full.

nonbonding, and antibonding assembly of orbitals is generated, and in this case the fully antibonding and fully bonding orbitals form the upper- and lower-energy bounds for an electron in the system, with the nonbonding orbital intermediate in energy. As the number of atoms and orbitals continues to increase, the number of molecular orbitals increases. For a very large number of atoms there is essentially an orbital at every energy between the two limits—a band of orbital energies comprised of mixed 2*s* orbitals. Because the valence orbital of lithium used to create the band was half full, $2s^1$, the band is half-filled, with, in the figure, the part filled with electrons shown in white and the unfilled part of the band black.

Note the energies of the orbitals as well. The fully bonding 3Li orbital is slightly lower in energy and "more bonding" than the 2Li fully bonding orbitals. Likewise, the fully antibonding 3Li orbital is higher in energy than the antibonding orbitals of 2Li. This does not continue to infinity, however. Once there are thousands, millions, or more atoms, adding one atom more makes no significant difference to the upper and lower bounds of the band.

Fig. 19.10 is a pictorial example, but a somewhat simpler way of generating such functions for periodic systems was given to us by Felix Bloch. The purpose of the equation is to determine the wavefunctions for a large array. This is achieved by using, as we did for molecular systems, a linear combination of atomic orbitals, but in this case we will need to make a fairly large modification to the way we think about the irreducible representations of the wavefunction. Since we have, essentially, an infinite number of atoms with an infinite number of atomic orbitals to combine, we need a way of getting at what is important about this very large array and its symmetry properties. It would not do to use an infinite basis of atomic orbitals to obtain an infinite number of wavefunctions as we would when using the methods presented in Chapter 6, and there is a simpler description.

For the moment, let us consider an infinite linear array of atoms with *s*-orbitals on them, like our string of lithium atoms with their 2*s* valence orbitals. There will be a constant spacing between atoms of a uniform distance, *a*. We will reference the atoms with a set of integers, starting somewhere in the string as 0, 1, 2, 3, and so on, and the wavefunction on each atom, such as 2*s* valence orbital, will be χ_1 on atom one, χ_2 on atom two, and so on.

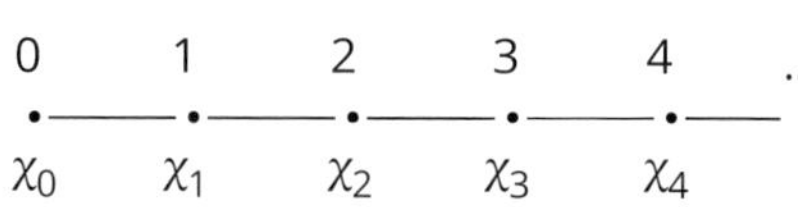

The wavefunction can be calculated with Eq. 19.11, which uses Euler's notation for a complex number (Appendix I). In this equation, *k* is a symmetry number that will run from 0 to π/a. The number *k*, in essence, is a mathematical stand-in for an irreducible representation for a wavefunction, such as a_{1g}, t_{2g}, e_g, and so on. The variable *n* is just an index that corresponds to integers: $n = 0, 1, 2, 3, \ldots$. The factor χ_n is the atomic wavefunction on each atom being mixed to obtain the overall wavefunction for linear extended array combination ψ_k.

$$\psi_k = \sum_n e^{ikna}\chi_n \qquad \textbf{19.11}$$

It is helpful to examine the limits of this function as applied to our string of lithium atoms mixing 2*s*-orbitals. As mentioned, $k = 0$ to π/a, so let us see what the wavefunction looks like at each of those values.

$$k = 0, \quad \psi_0 = \sum_n e^0\chi_n = \sum_n \chi_n$$

This equation for $k = 0$ simplifies to the sum of χ_n (because $e^0 = 1$), which means that all the wavefunctions on the atom have the same magnitude and the same phase. The accompanying figure shows the result with χ_n = *s*-orbital with their indexes, which is obviously the maximally bonding orbital.

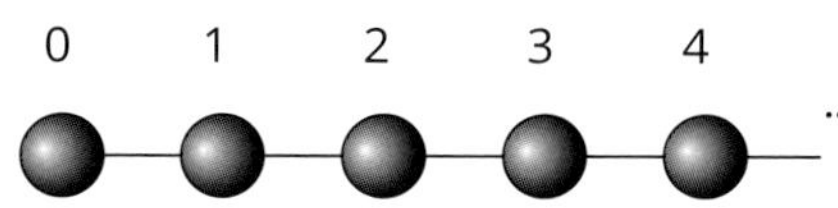

Solving the equation for $k = \pi/a$ produces the following equation. Keep in mind Euler's beautiful equation $e^{i\pi} + 1 = 0$, which can be rearranged to $e^{i\pi} = -1$. Consequently, $e^{i\pi n} = (-1)^n$.

$$k = \pi / a, \quad \psi_{\pi/a} = \sum_n e^{i\pi n}\chi_n = \sum_n (-1)^n \chi_n$$

Using this in the $k = \pi/a$ equation produces the result that the wavefunction will alternate phases. Again, using an *s*-orbital as χ_n for our string of lithium atoms produces a picture like the one here, which is obviously the maximally antibonding orbital.

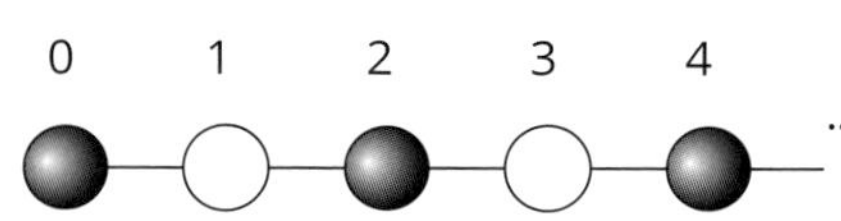

From these pictures, it is obvious that the $k = 0$ molecular orbital will be the lowest in energy and the $k = \pi/a$ molecular orbital will be the highest in energy when dealing with a string of *s*-orbitals. Naturally, all the other orbitals will be between these in energy, but there are a large number (practically infinite number) of them. Consequently, there is in essence a continuous group of orbitals bracketed by the energy of the $k = 0$ and $k = \pi/a$ molecular orbitals, which is referred to as a "band".

Note that the position of $k = 0$ or $k = \pi/a$ at either the top or bottom of the band depends on the type of orbital involved. For gerade orbitals such as *s*, $k = 0$ will be the most stable combination, and $k = \pi/a$ will be the highest in energy. For a string of *p*-orbitals (ungerade) oriented along the string, $k = \pi/a$ (phases alternating) will be the totally bonding combination and $k = 0$ (all phases the same) will be the totally antibonding combination.

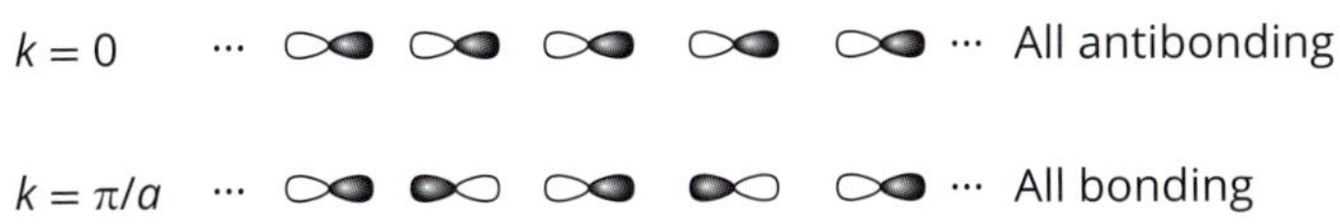

If one plots the energy of the molecular orbitals in the string versus the symmetry parameter k for a string of *s*-orbitals, the plot in Fig. 19.11 is obtained. Again, this large group of orbitals is practically continuous (but is in fact quantized) in energy from $k = 0$ to $k = \pi/a$, and is often viewed as a band of orbitals as shown on the left.

Every band will have a discrete beginning and end in energy at $k = 0$ and $k = \pi/a$, or vice versa. The difference in energy between these two extremes depends on the strength of the interactions between the atomic orbitals. If there is a strong interaction, the band may be very tall. If the orbitals interact poorly, the orbitals may be very isolated, making a very narrow band that at the extreme may be discrete orbitals on specific atoms with a single energy for the orbitals.

For example, if instead of looking at the 2*s*-valence orbital on lithium in our string, consider what the molecular orbitals would look like if we considered the core 1*s*-orbitals on each lithium. Because they are core orbitals, the interaction between each atom through this orbital will be extremely small, leading to a very narrow band compared to that made from the 2*s*-valence orbitals on each lithium. Naturally, the core orbitals are filled with electrons (white), while the valence 2*s*-orbitals are half-filled (Fig. 19.12).

The difference in energy between the $k = 0$ and $k = \pi/a$ molecular orbitals is called the "bandwidth", or sometimes "band dispersion". Consequently, the bandwidth is related to the difference in energy between the fully bonding MO and the fully antibonding MO; in other words, the bandwidth is related to orbital splitting in the MO. Just as in molecular systems, the

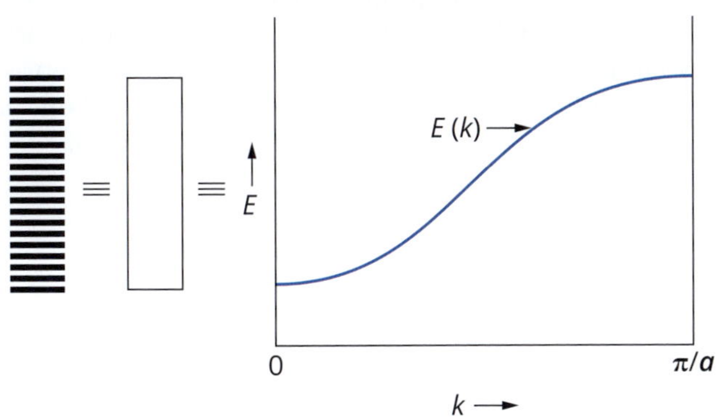

Fig. 19.11 As the number of orbitals of the same type increases (*left*) an essentially continuous band is created. (*right*) A plot of energy versus *k*, where *k* is a measure of the symmetry of the orbitals. In this case, the most bonding orbitals are on the left and least on the right. As the covalency in the system increases, the energy over which the band runs will increase. (Adopted from Hoffmann.[8])

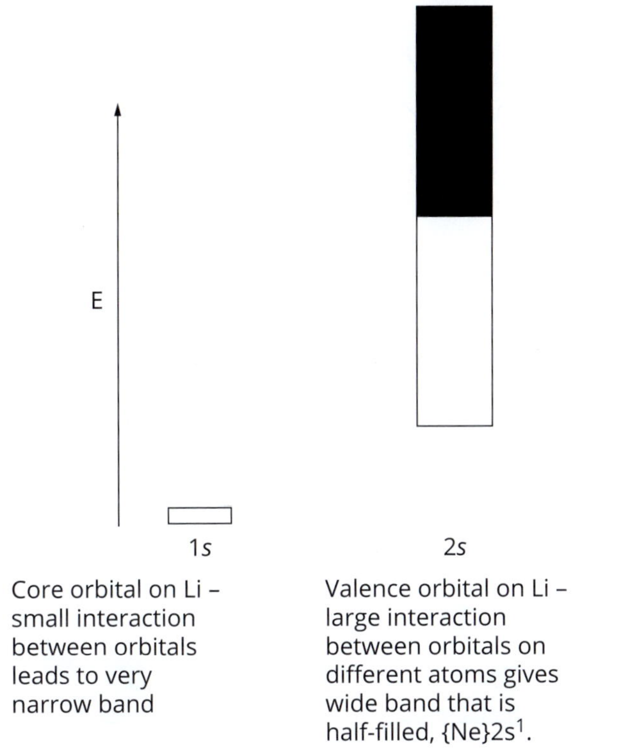

Fig. 19.12 Using the lithium atom as an illustration of the effect of orbital interactions on bands. The core 1s orbitals in a crystal of lithium interact very poorly with one another, which results in a very narrow band. The valence 2*s*-orbitals, which can interact, will have a broader, half-filled band.

bandwidth is then related to the energy difference between the interacting orbitals and the overlap, with the interaction being stronger when the energies are closer together and when the overlap is higher.

Naturally, these results can be extrapolated to a 3-dimensional framework, but the bands need not be the same in the *x*-, *y*-, and *z*-directions in the material. So how is this handled? The answer is, using symmetry. However, the exact details are beyond the scope of a general inorganic text, but can be found in more specialized texts such as Hoffman's excellent book on solid-state electronic structure.[8] In short, the symmetry of the system can be used to define axes that are related to the electronic structure, and this can be plotted as energy versus the vector k in the Bloch equation from $k = 0$ to $k = \pi/a$, just as before.

While the orbitals collect into bands, the number of levels in the band is not uniform throughout, and there are more energy levels of some energies than others. As a result, it is sometimes useful to think of a *density of states* (DOS), which is defined as the number of energy levels between some energy E and *d*E. For the 2*s*-band of lithium atoms discussed previously, the DOS plot takes the shape shown in Fig. 19.13.

For a 1-, 2-, and 3-dimensional solid, the shape of the DOS plot is as shown in Fig. 19.14. The energy labeled E_F is called the *Fermi level*, which in a metal is the maximum position of the band occupied at 0 K. In other words, if there is no thermal

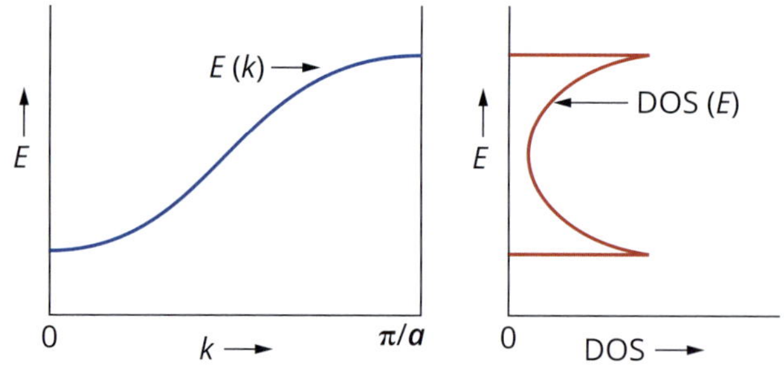

Fig. 19.13 Relationship between the E versus *k* plot and an E versus density of states (DOS) plot. (Adopted from Hoffmann.[8])

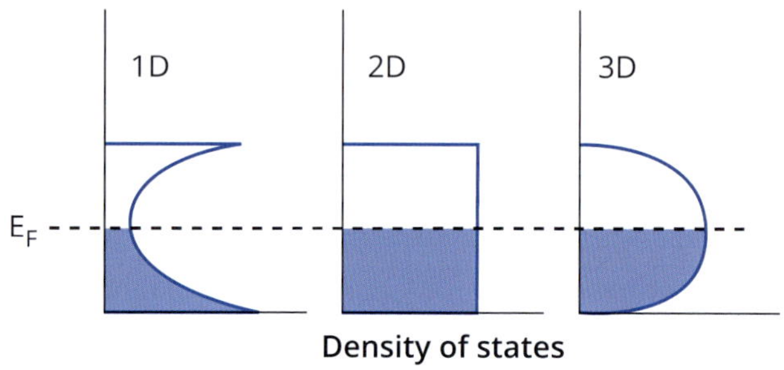

Fig. 19.14 Shapes of 1-, 2-, and 3-dimensional E versus DOS plots. (Adopted from Hoffmann.[8])

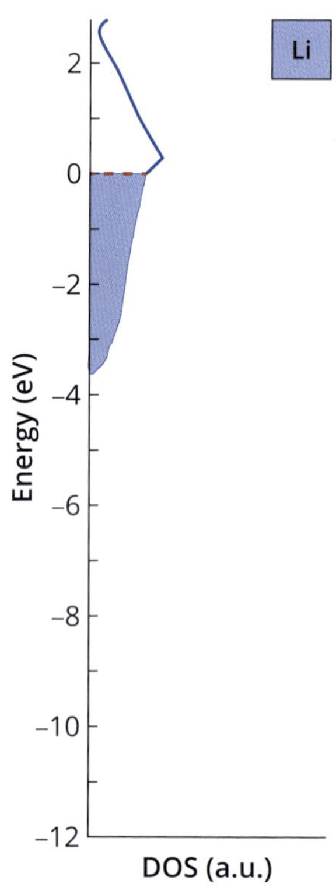

Fig. 19.15 Density of states plot for lithium metal, showing the valence band. The Fermi level of lithium is set to 0 eV.[9]

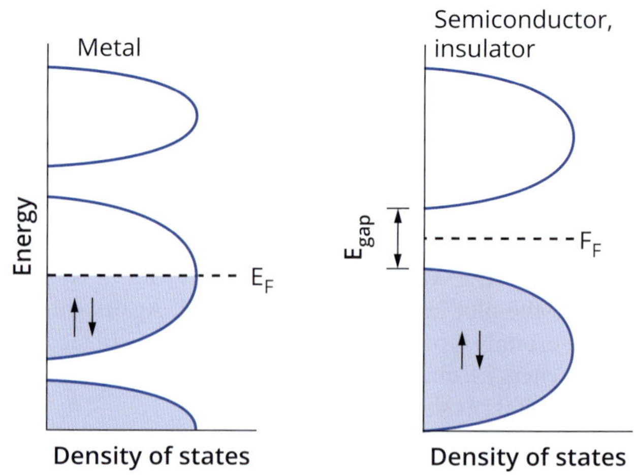

Fig. 19.16 In a conductor, the Fermi level (E_F) is typically within a band. In a semiconductor or insulator, the Fermi level will reside between the valence and conduction bands in the band gap.

promotion of electrons, it is the top MO of the solid that is occupied in a metallic system.

The DOS plot for Li metal is shown in Fig. 19.15. The plot shows both the 2*s* and 2*p* levels of Li, which happen to overlap. In a metallic system, the Fermi level will cut through a band, as shown for solid lithium.

The IUPAC definition of the Fermi level is: "The chemical potential of electrons in a solid (metals, semiconductors, or insulators) or in an electrolyte solution"; but perhaps better is the more physical definition as "the level in the material where the probability of finding an electron is 50%." In all definitions, the Fermi level is associated with the electrons at or near the highest occupied level in the material.

A metal (or generally any conductor) is a material in which the Fermi level cuts through a band and an infinitesimal amount of energy is required for electrons to move from filled orbitals into unfilled orbitals where they can move rapidly through the material. An insulator has a wide "band gap" (gap between filled bands and unfilled bands), with the Fermi level between (Fig. 19.16). A semiconductor has a small gap between filled and unfilled orbitals that is accessible under certain conditions, such as in the presence of an electric potential.

19.4 Phase Changes and Electronic Structure

A phase diagram shows changes in properties of gas, liquid, or solid as a function of conditions, such as pressure or temperature. Obviously, the structure of a solid may change with alteration of pressure and temperature, the most dramatic of which may be melting or sublimation. However, other changes are possible, including structural changes in the solid that may profoundly affect the properties, such as conductivity.

The phase diagram for tin is an interesting case. The tin metal with which we are familiar near room temperature and pressure is tetragonal white tin, which is also known as β-tin (Fig. 19.17). This allotrope is metallic. At low pressures there is also a cubic form, α-tin, that is nonmetallic and is a semiconductor with the diamond structure similar to its lighter congener carbon (Chapter 9). While the β-form is malleable and ductile, the α-form is brittle. Tin has a relatively low melting point, the lowest in Group 14, of 232 °C at 1 atm, which can also be seen in the diagram as the line separating the two different tetragonal phases and liquid. The structural change from tetragonal (white tin) to cubic (diamond) in the system profoundly impacts the physical properties, color, hardness, conductivity, and so on.

At cold temperatures, β-tin (white tin) can spontaneously transform into α-tin (cubic), which occurs around 13 °C (56 °F). Tin can be alloyed with other metals to lower the transition temperature, and for this reason commercial-grade tin often has small amounts of Bi, Sb, Pb, or Ag impurities.

In some cases this phase-change in tin has been ascribed historical significance. Some postulate that the tin buttons

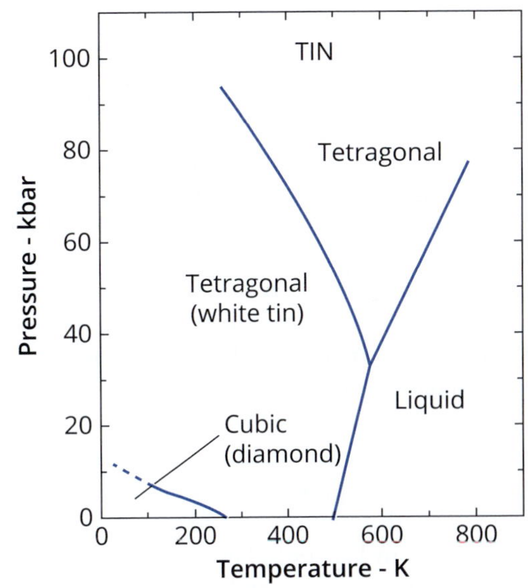

Fig. 19.17 Phase diagram for tin.[10]

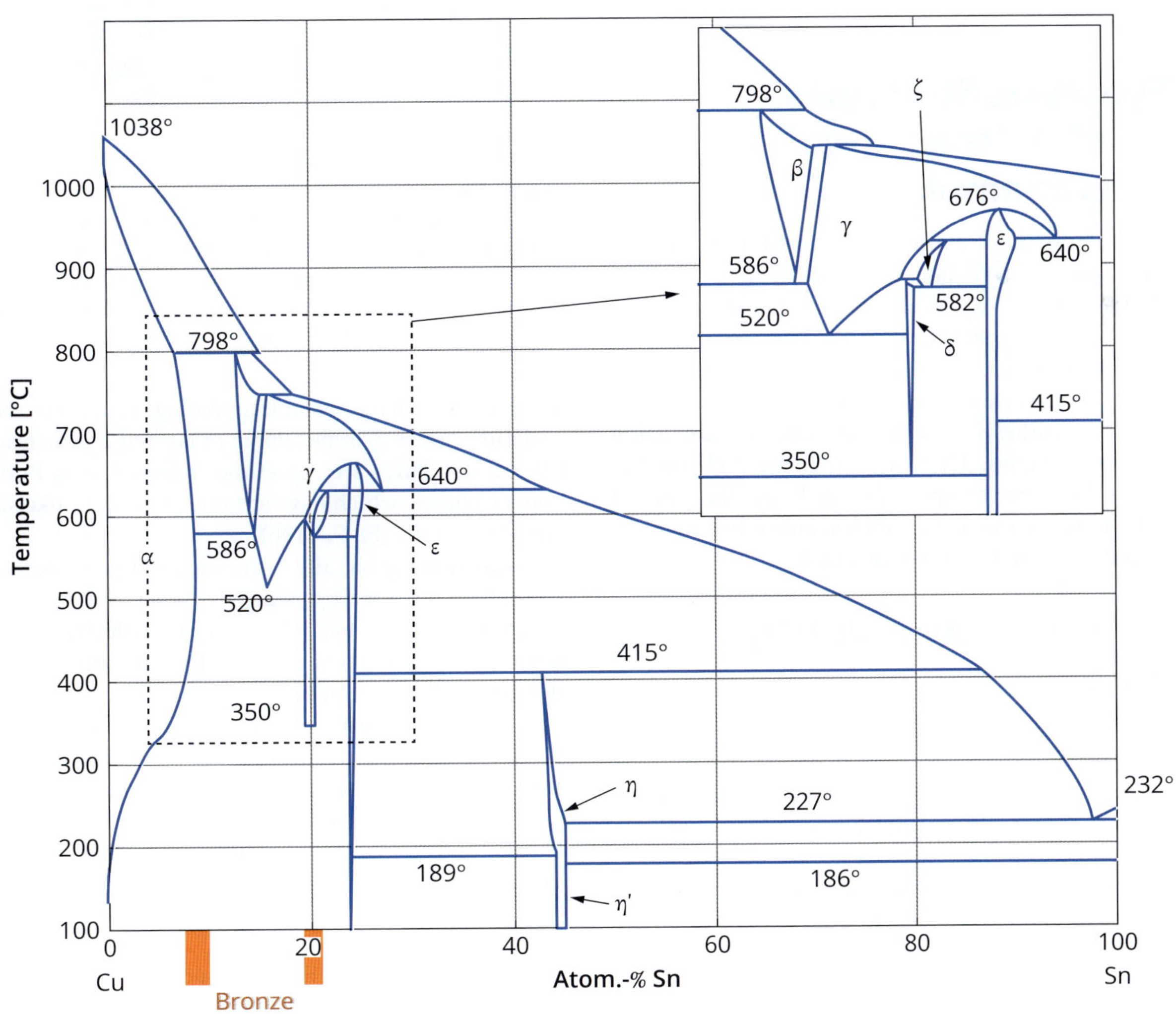

Fig. 19.18 Phase diagram for Cu/Sn alloys, showing the typical composition of bronze.[11]

used in French uniforms may have contributed to Napoleon's eventual defeat by the Russians. While the French won the Battle of Maloyaroslavets (October 24, 1812), their ranks were drastically reduced. It has been postulated that one major problem for the French may have been their tin buttons and the cold weather. The tin phase change can occur as high as 13 °C (depending on the purity of the tin), and the Russian winter temperatures can reach −30 °C. Consequently, all the French soldiers may have been suffering from wardrobe malfunctions on a massive scale as their buttons turned to dust. This degradation of tin with cold temperatures is sufficiently well known to have a common name: tin blight.

While the phase diagram as a pressure-versus-temperature plot, such as that for tin, is a common one, other types of phase diagrams are often employed. Also common are phase diagrams that vary with composition and a property, such as temperature. The one in Fig. 19.18 has composition of tin versus copper as the *x*-axis and temperature as the *y*-axis. Bronze is typically about 12% tin with the majority made up of copper. Different types of bronzes, with various properties, are made by adding a variety of different metals and metalloids such as aluminum, nickel, zinc, and arsenic.

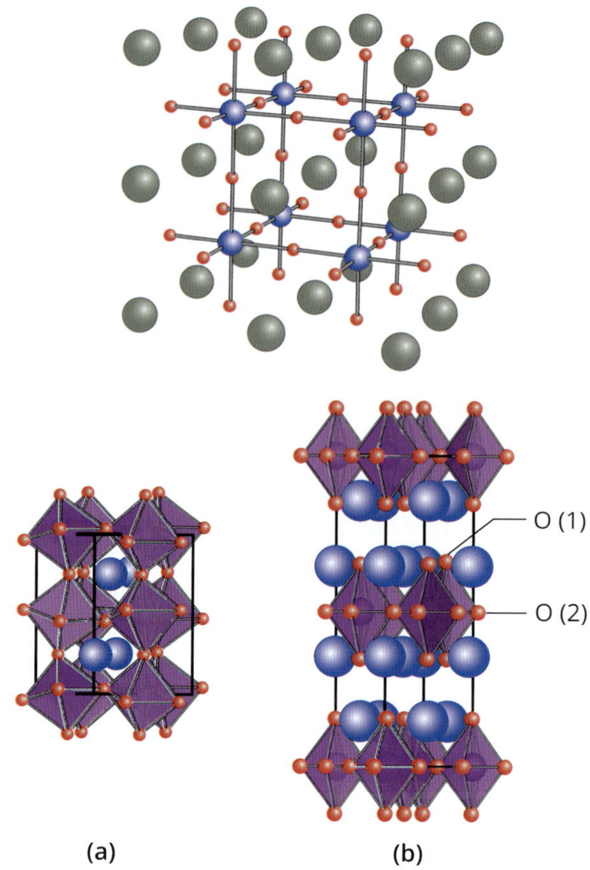

Fig. 19.19 (*top*) Structure of $CaTiO_3$, perovskite, showing the Ti (blue (medium)), O (red (small)), and Ca (light green (large)). (*bottom*) (a) A polyhedral representation of the perovskite structure for an ABO_3 system, showing cations A between BO_6 octahedra. (b) A Ruddlesden–Popper phase with a rock-salt-like O(1) and an equatorial perovskite-like site O(2).[13]

19.5 Synthetic Strategies for the Solid-State: Metal Oxides as Examples

Oxides constitute, perhaps at once, the most important and best-studied class of solid-state materials. Binary, ternary, and even more complex oxides are important for their electronic, magnetic, and physical properties. They are found everywhere, from electronic inductors to coatings on stealth aircraft. In this section we will briefly discuss a few common methods for the synthesis of solid-state compounds, using metal oxides as examples. There are other methods for the synthesis of solid-state compounds, such as the sol–gel method, but the following few examples suffice to provide a feel for the challenges and solutions available.

19.5.1 Direct Reaction: $PbTiO_3$ as a Case Study[12]

The simplest method for the synthesis of metal oxides involves taking commercially available metal oxides, mixing them in the desired proportions, and then heating them until a new phase is prepared. For example, consider the synthesis of lead titanate, $PbTiO_3$.

This lead titanate ceramic has a perovskite structure. Perovskite itself is a mineral of composition $CaTiO_3$; however, the perovskite structure type is a very common arrangement of solid-state compounds with the general formula ABE_3, where the usually metallic A and B bind to the E anions. The structure could be described as having octahedral B sites bound to six E with charge balancing A (Fig. 19.19, top and a). The structure can be 3-dimensional or layered 2-dimensional. The layered structures are sometimes referred to as Ruddlesden–Popper phases, which have anionic perovskite-like slabs separated by cations (Fig. 19.19b).

Lead titanate has the 3-dimensional perovskite structure and can be prepared by using traditional direct methods. The compound $PbTiO_3$ has interesting ferroelectric behavior and is used as a component in capacitors, thermistors, and optoelectronics. The compound can be synthesized from titanium dioxide and lead oxide (Eq. 19.12a). The oxide starting materials are put together and ground. This may be done with a mortar and pestle, but today it is more often done with a ball mill.[c] The importance of the grinding cannot be overstated. The method counts on diffusion of ions, often ions with multiple charges, through solid media. The process is difficult, and

[c] A ball mill is very much what it sounds like—a grinder that uses a group of balls that are in a rotated container to reduce the particle sizes in powders. The powder is ground by the action of the balls against the walls of the container and each other.

the diffusion lengths can be quite short. Consequently, the grinding is required to produce very small particles and excellent mixing of the starting materials.[14]

After grinding, the sample is often pressed into a pellet to limit the free space in the sample and lower diffusion distances. The reaction is heated to 800 °C for 2 hours, which is not a particularly high temperature for a solid-state reaction, as these often require >1,000 °C. After heating, the sample is ground again in the ball mill, reformed into a pellet, and heated again.[14] The reaction may be followed by powder X-ray diffraction, which produces a series of peaks characteristic of the compounds present. Consequently, X-ray diffraction can be used to determine whether the sample is fully converted to $PbTiO_3$ or if it still contains TiO_2 and PbO starting materials. If starting materials still remain, the mixture may be ground and heated any number of times until homogeneous. Once the $PbTiO_3$ is formed, the sample then may be sintered to cause the small crystallites to form larger crystals.

$$TiO_2 + PbO \xrightarrow{\substack{(1)\ \text{ball mill} \\ (2)\ 800\ ^\circ\text{C},\ 2\ \text{h} \\ (3)\ \text{ball mill} \\ (4)\ 800\ ^\circ\text{C},\ 2\ \text{h}}} PbTiO_3 \tag{19.12a}$$

Fig. 19.20 shows the powder X-ray diffraction pattern for pure $PbTiO_3$ as a function of sintering temperature.[12] The numbers on the peaks, such as (001) and (201), are the Miller indices, which correspond to planes through the crystal(s). For example, the (100) (read "one zero zero") index is coincident with face A of the unit cell. Indices like (110) run diagonally through the unit cells. The light can be thought of as diffracting off the electron density in these planes to give the peaks in the powder diffraction pattern. The angle of diffraction from the center of the beam and relative intensities are indicative of a particular structure. Note the increasing intensity of the peaks and increased ease of measuring peaks at a large angle from the beam (2θ) after sintering at higher temperatures, which increases the crystallinity of the sample.

While this brute-force method for making solid-state compounds is commonly employed, it has several drawbacks. Because of slow diffusion and repeated manipulations, the method can be expensive and time-consuming. In some cases, an incomplete reaction may result, giving an inhomogeneous product. Furthermore, there is little chance of making compounds that are not thermodynamic products, and interesting kinetic phases may be missed at the high temperatures required. The method is most likely to produce very fine powders, which then may have to be reprocessed to give the correct properties for many applications.

19.5.2 Precursor Method to Prepare $NiFe_2O_4$

The greatest difficulty with the direct reaction method is that the reactants cannot be mixed at the atomic level, and diffusion of the ions requires very high temperatures. This can only be combated with physical mixing and grinding to very small particle sizes, followed by hard packing of the starting materials into a pellet. An alternative is to make a molecular precursor with the correct composition, or nearly so, which would naturally produce an atomically mixed system from which to make the solid-state compound of interest. Ideally, all the other pieces of the molecule not found in the composition of the solid-state compound would volatilize away during heating.

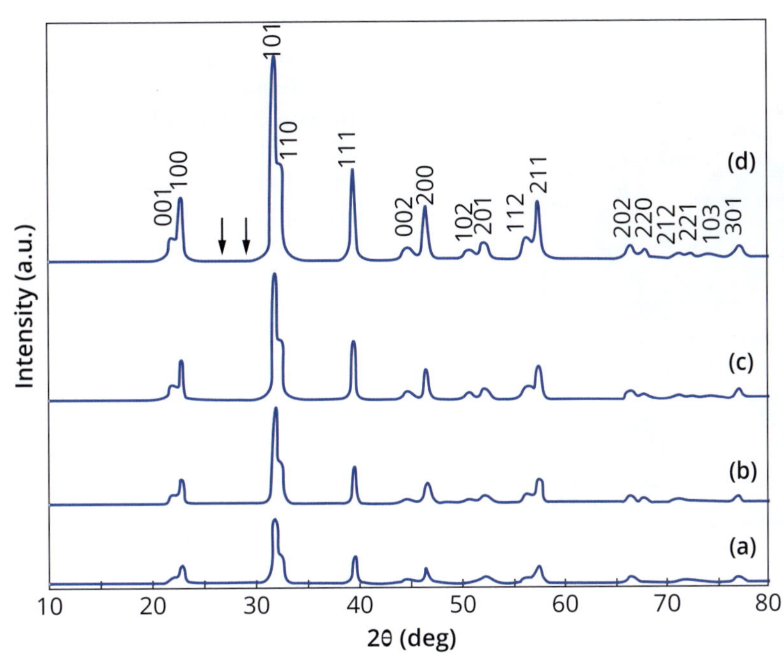

Fig. 19.20 Powder X-ray diffraction of samples of $PbTiO_4$ sintered at (a) 700 °C, (b) 750 °C, (c) 800 °C, and (d) 900 °C.[12]

This precursor method has been used to great effect in the production of ferrites of formula MFe_2O_4, where M = Mn^{2+}, Co^{2+}, Ni^{2+}, and Zn^{2+}. The strategy is to produce a "double acetate" in which two different metal centers are supported in the same acetate salt. In the case of the nickel salt, for example, a complex of formula $Fe_9Ni_4(OAc)_{26}(OH)_9{\cdot}23\ H_2O$ is prepared as the precursor to the solid-state complex by using the procedure of Weinland and Holtmeier.[15] This complex is prepared using the following reaction.

$$\begin{aligned}&Fe(NO_3)_2{\cdot}6H_2O\\&+5\ \text{equiv}\ Ni(NO_3)_2{\cdot}6H_2O \xrightarrow[H_2O]{} Fe_9Ni_4(OAc)_{19}(OH)_9{\cdot}23H_2O\\&+15.5\ \text{equiv}\ NaOAc{\cdot}3H_2O\end{aligned} \quad \textbf{19.12b}$$

This acetate salt was treated with hot pyridine, which liberates the water and provides a pyridine (py) complex of formula $Ni_3Fe_6(OAc)_{17}O_3OH{\cdot}12py$, which was used as the precursor to the nickel ferrate by Wickham, Whipple, and Larson.[16] The complex was slowly heated to begin, driving off the ligands without liquefying the complex, then the powder was heated to 1,000 °C in air. This provided a sample of $NiFe_2O_4$ with composition Fe 47.54 ± 0.03% and Ni 25.03 ± 0.01%—a mole ratio of Fe/Ni of 1.994 ± 0.004.

The ferrites to be synthesized are *spinels*—a class of minerals of general formula $A^{2+}B^{3+}{}_2O_4$. The mineral spinel itself is $MgAl_2O_4$, and the class includes many compounds including magnetite, which has the formula $Fe^{2+}Fe^{3+}{}_2O_4$ and is important for magnetic tapes, magnetic nanoparticles, and catalysis to produce ammonia from N_2 and H_2 (Haber–Bosch process, Chapter 1). The spinel structure, $MgAl_2O_4$, shown in Fig. 19.21, is comprised of tetrahedral Mg^{2+} sites (yellow) with octahedral Al^{3+} sites (blue) joined by oxygens (red). The nickel ferrite discussed previously has a cubic structure with *a* = 8.338, containing tetrahedral nickel sites with octahedral iron(III).[16]

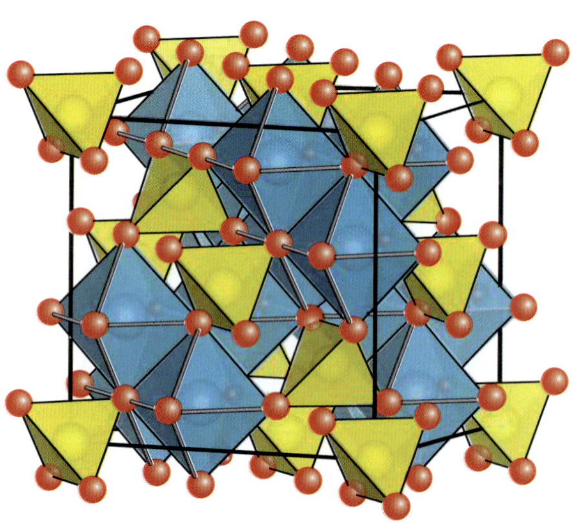

Fig. 19.21 The structure of spinel, $MgAl_2O_4$, comprised of tetrahedral Mg^{2+} sites (yellow) with octahedral Al^{3+} sites (blue) joined by oxygens (red spheres).[17]

19.5.3 Flux Method for Formation of $BaFe_{12}O_{19}$

A flux method typically involves the use of a relatively low melting salt to act as a solvent (flux) for the solid-state reaction. In many cases, the flux is not inert and provides materially to the product. Typical fluxes for solid-state reactions are metal halides or hydroxides such as $BaCl_2$ and KOH.

This method was used to prepare, for example, ternary oxides by Brixner. He found that $BaCl_2$ could be used as the flux for the reaction of Fe_2O_3 and BaO, where the barium oxide was formed *in situ* by reaction of the flux with water (Eq. 19.13). The reaction was heated to 1,300 °C in a platinum boat. The water was introduced by bubbling moist air through the reaction through a small hole in the bottom of the reaction vessel. The excess $BaCl_2$ is washed away with water, and the synthesis yields the product as ~2-mm long dark red thin plates. The compounds have a hexagonal structure ($P6_3/mmc$) with *a* = 5.88 Å and *c* = 23.20 Å.[18]

$$BaCl_2 + Fe_2O_3 + H_2O \xrightarrow[\substack{1300\ ^\circ C\\ -HCl}]{BaCl_2\ \text{flux}} BaFe_{12}O_{19} \quad \textbf{19.13}$$

19.5.4 Hydrothermal Methods to Prepare Vanadium Jarosites

While fluxes are often salts to allow their use at the very high reaction temperatures necessary for the formation of some solid-state compounds, lower boiling solvents can also be used at higher pressures, with by far the most popular being water. Reportedly, the first to use such a synthesis was Karl Emil von Schafhautl in 1845; however, Bunsen prepared Ba and Sr carbonates at 200 °C and 15 atm in sealed glass tubes from water, which was reported in 1848 in some musings on the geology of Iceland.[19] As one might expect, there is a significant contingent of those using hydrothermal methods from the Earth-sciences field, interested in how minerals may behave and form under pressure in the presence of water. Naturally, the method may be used to generate new phases for a variety of different applications and is particularly amenable to formation of metastable phases that are inaccessible when using higher-temperature syntheses.

An interesting example involves the formation of jarosites. Jarosite is an iron(III) mineral of composition $KFe_3(OH)_6(SO_4)_2$, which is a fairly common, yellow-orange mineral. In 2004, the mineral was also discovered on Mars by the *Opportunity* rover, using Mössbauer spectroscopy.

In 2002, Nocera and Grohol reported the synthesis of a series of vanadium jarosites, $AV_3(OH)_6(SO_4)_2$ where A = Na, K, Rb, Tl, and NH_4. The synthesis involves the reduction of V(IV) to V(III) under hydrothermal conditions. VCl_4 was added to water and HCl to produce a solution of "$VOCl_2$". Systems like this containing OV^{2+}, in which the vanadium is in the +4 oxidation state, are called "vanadyls" and are very common for aqueous conditions of the element. In a Teflon-lined container, Na_2SO_4, NaOH, NaCl, and water were added to the vanadyl solution, and the vessel was sealed and heated to 200 °C for 6 days. It was then cooled slowly to form dark-red crystals of the vanadium jarosite (Eq. 19.14).[20]

$$OVCl_2 + Na_2SO_4 + NaCl \xrightarrow[200\,^\circ C]{H_2O} NaV_3(OH)_6(SO_4)_2 \quad \textbf{19.14}$$

Jarosites adopt a 2-dimensional kagome lattice (Fig. 19.22(a)) featuring triangles shown in yellow and hexagons shown in red. In the jarosite at the top, the triangles are formed by hydroxide-bridged vanadium centers, and above and below these planes of vanadium atoms are sodium and sulfate ions.

Jarosite systems are of interest for their electronic properties, and the triangular clusters in (Fig. 19.23) show spin frustration. In an equilateral triangle occupied at the vertices by a single electron spin, if the spins want to align antiferromagnetically, spin 1 can align with spin 2 to cancel the overall spin. Spin 3 becomes "frustrated", however. If spin 3 aligns antiferromagnetically with spin 2, it will be ferromagnetically

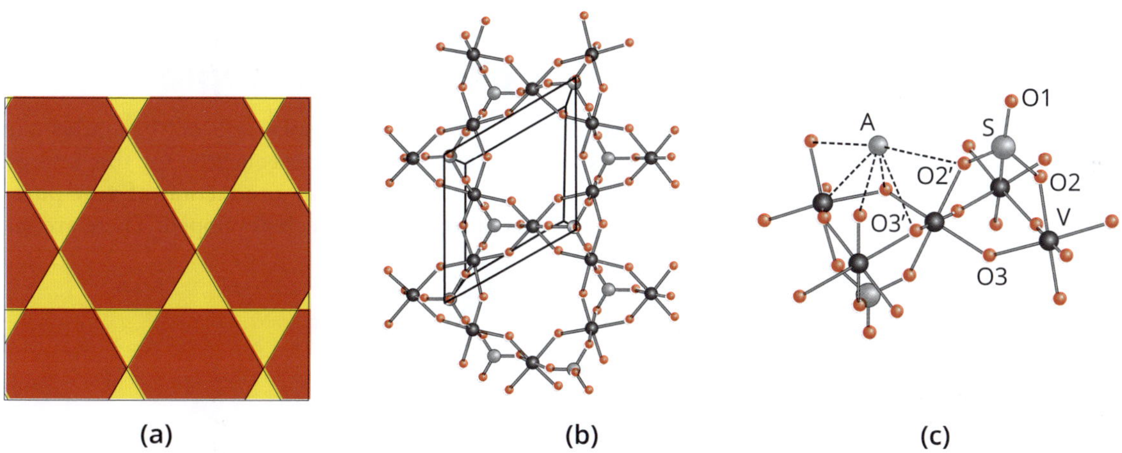

Fig. 19.22 (a) Kagomé lattice showing the triangles as yellow and hexagons as red. (b) The structure of vanadium jarosite $NaV_3(OH)_6(SO_4)_2$. The dark atoms of vanadium occupying the verticies of the triangles and hexagons are bridged by the white oxygen atoms. (c) A side-on view of the layer with OH (without their hydrogens) bridging vanadiums and the capping SO_4^{2-} and A^+ ions that separate the kagomé layers.[20–21]

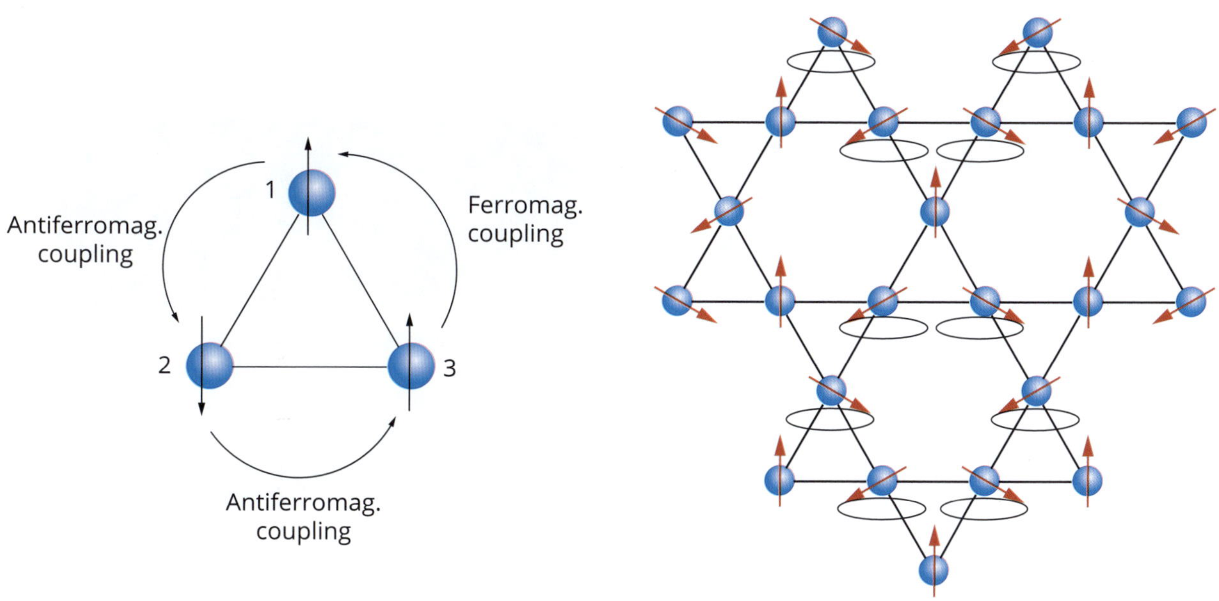

Fig. 19.23 (*left*) A spin-frustrated trigonal system. (*right*) One possible ground-state for a kagomé lattice. The dotted lines show that the spins can be rotated out of plane without changing the energy, which results in an infinite number of degenerate ground-states.[21]

aligned with spin 1, and vice versa. As a result, there is always a net electron spin in the system.

A kagomé lattice, which is comprised of similar triangular clusters, is a spin-frustrated system. In jarosite, the iron centers are high-spin d^5, and in the vanadium example the d^2 spins are examples of such magnetic systems with highly degenerate ground-states.

19.5.5 Case Study in Oxides Synthesis: $YBa_2Cu_3O_7$ (123 Superconductor)

In 1987, the first compound to exhibit superconductivity with a critical temperature above the boiling point of liquid nitrogen was reported by Maw-Keun Wu, Chu Ching-Wu, and coworkers. The compound has a defect perovskite structure and the formula $YBa_2Cu_3O_7$, sometimes referred to as "123-superconductor" or "YBCO". Superconductors, at temperatures below their critical temperature (T_c), conduct electricity without resistance and also display what is known as the *Meissner effect*, in which magnetic field lines are repelled by the material (Fig. 19.24).

This fascinating compound was originally prepared by using a mixture of carbonates and heating to 1,300 °C under O_2 (Eq. 19.15), but it has now been made by using a variety of different routes. The synthesis of the compound is sufficiently simple that the preparation has made its way into undergraduate classrooms, sometimes with observation of the Meissner effect in your material affecting your grade!

$$4\,BaCO_3 + Y_2(CO_3)_3 + 6\,CuCO_3 + \tfrac{1}{2}O_2 \longrightarrow 2\,YBa_2Cu_3O_7 + 13\,CO_2 \qquad 19.15$$

19.6 Crystallization, Impurities, and Doping: Silicon-Based Solar Cells as an Example

By far the most used solar cells on the market currently begin with pure silicon. Consequently, for this and other applications, methods for the production of very pure single crystals of silicon have been developed. There are many different

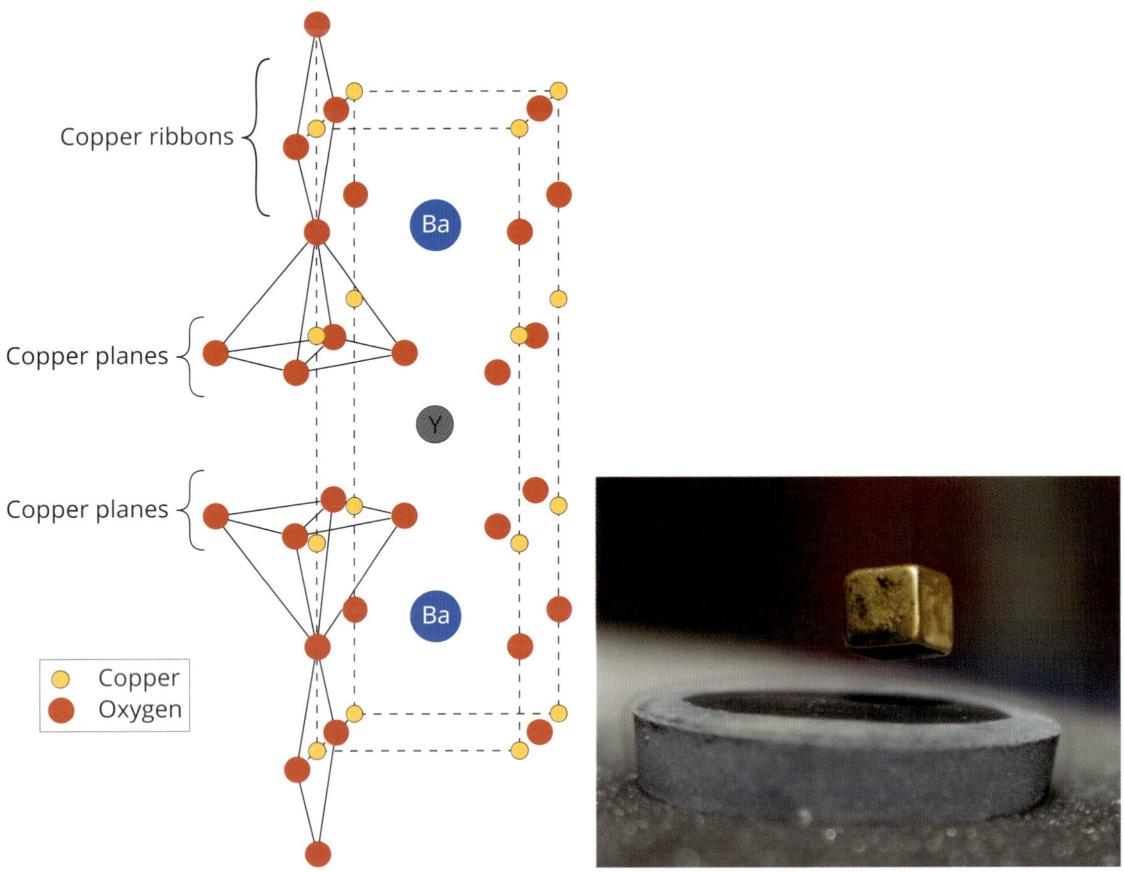

Fig. 19.24 (*left*) The structure of $YBa_2Cu_3O_7$. A small cube-shaped magnet levitating above a black disk of $YBa_2Cu_3O_7$ cooled by liquid nitrogen. The superconductor repels magnetic field lines, canceling the effect of gravity on the small magnet and causing it to levitate—an example of the Meissner effect.[22] Note that some students, on seeing a picture of this type, assume that the metal cube is "jumping" off the material, but the small cubic magnet is actually statically levitating and will do so as long as the material is superconducting (the temperature of the black disc is below T_c).

methods for crystallization of solid-state materials, and one that is very effective for silicon is the Czochroalski process.

The Czochroalski process simply involves creating a pool of molten silicon at about 1,500 °C, and a seed crystal is mounted on the end of a rod and dipped just below the surface of the liquid silicon. As the seed crystal is slowly lifted, it is rotated, and a single crystal of silicon is pulled from the pool. These massive, cylindrical single crystals may be a couple of meters in length and hundreds of millimeters in diameter, and one such ingot is shown in Fig. 19.25. These ingots are then typically sliced into wafers, which are polished flat for use in integrated circuits but are textured for solar cells in which a higher surface area is an advantage.

Fig. 19.25 An ingot of silicon purified with the Czochroalski process.

Solar cells convert light to electric current by using the energy in the photon to create charge separation. An electron is excited from a valence band containing electrons, creating a "hole" in the band (unbalanced positive charge of a nucleus) and an electron in the conduction band. The energy of the photon, therefore, goes to Coulombic potential in separating these charges. The hole and electron must then make their way to electrodes, and the electrons move through a wire back to the holes for recombination. In the process, they create an electric current. For this scheme to work well, the electrons and holes need an energetically favorable direction that draws them away from each other and towards their respective electrodes. The most common device for doing this is a p-n junction.

A p-type material is one that is relatively positive, and an n-type material is one that is relatively negative. In other words, the p-side has an excess of holes, and the n-side an excess of electrons. The p-n junction is created by "doping", adding specific impurities, to the silicon. A typical n-dopant for silicon is phosphorus, which has one additional electron over silicon; a typical p-dopant for silicon is boron, which has one fewer valence electron than silicon. These are typically diffused into either side of a single crystal of silicon to create the junction rather than making the doped material and creating the junction mechanically, as mechanical junctions create grain boundaries that reduce electron flow between the sides. Doping of silicon with phosphorus, for example, may be done with gas phase $POCl_3$ at temperatures of 800–900 °C with a nitrogen carrier gas.[23] A highly doped semiconductor may contain one dopant for every 1,000 silicon atoms, while low doping may involve one dopant for every billion silicon atoms.

Electronically, n-doped silicon can be viewed as in Fig. 19.26 (left). The electron-rich doping sites create donor levels higher in energy than the valence band of the silicon. This lowers the energy required to move an electron into the conduction band. Conversely, p-doping creates acceptor energy levels nearer in energy to the valence band (right).

A typical device includes several layers, as shown at the bottom of Fig. 19.26. A metal contact on the top is often made by silver "fingers" to collect the electrons but must allow most of the light through. A topcoat of Si_3N_4 may be used as an antireflection coating. Below this is an n-doped silicon on top of p-doped silicon. All of this is then backed with a metal contact, often aluminum.

19.7 Batteries and Solid-State Chemistry: Lithium-Ion Batteries as an Example[24–26]

Batteries are one of the most important commercial applications of advanced oxide materials, and this author currently sits working on a laptop powered by a Li-ion battery. In all likelihood, the reader has devices in their possession, if not on their person, that use these batteries, like most cellular

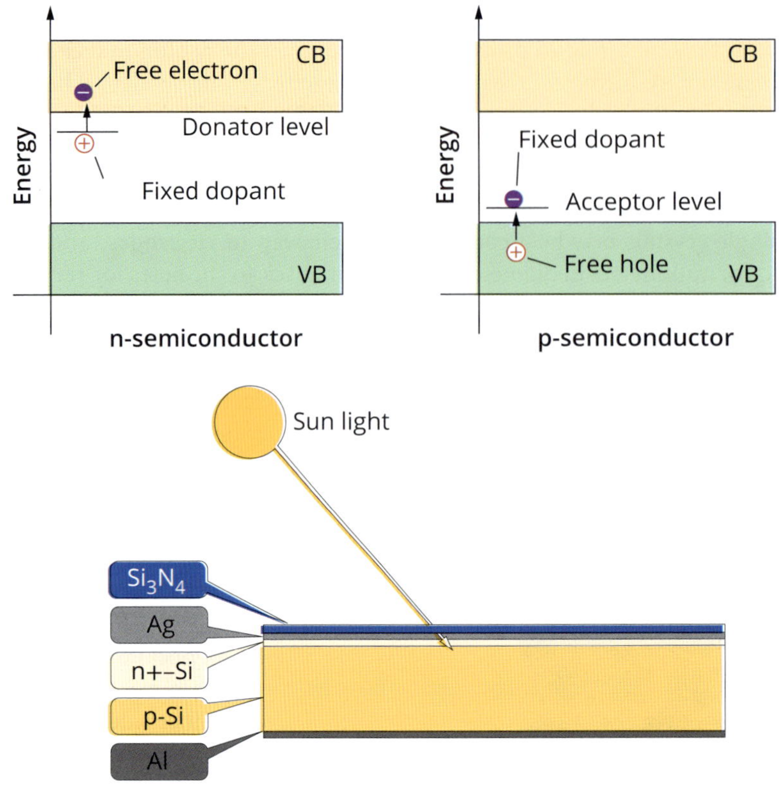

Fig. 19.26 (*top left*) n-doping creates electron-rich sites with energies above the valence band (VB) of silicon. This lowers the energy required to move an electron into the conduction band (CB). (*top right*) p-doping creates holes, electron-deficient sites, near the VB. (*bottom*) A typical silicon solar cell device with Si_3N_4 antireflection coating, silver "fingers" to collect electrons on the front face of the cell, n-doped silicon, p-doped silicon, and a back plate of aluminum.[23]

phones. Here we will briefly discuss the materials in some of the early lithium batteries, which are still some of the most common designs, although there are alternatives in commercial use as well.[27] Here we will discuss a few of the design features of this battery and some of the technological issues for production. Batteries are not terribly different from the Galvanic cell discussed in Section 13.2.2, and if necessary this section could be reviewed.

For any battery, the main concerns, and what sets the potential and amount of energy that can be stored, is the makeup of the cathode and anode. Lithium-ion batteries have a combination of (relatively) high energy and high power density that are currently unmatched for any system in large-scale production.[24] However, there are battery types that outperform them that may reach the market soon, such as lithium–air batteries.[28] While multivalent cations would allow for higher charge capacity per ion, the added charge reduces the mobility of the ions significantly making them more difficult to use in battery applications, although there are some applications using multivalent cations such as zinc–air batteries for hearing aids. Mobility is a (if not the) key issue in discharging and charging of the battery, so smaller monocationic materials are generally the norm. Furthermore, sodium-ion batteries have tremendous potential, both economically and technologically, and are under development.

For the lithium-ion battery in wide use today, lithium metal itself as the anode material would be the most weight-efficient design, but unfortunately this is technically not feasible, as lithium tends to crystallize into dendrites that short out the system. Instead, the anode is coated in graphite. The lithium atoms intercalate into the layers of graphite (Section 9.2). The intercalation is kinetically favorable and low-energy. The highest formula that is practical for storage of lithium in graphite has the formula LiC_6.

For the cathode, there are several metal oxides that can be used with one of the most common still being some materials proposed by Goodenough, such as $LiCoO_2$ and $LiNiO_2$. The cobalt octahedra are edge-shared into sheets separated by the lithium cations (Fig. 19.27). As the lithium is reduced on recharging, it is expelled from the $LiCoO_2$ with concomitant oxidation of the CoO-layers of the material for charge balance. As is required, the structure is relatively stable to removal and reinsertion of the lithium ions.

An electrolyte solution is necessary to complete the circuit. Goodenough used $LiBF_4$, but $LiPF_6$ is perhaps more common with a solvent like propylene carbonate.

A cross-section view of the battery that was used in Apple's iPhone 6 is shown in Fig. 19.28. To save space, the battery is composed of many equivalent layers; however, the materials are more or less as Goodenough and coworkers described, using $LiCoO_2$ at the cathode and graphite at the anode.

In 2019, John Goodenough, M. Stanley Whittingham, and Akira Yoshino shared the Nobel Prize in Chemistry for the

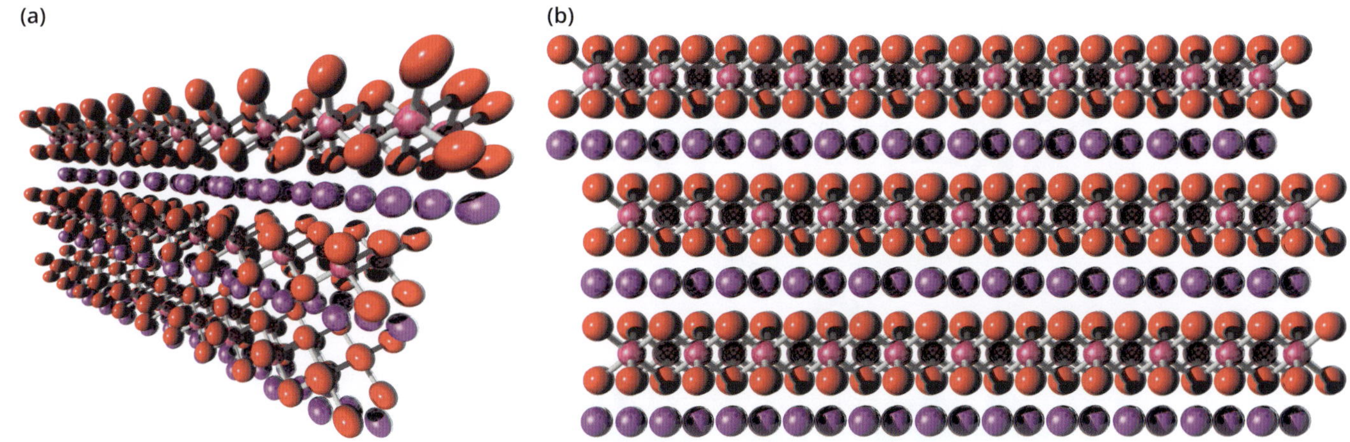

Fig. 19.27. Structure of $LiCoO_2$ showing the cobalt as purple octahedral centers, oxygen as red vertices, and lithium ions as violet spheres separating the layers.

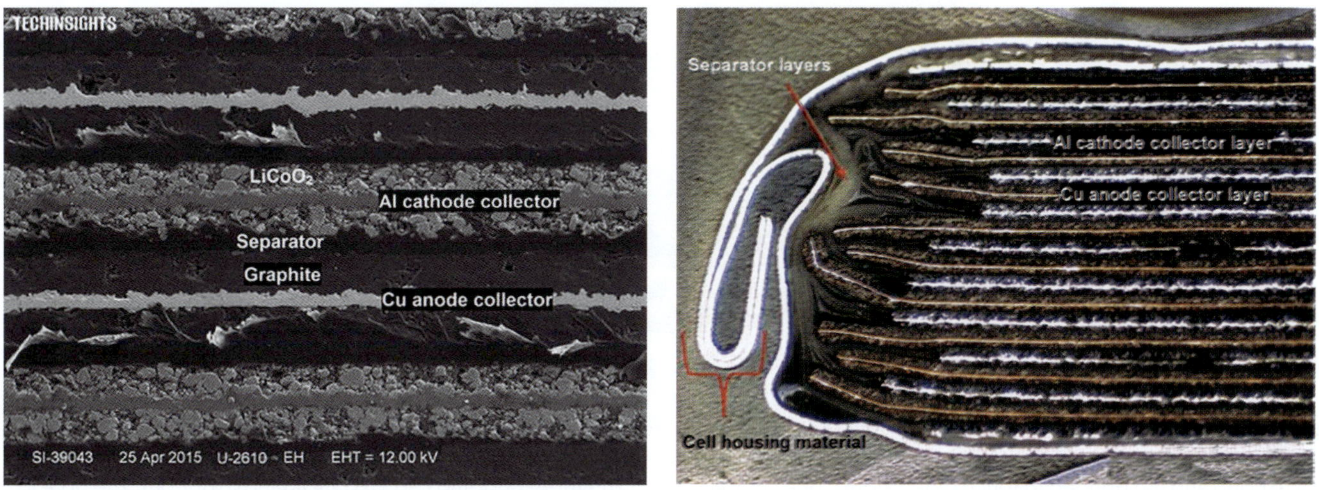

Fig. 19.28 (*left*) Cross-section of the battery in Apple's iPhone 6 showing the Cu anode and Al cathode conductors. In the SEM (*right*) one can see the graphite of the anode covering the copper conductor and the $LiCoO_2$ on the aluminum cathode collector.[29]

development of the lithium-ion battery. Goodenough developed the $LiCoO_2$, usually with small amounts of tin, used in the cathode. Yoshino made the batteries safer and developed tests to ensure safety. Whittingham developed the electrochemical intercalation of lithium into graphite to make a safe anode. Goodenough was 97 years of age, the oldest to ever receive the prize.

19.8 **Metal–Organic Frameworks (MOF)**

It is common in heterogeneous catalysis to use a porous material, often made of silica or alumina, as the support material for catalytic sites. These may be small-pore zeolites, such as the material ZSM-5, or mesoporous materials, often prepared using sol–gel conditions with various surfactants.[30] These materials offer great thermal stability and extraordinary surface areas, which give faster catalysis. These are by far the most common materials for supporting catalysts in industrial reactions, but are difficult to study because multiple catalyst-site types are the norm in these amorphous materials. While experienced industrial chemists can often make an educated guess as to the best catalyst composition for a transformation, the catalysts are optimized largely by trial and error.

Easier-to-study and control materials with high surface areas have been developed in the form of metal–organic frameworks (MOF). One of the best-studied materials in this class is MOF-5 reported by Yaghi and coworkers in 1999.[31] MOF-5 has a composition of $Zn_4O(BDC)_3$, where BDC = 1,4-benzenedicarboxylate. The structure was made by addition of trimethylamine to a solution of $Zn(NO_3)_2$ and H_2BDC with a small amount of hydrogen peroxide as O^{2-} donor. The joints of the structure have tetrahedrally disposed zinc atoms all bound to a central oxygen. Each pair of zinc atoms is bridged by a

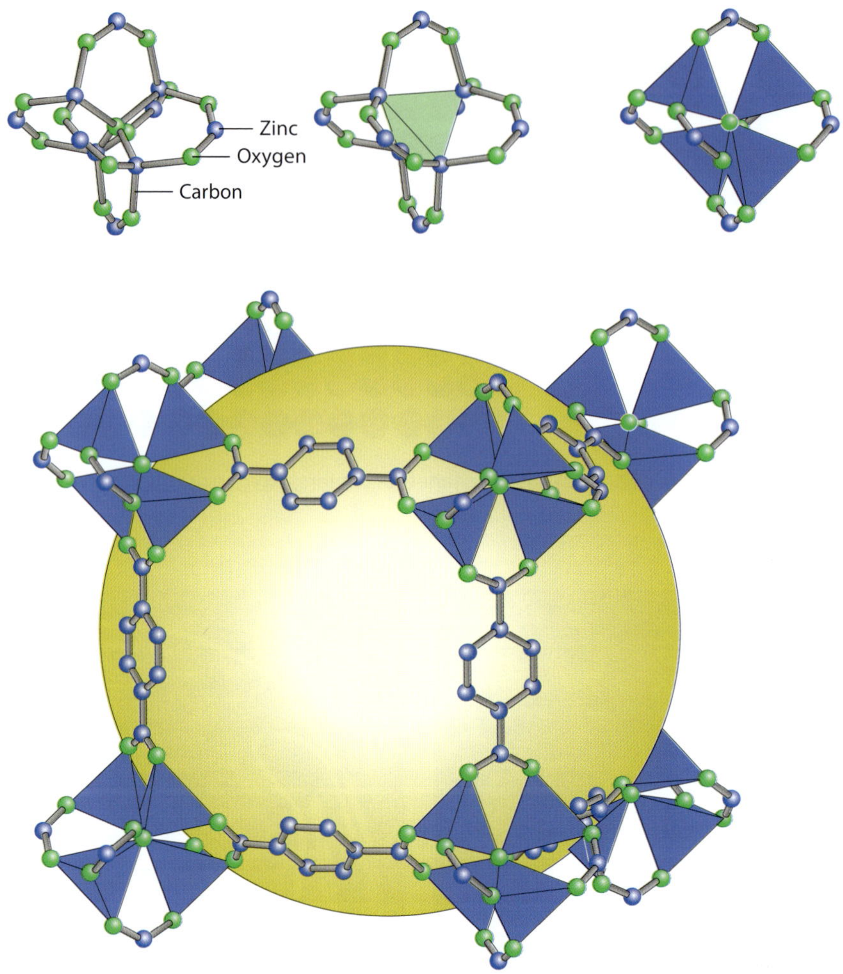

Fig. 19.29 Structure of MOF-5. (*top*) On the left is the cluster with zinc (blue), oxygen (green), and carbon (grey). The zinc atoms are tetrahedrally disposed around a central oxygen (green tetrahedron in the center top). The zinc atoms are tetrahedral (blue tetrahedron on the top right) and are bonded to four oxygen atoms: one central O atom and three bridging carboxylate ligands. (*bottom*) The pore in the center of the bridged zinc clusters is shown as a yellow sphere. (Figure adopted from the paper by Yaghi and coworkers.[31])

carboxylate arm (the organic linkers are often called "struts") that stretches to the next zinc cluster in the material. This produces a pocket, shown as a yellow sphere of diameter 18.5 Å in Fig. 19.29.

These materials have applications in areas such as gas storage, gas purification, and in catalysis.[32–33] One can make numerous modifications to the struts, change the structures of the bridging clusters, and so on.

Furthermore, it has been found that the metal centers in the clusters can be readily exchanged in some cases with other metals for applications in catalysis. This strategy allows the overall MOF synthesis to produce a known, stable structure and then be modified afterwards, while retaining the same overall arrangement. Dinca and Brozek have achieved these metal exchanges with MOF-5, for example.[34] Using this strategy, MOF-5 was modified to contain in place of a zinc atom: ClTi, V, ClV, Cr, ClCr, Mn, and Fe. This dramatically changes the properties of the MOF, while keeping the overall structure the same (Fig. 19.30). The syntheses of the complexes were, more or less, as simple as treating MOF-5 with various transition-metal chlorides in solvent.

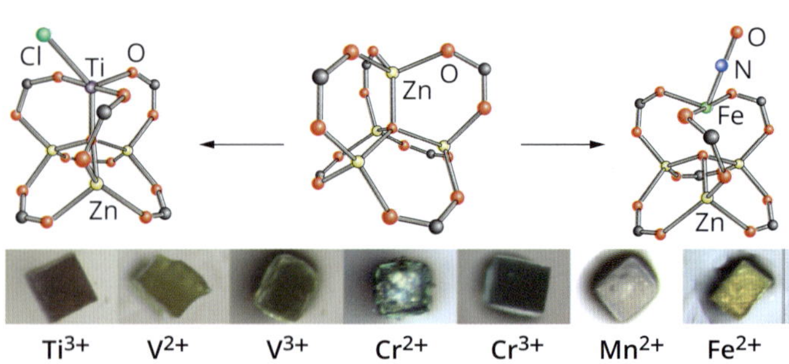

Fig. 19.30 Substitution of zinc in MOF-5 places metal centers in a trigonal bipyramidal environment. The structure is retained for a variety of different transition-metal ions while the properties (such as color) may vary widely.[34]

19.9 Intermetallics: From Metal Atoms to Materials (Advanced Topic)

One of the many very active areas of materials chemistry is in intermetallics, and we will briefly discuss this rich field here. While the word "alloy" is generally used for any mixture of two or more metallic elements, "intermetallics" are usually a subclass of alloy that has a well-defined stoichiometry and an ordered solid-state structure. Schulze defined the compounds as "solid phases containing two or more metallic elements, with optionally one or more nonmetallic elements, whose crystal structure differs from that of the other constituents."[35] As a result, the compound may include metalloids, such as aluminum, gallium, silicon, lead, and tellurium. Some researchers will even include certain carbides, such as Fe_3C, as intermetallic materials, largely because their electronic properties are similar.

An ordered structure generally implies covalency in the bonding, and intermetallics are often 3-dimensional structures built of polyhedra comprised of the constituent elements. It is therefore worthwhile to discuss metal clusters, molecules comprised of many metal atoms with metal–metal bonds and bonds to supporting ligands. In the next subsection, a method for understanding the structure and electronics of molecular metal clusters will be discussed, polyhedral skeletal electron pair theory (PSEPT), which are also known as Wade's rules, after Ken Wade.[36]

19.9.1 From Metal Atoms to Metal Clusters[36–42] (Advanced Topic)

PSEPT was originally devised as a method for anticipating the structure of boranes and is perhaps easiest taught by first using these as examples. Recall from Chapter 9 on main-group chemistry, that boron hydrides, "boranes", come in a wide variety with many bonding variations, which were discussed there. From the formulae, one can anticipate the types of bonds present and make guesses on the structure. PSEPT is a method for using the formula to discern the structure. This method was expanded by Mingos and others for use with metal clusters.

Let us start with a few examples using boron hydrides. First, here is the procedure (Wade's rules):

(1) Determine the total number of valence electrons by using the chemical formula. For each boron, three electrons are contributed, and, for each hydrogen, one electron is contributed.

(2) Some of these electrons will simply go to make B–H bonds and will not be involved in the formation of the skeleton. So, subtract two electrons for each B–H unit.

(3) Divide the number of remaining electrons by two to obtain the number of skeletal electron pairs (SEP).

(4) Determine the cluster type, which is used in conjunction with Fig. 19.31 and Table 19.5 to determine the base structure (B_n).

(5) If the number of boron atoms, n, is larger than $n + 1$ SEP, then the extra boron occupies a capping position on a triangular face.

The type of structure that occurs depends on the number of vertices and the number of skeletal electron pairs (SEP). The SEP are essentially the number of electrons involved in holding the cluster's core together. Fig. 19.31 shows some common polyhedral structures. The diagonal lines connect structures with equal numbers of SEP. Looking horizontally, the number of boron atoms, or more generally vertices, remains constant.

To illustrate, let us apply these rules to a couple of boron hydrides. First let us determine the structure of B_5H_{11}, which was discussed earlier in Chapter 9 on boranes. We found the compound to have a *styx* number = 3 2 0 3. While the *styx* number gives the types of bonds, such as the number of B–H–B τ-bonds, Wade's rules provide the structure of the boron core in which these bonds occur.

(1) Number of valence electrons = $(5 \times B) + (11 \times H) = (5 \times 3) + (11 \times 1) = 26$

(2) Number of electrons in B–H units = $(5 \times 2) = 10$

(3) Skeletal electron pairs (SEP): total number of skeletal electrons = 26–10 = 16; SEP = 16/2 = 8

(4) SEP = $n + 1 = 8$, and $n = 7$. The five boron atoms are vertices; therefore, there are $n - 2$ vertices (7–2)—an *arachno* structure based on a pentagonal bipyramid (B_7).

Table 19.5 Relationship between skeletal electron pairs (SEPs) and structure of the cluster.

Skeletal Electron Pairs	Structure Type	Cluster Geometry
$n - 1$		$(n - 2)$ polyhedron, 2 faces capped
n		$(n - 1)$ polyhedron, 1 face capped
$n + 1$	*closo*	(n)polyhedron
$n + 2$	*nido*	$(n + 1)$polyhedron, 1 vertex unoccupied
$n + 3$	*arachno*	$(n + 2)$polyhedron, 2 vertices unoccupied
$n + 4$	*hypho*	$(n + 3)$polyhedron, 3 vertices unoccupied

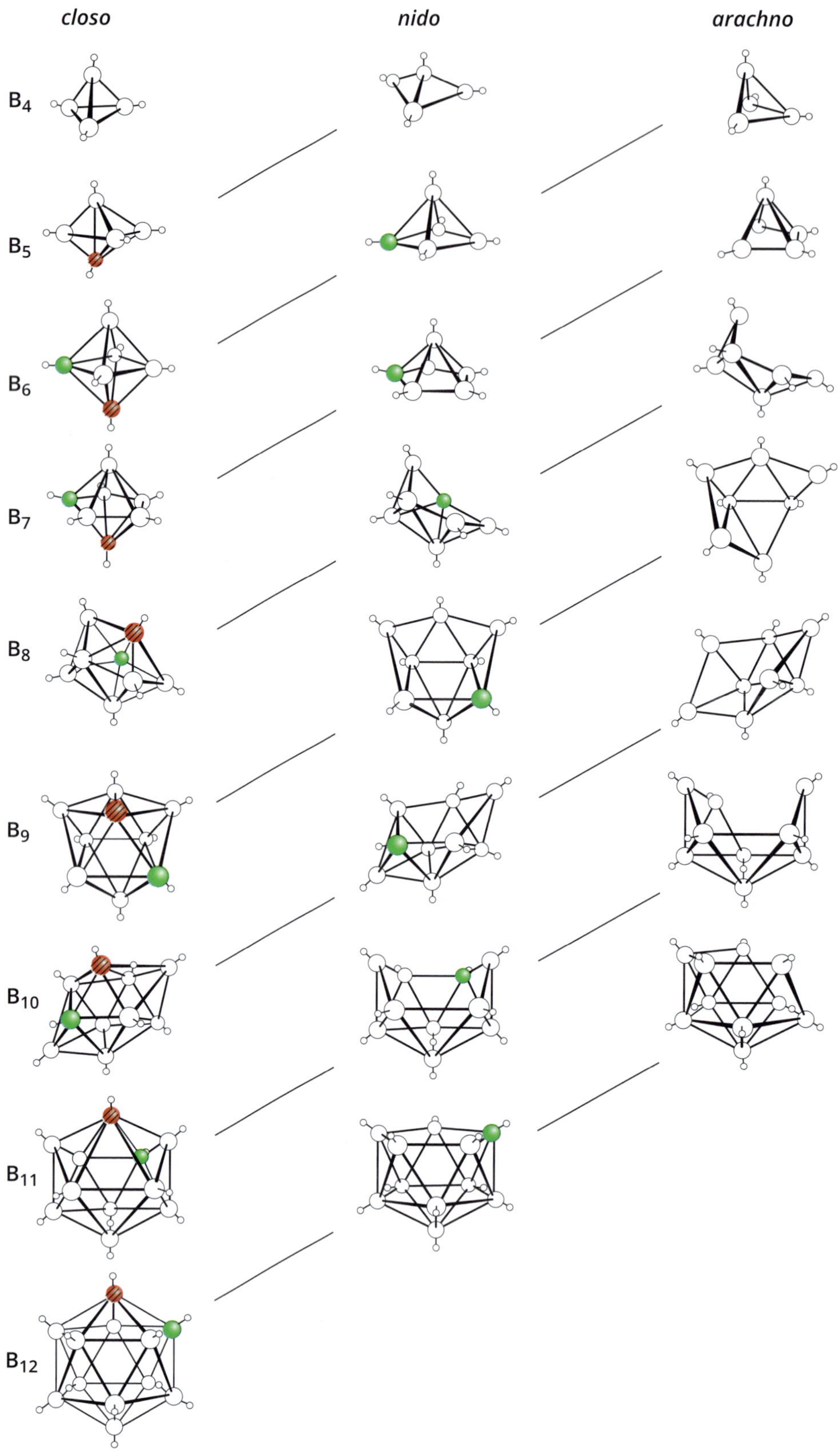

Fig. 19.31 Structural relationship between *closo*, *nido*, and *arachno* structures. The diagonal lines connect structures with equal numbers of skeletal electron pairs (SEP). H atoms, except for terminal B–H, are omitted. The red hatched vertex is removed from the related *closo* form first, followed by the green vertex, to form the *nido* and *arachno* structures, respectively. (Adapted from Rudolph.[43])

Now let us present an example with an atom different from boron in the cluster. Namely, we consider the neutral carborane, $C_2B_{10}H_{12}$, which is one of the most studied carboranes. Applying Wade's rules:

(1) Number of valence electrons = (2 × C) + (10 × 3) + (12 × H) = (2 × 4) + (10 × 3) + (12 × H) = 8 + 30 + 12 = 50.
(2) Number of electrons in B–H/C–H units = 12 × 2 = 24.
(3) SEP: total number of skeletal electrons = 50–24 = 26; SEP = 26/2 = 13.
(4) SEP = $n + 1$ = 13, and n = 12. Each carbon and boron can act as a vertex, and a *closo* B_{12}-type structure is expected.

In the case of $C_2B_{10}H_{12}$ there are three possible isomers (Fig. 19.32), which are often called *ortho*, *meta*, and *para*. The *ortho*-isomer is often made initially by reaction of acetylene and decaborane, $C_{10}H_{14}$, at high temperature. Heating the *ortho*-derivative initially produces the *meta*, and then the most thermally stable *para*.

The same method can be applied to metal clusters. In this case, one is concerned with the number of electrons donated by the metal center to the skeletal system. One can consider isolobal substitutions in the determination of the structures (Chapter 14).

First, consider the structure of tetrahedrane, C_4H_4, derivatives of which are known (Fig. 19.33).[44] As the name implies, tetrahedrane has a tetrahedral structure with each CH at a vertex; however, one can also show that CH is isolobal with $Co(CO)_3$. Consequently, $Co_4(CO)_{12}$ has a tetrahedral core of metal atoms surrounded by (terminal or bridging) carbonyls. Using the isolobal analogy, we can also show that CpNi is isolobal with $Co(CO)_3$; consequently, the cluster can also be prepared as part of this isolobal series. The Co_3Ni-cluster was prepared from $\{NiCp(CO)\}_2$ and $Co_4(CO)_{12}$ by Vahrenkamp and coworkers.[45]

An alternative method relates the number of valence electrons of a cluster to the "magic number" for polyhedral structures. The magic numbers can be generated by using the following equation, where it is assumed that each metal center

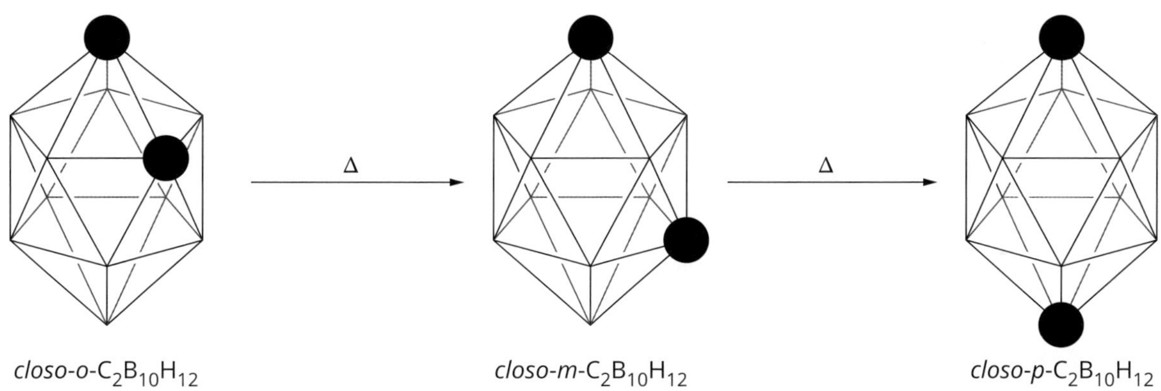

Fig. 19.32 Thermal rearrangement of *closo*-o-$C_2B_{10}H_{12}$ to *meta* and then *para*. The carbons are represented by black spheres, but H atoms are not shown.

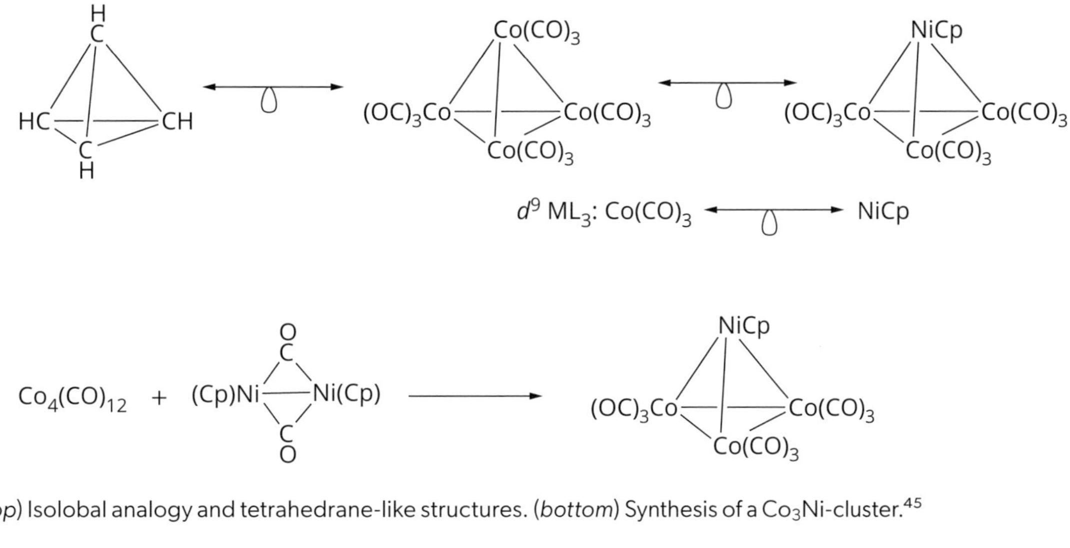

Fig. 19.33 (*top*) Isolobal analogy and tetrahedrane-like structures. (*bottom*) Synthesis of a Co_3Ni-cluster.[45]

will reach its noble gas configuration or 18-electrons. Each edge is considered to contain a metal–metal 2-electron bond as well.

$$\text{magic number} = (18 \times \text{number of vertices}) - (2 \times \text{number of edges})$$

Taking the number of electrons and shape into account produces Table 19.6.

Using this method, $Co_4(CO)_{12}$ would have $9 \times 4 = 36$ metal-based electrons and $12 \times 2 = 24$ electrons from the carbonyls, which gives it a magic number of 60 and a tetrahedral structure. What, then, is the structure of $Re_4(CO)_{16}$? This cluster has $7 \times 4 = 28$ metal-based electrons and $16 \times 2 = 32$ electrons from carbonyls, which also gives it a magic number of 60 and a tetrahedral structure again is expected.

In the case of $Re_4(CO)_{16}$, the cluster does not appear to have been structurally characterized; however, $Re_4(CO)_{16}^{2-}$ is known. The structure of this cluster can be anticipated by looking at Fig. 19.31. In order to go from the four vertex *closo* structure to the four vertex *nido* structure, two electrons must be added. In other words, $Re_4(CO)_{16}^{2-}$ should have the structure of two fused triangles shown in the top middle of Fig. 19.31; it is precisely this that was found by Churchhill and Bau (Fig. 19.34).[46]

The structures of many complex clusters can be understood by using these methods, including those containing interstitial main-group atoms like the carbide cluster $Ru_6(CO)_{17}C$ (Fig. 19.35), which has a *closo*-structure with 6-bonds to the carbon occupying the center of the octahedron, and the nitride cluster $\{Fe_5(CO)_{14}N\}^-$, which has a *nido*-structure with a central nitrogen atom. The interested reader is referred to the excellent text on organometallic chemistry by Elschenbroich and Salzer that covers such carbonyl clusters in detail.[47]

Table 19.6 The shape and magic number for transition-metal clusters.

Shape	Magic Number
triangle	48
tetrahedron	60
trigonal bipyramid	72
square pyramid	74
octahedron	86
trigonal prism	90
cube	120

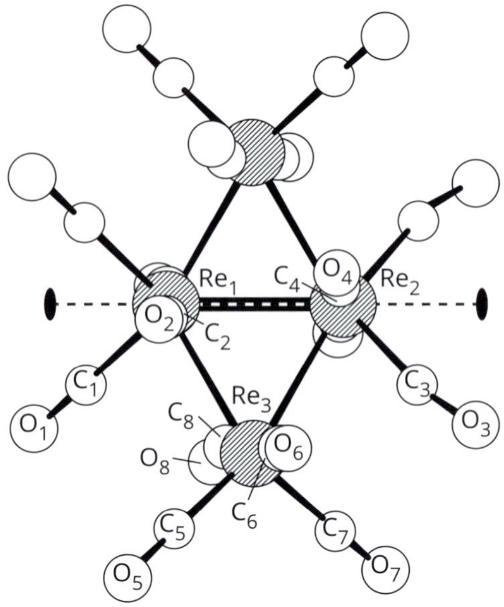

Fig. 19.34 Structure of $Ru_4(CO)_{16}^{2-}$. (From the paper by Churchhill and Bau.[46])

19.9.2 From Metal Clusters to Materials

One strategy for the access of intermetallics is to use exotic metallic fluxes.[48] For example, aluminum melts at 660 °C and can be used as a flux to form intermetallic phases. Aluminum also dissolves in nonoxidizing acids, such as HCl, which makes the excess Al easily removable and useful as a flux. Similar to the metallic clusters discussed briefly in the previous section, these materials are comprised of polyhedral structures with binary and ternary examples well known, such as $ReAl_6$ and $Ho_6Mo_4Al_{43}$. The geometry in $ReAl_6$, for example, was described by Neimann and Jietschko by using the

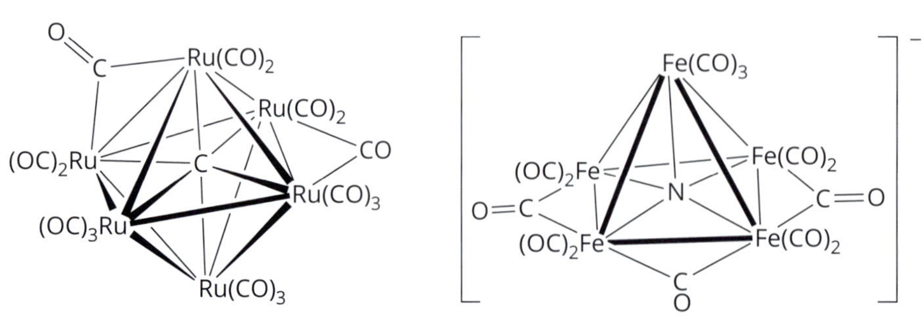

Fig. 19.35 Structures of $Ru_6(CO)_{17}C$ and $\{Fe_5(CO)_{14}N\}^-$.

polyhedra shown in Fig. 19.36.[d] The Re atoms are black, with the Al atoms as white spheres.[49–50]

A sense of how many of these reactions are possible can be gleaned from the relatively simple example of an intermetallic prepared from cobalt and tin in various ratios and at different temperatures. The phase diagram for the system is shown in Fig. 19.37. If one starts with β-Co_3Sn_2 at 1,180 °C, a congruous melt results in which the solid and liquid have the same composition. In contrast, a 1:1 mixture of Co:Sn at 966 °C is incongruous as solid β-Co_3Sn_2 is formed along with some liquid. If a melt of Co:Sn in a ratio 3:1 is cooled, the liquidous line on the right of the phase diagram is followed with solid CoSn formed until 571 °C, where $CoSn_2$ forms from the melt.

Instead of the peritectic reaction described previously, one may use a larger excess of a reactive flux to get a single phase. For example, $CoSn_3$ can be prepared from an excess of tin as flux and avoid the previous relatively slow reaction. The exact phase diagram of the mixture is rarely known in exploratory

[d] The symbols below the structures are the site symmetries of the central atoms. They are using *m* for mirror plane, a crystallographic convention, rather than σ, the group theory convention.

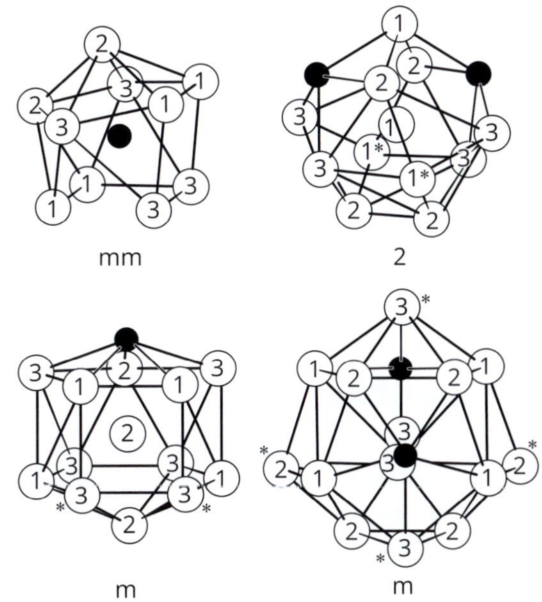

Fig. 19.36 The coordination polyhedra in $ReAl_6$ with the symmetry of the central atoms indicated. (Re, black; Al, white) The atoms marked with asterisks in some of the structures are at very long distances from the central atom, >3.4 Å.[49–50]

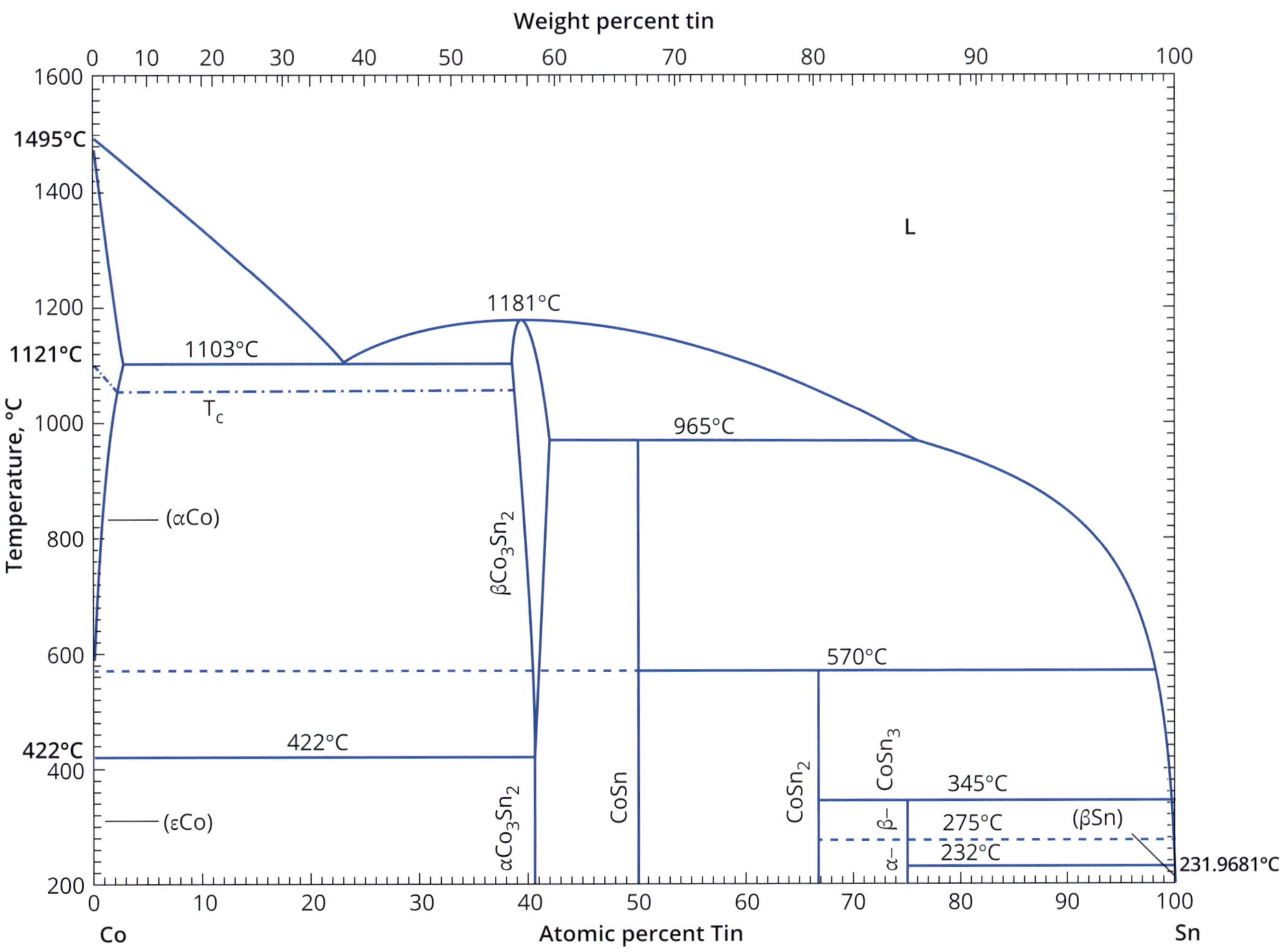

Fig. 19.37 Phase diagram for cobalt and tin.[51]

cases, of course, and researchers are searching compositions for interesting properties.

As mentioned, the structures of intermetallics are often 3-dimensional polymers of well-defined polyhedra. Understanding the structures of these intermetallics may be achieved by using models such as the Mott-Jones or Zintl model.[48] More recently, Yannello and Fredrickson presented a method for understanding the structures of intermetallics by using the isolobal analogy.[52–53]

19.10 Quantum Dots

Quantum dots are nanoscale particles comprised of some semiconductor material, often wrapped in stabilizing layers of material or ligands. These systems are synthesized in sizes varying from ~1 nm on the small end to hundreds of nanometers. On this scale they exhibit properties different from molecular species because of the large array of atoms present, but also different from bulk materials. Some of the differences from bulk materials are due to what is often referred to as "quantum confinement"—a phenomenon by which electrons in the conduction band are constrained to act much like the particle in a box problem often discussed in quantum mechanics courses. Changing the size of the particle (size of the box) changes the properties dramatically, even when the material is otherwise identical. The properties that are generally exploited are optical, especially photoluminescence (Fig. 19.38).

One difficulty is that as particles become smaller, the tendency is for there to be more defect sites created. Now, however, the production of quantum dots, which relies on a detailed understanding of the kinetics and thermodynamics of crystal formation along with lots of trial and error, is refined to the stage that they can be produced very reliably, and in some cases defect-free. This is often done by using a core-shell strategy, where a robust shell material such as a zinc chalcogen is used to cover the semiconductor. An example of emission from such a system is shown in Fig. 19.38 by Bawendi and coworkers. in which the CdSe core is coated in zinc sulfide.[54]

These core-shell quantum dots can have no nonradiative quenching processes, 100% quantum yield, very narrow emission, and emit in brilliant colors, which is why they are used in televisions and displays. The core-shell also helps to control photoluminescence intermittency in the quantum-dot emission, better known as "blinking". On excitation, quantum dots will seemingly self-quench, which can lead to loss of signal at low concentrations for brief periods of time. The phenomenon seems related to surface processes and can be controlled by carefully engineering the surface around the quantum dot.[55–57]

Quantum dots, because of their unusual electronic and optical properties, are being used in a host of different applications, including single-molecule detection in living cells,[58] fluorescence resonance energy transfer (FRET) for studying biological systems,[59] displays and lighting,[60] and a host of other electronic and device applications.[61–62]

19.10.1 History and Electronic Structure

Variations in color with colloidal particles of gold have been known for centuries and used in art such as stained glass windows, with spherical particles of ~25 nm producing the beautiful ruby reds found in many medieval works of art. One famous use of such particles is in the fourth-century Lycurgus cup, which contains both gold and silver nanoparticles. The cup appears ruby red when backlit, but is green like jade when lit from the front (Fig. 19.39).

Perhaps nanoparticles of this type were first studied in detail by Faraday, who realized that the particle color seemed related to particle size and shape. In 1857 he wrote: "It is clear

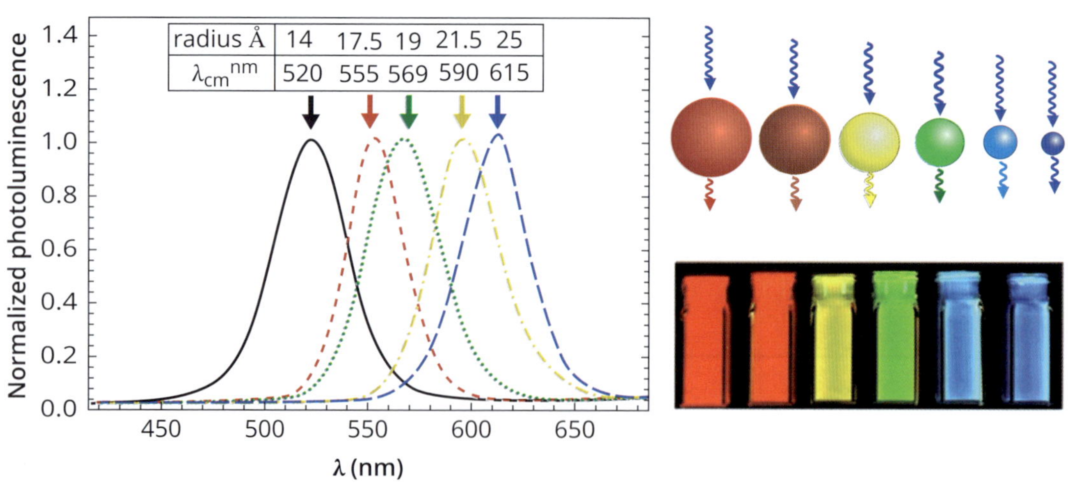

Fig. 19.38 Cadmium selenide emission spectra. In this case, to limit defect sites, Bawendi and coworker placed the CdSe in a shell of zinc sulfide, which is often written as (CdSe)ZnS. As shown, the emission varies from blue to red as the size of the CdSe nanoparticle changes from 1.4 to 2.5 nm, with smaller particles producing higher-energy emission.[54]

Figure 19.39 (*top*) The fourth-century Lycurgus cup, currently in the British Museum, incorporates silver and gold nanoparticles in the glass, producing a red color due to transmitted light and jade green due to reflected light.[64] (*bottom*) Colors of various-sized monodisperse gold nanoparticles.[65]

that [gold], reduced to small dimensions by mere mechanical means, can appear of two colours by transmitted light, whatever the cause of the difference may be."[63] Faraday described gold hammered thin as having a green hue, and very small particles as appearing red. Today one can purchase spherical gold nanoparticles from 5 nm to >100 nm (Fig. 19.39).

Semiconductor quantum dots were reported independently by Brus and Ekimov in the early 1980s. Ekimov, working in the USSR, grew microcrystals in a glassy media and noted changes in absorption spectrum with crystal size.[66–67] Brus was working on small crystallites of semiconductors at Bell Laboratories in New Jersey.[68–69] While these preliminary results were fascinating, they were difficult to reproduce because the syntheses of the particles were quite sensitive.

It is not too great an exaggeration to say that the field of quantum dots can be divided into work before Murray, Norris, and Bawendi published their synthesis of CdSe nanoparticle synthesis, and after this key paper in 1993.[70] Before this article, quantum dots were oddities that were difficult to reproduce. Afterwards, monodisperse quantum dots of almost any size were available for study.

The electronic structure of quantum dots has been studied in great detail by Brus and many others. One can begin to understand these systems in a top-down (starting with materials and looking at the consequences of confinement) or bottom-up (starting with the unit cell and building up to small clusters) fashion. Heath and Shiang published a qualitative electronic description in 1998 that is a little of both, which will serve as the basis of the brief discussion here.[71]

If we start at the "top", with a bulk material like CdSe, an energy versus k plot may look like that shown in Fig. 19.40. Recall from the previous discussion that k is a measure of the symmetry of the wavefunction, with $k = 0$ in this case being the totally bonding combination that spans the entire crystal of the bulk material. The $k = \pi/a$ will be the totally antibonding combination. The energy in the figure is based on Hückel theory, with E being the energy of a noninteracting p-orbital, and β being the energy from the resonance integral, which is related to the covalency in the system. If the system is covalent, the bands will be quite broad, covering the whole crystal, and if ionic, the bands will be narrow and localized.

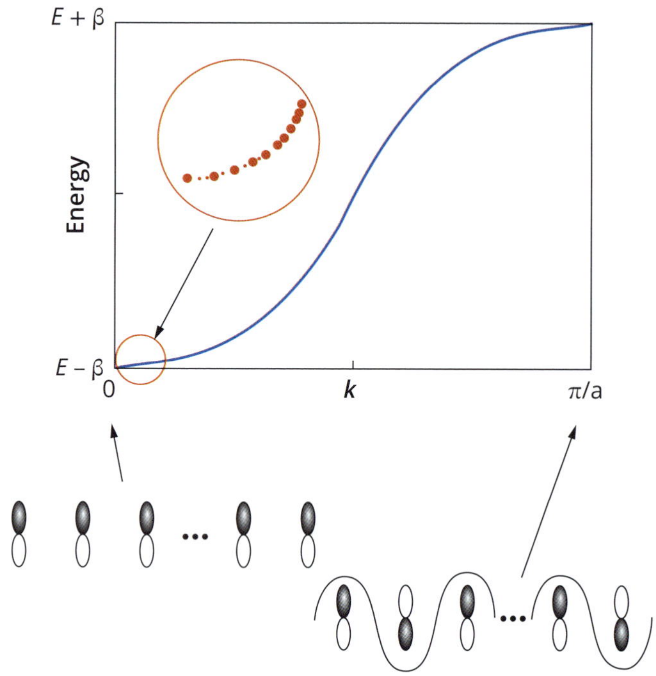

Fig. 19.40 Energy versus k plot of a 1-dimensional solid with a π-bonding *p*-orbital. The energy is measured in *E*, the energy of the isolated orbital, and *β*, the energy from the resonance integral, which is related to the covalency of the system in Hückel theory. The band is made of an array of *p*-orbitals, with *k* = 0 being the fully bonding combination and the *k* = π/a being the fully antibonding combination. For bulk material, the band is continuous. For a finite crystal, the number of points in the plot is equal to the number of unit cells in the crystal. So, as the crystal becomes smaller, the energy levels become more and more discrete. (Figure adapted from Heath and Shiang.[71])

If we take the bulk crystal with the broad covalent band and physically decrease the size, the continuous band will break into discrete levels with the number of points equal to the number of unit cells in the crystals (see the circular inset in Fig. 19.40). As the crystal becomes smaller and smaller, the "band" looks more and more like a collection of levels.

Consider the band in Fig. 19.40 as the valence band. A photon can be used to excite electrons in the valence band to the conduction band; these are called direct band gap materials, of which CdSe is an example. As the crystal becomes small, the molecular orbitals near $k = 0$ become too large for the crystal and must be removed from the diagram, localizing the wavefunctions and leaving higher energy levels for excitation. This, in conjunction with the granularization of the band itself as the number of unit cells is decreased, causes the absorption spectrum to exhibit higher-energy bands that are more discrete. The crystal absorbs light in transitions involving these more discrete bands to form excitons, electron-hole pairs that carry the energy of the photon, which will then emit a photon on a timescale dependent on their mobility. In other words, the photonic energy is stored as a charge-separated electron-hole pair (exciton), which eventually collapses to release light again. The light emitted is then (generally) of longer wavelength than adsorbed.[e]

For CdSe quantum dots, there is typically a nonlinear relationship between the extinction coefficient and the size of the particle. A simple relationship found experimentally between ΔE, the energy of the first absorption peak (in eV) and diameter, D, of the crystal is shown in Eq. 19.16,[72–73] but others have addressed this problem in more detail.[74]

$$\epsilon = 1600 \cdot \Delta E \cdot D^3 \tag{19.16}$$

The measurement typically used in this field for delocalization is the "effective mass" of the electrons and holes in the materials. These materials-based quantities can seem quite odd to a chemist's ear (for example, what is the effective mass of a hole?) but are related to delocalization. If one considers the electron to be a particle with a specific location, then it can move quickly or slowly through the material. If the electrons are viewed from a wave perspective, the wavefunction is very delocalized; from the electron-as-a-particle perspective, it must be moving very quickly. To compensate for this rapid movement, the theories will change the mass of the electron. Consequently, very delocalized electrons have low effective mass, m^*. In CdSe, the ratio of m^* to m_e (rest mass of the electron) has been measured as $m^*/m_e = 0.13$.

The theory for excitons uses the hydrogen atom as its basis, since both systems contain one positive and one negative charge, then adds corrections based on the differences in the two systems. For example, one can calculate the radius of the exciton (r_{exc}) from the Bohr picture (see Section 2.1.2), using Eq. 19.17, where ε^* = the relative permittivity of the material and a_0 = Bohr radius. Using this equation, one can find that excitons in CdSe have a radius of ~35 Å. When the size of the crystal approaches and drops below the exciton size, quantum confinement effects are observed.

$$r_{exc} = \{\varepsilon^*/(m^*/m_e)\}a_0 \tag{19.17}$$

Brus has published theoretical models for the behavior of quantum dots since the early 1980s, and the Brus equation (Eq. 19.18) describes the relationship between energy of the emitted light, $\Delta E(r)$, as a function of the radius, r, of the quantum dot, where E_{gap} is the band gap in the bulk material, m^*_e is the effective mass of the electron, and m^*_h is the effective mass of the hole. For CdSe, E_{gap} = 1.84 eV, $m^*_e = 0.13\ m_e$, and $m^*_h = 0.45\ m_e$. The Brus equation relies on the particle-in-a-box model, and derivations can be found in the primary literature.[68–69, 75]

$$\Delta E(r) = E_{gap} + \frac{h^2}{8r^2}\left(\frac{1}{m^*_e} + \frac{1}{m^*_h}\right) \tag{19.18}$$

[e] There are some special materials that will absorb two or more photons of lower energy and then emit a single, higher-energy photon, which is called "photon upconversion".

19.10.2 Synthesis of CdSe Quantum Dots

The synthetic procedures for the production of quantum dots have developed significantly since their discovery by Brus and Ekimov, with the paper by Bawendi and coworkers using the "hot injection method" being the start of modern syntheses.[70] The fabrication of the particles changes a great deal, depending on the type of material being studied, and only CdSe particle synthesis will be discussed here.

Murray, Norris, and Bawendi published two syntheses in their initial paper (Eq. 19.19).[70] Both syntheses use tri-*n*-octylphosphine (TOP) and tri-*n*-octylphosphine oxide (TOPO) as ligands for the surface. In Method A, $CdMe_2$, TOP, and tri-*n*-octylphosphine selenide (TOPSe) are loaded into a syringe. In one quick injection, the syringe is added to a solution of TOP/TOPO at 300 °C. Aliquots of the reaction are taken at 5–10-minute intervals and are interrogated by absorption spectroscopy. The reaction produces small quantum dots (~1.5 nm) at early times, and larger (11.5 nm) quantum dots at later times from the same preparation. By controlling the temperature and time, one can produce quantum dots of various sizes, with the optical spectra being used to follow reaction progress and particle size. Method B replaces TOPSe as the selenium source with $Se(SiMe_3)_2$. Since this study, detailed studies have been carried out on how precursor structure effects particle growth.[76]

$$\text{TOP} + \text{TOPO} \xrightarrow[\sim 300\,^\circ\text{C}]{\text{CdMe}_2,\ \text{TOP},\ \text{TOPSe or Se(SiMe}_3)_2} 1.5\text{–}11.5\,\text{Å CdSe quantum dots}$$

TOP = tri-*n*-octylphosphine

TOPO = tri-*n*-octylphosphine oxide **19.19**

The growth of quantum dots may be thermodynamically or kinetically controlled, and is generally understood by using LaMer's nucleation theory.[77] However, in 2015 Polte wrote: "No theory or theoretical model exists so far that is able to describe or predict the evolution of the particle size or size distribution comprehensively."[78]

Reactants are used to produce a "monomer", which is the actual precursor leading to crystal growth. In the case of CdSe nanoparticles, the TOPSe and $CdMe_2$ react to produce some type of compound, maybe with an Se–Cd bond, that can act as monomer. Perhaps needless to say, the exact monomer is not known for many nanoparticle syntheses.

In the LaMer model there are three distinct stages, defined based on the monomer concentration (Fig. 19.41). First (Phase I), the concentration of monomer must exceed the saturation level (C_S) and the level required to overcome the activation barrier (C_{min}). Then, near the maximum concentration of monomer (C_{max}), nucleation occurs. Second (Phase II), nucleation produces seeds for growth, and the concentration of monomer rapidly decreases. Third (Phase III), while the concentration of monomer is held between C_{min} and C_S, the particles undergo steady growth.

In order to obtain monodisperse nanoparticles, some effect must generally cause small particles to grow more quickly than large particles, which is sometimes referred to as "focusing" of the distribution. In 1998, Alivisatos and coworkers wrote: "At any given monomer concentration, there exists a critical size, which is at equilibrium. Nanocrystals smaller than the critical size have negative growth rates (dissolve), while larger ones grow at rates dependent strongly on size."[79] In other words, monodisperse nanoparticles are produced from dissolution of particles smaller than the critical size, along with particles just above some critical size growing quickly, while large particles grow more slowly.

If one allows the monomer concentration to begin to drop, the critical size becomes larger than the average size of the particles in solution. The result is that some particles are dissolved in order to make other particles larger, often with a loss in size uniformity. This process is called Ostwald ripening or defocusing. Consequently, having precursors that can decompose to produce a more or less constant concentration of monomer is one key to creating monodisperse nanoparticles.

While we have discussed quantum dots in only a cursory fashion here, this is a vast field with numerous and growing applications. These nanoparticle-sized materials are sometimes referred to as being "0-dimensional materials", point-like. Likewise, one can constrain dimensionality to make 1-dimensional, such as nanowires,[80–83] and 2-dimensional sheets, such as graphene.[84–87]

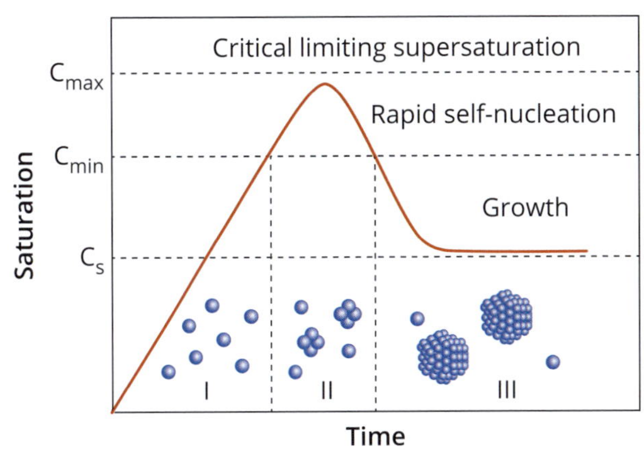

Fig. 19.41 A monomer concentration versus time plot to illustrate the LaMer nucleation model. (I) The concentration of monomer is increasing past that of saturation (C_S). (II) Finally, near C_{max}, the concentration of monomer reaches supersaturation levels that facilitate rapid nucleation. To do this, the solution must pass the concentration where the activation energy can be overcome, C_{min}. (III) Once nuclei are present, the particles grow while the monomer concentration is above saturation. (Figure from Polte.[78])

Chapter Summary

1. Solids can be divided into three different types: molecular solids, solid-state compounds, and low-dimensional materials.
2. Molecular solids have strong heterocovalent bonds only within the molecular species, and the solid is held together by weaker van der Waals forces.
3. Solid-state compounds can be understood at the extremes of ionic character by using Coulombic interactions only, using the Kapustinskii equation, which calculates the lattice energy using attraction between ions. More covalent systems are understood by using the Block equation, which allows MO-based techniques to be applied to extended systems.
4. Phase diagrams, of which there are several common types, are an important tool for showing important phase information as a function of conditions such as temperature, pressure, and composition. Phase changes have profound effects on electronic structure and properties.
5. There are many different synthetic protocols used to prepare solid-state compounds, and a few common ones have been discussed: direct reaction, precursor method, flux methods, and hydrothermal syntheses. For each, the synthesis of a specific compound was provided as an example.
6. The function of materials is oftentimes closely tied to the crystalline order and impurities within the crystals. In some cases, impurities are intentionally added in specific quantities to make them suitable for an application. Silicon-based solar cells were discussed as an example.
7. An important application of materials is to batteries, and the lithium-ion battery was used as an example.
8. Metal-based joints and organic "struts" are used to generate porous metal–organic frameworks (MOFs) for applications such as gas storage and catalysis.
9. Clusters of atoms can have specific structures based on the number of electrons in the framework. Intermetallics—covalently bound clusters of metals in extended frameworks—generate specific structures based on the number of electrons present.
10. Decreasing the dimensionality can change the properties of materials due to quantum confinement. For example, small chunks of semiconductors with chemically passivated surfaces called quantum dots have tunable properties based on the size of the particle. Quantum dots are sometimes referred to as 0-dimensional materials, due to their small size. Other materials, such as nanotubes (1-dimensional) and graphene (2-dimensional), are also studied for their unique electronic properties.

Practice Problems

1. Crystalline solids give characteristic X-ray powder patterns due to diffraction, and the pattern can be used as a "fingerprint" for the structure. For two similar solids, e.g., NaI and NaCl, why are the patterns distinguishable if the structures have the same symmetry? Include in your discussion Miller indices.
2. Discuss at least one method you could employ to understand the coordination environment of chromium(III) in ruby, despite the relatively low concentration of Cr? What information about the chromium center will be gathered from the method?
3. Use data on the NIST website to find the lattice energy of NaCl using a Born-Haber cycle.
4. In the table below are the experimental lattice energies of some silver and thallium halides along with the lattice energies from the Born-Mayer equation.[88] The difference is 11 kcal/mol for AgF but steadily grows to 21 kcal/mol down the halides. Based on the assumptions of the Born-Mayer equation, why may this be?

Salt	Experimental Lattice Energy (kcal/mol)	Lattice Energy from Born-Mayer Equation (kcal/mol)
AgF	228	219
AgCl	217	203
AgBr	215	197
AgI	211	190

5. What type of semiconductor (n or p) should a sample of silicon doped with antimony be and why? If one took that sample and began doping with gallium, how would the conducting properties of the solid change as a function of gallium concentration?
6. The electrical conductance of germanium is increased by a factor of 10^5 if a few parts per million of arsenic are added. Explain why?

7. Buckminsterfullerene, C_{60}, is a solid material comprised of a cage of carbon units in a truncated icosahedral structure. The C_{60} units rotate with three degrees of freedom in the solid structure, so there is no orientational order. The space group is cubic, a = 14.17 Å, in $Fm\bar{3}m$.[89] Name all of the forces in the crystal that are holding the molecules together. Why is rotation of the ball-like structures so facile?
8. Write possible formulas for a sulfide and a fluoride that might adopt the spinel structure.

Cited References

1. Dronskowski, R.; Kikkawa, S.; Stein, A., *Handbook of Solid State Chemistry*. Wiley-VCH: 2017.
2. Stout, G. H.; Jensen, L. H., *X-ray Structure Determination: A Practical Guide* 2nd ed.; Wiley: New York, 1989.
3. Shannon, R. D., Revised effective ionic-radii and systematic studies of interatomic distances in halides and chalcogenides *Acta Cryst. A* 1976, *32*, 751–767.
4. Hunt, K. L. C., Dispersion dipoles and dispersion forces: Proof of Feynman conjecture and generalization to interacting molecules of arbitrary symmetry *J. Chem. Phys.* 1990, *92*, 1180–1187.
5. Luo, F.; McBane, G. C.; Kim, G. S.; Giese, C. F.; Gentry, W. R., The weakest bond: experimental observation of helium dimer *J. Chem. Phys.* 1993, *98*, 3564–3567.
6. Weast, R. C., *CRC Handbook of Chemistry and Physics*. 70th ed.; CRC Press, Inc.: Boca Raton, FL, 1990.
7. Kapustinskii, A. F., Lattice energy of ionic crystals *Quar. Rev.* 1956, *10*, 283–294.
8. Hoffmann, R., *Solids and Surfaces: A Chemist's View of Bonding in Extended Structures* VCH Publishers: New York, NY, 1988; p x, 142 pages.
9. Meng, Y. S.; Arroyo-de Dompablo, M. E., First principles computational materials design for energy storage materials in lithium ion batteries *Energy & Env. Sci.* 2009, *2*, 589–609.
10. Young, D. A. *Phase Diagrams of the Elements*; U.S. Energy Research & Development Administration: Lawrence Livermore Laboratory, 1975.
11. Föll, H. Some Phase Diagrams. https://www.tf.uni-kiel.de/matwis/amat/iss/kap_6/illustr/i6_2_1.html (accessed March 28, 2018).
12. Rajpoot, J. S.; Mishra, A., Synthesis and study of the structural properties of pure $PbTiO_3$ ferroelectric ceramics. In *International Conference on Recent Trends in Physics*, Kane, S. N.; Mishra, A.; Gaur, A., Eds. 2014; Vol. 534.
13. Abrishami, M. E.; Risch, M.; Scholz, J.; Roddatis, V.; Osterthun, N.; Jooss, C., Oxygen Evolution at Manganite Perovskite Ruddlesden-Popper Type Particles: Trends of Activity on Structure, Valence and Covalence *Materials* 2016, *9*.
14. Hirakata, K.; Rhine, W. E.; Cima, M. J., Surface chemistry of lead titanate and its impact on binder removal *J. Am. Ceram. Soc.* 1996, *79*, 1002–1008.
15. Weinland, R.; Holtmeier, H., The ferri nickel etc. -acetate and on a very basic crystalised ferric acetate *Zeitschrift Fur Anorganische Und Allgemeine Chemie* 1928, *173*, 49–62.
16. Wickham, D. G.; Whipple, E. R.; Larson, E. G., The preparation of stoichiometric ferrites *J. Inorg. Nuc. Chem.* 1960, *14*, 217–224.
17. The world's most underappreciated gemstone - Red Spinel. https://crystallography365.wordpress.com/tag/spinel/ (accessed March 28, 2018).
18. Brixner, L. H., Preparation of the ferrites $BaFe_{12}O_{19}$ and $SrFe_{12}O_{19}$ in transparent form *J. Am. Chem. Soc.* 1959, *81*, 3841–3843.
19. Bunsen, R., Bemerkunge zu einigen Einwurfen gegen mehrere ansichten uber die chemisch-geologischen erscheinungen in island *Ann. Chem.* 1848, *65*, 70–85.
20. Grohol, D.; Nocera, D. G., Hydrothermal oxidation-reduction methods for the preparation of pure and single crystalline alunites: Synthesis and characterization of a new series of vanadium jarosites *J. Am. Chem. Soc.* 2002, *124*, 2640–2646.
21. Nocera, D. G.; Bartlett, B. M.; Grohol, D.; Papoutsakis, D.; Shores, M. P., Spin frustration in 2D kagome lattices: a problem for inorganic synthetic chemistry *Chem Eur. J.* 2004, *10*, 3850–9.
22. Yttrium barium copper oxide. https://en.wikipedia.org/wiki/Yttrium_barium_copper_oxide (accessed March 28, 2018).
23. Koyo Thermo Systems Co. Ltd., Crystalline silicon solar cells c-Si. https://www.crystec.com/klldope.htm (accessed July 1, 2019).
24. Nitta, N.; Wu, F.; Lee, J. T.; Yushin, G., Li-ion battery materials: present and future *Mater. Today* 2015, *18*, 252–264.
25. Goodenough, J. B.; Kim, Y., Challenges for Rechargeable Li Batteries *ChemMater.*. 2010, *22*, 587–603.
26. Goodenough, J. B., How we made the Li-ion rechargeable battery *Nature Elec.* 2018, *1*, 204.
27. Mizushima, K.; Jones, P. C.; Wiseman, P. J.; Goodenough, J. B., $LiXCoO_2$: A new cathode material for batteries of high-energy density *Mater. Res. Bull.* 1980, *15*, 783–789.
28. Asadi, M.; Sayahpour, B.; Abbasi, P.; Ngo, A. T.; Karis, K.; Jokisaari, J. R.; Liu, C.; Narayanan, B.; Gerard, M.; Yasaei, P.; Hu, X.; Mukherjee, A.; Lau, K. C.; Assary, R. S.; Khalili-Araghi, F.; Klie, R. F.; Curtiss, L. A.; Salehi-Khojin, A., A lithium-oxygen battery with a long cycle life in an air-like atmosphere *Nature* 2018, *555*, 502.
29. A Look Inside Apple iPhone 6 Plus Battery. *Tech Insights* 2015.
30. Cejka, J.; Corma, A.; Zones, S., *Zeolites and Catalysis Synthesis, Reactions and Applications* Wiley-VCH: Weinheim, 2010.
31. Li, H.; Eddaoudi, M.; O'Keeffe, M.; Yaghi, O. M., Design and synthesis of an exceptionally stable and highly porous metal-organic framework *Nature* 1999, *402*, 276–279.
32. Rosi, N. L.; Eckert, J.; Eddaoudi, M.; Vodak, D. T.; Kim, J.; O'Keeffe, M.; Yaghi, O. M., Hydrogen storage in microporous metal-organic frameworks *Science* 2003, *300*, 1127–1129.
33. Farrusseng, D., *Metal-Organic Frameworks: Applications from Catalysis to Gas Storage* Wiley-VCH: Weinheim, 2011.
34. Brozek, C. K.; Dinca, M., Ti^{3+}, $V^{2+/3+}$, $Cr^{2+/3+}$, Mn^{2+}, and Fe^{2+}-Substituted MOF-5 and Redox Reactivity in Cr- and Fe-MOF-5 *J. Am. Chem. Soc.* 2013, *135*, 12886–12891.
35. Schulze, G. E. R., *Metallphysik:Ein Lehrbuch* Springer-Verlag Wien: Berlin, 1974.
36. Wade, K., Structural significance of number of skeletal bonding electron-pairs in carboranes, higher boranes and borane anions, and various transition-metal carbonyl cluster compounds *J. Chem. Soc. D-Chem. Comm.* 1971, 792.
37. Housecroft, C. E., Clusters with interstitial atoms from the p-block: How do Wade's rules handle them? In *Structural and Electronic Paradigms in Cluster Chemistry*, Mingos, D. M. P., Ed. 1997; Vol. 87, pp 137–156.
38. Jemmis, E. D.; Balakrishnarajan, M. M.; Pancharatna, P. D., A unifying electron-counting rule for macropolyhedral boranes, metallaboranes, and metallocenes *J. Am. Chem. Soc.* 2001, *123*, 4313–4323.
39. Manning, M. C.; Trogler, W. C., Electronic-structures of transition-metal cluster complexes *Coord. Chem. Rev.* 1981, *38*, 89–138.
40. Mingos, D. M. P., General theory for cluster and ring compounds of main group and transition-elements *Nature-Physical Science* 1972, *236*, 99.
41. Mingos, D. M. P., Polyhedral skeletal electron pair approach *Acc. Chem. Res.* 1984, *17*, 311–319.
42. Mingos, D. M. P.; Slee, T.; Lin, Z. Y., Bonding models for ligated and bare clusters *Chem. Rev.* 1990, *90*, 383–402.
43. Rudolph, R. W., Boranes and heteroboranes - paradigm for electron requirements of clusters *Acc. Chem. Res.* 1976, *9*, 446–452.
44. Maier, G.; Pfriem, S.; Schafer, U.; Matusch, R., Small rings. 25. Tetra-tert-butyltetrahedrane *Angew. Chem. Int. Ed.* 1978, *17*, 520–521.
45. Beurich, H.; Blumhofer, R.; Vahrenkamp, H., Mixed metal-clusters: Metal exchange via dinuclear (cyclopentadienyl)metal carbonyls *Chem. Ber. Recl.* 1982, *115*, 2409–2422.

46. Churchill, M. R.; Bau, R., Crystallographic studies on polynuclear rhenium carbonyl hydrides and anions. 2. Characterization of $[(N\text{-}C_4H_9)_4N]_2Re_4(CO)_{16}$ *Inorg. Chem.* 1968, *7*, 2606.
47. Elschenbroich, C.; Salzer, A., *Organometallics: A Concise Introduction* 2nd ed.; VCH: Weinheim, Federal Republic of Germany ; New York, NY, USA, 1992.
48. Kanatzidis, M. G.; Pottgen, R.; Jeitschko, W., The metal flux: A preparative tool for the exploration of intermetallic compounds *Angew. Chem. Int. Ed.* 2005, *44*, 6996–7023.
49. Niemann, S.; Jeitschko, W., The crystal-structures of Re_2Al,Re_4Al_{11}, and $ReAl_6$ *Z. Naturforsch., B: Chem. Sci.* 1993, *48*, 1767–1773.
50. Wolff, M. W.; Niemann, S.; Ebel, T.; Jeitschko, W., Magnetic properties of rare-earth transition metal aluminides R_6TAl_{43} with $Ho_6Mo_4Al_{43}$-type structure *J. Magn. Magn. Mater.* 2001, *223*, 1–15.
51. Okamoto, H., Co-Sn (Cobalt-Tin) *Journal of Phase Equilibria and Diffusion* 2006, *27*, 308–308.
52. Miyazaki, K.; Yannello, V. J.; Fredrickson, D. C., Electron-counting in intermetallics made easy: the 18-n rule and isolobal bonds across the Os-Al system *Z. Kristal.* 2017, *232*, 487–496.
53. Yannello, V. J.; Kilduff, B. J.; Fredrickson, D. C., Isolobal Analogies in Intermetallics: The Reversed Approximation MO Approach and Applications to $CrGa_4$- and Ir_3Ge_7-Type Phases *Inorg. Chem.* 2014, *53*, 2730–2741.
54. Dabbousi, B. O.; Rodriguez-Viejo, J.; Mikulec, F. V.; Heine, J. R.; Mattoussi, H.; Ober, R.; Jensen, K. F.; Bawendi, M. G., (CdSe)ZnS core-shell quantum dots: Synthesis and characterization of a size series of highly luminescent nanocrystallites *J. Phys. Chem. B* 1997, *101*, 9463–9475.
55. Efros, A. L.; Nesbitt, D. J., Origin and control of blinking in quantum dots *Nature Nanotech.* 2016, *11*, 661–671.
56. Saniepay, M.; Mi, C.; Liu, Z.; Abel, E. P.; Beaulac, R., Insights into the Structural Complexity of Colloidal CdSe Nanocrystal Surfaces: Correlating the Efficiency of Nonradiative Excited-State Processes to Specific Defects *J. Am. Chem. Soc.* 2018, *140*, 1725–1736.
57. Yao, Y.; DeKoster, G. T.; Buhro, W. E., Interchange of L-, Z-, and Bound-Ion-Pair X-Type Ligation on Cadmium Selenide Quantum *Chem. Mater.* 2019, *31*, 4299–4312.
58. Alivisatos, P., The use of nanocrystals in biological detection *Nat. Biotechnol.* 2004, *22*, 47–52.
59. Medintz, I. L.; Mattoussi, H., Quantum dot-based resonance energy transfer and its growing application in biology *Physical Chemistry Chemical Physics* 2009, *11*, 17–45.
60. Supran, G. J.; Shirasaki, Y.; Song, K. W.; Caruge, J. M.; Kazlas, P. T.; Coe-Sullivan, S.; Andrew, T. L.; Bawendi, M. G.; Bulovic, V., QLEDs for displays and solid-state lighting *MRS Bull.* 2013, *38*, 703–711.
61. Pietryga, J. M.; Park, Y. S.; Lim, J. H.; Fidler, A. F.; Bae, W. K.; Brovelli, S.; Klimov, V. I., Spectroscopic and Device Aspects of Nanocrystal Quantum Dots *Chem. Rev.* 2016, *116*, 10513–10622.
62. Talapin, D. V.; Lee, J. S.; Kovalenko, M. V.; Shevchenko, E. V., Prospects of Colloidal Nanocrystals for Electronic and Optoelectronic Applications *Chem. Rev.* 2010, *110*, 389–458.
63. Faraday, M., *Phil. Trans. Royal Soc. London* 1857, *147*, 145.
64. Website of the British Museum. (accessed March 28, 2018).
65. Sigma-Aldrich online catalog. (accessed March 28, 2018).
66. Ekimov, A. I.; Efros, A. L.; Onushchenko, A. A., Quantum size effect in semiconductor microcrystals *Solid State Commun.* 1985, *56*, 921–924.
67. Ekimov, A. I.; Onushchenko, A. A., Quantum size effect in 3-dimensional microscopic semiconductor crystals *JETP Lett.* 1981, *34*, 345–349.
68. Brus, L. E., A simple-model for the ionization-potential, electron-affinity, and aqueous redox potentials of small semiconductor crystallites *J. Chem. Phys.* 1983, *79*, 5566–5571.
69. Brus, L. E., Electron electron and electron-hole interactions in small semiconductor crystallites: The size dependence of the lowest excited electronic state *J. Chem. Phys.* 1984, *80*, 4403–4409.
70. Murray, C. B.; Norris, D. J.; Bawendi, M. G., Synthesis and characterization of nearly monodisperse CDE (E = S, Se, Te) semiconductor nanocrystallites *J. Am. Chem. Soc.* 1993, *115*, 8706–8715.
71. Heath, J. R.; Shiang, J. J., Covalency in semiconductor quantum dots *ChemSoc. Rev.*. 1998, *27*, 65–71.
72. Yu, W. W.; Qu, L. H.; Guo, W. Z.; Peng, X. G., Experimental determination of the extinction coefficient of CdTe, CdSe, and CdS nanocrystals *Chem. Mater.* 2003, *15*, 2854–2860.
73. Yu, W. W.; Qu, L. H.; Guo, W. Z.; Peng, X. G., Experimental determination of the extinction coefficient of CdTe, CdSe and CdS nanocrystals: Addition *Chem. Mater.* 2004, *16*, 560–560.
74. Jasieniak, J.; Smith, L.; van Embden, J.; Mulvaney, P.; Califano, M., Re-examination of the Size-Dependent Absorption Properties of CdSe Quantum Dots *Journal of Physical Chemistry C* 2009, *113*, 19468–19474.
75. Kippeny, T.; Swafford, L. A.; Rosenthal, S. J., Semiconductor nanocrystals: A powerful visual aid for introducing the particle in a box *J. Chem. Educ.* 2002, *79*, 1094–1100.
76. Hendricks, M. P.; Campos, M. P.; Cleveland, G. T.; Jen-La Plante, I.; Owen, J. S., A tunable library of substituted thiourea precursors to metal sulfide nanocrystals *Science* 2015, *348*, 1226–1230.
77. LaMer, V. K.; Dinegar, R. H., Theory, production and mechanism of formation of monodispersed hydrosols *J. Am. Chem. Soc.* 1950, *72*, 4847–4854.
78. Polte, J., Fundamental growth principles of colloidal metal nanoparticles - a new perspective *Crystengcomm* 2015, *17*, 6809–6830.
79. Peng, X. G.; Wickham, J.; Alivisatos, A. P., Kinetics of II-VI and III-V colloidal semiconductor nanocrystal growth: "Focusing" of size distributions *J. Am. Chem. Soc.* 1998, *120*, 5343–5344.
80. Dasgupta, N. P.; Sun, J. W.; Liu, C.; Brittman, S.; Andrews, S. C.; Lim, J.; Gao, H. W.; Yan, R. X.; Yang, P. D., 25th Anniversary Article: Semiconductor Nanowires Synthesis, Characterization, and Applications *Adv. Mater.* 2014, *26*, 2137–2184.
81. Mai, L. Q.; Tian, X. C.; Xu, X.; Chang, L.; Xu, L., Nanowire Electrodes for Electrochemical Energy Storage Devices *Chem. Rev.* 2014, *114*, 11828–11862.
82. Rycenga, M.; Cobley, C. M.; Zeng, J.; Li, W. Y.; Moran, C. H.; Zhang, Q.; Qin, D.; Xia, Y. N., Controlling the Synthesis and Assembly of Silver Nanostructures for Plasmonic Applications *ChemRev.* 2011, *111*, 3669–3712.
83. Yan, R. X.; Gargas, D.; Yang, P. D., Nanowire photonics *Nature Photonics* 2009, *3*, 569–576.
84. Castro Neto, A. H.; Guinea, F.; Peres, N. M. R.; Novoselov, K. S.; Geim, A. K., The electronic properties of graphene *Rev. Mod. Phys.* 2009, *81*, 109–162.
85. Dreyer, D. R.; Park, S.; Bielawski, C. W.; Ruoff, R. S., The chemistry of graphene oxide *Chem. Soc. Rev.* 2010, *39*, 228–240.
86. Geim, A. K., Graphene: Status and Prospects *Science* 2009, *324*, 1530–1534.
87. Allen, M. J.; Tung, V. C.; Kaner, R. B., Honeycomb Carbon: A Review of Graphene *Chem.Rev.* 2010, *110*, 132–145.
88. Mayer, J. E., The lattice energies of the silver and thallium halides *J. Chem. Phys.* 1933, *1*, 327–334.
89. Dyachenko, O. A.; Graja, A., Crystal structure of C_{60} and C_{70} compounds *Fullerene Sci. Technol.* 1999, *7*, 317–385.

CHAPTER 20

Bioinorganic Chemistry

"In 1926, Sumner isolated from jack beans (Canavalia ensiformis) *the first crystalline enzyme, urease, and defined the proposition that enzymes could be proteins devoid of organic coenzymes and metal ions. It is therefore with some sadness that in this communication we adduce evidence which strongly indicates that urease is a nickel metalloenzyme."*

— **Dixon, Gazzola, Blakeley, and Zerner**[1]

CONCEPT COVERAGE

1. **Introduction and brief history**
2. **Ligands, metals, and structures found in biological systems**
3. **Common reaction types found in biology**
4. **Spectroscopic methods**
5. **Case studies on metal uptake and small molecule activation**

Bioinorganic chemistry is the subdiscipline of inorganic chemistry that seeks to understand the role of inorganic elements, usually metal ions, in biological systems. It is intricately linked with modern biochemistry/molecular genetics and the structure and function of proteins, but also includes the role of metal ions and coordination chemistry. The ligands in a biological system are more limited than those in a synthetic inorganic chemistry laboratory, as they are principally derived from amino-acid side chains from proteins and phosphodiester linkages and base-pair heteroatoms from DNA. However, especially for proteins, the specific orientation of the ligands can be precisely controlled by the overall structure of the protein allowing for fascinating chemical transformations that cannot yet be carried out by a synthetic chemist in the laboratory. One major reason for this is the intimate communication between different proteins that allow information to be transferred in a cooperative fashion across long distances. This chapter will begin with a brief discussion of both the ligands and the metals present in biological systems before moving on to reaction chemistry. The field relies on some fairly advanced or specialized spectroscopic techniques, and these will be introduced to provide a sense of both their role and importance. The chapter continues with a discussion of the question of how an organism selectively uptakes and uses a specific inorganic element with case studies on iron, copper and oxygen, and ends with three case studies that illustrate some of the spectroscopic techniques and show how they were used to elucidate structure and mechanism of action for several metalloenzymes that activate the small molecules N_2 to NH_3, CH_4 to CH_3OH, and H_2O to O_2. The field of bioinorganic chemistry is vast, and many important and interesting systems have been omitted. The interested reader is encouraged to consult in-depth specialized texts.[2–6]

20.1 Introduction and History

An enzyme is a chemical reagent that is produced by a living organism which acts to catalyze a specific chemical reaction. The first enzyme, amylase, was discovered in 1833 by the French chemist Anselme Payen. In 1835, the Swedish chemist Jon Jakob Berzelius studied enzymes and discovered that they had catalytic activity, but their biochemical makeup was unknown. The term "enzyme"—from the Greek ζυμόω (*zumoó*), "leaven"—was not coined for another forty years, by the German physiologist Wilhelm Kühne, and it was not until 1926 that the first enzyme was isolated and crystallized by the American chemist James Sumner,[7] whose experiments proved that enzymes were proteins, for which he was awarded the Nobel Prize in 1946. The protein he studied, jack bean urease, catalyzes the decomposition of urea into carbon dioxide and ammonia, according to Eq. 20.1.

$$(H_2N)_2C{=}O + H_2O \rightarrow 2\,NH_3 + CO_2 \qquad \textbf{20.1}$$

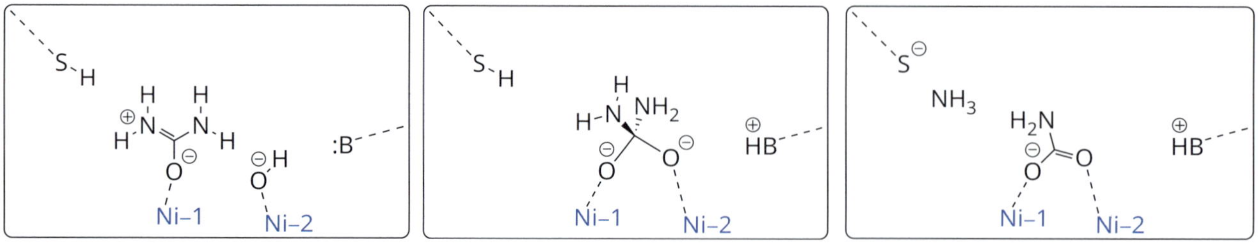

Fig. 20.1 The nickel-catalyzed hydrolysis of urea in jack bean urease. Nickel 2 activates a hydroxide to be more nucleophilic so that it can attack the urea carbon.

His observation, which was only relatively slowly accepted by the scientific community, eventually revolutionized and modernized the field of biochemistry, leading to the understanding that proteins are the species in biological systems that carry out chemical reactions. It was only in 1976 that Zerner[1] and coworkers discovered that Sumner's enzyme actually contained two nickel atoms per protein, and eventually a mechanism was proposed that involved both metal ions to activate urea (Fig. 20.1). An interesting thought experiment has been proposed to consider how the field of biochemistry might have developed differently if the discovery that enzymes contained metal ions had occurred earlier.[3] Nevertheless, it is now believed that half of proteins contain metal ions in any role, while somewhere between 25% and 35% of proteins *require* that metal ion for the activity of the enzyme to be observed. Enzymes in which the catalytic activity takes place at a metal center are termed *metalloenzymes*, some of which have metal binding sites within the protein structure, while others use metal-containing cofactors for their function. Cofactors are small chemical compounds that can be covalently or noncovalently incorporated into a protein to carry out a chemical transformation. Several metal containing cofactors will be discussed in more detail in Section 20.7.1.

What do metals do in biological systems? Although this chapter focuses on the catalytic activity of metal ions in active sites of protein enzymes, metals have other functions in biology as well. Metal ions have a positive charge and can be used both for charge balance and as signaling agents. The flux of sodium cations through a specific channel in the taste buds is how we are able to taste "salty". Metal ions are part of the mineralization process and are found in shells ($CaCO_3$), bones, and teeth (the mineral content of both being approximately that of hydroxyapatite, $Ca_{10}(PO_4)_6(OH)_2$). Metal ions are Lewis acids and can coordinate to biological structures (proteins and nucleic acids) to help stabilize or promote certain geometries. One such structure, the zinc finger, which is important for gene regulation, will be discussed in Section 20.7.4. Coordinated water molecules can be activated by Lewis acids to become much more acidic than free water. Water can be activated to become a potent nucleophile as well, as seen in the urease enzyme already described. Similarly, in the enzyme carbonic anhydrase, a zinc atom activates water which undergoes a nucleophilic attack on carbon dioxide to generate bicarbonate.

Why use transition metals in enzymes? Unlike almost all of the other elements found in a biological system, transition-metal ions are capable of having multiple oxidation states. This allows metal ions to serve as redox centers. The change in redox potential by coordination in even slightly different chemical environments allows the flow of electrons to be highly controlled to extract the maximum amount of useful energy. In addition, transition-metal ions have multiple coordination environments and different ionic radii in different oxidation states, allowing for structural flexibility. The flexibility in ionic radius, particularly, can be used to drive major conformational changes that assist with small molecule binding. One example is hemoglobin, where the cooperative binding of dioxygen is assisted by a significant structural change upon binding; this will be discussed in Section 20.8.4.

One last important use for transition metals in biological systems is in medical applications. Some of the most potent anti-cancer chemotherapy agents are the platinum drugs. The parent drug is known as *cis*-platin and will be discussed in Section 20.5.3. Other metal complexes are used as imaging agents. Gadolinium complexes are used as magnetic resonance imaging contrast agents and will be discussed in Section 21.3.4.1.

20.2 Ligands in Biological Systems

As you have read previously in this book, many small molecules can act as a ligand to a transition metal. In biological systems, these ligands are usually derived from sugars, proteins, and nucleic acids. The majority of the discussion that follows will focus on metalloenzymes, and as such, the focus will be on proteins binding to metals.

In metalloenzymes, the types of ligands that can bind to and support the transition metal are significantly more limited than those in the synthetic chemist's toolbox. Protein sidechains (described in detail in Section 20.3) contain variations on common ligating groups such as water, acetate, and ammonia, with a few additions such as thiol, thioether, guanidine, and imidazole. From the simplified perspective of an inorganic chemist, the purpose of the protein side chains is to hold the transition-metal ion in a fixed and carefully controlled geometry, while the rest of the protein serves to provide channels or pathways to allow substrates to enter and products to leave, or to provide docking sites so that two proteins can come into

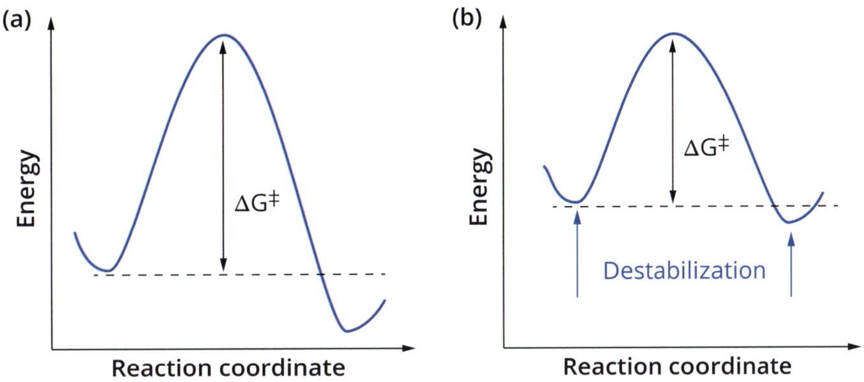

Fig. 20.2 (a) A reaction coordinate diagram for a generic chemical reaction. (b) The same chemical reaction by which both reactant and product are in an entatic state. This state is destabilized relative to the minimum energies of both the reactant and product, thus lowering the activation barrier for the chemical reaction.

close contact for an electron transfer event. While it is easier to focus only on the active site and the metal atom, this is a simplistic view of the overall biological system. It is clear that the larger structure of the protein is important for its function. The protein structure is also involved in enzyme regulation, binding, recognition, and many other important processes that will not be discussed at length in this chapter, though this does not diminish their obvious importance in biochemistry.

For metalloenzymes that carry out a chemical reaction, especially for redox reactions, the protein side-chains often orient the metal in an unusual or high-energy geometry, the entatic state. This word derives from the Greek *entasis*, which means "under tension". For the bioinorganic chemist, the state is defined as having a geometry that is intermediate between those of the starting material and product, thus lowering the activation barrier and increasing the rate of the corresponding reaction. As seen in Fig. 20.2, a protein that destabilizes both the reactant and product geometries would result in a reaction coordinate diagram with a lower free energy of activation, and thus a faster rate.

The classic example of the entatic phenomenon is the blue copper site found in plastocyanin. Most 4-coordinate Cu(I) complexes are tetrahedral due to the spherically symmetric d^{10} electronic configuration, while Cu(II) sites are typically Jahn–Teller distorted square–planar (see Chapter 15). The transition state that one might imagine for a redox event involving Cu(I)/Cu(II) that involved a geometrical change from tetrahedral to square planar would probably have a distorted square-planar or trigonal pyramidal geometry, and the geometry of copper in plastocyanin is in fact intermediate between the two limiting geometries (Fig. 20.3). The copper center is in an approximate trigonal pyramidal geometry with two nitrogen atoms from histidine amino acids and one sulfur atom from a cysteine bound in the trigonal plane with a sulfur atom from a methionine bound axially; amino acids are discussed in Section 20.3. Importantly, even the apoprotein (the protein without the metal bound) has the ligands oriented in this intermediate geometry, suggesting that the geometry is imposed by the protein itself and not caused by coordination to the metal ion. It is the protein environment and specific ligation of the metal by amino-acid side-chains in a specific orientation that causes the entatic state to be observed.

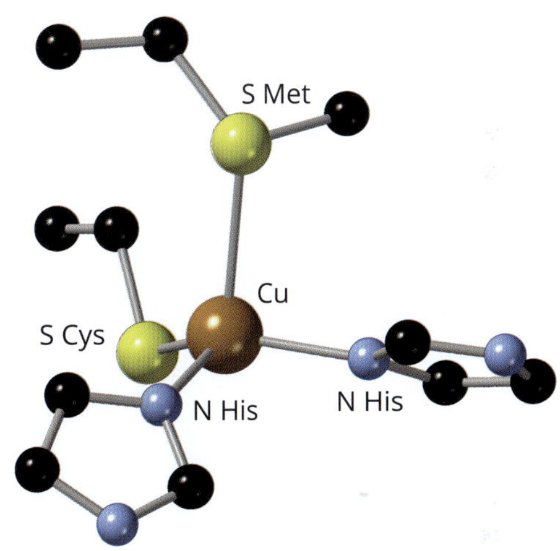

Fig. 20.3 Structure of the copper center in plastocyanin showing the distorted, almost trigonal pyramidal structure. Copper is bound by two histidine nitrogens: a cysteine sulfur and a methionine sulfur. Figure drawn using CrystalMaker[8] software with coordinates from the protein databank (PDB ID 1IUZ).[9–10]

While initially postulated to describe redox processes, chemists have described other biological systems by using these principles and have developed model complexes that have intermediate geometries. Two of these systems include phosphatases (an enzyme that cleaves a phosphodiester bond) and carbonic anhydrase (an enzyme that converts CO_2 into bicarbonate). In both of these reactions, the entatic state is proposed to activate a hydroxide substrate, making it more nucleophilic.

20.3 Protein Structure

Most proteins are made up of the twenty amino acids that are coded genetically by DNA. Of those twenty, ten are essential for human life, which means that there are no biosynthetic pathways available in humans and they must therefore be ingested. Other organisms, such as bacteria and plants, are capable of synthesizing these essential amino acids. The structure of a protein is highly regulated and governed by the "central dogma" of molecular biology—that DNA makes RNA, and that RNA makes proteins. The ability to alter DNA by molecular genetics techniques allows a scientist to alter protein structure in a controlled way.

An amino acid contains a central carbon about which an amine, a carboxylic acid, a hydrogen, and a side-chain R group are appended. All biological amino acids are α-amino acids, with the amine bound to the α-carbon from the carboxylic acid, and all that are found in proteins are found as the L-enantiomer (except the achiral glycine whose R group is a hydrogen atom). Some D-enantiomers can be found in biological systems, such as bacterial cell walls, but never in proteins. The side-chain R groups can be lipophilic or hydrophilic in nature, and among the hydrophilic side-chains there are neutral, acidic, and basic groups. The twenty amino acids are shown in Fig. 20.4. It is remarkable that nature is able to use

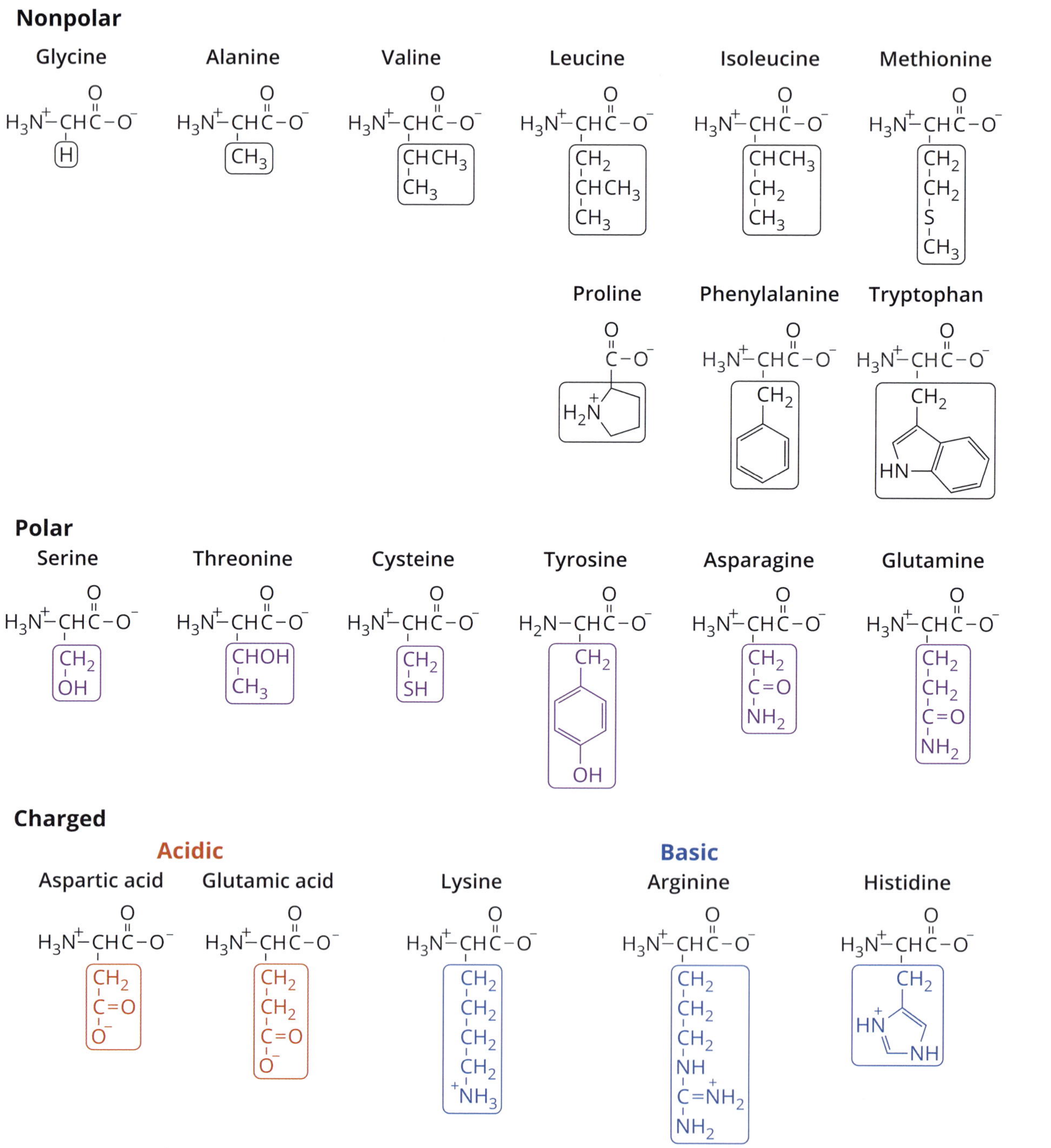

Fig. 20.4 The twenty naturally occurring genetically coded amino acids. Side chains are boxed, polar side chains are magenta, acidic side chains are red, and basic side chains are blue.

these amino acids to shape the overall protein structure, and then use others—typically aspartate, glutamate, cysteine, histidine, methionine, and tyrosine—as ligands for metals. Some amino acids are redox-active, especially tyrosine, which is implicated in a number of redox processes involving the tyrosyl radical (Section 20.9.1). Others, such as aspartic acid, glutamic acid, and lysine, are present to regulate the local pH and polarity of the active site, or to provide pathways for protons to enter an active site. There are two additional amino acids found in proteins (selenocysteine and pyrrolysine), but they are not coded directly by DNA.

Proteins are composed of a long chain of amino acids that condense the amine end of one molecule with the acid end of another forming a peptide bond. This chain of amino acids (commonly called residues) is known as the primary structure. The amine, or *N*-terminus, of the peptide is known by convention as the beginning of the protein, while the free carboxyl end, the *C*-terminus, is described as the end of the protein. The proteins are synthesized in this direction by the biological machinery of DNA translation. These protein chains can fold into additional structures, known as secondary, tertiary, and in some cases, quaternary structures.

The secondary structure results from hydrogen bonds that are both intramolecular (between different amino acid strands) and intermolecular (within the same amino acid strand). These hydrogen bonds link the N–H bond of an amide bond to the C=O bond of another amide bond and allow the protein chains to coil into more complex shapes, the most common being the α-helix and the β-sheet. The α-helix is a right-handed helical coil in which an intermolecular hydrogen bond forms between the carboxyl oxygen of one amino acid interacting with the amine hydrogen of the amino acid four residues away (Fig. 20.5). The helix completes one turn every 3.6 residues with an approximate 5.4 Å rise during each turn. There are additional, though less common, helicial structures. The β-sheet (or β-pleated sheet) is formed by intermolecular hydrogen bonds between a carboxyl oxygen of one amino acid interacting with the amine hydrogen on an adjacent peptide chain. These adjacent chains can be oriented parallel, with alignment of both *N*-terminal end to *C*-terminal directions, or antiparallel, where the adjacent chains are oriented in opposite directions, as shown in Fig. 20.5. The regions between the ordered α-helix and β-sheet are usually linked with short loops of less structured protein known as random coil. Atomic diagrams of both α-helices and β-sheets are difficult to visualize due to the large number of atoms. Ribbon diagrams are a cartoon-like representation of the structure that focus on the larger structure of the protein instead of the atomic structure more familiar to chemists.

Tertiary structures involve interactions between the various side-chain R groups within a single chain of the protein. These weak interactions, such as van der Waals forces (hydrophobic interactions), polar interactions, hydrogen bonds, and disulfide bonds, link together sections of secondary structure to form distinct 3-dimensional shapes of the protein (Fig. 20.6). Tertiary structure can also involve the incorporation of small prosthetic groups, which are nonprotein groups that are essential for the function of the protein. Prosthetic groups can be organic or can contain metal atoms, and are bound to the protein through either covalent or noncovalent interactions. Examples of prosthetic groups include vitamins, lipids, or metal-containing subunits such as heme, described in Section 20.7.2.

Some enzymes are not single proteins but are instead made up of protein complexes. The interaction of multiple peptide chains to form the overall aggregate structure is known as the quaternary structure. The interactions that form quaternary structures are the same as those that form tertiary structures, though the interactions are *interchain* instead of *intrachain*. Hemoglobin, (Fig. 20.6)—the protein responsible for carrying

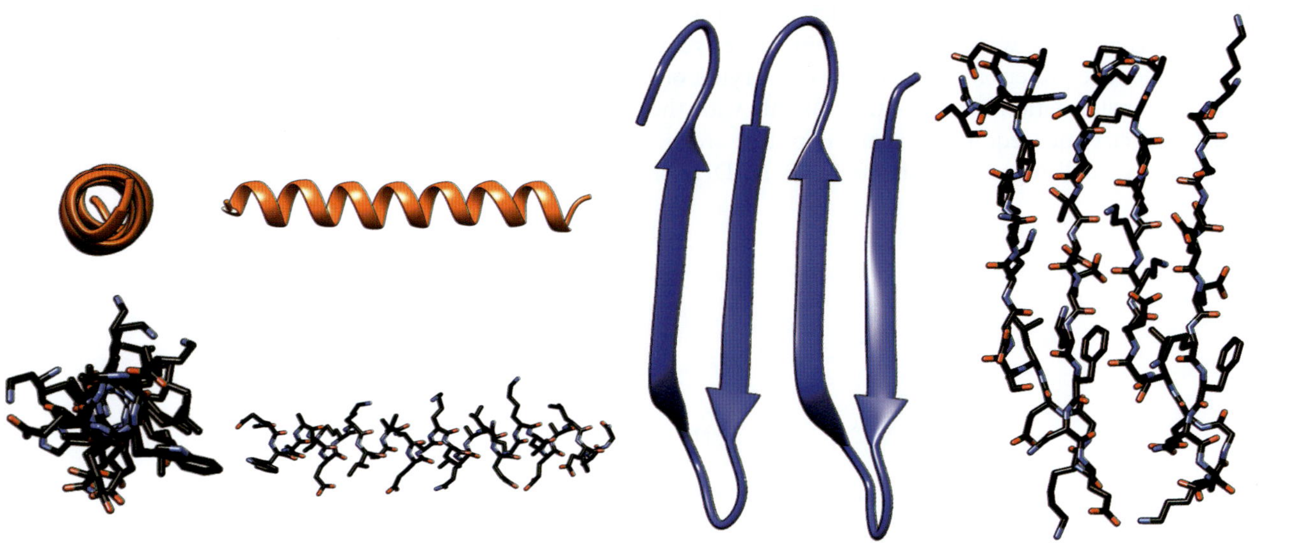

Fig. 20.5 An α-helix and an antiparallel β-sheet in both atomic and ribbon diagram form. Figure generated with UCSF Chimera[11] with coordinates from the protein databank (PDB ID 1COS and 2FKG).[9,12–13]

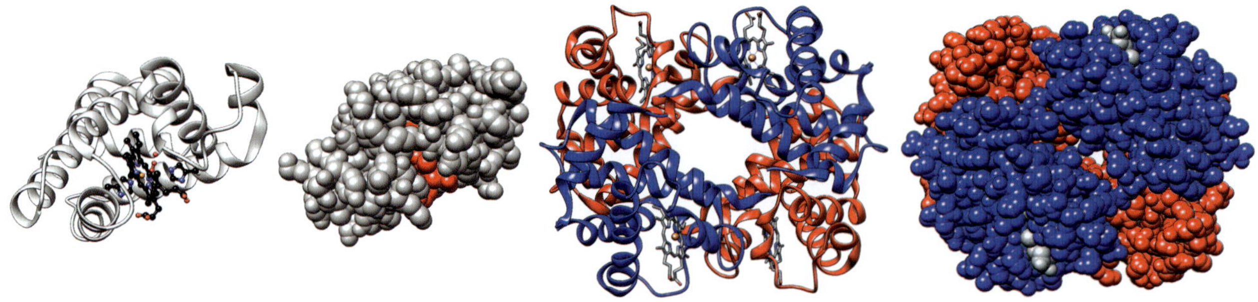

Fig. 20.6 Tertiary structure of protein chains in myoglobin (*left*), showing ribbon diagrams and atomic positions. Quaternary structure of protein chains in hemoglobin (*right*, made up of four almost identical myoglobin subunits), showing both ribbon diagrams and atomic positions. Figure generated with UCSF Chimera[11] with coordinates from the protein databank (PDB ID 1MBO and 2DHB).[9,14–15]

oxygen in mammalian blood—has a quaternary structure consisting of four almost identical myoglobin subunits, and will be discussed further in Section 20.7.2. The nitrogenase enzyme, responsible for the ability of some bacteria to split N_2 into NH_3, consists of two proteins: a homodimeric protein containing iron, and a heterotetrameric protein containing molybdenum and iron. Nitrogenase will be discussed in Section 20.9.6.

When represented visually, proteins are usually not depicted showing every single atom as is common in small-molecule chemistry. Instead, the proteins are shown in ribbon diagrams which more clearly show the gross structure at the expense of atomic precision. Both α-helix and β-sheets are relatively easy to depict in ribbon diagrams, making it easier to understand the larger interactions present that form the tertiary and quaternary structures.

Proteins serve many functions in biology, from structural (keratin is a key structural component in hair, horns, and skin) to movement (actin and myosin are the principal components of muscle tissue) to transport and storage (oxygen is carried in hemoglobin and stored in myoglobin) and to catalysts (cytochrome c oxidase is the enzyme in the last step in respiration where oxygen is reduced to water). When proteins serve as catalysts, sometimes the protein itself is capable of carrying out the chemical transformation by interacting directly with the substrate. Other enzymes require external groups that are not proteins, such as small organic fragments (ATP, NADH) or metal ions. For the purposes of this chapter, the most important function of protein structure is to orient the amino-acid side-chains so that they can serve as ligands for metal ions.

20.4 Nucleic Acids: RNA and DNA

In biology, the principal purpose of the nucleic acids is information storage. DNA in chromosomes is responsible for storing the genetic code that the organism uses to synthesize RNA and then proteins. DNA is a long polymer made up of four different nitrogenous bases: adenine, guanine, thymine, and cytosine (A, G, T, and C). These bases are linked to the pentose β-D-2-deoxyribose, and the pentoses are linked together by 3′-5′ phosphodiester bonds. The double helix of DNA is formed by hydrogen bonding between the base pairs A and T or G and C (Fig. 20.7). Although there are several types of hydrogen bonding interactions that can give rise to slightly different DNA structures, B-DNA is the most common form in cells. It forms a right-handed helix that is 23.7 Å wide, with the rise between base pairs of 3.5 Å. The DNA helix makes one complete turn every 10.4 base pairs. DNA has two grooves between the phosphate sugar backbone strand. One groove, the major groove, is significantly larger in both width and depth than the other, the minor groove. DNA can be selectively bound by proteins that fit into the different groove sizes. The major groove provides access for

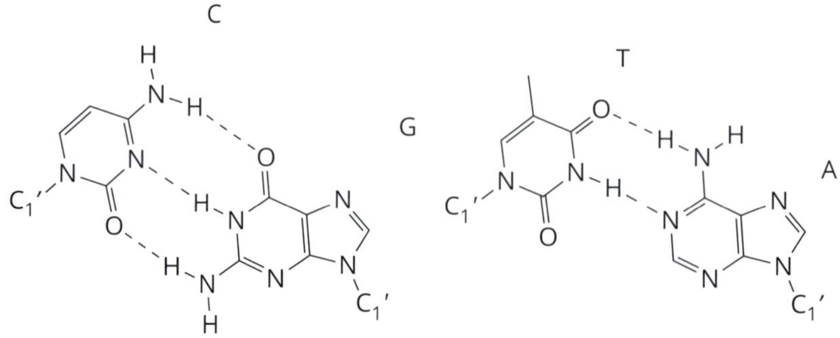

Fig. 20.7 The hydrogen bonding interactions between C–G and T–A base pairs in DNA.

binding to the DNA bases by transcription factors, metal cations, and regulatory proteins.

DNA bases can readily bind to metal cations, with the most favored binding sites to the endocyclic nitrogen atoms (the nitrogen atoms *in* the rings and not the appended amines). While important for life, the remainder of this chapter will not focus on DNA–metal interactions beyond the example of *cis*-platin in Section 20.5.3, but will instead focus on metalloenzymes.

20.5 Elements in Biological Systems

The chemist's principal tool is the periodic table, and you are probably quite familiar with it by this point in your academic career. Many of the elements of the periodic table are required for life, ranging from bulk elements (those that make up the structure of sugars, proteins, and nucleic acids, and structural features such as bones) to the trace elements (Fig. 20.8).[16] The ability for organisms to uptake and store trace elements is a huge area of modern research, but will not be discussed beyond that of iron and copper in subsequent sections.

20.5.1 Types of Biological Elements

Bulk biological elements: H, C, N, O, Na, Mg, P, S, Cl, K, Ca. The bulk biological elements are those that make up a large proportion of a biological organism. Since much of living things contain organic structures, it is no surprise that C, H, N, and O are bulk elements, but phosphorus is required to form the phosphate backbone of DNA and RNA, and sulfur is found in two of the essential amino acids. In addition to their charge-carrying role, to balance the net negative charge of DNA, Na, Mg, Ca and K are also commonly used for nerve signal transduction. The halides Cl and Br are also used as charge carriers and help to provide the high ionic strength found in blood plasma and tissues.

Essential elements: B, F, Si, V, Cr, Mn, Fe, Co, Ni, Cu, Zn, Se, Mo, I. The essential elements are those for which important biological roles are known for most organisms. The first-row transition elements from vanadium to zinc are those that will be considered the most in this chapter, as these are what make up bioinorganic enzymes.

Possibly essential (for some species): Ti, As, Br, Sr, Cd, Sn, W. These elements are not typically considered essential, and in fact, most would classify arsenic, cadmium, and tin as toxic or poisonous. However, in small doses these elements are in fact required for some species, while in larger doses they can interfere with other biological processes.

Poisonous (depends on dose): Be, Sb, Os, Hg, Tl, Pb, Po. All metals (and other chemicals) that are required for life have a dose-response curve. The curves for different metals can vary widely, depending on the need or concentration in the Earth's crust/hydrosphere. When the concentration of an essential element is too low within an organism, the organism can exhibit symptoms or diseases due to mild or severe deficiency (Fig. 20.9). On the other extreme, when the concentration is too high there can be symptoms due to mild to severe toxicity. Toxic metals have a differently shaped dose-response curve.

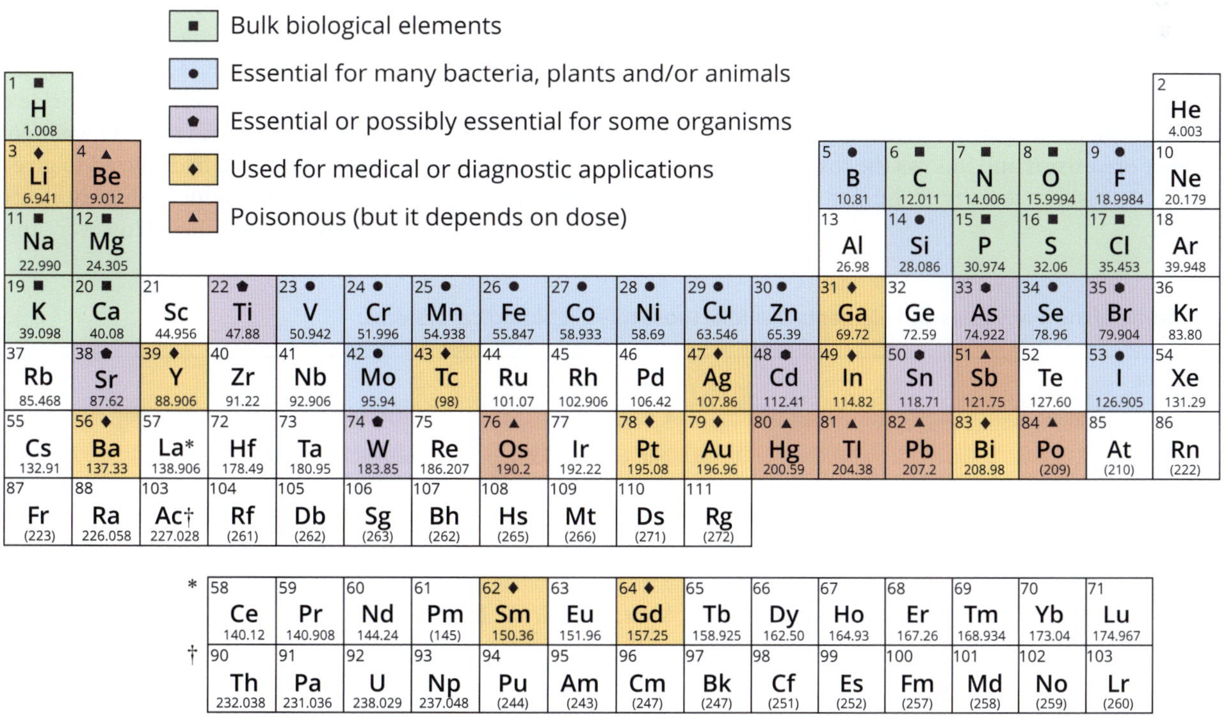

Fig. 20.8 The biological periodic table. Green, bulk biological elements; blue, essential for a wide number of organisms; gray, essential for some organisms; gold, medical or diagnostic; red, poisonous.

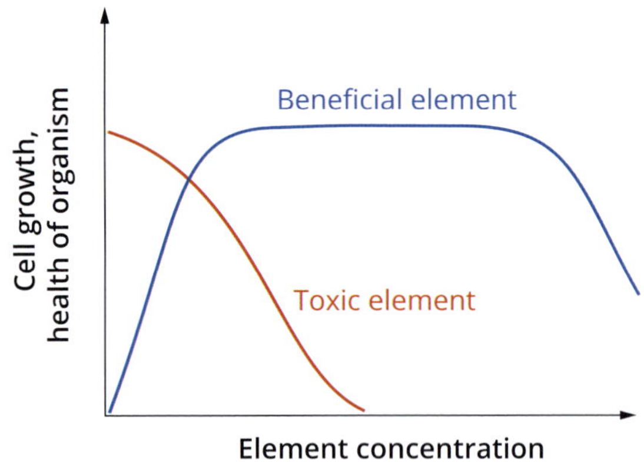

Fig. 20.9 Dose response curves for beneficial and toxic metal ions.

At low doses there is no effect, but as the concentration increases the organism responds unfavorably.

Iron is an essential mineral that has important functions. Some symptoms of low iron are anemia and tiredness due the importance of iron in transporting oxygen, among other roles. A typical person has on the order of 3–4 grams of iron in their body, while less than 1–2 mg is lost per day. On the other hand, iron overload results from the accumulation of too much iron, typically from a hereditary iron update disease or blood transfusions. Symptoms can range from pain and fatigue to heart failure, but most who have high iron levels are asymptomatic. The safe range for the consumption of iron is on the order of 10–20 mg/day, as not all of the iron ingested accumulates in the body. However, since the body is very good at iron uptake and storage, blood serum levels of iron are typically less than one ng/L. Increases of the serum level can be measured, as can precipitation of elemental iron in some tissues. In contrast, the element cadmium, while believed to be an essential micronutrient for marine phytoplankton, is toxic in higher organisms and follows the toxic dose response curve. It begins to show toxicity in humans above approximately 10 μM.

A brief illustration of the many roles that metals and other inorganic species serve in biology is presented in Table 20.1. Many of these roles are more biological or physical rather than chemical, and this chapter will focus on the use of transition metals in reaction chemistry.

20.5.2 Medicinal or Diagnostic Elements

Although there are many elements that have medical roles in biology, we will present one of them in more detail: the anti-cancer drug cisplatin and its derivatives. The suite of gadolinium complexes used as MRI contrast agents will be discussed in 22.3.4.1. The interested reader is referred to the excellent review article by Sadler that illustrates the use of metal elements for anti-cancer therapy, photodynamic therapy, antifungals, Alzheimer's disease, Huntington's disease, and diagnostic and therapeutic radiopharmaceuticals.[17] There is also an excellent review of nuclear medicine by Blower.[18]

20.5.3 The Platin Family of Chemotheraputics

Some of the most potent and widely used anti-cancer chemotherapy agents are the platinum drugs. Although cisplatin was discovered in the 1845 by Peyrone, the use of platinum drugs as pharmaceuticals was not developed until more than 100 years later after the serendipitous discovery that electrolysis of a platinum electrode inhibited cell division in *E. coli*. This class of drugs is known collectively as "platins", and approximately half of all people that are treated with chemotherapy for cancer receive a platinum drug.

The drug cisplatin is *cis*-$PtCl_2(NH_3)_2$. The drug is administered intravenously, and the relatively high chloride concentration in blood keeps it in its neutral state. Once it enters the cytoplasm, the chloride ligands are hydrolyzed to give the cationic *cis*-diaquo complex. This product is highly electrophilic and can react with many nucleophiles. However, analysis of

Table 20.1 Some of the many roles and elements that fulfil those roles in biological systems.

Charge balance	Na, K, Cl, Br
Structure and templating	Ca, Zn, Si
Signaling	Na, Ca, NO, Zn
Brønsted acid or base	P, Si, C
Lewis acid or base	Zn, Fe, Ni, Mn, Mg
Electron transfer	Fe, Cu, Mo
Group transfer (O, S)	V, Fe, Co, Ni, Cu, Mo, W
Redox	V, Mn, Fe, Co, Ni, Cu, W, S, Se
Energy storage	H, P, S, Na, K, Fe
Biomineralization (bones, shells)	Ca, Mg, Fe, Si, Sr, Cu, P
Medical or diagnostic	Li, Ba, Ti, Y, Tc, Ag, Au, Pt, Ga, In, Bi, Sm, Gd

treated cells shows that most of the platinum ends up bound to nitrogen donors on nucleic acid bases, especially guanine.[19] These complexes are called DNA adducts of platinum. Thus, after incorporation into the body, the platinum finds its way to the nucleus where it binds to DNA.

Intense efforts were made to study DNA adducts of platinum complexes in order to understand how the drug worked. Among many of the experiments, the binding of *cis*-$Cl_2Pt(NH_3)_2$ group to a 12 base pair palindromic DNA strand (d(CCTCTGGTCTCC), d(GGAGACCAGAGG)) was studied both in solution and in the solid state. The water ligands on the hydrolyzed *cis*-platin are lost and are replaced with guanine DNA base ligands, thus forming the complex *cis*-$[Pt(NH_3)_2(GG)]$. The structure of the adduct was determined by both solid-state X-ray and solution NMR methods (Fig. 20.10).[19] The binding of the platinum atom to the two GG bases causes a significant kink in the structure of approximately 50° (X-ray) to 70° (NMR). Most of the binding differences between the two techniques are ascribed to packing of the DNA in the solid state in the X-ray study.

In addition to the GG intrastrand cross-link shown here, which accounts for >60% of all observed adducts, there are also GA cross-links (20%). and there is also evidence for interstrand cross-linking, in which the Pt binds to guanine bases on opposite strands (5%). There are also some studies that show DNA–protein cross-linking by the platinum center.[21]

The formation of cisplatin-DNA interactions causes a significant distortion in the structure of DNA. This is due both to unwinding and shortening of the DNA strand. These effects result in a loss of helix stability by approximately 6 kcal/mol for each GG adduct. This significant distortion has clear implications for the ability to translate DNA. Site-specific binding proteins will not be able to find their targets, and the DNA replication machinery is disrupted at both the transcription and translation stages.

Since the discovery of cisplatin and the determination of its mechanism, there has been an intense effort to discover and use derivatives of cisplatin that have enhanced chemotherapeutic effects with reduced cytotoxicity, higher specificity for tumors, or enhanced stability *in vivo*. Several complexes are currently clinically approved, and they all are believed to operate by a similar mechanism to cisplatin (Fig. 20.11).[21]

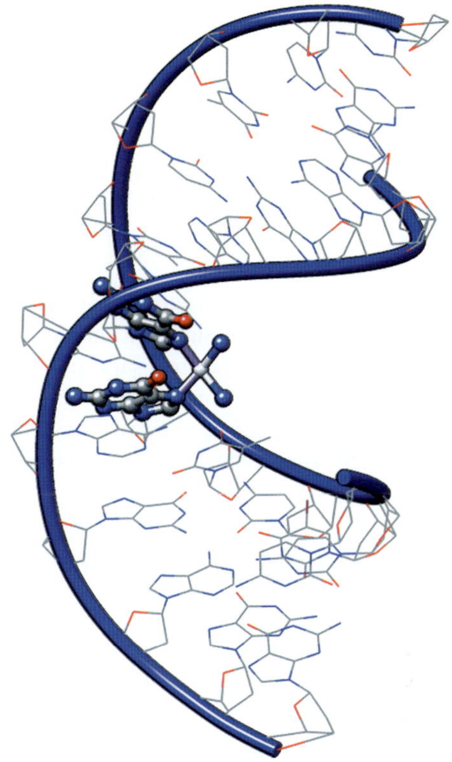

Fig. 20.10 View of a 12-mer of DNA where cisplatin binds to adjacent guanine residues. The structure from X-ray crystallography shows the $Pt(NH_3)_2$ group bound to two of the exocyclic nitrogen atoms in adjacent guanine residues (ball and stick view). The other base pairs are drawn as wireframe, and the ribbon indicates the phosphodiester backbone of the DNA. Figure generated with UCSF Chimera[11] with coordinates from the protein databank (PDB ID 1AIO).[9,20]

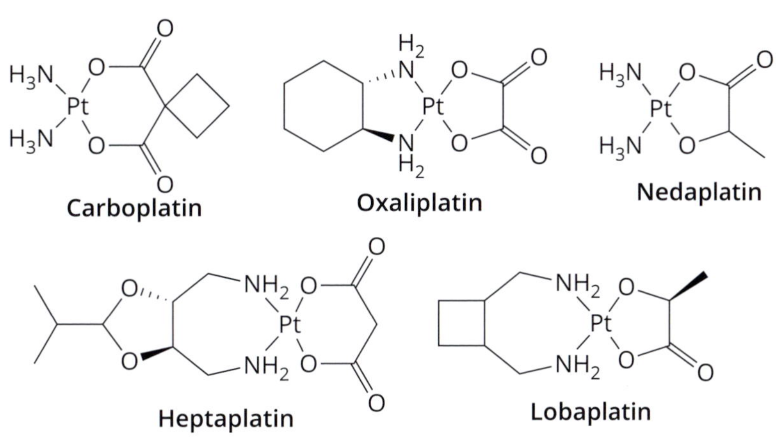

Fig. 20.11 Some modern platinum anti-cancer drugs derived from cisplatin.

20.6 Experimental Methods

This section will introduce a variety of experimental and spectroscopic techniques with an emphasis on techniques that are used in bioinorganic chemistry. The elucidation of either structure or function of metalloenzymes depends on piecing together many clues to build a more complete picture or model. Each experimental technique provides one piece of data, and these are all put together to make a consistent description of the overall structure. Model complexes—synthetic complexes that are designed to model one or more spectroscopic or reactivity features—are included in this section in order to illustrate and emphasize their role in structure elucidation. However, before one can carry out a spectroscopic measurement, one must first collect, synthesize, and isolate the protein.

20.6.1 Obtaining the Protein

The advent of modern biochemical and molecular biology techniques has made it possible to overexpress proteins in bacteria and other organisms. The general procedure for procuring proteins to study will be outlined in the barest detail only to provide a sense of the process and an appreciation for the steps and difficulty involved.

Once a protein is chosen to study, the DNA sequence that corresponds to the protein must be found. This can be achieved by sequencing the genome or parts of the genome of the organism. The DNA sequence is converted into a plasmid, a circular DNA strand that is used by bacteria. When the plasmid is incorporated into a bacterial cell line, the bacteria will process that DNA as its own, converting the DNA into RNA, and then into proteins. Ideally, the plasmid and the protein are overexpressed by the bacteria, so that it makes much more of that protein than other proteins it normally makes. The bacteria grow rapidly, and the protein can be isolated from the bacteria through relatively standard chemical and mechanical steps. Difficulties can arise, depending on the sensitivity of the protein to factors like pH or salinity. Purification of the protein can be carried out by chromatographic separation or crystallization. For use in solution studies such as spectroscopy or reactivity studies, single crystals of the protein are not required, but for X-ray structural studies, single crystals must be grown. In the special case of membrane-bound proteins, where the protein resides within a lipid bilayer, further complications can occur due to the added difficulty of simultaneously crystallizing the protein in a section of a lipid bilayer.

One aspect of molecular biology that has significantly aided bioinorganic chemistry is the ability to change the genetic code at will. It is often possible to specifically alter individual amino acids within a larger protein and determine the effects of this change on the chemical processes that ensue. For example, if a polar residue such as asparagine is thought to be important for protein function, it could be changed to a nonpolar but similar-sized amino acid such as leucine. Channels that substrates travel through in order to reach the active site from the protein exterior may be lined with small nonpolar amino acids such as alanine, for nonpolar substrates such as hydrocarbons or dioxygen, or with polar amino acids such as serine, for polar substrates such as water or protons. Substitution of that group with a bulkier one, such as phenylalanine, may slow down substrate access, change substrate specificity, or even shut down the activity all together. Changing the polarity of a channel can also impact substrate entrance or product exit.

One important method for solving the single-crystal X-ray structures of metalloenzymes, the Patterson method, is the incorporation of heavy atoms. These can be installed in the protein backbone with limited structural or functional changes by substitution of the amino acid selenocysteine for cysteine. Binding sites for heavy metal atoms can be added to the periphery of the protein using thiol-containing amino acids such as methionine and cysteine. Finally, multiple histidine residues can be attached to the *N*- or *C*-terminus of the protein. These histidine side-chains can bind to immobilized Ni^{2+} or Cu^{2+} on a chromatography column, allowing for relatively easy separation of the desired protein from other proteins in the mixture.

20.6.2 UV–Visible Methods

As was discussed in Chapter 16 for transition metal complexes, UV-visible spectroscopy is useful to identify the presence of transition-metal ions in proteins due to their often intense colors. As a reminder, the energy of UV–visible photons is in the order of several hundreds of kcal/mol, and the absorption of a photon results in an electronic transition from an electron in an occupied orbital to a higher-energy unfilled orbital. UV–visible spectroscopy can be used to determine Δ_o (the crystal field splitting energy) for a transition-metal complex, as well as the energies of other electronic transitions. As such, it provides valuable information about the electronic structure of a metal in an enzyme.

Many copper-containing proteins are blue due to absorption of a photon by the d^9 Cu(II) center. For this reason, a whole class of copper enzymes, including plastocyanin (Fig. 20.3), are called "blue copper proteins". These proteins absorb light around 600 nm due to an LMCT band from a cysteine sulfur ligand to an unoccupied copper orbital. Arthropods have blue blood because their oxygen carrier is based on copper. Mammalian blood is red due to the intense absorption of light at 450–550 nm by the iron-containing protein hemoglobin (see Sections 20.7.2 and 20.8.4). Metal-containing proteins can be characterized in much the same way that UV–visible spectroscopy is used to identify and describe the electronic structure of transition-metal complexes.

Circular dichroism (CD) spectroscopy is a UV–visible absorption experiment that relies on the fact that chiral molecules have differential absorption of circularly polarized light. A common technique for characterizing a chiral substance is to determine its specific optical rotation, $[\alpha]_D$, using the sodium D line (Chapter 2), while CD spectroscopy measures the optical rotation across a spectrum of UV and visible wavelengths.

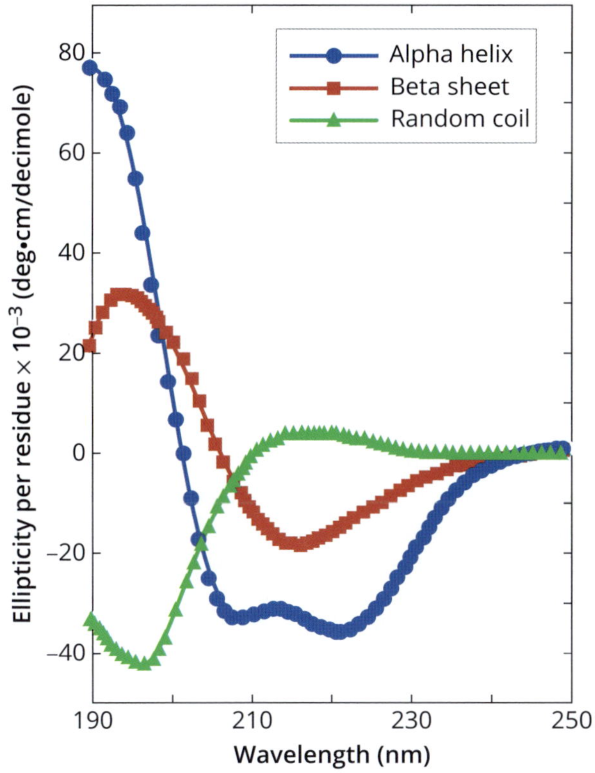

Fig. 20.12 Characteristic CD spectra in the UV region for protein secondary structures.

Proteins, being made up of chiral building blocks, are chiral and have strong CD signals in the UV region. The secondary structures of proteins (α-helices and β-sheets, discussed in Section 20.3) have characteristic signals in this region (Fig. 20.12), and CD is a common and rapid technique to verify that proteins are folded correctly before carrying out other experiments. These absorptions can be thought of as a fingerprint for the correctly folded protein, though more detailed analyses can be done. Metalloenzymes may have additional features in the visible region, due to LMCT and d-d transitions that can be observed.

The more sophisticated technique of magnetic circular dichroism is carried out in the presence of a strong magnetic field parallel to the direction of propagation of the light. Splitting of degenerate energy levels by the magnetic field (Zeeman splitting) gives rise to characteristic spectral patterns that can be assigned to metal-containing proteins. The technique can be used to characterize the oxidation state and spin state of metal centers.[22–23]

20.6.3 NMR

Nuclear magnetic resonance spectroscopy relies on the fact that some nuclei, such as ^{1}H, ^{13}C, and ^{15}N, have a nuclear spin. The important parameters obtained from NMR spectroscopy are the chemical shift (δ) and any couplings (J) due to nearby nuclei that can interact with the nucleus under study. The technique will be briefly reviewed here, as it is commonly covered in depth in the organic chemistry curriculum. Table 20.2 provides information obtainable by NMR techniques20.

In the presence of a strong external magnetic field, the nuclear spin can align either with or against the applied field. The energy gap has energy in the radio-frequency region of the electromagnetic spectrum, and the spin can be flipped by absorption of a photon. The relaxation of the proton back to its original position is the source of the signal in NMR spectroscopy. What makes the technique so powerful is the small changes in local magnetic field due to environment that can be detected. The NMR signals are reported with a chemical shift, δ, in parts per million relative to a standard. The normal region for ^{1}H NMR spectroscopy is 0–10 ppm: an aldehyde proton has its resonance frequency at around 10 ppm, while a methyl group in a hydrocarbon appears around 1 ppm. In addition, the J coupling and splitting of protons provides important connectivity information. The number of peaks into which a proton is split depends on 1 plus the number of protons attached to the adjacent carbon—the "n+1" rule—and the intensity of the signal is described by Pascal's triangle. For example, a CH_3 group next to a CH_2 group appears as a 1:2:1 triplet, while the CH_2 group appears as a 1:3:3:1 quartet (Fig. 20.13).

NMR is an incredibly powerful structural tool in synthetic inorganic chemistry—much as it is for organic chemistry—but its use in metalloenzyme systems is perhaps more limited. The metal center of interest is only one small part of the much larger protein environment, and the observed signals will be mostly due to the protein. In addition, many transition-metal centers are paramagnetic, and paramagnetic centers can lead

Table 20.2 Summary of major NMR methods and the information that can be obtained by that technique.

NMR Method	Brief Description
1D NMR (diamagnetic system)	Environment of nucleus and adjacent atoms through coupling
1D NMR (paramagnetic system)	Proximity to unpaired electron density
COSY or HETCOR	Connectivity due to transfer of magnetization through bonds to adjacent atoms
TOCSY	Connectivity due to magnetization through long range bonding
NOESY	Connectivity through space, even if not directly bonded

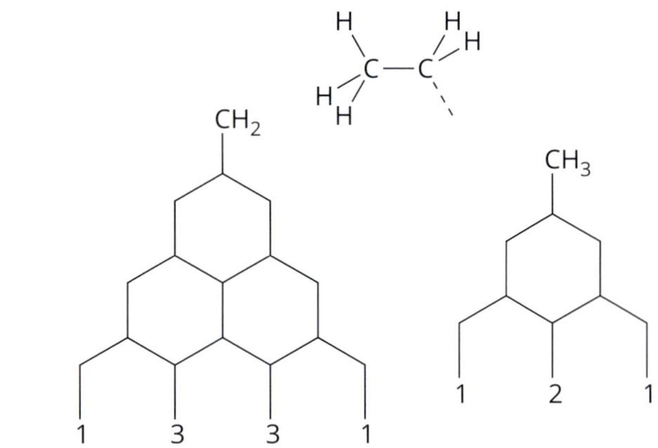

Fig. 20.13 Splitting of a CH_2 group by an adjacent CH_3 group produces a 1:3:3:1 quartet, while splitting of the CH_3 group by the adjacent CH_2 group produces a 1:2:1 triplet.

to significantly broadened and shifted resonances, which can limit the utility of NMR even further. However, the proximity of an amino-acid residue to a paramagnetic center can often be determined by whether the residue is affected by the paramagnetism of the metal center. The enzyme superoxide dismutase contains paramagnetic metal centers (usually copper, iron, or manganese), and the chemical shifts of the protons in histidine side-chains that bind to those metal centers are shifted well outside the normal region to a range of 20–85 ppm (Fig. 20.14). The peaks are also broadened relative to diamagnetic systems.

Two-dimensional spectroscopy can also be used to help assign protein structure. Three common examples are COSY (^{1}H-^{1}H COrrelation SpectroscopY), HETCOR (^{1}H-^{13}C HETeronuclear CORrelation spectroscopy), and TOCSY (TOtal Correlation SpectroscopY). In these techniques, magnetization is transferred through bonds in adjacent atoms. COSY and HETCOR reports only directly bonded atoms, while TOCSY can follow coupling through multiple bonds. As there is no hydrogen present on the carbonyl carbon of an amide bond, there is no coupling between adjacent amino-acid residues. Therefore, each TOCSY coupling system will correspond to a single residue in a protein. Strong COSY signals can be observed between adjacent amide N-H protons in both α-helices and β-sheets.

A more sophisticated NMR experiment relies on the nuclear overhauser effect: NOESY (NOE SpectroscopY). This technique can be used to determine spatial information about protein environments. The intensity of a proton signal can be modulated by the transfer of nuclear spin information from a nearby proton through space. Selective irradiation of one proton can cause signal enhancement of a proton that is nearby spatially even if not connected by a small number of bonds.

Protein structures can in some cases be determined in solution by using NMR. COSY, and TOCSY allow for specific peptides to be correlated and assigned using through bond-coupling. Since proteins fold, protons that are hundreds of bonds away can be brought into close proximity, and since the intensity of the NOESY peak is proportional to $1/r^6$, distance ranges can be assigned according to this intensity. More recent techniques include 3-dimensional and even 4-dimensional NMR. For example, a proton–carbon–nitrogen experiment can show a proton–carbon 2-dimensional experiment at each location along a nitrogen spectrum. Separating out signals in proteins is essential for NMR to be useful due to the large number of signals. The ^{1}H NMR spectrum of a relatively small protein fragment (fifty-five amino acids) is shown in Fig. 20.15a. The NMR spectrum of a 3-dimensional TOCSY-^{15}N experiment (Fig. 20.15b) shows how a TOCSY spectrum can

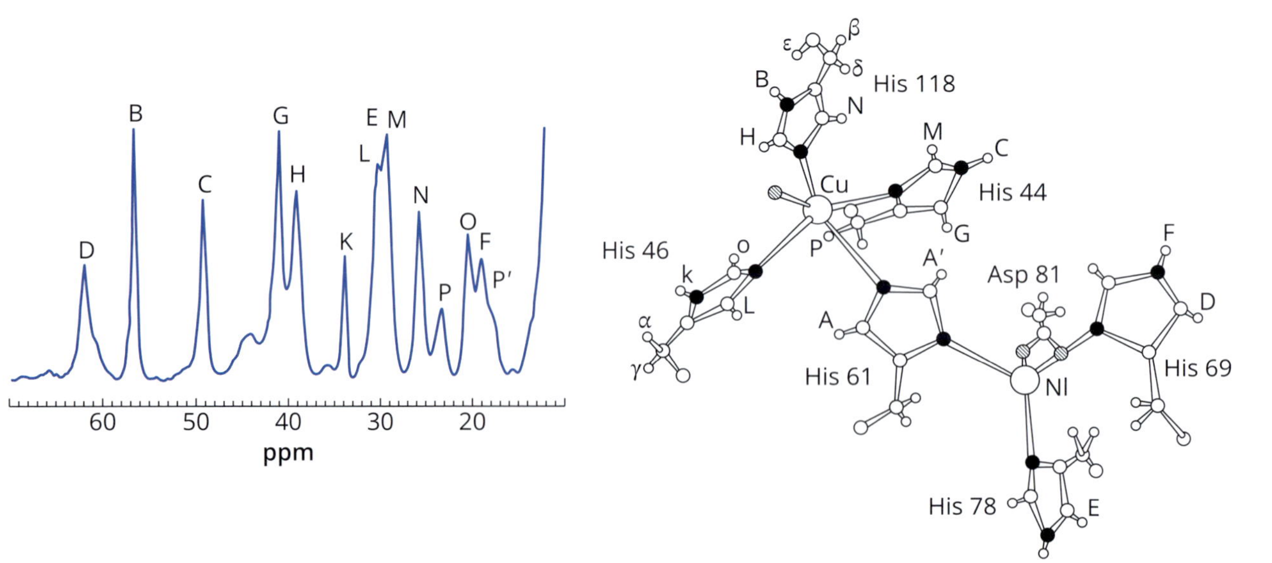

Fig. 20.14 The ^{1}H NMR spectrum of Cu,Ni-SOD. The Cu(II) center has one unpaired electron and causes a significant shift of bound histidine resonances (the signal due to proton A is off scale at 85 ppm; other assignments for B-P are as shown in structure). Figure modified from Figures 1 and 2 in Bertini.[24]

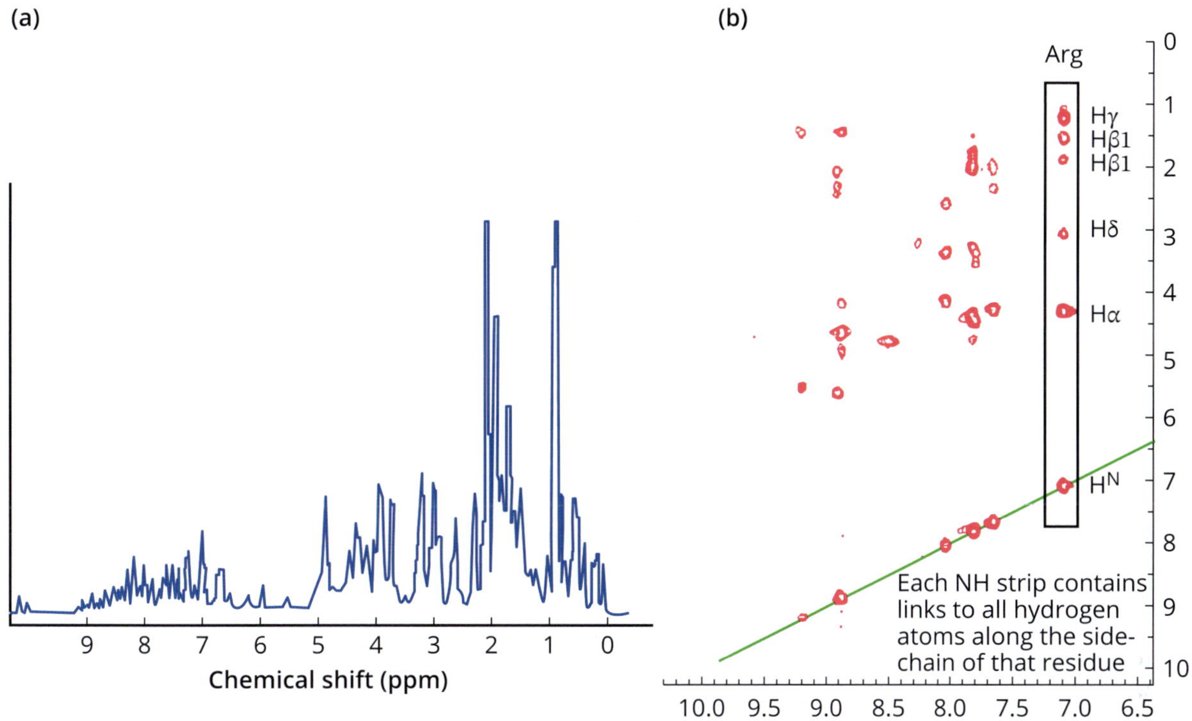

Fig. 20.15 (a) ^{1}H NMR spectrum of a fifty-five amino acid fragment of a protein.[25] (b) Several slices of a 3-dimensional TOCSY-^{15}N NMR spectrum of a protein.

be correlated with a single ^{15}N nucleus down a backbone, significantly simplifying the spectral analysis.

20.6.4 EPR

Electron paramagnetic resonance is a spectroscopic technique by which microwave radiation is absorbed by molecules containing one or more unpaired electrons. The physical basis for the phenomenon is similar to that of NMR. The presence of unpaired electrons makes the molecule a paramagnetic center. In the presence of an applied magnetic field, the unpaired electron spin magnetic moment (m_S, which can have values of ±½) splits into two energy levels—one aligned with the applied field (m_S = −½), and one anti-aligned (m_S = ½). This is known as the Zeeman effect. Each alignment has its own energy represented by Eq. 20.2, where B_0 is the applied magnetic field, μ_B is the Bohr magneton, and g_e is the dimensionless splitting factor or *g*-factor that is a proportionality constant relating the observed magnetic moment to its angular momentum quantum number.

$$E = m_S g_e \mu_B B_0 \qquad \textbf{20.2}$$

The presence of two energy levels for the unpaired electron implies that electromagnetic radiation of the proper wavelength—in this case, microwaves—will be absorbed simultaneously with promotion of the electron from the lower- to higher-energy state. The energy of the transition ($g_e\mu_B B_0$) is found by taking the difference between the two states, Fig. 20.16. If the frequency of light being absorbed and the applied magnetic field are both known, the dimensionless *g* factor (electron magnetic moment) can be readily calculated from this equation.

The value for g_e in a free electron is almost exactly 2. It is one of the most precisely known fundamental constants, with a value of 2.002 319 304 361 82(52), but it can deviate, in some cases significantly, for electrons in different environments. Thus, measured *g* values are a measure of the local chemical environment, much like chemical shifts in NMR. Organic compounds with unpaired electrons such as nitroxyl

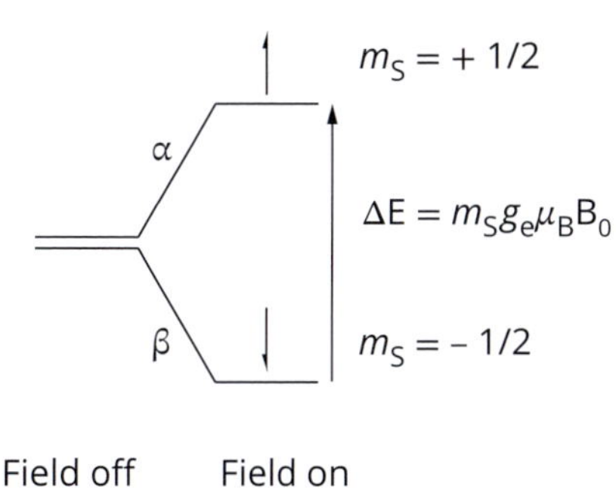

Fig. 20.16 The physical basis of EPR spectroscopy is due to the absorption of microwave radiation to promote an electron spin from m_S = −½ to +½ energy spin state.

radicals typically exhibit g values very similar to that of the free electron. However, in transition-metal complexes with unpaired electrons, spin–orbit coupling and interactions with local magnetic fields often result in significant deviations from this value. Additionally, the magnetic moment of an unpaired electron can be directional, or anisotropic, and the projections of the magnetic moment along x, y, and z axes need not be identical. Anisotropy in g can be used to help understand the type and symmetry of orbital the unpaired electron occupies. There are four limiting cases, shown in Fig. 20.17. For an isotropic case, all three g values are equivalent, and a single feature would be observed. For axial cases, the axis can be either along z or in the xy-plane, resulting in the shapes shown. In these cases, the g values are often called $g_{||}$(parallel to axis) and $g_{\perp}$ (perpendicular to axis). Finally, if all three g values are different, a rhombic spectrum would be obtained. Unlike NMR spectroscopy, which on modern instruments produces very sharp line shapes, EPR spectra are usually very broad, and thus are typically reported as first-derivative spectra to provide more obvious details. Absorption and derivative curves are also shown in Fig. 20.17.

If the unpaired electron is in an environment in which a nucleus with a nonzero spin is present, additional features will be observed due to coupling to other nuclei. A hyperfine interaction is one where the unpaired electron interacts with the atom on which it resides, while superhyperfine interactions are those in which the electron interacts with a different atom in the molecule. These couplings are the EPR equivalent to J couplings from NMR spectroscopy, and the coupling has the same relationship to Pascal's triangle as seen in NMR coupling (the n+1 rule). These interactions are typically reported in units of wavenumbers. Examples of hyperfine coupling are presented in Fig. 20.18, where the EPR spectrum of the superoxide dismutase is shown. This enzyme contains a copper atom, which has a 3/2 nuclear spin. As such, a four-line pattern is observed due to hyperfine coupling of the electron to the copper center. Note that since g and B are inversely proportional, the lower magnetic field corresponds to the larger g value for $g_{||}$ in this copper protein.

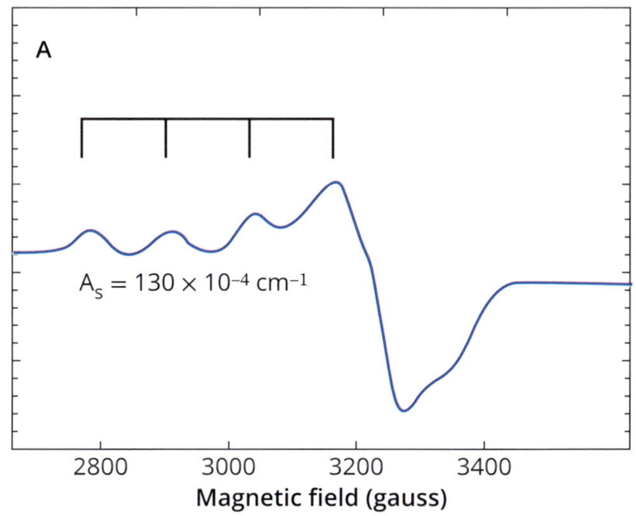

Fig. 20.18 The EPR spectrum of superoxide dismutase. This enzyme contains copper centers, and as Cu is a spin 3/2 nucleus, a four-line pattern is observed with A = 130 10^{-4} cm^{-1}. The spectrum is axial, with $g_{||} > g_{\perp}$.[6]

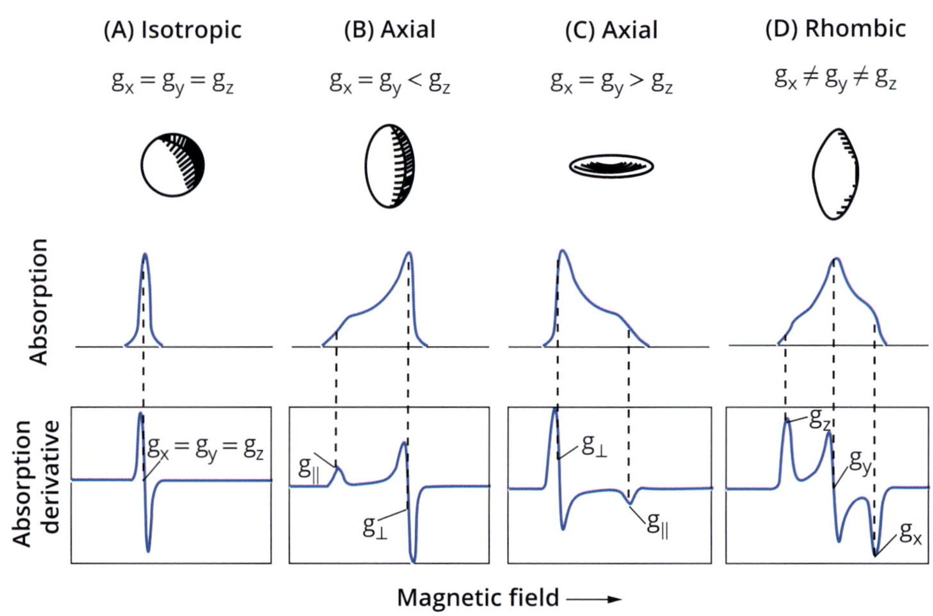

Fig. 20.17 The four limiting cases for anisotropy in EPR signals and the corresponding EPR spectra that would be observed for each.[6]

In addition to the relatively simple experiments described here, there are a number of pulsed-EPR experiments analogous to their NMR counterparts that result in more information. ENDOR (Electron Nuclear DOuble Resonance) and ESEEM (Electron Spin Echo Envelope Modulation) are two common pulsed-EPR techniques. These techniques measure interactions between unpaired electrons and atoms with nuclear spin and can be used to determine hyperfine coupling at a much higher resolution than with a continuous wave instrument. The HYSCORE technique (HYperfine Sub-level CORrElation) is a technique that is analogous to a 2-dimensional NMR, which shows correlations between the density of the electron spin on different nuclei.

20.6.5 X-ray Techniques

X-ray radiation has higher energy than ultraviolet, with a wavelength approximately that of the interatomic spacing in molecules. This similarity in wavelength allows for diffraction of the X-ray light by a crystalline lattice, and is important for determining the exact molecular structure of both large and small molecules, using a technique known as X-ray crystallography (See Box 10A). The higher energy of X-ray radiation allows its use in much the same way as UV–visible spectroscopy, but instead of looking at an energy gap from approximately the HOMO to the LUMO, electrons in core energy levels can be promoted either to a higher energy state or, more often, complete ejection of the electron to the continuum. The continuum is the set of energy levels for an electron whose gaps are negligibly small, and it represents the point where the electron is no longer bound to the atom. X-ray absorption spectroscopy (XAS) has two main forms of analysis: XANES (X-ray Absorption Near-Edge Structure) and EXAFS (Extended X-ray Absorption Fine Structure). Due to the need for tunable, and intense source of broad-spectrum X-rays for XAS experiments, this type of spectroscopy is typically carried out at a synchrotron source. A related technique, X-ray photoelectron spectroscopy was discussed in Section 15.3.2.

20.6.5.1 Crystallography

Protein structures are drawn with ribbon diagrams to simplify the structure when viewing a molecule that contains thousands of atoms. These ribbon diagrams are derived from atomic coordinates based on X-ray crystallographic data. One of the main differences between protein and small-molecule crystallography is due to the reduced resolution possible for most protein structures. While small molecule structures are obtained and published with sub-Ångstrom resolution, the reduced diffraction intensity, reduced regularity of the protein from position to position in the lattice, high number of disordered water molecules (up to 50% of the crystal by mass), and the significantly increased number of parameters required to describe the structure, all result in a reduced resolution. High-resolution structures can be obtained (and have been obtained, see Section 20.9.6 on nitrogenase), but in general, protein structures are obtained and reported with resolution far greater than 1 Å. This section will focus on the major differences between small molecule and protein crystallography with an emphasis on data-reporting.

The first X-ray diffraction studies of metalloenzymes were carried out in the 1960s and 1970s with the elucidation of the structure of both hemoglobin, the oxygen-carrying protein, and myoglobin, the oxygen storage protein. In 1962, Kendrew and Perutz shared the Nobel Prize for this work. Like small-molecule crystallography, the field has benefited enormously from the increase in computing power since that time. The techniques of molecular biology have also improved the ability to obtain structural information on metalloenzymes. These techniques allow for overexpression of the genes that code the protein in a bacterial or other microorganism host allowing for the production of larger (much larger) quantities of the protein than would be obtained from the naturally occurring organism. Protein crystallography has also benefited from the ability to carry out site-directed mutagenesis of native proteins to examine the effects of single amino acid side-chains on the structure and function of the resulting active site. The protein databank (www.wwpdb.org) is a repository for protein structures and their analysis, and as of 2024 there are more than 123,000 X-ray-determined structures reported.

Unlike small-molecule crystals, which are often hard or brittle and relatively easy to transfer, the consistency of protein crystals can be more like a soft gel-like material due to the high water content. Crystallization of proteins, especially membrane-bound proteins which exist spanning lipid bilayers, is more art than science, and hundreds of trials are often carried out before a suitable crystal can be obtained.

It has been stated that even for routine small-molecule structures the most difficult part is obtaining the crystal,[26] but another significant challenge remains: the phase problem. When X-ray light is diffracted by the crystal, the intensity information is collected but the phase information is not. To calculate the electron density at a point in space in a unit cell ($\rho(x, y, z)$) with volume V, both the phases (α_{hkl}) and amplitudes ($|F_{hkl}|$, the square root of the intensities, I_{hkl}) are required for each reflection (Eq. 20.3).[27] Without phase information, the structure cannot be solved!

$$\rho(x,y,z) = \left(\frac{1}{V}\right)\iiint |F_{hkl}| \cos[2\pi(hx + ky + lz - \alpha_{hkl})] \qquad \textbf{20.3}$$

The way that crystal structures are determined, then, is that a model is constructed that accounts for the observed electron density, and then the phases are *calculated* for that model. This phasing is applied to the observed data, resulting in an improved electron-density map, and the process repeats iteratively until the structure is determined. The scattering of X-rays is due to electron density on atoms, and as a hydrogen atom has only one electron and it is shared with its bonding partner, hydrogen atoms are usually not located with high precision during an X-ray diffraction experiment. During the refinement process, hydrogen atoms are added at calculated positions according to well-established geometrical criteria. While a gross simplification, this is generally the process by which X-ray crystal structures are determined. Interestingly, in the related technique of neutron diffraction, hydrogen atoms are strong diffractors and can be located. However, this technique is much less common due to the crystal size requirements and the requirement for a nuclear reactor.

A more complete discussion of X-ray diffraction data collection, structure solution, and refinement is left to more specialized texts,[28–30] and the remainder of this section will address data presentation. For small molecules—molecules containing up to approximately 100 nonhydrogen atoms—the resulting structures are usually obtained at atomic resolution, and the individual atoms can be readily identified. Small-molecule crystal structures are presented with atoms shown as thermal ellipsoids which reflect the vibrational motion present in the crystal lattice during the experiment. Bond lengths and angles are reported with their respective estimated standard deviations (e.s.d., the number in parentheses is the experimental error in the last digit) because there are sufficient data from the diffraction spots that are obtained that these values have error bars.

A representative crystal structure of a small molecule derived from the amino acid valine is shown in Fig. 20.19.[31] In this structure, the hydrogen atoms are shown for three of the atoms in order to explicitly show the chirality and the presence of the hydrogen bond; often, hydrogen atoms are omitted from

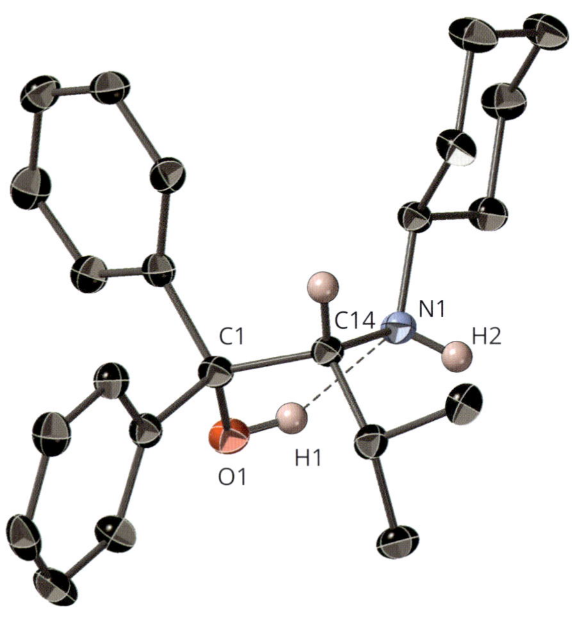

Fig. 20.19 A small molecule of a valine-derived organic molecule. Representative bond distances (Å) and angles (°), with e.s.d.: C1-O1, 1.4253(17); N1-C14, 1.4754(19); C1-C14, 1.5735(19); O1-H1, 0.87(2), N1-H2, 0.867(2); O1-C1-C14, 107.86(11); C1-C14-N1, 107.33(11); C1-O1-H1, 101.3(14). Figure drawn using CrystalMaker[8] software.

the presented structure for clarity unless the hydrogen atom is described in more detail.

The carbon, oxygen, and nitrogen atoms are shown as ellipsoids with their thermal displacement vectors, or the vibration of the atom, indicated by the size of the ellipse. Large thermal ellipsoids may also indicate two or more similar orientations of a substituent that are averaged across the whole crystal. The bond length between C1 and O1 is reported as 1.4253(17) Å, which corresponds to a length of 1.4253 Å with an e.s.d. of 0.0017 Å. Similarly, the O1-C1-C14 bond angle is reported as 107.86(11)°, which corresponds to an angle of 107.86° with an e.s.d. of 0.11°. In the structure, a hydrogen bond was located between O1 and N1 involving H1, and that is shown explicitly in the diagram.

On the other hand, protein crystal data is not usually collected at atomic resolution, and instead, blobs of electron density are obtained. Given that most of the atoms in a protein crystal structure are light atoms (C, N, O), it is very difficult to build a good model of the structure to solve the phase problem. This makes the determination of structure of large proteins much more difficult than for small molecules, and the data must be fit in an iterative fashion. Molecular biology and gene sequencing allow for easier determination of the correct model. If the gene for a particular protein is known and sequenced, the resulting amino acid sequence is also known, and these amino acids can be modeled and aligned with the measured electron density. There have, however, been cases where amino-acid sequencing has been proven incorrect by X-ray diffraction experiments (Section 20.9.4). Additionally, some proteins have post-translational modification, which makes this technique imperfect; but regardless, it allows for an initial model to be obtained. Often, heavy atoms are added to the structure, perhaps by binding a metal ion to cysteine sulfurs, or by replacing the sulfur atoms in methionine with selenium. The location of these atoms is more easily determined, allowing a better first estimate of phases to be determined. Other methods to solve the phase problem include collecting data at more than one wavelength; anomalous diffraction of these different wavelengths from heavier elements (sulfur and beyond) can be used to help assign phases.

A representative crystal structure of the particulate methane monooxygenase protein downloaded from the protein databank is shown in Fig. 20.20.[32] This is a membrane-bound protein; the bundles of α-helices in the structure are hydrophobic and allow the protein to sit within the lipid bilayer. In this structure, atoms are not indicated by using ellipsoids, as the data does not have enough resolution to precisely determine those values. Instead, the structure is presented as a ribbon diagram. The α-helices are drawn as coiled ribbons or sometimes cylinders, and β-sheets are drawn as arrows or straight ribbons, as shown in Section 20.3. The short amino-acid sequences between those major structural features are shown as an interpolated smooth curve containing the polypeptide backbone. This particular structure is a $\alpha_3\beta_3\gamma_3$ trimer, which means that there are three copies of the α, β, and γ subunits all within the main superstructure. The chemistry of a related MMO protein will be discussed in Section 20.9.2.

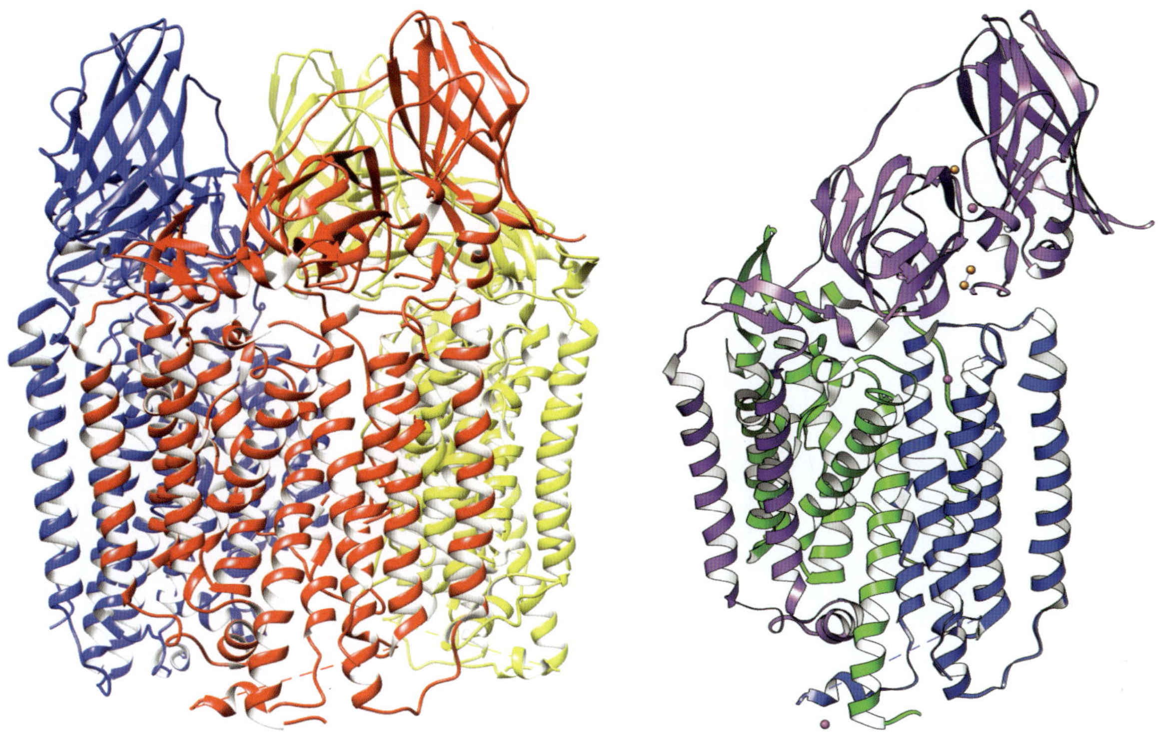

Fig. 20.20 pMMO X-ray structure. Left, complete structure of $\alpha_3\beta_3\gamma_3$ trimer showing α-helices as ribbons and β-sheets as arrows. Right, one copy of the αβγ monomer. The dicopper center (right, center top) is the active site for methane activation. Figure generated with UCSF Chimera[11] with coordinates from the protein databank (PDB ID 1YEW).[9, 32]

20.6.5.2 X-ray Absorption Spectroscopy

When high-energy X-ray radiation interacts with a sample, core electrons can be promoted to higher energy levels or even ejected to the continuum. Just as in UV–visible spectroscopy, the energy required to promote the electron from the core energy levels can be measured, and this technique is called X-ray absorption spectroscopy (XAS). A representative X-ray absorption spectrum is shown in Fig. 20.21. The spectrum has three main features: the pre-edge and the rising edge, which together correspond to the X-ray absorption near edge structure (XANES), and the extended X-ray absorption fine structure (EXAFS) region, after the electron has been promoted to the continuum. Each region provides details and information about the element(s) present and their electronic state.

In K-edge spectroscopy, the features that are observed represent absorption of X-ray photons by a 1*s* electron. The L edge and M edge spectroscopies measure the absorption by 2*s*/2*p* and 3*s*/3*p*/3*d* electrons, respectively. The pre-edge often contains weak transitions from the 1*s* to a bound state at the valence *d*-level. While this is a dipole forbidden transition (*s* and *d* orbitals are both centrosymmetric and the absorption operator has ungerade symmetry, see Section 16.6), these transitions can be observed due to *d*/*p* orbital mixing in complexes with an unfilled *d* shell. The region within 50 eV of the absorption edge is described as the XANES region by convention, and this region of the spectrum shows promotion of the 1*s* electron to a 2*p* level. The transition from 1*s* to 2*p* is symmetry-allowed, but for most elements the 2*p* levels are filled, so there would be no possible electronic transition. Strong ligand–metal 2*p*-3*d* hybridization results in stronger absorptions in the pre-edge region. The resulting spectra thus describe the differing electronic environment of the absorbing atom when ligands with their electronic structures are present. The edge is highly sensitive to formal oxidation state, ligand type, and coordination number, and as such there can be a complex structure to the absorption edge. The edge itself represents absorption of a photon that takes the 1*s* electron to the continuum.

The discrimination of elements possible using XAS is high, as each element has a very different energy that corresponds to the promotion of the 1*s* electron to the continuum. This energy ranges from 4 to 10 keV for first-row transition elements. The near-edge and pre-edge features are highly dependent on the oxidation state, coordination number, and electronic structure of the atom, such that different oxidation states of elements can be readily discriminated by using this technique. For example, Fig. 20.22 shows the XANES of two different Pu standards as well as several samples that were contaminated with Pu of unknown oxidation state. The edge moves to higher energy as the oxidation state of the Pu center increases, which is explained by the fact that the 1*s* electron is

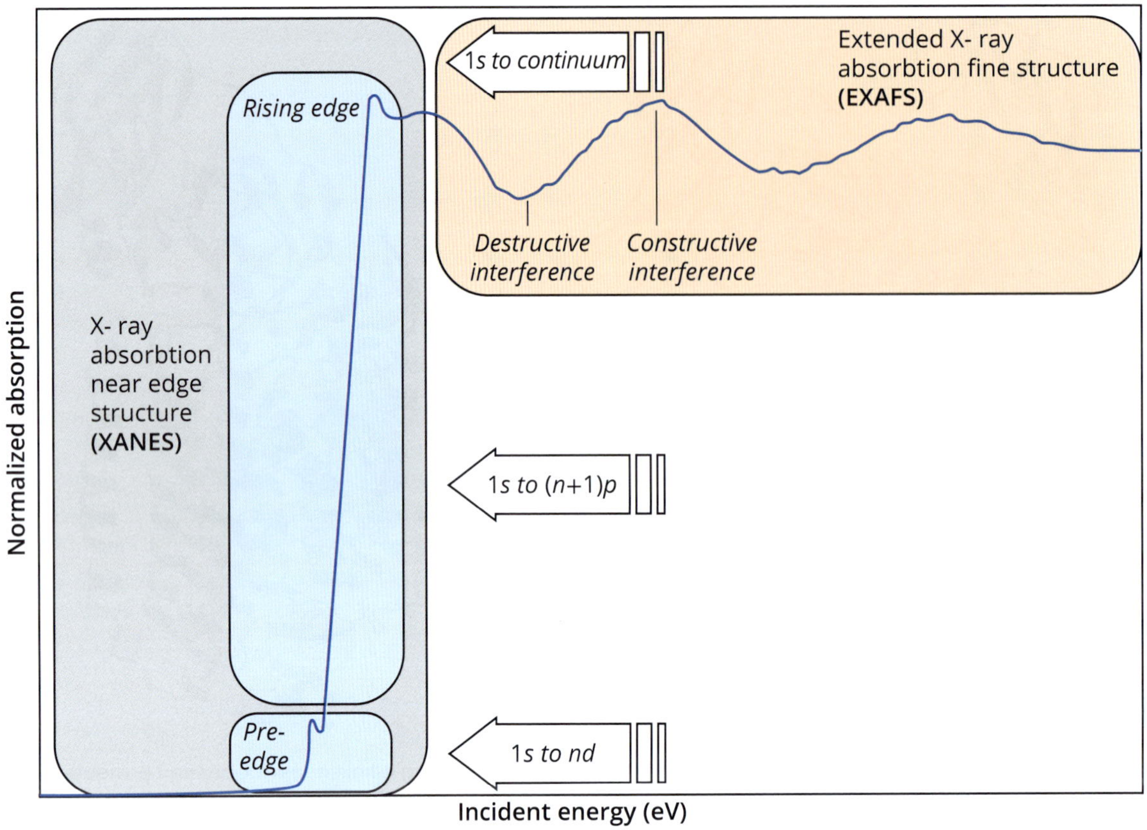

Fig. 20.21 X-ray absorption spectroscopy of an element, showing pre-edge features (1s to nd), near-edge structure (1s to (n+1)p), and EXAFS features. The energy of absorption is specific to the element studied, but ranges from 4 to 10 keV for first-row transition elements.

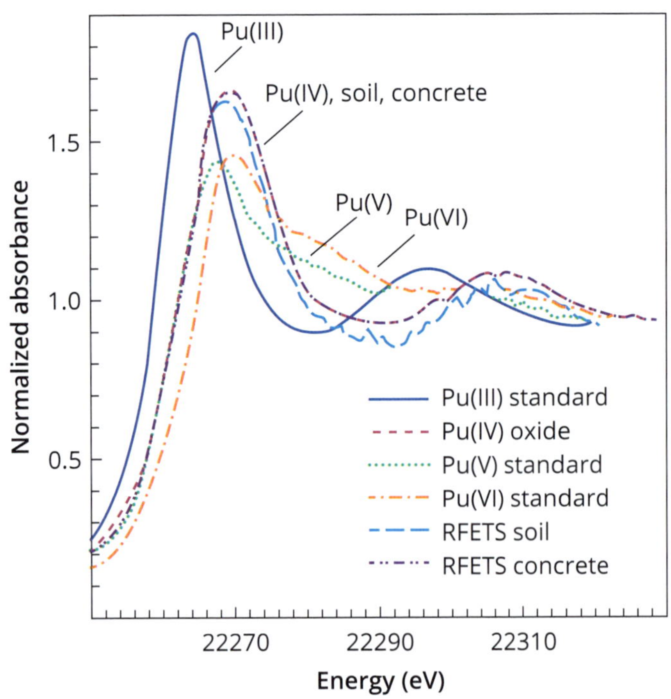

Fig. 20.22 The X-ray absorption near edge spectra (XANES) of several plutonium samples, showing the difference in pre-edge absorption by different oxidation states of the metal.

more tightly bound in the higher-oxidation-state metal. The Pu(IV) standard is a good match for the state of Pu found in the sample. Pattern-matching or fingerprinting of edge locations or linear combinations of known spectra are often used to determine the components present in an unknown sample.

20.6.6 EXAFS

The region of XAS beyond the edge is not featureless. One might think that taking the 1*s* electron to an unbound state would result in a flat-line absorption spectrum above the edge, but this is not the case. The electron that has been just barely excited to the unbound state now has a relatively small kinetic energy and can interact with electrons on neighboring atoms. This interaction results in a series of constructive and destructive interference patterns that are highly dependent on both the number and identity of the ligating atoms. Complete analysis of the EXAFS region requires sophisticated data reduction and iterative computer modeling. Fourier transformation of the observed frequencies of scattered radiation is converted into distances from the scattering atom. Each scattering atom contributes to the scattering intensity as a series of damped sine waves. The intensity and frequency of the scattering depend strongly on the atom type of the scatterer and distance from the central atom. Proper fitting of EXAFS data can provide very precise measurement of bond distance and atom type from the absorbing atom. A relatively simple example is the EXAFS of iron oxide is shown in Fig. 20.23. In this figure, the data and fit for scattering from six neighboring O atoms at 2.1 Å and from twelve next-neighboring Fe atoms at 3.1 Å are required to obtain even a reasonable fit, but these two scatterers end up accounting for most of the signal in the data. The structure is consistent with the known rock-salt structure of FeO. An excellent tutorial on EXAFS is available.[33]

EXAFS was used to help correctly determine the structure of the iron–sulfur protein rubredoxin, Section 20.7.3. The original X-ray structure was modeled to have one long Fe–S and three short Fe–S interactions, but a better fit to the EXAFS data was obtained when all four Fe–S bond lengths were the same length (Fig. 20.24).[34] Rerefining the original X-ray data with similar Fe–S bond lengths gave a better structural description. This use of EXAFS cemented its utility and credibility for the field of bioinorganic chemistry, and is a good example of the complementarity of the various spectroscopic techniques for self-checking data and structure.

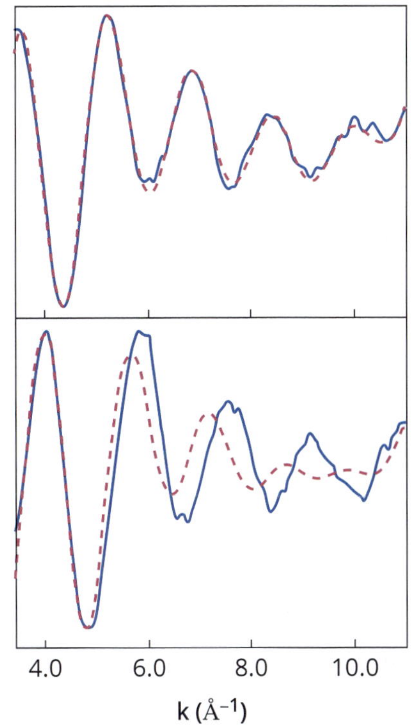

Fig. 20.24 EXAFS spectra for the iron center in rubredoxin (solid curve) and modeled with either the original Fe-S distances from X-ray crystallography (dashed curve, bottom) or with four identical Fe-S distances (dashed curve, top). Figure adapted from Shulman, Eisenberger and Blumberg.[34]

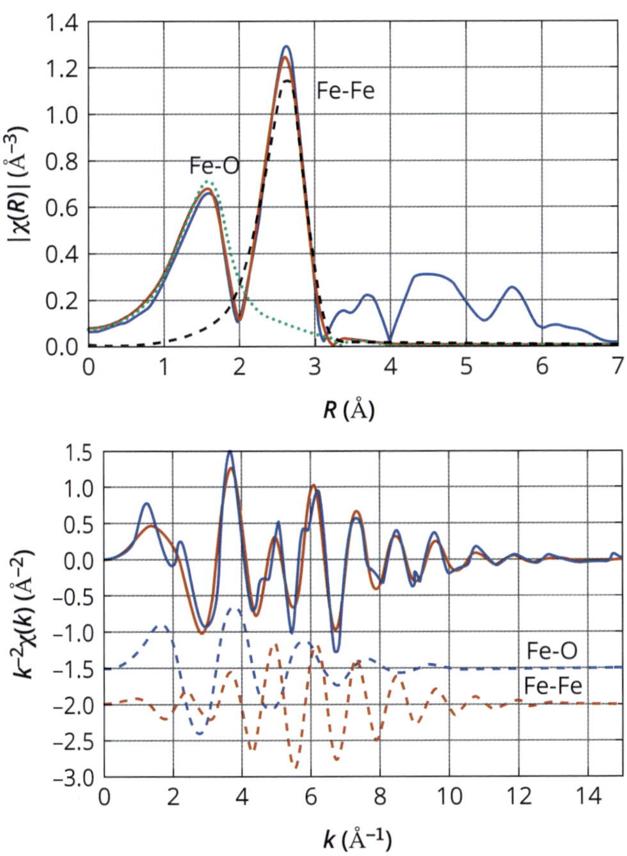

Fig. 20.23 EXAFS data for iron oxide. The Fourier transform (upper) of the baseline corrected absorption data (lower) result in Fe–Fe and Fe–O scattering interactions at the Fe–Fe and Fe–O bond distances. The sum of the two scattering interactions (dotted curves) matches the experimental data well. Figure adapted from Newvillle.[33]

20.6.7 Mössbauer

Mössbauer spectroscopy is a technique that reveals information about nuclear energy levels. Similar to NMR and EPR, nuclear energy levels can be probed with the appropriate electromagnetic radiation, which for nuclear transitions are gamma-rays. By far the most common element that is studied by using this technique is iron, which is conveniently also a very relevant nucleus for bioinorganic chemistry. The source for gamma-rays for the ^{57}Fe nucleus is ^{57}Co—a relatively stable isotope of cobalt with a half-life of about 270 days. The ^{57}Co nucleus decays to an excited state of ^{57}Fe, which then emits a gamma-ray as the nucleus relaxes to the ground state. For free ions or atoms, this relaxation would result in the emission of a gamma-ray that is lower in energy than that of the nuclear transition, because some of the energy would be lost to recoil of the nucleus. Similarly, in order for absorption of a gamma-ray to occur, it would have to have slightly higher energy than that of the nuclear transition, as the nucleus would gain momentum upon absorption of the photon. A recoilless energy absorption is essential to the utility of the technique, so the source of ^{57}Co is in a solid foil and the sample is either a solid or frozen solution. The lack of recoil results in a strong, sharp resonant transition instead of a more diffuse transition that would occur in a liquid or gas phase. The effect was discovered by Rudolf Mößbauer in 1958, and the resulting spectroscopic technique is named in his honor. He shared the Nobel prize in physics in 1961.

Because both the source and sample are the same nucleus, the energy of the nuclear transitions are the same, but are ever so slightly changed due to chemical environment. The source resides in a metal foil, while the sample, especially for bioinorganic samples, is in a very different environment, so its nuclear transitions will have slightly different energies. The energy of the transition can be adjusted using the Doppler effect (Eq. 20.4). The source is mounted on a holder that moves towards and away from the sample, so the cos(θ) term is equal to unity. As the source accelerates through a range of velocities, the energy of the emitted gamma-ray is altered slightly. This range of energies is small, and a typical sample exhibits an absorption near 1 mm/s which corresponds to an energy of 5 10^{-8} eV for the 14.4 keV source.

$$\Delta E = E_{\gamma}(v/c)\cos(\theta) \qquad \textbf{20.4}$$

Like NMR spectroscopy, the energy states of the iron nucleus are dependent on chemical environment, and the shift is reported as the isomer shift, δ. Isomer shifts are a measure of the s-electron density in the nucleus, which can be changed by hybridization, oxidation state, or spin state of the iron center. The isomer shift for iron centers range from 0–0.1 mm/s for Fe(IV) heme iron, to greater than 1 mm/s for Fe(II) with nitrogen and oxygen ligands. Increasing the iron oxidation state removes *d*-electrons which decreases the screening of the s electron density at the nucleus. This loss of screening increases the s electron density at the nucleus, giving a lower isomer shift. This technique can thus be used to determine oxidation state of iron centers much like XAS.

For ^{57}Fe spectroscopy, the absorption spectrum is observed as a doublet due to quadrupole splitting (ΔE_Q) where the excited state with $l = 3/2$ splits into two substates with $m_l = \pm 1/2$ and $\pm 3/2$ (Fig. 20.25). This gives rise to two distinct peaks in the spectrum, with typical quadrupole splitting of several mm/s. The magnitude of the ΔE_Q in general correlates with the degree of asymmetry of the electric field at the nucleus. Fe(III), with a spherically symmetric d^5 electron configuration has a lower ΔE_Q (0.5–1.5 mm/sec) than Fe(II) with similar ligands (1.5–3 mm/sec). Some representative spectra for biologically relevant iron–sulfur proteins are shown in Fig. 20.26. Iron–sulfur proteins are discussed in more detail in Section 20.7.3. Gamma-ray absorption is plotted as a function of the source velocity (mm/s). The Fe(III) center in a rubredoxin model complex $Fe(SR)_4$ exhibits a small quadrupolar splitting of 0.8 mm/s (A) while the reduced Fe(II) complex exhibits a much larger splitting of 3.5 mm/s (B). The spectra for oxidized ferredoxin (an Fe_2S_2 cluster) shows only Fe(III) centers (C) while the reduced form shows both Fe(III) and Fe(II) centers (D).

Additional magnetic hyperfine splitting can sometimes be observed. These are due to interactions of the nucleus with a

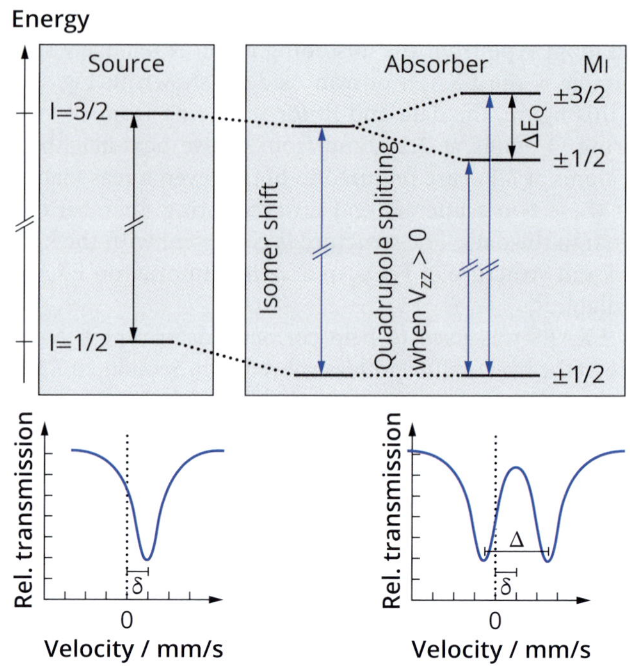

Fig. 20.25 Nuclear energy transitions for Mössbauer spectroscopy. The upper *I* = 3/2 energy level is split into two states so the observed spectrum is a doublet.

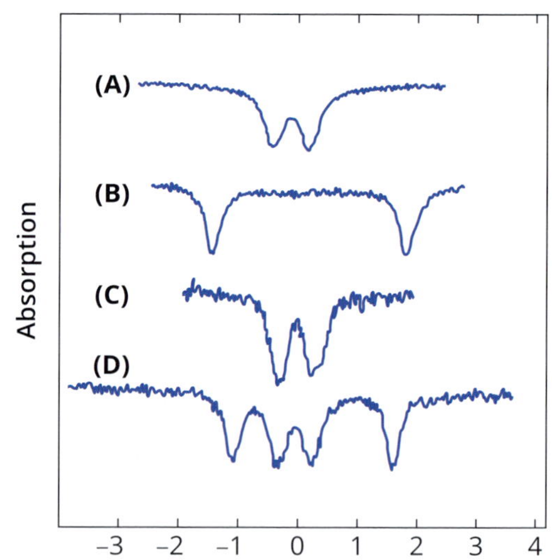

Fig. 20.26 Mössbauer spectra (absorption versus source velocity in mm/s) of oxidized (A) and reduced (B) $Fe(SR)_4$ rubredoxin model complexes, and oxidized (C) and reduced (D) ferredoxin proteins. (Adapted from Figure 7.7 in Bertini, Gray, Lippard and Valentine.)

nearby magnetic field, such as an unpaired electron. For diamagnetic samples, this splitting can be induced with an externally applied magnetic field. The two substates of the Fe nucleus are further split as shown in Fig. 20.27. Only six transitions are allowed: those with $|\Delta m_l| = 1$ (transitions 1, 3, 4 and 6), or $|\Delta m_l| = 0$ (transitions 2 and 5).

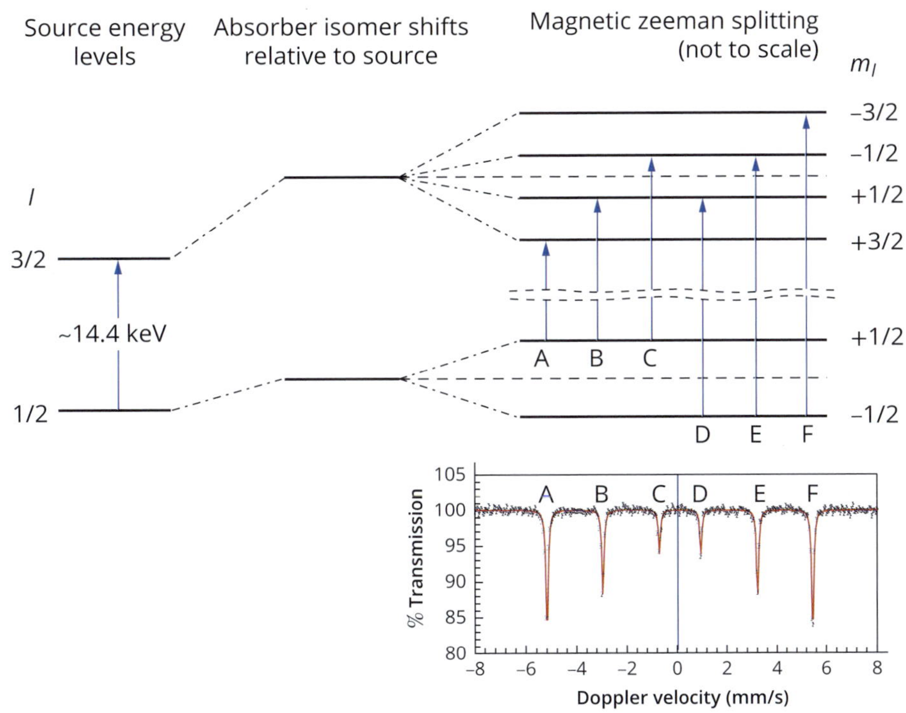

Fig. 20.27 Magnetic hyperfine splitting of the Mössbauer spectrum observed in the presence of a magnetic field.

20.6.8 Electrochemistry

The importance of electron flow in and through metalloenzymes can be seen by their use in photosynthesis and respiration. The presence of transition-metal centers capable of multiple stable oxidation states (Cu(I)/Cu(II), Fe(II)/Fe(III)) makes electron transfer possible. An important method for characterizing metal containing enzymes, especially those involved in redox processes is by measuring their redox potential by a technique such as cyclic voltammetry (CV, Section 13.2.3). Some proteins have their redox centers buried within the protein, and electrons are shuttled to the center by small organic molecules, but direct electrochemical measurements are not possible.

Although CV of proteins is essentially the same technique as that for soluble small molecules, it has some additional difficulties. It is typically done on a small scale due to the expense and difficulty of purifying the proteins, which is a technical complication. Proteins may deposit on the electrode surface or denature under the measurement conditions, which is a more of a physical complication. However, reduction potentials of metal centers in proteins, once determined, can be correlated with their environment. This allows for a highly tunable multistep electron transfer process to control the flow of electrons in a living organism.

20.6.9 Model complexes

The study and characterization of biological metalloenzymes is difficult and relies on piecing together characteristic information from a number of spectroscopic techniques. In order to verify that the proposed structure is consistent with all the observed information, synthetic chemists prepare model complexes. These are carefully designed small-molecule coordination compounds that have metal centers ligated by amino-acid side-chain mimics or other relevant functional groups. There are two broad classes of model complexes: spectroscopic, in which the spectroscopic features of the model compound match those of the biological one, and functional, in which the chemical behavior of the model compound matches that of the biological one. Within the subclass of spectroscopic models are those that mimic the coordination environment and geometry of a protein active site, and these are structural models.

An example of both spectroscopic and structural model compounds can be found in the work of Richard Holm, who spent most of his career studying model complexes predominantly at Harvard University. A classic example is that of the rubredoxin protein, which was postulated to exhibit the entatic state (Section 20.2). These complexes will be discussed in Section 20.7.3. The high-energy coordination environments were observed in biological systems and proposed as a model to explain the faster electron transfer rates. To explore this theory, Holm prepared a synthetic complex with simple thiolate ligands in order to mimic the coordination environments of the observed biological systems. The FeS_4 cluster in rubredoxin was modeled as an iron tetrathiolate, which had almost perfect tetrahedral coordination. A number of spectroscopic methods were employed to compare the model complex with that of the protein (Table 20.3).

Table 20.3 Comparison of spectroscopic and structural data for Rubredoxin and Holm's model complex.

Technique	Rubredoxin	Model
UV-visible (λ_{max}, ε)	380 (10.9), 490 (8.9) 580(sh)	354 (8.5), 486 (6.3), 640–680 (1.8)
X-ray: Fe-S (Å)	2.34, 2.32, 2.24, 2.05	2.268, 2.265, 2.279, 2.258
S-Fe-S (°)	101–115	105.8–112.6
Mössbauer δ	0.25 mm/sec	0.13 mm/s
Voltammetry	~0 V	−1.03 V

When comparing the relatively simple model complex and that of the rubredoxin protein, the UV-visible spectra had similar bands and intensities, and well as a similar Mössbauer isomer shift. A large difference in the reduction potential was observed relative to that of the native protein, and this was ascribed to "influences of protein tertiary structure and active site environment which are not yet interpretable."[35] The largest difference observed was in the X-ray crystal structures of the model versus the FeS_4 core, in which the native protein exhibits a severely distorted geometry from tetrahedral. The spectroscopic data were ascribed as belonging to a symmetrized version of the asymmetric protein active site.

More recently, Marcetta Darensbourg from Texas A&M University has studied a number of model complexes for another iron–sulfur metalloenzyme called hydrogenase.[36] These enzymes are found in bacteria and Archaea, and are involved in the reversible oxidation of dihydrogen. These organisms can use hydrogen as fuel, and the protein is of interest for the development of hydrogen fuel cells. Darensbourg's model complexes are functional models in that they are able to reversibly uptake and release dihydrogen. Her work has led to mechanistic insights that may also be present in the biological system.

The active site of iron hydrogenase consists of a diiron core coordinated to CO, CN, cysteine, and a bridging dithiolate ligand with a nearby Fe_4S_4 unit (Fig. 20.28). The model complex retains the diiron core and the CO ligands.

The model complex and others like it have not been able to mimic the bridging CO ligand or the mixed-valent Fe(II)Fe(I) oxidized form of the metalloenzyme. This lack of synthetic model complex is ascribed to the important role of the protein in stabilizing the cluster—another example of the entatic state. However, like the biological system, the model complex exhibits nonrigidity and rotation of the tricarbonyl turnstile with a barrier of 3.9 kcal/mol. The rotation of $Fe(L)_3$ units is postulated to be essential for ligand exchange in the biological system (Fig. 20.29).

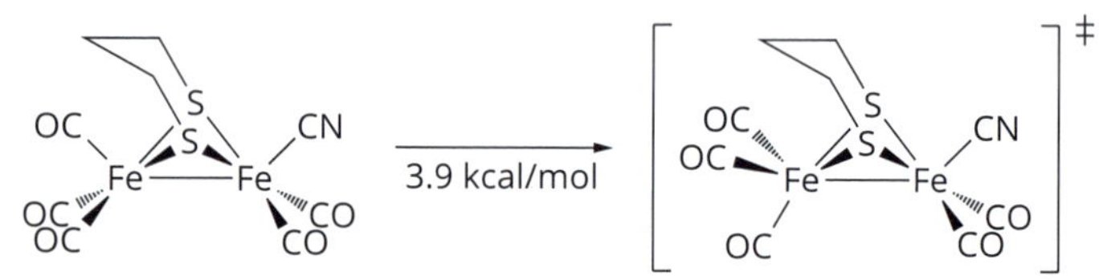

Fig. 20.29 The low rotational barrier observed in one derivative of the hydrogenase model complex. Figure adapted from Darensbourg.[36]

Since the model complexes matched the structure of the enzyme active site well, it was examined for functional activity. The all carbonyl derivative of the complex did not bind a proton or dihydrogen, but the *bis*-PMe_3 derivative (not pictured) was able to bind a proton to form a bridging hydride that was capable of scrambling H_2 and D_2 in the presence of water (Fig. 20.30). The authors postulated a mechanism for this exchange that involves a bound dihydrogen molecule. The authors state that this model is a functional model for update and reactivity of dihydrogen, but not necessarily the mechanism in the biological system.

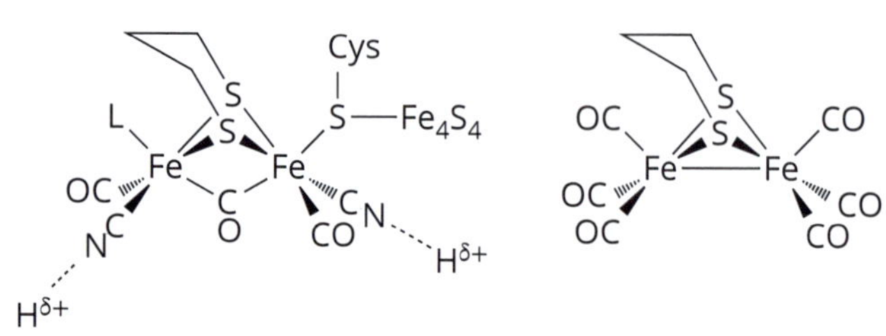

Fig. 20.28 The hydrogenase enzyme active site and a functional model complex.

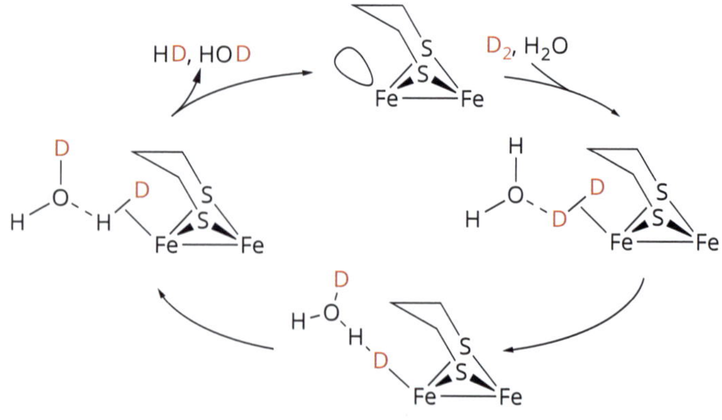

Fig. 20.30 Proposed mechanism for H_2/D_2 exchange by a diiron model complex. Figure adapted from Darensbourg.[36]

BOX 20A

Summary of physical techniques

Technique	Physical Property Measured	Information Obtained
UV–vis	electronic transitions	presence of transition metals, electronic energy levels
CD	differential absorption of polarized light	identification of α_helices and β_sheets
NMR	nuclear spin transitions	chemical environment of nucleus under study
EPR	electronic spin transitions	chemical environment of an unpaired electron
X-ray Crystallography	diffraction of X-rays by a crystal	atomic structure of molecule/protein
X-ray absorption	promotion of core electrons to the continuum	identity and electronic structure of metal atom
EXAFS	constructive/destructive interference of electrons with neighboring atoms	identity of and distance to nearby atoms
Mössbauer	nuclear energy level transitions	electronic environment, commonly for iron
Electrochemistry	ability to gain or lose electrons at a set external potential	redox potential of electron transport centers

20.7 Structure in Bioinorganic Chemistry

Now that we have examined how proteins can fold into different shapes and have seen the ways that metal ions interact with a biological system, we turn our attention to some of the more specific structures that metal ions exhibit in biological systems. Cofactors such as porphyrins and other related nitrogen macrocycles are important ligands that are used to store and/or transport small molecules but can also support redox centers. Iron–sulfur clusters are a class of prosthetic groups (Section 20.3) that can be thought of as small batteries or capacitors for the delivery of electrons at very specific and tunable redox potentials. Finally, zinc fingers are structures of DNA and zinc cations that are important for DNA regulation.

20.7.1 Cofactors

In addition to protein-bound metal ions, there are a number of metal-containing small molecules that are produced within a living system. These small molecules can either be covalently or noncovalently attached to a protein or metalloprotein in a post-translational step and assist the protein in carrying out chemical transformations. These small molecules are called cofactors or prosthetic groups. Some of these cofactors may be familiar to you, such as vitamin B_{12}, a cobalt containing molecule. There are a number of nonmetal-containing cofactors as well, such as NAD^+ and $NADP^+$, which are electron sources, and folate, which is used to deliver carbon atoms, but this section will discuss only metal cofactors.

20.7.2 Heme and Related Structures

The heme cofactor is an iron porphyrin complex. Porphyrin, which was introduced in Section 11.4.3, is an aromatic, nitrogen-containing heterocycle with four pyrrole rings. Biologically relevant porphyrins are functionalized derivatives of the parent molecule, porphyrin (Fig. 20.31a), such as protoporphyrin IX, which is found in hemoglobin and cytochrome c (Fig. 20.31b). Porphyrins have eighteen contiguous π electrons and are therefore aromatic by Hückel's rule ($4n+2$ where $n = 4$). As a result, these macrocycles are planar and rigid structures. Porphyrins are made from the amino acid glycine and succinyl coenzyme A in a complex pathway involving multiple enzymatic steps. Porphyrins can also be made under laboratory conditions and are used to make synthetic model complexes such as tetraphenylporphyrin (Fig. 20.31 (c)).

Heme is a prosthetic group derived from porphyrin that specifically coordinates an iron atom within the macrocyclic ring. Usually, a heme also contains one or two additional axial ligands. In the hemoglobin and myoglobin proteins which transport and store dioxygen in animals, one axial ligand is a histidine while the other is either vacant or the dioxygen molecule (Fig. 20.32a). The histidine binds to the iron, keeping the porphyrin in place inside the protein environment, while the other side is open to allow for reversible dioxygen binding. When the heme is accidentally released from a protein, which can happen under oxidative stress, it can float freely in the cytoplasm and damage cells by the uncontrolled production of free radicals. In cytochrome c, a protein that is involved in redox processes, the axial ligands are a histidine and a methionine (Fig. 20.32b). In this complex, the porphyrin is held in place by covalent bonds to the protein.

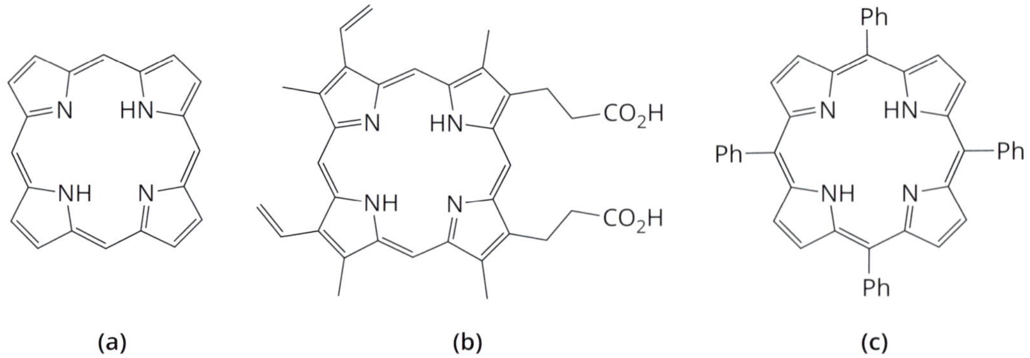

Fig. 20.31 (a) Structure of the parent molecule porphyrin, (b) protoporphyrin IX, found in heme, and (c) tetraphenylporphyrin, a synthetic porphyrin.

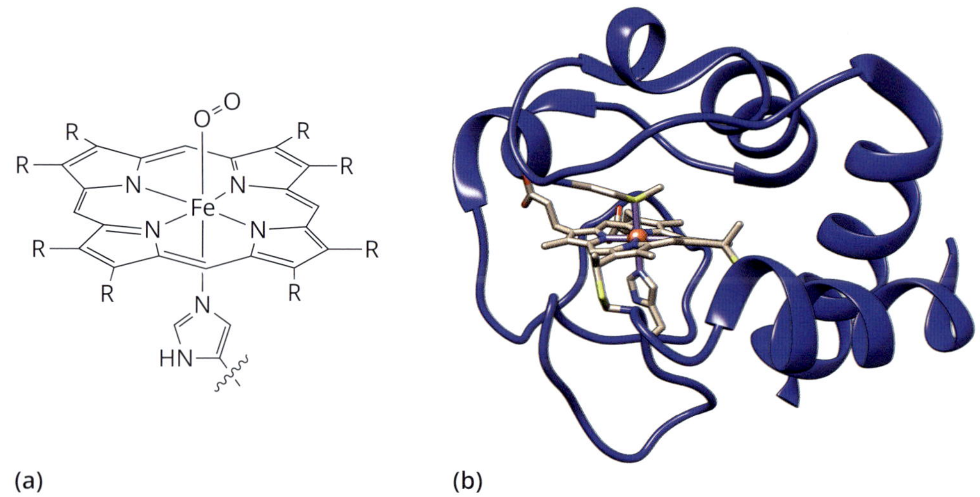

Fig. 20.32 Figure of the hemes found in (a) the protein hemoglobin, which binds dioxygen reversibly, and (b) in cytochrome c, which is involved in electron transport and redox processes. The heme in hemoglobin has an empty coordination site available to bind dioxygen, while the heme in cytochrome c is bound into the small globular protein by a methionine and a histidine. Figure generated with UCSF Chimera[11] with coordinates from the protein databank (PDB ID 1HRC).[9,37]

The functional diversity of heme groups is due to the structural diversity possible with different porphyrin rings as well as the highly specific amino acid environment in which they reside. There are many types of cytochromes, all with slightly different redox potentials due to the exact polarity and environment at the iron center. The binding of oxygen by hemoglobin and myoglobin is highly dependent on the partial pressure of oxygen and even the pH of the tissue. The functional aspects of these two proteins will be discussed in Section 20.8.4.

While a porphyrin is a specific organic functional group, nature uses other similar ligands to support a variety of metals and provide additional variability in redox potentials. One porphyrin-like ligand is the chlorin ring which resembles the porphyrin ring, but one of the double bonds is missing in one of the nitrogen-containing rings and there is an additional fused ring (Fig. 20.33 (a)). The magnesium complex of this ligand is found in chlorophyll. In this molecule, the long hydrophobic ester R group is a 20-carbon diterpene chain. This group is involved in hydrophobic interactions that anchor the complex within the thylakoid membrane found in chloroplasts. A second ligand is the corrin, which is similar to a porphyrin, but one of the linker methylene carbons is removed between two of the nitrogen containing rings (Fig. 20.33 (b)). A protein that has a corrole ring is coenzyme B_{12}. This cobalt complex is a cofactor that is used in carbon–carbon bond-forming and rearrangement reactions, and will be discussed in Section 20.8.5.

20.7.3 Iron Sulfur Clusters

Iron–sulfur clusters are extremely important cofactors that are involved in electron transport in biological systems; these clusters were introduced in Section 12.3.4.2. They are classified into four main groups based on the ligands and number of inorganic sulfur atoms (sulfide, S^{2-}) present in the cluster (Fig. 20.34). Rubredoxins are single iron atoms coordinated

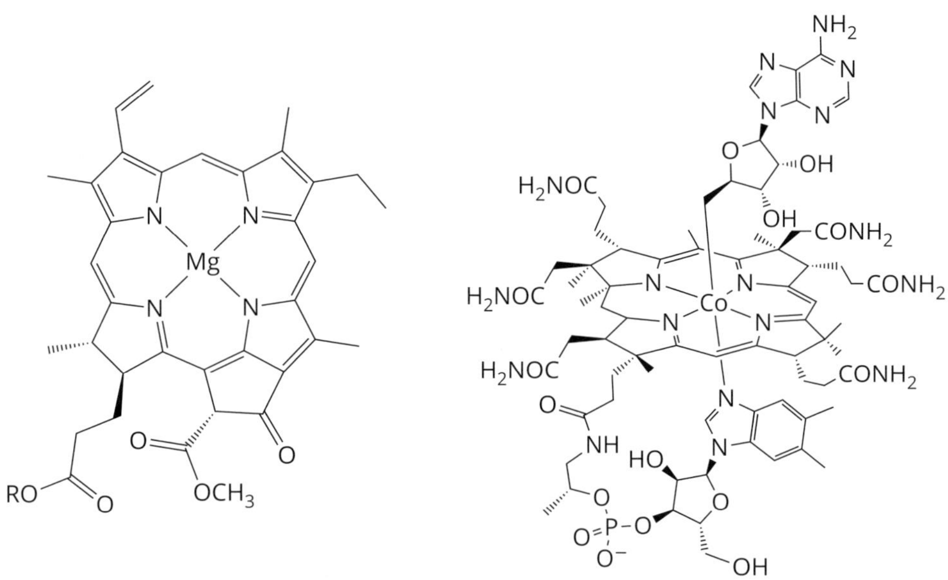

(a) (b)

Fig. 20.33 Two porphyrin-like complexes: (a) The magnesium chlorin complex found in chlorophyll, and (b) the cobalt corrin complex found in coenzyme B_{12}.

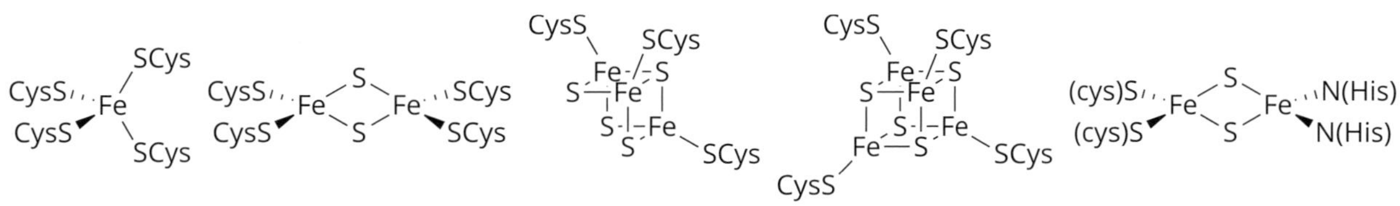

Fig. 20.34 Structures of rubredoxin ($Fe(SR)_4$), ferredoxin (Fe_2S_2, Fe_3S_4, and Fe_4S_4), and Rieske protein iron centers.

by four cysteine sulfurs and no sulfides. Ferredoxins consist of two, three, or four iron atoms, each with one or two cysteine sulfur atoms and two or four additional sulfides present in the cluster. Some ferredoxins have two nearby Fe_4S_4 or Fe_3S_4 centers, but they are believed to behave independent of one another. Rieske proteins contain a Fe_2S_2 core in which one iron is coordinated by two histidine nitrogen atoms and the other by two cysteine sulfur atoms. Finally, high-potential iron–sulfur proteins, or HiPIP centers, contain an Fe_4S_4 core in which each iron is coordinated by one cysteine sulfur, but the protein environment in which the core resides is significantly different from that of the corresponding ferredoxin environment. The structure of these major classes of iron–sulfur clusters will be discussed in this section, and several specific examples of iron–sulfur clusters in biological redox processes will be described in subsequent sections. A variety of experimental data for this class of proteins is shown in Table 20.4. The spectroscopic differences in these proteins allows for their identification in biological systems.

Rubredoxin proteins are small proteins found in bacteria. The structures of several rubredoxins are known, and they all share similar features, such as two small α-helices and a short two- to three-strand β-sheet. A consensus sequence is a strand of amino acids (or the corresponding DNA sequence) that is conserved across many different organisms. These sequences are used to identify metal-binding regions or similar protein-folding patterns that do not change substantially from organism to organism. More highly conserved sequences suggest a tighter control requirement for that metal-atom binding site. The consensus sequence for iron binding is shown in Eq. 20.5; X can be any residue, and the iron-binding residues are shown in bold. The coordinating ligands come from two different parts of the protein with a variable number of residues (X_n) between the two binding domains, and the binding of the iron atom helps keep the protein folded in the correct shape. The iron atom is coordinated in a tetrahedral environment with S–Fe–S bond angles between 102° and 114°. The biologically relevant oxidation states are Fe(III) and Fe(II). The redox potentials for rubredoxin proteins are between −60 and +40 mV.

cys-pro-X-**cys**-gly-X_n-**cys**-X-X-**cys** **20.5**

Ferredoxins are iron–sulfur proteins that are found in organisms ranging from bacteria to mammals. They consist of two

Table 20.4 Iron–sulfur clusters. Data from Lippard and Berg,[3] Bertini, Gray, Stiefel and Valentine,[2] and references cited therein.

Protein	Cluster	oxidation state	Potential (V)	EPR	Mössbauer isomer shift (δ)	$\lambda_{max}(\varepsilon \cdot 10^{-3}$, per Fe)	Fe-SRÅ	Fe-S$_i$Å
Rubredoxin	Fe_1S_0	Fe^{3+}	−0.061 – +0.042	4.3, 9	0.25	390(10.8), 490(8.8)	2.26–2.31	
		Fe^{2+}		none	0.65	310(10.8), 335(6.3)	2.24–2.29	
Ferredoxin	Fe_2S_2	$Fe^{3+}{}_2$	−0.428 – −0.239	none	0.26	325(6.4), 420(4.8), 465(4.9)	2.27–2.36	2.16–2.23
		$Fe^{2+}Fe^{3+}$		1.89, 1.95, 2.05	0.25, 0.55			
	Fe_3S_4	$Fe^{3+}{}_3$	−0.13	1.97, 2.00, 2.02	0.27	305(7.7), 415(5.2), 455(4.4)	2.26–2.30	2.23–2.33
		$Fe^{3+}{}_2Fe^{2+}$		None	0.30, 0.46	425(3.2)		
	Fe_4S_4	$Fe^{3+}{}_3Fe^{2+}$	−0.645 – −0.280	*2.04, 2.04, 2.12*	*0.31*	325(8.1), 385(5.0), 450(4.6)	2.29–2.32	2.29–2.34
		$Fe^{3+}{}_2Fe^{2+}{}_2$		None	0.42	305(4.9), 390(3.8)		
		$Fe^{3+}Fe^{2+}{}_3$		1.88, 1.92, 2.06	0.57			
		$Fe^{2+}{}_4$		Complex	0.68			
Rieske[38–39]	$N_2Fe_2S_2$	$Fe^{3+}{}_2$	−0.1 – +0.4	weak	0.32, 0.24	325(11.5), 458(60), 560(3)	2.22–2.34	2.24–2.35
		$Fe^{3+}Fe^{2+}$		1.8, 1.9, 2.02	0.31, 0.74	380(4.8), 425(3.8), 550(2.2)		
HiPIP[40–42]	Fe_4S_4	$Fe^{3+}{}_3Fe^{2+}$	0.093 – 0.458	2.14, 2.03	0.40, 0.29	283(11)	2.25–2.30	2.23–2.33
		$Fe^{3+}{}_2Fe^{2+}{}_2$		none	0.43	283(13), 385(9)		

main substructures, the Fe_2S_2 and Fe_4S_4 cores, and there are also examples of Fe_3S_4 cores that have essentially the same binding but with one less iron atom. All have a very similar binding sequence across their families with four cysteine residues within a short sequence of the protein (eq. 20.6); X can be any residue and the iron binding residues are shown in bold. The iron–sulfur clusters are near the surface of the protein. The biologically relevant clusters are Fe(III)$_2$/Fe(III)Fe(II) and Fe(III)$_2$Fe(II)$_2$/Fe(III)Fe(II)$_3$. The redox potentials for these proteins span a large range between −200 and −400 mV for the Fe_2S_2 class, and down to −650 mV for the Fe_4S_4 class.

$$\textbf{cys}\text{-}X_{2-4}\text{-}\textbf{cys}\text{-}X\text{-}X\text{-}\textbf{cys}\text{-}X_n\text{-}\textbf{cys} \qquad \textbf{20.6}$$

Rieske proteins are found in both prokaryotic and eukaryotic organisms. These proteins bind in a conserved structure consisting of three antiparallel β-sheets containing ten β-strands. There are no α-helixes in the protein, and the Fe_2S_2 unit is on the exterior of the protein, exposed to solvent. The major differentiation between these centers and ferredoxins are the unsymmetrical binding of the Fe_2S_2 core by two histidine nitrogens for one iron and two cysteine sulfurs for the other. The binding of the iron center by nitrogen gives these centers a different redox potential relative to ferredoxins, ranging from −100 to +400 mV. The consensus sequence for binding in Rieske proteins is shown in Eq. 20.7; X can be any residue and the iron binding residues are shown in bold. Unlike other binding domains where two residues binding one iron are close in sequence, the first and third binding residue are for one iron while the second and fourth are for the other iron. Rieske centers are typically found as part of a larger protein complex and provide electrons to centers within that complex instead of being redox carriers on their own.

$$\textbf{cys}\text{-}X\text{-}\textbf{his}\text{-}X_n\text{-}\textbf{cys}\text{-}X\text{-}X\text{-}\textbf{his} \qquad \textbf{20.7}$$

HiPIP iron–sulfur proteins are found in anaerobic photosynthetic bacteria. The proteins are small and have very little secondary structure. They incorporate an Fe_4S_4 cluster, but it is oxidized by one electron relative to the cluster found in ferredoxins. Three of the iron atoms are Fe(III), while one is Fe(II) in the oxidized form. HiPIP proteins have been studied by XAS in order to probe the oxidation states of the cluster.[43] Because the redox center is shifted by one electron, the redox potential of these proteins is shifted to higher potential, from 100 to 450 mV. The metal cluster is shielded from the aqueous environment, and this is a requirement for its function, as in the oxidized form the cluster is susceptible to hydrolysis. The consensus sequence for binding the metal cluster is highly conserved (Eq. 20.8), including a hydrophobic tyrosine or tryptophan residue (Z), and there are a number of hydrophobic side chains; X can be any residue and the iron binding residues are shown in bold. The distance between the second pair of cysteine residues is variable. The function of HiPIP in higher organisms is taken over by other types of iron–sulfur proteins, hemes, or copper proteins.

$$\mathbf{cys}\text{-X-X-}\mathbf{cys}\text{-X}_{8-16}\text{-}\mathbf{cys}\text{-X}_{10-13}\text{-gly-Z-}\mathbf{cys} \qquad \mathbf{20.8}$$

20.7.4 Zinc Fingers

A zinc finger is a protein sequence that is crucial for protein–DNA and protein–RNA interactions. In the 1960s and 1970s it was well known that zinc was essential for DNA and RNA replication, and in the 1980s a transcription factor called TFIIIA was isolated *Xenopus*, in which it was found that each mole of protein contained 2–3 mol of zinc. TFIIIA is a site-specific DNA binding protein that is involved in gene regulation by activating transcription. The protein binds tightly to a 45-base-pair sequence of DNA. Within a 30-amino-acid section of the protein there are two cysteine and two histidine residues that provide a tetrahedral coordination environment for a single zinc(II) cation. Zinc is an especially good choice for a DNA-binding protein, as it is both naturally abundant and does not have redox activity. Oxidative damage could lead to mutation of the DNA. The possibility of this damage is avoided with Zn relative to the use of Fe or Cu, both of which have multiple stable oxidation states in biological environments.

Unlike α-helices or β-sheets, there is significant structural diversity in zinc-finger proteins, since each has its own DNA sequence specificity. The discussion in this section is limited to the most common and well understood ββα domain, but other structures are known. However, even with the structural diversity, parts of the protein structure in zinc fingers are highly conserved. The consensus sequence of amino acids in zinc-finger proteins across many different species is shown in Eq. 20.9, in which X can be any residue, and with the zinc binding residues shown in bold.

$$\text{X}_2\text{-}\mathbf{cys}\text{-X}_{2,4}\text{-}\mathbf{cys}\text{-X}_{12}\text{-}\mathbf{his}\text{-X}_{3-5}\text{-}\mathbf{his} \qquad \mathbf{20.9}$$

In addition to the highly conserved 23–28-amino-acid repeat unit, there is a conserved protein structure consisting of two antiparallel β-sheets and an α-helix that includes the zinc finger and the metal-binding amino-acid residues (Fig. 20.35a). The two cysteine residues lie in the β-sheet, and the two histidine residues are towards the C-terminal end of the α-helix. The coordination of the zinc ion by two sulfur and two nitrogen atoms is essential, as it stabilizes the structure of the protein, and using gene editing to remove any of the cysteine or histidine residues results in the lack of ability for metal coordination and loss of structure and function.

The secondary structure of the zinc-finger protein provides a regular template that can be highly modified by the specific amino-acid residues. The α-helix is often called the "recognition helix" as it is the part of the structure responsible for making specific contacts to the DNA bases in the binding target (Fig. 20.35b). Most zinc fingers of the ββα type have at least two and often more zinc finger domains providing specific protein-DNA interactions in order for even more specific binding. The protein shown in Fig. 20.35 (b) consists of three zinc-finger domains linked together. When linked together, there is often a short amino acid residue spacer between the zinc-finger domains. Since these proteins typically play a gene

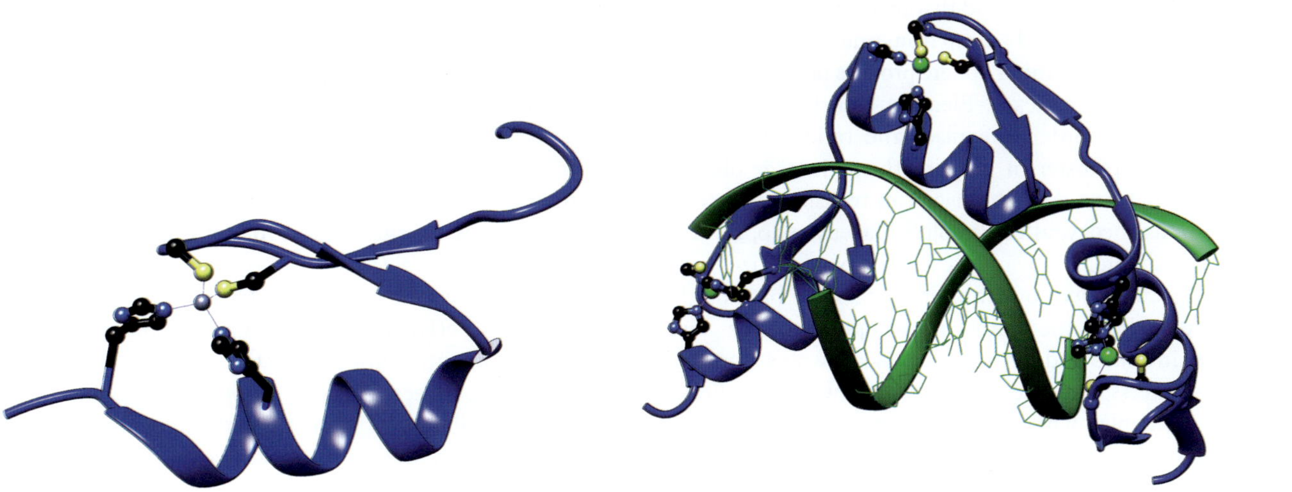

(a) (b)

Fig. 20.35 A) Zinc finger protein shown in ribbon form showing the zinc atom holding the fold in place forming the "finger," and the highly variable α-helix region that is responsible for site-specific binding. B) a protein containing three linked zinc fingers binding to a DNA strand. Figures generated with UCSF Chimera[11] with coordinates from the protein databank (PDB ID 3IUF and 1ZAA).[9, 44–45]

regulation role, specific DNA interactions are required. The specific amino acid residues in the nonconserved region allow for site-specific binding to DNA.

Other roles for zinc fingers include gene transcription and translation, cell adhesion, and protein-folding. The structure of the protein is so important for target-binding; thus, unlike many other proteins there is rarely a conformational change of the zinc finger upon binding the target. However, upon binding to DNA, the conformation of the DNA itself usually changes through opening up of the major groove.

20.8 Metalloenzymes as Group Transfer Reagents

We now turn our attention to the ability for metalloenzymes to function as group transfer reagents. This is but one role of this broad class of proteins. We have already seen examples of metalloenzymes acting as recognition agents (zinc fingers) and as activators of small molecules (jack bean urease). We will first discuss electron transfer—the process by which some living organisms harness light energy to form biological fuels during photosynthesis—and then controlled combustion of these fuels for life during respiration. Next, we will look at the uptake and storage of iron and copper—two of the most important elements for living things—followed by a discussion of the transport and storage of dioxygen. The section ends with a discussion of the cobalt alkylation coenzyme B_{12}.

20.8.1 Electron Transfer

Electron transfer (ET) in proteins is characterized by long-range processes by which electron donors and acceptors are separated in space. Iron–sulfur clusters are often buried deep within a protein environment, and close contact between a metal center in one protein and that in another is not possible. The electron cannot simply jump from one metal center to the next like it could in an inner sphere process between two solution phase metal complexes. Electron transfer theories have been developed in order to explain and rationalize the high rate of electron transfer in proteins through large distances. The theory that best describes electron transfer in proteins is called Marcus theory, after Rudolph Marcus, who, for its development (Chapter 17), was awarded the Nobel Prize in 1992.

A bimolecular electron transfer process can be thought of in three steps. First, the donor and acceptor must be close to each other in space and develop what is known as a precursor complex. Outer-sphere electron transfer follows the Franck–Condon principle, which means that during the transfer event, the nuclei are stationary. Thus, the nuclei must move to the transition-state geometry found in the precursor complex in order for the electron to transfer. In the second step the electron transfers, and in the third step the resulting successor complex (of the same geometry as the precursor complex, just with a different electronic configuration) dissociates.

In biological systems, the first step is severely limited due to the inability of the free metal ions to approach each other spatially. Electron-transfer centers are often (but not always) situated towards the outside of the protein to minimize the distance required for the electron-transfer event.

Marcus developed an equation to describe the rates of outer-sphere electron transfer (Eq. 20.10). The rate of the process depends on the nuclear frequency factor (ν_n), the transmission coefficient (κ), the $\Delta G°$ of the process, and λ, the reorganization energy. The transmission coefficient ranges from 0 to 1 and is a measure of how adiabatic the process is; an adiabatic electron transfer remains on the same electronic energy surface during the event. The reorganization energy is a parameter used to describe the bond lengths and geometry changes that must occur for the donor and acceptor to reach their respective transition state geometries. For electron transfer in small molecules, the reorganization energy and transmission coefficient play a large role in the rate. For metalloenzymes, the transmission coefficient can be quite small due to the long distance required for electron transfer and low orbital overlap between donor and acceptor. The reorganization energy is small in a protein that has an entatic geometry (see Section 20.2). For donors and acceptors separated by long distances and at a fixed orientation, such as metalloenzymes, a different model is used to describe the electron transfer (Eq. 20.11). In this model, the overlap integral between the donor and acceptor is explicitly parameterized as H_{AB}, as this parameter is the most important in describing the rate of electron transfer.

$$k_{ET} = \nu_n \kappa \exp\left(-\frac{(\lambda + \Delta G°)^2}{4\lambda RT}\right) \quad \textbf{20.10}$$

$$k_{ET} = \sqrt{\frac{4\pi^3}{h^2\lambda RT}} H_{AB}^2 \exp\left(-\frac{(\lambda + \Delta G°)^2}{4\lambda RT}\right) \quad \textbf{20.11}$$

The fastest possible electron transfer is when $\Delta G° = -\lambda$, and at this point the reaction is barrierless. The rate then depends only on the electronic coupling between donor and acceptor, H_{AB}. For reactions occurring with van der Waals contact between the donor and acceptor, the coupling parameter is large and electron transfer is fast. For metalloenzymes, the H_{AB} value is small; the donor and acceptor are not in close proximity, and the overlap of their wavefunctions is small. In these processes, the transition state must be reached many times before the electron will transfer.

In aqueous media, the transfer of an electron from an Fe(II) and Fe(III) center has an activation barrier of 0.66 eV. The major contributor to this barrier is the 0.14 Å bond length difference between the two oxidation states of iron, while another contributor of almost equal importance is the solvent reorganization that must occur around these two centers during the reaction. These large barriers would prevent electron transfer from occurring at a fast enough rate to support a living system. However, inside protein cavities, water can be excluded, and the iron centers can be held in a fixed environment, reducing the reorganization energy significantly.

Electron transfer occurs when the electron tunnels through the barrier between the donor and acceptor. At a reasonable distance of 20 Å, were the electron to tunnel through vacuum, the time required for an event to occur would be 10^{17} years. If the medium were water molecules, the time required drops to 10^4 years. However, if the electron can tunnel through hydrocarbon bonds like those found in proteins, the time required is on the order of micro- to milliseconds. The protein environment thus serves two vitally important functions. It organizes the donor and acceptor to reduce the barrier for electron transfer, and it provides an alternative pathway for electron tunneling to allow the electron to transfer at a reasonable rate.

There are several general features of electron transfer in metalloproteins that can be presumed by analysis of Marcus theory. The electron travels from a metal center through a medium (bonds or solvent water molecules) and ends up on a receptor. The receptor may be an organic small molecule such as a quinone or NADP, or a small protein such as cytochrome c or a ferredoxin. The receptor then moves to a new location and reduces the next center at a slightly lower reduction potential. Electron transfer proteins transfer a single electron at a time because metal redox centers change only by ±1 electron. The reduction potential of the electron transfer process is highly tunable. The range of reduction potential for iron–sulfur proteins ranges from −0.65 to +0.45 V. The reduction potentials (free energies of reduction) are designed to take advantage of higher rates when the thermodynamic driving force is not too large and not too small (ideal rate when $\Delta G = -\lambda$). The rates of ET processes are fast. The reorganization energy for ET processes is usually minimal due to the formation of entatic state geometries.

In addition to the iron–sulfur clusters and heme centers described previously in this chapter, there is a variety of copper proteins that can carry out electron transfer reactions. Copper proteins are found in a variety of organisms from plants to bacteria, and it is the only element besides iron that is used for redox reactions in nature. Copper proteins have much less structural diversity than iron-containing proteins. The most common of these consists of a small protein consisting of around 100 amino acid residues. The protein exhibits a structure known as a β-barrel, which has two linked β-sheets. Sometimes small α-helical segments are present. The copper binds close to the surface of the protein, though it is not exposed to the aqueous environment. It is bound by the conserved binding domain of a cysteine and two histidine residues. Structural variety is seen in the fourth axial ligand sometimes present, which may be either a methionine or glutamine. The bond present between the copper and cysteine sulfur gives rise to the strong LMCT band, and EPR studies suggest that the unpaired electron is shared almost equally between the copper and sulfur atoms.

Electron transfer is seen throughout living systems. Photosynthesis takes light energy to promote an electron to a higher energy state, just like visible absorption spectroscopy, but instead of thermal decay of the excited state, the electron is separated from the hole and brought through the photosynthetic reaction center to drive the formation of chemical bonds to store the light energy. Respiration is the controlled combustion of these fuel molecules with the net reduction of oxygen to water. Nitrogen-fixing bacteria take ATP molecules provided by a symbiotic host and use the energy they can obtain from those molecules to reduce dinitrogen to ammonia. Methane monooxygenase bacteria use their reducing equivalents to activate dioxygen, which then activates methane. The nitrogenase and methane monooxygenase protein centers will be discussed in detail in the last section of this chapter.

20.8.2 Iron Uptake

As has been seen in previous sections, iron is an essential and ubiquitous element in bioinorganic chemistry. However, the solubility of Fe(III) at biologically relevant pH is vanishingly small. Bacteria have developed specialized iron binding agents known as siderophores that are engineered to bind strongly to iron centers and can be specifically taken into the cell. The mammalian system for iron uptake and storage relies on two proteins: transferrin, which transports iron, and ferritin, which stores it.

20.8.2.1 Bacterial Siderophores

Siderophores are small molecules produced by bacteria and other microorganisms that are specifically designed to bind to Fe(III). Siderophores are produced under the stress condition of low iron levels, and the bacteria also produce receptor proteins on their outer cell membranes to bind to and take in the iron siderophore complexes. Through a feedback mechanism, once sufficient levels of Fe(III) are present in the organism, production of the siderophores and binding proteins ceases.

While there are hundreds of known siderophores, most contain one of two binding groups, catechol (dihydroxybenzene) and hydroxamic acid (RC(O)N(OH)R′). These two binding groups are shown in two common siderophores: enterobactin, found in bacteria such as *E. coli*, and desferrioxamine B, found in *Streptomyces* (Fig. 20.36). Enterobactin is a cyclic lactone with three catechol ligands joined by amide bonds that forms a hexadentate complex with iron; it has the highest formation constant of any known siderophore, at 10^{49}(!) for the chemical reaction shown in Eq. 20.12. Another measure of the binding affinity for siderophores is the pM, or the $-\log_{10}$ of the free metal ion concentration, which is 35.5 for enterobactin. This means that the concentration of free iron is $10^{-35.5}$ M! The three catechol binding groups form an octahedral iron center. The crystal structure for the iron enterobactin complex was determined in 2017,[46] while the crystal structure of a closely related iron siderophore complex was reported in 2003.[47]

$$Fe^{3+}(aq) + (ent)^{6-}(aq) = [Fe(ent)]^{3-}(aq);\ K_f = 10^{49} \qquad \textbf{20.12}$$

The free enterobactin ligand is oriented with its catechol group pointed "outwards", stabilized with a hydrogen bond between the ortho hydroxyl group and the amide oxygen (Fig. 20.37). Upon binding to iron or deprotonation (at physiological pH, approximately half of the enterobactin is deprotonated),

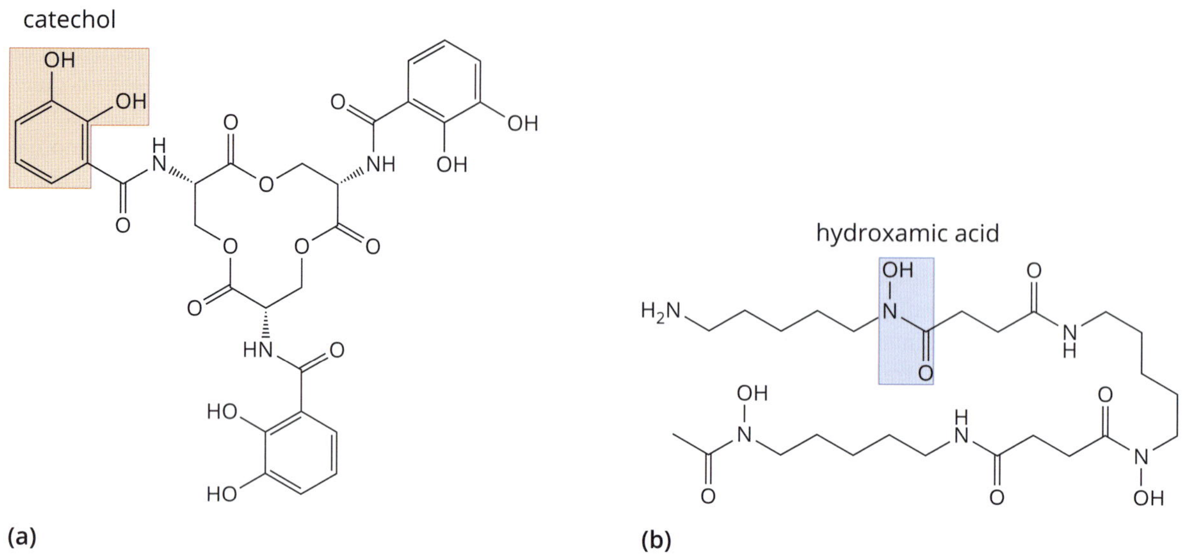

Fig. 20.36 The common siderophores (a) enterobactin, with a *tris*-catechol binding group (one catechol is shown in red), and (b) desferrioxamine B, with a hydroxamic acid binding group (one hydroxamic acid is shown in blue).

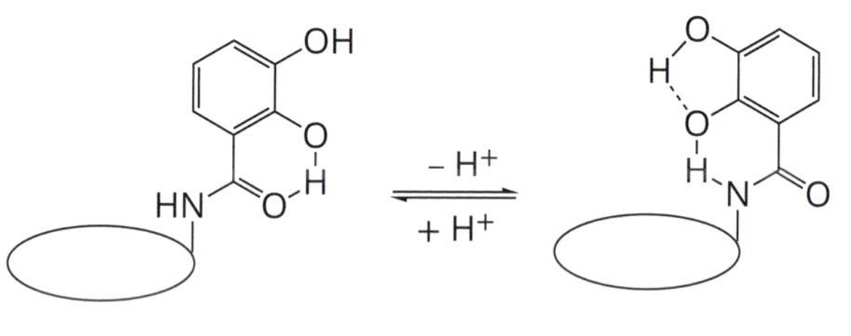

Fig. 20.37 Structural changes upon deprotonation of enterobactin that drive iron uptake and binding by the siderophore.

the catechol changes orientation to engage in hydrogen bonding between the amide proton and the *ortho*-phenoxide. This siderophore appears "optimally tuned to perform both tasks" of iron uptake by the outward-facing catechols, and storage by the inward facing ones.[47]

Bacteria have developed a complex multiprotein complex for the recognition and uptake of the iron enterobactin complex. An outer membrane receptor protein binds the complex in a rate-limiting step. Once bound, the complex is transported into the bacterial cytoplasm in an ATP dependent process. In a fascinating twist in biology, mammalian immune cells have developed binding proteins that target the iron enterobactin complex. These proteins sequester the bacterial iron sequestration agents, preventing the bacteria from taking up their own siderophores and iron and thus acting as a natural antibiotic. The crystal structure of this protein-Fe(ent) complex shows specific binding of the siderophore into the protein.[48]

20.8.2.2 Mammalian Transferrin

Multicellular organisms have developed complex proteins to transport iron from food sources into their cells. This family of proteins is known as transferrins. Human transferrin is a 679 amino-acid chain with two carbohydrate chains appended. The protein is an approximate dimer, with one iron binding site in each half of the protein (the N-terminal end and the C-terminal end). The binding of iron to the protein causes a significant conformational change from an "open" to a "closed" configuration (Fig. 20.38).[49] The iron is bound within a hydrophilic region, and the iron is ligated in an approximately octahedral geometry by two tyrosinates, an aspartate, a histidine, and a carbonate. The carbonate is located in a pocket with a positively charged arginine and is engaged in multiple hydrogen bonds with amide hydrogen atoms. It is believed that the synergistic binding of carbonate is important for the conformational change upon binding iron.

Transferrin stores iron for use by cells, and much like enterobactin, there are specific transferrin receptors in cell membranes. The receptor, at pH 7.4, specifically binds to the iron-loaded transferrin in the "closed" form. When transferrin binds to the receptor, the cell membrane undergoes endocytosis. A cartoon of this process is shown in Fig. 20.39. This process results in a small vesicle that contains the bound transferrin protein inside. The vesicle membrane has ATP driven proton pumps that lower the pH of the vesicle from 7.4 to 5.6. At this lower pH the receptor binds more strongly to the apoprotein, which helps to open the transferrin and release the iron atom inside the vesicle. Protonation of the carbonate and tyrosinate ligands helps with iron release. The vesicle then recombines with the cell membrane at pH 7.4, where the apotransferrin is released back outside the cell. Large differences in pH are used to regulate the iron binding as well as the binding of the transferrin to the receptor. The released iron is then incorporated into hemes, iron dependent proteins, or for long-term storage in the protein ferritin.

20.8.2.3 Mammalian Ferritin

Due to the importance of iron in living cells, iron is stored to make it available when needed to prevent both iron overload

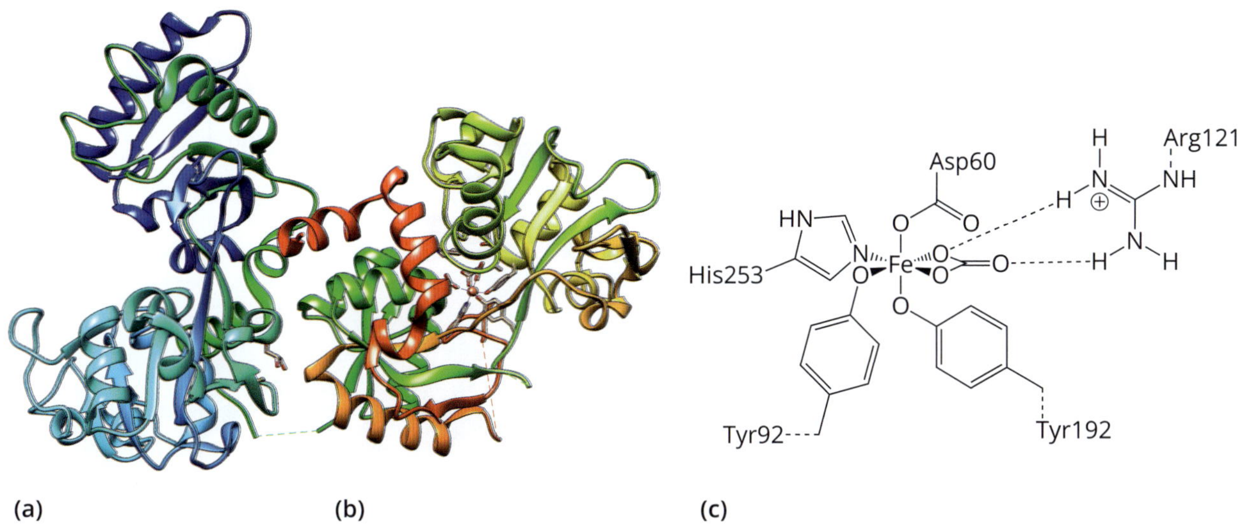

Fig. 20.38 The mammalian iron binding protein transferrin showing the (a) open and (b) closed forms in the same structure. The closed form binds to a single iron atom as well as a carbonate, and the binding pocket is shown in (c). Figures generated with UCSF Chimera[11] with coordinates from the protein databank (PDB ID 4X1B).[9,49]

and iron deficiency. The storage protein is called ferritin—a highly conserved protein found in organisms from algae to humans. The protein itself is highly symmetrical, consisting of twenty-four identical alpha-helical subunits that aggregate into a hollow nanocage with four-fold, three-fold, and two-fold axes with "O" point-group symmetry. The interior volume of ferritin is approximately 250 nm^3, and it can store around 4,500 iron atoms along with hydroxide ions as a single nanoparticle with a structure similar to ferrihydrite, otherwise known as rust. The ferritin structure is shown in Fig. 12.36 in the context of oxo ligands for iron.

The approximate chemical reactions that result in the formation of the iron oxide nanoparticle are shown in the following equation; all the species are hydrated in the aqueous environment. Iron(II) ions react with dioxygen to form the Fe(III) bridging peroxo complex. This complex decomposes and releases hydrogen peroxide, and the Fe(III) centers combine with each other to form hydrated Fe(III) oxide.

$$\mathrm{Fe(II)} + \mathrm{O_2} \rightarrow [\mathrm{Fe(III)\text{-}O\text{-}O\text{-}Fe(III)}]$$
$$[\mathrm{Fe(III)\text{-}O\text{-}O\text{-}Fe(III)}] \rightarrow 2\,\mathrm{Fe(III)} + \mathrm{H_2O_2}$$
$$\mathrm{Fe(III)} + 6\,\mathrm{H_2O} \rightarrow [\mathrm{Fe_2O_3(H_2O)_3}]_n + 6\,\mathrm{H^+}$$

The large number of protons released during the process must be translocated out of the interior of the ferritin particle, or acidic hydrolysis of the peptide bonds would occur. The mineralization of iron takes on the order of hours due to the relatively slow release of protons from the interior. In addition, the hydrogen peroxide that is released must be reduced by enzymes known as peroxidases. The storage of iron is an expensive but important process for the cell.

The pores through which iron atoms enter or leave the ferritin structure are believed to be the three-fold axes. Under iron-deficient conditions, the three-fold pore opens to allow access by reductants, protons, and chelators. This gating was modeled by the addition of urea or guanidine reagents to a folded ferritin protein. These denaturing agents specifically affected the alpha-helical regions surrounding the three-fold pore, disrupting the protein structure and opening the channel.[50] At low concentrations of reagent, the global ferritin structure was unchanged, but the alpha-helices at the pores were significantly disrupted and no ordered electron density was observed (Fig. 20.40).

20.8.3 Copper Uptake and Delivery

Like iron, copper is an essential transition metal required for its redox chemistry in a number of metalloenzymes. Copper imbalances can lead to Menkes' syndrome or Wilson's disease, which are both characterized by the diminished ability to distribute copper to where it is needed. The proteins responsible for copper uptake and transport were less well understood than those for iron until relatively recently. Ceruloplasmin, a glycosylated protein, stores more than 95% of copper in eukaryotic cells, but when the gene encoding for that protein was removed from a mouse model there was no impact on the absorption or uptake of copper.

Proteins from several species are now known to be copper transport proteins and have been classified into the Ctr family of high-affinity copper transport proteins. Their structural domains are highly conserved across yeast, plants, and animals. These proteins contain transmembrane regions with hydrophobic side-chains—a region rich in methionines in the N-terminal end with **met**-X_n-**met** ($n = 1, 2$) conserved regions, and a highly charged region at the C-terminal end that also contains cys and his amino acids. This conserved structural motif was modeled as a methionine-rich region that captures extracellular copper ions with cysteine and histidine side-chains to bind copper inside the cell (Fig. 20.41). In addition to

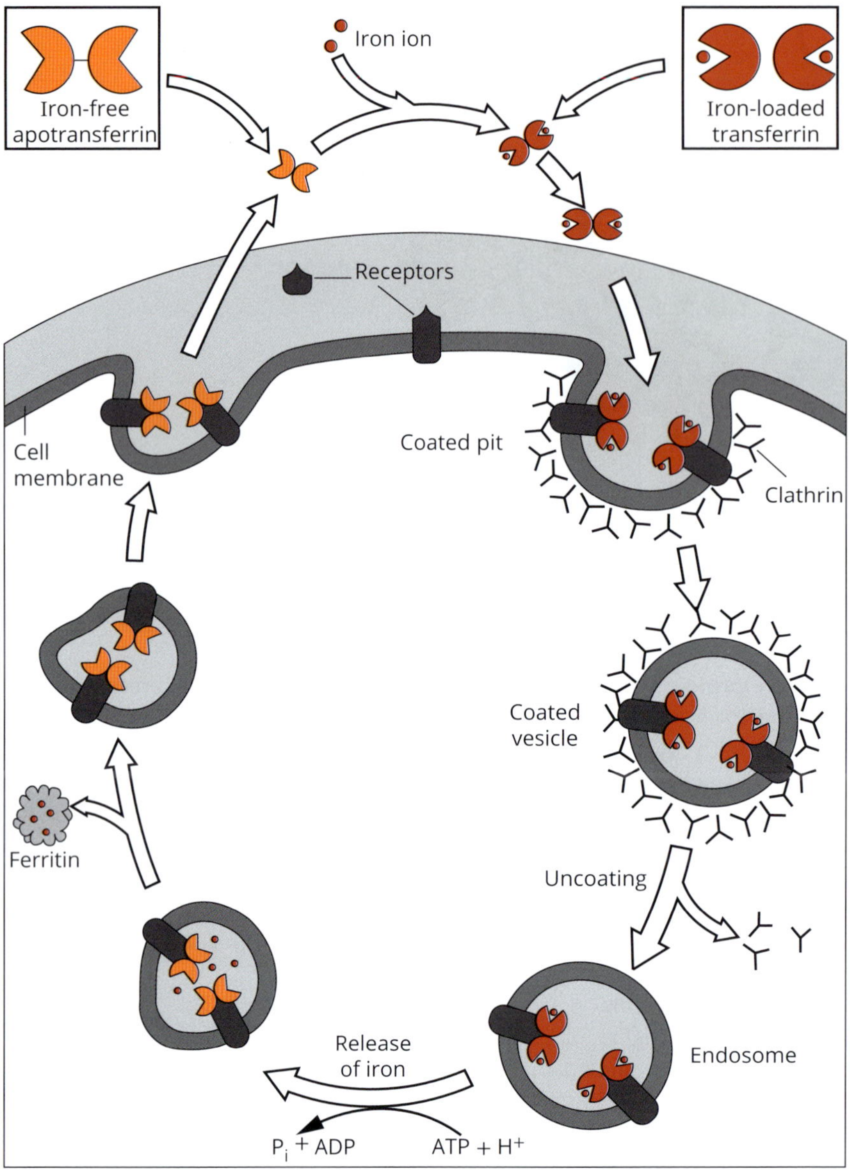

Fig. 20.39 The pH-dependent process by which iron is taken into a mammalian cell by the ferritin receptor. (Figure adapted from Lippard and Berg.[3])

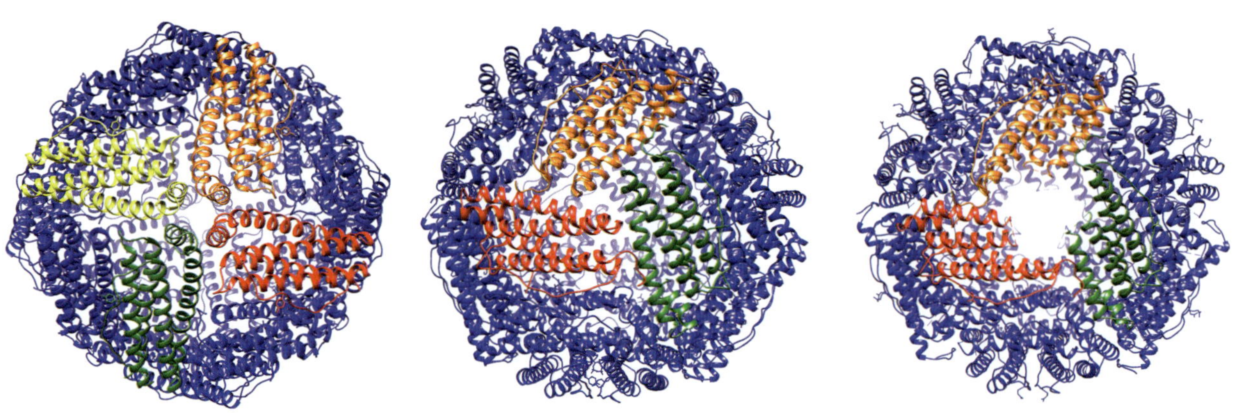

Fig. 20.40 Structure of the four-fold pore (left) and the three-fold pore of ferritin (center) in the closed form, and the three-fold pore of the open form (right) with disordered pore helices. Figure generated with UCSF Chimera[11] with coordinates from the protein databank (PDB ID 1BG7 and 1RCC).[9,51–52]

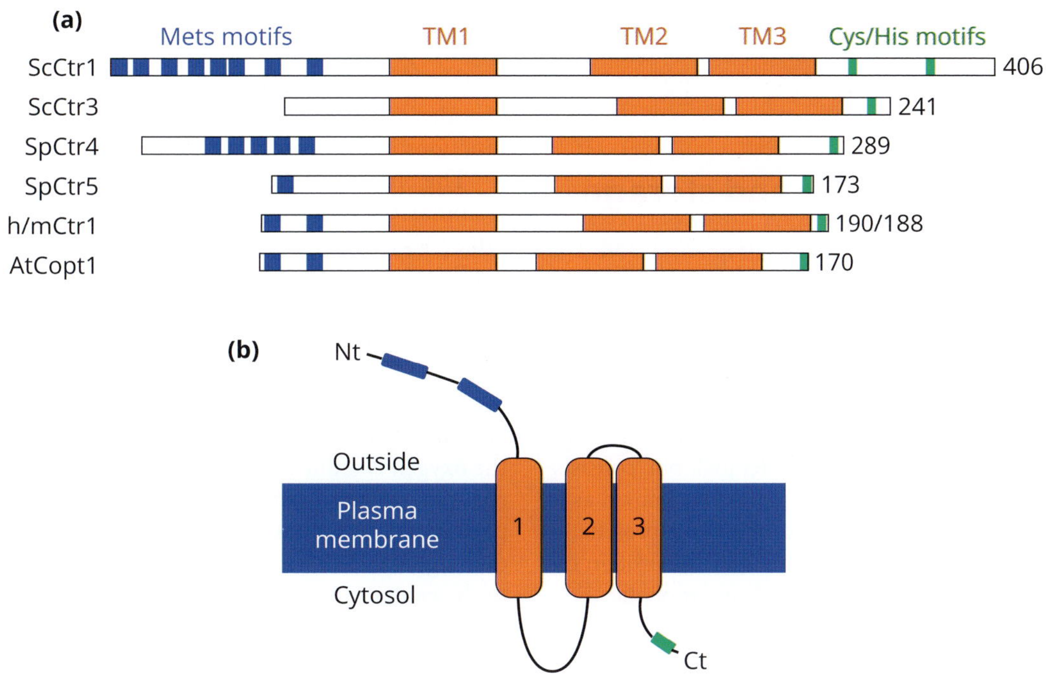

Fig. 20.41 The highly conserved structural motif in the Ctr class of high-affinity copper transport proteins. Figure adapted from Thiele.[53]

these high-affinity copper proteins, there are also low-affinity copper transport proteins. These proteins are activated when there is a high concentration of copper ions in the environment. These proteins, in some cases, can also transport iron.

Free copper ions are toxic, and therefore once the copper has been captured it must be carefully controlled and distributed to the cells where it is needed. A large number of proteins are known that can transport copper ions to different locations, including for secretion from the cell into the extracellular space. A protein where copper delivery is reasonably well understood is Cu, Zn superoxide dismutase (CuZn–SOD). This enzyme uses copper to catalyze the disproportionation of superoxide to hydrogen peroxide and dioxygen to reduce the threat of oxidative damage. Copper is delivered to this enzyme by a specific copper chaperone protein named CCS.

The structures of CuZn–SOD and the CCS protein are both known,[54–55] as is the heterodimer between CCS and a modified SOD that lacks a histidine and thus the terminal copper binding site near the zinc atom (Fig. 20.42).[56] The CCS protein has three domains. Domain 1, in purple, contains a **met**-x-**cys**-x-x-**cys** copper binding motif (not shown). Domain 2, in green, is a linking unit that binds to SOD. Domain 3, in blue, contains a **cys**-x-**cys** motif that is required to transport the copper from domain 1 to SOD but is not involved in binding to SOD. The evidence points to the following model for copper insertion into SOD from the storage and transport protein. First, the CCS protein receives a copper atom from a donor which binds in domain 1. Next, domain 2 is involved in

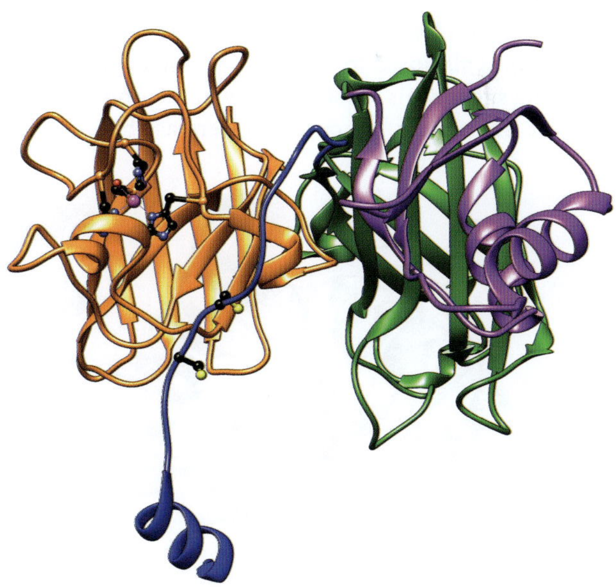

Fig. 20.42 The X-ray structure of the CCS protein (right) docked with a mutated CuZn-SOD that only binds to zinc (left). The zinc atom bound in SOD to three histidines and an aspartate is shown (center left) while the two cystine residues that bind to copper in Domain 3 are shown as ball and stick diagram. This structure was obtained with a mutant that has no Cu binding site in the SOD, but copper binds near the Zn. Figure generated with UCSF Chimera11 with coordinates from the protein databank (PDB ID 1JK9).[9,56]

an interaction with SOD to bring the two proteins into close proximity. Finally, the copper atom shifts form domain 1 to domain 3, which is flexible and can deliver the copper atom to the SOD active site.

20.8.4 Oxygen transfer and storage

The uptake, delivery, and storage of oxygen is critically important for aerobic organisms, and complex biochemical pathways have evolved to carry out this process. The oxygen storage protein found in muscle tissue of most vertebrate animals is myoglobin, while the oxygen transport protein found in red blood cells of most vertebrate animals is hemoglobin. Myoglobin was the first protein to be characterized by X-ray crystallography, determined by John Kendrew.[57] Hemoglobin consists of eight alpha-helices connected together in a fold known as a globin fold that creates a binding pocket for a heme prosthetic group (Fig. 20.43). The heme is held in place in the protein by a number of noncovalent interactions with hydrophobic side chains, and a single "proximal" histidine. A second "distal" histidine is positioned at the opposite face in a location that it can interact with a bound oxygen molecule. The binding affinity for myoglobin is higher than that of hemoglobin, which makes sense as it is designed to take the oxygen molecule from hemoglobin and store it in muscle tissue.

Hemoglobin, to a first approximation, has a tetrahedral noncovalently bound quaternary structure of four almost identical myoglobin-like molecules in an ααββ arrangement. The X-ray structure of hemoglobin was determined by Max Perutz, who, with Kendrew, shared the Nobel Prize in 1962.[58] Hemoglobin is also used to transport carbon dioxide, though the binding site for this substrate is not at the iron center. The affinity of the protein for oxygen lowers as the pressure of carbon dioxide increases, and the opposite trend of increased affinity for carbon dioxide as the pressure of oxygen decreases is also observed. This opposite affinity for oxygen and carbon dioxide is critical for describing the two different transport functions of the protein.

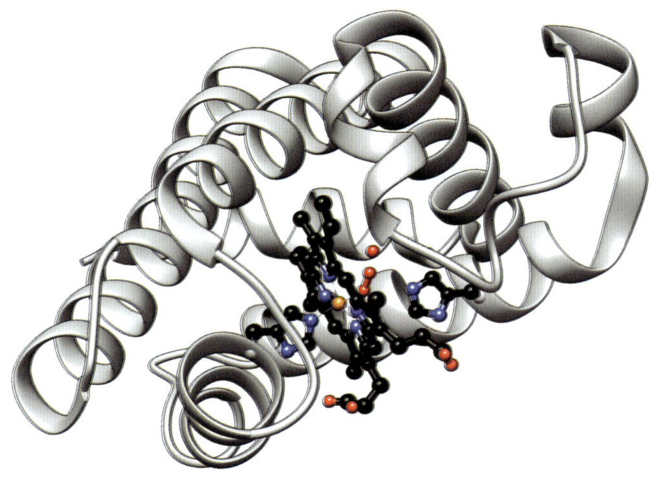

Fig. 20.43 The 8-alpha helix bundle that makes up the globin fold in the protein myoglobin. The heme group is held in place by a histidine bound to iron (orange sphere), while a second histidine is present to interact with the bound dioxygen (red spheres). One water molecule is shown near the dioxygen. Figure generated with UCSF Chimera[11] with coordinates from the protein databank (PDB ID 1MBO).[9,14]

The binding of dioxygen to the heme site in both proteins is similar. Considering the standard reduction potentials of O_2 (+0.82 V) and heme Fe(III) (+0.17 V), dioxygen is capable of oxidizing the iron center under reasonable concentrations and pH. The binding of oxygen to the heme iron causes a significant structural rearrangement of the iron–porphyrin–protein. The 5-coordinate Fe(II) center is high-spin and sits outside the porphyrin plane, bound to the proximal histidine. Upon binding of dioxygen, the iron center is oxidized to Fe(III), and the oxygen is reduced to superoxide (O_2^-). The stability of this reduced oxygen is enhanced by coordination to the distal, protonated, histidine. The 6-coordinate iron is low-spin and diamagnetic. The smaller radius of Fe(III) allows it to move to the center of the porphyrin pocket from its initial location, and the bound histidine moves with it. This movement of the histidine has profound implications for cooperative binding of dioxygen in hemoglobin, as will be discussed shortly.

There is substantial evidence for the presence of Fe(III) in hemoglobin, including XANES, where the edge energy is consistent with Fe(III), IR of the O–O stretch, which is consistent with an O–O bond order of 1.5, and the fact that the oxygen-bound system is diamagnetic. Binding of the paramagnetic superoxide anion to a low spin d^1 Fe(III) center results in strong antiferromagnetic coupling and a diamagnetic species. This model was proposed by Weiss,[59] resulting in a debate[60–61] between him and Pauling in the journal *Nature*. Detailed computational results support Weiss's original proposal (Fig. 20.44).[62]

In addition, the binding of oxygen is selective relative to other potential substrates such as CO. When dioxygen binds to the porphyrin, it binds in a bent fashion with an Fe–O–O angle of approximately 125°. The superoxide is involved in a hydrogen bonding interaction with the protonated distal histidine. The binding of CO to an iron porphyrin would preferentially result in a linear Fe–C–O arrangement with a binding affinity an order of magnitude above that of dioxygen. In the biological system, the distal histidine was initially postulated to provide a steric interaction that inhibited binding of CO. However, additional studies suggest that sterics are less important for the discrimination of binding than initially thought.

The presence of a water molecule near the binding site is now the best explanation of the high discrimination seen for dioxygen binding relative to CO. In deoxymyoglobin, a water molecule is coordinated to the distal histidine, which must exit the active site in order for dioxygen to bind (Fig. 20.45). When the cavity is vacant, dioxygen can enter and bind to the iron center. For CO, this binding is metastable, but for dioxygen the distal histidine tautomerizes to interact strongly with the superoxide ligand. The net energy cost for water loss is +6 kJ/mol. For dioxygen binding, the hydrogen bond is

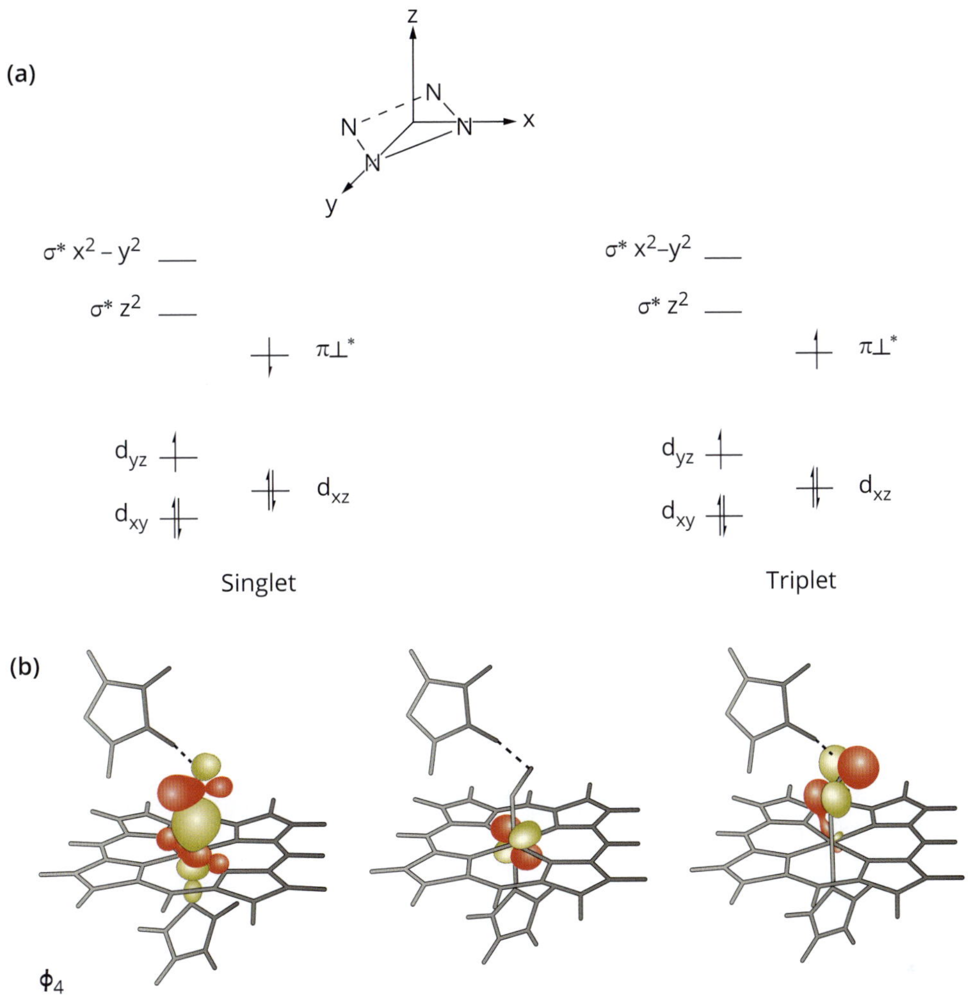

Fig. 20.44 (a) Frontier molecular orbital diagram for the binding of O_2 to the heme group in either hemoglobin or myoglobin. Shown is the lowest-energy open-shell singlet state and the slightly higher-energy triplet state. The O_2 molecule is in the *xz* plane. (b) The two orbitals responsible for Fe–O_2 bonding in heme oxygen complex. Orbital ϕ_4 is the sigma bond between Fe and O, while the other two orbitals, d_{yz} and O–O π*, each have a single electron. Figures modified from Chen.[62]

–17 kJ/mol for a net stabilization of –11 kJ/mol. For CO binding the hydrogen bond is only –3 kJ/mol for a net "stabilization" of +3 kJ/mol.

It has been known for more than a century that the binding of oxygen to hemoglobin follows a sigmoidal curve that could not be explained by a simple on-off binding of the dioxygen molecule. Instead, a cooperative binding model is required to explain the data. The binding of the first dioxygen molecule encourages binding of three additional ones, and the binding is inversely correlated with that of carbon dioxide. X-ray crystallography has shown that hemoglobin undergoes substantial conformational reorganization upon dioxygen binding. The movement of the proximal histidine, the histidine bound to iron, upon oxygen binding is critical for the explanation of the cooperative binding of oxygen in hemoglobin. Upon oxidation of the Fe(II) center, which sits outside the porphyrin ligand bound to the proximal histidine, to Fe(III), the iron moves into the plane of the porphyrin ligand. Movement of the iron atom is transferred via the bound histidine, which amplifies the relatively small atomic motion to significant structural changes in the ααββ tetrameric protein. There is a net rotation of the two pairs of αβ subunits by 15° relative to each other (Fig. 20.46). This movement results in an increase in oxygen binding constant for the three remaining hemes in the protein complex.[63] Mutants where the proximal histidine is changed to glycine do not exhibit the cooperative binding of dioxygen, because the motion of the iron into the plane of the porphyrin is not translated to the rest of the hemoglobin protein.

The tense, or T state, is the structure observed when there is no oxygen bound to the protein. When oxygen is bound, the new conformation is called the relaxed, or R state. As the hemoglobin changes conformation, protons are lost from the

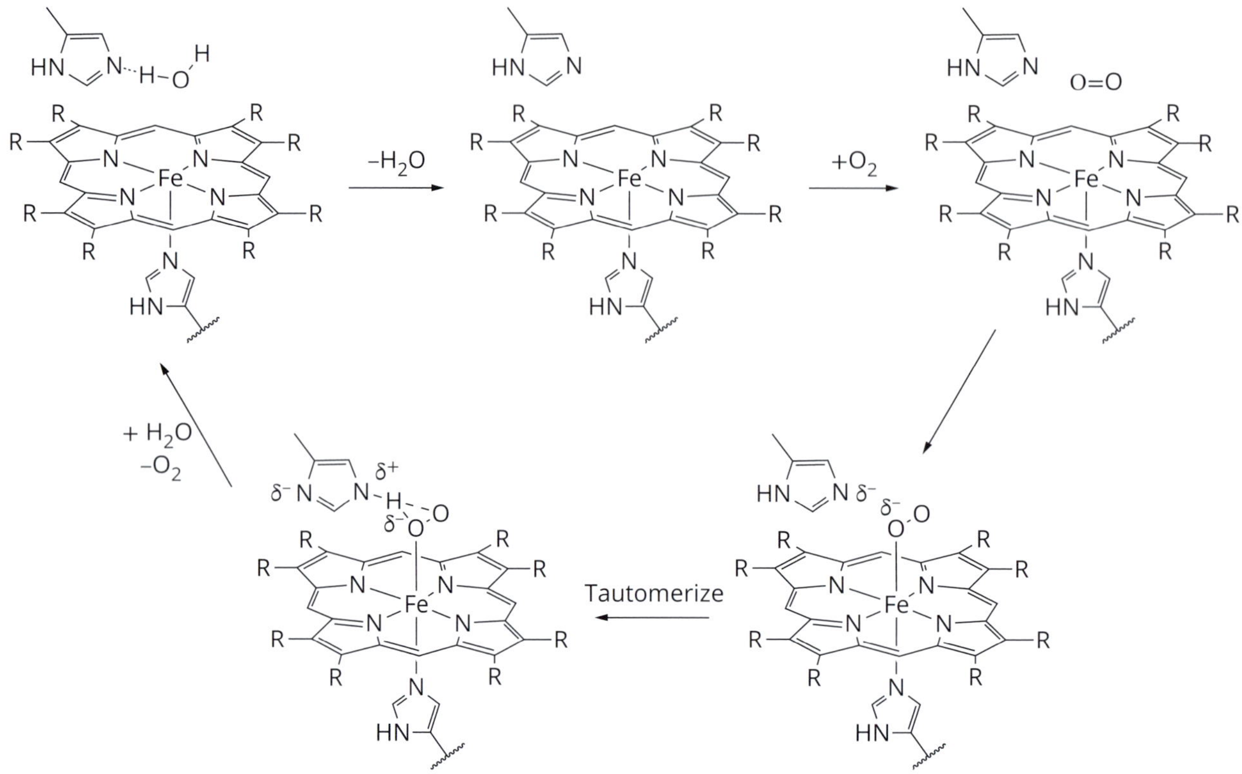

Fig. 20.45 The proposed mechanism for dioxygen binding to the heme in myoglobin showing the bent Fe–O–O geometry and the hydrogen bond to the distal histidine post-tautomerization.

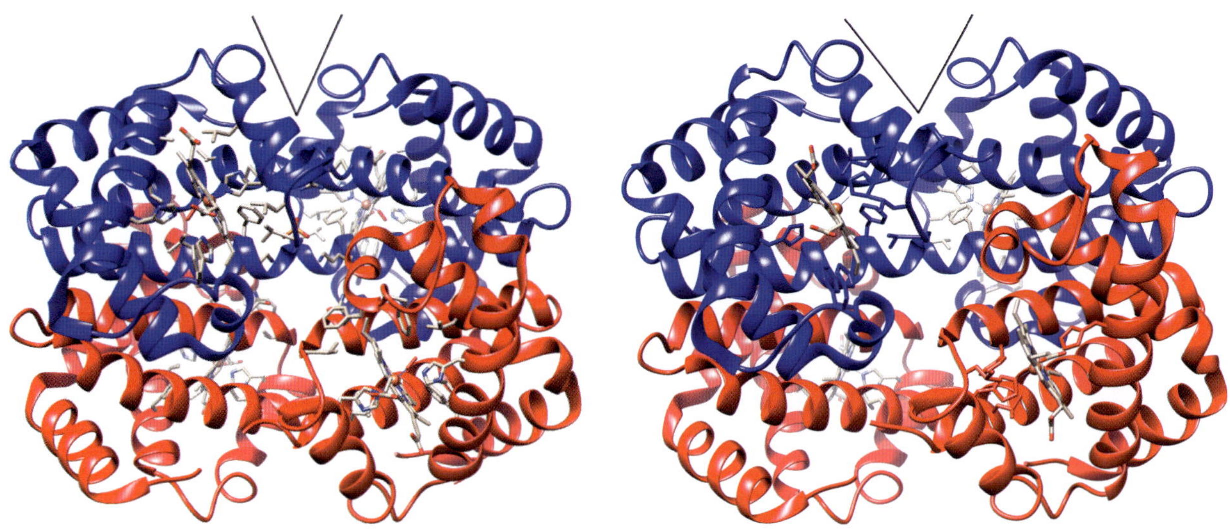

Fig. 20.46 The structural changes upon oxygen binding to hemoglobin. (*Left*) Oxyhemoglobin, (*right*) deoxyhemoglobin. Lines are provided to guide the eye to see the structural changes upon binding oxygen. Figure generated with UCSF Chimera[11] with coordinates from the protein databank (PDB ID 1HHO and 2HHB).[9,64–65]

amino-acid backbone. These protons are used to help maintain the conformation of the R state by forming hydrogen-bond networks. In lower pH regions, the T state, or oxygen free state, is preferred.

In regions of relatively low pressure of oxygen and high pressure of carbon dioxide, such as in muscles that are respiring at a rapid rate, the relative binding ability of hemoglobin drops below that of myoglobin, and as the oxygen molecules leave the protein it changes conformation back to the T state. The binding of oxygen is also correlated with the pH of the environment; in lower pH regions (a high concentration of carbonic acid), the oxygen is released more readily.

Approximately 25% of exhaled carbon dioxide is transported by hemoglobin that forms a carbamate with the terminal amine groups of each of the α and β monomers, while 70% is transported as bicarbonate and carried in blood plasma. The binding of the carbon dioxide reduces hemoglobin's affinity for oxygen. The work of hemoglobin is assisted by the zinc enzyme carbonic anhydrase, which catalyzes the conversion of CO_2 to bicarbonate in tissues, and then the conversion of bicarbonate to CO_2 in the lungs.

The net effect of all of these cooperative events are as follows. In an oxygen-rich, carbon dioxide-poor, and neutral-pH region, such as the lungs, hemoglobin releases protons and carbon dioxide and binds four dioxygen molecules. The oxygen is bound tightly in the red blood cells, which then travel to locations that are oxygen-poor, carbon dioxide-rich, and lower-pH, especially muscle tissue. In this region, the hemoglobin releases the bound oxygen to myoglobin, which has a higher binding affinity at these pressures, and binds protons and carbon dioxide. This process results in the delivery of oxygen to needed tissues, and the export of waste products in the lungs.

20.8.5 Carbon Group Transfer

Cobalamin is a cobalt complex of a corrin ring, which under physiological conditions can exist as the Co(I), Co(II), and Co(III) forms. The four nitrogen donors coordinate in the *xy*-plane and the axial ligands are an appended nucleotide, dimethylbenzimidazole (DMB), and an R group (Fig. 20.47). The X-ray crystal structure determination of Vitamin B_{12} was carried out by Dorothy Crowfoot Hodgkin in 1956.[66–67] The biological forms of the cofactor are methylcobalamin (MeCbl), where the R group is a methyl group, and adenosylcobalamin (AdoCbl), where the R group is an adenosyl group. Vitamin B_{12} is the form with which most people are familiar, and its R group is a cyano group due to the method of its synthesis.

The cobalamins are involved in a variety of carbon–carbon skeletal rearrangements and radical reactions, the metabolism of both fatty acids and amino acids, and is used in the synthesis of DNA and myelin.

The resting state of the cofactor contains a Co(III) ion with a low-spin d^6 electronic configuration. The Co(II) form has one unpaired electron in its d_{z^2} orbital. The EPR spectrum of this oxidation state exhibits g factors of 2.0 ($g_{\|}$) and 2.3 ($g_{\perp}$). The $g_{\|}$ component exhibits hyperfine coupling to both Co (spin = 7/2) of 110 G and the nitrogen of the DMB ligand (spin = 1) of 18 G. The Co(I) form has two paired electrons in its d_{z^2} orbital and is a strong nucleophile.

The cobalt–carbon bond is weak; the bond dissociation enthalpies for Co–C bond homolysis are 31 kcal/mol for AdoCbl and 37 kcal/mol for MeCbl. When AdoCbl releases its adenosyl radical it is trapped in the solvent cage and reforms with a rate constant of $1\ 10^9\ s^{-1}$. However, the resulting methyl radical escapes from the solvent cage and does not reform the Co–C bond. Other proteins recruit coenzyme B_{12} to carry out chemical reactions. In the case of methylmalonyl-CoA mutase, the coenzyme carries the isomerization of methylmalonyl-CoA to succinyl-CoA (Fig. 20.48). This reaction is part of the degradation of odd-number fatty acids and several amino acids in the kidneys.

The enzymes that use coenzyme B_{12} all carry out rearrangements of substrates that are facilitated by the formation of

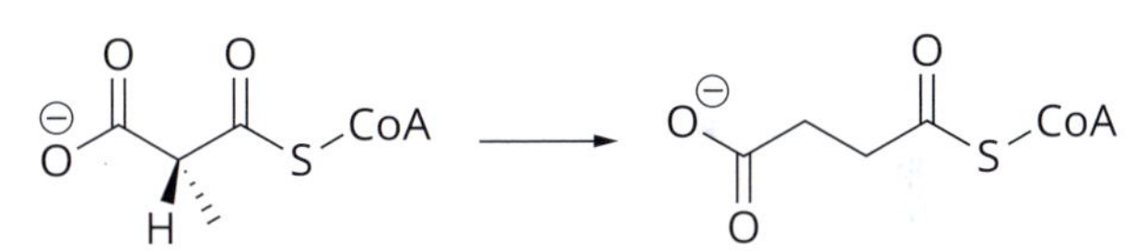

Fig. 20.48 The skeletal rearrangement of methylmalonyl-CoA to succinyl-CoA, carried out by the methylmalonyl-CoA mutase enzyme with the assistance of coenzyme B_{12}.

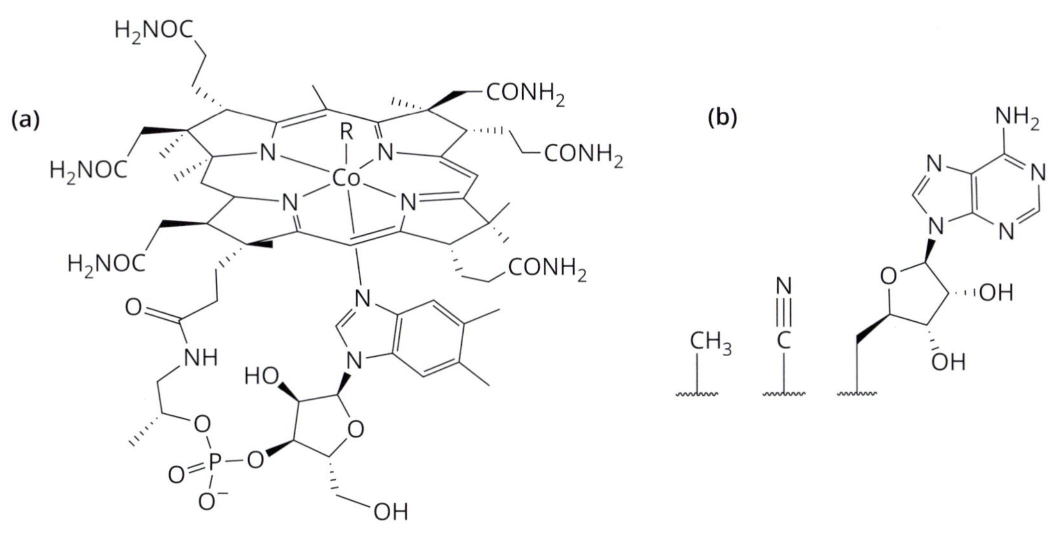

Fig. 20.47 (a) The structure of cobalamin. (b) R = Me, CN, or adenosyl (right).

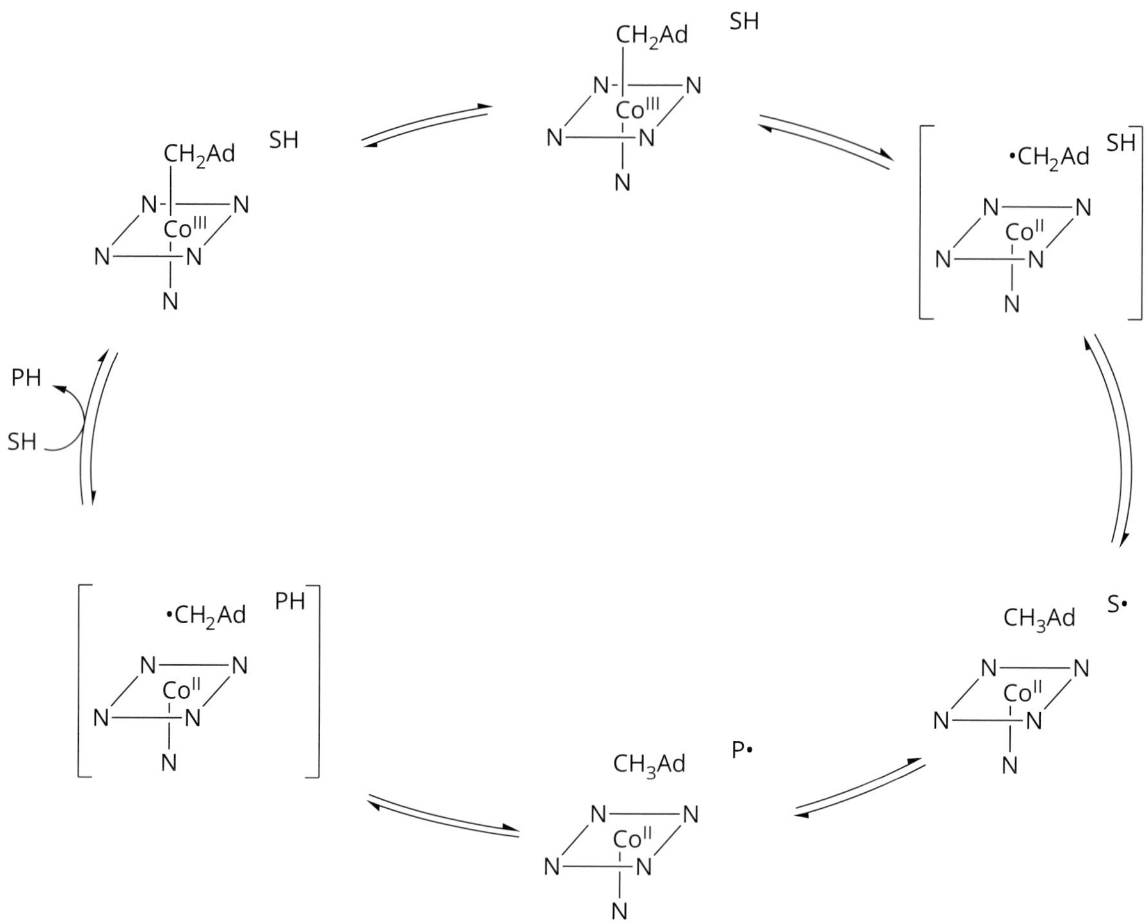

Fig. 20.49 The generic radical reactions catalyzed by AdoCbl. The corrin ring is represented by the square of four nitrogen atoms, and the axial DMB is the fifth nitrogen. The short-lived adenosyl radical can react with substrate (SH) to generate the substrate radical (S•). A radical rearrangement to product radical (P•) then occurs. Finally, the product radical abstracts a hydrogen from the adenosine to reform the adenosinyl radical which recombines with the cobalt.

the adenosinyl radical (Fig. 20.49). When the substrate is near the cofactor, the adenosinyl radical can abstract a hydrogen atom from the substrate generating a new radical. This radical rearranges to a product radical which then abstracts the hydrogen atom from the adenosine, regenerating the adenosinyl radical which can recombine with the cobalt, ending the catalytic cycle. The cobalt center remains a spectator during the reaction, and no metal–carbon bonds are formed except that with the regenerated adenosinyl radical. The stereochemistry of cobalamin-catalyzed reactions can exhibit inversion, retention, or even racemization depending on the system. Stereochemical control of the reaction is usually not necessary as the resulting products are nonchiral.

Rapid experimental methods, such as stopped flow or freeze quench, coupled with EPR spectroscopy have been used to identify the organic radical intermediates required for the proposed mechanism. The specifics of the mechanisms are not well known and may proceed by either cyclopropyl ring formation or radical fragmentation. Experimental data are often used to inform quantum mechanics/molecular mechanics calculations to suggest the most plausible mechanism. In the enzyme methylmalonyl-CoA mutase, the reaction is believed to proceed by a cyclopropane intermediate, while in glutamate mutase, a similar rearrangement is believed to proceed by a radical fragmentation and recombination (Fig. 20.50).[68]

The migration of the groups is not generally known to be either stepwise or concerted. In the rearrangement of propane diol to propanal by the diol dehydrase enzyme, the proposed transition state is a concerted process with hydrogen-bonding of the transferring group to a histidine (H143), a potassium, and an aspartic acid (D335) (Fig. 20.51). The substrate is held in place by additional hydrogen bonds to glutamic acid (E170) and glutamine (Q296).[69]

The use of MeCbl has been studied in the bacterial methionine synthase enzyme. This enzyme catalyzes the methylation of a zinc activated thiolate in homocysteine that is then converted to methionine. The source of the methyl group on MeCbl is the methyl transfer reagent *N*-methyl tetrahydrofolate. The nucleophilic Co(I) abstracts the methyl from this reagent by an S_N2 mechanism. The use of a chiral methyl group ($C^1H^2H^3H$) on MeCbl showed retention of configuration,

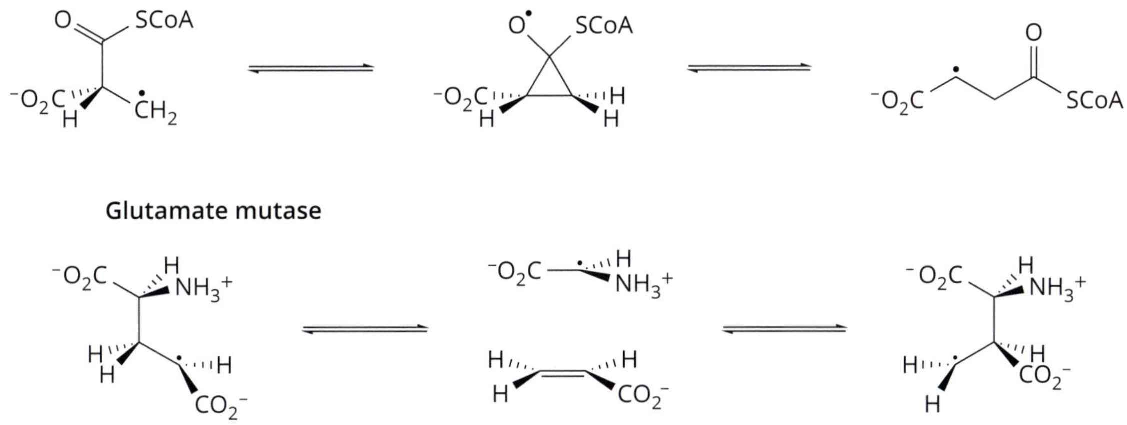

Fig. 20.50 Proposed rearrangement pathways that involve cyclopropyl groups or radical fragmentation.

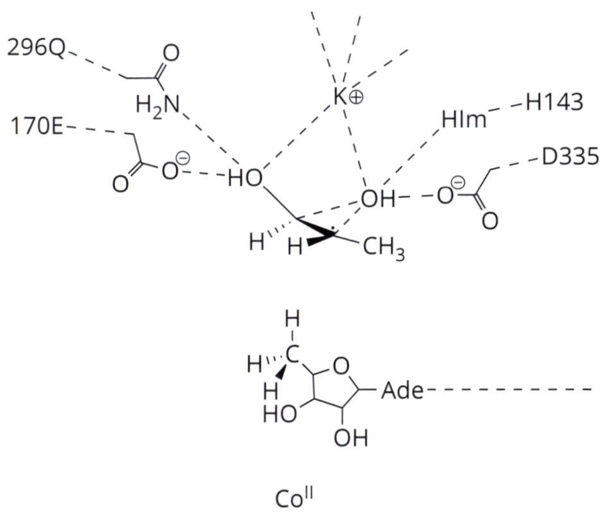

Fig. 20.51 The proposed transition state for hydroxy group migration in diol dehydrase.

which suggests that both methyl transfers occur by inversion by an S_N2 mechanism. Other mechanisms for the transfer of the methyl from MeCbl to the sulfur atom are also possible, so the exact details of this reaction are not known with certainty.

20.9 Biological Small-Molecule Activation

The remainder of the chapter will explore three important biological enzymes chosen for their importance, use of multiple spectroscopic methods for their functional elucidation, and of interest to a broad audience. Photosystem II includes the oxygen-evolving complex in which water is oxidized to oxygen. Methane monooxygenase is an enzyme that allows bacteria to use methane as their sole carbon source and can convert methane to methanol. Nitrogenase biologically converts inert dinitrogen into useful ammonia.

20.9.1 Oxygen Evolving Complex

Oxygen is essential for all aerobic life, as it is the final electron-acceptor in cellular respiration—the process used during the metabolism of foods that generates ATP. Oxygen is synthesized by plants, algae, and cyanobacteria in a process that began taking place approximately 3 billion years ago. A single dioxygen molecule is produced by the oxidation of two water molecules. A requirement for the process was the evolution of a metal complex that could store oxidizing equivalents until the water oxidation took place. The electrons that are removed from the water molecules are then made available in the organism for reductive reactions that turn carbon dioxide into sugars. The reduced compounds have a reduction potential equivalent to the reduction potential of dihydrogen.

The metal complex that releases oxygen is called the oxygen-evolving complex (OEC) and is located in photosystem II (PS II), a membrane bound protein complex where sunlight, harvested by pigment molecules, is used to photoexcite electrons that are then used as reducing equivalents in the organism. Oxidation of water also releases protons, setting up a proton gradient that can be used to synthesize ATP in the protein ATP synthase. Dioxygen is but a byproduct of the photosynthesis process. The structure of the OEC has been under study for decades. In 1996 a review article[70] described the state of the art structure based on spectroscopic studies, including EPR, NMR, XANES, and EAS. The first X-ray crystallographic study of PS II appeared in 2001, though that structure had only 3.8 Å resolution. Since that time, additional studies, principally X-ray spectroscopy and crystallographic characterization (up to a resolution of 1.9 Å),[71] have significantly increased our understanding of the structure and mechanism of PS II and the OEC.[72]

The X-ray structure of PS II reveals a pseudo-two-fold symmetrical structure consisting of thirty-five integral and three peripheral protein subunits. The protein complex has a significant number of α-helices (more than seventy) that span the thylakoid membrane in chloroplasts. The PS II protein complex resides within that membrane, allowing the proton gradient to be created. The complex contains light-harvesting molecules, including more than seventy chlorophyll molecules (Chl), pheophytin molecules (Pheo), which are simply chlorophyll molecules lacking the central Mg^{2+} ion, twenty-four trans β-carotene molecules, and several hemes, as well as electron-carrying groups, including both bound and free quinones, and, of course, the OEC.

The details of photosynthesis are complex, but the basic chemistry important for understanding the OEC is that sunlight is absorbed by light-harvesting molecules in the chloroplast, causing an electron to be excited in a dimer of chlorophyll molecules P_{D1}, called the special pair (Fig. 20.52). In order to use this electron for chemical synthesis, it must be separated from the corresponding hole that is generated rapidly, before it can relax to the ground state. A series of highly tuned electron-transfer reactions rapidly separate the electron–hole pair in space in less than 10 picoseconds. The electron travels 10.4 Å to Chl_{D1}, 10.5 Å to $Pheo_{D1}$ and 13.1 Å to plastoquinone Q_A, a tightly bound quinone molecule. Q_A transfers its electron to Q_B, a more loosely bound plastoquinone which can leave the binding pocket after being reduced by two electrons and gaining two protons to form plastoquinol, Q_BH_2. This molecule eventually transfers its electrons to the copper protein plastocyanin that is involved in further photosynthetic reactions. Interestingly, although the protein has pseudo-two-fold symmetry, electron transfer through PS II occurs only through one of the branches. The other, inactive, branch has slightly longer center-to-center bond distances severely reducing the rate of electron tunneling. It is hypothesized, however, that the tyrosine Y_D is involved in stabilization of the complex during electron transfer.

The hole that remains at P_{D1} is reduced by the neutral tyrosyl radical (Y_Z) that also serves as the oxidant for water oxidation in the nearby OEC. In order to provide these reducing equivalents to Y_Z, the OEC cycles through five oxidation states that were first proposed by Kok in 1970 (Fig. 20.53).[73] Electrons are removed from the OEC and are sequentially delivered to the tyrosine residue until four oxidizing equivalents have been stored in the manganese centers. At this time, water is spontaneously oxidized to dioxygen, regenerating the reduced from of the OEC (state S_0). The S_0 state is the most reduced form of the OEC, but the dark adapted resting state is oxidized by one electron by the nearby tyrosine to state S_1. The OEC can be cycled through all five oxidation states by using

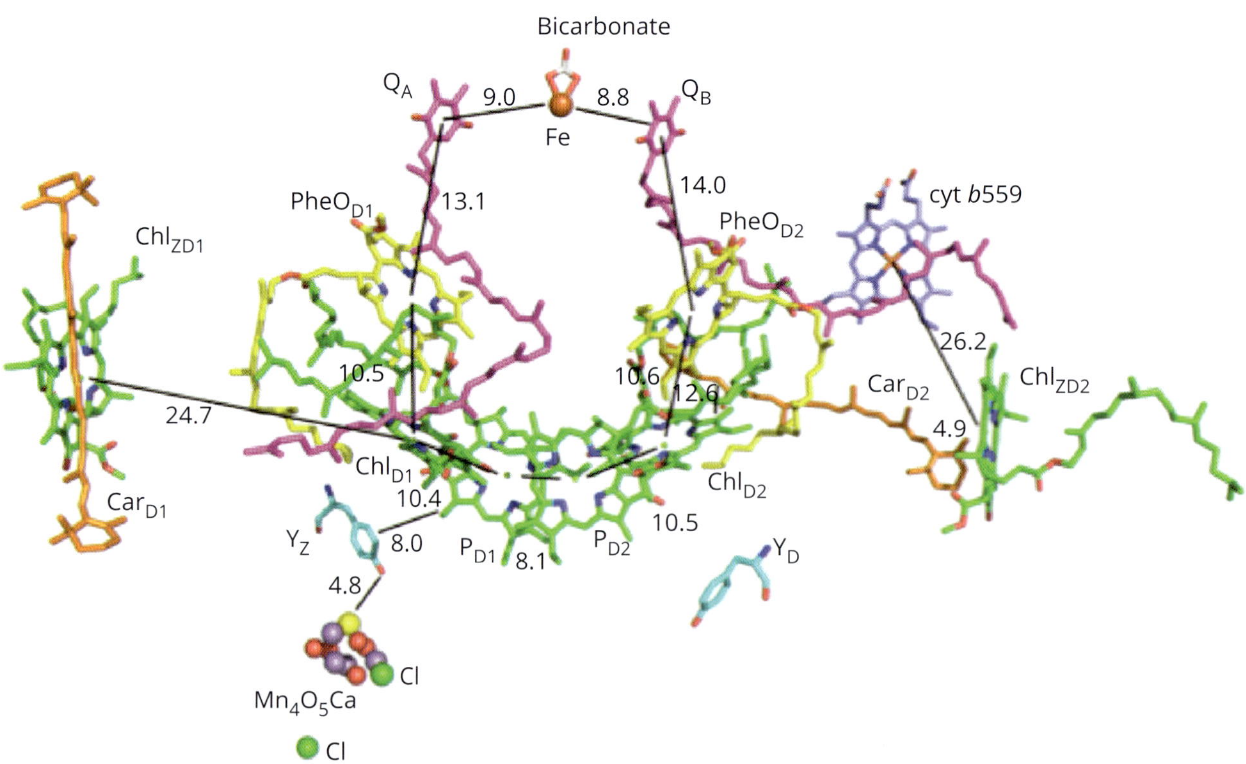

Fig. 20.52 Cofactors in both branches of photosystem II; distances between groups are shown in Å. Only the left branch consisting of Chl_{D1}, $Pheo_{D1}$ and Q_A is active. The special pair consists of P_{D1} and P_{D2}. The OEC is the Mn_4O_5Ca cluster located at the lower terminus of the left branch near Y_Z, a tyrosine. Figure adapted from Yano.[72]

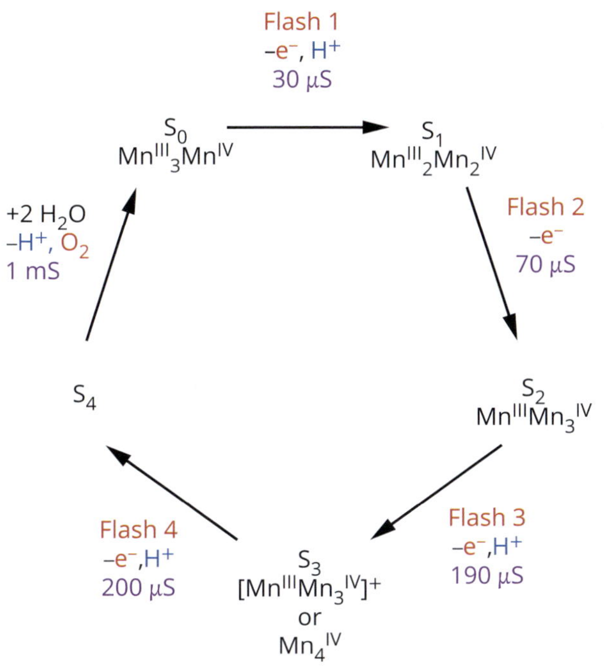

Fig. 20.53 The S-state Kok scheme for the OEC and the proposed Mn oxidation states during the synthesis of oxygen from water. Starting with the dark state, S_1, four flashes oxidize the OEC by four electrons causing a release of O_2 after the third flash, and then after each fourth subsequent flash.

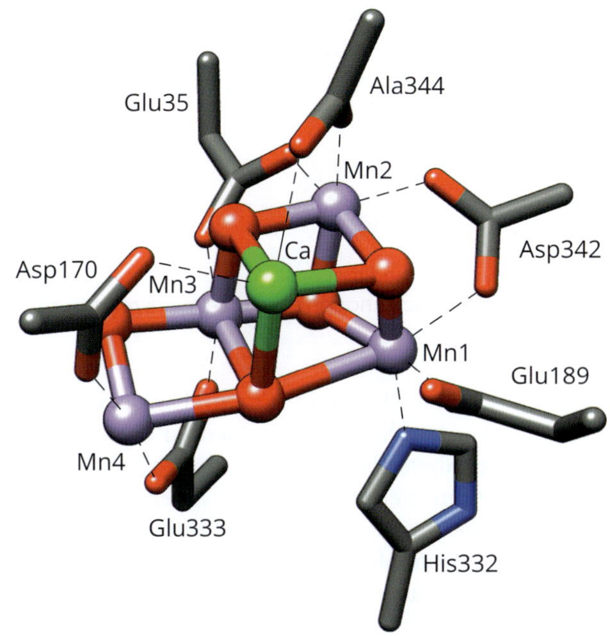

Fig. 20.54 The structure of the Mn_4CaO_5 cluster, showing the amino-acid residues that serve as ligands for the Mn and Ca atoms. Figure generated with UCSF Chimera[11] with coordinates from the protein databank (PDB ID 3WU2).[9,71]

light flashes, and dioxygen is released after every fourth flash. The OEC cluster provides a number of stable redox states, while the surrounding protein environment tunes the redox potentials and provides pathways for substrate (H_2O) entry, product (O_2) release, electrons, and protons.

The structure of the OEC was initially proposed based on EXAFS and other X-ray spectroscopic data.[74] One significant issue that hindered the structural determination of PS II and especially the OEC was that the high-intensity X-rays used for either structural determination or spectroscopy damaged the Mn_4Ca cluster. This is a well-known problem when using X-ray techniques to study metalloproteins. However, the structure based on EXAFS is extremely similar to that eventually obtained by the high-resolution X-ray diffraction study. The OEC consists of a Mn_3CaO_4 cubane core with a single additional Mn atom linked by two bridging oxygen atoms (Fig. 20.54). The complex contains four bound water molecules—two bound to Mn4 and two to the Ca atom. The cluster is held in place by direct contact with seven amino-acid residues that bind to the Mn and Ca atoms, including aspartate, glutamate, histidine, and the carboxylate terminus of an alanine residue in one of the proteins. There are a further three amino acids that are engaged in hydrogen-bonding interactions to the bridging oxygen atoms and bound water molecules (histidine, glutamate, and aspartate, not shown).

Significant structural changes (changes in the Mn–Mn bond lengths in the cluster by approximately 0.2 Å) have been observed in the OEC during the oxidation cycle by EXAFS. These changes make sense due to the different Mn–Mn bond distances expected for Mn(III) and Mn(IV), though additional structural reorganization of the cluster is required to completely explain the observed data. Flexible coordination geometry, in addition to multiple accessible oxidation states, is essential to maintain the function of the OEC through all the electron transfer reactions that occur.

The role of calcium in the cluster is not fully understood. Currently, it is believed to control the access of substrate water molecules to the Mn_4 cluster. Removal of the Ca(II) ion by chemical or biochemical processes leads to an inhibition of oxygen evolution, but the activity can be restored by addition of calcium ions. EXAFS of calcium depleted samples does not show a substantial change in the magnetic spectroscopic properties of the remaining Mn_4 cluster, suggesting that the structure of the remaining cluster does not change very much.

It is tempting to postulate that the two water molecules bound to Mn4 are those that eventually form the dioxygen molecule, but there is very limited experimental information about the mechanism of the O–O bond formation. Several mechanistic proposals have been postulated, including nucleophilic attack by a calcium-bound hydroxide on a terminal Mn=O group (Fig. 20.55a), direct O–O bond formation of two of the bridging oxygen atoms in the cubane (Fig. 20.55b), and coupling of the two terminal water/hydroxy ligands (Fig. 20.55 (c) and (d)). The mechanism with the most experimental evidence thus far is the coupling of a terminal manganese oxyl radical (MnOH·) with a bridging oxo ligand (Fig. 20.55e). Much work still needs to be done to determine the mechanism of the O–O bond formation reaction.

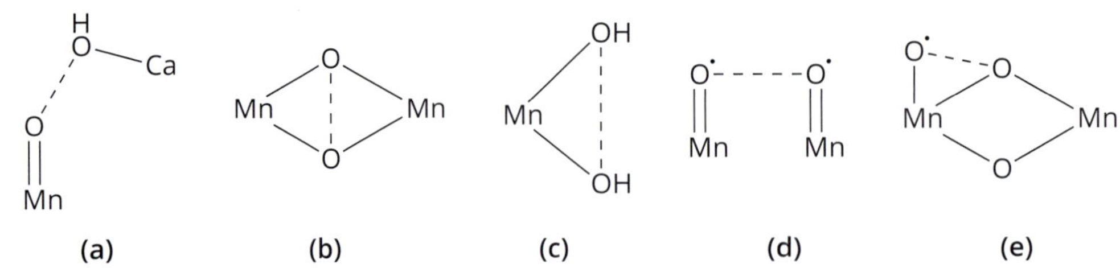

Fig. 20.55 Hypothesized mechanistic pathways for the conversion of two water molecules into one dioxygen molecule at the OEC during photosynthesis. Although significant work remains to be done, mechanism e currently has the most experimental support.

The study of the OEC and the natural synthesis of dioxygen during photosynthesis have influenced current research on synthetic catalysts for water oxidation. The transition to a renewable resource economy using solar energy to produce fuels requires both reducing and oxidizing equivalents. The artificial leaf—a device for splitting water into dihydrogen and dioxygen—is one such possible technology. This technique was first described in a patent in 1983, but more recent work that has stimulated much additional research is the report in 2011 by Nocera that uses nonprecious metal catalysts.[75] A thin semiconductor of silicon is coated with a Ni–Mo–Zn alloy on one side and a Co-based catalyst on the other. The cobalt side is the site of oxygen evolution, while the Ni–Mo–Zn side releases hydrogen. Simply dipping the device into water in the presence of sunlight causes the release of both dioxygen and dihydrogen. While technical hurdles remain, insights gained in the study of the OEC will almost certainly lead to the continued improvement of catalysts for splitting water, with potential applications in solar energy conversion.

20.9.2 Methane Monooxygenase

Methane is a relatively inert molecule formed naturally in anaerobic decomposition of organic material, and is the principal component of natural gas. It can be converted industrially to methanol, a liquid fuel, by first being decomposed to synthesis gas (CO and H_2) by a process known as steam reforming (Eq. 20.13), and subsequent reaction of the CO and H_2 can provide methanol (Eq. 20.14). There is no industrial process that can selectively oxidize methane to methanol on a scale sufficient to be practicable. Overoxidation of methane to formaldehyde, formic acid, and finally CO_2 is thermodynamically favored, and the strong C–H bonds in methane. The first bond of methane is the strongest at 104 kcal/mol (average for CH_4 is 99.3 kcal/mol), which makes stopping after a single activation difficult. Additionally, direct reaction of the triplet $^3\Sigma$ state of O_2 with the singlet CH_4 is spin-forbidden.

$$CH_4 + H_2O \rightarrow CO + 3\,H_2 \qquad \textbf{20.13}$$

$$CO + 2\,H_2 \rightarrow CH_3OH \qquad \textbf{20.14}$$

While organometallic chemists have struggled to activate the C–H bond of methane, there are approximately 130 species of bacteria whose sole carbon source is methane. Methanotrophic bacteria have been known to exist for over 100 years, and the first step of the metabolic pathway that converts methane to other substrates is shown in Eq. 20.15.

$$CH_4 + NADH + H^+ + O_2 \rightarrow CH_3OH + NAD^+ + H_2O \qquad \textbf{20.15}$$

These methanotrophic bacteria have fascinated structural and molecular biologists and synthetic chemists for decades. The complex protein aggregates that carry out the transformation of methane to methanol include both a particulate form of methane monooxygenase (pMMO), which contains copper (Fig. 20.20), and a soluble form (sMMO), which contains iron. Most methanotrophs can express pMMO, and those that also express sMMO usually only do so under conditions of low copper availability. This section will focus on only sMMO, which has been well-studied due to the relative ease of crystallography of the soluble protein relative to the membrane bound protein. The pMMO is significantly more active for selective oxidation of methane. Interested readers can find a good discussion of sMMO, pMMO and other hydrocarbon oxidases in a series of papers and review articles spanning the last twenty-five years.[76–83]

20.9.3 Early studies on the structure of MMO

The most studied form of sMMO consists of three proteins. The first is a reductase, MMOR, a <38.6-kDA> protein which transfers electrons to the hydroxylase using a Fe_2S_2 ferredoxin and flavin adenine dinucleotide, FAD, an organic electron carrier. The second is a coupling protein, MMOB, a 15.5-kDa protein which serves regulatory functions. Finally, MMOH is the hydroxylase, a 251-kDa protein that carries out the chemical transformation. The genes for the three proteins are contiguous and form a single MMO operon. Gel filtration or electrophoresis of the hydroxylase protein showed that the hydroxylase consisted of two copies each of three different proteins. The protein thus has an $\alpha_2\beta_2\gamma_2$ structure, and the three subunits are 54, 43, and 23 kDa in size. Molecular genetics was used to overexpress the three subunits so that the protein could be isolated in reasonable quantities for study.

The reagent EDC, (1-ethyl-3-(3-dimethylaminopropyl) carbodiimide, can couple carboxylate and primary amines

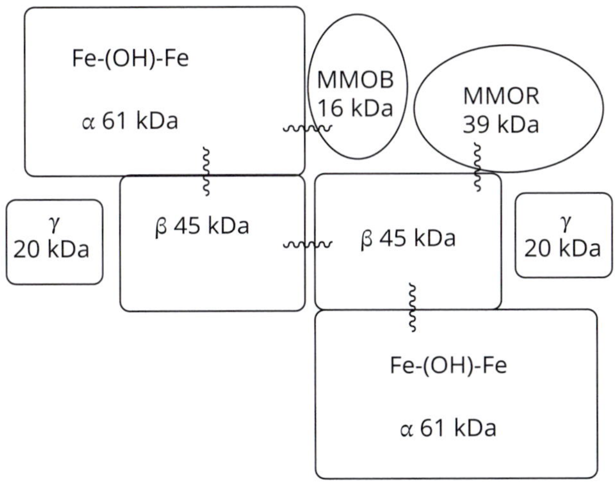

Fig. 20.56 A diagram of the sMMO protein complex resulting from chemical cross-linking studies, with cross-links indicated by wavy lines.

into an amide bond. Chemical cross-linking using EDC was carried out, and it showed chemical connectivity between the α-β and β-β subunits, with additional links from MMOH to both MMOB and MMOR. The results suggest that the whole protein complex is linked together in the native protein by inter subunit salt bridges, a relatively weak ionic bond between positively charged amino acids (arginine or lysine) and negatively charged amino acids (aspartate or glutamate) in a protein. Based on the results of these studies, a diagram of the hydroxylase was proposed (Fig. 20.56).

EPR studies of MMOH suggested a dinuclear iron center. Reduced samples showed signals consistent with Fe(II)Fe(III) ($g < 2$), while further reduction gave a broad resonance at $g \approx 16$, characteristic of an even spin Fe(II)Fe(II) state. A variety of species has been cloned, and there was high DNA-sequence homology, especially in the α subunit. Based on sequence alignment, it was thought that the diiron center in the hydroxylase protein was located in a four-helix bundle in subunit α. This structural motif is relatively common for a number of diiron centers found elsewhere in biological systems. Circular dichroism indicated a very high content of α-helices, so there was the possibility for these to form the four-helix bundle required to bind the diiron center.

The ligands that linked the two iron centers to each other were then discovered. The MMO proteins are essentially colorless, with no optical bands beyond 300 nm. Oxo-bridged Fe(III) proteins typically exhibit absorptions in the 300–800-nm range (LMCT bands), but these were not observed. Other proteins exhibit resonance Raman bands due to an Fe–O–Fe unit, but the MMOH protein did not exhibit this behavior. The EPR spectra showed strong antiferromagnetic coupling between the two iron centers, consistent with presence of a monatomic hydroxy, carboxylato, or alkoxo bridge. ENDOR spectroscopy produce a spectrum that was very similar to a known complex with a bridging water or hydroxide; coupling to a proton was observed. ESEEM and ENDOR spectroscopy showed that nitrogen atoms were ligated to each of the two iron atoms, suggesting histidine coordination. When the MMOH protein was treated with diethyl pyrocarbonate (a reagent that reacts with free histidine side chains), a total of twelve histidines were modified, but in the protein without the diiron cofactor, fourteen histidines were modified, further evidence suggesting histidine coordination of the iron centers.

EXAFS studies were carried out on the protein which produced information on both the ligands holding the iron centers into the four-helix bundle, and also the ligands that bridged the two iron atoms. The first shell coordination environment was modeled as having five or six oxygen or nitrogen atoms at 2.04 Å for the oxidized form (Fe(III)Fe(III)), 2.06–2.08 Å for the mixed valence species (Fe(III)Fe(II), and 2.15 Å for the fully reduced form (Fe(II)F(II)). Other proteins with diiron centers that are ligated with histidine ligands have longer bond distances, thus it was postulated that the Fe's are ligated mostly by oxygen atoms. An Fe–Fe distance of 3.42 Å was observed for both Fe(III)Fe(III) and Fe(II)Fe(III) hydroxylases. No short Fe–O (1.8 Å) distances were observed, ruling out a bridging oxo ligand, but the distance was consistent with the presence of a monatomic or additional bridging ligands such as a hydroxide, carboxylate, or alkoxide. The EXAFS data for the model compound $[Fe_2(OH)(OAc)_2(HB(pz)_3)_2]^+$, where $HB(pz)_3$ is trispyrazolyl borate, a histidine mimic, were found to be very similar to those for the biological cofactor suggesting that both a hydroxide and carboxylate from glutamate or aspartate were bridging the two iron atoms. The best structure for the diiron center that was consistent with all the spectroscopic data is shown in Fig. 20.57 (a). The diiron center is ligated to the protein by at least one histidine per iron atom, with additional oxygen-atom donors, a bridging hydroxide, a bridging carboxylate, and a terminal water or hydroxide on one iron center. Aligning the amino-acid sequence with the structurally characterized diiron protein ribonucleotide reductase suggested that the diiron center in MMO would also sit in a four-helix bundle, and there were nearby histidine and glutamic acid residues available for binding to the iron atoms (Fig. 20.57b).

20.9.4 X-ray Crystal Structure of sMMO

The crystallization and solution of the MMO structure by X-ray diffraction methods is a good case study for the difficulty of protein crystallography, and interested readers are referred to Lippard's account of the first crystal structure determinations that over a period of three years eventually resulted in the complete structural determination of the protein and metal binding sites.[76] Combination of all the available data led to a crystal structure that accounted for 97% of the amino acids in the α-subunit, 99% of the β-subunit, and 95% of the γ-subunit. The X-ray data indicated that the gene sequence for the MMO DNA was incorrect, and later resequencing of the gene was consistent this conclusion.

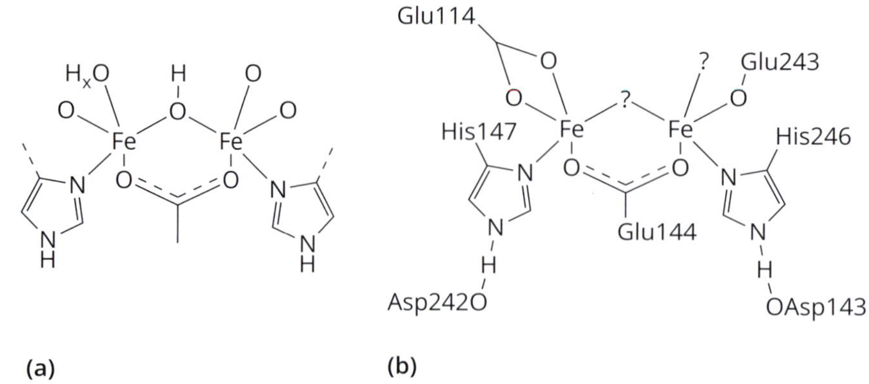

Fig. 20.57 (a) The model for the diiron center in MMOH based on EPR, ESEEM, ENDOR, Mössbauer, and EXAFS spectroscopy. (b) The model for the diiron center based on sequence homology with the structurally characterized diiron protein ribonucleotide reductase.

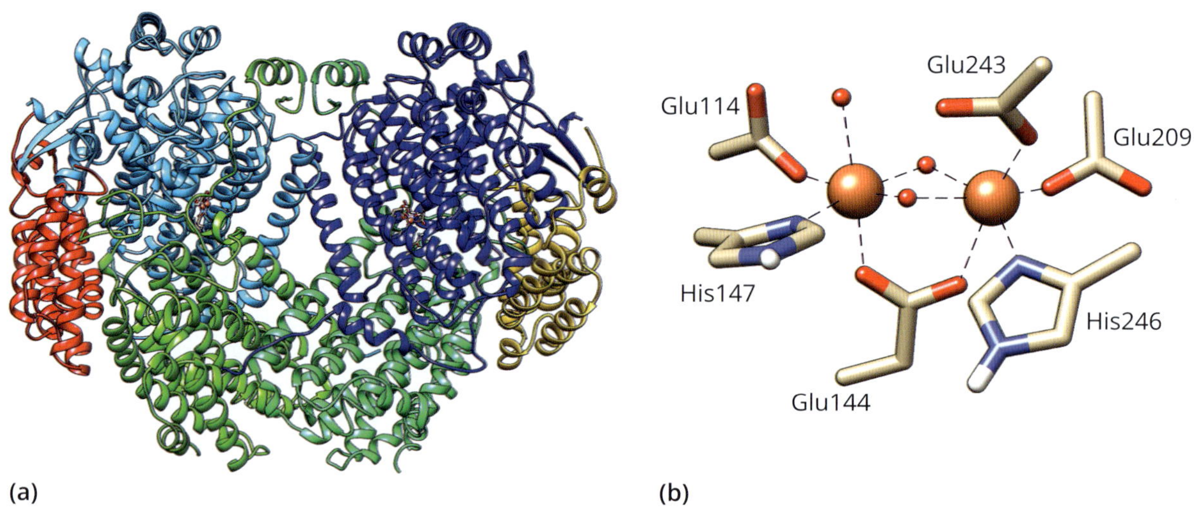

Fig. 20.58 The crystal structure of the MMOH protein as determined by X-ray crystallography. (a) the tertiary structure of the αβγ dimer. (b) The diiron center showing two bridging and one terminal water/hydroxides. Figure generated with UCSF Chimera[11] with coordinates from the protein databank (PDB ID 1MTY).[9,84]

The termini of the three subunits include protein chains that have significant flexibility, and no electron density for those residues was observed.

The final crystal structure was, as expected, given the spectroscopic model, an $\alpha_2\beta_2\gamma_2$ dimer (Fig. 20.58). The two αβγ monomers form a large "heart" shape, and they are related by a two-fold axis down the middle. The structure also shows possible sites for binding of the coupling protein and the reductase. There is a wide "canyon" between the αβ pairs in the dimer that is near the location of the iron centers, allowing an interaction of the coupling protein at that location to affect the redox potentials of the Fe centers.

The MMOH protein consists almost entirely of α-helices, with only small sections of β-sheet and random coil linking them together. This is consistent with the circular dichroism spectroscopy that had indicated a significant number of α-helices. The X-ray structure is also consistent with the observed chemical cross-linking.

The diiron center is (as predicted) in a four-helix bundle that is remarkably similar to that in the ribonucleotide reductase protein. There are two diiron sites in the hydroxylase, one in each α-subunit, separated by around 45 Å. The four-helix bundle donates the coordinating ligands. The structure of the diiron center is as shown in Fig. 20.58b. There are three ligands bridging the two irons, a carboxylate from Glu 243, which has relatively long Fe–O bonds of 2.5 and 2.7 Å, a hydroxide, as predicted from ENDOR, and an acetate ion, presumably from the ammonium acetate buffer from which the protein crystals were grown. The measured Fe–Fe distance of 3.4 Å is in agreement with the EXAFS data. Nonbridging ligands include a histidine, as expected by ESEEM, ENDOR EXAFS, and sequence alignment. Fe(1) is coordinated to one monodentate carboxylate (Glu 114), while Fe(2) is coordinated to two monodentate carboxylates (Glu 209 and Glu 243). Fe(1) also has a coordinated water, making the active site neutral overall. The water is engaged in hydrogen bonding interactions to two neighboring

Glu residues (243 and 114). In all, the X-ray crystal structure of the diiron center is in remarkable agreement with that predicted from the spectroscopic data.

20.9.5 Mechanism of Methane Oxidation by sMMO (Advanced Topic)

The mechanism by which MMO proteins selectively activate the CH bond in methane without subsequent overoxidation by activation of the resulting CH bonds in methanol has been under study for more than twenty-five years. The general features of the mechanism were first established in the mid 1990s, and an almost complete mechanism was reported. Since then, additional studies have been carried out to definitively determine the structure and formation of the most important intermediate, compound Q, the species that oxidizes methane.

Initial mechanistic work was carried out using stopped-flow techniques to identify kinetic intermediates in a single turnover reaction. Subsequent studies used rapid freeze–quench techniques. The catalytic cycle has been probed by a number of spectroscopic techniques, including EXAFS, EPR, ENDOR/ESEEM, MCD, and Mössbauer. The starting point for the catalytic cycle is $MMOH_{ox}$, the oxidized diiron(III) core (Fig. 20.59). This core has two bridging hydroxide ligands. At this stage of the cycle, the MMOB regulatory protein is not bound to MMOH, allowing the reductase, MMOR, to bind and transfer electrons to the diiron center. Two equivalents of MMOR must bind to MMOH, each delivering a single electron, with a net reduction to the $MMOH_{red}$, the reduced diiron(II) core. Protons are also transferred to the bridging hydroxide ligands, and two water molecules are lost. Reduction of Fe(III)Fe(III) to Fe(II)Fe(II) causes small protein structural changes that destabilize the $MMOH_{red}$-MMOR interaction, causing the MMOR protein to dissociate. This allows the MMOB protein to bind and form the $MMOH_{red}$-MMOB

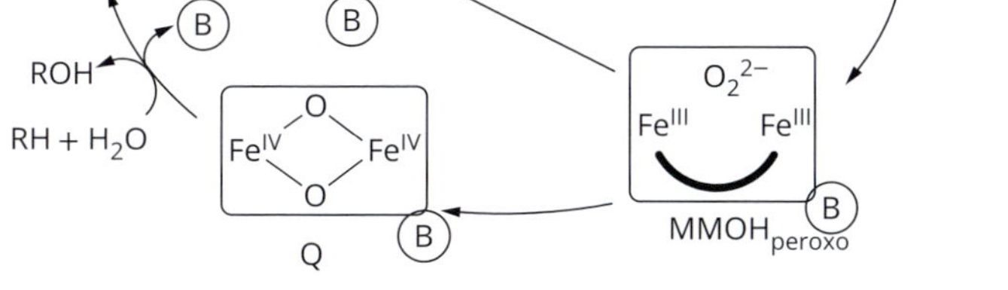

Fig. 20.59 The catalytic cycle for methane oxidation by sMMO. The regulatory protein MMOB is indicated by a B within a circle.

complex which then incorporates a molecule of dioxygen which comes in through a hydrophobic channel.

The next step is the reaction of $MMOH_{red}$ with dioxygen to form a transient O_2 adduct. Computational results suggest that the first intermediate upon addition of dioxygen is the mixed valence complex resulting from one electron reduction of dioxygen to the Fe(III)Fe(II) superoxide species. A kinetic isotope effect of 1.106 was observed for the reaction of $^{16}O_2$ versus $^{18}O_2$, which indicates significant decrease in O–O bond order before the first irreversible step. This argues against the equilibrium formation of an $MMOH_{red}$-O_2 complex, but instead a one-electron reduction to superoxide.

This superoxide complex is rapidly reduced to first the *cis*- and then the *trans*-μ-1,2-peroxo complexes P* and $MMOH_{peroxo}$, respectively. These intermediates are short-lived, but have been observed transiently by UV–visible (a band at 725 nm) and Mössbauer (two high spin Fe(III) centers) spectroscopy. The peroxo intermediate is diamagnetic at 4 K, suggesting strong antiferromagnetic coupling leading to a diamagnetic center. This spectroscopy is consistent with the peroxo complex, and their presence is required to fit the data to the kinetic model.

$MMOH_{peroxo}$ can oxidize electron rich substrates, but in the absence of those, it rearranges by cleavage of the O-O bond to form compound Q. This complex has a bright yellow color with UV–visible absorption bands at 350 and 420 nm. The structure of compound Q has been the subject of intense study and debate. Mössbauer spectroscopy suggested equivalent environments for both iron atoms with an antiferromagnetically coupled diamagnetic ground state. EXAFS showed a short Fe–Fe distance of 2.46 Å, and both short (1.77 Å) and long (2.0 Å) Fe-O interactions. The data was interpreted to suggest that the structure contained two single oxygen atom bridges and also a bridging carboxylate ligand, consistent with Fe(IV)(μ_2-O)$_2$Fe(IV), known as the diamond core. Several other structures have been proposed for compound Q, including the "open core" structures O=Fe(V)(μ-O)Fe(III) and O=Fe(IV)(μ-O)Fe(IV), and even a hydrogen bonded oxo–hydroxide species (Fig. 20.60). The current understanding of the structure of compound Q is under debate, with experimental evidence for both closed and open forms.[83]

Compound Q contains iron in a high oxidation state, consistent with its reactivity. The decay rate of compound Q increases in the presence of hydrocarbon substrates; this is the species that reacts with CH_4. Significant effort was undertaken to determine how methane reacts with compound Q. Reasonable mechanistic pathways included concerted, radical, and carbocation pathways. Many data were collected, and most of it is consistent with more than one mechanism. Isotopically substituted chiral alkanes such as CH_3CHDT were examined to probe the stereochemical outcome of C-H/D/T bond activation. A concerted mechanism would be expected to produce retention or inversion, while scrambling would occur with carbocation or radical. Reaction of CH_3CHDT produce 64–74% retention of configuration. The measurement of some racemization was taken to support a radical mechanism. However,

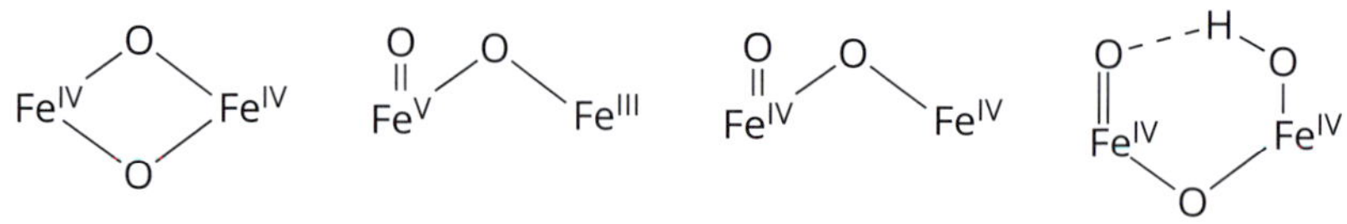

Fig. 20.60 Postulated structures for compound Q, including the bis-μ-oxo diiron(IV) diamond, and several with terminal iron oxo ligands known as the open-core structures.

70% retention would correspond to a rebound rate for the putative ethyl radical of 10^{13} s^{-1}, which is too fast to propose the formation of a discrete radical intermediate.

The deuterium kinetic isotope effect calculated for hydrogen atom abstraction was lower than experimentally observed values (ranging from 50 to 100 in some systems, and 4 to 20 for others) which are indicative of C–H bond-breaking at the transition state. However, the calculated values are consistent with the product distribution seen in experiments using CH_4 and CH_2D_2 when tunneling effects on the rate constants are taken into account.

The use of radical clock substrates (substrates that react differently as radicals or cations) were also used to help differentiate between the various mechanisms (Fig. 20.61). For example, 1,1-dimethylcyclopropane gives different products, depending on radical (top), carbocation (middle), or concerted (bottom). The observed product distribution is 81% concerted, 6% radical, and 13% carbocation. Related results from a different substrate place an upper limit of 150 fs on the lifetime for a radical species, again ruling out the formation of a discrete radical intermediate.

At this point of the catalytic cycle, the 2-electron oxidation of methane to methanol has taken place, resulting in a (μ-oxo)(μ-methanol) diiron(III) complex. The complex dissociates methanol and adds water. Proton transfer regenerates the bis-μ-hydroxy diiron(III) $MMOH_{ox}$ complex which undergoes a conformational change, releasing MMOB and allowing MMOR to bind, restarting the catalytic cycle.

While the functionalization of methane takes place at the diiron center, the larger protein environment is critical for the function of the enzyme. Amino-acid side-chains gain access to the active site and control the size of substrates that are allowed in, as well as both regio- and stereoselectivity. As in other redox centers, the local protein environment helps to control the oxidizing strength of the metal center. Even the MMOB protein can affect the regioselectivity of hydroxylation. Addition of only 10% of MMOB changed the product distribution relative to a functioning MMOH. This is all an example of the complexities of protein–protein interactions that inorganic chemists have yet to fully understand or replicate in the synthetic laboratory.

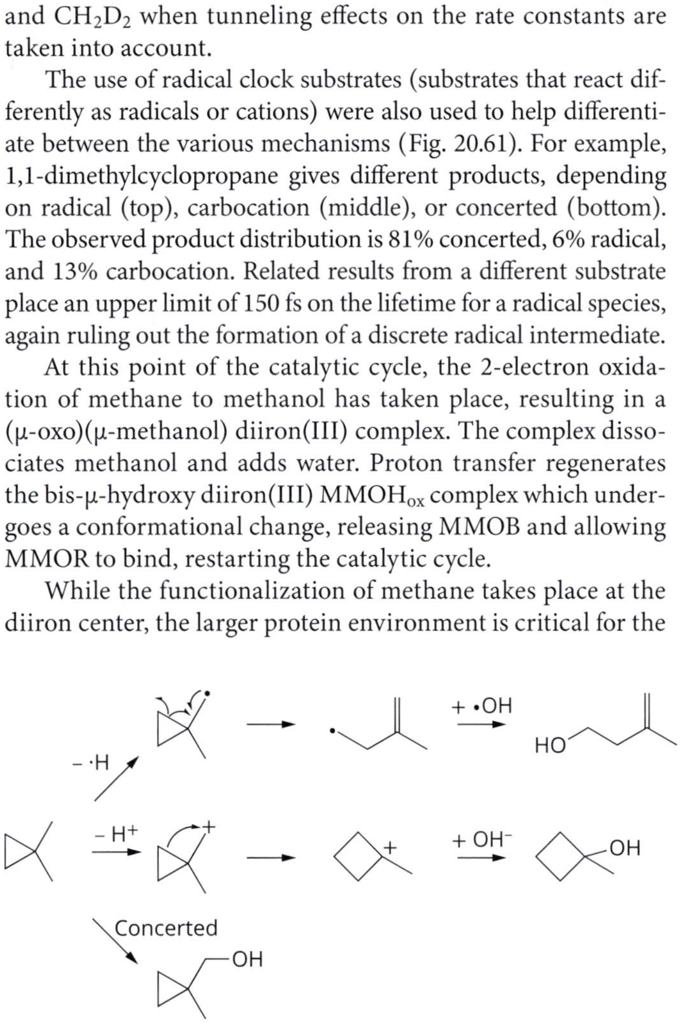

Fig. 20.61 Radical clock substrate 1,1-dimethylcyclopropane gives different products based on the mechanism of C–H bond activation. A radical reaction would be expected to produce 3-methyl-3-buten-1-ol, a carbocation reaction would produce 1-methylcyclobutanol, while a concerted process would produce (1-methylcyclopropyl)methanol.

20.9.6 Nitrogenase

It has been known since antiquity that rotation of crops leads to higher production. Alternation of planting legumes (peas and beans) with other plants results in higher crop yields. In the late 1800s it was discovered that legume plants could "fix" atmospheric dinitrogen to biologically relevant ammonia. Later, it was found that bacteria that lived in nodules in the plant roots were responsible for this nitrogen fixation. These bacteria were isolated early in the twentieth century, and by the 1960s cell-free extracts could be prepared that could carry out the chemical reaction. It is now known that only about a hundred species of bacteria and archaea have the genes necessary for the proteins that carry out this chemical reaction, and there are no examples of eukaryotic organisms capable of fixing dinitrogen. Since the 1960s, intense effort has been expended trying to understand the structure and mechanism of action of the protein complex nitrogenase that is responsible for the observed chemical reactivity.

The availability of fixed forms of nitrogen from the nitrogen cycle is crucially important for supporting life on Earth. The Haber–Bosch process for the industrial synthesis of ammonia from dinitrogen and dihydrogen was developed in response to the projected shortages of fertilizer for food at the turn of the last century. This process is energy intensive, using approximately 2% of global energy production to generate the high temperatures and pressures required (250 atm, 400–600° C) and for the synthesis of the dihydrogen that feeds into the reactors. Before this process found wide use, beginning in the 1950s, the majority of fixed nitrogen required for the biosynthesis of amino acids was a result of the nitrogenase reaction. Some geochemical processes, such as lighting, can also cleave the strong N–N bond, (226 kcal/mol) but even today it is estimated that more than half of fixed nitrogen derives from biological sources.

Nitrogenase fixes dinitrogen in its own energy-intensive process requiring eight electrons, eight protons, and sixteen MgATP molecules per dinitrogen substrate. Hydrolysis of the MgATP releases inorganic phosphate, P_i. Up to 40% of bacterial ATP production is devoted to the reduction of dinitrogen when the cell is actively undergoing the process. In addition, there is a required reduction of protons to dihydrogen—a step that seems uncharacteristically wasteful for a biological enzyme (Eq. 20.16). Nitrogen fixation is coupled to ferredoxin oxidation, and the source of electrons is a ferredoxin external to the nitrogenase protein complex.

$$\begin{aligned} &N_2 + 8\,H^+ + 8\,e^- + 16\,MgATP + 16\,H_2O \\ &\quad \rightarrow 2\,NH_3 + H_2 + 16\,MgADP + 16\,P_i \end{aligned} \qquad \textbf{20.16}$$

There are two major classes of nitrogenase: the Mo/V/Fe nitrogenases, which are all believed to have a common ancestor, and that from a thermophilic bacteria, which has a very different structure. The most information is known about the Mo nitrogenase, so it will be the focus of this discussion. More details can be found in the expansive review articles that cover this protein complex.[85–90] The function of nitrogenase is derived from two separate metal-containing proteins: the Fe protein and the FeMo protein. These proteins have no function separately. The function of the enzyme is lost in the presence of dioxygen. An external ferredoxin reduces the Fe protein, which then reduces the FeMo protein. There is a high level of amino-acid sequence conservation among a number of Mo nitrogenases, and the proteins encoded by different organisms are able to react with each other.

Beginning in the early 1990s, several single-crystal X-ray structures of the nitrogenase enzyme were obtained at ever-increasing resolution. The nitrogenase protein complex consists of both the iron protein and the MoFe protein (Fig. 20.62). The iron protein is a γ_2-dimer (the nomenclature in the nitrogenase field uses this notation for the iron protein for ease of comparison to historical data). The dimer has an MgATP binding site within each monomer and a bridging Fe_4S_4 cluster between the subunits. The iron cluster resides within a cleft of the iron molybdenum protein, allowing it to approach to within 14 Å of the P cluster within the MoFe protein. The MoFe protein is an $\alpha_2\beta_2$ tetramer with two different metal cofactors per $\alpha\beta$ subunit. The first is a Fe_8S_7 cluster, called the P-cluster, and the second is FeMoco, which was found to be an $MoFe_7S_9$ cluster.

The metal-containing cofactors responsible for the nitrogenase activity include an Fe_4S_4 cluster in the iron protein and two unique clusters in the MoFe protein. The first is the P-cluster, an Fe_8S_7-cluster, and the second is the iron molybdenum cofactor (FeMoco), which was found to be an $MoFe_7S_9$-cluster. Both of these cofactors were unprecedented in biology, which is perhaps not surprising as the reaction that the enzyme carries out is also unprecedented. The crystal structure launched an extensive exploration of iron–sulfur clusters as model complexes for nitrogenase activity.

In 2011 the crystal structure of the FeMo protein was redetermined at 1.16 Å resolution. This high-resolution structure indicated that FeMoco contained a central light atom, but initially it was not known whether it was carbon, oxygen, or nitrogen. The determination of the central atom required several studies beyond X-ray diffraction, including a statistical analysis of electron density from a subsequent 1.0 Å resolution structure, X-ray emission spectroscopy, and a pulsed NMR/EPR technique known as ESEEM spectroscopy. All three techniques demonstrated that the central atom was carbon.

With the determination of the central atom in FeMoco, the complete structure of the nitrogenase protein complex was known, but the mechanism of cleaving the N–N bond to form NH_3 was not known. The close proximity of the three metal-containing clusters in the nitrogenase protein complex suggested an electron-transfer pathway in the protein. However, ATP-driven electron transfer as seen in other types of enzymes was initially ruled out because of the long distance between the MgATP binding site and the Fe_4S_4 in the Fe protein. More studies were required to determine a mechanism for the enzyme. The mechanism would need to explain the observed rates and the function of the MgATP binding site, as well as the requirement that the enzyme generate one equivalent of dihydrogen during the dinitrogen reduction process.

During the 1970s and 1980s a detailed kinetic model was developed as a result of extensive experimental work on FeMoco. This model is known as the LT kinetic scheme, after the groups of Lowe and Thorneley (Fig. 20.63). A series of four proton and electron transfer steps reduce FeMoco from state E_0 to state E_4. The E_4 state is referred to as the Janus intermediate, after the ancient Roman mythical figure who looks forwards and backwards to the future and past. The Janus intermediate is the species that binds to dinitrogen, and it does so with an obligatory release of dihydrogen. The seemingly wasteful reduction of protons to dihydrogen is actually a mechanistic requirement. To generate the highly reduced cofactor state that is capable of binding to and reducing dinitrogen, a reductive elimination of two hydrides to one dihydrogen is required.

To reduce N_2 to two NH_3 molecules, a total of eight electrons and eight protons are required, as two additional protons and electrons are used to generate a single molecule of dihydrogen. Each electron-transfer step takes an electron from the Fe_4S_4-cofactor in one iron protein and delivers it to FeMoco, requiring that there be a total of eight separate docking and undocking steps that occur to fully reduce N_2. Each docking step results in the cleavage of both ATP molecules in the Fe protein, for a total of sixteen ATPs cleaved during the process.

The kinetic model accounts for the rates and chemical steps required for the cleavage of the N–N bond, but a detailed structural mechanistic model took significant additional experimental and computational work over an additional thirty years. While the exact details of the process are not known

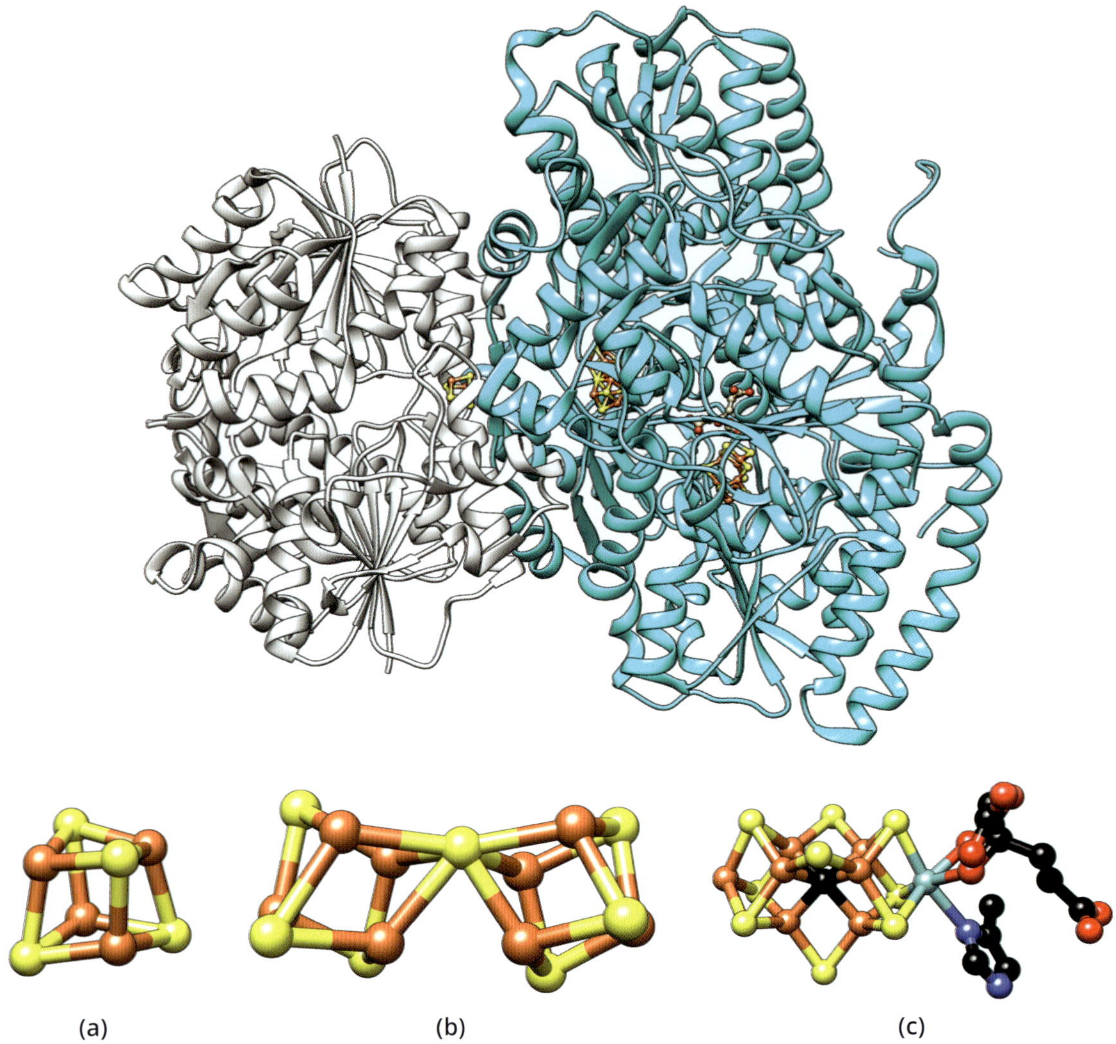

Fig. 20.62 Structure of the nitrogenase protein complex at 1.0 Å resolution (2011) consisting of the Fe protein (left), the FeMo protein (right) with the Fe_4S_4 center at the interface. Below are ball and stick models of a) the Fe_4S_4 center, b) the P-cluster, and c) FeMoco. Figure generated with UCSF Chimera[11] with coordinates from the protein databank (PDB ID 3U7Q and 1M34).[9, 91–92]

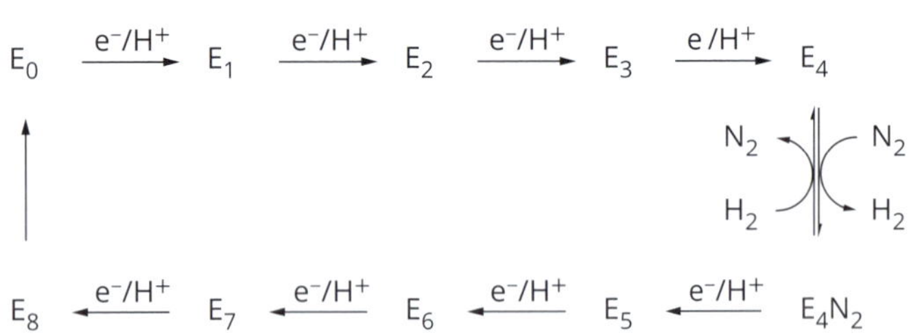

Fig. 20.63 The LT kinetic scheme for the reduction of dinitrogen to ammonia. Starting from the resting state, E_0, four separate correlated electron/proton transfer steps generate state E_4, known as the Janus intermediate, which is the halfway point of the mechanism and the point where obligatory dihydrogen reduction occurs as dinitrogen binds. Four subsequent electron/proton transfer steps result in the formation of two molecules of ammonia. Each electron/proton transfer step involves the docking of the Fe protein, cleavage of two ATP molecules, and a conformational change, electron transfer, and undocking of the Fe protein.

with certainty even now, a model has been proposed that is consistent with all the observed experimental data.[86]

The study of the nitrogenase enzyme is important for several reasons. Although it has been a tenacious protein to study, the reaction it carries out is difficult in a laboratory environment. The enzyme resisted study because it was not possible to isolate the enzyme in a state that was ready to react with dinitrogen. Without that substrate, the enzyme rapidly reduces protons to dihydrogen. Additionally, the enzyme itself exists as a protein complex made up of two separate proteins with different roles. However, the development of understanding of the reaction is hoped to someday allow for the fixation of dinitrogen on a large scale in a less energy-intensive fashion than the Haber–Bosch process. It also satisfies a deeply held intellectual curiosity as to how the reaction can occur, and its study has helped to refine our knowledge of inorganic chemistry.

20.9.7 Structural Details (Advanced Topic)

Two separate cofactors have been isolated from nitrogen-fixing bacteria: an iron-containing protein, the Fe protein, which contains an Fe_4S_4 center, and an iron/molybdenum-containing protein, the FeMo protein, which includes both the P-cluster and the iron molybdenum cofactor, FeMoco. Even before structural data from crystallography became available, the iron protein was believed to be an electron transfer protein that contained an ATP binding site, based on gene sequence similarity with other ATP binding proteins. Meanwhile, the iron/molybdenum protein was thought to be the site of dinitrogen reduction. FeMoco could be extracted from the FeMo protein into solvent, but the cofactor then lost all nitrogen fixing activity. However, when the extracted cofactor was added to cells where the genes necessary to produce the cofactor had been edited out, it could restore nitrogen-fixing activity to the cell.

The nitrogenase proteins were subjected to a number of spectroscopic investigations. Chemists wanted to know the structure of the enzyme so that they could develop model complexes that might mimic either the spectroscopic properties or reactivity of the protein. The resting state of FeMoco within the FeMo protein exhibits a distinctive EPR signal with g values of 4.3, 3.7, and 2.01 due to an $S = 3/2$ state. The isolated cofactor exhibits a similar spectrum. When the enzyme is actively reducing dinitrogen in the fully reduced state, this EPR signal disappears. Mössbauer spectroscopy was carried out on nitrogenase and FeMoco. These studies revealed that the fully reduced state has $S = 0$, the resting state has $S = \frac{1}{2}$, and the oxidized state has $S \geq 1$. When spectra were obtained in the presence of a magnetic field, four types of ^{57}Fe hyperfine coupling were observed, meaning that there were at least four iron atoms in FeMoco. Later ENDOR (Section 20.6.4) measurements showed that there was a minimum of five iron atoms in the cofactor.

In the 1970s, X-ray absorption spectroscopy (XAS and EXAFS) was carried out on both the iron molybdenum protein and isolated FeMoco; nitrogenase was the first enzyme to be studied by this technique. The results for both systems were essentially indistinguishable, meaning that the structure of the cofactor did not change when it was removed from the protein. The coordination environment around the Mo was found to have three or four sulfur atoms at 2.36 Å and two or three iron atoms at 2.72 Å. Two models were proposed for FeMoco: a $(cys\text{-}S)_2MoFe_3S_4$ cubane structure similar to a ferredoxin, and a square pyramidal $(cys\text{-}S)MoS_4Fe_2$ structure (Fig. 20.64). Model complexes were prepared that contained the $MoFe_3S_4$ cubane structure. and the X-ray spectroscopic studies of these model complexes produce results similar to that of FeMoco, suggesting that this model was potentially correct.

In the early 1990s, several complete single-crystal X-ray structures of the nitrogenase enzyme were obtained, first at 3.0 Å and subsequently at 2.2 Å resolution, allowing the protein

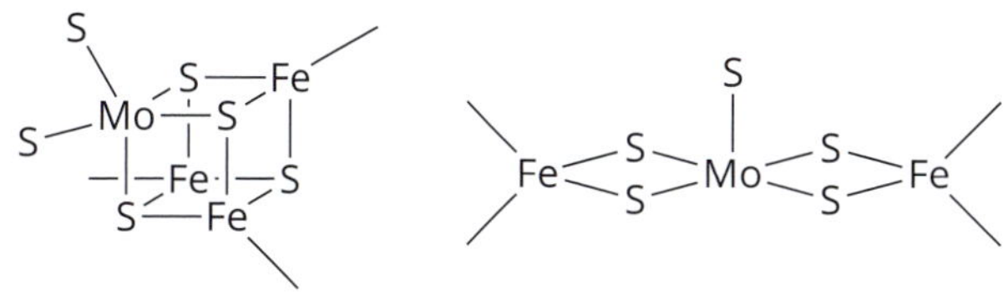

Model 1 Model 2

Fig. 20.64 Two models for the coordination environment around molybdenum in FeMoco based on X-ray absorption spectroscopy.

structure as well as the atomic structure of FeMoco to finally be described. The protein was isolated as the $MgADP \cdot AlF_4$ adduct. Adducts of this type are often used to stabilize the "cleaved ATP" structure of a protein as the aluminum tetrafluoride anion binds to the phosphate binding site without dissociating. The nitrogenase protein complex consists of both the iron protein and the MoFe protein (Fig. 20.62).

The P cluster in the reduced form, as initially crystallized, was found to have a highly symmetrical structure. This structure can be described as two Fe_4S_3 cubes sharing a vertex sulfur atom with two bridging cysteine sulfurs. The cluster also contains two bridging cysteine sulfur atoms. Later structures show that there is a substantial rearrangement of the FeS core of the P cluster upon oxidation to a more open and less symmetrical structure (Fig. 20.65).

FeMoco is buried approximately 10 Å below the protein surface. The molybdenum atom in FeMoco is bound to the protein by a histidine and has a homocitrate molecule attached to complete its octahedral coordination (Fig. 20.66a). It connects to the rest of the cofactor with three bridging sulfur atoms to a set of three equivalent iron atoms. The six iron atoms were each initially found to be three-coordinate in a trigonal planar environment, each slightly pyramidalized towards the center of the cluster. The last iron center is attached to the protein by a cysteine. The unprecedented structure of FeMoco, especially the 3-coordinate iron atoms, launched extensive synthetic investigations to try to model the structure or reactivity. The void in the center of FeMoco looks appealingly like a place where the N_2 molecule could bind to six iron atoms and be reduced by six electrons, but the cavity is too small to accommodate a diatomic molecule.

In 2002, a crystal structure was obtained at 1.16 Å (Fig. 20.66b). This higher-resolution allowed for the determination of a central atom in the middle of the cluster, but the data were not of sufficient quality to determine whether it was a nitrogen, oxygen, or carbon atom. The reason why the central atom was not observed in the 1992 structure is a result of artifacts of Fourier series termination due to the symmetric structure of six heavy iron atoms. This artifact results in a cancelation of electron density, reducing it to near zero for resolutions above 1.5 Å, and suggesting a void in the cavity instead of an atom.

Three separate lines of experimental evidence were obtained in order to determine the identity of the central atom. X-ray emission spectroscopy (X-ray fluorescence) is a

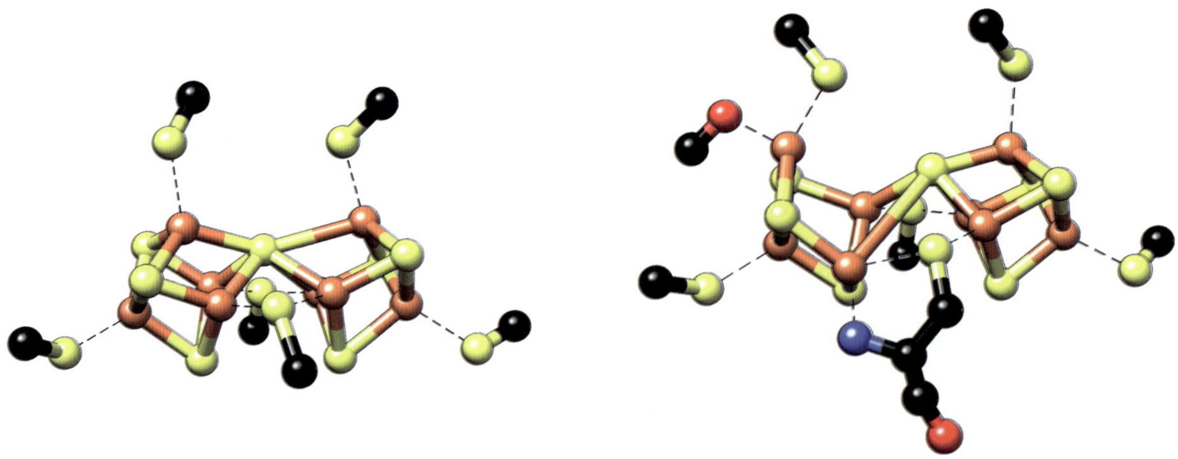

Fig. 20.65 Structures of the symmetrical P-cluster in the reduced form (left) and the unsymmetrical P-cluster in the oxidized form (right). In the reduced form, all iron atoms are coordinated by sulfur or cysteine sulfur atoms, while in the oxidized form, one iron is coordinated by a serine oxygen and another iron is coordinated by the cysteine nitrogen. Figure generated with UCSF Chimera[11] with coordinates from the protein databank (PDB ID 3MIN and 2MIN).[9,93]

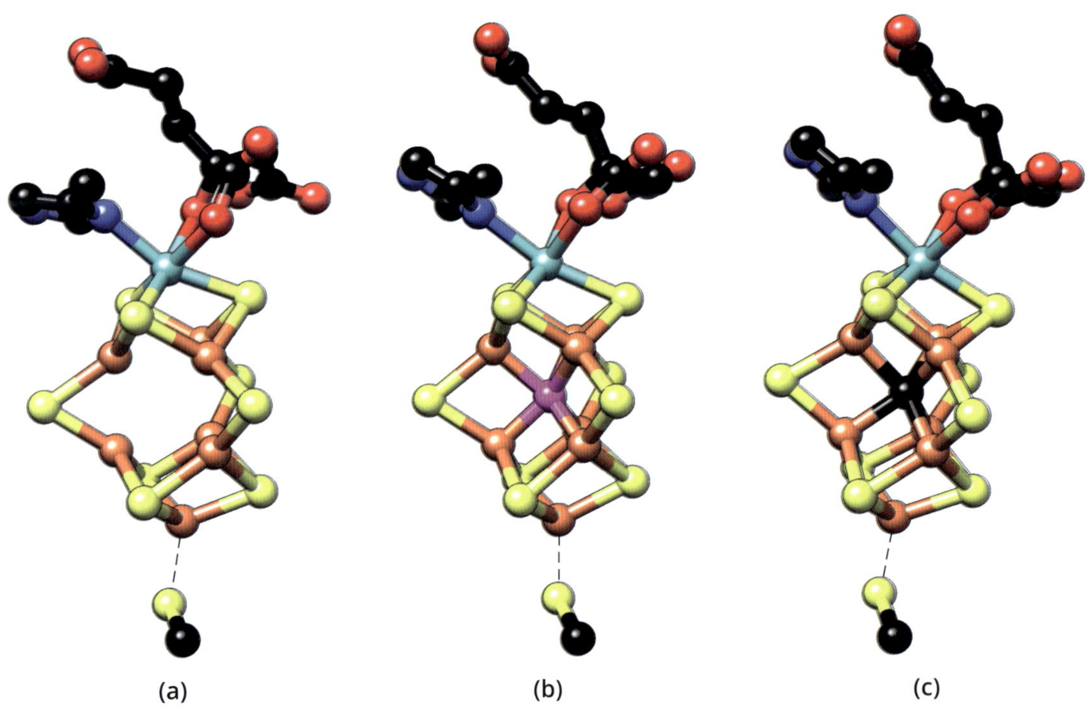

Fig. 20.66 The progression of the structure of FeMoco as X-ray diffraction data was obtained with higher resolution. (a) The original structure at 3.0 Å, (b) 1.16 Å resolution. and (c) 1.0 Å resolution. The cluster is bound to the protein by cysteine 275 (bottom) and histidine 442 (top), and the molybdenum atom is also coordinated by a homocitrate molecule. Figure generated with UCSF Chimera[11] with coordinates from the protein databank (PDB ID 3U7Q, 1M1N and 1MIO).[9,91,94–95]

technique that monitors the emission of X-ray photons after an X-ray absorption event. Although the difference between the 2p–1s transition was obstructed by emission from sulfides, the 2s–1s transition was clear and most consistent with the identity of the central atom as carbon (Fig. 20.67). In 2011, a follow-up X-ray diffraction study was carried out at 1.0 Å resolution. Although this increased resolution was insufficient to directly determine the identity of the central atom, the increased data available allowed for a statistical analysis of electron density in FeMoco. Measuring the electron density as a function of radius for more than 10,000 carbon, 2,700 nitrogen, and 5,600 oxygen atoms produced excellent calibration curves. When the electron density of the central atom was plotted, it matched that of carbon but not nitrogen or oxygen. A final experiment was the pulsed NMR/EPR spectroscopic technique known as ESEEM. This technique was used to detect coupling between unpaired electrons and isotopically labeled ^{13}C, ^{14}N, or ^{17}O. The technique had been carried out

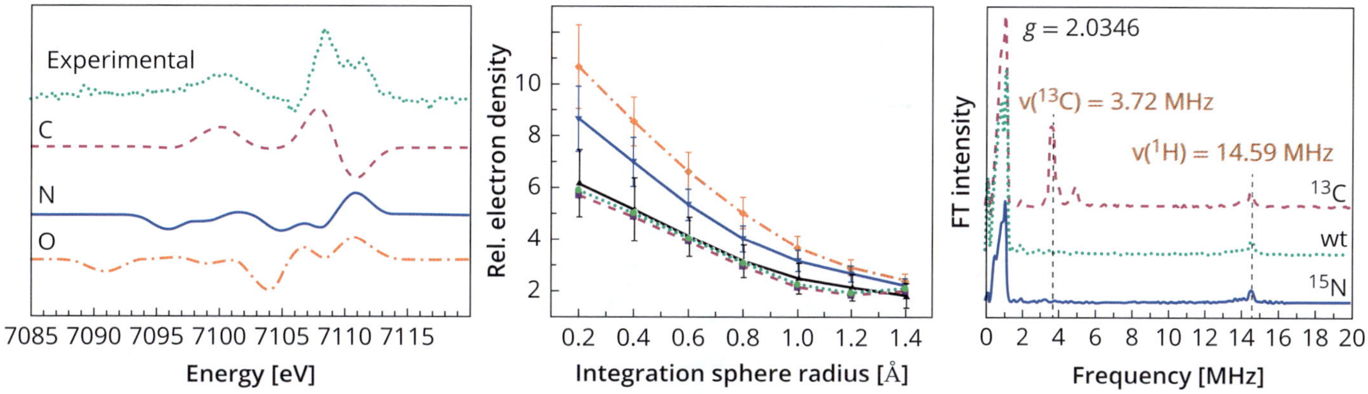

Fig. 20.67 Experimental evidence supporting the claim that the central atom in FeMoco is a carbon atom. (*left*) X-ray fluorescence of the 2s–1s transition; (*center*) relative electron density as a function of integration sphere; (*right*) ESEEM spectroscopy showing coupling of the electron to ^{13}C. All three techniques support the central atom being C and not N or O. Figure adapted from Einsle.[90]

previously but was inconclusive due to the high expense of fully labeling ^{13}C in the bacterial growth media. With only 5–10% labeling, no electron–carbon coupling was observed. After other results suggested that the central atom was in fact a carbon, the experiment was repeated. With higher labeling efficiency, a clear coupling between the electron and the labeled carbon was observed, while no coupling was observed with labeled oxygen or nitrogen, supporting the claim that the central atom was carbon.

20.9.8 Mechanism of Dinitrogen Reduction in FeMoco (Advanced Topic)

The LT kinetic scheme for the reduction of dinitrogen to ammonia was developed during the 1970s and 1980s as a result of extensive experimental work on FeMoco (Fig. 20.63). Very little is known about the first three reduced states in the kinetic model, states E_1–E_3. The states have been studied by EPR, and rates of the formation of three different species have been observed and measured. The addition of the first proton and electron results in the E_1 state, and Mössbauer spectroscopy suggested that this results in a reduced state of FeMoco. However, subsequent analysis suggests that instead the E_1 state consists of an oxidized core with an anionic hydride ligand. Addition of the second proton and electron produces a cluster with the same oxidation state as the initial E_0 state, with both a bridging hydride and a proton, presumably bound to a sulfur atom. It was proposed that the eight electron-transfer steps result in a cycling between only two oxidation states of FeMoco, with reducing equivalents stored as hydrides. Subsequent studies supported this hypothesis.

The E_4 state has been trapped and observed by EPR/ENDOR. This state shows coupling between the electrons and hydrogens with isotropic hyperfine coupling constants of 24 MHz, consistent with there being metal-bound hydrogens, and anisotropic coupling constants that suggest a rhombic rather than an axial geometry which rules out terminal hydrides and supports a hydride that bridges two paramagnetic metal centers. Subsequent ^{95}Mo ENDOR spectroscopy provided evidence that the hydrides were bridging iron atoms and not molybdenum atoms. Several structures for this intermediate with hydrides bridging several different iron atoms have been postulated, though quantum-mechanics calculations support only a subset of those. For simplicity, only one structure of the E_4 state is shown (Fig. 20.68). It was postulated that both hydrogen atoms were bound with one terminus on Fe-6. This stage has collected four protons and four electrons. The bridging hydrides account for two of the protons and all four electrons; the remaining two protons are placed on nearby sulfur atoms for electrostatic stabilization of the reduced dinitrogen that forms in a subsequent step. In 2019 the structure of the Janus intermediate was firmly established by high-resolution ENDOR and quantum-chemical calculations.[96] The intermediate does in fact have two bridging hydrides, though the hydrides bridge Fe-6/Fe-2 and Fe-7/Fe-3.

The state of the protein at this point is referred to as the Janus intermediate, referring to the Greek mythological figure who looks backwards and forwards simultaneously. It is this point that dinitrogen binds to the cluster. The early part of the mechanism, states E_0–E_4, have been studied independently from the later part of the mechanism, states E_5–E_8. The first part of the mechanism is an accumulation of reducing equivalents stored as hydrides, while in the later part of the mechanism the reducing equivalents are used in a productive fashion to generate ammonia. The Janus intermediate is the linker between the two halves of the mechanism. In the absence of dinitrogen substrate, the intermediate releases dihydrogen, regenerating the E_0 state.

The binding of dinitrogen to form diazene is the thermodynamically most challenging step of the entire nitrogenase process, uphill by 53 kcal/mol, and it is now believed that the enzyme uses the reductive elimination of dihydrogen to

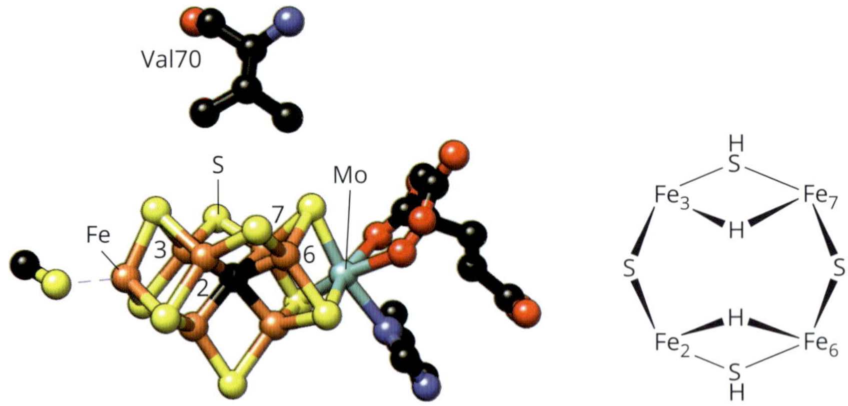

Fig. 20.68 The proposed structure of E_4, the Janus intermediate, which has collected a total of four reducing equivalents. On the left is the nitrogenase core. On the right is the structure of the Janus intermediate. Both bridging hydrides interact with the Fe(2)-Fe(3)-Fe(6)-Fe(7) face, while the protons are bound to nearby sulfides. Fe is orange, Mo is cyan, S is yellow. Figure generated with UCSF Chimera[11] with coordinates from the protein databank (PDB ID 3U7Q).[9,91]

generate a "superreduced" iron core involving iron atoms 2, 3, 6, and 7. The E_4 intermediate has collected four reducing equivalents and four protons. Reductive elimination of H_2 takes away only two of the reducing equivalents. Much like carbon monoxide and olefins bind more strongly to low-valent-transition metal centers, this reduced iron atom is a strong π-donor and formally reduces the dinitrogen to diazene. As dinitrogen binds to the reduced iron center a significant negative charge build-up would be thought to occur. This negative charge is counteracted by the two sulfur bound protons that provide electrostatic stabilization of the reduced intermediate. Rapid proton transfer to the bound dinitrogen results in a bound diazene intermediate.

The diazene binding is reversible, and the presence of dihydrogen can cause the release of dinitrogen and the regeneration of the E_4 state. Reactivity studies using deuterium and tritium (D_2 and T_2) were used to rule out an alternative mechanism that involved protonation of the hydride with the sulfur bound proton. If the process went by this mechanism, D^+/T^+ would be released into the solvent, but this was never observed.

Once the dinitrogen is bound to the protein, the mechanism of ammonia production was the subject of much debate. Both a distal (D) and alternating (A) pathway were proposed (Fig. 20.69). In the D pathway, once the dinitrogen is bound, the terminal nitrogen atom is removed by a series of three proton/electron transfers, and the second is removed after three additional transfers. In the A pathway, hydrogen atoms are added pairwise, forming first diazene, then hydrazine, and finally two ammonia molecules. Evidence for the A pathway came from kinetic and spectroscopic studies. The reduction of dinitrogen, diazene, and hydrazine all give rise to the same late-stage intermediate, suggesting that these substrates enter the catalytic cycle at their appropriate stages of the A pathway and are then trapped by hydrogen atom equivalents.

Fig. 20.70 shows a complete structural mechanism that accounts for all the observed spectroscopic data and is consistent with the reactivity and kinetics. This mechanism shows both hydrides bridging to Fe-6, but the overall mechanistic steps remain the same. The accumulation of reducing equivalents from states E_0 to E_4 results in an intermediate

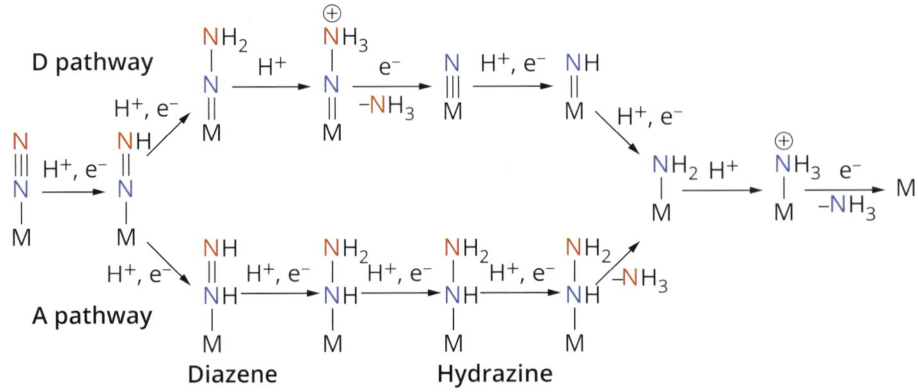

Fig. 20.69 The D and A pathways for the reduction of dinitrogen to ammonia.

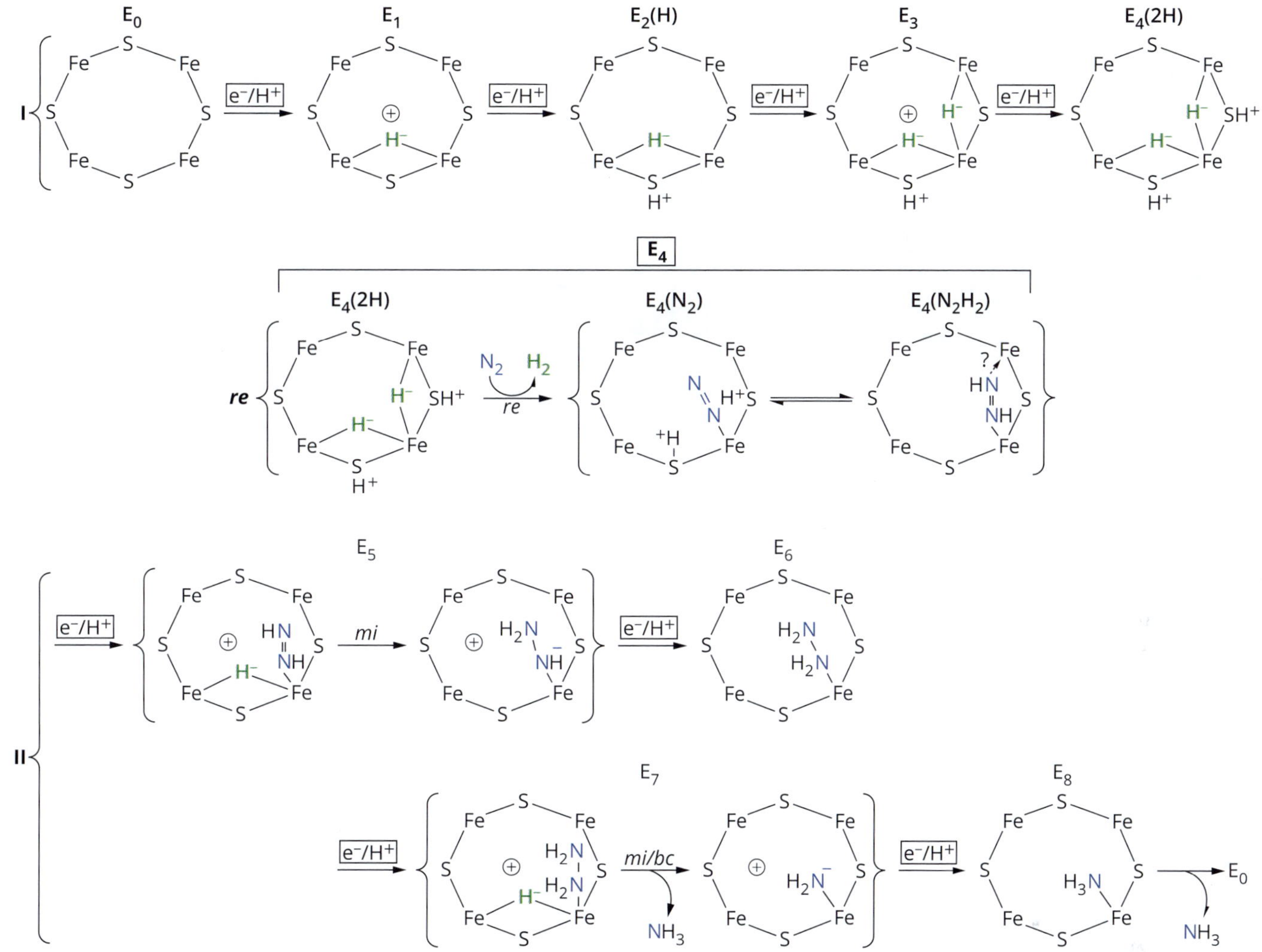

Fig. 20.70 A complete structural mechanistic model for all eight states in the LT kinetic model. This model accounts for the reactivity of diazene and hydrazine, is consistent with the kinetics and spectroscopic data available, and explains the obligatory production of dihydrogen during the nitrogen-reduction pathway. Figure from Seefeldt and Hoffman.[86]

poised to bind to dinitrogen. As dihydrogen is released in a reductive elimination step, the iron core becomes “superreduced” and is able to bind to dinitrogen. Proton transfer to the reduced dinitrogen forms the bound diazene complex. States E_5 to E_8 are productive hydrogen-atom transfer steps that generate hydrazine and then finally two equivalents of ammonia, regenerating state E_0.

20.9.9 Protein Complex Structure and Role of the Iron Protein (Advanced Topic)

While the detailed mechanism of the nitrogenase enzyme is consistent with observed data, the reducing equivalents (electrons) need to come from somewhere. The electrons that are delivered to FeMoco come from the iron protein. The relatively close proximity of the Fe_4S_4 cluster in the Fe protein to that of the P cluster and FeMoco had suggested that electron transfer from a reduced Fe protein into FeMoco was the source of reducing potential in the nitrogenase system. A series of Fe/MoFe protein clusters was crystallized in the presence of MgADP (adp), a nonhydrolyzable MgATP analog named MgAMPPCP (pcp), or in a nucleotide-free preparation (nf). The structure of the atp analog was very similar to that of the MgADP·AlF_4 structure shown in Fig. 20.62. In all three crystal structures there is no discernable structural change of the MoFe protein, but there are significant changes in both the structure of the Fe protein and its docking location on the MoFe protein. The MoFe protein contains multiple binding locations for the different geometries of the Fe protein (Fig. 20.71).

An angle was defined in the MgATP analog structure that related the orientation of two α-helices in the Fe protein. In the nf structure, the Fe protein is rocked by about −15° relative to the pcp structure, while the adp structure is rocked by about

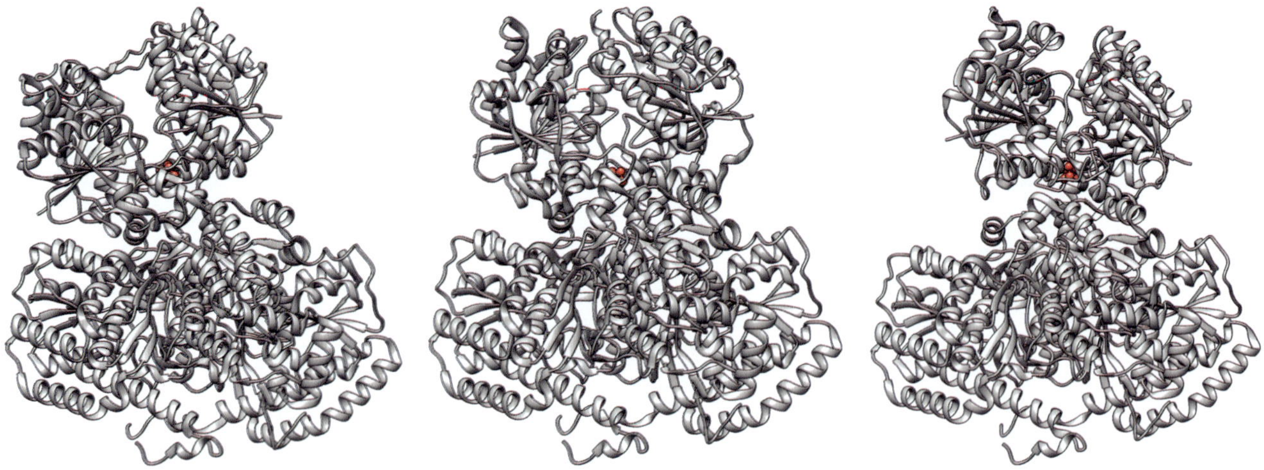

Fig. 20.71 Dramatic structural changes observed in the docking of the Fe protein to the MoFe protein during ATP hydrolysis. From left, the nucleotide free structure, the MgATP analog structure, and the MgADP structure. The Fe_4S_4 cluster in the iron protein is highlighted in red. Figure generated with UCSF Chimera[11] with coordinates from the protein databank (PDB ID 2AFH, 4WZB and 2AFI).[9,97]

+12° relative to the pcp structure. Additionally, the distance from the Fe_4S_4 cluster in the Fe protein changes significantly upon ATP hydrolysis. In the nf structure, the distance from the Fe_4S_4 centroid to the P cluster centroid is 23 Å. In the pcp structure, that distance drops to 18 Å, while in the adp structure, the distance again rises to 23–24 Å (Fig. 20.72). The net effect is that ATP binding to the nucleotide free Fe protein causes significant conformational changes resulting in a 5 Å contraction in the distance needed for the electron transfer process. While 18 Å is still an extremely long distance for an electron-tunneling process to occur, this change in distance corresponds to a two- to three-order of magnitude increase in the rate of electron transfer from the Fe_4S_4 cluster to the P cluster.[98] After the electron transfers, ATP hydrolysis causes another conformational change in the Fe protein, increasing the electron-transfer distance by 5–6 Å, preventing backwards flow of electrons.

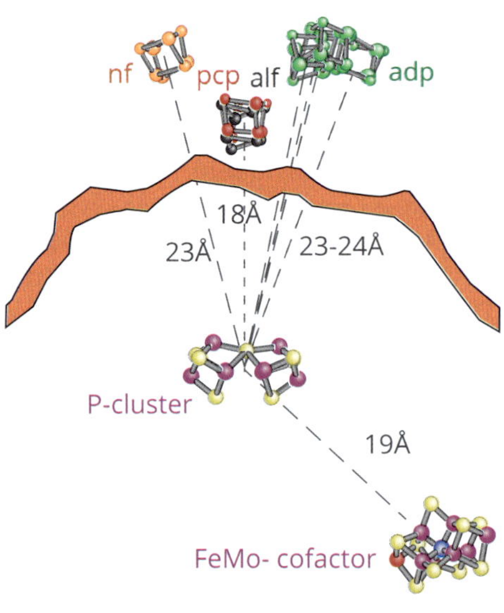

Fig. 20.72 Location of the Fe_4S_4 cluster relative to the P cluster and FeMoco. Distances shown are from the metal cofactor centroids. Figure from Rees.[97]

A more complete picture of the reduction of dinitrogen can now be described as follows. The reduced Fe protein docks to the MoFe protein in the nucleotide free state. As ATP binds, the Fe protein changes conformation and rocks by about 14°, bringing the Fe_4S_4 cluster 5 Å closer to the P cluster. The electron from the reduced Fe_4S_4 cluster tunnels through the protein amino acid backbone or nearby alpha helices and ends up on the P cluster. ATP hydrolysis causes another conformational change of the Fe protein, rocking by 12°, moving the Fe_4S_4 cluster 5 Å further away from the P cluster, trapping the electron in the MoFe protein. Electron transfer from the P cluster to the FeMoco occurs along with the transfer of a proton. This process repeats eight times during the reduction of dinitrogen, following the LT model from state E_0 to E_8.

20.9.10 The Site of Nitrogen-Binding and Implications for the Mechanism (Advanced Topic)

The site for the nitrogen-binding and ammonia-release has been identified by site directed mutagenesis and high-pressure X-ray crystallography under xenon. These studies help to confirm that the location of N_2 binding in FeMoco on the iron 2, 3, 6, 7 face. These two techniques have revealed probable pathways or channels for the ingress of N_2 and the egress of NH_3 and H_2 from the active site of the protein. A valine close to the iron face using irons 2, 3, 6, and 7, Valine 70 in Fig. 20.68, was replaced with either alanine or glycine. This was done to open up a potential channel, and when this was done the native nitrogenase enzyme was able to reduce larger substrates such as propargyl alcohol. This result suggests that the channel that terminates near Valine 70 is the pathway by which dinitrogen enters to bind to iron. The product of the reduction or propargyl alcohol was detected bound to iron, very suggestively indicating that the Fe(2)-Fe(3)–Fe(6)–Fe(7) face is important for catalysis.

High-pressure X-ray crystallography under a Xe atmosphere was also used to help identify channels for ingress and egress of reactants and products. The diameter of Xe is approximately the same as N_2, and when crystals of nitrogenase were pressurized at 14 atm under an atmosphere of Xe, these crystals took up Xe atoms in several different channels that linked FeMoco to the surface of the protein. It is probable that multiple channels are used by substrates and products, but at least one of the channels is also consistent with the site-directed mutagenesis data discussed previously. This channel begins at the protein surface, passes by Valine 70 and terminates at the Fe(2)–Fe(3)–Fe(6)–Fe(7) face of FeMoco (Fig. 20.73).

The mechanism of nitrogenase is now fairly well understood, and the proposed mechanism is in agreement with all experimental evidence. It took three decades and multiple crystal structures to resolve the structure of FeMoco and determine that there was a central carbon atom. This carbon atom is invoked to hold the iron centers as a rigid platform so that the hydrogen additions and reductive elimination take place on a stable surface. The nitrogenase enzyme has been affectionally claimed to have a "heart of steel", referring to the carbon and molybdenum additions to the iron atoms. Multiple spectroscopic techniques from X-ray fluorescence and EXAFS to EPR and high-pressure X-ray diffraction have all been used to determine the structure of the active site. What does the future hold for nitrogenase? Will it be possible to synthesize small-molecule analogs that have nitrogen-splitting activity? Will we be able to incorporate this complex dual protein cluster into plants that do not currently have a symbiotic bacterial relationship? Will we be able to control the nitrogen cycle without relying on bacteria for more than half of our fixed nitrogen? Our understanding of nitrogenase is still a work in progress.

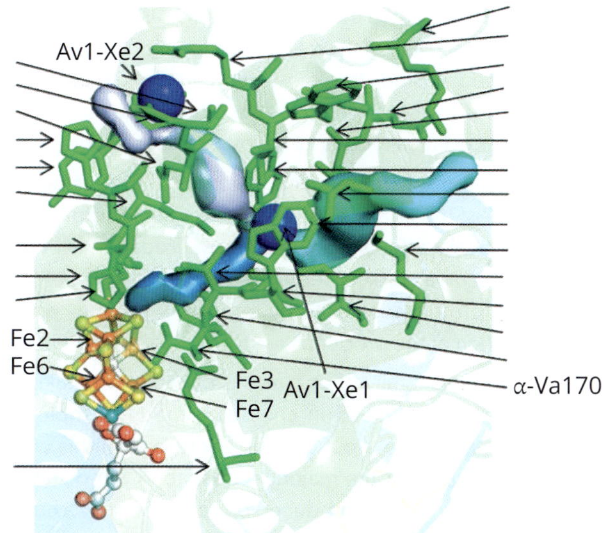

Fig. 20.73 Some of the channels located by high-pressure X-ray crystallography. Xenon atoms are shown in dark blue. In this figure, the light-blue channel starts at the protein surface and terminates at the 2367 face which is gated by Valine 70. Figure adapted from Rees.[99]

Chapter Summary

1. Inorganic elements in biological systems have a range of functions. Some elements are used in bulk, for example, in bones and shells. Some metal ions have structural roles and help proteins or DNA to fold properly, while others have functional roles, carrying out chemical reactions. Nonbiological metal complexes have use in medical imaging, diagnosis, and therapy.
2. Metalloenzymes are proteins that incorporate metal ions to carry out their chemical transformations. The metals can be coordinated by amino-acid side-chains, nucleic acids, or as part of a cofactor that includes other ligands such as porphyrins or inorganic sulfide.
3. A large number of experimental methods have been applied to metalloenzymes in order to understand their structure and chemical reactivity. Since the metal is only a small part of the protein, many spectroscopic techniques must be carried out to gather a complete picture. X-ray crystallography is perhaps the most important tool for the bioinorganic chemist because it gives a 3-dimensional atomic structure, though this technique is not without its own set of problems. Proteins are more difficult to crystallize than small molecules, and the large amount of disordered water molecules in the crystal can limit the resolution.
4. Model complexes are synthesized for two major purposes. Spectroscopic models are designed to mimic the spectroscopic features of a metalloenzyme, while functional models are designed to mimic the chemical behavior of the metalloenzyme. There are important and interesting biological proteins with chemical behavior that are yet to be synthesized successfully in the laboratory.
5. Metalloenzymes are involved in reactions that involve electron transfer, oxygen uptake and storage, and carbon group transfer. Many different proteins are involved in the acquisition and storage of important metal ions, such as copper and iron from the environment.
6. The oxygen evolving complex is a manganese cluster with a calcium atom appended in a distorted cubane structure. Water molecules are oxidized to dioxygen in the complex, and the resulting electrons are used to reduce a nearby tyrosine radical. This tyrosine reduces the photooxidized special pair, restarting the photosynthesis process.
7. The methane monooxygenase family of enzymes is capable of the selective oxidation of methane to methanol, breaking the strongest C–H bond in chemistry. This process converts a gaseous

hydrocarbon to a liquid that the bacteria use as their sole carbon source. The discovery of the mechanism for this process has been under continuous study for almost fifty years. Spectroscopic studies almost perfectly predicted the structure of the protein-active site that was eventually determined by X-ray crystallography.

8. The mechanism of the reaction is very well understood. The diiron core of the enzyme activates dioxygen to form complex Q, the most powerful oxidant in biology. Oxidation of methane takes place by a "radical rebound" mechanism. Careful control of the protein-active site by regulatory proteins prevents overoxidation of the product.
9. The nitrogenase family of enzymes is capable of breaking the strongest nonpolar bond in chemistry, the N–N triple bond in dinitrogen. Before around 1950, bacteria with this enzyme were responsible for the majority of useable, "fixed", nitrogen on Earth. This enzyme is produced by approximately a hundred species of bacteria and directly converts atmospheric dinitrogen into ammonia. The protein complex consists of several individual proteins that must dock and undock in order to transfer the reducing equivalents. The energy cost for the protein is large, with eight ATP molecules cleaved per ammonia molecule produced.
10. Several models for the nitrogenase active site were proposed based primarily on EXAFS data. The structure of the active site was determined through multiple iterations of increasingly higher-resolution X-ray crystallography experiments. These high-resolution pictures showed several unprecedented metal clusters that are required to carry out the chemical reaction.
11. The detailed mechanism for the production of ammonia by nitrogenase has been elucidated through multiple spectroscopic and crystallographic studies over the past thirty years. The reducing equivalents are stored as bridging hydride ligands in a the Fe_7S_9Mo core, the iron molybdenum cofactor. Reductive elimination of dihydrogen—a necessary component of the reaction—generates a "super-reduced" four-iron core that then has sufficient reducing power to take dinitrogen to diazene and then to hydrazine and ammonia.

Practice Problems

1. a. Cytochrome c oxidase is the protein that produces water in the last step of the electron transport chain. Cyanide is toxic because it binds to the iron center of this protein better than the biological substrate. What is the biological substrate? Explain why cyanide binds better than the biological substrate. A cyanide antidote is amyl nitrite ($(CH_3)_2CHCH_2CH_2ONO$). Propose a mechanism by which this compound could inhibit cyanide binding.

   b. Similarly, carbon monoxide is toxic because it binds to the iron center in hemoglobin better than the biological substrate. What is the biological substrate and why does CO bind better?

   c. Although both toxins bind to different proteins, what would the net effect of their binding be in terms of what biological process is affected? The same, or different processes?
2. Some diseases, such as Cooley's anemia, require regular blood transfusions and this can result in iron overload diseases. Why does the body not have good mechanisms for the removal of excess iron?
3. Many proteins are modified post translation (after conversion of RNA to protein) by being glycosylated which means that sugar chains are appended to the protein. The sugars can then be recognized by receptors. a) why are sugars good targets for receptors? b) what advantages are there for recognition of a post translated protein rather than the protein itself?
4. Chelators that are good for binding to iron are also generally good at binding to radioactive actinides since the radius/charge ratio is similar. What does this suggest about the toxicity of radionucleotides?
5. Chemists often use gene sequence similarity or identity to find metal binding sites in newly discovered proteins. What amino acid residues would you expect to be more conserved, and why might the spacing between those residues also be conserved?
6. It has been suggested that mitochondria in mammalian cells were bacteria that were not able to survive in the newly oxic atmosphere that formed after photosynthesis developed on the Earth. These bacteria found hosts in eukaryotic cells and have, over time, lost their own ability to reproduce. Compare and contrast this theory with the symbiotic nitrogen fixing bacteria that live in root nodules of legumes.
7. Verify that in a Mössbauer spectroscopy, a 1 mm/s velocity of the source relative to the sample using 14.4 keV gamma rays results in a $5 \cdot 10^{-8}$ eV difference in energy of the emitted photon.
8. For the chemical equation shown below (a) for the binding of Fe(III) by enterobactin, the equilibrium constant is 10^{49}. However, this chemical equation does not tell the whole story, as the favored ionization state for enterobactin at pH 7.4 is H_6ent. Six protons must be removed for the tris-catecholate to bind to the iron center. The pM ($-\log_{10}$[Fe]) for enterobactin was measured as 35.5 at pH 7.4 with a total siderophore concentration of 10 μM and a total Fe^{3+} concentration of 1 μM.

   i. Given this pM value, determine the effective equilibrium constant for the reaction b.

   $$Fe^{3+} + ent^{6-} \rightarrow \left[Fe(ent)\right]^{3-} \quad (a)$$

   $$Fe^{3+} + H_6ent \rightarrow \left[Fe(ent)\right]^{3-} + 6H^+ \quad (b)$$

   ii. use the following reaction (deprotonation of phenol) to rationalize your answer with the fact that enterobactin is the strongest iron chelator known.

   $$C_6H_5OH + H_2O \rightarrow C_6H_5O^- + H_3O^+ K_{eq} \sim 10^{-10}$$

Cited References

1. Dixon, N. E.; Gazzola, C.; Blakeley, R. L.; Zerner, B., Jack bean urease (EC 3.5.1.5). Metalloenzyme. Simple biological role for nickel *Journal of the American Chemical Society* 1975, *97*, 4131–4133.
2. Bertini, I.; Gray, H. B.; Stiefel, E. I.; Valentine, J. S., *Biological Inorganic Chemistry: structure and reactivity* University Science Books: Sausalito, California, 2007.
3. Lippard, S. J.; Berg, J. M., *Principles of Bioinorganic Chemistry* University Science Books: Mill Valley, California, 1994.
4. Bertini, I.; Gray, H. B.; Lippard, S. J.; Valentine, J. S., *Bioinorganic Chemistry* University Science Books: Mill Valley, CA, 1994.
5. Crabb, E.; Moore, E. A.; University, O., *Metals and Life* RSC Publishing: Cambridge, United Kingdom, 2010.
6. Roat-Malone, R. M., *Bioinorganic Chemistry: A Short Course* John Wiley & Sons: Hoboken, NJ, 2002.
7. Sumner, J. B., The isolation and crystallization of the enzyme urease: Preliminary paper *J. Biol. Chem.* 1926, 435–441.
8. *Crystalmaker*, 10.4.7; CrystalMaker Software Ltd, Oxford, England (www.crystalmaker.com): 2018.
9. Berman, H. M.; Westbrook, J.; Feng, Z.; Gilliland, G.; Bhat, T. N.; Weissig, H.; Shindyalov, I. N.; Bourne, P. E., The Protein Data Bank, rcsb.org *Nucleic Acids Research* 2000, *28*, 235–242.
10. Shibata, N.; Inoue, T.; Nagano, C.; Nishio, N.; Kohzuma, T.; Onodera, K.; Yoshizaki, F.; Sugimura, Y.; Kai, Y., Novel Insight into the Copper-Ligand Geometry in the Crystal Structure of Ulva pertusa Plastocyanin at 1.6-Å Resolution: Structural basis for regulation of the copper site by residue 88 *Journal of Biological Chemistry* 1999, *274*, 4225–4230.
11. Pettersen, E. F.; Goddard, T. D.; Huang, C. C.; Couch, G. S.; Greenblatt, D. M.; Meng, E. C.; Ferrin, T. E., UCSF Chimera--a visualization system for exploratory research and analysis *J. Comput. Chem.* 2004, *25*, 1605–12.
12. Lovejoy, B.; Choe, S.; Cascio, D.; McRorie, D. K.; DeGrado, W. F.; Eisenberg, D., Crystal structure of a synthetic triple-stranded alpha-helical bundle *Science* 1993, *259*, 1288.
13. Makabe, K.; McElheny, D.; Tereshko, V.; Hilyard, A.; Gawlak, G.; Yan, S.; Koide, A.; Koide, S., Atomic structures of peptide self-assembly mimics *Proceedings of the National Academy of Sciences* 2006, *103*, 17753.
14. Phillips, S. E. V., Structure and refinement of oxymyoglobin at 1·6 Å resolution *Journal of Molecular Biology* 1980, *142*, 531–554.
15. Bolton, W.; Perutz, M. F., Three Dimensional Fourier Synthesis of Horse Deoxyhaemoglobin at 2.8 Å Resolution *Nature* 1970, *228*, 551–552.
16. Frieden, E., New perspectives on the essential trace elements *Journal of Chemical Education* 1985, *62*, 917.
17. Barry, N. P. E.; Sadler, P. J., Exploration of the medical periodic table: towards new targets *Chem. Commun.* 2013, *49*, 5106–5131.
18. Blower, P. J., A nuclear chocolate box: the periodic table of nuclear medicine *Dalton Transactions* 2015, *44*, 4819–4844.
19. Jamieson, E. R.; Lippard, S. J., Structure, Recognition, and Processing of Cisplatin-DNA Aducts *Chem. Rev.* 1999, *99*, 2467–2498.
20. Takahara, P. M.; Rosenzweig, A. C.; Frederick, C. A.; Lippard, S. J., Crystal structure of double-stranded DNA containing the major adduct of the anticancer drug cisplatin *Nature* 1995, *377*, 649–652.
21. Brabec, V.; Hrabina, O.; Kasparkova, J., Cytotoxic platinum coordination compounds. DNA binding agents *Coordination Chemistry Reviews* 2017, *351*, 2–31.
22. Solomon, E. I.; Pavel, E. G.; Loeb, K. E.; Campochiaro, C., Magnetic circular dichroism spectroscopy as a probe of the geometric and electronic structure of non-heme ferrous enzymes *Coordination Chemistry Reviews* 1995, *144*, 369–460.
23. Zoppellaro, G.; Bren, K. L.; Ensign, A. A.; Harbitz, E.; Kaur, R.; Hersleth, H.-P.; Ryde, U.; Hederstedt, L.; Andersson, K. K., Review: Studies of ferric heme proteins with highly anisotropic/highly axial low spin (S = 1/2) electron paramagnetic resonance signals with bis-Histidine and histidine-methionine axial iron coordination *Biopolymers* 2009, *91*, 1064–1082.
24. Bertini, I.; Luchinat, C.; Ming, L. J.; Piccioli, M.; Sola, M.; Valentine, J. S., Two-dimensional proton NMR studies of the paramagnetic metalloenzyme copper-nickel superoxide dismutase *Inorganic Chemistry* 1992, *31*, 4433–4435.
25. Berg, J. M.; Tymoczko, J. L.; Stryer, L., *Biochemistry* 5th ed.; W. H. Freeman: New York, 2002.
26. Fanwick, P. E., Small molecule crystallography past, present, and future: a career long review of chemical crystallography *Crystallography Reviews* 2016, *22*, 250–279.
27. Pike, A. C. W.; Garman, E. F.; Krojer, T.; von Delft, F.; Carpenter, E. P., An overview of heavy-atom derivatization of protein crystals *Acta Crystallographica Section D* 2016, *72*, 303–318.
28. Clegg, W.; Blake, A. J.; Gould, R. O.; Main, P., *Crystal Structure Analysis* Oxford University Press: Oxford, 2001.
29. Glusker, J. P.; Trueblood, K. N., *Crystal Structure Analysis a Primer, 2nd ed.* Oxford University Press: Oxford, 1985.
30. Clegg, W., *Crystal Structure Determination* Oxford University Press: Oxford, 1998.
31. Reinheimer, E. W.; Hickman, A. J.; Moretti, J. E.; Ouyang, X.; Kantardjieff, K. A.; Johnson, A. R., Synthesis and Structure of 2-Cyclohexylamino-3-methyl-1,1-diphenyl-butan-1-ol (D-N-Cyclohexyl-diphenylvalinol), an Amino Alcohol Ligand Useful in Asymmetric Catalysis *J Chem Crystallogr* 2012, *42*, 911–915.
32. Lieberman, R. L.; Rosenzweig, A. C., Crystal structure of a membrane-bound metalloenzyme that catalyses the biological oxidation of methane *Nature* 2005, *434*, 177.
33. Newville, M. "EXAFS Analysis," Center for Advanced Radiation Sources, University of Chicago. https://docs.xrayabsorption.org/Workshops/IIT2013/Newville_Analysis.pdf (accessed 2019).
34. Shulman, R. G.; Eisenberger, P.; Blumberg, W. E.; Stombaugh, N. A., Determination of the iron-sulfur distances in rubredoxin by x-ray absorption spectroscopy *Proceedings of the National Academy of Sciences* 1975, *72*, 4003.
35. Lane, R. W.; Ibers, J. A.; Frankel, R. B.; Holm, R. H., Synthetic analogs of active sites of iron-sulfur proteins: bis (o-xylyldithiolato) ferrate (III) monoanion, a structurally unconstrained model for the rubredoxin Fe-S4 unit *Proceedings of the National Academy of Sciences* 1975, *72*, 2868.
36. Darensbourg, M. Y.; Lyon, E. J.; Zhao, X.; Georgakaki, I. P., The organometallic active site of [Fe]hydrogenase: Models and entatic states *Proceedings of the National Academy of Sciences* 2003, *100*, 3683.
37. Bushnell, G. W.; Louie, G. V.; Brayer, G. D., High-resolution three-dimensional structure of horse heart cytochrome c *Journal of Molecular Biology* 1990, *214*, 585–595.
38. Bönisch, H.; Schmidt, C. L.; Schäfer, G.; Ladenstein, R., The Structure of the Soluble Domain of an Archaeal Rieske Iron–Sulfur Protein at 1.1Å Resolution *Journal of Molecular Biology* 2002, *319*, 791–805.
39. Fee, J. A.; Findling, K. L.; Yoshida, T.; Hille, R.; Tarr, G. E.; Hearshen, D. O.; Dunham, W. R.; Day, E. P.; Kent, T. A.; Münck, E., Purification and characterization of the Rieske iron-sulfur protein from Thermus thermophilus. Evidence for a [2Fe-2S] cluster having non-cysteine ligands *Journal of Biological Chemistry* 1984, *259*, 124–133.
40. Bertini, I.; Campos, A. P.; Teixeira, M.; Luchinat, C., A mössbauer investigation of oxidized Fe4S4 HiPIP II from Ectothiorohodospira halophila *Journal of Inorganic Biochemistry* 1993, *52*, 227–234.
41. Middleton, P.; Dickson, D. P. E.; Johnson, C. E.; Rush, J. D., Interpretation of the Mössbauer Spectra of the High-Potential Iron Protein from Chromatium *European Journal of Biochemistry* 1980, *104*, 289–296.
42. Bartsch, R. G., [35] Purification of (4Fe-4S)1—2– ferrodoxins (high-potential iron-sulfur proteins) from bacteria. In *Methods in Enzymology*, Fleischer, S.; Packer, L., Eds. Academic Press: 1978; Vol. 53, pp 329–340.
43. Glaser, T.; Bertini, I.; Moura, J. J. G.; Hedman, B.; Hodgson, K. O.; Solomon, E. I., Protein Effects on the Electronic Structure of the [Fe4S4]2+ Cluster in Ferredoxin and HiPIP *Journal of the American Chemical Society* 2001, *123*, 4859–4860.
44. Zhang, W.; Xu, C.; Bian, C.; Tempel, W.; Crombet, L.; MacKenzie, F.; Min, J.; Liu, Z.; Qi, C., Crystal structure of the Cys2His2-type zinc finger domain of human DPF2 *Biochemical and Biophysical Research Communications* 2011, *413*, 58–61.
45. Pavletich, N. P.; Pabo, C. O., Zinc finger-DNA recognition: crystal structure of a Zif268-DNA complex at 2.1 A *Science* 1991, *252*, 809.
46. Johnstone, T. C.; Nolan, E. M., Determination of the Molecular Structures of Ferric Enterobactin and Ferric Enantioenterobactin Using Racemic Crystallography *Journal of the American Chemical Society* 2017, *139*, 15245–15250.

47. Raymond, K. N.; Dertz, E. A.; Kim, S. S., Enterobactin: An archetype for microbial iron transport *Proceedings of the National Academy of Sciences* 2003, *100*, 3584.
48. Goetz, D. H.; Holmes, M. A.; Borregaard, N.; Bluhm, M. E.; Raymond, K. N.; Strong, R. K., The Neutrophil Lipocalin NGAL Is a Bacteriostatic Agent that Interferes with Siderophore-Mediated Iron Acquisition *Molecular Cell* 2002, *10*, 1033–1043.
49. Wang, M.; Lai, T. P.; Wang, L.; Zhang, H.; Yang, N.; Sadler, P. J.; Sun, H., "Anion clamp" allows flexible protein to impose coordination geometry on metal ions *Chemical Communications* 2015, *51*, 7867–7870.
50. Liu, X.; Jin, W.; Theil, E. C., Opening protein pores with chaotropes enhances Fe reduction and chelation of Fe from the ferritin biomineral *Proceedings of the National Academy of Sciences* 2003, *100*, 3653.
51. Takagi, H.; Shi, D.; Ha, Y.; Allewell, N. M.; Theil, E. C., Localized Unfolding at the Junction of Three Ferritin Subunits: A mechanism for iron release? *Journal of Biological Chemistry* 1998, *273*, 18685–18688.
52. Trikha, J.; Theil, E. C.; Allewell, N. M., High Resolution Crystal Structures of Amphibian Red-Cell L Ferritin: Potential Roles for Structural Plasticity and Solvation in Function *Journal of Molecular Biology* 1995, *248*, 949–967.
53. Puig, S.; Thiele, D. J., Molecular mechanisms of copper uptake and distribution *Current Opinion in Chemical Biology* 2002, *6*, 171–180.
54. Lamb, A. L.; Wernimont, A. K.; Pufahl, R. A.; Culotta, V. C.; O'Halloran, T. V.; Rosenzweig, A. C., Crystal structure of the copper chaperone for superoxide dismutase *Nature Structural Biology* 1999, *6*, 724–729.
55. Bertini, I.; Mangani, S.; Viezzoli, M. S., Structure and Properties of Copper-Zinc Superoxide Dismutases *Adv. Inorg. Chem.* 1998, *45*, 127–250.
56. Lamb, A. L.; Torres, A. S.; O'Halloran, T. V.; Rosenzweig, A. C., Heterodimeric structure of superoxide dismutase in complex with its metallochaperone *Nature Structural Biology* 2001, *8*, 751–755.
57. Kendrew, J. C.; Bodo, G.; Dintzis, H. M.; Parrish, R. G.; Wyckoff, H.; Phillips, D. C., A Three-Dimensional Model of the Myoglobin Molecule Obtained by X-Ray Analysis *Nature* 1958, *181*, 662–666.
58. Perutz, M. F.; Bolton, W.; Diamond, R.; Muirhead, H.; Watson, H. C., Structure of Hæmoglobin: An X-ray Examination of Reduced Horse Hæmoglobin *Nature* 1964, *203*, 687–690.
59. Weiss, J. J., Nature of the Iron–Oxygen Bond in Oxyhæmoglobin *Nature* 1964, *202*, 83–84.
60. Pauling, L., Nature of the Iron–Oxygen Bond in Oxyhæmoglobin *Nature* 1964, *203*, 182–183.
61. Weiss, J. J., Nature of the Iron–Oxygen Bond in Oxyhæmoglobin *Nature* 1964, *202*, 183.
62. Chen, H.; Ikeda-Saito, M.; Shaik, S., Nature of the Fe–O_2 Bonding in Oxy-Myoglobin: Effect of the Protein *Journal of the American Chemical Society* 2008, *130*, 14778–14790.
63. Eaton, W. A.; Henry, E. R.; Hofrichter, J.; Mozzarelli, A., Is cooperative oxygen binding by hemoglobin really understood? *Nature Structural Biology* 1999, *6*, 351.
64. Shaanan, B., Structure of human oxyhaemoglobin at 2·1resolution *Journal of Molecular Biology* 1983, *171*, 31–59.
65. Fermi, G.; Perutz, M. F.; Shaanan, B.; Fourme, R., The crystal structure of human deoxyhaemoglobin at 1.74 Å resolution *Journal of Molecular Biology* 1984, *175*, 159–174.
66. Hodgkin, D. C.; Kamper, J.; Mackay, M.; Pickworth, J.; Trueblood, K. N.; White, J. G., Structure of Vitamin B12 *Nature* 1956, *178*, 64.
67. Lenhert, P. G.; Hodgkin, D. C., Structure of the 5,6-Dimethylbenzimidazolylcobamide Coenzyme *Nature* 1961, *192*, 937.
68. Banerjee, R., Radical Carbon Skeleton Rearrangements: Catalysis by Coenzyme B12-Dependent Mutases *Chemical Reviews* 2003, *103*, 2083–2094.
69. Toraya, T., Radical Catalysis in Coenzyme B12-Dependent Isomerization (Eliminating) Reactions *Chemical Reviews* 2003, *103*, 2095–2128.
70. Yachandra, V. K.; Sauer, K.; Klein, M. P., Manganese Cluster in Photosynthesis: Where Plants Oxidize Water to Dioxygen *Chemical Reviews* 1996, *96*, 2927–2950.
71. Umena, Y.; Kawakami, K.; Shen, J.-R.; Kamiya, N., Crystal structure of oxygen-evolving photosystem II at a resolution of 1.9 Å *Nature* 2011, *473*, 55–60.
72. Yano, J.; Yachandra, V., Mn_4Ca Cluster in Photosynthesis: Where and How Water is Oxidized to Dioxygen *Chemical Reviews* 2014, *114*, 4175–4205.
73. Kok, B.; Forbush, B.; McGloin, M., Cooperation of charges in photosynthetic O_2 evolution–I. A linear four step mechanism *Photochemistry and Photobiology* 1970, *11*, 457–475.
74. Yano, J.; Yachandra, V. K., Where Water Is Oxidized to Dioxygen: Structure of the Photosynthetic Mn_4Ca Cluster from X-ray Spectroscopy *Inorganic Chemistry* 2008, *47*, 1711–1726.
75. Reece, S. Y.; Hamel, J. A.; Sung, K.; Jarvi, T. D.; Esswein, A. J.; Pijpers, J. J. H.; Nocera, D. G., Wireless Solar Water Splitting Using Silicon-Based Semiconductors and Earth-Abundant Catalysts *Science* 2011, *334*, 645.
76. Rosenzweig, A. C.; Lippard, S. J., Determining the Structure of a Hydroxylase Enzyme That Catalyzes the Conversion of Methane to Methanol in Methanotrophic Bacteria *Accounts of Chemical Research* 1994, *27*, 229–236.
77. Wallar, B. J.; Lipscomb, J. D., Dioxygen Activation by Enzymes Containing Binuclear Non-Heme Iron Clusters *Chemical Reviews* 1996, *96*, 2625–2658.
78. Merkx, M.; Kopp, D. A.; Sazinsky, M. H.; Blazyk, J. L.; Müller, J.; Lippard, S. J., Dioxygen Activation and Methane Hydroxylation by Soluble Methane Monooxygenase: A Tale of Two Irons and Three Proteins *Angewandte Chemie International Edition* 2001, *40*, 2782–2807.
79. Balasubramanian, R.; Smith, S. M.; Rawat, S.; Yatsunyk, L. A.; Stemmler, T. L.; Rosenzweig, A. C., Oxidation of methane by a biological dicopper centre *Nature* 2010, *465*, 115.
80. Tinberg, C. E.; Lippard, S. J., Dioxygen Activation in Soluble Methane Monooxygenase *Accounts of Chemical Research* 2011, *44*, 280–288.
81. Nichol, T.; Murrell, J. C.; Smith, T. J., Controlling the Activities of the Diiron Centre in Bacterial Monooxygenases: Lessons from Mutagenesis and Biodiversity *European Journal of Inorganic Chemistry* 2015, *2015*, 3419–3431.
82. Wang, V. C. C.; Maji, S.; Chen, P. P. Y.; Lee, H. K.; Yu, S. S. F.; Chan, S. I., Alkane Oxidation: Methane Monooxygenases, Related Enzymes, and Their Biomimetics *Chemical Reviews* 2017, *117*, 8574–8621.
83. Banerjee, R.; Jones, J. C.; Lipscomb, J. D., Soluble Methane Monooxygenase *Annual Review of Biochemistry* 2019, *88*, 409–431.
84. Rosenzweig, A. C.; Brandstetter, H.; Whittington, D. A.; Nordlund, P.; Lippard, S. J.; Frederick, C. A., Crystal structures of the methane monooxygenase hydroxylase from Methylococcus capsulatus (Bath): Implications for substrate gating and component interactions *Proteins: Structure, Function, and Bioinformatics* 1997, *29*, 141–152.
85. Lee, S. C.; Holm, R. H., Speculative synthetic chemistry and the nitrogenase problem *Proceedings of the National Academy of Sciences* 2003, *100*, 3595.
86. Hoffman, B. M.; Lukoyanov, D.; Yang, Z.-Y.; Dean, D. R.; Seefeldt, L. C., Mechanism of Nitrogen Fixation by Nitrogenase: The Next Stage *Chemical Reviews* 2014, *114*, 4041–4062.
87. Hu, Y.; Ribbe, M. W., A journey into the active center of nitrogenase *JBIC Journal of Biological Inorganic Chemistry* 2014, *19*, 731–736.
88. Hu, Y.; Ribbe, M. W., Nitrogenases—A Tale of Carbon Atom(s) *Angewandte Chemie International Edition* 2016, *55*, 8216–8226.
89. Ribbe, M. W.; Hu, Y.; Hodgson, K. O.; Hedman, B., Biosynthesis of Nitrogenase Metalloclusters *Chemical Reviews* 2014, *114*, 4063–4080.
90. Einsle, O., Nitrogenase FeMo cofactor: an atomic structure in three simple steps *J. Biol. Inorg. Chem.* 2014, *19*, 737–745.
91. Spatzal, T.; Aksoyoglu, M.; Zhang, L.; Andrade, S. L. A.; Schleicher, E.; Weber, S.; Rees, D. C.; Einsle, O., Evidence for Interstitial Carbon in Nitrogenase FeMo Cofactor *Science* 2011, *334*, 940.
92. Schmid, B.; Einsle, O.; Chiu, H.-J.; Willing, A.; Yoshida, M.; Howard, J. B.; Rees, D. C., Biochemical and Structural Characterization of the Cross-Linked Complex of Nitrogenase: Comparison to the ADP-AlF4--Stabilized Structure *Biochemistry* 2002, *41*, 15557–15565.
93. Peters, J. W.; Stowell, M. H. B.; Soltis, S. M.; Finnegan, M. G.; Johnson, M. K.; Rees, D. C., Redox-Dependent Structural Changes in the Nitrogenase P-Cluster *Biochemistry* 1997, *36*, 1181–1187.
94. Einsle, O.; Tezcan, F. A.; Andrade, S. L. A.; Schmid, B.; Yoshida, M.; Howard, J. B.; Rees, D. C., Nitrogenase MoFe-Protein at 1.16 Å Resolution: A Central Ligand in the FeMo-Cofactor *Science* 2002, *297*, 1696.
95. Kim, J.; Woo, D.; Rees, D. C., X-ray crystal structure of the nitrogenase molybdenum-iron protein from Clostridium pasteurianum at 3.0-.ANG. resolution *Biochemistry* 1993, *32*, 7104–7115.
96. Hoeke, V.; Tociu, L.; Case, D. A.; Seefeldt, L. C.; Raugei, S.; Hoffman, B. M., High-Resolution ENDOR Spectroscopy Combined with Quantum Chemical Calculations Reveals the Structure of Nitrogenase Janus Intermediate E4(4H) *Journal of the American Chemical Society* 2019.
97. Tezcan, F. A.; Kaiser, J. T.; Mustafi, D.; Walton, M. Y.; Howard, J. B.; Rees, D. C., Nitrogenase Complexes: Multiple Docking Sites for a Nucleotide Switch Protein *Science* 2005, *309*, 1377.
98. Tezcan, F. A.; Crane, B. R.; Winkler, J. R.; Gray, H. B., Electron tunneling in protein crystals *Proceedings of the National Academy of Sciences* 2001, *98*, 5002.
99. Morrison, C. N.; Hoy, J. A.; Zhang, L.; Einsle, O.; Rees, D. C., Substrate pathways in the nitrogenase MoFe protein by experimental identification of small molecule binding sites *Biochemistry* 2015, *54*, 2052–2060.

CHAPTER 21

f-Element Chemistry

"I was a 28-year-old kid, and I didn't stop to ruminate about it. I didn't think, 'My God, we've changed the history of the world!'"

— **Glenn Seaborg on the discovery of plutonium**

"The rare earth elements perplex us in our researches, baffle us in our speculations, and haunt us in our very dreams. They stretch like an unknown sea before us mocking, mystifying and murmuring strange revelations and possibilities."

— **William Crookes**

"Benjamin: Come on, Mr. Simpson, you'll never pass this course without learning the periodic table.
Homer: I'll write it on my hand.
Benjamin: Hoh! Including all the known lanthanides and actinides? Good luck!

— **The Simpsons**

CONCEPT COVERAGE

1. **Radii and energies of the orbitals for the elements in the *f*-block**
2. **Abundance and synthesis of the elements**
3. **Lanthanide redox activity**
4. **Lanthanide covalency**
5. **Actinide chemistry**

The *f*-block, like the *s*-, *p*-, and *d*-blocks, has its own chemistry that is only partially informed by the science defining the other blocks of the periodic table. This book has thus far focused on the theories that describe the chemistry and properties of the *s*-, *p*-, and *d*-blocks, and it is these areas that the training of inorganic chemistry is usually focused. Unfortunately, it is not uncommon for those that venture into *f*-block chemistry to find their training, often largely based on *d*-block chemistry, to be partially if not wholly inadequate. For this inadequacy one cannot fault the chemist's instructors. The fact is that the theories to understand the chemistry of the actinides and lanthanides are not as developed as for the other regions of the periodic table.

In this chapter we will attempt to describe the trends within, and will present examples of, *f*-block reactivity and properties. The field is changing rapidly, however. Not long ago, people working in the field often espoused the idea that the bonding in this area of the periodic table was entirely electrostatic. Now, the new trend in the field seems to be the discussion of how much covalency is involved in bonding.

Lanthanides have numerous applications in magnets, imaging agents, catalysis, and so on. There are some applications where people encounter lanthanides on a regular basis, like the coating inside the oven in your kitchen. Until recently it was thought that the lanthanides basically existed in the +3 oxidation state, with very few exceptions. Now, the +2 oxidation state is known for virtually all of the lanthanides, often with what seems to be *d*-orbital occupation, as will be explained. Furthermore, there are some lanthanides that access the +4 oxidation state.

The actinides are the bottom of the periodic table, as arranged by Seaborg. They have relatively few direct applications to most people's daily lives as they are all radioactive. Their impact on the economy largely comes in their use in nuclear power. Actinide chemistry is

Lanthanum 57 La 138.90547(7)	Cerium 58 Ce 140.116(1)	Praseodymium 59 Pr 140.90766(2)	Neodymium 60 Nd 144.242(3)	Promethium 61 Pm [144.91]	Samarium 62 Sm 150.36(2)	Europium 63 Eu 151.964(1)	Gadolinium 64 Gd 157.25(3)	Terbium 65 Tb 158.92535(2)	Dysprosium 66 Dy 162.500(1)	Holmium 67 Ho 164.93032(2)	Erbium 68 Er 167.259(3)	Thulium 69 Tm 168.93421(2)	Ytterbium 70 Yb 173.054(5)
Actinium 89 Ac [227.03]	Thorium 90 Th 232.03806(2)	Protactinium 91 Pa 231.03588(2)	Uranium 92 U 238.0281(3)	Neptunium 93 Np [237.05]	Plutonium 94 Pu [244.06]	Americium 95 Am [243.06]	Curium 96 Cm [247.07]	Berkelium 97 Bk [247.07]	Californium 98 Cf [251.08]	Einsteinium 99 Es [252.08]	Fermium 100 Fm [257.10]	Mendelevium 101 Md [258.10]	Nobelium 102 No [259.10]

studied, not only for academic interest, but also to better understand their handling and chemical properties. Many of the actinides are strategic elements in defense and power generation, and, as a consequence, they are handled and stored often. In the case of the transuranic elements that do not occur naturally, we do not even have geological precedent for their chemical behavior, and understanding their chemistry is important for safe storage and handling. Furthermore, these elements may have much to teach about the entire periodic table as these are systems involving the extremes of relativistic, primogenic, and other effects discussed throughout this book.

As a quick note on nomenclature, there is a campaign by IUPAC to change the name of the "lanthanides and actinides" to "lanthanoids and actinoids". This was suggested to avoid confusion due to the "ide" ending typically being used for anions, such as chloride. However, the authors hear few in the field using these names, perhaps because they have never actually seen students confused by the ending. Nevertheless, some books and papers use "lanthanoid" and "actinoid", and one should be aware that this is simply an alternative name for these elements.

21.1 Orbital Energy, Shape, and Extension in the *f*-Block Elements

One of the major arguments among researchers in the field of *f*-block chemistry, and a major focus of the field, is in understanding the covalent and ionic character of the bonding in these elements. In fact, in some respects this will be a theme throughout this chapter. The degree of ionic character depends on the orbital sizes, their overlap with ligands, energies of the electrons in those orbitals relative to ligand orbitals, and so on. It is important to understand this as the energetic "rules" are a little different for covalent systems and purely ionic ones. In covalent systems, the geometry of the complex, for example, is determined by the nature of hybrids on the central atom and maximizing bonding interactions. In ionic systems, such as a potassium cation surrounded by water molecules, the angles are typically less well defined and simply determined by ligand–ligand interactions. As a result, very polar systems may have many more interactions to the same atom, such as the 16 Cs–F interactions in $Cs\{H_2N\text{-}B_2(C_6F_5)_6\}$ in the solid state.[1] The number and geometry of bonds in a very ionic system seems more determined by the strength of the interaction and the sterics of the groups than factors such as orbital makeup and hybridizations.

There is little doubt that the *f*-block elements form bonds to ligands that are typically less covalent than their transition-metal counterparts, but some of the argument, at one point, was whether there was *any* significant degree of covalent character in the bonds. Now, for the actinides, it seems that most in the field would agree that there is some covalency in the bonding. In the lanthanides, the amount of covalent character in the bonds is probably less than in the actinides and seems highly dependent on the ligand set. Another bonding controversy centered around the degree of *f*-orbital (see Appendix L for the f-orbital shapes) participation, and currently it is perhaps accepted by the community at large that the actinides can have significant *f*-orbital participation in bonds, while the lanthanides generally have less. The reasons behind these observations will now be discussed, but this requires numerous charts to present the simple points just discussed. If one wants to avoid the details and skip to Section 21.2 on abundance and occurrence of the elements, the following are the main points from the discussion.

(1) The covalency of bonding in the actinides is generally higher than the lanthanides. The lanthanides are often regarded as having largely ionic bonding.
(2) The participation of the *f*-orbitals in bonding is much higher for the actinides than the lanthanides in most cases.
(3) The *f*-orbitals are significantly smaller than some of the core orbitals for the lanthanides, making the participation of the *f*-orbitals in bonding very small. The *f*-orbital of the actinides are primogenically expanded to allow better involvement.
(4) The orbital energies of the lanthanides and actinides are higher than electronegative elements, but are not much higher than a typical transition element, such as molybdenum.
(5) The *f*-orbital energies for the actinides are higher for the left, naturally occurring side of the series, and drop significantly as one moves to the right, which may make the heavier actinides more ionic in their bonding.
(6) Because of accessibility, uranium and thorium compounds are far more common than the rest of the actinides.

21.1.1 The Sizes of the Orbitals in the *f*-Block Elements

The sizes of orbitals have been discussed a great deal with regards to hybridization and overlap when discussing the main group elements. Orbital size is of no less importance in other areas of the periodic table, and the sizes of the orbitals in the *f*-block seem to be quite important in determining chemical properties.

21.1.1.1 The Sizes of the Orbitals in the Lanthanides

The *f*-orbitals of the lanthanides can contract towards the nucleus because they are not restricted to be orthogonal to any lower-lying orbital set of the same angular momentum; that is, they have no primogenic repulsion (Section 8.2.1). As a result, the 4*f*-orbitals are exceedingly small relative to the other valence orbitals for the lanthanides, such as 5*d*- and 6*s*-orbitals. As shown in Fig. 21.1, the radius of maximum density (R_{max}) for the 4*f*-orbitals are all below 0.5 Å. Importantly, this is well below the size of many of the filled core orbitals, such as the 5*p*-orbitals (line shown in red). Orbitals smaller than the core orbitals will be very difficult to engage in bonding interactions with other atoms due to electron–electron repulsions with the core electrons.

The 6*s*-orbitals for the lanthanides are relatively diffuse at around 2 Å; this large size can lower overlap with many ligands, which are typically from the first row of the *p*-block with orbital sizes ~0.5 Å. A second-row transition element like Mo has *d*-orbital radii of ~0.7 Å and is well suited to overlap with ligands from the first row of the *p*-block. In contrast, the *d*-orbitals of the lanthanides (cyan in Fig. 21.1) are probably the best-suited valence orbitals to bond with incoming ligands at ~1.1 Å. These orbitals are typically empty in the neutral atoms, with the exception of cerium ({Xe}$6s^24f^15d^1$), gadolinium ({Xe}$6s^24f^75d^1$), and lutetium ({Xe}$6s^24f^{14}5d^1$). Also, these *d*-orbitals have orbital maxima that extend beyond the core electrons, making them accessible as acceptors of electron density.

21.1.1.2 The Sizes of the Orbitals in the Actinides

For the actinides, the exceedingly small size of the 4*f*-orbitals means that the 5*f* are also relatively small compared

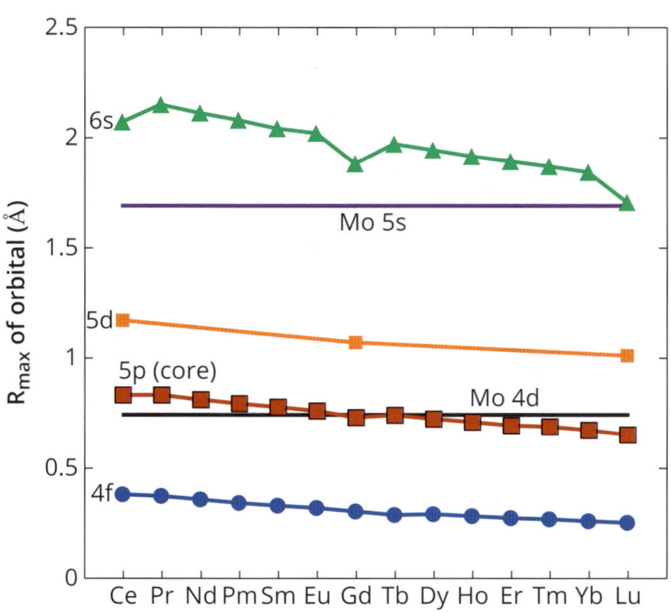

Fig. 21.1 The distance of the maximum density for pertinent orbital radii (R_{max}) of the lanthanides in Å. For comparison, the molybdenum 4*d* and 5*s* radii are also shown.[2]

to other valence orbitals; however, the 5*f*-orbitals are expanded relative to the 4*f*-orbitals of the lanthanides to maintain orthogonality. As a result, the 5*f* radii are only ~0.3 Å smaller than the 6*p* core orbitals (Fig. 21.2), rather than the ~0.45 Å found in the lanthanides (Fig. 21.1). Keep in mind that first-row transition elements often have 3*d*-orbitals similar in size to the core 3*p*-orbitals as well, and obviously these *d*-orbitals still participate in bonding. For example, the iron atom has 3*d* radii of 0.363 Å and 3*p* radii of 0.373 Å. In the case of actinides, the *f*-orbitals can certainly participate in bonding along with the 6*d* manifold, which is well beyond the core electrons and around 1.3 Å in size. The 7*s* valence orbital is accessible, but just as in the lanthanides it is fairly diffuse at around 1.9 Å in size, which probably results in poor overlap with many ligands. Nevertheless, *s*-hybridization with *d*- and *f*-orbitals is probably of importance in some systems.

For reference, the size and breadth of the wavefunctions for a neutral uranium atom are shown to the right in Fig. 21.2. Generally, core orbitals have somewhat narrower wavefunctions, and the 5*f*-orbitals, which are slightly smaller than the core 6*p*, do extend to about the same distance from the nucleus because the valence orbitals have somewhat broader radial functions (compare the solid black line of the 5f to the dotted green of the core 6p).

21.1.2 The Energies of the Orbitals of the *f*-Block Elements

The orbital sizes, which effect overlap, were discussed in the previous section. The other major factor in determining the strength and character of the bonding interaction is the energy difference between the metal and ligand orbitals. Orbitals close in energy with large overlap have a more covalent bonding interaction. If the orbital energies are very different, a more ionic bond is created, in which case the bond energy is determined by the amount of electron density exchanged and the distance (Chapter 3). For the ionic portion of a bond, the bond energy increases with short distance and when larger charge is transferred.

21.1.2.1 The Energies of the Orbitals of the Lanthanides

Fig. 21.3 is a graphic for the binding energy of electrons in the 4*f*, 5*d*, and 6*s* orbitals of the lanthanides. For reference, the binding energy of valence orbitals of nitrogen, the middle of the first-row *p*-block, are also included. If the ligand attaches to the metal at nitrogen, the orbital on the ligand used will be some *sp*-hybrid of energy, typically speaking, between the gray and red lines in the figure. Generally, the *f*-element orbitals are higher in energy than those of the electronegative nitrogen atom, as is to be expected, However, they are not far

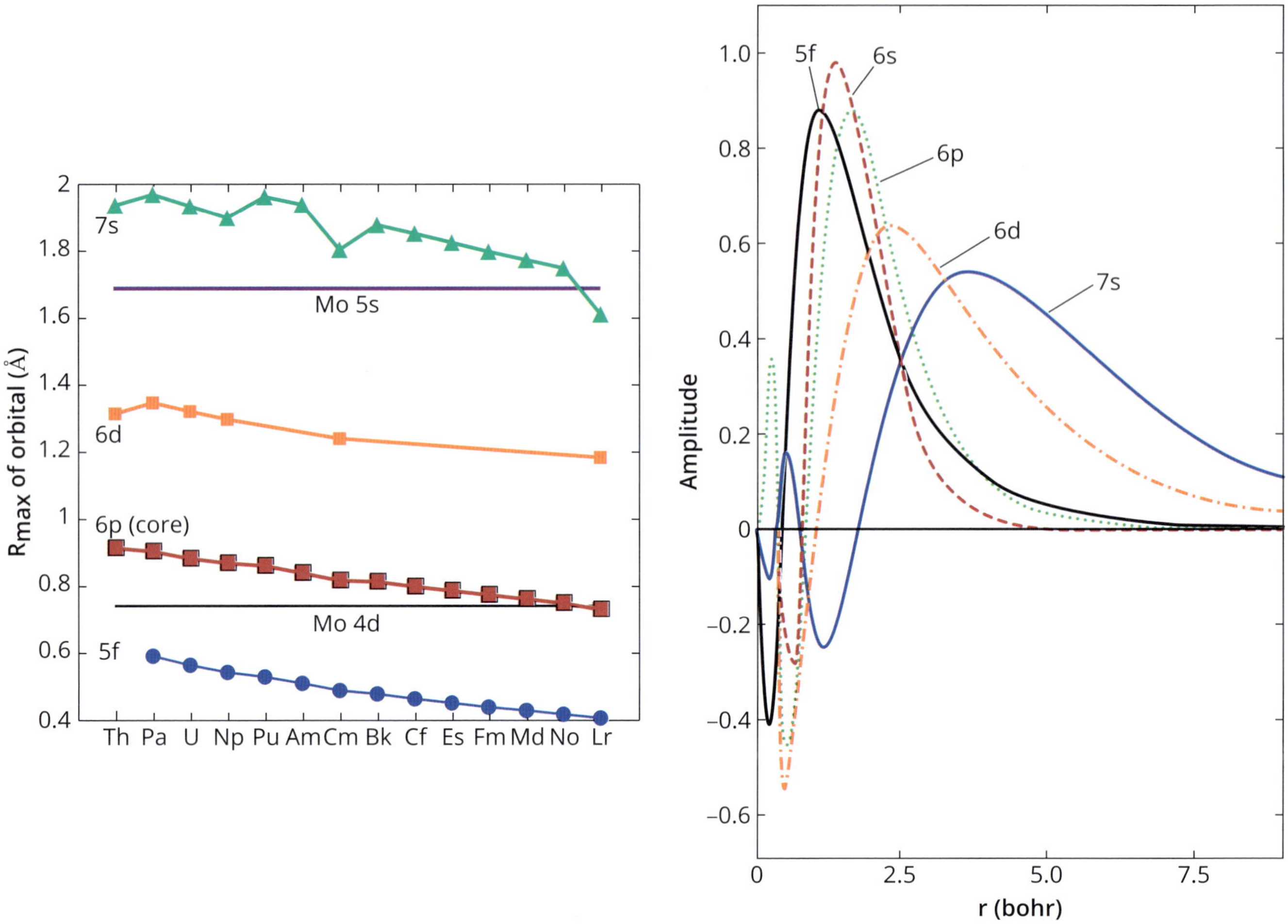

Fig. 21.2 (*left*) The distance of the maximum density for pertinent orbital radii (R_{max}) of the actinides in Å.[2] (*right*) The amplitude of the wavefunctions of neutral uranium versus radius. The *x*-axis uses Bohr radii as the unit.[3]

different from a typical transition metal, such as molybdenum, which is shown for reference.[a]

The molybdenum valence *s*- and *d*-orbitals are similar in energy to the lanthanides (Fig. 21.3), and since the molybdenum-element bonds are considered to be fairly covalent, this suggests that significant covalency in bonding is energetically possible for the lanthanides as well. The 4*f*-orbitals (dark blue) are an even better match to the nitrogen orbitals energetically. However, the limitation of the lanthanides with regards to covalency comes in the form of their orbital sizes and electronic configuration. The *d*- and *s*-orbitals of the lanthanides are generally more diffuse than a typical transition metal. The *f*-orbitals are generally an excellent size to overlap with the 1st row of the *p*-block, but this size leaves them smaller than some core orbitals, which makes the *f*-orbitals difficult to access. However, there is some indication that increasing the oxidation state of the lanthanide may increase *f*-orbital participation (*vide infra*), which is perhaps

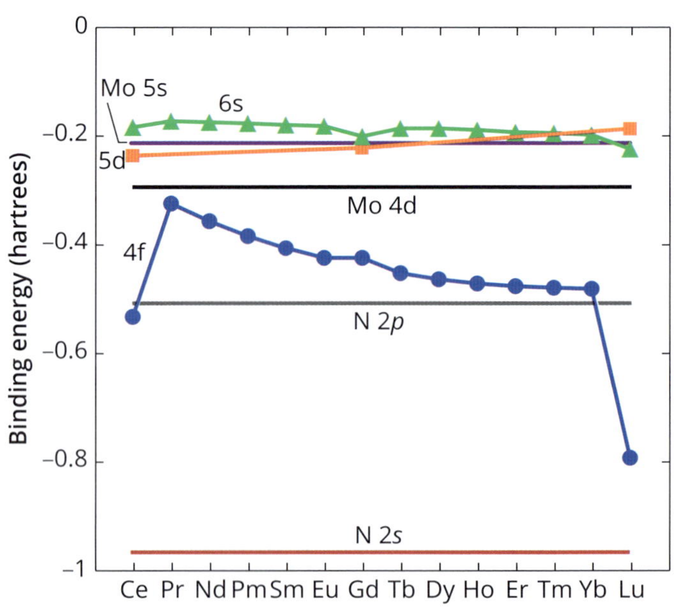

Fig. 21.3 Orbital energies in the lanthanides for the neutral, free atoms. The 2*s*/2*p* orbital energies for nitrogen and 4*d*/5*s* orbitals energies for molybdenum are shown for reference.

[a] Keep in mind that orbital energies change with charge and occupation. The energies given are for the neutral, free atom.

due to increasing *f*-orbital size and/or decreasing core orbital size with increasing Z_{eff}.

While the *f*-orbitals are poorly suited to overlap with incoming ligands due to their protection by the core electrons, they are the lowest-energy valence orbitals. Consequently, the *f*-orbital manifold acts as a vault, into which the metal puts its electron density, where these electrons are largely noninteracting with the environment around the atom. This electron density is accessible through redox processes but typically not through covalent bonding. Since covalent bonding, by definition, requires a sharing of electrons, the *f*-orbitals of the lanthanides are poorly suited to this task. As a result, it is believed that their bonding is by and large ionic in most complexes.

21.1.2.2 The Energies of the Orbitals of the Actinides

The actinide orbital binding energies are shown in Fig. 21.4. As shown, the 6*d*- and 7*s*-orbitals are very similar in energy to those of a typical transition metal such as molybdenum. The 5*f*-orbital energies change dramatically as the nuclear charge increases across the series, which mirrors the general shrinking of the *f*-orbitals across the series but is in some ways more dramatic.

For the actinides, because the *f*-orbitals are more energetically accessible on the left side of the series, it is thought that compounds of, for example, Th and U are more covalent than those further to the right. The smaller and lower-energy *f*-orbitals for the rightmost actinides may lead to their being more lanthanide-like with greater ionicity to their bonding. At the time of this writing there is sparse experimental evidence to support this claim, as the vast majority of actinide chemistry is with uranium and thorium.[4]

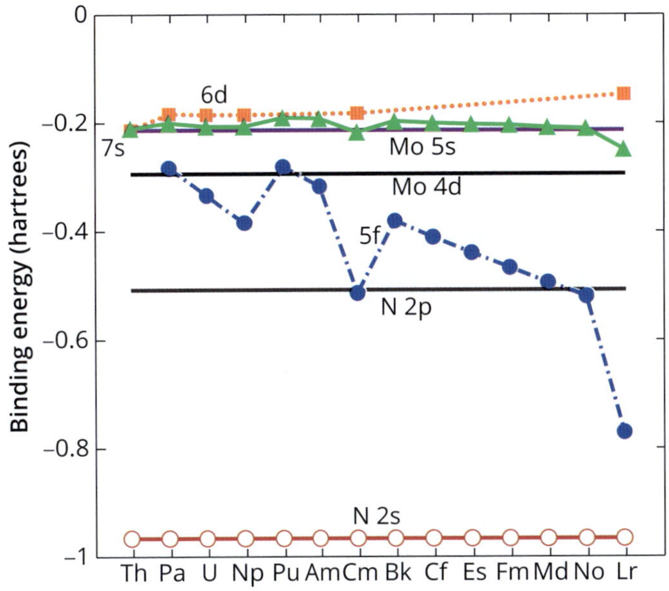

Fig. 21.4 Orbital energies in the actinides for the neutral, free atoms. The *2s/2p* orbital energies for nitrogen and *4d/5s* orbitals energies for molybdenum are shown for reference.

Perhaps this focus on U and Th chemistry in the literature is well-illustrated with Fig. 21.5, which shows the number of

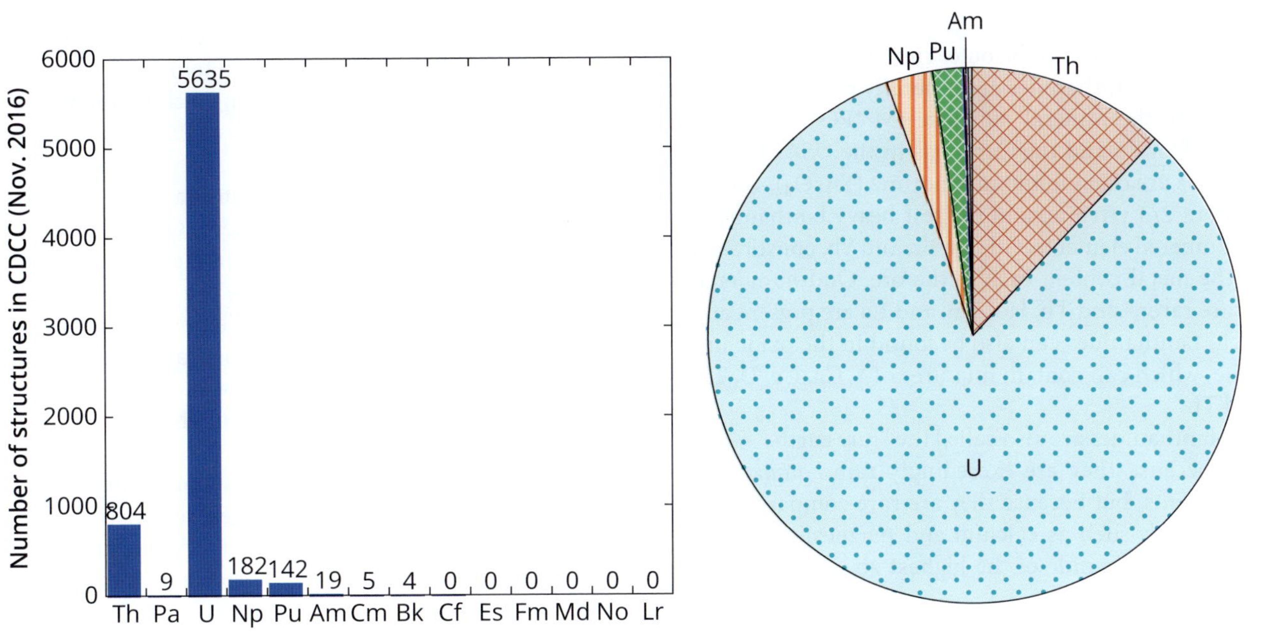

Fig. 21.5 Number of reported structures from X-ray diffraction from the CCSD for the actinide elements as of 2016. The same data were used for the pie chart on the right.

structures in the Cambridge Crystallographic Structural Database (CCSD) for all of the actinides as of November 2016. Out of 6,800 structures deposited, 6,439 (about 95%) are of thorium and uranium complexes.

21.2 Abundance and Preparation of the Elements

When the Earth was initially formed as a molten blob hurtling through space, the elements swirled and mixed. The old adage "like dissolves like" does, in fact, often have predictive power, and these elements collected into mixtures involving elements of similar properties. Because of their electronic structure (*vide supra*), all the lanthanides are relatively similar, and these collected into ribbons of ores containing mixtures of the lanthanide elements that were trapped in this spatial configuration on cooling of the Earth's crust. When these elements have been discovered near the planet's surface—for example, at Ytterby in Sweden—the metals are often found as an intimate mixture of various lanthanides in the same ore.

The lanthanide elements are gathered in this way in minerals because their properties are similar, but some ores do favor early (larger) lanthanides over later (smaller) lanthanides and vice versa, due to changes in atomic radii affecting the materials. Likewise, they are also fairly difficult to separate due to their chemical similarities, but the main property used in their separation is the ionic radius, which decreases gradually from left to right in the series.

Four different lanthanides are named after Ytterby, their place of initial discovery: yttrium, erbium, terbium, ytterbium. Three other elements—holmium, thulium, and gadolinium—were also discovered in the same location. Appropriately, Ytterby, which means "outer village" in Swedish, is quite rich in many of the elements on the outer rim of the periodic table.

The lighter lanthanides are generally more abundant and are found in the mineral monazite ($M^{3+}PO_4^{3-}$) in locations including Florida and the Carolinas in the USA. The heavier lanthanides are often found in xenotime ($YAsO_4$), which contains about 2/3 yttrium, and the rest heavy lanthanides. The even-numbered elements, Gd, Dy, Er, and Yb, constitute about 5% of the metal content each in the ore. Odd-numbered elements are rarer,[b] and Tb, Ho, Tm, and Lu are present at about 1% each. Lanthanides can also be found in the oxide mineral euxenite, which can contain mixtures of Y, Ca, Ce, U, Th, Nb, Ta, and Ti.

The lanthanides are also called the "rare earths"—not necessarily because they are exceedingly rare compared to all other elements, but because they are rare compared to many mineral metals such as calcium and magnesium (Fig. 21.6). Cerium, the lanthanide of the highest abundance, is present

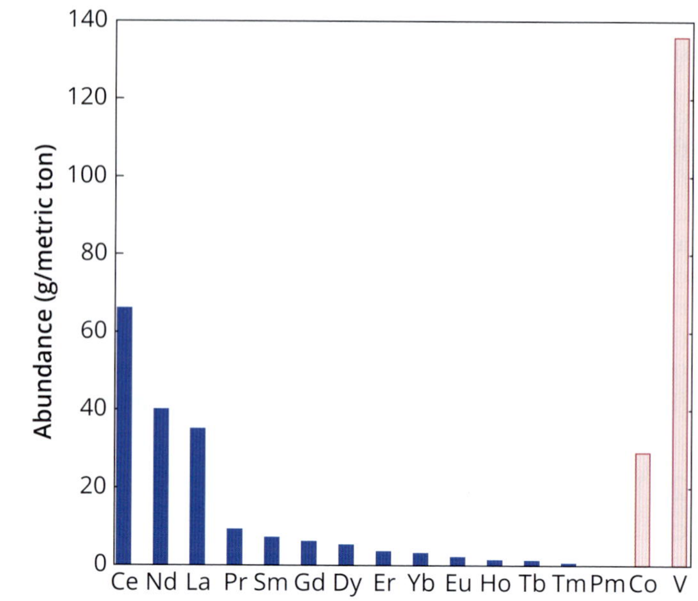

Fig. 21.6 Average abundance of the lanthanides in g/metric ton of crustal rock. The abundances for two first-row transition elements, cobalt and vanadium, are shown for reference. Data from Greenwood and Earnshaw.[5]

in the Earth's crust at an estimated 66 ppm, whereas the least abundant first-row transition-metal elements scandium and cobalt are present at 25 and 29 ppm, respectively. In fact, cerium is more abundant than all the transition metals except those most abundant: Ti, Zr, V, Cr, Mn, Fe, Ni, and Cu. The heavy, odd-numbered thulium (Tm), perhaps the rarest naturally occurring lanthanide,[c] has an abundance of about 0.5 ppm, which is still more prevalent than many transition elements such as Ru, Rh, Pd, Ag, Re, Os, Ir, Pt, and Au. In fact, Tm is more abundant than iodine.

Promethium (Pm) has the distinction of being only one of two elements lighter than uranium to not occur in significant quantities naturally, the other being technetium in the middle of the *d*-block. All of the isotopes of Pm are radioactive with fairly short half-lives.

The actinides are, from an abundance perspective, quite different from the row above them. Only thorium, protactinium, and uranium are found in significant quantities in the Earth's crust, and these were the only actinides known prior to 1940. However, the elements higher in atomic number than uranium (the transuranics, followed by the formally main-group and transition-metal transactinides) now constitute about a quarter of the entries in the periodic table.

Table 21.1 shows the actinides along with their synthetic route and the half-life of the longest-lived isotope or the one often used to study their chemistry. Applications of the elements often involve isotopes less stable than those listed. Californium-252, for example, is useful as a neutron emitter.

[b] This is generally true when looking at similar elements in a series and is called the Oddo–Harkins rule. It is simply a by-product of the nuclear reactions from which elements are made.

[c] This statement excludes radioactive promethium.

Table 21.1 The method of preparation, type of radiation, and half-life of the most stable isotope or commonly used isotope for the actinides.[a]

Most Common Isotope in Chemistry	Half-Life of Most Stable Isotope	Method of Preparation	Notes
^{232}Th	14.5×10^9 y	Naturally occurring.	Longest half-life of any actinide isotope.
^{231}Pa	32,760 y	Naturally occurring decay product of uranium.	Isotopes of Pa are "neutron poisons" that need to be removed from nuclear reactors.
^{238}U	4.5×10^9 y	Naturally occurring.	Weak α-emitter with many commercial applications including military projectiles. The metal is sometimes used to store radioactive materials, even though it is radioactive itself, because it is relatively inexpensive and denser than lead. Dangers are largely from ingestion and radioactivity of daughter isotopes. Perhaps the most studied actinide.
^{237}Np	2.1×10^6 y	Traces detected naturally. Multiple steps from ^{235}U.	First reported by E. M. McMillan and P. Abelson in 1940.
^{239}Pu/^{242}Pu	2.4×10^4 y/ 3.7×10^5 y	Traces detected naturally. Bombardment of ^{238}U with ^{1}n.	^{238}Pu was the first prepared by Seaborg, McMillan, Kennedy, and Wahl in 1940 by bombardment of ^{238}U with ^{2}H. This isotope has a very short half-life of 88 y, however. ^{239}Pu and ^{242}Pu have much longer half-lives and are more common for studying the chemistry of the element.
^{243}Am	7370 y	Bombardment of ^{239}Pu with ^{1}n.	First prepared by Seaborg, James, Morgan, and Ghiorso in 1944.
^{248}Cm	3.4×10^5 y	Decay of ^{252}Cf.	Various Cm isotopes can be prepared, but the most long-lived is often isolated by decay of ^{252}Cf, which has a half-life of 2.65 y. This allows milligram quantities of the element to be used for research purposes each year.
^{247}Bk	1380 y	Bombardment of ^{244}Cf with α.	There have been about twenty isotopes of Bk observed, and the longest-lived is ^{247}Bk, which has a $t_{1/2}$ of only 1,380 y. Very little chemistry has been achieved. A recent rare chemistry study on the element used ^{249}Bk ($t_{1/2}$ = 330 d, β^--emission), which can be made from Cf^{249} by β^--capture.[6]
^{249}Cf	351 y	β^--emission from ^{249}Bk.	^{251}Cf has a longer half-life (898 y) but is more difficult to produce.
^{253}Es	20 d	Bombarding ^{252}Cf with neutrons.	^{252}Es is more stable ($t_{1/2}$ = 472 d) but is more difficult to produce. Very little to nothing is known about its chemistry.
^{252}Fm	25 h	Microgram to picogram quantities have been reported from a few different routes. One such route is a thermonuclear explosion to make milligram quantities.	The most stable isotope is ^{257}Fm ($t_{1/2}$ = 100 d). In one of the few chemistry studies, a USSR team claimed to produce aqueous Fm(II).
^{258}Md	52 d	Bombardment of Es with α-particles.	Two other, less stable isotopes are currently known in ^{257}Md and ^{260}Md.
^{259}No	58 min	Bombardment of ^{248}Cm with oxygen nuclei.	The longest-lived isotope known has $t_{1/2}$ = 58 min. All others are 3 min or less. ^{261}No is predicted to have a $t_{1/2} \sim 170$ min but is currently unknown.
^{266}Lr	10 h	α-decay of ^{105}Db.	Most of the other isotopes have $t_{1/2} < 1$ min, and the element was isolated by fast extraction.

[a] The isotope used by researchers will necessarily depend on the source that they are using to access the elements. The isotope shown is either the most abundant (for Th and U), the one used most often in a survey of the literature at this time, or the most stable currently known isotope.

21.3 Some Selected Chemistry of the Lanthanides

Any discussion of lanthanide chemistry should probably begin with the electronic configuration of the atoms, shown in Table 21.2. According to the Aufbau principle, the order of filling should be *6s* → *4f* → *5d*, and, more often than not, this is what is found. However, there are a few exceptions: namely, La, Ce, and Gd. In a complex, the *s*- and *d*-orbitals, based on the previous discussions, will be those largely involved in bond-formation with incoming ligands. Once the metal is in the +3 oxidation state, the remaining electrons reside in the exceedingly small and energetically stabilized 4*f*-orbitals of the lanthanide. Further increases in formal oxidation state are more difficult. As a result, +3 is the most common formal oxidation state for the entire lanthanide series. The redox states of lanthanides will be discussed in more detail in the next section. Note that "Ln" is often used as an abbreviation for any lanthanide atom, while" La" is the element lanthanum.

Because the *f*-electrons shield the nucleus poorly, the increasing nuclear charge dramatically decreases the ionic radius of the ions. It is common to look at reactivity and structure of new lanthanide complexes with several different ions, all usually in the +3 oxidation state, to study the effect of changing radii.

Consider the effect of metal radius on the intramolecular hydroamination reported by Peter Roesky and coworkers.[8] In their study, they used bis(phosphinimino)methanide ancillary ligands, which in some respects are akin to acetylacetonates with hypervalent phosphines in place of carbonyls (Fig. 21.7). Placing one of these ancillary ligands on the lanthanides is accomplished by reaction with $Ln\{N(SiHMe_2)_2\}_3(THF)_2$, by which one of the basic amide ligands removes an acidic hydrogen from the methylene; this was done with Ln = La, Sm, Ho, Y, and Lu, spanning the lanthanide series. Hydroamination catalysis (Section 18.3.8.3) using the substrate shown in Fig. 21.7 results in clean cyclization, and the rate of reaction is dependent upon the ionic radius of the lanthanide employed with larger metal centers producing faster rates.

Note that the radii used by Roesky and coworkers provide an excellent linear correlation with the rate data as shown;[9] however, the radii shown in Table 21.2 also provide an excellent linear correlation. There are several sets of self-consistent radii tables that could be used. Mixing radii from a variety of different sources rarely produces good results, and one should try to find tables that contain all the radii, found using the same method, that are needed for the correlation.

21.3.1 Redox Chemistry of the Lanthanides[10]

Not many years ago, a discussion of the "redox chemistry of the lanthanides" would have been exceedingly short. As mentioned, the lanthanides tend to favor the +3 oxidation state regardless of *f*-electron count. In fact, in their 1971 book, Pimentel and Sprately comment: "Lanthanum has only one important oxidation state in aqueous solution, the +3 state. With few exceptions, that statement tells the whole boring

Table 21.2 Electronic configurations of the lanthanides and their ionic radii in the +3 formal oxidation state.

Element	Symbol	Electronic Configuration of Ln^0 all have a Xe core	Electronic Configuration of Ln^{3+} all have a Xe core	Ln^{3+} ionic radii[a]
lanthanum	La	$6s^25d^1$	$4f^0$	1.250
cerium	Ce	$6s^24f^15d^1$	$4f^1$	1.220
praseodymium	Pr	$6s^24f^3$	$4f^2$	1.200
neodymium	Nd	$6s^24f^4$	$4f^3$	1.175
promethium	Pm	$6s^24f^5$	$4f^4$	—
samarium	Sm	$6s^24f^6$	$4f^5$	1.140
europium	Eu	$6s^24f^7$	$4f^6$	1.120
gadolinium	Gd	$6s^24f^75d^1$	$4f^7$	1.105
terbium	Tb	$6s^24f^9$	$4f^8$	1.090
dysprosium	Dy	$6s^24f^{10}$	$4f^9$	1.075
holmium	Ho	$6s^24f^{11}$	$4f^{10}$	1.055
erbium	Er	$6s^24f^{12}$	$4f^{11}$	1.040
thulium	Tm	$6s^24f^{13}$	$4f^{12}$	1.025
ytterbium	Yb	$6s^24f^{14}$	$4f^{13}$	1.010
lutetium	Lu	$6s^24f^{14}5d^1$	$4f^{14}$	0.995

[a] Data from D'Angelo and coworkers.[7] Note that "Ln" is often used as an abbreviation for any lanthanide atom, while "La" is the symbol for the element lanthanum.

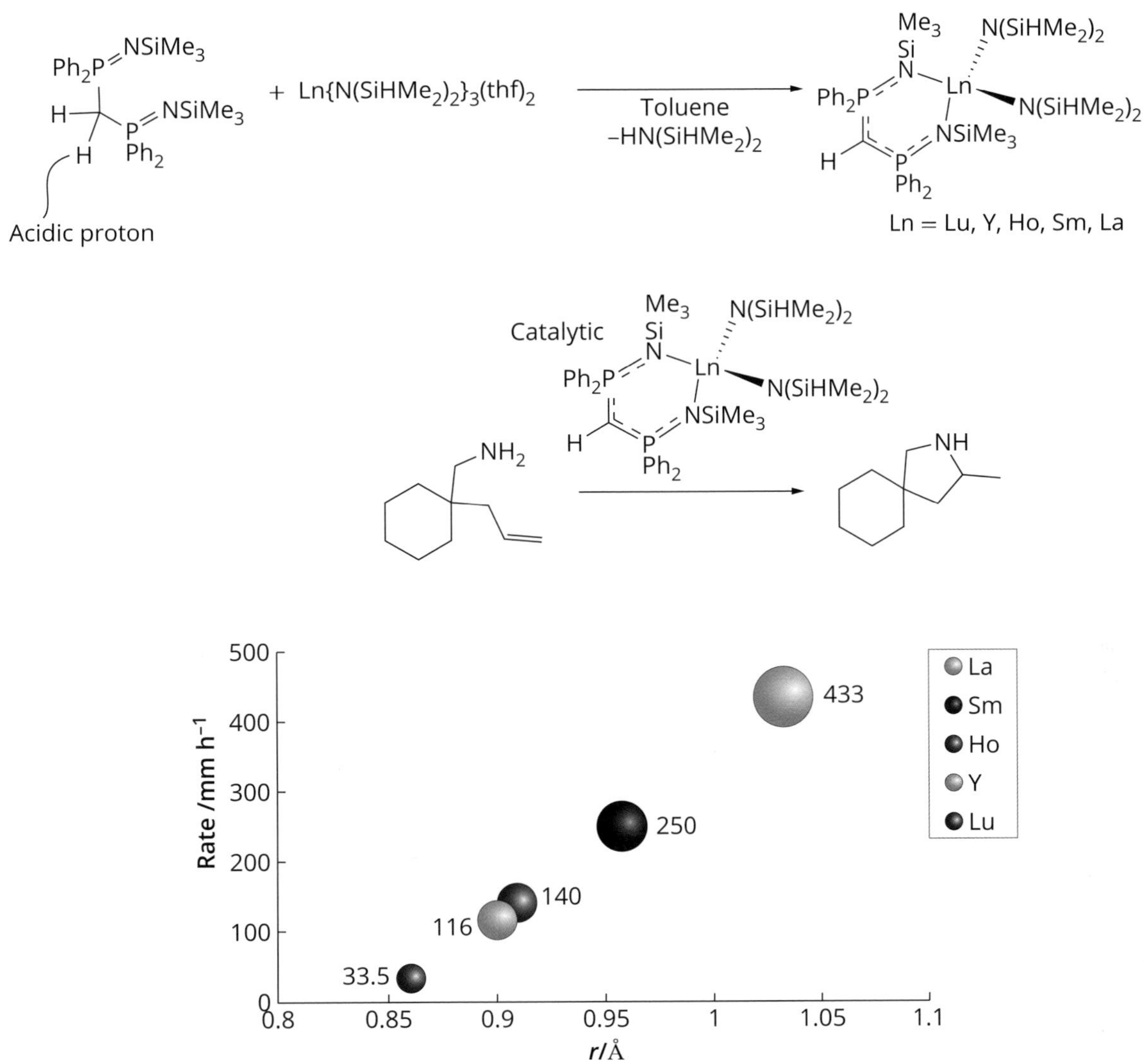

Fig. 21.7 An example of the effect of ionic radius of the lanthanides on reactivity. The isostructural complexes were prepared with five different lanthanides, and the rate of reaction for the cyclization shown is linearly related to the size of the lanthanide, with larger metals producing faster catalysis. (Figure from Roesky's paper.[8])

story about the other fourteen lanthanides."[11] This is by no means true now, especially outside of water! As will be shown shortly, virtually all the lanthanides have multiple known formal oxidation states.

As shown in Fig. 21.8, the lanthanides were almost exclusively known in the +3 formal oxidation state around the turn of the millennium. The only rare-earth elements with known +2 oxidation states were, not coincidentally, those with the smallest negative potentials for reduction: Eu, Yb, and Sm. These Ln^{2+} ions are stabilized by a half-filled shell (Eu^{2+}, f^7), filled-shell (Yb^{2+}, f^{14}), and a nearly-half-filled shell (Sm^{2+}, f^6).

Cerium is unusual in that cerium(IV) is readily accessible due to stabilization of the f^0-configuration for the ion and is a common oxidant in organic and inorganic chemistry. Certainly, cerium(III) compounds are also readily available, such as Ce(III) halides and nitrate. In contrast to the other lanthanides, cerium(IV) compounds are also commercially available, such as cerium(IV) oxides, sulfate, and fluoride. The hexanitrate, $(NH_4)_2Ce(NO_3)_6$, is perhaps the most popular reagent among these and is an inexpensive oxidant.

In all, it appears that at least five lanthanides may form stable complexes in the +4 formal oxidation state: Ce, Pr, Nd, Tb, and Dy. Cerium, as just mentioned, forms many stable complexes and has a reported $Ce^{III/IV}$ redox couple of +1.72 V in water. Praseodymium and terbium have been reported in acidic water to have redox couples of +3.20 and +3.10 V to give the +4 ions, respectively, which is strong enough to oxidize neutral water. Terbium(IV) and praseodymium(IV) carbonates have been reported in aqueous solutions.[12] Furthermore, the compounds of the general formula Cs_3LnF_7 are known, in which Ln = Pr, Nd, Tb, and Dy.

A molecular, fully characterized example of a terbium(IV) complex has also appeared from the Mazzanti group (Fig. 22.9). A salt-elimination reaction of TbI_3 with four

	La	Ce	Pr	Nd	Pm	Sm	Eu	Gd	Tb	Dy	Ho	Er	Tm	Yb	Lu
Valence electrons in element	3	4	5	6	7	8	9	10	11	12	13	14	15	16	17
Known oxidation states		4													
	3	3	3	3	3	3	3	3	3	3	3	3	3	3	3
						2	2							2	

Ln	Potential (V vs SHE)	Ln	Potential (V vs SHE)
Eu	−0.35	Pr	−2.7
Yb	−1.15	Y	−2.8
Sm	−1.55	Ho	−2.9
Tm	−2.3	Er	−3.1
Dy	−2.5	La	−3.1
Nd	−2.6	Ce	−3.2
Pm	−2.6	Tb	−3.7
Lu	−2.7	Gd	−3.9

Fig. 21.8 (*top*) Table of "traditional" oxidation states for the lanthanides, all of which are stable as aqueous solutions. (*bottom*) Estimated (—0.2 V) Ln(III)/Ln(II) standard reduction potentials of the lanthanides versus the standard hydrogen electrode (SHE).[10]

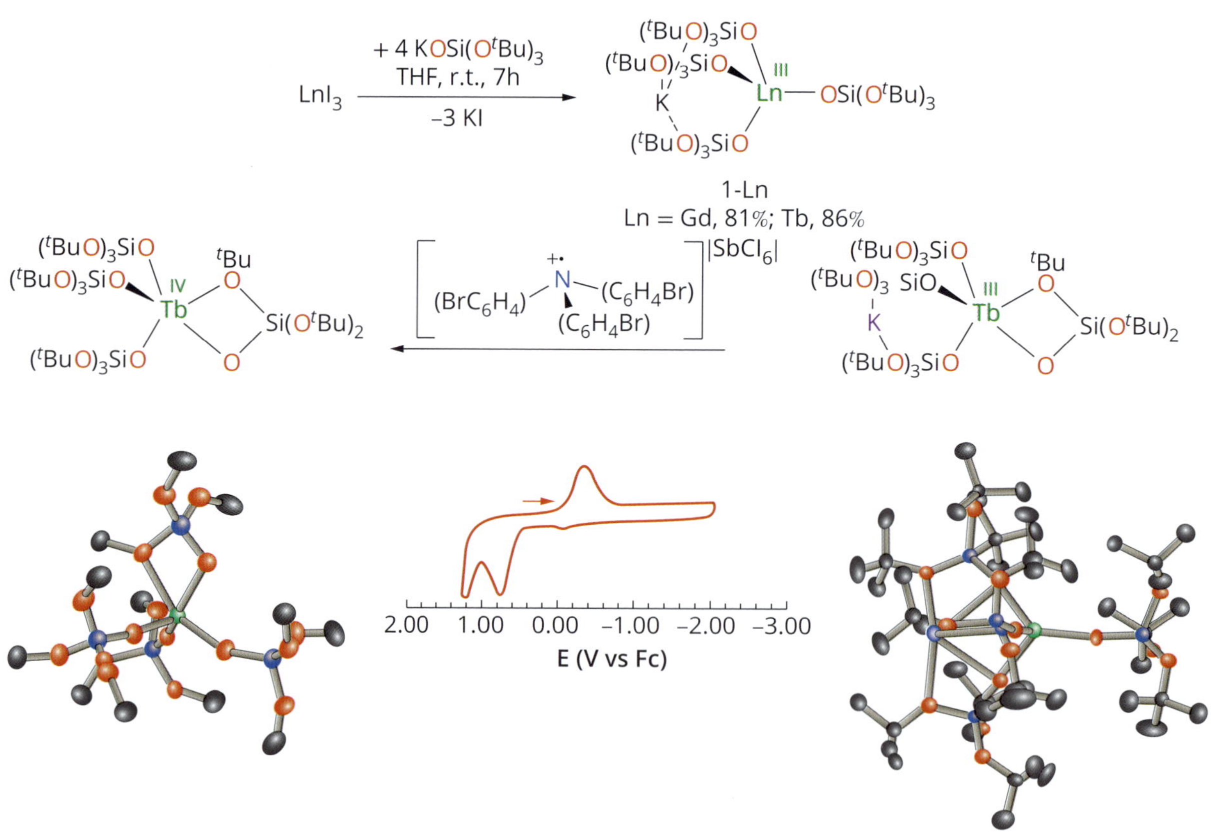

Fig. 21.9 (*top*) The reaction of TbI$_3$ with KOSi(O^{t}Bu)$_3$ produces K{Tb(OSi(O^{t}Bu)$_3$)$_3$}. The potassium salt of the Tb(III) complex can be oxidized chemically by {N(*p*-BrC$_6$H$_4$)$_3$}$^+$ to produce stable Tb{OSi(OBu)$_3$}$_3${κ^2*O*,*O*′- OSi(OBu)$_3$}, which is terbium(IV). The CV versus Fc/Fc$^+$ is shown at the bottom for the irreversible oxidation of Tb(III) to Tb(IV). (Figures from the JACS paper.[13])

equivalents of $KOSi(O^tBu)_3$ produces $K\{Tb(OSi(O^tBu)_3)_3\}$, which can be oxidized chemically or electrochemically to the isolable Tb(IV) complex $Tb\{OSi(OBu)_3\}_3\{\kappa^2O,O'\text{-}OSi(OBu)_3\}$, in which one of the siloxide ligands is now bidentate to the metal. At the time of writing, this is the only fully characterized lanthanide complex in the +4 oxidation state that is not of cerium(IV).[13]

The vast majority of lanthanides produce the +3 materials Ln_2O_3 on heating under O_2; the exceptions are cerium, praseodymium, and terbium. Cerium(IV) oxide, CeO_2, is the most important +4 lanthanide compound industrially. Praseodymium and terbium on heating with O_2 usually produce Pr_6O_{11} and Tb_4O_7, respectively. Heating in pure oxygen under pressure does allow access to PrO_2 and TbO_2.[14]

Recently, the first lanthanides in the +5 oxidation state were reported for praseodymium. Using matrix isolation methods, Pr atoms were found to react with O_2 in a helium matrix to produce PrO_2^+. The PrO_2^+ species was found to be linear and interacted datively with additional oxygen molecules to form complexes like $PrO_2(O_2)_6^+$. In argon, species like $PrO_2(Ar)_5^+$ were observed.[15] Reaction of NO with praseodymium atoms in a neon matrix reportedly provides linear NPrO.[16]

Under the proper circumstances and with the judicious choice of ligands, access to many lanthanide complexes of formal oxidation states greater than +3 seems achievable. Based on the current state of the field, it is expected that stable molecular lanthanide(IV) complexes are likely on the horizon outside of the Tb and Ce species currently known, especially of Pr, Nd, and Dy.

A few lanthanides in the +2 oxidation state are relatively easily accessed, especially compounds of europium, erbium, and samarium. From 1997 to 2001, Bochkarev, Evans, Schumann, and coworkers reported the syntheses and structures of the lanthanide(II) complexes $TmI_2(DME)_3$, $DyI_2(DME)_3$, and $NdI_2(DME)_3$, which were very unusual molecular compounds for these oxidation states. Notice, these are the next three elements with the lowest estimated reduction potentials in Fig. 21.8.

Using tris(cyclopentadienyl)lanthanide derivatives, the majority of the lanthanides are now known in the +2 formal oxidation state as molecular compounds, thanks largely to the efforts of the Evans and Lappert groups. The strategy for the preparation is a relatively simple one (Eq. 21.1); the lanthanide(III) cyclopentadienyl complex is reduced by potassium graphite (KC_8) in the presence of 2.2.2-cryptand under Ar.[d]

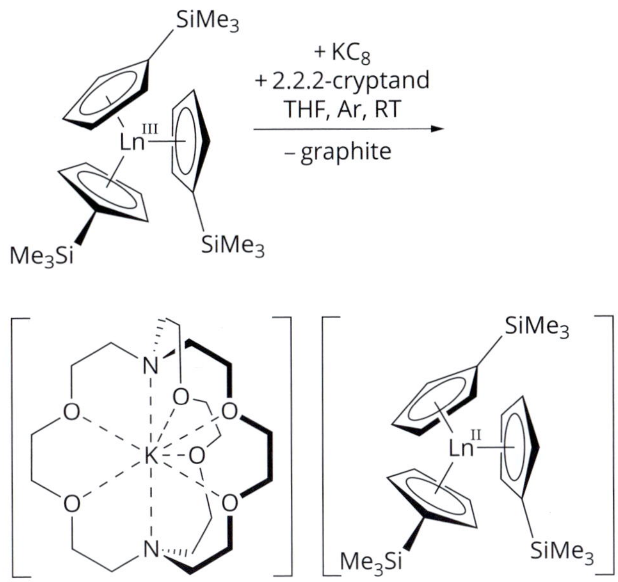

21.1

Using this strategy, virtually all of the Ln(II) ions have now been prepared and characterized. The only exception being radioactive promethium (Table 21.3).

One surprise that was found was that the electronic structures of these +2 lanthanides did not simply have an extra electron in an *f*-orbital compared to their +3 analogs in all cases. For the six systems easiest to reduce from +3 to +2 (Eu, Yb, Sm, Tm, Dy, and Nd), the new configuration on reduction of the Ln(III) simply has another electron in an *f*-orbital; for example, if the Ln^{3+} ion is $4f^n$, then the Ln^{2+} ion was $4f^{n+1}$. For the other lanthanides, the electron added to the +3 configurations do not occupy the 4*f*-manifold but instead goes into a 5*d*-orbital to produce configurations of $4f^n5d^1$. As mentioned previously, the six metals that are easier to reduce have some feature of their electronic structure that stabilizes the +2 oxidation state, such as a half-filled or filled *f*-shell; apparently, if

[d] Many of the Ln(II) complexes are reducing enough to activate dinitrogen, so argon is used as an inert atmosphere.

Table 21.3 Currently known oxidation states of the lanthanides as of 2017.

	La	Ce	Pr	Nd	Pm	Sm	Eu	Gd	Tb	Dy	Ho	Er	Tm	Yb	Lu
Valence electrons in element	3	4	5	6	7	8	9	10	11	12	13	14	15	16	17
Known oxidation states			5												
		4	4	4					4	4					
	3	3	3	3	3	3	3	3	3	3	3	3	3	3	3
	2	2	2	2		2	2	2	2	2	2	2	2	2	2

no such stabilization is present, it is more favorable to add the electron to the *d*-manifold than the *f*.[e]

21.3.2 Separation of the Lanthanides[17]

The properties of the lanthanides are generally quite similar with all in the +3 formal oxidation state and, generally speaking, little redox activity. They are often found as mixtures in the same ores, and their separation can be a challenge. As a result, identification of the lanthanides was quite difficult. For metals that are redox active, such as cerium(IV) and europium(II), they can be effectively separated by conversion to the different oxidation state to more greatly differentiate their properties from the other lanthanides. Otherwise, lanthanide separation depended on the properties in aqueous solutions of the +3 complexes.

The hydroxides of the heavy lanthanides precipitate before the lighter ones on gradual addition of base, which provided a method for separation. In some cases, various salts such as oxalates, sulfates, nitrates, and mixtures of anions were used in crystallizations. Because the ions have slightly different radii, their lattice energies, heats of solvation, and so on, are slightly different. Often, either or both of these procedures were repeated many times to obtain a single lanthanide sample. For example, in 1911 American chemist Charles James carried out 15,000 recrystallizations to obtain pure thulium bromate![5]

Industrially, lanthanides may be separated by extraction of the aqueous nitrates with a solution of P^nBu_3 in kerosene, for example, with the smaller ions binding more tightly. Ion-exchange chromatography with the EDTA complexes is sometimes used as well. Lanthanides of very high purity may be obtained by chromatography.

Because lanthanides are of strategic importance for the production of batteries, magnets, motors, catalysis, and energy-efficient lighting, a report of the US Department of Energy in 2011 identified several lanthanides as "critical" elements for the economy. These key elements for energy technology were Li, Mn, Co, Ni, Ga, Y, In, Te, La, Ce, Pr, Nd, Sm, Eu, Tb, and Dy. In other words, of these sixteen key elements, eight were lanthanides. As a result, new methods for separations and processing of lanthanides are still a very active area of research.

[e] The *d*-orbital occupation in these cases could be due to the ligand geometry, which has three Cp-derivatives in a belt around the metal in the *xy*-plane. This ligand geometry may allow d_{z^2} to be low-energy, leading to its occupation. The field awaits more extensive investigations with ligands other than cyclopentadienyl in these lower oxidation states.

21.3.3 Metal–Ligand Multiple Bonds of the Lanthanides as a Case Study in Covalency

There is perhaps no place that ionicity impacts the chemistry of a metal more than in metal–ligand multiple bonds. In a metal nitride, for example, the limiting covalent and ionic structures could be drawn as shown in Fig. 21.10a. In the purely covalent depiction of a nitride–metal bond, the three bonding electrons of the nitrogen are equally shared. In the purely ionic structure, the eight electrons and −3 charge must be supported by a single atom—a high-energy situation even for an electronegative atom.

As the complex becomes more ionic, the situation becomes closer (although never terribly close) to that exemplified on the right in the figure, and various ionic forms seem to contribute significantly for most lanthanide systems. As a result, the vast majority of known metal–ligand multiple bonds (such as oxo, imido, nitrido, alkylidene, alkylidyne, and so on) with lanthanide metals are typically bridged, which provides another metal atom with the opportunity to help support the electron density in ionic systems. The atom helping to support the electron density on the α-atom in the metal–ligand multiple bond group is often an alkali metal or even a hydrogen bond (Fig. 21.10b).

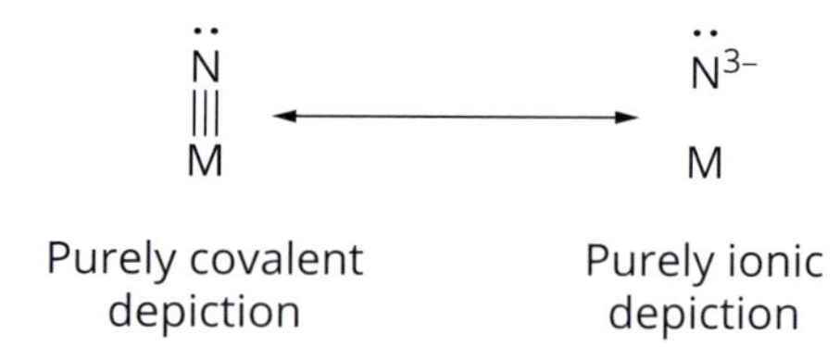

Fig. 21.10a Two extremes of nitride bonding.

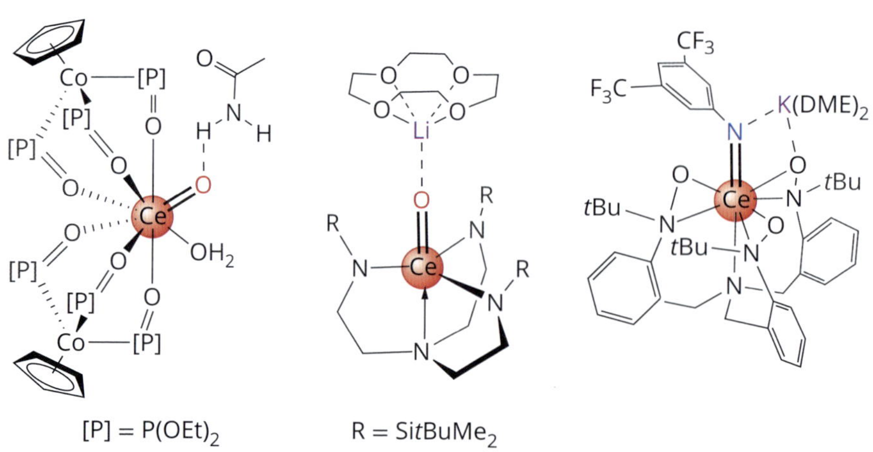

Fig. 21.10b Selected examples of metal–ligand multiple bonds with lanthanides. Note that all these systems have atoms to help support the high negative charge on the α-atom, presumably due to high polarity in the bonding. (Drawing from Schelter and coworker's paper.[18])

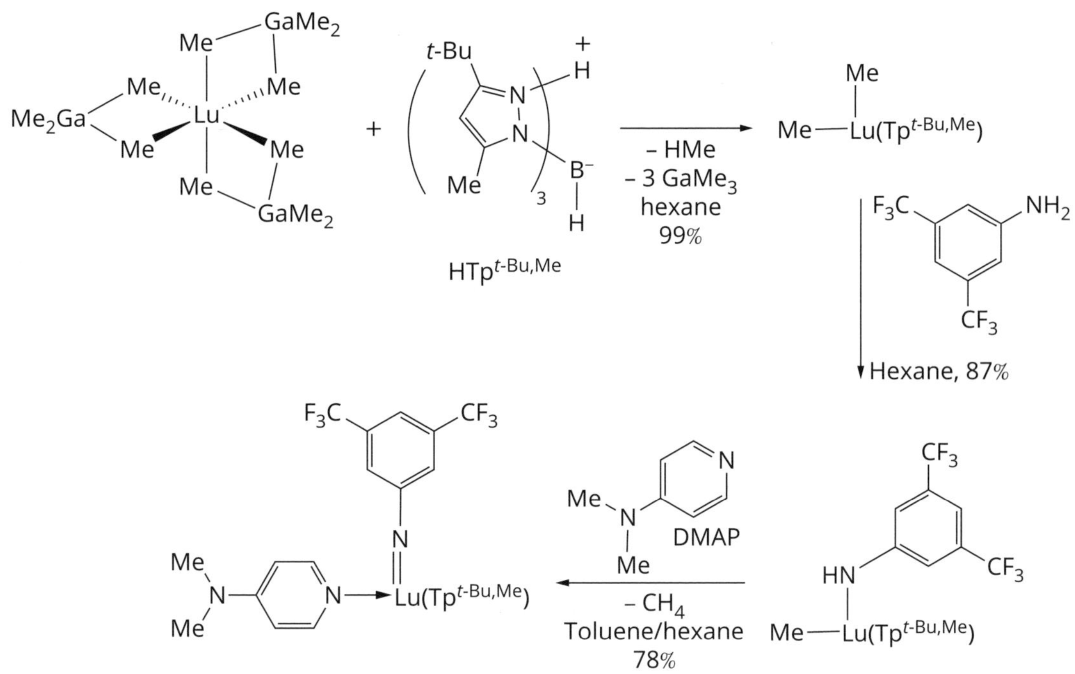

Fig. 21.10c Anwander and coworkers' synthesis of a lutetium terminal imide complex.

In 2015 an early structurally characterized example of a terminal imido of a rare-earth element was published by Anwander and coworkers, with lutetium (Fig. 21.10c).[19] The synthesis involves production of $Lu(GaMe_4)_3$ as a starting material, which has a hexacoordinate lutetium atom with τ-bonded methyl groups. The complex $Lu(GaMe_4)_3$ is prepared from $Li_3\{Lu(NMe_2)_3Cl_3\}$ and excess $GaMe_3$ and acts as a soluble, metastable form of "$LuMe_3$". The trigallate $Lu(GaMe_4)_3$ can be treated with $HTp^{t\text{-Bu,Me}}$ to generate stable $Lu(Tp^{t\text{-Bu,Me}})Me_2$.[20] One equivalent of methane is liberated from the dimethyl on addition of H_2N-3,5-$(CF_3)_2C_6H_3$, which provides the mono(amido) $Lu(Tp^{t\text{-Bu,Me}})Me(HN$-3,5-$(CF_3)_2C_6H_3)$. Addition of DMAP initiates the loss of a second equivalent of methane and provides the terminal imide. In this case, the terminal imide is prevented from dimerization, or other methods of supporting the negative charge on the imido nitrogen, by the very large $Tp^{t\text{-Bu,Me}}$ ligand.

In 2005, Clark and coworkers published an intriguing paper regarding covalency in the lanthanides.[21] In the article, calculations were reported that suggest cerium(IV) complexes would probably show much higher covalency than Ln(III) complexes, due to better orbital matching and, surprisingly, *f*-orbital extension. The *f*-orbital extension was attributed to hybridization of the *f*-manifold and occupancy of the orbitals.[22] The calculations suggested that terminal imides, for example, of higher-oxidation-state lanthanides such as cerium(IV) should be accessible with less steric hindrance required.

In 2017, Schelter and coworkers produced the first cerium(IV) terminal imido complex, and the complex only requires relatively small ligands near the imido to prevent bridging, which is consistent with the predictions of high covalency by Clark and coworkers.[18] The chosen ancillary was a tetradentate, deprotonated trihydroxylamine shown in Fig. 21.10d, which was placed on cerium by transamination of tris(hexamethyldisilazide)cerium(III) to produce the THF complex. Addition of I_2 oxidizes to cerium(IV), and the intermediate iodide was replaced with tetramethylsilazide, $^-N(SiMe_2H)_2$. Transamination of the tetramethylsilazide with 3,5-bis(trifluoromethyl)aniline produced the Ce(IV) anilide, which was deprotonated to provide a series of alkali metal salts with weak N(imido)–M interactions (M = Li, K, Rb, and Cs). The cesium atom can be removed from the imido nitrogen by encapsulation in cryptand to produce the terminal imido complex shown.

In computational studies on the cerium(IV) imides, it was found that 4*f*-orbital participation increased with negative charge on the imido nitrogen. Similarly, 4*f*-orbital participation is high for cerium(IV) $CeCl_6^{2-}$, but *f*-orbital participation is small in cerium(III) $CeCl_6^{3-}$, according to X-ray absorption spectroscopy and DFT studies by Clark and coworkers.[23] As a result, increasing the metal oxidation state and basicity of the ligands may increase covalency and *f*-orbital participation in the lanthanides. Consistent with this assertion, recent calculations on matrix isolable the praseodymium(V) species PrO_2^+ suggest a Pr–O Mayer bond order of 2.10, and NBO suggests the Pr–O σ-bond is 38% Pr and 62% O, with the praseodymium using 1:1 *d*/*f*-hybrids to form the bonds.[24] Considering all of this, higher-oxidation-state lanthanides are an intriguing field of study that may have significant covalency to their bonds.

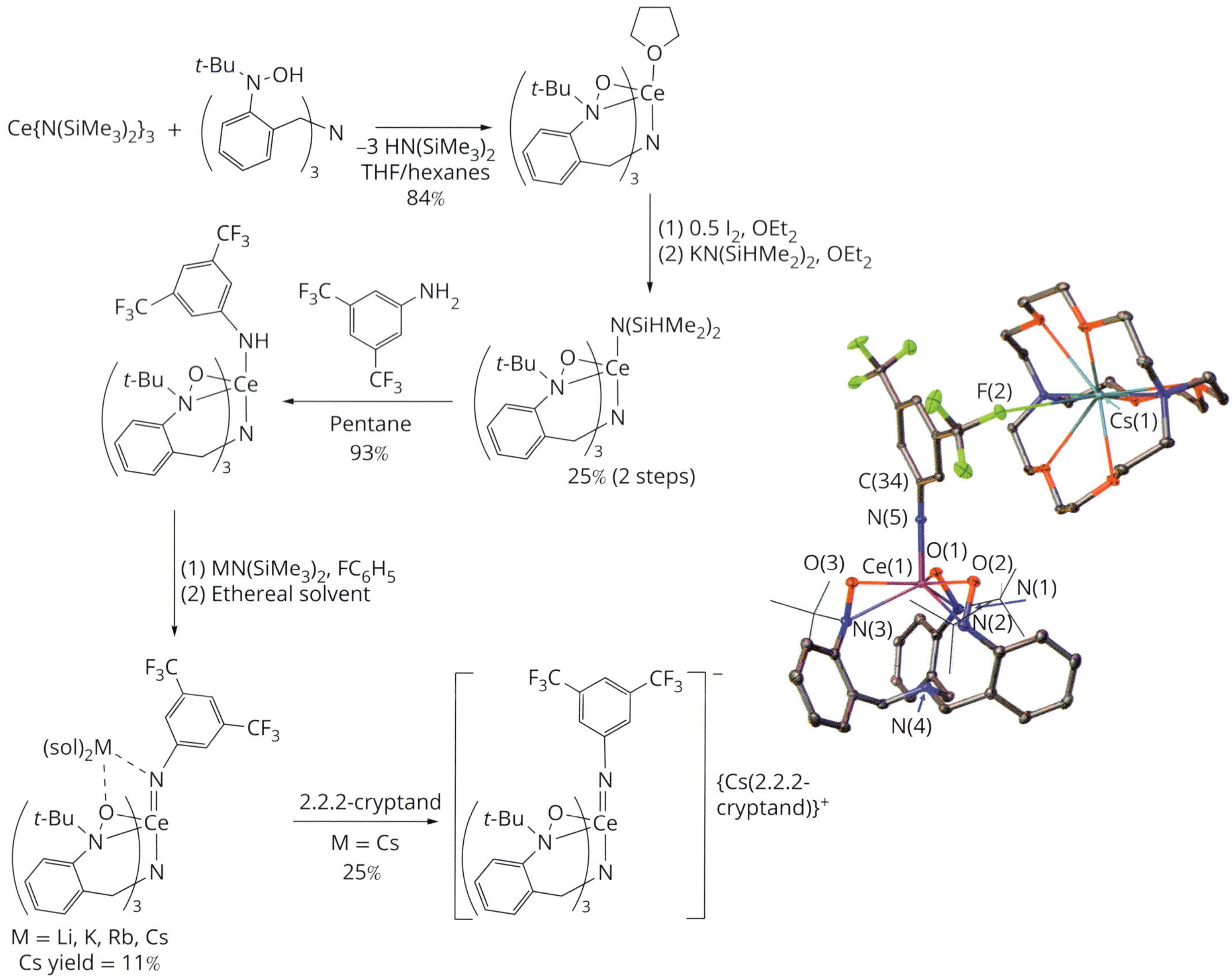

Fig. 21.10d (*left*) Schelter and coworkers' synthesis of a terminal cerium(IV) imide. (*right*) Structure of the cerium(IV) terminal imido from X-ray diffraction. There is a weak F–Cs interaction in the structure. (Structure adopted from the publication by Schelter.[18])

21.3.4 Applications of the Lanthanides

The rare-earth elements have a variety of different uses, though they are generally employed in all these applications in relatively small amounts. The elements are important for metallurgy, catalysts, glass and ceramic coloring, magnets, phosphors, and other applications (Fig. 21.11).[17] In the following sections a couple of select applications will be discussed briefly: applications in imaging and catalysis.

21.3.4.1 Imaging

An important therapeutic application of rare-earth elements is as MRI contrast agents. Gadolinium(III), which has the largest known electron spin at a single metal center of 7/2, is employed for this purpose in many currently used agents. The first to be approved by the FDA was Magnevist, with, at the time of writing, a total of nine such approved reagents overall. However, recently some complexes have fallen out of favor, and the number in use is dwindling. In every case, the metal center is encapsulated by a multidentate ligand in an attempt to prevent loss of gadolinium, as the metal is well-established as one that damages kidney function if not passivated. The "linear" chelates, like that used in Magnevist (Fig. 21.12), have been shown to release gadolinium more readily than the macrocyclic ligands, such as the DOTA ligand used in Dotarem.

In MRI, one is observing the same signal as in NMR, in which nuclear spins are aligned with a magnetic field and radiofrequency excitations are measured. In essence, the technique is a spatially resolved ^{1}H NMR experiment in which the protons in water are observed. Unlike X-ray diagnostics,

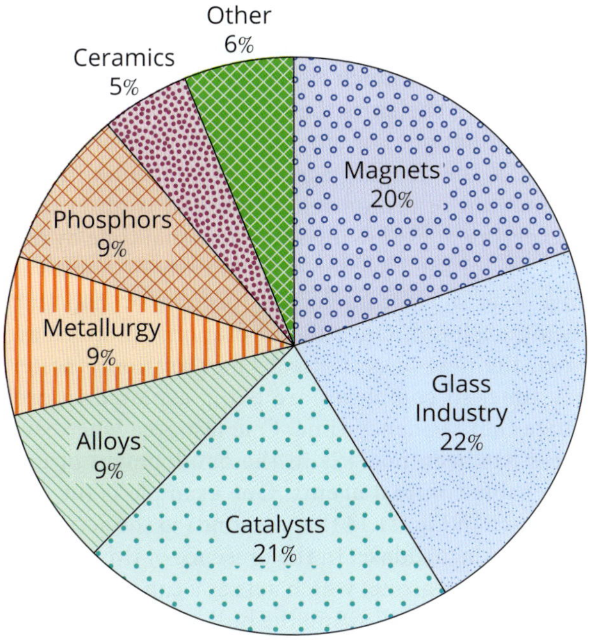

Fig. 21.11 Uses of the rare-earth elements. Data from the lanthanides section of Kirk–Othmer Encyclopedia of Chemical Technology.[25]

which are best for visualization of hard tissue, MRI is used for high-resolution imaging of soft tissue and can be used to identify brain tumors and diseases, cerebrovascular disease, spinal abnormalities, and joint problems. One important application of these agents is in imaging brain tumors. The contrast agent cannot pass the blood–brain barrier, and if increased contrast is observed in the brain after an intravenous injection it is indicative of a breakdown of that barrier due to stroke or tumor formation. The development of MRI resulted in the award of the Nobel Prize in 2003, and an account of the history and physics of the development of MRI was published soon thereafter.[27] A special issue of the journal *Inorganic Chemistry* was devoted to "Next-Generation Magnetic Resonance Imaging Contrast Agents."[28]

The random motion of paramagnetic metal centers in solution promotes relaxation of the excited nuclear spin state on the water molecules, and it is this property that is used for MRI contrast agents. An unpaired electron has, due to its spin and charge, a magnetic moment and can transfer its magnetic moment to a proton nucleus, which also has nuclear spin and charge. This can take place either through space, where it is called a dipolar or pseudocontact interaction, or through bond, where it is called a contact interaction. The magnitude of spin transfer falls off rapidly with distance or number of chemical bonds between the paramagnetic metal center and the water molecule, and directly coordinated water molecules experience the largest effect. The paramagnetic center significantly shortens the relaxation of proton spin. The protons in typical tissues relax on the order of 0.5–2 s^{-1}, but administration of the contrast agent shortens that rate by ten- to forty-fold at typical concentrations (2–5 mM).

The ability of a paramagnetic center to relax a nearby water molecule depends directly on the spin state of the center (related to the number of unpaired electrons), and inversely on distance to the water. The closest possible contact is when the water molecule is directly bound to the metal center, so the exchange rate of the water molecules must be fast (> 10^6 s^{-1}) for efficient transfer of the relaxation to the bulk water. There are a number of other factors that determine the ability of the paramagnetic center to relax the proton environments.

Gd(III) (and also high-spin Mn^{2+} and Fe^{3+}) are well suited for MRI contrast applications as they have either five (for Mn^{2+} or Fe^{3+}) or seven (for Gd^{3+}) unpaired electrons and thus a large magnetic susceptibility. Additionally, they exhibit rapid water exchange rates that allow a more rapid transfer of paramagnetic contrast enhancement into the bulk water pool. The larger number of unpaired electrons results in Gd(III)-based MRI contrast agents being used in more than 40% of all MRI diagnostics run per year. The increase in contrast possible with these agents is remarkable (Fig. 12.13).

While gadolinium is toxic, the chelating agent binds tightly to the ion, preventing its loss in the body and allowing it to be eventually excreted in urine. Recently, there has been concern of the development of nephrogenic systemic fibrosis that correlates strongly with Gd(III) in patients with reduced kidney function. There is also some evidence that gadolinium accumulates in brain tissue. The FDA has therefore

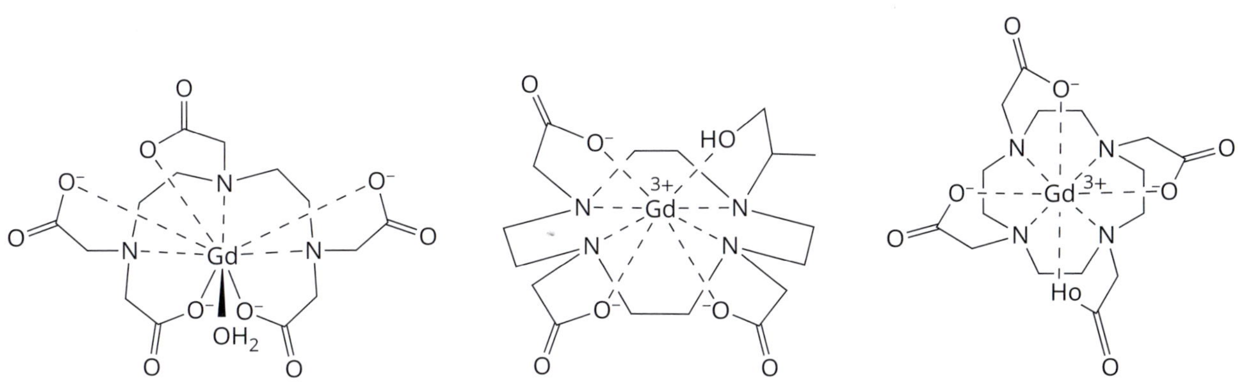

Fig. 21.12 Examples of gadolinium(III) MRI contrast agents.[26]

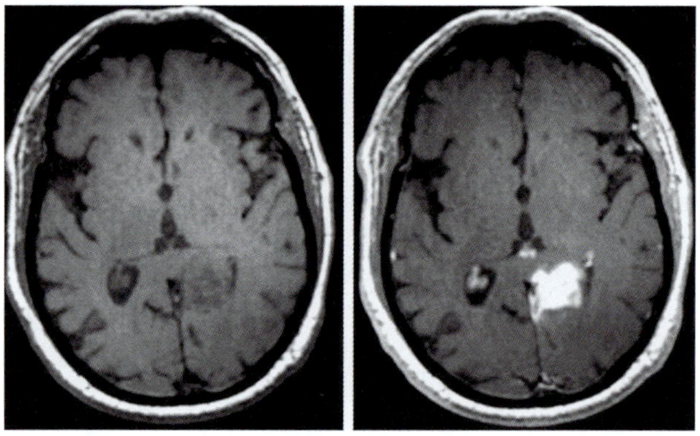

Fig. 21.13 An MRI image before (left) and after (right) administration of the Gd(III)-based MRI contrast agent Gd-HP-DO3A showing a gioblastoma, a type of brain tumor. Figure from Caravan.[26]

limited the use of the three contrast agents with the lowest formation constants. The formation constant is the equilibrium constant that favors the formation of a metal–ligand complex. Typical MRI contrast agents have K_f values above 10^{22}. Contrast agents based on iron nanoparticles have also been implicated in anaphylactic shock or extreme sensitivity in a small (<1%) portion of the patient population, resulting in the death of approximately 25% of the affected patients. As a result, there is continued interest in the development of new gadolinium as well as other transition metal-based MRI contrast agents.

Looking to the future, there are several areas of active research aimed at improving MRI contrast agents and reducing toxicity issues. All of the FDA-approved Gd(III) contrast agents are octadentate and thus have only the capacity for a single bound water molecule. The development of chelators that are only hepta- or hexadentate would allow for one or two additional waters to coordinate to the metal center. This would increase the rate of exchange with the solvent pool by two- to three-fold. However, a ligand with fewer coordinating atoms would give a less kinetically inert complex, so there is much effort in the development of more tightly binding chelators that still allow for additional free water coordination. For current contrast agents, rotational tumbling of the imaging agent is often the limiting factor in the amount of proton relaxation that is observed in an MRI experiment.[26] Slowing this rotational time by increasing the size of the chelator or covalently or noncovalently linking it to a macromolecule (protein or nanoparticle) are being explored. Once these rotational motions are optimized, the next largest effect on relaxation is water residency time, so complexes are being explored in order to optimize that residency time. One last area where exciting progress has been made is in the development of targeted or switchable contrast agents. One example is the contrast agent shown in Fig. 21.14 that responds to the activity of the enzyme β-galactosidase. In the presence of this enzyme, the galactopyranose group, which acts as a cap to prevent water binding, is cleaved, allowing the bulk water to interact with the Gd(III) center and brighten contrast.

Gadolinium-based imaging agents are a very active area of research. For example, nanoparticles adorned with Gd complexes are being applied to new *in vitro* and *in vivo* imaging.[29] New applications of known contrast agents are frequently reported from researchers across biological and medicinal science.

21.3.4.2 Catalysis

Rare-earths are used in industrial catalysts for many processes. They are employed in zeolites for oil-cracking during gasoline production. Cerium oxide is used in the enamel coating of self-cleaning ovens as a catalyst. Cerium oxide is also one of the major components of "three-way" catalysts used in the catalytic converters in cars; typically, the catalyst in most

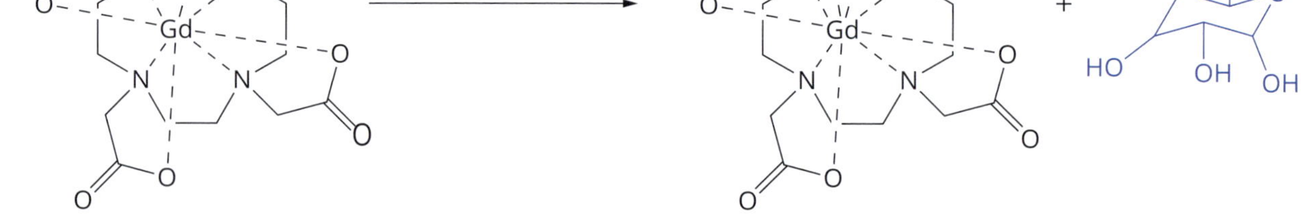

Fig. 21.14 A switchable MRI contrast agent that responds to the activity of the enzyme β-galactosidase. (Figure from Caravan.[26])

cars involves 100–3,000 ppm of Pd, Rh, or Pt dispersed in 20 wt percent CeO_2 with alumina or zirconium oxide. Lanthanides are used in a host of different catalysts for organic synthesis, often as Lewis acids, such as lanthanum triflate, but also in processes like diene polymerization by neodymium carboxylates.

There are many interesting homogeneous catalyses with lanthanides as well, and one is mentioned here as an example. Marks and coworkers have studied lanthanide-based catalysts for decades with applications in hydroelementation, such as hydroalkoxylation, hydroamination, hydrosulfurization, and so on. The group has also studied ether cleavage reactions in tandem with palladium-catalyzed hydrogenation. Generally speaking, cleavage of an ether to an alcohol and alkene is thermodynamically unfavorable; however, simple lanthanide catalysts will encourage the reaction in the presence of H_2 and Pd nanoparticles, which hydrogenate the olefin to make the reaction thermodynamically downhill. In this case, lanthanides of smaller radii were found to produce faster catalysis with turnover frequencies going as Yb > Sm > La. An example of the catalysis and the proposed catalytic cycle are shown in Fig. 21.15. The reactions used the ionic solvent 1-ethyl-3-methylimidazolium triflate (EMIM OTf).[30–31]

21.4 Some Selected Chemistry of the Actinides

By far the most explored actinide elements are uranium and thorium, which comes as no surprise as these are the only members of the series that occur naturally in significant quantities. Nevertheless, all the isotopes of the actinides are radioactive, and radiation safety along with chemical safety is always a concern. For example, the radioactivity in the case of the most common isotope of uranium, ^{238}U, is relatively benign with a half-life of 4.5 billion years. Depleted uranium, which has the more valuable ^{235}U removed and is therefore all ^{238}U, is a weak α-emitter, which means that the radiation is stopped by the dead skin cells on the outer layer of your epidermis. Depleted uranium has even been used in radiation shields, especially for neutron emitters, as the element is more effective for this purpose than lead, and the radiation from depleted uranium is readily controlled.

In the laboratory, the α-particles from ^{238}U are difficult to detect with most standard equipment. However, on emission of an α-particle, ^{234}Th is produced, which is a β-emitter with a half-life of about 24 days. The β-particles are easier to detect,

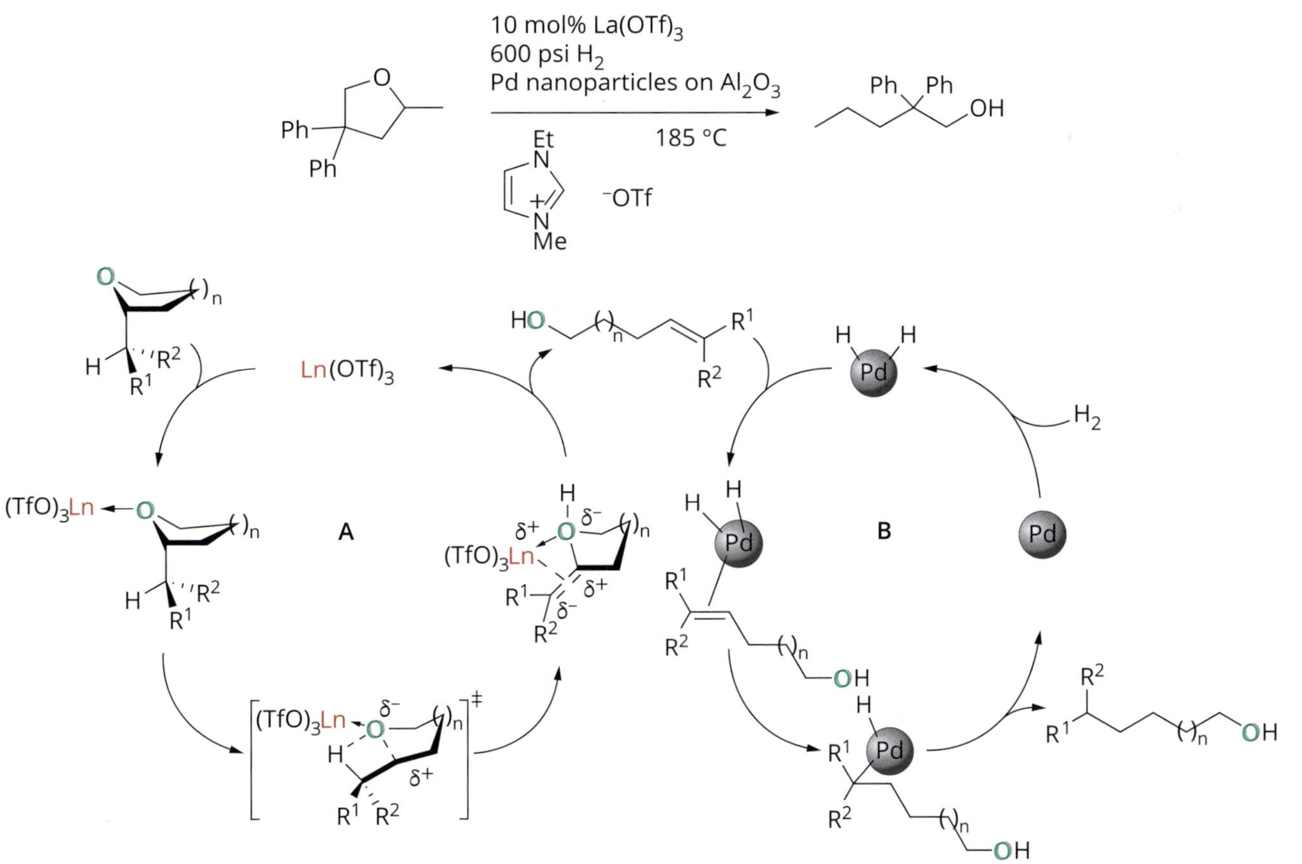

Fig. 21.15 Ether cleavage and hydrogenation by tandem lanthanide/palladium catalysis. (*top*) An example of the reaction. (*bottom*) The proposed catalytic cycle. (Bottom figure from the paper by Marks and coworkers.[30–31])

and, as a consequence, one may track ^{238}U contamination by the slow production of its primary daughter, ^{234}Th.

Thorium-232, the most common isotope of this element, has the exceedingly long half-life of 14 billion years and is a weak α-emitter. Unfortunately, the daughter isotopes of thorium, radium and radon, are quite dangerous, albeit slowly produced.

All the other elements of the actinides have relatively short lifetimes and require nuclear reactions to access. As a result, their chemistry is far less explored, and inorganic chemists must find access to these elements through collaborations with nuclear chemists. In recent years, some progress has been made in understanding how the chemistry of the actinides changes across the series. The conclusions are very much still in flux as there is relatively little data, but some of the current thinking will be discussed here. In addition, some chosen reactions involving the actinides will be given to illustrate the type of chemistry currently under investigation.

One intriguing recent finding regards favored oxidation states in the actinides. While lanthanide chemistry is predominantly in the +3 formal oxidation state, actinides are more redox active. Thorium has four valence electrons, and its chemistry is dominated by the +4 formal oxidation state. Even so, thorium compounds in lower oxidation states are known (*vide infra*). Many uranium complexes from the +3 to +6 formal oxidation states are all well known; furthermore, there are a few reports of the +2 formal oxidation state for uranium.[32–34] In general, the +4 formal oxidation state is perhaps most favored. There is evidence that the actinides become more lanthanide-like farther to the right, with the +3 oxidation state becoming favored. In addition, there is evidence that the +2 formal oxidation state may be favorable for californium and the heaviest actinides.[4]

In 1954, Seaborg and coworkers published a detailed study on the ion exchange elution times of many lanthanides and actinides versus HCl concentration.[35] From these studies, it was decided that the actinides under scrutiny, such as americium, displayed higher covalency than lanthanides at higher HCl concentrations where $AmCl_6^{3-}$ formation would be prevalent. Seaborg attributed the higher covalency in the actinides to 5*f*-orbital participation in the bonding. In a recent report, Kozimor and coworkers investigated the degree of covalency in $AmCl_6^{3-}$ by K-edge X-ray absorption spectroscopy (XAS). In the XAS of $AmCl_6^{3-}$, it was indeed seen that there is overlap between the Am 5*f*-orbitals and the chorine *p*-orbitals, confirming some degree of covalent bonding in these systems.[36]

21.4.1 Thorium

The chemistry of thorium is largely dominated by the +4 formal oxidation state. The most important mineral is ThO_2 (thorianite), but the metal is also found in silicates, phosphates, and other sources. Applications of thorium are generally small scale and include coatings and ceramics. Many of those applications are due to the exceedingly high melting point of ThO_2 of 3,370 °C.

There are a few known molecular Th(III) complexes and a Th(II) complex was reported in 2015.[37] The synthesis of the thorium(II) compound begins with a complex prepared by Lappert and coworkers, $Cl_2Th(Cp'')_2$, where Cp″ = 1,3-bis(trimethylsilyl)cyclopentadienyl (Fig. 21.16), which is prepared by salt

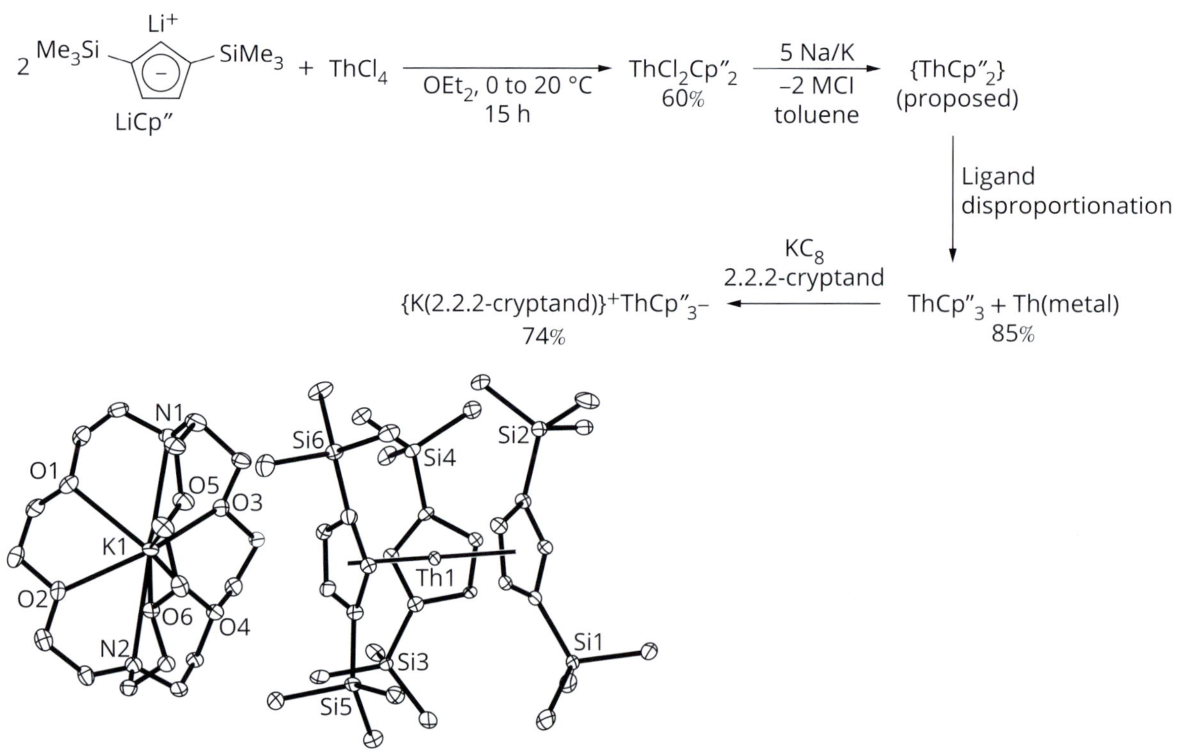

Fig. 21.16 Synthesis of thorium(III) $Th(Cp'')_3$ and reduction to Th(II). (Structure from the report by Evans and coworkers.[37])

elimination between $ThCl_4$ and LiCp″.[38] On addition of the very strong reductant NaK, it is proposed that the $Cl_2ThCp''_2$ is reduced to thorocene, $ThCp''_2$. It was further proposed that $ThCp''_2$ undergoes ligand disproportionation in a process almost quantitative in Cp″ ligands to produce $ThCp''_3$ and thorium metal.[39] Evans and coworkers reported that the $ThCp''_3$ complex can be reduced with potassium graphite to produce $K^+ \{ThCp''_3\}^-$, and the potassium was sequestered into 2.2.2-cryptand.[40]

21.4.2 Uranium

Some uranium salts and ores that are stable to air are derived from the uranyl ion, UO_2^{2+}, which has two *trans*-oxo groups and is in the +6 formal oxidation state. Uranium metal, often in the form of turnings, can also be used to generate uranium halides that are useful for producing a wide variety of complexes.

Prepared directly from uranium oxide, UCl_4 can be synthesized from $UO_3 \cdot 2H_2O$ by reaction with hexachloropropene at high temperature (Fig. 21.17).[41] The uranium(III) starting material $UI_3(THF)_4$ can be prepared by the action of I_2 on uranium metal in tetrahydrofuran. The uranium metal is made "oxide-free" by treatment with nitric acid to clean the surface followed by amalgamation with HgI_2. The oxide-free uranium metal is pyrophoric and should be handled and stored under purified nitrogen. The surface-protected and activated uranium is then dried and treated with a slight deficiency of I_2 to avoid over-oxidation.[42] Kiplinger and coworkers showed that the "oxide-free" uranium turnings can be used to prepare several different uranium(III and IV) halide/solvates, as shown in Fig. 21.17.[43]

21.4.2.1 Uranyl and the Inverse *trans*-Influence (ITI)

An effect being recognized as one of great importance in uranium chemistry, and perhaps elsewhere, is the inverse *trans*-influence (ITI). Recall that the *trans*-influence is a thermodynamic effect where ligands situated *trans* to a strongly donating ligand in a square planar or octahedral system may have their bonds weakened (Section 17.5.4); the weakening is then measurable by the metal–ligand bond distances, vibrational frequencies, NMR coupling constants between the metal and ligand, and so on, in comparison with reference values, ideally within the same complex but possibly to other structures where the *trans*-influence is expected to be negligible. As an explanation of *trans*-influence, one can consider the strongly donating ligand to occupy more of the lower-energy acceptor orbital on the metal center (see Bent's rule), while the weaker donor's bond is weakened by having to interact with higher-energy orbitals (Box 17C).

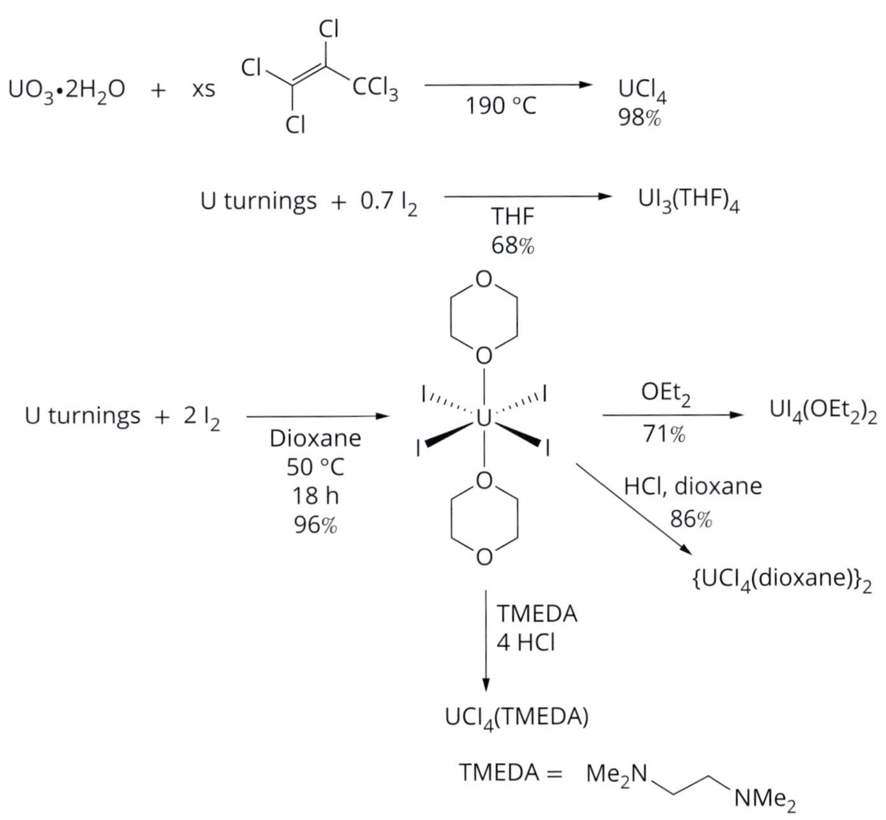

Fig. 21.17 Some representative examples of uranium halides and their preparation.

One result is that strongly donating ligands to transition metals typically are not found *trans* to one another, as their competition for the same orbitals on the metal would be energetically unfavorable. There are exceptions, however. For example, pseudo-octahedral d^2 dioxo complexes of the transition metals often have the two oxo groups *trans*. One explanation for this is that placing the two oxo ligands *cis* creates a situation where every *d*-orbital on the metal is either σ- or π-antibonding, and the two *d*-electrons on the metal must then go into antibonding orbitals. If the two oxo ligands are *trans*, the oxo ligands must compete for the same orbitals, but the d_{xy} orbital is nonbonding and can house the electrons on the metal.

For example, the complexes $MoE_2(dppe)_2$ were prepared and structurally characterized by Cotton and Schmid, where dppe = 1,2-bis(diphenylphosphino)ethane and E = O, S, Se, and Te.[44] The oxo complex, for example, was prepared from Hidai's $Mo(N_2)_2(dppe)_2$ by addition of SO_3 (Fig. 21.18). For all E, the double-bonded groups are *trans* to one another in the d^2 system, which may be analyzed by using a crystal field theory approach as shown in Fig. 21.14 with a nonbonding, filled d_{xy}. As noted, this is an exception that proves the rule. Typically, strong donors are *cis* in transition metal systems, except when there are overriding orbital effects to favor the *trans*-configuration.

Some *f*-block compounds, from a phenomenological perspective, display an inverse trans-influence (ITI). For example, dioxo–metal complexes strongly prefer a *trans*-arrangement for uranium(VI). In fact, the UO_2^{2+} unit is so common that it is given a trivial name, "uranyl", and the oxo groups are always *trans* in such compounds. The *trans*-arrangement is not restricted to just oxo groups, and there seems to be a strong energetic preference for imide (=NR) and other strong donors to be *trans* as well. Some examples are shown in Fig. 21.19, but there are hundreds of such uranyl structures reported. In the top two structures in the figure are the simple dianion $UO_2Br_4^{2-}$ with a O–U–O angle of 178(1)° and the hexagonal bipyramidal tricarboxylate $UO_2(O_2CEt)_3^-$ with an O–U–O angle of 179.5(3)°.[45–46] One can use the other ligands to try to lower the O–U–O angle; for example, Hayton and coworkers used a macrocyclic tetradentate ligand to occupy one side of the uranium coordination sphere and force a smaller uranyl angle of 164°.

Bis(imido) complexes typically prefer a linear N–U–N angle as well, but there has been substantial work in lowering this angle by using Cp and Cp-like ligands. For example, Burns and coworkers prepared $Cp^*_2U(NPh)_2$ with a N–U–N angle of 98.7°.[47] Boncella and coworkers showed that the large Cp* ligands were not required for this chemistry and prepared $Cp_2U(N\text{-}t\text{-}Bu)_2$ with a slightly larger N–U–N angle of 103°.[48] Boncella and coworkers also reported a methylene-bridged pyrrolyl ligand on bis(imido) uranium, which gives a bis(Cp)-like structure with an N–U–N angle of 116°. On addition of a bidentate phosphine ligand (dmpe), the η^5-pyrrolyl convert to η^1-pyrrolyl and give a uranyl-like, pseudo-octahedral structure with an N–U–N angle of 166°.[49]

As a means of comparison, consider the complexes $MoO_2Cl_2(OPPh_3)_2$ and $UO_2Cl_2(OPPh_3)_2$ with identical ligand

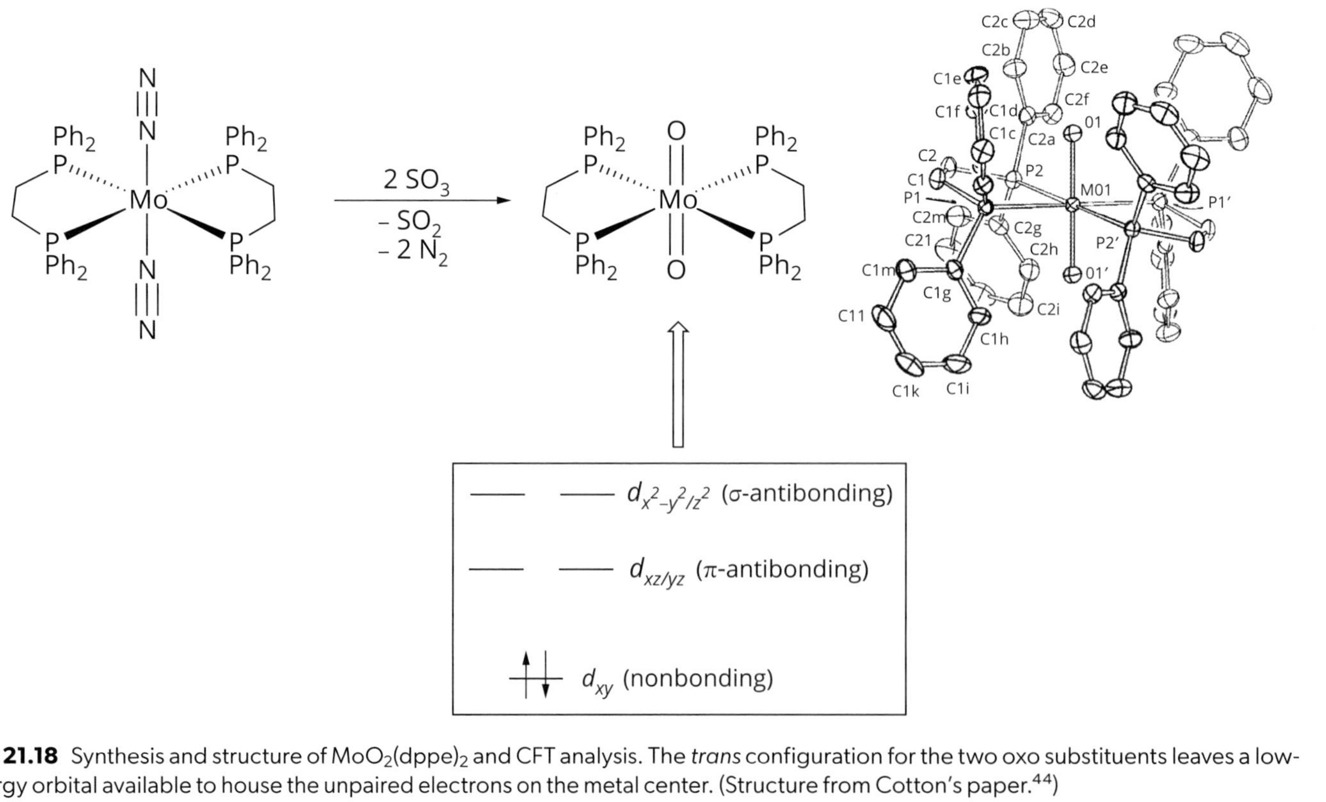

Fig. 21.18 Synthesis and structure of $MoO_2(dppe)_2$ and CFT analysis. The *trans* configuration for the two oxo substituents leaves a low-energy orbital available to house the unpaired electrons on the metal center. (Structure from Cotton's paper.[44])

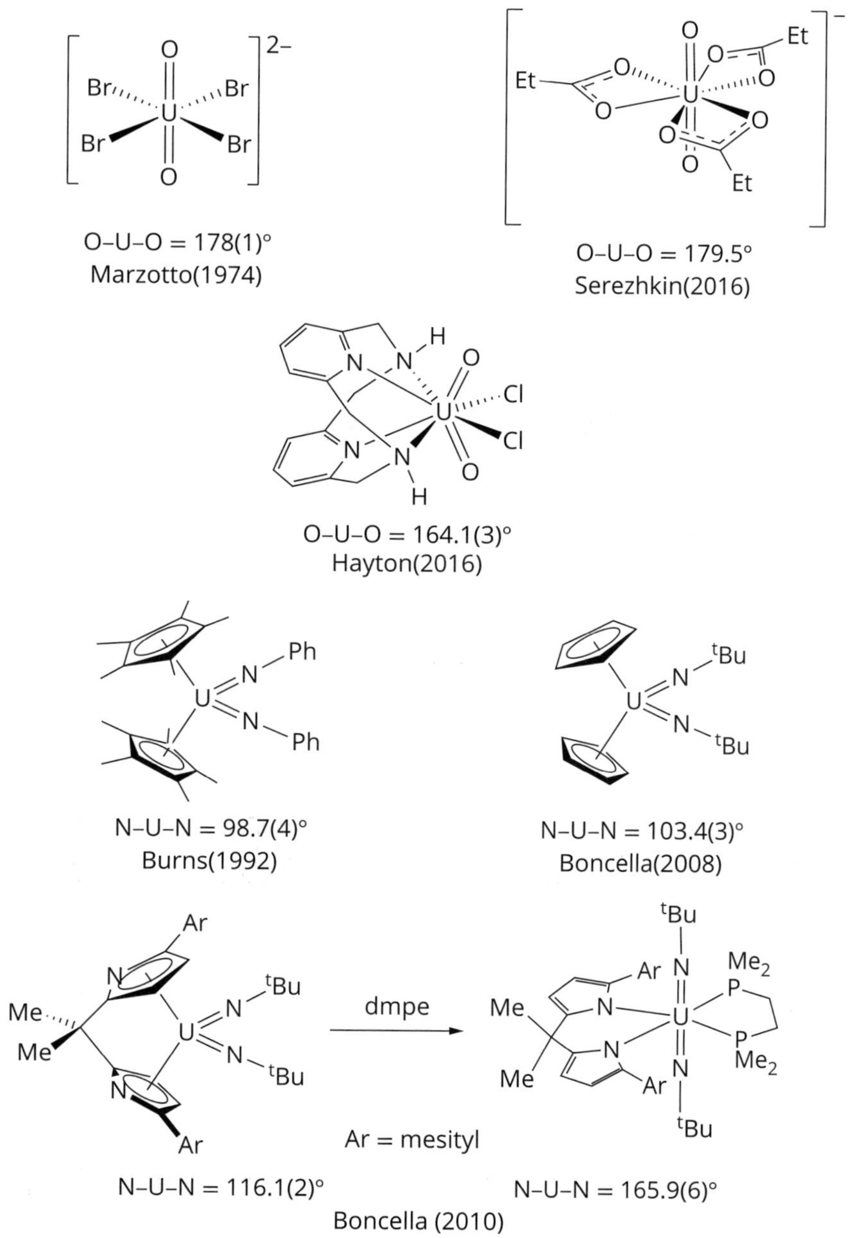

Fig. 21.19 Some selected examples of uranium(VI) dioxo and bis(imido) complexes.

sets shown in Fig. 21.20. The molybdenum complex has the expected geometry for an $MX_2Y_2Z_2$ complex, with the two strongest donors *cis* to one another and *trans* to the weakest donors (in this case, dative phosphine oxides) with the medium donors (chlorides) mutually *trans*. The uranyl complex has

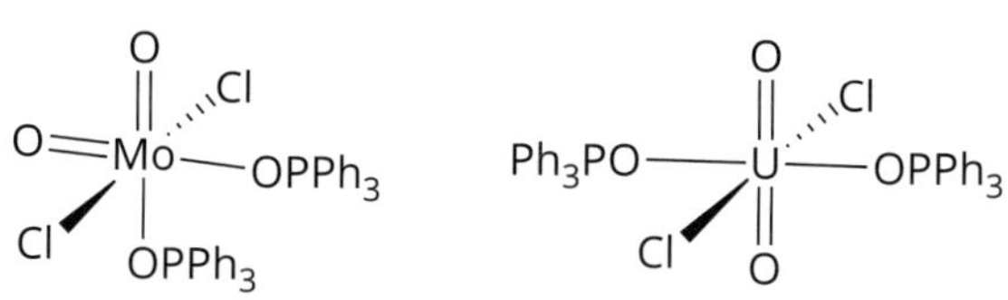

Fig. 21.20 Structures of $MoO_2Cl_2(OPPh_3)_2$ and $UO_2Cl_2(OPPh_3)_2$.

the atypical geometry, for transition metals at least, shown to the right where the strongest donors, the oxo ligands, are *trans*. There does not seem to be a strong preference for *cis*- or *trans*-chlorides in the uranium complex as they are often prepared as a mixture of isomers and both isomers have been structurally characterized.[50] Why do uranium complexes prefer to place these strong donors along the same axis, especially when the metal is in a high oxidation state?

The origin of the ITI seems to be complex, but it is well established that one major contributor to the effect is a mixing of the core 6*p*-orbitals. The role of 6*p*-orbitals in the bonding of uranyl was clearly established in 1980 by Tatsumi and Hoffmann, using extended Hückel calculations. In the manuscript, the authors examined the orbital energies as a function of O–M–O angle for M = Mo and U in MO_2^{2+}. For molybdenum, there is a strong energetic minimum around ~110° that dominates and leads to the strong preference for the bent geometry. Interestingly, when the valence 5*f*-, 6*d*-, 7*s*-, and 7*p*-orbitals are examined for UO_2^{2+}, the valence *p*-, *d*-, and *f*-orbitals all have minima near ~110°, similar to the transition-metal case (Fig. 21.21a), although the potentials are relatively flat. If the inner 6*p*-orbitals are included with the valence *fdsp*-set of the uranium, there is a strong energetic preference for the 180° configuration. Note that this is *not* simply due to a preference of the 6*p*-orbital alone, which has a minimum at the 90° O–U–O bond angle. Consequently, the ITI is due to orbital interactions between the 6*p* and valence orbitals of the uranium.[51]

Recall from Chapter 5 on group theory that orbitals with an even angular momentum quantum number (*s*, *d*, and so on) have gerade symmetry and odd angular momentum orbitals (*p*, *f*, and so on) have ungerade symmetry; the parity of the *d*- and *f*-orbitals are different, with *f*-orbitals having the same parity as *p*-orbitals. For interactions of core electrons with valence electrons, this is quite significant, as *p*-orbitals are often the highest-energy core orbitals and most likely to interact with the valence set. Hoffmann and Tatsumi interpreted their results as mixing of the filled-core 6*p*-orbitals with an *f*-orbital to stabilize the 180° geometry, where this *fp*-hybrid on uranium is composed of about 21% 6*p*-character and 79% 5*f*-character and interacts with the in-phase combination of 2*p*-orbitals on the oxygens as shown in Fig. 21.21b.

In a related but alternative model for understanding the ITI, Denning described "a naïve, but pleasingly simple view of the source of the *trans*-influence for metal ions with a formally empty valence shell."[3] In this model, the polarization of the core electrons by one strongly donating ligand is expressed in terms of dipolar (along the same axis as the strongly donating ligand) and quadrupolar (in the plane perpendicular

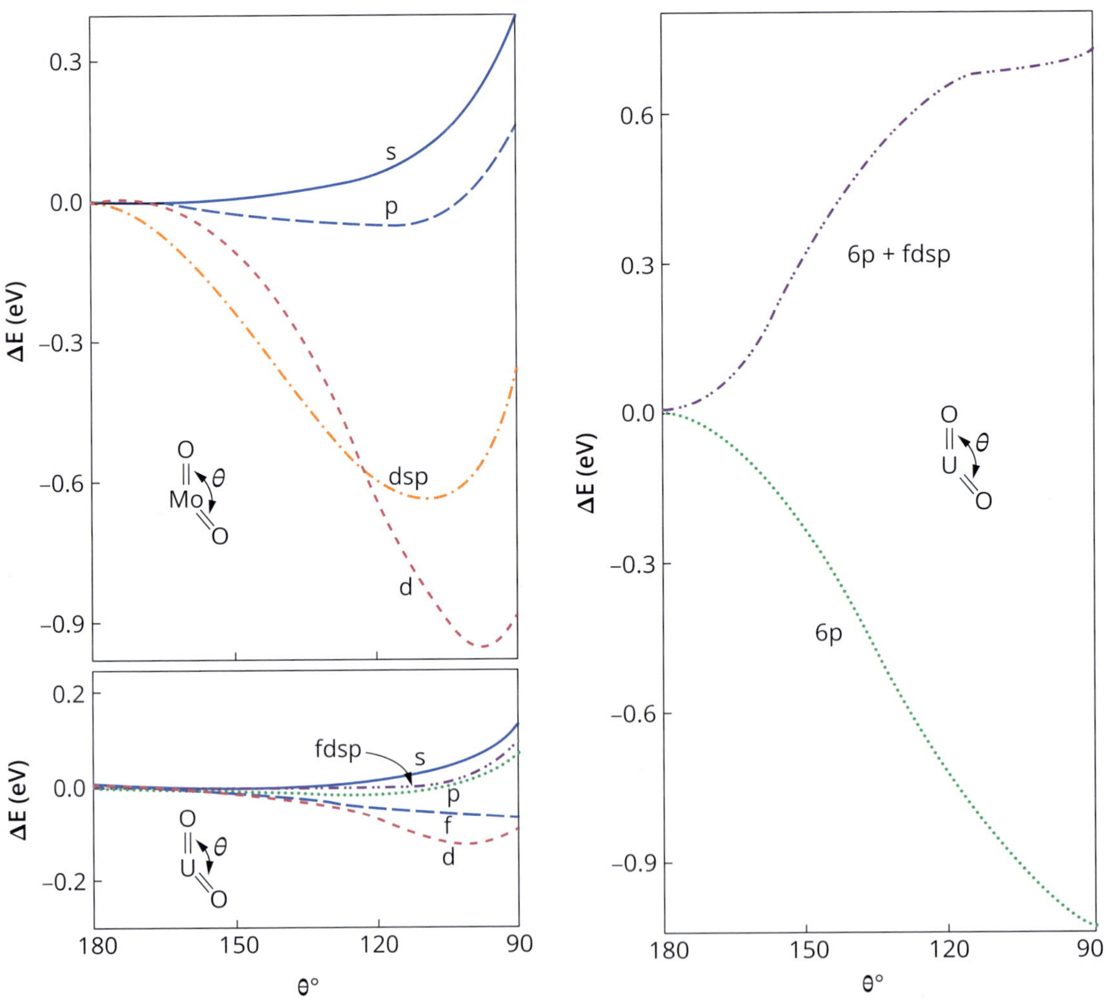

Fig. 21.21a (*left, top*) Relative total energy as a function of bond angle for Mo 4*d*, 5*s*, 5*p* (separately), and *dsp* (all together) basis sets in MoO_2^{2+}. (*left, bottom*) Relative total energy as a function of bond angle for U 5*f*, 6*d*, 7*p* (separately) and *fdsp* (all together) basis sets in UO_2^{2+}. (*right*) relative total energy as a function of O–U–O bond angle for U 6*p* and 6*p* + valence *fdsp* basis sets in UO_2^{2+}. (Figures from the study by Tatsumi and Hoffmann.[51])

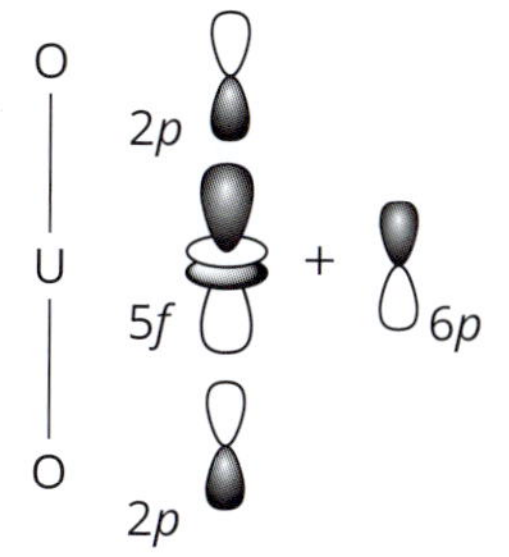

Fig. 21.21b Proposed mixing of the 5*f* and 6*p* orbitals on uranium.

to the bond of the strongly donating ligand) polarizations. In essence, the ligand interaction perturbs the core electrons and leads to their mixing with the valence orbitals.

If the highest-energy core orbitals have opposite parity to the lowest-energy valence orbitals, the dominant polarization will be dipolar, leading to an accumulation of negative charge *trans* to the strongest donating ligand. The next strongest donating ligand added to the metal will then energetically prefer the *cis*-position away from this negative charge. This is the case for the transition metals where the core orbital of highest energy is a *p*-orbital (ungerade) and the lowest-energy valence orbitals are *d*-orbitals (gerade).

If the highest-energy core orbitals of an atom have the same parity as the lowest-energy valence orbitals, the dominant polarization will be quadrupolar, leading to accumulation of negative charge *cis* to the strongest donating ligand. The 6*p* and 5*f* orbitals of uranium have the same parity, which causes negative charge build-up *cis* to the strongly bonding ligand. The next strongest bonded ligand therefore energetically prefers the *trans* position. This is the case for the actinides, where the core orbital of highest energy is the *p*-orbital (ungerade) and the lowest-energy valence orbitals are *f*-orbitals (ungerade).

Does the ITI occur elsewhere in chemistry? It seems that the effect is fairly pervasive among systems with large central atoms having relatively high-energy core orbitals; however, other MO or VB explanations can be used in many cases rather than the previous polarization explanation. For example, 2-coordinate Ag(I), Au(I), and Hg(II) systems are typically *trans*, which can be explained as being due to filled *d*-subset (gerade) pseudo-core interactions with the *s*-orbital valence (gerade), leading to a quadrupolar interaction that favors *trans*-arrangement of the ligands. In BaF_2, the filled 5*p*-orbitals (ungerade) interact with the 6*s* (gerade) valence orbitals, resulting in a dipole that leads to the observed bending in the system, in violation of VSEPR.

The *trans*-geometry of d^2 transition-metal dioxo complexes (*vide supra*) can be interpreted using this polarity model as well. The first oxo acts to induce quadrupolar charge distributions in the valence shell; that is, a filled d_{xy} orbital, and the second oxo then prefers the *trans* position. The ITI has also been observed in lanthanides in bis(carbene) complexes of Ce(IV) and in the +4 oxidation states of U and Th.[52]

The ITI is one of the few overarching principles discovered thus far for *f*-block chemistry, at least in the middle to high oxidation states. The effect is probably whole or in part the result of core *p*-orbital mixing with the valence *f*-orbitals.

21.4.2.2 Organometallic Chemistry

The organometallic chemistry of uranium and thorium has been investigated, relatively speaking, for actinide chemistry, in great detail, and a few such complexes were shared previously, such as cyclopentadienyl bis(imido) complexes. In fact, cyclopentadienyl derivatives are in the Cambridge Crystallographic Database for U, Th, Np, Cm, Bk, and Cf, with the later transuranics often being represented exclusively by MCp_3 structures. Cyclopentadienyl derivatives are "go to" ancillary ligands in uranium and thorium chemistry, in fact, as there are many varieties available with different steric profiles, they bind strongly to most metal centers, and the ligands occupy a significant percentage of the space around the large radii metal centers, which can simplify the chemistry.

One ligand of particular interest for the actinides, but also employed for the early transition metals, is cyclooctatetrene dianion (COT). This eight-membered ring is too large to coordinate in an η^8-configuration with many of the later transition metals, but binds in an aromatic fashion to metals such as uranium and thorium. The $U(COT)_2$ complex is simply known as "uranocene". The bis(COT) complexes of U, Th, Pa, Pu, and Np can currently be found in the CCSD (Fig. 21.22),[53–54] although the Pu complex appears to only be structurally characterized as the Pu(III) anion.[55] The chemistry of the ligand with U and Th is extensive. For example, Cloke and coworkers have published a series of uranium bis(triisopropylsilyl)cycloctatetraenyl pentamethylcyclopentadienyl complexes including the hydride and alkyl derivatives.[56]

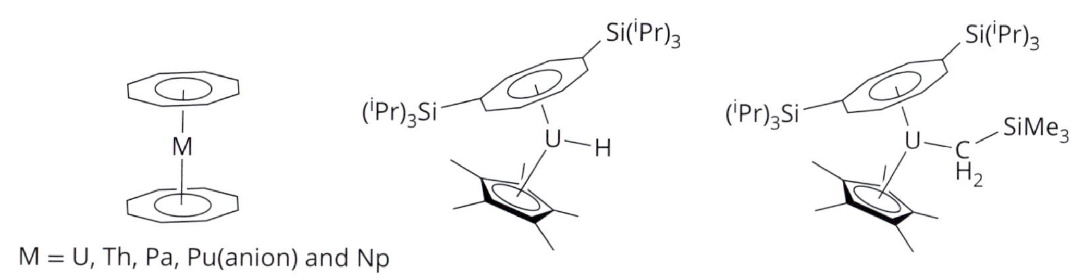

Fig. 21.22 Structure of uranocene and other cycloctatetraenyl (COT) complexes of the actinides.

There has been some success in the production of organometallic complexes of uranium not involving π-aromatic ligands like COT or Cp. For example, Hayton and coworkers have reported UMe_6^{2-} as an "ate" complex with lithium(*N,N,N′,N′*-tetramethyl-1,2-ethylenediamine), Li(TMEDA).[57] Bart and coworkers have prepared a series of $U(benzyl)_4$ derivatives as neutral monomeric compounds, such as the homoleptic 4-*t*-Bu-benzyl complex U(4-*t*-Bu-benzyl)$_4$.[58]

There are few "rules" currently known for organometallic reactivity in the actinides, but there is a tendency for actinides to do single-electron chemistry rather than familiar 2-electron reactions such as reductive elimination. Many of the transition metals behave similarly, and one avenue to do 2-electron chemistry with such species is to employ redox noninnocent ligands. Bart has shown that uranium can achieve clean reductive elimination chemistry in the presence of a redox noninnocent ligand (bottom, Fig. 21.23).[59]

21.4.2.3 Catalysis

Catalytic reactions within the actinides are a growing field. Like all of actinide chemistry, catalysis has not seen the type of development which the transition metals have enjoyed; however, uranium has an intriguing history. In 1909, Haber wrote about the metal for N_2 reduction to NH_3 conversion: "Uranium as a catalyst exceeds all literature-reported systems in efficiency."[60] Furthermore, uranium has seen extensive investigation as a catalyst for the Fischer–Tropsch process. Uranium's large radius and orbital make-up result in systems that are attractive for catalytic applications, and the metal will probably yield exciting new reaction chemistry in the future. Currently there are relatively few catalytic reactions that are well characterized and mechanistically well described, but a couple of examples will be discussed briefly.[61]

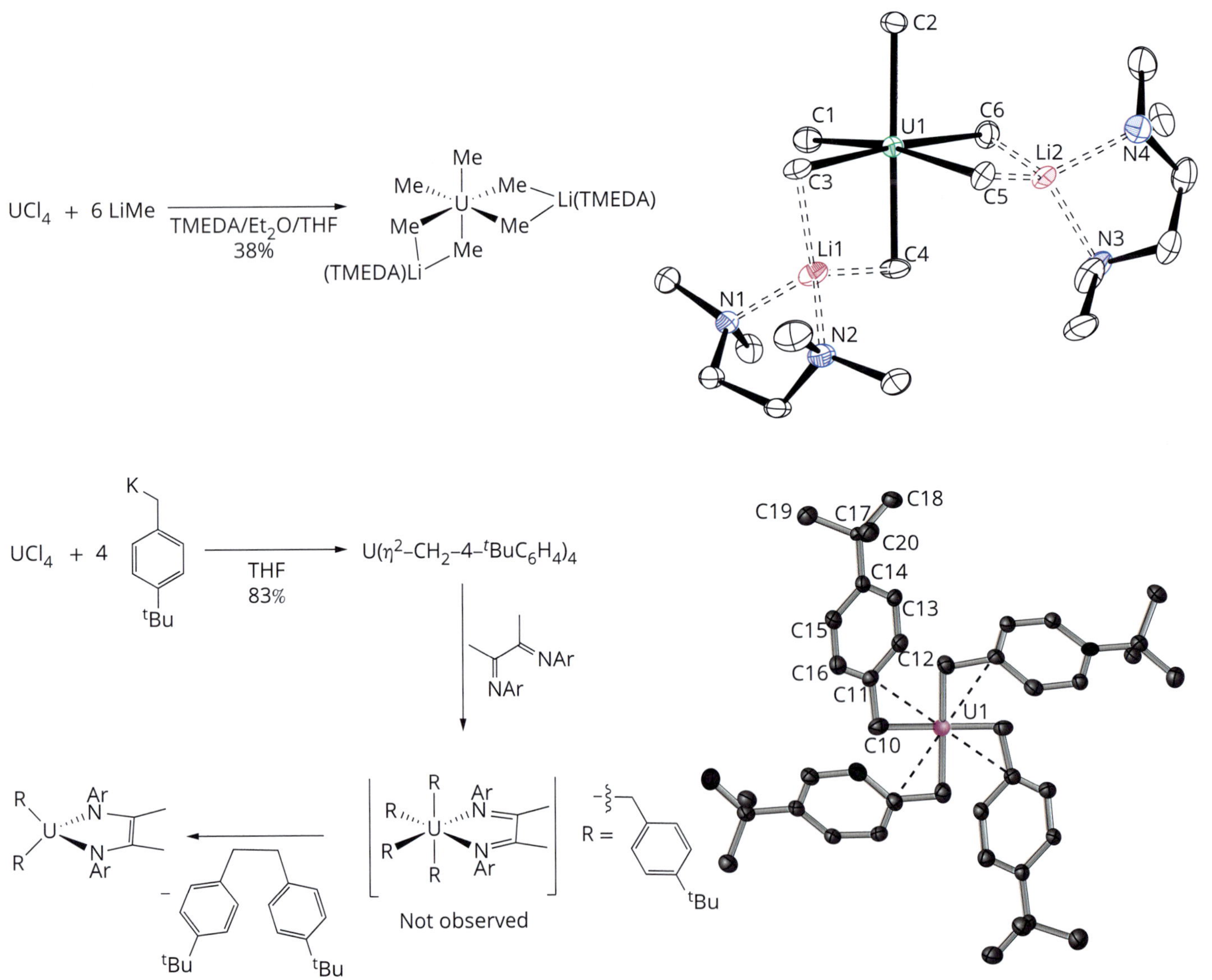

Fig. 21.23 (*top*) Hayton and coworkers' synthesis and structure of UMe_6^{2-}.[57] (*middle*) Bart and coworkers' U(4-*t*-Bu-benzyl)$_4$ complex.[58] (*bottom*) Bart and coworkers reductive elimination from uranium, utilizing a redox noninnocent ligand.[59]

Olefin hydrogenation reactions catalyzed by thorium- and uranium-cyclopentadienyl complexes have been reported by Marks and coworkers. They were able to show that the hydrogenolysis of $Cp^*_2MR_2$, where M = Th or U, does not involve oxidative addition and reductive elimination. Instead, a four-center reaction is proposed where the metal stays in the +4 oxidation state throughout. The four-center transition state was suggested, partially, based on activation parameters derived from variable-temperature kinetics; the experiments for $Cp^*_2U(Np)\{OCH(t\text{-}Bu)_2\}$ gave $\Delta H^\ddagger = 9(2)$ kcal/mol and $\Delta S^\ddagger = -45(5)$ kcal/mol, which suggests a very ordered transition state like the polar-cyclic structure in Fig. 21.24.[62]

The dihydride of the thorium-based system is quite stable but seems to exist largely or exclusively as the dimer $\{H(\mu\text{-}H)ThCp^*_2\}_2$. There are many variations of mechanisms that fit the reported data, but a plausible mechanism is shown in Fig. 21.24. The catalyst proposed in the scheme is the monoalkyl hydride complex **A**, which could be generated either through a dimeric route (Activation Pathway 1) utilizing olefin 1,2-insertion by the dimeric hydride or by breakup of the dimer to a monomeric dihydride (Activation Pathway 2). After either activation pathway, the monohydride alkyl complex **A** would then insert another olefin to give the monomeric dialkyl complex, which can undergo hydrogenolysis of one of the two alkyls by the polar four-center mechanism. This mechanism would presume fast olefin insertion and relatively slow hydrogenolysis; that is, once a terminal hydride is present an olefin quickly inserts to go back to the dialkyl.[63]

The hydrogenation catalysis using $\{H(\mu\text{-}H)UCp^*_2\}_2$ was much faster than Th at 70 turnovers/h for U, versus 0.5 h^{-1} for the thorium analog. However, the uranium reaction is more complex due to possible redox processes. Uranium(IV) $\{H(\mu\text{-}H)UCp^*_2\}_2$ is converted under heating to the uranium(III) $\{HUCp^*_2\}_x$ and H_2 as shown in Eq. 21.2. Marks and coworkers state that involvement of a uranium(III) $HUCp^*_2$ complex, or some oligomer thereof, "seems likely."[63]

$$\{H(\mu\text{-}H)UCp^*_2\}_2 \rightarrow (2/x)\{HUCp^*_2\}_x + H_2 \qquad \textbf{21.2}$$

Karsten Meyer and coworkers have developed a uranium-based water-oxidation catalysis that electrocatalytically produces O_2 and H_2. The catalyst employed (Fig. 21.25) is a pseudo-C_{3v} uranium(III) tris(aryloxide) complex (**1**, U{mes$(OAr^{Ad,Me})_3$}) with a metal–arene interaction in an axial position. The proposed mechanism involves coordination of water *trans* to the axial arene ring and oxidative addition to form a hydroxo hydride uranium(V) intermediate, which then eliminates H_2 unimolecularly to form a terminal oxo. The arene bound to the uranium is critical as a redox noninnocent ligand for this oxidative addition step; compare the reductive elimination by Bart in Fig. 21.23 using a redox noninnocent ligand.[64]

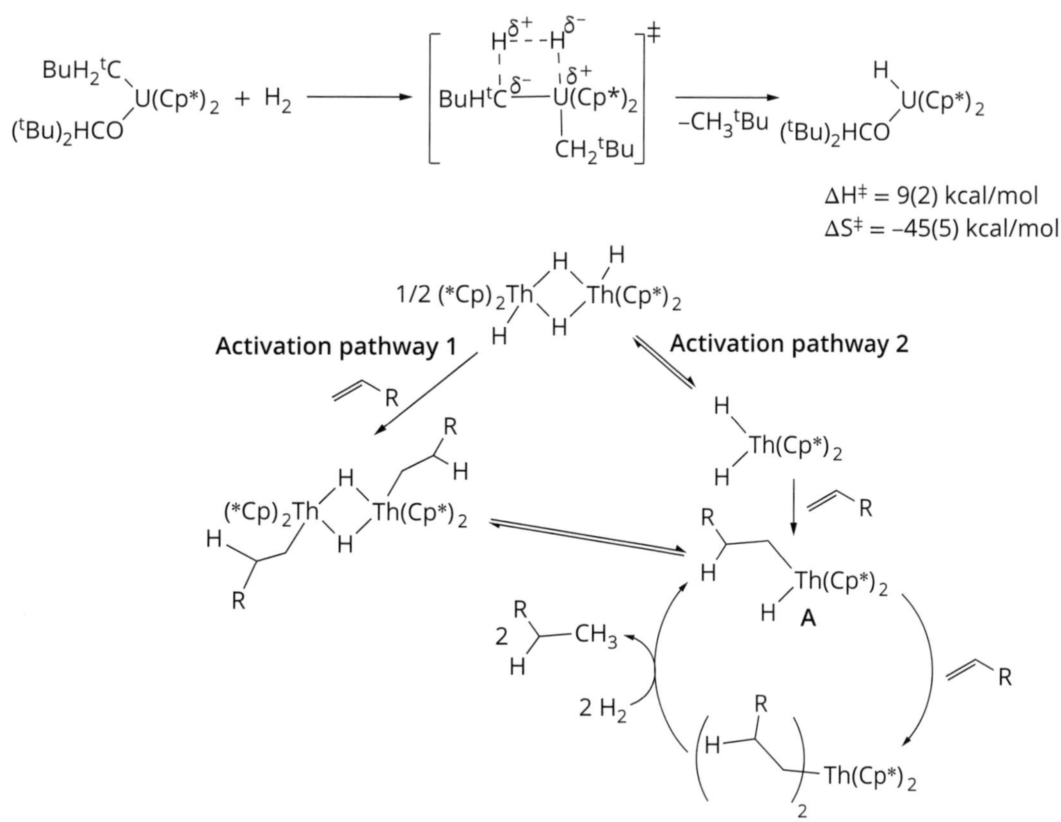

Fig. 21.24 Proposed mechanisms for (*top*) hydrogenolysis of a uranium alkyl and (*bottom*) catalytic hydrogenation of an olefin.

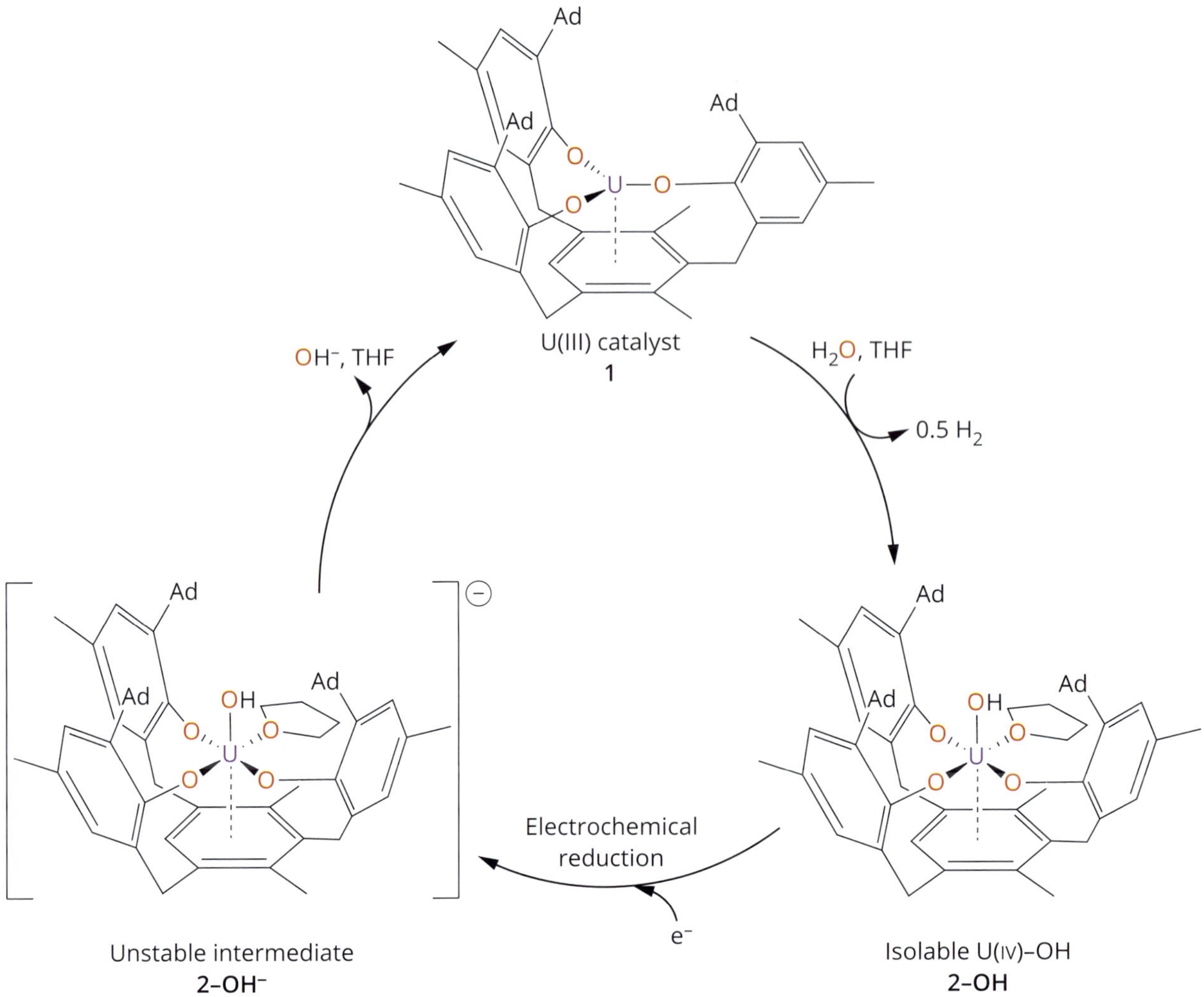

Fig. 21.25 Meyer and coworkers proposed mechanism for electrochemical oxidation of water by a uranium catalyst. (Figure from their publication.[65])

One equivalent of the U(V) terminal oxo species valence comproportionates with a U(III) aquo complex to produce two equivalents of U(IV) hydroxide (Eq. 21.3). The U(IV)-OH complex is reduced electrochemically, which initiates dissociation of hydroxide and reforms the U(III) catalyst. Using 0.22 M H_2O in THF, the catalyst produces 10^6 turnovers/h at an overpotential of 1.3 V.[65]

$$\begin{aligned} &O{=}U\{mes(OAr^{Ad,Me})_3\} + H_2O{-}U\{mes(OAr^{Ad,Me})_3\} \\ &\quad \rightarrow 2HO{-}U\{mes(OAr^{Ad,Me})_3\} \end{aligned} \qquad \textbf{21.3}$$

There is a tendency by some to view all actinides as elements only fit for use in nuclear reactors and to be buried in vaults out of sight and out of mind. However, uranium in particular has a unique combination of orbital make-up and redox properties that conjure in the minds of inorganic chemists intriguing possibilities for catalysis, and ^{238}U is relatively easily handled with a very long (by any geologist's standard) half-life. Uranium is not terribly rare, with an abundance similar to bromine, and, considering all this, is an attractive target for new catalytic chemistry.

21.5 Brief Notes on Absorption Spectroscopy of the Actinides and Lanthanides

The magnetism and spectroscopy of the actinides and lanthanides is an active and complex field. Nevertheless, it is also an important field with many applications. A full discussion is well beyond the scope of this text and is a highly specialized area; however, a few broad strokes about absorption spectroscopy in this field are appropriate.

Unpaired electron density in the lanthanides is generally in the *f*-orbitals, just discussed extensively, protected by core electrons from interactions with the ligands in most cases. In the *d*-orbitals, the strength of the energies perturbing the electrons for a first-row octahedral metal complex can be placed in the order

$$1^{st}\ \textit{Row Transition Metal}\text{: e-e repulsions} \sim \text{octahedral ligand field} > \text{spin-orbit} > k_B\text{T}$$

with electron–electron repulsions that lead to free ion terms of similar energetic magnitude to the overall octahedral ligand field. For the transition metals, the spin–orbit coupling is then smaller than these effects, which are smaller than the thermal energy under typical conditions (k_BT).

21.5.1 Spectroscopy of the Lanthanides

For the lanthanides, the spin–orbit coupling term is typically much larger than the ligand-field term due to the poor interaction between the *f*-orbitals and the ligands. In addition, the interelectron repulsions are about one-quarter as large as in a first-row transition element, while the spin–orbit coupling (ζ) is much larger. As a result, the following order is generally found for the lanthanides.

$$\textit{Lanthanides}\text{ : e-e repulsions} \sim \text{spin-orbit}\,(\zeta) > \text{ligand field} \sim k_B\text{T}$$

As for the transition metals where *d*–*d* transitions are Laporte forbidden and have relatively weak intensities, the *f*–*f* transitions of the lanthanides are similarly of low intensity with typical ε <300 $cm^{-1}M^{-1}$. The spectra, due to the weak ligand field effects, are typically affected very little by the environment, with gas-phase spectra of the ions being similar to the solution or solid-phase spectra of compounds.

The crystal-field splittings are small, but it is possible to assign the energy levels of some complexes fully. These experiments often involve cooling samples to near absolute zero with liquid helium to make the measurements because, as mentioned, k_BT is of similar energetic magnitude to the ligand field. The assignments, even beyond the experimental difficulties just mentioned, are a Herculean task, as there can be literally hundreds of energy levels and dozens of crystal-field parameters for a complex. For example, a crystal-field description of $Eu(O_2CCH_3)_3(OH_2)_2{\cdot}2H_2O$ has been developed based on X-ray diffraction, absorption spectroscopy, emission spectroscopy, and calculations which describes the 130 energy levels of the C_1 complex in terms of twenty-seven crystal-field parameters![66] The low-temperature absorption spectra for the europium acetate hydrate are shown in Fig. 21.26, along with their assignments. Note that the sharpness of the absorption bands in the spectra are typical for lanthanides due to the lack of ligand vibrational effects on the bands.

Continuing with Eu^{3+} as the example, the complexes of this ion will be f^6, and Hund's rules can be used to determine the ground electronic configurations. The state with the highest possible multiplicity will be favored, which will give all six electrons the same spin and 2S + 1 then equals 2(6/2) + 1 = 7. In other words, the ground state will be a septet. Hund's rules then state that the term with the highest spin multiplicity along with the highest angular momentum (*L*) is favored. Placing all the electrons into the *f*-orbitals with the same spin, which of course requires all of the electrons to be in different orbitals, is done while maximizing the angular momentum, shown as follows. This produces a ground state of (3 + 2 + 1 + 0–1–2 = 3→ an F term) ^{7}F.

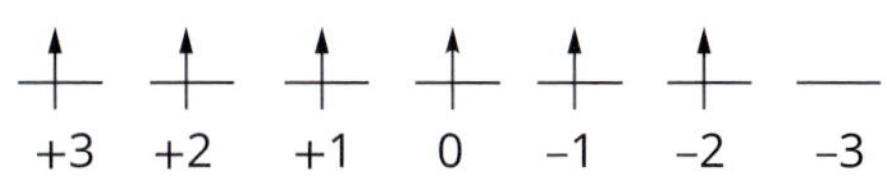

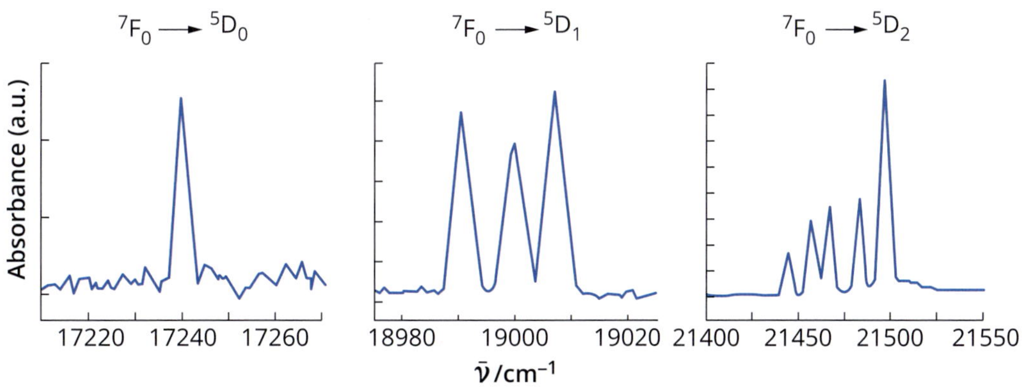

Fig. 21.26 Absorption spectra for $Eu(O_2CCH_3)_3(OH_2)_2{\cdot}2\,H_2O$ at 4.2 K. (Figure from the article by Karbowiak and Mondry.[66])

Hund's rules then state that for shells half-filled or less, the ground state has the lowest possible J value, and for states more than half-filled the ground state has the highest possible J value. The $4f^6$ configuration is less than half-filled, and the possible J terms are $L+S, L+S-1, L+S-2, \ldots, |L-S|$ or (for L = 3 and S = 3) $J = 6, 5, 4, 3, 2, 1, 0$. Consequently, the ground state of europium(III) is $^7F_0$. While this is the ground state, there are 119 terms for the $4f^6$ configuration of Eu(III)![67] A partial crystal-field splitting diagram for the ion is shown in Fig. 21.27.[68]

In contrast, cerium(III) and terbium(III) have very intense bands in the ultraviolet region due to transitions between *f*- and *d*-orbitals. The transitions involve excitation from the ground state f^1 of cerium(III), for example, to the *d*-manifold, and are therefore Laporte allowed. The terbium system is a (half-filled shell + 1) configuration, and the half-filled *f*-orbitals are stabilizing on excitation, just like the empty *f*-shell is for cerium.

The lanthanides are useful for their fluorescent and phosphorescent properties. For example, lanthanides have been used in color television tubes where red was from Eu(III) doped into Y_2O_3, green was from Ce(III) in CaS, and blue was from Eu(II) in $(Sr,Mg)_2P_2O_7$. The mixture produces white light. Furthermore, neodymium is used in one of the most common high-power laser sources, $Y_3Al_5O_{12}$ (yttrium aluminum garnet, YAG) doped with Nd(III), Nd:YAG. Furthermore, lanthanides have been explored of late in photoredox catalysis for organic transformations.[69–71]

21.5.2 Spectroscopy of the Actinides

The actinides have optical absorption properties similar to their lanthanide cousins, but are complicated by higher ligand-field perturbations on the energies. The absorption bands look slightly more transition-metal-like in that they are broader due to interactions with the ligands. As an example, the UV-visible–NIR absorption spectrum is shown in Fig. 21.28 for $\{UCl_5(THF)\}^-$ (red line) and $\{UBr_5(THF)\}^-$ (black line).[72]

Uranium(IV), for example, has an f^2 configuration and therefore a $^3H_4$ ground state. The band at 277 nm that does not shift with halide ligand was assigned to the $^3H_4$ to $^3F_2$ transition—an *f-d* transition that is Laporte allowed. The more intense band that shifts from 303 nm in the chloride to 322 nm in the bromide (*) was assigned to a charge-transfer transition.

Further into the visible and near-infrared (NIR), there are typically a series of narrow lanthanide-like transitions corresponding to Laporte forbidden *f–f* transitions (Fig. 21.29). These transitions, despite being *f*-orbital-centered, change somewhat from complex to complex (unlike what is usually found for the lanthanides) consistent with significant ligand interaction with the *f*-orbitals. As examples, the absorption spectra in this region in THF for $\{UCl_5(THF)\}^-$ (*top*) and a tris(aryloxy)amine azide by Meyer and coworkers are shown.[72–73] Even so, for complexes with similar ligand sets, these *f–f* transitions can serve as a "fingerprint" for the type of complex obtained. Comparing the tris(aryloxy)amine azide (*middle*) with the tris(aryloxy)tacn azide (*bottom*) shows that the two uranium(IV) 7-coordinate azide tris(aryloxide) complexes do have similar spectra in this region.[74]

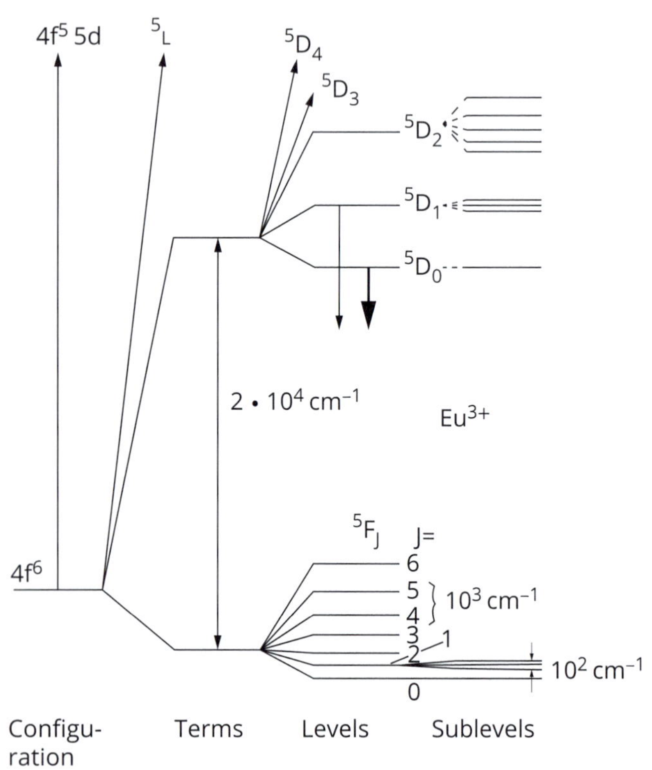

Fig. 21.27 Partial energy diagram for Eu(III) showing the relative energies of the electron–electron repulsions (terms), spin-orbit coupling, and the crystal-field splittings. The arrows pointing down indicate the levels from which luminescence typically occurs, $^5D_0$ and $^5D_1$. (Figure from Bünzli's article.[68])

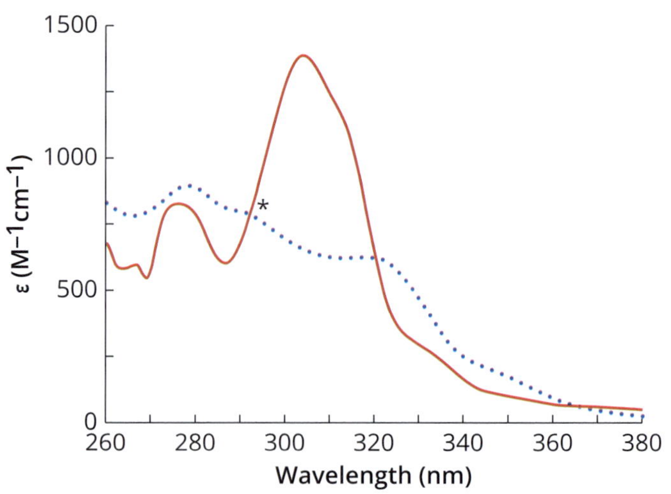

Fig. 21.28 UV-visible absorption spectrum of $\{UCl_5(THF)\}^-$ (red line) and $\{UBr_5(THF)\}^-$ (blue line) in THF. (The * indicates a feature due to a small uranyl impurity.) (Spectrum from the article by Hashem and coworkers.[72])

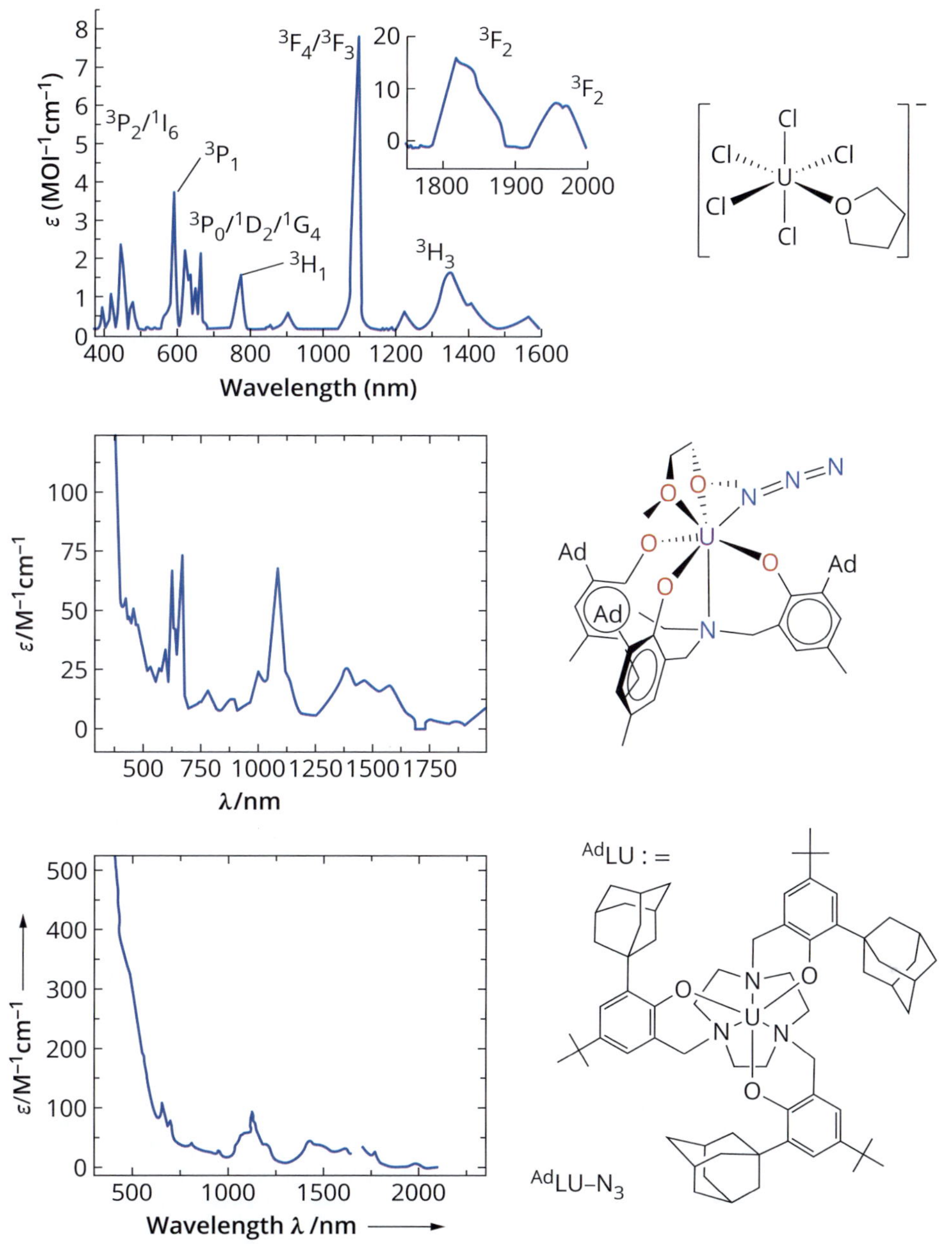

Fig. 21.29 (*top*) Visible and near-infrared (NIR) absorption spectrum for $\{UCl_5(THF)\}^-$. (Spectrum from the article by Hashem and coworkers.[72]) (*middle*) Visible–NIR absorption spectrum of the tris(phenoxyl)amine azide complex shown. (Figures from the article by Meyer and coworkers.[73]) (*bottom*) Visible–NIR absorption spectrum of the tris(phenoxyl)tacn azide complex shown. (Figures from the article by Meyer and coworkers.[74])

Chapter Summary

1. The sizes of the valence 4*f*-orbitals in the lanthanides leads to them typically being far too small to participate in bonding. As a result, they often act as a storehouse for electron spin.
2. The actinide 5*f*-orbitals are larger due to primogenic repulsion and often participate in bonding.
3. Many of the lanthanides are not rare in the Earth's crust, with Ce, Nd, and La all being more abundant than cobalt.
4. One application of gadolinium(III) is as an MRI contrast agent, where its high spin (7/2) and fast exchange rate of coordinated water is advantageous.
5. There are some investigations of the lanthanides, thorium, and uranium complexes as catalysts, including fairly extensive investigations into hydroelementation, where their large radial size is likely an advantage in achieving rapid reaction rates.

6. The lanthanides and actinides have very large spin–orbit coupling, and their electronic structure is understood in the context of a dominant *J*-term.
7. The *f–f* transitions in the actinides and lanthanides are Laporte-forbidden, which leads to fairly weak bands. For the lanthanides, where the *f*-orbitals rarely participate in bonding, this produces a signature of the metal center that changes little from complex to complex. The *f–f* transition for the actinides often change somewhat with ligand structure.

Practice Problems

1. A common material for a solid-state laser is to use a crystal of neodymium-doped ytterium aluminum garnet, $Y_3Al_5O_{12}$, (Nd:YAG). The Nd:YAG laser emits most intensely at 1,064 nm. Changing the host material to yttrium orthovanadate, YVO_4, does not shift this transition at all (1,064 nm), and using yttrium lithium fluoride only shifts it to 1,053 nm. Why is the emission so insensitive to the nature of the material into which neodymium is doped?
2. Explain why the lanthanide orbitals are contracted and rarely participate in bonding. Provide at least two causes for the effect.
3. All of the actinides are radioactive.
   a. What types of emitters are Th and U, and what are their half-lives?
   b. Why are thorium and uranium the only naturally occurring actinides?
   c. Subatomic particles can be viewed by using a cloud chamber. Search for a video using "uranium ore cloud chamber". Describe how a cloud chamber works, and what is happening in the video. Remember to give the link for your video.
4. Answer the following questions regarding uranocene, $U(C_8H_8)_2$.
   a. Draw the eclipsed structure of uranocene. What is the point group?
   b. For free $C_8H_8^{2-}$, draw the molecular orbitals for the π-system. Refer to Chapter 12 if necessary.
   c. For your answer in part b, place two $C_8H_8^{2-}$ rings parallel, and make in-phase and out-of-phase combinations.
   d. By inspection, what are the symmetries of the SALCs from part c in the point group?
   e. What orbitals on uranium would overlap with the SALCs?
5. In 2017, Kozimor and coworkers prepared and characterized $AmCl_6^{3-}$, a homoleptic americium complex with an octahedral geometry. The SALCs for the system should be well known to you at this point. What orbitals on Am can overlap with the SALC's? Discuss the differences between the bonding in O_h geometry for a transition metal and actinide.[36]

Cited References

1. Pollak, D.; Goddard, R.; Poerschke, K.-R., $Cs[H_2NB_2(C_6F_5)_6]$ Featuring an Unequivocal 16-Coordinate Cation *J. Am. Chem. Soc.* 2016, *138*, 9444–9451.
2. Desclaux, J. P., Relativistic Dirac-Fock Expectation Values for Atoms with Z = 1 to Z = 120 *At. Data Nucl. Data Tables* 1973, *12*, 311–406.
3. Denning, R. G., Electronic-structure and bonding in actinyl ions *Struct. Bond.* 1992, *79*, 215–276.
4. Cary, S. K.; Vasiliu, M.; Baumbach, R. E.; Stritzinger, J. T.; Green, T. D.; Diefenbach, K.; Cross, J. N.; Knappenberger, K. L.; Liu, G.; Silver, M. A.; DePrince, A. E.; Polinski, M. J.; Van Cleve, S. M.; House, J. H.; Kikugawa, N.; Gallagher, A.; Arico, A. A.; Dixon, D. A.; Albrecht-Schmitt, T. E., Emergence of californium as the second transitional element in the actinide series *Nature Comm.* 2015, *6*.
5. Greenwood, N. N.; Earnshaw, A., *Chemistry of the Elements* 1st ed.; Pergamon Press: Oxford 1984.
6. Silver, M. A.; Cary, S. K.; Johnson, J. A.; Baumbach, R. E.; Arico, A. A.; Luckey, M.; Urban, M.; Wang, J. C.; Polinski, M. J.; Chemey, A.; Liu, G.; Chen, K.-W.; Van Cleve, S. M.; Marsh, M. L.; Eaton, T. M.; de Burgt, L. J. v.; Gray, A. L.; Hobart, D. E.; Hanson, K.; Maron, L.; Gendron, F.; Autschbach, J.; Speldrich, M.; Koegerler, P.; Yang, P.; Braley, J.; Albrecht-Schmitt, T. E., Characterization of berkelium(III) dipicolinate and borate compounds in solution and the solid state *Science* 2016, *353*.
7. D'Angelo, P.; Zitolo, A.; Migliorati, V.; Chillemi, G.; Duvail, M.; Vitorge, P.; Abadie, S.; Spezia, R., Revised Ionic Radii of Lanthanoid(III) Ions in Aqueous Solution *Inorg. Chem.* 2011, *50*, 4572–4579.
8. Rastaetter, M.; Zulys, A.; Roesky, P. W., Bis(phosphinimino)methanide rare earth amides: Synthesis, structure, and catalysis of hydroamination/cyclization, hydrosilylation, and sequential hydroamination/hydrosilylation *Chem Eur. J.* 2007, *13*, 3606–3616.
9. Shannon, R. D.; Prewitt, C. T., Effective ionic radii in oxides and fluorides *Acta Cryst. B: Struc. Sci.* 1969, *B 25*, 925.
10. Evans, W. J., Tutorial on the Role of Cyclopentadienyl Ligands in the Discovery of Molecular Complexes of the Rare-Earth and Actinide Metals in New Oxidation States *Organometallics* 2016, *35*, 3088–3100.
11. Pimentel, G. C.; Spratley, R. D., *Understanding chemistry* Holden-Day: San Francisco, 1971.
12. Hobart, D. E.; Samhoun, K.; Young, J. P.; Norvell, V. E.; Mamantov, G.; Peterson, J. R., Stabilization of praseodymium(IV) and terbium(IV) in aqueous carbonate solution *Inorg. Nuc. Chem. Lett.* 1980, *16*, 321–328.
13. Palumbo, C. T.; Zivkovic, I.; Rosario, S.; Mazzanti, M., Molecular Complex of Tb in the +4 Oxidation State *J. Am. Chem. Soc.* 2019, *141*, 9827–9831.
14. Sroor, F. M. A.; Edelmann, F. T., Lanthanides: Tetravalent Chemistry: Inorganic In *Encyclopedia of Inorganic and Bioinorganic Chemistry*, Scott, R. A., Ed. Wiley: Hoboken, N.J., 2012.
15. Zhang, Q.; Hu, S.-X.; Qu, H.; Su, J.; Wang, G.; Lu, J.-B.; Chen, M.; Zhou, M.; Li, J., Pentavalent Lanthanide Compounds: Formation and Characterization of Praseodymium(V) Oxides *Angew. Chem. Int. Ed.* 2016, *55*, 6896–6900.
16. Hu, S.-X.; Jian, J.; Su, J.; Wu, X.; Li, J.; Zhou, M., Pentavalent lanthanide nitride-oxides: NPrO and $NPrO^-$ complexes with N equivalent to Pr triple bonds *Chem. Sci.* 2017, *8*, 4035–4043.
17. Bohnet, M., *Ullmann's encyclopedia of industrial chemistry* 6th ed.; Wiley-VCH: Weinheim, 2003.
18. Solola, L. A.; Zabula, A. V.; Dorfner, W. L.; Manor, B. C.; Carroll, P. J.; Schelter, E. J., Cerium(IV) Imido Complexes: Structural, Computational, and Reactivity Studies *J. Am. Chem. Soc.* 2017, *139*, 2435–2442.

19. Schadle, D.; Meermann-Zimmermann, M.; Schadle, C.; Maichle-Mossmer, C.; Anwander, R., Rare-Earth Metal Complexes with Terminal Imido Ligands *Eur. J. Inorg. Chem.* 2015, 1334–1339.
20. Zimmermann, M.; Litlabo, R.; Tornroos, K. W.; Anwander, R., "Metastable" $Lu(GaMe_4)_3$ Reacts Like Masked $LuMe_3$: Synthesis of an Unsolvated Lanthanide Dimethyl Complex *Organometallics* 2009, *28*, 6646–6649.
21. Clark, D. L.; Gordon, J. C.; Hay, P. J.; Poli, R., Existence and stability of lanthanide-main group element multiple bonds. New paradigms in the bonding of the 4f elements. A DFT study of Cp_2CeZ (Z = F^+, O, NH, CH^-, CH_2 and the ligand adduct $Cp_2Ce(CH_2)(NH_3)$ *Organometallics* 2005, *24*, 5747–5758.
22. However, there is a "chicken and egg" problem with these types of arguments. A better match in size between orbitals may occur if there is better hybridization, but hybridization typically increases for orbitals of similar size. Perhaps the f-orbital size is increasing at higher oxidation state leading to the better hybridization.
23. Loble, M. W.; Keith, J. M.; Altman, A. B.; Stieber, S. C. E.; Batista, E. R.; Boland, K. S.; Conradson, S. D.; Clark, D. L.; Pacheco, J. L.; Kozimor, S. A.; Martin, R. L.; Minasian, S. G.; Olson, A. C.; Scott, B. L.; Shuh, D. K.; Tyliszczak, T.; Wilkerson, M. P.; Zehnder, R. A., Covalency in Lanthanides. An X-ray Absorption Spectroscopy and Density Functional Theory Study of $LnCl_6^{x-}$ (x = 3, 2) *J. Am. Chem. Soc.* 2015, *137*, 2506–2523.
24. Zhang, Q. N.; Hu, S. X.; Qu, H.; Su, J.; Wang, G. J.; Lu, J. B.; Chen, M. H.; Zhou, M. F.; Li, J., Pentavalent Lanthanide Compounds: Formation and Characterization of Praseodymium(V) Oxides *Angew. Chem. Int. Ed.* 2016, *55*, 6896–6900.
25. Kirk-Othmer, R. E., *Kirk-Othmer Encyclopedia of Chemical Technology*. 5th ed.; Wiley-Interscience Imprint, John Wiley & Sons: Hoboken, 2004.
26. Wahsner, J.; Gale, E. M.; Rodríguez-Rodríguez, A.; Caravan, P., Chemistry of MRI Contrast Agents: Current Challenges and New Frontiers *Chemical Reviews* 2019, *119*, 957–1057.
27. Fry, C. G., The Nobel Prize in Medicine for Magnetic Resonance Imaging *Journal of Chemical Education* 2004, *81*, 922.
28. Morrow, J. R.; Tóth, É., Next-Generation Magnetic Resonance Imaging Contrast Agents *Inorganic Chemistry* 2017, *56*, 6029–6034.
29. Hu, F. Q.; Joshi, H. M.; Dravid, V. P.; Meade, T. J., High-performance nanostructured MR contrast probes *Nanoscale* 2010, *2*, 1884–1891.
30. Assary, R. S.; Atesin, A. C.; Li, Z.; Curtiss, L. A.; Marks, T. J., Reaction Pathways and Energetics of Etheric C-O Bond Cleavage Catalyzed by Lanthanide Triflates *ACS Catalysis* 2013, *3*, 1908–1914.
31. Atesin, A. C.; Ray, N. A.; Stair, P. C.; Marks, T. J., Etheric C-O Bond Hydrogenolysis Using a Tandem Lanthanide Triflate/Supported Palladium Nanoparticle Catalyst System *J. Am. Chem. Soc.* 2012, *134*, 14682–14685.
32. Meyer, G., All the Lanthanides Do It and Even Uranium Does Oxidation State +2 *Angew. Chem. Int. Ed.* 2014, *53*, 3550–3551.
33. Vlaisayljevich, B.; Diaconescu, P. L.; Lukens, W. L.; Gagliardi, L.; Cummins, C. C., Investigations of the Electronic Structure of Arene-Bridged Diuranium Complexes *Organometallics* 2013, *32*, 1341–1352.
34. Billow, B. S.; Livesay, B. N.; Mokhtarzadeh, C. C.; McCracken, J.; Shores, M. P.; Boncella, J. M.; Odom, A. L., Synthesis and Characterization of a Neutral U(II) Arene Sandwich Complex *J. Am. Chem. Soc.* 2018, *140*, 17369–17373.
35. Diamond, R. M.; Street, K.; Seaborg, G. T., An ion-exchange study of possible hybridized 5f-bonding in the actinides *J. Am. Chem. Soc.* 1954, *76*, 1461–1469.
36. Cross, J. N.; Su, J.; Batista, E. R.; Cary, S. K.; Evans, W. J.; Kozimor, S. A.; Mocko, V.; Scott, B. L.; Stein, B. W.; Windorff, C. J.; Yang, P., Covalency in Americium(III) Hexachloride *J. Am. Chem. Soc.* 2017, *139*, 8667–8677.
37. Langeslay, R. R.; Fieser, M. E.; Ziller, J. W.; Furche, F.; Evans, W. J., Synthesis, structure, and reactivity of crystalline molecular complexes of the $\{[C_5H_3(SiMe_3)_2]_3Th\}^-$ anion containing thorium in the formal +2 oxidation state *Chem. Sci.* 2015, *6*, 517–521.
38. Blake, P. C.; Lappert, M. F.; Taylor, R. G.; Atwood, J. L.; Hunter, W. E.; Zhang, H. M., Synthesis, spectroscopic properties and crystal-structures of $[ML_2Cl_2]$ [M = Th or U; L = η-$C_5H_3(SiMe_3)_2$-1.3] and $[UL_2X_2]$ (X = Br, I or BH_4) *Dalton Trans.* 1995, 3335–3341.
39. Blake, P. C.; Lappert, M. F.; Atwood, J. L.; Atwood, J. L.; Zhang, H. M., The synthesis and characterization, including X-ray-diffraction study, of Th(η-$C_5H_3(SiMe_3)$): The 1st thorium(III) crystal-structure *Chem. Comm.* 1986, 1148–1149.
40. In actuality, Evans's group used a slight modification of the Lappert procedure to produce ThCp"3, utilizing a bromide precursor instead of chloride with KC8 as the reductant.
41. Kiplinger, J. L.; Morris, D. E.; Scott, B. L.; Burns, C. J., Convenient synthesis, structure, and reactivity of $(C_5Me_5)U(CH_2C_6H_5)_3$: a simple strategy for the preparation of monopentamethylcyclopentadienyl uranium(IV) complexes *Organometallics* 2002, *21*, 5978–5982.
42. Clark, D. L.; Sattelberger, A. P., Synthesis of $UI_3(THF)_5$ *Inorg. Synth.* 1997, *31*, 307.
43. Monreal, M. J.; Thomson, R. K.; Cantat, T.; Travia, N. E.; Scott, B. L.; Kiplinger, J. L., UI_4(1,4-dioxane)$_2$, UCl_4(1,4-dioxane)$_2$, and UI_3(1,4-dioxane)$_{1.5}$: Stable and Versatile Starting Materials for Low- and High-Valent Uranium Chemistry *Organometallics* 2011, *30*, 2031–2038.
44. Cotton, F. A.; Schmid, G., Mononuclear molybdenum(IV) complexes with two multiply bonded chalcogen ligands in trans configuration and chelating biphosphine ligands *Inorg. Chem.* 1997, *36*, 2267–2278.
45. Marzotto, A.; Graziani, R.; Bombieri, G.; Forsellini, E., Preparation, properties and crystal structure of bis(acetylcholine)tetrabromodioxouranium(VI) *J. Cryst. Mol. Struct.* 1974, *4*, 253–262.
46. Serezhkin, V. N.; Grigoriev, M. S.; Abdulmyanov, A. R.; Fedoseev, A. M.; Savchenkov, A. V.; Serezhkina, L. B., Synthesis and X-ray Crystallography of $Mg(H_2O)_6$ $[AnO_2(C_2H_5COO)_3]_2$ (An = U, Np, or Pu) *Inorg. Chem.* 2016, *55*, 7688–7693.
47. Arney, D. S. J.; Burns, C. J.; Smith, D. C., Synthesis and structure of the 1st uranium(VI) organometallic complex *J. Am. Chem. Soc.* 1992, *114*, 10068–10069.
48. Spencer, L. P.; Gdula, R. L.; Hayton, T. W.; Scott, B. L.; Boncella, J. M., Synthesis and reactivity of bis(imido) uranium(VI) cyclopentadienyl complexes *Chem. Commun.* 2008, 4986–4988.
49. Swartz, D. L.; Spencer, L. P.; Scott, B. L.; Odom, A. L.; Boncella, J. M., Exploring the coordination modes of pyrrolyl ligands in bis(imido) uranium(VI) complexes *Dalton Trans.* 2010, *39*, 6841–6846.
50. Akona, S. B.; Fawcett, J.; Holloway, J. H.; Russell, D. R.; Leban, I., Structures of *cis* and *trans*-dichlorodioxobis(triphenylphosphine oxide)-uranium(VI) *Acta Cryst. C* 1991, *47*, 45–48.
51. Tatsumi, K.; Hoffmann, R., Bent cis d^0 MoO_2^{2+} vs linear trans d^0f^0 UO_2^{2+}: A significant role for non-valence 6p orbitals in uranyl *Inorg. Chem.* 1980, *19*, 2656–2658.
52. Gregson, M.; Lu, E.; Mills, D. P.; Tuna, F.; McInnes, E. J. L.; Hennig, C.; Scheinost, A. C.; McMaster, J.; Lewis, W.; Blake, A. J.; Kerridge, A.; Liddle, S. T., The inverse-trans-influence in tetravalent lanthanide and actinide bis(carbene) complexes *Nature Comm.* 2017, *8*.
53. Avdeef, A.; Zalkin, A.; Raymond, K. N.; Hodgson, K. O., Two isostructural actinide π-complexes - crystal and molecular structure of bis(cyclooctatetraenyl)uranium(IV), $U(C_8H_8)_2$, and bis(cyclooctatetraenyl)thorium(IV), $Th(C_8H_8)_2$ *Inorg. Chem.* 1972, *11*, 1083.
54. DeRidder, D. J. A.; Rebizant, J.; Apostolidis, C.; Kanellakopulos, B.; Dornberger, E., Bis(cyclooctatetraenyl)neptunium(IV) *Acta Cryst. C* 1996, *52*, 597–600.
55. Karraker, D. G.; Stone, J. A., Bis(cyclooctatetraenyl)neptunium(III) and bis(cyclooctatetraenyl)plutonium(III) compounds *J. Am. Chem. Soc.* 1974, *96*, 6885–6888.
56. Higgins, J. A.; Cloke, F. G. N.; Roe, S. M., Synthesis and CO2 Insertion Chemistry of Uranium(IV) Mixed-Sandwich Alkyl and Hydride Complexes *Organometallics* 2013, *32*, 5244–5252.
57. Fortier, S.; Melot, B. C.; Wu, G.; Hayton, T. W., Homoleptic Uranium(IV) Alkyl Complexes: Synthesis and Characterization *J. Am. Chem. Soc.* 2009, *131*, 15512–15521.
58. Johnson, S. A.; Kiernicki, J. J.; Fanwick, P. E.; Bart, S. C., New Benzylpotassium Reagents and Their Utility for the Synthesis of Homoleptic Uranium(IV) Benzyl Derivatives *Organometallics* 2015, *34*, 2889–2895.
59. Kraft, S. J.; Fanwick, P. E.; Bart, S. C., Carbon-Carbon Reductive Elimination from Homoleptic Uranium(IV) Alkyls Induced by Redox-Active Ligands *J. Am. Chem. Soc.* 2012, *134*, 6160–6168.
60. Lam, O. P.; Meyer, K., Hydrogenation of CO at a Uranium(III) Center *Angew. Chem. Int. Ed.* 2011, *50*, 9542–9544.
61. Fox, A. R.; Bart, S. C.; Meyer, K.; Cummins, C. C., Towards uranium catalysts *Nature* 2008, *455*, 341–349.
62. Lin, Z. R.; Marks, T. J., Metal, bond-energy, and ancillary ligand effects on actinide-carbon sigma-bond hydrogenolysis: A kinetic and mechanistic study *J. Am. Chem. Soc.* 1987, *109*, 7979–7985.
63. Fagan, P. J.; Manriquez, J. M.; Maatta, E. A.; Seyam, A. M.; Marks, T. J., Synthesis and properties of bis(pentamethylcyclopentadienyl) actinide hydrocarbyls and hydrides: A new class of highly reactive f-element organo-metallic compounds *J. Am. Chem. Soc.* 1981, *103*, 6650–6667.
64. Halter, D. P.; Heinemann, F. W.; Maron, L.; Meyer, K., The Role of Uranium-Arene Bonding in H2O Reduction Catalysis *Nature Chem.* 2018, *10*, 259–267.

65. Halter, D. P.; Heinemann, F. W.; Bachmann, J.; Meyer, K., Uranium-mediated electrocatalytic dihydrogen production from water *Nature* 2016, *530*, 317.
66. Karbowiak, M.; Mondry, A., Electronic energy-level structure of $4f^6$ configuration in europium(III) triacetate tetrahydrate *Chem. Phys.* 2008, *354*, 86–93.
67. Binnemans, K., Interpretation of europium(III) spectra *Coord. Chem. Rev.* 2015, *295*, 1–45.
68. Bunzli, J. C., The europium(III) ion as spectroscopic probe in bioinorganic chemistry *Inorg. Chim. Acta* 1987, *139*, 219–222.
69. Guo, J.-J.; Hu, A.; Chen, Y.; Sun, J.; Tang, H.; Zuo, Z., Photocatalytic C-C Bond Cleavage and Amination of Cycloalkanols by Cerium(III) Chloride Complex *Angew. Chem. Int. Ed.* 2016, *55*, 15319–15322.
70. Yin, H.; Carroll, P. J.; Manor, B. C.; Anna, J. M.; Schelter, E. J., Cerium Photosensitizers: Structure-Function Relationships and Applications in Photocatalytic Aryl Coupling Reactions *J. Am. Chem. Soc.* 2016, *138*, 5984–5993.
71. Yin, H.; Jin, Y.; Hertzog, J. E.; Mullane, K. C.; Carroll, P. J.; Manor, B. C.; Anna, J. M.; Schelter, E. J., The Hexachlorocerate(III) Anion: A Potent, Benchtop Stable, and Readily Available Ultraviolet A Photosensitizer for Aryl Chlorides *J. Am. Chem. Soc.* 2016, *138*, 16266–16273.
72. Hashem, E.; Swinburne, A. N.; Schulzke, C.; Evans, R. C.; Platts, J. A.; Kerridge, A.; Natrajan, L. S.; Baker, R. J., Emission spectroscopy of uranium(IV) compounds: a combined synthetic, spectroscopic and computational study *RSC Adv.* 2013, 3, 4350–4361.
73. Lam, O. P.; Franke, S. M.; Nakai, H.; Heinemann, F. W.; Hieringer, W.; Meyer, K., Observation of the Inverse Trans Influence (ITI) in a Uranium(V) Imide Coordination Complex: An Experimental Study and Theoretical Evaluation *Inorg. Chem.* 2012, *51*, 6190–6199.
74. Castro-Rodriguez, I.; Nakai, H.; Zakharov, L. N.; Rheingold, A. L.; Meyer, K., A linear, O-coordinated η^1-CO_2 bound to uranium *Science* 2004, *305*, 1757–1759.

APPENDIX A

Polar Covalence

Polar covalence theory is not commonly covered in inorganic courses but can be used to illustrate many general features in bonding, calculate charges by hand, estimate bond energies, and so on. In this appendix, a few additional features of the theory are covered to supplement the large amount of information in Chapters 3 and 8.

The Constant 332 in the Equation to Determine E_i

The energy of the ionic interaction is estimated as simply being due to the Coulombic energy. More specifically, the ionic interaction is the electric potential energy (U_E) due to the interaction of the cation and anion at distance R_0. The equation for the potential energy is as follows, where q_1 and q_2 are the magnitudes of the charges and ε_0 is the permittivity of a vacuum.

$$U_E = \frac{1}{4\pi\varepsilon_0}\frac{q_1 q_2}{R_0}$$

The charges are +1 and −1 in all cases (Fig. 3.16). The unit electric charge is 1.602×10^{-19} A · s and $\varepsilon_0 = 8.854 \times 10^{-12}$ A^2 · s^4/kg · m^3.

$$U_E = \frac{1}{4\pi(8.854 \times 10^{-12}\,A^2 s^4/Kg \cdot m^3)}\frac{(1.602 \times 10^{-19}\,A \cdot s)(-1.602 \times 10^{-19}\,A \cdot s)}{R_0}$$

$$U_E = \frac{-2.307 \times 10^{-28}\,Kg \cdot m^3/s^2}{R_0}$$

One joule = 1 Kg · m^2/s^2. Consequently, the numerator is in J · m in this expression. Converting to kcal/mol is achieved by multiplying by 2.39×10^{-4} and by Avogadro's number, 6.022×10^{23} mol^{-1}, which produces the following expression.

$$U_E = \frac{-3.32 \times 10^{-8}\,kcal \cdot m/mol}{R_0}$$

Here the electric potential energy is negative, signifying an attractive interaction between the charges. In terms of bond energy we would consider the value to be a positive energy contributing to the strength of the bond. Making the value positive, substituting E_i for U_E in the equation, and converting from meters to Ångstroms, produces Eq. 3.19, where the ionic bond energy in kcal/mol is found if the distance, R_0, is in Å.

$$E_i = \frac{332\ kcal \cdot \text{Å}/mol}{R_0\ \text{Å}}$$

The equation in SI units of kilojoules with R_0 in picometers would be as follows.

$$E_i = \frac{1.39 \times 10^5 kJ \cdot pm/mol}{R_0 pm}$$

Bond Dissociation Enthalpies for More Complex Molecules

The same principles may be applied to more complex molecules that were applied to binary (EX_n) compounds in the text. If we propose that electronegativity equalization holds for all molecules, the ideas behind polar covalence theory can be applied consistently to any system. However, there are some complicating factors when applying the theory to more complex systems, which will be expanded here.

As an alternative to treating the entire molecule as previously, the Sanderson method can be used to calculate the electronegativity of a group. We could assume that the methyl group, for example, can be treated using the equivalent atom method (EAM).[1] In other words, we could postulate that we can produce an equivalent single atom with the electronegativity and bonding properties of the methyl group, and use that in the calculations.

The EAM method is of greatest utility in cases where the charge scheme for polar covalence breaks down, as in cases where there are multiple environments for the same element. One example presented in Chapter 3 is $F_3C{-}CH_3$, where the usual charge calculation method will show that there is no polarity in the C–C bond, which is obviously incorrect.

We can instead treat $F_3C{-}CH_3$ as being a bond between two pseudoatoms (F_3C)–(CH_3), which have different electronegativities derived by using Eq. 3.23. To find the electronegativity of the CH_3 $\left(\chi_{CH_3}\right)$ and CF_3 $\left(\chi_{CF_3}\right)$ groups, the following calculations are carried out.

$$\chi_{CF_3} = \sqrt[4]{2.746 \times 4^3} = 3.641$$
$$\chi_{CH_3} = \sqrt[4]{2.746 \times 2.592^3} = 2.630$$

The molecular electronegativity is

$$\chi_m = \sqrt[8]{2.746^2 \times 4^3 \times 2.592^3} = 3.094.$$

The charges on the (CH_3) and (CF_3) pseudoatoms are then calculated as before.

$$\delta_{CH_3} = \frac{\left(\chi_m - \chi_{CH_3}\right)}{1.57\sqrt{\chi_{CH_3}}} = \frac{(3.094 - 2.630)}{1.57\sqrt{2.630}} = +\,0.182$$
$$\delta_{CF_3} = \frac{\left(\chi_m - \chi_{CF_3}\right)}{1.57\sqrt{\chi_{CF_3}}} = \frac{(3.094 - 3.641)}{1.57\sqrt{3.641}} = -\,0.183$$

The ionic blending coefficient is then $t_i = 0.183$.

E_c and E_i are calculated as described in Chapter 3. The C–C distance in $F_3C{-}CH_3$ is 1.497 Å (= R_o). The slight shortening of the C–C bond from ethane and hexafluoroethane will cause slight increases in both the covalent and ionic bond energies.

$$E_c = \frac{R_c}{R_0}\sqrt{E_{CC}E_{CC}} = \frac{2 \times 0.772}{1.497} \times 85.4 = 88.1$$
$$E_i = \frac{332}{R_0} = \frac{332}{1.497} = 221.8$$

The BDE in the polar compound using the pseudoatom method with polar covalence theory is then

$$E_{(F_3C-CH_3)} = t_c E_c + t_i E_i$$
$$\text{using}\left(1 = t_c + t_i\right)$$
$$E_{(F_3C-CH_3)} = \left(1 - t_i\right)E_c + t_i E_i$$
$$E_{(F_3C-CH_3)} = (1 - 0.183)88.1 + (0.183)221.8$$
$$E_{(F_3C-CH_3)} = 112.6$$

and the value of 112.6 kcal/mol is a reasonable estimate of the G3 energy for the C–C bond $H_3C{-}CF_3$ of 103.1 kcal/mol—a 9% error. This error is higher than typical for polar covalence on binary compounds, perhaps due to the necessity of employing the EAM.

We can apply this method to any molecule, but it may increase errors somewhat. In a few of the following cases, EAM will be used where it would ordinarily not be necessary just to illustrate.

The usual method to find the B–C (or C–H if desired) bond energies in $B(CH_3)_3$ can be carried out much as before. The electronegativity of the molecule is based on the geometric mean of all the atoms in the molecule. For BMe_3, which has the formula BC_3H_9, the equation and results are as follows.

$$\chi_M = \sqrt[13]{\chi_H^9 \chi_C^3 \chi_B} = \sqrt[4]{(2.592)^9(2.592)^3 2.275} = 2.601$$

The charge on a particular atom in a molecule is dependent upon the electronegativity difference between the individual atoms and the molecular electronegativity. This difference is divided by the electronegativity of the atom with unit change in charge. The results mathematically and pictorially are shown as follows. As would be expected for this case, all the hydrogens, for example, bear the same charge.

$$\delta_B = \frac{\chi_M - \chi_B}{1.57\sqrt{\chi_B}} = \frac{2.601 - 2.275}{1.57\sqrt{2.275}} = +0.138$$
$$\delta_C = \frac{\chi_M - \chi_C}{1.57\sqrt{\chi_C}} = \frac{2.601 - 2.746}{1.57\sqrt{2.746}} = -0.056$$
$$\delta_H = \frac{\chi_M - \chi_H}{1.57\sqrt{\chi_H}} = \frac{2.601 - 2.592}{1.57\sqrt{2.592}} = +0.004$$

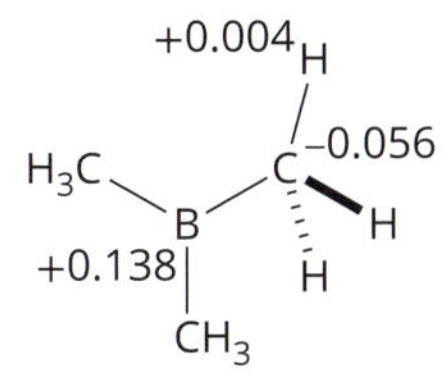

To calculate the BDE of the B–C bond in the molecule, the ionic blending coefficient, t_i, is estimated as the average of the absolute values of the charges across that bond. The covalent and ionic bond energies are calculated by using the B–C bond distance in the molecule of 1.583 Å. (Distance found using CCSD with 6-311++G** on Gaussian09.) The BDE obtained is 93.9 kcal/mol, which has an error of 7% from the experimental value of 87.5 kcal/mol for the first B–Me bond cleavage.

$$t_i = \frac{0.138 + 0.056}{2} = 0.097$$

$$E_c = \frac{0.772 + 0.822}{1.582}\sqrt{76.7 \times 85.4} = 81.5$$

$$E_i = \frac{332}{1.583} = 209.7$$

$$\begin{aligned} E_{B-Me} &= E_c - t_i E_c + t_i E_i \\ &= 81.5 - (0.097)81.5 + (0.097)209.7 = 93.9 \end{aligned}$$

The results for more examples relative to their experimental and theoretical BDEs are shown in Table 3.3. In all cases, the experimental values are for the cleavage of the first bond, and the polar covalence theory estimates are for the average of all the bonds, which probably explains most of the (generally fairly small) differences calculated and experimental BDEs.

Charge Dissipation Through Groups

One consequence of the principle of electronegativity equalization is that charge tends to be dissipated throughout a group instead of localizing on particular atoms. The theory of polar covalence can be used to determine the electronegativity of charged groups and the charges on the individual atoms within the group.

Eq. 3.21 relates charge and Sanderson electronegativity under the theory of polar covalence. If we rearrange the expression in Eq 3.21, it is seen that the equation relates the electronegativity of an atom with some partial charge, χ_δ, to the electronegativity of the free atom A, χ_A, and the partial charge, δ. The electronegativity of an atom is linearly related to the partial charge on the atom with the intercept as the electronegativity at

Table 3.3 Examples of bond dissociation enthalpies (kcal/mol) calculated by using polar covalence theory versus experimental or theoretical values for the first bond-breaking event.

Compound[b]	Bond[c]	Bond Length[a] (Å)	Calc fromTheory of Polar Covalence[d]	Experimental or Theoretical (kcal/mol)
BMe_3	$B{-}CH_3$	1.583	93.9	87[2], 87.5[3], 102.7[e]
$AlMe_3$	$Al{-}CH_3$	1.971	90.8	62.9[3], 82.7[e]
$AlEt_3$	Al–Et	1.979	90.5	77.0[e]
$GaMe_3$	$Ga{-}CH_3$	1.991	66.8	57.5[3], 63 ± 4[4], 76.7[e]
$B(OH)_3$	B–OH	1.370	133.6[g]	132.7 ± 1.7[5,6], 149.1[e]
$B(OMe)_3$	B–OMe	1.371	135.3[g]	118.0 ± 2[6], 138.2[e]
$B(OPr^n)_3$	$B{-}OPr^n$	1.369	134.8[g]	119.0 ± 2[6]
CMe_4	C–{Me}	1.528	86.3	85.3[3], 84 ± 1[4], 87.7[e]
OMe_2	O–Me	1.413	82.9[f]	84.5[3], 83.3 ± 2[4], 83.3[e]
NMe_3	N–Me	1.455	75.6[f]	71.5[3], 82.2 ± 2[4], 78.5[e]
PMe_3	P–Me	1.863	73.3[f]	65.3[3], 73.2[e]
$AsMe_3$	As–Me	1.989	58.5	51.5[3], 67 ± 4[4], 65.4[e]
$SbMe_3$	Sb–Me	2.201	57.7	49.7[3], 61 ± 4[4]
$BiMe_3$	Bi–Me	2.272	59.3	33.7[3], 52 ± 4[4]
$SnMe_4$	$Sn{-}CH_3$	2.169	59.7	53.3[3], 52.6 ± 2[7], 43.0[8], 71 ± 4[9]
$HgMe_2$	$Hg{-}CH_3$	2.083	44.4	30[3], 61 ± 4[9]

[a] Bond distances (R_0) were found by using gas-phase calculations B3LYP/6-311++G** or CCSD/6-311++G**, except for $SnMe_4$, $BiMe_3$, and $SbMe_3$, where CCSD/LANL2DZ was used. For $HgMe_2$ the gas-phase value, 2.083 Å, from electron diffraction was used.
[b] All values are for the monomers in the gas phase.
[c] Bond investigated for its BDE.
[d] Reorganization energies are not compensated for in the values given.
[e] Value from the G3 method using Gaussian09.
[f] Fully weakened value.
[g] Fully unweakened value.

zero charge and the slope as the change in electronegativity with unit charge, $1.57\sqrt{\chi_A} = \Delta\chi_A$.

$$\chi_\delta = \chi_A + \left(1.57\sqrt{\chi_A}\right)\delta$$
$$= \chi_A + \left(\Delta\chi_A\right)\delta$$

The electronegativity of any charged group—say an isolated anion such as NO_3^-—can be found by assigning an arbitrary electronegativity to the group and then calculating all the charges for the atoms in the group. Summing the charges on the atoms produces the charge required to be on the group for it to have the electronegativity assigned. The electronegativity is then assigned a different value, and the charge of the group is recalculated. Once the electronegativity and charge are known for two different points for that group, the line relating charge to electronegativity can be found. Once the equation for that group is known, its electronegativity with any charge can be calculated.

As an illustration, let us find the electronegativity of the hexafluoroantimonate monoanion, SbF_6^-, and the charges on the atoms. If we estimate the electronegativity of the SbF_6 assembly to be 3.00, the charges on the antimony and fluorine atoms are ($\chi_M = 3.00$) +0.220 and −0.318, respectively.

$$\delta_A = \frac{(\chi_M - \chi_A)}{1.57\sqrt{\chi_A}} = \frac{(\chi_M - \chi_A)}{\Delta\chi_A}$$
$$\delta_{Sb} = \frac{(3.00 - 2.458)}{2.461} = +0.220$$
$$\delta_{Sb} = \frac{(3.00 - 4.000)}{3.140} = -0.318$$

The total charge on the molecule corresponding to a molecular electronegativity of 3.00 is then $\delta_{Sb} + 6\delta_F = 0.22 + 6(-0.318) = -1.691$. Obviously, this value for the electronegativity is incorrect because the actual charge is −1. So, let us try a higher electronegativity value.

If we estimate the electronegativity of the SbF_6 group to be 3.50, the charges on the antimony and fluorine atoms are +0.423 and −0.159, respectively.

$$\delta_A = \frac{(\chi_M - \chi_A)}{1.57\sqrt{\chi_A}} = \frac{(\chi_M - \chi_A)}{\Delta\chi_A}$$
$$\delta_{Sb} = \frac{(3.50 - 2.458)}{2.461} = +0.436$$
$$\delta_{Sb} = \frac{(3.50 - 4.000)}{3.140} = -0.159$$

The total charge on the molecule corresponding to an electronegativity of 3.50 is then $\delta_{Sb} + 6\delta_F = 0.436 + 6(-0.159) = -0.518$. This value for the total charge is too low. Considering that our previous calculation produced a value that was too high, a line can be found relating the total charge to electronegativity in the system where the actual value, −1, is between these two calculated values, −0.518 and −1.691.

We determine the line including these two points has the form $\chi = 0.426\,\delta + 3.721$, where δ is the charge on SbF_6 and χ is the electronegativity of the group. The intercept of the line is the electronegativity of SbF_6 without a charge, which is about the same value found from using Eq. 3.21. The electronegativity of SbF_6^-, with the actual charge $\delta = -1.0$, is 3.295 on the Sanderson electronegativity scale.

The electronegativity of various groups relative to total charge on the collection of atoms can be found. Then, because every atom in a molecule has the same electronegativity, the partial charge on each atom can be calculated. In this way, the dissipation of the charge through the entire group, driven by electronegativity equalization, can be seen.

How does having a group of atoms to dissipate charge differ from a single atom? If Eq. 3.21 is used to plot the Sanderson electronegativity of atoms, χ_A^S, versus the partial charge, δ, the slopes of the lines are $1.57\sqrt{\chi_A}$.

While the electronegativity of an individual atom changes sharply with charge, a group will dissipate the charge throughout all of its atoms, and the partial charge on the bonded atom does not change dramatically in most cases (Fig. A1).

Let us consider how an atom or group is affected by a local charge, such as formation of a cation or anion on the group. An important factor in the stabilization is how the charge is dissipated through the atoms of the group. In general, isolating large charges on one or a few atoms is energetically costly and is disfavored; this isolation of charge on one atom causes its electronegativity to rise or fall sharply as well. In Fig. A1, steep lines are found for the individual atoms (H, F, Cl) with changes in charge. However, for the atoms in a group the electronegativity changes much more slowly with charge. In other words, groups support the charge without large, and energetically costly, changes to electronic structure by dissipating the electron density change throughout the group—a result

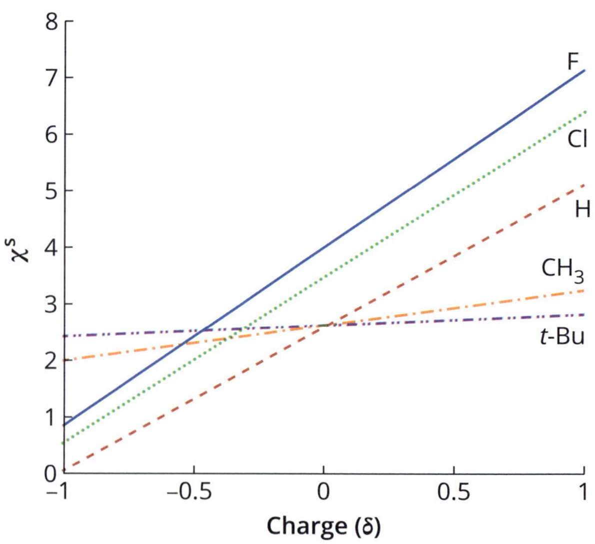

Fig. A1 Sanderson electronegativity (χ^S) of the atom or group versus partial charge (δ).

perhaps obvious from electronegativity equalization but very important for understanding reactivity.

This explains the chemical behavior of groups relative to individual atoms. For example, alkyl groups are slightly more electronegative (methyl $\chi^S_{CH_3} = 2.630$) than hydrogen ($\chi^S_H = 2.592$) due to the higher electronegativity of carbon. However, alkyl groups are often *more* electron-donating than H in molecules when attached to electronegative groups or atoms such as N, O, aryl, or halogens. Because hydrogen's electronegativity rises much more steeply with charge, the alkyl groups are actually more electron-releasing when attached to electronegative groups, because they dissipate the charge throughout the atoms of the group. This explains, for example, why alkyl groups are more activating than H in electrophilic aromatic substitution, even though the calculated electronegativity of neutral alkyl groups suggests that they are slightly more electron-withdrawing than hydrogen. In fact, the charge-dissipating effect of groups explains many chemical phenomena, such as why basicity of amines decreases in the order $NR_3 > HNR_2 > H_2NR > NH_3$ (R = alkyl group such as methyl) as another example.[10]

Cited References

1. Mortier, W. J., Electronegativity equalization and its applications *Struct. Bond.* 1987, *66*, 125–143.
2. Bamford, C. H.; Tipper, C. F. H., *Decomposition of Inorganic and Organometallic Compounds* Elsevier: Amsterdam, New York, 1972.
3. Long, L. H., Dissociation Energies of Metal-Carbon bonds and the Excitation Energies of Metal Atoms in Combination *Pure Appl. Chem.* 1961, *2*, 61–69.
4. McMillen, D. F.; Golden, D. M., Hydrocarbon bond-dissociation energies *Annu. Rev. Phys. Chem.* 1982, *33*, 493–532.
5. Grant, D. J.; Dixon, D. A., Heats of Formation and Bond Energies of the $H_{3-n}BX_n$ Compounds for (X = F, Cl, Br, I, NH_2, OH, and SH) *J. Phys. Chem. A* 2009, *113*, 777–787.
6. Fenwick, J. T. F.; Wilson, J. W., Thermochemistry of triphenoxyborane and lewis acidity of boron atom *Dalton Trans.* 1972, 1324.
7. Davies, J. V.; Skinner, H. A., The Thermochemistry of Tin Alkyls: The Heats of Combustion of Tin tetramethyl, Tin tetraethyl and Trimethylethyl tin *Contract Report for U.S. Army* 1962.
8. Balandin, A. A.; Klabunovskii, E. I.; Kozina, M. P.; Ul'ianova, O. D., Thermochemical Determination of Bond Energies. Communication 1. Sn–C Bond Energies in Tetramethyl- and Tetraethyltin *Russ. Chem. Bull.* 1957, *7*, 10–15.
9. Weast, R. C., *CRC Handbook of Chemistry and Physics.* 70th ed.; CRC Press, Inc.: Boca Raton, FL, 1990.
10. Bratsch, S. G., A Group Electronegativity Method With Pauling Units *J. Chem. Educ.* 1985, *62*, 101–103.

APPENDIX B

Some Representative Bond Distances and Bond Energies

Some further experimental values for comparison with simple polar covalence bond energies. All these values are straightforward to calculate by using the descriptions in the text.

Compound	Bond[a]	Bond Distance (Å)[b]	Bond Energy (kcal/mol)[b]
Simple Gases			
CO	C≡O‴	1.128	257.3
N_2	N‴≡N‴	1.098	225.9
P_2	P‴≡P‴	1.89	125.1
As_2	As‴≡As‴	2.17	91
CO_2	C=O″	1.16	192.3
Representative Main Group Halides			
BeF_2	Be–F″	1.40	152.5
CaF_2	Ca–F‴	2.10	134
SrF_2	Sr–F‴	2.20	129.8
BaF_2	Ba–F‴	2.32	137.9
BF_3	B–F‴	1.31	141.5
CF_4	C–$F^{2\prime/2\prime\prime}$	1.32	117.0
SiF_4	Si–F‴	1.54	142.4
GeF_4	Ge–F‴	1.68	112.5
PF_3	P–F‴	1.57	120.3
AsF_3	As–F‴	1.712	105.6
SbF_3	Sb–F‴	1.90	103.0
$CaCl_2$	Ca–Cl‴	2.51	106.7
$SrCl_2$	Sr–Cl‴	2.67	106.6
$BaCl_2$	Ba–Cl‴	2.892	113.1
BCl_3	B–Cl‴	1.75	105.7
$AlCl_3$	Al–Cl‴	2.06	101.8
$InCl_3$	In–Cl‴	2.46	73.8
$SiCl_4$	Si–Cl‴	2.019	95.3
$SnCl_4$	Sn–Cl‴	2.31	75.2
$SbCl_3$	Sb–Cl‴	2.325	75.1
$BiCl_3$	Bi–Cl‴	2.48	66.7
$CaBr_2$	Ca–Br‴	2.63	95
$SrBr_2$	Sr–Br‴	2.82	95.3

Compound	Bond[a]	Bond Distance (Å)[b]	Bond Energy (kcal/mol)[b]
$BaBr_2$	Ba–Br‴	2.99	98.8
BBr_3	B–Br″	1.87	87.6
$AlBr_3$	Al–Br‴	2.27	85.7
$InBr_3$	In–Br‴	2.58	64.2
$SiBr_4$	Si–Br‴	2.30	78.8
$SnBr_4$	Sn–Br‴	2.44	63.6
$AsBr_3$	As–Br‴	2.33	61.2
$SbBr_3$	Sb–Br‴	2.51	63.3
$BiBr_3$	Sb–Br‴	2.63	56.0
CaI_2	Ca–I‴	2.88	78.3
SrI_2	Sr–I‴	3.03	77.6
BaI_2	Ba–I‴	3.20	85.8
AlI_3	Al–I‴	2.44	68.1
InI_3	In–I‴	2.458	50.2
GeI_4	Ge–I‴	2.49	51.4
BiI_3	Bi–I‴	2.84	43.3

[a] These are given for comparison with polar covalence theory. For systems with E′ the fully weakened covalent bond value works best. For systems with E″ the partially unweakened covalent bond value works best. For systems with E‴ the fully unweakened covalent bond value works best.
[b] Bond distances and energies are from Sanderson, R. T. *Polar Covalence*, Academic Press: New York, 1983.

APPENDIX C

Tabulated Values for Determination of Bond Order: r_0 and b

Pauling[1] suggested a simple function to relate bond order to bond length. The bond order between atoms A and B, B_{AB}, is a function of the bond distance (r), single bond distance (r_0), and an empirical fitting parameter (b). The bond order is then expected to change exponentially with differences in the A–B bond distance.

$$B_{AB} = \exp\left(\frac{-(r - r_0)}{b}\right)$$

Below are some representative values for r_0 and b for many elemental combinations taken from Brown's review on bond valence.[2] Taking all the values from a database (1,600 values for a variety of molecules) and averaging to find b produced average = 0.377, st. dev. = 0.04, lowest value = 0.26, and highest value = 0.94.

Metal Center	Type of Ligand	r_0 (Å)	b (Å)
Li(I)	O	0.59	1.174
Na(I)	O	0.483	1.56
K(I)	O	0.422	1.973
Cs(I)	O	0.403	2.298
B(III)	O	0.37	1.371
B(III)	F	0.37	1.289
B(III)	S	0.37	1.815
B(III)	N	0.37	1.482
B(III)	P	0.37	1.920
P(V)	O	0.37	1.615
Si(IV)	O	0.37	1.622
Cu(I)	O	0.37	1.567
Cr(II)	O	0.37	1.739
Cr(III)	O	0.37	1.708
Sn(II)	S	0.37	2.423
Sn(IV)	S	0.37	2.391
Mo(IV)	O	0.37	1.856
Tl(I)	O	0.50	1.927
Mn(III)	F	0.36	1.666

Metal Center	Type of Ligand	r_0 (Å)	b (Å)
Cu(II)	O	0.37	1.655
Cu(I)	Cl	0.37	1.840
Cu(II)	Cl	0.37	1.999
La(III)	O	0.37	2.148
Ce(III)	O	0.37	2.116
Nd(III)	O	0.37	2.086
Eu	Cl	0.37	2.457
Dy(III)	S	0.37	2.475

Cited References

1. Pauling, L., Atomic radii and interatomic distances in metals *J. Am. Chem. Soc.* 1947, *69*, 542–553.
2. Brown, I. D., Recent Developments in the Methods and Applications of the Bond Valence Model *Chem. Rev.* 2009, *109*, 6858–6919.

APPENDIX D

Character Tables

Groups C_1, C_s, C_i

C_1	E	$h=1$
A	1	

C_s	E	σ_h	Linear functions	Quadratic functions	Cubic functions
A′	1	1	x, y, R_z	x^2, y^2, z^2, xy	$xz^2, yz^2, x^2y, xy^2, x^3, y^3$
A″	1	−1	z, R_x, R_y	xz, yz	z^3, xyz, y^2z, x^2z

C_i	E	i	Linear functions	Quadratic functions	Cubic functions
A_g	1	1	R_x, R_y, R_z	$x^2, y^2, z^2, xy, xz, yz$	–
A_u	1	−1	x, y, z		all cubic functions

Groups C_n

C_2	E	C_2	Linear functions	Quadratic functions	Cubic functions
A	1	1	z, R_z	x^2, y^2, z^2, xy	z^3, xyz, y^2z, x^2z
B	1	−1	x, y, R_x, R_y	xz, yz	$xz^2, yz^2, x^2y, xy^2, x^3, y^3$

C_3	E	C_3	C_3^2	Linear functions	Quadratic functions	Cubic functions
A	1	1	1	z, R_z	x^2+y^2, z^2	$z^3, y(3x^2-y^2), x(x^2-3y^2), z(x^2+y^2)$
E	$\left\{\begin{matrix}1 \\ 1\end{matrix}\right.$	$\begin{matrix}\varepsilon \\ \varepsilon^*\end{matrix}$	$\left.\begin{matrix}\varepsilon^* \\ \varepsilon\end{matrix}\right\}$	$x+iy; R_x+iR_y$ $x-iy; R_x-iR_y$	$(x^2-y^2, xy)\ (xz, yz)$	$(xz^2, yz^2)\ [xyz, z(x^2-y^2)]\ [x(x^2+y^2), y(x^2+y^2)]$

$\varepsilon=\exp(2\pi i/3)$

C_4	E	C_4	C_2	$(C_4)^3$	linear functions, rotations	quadratic functions	Cubic functions
A	+1	+1	+1	+1	z, R_z	x^2+y^2, z^2	$z^3, z(x^2+y^2)$
B	+1	−1	+1	−1	–	x^2-y^2, xy	$xyz, z(x^2-y^2)$
E	+1 +1	+i −i	−1 −1	−i +i	$x+iy; R_x+iR_y$ $x-iy; R_x-iR_y$	(xz, yz)	$(xz^2, yz^2)\ (xy^2, x^2y)\ (x^3, y^3)$

C_5	E	C_5	$(C_5)^2$	$(C_5)^3$	$(C_5)^4$	linear functions, rotations	quadratic functions	cubic functions
A	+1	+1	+1	+1	+1	z, R_z	x^2+y^2, z^2	$z^3, z(x^2+y^2)$
E_1	+1	$+\varepsilon$	$+\varepsilon^2$	$+\varepsilon^{2*}$	$+\varepsilon^*$	$x+iy, R_x+iR_y$	(xz, yz)	(xz^2, yz^2) $[x(x^2+y^2), y(x^2+y^2)]$
	+1	$+\varepsilon^*$	$+\varepsilon^{2*}$	$+\varepsilon^2$	$+\varepsilon$	$x-iy, R_x-iR_y$		
E_2	+1	$+\varepsilon^2$	$+\varepsilon^*$	$+\varepsilon$	$+\varepsilon^{2*}$	–	(x^2-y^2, xy)	$[xyz, z(x^2-y^2)]$ $[y(3x^2-y^2), x(x^2-3y^2)]$
	+1	$+\varepsilon^{2*}$	$+\varepsilon$	$+\varepsilon^*$	$+\varepsilon^2$			

$\varepsilon=\exp(2\pi i/5)$

C_6	E	C_6	C_3	C_2	$(C_3)^2$	$(C_6)^5$	linear functions, rotations	quadratic functions	cubic functions
A	+1	+1	+1	+1	+1	+1	z, R_z	x^2+y^2, z^2	$z^3, z(x^2+y^2)$
B	+1	−1	+1	−1	+1	−1	–	–	$y(3x^2-y^2), x(x^2-3y^2)$
E_1	+1	$+\varepsilon$	$-\varepsilon^*$	−1	$-\varepsilon$	$+\varepsilon^*$	$x+iy; R_x+iR_y$	(xz, yz)	(xz^2, yz^2) $[x(x^2+y^2), y(x^2+y^2)]$
	+1	$+\varepsilon^*$	$-\varepsilon$	−1	$-\varepsilon^*$	$+\varepsilon$	$x-iy; R_x-iR_y$		
E_2	+1	$-\varepsilon^*$	$-\varepsilon$	+1	$-\varepsilon^*$	$-\varepsilon^*$	–	(x^2-y^2, xy)	$[xyz, z(x^2-y^2)]$
	+1	$-\varepsilon$	$-\varepsilon^*$	+1	$-\varepsilon$	$-\varepsilon$			

$\varepsilon=\exp(2\pi i/6)$

C_7	E	C_7	$(C_7)^2$	$(C_7)^3$	$(C_7)^4$	$(C_7)^5$	$(C_7)^6$	linear functions, rotations	quadratic functions	cubic functions
A	+1	+1	+1	+1	+1	+1	+1	z, R_z	x^2+y^2, z^2	$z^3, z(x^2+y^2)$
E_1	+1	$+\varepsilon$	$+\varepsilon^2$	$+\varepsilon^3$	$+\varepsilon^{3*}$	$+\varepsilon^{2*}$	$+\varepsilon^*$	$x+iy, R_x+iR_y$	(xz, yz)	(xz^2, yz^2) $[x(x^2+y^2), y(x^2+y^2)]$
	+1	$+\varepsilon^*$	$+\varepsilon^{2*}$	$+\varepsilon^{3*}$	$+\varepsilon^3$	$+\varepsilon^2$	$+\varepsilon$	$x-iy, R_x-iR_y$		
E_2	+1	$+\varepsilon^2$	$+\varepsilon^{3*}$	$+\varepsilon^*$	$+\varepsilon$	$+\varepsilon^3$	$+\varepsilon^{2*}$	–	(x^2-y^2, xy)	$[xyz, z(x^2-y^2)]$
	+1	$+\varepsilon^{2*}$	$+\varepsilon^3$	$+\varepsilon$	$+\varepsilon^*$	$+\varepsilon^{3*}$	$+\varepsilon^2$			
E_3	+1	$+\varepsilon^3$	$+\varepsilon^*$	$+\varepsilon^2$	$+\varepsilon^{2*}$	$+\varepsilon$	$+\varepsilon^{3*}$	–	–	$[y(3x^2-y^2), x(x^2-3y^2)]$
	+1	$+\varepsilon^{3*}$	$+\varepsilon$	$+\varepsilon^{2*}$	$+\varepsilon^2$	$+\varepsilon^*$	$+\varepsilon^3$			

$\varepsilon=\exp(2\pi i/7)$

C_8	E	C_8	C_4	$(C_8)^3$	C_2	$(C_8)^5$	$(C_4)^3$	$(C_8)^7$	linear functions, rotations	quadratic functions	cubic functions
A	+1	+1	+1	+1	+1	+1	+1	+1	z, R_z	x^2+y^2, z^2	$z^3, z(x^2+y^2)$
B	+1	−1	+1	−1	+1	−1	+1	−1	–	–	–
E_1	+1	$+\varepsilon$	+i	$-\varepsilon^*$	−1	$-\varepsilon$	−i	$+\varepsilon^*$	$x+iy, R_x+iR_y$	(xz, yz)	(xz^2, yz^2) $[x(x^2+y^2), y(x^2+y^2)]$
	+1	$+\varepsilon^*$	−i	$-\varepsilon$	−1	$-\varepsilon^*$	+i	$+\varepsilon$	$x-iy, R_x-iR_y$		
E_2	+1	+i	−1	−i	+1	+i	−1	−i	–	(x^2-y^2, xy)	$[xyz, z(x^2-y^2)]$
	+1	−i	−1	+i	+1	−i	−1	+i			
E_3	+1	$-\varepsilon$	+i	$+\varepsilon^*$	−1	$+\varepsilon$	−i	$-\varepsilon^*$	–	–	$[y(3x^2-y^2), x(x^2-3y^2)]$
	+1	$-\varepsilon^*$	−i	$+\varepsilon$	−1	$+\varepsilon^*$	+i	$-\varepsilon$			

$\varepsilon=\exp(2\pi i/8)$

Groups D_n

D_2	E	$C_2(z)$	$C_2(y)$	$C_2(x)$	Linear functions	Quadratic functions	Cubic functions
A	1	1	1	1		x^2, y^2, z^2	xyz
B_1	1	1	−1	−1	z, R_z	xy	z^3, y^2z, x^2z
B_2	1	−1	1	−1	y, R_y	xz	yz^2, x^2y, y^3
B_3	1	−1	−1	1	x, R_x	yz	xz^2, xy^2, x^3

D_3	E	$2C_3$	$3C_2$	Linear functions	Quadratic functions	Cubic functions
A_1	1	1	1		x^2+y^2, z^2	$x(x^2-3y^2)$
A_2	1	1	−1	z, R_z		$z^3, y(3x^2-y^2), z(x^2+y^2)$
E	2	−1	0	(x, y) (R_x, R_y)	$(x^2-y^2, xy)(xz, yz)$	$(xz^2, yz^2)[xyz, z(x^2-y^2)]$ $[x(x^2+y^2), y(x^2+y^2)]$

D_4	E	$2C_4$ (z)	C_2 (z)	$2C'_2$	$2C''_2$	linear functions, rotations	quadratic functions	cubic functions
A_1	+1	+1	+1	+1	+1	–	x^2+y^2, z^2	–
A_2	+1	+1	+1	−1	−1	z, R_z	–	$z^3, z(x^2+y^2)$
B_1	+1	−1	+1	+1	−1	–	x^2-y^2	xyz
B_2	+1	−1	+1	−1	+1	–	xy	$z(x^2-y^2)$
E	+2	0	−2	0	0	(x, y) (R_x, R_y)	(xz, yz)	(xz^2, yz^2) (xy^2, x^2y) (x^3, y^3)

D_5	E	$2C_5$ (z)	$2(C_5)^2$	$5C'_2$	linear functions, rotations	quadratic functions	cubic functions
A_1	+1	+1	+1	+1	–	x^2+y^2, z^2	–
A_2	+1	+1	+1	−1	z, R_z	–	$z^3, z(x^2+y^2)$
E_1	+2	$+2\cos(2\pi/5)$	$+2\cos(4\pi/5)$	0	(x, y) (R_x, R_y)	(xz, yz)	(xz^2, yz^2) $[x(x^2+y^2), y(x^2+y^2)]$
E_2	+2	$+2\cos(4\pi/5)$	$+2\cos(2\pi/5)$	0	–	(x^2-y^2, xy)	$[xyz, z(x^2-y^2)]$ $[y(3x^2-y^2), x(x^2-3y^2)]$

D_6	E	$2C_6$ (z)	$2C_3$ (z)	C_2 (z)	$3C'_2$	$3C''_2$	linear functions, rotations	quadratic functions	cubic functions
A_1	+1	+1	+1	+1	+1	+1	–	x^2+y^2, z^2	–
A_2	+1	+1	+1	+1	−1	−1	z, R_z	–	$z^3, z(x^2+y^2)$
B_1	+1	−1	+1	−1	+1	−1	–	–	$x(x^2-3y^2)$
B_2	+1	−1	+1	−1	−1	+1	–	–	$y(3x^2-y^2)$
E_1	+2	+1	−1	−2	0	0	(x, y) (R_x, R_y)	(xz, yz)	(xz^2, yz^2) $[x(x^2+y^2), y(x^2+y^2)]$
E_2	+2	−1	−1	+2	0	0	–	(x^2-y^2, xy)	$[xyz, z(x^2-y^2)]$

D_7	E	$2C_7$	$2(C_7)^2$	$2(C_7)^3$	$7C'_2$	linear functions, rotations	quadratic functions	cubic functions
A_1	+1	+1	+1	+1	+1	–	x^2+y^2, z^2	–
A_2	+1	+1	+1	+1	−1	z, R_z	–	$z^3, z(x^2+y^2)$
E_1	+2	$+2\cos(2\pi/7)$	$+2\cos(4\pi/7)$	$+2\cos(6\pi/7)$	0	(x, y) (R_x, R_y)	(xz, yz)	(xz^2, yz^2) $[x(x^2+y^2), y(x^2+y^2)]$
E_2	+2	$+2\cos(4\pi/7)$	$+2\cos(6\pi/7)$	$+2\cos(2\pi/7)$	0	–	(x^2-y^2, xy)	$[xyz, z(x^2-y^2)]$
E_3	+2	$+2\cos(6\pi/7)$	$+2\cos(2\pi/7)$	$+2\cos(4\pi/7)$	0	–	–	$[y(3x^2-y^2), x(x^2-3y^2)]$

D_8	E	$2C_8$	$2C_4$	$2(C_8)^3$	C_2	$4C'_2$	$4C''_2$	linear functions, rotations	quadratic functions	cubic functions
A_1	+1	+1	+1	+1	+1	+1	+1	–	x^2+y^2, z^2	–
A_2	+1	+1	+1	+1	+1	−1	−1	z, R_z	–	$z^3, z(x^2+y^2)$
B_1	+1	−1	+1	−1	+1	+1	−1	–	–	–
B_2	+1	−1	+1	−1	+1	−1	+1	–	–	–
E_1	+2	$+(2)^{1/2}$	0	$-(2)^{1/2}$	−2	0	0	(x, y) (R_x, R_y)	(xz, yz)	(xz^2, yz^2) $[x(x^2+y^2), y(x^2+y^2)]$
E_2	+2	0	−2	0	+2	0	0	–	(x^2-y^2, xy)	$[xyz, z(x^2-y^2)]$
E_3	+2	$-(2)^{1/2}$	0	$+(2)^{1/2}$	−2	0	0	–	–	$[y(3x^2-y^2), x(x^2-3y^2)]$

Groups C_{nv}

C_{2v}	E	C_2	$\sigma_v(xz)$	$\sigma'_v(xz)$	Linear functions	Quadratic functions	Cubic functions
A_1	1	1	1	1	z	x^2, y^2, z^2	z^3, x^2z, y^2z
A_2	1	1	−1	−1	R_z	xy	xyz
B_1	1	−1	1	−1	x, R_y	xz	xz^2, x^3, xy^2
B_2	1	−1	−1	1	y, R_x	yz	yz^2, y^3, x^2y

C_{3v}	E	$2C_3$	$3\sigma_v$	Linear functions	Quadratic functions	Cubic functions
A_1	1	1	1	z	x^2+y^2, z^2	$z^3, x(x^2-3y^2), z(x^2+y^2)$
A_2	1	1	−1	R_z		$y(3x^2-y^2)$
E	2	−1	0	(x, y) (R_x, R_y)	$(x^2-y^2, xy)(xz, yz)$	(xz^2, yz^2) $[xyz, z(x^2-y^2)]$ $[x(x^2+y^2), y(x^2+y^2)]$

C_{4v}	E	$2C_4$	C_2	$2\sigma_v$	$2\sigma_d$	Linear functions	Quadratic functions	Cubic functions
A_1	1	1	1	1	1	z	x^2+y^2, z^2	$z^3, z(x^2+y^2)$
A_2	1	1	1	−1	−1	R_z		–
B_1	1	−1	1	1	−1		x^2-y^2	$z(x^2-y^2)$
B_2	1	−1	1	−1	1		xy	xyz
E	2	0	−2	0	0	(x, y) (R_x, R_y)	(xz, yz)	(xz^2, yz^2) (xy^2, x^2y) (x^3, y^3)

C_{5v}	E	$2C_5(z)$	$2(C_5)^2$	$5\sigma_v$	linear functions, rotations	quadratic functions	cubic functions
A_1	+1	+1	+1	+1	z	x^2+y^2, z^2	$z^3, z(x^2+y^2)$
A_2	+1	+1	+1	−1	R_z	–	–
E_1	+2	$+2\cos(2\pi/5)$	$+2\cos(4\pi/5)$	0	(x, y) (R_x, R_y)	(xz, yz)	(xz^2, yz^2) $[x(x^2+y^2), y(x^2+y^2)]$
E_2	+2	$+2\cos(4\pi/5)$	$+2\cos(2\pi/5)$	0	–	(x^2-y^2, xy)	$[xyz, z(x^2-y^2)]$ $[y(3x^2-y^2), x(x^2-3y^2)]$

C_{6v}	E	$2C_6$	$2C_3$	C_2	$3\sigma_v$	$3\sigma_d$	Linear functions	Quadratic functions	Cubic functions
A_1	1	1	1	1	1	1	z	x^2+y^2, z^2	$z^3, z(x^2+y^2)$
A_2	1	1	1	1	−1	−1	R_z		–
B_1	1	−1	1	−1	1	−1			$x(x^2-3y^2)$
B_2	1	−1	1	−1	−1	1			$y(3x^2-y^2)$
E_1	2	1	−1	−2	0	0	(x, y) (R_x, R_y)	(xz, yz)	(xz^2, yz^2) $[x(x^2+y^2), y(x^2+y^2)]$
E_2	2	−1	−1	2	0	0		(x^2-y^2, xy)	$[xyz, z(x^2-y^2)]$

C_{7v}	E	$2C_7$	$2(C_7)^2$	$2(C_7)^3$	$7\sigma_v$	linear functions, rotations	quadratic functions	cubic functions
A_1	+1	+1	+1	+1	+1	z	x^2+y^2, z^2	$z^3, z(x^2+y^2)$
A_2	+1	+1	+1	+1	−1	R_z	–	–
E_1	+2	$+2\cos(2\pi/7)$	$+2\cos(4\pi/7)$	$+2\cos(6\pi/7)$	0	(x, y) (R_x, R_y)	(xz, yz)	(xz^2, yz^2) $[x(x^2+y^2), y(x^2+y^2)]$
E_2	+2	$+2\cos(4\pi/7)$	$+2\cos(6\pi/7)$	$+2\cos(2\pi/7)$	0	–	(x^2-y^2, xy)	$[xyz, z(x^2-y^2)]$
E_3	+2	$+2\cos(6\pi/7)$	$+2\cos(2\pi/7)$	$+2\cos(4\pi/7)$	0	–	–	$[y(3x^2-y^2), x(x^2-3y^2)]$

C_{8v}	E	$2C_8$	$2C_4$	$2(C_8)^3$	C_2	$4\sigma_v$	$4\sigma_d$	linear functions, rotations	quadratic functions	cubic functions
A_1	+1	+1	+1	+1	+1	+1	+1	z	x^2+y^2, z^2	$z^3, z(x^2+y^2)$
A_2	+1	+1	+1	+1	+1	−1	−1	R_z	–	–
B_1	+1	−1	+1	−1	+1	+1	−1	–	–	–
B_2	+1	−1	+1	−1	+1	−1	+1	–	–	–
E_1	+2	$+(2)^{1/2}$	0	$-(2)^{1/2}$	−2	0	0	(x, y) (R_x, R_y)	(xz, yz)	(xz^2, yz^2) $[x(x^2+y^2), y(x^2+y^2)]$
E_2	+2	0	−2	0	+2	0	0	–	(x^2-y^2, xy)	$[xyz, z(x^2-y^2)]$
E_3	+2	$-(2)^{1/2}$	0	$+(2)^{1/2}$	−2	0	0	–	–	$[y(3x^2-y^2), x(x^2-3y^2)]$

Groups C_{nh}

C_{2h}	E	$C_2(z)$	i	σ_h	linear functions, rotations	quadratic functions	cubic functions
A_g	+1	+1	+1	+1	R_z	x^2, y^2, z^2, xy	–
B_g	+1	−1	+1	−1	R_x, R_y	xz, yz	–
A_u	+1	+1	−1	−1	z	–	z^3, xyz, x^2z, y^2z
B_u	+1	−1	−1	+1	x, y	–	$xz^2, yz^2, x^2y, xy^2, x^3, y^3$

C_{3h}	E	$C_3(z)$	$(C_3)^2$	σ_h	S_3	$(S_3)^5$	linear functions, rotations	quadratic functions	cubic functions
A′	+1	+1	+1	+1	+1	+1	R_z	x^2+y^2, z^2	$y(3x^2-y^2), x(x^2-3y^2)$
E′	+1	$+\varepsilon$	$+\varepsilon^*$	+1	$+\varepsilon$	$+\varepsilon^*$	$x+iy$	(x^2-y^2, xy)	(xz^2, yz^2) $[x(x^2+y^2), y(x^2+y^2)]$
	+1	$+\varepsilon^*$	$+\varepsilon$	+1	$+\varepsilon^*$	$+\varepsilon$	$x-iy$		
A″	+1	+1	+1	−1	−1	−1	z	–	$z^3, z(x^2+y^2)$
E″	+1	$+\varepsilon$	$+\varepsilon^*$	−1	$-\varepsilon$	$-\varepsilon^*$	R_x+iR_y	(xz, yz)	$[xyz, z(x^2-y^2)]$
	+1	$+\varepsilon^*$	$+\varepsilon$	−1	$-\varepsilon^*$	$-\varepsilon$	R_x-iR_y		

$\varepsilon=\exp(2\pi i/3)$

C_{4h}	E	$C_4(z)$	C_2	$(C_4)^3$	i	$(S_4)^3$	σ_h	S_4	linear functions, rotations	quadratic functions	cubic functions
A_g	+1	+1	+1	+1	+1	+1	+1	+1	R_z	x^2+y^2, z^2	–
B_g	+1	−1	+1	−1	+1	−1	+1	−1	–	x^2-y^2, xy	–
E_g	+1	+i	−1	−i	+1	+i	−1	−i	R_x+iR_y	(xz, yz)	–
	+1	−i	−1	+i	+1	−i	−1	+i	R_x-iR_y		
A_u	+1	+1	+1	+1	−1	−1	−1	−1	z	–	z^3, $z(x^2+y^2)$
B_u	+1	−1	+1	−1	−1	+1	−1	+1	–	–	xyz, $z(x^2-y^2)$
E_u	+1	+i	−1	−i	−1	−i	+1	+i	$x+iy$	–	(xz^2, yz^2) (xy^2, x^2y) (x^3, y^3)
	+1	−i	−1	+i	−1	+i	+1	−i	$x-iy$		

C_{5h}	E	C_5	$(C_5)^2$	$(C_5)^3$	$(C_5)^4$	σ_h	S_5	$(S_5)^7$	$(S_5)^3$	$(S_5)^9$	linear functions, rotations	quadratic functions	cubic functions
A′	+1	+1	+1	+1	+1	+1	+1	+1	+1	+1	R_z	x^2+y^2, z^2	–
E'_1	+1	$+\varepsilon$	$+\varepsilon^2$	$+\varepsilon^{2*}$	$+\varepsilon$	+1	$+\varepsilon$	$+\varepsilon^2$	$+\varepsilon^{2*}$	$+\varepsilon^*$	$x+iy$	–	(xz^2, yz^2) $[x(x^2+y^2), y(x^2+y^2)]$
	+1	$+\varepsilon^*$	$+\varepsilon^{2*}$	$+\varepsilon^2$	$+\varepsilon$	+1	$+\varepsilon^*$	$+\varepsilon^{2*}$	$+\varepsilon^2$	$+\varepsilon$	$x-iy$		
E'_2	+1	$+\varepsilon^2$	$+\varepsilon^*$	$+\varepsilon$	$+\varepsilon^{2*}$	+1	$+\varepsilon^2$	$+\varepsilon^*$	$+\varepsilon$	$+\varepsilon^{2*}$	–	(x^2-y^2, xy)	$[y(3x^2-y^2), x(x^2-3y^2)]$
	+1	$+\varepsilon^{2*}$	$+\varepsilon$	$+\varepsilon^*$	$+\varepsilon^2$	+1	$+\varepsilon^{2*}$	$+\varepsilon$	$+\varepsilon^*$	$+\varepsilon^2$			
A″	+1	+1	+1	+1	+1	−1	−1	−1	−1	−1	z	–	z^3, $z(x^2+y^2)$
E''_1	+1	$+\varepsilon$	$+\varepsilon^2$	$+\varepsilon^{2*}$	$+\varepsilon$	−1	$-\varepsilon$	$-\varepsilon^2$	$-\varepsilon^{2*}$	$-\varepsilon^*$	R_x+iR_y	(xz, yz)	–
	+1	$+\varepsilon^*$	$+\varepsilon^{2*}$	$+\varepsilon^2$	$+\varepsilon$	−1	$-\varepsilon^*$	$-\varepsilon^{2*}$	$-\varepsilon^2$	$-\varepsilon$	R_x-iR_y		
E''_2	+1	$+\varepsilon^2$	$+\varepsilon^*$	$+\varepsilon$	$+\varepsilon^{2*}$	−1	$-\varepsilon^2$	$-\varepsilon^*$	$-\varepsilon$	$-\varepsilon^{2*}$	–	–	$[xyz, z(x^2-y^2)]$
	+1	$+\varepsilon^{2*}$	$+\varepsilon$	$+\varepsilon^*$	$+\varepsilon^2$	−1	$-\varepsilon^{2*}$	$-\varepsilon$	$-\varepsilon^*$	$-\varepsilon^2$			

$\varepsilon=\exp(2\pi i/5)$

C_{6h}	E	$C_6(z)$	C_3	C_2	$(C_3)^2$	$(C_6)^5$	i	$(S_3)^5$	$(S_6)^5$	σ_h	S_6	S_3	linear functions, rotations	quadratic functions	cubic functions
A_g	+1	+1	+1	+1	+1	+1	+1	+1	+1	+1	+1	+1	R_z	x^2+y^2, z^2	–
B_g	+1	−1	+1	−1	+1	−1	+1	−1	+1	−1	+1	−1	–	–	–
E_{1g}	+1	$+\varepsilon$	$-\varepsilon^*$	−1	$-\varepsilon$	$+\varepsilon^*$	+1	$+\varepsilon$	$-\varepsilon^*$	−1	$-\varepsilon$	$+\varepsilon^*$	R_x+iR_y	(xz, yz)	–
	+1	$+\varepsilon^*$	$-\varepsilon$	−1	$-\varepsilon^*$	$+\varepsilon$	+1	$+\varepsilon^*$	$-\varepsilon^*$	−1	$-\varepsilon^*$	$+\varepsilon$	R_x-iR_y		
E_{2g}	+1	$-\varepsilon^*$	$-\varepsilon$	+1	$-\varepsilon^*$	$-\varepsilon$	+1	$-\varepsilon^*$	$-\varepsilon$	+1	$-\varepsilon^*$	$-\varepsilon$	–	(x^2-y^2, xy)	–
	+1	$-\varepsilon$	$-\varepsilon^*$	+1	$-\varepsilon$	$-\varepsilon^*$	+1	$-\varepsilon$	$-\varepsilon^*$	+1	$-\varepsilon$	$-\varepsilon^*$			
A_u	+1	+1	+1	+1	+1	+1	−1	−1	−1	−1	−1	−1	z	–	z^3, $z(x^2+y^2)$
B_u	+1	−1	+1	−1	+1	−1	−1	+1	−1	+1	−1	+1	–	–	$y(3x^2-y^2)$, $x(x^2-3y^2)$
E_{1u}	+1	$+\varepsilon$	$-\varepsilon^*$	−1	$-\varepsilon$	$+\varepsilon^*$	−1	$-\varepsilon$	$+\varepsilon$	+1	$+\varepsilon$	$-\varepsilon^*$	$x+iy$	–	(xz^2, yz^2) $[x(x^2+y^2), y(x^2+y^2)]$
	+1	$+\varepsilon^*$	$-\varepsilon$	−1	$-\varepsilon^*$	$+\varepsilon$	−1	$-\varepsilon^*$	$+\varepsilon^*$	+1	$+\varepsilon^*$	$-\varepsilon$	$x-iy$		
E_{2u}	+1	$-\varepsilon^*$	$-\varepsilon$	+1	$-\varepsilon^*$	$-\varepsilon$	−1	$+\varepsilon^*$	$+\varepsilon^*$	−1	$+\varepsilon^*$	$+\varepsilon$	–	–	$[xyz, z(x^2-y^2)]$
	+1	$-\varepsilon$	$-\varepsilon^*$	+1	$-\varepsilon$	$-\varepsilon^*$	−1	$+\varepsilon$	$+\varepsilon^*$	−1	$+\varepsilon$	$+\varepsilon^*$			

$\varepsilon=\exp(2\pi i/6)$

Groups D_{nh}

D_{2h}	E	$C_2(z)$	$C_2(y)$	$C_2(x)$	i	$\sigma(xy)$	$\sigma(xz)$	$\sigma(yz)$	Linear functions	Quadratic functions	Cubic functions
A_g	1	1	1	1	1	1	1	1		x^2, y^2, z^2	–
B_{1g}	1	1	−1	−1	1	1	−1	−1	R_z	xy	–
B_{2g}	1	−1	1	−1	1	−1	1	−1	R_y	xz	–
B_{3g}	1	−1	−1	1	1	−1	−1	1	R_x	yz	–
A_u	1	1	1	1	−1	−1	−1	−1			xyz
B_{1u}	1	1	−1	−1	−1	−1	1	1	z		z^3, y^2z, x^2z
B_{2u}	1	−1	1	−1	−1	1	−1	1	y		yz^2, y^3, x^2y
B_{3u}	1	−1	−1	1	−1	1	1	−1	x		xz^2, xy^2, x^3

D_{3h}	E	$2C_3$	$3C_2$	σ_h	$2S_3$	$3\sigma_v$	Linear functions	Quadratic functions	Cubic functions
A'_1	1	1	1	1	1	1		x^2+y^2, z^2	$x(x^2-3y^2)$
A'_2	1	1	−1	1	1	−1	R_z		$y(3x^2-y^2)$
E'	2	−1	0	2	−1	0	(x, y)	(x^2-y^2, xy)	(xz^2, yz^2) $[x(x^2+y^2), y(x^2+y^2)]$
A''_1	1	1	1	−1	−1	−1			–
A''_2	1	1	−1	−1	−1	1	z		$z^3, z(x^2+y^2)$
E''	2	−1	0	−2	1	0	(R_x, R_y)	(xz, yz)	$[xyz, z(x^2-y^2)]$

D_{4h}	E	$2C_4$	$C_2(=C_4^2)$	$2C'_2$	$2C''_2$	i	$2S_4$	σ_h	$2\sigma_v$	$2\sigma_d$	Linear functions	Quadratic functions	Cubic functions
A_{1g}	1	1	1	1	1	1	1	1	1	1		x^2+y^2, z^2	–
A_{2g}	1	1	1	−1	−1	1	1	1	−1	−1	R_z		–
B_{1g}	1	−1	1	1	−1	1	−1	1	1	−1		x^2-y^2	–
B_{2g}	1	−1	1	−1	1	1	−1	1	−1	1		xy	–
E_g	2	0	−2	0	0	2	0	−2	0	0	(R_x, R_y)	(xz, yz)	–
A_{1u}	1	1	1	1	1	−1	−1	−1	−1	−1			–
A_{2u}	1	1	1	−1	−1	−1	−1	−1	1	1	z		$z^3, z(x^2+y^2)$
B_{1u}	1	−1	1	1	−1	−1	1	−1	−1	1			xyz
B_{2u}	1	−1	1	−1	1	−1	1	−1	1	−1			$z(x^2-y^2)$
E_u	2	0	−2	0	0	−2	0	2	0	0	(x, y)		(xz^2, yz^2) (xy^2, x^2y) (x^3, y^3)

(x axis coincident with C'_2 axis)

D_{5h}	E	$2C_5$	$2(C_5)^2$	$5C'_2$	σ_h	$2S_5$	$2(S_5)^3$	$5\sigma_v$	linear functions, rotations	quadratic functions	cubic functions
A'_1	+1	+1	+1	+1	+1	+1	+1	+1	–	x^2+y^2, z^2	–
A'_2	+1	+1	+1	−1	+1	+1	+1	−1	R_z	–	–
E'_1	+2	+2cos(2π/5)	+2cos(4π/5)	0	+2	+2cos(2π/5)	+2cos(4π/5)	0	(x, y)	–	(xz^2, yz^2) $[x(x^2+y^2), y(x^2+y^2)]$
E'_2	+2	+2cos(4π/5)	+2cos(2π/5)	0	+2	+2cos(4π/5)	+2cos(2π/5)	0	–	(x^2-y^2, xy)	$[y(3x^2-y^2), x(x^2-3y^2)]$
A''_1	+1	+1	+1	+1	−1	−1	−1	−1	–	–	–
A''_2	+1	+1	+1	−1	−1	−1	−1	+1	z	–	$z^3, z(x^2+y^2)$
E''_1	+2	+2cos(2π/5)	+2cos(4π/5)	0	−2	−2cos(2π/5)	−2cos(4π/5)	0	(R_x, R_y)	(xz, yz)	–
E''_2	+2	+2cos(4π/5)	+2cos(2π/5)	0	−2	−2cos(4π/5)	−2cos(2π/5)	0	–	–	$[xyz, z(x^2-y^2)]$

(x axis coincident with C'_2 axis)

D_{6h}	E	$2C_6$	$2C_3$	C_2	$3C'_2$	$3C''_2$	i	$2S_3$	$2S_6$	σ_h	$3\sigma_d$	$3\sigma_v$	Linear functions	Quadratic functions	Cubic functions
A_{1g}	1	1	1	1	1	1	1	1	1	1	1	1		x^2+y^2, z^2	–
A_{2g}	1	1	1	1	−1	−1	1	1	1	1	−1	−1	R_z		–
B_{1g}	1	−1	1	−1	1	−1	1	−1	1	−1	1	−1			–
B_{2g}	1	−1	1	−1	−1	1	1	−1	1	−1	−1	1			–
E_{1g}	2	1	−1	−2	0	0	2	1	−1	−2	0	0	(R_x, R_y)	(xz, yz)	–
E_{2g}	2	−1	−1	2	0	0	2	−1	−1	2	0	0		(x^2-y^2, xy)	–
A_{1u}	1	1	1	1	1	1	−1	−1	−1	−1	−1	−1			–
A_{2u}	1	1	1	1	−1	−1	−1	−1	−1	−1	1	1	z		$z^3, z(x^2+y^2)$
B_{1u}	1	−1	1	−1	1	−1	−1	1	−1	1	−1	1			$x(x^2-3y^2)$
B_{2u}	1	−1	1	−1	−1	1	−1	1	−1	1	1	−1			$y(3x^2-y^2)$
E_{1u}	2	1	−1	−2	0	0	−2	−1	1	2	0	0	(x, y)		(xz^2, yz^2) $[x(x^2+y^2), y(x^2+y^2)]$
E_{2u}	2	−1	−1	2	0	0	−2	1	1	−2	0	0			$[xyz, z(x^2-y^2)]$

Groups D_{nd}

D_{2d}	E	$2S_4$	C_2	$2C'_2$	$2\sigma_d$	Linear functions	Quadratic functions	Cubic functions
A_1	1	1	1	1	1		x^2+y^2, z^2	xyz
A_2	1	1	1	−1	−1	R_z		$z(x^2-y^2)$
B_1	1	−1	1	1	−1		x^2-y^2	–
B_2	1	−1	1	−1	1	z	xy	$z^3, z(x^2+y^2)$
E	2	0	−2	0	0	$(x, y)(R_x, R_y)$	(xz, yz)	$(xz^2, yz^2)(xy^2, x^2y)$ (x^3, y^3)

D_{3d}	E	$2C_3$	$3C_2$	i	$2S_6$	$3\sigma_d$	Linear functions	Quadratic functions	Cubic functions
A_{1g}	1	1	1	1	1	1		x^2+y^2, z^2	–
A_{2g}	1	1	−1	1	1	−1	R_z		–
E_g	2	−1	0	2	−1	0	(R_x, R_y)	(x^2-y^2, xy) (xz, yz)	–
A_{1u}	1	1	1	−1	−1	−1			$x(x^2-3y^2)$
A_{2u}	1	1	−1	−1	−1	1	z		$y(3x^2-y^2)$, z^3, $z(x^2+y^2)$
E_u	2	−1	0	−2	1	0	(x, y)		(xz^2, yz^2) $[xyz, z(x^2-y^2)]$ $[x(x^2+y^2), y(x^2+y^2)]$

D_{4d}	E	$2S_4$	$2C_4$	$2S_5^3$	C_2	$4C_2'$	$4\sigma_d$	Linear functions	Quadratic functions	Cubic functions
A_1	1	1	1	1	1	1	1		x^2+y^2, z^2	–
A_2	1	1	1	1	1	−1	−1	R_z		–
B_1	1	−1	1	−1	1	1	−1			–
B_2	1	−1	1	−1	1	−1	1	z		z^3, $z(x^2+y^2)$
E_1	2	$\sqrt{2}$	0	$-\sqrt{2}$	−2	0	0	(x, y)		(xz^2, yz^2) $[x(x^2+y^2), y(x^2+y^2)]$
E_2	2	0	−2	0	2	0	0		(x^2-y^2, xy)	$[xyz, z(x^2-y^2)]$
E_3	2	$-\sqrt{2}$	0	$\sqrt{2}$	−2	0	0	(R_x, R_y)	(xz, yz)	$[y(3x^2-y^2), x(x^2+3y^2)]$

D_{5d}	E	$2C_5$	$2(C_5)^2$	$5C_2'$	i	$2(S_{10})^3$	$2S_{10}$	$5\sigma_d$	linear functions, rotations	quadratic functions	cubic functions
A_{1g}	+1	+1	+1	+1	+1	+1	+1	+1	–	x^2+y^2, z^2	–
A_{2g}	+1	+1	+1	−1	+1	+1	+1	−1	R_z	–	–
E_{1g}	+2	+2cos(2π/5)	+2cos(4π/5)	0	+2	+2cos(2π/5)	+2cos(4π/5)	0	(R_x, R_y)	(xz, yz)	–
E_{2g}	+2	+2cos(4π/5)	+2cos(2π/5)	0	+2	+2cos(4π/5)	+2cos(2π/5)	0	–	(x^2-y^2, xy)	–
A_{1u}	+1	+1	+1	+1	−1	−1	−1	−1	–	–	–
A_{2u}	+1	+1	+1	−1	−1	−1	−1	+1	z	–	z^3, $z(x^2+y^2)$
E_{1u}	+2	+2cos(2π/5)	+2cos(4π/5)	0	−2	−2cos(2π/5)	−2cos(4π/5)	0	(x, y)	–	(xz^2, yz^2) $[x(x^2+y^2),$ $y(x^2+y^2)]$
E_{2u}	+2	+2cos(4π/5)	+2cos(2π/5)	0	−2	−2cos(4π/5)	−2cos(2π/5)	0	–	–	$[xyz, z(x^2-y^2)]$ $[y(3x^2-y^2),$ $x(x^2-3y^2)]$

(x axis coincident with C'_2 axis)

D_{6d}	E	$2S_{12}$	$2C_6$	$2S_4$	$2C_3$	$2(S_{12})^5$	C_2	$6C_2'$	$6\sigma_d$	linear functions, rotations	quadratic functions	cubic functions
A_1	+1	+1	+1	+1	+1	+1	+1	+1	+1	–	x^2+y^2, z^2	–
A_2	+1	+1	+1	+1	+1	+1	+1	−1	−1	R_z	–	–
B_1	+1	−1	+1	−1	+1	−1	+1	+1	−1	–	–	–
B_2	+1	−1	+1	−1	+1	−1	+1	−1	+1	z	–	z^3, $z(x^2+y^2)$
E_1	+2	$+(3)^{1/2}$	+1	0	−1	$-(3)^{1/2}$	−2	0	0	(x, y)	–	(xz^2, yz^2) $[x(x^2+y^2), y(x^2+y^2)]$
E_2	+2	+1	−1	−2	−1	+1	+2	0	0	–	(x^2-y^2, xy)	–
E_3	+2	0	−2	0	+2	0	−2	0	0	–	–	$[y(3x^2-y^2), x(x^2-3y^2)]$
E_4	+2	−1	−1	+2	−1	−1	+2	0	0	–	–	$[xyz, z(x^2-y^2)]$
E_5	+2	$-(3)^{1/2}$	+1	0	−1	$+(3)^{1/2}$	−2	0	0	(R_x, R_y)	(xz, yz)	–

(x axis coincident with C'_2 axis)

D_{7d}	E	$2C_7$	$2(C_7)^2$	$2(C_7)^3$	$7C_2'$	i	$2(S_{14})^5$	$2(S_{14})^3$	$2S_{14}$	$7\sigma_d$	linear functions, rotations	quadratic functions	cubic functions
A_{1g}	+1	+1	+1	+1	+1	+1	+1	+1	+1	+1	–	x^2+y^2, z^2	–
A_{2g}	+1	+1	+1	+1	–1	+1	+1	+1	+1	–1	R_z	–	–
E_{1g}	+2	$+2\cos(2\pi/7)$	$+2\cos(4\pi/7)$	$+2\cos(6\pi/7)$	0	+2	$+2\cos(2\pi/7)$	$+2\cos(4\pi/7)$	$+2\cos(6\pi/7)$	0	(R_x, R_y)	(xz, yz)	–
E_{2g}	+2	$+2\cos(4\pi/7)$	$+2\cos(6\pi/7)$	$+2\cos(2\pi/7)$	0	+2	$+2\cos(4\pi/7)$	$+2\cos(6\pi/7)$	$+2\cos(2\pi/7)$	0	–	(x^2-y^2, xy)	–
E_{3g}	+2	$+2\cos(6\pi/7)$	$+2\cos(2\pi/7)$	$+2\cos(4\pi/7)$	0	+2	$+2\cos(6\pi/7)$	$+2\cos(2\pi/7)$	$+2\cos(4\pi/7)$	0	–	–	–
A_{1u}	+1	+1	+1	+1	+1	–1	–1	–1	–1	–1	–	–	–
A_{2u}	+1	+1	+1	+1	–1	–1	–1	–1	–1	+1	z	–	$z^3, z(x^2+y^2)$
E_{1u}	+2	$+2\cos(2\pi/7)$	$+2\cos(4\pi/7)$	$+2\cos(6\pi/7)$	0	–2	$-2\cos(2\pi/7)$	$-2\cos(4\pi/7)$	$-2\cos(6\pi/7)$	0	(x, y)	–	(xz^2, yz^2) $[x(x^2+y^2), y(x^2+y^2)]$
E_{2u}	+2	$+2\cos(4\pi/7)$	$+2\cos(6\pi/7)$	$+2\cos(2\pi/7)$	0	–2	$-2\cos(4\pi/7)$	$-2\cos(6\pi/7)$	$-2\cos(2\pi/7)$	0	–	–	$[xyz, z(x^2-y^2)]$
E_{3u}	+2	$+2\cos(6\pi/7)$	$+2\cos(2\pi/7)$	$+2\cos(4\pi/7)$	0	–2	$-2\cos(6\pi/7)$	$-2\cos(2\pi/7)$	$-2\cos(4\pi/7)$	0	–	–	$[y(3x^2-y^2), x(x^2-3y^2)]$

D_{8d}	E	$2S_{16}$	$2C_8$	$2(S_{16})^3$	$2C_4$	$2(S_{16})^5$	$2(C_8)^3$	$2(S_{16})^7$	C_2	$8C_2'$	$8\sigma_d$	linear functions, rotations	quadratic functions	cubic functions
A_1	+1	+1	+1	+1	+1	+1	+1	+1	+1	+1	+1	–	x^2+y^2, z^2	–
A_2	+1	+1	+1	+1	+1	+1	+1	+1	+1	–1	–1	R_z	–	–
B_1	+1	–1	+1	–1	+1	–1	+1	–1	+1	+1	–1	–	–	–
B_2	+1	–1	+1	–1	+1	–1	+1	–1	+1	–1	+1	z	–	$z^3, z(x^2+y^2)$
E_1	+2	$+2\cos(1\pi/8)$	$+(2)^{½}$	$+2\cos(3\pi/8)$	0	$-2\cos(3\pi/8)$	$-(2)^{½}$	$-2\cos(1\pi/8)$	–2	0	0	(x, y)	–	(xz^2, yz^2) $[x(x^2+y^2), y(x^2+y^2)]$
E_2	+2	$+(2)^{½}$	0	$-(2)^{½}$	–2	$-(2)^{½}$	0	$+(2)^{½}$	+2	0	0	–	(x^2-y^2, xy)	–
E_3	+2	$+2\cos(3\pi/8)$	$-(2)^{½}$	$-2\cos(1\pi/8)$	0	$+2\cos(1\pi/8)$	$+(2)^{½}$	$-2\cos(3\pi/8)$	–2	0	0	–	–	$[x(x^2-3y^2), y(3x^2-y^2)]$
E_4	+2	0	–2	0	+2	0	–2	0	+2	0	0	–	–	–
E_5	+2	$-2\cos(3\pi/8)$	$-(2)^{½}$	$+2\cos(1\pi/8)$	0	$-2\cos(1\pi/8)$	$+(2)^{½}$	$+2\cos(3\pi/8)$	–2	0	0	–	–	–
E_6	+2	$-(2)^{½}$	0	$+(2)^{½}$	–2	$+(2)^{½}$	0	$-(2)^{½}$	+2	0	0	–	–	$[xyz, z(x^2-y^2)]$
E_7	+2	$-2\cos(1\pi/8)$	$+(2)^{½}$	$-2\cos(3\pi/8)$	0	$+2\cos(3\pi/8)$	$-(2)^{½}$	$+2\cos(1\pi/8)$	–2	0	0	(R_x, R_y)	(xz, yz)	–

Groups S_n

S_4	E	S_4	C_2	$(S_4)^3$	linear functions, rotations	quadratic functions	cubic functions
A	+1	+1	+1	+1	R_z	x^2+y^2, z^2	$xyz, z(x^2-y^2)$
B	+1	−1	+1	−1	z	x^2-y^2, xy	$z^3, z(x^2+y^2)$
E	+1	+i	−1	−i	$x+iy; R_x+iR_y$	(xz, yz)	(xz^2, yz^2) (xy^2, x^2y) (x^3, y^3)
	+1	−i	−1	+i	$x-iy; R_x-iR_y$		

S_6	E	$C_3(z)$	$(C_3)^2$	i	$(S_6)^5$	S_6	linear functions, rotations	quadratic functions	cubic functions
A_g	+1	+1	+1	+1	+1	+1	R_z	x^2+y^2, z^2	–
E_g	+1	$+\varepsilon$	$+\varepsilon^*$	+1	$+\varepsilon$	$+\varepsilon^*$	R_x+iR_y	(x^2-y^2, xy) (xz, yz)	–
	+1	$+\varepsilon^*$	$+\varepsilon$	+1	$+\varepsilon^*$	$+\varepsilon$	R_x-iR_y		
A_u	+1	+1	+1	−1	−1	−1	z	–	$z^3, y(3x^2-y^2), x(x^2-3y^2), z(x^2+y^2)$
E_u	+1	$+\varepsilon$	$+\varepsilon^*$	−1	$-\varepsilon^*$	$-\varepsilon^*$	$x+iy$	–	(xz^2, yz^2) $[xyz, z(x^2-y^2)]$ $[x(x^2+y^2), y(x^2+y^2)]$
	+1	$+\varepsilon^*$	$+\varepsilon$	−1	$-\varepsilon^*$	$-\varepsilon$	$x-iy$		

$\varepsilon=\exp(2\pi i/3)$

S_8	E	S_8	$C_4(z)$	$(S_8)^3$	C_2	$(S_8)^5$	$(C_4)^3$	$(S_8)^7$	linear functions, rotations	quadratic functions	cubic functions
A	+1	+1	+1	+1	+1	+1	+1	+1	R_z	x^2+y^2, z^2	–
B	+1	−1	+1	−1	+1	−1	+1	−1	z	–	$z^3, z(x^2+y^2)$
E_1	+1	$+\varepsilon$	+i	$-\varepsilon^*$	−1	$-\varepsilon$	−i	$+\varepsilon^*$	$x+iy$	–	(xz^2, yz^2) $[x(x^2+y^2), y(x^2+y^2)]$
	+1	$+\varepsilon^*$	−i	$-\varepsilon$	−1	$-\varepsilon^*$	+i	$+\varepsilon$	$x-iy$		
E_2	+1	+i	−1	−i	+1	+i	−1	−i	–	(x^2-y^2, xy)	$[xyz, z(x^2-y^2)]$
	+1	−i	−1	+i	+1	−i	−1	+i			
E_3	+1	$-\varepsilon$	+i	$+\varepsilon^*$	−1	$+\varepsilon$	−i	$-\varepsilon^*$	R_x+iR_y	(xz, yz)	$[y(3x^2-y^2), x(x^2-3y^2)]$
	+1	$-\varepsilon^*$	−i	$+\varepsilon$	−1	$+\varepsilon^*$	+i	$-\varepsilon$	R_x-iR_y		

$\varepsilon=\exp(2\pi i/8)$

S_{10}	E	C_5	$(C_5)^2$	$(C_5)^3$	$(C_5)^4$	i	$(S_{10})^7$	$(S_{10})^9$	S_{10}	$(S_{10})^3$	linear functions, rotations	quadratic functions	cubic functions
A_g	+1	+1	+1	+1	+1	+1	+1	+1	+1	+1	R_z	z^2, x^2+y^2	–
E_{1g}	+1	$+\varepsilon$	$+\varepsilon^2$	$+\varepsilon^{2*}$	$+\varepsilon^*$	+1	$+\varepsilon$	$+\varepsilon^2$	$+\varepsilon^{2*}$	$+\varepsilon^*$	R_x+iR_y	(xz, yz)	–
	+1	$+\varepsilon^*$	$+\varepsilon^{2*}$	$+\varepsilon^2$	$+\varepsilon$	+1	$+\varepsilon^*$	$+\varepsilon^{2*}$	$+\varepsilon^2$	$+\varepsilon$	R_x-iR_y		
E_{2g}	+1	$+\varepsilon^2$	$+\varepsilon^*$	$+\varepsilon$	$+\varepsilon^{2*}$	+1	$+\varepsilon^2$	$+\varepsilon^*$	$+\varepsilon$	$+\varepsilon^{2*}$	–	(x^2-y^2, xy)	–
	+1	$+\varepsilon^{2*}$	$+\varepsilon$	$+\varepsilon^*$	$+\varepsilon^2$	+1	$+\varepsilon^{2*}$	$+\varepsilon$	$+\varepsilon^*$	$+\varepsilon^2$			
A_u	+1	+1	+1	+1	+1	−1	−1	−1	−1	−1	z	–	$z^3, z(x^2+y^2)$
E_{1u}	+1	$+\varepsilon$	$+\varepsilon^2$	$+\varepsilon^{2*}$	$+\varepsilon^*$	−1	$-\varepsilon$	$-\varepsilon^2$	$-\varepsilon^{2*}$	$-\varepsilon^*$	$x+iy$	–	(xz^2, yz^2) $[x(x^2+y^2), y(x^2+y^2)]$
	+1	$+\varepsilon^*$	$+\varepsilon^{2*}$	$+\varepsilon^2$	$+\varepsilon$	−1	$-\varepsilon^*$	$-\varepsilon^{2*}$	$-\varepsilon^2$	$-\varepsilon$	$x-iy$		
E_{2u}	+1	$+\varepsilon^2$	$+\varepsilon^*$	$+\varepsilon$	$+\varepsilon^{2*}$	−1	$-\varepsilon^2$	$-\varepsilon^*$	$-\varepsilon$	$-\varepsilon^{2*}$	–	–	$[xyz, z(x^2-y^2)]$
	+1	$+\varepsilon^{2*}$	$+\varepsilon$	$+\varepsilon^*$	$+\varepsilon^2$	−1	$-\varepsilon^{2*}$	$-\varepsilon$	$-\varepsilon^*$	$-\varepsilon^2$			$[y(3x^2-y^2), x(x^2-3y^2)]$

$\varepsilon=\exp(2\pi i/5)$

S_{12}	E	S_{12}	C_6	S_4	C_3	$(S_{12})^5$	C_2	$(S_{12})^7$	$(C_3)^2$	$(S_4)^3$	$(C_6)^5$	$(S_{12})^{11}$	linear functions, rotations	quadratic functions	cubic functions
A	+1	+1	+1	+1	+1	+1	+1	+1	+1	+1	+1	+1	R_z	z^2, x^2+y^2	–
B	+1	−1	+1	−1	+1	−1	+1	−1	+1	−1	+1	−1	z	–	$z^3, z(x^2+y^2)$
E_1	+1	$+\varepsilon$	$+\varepsilon^2$	+i	$-\varepsilon^{2*}$	$-\varepsilon^*$	−1	$-\varepsilon$	$-\varepsilon^2$	−i	$+\varepsilon^{2*}$	$+\varepsilon^*$	$x+iy$	–	(xz^2, yz^2) $[x(x^2+y^2),$
	+1	$+\varepsilon^*$	$+\varepsilon^{2*}$	−i	$-\varepsilon^2$	$-\varepsilon$	−1	$-\varepsilon^*$	$-\varepsilon^{2*}$	+i	$+\varepsilon^2$	$+\varepsilon$	$x-iy$		$y(x^2+y^2)]$
E_2	+1	$+\varepsilon^2$	$-\varepsilon^{2*}$	−1	$-\varepsilon^2$	$+\varepsilon^{2*}$	+1	$+\varepsilon^2$	$-\varepsilon^{2*}$	−1	$-\varepsilon^2$	$+\varepsilon^{2*}$	–	(x^2-y^2, xy)	–
	+1	$+\varepsilon^{2*}$	$-\varepsilon^2$	−1	$-\varepsilon^{2*}$	$+\varepsilon^2$	+1	$+\varepsilon^{2*}$	$-\varepsilon^2$	−1	$-\varepsilon^{2*}$	$+\varepsilon^2$			
E_3	+1	+i	−1	−i	+1	+i	−1	−i	+1	+i	−1	−i	–	–	$[x(x^2-3y^2),$
	+1	−i	−1	+i	+1	−i	−1	+i	+1	−i	−1	+i			$y(3x^2-y^2)]$
E_4	+1	$-\varepsilon^{2*}$	$-\varepsilon^2$	+1	$-\varepsilon^{2*}$	$-\varepsilon^2$	+1	$-\varepsilon^{2*}$	$-\varepsilon^2$	+1	$-\varepsilon^{2*}$	$-\varepsilon^2$	–	–	$[xyz, z(x^2-y^2)]$
	+1	$-\varepsilon^2$	$-\varepsilon^{2*}$	+1	$-\varepsilon^2$	$-\varepsilon^{2*}$	+1	$-\varepsilon^2$	$-\varepsilon^{2*}$	+1	$-\varepsilon^2$	$-\varepsilon^{2*}$			
E_5	+1	$-\varepsilon^*$	$+\varepsilon^{2*}$	+i	$-\varepsilon^2$	$+\varepsilon$	−1	$+\varepsilon^*$	$-\varepsilon^{2*}$	−i	$+\varepsilon^2$	$-\varepsilon$	R_x-iR_y	(xz, yz)	–
	+1	$-\varepsilon$	$+\varepsilon^2$	−i	$-\varepsilon^{2*}$	$+\varepsilon^*$	−1	$+\varepsilon$	$-\varepsilon^2$	+i	$+\varepsilon^{2*}$	$-\varepsilon^*$	R_x+iR_y		

$\varepsilon=\exp(2\pi i/12)$

Cubic Groups

T	E	$4C_3$	$4(C_3)^2$	$3C_2$	linear functions, rotations	quadratic functions	cubic functions
A	+1	+1	+1	+1	–	$x^2+y^2+z^2$	xyz
E	+1	$+\varepsilon$	$+\varepsilon^*$	+1	–	$(x^2-y^2, 2z^2-x^2-y^2)$	–
	+1	$+\varepsilon^*$	$+\varepsilon$	+1			
T	+3	0	0	−1	(x, y, z) (R_x, R_y, R_z)	(xy, xz, yz)	(x^3, y^3, z^3) (xy^2, x^2z, yz^2) (xz^2, x^2y, y^2z)

$\varepsilon=\exp(2\pi i/3)$

T_h	E	$4C_3$	$4(C_3)^2$	$3C_2$	i	$4(S_6)^5$	$4S_6$	$3\sigma_h$	linear functions, rotations	quadratic functions	cubic functions
A_g	+1	+1	+1	+1	+1	+1	+1	+1	–	$x^2+y^2+z^2$	–
E_g	+1	$+\varepsilon$	$+\varepsilon^*$	+1	+1	$+\varepsilon$	$+\varepsilon^*$	+1	–	$(x^2-y^2, 2z^2-x^2-y^2)$	–
	+1	$+\varepsilon^*$	$+\varepsilon$	+1	+1	$+\varepsilon^*$	$+\varepsilon$	+1			
T_g	+3	0	0	−1	+3	0	0	−1	(R_x, R_y, R_z)	(xy, xz, yz)	–
A_u	+1	+1	+1	+1	−1	−1	−1	−1	–	–	xyz
E_u	+1	$+\varepsilon$	$+\varepsilon^*$	+1	−1	$-\varepsilon^*$	$-\varepsilon^*$	−1	–	–	–
	+1	$+\varepsilon^*$	$+\varepsilon$	+1	−1	$-\varepsilon$	$-\varepsilon$	−1			
T_u	+3	0	0	−1	−3	0	0	+1	(x, y, z)	–	(x^3, y^3, z^3) (xy^2, x^2z, yz^2) (xz^2, x^2y, y^2z)

$\varepsilon=\exp(2\pi i/3)$

T_d	E	$8C_3$	$3C_2$	$6S_4$	$6\sigma_d$	Linear functions	Quadratic functions	Cubic functions
A_1	1	1	1	1	1		$x^2+y^2+z^2$	xyz
A_2	1	1	1	−1	−1			–
E	2	−1	2	0	0		$(2z^2-x^2-y^2, x^2-y^2)$	–
T_1	3	0	−1	1	−1	(R_x, R_y, R_z)		$[x(z^2-y^2), y(z^2-x^2), z(x^2-y^2)]$
T_2	3	0	−1	−1	1	(x, y, z)	(xy, yz, zx)	(x^3, y^3, z^3) $[x(z^2+y^2), y(z^2+x^2), z(x^2+y^2)]$

O	E	$8C_3$	$6C_2'$	$6C_4$	$3C_2=(C_4)^2$	linear functions, rotations	quadratic functions	cubic functions
A_1	+1	+1	+1	+1	+1	–	$x^2+y^2+z^2$	–
A_2	+1	+1	−1	−1	+1	–	–	xyz
E	+2	−1	0	0	+2	–	$(x^2-y^2, 2z^2-x^2-y^2)$	–
T_1	+3	0	−1	+1	−1	(x, y, z) (R_x, R_y, R_z)	–	(x^3, y^3, z^3) $[x(z^2+y^2), y(z^2+x^2), z(x^2+y^2)]$
T_2	+3	0	+1	−1	−1	–	(xy, xz, yz)	$[x(z^2-y^2), y(z^2-x^2), z(x^2-y^2)]$

O_h	E	$8C_3$	$6C_2$	$6C_4$	$3C_2(=C_4^2)$	i	$6S_4$	$8S_6$	$3\sigma_h$	$6\sigma_v$	Linear functions	Quadratic functions	Cubic functions
A_{1g}	1	1	1	1	1	1	1	1	1	1		$x^2+y^2+z^2$	–
A_{2g}	1	1	–1	–1	1	1	–1	1	1	–1			–
E_g	2	–1	0	0	2	2	0	–1	2	0		$(2z^2-x^2-y^2, x^2-y^2)$	–
T_{1g}	3	0	–1	1	–1	3	1	0	–1	–1	(R_x, R_y, R_z)		–
T_{2g}	3	0	1	–1	–1	3	–1	0	–1	1		(xy, yz, zx)	–
A_{1u}	1	1	1	1	1	–1	–1	–1	–1	–1			–
A_{2u}	1	1	–1	–1	1	–1	1	–1	–1	1			xyz
E_u	2	–1	0	0	2	–2	0	1	–2	0			–
T_{1u}	3	0	–1	1	–1	–3	–1	0	1	1	(x, y, z)		(x^3, y^3, z^3) $[x(z^2+y^2), y(z^2+x^2), z(x^2+y^2)]$
T_{2u}	3	0	1	–1	–1	–3	1	0	1	–1			$[x(z^2-y^2), y(z^2-x^2), z(x^2-y^2)]$

I	E	$12C_5$	$12C_5^2$	$20C_5$	$15C_2$	Linear functions	Quadratic functions	Cubic functions
A_1	1	1	1	1	1		$x^2+y^2+z^2$	–
T_1	3	$\frac{1}{2}(1+\sqrt{5})$	$\frac{1}{2}(1-\sqrt{5})$	0	–1	(x, y, z) (R_x, R_y, R_z)		$[x(z^2+y^2), y(z^2+x^2), z(x^2+y^2)]$
T_2	3	$\frac{1}{2}(1-\sqrt{5})$	$\frac{1}{2}(1+\sqrt{5})$	0	–1			$[x^3, y^3, z^3]$
G	4	–1	–1	1	0			$[x(z^2-y^2), y(z^2-x^2), z(x^2-y^2), xyz]$
H	5	0	0	–1	1		$(2z^2-x^2-y^2, x^2-y^2, xy, yz, zx)$	–

I_h	E	$12C_5$	$12(C_5)^2$	$20C_3$	$15C_2$	i	$12S_{10}$	$12(S_{10})^3$	$20S_6$	15σ	linear functions, rotations	quadratic functions	cubic functions
A_g	+1	+1	+1	+1	+1	+1	+1	+1	+1	+1	–	$x^2+y^2+z^2$	–
T_{1g}	+3	$-2\cos(4\pi/5)$	$-2\cos(2\pi/5)$	0	–1	+3	$-2\cos(2\pi/5)$	$-2\cos(4\pi/5)$	0	–1	(R_x, R_y, R_z)	–	–
T_{2g}	+3	$-2\cos(2\pi/5)$	$-2\cos(4\pi/5)$	0	–1	+3	$-2\cos(4\pi/5)$	$-2\cos(2\pi/5)$	0	–1	–	–	–
G_g	+4	–1	–1	+1	0	+4	–1	–1	+1	0	–	–	–
H_g	+5	0	0	–1	+1	+5	0	0	–1	+1	–	$[2z^2-x^2-y^2, x^2-y^2, xy, xz, yz]$	–
A_u	+1	+1	+1	+1	+1	–1	–1	–1	–1	–1	–	–	–
T_{1u}	+3	$-2\cos(4\pi/5)$	$-2\cos(2\pi/5)$	0	–1	–3	$+2\cos(2\pi/5)$	$+2\cos(4\pi/5)$	0	+1	(x, y, z)	–	$[x(z^2+y^2), y(z^2+x^2), z(x^2+y^2)]$
T_{2u}	+3	$-2\cos(2\pi/5)$	$-2\cos(4\pi/5)$	0	–1	–3	$+2\cos(4\pi/5)$	$+2\cos(2\pi/5)$	0	+1	–	–	$[x^3, y^3, z^3]$
G_u	+4	–1	–1	+1	0	–4	+1	+1	–1	0	–	–	$[x(z^2-y^2), y(z^2-x^2), z(x^2-y^2), xyz]$
H_u	+5	0	0	–1	+1	–5	0	0	+1	–1	–	–	–

Linear groups

$C_{\infty v}$	E	$2C_\infty$	···	$\infty\,\sigma_v$	linear functions, rotations	quadratic functions	cubic functions
$A_1{=}\Sigma^+$	+1	+1	···	+1	z	$x^2{+}y^2$, z^2	z^3, $z(x^2{+}y^2)$
$A_2{=}\Sigma^-$	+1	+1	···	−1	R_z	–	–
$E_1{=}\Pi$	+2	+2cos(φ)	···	0	(x, y) (R_x, R_y)	(xz, yz)	(xz^2, yz^2) $[x(x^2{+}y^2), y(x^2{+}y^2)]$
$E_2{=}\Delta$	+2	+2cos(2φ)	···	0	–	$(x^2{-}y^2, xy)$	$[xyz, z(x^2{-}y^2)]$
$E_3{=}\Phi$	+2	+2cos(3φ)	···	0	–	–	$[y(3x^2{-}y^2), x(x^2{-}3y^2)]$
···	···	···	···	···	–	–	–
E_n	+2	+2cos(nφ)	···	0	–	–	–

$D_{\infty h}$	E	$2C_\infty$	···	$\infty\sigma_v$	i	$2S_\infty$	···	$\infty C_2'$	Linear functions, rotations	Quadratic functions	Cubic functions
$A_{1g}{=}\Sigma^+{}_g$	+1	+1	···	+1	+1	+1	···	+1	–	$x^2{+}y^2$, z^2	–
$A_{2g}{=}\Sigma^-{}_g$	+1	+1	···	−1	+1	+1	···	−1	R_z	–	–
$E_{1g}{=}\Pi_g$	+2	+2cos(ϕ)	···	0	+2	−2cos(ϕ)	···	0	(R_x, R_y)	(xz, yz)	–
$E_{2g}{=}\Delta_g$	+2	+2cos(2ϕ)	···	0	+2	+2cos(2ϕ)	···	0	–	$(x^2{-}y^2, xy)$	–
$E_{3g}{=}\Phi_g$	+2	+2cos(3ϕ)	···	0	+2	−2cos(3ϕ)	···	0	–	–	–
E_{ng}	+2	+2cos(nϕ)	···	0	+2	$(-1)^n$2cos(n ϕ)	···	0	–	–	–
···	···	···	···	···	···	···	···	···	–	–	–
$A_{1u}{=}\Sigma^+{}_u$	+1	+1	···	+1	−1	−1	···	−1	z	–	z^3, $z(x^2{+}y^2)$
$A_{2u}{=}\Sigma^-{}_u$	+1	+1	···	−1	−1	−1	···	+1	–	–	–
$E_{1u}{=}\Pi_u$	+2	+2cos(ϕ)	···	0	−2	+2cos(ϕ)	···	0	(x, y)	–	(xz^2, yz^2) $[x(x^2{+}y^2), y(x^2{+}y^2)]$
$E_{2u}{=}\Delta_u$	+2	+2cos(2ϕ)	···	0	−2	−2cos(2ϕ)	···	0	–	–	$[xyz, z(x^2{-}y^2)$
$E_{3u}{=}\Phi_u$	+2	+2cos(3ϕ)	···	0	−2	2cos(3ϕ)	···	0	–	–	$[y(3x^2{-}y^2), x(x^2{-}3y^2)]$
E_{nu}	+2	+2cos(nϕ)	···	0	−2	$(-1)^{n+1}$2cos(nϕ)	···	0	–	–	–
···	···	···	···	···	···	···	···	···	–	–	–

APPENDIX E

Review of Matrix Manipulations

A matrix is an array of numbers that can be used to manipulate systems of linear equations. Consider the following system of equations:

$$
\begin{aligned}
1x + 2y + 3z &= 0 \\
3x + 4y + 7z &= 2 \\
6x + 5y + 9z &= 11
\end{aligned}
$$

We can define two matrices based on this system:

$$
A = \begin{bmatrix} 1 & 2 & 3 \\ 3 & 4 & 7 \\ 6 & 5 & 9 \end{bmatrix}, \ B = \begin{bmatrix} 0 \\ 2 \\ 11 \end{bmatrix}
$$

In general, a matrix may have any number of elements within it. We can designate a specific element (a) by its row (i) and column (j) as follows, where each entry is given a symbol "a_{ij}".

$$
\begin{bmatrix} a_{11} & \cdots & a_{1j} \\ \vdots & \ddots & \vdots \\ a_{i1} & \cdots & a_{ij} \end{bmatrix}
$$

Matrix Multiplication

One operation we will need is matrix multiplication. The thing to remember is "row times column" or the RC Rule.

To illustrate, here is a simple example, which is probably more difficult than any we will need in our group theory examples. However, some of the group theory examples can involve quite large matrices. Let us multiply the matrices A and B.

$$
\mathrm{A} = \begin{bmatrix} 2 & -1 \\ 1 & 0 \\ -3 & 4 \end{bmatrix} \quad \mathrm{B} = \begin{bmatrix} 1 & -2 & -5 \\ 3 & 4 & 0 \end{bmatrix}
$$

In order to be able to multiply two matrices, the number of columns in the first matrix must equal the number of rows in the second.

Let us multiply A × B. This is done by taking the first entry in the first matrix (a_{11}) and multiplying by the first element in the second (b_{11}); then this is summed with the product of a_{12} and b_{21}. In other words, the row $[2 \quad -1]$ is multiplied by the column $\begin{bmatrix} 1 \\ 3 \end{bmatrix}$, which becomes the first entry in our answer (ab_{11} = 2−3 = −1). The first row of A, [2 −1], is then multiplied by the next column of B, $\begin{bmatrix} -2 \\ 4 \end{bmatrix}$, to produce −4 −4 = −8. Finally, the first row of A is multiplied by the last column of B to produce entry ab_{13} = −10 + 0 = −10. The other two rows of A are treated similarly until the 3 × 3 square matrix AB is obtained. Notice that Row 1 is multiplied by Column 1, which produce the answer for the term (1, 1) in our product matrix. Similarly, multiplying Row 3 by Column 1 will produce the term (3, 1) in our product matrix, and so on.

$$\begin{bmatrix} 2 & -1 \\ 1 & 0 \\ -3 & 4 \end{bmatrix} \begin{bmatrix} 1 & -2 & -5 \\ 3 & 4 & 0 \end{bmatrix} = \begin{bmatrix} 2-3 & -4-4 & -10+0 \\ 1+0 & -2+0 & -5+0 \\ -3+12 & 6+16 & 15+0 \end{bmatrix}$$

$$AB = \begin{bmatrix} -1 & -8 & -10 \\ 1 & -2 & -5 \\ 9 & 22 & 15 \end{bmatrix}$$

Now let us multiply B × A. You will notice that BA ≠ AB and multiplication of matrices is not, in general, communicative. In many cases, one says, for example, that "A is left multiplied by B" to designate.

$$\begin{bmatrix} 1 & -2 & -5 \\ 3 & 4 & 0 \end{bmatrix} \begin{bmatrix} 2 & -1 \\ 1 & 0 \\ -3 & 4 \end{bmatrix} = \begin{bmatrix} 2-2+15 & -1+0-20 \\ 6+4+0 & -3+0+0 \end{bmatrix}$$

$$BA = \begin{bmatrix} 15 & -21 \\ 10 & -3 \end{bmatrix}$$

Another operation that will be used in matrix algebra is calculation of the character, which comes from the trace of a matrix. The character of a matrix, assigned the symbol χ, is the sum of all the diagonal elements from top left to bottom right.

$$\chi = \sum_{i=j} a_{ij}$$

As examples, we can calculate the trace of AB found previously. The diagonal elements are a_{11} = −1 (top row, left), a_{22} = −2 (second row, middle), and a_{33} = 15 (bottom row, right). The simple sum of these is −1 −2 + 15 = 12.

$$AB = \begin{bmatrix} -1 & -8 & -10 \\ 1 & -2 & -5 \\ 9 & 22 & 15 \end{bmatrix}$$

$$\chi_{AB} = \sum_{i=j} a_{ij} = -1 + -2 + 15$$

$$\chi_{AB} = 12$$

Finally, we will define one other linear algebra term: inverse. The inverse of the matrix is the matrix that, if multiplied by the original matrix, produces an identity matrix (all those along the diagonal with zeros everywhere else). So, $AA^{-1} = A^{-1}A = I$.

There are several mathematical methods for finding the inverse of a matrix, which will generally be unnecessary for our discussions here, but we can illustrate inverse by using a very simple case. If we start with a simple 2 × 2 matrix and its inverse:

$$A = \begin{bmatrix} 0.8 & -0.6 \\ 0.6 & 0.8 \end{bmatrix}, \quad A^{-1} = \begin{bmatrix} 0.8 & 0.6 \\ -0.6 & 0.8 \end{bmatrix}$$

We can multiply them:

$$AA^{-1} = \begin{bmatrix} 0.8 & -0.6 \\ 0.6 & 0.8 \end{bmatrix} \begin{bmatrix} 0.8 & 0.6 \\ -0.6 & 0.8 \end{bmatrix}$$

This produces (row times column):

$$AA^{-1} = \begin{bmatrix} 0.8^2 + (-0.6)^2 & 0.8 \times 0.6 - 0.6 \times 0.8 \\ 0.6 \times 0.8 - 0.8 \times 0.6 & 0.6^2 + 0.8^2 \end{bmatrix} = \begin{bmatrix} 1 & 0 \\ 0 & 1 \end{bmatrix}$$

If you look back at the matrix A just used, and its inverse A^{-1}, you will see that there is something special about their relationship. In this case, A^{-1} is the "transpose" of A—the result of swapping the columns and rows of a matrix. The transpose is often assigned a superscripted "T", so the transpose of A = A^T. In this case, $A^T = A^{-1}$, which is not the usual case for matrices. Matrices that have this property are called "orthonormal", which means that the vectors in it are orthogonal. While this is not really the norm for matrices in general, it will not be unusual in our applications of group theory to chemistry for the transpose to be the inverse.

The square identity matrix is, to some extent, the matrix equivalent of the number "1" in regular algebra. Just like $a \times a^{-1} = 1$, $AA^{-1} = I$, the identity matrix of the appropriate size.

Determinants

The "determinant" of a matrix *A* is sometimes written as det(*A*) or |*A*|, with straight lines on either side of the matrix. This is a scalar (a vector length or number) associated with a matrix. The determinant has special properties for the matrix and is found by multiplying along various diagonals from top left to bottom right, and then subtracting the diagonals from top right to bottom left.

The easiest way to see this is with an illustration. First let us look at a 2 × 2 matrix. Let us look at the simple matrix as follows. First we multiply the top left by the bottom right to obtain 15, and then we take the negative of the top right to bottom left to obtain −8. The result is then 7.

$$\begin{vmatrix} 5 & 4 \\ 2 & 3 \end{vmatrix} = (5 \times 3) - (4 \times 2) = 15 - 8 = 7$$

Higher-order matrices are no more difficult but can be much more time-consuming. To illustrate the process, let us do a 3 × 3 matrix. This problem must include all the diagonals of the matrix. We can start with the first diagonal at the top left to bottom right (2, 5, 4).

$$\begin{vmatrix} 2 & 1 & 8 \\ 0 & 5 & -2 \\ 6 & -3 & 4 \end{vmatrix} = (2 \times 5 \times 4) + \ldots$$

To find the next diagonal we move to the next element of the first row, which has a "1", and then move down and to the right, "–2". Just like on the main diagonal, we need three elements, so we jump to the first element of the bottom row, "6". Now, our calculation looks like this:

$$\begin{vmatrix} 2 & 1 & 8 \\ 0 & 5 & -2 \\ 6 & -3 & 4 \end{vmatrix} = (2 \times 5 \times 4) + (1 \times -2 \times 6) + \ldots$$

Our last positive diagonal starts with the third element of the first row, "8". Then it goes to the first element of the second row, "0", and second element of the third row, "–3".

$$\begin{vmatrix} 2 & 1 & 8 \\ 0 & 5 & -2 \\ 6 & -3 & 4 \end{vmatrix} = (2 \times 5 \times 4) + (1 \times -2 \times 6) + (8 \times 0 \times -3) - \ldots$$

Next, we turn to the diagonals that run from top right to bottom left, which get subtracted. The main diagonal here is (8, 5, 6).

$$\begin{vmatrix} 2 & 1 & 8 \\ 0 & 5 & -2 \\ 6 & -3 & 4 \end{vmatrix} = (2 \times 5 \times 4) + (1 \times -2 \times 6) + (8 \times 0 \times -3) - (8 \times 5 \times 6) - \ldots$$

Moving to the left, the next diagonal goes (1, 0, 4).

$$\begin{vmatrix} 2 & 1 & 8 \\ 0 & 5 & -2 \\ 6 & -3 & 4 \end{vmatrix} = (2 \times 5 \times 4) + (1 \times -2 \times 6) + (8 \times 0 \times -3) - (8 \times 5 \times 6) - (1 \times 0 \times 4) - \ldots$$

Finally, the last diagonal would be (2, –2, –3).

$$\begin{vmatrix} 2 & 1 & 8 \\ 0 & 5 & -2 \\ 6 & -3 & 4 \end{vmatrix} = (2 \times 5 \times 4) + (1 \times -2 \times 6) + (8 \times 0 \times -3) - (8 \times 5 \times 6) - (1 \times 0 \times 4) - (2 \times -2 \times -3)$$

$$= 40 - 12 + 0 - 240 - 0 - 12 = -224$$

APPENDIX F

Expansion of the AOM to Include S^4 Terms

Following Albright, Burdett, and Whangbo,[1] it can be shown that the energy of interaction between a metal d-orbital and a ligand-based donor orbital with the usual assumptions of the angular overlap model can be represented as follows

$$E = A \pm (B^2 + k^2 S_{ML}^2)^{1/2}$$

where $A = ½(H_{MM} + H_{LL})$ and $B = ½(H_{MM} - H_{LL})$. This equation can be subjected to a Taylor expansion based on the relation

$$\sqrt{1+x} = 1 + \frac{x}{2} - \frac{x^2}{8} + \frac{x^3}{16} - \ldots$$

which we truncated after the quadratic term. The result is shown as follows, where we have chosen the negative term in ±.

$$E = A - B\left[1 + \frac{k^2 S_{ML}^2}{2B^2} - \frac{k^4 S_{ML}^4}{8B^4}\right]$$

The stabilization energy of the orbital (ε) can then be estimated as

$$\varepsilon \approx \beta e_\sigma F^2(\theta,\phi) - \gamma e_\sigma F^4(\theta,\phi)$$

where the expression above is appropriate for evaluation of different geometries where the analysis based on F^2 does not differentiate the geometries in energy. The coefficients are such that $\beta > \gamma$, and both are always positive.

In Chapter 15, the total stabilization energy of various 4-coordinate complexes were calculated with three geometries: tetrahedral (T_d), square-planar (D_{4h}), and *cis*-divacant octahedral (C_{2v}). The energies of the orbitals truncated at the F^2-term are shown in Table 15.5. The energies expanded to include the quartic terms are as follows.

For a tetrahedral geometry, the orbital stabilization energies are then:

$$\varepsilon_{xz,yz,xy} = \beta\left(\frac{4}{3}e_\sigma + \frac{8}{9}e_\pi\right) - \gamma\left(\frac{16}{9}e_\sigma + \frac{64}{81}e_\pi\right)$$

$$\varepsilon_{x^2-y^2,z^2} = \beta\frac{8}{3}e_\pi - \gamma\frac{64}{9}e_\pi$$

For a square-planar geometry:

$$\begin{aligned}\varepsilon_{x^2-y^2} &= \beta 3e_\sigma - \gamma 9e_\sigma\\ \varepsilon_{z^2} &= \beta e_\sigma - \gamma e_\sigma\\ \varepsilon_{xy} &= \beta 4e_\pi - \gamma 16e_\pi\\ \varepsilon_{xz,yz} &= \beta 2e_\pi - \gamma 4e_\pi\end{aligned}$$

For a *cis*-divacant octahedral geometry:

$$\begin{aligned}\varepsilon_{z^2} &= \beta\frac{5}{2}e_\sigma - \gamma\frac{25}{4}e_\sigma\\ \varepsilon_{x^2-y^2} &= \beta\frac{3}{2}e_\sigma - \gamma\frac{9}{4}e_\sigma\\ \varepsilon_{xz} &= \beta 3e_\pi - \gamma 9e_\pi\\ \varepsilon_{xy,yz} &= \beta 2e_\pi - \gamma 4e_\pi\end{aligned}$$

For a low spin d^6 complex, it was determined that the system truncated at F^2-terms produced equal energies for the *cis*-divacant octahedral and square-planar geometries. We can use the previous expressions to determine if this was simply due to truncation of the Taylor expansion at the quadratic terms, and if the energies for the two geometries are actually different when quartic terms are included.

The total stabilization energy of the system is determined by the holes, h, in the two highest-energy orbitals of the two geometries and the orbital energies. Using these values in Eq. 15.12 produces the following equations.

$$\begin{aligned}\Phi_{22200}^{D_{4h}} &= \sum_i h_i\varepsilon_i^*\\ &= 2\times\left(\beta 3e_\sigma - \gamma 9e_\sigma\right) + 2\times\left(\beta e_\sigma - \gamma e_\sigma\right)\\ &= 2e_\sigma(4\beta - 10\gamma)\end{aligned}$$

$$\begin{aligned}\Phi_{22200}^{C_{2v}} &= \sum_i h_i\varepsilon_i^*\\ &= 2\times\left(\beta\frac{5}{2}e_\sigma - \gamma\frac{25}{4}e_\sigma\right) + 2\times\left(\beta\frac{3}{2}e_\sigma - \gamma\frac{9}{4}e_\sigma\right)\\ &= 2e_\sigma\left(4\beta - 8\frac{1}{2}\gamma\right)\end{aligned}$$

Since the C_{2v} geometry has a larger total stabilization energy (remember, β and γ are always positive), *cis*-divacant octahedral is the more stable geometry for low spin d^6 (22200).

Cited Reference

1. Albright, T. A., Burdett, J. K., and Whangbo, M.-H. *Orbital Interactions in Chemistry*, 2nd edn. Hoboken, NJ: Wiley, 2013.

APPENDIX G

Matrices Associated with Strong Field Parameters

Table G1 Strong-field energy matrices for d^2, in O_h.
A constant value of *A* should be added to all diagonal elements to properly represent the actual eigenvalues of the electron–electron repulsion terms.

$^1A_{1g}$	t_{2g}^2	e_g^2
t_{2g}^2	$-8Dq + 10B + 5C$	$(2B + C)\sqrt{6}$
e_g^2	$(2B + C)\sqrt{6}$	$12Dq + 8B + 4C$

^1E_g	t_{2g}^2	e_g^2
t_{2g}^2	$-8Dq + B + 2C$	$2B\sqrt{3}$
e_g^2	$2B\sqrt{3}$	$12Dq + 2C$

$^1T_{1g}$	$t_{2g}^1e_g^1$
$t_{2g}^1e_g^1$	$2Dq + 4B + 2C$

$^1T_{2g}$	t_{2g}^2	$t_{2g}^1e_g^1$
t_{2g}^2	$-8Dq + B + 2C$	$-2B\sqrt{3}$
$t_{2g}^1e_g^1$	$-2B\sqrt{3}$	$2Dq + 2C$

$^3A_{2g}$	e_g^2
e_g^2	$12Dq - 8B$

$^3T_{1g}$	t_{2g}^2	$t_{2g}^1e_g^1$
t_{2g}^2	$-8Dq - 5B$	$-6B$
$t_{2g}^1e_g^1$	$-6B$	$2Dq + 4B$

$^3T_{2g}$	$t_{2g}^1e_g^1$
$t_{2g}^1e_g^1$	$2Dq - 8B$

Table G2 Strong-field energy matrices for d^3, in O_h.
A constant value of 3*A* should be added to all diagonal elements to properly represent the actual eigenvalues of the electron–electron repulsion term.

$^2A_{1g}$	$t_{2g}^2(^1E_g)e_g^1$
$t_{2g}^2(^1E_g)e_g^1$	$-2Dq - 11B + 3C$

$^2A_{2g}$	$t_{2g}^2(^1E_g)e_g^1$
$t_{2g}^2(^1E_g)e_g^1$	$-2Dq + 9B + 3C$

Table G2 Continued

^2E_g	t_{2g}^3	$t_{2g}^2(^1A_{1g})e_g^1$	$t_{2g}^2(^1E_g)e_g^1$	e_g^3
t_{2g}^3	$-12Dq-6B+3C$	$-6B\sqrt{2}$	$3B\sqrt{2}$	0
$t_{2g}^2(^1A_{1g})e_g^1$	$-6B\sqrt{2}$	$-2Dq+8B+6C$	$-10B$	$(2B+C)\sqrt{3}$
$t_{2g}^2(^1E_g)e_g^1$	$3B\sqrt{2}$	$-10B$	$-2Dq-B+3C$	$-2B\sqrt{3}$
e_g^3	0	$(2B+C)\sqrt{3}$	$-2B\sqrt{3}$	$18Dq-8B+4C$

$^2T_{1g}$	t_{2g}^3	$t_{2g}^2(^3T_{1g})e_g^1$	$t_{2g}^2(^1T_{2g})e_g^1$	$t_{2g}^1e_g^2(^3A_{2g})$	$t_{2g}^1e_g^2(^1E_g)$
t_{2g}^3	$-12Dq-6B+3C$	$3B$	$-3B$	0	$-2B\sqrt{3}$
$t_{2g}^2(^3T_{1g})e_g^1$	$3B$	$-2Dq+3C$	$-3B$	$3B$	$-3B\sqrt{3}$
$t_{2g}^2(^1T_{2g})e_g^1$	$-3B$	$-3B$	$-2Dq-6B+3C$	$-3B$	$B\sqrt{3}$
$t_{2g}^1e_g^2(^3A_{2g})$	0	$3B$	$-3B$	$8Dq-6B+3C$	$-2B\sqrt{3}$
$t_{2g}^1e_g^2(^1E_g)$	$-2B\sqrt{3}$	$-3B\sqrt{3}$	$B\sqrt{3}$	$-2B\sqrt{3}$	$8Dq-2B+3C$

$^2T_{2g}$	t_{2g}^3	$t_{2g}^2(^3T_{1g})e_g^1$	$t_{2g}^2(^1T_{2g})e_g^1$	$t_{2g}^1e_g^2(^3A_{1g})$	$t_{2g}^1e_g^2(^1E_g)$
t_{2g}^3	$-12Dq+5C$	$-3B\sqrt{3}$	$5B\sqrt{3}$	$4B+2C$	$2B$
$t_{2g}^2(^3T_{1g})e_g^1$	$-3B\sqrt{3}$	$-2Dq-6B+3C$	$-3B$	$-3B\sqrt{3}$	$-3B\sqrt{3}$
$t_{2g}^2(^1T_{2g})e_g^1$	$5B\sqrt{3}$	$-3B$	$-2Dq+4B+3C$	$B\sqrt{3}$	$-B\sqrt{3}$
$t_{2g}^1e_g^2(^3A_{1g})$	$4B+2C$	$-3B\sqrt{3}$	$B\sqrt{3}$	$8Dq+6B+5C$	$10B$
$t_{2g}^1e_g^2(^1E_g)$	$2B$	$-3B\sqrt{3}$	$-B\sqrt{3}$	$10B$	$8Dq-2B+3C$

$^4A_{2g}$	t_{2g}^3
t_{2g}^3	$-12Dq-15B$

$^4T_{2g}$	$t_{2g}^2(^3T_{1g})e_g^1$
$t_{2g}^2(^3T_{1g})e_g^1$	$-2Dq-15B$

$^4T_{1g}$	$t_{2g}^2(^3T_{1g})e_g^1$	$t_{2g}^1e_g^2(^3A_{2g})$
$t_{2g}^2(^3T_{1g})e_g^1$	$-2Dq-3B$	$6B$
$t_{2g}^1e_g^2(^3A_{2g})$	$6B$	$8Dq-12B$

Table G3 Strong-field energy matrices for d^4, in O_h.
A constant value of 6*A* should be added to all diagonal elements to properly represent the actual eigenvalues of the electron–electron repulsion terms.

$^1A_{1g}$	t_{2g}^4	$t_{2g}^3(^2E_g)e_g^1$	$t_{2g}^2(^1A_{1g})e_g^2(^1A_{1g})$	$t_{2g}^2(^1E_g)e_g^2(^1E_g)$	e_g^4
t_{2g}^4	$-16Dq + 10C$	$-12B\sqrt{2}$	$(4B + 2C)\sqrt{2}$	$-2B\sqrt{2}$	0
$t_{2g}^3(^2E_g)e_g^1$	$-12B\sqrt{2}$	$-6Dq + 6C$	$-12B$	$6B$	0
$t_{2g}^2(^1A_{1g})e_g^2(^1A_{1g})$	$(4B + 2C)\sqrt{2}$	$-12B$	$4Dq + 14B + 11C$	$-20B$	$(2B + C)\sqrt{6}$
$t_{2g}^2(^1E_g)e_g^2(^1E_g)$	$-2B\sqrt{2}$	$6B$	$-20B$	$4Dq - 3B + 6C$	$-2B\sqrt{6}$
e_g^4	0	0	$(2B + C)\sqrt{6}$	$-2B\sqrt{6}$	$24Dq - 16B + 8C$

$^1A_{2g}$	$t_{2g}^3(^2E_g)e_g^1$	$t_{2g}^2(^1E_g)e_g^2(^1E_g)$
$t_{2g}^3(^2E_g)e_g^1$	$-6Dq - 12B + 6C$	$-6B$
$t_{2g}^2(^1E_g)e_g^2(^1E_g)$	$-6B$	$4Dq - 3B + 6C$

^1E_g	t_{2g}^4	$t_{2g}^3(^2E_g)e_g^1$	$t_{2g}^2(^1E_g)e_g^2(^1A_{1g})$	$t_{2g}^2(^1A_{1g})e_g^2(^1E_g)$	$t_{2g}^2(^1E_g)e_g^2(^1E_g)$
t_{2g}^4	$-16Dq - 9B + 7C$	$6B$	$-(2B + C)\sqrt{2}$	$-2B$	$4B$
$t_{2g}^3(^2E_g)e_g^1$	$6B$	$-6Dq - 6B + 6C$	$3B\sqrt{2}$	$-12B$	0
$t_{2g}^2(^1E_g)e_g^2(^1A_{1g})$	$-(2B + 2)\sqrt{2}$	$3B\sqrt{2}$	$4Dq + 5B + 8C$	$-10B\sqrt{2}$	$-10B\sqrt{2}$
$t_{2g}^2(^1A_{1g})e_g^2(^1E_g)$	$-2B$	$-12B$	$-10B\sqrt{2}$	$4Dq + 6B + 9C$	0
$t_{2g}^2(^1E_g)e_g^2(^1E_g)$	$4B$	0	$-10B\sqrt{2}$	0	$4Dq - 3B + 6C$

$^1T_{1g}$	$t_{2g}^3(^2T_{1g})e_g^1$	$t_{2g}^3(^2T_{2g})e_g^1$	$t_{2g}^2(^2T_{2g})e_g^2(^1E_g)$	$t_{2g}^1e_g^3$
$t_{2g}^3(^2T_{1g})e_g^1$	$-6Dq - 3B + 6C$	$5B\sqrt{3}$	$-3B$	$B\sqrt{6}$
$t_{2g}^3(^2T_{2g})e_g^1$	$5B\sqrt{3}$	$-6Dq - 3B + 8C$	$5B\sqrt{3}$	$(B + C)\sqrt{2}$
$t_{2g}^2(^2T_{2g})e_g^2(^1E_g)$	$-3B$	$5B\sqrt{3}$	$4Dq - 3B + 6C$	$B\sqrt{6}$
$t_{2g}^1e_g^3$	$B\sqrt{6}$	$(B + C)\sqrt{2}$	$B\sqrt{6}$	$14Dq - 16B + 7C$

Table G3 Continued

$^1T_{2g}$	t_{2g}^4	$t_{2g}^3(^2T_{1g})e_g^1$	$t_{2g}^3(^2T_{2g})e_g^1$	$t_{2g}^2(^3T_{1g})e_g^2(^3A_{2g})$	$t_{2g}^2(^1T_{2g})e_g^2(^1E_g)$	$t_{2g}^2(^1T_{2g})e_g^2(^1A_{1g})$	$t_{2g}^1e_g^3$
t_{2g}^4	$-16Dq-9B$ $+7C$	$-3B\sqrt{2}$	$-5B\sqrt{6}$	0	$2B\sqrt{2}$	$-(2B+C)\sqrt{2}$	0
$t_{2g}^3(^2T_{1g})e_g^1$	$-3B\sqrt{2}$	$-6Dq$ $-9B+6C$	$5B\sqrt{3}$	$3B$	$-3B$	$-3B$	$B\sqrt{6}$
$t_{2g}^3(^2T_{2g})e_g^1$	$-5B\sqrt{6}$	$5B\sqrt{3}$	$-6Dq+3B+8C$	$3B\sqrt{3}$	$-5B\sqrt{3}$	$5B\sqrt{3}$	$(3B+C)\sqrt{2}$
$t_{2g}^2(^3T_{1g})e_g^2(^3A_{2g})$	0	$3B$	$3B\sqrt{3}$	$4Dq-9B+6C$	$-6B$	0	$3B\sqrt{6}$
$t_{2g}^2(^1T_{2g})e_g^2(^1E_g)$	$2B\sqrt{2}$	$-3B$	$-5B\sqrt{3}$	$-6B$	$4Dq-3B+6C$	$-10B$	$-B\sqrt{6}$
$t_{2g}^2(^1T_{2g})e_g^2(^1A_{1g})$	$-(2B+C)\sqrt{2}$	$-3B$	$5B\sqrt{3}$	0	$-10B$	$4Dq+5B+8C$	$-B\sqrt{6}$
$t_{2g}^1e_g^3$	0	$B\sqrt{6}$	$(3B+C)\sqrt{2}$	$3B\sqrt{6}$	$-B\sqrt{6}$	$-B\sqrt{6}$	$14Dq+7C$

$^3A_{1g}$	$t_{2g}^3(^2E_g)e_g^1$
$t_{2g}^3(^2E_g)e_g^1$	$-6Dq-12B+4C$

$^3A_{2g}$	$t_{2g}^3(^2E_g)e_g^1$	$t_{2g}^2(^1A_{1g})e_g^2(^3A_{2g})$
$t_{2g}^3(^2E_g)e_g^1$	$-6Dq-8B+4C$	$-12B$
$t_{2g}^2(^1A_{1g})e_g^2(^3A_{2g})$	$-12B$	$4Dq-2B+7C$

^3E_g	$t_{2g}^3(^2E_g)e_g^1$	$t_{2g}^3(^2E_g)e_g^1$	$t_{2g}^2(^1E_g)e_g^2(^3A_{2g})$
$t_{2g}^3(^4A_{2g})e_g^1$	$-6Dq-13B+4C$	$-4B$	0
$t_{2g}^3(^2E_g)e_g^1$	$-4B$	$-6Dq-10B+4C$	$3B\sqrt{2}$
$t_{2g}^2(^1E_g)e_g^2(^3A_{2g})$	0	$3B\sqrt{2}$	$4Dq-11B+4C$

$^3T_{1g}$	t_{2g}^4	$t_{2g}^3(^2T_{1g})e_g^1$	$t_{2g}^3(^2T_{2g})e_g^1$	$t_{2g}^2(^3T_{1g})e_g^2(^1A_{1g})$	$t_{2g}^2(^3T_{1g})e_g^2(^1E_g)$	$t_{2g}^2(^1T_{2g})e_g^2(^3A_{2g})$	$t_{2g}^1e_g^3$
t_{2g}^4	$-16Dq-15B$ $+5C$	$-B\sqrt{6}$	$-3B\sqrt{2}$	$-(2B+C)\sqrt{2}$	$2B\sqrt{2}$	0	0
$t_{2g}^3(^2T_{1g})e_g^1$	$-B\sqrt{6}$	$-6Dq-11B$ $+4C$	$5B\sqrt{3}$	$-B\sqrt{3}$	$B\sqrt{3}$	$3B$	$B\sqrt{6}$
$t_{2g}^3(^2T_{2g})e_g^1$	$-3B\sqrt{2}$	$5B\sqrt{3}$	$-6Dq-3B$ $+6C$	$3B$	$3B$	$5B\sqrt{3}$	$(B+C)\sqrt{2}$
$t_{2g}^2(^3T_{1g})e_g^2(^1A_{1g})$	$-(2B+C)\sqrt{2}$	$-B\sqrt{3}$	$3B$	$4Dq-B+6C$	$-10B$	0	$-3B\sqrt{2}$
$t_{2g}^2(^3T_{1g})e_g^2(^1E_g)$	$2B\sqrt{2}$	$B\sqrt{3}$	$3B$	$-10B$	$4Dq-9B+4C$	$2B\sqrt{3}$	$3B\sqrt{2}$
$t_{2g}^2(^1T_{2g})e_g^2(^3A_{2g})$	0	$3B$	$5B\sqrt{3}$	0	$2B\sqrt{3}$	$4Dq-11B+4C$	$B\sqrt{6}$
$t_{2g}^1e_g^3$	0	$B\sqrt{6}$	$(B+C)\sqrt{2}$	$-3B\sqrt{2}$	$3B\sqrt{2}$	$B\sqrt{6}$	$14Dq-16B$ $+5C$

(continued)

Table G3 Continued

$^3T_{2g}$	$t_{2g}^3(^2T_{1g})e_g^1$	$t_{2g}^3(^2T_{2g})e_g^1$	$t_{2g}^2(^3T_{1g})e_g^2(^3A_{2g})$	$t_{2g}^2(^3T_{1g})e_g^2(^1E_g)$	$t_{2g}^1e_g^3$
$t_{2g}^3(^2T_{1g})e_g^1$	$-6Dq-9B+4C$	$5B\sqrt{3}$	$B\sqrt{6}$	$-B\sqrt{3}$	$B\sqrt{6}$
$t_{2g}^3(^2T_{2g})e_g^1$	$5B\sqrt{3}$	$-6Dq-5B+6C$	$3B\sqrt{2}$	$3B$	$(3B+C)\sqrt{2}$
$t_{2g}^2(^3T_{1g})e_g^2(^3A_{2g})$	$B\sqrt{6}$	$3B\sqrt{2}$	$4Dq-13B+4C$	$2B\sqrt{2}$	$6B$
$t_{2g}^2(^3T_{1g})e_g^2(^1E_g)$	$-B\sqrt{3}$	$3B$	$2B\sqrt{2}$	$4Dq-9B+4C$	$3B\sqrt{2}$
$t_{2g}^1e_g^3$	$B\sqrt{6}$	$(3B+C)\sqrt{2}$	$6B$	$3B\sqrt{2}$	$14Dq-8B+5C$

^5E_g	$t_{2g}^3(^4A_{2g})e_g^1$
$t_{2g}^3(^4A_{2g})e_g^1$	$-6Dq-21B$

$^5T_{2g}$	$t_{2g}^2(^3T_{1g})e_g^2$
$t_{2g}^2(^3T_{1g})e_g^2(^3A_{1g})$	$4Dq-21B$

Table G4 Strong-field energy matrices for d^5, in O_h.
A constant value of 10*A* should be added to all diagonal elements to properly represent the actual eigenvalues of the electron–electron repulsion terms.

$^2A_{1g}$	$t_{2g}^4(^1E_g)e_g^1$	$t_{2g}^3(^2E_g)e_g^2(^1E_g)$	$t_{2g}^3(^4A_{2g})e_g^2(^3A_{2g})$	$t_{2g}^2(^1E_g)e_g^3$
$t_{2g}^4(^1E_g)e_g^1$	$-10Dq-3B+9C$	$-3B\sqrt{2}$	0	$-(6B+C)$
$t_{2g}^3(^2E_g)e_g^2(^1E_g)$	$-3B\sqrt{2}$	$-12B+8C$	$4B\sqrt{3}$	$-3B\sqrt{2}$
$t_{2g}^3(^4A_{2g})e_g^2(^3A_{2g})$	0	$4B\sqrt{3}$	$-19B+8C$	0
$t_{2g}^2(^1E_g)e_g^3$	$-(6B+C)$	$-3B\sqrt{2}$	0	$10Dq-3B+9C$

$^2A_{2g}$	$t_{2g}^4(^1E_g)e_g^1$	$t_{2g}^3(^2E_g)e_g^2(^1E_g)$	$t_{2g}^2(^1E_g)e_g^3$
$t_{2g}^4(^1E_g)e_g^1$	$-10Dq-23B+9C$	$3B\sqrt{2}$	$2B-C$
$t_{2g}^3(^2E_g)e_g^2(^1E_g)$	$3B\sqrt{2}$	$-12B+8C$	$-3B\sqrt{2}$
$t_{2g}^2(^1E_g)e_g^3$	$2B-C$	$-3B\sqrt{2}$	$10Dq-23B+9C$

Table G4 Continued

^2E_g	$t_{2g}^4(^1A_{1g})e_g^1$	$t_{2g}^4(^1E_g)e_g^1$	$t_{2g}^3(^2E_g)e_g^2(^1A_{1g})$	$t_{2g}^3(^2E_g)e_g^2(^3A_{2g})$	$t_{2g}^3(^2E_g)e_g^2(^1E_g)$	$t_{2g}^2(^1E_g)e_g^3$	$t_{2g}^2(^1A_{1g})e_g^3$
$t_{2g}^4(^1A_{1g})e_g^1$	$-10Dq-4B+12C$	$10B$	$6B$	$6B\sqrt{3}$	$6B\sqrt{2}$	$2B$	$4B+2C$
$t_{2g}^4(^1E_g)e_g^1$	$10B$	$-10Dq-13B+9C$	$-3B$	$3B\sqrt{3}$	0	$-2B-C$	$2B$
$t_{2g}^3(^2E_g)e_g^2(^1A_{1g})$	$6B$	$-3B$	$-4B+10C$	0	0	$3B$	$-6B$
$t_{2g}^3(^2E_g)e_g^2(^3A_{2g})$	$6B\sqrt{3}$	$3B\sqrt{3}$	0	$-16B+8C$	$2B\sqrt{6}$	$3B\sqrt{3}$	$6B\sqrt{3}$
$t_{2g}^3(^2E_g)e_g^2(^1E_g)$	$6B\sqrt{2}$	0	0	$2B\sqrt{6}$	$-12B+8C$	0	$6B\sqrt{2}$
$t_{2g}^2(^1E_g)e_g^3$	$2B$	$-2B-C$	$3B$	$3B\sqrt{3}$	0	$10Dq-13B+9C$	$10B$
$t_{2g}^2(^1A_{1g})e_g^3$	$4B+2C$	$2B$	$-6B$	$6B\sqrt{3}$	$6B\sqrt{2}$	$10B$	$10Dq-4B+12C$

$^2T_{1g}$	$t_{2g}^4(^3T_{1g})e_g^1$	$t_{2g}^4(^1T_{2g})e_g^1$	$t_{2g}^3(^2T_{1g})e_g^2(^1A_{1g})$	$t_{2g}^3(^2T_{1g})e_g^2(^1E_g)$	$t_{2g}^3(^2T_{2g})e_g^2(^3A_{2g})$	$t_{2g}^3(^2T_{2g})e_g^2(^1E_g)$	$t_{2g}^2(^1T_{2g})e_g^3$	$t_{2g}^2(^3T_{1g})e_g^3$
$t_{2g}^4(^3T_{1g})e_g^1$	$-10Dq-22B+9C$	$-3B$	$-\frac{3}{2}B\sqrt{2}$	$\frac{3}{2}B\sqrt{2}$	$\frac{3}{2}B\sqrt{2}$	$-\frac{3}{2}B\sqrt{6}$	0	$-C$
$t_{2g}^4(^1T_{2g})e_g^1$	$-3B$	$-10Dq-8B+9C$	$\frac{3}{2}B\sqrt{2}$	$\frac{3}{2}B\sqrt{2}$	$-\frac{15}{2}B\sqrt{2}$	$\frac{5}{2}B\sqrt{6}$	$-4B-C$	0
$t_{2g}^3(^2T_{1g})e_g^2(^1A_{1g})$	$-\frac{3}{2}B\sqrt{2}$	$\frac{3}{2}B\sqrt{2}$	$-4B+10C$	0	0	$10B\sqrt{3}$	$-\frac{3}{2}B\sqrt{2}$	$\frac{3}{2}B\sqrt{2}$
$t_{2g}^3(^2T_{1g})e_g^2(^1E_g)$	$\frac{3}{2}B\sqrt{2}$	$\frac{3}{2}B\sqrt{2}$	0	$-12B+8C$	0	0	$\frac{3}{2}B\sqrt{2}$	$\frac{3}{2}B\sqrt{2}$
$t_{2g}^3(^2T_{2g})e_g^2(^3A_{2g})$	$\frac{3}{2}B\sqrt{2}$	$-\frac{15}{2}B\sqrt{2}$	0	0	$-10B+10C$	$-2B\sqrt{3}$	$\frac{15}{2}B\sqrt{2}$	$-\frac{3}{2}B\sqrt{2}$
$t_{2g}^3(^2T_{2g})e_g^2(^1E_g)$	$-\frac{3}{2}B\sqrt{6}$	$\frac{5}{2}B\sqrt{6}$	$10B\sqrt{3}$	0	$-2B\sqrt{3}$	$-6B+10C$	$-5/2B\sqrt{6}$	$\frac{3}{2}B\sqrt{6}$
$t_{2g}^2(^1T_{2g})e_g^3$	0	$-4B-C$	$-\frac{3}{2}B\sqrt{2}$	$\frac{3}{2}B\sqrt{2}$	$\frac{15}{2}B\sqrt{2}$	$-5/2B\sqrt{6}$	$10Dq-8B+9C$	$-3B$
$t_{2g}^2(^3T_{1g})e_g^3$	$-C$	0	$\frac{3}{2}B\sqrt{2}$	$\frac{3}{2}B\sqrt{2}$	$-\frac{3}{2}B\sqrt{2}$	$\frac{3}{2}B\sqrt{6}$	$-3B$	$10Dq-22B+9C$

(continued)

Table G4 Continued

$^2T_{2g}$	t_{2g}^5	$t_{2g}^4(^3T_{1g})e_g^1$	$t_{2g}^4(^1T_{2g})e_g^1$	$t_{2g}^3(^2T_{1g})e_g^2(^3A_{2g})$	$t_{2g}^3(^2T_{1g})e_g^2(^1E_g)$	$t_{2g}^3(^2T_{2g})e_g^2(^1A_{1g})$	$t_{2g}^3(^2T_{2g})e_g^2(^1E_g)$	$t_{2g}^2(^1T_{2g})e_g^3$	$t_{2g}^2(^3T_{1g})e_g^3$	$t_{2g}^1e_g^4$
t_{2g}^5	$-20Dq-20B+10C$	$-3B\sqrt{6}$	$B\sqrt{6}$	0	$2B\sqrt{3}$	$4B+2C$	$2B$	0	0	0
$t_{2g}^4(^3T_{1g})e_g^1$	$-3B\sqrt{6}$	$-10Dq-8B+9C$	$-3B$	$-\frac{1}{2}B\sqrt{6}$	$-\frac{3}{2}B\sqrt{2}$	$-\frac{3}{2}B\sqrt{6}$	$-\frac{3}{2}B\sqrt{6}$	0	$-4B-C$	0
$t_{2g}^4(^1T_{2g})e_g^1$	$B\sqrt{6}$	$-3B$	$-10Dq-18B+9C$	$\frac{3}{2}B\sqrt{6}$	$\frac{3}{2}B\sqrt{2}$	$\frac{5}{2}B\sqrt{6}$	$-\frac{5}{2}B\sqrt{6}$	$-C$	0	0
$t_{2g}^3(^2T_{1g})e_g^2(^3A_{2g})$	0	$-\frac{1}{2}B\sqrt{6}$	$\frac{3}{2}B\sqrt{6}$	$-16B+8C$	$-2B\sqrt{3}$	0	0	$\frac{3}{2}B\sqrt{6}$	$-\frac{1}{2}B\sqrt{6}$	0
$t_{2g}^3(^2T_{1g})e_g^2(^1E_g)$	$2B\sqrt{3}$	$-\frac{3}{2}B\sqrt{2}$	$\frac{3}{2}B\sqrt{2}$	$-2B\sqrt{3}$	$-12B+8C$	$10B\sqrt{3}$	0	$\frac{3}{2}B\sqrt{2}$	$-\frac{3}{2}B\sqrt{2}$	$2B\sqrt{3}$
$t_{2g}^3(^2T_{2g})e_g^2(^1A_{1g})$	$4B+2C$	$-\frac{3}{2}B\sqrt{6}$	$\frac{5}{2}B\sqrt{6}$	0	$10B\sqrt{3}$	$2B+12C$	0	$\frac{5}{2}B\sqrt{6}$	$-\frac{3}{2}B\sqrt{6}$	$4B+2C$
$t_{2g}^3(^2T_{2g})e_g^2(^1E_g)$	$2B$	$-\frac{3}{2}B\sqrt{6}$	$-\frac{5}{2}B\sqrt{6}$	0	0	0	$-6B+10C$	$\frac{5}{2}B\sqrt{6}$	$\frac{3}{2}B\sqrt{6}$	$-2B$
$t_{2g}^2(^1T_{2g})e_g^3$	0	0	$-C$	$\frac{3}{2}B\sqrt{6}$	$\frac{3}{2}B\sqrt{2}$	$\frac{5}{2}B\sqrt{6}$	$\frac{5}{2}B\sqrt{6}$	$10Dq-18B+9C$	$-3B$	$B\sqrt{6}$
$t_{2g}^2(^3T_{1g})e_g^3$	0	$-4B-C$	0	$-\frac{1}{2}B\sqrt{6}$	$-\frac{3}{2}B\sqrt{2}$	$-\frac{3}{2}B\sqrt{6}$	$\frac{3}{2}B\sqrt{6}$	$-3B$	$10Dq-8B+9C$	$-3B\sqrt{6}$
$t_{2g}^1e_g^4$	0	0	0	0	$2B\sqrt{3}$	$4B+2C$	$-2B$	$B\sqrt{6}$	$-3B\sqrt{6}$	$20Dq-20B+10C$

Table G4 Continued

$^4A_{1g}$	$t_{2g}^3(^4A_{2g})e_g^2(^3A_{2g})$
$t_{2g}^3(^4A_{2g})e_g^2(^3A_{2g})$	$-25B + 5C$

$^4A_{2g}$	$t_{2g}^3(^4A_{2g})e_g^2(^1A_{1g})$
$t_{2g}^3(^4A_{2g})e_g^2(^1A_{1g})$	$-13B + 7C$

^4E_g	$t_{2g}^3(^2E_g)e_g^2(^3A_{2g})$	$t_{2g}^3(^4A_{2g})e_g^2(^1E_g)$
$t_{2g}^3(^2E_g)e_g^2(^3A_{2g})$	$-22B + 5C$	$-2B\sqrt{3}$
$t_{2g}^3(^4A_{2g})e_g^2(^1E_g)$	$-2B\sqrt{3}$	$-21B + 5C$

$^4T_{1g}$	$t_{2g}^4(^3T_{1g})e_g^1$	$t_{2g}^3(^2T_{2g})e_g^2(^3A_{2g})$	$t_{2g}^2(^3T_{1g})e_g^3$
$t_{2g}^4(^3T_{1g})e_g^1$	$-10Dq - 25B + 6C$	$3B\sqrt{2}$	$-C$
$t_{2g}^3(^2T_{2g})e_g^2(^3A_{2g})$	$3B\sqrt{2}$	$-16B + 7C$	$-3B\sqrt{2}$
$t_{2g}^2(^3T_{1g})e_g^3$	$-C$	$-3B\sqrt{2}$	$10Dq - 25B + 6C$

$^4T_{2g}$	$t_{2g}^4(^3T_{1g})e_g^1$	$t_{2g}^3(^2T_{1g})e_g^2(^3A_{2g})$	$t_{2g}^2(^3T_{1g})e_g^3$
$t_{2g}^4(^3T_{1g})e_g^1$	$-10Dq - 17B + 6C$	$-B\sqrt{6}$	$-4B - C$
$t_{2g}^3(^2T_{1g})e_g^2(^3A_{2g})$	$-B\sqrt{6}$	$-22B + 5C$	$-B\sqrt{6}$
$t_{2g}^2(^3T_{1g})e_g^3$	$-4B - C$	$-B\sqrt{6}$	$10Dq - 17B + 6C$

$^6A_{1g}$	$t_{2g}^3(^4A_{2g})e_g^2(^3A_{2g})$
$t_{2g}^3(^4A_{2g})e_g^2(^3A_{2g})$	$-35B$

Table G5 Strong-field energy matrices for d^6, in O_h.
A constant value of 15*A* should be added to all diagonal elements to properly represent the actual eigenvalues of the electron–electron repulsion terms.

$^1A_{1g}$	$t_{2g}^2(^1A_{1g})e_g^4$	$t_{2g}^3(^2E_g)e_g^3$	$t_{2g}^4(^1A_{1g})e_g^2(^1A_{1g})$	$t_{2g}^4(^1E_g)e_g^2(^1E_g)$	$t_{2g}^6(^1A_{1g})$
$t_{2g}^2(^1A_{1g})e_g^4$	$16Dq - 14B + 17C$	$-12B\sqrt{2}$	$(4B + 2C)\sqrt{2}$	$2B\sqrt{2}$	0
$t_{2g}^3(^2E_g)e_g^3$	$-12B\sqrt{2}$	$6Dq - 14B + 13C$	$-12B$	$-6B$	0
$t_{2g}^4(^1A_{1g})e_g^2(^1A_{1g})$	$(4B + 2C)\sqrt{2}$	$-12B$	$-4Dq + 18C$	$20B$	$(2B + C)\sqrt{6}$
$t_{2g}^4(^1E_g)e_g^2(^1E_g)$	$2B\sqrt{2}$	$-6B$	$20B$	$-4Dq - 17B + 13C$	$2B\sqrt{6}$
$t_{2g}^6(^1A_{1g})$	0	0	$(2B + C)\sqrt{6}$	$2B\sqrt{6}$	$-24Dq - 30B + 15C$

(continued)

Table G5 Continued

$^1A_{2g}$	$t_{2g}^3(^2E_g)e_g^3$	$t_{2g}^4(^1E_g)e_g^2(^1E_g)$
$t_{2g}^3(^2E_g)e_g^3$	$6Dq-26B+13C$	$6B$
$t_{2g}^4(^1E_g)e_g^2(^1E_g)$	$6B$	$-4Dq-17B+13C$

^1E_g	$t_{2g}^2(^1E_g)e_g^4$	$t_{2g}^3(^2E_g)e_g^3$	$t_{2g}^4(^1E_g)e_g^2(^1A_{1g})$	$t_{2g}^4(^1A_{1g})e_g^2(^1E_g)$	$t_{2g}^4(^1E_g)e_g^2(^1E_g)$
$t_{2g}^2(^1E_g)e_g^4$	$16Dq-23B+14C$	$6B$	$-(2B+C)\sqrt{2}$	$2B$	$-4B$
$t_{2g}^3(^2E_g)e_g^3$	$6B$	$6Dq-20B+13C$	$3B\sqrt{2}$	$12B$	0
$t_{2g}^4(^1E_g)e_g^2(^1A_{1g})$	$-(2B+C)\sqrt{2}$	$3B\sqrt{2}$	$-4Dq-9B+15C$	$10B\sqrt{2}$	$10B\sqrt{2}$
$t_{2g}^4(^1A_{1g})e_g^2(^1E_g)$	$2B$	$12B$	$10B\sqrt{2}$	$-4Dq-8B+16C$	0
$t_{2g}^4(^1E_g)e_g^2(^1E_g)$	$-4B$	0	$10B\sqrt{2}$	0	$-4Dq-17B+13C$

$^1T_{1g}$	$t_{2g}^3(^2T_{1g})e_g^3$	$t_{2g}^3(^2T_{2g})e_g^3$	$t_{2g}^4(^2T_{2g})e_g^2(^1E_g)$	$t_{2g}^5e_g^1$
$t_{2g}^3(^2T_{1g})e_g^3$	$6Dq-17B+13C$	$-5B\sqrt{3}$	$-3B$	$B\sqrt{6}$
$t_{2g}^3(^2T_{2g})e_g^3$	$-5B\sqrt{3}$	$6Dq-17B+15C$	$-5B\sqrt{3}$	$-(B+C)\sqrt{2}$
$t_{2g}^4(^2T_{2g})e_g^2(^1E_g)$	$-3B$	$-5B\sqrt{3}$	$-4Dq-17B+13C$	$B\sqrt{6}$
$t_{2g}^5e_g^1$	$B\sqrt{6}$	$-(B+C)\sqrt{2}$	$B\sqrt{6}$	$-14Dq-30B+14C$

$^1T_{2g}$	$t_{2g}^2(^1T_{2g})e_g^4$	$t_{2g}^3(^2T_{1g})e_g^3$	$t_{2g}^3(^2T_{2g})e_g^3$	$t_{2g}^4(^3T_{1g})e_g^2(^3A_{2g})$	$t_{2g}^4(^1T_{2g})e_g^2(^1E_g)$	$t_{2g}^4(^1T_{2g})e_g^2(^1A_{1g})$	$t_{2g}^5e_g^1$
$t_{2g}^2(^1T_{2g})e_g^4$	$16Dq-23B+14C$	$3B\sqrt{2}$	$-5B\sqrt{6}$	0	$-2B\sqrt{2}$	$-(2B+C)\sqrt{2}$	0
$t_{2g}^3(^2T_{1g})e_g^3$	$3B\sqrt{2}$	$6Dq-23B+13C$	$-5B\sqrt{3}$	$3B$	$-3B$	$3B$	$B\sqrt{6}$
$t_{2g}^3(^2T_{2g})e_g^3$	$-5B\sqrt{6}$	$-5B\sqrt{3}$	$6Dq-11B+15C$	$-3B\sqrt{3}$	$5B\sqrt{3}$	$5B\sqrt{3}$	$-(3B+C)\sqrt{2}$
$t_{2g}^4(^3T_{1g})e_g^2(^3A_{2g})$	0	$3B$	$-3B\sqrt{3}$	$-4Dq-23B+13C$	$-6B$	0	$3B\sqrt{6}$
$t_{2g}^4(^1T_{2g})e_g^2(^1E_g)$	$-2B\sqrt{2}$	$-3B$	$5B\sqrt{3}$	$-6B$	$-4Dq-17B+13C$	$10B$	$-B\sqrt{6}$
$t_{2g}^4(^1T_{2g})e_g^2(^1A_{1g})$	$-(2B+C)\sqrt{2}$	$3B$	$5B\sqrt{3}$	0	$10B$	$-4Dq-9B+15C$	$B\sqrt{6}$
$t_{2g}^5e_g^1$	0	$B\sqrt{6}$	$-(3B+C)\sqrt{2}$	$3B\sqrt{6}$	$-B\sqrt{6}$	$B\sqrt{6}$	$-14Dq-14B+14C$

Table G5 Continued

$^3A_{1g}$	$t_{2g}^3(^2E_g)e_g^3$
$t_{2g}^3(^2E_g)e_g^3$	$6Dq - 26B + 11C$

$^3A_{2g}$	$t_{2g}^3(^2E_g)e_g^3$	$t_{2g}^4(^1A_{1g})e_g^2(^3A_{2g})$
$t_{2g}^3(^2E_g)e_g^3$	$6Dq - 22B + 11C$	$12B$
$t_{2g}^4(^1A_{1g})e_g^2(^3A_{2g})$	$12B$	$-4Dq - 16B + 14C$

^3E_g	$t_{2g}^3(^2E_g)e_g^4$	$t_{2g}^3(^2E_g)e_g^4$	$t_{2g}^4(^1E_g)e_g^2(^3A_{2g})$
$t_{2g}^3(^2E_g)e_g^4$	$6Dq - 27B + 11C$	$-4B$	0
$t_{2g}^3(^2E_g)e_g^4$	$-4B$	$6Dq - 24B + 11C$	$-3B\sqrt{2}$
$t_{2g}^4(^1E_g)e_g^2(^3A_{2g})$	0	$-3B\sqrt{2}$	$-4Dq - 25B + 11C$

$^3T_{1g}$	$t_{2g}^2(^3T_{1g})e_g^4(^1A_{1g})$	$t_{2g}^3(^2T_{1g})e_g^3$	$t_{2g}^3(^2T_{2g})e_g^3$	$t_{2g}^4(^3T_{1g})e_g^2(^1A_{1g})$	$t_{2g}^4(^3T_{1g})e_g^2(^1E_g)$	$t_{2g}^4(^1T_{2g})e_g^2(^3A_{2g})$	$t_{2g}^5e_g^1$
$t_{2g}^2(^3T_{1g})e_g^4(^1A_{1g})$	$16Dq - 29B + 12C$	$B\sqrt{6}$	$-3B\sqrt{2}$	$-(2B + C)\sqrt{2}$	$-2B\sqrt{2}$	0	0
$t_{2g}^3(^2T_{1g})e_g^3$	$B\sqrt{6}$	$6Dq - 25B + 11C$	$-5B\sqrt{3}$	$B\sqrt{3}$	$B\sqrt{3}$	$3B$	$B\sqrt{6}$
$t_{2g}^3(^2T_{2g})e_g^3$	$-3B\sqrt{2}$	$-5B\sqrt{3}$	$6Dq - 17B + 13C$	$3B$	$-3B$	$-5B\sqrt{3}$	$-(B + C)\sqrt{2}$
$t_{2g}^4(^3T_{1g})e_g^2(^1A_{1g})$	$-(2B + C)\sqrt{2}$	$B\sqrt{3}$	$3B$	$-4Dq - 15B + 13C$	$10B$	0	$3B\sqrt{2}$
$t_{2g}^4(^3T_{1g})e_g^2(^1E_g)$	$-2B\sqrt{2}$	$B\sqrt{3}$	$-3B$	$10B$	$-4Dq - 23B + 11C$	$2B\sqrt{3}$	$3B\sqrt{2}$
$t_{2g}^4(^1T_{2g})e_g^2(^3A_{2g})$	0	$3B$	$-5B\sqrt{3}$	0	$2B\sqrt{3}$	$-4Dq - 25B + 11C$	$B\sqrt{6}$
$t_{2g}^5e_g^1$	0	$B\sqrt{6}$	$-(B + C)\sqrt{2}$	$3B\sqrt{2}$	$3B\sqrt{2}$	$B\sqrt{6}$	$-14Dq - 30B + 12C$

$^3T_{2g}$	$t_{2g}^3(^2T_{1g})e_g^3$	$t_{2g}^3(^2T_{2g})e_g^3$	$t_{2g}^4(^3T_{1g})e_g^2(^3A_{2g})$	$t_{2g}^4(^3T_{1g})e_g^2(^1E_g)$	$t_{2g}^5e_g^1$
$t_{2g}^3(^2T_{1g})e_g^3$	$6Dq - 23B + 11C$	$-5B\sqrt{3}$	$B\sqrt{6}$	$-B\sqrt{3}$	$B\sqrt{6}$
$t_{2g}^3(^2T_{2g})e_g^3$	$-5B\sqrt{3}$	$6Dq - 19B + 13C$	$-3B\sqrt{2}$	$-3B$	$-(3B + C)\sqrt{2}$
$t_{2g}^4(^3T_{1g})e_g^2(^3A_{2g})$	$B\sqrt{6}$	$-3B\sqrt{2}$	$-4Dq - 27B + 11C$	$2B\sqrt{2}$	$6B$
$t_{2g}^4(^3T_{1g})e_g^2(^1E_g)$	$-B\sqrt{3}$	$-3B$	$2B\sqrt{2}$	$-4Dq - 23B + 11C$	$3B\sqrt{2}$
$t_{2g}^5e_g^1$	$B\sqrt{6}$	$-(3B + C)\sqrt{2}$	$6B$	$3B\sqrt{2}$	$-14Dq - 22B + 12C$

(continued)

Table G5 Continued

^5E_g	$t_{2g}^3(^4A_{2g})e_g^3$
$t_{2g}^3(^4A_{2g})e_g^3$	$6Dq - 35B + 7C$

$^5T_{2g}$	$t_{2g}^4(^3T_{1g})e_g^2$
$t_{2g}^4(^3T_{1g})e_g^2$	$-4Dq - 35B + 7C$

Table G6 Strong-field energy matrices for d^7, in O_h.
A constant value of 21*A* should be added to all diagonal elements to properly represent the actual eigenvalues of the electron–electron repulsion term.

$^2A_{1g}$	$t_{2g}^4(^1E_g)e_g^3$
$t_{2g}^4(^1E_g)e_g^3$	$2Dq - 39B + 17C$

$^2A_{2g}$	$t_{2g}^4(^1E_g)e_g^3$
$t_{2g}^4(^1E_g)e_g^3$	$2Dq - 19B + 17C$

^2E_g	$t_{2g}^3(^2E_g)e_g^4$	$t_{2g}^4(^1A_{1g})e_g^3$	$t_{2g}^4(^1E_g)e_g^3$	$t_{2g}^6e_g^1$
$t_{2g}^3(^2E_g)e_g^4$	$12Dq - 34B + 17C$	$6B\sqrt{2}$	$-3B\sqrt{2}$	0
$t_{2g}^4(^1A_{1g})e_g^3$	$6B\sqrt{2}$	$2Dq - 20B + 20C$	$-10B$	$(2B + C)\sqrt{3}$
$t_{2g}^4(^1E_g)e_g^3$	$-3B\sqrt{2}$	$-10B$	$2Dq - 29B + 17C$	$-2B\sqrt{3}$
$t_{2g}^6e_g^1$	0	$(2B + C)\sqrt{3}$	$-2B\sqrt{3}$	$-18Dq - 36B + 18C$

$^2T_{1g}$	$t_{2g}^3(^2T_{1g})e_g^4$	$t_{2g}^4(^3T_{1g})e_g^3$	$t_{2g}^4(^1T_{2g})e_g^3$	$t_{2g}^5e_g^2(^3A_{2g})$	$t_{2g}^5e_g^2(^1E_g)$
$t_{2g}^3(^2T_{1g})e_g^4$	$12Dq - 34B + 17C$	$3B$	$-3B$	0	$2B\sqrt{3}$
$t_{2g}^4(^3T_{1g})e_g^3$	$3B$	$2Dq - 28B + 17C$	$-3B$	$-3B$	$3B\sqrt{3}$
$t_{2g}^4(^1T_{2g})e_g^3$	$-3B$	$-3B$	$2Dq - 34B + 17C$	$3B$	$-B\sqrt{3}$
$t_{2g}^5e_g^2(^3A_{2g})$	0	$-3B$	$3B$	$-8Dq - 34B + 17C$	$-2B\sqrt{3}$
$t_{2g}^5e_g^2(^1E_g)$	$2B\sqrt{3}$	$3B\sqrt{3}$	$-B\sqrt{3}$	$-2B\sqrt{3}$	$-8Dq - 30B + 17C$

$^2T_{2g}$	$t_{2g}^3(^2T_{2g})e_g^4$	$t_{2g}^4(^3T_{1g})e_g^3$	$t_{2g}^4(^1T_{2g})e_g^3$	$t_{2g}^5e_g^2(^3A_{1g})$	$t_{2g}^5e_g^2(^1E_g)$
$t_{2g}^3(^2T_{2g})e_g^4$	$12Dq - 28B + 19C$	$3B\sqrt{3}$	$-5B\sqrt{3}$	$-4B - 2C$	$2B$
$t_{2g}^4(^3T_{1g})e_g^3$	$3B\sqrt{3}$	$2Dq - 34B + 17C$	$-3B$	$-3B\sqrt{3}$	$3B\sqrt{3}$
$t_{2g}^4(^1T_{2g})e_g^3$	$-5B\sqrt{3}$	$-3B$	$2Dq - 24B + 17C$	$B\sqrt{3}$	$B\sqrt{3}$
$t_{2g}^5e_g^2(^3A_{1g})$	$-4B - 2C$	$-3B\sqrt{3}$	$B\sqrt{3}$	$-8Dq - 22B + 19C$	$-10B$
$t_{2g}^5e_g^2(^1E_g)$	$2B$	$3B\sqrt{3}$	$B\sqrt{3}$	$-10B$	$-8Dq - 30B + 17C$

Table G6 Continued

$^4A_{2g}$	$t_{2g}^3(^4A_{2g})e_g^4$
$t_{2g}^3(^4A_{2g})e_g^4$	$12Dq - 43B + 14C$

$^4T_{2g}$	$t_{2g}^4(^3T_{1g})e_g^3$
$t_{2g}^4(^3T_{1g})e_g^3$	$2Dq - 43B + 14C$

$^4T_{1g}$	$t_{2g}^4(^3T_{1g})e_g^3$	$t_{2g}^5e_g^2(^3A_{2g})$
$t_{2g}^4(^3T_{1g})e_g^3$	$2Dq - 31B + 14C$	$-6B$
$t_{2g}^5e_g^2(^3A_{2g})$	$-6B$	$-8Dq - 40B + 14C$

Table G7 Strong-field energy matrices for d^8, in O_h.
A constant value of 28*A* should be added to all diagonal elements to properly represent the actual eigenvalues of the electron–electron repulsion terms.

$^1A_{1g}$	$t_{2g}^6e_g^2(^1A_{1g})$	$t_{2g}^4(^1A_{1g})e_g^4$
$t_{2g}^6e_g^2(^1A_{1g})$	$-12Dq - 34B + 25C$	$(2B + C)\sqrt{6}$
$t_{2g}^4(^1A_{1g})e_g^4$	$(2B + C)\sqrt{6}$	$8Dq - 32B + 26C$

^1E_g	$t_{2g}^6e_g^2(^1E_g)$	$t_{2g}^4(^1E_g)e_g^4$
$t_{2g}^6e_g^2(^1E_g)$	$-12Dq - 42B + 23C$	$-2B\sqrt{3}$
$t_{2g}^4(^1E_g)e_g^4$	$-2B\sqrt{3}$	$8Dq - 41B + 23C$

$^1T_{1g}$	$t_{2g}^5e_g^3$
$t_{2g}^5e_g^3$	$-2Dq - 38B + 23C$

$^3A_{2g}$	$t_{2g}^6e_g^2(^3A_{2g})$
$t_{2g}^6e_g^2(^3A_{2g})$	$-12Dq - 50B + 21C$

$^1T_{2g}$	$t_{2g}^4(^1T_{2g})e_g^4$	$t_{2g}^5e_g^3$
$t_{2g}^4(^1T_{2g})e_g^4$	$8Dq - 41B + 23C$	$2B\sqrt{3}$
$t_{2g}^5e_g^3$	$2B\sqrt{3}$	$-2Dq - 42B + 23C$

$^3T_{1g}$	$t_{2g}^4(^3T_{1g})e_g^4$	$t_{2g}^5e_g^3$
$t_{2g}^4(^3T_{1g})e_g^4$	$8Dq - 47B + 21C$	$6B$
$t_{2g}^5e_g^3$	$6B$	$-2Dq - 38B + 21C$

$^3T_{2g}$	$t_{2g}^5e_g^3$
$t_{2g}^5e_g^3$	$-2Dq - 50B + 21C$

APPENDIX H

Changes in Atomic Radius with Charge and Coordination Number

Radii (Å) of Some Element in Different Oxidation States[a]

Element	Radius (Å) in Oxidation State[b]			
	I	II	III	IV
Ti		1.00	0.81	0.745
V		0.93	0.78	0.72
Cr		0.94 0.87[LS]	0.755	0.69
Mn		0.97 0.81[LS]	0.785 0.72[LS]	0.67
Fe		0.92 0.75[LS]	0.785 0.69[LS]	0.725
Co		0.885 0.79[LS]	0.75 0.685[LS]	0.67[LS]
Ni		0.69 0.63[SP]	0.74 0.70[LS]	0.62[LS]
Cu		0.71[SP] 0.87	0.68[LS]	
Zn		0.74 0.88		
Al			0.68	
Ga			0.76	
Ge		0.87		0.67
As			0.62	
Se				0.64
Zr				0.86
Nb			0.86	0.82
Mo			0.83	0.79
Tc				0.785
Ru			0.82	0.76
Rh			0.805	0.74
Pd		0.78	0.90	0.755
Ag		0.93	0.81	
Cd		1.09		
In			0.94	
Sn		1.22		0.83
Sb			0.94	
Te				1.11
Hf				0.85
Ta			0.86	0.82
W				0.80
Re				0.77
Os				0.77
Ir			0.82	0.765
Pt		0.74		0.765
Au	1.51		0.82	
Tl	1.64		1.025	
Pb		1.33		0.915
Bi			1.17	
Po				1.08

[a] Shannon, R.D. *Acta Cryst.* 1976, *A32*, 751.
[b] LS, low spin. SP, square-planar.

Ionic Radii (Å) of Some Complex Anions

Anion	Radius (Å)
CO_3^{2-}	1.85
NO_3^{-}	1.89
BF_4^{-}	2.28
PO_4^{3-}	2.38
SO_4^{2-}	2.30
ClO_4^{-}	2.36
SiF_6^{2-}	1.94
AsO_4^{3-}	2.48
CrO_4^{2-}	2.30
MnO_4^{-}	2.40
TeO_4^{2-}	2.54
IO_4^{-}	2.49
MoO_4^{2-}	2.54
$TiCl_6^{2-}$	2.48
$IrCl_6^{2-}$	2.54
$ZrCl_6^{2-}$	2.47
$PtCl_6^{2-}$	2.59

Douglas, B. E., McDaniel, D. H., and Alexander, J. J. *Concepts and Models of Inorganic Chemistry*, 3rd edn. New York, NY: John Wiley & Sons, 1994.

APPENDIX I

Imaginary Numbers and Complex Conjugates

Wave mechanics often uses imaginary numbers, which are used to mathematically describe waves of all types. An imaginary number is, quite simply, any number that includes *i*, where $i = \sqrt{-1}$. Obviously, any real number squared is always positive, but $i^2 = -1$. However, imaginary numbers can easily be understood as being part of a 2-dimensional space, and all we really need to understand them is a little geometry, such as the Pythagorean theorem.

All the real numbers form a number line from −infinity to +infinity, which we will place on the *x*-axis. If we say that all imaginary numbers then form a number line along the *y*-axis from −infinity(*i*) to +infinity(*i*), we obtain a 2-dimensional plane with real numbers in one direction and imaginary numbers in the other direction. All imaginary numbers are then vectors on this plane.

For example, if we have the imaginary number $4 + 3i$, this means that we are at a point that is 4 along the *x* axis and 3 along the *y* axis. It looks something like this.

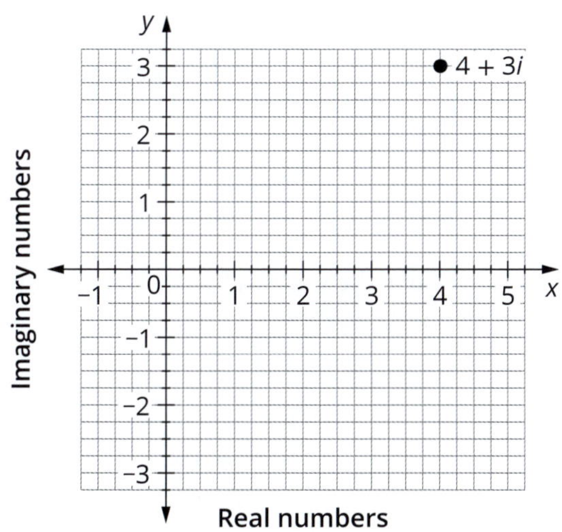

Any point $a + bi$ has a magnitude and an angle. The magnitude we can find by using the Pythagorean theorem, $c^2 = a^2 + b^2$. Inserting our point $4 + 3i$ produces

$$|a + bi| = \sqrt{a^2 + b^2}$$

Using this equation with our example of $4 + 3i$ produces

$$|4 + 3i| = \sqrt{4^2 + 3^2}$$
$$|4 + 3i| = \sqrt{16 + 9} = \sqrt{25} = 5$$

The magnitude of our complex number, the length of the vector from the origin, is then 5. The absolute-value sign was added with the complex number as meaning "magnitude", for good reason. Absolute value means "the distance from the origin" if you are talking about real numbers. It means the same thing here for complex numbers.

We have all the sides of a triangle, so we can use any trigonometric relationship to find the angle between the *x*-axis and the point. For example, if we use the definition of tangent, we obtain

$$\tan\theta = \frac{opposite}{adjacent} = \frac{3}{4} = 0.75$$

The angle is then $\tan^{-1} 0.75 = 36.87°$.

Multiplication of Complex Numbers

If you multiply complex numbers (numbers with real and imaginary components), you

- Add the angles.
- Multiply the magnitudes.

Let us take our previous example and multiply it by itself $(4 + 3i)^2$. Using the two bullet points, the angle of the answer should be 73.74° with a magnitude of 25.

$$(4 + 3i)(4 + 3i) = 16 + 9i^2 + 24i = 16 + (-9) + 24i = 7 + 24i$$

The magnitude of $7 + 24i$ is then

$$|a + bi| = \sqrt{a^2 + b^2} = \sqrt{7^2 + 24^2} = \sqrt{49 + 576} = 25.$$

The angle using the definition of tangent again is

$$\tan\theta = \frac{opposite}{adjacent} = \frac{24}{7} = 3.43$$

$$\tan^{-1}(3.43) = 73.74°$$

which is exactly twice the angle of the original $4 + 3i$ function!

The Complex Conjugate

Finding the complex conjugate of an imaginary function is very simple; you just change the sign on the imaginary part of the function. For our example we have been using $4 + 3i$, by which the complex conjugate is then simply $4-3i$. You leave the real part of the function (the "4") alone and change the sign on the imaginary part ("$3i$"). In quantum mechanics the complex conjugate is often written with an asterisk (*). So, if we have a function $t = 1-4i$, then $t^* = 1 + 4i$; just change the sign of the part with the *i*.

The more important question is, what does multiplying by the complex conjugate do? When we change the sign on the imaginary part of the function, the point is moved to its reflection along the *y*-axis. If the angle to the original point $4 + 3i$ is θ, then the complex conjugate is in the opposite direction from the *x*-axis at −θ.

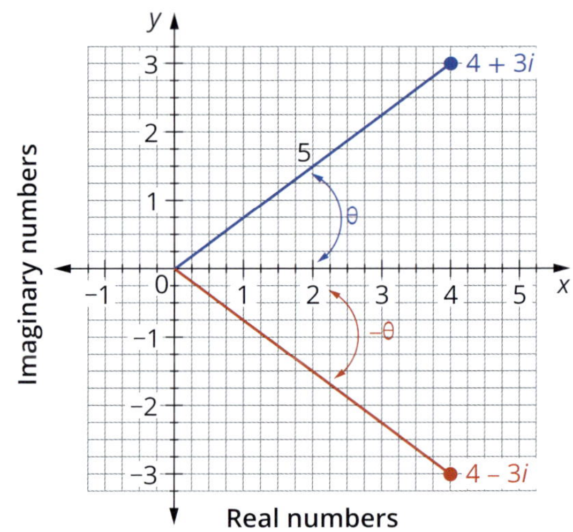

When we multiply two complex numbers we add the angles and multiply the magnitudes. As a result, multiplying a complex number by its complex conjugate produces a real number (along the *x*-axis at θ = 0). The magnitudes of the two are then multiplied.

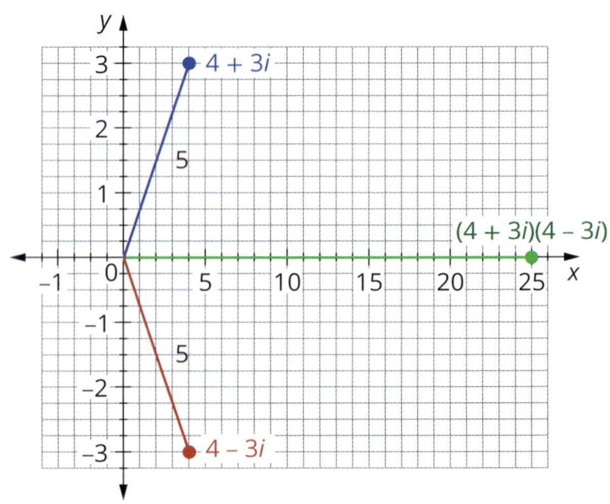

If we do this with our example, $4 + 3i$:

$$(4 + 3i)(4 - 3i) = 16 - 9i^2 = 16 - (-9) = 25.$$

Consequently, multiplying a complex number by its complex conjugate is a way of squaring a number so that the result is a real number. As you can see from the previous, multiplying some function $z = a + bi$ by its complex conjugate $z^* = a-bi$ produces

$$zz^* = a^2 + b^2$$

In quantum mechanics, the probability of finding the electron at a specific point (and time if using the time-dependent function) is ψ^2, which is found by multiplying a wavefunction by its complex conjugate, $\psi\psi^*$, and integrating over all space. In this process, the angular (phase) information in the wavefunction is lost, and the result is a real number (one having no imaginary component).

Euler's Formula for Complex Numbers

The following equation

$$e^{i\pi} + 1 = 0$$

is a beautifully simple, although perhaps not obvious, method for writing complex numbers. In the equation, e = Euler's number, which is an irrational number that can be approximated as 2.718. The number is named after Swiss-born Leonhard Euler (pronounced "oiler") who lived from 1707 to 1783 and is generally considered one of the greatest mathematicians of all time. The number e pops up a great deal in very simple equations. For example, $(1 + 1/n)^n$ approaches e. The sums of $1/(n!)$, where n = positive integers from 0 to infinity, also approaches e.

$$e = \frac{1}{0!} + \frac{1}{1!} + \frac{1}{2!} + \frac{1}{3!} + \frac{1}{4!} + \frac{1}{5!} + \ldots$$

Although the constant bears Euler's name, it is believed to have been discovered by Jacob Bernoulli in 1687, while studying compound interest.

The equation $e^{i\pi} + 1 = 0$ derives from Euler's formula.

$$e^{ix} = \cos x + i(\sin x)$$

The derivation of this very elegant equation uses the Taylor series, which was already known in Euler's day, as follows.

$$e^x = \sum_{n=0}^{\infty} \frac{x^n}{n!} = 1 + x + \frac{x^2}{2!} + \frac{x^3}{3!} + \frac{x^4}{4!} + \ldots$$

If you put ix instead of x:

$$e^{ix} = 1 + ix + \frac{(ix)^2}{2!} + \frac{(ix)^3}{3!} + \frac{(ix)^4}{4!} + \frac{(ix)^5}{5!} + \ldots$$

this becomes the following expression on simplification ($i^2 = -1$), and some terms become negative.

$$e^{ix} = 1 + ix - \frac{x^2}{2!} - \frac{ix^3}{3!} + \frac{x^4}{4!} + \frac{(ix)^5}{5!} - \ldots$$

Euler then gathered the term into real and imaginary parts and recognized the nature of the functions that results.

$$e^{ix} = \left(1 - \frac{x^2}{2!} + \frac{x^4}{4!} - \ldots\right) + i\left(x - \frac{x^3}{3!} + \frac{x^5}{5!} - \ldots\right)$$

The real part is the Taylor series for cosine, and the imaginary part is the Taylor series for sine.

$$\cos x = \left(1 - \frac{x^2}{2!} + \frac{x^4}{4!} - \ldots\right)$$
$$\sin x = \left(x - \frac{x^3}{3!} + \frac{x^5}{5!} - \ldots\right)$$

Consequently,

$$e^{ix} = \cos x + i(\sin x)$$

For the special case of $x = \pi$, and remembering that $\cos \pi = -1$ and $\sin \pi = 0$;

$$e^{i\pi} = \cos x + i(\sin x)$$
$$e^{i\pi} = -1 + i(0)$$
$$e^{i\pi} = -1$$
$$e^{i\pi} + 1 = 0$$

Feynman called $e^{i\pi} + 1 = 0$ the "most remarkable formula in mathematics" for its single use of 0, 1, e, i, and π. In 1988, the readers of *Mathematical Intelligencer* voted it the "Most beautiful mathematical formula ever"; but how is this function used to represent complex numbers?

One way of thinking about the function $e^{i\pi} + 1 = 0$ is as a circle with radius 1 in our y = *imaginary numbers* and x = *real numbers* 2-dimensional plot. We can change the radius of the circle by multiplying $e^{i\pi}$ by some number; let us use 5, $5e^{i\pi}$, which just makes the radius of the circle 5 instead of 1. Note that whatever is multiplied by Euler's number is exactly the magnitude that we were using previously! The magnitude of $5e^{i\pi}$ is 5.

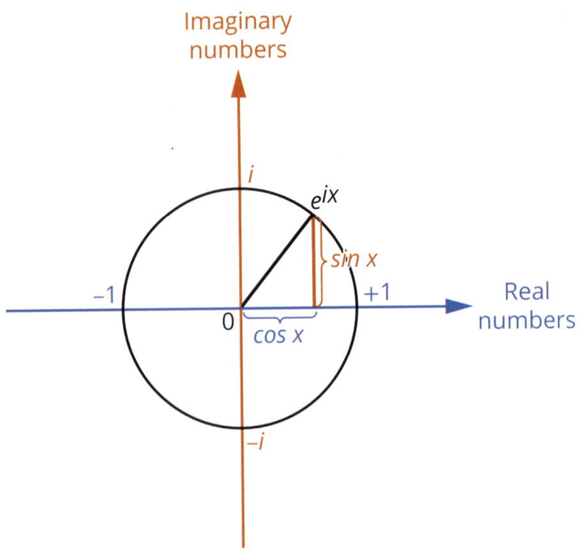

The angle is then in the superscript of Euler's number. If we are using e^{ix} and our complex number is 4 + 3*i*, then

$$x = tan^{-1}(imaginary\ part/real\ part) = tan^{-1}(3/4) = 36.87°$$

the angular part at which we arrived previously for the expression 4 + 3*i*! In other words, $5e^{i(36.87°)} = 4 + 3i$. More often, the term in the superscript is put in terms of radians; the conversion is degrees = (180° · radians)/π or 1 radian = 180°/π = 57.296°. In this case, in radians, $\tan^{-1}(3/4) \sim 0.644$. So, an equivalent way of writing 4 + 3*i* is as $5e^{0.644i}$. Euler's equation can be used directly to turn $5e^{0.644i}$ back into the type of expression used previously.

$$\begin{aligned} e^{ix} &= cos\,x + i(sin\,x) \\ 5e^{0.644i} &= 5[\cos(0.644)] + 5i[sin(0.644)] \\ 5e^{0.644i} &= 4 + 3i \end{aligned} \quad (1)$$

Euler's formula directly produces the magnitude and the angle (in radians) of the complex number—a very convenient expression.

Just as before, to make the complex conjugate, we just change the sign on the imaginary part of the function, and in this case the sign of the superscript containing *i* is changed. The complex conjugate of $5e^{0.644i}$ is then $5e^{-0.644i}$. Multiplying the two produces $(5e^{0.644i})(5e^{-0.644i}) = 25e^0 = 25$; compare $(4 + 3i)(4 - 3i) = 25$.

APPENDIX J

Standard Reduction Potentials

Acidic Solution	E° (V)
$F_2(g) + 2\,e^- = 2\,F^-(aq)$	+2.87
$Ag^{2+} + e = Ag^+$	+1.98
$Co^{3+}(aq) + e^- = Co^{2+}(aq)$	+1.92
$H_2O_2(aq) + 2\,H^+(aq) + 2\,e^- = 2\,H_2O$	+1.78
$N_2O(g) + 2\,H^+(aq) + 2\,e^- = N_2 + H_2O$	+1.766
$NiO_2(s) + 4\,H^+(aq) + 2\,e^- = Ni^{2+}\,(aq) + 2\,H_2O$	+1.68
$PbO_2(s) + SO_4^{2-}(aq) + 4\,H^+(aq) + 2\,e^- = PbSO_4(s) + 2\,H_2O$	+1.685
$Au^+(aq) + e^- = Au(s)$	+1.52
$HClO(aq) + H^+(aq) + e^- = 1/2\,Cl_2(g) + H_2O$	+1.61
$Ce^{4+}(aq) + e^- = Ce^{3+}(aq)$	+1.72
$MnO_4^-(aq) + 8\,H^+(aq) + 5\,e^- = Mn^{2+}(aq) + 4\,H_2O$	+1.51
$Au^{3+}(aq) + 3\,e^- = Au(s)$	+1.50
$ClO_3^-(aq) + 6\,H^+(aq) + 5\,e^- = 1/2\,Cl_2(g) + 3\,H_2O$	+1.47
$PbO_2 + 2\,e^- = Pb^{2+}(aq)$	+1.45
$BrO_3^- + 6\,H^+(aq) + 6\,e^- = Br^-(aq) + 3\,H_2O$	+1.42
$Cl_2(g) + 2\,e^- = 2\,Cl^-(aq)$	+1.358
$HCrO_4^- + 7\,H^+(aq) + 3\,e^- = Cr^{3+}(aq) + 4\,H_2O$	+1.35
$PtCl_6^{2-}(aq) + 2\,e^- = PtCl_4^{2-}(aq) + 2\,Cl^-(aq)$	+1.288
$MnO_2(s) + 4\,H^+(aq) + 2\,e^- = Mn^{2+}(aq) + 2\,H_2O$	+1.22
$O_2(g) + 4\,H^+(aq) + 4\,e^- = 2\,H_2O$	+1.229
$Pt^{2+}(aq) + 2\,e^- = Pt(s)$	+1.188
$IO_3^-(aq) + 6\,H^+(aq) + 5\,e^- = 1/2\,I_2(aq) + 3\,H_2O$	+1.195
$ClO_4^-(aq) + 2\,H^+(aq) + 2\,e^- = ClO_3^-(aq) + H_2O$	+1.189
$Br_2(l) + 2\,e^- = 2\,Br^-(aq)$	+1.066
$Pd^{2+}(aq) + 2\,e^- = Pd(s)$	+0.951
$NO_3^-(aq) + 4\,H^+(aq) + 3\,e^- = NO(g) + 2\,H_2O$	+0.957
$NO_3^-(aq) + 3\,H^+(aq) + 2\,e^- = HNO_2(aq) + H_2O$	+0.934
$2\,Hg^{2+}(aq) + 2\,e^- = Hg_2^{2+}(aq)$	+0.920
$Hg^{2+}(aq) + 2\,e^- = Hg(l)$	+0.851
$Ag^+(aq) + e^- = Ag(s)$	+0.7996
$Hg_2^{2+}(aq) + 2\,e^- = 2\,Hg(l)$	+0.797

(*continued*)

Acidic Solution	E° (V)
$Fe^{3+}(aq) + e^- = Fe^{2+}(aq)$	+0.771
$PtCl_4^{2-}(aq) + 2\ e^- = Pt(s) + 4\ Cl^-(aq)$	+0.755
$O_2(g) + 2\ H^+(aq) + 2\ e^- = H_2O_2(aq)$	+0.695
$TeO_2(s) + 4\ H^+(aq) + 4\ e^- = Te(s) + 2\ H_2O$	+0.593
$I_2(s) + 2\ e^- = 2\ I^-(aq)$	+0.5355
$Cu^+(aq) + e^- = Cu(s)$	+0.521
$RhCl_6^{3-}(aq) + 3\ e^- = Rh(s) + 6\ Cl^-(aq)$	+0.431
$Cu^{2+}(aq) + 2\ e^- = Cu(s)$	+0.342
$HgCl_2(s) + 2\ e^- = 2\ Hg(l) + 2\ Cl^-(aq)$	+0.268
$AgCl(s) + e^- = Ag(s) + Cl^-(aq)$	+0.222
$SO_4^{2-}(aq) + 4\ H^+(aq) + 2\ e^- = H_2SO_3(g) + H_2O$	+0.172
$Cu^{2+}(aq) + e^- = Cu^+(aq)$	+0.153
$Sn^{4+}(aq) + 2\ e^- = Sn^{2+}(aq)$	+0.151
$S(s) + 2\ H^+(aq) + 2\ e^- = H_2S(aq)$	+0.142
$AgBr(s) + e^- = Ag(s) + Br^-(aq)$	+0.0713
$2\ H^+(aq) + 2\ e^- = H_2(g)$ (reference electrode)	0 (by definition)
$Pb^{2+}(aq) + 2\ e^- = Pb(s)$	−0.126
$Sn^{2+}(aq) + 2\ e^- = Sn(s)$	−0.138
$AgI(s) + e^- = Ag(s) + I^-(aq)$	−0.152
$CO_2(g) + 2\ H^+(aq) + 2\ e^- = HCOOH$	−0.199
$Ni^{2+}(aq) + 2\ e^- = Ni(s)$	−0.257
$Co^{2+}(aq) + 2\ e^- = Co(s)$	−0.28
$Tl^+(aq) + e^- = Tl(s)$	−0.336
$PbSO_4(s) + 2\ e^- = Pb(s) + SO_4^{2-}(aq)$	−0.359
$Se(s) + 2\ H^+(aq) + 2\ e^- = H_2Se(aq)$	−0.399
$Cd^{2+}(aq) + 2\ e^- = Cd(s)$	−0.403
$Cr^{3+}(aq) + e^- = Cr^{2+}(aq)$	−0.407
$Fe^{2+}(aq) + 2\ e^- = Fe(s)$	−0.447
$Ga^{3+}(aq) + 3\ e^- = Ga(s)$	−0.560
$Cr^{3+}(aq) + 3\ e^- = Cr(s)$	−0.744
$Zn^{2+}(aq) + 2\ e^- = Zn(s)$	−0.762
$Cr^{2+}(aq) + 2\ e^- = Cr(s)$	−0.913
$Mn^{2+}(aq) + 2\ e^- = Mn(s)$	−1.185
$V^{2+}(aq) + 2\ e^- = V(s)$	−1.175
$ZrO_2 + 4\ H^+ + 4\ e^- = Zr(s) + 2\ H_2O$	−1.553
$Al^{3+}(aq) + 3\ e^- = Al(s)$	−1.662
$H_2(g) + 2\ e^- = 2\ H^-(aq)$	−2.23
$Mg^{2+}(aq) + 2\ e^- = Mg(s)$	−2.372
$Na^+(aq) + e^- = Na(s)$	−2.71

(continued)

Acidic Solution	E° (V)
$Ca^{2+}(aq) + 2\ e^- = Ca(s)$	−2.868
$Sr^{2+}(aq) + 2\ e^- = Sr(s)$	−2.89
$Ba^{2+}(aq) + 2\ e^- = Ba(s)$	−2.912
$Rb^+(aq) + e^- = Rb(s)$	−2.98
$K^+(aq) + e^- = K(s)$	−2.93
$Li^+(aq) + e^- = Li(s)$	−3.04
Basic Solution	
$ClO^-(aq) + H_2O + 2\ e^- = Cl^-(aq) + 2\ OH^-(aq)$	+0.89
$OOH^-(aq) + H_2O + 2\ e^- = 3\ OH^-(aq)$	+0.878
$ClO_3^-\ (aq) + 3\ H_2O + 6\ e^- = Cl^-(aq) + 6\ OH^-(aq)$	+0.62
$MnO_4^-(aq) + 2\ H_2O + 3\ e^- = MnO_2(s) + 4\ OH^-(aq)$	+0.595
$O_2(g) + 2\ H_2O + 4\ e^- = 4\ OH^-(aq)$	+0.401
$ClO_4^-(aq) + H_2O + 2e^- = ClO_3^-(aq) + 2\ OH^-(aq)$	+0.36
$Ag_2O(s) + H_2O + 2e^- = 2\ Ag(s) + 2\ OH^-(aq)$	+0.342
$2\ NO_2^-(aq) + 3\ H_2O + 4\ e^- = N_2O(g) + 6\ OH^-(aq)$	+0.15
$Co(NH_3)_6^{3+}(aq) + e^- = Co(NH_3)_6^{2+}(aq)$	+0.108
$HgO(s) + H_2O + 2e^- = Hg(l) + 2\ OH^-(aq)$	+0.0977
$NO_3^-(aq) + H_2O + 2\ e^- = NO_2^-(aq) + 2\ OH^-(aq)$	+0.01
$O_2(g) + H_2O + 2\ e^- = OOH^-(aq) + OH^-(aq)$	−0.076
$CrO_4^{2-}(aq) + 4\ H_2O + 3\ e^- = Cr(OH)_3(s) + 5\ OH^-(aq)$	−0.13
$Cu(OH)_2(s) + 2\ e^- = Cu(s) + 2\ OH^-(aq)$	−0.222
$NiO_2(s) + 2\ H_2O + 2\ e^- = Ni(OH)_2(s) + 2\ OH^-(aq)$	−0.490
$2\ H_2O + 2\ e^- = H_2(g) + 2\ OH^-(aq)$	−0.8277
$SO_4^{2-}(aq) + H_2O + 2\ e^- \rightarrow SO_3^{2-}(aq) + 2\ OH^-(aq)$	−0.93
$ZnO_2^{2-}(aq) + 2\ H_2O + 2\ e^- \rightarrow Zn(s) + 4\ OH^-(aq)$	−1.215
$Cr(OH)_3(s) + 3\ e^- \rightarrow Cr(s) + 3\ OH^-(aq)$	−1.48
$SiO_3^{2-}(aq) + 3\ H_2O + 4\ e^- \rightarrow Si(s) + 6\ OH^-(aq)$	−1.697

Weast, R. C. (ed.). *CRC Handbook of Chemistry and Physics*, 70th edn. Boca Raton, FL: CRC Press, 1989.

APPENDIX K

Valence Orbital Energies

Atomic Number	Element	Configuration	Valence Orbitals	Ionization Energy (eV)[a]	Calculated Orbital Energies (Hartree)[b]
1	H	$1s^1$	1s	13.6	−0.499
2	He	$1s^2$	1s	24.6	−0.914
3	Li	$[He]2s^1$	2s	5.39	−0.086
			2p		0.032
4	Be	$[He]2s^2$	2s	9.32	−0.309
			2p		0.050
5	B	$[He]2s^22p^1$	2s	14.0	−0.495
			2p	8.30	−0.002
6	C	$[He]2s^22p^2$	2s	19.5	−0.706
			2p	10.7	−0.088
7	N	$[He]2s^22p^3$	2s	25.5	−0.943
			2p	13.1	−0.213
8	O	$[He]2s^22p^4$	2s	32.3	−1.24
			2p	15.9	−0.390
9	F	$[He]2s^22p^5$	2s	46.4	−1.572
			2p	18.7	−0.596
10	Ne	$[He]2s^22p^6$	2s	48.5	−1.93
			2p	21.5	−0.845
11	Na	$[Ne]3s^1$	3s	5.21	−0.083
			3p		0.163
12	Mg	$[Ne]3s^2$	3s	7.68	−0.253
			3p		0.037
13	Al	$[Ne]3s^2p^1$	3s	11.3	−0.394
			3p	5.95	−0.006
14	Si	$[Ne]3s^2p^2$	3s	15.0	−0.541
			3p	7.81	-0.075
15	P	$[Ne]3s^2p^3$	3s	18.7	−0.696
			3p	10.2	−0.168
16	S	$[Ne]3s^2p^4$	3s	20.7	−0.881
			3p	11.7	−0.289
17	Cl	$[Ne]3s^2p^5$	3s	25.3	−1.074
			3p	13.8	−0.443
18	Ar	$[Ne]3s^2p^6$	3s	29.2	−1.28
			3p	15.9	−0.590
19	K	$[Ar]4s^1$	4s	4.34	−0.068[c]
			4p		0.067[c]
20	Ca	$[Ar]4s^2$	4s	6.07	−0.196
			4p		0.023

Atomic Number	Element	Configuration	Valence Orbitals	Ionization Energy (eV)[a]	Calculated Orbital Energies (Hartree)[b]
21	Sc	$[Ar]4s^23d^1$	3d	4.71	0.051
			4s	5.70	−0.210
			4p	3.22	0.239
22	Ti	$[Ar]4s^23d^2$	3d	5.58	0.0045
			4s	6.07	−0.22
			4p	3.35	0.027
23	V	$[Ar]4s^23d^3$	3d	6.32	−0.058
			4s	6.32	−0.232
			4p	3.47	0.161
24	Cr	$[Ar]4s^13d^5$	3d	7.19	−0.080
			4s	6.57	−0.096
			4p	3.47	0.050
25	Mn	$[Ar]4s^23d^5$	3d	7.93	−0.137
			4s	6.82	−0.103
			4p	3.59	0.055
26	Fe	$[Ar]4s^23d^6$	3d	8.68	−0.207
			4s	7.07	−0.104
			4p	3.72	0.058
27	Co	$[Ar]4s^23d^7$	3d	9.42	−0.289
			4s	7.32	−0.107
			4p	3.84	0.063
28	Ni	$[Ar]4s^23d^8$	3d	10.0	−0.286
			4s	7.56	−0.110
			4p	3.84	0.065
29	Cu	$[Ar]4s^13d^{10}$	3d	10.7	−0.642
			4s	7.69	−0.285
			4p	3.97	0.051
30	Zn	$[Ar]3d^{10}4s^2$	4s	9.42	−0.292
			4p		0.054
31	Ga	$[Ar]3d^{10}4s^24p^1$	4s	12.6	−0.425
			4p	5.95	−0.003
32	Ge	$[Ar]3d^{10}4s^24p^2$	4s	15.6	−0.554
			4p	7.56	−0.076
33	As	$[Ar]3d^{10}4s^24p^3$	4s	17.6	−0.686
			4p	9.05	−0.163
34	Se	$[Ar]3d^{10}4s^24p^4$	4s	20.8	−0.840
			4p	10.8	−0.255
35	Br	$[Ar]3d^{10}4s^24p^5$	4s	24.0	−0.994
			4p	12.5	−0.391
36	Kr	$[Ar]3d^{10}4s^24p^6$	4s	27.5	−1.15
			4p	14.3	−0.523

[a] Average energies of the ground-state and ionized-state configurations, from Gray, H. *Electrons and Chemical Bonding*, Benjamin, 1964. Orbitals that are higher in energy have smaller values.

[b] Calculated using CCSD/ccPVDZ with all electrons. In most cases not all orbital energies of a given type (such as 2p) are the same, and the number given is the average energy of all the orbitals. Single-point calculations were used with the correct spin state of the ground-state electronic configuration, and other effects were not considered. The values are provided as starting points for by-hand calculations of molecular orbitals. Orbitals that have smaller values are lower in energy.

[c] The basis set used was TZVP.

APPENDIX L

Shapes of the *f*-Orbitals

The *f*-orbitals all have three angular nodes and ungerade symmetry. Here, in analogy to the orbitals for *p* and *d* discussed previously, we use the Cartesian-based cubic set of *f*-orbitals. There are three orbitals, somewhat analogous to d_{z^2} in shape, directed along the three coordinate axes: $4f_{x^3}$, $4f_{y^3}$, and $4f_{z^3}$. There are then four orbitals with a double four-leaf-clover shape, and in these, the lobes reside between the axes. These orbitals are designated as $4f_{x(z^2-y^2)}$, $4f_{y(z^2-x^2)}$, $4f_{z(x^2-y^2)}$, and $4f_{xyz}$.

It should be noted that for *f*-orbitals the lobes 180° through the origin in every case have the opposite phase, quite different from *d*-orbitals. The seven 4*f*-orbitals are shown.

For three of these orbitals with the double four-leaf-clover shape—$4f_{x(z^2-y^2)}$, $4f_{y(z^2-x^2)}$, and $4f_{z(x^2-y^2)}$—the lobes reside between axes but in the planes. For example, $4f_{z(x^2-y^2)}$ has four lobes in the *xz*-plane and four in the *yz*-plane. The unique direction for the $4f_{z(x^2-y^2)}$-orbital is along the *z*-axis, which is orthogonal to the plane of the stacked four-leaf clovers (see the diagram). The unique directions for the $4f_{x(z^2-y^2)}$ and $4f_{y(z^2-x^2)}$-orbitals are the *x*- and *y*-axes, respectively.

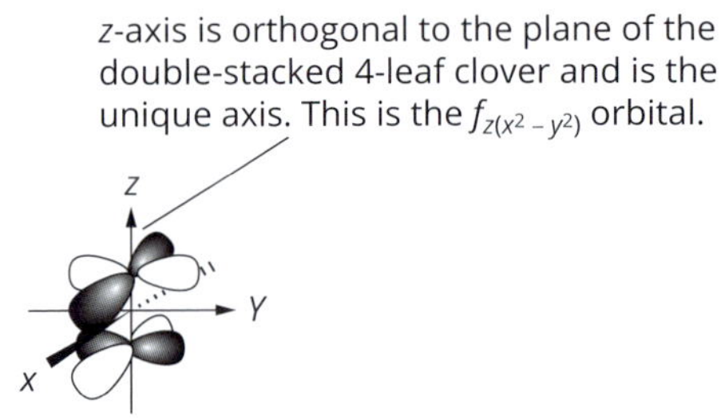

The fourth orbital with the double four-leaf-clover shape, $4f_{xyz}$, has no lobes that sit on any of the three coordinate planes; that is, it has no lobes in the *xz*-, *yz*-, or *xy*-planes. The eight lobes extend into the eight quadrants around the origin and between axes.

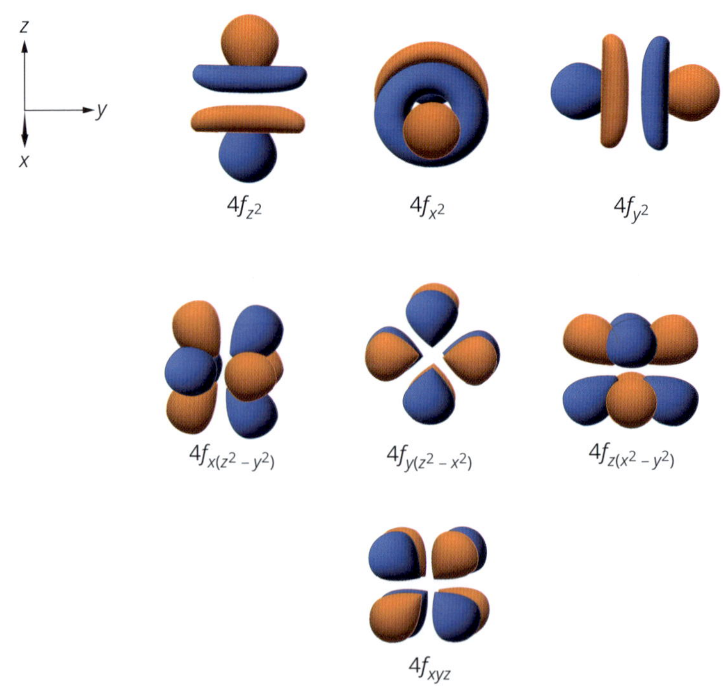

The *f*-orbitals are a little more complex than their *s*-, *p*-, and *d*-counterparts. Adding to the complexity is that there is a completely different set of linear combinations in common usage (f_{z^3}, f_{xz^2}, f_{xyz}, $f_{x(x^2-3y^2)}$, f_{yz^2}, $f_{z(x^2-y^2)}$, and $f_{y(3x^2-y^2)}$). Both sets are common in the *f*-element community, and they have decidedly different shapes. We recommend learning one set in at least a cursory fashion as an inorganic chemist, especially if you are not an *f*-element chemist. As far as which should be learned, any preferences are due to aesthetics, as they are linear combinations of one another.

Orbitals with $l > 4$ can readily be drawn mathematically but are of little chemical importance, as there are no elements (yet) with these orbitals occupied. However, they have the expected mathematical properties, such as the *g*-orbitals like the two shown here, which have three angular nodes.

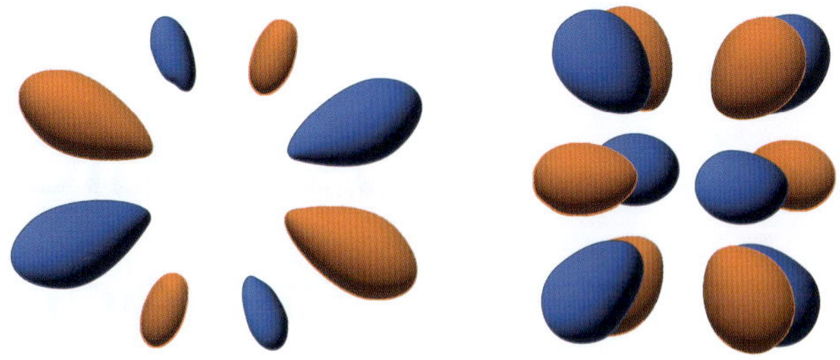

APPENDIX M

Linear Combination of Atomic Orbitals (LCAO) Molecular Orbital Method Review

The LCAO method is a common approximation for the determination of molecular orbitals (MO) where it is assumed that the orbitals in the molecule can be found by combining the valence orbitals of the individual atoms. Improvements on this approximation are commonly employed computationally, which will be discussed elsewhere. The efficacy of the improvements can be monitored for some computational methods, those that are "variationally bound", meaning that the value for the energy calculated will always be either higher or equal to the actual energy. As a result, one can improve the computed estimate of the energy by removal of approximations, increasing the size of the basis set, and so on, and the energy will decrease as it approaches the real (lower) value.

The *variational principle* can be expressed as shown in the following equation, stating that the energy of an orbital can be estimated from the wavefunction, ψ, and the Hamiltonian, H. The term ψ^* is the complex conjugate of ψ. (See Appendix I for a discussion of complex conjugates.) If the wavefunction is entirely real—if it does not include i at all—then $\psi = \psi^*$. The integrals are over all space, $d\tau$.

$$E \leq \frac{\int \psi^* H\psi \, d\tau}{\int \psi^* \psi \, d\tau}$$

An alternative method for writing integrals of this type is Dirac's "bra-ket" notation by which the integrated functions are placed in brackets separated by a vertical line or lines. An alternative writing of the previous equation using this notation is shown as follows, and this notation will be used on occasion in this book where it is convenient. Note that in the notation, the complex conjugate is implied by the function being at the left side of the bracket.

$$E \leq \frac{\langle \psi \mid H \mid \psi \rangle}{\langle \psi \mid \psi \rangle}$$

If we assume that the wavefunctions are real—that is, they have no imaginary components—then $\psi = \psi^*$, and the previous expression becomes

$$E = \frac{\int \psi H\psi \, d\tau}{\int \psi^2 d\tau}$$

If we use the linear combination of atomic orbitals (LCAO), the nature of the wavefunction, ψ, can be approximated by using the following equation, where c_1 and c_2 are coefficients, ϕ_1 is a wavefunction on atom 1, and ϕ_2 is a wavefunction on a different atom 2.

$$\psi = c_1\phi_1 + c_2\phi_2$$

Putting this into our expression for E:

$$E = \frac{\int \psi H \psi d\tau}{\int \psi^2 d\tau} = \frac{\int (c_1\phi_1 + c_2\phi_2)H(c_1\phi_1 + c_2\phi_2)d\tau}{\int (c_1\phi_1 + c_2\phi_2)^2 d\tau}$$

The numerator can then be expanded.

$$\int \psi H \psi d\tau = c_1^2 \int \phi_1 H \phi_1 d\tau + c_1 c_2 \int \phi_1 H \phi_2 d\tau + c_2 c_1 \int \phi_2 H \phi_1 d\tau + c_2^2 \int \phi_2 H \phi_2 d\tau$$

For simplicity, the integral will be abbreviated. The integral involving the orbital on atom 1 only will be labeled as α_1, and the orbital involving the orbital on atom 2 only will be α_2. There are also two integrals involving both atoms, β_{12} and β_{21}. The integral α is called a "Coulomb integral" and is a measure of the energy of the atomic orbital in the environment of the molecule. The integral β is called a "resonance integral" and is a measure of the strength of the bonding interaction.

$$\alpha_1 = \int \phi_1 H \phi_1 d\tau$$
$$\alpha_2 = \int \phi_2 H \phi_2 d\tau$$
$$\beta_{12} = \int \phi_1 H \phi_2 d\tau$$
$$\beta_{21} = \int \phi_2 H \phi_1 d\tau$$

We can then rewrite the numerator, keeping in mind that since we said that the wavefunctions are all real, $\beta_{12} = \beta_{21}$. We drop the subscripts and use just β, since they are the same. The coefficients are generally simple decimals or fractions, and $c_2c_1 = c_1c_2$.

$$\int \psi H \psi d\tau = c_1^2\alpha_1 + 2c_{12}\beta + c_2^2\alpha_2$$

Turning to the denominator and expanding:

$$\int \psi^2 d\tau = c_1^2 \int \phi_1^2 d\tau + c_1 c_2 \int \phi_1\phi_2 d\tau + c_2 c_1 \int \phi_2\phi_1 d\tau + c_2^2 \int \phi_2^2 d\tau$$

Since the orbitals are normalized, the overlap of an orbital with itself is unity. So, the first and last integrals are equal to the square of the coefficients.

$$c_1^2 \int \phi_1^2 d\tau = c_1^2$$
$$c_2^2 \int \phi_2^2 d\tau = c_2^2$$

The other integrals (such as $c_1c_2 \int \phi_1\phi_2 d\tau$) involving both atoms in the bond (such as a metal and its ligand) are the overlap between the orbitals, which we will abbreviate S_{12}. Our expression for the energy of the MO can now be written as

$$E = \frac{c_1^2\alpha_1 + 2c_1c_2\beta + c_2^2\alpha_2}{c_1^2 + c_2^2 + 2c_1c_2S_{12}}$$

If we rearrange the expression above we obtain

$$E(c_1^2 + c_2^2 + 2c_1c_2S_{12}) = c_1^2\alpha_1 + 2c_1c_2\beta + c_2^2\alpha_2$$

We wish to find where the coefficients make this expression a minimum. We differentiate with respect to c_1 and c_2, and then set $\frac{\partial E}{\partial c_1} = 0$ and $\frac{\partial E}{\partial c_2} = 0$. Differentiation with respect to c_1 produces the following expression, which is set equal to zero to find the minimum. (The differentiation at the left side involves the "product rule".)

$$c_1(\alpha_1 - E) + c_2(\beta - ES_{12}) = 0$$

Differentiation with respect to c_2 produces the following expression, which is also set equal to zero because we are trying to find a minimum.

$$c_1(\beta - ES_{12}) + c_2(\alpha_2 - E) = 0$$

These are two simultaneous equations in c_1 and c_2 set equal to zero, also called *secular equations*. We write the following secular determinant. (The matrix is formed by using the previous two equations of form $Ac_1 + Bc_2 = 0$ and using the As and Bs as the rows. The first row is the equation found by differentiation with respect to c_1, and the second row was found by differentiation with respect to c_2.)

$$\begin{vmatrix} \alpha_1 - E & \beta - ES_{12} \\ \beta - ES_{12} & \alpha_2 - E \end{vmatrix} = 0$$

Solutions of the secular determinant can be achieved in several ways to find the orbital energies, and one solution will be used in the angular overlap model in Chapter 15. Here we can look at some special cases.

If the two atoms bonding are identical, such as the two hydrogen atoms in H_2, then $\alpha_1 = \alpha_2 = \alpha$, and the secular determinate can be simplified as follows. The overlap integral is also simplified as $S_{12} = S$.

$$\begin{vmatrix} \alpha - E & \beta - ES \\ \beta - ES & \alpha - E \end{vmatrix} = 0$$

The determinant is then shown as follows, which leads to the solution shown.

$$(\alpha - E)^2 - (\beta - ES)^2 = 0$$
$$(\alpha - E)^2 = (\beta - ES)^2$$
$$(\alpha - E) = \sqrt{(\beta - ES)^2}$$
$$(\alpha - E) = \pm(\beta - ES)$$

Solving this equation for the orbital energy when the sign is positive (E_+) and negative (E_-) produces the following equation.

$$E_+ = \frac{(\alpha + \beta)}{(1 + S)}$$
$$E_- = \frac{(\alpha - \beta)}{(1 - S)}$$

The resonance integral (β) will be negative (stabilizing); consequently, the lower-energy orbital is E_+, the bonding orbital. The root E_- is then the antibonding orbital energy. A better presentation for LCAO MO theory is to rewrite the previous expressions as a perturbation on the atomic orbital energy, α, as follows.[16] In the equation, α is the atomic orbital energy in the environment of the molecule and a second term, $\frac{(\beta - \alpha S)}{(1 + S)}$, stabilizes or destabilizes that orbital.

$$E_+ = \alpha + \frac{(\beta - \alpha S)}{(1 + S)}$$
$$E_- = \alpha - \frac{(\beta - \alpha S)}{(1 - S)}$$

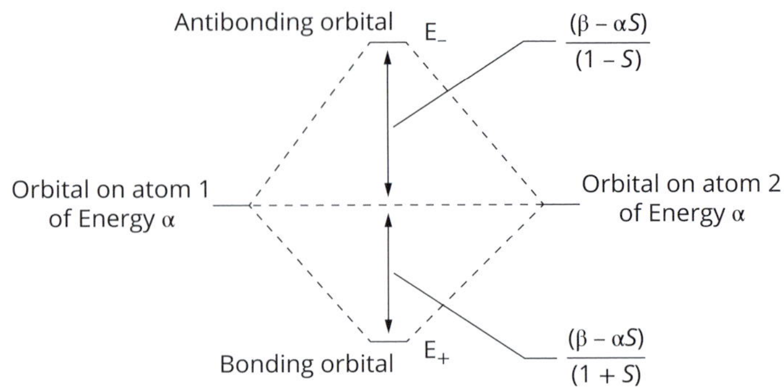

Fig. M1 LCAO MO picture with $\alpha_1 = \alpha_2 = \alpha$, illustrating the greater destabilization of the antibonding orbital relative to the bonding orbital.

Because S is positive, the antibonding orbital E_- is destabilized more than the bonding orbital is stabilized, as shown in Fig. M1.

It is a common approximation to assume that described in Chapter 6, to assume that the destabilization of the antibonding orbital is about the same as the stabilization of the bonding orbital. However, this is not exactly true unless the overlap is zero, $S = 0$, in which case the MO orbital energies are readily seen to be as shown in the following equation. (Again, keep in mind that β is negative. So, E_+ is lower in energy, and E_- is higher.)

$$E_+ = \alpha + \beta$$
$$E_- = \alpha - \beta$$

APPENDIX N

Pascal's Constants and the Diamagnetic Susceptibility

The observed susceptibility, χ_{obs}, is the parameter that provides the chemical information from the effect of a magnetic field on a substance, which is the sum (Eq. N1) of the diamagnetic (χ_{dia}) and paramagnetic susceptibilities (χ_{para}). The diamagnetic susceptibility of a substance is related to having pairs of electrons in specific environments. Since virtually all species have some number of paired electrons (with things like the hydrogen atom being one of the few exceptions), almost everything requires some accounting of the diamagnetic susceptibility. For any paramagnetic atom, the paramagnetic susceptibility, χ_{para}, is much larger than χ_{dia}. However, χ_{dia} is additive, with larger molecules generally having larger χ_{dia}. As a result, for most molecules some accounting of χ_{dia} is worthwhile, and, for very large molecules, χ_{dia} may dominate the observed susceptibility, χ_{obs}.

$$\chi_{obs} = \chi_{dia} + \chi_{para} \qquad \textbf{N1}$$

Pierre Curie (1859–1906) divided the diamagnetic susceptibility into an effect due to the atom (χ_A) and an effect due to electrons in bonds (λ_i). Again, the whole effect is a sum over the atoms and bonds in a molecule (Eq. N2). The values of χ_A and λ_i are now known as "Pascal's constants".

$$\chi_{dia} = \sum_i \chi_{A_i} + \sum_i \lambda_i \qquad \textbf{N2}$$

Table N1 lists some common values for χ_A and λ_i, most of which are from the article by Bain and Berry, in which other values may be found.[1]

Instead of summing each bond and atom in a molecule individually, some χ_{dia} values for commonly encountered cations, anions, and ligands have been collected in Table N2. These can be summed with other atoms or groups in molecules to derive the total diamagnetic susceptibility.

Table N1 Diamagnetic susceptibility of atoms (χ_{A_i}) in covalent molecules and susceptibilities for some bond types (λ_i). These values can be used in Eq. N2.

Atom	$\chi_{A_i}\left(1\times10^{-6}\frac{emu}{mol}\right)$	Bond	$\lambda_i\left(1\times10^{-6}\frac{emu}{mol}\right)$
Ag	−31.0	C=C	+5.5
Al	−13.0	C≡C	+0.8
As(III)	−20.9	C=O	+6.3
B	−7.0	COOH	−5.0
Bi	−192.0	$C(O)NH_2$	−3.5
Br	−30.6	C(O)OH/ C(O)OR	−5.0
C	−6.00	C=N	+8.15
Li	−4.2	−C≡N	+0.8
Mg	−10.0	−N≡C	0.0
N(ring)	−4.61	$-NO_2$	−2.0
N(open chain)	−5.57	C–Cl	+3.1
Na	−9.2	$RCHCl_2$	+6.43
O	−4.6	C–Br	+4.1
P	−26.3	C–I	+4.1
Pb(II)	−46.0	Ar–OH/Ar–OR	−1
S	−15.0	$Ar-NR_2$	+1
Sb(III)	−74.0	Ar–C(=O)	−1.5
Se	−23.0	Ar–C=C	−1.00
Si	−13	Ar–C≡C	−1.5
Sn(IV)	−30	Ar–Ar	−0.5
Te	−37.3	$Ar-NO_2$	−0.5
Tl(I)	−40.0	Ar–Br	−3.5
Zn	−13.5	Ar–Cl	−2.5
		Ar–I	−3.5
		Benzene	−1.4
		Cyclohexane	+3.0
		Cyclopentane	0.0
		Furan	−2.5
		Pyridine	+0.5
		Pyrrole	−3.5
		Tetrahydrofuran	0.0
		Thiophene	−7.0

Table N2 Some values for common cations, anions, and ligands.

Species	$\chi_{dia}\left(1\times10^{-6}\frac{emu}{mol}\right)$	Species	$\chi_{dia}\left(1\times10^{-6}\frac{emu}{mol}\right)$
Ag^{+}	−28	Rh^{3+}	−22
Al^{3+}	−2	Pb^{2+}	−32.0
As^{3+}	−9	Pb^{4+}	−26
As^{5+}	−6	Pd^{2+}	−32.0
Au^{3+}	−32	Pm^{3+}	−27
B^{3+}	−0.2	Pr^{3+}	−20
Ba^{2+}	−26.5	Pt^{2+}	−40
Ca^{2+}	−10.4	Pt^{4+}	−28
Cd^{2+}	−24	Re^{3+}	−36
Ce^{3+}	−20	Sb^{3+}	−17
Ce^{4+}	−17	Sb^{5+}	−14
Co^{2+}	−12	Sc^{3+}	−6
Co^{3+}	−10	Si^{4+}	−1
Cr^{3+}	−11	Sm^{2+}	−23
Cr^{4+}	−8	Sm^{3+}	−20
Cr^{5+}	−5	Sn^{2+}	−20
Cr^{6+}	−3	Sn^{4+}	−16
Cs^{+}	−35.0	Sr^{2+}	−19.0
Cu^{+}	−12	Ta^{5+}	−14
Cu^{2+}	−11	Tb^{3+}	−19
Dy^{3+}	−19	Ti^{3+}	−9
Er^{3+}	−18	Ti^{4+}	−5
Eu^{2+}	−22	Tl^{+}	−35.7
Eu^{3+}	−20	Tl^{3+}	−31
Fe^{2+}	−13	Tm^{3+}	−18
Fe^{2+}	−10	U^{3+}	−46
Ga^{3+}	−8	U^{4+}	−35
Ge^{4+}	−7	U^{5+}	−26
Gd^{3+}	−20	U^{6+}	−19
H^{+}	0	V^{3+}	−10
Hf^{4+}	−16	V^{4+}	−7
Hg^{2+}	−40	V^{5+}	−4
Ho^{3+}	−19	VO^{2+}	−12.5
In^{3+}	−19	W^{3+}	−36
Ir^{+}	−50	W^{6+}	−13
Ir^{3+}	−35	Y^{3+}	−12

Table N2 Continued

Species	$\chi_{dia}\left(1\times10^{-6}\frac{emu}{mol}\right)$	Species	$\chi_{dia}\left(1\times10^{-6}\frac{emu}{mol}\right)$
K^+	−14.9	Yb^{3+}	−18
La^{3+}	−20	Zn^{2+}	−15.0
Li^+	−1.0	Zr^{4+}	−10
Lu^{3+}	−17	BF_4^-	−37
Mg^{2+}	−5.0	Br^-	−34.6
Mn^{2+}	−14	Cl^-	−23.4
Mn^{3+}	−10	ClO_4^-	−32.0
Mn^{4+}	−8	CN^-	−13.0
Mn^{6+}	−7	$C_5H_5^-$	−65
Mo^{3+}	−23	CO_3^{2-}	−28.0
Mo^{6+}	−7	F^-	−9.1
NMe_4^+	−52	I^-	−50.6

Species	$\chi_{dia}\left(1\times10^{-6}\frac{emu}{mol}\right)$	Species	$\chi_{dia}\left(1\times10^{-6}\frac{emu}{mol}\right)$
NEt_4^+	−101	NO_3^-	−18.9
Na^+	−6.8	O^{2-}	−12.0
Nb^{5+}	−9	S^{2-}	−30
Nd^{3+}	−20	SO_4^{2-}	−40.1
Ni^{2+}	−12	acac⁻	−52
Os^{3+}	−36	bipy	−105
Os^{4+}	−29	CO	−10
Os^{6+}	−18	en	−46.5
Os^{8+}	−11	H_2O	−13
P^{3+}	−4	NH_3	−18
P^{5+}	−1	PPh_3	−167
Re^{7+}	−12	pyridine	−49

Cited Reference

1. Bain, G. A. and Berry, J. F. Diamagnetic corrections and Pascal's constants. *J. Chem. Ed.* 2008, *85* (4), 532–536.

Index

T

U